HANDBUCH DER PFLANZENERNÄHRUNG UND DÜNGUNG

BEGRÜNDET VON
KARL SCHARRER UND HANS LINSER

HERAUSGEGEBEN VON
DR. PHIL. HANS LINSER
O. UNIVERSITÄTSPROFESSOR FÜR PFLANZENERNÄHRUNG
DIREKTOR DES INSTITUTS FÜR PFLANZENERNÄHRUNG
DER JUSTUS-LIEBIG-UNIVERSITÄT IN GIESSEN

IN DREI BÄNDEN

ZWEITER BAND
BODEN UND DÜNGEMITTEL
ERSTE HÄLFTE

1966
SPRINGER-VERLAG
WIEN · NEW YORK

BODEN
UND DÜNGEMITTEL

BEARBEITET VON

E. ABRAHAMCZIK · J. M. ALBAREDA HERRERA · H.-J. ALTEMÜLLER
A. AMBERGER · N. ATANASIU · J. BAEYENS · H. BANTHIEN · G. BARBIER
P. BOEKEL · H. BORTELS · G. BRAUNE · M. DE BOODT · L. DE LEENHEER
E. ERIKSSON · W. FLAIG · H. FRANZ · H. FRESE · A. FRUHSTORFER
S. GERICKE · G. GLATHE · H. GLATHE · W. HÖFNER · K. KAINDL
E. KNICKMANN · H. KOEPF · J. J. LEHR · H. LINSER · H. LÖCKER
E. MÜCKENHAUSEN · H. MÜLLER · D. J. D. NICHOLAS · T. NIEDERMAIER
P. K. PEERLKAMP · V. RANK · K. RAUHE · E. RAUTERBERG · G. RIEHLE
J. CHR. SALFELD · H. K. SCHÄFER · K. C. SCHEEL · U. SCHWERTMANN
H. SÖCHTIG · S. TROCMÉ · B. WOHLRAB · F.-H. ZIEMER

ERSTE HÄLFTE

MIT 188 ZUM TEIL FARBIGEN ABBILDUNGEN
IM TEXT UND AUF EINER AUSSCHLAGTAFEL

1966
SPRINGER-VERLAG
WIEN · NEW YORK

ISBN-13: 978-3-7091-8198-0 e-ISBN-13: 978-3-7091-8197-3
DOI: 10.1007/978-3-7091-8197-3

Softcover reprint of the hardcover 1st edition 1966

Library of Congress Catalog Card Number 66-16426

Titel-Nr. 8323

Vorwort

Seit dem Erscheinen des zweibändigen Standardwerkes von F. HONCAMP, „Handbuch der Pflanzenernährung und Düngerlehre", sind mehr als dreißig Jahre vergangen. Bei den stürmischen Fortschritten auf allen Gebieten der Naturwissenschaften, insbesondere der Chemie, Pflanzenphysiologie, Biochemie, Bodenkunde und Technologie, ist dieses zur Zeit seines Erscheinens so ausgezeichnete Handbuch natürlich längst veraltet.

Zweifellos besteht heute ein dringendes Bedürfnis nach einem ähnlichen Werk. Wenn wir von dem ursprünglichen Plan abgekommen sind, das HONCAMPsche Handbuch in neuer Auflage zu bearbeiten, so vor allem deswegen, weil die tiefgreifende Entwicklung auf diesem Gebiet eine völlig neue Gestaltung des Werkes geboten erscheinen ließ. Es gliedert sich nun in drei Bände.

Band I behandelt im wesentlichen die physiologischen Grundlagen der Pflanzenernährung. Er enthält eine ausführliche und systematische Darstellung der eigentlichen Ernährungsphysiologie. Auf eine Erörterung der allgemeinen pflanzenphysiologischen Grundlagen konnte dabei verzichtet werden, weil im RUHLANDschen „Handbuch der Pflanzenphysiologie" ein umfassendes und modernes Nachschlagewerk zur Verfügung steht.

Band II beschäftigt sich mit dem Boden und den Düngemitteln. Unter Berücksichtigung der modernen Agrikulturchemie haben wir das heutige Wissen über den Boden als Standort und Nährstoffquelle der Pflanze unter den Verhältnissen der landwirtschaftlichen Praxis und des Gartenbaues bzw. des Forstwesens systematisch zusammengefaßt. Die folgenden Kapitel befassen sich mit der Beschreibung und Herstellung der verschiedenen Wirtschafts- und Handelsdüngemittel. Gerade auf diesem Gebiet bestehen in der wissenschaftlichen Literatur erhebliche Lücken, da keine größeren Lehr- oder Handbücher vorhanden sind, die sich eingehender mit der Technologie der Düngemittel und der Düngung beschäftigen.

Band III behandelt schließlich die Ausbringungsverfahren für Düngemittel sowie die Düngung aller wichtigen landwirtschaftlichen, gärtnerischen und forstlichen Kulturpflanzen einschließlich wichtiger tropischer Arten. Sondergebiete, wie die Düngung von Moor und Anmoor und die Düngung von Teichen, sind in eigenen Abschnitten behandelt. Besonderes Augenmerk ist in ausführlichen Kapiteln der Rentabilität und der wirtschaftlichen Bedeutung der Düngung wie auch ihren ernährungsphysiologischen Auswirkungen gewidmet. Ein Anhang gibt einen Überblick über die während der letzten Jahre in verschiedenen Ländern hauptsächlich verwendeten Nährstoffverhältnisse in Düngemitteln.

Das vorliegende Werk soll der agrikulturchemischen Forschung einen Überblick über den heutigen Entwicklungsstand auf dem Gebiet der Pflanzenernährung und Düngung vermitteln. Ist uns dies gelungen, so war die dabei aufgewandte Mühe nicht vergeblich. Die Erfordernisse der praktischen Landwirtschaft bringen es mit sich, daß einzelne Fragen recht unterschiedliches Gewicht haben. Die Gliederung des Stoffes konnte daher nicht nur nach logischen und theoretisch-sachlichen Gesichtspunkten vorgenommen werden. Wir hielten es auch für richtig, einzelne Gegenstände unter verschiedenen Gesichtspunkten darzustellen, wie sie sich aus den unterschiedlichen Betrachtungsweisen verschiedener Forscher und Autoren ergeben. Wenn dadurch gewisse Überschneidungen verursacht wurden, so halten wir dies nicht für einen Nachteil.

Professor Dr. Dr. h. c. Dr. h. c. KARL SCHARRER, der langjährige Direktor des agrikulturchemischen Instituts der Justus-Liebig-Universität in Gießen, gab den Anstoß zum Entstehen dieses Werkes, mit dem er seinen grundlegenden Arbeiten auf dem Gebiete der Agrikulturchemie und insbesondere auf dem des Düngungswesens einen zusammenfassenden Abschluß geben wollte. Seine letzten Bemühungen vom Krankenbett aus galten diesem Buch. Sein früher Tod ließ ihn das Entstehen des Gesamtwerkes nicht mehr erleben. Der Herausgeber hofft, es in seinem Sinne weitergeführt zu haben.

Über hundert Wissenschaftler und Techniker aus aller Welt haben an dem Werk mitgearbeitet. Ihnen allen sei an dieser Stelle besonders gedankt. Zu danken ist aber auch jenen, die bei der umfangreichen redaktionellen Arbeit tatkräftig mitgewirkt haben, vor allem Frau Dr. ELFRIEDE PRESOLY (Linz), Herrn Dr. KARL-HERMANN NEUMANN (Gießen), der auch die Register angefertigt hat, sowie Frau EDITH HARNDT, Frau CHRISTEL FRIEDRICH und Frau ELLEN NEUMANN (Gießen) sowie den Herren Dr. JOACHIM BOHRING und Dipl.-Landw. JÜRGEN DRESSEL.

Dem Springer-Verlag in Wien, der sich dem Plan von Anfang an aufgeschlossen gezeigt hat, gebührt Dank für alle die Sorgfalt, die er dem Werk hat angedeihen lassen.

Gießen, im Frühjahr 1966

Hans Linser

Inhaltsverzeichnis der ersten Hälfte

Seite

Inhaltsübersicht der zweiten Hälfte

II. Düngemittel

Mitarbeiter von Band II

ABRAHAMCZIK, Dr. E., Badische Anilin- & Soda-Fabrik A. G., Analytisches Laboratorium, *Ludwigshafen/Rhein*, Bundesrepublik Deutschland.

ALBAREDA HERRERA, Professor Dr. J. M., Instituto de Edafología y Fisiología Vegetal, Serrano 113, *Madrid*, España.

ALTEMÜLLER, Dr. H.-J., Institut für Bodenbearbeitung der Forschungsanstalt für Landwirtschaft, Bundesallee 50, *Braunschweig-Völkenrode*, Bundesrepublik Deutschland.

AMBERGER, Professor Dr. A., Leiter der Abteilung Agrikulturchemie und Versuchswesen an der Landesanstalt für Bodenkultur, Pflanzenbau und Pflanzenschutz, Menzingerstraße 54, *München*, Bundesrepublik Deutschland.

ATANASIU, Professor Dr. N., Leiter der Abteilung Pflanzenbau und Pflanzenzüchtung der Sektion Pflanzliche Produktion des Tropen-Instituts der Justus-Liebig-Universität, Schottstraße 2—4, *Gießen*, Bundesrepublik Deutschland.

BAEYENS, Professeur Dr. J., Directeur de l'Institut Pédologique, Université de Louvain, Av. Card. Mercier 92, *Heverlee*, Belgique.

BANTHIEN, Dr. H., Hoechst-Uhde International G. m. b. H., Brüningstraße 45, *Frankfurt/Main-Höchst*, Bundesrepublik Deutschland.

BARBIER, Professeur Dr.-Ing. G., Directeur de Recherches à l'Institut National de la Recherche Agronomique, 18, rue de l'Orangerie, *Versailles* (Yvelines), France.

BOEKEL, Ir. P., Instituut voor Bodemvruchtbaarheid, van Hallstraat 3, *Groningen*, Niederlande.

BORTELS, Professor Dr. H., Institut für Bakteriologie der Biologischen Bundesanstalt für Land- und Forstwirtschaft, Messeweg 11—12, *Braunschweig-Gliesmarode*, Bundesrepublik Deutschland.

BRAUNE, Dr. G., Kaliforschungsinstitut e. V., Georgstraße 29, *Hannover*, Bundesrepublik Deutschland.

DE BOODT, Professor Dr.-Ing. M. F., M. S., Fakultät für Landwirtschaftswissenschaften, Coupure Links 235, *Gent*, Belgien.

DE LEENHEER, Professor Dr. L., Leerstoel voor Bodemkunde en Geologie, Rijkslandbouwhogeschool, Coupure Links 235, *Gent*, Belgien.

ERIKSSON, Professor Dr. E., International Meterological Institute, Tulegatan 41, *Stockholm*, VA, Sweden.

FLAIG, Professor Dr. W., Direktor des Instituts für Biochemie des Bodens der Forschungsanstalt für Landwirtschaft, Bundesallee 50, *Braunschweig-Völkenrode*, Bundesrepublik Deutschland.

FRANZ, Professor Dr.-Ing. H., Vorstand des Instituts für Bodenforschung der Hochschule für Bodenkultur, Gregor-Mendel-Straße 33, *Wien* XVIII, Österreich.

FRESE, Professor Dr. H., Direktor des Instituts für Bodenbearbeitung der Forschungsanstalt für Landwirtschaft, Bundesallee 50, *Braunschweig-Völkenrode*, Bundesrepublik Deutschland.

FRUHSTORFER, Professor Dr. A., *Weißensee-Oberried* über Füssen/Allgäu, Bundesrepublik Deutschland.

GERICKE, Dr. S., Landwirtschaftliche Versuchsanstalt der Thomasphosphatfabriken GmbH., Prinz-Adolf-Straße 2, *Essen-Bredeney*, Bundesrepublik Deutschland.

GLATHE, Dr. GERTRAUDE, Goethestraße 30, *Gießen*, Bundesrepublik Deutschland.

GLATHE, Professor Dr. H., Direktor des Instituts für landwirtschaftliche Mikrobiologie der Justus-Liebig-Universität, Landgraf-Philipp-Platz 4, *Gießen*, Bundesrepublik Deutschland.

HÖFNER, Dr. W., Wissenschaftlicher Rat am Institut für Pflanzenernährung der Justus-Liebig-Universität, Braugasse 7, *Gießen*, Bundesrepublik Deutschland.

KAINDL, Univ.-Doz. Dr. K., Leiter des Instituts für Biologie und Landwirtschaft des Reaktorzentrums Seibersdorf der Österreichischen Studiengesellschaft für Atomenergie, Lenaugasse 10, *Wien* VIII, Österreich.

KNICKMANN, Professor Dr. E., Hessische Lehr- und Forschungsanstalt für Wein-, Obst- und Gartenbau, *Geisenheim/Rheingau*, Bundesrepublik Deutschland.

KOEPF, Professor Dr. H., Institut für Bodenkunde der Landwirtschaftlichen Hochschule Hohenheim, *Stuttgart-Hohenheim*, Bundesrepublik Deutschland, z. Zt. Threefold Farm, *Spring Valley*, N.Y., USA.

LEHR, Direktor Dr. J. J., Laboratorium voor Bemestingsonderzoek, Landbouwkundige Dienst voor Chilisalpeter, Diedenweg 20, *Wageningen*, Niederlande.

LINSER, Professor Dr. H., Direktor des Instituts für Pflanzenernährung der Justus-Liebig-Universität, Braugasse 7, *Gießen*, Bundesrepublik Deutschland.

LÖCKER, Dr.-Ing. H., Carl-Bosch-Weg 3, *Linz*, Österreich.

MÜCKENHAUSEN, Professor Dr. Dr. E., Direktor des Instituts für Bodenkunde der Rheinischen Friedrich-Wilhelms-Universität, Nußallee 13, *Bonn*, Bundesrepublik Deutschland.

MÜLLER, Dr. H., Deutsche Hyperphosphat Gesellschaft m. b. H., Hofgut Oppelshausen, *Altenstadt/Hessen*, Bundesrepublik Deutschland.

NICHOLAS, Professor Dr. D. J. D., Head, Department of Agricultural Biochemistry and Soil Science, Waite Agricultural Research Institute, The University of Adelaide, *Glen Osmond*, South Australia.

NIEDERMAIER, Dr. T., Badische Anilin- & Soda-Fabrik A. G., Analytisches Laboratorium, *Ludwigshafen/Rhein*, Bundesrepublik Deutschland.

PEERLKAMP, Dr. P. K., Instituut voor Bodemvruchtbaarheid, van Hallstraat 3, *Groningen*, Niederlande.

RANK, Dr. V., Carl-Bosch-Weg 2 b, *Linz*, Österreich.

RAUHE, Professor Dr. K., Institut für Acker- und Pflanzenbau der Karl-Marx-Universität, Johannisallee 19, *Leipzig*, DDR.

RAUTERBERG, Professor Dr. E., Direktor des Instituts für Pflanzenernährung, Bodenchemie und Bodenbiologie der Technischen Universität Berlin, Lentze-Allee 55—57, *Berlin-Dahlem*, Deutschland.

RIEHLE, Dr. G., Badische Anilin- & Soda-Fabrik A. G., Landwirtschaftliche Versuchsstation, Postfach 220, *Limburgerhof/Pfalz*, Bundesrepublik Deutschland.

SALFELD, Dipl.-Chem. Dr. J. CHR., Institut für Biochemie des Bodens der Forschungsanstalt für Landwirtschaft, Bundesallee 50, *Braunschweig-Völkenrode*, Bundesrepublik Deutschland.

SCHÄFER, Dr. H. K., Farbwerke Hoechst A. G., *Frankfurt/Main-Höchst*, Bundesrepublik Deutschland.

SCHEEL, Dr. K. C., Guano-Werke A. G., Spitalerstraße 10, *Hamburg*, Bundesrepublik Deutschland.

SCHWERTMANN, Professor Dr. U., Direktor des Instituts für Bodenkunde der Technischen Universität Berlin, Königin-Luise-Straße 22, *Berlin-Dahlem*, Deutschland.

SÖCHTIG, Dipl.-Landw. Dr. H., Institut für Biochemie des Bodens der Forschungsanstalt für Landwirtschaft, Bundesallee 50, *Braunschweig-Völkenrode*, Bundesrepublik Deutschland.

TROCMÉ, S., Maître de Recherches à l'Institut National de la Recherche Agronomique, Route de Saint-Cyr, *Versailles* (Yvelines), France.

WOHLRAB, Priv.-Doz. Dr. B., Regierungsdirektor an der Landesanstalt für Immissions- und Bodennutzungsschutz des Landes Nordrhein-Westfalen, Am Hohwege 5, *Bochum-Linden*, Bundesrepublik Deutschland.

ZIEMER, Dipl.-Landw. Dr. F.-H., Düngekalk-Hauptgemeinschaft, „Kalkdienst", Land- und forstwirtschaftliche Abteilung, Kaiser-Wilhelm-Ring 26, *Köln*, Bundesrepublik Deutschland.

Introduction

The General Problem of Interaction Between Soils and Fertilizers in Their Action on Plant Yield

By

J. M. Albareda Herrera

Soil is the culture medium of plants. The larger modifications which cultivation—with the great diversity of its operations—introduces in natural soils result in agricultural development of the crops of cultivated plants.

Soil has a structure and composition which enables it to provide for the nutrition and growth of plants. As a convergence of mineralogical and biological phenomena, it is a result both of rock desintegration and decomposition processes and of the transformation of organic matter. Finally being a disperse and dynamic system, soil reveals the different stages of its development in mineral and organic fractions which are different as far as their composition particle size and shape are concerned.

It may seem surprising that soil, which has been studied from the most remote past and which has been the object of continuous investigations during the last century, should present so many aspects which have only been investigated during recent years. Soil is a heterogenous system and up to now it has been impossible to consider it from the many different angles which its study requires. The extraordinary scientific development of our day reveals common points on different fronts: after the analytical and extensive study which yields its chemical composition, an intensive investigation pertaining to everything related to structure must be carried out; there is a dynamic consideration, an investigation of processes; a study of the groupings, of associations is also necessary and as a result of all this, we have a convergence which results in the projection of different sciences on each other.

The term structure holds sway in the physics, chemistry and biology of our time and permeates the lexicon of social sciences. The sweeping development of nuclear physics is due exclusively to the fact that it has been able to penetrate into atomic structure with instruments undreamed of a few years ago. The renewed freshness of chemistry, opening new and extensive fields which seemed forever exhausted, has its roots in that structural character which prevails in the best treatises on chemistry. In biology, after colloidal chemistry had overshadowed the importance of structure, under the sway of the exclusive and independent idea of surface development, we are confronted with the knowledge of submicroscopic structures, of increasing complexity and harmony. And it is not only that this common term has a formal identity and a widely different significance in the realm of the different sciences. The same techniques have allowed us to apprehend in detail the crystalline lattice of minerals, establishing the foundation of clay mineralogy and unravelling the architecture of what in the past were considered as very complex molecules, such as those defined by the spirals of deoxyribonucleic acids, which, at the same time, afforded an exploration of character diversification and hereditary continuity.

But apart from the structural design of the beings which biology and geology offer us or of the bodies which technology produces, we have to dwell on the relevant consideration of their dynamic processes. Nuclear reactions, with their fissions permit us to begin to understand solar energy, the radioactivity of the elements on which we can base datings, extending from precambrian geology with its billions of years, to the organic residues which the Quaternary offers to our contemplation and to the growth of crystals with their dislocations and reticular effects. Biochemistry has also begun to outgrow the static structure stage and modern research investigates the metabolic and dynamic aspects of living beings through enzymatic routes. Its mission is cellular unity and analysis; the intimate interpenetration of biochemistry and genetics is one of the great developments of scientific speculation during recent years.

In the natural sciences, after the development of systematics, in other words, of the establishment of an order according to form, we have passed to the study of associations, of mineral or biological assemblies and, eventually, to that of the reasons which explain their existence. As Jakob (1945) declares, the idea of a chaotic state of the Earth is basically false: the Earth is constructed with an unmistakable regularity from its very foundations. For the uninitiated it is hard to see any regularity in the constitution of the Earth. In the measure in which we proceed to further our observations of what is taking place in the surface of our planet we begin to acquaint ourselves with a greater number of regularities and apprehend a clearer picture of the Earth. The effect of foods—says Linser (1957)—is always achieved only by their common action with other foods and we know that the action of one nutrient material does not only depend on its amount but also on the amounts of other nutrient materials which are supplied simultaneously or had been supplied before.

Our time is also characterised by the development of bridge and integration sciences, for in attempting to study a being or a process, we need to refer to the different sciences. This has led to the formation of composite sciences, such as geochemistry or biophysics and of integrated sciences, i.e. ecology.

Soil science is both an integrated and a bridge science: it is a bridge science which bonds geology and biology when the study of the mineral nutrition of plants links the colloidal system of soil with the root system of plants. It is also an integrated science, for soil is a natural complex, the study of which requires the convergence of physical, chemical, geological and biological knowledge. Soil science is one of those sciences which favour a return to the dynamic unity of the natural sciences, divided and subdivided from the moment they abandoned the systematics of classical natural history for the realm of the diverse geological and biological processes.

Here soil appears as a solid link between geology and biology, two realms, which are drifting further apart in many fundamental aspects as time goes by due to the study of structures and processes, a study which has given birth to new sciences—crystalochemistry, enzymology, etc.—in which the old naturalisticsystematic unity has been extinguished. Face to face with the regressive process through which biology leads to geology, giving birth to enormous paleontological formations (Twenshofel 1932), soil acts as a sort of geological-phytobiological bridge through the chemical nutrition of plants.

Soil Science, Sedimentology and Geochemistry

The dispersive process of primary rocks, the study of mineral transfer and deposition are studied in sedimentology, which is essential in the study of surface geology. At the same time the study of the mineral nutrition of plants requires a knowledge of the frequency and distribution of chemical elements. This is the task of geochemistry. The studies of RANKAMA (1950) which pursued the location of mineral deposits through the incorporation of trace cations into the soil, and from it into the plant, offer us an idea of the connexion and intimacy of processes in which soil is the basic link. Geochemistry describes the cycle of each of those elements, the position of which we consider in mineral structure, and their dynamism in the geological process. Thus, both sedimentology and geochemistry are sciences which contribute with important knowledge to the realm of soil science.

These considerations on the present scientific development are of particular application to a system so heterogeneous and dynamic as soil. With the structure and composition and dynamism of its components, it has a capacity to nourish plant life. This capacity is increased through the addition of fertilizers.

The weathering processes by which igneous rocks (which result from magmatic crystallization) are transformed into an assembly of secondary minerals, in equilibrium with surface conditions, are varied. Many of its phases lead to the formation of soil, where detritic minerals of the primary rock and newly formed minerals are present. The coincidence of a mechanical milling process with the chemical reactions which lead to the new minerals supplied soil with an inorganic material, which in part has been produced in its medium, and, therefore, serves to define it. Starting from a raw organic material, the presence of an intense biological activity in the soil forms products elaborated by a flora and a fauna which are integrated into it and, at the same time, define it. JENNY (1941) has studied soil formation in terms of the following factors: primary rock, climate, topography, time, vegetation and has tried to find a representative equation of the influence exerted by each of these factors.

The present development of sedimentology offers a reciprocal aid to the progress of edaphic knowledge. Sedimentogenesis systematises a group of processes among which we find, in part, the soil forming processes.

RANKAMA and SAHAMA (1950) have described the exogenous cycle of geological material and the desintegration sequence of rocks into their different components. There is a coincidence between the magmatic crystallization sequence and that of the stability of minerals in respect to decomposition agents. This coincidence is of a physicochemical character (KELLER 1955, RAMBERG 1952). STRAHKOV (1958) speaks of four different sedimentary processes: glaciar, humid, arid and effusive-sedimentary. The first named is essentially mechanical in nature, milling and transfer, but in the humid type the differences in solubility call for a different distribution of materials. Those which are easily soluble: sodium, magnesium and calcium chlorides and sulfates; those which are less soluble going into suspensions or solutions: calcium carbonate, silica, organic substances; those which are even less soluble, such as iron, manganese, aluminium and phosphorus, in part as dominantly colloidal solutions; and, finally silicates, which form suspensions and the greatest portion of deposited material.

To this modality of separation, based on the different solubility of the materials formed as well as for the products of their chemical reactions, we can add the one established by biological actions. These can attain a considerable diversity, considering the wide variety of living organisms capable of exerting their activity and the different action implied by the different medium conditions.

In these superficial processes—coexistence of inanimate factors with living matter—we can attain a different distribution of the elements and, in specific areas, strong concentrations are formed which give place to mineral deposits originating by hydrolitic action—iron, manganese and aluminium oxides and hydroxides—elements which are scarce in other areas where they have been washed away. This elution decreases in the arid type, where chemical decomposition processes are of lesser importance and accumulations of minerals formed by oxides and hydroxides of iron, manganese and aluminium are not found. The intense evaporation, which is present in areas of basin drainage in arid regions results in the concentration and deposition of sodium, magnesium, potassium and calcium salts: halogenesis areas. A lack of water reduces vegetation to a few communities adapting themselves to this medium, and leads to a scarcity of soil organic matter.

There exists a contrast between the types of soils which correspond to the dry and humid surface processes, as well as between their respective agricultural problems.

The degree of transformation of the originary geological material of soils is very different and has been the subject of intense research the aim of which was the determination of the degree of weathering of the different minerals. As the soil is a dynamic system, both from the point of view of its genesis as from that of its plant nutrition capacity, the stability of its components is of decisive significance. In the inorganic fraction the decomposition facility of primary minerals, from which soil is formed, as well as the degree of correlation between the secondary minerals produced in the soil and the different types of it, have been studied (JACKSON 1953, FIELDES 1954). Furthermore, a division betwen nutrient humus and stable humus has been drawn (SCHEFFER 1941), which clearly enhances this idea of stability in opposition to that of nutrient dynamism.

GOLDSCHMIDT (1954) has classified sediments on the basis of geochemical principles and RANKAMA (1950) has developed and outlined this classification. Thus the heterogeneity of these materials is systematised in the following groups: unattacked (Si), hydrolysed (Si, Al, Fe, K), oxidised (Fe, Mn), reduced precipitated (Ca, Mg), evaporated (K, Na, Ca, Mg) and biolytic. This geochemical concept is very important in the study of soil. Edafology has contributed in a large scale to the development of sedimentary geochemistry.

The passage of primitive geological material into sedimentary material, when the soil is formed, results in a loosening of the chemical bonds in such a way as to leave the nutrient elements of plants in a considerably more favourable condition for their assimilation. Each element is linked in the crystal by an electrostatic bond which holds it in a fixed position in the lattice. The chemical decomposition processes which take place in the soil furthers its disintegration and liberate it up to a point in which its biological assimilation is possible. Inversely, these processes can lead to a fixation of the element, leaving it outside the scope of the assimilative capacity of the plant.

The greatest portion of the rocks which make up the surface of the earth are siliceous. If we consider the frequency of lithological elements, we see that oxygen and silicon account for three fourths of the composition of the earth's crust (CLARKE 1924). This siliceous material reveals a different tendency to decompose under conditions prevailing in the surface. This tendency becomes greater through the sequence three dimensional lattices to lattices with independent SiO_4 tetrahedrons; that is to say, as the degree of polymerization of the tetrahedrons decreases.

The weathering series of the heavy minerals has been the object of a detailed study. PETTIJOHN (1941) considers this sequence from the standpoint of the disappearance of the minerals in definite periods. Thus, olivine, which easily decomposes, is found in rocks formed before the Pleistocene, augite, which occupies an intermediate position, is not to be found or, rarely so, in Precambrian rocks, but is present profusely in younger ones. Apart from these considerations on the greater or lesser decomposition tendency of primitive minerals in order to become soluble or assimilable elements, we have the reverse process by which soluble or assimilable elements can regress to their unassimilable form through fixation. It is not that an element should traverse a route through which, by successive decreases of its bonding energy, it can attain the soluble ion stage. JENNY (1939, 1953) has shown the possibility of contact adsorption by which the soil ion passes directly into the root without any intervention whatsoever of the nutrient solution[1].

Soil Minerals

Soil mineralogy shows each chemical element in the chemical and morphological entity of which it forms a part. MARSHALL (1940) has shown "the very powerful aid which petrological methods can render first in deciding questions of origin and second in enabling us to form a clearer picture of soil-forming processes."

The mineral content of sedimentary deposits which exist in the earth is made up of very different materials, which are derived from igneous rocks or have been produced by their transformation: quartz alone accounts for more than a third part: mica and carbonates for a fifth each; clays and feldspars, taken together for somewhat more than a sixth; limonite is found in lesser amounts. Apart from these main minerals there are other constituents which are found in smaller amounts and belong to the group of the so-called accessory minerals.

The mineralogical study of soil is carried out by the determination of the mineral constituents as well as of that of their nature, distribution and characteristics. In problems of soil classification knowledge of the constituent minerals is important. The correlation of characteristics of minerals with field observations—says JEFFRIES (1950)—is a branch of study in which little research work has been undertaken. Correlations between soil field observations and primary minerals, that is to say underlying rocks, have been undertaken in a more extensive scale, but the idea that those minerals of which the soil is made up are derived from the rocks on which they lie has not been established as a certainty.

Soil mineralogy reveals both unity and diversity from different points of view. It has the continuity of a disperse system and WIEGNER (1926) was able to draw a picture of the variations of the chemical and physical properties of the soil, with their influence on agricultural fertility or forestry development. ENGELHARDT (1938) has shown the distribution of different soil minerals in terms of particle size. The so-called fine and heavy portion are separated on the basis of their respective richness in nutrient elements, physical properties and determination techniques. The enormous diversity of the material produced by the

[1] This theory has been the object of several investigations and discussions (LAGERWERFF 1958, 1960, BLANCHET, CHAUMONT and STUDER 1962).

dispersion process has been the object of a first dispersoid-chemical systematization, which considers particle size, surface area. Even though the study of the heavy fractions of the soil progresses in close relation with the development of the petrographic microscope, that of the fine, active, portionsrequires the development of physical methods adequate for small size particles. Together with other supplementary methods X-ray-analysis appears to be the basic method of identification. To this we have still to add that material which is to be found in the gel state, the adsorbing properties of dissolved ions which have been the subject of fundamental studies carried out by Mattson (1929).

These differences in the techniques employed in the mineralogical study of the fine and heavy fractions divided soil mineralogy into two fields which are too for apart.

The adsorbing properties of clays and the diversity of their applications in the soil and in industry have resulted in considerable development of clay mineralogy during the last few years (Schachtschabel 1940, Marshall 1949, Jasmund 1951, Grim 1953, Proceedings of the National Conferences on Clays and Clay Minerals, Eitel 1954).

Sand, which has a low rate of decomposition as regards the supply of nutrient elements, has been neglected as a factor, for its slow rate of decomposition excludes its consideration in view of the requirements of the production of crops by intensive culture techniques. Sand also plays the role of an indicator of the primary rock in the sedimentological processes.

Mineral Stability and Fertility

The rate of decomposition changes with climate as has been evidenced in tropical agriculture (Vageler 1930, Mohr 1954, Tamm 1934, Perez Mateos 1957) has attempted to apply the basic mineral number to tropical soils. There is a group of numbers the object of which is to measure the degree of decomposition of the mineral on the basis of the ease with which the different participants can be separated and mobilised; from the $SiO_2/Al_2O_3+Fe_2O_3$ to the K_2O/Na_2O molecular relations. These values measure the various degree of mobilization of the different components.

The richness of rocks or sands in chemical bases is important, for the presence of bases favours chemical decomposition: if alkalis are lacking only kaolinitic or lateritic decomposition products are formed (Grim 1953).

These two characteristics of modern scientific progress, studies of structure and dynamics, converge very clearly in this case. Jackson (1953) has studied in detail the decomposition or resistance sequence of the difference minerals. The intensity of decomposition depends on weathering agents (intensity factors) and on the nature and particle size of the minerals (capacity factors).

Mohr (1954) has established the weathering intensity in tropical soil in Indonesia dividing them into: fresh, juvenile, virile, senile and final or lateritic. Fertility attains an optimum in virile soils. Jackson (1953) has established a mineral series according to their increasing resistance to the appreciation of their degree of decomposition. He indicates that in this sequences these mineralogical decompositions of the soil in no way implies a simple process evolving from primary and secondary silicates, but that in some cases a colloidal mineral may be the starting material for successive colloidal products. At the same time as these decomposition reactions are reversible in character one must examine the mineral content of the clay fraction as a whole and also study as a whole the mineralogy of soils and sediments and not only clay silicate minerals.

JACKSON (1948) has undertaken a crystalochemical study of these associations, searching for the explanation of the relative stability in the structural factors of the crystal and in the chemical transformations which its formation implies; we can thus link dynamism and structure. We can adduce the following example; albite has a greater stability due to the threedimensional arrangement of its bonds, but its base content furthers hydrolysis and even places feldspars in a lower stability level than quartz, which has also threedimensional lattice bonds, but is lacking in bases. Its solubility is a linear function of the specific surface and, therefore, the amount decreases with particle size. The situation of micas in a more stable level than quartz is in opposition to Goldich's series. Thus the stability factor of clay silicates is a consequence of the crystalline structure, but can be modified and even reversed according to the ionic substitutions which take place. JACKSON (1948) applied this physicochemical study to soil forming processes. From this angle it may be a case of interpreting those studies which relate soil types and mineral associations.

MOHR (1954) relates clay minerals to fertility and finds that virile soils have a large amount of clay minerals with three layer lattices and that kaolinisation marks the end of the virile stage and the passage into free oxides in which fertility decreases rapidly. As regards potassium from the viewpoint of fertility the mineralogical composition of soil, even of the heavy fraction, is important, more so if one is dealing with tropical areas. Also from the standpoint of potassium fixation and liberation dynamics the mineralogical composition of clay and the possible transformations of these minerals (the associations of which have been related by several authors with soil types in the mediterranean area) are of decisive importance. An example of this correlation, which includes humid and arid zone soils, is given by MUÑOZ TABOADELA (1952–53). Therefore, the degree of decomposition of a soil is closely related to fertility. According to LAATSCH (1954) the natural fertility of a soil depends, on the one hand, on its content of easily erodable minerals and the rate of decomposition and, on the other, on the degree of accumulation of the most important nutrient substances of plants by biological processes. But only a proportion of the soil falls within the scope of soil mineralogy, for soil also contains organic matter.

The Various Actions of Humus

The study uf humus reveals the standpoint of present scientific progress and the enormous volume and diversity of the routes employed in humus research during the different historical periods of the progress of chemistry can be inferred from general treatises—from WAKSMAN's (1936) to KONONOWA's (1958).

The analytical conception, the strictly chemical consideration of plant nutrition minimizes the value of humus when measured by the mineralizable elements which it can supply to the soil. But when we were able to focus our study on humus from the viewpoint of great diversity represented by a group of different sciences of our days, its importance has grown larger and larger and today humus is not considered as necessarily linked with a backward agriculture, as it may have seemed in the day when agricultural chemistry prevailed exclusively, but as a very up to date subject of research, because it influences structure and, in a very varied way, soil dynamics. Therefore, the study of humus decomposition is not the exclusive object of modern investigation, for it has enlarged its field of activity to consider the components of humus from the standpoint of their physical action, which leads to a good soil structure.

In this sense, Flaig's studies (1955) with the electron-microscope, as well as determining the physico-chemical properties of humus have shown that humic acids have a spherical form and a size which lies between 60 and 80 Å, forming bunched aggregates. We consider them as the colloidal organic fraction and many physical properties of the soil can be attributed to them. But humus is also examined from the standpoint of its reciprocal actions on soil minerals. The micromorphological studies of Kubiëna (1953) have revealed the different stages of the process which is initiated in plant residues, in the litter and undergoes a fermentation and a wide modification, as a result of the living organisms that it houses. This modification includes the activity of bacteria, fungi and lower animals (Franz 1951). It can thus lead to the integration of the organic and mineral portions and to the formation of the clay-humus complex which prevails in the mull (Tjulin 1948).

Duchaufour (1956) has undertaken a biochemical and microbiological study of humus which can be applied to the formation and evolution of forestry humus types and to the possible transformation of moor into mull.

Humus does not only act morphologically upon mineral material, but it also contributes to the mobilization of the nutrient elements of minerals, increases the CO_2 concentration and furthers the solubility of phosphates, and its organic acids form complexes with metallic cations which can prevent its immobilization such as in the case of iron (Lavollay 1955). These different activities of humus can be carried out directly by its components and also through the development of those microorganisms living on humus. A large part of the N-cycle takes place in the soil in intimate relation with organic matter. Humification is a nitrogen enrichment process: its mineralization makes the organic nitrogen reserve available to the plant. Of course, the importance of this element is not limited to humus: nitrogen takes upon itself the most varied problems of plant nutrition: its fixation from air, its biochemical cycle in the soil, the diversity of organic and inorganic fertilizers, the different degree of influence on the different plant organs and the composition of these (Russell 1950).

But humus also promotes the absorption of mineral elements, as Chaminade and Blanchet (1951) have shown, exerting its activity on the root adsorption of mineral ions and, besides, on the permeability of the cytoplasm and the absorption, in the strict sense of the word, of mineral elements (Blanchet 1958). It also exerts its influence on plant nutrition and growth by means of growth substances. A large research effort has been undertaken starting with Bottomley (1914) and Rocasolano (1915–16), and stretching to Flaig (1955–56), Hofman (1956), Chaminade (1956) and Kononowa (1959).

All these aspects cannot be considered from a purely chemical standpoint of the mineral nutrition of plants, but the development of all these ample perspectives includes, as an important item, the chemical study of humus on which both chemical (Laatsch 1952, Flaig 1955, Bremmer 1955) and biological research converge. Thus the problem of humus constitution continues to be a subject of modern research. Despite the variations inherent in different soils, peats and coals, humic substances (including humic and fulvic acids) are characterized by the presence of an aromatic ring, which has the form of a two-dimensional atomic carbon lattice linked to side chains carrying functional groups (among which carboxyl is to be found) and by that of side chain cyclic nitrogen (Kononowa 1959).

From the ideas of Waksman (1936) who considers humus as a lignoproteinic complex, in which lignine, on the one hand, and the microbial synthesis of protein on the other appear as the dominant factors in humus formation, we have evolved to a wider conception of the initial products of humus, considering both carbohydrates and, specifically, cellulose as the primary material. The twin routes of chemical synthesis and microbial metabolism join in the production of these natural products, the permanence of which is an unequivocal evidence of their stability. In the frame of the diverse influences exerted by humus we must not forget its direct nutrient value for plants; its content in such important elements as nitrogen and phosphorus, which are concentrated both in lower and higher plants.

The rehabilitation of humus is a consequence of the study of the complex natural facts by means of the convergence of different sciences. One realises the present value which is attributed to humus when modern research expresses its viewpoints as in the case of SCHARRER (1953): modern plant nutrition conscientiously avoids the unilateral theory of humus sustained by THAER, as also that which postulates an exclusively mineral nutrition. The presence of humic substances in commercial organic fertilizer is of paramount importance in the conservation and development of soil fertility. In the same way one has to supply the soil the nutrient elements of which it is lacking through the addition of mineral fertilizer. Therefore, a sort of synthesis between THAER's humus theory and LIEBIG's mineral theory takes place.

Mineralogy and soil biochemistry reveal the soil content of those elements required by plants as well as the position of each element in the different mineral and organic combinations. But there is no clear relation between the structural position of an element and its assimilability.

The Assimilability of the Elements

Potassium is an outstanding case of the structural knowledge of minerals in which it participates, of its different reactions and at the same time, of the difficulty of establishing the physico-chemical delimitation of assimilable potassium; in other words, the correspondence between the different situations of potassium and its degree of assimilability by the plant. We know the most important potassium minerals existing in the soil which different authors (ANDREATTA 1955, REITEMEIER 1951, REITEMEIER, BROWN and HOLMES 1951) have considered. The work of CORRENS and ENGELHARDT (1939—40) on the decomposition of silicates and the liberation of potassium in relation to silica and aluminic oxide is already a classic in this field. We are acquainted with the adsorption of potassium and the mechanism of its fixation to mica, illite and vermiculite in clay minerals. We can thus establish the equilibrium reactions extending from the dissolved potassium ion to the adsorbed one and to that fixed either interlaminarly or in three-dimensional lattices. In view of the existence of assimilable potassium from these different sources it is difficult to closely relate natural fertility and mineralogical composition. Potassium does not have to attain the soluble ion stage, not even that of the exchange ion for the plant to be able to assimilate it, for the former can take it up from potassium minerals, such as ortose or muscovite. Furthermore, fixation is a relative problem, an adsorption against removal by pure water but not against the aqueous solution of an electrolyte or root absorption. A fixation retaining the elements in an ion exchange process carried out with an electrolyte is possible, but not with an exchange resin. The equilibria which have been established between fixed and exchanged potassium, between the last named and dissolved potassium, can displace themselves with different intensities according to the reagent employed when we assign a relative value both to fixation and exchange concepts. That is why this equilibrium is considered of lesser importance from the standpoint of potassium supply to plants. Chaves (1958) and Gonzalez (1958) have studied potassium fixation in soil types which are frequent found in the mediterranean regions and have attempted to interpret this function in terms of the mineralogical composition of the clays in the soils. Schachtschabel (1937) has shown an example of these processes, which, as Laatsch (1954), declares, can be explained by the transformations which take place in clay minerals. Thus, composition, structure, stability or dynamism and nutrient capacity appear to be intimately linked as different aspects of the same problem.

In connexion with the structural changes experiences by potassium minerals and the biological diversity of the assimilation cycles of nitrogen, we have the complex biogeochemistry of phosphorus (Hutchinson 1952).

Mobilization, fixation and assimilation processes and reactions take place between the phosphorus which is found in the lithosphere and the soil and that existing in animals and plants. The fixation or assimilation state which phosphorus presents in soil is influenced by many factors. Inorganic phosphorus is fixed either by the calcium ion or by the ferric and aluminic ions. This explains the different behaviour of acid soils and neutral and basic soils (Kurtz 1953, Olsen 1953).

The solubility of phosphorus, its reaction with clay minerals, oxides and carbonates (Wey 1955, Truog 1946, Barbier 1949); the different adsorption phenomena, their different forms (Hernando 1954); the existence also of organic phosphorus (Black 1953) results in a set of equilibria which can be displaced either in the direction of solubilization and mobility or in that of insolubilization and fixation (Dean 1949, Williams 1950, Barbier 1948). The different classes of phosphorus are distributed in diverse ways in the different mechanical fractions of soil (Hoyos 1958). Organic combinations by microorganisms and higher plants can reach a high proportion up to one half of the soil phosphorus.

Truog (1953) has shown the intimate relationship which links the assimilation degree of phosphorus and lime treatment.

As Schuffelen (1958) asserts, the science of soil fertility is mostly concerned with the mineral nutrition of plants and, due to this, it deals, on the one hand, with the supply capacity of nutrient elements which the soil offers and, on the other, with the different facility with which the plant takes up these elements. That is why we have to consider the provision of nutrient elements and their uptake by plants. Only the first named depends on the soil. That is why it has a considerable importance and why the study of the concentration and ionic activity of the soil is an outstanding feature of present day research.

Sulfate, nitrate and chloride ions are generally considered not to be fixed. The fixation of phosphorus, potassium and ammonium nitrogen, as well as that of minor-elements, has been the object of profound study.

The fixation and liberation or "activity" mechanisms of the different ions are very diverse in nature. They differ according to the adsorbent material, to the adsorbed, to the conditions of the process; pH, for example, has a multiple activity. It exerts a strong influence on the nature of the clay minerals formed, in deciding their adaptation of a kaolinitic or montmorillonitic constitution, so different from each other, as regards their adsorbent properties. It is also influential in deciding the existence of $Al^{\cdots}$ and $Fe^{\cdots}$ or $Ca^{\cdot\cdot}$ cations in soil solution, on which the fixation of phosphoric anions depends.

Wiklander (1958) has systematically arranged the different ways in which the elements appear in soil materials. They can be found in solution, adsorbed, as exchange ions, linked in a non-exchangeable form and in soluble or insoluble form in organic matter. This outline suffices to show the diverse complexity in which they are found, as their position in organic matter can be very different and the other three situations merely indicate that the adsorbed material appears between the dissolved and solid state materials. Though this adsorbed material is present in the solid state, it can very easily pass into solution. It must be indicated that it was in the soil where this natural state of the adsorbed elements could be initially studied and during more than a century it was soil scientists who dedicated themselves with great intensity to this research.

A liquid phase, a solution, exists in the soil, for water is an essential soil component. But, by ionic exchange reactions, this solution is in equilibrium with the colloidal material—clay minerals, alumini-ferric hydroxides, humic materials—of the soil. At the same time this finely dispersed material is being formed by the transformation of the original material, as we have just seen before. On the other hand, vegetation can assimilate these elements, taking them up in very different dispersion degrees, as it is able to assimilate directly elements of the original silicates or those obtained by chemical transformations produced by base exchange or from solution. The state of the soil solution has a considerable interest, but the techniques which guarantee the extraction of this solution are confronted sometimes with unsurmountable difficulties. Thus, if it is true that the study of plant nutrition with solution cultures is rapidly expanding, and that we can avail ourselves of a proficient technique for the preparation of these solutions which are employed, this does not apply to the possibility of extracting those solutions which are present in the soil.

Wiklander (1958) considers the anions and cations which appear in soil solutions and studies the factors which influence their different concentrations, as well as their influence of this concentration on the solubility of some nutrient elements, such as the so-called saline effect. Absorbed and exchangeable cations or anions can not be considered as foreign to a solution into which they are able to introduce themselves so easily and in so many ways. As Schuffelen (1958) say the supply and capacity of nutrient materials which a given system, the soil, presents, is characterized by the total amount of nutrient materials present and by the energy level to which they are supplied. Therefore, a complete characterization would require for each nutrient element the functional relationship of the amount to the energy level. The characterization of a definite soil would require a parameter, such as the thermodynamic potential or some thing like that. In some simple systems the concentration can be approximately equivalent to the said potential, but in all other cases factors, such as the adsorption forces, have to be considered.

Schofield (1955) has attempted to apply to available phosphorus the pF conception, which he had already developed to the end of measure the amount of available water, assuming that essentially the same considerations are of application to phosphate in soil. As a working hypothesis it seems reasonable to assume that phosphorus availability in soil is determined by an appropriate chemical potential and its rate of decrease with the absorption of phosphorus. The numerical values depend on the species and age of the plants and the conditions of the culture, such as soil temperature and pH. But despite these complexities, this hypothesis should be explored with the hope of transforming present chaos into something resembling order. Along the same lines of the phosphate nutrition of plants Barbier (1956) and Blanchet (1958) have studied the approximate determination of phosphate ions which can exchange or diffuse in the soil by isotopic dilution, as well as the bonding energy and rate of the said isotopic dilution.

Adsorption and Ionic Exchange

The problem of ion adsorption and exchange pervades the study of agricultural chemistry, for the most characterized soil components are those which form their inorganic and organic colloidal fraction, and these act and react with each and every ion in soil.

The quantitative study of adsorption and exchange equilibrium has been the object of widespread investigations in which types of equations based on absorption isotherms or on thermodynamic laws, through the application of the mass law, or the kinetic theory, have been employed. Eriksson (1952) compares several equations which assume that the adsorbent and solution behave as a Donnan system. Vanselow's (1932) equation introduces the activities of the adsorbed cations and of those appearing in solution which take part in the reaction.

The determination of the activity of the cations in solution is no problem; it can be carried out starting from the respective activity coefficients, though duly taking into account that we are not dealing with a mixture of electrolytes. The difficulty lies in finding an expression for the activity of the adsorbed cations. For the present, there is no direct experimental method for the determination of this magnitude and the procedure employed started by assuming the state of the cations present in the adsorbent surface, and then proceeded to compute theoretically the desired magnitude. The confirmation of the validity of this assumption results by proving that we obtain constant values when we introduce the values thus obtained in the expression of the equilibrium constant.

We are indebted to Mehlich (1952) for a comprehensive study of ionic activity, on which Marshall's (1949, 1950) work has such a direct bearing.

Vanselow (1932) believes that the activity of the adsorbed cations can be expressed in terms of the respective molar fractions, due to the fact that the cations are statistically distributed forming a single solid phase. In this case constant potassium values are obtained in dilute solution, for ions of the same charge and in the case of those equilibria in which the hydrogen ion does not participate.

On the basis of kinetics Gutierrez Rios (1957) deduces that the equilibrium constant is a function of the hydrated radii of the cations involved in the exchange process. According to this the quotient which represents the equilibrium constant would, as a fact, have a constant value when the hydration of the cation is not modified during the exchange process. The constancy which in the value of the said quotient is observed for equilibria between two, alkaline or alkaline earth cations, at ordinary temperature and in dilute solutions, must be attributed to the fact that, under these conditions, the hydration of the cations is not subjected to any modification.

It seems logical to assume that the properties of the adsorbing phase vary when those of the adsorbed cations do, which depend on the solution and, therefore, that they vary when the ionic strength of the solution varies. This is due to the fact that both phases, adsorbent and solution, are in equilibrium. This observation has been the principal objection of a theoretical character which has been raised against Vanselow's formulation (1932). In agreement with this the activities of the adsorbed cations will be correctly expressed by multiplying the respective molar fractions by appropriate activity coefficients.

Strasman, Quidet and Blanchet (1958) studied the form of potassium in the soil in search of a fertility criterion and drew up reaction numbers of some cultures in which potassium fertilizer is employed by establishing the variation of the potassium extraction in terms of the concentration of the potassium solution in the soil. But it seems that more than to the concentration of the potassium ions, nutrition is linked to the ratio of these to all the cations of the colloidal complex.

Soil is a heterogenous medium in which the presence of particles may strongly hinder the assimilation of nutrients by the plant roots. The researchers of the University of Ohio (McLean and Bartlett 1959) have attempted to assess this influence by obtaining differences in the growth of the plant and in the absorption of potassium between suspension and solution, although activity has the same level in both. This activity is measured with electrodes made of fine clay films designed to measure directly the activity of the potassium and calcium ions. But we know of other experiments of root absorption which reveal no differences at all between solutions and suspensions.

The proportion of the elements which will pass into the root system can not be determined with certainty by the study of soil and the physico-chemical determinations which are carried out in it. We can not do without plant experiments.

The study of the yield which vegetation can obtain from an existing element in the soil has been approached through the ample route of agricultural chemistry by using physiological and agricultural experimentation. These are realistic criteria, which have certainly been complemented and explained in the main through physico-chemical studies and the development of mineralogy and soil biology.

But the progress of agricultural technology, the increase in crops through the rational application of fertilizers, the state of the soil from the viewpoint of its nutritional strength in each element have advanced considerably due, especially, to the development registered by analytical chemistry. There are also biological methods applied to the determination of the fraction of each element which is considered assimilable. And this in close coordination with field experimentation techniques, under conditions which, owing to their variable character, require a precise statistical approach.

Nutrient Elements

Of the essential nutrient elements in the growth of the plant (C, H, O, N, S, P, K, Ca, Mg, Mn, B, Fe, Cu, Zn, Mo, Cl) the three named in first place come principally from air and water; the remainder come from the soil, except atmospheric nitrogen. Ash analysis has enabled us to fix the composition of plants in terms of these mineral elements. Leroux (1951) describes Bertrand's work on the composition of the ash of plants. These analyses have been carried out by different authors, i.e. Baumeister (1952). But the existence of a permanent element in the composition of plants is not sufficient to assert its essential character. Up to 1869, in which year Raulin discovered the role of zinc in Aspergillus niger nutrition, besides carbon, hydrogen and oxygen only the following elements were considered to be essential: nitrogen, phosphorus, potassium, calcium, magnesium, sulphur and iron.

The problems of fixation and liberation are of a special interest for minor elements, which are the source of nutritional trouble both by defect and excess.

If in the course of the development of the study of the soil the passage of the total determination of an element into the determination of the amount of its possible forms is observed, the determination of the form or state is perhaps never so important as in the study of minorelements. Generally speaking deficiency levels appear: 1) due to the absence of the relevant microelement in the soil;

2) due to conditions which result in excessive lacking and its absence at a given moment in the soil; 3) due to conditions which permit its fixation.

Several authors have studied the presence of minorelements in relation to the geological genealogy of the soil (HASEMAN and MARSHALL 1945, SWABY and PASSEY 1953). In this light, soil appears as a natural entity which is able to concentrate certain microelements. MITCHELL (1955) has studied its geochemical distribution; SWAINE (1955) has collected the results of the work carried out on this subject.

Even though in the first two cases adequate correction requires the incorporation of the micronutrient in the soil, in the third the cause of immobilization may be due to unfavourable conditions of the medium, which duly corrected, may further immobilization of the minorelement as it has been seen to be the case after an intensive lime treatment. Thus LAATSCH (1950) has found copper, manganese and boron deficiencies in Schleswig-Holstein due to the original scarcity of these elements or to their immobilization by an excessive application of lime. In the humid and acid areas of Australia ANDERSON (1956) has studied the dominating role of molybdenum and the correction of its amounts has produced important agricultural developments.

Arid zone deficiencies are of another type: thus in Spain RODRIGUEZ MUÑOZ (1955) has studied chlorosis phenomena in relation to soil and light. In principle, the double consideration of soil genesis as a starting material and of its degree of acidity can focus our attention on areas of possible deficiencies or of toxic levels in the soil.

In this sense the practice of adding lime to acid soils in an empirical way—which today has given place to better methods—may have been the cause of deficiency effects in the initial stages of that correction.

SCHARRER's (1955) fundamental work explains the diversity of processes in which these elements participate. The problem is of a complex nature, for a different behaviour appears not only in animal and plant media but between higher and lower plants and different plant species.

An element can be considered as universally essential if it meets the following conditions: "(a) a deficiency of it makes it impossible for the plant to complete the vegetative or reproductive stage of its life cycle; (b) such deficiency is specific to the element in question and can be prevented or corrected only by supplying this element, and (c) the element is directly involved in the nutrition of the plant quite apart from its possible effects in correcting some unfavourable microbiological or chemical condition of the external medium" (ARNON 1958). Different plants have different requirements. Thus ARNON shows that beans have greater boron requirements than barley. Alfalfa is able to absorb sufficient zinc in a medium in which maize reveals acute deficiency symptoms. There is no evidence whatsoever that vanadium is indispensable to plants excepting green algae, but this element is widely distributed throughout the plant realm.

The investigation of minorelements in relation to the culture of microorganisms has been specially undertaken by LAVOLLAY (1955), NICHOLAS (1956), GERETSEN (1954). In the domain of higher plants WALLACE's studies (1951) are fundamental in character. Following the spectacular description of these deficiency diseases and the study of the relation of minorelements in the soil and the different plants cultivated, there is a line of work which studies the causality of these effects in the field of enzymology.

In this sense we quote BALDWIN's (1952) words: "Many different metallic ions are now known to function as activators of one enzyme or another and, as a general rule, the relationship between the enzyme and the metallic ions is very specific. As is well known, minute quantities of numerous trace elements including Mn, Mo, Zn and Co for example, are indispensable nutritional factors for living organisms, ranging from bacteria to mammals. It is becoming increasingly clear that they play the part of specific activators for particular enzymes. In general they appear to be parts of the activating system, i.e. the enzyme protein fails to activate the substrate unless the appropriate metallic ion is also present."

Fertilizers

The fertility of a soil and its productive capacity is a function of many factors which interfere with each other. On the other hand the outstanding problems which the study of fertility in different regions of the earth presents, differ considerably among themselves. While in humid regions the determination of acidity, lime supply, the nature and forms of humus, with its influence on the whole of mineral nutrition—macroelements and minor elements—are the subjects which awaken a greater interest, in arid zones of less dynamic soils, another form of investigation is prevalent: as well as water, salinity and the principal fertilizer elements, there is the fixation of a great portion of minorelements. Apart from the activity of each element, we are confronted with the different forms and ways in which each element can be applied, giving birth to the great diversity of fertilizers, both as regards composition and structure. A combination of fertilizer manufacturing techniques and agricultural experimentation has opened new and great possiblities for the preparation of new or improved fertilizers and methods of their application (Jones and Rogers 1949). During recent years these methods of nutrient supply have been considerably extended. The diffusion and concentration processes in the soil are influenced by the position of the fertilizer, by its degree of pulverization or granulation; attempts are made to surmount the fixing action of the soil, in some cases of minorelements by substituting the ionic form by complexes, such as chelates, or by applying the product directly to the plant by injection or spraying, dispensing with any intervention of soil whatsoever.

Even the approach to the fertilizer requirements of the soil, as far as the three most essential and employed elements—nitrogen, phosphorus, potassium—are concerned, has been modified with the appearance of composite fertilizers. Ten years ago the mixed fertilizer consumption of the U.S.A. amounted to 75% of the total, and furthermore this method of fertilization was growing (Jones and Rogers 1949). Ferrari (1959) has studied the systematization of fertilizers, as far as their definition is concerned and he distinguishes the following types: simple, composite, mixed and complex. Furthermore, calcium has been specifically considered and magnesium deficiencies are also taken into account due to the fact that it is regarded as the fifth essential element (Jacob 1955).

We have also to consider the addition of organic matter and lime, which can be considered as first class soil modifiers.

Fertilizers make possible large changes in the supply of plant nutrients in the soil. The production of fertilizers is one of the great industries of modern technology. Russell (1933) describes the history of the discovery of fertilizers starting with Home in 1755, who, in the Edinburgh Society for the Improvement of Arts and Manufacture, attempted to apply chemistry to agriculture. There are outstanding personalities in the history of these two centuries in which the work of plant physiologists was undertaken in close connection with that of chemists. The scientific method based on field experiments was initiated by Boussingault and continued, in Rothamsted, by Lawes and Gilbert. From the beginning of this epoch, in which bones were the most important source of phosphorus fertilization to our days, the chemical industry has made important advances which have found immediate application in the solution of the most difficult problems of population growth. ROEMER (1949) has shown the potential contribution of fertilizers to the increase of food production required by the rate of growth of world population. With the application of modern technology the world is in a position to produce a much larger amount of food, although the problem of its proportionate distribution has yet to be solved (CUMMINGS 1953). Today fertilizers, both organic (from natural sources) and chemical fertilizers, are produced in ever growing amounts. The production of and demand for commercial fertilizers have reached their highest level and still continue to increase. For the year ended June 30, 1950, the Food and Agriculture Organization of the United Nations, 1951 reports that, excluding the Soviet-Union, the world's output of the three primary plant nutrients was 12.8 million metric tons as compared with 8.9 million tons in 1938—1939. Of the production in 1949—1950, nitrogen (N) comprised 3.6 million tons, phosphoric oxide (P_2O_5) 5.4 million tons, and potash (K_2O) 3.8 million tons. Without this production to supplement the nutrients provided from noncommercial sources, much of the arable land of the world could not be farmed economically and world crop yields could not by maintained (CUMMINGS 1953). But this history of two centuries ends with several years which are marked by a parallel growth in the development of both scientific research and industrial production technology. Thus, as PIERRE (1953) says, when speaking of phosphoric fertilization in the period extending from 1935 to 1950 only in the United States 673 important articles were published. Of these 240 concerned themselves with studies on phosphate fertilizer sources, mining technology, fertilizer manufacture and the largest aspects of consumption, while the remainder, 433, deal with phosphorus in soil, the soil-plant relation and phosphate fertilization. Approximately the same could be said of other countries and of the other main fertilizers. During these last few years a particularly intensive investigation on minorelements which as SCHARRER (1958) has said, are dwarfs with the strength of giants, has been taking place.

Their production capacity is extraordinarily large, and the same can be said of their influence on the quality of products. Small amounts are enough to solve growth difficulties, but the difference between sufficiency and excess may not be very great.

Besides their influence on production fertilizers also bear on the quality and chemical composition of the crop, on the floristic composition of meadows (MULDER 1949, STEWART and HOLMES 1953, DEAN 1953) and on the soil fauna (FRANZ 1953). The influence of the different fertilizers on the amount of active ingredients of medicinal plants has been also studied (HUTER 1947, BOSHART 1938, COURT and ALLEMAN 1943).

Growth Factors and Plant Yield

We can express quantitatively the relation of crop size to various factors. There are a series of laws of which we can consider the different essential changes. These laws are intimately linked to the different dominant ideas on the character of plant nutrition.

The study of plant nutrition brings to the fore representative names of periods controlled by the then dominant conception. THAER beliefes that the organic complex humus was the nutrient of plants. Then LIEBIG (1840) formulated his mineral theory of plant nutrition according to which plants assimilated inorganic substances. But, furthermore, certain ions e.g. potassium, iron, promote and regulate assimilation in their role as catalysts. Others must exist in perfect equilibrium, preserving the cellular membrane and protoplasm in those physico-chemical states on which the permeability of solutions and the harmonious development of root absorption depend. The health of plants is linked to a mineral equilibrium in cells (GALLAY 1948).

But LIEBIG has also enunciated the so-called minimum law. The size of the crop is related to that nutrient material which is essential for the complete development of the plant found in minimum amount—in appropriate conditions and form—in the soil. MAYER enunciated it as follows: "The crop depends on that factor which is present in minimum amount and is proportional to the same". In the development of the minimum law we come upon quotations by LIEBIG which are not in agreement as to the proportionality of that relation. As far back as 1859 LIEBIG had introduced the idea of resistance to increase, defining crop as the difference between nutrition and the said resistance, for the experiments carried out in pots did not result in that proportionality. Thus the idea of a depression promoted by an excess of the plant factor is introduced. In 1883 WAGNER showed the existence of a lowering of crop increase in contrast to the idea of proportional crop.

However, the most adequate enunciation from the quantitative standpoint was given in 1909 by MITSCHERLICH, when he published his law of "the action of growth factors". With this law MITSCHERLICH surpassed LIEBIG's study, on the basis that the crop yield—even when a factor increases—is always influenced by all the growth factors. Starting from a solid experimental foundation MITSCHERLICH enunciated that: "the plant yield increases as each factor increases, and in direct proportion to the difference between the maximum attainable yield and that obtained in each case". MITSCHERLICH's contribution to the study of quantitative relations between yield and growth factors has offered us the possibility of determining and estimating in a soil the active amount of each nutrient element. The proportionality factor of MITSCHERLICH called "Wirkungswert" shows up in this formulation. This "Wirkungsfaktor" ("activity value") is constant and independent of the interaction between other growth factors as well as of the maximum yield.

Other studies by MITSCHERLICH as well as those by BOGUSLAWSKI and collaborators and others were aimed at the determination of this activity value through a large number of experiments. In this way the activity values for N, K, P, Ca, Mg, etc. were determined. MITSCHERLICH's law has been extendet to embrace every aspects which influences plant growth. His book, "Bodenkunde" (1923) is an application of this law to the most diverse plant factors. This large scope extends from the interpretation of WOLLNY's classical studies on soil physics to organic matter and its humic constituents, climatic factors and, lately, minor-elements themselves, for which MITSCHERLICH assumes a high activity value if the law is valid. Furthermore, the crop yield of plant portions also obeys this law with slight modifications (BOGUSLAWSKI 1936). The application of this law to the field of plant nutrition is of great value. Therefore if we consider the yield which is to be obtained in each case, the immobilization of any nutrient element, as is the case of P by $Ca^{\cdot\cdot}$, can be indirectly determined.

Though the revisions of this law show that the activity value given by MITSCHERLICH is not constant in every case, this does not imply any contradiction, for the studies we based on experimental data and the said activity value was considered as constant under definite conditions. The yield law must rather be interpreted along the lines of the relation of nutrient elements when their absorption and apparent utilization are investigated. Physical and climatic factors play a very important role when the action of chemical growth factors is investigated, owing to their immediate activity upon them.

The studies of Rippel (1926—27), as well as those of Linser and Kaindl (1951), Kaindl (1953) and Bertalanffy (1942) have confirmed Mitscherlich's yield law. Recent work carried out by Baule (1953—54, 1956) deals with new formulations of Mitscherlich's law. Boguslawski (1958) has undertaken a global study of these laws.

Considering the complicated character of the physiological growth processes, any mathematical formulation can only attain a relative approximation to reality. Nevertheless, the reason why up to the present moment Mitscherlich's simple formulation represents the greatest approximation to reality lies in the affirmation of the asymptotic increment of yield in terms of growth factors.

Mitscherlich's work marks the beginning of an era in the study of plant production. Now it is not the case of a single factor, for every factor takes part in the said production and the crop increment is related to the difference which separates it from the maximum possible crop. Mitscherlich considers that each of the plant factors exerts a constant effect. In the present conception of the problem there are two factors which have an outstanding interest: the first concerns the interaction of elements. The dependence of the crop on a single factor has been superseded, for we have an interaction of the elements which, as in many sciences of our time, compels one to go from the specific and individual study of each factor to that of its different groupings. In the whole realm of science the study of assemblies or groups is evident. Despite its complexity and variability, which was examined in the introduction to this chapter, considering that soil represents the link between inanimate matter and organisms it is of fundamental importants to study and understand its properties and functions: "Gesunder Boden, gesunde Pflanze, gesundes Tier, gesunder Mensch" (Scharrer 1953).

Literature

Anderson, A. J.: Molybdenum as a fertilizer. Advances in Agronomy **8**, 163–202 (1956). — Andreatta, C.: L'alterazione dei minerali delle rocce in rapporto alla messa in circolazione del potassio. Potassium Symposium, Roma, p. 107–119. 1955. — Arnon, D. I.: The role of micronutrients in plant nutrition with special references to photosynthesis and nitrogen assimilation. *In*: Trace Elements, p. 1–32. New York: Academic Press 1955.

Baldwin, E.: Dynamic aspects of biochemistry, 2nd ed., p. 41. Cambridge: University Press. 1952. — Barbier, G.: Introduction à l'étude des phosphatés du sol. VIe Congrès International de la Science du Sol, Paris, B 387–402 (1956). — Barbier, G., J. Chabannes et A. Marquis: Utilisation des engrais phosphatés par les plantes, après leur absorption dans le sol. Plant a. Soil **1**, 11–17 (1948). — Barbier, G., et J. Chabannes: Sur le mécanisme de l'adsorption des phosphates par divers constituants des argiles des sols. Ann. Agronom. **19**, 343–379 (1949). — Barshad, I.: Soil development. *In*: Chemistry of the Soil, Ed. by F. E. Bear, p. 1–52. New York: Reinhold. 1955. — Baule, B.: Über die Weiterentwicklung der Ertragsgesetze von Liebig und Mitscherlich. Z. Acker- u. Pflanzenbau **96**, 173–186 (1953). — Eine physikalische Analogie zum Pflanzenertragsgesetz. Z. Acker- u. Pflanzenbau **98**, 265–278 (1954). — Das Ertragsgesetz als Mittel zur Bestimmung der Bodennährstoffe. Z. Acker- u. Pflanzenbau **100**, 261–272 (1956). — Baumeister, W.: Mineralstoffe und Pflanzenwachstum. Jena: G. Fischer. 1952. — Die Aschenstoffe. *In*: Handbuch der Pflanzenphysiologie, Bd. IV, S. 5–36. Berlin: Springer. 1958. — Bertalanffy, L. v.: Theoretische Biologie, Bd. II, S. 239–253. Berlin: Borntraeger. 1942. — Blanchet, R.: Energie de liaison et vitesse de dilution isotopique des ions phosphoriques dans le sol; application à l'alimentation phosphatée des plantes. II Simposio Internazionale di Agrochimica, Procchio, 8–13 Sett. 1958. — Actions directes et indirectes de la matière organique humufiée sur la nutrition des végétaux vasculaires. Ann. Agronom. **9**, 449–532 (1958). — Black, C. A., and C. A. I. Goring:

Organic phosphorus in soils. *In*: Soil and Fertilizer Phosphorus in Crop Nutrition, Ed. by W. H. PIERRE und A. G. NORMAN, p. 123–144. New York: Academic Press. 1953. — BLANCHET, R., C. CHAUMONT und R. STUDER: Alimentation phosphorique des plantes et phénomènes de contact racines-particles de sol. Ann. Agron. **13**, 21–29 (1962). — BOGUSLAWSKI, E. v.: Untersuchungen über das Bodenkali sowie seine Aufnahme und Verwertung durch Hafer. Habil.-Schr. Dessau 1936. — Das Ertragsgesetz. *In*: Handbuch der Pflanzenphysiologie, Bd. IV, S. 943–971. Berlin: Springer. 1958. — BOSHART, K.: Düngungsversuche mit Heil- und Gewürzpflanzen. Forschungsdienst, Sonderheft 8, 428–431 (1938). — BOTTOMLEY, W. B.: Some accessory factors in plant growth and nutrition. Proc. Roy. Soc. London B **88**, 237–247 (1914). — BREMNER, J. M.: Recent work on soil organic matter of Rothamsted. Z. Pflanzenernähr., Düng., Bodenkde. **69**, 32–38 (1955).

CHAMINADE, R.: Action de l'acide humique sur le dévelopment et la nutrition minérale des végétaux. VIe Congrès International de la Science du Sol, Paris, D, 443–448 (1956). — CHAMINADE, R., et R. BLANCHET: Action de l'humus colloidal sur la nutrition minérale des végétaux. C. R. Acad. Sci. de Paris **233**, 1486–1488 (1951). — CHAVES, M.: Fijación de potasio en suelos de Andalucia Occidental. An. Edafología y Fisiología Vegetal (in Druck). — CLARKE, F. W., and H. S. WASHINGTON: The composition of the earth's crust. United States Geological Survey, Professional Paper n° 127, 1924. — Clays and Clay Minerals. Proceedings of the VI National Conference on Clays and Clay Minerals, 1957. London: Pergamon Press. 1959. — CORRENS, C. W., und W. ENGELHARDT: Neue Untersuchungen über die Verwitterung des Kalifeldspates. Chemie der Erde **12**, 1–22 (1939–40). — COURTH, G. Th., und O. ALLEMANN: Einige Erfahrungen über die Kultur von Digitalis lanata in der Schweiz. Pharmaceut. Acta Helvet. **18**, 369 (1943). — CUMMINGS, R. W.: Fertilizer technology in a changing world. *In*: Fertilizer technology and resources in the United States, Ed. by K. D. JACOB, p. 1–14. New York: Academic Press. 1953.

DEAN, L. A.: Fixation of soil phosphorus. Advances in Agronomy **I**, 391–411 (1949). — DEAN, L. A., and M. FRIED: Soil-plant relationships in the phosphorus nutrition of plants. *In*: Soil and Fertilizer Phosphorus in Crop Nutrition, Ed. by W. H. PIERRE and A. G. NORMAN, p. 43–58. New York: Academic Press. 1953. — DUCHAUFOUR, Ph.: Formation et evolution des humus forestiers. Bull. Ass. Franc. Etude du Sol **75**, 457–463 (1956). — DUCHAUFOUR, PH., et F. MANGENOT: Recherches sur l'évolution expérimentale de certains humus. Ann. Agronom. **2**, 159–181 (1956).

EITEL, W.: The physical chemistry of the silicates. Chicago: The University of Chicago Press. 1954. — ENGELHARDT, W.: Mineralogische Beschreibung eines mecklenburgischen Bodenprofils. Chemie der Erde **11**, 17–37 (1938). — ERIKSSON, E.: Cation exchange equilibria on clay minerals. Soil Sci. **74**, 103–113 (1952).

FERRARI, C.: Aspetti chimici dei concimi complessi. Concimi e Concimazioni **3**, 9–16 (1959). — FIELDES, M., and L. D. SWINDALE: Chemical weathering of silicates in soil formation. New Zealand J. Sci. Technology **36**, 140–154 (1954–55). — FLAIG, W.: Contribución al estudio de los ácidos humínicos. Monografías de Ciencia Moderna, n° 46. Madrid: C.S.I.C. 1955. — FLAIG, W., und E. SAALBACH: Zur Kenntnis der Huminsäure, X, XI, XII und XIII. Z. Pflanzenernähr., Düng., Bodenkde. **71**, 208–215, 215–224 (1955); **72**, 1–7, 8–15 (1956). — FRANZ, H.: Über die Bedeutung terricoler Kleintiere für den Stickstoff- und Humushaushalt des Bodens. Z. Pflanzenernähr., Düng., Bodenkde. **55**, 44–52 (1951). — Der Einfluß verschiedener Düngungsmaßnahmen auf die Bodenfauna. Angew. Pflanzensoziologie **21**, 1–50 (1953).

GALLAY, R., J. RAPIN et J. STALE: Les Engrais. Lausanne: Librairie Payot. 1948. — GERRETSEN, F. C., and H. DE HOOP: Boron, an essential microelement for Azotobacter croococcum. Plant a. Soil **5**, 349–367 (1954). — GOLDSCHMIDT, V. M.: Geochemistry. Oxford: Clarendon Press. 1954. — GONZALEZ GARCIA, G.: Estudio de la naturaleza, constitución y propiedades de la arcillas de los principales tipos de suelos de Andalucía Occidental. Agrochimica (in Druck). — GRIM, R. E.: Clay mineralogy, p. 331. New York: Mac Graw-Hill. 1953. — GUTIERREZ RIOS, E.: Equilibrio de cationes entre la superficie de sólidos y soluciones acuosas. Trabajos de la III Reunión Internacional sobre Reactividad de los sólidos **1**, 21–56 (1957).

HASEMAN, J. F., and C. E. MARSHALL: The use of heavy minerals in studies of the origin and development of soils. Missouri Agricultural Experiment Station, Res. Bull. 387 (1945). — HERNANDO, V., y M. LOSADA: Estudio de las formas de fósforo presentes en una serie de suelos. An. Edafología y Fisiología Vegetal **13**, 1–28 (1954). — HOFMANN, E.: Der Enzymspiegel im Boden. Landwirtsch. Forsch. 7. Sonderheft, **8**, 80–82 (1956). — HOYOS, A., y R. GARCIA: Distribución de las distintas clases de fósforo en las fracciones de suelos de varios cortijos. II Simposio Internazionale di Agrochimica, Procchio, 8–13 Sett. 1958. — HUTER, R.: Die Stickstoffernährung der

Tabakpflanze. Diss. ETH. Zürich. 1947. — HUTCHINSON, G. E.: The biogeochemistry of phosphorus. *In*: The biology of phosphorus, p. 1–35. Michigan: State College Press. 1952.

JACKSON, M. L., *et al.*: Weathering sequence of clay-size minerals in soils and sediments, I. J. Phys. Colloid Chemistry **52**, 1237–1260 (1948). — JACKSON, M. L., and G. D. SHERMAN: Chemical weathering of minerals in soils. Advances in Agronomy **5**, 219–318 (1953). — JACOB, A.: Magnesia, der fünfte Pflanzenhauptnährstoff. Stuttgart: F. Enke. 1955. — JAKOB, J.: Der chemische Aufbau unseres Planeten, S. 160. Zürich: Büchergilde Gutenberg. 1945. — JASMUND, K.: Die silicatischen Tonminerale. Berlin: Verlag Chemie. 1951 .— JEFFRIES, C. D.: The minerals of the soil. *In*: Agricultural Chemistry, By D. E. H. FREAR, Vol. I, p. 745–762. New York: Van Nostrand. 1950. — JENNY, H.: Factors of soil formation. New York: MacGraw-Hill. 1941. — Mineral nutrition of plants. Madison: University of Wisconsin Press. 1951. — Mineral nutrition of plants. Contact Phenomena between Adsorbents and their Significance in Plant Nutrition, p. 107–132. Wisconsin: The University of Wisconsin. 1953. — JENNY, H., and R. OVERSTREET: Cation interchange between plant roots and soil colloids. Soil Sci. **47**, 257–272 (1939). — JONES, R. J., and H. T. ROGERS: New fertilizers and fertilizer practices. Advances in Agronomy **1**, 39–76 (1949).

KAINDL, K.: Versuch einer biophysikalischen Deutung des Pflanzenwachstums. Biochim. Biophys. Acta **10**, 241–255 (1953). — KELLER, W. D.: The principles of chemical weathering. Columbia: Lucas Brothers. 1955. — KONONOWA, M. M.: Die Humusstoffe des Bodens, S. 142. Berlin: Deutscher Verlag der Wissenschaften. 1958. — KONONOWA, M. M., and F. V. ALEKSANDROVA: The biochemistry of humus and some problems of plant nutrition. Soils a. Fertilizers **22**, 77–83 (1959). — KUBIËNA, W.: Suelo y formación del suelo desde el punto de vista biológico. Madrid: Instituto de Edafología y Fisiología Vegetal. 1944. — The soils of Europe. London: Th. Murby. 1953. — KUEHNELT, W.: Bodenbiologie. Wien: Herold. 1950. — KURTZ, L. T.: Inorganic phosphorus in acid and neutral soils. *In*: Soil and Fertilizer Phosphorus in Crop Nutrition, Ed. by W. H. PIERRE and A. G. NORMAN, p. 59–80. New York: Academic Press. 1953.

LAATSCH, W.: Die Kupfer-, Mangan- und Borernährung unserer Kulturpflanzen. Schriftenreihe der Landwirtschaftlichen Fakultät der Universität Kiel **3**, 31–44 (1950). — Modernos conceptos sobre el humus. Monografías de Ciencia Moderna, nº 35. Madrid: C.S.I.C. 1952. — Dynamik der mitteleuropäischen Mineralböden, 3. Aufl., S. 45 u. 92. Dresden: Th. Steinkopff. 1954. — LAGERWERFF, J. V.: Comparable effects of adsorbed and dissolved cations on plant growth. Soil Sci. **86** 63–74 (1958). — The contact-exchange theory amended. Plant Soil **13** 253–264 (1960).— LAVOLLAY, J.: Trois aspects de la nutrition minérale. Bull. Soc. Franç. Physiol. Végétale **3**, 67–74 (1955). — LEROUX, D.: Engrais amendements produits pour la protection des cultures, p. 1. Paris: Gauthier-Villars. 1951. — LIEBIG, J.: Die organische Chemie in der Anwendung auf Agrikultur und Physiologie. Braunschweig: F. Vieweg. 1840. — LINSER, H.: Die pflanzenphysiologische Wirkung der Makronährstoffe (N, P, K, Ca, Mg) im Lichte der neuesten wissenschaftlichen Erkenntnisse. 100 Jahre erfolgreiche Düngewirtschaft. Heidelberg, September 1957. — LINSER, H., und K. KAINDL: Versuch einer trefferstatistischen Deutung des Mitscherlich'schen Ertragsgesetzes. Z. Pflanzenernähr. Düng., Bodenkde. **53** (98), 47–63 (1951). — Versuch einer biophysikalischen Deutung des Mitscherlich'schen Ertragsgesetzes. Biologia generalis **19**, 349–367 (1951). — LUNDEGÅRDH, H.: Physiological and biochemical aspects on microelements. Semaine d'Etude sur le Problème des oligoéléments dans la vie végétale et animale, Roma, p. 172–192. Pontificia Academia Scientiarum. 1956.

MAC LEAN, E. O., and R. J. BARTLETT: Soil particles affect nutrient uptake. Crops a. Soils **11**, 18 (1959). — MARSHALL, C. E.: A petrographic method for the study of soil formation processes. Proc. Soil Sci. Soc. Amer. **5**, 100–103 (1940). — The colloid chemistry of the silicate minerals. New York: Academic Press. 1949. — The electrochemistry of the clay minerals in relation to pedology. Transactions of the Fourth International Congress of Soil Science, Amsterdam, **1**, 71–82 (1950). — MARSHALL, C. E., and S. A. BARBER: The calcium-potassium relationships of clay minerals as revealed by activity measurements. Proc. Soil Sci. Soc. Amer. **14**, 86–89. 1949. — MATTSON, S.: The laws of soil colloidal behavior. I. Soil Sci. **28**, 179–220 (1929) and follow. — MEHLICH, A., and N. T. COLEMAN: Type of soil colloid and the mineral nutrition of plants. Advances in Agronomy **4**, 67–99 (1952). — MITCHELL, R. L.: Trace elements. *In*: Chemistry of the Soil, Ed. by F. E. BEAR, p. 253–285. New York: Reinhold. 1955. — MITSCHERLICH, E. A.: Bodenkunde für Land- und

Forstwirte, 4. Aufl. Berlin. P. Parey. 1923. — MITSCHERLICH, E. A., E. v. BOGUSLAWSKI und A. GUTMANN: Schr. Königsberger Gelehrte Ges. Naturwissenschaftl. Kl. **12**, H. 2 (1935). — MITSCHERLICH, E. A., und N. ATANASIU: Der Wirkungswert des Stickstoffs (C = 0, 20°). Z. Acker- u. Pflanzenbau **94**, 253–280 (1952). — MOHR, E. C. J., and F. A. v. BAREN: Tropical soils. Amsterdam: W. van Hoeve. 1954. — MULDER, E. G.: Mineral nutrition in relation to the Biochemistry and Physiology of potatoes. I. Effect of nitrogen phosphate, potassium, magnesium and copper nutrition on the tyrosine content and tyrosine activity with particular reference to blackening of the tubers. Plant a. Soil **2**, 59–121 (1949). — MUÑOZ TABOADELA, M.: Composición mineralógica de las arcillas de los podsoles asturianos y su variación a través del perfil. An. Edafología y Fisiología Vegetal **11**, 35–58 (1952) and J. Soil Sci. **4**, 48–55 (1953).

NICHOLAS, D. J. D.: Trace metal requirements and some enzyme systems in a riboflavin-requiring mutant of Neurospora crassa. J. Gen. Microbiol. **15**, 470–477 (1956).

OLSEN, S. R.: Inorganic phosphorus in alkaline and calcareous soils. *In*: Soil and Fertilizer Phosphorus in Crop Nutrition, Ed. by W. H. PIERRE and A. G. NORMAN, p. 89–118. New York: Academic Press. 1953.

PEREZ MATEOS, J.: Aplicación del "índice de los minerales básicos" para deducir la fertilidad de los suelos, en los trópicos. An. Edafología y Fisiología Vegetal **16**, 29–41 (1957). — PETTIJOHN, F. J.: Resistance of heavy minerals and geologic age. J. Geol. **49**, 610–625 (1941). — PIERRE, W. H.: Introduction. *In*: Soil and Fertilizer Phosphorus in Crop Nutrition, Ed. by W. H. PIERRE and A. G. NORMAN, p. X. New York: Academic Press. 1953.

RAMBERG, H.: Chemical bonds and distribution of cations in silicates. J. Geol. **60**, 331–355 (1952). — RANKAMA, K., and TH. G. SAHAMA: Geochemistry, p. 189 and 344. Chicago: University of Chicago Press. 1950. — REITEMEIER, R. F.: The chemistry of soil potassium. Advances in Agronomy **3**, 113–164 (1951). — REITEMEIER, R. F., I. C. BROWN and R. S. HOLMES: Release of native and fixed nonexchangeable potassium of soils containing hydrous mica. United States Department of Agriculture, Techn. Bull. nº 1049. 1951. — RIPPEL, A.: Zur experimentellen Widerlegung der Mitscherlich-Bauleschen Wirkungsgesetze der Wachstumsfaktoren. (Weiterer Beitrag zur Kenntnis physiologischer Gleichgewichtszustände und der Regel der konstanten Verschiebung der Pflanzen.) Z. Pflanzenernähr., Düng., Bodenkde. **8 A**, 65–80 (1926–27). — ROCASOLANO, G. A.: El manganeso como catalizador de las reacciones bioquímicas, por las cuales el nitrógeno atmosférico, por vía bacteriana, es asimilado por las plantas. Rev. Real Acad. Ci. Madrid **14**, 681–693 (1916). — ROCASOLANO, G. A., y S. FERNÁNDEZ: Investigaciones sobre la alimentación nitrogenada de las plantas por vía bacteriana. Bol. Real Soc. Española de Historia Natural **15**, 112–117; 141–199; 244–252 (1915). — RODRIGUEZ MUÑOZ, C., et C. DIEZ ALTARÉS: Chlorose dans les cultures des zones semi-arides: influence du sol et de la lumière. Plant Ecology. Proceedings of the Montpellier Symposium, Montpellier, p. 89–91, 1955. — ROEMER, T. H.: Wird die Lehre von Robert Malthus 1789–1805 in der zweiten Hälfte des 20. Jahrhunderts doch noch Wirklichkeit? Stiftung F.V.S. zu Hamburg. Justus-von-Liebig-Preis 1949. — RUSSELL, E. J.: Artificial fertilizers in modern agriculture. London: His Majesty's Stationery Office. 1933. — Soil conditions and plant growth, 8th Ed., p. 30. London: Longmans, Green. 1950.

SCHACHTSCHABEL, P.: Aufnahme von nicht-austauschbarem Kali durch die Pflanzen. Bodenkunde u. Pflanzenernähr. **3**, 107–133 (1937). — Mikroskopische und röntgenographische Untersuchungen von Böden. Bodenkunde u. Pflanzenernähr. **5**, 375–389 (1937). — Untersuchungen über die Sorption der Tonmineralien und organischen Bodenkolloide und die Bestimmung des Anteils dieser Kolloide an der Sorption im Boden. Kolloid-Beihefte **51**, 199–276 (1940). — SCHARRER, K.: Die Bedeutung der agrikulturchemischen Forschungen Justus Liebigs für die Landwirtschaft. Ber. Landwirtsch. **31**, 1–128 (1953). — Gesunder Boden, gesunde Pflanze, gesundes Tier, gesunder Mensch. Landwirtsch. Forsch. **5**, 3–21 (1953). — Biochemie der Spurenelemente, 3. Aufl. Berlin: P. Parey. 1955. — III. Congrès Mondial des fertilisants. Heidelberg 9–12 septembre 1957. Bull. Centre International des Engrais Chimiques **5**, 3–5 (1958). — SCHEFFER, F.: Agrikulturchemie. Teil C: Humus und Humusdüngung. Stuttgart: F. Enke. 1941. — SCHOFIELD, R. K.: Can a precise meaning be given to "available" soil phosphorus ?. Soils a. Fertilizers **18**, 373–375 (1955). — SCHUFFELEN, A. C.: Ionic equilibria in soils. Transactions of the II and IV Commission of the International Society of Soil Science, Hamburg, **2**, 132–146 (1958). — SMITHSON, F.: Grass opal in British soils. J. Soil Sci. **9**, 148–154 (1959). — STEWART, A. B., and W. HOLMES: Manuring of grassland. I. Some effects of heavy dressings of nitrogen on the mineral composition of grassland herbage. J. Sci. Food a. Agriculture **9**, 401–408 (1953). — Strakhov, N. M.: Méthodes d'études des roches sédimentaires, Vol. II, p. 12. Paris: Service d'Information Géologique. 1958. — Strasman, A., P. Quidet et R. Blanchet: Dynamique du potassium dans le sol et alimentation potassique des plantes. Ann. Agronom. **9**, 637–660 (1958). — Swaby, R. J., and B. I. Passey: Availability of trace elements from rocks and minerals. Austral. J. Agric. Res. **4**, 292–304 (1953). — Swaine, D. J.: The trace element content of soils. Bucks: Commonwealth Agricultural Bureaux. 1955.

Tamm, O.: Om de lågproduktiva sandmarkerna a Hökensäs och i övre Lagadalen. Meddelanden från Statens skogsförsöksanstalt **30**, 1 (1934). — Tjulin, A. F.: Kolloidchemische Erforschung der Böden mit agronomischen Zielen. Arbeiten des Unionsinstituts für Dünger- und Bodenkunde, Heft 27 (1948). — Truog, E.: Soil reaction influence on availability of plant nutrients. Proc. Soil Sci. Soc. Amer. **11**, 305–308. 1946. — Liming in relation to availability of native and applied phosphates. *In*: Soil and Fertilizer Phosphorus in Crop Nutrition, Ed. by W. H. Pierre and A. G. Norman, p. 281–296. New York: Academic Press. 1953. — Twenhofel, W. H.: Treatise on sedimentation, p. 189. Baltimore: Williams and Wilkins. 1932.

Vageler, P.: Grundriß der tropischen und subtropischen Bodenkunde. Berlin: Verlagsges. für Ackerbau. 1930. — Vasenlow, A. P.: Equilibria of the base-exchange reactions of bentonites, permutites, soil colloids and zeolites. Soil Sci. **33**, 95–113 (1932).

Waksman, S. A.: Humus. Baltimore: Williams and Wilkins. 1936. — Wallace, T.: The diagnosis of mineral deficiencies in plants by visual symptoms. London: His Majesty's Stationery Office. 1951. — Wey, R.: Etude de la rétention des anions phosphoriques par les argiles: montmorillonite et kaolinite. Thèse. Université de Strasbourg. 1955. — Wiegner, G.: Boden und Bodenbildung. Dresden: Th. Steinkopff. 1926. — Wiklander, L.: The soil. *In*: Handbuch der Pflanzenphysiologie, Bd. IV, S. 118–169. Berlin: Springer. 1958. — Williams, E. G.: Phosphate fixation and availability. J. Sci. Food a. Agriculture **1**, 244–248 (1950).

I. Der Boden als Standort der Pflanze

A. Bodentypen und Bodenarten

a) Verwitterung

Von

E. Mückenhausen

1. Begriffsbestimmung

Der Begriff „Verwitterung" war schon in der Geologie vorhanden, als es noch keine Bodenkunde als selbständige Naturwissenschaft gab. Ursprünglich wurde der Begriff „Verwitterung" von „Witterung" abgeleitet. In diesem Sinne sollte „Verwitterung" den Einfluß der Witterung, d. h. des Wetters mit allen seinen physikalischen und chemischen Wirkungen, auf die Umbildung der Gesteine ausdrücken. Heute wird der Begriff weiter gefaßt; er schließt auch Prozesse der Gesteinszersetzung ein, die mit der Witterung unmittelbar nichts zu tun haben, so u. a. die zerstörende Wirkung der Pflanzen. Andererseits wurden früher vielfach die exogenen geologischen Kräfte, welche die Gesteine an der Erdoberfläche zerstören, auch als Kräfte der Verwitterung aufgefaßt. Wenn das heute auch bisweilen noch geschieht, so wollen wir hier eine klare Trennung zwischen diesen geologischen Kräften und den Prozessen der Verwitterung im engeren Sinne machen.

Dementsprechend wird die Verwitterung folgendermaßen definiert: Unter Verwitterung versteht man die Gesteinszerstörung an der Erdoberfläche, hervorgerufen durch die direkte und indirekte Einwirkung der Atmosphäre, der Pflanzen und der Tiere. Ausgenommen sind jedoch die exogenen geologischen Vorgänge der mechanischen Gesteinszerstörung, nämlich die Erosion, die Exaration und die Korrasion.

2. Die Gesteinszerstörung durch exogene geologische Kräfte

Wenn auch die exogenen geologischen Kräfte der mechanischen Gesteinszerstörung heute meistens nicht mehr zu dem Begriff „Verwitterung" gehören, so sollen dennoch diese Kräfte hier kurz behandelt werden, um ein geschlossenes Bild von der Gesteinszerstörung an der Erdoberfläche zu geben. Für ein eingehenderes Studium dieser Vorgänge werden folgende Werke empfohlen: Wagner 1950, Machatschek 1955, Brinkmann 1956, Derruau 1958, Louis 1960, Scheidegger 1961.

A. Die Erosion

Unter Erosion (von lateinisch erodere = ausnagen) versteht man die einnagende Wirkung des fließenden Wassers (Bäche, Flüsse). Dabei ist es gleich, ob das Einschneiden in lockeres oder festes Material erfolgt. Der flächenhafte Abtrag wird in der Geologie Denudation (von lateinisch denudare = entblößen) genannt. Im Gegensatz zum rinnenartigen Einschneiden des Wassers wird durch die Denudation auf breiter Fläche loses Material abgespült. Der Begriff „Bodenerosion“ (durch Wasser) umfaßt den flächenhaften und rinnenartigen Abtrag des *Bodens*. Dagegen ist der geologische Begriff „Erosion“ umfassender; er bezieht sich auf den Abtrag aller Lockermassen (Boden und Lockergesteine) und das Einschneiden des fließenden Wassers in das feste Gestein. Letzteres zeigen am deutlichsten die Gebirgsbäche und -flüsse.

Das fließende Wasser vermag loses Material aufzunehmen und fortzutragen, das um so mehr, je größer die Wassermenge und je stärker die Strömung ist. Um jedoch in festes Gestein einzuschneiden, muß das fließende Wasser ein Schleifmittel führen; dieses sind die vom Wasser mitgeführten Gesteinsstücke verschiedener Größe (einschließlich Sand). Der reißende Gebirgsbach transportiert Blöcke bis zu etwa 1 m Durchmesser. Die vom Wasser geschleppten Steine schleifen die Bach- bzw. Flußsohle ein und reiben sich dabei außerdem gegenseitig ab. Weiche Gesteine (nicht verkieselter Sandstein, Kalkstein) erfahren bei 10 bis 50 km Transportweg einen Abrieb auf die Hälfte ihres Durchmessers, harte Gesteine (Silikatgesteine, Quarzite) benötigen dafür 100 bis 200 km. Bei diesem Vorgang wird das Gestein mechanisch bis zum feinsten Korn zerkleinert, ohne aber dabei chemisch verändert zu werden.

Die Erosion ist um so stärker, je höher der Niederschlag, je steiler das Gelände und je weniger die Bodenoberfläche durch Vegetation geschützt wird. Im Gebirge ist also der Erosionseffekt am größten, was sich eindrucksvoll in der Fracht der Gebirgsbäche und -flüsse kundtut.

B. Die Exaration

Unter der Exaration (von lateinisch exarare = auspflügen) wird die schleifende und auspflügende Wirkung der Gletscher verstanden. Die Gletscher schieben sich langsam von höheren in tiefere Lagen, wobei hoher Druck und große Reibung auf die überfahrene Sohle einwirken. Lose und nicht sehr fest im Gesteinsverband sitzende Gesteinsstücke (ebenso loses Feinmaterial) werden vom Eis mitgenommen. Die an der Gletschersohle im Eis eingepackten Gesteinsstücke schrammen die Gesteinsunterlage (Gletscherschrammen). Dabei wird nicht nur vom Untergrund Gesteinsmasse abgeschliffen, sondern es reiben sich auch die als Hobel wirkenden Gesteine ab. Die von einem Gletscher abgeschliffenen sogenannten Rundhöcker und geglätteten Gesteinswände sowie die kantengerundeten Gesteinsstücke in den Moränen zeugen eindrucksvoll von der Gesteinszerstörung durch die Exaration. Die ausgepflügten Trogtäler, auch U-Täler genannt, geben einen Begriff von dem Ausmaß der Gletscherwirkung.

C. Die Korrasion

Mit Korrasion (von lateinisch corradere = zusammenscharren) bezeichnet man die schleifende Wirkung des vom Wind getragenen Sandes. Der Begriff „Korrasion“ soll zunächst ausdrücken, daß der Wind das durch ihn tragfähige Material zusammenträgt = zusammenscharrt. Nur dort kann der Wind diese Tätigkeit entfalten, wo der Boden nicht von Vegetation bedeckt ist. In erster Linie sind es die trockenen, warmen Wüsten, in denen das Gestein durch die physikalische Verwitterung bis zu Sandkorngröße zerkleinert wird und der Wind diesen Sand in großen Mengen verweht. Der Entfaltung größerer Windgeschwindigkeit am Boden stehen weder Baum noch Strauch entgegen. Der Wind mit dem Sand wirkt wie ein Sandstrahlgebläse; er schleift alles an, was er trifft. So werden an den lose am Boden liegenden Steinen Facetten geschliffen (Schliffacetten, Windkanter). Aufragende Felsen werden am Boden angeschliffen, so daß pilzartige Formen entstehen können (Pilzfelsen). An größeren Felsen wird ständig „genagt“; sie stehen mit kahlen, zerschliffenen, ziemlich steilen Hängen als stumme Zeugen dieser Zerstörung in der Wüstenlandschaft (Zeugenberge). Die in diesem Klimabereich stark wirksame physikalische Verwitterung ist an dieser Gesteinszerstörung stark mitbeteiligt; meistens ist in der Wüste die Gesteinszerkleinerung durch die physikalische Verwitterung größer als durch die Korrasion (s. Kap. a, 3). Der Anblick einer Wüstenlandschaft vermittelt das Ausmaß dieser beiden gesteinszerstörenden Prozesse.

3. Die physikalische Verwitterung

Die physikalische Verwitterung umfaßt die Prozesse an der Erdoberfläche, welche die Gesteine mechanisch zerkleinern, ohne daß dabei eine substantielle Veränderung stattfindet; im strengen Sinne ist die Gesteinszerstörung durch geologische Kräfte (Wasser, Gletscher, Wind) ausgenommen (s. Kap. a, 1 und a, 2). Die Kräfte der physikalischen Verwitterung sind: Insolation, Frostsprengung, Salzsprengung, Wurzeldruck.

Zur Vertiefung des Studiums der physikalischen Verwitterung werden folgende Arbeiten empfohlen: Lang 1920, 1922, Blanck 1929a, Meinecke 1957, Wilhelmy 1958, Hempel 1958, 1960, Murawski 1958, Knetsch 1960.

A. Die Insolation

Insolation bedeutet die physikalische Verwitterung infolge der Sonnenbestrahlung, wodurch mehr oder weniger große Temperaturdifferenzen im Gestein entstehen (Abb. 1). Wie andere Körper dehnt sich das Gestein bei Erwärmung aus und mindert bei Abkühlung sein Volumen. Bei Granit und Sandstein beträgt die Ausdehnung auf 1 m Gesteinsmasse und bei 50° C Temperaturerhöhung 0,25 bis 0,6 mm (Blanck 1929a). Da das Wärmeleitvermögen der Gesteine klein ist, wird die Wärme von außen nach innen nur langsam weitergeleitet (etwa 3 cm/Stunde). Die Folge davon ist, daß die bestrahlte Seite des Gesteins eine Ausdehnung erfährt, während im übrigen das Gestein sein Volumen nicht verändert. Ist das Gestein durchgehend erwärmt und erfolgt nun (während der Nacht) eine Abkühlung, so verläuft der Prozeß umgekehrt, d. h. außen wird das Bestreben der Volumenverkleinerung bestehen, während das im Innern noch nicht der Fall ist. Die partielle Volumenvergrößerung oder -verkleinerung erzeugt außerordentlich hohe Spannungen im Gestein, und diese führen zu schalen- oder schuppenförmigem Abplatzen von Gesteinsstücken. Schroffer Temperaturwechsel mit hoher Temperaturdifferenz kann ein Zersprengen großer Gesteinsblöcke (Kernsprünge) herbeiführen.

Neben diesen Spannungen, die im Gestein infolge seines geringen Wärmeleitvermögens entstehen, verursachen auch noch andere physikalische Erscheinungen die gleiche Wirkung, wenn auch in weit geringerem Maße. Die Ausdehnung der einzelnen Mineralarten der Gesteine ist verschieden. Reguläre und amorphe Minerale zeigen bei Temperaturerhöhung zwar eine verschiedene, jedoch jede Mineralart nach allen Seiten die gleiche Ausdehnung. Dagegen ist die Ausdehnung bei weniger symmetrisch gebauten Mineralen nach den kristallographisch verschiedenen Kristallflächen unterschiedlich. Diese Tatsache erzeugt im Gestein Spannungen auf kleinem Raum. Ferner erwärmen sich dunkle Minerale schneller als helle. Mithin werden in den vielfarbigen Gesteinen auch dadurch Spannungen entstehen. Gesteine mit rauher Oberfläche nehmen die Wärme schneller auf und geben sie auch wieder schneller ab als glatte. — Alle diese Ursachen einer ungleichen Erwärmung des Gesteins führen gleichfalls zu ungleichen Spannungen, und diese können das feste Gesteinsgefüge schwächen und schließlich den Zerfall des Gesteins herbeiführen.

Abb. 1. Zerfall des Gesteins (Gneis), vor allem schaliges Abplatzen, durch Temperaturwitterung (Insolation) im warmen, trockenen Klima Colorados, USA (östliche Rocky Mountains)

Die Wüste, in der das Gestein der unmittelbaren, starken Sonnenbestrahlung ausgesetzt ist und ein starker Temperaturwechsel mit großer Schwankungsbreite erfolgt, ist der Schauplatz dieser intensiven mechanischen Gesteinszerstörung. Auch das vegetationsfreie Hochgebirge erfährt große Temperaturschwankungen; deshalb ist auch hier die Insolation stark. Wo Vegetation und Boden das Gestein abdecken, hat dieser Verwitterungsprozeß keine Wirkung. — Schichtige Gesteine und Minerale mit guter Spaltbarkeit werden durch die Insolation schneller zerlegt als massige (kompakte) Gesteine und schlecht spaltbare Minerale.

B. Die Frostsprengung

Die Frostsprengung, auch Frostverwitterung oder Spaltenfrost genannt, beruht auf der Ausdehnung des Wassers beim Gefrieren (etwa 9% seines Volumens), wobei das werdende Eis einen hohen Druck entfaltet. Sind Spalten, Klüfte und kleine, rundliche Hohlräume des Gesteins ganz mit Wasser gefüllt und gefriert dieses, so übt das Eis starken Druck auf die Gesteinswandung aus. Dieser Druck erreicht bei —22° C das Maximum mit 2100 kg/cm² (Blanck 1929a). Nur dann kann der volle Druck zur Wirkung kommen, wenn die Gesteinshohlräume *ganz* mit Wasser ausgefüllt sind und wenn der Ausgang der Hohlräume klein ist, so daß das Eis sich nicht oder kaum aus den Hohlräumen herausquetschen kann. Daher erzeugen flaschenförmige Hohlräume („Flaschenhals" = Hohlraumausgang) den größten Effekt. Der vom Eis ausgeübte Druck schwächt zum mindesten den Gesteinsverband und kann ebenfalls zum Absprengen von Gesteinsteilen führen. Was ein einmaliges Gefrieren nicht vermag, kann aber bei wiederholtem erreicht werden. Nicht nur größere Gesteinsstücke werden losgesprengt, sondern es erfolgt eine intensivere Zerkleinerung der Gesteinsmasse bis zu den feinen Korngrößen des Lösses, wie DÜCKER 1937 gezeigt hat.

Die Orte dieses Verwitterungsvorganges sind die Tundra und das Hochgebirge; die Frostsprengung ist aber ferner — wenn auch schwächer — überall wirksam, wo Gestein befeuchtet wird und die Feuchtigkeit gefriert. — Am meisten werden solche Gesteine von der Frostsprengung betroffen, die viele kleine Hohlräume mit engem Ausgang für die Aufnahme von Wasser besitzen.

C. Die Salzsprengung

Die Salzsprengung stellt im Prinzip den gleichen Vorgang dar wie die Frostsprengung, d. h. bei der Kristallisation eines Stoffes wird Druck erzeugt (MORTENSEN 1933). Es sind jedoch drei Vorgänge zu unterscheiden:

1. Ein Salz kann auskristallisieren, wobei eine Vermehrung des Volumens (gegenüber der Lösung) eintritt. Dabei wird ein Druck erzeugt, der z. B. beim Auskristallisieren von Alaun 130 kg/cm^2 beträgt.

2. Die Hydratisierung eines Salzes ist oft mit großer Druckentfaltung verbunden, z. B. wird bei der Bildung von $CaSO_4 \cdot 2\,H_2O$ (Gips) aus $CaSO_4$ (Anhydrit) ein Druck von 1100 kg/cm^2 entwickelt, wobei gleichzeitig eine Volumenzunahme von etwa 60% erfolgt. Der Druck auf die Umgebung ist sehr hoch.

3. Wachsende Kristalle können ebenfalls einen gewissen Druck ausüben. Wächst z. B. der Alaun auf Oktaederflächen, so wird von diesen Flächen aus ein Druck von 40 kg/cm^2 entwickelt. Eine Glasplatte wird davon angehoben, ohne daß der Kristallisationsraum ausgefüllt zu sein braucht, ein Beweis dafür, daß der Druck allein von den wachsenden Kristallflächen ausgeht. Ein zum Würfel kristallisierender Alaunkristall erzeugt in diesem Falle keinen Druck.

Die Salzsprengung hat als physikalische Verwitterungsform nur eine geringe Bedeutung. Naturgemäß wird dort die Salzsprengung auftreten, wo entsprechende Salzmengen vorkommen, sei es, daß diese in einem Gestein enthalten sind, oder sei es, daß sie sich in einem ariden Klima an oder nahe der Oberfläche anreichern. In erster Linie wird sie in der Wüste wirksam sein und am ehesten porige, weniger feste Gesteine betreffen.

D. Die Kapillarwirkung des Wassers

Wenn ein trockenes Gestein Wasser in seine kapillaren Hohlräume aufnimmt, so kann es sich dabei ausdehnen, wie beobachtet worden ist. Eine solche Ausdehnung kann eine Lockerung des Gesteinsgefüges zur Folge haben. Der dabei entfaltete Druck ist nicht bekannt; wahrscheinlich ist die Bedeutung dieser physikalischen Einwirkung auf das Gestein gering.

E. Der Druck durch Pflanzenwurzeln

Die Zellen der Pflanzenwurzeln entwickeln einen Turgordruck von etwa 10 at. Dementsprechend bedarf es eines Gegendruckes von 10 bis 15 kg/cm^2, um das Dickenwachstum einer Baumwurzel aufzuhalten. Bei der Ausbreitung der Wurzeln von Bäumen, die eine Straße säumen, werden nicht selten Bord- und Pflastersteine angehoben, sogar eine kleine Vorgartenmauer kann dabei entzweigehen. Diese Beispiele zeugen von dem Druck, der hierbei entfaltet wird. Die Bäume, vor allem Tiefwurzler (Stieleiche, Hainbuche, Tanne, Rebe) sind bestrebt, ihre Wurzeln möglichst tief in den Boden zu senken, um die Versorgung mit Wasser und Nährstoffen zu sichern. Auf Standorten mit nahem Gesteinsuntergrund zwängen sich die Wurzeln der Bäume in Gesteinsspalten und Schichtfugen, um die Existenz zu sichern. Dabei wird ein starker Druck (je nach Baumart mehr oder weniger) auf die Gesteinswände ausgeübt, und es kann zur Absprengung von Gesteinsteilen oder wenigstens zur Aufweitung bestehender Hohlräume kommen. Wo infolge des Pflanzenbestandes die übrigen physikalischen Verwitterungsprozesse nichts oder nur wenig mehr vermögen, kann der Wurzeldruck für die mechanische Auflockerung bzw. Zerlegung des Gesteins Bedeutung haben.

F. Die mechanische Gesteinszerkleinerung durch Tiere

Die Tierwelt beteiligt sich nur wenig an der mechanischen Zerstörung des Gesteins. An der Meeresküste gibt es Tiere, die sich in das Gestein einbohren (Bohrmuscheln, Bohrschwämme, Bohrwürmer). Dagegen ist auf dem Lande (Palästina) nur eine Schneckenart (*Helix lithophaga*) bekannt geworden, die das Gestein anbohrt. Im übrigen greifen die Tiere mechanisch das feste Gestein nicht unmittelbar an, wohl aber wirken sie auf Lockergesteine und Verwitterungsmassen (Böden) ein, indem diese von Bodenwühlern und Bodenbewohnern mehr oder weniger stark durchmischt werden, wodurch die Verwitterung insgesamt gefördert wird. Solche Tiere sind: Wildschweine, Füchse, Kaninchen, Steppenbewohner (Ziesel u. a.), Hamster, Ratten, Mäuse, Maulwürfe, Käfer, Ameisen, Termiten, Würmer, Spinnen, Larven u. a. Sie unterstützen zwar mit ihrer Arbeit die Verwitterung, fördern sie aber nicht direkt.

4. Die chemische Verwitterung

Unter der chemischen Verwitterung verstehen wir die chemische Zersetzung, d. h. die *stoffliche Umbildung der Gesteine* an der Erdoberfläche. Während bei der physikalischen Verwitterung nur eine mechanische Zerkleinerung des Gesteins erfolgt, wird bei der chemischen Verwitterung auch die Substanz umgebildet, d. h. die Minerale werden größtenteils zerstört, und es entstehen neue Substanzen (Tonminerale und amorphe Verbindungen).

Dieser Prozeß ist für die Ernährung der Pflanzen von größtem Wert. Mechanisch zerfallenes Gestein, also Gesteinsgrus, bildet noch keinen Standort für anspruchsvolle Pflanzen; erst durch die chemische Verwitterung werden Nährstoffe aus den Mineralen herausgelöst und pflanzenaufnehmbar. Die bei der chemischen Verwitterung gebildete Tonsubstanz bildet die anorganische Kolloidsubstanz des Bodens, welche neben der organischen Masse die Speicherung der Nährstoffe und des Wassers übernimmt. Die entscheidendsten Eigenschaften des Pflanzenstandortes werden somit durch die chemische Verwitterung geschaffen.

Die chemische Verwitterung stellt eine Reihe von chemischen Prozessen dar, die teils als Einzelvorgänge erkannt, teils in ihrem Zusammenwirken nur in großen Zügen übersehbar sind. Um von den schwierigen Vorgängen ein einigermaßen klares Bild zu entwerfen, müssen wir zunächst die Wirkung der Verwitterungsagenzien untersuchen und dann den Gesamtprozeß in einzelne Vorgänge aufgliedern. Die Betrachtung der einzelnen chemischen Reaktionen wird eine Vorstellung von dem komplizierten chemischen Zusammenspiel dieser Verwitterungsart vermitteln. Für eine Vertiefung in das Gebiet der chemischen Verwitterung werden die Arbeiten folgender Autoren vorgeschlagen: Lang 1920, Ramann 1921, Niggli 1926, Behrend und Berg 1927, Blanck 1929b, Butler 1954, Bespalow und Sokolow 1957, Laatsch 1957, Anderson und Hawkes 1958, Barth, Correns und Eskola 1960.

A. Die Agenzien der chemischen Verwitterung

Überwiegend sind die Agenzien der chemischen Verwitterung Bestandteile der Atmosphäre, die als Atmosphärilien bezeichnet werden. Außer den Atmosphärilien beteiligen sich aber auch bodenbürtige Stoffe (Kohlensäure, Schwefelsäure, organische Säuren) an der chemischen Verwitterung.

1. Das Wasser. *Reines (destilliertes) Wasser*, das es zwar in der Natur nicht gibt, vermag lösend auf bestimmte Minerale und Gesteine einzuwirken. So lösen 100 Teile reinen Wassers bei etwa 8 bis 9° C 36 Teile NaCl, 0,25 Teil $CaSO_4 \cdot 2\,H_2O$ und 0,003 Teil $CaCO_3$. Die Lösung von Salzgestein an der Erdoberfläche ist auf geringe Vorkommen beschränkt (z. B. am Toten Meer und bei Barcelona), während unterirdische Salzauflösung einen größeren Umfang hat. Durch die Auflösung des Gipsgesteins entstehen Spalten und Höhlen (Barbarossahöhle im Kyffhäuser) sowie Dolinen wie bei der Verkarstung der Kalksteine. Reines Wasser bewirkt ferner die Hydratbildung; z. B. entsteht aus Anhydrit durch Aufnahme von Wasser Gips, wobei unter starker Druckentfaltung eine Volumenzunahme von 60% erfolgt. Überlagerndes Gestein kann dadurch zerbrochen werden. Auch mit Eisenverbindungen kann reines Wasser reagieren, z. B. kann aus Roteisen (Fe_2O_3) Brauneisen ($Fe_2O_3 \cdot 1$—$3\,H_2O$) entstehen. Die stärkste Wirkung hat — regional gesehen — das Wasser durch die Hydratation und die Hydroylse bei der Silikatverwitterung, die weiter unten behandelt wird (4, C).

Das *Regenwasser* besitzt eine weit stärkere Lösungsenergie als chemisch reines Wasser; es enthält Stoffe der Atmosphäre, welche seine Lösungskraft erhöhen. Ein Liter Regenwasser enthält 25 cm^3 Luft von der ungefähren Zusammensetzung: 62% N, 30% O und 8% CO_2 (Vol.-%) (Wagner 1950); Correns (Barth, Correns und Eskola 1960) gibt allerdings nur rund 3 Vol.-% CO_2 in der Luft des Regenwassers an. Der Gehalt an Sauerstoff und Kohlensäure liegt also in der Luft des Regenwassers wesentlich höher als in der normalen atmosphärischen Luft. Aus der Luft kann das Regenwasser ferner aufnehmen: SO_2, NH_3, HNO_3 und NaCl. Durch alle diese Stoffe wird das Regenwasser zu einem wesentlich stärkeren Verwitterungsagens als das reine Wasser. Die Wirkung der genannten Stoffe soll einzeln erörtert werden.

2. Die Kohlensäure. Oben wurde schon erwähnt (Kap. 4, A, 1), daß die Luft des Regenwassers wenigstens 100mal mehr Kohlensäure enthält als die atmosphärische Luft (wenigstens 3 gegenüber 0,03 Vol.-%). Dabei darf nicht vergessen werden, daß von der im Wasser gelösten Kohlensäure nur etwa 1% dissoziiert ist, und nur diese ist „aktiv“.

Während im reinen Wasser nur 14 mg/l $CaCO_3$ löslich sind, löst das mit CO_2 beladene Regenwasser 40 mg/l und mit CO_2 gesättigtes Wasser bei normalem Druck 900 mg/l. Die starke Lösungskraft der Kohlensäure auf Kalkgestein wird durch diese Zahlen offenkundig. Die Kohlensäure löst das Calciumcarbonat unter Bildung von löslichem Calciumbicarbonat ($CaCO_3 + H_2CO_3 = Ca(HCO_3)_2$), das mit dem Wasser weggetragen wird und an anderer Stelle unter Abgabe von Kohlensäure wieder als Calciumcarbonat ausfallen kann. Mit steigender Temperatur und fallendem Druck sinkt die Aufnahmefähigkeit des Wassers für CO_2, und mit abnehmender Temperatur und steigendem Druck nimmt sie zu.

Die Lösung des Calciumcarbonates ist auf der Erde am wichtigsten, weil dieses Carbonat, also der Kalkstein, die größte Verbreitung besitzt. Dann folgt der Dolomit ($CaMg(CO_3)_2$), wovon in kohlensäuregesättigtem Wasser 310 mg/l löslich sind. Im gleichen Medium werden 115 mg/l Bitterspat ($MgCO_3$), 500 mg/l Manganspat ($MnCO_3$) und 720 mg/l Eisenspat ($FeCO_3$) gelöst.

3. Der Sauerstoff. Der Sauerstoff kann in der Luft des Regenwassers bis zu 34 Vol.-% betragen. Dadurch besitzt das Regenwasser eine relativ stark oxydierende Wirkung. Das zweiwertige Eisen wird davon besonders betroffen; es wird in dreiwertiges Eisen übergeführt. Eisencarbonat z. B. wird zu Eisenhydroxyd (Brauneisen) oxydiert: $2\,FeCO_3 + {}^1/_2\,O_2 + 3\,H_2O = 2\,Fe(OH)_3 + 2\,CO_2$. Meistens entsteht ein Eisenoxydhydrat und weniger die wasserreiche Form Eisenhydroxyd. Sulfide können in Sulfate und Schwefelsäure umgewandelt werden: $4\,FeS_2 + 4\,H_2O + 14\,O_2 = 4\,FeSO_4 + 4\,H_2SO_4$.

4. Das Schwefeldioxyd. Die Luft erhält SO_2 durch die Verbrennung organischer Stoffe, vor allem der Steinkohle. Das dabei frei werdende SO_2 verbindet sich mit dem Wasser der Luft zu schwefeliger Säure (H_2SO_3), die durch Sauerstoffaufnahme in Schwefelsäure (H_2SO_4) übergeht. Diese Säuren begünstigen die chemische Verwitterung, da sie stark dissoziieren und die Wasserstoffionen die aktiven Teilchen der Silikatverwitterung darstellen. Den Beweis dafür liefert die starke Verwitterung der Natursteine (mit Silikaten) an Bauten großer Städte, vor allem solcher mit großen Industrieanlagen, in denen große Mengen SO_2 in die Luft gehen (Beispiele: Kölner Dom, Genter Rathaus). Wo Gesteine in natürlicher Lagerung an der Erdoberfläche liegen (Hochgebirge, Wüste), befinden sich im allgemeinen keine großen Siedlungen, so daß dort der SO_2-Gehalt der Luft geringer und daher auch die chemische Gesteinszersetzung durch diesen Stoff nur schwach ist. Möglicherweise sind bei der Lochverwitterung am Granit in der Namib unweit der südwestafrikanischen Küste (bei Walvis Bay) schwefelige Säure und Schwefelsäure beteiligt. Die Küstennebel tragen hier den Schwefel als Schwefelwasserstoff (Zersetzungsprodukt des Planktons) vom Meere heran (Scholz 1963).

5. Die Salpetersäure. Salpetersäure (HNO_3) und salpetrige Säure entstehen in der Atmosphäre durch elektrische Entladung (Gewitter). Wo also größere elektrische Entladungen stattfinden (Tropen), wird auch eine entsprechend große Menge dieser Säuren in der atmosphärischen Luft gebildet. So enthält das Regenwasser in England nur 0,2 mg/l dieser Säuren, dagegen in tropischen Gebieten das Zehnfache. Diese aggressiven Säuren wirken bei der chemischen Verwitterung wie die Schwefelsäure.

6. Das Ammoniak. Ammoniak entsteht bei der Zersetzung organischer Stoffe unter weitgehendem Abschluß von Sauerstoff (Fäulnis). Größere Mengen gehen aus Großstädten in die Atmosphäre. In Deutschland besitzt das Regenwasser etwa 1 bis 2 mg/l NH_3; mit dem Regenwasser kommen hier im Jahr 10 kg/ha NH_3 in den Boden. In den lezten Jahrzehnten hat dieser Betrag wahrscheinlich zugenommen. Der Vollständigkeit wegen wird das Ammoniak als eines der Atmosphärilien erwähnt, ist aber für die Verwitterung ohne Bedeutung.

7. Das Steinsalz. Das Wasser der Luft erhält das NaCl vom Meerwasser; vom Meer her wird NaCl-haltiges Wasser landeinwärts getragen. In den Alpen beträgt der Steinsalzgehalt des Regenwassers 2 mg/l, in Paris 8 mg/l und an der englischen Küste 90 mg/l.

NaCl im Wasser erhöht die Löslichkeit bestimmter Verbindungen, z. B. des Gipses. Im übrigen hat indes das Steinsalz wenig Einfluß auf die chemische Verwitterung. Es hat aber Bedeutung für die Anreicherung des Bodens mit Na-Ionen. Während das Na-Ion in humiden Bereichen schnell ausgewaschen wird, kann es in ariden Gebieten weitgehend im Boden verbleiben (Salzkrustenbildung, Salzsprengung); allerdings kann hier infolge des geringen Niederschlages durch diesen dem Boden nicht viel Salz zugeführt werden. Andererseits tragen die Nebel der Nebelwüste (Namib) viel NaCl in diese aride Landschaft (Scholz 1963).

8. Säuren des Bodens. Im Boden wird Kohlensäure bei der Zersetzung der organischen Substanz gebildet. Deren Wirkung ist bereits im Kap. 4, A, 2 beschrieben. Ferner kann aus dem Schwefel des Eiweißes der Pflanzen Schwefelsäure entstehen. Die darüber hinaus aus der pflanzlichen Masse entstehenden organischen Säuren (Ameisen-, Essig-, Propion-, Äpfel-, Lävulin- und Oxalsäure), besonders Humussäuren, sind mannigfaltig. Diese Säuren sind mehr oder weniger aggressiv und fördern somit die chemische Verwitterung.

B. Die Lösungsverwitterung

Im Kap. 4, A, 1 und 4, A, 2 wurde bereits die lösende Wirkung des reinen Wassers und des kohlensäurehaltigen Wassers beschrieben; sie wird als Lösungsverwitterung bezeichnet. Wenn wir die der Lösungsverwitterung anheimfallenden Gesteine nach Art und Verbreitung betrachten, so sind Kalksteine, Mergel und Dolomite die wichtigsten. Kalksteine werden am intensivsten angegriffen; sie lösen sich fast dreimal so schnell auf wie Dolomit. Mergelgesteine lassen bei der Verwitterung eine große Menge Tonsubstanz (oder sandig-tonige Substanz) zurück. Hat sich eine mächtige Schicht dieses Rückstandes über dem frischen Mergelgestein angesammelt, so wird die weitere Verwitterung gebremst. — Das Calciumcarbonat in der Modifikation des Aragonits ist leichter löslich als das in der Modifikation des Kalkspates; letzteres überwiegt aber bei weitem.

Abb. 2. Lösungsverwitterung auf dem Kalkstein des Wallberges bei Tegernsee, Oberbayern. Die sichtbaren Spalten im Kalkstein, die sogenannten Karren oder Schratten, entstanden durch Auflösung des Kalksteins infolge der Einwirkung von kohlensäurehaltigem Wasser. Klüfte und Schichtfugen erweitern sich durch diesen Auflösungsprozeß zu Karren

Die charakteristischsten Erscheinungen der Lösungsverwitterung zeigt der Kalkstein. Von der Oberfläche her wird zunächst der Kalk unter Bildung von Calciumbicarbonat abgelöst. Durch diesen Vorgang verliert der Kalkstein in Mitteleuropa bei durchschnittlichem Jahresniederschlag (650 mm) eine Schicht von 0,01 mm Mächtigkeit jährlich. Das macht allein für den Weißen Jura der Württembergischen Alb jährlich einen Abtrag von 60000 m³ Kalkstein aus (WAGNER 1950). In Kluft- und Schichtfugen greift die Lösungsverwitterung am stärksten an, weil sich in diesen Fugen mehr Wasser sammelt und daher auch mehr Kohlensäure mit dem Kalkstein in Berührung kommt. Die Fugen vertiefen und verbreitern sich; die Oberfläche des Kalksteins erhält eine unregelmäßige, starke Zerfurchung; diese Furchen nennt man *Karren* oder *Schratten* (Abb. 2).

Das kohlensäurehaltige Wasser löst sich gleichsam selbst seinen Weg in die Tiefe des Kalksteins. Klüfte und kleine Spalten werden mehr und mehr erweitert; schließlich entstehen kleine und große Hohlräume, sogar große, weitverzweigte Höhlen, die sogenannten Karsthöhlen (z. B. die Dechenhöhle bei Iserlohn/Westfalen). Im Westen des Staates Kentucky, USA, wurden rund 60000 Karsthöhlen gefunden, wovon allein eine 48 km und mit allen Verzweigungen sogar etwa 200 km lang ist. Von der Decke der Höhlen tropft das kalkhaltige Wasser. An den Tropfstellen der Höhlendecke entstehen Kalkzapfen, die sogenannten *Stalaktiten* (von griechisch stalaktos = tröpfelnd), und auf dem Höhlenboden, wo das Wasser auffällt, bilden sich kegelartige Kalkgebilde, die sogenannten *Stalagmiten* (von griechisch stalagmos = Tropfen). In kleinen Hohlräumen des Kalksteins kann ebenfalls Kalk als Calcit ausgeschieden werden (*Kalkdrusen*). Die Decken der Karsthöhlen können bei fortschreitender Lösungsverwitterung so dünn und im Gesteinsverband so schwach werden, daß sie einstürzen. So entstehen *Erdfälle*, auch *Dolinen* genannt. Bricht eine längere Höhle an mehreren Stellen ein, so entstehen mehrere Dolinen hintereinander, eine *Dolinenkette*. Durch diese Lösungserscheinungen entsteht der sogenannte *Karst*. Das charakteristischste Gebiet der Verkarstung als Ergebnis der Lösungsverwitterung ist der klassische Karst Dalmatiens. Hier wurde auch der Begriff „Karst" geprägt. Wahrscheinlich wird er von dem slawischen „krs" = Fels oder „krsti" = (soviel wie) zerbrochen (da der Kalkstein an vielen Klüften zerbrochen ist) entlehnt. Nach JANEKOVIĆ (persönliche Mitteilung 1961) bedeutet „Karst" etwa „Durcheinander", womit die wechselvolle unruhige Bodenoberfläche gekennzeichnet werden soll. In diesem Gebiet wird auch besonders augenfällig, daß das schnelle Eindringen des Niederschlagswassers in den tiefen Untergrund zu einer schnellen Austrocknung des Bodens führt. Weil bei der Lösungsverwitterung des Kalksteins wenig Rückstand verbleibt, sammelt sich selbst in langer Zeit nur wenig Bodenmasse, so daß auch die Wasserkapazität der Böden aus Kalkstein in der Regel gering ist. Auch dieser Umstand bedingt einen trockenen Standort.

Das Wasser kann im verkarsteten Kalk gut zirkulieren; in den zusammenhängenden Spalten bildet sich ein Grundwasserspiegel. Nach den Gesetzen der Quellenbildung tritt dieses Karstwasser als *Karstquelle* zu Tage (z. B. Quelle der Blau und der Brenz). Das Karstquellwasser gibt in der Regel beim Austreten aus dem Gestein Kohlensäure aus dem mitgeführten Bicarbonat des Calciums ab; es fällt Calciumcarbonat als *Kalksinter* aus, der in schichtiger Ausbildung auch *Travertin* genannt wird. Die mächtigen Ablagerungen von Alm (Alm = örtliche Bezeichnung für weichen Kalksinter in Bayern) nördlich von München und von Travertin in Tivoli bei Rom zeugen von dem gewaltigen Ausmaß der Lösungsverwitterung der benachbarten Kalkgebirge (Alpen, Apennin).

C. Die Silikatverwitterung

Da die heutige *Gesteinshülle der Erde* zu *drei Vierteln aus Silikaten* besteht, so muß der Verwitterung dieser Minerale besondere Bedeutung zukommen, nicht zuletzt auch deshalb, weil durch diesen Verwitterungsprozeß der weit überwiegende Teil der Tonsubstanz unserer Böden entsteht; vor allem die Tonsubstanz speichert das Wasser und die Pflanzennährstoffe. Wenn wir den Ablauf der ganzen Erdgeschichte hierbei in Betracht ziehen, so entstammt aller Ton der Erde ursprünglich den Silikaten, und zwar entstand er durch die Silikatverwitterung. Bei den Silikaten der Erdrinde machen die Feldspäte insgesamt 58% (Gew.-%) aus; davon sind 40,3% Kalknatronfeldspäte (Plagioklase) und 17,7% Kalifeldspat (Orthoklas).

Die Silikatverwitterung stellt eine Kette von komplizierten Prozessen dar, deren Hauptphasen wir nur chemisch überblicken können. Wir wollen die wichtigsten Prozesse der Silikatverwitterung an einem Vertreter der Feldspäte darstellen, und zwar wählen wir dafür den einfach zusammengesetzten *Kalifeldspat*, weil hierbei die chemischen Vorgänge gut zu verfolgen sind (Abb. 3). Grundsätzlich verläuft die Verwitterung der übrigen Silikate in der gleichen Weise, allerdings ist dieser Vorgang um so komplizierter, je mehr Elemente bzw. Ionen mit verschiedenen Eigenschaften am Aufbau des Kristallgitters beteiligt sind. Bei der Darlegung der Verwitterung des Kalifeldspates wird den Vorstellungen von BEHREND und BERG (1927), BLANCK (1929b, 1949) und LAATSCH (1957) gefolgt. Außer diesen Werken werden die von CORNU (1909), RAMANN (1921), NIGGLI (1926) und BARTH, CORRENS und ESKOLA (1960) zum weiteren Studium der Silikatverwitterung empfohlen.

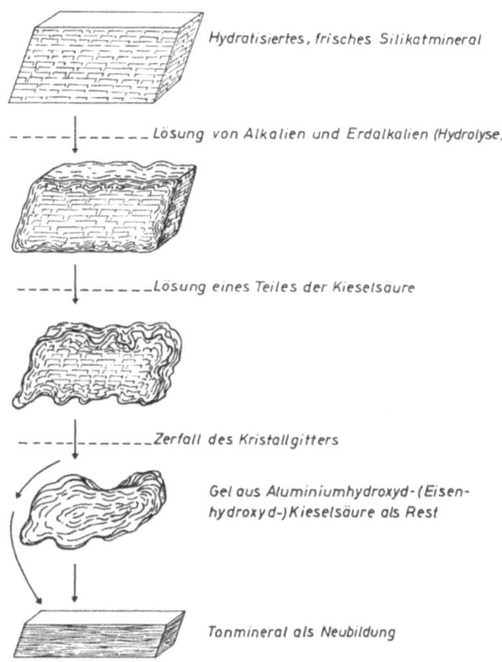

Abb. 3. Schematische Darstellung der chemischen Verwitterung eines Silikatminerals (Feldspäte, Glimmer, Pyroxene, Amphibole u. a.). Der Prozeß verläuft etappenweise

1. Die Mineralauflockerung durch Hydratation. Die Hydratation eines Minerals besteht darin, daß sich Wassermoleküle als Dipole dicht um die Kationen des Minerals anhäufen (Abb. 3). Bei einem *Kalifeldspat* wird sich die Hydratation an den Kaliumionen der Kristalloberfläche vollziehen. Die Dipole des Wassers binden elektrische Feldkräfte der in der Außenzone des Kristallgitters stehenden Kaliumionen. Im Innern der Kristalle gehen von jedem Massenpunkt elektrische Kraftlinien (der Anziehung) in jeder Raumrichtung aus zu den benachbarten, entgegengesetzt geladenen Ionen. An den Grenzflächen, Kanten und Ecken eines Kristalls sind die Ionen nicht mehr nach allen Raumrichtungen gebunden; zum freien Raum hin strahlen also die Ionen der Randstellung elektrische Anziehungskräfte aus, welche durch die Wasserdipole abgesättigt werden. So vollzieht sich die Hydratation.

Die dichten Hydratbeläge wirken sich als Isolatoren aus, welche die Ionen im Gitter der Kristallgrenzflächen gegeneinander abschirmen und damit ihren Zusammenhalt schwächen. In den so entstehenden Schwächezonen können sich Risse bilden. Die Oberfläche dieser Risse besitzt ebenfalls freie Anziehungskräfte, so daß nun auch hier eine Hydratation erfolgt. Das hat wiederum eine Schwächung innerhalb der Risse zur Folge. So schreitet die Kristallauflockerung fort, und das geschieht um so schneller, je mehr Angriffspunkte die Kristalle bieten, d. h. je kleiner sie sind und je mehr Risse sie haben.

2. Die Oxydation. Im Wasser gelöster Sauerstoff dringt in die feinen Risse, die durch die oben beschriebene Hydratation oder durch physikalische Verwitterungsvorgänge entstanden sein können, ein und kann oxydierbare Stoffe zur Oxydation bringen. Im Kalifeldspat fehlen solche Stoffe, während die zweiwertigen Eisen- und Manganionen im Biotit, Augit und in der Hornblende zu dreiwertigen Verbindungen oxydiert werden können, z. B. kann dadurch bei gleichzeitiger Wasseraufnahme α — FeOOH (Goethit), γ — FeOOH (Lepidokrokit) oder ein wasserreicheres Fe(III)oxydhydrat entstehen. Äußerlich ist dieser Vorgang an einer rostgelben oder rostbraunen Verfärbung der Minerale bzw. der Gesteine erkenntlich. Durch die Oxydation findet ebenfalls eine Kristallauflockerung statt.

3. Die Hydrolyse. Unter Hydrolyse versteht man die Umsetzung eines Salzes mit den Ionen des Wassers. Bei den *löslichen* Salzen geht dieser chemische Prozeß schnell vor sich, dagegen ist er bei einem in Wasser unlöslichen Silikat nicht so offenkundig. Da die Silikate aus einer schwachen Säure und stark dissoziierbaren Basen aufgebaut sind, haben ihre Ionen ebenfalls das Bestreben, sich mit den Ionen des Wassers umzusetzen. Diese Reaktion können indes nur einzelne Ionen der Grenzflächen des unlöslichen Kristalls vollführen. In erster Linie sind das Na, K, Ca, Mg, Fe(II) und Mn(II), die zuvor durch die Hydratation im Kristallgitter gelockert worden sind. Die Verdrängung dieser hydratisierten Grenzflächenionen aus dem Kraftfeldbereich des Kristallgitters in die Bodenlösung kann durch die Wasserstoffionen erfolgen, die ihrer geringen Masse wegen eine hohe Bewegungs- und damit Verdrängungsenergie besitzen. Beim *Kalifeldspat* werden durch diesen Vorgang K-Ionen gegen H-Ionen ausgetauscht (Abb. 3). Das Kalium verbindet sich mit OH-Ionen der Bodenlösung zu KOH; damit wird die Bodenlösung alkalisch. Diese Reaktion kann man mit fein gemahlenem Kalifeldspat und destilliertem Wasser unter Hinzufügung eines Indikators demonstrieren; schon nach einem Tag ist die alkalische Reaktion der Lösung am Farbumschlag sichtbar.

4. Herauslösung von Kieselsäure. Durch den Austritt von Kaliumionen aus den Grenzflächen des Kalifeldspatgitters wird der Zusammenhalt der Gittergrenzflächen so geschwächt, daß neben den K-Ionen auch *ein Teil* der Kieselsäure aus der Grenzzone des Kristalls austritt (Abb. 3). Sie hydratisiert in der Bodenlösung oder reagiert mit Al-Ionen.

5. Zerfall des Kristallgitters. Mit dem Austritt eines Teiles der Kieselsäure wird der Gitterverband in der Grenzzone des Kristalls so gelockert, daß sich nun weitere Kieselsäure und Al-Ionen ablösen. Das bedeutet die Auflösung des Gitters in den Grenzflächen (Abb. 3).

Nach CORRENS (1940, 1949) bildet sich um die verwitternden Feldspatkörnchen eine sehr dünne Rinde aus Kieselsäure und Aluminiumhydroxyd, durch welche nun die vom Gitter abgespaltenen Ionen hindurchtreten müssen. Damit erklärt sich, daß die Verwitterung frischer Kristalle schneller verläuft als die der bereits in Verwitterung befindlichen Kristalle.

6. Aufbau neuer Verbindungen. Die aus dem Gitter ausgetretene Kieselsäure kann hydratisieren, und die Al-Ionen können sich mit OH-Ionen zu Aluminiumhydroxyd verbinden. Großenteils vereinigen sich Kieselsäure und Al zu amorpher Tonsubstanz *(Allophan)* oder je nach dem Verwitterungsmilieu zu bestimmten *Tonmineralen* (Abb. 3).

7. Silikatverwitterung durch Säuren. Der Prozeß der Hydrolyse durch das Wasser zeigt, daß selbst die relativ geringe Menge der H-Ionen des reinen Wassers die Abspaltung von K-Ionen (und auch anderer Kationen) aus den Grenzzonen eines Kalifeldspates bewirken kann. In weit größerem Umfang kann der Austausch von K-Ionen (und anderen) gegen H-Ionen vor sich gehen, wenn viele H-Ionen in der Bodenlösung sind. Die Gegenwart von Säuren schafft diese Voraussetzung. Säuren sind in der Bodenlösung stets vorhanden, zumindest Kohlensäure, die laufend im Boden produziert wird und die mit dem Regenwasser in den Boden bzw. auf das Gestein gelangt. In vielen Böden der humiden Klimate sind aber auch andere Säuren in der Bodenlösung, die oben schon genannt wurden (Kap. 4, A, 8). Diese Säuren geben eine Menge H-Ionen in die Bodenlösung, und dadurch wird die Silikatverwitterung schnell vorwärtsgetrieben. In einem stark sauren Boden (pH 3 bis 4) sind etwa 1000mal mehr H-Ionen vorhanden als in reinem Wasser. Man kann erwarten, daß die Humussäuren einer Rohhumusdecke in etwa 100 Jahren eine stärkere Verwitterung bewirken, als die Kohlensäure einer normalen Bodenlösung es in 5000 Jahren vermag (LAATSCH 1957). In den humiden Klimaten sind also die H-Ionen die wichtigste treibende Kraft der Silikatverwitterung.

Nach CORRENS und ENGELHARDT (1938) greifen nicht nur saure, sondern auch alkalische Lösungen die Silikate an, so daß auch in alkalisch reagierenden Böden die Silikatverwitterung stattfindet. In humiden Klimabereichen treten diese Böden gegenüber den sauren zurück, und wo auf der Erde vorwiegend alkalische Böden vorliegen, fehlt meistens das für die chemische Verwitterung notwendige Wasser; hier herrscht die physikalische Verwitterung vor.

5. Die biologisch-chemische Verwitterung

Unter der biologisch-chemischen Verwitterung wird die chemische Zersetzung der Gesteine (feste und lockere) durch die Einwirkung von Pflanzen und Tieren (lebende und abgestorbene) verstanden. Der physikalische Einfluß der Pflanzen wurde oben schon behandelt (Kap. 3, E). Die biologisch-chemische Verwitterung durch niedere und höhere Pflanzen sowie durch deren abgestorbene Massen ist zum mindesten quantitativ sehr verschieden und muß deshalb getrennt behandelt werden. Die chemische Einwirkung durch Tiere ist so geringfügig, daß auf eine besondere Darstellung verzichtet werden kann. — Die biologisch-chemische Verwitterung hat in allen Bereichen der Erde große Bedeutung, in denen eine dichte Vegetationsdecke vorhanden ist und besonders dann, wenn infolge Flachgründigkeit die Vegetation unmittelbar mit dem Gestein in Berührung steht.

Für ein eingehendes Studium der biologisch-chemischen Verwitterung werden die Arbeiten folgender Autoren empfohlen: BLANCK 1929c, 1929d, SCHELLENBERG 1929. In diesen Arbeiten befindet sich eine große Zahl von Spezialschriften.

A. Die biologisch-chemische Verwitterung durch niedere Pflanzen

Während höhere Pflanzen auf dem unverwitterten Gestein nicht leben können, vermögen niedere Pflanzen trotz mangelnden Bodens und oft langer Trockenheit hierauf zu existieren (SCHELLENBERG 1929).

Die einfachsten pflanzlichen Organismen, die das Gestein besiedeln können und in Spalten und Klüften bisweilen metertief eindringen, sind die *Bakterien*. Die von ihnen ausgeschiedene Kohlensäure vermag nicht nur Carbonatgesteine anzugreifen, sondern auch Silikatgesteine.

Die ersten sichtbaren Besiedler der nackten Gesteine sind weit überwiegend die *Flechten*. Die Algen können als vorwiegende Wasserpflanzen auf stark austrocknendem Gestein nur in seltenen Fällen existieren, und Pilze finden als nicht autotrophe niedere Pflanzen auf dem Gestein keine Nahrung. Aus diesem Grunde finden wir auf dem unverwitterten Gestein fast nur die Flechten, die eine Lebensgemeinschaft von Pilzen und Algen darstellen. Zuerst siedeln sich meistens Krustenflechten an, dann folgen Laub- und Strauchflechten. Erst wenn das Gestein durch den chemischen Angriff aufgerauht ist, folgen die Moose. Die

Flechten leben meistens auf dem Gestein (epilithische Flechten), teils dringen sie in das Gestein ein (endolithische Flechten). Ein Teil der Flechten gedeiht besser auf Kalkgesteinen und kalkhaltigen Gesteinen (calcisede Flechten), andere ziehen als Standort die Kieselgesteine oder kieselsäurehaltigen Gesteine vor (silizisede Flechten). Die epilithischen Flechten zeigen auf dem Gestein verschiedene Farben, die häufigsten sind grau, ocker, gelb, braunrot und schwarz. Am auffälligsten ist die zitronengelbe, sogenannte Landkartenflechte (*Rhizocarpon geographicum*) mit schwarzem Saum. Meistens bilden die Flechten auf dem Gestein rundliche Lager und wachsen im Kreise nach allen Seiten weiter.

Die Flechten dringen mit ihren Hyphen in das Gestein ein. Die dabei entfaltete physikalische Wirkung ist gering, viel stärker ist die chemische Ein-

Abb. 4. Biologisch-chemische Verwitterung von Schieferlagen im Gesteinsverband von Schiefer und Grauwacke (U. Devon) in der nördlichen Eifel. Die von links unten nach rechts oben verlaufenden, dünnen Schieferlagen sind durch die Einwirkung der Gerbsäure (u. a. Säuren) der Eichenwurzeln weißgrau gebleicht, d. h. das Eisen ist hierbei herausgelöst worden

wirkung, die in erster Linie von der ausgeschiedenen Kohlensäure, aber auch von anderen Säuren ausgeht; diese Säuren entstehen weitgehend bei der Zersetzung der abgestorbenen Flechtenmasse (s. Kap. 4, A, 8). Die biologisch-chemische Verwitterung durch niedere Pflanzen spielt in Bereichen eine große Rolle, in denen höhere Pflanzen nicht mehr leben können, nämlich in Wüsten mit großer Wärme und langer Trockenheit und in hohen Breiten sowie in Hochgebirgen mit großer Kälte.

Die Flechten bereiten den Boden für die *Moose*. Durch die Einwirkung der Flechten wird das Gestein aufgerauht und angewittert, so daß die Moose Fuß fassen und aus der vergrößerten Teilchenoberfläche Nährstoffe aufnehmen können. Nach längerer Zeit biologischer Verwitterung durch niedere Pflanzen werden Gesteinspartikel aus dem Gesteinsverband gelockert. Diese mehr oder weniger verwitterten Mineralteilchen und abgestorbenes pflanzliches Material sammeln sich in Gesteinsvertiefungen, in denen damit die Voraussetzung für das Wachstum höherer Pflanzen geschaffen ist.

B. Die biologisch-chemische Verwitterung durch höhere Pflanzen

Wie oben gesagt, müssen erst bestimmte Verwitterungsvorgänge den Standort für die höheren Pflanzen vorbereiten. Haben sie Fuß gefaßt, so dringen sie mit ihren Wurzeln in Spalten, Klüfte und Schichtfugen des Gesteins ein und vollziehen hier neben dem Wurzeldruck (s. Kap. 3, E) einen starken chemischen Angriff auf das Gestein (Abb. 4). Je tiefer sich eine Pflanze in das Gestein zwängen kann, je weiter sie ihr Wurzelsystem auszubreiten vermag, je mehr Säuren ihre Wurzeln ausscheiden und je aggressiver diese sind, um so stärker ist der biologisch-chemische Angriff auf das Gestein. In den aufgezählten Wirkungen unterscheiden sich die höheren Pflanzen sehr. Die Vegetation des Waldes entfaltet einen ungleich stärkeren Angriff auf den Gesteinsuntergrund als die landwirtschaftlichen Kulturpflanzen und die Gartenpflanzen. Es bestehen aber auch innerhalb dieser Pflanzengruppen große Unterschiede. Bekannt ist, daß z. B. die Stieleiche und die Hainbuche eine große Wurzelenergie besitzen, während die Fichte nur auf sehr lockeren Böden in die Tiefe strebt (Köstler 1950). Die Stieleiche scheidet im Vergleich zur Hainbuche wesentlich mehr Säuren aus und wirkt damit chemisch weit intensiver auf das Gestein und die Veränderung im Boden ein. Der Rebstock hat eine ähnliche Wurzelenergie wie die wurzelkräftigen Waldbäume. Auch die landwirtschaftlichen Kulturpflanzen entfalten ein sehr verschiedenes Wurzelsystem (Klapp 1958, Kutschera 1960) und zeigen ein verschiedenes Aneignungsvermögen für Nährstoffe. Letzteres bedeutet eine verschieden starke Lösungsenergie, und diese bewirkt einen Angriff auf Gestein und Bodenmasse. Die Lösungskraft der landwirtschaftlichen Kulturpflanzen auf Gesteinspartikel allgemein und die unterschiedliche Lösungsenergie der einzelnen Arten werden durch einen Versuch von Haselhoff (1909) gut demonstriert. Nach den Untersuchungen dieses Autors wurden von folgenden Kulturpflanzen aus den als Düngung gegebenen Gesteinsmehlen folgende Mengen an CaO und K_2O entzogen:

Tabelle 1

Pflanzenart	Aufnahme an CaO aus				Aufnahme an K_2O aus			
	Buntsandstein	Basalt	Grauwacke	Muschelkalk	Buntsandstein	Basalt	Grauwacke	Muschelkalk
Bohne	0,478	0,130	0,380	0,937	0,197	0,115	0,175	0,012
Erbse	0,572	0,336	0,547	0,548	0,247	0,253	0,189	0,114
Lupine	0,984	0,224	0,539	0,114	0,410	0,084	0,118	0,019
Gerste	0,019	0,095	0,129	0,031	0,019	0,017	0,020	0,003
Weizen	0,026	0,010	0,010	0,027	0,034	0,021	0,032	0,027

Der Versuch zeigt, daß das Aneignungsvermögen der Lupine am größten ist, dann folgen die übrigen Leguminosen und zuletzt die Gramineen. Vor etwa 50 Jahren sind viele Versuche über die Düngewirkung von Gesteinsmehlen gemacht worden. Darüber hat Blanck (1929c) zusammenfassend berichtet, wobei er sich auf viele eigene Untersuchungen und solche seiner Mitarbeiter stützt. Dabei hat sich ergeben, daß die Minerale des Phonoliths (Nephelin und andere) sowie Feldspäte (Plagioklas leichter als Orthoklas) und Glimmer (Biotit leichter als Muskowit) relativ leicht angreifbar sind. Vor allem werden Kalium und Calcium verhältnismäßig leicht aus dem Kristallverband gelöst. Der Phosphor wird dagegen nur langsam aus den Apatiten freigemacht.

C. Die biologisch-chemische Verwitterung durch abgestorbene Pflanzenmasse

Der guten Übersicht über das verwickelte Gebiet der Verwitterung wegen wird in einem gesonderten Kapitel die Wirkung der abgestorbenen Pflanzenmasse auf die Zerstörung des Gesteins herausgestellt. Blanck (1929d) hat von diesen Prozessen unter Auswertung vieler Spezialarbeiten eine eingehende Darstellung gegeben (s. auch NIKLAS 1912).

Wenn wir die längere Wirkung einer Rohhumusauflage auf den Boden betrachten, so wird die davon ausgehende intensive Lösungsenergie deutlich (s. Kap. 4, A, 8). Dabei sind vor allem aggressive, niedrigmolekulare Humussäuren (Fulvosäuren) wirksam. Eisen-, Aluminium- und Kieselsäureverbindungen gehen dabei in Lösung; es entsteht unter dem Rohhumus eine Bleichzone. Die gleiche Wirkung, die bei dieser Bleichung (Podsolierung) auf die Bodenmasse ausgeübt wird, wirkt selbstverständlich auch auf das Gestein ein, wenn dieses so hoch ansteht, daß die Säuren des Rohhumus damit in Berührung kommen. In Bereichen flachgründiger Böden (Ranker) trifft das zu. — Der saure Moder hat eine ähnliche, aber schwächere Wirkung als der Rohhumus.

Besonders intensiv ist die Zersetzung des Gesteins unter saurem Anmoor und Hochmoor. In Hochgebirgslagen, wo infolge des kühlen und niederschlagsreichen Klimas eine saure Moorbildung leicht einsetzt, liegt diese intensive biologisch-chemische Einwirkung auf den Gesteinsuntergrund vor. Eindrucksvoll zeigt sich diese Wirkung auch im Untergrund des nassen Podsols mit starker, fast anmooriger Rohhumusauflage (Eisenbändchen-Podsol) in hohen Mittelgebirgslagen (hoher Buntsandstein-Schwarzwald, Schottland).

Wo die Zersetzung der organischen Masse infolge günstiger Temperatur- und Basenverhältnisse schnell vonstatten geht, entstehen zwar keine großen Mengen organischer Säuren, jedoch viel Kohlensäure, die aber eine wesentlich geringere Zerstörungsenergie aufweist; zudem entweicht davon ein großer Teil in die Luft und wird für die Verwitterung gar nicht wirksam. In solchen Böden mit intensiver Zersetzung der organischen Substanz sammelt sich nur wenig organische Masse an. Wird andererseits in basenreichen Böden organische Substanz infolge klimatischer Umstände (kalte Winter, trockene Sommer) angesammelt (Schwarzerde), so ist von einem solchen säurearmen Humusmaterial ebenso nur eine schwache Verwitterungswirkung zu erwarten.

6. Die Wabenverwitterung

Unter der Wabenverwitterung versteht man die lochartige Auswitterung von Gesteinswänden (meistens Sandsteine), die den Eindruck einer „Wabe" vermitteln. Die Herauswitterung dieser Wabenstruktur beruht in erster Linie auf einer ungleichen Verfestigung (Diagenese) des Gesteins, und zwar ist diese zellenartig erfolgt, d. h. die „Zellenwände" sind stärker verfestigt als das „Zelleninnere". Dementsprechend wird das „Zelleninnere" schnell von der Verwitterung erfaßt, während die „Zellenwände" weniger schnell verwittern und deshalb erhaben aus der Gesteinswand ragen. An dem Zustandekommen dieser Verwitterungsform soll auch das Lokalklima beteiligt sein. Die Verfestigungspartien können auch streifig sein. Das Phänomen der Wabenverwitterung ist eindrucksvoll und hat vor allem die Geomorphologen stets beschäftigt. Sie ist aber für das gesamte Ausmaß der Verwitterung nicht bedeutsam (WILHELMY 1958, LOUIS 1960).

7. Die submarine und sublakustrische Verwitterung

Der Vollständigkeit wegen wird auf die submarine und sublakustrische Verwitterung hingewiesen. Die Frage ist, ob die Vorgänge am Boden des Meeres und der Seen als Verwitterungsprozesse gelten können oder ob sie als Prozesse der Sedimentbildung anzusehen sind (BARTH, CORRENS und ESKOLA 1960).

Die vom Lande in das Meer und in die Seen gelangenden Stoffe sind überwiegend schon verwittert, somit also den Bedingungen an der Erdoberfläche bereits angepaßt. Nur die vom Wind in das Meer und in die Seen getragenen vulkanischen Aschen können im Wasser eine gewisse Umsetzung erfahren. Der Angriff auf die Silikate ist aber geringfügig, wie das Vorkommen frischer Feldspäte, Augite und Olivine in den Meeressedimenten zeigt. Das Medium ist für die Zersetzung der Silikate wenig günstig, weil es reich an Ionen von Na, K, Ca und Mg, dagegen arm an pflanzlichen Abbauprodukten ist. Allerdings kann der in der Nähe der Meeresoberfläche gebildete Kalk am Meeresboden wieder aufgelöst werden, wenn hier das Wasser an Kalk ungesättigt ist.

8. Das Verhalten der bei der Verwitterung gelösten Stoffe

Getrennt von den vielseitigen Verwitterungsprozessen soll das Verhalten der bei der Verwitterung gelösten Stoffe geschlossen dargestellt werden, da Bindungsart, Verbleib und Wanderung dieser Stoffe für die Bodenbildung und für die Eigenschaften des Bodens sehr bedeutsam sind (Barth, Correns und Eskola 1960).

A. Das Verhalten der Alkalien und Erdalkalien

Die Alkalisalze (Chloride, Sulfate) werden mehr oder weniger schnell aus der Verwitterungsmasse (Boden) durch das Regenwasser ausgewaschen, da sie leicht löslich sind. Dagegen sind Erdalkalisalze (fast nur solche des Ca und Mg) weniger schnell in Lösung zu setzen. Überwiegend liegen Ca und Mg als schwer lösliche Carbonate im Boden vor und werden in Verbindung mit Kohlensäure als Bicarbonate leicht löslich. Als relativ leicht lösliche Sulfate treten die Erdalkalien, mit Ausnahme in semiariden und in ariden Gebieten, selten auf. Die genannten ausgewaschenen Salze wandern im humiden Klimabereich in das Grundwasser und in das Flußwasser. In semiariden und ariden Bereichen kann es zur Ansammlung von Alkali- und Erdalkalisalzen an oder nahe an der Erdoberfläche kommen.

B. Das Verhalten der Kieselsäure

Bei der Silikatverwitterung geht Kieselsäure in Lösung (s. Kap. 4, C, 4). Bei der Austrocknung des Bodens kann sie als wasserhaltige Gallerte, also als Gel ausgeschieden werden und schließlich nach Wasserabgabe Opal bilden. In saurer Verwitterungslösung (Bodenlösung) ist die Kieselsäure nur wenig löslich; sie wird dann vorwiegend als Gel (später Opal) ausgeschieden. Mit steigendem pH nimmt im allgemeinen die Löslichkeit der Kieselsäure im Boden zu. In warmen, ariden Räumen kann die Kieselsäure an der Oberfläche als Kruste abgeschieden und bei großer Hitze sogar kristallin (Quarz) werden. Aus tropischen Böden dagegen erfolgt eine starke Abfuhr der Kieselsäure.

C. Das Verhalten des Aluminiums

Das Al-Ion bleibt unter pH 4 und über pH 9 in der Bodenlösung gelöst. Als freies Ion kann Al sorbiert und ausgetauscht werden; das scheint bei der Flockung der Feinsubstanz des Bodens wichtig zu sein. Andererseits stellt es ein Pflanzengift dar. Um pH 7 bildet sich Aluminiumhydroxyd. Da bei etwa pH 7 Kieselsäure löslich ist und bei entsprechendem Klima stark weggeführt wird, kann es zu starker Anreicherung von Aluminiumhydroxyd kommen (Bauxit). Die hierdurch gekennzeichnete Verwitterungsart nennt man *allitisch* im Gegensatz zur *siallitischen*, bei der die Kieselsäure ganz oder weitgehend in der Verwitterungsmasse verbleibt. Bei Verdunstung der Bodenlösung kann ebenfalls das Al ausgeschieden werden.

D. Das Verhalten des Eisens

Bei der Verwitterung bildet sich dreiwertiges Eisen, das eine Reihe von Verbindungen mit Wasser eingeht, die unter dem Sammelbegriff „Brauneisen" zusammengefaßt werden; es verleiht der Verwitterungsmasse gelbe und rostbraune Farbtöne. Teils sind diese Verbindungen kristallin ($\alpha-FeOOH=$Goethit, $\gamma-FeOOH=$Lepidokrokit), teils auch amorph. In warmen, ariden Gebieten sind die dreiwertigen Eisenverbindungen arm an Wasser und daher von leuchtend gelber, braunroter und roter Farbe. Im Extrem entstehen Hydrohämatit und Hämatit. In warmen, ariden Gebieten können diese an der Erdoberfläche dünnkrustig abgeschieden werden (Wüstenlack).

Das dreiwertige Eisenion ist nur unter pH 3 beständig. Dann ist es auch sorbierbar sowie austauschbar und trägt zur Flockung der Feinsubstanz des Bodens, wie das Al-Ion, entscheidend bei. Eisen ist als Ferrat im basischen Milieu löslich, aber nicht beständig. In den anderen pH-Bereichen fallen die reinen Eisen(III)verbindungen aus. Sie können indes als Eisenoxydhydrat in Kolloidform (unter Schutzwirkung von Kieselsäure und Humussäure), als Eisenchelat oder in der Koppelung mit anderen organischen Substanzen wanderungsfähig sein.

Im Reduktionsmilieu, verursacht durch Luftabschluß und die Gegenwart von organischer Substanz, wird dreiwertiges Eisen in die zweiwertige Form übergeführt, was an grauen, grünlichen oder bläulichen Farben erkennbar ist. Das zweiwertige Eisenion ist wanderungsfähig, oft bildet es mit Kohlensäure lösliches Bicarbonat.

E. Bildung von Aluminiumsilikaten

Aluminiumhydroxyd und Kieselsäure können sich in der Verwitterungsmasse (Boden) zu Aluminiumsilikaten vereinigen. Je nach den vorliegenden Bedingungen entsteht eine bestimmte Silikatart oder eine Artengruppe von Silikaten. Die Aluminiumsilikate sind teils kristallin und stellen als solche die *Tonminerale* dar, teils sind sie amorph und heißen dann *Allophane*. Die Tonminerale der Montmorillonitgruppe scheinen sich vorwiegend in basischen Böden des semihumiden und noch trockeneren Klimas, jedoch auch im Ca- und Mg-reichen, feuchten Milieu zu bilden; der Kaolinit entsteht dagegen in erster Linie im feuchten tropischen und subtropischen Klima, während im gemäßigt warmen, humiden Bereich der Illit bei weitem überwiegt (Näheres s. bei CORRENS 1949, JASMUND 1955 und bei v. ENGELHARDT 1961).

9. Die Intensität und die Eigenart der Verwitterung in den wichtigsten Klimazonen

Die Verwitterung ist stark klimaabhängig. Das Klima wirkt direkt durch Niederschlag, Luftfeuchtigkeit und Temperatur und indirekt über die Vegetation. Bei der Intensität der Verwitterung muß neben der Stärke der Verwitterungseinflüsse der Faktor „Zeit" stets in Rechnung gestellt werden. — Nur die Hauptzüge der Verwitterung in den vier Hauptklimabereichen sollen kurz dargestellt werden; selbstverständlich gehen diese Klimabereiche auch hinsichtlich der

Verwitterungsart allmählich ineinander über (Näheres s. bei BRINKMANN 1956, LAATSCH 1957, DERRUAU 1958, WILHELMY 1958, LOUIS 1960, SCHEIDEGGER 1961).

A. Der nivale Klimabereich

Im nivalen Klimabereich erlaubt die niedrige Temperatur nur eine *schwache* chemische Verwitterung; die Bildung von Feinsubstanz ist gering. Dagegen bewirkt die physikalische Verwitterung (in erster Linie Frostsprengung, teils auch in geringem Maße Insolation) eine starke mechanische Zerlegung des Gesteins. An der Oberfläche, wo sich der Wechsel des Gefrierens und Auftauens im Ablauf von Tag und Nacht abspielt, ist die Zerstörung am stärksten; sie wirkt sich aber auch bis zur jährlichen Auftaugrenze der stets in der Tiefe gefrorenen Erdrinde aus.

Unter Rohhumus, saurem Anmoor und saurem Moor (Palsen) kann eine intensive biologisch-chemische Zersetzung des Gesteins stattfinden.

B. Der gemäßigt warme, humide Klimabereich

Für eine mittelmäßige, chemische Verwitterung bietet dieser Klimabereich die notwendige Feuchtigkeit und Wärme. Eine geschlossene Pflanzendecke stellt sich ein, sobald eine auch nur schwache Verwitterungshaut anspruchslosen Pflanzen einen Standort bietet. Die biologisch-chemische Verwitterung ist mithin in diesem Klima von großer Bedeutung. Dagegen treten alle Formen der physikalischen Verwitterung stark zurück. — Die Verwitterungstiefe des gemäßigt warmen, humiden Klimaraumes überschreitet selten 2 m.

C. Der warme, humide Klimabereich

Viele und regelmäßige Niederschläge sowie fast stets relativ hohe Temperaturen sind in diesem Klimabereich die Vorbedingungen für eine starke chemische Verwitterung (SCHAUFELBERGER 1950, 1953, SIMONS 1951). Die mächtigen Verwitterungsdecken der Tropen von zum Teil über 30 m sind Zeugen dieser intensiven Gesteinszerstörung. Selbstverständlich müssen wir für die Bildung solcher Decken lange Zeiträume annehmen. Die Bildungszeit der mächtigen Verwitterungsschichten in Zentralafrika und im äquatorialen Südamerika reicht z. B. großenteils in das Mesozoikum zurück. Für die biologisch-chemische Verwitterung sind in diesem Klima auch die Vorbedingungen günstig, indes wird sie von der rein chemischen Verwitterung weit übertroffen; davon zeugt die chemische Verwitterung in relativ großer Tiefe unter der Erdoberfläche, wohin keine Pflanzenwurzel gelangt. Die physikalische Verwitterung tritt in diesem Klimabereich ganz zurück.

D. Der warme, aride Klimabereich

In warmen, ariden Klimagebieten fehlt für eine intensive chemische Verwitterung die Feuchtigkeit; die chemische Verwitterung ist also gering. Wohl werden an der Erdoberfläche die Gesteine langsam zersetzt. Das zeigt sich sogar an Steinen, die mit „Wüstenlack" (=Eisenkruste) oder einer anderen „Schutzrinde" überzogen sind; im Innern zermürben sie (LOUIS 1960). Im ganzen gesehen ist das jedoch nicht als starke chemische Verwitterung aufzufassen. Dieser warme, aride Klimabereich ist der Schauplatz der Insolation (s. Kap. 3, A), die einen starken mechanischen Gesteinszerfall herbeiführt; dieser spielt sich indes nur an und nahe an der Erdoberfläche ab.

Literatur

Anderson, D. H., und H. W. Hawkes: Relative mobility of the common elements in weathering of some schist granite areas. Geochim. Cosmochim. Acta **14**, H. 3, 204–210 (1958).

Barth, T. F. W., C. W. Correns und P. Eskola: Entstehung der Gesteine, S. 118–130, 206, 212–247. Berlin-Göttingen-Heidelberg: Springer. 1960. — Behrend, F., und G. Berg: Chemische Geologie, S. 220–389. Stuttgart: Enke. 1927. — Bespalow, W. F., und W. A. Sokolow: Perioden chemischer Verwitterung auf dem Territorium von Kasachstan. Utschenye Sapiski Kasach. Uniw. **27**, No. 3, 55–66 (1957). — Blanck, E.: Physikalische Verwitterung. Handb. d. Bodenl., Bd. II. Berlin: Springer. 1929 a. — Chemische Verwitterung. Handb. d. Bodenl., Bd. II. Berlin: Springer. 1929 b. — Biologische Verwitterung durch lebende Organismen (höhere Pflanzen). Handb. d. Bodenl., Bd. II. Berlin: Springer. 1929 c. — Die biologische Verwitterung als Ausfluß der in Zersetzung begriffenen organischen Substanz. Handb. d. Bodenl., Bd. II. Berlin: Springer. 1929 d. — Einführung in die genetische Bodenlehre, S. 89–281. Göttingen: Vandenhoeck & Ruprecht. 1949. — Brinkmann, R.: Abriß der Geologie, Bd. I, 8. Aufl., S. 7–14. Stuttgart: Enke. 1956. — Butler, J. R.: The geochemistry and mineralogy of rock weathering. Geochim. Cosmochim. Acta **6**, H. 5/6, 268–282 (1954).

Cornu, F.: Die heutige Verwitterungslehre im Lichte der Kolloidchemie. Kolloid-Z. **4**, 291–295 (1909). — Correns, C. W., und W. v. Engelhardt: Neue Untersuchungen über die Verwitterung des Kalifeldspates. Naturwiss. **26**, 137 (1938). — Correns, C. W.: Die chemische Verwitterung der Silikate. Naturwiss. **28**, 369 (1940). — Einführung in die Mineralogie. Berlin-Göttingen-Heidelberg: Springer. 1949.

Derruau, M.: Précis des Géomorphologie. Paris: Masson. 1958. — Dücker, A.: Über Strukturböden im Riesengebirge, ein Beitrag zum Bodenfrost- und Lößproblem. Z. Dtsch. Geol. Ges. **89**, 113–129 (1937).

Engelhardt, W. v.: Neuere Ergebnisse der Tonmineralienforschung. Geol. Rdsch. **51**, H. 2, 457–477 (1961).

Haselhoff, E.: Untersuchung über die Zersetzung bodenbildender Gesteine. Landw. Versuchsstat. **70**, 53–143 (1909). — Hempel, L.: Über Verwitterung und Abtragung im Granit Nordostspaniens. N. Jb. Geol. Paläont., Mh. H. 5, 227–233 (1958). — Bilanzen zur Reliefgestaltung der Erde. Geogr. Ber. **5**, Nr. 15, H. 2, 97–107 (1960).

Jasmund, K.: Die silikatischen Tonminerale, 2. Aufl. Weinheim/Bergstr.: Verlag Chemie. 1955.

Klapp, E.: Acker- und Pflanzenbau, 5. Aufl. Berlin-Hamburg: Parey. 1958. — Knetsch, G.: Über aride Verwitterung unter besonderer Berücksichtigung natürlicher und künstlicher Wände in Ägypten. Z. Geomorph., Suppl. **1**, 190–205 (1960). — Köstler, J: Waldbau. Berlin-Hamburg: Parey. 1950. — Kutschera, L.: Wurzelatlas mitteleuropäischer Ackerunkräuter und Kulturpflanzen. Frankfurt/M.: DLG. 1960.

Laatsch, W.: Dynamik der mitteleuropäischen Mineralböden, 4. Aufl. Dresden u. Leipzig: Steinkopff. 1957. — Lang, R.: Verwitterung und Bodenbildung als Einführung in die Bodenkunde. Stuttgart: Schweizerbart. 1920. — Die Verwitterung. Fortschr. Min. **7**, 176 (1922). — Louis, H.: Allgemeine Geomorphologie. Berlin: de Gruyter. 1960.

Machatschek, F.: Das Relief der Erde, Bd. I und II. Berlin-Nikolassee: Borntraeger. 1955. — Meinecke, F.: Granitverwitterung, Entstehung und Alter der Granitklippen. Z. Dtsch. Geol. Ges. **109**, 483–498 (1957). — Mortensen, H.: Die Bedeutung der Salzsprengung. Peterm. Geogr. Mitt. 1933. — Murawski, H.: Beispiele für die Wirkungsabhängigkeit mechanischer Kräfte von der Gesteinsbeschaffenheit. Z. Geomorph. **2**, 1–12 (1958).

Niggli, P.: Die chemische Gesteinsverwitterung in der Schweiz. Schweiz. Min. Petr. Mitt. **5**, H. 2, 322 (1926). — Niklas, H.: Untersuchungen über den Einfluß von Humusstoffen auf die Verwitterung von Silikaten. Intern. Mitt. Bodenkde. **2**, 214 (1912).

Ramann, E.: Bodenkunde, 3. Aufl. Berlin: Springer. 1911. — Kohlensäure und Hydrolyse bei der Verwitterung. Cbl. Min., Geol. u. Paläont. 233—265 (1921).

Schaufelberger, P.: Wie verläuft die Gesteinsverwitterung und Bodenbildung in den Tropen? Schweiz. Min. Petr. Mitt. **30**, 238–499 (1950). — Tropische Verwitterung und Bodenbildung über Andesit und Diorit. Schweiz. Min. Petr. Mitt. **33**, 127–143 (1953). — Scheidegger, A. E.: Theoretical Geomorphology. Berlin-Göttingen-Heidelberg: Springer. 1961. — Schellenberg, G.: Biologische Verwitterung durch lebende Organismen (niedere Pflanzen). Handb. d. Bodenl., Bd. II. Berlin: Springer. 1929. — Scholz, H.: Studien über die Bodenbildung zwischen Rehoboth und Walvis Bay.

Diss. Bonn 1963. — Simons, H. S.: Zur Frage des Verlaufs der Gesteinsverwitterung in den Tropen. Schweiz. Min. Petr. Mitt. **31**, 589–591 (1951).

Wagner, P.: Einführung in die Erd- und Landschaftsgeschichte, S. 17–249. Öhringen: Hohenlohesche Buchhandlung F. Rau. 1950. — Wilhelmy, H.: Klimamorphologie der Massengesteine. Braunschweig: Westermann. 1958.

b) Soil Texture

By

L. De Leenheer

1. Definition

Most soils are mineral soils, in which the mineral component largely predominates. Moreover, in these soils the mineral particles are generally smaller than 2 mm. The size distribution of the particles or grains is different from soil to soil, but each soil sample contains particles ranging in size from those easily visible to very small ones, discernible only under an ultramicroscope.

By convention, the mineral grains have been divided into size groups, such as clay, silt and sand (Ton, Schluff und Sand), each group having a lower and an upper limit of the particle diameter. Thus, each soil has a characteristic particle size distribution, which is called the texture. The analytical method used to determine the texture, that is, the content of the individual size groups or separates in the soil, is called the mechanical analysis[1].

The texture of the soil is one of its most important characteristics. Therefore knowledge of texture is basic for all soil research work and soil classification. Certainly a given soil class, for example a sand soil, may show differences in physical and chemical properties, but these changes are limited and a relationship between soil texture and such soil properties as permeability, plasticity, water-holding capacity, base-exchange capacity, etc. is evident. Soil texture is also one of the most permanent properties, not being changed by man using normal methods of land cultivation.

2. Soil Separates

The particle size groups or separates, as they are conventional, are not the same all over the world.

There are three schemes of classification in common use (Table 2): the first in the table was established by the United States Department of Agriculture (U. S. D. A.); the second was proposed by the International Society of Soil Science (I. S. S. S.), and the third is that accepted by modern European soil scientists.

Nowadays the second is only in use in soil testing laboratories (for fertilizer requirements of soils), being too incomplete for soil survey and soil research work.

The diameter limit 0.25 mm in use in the United States is not used in Europe, where 0.2 mm was chosen; consequently the classes medium sand and fine sand do not have the same meaning.

A very important observation is that a soil sample never consists of one single separate; thus a clay soil has a high content of the clay fraction and a sandy soil a high proportion of the sand fraction, but both always contain a certain pro-

[1] The term "mechanical analysis" is internationally used, although the French term "analyse granulométrique" (granulometric analysis) is much better. Many English writing authors prefer "particle size analysis" over "mechanical analysis".

Table 2. *Soil texture classification*

U.S.D.A. scheme[1]			I.S.S.S.[2] or Atterberg scheme			Modern European scheme			(10) Approximate surface area in cm² of 1 g material
(1) Name	(2) German French	(3) Diameter limits in mm	(4) Name	(5) German French	(6) Diameter limits in mm	(7) Name	(8) German French	(9) Diameter limits in mm	
very coarse sand	sehr grober Sand; sable très grossier	2.00—1.00				very coarse sand	see column (2)	2.00—1.00	12
coarse sand	grober Sand; sable grossier	1.00—0.50	coarse sand	Grobsand; sable grossier	2.00—0.20	coarse sand	see column (2)	1.00—0.50	23
medium sand	mäßig feiner Sand; sable moyen	0.50—0.25				medium sand	see column (2)	0.50—0.20	47
fine sand	feiner Sand; sable fin	0.25—0.10	fine sand	Feinsand; sable fin	0.20—0.02	fine sand	see column (2)	0.20—0.10	120
very fine sand	sehr feiner Sand; sable très fin	0.10—0.05				very fine sand	see column (2)	0.10—0.05	227
silt	(no appropriate translation)[3]	0.05—0.002	silt[4]	Schluff; limon	0.02—0.002	pulver[5] or coarse silt	Pulver; pulver	0.05—0.02	470
						fine silt	Schluff; limon[6]	0.02—0.002	1820
clay	Ton; argile	<0.002	clay	Ton; argile	<0.002	clay	Ton; argile	<0.002	17000

[1] Scheme of the U.S. Department of Agriculture, not to be confounded with the older classification of the American Bureau of Soils, which is now abandoned.

[2] I.S.S.S. = International Soil Science Society, now in general abandoned too, except in soil testing laboratories.

[3] The 0.05–0.002 mm fraction is now called "fraction limoneuse" by the Belgian Soil Mapping Centre; this however is wrong as the French meaning for "limon" and "fraction limoneuse" has always been the 0.02–0.002 mm fraction.

[4] The term silt, sometimes used for the 0.02–0.002 mm fraction, should be rejected, as the common meaning is now: 0.05–0.002 mm fraction.

[5] The latin word "Pulver" (powder), which is used in German and Dutch with the same meaning, and which is known as a verb in English (pulverise) and French (pulvériser), was recently proposed for this fraction (De Leenheer and Van Ruymbeke 1960).

[6] The Netherlands are the only country in the world taking 0.016 instead of 0.020 mm as the upper limit of the "slib-fractie" ("limon" fraction). This was a consequence of an erroneous calculation, made several years ago using the law of Stokes. Instead of correcting the time of sedimentation, the size limit was changed after the error was detected, so that all the analyses made before, were comparable with those made subsequently.

portion of the other size fractions. This means that the word "clay" considered as a soil separate has a different sense from the word "clay" used by a farmer, to whom it means a "clay soil".

Therefore, to avoid misunderstanding, the word "fraction" should be used after the name of a soil separate; this is particularly important for European soil scientists. For example the French word "limon" and the German word "Lehm" are used for loams in which the predominant soil separate is not the "limon" fraction (0.02 — 0.002 mm) but the "pulver" fraction (0.02 — 0.05 mm).

Before making a soil analysis, the coarse fragments in the soil sample are eliminated by passing it through a 2 mm sieve (with circular sieve openings). The coarse fragments are called fine gravel (2 — 20 mm), coarse gravel (2 — 20 cm) and stony material (larger than 20 cm). Soils are usually grouped into soil textural classes, which are named according to the fraction which is most important.

3. Mechanical Analysis

In a soil sample the ultimate particles are closely associated with soil organic matter, carbonates, hydrated iron oxides and other cementing materials, to build aggregated particles such as clods, crumbs and concretions. The term structure is used to express these grouping into compound soil particles.

A mechanical analysis, i. e. the determination of the percentage of the grain size fractions of a soil sample, is not possible until the cementing materials have been destroyed or broken down; all the mineral soil particles must be present as individuals.

A. Pre-treatment of the Soil Sample

The usual procedure consists in removing first the gravelly and stony material, root fragments, etc., by passing the air-dry sample through a 2 mm sieve; aggregates larger than 2 mm should be crushed until they pass the 2 mm sieve.

Soil organic matter is then destroyed with hydrogen peroxide. Next the soil sample is treated with hydrochloric acid to destroy the carbonates and the easily soluble hydrated iron oxides. The soil components dissolved by these treatments must be eliminated either by filtration or decantation, to prevent the small soil particles from sticking together. Finally a dispersing agent is added to make sure that all the soil particles will stay in suspension as individuals, without flocculation.

It is evident that such a pre-treatment is conventional and several methods have been proposed. In the footnote is described the pre-treatment, which is in use in the Soil Research Centre of Ghent, where an average of 400 mechanical analyses are carried out monthly[1].

---

[1] 100 ml H_2O_2, 30 percent are added 2 to 3 times, to 20 g of soil prepared by sieving through a 2 mm sieve and crushing the aggregates in a mill. Each time, but only after a few hours, the solution is brought to boiling temperature. To the warm suspension 0.33 ml of concentrated HCl is added from a burette for each percentage unit of $CaCO_3$ present in the sample (a result known from a separate carbonate analysis), followed by an extra addition of 100 ml 0.2 N HCl. Enough water is added to double the suspension volume, which is then boiled for 10 minutes. After sedimentation of the suspension, the clear supernatant liquid is removed by suction and about 400 ml of distilled water is added. After a certain time the clear supernatant liquid is removed again. The washing and sedimentation process of the soil suspension is repeated 3 to 4 times (until the supernatant liquid stops staying clear after 2 to 3 hours of sedimentation). Finally 50 ml of a dispersing agent is added (obtained by dissolving 3.17 g $Na_2CO_3.2H_2O$ and 14.28 g sodium hexametaphosphate per liter) (De Leenheer and Van Hove 1956, 1957).

B. Separation of the Sand- from the Silt- and Clay Fraction

The grain size lower limit for the sand fraction is 0.05 mm or 50 microns. The separation of the sand size groups from the silt- and clay fractions is important, because the further separation of the different sized sand groups on one hand and the separation of the clay-, "limon"- and "pulver" fractions on the other hand is done using different techniques.

The differentiation of the sand particles is carried out by using calibrated sieves (standard sieves); the determination (it may not of course be a real separation) of the percent of the finer size intervals is based on the settling velocity of grains in a vertical water column at a constant temperature, following Stokes' law. Two problems arise for grains with a diameter of about 50 microns: they have such a high velocity of sedimentation that a good determination in a water column of normal dimensions (30 cm high) is not feasible, and the dry grains are too small to be sieved conveniently in a mechanical way because of effect of "jumping" on the sieve wires.

A good way for separating sand from silt (silt meaning the "limon" fraction plus "pulver" fraction) and clay is to make use of the method called "sieving under distilled water". The water passing through the sieve openings carries along the finer particles; a gentle rubbing of the soil sample on the sieve with the finger while washing with distilled water insures a quantitative separation in a short time.

C. Separation of the Sand Separates

The mineral soil fraction which remains on the 50-micron sieve is collected, dried and then allowed to pass through a set of standard sieves. The operation of sieving is done mechanically for 10 minutes, the sieve with the largest openings (1 mm) being put on the top of the set.

D. Separation of the Finer Separates

The formula of Stokes (Fig. 5) is the starting point for the calculation of the velocity of sedimentation in a fluid column, the grains being supposed spherical.

$$V = \frac{2}{9} \cdot \frac{g\, r^2 \,(D_m - D_w)}{\eta} \quad [1]$$

where V = velocity of the settling particles.
g = acceleration of gravity being 981 cm/sec^2 for a latitude of $\pm 51°$ (Western Europe).

[1] It is important to know the underlying assumptions in Stokes' law, so that the limitations of the formula will be recognized. The most important of these assumptions are:

1. The spherical particles (grains) must be large compared to the size of the molecules of the liquid, because of Brownian motion. This puts a lower limit on the measurable particles.

2. No turbulent currents or slipping between particles and fluids may occur. This puts an upper limit on the measurable particles of 0.060–0.090 mm.

3. The concentration of the particles in the liquid must not be too large. It must be smaller than 5 percent to avoid edge effect and subsequent capillary forces.

For the detailed discussion of all the assumptions, it is referred to Gessner (1931) and Hooghoudt (1945).

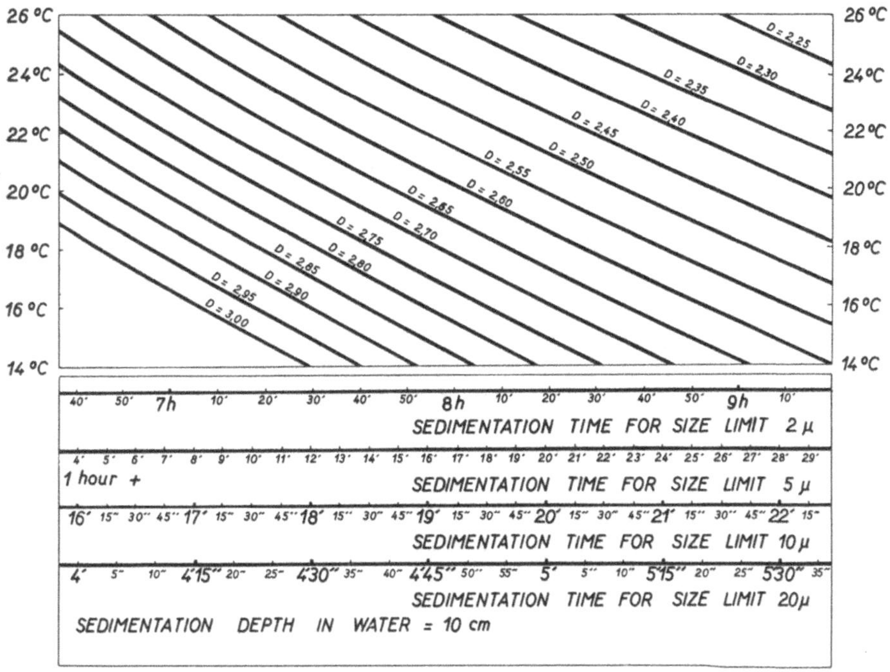

Fig. 5. Graphical presentation of STOKES' law

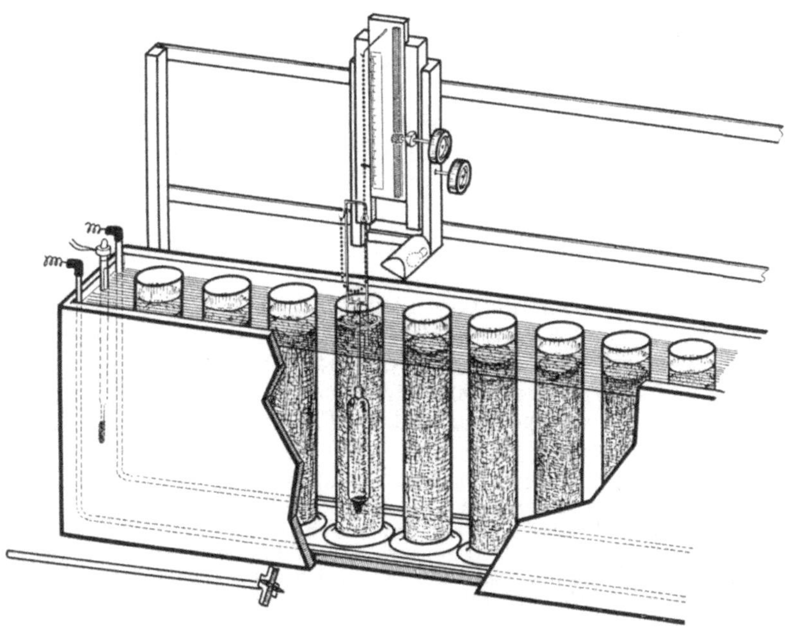

Fig. 6. Large water container for sedimentation cylinders at constant temperature (for chain-hydrometer determinations; see Fig. 8)

r = radius of the settling spherical grains.

$D_m = 2.675$ = average density of finer particles[1].

D_w = density of water, after the addition of the dispersing agent, at the temperature of the determination.

η = viscosity in poises of the water at the temperature of the determination.

The value of η is highly influenced by changes in temperature during the determination, so it is very important to keep this temperature constant (using a large water container at constant temperature with the sedimentation cylinders in it) (Fig. 6).

Under these circumstances r is the only variable in the experiment. This means that for a particle size of $2\,\mu$ (upper limit of the size of clay particles) the settling velocity is so small that a sedimentation over a depth of 10 cm requires about 8 hours at 20° C and about 7 hours at 25° C; for a particle with a diameter of $20\,\mu$ the sedimentation time for the same depth is only 4′40″ (4 min. 40 sec.) at 20° C and 4′10″ at 25° C (Fig. 5).

Several techniques have been developed for the determination of the fine separates. In the first methods a real separation of the fractions was realized by the decantation or the drawing off of the remaining suspension after a given time of settling, such as in the Atterberg method (Fig. 7A).

An important improvement in mechanical analysis came with the pipette method in 1922-23[2].

In this method a small volume of the suspension (as 25 or 50 ml) is taken with a pipette at a given depth and after a given time of sedimentation. The volume withdrawn is evaporated and the dry material weighed. Knowing the initial total volume of the suspension, it is possible to calculate the amount of the finer separates.

The most satisfactory apparatus for pipette analysis is that of Köhn (1928), which is still in use (Fig. 7B).

Other methods have been proposed, based on the principle of the rising current elutriation (in German: Spülmethoden); this means a separation of particles into size groups by a rising water current. Fig. 7C shows the Kopecky's (1901) elutriator as example.

A general discussion of these methods, which are practically out of use now, is made by Gessner (1931).

A much easier but unsatisfactory method in precise works is the hydrometer or, so called, Bouyoucos method (1927), which is still used (mostly by civil engineers studying soil mechanics (Krynine 1947). It consists in the determination of the density of a suspension after a given time of settling. A hydrometer, calibrated to read grams of suspension per liter (Fig. 7D), is put into the suspension

---

[1] D_m is generally put = 2.65 (density of quartz particles); a large number of determinations (De Leenheer and Van Hove 1956) of D_m of the finer particles ($< 50\,\mu$) in Belgian soils gave a value = 2.675, which may be explained as the influence of a small film of undissolved iron oxides surrounding the silt and clay particles. For very small clay particles, e.g. between 0.1–0.2 μ, the density becomes smaller, about 2.39, because of the water hull surrounding the clay.

[2] The pipette method was developed independently in 1922–23 in England (Robinson 1922), in the United States (Jennings, Thomas and Gardner 1922) and in Germany (Krauss 1923).

at given intervals and from the readings a cumulative curve may be drawn directly. In this method the moment of the determination and thus the period of settling is well known, but the depth of the determination is never constant, as the depth to which the hydrometer sinks changes with the density of the suspension, that is, with the clay- or clay- and silt content. A constant depth of determination, however, is one of the first conditions to be fulfilled. For a discussion of the hydrometer methods we refer to CASAGRANDE (1934).

A satisfactory and rapid technique is the chain-hydrometer method, which combines the speed of the hydrometer method (without evaporations or weighings) and the accuracy of a weighing. This method is now in use in many laboratories of Austria, Belgium, France, Germany, etc. For this efficient method some details follow (DE LEENHEER *et al.* 1952, 1954, 1955, 1956, 1957).

In a chain-hydrometer, such as illustrated in Fig. 8, the distance between the point of the needle b' and the middle of the immersion body a is a constant value; by slowly immersing the chain-hydrometer until the point of the needle touches the surface of the water column, one makes sure that the depth of the measurement is always constant.

For each chain-hydrometer the following values must be known:

P = the weight of the hydrometer without chain,

cm = the weight of one cm of chain,

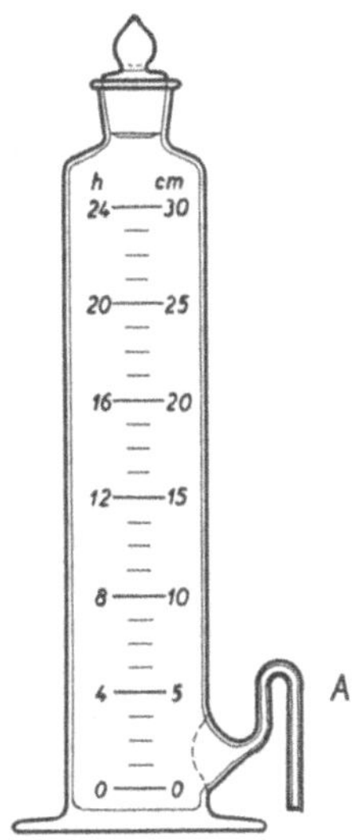

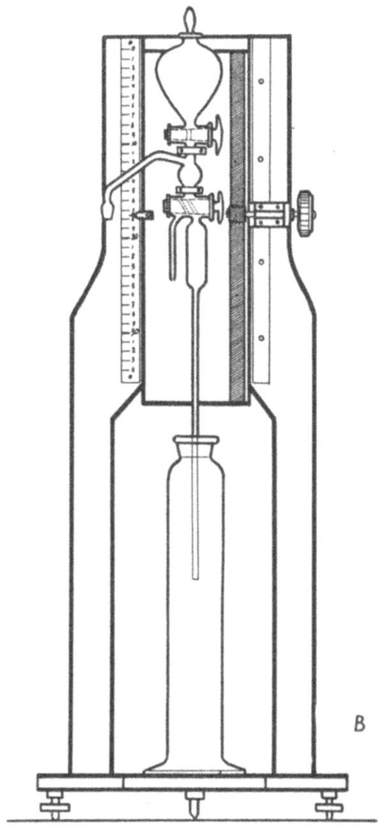

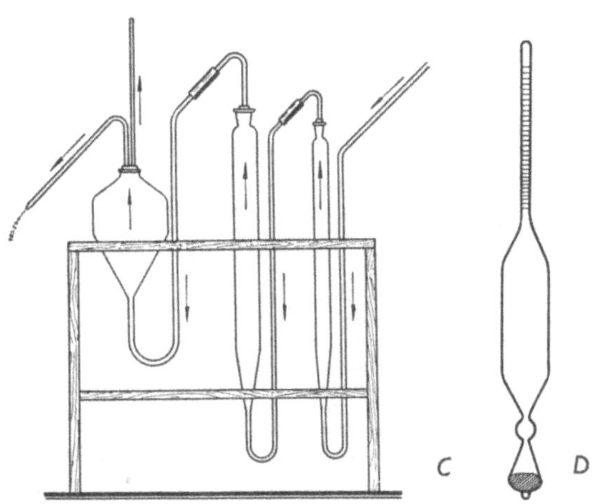

Fig. 7. *A* ATTERBERG cylinder for drawing off the finer soil separates. At the right the height of the water column in cm, at the left the sedimentation time in h. — *B* Pipette apparatus of KÖHN. — *C* KOPECKY's elutriator. The limits of the size groups obtained depend upon the diameters of the tubes and the volume of water used per second. — *D* Improved hydrometer of BOUYOUCOS; the calibrated scale on the stem gives the amount of suspension in grams per liter

V = the volume of the hydrometer at the moment the point of the needle touches the water surface, $V = \dfrac{P + p' + p''}{D}$,

p' = the amount of supplementary weights (p' in Fig. 8),

p'' = the weight given by the chain to sink the hydrometer in distilled water at a given temperature (the temperature chosen for the determinations) until the needle point b' touches the water surface,

L = the theoretical depth of the determination (distance between the point of the needle and the centre of the immersion body),

L' = the real depth of the determination; by sinking the hydrometer with a volume V the water level in the sedimentation cylinder (with radius $= R$) will rise, so a correction is necessary:

$$L' = L - \frac{V}{2\pi R^2}$$ [1]

(De Leenheer and Maes 1952).

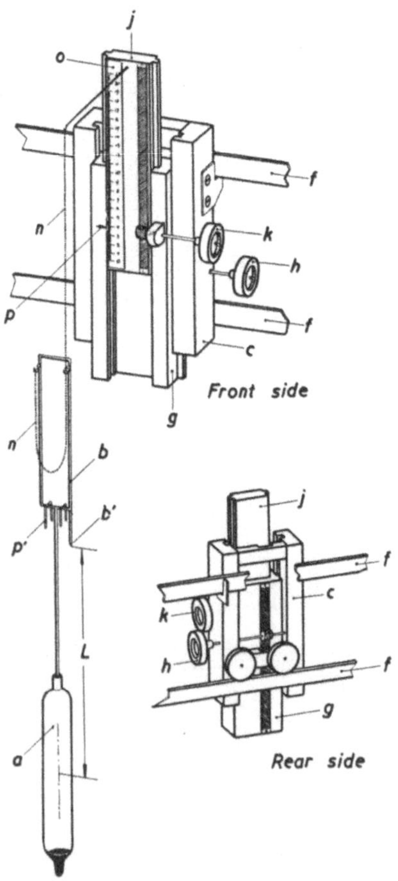

Fig. 8. Chain-hydrometer equipment (see also Fig. 6). *a* Immersion body, partly filled with mercury. — *b* Hydrometer stem, with *b'* point of the needle which touches the water surface, *p'* supplementary weight. — *L* Theoretical depth of determination. — *n* Chain with a well-known weight per cm. — *p* Zero dash for the cm and mm scale *o* (reading of the charge of the chain). The support *c* is movable along the rails *f*; it allows a general adjustment through the sliding body *g* (revolving button *h*) and an adjustment for the scale *o* through the sliding body *j* (revolving button *k*)

Knowing these characteristics, the sedimentation time t for a determination is calculated easily

$$t = L'/v$$

where v is the settling velocity, deduced from the law of Stokes.

A most important observation must be made. It is common practice in mechanical analysis to make the determination of all the separates or fractions, except one, which is generally the fraction $20-50\,\mu$. The latter is then calculated by difference, a procedure which may be very misleading, because all the errors are incorporated in one separate. Therefore it is extremely important to start the determination of the finer fractions by sucking off with a pipette a given volume (say 50 ml) of the suspension, after a vigorous shaking and a sedimentation time equal to zero. The withdrawn suspension, after evaporation and weighing, gives the total amount of the finer separates ($0-50\,\mu$). With

[1] The water level rises $\dfrac{V}{\pi R^2}$ cm, but the correction for the centre of the immersion body must be calculated for $\dfrac{V}{2}$.

Table 3. *Mechanical analysis and calculation of some characteristics of the particle size distribution*

Loam soil

(1) Class intervals in mm	(2) Percent of frequency n	(3) Cumulative percent c	(4) Midpoint of each size grade m	(5) Product of col. (2) and (4) nm	(6) Deviations midp. from arith. mean d	(7) Deviations squared. d^2	(8) Product of col. (2) and (7) nd^2	(9) Log of the midp. of each size grade $\log m$	(10) Product of col. (2) and (9) $n \log m$
0. —0.002	18.6	18.6	0.001	0.0186	—0.025	0.000625	0.011625	—3.00000	—55.800000
0.002—0.010	6.8	25.4	0.006	0.0408	—0.020	0.000400	0.002720	—2.22185	—15.108580
0.010—0.020	13.1	38.5	0.015	0.1965	—0.011	0.000121	0.001585	—1.82391	—23.893221
0.020—0.050	56.8	95.3	0.035	1.9880	+0.009	0.000081	0.004601	—1.45593	—82.696824
0.050—0.100	4.5	99.8	0.075	0.3375	+0.049	0.002401	0.010805	—1.12494	—5.062230
0.100—0.200	0.2	100.0	0.150	0.0300	+0.124	0.015376	0.003075	—0.82391	—0.164782
0.200—0.500	0	100.0	0.350	0	+0.324	0.104976	0	—0.45593	0
0.500—1.000	0	100.0	0.750	0	+0.724	0.524176	0	—0.12494	0
				Sum $= \sum(nm)$ $= 2.6114$ $AM = \frac{\Sigma(nm)}{100}$ $AM = 0.026$ mm			Sum $= \sum(nd^2) =$ $= 0.034411$ $\sigma = \sqrt{\frac{\Sigma(nd^2)}{100}}$ $\sigma = \sqrt{\frac{0.034411}{100}}$ $\sigma = 0.019$ mm		Sum $=$ $\sum(n \log m) =$ -182.725637 $GM =$ antilog of $\frac{\Sigma(n \log m)}{100}$ $GM = 0.015$ mm

the chain-hydrometer the fractions 0—2, 0—10 and 0—20 μ are then determined and a check of the results is very easy.

When the total of the soil components (percent organic matter + percent $CaCO_3$ + percent moisture at 105° C + all the separates of the mechanical analysis) is not equal to 100 ±3, one (or more) of the determinations is incorrect and the error has to be detected. This is the criterium used in the Soil Research Centre at Ghent.

4. Representation of Results of Mechanical Analysis

In every textbook on soils, texture is described as an important physical property, but it is unusual to find a textbook in which the representation of the results of the mechanical analysis is given the place it deserves.

The results of the mechanical analysis are given as percentages by weight of the dry mineral soil sample (after pre-treatment). Each percentage refers to a certain fraction of which the upper and lower size limits are always given. In that form, the data are frequently unwieldy and sometimes difficult to interpret and a comparison of a great number of soils is then impossible. Consequently other methods have been tried; in this respect the contribution of sedimentary petrologists was very important (KRUMBEIN and PETTIJOHN 1938).

A. Histograms

The simplest way of depicting the results of the mechanical analysis is the histogram. Rather than giving a definition, we will describe how to make a correct

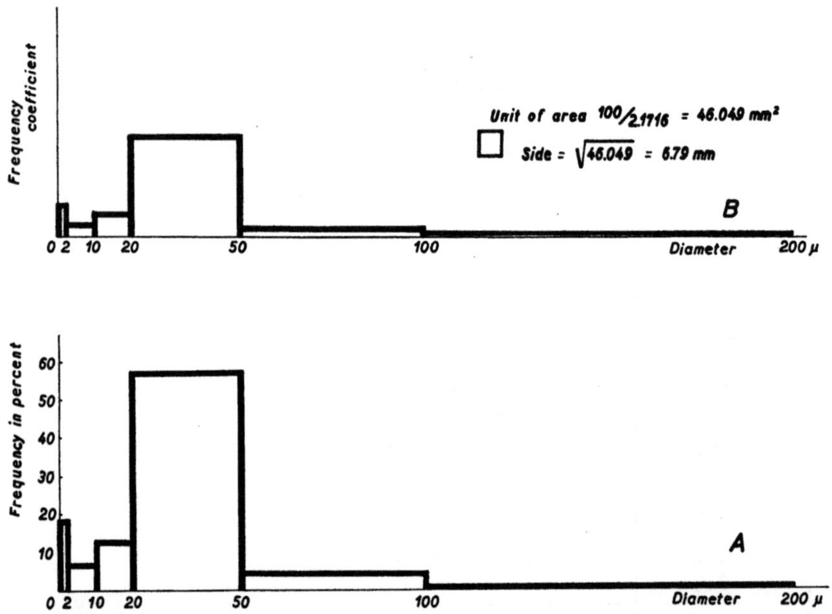

Fig. 9. Incorrect (*A*) and correct (*B*) way of drawing a histogram

histogram. Table 3 shows the results of the mechanical analysis for a loam soil. In column (1), the class intervals are given in mm and in column (2) are the percentages of sample weight, referred to as the frequencies.

On a graph the diameters in millimeters are put as the independent variable and the frequency as the dependent variable (Fig. 9). The class intervals are laid off along the horizontal axis and above each of the classes a vertical rectangle is drawn with a width equal to the class interval and a height, *not equal* to the frequency *of* the class, *but proportionate* to the frequency coefficient *in* the class, as described below.

The first way of drawing a histogram, in which the heights of the vertical rectangles are drawn directly in terms of the frequency, is the most common, but it is incorrect (Fig. 9A). The height of the vertical blocks should be drawn in terms of the area and a *unit area* always should be given for comparison (Fig. 9B). The two drawings of Fig. 9 illustrate the correct and the incorrect way. In the correct method the area but not the heights of the rectangles represent the frequencies. As area involves both width and height, the height of the rectangle will depend on the class interval. Furthermore, the total area of all the rectangles should be equal to unity. How this calculation is carried out, becomes evident from Table 4 and Fig. 9B.

Table 4. *The correct way to draw a histogram (see also Fig. 9)*

Loam soil

(1)	(2)	(3)	(4)	(5)	(6)
Class interval in mm	Difference between upper and lower limit	Frequency	Area (2) × (3)	Proport. area being (4) : $\frac{1}{2.1716}$ = 0.46049	New height of each area (5) : (2)
0. —0.002	0.002	18.6	0.0372	0.01713	8.565
0.002—0.010	0.008	6.8	0.0544	0.02505	3.131
0.010—0.020	0.010	13.1	0.1310	0.06032	6.032
0.020—0.050	0.030	56.8	1.7040	0.78468	26.156
0.050—0.100	0.050	4.5	0.2250	0.10361	2.072
0.100—0.200	0.100	0.2	0.0200	0.00921	0.092
totals	—	100.0	2.1716	1.00000	

A histogram is not always satisfactory, as it is influenced by the class intervals used in the analysis and its shape varies according to the particular class limits chosen. Therefore this correct way of graphic presentation is seldom used.

B. The Frequency Curves

Histograms give step by step the variation of the dependent variable as a function of the independent variable, while frequency curves are smooth curves, which show the variation of the dependent variable, as a continuous function of the independent variable (Fig. 10A). Histograms, cumulative curves and frequency curves are related mathematically and any one may be obtained from any other (KRUMBEIN and PETTIJOHN 1938).

One can say that a frequency curve is equal to a histogram when the class intervals become smaller and smaller and finally reach zero (Fig. 10B).

It is easy to draw a smooth curve over the histogram. The smooth curve and the step function will be equivalent if the shaded areas above and below each horizontal line of the histogram and the frequency curve are drawn equal for each step in the histogram (Fig. 10A).

The grain size distribution of a sediment also can be calculated to enable one to draw directly a frequency distribution. Therefore the distribution should follow a *normal curve* of which the formula is:

$$y = \frac{iN}{s}\left(\frac{1}{\sqrt{2\pi}}\, e^{-\frac{1}{2}\left(\frac{x-\bar{x}}{s}\right)^2}\right) \text{ (DIXON and MASSEY 1957).}$$

where y = ordinate, height of curve corresponding to an assigned value of x,
x = abscissa (values on the horizontal axis),
N = number of cases observed,
i = length of class interval used to draw the histogram,
s = a parameter being equal to the standard deviation of the distribution,
π = a constant (approximately 3.1416),
e = a constant (approximately 2.7183),
$\bar{x}$ = a parameter being equal to the mean of the distribution.

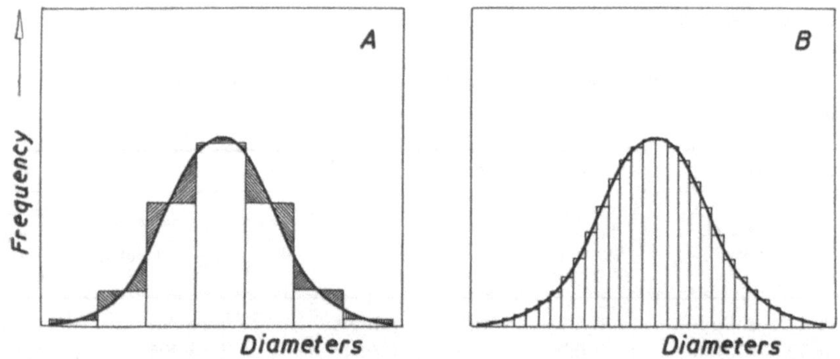

Fig. 10. Graphical way of fitting a normal curve to a histogram

A frequency curve cannot be drawn in this way for every grain size distribution, because the curve should be normal. To check this, a statistical method can be used by which a "chi square test" is carried out. As this can be found in textbooks on statistics (see for example OSTLE 1958, p. 54) and as we will discuss a graphical way of checking normality a few paragraphs further, we will not discuss the "chi square test" here.

A limitation in the application of the above stated formula is that i, the class interval, should be the same for the whole granulometric analysis. This in general is not the case. However, it can be deduced from the cumulative curve, as frequency values for equal class intervals can be read off. Once $i, N, s, \bar{x}$ are known, it is possible by reading off the value $\left(\frac{1}{\sqrt{2\pi}}\, e^{-\frac{1}{2}\left(\frac{x-\bar{x}}{s}\right)^2}\right)$ in a table to draw the frequency curve.

As this mathematical procedure is laborious, it is more convenient to use a quick graphical way by which it is possible not only to check the normality of the distribution, but also to read off the mean and the standard deviation, which completely characterize the distribution.

C. Summation Curves and Deduced Characteristics

A method of representing the analytical data, independently of the class intervals used, is the cumulative curve or summation curve. The summation curve is made by plotting ordinates which represent the total amount of material of which the diameter is given in the abscissa. These curves allow one to read the content of particles below any given diameter. Theoretically two types of cumulative curves are possible. They are of equal validity: the ordinates give the percentage of material with a diameter greater or smaller than the given diameter. In soil science the second type of summation curve is commonly used. An example of how such a curve is calculated is given in Table 3, columns (1), (2) and (3) and is represented in Fig. 11. The horizontal scale may be either arithmetic or logarithmic. For convenience, the logarithmic scale is ordinarily used in soil science, as it is possible in this way to put all size classes in a relatively small distance on the horizontal axis.

A summation curve allows one to compare the results from two laboratories using different diameter limits, and to correct the result of a dry sieving when the sieve opening does not correspond to the limits of a particular size range, e. g. using an American Taylor sieve with openings of 105 and 210 microns instead of 100 and 200.

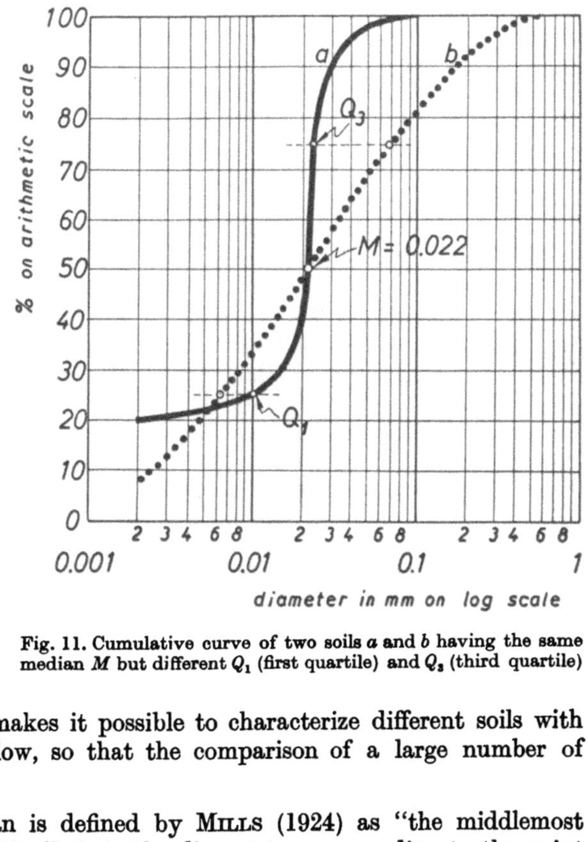

Fig. 11. Cumulative curve of two soils *a* and *b* having the same median *M* but different Q_1 (first quartile) and Q_3 (third quartile)

The great value of the summation curve is that it makes it possible to characterize different soils with a few numbers as given below, so that the comparison of a large number of samples is possible.

The Median. The median is defined by MILLS (1924) as "the middlemost member of the size distribution"; it is the diameter corresponding to the point where the cumulative curve is crossed by the 50 percent line: thus 50 percent of the material present has a smaller and 50 percent a greater diameter than the median. In the examples shown in Fig. 11, the median is 0.022 mm.

The median can be determined even when one does not have the complete analysis, as it is not influenced by the extreme limits in particle size at each end of the distribution. The disadvantage is that it cannot be manipulated algebraically: for example one cannot take averages. Furthermore the median does not characterize a sediment or a soil: two loamy soils may have the same median, but be of different texture (Fig. 11).

The Quartile Deviation. To measure the average spread in respect to the median, the quartile deviation is used. One finds it on either side of the median by reading the diameters which correspond to frequencies of 25 and 75 percent. From the way the cumulative curve is constructed, it is evident that the quartile value at 25 percent (or first quartile Q_1) is the diameter for which 25 percent of the distribution is smaller and 75 percent larger. The third quartile is designated as Q_3 and is the diameter for which 75 percent of the distribution is smaller and 25 percent larger. In the example (soil *a*, Fig. 11) Q_1 and Q_3 are respectively 0.009 and 0.023 mm. Both deviations can be given as a geometric quartile deviation S_0, which is called the "sorting coefficient", and which is given by the equation $S_0 = \sqrt{Q_1/Q_3}$, but is always such that the larger quartile is the numerator and $S_0 > 1$. In our example $S_0 = \sqrt{\frac{0.023}{0.009}} = \sqrt{2.56} = 1.6$.

The sorting coefficient S_0 is a measure of the spread of the particle size distribution around the median. The coarseness of the material has no bearing on it, nor would it make any difference if centimeters were used instead of millimeters as the units to express the diameters.

For the study of pedogenetic processes the knowledge of the sorting coefficient of the sedimentary parent material is often important. From a long experience some critical values of S_0 are known. A value of S_0 less than 2.5 indicates a well sorted sediment, a value of about 3.0 means a normally sorted sediment and a value greater than 4.5 a poorly sorted sediment (Trask 1932).

An important remark must be made here. The S_0 values cannot be compared directly with one another on a purely arithmetic basis. When one soil parent material has a value $S_0=3$ and another $S_0=1.5$, one cannot say that the latter is twice as well sorted as the first. However, since S_0 increases geometrically, the logs of S_0 will form an arithmetic series, so that the logs of the S_0 values may be directly compared. Log_{10} of $S_0=3$ is equal to 0.477 and $\log_{10}$ of $1.5=0.176$; so the spread in the grain size distribution is $0.477:0.176=2.71$ or nearly three times as great in the first as in the second material.

D. Some Arithmetic and Statistical Values to Characterize the Grain Size Distribution

The Arithmetic Mean Diameter. The arithmetic mean diameter represents the value of the particle diameter at the centre of the frequency distribution, taking into account the centre of every size interval. The arithmetic mean diameter is affected by every particle in the distribution and in some respects it is more typical than the median. Besides this, the arithmetic mean can be manipulated algebraïcally.

The way the arithmetic mean *AM* is calculated is given in Table 3. In the first column the class intervals are given in millimeters. The percentages by weight (frequency *n*) are placed in column (2). Column (4) has the midpoints *m* of each size interval, regardless of whether the intervals are equal or not. In column (5) the frequency is multiplied by the midpoint value and the sum of the products $\Sigma\,(nm)$ is at the foot of the column. This sum divided by the total frequency (100) gives the arithmetic mean diameter in millimeters: $AM = \frac{\Sigma\,(nm)}{100}$. In the example of Table 3 it is 0.026 mm.

This number does not characterize completely the distribution of the particles. It is quite possible that two samples with the same mean size may differ because of the spread. Consequently it is necessary to know how the various sizes are distributed around the mean; therefore the standard deviation is calculated.

The Standard Deviation. If the frequency curve is perfectly symmetrical, the standard deviation is a measure of the spread, such that about 68 percent of the distribution is contained within the 2 values deduced from the arithmetic mean plus and minus the standard deviation. This calculation also is illustrated in Table 3. Once the arithmetic mean is found, the deviation *d* of the midpoint from the arithmetic mean is entered in column (6). In the following column these deviations are squared and in column (8) the squared deviations are multiplied by the frequency in each grade. This total sum at the foot of the column is devided by 100 and the square root is taken, giving the standard deviation σ.

$$\sigma = \sqrt{\frac{\Sigma\,(nd^2)}{100}}$$

In our example the standard deviation equals 0.019 mm. In the example of Table 3 this means that about 68 percent of the particles must be found between $AM \pm \sigma = 0.026 \pm 0.019 =$ between 0.045 and 0.007 mm. Column (2) of Table 3 confirms it: the total amount between 0.050 and 0.010 mm = 69,9 percent.

The Mean Weight Diameter. In calculating the arithmetic mean diameter, one assumes a linear distribution of the particles in each size class. This can be true if a very large number of size intervals are used, but normally this is not the case. In order to use a more accurate method, the mean weight diameter can be calculated graphically. By doing so, the actual centre of the size interval is better observed. In order to find the mean weight diameter a graph is drawn with the accumulated frequency in ordinate and the upper limit of the particles on an arithmetic scale in abscissa (Fig. 12). The shaded area in the figure is measured with the planimeter. As one axis reads in percent and the other axis is 10 times magnified (1 mm = 1 cm) the measured area should be divided by $100 \times 10 = 1000$. In our given example the mean weight diameter is 22:1000 = 0.022 mm.

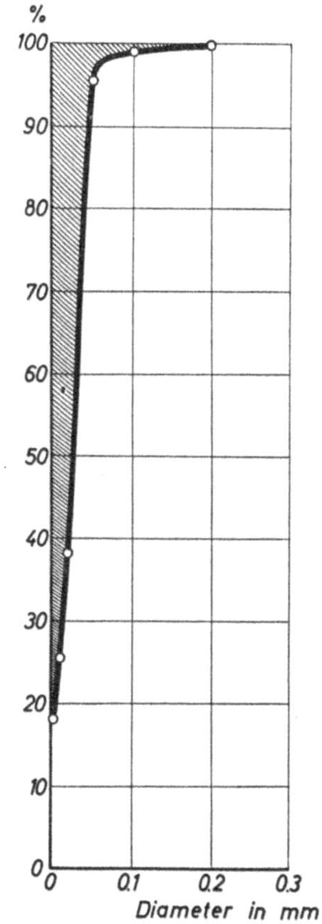

Fig. 12. Graphical determination of the mean weight diameter. The shaded area is measured with a planimeter

The Mode. In a given particle size distribution of soils or sediments the influence of the larger particles may be great. The mean weight diameter discussed above gives a disproportionate (to high) weight to the larger particles, so that a number of research workers feel that the emphasis should be on the particles which are most numerous. Two parameters are mainly used in this respect: the geometric mean and the mode.

The mode is the particle size which is the most abundant, and therefore the most typical in the distribution. It has its most precise meaning in a unimodal distribution with a single peak. The mode is then the mean diameter of the most abundant size interval and can be deduced from an ordinary frequency curve or from the data given in Table 3. In our example the mode equals 0.035 mm, this value being the midpoint of the most abundant size interval 0.02 — 0.05 mm (Table 3, column (4)). The midpoint can be influenced by many factors, such as size interval. Therefore a better characteristic is obtained by the geometric mean diameter.

The Geometric Mean Diameter and Deduced Characteristics. The mathematical definition is as follows: it is the nth root of the product of n items. Its direct computation is tedious, but by using the logarithm of the midpoint of each class interval the method can be made similar to that used in computing the arithmetic mean (Table 3, columns (9) and (10)). The algebraic sum of the products of the frequency n multiplied by the logarithm of the midpoints m (column 10) is divided by the total frequency 100 and the antilogarithm is calculated; this value is the geometric mean diameter GM, which is 0.015 mm in our example.

$$GM = \text{antilog of } \frac{\Sigma\,(n \log m)}{100}$$

More convenient is the use of a graphical method by which the geometric mean and standard deviation are obtained directly if the sediments follow a normal distribution. It is the experience that undisturbed sediments follow a log-normal distribution; consequently a log-normal distribution plotted on log-probability paper[1] produces a straight line (Krumbein and Pettijohn 1938) as can be seen

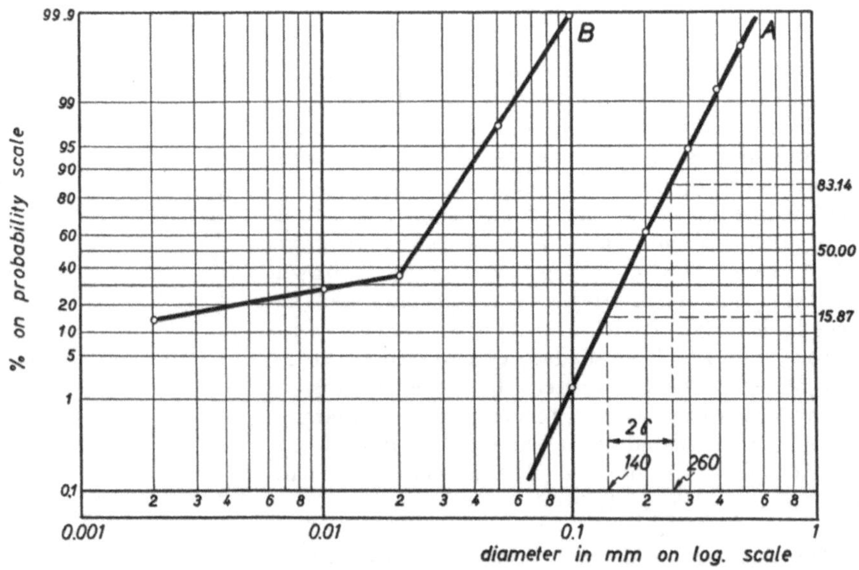

Fig. 13. Undisturbed sediments follow a log-normal distribution, with a straight line (*A*) on log-probability paper. A soil (*B*) may show a broken line (as a consequence of alteration). *A* represents a sandy sediment, *B* a subsoil sample (B-horizon of a gray brown podsolic profile on loam)

in Fig. 13, the right hand curve *A* of a sand sediment. This is a slightly different opinion of the one expressed by Doeglas (1952), who believes that undisturbed sediments should give a straight line when plotted on arithmetic-probability paper. The difference between the two opinions concerns the extremes of the diameter range. Indeed, it is found that for any real distribution, some deviation from a straight line is to be expected at the extremes of the diameter range, as well on log-probability paper as on arithmetic-probability paper. On the first however the deviation is smaller than on the second, and therefore preference is given to the use of log-probability paper. The deviation from the straight line at the extremes of the diameter range is not a serious limitation, since these extreme sizes account for only a small portion of the total sample weight.

[1] On log-probability paper the diameter of the particle is put on logarithmic scale in the abscissa and the frequency on probability scale in ordinates.

Using a log-normal graph, the mean diameter of the log-distribution is the size corresponding to 50 percent and the standard deviation of this distribution is numerically equal to the difference between the values of the size of the particles at 50 percent and 84.13 or 15.87 percent $\left(50 \pm \frac{68.26}{2}\right)$.

Thus σ = size at 84.13 percent, minus the size at 50 percent = size at 50 percent, minus the size at 15.87 percent.

If the data of Table 3 are plotted on a log-probability curve, we obtain a broken line as can be seen in Fig. 13 (the left-hand curve *B*). This shows that the loam soil material is no longer undisturbed; indeed, the loessial deposit which was the parent material of the soil was subject to alterations and changed into a gray-brown podsolic profile. The sample is taken from the *B*-horizon in which clay accumulation took place during the soil development process.

5. Physical Nature and Properties of Soil Separates

As a soil never is composed of only one separate, it never will have the properties of one single textural fraction. The understanding of the agricultural properties of a soil therefore requires a knowledge of the properties of the major particle size groups: sand, silt and clay.

The Sand Fraction. A given weight (say 10 g) of sand grains has a small specific surface[1] compared to that of the same weight of clay particles. Owing to this relatively small surface a sand particle has low physico-chemical activity; a sandy soil shows practically no plasticity, works easily and is therefore called a "light" soil (which has nothing to do with the real weight of the soil).

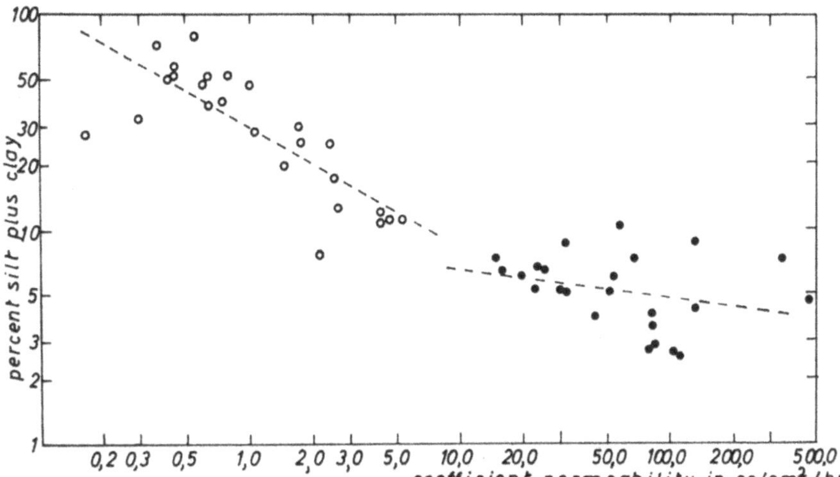

Fig. 14. The correlation of permeability coefficient with percentage of silt plus clay changes according to the dominant sand grain size. ○ the predominant sand fraction is smaller than 105 μ; ● the predominant sand fraction is larger than 105 μ but smaller than 250 μ (Aronovici 1946)

The sand grains, having a relatively large diameter, will also have large spaces between the grains; these pores will facilitate not only the movement of water but also of air.

[1] The specific surface is the surface of material per unit weight; it is expressed in cm^2/g. See Table 3, column (10).

Sandy soils are well aerated and drain well (possibly too rapidly); their water-holding capacity is low. The larger the average grain size of the sand the smaller the water-holding capacity and the less desirable its agricultural value under normal conditions of farming.

The influence of the average grain size of the fraction on physical soil properties is illustrated in Fig. 14. This figure makes evident that the correlation of permeability coefficient with percentage of silt plus clay changes according to the dominant sand grain size (Aronovici 1946).

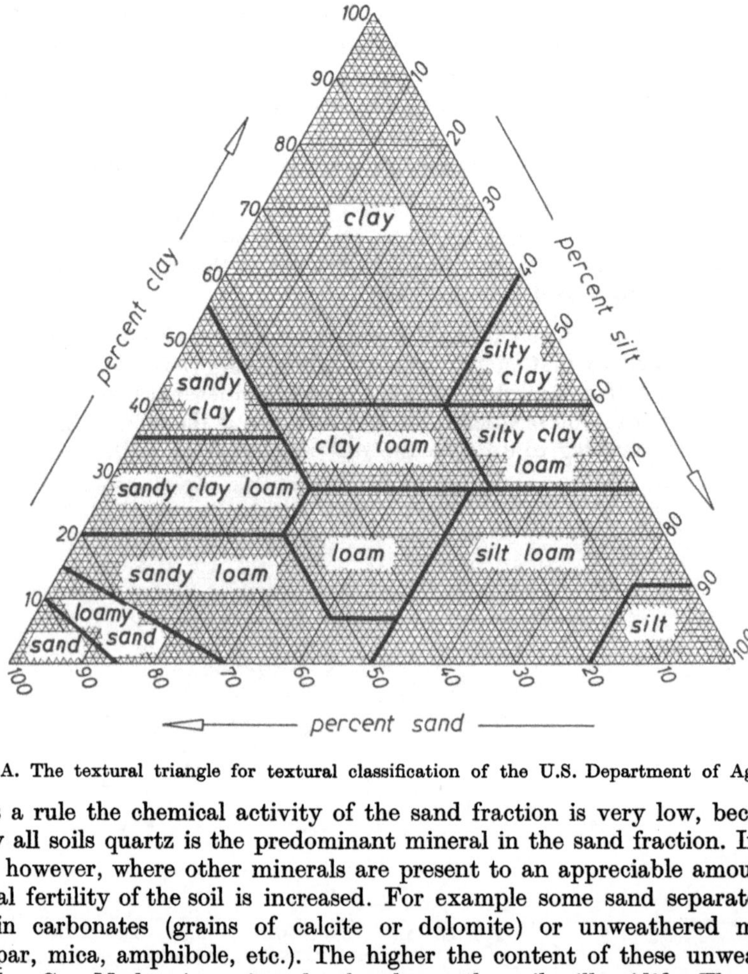

Fig. 15A. The textural triangle for textural classification of the U.S. Department of Agriculture

As a rule the chemical activity of the sand fraction is very low, because in nearly all soils quartz is the predominant mineral in the sand fraction. In those cases, however, where other minerals are present to an appreciable amount, the natural fertility of the soil is increased. For example some sand separates may contain carbonates (grains of calcite or dolomite) or unweathered minerals (feldspar, mica, amphibole, etc.). The higher the content of these unweathered K-, Na-, Ca-, Mg-bearing minerals, the slower the soil will acidify. Therefore a microscopic examination of the sand fraction may be of great interest.

The Clay Separate. The sand particles are considered as the residual grains of physical weathering; the clay particles, on the other hand, are generally the newly formed products of chemical weathering and these synthesized particles have the shape of very small plates. Moreover these plates are not compact but show a lamellar porosity, and some of them are swellable under the influence of water uptake. All these properties explain the very high active surface of the clay fraction: therefore a clay soil (supposing poor structure conditions) has very bad aeration and drainage, a high plasticity when wet, but hard and compact through shrinkage when dry. These soils being hard to work are called heavy soils by the farmers.

The large active surface confers on clay soils a high adsorbing power for nutritive elements and a high water-holding capacity. As the pores between the clay particles are extremely small, the water movement is extraordinarily slow and the water available for plant growth is only a fraction of the water-holding capacity.

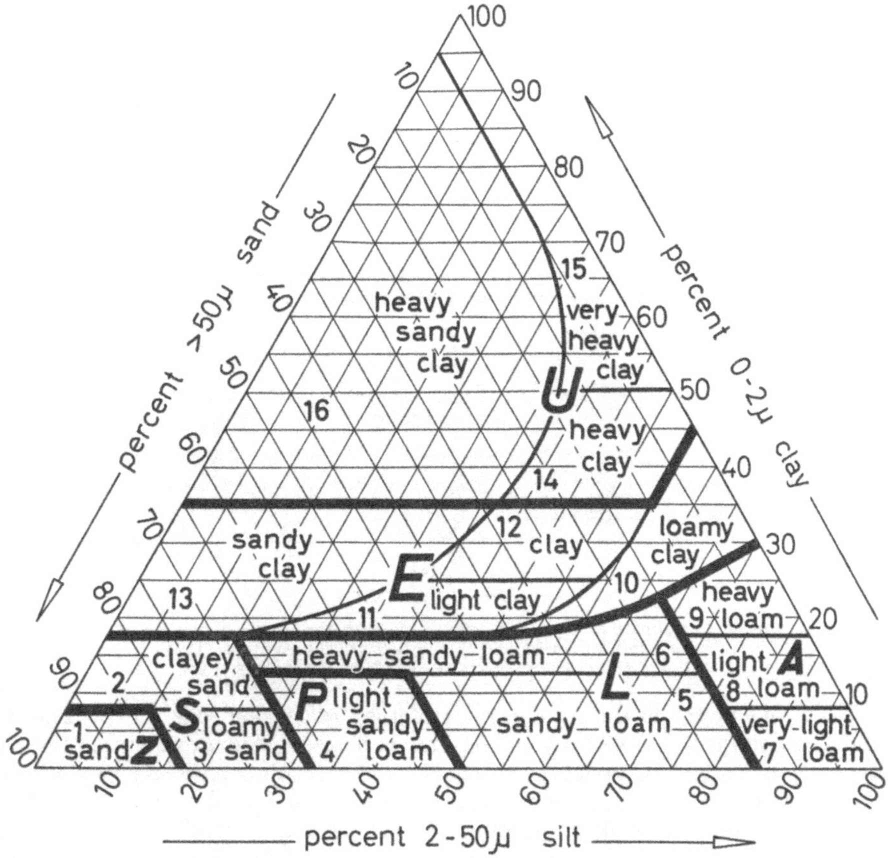

Fig. 15 B. The textural triangle in use in Belgium (Soil Survey Committee). The letters are the textural symbols used in the formulae of the soil types (limits in thick lines), the numbers represent the different soil textural classes (limits in thin lines). (Tavernier *et al.* 1958)

The chemical activity of the soils, especially clay soils, changes very much with the mineralogical composition of the clay fraction. In this respect mention must be made of the clay mineral illite, which is known for its important "potassium supplying power"; therefore the knowledge of the illite-content of clay soils is an important condition to be fulfilled for a good potassium fertilizing advice (Van Ruymbeke 1964).

The Silt Separate. The particles of the silt size interval are similar in shape to those of the sand fraction, irregularly rounded but seldom flat. The grains of the silt separate are in fact microsand particles, and quartz is normaly the dominant mineral.

In the silt fraction, however, the content of unweathered minerals is generally higher, so that a microscopic examination of this separate is often valuable for getting an idea of the natural reserve of plant nutrient elements. This examination is not easy and only the "pulver" fraction (20—50 μ) can be studied under the normal polarising microscope. The other properties, such as plasticity and stickiness, active surface, drainage, aeration and water-holding capacity, are midway between those of the sand and clay separates.

In respect to the water economy of the soil, the fraction 20—50 μ seems to play an important part, as far as water content at field capacity is concerned, but has only a low influence on the water content at wilting point.

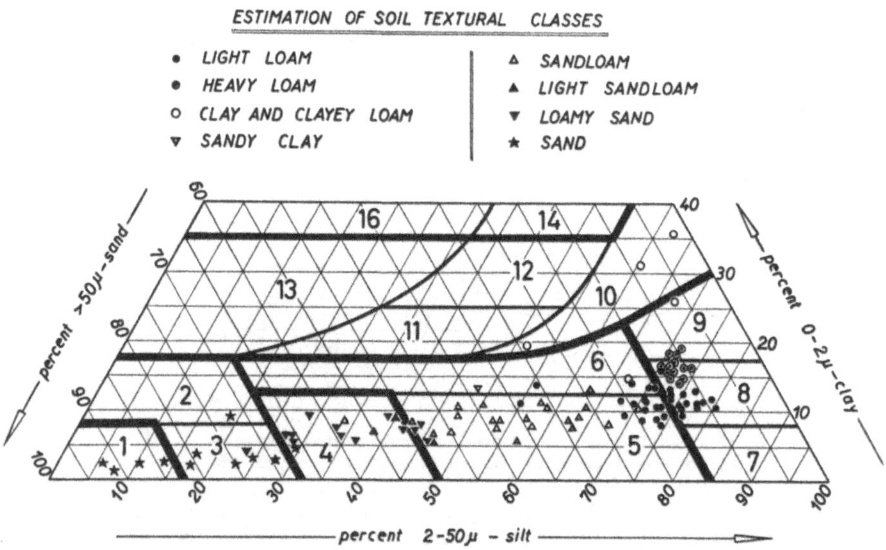

Fig. 16. Texture field estimations and their comparison with the results of the mechanical analysis. The numbers correspond to the textural classes of Fig. 15 B (DE LEENHEER, APPELMANS and VAN RUYMBEKE 1958)

This was observed during a regional survey of marine soils (DE LEENHEER and VAN RUYMBEKE 1960) with a large range in size distribution, and in $CaCO_3$- and humus content. Correlation calculations were made by the electronic brain (I.B.M. Ordinator 650) for more than hundred soil samples from arable land and about one hundred from meadows. As independent variables were taken the following soil components expressed in percent of weight: clay fraction (0—2 μ), "limon" fraction (2—20 μ), "pulver" fraction (20—50 μ), the percent $CaCO_3$, the percent organic matter and the water content when the soil sample was taken (which was practically always at field capacity).

As dependent variables were taken the water content at field capacity, the water content at wilting point, the volume weight, the volume percent of drainage pores, the volume percent of useful water retaining pores and the stability of the aggregates.

For the 117 soil samples from arable land the I.B.M. ordinator calculated the following formulae for water content at field capacity (F.C.) and at wilting point (W.P.).

$$\text{F.C.} = +0.290655 + 0.400696\ (\%\ 0\text{—}2\ \mu) + 0.262886\ (\%\ 20\text{—}50\ \mu) + 0.426860\ (\%\ CaCO_3) + 3.224689\ (\%\ \text{org. mat.})$$

$$W.P. = +0.666156+0.363627\ (\%\ 0\text{—}2\ \mu)+0.044663\ (\%\ 20\text{—}50\ \mu)+0.199028\ (\%\ CaCO_3)+0.849308\ (\%\ \text{org. mat.}).$$

In both cases the "limon" fraction (2—20 μ) shows no specific influence on the water content; the same was true for the samples of meadows.

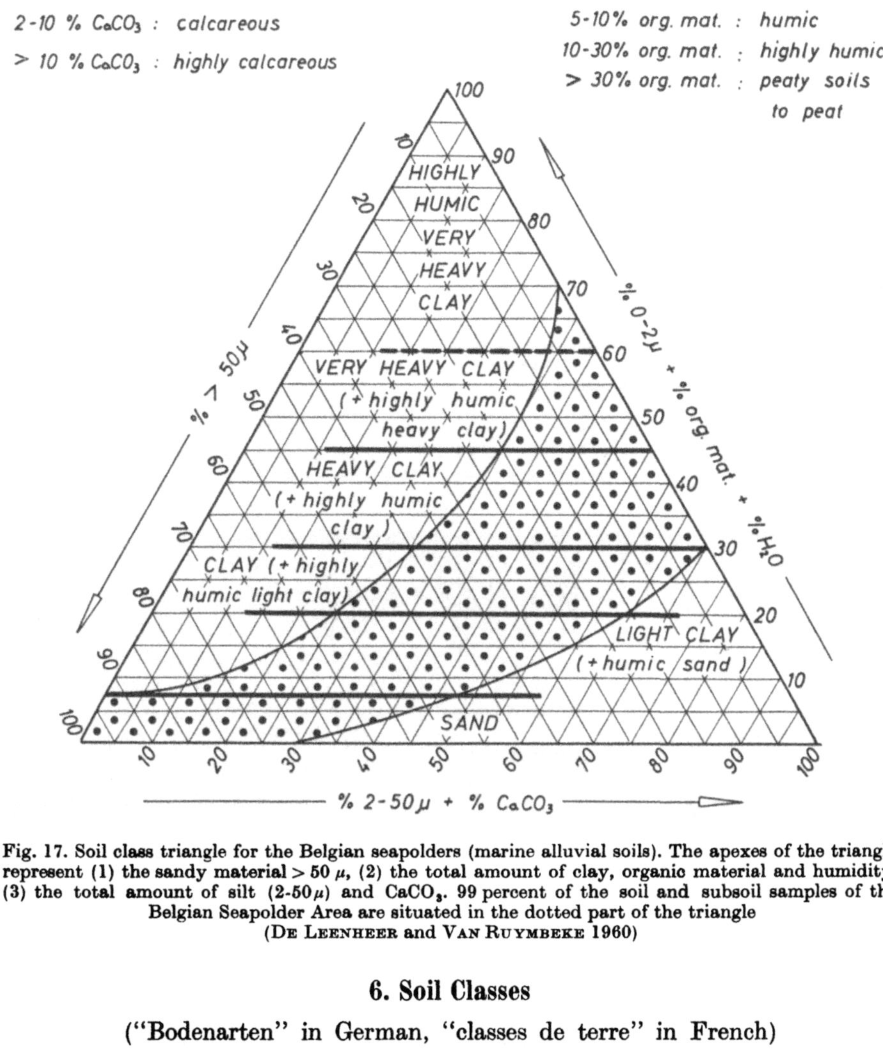

Fig. 17. Soil class triangle for the Belgian seapolders (marine alluvial soils). The apexes of the triangle represent (1) the sandy material > 50 μ, (2) the total amount of clay, organic material and humidity, (3) the total amount of silt (2-50 μ) and $CaCO_3$. 99 percent of the soil and subsoil samples of the Belgian Seapolder Area are situated in the dotted part of the triangle (De Leenheer and Van Ruymbeke 1960)

6. Soil Classes

("Bodenarten" in German, "classes de terre" in French)

There is some confusion in the use of soil textural class names (derived from laboratory analysis) and soil classes (subjective terms given by agronomists and farmers); some authors consider them as being synonymous.

The use, however, of these terms in a synonymous manner is not correct. Another point of confusion is the fact that soil fertility and productivity are often put in direct relation to the textural class of the arable layer. This also is an erroneous correlation. Only in a given area or country, and as long as the soil profile may be considered as homogeneous, can the foregoing correlations be considered approximately correct. Consequently an error is made if these correlations are presented for universal application.

On the other hand it is important to introduce a standardization of soil textural class names to be able to make a comparison of soil descriptions from different countries. A suitable way of doing this is not to try to come to an international standardization of terms, which will never be realized, but to refer to the position of the sediment in a textural triangle. An experienced field pedologist,

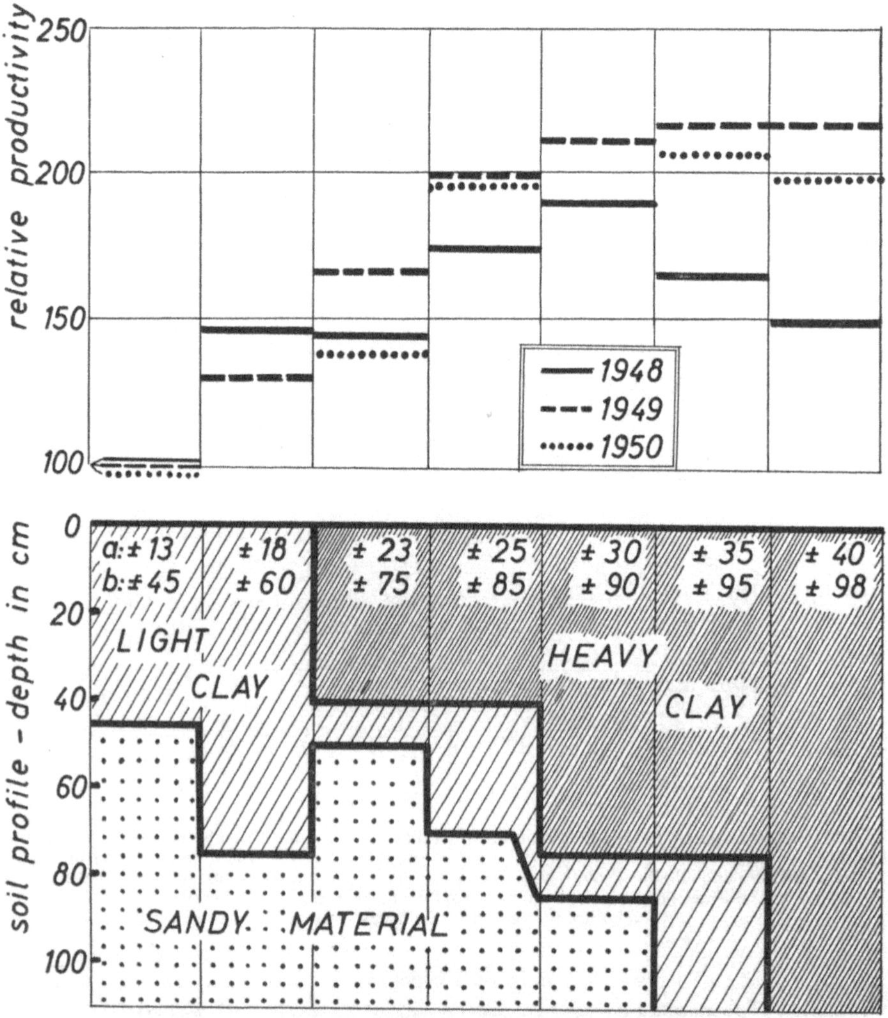

Fig. 18. Soil productivity, soil texture and soil profile in the "Old Seapolder Area" of Belgium. A relative productivity of 100 is given to the lightest soil profile; in the arable layer *a* represents the clay content (0-2 μ) and *b* the clay- and silt content (0-50 μ). In a relatively wet year (1948) an undeep clay profile with an arable layer containing about 20 percent clay, produces as much as a deep clayey profile with 40 percent clay content; in a very dry year (1949) the deep clayey profile produces nearly twice as much

by feeling the soil (when wet at field capacity) between his fingers will be able to distinguish ten or more soil textural classes, such as sand, loamy sand, sandy loam, loam, silt loam, clay loam, etc., but this is only true for soils not rich in organic matter or $CaCO_3$, when the weather is not too rainy and cold, and for field pedologists who are well trained and whose field estimations are regularly checked.

Ten years experience of the author has proved that the estimations of a given field pedologist are not constant all the year through (De Leenheer, Appelmans, De Caestecker 1954). The best way to come to an international understanding is to give the position of the soil textural names used in a textural triangle. This makes possible the reading of a simplified grain size distribution with three separates (usually sand, silt and clay).

The textural triangle in use in the U.S.A. (Soil Survey) and the one used by the Belgian Soil Survey are given here as an example (Fig. 15A and 15B).

A "soil class" (Bodenart, Classe de terre) may be considered synonymous with "soil textural class" when the total of the mineral components (not including the calcium salts present as carbonate or sulphate) is more than 90 percent of the

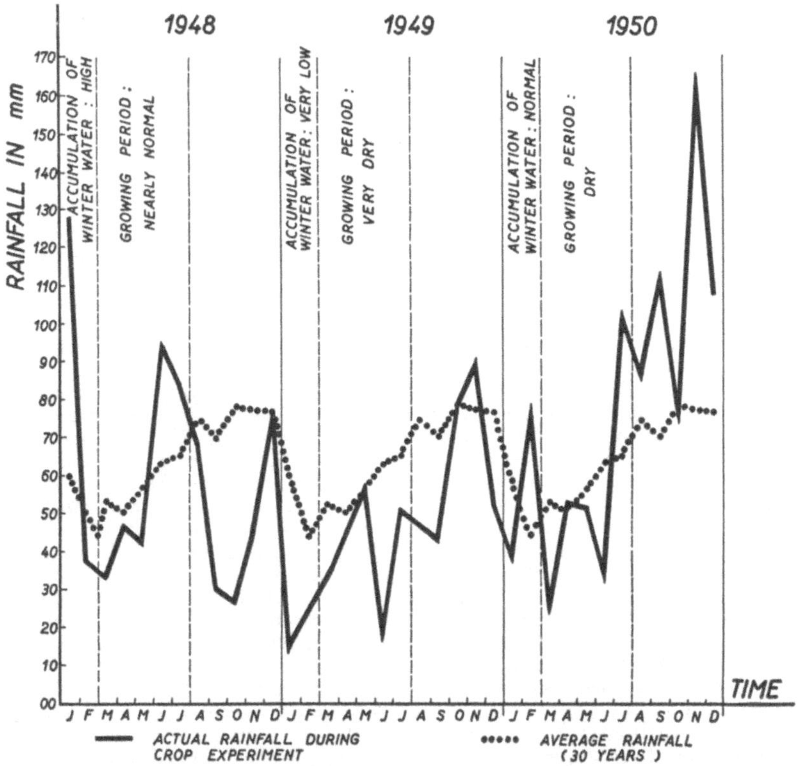

Fig. 19. This figure is to be confronted with Fig. 18 to show the influence of rainfall distribution on crop production

untreated air dried soil sample. In such cases there is a good correlation between the evaluation of the soil class in the field and the soil textural classes in the soil class triangle, as can be seen in Fig. 16 for about 100 soil samples in the loamy region of Belgium (De Leenheer, Appelmans, Van Ruymbeke 1958). This, however, is not true for alluvial soils in general, nor for soils which are the weathering product of a compact (rocky) parent material, and certainly not for tropical and subtropical soils containing hard concretions.

For alluvial soils the soil class triangle is entirely different from the soil textural class triangle, as is illustrated for the soil classes of the Belgian marine alluvial soils (Fig. 17) (De Leenheer and Van Ruymbeke 1960). The estimation in the field of a soil class is greatly influenced by the soil organic matter and carbonate content, but it is evident that a farmer is more interested in a good soil class evaluation rather than in a soil textural class estimation. *Therefore a good soil class triangle will always have a regional significance* and will be the result of an intense agro-pedological survey.

Finally, it must be stressed that in a given area it is dangerous to correlate directly soil texture with soil productivity. In some landscapes with homogeneous soil profiles it may appear that there is some direct correlation, but then the productivity differences will normally not be important, as the changes in texture themselves are limited. In Fig. 18 it is evident that the correlation between the soil texture of the arable soil (e. g. the clay content or the clay and silt content) and productivity is only apparent and is really misleading, the productivity being first the result of the properties of the soil profile (De Leenheer and De Caestecker 1950) which is greatly influenced by the rainfall distribution during the year of the observation (Fig. 19). The concept of the "available water reserve" of the soil profile at the moment that crop growth starts in the spring seems to be, in temperate climates, one of the most important in moden pedology (De Leenheer and Van Ruymbeke 1960).

Literature

Aronovici, V. S.: The mechanical analysis as an index of subsoil permeability. Soil Sci. Soc. Amer. Proc. **11**, 137–141 (1946).

Bouyoucos, G. J.: The hydrometer as a new method for the mechanical analysis of soils. Soil Sci. **23**, 343–353 (1927).

Casagrande, A.: Die Aräometer-Methode zur Bestimmung der Kornverteilung von Böden. Berlin: Springer. 1934.

De Leenheer, L., and K. De Caestecker: De proefoogsten, hun economische betekenis en wetenschappelijke waarde bij Bodemkartering. Comptes rendus de recherches; Institut pour l'Encouragement de la Recherche Scientifique dans l'Industrie et l'Agriculture, **4**, 189–209 (1950). — De Leenheer, L., and L. Maes: Analyse granulométrique avec l'hydromètre à chaîne. Bull. Soc. belge de Géologie **61**, 138–164 (1952). — De Leenheer, L., F. Appelmans and K. De Caestecker: Sur la valeur objective des évaluations de la texture sur le terrain. Transact. 5th Intern. Congr. Soil Sci., Léopoldville, **4**, 374–382 (1954). — De Leenheer, L., and M. Van Ruymbeke: L'analyse granulométrique par l'hydromètre à chaîne de terres riches en matières organiques. Transact. 5th Intern. Congr. Soil Sci., Léopoldville, **2**, 390–400 (1954). — De Leenheer, L., M. Van Ruymbeke and L. Maes: Die Kettenaräometer-Methode für die mechanische Bodenanalyse. Z. Pflanzenernähr., Düng., Bodenkde. **68** (**113**), 10–19 (1955). — De Leenheer, L., and J. Van Hove: Werkwijze voor de mechanische analyse met de kettinghydrometer. Mededel. Landbouwhogeschool, Gent, **21**, 249–274 (1956). — Vergelijkende studie over het gebruik van natriumoxalaat-carbonaat, natriumpyrofosfaat en natriumhexametafosfaat als peptisatiemiddel voor de mechanische analyse van gronden. Mededel. Landbouwhogeschool, Gent, **22**, 225–242 (1957) — De Leenheer, L., F. Appelmans and M. Van Ruymbeke: Etude de la Planchette 116 E (La Hulpe) de la carte pédologique. Non published report of the Soil Research Centre Ghent, 1958. — De Leenheer, L., and M. Van Ruymbeke: Monografie der Zeepolders. Mededel. Landbouwhogeschool, Gent, **25**, 1–416 (1960). — Dixon, W. J., and F. J. Massey: Introduction to statistical analysis, p. 61. New York: McGraw Hill. 1957. — Doeglas, D. J.: Afzettingsgesteenten. Den Haag: Servire. 1952.

Gessner, H.: Die Schlämmanalyse. Leipzig: Akademische Verlagsgesellschaft. 1931.

Hooghoudt, S. B.: Een gecombineerde zeef- en pipetmethode voor de bepaling van de granulaire samenstelling van gronden. Verslagen van Landbouwkundige Onderzoekingen, 's Gravenhage (Nederland), **50** (**13**) A, 671–993 (1945).

Jennings, D. S., M. D. Thomas and W. Gardner: A new method of mechanical analysis of soils. Soil Sci. **14**, 485–499 (1922).

Köhn, M.: Beiträge zur Theorie und Praxis der mechanischen Bodenanalyse. Landwirtsch. Jahrb. **67**, 485–546 (1928). — Krauss, G.: Über eine . . . neue Methode der mechanischen Bodenanalyse. Intern. Mitt. Bodenkde. **13**, 147–160 (1923). — Krumbein, W. C., and F. J. Pettijohn: Manual of sedimentary petrography.

New York: Appleton-Century-Crofts. 1938. — KRYNINE, D. P.: Soil mechanics, 2nd Ed., New York and London: McGraw Hill. 1947.

MILLS, F. C.: Statistical methods, p. 112. New York: H. Holt. 1924.

OSTLE, B.: Statistics in research, 3rd Ed. Ames, Iowa: Iowa State College Press. 1958.

ROBINSON, G. W.: A new method for the mechanical analysis of soils. J. Agric. Sci. **12**, 306–321 (1922).

TAVERNIER, R., *et al.* (Soil Survey Staff): Carte des sols de la Belgique. Texte explicatif de la planchette de... (p. ex. Herderen 107/E, p. 13). Comité pour l'établissement de la carte des sols et de la végétation de la Belgique. Rozier, 6, Gand. 1958. — TRASK, P. D.: Origin and environment of source sediments of petroleum. Houston, Texas, 1932.

VAN HOVE, J., and L. DE LEENHEER: Studiën van de ijkingsfouten van de kettinghydrometer voor de mechanische analyse van gronden. Mededel. Landbouwhogeschool, Gent, **22**, 305–333 (1957). — VAN RUYMBEKE, M.: Bydrage tot de studie van de kleimineralen in de Zeepolders in verband met de kalihuishouding van deze bodens. Thesis, A.H.O. Landbouwhogeschool, Gent 1964. — VAN RUYMBEKE, M., and L. DE LEENHEER: Etude comparative d'analyses granulométriques par décantations successives et par l'hydromètre à chaîne. Transact. 5th Intern. Congr. Soil Sci. Léopoldville, **2**, 322–328 (1954).

c) Bodentypen und Bodensystematik

Von

E. Mückenhausen

1. Zweck und Umfang des Kapitels „Bodentypen"

Die Bedeutung des Teilgebietes „Bodentypen" der Naturwissenschaft „Bodenkunde" geht anschaulich aus dem Anteil hervor, den die Bodentypen im Handbuch der Bodenlehre einnehmen (BLANCK 1929/39), nämlich den ganzen Band III mit 521 Seiten. Seit dem Erscheinen dieses III. Bandes 1930 wurde eine große Anzahl von Publikationen über Bodentypen herausgebracht. Diese alle in diesem Kapitel zu erwähnen, würde den gegebenen Rahmen überschreiten. Vielmehr muß eine Auswahl getroffen werden, die auf das Thema dieses Handbuches ausgerichtet ist. Damit ist die Richtschnur für dieses Kapitel aufgezeigt, d. h. die Behandlung der Bodentypen soll nicht in gleicher Breite bis in alle Einzelheiten erfolgen, vielmehr soll dabei stets das Thema dieses Handbuches im Vordergrund stehen. Andererseits soll dieses Kapitel den Leser in die Lage versetzen, sich auf Grund der Beschreibung der Bodentypen und der Aufzeigung der einschlägigen Literatur in die eine oder andere Teilfrage zu vertiefen.

Im ersten Abschnitt dieses Kapitels werden wichtige Definitionen und Erläuterungen gegeben, die für das Verständnis des Folgenden notwendig sind. Anschließend werden das Wirken der Bodenbildungsfaktoren und die wichtigsten Vorgänge der Bodenbildung geschildert. Der folgende größere Abschnitt behandelt die einzelnen Bodentypen nach Entstehung, Aufbau, Eigenschaften und Verbreitung. Eine zusammengedrängte Aufzählung der wichtigsten Bücher, die Auskunft über die Bodentypen größerer Räume geben, sowie von Bodentypenkarten sollen die Möglichkeit einer Vertiefung des in diesem Buche gegebenen Überblicks geben. Die zum Schluß gebotene Bodensystematik kann nur ein Gegenwartsbild vermitteln, und dieses nur von einem Standpunkt, der zwar in Mitteleuropa vielen Fachleuten zusagt, jedoch nicht weltweite Gültigkeit beanspruchen kann, weil in anderen Ländern viele Ursachen (Bodenverhältnisse, Entwicklungsweg und Stand der bodenkundlichen Forschung u. a.) zu einer anderen Klassifikation oder Systematik der Böden der betreffenden Räume geführt haben.

2. Definitionen und Erläuterungen zum Kapitel „Bodentypen“

In der Geschichte der bodenkundlichen Forschung sind immer wieder neue Definitionen für den Boden gegeben worden, eine löste die andere ab, weil jeweils die ältere nicht befriedigte oder weil der Boden nur von einem bestimmten Gesichtspunkt aus definiert wurde und deshalb die Definition einseitig war. Die Definition des Bodens muß *allgemein* und zugleich *umfassend* sein. Diese Forderung erfüllt die Bodendefinition von Laatsch und Schlichting (1959): „Unter ‚Böden‘ verstehen wir beliebige dreidimensionale Ausschnitte aus der Pedosphäre von der Streu bis zum Gestein.“

Die genannten Autoren betrachten die *Pedosphäre* als den Bereich der Erdrinde, in welchem die *Lithosphäre* (=Gesteinsmantel der Erde) durch Atmosphärilien und Organismen umgewandelt wird. Die Umgestaltung der Lithosphäre vollzieht sich am jeweiligen Ort der Erdoberfläche unter der Wirkung einer bestimmten Konstellation von Bodenbildungsfaktoren (s. Kap. I, A, a), wodurch charakteristische *Umbildungsformen der Lithosphäre* entstehen, die wir *Bodentypen* nennen. Jeder Bodentyp wird durch eine bestimmte Gruppe von Merkmalen charakterisiert, die Laatsch und Schlichting (1959) als *korrelative Merkmalsgruppe* bezeichnet haben.

Bei der Umbildung der Lithosphäre entstehen im Bodenprofil (=Bodeneinschnitt=Bodenwand) vor allem charakteristische Lagen, die *Horizonte* genannt werden. Bodenhorizonte sind also pedogenetisch bedingt, dagegen sind *Bodenschichten* geologischer Entstehung. Bodenschichten und Bodenhorizonte können unter den vereinigenden Begriff „*Lagen*“ gestellt werden.

Der Begriff „Verwitterung“ als Vorgang der Gesteinszerstörung wurde bereits im Kap. I, A, a näher erläutert. Demgegenüber beinhaltet der Begriff „*Bodenbildung*“ mehr, nämlich die Bildung der Böden in ihrer heute noch nicht übersehbaren Formenfülle. Die Bodenbildung schafft das *Solum*, d. h. die von der Bodenbildung umgeformte Schicht der Erdrinde. Die *Bodenbildungstiefe* entspricht dem Solum.

Die *Bodenentwicklung* ist begrifflich umfassender als die „Bodenbildung“. Der Begriff Bodenentwicklung beinhaltet den Ablauf der Bodenbildung von Stadium zu Stadium; z. B. entwickelt sich der stark ausgeprägte Podsol vom Initialstadium bis zur starken Ausprägung.

Die im Text, vor allem in den Profilbeschreibungen, benutzten Bezeichnungen für das *Makrogefüge* (=*Struktur*) des Bodens sind in einer Arbeit von Mückenhausen (1958b) näher beschrieben und auch mit den in der englischen Literatur gebräuchlichen parallelisiert worden. Die Bezeichnungen des *Mikrogefüges* werden in Anlehnung an Kubiëna (1935) vorgenommen. — Hinsichtlich der Benennung der Bodenarten (=Texturen) und des Humusgehaltes wird auf die entsprechenden Beiträge dieses Handbuches verwiesen.

Die Bezeichnung der Bodenhorizonte geschieht mit den zur Zeit in Deutschland und überwiegend auch im Ausland gebräuchlichen *Horizontsymbolen*. In vielen Ländern ist man bestrebt, die Horizontsymbole zu vermehren, vor allem in genetischer Hinsicht zu spezifizieren, und als Indizes Kleinbuchstaben statt Zahlen zu verwenden. Da dieses Bemühen noch divergent verläuft, wird zunächst bei der alten Bezeichnung verblieben, jedoch werden die neuen Symbole in den folgenden Definitionen hinzugefügt.

Übergangshorizonte enthalten die Merkmale von zwei Horizonten; sie werden durch die Kombination der entsprechenden Symbole gekennzeichnet. Dabei steht das Symbol am Schluß, welches das in dem betreffenden Übergangshorizont Vorherrschende anzeigen soll, z. B. wird mit gB ein pseudovergleyter B-Horizont und mit Bg ein Pseudogley-Horizont mit den untergeordneten Eigenschaften eines B-Horizontes gekennzeichnet.

Die *wichtigsten Horizontsymbole* sind:

A = allgemein ein humoser, an der Oberfläche gebildeter Mineralbodenhorizont;

(A) = A-Horizont ohne sichtbaren Humus, jedoch belebt und mit beginnender Bodenbildung; auch A_i genannt;

A_{00} = unzersetzte Pflanzenteile, z. B. Blätter und Nadeln; dieser Horizont wird auch L oder O_L (von englisch litter) genannt;

A_0 = Auflage von Rohhumus oder Moder; auch O genannt;

A_{01} = Vermoderungshorizont; er wird auch F oder O_F (von englisch fermentation) genannt;

A_{02} = Humusstoffhorizont; er wird auch H oder O_H (von Humus) genannt;

A_1 = humoser, an der Oberfläche gebildeter Mineralbodenhorizont. Der Index 1 soll hier nur besagen, daß der A-Horizont untergliedert ist, daß also unter dem A_1-Horizont noch ein weiterer A-Horizont folgt oder ein A_0-Horizont darüber liegt; er wird auch A_h (h von Humus) genannt;

A_2 = verarmter, gebleichter, hellgrauer (holzaschefarbener) Horizont des Podsols und podsoliger Böden; er wird auch A_e (e von eluvial) genannt;

A_3 = hellerer, durch Eisenhydroxyd noch gleichmäßig gefärbter (meist ocker) Verarmungshorizont; charakteristisch für die Parabraunerde; der A_3g des sekundären Pseudogleyes hat eine hellgraue Farbe; in anderen Ländern wird der A_3 ebenfalls A_2 genannt; neuerdings A_l;

(B) = verlehmter Horizont zwischen dem A- und C-Horizont ohne oder ohne nennenswerte Illuviation, charakteristisch für die typische Braunerde; er wird auch B_v (v von Verwitterung) genannt;

B = Anreicherungshorizont, d. h. Illuvialhorizont terrestrischer Böden, wobei die Anreicherung aus Sesquioxyden (auch B_s genannt), aus unzerstörtem Ton (auch B_t genannt), aber auch bzw. außerdem aus Humusstoffen (auch B_h bzw. B_{sh} genannt) bestehen kann. Der B-Horizont kann unterteilt sein, so daß man dann vom B_1- und B_2-Horizont spricht. B_1 ist zum Beispiel beim Podsol der Orterde- oder Ortsteinhorizont, der B_2 der rostbraune Horizont unter dem B_1;

C = allgemein Gestein, aus dem der Boden entstand;

C_1 = schwach verwitterter Übergangshorizont zum frischen Gestein; auch C_v;

C_2 = gänzlich unverwittertes Gestein; auch C_n;

Ca = (Ca chemisches Symbol für Calcium) ein mit Calziumcarbonat angereicherter Horizont, der sich in den betreffenden Böden z. B. zwischen B und C einschieben kann;

P = (von Pelosol) Verwitterungshorizont unter dem A-Horizont des Pelosols, entkalkt (betrifft Pelosole aus Tonmergel); die Verkittung der Tonteilchen im Ausgangsgestein, also das Gesteinsgefüge, ist aufgelöst; meist schwache Illuviation;

G = (G von Gley) durch das Grundwasser beeinflußter Horizont. Ist ein anderer Haupthorizont vom Grundwasser beeinflußt, so tritt das G zu dem Symbol des Haupthorizontes, z. B. GB;

G_o = (o von Oxydation) Oxydationshorizont eines Grundwasserbodens, der im allgemeinen mit dem Spiegelschwankungsbereich einschließlich des Kapillarraumes zusammenfällt;

G_r = (r von Reduktion) Reduktionshorizont eines Grundwasserbodens, der ein reduzierendes Milieu in dem betreffenden Horizont anzeigt;

g = (g von gleyartig) staunasse Horizonte der pseudovergleyten (früher = gleyartigen) Böden. Sind andere Haupthorizonte pseudovergleyt (meist schwächere Ausprägung), so tritt das g zu dem Symbol des Haupthorizontes, z. B. gB; neuerdings wird S für g gebraucht;

Sa = (Sa von Salz) Salzanreicherungshorizont der Salzböden; jetzt auch Z;

M = (M von lateinisch migrare = wandern, d. h. das Bodenmaterial ist gewandert) Bodensediment, d. h. vom Wasser in den Tälern sedimentiertes Material erodierter Böden;

D = Material (Gestein oder Verwitterungsmaterial) des Untergrundes, aus dem der darüber liegende Boden selbst nicht entstand;

T = (T von russisch und deutsch Torf) allgemein für eine Torfschicht ohne nähere Kennzeichnung der Torfart. Verschiedene Torfschichten übereinander werden durch T_1, T_2, T_3 usw. gekennzeichnet. Neuerdings wird auch das Symbol H dafür verwendet, das allerdings bereits für den A_{02}-Horizont vergeben ist.

3. Die Wirkung der Bodenbildungsfaktoren

Vor etwa 80 Jahren erkannte der russische Bodenforscher Dokutschajew (1879b), daß die Bodenbildung durch äußere Kräfte gesteuert wird. In den weiten Ebenen Rußlands fand er in Abhängigkeit von den Klimazonen spezifisch geformte Böden. Dabei entdeckte er gleichzeitig, daß auch andere Faktoren die Bodenbildung beeinflussen können. Die wichtigsten Faktoren der Bodenbildung waren damit aufgedeckt. Später wurden die grundlegenden Erkenntnisse von Dokutschajew (1879a, 1879b) durch Hilgard (1892), Sibirzew (1895) und andere erweitert, jedoch änderte sich an den grundsätzlichen Gedankengängen Dokutschajews nichts. In erster Linie hat Stremme (1926a) das Gedankengut der russischen Bodenkundler über die Bodenbildungsfaktoren nach Westeuropa übertragen.

Folgende Faktoren der Bodenbildung werden heute als maßgeblich für die Entstehung der Böden aufgefaßt: Klima, Vegetation, Wasseransammlung, Relief, Tier, Mensch, Ausgangsmaterial und Zeit (Dauer des Vorgangs). Die ersten sechs Faktoren sind als äußere Kräfte zu betrachten, die auf das Gestein als Ausgangsmaterial in der „Zeit“ einwirken. Die genannten Faktoren wirken nie allein; es sind stets mehrere an der Bildung eines bestimmten Bodens beteiligt. Dabei wechseln sowohl die jeweilige Konstellation der Faktoren als auch die anteilige Intensität ihrer Wirkung. Daraus ergibt sich, daß die Faktoren der Bodenbildung in mannigfacher Weise auf das Gestein einwirken können, und darauf beruht die Entstehung der vielen Bodentypen und deren Übergangsbildungen auf der Erde.

In diesem Kapitel soll die Wirkung der einzelnen Faktoren gesondert dargestellt werden, um einerseits die Hauptlinien der Bodenbildung aufzuzeigen und andererseits das komplizierte Zusammenwirken der Faktoren besser verstehen zu können.

A. Das Klima

Die Wirkung des Klimas auf die Bodenbildung wird in erster Linie durch die Menge des Niederschlags, die Niederschlagsverteilung, die Temperatur und die Luftfeuchtigkeit bestimmt. Die wesentlichen Bodenbildungstendenzen auf der Erde werden durch diese Klimafaktoren hervorgerufen. Sehr eindrucksvoll läßt sich das bei einem Vergleich der Klimazonen und der Bodenzonen der Sowjetunion und Nordamerikas erkennen. Mit steigendem Niederschlag und fallender Temperatur wachsen die Auswaschung und die Verarmung der Böden. Die Profile der Bodentypen auf den Tafeln I und II (S. 101 und 103) zeigen diesen starken Effekt des Klimas auf den Aufbau und die damit gegebenen Eigenschaften der Böden aus gleichem Ausgangsmaterial (Löß). Vom Sierosem zur Braunerde und Parabraunerde steigen Niederschlag und Versickerungsmenge. Mit einer hohen Niederschlagsmenge ist normalerweise eine große Versickerungsmenge verbunden, letztere wird jedoch mit zunehmender Temperatur geringer. Andererseits wird mit steigender Luftfeuchtigkeit der Verdunstungswert herabgesetzt. Diese funktionellen Zusammenhänge der Klimafaktoren variieren stark und damit auch der Einfluß des Klimakomplexes auf die Bodenbildung. Nicht nur der Durchschnittswert von Niederschlagsmenge und Temperatur sind entscheidend, sondern auch deren Werte im Ablauf des Jahres. Eine gleichmäßige Regenverteilung wirkt sich anders aus als die Abwechslung von Regen- und Trockenzeiten. Lange Trockenzeiten und hohe Temperaturen führen die anorganischen Kolloide in einen mehr oder weniger irreversiblen Zustand über. Starke Austrocknung des Bodens erzeugt einen Benetzungswiderstand der Bodenmasse, so daß der Niederschlag oberflächlich abläuft oder in den durch Trockenheit entstandenen Bodenspalten schnell in den Untergrund absinkt. Dadurch wird Bodenmaterial oberflächlich abgespült und (oder) Bodenfeinsubstanz in den Unterboden mitgenommen.

Während das Großklima im allgemeinen die großen Bodenzonen entstehen läßt, kann das Kleinklima die Bodenbildung in kleineren Räumen stark beeinflussen. Die besten Beispiele dafür bieten die verschiedenen Expositionen im Gebirge. Im Hochgebirge können auf Süd- und Nordhang extreme Bodenbildungen auftreten, wie das z. B. von den Schweizer Bodenkundlern (Frei, Pallmann, Richard) im Engadin auf einem kleinen, kegelförmigen Berg aus Kalkstein gezeigt worden ist. Auf der warmen Südseite ist die Mullrendzina und auf der kalten Nordseite die Tangelrendzina (Rendzina mit rohhumusartiger Auflage) ausgebildet. Selbst im Mittelgebirge finden wir auf der Südseite oft den Ranker und auf der verwitterungsintensiveren Nordseite die Braunerde oder eine andere stärkere Profilentwicklung. Überraschend ist, daß unter Umständen nur geringe Unterschiede des Klimas die Bildung verschiedener Bodentypen veranlassen. Nur eine geringe Menge mehr an Niederschlag kann unter Umständen den Pseudogley statt der Braunerde oder Parabraunerde entstehen lassen.

Auch die durch die Vegetation bedingten Unterschiede im Kleinklima sind für die Bodenbildung bedeutend. Z. B. erzeugen die Bäume des Waldes und die Pflanzen des Ackers ein verschiedenes Kleinklima, das bei längerer Einwirkung die Bodenbildung verschieden beeinflußt. Klima und Vegetation wirken in diesem Falle zusammen; ihr Anteil an der Bodenbildung läßt sich dabei nicht erfassen.

B. Die Vegetation

Die Vegetation, einschließlich der Mikroflora, wirkt in verschiedener Weise mehr oder weniger auf die Bodenbildung ein. Die Abfälle der Vegetation bilden den wichtigsten Teil der Ausgangssubstanz für die Humusbildung. Ohne den Humus und seine Wirkungen wäre der Boden etwas Unvollkommenes. Die Abfälle der Pflanze sind indes im Hinblick auf die Bildung des Humus sehr unterschiedlich, teils sind sie leicht zersetzbar, andererseits gibt es pflanzliche Substanz, die nur langsam zersetzt wird und deshalb zur Rohhumusbildung neigt.

Die Wurzeln der Pflanzen sind zunächst stofflich sehr verschieden. Die feinen Wurzeln von Gräsern und Kräutern werden schnell zersetzt, während die Wurzeln von Zwergsträuchern, Sträuchern und Bäumen holzig sind und nur langsam nach dem Absterben der Pflanzen im Boden vermodern. Das Wurzelsystem der Pflanzen durchdringt den Boden je nach Pflanzenart und Bodenaufbau mehr oder weniger stark. Tiefwurzelnde Bäume und Weinreben durchsetzen den Boden bis in größere Tiefe, lockern ihn dabei auf und schaffen ein Kanalsystem für Wasser und Luft. Andererseits vermögen im allgemeinen flachwurzelnde Pflanzen (Fichte) diese Auflockerungsarbeit nur in den oberen Bodenhorizonten zu leisten. Die Pflanzen, besonders die Bäume, nehmen Nährstoffe aus Unterboden und Untergrund auf und geben diese über den Bestandesabfall dem Oberboden wieder.

Während auf der einen Seite die pflanzliche Substanz zur Ernährung der Bodenorganismen unbedingt erforderlich ist und damit für einen ökologisch günstigen Bodenaufbau eine unentbehrliche Ausgangssubstanz darstellt, kann andererseits die pflanzliche Masse die Verarmung des Bodens beschleunigen. Die Nadeln der Koniferen und die Rückstände der Zwergsträucher (Heide und Heidelbeere) erzeugen den an niedrigmolekularen Humussäuren (Fulvosäuren) reichen Moder und Rohhumus. Diese Säuren wirken zerstörend auf die Feinsubstanz des Bodens und fördern damit den sogenannten Podsolierungsprozeß. Die Sesquioxyde werden aus der Tonsubstanz frei gemacht und wandern im Bodenprofil abwärts; dabei üben die Humussäuren eine Schutzkolloidwirkung beim Transport aus. Ferner können organische Abbaustoffe der pflanzlichen Substanz mit anorganischen Substanzen wanderungsfähige Komplexverbindungen bilden. In welcher unterschiedlichen Weise die Vegetation mit ihren Abfällen auf die Entwicklung des Bodens einwirkt, veranschaulicht sehr gut die Vegetationsabfolge in Nordeuropa: Unter dem ehemaligen Eichen-Birken-Wald entstand die Parabraunerde oder die Podsol-Parabraunerde; es folgte die Heide, unter der sich das typische Podsolprofil entwickelte. Eine besonders starke Ausbleichung des Bodens erzeugen die organischen Säuren der Rinde und der Wurzeln des Kauribaumes im Norden von Neuseeland. Unter dem Wurzelstock und um den Stamm (hier Ansammlung von abgefallener Rinde) entwickelt sich ein mächtiger A_2-Horizont von 1 m und mehr Mächtigkeit. Die napfartige Verdickung des A_2-Horizontes unter dem Wurzelstock führte zu dem Namen „Eierbecher-Podsol" (BLOOMFIELD 1953b).

Ein weiteres Beispiel für den großen Einfluß der Vegetation bietet die Bildung der Schwarzerde. Ohne die hohe Produktion an organischer Substanz käme der Humusreichtum in der Schwarzerde nicht zustande. Dabei spielt die leichte Zersetzlichkeit der hier gedeihenden Gräser und Kräuter eine wichtige Rolle. Selbstverständlich sind in erster Linie die Klimabedingungen für die Bildung der Schwarzerde entscheidend.

In diesem Zusammenhang muß auch die große Bedeutung erwähnt werden, welche die Einschaltung von mehrjährigem Kleegras in die Fruchtfolge besitzt. Der Ackerbau fördert den Abbau der organischen Substanz des Bodens, vor allem in Klimabereichen mit milden Wintern. Das mehrjährige Kleegras erhöht nicht nur den Gehalt an Humus, sondern verstärkt ebenfalls die Tätigkeit der Bodenorganismen, weil diese Vegetation ihr eine ständige und zusagende Futterquelle bietet. Bekannt ist diese Bodenverbesserung aus Mitteleuropa, aber auch aus England und der Sowjetunion; sie hat sich ebenfalls in den USA in den letzten Jahren bewährt.

Die ausgleichende Wirkung des Waldes auf den Wasserhaushalt und auf die Bodentemperatur ist bekannt. Ein alter Fichtenbestand vermag einen staunassen Boden (Pseudogley), auch in der feuchten Jahreszeit, weitgehend trocken zu halten. Dadurch wird in dem Pseudogley eine andere Wasserdynamik erzeugt, so daß dadurch Reduktionsvorgänge zurückgehalten und Oxydationsvorgänge begünstigt werden. Diese günstige Wirkung des Fichtenbestandes wird andererseits jedoch durch die Lieferung von schwer zersetzlichen Bestandesabfällen beeinträchtigt.

Eine weitere wichtige Wirkung der Vegetation auf den Boden liegt darin, daß die Vegetation in hängigen Lagen den Boden vor dem Abtrag schützt. Selbst an steilen Hängen verankern Bäume die Verwitterungsschicht (Solum) weitgehend, indem die Wurzeln durch das Solum bis in Gesteinsfugen vordringen und sich auf diese Weise selbst und den Boden verankern. Der Wald ermöglicht somit auch in steilen Gebirgslagen eine mehr oder weniger ungestörte Bodenbildung, wenn wir von dem sogenannten Bodenkriechen, d. h. einer sehr langsamen Bodenbewegung in Hanglage, absehen. Auch die Grünlandnarbe schützt den Boden weitgehend vor dem Abtrag durch das Niederschlagswasser, weil sie den Boden dicht bedeckt und die oberste Schicht intensiv mit einem Wurzelnetz durchzieht. Dagegen birgt die Ackerkultur stets die Gefahr des Bodenabtrags, wenn auch die Getreidearten den Boden etwas besser schützen als die schlecht abdeckenden Hackfrüchte. Damit sind nur die wichtigsten Einwirkungen der Vegetation auf die Bodenbildung skizziert; in Wirklichkeit ist ihr Einfluß viel mannigfaltiger.

C. Die Wasseransammlung als Staunässe und Grundwasser

Das Niederschlagswasser ist in dem Bodenbildungsfaktor „Klima" einbezogen. Die Wasseransammlungen als Staunässe und Grundwasser wirken spezifisch auf die Bodenbildung ein und schaffen besondere Bodentypen (Pseudogley, Auenboden, Gley, Marsch und Moor).

Steht das Wasser in tieferen Geländelagen stets über der Bodenoberfläche, so entstehen die subhydrischen Böden. In einem seichten Wasser gedeihen Wasserpflanzen, wie Rohrkolben, Schilf und Großseggen, deren abgestorbenes Material das Niedermoor aufbaut. Bei einem Grundwasserstand in der Nähe der Bodenoberfläche bilden sich die Gleye, deren Profile mit G_o- und G_r-Horizonten charakteristische Bildungen des Grundwassers sind. Wenn das Grundwasser zeitweise den Oberboden erfüllt und luftarm macht, wird die Zersetzung der organischen Substanz gehemmt; es kommt zur Bildung des Anmoorgleys. Das Grundwasser der Marschen schwankt in kurzen Abständen mit der Tide, wenn nicht der Mensch durch kulturtechnische Maßnahmen dieses verhindert. Die Staunässe bildet sich, indem das Niederschlagswasser an der normalen Versickerung infolge eines dichten Unterbodens oder Untergrundes gehindert wird; dadurch entsteht der Pseudogley.

Das sind die wesentlichsten Züge des Wassereinflusses auf die Bodengestaltung; die Intensität des Wassereinflusses ist sehr variabel und bedingt mannigfache Abstufungen der erwähnten Bodenbildungen.

D. Das Relief

Von der Oberflächengestaltung hängt es ab, ob die Bodenbildung ohne Störung ablaufen kann oder nicht. Unebenheiten des Geländes veranlassen Bodenabtrag und Bodenauftrag, wenn nicht eine Pflanzendecke dieses verhindert. Die verstümmelten Profile auf Kuppen und Hängen sowie die Kolluvien am Hangfuß und in den Senken und Tälern zeugen davon. Die Entblößung der Hänge des Gebirges, vor allem in Südeuropa, sind ebenso überzeugende wie auch abschreckende Beispiele für den Einfluß des Bodenabtrags im hängigen Gelände. Aber auch im Berg- und Hügelland stört dieser Vorgang die Entwicklung und Erhaltung der Bodenprofile, wenn das Wasser ungehindert das Bodenmaterial abschwemmen kann. Das beste Beispiel dafür bieten viele Ackerflächen Nordamerikas. Aber auch in wenig hängigen Lagen und in sehr porösem Boden, wo der Bodenabtrag keine Wirkung zu haben scheint, summieren sich doch im Laufe jahrhundertelanger Ackerkultur die geringen Mengen der Bodenverlagerung. Dieses beweist die Einebnung der flachwelligen Landschaft des Neuwieder Beckens mit seinen wasserdurchlässigen Böden aus Trachyttuff. Die Folge des Bodenabtrags hat hier auf kurze Entfernung Ranker, gekappte Braunerde bzw. Parabraunerde und Kolluvien dieser Böden nebeneinander geschaffen. Der Bodenabtrag hügeliger Löß- und Moränenlandschaften Mitteleuropas ist ebenso ein überzeugendes Beispiel für die starke Einwirkung des Reliefs in Verbindung mit dem Niederschlag auf die Bodenbildung.

Bei jeder Einwirkung des Reliefs auf die Bodenbildung müssen wir Niederschlag und Vegetation einbeziehen. Das bewegte Relief bedingt nur Niveauunterschiede und erzeugt damit die Schwerkraftenergie für den Bodentransport. Indes wird der Abtrag selbst vom Wasser bewirkt, und das Vorhandensein oder Nichtvorhandensein der Vegetation entscheidet darüber, ob der Boden der abtragenden Wirkung des Wassers anheimfällt. Hier liegt ein typisches Beispiel vom Zusammenwirken der Faktoren der Bodenbildung vor.

E. Die Tiere

Die Tätigkeit der Tiere bei der Bodenbildung erstreckt sich in erster Linie auf das Durchwühlen der Bodenmasse und auf die Zerkleinerung der organischen Substanz sowie deren Vermengung mit dem Mineralboden.

Es handelt sich je nach Klima um verschiedene Tierarten. Im Steppenklima sind es eine Reihe von Nagetieren (Ziesel, Erdhörnchen, Steppenmurmeltier, Pferdespringer u. a.), die größere Gänge in den Boden graben, den Boden durchmischen und vor allem den A-Horizont durch die wühlende Tätigkeit stets vertiefen. Der humide Klimaraum hat zwar auch eine Reihe von Erdbewohnern (Maulwurf, Feldmaus, Wühlmaus, teils auch Hamster u. a.), die den Boden durchmischen; es sind aber zahlenmäßig bei weitem nicht so viele wie in der Steppe. Wichtiger sind im gemäßigten Klimaraum die Regenwürmer und andere Bodenkleintiere, die sich an der Zerkleinerung der organischen Substanz beteiligen und den Boden durchmischen. Die Regenwürmer durchziehen den Boden in vertikaler Richtung und geben ihm dadurch ein siebartiges Hohlraumsystem, das für den Wasser- und Lufthaushalt wichtig ist. Dabei verschleppen diese Kleintiere Bodenmaterial sowohl von oben nach unten als auch von unten nach oben, ein Vermischungsvorgang, den die niederländischen Bodenkundler als „Homogenisierung" bezeichnet haben. Neben den Regenwürmern sind besonders wichtig: Enchytraeiden (kleine Borstenwürmer), Tausendfüßler, Asseln, Larven, Springschwänze, Milben, Spinnen, Ameisen und andere. In den Tropen vollbringen die Termiten eine tiefgreifende, innige Durchmischung des Bodens, wobei oftmals die naturgegebenen Bodenhorizonte verschwinden; nur die oberste Schicht mit junger organischer Substanz hebt sich etwas ab.

An der Humusbildung sind die Kleintiere ebenfalls beteiligt, indem sie organische und mineralische Substanz aufnehmen und durch ihren Verdauungstrakt passieren lassen. Dabei werden die Substanzen zu stabilen Bodenaggregaten verbunden, die organische Substanz wird für die weitere Humusbildung durch Zerkleinerung und chemischen Aufschluß vorbereitet. Dieser Vorgang ist nicht ohne weiteres sichtbar; für die Bodenbildung im gemäßigten Klima ist er von großer Bedeutung.

F. Der Mensch

Seit der Mensch den Boden bewirtschaftet, stellt er in verschiedener Weise den intensivsten Faktor der Bodenbildung dar. Anschauliche Beispiele dafür sind das Rigolen der Weinberge, das sogenannte Kuhlen der verarmten Marsch, das Aufbrechen des Ortsteins, die Absenkung des Grundwassers (Flußregulierung), Dränung, tiefes Umpflügen von Marsch und Moor, die Besandung des Moores und anderes mehr. Alle diese Maßnahmen ändern das Bodenprofil mit seiner natürlichen Horizontierung von heute auf morgen. Es ist Auffassungssache, ob man diese Eingriffe zur Bodenbildung rechnet. Demgegenüber wird die durch die intensive Gartenkultur bedingte Veränderung des Bodens, wodurch ein schwarzerdeartiger, mächtiger A-Horizont entsteht, allgemein als ein vom Menschen bedingter Bodenbildungsvorgang aufgefaßt. Auch die Entstehung eines mächtigen A-Horizontes durch die Plaggendüngung stellt ein Werk des Menschen dar. Eine fast weltweite Einwirkung auf den Boden übt der Mensch aus, indem er den Boden bearbeitet und ihm Kalk sowie (organische) Wirtschaftsdünger und Handelsdünger zuführt. Dabei wird im allgemeinen das Profil nur in den obersten 15 bis 30 cm durch die Bearbeitung umgewandelt. Unter dieser bearbeiteten Schicht hält sich aber, sofern nicht Untergrundlockerung erfolgt, das natürliche Bodenprofil über lange Zeit. Aus diesem Grunde werden solche Böden nicht als anthropogen bezeichnet. Dabei darf nicht vergessen werden, daß die chemischen Eigenschaften durch Kalk und Dünger, vor allem in der obersten, bearbeiteten Schicht, verändert werden; nach und nach ändern sich auch die chemischen Eigenschaften der Unterbodenhorizonte, sofern diese Stoffe mit dem Sickerwasser in den Unterboden getragen werden.

Der Einfluß des Menschen auf die Bodengestaltung wurde in dem Augenblick sehr stark, als er den Naturwald rodete. Unter dem Walde war in Jahrtausenden ein Bodenprofil entstanden, das selbst im Gebirge vor der Bodenerosion geschützt war. Nachdem aber der Mensch die schützende Decke dem Boden nahm, war das Bodenprofil der Verstümmelung durch die Bodenerosion ausgesetzt. Auf Sandböden veranlaßte der Kahlschlag oftmals die Winderosion, wodurch Boden- und Lockergesteinsmaterial von einer Stelle zur anderen transportiert wurden; auf der neuen Ablagerung begann die Bodenbildung aufs neue.

Der Kahlschlag ließ in vielen Fällen eine andere Vegetation aufkommen, so im deutschen Mittelgebirge in den vergangenen Jahrhunderten die Heide, welche die Podsolierungstendenz mit sich brachte. Die menschenbedingte Ausdehnung des Nadelwaldes auf Kosten des Laubwaldes führte ebenso zu einer schlechteren Humusform und zu Podsolierungserscheinungen.

Die eindrucksvollsten Beispiele für den Bodenbildungsfaktor „Mensch“ sind der schon erwähnte Plaggenesch und der alte Gartenboden. Beim Plaggenesch wächst durch die Zufuhr von Plaggen und Stalldung ein mächtiger A-Horizont über dem ehemaligen Bodenprofil auf. Der mächtige A-Horizont alter Gartenböden entsteht in erster Linie durch starke Düngung mit organischem Material, die Grabarbeit, die Zufuhr von Wasser in Trockenzeiten (Begießen) und die Beschattung. Dadurch werden für die Bodentiere, vor allem für die Regenwürmer, beste Lebensbedingungen geschaffen; sie vertiefen mehr und mehr den A-Horizont. Bei starker Zufuhr von organischer Substanz wächst der A-Horizont sogar über das ursprüngliche Oberflächenniveau hinaus. Mensch und Tier sind bei dieser Bodenbildung gleichermaßen beteiligt. Das ist wieder ein Beispiel dafür, daß die Bodenbildungsfaktoren ineinandergreifen.

G. Das Ausgangsmaterial

Die oben beschriebenen *sechs* Faktoren der Bodenbildung stellen äußere Kräfte dar, die auf das Gestein als Ausgangsmaterial des Bodens und auch auf die bereits gebildete Bodensubstanz einwirken. Gegenüber den anderen Bodenbildungsfaktoren nimmt das Ausgangsmaterial als Bodenbildungsfaktor eine Sonderstellung ein, indem es gleichsam von „innen heraus" auf die Bodengestaltung Einfluß nimmt.

Das Ausgangsmaterial des Bodens bildet in den meisten Fällen ein Gestein (festes Gestein oder Lockergestein), teils auch älteres, umgelagertes Bodenmaterial.

Die Beeinflussung der Bodenbildung durch die Eigenart des Gesteins ist sehr verschieden. Schon lange bekannt ist der durchschlagende Einfluß der Carbonatgesteine auf die Bodenbildung. In weiter Klimaspanne, von Nordeuropa bis zum Mittelmeergebiet, entsteht aus den Carbonatgesteinen die Rendzina, die sich allerdings im Mediterranklima relativ schnell zur Terra fusca oder Terra rossa weiterentwickelt. In letzterem Falle wirkt das Klima in starkem Maße bei der gesteinsbedingten Bodenbildung mit. — Tonige Gesteine geben der Bodenbildung eindeutig die Richtung. Die tonige Verwitterungsmasse zeigt eine starke Resistenz gegen jede Umbildung des Materials und bedingt einen charakteristischen, den übrigen Böden fremden Wasserhaushalt. Der tonige Boden (Pelosol) quillt bei Befeuchtung zu einer weitgehend kohärenten und dichten Masse und schrumpft bei der Austrocknung unter Rissebildung stark zusammen.

Im gemäßigt warmen, humiden Klima verläuft die chemische Verwitterung mit mäßiger Energie. Die Eigenschaften der hier entstehenden Braunerde und Parabraunerde sind in erster Linie durch das Ausgangsmaterial bestimmt, wobei nicht zu übersehen ist, daß eine lange Einwirkung des hier herrschenden Klimas und anderer Faktoren bedeutende Unterschiede, d. h. Entwicklungsgrade schaffen. Die gesteinsbedingte Verschiedenheit der Korngrößen und der Mineralzusammensetzung bleibt in jeder Entwicklungsreihe der Böden ein grundlegender Faktor. — Selbst bei der energischen Bodenbildung des Podsols bildet das Ausgangsgestein noch ein stark variierendes Moment. Die Podsolierung kann die spezifischen Charaktereigenschaften der Gesteine nicht verwischen. Ein stark ausgeprägter Podsol aus jungdiluvialem Schmelzwassersand, aus tertiärem Quarzsand, aus Neokomsandstein oder aus Konglomerat des Hauptbuntsandsteins ist in seinen Eigenschaften sehr verschieden, und das beruht auf der Verschiedenartigkeit des Gesteins.

Als in der zweiten Hälfte des vergangenen Jahrhunderts die Zonen der Bodentypen auf den ausgedehnten Lößflächen Rußlands entdeckt wurden, rückte angesichts dieser durch äußere Kräfte bedingten Bodenunterschiede das Ausgangsmaterial zu sehr in den Hintergrund. Das wäre wahrscheinlich nicht geschehen, wenn die ersten grundlegenden bodengenetischen Forschungen in einem Gebiet mit starkem Gesteinswechsel durchgeführt worden wären.

Nach den vorstehenden Überlegungen müssen wir sowohl bei der allgemeinen Charakterisierung als auch bei der Systematik des Bodens dem Ausgangsmaterial einen bevorzugten Platz einräumen. Zu jeder ausreichenden Bodenbeurteilung gehört auch das Gestein, aus dem die Textur entstand, und in letzterer vollzog sich die bodentypologische Horizontdifferenzierung.

H. Die Zeit

Die Zeit, d. h. die *Dauer des Bodenbildungsvorganges*, wird nicht von allen Bodenkundlern als Faktor der Bodenbildung angesehen, weil die Zeit *kein direkt* einwirkender Faktor ist, sondern die Bodenbildungsvorgänge nur *in ihr* ablaufen.

Diese Begründung hat ihre Berechtigung. Jedoch werden durch die Dauer der Bodenbildung unter gleichen Faktoren große Unterschiede im Bodenaufbau geschaffen, die bodengenetisch nur unter Einbeziehung der „Zeit" erklärt werden können.

Im mitteleuropäischen Raum können aus Löß unter gleichbleibenden Faktoren der Bodenbildung nacheinander Gesteinsrohboden, Pararendzina, Braunerde und schließlich verschiedene Entwicklungsstadien der Parabraunerde entstehen. Diese Abfolge ist nur durch die Einbeziehung des Faktors „Zeit" in die bodengenetischen Überlegungen zu verstehen. Ebenso ist das Durchlaufen verschiedener Stadien der Podsolierung nur eine Frage der „Zeit". Aus diesen Gründen der universellen bodengenetischen Betrachtungsweise erscheint es trotz begrifflicher Inkonsequenz zweckmäßig, die Zeit als einen Faktor der Bodenbildung aufzufassen, wobei wir uns stets bewußt sind, daß es sich um die *Dauer der Wirkung* der sieben unmittelbar wirksamen Bodenbildungsfaktoren handelt.

Sehr eindrucksvoll wirkt sich die Dauer der Bodenbildung auf den verschieden alten Moränenablagerungen in Nord- und Süddeutschland aus, wodurch nicht nur eindeutige Unterschiede im Profilaufbau (verschiedene Bodentypen), sondern auch große wertmäßige Unterschiede des Pflanzenstandortes entstanden. Das gleiche gilt für die verschieden alten Marschen an der Nordsee. Noch mehr fällt der Faktor „Zeit" ins Auge, wenn ein Bodenbildungsvorgang so schnell verläuft, daß eine wesentliche Veränderung des Bodenaufbaues und der Bodeneigenschaften im Ablauf eines Menschenlebens beobachtet werden kann (Bodenerosion).

4. Die wichtigsten Vorgänge bei der Bodenbildung

Die Bodenbildung wird nie durch einen Vorgang bewirkt; meistens laufen mehrere Prozesse gleichzeitig ab. Jedes Stadium der Bodenentwicklung wird durch eine bestimmte Kombination von Bildungsprozessen beherrscht. Erreicht die Bodenbildung ein neues Stadium, so ändert sich irgendwie die Kombination der Bildungsprozesse. Dieses Nebeneinander und die zeitliche Ablösung der Bodenbildungsprozesse lassen sich besser übersehen, wenn die Einzelwirkungen bekannt sind; sie werden daher im folgenden kurz beschrieben. Als ergänzende Literatur werden die zusammenfassenden Darstellungen folgender Autoren empfohlen: Jenny (1941), Mohr und van Baren (1954), Buckman und Brady (1957), Laatsch (1957) und Rode (1961).

A. Mechanische Zerkleinerung des Gesteins

Die *festen Gesteine* bilden selten eine kompakte Masse. Vielmehr sind sie durch Schichtfugen, Spalten, Klüfte und Absonderungsfugen (in Magmatiten) mehr oder weniger stark zerteilt. Die exogenen geologischen Prozesse (Erosion, Exaration und Korrasion) zerkleinern das Gestein mechanisch. Die Prozesse der physikalischen Verwitterung (Sonnenbestrahlung, Spaltenfrost, Salzsprengung, Wurzelsprengung) zerstören ebenfalls das feste Gestein mechanisch. Dabei entstehen Gesteinstrümmer verschiedener Größe, ohne daß eine substantielle Änderung des Gesteins vor sich geht (s. Kap. I, A, a).

Die *Lockergesteine* können durch mechanische Kräfte weiter zerlegt werden. Gesteinsstücke, die vom Wasser transportiert werden, erfahren durch Abrieb eine Zerkleinerung. Im Gletschereis erfolgt ebenso ein Abschleifen der mitgenommenen Gesteinsstücke. Die vom Wind getragenen Gesteins- und Mineralteilchen reiben sich gegenseitig ab. Sonneneinstrahlung und Frost zerkleinern die Lockergesteine bis zu Größen des Staubsandes.

B. Chemische Zerstörung der primären Minerale

Die chemische Zersetzung der primären Minerale der Gesteine ist äußerst verwickelt; sie wird von vielen Faktoren gesteuert, in erster Linie durch die Widerstandskraft und die Größe der Minerale, die Stärke des zersetzenden Agens, die Temperatur und die Feuchtigkeit. Carbonatische Minerale bzw. Gesteine werden von reinem Wasser nur wenig, dagegen von kohlensäurehaltigem schneller aufgelöst. Die Sulfate der Alkalien und Erdalkalien (Minerale bzw. Gesteine) lösen sich ebenfalls leicht.

Widerstandsfähig gegen chemische Zersetzung (Verwitterung) sind die Silikate, die den größten Anteil der Gesteine ausmachen. In großen Zügen laufen hierbei folgende Prozesse ab: Die Hydratation schwächt den Zusammenhalt der Atome in der Randzone der Minerale. Durch Hydrolyse löst das Wasserstoffion Alkalien und Erdalkalien aus der Randzone des Silikatgitters. Die Folge ist die Freisetzung eines Anteils der Kieselsäure. Dadurch zerfällt das Silikatgitter in die restlichen Teile, nämlich Aluminiumhydroxyd und Kieselsäure. Etwa vorhandenes Eisen wird oxydiert zu einer Form des Brauneisens, u. a. auch vorübergehend zu Eisenhydroxyd. Der stufenweise Zerfall geht in der Außenzone der Minerale bzw. Gesteine vor sich und schreitet nach dem Inneren bis zur völligen Auflösung fort (s. Kap. I, A, a).

C. Bildung der Tonsubstanz

Aus den bei der Silikatverwitterung entstehenden Produkten (Aluminiumhydroxyd und Kieselsäure) können sich je nach Milieu verschiedene Tonminerale bilden. Allgemein entsteht im calcium- und magnesiumreichen Milieu mit schwacher Kieselsäureabfuhr eine große Menge der Tonmineralgruppen Montmorillonit, Illit und Vermiculit. Semihumides und semiarides Klima mit schwacher Perkolation scheinen die Bildung dieser Tonminerale zu begünstigen. Der große Anteil von Illit in den Böden des gemäßigt warmen, humiden Klimas spricht für eine spezifische Bildungstendenz von Illit in diesem Klimaraum. Im tropischen und subtropischen Klima mit starker Kieselsäureabfuhr entsteht bei der Bodenbildung eine große Menge von Kaolinit und verwandter Tonminerale. Diese klimaabhängigen Tendenzen werden in vielen Fällen durch örtliche Besonderheiten des Verwitterungsmilieus abgewandelt. Wenn Aluminiumhydroxyd und Kieselsäure sich vereinigen, so entstehen nicht in jedem Falle Tonminerale, zum Teil wird auch amorpher Allophan gebildet, der mit der Zeit kristallin, d. h. zu Tonmineralen umgewandelt werden kann (Näheres siehe bei JASMUND 1955, v. ENGELHARDT 1961).

D. Bildung von Eisenverbindungen

Bei der chemischen Verwitterung wird das in die primären Silikate und Oxyde eingebaute Eisen frei. Dabei verbindet es sich im gemäßigt warmen, humiden Klima mit Sauerstoff und Wasser vorwiegend zu Eisenoxydhydraten, die teils amorph, teils aber auch kristallin (Goethit, Lepidokrokit) sind. Auch stärkere Wasseranlagerungen an Fe kommen vor. Diese Eisenverbindungen mit unterschiedlichen Mengen an Wasser werden unter dem Begriff Brauneisen zusammengefaßt. In den Böden des gemäßigt warmen, humiden Klimas liegt das Eisen stets als Brauneisen in homogener (Braunerde) oder ungleicher Verteilung (Pseudogley) vor. Seine Farbe schwankt je nach Menge und Verteilungsart von Gelb bis Braunschwarz. Bei sehr dunkler Farbe sind meistens Manganverbindungen beteiligt.

In warmen und (wenigstens zeitweise) trockenen Gebieten entstehen wasserarme und wasserfreie Eisenverbindungen von leuchtend gelber, roter und braunroter Farbe (meistens Hämatit und Hydrohämatit); die Böden sind entsprechend gefärbt.

E. Stabilisierung der Tonsubstanz

Die Stabilisierung der Tonsubstanz im Boden geht auf verschiedene Ursachen zurück, oft sind mehrere Ursachen komplex beteiligt. Die größte Stabilisierung der Tonsubstanz liegt in der Roterde vor. Das periodisch trockene und warme Klima der Roterdegebiete führt die Aluminium-Eisenverbindungen in einen irreversiblen, sehr stabilen Zustand über. Dabei können diese Verbindungen eine so große Stabilität erreichen, daß sie nicht oder kaum mehr quellbar, ferner unplastisch sind und ein schorfig-krümeliges Makrogefüge zeigen; diese tonige Substanz verhält sich bodenphysikalisch ähnlich wie Mineralkörner. In dem calciumreichen Tschernosem verleiht in erster Linie die große Menge an Ca-Ionen der Tonsubstanz ihre Stabilität; letztere wird bereits beeinträchtigt, wenn die volle Basensättigung unterschritten wird. Die Degradation der Schwarzerde ist die Folge.

Die typische Braunerde des gemäßigt warmen, humiden Klimas besitzt nicht, wie früher angenommen wurde, eine vollkommen stabilisierte Tonsubstanz; nur ein Teil davon zeigt geflockte, stabilisierte Aggregate. Die saure Braunerde, die infolge der Calciumarmut ihres Ausgangsmaterials stets sauer war, besitzt eine ziemlich stabile Tonsubstanz. Wahrscheinlich beruht in diesem Falle die Stabilisierung auf der Wirkung von Eisen(III)-Ionen, die im sauren Milieu frei und reaktionsfähig werden (BARGON 1960, B. MEYER, mündliche Mitteilung 1960). Diese Auffassung wird bezeugt durch die Braunerdebildung aus dem Eisensandstein des Doggers. Die dabei entstehende große Menge an Eisenionen flocken und stabilisieren die Feinsubstanz des Bodens derart, daß ein fast unzerstörbares Krümelgefüge entsteht. Eine Braunerde dieser Art entstand auf dem mitteljurassischen Eisensandstein Englands.

Zusammenfassend kann man sagen, daß in erster Linie Calciumionen, Eisenionen (wahrscheinlich auch Aluminiumionen) und Alterung der Al-Fe-Verbindungen die wichtigsten Stabilisatoren der Tonsubstanz (= mineralische Feinsubstanz) im Boden sind.

F. Humusbildung

Die Humusbildung wird hier nur so weit besprochen, wie es zum Verständnis der Bodenbildungsvorgänge erforderlich ist. Für eine eingehende Orientierung über die Humusbildung werden die umfassenden Arbeiten von KONONOWA (1958) sowie von SCHEFFER und ULRICH (1960), ferner die Spezialarbeiten von FLAIG, LAATSCH und WITTICH empfohlen.

Die Zersetzung der auf und in den Boden gelangenden organischen Substanz wird durch folgende Faktoren beschleunigt: leichte Zersetzbarkeit der organischen Masse, um den Neutralpunkt liegende Bodenreaktion, mäßige Feuchtigkeit, Wärme, Nährstoffe für die Mikroben (die gleichen wie für höhere Pflanzen). Die organische Substanz wird weitgehend abgebaut, wenn alle diese Faktoren optimal sind. Weicht dagegen einer (oder mehrere) dieser Faktoren weit vom Optimum ab, so ist der Abbau der organischen Substanz gehemmt. Beispielsweise führt Wasserüberstauung zur Moorbildung, stark saure Reaktion und schwer zersetzbare organische Masse (Beersträucherrückstände, Bestandesabfall von Koniferen) zur Rohhumusbildung; andererseits kann bei alkalischer Reaktion, aber starker

Austrocknung in Südwestexposition, aus schwer zersetzbarem Abfall der Schwarzkiefer ein Trockenmoder entstehen. Bei niedrigen Temperaturen bildet sich in nördlichen Breiten und im Hochgebirge ebenfalls eine rohhumusähnliche Auflage.

Die Bildung einer stickstoffreichen, stabilen Humussubstanz hängt einerseits von zeitweilig günstigen Abbaubedingungen und andererseits von zeitweiliger Hemmung eines zu stürmischen Abbaues ab (kontinentales Klima mit Unterbrechung der biologischen Arbeit durch Wärme, Austrocknung und Kälte), ferner von der Gegenwart stabilisierender Substanz, nämlich Calcium und aktiver Tonsubstanz, vielleicht auch Eisen. Die Bildung und Konservierung großer Mengen bester Humusstoffe vollzieht sich im Tschernosem des semihumiden Kontinentalklimas.

Im gemäßigt warmen, humiden Klima findet bei nicht zu saurer Reaktion eine relativ geringe Humusbildung statt, weil unter günstigen Feuchte- und Wärmebedingungen die Mikroorganismen eine lange Arbeitszeit haben, während der ein starker Abbau der organischen Masse vor sich geht. Im allgemeinen fehlt auch das stabilisierende Calcium. Es wird in diesem Falle viel organische Masse abgebaut und wenig Humus konserviert.

In warmen, feuchten Klimaten (Subtropen und Tropen) besitzen die Böden allgemein wenig Humus. Hier sind die Voraussetzungen für die Zellulosezersetzer so günstig, daß die reichlich anfallende organische Masse einem rapiden Abbau anheimfällt. Auf dem Boden des tropischen Regenwaldes sammelt sich infolgedessen trotz hoher Produktion an organischer Masse kein Bestandesabfall an. Auch im tropischen Wechselklima mit feuchten und trockenen Perioden läßt der Abbau der pflanzlichen Substanz in der feuchten Zeit keine Ansammlung von Humus zustande kommen. Andererseits beruht das Wachstum auf tropischen Böden in erster Linie auf der schnellen Mineralisierung der organischen Masse (schneller Humusumsatz). — Das sind nur die Hauptzüge der Humusbildung in Abhängigkeit von wichtigen Klima- und Bodenfaktoren.

G. Entbasung

In jedem Klima, in dem mehr Niederschlag fällt als verdunstet, sickert ein Teil des Wassers in den Untergrund ab, wenn nicht undurchlässige Schichten das verwehren. Das Sickerwasser nimmt lösliche Bestandteile und austauschbare Ionen mit in den Untergrund. Diesem Auswaschungsprozeß unterliegen zunächst die wasserlöslichen Salze der Alkalien und Erdalkalien, dann die leicht im kohlensäurehaltigen Wasser löslichen Carbonate. Schließlich werden auch die sorbierten Ionen mehr oder weniger schnell, je nach Ionenart, ausgewaschen, wobei weitgehend H-Ionen die basisch wirkenden Kationen ersetzen. Das bedeutet Basenverarmung und gleichzeitig Versauerung mit allen Folgeerscheinungen. Dieser Prozeß verläuft um so schneller, je mehr Wasser versickert (hoher Niederschlag, geringe Verdunstung, gute Durchlässigkeit) und je ärmer der Boden an Basen ist.

H. Tondurchschlämmung

Der Prozeß der Tondurchschlämmung wird eine eingehende Erörterung in dem Abschnitt „Parabraunerde" erfahren, so daß hier nur das Grundsätzliche der Vollständigkeit wegen zu sagen ist.

Unter Tondurchschlämmung wird die Verlagerung toniger Substanz (einschließlich Eisen) aus einem höheren in einen tieferen Bodenhorizont verstanden. Dabei wird die Tonsubstanz nicht wie bei der Podsolierung zerstört, da für eine Tonzerstörung die Wasserstoffionenkonzentration zu gering ist. Der Prozeß beschränkt sich nicht auf die Parabraunerde und äquivalente Böden (sol brun lessivé, sol lessivé, gray-brown podzolic soils, Derno-Podsol), sondern gehört auch zur Dynamik vieler anderer Bodentypen. Er wird bei der Degradation des Tschernosems, bei Mediterranböden, bei Rotlehm und sogar bei Böden des semiariden Klimas beobachtet. Selbst auf sehr schweren Böden (Pelosolen) findet dieser Vorgang statt (Diez 1959). Deshalb erscheint die Bezeichnung „Lessivierung", abgeleitet vom Sol lessivé, für diesen Vorgang nicht zweckmäßig zu sein.

Die Tonverarmung der oberen Bodenhorizonte und die Tonanreicherung im Unterboden als Effekt einer Tondurchschlämmung wurden schon vor mehreren Jahrzehnten beobachtet, aber es wurde zunächst keine Unterscheidung zwischen Tondurchschlämmung und Podsolierung getroffen.

In den Erläuterungen zu den geologischen Karten im Maßstab 1:25000 im Bereich des Jungdiluviums in Norddeutschland wurde schon auf die Entkalkung und Tonverarmung der oberen Horizonte der Geschiebemergelböden hingewiesen (Behrend und Schmidt 1938). Für bestimmte Braunerden der Schweiz hat Pallmann u. a. (1943) das gleiche früh erkannt. Um den Vorgang der Tondurchschlämmung von der Podsolierung zu unterscheiden, wurden Begriffe gebraucht, wie „podsolig", „podsoliert", „gebleicht" und „podzolic"; z. B. sagte man gebleichter brauner Waldboden, welcher der Parabraunerde entspricht, oder gray-brown podzolic, eine große Bodengruppe, welche die Parabraunerde einschließt. Die sowjetischen Bodenkundler haben bisher nur von Podsolierung gesprochen, jedoch hat neuerdings Fridland (1958) die Tondurchschlämmung in klarer Gegenüberstellung zur Podsolierung als Illimerisation bezeichnet. Der Mechanismus der Tondurchschlämmung kann gewiß nicht auf eine Ursache zurückgeführt werden. Die Tatsache, daß dieser Prozeß in sehr verschiedenen Bodentypen abläuft, spricht für mehrere Ursachen. Wenn wir alle betroffenen Bodentypen ins Auge fassen, so scheinen folgende Möglichkeiten für die Tonwanderung zu bestehen: 1. mechanische Durchspülung und Absatz bei Verlangsamung des Sickerwasserstromes; 2. Verbindung der Tonsubstanz mit bestimmten organischen Verbindungen unter Bildung von Chelaten oder anderer organisch-anorganischer Komplexe, die wanderungsfähig sind (Bloomfield 1953a, 1954b, c, 1955b); 3. Beweglichmachung der Tonsubstanz im Zuge der Entbasung, wobei die stabilisierende Wirkung des Calciums weitgehend verlorengeht (B. Meyer, mündliche Mitteilung 1960); 4. Wanderung der Tonsubstanz durch die Schutzkolloidwirkung der Kieselsäure (Kubiëna 1953).

I. Podsolierung

Wie schon im vorigen Kapitel gesagt, wurden bisher Tondurchschlämmung und Podsolierung nicht scharf getrennt. Heute begrenzen die meisten Bodenkundler den Prozeß der Podsolierung auf die Tonzerstörung, die Wanderung der Sesquioxyde des Eisens und des Aluminiums und deren Absatz in einem tieferen Horizont (B-Horizont), wobei in der Regel Humus mitwandert (Rode 1937). Die Ausfällung der verfrachteten Sesquioxyde kann durch eine höhere Konzentration von Ca-Ionen, durch Überschreiten des isoelektrischen Punktes und durch Konzentration der Lösung im Unterboden erfolgen. In neuerer Sicht wird für die Wanderung des Eisens bei der Podsolierung die Chelatbildung in größerem Umfange für wahrscheinlich gehalten (Bloomfield 1955b, Stobbe und Wright 1959). Bei niedrigem pH vermögen auch freie Fe(III)- und Al-Ionen im Bodenprofil zu wandern.

Durch den Podsolierungsprozeß verschiebt sich das Verhältnis von Kieselsäure zu Sesquioxyden im Auswaschungs- und Einwaschungshorizont; im Auswaschungshorizont wird das Verhältnis SiO_2/R_2O_3 zu Gunsten der Kieselsäure verschoben, und im Einwaschungshorizont ist es umgekehrt. So ist ein Nachweis der Podsolierung sicher möglich, vorausgesetzt, daß einheitliches Ausgangsmaterial vorliegt.

In den Böden des gemäßigt warmen, humiden Klimabereiches werden bei einem pH von 4,5 bis 5 Tondurchschlämmung und Podsolierung *gleichzeitig* stattfinden.

K. Naßbleichung

Nässe, Luftabschluß, hohe Wasserstoffionen-Konzentration und organische Substanz bedingen eine energische Lösung des Eisens im Boden. Wenn bei diesen Bedingungen eine Wegfuhr des Eisens möglich ist, so findet damit eine starke Ausbleichung des Bodens statt, die sich als hellgraue Bodenfärbung kenntlich macht. Der Mineralboden unter saurer Moorauflage, ferner unter dem A-Horizont eines sauren Anmoorgleyes, eines sauren Pseudogleyes und eines Stagnogleyes zeigt die Naßbleichung deutlich.

Die Naßbleichung unterscheidet sich von der Bleichung, die im Zuge der Podsolierung erfolgt, durch die Beteiligung der Nässe (Luftabschluß) und den andersartigen Transport und Wiederabsatz des Eisens. In Grundwasserböden wird das Eisen teils weit verfrachtet. Dagegen ist der überwiegend seitliche Transport des Eisens im Pseudogley und Stagnogley gering. Es wird partiell gelöst (helle Flecken) und nach kurzem Transport von oft nur wenigen cm wieder ausgefällt (rostbraune Flecken und Streifen). — Das Mangan verhält sich bei diesem Vorgang ähnlich wie das Eisen.

L. Vergleyung

Unter dem Begriff Vergleyung im weiteren Sinne werden die vom Grundwasser bewirkten Veränderungen im Boden verstanden. In erster Linie stellen die vom Grundwasser abgesetzten Stoffe die Vergleyung vor; vor allem sind es im gemäßigt warmen, humiden Klimaraum die Eisenverbindungen.

Im sauerstoffarmen Grundwasserbereich wird Eisen reduziert (graue, grünliche und bläuliche Färbung); es geht als zweiwertiges Ion in Lösung. Kommt das eisenführende Grundwasser mit der Luft in Berührung, so oxydiert das Eisen und fällt als Brauneisen in Form von rostgelben und rostbraunen Flecken, Streifen oder Konkretionen aus. Dieses so entstehende Fleckenbild stellt die bekannteste Vergleyungserscheinung dar. Daneben kann auch Calcium als Calciumbicarbonat vom Grundwasser transportiert und unter Freiwerden von CO_2 als Calciumcarbonat in feiner Verteilung, in Form von Kalkknoten, von Wiesenkalk oder von Kalksinter ausgefällt werden. Tonsubstanz wird weniger verfrachtet. In warmen Klimaten kann auch Kieselsäure vom Grundwasser transportiert und wieder ausgeschieden werden. Alle diese Wirkungen gehören in weiterem Sinne zur Vergleyung.

Im allgemeinen bringt die Vergleyung eine Bodenverdichtung mit sich. Gleichzeitige Vergleyung und Podsolierung führen zu besonders dichten und bei Austrocknung harten Unterboden-Horizonten.

M. Pseudovergleyung

Die Staunässe, d. h. das im Boden auf einer mehr oder weniger dichten Unterbodenschicht gestaute Niederschlagswasser, bewirkt ebenfalls eine Lösung des Eisens. Bei längerer Einwirkung, tiefem pH und Gegenwart von organischer Substanz erfolgt eine Reduktion des Eisens. Nach kurzem Transport wird es bei

Austrocknung des Bodens wieder als Fe(III) in Flecken, Streifen und Konkretionen ausgefällt. Eine Verlagerung von dreiwertigem Eisen kann ebenfalls erfolgen, und zwar bei sehr tiefem pH als Ion oder bei weniger tiefem pH unter Beteiligung von Schutzkolloiden (meistens saure organische Stoffe). Schließlich kommt auch noch die Möglichkeit der Bildung von Chelaten oder anderer wanderungsfähiger organisch-mineralischer Komplexverbindungen in Frage.

Mangan verhält sich ähnlich wie das Eisen. Auch die tonigen Substanzen, in erster Linie Al-Verbindungen, nehmen an der kurzstreckigen Verlagerung teil, wie tonreichere und tonärmere Stellen im Pseudogley-Profil bezeugen. Im allgemeinen erzeugt der Prozeß der Pseudovergleyung eine dichte Lagerung des Bodens, die vielfach durch ein Plattengefüge äußerlich erkennbar ist.

Das durch den geschilderten Prozeß entstehende Fleckenbild des Bodenprofiles ist dem des Gleyes ähnlich; diese Tatsache führte zu dem Namen Pseudogley (Kubiëna 1953).

In Hanglagen zieht die Staunässe je nach dem Porensystem mehr oder weniger schnell hangabwärts, so daß die Möglichkeit eines weiteren Transportes gelöster Stoffe besteht. Im Pseudogley der Hanglagen nimmt das bewegte Wasser mehr Luft auf, so daß Reduktionsvorgänge zurücktreten. Ökologisch ist diese Wasserform günstiger.

Der Stagnogley unterliegt ebenfalls dem Prozeß der Pseudovergleyung, jedoch in wesentlich stärkerem Maße als der Pseudogley, vor allem in bezug auf Reduktionsvorgänge. Oft werden im extrem sauren Stagnogley toxisch wirkende Al-Ionen frei; aber auch im Bodenwasser gelöste, niedrigmolekulare Humussäuren schädigen die Pflanzen.

N. Versalzung

Der breite Raum vom semihumiden bis zum ariden Klima beherbergt die sogenannten Salzböden, die sich durch einen hohen Salzgehalt (oder die Kationen ehemals vorhandener Salze) in den oberen Horizonten auszeichnen. Überwiegend sind es Natrium- und Calciumsalze. Die Versalzung beruht auf folgenden Voraussetzungen: 1. dichter Unterboden oder Untergrund, so daß die Versickerung gehemmt oder ganz unterbunden ist; 2. hochstehendes Grundwasser, das Salze enthält; indes kann der Salzgehalt sehr gering sein. Das Salz kann auch aus salzhaltigen, oberflächennahen Schichten stammen, wie es am Neusiedler See in Österreich der Fall ist (Franz 1960). Durch die starke Verdunstung in den warmen Verbreitungsgebieten der Salzböden wird der Kapillarzug zur Bodenoberfläche stark gefördert. Das Wasser verdunstet, und die Salze werden ausgeschieden; so entsteht der Solontschak. Wird das Grundwasser irgendwie abgesenkt, so daß sein Kapillarwasser die Oberfläche nicht mehr erreicht, so schreitet die Versalzung nicht fort, im Gegenteil wird das vorhandene Salz allmählich mit dem Niederschlagswasser in den Untergrund abgeführt; indes bleiben die Kationen der Salze noch lange im Boden sorbiert; so entsteht der Solonetz.

O. Krustenbildung

In ariden Gebieten entstehen die Wüstenkrusten. Der Boden wird in diesem Klima nur selten durch Niederschlag befeuchtet. In den langen Trockenzeiten mit starker Verdunstungsenergie steigt die geringe Bodenfeuchtigkeit an die Oberfläche, und die in ihr gelösten Stoffe werden nahe der Oberfläche ausgeschieden. Hält dieser Vorgang lange an, so entwickelt sich mit der Zeit eine krustenartige Schicht, die bei starker Anlieferung von Salzen bis etwa 2 m

Mächtigkeit erreichen kann; solche Krusten sind sehr alt (fossil). Die häufigsten Krusten sind die aus Calciumcarbonat, Gips und Natriumsalzen sowie aus Gemischen der genannten Salze.

Zu diesen Bildungen gehören auch die sogenannten Schutzrinden, wovon der Wüstenlack aus Limonit und die Kieselsäurerinde am häufigsten sind. Die Schutzrinden umkleiden das Gestein und hemmen die chemische Verwitterung. Der Bildungsvorgang ist im grundsätzlichen der gleiche wie bei den Krusten.

P. Lateritisierung

Der Vollständigkeit wegen wird auch die Lateritisierung erwähnt. Die vielseitigen Deutungen dieses Vorganges und seine ganze Problematik können hier nicht diskutiert werden (Harrassowitz 1926, Mohr und van Baren 1954). Die Lateritisierung gilt nicht uneingeschränkt als ein Vorgang der Bodenbildung, sondern wird von geologischer Seite auch als Lagerstättenbildung aufgefaßt.

Der Laterit, eine Bildung der Tropen, fällt besonders durch seine krusten- oder knollenartige Eisenanreicherung (Lateritpanzer) an der Oberfläche auf. Diese Anreicherung von Eisen kann auf folgenden Vorgängen beruhen: 1. Viele Laterite entstanden durch Eisenanreicherung an der Oberfläche eines grundwassernahen Bodens (Grundwasser-Laterit). Das kapillar gehobene Grundwasser verdunstet, und das in ihm gelöste Eisen fällt an oder nahe der Oberfläche aus. 2. In Hangfußlagen und Senken ohne ständiges Grundwasser kann langsam zuziehendes Wasser aus eisenhaltigen Verwitterungsschichten Eisen zusammentragen; das Wasser verdunstet an den bezeichneten Stellen, und Eisen scheidet sich an oder nahe der Oberfläche aus. 3. In der Verwitterungsschicht eisenreicher Gesteine (besonders basischer Magmatite) trägt die durch starke Verdunstungsenergie nach oben steigende Bodenfeuchtigkeit das Eisen an die Oberfläche; auf diesen eisenreichen Verwitterungsschichten können mächtige Panzer entstehen. In diesem dritten Falle entsteht der Laterit auch ohne seitliche Zufuhr des Eisens.

Mit der Lateritbildung sind starke Kieselsäureabwanderung und Kaolinitbildung verbunden. Teils wird die Kieselsäure auch nur von Eisen und Aluminium getrennt, nur wenig transportiert und schalig-konkretionär abgeschieden. Unter dem Lateritpanzer folgt ein rot und grau gefleckter Horizont (Fleckenzone), der allmählich in einen mächtigen, hellgrauen Horizont übergeht (Bleichzone).

In den Tropen wird auch zusammengetragenes Lateritmaterial in Form kleiner Knöllchen gefunden. Diese Umlagerungsprodukte fossiler Laterite zeugen davon, daß dieser Prozeß in den Tropen über verschiedene erdgeschichtliche Epochen zurückreicht.

5. Beschreibung der Bodentypen

Die *Bodentypen* wurden in Kap. c, 2 als charakteristische *Umbildungsformen der Lithosphäre* definiert. In diesem Sinne wird in den meisten Ländern Europas und Asiens der Begriff „Bodentyp" aufgefaßt. Dieser Begriff entspricht auch der ursprünglichen Vorstellung der russischen Bodenforscher, besonders der von Dokutschajew. In den USA versteht man unter „soil type" eine niedrige Einheit der Bodengliederung, die in erster Linie durch den variierenden Faktor der Bodenart (Textur) gebildet wird.

Die Beschreibung der Bodentypen zwingt vorab zu der Beantwortung von zwei Fragen: 1. In welcher Reihenfolge sollen die Bodentypen abgehandelt werden? 2. Welche Bezeichnungen sollen verwendet werden? Die erste Frage erledigt sich bei der Abwägung der zweckmäßigsten Gruppierung. Demgegenüber ist aber die zweite Frage nicht ohne weiteres lösbar, weil die Bezeichnungen der Bodentypen nicht einheitlich sind. Die verschiedenen Bezeichnungen bereiten aber an sich nicht die entscheidenden Schwierigkeiten, denn man könnte die Synonyme nebeneinanderstellen. Das setzt aber voraus, daß die Synonyme den gleichen sachlichen Inhalt haben, d. h. daß die mit verschiedenen Namen belegten Bodentypen dem Wesen nach gleich sind. Das trifft aber in den meisten Fällen nicht zu. Vielmehr sind zunächst die mit gleichen Namen bezeichneten Bodentypen weit auseinander liegender Räume nicht in Entstehung, Aufbau und Eigenschaften gleich, z. B. die Schwarzerde Mitteldeutschlands und Südwestsibiriens. Ferner werden in einem Land bestimmte Bodentypen herausgestellt, die im anderen Lande anders abgegrenzt werden, z. B. ist der Rasenpodsol (=Derno-Podsol) der Sowjetunion nicht identisch mit unserer Parabraunerde; der Rasenpodsol verkörpert nämlich eine größere Bodenbildungsbreite (Pseudogley ist eingeschlossen) als unsere Parabraunerde. Dieser Umstand erschwert die Darstellung der Bodentypen der Erde erheblich. Die Beschreibung der Bodentypen eines kleinen Raumes ist schon schwierig und erfordert viele Seiten in einem Buch (s. Mückenhausen u. a. 1962).

Aus den aufgezeigten Schwierigkeiten, die bei der Beschreibung der Bodentypen bestehen, darf nicht geschlossen werden, daß es am Bemühen der bodentypologischen (=pedosystematischen) Forschung gefehlt hat. Am besten wird die Situation in der bodentypologischen Forschung klar, wenn man damit die stratigraphische Forschung der Geologie vergleicht; denn Bodentypologie und geologische Stratigraphie sind im Wesen ähnlich. Obgleich die Geologie als Naturwissenschaft wesentlich älter als die Bodenkunde ist, können heute bei weitem nicht alle Gesteinsschichten parallelisiert und mit gleichen Namen belegt werden. Wie in der Bodentypologie die Abgrenzung der Typen über große Räume Schwierigkeiten bereitet, so ist auch in der geologischen Stratigraphie die Abgrenzung der Schichten schwierig, oft unmöglich. Die Ursache dafür liegt bei beiden Fachgebieten in dem oft nicht zu entwirrenden Komplex der Entstehungsumstände und den wenigen Indizien für einen eindeutigen Vergleich und für eine sichere Einordnung.

Angesichts dieser Situation müssen wir uns bei der Darstellung der Bodentypen in diesem Handbuch auf einen Überblick beschränken, wobei die Bodenbildungen in erster Linie nach Aufbau und Eigenschaften beschrieben werden, ohne dabei das Gesamtbodenbild der Erde durch zu viele Bodennamen zu verwirren. Die mitgeteilte Literatur gibt genügend Ansatzpunkte zur Vertiefung in die einzelnen pedogenetischen Umbildungsformen der Lithosphäre.

Die Bodentypen folgender Räume sollen behandelt werden:

A. Die Bodentypen des gemäßigt warmen, humiden Klimas, besonders Mitteleuropas.

B. Die Bodentypen der übrigen Klimaräume der Erde.
   - Die Bodentypen des kalten, feuchten (arktischen) Klimas (arktische Böden).
   - Die Bodentypen des kühlen bis gemäßigt warmen, feuchten Klimas (Podsolregion).
   - Die Bodentypen des mediterranen Klimas und ähnlicher Klimate.
   - Die Bodentypen der semihumiden und semiariden Steppe.
   - Die Bodentypen der Halbwüste und der Wüste.
   - Die Salzböden.
   - Die Bodentypen der feuchten Subtropen und Tropen.
   - Die Bodentypen des Hochgebirges.

Die Bodentypen des gemäßigt warmen, humiden Klimas, besonders Mitteleuropas, werden eingehender beschrieben als die der übrigen Räume der Erde. Sie sind am besten bekannt wichtigsten Bodentypen. Es läßt sich nicht vermeiden, verschiedene Nomenklaturen anzuwenden, denn die Parallelisierung der Synonyme ist nicht ohne weiteres möglich, weil der Inhalt der Namen nicht immer der gleiche ist, wie oben schon des näheren ausgeführt wurde.

Die Bodentypen der Erde außer denen Mitteleuropas, werden in der Bodenlehre überwiegend entweder nach der US-amerikanischen oder der sowjetischen Nomenklatur bezeichnet, und zwar wird in Nord- und Südamerika sowie Australien meistens die US-amerikanische, in Osteuropa und Asien meistens die sowjetische angewendet. In den einzelnen europäischen Ländern hat sich eine Nomenklatur herausgebildet, die teils mehr US-amerikanische, teils mehr sowjetische Elemente enthält. Die Erforschung und Benennung der Bodentypen Deutschlands waren ursprünglich überwiegend von der russischen Bodenforschung angeregt (STREMME 1926a), später hat aber dieser Forschungszweig eigene Wege gesucht (KUBIËNA 1953). Die beiden neueren Bodenkarten der Erde zeigen die Nomenklaturschwierigkeiten sehr klar. Die eine Bodenübersichtskarte der Erde wurde vom Soil Survey der USA in Zusammenarbeit mit der FAO mit der US-amerikanischen Nomenklatur (Abb. 20) und die andere vom *Dokutschajew-Institut* der UdSSR (1957) mit der sowjetischen Nomenklatur herausgegeben. Bei dieser starken Zusammenfassung der Böden, wobei nur die allgemeinen, durch die Klimazonen bedingten Entwicklungszüge mit nur sehr grober Abgrenzung wiedergegeben werden können, sind ohne Schwierigkeit beide Nomenklaturen anwendbar. Dabei müssen in jedem Falle Begriffe gebraucht werden, die allgemein verständlich sind. Je mehr die Böden auf Grund einer vom Klima bestimmten Entwicklungstendenz zusammengefaßt werden, um so leichter ist ihre allgemein verständliche Bezeichnung, die in der Regel allgemeine Anerkennung findet. Je mehr die Bodenbildungen differenziert werden, desto schwieriger werden Abgrenzung und Benennung.

Damit sind wir bei der Frage der *Übergangsbildungen* zwischen zwei oder mehr Bodentypen angelangt. Wenn zwei Bodenbildungsfaktoren mit etwa gleicher Intensität auf ein Ausgangsmaterial einwirken, so entsteht kein „reiner" Bodentyp, sondern eine Umbildungsform der Lithosphäre, welche die Merkmale zweier Bodentypen zeigt. Sind z. B. die Bildungsumstände für einen Podsol gegeben, steht aber an dem betreffenden Ort das Grundwasser hoch, so entsteht ein Boden, der oben die Merkmale des Podsols und unten die des Gleyes besitzt; es liegt ein Gley-Podsol vor, der schon lange bekannt ist. Ein weiteres, weit verbreitetes Beispiel bildet die Pseudogley-Parabraunerde. Diese Übergangsbildung zeigt im oberen Profilteil Horizonte der Parabraunerde und im tieferen Unterboden (infolge starker Staunässe) die Kennzeichen des Pseudogleyes. Ist der Staunässeeinfluß stärker, so daß dieser die Profilausbildung stark beeinflußt und die Horizontbildung der Parabraunerde zurücktritt, so steht dieser Boden dem Pseudogley näher; es liegt ein Parabraunerde-Pseudogley vor. Das *Vorherrschende* steht also immer betont *am Schluß* und das *Untergeordnete* wie ein Adjektiv *voran.*

Statt der Vereinigung von zwei Bodentypennamen zur Bezeichnung der Übergangsbildungen kann man auch den Namen des untergeordneten Bodentyps mit der Endung „artige(r)" versehen, z. B. pseudogleyartige Parabraunerde, parabraunerdeartiger Pseudogley, gleyartiger Podsol, podsolartiger Gley. Diese Bezeichnungsweise kann fakultativ neben der oben genannten (Vereinigung der Bodentypennamen) angewendet werden; beides ist richtig und treffend. Im folgenden Abschnitt sollen nur weit verbreitete Übergangsbildungen genannt werden; eine eingehende Darstellung ist bei MÜCKENHAUSEN u. a. (1962) zu finden.

A. Die Bodentypen des gemäßigt warmen, humiden Klimas, besonders Mitteleuropas

Die gewählte Reihenfolge für die Abhandlung der Bodentypen Mitteleuropas folgt nicht starr einem Prinzip, sie beachtet jedoch in großen Zügen die in Deutschland weitgehend übliche bodensystematische Gliederung. Je nach Verbreitung und Bedeutung werden die einzelnen Bodentypen mehr oder weniger eingehend beschrieben.

Syrosem oder Gesteinsrohboden. Der Syrosem (KUBIËNA 1953) stellt die beginnende Bodenbildung mit geringer chemischer Verwitterung in der *gemäßigten* Klimazone dar. Der Boden ist sehr flachgründig und steinig, sofern er aus festem Gestein entsteht, und besitzt sehr wenig Humus. Das Profil ist sehr einfach; es besitzt nur die beiden Horizonte (A) und C. In den USA nennt man diesen Bodentyp *Lithosol,* wenn er aus festem Gestein entsteht, und *Regosol,* wenn er sich aus einem Lockergestein gebildet hat. Dabei macht man keine scharfe Grenze zu dem fortgeschrittenen Stadium der Bodenentwicklung mit einem A-Horizont, das in Mitteleuropa Ranker genannt wird. Die Unterscheidung der Anfangsbodenbildungen auf festem und lockerem Gestein werden wir in Deutschland in Zukunft auch vornehmen müssen (MÜCKENHAUSEN u. a. 1962). Für den Bodentyp Syrosem existieren auch noch andere Namen, z. B. *Silikatrohboden* und *Karbonatrohboden* (PALLMANN 1942), *Skelettboden* wegen des hohen Steinanteiles, *Schuttboden,* wenn er auf Gesteinstrümmern liegt, und noch andere (JENNY 1930). Der Syrosem aus festem Gestein bietet nur anspruchslosen Pflanzen ein bescheidenes Dasein. Ist er auf einem mineralkräftigen Lockergestein entwickelt, so gibt dieses den Pflanzen bessere, teils sogar relativ gute Standortsbedingungen, z. B. roher Löß. Im ganzen gesehen bestimmt beim Syrosem nicht der Bodentyp, sondern das Ausgangsmaterial den Pflanzenstandort (MÜCKENHAUSEN 1959).

Ranker (Tafel III). Der Name Ranker wurde von KUBIËNA (1948, 1953) vorgeschlagen; er hat sich in Deutschland und Österreich eingeführt und auch in anderen Ländern Anklang gefunden. Die Namengebung war von der Vorstellung geleitet, daß dieser flachgründige Boden an den Berghängen gleichsam „rankt" (= sich festhält). In den USA und auch in anderen Ländern gebraucht man für diesen flachgründigen Boden die Bezeichnung *Lithosol.* Heute wird der Name Ranker in Mitteleuropa auch für A—C-Böden aus silikatischen Lockergesteinen verwendet, während diese in den USA und anderen Ländern *Regosol* genannt werden. In Zukunft werden wir in Deutschland zweckmäßig auch die A—C-Böden aus festen und lockeren Silikatgesteinen trennen müssen.

Entstehung. Der Ranker entsteht im gemäßigt warmen, humiden Klimabereich aus Silikat- und Kieselgesteinen durch fortschreitende Bodenbildung aus dem Syrosem. Diese Entwicklung verläuft langsam über die Übergangsbildungen Ranker-Syrosem und Syrosem-Ranker. Dabei schreitet infolge chemischer Verwitterung die Tonbildung fort; der etwas verbesserte Standort erlaubt mehr Pflanzenwuchs, und dadurch wird die organische Substanz im A-Horizont vermehrt. Je trockener das Klima, um so langsamer entwickelt sich aus dem Syrosem der Ranker. Das ist an trockenen Südwesthängen des Gebirges im Gegensatz zu den Nordhängen zu beobachten. In Südeuropa verweilt die Bodenbildung in vielen Fällen lange im Rankerstadium (FRANZ 1956). Die Humusform des A-Horizontes ist sehr verschieden; alle Formen vom Mull bis zum Rohhumus können entwickelt sein. Die verschiedene Mineralzusammensetzung des Gesteins, das Klima und die Vegetation sind die variierenden Faktoren der Humusbildung. In warmen, feuchten Lagen bildet sich unter Laubwald der Mull, auf Ca-armen Gesteinen unter Nadelwald der Rohhumus und im kühlfeuchten Bereich des Hochgebirges, selbst auf mineralkräftigem Gestein der Tangel (Tangelranker), eine bessere Form des Rohhumus.

Normalerweise bildet sich in Mitteleuropa aus dem Syrosem-Ranker ein brauner Ranker, der bei weiterer Bodenbildung in den Braunerde-Ranker mit beginnender Bildung eines (B)-Horizontes übergeht. Auf basen- und silikatarmen Gesteinen entwickelt sich ein Ranker mit filzigem Rohhumus, den KUBIËNA (1953) dystrophen Ranker und PALLMANN und HAFFTER (1933) A_0-Silikatrohboden genannt haben. Aus diesem armen Ranker bildet sich leicht der Podsol-Ranker. Der graue Ranker entsteht bei verzögerter chemischer Verwitterung in trockenen Lagen, besonders aus eisenarmen Gesteinen, und ist vor allem in Südeuropa verbreitet (KUBIËNA 1953).

Aufbau und Eigenschaften. Der Ranker zeigt ein A—C-Profil. Die Varianz in den Eigenschaften wird in erster Linie durch die Art des Gesteins gegeben. Ranker aus festem und lockerem Gestein zeigen recht verschiedene Eigenschaften, vor allem in ökologischer Hinsicht. Andererseits macht sich die mineralogische Zusammensetzung des Gesteins unmittelbar geltend, und zwar sowohl in der Bodenbildung, besonders in der Humusbildung, als auch in der Ernährung der Pflanzen.

Verbreitung. Der Ranker mit seinen Subtypen ist vorwiegend im Mittel- und Hochgebirge auf Silikat- und Kieselgesteinen zu finden, wo die Erosion keine ungestörte, lange Bodenbildung erlaubt hat, also vorwiegend in Rücken-, Kuppen- und Steilhanglagen. Wärme und trockene Lagen erhalten das Rankerstadium länger. Auch im Hügelland tritt der Ranker auf erodierten Flächen auf, ferner auch auf jungen Dünen.

Rendzina (Tafel III). Der Name Rendzina ist inzwischen allgemein in die internationale Literatur aufgenommen worden (MIKLASZEWSKI 1922, NIDDA 1932, BLANCK 1939b, FREI 1944, DOBRZANSKI und SKARZYNSKA 1950, GOGOLEW 1952, LIONETT 1952, LÅG 1959). In erster Linie beruht die schnelle und allgemeine Übernahme dieses Namens auf seiner Kürze und guten Einprägsamkeit, aber auch auf der Möglichkeit, die Böden aus Ca-reichen Gesteinen klar abzugrenzen. „Rendzina" ist eine volkstümliche polnische Bezeichnung für flachgründige Böden aus Karbonat- und Gipsgesteinen. Der Name soll von „rauschen" hergeleitet sein; denn das Pflügen dieser Böden verursacht wegen des hohen Steingehaltes und des meist nahen Gesteinsuntergrundes ein „Geräusch". In Süddeutschland ist „Fleinserde" eine alte, volkstümliche Bezeichnung für den flachgründigen Kalkboden. In der Schweiz wurde der Begriff „Humuskarbonatboden" dafür geprägt. Die Rendzina aus Gipsgestein trägt im Kaukasus die Namen Gadscha, Gaza und Gaschi (KUBIËNA 1953), in anderen Gebieten wird die Bezeichnung Humussulfatboden gebraucht. Die braune Rendzina (und Pararendzina) trägt in Frankreich den Namen „sol brun calcaire" und in England „brown calcarious soil". Dabei sind auch Übergänge zur Braunerde und Terra fusca einbezogen.

Entstehung. Hoher Calciumgehalt des Gesteins bildet die wichtigste Voraussetzung für die Bildung der Rendzina über eine weite Klimaspanne hinweg. Sie tritt in Norwegen auf (LÅG 1959) und scheint im Mittelmeerraum ebenfalls die erste Bodenbildung auf dem Kalkgestein zu sein. Auf den Carbonat- und Gipsgesteinen herrscht die *Lösungsverwitterung*, d. h. Calcium und Magnesium gehen unter Aufnahme von Kohlensäure in Bicarbonat über und werden weggeführt; Gipsgestein löst sich ebenfalls. Nur ein sogenannter Lösungsrest, überwiegend Aluminiumsilikate, Quarz und Kieselskelettbruchstückchen, bleibt für die Bildung der Bodenmasse zurück. Diese Teilchen besitzen meist $< 2\mu$ Durchmesser und gehören damit der Tonfraktion an. Somit sind die Rendzinen stets schwere Böden, was jedoch bei der Bodenbearbeitung wegen des meist hohen Calcium- und Humusgehaltes nicht stark ins Gewicht fällt. Durch Pflanzenwurzeln, Bodenfrost, Tiere und Bodenbearbeitung gelangen Steine aus dem Untergrund in den A-Horizont, der mithin stets mehr oder weniger steinig ist. Durch die Lösungsverwitterung entstehen im Carbonat- bzw. Gipsgestein Spalten und Höhlen, die sich mehr und mehr erweitern; das Gestein verkarstet.

Ein weiteres Charakteristikum der Rendzinabildung ist die Humusform. Der hohe Calciumgehalt des Solums und des Untergrundes läßt eine hochwertige, stabile, grauhuminsäurereiche Humusform entstehen, die der Rendzina die dunkle, unter Wald meist grauschwarze Farbe verleiht. Die Trockenheit im Sommer hemmt den Abbau der organischen Substanz, so daß die Rendzina normalerweise mehr als 4% Humus enthält. Meistens besitzt der A-Horizont Calciumcarbonat, zum mindesten in Form von Steinen und Gesteinssplittern.

Die Rendzinabildung geht von der Syrosem-Rendzina aus, die KUBIËNA (1953) *Protorendzina* genannt hat. Die typische Ausbildung ist die Mullrendzina mit einer ausgezeichneten Krümelbildung. Die verschiedenen Stadien der Rendzinaentwicklung hat KUBIËNA (1943a) eingehend beschrieben.

Das humide Klima treibt die Lösungsverwitterung und Entbasung schnell weiter, die Humusform verschlechtert sich und damit läuft die Entwicklung in Richtung zur Braunerde oder zu einem der Terra fusca ähnlichen Boden, im wärmeren Klima (Mediterranklima) zur Terra fusca und Terra rossa. So entstehen die mannigfaltigen Übergänge zwischen der Rendzina und den aufgezeigten Bodentypen. In trockenen, warmen Süd- und Südwestlagen kann auch in Mitteleuropa eine „trockene" Rendzinaart von grauer Farbe entstehen, bei der nicht selten der Abbau der organischen Substanz stockt und sogenannter Kalkmoder entsteht. Im Mediterranklima entsteht bei schwachem Pflanzenwuchs die humusarme Xerorendzina (KUBIËNA 1953). Das kühl-feuchte Klima des Hochgebirges verlangsamt selbst auf den Carbonatgesteinen den Abbau der organischen Substanz, es kommt zur Bildung einer rohhumusartigen Humusauflage, des sogenannten Tangels (KUBIËNA 1953). Unter Ackerkultur wandelt sich im allgemeinen die grauschwarze Rendzina in eine braune um, womit Humusabbau verbunden ist. In vielen Rendzinagebieten Mitteleuropas ist dem A-Horizont etwas Lößmaterial beigemischt, wodurch die Farbe stets braun und die Textur leichter wird. — Aus Mergel entsteht die Rendzina schneller, weil bei der Verwitterung viel Tonsubstanz entsteht.

Aufbau und Eigenschaften. Der Aufbau der Rendzina ist durch das A—C-Profil gekennzeichnet. Der A-Horizont wird selten über 30 cm mächtig. Die Textur stellt meistens einen steinigen Lehm bis steinigen Ton mit 3 bis 8% Humus dar. In der Rendzina wird das Niederschlagswasser schnell in den Untergrund abgeführt, was die Austrocknung des flachgründigen und daher nur wenig wasserhaltenden Bodens beschleunigt. Die schnelle Austrocknung ist ökologisch der schwache Punkt der Rendzina. Hinzu kommt, daß der meist dunkle Boden sich schnell erwärmt, der Kalkuntergrund ein warmes Bodenklima schafft und der Kalkgehalt ökologisch gleichsam Bodenwärme kompensiert. Darauf ist es zurückzuführen, daß anspruchsvolle Kulturpflanzen (Weizen) auf der Rendzina noch in Höhenlagen wachsen, in denen andere Böden dafür versagen.

Die Rendzina aus Mergel besitzt meistens einen mächtigeren A-Horizont sowie einen besseren Wasserhaushalt (trocknet nicht so schnell aus), und damit werden auch die ökologischen Bedingungen günstiger.

Meistens enthält der A-Horizont noch Calciumcarbonat, zum mindesten Kalksteine und Kalkgesteinssplitter, die zur Lockerung des Bodens beitragen.

Sollte fein verteilter Kalk fehlen, so ist aber stets der Ca-Ionen-Anteil hoch (hoher V-Wert). Die Reaktion liegt zwischen basisch und schwach sauer. Alle diese physikalischen und chemischen Eigenschaften bedingen ein gut ausgeprägtes Krümelgefüge oder bei weniger Humus ein Korngefüge. Die guten Bedingungen für Bodenorganismen werden nur durch zeitweilige Trockenheit eingeschränkt.

Verbreitung. Die Verbreitung der verschiedenen Arten von Rendzina ist an Kalke, Dolomite, Mergel und Gipsgesteine gebunden. Diese Gesteine finden wir im Mittel- und Oberdevon sowie Unterkarbon des variszischen Gebirges, das Teile Mitteleuropas einnimmt; ferner auf dem Muschelkalk sowie auf den Kalken des Juras und der Kreide. Die Kalke sind flächenweise in ganz Mitteleuropa — mit Ausnahme des nördlichen Teiles — verbreitet, vor allem auf dem Jurabogen vom Fränkischen bis zum Schweizer Jura (HUMMEL 1957). Die großen Kalkmassive der Alpen tragen auf den nicht zu steilen Lagen die Rendzina.

Pararendzina. Der Name Pararendzina wurde von KUBIËNA (1953) geprägt, soll „Nebenrendzina" bedeuten und damit die Verwandtschaft zur Rendzina zum Ausdruck bringen. Wegen abweichender Eigenschaften von der Rendzina nannte MIECZINSKI (1932) sie „gemischte Rendzina". In Frankreich steht sie beim „sol brun calcaire". In Deutschland hat sich die Trennung von Rendzina und Pararendzina weitgehend eingeführt, im Ausland (außer Österreich) kaum (LAATSCH 1957, FRANZ 1960, SCHEFFER und SCHACHTSCHABEL 1960).

Entstehung. Die Pararendzina entsteht aus calciumcarbonathaltigen Silikatgesteinen (festen und lockeren); anteilmäßig kann auch Magnesiumcarbonat vorhanden sein. Im Gegensatz zu den Rendzinen besitzt die Pararendzina mehr Quarzsand, und es fehlt ihr der Kalksteinuntergrund mit allen seinen Wirkungen auf den Gesamtboden. Die häufigsten Ausgangsgesteine sind: Kalksandstein, Geschiebemergel, kalkhaltiger Moränenschutt, Löß, Mischungen von Silikat- und Kalkschottern, im trockenen Klima auch basische Eruptiva. Die Pararendzina entwickelt sich aus der Syrosem-Pararendzina und entwickelt sich weiter über die verbraunte Pararendzina zur Braunerde-Pararendzina. Dieser Entwicklungsweg ist zwangsläufig im humiden Klima mit der Basenauswaschung und zunehmender chemischer Verwitterung verbunden.

Aufbau und Eigenschaften. Die Pararendzina zeigt ebenso ein A—C-Profil. Sie ist kalkhaltig bis zur Oberfläche, oder der A-Horizont zeigt wenigstens eine hohe Ca-Sättigung (hoher V-Wert). Im allgemeinen ist der Humusgehalt geringer als in der Rendzina, jedoch ist die Humusform gleich gut. Das quarz- und silikathaltige Ausgangsmaterial bedingt eine leichtere Textur als bei der Rendzina. Die kalkhaltigen, sandig-lehmigen Ausgangsgesteine (lehmiger Moränenschutt, Geschiebemergel, Löß) verursachen günstige, dagegen Kalksandstein, kalkhaltiger Sand und kalkhaltiger Schotter ungünstige bodenphysikalische Bedingungen, so daß eine große Varianz in den Eigenschaften besteht, die bei schwacher Bodenbildung in erster Linie gesteinsabhängig sind. Die Lebensbedingungen für die Bodenorganismen richten sich vornehmlich nach den geschilderten physikalischen und chemischen Eigenschaften.

Verbreitung. In hügeligen, erodierten Löß- und Geschiebemergellandschaften finden wir die Pararendzina auf Kuppen und in Hanglagen, wo der ehemalige Boden abgetragen ist, die Bodenbildung neu begonnen hat und sich im Stadium der Pararendzina befindet. Vor allem finden wir sie in etwas trockenerem Klima, z. B. in Mitteldeutschland und im Mainzer Becken. Auf dem Kalksandstein der Kreideformation (Ostwestfalen, Südniedersachsen) und auf den kalkigen und mergeligen Sandsteinen Südwestdeutschlands tritt sie auch kleinflächig auf. Größere Verbreitung hat sie auf den kalkhaltigen Schottern des Alpenvorlandes und auf dem kalkhaltigen Sand im nördlichen Oberrheintal.

Tschernosem (Schwarzerde) (Tafel I). Der Name Tschernosem ist die russische, volkstümliche Bezeichnung für die Schwarzerde (= schwarze Erde) der Ukraine. Diese Schreibweise ist wegen Verschiedenheiten der deutschen und russischen Schriftzeichen gewählt, die phonetische Schreibweise des russischen Wortes ist „Tschernosiom". Der Name Tschernosem (englisch Chernozem) wird heute international verwendet. Schon 1763 wurde dieser Name von LOMONOSSOFF (in GLINKA 1914) gebraucht, und 1879 und 1883 schrieb DOKUTSCHAJEW (1879b, 1879c und 1883) die ersten größeren Arbeiten über den Tschernosem. Später erschienen von russischen Bodenkundlern eine Reihe von guten Arbeiten über den Tschernosem (TANFILIEW 1928, KONAKOW 1930, PANKOW 1930, BOLISCHEW 1947, GORSCHENIN 1947, WOLOBUJEW 1958, POPOWA 1959). Ferner sollen noch die umfangreichen Arbeiten von STREMME (1930a, 1930b) und von FREHE (1931) genannt werden.

Entstehung. Zunächst ist hervorzuheben, daß der Tschernosem kein typischer Vertreter der gemäßigten, humiden Klimazone ist. Er wurde hier in der postglazialen Wärmezeit, hauptsächlich wohl im Boreal, gebildet und erhielt sich trotz Klimaänderung bis heute, wenn auch durch vermehrten Niederschlag und die Ackerkultur eine mehr oder weniger starke Degradation erfolgte. In der postglazialen Wärmezeit war das Klima des großen mitteldeutschen Schwarzerdegebietes kontinentaler, da die Nordseeküste weiter nördlich (nördlich der Doggerbank) lag.

Der Tschernosem stellt einen charakteristischen Bodentyp des kontinentalen, semihumiden Klimas mit etwa 350 bis 500 mm Jahresniederschlag dar. Die Grassteppe ist seine typische Vegetation; an feuchteren Stellen (Täler und Senken) siedeln sich auch Bäume an, ebenso erscheinen Baumgruppen zum humideren Bereich hin (Waldsteppe). Im Frühjahr sind alle Wachstumsfaktoren optimal: Wärme, Feuchtigkeit, Nährstoffe. Die Vegetation entwickelt sich schnell und üppig; auf diese Weise wird viel organische Masse für die Humusbildung geschaffen. Der trockene Sommer und der lange, kalte Winter lassen das Bakterienleben im Boden erstarren; nur eine relativ kurze Zeit des Jahres bleibt für die Tätigkeit der bakteriellen Zersetzung übrig. Diese reicht zwar aus, um die organische Substanz für die Humusbildung vorzubereiten, aber sie führt nicht zu einem stürmischen Abbau, so daß es zur Ansammlung von Humus kommt. Die abgestorbene Bakterienmasse bietet reichlich Stickstoff für den Aufbau einer stickstoffreichen, stabilen Humusform (FLAIG, SCHEFFER und KLAMROTH 1955). Bei der Stabilisierung des Humus spielen auch die Basen (vor allem Calcium) des Ausgangsgesteins, die in dem semihumiden Klima einer nur schwachen Auswaschung unterliegen, eine große Rolle. Die Bodenkleintiere (besonders Regenwürmer) und die Bodenwühler der Steppe durchmischen den A-Horizont intensiv, wobei ständig Bestandteile des C-Horizontes einbezogen werden. So vertieft sich der A-Horizont, der 80 cm und mächtiger werden kann.

Eine wichtige Entstehungsbedingung für die Schwarzerde stellt das Ausgangsmaterial dar. Der kalkreiche, lockere Löß bietet die beste Voraussetzung. Aber auch auf anderen kalkhaltigen Lockergesteinen ist sie entstanden, z. B. auf sandreichem Mergel und auf Geschiebemergel. Auf basenreichen festen Gesteinen kann die Schwarzerde auch gebildet werden (Nordkaukasus), aber ihre Entwicklung verläuft in dem semihumiden Klima mit relativ geringer chemischer Verwitterung langsam. Sogar auf dem Granit entsteht sie (Kasachstan). Aus kalkhaltigen Sanden bildet sich ein tschernosemähnlicher Boden, den KUBIËNA

(1953) und FRANZ (1955) Paratschernosem genannt haben. Östlich von Budapest tritt dieser Boden auf kalkhaltigem, pliozänem Sand auf.

Nach dem Bildungsoptimum der Schwarzerde in Mitteleuropa wurde das Klima etwas kühler und regenreicher. Der Wald rückte in die Steppe vor; außerdem wurde das mitteleuropäische Steppengebiet bereits vor einigen tausend Jahren vom Menschen in Kultur genommen. Diese Umstände führten zur Degradation der Schwarzerde, die sich in folgendem äußert: Entbasung, Abbau und Verschlechterung des Humus, Tondurchschlämmung, Bildung eines B-Horizontes, bei dichtem Untergrund sogar Vernässung. Die kleinen Schwarzerdegebiete Deutschlands, z. B. die Soester Börde und das Leinetal bei Göttingen (SCHEFFER und MEYER 1958), fielen der Degradation vollends anheim; nur Relikte blieben als Tschernosem-Parabraunerde erhalten. Im übrigen entwickelte sich aus dem Tschernosem eine Parabraunerde. Wo die Zunahme der Niederschläge zur Vernässung führte, entstanden Übergänge zum Pseudogley und Gley, die SELLKE (1935) „feuchte Steppenböden" genannt hat, die neuerdings auch Pseudotschernoseme genannt werden (SCHEFFER und MEYER 1958, SCHEFFER und SCHACHTSCHABEL 1960). In Osteuropa nennt man sie Wiesentschernoseme. Zwischen dem Tschernosem und der Braunerde entstehen im Übergangsgebiet zur Braunerde ebenfalls Übergangsbildungen (Braunerde-Tschernosem). Wo der A-Horizont des entbasten Tschernosems teils erodiert wurde und die Bodentiere kalkhaltiges Untergrundmaterial mit dem Krumenrest vermischten, entstand der kalkhaltige Tschernosem (die übrigen Tschernoseme besitzen keinen Kalk im A-Horizont). Kalkhaltiger Tschernosem kann aber auch in den trockenen Bereichen der Steppe vorkommen, so im Übergangsgebiet zum Kastanosem. Selbst in dem Trockengebiet des Mainzer Beckens finden wir die kalkhaltige Schwarzerde.

In diesem lokal trockenen Raum des Mainzer Beckens hat sich ein „*brauner Steppenboden*" entwickelt, der kalkhaltig bis zur Oberfläche ist; er kommt auch in Rußland vor und wird dort „brauner Tschernosem" genannt. Die braunen Steppenböden der Pfalz, die als pedogenetisches Phänomen das hier herrschende Klima (Weinklima) kennzeichnen, haben die Bodenforschung stets sehr beschäftigt (STREMME und SCHLACHT 1927, SCHLACHT 1933, STREMME 1953, ZAKOSEK 1956, 1962).

Schwarzerdeähnliche Böden entstanden in kleineren Vorkommen, wo das Klima zwar nicht trocken genug war, um die Steppe zu erzeugen, wo aber die Bodenfeuchte den Abbau der organischen Masse hemmte. Ein solcher Boden scheint z. B. auf der Insel Fehmarn vorzuliegen (SCHLICHTING 1953, STREMME 1955, MÜCKENHAUSEN 1959).

Aufbau und Eigenschaften. Das normale Tschernosemprofil zeigt den Aufbau A—C, wobei oft zwischen A und C die Übergangshorizonte AC und (oder) CaC eingeschaltet sind. Der degradierte Tschernosem besitzt einen B-Horizont; bei starker Degradation liegt auch ein A_3-Horizont vor. Die Mächtigkeit des A-Horizontes schwankt zwischen 30 und 80 cm, kann aber auch darüber hinausreichen. Im allgemeinen ist die Textur sandig-lehmig; der Humusgehalt übersteigt in Mitteleuropa selten 5%, kann in Rußland aber 15% erreichen.

Physikalische Eigenschaften: Ausgezeichnetes Krümelgefüge, hohes Porenvolumen, hohe Wasserkapazität, guter Wasser- und Lufthaushalt. Die Degradation verschlechtert alle diese Eigenschaften.

Chemische Eigenschaften: Nur die „kalkhaltige Schwarzerde" besitzt Calciumcarbonat im A-Horizont, sonst ist der A-Horizont kalkfrei, besitzt aber eine hohe Sättigung an basischen Kationen, vor allem an Ca-Ionen. Der V-Wert liegt infolgedessen hoch und die Reaktion bewegt sich um den Neutralpunkt.

Der T-Wert ist infolge des hohen Humusgehaltes und eines höheren Anteils an Tonmineralen der Montmorillonitgruppe hoch (ABRUKOWA 1960).

Biologische Eigenschaften: Großer Besatz an Bodenorganismen, vor allem an Bakterien, Bodenkleintieren und Bodenwühlern, daher die große biologische Aktivität.

Verbreitung. In Mitteleuropa gibt es größere Tschernosemvorkommen nur in Mitteldeutschland (Magdeburger Börde). Kleine Vorkommen treten im Oberrheintal auf (ZAKOSEK 1956, GANSSEN und HARTH 1957). Die kleinen Vorkommen von degradiertem Tschernosem (Soester Börde, Leinetal, Limburger Becken, Wetterau) sind unbedeutend. Größere Flächen von Tschernosem liegen in Südosteuropa, Osteuropa, Südwestsibirien, Argentinien und in den USA.

Braunerde (Tafeln II bis IV). Der Begriff Braunerde hat sich seit seiner Prägung durch RAMANN (1905) stark gewandelt. In der Sowjetunion war statt dessen stets der Name „brauner Waldboden" in Gebrauch und ist auch heute dort noch üblich (ANTIPOW-KARATAEW 1947). STREMME (1930a) führte diese russische Bezeichnung in Deutschland ein und schloß unter dieser Bezeichnung auch die heute abgetrennte Parabraunerde ein. In den USA, England und noch anderen europäischen Ländern werden „brown forest soil" und „brown earth" als identische Bezeichnungen gebraucht, und man versteht darunter die braunen Böden mit A – (B) – C-Profil *ohne* Tondurchschlämmung, während die graubraunen Böden mit Tondurchschlämmung als „gray-brown podzolic soils" abgetrennt wurden (CLINE 1949). In Frankreich und Belgien wird diese Unterscheidung auch seit über zehn Jahren gemacht, aber es bestehen dort andere Namen, nämlich „sol brun" für die Braunerde, „sol brun lessivé" für die schwach durchschlämmte Parabraunerde und „sol lessivé" für die stark durchschlämmte Parabraunerde (DUCHAUFOUR 1951, AUBERT 1952, DUDAL 1953). Heute wird allgemein unter Braunerde und synonymen Begriffen ein brauner Boden des gemäßigt warmen, humiden Klimaraumes mit A – (B) – C-Profil verstanden. Entscheidend dabei ist die Ausbildung des (B)-Horizontes. Dieser Verwitterungshorizont ist nur durch Verwitterung entstanden, es fehlt die Illuviation; das Makrogefüge ist subpolyedrisch oder bröckelig, der Horizont ist porenreich und durchlässig (MANCINI 1955, HOYOS DE CASTRO und SOLER 1957, KUNDLER 1957, TAVERNIER und SMITH 1957, EHWALD 1958).

Entstehung. Die Braunerde gehört zu den charakteristischen Bodentypen des gemäßigt warmen, humiden Klimas. Der Laubwald ist ihre natürliche Vegetation. Sehr entscheidend für die Entstehung der Braunerde in einem Klimaraum, der auch die Parabraunerde beheimatet, ist in erster Linie das Ausgangsmaterial, und zwar die bei der Verwitterung freiwerdenden Ca-Ionen einerseits sowie Fe- und Al-Ionen andererseits. Die bisherigen Beobachtungen sprechen dafür, daß die kalkhaltigen Ausgangsgesteine folgende Entwicklungsreihe von Bodentypen bedingen: Syrosem — Pararendzina — Braunerde — Parabraunerde. Dabei ist das Entwicklungsstadium der Braunerde kurz; ihr Entbasungszustand ist neutral bis schwach sauer und meistens nicht ohne Spuren von Illuviation, die jedoch für die Braunerde noch toleriert werden. Dieses Stadium der Ca-Ionen-reichen (= basenreichen) Braunerde scheint sich in einem etwas trockeneren, vor allem sommertrockeneren Klima länger zu halten, zudem ist in diesem Klima die bröckelige Ausprägung des (B)-Horizontes besser und die Neigung zur Tondurchschlämmung geringer. Das scheint seine Begründung darin zu finden, daß die Bodenfeinsubstanz durch die Trockenheit stabiler gegen Verlagerung wird. Die Braunerde aus Löß im Burgenland und Ungarn liefert dafür den Beweis.

Bei fortschreitender Entbasung der Braunerde, wobei Ca- durch H-Ionen ersetzt werden, folgt ein Stadium der Beweglichkeit der feinen Tonsubstanz;

die Durchschlämmung (= Illuviation) beginnt, es entsteht die Parabraunerde, die als nächster Typ behandelt wird. Ist aber die Entbasung so weit fortgeschritten, daß Al- und Fe-Ionen frei werden, so erfolgt wieder eine Stabilisierung der Tonsubstanz. Dafür spricht die Beobachtung, daß in dem entbasten, verlagerten Material der oberen Horizonte der Parabraunerde keine Illuviation mehr stattfindet. Die von SCHÖNHALS (1957) und von BARGON (1960) beschriebene Lockerbraunerde, die aus entbastem, umgelagertem Löß entstand, ist ein Beispiel dafür. Ebenso entwickelt sich aus calciumarmen Gesteinen unmittelbar die basenarme oder saure Braunerde (Tafel III), die keine Tondurchschlämmung zeigt (SCHEFFER, MEYER und FÖLSTER 1961). MANIL und HANOTIAUX (1957) haben ebenfalls die saure Braunerde ohne Illuviation in den Ardennen studiert und als „sol brun acide" bezeichnet; sie wurde auch in den nordöstlichen USA gefunden (BAUR und LYFORD 1957). Eine saure Braunerde aus sandigem Ausgangsgestein hat AVERY (1956) in England beschrieben und als sandige Braunerde bezeichnet. Bei der Entstehung der Braunerde übt die natürliche Laubwaldvegetation großen Einfluß auf die Bodenbildung aus, indem die Wurzeln den Boden durchdringen, ihn locker halten und Nährstoffe aus dem Untergrund aufnehmen, die mit dem Bestandesabfall wieder der Krume zugeteilt werden. Die Mullbildung wird dadurch gefördert.

Die basenreiche Braunerde kommt von der Pararendzina her und die primär basenarme (saure) Braunerde vom Ranker. Dementsprechend gibt es die Übergänge Pararendzina-Braunerde und Ranker-Braunerde. Die basenreiche Braunerde entwickelt sich unter Einschaltung von Übergängen zur Parabraunerde; die saure Braunerde kann durch Basenverlust und Rohhumusentwicklung podsolig werden und schließlich den Übergang Podsol-Braunerde (Tafel IV) vorstellen. Sofern keine Durchschlämmung vorliegt, ist der „brown podsolic soil" der USA damit identisch (LYFORD 1947, TAMURA und SWANSON 1954). Staunässe im tieferen Unterboden erzeugt den Übergang Pseudogley-Braunerde, Grundwasser die Gley-Braunerde. Das sind die wichtigsten Übergangsbildungen; sie zeigen, welche Mannigfaltigkeiten die Pedogenese bei dem Bodentyp „Braunerde" haben kann. Eine noch weitere pedogenetische Differenzierung ist der Arbeit von MÜCKENHAUSEN u. a. (1962) zu entnehmen.

Aufbau und Eigenschaften. Die Braunerde ist stets durch ein A – (B) – C-Profil gekennzeichnet. Das zweite charakteristische Merkmal stellt das *porenreiche, bröckelige* oder *subpolyedrische Makrogefüge* des *(B)-Horizontes* dar, der ein reiner Verwitterungshorizont *ohne* Toneinschlämmung ist; allerdings darf eine geringe Illuviation toleriert werden, wenn im übrigen das Gefüge für die Braunerde charakteristisch ist. Der A-Horizont ist 10 bis 25 cm mächtig und ist überwiegend als Mull ausgebildet; die saure Braunerde kann Moder besitzen und die Podsol-Braunerde sogar Rohhumus.

Die physikalischen Eigenschaften sind stark von der Textur abhängig, vor allem die Wasserkapazität. Mit Ausnahme der Pseudogley-Braunerde besitzen aber alle sandig-lehmigen Subtypen der Braunerde einen guten Wasser- und Lufthaushalt. In den chemischen Eigenschaften liegt eine große Breite vor, vor allem im Basensättigungsgrad, der zwischen gering bis hoch schwankt. Dementsprechend liegen die Reaktionswerte etwa zwischen 3,5 und 7. Diese große Varianz in den physikalischen und chemischen Eigenschaften bedingt auch große Unterschiede in der biologischen Aktivität, die in den verschiedenen Humusformen sichtbar werden.

Verbreitung. Die Braunerde besitzt im gemäßigt warmen, humiden Klima nicht die große Verbreitung, wie ursprünglich angenommen wurde. Nachdem der Braunerde das A – (B) – C-Profil gleichsam zugewiesen wurde und alle braunen

Böden mit Tonilluviation dieses Klimaraumes als Parabraunerden bezeichnet wurden, sind die Flächen stark dezimiert. Die basenreiche Braunerde tritt kleinflächig im Gebiet kalkhaltiger, sandig-lehmiger Substrate in Gesellschaft der Pararendzina und Parabraunerde auf, z. B. in den Gebieten des Löß, des jungen Geschiebemergels und junger, kalkhaltiger Moränen und Schotter. Größere Flächen nimmt die saure Braunerde auf kalkfreien Silikatgesteinen ein, z. B. auf Grauwacke, Schiefer, Granit, Gneis und bestimmten Sandsteinen.

Parabraunerde (Tafeln II und IV). Bei der Besprechung der Braunerde wurde mehrmals die Parabraunerde als brauner Boden mit Tondurchschlämmung der gemäßigt warmen, humiden Klimazone genannt. Der Name Parabraunerde wurde gewählt, um die Verwandtschaft zur Braunerde aufzuzeigen, aber auch deshalb, weil die Parabraunerde synonym den Bezeichnungen „gebleichter (oder podsolierter) brauner Waldboden“ und „podsolierte Braunerde“ ist.

In Frankreich und Belgien ist die Parabraunerde synonym mit dem „sol brun lessivé“ (schwach durchschlämmt) und dem „sol lessivé“ (stark durchschlämmt) (AUBERT 1952, DUDAL 1953). In Österreich scheint sich die Bezeichnung „sol lessivé“ auch einzuführen (FRANZ 1960). In den USA werden alle graubraunen Böden mit Tondurchschlämmung dieser Klimazone in der großen Bodengruppe „gray-brown podzolic soils“ zusammengefaßt. Dazu gehören auch die Parabraunerden, daneben jedoch auch die vielen Übergänge zum Pseudogley (CLINE 1949, MCCALEB und CLINE 1950). In Rußland wird die Parabraunerde als podsolierter brauner Waldboden bezeichnet, während die Übergänge von der Parabraunerde zum Pseudogley zum Derno-Podsol gestellt werden. Im ganzen gesehen sind die Benennung und die Trennung der Typen im Bereich der graubraunen Böden mit Tondurchschlämmung nicht einheitlich und schwierig zu übersehen (MÜCKENHAUSEN u. a. 1962).

Entstehung. Die Parabraunerde, oder allgemein gesprochen der braune Boden mit Tondurchschlämmung, ist an den gemäßigt warmen, humiden Klimaraum gebunden. Bei der Tondurchschlämmung wird die Tonsubstanz, d. h. die verschiedenartigen Aluminiumsilikate, nicht zerstört im Gegensatz zur Podsolierung, bei der eine Trennung der Sesquioxyde von der Kieselsäure erfolgt und vorzugsweise die Sesquioxyde im Profil abwandern. Obschon die Tondurchschlämmung schon lange festgestellt worden ist, kann sie bis heute pedogenetisch noch nicht sicher gedeutet werden. Viele Autoren haben das Phänomen untersucht und teils auch eine Deutung versucht (DOBRZANSKI 1947, CLINE 1949, FREI und CLINE 1949, MCCALEB und CLINE 1950, STOBBE 1952, GARDNER und WHITESIDE 1952, DUDAL 1953, VAN DEN BROEK 1956, KOHL 1958, BRUNNACKER 1959, HOLLSTEIN 1959, SCHLICHTING und BLUME 1959).

Die Tonwanderung war schon früher erkannt, aber mit der Podsolierung gleichgestellt worden (STREMME 1930a, SELLKE 1935), wie es auch heute noch vielfach geschieht. Die im weiten Raum der jüngeren Grundmoräne in Norddeutschland beobachtete „Versandung“ des oberen Bodenprofils wurde in den Erläuterungen zur Geologischen Spezialkarte von Preußen im Maßstab 1:25000 schon lange als Tonverarmung der oberen Horizonte gedeutet. Die russischen Bodenkundler haben bisher meistens nicht zwischen Tondurchschlämmung und Podsolierung unterschieden, fordern das aber neuerdings (FRIDLAND 1958).

Bei vielen pedogenetischen Deutungen der Tonwanderung schälen sich die folgenden vier Vorstellungen heraus: 1. Die belgischen Bodenkundler (TAVERNIER, mündliche Mitteilung 1958) nehmen eine rein mechanische Einspülung der feinen Tonsubstanz durch den Niederschlag nach Bodenaustrocknung an. 2. Die Untersuchungen von BLOOMFIELD (1953/54/55) über die Verlagerung von anorganischen Kolloiden im Bodenprofil durch Extrakte aus Laub ergaben, daß die Verlagerung der Tonsubstanz im Boden unter Beteiligung organischer Verbindungen vor sich geht, daß diese organischen Stoffe im Unterboden bakteriell abgebaut werden und damit die anorganische Komponente wieder unbeweglich wird. Über die Chelatbildung, die dabei eine große Bedeutung haben kann, haben kürzlich SCHEFFER, ULRICH und HIESTERMANN (1957) berichtet. 3. KUBIËNA (1953, 1956) macht die Kieselsäure (Schutzkolloidwirkung) für die Mobilisierung der Tonsubstanz verantwortlich. Das dürfte für das warm-feuchte Klima zutreffen, aber weniger für das gemäßigt warme. 4. Eine sehr einleuchtende Erklärung vertritt B. MEYER (mündliche Mitteilung 1960), indem er die Fließgrenze der Tonsubstanz in Abhängigkeit vom Anteil der Ca- und H-Ionen bringt, die das elektrokinetische Potential der Kolloidsubstanz bestimmen. Steigt letzteres nämlich im Zuge des Ca-Verlustes, so wird die Tonsubstanz beweglich, um bei noch stärkerer Versauerung wieder unter dem Einfluß von Al- und Fe-Ionen stabiler zu werden. Auffallend ist, daß die Parabraunerde aus Substraten hervorgeht, die den „kritischen Bereich der Ca-Verarmung" durchlaufen, während aus Ca-armem Gestein die saure Braunerde entsteht (s. bei Braunerde). So überzeugend diese Deutung ist, wir müssen auch die übrigen Deutungen wenigstens anteilmäßig für möglich erachten.

Die Bildung der Parabraunerde, d. h. die Durchschlämmung einer nennenswerten Tonmenge, benötigt nicht lange Zeit (etwa 1000 bis 2000 Jahre), wie Beobachtungen von HEIDE, S. MÜLLER und v. ZEZSCHWITZ (mündliche Mitteilung 1959/60) gezeigt haben.

Die Parabraunerde bildet oft — wie die Braunerde — Übergänge zu anderen Bodentypen, vor allem zur *Braunerde*, zum *Pseudogley* (Tafel IV) und zum *Podsol*. Den Übergang der sandigen Parabraunerde zum Podsol nennt KUNDLER (mündliche Mitteilung 1961) neuerdings Braunpodsol; in den USA entspricht diese Übergangsbildung dem „brown podzolic soil", sofern dieser durchschlämmt ist (FREI und CLINE 1949). Es besteht die Tendenz, die stark durchschlämmte Parabraunerde als *Fahlerde* zu bezeichnen.

Aufbau und Eigenschaften. Die Parabraunerde besitzt die Horizontfolge $A_1 - A_3 - B - C$ oder $A_1 - A_3 - B - (B) - C$. Der B-Horizont kann pseudovergleyt sein und trägt dann das Symbol gB. Infolge des Tonverlustes haben A_1- und A_3-Horizont einen graubraunen Farbton erhalten; daher kommt die Bezeichnung „gray-brown podzolic soil". Die Toneinschlämmung im B-Horizont hat eine etwas rötlichbraune Farbe dieses Horizontes veranlaßt. Der Unterschied im Tongehalt zwischen A_1- und A_3-Horizont einerseits und B-Horizont andererseits ist deutlich mit der Fingerprobe feststellbar; der Unterschied beträgt wenigstens 4% Ton, oft hat der B-Horizont die doppelte Menge Ton wie der A_1- und der A_3-Horizont. Das Makrogefüge im A_1- und A_3-Horizont ist bröckelig und stark porös, im B-Horizont ist es *ausgeprägt polyedrisch*, wobei die Gefügeaggregate mehr oder weniger porös sind. Bei starker Durchschlämmung kann der B-Horizont mehr oder minder verdichtet sein, wodurch sich Staunässe im Unterboden bildet.

Unter Wald ist der A_1 selten über 20 cm mächtig, meist sind Moder oder Mull entwickelt oder Übergänge zwischen diesen Humusformen. Dagegen kommt Rohhumus selten vor, jedoch ist er bei der Podsol-Parabraunerde bisweilen vorhanden.

Der B-Horizont der Parabraunerde aus Sand besteht aus Bändern, zwischen denen das Ausgangsmaterial unverwittert ist (C-Horizont). Die B-Bänder, die oben breiter und dichter als unten im Profil sind, reichen oft tief in das Profil hinab (bis 4 m etwa), und dann folgt erst der reine C-Horizont.

Die Wasserkapazität hängt von der Textur ab. Während der B-Horizont mit starker Tonilluviation Staunässe veranlassen und dann den Wurzeltiefgang hemmen kann, verbessern die B-Bänder den Wasserhaushalt der Sand-Parabraunerde. Der B-Horizont der Parabraunerde ist stets schwerer durchwurzelbar als der (B)-Horizont der Braunerde, was für die ökologische Wassernutzung wichtig ist.

Die chemischen Eigenschaften der einzelnen Parabraunerden sind verschieden; das pH schwankt etwa zwischen 6,5 und 3,5. Dementsprechend schwankt auch der V-Wert etwa zwischen 70 und 20. Das pH nimmt vom A_1- zum C-Horizont zu. Mit der Verarmung an Ca und Mg geht ebenso eine Verarmung an den übrigen Pflanzennährstoffen einher.

Den stark verschiedenen physikalischen und chemischen Eigenschaften entspricht ein ziemlicher Unterschied im Besatz mit Bodenorganismen. Dichte B-Horizonte werden gemieden.

Verbreitung. Die Parabraunerde oder, weiter gefaßt, die große Bodengruppe „gray-brown podzolic soils" hat im Bereich des gemäßigt warmen, humiden Klimas eine weite Verbreitung, wie die neuen Bodenübersichtskarten zeigen. Besonders hat dieser braune Boden mit Tonverlagerung sich in den Gebieten mit folgenden Lockergesteinen entwickelt: Löß, Geschiebemergel, kalkhaltige Moräne, kalkhaltige Glazialschotter. Kalkhaltige, sandig-lehmige Flußabablagerungen außerhalb des Überflutungs- und Grundwasserbereiches neigen zu der gleichen pedogenetischen Entwicklung. Auch kann dieser Boden auf kalkhaltigen oder (und) calciumhaltigen (Silikate) Gesteinen entstehen. Die sandige Parabraunerde mit B-Bändern entsteht aus silikatreichen Sanden, wie sie z. B. die Schmelzwässer des Inlandeises aufgeschüttet haben.

Podsol (Tafel IV). Der Name Podsol kommt aus dem Russischen und bedeutet frei übersetzt „Boden unter Asche". Damit wird ausgedrückt, daß unter dem holzaschefarbenen, völlig verarmten Oberboden ($A_1 + A_2$) der etwas nährstoffreichere Mineralboden folgt. „Podsol" ist auch einfach als „Ascheboden" aus dem Russischen übersetzt worden. Von „Podsol" ist „Podsolierung" abgeleitet, worunter der bodendynamische Prozeß im Podsol verstanden wird. Davon sind wieder „podsolieren" und „podsolig" abgeleitet.

Der Bodentyp Podsol ist schon lange bekannt. Trotzdem wird seine Abgrenzung zu anderen Bodentypen nicht einheitlich vorgenommen. In Deutschland möchten wir nur die Böden als Podsol ansehen, die infolge starker Versauerung eine Tonzerstörung zeigen und in denen ein klar ausgeprägtes $A_0 - A_1 - A_2 - B_1 - B_2 - C$-Profil vorliegt. Dieses Profil zeigt eindeutig die Eluviation und die Illuviation ohne andere pedogenetische Einwirkungen. Z. B. erzeugt der gleichzeitige Einfluß von Staunässe und Grundwasser eine andere Horizontfolge. Auf sandigen Substraten entstehen im allgemeinen die typischen Podsolprofile, wenn das Grundwasser tief steht. In den USA bezeichnet man den Podsol mit der oben gekennzeichneten Horizontfolge als „typical podzol". In der Sowjetunion wird dieser Boden, der sich aus sandigen Substraten entwickelt, „Sand-Podsol" genannt.

Im Norden der Sowjetunion treten Böden mit Tonzerstörung und einem mächtigen Bleichhorizont auf, die gleichzeitig staunaß sind, jedoch zu den Podsolen gestellt werden (Panomarewa 1956). In Deutschland würden wir diesen staunassen Podsol als Übergang zwischen Podsol und Pseudogley auffassen. Die sowjetischen Bodenkundler haben südlich der Podsolregion die breite Zone des Rasenpodsols (= Derno-Podsol) ausgeschieden. Teils sind dieses stark versauerte Böden, in denen eine Tonzerstörung stattfindet, teils liegt aber *nur* eine Tonverlagerung vor. Meist ist auch Staunässeeinfluß vorhanden (Katschinskaya 1937). In Frankreich faßt man ebenfalls den Podsolbegriff weiter (Duchaufour 1956). In der Tschechoslowakei wird der stark ausgeprägte Pseudogley mit hellem g_1-Horizont als Podsol aufgefaßt (Pelišek 1955a, 1955b).

so auch bisher in Jugoslawien (Ćirić 1956). Das möge genügen, um die unterschiedliche Auffassung und die Abgrenzung des Podsols darzulegen.

Für eine großzügige Abgrenzung des Podsols könnte man lediglich die Tonzerstörung fordern, wie es geschieht, wenn man von „podsolierten Böden" spricht. Dabei ist aber zu bedenken, daß andere pedogenetische Einflüsse die Ausbildung der Profile, die Dynamik und auch die Eigenschaften stark differenzieren können, z. B. Staunässe, Grundwasser oder die für den Podsol sonstwie artfremden Horizonte im Unterboden, die dann vorhanden sind, wenn der Podsol sekundär aus einem anderen Bodentyp entstand.

Entstehung. Die für die Podsol-Genese erste Bedingung ist ein humides Klima. Dabei ist weniger die Menge des Niederschlages entscheidend als vielmehr die Wassermenge, die den Boden durchsickert. Wenn die Verdunstung durch niedrige Jahrestemperatur, vor allem niedrige Sommertemperatur, und durch hohe Luftfeuchtigkeit stark herabgesetzt wird, so nimmt damit die Versickerungsmenge zu, und die Tendenz der Podsolierung wird stärker. Niedrige Temperaturen hemmen außerdem den Abbau der organischen Substanz; dadurch wird die Moder- und Rohhumusbildung gefördert und dadurch wiederum die Podsolierung. Die Verbreitung der Podsole im kühlen, feuchten Klimagebiet bestätigen diese Überlegungen. Dabei ist es nicht ausgeschlossen, daß kleinere Flächen mit dem Podsoltyp selbst in den Tropen vorkommen, wenn günstige Bedingungen für seine Entstehung vorliegen, vor allem ein basen- und silikatarmer Sand oder Sandstein.

Im kühlen, feuchten Klima finden Nadelbäume und Zwergsträucher (Heide, Heidelbeere) gute Wachstumsbedingungen. Ihr Bestandesabfall neigt zur Bildung von Moder und Rohhumus, deren niedrigmolekulare Humussäuren (Fulvosäuren) die Podsolierung stark fördern.

Beim Podsolierungsprozeß wird die Tonsubstanz im A_1- und A_2-Horizont zerstört, d. h. die Sesquioxyde trennen sich von der Kieselsäure und wandern mit Humus in einen tieferen Bereich des Profils, wo sie wieder ausfallen und damit den B-Horizont bilden. An dem Verhältnis von Sesquioxyden zu Kieselsäure in den oberen (A_1 und A_2) und den unteren Horizonten (B_1 und B_2) läßt sich analytisch klar erkennen, ob eine Wanderung der Sesquioxyde im Sinne der Podsolierung stattgefunden hat. Dieser Vorgang der Sesquioxyd-Verlagerung wird verschieden gedeutet.

Mattson und seine Mitarbeiter haben in mehreren experimentellen Arbeiten vor einigen Jahrzehnten nachweisen können, daß Fe und Al mit organischen Säuren und einer geringen Menge von Kieselsäure als elektropositive Sole im Podsolprofil wandern können. Die Ladung und die Stabilität der Sole sind um so größer, je höher die Wasserstoffionenkonzentration der Bodenlösung ist. Da die Wasserstoffionenkonzentration mit der Tiefe im Podsolprofil geringer wird, verlieren die Sole beim Abwandern im Profil ihre Ladung. Wenn die Sole schließlich ihr isoelektrisches pH erreichen, so koagulieren sie zu einem meist rostfarbenen Gel. Auch humusreiche Sole mit Fe und Al, die nur im extrem sauren Bereich koagulieren, werden im B-Horizont des Podsols geflockt, da die hier angereicherten Fe- und Al-reichen Gele einen hohen isoelektrischen Punkt erzeugt haben, bei dem auch die humusreichen Sole zur Koagulation gebracht werden (Laatsch 1957).

Später wurde die Möglichkeit aufgezeigt, daß organische Säuren bei dem Bleichungs- und Verlagerungsvorgang im Podsolprofil mitwirken können. Besonders aufschlußreich sind in diesem Zusammenhang die Experimente von Bloomfield (1953/55). Mit wasserlöslichen, organischen Stoffen (Wasserextrakten) von Laub bzw. Nadeln verschiedener Baumarten konnte er Fe und

Al aus dem Boden lösen und in komplexe Bindungen überführen. Das Fe bildet dabei einen stabilen, aber wanderungsfähigen Fe(II)-Komplex. Der Autor nimmt an, daß nach einer gewissen Wanderung der organo-mineralischen Komplexe im Bodenprofil der organische Anteil abgebaut wird und dadurch die Ausfällung erfolgt. Die Arbeit von STOBBE und WRIGHT (1959) weist auf die gleichen Möglichkeiten hin.

Die Ausfällung der wanderungsfähigen Stoffe im Unterboden des Podsols kann auch durch Wasserverlust und Austrocknung hervorgerufen werden, worauf LAATSCH (1957) und KURON (mündliche Mitteilung, 1954) hingewiesen haben. Der Wasserverlust der gewanderten Kolloidsubstanz wird im Unterboden durch ausgetrocknete, hydratationskräftige Kolloide oder auch durch die Pflanzen bewirkt.

Nur einige zur Zeit im Vordergrund stehende Erklärungen der Podsolierung sind damit gegeben. Daneben sind viele Arbeiten über diesen Gegenstand erschienen, die aber nicht grundsätzlich von den geschilderten Konzeptionen abweichen (HARRASSOWITZ 1930d, JOFFE 1931, REMEZOW 1937, RODE 1937, FREI 1944, DEB 1949/50, YARKOW 1954, PELIŠEK 1955a, 1955b, ĆIRIĆ 1956, MARTIN und REEVE 1957, 1958).

Nach den Anteilen des Eisens und des Humus im B-Horizont unterscheidet man Eisenpodsol (B ist rostbraun), Eisenhumuspodsol (B ist kaffeebraun) und Humuspodsol (B ist schwarz). Zwischen dem Podsol und einigen anderen Bodentypen gibt es Übergänge; diejenigen zur Braunerde, zur Parabraunerde, zum Gley und zum Pseudogley sind besonders häufig (MÜCKENHAUSEN u. a. 1962).

Aufbau und Eigenschaften. Die typische Horizontfolge des Podsols ist: A_0—A_1—A_2—B_1—B_2—C. Meistens besteht der A_0-Horizont aus Rohhumus, seltener aus Moder. A_1- und A_2-Horizont unterliegen starker Zersetzung der Silikate und der Verarmung an Basen, Sesquioxyden und noch anderen Stoffen (Phosphorverbindungen). Durch den Eisenverlust verlieren diese beiden Horizonte die durch Eisen bedingte Braunfärbung; sie werden grau wie Holzasche. Der A_1 ist infolge stärkerer Humusbeimengung stets dunkler als der meistens weißlichgraue A_2. Die Illuvialhorizonte B_1 und B_2 sind je nach den Anteilen von Eisen und Humus rostbraun bis tiefschwarz, wobei der Humus die Schwarzfärbung bewirkt. Den Anteilen dieser Stoffe entsprechend werden Eisenpodsol, Eisenhumuspodsol und Humuspodsol unterschieden. Der B_1-Horizont kann dicht und hart sein, wenn ein ausgeprägtes Hüllengefüge vorhanden ist; in diesem Falle liegt Ortstein vor. Bei nicht ganz verfestigtem B_1-Horizont spricht man von Orterde. Der dem B_1- folgende B_2-Horizont ist durch eine geringere Illuviation ausgezeichnet, oft besteht er aus aufeinanderfolgenden Bändern, die im allgemeinen in Stoff und Farbe dem B_1-Horizont ähneln. Bei gleichmäßiger Ausbildung des B_2-Horizontes enthält dieser im allgemeinen mehr Eisen und weniger Humus als der B_1-Horizont. Der B_2-Horizont geht allmählich in den C-Horizont über. Die Übergänge zwischen Podsol und anderen Bodentypen sind stets am Aufbau der Unterboden-Horizonte zu erkennen (MÜCKENHAUSEN u. a. 1962). Die physikalischen Eigenschaften des Podsols aus Sand sind: geringe Wasserkapazität, Trockenheit, Wasserstau im Podsolprofil nur bei dichtem Ortstein, der auch ein Wurzelhemmnis darstellt, Einzelkorngefüge, das im A_1 schwach durch Humus verkittet ist, im B Hüllengefüge. Bei sandig-lehmigem Unterboden verbessert sich der Wasserhaushalt. Der Pseudogley-Podsol ist staunaß.

Chemische Eigenschaften: im ganzen basenarm, versauert, niedriger V-Wert, geringer T-Wert, im B Anreicherung von Sesquioxyden, Humus und Phosphor.

Profil 1. *Sierosem* aus Löß

Profile 1. *Sierozem*, developed from loess

A_{11}	0— 4 cm	Gelblichgrauer, sehr schwach humoser, kalkhaltiger lehmiger Feinsand (Löß), schwaches Bröckelgefüge, locker, sehr porös, einige Wurzeln.	Yellowish-gray, very low in humus, calcareous loamy fine sand (loess), very weak subangular blocky structure[1], friable[2], very porous, some roots.
A_{12}	4— 22 cm	Dunkelgelblichgrauer, schwach humoser, kalkhaltiger lehmiger Feinsand (Löß), Bröckelgefüge, locker, sehr porös, einige Wurzeln.	Dark-yellowish-gray, low in humus, calcareous loamy fine sand (loess), weak subangular blocky structure, friable, very porous, some roots.
ACa	22— 28 cm	Grauweißlicher (etwas dunkler als CCa), stark kalkhaltiger lehmiger Feinsand (Löß), Spuren von Humus, etwas kohärent, sonst bröckelig, porös.	Gray-whitish (somewhat darker than CCa), strongly calcareous, loamy fine sand (loess), traces of humus, somewhat coherent, otherwise subangular blocky structure, porous.
CCa	28— 40 cm	Grauweißlicher, stark kalkhaltiger lehmiger Feinsand (Löß), etwas verfestigt, zerbricht in Bröckel, porös.	Gray-whitish, strongly calcareous loamy fine sand (loess), somewhat cemented, breaks down to weak subangular blocky structure, porous.
C	40—120 cm +	Hellgrauer, stark feinsandiger Löß, schwach kohärent, schwaches Bröckelgefüge.	Light-gray, very fine sandy loess, weakly coherent, very weak subangular blocky structure[3].

[1] Weak subangular blocky structure (= Bröckelgefüge), e.g. the soil breaks into poorly formed, indistinct, weakly coherent porous aggregates.

[2] Friable (= locker): soil material crushes easily under gentle to moderate pressure between thumb and forefinger and coheres when pressed together.

[3] The symbol + means that the horizon continues beyond the designated lower boundary.

Profil 2. *Kastanosem* aus Löß

Profile 2. *Chestnut Soil* (Kastanosem), developed from loess

A_{11}	0— 15 cm	Hellkastanienfarbiger, schwach humoser, kalkhaltiger feinsandiger Lehm (Löß), Bröckelgefüge, locker, porös, einige Wurzeln.	Light-chestnut coloured, low in humus, calcareous fine sandy loam (loess), weak subangular blocky structure, friable, porous, some roots.
A_{12}	15— 45 cm	Kastanienfarbiger (dunkler als A_{11}, etwas rötlichbraun), schwach humoser, kalkhaltiger feinsandiger Lehm (Löß), Bröckelgefüge, weniger locker als A_{11}, hohlraumreich, einige Wurzeln.	Chestnut coloured (darker than A_{11}, somewhat reddish-brown), low in humus, calcareous fine sandy loam (loess), weak subangular blocky structure, less friable than A_{11}, very porous, some roots.
ACa	45— 60 cm	Übergang von A_{12} zu CCa, dementsprechend gehen auch die Farben verzahnend ineinander über, gelblichweiße Kalkaugen.	Transition from A_{12} to CCa, colours intergrading from A_{12} to CCa, yellowish-white lime-nodules.
CCa	60— 70 cm	Grauweißlicher, kalkhaltiger feinsandiger Lehm (Löß), etwas verfestigt durch den Kalk, bricht kantig-bröckelig, porös.	Gray-whitish, calcareous fine sandy loam (loess), somewhat cemented by lime, breaks down to angular to subangular blocky structure, porous.
C	70—120 cm +	Gelblichgrauer Löß, schwach kohärent, schwaches Bröckelgefüge.	Yellowish-gray loess, weakly coherent, very weak subangular blocky structure.

Profil 3. *Tschernosem* aus Löß

Profile 3. *Chernozem*, developed from loess

A_{11}	0— 40 cm	Braunschwarzer, mittelhumoser feinsandiger Lehm (Lößlehm), Krümelgefüge und Wurmlosungsgefüge, locker, hohlraumreich, Regenwurmröhren, von feinen Wurzeln durchzogen.	Brown-black, medium in humus, fine sandy loam (loess-loam), crumbstructure and worm casts, friable, very porous, worm-holes, many fine roots.
A_{12}	40— 60 cm	Braunschwarzer (etwas heller als A_{11}), mittelhumoser feinsandiger Lehm (Lößlehm), nach unten zu weißliche Kalkfäden (Pseudomyzel), Krümelgefüge, locker, hohlraumreich, Regenwurmröhren, von feinen Wurzeln durchzogen.	Brown-black (somewhat lighter coloured than A_{11}), medium in humus, fine sandy loam (loess-loam), with increasing depth whitish threads of lime (pseudomycelium), crumb, fluffy, many pores, worm-holes, many fine roots.
ACa	60— 70 cm	Übergang vom A_{12} zu CaC, d. h. A_{12} greift streifig und hellerwerdend in CaC hinein, Bröckelgefüge, porös, Regenwurmröhren.	Transition from A_{12} to CaC, e.g. A_{12} tongues of organic matter cause a streaked colour pattern, weak subangular blocky structure, porous, worm-holes.
CaC	70— 80 cm	Weißgelblicher, stark kalkhaltiger Löß, etwas kohärent, zerbricht bröckelig.	White-yellowish, strongly calcareous loess, somewhat coherent, breaks down to weak subangular blocky structure.
C	80—120 cm +	Graugelber Löß, schwach kohärent, schwaches Bröckelgefüge.	Gray-yellow loess, weakly coherent, very weak subangular blocky structure.

Tafel I

Profil 1 *Sierosem* | Profil 2 *Kastanosem* | Profil 3 *Tschernosem*

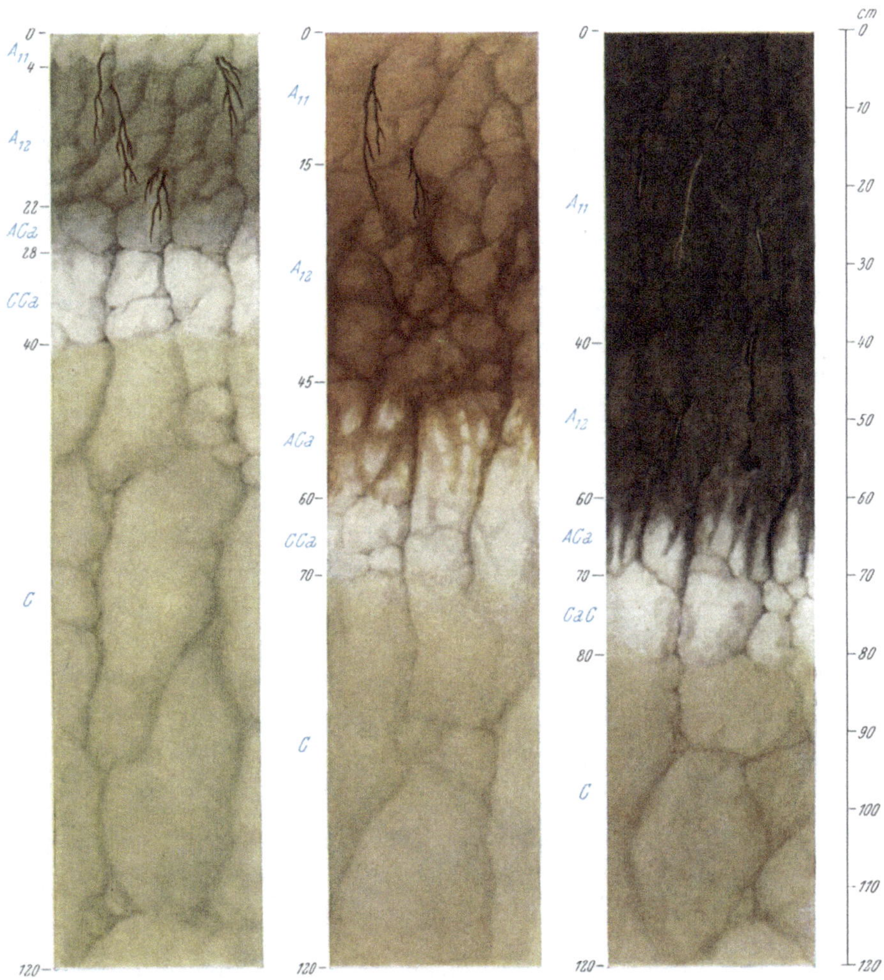

(Nach einem Entwurf von E. Mückenhausen, gemalt von C. Krahberg)

Profil 4. *Braunerde* aus Löß

Profile 4. *Brown earth* (= brown forest soil), developed from loess

A	0— 18 cm	Dunkelsepiabrauner, mittelhumoser feinsandiger Lehm (Lößlehm), Bröckelgefüge, teils Krümelgefüge, porös, Regenwürmer, stark durchwurzelt, Übergang zu $(B)_1$ allmählich.	Dark-sepiabrown, medium in humus, fine sandy loam (loess-loam), weak subangular blocky structure, partly crumb structure, porous, many roots, transition to $(B)_1$ gradual.
$(B)_1$	18— 65 cm	Sepiabrauner feinsandiger Lehm (Lößlehm), ausgeprägtes Bröckelgefüge, porös, Regenwurmröhren, gut durchwurzelt, Übergang zu $(B)_2$ allmählich.	Sepia-brown, fine sandy loam (loess-loam), moderate subangular blocky structure, porous, worm-holes, many roots, transition to $(B)_2$ gradual.
$(B)_2$	65—110 cm	Wie $(B)_1$, aber heller.	Similar to $(B)_1$, but lighter coloured.
C	110—120 cm +	Gelbbrauner Löß, viele weiße Kalkflecken (Ca-Zone), die nach unten verschwinden, schwach kohärent, sehr porös.	Yellow-brown loess, many white mottles of lime (Ca-zone), which disappear with increasing depth, weakly coherent, very porous.

Profil 5. *Parabraunerde* aus Löß

Profile 5. *Parabraunerde* (gray-brown podzolic soil), developed from loess

A_1	0— 16 cm	Graubrauner, mittelhumoser, stark lehmiger Feinsand (Lößlehm), Bröckelgefüge, wenig Krümelgefüge, locker, stark porös, einige Regenwurmröhren, stark durchwurzelt.	Gray-brown, medium in humus, very loamy fine sand (loess-loam), weak subangular blocky structure, some parts with crumb structure, friable, very porous, some worm-holes, many roots.
A_3	16— 40 cm	Ockerbrauner, stark lehmiger Feinsand (Lößlehm), Bröckelgefüge, locker, stark porös, einige Regenwurmröhren, gut durchwurzelt.	Ochre-brown, loamy fine sand (loess-loam), weak subangular blocky structure, very porous, some worm-holes, friable, many roots.
A_3B	40— 50 cm	Übergang von A_3 zu B.	Transition from A_3 to B.
B	50—110 cm	Rötlichbrauner feinsandiger Lehm (Lößlehm), ausgeprägtes Polyedergefüge, Tonhäutchen über den Gefügeaggregaten (Polyeder), Polyeder etwas porös, sehr wenige Regenwurmgänge, nach unten abnehmend durchwurzelt.	Reddish-brown, fine sandy loam (loess-loam)-strong angular blocky structure, peds with clay-coatings and somewhat porous, very few worm, holes, roots decreasing with increasing depth.
B)	110—120 cm +	Sepiabrauner feinsandiger Lehm (Lößlehm), ausgeprägtes Bröckelgefüge, porös. Der gelbbraune Löß als C-Horizont folgt tiefer.	Sepia-brown fine sandy loam (loess-loam), moderately weak subangular blocky structure, porous. Yellow-brown loess (C-horizon) follows underneath.

Profil 6. *Pseudogley* aus Löß

Profile 6. *Pseudogley*, developed from loess

A_0	0— 3 cm	Braunschwarzer Moder, stark durchwurzelt[1].	Brown-black mould, many roots.
A_1	3— 8 cm	Schwarzgrauer, mittelhumoser feinsandiger Lehm (Lößlehm), Bröckelgefüge, etwas plattig und dicht, gut durchwurzelt.	Black-gray, medium in humus, fine sandy loam (loess-loam), weak subangular blocky structure, somewhat platy and dense[1], many roots.
A_2g	8— 28 cm	Weißlichgrauer, schwach rostbraun gefleckter feinsandiger Lehm (Lößlehm), einige schwarzbraune Konkretionen, Bröckelgefüge, wenig plattig, etwas dicht gelagert, durchwurzelt.	Whitish-gray, faintly rusty brown mottled, fine sandy loam (loess-loam), some blackbrown concretions, weak subangular blocky structure, few parts with platy structure, somewhat dense, some roots.
g_1	28— 55 cm	Weißlichgrau, rostgelb und rostbraun gefleckter feinsandiger Lehm (Lößlehm), Polyedergefüge, schwache Tonhäutchen über den Gefügeaggregaten, teils etwas plattig, Gefügeaggregate schwach porös, nach unten abnehmend durchwurzelt.	Whitish-gray, rusty yellow and rusty brown mottled fine sandy loam (loess-loam), angular blocky structure, partly somewhat platy structure, weak clay skins on the peds, peds weakly porous, roots decreasing with increasing depth.
g_2	55—120 cm +	Rötlichbrauner, stark mittelgrau und rostbraun gefleckter feinsandiger Lehm (Lößlehm), nach unten weniger gefleckt, Polyedergefüge, nur schwache Tonhäutchen über den Gefügeaggregaten, die wenig porös sind.	Prominently mottled (colours: reddish brown, medium gray and rusty brown) fine sandy loam (loess-loam), mottling decreasing in the lower horizon, angular blocky structure, only weak clay skins on the peds, which are faintly porous.

[1] Mit der Humusform kann man auch den gesamten humosen Oberboden kennzeichnen (Kubiena 1953).

[1] Dense (= dicht): the soil is rather impermeable for roots and hinders the water movement.

Tafel II

Profil 4	Profil 5	Profil 6
Braunerde	*Parabraunerde*	*Pseudogley*

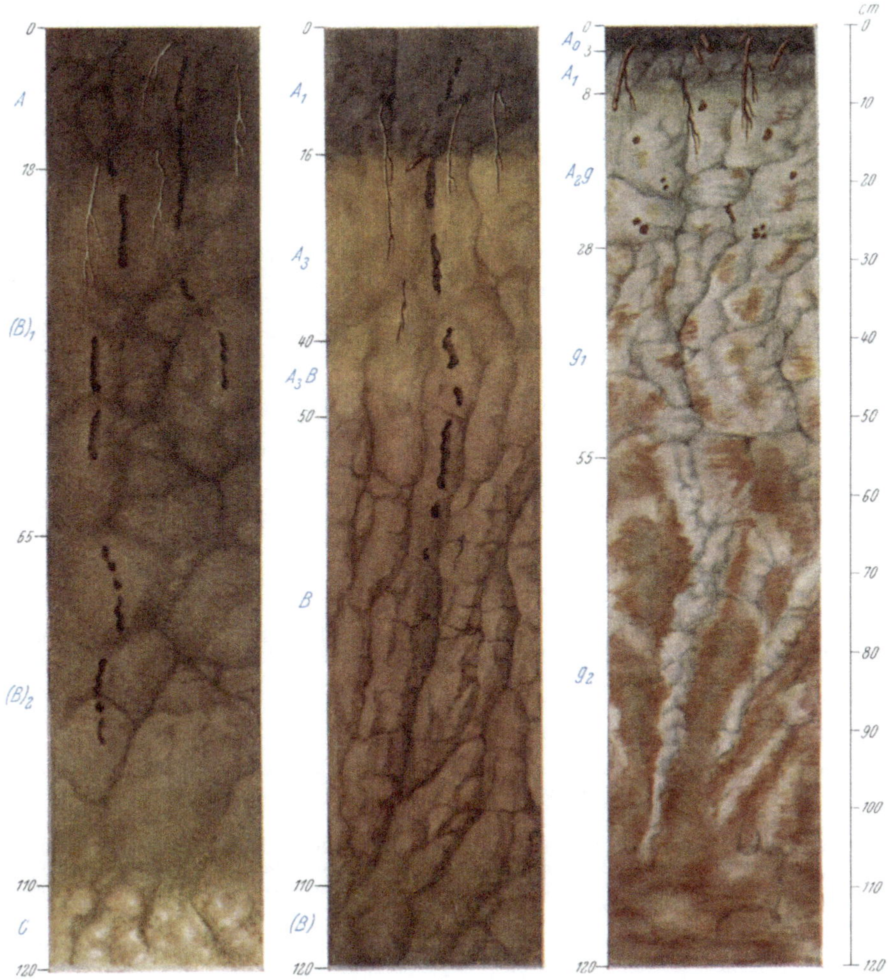

(Nach einem Entwurf von E. MÜCKENHAUSEN, gemalt von C. KRAHBERG)

Profil 7. *Rendzina* aus Kalkstein

Profile 7. *Rendzina*, developed from limestone

A	0— 24 cm	Braunschwarzer, sehr stark humoser, schluffiger toniger Lehm (Mull), Krümelgefüge, locker, stark porös, sehr stark durchwurzelt.	Brown-black, very strong in humus, silty clay loam (mull), crumb structure, friable, very porous, very many roots.
C_1	24— 45 cm	An den Außenflächen verwitterter, grauweißlicher, plattiger Kalkstein, teils durch Brauneisen etwas rostgelb gefärbt, Gesteinsverband gelockert, auf den Schichtfugen etwas braune, tonige Verwitterungssubstanz, oben noch einige Wurzeln.	On the surface weakly weathered, gray-whitish, platy limestone, partly somewhat rusty yellow coloured by hydrated oxides of iron, bedrock loosened, on bedding joints somewhat brown clayey weathered material, in the upper part some roots.
C_2	45—120 cm +	Grauweißlicher, plattiger Kalkstein, teils von Brauneisen rostgelb überzogen (U. Muschelkalk).	Gray-whitish, platy limestone, partly covered with rusty yellow hydrated oxides of iron (Lower Muschelkalk).

Profil 8. *Ranker* aus Schiefer und Grauwacke

Profile 8. *Ranker* (similar lithosol), developed from schist and graywacke

A_{00}	0— 2 cm	Bestandesabfall von Eichen (Niederwald).	Litter of oaks (shrub).
A_1	2— 23 cm	Graubrauner, mittelhumoser, schwach steiniger, mittelgrusiger sandiger Lehm, Subpolyedergefüge und Bröckelgefüge, locker, porös, stark durchwurzelt.	Gray-brown, medium in humus, weakly stony, medium cherty sandy loam, strong subangular blocky structure and weak subangular blocky structure, friable, porous, many roots.
C_1	23— 75 cm	Schiefer mit Grauwackenlagen von etwa 10 bis 20 cm Mächtigkeit, schwach verwittert, teils durch etwas Brauneisen rostgelb gefärbt, Gesteinsverband gelockert.	Schist with layers of graywacke of about 10 to 20 cm thickness, weakly weathered, partly rusty yellow coloured by hydrated oxides of iron, bedrock loosened.
C_2	75—120 cm +	Schiefer mit Grauwackenlagen wie bei C_1, aber unverwittert (U. Devon).	Schist with layers of graywacke as described under C_1, but unweathered (Lower Devonian).

Profil 9. *Saure Braunerde* aus Schiefer

Profile 9. *Acid brown earth* (=acid brown forest soil = sol brun acide), developed from schist

A_{00}	0— 3 cm	Bestandesabfall von Rotbuchen.	Litter of beeches.
A_{0+1}	3— 10 cm	Schwärzlichbrauner, schwach grusiger, feinsandig-lehmiger Moder, locker, stark porös, stark durchwurzelt.	Blackish-brown, weakly cherty fine sandy mould, friable, very porous, many roots.
$(B)_1$	10— 35 cm	Graubrauner (etwas ockerfarbig), schwach grusiger feinsandiger Lehm, Subpolyedergefüge, Gefügeaggregate rauh und porös, locker, gut durchwurzelt.	Gray-brown (somewhat ochre), weakly cherty fine sandy loam, strong subangular blocky structure, peds porous and with rough surfaces, friable, many roots.
$(B)_2$	35— 60 cm	Wie $(B)_1$, aber etwas dunkler, mehr Grusgehalt und weniger locker.	Like $(B)_1$, but somewhat darker, more cherty fragments and less friable.
(B)C	60— 80 cm	Sepiabraun und rostbraun unregelmäßig gefärbter Grus mit gleichfarbigem feinsandigem Lehm dazwischen, etwas dicht, aber noch etwas durchwurzelt, allmählicher Übergang zu C_1.	Sepia-brown and rusty brown irregularly coloured chert, mixed with similarly coloured fine sandy loam, somewhat dense, but still some roots, gradual transition to C_1.
C_1	80—105 cm	Mittelgrauer (etwas bläulich) Schiefer, schwach verwittert, teils durch Brauneisen etwas rostgelb und rostbraun gefärbt, Gesteinsverband gelockert.	Medium gray (somewhat bluish) schist, weakly weathered, partly somewhat rusty yellow and rusty brown coloured by hydrated oxides of iron, bedrock loosened.
C_2	105—120 cm +	Mittelgrauer (etwas bläulich) Schiefer wie bei C_1, aber unverwittert (U. Devon).	Medium gray (somewhat bluish) schist as described under C_1, but unweathered (Lower Devonian).

Tafel III

Profil 7 *Rendzina* | Profil 8 *Ranker* | Profil 9 *Saure Braunerde*

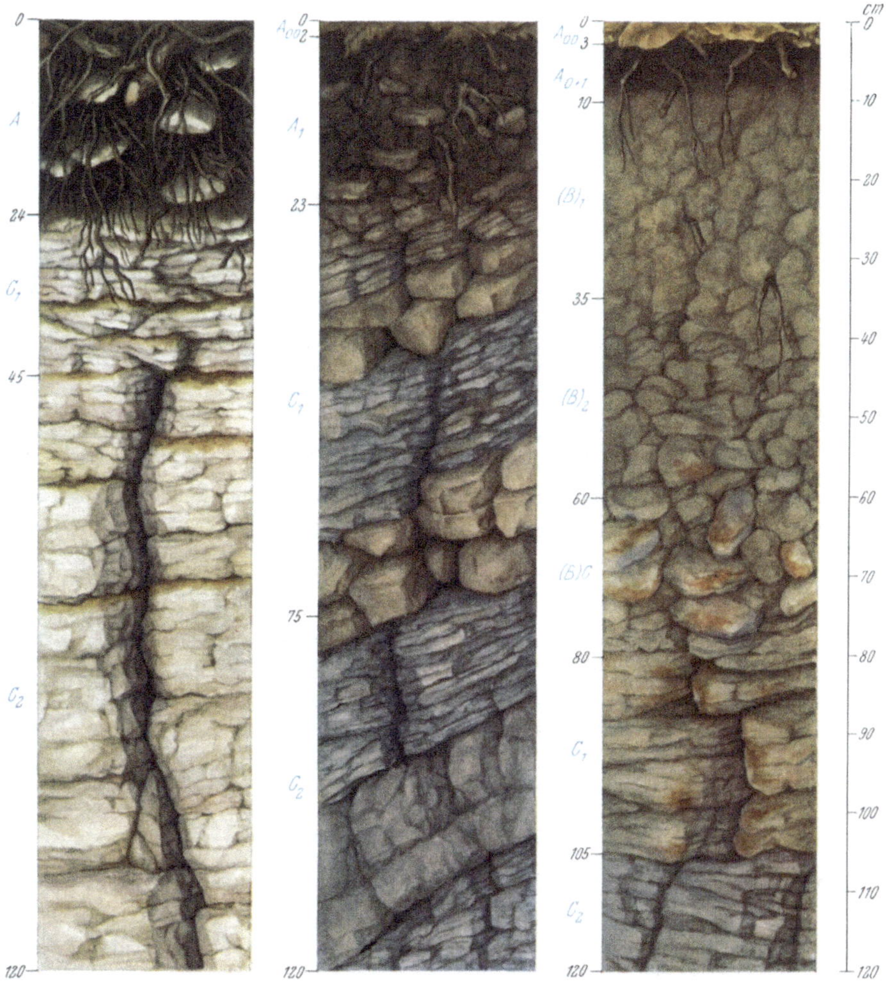

(Nach einem Entwurf von E. MÜCKENHAUSEN, gemalt von C. KRAHBERG)

(Nach einem Entwurf von E. MÜCKENHAUSEN, gemalt von C. KRAHBERG)

Profil 10. *Pseudogley-Parabraunerde* aus Löß

Profile 10. *Pseudogley-Parabraunerde* (gray-brown podzolic soil with signs of wetness in the lower part of the solum), developed from loess

A_0	0— 4 cm	Schwärzlichbrauner Moder.	Blackish-brown mould.
A_1	4— 15 cm	Dunkelgraubrauner, mittelhumoser, stark feinsandiger Lehm (Lößlehm), Bröckelgefüge, locker, stark porös, stark durchwurzelt.	Dark gray-brown, medium in humus, very fine sandy loam (loess-loam), weak subangular blocky structure, friable, very porous, many roots.
A_3	15— 42 cm	Ockerbrauner, stark feinsandiger Lehm (Lößlehm), Bröckelgefüge, locker, porös, gut durchwurzelt.	Ochre-brown, very fine sandy loam (loess-loam), weak subangular blocky structure, friable, porous, many roots.
gB	42— 85 cm	Etwas rötlichbrauner, mittelgrau, rostgelb und rostbraun gefleckter feinsandiger Lehm, einige hellgraue, vertikale Streifen mit rostbraunem Saum, einige schwarzbraune, weiche Konkretionen, Polyedergefüge, Gefügeaggregate mit Tonhäutchen überzogen und wenig porös, ziemlich dicht gelagert, oben noch einige Wurzeln.	Somewhat reddish-brown, medium gray, rusty yellow and rusty brown mottled fine sandy loam, some light gray, vertical tongues with rusty brown seams, some black-brown, soft concretions, angular blocky structure, peds covered with coatings and faintly porous, rather dense, in the upper part still some roots.
Bg	85—120 cm +	Wie gB, aber weniger stark gefleckt und gestreift, aber ebenso dicht. Ein dichter D-Horizont folgt tiefer.	Like gB, but less mottled and fewer tongues, but as dense as the gB-horizon. A dense D-horizon follows underneath.

Profil 11. *Podsol-Braunerde* aus silikatreichem Sand

Profile 11. *Podzol-brown earth* (Brown podzolic soil), developed from sand, rich in silicates

A_0	0— 3 cm	Schwärzlichbrauner, faseriger Rohhumus aus Bestandesabfall von Kiefer und Heidelbeere, gut durchwurzelt.	Blackish-brown, fibrous raw humus of litter of pine and blueberry, many roots.
A_{1+2}	3— 9 cm	Hellgrauer (schimmert etwas violett), oben etwas schwärzlicher, schwach humoser, schwach kiesiger Sand, Einzelkorngefüge, locker, gut durchwurzelt.	Light gray (shines somewhat violet), in the upper part somewhat more blackish, weak in humus, somewhat gravelly sand, single grain structure, loose, many roots.
$(B)_1$	9— 55 cm	Rostbrauner (etwas nach sepiabraun), schwach kiesiger anlehmiger Sand, schwach ausgeprägtes Bröckelgefüge bis Einzelkorngefüge, locker, porös, ziemlich gut durchwurzelt, Übergang von $(B)_1$ zu $(B)_2$ kontinuierlich.	Rusty brown (intergrading to sepia-brown), somewhat gravelly loamy sand, very weak subangular blocky structure intergrading to single grain structure, very friable, porous, many roots, gradual transition from $(B)_1$ to $(B)_2$.
$(B)_2$	55—105 cm	Wie $(B)_1$, aber nach unten zunehmend heller und weniger Feinsubstanz, auch die Durchwurzelung nimmt ab, Übergang zu C allmählich.	Like $(B)_1$, but with increasing depth growing lighter coloured and less clayey, roots decreasing, gradual transition to C.
C	105—120 cm +	Hellgelblichgrauer kiesiger Sand, Einzelkorngefüge (glazigener Sand, Würm).	Light yellowish-gray gravelly sand, single grain structure (glacial sand, Würm = Wisconsin).

Profil 12. *Eisenhumuspodsol* aus silikatarmem Sand

Profile 12. *Iron-humus podzol*, developed from sand, poor in silicates

A_0	0— 6 cm	Schwärzlichbrauner, faseriger Rohhumus aus Bestandesabfall von Kiefer und Heide, durchwurzelt.	Blackish-brown, fibrous raw humus of litter of pine and heath, roots.
A_1	6— 13 cm	Schwarzgrauer, stark humoser Sand, Einzelkorngefüge, durch Humus sind die Körner etwas verklebt, durchwurzelt.	Black-gray, strong in humus, sand, single grain structure, the grains are somewhat coherent due to humus, roots.
A_2	13— 28 cm	Hellgrauer, schwach humoser Sand, Einzelkorngefüge, durchwurzelt.	Light gray, weak in humus, sand, single grain structure, roots.
B_1	28— 37 cm	Braunschwarzer Eisenhumusortstein (in Sand), Hüllengefüge, einige Wurzeln wachsen horizontal hinein.	Brown-black iron humus ortstein (in sand), "Hüllengefüge"[1], moderately cemented, weakly platy, moderately dense, horizontally growing roots.
B_2	37— 85 cm	Oben schwärzlichbrauner Sand, nach unten heller werdend bis zu gelblichgrauem Sand, von vielen braunschwarzen horizontalwellig verlaufenden Orterdebändern durchzogen, einzelne Ortsteintöpfe greifen von B_1 in B_2 hinein, Einzelkorngefüge, teils schwaches Hüllengefüge (darum schwach verfestigt), nicht durchwurzelt.	In the upper part blackish-brown sand, the sand is growing lighter to yellowish-gray with increasing depth, interrupted by some brown-black horizontal wavy bands of orterde, some isolated ortstein-pots reaching from B_1 into B_2, single grain structure, partly weak "Hüllengefüge" (therefore weakly cemented), no roots.
C	85—120 cm +	Gelblichgrauer Sand, Einzelkorngefüge (glazigener Sand, Riß).	Yellowish-gray sand, single grain structure (glacial sand, Riß = Illinoian).

[1] Cementation between mineral grains, formed by connecting sesquioxide-organic matter coatings.

Tafel IV

Profil 10	Profil 11	Profil 12
Pseudogley-Parabraunerde	*Podsol-Braunerde*	*Eisenhumuspodsol*

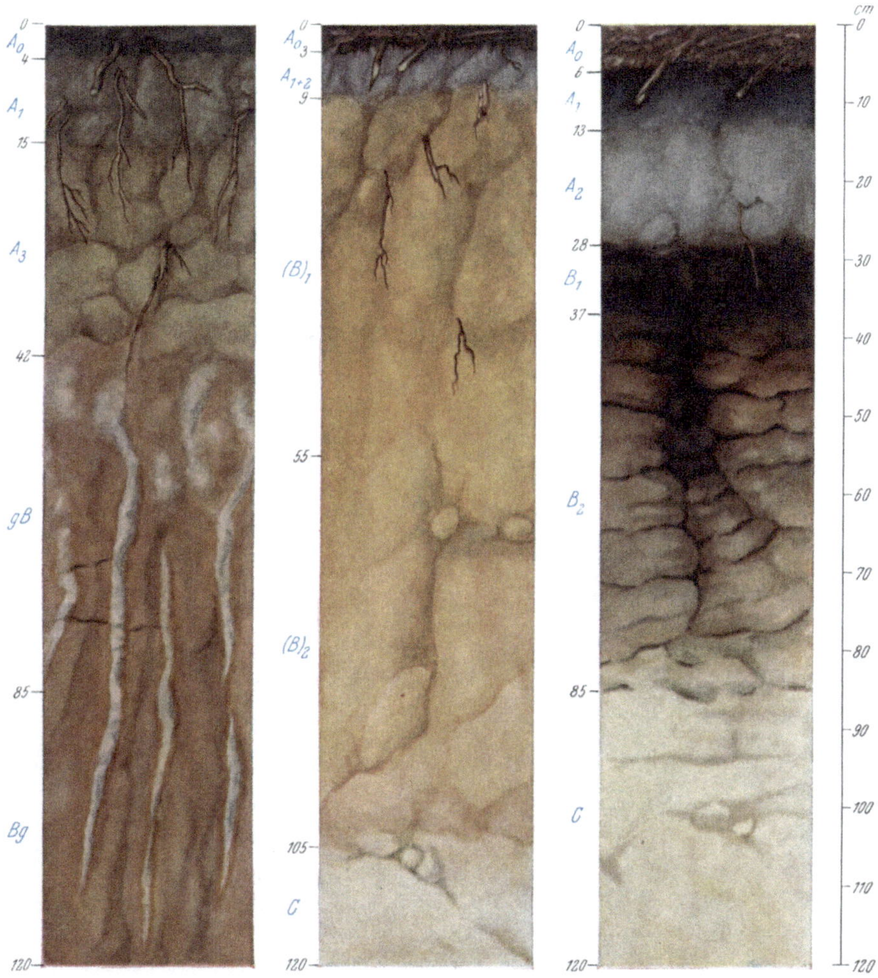

(Nach einem Entwurf von E. Mückenhausen, gemalt von C. Krahberg)

Profil 13. *Gebleichter Rotlehm* aus Gneis

Profile 13. *Red-yellow podzolic soil* (= Gebleichter Rotlehm), developed from gneiss

A_1	0— 18 cm	Dunkelgrauer (etwas bräunlich), schwach humoser, stark sandiger Lehm, Bröckelgefüge, teils schwach ausgeprägtes Krümelgefüge, ziemlich locker und porös, gut durchwurzelt, allmählicher Übergang zu A_3.	Dark gray (somewhat brownish), weak in humus, very sandy loam, weak subangular blocky structure, partly weak crumb structure, rather friable and porous, many roots, gradual transition to A_3.
A_3	18— 34 cm	Hellgelblichgrauer, stark sandiger Lehm, Subpolyeder-, teils auch Bröckelgefüge, etwas plattig, ziemlich locker und porös, durchwurzelt, geringmächtiger, etwas verzahnter Übergang zu B.	Light yellowish-gray, very sandy loam, subangular blocky, partly also weak subangular blocky structure, somewhat platy, rather friable and porous, roots, thin somewhat irregular transition to B.
B	34— 90 cm	Braunrötlicher sandiger Lehm, plastisch, ausgeprägtes Polyedergefüge, Gefügeaggregate > 2 cm ⌀, mit Tonhäutchen überzogen und ziemlich dicht, schwach durchwurzelt, allmählicher Übergang zu (B).	Brown-reddish sandy loam, plastic, strong angular blocky structure, peds > 2 cm in diameter, covered with coatings and rather dense, some roots, gradual transition to (B).
(B)	90—110 cm	Braunrötlicher sandiger Lehm, etwas heller und weniger plastisch (weniger Feinsubstanz) als B, Polyedergefüge nicht so ausgeprägt wie in B, Gefügeaggregate besitzen keine Tonhäutchen und sind porös.	Brown-reddish sandy loam, somewhat lighter and less plastic (less clay) than B, angular blocky structure not as strong as in the B, peds have no coatings and are porous.
C_1	110—120 cm +	Gelbrötlich und graugelb (Farben gehen ineinander über) kleingefleckter, vergruster Gneis, horizontale, blätterige Absonderung.	Yellow-reddish and gray-yellow (diffuse colour pattern), fine mottles, cherty gneiss, very thin platy structure.

Profil 14. *Roterde* aus Basalt

Profile 14. *Latosol* (=Roterde), developed from basalt

A	0— 16 cm	Braunzinnoberroter, mittelhumoser sandiger Lehm, schorfig-krümeliges Gefüge, locker, stark porös, stark durchwurzelt, unscharfer Übergang zu $(B)_1$.	Brown-vermilion, medium in humus, sandy loam, scabby crumb structure, friable, very porous, many roots, gradual transition to $(B)_1$.
$(B)_1$	16— 75 cm	Zinnoberroter sandiger Lehm, schorfig-krümeliges Gefüge, locker, stark porös, durchwurzelt, allmählicher Übergang zu $(B)_2$.	Vermilion sandy loam, scabby crumb structure, friably, very porous, roots, gradual transition to $(B)_2$.
$(B)_2$	75—110 cm	Wie $(B)_1$, aber etwas dunkler zinnoberrot, etwas bröckelig neben schorfig-krümelig, wenig plattig, weniger locker als $(B)_1$, aber noch stark porös, weniger durchwurzelt als $(B)_1$.	Like $(B)_1$, but somewhat darker vermilion, some weak subangular blocky besides scabby crumb structure, also some platy structure, less friable than $(B)_1$, but still very porous, less roots than in the $(B)_1$.
(B)C	110—120 cm +	Wie $(B)_2$, aber mit rundlichen Basaltsteinen stark durchsetzt, die eine hellzinnoberrote Verwitterungsrinde tragen. Der C-Horizont als Basalt (Tertiär) folgt tiefer.	Like $(B)_2$, but very much mixed with stones of basalt which are covered with a light vermilion weathering crust. The C horizon, basalt bedrock (tertiary), follows underneath.

Profil 15. *Kalkkrustenboden* aus Kalksandstein

Profile 15. *Calcareous crust soil* (Calcareous caliche), developed from calcareous sandstone

Ca_1	0— 15 cm	Cremefarbige, teils etwas rötlich und mittelgrau gefleckte, leicht gebänderte, harte Kalkkruste, von endolithischen Flechten schwach besiedelt[1].	Light yellowish coloured, partly somewhat reddish and medium gray mottled, slightly banded, cemented calcareous crust, weakly penetrated by endolithic lichen[1].
Ca_2	15— 65 cm	Cremefarbige, etwas rötliche, sandigkalkige Masse, nur wenig verfestigt, läßt sich zerdrücken, stark porös.	Light yellowish coloured, somewhat reddish, sandy-limy mass, only little indurated, crushable, very porous.
C_1	65—105 cm	Gelblichgrauer, schwach rötlich gefleckter, schwach verwitterter, dünnbankiger Kalksandstein in schwach geneigter Lagerung, Gesteinsverband gelockert.	Yellowish-gray, faint reddish mottles, weakly weathered, thin layers of calcareous sandstone weakly inclined, bedrock loosened.
C_2	105—120 cm +	Gelblichgrauer, unverwitterter, dünnbankiger Kalksandstein in schwach geneigter Lagerung.	Yellowish-gray, unweathered, thin layers of calcareous sandstone, weakly inclined.

[1] Der ehemalige Horizont über der Kruste wurde durch Winderosion beseitigt.

[1] The former horizon above the crust is removed by Winderosion.

Tafel IV

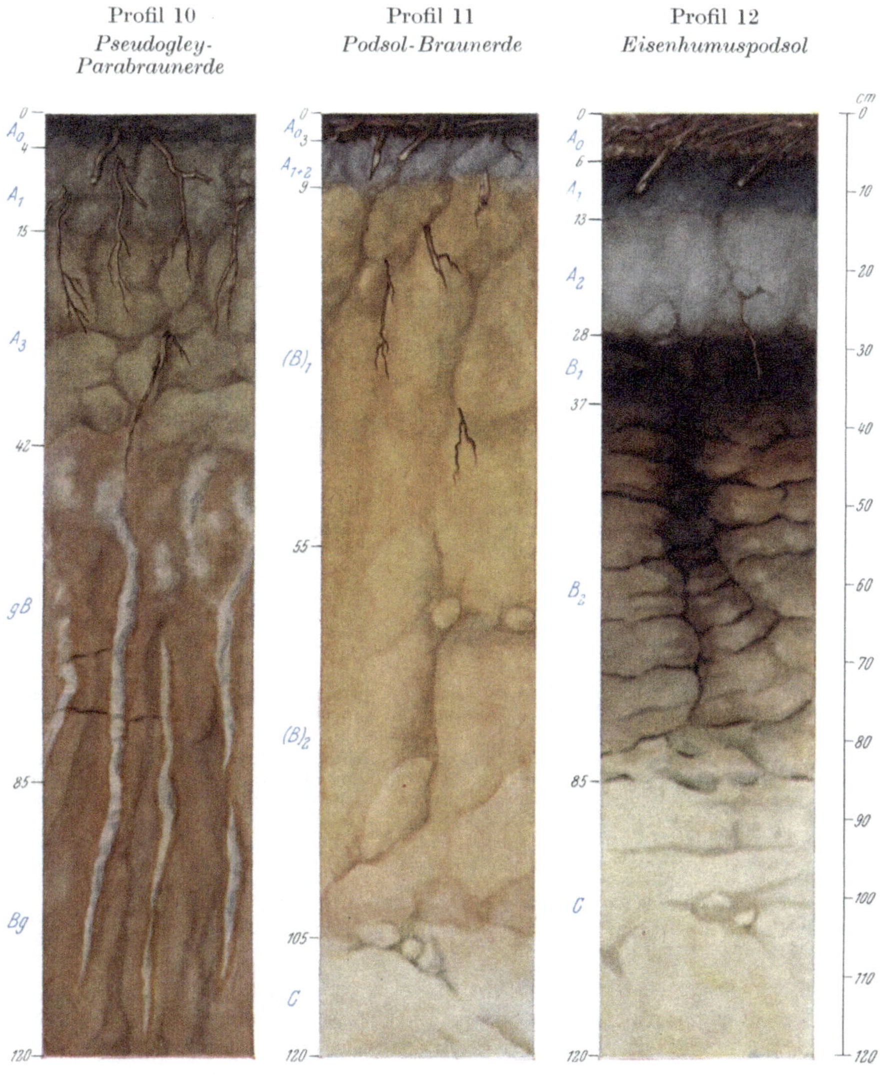

(Nach einem Entwurf von E. MÜCKENHAUSEN, gemalt von C. KRAHBERG)

Profil 16. *Solontschak* aus lehmigem Alluvium

ASa 0— 4 cm — Hellgrauer, sehr schwach humoser sandiger Lehm, kantiges Bröckelgefüge, im ausgetrockneten Zustand eine weißliche Salzkruste tragend.

A 4— 13 cm — Mittelgrauer, schwach humoser, schwach sandiger Lehm, großprismatisches Gefüge, Gefügeaggregate ziemlich dicht, in den Schwundrissen viele, dagegen in den Gefügeaggregaten wenige Wurzeln.

G_o 13— 34 cm — Hellgrauer, schwach sandiger Lehm, wenige schwach rostbraune Flecken, großprismatisches Gefüge, Gefügeaggregate dicht, in den Schwundrissen gehen Wurzeln in die Tiefe und wird etwas Humus eingespült; alle G-Horizonte gehen allmählich ineinander über.

G_{ro} 34— 75 cm — Hellgrauer, rostbraun gefleckter, schwach sandiger Lehm, großpolyedrisches (blockiges) Gefüge, dicht, nur oben noch einige Wurzeln auf den Schwundrissen.

G_{or} 75— 95 cm — Hellgrauer (etwas grünlich), mäßig rostbraun gefleckter, schwach sandiger Lehm, schwach ausgeprägtes großpolyedrisches Gefüge, dicht.

G_r 95—120 cm + — Wie G_{or}, aber nicht rostbraun gefleckt, grünliche Reduktionsfarben.

Profile 16. *Solonchak*, developed from loamy alluvial deposits

ASa 0— 4 cm — Light gray, very weak in humus, sandy loam, subangular blocky structure, somewhat sharp angles, covered with a whitish salt crust when dry.

A 4— 13 cm — Medium gray, weak in humus, very sandy loam, coarse prismatic structure, peds rather dense, along the cracks many but in the peds very few roots.

G_o 13— 34 cm — Light gray, very sandy loam, some weak rusty brown mottles, coarse prismatic structure, peds dense, roots and some humus-illuviation along the cracks; all G-horizons with diffuse boundaries.

G_{ro} 34— 75 cm — Light gray, rusty brown mottled, very sandy loam, coarse angular blocky structure, dense, only in the upper part in the cracks some roots left.

G_{or} 75— 95 cm — Light gray (somewhat greenish), moderately rusty brown mottled, very sandy loam, weak coarse angular blocky structure, dense.

G_r 95—120 cm + — Like G_{or}, but not rusty brown mottled, greenish colours indicating reduction.

Profil 17. *Typischer Gley* aus sandig-lehmigem Alluvium

A 0— 19 cm — Dunkelgraubrauner, mittelhumoser sandiger Lehm, einige rostbraune Flecken, Bröckelgefüge, Gefügeaggregate etwas kantig und fest, aber porös, stark durchwurzelt von Gräsern.

G_o 19— 55 cm — Mittelgrau und rostbraun stark gefleckter sandiger Lehm, Bröckel- und Polyedergefüge, Prismengefüge oben angedeutet und bei stärkerer Austrocknung vorhanden, Gefügeaggregate rauh und wenig kantig, porös, Durchwurzelung nach unten abnehmend, allmählicher Übergang zu G_{ro}.

G_{ro} 55— 75 cm — Mittelgrau (etwas grünlich) und mäßig rostbraun gefleckter sandiger Lehm, Gefüge nicht deutlich ausgeprägt, jedoch zu Polyedern neigend, ziemlich dicht, allmählicher Übergang zu G_{or}.

G_{or} 75— 97 cm — Wie G_{ro}, aber nur wenig rostbraun gefleckt und noch weniger Gefügeabsonderung, allmählicher Übergang zu G_r.

G_r 97—120 cm + — Grünlichgrauer sandiger Lehm, kohärente Masse, ziemlich dicht.

Profile 17. *Typical gley*, developed from sandy-loamy alluvial deposits

A 0— 19 cm — Dark gray-brown, medium in humus, sandy loam, some rusty brown mottles, weak subangular blocky structure, peds somewhat sharp angles and firm, but porous, many roots of grasses.

G_o 19— 55 cm — Medium gray and rusty brown prominently mottled sandy loam, subangular to angular blocky structure, weak prismatic structure in the upper part, distinct when dry, peds with rough surfaces and weakly pronounced angles, porous, roots decreasing with increasing depth, gradual transition to G_{ro}.

G_{ro} 55— 75 cm — Medium gray (somewhat greenish) and moderately rusty brown mottled sandy loam, structure not distinct, tending to angular blocky, rather dense, gradual transition to G_{or}.

G_{or} 75— 97 cm — Like G_{ro}, but only faintly rusty brown mottled and still less aggregation, gradual transition to G_r.

G_r 97—120 cm + — Greenish-gray sandy loam, massiv, rather dense.

Profil 18. *Hochmoor* vorwiegend aus Torfmoos und Wollgras

T_1 0— 22 cm — Tiefbrauner, gut zersetzter jüngerer Moostorf, gut gekrümelt, sehr locker und stark porös, stark durchwurzelt (Bunkerde).

T_2 22— 65 cm — Umbrabrauner (gelblichbraun-fuchsig) jüngerer Moostorf, schwach zersetzt, schichtig, stark porös, oben noch durchwurzelt, mit Wasser übersättigt.

T_3 65—120 cm + — Schwarzbrauner älterer Moos- und Wollgrastorf, mäßig zersetzt, schmierig, schichtig, teils faserig (von Wollgras), ziemlich dicht gelagert, mit Wasser übersättigt.

Profile 18. *Acid bog soil* (Hochmoor = high moor), developed from *Sphagnum* and *Eriophorum*

T_1 0— 22 cm — Strong brown, well decomposed young *Sphagnum*-peat, well crumbly, very friable and very porous, many roots (tirr = Bunkerde).

T_2 22— 65 cm — Umbra-brown (yellowish-brown) young *Sphagnum*-peat, weakly decomposed, occurring in layers, very porous, in the upper part still roots, oversaturated by water.

T_3 65—120 cm + — Black-brown older *Sphagnum*- and *Eriophorum*-peat, moderately decomposed, smeary, occurring in layers, partly fibrous (from *Eriophorum*), rather dense, oversaturated by water.

Tafel VI

Profil 16	Profil 17	Profil 18
Solontschak	*Typischer Gley*	*Hochmoor*

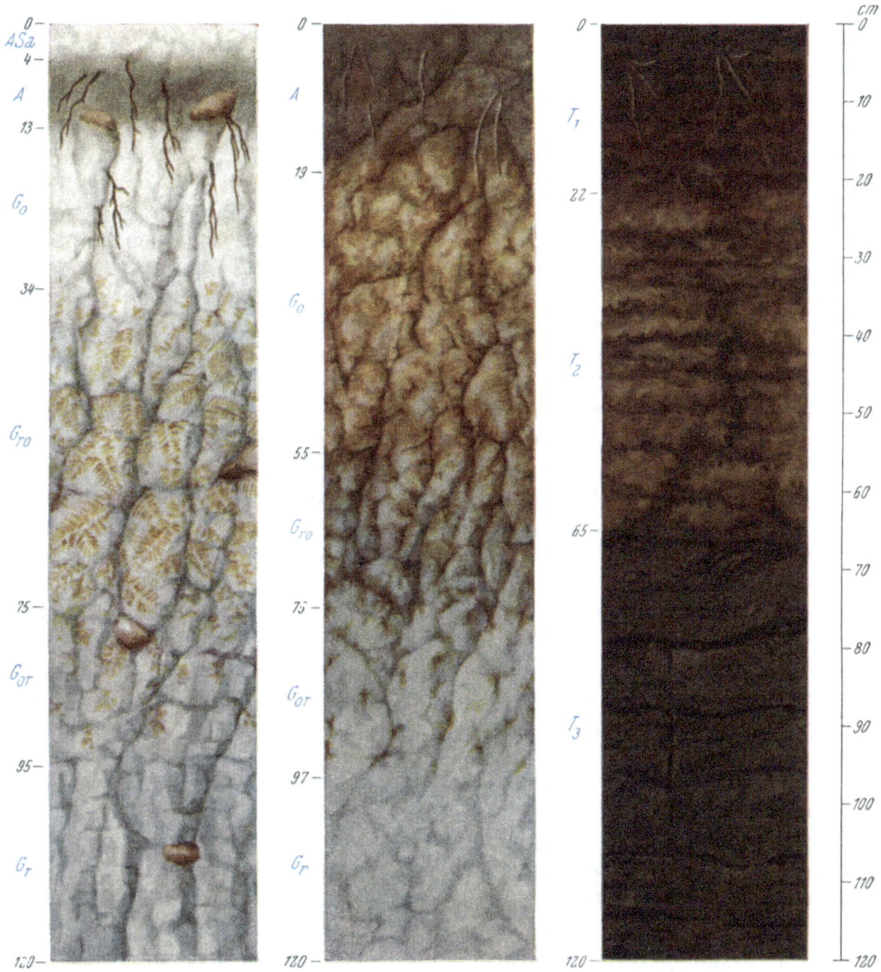

(Nach einem Entwurf von E. MÜCKENHAUSEN, gemalt von C. KRAHBERG)

Biologische Eigenschaften: geringer Besatz an Bodenbakterien, daher ungenügende Zersetzung der organischen Substanz; die vorhandenen Pilze bewirken diese Zersetzung nur unvollkommen. Bodentiere fehlen weitgehend, weil das Milieu zu sauer ist.

Die Podsole verändern sich unter dem Einfluß der landwirtschaftlichen Bewirtschaftung, wobei der Stickstoff eine bedeutende Rolle spielt (TAVERNIER und AMERYCKX 1957, ADERIKHIN u. a. 1960). Die Schweizer Bodenkundler haben gezeigt, daß der Ortstein unter Weide allmählich abgebaut wird.

Für das nähere Studium über den Podsol, seine Entstehung und Eigenschaften wird noch auf folgende größere Arbeiten hingewiesen: RATHLEF 1930, STREMME 1930c, JOFFE 1932, LUNDBLAD 1934, BLANCK 1939b, DUTT 1947, BARSHAD und ROJAS-CRUZ 1950, HALLSWORTH u. a. 1953, SCHEYS u. a. 1954, MUSIEROWICZ 1954b, WORTMANN und MAAS 1954, DIMBLEBY und GILL 1955, VAN DIEPEN 1956, SCHELLING 1956. HAMBLOCH 1958, SABOJEWA 1958. Bezüglich der landwirtschaftlichen Nutzung der Podsole wird noch auf folgende Arbeiten hingewiesen: BADEN 1951, FINGER 1951, WEHRMANN 1955, KÖHNLEIN und SCHLICHTING 1959, MUIR 1961.

Verbreitung. Das Verbreitungsgebiet der Podsole ist das kühle, feuchte Klima Nordeuropas, der nördlichen Sowjetunion und des nördlichen Nordamerikas. Aber stets entsteht er am schnellsten auf basenarmen, quarzreichen Gesteinen. In Mitteleuropa kommt er in Nordwestdeutschland, im westlichen Schleswig-Holstein, in den nördlichen Niederlanden und im nördlichen Belgien vor. In Frankreich ist er auf den Sanden der Landschaft Landes verbreitet. Auf basenarmem Gestein des kühlen, feuchten Hochgebirges tritt er auch oft auf. In Deutschland findet er sich noch auf den armen Sandsteinen des Buntsandsteins, des Keupers (Südwestdeutschland), des Juras (bei Trier) und der Kreide sowie auf dem Quarzit des Hunsrücks und des Taunus. Seltener hat er sich auf Granit und Gneis des Hochschwarzwaldes und des Bayerischen Waldes entwickelt.

Pseudogley (Tafel II). Der Pseudogley wird unter der Einwirkung von Staunässe gebildet. Unter Staunässe versteht man in der Bodenkunde das am Ort gefallene und im nahen Untergrund durch einen mehr oder weniger dichten Horizont gestaute Niederschlagswasser. Sie verhält sich hydrostatisch wie das Grundwasser und wird deshalb in der Hydrologie nicht von diesem getrennt (ZAKOSEK 1957). Vom Grundwasser unterscheidet sich die Staunässe in ihrer bodendynamischen Wirkung: Sie wird innerhalb der obersten 1,5 m von dem örtlichen Niederschlag gebildet, verbleibt am Ort oder bewegt sich nur wenig seitlich und verschwindet in der Regel im Sommer, so daß dann der Boden austrocknet.

Die Staunässe erzeugt ein grau und rostbraun geflecktes Profil, das zuerst SIBIRZEW (1901, mündliche Mitteilung von TIURIN 1960) und TREITZ (1912) erwähnen. SIBIRZEW nannte diesen Boden „gleyiger Podsol" (russisch Ilowka). Vor einigen Jahrzehnten wußte man bereits, daß Nässe dieses Profil prägt; deshalb sprach man von *Naßboden,* ohne einen Unterschied zwischen Staunässe- und Grundwasserböden zu machen (STREMME 1930c). KRAUSS und Mitarbeiter (1928, 1939) erkannten in vollem Umfange die spezifische bodendynamische Wirkung der Staunässe. Sie nannten den Staunässeboden *„gleyartigen Boden"*, da das Profil dem des Gleyes ähnlich ist. Später hat KUBIËNA (1953) dafür den treffenden Namen Pseudogley vorgeschlagen.

Entstehung. Ein genügend hoher Niederschlag und eine mehr oder weniger verdichtete Schicht im *nahen* Untergrund sind die entscheidenden Voraussetzungen für die Entstehung des Pseudogleyes. Der typische Pseudogley gehört in die Bodentypengesellschaft des gemäßigt warmen, humiden Klimas. Das

Niederschlagswasser sickert in den obersten Horizonten (= Stauzone oder Staunässeleiter) ungehindert ab, wird aber auf und in einem folgenden, mehr oder weniger dichten Horizont (= Staukörper oder Staunässesohle) aufgehalten. Bei anhaltendem Niederschlag füllen sich die Bodenporen mehr und mehr, das Wasser staut vom Staukörper nach oben in die Stauzone hinein. Alle Momente, welche die Vernässung begünstigen, verstärken die Tendenz der Pseudogley-Bildung (hoher Niederschlag, wenig Abfluß, geringe Verdunstung durch niedrige Temperatur und hohe Luftfeuchtigkeit). Wasserabfluß in Kuppen- und Hanglagen mindert die Bildungsenergie, dagegen wird sie verstärkt, wenn in schwach muldigen Lagen etwas Wasser zuzieht (ohne daß ein Gley entsteht).

Stauzone und Staukörper können verschiedenen Ursprungs sein. Es kann eine primäre geologische Schichtung von Durchlässigem über Undurchlässigem vorliegen, z. B. von Lößlehm über Ton. Oft wird der Staukörper durch einen älteren Bodenhorizont gebildet, der durch eine lange Zeit der Verwitterung und oft auch durch Illuviation dicht geworden ist (MÜCKENHAUSEN 1954b, BRUNNACKER 1957, LOŽEK und KULKA 1959, SMOLÍKOVÁ 1960b). Nur selten hat die rezente Tondurchschlämmung zur Bildung des Pseudogleyes geführt, wohl zu Übergängen zwischen Parabraunerde und Pseudogley.

Zwischen dem Pseudogley und einigen anderen Bodentypen gibt es viele gleitende Übergangsbildungen, vor allem zur Parabraunerde, zur Braunerde, zum Podsol, zum Pelosol, zum Stagnogley und zum Gley (SCHNEIDER 1950).

Aufbau und Eigenschaften. Den typischen Pseudogley bauen die Horizonte A—g_1—g_2 auf. A- und g_1-Horizont ergeben zusammen die durchlässige Stauzone, während der g_2-Horizont den mehr oder minder dichten Staukörper vorstellt. Diese drei Horizonte können infolge von Unterschieden in der Textur, Fleckung usw. in Subhorizonte und (oder) Übergangshorizonte aufgeteilt sein, aber das ändert nichts am grundsätzlichen, dreigeteilten Profilbau. Während der A-Horizont gleichmäßig graubraun oder braungrau gefärbt ist, zeigt der g_1-Horizont meistens einen fahlgrau, rostgelb und rostbraun gefleckten Aspekt, oft mit dunkelbraunen und schwarzbraunen Konkretionen durchsetzt. Der g_2-Horizont ist meist weniger fleckig (s. Tafel II).

Für die spezifische Wasserdynamik ist es entscheidend, welchen Dichtegrad der Staukörper besitzt und wie tief er unter der Oberfläche ansteht; ferner ist dafür noch die Oberflächengestaltung wichtig. Ein dichter, tiefliegender Staukörper und eine etwas muldige Geländelage (Wasserzuzug) erzeugen eine lange Naßphase (=Nässezeit) und damit die Ausbildung eines *stark ausgeprägten* Pseudogleyes. Andererseits wird die Naßphase verkürzt und die Trockenphase verlängert, wenn der Staukörper hoch ansteht, ebenso dicht ist und das Wasser Abzug hat (MÜCKENHAUSEN 1950, 1954a, HABIG 1953, ZAKOSEK 1954a, 1954b, KERPEN 1960, RÜCKERT 1960); in diesem Falle trocknet der Pseudogley schnell aus und ist dann sehr hart. Wenn die Oberfläche einer dichten Bodenschicht tiefer als 1,5 m unter Flur ansteht, so kommt es selten zur Bildung des Pseudogleyes, weil in der darüber liegenden, durchlässigen Schicht das Niederschlagswasser im allgemeinen einen genügend großen Verteilungsraum hat.

Für den ökologischen Wert des Pseudogleyes ist entscheidend, wie lange seine *Feuchtphase* in der Vegetationszeit andauert. In der Feuchtphase besitzt der Pseudogley einen Feuchtgehalt, welcher der Pflanze eine ausreichende Versorgung mit Wasser und Luft (im Wurzelraum) gewährt (MÜCKENHAUSEN 1958a). In Hanglagen zieht das im Boden gestaute Wasser stets etwas hangabwärts und ist dadurch ökologisch wertvoller; es ist die sogenannte Hangnässe, die zwar ebenfalls einen Pseudogley entstehen läßt, aber mit der spezifischen Wasserform „Hangnässe" (RÜCKERT 1960).

Der Wechsel von Vernässung und Austrocknung bildet das bodenphysikalische Charakteristikum des Pseudogleyes. Davon wird der Bodentyp in allen seinen sichtbaren Merkmalen und seinen Eigenschaften weitgehend geprägt. Darum, aber auch wegen der ökologischen Bedeutung der Vernässung und Austrocknung, sollte der Beurteilung dieser bodenphysikalischen Eigenschaft besondere Aufmerksamkeit gewidmet werden (ARENS, KRÄMER und LANGNER 1958, RID 1958, SCHMIEDEL 1959, KERPEN 1960). Vor allem ist von dieser Wasserdynamik der Lufthaushalt abhängig.

Das Wechselspiel von Vernässung und Austrocknung erzeugt das fleckige Profilbild. In der Vernässungszeit gehen Eisen und auch Aluminium in Lösung und werden bei der Austrocknung wieder konzentriert ausgeschieden. So entsteht die partielle Eisen- und teils auch Aluminiumverarmung, sichtbar an den hellgrauen Flecken, und die partielle Anreicherung dieser Stoffe, kenntlich an den rostgelben und rostbraunen Flecken sowie braunschwarzen Konkretionen. Die Profilbildung verläuft teils bei starker Wasserstoffionenkonzentration ab, bei der Fe und Al als freie Ionen vorliegen und wandern können. Bei langer Vernässung ist aber auch eine Reduktion des Eisens in die zweiwertige Form möglich. Dabei sind an dem Lösungs- und Verlagerungsvorgang der genannten Stoffe organische Verbindungen sehr wahrscheinlich beteiligt, worüber beim „Podsol" Näheres gesagt ist. Neben der Fulvosäure spielt die Gerbsäure bei der Bildung des fleckigen Profils, vor allem bei der Ausbleichung in Wurzelröhren und Schwundrissen eine Rolle, wodurch das Profil hellgraue Streifen erhält (LAATSCH 1957). Die Umlagerung von Tonsubstanz im Profil (Tondurchschlämmung) und damit die Bildung eines A_3g-Horizontes ist die Regel. Bei niedrigem pH ist gleichzeitig auch mit einer Tonzerstörung zu rechnen (LAATSCH 1957). In diesem Falle wird der g_1-Horizont sehr hell. Diese Ausbildung ist im Norden der Sowjetunion häufig und wird dort als Podsol oder vergleyter Podsol bezeichnet.

In den meisten Fällen besitzt der Pseudogley eine saure oder stark saure Reaktion, entsprechend niedrig ist der V-Wert. Diese und die geschilderten physikalischen Eigenschaften lassen nur ein schwaches Bodenleben zu, das sich fast nur auf den A-Horizont konzentriert. Bei diesen Gegebenheiten gelingt den Pflanzenwurzeln nur in geringem Maße ein Vordringen in den Unterboden. In seltenen Fällen besitzt der Pseudogley einen höheren Basengehalt. Damit verbessern sich viele andere Eigenschaften, vor allem wirkt sich eine Dränung bei diesem Pseudogley besonders schnell und günstig aus.

Für die ackerbauliche Nutzung muß der ausgeprägte Pseudogley dräniert werden. Das Bodenprofil und die Beobachtung der Vernässungszeiten zeigen an, ob diese Maßnahme notwendig ist. Bei der Ausdeutung des Pseudogleyprofiles für Meliorationsmaßnahmen sollten alle Umstände berücksichtigt werden, wie das ZAKOSEK (1951) dargelegt hat. Nach der Dränung wirkt eine zeitgerechte (nicht bei nassem Bodenzustand) Unterboden-Lockerung im allgemeinen gut, während eine Krumenvertiefung mit äußerster Vorsicht (wenn überhaupt, dann nur langsam) vorgenommen werden darf (KLAPP 1958).

Verbreitung. Im Bereich der Braunerde und vor allem der Parabraunerde ist der Pseudogley in meist kleineren, selten größeren Flächen verbreitet, vor allem in den Löß- und Geschiebelehmgebieten des ganzen mäßig warmen, humiden Klimaraumes. Seine Entstehung ist dort begünstigt, wo durchlässiges Material über undurchlässigem abgelagert wurde, z. B. Löß über tonigen Substraten, wie z. B. in Südwestdeutschland.

Stagnogley. Eine auffällige und daher schon lange bekannte spezifische Bodenbildung stellt der Stagnogley dar, der von VOGEL (mündliche Mitteilung

1956) wegen der Stagnation der fast stets vorhandenen Bodennässe so genannt wurde. Die weißliche Farbe (wie Molke) des Unterbodens hat diesem nassen Boden der ebenen Mittelgebirgshochlagen Mitteldeutschlands den Namen Molkenboden eingetragen (GRUPE 1909, 1923, VOGEL v. FALCKENSTEIN und v. ROMBERG 1915, ALBERT und KÖHN 1930, 1932). Im Schwarzwald hat der gleiche Boden die volkstümliche Bezeichnung Misseboden oder Missenboden (RAMM 1908). Wegen der starken Bleichung nannte ihn KUBIËNA (1953) Molkenpodsol und stellte ihn zu den Podsolen.

Entstehung. Der typische Stagnogley hat seine günstigsten Entstehungsbedingungen in dem kühlen, feuchten Klima der höheren, ebenen Mittelgebirgslagen mit dichtem, sandig-lehmigem bis schluffig-lehmigem oder schluffigtonigem Untergrund, der meistens umgelagertes, fossiles Bodenmaterial darstellt. Von den ziemlich hohen Niederschlägen verdunstet in dem kühlen Lokalklima relativ wenig, die ebene Lage gestattet wenig Abfluß und der dichte Untergrund nur eine geringe oder keine Versickerung. Das sind die Umstände, die zu einer starken Vernässung führen. Diese verschwindet in den meisten Sommern nicht und hemmt im Verein mit der Bodensäure den Abbau der organischen Substanz stark. Die Bildung von nassem Rohhumus und schließlich einer Torfauflage ist die Folge. Aus dieser sauren Humusauflage gelangen niedrigmolekulare Humussäuren in den Unterboden und bleichen diesen stark aus; so entsteht der „Molkenhorizont“. Bei gleichbleibenden Bedingungen wächst an besonders feuchten Stellen die Torfauflage zu einem Hochmoor auf.

Neben dem typischen Stagnogley mit Rohhumus und „Molkenhorizont“ gibt es einen Übergang zum Hochmoor, den Torf-Stagnogley, und einen Übergang zum Pseudogley mit geringerer Vernässung. In niedrigeren Lagen, wo die Humuszersetzung infolge höherer Temperatur besser ist, entsteht ein rohhumusfreier Stagnogley. In seltenen Fällen gibt es sogar einen basenreichen Stagnogley, der aber die fast ständige Vernässung mit der typischen Ausprägung gemein hat.

Aufbau und Eigenschaften. Der Rohhumus als oberster Horizont des typischen Stagnogleyes zeigt zwei Subhorizonte, einen weniger zersetzten oberen und einen stärker zersetzten, in nassem Zustand schlammigen unteren. Der darunter folgende, naßgebleichte, helle Horizont zeigt oben eine schmutziggraue Farbe, die durch Einschlämmung von sauren Humusstoffen dunkler geworden ist als der folgende weißliche „Molkenhorizont“. Diesen unterlagert der meist rostgelb und rostbraun, oft auch rötlichbraun gefleckte Staukörper. Der Aufbau der oben genannten Übergangsbildungen weicht entsprechend ab.

Die bodenphysikalischen Eigenschaften des Stagnogleyes werden ganz von der Nässe bestimmt; Luftarmut des Bodens ist die übelste Folge. Starke Versauerung und damit im Gefolge starker Angriff auf die Bodensubstanz kennzeichnen die chemischen Eigenschaften. Al-Ionen und saure Humusstoffe befinden sich in der Bodenlösung und wirken toxisch auf die Pflanzen. Bodenorganismen finden schlechteste Bedingungen. Wenn es nicht gelingt, diesen Bodentyp zu entwässern, so ist keine erfolgbringende Nutzung möglich. Die Fichte erreicht auf dem Stagnogley in 100 Jahren bisweilen nur 5 bis 10 m Höhe.

Verbreitung. Auf den ebenen Hochlagen der Mittelgebirge mit kühlem, feuchtem Lokalklima, in Deutschland im Solling, Reinhardswald, Bramwald, Schwarzwald, Kaufunger Wald. Nur selten tritt er kleinflächig im feuchteren Bereich des gemäßigt warmen, humiden Klimas auf, z. B. in Nordwestdeutschland.

Pelosol. Der Name „Pelosol“ wurde 1954 von VOGEL (mündliche Mitteilung) für den Boden aus tonigen Substraten vorgeschlagen. Die spezifische Dynamik dieses Bodens hatte schon vor einigen Jahrzehnten Veranlassung zu einer bodentypologischen Sonderstellung gegeben. Er wurde wegen seiner verschiedenen, meist lebhaften Färbung „Bunter Ton- und Mergelboden" (v. HOYNINGEN-HUENE 1930a, STREMME 1930c) und „Bunter Gesteinsboden" (STREMME 1936), ferner auch „Buntton" (LAATSCH 1957) oder Lettenboden genannt. Die Bodenkundler der USA nennen diese Tonböden „low humus clay".

Entstehung. Tonreiche Ausgangsgesteine bilden die entscheidende Entstehungsbedingung für den Pelosol. Meistens liefern Ton und Tonmergel das Ausgangsmaterial, aber er kann ebenso aus Tonschiefer, Schieferton und festem Mergel entstehen, wenn diese (festen) Gesteine intensiv verwittert sind. Der hohe Kolloidgehalt der Bodenmasse quillt bei Befeuchtung, die Luft wird abgeschlossen, so daß dann die Verwitterung bzw. Bodenbildung weitgehend ruht. Wenn bei der Austrocknung der Boden aufreißt, kann die Luft zwar eindringen, jedoch bleibt mangels Feuchtigkeit die Bodenbildung weitgehend gehemmt. Im gemäßigt warmen, humiden Klima verläuft die Bodenbildung aus tonigen Substraten im ganzen gesehen langsam. Nach den Feststellungen von DIEZ (1959) erfolgt dabei keine nennenswerte Umformung der Tonsubstanz, d. h. des Tonmineralbestandes. Sofern es sich um ein festes, toniges Gestein (Tonschiefer, Schieferton) handelt, so löst sich zwar das Gestein an vielen Gesteinsfugen durch Verwitterungsvorgänge auf, und es entsteht eine tonige Bodenmasse, die jedoch der weiteren Bodenbildung gegenüber resistent ist. Bei einem durch diagenetische Vorgänge feinaggregierten Ton oder Tonmergel (Lockergestein) lösen sich die stabilisierten Teilchen zu plastischer Tonmasse auf, ohne dabei eine substantielle Umformung zu erfahren; die nun folgende Bodenbildung verläuft langsam (DIEZ 1959).

Die tiefe Bodenbildung der Pelosole Mitteleuropas kann nicht nur in der Nacheiszeit erreicht worden sein. Die pleistozäne, solifluktive Umlagerung der tonigen Verwitterungsmasse des Pelosols beweist, daß im Pleistozän bereits eine beachtliche Verwitterungsdecke bestanden haben muß. Auf dem umgelagerten Material ging die Bodenbildung weiter. Aber auch der unverwitterte Ton und Tonmergel wurde teils im Pleistozän von der Solifluktion erfaßt, wobei ganze Schollen abrutschten. Selbst im jetzigen Klima rutschen stark quellfähige Tone leicht (OSTENDORFF 1951).

Die bei der Besprechung der Parabraunerde erläuterte Tondurchschlämmung erfolgt auch im oberen Profilteil des Pelosols, was bei dem tonreichen Filtergerüst nicht erwartet werden sollte. DIEZ (1959) hat diesen Prozeß für den Feuerletten des Keupers nachgewiesen und BURNHAM auf einem Pelosol aus Tonmergel des Keupers im mittleren England (Vortrag auf der Tagung der Britischen Bodenkundlichen Gesellschaft 1959).

Aufbau und Eigenschaften. Die junge Bodenbildung beim Pelosol wird vom A—C-Profil präsentiert. Schreitet die Bodenbildung fort, so schaltet sich zwischen A- und C- ein P-Horizont ein; dieser unterliegt der Bodenbildung und besitzt andere Eigenschaften als der C-Horizont, was sich vor allem in einer stärkeren Plastizität und meistens auch in einer Fleckenbildung äußert. Am besten zeigt sich diese Horizontfolge A—P—C in Pelosolen aus Tonschiefer und Schieferton.

Der Entwicklungsgrad des Pelosolprofils bedingt mehrere Subtypen; Syrosem-Pelosol, geringmächtiger Pelosol (kalkfrei, mit A—C-Profil), Kalkpelosol (kalkhaltig, mit A—C-Profil), typischer Pelosol (mit A—P—C-Profil), durchschlämmter Pelosol (A_{1+3}—BP—C-Profil). Der P- oder BP-Horizont kann verbraunen, indem sich Brauneisen gleichmäßig im oberen Profil verteilt (verbraunter Pelosol), ferner kann er pseudovergleyt sein (Pseudogley-Pelosol).

Die spezifischen bodenphysikalischen Eigenschaften des Pelosols bestimmen seine bodentypologische Sonderstellung: Infolge des hohen Kolloidgehaltes erfolgt starkes Quellen bei Befeuchtung; in diesem Zustand ist der Boden luftarm und weitgehend undurchlässig, bei Austrocknung erfolgt starkes Schrumpfen mit gleichzeitiger Rissebildung; dann kann die Luft wieder eindringen, vermag aber wegen der Trockenheit die sonst üblichen Prozesse nur stark vermindert zu vollbringen (Oxydation, O-Versorgung der Pflanzenwurzeln). Der bei der Austrocknung stattfindende Schwundprozeß der Bodenmasse erzeugt prismatische und polyedrische Gefügeaggregate (RID, BRUNNACKER u. a. 1960). Im Zustand günstiger Befeuchtung beschränkt sich der Lebensraum der Bodenorganismen auf einige cm an der Oberfläche, eine tiefergreifende biologische Aktivität ist nicht möglich; das ist ein wichtiger Hinweis für die zweckmäßigste Bearbeitungstiefe. Wenn infolge Austrocknung und Schrumpfung Luft in den Boden kommt, fehlt für eine biologische Entfaltung die Feuchtigkeit. Kalkgehalt verbessert den bodenphysikalischen Zustand.

Je nach dem Gehalt an Kalk (und anderen Basen) im Ausgangsmaterial und je nach dem Entwicklungsgrad sind die chemischen Eigenschaften gestaltet. Im allgemeinen zeigen die Pelosole aus Tonmergel und Mergelgestein eine günstige Reaktion und dementsprechend einen hohen V-Wert; in diesem Falle sind auch andere Pflanzennährstoffe in nennenswertem Maße vorhanden, während die Pelosole aus kalkfreien Tonen ungünstige chemische Eigenschaften im Hinblick auf Pflanzennährstoffe und Ökologie aufweisen. Letztere können deshalb nur der Grünland- und Waldwirtschaft dienen, während die Pelosole aus kalkhaltigen Substraten auch beackert werden können; allerdings erfordert das Erfahrung (VOGEL 1957).

Verbreitung. In Deutschland finden wir den Pelosol auf den tonigen Substraten des Oberen Buntsandsteins, des Mittleren Muschelkalkes (Mergel), des Keupers, des Juras, der Kreide und des Tertiärs. Sie stehen in Südwestdeutschland, Mitteldeutschland, Ostwestfalen, im Münsterland und in der Südeifel an. Starke Verbreitung hat der Pelosol auf den tonigen Keuper-, Jura- und Kreideschichten Süd- und Mittelenglands.

Plaggenesch. Der Plaggenesch ist ein Repräsentant der anthropogenen Bodenklasse. Auf der gegenüber der Umgebung etwas erhöht liegenden Feldflur Nordwestdeutschlands, Esch genannt, kommt er vielfach vor, und danach erhielt er früher den Bodentypennamen „Esch" (v. HOYNINGEN-HUENE 1930a, STREMME 1930c). Die Absicht des Menschen, den Acker fruchtbar zu erhalten, formte diesen Boden, der in den letzten Jahrzehnten durch mehrere aufschlußreiche Forschungsarbeiten in Entstehung und Eigenschaften gut bekannt wurde (NIEMEIER 1939, JOHANNSEN und STREMME 1954, v. RAUPACH 1955, FASTABEND und v. RAUPACH 1961, 1962, HEINEMANN 1958, 1960). Da der Plaggenesch auch außerhalb der Esch-Feldflur auftritt, nannten ihn NIEMEIER und TASCHENMACHER (1939) Plaggenboden. Dem Vorschlag von KUBIËNA (1953) folgend, in Bodentypennamen möglichst den Bestandteil „boden" zu vermeiden, hat FASTABEND „Plaggenesch" vorgeschlagen (mündliche Mitteilung, 1956).

Entstehung. Der Plaggenesch entstand durch eine jahrhundertelange Düngung des Ackerbodens mit einem Gemisch von Heideplaggen oder (und) Grasplaggen und Stalldung, das meistens kompostiert wurde, wobei auch andere organische Substanzen (Küchen- und Gartenabfälle, Grabenaushub u. a.) beigemischt wurden. Plaggen oder Soden sind flach abgehackte Krumenstücke, in diesem Falle der Heide oder des Graslandes. Die Plaggen enthielten auch einen hohen Anteil an anorganischem Material (meist Quarzkörnchen). Dadurch, aber vor allem durch das niedrige pH, wurde die dem Acker zugeführte Düngermasse nur zum Teil abgebaut, sie reicherte sich an, die Plaggenauflage wuchs und erreichte mit der Zeit 80 cm und mehr über dem ehemaligen Boden, der meistens ein unfruchtbarer Podsol ist. Die Plaggenwirtschaft ist alt; gemäß den Untersuchungen von

NIEMEIER (1939, 1959) reicht sie 1000, vielleicht sogar 2000 Jahre zurück, gemäß denen von FASTABEND und v. RAUPACH (1962) wird sie seit etwa 600 bis 800 n. Chr. praktiziert.

Wurden nur Heideplaggen verwendet, so entstand der *graue* Plaggenesch, dagegen erzeugten die Grasplaggen den *braunen* Plaggenesch; Mischungen dieser Plaggen führten zu Übergangsbildungen (MÜCKENHAUSEN 1936a, 1936b). Überwiegend baut die Plaggenauflage auf dem Podsol auf, daneben, wenn auch seltener, auf dem Gley-Podsol, dem Podsol-Gley und dem typischen Gley, nur vereinzelt finden wir ihn auf anderen Bodentypen. Unter einer guten Plaggenauflage zerfällt mit der Zeit der Ortstein.

Aufbau und Eigenschaften. Der mächtige A_p-Horizont von etwa 40 bis 80 cm kennzeichnet den makromorphologischen Aufbau; bisweilen hebt sich der beackerte Horizont (A_{p1}) schwach von dem unteren Teil (A_{p2}) ab. Auf die Farbe der verschiedenen Plaggen wurde oben bereits hingewiesen. Die Humusform ist ebenfalls abhängig von der verwendeten Plaggenart und den beim Kompostieren sonst noch verwendeten organischen Substanzen (NIERMANN 1940). Die anorganische Beimengung der Plaggenauflage enthält wenig Kolloidsubstanz, sondern überwiegend Sand. Infolgedessen verbindet nur der Humus die Bodenteilchen zu bröckeligen Aggregaten; Teilbereiche haben Einzelkorngefüge. Unter der Plaggenauflage folgen die Horizonte des ursprünglichen natürlichen Bodentyps, wie oben bereits ausgeführt wurde.

Bodenphysikalische Eigenschaften: Porenreichtum, guter Lufthaushalt, relativ geringe Wasserkapazität, gute Durchwurzelbarkeit, im ausgetrockneten Zustand hoher Benetzungswiderstand.

Chemische Eigenschaften: niedriger pH-Wert, geringer V-Wert, geringe Mengen an Pflanzennährstoffen (GEILMANN 1923). Zu starke Kalkung führt zu heftigem Humusabbau und spürbarer Festlegung von Spurenelementen.

Biologische Eigenschaften: infolge der hohen Wasserstoffionenkonzentration und der Austrocknung geringer Besatz an Bodenorganismen.

Beim braunen Plaggenesch sind alle Eigenschaften und auch der Nutzwert günstiger.

Verbreitung. Der Plaggenesch wurde in erster Linie auf den silikatarmen Schmelzwasser- und Talsanden der Riß-Eiszeit und teils auch auf anderen armen Sanden Nordwesteuropas geschaffen (Dänemark, Nordwestdeutschland, Niederlande, Flandern). Der graue Subtyp herrscht bei weitem vor. Der braune Subtyp tritt nur da auf, wo Grasplaggen zur Verfügung standen, also in der Nähe von Niederungen mit Grasland; ferner finden wir diesen Subtyp in der Nähe der Flottlehm- und Flottsandgebiete (Oldenburg).

Anthropogene Böden (außer Plaggenesch). Die anthropogenen Böden erfahren eine kurze, zusammenfassende Darstellung, da sie nur kleine Flächen einnehmen und ihre Einzeldarstellung viel Raum erfordern würde (MÜCKENHAUSEN u. a. 1962). Der verbreitete Plaggenesch wurde oben eingehender beschrieben.

Zu den anthropogenen Böden zählen nicht unsere Ackerböden, obwohl sie auch durch die Düngung eine Änderung der chemischen Eigenschaften in den oberen Horizonten erfahren. Zwar wird durch die Bearbeitung ein Profilbereich von etwa 15 bis 35 cm umgestaltet, jedoch bleibt im übrigen der *natürliche Bodenaufbau* erhalten. Als anthropogene Böden werden nur die aufgefaßt, deren Profil durch den Menschen eine grundlegende Umgestaltung erfährt.

Ein kleinflächig und relativ häufig auftretender Bodentyp der anthropogenen Bodenklasse ist ein alter Gartenboden, der 1954 von VOGEL (mündliche Mitteilung) *Hortisol* genannt wurde. Er entsteht durch jahrzehnte- bzw. jahrhundertelange, intensive Gartenkultur. Ständige und starke Zufuhr von organischer Substanz, intensive Bearbeitung, insbesondere das tiefe Umgraben, die meist zusätzliche und dadurch geregelte Wasserversorgung (Begießen) und die weitgehende Beschattung sind die wichtigsten genetischen Vorbedingungen für seine Entstehung. Diese geben den Bodentieren, vor allem den Regenwürmern, beste Lebensmöglichkeiten; sie vertiefen mehr und mehr die Krume, die nicht selten 50 cm und mehr erreicht. In alten Klostergärten, in alten Gärtnereien und überhaupt in alten Gärten findet man den Hortisol häufig. Alte, verlassene Siedlungen lassen sich genau auffinden, indem man die Verbreitung des Hortisols kartiert (EDELMAN, mündliche Mitteilung 1958). Die nachhaltige Fruchtbarkeit des Hortisols wird dadurch bezeugt, daß in früherer Zeit sein A-Horizont abgehoben und zur Verbesserung der verarmten Ackerflur verwendet wurde. Der Hortisol findet sich nicht nur auf terrestrischen Bodenflächen, sondern auch in den Flußauen, also im semiterrestrischen Bereich, z. B. bei Schweinfurt, Bamberg und Mainz.

Der Typ *Rigosol* umfaßt die durch den Menschen tief umgewendeten Böden, z. B. die rigolten Weinbergböden und Baumscheiben, tief gewendete Marschen und Moore, durch Vergraben von Sand und Heraufbringen besseren Materials meliorierte Böden usw. Zu den Rigosolen gehören nicht die Böden, die durch Bewegung größerer Bodenmassen entstehen, z. B. durch Planieren, Aufschüttung von Halden u. ä.

Auenböden

Die Auenböden, Gleye und Marschen, teils auch die Moore, gehören im Sinne von KUBIËNA (1948, 1953) zu den semiterrestrischen Böden; sie stellen in seiner Bodensystematik (1953) eine Abteilung dar. In der Klasse der Auenböden gibt es mehrere Bodentypen, die aber hier kurz behandelt werden können, wenn das Verbindende aller Auenbodentypen, das Wasserregime, voraufgeschickt wird.

Die Auenböden sind die Böden der meist breiteren Täler (=Auen), die unter natürlichen Bedingungen jährlich ein- oder mehrmals überflutet werden. In der sogenannten Kulturlandschaft verhindert der Mensch dieses in großem Umfange durch Deichbau. Deshalb sind überflutete und nicht mehr überflutete Auenböden zu unterscheiden; denn die Überflutung bringt Auflandungsstoffe, die bei den nicht mehr überfluteten ausbleiben. Das Auflandungsmaterial ist in seiner stofflichen Zusammensetzung vom Einzugsgebiet des Flusses abhängig. Wenn der Deichschutz die Überflutung verhindert, drückt das Hochflutwasser in vielen Fällen *unter* dem Deich durch und überstaut als Druck- oder Qualmwasser den näheren Bereich des Flußlaufes. Aber auch dann bleiben die starken Grundwasserschwankungen und der sommerliche Grundwassertiefstand charakteristisch für den Wasserhaushalt der Auenböden. Der Tiefstand des Grundwassers im Sommer gestattet ein Eindringen der Luft und damit Oxydationsvorgänge bis tief ins Profil, oft bis in 2 m Tiefe und darüber hinaus. Das unterscheidet die Profile der Auenböden grundsätzlich von denen der Gleye. Abgesehen von der Einwirkung des Wassers an sich, führt das Wasser gelöste Stoffe mit, vor allem Calcium als Calciumbicarbonat, das im durchtränkten Auenboden zum Teil unter Abgabe von Kohlensäure als Calciumcarbonat ausgefällt wird. Der Auenboden erfährt also auch noch eine Bereicherung, wenn er zwar nicht mehr überflutet wird, jedoch noch vom Hochwasser (als Druckwasser) von unten her durchtränkt wird.

Das den Auenboden aufbauende Material variiert sehr stark. Die Textur zeigt alle Körnungen vom Geröll bis zum Ton, und zwar in der horizontalen Verteilung und auch im vertikalen Aufbau des einzelnen Profils. Allerdings besteht insofern eine gewisse Gesetzmäßigkeit, als die Auenböden im strömungsstarken Oberlauf meist grobe Körnungen und im strömungsschwachen Unterlauf meist feine Körnungen aufweisen. Im Ablagerungsraum stark erodierender Flüsse (z. B. Alpenflüsse) geht die Auenbodenbildung vom unverwitterten Gesteinszerreibsel aus; hier finden wir vielfach *junge* Auenböden, aus denen sich, wenn die weitere Auflandung ausbleibt, ältere, reifere *autochthone* Auenböden entwickeln, deren Zusammensetzung vom Gestein des Flußeinzugsgebietes abhängt. Dagegen kommt im Unterlauf der Flüsse vorherrschend — zum mindesten anteilmäßig — erodiertes Bodenmaterial zur Ablagerung, aus dem sich *allochthone* Auenböden bilden. Diese stehen in ihrer stofflichen Zusammensetzung in starker Abhängigkeit von den (erodierten) Böden des Einzugsgebietes. Auf Grund dieser allgemeinen Leitlinien der Auenbodenbildung können die Bodentypen der Auenbodenklasse kurz behandelt werden.

Auenrohboden. Der Auenrohboden, von KUBIËNA (1953) *Rambla* (spanisch, abgeleitet vom arabischen Wort „ramla“ = grober Sand) genannt, ist ein junges Flußsediment mit sehr geringer Verwitterung und (A)—C-Profil. Die frischen, gröberen oder feineren, kalkfreien oder kalkhaltigen Gesteinstrümmer sind noch „roh“, der (A)-Horizont enthält kaum Humus, aber es sind schon Pflanzen (wenn auch wenige) vorhanden.

Junger Auenboden. Der junge Auenboden, von KUBIËNA (1953) *Paternia* genannt (von Rio Paterna der Sierra Nevada Spaniens), bildet das Fortentwicklungsstadium des Auenrohbodens. Durch Verwitterung und Humusbildung ist bereits ein A- über dem C-Horizont entstanden. Der junge Auenboden zeigt einen starken Wechsel in der Textur, er kann grau oder braun, kalkfrei oder kalkhaltig sein, womit er sehr unterschiedliche Standorteigenschaften aufweist.

Rendzinaartiger Auenboden. Dem rendzinaartigen Auenboden wurde von KUBIËNA (1953) der Name *Borowina* (volkstümliche polnische Bezeichnung für die „Rendzina der Aue“) gegeben. Er entwickelt sich aus kalkigen, grobkörnigen Flußsedimenten (C-Horizont). Im ganzen gesehen ist dieser Bodentyp relativ jung und chemisch schwach verwittert. Der A-Horizont zeigt eine schwärzlichgraue bis grauschwarze Farbe, ist kalkhaltig und meist stark humos (gute Humusform). Mit diesen Eigenschaften gleicht er dem A-Horizont der Rendzina (VOGEL 1952, 1953).

Tschernosemartiger Auenboden. Der tschernosemartige Auenboden ist an das gemäßigt warme, humide Klima gebunden und darf nicht mit *feuchten, schwarzen* Bodenbildungen anderer Klimate gleichgestellt werden, z. B. mit Smonitza [= Smonica (serbisch), Smolnica (bulgarisch), smonicy oder smolnicy (albanisch), Sionica (bosnisch)], Wiesenboden (Sowjetunion), Tirs (Nordafrika), black cotton soil (Afrika), regur soil (Indien), margalite soil (Java), Grumusol (USA), sol noir (Frankreich) u. a. (OSTENDORFF 1930, STEBUTT 1930, KRAUSE und UTESCHER 1934, STREMME 1936, KOHL, VOGEL und WACKER 1954, BOGATYREW 1958, SCHEFFER, FÖLSTER und MEYER 1960, NEUGEBAUER und ZAKOSEK 1962). Für die Bildung des tschernosemartigen Auenbodens müssen wir feuchtere Bedingungen in den Flußtälern annehmen als heute. Er ging aus einem kalkhaltigen, meist sandig-lehmigen Anmoorgley hervor, als im Zuge der Flußregulierung und Entwässerung oder auch durch natürliche Flußerosion das Grundwasser abgesenkt wurde. Bodenorganismen hatten nun beste Lebensbedingungen; sie durchwühlten den Boden tief und halfen bei der Stabilisierung des stickstoffreichen Humus mit. Unter dem mächtigen A-Horizont liegt im allgemeinen ein G_o-Horizont als Zeichen des ehemals hochstehenden Grundwassers. Trockeneres Lokalklima förderte diese Bodenbildung, welche alle guten Eigenschaften für die Pflanzenproduktion schaffte.

Brauner Auenboden. Der braune Auenboden erhielt von DEL VILLAR (1927, 1929) den Namen „*Vega*“, der auch von KUBIËNA (1953) übernommen wurde.

Stremme (1936) nannte diesen Boden „brauner Auenwaldboden". Dieser braune Auenbodentyp hat die weiteste Verbreitung im gemäßigt warmen, humiden Klimaraum, und daher wurde ihm in der bodenkundlichen Forschung mehr Beachtung geschenkt als den anderen Auenbodentypen (Dobrowolski 1956, Ostendorff 1930, Mückenhausen 1934, Kubiëna 1953, Schneider 1954, Franz, Gunhold und Pschorn-Walcher 1959, Jaklitsch 1959, Roeschmann 1960).

Der braune Auenboden kann sich außerhalb des Überflutungsbereiches durch chemische Verwitterung autochthon aus frischem Gesteinstrümmermaterial bilden, wobei sich meistens ein A—(B)—G_o-Profil entwickelt; zwischen (B) und G_o können die Übergangshorizonte G_o(B) und (B)G_o eingeschaltet sein. Meistens ist er kalkfrei, er kann jedoch im Untergrund große Kalkreserven besitzen.

Im Unterlauf der Flüsse geht der braune Auenboden in der Regel aus braunem Bodensediment, das überwiegend aus erodierter Braunerde und Parabraunerde besteht, hervor (allochthon); er kann kalkfrei oder kalkhaltig sein und zeigt meistens ein A—M—G_o-Profil. In diesem bildet sich mit der Zeit ein (B)-Horizont (später ein B-Horizont), wenn die Überflutung ausbleibt, so daß dann die Horizontfolge A—(B)—M—G_o vorliegt. In Tälern von Gebieten mit fossilen Böden können auch diese durch Abtrag in die Auen gelangt sein, z. B. Terra-fusca-Material in den Tälern der Alb und den Tälern des Muschelkalkgebietes der Südeifel.

Allgemein befinden sich naturgemäß die verschiedenen Bodentypen der Auenbodenklasse in den breiteren Flußtälern mit starken Grundwasserschwankungen und tiefem Grundwasserspiegel im Sommer.

Gleye

„Gley" ist eine volkstümliche russische Bezeichnung für eine schlammige Bodenmasse. Im russischen Schrifttum finden wir den Namen „Gley" bereits bei Wyssotzki (1900, 1905) und bei Glinka (1914). Wyssotzki bezeichnete mit diesem Wort den Grundwasser-Reduktionshorizont. Später wurde „Gley" als Bezeichnung für den Mineralboden mit hohem Grundwasser international eingeführt. Von „Gley" ist „Vergleyung" abgeleitet, womit die Ausfällung von Stoffen (Brauneisen, Braunstein, Kalk, Vivianit) im Boden durch das Grundwasser gekennzeichnet wird. Davon wird das Adjektiv „vergleyt" abgeleitet. Als junge Bildungen der Täler und Senken werden Auenböden und Gleye in der englischsprachigen Literatur mit der Bezeichnung „alluvial soils" oder „hydromorphic soils" zusammengefaßt.

In Entstehung, Aufbau, Dynamik und Eigenschaften bestehen bei den Bodentypen der Gleyklasse gemeinsame Züge, die im Zusammenhang erörtert werden, bevor die einzelnen Typen der Gleyklasse kurz beschrieben werden. Den wichtigsten Bodenbildungsfaktor bildet hochstehendes Grundwasser, welches das A—G-Profil mit einer vielfältigen Horizontdifferenzierung formt. Das Grundwasser zeigt Unterschiede in folgendem: Dauer des Hoch- und Tiefstandes, Schwankungsamplitude, Fließgeschwindigkeit (unter Umständen Stagnation), Gehalt an organischen Stoffen sowie O-, Ca-, CO_2- und SO_4-Gehalt (Dietrich 1958, Mückenhausen 1958c).

Im Gley steigt das Grundwasser mehr oder weniger hoch (wichtiges Unterscheidungsmerkmal) in den normalen Wurzelraum des Bodens hinein, verändert den Boden und beeinflußt das Pflanzenwachstum. Einerseits gelangen organische Stoffe mit dem Sickerwasser in das Grundwasser, andererseits steigt das Grundwasser bis in die humushaltigen Horizonte und nimmt dabei auch organische Stoffe auf, welche insgesamt die Reduktionskraft des Grundwassers verstärken. Die Reduktion im Grundwasserbereich wird an der Bildung von Eisen(II)-Verbindungen (graue, teils grünliche oder bläuliche Farbtöne) aus Eisen(III)-Verbindungen (rostgelbe und rostbraune Farben) sichtbar. Der Berührungszone mit der Luft geben die Eisen(III)-Verbindungen ihr Gepräge; sie umfaßt den G_o-Horizont (o von Oxydation). Nach der Tiefe nimmt das *typische* Gleyprofil mit abnehmendem Sauerstoffgehalt mehr und mehr die obengenannten Reduktionsfarben an; hier liegt der G_r-Horizont (r von Reduktion). Zwischen dem G_o- und G_r-Horizont sind mehr oder minder mächtige Übergangshorizonte G_{ro} und G_{or} ausgebildet, in denen die Prozesse der Oxydation und Reduktion abwechselnd ablaufen; dementsprechend gestaltet sich die Farbe. Der A-Horizont kann auch vom aufsteigenden Grundwasser erreicht werden und rostfleckig sein; dann sprechen wir vom AG_o-Horizont.

Die Eisen(III)-Verbindungen umhüllen die Mineralkörnchen und sind im Gley praktisch nicht wanderungsfähig; nur bei tiefem pH kann Fe(III) als Ion wandern. Demgegenüber wandern die Ionen der Eisen(II)-Verbindungen leicht. Nach BLOOMFIELD (1951, 1952) kann dieses Ion wahrscheinlich komplex an wasserlösliche organische Stoffe gebunden und leicht transportiert werden. Das Fe(II)-Ion kann teils auch an Kolloide sorptiv gebunden und ausgetauscht werden. Eisenbicarbonat, Eisen(II)-Sulfat und Eisen(II)-Silikat können gebildet werden; die beiden letzteren scheinen den grünlichen Farbton im G_r-Horizont zu verursachen. Das Verhalten des Fe(II) ist nicht in allen Teilen geklärt.

Das durch Oxydation im G_o-Horizont ausgefällte Eisenoxydhydrat ist nach BROWN (1953, 1954) überwiegend Lepidokrokit; BECKEL (1958) fand auch Goethit. GEWEHR (mündliche Mitteilung 1954) stellte fest, daß Eisenausfällungen des G_o-Horizontes amorph waren. Mit dem Vergleyungsprozeß, besonders mit den dabei entstehenden Eisenformen haben sich ferner JOFFE (1935), DEB (1950), HRUSKA (1956), LUDWIEG (1956), YAMANAKA und MOTOMURA (1959) befaßt, speziell mit der Entstehung von Eisenkonkretionen SCHROEDER und SCHWERTMANN (1955).

Die Reduktion im Gley kann aus höherwertigen Manganverbindungen Mangan(II)-Ionen bilden, die in Berührung mit Luft beständiger sind als zweiwertiges Eisen; daher ist in diesen Böden leicht zu viel pflanzenaufnehmbares Mangan vorhanden, das pflanzenschädigend wirken kann (LÖHNIS 1951). In schwarzbraunen und schwarzen Konkretionen sind meistens die Oxydationsformen des Eisens und des Mangans enthalten, die nach längerer Alterungszeit gegen Reduktion ziemlich beständig und wenig reaktionsfähig sind (BECKEL 1958).

Unter Luftabschluß entsteht in stark reduzierendem Milieu aus Schwefel und Eisen schwarzes Eisen(II)-Sulfid, das in geringer und feiner Verteilung dem Reduktionshorizont eine graublaue Farbe verleiht. Ferner entsteht unter gleichen Bedingungen bei Vorhandensein von Phosphorverbindungen das farblose oder weißliche Vivianit, ein Eisen(II)-Phosphat, das an der Luft intensiv blau wird; es kommt selten und zwar nesterweise vor.

Nach Absenkung des Grundwassers im Gley bleibt eine weitere Vergleyung der nun grundwasserfreien Horizonte aus, aber das Fleckenbild des Vergleyungshorizontes hält sich noch sehr lange. Bei guter Belüftung durch Bodenbearbeitung und starker biologischer Aktivität des A-Horizontes beginnt von diesem aus allmählich nach der Tiefe hin eine Umbildung in Richtung eines terrestrischen Bodens (MÜCKENHAUSEN 1936, SCHNEIDER 1954).

Die Bodentypen der Gleyklasse werden in erster Linie durch die Höhe und die Schwankung des Grundwasserspiegels geschaffen. KUBIËNA (1953) unterscheidet Syrogley, Modergley, Mullgley und verbraunten Gley, wobei weniger der unmittelbare Einfluß des Grundwassers im Vordergrund steht. Steht das Grundwasser so tief, daß im oberen Profilteil noch die für terrestrische Böden typischen Prozesse stattfinden können, so entstehen die sogenannten *Semi-* oder *Halbgleye* (MÜCKENHAUSEN u. a. 1962). Die Typen der Gleyklasse brauchen nach der Beschreibung dieser allgemein geltenden Entstehungsbedingungen, des Chemismus und der Eigenschaften nur kurz behandelt zu werden.

Gley (Tafel VI). Der Typus „Gley" entsteht bei höherem Grundwasserspiegel (höher als etwa 80 cm u. Fl.) und weniger schwankendem Grundwasserspiegel als bei den Auenböden. Das hochstehende Grundwasser formt das A—G-Profil, das je nach Stand und O-Gehalt des Grundwassers differenziert ist. Als Subtypen treten auf: *Typischer Gley* mit A—G_o—G_r-Profil (s. oben allgemeine Ausführungen); *sauerstoffreicher Gley* mit A—G_o-Profil, also ohne G_r-Horizont, weil der vorhandene Sauerstoff keine Reduktion zuläßt (z. B. bei stark fließendem Grundwasser); *eisenreicher Gley*, der bei hoher Eisenführung des Grundwassers und guten Oxydationsbedingungen entsteht, im Extremfall bildet sich dabei Raseneisenstein; *Kalkgley*, der bei kalkreichem Grundwasser entstehen kann; *gebleichter Gley*, der aus basenarmem Material (silikatarme Sande) unter Moder und Rohhumus hervorgeht; *Auengley*, der im Bereich der Überflutung in der Aue liegt, aber das Grundwasserregime des Gleyes besitzt. Ferner gibt es Übergänge, vor allem zu nachstehenden Bodentypen: Braunerde, Parabraunerde, Podsol, Pseudogley und Pelosol. Diese Übergänge, z. B. der Braunerde-Gley, sind die *Semi-* oder *Halbgleye*.

Naßgley. Der Naßgley entsteht bei im Ablauf des Jahres meist hohem Grundwasser und besitzt wegen dieser fast ständig starken Nässeeinwirkung ein AG_o—G_r-Profil. Der Kapillarsaum steigt meistens höher als 0,2 m u. Fl., und die Schwankung des Grundwasserspiegels spielt sich im Laufe des Jahres etwa zwischen 0,0 und 0,6 m u. Fl. ab. Nur selten ist bei diesem Typ das Grundwasser luftreich; ist dies jedoch der Fall, so bedeutet das Oxydation und bessere ökologische Bedingungen. Meistens beschränkt sich der Naßgley auf kleinere Flächen. Er bedarf dringend der Dränung, wenn er einen brauchbaren Grünlandstandort abgeben soll.

Anmoorgley. Der Anmoorgley, auch einfach *Anmoor* genannt, entsteht unter fast ständig hohem Grundwasser, so daß infolge Luftabschlusses die Zersetzung der organischen Substanz gemindert wird. Wir zählen diejenigen nassen, humusreichen Böden zum Anmoorgley, die nicht mehr als 20% organische Masse (vom Bodentrockengewicht) im Oberboden aufweisen. Semiterrestrische Böden mit mehr organischer Substanz gehören zu den Mooren. Der Anmoorgley kann durch die Verschiedenheit in der Textur sowie im Calcium- und Nährstoffgehalt (Stickstoff) einen sehr verschiedenen ökologischen Wert aufweisen. Sachgerechte Grundwasserabsenkung verbessert den Wert des Anmoorgleyes. Er kann dann auch der Ackerkultur dienen; im übrigen ist der Anmoorgley ein naturgegebener Grünlandstandort.

Moorgley. Unter Moorgley verstehen wir einen Boden mit hohem Grundwasser und einer Torfauflage bis zu 20 cm Mächtigkeit. Ist die Torfauflage höher, so stellen wir diesen Boden zum Moor. Die Torfauflage des Moorgleyes kann in Entstehung, Eigenschaften und ökologischem Wert dem Niedermoor, Übergangsmoor oder Hochmoor nahe stehen (s. bei „Moore").

Die Typen der Gleyklasse sind in (meist kleineren) Tälern und Senken mit hohem Grundwasser des ganzen gemäßigt warmen, humiden Klimaraumes zu finden. Da das Grundwasser der entscheidende Entstehungsfaktor ist, treten sie auch in anderen Klimaten auf; selbstverständlich gibt es hier außer den genannten auch andere Subtypen, und vor allem gibt es auch andere Halbgleye als Übergänge zu den Bodentypen des betreffenden Raumes. Schließlich bedingt auch die Eigenart des in Tälern und Senken zusammengetragenen Ausgangsmaterials in den einzelnen Klimaräumen eine gewisse Differenzierung.

Marschen

„Marsch“ ist eine Bezeichnung für die ebene Landschaft der Nordseeküste mit überwiegend schweren Böden aus Schlick und mit hohem Grundwasserspiegel, dessen Schwankungen mit der Tide konform gehen, wenn nicht der Mensch durch Deich- und Wehrbau den Grundwassergang anders gestaltet hat. Die Landschaftsbezeichnung „Marsch“ wird schon lange für die hier vorkommenden Böden mit spezifischen Eigenschaften gebraucht und ist später als Bodentypenname vorgeschlagen worden (Kubiëna 1953, Mückenhausen 1936, 1937). Die Bezeichnung „Marsch“ kommt aus der volkstümlichen norddeutschen Sprache; in der englischen Fachsprache finden wir ebenso den Begriff „marsh“. Früher wurden noch folgende Namen für die Marsch gebraucht, die teils und gelegentlich auch heute noch in Anwendung sind: Kleiboden, Koogsboden, Polderboden (soweit im Polder befindlich), Mar, Seemarschboden, Seemarscherde.

Bis vor einigen Jahren wurden die Marschen bodentypologisch in Jungmarsch, Altmarsch und Knickmarsch gegliedert. Diese Auffassung resultierte aus der Unterstellung der bis dahin bekannten pedogenetischen Prozesse, die allerdings neben den neueren Erkenntnissen ihre Gültigkeit behalten. Dieses kann bei der Benutzung der älteren Literatur, die auch heute noch eine gute regionale Orientierung bietet, unterstellt werden (Schucht 1913, v. See 1920, Stremme 1926a, Wildvang 1929, v. Hoyningen-Huene 1930b, Schütte 1935, Mückenhausen 1937). Diese Arbeiten, die eine umfangreiche regionale Darstellung geben und die Zusammenhänge zwischen Alter, Entkalkung und Knickbildung behandeln, wurden durch spätere Forschungen, welche die Eigenart der Sedimentation und des Sedimentationsraumes einschließen, erweitert (Edelman 1950, Lüttmer 1952, Ameryckx 1954, Veenenbos 1955). An diese Arbeitsergebnisse anknüpfend, entwickelte Müller (1954, 1955, 1957, 1959a, 1959b) auf Grund umfangreicher Feld- und Laboruntersuchungen einen neuen Vorschlag für die typologische Gliederung der Marschen der Nordseeküste.

Gemäß diesen neueren pedogenetischen Vorstellungen müssen die Marschen als eine Gruppe (Bodenklasse) aufgefaßt werden, die mehrere (vier) Bodentypen umfaßt (Müller 1959a, 1959b). Das Gemeinsame dieser Marsch-Bodentypen wollen wir vorausschicken, um dann die einzelnen Typen kürzer darstellen zu können. Folgende Merkmale kennzeichnen die Marschen und rechtfertigen eine Abtrennung von den Grundwasserböden des Binnenlandes: 1. die Eigenart der Sedimentation im Zuge der Trans- und Regression des Meeres; 2. das meist feinkörnige Sediment, aufgespült durch die See im Küstenbereich und (oder) den Fluß im Deltabereich; 3. der unter natürlichen Bedingungen mit der Tide konform gehende, meist hohe Grundwasserstand; 4. der durch See, Brack- oder Flußwasser bestimmte Kationenbelag der Bodenkolloide und die dadurch erzeugten kolloidchemischen und physikalischen Bodeneigenschaften. Daraus resultiert die Gliederung der Marschen in die Bodentypen Seemarsch, Brackmarsch, Flußmarsch und Moormarsch.

Der die Marschen aufbauende Schlick ist von Natur aus fruchtbar (Fettweiden). Aus diesem Grunde werden seit Jahrhunderten an der belgischen, niederländischen und deutschen Küste umfangreiche wasserwirtschaftliche Arbeiten (Deichbauten) unternommen, um die Marschen der landwirtschaftlichen Nutzung zu erschließen. Darüber hinaus werden seit Jahrzehnten von bodenkundlicher, grünlandwirtschaftlicher und pflanzenbaulicher Seite große Anstrengungen gemacht, einerseits frischen Schlick für die landwirtschaftliche Nutzung vorzubereiten und andererseits die weniger fruchtbare (meist kalkfreie, ältere) Marsch zu verbessern (Hebestreit 1955a, 1955b, Nieschlag 1955a,

1955b, 1956a, 1956b, KÖHNLEIN 1957, KUNTZE 1958, SCHACHTSCHABEL und Mitarbeiter 1958a, 1958b, 1958c, 1959a, 1959b). Besonders hervorzuheben ist die sogenannte Blausandmelioration, bei welcher der kalkreiche Marschuntergrund (Kuhlerde) von Hand (früher) oder mit einer Spezialmaschine (Kuhlmaschine mit Förderschnecke) an die Oberfläche gebracht wird (NAGEL 1935, RATHJENS 1936, BRÜNE und HUSEMANN 1937, SCHEFFER 1938, IVERSEN 1942). SOMMERKAMP, GALENSA und KUNTZE (1960) haben in einem Überblick über den derzeitigen Stand der Erforschung der Marschen an der deutschen Nordseeküste eine ausführliche Literaturzusammenstellung gegeben, auf die besonders hingewiesen sei. KUNTZE (1960b) stellte außerdem in einer anderen Arbeit die neueste Auffassung über Entstehung, Eigenschaften, Nutzungs- und Verbesserungsmöglichkeiten der Marschen dar. Der weiteren Erforschung der Marschen gelten die im Gang befindlichen, großmaßstäblichen Kartierungen durch das Niedersächsische Landesamt für Bodenforschung (RICHTER 1957, MÜLLER und RUYTHER 1957/58).

Seemarsch. Der graue, kalkreiche Schlick der Seemarsch wurde, wie der Name zum Ausdruck bringt, vom Meere angespült. Der grundsätzliche Profilaufbau zeigt einen A-Horizont und infolge Schichtung und verschieden tiefgreifender Oxydation mehrere G-Horizonte. Das marine Sediment ist überwiegend ton- und schluffreich; Sturmfluten brachten feinsandreichere Schichten. Bei der Ablagerung besitzt das Sediment einen Salzgehalt von etwa 20‰. Wird es der Überflutung entzogen, so wird das NaCl durch die Niederschläge in den Untergrund gewaschen; sorbierte Na-Ionen werden gegen Ca ausgetauscht. Der Seeschlick ist von Natur kalkreich (teils Schalen von Meerestieren), worauf das günstige Gefüge der jungen Seemarsch (Jungmarsch) beruht. Demgegenüber sind Mg- und Na-Gehalt relativ gering; das Verhältnis von Ca/Mg liegt über 4, schwankt aber in gewissen Grenzen. Das feuchte Küstenklima bewirkt einen ständigen Verlust an basisch wirkenden Kationen, so daß die Seemarsch mit der Zeit den ehemals vorhandenen Kalk und auch die sorbierten Ca-Ionen mehr und mehr verliert (Altmarsch), wodurch die chemischen, aber auch wesentliche physikalische Eigenschaften, besonders das Gefüge, verschlechtert werden. Andererseits kann der über *flache* Küstenbereiche sedimentierte Schlick zum Teil von vornherein kalkfrei sein (LÜTTMER 1952). Ältere Marschen „verbraunen" in der Regel in der Krume, später auch darunter.

Folgende Subtypen der Seemarsch lassen sich unterscheiden: Die *Vorland-Seemarsch* liegt im noch überfluteten und überschlickten Deichvorland und zeigt rohbodenartigen Charakter. Die *kalkhaltige Seemarsch* zeichnet sich durch beachtlichen Kalk- und Nährstoffgehalt sowie gutes Gefüge aus, im Gegensatz zur *kalkfreien Seemarsch.* Die *Knick-Seemarsch* ist ebenfalls — wenigstens im oberen Profilteil — kalkfrei und besitzt einen dichten Horizont im Unterboden (=Knick), der sedimentationsbedingt ist oder einen ehemaligen Oberflächenhorizont darstellt. Zur Brackmarsch gibt es gleitende Übergänge (*Brackmarsch-Seemarsch*).

Brackmarsch. Die Brackmarsch entsteht in der Brackwasserzone (Mischung von Salz- und Süßwasser) aus grauem, kalkhaltigem bis kalkfreiem, meist schluff- und tonreichem Schlick mit primär hohem Gehalt an Mg-, Na- und K-Ionen. Innerhalb der breiten Zone des Brackwasserbereiches werden oberes (flußaufwärts), mittleres und unteres (seewärts) Brackwasser unterschieden, womit der Grad der Versalzung des Flußwassers zum Ausdruck gebracht wird. Durch Transgression und Regression des Meeres verschieben sich die Aussüßungsbereiche. Dadurch entstehen Profile, in denen brackischer und mariner Schlick *übereinander* vorkommen; diese Tatsache erschwert die bodentypologische

Einordnung der Marsch sehr (MÜLLER und RUYTHER 1957/58). Am Geestrand entsteht dann die Brackmarsch, wenn von der Geest Süßwasser in die vom Meere überspülte Marsch eintritt.

Im Brackwasserbereich kommt im allgemeinen schwerer Schlick mit etwa 60 bis 90% Schluff und Ton zum Absatz, weil Meer- und Flußwasser bis zu diesem Bereich einen weiten Weg zurückgelegt und die größeren Bestandteile bereits verloren haben. Die Ausfällung der feinen Schwebstoffe wird in dem Gemisch „Salz-Süßwasser“ durch kolloidchemische Vorgänge gefördert. Das Seewasser hat bereits vor dem Erreichen des Flußwassers die Kalkschalentrümmer abgesetzt und einen großen Teil des Calciumcarbonates unter Abgabe von CO_2 ausgefällt, so daß der in der Brackwasserzone abgesetzte Schlick relativ wenig Kalk besitzt, dafür aber primär reich an Mg-, Na- und K-Ionen ist. Der Salzgehalt des frischen Schlickes beträgt etwa 2,0 bis 0,25‰. Das Verhältnis Ca:Mg (sorbierte Ionen) schwankt zwischen 3,5 und 1,0 je nach Lage des Sedimentationsraumes, das heißt je nach den Mengenanteilen von Meer- und Flußwasser. Der hohe Anteil an Schluff und Ton sowie der relativ hohe Gehalt an sorbierten Mg- und Na-Ionen bedingt ein dichtes Bodengefüge und damit den gefürchteten „*Knick*“, d. i. ein verdichteter Horizont. Wenn das auch die häufigste Entstehungsweise von Knickhorizonten ist, so kann ein solcher auch eine andere Entstehungsursache haben. Ein durch Überschlickung „begrabener“, ehemaliger Oberflächenhorizont kann z. B. sehr dicht sein (s. bei Knick-Seemarsch). Ebenfalls der Verlust an Calcium im Laufe der Alterung, ferner häufige Schwankungen des Grundwassers und damit Vergleyung können zur Dichtlagerung beigetragen haben (STREMME 1926a, MÜCKENHAUSEN 1937, KUNTZE 1960b). Mit einer Tondurchschlämmung muß in der an Mg- und Na-Ionen reichen Brackmarsch gerechnet werden, da diese Ionen die Peptisation und damit die Verlagerungsbereitschaft der Kolloide fördern (VEENENBOS 1955). Das Makrogefüge der an Mg- und Na-Ionen reichen Verdichtungshorizonte zeigt großprismatische oder großsäulige, dichte Aggregate; im gequollenen Zustand bilden sie eine dichte, kohärente Masse mit teils fehlenden dränfähigen Poren. Über den Nährstoffgehalt orientieren die Arbeiten von SCHACHTSCHABEL und Mitarbeiter (1958a, 1958b).

Die Subtypen der Brackmarsch sind (analog zur Seemarsch) folgende: *Vorland-Brackmarsch, kalkhaltige Brackmarsch, kalkfreie Brackmarsch, Knick-Brackmarsch, Seemarsch-Brackmarsch, Flußmarsch-Brackmarsch.* Die Knick-Brackmarsch bereitet bei der Nutzung die größten Schwierigkeiten (NIESCHLAG 1955b, 1956b, KUNTZE 1958, 1960a).

Flußmarsch. Die Flußmarsch entsteht im Gezeitenbereich des Flußunterlaufes, d. h. sie wurde im Zustand der Aufschlickung täglich zweimal überflutet. Das unterscheidet sie genetisch deutlich vom Auenboden. Sie besteht aus meist schluffig-tonigem, kalkhaltigem oder kalkfreiem Flußschlick, dessen Zusammensetzung weitgehend von den im Einzugsgebiet des jeweiligen Flusses anstehenden Böden und Gesteinen abhängig ist. Darin stimmen wieder Flußmarsch und Auenboden überein. Der Salzgehalt des frischen Flußschlickes liegt unter 0,5‰, ist also gering. Dabei überwiegen die Ca-Ionen die Mg- und Na-Ionen stark. Dementsprechend sind die Eigenschaften der Flußmarsch wesentlich besser als die der Brackmarsch; sie stimmen mehr mit denen der Seemarsch und der braunen Auenböden überein. Die Subtypen der Flußmarsch werden analog zur See- und Brackmarsch gebildet; es wird daneben noch eine *verbraunte Flußmarsch* ausgeschieden.

Moormarsch. Die Moormarsch besteht aus einer dünnen schluff- und tonreichen, grauen, rostbraun gefleckten, basenarmen Schlickdecke mit brackischem

Charakter über meist mit Schlick durchsetztem Torf des Niedermoores, selten des Übergangs- und Hochmoores. Sie entsteht am Geestrand, meistens weit von der Küste entfernt, wo die Bedingungen für die Moorbildung günstig sind, aber nur wenig Schlick angelandet wird. Auch während der Moorbildung ist eine geringe Schlickaufspülung erfolgt, die aber immer wieder von den Moorpflanzen durchwachsen wurde (Darg). Durch Sturmfluten wurde das geestnahe Moor zugeschlickt; so kam die Schichtenfolge „Marsch über Moor" zustande. Das wechselvolle Geschehen an der Küste führte zu einem starken Wechsel in der Horizontfolge der Profile der Moormarsch, worüber KUNTZE (1960b) berichtet hat. Die Schlickschicht der Moormarsch ist zum Teil auch vom Menschen aufgetragen worden, um tiefer gelegenes Moorgelände zu erhöhen, zu festigen und zu nutzen. Das macht die genetische Behandlung der Moormarsch noch komplizierter.

Die Eigenschaften der Moormarsch sind ähnlich schlecht wie bei der dichten Brackmarsch: pH-Werte um 4, V-Werte oft zwischen 20 und 30%, enges Ca-Mg-Verhältnis, Kalium ist etwas vorhanden, Phosphor sehr wenig, dichtes Gefüge.

Moore

„Moor" ist eine Landschaftsbezeichnung (wie Marsch), die ebenso für die Benennung der Moor-Bodentypen übernommen wurde. Die Moore können hier kurz abgehandelt werden, da sie an anderer Stelle im Zusammenhang mit der Düngung ebenfalls beschrieben werden; sie müssen aber auch hier erwähnt werden, um die Bodentypen-Zusammenstellung vollständig zu machen.

Das Moor wird teils subhydrisch gebildet und gehört dann nach dem Wasserregime zu den subhydrischen Böden. Andere Moore entstehen unter semiterrestrischen Bedingungen und wieder andere über dem Grundwasser unter stark humidem Einfluß. Indes haben alle Moore gemeinsam, daß sie rein organische Bildungen sind, nicht aus „fremdem" Material bestehen, sondern bei ihrer Bildung gleichsam ihre „eigene" Muttersubstanz aufbauen. Das unterscheidet sie pedogenetisch klar von allen anderen Böden.

Wie schon früher (beim Moorgley) gesagt, gehören alle Böden mit mehr als 20 cm Torf über dem Mineralboden zu den Mooren. Bei der bodentypologischen Gliederung werden die alten, bewährten Bezeichnungen Niedermoor, Übergangsmoor und Hochmoor gebraucht. Es ist nicht die Aufgabe dieser kurzen Darstellung, die ganze Problematik der Entstehung, der Eigenschaften und der verschiedenen Einteilungsmöglichkeiten der Moore aufzuzeigen. Nur die allgemeinen Wesenszüge können erörtert werden, ohne auf die lokalen Verschiedenheiten einzugehen. Zur Vertiefung in das Stoffgebiet werden die Arbeiten der Moorspezialisten empfohlen; aber auch aus diesen Arbeiten kann nur eine Auswahl erfolgen: WEBER 1903, v. POST 1924, TACKE 1930, HUSEMANN 1947, BRÜNE 1948, OVERBECK 1950, KIVINEN 1954, BADEN 1958, GÖTTLICH 1958, LUTZ 1936 u. a. Ferner wird zur Ergänzung und Vertiefung noch auf folgende außerdeutsche Spezialarbeiten hingewiesen: v. RATHLEF 1930, DOKTUROWSKI 1932, ERDTMAN 1932, GRUNLAND 1932, WAKSMAN und SCHULHOFF 1942, WIMMER 1952, HEIKUREINEN 1953, BARRY 1954, FRASER 1943, 1954, DAWSON 1956, KATZ 1957, 1958, TOMASZEWSKI 1957, SMITH, BRYSON und THOMSON 1958.

Niedermoor. Das Niedermoor entsteht in flachen, stehenden Gewässern aus abgestorbenen Wasserpflanzen, hauptsächlich Rohrkolben (*Typha*), Schilf (*Phragmites*) und Großseggen (*Carex*), untergeordnet auch Astmoos (*Hypnum*), Erle (*Alnus*) und Weide (*Salix*). Der Name Niedermoor wird von der „niederen Lage" (zur Umgebung) abgeleitet; wegen der flachen Oberfläche (im Gegensatz zum Hochmoor) wird dieses Moor auch Flachmoor genannt. Da es geländeabhängig ist, wurde auch die Bezeichnung „topogenes Moor" und wegen des Vorkommens in Niederungen „Niederungsmoor" gebraucht. Die englische Bezeichnung „Fen" für Niedermoor (auch teils für andere organische Böden gebraucht) entspricht dem niederländischen „Veen" und dem deutschen „Fehn". Das nährstoffreiche Niedermoor wird im Englischen „rich fen", im Schwedischen „rikkör" genannt. Ferner werden mitunter folgende Bezeichnungen für das Niedermoor gebraucht: Moos, Wasenmoos, Wampenmoos, Talmoos, Verlandungsmoor, limnisches Moor, subaquatisches Moor, eutrophes Moor, Ried.

Der Kalk- und Nährstoffgehalt des Niedermoores hängt weitgehend vom Grundwasser ab, in dem das Moor entsteht. Ist das Grundwasser an diesen Stoffen reich, so entsteht auch ein entsprechend günstiges Niedermoor, das insbesondere calcium- und stickstoffreich ist. In einem armen Grundwasser dagegen kann ein Niedermoor von fast der gleichen Nährstoffarmut entstehen, wie sie für das Hochmoor typisch ist. Große Verbreitung haben die Niedermoore in Bayern, Nordwestdeutschland, den Niederlanden, Skandinavien, Polen und der westlichen Sowjetunion.

Übergangsmoor. Das Übergangsmoor, auch Zwischenmoor genannt, nimmt eine Übergangs- oder Zwischenstellung zwischen Nieder- und Hochmoor ein. Es besitzt noch Pflanzen des Niedermoores, aber auch bereits solche des Hochmoores. Das Übergangsmoor wächst über dem Niedermoor auf, wenn die Wasser- und Nährstoffversorgung der Pflanzen infolge der Aufhäufung von Pflanzenmaterial oder Grundwasserabsenkung schlechter wird, oder es entsteht auf nassem, nährstoffarmem Anmoorgley. Der genetischen Zwischenstellung zwischen Nieder- und Hochmoor entspricht auch in der Regel ein mittelmäßiger Gehalt an Pflanzennährstoffen.

Hochmoor (Tafel VI). Das Hochmoor entsteht in niederschlagsreichem, luftfeuchtem Klima *über* dem Grundwasserspiegel (daher die ältere Bezeichnung supraaquatisches Moosmoor) und erhebt sich uhrglasförmig aus der Landschaft (daher Hochmoor). Da es ein niederschlagsbedingtes Moor ist, wird es auch ombrogenes Moor oder Weichwassermoor (da das Niederschlagswasser kein Ca enthält) genannt. Das Torfmoos, auch Bleichmoos oder Weißmoos genannt, bildet den Hauptbestandteil des Hochmoors; davon sind die Bezeichnungen Bleichmoostorfmoor und Weißmoostorfmoor abgeleitet. In Oberbayern trägt das latschen- und spirkenbestockte Hochmoor die volkstümliche Bezeichnung Filze (in Niederbayern Filz).

Da das Wasser des Hochmoores aus dem Niederschlag stammt, somit ein nährstoffarmes Milieu vorliegt, können nur anspruchslose, typische Hochmoorpflanzen gedeihen. In erster Linie sind es Torfmoose (*Sphagnum*), Wollgras (*Eriophorum*) und die Sumpfbeise (*Scheuchzeria palustris*). Andere für das Hochmoor charakteristische Pflanzen spielen mengenmäßig keine bedeutende Rolle am Aufbau des Hochmoores. In Ostdeutschland und Osteuropa ist auch die Kiefer, in Oberbayern die Latsche (bzw. Spirke) auf dem Hochmoor zu finden. Nach der Entwässerung siedelt sich als Sekundärvegetation gern die Heide an.

Die typischen Hochmoorpflanzen, vor allem die Torfmoose, besitzen ein schwammartiges Porensystem, in dem das Niederschlagswasser gespeichert wird. Dadurch wird das Moor stark vernäßt und luftarm. Sauerstoffarmut und Humussäure hemmen die Zersetzung der pflanzlichen Masse; sie vertorft.

Das Hochmoor wächst vielfach über dem Übergangsmoor auf, wenn die Pflanzen die Verbindung mit dem Grundwasser verloren haben. Auch kann es unmittelbar auf sehr nährstoffarmem Anmoorgley aufwachsen. Viele Hochmoore Deutschlands entstanden in der Nacheiszeit bei zunehmender Humidität durch

Versumpfung von Wäldern. Den oberen Teil des Hochmoores baut der wenig zersetzte, umbrabraune sogenannte Weißtorf auf, darunter folgt der gut zersetzte, schwärzlichbraune sogenannte Schwarztorf; getrennt sind beide meistens durch den mehr oder minder gut erkennbaren „Grenzhorizont", der eine Zeit des Stillstandes der Torfbildung und Torfzersetzung anzeigt (OVERBECK und SCHNEIDER 1940).

Im Entstehungszustand herrscht im Hochmoor große Nässe; wo dieses Moor stärker wächst, bilden sich Bülten, dazwischen ziehen sich nasse Schlenken hin. Die Nässe beherrscht den bodenphysikalischen Zustand. Ferner kennzeichnen Armut an allen Pflanzennährstoffen und biologische Inaktivität das Hochmoor. Nach der Entwässerung werden die Poren des Hochmoortorfes mehr oder weniger wasserfrei; dadurch wird er luftreich und gut erwärmbar; er ist nach längerem Austrocknen schwer benetzbar und wird leicht puffig.

Der größte Teil der Hochmoore Deutschlands liegt in den klimafeuchten Gebieten Nordwestdeutschlands und am nördlichen Alpenrand. Kleinere Hochmoore treten in den höheren Mittelgebirgslagen Mitteleuropas auf. Große Hochmoore sind in ganz Nordeuropa, vor allem in Finnland und der nördlichen Sowjetunion verbreitet.

Subhydrische Böden

Die subhydrischen Böden umfassen einige Bodentypen, deren gemeinsame Wesenszüge vorausgeschickt werden. Diese Bodentypen werden *unter* dem Wasser gebildet und von diesem *allseitig* durchdrungen und beeinflußt. Sie haben in Mitteleuropa keine große Verbreitung, wohl aber in Skandinavien (v. POST 1862, NAUMANN 1917, 1930), wo sie inzwischen großflächig unter terrestrische Bedingungen gekommen sind und in landwirtschaftlicher Nutzung wegen ihrer Eigenart eine besondere Behandlung erfahren müssen. Auch der Vollständigkeit wegen müssen sie kurz erörtert werden.

Die Auffassungen darüber, ob die subhydrischen Böden wirkliche Böden im bodenkundlichen Sinne sind oder ob sie lediglich Sedimente darstellen, gehen auseinander. Von geologischer und biologischer Seite neigt man teilweise zu letzterer Auffassung (POTONIÉ 1903, POTONIÉ 1938, MORTIMER 1949), während andere Autoren eindeutig von Böden sprechen (LUNDQUIST 1927, TACKE 1930, WASMUND 1930, KUBIËNA 1953). Die Benennung der subhydrischen Böden hat im Laufe der Zeit stark gewechselt, so daß eine einheitliche und eindeutige Kennzeichnung schwierig ist. In einer neuen Arbeit hat GROSSE-BRAUCKMANN (1961) sich eingehend mit den bestehenden Bezeichnungen für die subhydrischen Bildungen auseinandergesetzt. Hier wird der Nomenklatur von KUBIËNA (1953) gefolgt, die der heute geltenden weitgehend entspricht.

Protopedon. Darunter versteht KUBIËNA (1953) einen subhydrischen Rohboden, d. h. ein Unterwassersediment mit beginnender Bodenbildung, ohne mikroskopisch sichtbaren Humus, aber besiedelt durch Organismen (kalkarmes, klastisches Sediment, See-Erz, Seemergel, Seekreide).

Gyttja. Mit Gyttja, einer volkstümlichen schwedischen Bezeichnung (v. POST 1862), wird ein subhydrischer, grauer, graubrauner oder schwärzlicher, an organischen Stoffen reicher, oberflächlich *organismenreicher* Schlamm mehr oder minder sauerstoffreicher Gewässer bezeichnet; auch ist dafür der Name Mudde gebraucht worden (GROSSE-BRAUCKMANN 1961). KUBIËNA (1953) unterscheidet sieben limnische bzw. marine Gyttjen. In Schweden und Finnland befinden sich trocken gefallene Gyttjen ehemaliger großer Seen in landwirtschaftlicher Kultur (KIVINEN 1938); starke Festlegung der Phosphorsäure fällt besonders auf.

Sapropel. Unter Sapropel (Lauterborn 1901, 1915, Solowjew 1933) versteht man einen übelriechenden Faulschlamm mit reduzierendem Milieu und meist (von Schwefeleisen) schwärzlicher Farbe. Als Bildung sauerstoffarmer Gewässer ist das Sapropel *organismenarm*, enthält jedoch anaerobe Bakterien; es ist meistens reich an organischer Substanz. Kubiëna (1953) unterscheidet limnisches und marines Sapropel, bei letzterem zwei Varianten. Bei Trockenlegung oxydiert das Schwefeleisen des Sapropels durch Luftzutritt; es entsteht Schwefelsäure, und der Boden versauert stark, wenn nicht genügend Kalk vorhanden ist.

Dy. Dy, eine volkstümliche schwedische Bezeichnung für Torfschlamm (v. Post 1862), stellt eine leberbraune bis schwarzbraune, saure, biologisch träge Humusmasse von Binnenseen dar, denen braunes, saures Moorwasser zufließt; aus letzterem wird die Humusmasse größtenteils ausgefällt. Dy ist bei Trockenlegung als Kulturboden zunächst nicht brauchbar.

Die *subhydrischen Böden* entwickeln sich nach Entfernung des darüberstehenden Wassers (Trockenlegung) in Richtung der semiterrestrischen oder terrestrischen Böden. Oft besitzen sie dann die Wasserverhältnisse des Gleyes oder Naßgleyes und werden dann fälschlicherweise mit diesen Namen bezeichnet. Auch nach der Trockenlegung behalten die subhydrischen Böden viele ihrer Eigenschaften, die nur aus ihrer Genese erklärbar sind.

Fossile Böden (Paläoböden)

In der gemäßigt warmen, humiden Klimaregion treten Böden auf, die nicht in der Nacheiszeit, sondern früher gebildet worden sind und daher fossile Böden genannt werden (Harrassowitz 1930b, 1930c). Die präpleistozänen Böden entstanden in tropischem oder subtropischem, teils auch in einem warmen, trockenen Klima. Wie weit die Bildung der präpleistozänen Böden in die Erdgeschichte zurückreicht, ist in den meisten Fällen nicht zu entscheiden; jedenfalls geht sie in einzelnen Fällen nachweislich in das Prätertiär zurück (Schmidt und Wolters 1952). Von den präpleistozänen Böden sind bisher noch keine ungestörten Profile gefunden worden, meistens sind es verstümmelte Profile oder umgelagertes Bodenmaterial. Häufig ist fossiles Bodenmaterial bei der Umlagerung mit anderem Gesteins- oder Bodenmaterial vermischt worden, übt aber auch dann noch seinen spezifischen Einfluß aus. Von den fossilen Böden sind also nur noch „Relikte“ vorhanden; deshalb spricht man in diesem Falle auch von Reliktböden. Die pleistozänen Böden gehören teils den Kaltzeiten (arktische Böden) und teils den Warmzeiten mit einem ähnlichen Klima wie heute an.

Die im gemäßigt warmen, humiden Klima auftretenden „fossilen“ Bodentypen kommen in anderen Klimazonen als „rezente“ Bodentypen vor und werden in dem entsprechenden Abschnitt näher behandelt. Deshalb brauchen hier die fossilen Böden nur kurz abgehandelt zu werden. (Näheres s. in Mückenhausen u. a. 1962.)

Die fossilen Böden des Präpleistozäns gehören zu den *Terrae calcis*, den *Plastosolen* und den *Latosolen* (Kubiëna 1953).

Die *Terrae calcis* entstanden unter einem warmen, feuchten Klima aus Kalkstein, Dolomit und Mergel und sind vertreten durch die Bodentypen *Terra fusca* und *Terra rossa*, wovon erstere die größere Verbreitung zeigt (Ložek und Prosek 1956, Ložek und Tyracek 1958, Werner 1958, Smolíkowá 1959, 1960a). Oft ist Material dieser Bodentypen (vor allem von der Terra fusca) mit jüngerem Bodenmaterial (Rendzina) bei der Umlagerung vermischt worden (Mückenhausen 1953). In Spalten und Dolinen blieb meistens das Material dieser Bodentypen vor der Erosion bewahrt; nur selten bilden Terrae größere Flächen, wie das auf dem Jurakalk der Alb und auf dem Muschelkalk der Südeifel der Fall ist. Auf Kalkstein, Dolomit und Mergel aller Formationen finden sich Reste dieser Bodentypen sporadisch in Mitteleuropa.

Die *Plastosole*, vertreten durch die Bodentypen *Braunlehm*, *Graulehm* und *Rotlehm*, entstanden unter tropischem und subtropischem Klima aus Silikatgesteinen; sie haben größere Verbreitung als die Terrae. Während Braunlehm und Rotlehm eine geringe Verbreitung haben (MÜCKENHAUSEN u. a. 1962), kommen größere Flächen des Graulehms in den Mittelgebirgen vor (PFEFFER 1938, MÜCKENHAUSEN 1958d). Gegenüber den jüngeren Böden besitzen sie vor allem wegen der beweglichen Feinsubstanz und des hohen Kaolinitgehaltes ungünstige bodenphysikalische Eigenschaften, außerdem sind sie meistens nährstoffarm. Die schwierigen Standortseigenschaften haben zu intensiven Untersuchungen Anlaß gegeben (ARENS, KRÄMER und LANGNER 1958, MÜCKENHAUSEN, GERKHAUSEN und KERPEN 1959, KERPEN 1960).

Den *Latosol*, und zwar fast nur den roten Typ, finden wir selten, in etwas größerer Verbreitung im Gebiet des Vogelsberges (HARRASSOWITZ 1930b, 1930c). Dieser Bodentyp entstand auch unter tropischen Bedingungen, allerdings ist für seine Bildung ein Wechsel von Niederschlags- und Trockenperioden anzunehmen. Dabei wird die Kieselsäure teils weggeführt, und die Fe- und Al-Verbindungen gehen weitgehend in stabile, irreversible, schorfige Aggregate über. Trotz seines hohen Tongehaltes ist dadurch der Latosol durchlässig, aber auch schnell trocken.

Die *Kaltzeiten des Pleistozäns* haben eine Fülle von Formen *arktischer Böden* hinterlassen. Die Kryoturbationen haben oft starke Bodenunterschiede auf kleinem Raum geschaffen. Würge- und Taschenböden kommen am häufigsten vor (TROLL 1944, STEEGER 1948, SCHÖNHALS 1951b, 1956, WOLDSTEDT 1954, KAISER 1958), dagegen sind Steinnetzboden, Tropfenboden und Tundragley seltener (WORTMANN 1956, STEUSLOFF 1957).

Die *Warmzeiten und Interstadialzeiten des Pleistozäns* hinterließen verschiedenartige Böden. In den zum Teil etwas wärmeren, vor allem aber in den langen Warmzeiten entstanden Böden, die eine Übergangsbildung zwischen Braunlehm und Parabraunerde darstellen. Vollständige Profile blieben selten erhalten, aber die verstümmelten Profile dieser Böden haben eine große Verbreitung auf den altpleistozänen Eis- und Flußablagerungen (MÜCKENHAUSEN 1954b, 1959). Da das gekappte Profil meistens relativ dicht ist, hat es Anlaß zur Staunässe gegeben, selbst in den Fällen, in denen ein später aufgetragenes, durchlässigeres Material darüber liegt. In den Interstadialzeiten der letzten Vereisung entwickelten sich Böden, die unserer Parabraunerde, dem Pseudogley und den Übergängen zwischen diesen Typen entsprechen. Je nach der Dauer des Interstadials ist die Bodenbildung stärker oder schwächer (SCHÖNHALS 1951a, WOLDSTEDT 1954, FINK 1954, BRUNNACKER 1957, SMOLÍKOWÁ 1960b). Meistens sind diese Böden durch jüngere Ablagerungen (meistens Löß) überdeckt worden. Wir müssen indes annehmen, daß sie in manchen Fällen an der Oberfläche liegen. Die starke Bodenbildung, oft auch Verdichtung und Pseudovergleyung, läßt in diesen Böden eine ältere Bodenbildung vermuten. Da jedoch keine jüngere Ablagerung darüber liegt, die eine Alterseinstufung erlauben würde, kann in diesen Fällen nichts über den Beginn der Bodenentwicklung gesagt werden (BRUNNACKER 1957).

Diese kurze Darstellung der wichtigsten fossilen Böden Mitteleuropas läßt erkennen, daß solche Böden flächenmäßig keine große Bedeutung haben. Die örtlichen fossilen Bildungen weichen ganz von den rezenten Böden ab; die Relikte fossiler Böden können die jüngere Bodenbildung örtlich in mancher Weise beeinflußt haben.

B. Die Bodentypen der übrigen Klimaräume der Erde

Nachdem die Bodentypen des gemäßigt warmen, humiden Klimas, die am besten bekannt und für die landwirtschaftliche Nutzung am wichtigsten sind, eingehender behandelt wurden, sollen nun die Bodentypen der übrigen Erde eine kurze Darstellung erfahren. Dabei werden die Böden in acht bodenkundliche Räume aufgeteilt, womit einer üblichen Einteilung gefolgt wird, die z. B. bei der Darstellung der Böden auf Übersichtskarten Anwendung findet. Schwierig ist wieder die Wahl der Namen für die betreffenden Bodentypen, weil, wie oben schon geschildert, keineswegs eine einheitliche Nomenklatur besteht. Soweit als möglich, werden die Synonyme aufgeführt.

Um einen Überblick über die Bodentypen der Erde zu gewinnen, wird das Studium der auf der Tagung der Internationalen Bodenkundlichen Gesellschaft in Madison (USA) 1960 vorgelegten neuen Bodenübersichtskarten folgender Räume empfohlen: Westeuropa im Maßstab 1:2,5 Mill. (TAVERNIER u. a. 1963), Osteuropa im Maßstab 1:2,5 Mill. (*Dokutschajew-Institut* 1960, nicht veröffentlicht), Asien im Maßstab 1:6 Mill. (KOWDA und LOBOWA 1960), Australien im Maßstab 1:5 Mill. (STEPHENS u. a. 1960), Afrika im Maßstab 1:25 Mill. (D'HOORE 1960), Südamerika im Maßstab 1:10 Mill. (BRAMAO und LEMOS 1960). Von den USA liegt eine Bodenübersichtskarte kleinen Maßstabes von KELLOGG und KNIGHT vor, die in „Soils and Men“ (U.S. Dept. Agr. 1938) veröffentlicht worden ist. Eine Bodenkarte der Sowjetunion im Maßstab 1:4 Mill. gaben Dokutschajew-Institut und GUKG als Gemeinschaftsarbeit 1954 heraus. Ferner existieren zwei neue Weltbodenkarten kleinen Maßstabes, eine mit der Nomenklatur der USA (Abb. 20) und eine (im Maßstab 1:20 Mill.) mit der Nomenklatur der Sowjetunion (WILENSKI 1957). Von den vielen Lehrbüchern über regionale Bodenkunde können nur wenige genannt werden: U.S. Dept. Agr. 1938, ROBINSON 1939, KUBIËNA 1953, GANSSEN 1957, GERASSIMOW und GLASOWSKAJA 1960, GANSSEN und HÄDRICH 1965.

Die Bodentypen des kalten, feuchten (arktischen) Klimas (arktische Böden)

Die Bodentypen des kalten, feuchten Klimas (Abb. 20) besitzen folgende gemeinsame Wesenszüge:

1. Die Bodenbildung erfolgt unter meist niedriger Temperatur und bei fast stetiger Feuchtigkeit. Die Niederschlagsmenge ist zwar meistens nicht groß, aber die Luftfeuchtigkeit ist relativ hoch und die Verdunstung gering.

2. Der Dauerfrost oder die ewige Gefrornis (englisch permafrost, schwedisch Tjäle, russisch Merslota) befindet sich stets ab einer gewissen Tiefe. Im Sommer taut der Boden je nach geographischer Lage mehr oder minder tief (0,4 bis 6 m) auf, während unter der Auftauzone der ewige Frost bleibt. Im Herbst beginnt die Auftauzone von oben her wieder zu gefrieren. Während dieser Zeit des erneuten Einfrierens ist der Boden unten und oben erstarrt; dazwischen befindet sich zunächst noch die weiche Auftauschicht (POHLE 1924).

3. Starke physikalische, geringe chemische Verwitterung infolge des kühlen Klimas; deshalb sind grobkörnige, steinhaltige Bodenarten verbreitet.

4. Starke Vernässung der Böden; denn die Gefrornis läßt das Sickerwasser nicht in den Untergrund abziehen, und die Verdunstung ist infolge der niedrigen Temperatur und des spärlichen Pflanzenwuchses gering.

5. Spärliche Pflanzendecke und schwaches Pflanzenwachstum infolge der Kälte.

6. Schwacher Besatz mit Bodenorganismen und geringe biologische Aktivität infolge der Kälte, der Bodennässe und oft auch der Bodensäure. Die Zersetzung der pflanzlichen Abfälle ist deshalb stark gehemmt; daher besteht die Tendenz zur Ansammlung von Rohhumus und Torf.

7. Infolge der starken Vernässung des Bodens zu einer schlammigen Masse und infolge des fehlenden oder geringen Vegetationsschutzes kommt der Boden selbst in Hanglagen mit nur schwacher Neigung in langsames Fließen (Bodenfließen oder Solifluktion).

Für ein eingehendes Studium der arktischen Böden werden folgende Arbeiten empfohlen: DE LEFFINGWELL 1915, GRIPP 1926, MEINARDUS 1930, BLANCK 1939a, TROLL 1944, KUBIËNA 1953, WOLDSTEDT 1954, SCHENK 1955, BÜDEL 1959.

In den Kaltzeiten des Pleistozäns war die kalte, feuchte Klimaregion weit nach Süden gerückt. In Mitteleuropa und in den übrigen eisnahen (jedoch eisfreien) Gebieten der Erde, Periglazial genannt, herrschte die Tundra. Hier entstanden damals arktische Böden, vor allem Fließerden, Strukturböden und Eisspalten, die erhalten blieben und heute als fossil anzusehen sind (TROLL 1944, STEEGER 1948, GALLWITZ 1949, SCHÖNHALS 1951a, 1951b, 1956, WORTMANN 1956, STEUSLOFF 1957, KAISER 1958, DAHM 1959). Diese fossilen Bodenbildungen wurden später großenteils durch jüngere (meist äolische) Ablagerungen zugedeckt, sie blieben aber im Unterboden bis heute erhalten und bedingen infolge der starken Störung der ehemaligen Bodenschichtung große Bodenunterschiede auf kleinem Raum. Das ist z. B. bedeutsam bei der Auswahl von Versuchsfeldern.

Die *wichtigsten Bodentypen* des kalten, feuchten Klimas können nur kurz dargestellt werden. Dabei wird in großen Zügen der von KUBIËNA (1953) vorgeschlagenen Einteilung gefolgt, weil sie mehr pedologisch ausgerichtet ist, als es die Vorschläge anderer Fachkreise sind.

Arktischer Rohboden. Der arktische Rohboden tritt in kalten Regionen mit schwachem Pflanzenwuchs auf. Wie für alle Rohböden, so ist auch für diesen das (A)—C-Profil charakteristisch. Humus ist also kaum vorhanden. Geringe chemische, aber starke physikalische Verwitterung bedingen grobkörnige Bodenarten. Infolge des weitgehenden Fehlens der Vegetation erfolgt das Bodenfließen bereits bei geringer Hangneigung. Der arktische Rohboden gehört mit zu den am weitesten in die Arktis und Antarktis hineinreichenden Böden.

Arktischer Steinpflaster-Rohboden. KUBIËNA (1953) nennt den arktischen Steinpflaster-Rohboden „arktischer Hamada-Rohboden". Infolge der geringen chemischen und der starken physikalischen Verwitterung kommen steinige Bodenarten zustande. Wird die wenige Feinerde von heftigen Winden ausgeblasen, so entsteht an der Oberfläche das für diesen Boden charakteristische Steinpflaster.

Arktischer Strukturboden. Der arktische Strukturboden tritt in einer außerordentlichen Mannigfaltigkeit auf, d. h. die ihn kennzeichnenden Strukturen sind sehr zahlreich. Sie kommen zustande, indem das Bodenmaterial der Auftauzone eine bestimmte Sortierung und Anordnung der verschieden großen Bestandteile erfährt. Die mannigfaltigen Erscheinungen können heute nicht in jedem Falle eindeutig genetisch gedeutet werden.

Häufig ist der Strukturboden durch Verknetungen, sogenannte Kryoturbationen, des Bodenmaterials gekennzeichnet. Sie kommen in der Auftauzone zustande, wenn diese im Herbst von oben wieder gefriert und dabei die noch nicht gefrorene, wassererfüllte Schicht unter einen ungleichmäßigen Druck gesetzt wird. Das noch nicht gefrorene, breiige Bodenmaterial weicht dabei nach der Stelle geringeren Druckes aus. Dabei kommt es zu Verwürgungen oder Verknetungen (Würgeboden, Wickelboden, Kryoturbationen) des Materials, die oft taschenartig ausgebildet sind (Taschenboden).

Abb. 20. Bodenkarte der Welt.

Herausgegeben vom United States Department of Agriculture, Soil Conservation Service, Soil Survey, und der Food and Agriculture Organization of the United Nations, Soil Survey, 1960.

1 Prärieböden, degradierte Schwarzerde

2 Schwarzerde und rötlich-kastanienfarbige Böden (mit einigen Flächen von dunkelgrauen und schwarzen Böden der Subtropen und Tropen)

3 Dunkelgraue und schwarze Böden der Subtropen und Tropen (mit einigen Flächen von Schwarzerden, rötlich-kastanienfarbigen Böden und hydromorphen Böden)

4 Kastanienfarbige Böden, braune und rötlichbraune Böden

5 Sieroseme, Wüstenböden und rote Wüstenböden (mit Flächen von Lithosolen, Regosolen und Salzböden)

6 Podsole und schwach podsolierte Böden

7 Graubraune podsolige Böden, braune Waldböden (= Braunerden) usw.

8 Latosole, rotgelbe podsolige Böden (mit Flächen von hydromorphen Böden, Lithosolen und Regosolen)

9 Rotgelbe Mittelmeerböden (einschließlich Terra rossa), meist gebirgig (einschließlich vieler Gebiete mit Rendzinen)

10 Böden der Gebirge und Gebirgstäler (Komplex)

11 Tundra

12 Alluviale Böden (viele kleine, aber wichtige Gebiete dieser Böden kommen in allen Teilen der Welt vor, sind aber nicht auf der Karte dargestellt)

1 Prairie soils, degraded chernozem

2 Chernozems and reddish chestnut soils (with some inclusions of dark gray and black soils of the subtropics and tropics)

3 Dark gray and black soils of the subtropics and tropics (with some inclusions of chernozems, reddish chestnut soils and hydromorphic soils)

4 Chestnut, brown and reddish-brown soils

5 Sierozems, desert and red desert soils (with inclusions of lithosols, regosols and saline soils)

6 Podzols and weakly podzolized soils

7 Gray-brown podzolic soils, brown forest soils, etc.

8 Latosols, red-yellow podzolic soils (with inclusions of hydromorphic soils, lithosols and regosols)

9 Red-yellow mediterranean soils (including Terra rossa), mostly mountainous (including many areas of rendzina soils)

10 Soils of mountains and mountain valleys (complex)

11 Tundra

12 Alluvial soils (many small but important areas, not shown on map, occur in all parts of the world)

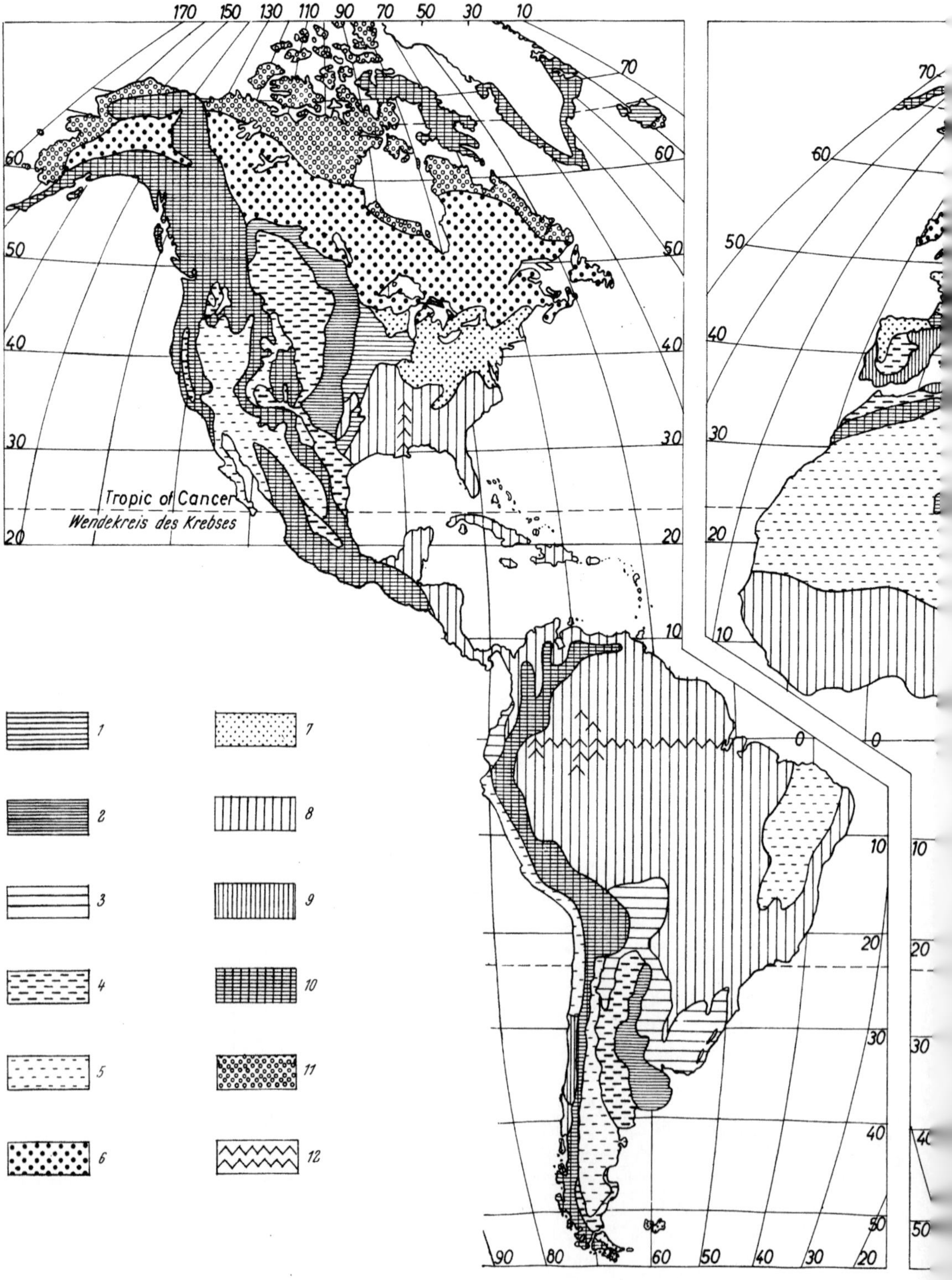

170 150 130 110 90 70 50 30 10
70
60
50
40
30
Tropic of Cancer
Wendekreis des Krebses
20
10
0
90 80 60 50 40 30 20
1
2
3
4
5
6
7
8
9
10
11
12

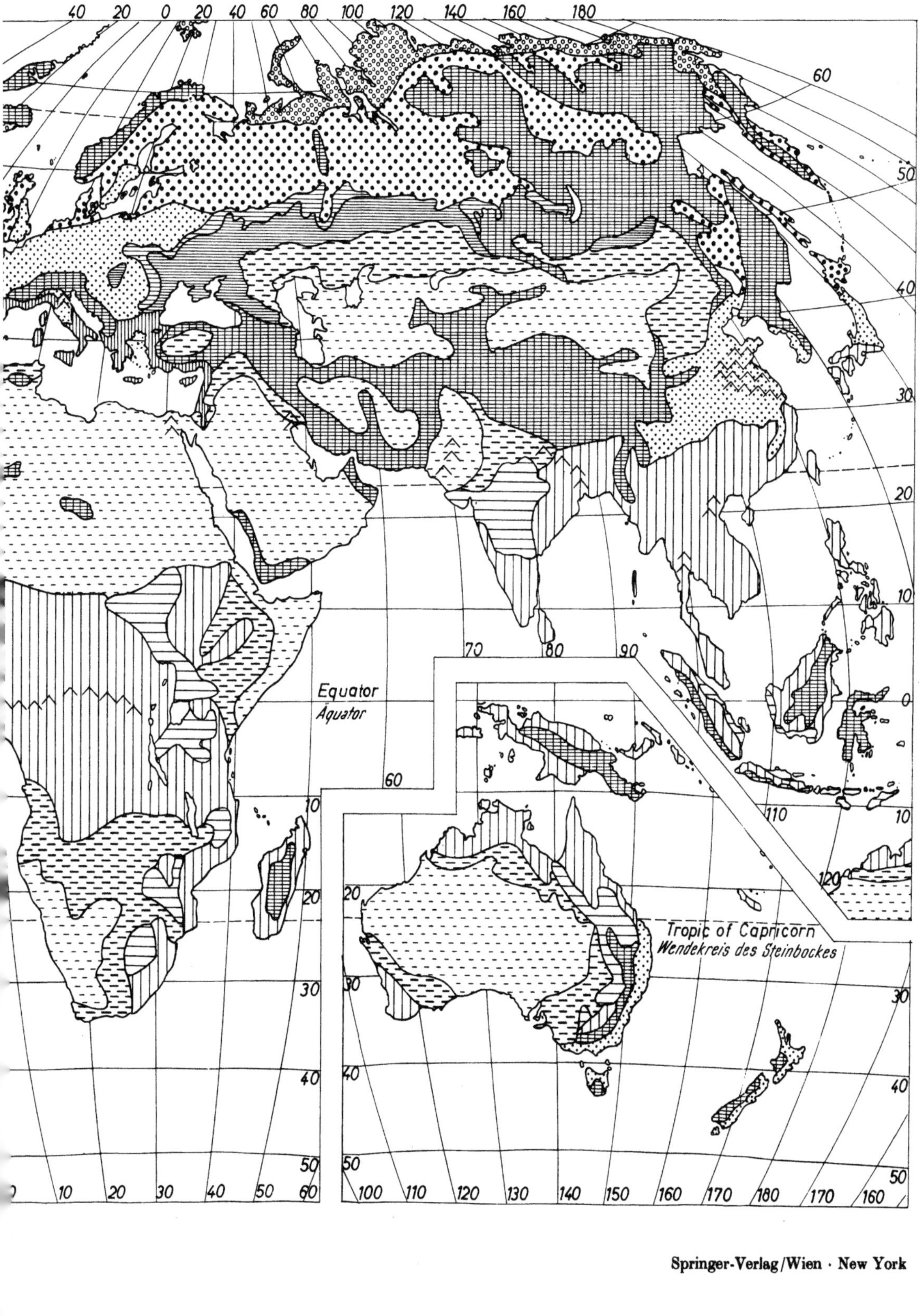

Springer-Verlag/Wien · New York

Durch Schwund, der nach SCHENK (1955) durch die Dehydratation der Bodenteilchen im Gefolge der Eisbildung im Boden hervorgerufen wird, entstehen Spalten, die den Boden in Polygone aufteilen (Polygonboden, Rautenboden, Quarréboden, Facettenboden). Von oben zeigt dieser Boden ein wabenartiges Aussehen. Die Polygone haben einen Durchmesser von einigen Dezimetern bis zu einigen Metern; die größeren Polygone teilen sich in kleinere auf. Die etwas nach oben gewölbten Polygonkerne bestehen aus feinerdereicherem und die Streifen („Spalten") zwischen den Polygonen aus gröberem Material. Bei steinhaltigen Bodenarten reichern sich mit der Zeit die Steine auf den „Spalten" zwischen den Polygonen an; so entstehen an der Oberfläche Steinringe (Steinringboden, Steinnetzboden). In Hanglagen bewegen sich die Steinmassen hangabwärts; es bilden sich Steinstreifen (Steinstreifenboden). Die ringartige Anreicherung der Steine an der Oberfläche ist nicht sicher geklärt. Wahrscheinlich gelangen die Steine durch Frosthebung (Bildung von Eis im Lagebett der Steine) allmählich an die Oberfläche. Beim Auftauen gleiten sie von den gewölbten Polygonen in die tiefer liegenden Streifen zwischen diesen. Vereinzelt wurde auch dieser Steinringboden fossil gefunden (WORTMANN 1956). Gemäß den Forschungen von SCHENK (1955) und in neuester Zeit von BÜDEL (1961) sind diese Polygonböden zum Teil sehr alt und teils nicht mehr „aktiv", d. h. ihre Bildung geht infolge veränderter Wasserverhältnisse (trockener) nicht weiter. Der Polygonboden Alaskas, bei dem die großen Polygone nicht durch gröberes Material, sondern lediglich durch Eisspalten getrennt sind, stellen grundsätzlich nichts anderes dar; sie werden auch Tetragonal- oder Schachbrettboden genannt (SCHENK 1955).

Der sogenannte Tropfenboden entsteht, wenn in der Auftauzone spezifisch schwereres über spezifisch leichterem Bodenmaterial liegt. Im Zustand völliger Wasserdurchtränkung sinkt das spezifisch schwerere Material in das leichtere *tropfenartig* hinein (STEUSLOFF 1957).

Eisspalten oder Eiskeile sind eine allgemeine Erscheinung im periglazialen Raum. Sie entstehen wahrscheinlich ebenfalls durch Bodenschwund im Zuge der Dehydratation der Bodenmasse beim Gefrierprozeß im Sinne von SCHENK (1955). Später werden die Eisspalten mit Fremdmaterial von oben gefüllt und so konserviert (GALLWITZ 1949). Im pleistozänen Periglazial Mitteleuropas und entsprechender anderer Räume werden fossile Eisspalten oder Eiskeile oft gefunden.

KUBIËNA (1953) hat noch eine „arktische Rutmark" ausgeschieden, die er im Gegensatz zu den terrestrischen arktischen Strukturböden zu den semiterrestrischen Böden stellt, weil sie im feuchten Milieu des Schwemmlandes der Arktis liegt.

Tundraranker. Wie die übrigen Ranker kennzeichnet den Tundraranker ein A—C-Profil. Mit den arktischen Böden hat er die zu Anfang des Kapitels erwähnten Eigenschaften gemeinsam. Er entsteht meistens auf den Felsfluren oder auf dem Gesteinsschutt im kalten Klimabereich. Die ewige Gefrornis spielt hierbei keine entscheidende Rolle, da Fels oder Gesteinstrümmer hoch anstehen. Charakteristisch ist ferner im Gegensatz zu den meisten Böden des kalten Klimabereiches die geringere Vernässung. Vielfach hat sich ein Rohhumus (terrestrische Form) gebildet.

Nordischer Zwergpodsol. Eine „trockene" Bildung des kalten Klimas ist ein Podsol mit einer Entwicklungstiefe, die selten mehr als 25 cm beträgt; nur auf Lockergesteinen reicht die Entwicklung des Profils tiefer. Alle charakteristischen Horizonte des Podsols sind vorhanden, aber meistens nur einige cm mächtig. Die Geringmächtigkeit der Profile führte zu dem Namen „Zwergpodsol"; von

KUBIËNA (1953) wird dieser Podsol auch „nordischer Nanopodsol" genannt. Im übrigen besitzt dieser nordische Zwergpodsol alle charakteristischen Eigenschaften des Podsols. Er entwickelt sich im Norden Europas vorwiegend auf trockeneren Standorten unter bultenartigem Rohhumuspolster auf calciumarmen Gesteinen oder deren Schutt (saure Eruptiva, Kristalline Schiefer). Die Podsolbildung wurde auch in Alaska beobachtet (KUBOTA und WHITTIG 1960).

Tundragley. In tieferen Lagen, die für die Wasseransammlungen günstig sind, bildet sich der Tundragley. Vom Gley des gemäßigt warmen, humiden Klimas unterscheidet er sich durch die Dynamik im arktischen Frostboden. Im Tundragley herrschen nur in der Auftauzeit ähnliche Bedingungen wie im normalen Gley. Beginnt die Auftauzone im Herbst zu gefrieren, so spielen sich in der noch nicht gefrorenen Zwischenschicht ebenfalls Verknetungen des Bodenmaterials ab. Die noch schwach gefrorene Oberschicht kann auch aufbrechen, und der Bodenbrei wird an die Oberfläche gedrückt (sogenannte Erdquellen). Die Farbe des Tundragleyes ist mittel- bis hellgrau. Wir finden den Tundragley auch fossil im pleistozänen Periglazial Mitteleuropas, wo die Bezeichnung „Tundra-Naßboden" üblich ist.

Tundraanmoor. Die Entstehung des Tundraanmoores erfordert noch größere Vernässung als beim Tundragley; sumpfiges Gelände bildet den Entstehungsort. Vom Tundragley unterscheidet es sich durch einen humusreichen A-Horizont. Der Humus ist schwärzlichbraun oder schwärzlich, dystroph und wenig zersetzt. Unter dem A-Horizont folgt der mittel- bis hellgraue Gleyhorizont.

Tundramoor. KUBIËNA (1953) nannte das Tundramoor „Tundra-Moos" oder „Tundra-Torfmoor". Von den Mooren des gemäßigt warmen, humiden Klimas unterscheidet es sich durch die Gefrornis und die starke Oberflächengliederung in Torfhügel und dazwischen gelegene Schlenken. Das Tundramoor entwickelt sich durch Vermoosen der Rohhumuslage des Tundrarankers, oder es baut sich auf Verlandungstorfen auf; auch ein Tundraanmoor kann den Anfang darstellen. Überwiegend baut sich das Tundramoor aus *Sphagnaceen* auf, daneben sind auch andere Moosarten beteiligt (*Dicranum*- und *Polytrichum*-Arten). Ferner beleben Krähenbeere, Moltebeere, Bärentraube, Weißbirke, Zwergbirke und Flechten die hügelige Oberfläche des Tundramoores. Die eine Eislinse einschließenden größeren Hügel nennt man Palsen; sie können mehrere Meter hoch werden. Weniger nasse Formen werden Thufur genannt. Neben den Torfhügeln gibt es wallartige Formen, die auch Ringmoore genannt werden. In Hanglagen können sich strangartige Torfrücken bilden (Strangmoore).

Die Bodentypen des kühlen bis gemäßigt warmen, feuchten Klimas (Podsolregion)

Zwischen dem arktischen und dem gemäßigt warmen, feuchten Klima (Abb. 20) befindet sich eine relativ breite Klimazone, die sich durch die Verbreitung des Podsols hervorhebt. Diese Podsolzone greift mit schmalem Saum in die gemäßigt warme, feuchte Klimazone Mitteleuropas über, nämlich wo basenarme, quarzreiche Ausgangsgesteine die Entwicklung des Podsols begünstigen.

Die typische Podsolregion im Norden Europas, Asiens und Nordamerikas (hauptsächlich Kanada) trägt folgende gemeinsame Wesenszüge:

1. Die Bodenbildung läuft im überwiegend kühlen bis gemäßigt warmen, feuchten Klima ab. Hohe Luftfeuchtigkeit und niedrige Temperaturen bewirken eine geringe Verdunstung; infolgedessen ist die Sickerwassermenge groß.

2. Die natürliche Pflanzendecke ist der Wald; vielfach ist es Nadelwald mit schlecht zersetzbarer Streu. Teilweise sind auch rohhumusfördernde Zwergsträucher vorhanden.

3. Normalerweise entwickelt sich das A—B—C-Profil, das meistens folgendermaßen differenziert ist: A_0—A_1—A_2—B_1—B_2—C. Staunässe und Grundwasser verändern diese Horizontfolge unter Einschaltung vernäßter Horizonte, z. B. eines Bg oder BG.

4. Starke Verluste an Basen und Pflanzennährstoffen, oft auch an Spurenelementen; starke Versauerung ist die Folge.

5. Infolge der Versauerung und der stark verminderten biologischen Aktivität bildet sich aus schwer zersetzbarer Pflanzensubstanz der Rohhumus.

6. Zerstörung der Tonsubstanz und Wanderung der Sesquioxyde des Eisens und des Aluminiums in einen tieferen Horizont. Im Gefolge dieses Podsolierungsprozesses entwickeln sich der hellgraue A_2-Horizont und der dichte, meist kaffeebraune oder rostbraune (wenn trocken) B-Horizont.

In Abschnitt A über die Bodentypen Mitteleuropas ist die Entstehung des Podsols dargestellt und die wichtigste Literatur zitiert, worauf hingewiesen sei. Darüber hinaus werden noch folgende Spezialarbeiten aufgeführt, die sich besonders mit den Podsolen des Nordens befassen: Stremme 1930c, Ekström 1950, Iwanowa 1956, Låg 1960, Muir 1961.

Typischer Podsol. Der typische Podsol wurde schon bei den Böden Mitteleuropas besprochen. Er besitzt ein A_0—A_1—A_2—B_1—B_2—C-Profil; Rohhumus, Bleichhorizont, Orterde oder Ortstein sind entwickelt. Überwiegend entsteht dieser Bodentyp aus Sanden und Sandsteinen, und zwar um so eher, je silikatärmer diese Gesteine sind. Rohhumusliefernde Pflanzen, kühle Witterung und hoher Niederschlag fördern den Prozeß. Die naturgegebene Armut des Ausgangsmaterials wird durch den Podsolierungsprozeß noch verstärkt; andererseits schließt er die wenigen Silikate energisch auf. In Niederungen stört oft das Grundwasser die Ausbildung des typischen Podsolprofils; es verbessert aber den Pflanzenstandort (Wasserreserve).

Podsolierte Böden (podzolic soils). Unter podsolierten Böden sollen alle Böden mit deutlicher Entwicklung eines A_2-Horizontes, in dem sich Tonzerstörung anzeigt, verstanden werden, wobei sich aber kein nennenswerter Orterde- oder Ortsteinhorizont gebildet hat. Der B-Horizont ist vielmehr homogen rostfarbig (bei normalem Wasserhaushalt) oder rostbraun und grau gefleckt (bei Wassereinfluß). Diese podsolierten Böden entstehen meistens aus silikatreicheren Gesteinen (lockeren und festen). Auch die *Derno-Podsole* (=Rasenpodsole) der sowjetischen Bodenkundler sind hierzu zu zählen, sofern sie einen hellgrauen A_2-Horizont besitzen. — Die podsolierten Böden, mannigfaltig im Profilaufbau, in Textur und Ausgangsmaterial, nehmen einen breiten Raum in Nordeuropa, Nordasien und in Kanada ein.

Der Derno-Podsol mit ockerfarbigem A_3-Horizont und ohne nennenswerte Staunässe entspricht der Parabraunerde und dem trockenen gray-brown podzolic soil. Die hellgelben podsoligen Böden Rußlands gehören auch dazu (Nogina 1952)

Nordischer Zwergpodsol. In nördlichen Breiten und in hohen Lagen entwickelt sich ein Podsol mit auffallend geringmächtigem Profil (selten mächtiger als 25 cm), den Kubiëna (1953) nordischer Zwerg- oder Nanopodsol genannt hat. Die geringmächtige Profilentwicklung beruht auf der geringen chemischen Verwitterung in diesem kühlen Klima (s. Weiteres im Kapitel über arktische Böden).

Podsol mit Eisenbändchen. Im Gebirge Nordeuropas wurde ein nasser Podsol mit auffallend dünnem (etwa 2 cm), hartem B_1-Horizont (Eisenbändchen) gefunden. Dieses Eisenbändchen staut das Sickerwasser, so daß in dem feuchten, verdunstungsschwachen Klima die darüber liegenden Horizonte über einen großen Teil des Jahres naß oder feucht sind. Aus einer meist mächtigen und feuchten Rohhumusauflage wandern saure Humussole in den A_2-Horizont, der dadurch eine schwärzlichgraue Farbe erhält. Feuchte, Bodensäure und Nährstoffarmut bedingen einen dürftigen Pflanzenstandort. Oft geht dieser Podsol in Hochmoor über. Er tritt auch im Hochgebirge Neuseelands auf.

Staunasse Podsole. Die im Norden, vor allem in der Taiga, aus sandig-lehmigen Substraten sich bildenden Podsole besitzen meistens Staunässe. Man kann sie als Übergänge zwischen Podsol und Pseudogley (Pseudogley-Podsol und Podsol-Pseudogley) auffassen. Charakteristisch für diese Podsole sind der meist feuchte Rohhumus, der hellgraue (teils hellgelbliche) und meistens mit rostbraunen Flecken und Konkretionen durchsetzte A_2g-Horizont und der stark rostbraun und fahlgrau gefleckte Bg-Horizont. Diese verarmten, verdichteten und vernäßten Podsole haben eine große Verbreitung im Norden der Sowjetunion und wahrscheinlich auch in Kanada.

Die Bodentypen des mediterranen Klimas und ähnlicher Klimate

Die Böden des mediterranen Klimabereiches (Abb. 20) haben folgende Wesens züge gemeinsam:

1. Die Bodenbildung läuft unter stark gegensätzlichen Klimabedingungen ab, d. h. eine relativ milde, feuchte Periode (Jahresniederschlag etwa zwischen 400 und 1000 mm) vom Herbst bis zum Frühjahr wechselt mit einer warmen, trockenen Periode vom Frühjahr bis zum Herbst ab. Dieser periodische Klimawechsel ist für die Bodenbildung einschneidend.

2. Die Wärme vom Frühjahr bis zum Herbst führt zur Dehydratisierung der Eisenverbindungen und daher vielfach zu roten oder rötlichen Bodenfarben.

3. Die Tendenz der Bildung eines A—B—C-Profils ist, im ganzen gesehen, vorhanden.

4. Die *Voraussetzungen* für die Bodenerosion durch Wasser sind günstig: vorwiegend hängiges Gelände, Niederschläge periodisch und oftmals stark, Niederschläge in der Jahreszeit (Herbst bis Frühjahr), in welcher die ackerbaulich genutzten Flächen keinen oder nur geringen Vegetationsschutz haben.

5. Die *erfolgte* Bodenerosion durch Wasser hat im ganzen Mittelmeerraum ein großes Ausmaß erreicht, ausgelöst durch die unter 4. genannten natürlichen Voraussetzungen, ferner durch die großflächigen Entwaldungen und die Vernichtung der Waldverjüngung durch Schafe und Ziegen. Der Ab- und Auftrag von Bodenmaterial ist infolgedessen fast allerorts sehr stark. — Die Winderosion hat dagegen keine wesentliche Bedeutung.

6. Die Böden sind teilweise sehr alt; ihre Entstehung reicht ins Pleistozän, teils ins Präpleistozän zurück (NEVROS und ZVORYKIN 1939). Vor allem enthalten die Umlagerungsprodukte in den Tälern fossiles Bodenmaterial; darauf ging die Bodenbildung weiter.

Die Nomenklatur der Böden des Mediterranklimas ist leider nicht einheitlich. Hier werden die Namen gebraucht, die heute im allgemeinen in den Mittelmeerländern üblich sind und die auch auf der neuen Bodenkarte von Westeuropa verwendet werden (TAVERNIER u. a. 1963).

Die Bodenforschung war in den letzten Jahrzehnten in den Mittelmeerländern sehr aktiv, so daß die wichtigsten Bodentypen dieses Raumes bekannt sind. Dabei hat der rote Mediterranboden aus Kalkstein die meiste Beachtung gefunden. Leider läßt sich zur Zeit noch nicht sagen, welche Räume der übrigen Erde gleiche oder sehr ähnliche Bodentypen aufweisen. So scheint der zimtfarbige Boden Bulgariens und des Südkaukasus mit dem rötlichbraunen Mediterranboden übereinzustimmen. Der gleiche Boden tritt auch in dem halbtrockenen Klima der westlichen USA auf. Eine Korrelation der Böden dieser Räume steht noch aus.

Für ein eingehendes Studium der Böden des mediterranen Klimas werden folgende Schriften und Bodenkarten empfohlen: REIFENBERG 1929, BLANCK 1930, DEL VILLAR 1930, BLANCK und MUSIEROWICZ 1931, LIATSIKAS 1935, AGAFONOFF 1937, HOLLSTEIN 1938, NEVROS und ZVORYKIN 1939, LIATSIKAS 1942, PENNINGSFELD 1942, GRAČANIN 1947, ALBAREDA und ALVIRA 1950, COMEL 1952, KUBIËNA 1953, OAKES 1957, TAMÉS 1958, CATACOUSINOS 1959, MANCINI 1960.

Die wichtigsten Bodentypen des mediterranen Klimas und ähnlicher Klimate werden nunmehr kurz beschrieben.

Ranker. Während im gemäßigt warmen, feuchten Klima überwiegend der braune Ranker auftritt, entwickelt sich im sommertrockenen Mediterranklima häufig ein *grauer Ranker*, der vorwiegend durch mechanische und geringe chemische Verwitterung gebildet wird (KUBIËNA 1953, FRANZ 1956). Dieses Rankerstadium hält sich klimabedingt sehr lange. Wie für die Ranker allgemein, so ist auch für diesen das A—C_1—C_2-Profil charakteristisch. Der A-Horizont besteht aus mullartigem Moder oder Mull. Den Ranker trockener Lagen mit wenig Humus, aber Mullbildung hat KUBIËNA (1953) *Xeroranker* genannt, den grauen Ranker mit feinerdereichem C_1-Horizont bezeichnet er als *Gantranker* und in dem Falle, in dem der C_1-Horizont leicht verschlämmbar und dicht ist, als *Launaranker*. Im Hochgebirge des Mittelmeeres fand KUBIËNA (1953) ferner einen sehr flachgründigen Ranker an windexponierten Stellen, der sich durch einen filzigen, nicht dystrophen A-Horizont auszeichnet (*Eilagranker*).

Die Ranker des Mittelmeergebietes haben im allgemeinen die große Austrocknung im Sommer, geringe chemische Verwitterung, geringen Humusgehalt und langes Verharren in diesem Entwicklungsstadium gemeinsam. Man könnte sie als Mediterran-Ranker von den übrigen abtrennen. Sie werden vielfach mit dem braunen Mediterranboden (TAVERNIER u. a. 1963) zusammengefaßt; früher zählte man sie häufig zu dem mediterranen Trockenwaldboden (LIATSIKAS 1935).

Rendzina. Die Rendzinen des Mittelmeergebietes unterscheiden sich von denen des gemäßigt warmen, humiden Klimas durch die Sommertrockenheit und meistens auch durch einen geringen Humusgehalt, ferner noch durch das weite Farbspektrum. Von Hellgrau bis Dunkelgrau, von Gelblich bis Rot kommen alle Farbnuancen vor. Die Frage ist, ob der flachgründige, kalkhaltige, rote Boden aus Kalkstein als rote Rendzina (ZVORYKIN und STOGIANNIS 1958) oder als roter Mediterranboden (TAVERNIER u. a. 1963) zu bezeichnen ist. Gemeinsam ist allen das A—C-Profil, die Bildung aus Carbonatgesteinen, der Kalkgehalt, der A-Horizont, die gute Humusform, die schnelle Austrocknung, das bodenbedingte, warme Bodenklima sowie der mit Spalten und Dolinen durchsetzte Untergrund (Karst). Für die hellgraue, humusarme, kalkreiche Rendzina der besonders trockenen Gebiete bzw. Lagen des Mittelmeerraumes hat KUBIËNA (1953) den Namen *Xerorendzina* vorgeschlagen, in Spanien auch Rendzina cinérea genannt. Bei fortschreitender Bodenbildung entwickeln sich die Rendzinen im Mediterranklima zum roten Mediterranboden. Wie die roten Verwitterungskrusten an Kalkbausteinen alter Bauten zeigen, kann der rote Mediterranboden auch direkt aus dem Gestein entstehen (NEVROS und ZVORYKIN 1939).

Auf großen Flächen sind Rendzinen entstanden, wo der alte Boden abgetragen worden ist. Da Carbonatgesteine (Kalk, Dolomit, Mergel) in den Gebirgen des Mittelmeerraumes eine große Verbreitung haben, so sind auch die Rendzinen entsprechend häufig neben dem roten Mediterranboden zu finden.

Roter und brauner Mediterranboden aus Carbonatgesteinen. Roter und brauner Mediterranboden oder red and brown mediterranean soil (aus Carbonatgestein) ist im Mediterrangebiet und ähnlichen Klimaten die neue und übliche Bezeichnung der Terrae calcis (Terra rossa und Terra fusca) KUBIËNAS (1948, 1953). Die Farben gehen bis zum Gelb hin, so daß es besser wäre, „roter *bis gelber* Mediterranboden" zu sagen, wie es in Portugal geschieht. Der rote Typ wurde früher meistens Mediterran-Roterde oder Terra rossa genannt, womit man meistens, aber nicht immer, die roten Bildungen aus Carbonatgesteinen verstanden wissen wollte (REIFENBERG 1929, BLANCK 1930, PRINCIPI 1937, GRAČANIN 1947, COMEL 1952).

Der braune bis gelbe Mediterranboden (aus Carbonatgesteinen) wurde früher auch Gelberde genannt (BLANCK 1939c); er ist identisch mit der Terra fusca KUBIËNAS (1953).

Der rote Mediterranboden tritt auf den reineren Kalken und in besonders warmen Lagen auf. Die reinen Kalke hinterlassen wenig Lösungsrückstand, so daß zunächst ein flachgründiger Boden mit warmem Bodenklima entsteht. Auf Mergel, aber auch auf Kalkstein in feuchtigkeitsbegünstigten Lagen (Hangfuß, Mulden) tritt dagegen der braune (oder gelbe) Mediterranboden auf. In jedem Falle scheint die etwas größere Bodenfeuchte gegenüber dem Standort des roten Mediterranbodens Anlaß zur Entstehung des braunen Typs zu geben. Wegen der besseren Feuchtigkeitsverhältnisse ist letzterer als Pflanzenstandort günstiger zu bewerten.

Der rote und der braune Mediterranboden aus Carbonatgesteinen entwickeln normalhin ein A—B—C-Profil; dem geht selbstverständlich ein A—C-Stadium vorauf, das teils als rote Rendzina bezeichnet wird, wie oben schon gesagt wurde. Die Bodenmasse des roten und braunen Mediterranbodens ist entkalkt; sie kann aber noch Kalksplitter und Kalksteinstücke enthalten, die meistens durch Umlagerung hineingelangt sind. Je tiefer die Profilentwicklung reicht, desto stärker sind Entkalkung und Entbasung fortgeschritten (Verlust an basischen Kationen). Damit wird auch das Makrogefüge ungünstiger; meistens sind es schwere, plastische Böden. KUBIËNA (1953) hat das Mikrogefüge dieser Böden mit Hilfe von Dünnschliffen mikroskopisch genauer untersucht und stellte ein sogenanntes „Lehmgefüge" fest; deshalb sprach er auch von Kalksteinrotlehm (=Terra rossa) und Kalksteinbraunlehm (=Terra fusca). Das „Lehmgefüge" wird durch eine leicht peptisierbare, bewegliche Feinsubstanz des Bodens bedingt, die mit dem Sickerwasser aus dem Ober- in den Unterboden verlagert (durchschlämmt) werden kann und die ein Fließgefüge (Teilcheneinregelung) erkennen läßt. So erfolgt die Bildung des B-Horizontes wie in der Parabraunerde (s. dort). KUBIËNA (1953) nennt die stärker durchschlämmten Terrae calcis „gebleichte Terra rossa" und „gebleichte Terra fusca".

Die lange, warme Sommerperiode führt vor allem in den stark erwärmbaren Lagen auf Kalkgesteinen zu einer Alterung und Stabilisierung der Fe- und Al-Verbindungen, wobei in der Regel Kieselsäure abgeführt wird. Dadurch wird das Gefüge „erdig" im Sinne KUBIËNAS (1953), d. h. die Feinsubstanz wird nicht mehr peptisiert, sie bleibt geflockt, der Boden behält seine Krümelung. KUBIËNA (1953) spricht im Anfangsstadium dieser Bodenbildung von „vererdet" und im fortgeschrittenen von „allitisch" (letzteres nur bei Terra rossa). Eine nachweisbare Allitisierung stellten auch NEVROS und ZVORYKIN (1939) auf Kreta durch Bauschanalysen fest. Bezüglich einer weiteren Unterteilung der Terrae durch KUBIËNA (1953) wird auf sein Werk verwiesen.

Aus diesen Ausführungen ist zu entnehmen, daß der rote und der braune Mediterranboden eine Tondurchschlämmung und in besonders warmen Lagen auch eine „Vererdung“ und Allitisierung erfahren können. In jedem Falle werden diese Mediterranböden durch starke chemische Verwitterung und Verarmung an Pflanzennährstoffen charakterisiert. Den Anfang ihrer Entwicklung müssen wir teils bis ins Präpleistozän zurückverlegen (KLINGE 1957, 1958). Sie haben eine große Verbreitung im Mittelmeerraum entsprechend dem großflächigen Auftreten der Carbonatgesteine.

Roter und brauner Mediterranboden aus kalkfreien Gesteinen (Silikat- und Kieselgesteinen). Auf kalkfreien Gesteinen vollzieht sich im Prinzip die gleiche Bodenbildung wie auf den Carbonatgesteinen, und doch ist der Unterschied infolge des verschiedenen Ausgangsmaterials und der damit gegebenen Eigenschaften sehr groß. Bei diesen Böden fehlt die Carbonatreserve des Untergrundes; ferner ist ihr Basengehalt im allgemeinen geringer. Auch zeigen die Silikat- und Kieselgesteine nie das Karstphänomen. Es ist deshalb begründet, die Mediterranböden nach dem Gestein in zwei Gruppen zu teilen.

Wieweit der rote Mediterranboden (aus Silikat- und Kieselgesteinen) mit dem Rotlehm KUBIËNAS (1953) und der braune Mediterranboden mit dem Braunlehm des gleichen Autors übereinstimmen, kann heute nicht klar entschieden werden. KUBIËNA (1953) hat die genannten Bodentypen klar als plastische Böden mit „Lehmgefüge“, entstanden aus Silikat- und Kieselgesteinen, definiert, aber es fehlen ausreichende regionale Untersuchungen, welche die Entscheidung ermöglichen, ob die von KUBIËNA herausgestellten Bodentypen mit den genannten Mediterranböden identisch sind. Wir können aber mit ziemlicher Sicherheit annehmen, daß ein großer Teil der roten und der braunen Mediterranböden fossil ist und daß diese fossilen Böden mit den genannten Bodentypen KUBIËNAS übereinstimmen. Diese Mediterranböden sind chemisch stark verwittert und entbast; ihr Makrogefüge ist im Hinblick auf die Bearbeitbarkeit ungünstig.

Normalerweise besitzen der rote und der braune Mediterranboden aus Silikat- und Kieselgesteinen ein A—B—C-Profil. Die Entwicklung dieses Profils scheint vom Ranker auszugehen und noch ein Zwischenstadium mit einem A—(B)—C-Profil zu durchlaufen. Dieses Zwischenstadium wird heute mit dem braunen Mediterranboden zusammengefaßt. Die von KUBIËNA (1953) ausgeschiedene meridionale Braunerde entspricht diesem Profil. Dieser Boden wurde früher auch mediterrane Braunerde oder südliche Braunerde und von DEL VILLAR (1930) brauner Trockenwaldboden genannt. KUBIËNA (1953) fand in diesem Boden mikromorphologische Merkmale der Braunerde und des Braunlehms, was für seine Verwandtschaft mit dem braunen Mediterranboden spricht.

Zwischen dem roten und dem braunen Mediterranboden gibt es Böden mit vielen Farbübergängen. Der rötlichbraune (manchmal auch andersfarbige) Mediterranboden scheint mit dem zimtfarbigen Boden im Sinne der sowjetischen Bodenkundler übereinzustimmen, der südlich des Kaukasus und in Bulgarien gefunden wurde. Auch in dem sommertrockenen Gebiet der westlichen USA wurde er gefunden. Demnach scheinen die Mediterranböden nicht auf das Mittelmeergebiet beschränkt zu sein.

Weitere Böden des mediterranen Raumes. In den *Hochgebirgen* des Mittelmeerraumes treten die gleichen Böden auf wie in den Hochgebirgen Mitteleuropas, nur in entsprechend höherer Lage. Die Hochgebirgsböden werden später zusammenhängend besprochen, um Wiederholungen zu vermeiden.

Im mediterranen Raum treten *Steppenböden*, *Wüstenböden* und *Salzböden* auf. Diese werden ebenfalls in den folgenden Kapiteln besprochen.

Die starke Erosion durch das Wasser hat im ganzen mediterranen Raum gewaltige Mengen von Verwitterungsmaterial in die *Täler* gebracht (Vageler 1955). Dabei handelt es sich teils um steinigen Verwitterungsschutt, teils aber auch um Bodenmaterial; oft ist beides gemischt. Zum großen Teil füllen diese Abtragsmassen schon lange die Täler aus; teilweise bilden sie pleistozäne Terrassen. Die Bodenentwicklung ging hier weiter, und zwar je nach Grundwasser- und Überflutungsverhältnissen in Richtung terrestrischer oder semiterrestrischer Böden. Die breiten Talauen sind für die landwirtschaftliche Nutzung besonders wertvoll.

In den Niederungen des Mittelmeerraumes gibt es an einigen Stellen *schwarze Böden*, die Ähnlichkeit mit dem Tschernosem oder noch mehr mit dem tschernosemartigen Auenboden Mitteleuropas haben. Einen Boden dieser Art gibt es in Italien (Mancini 1960) und auch in Spanien; er ist dort als Andalusische Schwarzerde bekannt (del Villar 1930, Albareda, Burriel und Taboadela 1950). Er ist sehr fruchtbar und daher bei den allgemein ungünstigen Bodenverhältnissen des Mittelmeerraumes sehr bedeutungsvoll.

Die Bodentypen der semihumiden und semiariden Steppe

Das Gemeinsame der Bodentypen der semihumiden und semiariden Steppe (Abb. 20) wird in folgenden Punkten zusammengefaßt:

1. Sie liegen meistens in einem kontinentalen Klima mit relativ geringer Niederschlagsmenge (meist zwischen 250 und 500 mm), geringer Luftfeuchtigkeit, starker Verdunstung, warmen Sommern und kalten Wintern. Perkolation, Pflanzenwachstum und Bodenorganismen sind durch diese klimatischen Bedingungen stark beeinflußt.
2. Infolge geringer Niederschläge und starker Verdunstung ist die Auswaschung an löslichen Bodenstoffen (vor allem an Basen) aus diesen Böden, mit Ausnahme des degradierten Tschernosems, gering. Überwiegend sind diese Bodentypen kalkhaltig, oder sie besitzen zumindest eine hohe Basensättigung.
3. Mit Ausnahme der Salzböden und anderer Niederungsböden zeigen diese Bodentypen ein A—CaC—C-Profil; die Mächtigkeit des A-Horizontes ist bei den einzelnen Typen verschieden.
4. Die Humusform ist bei allen diesen Bodentypen, mit Ausnahme der Salzböden, gut; sie ist stabil und stickstoffreich.
5. Das Makrogefüge zeichnet sich allgemein, mit Ausnahme der Salzböden, durch gute Krümelung aus; das Mikrogefüge ist überwiegend „erdig" im Sinne Kubiënas (1953).
6. Die pflanzliche Produktion wird nicht durch die Bodeneigenschaften (ausgenommen Salzböden), sondern durch das Klima mehr oder weniger stark eingeschränkt.
7. In großen Gebieten mit diesen Bodentypen herrscht die Bodenerosion durch Wasser, wodurch die Böden schluchtenartig zerschnitten werden.
8. Im allgemeinen trocknet der Wind diese Bodentypen stark aus; dem wird durch die Anlagen von Windschutzstreifen entgegengewirkt.

Folgende Werke zur Vertiefung in das Wesen dieser Bodentypengruppe werden empfohlen: Glinka 1914, Stebutt 1930, Kubiëna 1953, Kellogg und Smith 1960, Gerassimow und Glasowskaja 1960.

Brunizem oder Prärieboden. Der Brunizem, früher Prärieboden genannt, fügt sich in den USA räumlich zwischen den gray-brown podzolic soil und den

Tschernosem ein. Die Niederschlagsmenge liegt bei etwa 700 mm und steigt damit über die für den Tschernosem übliche hinaus. Im Norden zeigt der Brunizem eine dunkelbraune und im Süden eine rötlichbraune Farbe (Temperaturunterschied). Mit seinem A—B—C-Profil weicht der Brunizem deutlich vom Tschernosem ab. Der A-Horizont, selten mächtiger als 50 cm, ist vollkommen entkalkt, die Basensättigung liegt vorwiegend im mittleren bis hohen Bereich, das Makrogefüge zeigt hauptsächlich Krümelung, Wasser- und Lufthaushalt sind meistens sehr gut. In ebenen und muldigen Lagen liegt fast stets Staunässe vor. Der B-Horizont ist in der Regel braun gefärbt, bei Staunässeeinfluß ist der Bg-Horizont rostbraun und grau gefleckt.

In der Sowjetunion schiebt sich zwischen den *grauen Waldboden* (Bodenzone zwischen Derno-Podsol und Tschernosem) und den ausgelaugten Tschernosem der podsolierte Tschernosem mit B-Horizont ein. Ob dieser und der Brunizem der USA identisch sind, kann vermutet werden, jedoch liegen wahrscheinlich geringe Unterschiede vor. Jedenfalls kann der Brunizem als ein tschernosemartiger Boden mit B-Horizont angesehen werden.

Wegen seiner Fruchtbarkeit (höherer Niederschlag als im Tschernosemgebiet) wird er besonders hervorgehoben. In den USA bildet er zum großen Teil den Maisgürtel (FRESE, GADE und MÜCKENHAUSEN 1952).

Im Randgebiet des Brunizems zum gray-brown podzolic soil tritt auf kleinen, staunassen Flächen der *Planosol* auf, der sich durch einen mächtigen, hellgrauen Tonverarmungshorizont (A_3g) und einen dichten Illuvialhorizont (Bg) auszeichnet.

Tschernosem (Schwarzerde) (Tafel I). Der Tschernosem wurde bereits bei den Böden Mitteleuropas besprochen, so daß hier eine kürzere Fassung möglich ist. Der Tschernosem besitzt ein A—C-Profil; vielfach schaltet sich ein CaC-Horizont zwischen A und C ein. Der 40 bis 100 cm mächtige A-Horizont zeichnet sich durch folgende Eigenschaften aus: grauschwarze Farbe, ideale Krümelung, guter Wasser- und Lufthaushalt, hohe Basensättigung, 4 bis 16% Humus, beste Humusform, große Stickstoffreserve, starke biologische Aktivität.

In Rußland wurde diesem Boden schon früh Aufmerksamkeit gewidmet (DOKUTSCHAJEW 1879, 1883), weil er eine fast unerschöpfliche natürliche Fruchtbarkeit und daher eine große wirtschaftliche Bedeutung besitzt (FREHE 1931). In den vergangenen Jahrzehnten wurde vor allem in der Sowjetunion die Schwarzerde intensiv durchforscht, so daß heute dort die Subtypen dieses Bodentyps gut bekannt sind (MISUSTIN 1928, TANFILIEW 1928, KONAKOW 1930, PANKOW 1930, BOLISCHEW 1947, GORSCHENIN 1947, TATARINKOW 1949, DURASOW 1955, SSELJAKOW 1955, WOLOBUJEW 1958, POPOWA 1959). Nach diesen umfangreichen Arbeiten werden in der Sowjetunion folgende wichtige Subtypen des Tschernosems vom feuchteren zum trockneren Klimabereich unterschieden:

Der *graue Waldboden* mit A—B—C-Profil begrenzt die Tschernosemzone im Norden, südlich folgen: *Podsolierter Tschernosem* mit ziemlich starker Auswaschung der Basen und Tondurchschlämmung (etwa 600 mm Niederschlag), so daß ein A—B—C-Profil vorliegt. Er hat viel Ähnlichkeit mit dem Brunizem der USA. *Ausgelaugter Tschernosem* mit Basenverarmung und Entwicklung eines schwachen (B)-Horizontes zwischen A un C. *Typischer* (oder mächtiger) *Tschernosem* mit einem A-Horizont (pH etwa 6,5) von 80 bis 100 cm und 10 bis 16% Humus; nur der untere Teil des A-Horizontes ist kalkhaltig (fadenartige Kalkausblühungen = Pseudomyzelien). *Gewöhnlicher Tschernosem* mit einem 60 bis 80 cm mächtigen A-Horizont und 6 bis 10% Humus; ab etwa 50 cm kalkhaltig. *Südlicher Tschernosem* mit einem 40 bis 60 cm mächtigen A-Horizont und 4 bis 6% Humus;

ab etwa 20 bis 40 cm kalkhaltig. Dieser Subtyp geht allmählich in den dunklen kastanienfarbigen Boden über.

Der Tschernosem besitzt teilweise eine braune oder dunkelgraue Farbe in Gebieten mit milderem Winter; das zeigt an, daß der Abbau der organischen Substanz stärker und der Humusgehalt geringer (1,5 bis 4%) ist. Dazu gehört der *Krim-Tschernosem* (Abrukowa 1960) und der *Woiwodina-Tschernosem* (Neugebauer 1951) und schließlich auch der *braune Steppenboden der Rheinpfalz* (Zakosek 1956).

Der Tschernosem zieht sich weit nach Sibirien hinein bis fast nach Alma Ata, in westlicher Richtung reicht er bis nach Polen (Dobrzanski und Zbysaw 1955), Rumänien, Ungarn, Jugoslawien und Österreich (Neugebauer 1951, Franz 1955).

Der Tschernosem der USA schließt sich westlich an den Brunizem an und besitzt häufig ein A—B—C-Profil; der A-Horizont ist stets entkalkt und etwa 40 bis 70 cm mächtig (Larson u. a. 1948, McClelland u. a. 1959). Im Norden zeigt die Schwarzerde der USA schwärzliche und dunkelgraubraune Farbtöne, dagegen im Süden infolge höherer Temperatur dunkelbraune bis rötlichbraune Farben. Als starke Degradationsstufe des Tschernosems gilt im Norden der USA und in Kanada der gray wooded soil.

Im Klimagebiet der Schwarzerde entstehen aus Sanden und ähnlichen kalkarmen oder kalkfreien Sedimenten schwarzerdeähnliche Böden, die Kubiëna (1953) und Franz (1955) *Para-Tschernosem* genannt haben. Sie treten in allen Tschernosemgebieten auf.

Kastanosem (Kastanienfarbiger Boden) (Tafel I). Der kastanienfarbige Boden, von Kubiëna (1953) Kastanosem genannt, fügt sich in der Sowjetunion der Region des südlichen Tschernosems an. Das für den Kastanosem charakteristische Klima ist also arider als das des Tschernosems. Auch in den USA liegt die Zone des kastanienfarbigen Bodens zwischen den Tschernosemen und den Böden der Halbwüste. — Der Name „kastanienfarbiger Boden“ wurde von seiner Farbe abgeleitet; diese ist intensiv dunkelbraun oder rötlichbraun wie die Farbe der Kastanie (Tafel I).

Folgende Eigenschaften charakterisieren den Kastanosem: A–C- oder A–CaC–C-Profil; A-Horizont rötlichbraun (kastanienbraun), 30 bis 60 cm mächtig, Humusgehalt 3 bis 5%, gute, stickstoffreiche Humusform, gute Krümelung, guter Luft- und Wasserhaushalt, oft schwach kalkhaltig bis zur Oberfläche, in jedem Falle hohe Basensättigung, Bodenleben etwas durch Trockenheit beeinträchtigt, Pflanzendecke infolge Trockenheit nicht so stark entwickelt wie bei dem Tschernosem (Trockensteppe). Der A-Horizont ist meistens unterteilt in einen etwas helleren A_{11} und einen etwas dunkleren A_{12}. Auch ein B-Horizont ist möglich.

Der Kastanosem ist fruchtbar, soweit die Salzkonzentration im A-Horizont nicht zu hoch ist. Es bestehen Übergänge zum Solonetz (*solonetzartiger Kastanosem*), deren Standortseigenschaften infolge Salzanreicherung und schlechterer physikalischer Eigenschaften (Säulenstruktur, Verschlämmungsneigung, Härte im ausgetrockneten Zustand) ungünstiger sind.

In der Sowjetunion werden ein dunkler und ein heller kastanienfarbiger Boden unterschieden (Merschnin 1956). Ersterer schließt sich unmittelbar dem südlichen Tschernosem an und geht allmählich in den helleren Subtyp über, der zum arideren Raum hin in die Halbwüstenböden überleitet. Der dunkle Kastanosem kann trotz geringen Niederschlages unter Einschaltung der Brache noch weitgehend dem Getreideanbau dienen, dagegen ist das auf dem hellen Kastanosem nur eingeschränkt möglich. Hier ist die Wasserzufuhr am Platze. — In den USA schließt sich der kastanienfarbige Boden nach Westen (zum arideren

Bereich hin) an den Tschernosem an. Im Norden ist dieser Kastanosem dunkelbraun und im Süden infolge höherer Sommertemperatur rötlichbraun gefärbt.

Kastanienfarbige Böden treten ferner in Argentinien, Kleinasien und in Spanien auf. Diese verschiedenen Vorkommen sind nicht in allen Eigenschaften gleich. Im Osten Mittelafrikas, im Süden Afrikas und in Australien gibt es dem Kastanosem ähnliche Böden, die aber mehr Verwandtschaft mit dem braunen Halbwüstenboden zu haben scheinen.

Im Gebiet des Tschernosems und des kastanienfarbigen Bodens treten in Tälern und Senken grundwasserbeeinflußte Übergangsbildungen auf, die von den sowjetischen Bodenkundlern Wiesen-Tschernosem bzw. Wiesen-Kastanosem genannt werden (Rode 1959).

Im Bereich der kastanienfarbigen Böden, vor allem in den arideren Tälern, finden sich ferner große Flächen von Salzböden, die in dem Kapitel „Die Salzböden" beschrieben werden.

Die Bodentypen der Halbwüste und der Wüste

Die Bodentypen der Halbwüste und der Wüste (Abb. 20) sind nicht so eingehend erforscht wie die Böden mit höherem wirtschaftlichem Wert. Leider haben die Forschungen in sehr verschiedenen Räumen, hauptsächlich in Nord- und Südafrika, im Westen der USA, in Zentralasien und Australien, zu einer verschiedenen Benennung dieser Bodentypen geführt. Im Augenblick ist es nur möglich, die gebräuchlichsten Namen zu verwenden. Zum eingehenderen Studium dieser Bodentypen werden folgende Schriften empfohlen: Linck 1930, Mortensen 1930, Schokalskaja 1953, Durand 1954a, 1954b, 1959, Stephens 1954, Jewitt 1955, Stephens 1956, Boulaine 1957 (mit umfangreichem Schrifttum), Yaalon 1959, Buringh 1960, Ganssen 1960, d'Hoore 1960, Finck 1963.

Die Bodentypen der Halbwüste und der Wüste haben folgende Charakterzüge gemeinsam:

1. Sie bilden sich im niederschlagsarmen (weniger als etwa 250 mm jährlich), lufttrockenen, warmen, verdunstungsstarken Klima.

2. Die physikalische Verwitterung überwiegt die chemische infolge der Trockenheit und des starken Temperaturwechsels bei weitem.

3. Infolge des geringen Niederschlages und der starken Verdunstung werden die Salze des Bodens gebietsweise schwach ausgewaschen, im übrigen werden sie in den oberen Bodenschichten angereichert; im Extrem bilden sich Krusten (caliche).

4. Zu der Bodengesellschaft der Halbwüste und der Wüste gehören ebenfalls Salzböden.

5. Der A-Horizont ist geringmächtig, der Humusgehalt niedrig und der Besatz an Bodenorganismen gering.

6. Der Pflanzenwuchs ist schwach oder fehlt ganz. Nur Xerophyten können existieren, auf den Salzböden nur Halophyten.

7. Weitgehende oder völlige Vegetationsfreiheit führt bei plötzlichen, starken Niederschlägen zu verheerender Bodenerosion (Schluchtenbildung).

Burosem. Für diesen Bodentyp werden bzw. wurden auch folgende Namen verwendet: Brauner Boden, brauner Boden der Halbwüste, hellbrauner Steppenboden, brown desert soil.

Der Burosem, von Kubiëna (1953) so genannt, folgt zonal zum Arideren hin dem hellen kastanienfarbigen Boden. Meistens überzieht ihn eine Pflanzendecke von *Artemisia*-Arten, die typisch für die Trockensteppe ist. Der A-Horizont teilt sich in einen helleren A_{11} und einen etwas dunkleren A_{12}, darunter folgt ein ACa oder CaC, der in den C übergeht. Gebietsweise hat sich auch ein B-Horizont entwickelt, z. B. flächenweise im Westen der USA. Sonstige Eigenschaften: braune oder graubraune Farbe, meist kalkhaltig bis zur Oberfläche, teils aber auch kalkfrei, aber dann hohe Basensättigung, krümeliges, lockeres Gefüge, der ACa und CaC teils etwas verfestigt, 2 bis 3% stickstoffreicher Humus, geringes Bodenleben. Der an sich einfache Profilaufbau zeigt viele Variationen. Das aride Klima bedingt Übergänge zum Solontschak und zum Solonetz.

Der Burosem tritt im Übergangsgebiet von der Steppe zur Halbwüste, in der sogenannten Trockensteppe, auf, vor allem auf großen Flächen in Südostrußland und im Westen der USA.

Sierosem (Tafel I). Sierosem oder Sierosiom ist die russische Bezeichnung für einen grauen Boden der Halbwüste, auch Serosem, Grauerde, grauer Wüstensteppenboden, grauer Boden der Halbwüste und gray desert soil genannt.

Der Sierosem tritt in noch arideren Gebieten auf als der Burosem. Dementsprechend spärlich ist die xerophytische Pflanzendecke. Die wichtigsten Eigenschaften sind: Horizontfolge A-CaC-C, geringmächtiger, schwach humoser (0,5 bis 2%), grauer, meist kalkhaltiger A-Horizont, geringes Bodenleben. Gleitende Übergänge bestehen vom Sierosem zum Solontschak sowie zum Solonetz.

Der Sierosem kommt am Rande der Trockenwüsten vor, besonders große Flächen befinden sich in Turkestan, Südostrußland und im Westen der USA. Auch in Spanien und in Kleinasien treten größere Flächen auf.

Rötlichbrauner Halbwüstenboden. Der rötlichbraune Typ der Halbwüstenböden hat vieles gemeinsam mit dem Burosem. Er kann kalkhaltig und kalkfrei sein. Das Profil zeigt die Horizontfolge A—C oder A—B—C. Eine über 50 cm tiefe Profilentwicklung, die im Westen der USA beobachtet wird, weist auf eine lange Bodenbildungszeit hin. Für die Entstehung der rötlichbraunen Halbwüstenböden scheinen hohe Sommertemperaturen nötig zu sein, die auch in früherer Zeit wirksam gewesen sein können. Nur eine spärliche xerophytische Pflanzendecke bedeckt diesen Bodentyp, der im Westen der USA beachtliche Flächen einnimmt, aber auch in anderen Randgebieten der Wüste (Halbwüste) auftritt. Auch dieser Halbwüstenboden bildet Übergänge zum Solontschak und zum Solonetz.

Als eine Eigenart dieses Bodentyps kann der durch Kieselsäure verfestigte, harte Horizont (hardpan) angesehen werden, der flächenweise auftritt, durch Kieselsäureabwanderung vom Ober- in den Unterboden entstand und dem Wachstum der Pflanzenwurzeln sehr hinderlich ist.

Lockere Rohböden der Wüste. Die lockeren Rohböden der Wüste werden als *lockere* Bodenbildungen den „rohen" Wüstenkrusten gegenübergestellt, um die Rohböden der Trockenwüste nach einem wichtigen Merkmal in zwei Bodentypen zu gliedern. Die lockeren Rohböden der Wüste können unweit unter der Oberfläche ebenfalls eine leichte Verkittung aufweisen, aber diese ist im Vergleich zu der Verhärtung der Wüstenkrusten gering. Zudem gibt es Übergänge zwischen den lockeren und „festen" Rohböden (Krusten) der Wüste. In erster Linie gliedert man diese lockeren Rohböden nach den Korngrößen und einem etwaigen Salzgehalt, wie es KUBIËNA (1953) getan hat.

Steinpflaster-Wüstenboden. KUBIËNA (1953) nannte den Subtyp „Steinpflaster-Wüstenboden" treffend „Hamada-Wüstenboden" oder „Hamada-Yerma". Die vorwiegend physikalische Verwitterung zerlegt in der Wüste das Gestein in Fragmente vom Block bis zum Staub. Der Wind, der sich auf den ganz oder fast vegetationsfreien Flächen entfalten kann, bläst von der Oberfläche fortwährend das feine Bodenmaterial aus, und es bleibt ein Steinpflaster zurück. Darunter folgt in der Regel Verwitterungsmaterial, das auch feine Bestandteile enthält. Die Steine des Pflasters zeigen oft Überzüge, z. B. den bekannten Wüstenlack aus Limonit, ferner auch andere sogenannte Schutzrinden (LINCK 1930).

Sandwüstenboden. Der Sandwüstenboden stellt als Subtyp des lockeren Rohbodens den Inbegriff der Wüste dar. Die volkstümliche arabische Bezeichnung *Erg* ging auch in die wissenschaftliche Literatur ein; KUBIËNA (1953) nannte ihn Sand-Yerma oder auch Sandwüstenboden. Er stellt einen fast humusfreien, feinerdearmen, lockeren, im Einzelkorngefüge befindlichen Rohboden der Wüste dar mit überwiegend Korngrößen des Sandes und des Feinkieses. Bekannt sind die weißlichen, grauen und ockergelben Farben, daneben treten rotgelbe bis intensiv rote auf. In den Rissen der Sandkörner befindet sich oft verhärtete, feine Substanz, die meist rot oder gelblich gefärbt ist und dem Boden seine Farbe verleiht (DURAND 1959).

Salzstaubboden. Der Subtyp Salzstaubboden, von KUBIËNA (1953) auch Salzstaub-Yerma genannt, bildet eine lockere, staubige, sehr humusarme Masse, die mit bloßem Auge erkennbare Ausscheidungen von wasserlöslichen Salzen enthält. Extrem trockene, sommerheiße Gebiete der Wüste bilden das Verbreitungsgebiet.

Wüstenkrusten (Tafel V). Der Bodentyp Wüstenkruste unterscheidet sich von den lockeren Rohböden der Wüste durch eine steinähnliche Kruste (englisch caliche) wechselnder Mächtigkeit (von 5 bis 200 cm) nahe der Oberfläche. Die Kruste entsteht durch die Ausscheidung von Salzen im oberen Bodenbereich; die Salze gelangen mit aufsteigender Bodenlösung in diesen Bereich. Gelegentliche Niederschläge, die in den Boden dringen, werden durch die starke Verdunstungskraft des heißen Klimas wieder gleichsam nach oben gezogen. Alkalisalze sind selten an der Krustenbildung beteiligt, da diese leicht löslich sind und mit der Zeit doch von den gelegentlichen Niederschlägen abgeführt werden. Doch gibt es in extremen Trockengebieten auch *Alkalisalzkrusten.* Bei den in der Wüste herrschenden Sandverwehungen werden die Krusten häufig mehr oder weniger mit Sand, Staub oder Salzstaub überdeckt; andererseits kann der Wind den oberen lockeren Profilteil abtragen und so die Kruste freilegen. Infolge dieser Winderosion bildet heute die Kruste weitverbreitet die Oberfläche (SCHOLZ 1963). Für die Bildung mächtiger Krusten sind lange Zeiträume nötig (BRINKMANN, mündliche Mitteilung 1960, STUART u. a. 1961).

Gipskrustenboden. Beim Gipskrustenboden oder Gipskrusten-Yerma (KUBIËNA 1953) besteht die Kruste überwiegend aus Gips; es kann auch Calciumcarbonat beteiligt sein. Die Gipskruste zeigt folgende Merkmale: meistens fahlweiße Farbe, matte und gerauhte Oberfläche, manchmal einzelne, durch Rekristallisation gebildete, größere Kristalle, manchmal sinterartige, knollige Bildungen; sie läßt sich leicht ritzen.

Kalkkrustenboden (Tafel V). Die Kalkkruste, auch Kalkkrusten-Yerma (KUBIËNA 1953) oder Nari (volkstümliche Bezeichnung in Palästina) genannt, besteht überwiegend aus Calciumcarbonat. Gips und Kieselsäure können beteiligt sein (STUART u. a. 1961). Reine *Kieselsäurekrusten* sind selten. Das Calcium wird mit der aufsteigenden Bodenlösung als Calciumbicarbonat transportiert und nahe der Oberfläche als Calciumcarbonat unter Abgabe von Kohlensäure ausgeschieden (SCHOLZ 1963). Eigenschaften der Kruste: meist cremefarbig, teils rötlich, feinkristallin, dicht, sehr hart, oft schichtig.

Die Salzböden

Die Salzböden sind an folgende Entstehungsbedingungen gebunden: geringer Niederschlag, starke Verdunstung, dichte Untergrundschicht, Vorhandensein von Salzen im Boden oder deren Zuführung durch das Grund- oder Oberflächenwasser. Die Bodenregionen der Steppe, Halbwüste und Wüste schließen die meisten Vorkommen ein. Auch im Mediterranklima finden sich relativ kleine Flächen dieser Bodentypen (Spanien, Griechenland, Jugoslawien). Salzböden der Küstenregion, deren Salze aus dem Meerwasser stammen und deshalb eine andere Zusammensetzung aufweisen als die Salzböden der Trockengebiete, sind begrenzt und können hier außer acht bleiben. Die Salzböden zeigen folgende gemeinsame Merkmale:

1. Sie entstehen in niederschlagsarmen, verdunstungsstarken Klimaräumen.
2. Eine mehr oder weniger dichte Untergrundschicht hemmt oder unterbindet die Versickerung.
3. Aktuelle oder frühere, größere oder kleinere Salzanreicherung in den oberen Bodenschichten.
4. Ungünstiger bodenphysikalischer Zustand (dicht und kohärent oder große, kompakte Gefügeaggregate).
5. Das Bodenleben kann sich nicht entfalten.
6. Das Pflanzenwachstum ist eingeschränkt, bei starker Salzkonzentration können nur Halophyten existieren; starke Salzausscheidungen an der Oberfläche verhindern jegliches Pflanzenwachstum.

Wie bei vielen anderen Bodentypen, so ist auch bei den Salzböden die Namengebung nicht einheitlich. Hier sollen die ursprünglichen russischen Bezeichnungen gebraucht werden, die auch KUBIËNA (1953) vorgeschlagen hat und die am bekanntesten sind. Soweit wie möglich werden die Synonyme angegeben. Für ein eingehendes Studium der Salzböden werden die Arbeiten folgender Autoren empfohlen: HILGARD 1910, GLINKA 1915, HARRIS 1920, GANGULEE 1924, v. SIGMOND 1930, KUBIËNA 1953, KOWDA 1964, SZABOLCS 1966.

In den zwei letzten Jahrzehnten, besonders in den beiden bodenkundlichen Zeitschriften „Potschwowedenie" und „Soil Science", sind eine Anzahl von Spezialarbeiten über Entstehung, Eigenschaften, Verbesserung und Nutzung der Salzböden erschienen.

Solontschak (Tafel VI). Die Bezeichnung Solontschak ist der russischen Fachsprache entlehnt. Synonyme sind: Szikboden oder Székboden (ungarisch), Zickboden (burgenländisch); diese Namen werden örtlich auch für Solonetz gebraucht; ferner white alkali soil (HILGARD 1910); heute unterscheidet man in den USA saline soil (mäßig alkalisch) und alkaline soil (stark alkalisch); Kebir (kirgisisch); strukturloser Salzboden (weil Säulenstruktur fehlt); Reh, Kalar und Usar (indisch, gemäß GANGULEE 1924); Trona (ägyptisch).

Eigenschaften des Solontschaks: hellgraue, seltener dunkelgraue Farben wie beim Gley, überwiegend Eisen(II), nur wenig Eisen(III) als rostgelbe und rostbraune Flecken (Tafel VI), A-Horizont mehr oder weniger humushaltig, Salzausscheidung an der Oberfläche oder in einem tieferen Horizont (überwiegend Sulfat, Chlorid und Carbonat des Natriums, Calciumsulfat auch fast immer vorhanden, teils auch Magnesiumsulfat und Calciumchlorid), bei Austrocknung in allen Horizonten Salze, im Frühjahr meist naß, im Sommer austrocknend, durch den hohen Na-Gehalt kohärente Masse, im Frühjahr Begrünung, im Sommer verdorren die Pflanzen.

Nach der Salzart können unterschieden werden (KUBIËNA 1953): *Sodasolontschak* mit großer Menge von Natriumcarbonat, *Gipssolontschak* mit vorwiegend Calciumsulfat, *Calcium-Natriumsolontschak* mit reichlich Calciumsalzen neben Natriumsalzen, *Kalksolontschak* mit viel Calciumcarbonat; ferner scheidet KUBIËNA (1953) noch aus: *Kryptosolontschak*, bei dem die Salzausscheidung unter der Oberfläche (verdeckt) erfolgt; *Trockensolontschak*, der wenig durchfeuchtet wird und im ganzen Profil Salzausscheidungen zeigt; *nasser Solontschak*, der selten völlig austrocknet und daher nur an der Oberfläche Salze auskristallisiert.

Die salzreichen Solontschake finden sich meistens in Gesellschaft der Halbwüsten- und Wüstenböden. In der Steppe sind sie ebenfalls zu finden; es sind aber hier meistens jüngere Bildungen (KUBIËNA 1953). Vom Solontschak gibt es viele Übergänge zum Solonetz und zu den vergesellschafteten terrestrischen Bodentypen.

Die Kultivierung der Solontschake stellt ein altes, schwieriges Problem dar. Die für den Anbau von Kulturpflanzen notwendige Wasserzufuhr bringt oft neues Salz. Die fast stets vorhandene, mehr oder weniger dichte Untergrundschicht erlaubt nicht die Durchspülung und Entsalzung des Bodens. Ein durchlässiger Untergrund und die Zufuhr von salzarmem (süßem) Wasser vermögen sowohl die Versalzung zu verhindern als auch die Entsalzung zu bewirken und damit eine dauernde Nutzung zu gewährleisten.

Solonetz. Auch der Name Solonetz kommt aus dem Russischen. Andere Namen sind: Ton-Szikboden (v. SIGMOND 1930), black alkali soil (HILGARD 1910), Schokat (kirgisisch), Struktur-Salzboden (wegen der Säulenstruktur).

Der Solonetz geht in der Regel aus Solontschaken durch weitgehende oder vollständige Auswaschung der Salze hervor, ohne daß dabei jedoch auch die am Sorptionskomplex gebundenen Na-Ionen verdrängt werden. Das Profil zeigt drei charakteristische Horizonte: der mittelgraue, humose, dichte A-Horizont; der ebenfalls mittelgraue, humushaltige, in Säulenstruktur abgesonderte, dichte AB-Horizont; der hellgraue, schwach rostgefleckte, kohärente, oft noch salzhaltige G-Horizont. A- und AB-Horizont sind schwach alkalisch bis neutral. Das Grundwasser steht tiefer als im Solontschak, meistens wurde es auf natürlichem oder künstlichem Wege abgesenkt. Infolge des hohen Gehaltes an Na-Ionen verschlämmt die Bodenmasse leicht und wird dicht, der Humus wird teils löslich und wandert mit dem Sickerwasser in den AB-Horizont. Im Frühjahr bildet der Boden eine breiige Masse; bei der Austrocknung wird er rissig und hart. Die Bedingungen für Bodenorganismen und höhere Pflanzen sind ungünstig. Deshalb bereitet auch die Kultivierung des Solonetz große Schwierigkeiten.

Zwischen Solonetz und Solontschak gibt es Übergänge, die *Solontschak-Solonetz* genannt werden. Diese besitzen den hohen Salzgehalt des Solontschaks und die Struktur des Solonetzes. Ferner werden *kalkfreier* und *kalkhaltiger Solonetz* unterschieden (KUBIËNA 1953).

Solod. Die Bezeichnung Solod (volkstümlicher russischer Name) wurde ebenfalls von russischen Bodenforschern vorgeschlagen. Weitere Namen für diesen Boden sind Solodi, Soloti, Steppenbleicherde, salzerdiger Podsol.

Der Solod entsteht aus dem Solonetz durch die Durchschlämmung von Bodenfeinsubstanz aus dem Ober- in den Unterboden. Diese Tonverlagerung wird in erster Linie durch die starke Dispergierung der Na-reichen Kolloidsubstanz und das Vorhandensein von Kieselsäuresolen veranlaßt. So entsteht ein mittelgrauer (durch Humus dunkler), schwach humoser Oberboden und ein darunter folgender hellgrauer Bleichhorizont, der dem A_2-Horizont des Podsols sehr ähnlich ist. Diesem Horizont folgt ein stark rostbraun und grau gefleckter BG-Horizont. Im Gegensatz zu der im stark sauren Milieu und in der Regel unter einer Rohhumusdecke ablaufenden Podsolierung erfolgt die Durchschlämmung im Solod durch die Wirkung von Na-Ionen und Kieselsäuresolen. Neuerdings wird von einigen Bodenkundlern angenommen, daß gebietsweise die oberen Horizonte des Solods von einer jüngeren Sedimentation stammen (Theiß-Niederung). — Der Solod findet sich in Gesellschaft des Solonetzes. Zu letzterem bildet der Solod Übergänge.

Die Bodentypen der feuchten Subtropen und der Tropen

Keine Bodenregion läßt sich heute so schlecht überblicken wie die der Subtropen und Tropen. Darum kann das Bild von den Bodentypen dieses Klimaraumes nicht befriedigend dargeboten werden. Zur Zeit sind wissenschaftliche Diskussionen der in den Subtropen und Tropen erfahrenen Bodenkundler im Gange mit dem Bestreben, eine Gliederung, ferner eine einheitliche Abgrenzung und Benennung der Bodentypen dieses Klimagebietes zu finden. Da sich dabei Neues ergeben wird, ist im Augenblick Zurückhaltung geboten. In erster Linie sollen hier die Wesenszüge der wichtigsten subtropischen und tropischen Böden dargestellt werden. Die Namen für diese Bodentypen haben in den letzten Jahrzehnten gewechselt, und es ist nicht klar, welche sich in der Zukunft durchsetzen. Hier sollen die älteren Bezeichnungen gebraucht werden; auf die neueren wird hingewiesen.

Für die subtropischen und tropischen Böden (Abb. 20) der feuchten Region gelten folgende allgemeine Merkmale:

1. Sie werden im niederschlagsreichen, mehr oder weniger luftfeuchten, warmen Klima gebildet. Klimatisch sind zu unterscheiden die stets feuchten Gebiete und die mit periodischem Wechsel von Niederschlag und Trockenheit.

2. Feuchtigkeit und Wärme verursachen eine intensive *chemische* Verwitterung und Verarmung an Basen und Pflanzennährstoffen.

3. Die spezifische Verwitterung in diesem Klima bewirkt die Abfuhr eines großen Teiles der Kieselsäure.

4. Dehydratisierung und Stabilisierung von Al- und Fe-Verbindungen.

5. Intensiv gelbe bis rote Färbung des Bodens durch wasserarme und wasserfreie Fe-Verbindungen.

6. Starker Abbau der organischen Substanz, so daß die gebietsweise hohe Produktion an organischer Masse nicht zu Humusreichtum führt.

7. Die Böden sind in tektonisch ruhigen Gebieten sehr alt.

8. Mächtige Boden-Kolluvien treten örtlich auf und zeugen von der Umlagerung alter Böden; auf den Kolluvien geht die Bodenbildung weiter.

9. Bei Wegnahme der Vegetationsdecke ist im hängigen Gelände die Erosionsgefahr groß.

10. Die klimatischen Bedingungen für das natürliche Pflanzenwachstum sind überwiegend gut (Feuchtigkeit, Wärme), jedoch ist bei der Nährstoffarmut des Bodens das Wachstum weitgehend vom Humusumsatz abhängig. Lange periodische Trockenheit (Savanne) mindert den Wert des Pflanzenstandortes.

Für das nähere Studium der Bodentypen der Subtropen und Tropen werden folgende Arbeiten empfohlen: Harrassowitz 1926, 1930a, 1930b, Vageler 1930, Vageler und Alten 1932, Thorp 1936, Mohr 1948, Kellogg 1950, Kubiëna 1943b, 1953, van der Merve 1950, Schokalskaja 1953 (mit umfangreichem Schrifttum), Mohr und van Baren 1954, d'Hoore 1960, Sys 1960, Ochse u. a. 1961, Finck 1963.

Rotlehm aus Silikatgesteinen (Tafel V). Rotlehm und Roterde sind die auffälligsten und bekanntesten Bodenbildungen der Tropen und Subtropen.

Die ältere Bezeichnung in den USA ist red loam, der heute übliche Name für den gebleichten Subtyp red podzolic soil (gehört zu der great soil group „red-yellow podzolic soils"), in der Sowjetunion wird Krasnosem gebraucht, HARRASSOWITZ (1926) nannte ihn Kresslehm (wird nicht mehr gebraucht), und neuerdings hat D'HOORE (1960) auf der Bodenkarte Afrikas die Bezeichnung Ferrisols gebraucht. KUBIËNA (1953) knüpfte an den alten Vorschlag von WOHLTMANN (1892) an und definiert den Rotlehm als einen plastischen, stark verwitterten, kieselsäurereichen Boden aus Silikatgesteinen, entstanden im subtropischen und tropischen Klima. Damit ist eine klare Abtrennung von der Roterde vorgenommen. Der Rotlehm und die anderen „Lehme" gehören zu der Klasse der *Plastosole*, die sich alle durch eine hohe Plastizität auszeichnen.

Der Rotlehm besitzt ein A—(B)—C- oder A_1—A_3—B—C-Profil. Wichtigste Eigenschaften: gelbrote und ziegelrote bis braunrote Farbe, A-Horizont geringmächtig und humusarm, er hebt sich nur wenig vom Unterboden ab (beim gebleichten Subtyp mehr), stark chemisch verwittert, basen- und nährstoffarm, bewegliche Feinsubstanz („Lehmgefüge" im Sinne KUBIËNAS 1953), Durchschlämmungstendenz im Profil, plastisch, leicht dicht, hart bei Austrocknung, enthält viel Kaolinit, erodiert leicht. Neben dem typischen Rotlehm (ohne Bleichung) gibt es vielfach den *gebleichten Rotlehm* (Tafel V), der dem red podzolic soil der USA entspricht (U.S. Dept. Agr. 1938); letzterer ist meist fossil. Während sich der Rotlehm in feuchteren Gebieten lange hält, kann er in sommerheißen Gebieten in den *erdigen Rotlehm* übergehen, der einen Übergang zur Roterde darstellt (KUBIËNA 1953). Die kleinen Rotlehmvorkommen in der gemäßigt warmen Klimaregion sind fossil (Präpleistozän).

Braunlehm aus Silikatgesteinen. Die Bezeichnung Braunlehm wurde schon von HARRASSOWITZ (1926) gebraucht. Von KUBIËNA (1953) ist Braunlehm als analoge Bildung des Rotlehms herausgestellt worden, nämlich als plastischer, kieselsäurereicher Boden der Subtropen und Tropen mit leuchtendgelber bis rotbrauner Farbe. Das Farbspektrum ist also ziemlich weit. Der Gelblehm von HARRASSOWITZ (1926), die tropische Gelberde von WOHLTMANN (1892) und der Jeltozem der sowjetischen Bodenkundler sind mit dem Braunlehm identisch. Bei D'HOORE (1960) fällt der Braunlehm unter die Ferrisols.

Der Braunlehm, ein Typ der Plastosole, besitzt weitgehend die gleichen Eigenschaften wie der Rotlehm; diese „Lehme" unterscheiden sich in der durch Menge und Form des Eisens bedingten Farbe. Der Dünnschliff bringt dieses eindrucksvoll zum Ausdruck (KUBIËNA 1953). Für die Entstehung des Braunlehms neben dem Rotlehm scheinen folgende Bedingungen entscheidend zu sein: eisenärmeres Gestein, feuchtere Bildungsumstände und im Gefolge eine weniger starke Dehydratisierung der Eisenverbindungen, kürzere Dauer der Bodenbildung. Bezüglich der übrigen Eigenschaften wird auf den Rotlehm hingewiesen.

Zwischen Rot- und Braunlehm gibt es gleitende Übergänge. Bei stärkerem Einfluß des Bodenwassers infolge Bodendichte sowie ebener oder muldiger Lage kommt ein mehr oder minder rot, gelb und grau geflecktes Profil zustande. Dafür ist bereits der Name *Buntlehm* verwendet worden, indes scheint diese Sonderstellung nicht erforderlich zu sein. Der *gebleichte Braunlehm* ist sehr verbreitet. Auch er gehört in die great soil group „red-yellow podzolic soils" der USA, und zwar ist er der hellere Subtyp. Wenn bei der Bleichung vorzugsweise das Eisen gewandert ist, so wendet KUBIËNA (1953) die Bezeichnung *enteisenter Braunlehm* an, während er dem im Oberboden an Feinsubstanz stark verarmten Braunlehm den Namen *sandiger Braunlehm* zulegt; dieser entsteht vorwiegend aus sandigen Substraten. In sommerheißen Lagen können die Al- und

Fe-Verbindungen eine „erdige" Form im Sinne KUBIËNAS (1953) annehmen. Dann liegt der *vererdete Braunlehm* vor, der den Übergang zur Gelberde (Gelber Latosol) darstellt. Ferner gibt es Übergänge zur Braunerde und zur Parabraunerde sowie zu den Mediterranböden.

In den Subtropen und Tropen findet sich der Braunlehm an Stellen, wo die oben genannten Bildungsbedingungen erfüllt sind. Der große Komplex von Braun- und Rotlehmen (red-yellow podzolic soils) im SO der USA gibt einen Hinweis auf die regionale und örtliche Verteilung der beiden Bodentypen. Nach Norden, d. h. mit abnehmender Temperatur, tritt mehr der yellow-podzolic soil auf, ebenso finden wir diesen in südlichen Bereichen (mit höherer Temperatur) an feuchteren Stellen. Höhere Temperaturen und geringe Feuchtigkeit fördern andererseits die Bildung des Rotlehms (red podzolic soil).

Relikte von Braunlehm werden auch im gemäßigten Klima gefunden; sie gelten als präpleistozän. Fossile Übergänge von Braunlehm zur Parabraunerde sind häufiger; diese können auch dem Pleistozän angehören.

Rotlehm und Braunlehm aus Carbonatgesteinen. Im Mediterrangebiet werden die roten und braunen Mediterranböden in zwei Gruppen nach den beiden Gesteinsgruppen Silikatgesteine und Carbonatgesteine aufgeteilt. Aus den letzteren entwickeln sich Terra rossa und Terra fusca. Diese Zweiteilung nach dem Gestein scheint auch für Rot- und Braunlehm der Subtropen und Tropen zweckmäßig zu sein, wie es HORDY (1933) bereits vorgeschlagen hat. Wenigstens ist das in den Fällen angezeigt, wo das Solum (= Verwitterungsschicht) 2 m nicht übersteigt. Bei großer Verwitterungstiefe (>2 m) und der damit verbundenen starken chemischen Verwitterung wirkt sich das Ausgangsmaterial nicht mehr wesentlich aus, so daß in diesen Fällen der Hinweis auf das Ausgangsmaterial zurücktreten kann.

Die Eigenschaften des Solums des Rot- und des Braunlehms aus Carbonatgesteinen entsprechen im wesentlichen denen der entsprechenden Bodentypen aus Silikatgesteinen. Deshalb kann auf eine nähere Charakterisierung verzichtet werden. Den entscheidenden Unterschied bildet der Carbonatuntergrund.

Wieweit die roten und braunen Mediterranböden aus Carbonatgesteinen mit dem Rot- und Braunlehm (aus Carbonatgesteinen) der Subtropen und Tropen gleiche oder ähnliche Eigenschaften besitzen, ist nicht genügend geklärt. Sicher ist, daß Unterschiede bestehen und deshalb eine Trennung dieser Böden nötig ist.

Eigentümlich ist, daß die red-yellow podzolic soils in den USA auf den Carbonatgesteinen am weitesten nach Norden vorgreifen; dies ist ein Zeichen dafür, daß das durch die Carbonatgesteine erzeugte warme Bodenklima die Bildung dieser Böden fördert.

Graulehm. Der Graulehm zeichnet sich durch eine mittelgraue Farbe aus; rostgelbe und rostbraune Flecken sowie Konkretionen durchsetzen das Profil. Er entsteht in der ebenen oder muldigen Landschaft unter feuchten Bedingungen der Tropen. Wie die anderen „Lehme" zeichnet er sich durch einen hohen Kaolinitgehalt aus, wodurch ein dichtes Gefüge zustande kommt. Im übrigen entsprechen die Eigenschaften des Graulehms denen des Rot- und Braunlehms. Im Tertiär und früher entstand auf der Fastebene der mitteleuropäischen Mittelgebirge eine mächtige Decke von Graulehm, auch Weißlehm oder Weißverwitterung genannt (PFEFFER 1938, SCHMIDT und WOLTERS 1952). Beachtliche Flächen blieben auf tektonischen Senkungsfeldern erhalten (MÜCKENHAUSEN u. a. 1959, KERPEN 1960).

Roterde (Roter Latosol) (Tafel V). Die Roterde wurde von KUBIËNA (1953) als rotbrauner bis roter Boden der Tropen mit „erdigem" Gefüge und geringer oder fehlender Plastizität definiert. KELLOGG (1948) gab den Böden mit diesen Eigenschaften (einschließlich gelbe Farben) den treffenden Namen *Latosols*. Im Hinblick auf den hohen Gehalt an Al- und Fe-Verbindungen nannte D'HOORE (1960) sie auf der Bodenkarte Afrikas sols ferrallitiques. D'HOORE (1960) hat auf dieser neuen Bodenkarte von Afrika in etwas trockeneren Bereichen (z. B. südlich der Sahara) die *sols ferrugineux tropicaux* ausgeschieden, die Übergänge zu Böden trockener Gebiete darstellen, was allein schon der höhere Basensättigungsgrad bezeugt; die Kieselsäureabfuhr ist in ihnen wesentlich geringer als in den sols ferrallitiques.

Die normal ausgebildete Horizontfolge der Roterde ist: A—(B)—C. Oft liegt infolge der langen Bodenbildung ein mächtiges Solum mit einem allmählichen Übergang (Zersatzzone) in das Gestein vor. Teils ist Solummaterial akkumuliert, teils sind die Horizonte durch Termiten vermischt. Weitere Eigenschaften: überwiegend rote und braunrote Farbe, starke chemische Verwitterung, verarmt an Basen und Pflanzennährstoffen, relativ geringer Kieselsäuregehalt (Kieselsäureabfuhr); hoher Gehalt an H_2O-armen Al- und Fe-Verbindungen, die weitgehend irreversibel stabilisiert sind, daher schorfig-krümeliges, stabiles Gefüge; geringer Humusgehalt, wenig oder nicht plastisch, hohlraumreich, durchlässig, teils mit kleinen Lateritknöllchen durchsetzt. Die mit Lateritknöllchen durchsetzte Roterde leitet über zum Laterit. Andererseits gibt es Übergänge zum Rotlehm. Auf folgende Arbeiten über die Latosole wird noch hingewiesen: SHERMAN u. a. 1949, 1954, 1959, RUHE und CADY 1954.

Gelberde (Gelber Latosol). Die Gelberde unterscheidet sich durch Farbe, Eisengehalt und Eisenform von der Roterde; im übrigen besitzt sie die gleichen Eigenschaften wie die Roterde. Sie entsteht aus eisenärmeren Gesteinen und — was wohl das wichtigste ist — unter feuchteren Bildungsbedingungen. Letzteres läßt sich in der tropischen Landschaft aus der Verteilung der Rot- und Gelberde in Abhängigkeit vom Kleinrelief feststellen. Auch kann die Gelberde eine relativ junge Bodenbildung, die Vorstufe der Roterde, sein. Es bestehen gleitende Übergänge zwischen der Rot- und der Gelberde. Die beiden Latosoltypen und deren Übergänge kommen vergesellschaftet vor.

Laterit. Der Laterit hat stets ein großes Interesse der Geologen und Bodenkundler erregt, nicht zuletzt, weil die mächtigen Panzer als Eisenlagerstätte gelten (HARRASSOWITZ 1926, VAGELER 1930, SCHOKALSKAJA 1953, MOHR und VAN BAREN 1954). Es kann hier nicht die Aufgabe sein, den Laterit in seinem vielseitigen Auftreten und die verschiedenen Deutungen seiner Entstehung zu diskutieren. Die angeführte Literatur gibt die Möglichkeit der Orientierung.

Das typische Lateritprofil besteht aus dem Eisenpanzer an oder nahe der Oberfläche. Darunter befindet sich die rot und grau gefärbte Fleckenzone, die in eine weitgehend hellgraue, kaolinitreiche Zone meist großer Mächtigkeit übergeht.

Ein großer Teil des Laterites der Tropen ist genetisch ein sogenannter Grundwasser-Laterit, d. h. er entstand unter maßgeblicher Beteiligung des Grundwassers in muldigen Geländelagen. Das Grundwasser solcher Mulden wird von Wasser ergänzt, das auf und in den Verwitterungsschichten der Umgebung den tieferen Lagen zufließt. Dabei nimmt das Wasser Eisen aus dem Verwitterungsmaterial mit in das Grundwasserbecken. Die starke Verdunstungsenergie führt zu einer schnellen Verdunstung des aufgestiegenen kapillaren Grundwassers. Dabei wird das Eisen an oder nahe an der Oberfläche ausgeschieden, wobei mit der Zeit eine mächtige Eisenkruste entstehen kann. Eine Eisenkonzentration kann auch in niedrigeren Lagen durch langsam zusammenziehendes Wasser (ohne Grundwasserbildung) zustande kommen.

Aus eisenreichen Gesteinen (z. B. basenreiche Magmatite) kann bei intensiver, tiefgründiger chemischer Verwitterung im tropischen Klima auch ein Laterit entstehen, indem die eisenreichen Verwitterungslösungen infolge der starken Verdunstungsenergie an die Oberfläche steigen, verdunsten und Eisen abscheiden.

Der Laterit ist arm an Basen und aufnehmbaren Pflanzennährstoffen, überdies verhindert er durch seine Festigkeit den Pflanzenwuchs, wenn er als geschlossener Panzer vorliegt. — In allen tropischen Regionen ist er vorhanden.

Schwarze Böden der Subtropen und Tropen. Unter verschiedenen Namen sind die im Aussehen dem Tschernosem ähnlichen, schwarzen Böden der Subtropen und Tropen bekannt geworden. Die Namen Regur (Indien), black cotton soil (Afrika) und margalite soil (Java) sind am bekanntesten (GIESECKE 1930). Die mittelafrikanischen Vorkommen hat D'HOORE (1960) auf seiner neuen Bodenkarte Afrikas neutral „sols à argiles foncées" genannt, um die bestehenden vieldeutigen Begriffe zu vermeiden. In anderen Klimagebieten gibt es ebenfalls dem Tschernosem ähnliche Böden, auf die bereits bei den Böden des gemäßigten und bei denen des mediterranen Klimas aufmerksam gemacht wurde.

Die Entstehung, der Aufbau und die Eigenschaften dieser schwarzen Böden sind verschieden; überdies sind sie unzureichend erforscht. Immerhin kann gesagt werden, daß die meisten dieser schwarzen Böden der Subtropen und Tropen unter dem Einfluß von wenigstens zeitweilig gestautem Wasser stehen, das für die Bildung des mächtigen, schwarzen A-Horizontes in erster Linie verantwortlich ist (SCHEFFER, FÖLSTER und MEYER 1960). Die zeitweilige Nässe, den mächtigen, meist mäßig humosen A-Horizont mit überwiegend schwerer Textur haben diese Böden gemeinsam. Sie gehören zu den fruchtbarsten Böden dieser Klimaregion.

Braune Böden der Subtropen und Tropen. In etwas höheren Gebirgslagen entsteht in den Subtropen und Tropen ein der Braunerde ähnlicher Boden. Die Flächen sind nicht groß. Meistens bildet sich dieser Boden aus Magmatiten und besonders aus deren Tuffen. Der A-Horizont geht ohne scharfe Grenze in den (B)-Horizont über. Weitere Eigenschaften: lockeres Gefüge, Tonfraktion oft reich an Allophan, guter Wasser- und Lufthaushalt, meist gute Versorgung mit Basen und Pflanzennährstoffen, fruchtbar.

Böden der Täler in den Subtropen und Tropen. In den Tälern, Senken und Deltaräumen der Subtropen und Tropen treten ausgedehnte hydromorphe Böden auf, die überwiegend von schwerer Textur sind. Sie bestehen aus dem umgelagerten Material der geschilderten Böden und wechseln je nach Materialanteil des einen oder anderen Bodentyps. Überwiegend sind sie relativ fruchtbar.

Die Bodentypen des Hochgebirges

Zu den Böden des Hochgebirges (Abb. 20) zählen alle Bodenbildungen, die unter den dort herrschenden Bedingungen entstehen. Diese Entstehungsfaktoren sind vielfältig und wechseln auf kleinem Raum. Es ist nicht möglich, eine untere Höhengrenze für die Hochgebirgsböden anzugeben, da deren Auftreten nicht streng an eine bestimmte Höhe gebunden ist. Nicht nur der Breitengrad ist zu berücksichtigen, sondern auch die örtlich gegebenen Einflüsse der Hangneigung, der Exposition u. a.

Die Böden des Hochgebirges sind mannigfaltig, weil sehr verschiedene Bildungsfaktoren an ihrer Formung beteiligt sind. In einigen Hochgebirgen sind sie gut erforscht, z. B. in den Alpen (JENNY 1926, PALLMANN u. a. 1933,

1938, KUBIËNA 1953, FRANZ 1960), in vielen Hochgebirgen wurden Einzeluntersuchungen unternommen. Der praktische Anreiz, umfangreiche und kostspielige Forschungen anzustellen, ist nur gering, weil die Nutzung dieser Böden auf extensive Weide und Waldbau beschränkt ist.

Die Vielfalt der Böden des Hochgebirges erfordert hier eine starke Zusammenfassung. Dabei soll an die Begriffe angeknüpft werden, die bereits in früheren Abschnitten verwendet wurden; sie stimmen weitgehend mit den Bezeichnungen KUBIËNAS (1953) überein. Für das Studium der Böden des Hochgebirges werden angegeben: JENNY 1926, 1930, PALLMANN und HAFFTER 1933, BLANCK 1939a, BOGATYREW 1953, KUBIËNA 1953, RETZER 1956, GANSSEN 1957, FRANZ 1960.

Trotz der Vielfalt ihrer Erscheinungsformen lassen sich folgende gemeinsame Wesenszüge der Böden des Hochgebirges herausstellen:

1. Sie werden in dem meist regen- und schneereichen sowie kühlen, jedoch mit der Exposition stark wechselnden Klima gebildet. Andererseits gibt es auch relativ trockene Hochgebirgslagen (Westalpen, westliche USA).

2. Mit steigender Höhe steigt die Niederschlagsmenge, und die Temperatur sinkt. Dementsprechend ändert sich die Bodenbildung.

3. Mit steigender Höhe (abgesehen vom Einfluß der Exposition) nimmt infolge der Temperaturerniedrigung die chemische Verwitterung ab und die physikalische Verwitterung zu.

4. Fast überall herrscht mehr oder weniger starke Erosionsgefahr durch Wasser.

5. Flachgründige Böden in Kuppen- und Hanglagen sowie Ansammlung von Bodenmaterial in ebeneren Lagen.

6. Mit steigender Höhe (unter Berücksichtigung der Exposition) vermindert sich die Humuszersetzung, der Humusgehalt der Böden nimmt zu, schließlich entsteht eine Humusauflage. In hohen und steilen Lagen verhindert der geringe Pflanzenwuchs die Ansammlung von Humus.

7. Die niedrigen Temperaturen hemmen die Freilegung von Stickstoff aus der organischen Substanz.

8. Starke Tendenz der Bodenauswaschung, d. h. Verlust an Basen und Pflanzennährstoffen.

9. Mit steigender Höhe Verkürzung der Vegetationszeit und Abnahme des Pflanzenwuchses.

Alpine Frostböden. In der Nähe des ewigen Schnees im Hochgebirge entstehen Frostböden, die sehr ähnlich denen des arktischen Raumes sind. Die letzteren sind bereits beschrieben, so daß hier nur die Abwandlungen der alpinen Frostböden aufzuzeigen sind.

Die Frostböden des Hochgebirges kommen durch die Hängigkeit, das Überfahren durch die Gletscher und durch die großen Schmelzwassermengen nicht zu der ruhigen Entwicklung wie die meisten arktischen Frostböden. Das wulstartige Abgleiten der alpinen Frostböden ist abhängig von Hanglage und Nässe. Auch die Bildung von Steinstreifen wurde beobachtet, während in geschützteren, weniger hängigen Lagen auch Polygonböden entstehen können (BESKOW 1930, BLANCK 1939a, CAILLEUX und HUPÉ 1947, ZEUNER 1949). Wesentlich ist bei den alpinen Frostböden die starke Durchknetung des Bodenmaterials, die noch größer ist als bei den arktischen Frostböden.

Sie besitzen infolge der geringen chemischen Verwitterung wenig Feinerde und sind steinreich. Teils hat sich eine mächtige Schuttdecke gebildet, teils kann nur von einer dichten Steinbestreuung gesprochen werden; in beiden Fällen wendet KUBIËNA (1953) die Bezeichnung *alpiner Hamada-Rohboden* an. In

Reliefdepressionen des Hochgebirges kommt es zur Ansammlung von rohem Schwemmboden, der meist feucht ist und in dem sich Polygone bilden können; KUBIËNA (1953) nennt diesen Subtyp Schneetälchen-Rutmark oder *nivaler Schneetälchenboden*.

Alpine Gesteinsrohböden. Die alpinen Gesteinsrohböden (ohne Froststrukturen), in den USA lithosols genannt, gleichen im Erscheinungsbild den Gesteinsrohböden des gemäßigt warmen, humiden Klimas, allerdings sind die alpinen den aufgeführten Einflüssen des Hochgebirges ausgesetzt; daher ist eine Abtrennung angezeigt. Auch bei den alpinen Gesteinsrohböden wird eine Unterteilung nach dem Gestein vorgenommen: *alpiner Silikat-Rohboden* und *alpiner Karbonat-Rohboden*. Als Anfangsstadium der Bodenbildung mit (A)—C-Profil stellen sie den Übergang zwischen den Felsen der Gipfelregion, ferner den nackten Felsen schroffer Hanglagen und dem Protoranker sowie der Protorendzina dar.

Alpine Ranker. Der Ranker als flachgründige Bildung aus Silikatgesteinen mit A—C-Profil wurde bereits bei den Böden des gemäßigt warmen, humiden Klimas beschrieben. Im Hochgebirge bilden sich außerdem noch einige besondere Rankerformen. Der Übergang zum alpinen Gesteinsrohboden ist häufig (nach KUBIËNA [1953] *Protoranker*). Den sehr flachgründigen, biologisch untätigen, filzigen Ranker nennt KUBIËNA (1953) *dystropher Ranker* (A_0-Silikatrohboden von PALLMANN und HAFFTER 1933), die besondere Form des windexponierten Hochgebirges *Eilagranker*. *Brauner* und *grauer Ranker* mit *Mull* oder *mullartigem Moder* sind häufig (alpiner mullartiger Ranker nach KUBIËNA 1953; Humussilikatboden nach JENNY 1926 und PALLMANN und HAFFTER 1933). Je geringer die chemische Verwitterung wirksam ist, je mehr entwickelt sich der graue Ranker mit geringerer Eisenfreilegung. Nur in wärmebegünstigten Lagen kann sich die Mullform entwickeln. In höheren Lagen südlicher Hochgebirge bildet sich oft eine rohhumusartige Decke, die KUBIËNA (1953) als Tangel bezeichnet hat; demnach spricht er von *Tangelranker*. Früher wurde die Humusauflage in den Alpen und ähnlichen Gebirgen als Alpenhumus, Alpenmull, alpiner Humusboden, Alpenmodererde, Rasenerde, Humuspolsterboden, Gebirgstrockentorf usw. bezeichnet (JENNY 1930). Dabei waren die entsprechenden Bildungen auf Carbonatgesteinen (Tangelrendzina) eingeschlossen. Die Übergänge zum alpinen Podsol sind nicht selten (*alpiner Podsol-Ranker* und *alpiner Ranker-Podsol*). Allen Rankern des Hochgebirges sollte in der Benennung das Wort „alpiner“ vorgesetzt werden, damit die besondere Geländelage angezeigt wird.

Alpine Rendzinen. Die Rendzina wurde bereits bei den Bodentypen des gemäßigt warmen, humiden Klimas beschrieben. Den Namen der Rendzinen des Hochgebirges sollte in jedem Falle — wie bei den Rankern — das Wort „alpine“ vorausgestellt werden. In niedrigeren Lagen und sonnigen Expositionen des Hochgebirges kommen ähnliche Rendzinen wie im gemäßigt warmen, humiden Klima vor: *Alpine Protorendzina*, *alpine mullartige Rendzina*, *alpine Mullrendzina*. Im Kalkhochgebirge tritt unter Polsterpflanzen eine sehr flachgründige Rendzina auf, die einen zweigeteilten A-Horizont besitzt, nämlich einen rohhumusähnlichen oberen und einen gut humifizierten unteren. KUBIËNA (1953) nannte sie *alpine Polsterrendzina* und unterschied eine *graue* und eine *schwarze* Varietät. In der Zwergstrauch- und heidekrautreichen Nadelwaldstufe bildet sich über einem gut zersetzten A_1-Horizont eine ziemlich mächtige, rohhumusartige Auflage, die KUBIËNA (1953) Tangel genannt hat (*Tangelrendzina*). Die Humusauflage der Rendzinen wurde früher ebenfalls Alpenhumus usw. genannt (s. bei alpiner Ranker). Einen weiteren Rendzina-Subtyp des Kalkhochgebirges stellt die *alpine Pechrendzina* (KUBIËNA 1953) dar. Der mineralarme, pechartig aussehende A-Horizont gliedert sich in einen von Wurzeln filzartig durchsetzten oberen (10 bis 20 cm) und einen tiefschwarzen, gut zersetzten unteren Teil (10 bis 15 cm).

Während die Rendzinen des gemäßigt warmen, humiden Klimas mehrere Monate im Jahre trocken sind, bleiben die meisten alpinen Subtypen lange Zeit feucht; in muldigen Lagen vernässen sie sogar zeitweilig.

Die verschiedenen Subtypen der Rendzina entwickeln sich oft in Abhängigkeit von der Exposition auf kleinem Raum, z. B. auf der Südseite eines Kalkmassivs die alpine Mullrendzina und auf der Nordseite die Tangelrendzina.

Die alpinen Rendzinen stellen im Kalkhochgebirge die normale Bodenbildung dar, denn die Bildung ist von den Carbonatgesteinen abhängig. Wenn sie nicht zu flachgründig sind, gelten sie als gute Böden für die Almweide und für den Wald.

Alpine Terrae calcis. Im Hochgebirge treten Relikte von Terra fusca und Terra rossa auf, die als fossil gelten müssen und wegen ihrer Lage den Zusatz „alpine" erhalten sollten.

Alpine Braunerde. Auf den Silikatgesteinen, vorzugsweise auf den basenarmen, bildet sich im Hochgebirge eine Braunerde mit gehemmter chemischer Verwitterung und geringmächtigem Profil. Der lockere A-Horizont ist dunkelgraubraun und besitzt viel Humus. Wegen des Vorkommens in der Grasheidestufe des Hochgebirges nennt KUBIËNA (1953) diesen Boden *alpine Rasenbraunerde.*

Alpine Podsole. Auf basenarmen Silikatgesteinen (besonders Sandstein) bildet sich im Hochgebirge oftmals ein Podsol, der sich als alpine Form durch ein geringmächtiges Profil auszeichnet. Der alpine Podsol, von KUBIËNA (1953) *alpiner Rasenpodsol* genannt, kann einen rostfarbigen B-Horizont (*Eisenpodsol*) oder auch einen durch Eisen und Humus kaffeebraun gefärbten B-Horizont (*Eisenhumuspodsol*) aufweisen (PALLMANN u. a. 1938). In kühlen, feuchten Lagen (Nordhänge) ist auf diesen alpinen Podsolen die Rohhumusauflage ziemlich mächtig. Unter Beweidung wird der Rohhumus abgebaut, und der Ortstein zerfällt. — Den in höheren Lagen des Hochgebirges unter Grasvegetation vorkommenden Podsol mit besonders geringmächtigem Profil hat KUBIËNA (1953) *alpiner Nanopodsol* (alpiner Zwergpodsol) genannt.

Zwischen dem alpinen Podsol und der alpinen Braunerde gibt es gleitende Übergänge. KUBIËNA (1953) hat eine *stesopodsolige Braunerde* (*=stesomorpher Semipodsol*) ausgeschieden, die einen solchen Übergang vorstellt. Ebenso kann man von *alpiner Podsol-Braunerde* sprechen, bei deutlich sichtbarer Bleichung von *alpinem Braunerde-Podsol.*

Weitere alpine Böden. Mit den aufgezählten Typen und Subtypen erschöpft sich keineswegs das Mosaik der Hochgebirgsböden. Vor allem gibt es viele Übergänge zwischen den genannten Typen und Subtypen. Ferner haben sich in feuchten Lagen Übergänge zum Gley und Pseudogley gebildet; auch der *Pseudogley mit Hangnässe* ist vorhanden. Andererseits treten in den trockenen Hochgebirgen Böden des trockenen Klimas auf, z. B. haben die sowjetischen Bodenforscher auf ihren Bodenkarten den Gebirgs-Tschernosem, den Gebirgs-Kastanosem und den braunen Halbwüstenboden des Gebirges ausgeschieden, die aber großenteils keine großen Höhen erreichen. Immerhin bezeugt dieses allgemein die Vielfalt der Bodenbildungen des Gebirges. In den westlichen Alpen hat GANSSEN (Vortrag 1959) Böden untersucht, die ebenfalls einem relativ trockenen Klima zuzuordnen sind. Das ist ein Beispiel dafür, daß auch in einem geschlossenen Hochgebirge große klimatische Unterschiede bestehen, die eine entsprechende Wirkung auf die Bodenbildung ausüben.

6. Bodensystematik

Eine allgemein anerkannte Bodensystematik gibt es bisher nicht. Andererseits gibt es einige brauchbare Vorschläge für eine Systematik, jedoch fehlt die Anerkennung einer *bestimmten* Systematik zur *einheitlichen Anwendung*. Aus diesem Grund sollen hier in erster Linie die *Grundzüge einer Bodensystematik* dargestellt und *ein* Beispiel für eine Systematik der Böden aufgezeigt werden. Wenn als Beispiel eine in Deutschland ausgearbeitete Bodensystematik gewählt wird, so möge darin nicht eine Hervorhebung dieser Systematik gesehen werden; denn andere Vorschläge werden nicht als geringwertiger angesehen. Ferner ist es hier nicht möglich, die vielen Versuche der Bodenklassifizierung der vergangenen Jahrzehnte eingehend zu diskutieren; wir müssen uns diesbezüglich auf einige Hinweise beschränken.

A. Wesen und Zweck der Bodensystematik

Zwischen Systematik und Klassifikation besteht ein grundsätzlicher Unterschied. Eine Klassifikation von Naturobjekten erfolgt auf Grund eines gewählten Einteilungsprinzips, d. h. auf Grund eines Merkmals oder auch einiger. Die Systematik verlangt mehr, nämlich eine *stufenartige* Gliederung unter Berücksichtigung *aller* für die Systematik wesentlichen, den betreffenden Objekten innewohnenden, bekannten Eigenschaften.

Das *Wesen jeder Systematik* beruht in einer stufenartigen Gliederung einer großen Zahl zusammengehöriger Objekte. Dabei werden die Objekte auf Grund gemeinsamer Merkmale in höhere und niedere Einheiten, sogenannte *Kategorien*, aufgeteilt. Die Einordnung in höhere Kategorien erfolgt auf Grund stark hervortretender Merkmale; hierbei werden die Objekte in wenigen Kategorien zusammengefaßt. Zu den niedrigeren Kategorien hin erfolgt eine zunehmende Aufgliederung der Objekte, wobei die Merkmale in dieser Richtung weniger hervortretend sind.

Das *Wesen speziell der Bodensystematik* besteht demnach darin, die große Zahl verschiedenartiger Böden nach einem bestimmten systematischen Prinzip auf Grund bestimmter bodeneigener Kriterien in höhere und niedere Kategorien einzuordnen. Gemeinsame, besonders hervortretende Merkmale gestatten es, die Böden in relativ wenigen höheren Kategorien unterzubringen. Die weniger hervortretenden Merkmale ermöglichen es, die Böden weitgehend in niedrigere Kategorien aufzugliedern.

Der *Zweck jeder Systematik* besteht in der sinnvollen Ordnung für bestimmte Objekte, wodurch eine geordnete Übersicht geschaffen und die verbindenden Beziehungen zusammengehöriger Objekte aufgedeckt werden sollen. Die Systematik ermöglicht also das Erkennen wesensmäßiger Zusammenhänge der Objekte und erleichtert die Verständigung über das Gemeinsame und das Trennende der vielen Objekte. Die systematische Ordnung gleicht einem Regal mit gestaffelter Einteilung, d. h. die großen Fächer sind in kleinere unterteilt und diese wieder in kleinere bis zur kleinsten Einheit.

Der *Zweck speziell der Bodensystematik* ist darin zu sehen, die Böden logisch und übersichtlich so zu ordnen, daß eine leicht einprägsame Übersicht entsteht und daß das Verbindende und Trennende der vielgestaltigen Böden erkennbar wird. Wenn von einem Großraum der Erde eine Übersicht über dessen Böden gegeben werden soll, so wird dieses wesentlich durch die Anwendung einer systematisch fundierten Bodeneinteilung erleichtert. Liegt einer solchen Bodenübersicht (etwa einer Bodenkarte) eine bodensystematische Gliederung zugrunde, so bedeutet das nicht nur für den Hersteller, sondern auch für den Benutzer eine wesentliche Erleichterung; denn logische Zusammenhänge sind in jedem Falle einprägsam. Bei der bodenkundlichen Aufgliederung kleiner Räume können naturgemäß nur niedrigere Kategorien herausgestellt werden. Auch in diesem Falle empfiehlt es sich, die Bodeneinteilung nach einer systematischen Ordnung vorzunehmen. Aus dieser Überlegung der Zweckmäßigkeit folgt, daß jede wissenschaftliche Bodenkartierung bodensystematisch fundiert sein soll, indem die Kartenlegende mit der bodensystematischen Gliederung in Einklang steht. Selbstverständlich kann nicht jede Kartierungseinheit einer detaillierten Bodenkarte mit einer bodensystematischen Kategorie übereinstimmen, jedoch soll auch im Aufbau der Legende einer großmaßstäblichen Bodenkarte die systematische Ordnung wenigstens in großen Zügen enthalten sein. Dagegen müssen Bodenkarten für einen begrenzten praktischen Zweck in ihrem Aufbau (Legende) auf den betreffenden Zweck ausgerichtet sein.

B. Bodensystematische Kategorisierung

Die erste Frage ist, wie viele Kategorien für eine systematische Gliederung der Böden notwendig sind. In den Vorschlägen für eine Systematik oder Klassifikation der Böden der letzten 20 Jahre sind in der Regel 5 bis 7 Kategorien vorgesehen. Wenn sich diese Vorschläge in der Konzeption auch sehr unterscheiden, so läßt aber doch die weitgehende Übereinstimmung in der Anzahl der Kategorien vermuten, daß etwa 6 bis 7 Kategorien für eine Systematik der Böden ausreichend sein müssen. Daraus kann natürlich nicht gefolgert werden, daß diese ungefähre Anzahl der Kategorien auch in der Zukunft richtig sein wird. Jedenfalls ist sicher, daß mit der Vermehrung der Kategorien die Schwierigkeiten der Abgrenzung wachsen und die Anzahl der in den niederen Kategorien auszuscheidenden Bodeneinheiten kaum mehr zu übersehen ist, wodurch leicht Verwirrung entsteht. Darum sollte die Zahl der Kategorien beschränkt bleiben und die *feinere Differenzierung der Böden* sollte man jeweils der *Beschreibung* überlassen.

Die zweite, sehr schwierige und noch sehr umstrittene Frage ist, welche bodeneigenen Kriterien (=sichtbare und unsichtbare Merkmale) für die Bildung der einzelnen Kategorien maßgebend sein sollen. Daß dafür nur bodeneigene, d. h. dem Boden innewohnende Merkmale in Frage kommen, wird kaum mehr von Fachleuten bestritten. Ferner besteht auch darüber weitgehende Einigkeit, daß bei der Kategorisierung von den *Bodentypen als den Umbildungsformen der Lithosphäre* ausgegangen werden muß. Sie stellen gleichsam die *Ausgangsposition* für die Bildung höherer und niedrigerer Kategorien dar, d. h. die Bodentypen werden in höhere Kategorien zusammengefaßt und in niedrigere aufgeteilt.

Schwieriger ist aber die Entscheidung, *nach welchen Merkmalen* die Zusammenfassung der Bodentypen in höhere Kategorien erfolgen soll und welches die entscheidenden Merkmale für die Bildung niedrigerer Kategorien sein sollen. Es besteht die Frage, ob in jedem Falle ein bestimmtes Merkmal oder eine Merkmalsgruppe für die Bildung einer Kategorie herangezogen werden soll oder ob bei den einzelnen Bodentypen jeweils das am meisten hervorstechende Merkmal für die Bildung der Kategorien entscheidend sein soll. Darin scheiden sich noch die Auffassungen. Eine Gruppe von Fachleuten vertritt den Standpunkt, daß jeder Kategorie stets die gleichen Bodenmerkmale zuzuordnen sind. Damit wird eine strenge, der Willkür entrückte Systematisierung gewährleistet. Eine andere Gruppe möchte die Systematik elastischer gestalten, indem jeweils die am meisten hervortretenden Merkmale den Vorrang haben. Z. B. könnte es zweckmäßig erscheinen, bei stark wasserbeeinflußten Böden (Auenboden, Gley, Pseudogley) die genaue Charakterisierung der Wasserdynamik in einer höheren Kategorie vorzunehmen als bei nicht wasserbeeinflußten Böden. Ferner ist die Ausbildung des A-Horizontes (Mächtigkeit, Humusgehalt, Humusform) bei dem einen Bodentyp wichtiger als bei dem anderen, und dementsprechend könnte die Berücksichtigung dieser Merkmale in dem einen Falle in einer höheren Kategorie, in dem anderen in einer niedrigeren zweckmäßig erscheinen. Bei einem solchen Vorgehen wäre dem Ermessen des Bodenkundlers zu viel in die Hand gegeben, und eine gewisse, in der Subjektivität begründete Willkür wäre nicht zu umgehen.

Das Ermessen muß in der Systematik ausgeschaltet sein. Darum sollten jeder Kategorie bestimmte Bodenmerkmale (sichtbare und unsichtbare) zugeordnet werden. Jedoch erscheint es zur Zeit geboten, mit einer zu starren Systematisierung in den niedrigsten Kategorien zurückhaltend zu sein. Zunächst ist es notwendig, in den höheren und mittleren Kategorien, vor allem bei den Bodentypen und Subtypen, mit festgelegten systematischen Regeln Ordnung zu schaffen und die bestehenden Vorschläge für die Bodensystematik zu parallelisieren. Ist das geschehen, so werden sich zwangsläufig die Wege für die Bildung der niedrigeren Kategorien weisen. Mit Recht hat KUBIËNA (1953, 1958) vor einer zu starren Kategorisierung in der niedrigeren Ebene gewarnt.

C. Grundsätze für eine Systematik der Böden der Bundesrepublik Deutschland

Allgemeine Voraussetzungen

Die Deutsche Bodenkundliche Gesellschaft hat im Jahre 1952 eine Kommission für Bodensystematik gewählt, die eine Systematik der Böden Deutschlands ausarbeiten soll. Inwischen wurde diese Systematik, ergänzt durch „Entstehung und Eigenschaften" der entsprechenden Böden, herausgebracht (MÜCKENHAUSEN mit HEINRICH, LAATSCH und VOGEL 1962). Zwar ist diese Systematik im Titel des Buches auf die Böden der Bundesrepublik Deutschland beschränkt; jedoch ist sie darüber hinaus für alle Böden Mitteleuropas anwendbar. Mit diesem Vorschlag wollen die Autoren nichts Endgültiges vorstellen, vielmehr wird dieser Vorschlag als ein *Gegenwartsbild* aufgefaßt. Man kann also in der vorgeschlagenen Weise systematisieren, muß es aber nicht.

KUBIËNA hat 1953 eine Systematik der Böden Europas vorgelegt, die weitgehend auf dem in mehreren Jahrzehnten in Europa und darüber hinaus gesammelten bodenkundlichen Wissensgut aufbaut. Dieser Autor setzt seine Systematik in enge Beziehung zu der Entstehung und zu den Eigenschaften der Böden und verwendet soweit wie möglich die eingeführten Begriffe. Es lag nahe, an diese wichtige Vorarbeit anzuknüpfen.

KUBIËNA (1953, 1958) hat seine Systematik als natürliches System der Böden bezeichnet, weil er *alle* bekannten Bodeneigenschaften bei der systematischen Gliederung einbezogen hat. Unter einem künstlichen System versteht er eine Gliederung auf Grund *einzelner* Merkmale, und dieses Vorgehen entspricht dem Wesen einer Klassifikation.

Bildung und Kennzeichnung der Kategorien

In seiner Bodensystematik hat KUBIËNA (1953) die folgenden Kategorien vorgesehen: Abteilung, Klasse, Unterklasse, Typ, Subtyp, Varietät, Subvarietät. Davon sind MÜCKENHAUSEN u. a. (1962) insofern abgewichen, als sie auf die Unterklasse verzichtet und die „Form" hinzugefügt haben. Die Form ist selbst keine Kategorie, sondern stellt die geogene Ergänzung zu den pedogenen Kategorien 1 bis 6 dar.

In Anlehnung an KUBIËNA (1953) werden von der Kommission für Bodensystematik der Deutschen Bodenkundlichen Gesellschaft folgende sechs pedogene Kategorien und die geogene Form aufgestellt, die mit Buchstaben bzw. Ziffern gekennzeichnet werden:

Abteilungen:	Großbuchstaben, z. B. A
Klassen:	Kleinbuchstaben, z. B. a
Typen:	Römische Ziffern, z. B. I
Subtypen:	Römische Ziffern in (), z. B. (I)
Varietäten:	Arabische Ziffern, z. B. 1
Subvarietäten:	Arabische Ziffern in (), z. B. (1)
Formen:	Arabische Ziffern mit Stern, z. B. 1*

Die „Form" berücksichtigt Ausgangsmaterial und Textur. Sie nimmt eine Sonderstellung ein, weil sie grundsätzlich mit allen pedogenetischen Kategorien gebildet werden kann, jedoch sinnvoll nur mit dem Typ bis zur Subvarietät. Ein Beispiel für eine Formbildung mit einem Typ ist: Rendzina, steiniger lehmiger Ton, entstanden aus dickplattigem Kalkstein. Ein Beispiel für eine Formbildung mit einer Subvarietät ist: stark ausgeprägte Mullrendzina, A-Horizont 37 cm mächtig, steiniger toniger Lehm, entstanden aus dünnplattigem Kalkstein (Turon-Pläner).

In Übereinstimmung mit vielen Bodensystematikern des In- und Auslandes sind folgende Merkmale (=Eigenschaften) des Bodens für seine systematische Einordnung maßgebend:

1. Richtung und Ausmaß der Perkolation, d. h. der Wanderung echt- und kolloidgelöster Stoffe sowie anderer wanderungsfähiger Stoffe im Boden.

2. Die Profildifferenzierung (einschließlich der Humusdecke), soweit diese ein Ergebnis der Bodenentwicklung ist.

3. Das durch das Ausgangsmaterial bedingte Filtergerüst des Bodens; denn davon hängen weitgehend Bodenentwicklung und Wasserhaushalt ab.

4. Die spezifische Bodendynamik, die sich aus den bei den Punkten 1 bis 3 gegebenen Voraussetzungen ergibt.

Aus den unter 1 bis 4 genannten *bodeneigenen Kriterien* resultieren die wichtigsten physikalischen, chemischen und biologischen Eigenschaften des Bodens, die mithin bei der systematischen Kategorisierung einbezogen sind. Mit Hilfe dieser bodeneigenen Kriterien 1 bis 4 werden die einzelnen pedogenetisch bedingten, oben genannten Kategorien wie folgt gebildet:

1. Die *Abteilungen* umfassen die Böden mit gleicher Hauptrichtung der Perkolation, d. h. mit der gleichen Einwirkungskraft des Wassers. — Die Moorböden werden wegen ihrer Eigenart als Abteilung herausgestellt, obwohl das im Hinblick auf die Perkolation nicht gerechtfertigt ist.

2. Die *Klassen* umfassen Böden mit gleicher oder ähnlicher Horizontfolge, unter Umständen auch gleicher spezifischer Dynamik, die eine Sonderstellung als zweithöchste Kategorie rechtfertigt.

3. Die *Typen* vereinigen Böden mit einer charakteristischen Horizontfolge und spezifischen Eigenschaften der einzelnen Horizonte. Sie stellen die *Umbildungsformen der Lithosphäre* dar, bedingt durch spezifische Bodenbildungsprozesse und spezifische Eigenarten des Ausgangsmaterials.

4. Die *Subtypen* sind qualitative Modifikationen der Typen. Sie sind gegeben, wenn ein *artfremdes* Merkmal zu einem Typ tritt, z. B. Calziumcarbonat im A- und (B)-Horizont einer Braunerde (=kalkhaltige Braunerde). Vielfach entstehen Subtypen dadurch, daß zu den Merkmalen reiner Typen Merkmale anderer Typen hinzutreten; die Subtypen bilden in diesen Fällen die Übergangsformen von einem Typ zum anderen, wobei sich die Zuordnung nach den *vorherrschenden* Merkmalen eines Typs richtet. Z. B. ist ein Gley-Podsol ein Podsol mit Vergleyung im Unterboden; ein Podsol-Gley ist dagegen ein Gley mit Podsolierung in den oberen Horizonten.

5. Die *Varietäten* sind die *quantitativen* Abarten (Modifikationen) eines Subtyps; sie werden durch graduelle Unterschiede, also durch den Ausbildungsgrad bestimmter bodentypologischer Merkmale gebildet. Beispiele: schwach kalkhaltige Braunerde; stark ausgeprägter Gley-Podsol, schwach vergleyt („stark ausgeprägter" bezieht sich auf Podsol, also auf das pedogenetisch Vorherrschende, dagegen „schwach vergleyt" auf Gley, also auf das pedogenetisch Untergeordnete).

6. Die *Subvarietäten* berücksichtigen alle *qualitativen und quantitativen* pedogenetischen Besonderheiten der Varietäten; sie sind also die Modifikationen der Varietäten. Beispiele: schwach kalkhaltige Braunerde, 27 cm mächtiger A-Horizont; ferner stark ausgeprägter Gley-Podsol, schwach vergleyt, 25 cm mächtige Rohhumusdecke. In der Subvarietät soll u. a. auch der Faktor „Wasser" näher differenziert werden.

Die *Formen* (Bezeichnung von „Bodenlokalform") werden mit den Kategorien 3 bis 6 (Typ, Subtyp, Varietät, Subvarietät) gebildet, indem neben den pedogenen auch die geogenen Merkmale, nämlich Bodenart (Textur) und Ausgangsgestein (bzw. -material), einbezogen werden. Pedogene und geogene Merkmale zusammen ergeben die unübersehbare Fülle der *Bodenformen* der Pedosphäre. Die Form der Subvarietät entspricht etwa der Serie der USA. Beispiele für Formen: stark ausgeprägter Gley-Podsol, schwach vergleyt, aus kiesigem fluvioglazialem Sand der Riß-Eiszeit (=Form der Varietät); Braunerde aus Basalt, steiniger toniger Lehm (=Form des Typs).

Als Beispiel wird nach diesen Grundregeln die systematische Stellung eines Bodens der Braunerdeklasse gezeigt:

Abteilung:	A	Terrestrische Böden.
Klasse:	a	Braunerden.
Typ:	I	Parabraunerde.
Subtyp:	(I)	Basenarme Parabraunerde.
Varietät:	1	Stark durchschlämmte basenarme Parabraunerde.
Subvarietät:	(1)	Stark durchschlämmte basenarme Parabraunerde, 12 cm mullartiger Moder, schwach pseudovergleyt ab 80 cm Tiefe.
Form (mit Subvarietät):	1*	Stark durchschlämmte basenarme Parabraunerde, 12 cm mullartiger Moder, schwach pseudovergleyt ab 80 cm Tiefe, feinsandiger Lehm, entstanden aus Löß (Würm).

Bei der *Formbildung mit dem Typ* heißt es kurz: Parabraunerde aus Löß (Würm).

D. Systematische Kategorien „Abteilung bis Subtyp" der Böden der Bundesrepublik Deutschland

Um wenigstens einen Überblick über die systematische Ordnung der Böden der Bundesrepublik Deutschland zu geben, die auch gültig für ganz Mitteleuropa ist, werden nachstehend die von der Kommission für Bodensystematik der Deutschen Bodenkundlichen Gesellschaft (Mückenhausen u. a. 1962) beschriebenen Einheiten der systematischen Kategorien von der Abteilung bis zum Subtyp aufgeführt. Von den zahlreichen Einheiten niederer Kategorien sind in dem genannten Buch nur Beispiele gegeben.

A Terrestrische Böden
- a Terrestrische Rohböden
  - I Alpiner Rohboden
  - II Arktischer Strukturboden
    - (I) Kryoturbater Boden
    - (II) Tropfenboden
    - (III) Steinringboden
    - (IV) Eiskeilboden
  - III Syrosem oder Gesteinsrohboden
    - (I) Typischer Syrosem
    - (II) Rendzina-Syrosem
    - (III) Ranker-Syrosem
- b A—C-Böden ohne verlehmten Unterboden
  - I Ranker
    - (I) Syrosem-Ranker
    - (II) Grauer Ranker
    - (III) Brauner Ranker
    - (IV) Braunerde-Ranker
    - (V) Podsol-Ranker
    - (VI) Dystropher Ranker
    - (VII) Tangelranker
    - (VIII) Alpiner Ranker
  - II Rendzina
    - (I) Syrosem-Rendzina
    - (II) Mullartige Rendzina
    - (III) Mullrendzina (typische Rendzina)
    - (IV) Alpine Polsterrendzina
    - (V) Moderrendzina
    - (VI) Tangelrendzina
    - (VII) Verbraunte Rendzina
    - (VIII) Braunerde-Rendzina
    - (IX) Terra fusca-Rendzina
  - III Pararendzina
    - (I) Syrosem-Pararendzina
    - (II) Mullartige Pararendzina
    - (III) Mullpararendzina (typische Pararendzina)
    - (IV) Verbraunte Pararendzina
    - (V) Braunerde-Pararendzina
    - (VI) Calziumsilikat-Pararendzina
- c Steppenböden
  - I Tschernosem (Steppenschwarzerde)
    - (I) Typischer Tschernosem
    - (II) Carbonathaltiger Tschernosem
    - (III) Degradierter Tschernosem
    - (IV) Braunerde-Tschernosem
    - (V) Parabraunerde-Tschernosem
    - (VI) Pseudogley-Tschernosem
    - (VII) Gley-Tschernosem

II Brauner Steppenboden
(I) Typischer brauner Steppenboden
(II) Degradierter brauner Steppenboden

d Pelosole
I Pelosol
(I) Syrosem-Pelosol
(II) Kalkhaltiger Pelosol
(III) Geringmächtiger Pelosol
(IV) Typischer Pelosol
(V) Durchschlämmter Pelosol
(VI) Verbraunter durchschlämmter Pelosol
(VII) Pseudogley-Pelosol
(VIII) Verbraunter Pseudogley-Pelosol

e Braunerden
I Braunerde
(I) Ranker-Braunerde
(II) Pararendzina-Braunerde
(III) Rendzina-Braunerde
(IV) Terra fusca-Braunerde
(V) Pelosol-Braunerde
(VI) Eutrophe Braunerde
(VII) Oligotrophe Braunerde
(VIII) Kalkbraunerde
(IX) Basenreiche Braunerde
(X) Basenarme Braunerde
(XI) Ferritische Braunerde
(XII) Tschernosem-Braunerde
(XIII) Parabraunerde-Braunerde
(XIV) Podsol-Braunerde
(XV) Pseudogley-Braunerde
(XVI) Gley-Braunerde
(XVII) Tundragley-Braunerde
II Parabraunerde
(I) Braunerde-Parabraunerde
(II) Tschernosem-Parabraunerde
(III) Basenreiche Parabraunerde
(IV) Basenarme Parabraunerde
(V) Fahlerde-Parabraunerde
(VI) Podsol-Parabraunerde
(VII) Pseudogley-Parabraunerde
(VIII) Gley-Parabraunerde
(IX) Tundragley-Parabraunerde
III Fahlerde
(I) Typische Fahlerde
(II) Parabraunerde-Fahlerde
(III) Pseudogley-Fahlerde
(IV) Gley-Fahlerde
(V) Podsol-Fahlerde

f Podsole
I Podsol
(I) Eisenpodsol
(II) Eisenhumuspodsol
(III) Humuspodsol
(IV) Ortsteinstaupodsol
(V) Braunerde-Podsol
(VI) Fahlerde-Podsol
(VII) Pseudogley-Podsol
(VIII) Gley-Podsol
(IX) Plaggenesch-Podsol
(X) Alpiner Podsol
II Bändchenpodsol

g Terrae calcis (plastische, dichte Böden aus Carbonatgesteinen)

I Terra fusca
(I) Typische Terra fusca
(II) Braunerde-Terra fusca
(III) Parabraunerde-Terra fusca
(IV) Pseudogley-Terra fusca
(V) Terra rossa-Terra fusca
(VI) Kalkhaltige Terra fusca

II Terra rossa

h Plastosole (plastische Böden aus Silikatgestein oder bolusartige Silikatbö den

I Braunlehm (Braunplastosol)

II Graulehm (Grauplastosol)

III Rotlehm (Rotplastosol)

i Latosole

I Roterde (Rotlatosol)

II Gelberde (Gelblatosol)

k Staunässeböden (d. h. Böden mit vorwiegend lateralem Soltransport)

I Pseudogley
(I) Typischer Pseudogley
(II) Tschernosem-Pseudogley
(III) Braunerde-Pseudogley
(IV) Parabraunerde-Pseudogley
(V) Fahlerde-Pseudogley
(VI) Podsol-Pseudogley
(VII) Pelosol-Pseudogley
(VIII) Braunlehm-Pseudogley
(IX) Graulehm-Pseudogley
(X) Gley-Pseudogley
(XI) Stagnogley-Pseudogley

II Stagnogley
(I) Typischer Stagnogley
(II) Torf-Stagnogley
(III) Rohhumusfreier Stagnogley
(IV) Basenreicher Stagnogley
(V) Pseudogley-Stagnogley

l Terrestrische anthropogene Böden (Terrestrische Kultosole)

I Plaggenesch
(I) Grauer Podsol-Plaggenesch
(II) Grauer Gley-Plaggenesch
(III) Grauer Pseudogley-Plaggenesch
(IV) Grauer Rendzina-Plaggenesch
(V) Brauner Podsol-Plaggenesch
(VI) Brauner Gley-Plaggenesch

II Hortisol (Gartenboden)

III Rigosol (Rigolter Boden)

B Semiterrestrische Böden

a Auenböden

I Rambla
(I) Kalkrambla
(II) Paternia-Rambla

II Paternia
(I) Typische Paternia (grauer Auenboden)
(II) Braune Paternia (brauner junger Auenboden)
(III) Graue Kalkpaternia (grauer Kalkauenboden; grauer pararendzinaartiger Auenboden)
(IV) Braune Kalkpaternia (brauner junger Kalkauenboden; brauner pararendzinaartiger Auenboden)

III Borowina
IV Tschernosemartiger Auenboden
V Vega (brauner Auenboden)
(I) Autochthone braune Vega
(II) Autochthone braune Kalkvega
(III) Allochthone braune Vega
(IV) Allochthone braune Kalkvega
(V) Allochthone Terra-Vega
(VI) Pelosol-Vega

b Gleye

I Gley
(I) Typischer Gley
(II) Sauerstoffreicher Gley
(III) Eisenreicher Gley
(IV) Kalkgley
(V) Gebleichter Gley
(VI) Auengley
(VII) Braunerde-Gley
(VIII) Parabraunerde-Gley
(IX) Podsol-Gley
(X) Pseudogley-Gley
(XI) Pelosol-Gley
II Naßgley
III Anmoorgley
IV Moorgley
V Tundragley

c Marschen

I Seemarsch
(I) Vorland-Seemarsch
(II) Kalkhaltige Seemarsch (typische Seemarsch)
(III) Kalkfreie Seemarsch
(IV) Knick-Seemarsch
(V) Brackmarsch-Seemarsch
II Brackmarsch
(I) Vorland-Brackmarsch
(II) Kalkhaltige Brackmarsch
(III) Kalkfreie Brackmarsch
(IV) Knick-Brackmarsch (typische Brackmarsch)
(V) Seemarsch-Brackmarsch
(VI) Flußmarsch-Brackmarsch
III Flußmarsch
(I) Vorland-Flußmarsch
(II) Kalkhaltige Flußmarsch (typische Flußmarsch)
(III) Kalkfreie Flußmarsch
(IV) Verbraunte Flußmarsch
(V) Brackmarsch-Flußmarsch
IV Moormarsch

d Semiterrestrische anthropogene Böden

C Subhydrische Böden
I Protopedon
II Gyttja
III Sapropel
IV Dy

D Moore
I Niedermoor
II Übergangsmoor
III Hochmoor

E. Andere Vorschläge für eine Bodensystematik

Wenn in diesem Abschnitt nur auf einige Vorschläge für eine Bodensystematik hingewiesen wird, so soll daraus nicht geschlossen werden, daß andere Vorschläge als weniger brauchbar angesehen werden. *Jeder* Versuch, die Böden eines beschränkten Raumes systematisch zu ordnen, wird dazu beitragen, zu gegebener Zeit eine Bodensystematik für die Erde zu erarbeiten (MÜCKENHAUSEN u. a. 1962).

Die Kommission der Deutschen Bodenkundlichen Gesellschaft gab einen Überblick über die geschichtliche Entwicklung der Bodenklassifikation im Rahmen einer neuen Bodensystematik der Böden der Bundesrepublik Deutschland; dabei ist auch die wichtigste Literatur angegeben (MÜCKENHAUSEN u. a. 1962).

In der Sowjetunion wurden in den letzten Jahren viele Vorschläge für die Klassifikation der Böden ausgearbeitet. BASINSKI (1959) hat die sowjetischen Bodenklassifikationen mit einem umfangreichen Literaturnachweis dargestellt. In neuerer Zeit haben die russischen Bodenforscher einen größeren Arbeitsstab ausgewählt, der in konsequenter Arbeit eine Systematik der Böden erarbeiten soll (ROSSOW, KARAWAJEWA und RODE 1957, 1958). Mit der Definition der einzelnen Bodentypen und der Aufstellung der systematischen Kategorien wurde die mühevolle Arbeit mit kritischer Vorsicht und Gründlichkeit begonnen.

Vor einigen Jahren hat die Systematik der Böden Europas von KUBIËNA (1953) mit ihrer umfassenden Konzeption Aufsehen erregt. Der Autor vertritt darin die Auffassung, daß die Bodensystematik alle bekannten Bodenmerkmale berücksichtigen muß. Andere Bodenforscher möchten die Bodensystematik auf wenigen Bodenmerkmalen aufbauen (LEEPER 1952, 1954, 1956).

KELLOGG und SMITH (1960) haben inzwischen den 7. Entwurf einer Bodenklassifikation vorgelegt, der nicht als endgültig betrachtet wird, sondern noch ausgebaut werden soll. Dieser Vorschlag enthält viele neue Gedanken und ist im Aufbau sehr konsequent. Sieben Kategorien sind vorgesehen: Orders, Suborders, Groups, Subgroups, Families, Series, Typs. Für die neue Nomenklatur werden Wortbestandteile der griechischen und lateinischen Sprache verwendet, und zwar werden die Namen so gebildet, daß die erste Silbe die Order, die zweite die Suborder, die dritte die Group usw. kennzeichnet. Aus dem Namen können also die höheren Kategorien abgeleitet werden. Leider entstehen bei dieser Namensbildung schlecht einprägsame Bezeichnungen, und es ist deshalb die Frage, ob sich dieser Vorschlag trotz des konsequenten Aufbaues einführen wird. Wünschenswert wäre es, wenn er für großräumige Bodenvergleiche Anerkennung finden würde. Außer dieser grundlegenden Arbeit sind in den USA viele Beiträge zur Klassifikation der Böden geliefert worden.

Außer diesen umfangreicheren, für größere Räume (UdSSR und USA) gedachten Vorschlägen für eine Bodensystematik sind eine Reihe weiterer sehr beachtlicher bodensystematischer Arbeiten erschienen, auf die im einzelnen MÜCKENHAUSEN u. a. (1962) hingewiesen haben. Alle diese Arbeiten liefern so oder so einen Beitrag für eine spätere Systematik der Böden der Erde. Nur auf einige dieser Arbeiten (aufgeführt in alphabetischer Reihenfolge der Länder) kann hier hingewiesen werden, wobei nur das Land, für dessen Böden der Gliederungsvorschlag gilt, und der Autor genannt werden: Algerien (DURAND 1954), Angola (BOTELHO DA COSTA u. a. 1956), Australien (LEEPER 1954, STEPHENS 1950, 1954), Belgien (MANIL 1959, TAVERNIER und MARÉCHAL 1958), China (Anonym 1956, HSEUNG 1945, THORP 1936), England (AVERY 1956,

ROBINSON 1949/50), Finnland (AALTONEN u. a. 1949), Frankreich (AUBERT und DUCHAUFOUR 1956), Indien (SHIVA RAU und KASINATHAN 1951), Israel (YAALON 1959), Japan (KANNO 1961), Jugoslawien (GRAČANIN 1951, STEBUTT 1930), Kanada (STOBBE 1960), Neuseeland (BAUMGART 1953, TAYLOR und POHLEN 1962), Niederlande (BENNEMA 1956, SCHELLING 1959), Niederländisch-Neuguinea (BURINGH 1954), Ostafrika (CALTON 1954), Österreich (FINK 1956, FRANZ 1960), Polen (DOBRZANSKI u. a. 1959, MUSIEROWICZ 1954a, TOMASZEWSKI 1954), Rumänien (CHERNESCU 1956, CHIRITA 1954), Schweden (EKSTRÖM 1950), Spanien (DEL VILLAR 1932), Tschechoslowakei (JURČA 1956, PELIŠEK 1957), Ungarn (STEFANOWITS und SZUCS 1956, STEFANOWITS 1959). In dem Literaturnachweis der Arbeit von MÜCKENHAUSEN u. a. (1962) ist eine größere Zahl von Veröffentlichungen über Bodenklassifikation bzw. Bodensystematik zu finden, vor allem solche von Bodenforschern der UdSSR und der USA.

Literatur

AALTONEN, V. T., u. a.: A critical review of soil terminology and soil classification in Finland in the year 1949. J. Sci. Agric. Soc. Finland **2** (1949). — ABRUKOWA, L. P.: Mineralogische Zusammensetzung der Tonfraktion des Krim-Tschernosems (russ.). Potschwowedenie Nr. 12 (1960). — ADERIKHIN, P. G., E. P. TIKHOWA, P. V. KULAKOWA und N. J. DEGTIAREWA: Veränderungen in Podsolböden unter Bewirtschaftung im Gebiet von Murmansk (russ.). Potschwowedenie Nr. 4 (1960). — AGAFONOFF, V.: Sols types de Tunisie. Avec carte pédologique. Tunis 1937. — ALBAREDA, J. M., und T. ALVIRA: Mediterranean soils of the Spanish Levant and North-Africa. Trans. 4. Int. Congr. Soil Sci., Vol. 2, S. 185–186. Amsterdam 1950. — ALBAREDA, J. M., F. BURRIEL und M. M. TABOADELA: Study of the Andalusian Black soils and the reason for their colour. Trans. 4. Int. Congr. Soil Sci., Vol. 2, S. 186–187. Amsterdam 1950. — ALBERT, R., und M. KÖHN: Beitrag zur Kenntnis der Molkenböden, I. Z. Forst- und Jagdwesen **62**, 411 (1930). — Beitrag zur Kenntnis der Molkenböden, II. Z. Forst- und Jagdwesen **64**, 352 (1932). — AMERYCKX, J.: Die Bodenbeschaffenheit der belgischen Marsch östlich vom Blankenberger Deich. Diss. Bonn 1954. — Anonym: Systematik der Böden Chinas (russ.). Potschwowedenie Nr. 1, 89–101 (1956). — ANTIPOW-KARATAEW, I. N.: Braune und bräunliche Waldböden (russ.). Potschwowedenie **1947**, 697–703. — ARENS, R., F. KRÄMER und G. LANGNER: Über die Bodenstruktur, den Wasserhaushalt und die Ertragsfähigkeit von Pseudogleyen und gleyartigen Braunerden. Z. Acker- u. Pflanzenbau **107**, 67–98 (1958). — AUBERT, G.: Classification pédologique des sols de France. In: DEMOLON, A. Dynamique du sol, 5. Aufl. Paris 1952. — AUBERT, G., und P. DUCHAUFOUR: Projet de classification des sols. Rapp. 6. Congr. Int. Sci. Sol, Vol. E, S. 597–604. Paris 1956. — AVERY, B. W.: A classification of British soils. Rapp. 6. Congr. Int. Sci. Sol, Vol. E, S. 285–292. Paris 1956.

BADEN, W.: Untersuchungen an nordwestdeutschen Podsolprofilen und podsoligen Moorbildungen im Hinblick auf das jeweils zweckmäßige Urbarmachungsverfahren. Z. Pflanzenernähr., Düng., Bodenkde. **52** (97), 120–150 (1951). — Einige Besonderheiten der nordwestdeutschen Moorvorkommen. Trans. Int. Soc. Soil Sci. II. and IV. Comm., Vol. 2, S. 125–131. 1958. — BARGON, E.: Über die Entwicklung von Lockerbraunerden aus Solifluktionsmaterial im vorderen Odenwald. Z. Pflanzenernähr., Düng., Bodenkde. **90** (135), 229–243 (1960). — BARRY, T. A.: Some considerations affecting the classification of the bogs of Ireland, and their peats. Int. Peat Symp., Sect. B 2, 16 S. Dublin 1954. — BARSHAD, I., und L. A. ROJAS-CRUZ: A pedological study of a podzol soil profile from the equatorial region of Columbia, South America. Soil Sci. **70**, 221–236 (1950). — BASINSKI, J. J.: The russian approach of soil classification and its recent development. J. Soil Sci. **10**, 14–26 (1959). — BAUMGART, I. L.: Outline of New Zealand soils and their classification. Aust. Conf. Soil Sci., Bd. 1, S. 6. Adelaide 1953. — BAUR, A. J., und W. H. LYFORD: Saure Braunerden in den nördlichen USA. Proc. Soil Sci. Soc. Amer. **21**, 533–536 (1957). — BECKEL, A.: Bildungsbedingungen und Eigenart der Gleye in der Erftniederung. Diss. Bonn 1958. — BEHREND, F., und W. E. SCHMIDT: Erl. z. Bl. Nauen. Preuß. Geol. L.-A. Berlin 1938. — BENNEMA, J.: Klassifizierungsprinzipien der niederländischen Böden. Rapp. 6. Congr. Int. Sci. Sol., Vol. E, S. 331–336. Paris 1956. — BESKOW, G.: Erdfließen und Strukturböden der Hochgebirge im Lichte der Frosthebung. Geol. Fören. i Stockholm Förhandl. **52**, 622–638 (1930). — BLANCK, E.: Handbuch der Bodenlehre. Berlin:

Springer. 1929/39. — Die Mediterran-Roterde (Terra rossa). Handb. d. Bodenlehre v. E. Blanck, Bd. III, Berlin 1930. — Blanck, E., und A. Musierowicz: Nochmals zur Kenntnis der Roterden der Mittelmeerländer. Chem. d. Erde **6**, 381–389 (1931). — Blanck, E.: Die Böden der kalten Regionen (arktische, subarktische und Hochgebirgsböden). Handb. d. Bodenlehre v. E. Blanck, Erg.-Bd. I, S. 70–76. Berlin 1939 (1939a). — Böden der kühlen gemäßigten Regionen: Bleicherdewaldböden und podsolige Böden, Braunerde einschließlich Rendzina, nebst Bodenverteilung innerhalb Deutschlands. Handb. d. Bodenlehre v. E. Blanck, Erg.-Bd. I, S. 77–99. Berlin 1939 (1939b). — Böden der feuchtwarmen gemäßigten Regionen: Gelb- und Roterden (Terra rossa). Handb. d. Bodenlehre v. E. Blanck, Erg.-Bd. I, S. 99–117. Berlin 1939 (1939c). — Bloomfield, C.: Experiments on the mechanism of Gley formation. J. Soil Sci. **2**, 196–211 (1951). — The distribution of iron and aluminium oxides in Gley soils. J. Soil Sci. **3**, 167–171 (1952). — A study of podzolization, part I. The mobilization of iron and aluminium by Scots Pine needles. J. Soil Sci. **4**, 5–16 (1953) (1953a). — A study of podzolization, part II: The mobilization of iron and aluminium by the leaves and bark of Agathis Australis (Kauri). J. Soil Sci. **4**, 17–23 (1953) (1953b). — Sesquioxide immobilization and clay movement in podzolized soils. Nature **172**, 958 (1953) (1953c). — A study of podzolization, part III: The mobilization of iron and aluminium by Rimu (Dacrydium cupressinum). J. Soil Sci. **5**, 39–45 (1954) (1954a). — A study of podzolization, part IV: The mobilization of iron and aluminium by picked and fallen Larch needles. J. Soil Sci. **5**, 46–49 (1954) (1954b). — A study of podzolization, part V: The mobilization of iron and aluminium by Aspen and Ash leaves. J. Soil Sci. **5**, 50–56 (1954) (1954c). — A study of podzolization, part VI: The immobilization of iron and aluminium. J. Soil Sci. **6**, 284–292 (1955) (1955a). — Experimental production of podzolization. Chem. a. Ind. **1955**, 1596–1597 (1955b). — Bogatyrew, K. P.: Zur Kenntnis der Hochgebirgsböden in den oberen Gebirgsregionen (russ.). Potschwowedenie Nr. 5, 1–13 (1953). — Die Smonitza-Böden Albaniens (russ.). Potschwowedenie Nr. 4, 14–22 (1958). — Bolischew, N. N.: Die Genese gewöhnlicher Tschernoseme in einem Teil der westsibirischen Ebene (russ.). Potschwowedenie **1947**, 660–668. — Botelho da Costa, J. V., L. Ario Azevedo, R. Pinto Ricardo und E. P. Cardoso Franco: Klassifizierungs- und kartographische Probleme der Böden von Huila (Angola). Rapp. 6. Congr. Int. Sci. Sol, Vol. E, S. 227–232. Paris 1956. — Boulaine, J.: Étude des sols des plaines du Chélif. Ministère de l'Algérie. Birmandreis 1957. — Bramao, D. L., und P. Lemos: Broad soil associations of South America. Food and Agriculture Organization of the United Nations. Rome 1960. — Broek, J. van den: Some considerations on gray brown podzolic soils in the Netherlands. Trans. 6. Int. Congr. Soil Sci., Vol. E, S. 331–335. 1956. — Brown, G.: The occurence of lepidocrocite in some British soils. J. Soil Sci. **4**, 220–228 (1953). — Soil morphology and mineralogy, a qualitative study of some gleyed soils from North-Western England. J. Soil Sci. **5**, 145–154 (1954). — Brüne, F., und C. Husemann: Feldversuche zur Frage nach der praktischen Bedeutung der Kuhlerdemelioration in den Nordsee-Marschen. Landwirtsch. Jb. **84**, H. 1 (1937). — Brüne, F.: Die Praxis der Moor- und Heidekultur. Berlin und Hamburg 1948. — Brunnacker, K.: Die Geschichte der Böden im jüngeren Pleistozän in Bayern. Geol. Bavarica **34** (1957). — Bemerkungen zur Parabraunerde (Ergebnisse der Bodenkartierung in Bayern). Geol. Jb. **76**, 25–36 (1959). — Buckmann, H. O., und N. C. Brady: The nature and properties of soils. New York: Macmillan. 1957. — Büdel, J.: Periodische und episodische Solifluktion im Rahmen der klimatischen Solifluktionstypen. Erdkunde (Bonn) **13**, 297–314 (1959). — Die Abtragungs-Vorgänge auf Spitzbergen im Umkreis der Barents-Insel. Tagungsber. Deutsch. Geographentag 1961, S. 337–375. Wiesbaden: Steiner. 1961. — Buringh, P.: Soil and land classification in Dutch New Guinea. Rapp. 5. Congr. Int. Sci. Sol, Vol. IV, S. 93–95. Léopoldville 1954. — Soils and soil conditions in Iraq. Republic of Iraq, Ministry of Agriculture. Bagdad 1960.

Cailleux, A., und P. Hupé: Présence des sols polygonaux et striés dans les Pyrénées françaises. C. R. **225**, 135?–1355 (1947). — Calton, W. E.: The catena in relation to the classification of East African soils. Rapp. 5. Congr. Int. Sci. Sol., Vol. IV, S. 58–61. Léopoldville 1954. — Catacousinos, D. S.: Les sols de Grèce. Inst. Central pédologique du Ministère de l'Agriculture. Athènes 1959. — Chernescu, N.: Les sols de la région située entre le Danube, les Carpathes et la Mer Noire (République Populaire Roumaine). Rapp. 6. Congr. Int. Sci. Sol., Vol. E, S. 479–485. Paris 1956. — Chirita, C. D.: Genetische Klassifizierung nach einem einzigen bodenbildenden Prozeß der Böden der Rumänischen Volksrepublik (rumän.). Agrokém. Talajt. **3**, 416–418 (1954). — Ćirić, M.: Merkmale des Podsolierungsprozesses in Serbien (jugosl.). Jugosl. Druš to Pronč. Zemlj. Nr. 4, S. 9–73 (1956). — Cline, M. G.: Profile studies of normal soils of New York, I: Soil profile sequences involving brown forest, gray brown podzolic and brown podzolic soils. Soil Sci. 68, 259–272 (1949). — Comel, A.: Italian red soils. Ist. Chim. Agr. Sper. Gorizia, Nuovi Ann. 3, 3–110 (1952).

Dahm, H.-D.: Merkmale und Verbreitung periglazialer Fließerden im südlichen Egge-Gebirge. Geol. Jb. 76, 25–36 (1959). — Dawson, J. E.: Organic soils. Adv. Agron. 8, 377–401 (1956). — Deb, B. C.: The movement and precipitation of iron oxides in podzol soils. J. Soil Sci. 1, 112 (1949/50). — Estimation of free iron oxides in soils and clays and their removal. J. Soil Sci. 1, 212–220 (1950). — Diepen, D. van: Hydromorphological and morphogenetic profile caracteristics of humus podzols of the Netherlands. Rapp. 6. Congr. Int. Sci. Sol., Vol. E, S. 453–461. Paris 1956. — Dietrich, H.: Untersuchungen zur Morphologie und Genese grundwasserbeeinflußter Sandböden im Gebiet des nordostdeutschen Diluviums. Arch. Forstwes. 7, 577–640 (1958). — Diez, T.: Entstehung und Eigenschaften von Böden aus tonigen Substraten. Diss. Bonn 1959. — Dimbleby, G. W., und J. M. Gill: The occurence of podzols under deciduous woodland in the new forest. Forestry 28, 95–106 (1955). — Dobrowolski, G. W.: Auenböden am Unterlauf des Flusses Oka (russ.). Potschwowedenie Nr. 4, 47–58 (1956). — Dobrzanski, B.: Regradation podsoliger Lößböden. Roczn. Univ. Marii Curie-Sklodowska 2, 27–46 (1947). — Dobrzanski, B., und J. Skarzynska: Jurassische Rendzinen in der Umgebung von Przemysl (poln.). Ann. Univ. Marii Curie-Sklodowska 5B, 97–107 (1950). — Dobrzanski, B., und B. Zbysaw: Tschernosemböden auf Löß im Karpatenvorland (poln.). Ann. Univ. Marii Curie-Sklodowska, Sekt. E 10, 287–299 (1955). — Dobrzanski, B., A. Musierowicz, M. Strzemski, S. Uziak u. a.: Genetische Klassifikation der polnischen Böden (poln.). Odbitka z Roczn. Glebozn. T. VII, S. 103. Warschau 1959. — Dokturowski, V. S.: Niederungsmoore im Süden der UdSSR. Proc. and Pap. 2. Int. Bodenkdl. Kongr. Leningrad-Moskau, 6. Komm., S. 292. 1932. — Dokutschajew, W. W.: Vorläufiger Bericht über die Aufnahme des südöstlichen Teiles des Tschernosemgebietes in Rußland (russ.). Trudi Imp. Vol. Ekonom. Obsches 1879 (1879a). — Die Kartographie der russischen Böden (russ.). St. Petersburg 1879 (1879b). — Der russische Tschernosem (russ.). St. Petersburg 1879 (1879c) und 1883. — Dokutschajew-Institut und GUKG: Bodenkarte der Sowjetunion im Maßstab 1:4 Mill. Dokutschajew-Inst. Moskau 1954. — Duchaufour, P.: Lessivage et podzolisation. Rev. Forêt. Franç. 10, 647–652 (1951). — Note sur les phases de la podzolisation sur grès vosgiens. Rapp. 6. Congr. Int. Sci. Sol., Vol. E, S. 373. Paris 1956. — Dudal, R.: Étude morphologique et génétique d'une séquence de sol sur limon loessique. Agricultura I (2 Sér.) 2, 119–163 (1953). — Durand, J. H.: Les sols d'Algérie. S. C. H. Pédologie, Clairbois; Birmandreis. 1954 (1954a). — Durand, J. H., u. a.: Carte des sols d'Algérie, Gouvernement Général de l'Agriculture. Alger 1954 (1954b). — Durand, J. H.: La classification des sols utilisés pour les cartes pédologiques d'Algérie. Rapp. 5. Congr. Int. Sci. Sol., Vol. IV, S. 168–174. Léopoldville 1954 (1954c). — Les sols rouges et les croûtes en Algérie. Service des Études Scientifiques. Alger 1959. — Durasow, A. M.: Das Problem der Klassifizierung der Tschernoseme (russ.). Potschwowedenie Nr. 3, 37–38 (1955). — Dutt, A. K.: The influence of forest types on the degree of podzolization. Indian Forester 73, 495–498 (1947).

Edelman, C. H.: Soils of the Netherlands. Amsterdam 1950. — Ehwald, E.: Bemerkungen zur Abgrenzung und Gliederung der wichtigsten Bodentypen Mitteleuropas unter dem Gesichtspunkt einer internationalen Annäherung in der Bodensystematik. Z. Pflanzenernähr., Düng., Bodenkde. 80 (125), H. 1 (1958). — Ekström, G.: Soil classification in Scandinavia—Finland, especially in Sweden. Trans. 4. Int. Congr. Soil Sci., Vol. 2, S. 211–214. Amsterdam 1950. — Engelhardt, W. v.: Neuere Ergebnisse der Tonmineralienforschung. Geol. Rdsch. 51, H. 2, 457–477 (1961). — Erdtman, G.: Britische Moortypen. Erwiderung auf E. Grunland. Sveriges Unders. Årsbok 26, 380 (1932).

Fastabend, H., und F. v. Raupach: Zur Kenntnis der Plaggenböden in Nordwestdeutschland. Geol. Jb. 78, 139–172 (1961). — Ergebnisse der C^{14}-Untersuchung an einigen Plaggenböden des Emslandes. Geol. Jb. 79, 863–866 (1962). — Finger, H.: Die Wirkung von Bor, Mangan und Kupfer und steigenden Kalkgaben auf rohhumushaltigem Heidesandboden. Landwirtsch. Forsch. 6, 89 (1951). — Finck, A.: Tropische Böden. Hamburg u. Berlin: Parey. 1963. — Fink, J.: Die fossilen Böden im österreichischen Löß. Quartär (Bonn) 6 (1954). — Zur Systematik fossiler und rezenter Lößböden in Österreich. Rapp. 6. Int. Congr. Sci. Sol., Vol. E, S. 585–592. Paris 1956. — Flaig, W., F. Scheffer und B. Klamroth: Zur Kenntnis der Huminsäuren, VIII. Mitt.: Zur Charakterisierung der Huminsäuren des Bodens. Z. Pflanzenernähr., Düng., Bodenkde. 71, (116), 33 (1955). — Franz, H.: Zur Kenntnis der „Steppenböden" im pannonischen Klimagebiet Österreichs. Bodenkultur 8, 125–132 (1955). — Drei klimabedingte Ranker-Subtypen Europas. 6. Int.

Congr. Soil Sci., Vol. E, S. 135–141. Paris 1956. — FRANZ, H., P. GUNHOLD und H. PSCHORN-WALCHER: Die Kleintiergemeinschaften der Auwaldböden der Umgebung von Linz und benachbarter Flußgebiete. Naturkundl. Jb. d. Stadt Linz 1959. — FRANZ, H.: Feldbodenkunde. Wien und München 1960. — FRASER, G. K.: Peat deposits of Scotland. Dept. of Sci. and Indust. Research, Wartime Pamphlet Nr. 36. 1943. — Classification and nomenclature of peat and peat-deposits. Int. Peat. Symp., Sect. B 2, 8 S. Dublin 1954. — FREHE, K.: Die Schwarzerde (Tschernosem) und ihre wirtschaftsgeographische Bedeutung. Berlin 1931. — FREI, E.: Morphologische, chemische und kolloidchemische Untersuchungen subalpiner Weide- und Waldböden der Rendzina- und Podsol-Serie. Diss. E.T.H. Zürich, Landw. Abt. 1944. — FREI, E., und M. G. CLINE: Profile studies of normal soils of New York, II: Micromorphological studies of the gray brown podzolic—brown podzolic soil sequence. Soil Sci. **68**, 333–344 (1949). — FRESE, H., H. G. GADE und E. MÜCKENHAUSEN: Studienreise nach USA. Land- und Hauswirtschaftl. Auswertungs- und Informationsdienst. Bad Godesberg 1952. — FRIDLAND, V. M.: Zu den Problemen der Podsolierung und „Illimerisation" (russ.). Potschwowedenie Nr. 1, 27–38 (1958).

GALLWITZ, H.: Eiskeile und glaziale Sedimentation. Geologica **2** (1949). — GANGULEE, N.: The alkali soils of India. Actes IV. Conf. Intern. de Pédol., Vol. 2, Rome 1924. — GANSSEN, R.: Bodengeographie. Stuttgart: Köhler. 1957. — GANSSEN, R., und H. HARTH: Beiträge zur Kenntnis der Böden des Oberrheintalgrabens und der angrenzenden Gebiete, II. Mitt.: Versuch einer Kennzeichnung wichtiger Böden im trockensten Teil des nördlichen Oberrheingrabens. Z. Pflanzenernähr., Düng., Bodenkde. **76** (121), 213–223 (1957). — GANSSEN, R.: Landschaft und Böden in Südwestafrika. Erde **91**, H. 2, 115–131 (1960). — GARDNER, D. R., und E. P. WHITESIDE: Zonal soils of the transition region between the podzol and gray brown podzolic regions in Michigan. Proc. Soil Sci. Soc. Amer. **16**, 137–141 (1952). — GEILMANN, W.: Der Eschboden und seine Düngebedürftigkeit. Arb. aus d. landw. Versuchsfeld d. Univ. Göttingen. J. Landw. (Berlin) **71** (1923). — GERASSIMOW, J. P., und M. A. GLASOWSKAJA: Grundlagen der Bodenkunde und der Bodengeographie. Moskau 1960. — GIESECKE, F.: Subtropische Schwarzerden. Handb. d. Bodenlehre v. E. BLANCK, Bd. III, S. 341–352. Berlin 1930. — GLINKA, K. D.: Die Typen der Bodenbildung, ihre Klassifikation und geographische Verbreitung. Berlin 1914. — Bodenkunde, 1. Aufl. St. Petersburg 1915. — GÖTTLICH, K. H.: Beurteilungsrahmen II für Moor- und Anmoorböden. Landw. Forsch. **11**, 156–161 (1958). — GOGOLEW, I. N.: Die Genese dunkelfarbener Böden (Rendzinen) unter Wald (russ.). Potschwowedenie **1952**, 241–250. — GORSCHENIN, K.: Sibirische Tschernosemböden (russ.). Potschwowedenie Nr. 6, 330 (1947). — GRAČANIN, M.: Contribution to the genesis of terra rossa, I: Does terra rossa represent an illuvial horizon of forest soils? Poljopriv. Znanst. Smot. H. 10/11, 99–106 (1947). — Pedologija III: Sistematika tala (Systematik der Böden). Zagreb 1951. — GRIPP, K.: Über Frost- und Strukturböden auf Spitzbergen. Z. Ges. f. Erdk. **1926**, 7–8. — GROSSE-BRAUCKMANN, G.: Zur Terminologie organogener Sedimente. Geol. Jb. **79**, 117–144 (1961). — GRUNLAND, E.: Die Geologie der schwedischen Hochmoore, ihre Bildungsbedingungen, Entwicklungsgeschichte und Verbreitung, auch der Zusammenhang von Hochmoorbildung und Versumpfung. Sveriges Geol. Unders. Årsbok **26**, H. 1 (1932). — GRUPE, O.: Die Brücher des Sollings, ihre geologische Beschaffenheit und Entstehung. Z. Forst- u. Jagdwesen **3** (1909). — Zur Entstehung des Molkenbodens. Int. Mitt. f. Bodenkde. Berlin und Wien 1923.

HABIG, F.: Beitrag zur Dynamik und Genetik stauwasserbeeinflußter Böden. Diss. Hann.-Münden 1953. — HALLSWORTH, E. G., A. B. COSTIN und F. R. GIBBONS: Studies on pedogenesis in New South Wales, VI: On the classification of soils showing features of podzol morphology. J. Soil Sci. **4**, 241–256 (1953). — HAMBLOCH, H.: Über das Alter und die Bildungsdauer von Eisenhumuspodsolen. Z. Pflanzenernähr., Düng., Bodenkde. **83**, (128), 134–139 (1958). — HARRASSOWITZ, H.: Laterit. Fortschr. Geol. Paläontol. IV, H. 14, 153–516 (1926). — Böden der tropischen Region. Handb. d. Bodenlehre v. E. BLANCK, Bd. III. Berlin 1930 (1930a). — Fossile Verwitterungsdecken. Handb. d. Bodenlehre v. E. BLANCK, Bd. IV. Berlin 1930 (1930b). — Fossile Roterden, besonders auf Kalken. Handb. d. Bodenlehre v. E. BLANCK, Bd. IV, S. 267–275. Berlin 1930 (1930c). — Bleichungsvorgänge. Chem. d. Erde **5**, 146 (1930) (1930d). — HARRIS, S. A.: Soil alkali, its origin, nature and treatment. New York: Wiley. 1920. — HEBESTREIT, M. H.: Was ist Knick? Wasser u. Boden **7**, H. 10, 308 (1955) (1955a). — Zur Bewirtschaftung der Knickböden in der Marsch. Betriebsw. Mitt. f. d. Wirtschaftsberater. LK Schleswig-Holstein Nr. 16 (1955) (1955b). — HEIKUREINEN, L.: Die kiefernbewachsenen eutrophen Moore Nordfinnlands. Eine Moortypenstudie aus dem Gebiet des Kivalo-Höhenzuges. Ann. bot. Vanamo **26**, Nr. 2, 189 S. (1953). — HEINEMANN, B.: Über Aufbau und Verbreitung der Plaggenböden im Emsland. Jb. Emsländ. Heimatvereins (Meppen) **6**, 62–76 (1958). — Gräben und Grabensysteme unter Plaggenböden des Emslandes. Jb. Emsländ. Heimatvereins (Meppen) **8**, 19–32 (1960). — HILGARD, W. E.: A report on the relation of soil and climate. U.S. Dept. Agr. Weather Bureau Bull. 3. 1892. — Soils. New York 1910. — HOLLSTEIN, W.: Beiträge zur Bodenkunde des Mittelmeergebietes. Bodenk. Forsch. **6**, 91–121 (1938). — Bodenbildungsprobleme in Nordwestdeutschland. Geol. Jb. **76**, 1–10 (1959). — HOORE, J. D': La carte des sols d'Afrique au sud du Sahara. Pédologie **10**, H. 1, 191–204 (1960). — HORDY, F.: Cultivation properties of tropical red soils. Empire J. Exper. Agric. (Oxford) **1**, H. 2, 103 (1933). — HOYNINGEN-HUENE, P. F. v.: Die Bodentypen Nord- und Mitteldeutschlands. Jb. Preuß. Geol. L.-A. **51** (1930) (1930a). — Knick = geologische Aufschlickung und Durchschlämmung. Jb. Preuß. Geol. L.-A. **51** (1930) (1930b). — HOYOS DE CASTRO, A., und V. SOLER: A study of Canary-Island brown earths and rankers in relation to the parent rock. An. Edafol. Fisiol. vegetal **16**, 297–386 (1957). — HRUSKA, B.: Charakteristik des Gleyprozesses (tschech.). Sborn. výs. Skol. Zěměd. Lesn. Brne. Nr. 1C, 55–60 (1956). — HSEUNG, Y.: A suggested system of soil classification in China. Soils Quart. **4**, 65–74 (1945). — HUMMEL, P.: Rendzinen auf der Hohen Schwabenalb. Z. Pflanzenernähr., Düng., Bodenkde. **77** (122), 230–242 (1957). — HUSEMANN, C.: Die landwirtschaftliche Bewertung der Moorböden und ihre natürlichen Grundlagen. Lüneburg 1947.

IVERSEN, J.: Beitrag zur Frage der Blausandwirkung. Landwirtsch. Jb. **91** (1942). — IWANOWA, J. N.: Systematik der Böden des nördlichen Teils des europäischen Territoriums der UdSSR (russ.). Potschwowedenie Nr. 1, 70–88 (1956).

JAKLITSCH, L.: Zur Untersuchung von Aulehmböden der Oststeiermark, insbesondere auf den Terrassen des Ritscheintales. Diss. Hochschule f. Bodenkultur in Wien 1955 (gekürzt in Mitt. Österr. Bodenkdl. Ges. **3**, 15–30 (1959). — JASMUND, K.: Die silicatischen Tonminerale. Monographien zu „Angewandte Chemie" und „Chemie-Ingenieur-Technik". Weinheim/Bergstr.: Verlag Chemie. 1955. — JENNY, H.: Die alpinen Böden. In: BRAUN-BLANQUET, J., Vegetationsentwicklung und Bodenbildung usw. Denkschr. Schweiz. Naturforsch. Ges. (Zürich) **63** (1926). — Hochgebirgsböden. Handb. d. Bodenlehre v. E. BLANCK, Bd. III, S. 96–118. Berlin 1930. — Factors of soil formation. New York 1941. — JEWITT, T. N.: Gezira soil. The Republic of the Sudan, Ministry of Agriculture, Bull. 12. Khartoum 1955. — JOFFE, J. S.: Soil profile studies, III: The process of podzolization. Soil Sci. **32**, 303–323 (1931). — Soil profile studies, IV: Morphological and chemical evidence of podzolization. Soil Sci. **33**, 217–237 (1932). — Soil profile studies, VII: The Gley process. Soil Sci. **39**, 391–401 (1935). — JOHANNSEN, A., und H. E. STREMME: Plaggenböden auf För. Z. Pflanzenernähr., Düng., Bodenkde. **65** (110), 170–180 (1954). — JURČA, V.: Ein Beitrag zur Bodensystematik und Bodenklassifikation (tschech.). Acta Univ. Agric. Praha 1956.

KAISER, K.: Wirkungen des pleistozänen Bodenfrostes in den Sedimenten der Niederrheinischen Bucht. Eiszeitalter u. Gegenwart **9**, 110–129 (1958). — KANNO, J.: Genesis and classification of main genetic soil types in Japan. II. Red-yellow soils. Bull. Kyushu Agric. Exper. Stat. **7**, 187–306. Fukuoka/Japan 1961. — KATSCHINSKAYA, E. S.: Zur Frage der Geschwindigkeit des Verlaufs des Bodenprozesses unter den Bedingungen der podsoligen Zone (russ.). Potschwowedenie **23**, Nr. 7, 721 (1937). — KATZ, N. J.: Über die Grenzen der Torf-Moorböden, über die Verbreitungsgebiete der Pflanzen im Zusammenhang mit dem Klima und mit der geologischen Geschichte (russ.). Potschwowedenie Nr. 5, 12–21 (1957). — Moortypen und ihre Verbreitung in den kalten und gemäßigten Zonen der nördlichen Hemisphäre (russ.). Potschwowedenie Nr. 6, 13–20 (1958). — KELLOGG, C. E.: Preliminary suggestions for the classification and nomenclature of great soil groups in tropical and equatorial regions. Proc. First Commonwealth Bureau of Soil Sci. **1948**, 78–85. — Tropical soils. Trans. **4**. Int. Congr. Soil Sci., Vol. 1, 266–276 (1950). — KELLOGG, C. E., und G. D. SMITH: Soil classification. A comprehensive system. 7th approximation. Soil Survey Staff, Soil Conservation Service, U.S. Dept. Agr. Washington 1960. — KERPEN, W.: Die Böden des Versuchsgutes Rengen, Kartierung, Eigenschaften und Standortswert. Forsch. u. Beratung. Reihe B, H. 5. Hiltrup: Landwirtschaftsverlag. 1960. — KIVINEN, E.: Über die Eigenschaften der Gyttjaböden. Bodenkde. u. Pflanzenernähr. **9/10** (54/55), 122–134 (1938). — The classification and chemical composition of peat soils in Finland. Repr. Int. Peat Symp. Dublin, Sect. B2, 4 S. (1954). — KLAPP, E.: Acker- und Pflanzenbau, 5. Aufl. Berlin u. Hamburg: Parey. 1958. — KLINGE, H.: Über spanische Terra rossa-Vorkommen und die Möglichkeiten ihrer zeitlichen Einordnung auf Grund bodengeographischer Studien. Z. Pflanzenernähr., Düng., Bodenkde. **76** (121), 223–231 (1957). — Eine Stellungnahme zur Altersfrage von Terra rossa-Vorkommen (unter besonderer Berücksichtigung der Iberischen Halbinsel, der Balearischen Inseln und Marokkos). Z. Pflanzenernähr., Düng., Bodenkde. **81** (126), 56–63 (1958). — KÖHN-

LEIN, J.: Zu den pflanzenbaulichen Schwierigkeiten in den Marschen. Landwirtsch. Forsch. **10**, H. 10, 8 (1957). — KÖHNLEIN, J., und E. SCHLICHTING: Eigenschaften und Nutzung kultivierter Heidepodsole in Nordwestdeutschland. Z. Pflanzenernähr., Düng., Bodenkde. **84** (129), 296–308 (1959). — KOHL, F., F. VOGEL und F. WACKER: Vergleich zwischen bodenkundlicher und pflanzensoziologischer Kartierung. Landw. Jb. Bayern **31**, H. 9/10 (1954). — KOHL, F.: Durchwaschungs- und Durchschlämmungserscheinungen an Böden aus Bayern. Ein Beitrag zur Frage der Parabraunerden. Z. Pflanzenernähr., Düng., Bodenkde. **80** (125) 237–244 (1958). — KONAKOW, M.: Zur Dynamik der Tschernoseme (russ.). Arb. Wiss. Forsch.-Inst. Bodenkde. Charkow **1**, 145 (1930).— KONONOWA, M. M.: Die Humusstoffe des Bodens. Berlin: Deutscher Verlag der Wissenschaften. 1958. — KOWDA, V. A., und E. V. LOBOWA: Carte des sols de l'Asie i. M. 1:6 Mill. 7. Congr. Intern. Society of Soil Sci., Madison 1960. — KRAUSE, P. G., und K. UTESCHER: Über ein neues „Schwarzerde"-Vorkommen in Ostpreußen und die Kennzeichnung von Schwarzerde und Moorerde. Jb. Preuß. Geol. L.-A. **55**, 526–538 (1934). — KRAUSS, G.: Die sogenannten Bodenerkrankungen. Jahresber. Dtsch. Forstver. 1928. — KRAUSS, G., F. MÜLLER, G. GÄRTNER und H. SCHANZ: Standortsgemäße Durchführung der Abkehr von der Fichtenwirtschaft im nordwestsächsischen Niederland (mit grundsätzlichen Bemerkungen über „gleyartige" Böden). Thar. Forstl. Jb. H. 8 (1939). — KUBIËNA, W. L.: Über das Elementargefüge des Bodens. Bodenkdl. Forsch. (Berlin) **4**, 380–412 (1935). — Beiträge zur Bodenentwicklungslehre: Entwicklung und Systematik der Rendzinen. Bodenkde. u. Pflanzenernähr. **29** (74), 108–109 (1943) (1943a). — Gefügeuntersuchungen an tropischen und subtropischen Rotlehmen. Beitr. z. Kolonialforsch. **III**, 48–58. Berlin: Reiner/Andrews & Steiner. 1943 (1943b). — Entwicklungslehre des Bodens. Wien: Springer. 1948. — Bestimmungsbuch und Systematik der Böden Europas. Stuttgart 1953. — Zur Mikromorphologie, Systematik und Entwicklung der rezenten und fossilen Lößböden. Eiszeitalter u. Gegenwart **7**, 102–112 (1956). — The classification of soils. J. Soil Sci. **9**, Nr. 1, 9–19 (1958). — KUBOTA, J., und L. D. WHITTIG: Podzols in the vicinity of the Nelchina and Tazlina glaciers, Alaska. Proc. Soil Sci. Soc. Amer. **24**, 133 (1960). — KUNDLER, P.: Zur Charakterisierung und Systematik der braunen Waldböden. Z. Pflanzenernähr., Düng., Bodenkde. **78** (123), 109–232 (1957). — KUNTZE, H.: Die Meliorationsfähigkeit der Marschböden in bodenkundlicher Hinsicht. Wasser und Boden **10**, H. 11 (1958). — Beobachtungen und Untersuchungen zur Melioration des Grünlandes in der Marsch. Kalibriefe, Fachgebiet 4, Folge 1, 1960 (1960a). — Entstehung, Eigenschaften, Nutzungs- und Verbesserungsmöglichkeiten von Marschböden. Phosphorsäure **20**, Folge 5 (1960) (1960b).

LAATSCH, W.: Die Dynamik der mitteleuropäischen Mineralböden, 4. Aufl. Dresden u. Leipzig 1957. — LAATSCH, W., und E. SCHLICHTING: Bodentypus und Bodensystematik. Z. Pflanzenernähr., Düng., Bodenkde. **87** (132), H. 1, 97–108 (1959). — LÅG, J.: Rendzinaähnliche Böden in Norwegen (engl.). Meld. Norg. Landbr. Högsk. **38**, Nr. 4, 7 (1959). — Observations on special colours of the A_2-layer of podzol soils in Norway. Acta Agric. Scand. **10**, H. 1, 45–49 (1960). — LARSON, W. E., W. H. ALLAWAY und H. F. RHOADES: Characteristics of three soils from the chernozem and chestnut soil regions of Nebraska. Proc. Soil Sci. Soc. Amer. **12**, 420–423 (1948). — LAUTERBORN, R.: Die sapropelische Lebewelt. Zool. Anz. **24**, (1901). — Die sapropelische Lebewelt. Ein Beitrag zur Geologie des Faulschlamms natürlicher Gewässer. Verh. Naturhist. Med. Ver. Heidelberg. N. F. **13**, 395–481 (1915). — LEEPER, G. W.: On classifying soils. J. Aust. Inst. Agric. Sci. **18**, 77–80 (1952). — The classification of soils — an Australian approach. Rapp. 5. Congr. Int. Sci. Sol, Vol. IV, S. 217–226, Léopoldville 1954. — The classification of soils. J. Soil Sci. **7**, 59–64 (1956). — LEFFINGWELL, E. K. DE: Ground-ice wedges. The dominant form of ground-ice on the north-coast of Alaska. J. Geol. (Chicago) **23** (1915). — LIATSIKAS, N.: Die Verbreitung der Bodentypen in Griechenland. Bodenkundl. Forsch. **4**, H. 3, 413–441 (1935). — Bodenkarte von Griechenland im Maßstab 1:1 Mill. Athen 1942. — LINCK, G.: Die Schutzrinden. Handb. d. Bodenlehre v. E. BLANCK, Bd. III. Berlin 1930. — LIONETT, J. F. G.: Rendzina soils of coastal flats of the Seychelles. J. Soil Sci. **3**, 172–181 (1952). — LÖHNIS, M. P.: Manganese toxicity in field and market garden crops. Plant a. Soil **3**, 193 (1951). — LOŽEK, V., und F. PROSEK: Karstphänomene im Travertin und ihre stratigraphische Bedeutung (tschech.). Cs. Kras **10**, cis. 4, 145–158 (1956). — LOŽEK, V., und J. TYRACEK: Die stratigraphische Erforschung des Travertins in Tucin bei Prerau. Anthropozoikum (Prag) **7**, 261–286 (1958). — LOŽEK, V., und J. KULKA: Das Lößprofil von Leitmeritz an der Elbe, Nordböhmen. Eiszeitalter u. Gegenwart **10**, 81–104 (1959). — LUDWIEG, F.: Über die Eisenoxydhydrate im Boden unter besonderer Berücksichtigung des Einflusses auf den T-Wert der Tonminerale. Diss. Göttingen 1956. — LÜTTMER, J.: Die Böden des Rheiderlandes unter besonderer Berücksichtigung der Knickböden. Diss. Gießen 1952. — LUNDBLAD, K.: Studies on podzols and brown forest soils, I. Soil Sci. 37, 137–155 (1934). — LUNDQUIST, G.: Bodenablagerungen und Entwicklungstypen der Seen. Binnengewässer (Stuttgart) 2 (1927). — LUTZ, J. L.: Die Moore der Oberpfalz. Oberpfalz (Kallmünz) 30, H. 10/11 (1936). — LYFORD, W. H.: The morphology of the brown podzolic soils of New England. Proc. Soil Sci. Soc. Amer. 11, 486–492 (1947).

MANCINI, F.: Delle terre brune d'Italia. Ann. Acad. Ital. Sci. For. 3 (1955). — Carta dei Suoli d'Italia. Firenze 1960. — MANIL, G., und G. HANOTIAUX: Données descriptives sur les sols bruns acides et les sols bruns ocreaux (Brown podzolic soils) des Ardennes belges. Pédologie 7, 239–245 (1957). — MANIL, G.: General considerations on the problem of soil classification. J. Soil Sci. 10, 5–13 (1959). — MARTIN, A. E., und R. J. REEVE: Chemical studies of podzolic illuvial horizons, I. und II. J. Soil Sci. 8, 268–279 (1957). — Chemical studies of podzolic illuvial horizons, III. J. Soil Sci. 9, 89–100 (1958). — MCCALEB, S. B., und M. G. CLINE: Profile studies of normal soils of New York, III: Physical and chemical properties of brown forest and gray brown podzolic soils. Soil Sci. 70, 315–328 (1950). — MCCLELLAND, J. E., C. A. MORGAN, W. M. JOHNSON, F. W. SCHROER und J. S. ALLAN: Chernozems and associated soils of eastern North-Dakota: Some properties and topographic relationships. Proc. Soil Sci. Soc. Amer. 23, 51–56 (1959). — MEINARDUS, W.: Arktische Böden. Handb. d. Bodenlehre v. E. BLANCK, Bd. III, S. 27–96. Berlin 1930. — MERSCHNIN, A. P.: Physikalisch-chemische und landwirtschaftliche Eigenschaften jungfräulicher Schwarzerden und dunkel-kastanienfarbener Böden im Akmolinsker Gebiet Kasachstans (russ.). Nachr. landw. Timirjasew. Akad. 1, 117–134 (1956). — MERVE, C. R. VAN DER: Red and yellow earth. Trans. Fourth Int. Congr. Soil Sci., Vol. 2, S. 169–172. 1950. — MIECZINSKI, T.: Die Böden der Wojwodschaft Lublin. Materialien zur Erforschung polnischer Böden, H. 2 (1932). — MIKLASZEWSKI, S.: Contribution à la connaissance des sols nommés „Rendzinas". C. R. III. Conf. Agropédologique Prag 1922. — MISUSTIN, D.: Das Problem der Versauerung südlicher Tschernoseme (russ.). Potschwowedenie Nr. 1–2, 5–18 (1928). — MOHR, E. C. J.: The soils of equatorial regions with special reference to the Netherlands East Indies. Michigan: Edwards. 1948. — MOHR, E. C. J., und F. A. VAN BAREN: Tropical soils. The Hague and Bandung: Van Hoeve. 1954. — MORTENSEN, H.: Die Wüstenböden. Handb. d. Bodenlehre v. E. BLANCK, Bd. III. Berlin 1930. — MORTIMER, C. H.: Underwater 'soils': a review of lake sediments. J. Soil Sci. 1, 63–73 (1949). — MÜCKENHAUSEN, E.: Die Böden der weiteren Umgebung von Landsberg/Warthe und spezielle Untersuchungen an Grundwasserböden. Landw. Jb. 79, H. 2 (1934). — Die Bodentypenwandlungen des norddeutschen Flachlandes und besondere Beobachtungen von Bodentypenwandlungen in Nordniedersachsen. Geol. Jb. (Berlin) 56 (1936) (1936a). — Die deutschen Bodentypen nach dem heutigen Stand der Bodentypenlehre. Geol. Rdsch. 27, H. 2, 129–156 (1936) (1936b). — Die Bodenbildung der Nordseemarschen. Abh. Nat. Ver. Bremen 30, H. 1/2, 66–77 (1937). — Über gleyartige Böden im Rheinland. Z. Pflanzenernähr., Düng., Bodenkde. 50 (95), 113–134 (1950). — Fossile Böden in der nördlichen Eifel. Geol. Rdsch. 41 (1953). — Die Beurteilung des Faktors „Wasser" bei der bodenkundlichen Kartierung. Forstarch. 25, 12, 269–273 (1954) (1954a). — Fossile Böden im nördlichen Rheinland. Z. Pflanzenernähr., Düng., Bodenkde. 65 (110), 81–103 (1954) (1954b). — Der Wasserhaushalt der Pseudogleye und dessen Bedeutung für die Pflanze. Verh. d. II. u. IV. Komm. d. Int. Bodenkundl. Ges., II, S. 105–111. Hamburg 1958 (1958a). — Die Beurteilung des Bodengefüges bei der forstlichen Standortsaufnahme. Landwirtschaft – Angewandte Wissenschaft. Hiltrup/W.: Landwirtschaftsverlag. 1958 (1958b). — Die Beurteilung des Grundwassers bei der forstlichen Standortsaufnahme. Landwirtschaft – Angewandte Wissenschaft. Hiltrup/W.: Landwirtschaftsverlag. 1958 (1958c). — Entstehung und Umlagerung der Graulehme der Eifel. Fortschr. Geol. Nordrhein-Westfalen 2, 495–502 (1958) (1958d). — Die wichtigsten Böden der Bundesrepublik Deutschland. Frankfurt/M. 1959. — MÜCKENHAUSEN, E., W. GERKHAUSEN und W. KERPEN: Entstehung und Eigenschaften der Böden auf den fossilen Verwitterungsdecken der Eifel. Z. Acker- u. Pflanzenbau 108, H. 1/2, 201–222 (1959). — MÜCKENHAUSEN, E., mit F. HEINRICH, W. LAATSCH und F. VOGEL: Entstehung, Eigenschaften und Systematik der Böden der Bundesrepublik Deutschland. Frankfurt/M.: DLG-Verlag. 1962. — MÜLLER, W.: Untersuchungen über die Bildung und Eigenschaften von Knickschichten in Marschböden. Diss. Gießen 1954. — Verdichtungserscheinungen in Marschböden. Z. Pflanzenernähr., Düng., Bodenkde. 71 (116), 225 (1955). — Grundsätzliche Betrachtungen bei der Ansprache der Marschböden. Praxis u. Forschung 9, H. 2, 17 (1957). — MÜLLER, W., und J. L. RUYTHER: Geol.-bodenkundliche Karte der niedersächsischen Marschen. Amt f. Bodenforschung.

Hannover 1957/58. — Besonderheiten der Marschböden durch den Einfluß des Meeres während ihrer Entstehung. Z. Pflanzenernähr., Düng., Bodenkde. **84** (129), 289 (1959) (1959a). — Grundsätzliche Betrachtungen zur systematischen Gliederung der Marschböden. Geol. Jb. **76**, H. 11 (1959) (1959b). — MUIR, A.: The podzol and podzolic soils. Soil Survey of England and Wales. Adv. Agron. Vol. 13, 56 S. 1961. — MUSIEROWICZ, A.: Klassifizierung der polnischen Böden von der Polnischen Bodenkundlichen Gesellschaft (poln.). Roczn. Glebozn. **3**, 3–24 (1954) (1954a). — Sandige Podsolböden der Wojwodschaft Warszawa (poln.). Roczniki Nauk. rolniczych. Ser. A 70, 1–15 (1954) (1954b).

NAGEL, W.: Physikalische und chemische Zustandsänderungen des Sorptionskomplexes der Nordseemarschböden durch Blausandmelioration. Kühn-Arch. **39** (1935). — NAUMANN, E.: Untersuchungen über Phytoplankton und unter der pelagischen Region vor sich gehende Gyttja- und Dybildungen in gewissen süd- und mittelschwedischen Urgebirgswässern. K. Sv. Vet. Akad. Handb., Bd. 56. 1917. — Einführung in die Bodenkunde der Seen. Die Binnengewässer. Stuttgart 1930. — NEUGEBAUER, V.: Der Woiwodina-Tschernosem, seine Beziehung zum Tschernosem in Ost- und Südosteuropa und seine Degradierungsrichtung (jugosl.). Nauk. Zborn. Mat. Srp. **I**, 78 S. (1951). — NEUGEBAUER, V., und H. ZAKOSEK: Die Smonica. Notizbl. Hess. L.-A. Bodenforsch. **90**, 341–353 (1962). — NEVROS, K. I., und I. A. ZVORYKIN: Zur Kenntnis der Böden der Insel Kreta (Griechenland). Bodenkdl. Forsch. **4**, H. 4/5, 242–308 (1939). — NIDDA, S.: Die Rendzinen von Polessie (poln.). Materialien z. Erforschung poln. Böden, H. 2, 124–156 (1932). — NIEMEIER, G., und W. TASCHENMACHER: Plaggenböden. Westf. Forsch. **1**, 124 (1939). — NIEMEIER, G.: Die Altersbestimmung der Plaggenböden als kulturgeographisches Problem. Geogr. Anz. (Gotha), H. 9/10 (1939). — Datierungen der Kulturlandschaftsgeschichte Nordwestdeutschlands. Abh. Braunschw. Wiss. Ges. **11** (1959). — NIERMANN, W.: Untersuchungen über den Humuszustand der Eschböden. Bodenkde. u. Pflanzenernähr. **21/22** (1940). — NIESCHLAG, F.: Welche Besonderheiten weist der Marschboden auf? Landwirtschaftl. Bl. Weser-Ems **102** (1955) (1955a). — Gedanken zur Bewirtschaftung des Marschbodens. Betriebsw. Mitt. LK Schleswig-Holstein **17** (1955) (1955b). — Probleme der Bodenfruchtbarkeit der nordwestdeutschen Landschaft. Landw. Forsch. **9**, Sonderh. 32 (1956) (1956a). — Aufkalkungsversuche auf Knickböden (verdichtete Marschböden). Praxis u. Forsch. **8**, H. 1, 1 und 17 (1956) (1956b). — NOGINA, N. A.: Die hellgelben podsoligen Böden Weißrußlands (russ.). Potschwowedenie **1952**, 132–144.

OAKES, H.: The soils of Turkey. Ministry of Agriculture, Soil Conservation and Farm Irrigation Division Publication 1, Ankara 1957. — OCHSE, J. J., J. M. SOULE, M. J. DIJKMAN und G. WEHLBERG: Tropical and subtropical agriculture. New York: Macmillan. 1961. — OSTENDORFF, E.: Die Grundwasserböden des Weichseldeltas. Diss. Danzig 1930. — Schrumpfungen und Rutschungen in bindigen, quellbaren Böden. Z. Dtsch. Geol. Ges. **103**, 387–399 (1951). — OVERBECK, F., und S. SCHNEIDER: Torfzersetzung und Grenzhorizont, ein Beitrag zur Frage der Hochmoorentwicklung in Niedersachsen. Angew. Bot. **22**, 321–379 (1940). — OVERBECK, F.: Die Moore. Geologie und Lagerstätten Niedersachsens, Bd. III. Bremen 1950.

PALLMANN, H., und P. HAFFTER: Pflanzensoziologische und bodenkundliche Untersuchungen im Oberengadin. Ber. Schweiz. Bot. Ges. **42**, 337 (1933). — PALLMANN, H., A. HASLER und A. SCHMUTZIGER: Beitrag zur Kenntnis der alpinen Eisen- und Humuspodsole. Bodenkde. u. Pflanzenernähr. **9/10** (54/55), 94–122 (1938) (Berlin). — PALLMANN, H.: Grundzüge der Bodenbildung. Schweiz. Landw. Mh. 1942. — PALLMANN, H., E. FREI und H. HAMDI: Die Filtrationsverlagerungen hochdisperser Verwitterungsprodukte im Bodenprofil einiger Glieder der schweizerischen Braunerdeserie. Ber. Schweiz. Bot. Ges. **53** (1943). — PANKOW, A.: Versuch einer Untersuchung des Bodenkomplexes im Gebiet der mächtigen Schwarzerde (russ.). Potschwowedenie Nr. 3–4, 405 (1930). — PANOMAREWA, W. W.: Über Wesen und geographische Gesetzmäßigkeiten der Podsolbildung (russ.). Potschwowedenie Nr. 3, 31–47 (1956). — PELIŠEK, J.: Neue Forschungen über die Entstehung podsolierter Böden in den Waldgebieten Mitteleuropas in Beziehung zur Waldbiologie (tschech.). Sbornik. českoslav. Akad. zěměd. Věd. Lesnictvi **28**, 479–492 (1955) (1955a). — Bildung und Charakteristik der Podsolböden in den Waldgebieten Mittelböhmens (tschech.). Sbornik českoslav. Akad. zěměd. Věd. Lesn. **28**, 793–824 (1955) (1955b). — Übersicht der typologischen Klassifikation der Waldböden in der ČSSR (tschech.). Sbornik českoslav. Akad. zěměd. Věd. Lesn. **30**, 479–501 (1957). — PENNINGSFELD, F.: Zur Kenntnis der Böden Albaniens. Reichsamt für Bodenforschung, H. 205. Berlin 1942. — PFEFFER, P.: Verwitterungsstudien an Bodenprofilen auf alten Landoberflächen im Gebiet des Rheinischen Schiefergebirges.

Jb. Preuß. Geol. L.-A. **59** (1938). — POHLE, R.: Frostböden in Asien und Europa. Petermanns geogr. Mitt. **1924**, 86–89. — POPOWA, T. W.: Merkmale der Schwarzerden im Steppengebiet der Krim (russ.). Potschwowedenie Nr. 4, 76–86 (1959). — POST, H. v.: Studier öfver nutidens Koprogenajordbildninger, gyttja, dy och mull. K. Sv. Vet. Akad. Handb., Bd. IV. 1862. — POST, L. v.: Das genetische System der organogenen Bildungen Schwedens. Act. 4. Conf. Int. Pédologie, Comm. 4, Vol. III, S. 495. Rom 1924. — POTONIÉ, H.: Über Kalkgyttja aus dem Bäketal, aufgeschlossen durch den Bau des Teltowkanals bei Berlin. Engl. Bot. Jb. **33**, H. 3 (1903). — POTONIÉ, R.: Die Nomenklatur der Unterwasserablagerungen unter besonderer Berücksichtigung derjenigen vorwiegend organischer Natur. Erläuterungen zur Normung der Terminologie der subaquatischen Sedimente. Jb. Preuß. Geol. L.-A. **58**, 426–438 (1938). — PRINCIPI, P.: Osservazioni in torno ad alcuni terreni rossi della Republica di S. Marino. Bull. Soc. Geol. Italiana **56**. Roma 1937.

RAMANN, E.: Bodenkunde. Berlin 1905. — RAMM, K.: Bestandesverjüngung auf den vielfach zur Rohhumus- und Ortsteinbildung neigenden Böden der Buntsandsteinformation der württembergischen Erzgebiete. Ber. Württ. Forstver. **1908**. — RATHJENS, H.: Dreijährige Versuche über die Blausandmelioration in den Nordseemarschen. Kühn-Arch. **42**, 61 (1936). — RATHLEF, H. v.: Die Podsol- und Moorböden Rußlands und der baltischen Staaten. Sammelref. Bodenkde. u. Pflanzenernähr. **17** (62), 201–226 (1930). — RAUPACH, F. v.: Die Plaggenböden des südwestlichen Ammerlandes. Oldenb. Jb. **55** (1955). — REIFENBERG, A.: Die Entstehung der Mediterran-Roterde (Terra rossa). Kolloidchem. Beihefte **5** (1929). — REMEZOW, N. P.: Theorie des Podsolierungsprozesses (russ.). Potschwowedenie **23**, Nr. 8, 1139 (1937). — RETZER, J. L.: Alpine soils of the Rocky Mountains. J. Soil Sci. **7**, 22–32 (1956). — RICHTER, K.: Geol.-bodenkundliche Kartierung der niedersächsichen Marschen im Maßstab 1:5000 als Planungsgrundlage für die Praxis. Wasser u. Boden **9**, H. 2, 51 (1957) — RID, H.: Die Stellung der pseudovergleyten Böden im Acker- und Pflanzenbau. Trans. Int. Soc. Soil Sci. II. u. IV. Komm., Vol. 2, S. 99–104. Hamburg 1958. — RID, H., K. BRUNNACKER, H. ZEIDLER, A. WAGNER, E. EBERHARDT, H. J. ALTEMÜLLER, H. POSCHENRIEDER und TH. BECK: Untersuchung zur Dynamik schwerer Böden, I: Böden des Steigerwaldes. Bayer. Landw. Jb., Sonderh. 4, 3–102 (1960). — ROBINSON, G. W.: Die Böden. Berlin: Verlagsges. f. Ackerbau. 1939. — Some considerations on soil classification. J. Soil Sci. **1**, 150–155 (1949/50). — RODE, A. A.: Die Grundzüge des Podsolierungsprozesses (russ.). Potschwowedenie **23**, Nr. 6, 849 (1937). — Zur Frage des Begriffs der Hydromorphie der Böden in Anwendung auf die Klassifikation der „hydromorphen“ Böden der Steppen-, der Trockensteppen- und der Halbwüstenzone (russ.). Potschwowedenie Nr. 10, 1–13 (1959). — The soil forming prozess and soil evolution. Translated from russian. Published by the Israel Programme for Scientific Translations. Jerusalem 1961. — ROESCHMANN, G.: Die Grundwasserböden des Emstales zwischen Rheine und Papenburg und ihre Beziehung zur Geschichte des Emslaufes. Geol. Jb. **77**, 741–820 (1960). — ROSSOW, N. N., N. A. KARAWAJEWA und A. A. RODE: Erste Plenarsitzung der Kommission für Bodennomenklatur, -systematik und -klassifikation in der Akademie der Wissenschaften der UdSSR (russ.). Potschwowedenie Nr. 8, 60–65 (1957). — Zweite Plenarsitzung der Kommission für Bodennomenklatur, -systematik und -klassifikation in der Akademie der Wissenschaften der UdSSR (russ.). Potschwowedenie Nr. 9, 109–115 (1958). — RÜCKERT, G.: Der Einfluß der Hangnässe auf die Bodenbildung, dargestellt am Beispiel der Böden des Forstamtsbezirks Schieder. Diss. Bonn 1960. — RUHE, R. V., und J. G. CADY: Latosolic soils of the Central African interior high plateaus. Trans. 5. Int. Congr. Soil Sci., Vol. IV, S. 401–407, 1954.

SABOJEWA, I. W.: Gley-Podsolböden (russ.). Potschwowedenie Nr. 3, 24–33 (1958). — SCHACHTSCHABEL, P., R. HAAR und W. KÖSTER: Chemische Untersuchungen an Marschen, I: Böden des Niederelbegebietes. Z. Pflanzenernähr., Düng., Bodenkde. **83** (128), 103–124 (1958) (1958a). — SCHACHTSCHABEL, P., und K. HARTGE: Untersuchungen über die Porengrößenverteilung in drei Marschprofilen. Z. Pflanzenernähr., Düng., Bodenkde. **81** (126), 141 (1958) (1958b). — SCHACHTSCHABEL, P., und W. KÖSTER: Chemische Untersuchungen an Marschen, II: Böden des Ems- und Weser-Jade-Gebietes. 1. Korngrößenverteilung, Sorptionseigenschaften, organische Substanz und Phosphor. Z. Pflanzenernähr., Düng., Bodenkde. **83** (128), 203 (1958) (1958c). — Chemische Untersuchungen an Marschen, II: Böden des Ems- und Weser-Jade-Gebietes. 2. Lösliche Kieselsäure und Oxyde von Aluminium, Eisen und Mangan. Z. Pflanzenernähr., Düng., Bodenkde. **85** (130), 140–151 (1959) (1959a). — SCHACHTSCHABEL, P., und K. HARTGE: Die Messung der Bodenstruktur durch Bestimmung der pF-Kurve und der Strukturstabilität. Landw. Forsch. **12**, Sonderh. 30 (1959) (1959b). — SCHEFFER, F.: Die Bedeutung der Blausandmelioration in den Nordseemarschen.

Landw. Jb. 85, 891 (1938). — SCHEFFER, F., B. ULRICH und P. HIESTERMANN: Die Bedeutung der Chelatisierung in der Agrikulturchemie und Bodenkunde. Z. Pflanzenernähr., Düng., Bodenkde. 76 (121), 146–155 (1957). — SCHEFFER, F., und B. MEYER: Bodenkundliche Untersuchungen an neolithischen Siedlungsprofilen des Göttinger Leinetalgrabens. Gött. Jb. 1958, 3–19. — SCHEFFER, F., und P. SCHACHTSCHABEL: Lehrbuch der Agrikulturchemie und Bodenkunde, I. Teil: Bodenkunde. Stuttgart: Enke. 1956 und 1960. — SCHEFFER, F., und B. ULRICH: Lehrbuch der Agrikulturchemie und Bodenkunde, III. Teil: Humus und Humusdüngung. 1. Morphologie, Biologie, Chemie und Dynamik des Humus. Stuttgart: Enke. 1960. — SCHEFFER, F., H. FÖLSTER und B. MEYER: Zur Entstehung von Schwarzerden und schwarzerdeartigen Böden. Chem. d. Erde 20, H. 4, 302–330 (1960). — SCHEFFER, F., B. MEYER und H. FÖLSTER: Dreischicht-Tonminerale mit Aluminium-Zwischenschichtbelegung in mitteldeutschen sauren Braunen Waldböden. Z. Pflanzenernähr., Düng., Bodenkde. 92 (137), H. 3, 201–207 (1961). — SCHELLING, J.: The well-drained and excessively drained podzols and brown podzolics of the Netherlands. Trans. 6. Int. Congr. Soil Sci., Vol. E, S. 349-352. Paris 1956. — Doel en principe van de moderne bodemclassificatie. Landbouwkund. Tijdschr. (Den Haag) 71, 737–743 (1959). — SCHENK, E.: Die periglazialen Strukturbodenbildungen als Folge von Hydratationsvorgängen im Boden. Eiszeitalter u. Gegenwart 6, 170–184 (1955). — SCHEYS, G., R. DUDAL und L. BAEYENS: Une interprétation de la morphologie de podzols humo-ferriques. Rapp. 5. Congr. Int. Sci. Sol, Vol. IV, S. 274–281. Léopoldville 1954. — SCHLACHT, K.: Bodenstruktur und Sohlenbildung mit Kartierungsbeispielen aus der Vorderpfalz. Z. Pflanzenernähr., Düng., Bodenkde. 27 (1933). — SCHLICHTING, E.: Die Böden der Insel Fehmarn. Naturw. Ver. f. Schleswig-Holstein 26, H. 2 (1953). — SCHLICHTING, E., und H. P. BLUME: Nachweis einer Tonverlagerung in Bodenprofilen. Z. Pflanzenernähr., Düng., Bodenkde. 85 (130), 227 (1959). — SCHMIDT, W., und R. WOLTERS: „Basiston" der Aachener Kreide, Alttertiär und fossile Verwitterung am Nordrand der Eifel. Geol. Jb. 66 (1952). — SCHMIEDEL, H.: Beiträge zur Kenntnis pseudovergleyter Böden Sachsens. Diss. Dresden 1959. — SCHNEIDER, F.: Übergänge zwischen mineralischen Grundwasserböden und gleyartigen Böden. Diss. Bonn 1950. — Übergangsbildungen von mineralischen Grundwasserböden zu gleyartigen Böden in Westdeutschland. Z. Pflanzenernähr., Düng., Bodenkde. 65 (110), 32–52 (1954). — SCHÖNHALS, E.: Über fossile Böden im nicht vereisten Gebiet. Eiszeitalter u. Gegenwart 1, 109–130 (1951) (1951a). — Riß- und würmeiszeitliche Frostbodenstrukturen aus der Magdeburger Börde. Geol. Jb. f. d. Jahr 1949, 65, 589–602 (1951) (1951b). — Ein fossiler periglazialer Strukturboden bei Appenrode am Nordrand des Vogelsberges. Notizbl. Hess. L.-A. Bodenforsch. (Wiesbaden) 84, 374–379 (1956). — Spätglaziale äolische Ablagerungen in einigen Mittelgebirgen Hessens. Eiszeitalter u. Gegenwart 8, 5–17 (1957). — SCHOKALSKAJA, S. J.: Die Böden Afrikas. Berlin: Akademie-Verlag. 1953. — SCHOLZ, H.: Studien über die Bodenbildung zwischen Rehoboth und Walvis Bay. Diss. Bonn 1963. — SCHROEDER, D., und U. SCHWERTMANN: Zur Entstehung von Eisenkonkretionen im Boden. Naturwiss. 42, H. 9, 255 (1955). — SCHUCHT, F.: Über das Vorkommen von Bleicherde und Ortstein in den Schlickböden der Nordseemarschen. Int. Mitt. Bodenkde. 3, 104 (1913). — SCHÜTTE, H.: Das Alluvium des Weser-Jade-Gebietes. Wirtschaftswiss. Ges. z. Studium Niedersachsens, Reihe B, H. 13. Oldenburg 1935. — SEE, K. v.: Über den Profilaufbau der Marschböden. Int. Mitt. Bodenkde. 10, H. 5, 6 (1920). — SELLKE, M.: Die Böden Südniedersachsens. Wirtschaftswiss. Ges. z. Studium Niedersachsens, Reihe B, H. 12. Oldenburg 1935. — SHERMAN, G. D., Z. C. FOSTER und C. K. FUJIMOTO: Some of the properties of the ferruginous humic latosols of the Hawaiian Islands. Proc. Soil Sci. Soc. Amer. 13, 471–476 (1949). — SHERMAN, G. D., und Y. KANEHIRO: Origin and development of ferruginous concretions in Hawaiian latosols. Soil Sci. 77, 1–8 (1954). — SHERMAN, G. D., und L. T. ALEXANDER: Characteristics and genesis of low humic latosols. Proc. Soil Sci. Soc. Amer. 23, 168–170 (1959). — SHIVA RAU, H., und S. KASINATHAN: Studies of soil systematics: The black soils of the Madras Deccan. J. Soil Sci. 2, 61–66 (1951). — SIBIRZEW, N. M.: Die Grundlage der genetischen Bodenklassifikation (russ.). Mem. Inst. Agron. Novo-Alex. 9 (7) (1895). — SIGMOND, A. A. J. v.: Die Böden trockener Gebiete. Handb. d. Bodenlehre v. E. BLANCK, Bd. III. Berlin 1930. — SMITH, G. D., C. U. BRYSON und E. M. THOMSON: Chemical composition of the peat bogs of the maritime provinces. Canad. J. Soil Sci. 38, 120–127 (1958). — SMOLÍKOWÁ, L.: Die Bodenverhältnisse des südslowakischen Karstes. Acta Univ. Carolinae, Geologia (Prag) Nr. 3, 273–300 (1959). — Zur Altersfrage der untersuchten Terra fusca. (tschech.). Casopis pro Mineralogii a Geologii (Prag) 3, Ročnik V (1960) (1960a). — Fossile gefleckte Böden in der Tschechoslowakei (tschech.).

Věstník Ustredního Ustavu Geologického, Ročník (Srpen) **25**, C. 5 (1960) (1960b). — SOLOWJEW, M. M.: Das Sapropelproblem in der UdSSR (russ.). Leningrad: Verh. d. Akad. 1933. — SOMMERKAMP, G., F. GALENSA und H. KUNTZE: Probleme deutscher Marschen. Ein Überblick über den derzeitigen Stand der Forschung. Z. Kulturtechnik **1**, H. 5, 257–288 (1960). — SSELJAKOW, S. N.: Die Klassifikation der Tschernosemböden (russ.). Potschwowedenie Nr. **3**, 34–36 (1955). — STEBUTT, A.: Lehrbuch der allgemeinen Bodenkunde. Berlin 1930. — STEEGER, A.: Neue Beobachtungen über Frostspalten und Würgeböden am Niederrhein. Erdkunde (Bonn) **2**, (1948). — STEFANOWITS, P., und L. SZUCS: Die Prinzipien der Klassifizierung, angewandt bei der Erstellung der genetischen Karte ungarischer Böden. Rapp. 6. Congr. Int. Sci. Sol., Vol. E, S. 323–330. Paris 1956. — STEFANOWITS, P.: Genetisch-bodengeographische Klassifikation der Waldböden Ungarns (ungar.). Agrokémia és Talajtan **8**, 163–184 (1959). — STEPHENS, C. G.: The elements of soil classification. J. Dept. Agric. S. Aust. **54**, 141–142 (1950). — The classification of australian soils. Rapp. 5. Congr. Int. Sci. Sol. Vol. IV, S. 155–160. Léopoldville 1954. — A manual of Australian soils. Commonwealth Scientific and Industrial Research Organisation, Division of Soils. Melbourne 1956. — Soil map of Australia. Commonwealth Scientific and Industrial Research Organization, Division of Soils. Melbourne 1960. — STEUSLOFF, U.: Periglazialer „Tropfen"- und Taschenboden im südlichen Münsterland bei Haltern. Geol. Jb. **66** (1957). — STOBBE, B. C.: The morphology and genesis of the gray brown podzolic and related soils of Eastern Canada. Proc. Soil Sci. Soc. Amer. **16**, 81–84 (1952). — STOBBE, B. C., und J. A. WRIGHT: Modern concepts of the genesis of podzols. Proc. Soil Sci. Soc. Amer. **23**, 161–164 (1959). — STOBBE, B. C.: The great soil groups of Canada, their characteristics and distribution. Agric. Inst. Rev. **15**, Nr. 2, Dept. of Agric., Soils Res. Inst. Ottawa (1960). — STREMME, H.: Grundzüge der praktischen Bodenkunde. Berlin 1926 (1926a). — Über einige Systeme der natürlichen Bodeneinteilung nebst dem Vorschlage einer für Feldpedologen verwendbaren Klassifikation. Actes de la IVième Conférence Intern. de Pédologie. Rome 1926 (1926b). — STREMME, H., und K. SCHLACHT: Über Steppenböden des Rheinlandes. Chem. d. Erde **3**, 28 (1927). — STREMME, H.: Die Steppenschwarzerden. Handb. d. Bodenlehre v. E. BLANCK, Bd. III, S. 257–283. 1930 (1930a). — Die Braunerden. Handb. d. Bodenlehre v. E. BLANCK, Bd. III, S. 160–182. 1930 (1930b). — Die Böden Deutschlands. Handb. d. Bodenlehre v. E. BLANCK, Bd. V. Berlin 1930 (1930c). — Die Böden des deutschen Reiches und der Freien Stadt Danzig. Petermanns Mitt., Erg.-H. 226 (Gotha 1936). — STREMME, H. E.: Zum Vorkommen brauner Steppenböden im Oberrheingebiet. Z. Pflanzenernähr., Düng., Bodenkde. **60** (105), 273–278 (1953). — Bodenkarte der Insel Fehmarn im Maßstab 1:50000. Geol. L.-A. Schleswig-Holstein. Kiel 1955. — STUART, D. M., M. A. FOSBERG und G. C. LEWIS: Caliche in southwestern Idaho. Proc. Soil Sci. Soc. Amer. **25**, H. 2, 132–135 (1961). — SYS, C.: La carte des sols du Congo belge et du Ruanda-Urundi. Pédologie **10**, 48–116 (1960).

TACKE, B.: Die Humusböden der gemäßigten Breiten. Handb. d. Bodenlehre v. E. BLANCK, Bd. IV, S. 143–156. Berlin 1930. — TAMÉS, G.: Los grupos principales de Suelos de la España Peninsular. Ministerio de Agricultura, Mapa Agronómico Nacional. Madrid 1958. — TAMURA, T., und C. L. W. SWANSON: Chemical and mineralogical properties of a brown podzolic soil. Proc. Soil Sci. Soc. Amer. **18**, 148–153 (1954). — TANFILIEW, G.: Zur Entstehung des Tschernosembodens und der südrussischen Steppen (russ.). Potschwowedenie Nr. 1–2, 5–18 (1928). — TATARINKOW, S. F.: Merkmale der Schwarzerden im südlichen Transuralien (russ.). Potschwowedenie **1949**, 387–393. — TAVERNIER, R., und G. D. SMITH: The concept of Braunerde (Brown Forest Soil) in Europe and in the United States. Adv. Agron. **9**, 217–289 (1957). — TAVERNIER, R., und J. AMERYCKX: Le postpodzol en Flandre sablonneuse. Pédologie **7**, 89–96 (1957). — TAVERNIER, R., und R. MARÉCHAL: Carte des associations de sols de la Belgique. Pédologie **8**, 134–182 (1958). — TAVERNIER, R., u. a.: Soil map of Europe. Gent 1963. — TAYLOR, N. H., und I. J. POHLEN: Soil Survey Method. New Zealand Soil Bur. Bull. 25. 1962. — THORP, J.: Geography of the soils of China. Nanking: The National Geological Survey of China. 1936. — TOMASZEWSKI, J.: Kritischer Überblick über die polnische Bodenklassifikation (poln.). Roczn. Glebozn. **3**, 106–116 (1954). — Moorböden und ihre Bildungsverhältnisse (russ.). Potschwowedenie, Nr. 5, 1–11 (1957). — TREITZ, P.: Die Bildungsprozesse des Bodens im Osten des pannonischen Beckens. Jber. Kgl. Ungar. Geol. Reichsanst. 1912. — TROLL, C.: Strukturböden, Solifluktion und Frostklimate der Erde. Geol. Rdsch. **34**, 545–694 (1944).

United States Department of Agriculture: Yearbook of Agriculture – Soils and Men. Washington, D.C. 1938.

VAGELER, P.: Grundzüge der tropischen und subtropischen Bodenkunde. Berlin: Verlagsges. f. Ackerbau. 1930. — VAGELER, P., und F. ALTEN: Böden des Nil und Gash. Z. Pflanzenernähr. Düng., Bodenkde. Teil A, 23, H. 3/4, 1–191 (1932). — VAGELER, P.: Zur Bodengeographie Algiers. Petermanns Geogr. Mitt., Erg. H. 258 (1955). — VEENENBOS, J. S.: Gedanken zum Knickproblem. Z. Pflanzenernähr., Düng., Bodenkde. 68 (113), 141–157 (1955). — VILLAR, E. H. DEL: España en el Mapa internacional de suelos. Ministerio de Fomento. Publicaciones agrícolas. Madrid 1927. — Suelos de España. Madrid 1929. — Les sols méditerranéens. Madrid 1930. — Classification générale des sols. Proc. and Papers of the Second Intern. Congr. of Soil Sci. Comm. V. Moscow 1932. — VOGEL V. FALCKENSTEIN, K., und G. Frhr. v. ROMBERG: Die Molkenböden des Bram- und Reinhardswaldes im Buntsandsteingebiet der Oberweser. Int. Mitt. Bodenkde. (Berlin und Wien) 5, H. 2, 77 (1915). — VOGEL, F.: Bodenbildung und Bodenfruchtbarkeit, aufgezeigt an südbayerischen Bodentypen unter besonderer Berücksichtigung von Ackerböden. Z. Pflanzenbau u. Pflanzenschutz, Sonderh. 3, 11–43 (1952). — Erklärung von Fachausdrücken für die Bodentypen. Landw. Jb. Bayern, Sonderh. „50 Jahre Bayerische Pflanzenzüchtung". 1953. — Boden und Landschaft. München: Landw. Bildberatung. 1957.

WAKSMAN, S. A., und H. B. SCHULHOFF: The nature and properties of peats in New Jersey. Soil Sci. 54, 447 (1942). — WASMUND, E.: Lakustrische Unterwasserböden (Seeablagerungen des Bleicherdegebietes der nördlich gemäßigten Zone). Handb. d. Bodenlehre v. E. BLANCK, Bd. V. Berlin 1930. — WEBER, C. A.: Über Torf, Humus und Moor. Versuch einer Begriffsbestimmung mit Rücksicht auf die Kartierung und die Statistik der Moore. Abh. Naturwiss. Ver. Bremen 17, H. 2 (1903). — WEHRMANN, J.: Mangan, Kupfer und Kobalt in Pflanzen und Böden Schleswig-Holsteinischer Weidegebiete. Plant a. Soil 6, 61 (1955). — WERNER, J.: Zur Kenntnis der Braunen Karbonatböden (Terra fusca) auf der Schwäbischen Alb. Diss. Stuttgart 1958. — WILDVANG, D.: Der Boden Ostfrieslands. Aurich 1929. — WILENSKI, A. G.: Bodenkarte der Erde im Maßstab 1:20 Mill. Dokutschajew-Institut. Moskau 1957. — WIMMER, C.: Die Moore Vorarlbergs. Natur u. Land 38, 140–143 (1952). — WOHLTMANN, F.: Handbuch der Agrikultur. Leipzig: Duncker & Humblot. 1892. — WOLDSTEDT, P.: Das Eiszeitalter. Stuttgart 1954. — WOLOBUJEW, W. R.: Vergleichende geographische Untersuchungen der klimatischen Bedingungen der Schwarzerdebildung (russ.). Potschwowedenie Nr. 8, 84–93 (1958). — WORTMANN, H., und H. MAAS: Außergewöhnlich starke Humuspodsole bei Haltern/Westfalen. Z. Pflanzenernähr., Düng., Bodenkde. 65 (110), 15–26 (1954). — WORTMANN, H.: Ein erstes sicheres Vorkommen von periglazialem Steinnetzboden im Norddeutschen Flachland. Eiszeitalter u. Gegenwart 7, 119–126 (1956). — WYSSOTZKI, G. N.: Hydrologische und geobiologische Beobachtungen in Veliko-Anadol, V: Grundwasser (russ.). Potschwowedenie Nr. 22 (1900). — Gley (russ.). Potschwowedenie Nr. 7 (1905).

YAALON, D. H.: Classification and nomenclature of soils in Israel. Taxonomic comparison and genetic relationship with soils from other countries. Bull. Res. Coun. Israel 8 G, 91–118 (1959). — YAMANAKA, K., und S. MOTOMURA: Studies of the Gley formation of soils, I: On the mechanism of the formation of active ferrous iron in soils. Soil a. Plant Food 5, 134–140 (1959). — YARKOW, S. P.: The formation of podzolic soils. Rapp. 5. Congr. Int. Sci. Sol, Vol. IV, S. 83. Léopoldville 1954.

ZAKOSEK, H.: Über die Deutung des Profilgepräges gleyartiger Böden. Diss. Bonn 1951. — Über Pseudogleye in der Rheinpfalz. Notizbl. Hess. L.-A. Bodenforsch. 82, 258–268 (1954) (1954a). — Zur Beurteilung von Pseudogleyen. Z. Pflanzenernähr., Düng., Bodenkde. 65 (110), 27–31 (1954) (1954b). — Die Böden der Rheinpfalz. Notizbl. Hess. L.-A. Bodenforsch. 84, 356 (1956). — Grundwasser und Staunässe. Z. Pflanzenernähr., Düng., Bodenkde. 74 (119), 240–242 (1957). — Zur Genese und Gliederung der Steppenböden im Oberrheintal. Abh. Hess. L.-A. Bodenforsch. 37, 46 S. (1962). — ZEUNER, F. E.: Frost soils on Mount Kenya, and the relation of frost soils to aeolian deposits. J. Soil Sci. 1, 20–30 (1949). — ZVORYKIN, L. A., und G. R. STOGIANNIS: Soil map of Peloponnesus. Athens 1958.

Nachtrag:

GANSSEN, R., und F. HÄDRICH: Atlas zur Bodenkunde. Mannheim: Bibliographisches Institut. 1965. — KOWDA, V. A.: Problems of Soil Science and Reclamation. Paris: UNESCO. 1964. — SZABOLCS, I.: Proceedings of the Symposium on Sodic Soils. Budapest: Publishing-House of the Hungarian Academy of Science. 1966.

B. Soil Structure

a) Factors and Forces in Soil Structure

By

L. De Leenheer and M. De Boodt

1. Definition of Soil Structure

The primary soil particles do not exist as individuals in the field, but are more or less aggregated into compound soil particles. The term "structure" refers to this aggregation.

Soil structure is an important property, because it influences not only the physical conditions, aeration and water economy, but also the availability of plant nutrients, the decomposition of soil organic matter and all the microbiological activities. Speaking of soil structure as the expression of the spatial arrangement of soil particles, we must distinguish:

a) the spatial arrangement of elementary soil particles into aggregates and the study of the forces causing or destroying these aggregations. This is the more theoretical aspect of soil structure investigation.

b) the spatial arrangement of the compound soil particles and the pore size distribution resulting from it. This is the more practical aspect of soil structure investigation. It includes not only the study of the influence of pore size distribution on soil air and water economy and on plant growth, but also the methods of finding out the best conditions of the soil physical environment.

c) the spatial arrangement and genesis of the compound soil particles as it can be described either in the field (macrostructure) or in the laboratory with the aid of the microscope (microstructure): soil structure morphology. The latter is the meaning given by scientists interested in soil survey, soil classification and soil genesis.

In view of the distinctions made above it is evident that there must be confusion in the use of the term "structure".

2. Arrangement of Elementary Soil Particles into Aggregates

Soil aggregates are divided into micro-aggregates or micro-elements when their diameter is smaller than 250 μ and macro-aggregates when their diameter is smaller than 10 mm but larger than 250 μ. Structural elements larger than 10 mm are called clods (Werschinin 1958). Different forces keep together the components of these micro-aggregates, which components are the colloidal soil organic matter, the mineral soil particles (clay, silt, very fine to fine sand), anions, cations and water. The forces are mainly of three different kinds:

1) capillary forces or forces resulting from the existence of a meniscus,
2) intermolecular forces or forces of van der Waals,
3) chemical forces.

The intermolecular forces are the most important; for such the intensity of the attraction diminishes with the sixth power of the distance between the particles (Freundlich 1930). This explains why the soil particles must be put very close to each other in order to build micro-aggregates.

In soils, intermolecular forces originate either in the flocculation of soil colloids or in the shrinkage of a drying soil.

A. The Flocculation of Soil Particles[1]

Direct aggregation of soil particles takes place when they are smaller than 1 μ, and the micro-aggregates built this way are not larger than 250 μ (SMOLUCHOWSKI, WIEGNER and MULLER, cited by WERSCHININ 1958). When the elementary particles are larger than 5 μ, the aggregation does not take place directly, but through the presence of colloidal particles.

Direct flocculation is made possible when the electric charges of the colloidal particles are neutralized; this does not happen for 100 percent of the charge, so a flocculating micro-aggregate will have a residual electric charge, which is important for the flocculation of larger aggregates (KATSCHINSKI 1958), as it is illustrated in Fig. 21.

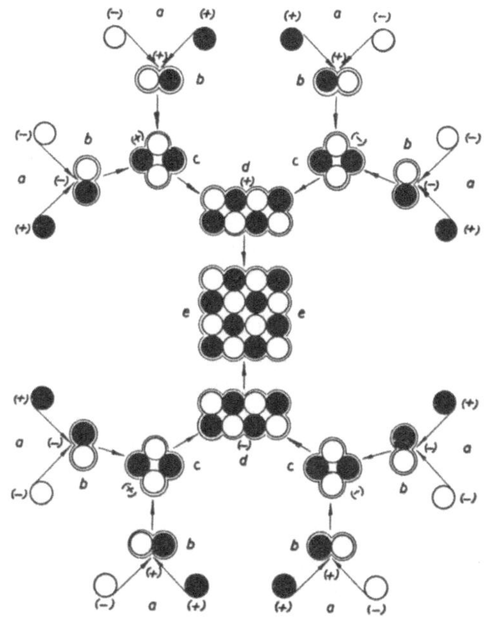

Fig. 21. Scheme of the formation of micro-aggregates through flocculation (KATSCHINSKY 1958)
a colloidal elementary particles; *b*, *c*, *d*, *e* micro-aggregates of lst, 2nd, 3rd, 4th order

The addition of salts to a colloidal solution is a well known way for reducing the electrokinetic potential difference in the electric double layer or GOUY-layer enveloping every charged particle, and in some cases it is possible to change the sign of the initial electric charge of the colloid (FREUNDLICH 1930).

The stability of the flocculation of the elementary particles depends mainly on three factors:

— the size and structure of the elementary particles,

— the physico-chemical properties of the electric double layer (GOUY-layer) of the colloids,

— the chemical stabilization of the aggregates.

Size and Structure of Elementary Particles as Factor of Flocculation. The smaller the particles are, the easier it is for the Brownian movement to bring them together and the larger the possible number of points of contact per unit surface; this means: the larger the specific surface, the higher the total surface energy. The alteration of the particles is important too: a mineral particle is composed of a crystalline nucleus and an amorphous swellable envelopment. The thickness of the latter at a given moment is in equilibrium with its own solubility in soil water (CORRENS 1940). The smaller the clay particle, the greater the relative importance of the swellable envelopment, the higher the active surface, and the more stable the aggregation. The structure of the crystalline nucleus also is very important (VAGELER 1932, WIEGNER 1935, SCHACHTSCHABEL 1940) and influences directly the physico-chemical properties of the electric double layer. The clay

[1] In English publications (e.g. BAVER 1948) the word "coagulation" is seldom used; French authors (e.g. DEMOLON 1938) prefer the word flocculation to coagulation, although they are considered as synonymous; in German (see BLANCK, Handbuch der Bodenlehre, 10 Volumes, 1928–1932) only the word "Koagulation" is used.

minerals generally acquire a negative charge through the substitution in the crystal lattice (called isomorphous replacement) of divalent for trivalent cations or of trivalent for quadrivalent cations (e. g. Mg taking the place of Al or Al taking the place of Si in the lattice).

This type of charge is active on the horizontal surface of the lamellar clay minerals.

The charge on the surface of the clay minerals may also be caused by the ionisation of the AlOH- and SiOH-groups, mainly on the broken lateral surface (broken bonds). These groups have an amphoteric character and their electric charge, both the sign and the magnitude of the charge, depends on the pH. In acid environment the exposed OH-groups tend to become positive. At high pH these OH-groups become negatively charged.

A third cause of the electrical charge on the clay surface are the cleavage planes. The elementary cell of some clay minerals, such as muscovite, including the potassium which is located between the "potential" cleavage planes, may be considered neutral. The increase of the distance between the cleavage planes, as a consequence of weathering, means a destruction of the neutrality of the elementary cell, because broken bonds originate. The elementary cell becomes electrically charged.

It is known (Grim 1953) that in kaolinite the cleavage is not so pronounced as in other clay minerals. It is also known that the isomorphous replacement does not occur in kaolinite, but is very important in montmorillonite. For this reason the basal charge (charge of the horizontal surface of the lamellar particles) of montmorillonite is so important that the charge caused by broken bonds becomes negligible in this mineral. The kaolinite is the other type of clay mineral, where the electrical charge is essentially the consequence of broken bonds, the total charge being about 15 times smaller than that of montmorillonite and the sign of the charge being undetermined (amphoteric character). This makes it evident why the stability of a montmorillonite clay suspension is better than that of a kaolinite clay suspension.

Other colloids in soils, which are subject to flocculation, are silica, sesquioxides of iron and aluminium, oxides of manganese, together with organic matter. These colloidal substances have amphoteric properties. Some of these components, such as iron oxides, are known for their positive electrical charge under normal soil conditions. To understand soil structure and soil fertility properties not only the physico-chemical properties of the colloidal particle are important, but also pure physical surface activities, as well as chemical reactions, including cementation, hydration and dehydration, binding forces between colloids and larger soil particles, and binding force between organic and mineral soil components.

The Physico-chemical Properties of the Electric Double Layer. The existence of a surface charge on the interface between the solid and the liquid phase causes the formation of an electric double layer in which the orientation, activity and potential of the ions are different from those in the same solution, but which are not in the immediate vicinity of the electrically charged solid surface. The concept of the electric double layer is very important to a clear understanding of clay flocculation. The first substantial theory of the electrical double layer was proposed in 1879 by von Helmholtz, who suggested that the double layer is virtually an electrical condenser with parallel plates at an extremely small (molecular) distance; the particle surface with its charge forms the inner layer and the counterions form the outer layer (very close to the former) of the electrical double layer (see Fig. 22, left part of the figure). The so-called new theory of the diffuse double layer at a plane surface is in fact an improvement of HELMHOLTZ's idea, made by GOUY in 1910. Later, CHAPMAN, although independently, developed the same idea.

So the same theory is meant when, according to the country, one is speaking of the improved HELMHOLTZ, the GOUY or the GOUY-CHAPMAN theory of the electric double layer. The distribution of the counterions in the double layer is determined by the electrostatic forces on the clay surface, the valence of the counterions, the total electrolyte concentration of the solution and the thermal movement of the ions. As at room-temperature the here mentioned levels of energies are of the same order of magnitude (BOLT 1962), the distribution of the ions is called "atmospheric" or diffuse (see Fig. 22, central part).

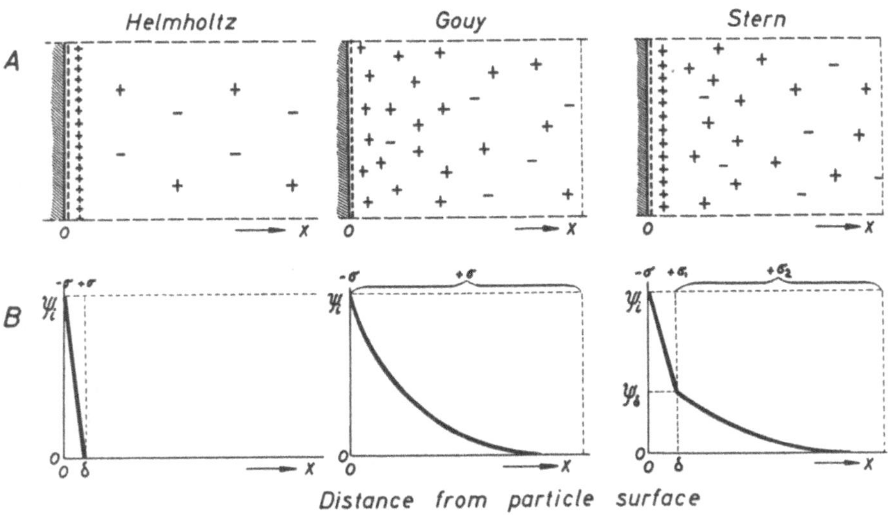

Fig. 22. Schematic representation of the ion distribution (A) and the decreasing potential (B) in the double layer, according to the theories of HELMHOLTZ, GOUY and STERN (ψ potential; σ surface charge density and x distance from particle surface) (WIKLANDER 1955)

To-day it is recognized that the true picture of the electric double layer is a bit more complicated than the one proposed by GOUY. Indeed, in the latter no mention is made of attractive forces, other than electrostatic, between the ions and the surface. Moreover the ion behaviour is considered ideal and the surface charges of clay minerals are considered homogeneously distributed over the surface. In fact they are spaced at considerable distances from each other; e. g. 14 Å for montmorillonite (SCHUFFELEN and BOLT 1958). In any case it is evident that a considerable portion of the exchangeable ions is held very close to the adsorbing surface. This was first recognized by STERN. In this way it may be said that STERN combines the theory of HELMHOLTZ with the theory of GOUY; the inner part of the ion swarm is indeed similar to HELMHOLTZ's layer and the outer part is similar to the GOUY layer (see Fig. 22, right part of the figure).

Up to now the attention has been drawn too much to the fact that the clay surface is negatively charged and that the counterions consequently are positive. In fact also negative counterions occur simultaneously with positive ones in the electric double layer. Let us suppose a negative charge of a clay particle in suspension; such a charge leads to a deficit in the anion-concentration close to the particle surface, in comparison to the concentration in the intermicellar liquid (see Fig. 23). This is called the negative adsorption of anions (BOLT 1962).

A theory which is different from the GOUY theory is the membrane equilibrium theory of DONNAN. This theory, developed in 1911, was applied in soil science by MATTSON and his school since 1929. DONNAN considers the boundary between the electrical double layer and the intermicellar liquid as a membrane which is not permeable to anions, assuming the electrical charge of the colloid is negative. MATTSON and his school, particularly WIKLANDER (1955), apply the DONNAN membrane theory to explain the ion exchange processes on clay minerals, which in the opinion of BOLT and SCHUFFELEN (1958) is of limited value.

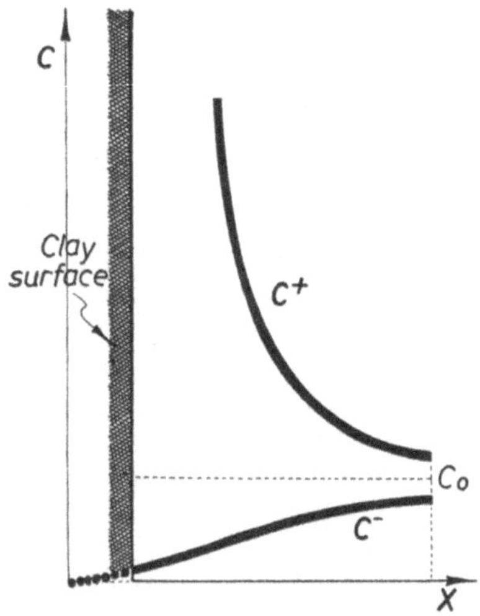

Fig. 23. Kationic (C^+) and anionic (C^-) distribution in a planar electric double layer, the concentration of the solution outside the double layer being Co (x represents the distance from the clay surface) (BOLT 1962)

Summarizing it may be said that, up to now, the theory of STERN explains very well the adsorption and exchange phenomenon in soils, as far as the explanation is limited to the experimental aspect of the problem. The more simple GOUY layer theory, however, is the only one which has been put already in a mathematical formula, so it can be used to calculate in advance some quantitative aspects of the adsorption and exchange phenomena.

Although some shortcomings of the GOUY theory are recognized, the calculated ion distribution in the diffuse double layer was found to be in good agreement with the observed ion distribution. A comprehensive study on this matter was published by SCHUFFELEN and BOLT (1958).

BOLT (1962), who applied the POISSON-BOLTZMANN equation (KRUYT 1947) to calculate the "atmospheric distribution" of the ions in the double layer, found that:

a) The increase of the concentration of the electrolyte with a factor n compresses the potential curve in the direction of the surface of the clay mineral with a factor proportional to n.

b) The increase of the valency of the ions with a factor n compresses the potential curve in the direction of the x and y axes with a factor proportional to n.

c) The change of the density of charge by isomorphous replacement gives rise to a translation of the potential curve over a distance inversely proportional to the density of charge. However, at a fairly high density of charge the translation of the potential curve is small.

d) Taking into account that the charge density of most clay minerals varies between 1 to 3×10^{-7} milliequivalents per cm^2, the thickness of the double layer is about 200 Å in a 10^{-3} molar mono-monovalent salt solution. The concentration of the counterions against the surface of the clay particle is then 2 to 10 molar. Using the BOLTZMANN equation (KRUYT 1947) the potential difference between the solid and the liquid phase in the electric double layer can be estimated.

This quantitative approach to the problem is valid whatever the kind of clay mineral and the ion composition in the diffuse double layer may be.

This does not mean that every adsorption phenomenon can already be put in an equation; for several practical aspects, e. g. of the flocculation of a soil suspension, one has still to go back to the earlier workers who derived exchange equations from laboratory experiments, their results being statistically checked and put in agreement with classic physico-chemical conceptions. This might be called the more experimental approach of the phenomenon.

The Electrokinetic Potential or Zeta Potential. To introduce the more experimental study of the flocculation phenomena which will be discussed now, the picture of the potential curve in an electric double layer is given in Fig. 24.

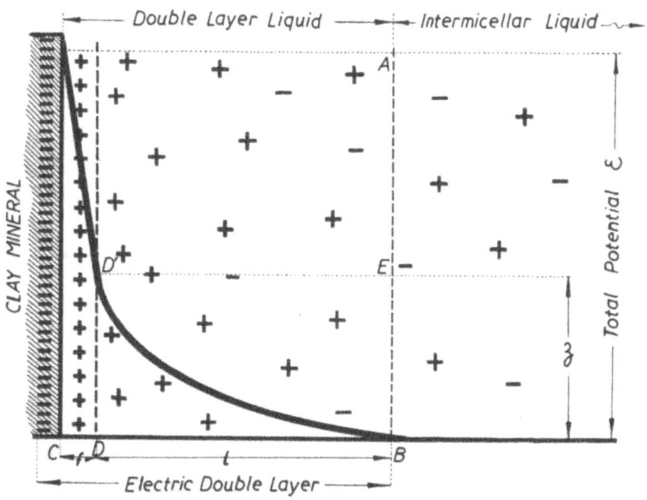

Fig. 24. The thermodynamic potential ε and the electric double layer. For a moving clay particle, the liquid coat *CD* strongly adsorbed to the particle moves with it. The electrokinetic potential *AB*—*AE*=*EB* is called the zeta (ζ) potential. *CD* is the fixed fluid coat (*f*), *DB* is the layer of losely fixed ions (*l*) (please, replace the wrong symbol ẓ in the figure by ζ)

Two different parts can be distinguished in this figure, i. e. the fixed fluid coat (*CD*) and the diffuse part *D'E*. When a clay particle is moving in an electric field, it is observed that from the whole double layer (this is from the clay surface down to the point *B*) only a part of it, *CD*, is moved with the clay mineral; this part is the fixed fluid coat. The movement of the layer *DB* decreases with the distance to the clay particle. Consequently, between the two layers of the double layer an electrokinetic potential, with a magnitude equal to *EB*, is originated. Looking at Fig. 24, the following definition may be given: the electrokinetic potential difference ζ, also called the zeta potential, is the difference in potential or magnitude of charge (when a clay particle is moving in a liquid) between two points in the fluid phase, one at the limit *D'* of the fixed fluid coat coat *CD* of the double layer, the other at or outside the limit *B* of the diffuse part of the double layer.

It is the general experience that there exists a close relationship between the ζ potential (measurable e. g. as the moving speed of particles in an electric field) and the amount of an electrolyte needed to flocculate or to disperse a suspension. So, depending on the ions adsorbed and the amount of NaCl added, a suspension will disperse or flocculate. It is evident that with increasing values of the ζ potential, the repulsive forces between two equally charged particles increase and no coagulation takes place. On the contrary, when the ζ potential is lowered, the repulsive forces are no longer strong enough to keep the particles apart and flocculation occurs. Clay particles in a suspension will flocculate as soon as the ζ potential is lowered, not necessarily to zero, but to a certain value, called the critical value or critical potential (Kruyt 1947). Consequently attention is paid to the factors which influence the ζ potential.

Experiments conducted to study the movement of colloidal particles in an electric field, have shown that the magnitude of the zeta potential, ζ, is equal to

$$\zeta = \frac{e \cdot d}{D \cdot r^2},$$

in which: e = the charge of the particles (total potential),
d = the thickness of the fixed fluid coat,
D = the dielectric constant of the liquid,
r = the radius of the particles (Scheffer and Schachtschabel 1960).

The charge of the particles of humic acids and clay minerals, which form the majority of the "colloidal fraction" of soils, cannot be changed very much. Theoretically it must be possible, by lowering the pH, to bring a soil colloid to its isoelectric point. However, as the isoelectric point of the soil colloids (except kaolinite) is much lower than the pH value of the hydrogen saturated soil col-

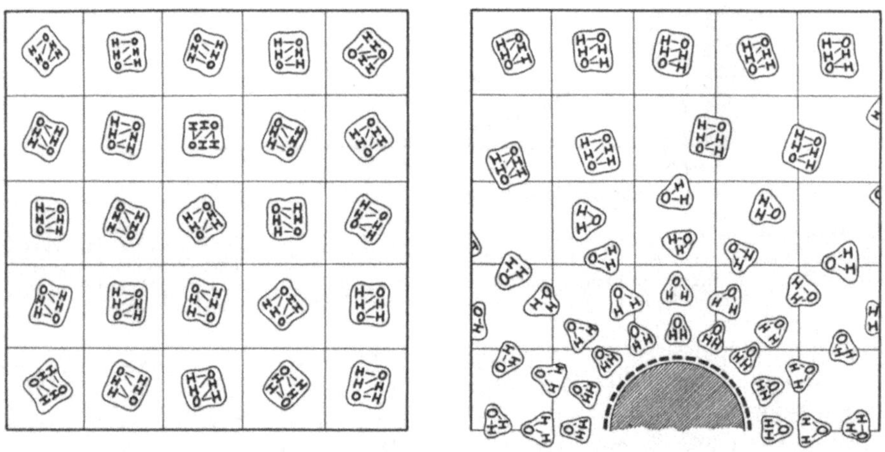

Fig. 25. The double water molecules in the liquid state (two dipoles with the opposite poles adjacent) are deformed under the influence of a clay particle or ion. The result is an apolar adsorbed water hull composed of single dipoles (de Sigmond 1938)

loids, important changes of charge, within the normal pH range which is found in soils, do not occur. This conclusion is not valid for free oxides of iron, aluminium and manganese, which have an isoelectric point at slightly acid pH values. Therefore the change of the particle charge of the free sesquioxides as a consequence of pH-changes in soils is more readily to be expected (Scheffer and Schachtschabel 1960).

It should be remarked that the value of the particle charge gets its real significance when considered per unit surface, which is expressed in the formula above as the ratio e/r^2. The thickness of the fixed fluid coat depends on the hydration, the concentration and valence of the ions in the electric double layer.

A very important phenomenon is the hydration of ions and colloidal particles. In 1925 Wiegner proved that one equivalent of the following monovalent ions

H, K, Na, Li, when completely hydrated, has the following hydration-values (in Mol. H_2O): H = 1, K = 11, Na = 18 and Li = 27. Ions in solution have therefore a larger effective radius than in a dehydrated state (such as in a crystal lattice). The water hull surrounding the ions in solution is composed of dipoles of water

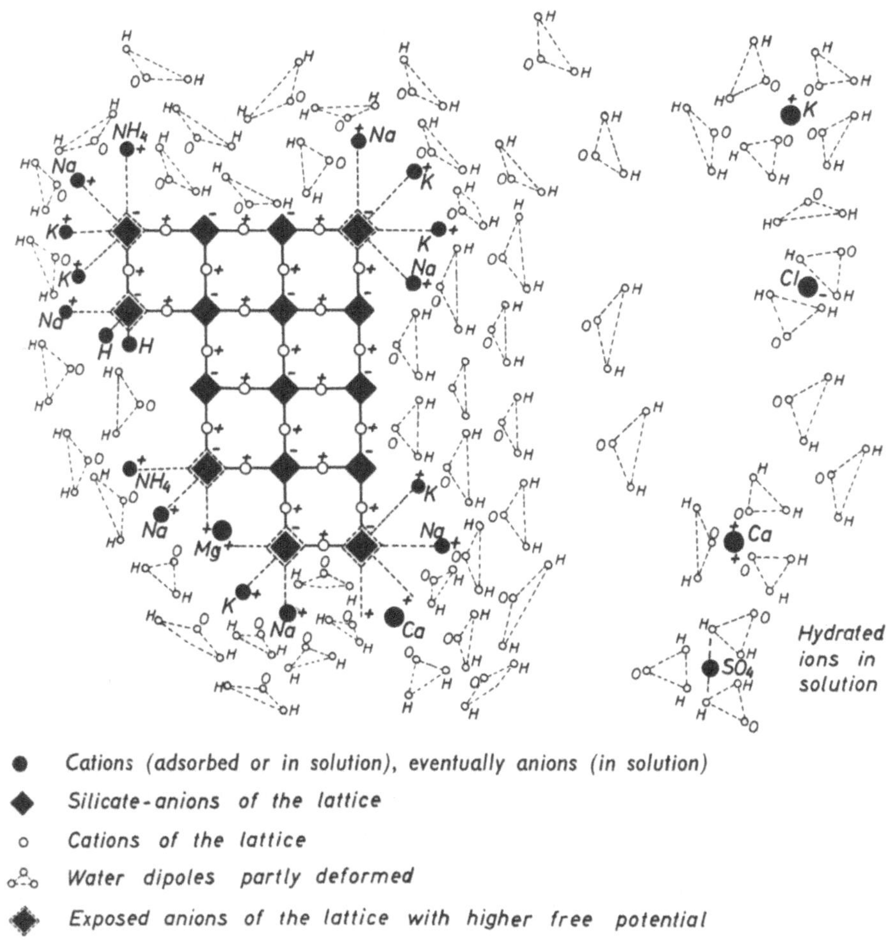

Fig. 26. Hydrated colloidal clay particle (with schematic cristal lattice): part of the water dipoles is attracted directly to the surface, part of the dipoles surrounding the adsorbed cations (VAGELER 1932)

molecules (under the influence of an electric field the water molecules — which are double molecules in the liquid state — are deformed, the two positively charged hydrogen atoms approaching one another and turning towards the negatively charged particle or ion) (Fig. 25).

The intensity of the electric field of an ion increases with its charge, but decreases with its radius squared, so small ions attract the largest number of water dipoles and hence have the thickest water hull.

A clay particle in soils therefore always is hydrated, part of the water dipoles being attracted directly to the surface, part of the dipoles surrounding the adsorbed cations (Fig. 26). In general the larger the amount of adsorbed cations (or exchange capacity) and the more highly hydrated the cations are, the greater is the hydration of the clay. Moreover, some clays with swellable crystal lattice have also water layers between the sheets of the lattice.

Jenny and Reitemeyer (1935) studied the relation of the potential of moving colloidal particles in suspension (the electrokinetic or zeta potential) to the charge and size of the adsorbed ions. They proved that the magnitude of charge (potential) of the clay particles is lower for bivalent than for monovalent adsorbed ions having equal radius (Fig. 27: compare the charges for adsorbed Mg^{++} and Li^{+}).

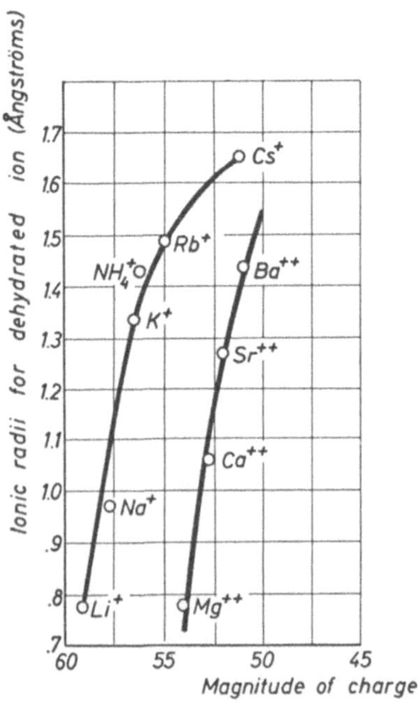

Fig. 27. Magnitude of charge on clay particles in relation to the radius of the adsorbed ion (Jenny and Reitemeyer 1935)

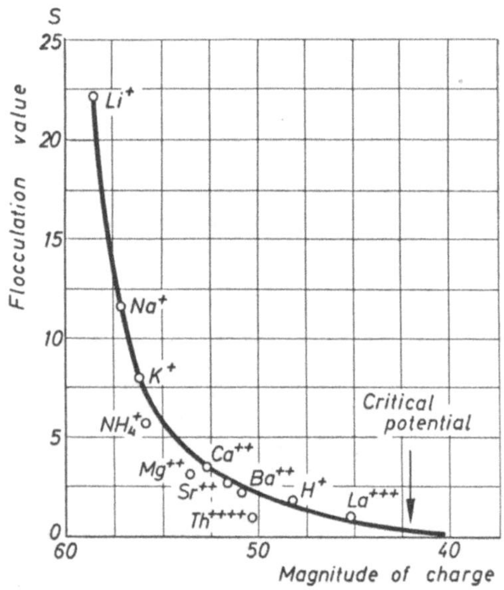

Fig. 28. The effect of the magnitude of the charge on colloidal particles to the flocculation value for symmetry concentration (Jenny and Reitemeyer 1935)

However, as a consequence of their greater activity, the divalent ions lower faster the zeta potential than the monovalent ions do, although the hydration of the former is greater. The divalent ions, when of the same size as the monovalent ions, will absorb about the double amount of water molecules; e.g. Li^{+} and Mg^{++} both have 1.56 Å as ion diameter, but the first one has only 6, and the second has an average of about 11 as hydration number[1] (Scheffer-Schachtschabel 1960). Jenny and Reitemeyer studied also the relation of the energy of adsorption and release of the adsorbed ions to the zeta potential. Here the radius of the hydrated ions is important. The magnitude of the zeta (ζ) potential is function of the average distance between the adsorbed ions and the ions directly located at the particle surface. But highly hydrated ions cannot approach as closely to the clay particle as ions with a thin water hull. The adsorption of highly hydrated ions therefore increases the thickness of the double electric layer and increases the ζ potential. The greater the thickness of the double layer, the greater is the average distance of the oscillation, which an adsorbed ion can make through

[1] Hydration number: number of water molecules per ion.

the influence of heat motion and Brownian movement; this means the more loosely an adsorbed ion is retained, the easier also the cation exchange. Therefore clays with high ζ potentials have easily exchangeable cations. Fig. 28 gives the flocculation values (obtained by adding KCl to clays saturated with different cations) as functions of the ζ potential of clay particles. The flocculation values are given for symmetry concentration, which is the percent release of an adsorbed ion when the replacing ion is added in equivalent amounts to the adsorbed ion.

Looking to Fig. 28 it is evident that the large ζ potential of the Na-saturated clay is the reason of its high state of dispersion, compared to the Ca- and H-saturated clays.

An exceptional position is taken by H^+: this "naked" ion (only surrounded with one Mol. of water) has a high electrokinetic potential, making its flocculation effect even a bit higher than of Ca^{++}.

The last factor to be explained in the formula determining the magnitude of the ζ potential is D, which is the dielectric constant of the liquid. In a normal clay suspension water has a value of $D=$ about 81,07. By adding ethylalcohol water is withdrawn from the double layer by which ζ diminishes.

In summary, when an electrolyte is added to a colloidal clay system,

1) there is an exchange of the ions present on the solution side of the fixed liquid coat of the double layer for those of the same sign added with the electrolyte. If the added ions are more strongly adsorbed than those which are displaced, the thickness of the double layer will be decreased; consequently the ζ potential will be lowered and flocculation occurs (whereas those ions which produce a high ζ potential will cause dispersion);

2) the free energy of the water is lowered in the bulk of the solution and water moves from the region of the double layer into the solution, because of the free energy gradient established. The result is a contraction of the double layer lowering also the ζ potential. The amount of elecrolytes to be added depends on the original ζ potential of the system. It is obvious that the addition of a large amount of an electrolyte causes always flocculation;

3) every collision between particles produces adhesion as soon as the ζ potential is lowered to or below the critical value; this is the case of a rapid flocculation. When two particles approach sufficiently close, the VAN DER WAALS' attractive forces cause them to stick together. These forces are proportional to $1/r^6$ and are consequently active only at very close range. If the particles are similarly charged, work has to be done in bringing up one through the electrostatic (COULOMB) repulsion of the other. The repulsive force is proportional to $1/r^2$ and hence is active at relatively long range. This force hinders the approach of the particles to the critical distance at which they stick together.

If the potential is higher than the critical ζ potential, some flocculation may still occur, for the momentum of the faster moving particles can carry them through the repulsive electrostatic fields within the range of the VAN DER WAALS' attractive forces.

In agricultural practice we see that aggregate stability is increased by adding Ca-ions and Fe-ions to the soil. Since more than 30 years (TIULIN 1928) it is known that liming is a good method to improve soil structure. The increase in water stability of the aggregates by saturating the clay with calcium has been proved several times (e.g. at the International Symposium on Soil Structure, Ghent, 1958, three communications on this subject were presented, two by BOEKEL and one by DE LEENHEER and DE BOODT).

Recently the addition of Fe-ions was also introduced in practice, e. g. the use of Flotal (Ferriammonium alum) to improve structure deteriorated soils (KALOVOULOS 1955, HANKE 1956, KNICKMANN 1959).

The easy dispersion of sodium saturated clay in newly reclamed polder soils and in semidesert soils is explained by the large waterhull around the sodium-ion. Consequently the salt content of irrigation water is to be watched very carefully. Another feature from practice to be explained is the high stability of acid soils; this stability is not only due to the H^+ion adsorption but also to the strong flocculation value of Al^{+++}ions in acid clay soils. In general, however, the aggregation in acid soils is less stable than in calcium saturated soils; one of the reasons is that in acid soils the organic matter does not have the sticking property which is a characteristic of Ca-saturated organic matter.

To get a better idea of the soil components which are active in the formation of stable aggregates KOLASKOVA and AKBERDINA (1959) examined aggregates of different sizes from the Volga-Kam wooded steppe. They found that the aggregates smaller than 0,25 mm, the so-called micro-aggregates, contained more clay, humus, mobile humus acids, SiO_2, R_2O_3, P_2O_5 and mobile Fe than the larger aggregates.

These results, however, could not be confirmed in the study of ANTIPOV-KARATAEV and KELLERMAN (1962) who compared aggregates from tschernosems, podsolic, gray-wooded and red loamy (Krasnosem) soils. It is believed that part of the explanation why there is a difference in composition in some but not in all soils between the stable and the unstable aggregates, may be found in the different genesis and consequently different composition of the colloidal material of the studied soils.

The Chemical Stabilization of the Aggregation. In soil survey one knows that aggregates in soils periodically suffering from an excess of water are more stable than the normal calcium-saturated aggregates in well-drained profiles. The explanation is that during periods of water excess, through the reduction processes in the soil profile, an amount of soluble bivalent iron ions is liberated which can penetrate the aggregates. During the drier periods, when the water table goes down, through oxidation the Fe^{++} changes into Fe^{+++} and is precipitated as colloidal $Fe(OH)_3$, which cements the aggregates. Dehydration and crystallization of the cementing $Fe(OH)_3$ may be the origine of very hard iron concretions or even iron pans. Not only iron, but also calcium, manganese and phosphorus may be important as chemical stabilizers of aggregates, when they are changing from the soluble into the insoluble state (carbonates, oxides, phosphates).

B. Formation of Soil Crumbs by Dessication

Dessication brings the colloidal soil particles closer together. When the electric forces, following the law of COULOMB, are inversely proportional to the second power of the distance, the VAN DER WAALS' forces, however, are inversely proportional to the sixth power of this distance. In crumb formation the rôle of water is really important.

In a dry and dusty soil the air surrounding the elementary soil particles prevents them from coming close together, so that no intermoleculary attractive forces originate. The moistening of such a soil causes the formation of a water hull around the particles; this way the small altered amorphous coat of the mineral particles swells and simultaneously a meniscus is formed between the particles (SCHOFIELD 1946). The force p of the meniscus (interfacial curvature) to keep particles together may be put equal to the capillary tension, for which the equation of the height of capillary rise is valid[1] (RUSSELL 1949).

$$h = p = \frac{2\,T \cos\theta}{r\,d\,g}$$

where h = height of capillary rise in cm,
p = force in dynes/cm^2,
r = radius of the pores in cm,
T = surface tension of the water in poise,
θ = angle of contact between water and soil (angle = 0; $\cos\theta = 1$),
d = density of water g/cm^3,
g = acceleration of gravity (981 cm/sec^2 in Western Europe).

The mechanical stability caused by capillary forces increases rapidly with water content until an optimum is reached, the forces then decreasing rapidly as more water is added.

This aspect of keeping soil particles together under the influence of a water miniscus is only the starting point for the formation of soil crumbs. If capillary forces alone were responsable for crumb stability, the crumbs would disaggregate completely when becoming dry or very wet (water saturated). This, however, happens only in sand soils, not in loam or clay soils. In the latter two kinds of soil, dessication increases very markedly the resistance against compression or penetration (WERSCHININ 1958). The original poor aggregation in loam and clay soils, under the influence of capillary forces goes over, through dessication, into a much more intense fixation in which the VAN DER WAALS' forces play the important rôle. The forces seem to be located on the spots which were originally wetted. It has been observed that the higher the original moistening, the better becomes the crumb stability after the loam or clay soil has dried out.

C. Formation of Macro-aggregates

The coagulation of colloidal particles makes also possible the cementation or linking of larger soil particles, such as sandy grains. Larger soil particles and micro-aggregates may be joined with a thin layer, a film of the dispersed colloid, which then acts not as a cement, but as a glue or a sticking product[2].

Crumbs may also stick together and form macro-aggregates under the influence of a drying-out process, as has been previously stated. The coagulation of micro-aggregates from a colloidal solution is a first step, but the actual aggregation of the floccules into larger aggregates starts with the loss of water of these floccules: climate, plant growth, soil fauna and men cause the formation of macro-aggregates through dessication, frost influence, pressure of the growing root-system of the plants, action of soil fauna, tillage, etc.

ROGOWSKI and KIRKHAM (1962) studied the influence of pressure and moisture treatments on silty clay loam material. Values of pressure up to 1000 psi (70,3 kgr per cm^2) and of moisture up to 30,5 percent failed to produce an aggregate water-stability comparable to that of normal soil aggregates taken in the field.

---

[1] Equation of the capillary rise, also known as "law of KELVIN" (in Anglo-American publications), as "law of BECHOLD" (in German studies), or as "law of JURIN" (in French publications) — (DE BOODT 1957).

[2] The words cement and cementation should be used when the sticking product undergoes a change in chemical composition. It is well known that soluble Ca-salt under influence of aeration changes into insoluble $CaCO_3$, which crystallizes with 6 H_2O. Iron in the soil must be considered also as a cementing element when chemical changes involving it take place, as it has already been described.

Since pressures exerted by farm machinery, plant roots and soil fauna are smaller than 1000 psi at any moisture content, the authors conclude that water-stable aggregation is largely due to factors other than applied pressure.

However, a distinction must be made between the formation of water-stable aggregates and of macro-aggregates which may be observed in every soil.

It is the every day farmer's experience that rather low tensions and pressures act in the soil to form macro-aggregates. These tensions and pressures explain the fact that in a given large soil mass places occur with higher density than on other places. Very important is the water content of the soil at the moment pressure acts on the aggregates.

For the different soil classes there is an optimum water content at which the soil density is maximum. In soil physics this water content is sometimes called the "structure forming water content", because with that exact amount of water the water meniscus surrounding the micro-aggregates is supposed to form an uninterrupted water film around the macro-aggregate (Fig. 21) (KATSCHINSKI 1958).

Even under a light external pressure, the soil particles may be moved within the macro-aggregate to present their largest surface to the direction of the pressure and hence to build up a mass with higher density (HILGARD 1910). This way the soil material gets a new property, plasticity.

When speaking about soil plasticity, one usually distinguishes an upper and a lower plasticity limit. The upper limit is the water content at which the clay mass becomes semi-liquid (syrupy), the lower limit being the water content at which a small soil cylinder starts to break by rolling. This lower limit should correspond to the "structure forming water content" and some pedologists believe that at this water content soil tillage has the best influence on structure improvement (crumb and aggregate formation). This opinion, however, is not generally admitted (see chapter on "Methods for Soil Structure Determination").

In agricultural practice the farmer knows that the soil water content at the moment of tillage is highly important. The optimum water content changes with soil class, soil structure, type of soil profile (internal lateral and vertical drainage), soil use, speed of plowing, etc.

In sandy material the plowing may be done at a moisture content, which is nearly equal to field capacity; the capillary forces are then at their optimum value to provide a mechanically stable aggregation. In heavy soil the optimum water content for plowing is at 60 to 90 percent of the field capacity. The better the soil structure, the higher the optimum water content may be for tillage. Experiments also proved that at relatively high water content a fairly good work can be done when the plowing speed is accelerated; the speed limit however seems to be at 7—8 km/hour (KATSCHINSKI 1958).

Frost influences on soil structure are important too, but the circumstances and the kind of soil must be taken into consideration. Free water freezes at 0° C, capillary water freezes between —0.5° C and —10° C, depending upon the diameter of the pores (capillary forces); for hygroscopic water a frost temperature of at least —70° C is needed. In terms of agricultural practice, this means that in a dry soil and for a normal frost in the temperate climate, no influence on soil structure is observed; however, when frost suddenly starts on a soil at field capacity, the growing ice crystals in the pore cavities will develop an important compression (100 volumes of water changing into 109 volumes of ice), which brings the soil particles closer together and increases the VAN DER WAALS' forces. Another influence of the frost is the dehydration of the colloids, which increases the aggregate stability.

When the soil is not at field capacity, but entirely saturated with water, the numerous ice crystals growing in all directions in the soil mass will be a cause of the disjunction of crumbs and aggregates. Thawing will liberate a large amount of melting water and the soil will change into a plastic paste.

D. Binding Forces of Organic and Mineral Particles

When aggregates are put in water, some will stay together, others will collapse. The first are called water stable and one can observe that their organic matter content for the same kind of soil is higher than that of the water unstable aggregates. This has been made evident by TIULIN (1928), who compared the stability of grassland crumbs with those of corresponding arable soils. The difference between mechanically stable and water stable aggregates is important.

Little is known about the way organic and inorganic soil material is bound together. The actual nature and colloidal activity of what is called soil organic matter or humus is not sufficiently known.

That montmorillonite is swellable can be proved easily by putting a pure montmorillonite clay in a saturated solution of NaCl and by gradually diluting this solution; the clay volume will increase (HOFMANN and BILKE 1936). The same experiment made with a montmorillonite clay soil, rich in organic matter, does not show swellability. Therefore, there must be a binding between the clay and the soil organic matter. Scientists recently tried to investigate this binding by studying what happens when a mineral soil is mixed with well known organic substances, such as "krilium" or gelatin.

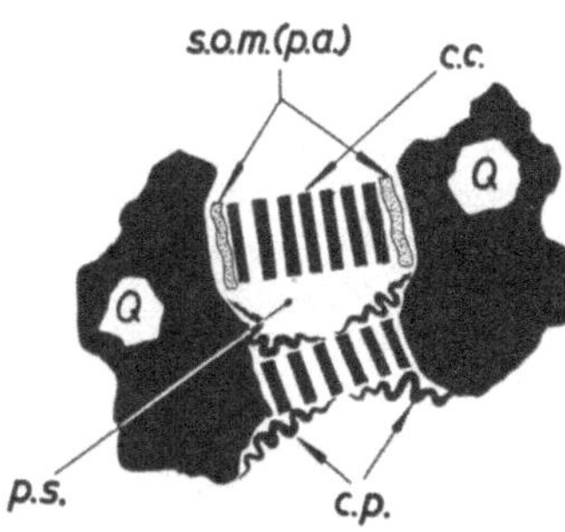

Fig. 29. Stability of soil crumbs and organic matter (EMERSON 1959) *Q* quartz; *c.p.* carboxylated polymer; *s.o.m.* (*p.a.*) soil organic matter or polyvinyl alcohol; *c.c.* clay crystals forming clay domain; *p.s.* pore space

The complex organic substances which may bind clay particles together are divided into two groups (EMERSON 1958 and 1959):

1) the ionizing polymers, such as those obtained with polyacrylic acid, for example the hydrolized polyacrylonitril (H. P. A. N.) or krilium. The aggregates are stabilized with this product, because the borders of the clay minerals or the borders of clay with fine sandy material are bound with each other through hydrogen bridges (or edge-linkages).

2) the non ionic-polymers, such as polyvinyl-alcohol and also in part anionic polymers, such as gelatin, which products build lamellar complexes (or basal linkages) with clay minerals, or between the clay and the fine sandy material.

Both kinds of linkages are presented in Fig. 29 (EMERSON 1959). Presenting the two kinds of linkages this way, it is easy to explain why the grassland crumbs, the arable land crumbs and the soil conditioner treated crumbs give similar moisture-release curves (JAMISON and KROTH 1958). Both kinds of bindings may be distinguished using a solution of sodiumpyrophosphate. The lamellar complexes are not changed in this solution; the others are dispersed.

The lamellar complexes themselves may be distinguished, using an alkali-solution; the complexes composed with non ionic-polymers, such as polyvinyl-alcohol, are not dispersed with alkali; those built with organic compounds, such as gelatin (amino-groups) undergo dispersion, because the positively charged groups of the polymers which are bound on the negatively charged clay particles, undergo dispersion when the pH-value of the solution is higher than the pK-value of the anionic group (about 9 for the amino-groups) (Emerson 1958).

In order to know with what kind of binding forces one is dealing in a soil, let us take as an example aggregates of an old pasture land (Emerson 1956) and put them in a solution of sodium-pyrophosphate; no dispersion will be discernible, so in this case there exist no bindings through hydrogen bridges. When the same aggregates are put in an alkaline solution of sodium-hydroxide, they will lose their stability, which proves that there was an electrostatical binding between positively and negatively charged components.

Soil organic matter seems to be a polymer, which is sensitive to the pH. The same happens with some organic soil conditioners, which are very sensitive to acidity and show an optimum activity at a given pH (Richard 1959).

In agricultural practice it is observed that the increase of the organic matter content in the soil means the improvement of structure stability; this however is limited to about 6 percent organic matter content, both for Central Europe and for colder climates (Schachtschabel 1950, Heinonen 1958). A larger input of organic material does not increase aggregate stability, because there is a limit for the linkage possibility between organic and inorganic soil components. Some pedologists explain this limit, saying that at that limit the sorptive capacity of the organic material should reach 50 percent of the total soil sorptive capacity, the 50 percent value being believed a critical value, exception being made for acid soils rich in organic matter, such as peat soils, heath soils, some forest soils, etc. (Scheffer and Schachtschabel 1956). Higher amounts of organic material are unable to make electrostatical bindings with clay minerals.

In arable land of non-black soils in Western and Central Europe with a clay content larger than 25 percent, the normal organic matter content is relatively low (2—3 percent), so the structural stability depens on the binding forces between clay-clay or clay-silt (or sand), which are not strong. In these soils the amount of active lime and the $CaCO_3$ reserve are very important for the aggregate stability.

In order to get a better information about the kind of organic material, which plays an active role in the binding with mineral soil particles, Greenland, Lindstrom and Quirk (1962) carried out the following experiment. Aggregates of Red Brown Earths were treated with periodate ($NaIO_4$) and compared with the behaviour of aggregates treated with sodiumchloride (NaCl). The reduction in stability after the treatment with periodate was small for samples taken from plots which had been under pasture for many years, but very large for samples which had been under pasture for only 4 years or less or from soils which were continuously cropped. Since periodate causes a breakdown of the polysaccharides and polyuronides in the soil, it is likely that these organic products are the primary factors of aggregate stability in cropped soils and in soils under young pasture. The smaller decrease in aggregate stability (after the periodate treatment) of samples from old pasture land or from soils containing free $CaCO_3$, such as solonized brown soil and especially rendzina, indicates that aggregate breakdown in soils is prevented by soil components already mentioned, such as the chemically precipitated $CaCO_3$ or the irriversibly flocculated active humus.

The formation of active humus should essentially take place in the rhizosphere of plant roots (Katschinski 1958, Werschinin 1958), where complex products are built up. Some of their components originate from the uronic acids liberated by young plant roots, some components contain nitrogen as they are autolysis products of the secretions of bacteria. As an example of this kind of humification, the water stable aggregates found under permanent pasture land may be cited. In these highly porous aggregates the irreversible flocculation of the colloids under the effect of dehydration during the warm summer period is another factor which increases the stability. In this way, the perfect water stability of the crumbs from the A horizon of the Chernosems and of the Mid-West prairie soils of the U.S.A. can be explained.

To keep arable land in a good structural status, the microbiological activity has to be taken care of too. Fig. 30 brings the illustration of the change in aggregate stability of two soils in function of time (PEERLKAMP 1950). The changes are nearly negligeable if no organic matter is added to the soil, but they are very important, due to microbiological activity, if easily decaying organic matter is incorporated into the soil. According to the temperature, the moisture content and aeration, and the C/N-ratio of the organic matter, the speed of the organic matter decomposition will be completely different, so the original situation will be reached within some weeks or only after several months. The durability of the created aggregate stability under influence of the different kinds of organic matter, may be ranked as follows: green manure, subsoil roots and leaves of crops, stable manure, compost (SCHEFFER and SCHACHTSCHABEL 1960). The part biological activity plays in a soil is synthesized by SEKERA and BRUNNER (1943) in the expression "Lebendverbauung", which means the biologically created stable aggregation through the mycelia and metabolic products of fungi and bacteria and through the rhizosphere activity of higher plant roots.

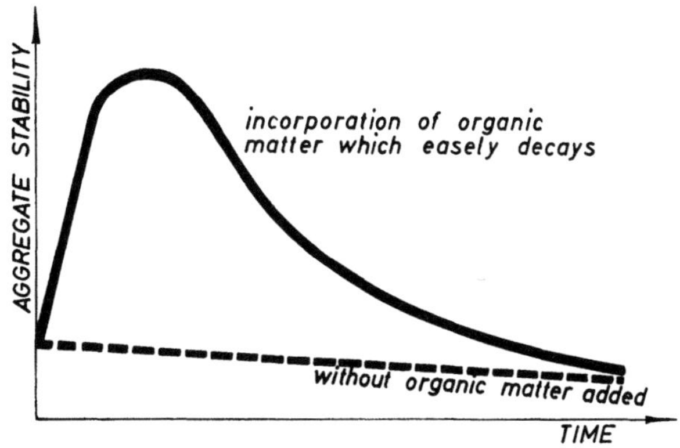

Fig. 30. Change of aggregate stability with time (PEERLKAMP 1950)

To come to an overall picture of the colloidal substances acting as binding forces in aggregates, ANTIPOV-KARATAEV and KELLERMAN (1962) worked out a method by which they could detect the different types of bindings (organic and inorganic), acting on water stable aggregates. (This method will be discussed in the following chapter: "Methods for Soil Structure Determination".)

In conclusion and irrespective of the colloidal, the biological or the organo-chemical point of vue one is dealing with, it may be said that all mesures which favour the biological activity of the soil create a better aggregate stability, not only through a better binding of the soil particles, but also through the increase of the water-repelling forces of the metabolic products.

Literature

ANTIPOV-KARATAEV, I. N., and V. V. KELLERMAN: Die Bodenaggregate und ihre Kolloid-chemische Analyse. Z. Pflanzenernähr., Düng., Bodenkde. 98 (143), 193–204 (1962).

BAVER, L. D.: Soil physics. New York: Wiley, and London: Chapman & Hall. 1948. — BLANCK, E.: Handbuch der Bodenlehre, Vol. I–X. Berlin: Springer. 1928–1932. — BOEKEL, P.: Some remarks on the influence of soil structure on plant growth. Intern. Symposium on Soil Structure (1958); Mededel. Landbouwhogeschool, Gent, 24, 52–57 (1959). — The amelioration of the structure of clay soils by liming. Intern. Symposium on Soil Structure (1958); Mededel. Landbouwhogeschool, Gent, 24, 122–127 (1959). — Bolt, C. H.: Fysisch-chemische aspekten van het vochttransport in poreuze media. De Ingenieur 28, 1–6 (1962).

Childs, E. C.: The space charge in the Gouy-layer between two plane parallel non-conducting particles. Trans. Faraday Soc. 50, 1356–1362 (1954). — Correns, C. W.: Mineralogische Untersuchungen an Böden. Bodenkde. u. Pflanzenernähr. 21–22, 656–666 (1940).

De Boodt, M.: Bijdrage tot de kennis van de bodemstructuur. Doctorale Thesis, Landbouwhogeschool, Gent, 1957. — De Leenheer, L., and M. De Boodt: Signification pratique de l'emploi de doses massives d'écume de sucrerie pour l'amélioration de la structure du sol. Intern. Symposium on Soil Structure (1958); Mededel. Landbouwhogeschool, Gent, 24, 128–136 (1959). — Demolon, A.: Principes d'Agronomie, Vol. I. La Dynamique du Sol. Paris: Dunod. 1938. — De Sigmond, A. A. I.: The principles of Soil Science. London: Thomas Murby. 1938.

Emerson, W. W.: Synthetic soil conditioners. J. Agric. Sci. 47, 117–121 (1956). — A comparison between the mode of action of organic matter and synthetic polymers in stabilizing soil crumbs. J. Agric. Sci. 47, 350–353 (1956). — Über die Kräfte in Bodenaggregaten. Probleme der Krümelstabilitätsmessung und der Krümelbildung; Deutsche Akademie der Landwirtschaftswissenschaften, Berlin, Tagungsberichte 13, 287–290 (1958). — Stability of soil crumbs. Nature 183, 538 (1959).

Freundlich, H.: Kapillarchemie, Vol. I. Leipzig: Akademische Verlagsgesellschaft. 1930.

Greenland, D. I., G. R. Lindstroh, and I. P. Quirk: Organic materials which stabilize natural soil aggregates. Soil Sci. Soc. Amer. Proc. 26, 366–371 (1962). — Grim, R.: Clay mineralogy. New York: Mc Graw-Hill. 1953.

Hanke, F.: Über die Wirkung von dreiwertigem Eisen und dem Bodenverbesserungsmittel Flotal auf die Struktur schwerer Böden. Dissertation, Univ. Bonn. 1956. — Hartman, R. J.: Colloid Chemistry. Cambridge: The Riverside Press. 1947. — Heinonen, R.: Die Bedeutung des Humusgehaltes im Boden für die Krümelbildung. Probleme der Krümelstabilitätsmessung und der Krümelbildung; Deutsche Akademie der Landwirtschaftswissenschaften, Berlin, Tagungsberichte 13, 279–285 (1958). — Hilgard, E. W.: Soils. New York: Macmillan. 1910. — Hofmann, U., and W. Bilke: Über die innerkristalline Quellung und das Basenaustauschvermögen des Montmorillonits. Kolloid-Z. 77, 283–285 (1936).

Jamison, U. C., and E. M. Kroth: Available moisture storage capacity in relation to textural composition and organic matter content of several Missouri soils. Soil Sci. Soc. Amer. Proc. 22, 189–192 (1958). — Jenny, H., and R. F. Reitemeyer: Ionic exchange in relation to the stability of colloidal systems. J. Phys. Chem. 39, 593–604 (1935).

Kalovoulos, J. M.: The italian soil conditioner "Flotal". Soil Sci. 80, 463–467 (1955). — Katschinski, N. A.: Die Natur der mechanischen Stabilität und Wasserstabilität der Bodenstruktur. Probleme der Krümelstabilitätsmessung und der Krümelbildung; Deutsche Akademie der Landwirtschaftswissenschaften, Berlin, Tagungsberichte 13, 139–150 (1958). — Knickmann, E.: Ergebnisse von Weinbergversuchen mit Ferriammonalaun (Flotal). Intern. Symposium on Soil Structure (1958), Mededel. Landbouwhogeschool, Gent, 24, 117–121 (1959). — Koloskova, A. V., and R. K. Akberdina: Qualitative composition of aggregates of some soils of the Volga-Kam wooded steppe. Pochvovedenie 10, 100–104 (1959). — Kruyt, H. R.: Colloid Science, I. Amsterdam: H. J. Paris. 1947.

Peerlkamp, P. K.: The influence on soil structure of the natural organic manuring by roots and stubbles of crops. Trans. 4th Intern. Cong. Soil Sci., Amsterdam, 1, 50–54 (1950).

Richard, F.: Die Wirkung von "Krilium" in neutralen und sauren Ton- und Staubböden. Intern. Symposium on Soil Structure (1958); Mededel. Landbouwhogeschool, Gent, 24, 97–102 (1959). — Rogowski, A. S., and D. Kirkham: Moisture, pressure and formation of water stable soil aggregates. Soil Sci. Soc. Amer. Proc. 26, 213–216 (1962). — Russell, M. B.: Methods of measuring soil structure and aeration. Soil Science 68, 25–35 (1949).

Schachtschabel, P.: Untersuchungen über die Sorption der Tonmineralien und organischen Bodenkolloide, und die Bestimmung des Anteils dieser Kolloide an der Sorption im Boden. Kolloid-Beihefte 51, 199–276 (1940). — Einfluß von Kalk auf Boden und Humus. Landwirtsch. Forsch. 2, 57–64 (1950). — Scheffer, F., and P. Schachtschabel: Bodenkunde. Stuttgart: F. Enke. 1956 und 1960. — Schofield, R. K.:

Ionic forces in thick films of liquid between charged surfaces. Trans. Faraday Soc. **42 B**, 219–225 (1946). — Schuffelen, A. C., and G. H. Bolt: Some equilibria in Soils. Vortragsveröffentlichungen der zweiten und vierten Kommission der I. B. G., Hamburg, **II**, 131–146 (1958). — Sekera, F., and A. Brunner: Beiträge zur Methodik der Gareforschung. Bodenkde. u. Pflanzenernähr. **29** (**74**), 169–212 (1943).

Tiulin, A. F.: Results of investigation. Agric. Exper. Station of Perm; Div. of Agric. Chem., **2**, 77–87 (1928).

Vageler, P.: Der Kationen- und Wasserhaushalt des Mineralbodens. Berlin: Springer. 1932.

Werschinin, P. W.: Die physikalischen Grundlagen der Strukturbildung. Probleme der Krümelstabilitätsmessung und der Krümelbildung; Deutsche Akademie der Landwirtschaftswissenschaften, Berlin, Tagungsberichte **13**, 129–138 (1958). — Wiegner, G.: Dispersität und Basenaustausch. Zsigmondy-Festschrift. Kolloid-Z. Ergänzung zu **36**, 341–369 (1925). — Ionenumtausch und Struktur. Trans. 3rd Int. Congr. Soil. Sci. **3**, 5–28 (1935). — Wiklander, L.: Cation and anion exchange phenomena. Chemistry of the soil. New York: Reinhold. 1955.

b) Methods for Soil Structure Determination

By

L. De Leenheer and M. De Boodt

1. Introduction

In the foregoing chapter on "Factors and Forces in Soil Structure" three kinds of definitions of soil structure were given. In this chapter soil structure should be understood in its second meaning: the spatial arrangement of the compound soil particles, the pore size distribution and other physical properties resulting from it, and their influence on plant growth.

Soil structure as defined here can be studied either from a more static aspect, with soil properties as direct consequences of a soil aggregate distribution at a given moment; or from a more dynamic aspect, with soil properties under changing circumstances or potentialities of soil structure. Some examples will make this clear:

a) Static properties or direct consequences of a soil aggregate distribution at a given moment are: the total pore space and the pore size groups; the air permeability; the infiltration and the water permeability; the water and air content, the latter eventually being specified in percentages of the soil air components; the aggregate distribution; etc.

b) Dynamic properties or potentialities of soil structure are: the water tension curve of a soil; the hydraulic conductivity in the saturated and unsaturated phase; the aggregate stability; the Atterberg plasticity limits and some other mechanical properties, such as shrinkage and swelling; the oxidation-reduction potential; the bio-electrical potential; etc.

Between the groups (a) and (b) there is no sharp limit, because when one of the properties under (a) is measured at different times during the growing season, with known environmental conditions, one already has an idea of the potentiality of the soil for the measured property.

Generally speaking, when using the different methods to measure soil structure properties, research workers are not always conscious if their method fits in one of the above stated subdivisions of measurements. In a method one sometimes finds a mixture of both; for example when the distribution of the wet stable aggregates is measured through sieving under water, without measuring the initial dry aggregate distribution, then one measures something more than the actual aggregate distribution, as the aggregates are sieved under water and consequently the stability is involved; but on the other hand, it is not a complete stability determination, as the initial aggregate distribution as taken from the field is not compared with what is left when the aggregates are wet sieved.

In this discussion, we cannot strictly maintain the above mentioned subdivisions in (a) and (b), but we will start with the methods which are always needed when measuring soil structure, such as taking undisturbed soil samples and determining the moisture content of the fresh samples. We will call this the indispensable preliminary procedure. Then follows a group of determinations, which aims to characterize soil structure in the way the soil is built up, as the bulk or apparent density, the total pore space, the specification of the soil porosity, and the aggregation subdivided in its actual distribution and in its stability. This group we call the direct determinations of soil structure.

Another group of determinations deals with the properties which are the consequences of the way the soil is built up, such as air and water permeability, sound absorption, etc. Here the subdivision between determinations to characterize soil structure at a given moment and the determinations of the potentialities of the soil structure can be carried out completely. This last group we will call the indirect determinations of soil structure.

It is evident that this outline will be incomplete, as it is impossible to discuss all the methods, but we will try to make a synthesis of the used methods, referring each time to representative papers.

Up to now, a few synoptical studies on methods used in soil structure research have been written. We will mention:

a) E. W. Russell, 1938: "Soil structure". Technical communication no. 37, published by the Imperial Bureau of Soil Science, Harpenden, England.

b) M. B. Russell, 1949: "Methods of measuring soil structure and aeration". Soil Science 68, 25—35.

c) "The national soil structure—organic matter work group of the United States" (1951). A survey without comments of all papers published in the U.S.A. on this subject; a few papers from abroad also are listed.

d) L. D. Baver, 1956: "Soil physics" (third edition). New York: Wiley, and London: Chapman & Hall.

e) "The method book on soil structure", which is being prepared by the West-European Working Group on Soil Structure.

For students it may be interesting to know that the Commonwealth Bureau of Soils, Rothamsted Experimental Station, Harpenden, England, prepares bimonthly the abstracts of all papers dealing with soil structure and soil physical research; these abstracts are published in the periodical "Soils and Fertilizers".

2. Indispensable Preliminary Procedure in Soil Structure Research

If one wants to determine soil structure at a given moment or if one wants to determine the potentialities of the soil structure, some preliminary activities are indispensable, which we will discuss first. They are:

a) Taking undisturbed soil samples.

b) Determining the moisture content of the fresh sample.

A. Taking Undisturbed Soil Samples

The great difference between sampling for chemical and for physical soil analysis, is that for chemical analysis one takes disturbed samples which can be mixed together to get a *representative sample* on which the determinations can be carried out. For physical soil analysis (except possibly for stability determi-

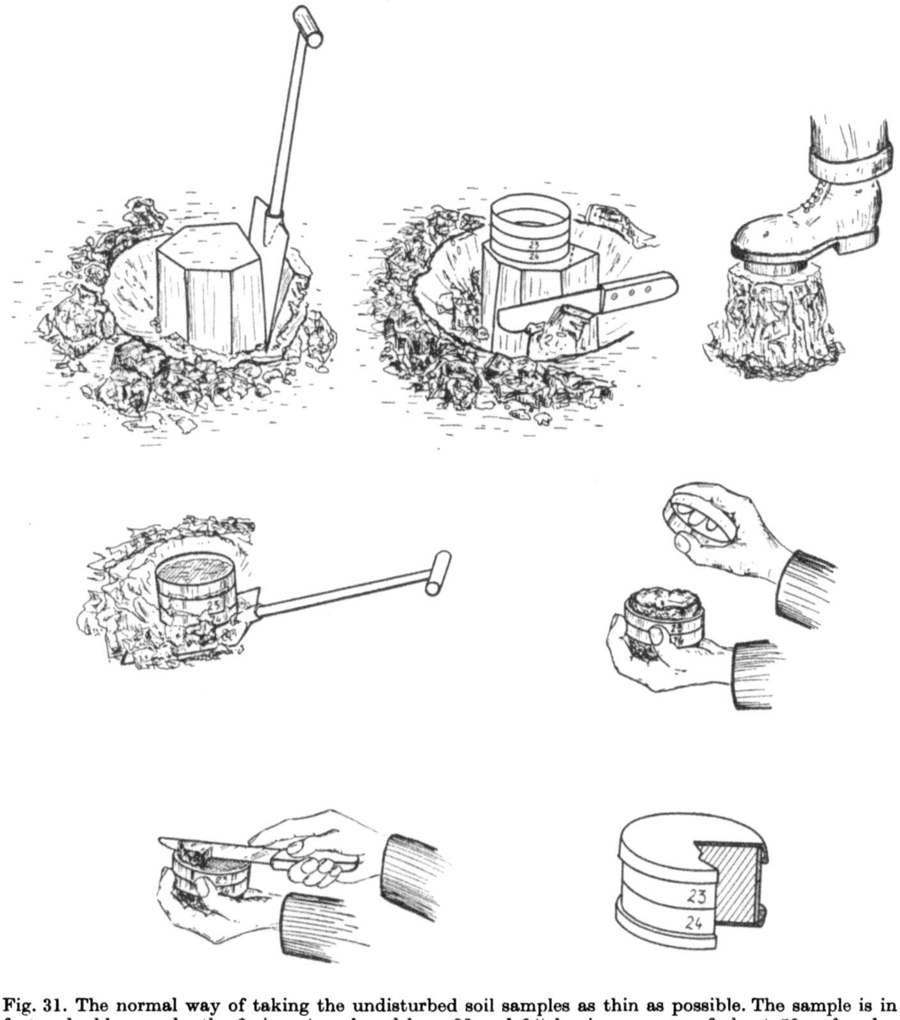

Fig. 31. The normal way of taking the undisturbed soil samples as thin as possible. The sample is in fact a double sample, the 2 rings (numbered here 23 and 24) having an area of about 50 cm² and a height of 2 cm. The third ring put on top is necessary to avoid soil compaction, whilst pushing the rings into the soil by the foot (De Leenheer and De Boodt 1959a)

nations) extreme care must be taken to get undisturbed samples; the determinations must be carried out on each sample individually; the results then are averaged to get a *representative value* of the measured property.

The primary aim of taking undisturbed soil samples was originally to determine the bulk or apparent density of a fixed volume of soil. Nowadays, undisturbed soil samples are often taken for other purposes, in particular for all determinations in the laboratory where the pore geometry of the soil is involved, such as the permeability determination. However, prior to such determinations, the bulk density must be known, this requiring only a few data.

Undisturbed soil samples are always taken in a container of a constant and known volume. By weighing the soil sample (the weight of the container is known in advance) and determining the moisture content, the apparent or bulk density (B. D.) can be calculated with the formula:

$$\text{B. D.} = \frac{\text{weight of soil, dried at } 105^\circ\,\text{C}}{\text{volume of the undisturbed soil}}$$

The earliest methods, as that used by SCHUMACHER (1904), for the determination of the B.D. consisted in pushing a metal cylinder into the soil. Then the container was dug out and the soil surface was cut flush with the cylinder top and bottom. Two difficulties arise in using this method:

1. Hammering the tool into the soil tends to shatter the soil and to disturb the original structure.

2. Pushing the cylinder in the soil may tend to induce compression.

These troubles can be lessened to some extent by:

a) choosing the right moisture content which should be close to field capacity[1]. When the soil is not at the right moisture content, a soil column should be prepared so the metal cylinders can slip over it, as shown in Fig. 31 (DE LEENHEER and DE BOODT 1959a);

b) by suitably designing the metal cylinder, so the friction between the cylinder and the soil can be considerably reduced. A large number of designs

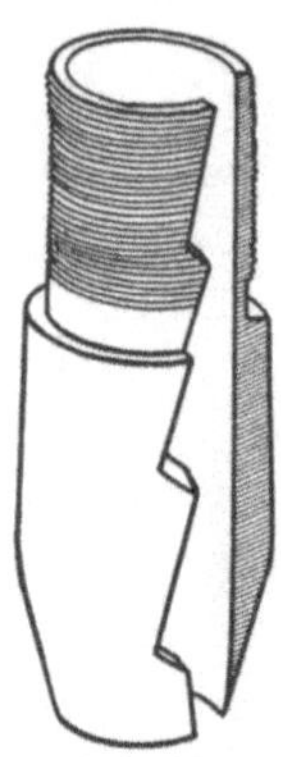

Fig. 32. Sampling tube of VEIHMEYER (1929)

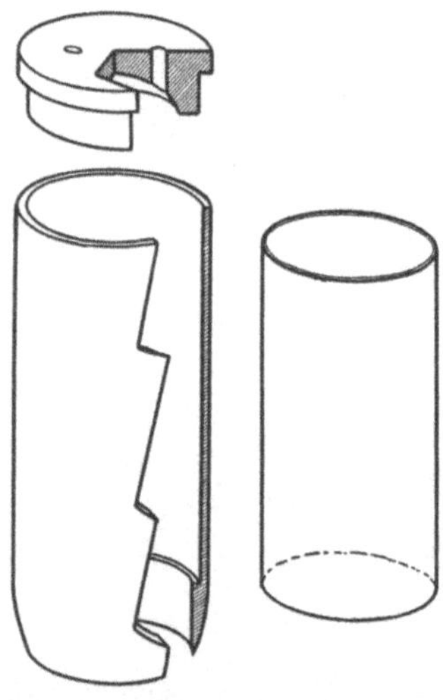

Fig. 33. Soil core sampler of UHLAND and O'NEAL (1951) which serves as container for an aluminium or carton box in which the soil sample enters

have been suggested in this respect, of which we will mention the one proposed by VEIHMEYER (1929), which has the advantage of avoiding lateral compression when the soil sample slides into the tube, but the inconvenience of difficulty in removing the sample (Fig. 32).

[1] Field capacity is the water content when all the drainage water has moved down in a soil with free drainage possibilities.

Lutz (1947), and later Uhland and O'Neal (1951), proposed a soil core sampler which serves as cylindrical container for an aluminum or carton box in

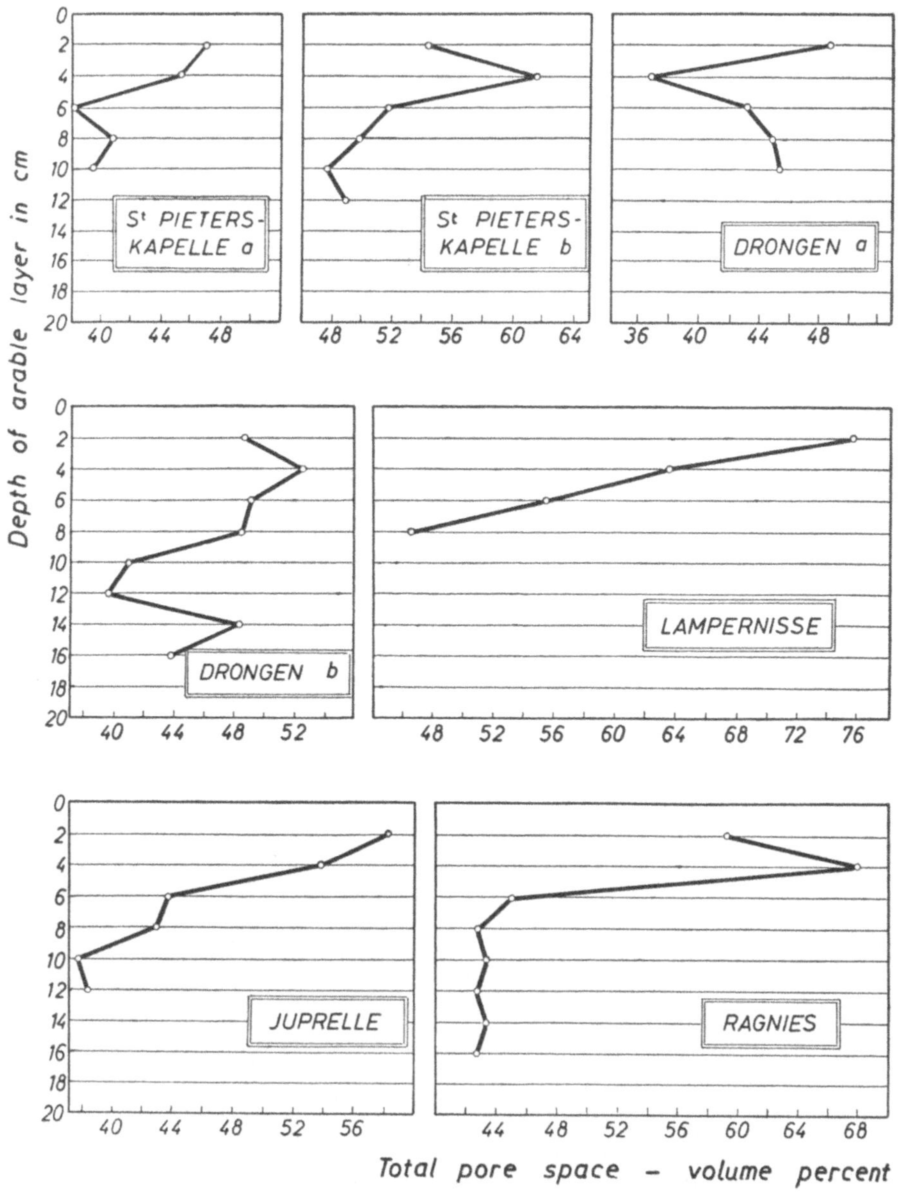

Fig. 34. The soil structure of the arable layer changes from cm to cm. The samples St. Pieterskapelle a and b and Lampernisse represent marine alluvial (Polder) soils; the samples Drongen a and b represent two sandy soils near Ghent. The samples b of St. Pieterskapelle and Drongen have a better structure than the samples a (see the vol percent). The samples of Juprelle and Ragnies represent two loamy soils, one (Juprelle) from the eastern and the other from the south western part of the Belgian loamy region

which the soil sample enters when the cylinder is driven into the soil. Together with the aluminum or carton board box the soil sample is removed from the sampler (Fig. 33). When sampling, one should always try to keep mechanical disturbance at a minimum to avoid lateral compression and to get the samples as thin as possible, because soil structure changes so quickly in the vertical direction, particularly in the plow-layer (Fig. 34). To meet these requirements De Leenheer and De Boodt (1959a) proposed to take samples 3 inches in diameter and only 2 cm thick. To sample longer soil columns, 2 or 3 brass rings with 1.5 mm wall thickness, can be put one above the other and in the laboratory the soil can be cut flush with the rings every 2 cm.

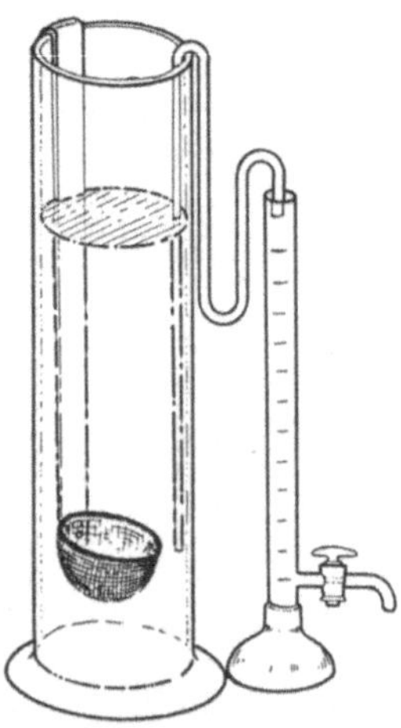

Fig. 35. Bulk density determination of weighed and waxed soil lumps. The volume is known with the accuracy of one waterdrop using an outflow glass tube with a double S-shape

The height of the ring samples is taken as 2 cm, because by experience it is known that this is the minimum height required to prevent the sample from falling out of the ring, especially for light soil. The soil is easily pushed out of these rings by using a full rubber cylinder somewhat smaller in diameter than the brass ring itself.

When the moisture content of the soil is near field capacity, the rings can be pushed gently in the soil by the foot. With a spade the rings are dug out, on top and bottom, the superfluous soil is trimmed flush, and closely fitting plastic snap caps are put on. Such a packing prevents crumbling of the sample during transportation and loss of moisture.

When the soil is not at the right moisture content, a little bit different method is followed (Fig. 31). At a short distance from the brass ring, the soil is cut away with a spade to a depth of about 10 cm. After the use of the spade, the soil is further trimmed away with a cheese-cutter, so that a column of cross-section about equal to that of the ring is obtained. By pushing gently, either by hand or by foot, on the rings, they will slip over the cut soil column.

The lateral compression can be expressed by a mathematical formula called the "area ratio", A, which is:

$$A = (D_e^2 - D_i^2)/D_i^2,$$

where D_e = external diameter and D_i = inside diameter of the sampling ring. A should be less than 0.1. In the proposed method by De Leenheer and De Boodt (1959a) $A = (7.93^2 - 7.63^2)/7.63^2 = 0.0802$, which is very satisfactory.

When obstructions of different kinds, such a stones, gravel, roots, etc. are present, sampling with rings or cylinders is impossible. In many cases a lump of soil can be carefully removed from the field, be coated with wax or paraffin and its volume thus determined. The accuracy here is limited by the unknown thickness of the wax coating on the outside of the lump. This, however, is of minor importance when the lumps used are large. The lump can be introduced in a tube filled with water and the rise in level is equal to the volume of the lump. The rise in level can be measured with the accuracy of one waterdrop using an outflow glass tube with a double S-shape (Fig. 35).

Vilensky (1945) introduced a method to determine the bulk density of aggregates in which no waxing was necessary. The aggregates are put in a graduated glass tube, partly filled with mercury. As mercury does not enter the pores, the rise in the mercury level is due to the volume of the aggregates. A variation on this method was introduced by STRICKLING (1955) using a vacuum on the aggregates before adding the mercury.

Another method of volume weight determination in stony soils consists of excavating carefully a volume of soil without disturbing the walls of the small pit. This volume is weighed. To determine exactly the volume, the excavated hole is filled up with a fine uniform sand (or even a thick lubricating oil) from a graduated cylinder under constant conditions. This field method is often used by civil engineers, when studying soil mechanics.

SIDERI (1936) compared different methods of determining the apparent density and found no uniformity in the results. The soil cylinder methods gave the lowest bulk densities; the paraffin method gave a value of about 10 percent higher, while the individual clods gave the highest results.

In addition to the mechanics of taking undisturbed soil samples, a second problem is the number of soil sample replications needed in soil structure research, for which we refer to DE BOODT (1959).

The third problem related to sampling is the difficulty of suitable storage when a large number of samples is taken, as for regional studies of the water economy properties of the different soil types in a mapped area. About this problem details are given by DE LEENHEER and DE BOODT (1959a).

B. Measurement of Moisture Content

In the Laboratory. The measurement of soil moisture can be done in the laboratory on disturbed samples. Usually a small amount of the sample is taken, the moisture content of which is determined by loss of water determination through heating at 105° C.

The moisture percent is calculated by the formula:

$$\frac{(\text{weight of moist soil—weight of dry soil}) \cdot 100}{\text{weight of dry soil}} = \text{percent moisture content.}$$

It is to be noted that in soil science the moisture percent is calculated on the dry soil (105° C) whereas in many other fields, such as the meal and flour industry, the moisture percent is calculated with reference to the weight of the moist material. Percent moisture is sometimes referred to the bulk volume of the dry soil.

The speed of moisture determination can be increased by putting the moist soil in an oven with warm air (105° C) circulation and with a built in balance, on which the water content percent can be read off directly, so it is not necessary to let the heated soil sample cool off in a dry atmosphere before weighing. Such determinations take only one hour or so, depending on the kind of material. As this method was originally developed for the meal and flour industry, the reading of the moisture content should be converted to percent of dry material.

Among the other quick methods recently developed, we mention the one described by VAN DER MAREL (1959). The water of the soil is extracted by dioxane and the change of the dielectric constant of the solution versus the pure dioxane is measured. From a chart the corresponding water content is read.

In the laboratory, soil moisture can also be determined in a quick way without expensive instruments by applying a method often used by civil engineers, in which soil moisture is calculated from the difference between the weights of a moist soil sample weighed in the air and under water (application of the law of ARCHIMEDES). A known weight of moist soil is therefore put in a graduated cylinder, which is filled with water to a given volume and weighed again. The moisture content w of the soil is then

$$w = \frac{G_0 (1 - d_w/d_k)}{G_1 - V \cdot d_w} - 1,$$

where G_0 = weight of moist soil,
G_1 = weight of moist soil plus the water in the cylinder,
V = volume of the soil plus the water,
d_w = density of the water,
d_k = density of the soil particles (2.65).

Example: a moist sample with a volume of 100 ml and a weight G_0 of 185 g is put in a cylinder and distilled water is added to a total volume v of 200 ml, G_1 being then 285 g.

$$w = \frac{185\,(1-0.38)}{285 - 200} - 1 = 35 \text{ percent.}$$

The above mentioned law of Archimedes has been worked out in a simpler formula by Maiboroda (in Russian, 1957), recently discussed in English by Wilde and Spyridakis (1962). The moisture content can easily be determined by making an immersion in water and a few weighings. The formula is:

$$w = \frac{Go - Gw \times K}{Gw \times K} \times 100, \text{ in which}$$

w = moisture percentage,
Go = weight of moist soil,
Gw = weight of moist soil in water,
K = ratio of the density of the soil particles to the density of the soil particles in water,
being: 2,65/(2,65—1) = 1,6.

Taking the same example as above, in which the weight of the moist soil sample in 185 g and the weight of the moist soil in water is 85 g, the calculated water content is:

$$w = \frac{185 - (85 \times 1.6)}{85 \times 1.6} \times 100 = 36 \text{ percent.}$$

Field Methods. Field methods that measure the moisture content of a soil sample near the surface or that measure soil samples in deeper layers without removing each time the upper soil layers are interesting. Among the earliest field methods is the procedure by which the moist soil is burned repeatedly with methyl- or ethylalcohol to constant weight. With the same formula as used when drying the soil in an oven at 105° C, the moisture content can be calculated. Another method, which is currently used and which gives good results, consists in bringing moist soil in contact with CaC_2, or even with CaH_2, in a closed container (Berg and Sitnikov 1958). With the first product acetylene (C_2H_2) with the second hydrogen (H_2) is produced and the resulting increase in pressure, which is in direct relationship with the moisture content of the soil, is measured. It is unnecessary to point out that both gasses are inflammable.

Another way of measuring soil moisture in the field can be found in the application of electricity, as the electrical conductivity of a moist sample is a function of the water content. The procedure described by Kirkham and Taylor (1950) may be used for direct measurement of moisture in the soil surface; Bouyoucos

(1949) describes an apparatus for continuous measurement of soil moisture of deeper layers using buried gypsum or nylon elements. The influence of the salt content in the soil is not always negligible in either of these methods. Therefore the thermal method of soil moisture measurement offers better possibilities to get accurate results (DE VRIES 1953; VAN DUIN and DE VRIES 1954), as the method is unaffected by salt concentration and covers a large moisture range with satisfactory sensitivity. BLOODWORTH and PAGE (1957) described an interesting instrument for this purpose using thermistors, while BOLLMEYER (1958) improved the embedding of the thermo-elements, using a mixture of sand-cement and asbestos. Measurements can be carried out 5 to 10 minutes after the electric current is switched on to heat up the elements.

A completely different way of measuring moisture content became possible after World War II, when nuclear science was applied for peaceful purposes.

The pioneer work concerning the method of neutron scattering for soil moisture determination in agriculture has been carried out by GARDNER and KIRKHAM (1952). Even earlier BELCHER *et al.* (1950—1953) had worked on the same problem and developed an equipment for soil tests in airport runway design. At present it is no problem to buy the ready-for-use equipment (soil moisture neutron scattering meters). The device works as follows: a source of fast neutrons, usually a mixture of radium and beryllium and a detector of slow neutrons are lowered into a pipe (called the access pipe), which is driven into the soil. At a given depth the fast neutrons emitted by the source penetrate over a certain distance into the soil and scan a spherical volume around the source. This scanned volume is called the sphere of influence (STONE, KIRKHAM and READ 1955, MORTIER *et al.* 1960). As a consequence of the collisions with all kinds of nuclei present in the soil, but mainly with hydrogen nuclei, the fast neutrons lose their energy until it is lowered to the energy of an ordinary gas molecule; those neutrons are called slow or thermal neutrons. The thermal neutrons are detected with a special device, which is located close to the source. Since nearly all the hydrogen atoms in the soil are associated with soil water, the number of slow neutrons detected per second is a measure of the soil water content per volume unit of the soil. The method is rapid; the measurement only takes half a minute to one minute, depending on the moisture content, for a determination at one single depth in the soil.

The reading of the moisture meter gives the average moisture of a 18 cm diameter sphere of influence, assuming the medium which slows down the neutrons (the moderator) is pure water. When the moisture content is only one percent, the diameter of the sphere of influence increases to about 80 cm. In some cases a smaller diameter is desired. MORTIER *et al.* (1959—1960) have shown how a paraffin sphere surrounding the radium-beryllium source may be used to reduce the sphere of influence about 30 percent. DE BOODT, DE LEENHEER and MORTIER (1963) have demonstrated that an analogous effect can be obtained by using a polyethylene access pipe instead of a steel or aluminium tube. However, when setting up a calibration curve for inorganic soils, accurate results can only be obtained if the bulk density (DE BOODT, DE LEENHEER and MORTIER 1963) and the amount of cristal water in the clay lattice are taken into account (MORTIER and DE BOODT 1956).

Another possibility for the determination of the soil moisture content by a radio-active method is the use of gamma-rays (KOROBOCKIN 1958). In fact it is a bulk density measurement (see below) which is made both on a dry and on a moist soil. The moisture content is calculated from the difference between the two.

3. The Direct Measurements of Soil Structure

In theory the so called "direct measurements of soil structure" should provide the necessary data, eventually complemented with the data of the soil components, such as the granulometric analysis, the kind of clay minerals, etc. to deduce other physical data, which characterize the properties of a given structure. This way of approach is not yet completely developed and in many cases it still is preferable to carry out direct measurements, such as temperature propagation, instead of deducing it from the porosity, moisture content, granulometric analysis, etc. However, the obtained results are notable.

The effort is done in two ways: one through pure physical reasoning and experimentation, and the other through correlation studies which recently got a new impetus through the possibilities offered by the electronic ordinator.

In the first group of papers let us mention the synoptical work of Marshall (1959), who reviews more than 300 papers to make an inventory of the actual knowledge of the relations between water and soil. There the equations are given relating permeability and gas diffusion to the size distribution of pores. He also shows how the permeability can be calculated from the curve relating water content to suction.

The laws governing the migration of water under the influence of salt concentration, temperature and moisture gradients, besides these of the air and vapor movement in soil due to wind, in function of the soil physical characteristics, are more or less well established, when dealing with isotropic material; but for anisotropic media, as soils often are, complementary research is necessary.

In the second group of papers let us mention the work by Burrows and Kirkham (1958), relating the field capacity of the soil to the clay content and the bulk density. On a much larger scale De Leenheer and Van Ruymbeke (1960b) for a given region correlated directly several soil structure properties such as the bulk density, the moisture content at field capacity and at the wilting point, the critical pore groups (the drainage and water holding pore space), and the stability of the aggregates to the most important soil components, such as clay and silt, organic matter, lime, and water content at the moment of sampling. Thus knowing the soil components, it is possible to deduce by calculation several important soil structure characteristics. The formulae of these correlations were calculated with an electronic ordinator (see chapter on Soil Texture).

A. The Apparent or Bulk Density (B.D.)

This aspect is already partly discussed in the description of the methods of taking soil samples. Here particular interest is paid to the quick field methods based on gamma ray emission.

The development of the technique of bulk density measurement by gamma radiation is closely related to the development of the moisture determination method by neutron scattering. Two reasons explain this parallelism in development:

1) The same counting unit (scaler) serves for the measurement of the backscattered or transmissed gamma rays and for the measurement of the slowed down neutrons. Just a different probe (for bulk density measurement or for moisture measurement) has to be connected to the scaler. The same commercial firms are bringing out both instruments.

2) The bulk density device measures the moisture plus the solids density; therefore one has to know the moisture content (using the neutron scattering

method) in order to calculate the soil-solids density from the total density, or one has to measure first the dry soil-solids density to calculate the moisture content from the bulk density[1].

It must be pointed out that the soil density measurements by gamma radation are elaborated at much greater length than the radioactive measurements of soil moisture. For the gamma-radiation technique two distinct procedures may be applied:

a) One is the scattering well-type procedure for which only one access pipe is needed (pipe in which also the neutron probe can be lowered).

b) The other is the transmission measurement in which the gamma source is lowered in one access pipe whilst in a second pipe, but at a fixed distance, the detector is held.

The pioneer work concerning soil density determination by the back scattering method was developed for soil tests in airport runway design by BELCHER *et al.* (1950). Gamma rays issued by a radium or cobalt 60 source are absorbed and scattered back by the outer orbit electrons of all atoms present in the soil, including the atoms of water molecules. Mostly two types of probes are used, i.e. a gauge to measure the depth density and one to measure the surface density.

The spherical volume scanned by the gamma rays has a diameter between 13 to 40 cm, the magnitude of the diameter decreasing with an increase in wet density.

The stronger the scattering, the more electrons are present in a given volume, which is in direct proportion to the compaction of the material, providing the same group of atoms is always present, as is the case in soils. The counting rates should be compared with a calibration curve to find the wet bulk density.

The initial work applying the transmission technique to measure soil bulk density was done by VOMOCIL (1954).

He postulated that the log. of the transmitted gamma rays should follow a straight calibration curve when plotted against the wet bulk density as predicted by the LAMBERT-BEER law (absorption law for monochromatic waves). As this was not the case, he pointed out that the presence of a secondary radiation was the reason for not obtaining the desired curve.

VAN BAVEL (1959) describes a technique whereby a separation of a primary and a secondary radiation could be achieved by scintillation counting and electronic discrimination. In field measurements VAN BAVEL (1959) obtained bulk density values, the precision of which was of the order of 1 procent and for which the scanned layer was about 2,5 cm thick (source and detector being placed in a different access tube at a fixed distance from each other). Consequently measurements with this device can be carried out closely to the surface. Individual counts can be made in 1 minute; in very dense soils a period of 2 minutes is preferable.

B. The Total Pore Space or Total Porosity

Once the bulk density is known, the total pore space is calculated using the following formula:

$$\text{percent pore space} = \left(1 - \frac{\text{bulk density}}{\text{solids density}}\right) \times 100\,.$$

[1] From a private communication (Dr. F. BERTHOLD, Wildbad, W. Germany) it was learned that through scintillation counting and electronic discrimination, one and the same probe may now contain both the neutron and the gamma source; depending on the position of the switch, the scaler measures the counts either for the moisture content or for the wet bulk density.

The difficulty consists in finding the solids density, as this is a function of the ratio between the mineral and the organic components of the soil. On an average, the solids density for the mineral components, without appreciable amounts of iron or other heavy minerals in it, is 2.65; for the organic material of normal soils the average value of 1.45 is admitted. When the amount of organic material in the soil is small, this is less than 1 percent, 2.65 is taken as solids density; if the amount of organic material is higher, the density is corrected and for each percent of organic matter the value 2.65 is diminished by 0.02. This rule, experimentally found, is not applicable for peat soils, where a direct measurement of the solids density is necessary (De Leenheer and Van Ruymbeke 1960).

Example: when the bulk density of a soil is 1.424 and its organic matter content is 3 percent, the total pore space is

$$1 - \frac{1.424}{2.65 - (0.02 \times 3)} = 45 \text{ percent.}$$

Attention is drawn to the fact that the total porosity (T. P.) is expressed on a *volume basis*. Another difficult point for calculating the apparent or bulk density is the swelling or shrinkage, which can be observed on soils having a certain amount of colloidal material, when the soil moisture content changes.

At present there is discussion on what basis the bulk density should be expressed. Perrier, Nielsen and Doan (1959) propose to do it on an oven-dry volume basis, although they claim that the samples should be taken at 1/3 bar moisture content, being equivalent to the field capacity[1]. De Leenheer and De Boodt (1959a) also proposed that the soil samples should be taken at field capacity, but then it is more logical that the bulk density should be expressed for that moisture content.

Besides the calculation of the total volume space, direct measurements also are possible. It is said that Say in 1797 was the first to describe an instrument, using Boyle's law, P. V. = constant, to measure the volume of non geometric spaces. For practical soil research work, Torstensson and Eriksson (1936), Visser (1937), von Nitzsch (1938) were among the first to describe a handy tool for this purpose, called an air picnometer. Since the war M. B. Russell (1949) made the instrument much more simple, so it could be used in the field; unfortunately the results are not sufficiently reproducible. Loebell (1953), who thought that the readings depend too much on the atmospheric pressure, made the instrument independent of the atmospheric pressure. However, Jamison (1953) points out that the air absorption on the soil colloidal material is the reason why these instruments cannot give satisfactory results.

It should be remarked that the word picnometer often is used in another sense, namely, in its original meaning, being an instrument to measure the real density of grains or irregular fragments of solids (minerals, ores, rocks, etc.).

C. The Specification of Soil Porosity. The Pore Size Distribution at a Given Moment

One of the earliest interests in soil physical characterization was in the specification of the pores, which were filled either with water or with air. This splitting up of the total pore volume became more definite when Schumacher (1864) introduced the terminology capillary and non-capillary pore space. This division was purely arbitrary until 1935, when Schofield clarified the suction mechanism

[1] 1 bar = 0.98 atmosphere.

in soils, and units of suction forces could be introduced to measure capillary and non-capillary pores. CHILDS (1940) showed how the diameter of the pores could be calculated from suction data. Originally there was a tendency to use a continuous pore size distribution curve as an expression of soil structure, as did CHILDS (1942) and FENG and BROWNING (1947). As a curve is not convenient for statistical calculations, it was found better to re-introduce volumes of characteristic pore size groups as a specification of soil porosity. The earlier denomination capillary and non-capillary pores, although still in use by many authors, was disapproved in 1946 by the Soil Science Society of America, Committee on

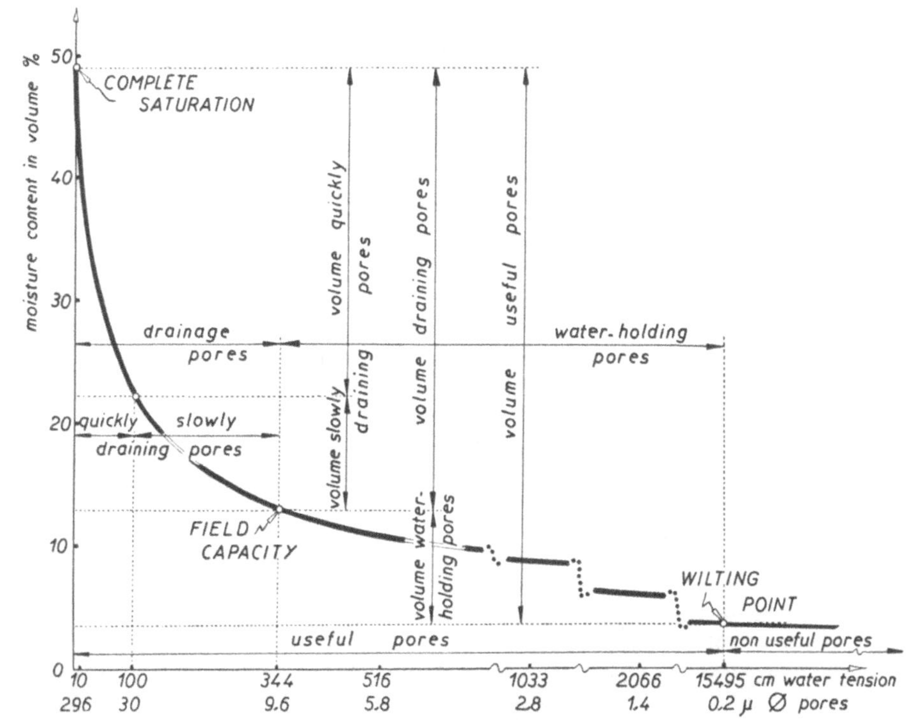

Fig. 36. The different soil pore groups.
The wilting point, the field capacity and the complete saturation are three critical points dividing the useful pore group in the group of the water-holding pores and the group of the drainage pores

Terminology (sub-committee for section I, soil physics), as the so called non-capillary pores still have capillary properties. Other terms were proposed, such as aeration pores or macropores and micropores (KÖHNKE 1947). This, however, is also to be rejected, because of ambiguity with the word microstructure, as there is no relationship between micropores and microstructure.

The above named committee proposed to specify the diameters of the pores in question, instead of using indefined terms. The subdivision of the pores in two main groups is all but uniform, although the distinction between small and large pores cannot be neglected. The lower limit of the large pore group differs very much, in America as well as in Europe, but the 15 μ radius (30 μ diameter), corresponding to 100 cm water suction, is frequently chosen for separating "large" from "small" pores.

To give a more practical meaning to the different soil pore groups, DE BOODT and De Leenheer (1959a) introduced the following grouping of the pores (see Fig. 36), the tensions characterizing the diameters being carefully specified.

- A. Useful pores
  - 1) drainage pores
    - a) quickly draining pores
    - ... limit between (a) and (b): pore diameter = 28.8 μ; 100 cm water or 1/10 bar tension
    - b) slowly draining pores
  - ... limit between (1) and (2): pore diameter = 8.6 μ; 0.337 bar tension
  - 2) water holding pores
- ... limit between A and B: pore diameter = 0.2 μ; 15 bar tension
- B. Non-useful pores.

The reason for this subdivision at the given limits of pore size and tension appears from the following.

(1) *Useful and non-useful pores:* all the water held in the soil at a tension larger than 15 bars is no longer useful for plant growth, because in temperate humid climates the average root suction of crops is about 15 bars. This corresponds to the water tension in voids of maximum 0.2 μ diameter. These voids are therefore called non-useful pores, the larger pores, which can hold water and air for plant growth, being called the useful pores.

(2) *Drainage and water-holding pores:* it was Veihmeyer and Hendrickson (1927) who first defined the term "field capacity" as the moisture content remaining in the soil after a heavy rainfall when free drainage has ceased. When looking at a water tension curve (see Fig. 36), it is obvious that there is no sudden break in the curve at the point corresponding to field capacity. Although we have a practical notion of this definition, theoretically is it hard to apply it to all soil types. Following the paper by Burrows and Kirkham (1958), there are soils which never come to field capacity, but may continue to lose water slowly until the plant wilts.

In general, however, one to two days after a soil is water saturated, free drainage under the pull of gravity ceases. The moisture content at field capacity can for certain soils be estimated in the laboratory by applying a suction on the moist soil equal to 1/3 atmosphere or 0.337 bar. This relationship was experimentally established for a number of Belgian soils (De Boodt 1958). It should be remarked, however, that there is no general ag reement on the equivalent tension of the field capacity. In America also the 1/3 atmosphere suction is generally used, while in Europe many research workers use the 100 cm tension as the equivalent to the field capacity. In stratified soils the field capacity as actually measured in the field will not agree with a laboratory result, because the latter depends on the structure of a sample withdrawn from the soil profile; whereas the field capacity, as measured directly at a certain depth in the field, will depend on the pore size conditions in all the strata above and below the point of measurement. For example if a clay layer underlies a sandy layer, the sandy layer in the field, a few days after a rain, will have a much higher moisture content than would be measured for the field capacity of the same sample in the laboratory. If the clay layer below the sand was thick and relatively dry before the rain on which the field capacity was based, then the clay after a few more days might suck considerably more water from the sand than would be done by 1/3 bar tension in the laboratory. Nevertheless the division of the group of useful pores in drainage pores (normally drained within 24—48 hours) and water holding pores (which can hold water against the pull of gravity) makes sense for practical use. The calculated upper diameter of the water-holding pores (using the formula of the capillary rise, given on this page) is 8.6μ, often rounded off to 10μ, and the lower diameter is 0.2μ. The drainage pores are those which have a diameter larger than 8.6μ.

(3) *Quickly draining pores and slowly draining pores:* in humid temperate climates the drain pipes in heavy soils generally are put at a depth of about 100 cm. The water in the pores of the arable layer, supposing a water saturated soil, is then subject to the pull of gravity of a water column of 100 cm length. Pores of a larger size, which can be emptied under a pull of maximum 100 cm water, will drain much quicker than the smaller pores, which need a pull of between 100 cm water up to nearly 0.337 bar to be emptied. The lower diameter of the quickly draining pores, corresponding to a maximum suction of 100 cm water, is 28.8 μ, often rounded off to 30 μ. The slowly draining pores have a diameter comprised between 28.8 and 8.6 μ (the latter being the upper limit of the water-filled pores at field capacity).

Present Method to Determine the Different Pore Size Groups. From the above stated data it is evident that the study of the different pore size groups is a matter of water content determination in a soil sample, when the water is in equilibrium with a known suction force exerted on the soil sample. The practical critical tension values which are of interest, according to the statements above, are 100 cm water, 0.337 bar and 15 bar.

L. A. RICHARDS (1949) played a leading part in the development of the needed instrument for serial analysis of this kind and reviewed the different methods. Only the present methodology with emphasis on some practical points will be discussed here.

In the laboratory method proposed by SCHOFIELD (1935), the suction needed to empty a water-filled soil pore was created by a difference in height between the moist soil sample and a free water level, having a water-filled connection without any air bubble. Only tension values smaller than one atmosphere could be applied this way. Completing a study of S. J. RICHARDS (1939), L. A. RICHARDS and FIREMAN (1943) constructed an apparatus, which is still up to date, by which a positive pressure could replace the suction on the soil samples to obtain the moisture equilibrium. The equipment, of which a detailed view is given in Fig. 37, is called pressure plate apparatus.

The porosity of the porous plate on which the soil samples will be put, is critical; the pores may not be too large; otherwise the air will blow through it when pressure is applied, but they should be large enough to allow a reasonable rate of drainage.

Using the formula of the capillary rise, one can calculate how large the diameter of the pores may be before air can blow through the plate, when it is covered with water and a known pressure is applied. This formula is $r = \frac{2\,\sigma \cos\theta}{pdg}$,

in which r = the radius of the pores,
p = the pressure expressed in cm water,
d = the mass of water,
g = the force of gravity (981 cm/sec^2 in Western Europe),
σ = the surface tension of the water (72.75 dynes/cm at 20° C),
θ = the angle of contact between the water and the wall of the pores.

To withstand a pressure of 2 bar (twice what is needed to have security), the diameter of the pores should be 1.5 μ[1]. To allow the water surplus to drain

[1] For the underlying assumptions to make this calculation, see the paragraph "Calculation of the Volume of the Different Pore Groups".

away quickly enough, the saturated permeability of the porous plate itself should be 0.01 to 0.1 ml cm²/min/bar.

The mechanism of the determination is as follows: wet soil is placed on the water-saturated porous plate, compressed air or nitrogen is introduced via the pressure inlet into the chamber and brought to a chosen pressure, say 0.1 or 1/3 atmosphere. Water passes out of the soil through the porous plate and continues to do so, until the forces holding the water in the soil just balance the gas pressure tending to force water out of it. This water is for the same reason pressed through the porous plate and is collected in a rubber bulge under the porous plate. The small pressure excess, which comes through the pores, forces the water through a tube connecting the space in the bulge with the rubber tubing that comes through the walls of the container to form the water outlet (see Fig. 37).

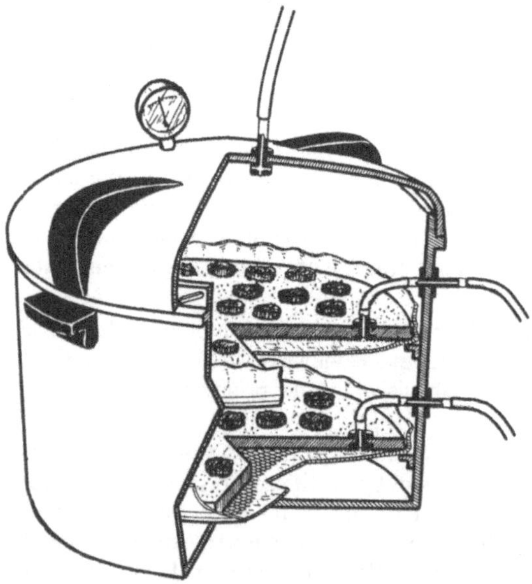

Fig. 37. Pressure plate apparatus for the determination of the volume of the slowly and quickly draining pores. Equipment with two porous plates

When the water flow ceases, the soil moisture is in equilibrium with the water outside the porous plate at atmospheric pressure. This equilibrium on a non-disturbed soil sample, 1 cm thick, is obtained, on an average, after 12 to 24 hours, depending on the kind of soil material.

If the soil in equilibrium is brought in direct contact again with water, the soil would quickly absorb it. To prevent this, a suction of the same magnitude of the original gas pressure needs to be applied to the water, to prevent its re-entering the soil. This proves that the positive pressure on the soil is equivalent to the negative pressure or suction of the soil at the equilibrium water content. Consequently the applied gas pressure may be interpreted as being equivalent to the same amount of tension (negative pressure).

For more details on an operational method to carry out this kind of measurements, see L. A. Richards and Fireman (1943) or De Boodt (1958).

The practical limitation for measurements with the porous plate apparatus is 2 atmospheres. On the other hand, we know through experience that the average moisture content at which the plants wilt, may be considered, for practical purposes, to be the same as the moisture content in the soil in equilibrium with a 15 bar tension. Woodruff (1940) and L. A. Richards (1941) gave at nearly the same time, but independently from one another, the practical solution for applying such high pressures on soil samples by using a cellophane membrane replacing the porous plate. An ultra-filtration apparatus was the source of inspiration to build it.

L. A. Richards (1947) has published full details and drawings of the pressure membrane apparatus. As can be seen in Fig. 38, the device consists of a closed chamber, the floor of which is a membrane of cellophane, suitably supported to withstand high pressure from inside. Water can pass through this membrane, but it will hold back air when wet. Practically the membrane with the very fine pores must be as thin as possible, to allow a reasonable permeability, but otherwise it should be thick enough to withstand high pressure. The results obtained in superposing two cellophane membranes, each 0.02 mm thick, are more satisfactory than when only one membrane of 0.04 mm is used (De Boodt 1958). In order to get the right swelling of the pores and to prevent air from coming through, the membrane should be wetted 24 hours in advance in distilled or rainwater at 20° C.

However, no matter what precautions are taken, there will be small leaks in the membrane when the pressure is up to 15 bars. In order to maintain the pressure, dry gas must be introduced into the chamber, which induces a small air current, with the consequence that there is a drying-out effect on the soil samples having nothing to do with the equilibrium pressure-water content. Therefore, the maximum leakage allowed, before error occurs, should be known. The upper limit of leakage allowed is 2.10^{-6} ml/cm²/second/atmosphere (De Boodt 1958). This leakage has to be controlled from time to time; therefore the outlet of the pressure membrane apparatus is connected with the Scheibler apparatus, shown in Fig. 39, which is well known for the volumetric determination of carbonates in soils. This way the gas volume, which goes through the membrane within a given time, can be read off directly. For a pressure-membrane apparatus with dimensions as described by L. A. Richards (1947) and as shown in Fig. 38, the allowed gas diffusion is maximal 67.8 ml per hour.

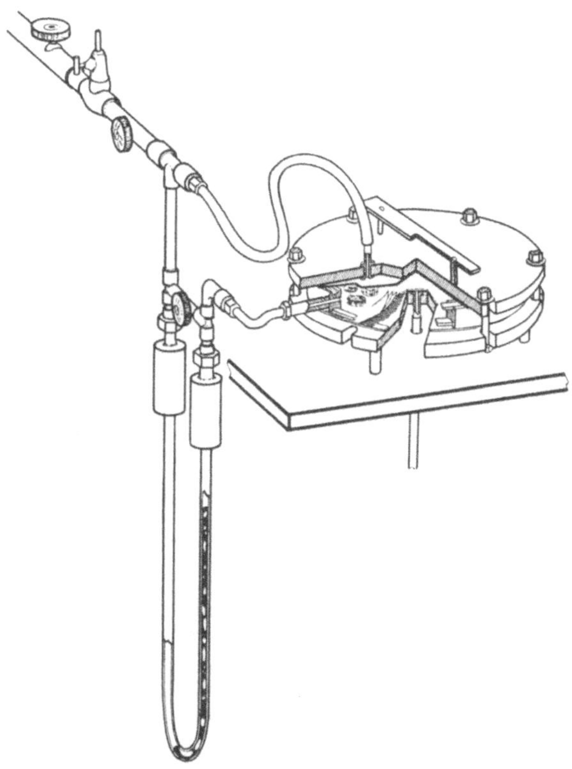

Fig. 38. Pressure-plate apparatus for the determination of the volume of the water-holding pores and the non-useful pores. In this equipment circular soil samples about 1 cm thick are put on two membranes (each 0.02 mm thick) having a sheet of blotting paper between them and suitably supported

The mechanism of the determination with the pressure-membrane apparatus is the same as explained for the porous-plate apparatus. Equilibrium with the pressure-membrane apparatus is reached after 24 to 48 hours for a sample 1 cm thick, depending on the nature of the soil material.

The remark should be made that the water content at 15 bar pressure is a function of the material itself and not of the soil structure as such, as the pores which can retain water at such a pressure are less than 0.2 μ in diameter. Therefore, the soil material used here should be dried and put through a mill in such a way that the largest particles are about 1 mm in diameter. After wetting again for 24 hours, the samples of 1 cm thick are put on the membrane already covered with a small layer of water to assure good contact between the soil sample and the membrane and then the water losing process of the soil under influence of the applied pressure is started. Increasing pressure values are applied on one and the same sample; each time that equilibrium is reached, a sub-sample is taken for the water content determination.

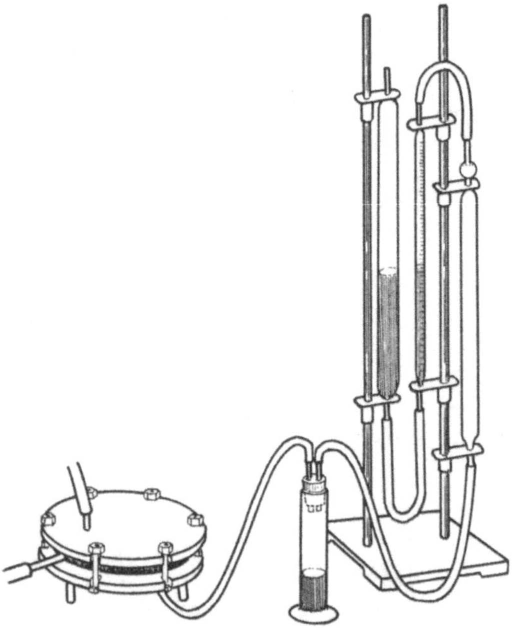

Fig. 39. Leakage control of the pressure plate equipment, using a SCHEIBLER apparatus (right-hand side of the figure)

Calculation of the Volume of the Different Pore Groups. Applying the formula of the capillary rise, assuming the contact angle between water and soil to be zero, the pores to be of circular section, and non-shrinking of the soil when moisture content changes, it is possible to make the following table, in which the inter-relations are given between the units, wherewith the suction force p can be expressed, and the equivalent radius of the largest pores which could be full of water for that pressure.

Table 5. *Equivalent diameter of the largest pores which could be full of water under pressure forces expressed in various units*

cm water pressure or cm capillary rise	pF or log. of cm water	Energy equiv. in ergs/g	Pressure in dynes per cm²	Pressure in millibar[1]	Diameter of the pores in microns
10	1	9.81×10^3	9.81×10^3	9.8	296.64
60 (4.4 cm Hg)	1.78	5.89×10^4	5.89×10^4	58.9	49.44
103 (1/10 atm.)	2.01	1.01×10^5	1.01×10^5	101.0	28.80
344 (25.3 cm Hg or 1/3 atm.)	2.54	3.38×10^5	3.38×10^5	338.0	8.66
516 (1/2 atm.)	2.71	5.06×10^5	5.06×10^5	506.0	5.75
1033 (1 atm.)	3.01	1.01×10^6	1.01×10^6	1013	2.87
2066 (2 atm.)	3.32	2.03×10^6	2.03×10^6	2027	1.44
3099 (3 atm.)	3.49	3.04×10^6	3.04×10^6	3040	0.96
5165 (5 atm.)	3.71	5.07×10^6	5.07×10^6	5076	0.57
10330 (10 atm.)	4.01	1.01×10^7	1.01×10^7	10134	0.29
15495 (15 atm.)	4.19	1.52×10^7	1.52×10^7	15201	0.19

[1] 1 millibar = 1.000 dynes/cm².

To calculate the volume of a group of soil pores having a diameter between two given values, one needs to know the difference between two moisture contents of the soil sample for two equilibriums, the pressures corresponding to the two chosen values of diameter. In other words: the volume of water withdrawn from a soil sample, when going from a first equilibrium to a second with higher suction or pressure, represents the volume of the pores with a diameter between the two corresponding limiting sizes, assuming the soil does not shrink through dessication and the pores are of circular section. To make this conversion, gram of water to ml volume, the difficulty arises that the moisture content of the soil is expressed in weight percentage and the volume of pores in volume percentage.

To convert weight percentage of the soil to volume percentage, one must multiply the weight percentage by the bulk density (B. D.) of the soil.

Example: a fine loamy sand has a moisture content at field capacity (1/3 bar tension) of 15.2 percent and at the wilting point (15 bar tension) of 5.2 percent, the bulk density being 1.24. The volume of water-holding pores is then $(15.2—5.2) \times \times 1.24 = 12$ percent. Let us suppose, that this useful water-holding pore volume should be a mean value for the different horizons of this fine loamy sand profile. When one assumes that the useful water-holding pore volume is water saturated at the beginning of spring time (which is true in humid temperate climates) the water reserve in the soil profile to a depth of 1 m equals 12 cm rainfall. This water reserve expressed in cm or mm rain at the beginning of the plant growth period is very interesting to compare with different soil profile types in a soil mapped area (De Leenheer and Van Ruymbeke 1960 a). The relationship between the suction forces, the pore diameters and the pore groups is given in Fig. 36.

D. Visual Method to Evaluate Total Porosity

The visual evaluation of soil structure is common practice by the soil surveyors, but they refer mostly to the aggregate morphology and size (*Soil Survey Manual* 1951). Visual and microscopical methods to study pore size distribution have been developed by Kubiëna (1938) and applied by Redlich (1940) and Day (1949). Impetus was given to it by Altemüller (1956), who wrote a special chapter on the subject in this book, to which is referred. A macroscopical method was developed by De Boodt and De Leenheer (1959 b), who used two schemes A and B, depending on the moisture content of the soils. If the soil is at field capacity, scheme A, valuable for clay, loam and sand soils is used. The evaluation is based on the frequency and size of the pores within a fixed area, namely the surface of a soil sample ring, 3 inches in diameter (7.62 cm) and a height of 2 cm (see beginning of this chapter). When the soil is dry, or when the moisture content is inferior to field capacity, scheme B is used, based on the morphology of the aggregates and valuable for clay and loam soils. Dry sand soils are also evaluated with scheme A.

Five different porosity classes are recognized, each class being subdivided in two sub-classes. The difference with other methods of determining visual soil structure, such as the "Spatenstich Diagnosis", developed by Görbing and discussed among others by Roemer and Frese (1952), is the fact that the evaluated classes and sub-classes can be checked by a bulk density determination, as each class differs from the foregoing one by 0.2 of a bulk density unit and the sub-classes by 0.1 of a bulk density unit.

A procedure which can be estimated as being half way between the "Spatenstich Diagnosis" and the method proposed by De Boodt and De Leenheer (1959 b), was described by Peerlkamp (1959). His estimations are based on the observation of the size and shape of the aggregates, the crushing of aggregates between the fingers, the porosity of the aggregates and of the entire plowlayer, the root development and the dispersion on the soil surface. He concludes that a 12 year experience period showed the suitability of the method for quantitative characterization of the structure of arable land, especially if many evaluations are required.

E. Determination of Aggregate Distribution and Aggregate Stability

Besides the study of the pore size distribution at a given moment, the determination of the aggregate distribution and of the aggregate stability is another important method of characterizing directly soil structure.

The two aspects of structure investigation mentioned above can clearly be recognized here: the determination of a structure property at a given moment (aggregate distribution) and the study of the effect or potentiality of a structural status (aggregate stability).

Sometimes, as we have seen in the introduction to this chapter, the determination carried out is actually a mixture of both tendencies.

The Aggregate Distribution. The measurement of the aggregate distribution has interested the earliest soil physicists, as can be concluded from the work of Puchner (1911), Mangelsdorff (1929) and Keen (1933). They took soil samples in the field and shaking by hand a nest of sieves of apertures up to about 40 mm and down to 3 or 4 mm, they determined on the spot the aggregate distribution. The shaking of soil aggregates in moist conditions offers, however, difficulties and therefore Volkov (1933) proposed to do the sieving on dry material. This method as a way of characterizing soil structure was aided by Chepil (1952), who uses it mainly to describe the sensitivity of a soil for wind erosion.

Aside from work of De Leenheer and De Boodt (1959 b), who use a dry aggregate as well as a wet aggregate distribution for stability determination, wet sieving of aggregates has, in the past, been emphasized. Instead of water, a liquid with non-dipole character was sometimes used. Russell (1938) reports that sieving dry aggregates in the air or in benzene yields the same results.

Not only the macro-aggregate distribution, but also the micro-aggregate distribution, has been studied, using the same method as for the granulometric analyses (Paschall, Burke and Baver 1935 and Hooghoudt 1945).

This kind of determination has lost its interest in the last decade.

The Aggregate Stability Measurements. Besides the determination of the pore size distribution, the aggregate stability measurement is one of the most common determinations in soil structure studies. To judge of the different methods, which are used now for this purposes, the evolution of the method should be traced briefly.

The most widely spread technique consists in sieving aggregates under water. This method became well established through the papers by Tiulin (1933), Yoder (1936) and Meyer and von Rennenkampff (1936). The method consists in obtaining a relative movement of the water versus the aggregates. Yoder (whose procedure is the origin of many stability determinations) used his method to study the erodibility of the soil and not the aggregate stability.

The *Committee on Physical Analysis* of the Soil Science Society of America (1953), described the essential features of the aggregate stability determination through wet sieving (more or less the same as the YODER method), leaving the possibilities of a great number of minor variations in the procedure. Although called an aggregate stability method, it is only a procedure to measure the aggregate distribution of a soil after wet sieving. As there is no comparison made between the obtained wet stable aggregates and the initial dry aggregate distribution, there is, in the opinion of DE LEENHEER and DE BOODT (1959b), not an actual aggregate stability measurement, although a stability determination is involved through the wet sieving. To make a real aggregate stability determination, from the under-water sieving, these authors measure first the dry aggregate distribution and then wet sieve the dry aggregates; finally they compare the obtained wet stable aggregate distribution with the initial one. The stability is expressed as the change in mean weight diameter of the two aggregate

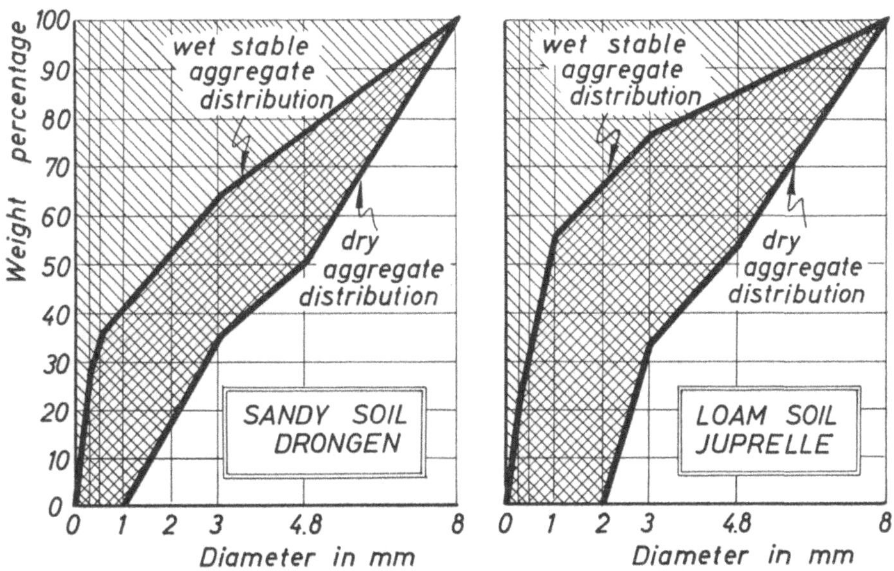

Fig. 40. Graphical determination of the change in mean weight diameter between the dry and wet aggregate distribution. The area between the two curves is measured with a planimeter (the sandy soil is a humic sandy soil)

distributions (Fig. 40). The discussion of the practical value of this method is given by DE BOODT, DE LEENHEER and KIRKHAM (1960) in a paper where their stability index is compared as to reproducibility and correlation with yields with other indexes, such as the geometric mean size of the water stable aggregates and the mean weight diameter. Their stability index gave better reproducibility and better correlation with yields than the others tested.

In assessing soil structure stability by which also an initial state is compared with an obtained new one, two papers should be mentioned. CHILDS and YOUNGS (1959) compared two moisture tension curves of 1 to 2 mm crumb fractions, the first one obtained with slowly wetted, and the second with quickly wetted crumbs. Depending on the area between the two curves, deductions were made as to different degrees of stability; the soil with no change had a maximal stability. Their results were confirmed by the types of sward established on the various sites studied. PAYNE (1959) uses another approach. He measured the surface area of the aggregates (2-1 mm crumbs) from the moisture tension curve. A first time the aggregates were slowly wetted to full saturation and a second time they were rapidly wetted. The difference in surface areas of the aggregates was a measure of the stability.

Another opinion of what stability measurements should be, is defended by Hénin, Robrichet and Jongerius (1955). They say that soil structure stability cannot be studied using only one method. So, for example, one can study the resistance which soil aggregates offer to the bursting effect of water entering the pores; to do this a pretreatment with carbon-tetrachloride (CCl_4) or chloroform ($CHCl_3$) is necessary to avoid any modification in the initial aggregate status. But a soil structure study may be conducted as a determination of the aggregate resistance against dispersion under the influence of water uptake, and then a pretreatment with alcohol or benzine is necessary.

Emerson (1954) and Dettmann and Emerson (1959) think that the stability of aggregates should be measured by finding the concentration of a NaCl solution, at which the permeability of a bed of aggregates, which have been Na-saturated, falls almost to zero, when a solution of decreasing concentration has been passed through.

Another way to determine the aggregate stability is the William-Fadejew method, as described by Faust (1959). 10 g of dry aggregates of defined size are put into a funnel and percolated with water. The amount of water which flows trough the soil is measured every minute during 15 minutes; the observed change in water permeability is considered as a measure of the aggregate breakdown, which, of course, is the opposite of the aggregate stability. Also inspired by Russian scientists is the method of Werschinin, as modified by Doblies and Faust (1963). The method is called the determination of the „Zerwaschungszahl" (disintegration number) and consists in a nephelometric measurement of the change in concentration of the suspension when aggregates are destroyed through a disperging agent (0,004 n $NaHP_2O_7$) under standardised stirring. More or less similar is the method by Davidson and Evans (1960), which is a turbidimeter technique, the stability of the soil aggregates being tested in a water-glycerol mixture after different mechanical mixings.

As can be concluded from a discussion by Kullmann (1958), the wet sieving method is still most used in Europe, while Gliemeroth (1958) finds it better to determine the resistance to washing, in using the flowing method with the aid of Kopecky's elutriating apparatus.

A simpler measurement is the flowing method proposed by Ebert (1958) where the water resistant aggregates are separated from the non-resistant aggregates by a constant waterstream. Further as a representative method to measure aggregate stability by simulating rainfall in the laboratory, the work by Koepf (1958) can be mentioned. Frese, Czeratzki and Schladerbusch (1956) studied the influence of different rain densities on aggregate deterioration and soil crust formation.

The reason why there are so many different methods on stability measurements, is that none of them gives general satisfaction. A main objection against the wet sieving technique, that it is not able to detect difference in stability when sodium or calcium ions are added to the soil, can be solved when the initial dry aggregate distribution is compared with the wet stable one. However, stability measurements are not a full structure determination and therefore one should look for something more, for example a soil structure index as proposed by De Boodt and De Leenheer (1959a) (see chapter on "Soil Structure and Soil Fertility").

Before ending this paragraph, it should be pointed out that at the International Symposium on Soil Structure, Ghent 28-31 May 1958, (p. 434 of its Proceedings[1]) general agreement was obtained on the following:

1) The term "structure stability" (with the meaning of "aggregate stability") should be avoided. It should be replaced in our papers by an explanation of what really was determined in the laboratory (for example, diameter of water stable aggregates after so many minutes of sieving and such a way of moistening, etc.) or observed in the field (structure morphology at different depths or changes in morphology under influence of rain, soil management, etc.).

2) Aggregate stability, whatever the method of measurement may be, should be expressed as a difference between an initial or a reference status and a final status; when a wet sieve method is used and when a wet aggregate distribution is given, this distribution should be compared with the initial dry aggregate distribution.

Quick Methods to Determine Aggregate Stability. In agricultural practice attention is only paid to the structure by the farmer when on a given plot the surface layer is really in a poor physical condition. At that moment it is often expensive to restore good structure or great effort is necessary.

In order to be able to warn farmers when their soil structural status is diminishing in a more detailed way than by the appreciation of a good, a medium or a poor structure, De Boodt and De Leenheer (1959 c) worked out a quick method based on the behavior of the aggregates in water-alcohol, or water-dioxane mixtures of different proportions, to evaluate the stability. The aggregates are put on a white plate with a number of small cavities (plates used in qualitative analytical chemistry for drop tests) and alcohol-water or dioxane-water mixtures are gently poored in the cavities. In the water-free liquid, the aggregate will not puddle, while a slacking effect will be observed as the water percentage increases. The percentage of water in alcohol or dioxane that an aggregate can stand, before it puddles, is a function of the stability of the aggregates. The theoretical background of the physical processes involved is also given.

The Soil Structural Index. Although it is possible to correlate some particular physical measurements with crop yields, it is known from experience that a more composite figure should be tried to express the value of the soil structure.

De Boodt and De Leenheer (1959 a) attempted to compose such an index, deduced from a critical pore space, the stability index and the kind (texture) of soil (see chapter on "Soil Structure and Fertility").

Another attempt in this direction was made by Dabin (1960) who calculated a structural index for tropical soils based on aggregate stability, useful porosity and available moisture.

Lemon and Kirstensen (1960) worked out what they call an edaphic expression of soil structure. Hereby the soil, as a three-phase system, is quantitatively defined as a geometric term in a steadystate radial flow equation, applicable to plant roots. In soils with high moisture range the oxygen flux to a simulated plant root, and in soils with low moisture content the hydroxyl-ion flux away from the simulated root surface are taken as the main factors of the edaphic expression of soil structure.

More information about soil structure as a factor of soil fertility is given in the chapter "Soil Structure and Soil Fertility".

---

[1] Discussions and conclusions; Proceedings Intern. Symposium on Soil Structure, Ghent, 1958. Mededelingen Landbouwhogeschool, Gent, 24, 422–434 (1959).

4. The Indirect Measurements of Soil Structure

It is difficult to draw a line between the so-called direct and indirect methods, because all depends on the definition one gives of soil structure and, as mentioned before, this can vary greatly.

Here we consider as indirect the methods to measure soil structure, in which pore size distribution or aggregate stability are indirectly, not directly, involved.

Under the indirect determinations of soil structure we consider the methods to measure the saturated and unsaturated water permeability, the air permeability and gas diffusion, the oxydation-reduction and the bio-electrical potential, the sound absorption, the modulus of rupture and the plastic limits.

Some of these measurements will be discussed in the chapter "Soil Structure and Soil Fertility", as many data concerning plant growth are available; the others are discussed below.

A. The Measurement of Permeability in Saturated and Unsaturated Soil

The best known and most common way to characterize the potentialities of the pore size distribution in the soil, is to obtain a curve of soil moisture content versus water tension. This is the curve relating water content to suction, as shown in Fig. 36. As the already cited assumptions concerning the circular section of the pores and the absence of shrinkage through dessication are mostly not fulfilled in soils, it is much better to take the data just as they can be read directly from the curve, rather than calculating the theoretical pore size distribution for further interpretation. In this respect, we refer to Marshall (1959), who discusses the limitations in the use of suction data.

The drainage capacity of the soil is one of the great points of interest in agriculture. The pore size distribution of the soil, through which water is passing, is of primary influence, as can be seen from Darcy's law, empirically found in 1856. The permeability measurements should be done on saturated as well as on unsaturated soil, in order to know what the practical significance will be of the structure in a given soil in draining off the water under different circumstances.

The measurement of water drainage in saturated soil has been mainly worked out by Hooghoudt (the Netherlands), Kirkham (U.S.A.), Childs (England) and their co-workers. The knowledge of the rate of seepage of water into an auger hole enables the determination of hydraulic conductivity. By this method an auger hole is made at a depth of about 15 cm below the water table; the water is pumped out and one measures the rate that the water enters the hole. This method, from which the hydraulic conductivity of the soil in place can be deduced (van Bavel and Kirkham 1949), was complemented by Childs (1952). The latter uses two auger holes, 15 cm in diameter and 1 meter apart. In one auger hole the water is pumped out and poured into the other, until a constant waterflow is observed in the emptied auger hole. This offers the possibilities of calculating the horizontal permeability which can add to the results obtained by the Kirkham method in obtaining the vertical permeability (Childs, Cole and Edwards, 1953). The results show that at more than half their sites, the soils were isotropic and Childs (1957) considers that appreciable anisotropy may not be as common as is sometimes believed.

Originally the calculation of the conductivity coefficient k from the data of the auger-hole method was very complicated and time consuming, until van Beers (1958) developed a graphical method and reduced the whole calculation of the conductivity coefficient to a formula as simple as $k=c\cdot\Delta Y/\Delta T$. The c value can be read from a graph (knowing the dimensions of the auger-hole); ΔY is the rise of water in the hole, expressed in cm, during a time ΔT, expressed in seconds.

A critical review of the methods used in this respect is given by Kirkham (1955). The principal disadvantage of the field methods is that they are limited to sites and horizons below a water table and that they are of value mainly in relation to drainage design in areas already troubled with waterlogging, structure not being the first point of interest.

If the latter is the case, the laboratory procedures can be used. The saturated permeability determinations at the laboratory are useful for structure studies as they give indirect informations on pore size distribution and structure stability of soils, which may not commonly be saturated under field conditions.

Permeability measurements in the laboratory are simple: water passes through soil held usually in a cylindrical container, which fits closely to the soil or is sealed to it by paraffin wax to prevent passage of water down along the wall of the soil column. From the difference in head between the two water levels (taking into account the section and the height of the soil column) and the measured rate of water flow, the permeability can then be calculated, following Darcy's law. Either undisturbed cores or artificially packed soil can be used. Methods for taking cores and using them for permeability measurements are reviewed by Reeve (1957).

In recent years simple but practical devices have been set up to measure water-saturated permeability on undisturbed soils. When large series of samples must be studied under a constant water head, the method of Leyton and Yadav (1960) may be cited; when only one sample is studied but different water pressures have to be tried out, the method of Hartge (1961) is interesting; the recent procedure of Flannery and Kirkham (1963) may be mentioned for very simple measurements in series, yielding, nevertheless, good results. This method is recommended if the laboratory equipment is elementary and one only has the disposal of unskilled people.

Using this kind of percolation methods in the laboratory, one normally observes that the rate of outflow will decrease at first, then it will increase, and finally it will slowly decrease. Greacen and Huon (1953) explain the phenomena in the first stage through progressive increase in the volume of the aggregates and the accompanying reduction in size of the large pores due to swelling; the second stage is accounted for through the gradual removal of air bubbles; and finally the decrease in the permeability at the third stage is a function of the aggregate stability. The two first steps, depending on the kind of soil and the applied hydraulic head, last a few hours, therefore permeability determinations should be followed hour per hour for the first eight or ten hours, at least.

It should be remarked that Uhland and O'Neal (1951) proposed a practical system of classifying soil permeability with seven permeability classes. O'Neal (1952) presented a visual method to evaluate soil permeability, by which these seven classes are distinguished. These classes are the same as described in "*Soil Survey Manual*" (1951) and are more or less universally accepted. These tentatively suggested classes of soil permeability are given below, with the corresponding possible rates of water flow.

To avoid misunderstanding it is better to use the term "hydraulic conductivity, K", known from the law of Darcy ($v = K \Delta H / \Delta 1$) instead of "permeability". The value K which is given in cm/sec is then calculated as rates of flow in mm or in inches per hour.

Table 6. *The seven permeability classes of Uhland and O'Neal*

	possible rates of flow per hour	
	in mm	in inches
Slow		
1. very slow	less than 1.25	less than 0.05
2. slow	1.25—5.0	0.05—0.2
Moderate		
3. moderately slow	5.0—20.3	0.2—0.8
4. moderate	20.3—63.5	0.8—2.5
5. moderately rapid	63.5—127	2.5—5.0
Rapid		
6. rapid	127—254	5.0—10.0
7. very rapid	over 254	over 10.0

Movement of water in the unsaturated phase occurs much more often than it does in the saturated phase. For the unsaturated phase, however, the theoretical equations are more complex than for the saturated phase, and accordingly, the knowledge of the movement of water in the unsaturated is less well known (Richards and Wadleigh 1952).

In a short but important communication Hallaire and Hénin (1958) proved that the conductivity equation (Darcy's law) and other improved formulae do not always hold in studying water movement in unsaturated soil. In four soil profiles they found that a relatively dry soil layer lost water through a more wet layer situated at the soil surface. In these cases Hallaire and Hénin believed that water cohesion and water movement in the adsorbed water films were a suitable explanation for water movement in the unsaturated phase.

The same year, however, the work of Gardner *et al.* (1958) (and other references cited there), shows that Darcy's law has applied in a number of cases in applied agricultural problems involving unsaturated flow.

To measure capillary conductivity in the unsaturated phase, Gardner (1956) introduced a great simplification over previous methods, using standard soil physics laboratory equipment. By measuring the pressure plate outflow from the soil sample as a function of time and using the solution of the flow equation, the capillary conductivity was calculated for each time interval. It was found that the conductivity decreases extremely rapidly by decreasing water content or suction.

A practical way to measure water permeability at different suction values in the low range, using fritted glass beads as porous plates, is described by Nielsen and Phillips (1958).

In the same line of thinking Kunze and Kirkham (1962) proposed a simplified device together with a simplified method to calculate the results of capillary conductivity determinations.

Theoretically, these methods are much directer to characterize soil water economy than the ones discussed before which are based on the pore-size distribution as deduced from pF curves; the properties deduced from a pore-size distribution curve are lacking any dynamic property. However, it is the general feeling that at this moment no laboratory method which measures the unsaturated water movement is completely satisfactory, as the reproducibility of the results is not good enough.

Real hope exists that good results will be obtained soon by field methods using both the neutron diffusion technique to measure the moisture content and the tensiometer method to measure moisture potentials (KIRKHAM and KUNZE 1962).

Because of the difficulties encountered in measuring permeability in the unsaturated phase, much attention has been given to the ways of calculating it from other data. The general principles governing the relation of permeability of saturated and unsaturated materials to pore space are discussed by MARSHALL (1959).

B. The Measurement of Air Permeability and Diffusion

This problem will be discussed from a more practical standpoint. See the chapter on "Soil Structure and Soil Fertility".

C. The Potential Measurements

The determination of the oxidation and reduction potential in soils and the measurement of the electrical potential between plant roots and soil will also be discussed in the chapter on "Soil Structure and Soil Fertility".

D. Sound Absorption Measurement

JANSE (1959) described a method using sound absorption to evaluate soil structure. This is possible because sound absorption is strictly concerned with the geometry of the air space. In this respect the method, aside from certain shortcomings with regard to the theoretical background, provides a rather direct approach to the study of the soil as a porous medium.

E. Modulus of Rupture Determination

The short description of the method and its discussion from practical standpoint will be given in the chapter on "Soil Structure and Soil Fertility".

F. The Plasticity Limits Determination

When water is added to dry soil, each particle is covered with a film of adsorbed water. If the addition of water is continued, the thickness of the water film on a particle increases. Increasing the thickness of the water films permits the particles to slide past one another more easily. So depending on the moisture content, the soil will behave differently towards gravity and pressure.

It was ATTERBERG (1911, 1912, 1913) who studied systematically the boundaries of the three states of the soil in terms of limits, later called the plasticity or ATTERBERG's limits. He observed a liquid limit or upper plasticity limit (sometimes called limit of flowability; German: Fliessgrenze; French: limite supérieure de plasticité) as being the boundary between the semi-liquid and the plastic state, the friability limit or lower plasticity limit (sometimes called the rolling-out limit; German: Ausrollgrenze; French: limite inférieure de plasticité) as the boundary between the plastic and the semi-solid state; the boundary between the semi-solid and solid states is called the shrinkage limit (German: Schrumpfungsgrenze; French: limite de rétrécissement).

A particular point of the soil between the upper and lower plasticity limit is the sticky point (sometimes called the scouring point; German: Klebepunkt; French: point d'adhésivité).

These limits have been better defined by Casagrande (1932) as the water content which exists under different experimental conditions. The plasticity number or plasticity range is defined as the difference in water content at the liquid limit and the friability limit. Is the plasticity number between 20 and 30, the soil is highly plastic; between 10 and 20 the soil is moderately plastic and the plasticity is low for a plasticity number smaller than 10.

The liquid limit is obtained when a groove drawn in a soil paste is closed after 25 standardized blows from a given height against the cup containing the soil. The friability limit is reached when a soil thread of 3 mm (about 1/8 inch) diameter crumbles, being rolled on a glass plate with hand. The shrinkage limit is the moisture content at which the volume of the soil does not shrink when the soil loses more water and the sticky point is the moisture content when the soil will no longer stick to a smooth metal spatula rubbed across it.

The sticky point is also the water content at which a kneaded soil ceases adhering to the fingers. The point can be determined with better accuracy than the lower plasticity limit (Hardy 1923, Keen and Coutts 1928) and should be more important for the study of the best water content for plowing than the lower plasticity limit.

Kuron (1948) proposed another limit, the crack limit (German: Rissgrenze; French: limite de fissuration) as being the water content at which a cylindrical soil sample under a given radial compression shows the first cracks.

Of course, all these procedures are standardized; and it is amazing to see that soil engineers rather than agronomists played in their development the leading part (Lambe: Soil Testing for Engineers, 1951). Atterberg (1912) introduced these limits for agricultural work, but for many years the agronomist's interest in this study was negligible, while in building and related fields interest was great.

Another point is that, although these limits all are determined on soil pastes, they differ according to the soil structure (Boekel and Peerlkamp 1955), and on the other hand they provide the necessary scientific background to the question: in what moist conditions may structure be improved or damaged when working the soil?

When the soil reaches its liquid limit, soil particles will orientate themselves with their largest surface perpendicular to the direction of the pressure on the soil, breaking all the links between the soil particles, which provided the aggregate stability and the pore size distribution.

Starting with the findings of Terzaghi (1926), who found a direct relation between moisture content and soil compaction through compression, Boekel (1959) developed a soil "structure index", based on the Atterberg consistency limits and on the field capacity (at pF 2.0). In sandy soils the structure is good if the quotient upper plasticity limit over field capacity is much greater than 1.0 (for example 1.4) and the soil will disperse easily if this quotient is only 1.0 or less. For heavy clay soils, however, it is the quotient of the lower plasticity limit over field capacity which is critical.

Haines (1923), Keen and Coutts (1928) and Keen and Scott Blair (1929) were among the earlier soil scientists who understood the sense of the plasticity limits for soils. The latter introduced a new concept, "the static rigidity", by which they mean the energy which would be needed to make a soil paste move in plastometric studies, and which should be a measure of the cohesion forces of the soil. HÉNIN (1936) also brought the meaning of the sticky point to the attention of the agronomist on plowing problems, while KURON (1948) discussed the practical significance of the plasticity limits on soil management in general.

Thanks to mechanisation in agriculture, the study of soil strength received new interest from soil physicists during the last decennia. Particularly the deformation of the soil under a certain stress has been studied in the laboratory. In Germany a so called "Schlagpendel" device (blow pendulum) was described by JACOB (1941) by which the mechanical resistance was measured on a dry soil briquet. A pendulum of a certain weight was used to determine the momentum needed to break the briquet.

As water content plays an important role in soil strength, PFEFFERKORN (1942) developed a simple apparatus which made possible the measurement of the deformation through a falling load on a soil paste, prepared at different water contents.

RHODES (1938) described an apparatus, called plastograph, by which the shear stress instead of the vertical stress could be measured through the torque needed to shear a soil paste when increments of water are added.

A new impulse to the more fundamental study of soil plasticity was given by BODMAN and RUBIN (1948) who constructed an apparatus which permitted the application of both the normal and shearing stresses on a mass of soil confined in a metal ring. The volume changes produced by these stresses could not only be measured, but it was also possible to determine the amount of work done on the soil by the applied stresses.

They found that the behaviour of the soil was not only influenced by the moisture content but also by the energy status of the solid phase.

It is the merit of GREACEN (1960a) that, using nearly the same apparatus as mentioned above, he could separate the influence of the water content from the inherent soil strength when measuring the deformation of a soil. He found that when a loose soil is compressed and sheared in the water saturated condition, a maximum resistance to shearing develops which is linearly proportional to the normal load; he also found that there is a unique load-water content-strength relation for straining watersaturated clay. For both the saturated and unsaturated state, agricultural soil compress under normal load to a density which appears to be a function of the aggregate strength and of the initial state of the bed. When a shear force is subsequently applied, the soil compresses further to a degree which is also determined by the moisture content and the aggregate strength. In another publication GREACEN (1960b) proves that the aggregate strength can be measured through the ratio of the water content of structured clay to the plasticity or liquid limit of the same clay sample when strained.

A method to measure the shearing strength of arable soils for assessing the stability of their structure was developed by SCHAFFER (1959). It is a simple device by which a 80 cm shaft carries at its base a crosswise arrangement of 10 cm long and 2,5 cm wide metal plates. After insertion to the desired depth into the soil, the tool is turned around its vertical axis and the torque is read on a graduated cm/kg scale. The shear strength, expressed in kg/cm^2 is calculated after elimination of the friction components and by taking soil-moisture content into consideration.

As mechanization for all kinds of plowing, management and crop harvesting brings constantly heavier machinery on the fields, it is expected that studies based on the plasticity limits and inherent soil strength will be more and more necessary, if the soil physicist is to be of help to the farmers, in advising them how to keep their soil structure in good conditions, despite mechanization.

Literature

ALTEMÜLLER, H. J.: Neue Möglichkeiten zur Herstellung von Bodendünnschliffen. Z. Pflanzenernähr., Düng., Bodenkde. **72**, 56–62 (1956). — ATTERBERG, A.: Die Plasticität der Tone. Internat. Mitt. Bodenkde. **1**, 10–44 (1911). — Die Konsistenz und die Bindigkeit der Böden. Internat. Mitt. Bodenkde. **2**, 149–189 (1912). — Die Plastizität und Bindigkeit liefernden Bestandteile der Tone. Internat. Mitt. Bodenkde. **3**, 291–324 (1913).

BAVER, L. D.: Soil physics, 3rd ed. New York: J. Wiley, and London: Chapman & Hall. 1956. — BELCHER, D. J., T. R. CUYKENDALL and H. S. SACK: The measurement of soil moisture and density by neutron and gamma ray scattering. Civil Aeronautics Administration; Technical Development and Evaluation Centre, Indianapolis, Indiana, U.S.A., 1950. — BELCHER, D. J., R. C. HERNER, T. R. CUYKENDALL and H. S. SACK: Use of radioactive material to measure soil moisture and density. Symposium on Radioisotopes, Special Technical Publication, American Society for Testing Materials, 10–22, 1953. — BERG, L. G., and N. V. SITNIKOV: Rapid determination of soil moisture by the hydride method. Soviet Soil Sci. **5**, 78–80 (1958). — BLOODWORTH, M. E., and J. B. PAGE: Use of thermistors for the measurement of soil moisture and temperature. Soil Sci. Soc. Amer. Proc. **21**, 11–15 (1957). — BODMAN, G. B., and J. RUBIN: Soil puddling. Soil Sci. Soc. Amer. Proc. **13**, 27–36 (1948). — BOEKEL, P.: Evaluation of the structure of clay soils by means of soil consistency. Proceedings of the International Symposium on Soil Structure. Mededelingen van de Landbouwhogeschool, Gent, **24**, 363–368 (1959). — BOEKEL, P., and P. K. PEERLKAMP: Soil consistency as a factor determining the soil structure of clay soils. Proceedings of the Informal Meeting of Physics in Agriculture, Wageningen, Netherlands. J. Agricult. Sci. **4**, 122–125 (1955). — BOLLMEYER, K.: Bestimmung der Bodenfeuchte aus der Messung der Wärmeleitfähigkeit. Z. Pflanzenernähr., Düng., Bodenkde. **80**, 245–253 (1958). — BOUYOUCOS, G. J.: Nylon electrical resistance unit for continuous measurement of soil moisture in the field. Soil Sci. **67**, 319–330 (1949). — BURROWS, W. C., and D. KIRKHAM: Measurement of field capacity with a neutron meter. Soil Sci. Soc. Amer. Proc. **22**, 103–106 (1958).

CASAGRANDE, A.: Research on the Atterberg limits of soils. Public Roads **13**, 121–130 (1932). — CHEPIL, W. S.: Improved rotary sieve for measuring state and stability of dry soil structure. Soil Sci. Soc. Amer. Proc. **16**, 113–117 (1952). — CHILDS, E. C.: The use of soil moisture characteristics in soil studies. Soil Sci. **50**, 239–252 (1940). — Stability of clay soils. Soil Sci. **53**, 79–92 (1942). — The measurement of the hydraulic permeability of saturated soil in situ, I, Principles of the proposed method. Proc. Roy. Soc. A **215**, 525–535 (1952). — The anisotropic hydraulic conductivity of soil. J. Soil Sci. **8**, 42–47 (1957). — CHILDS, E. C., A. H. COLE and D. H. EDWARDS: The measurement of the hydraulic permeability of saturated soil in situ. Proc. Roy. Soc. A **216**, 72–89 (1953). — CHILDS, E. C., and E. G. YOUNGS: The use of moisture characteristic curves in assessing the soil stability of open cast coal sites. Proceedings of the International Symposium on Soil Structure. Mededelingen van de Landbouwhogeschool, Gent, **24**, 415–421 (1959). — COLLIS-GEORGE, N.: Relationship between air and water permeabilities in porous media. Soil Sci. **76**, 232–250 (1953). — Committee on Physical Analysis: Report of Committee on physical analysis 1951–1953. Soil Sci. Soc. Amer. Proc. **17**, 416–418 (1953).

DABIN, B.: Utilisation d'un indice de structure pour la détermination de la qualité physique des sols tropicaux. Trans. 7th Intern. Congr. Soil Sci. **1**, 311–320 (1960). — DAVIDSON, J. M., and D. D. EVANS: Turbidimeter technique for measuring the stability of soil aggregates in a water-glycerol mixture. Soil Sci. Soc. Amer. Proc. **24**, 75–79 (1960). — DAY, P. R.: Experiments in the use of microscope for study of soil structure. Soil Sci. Soc. Amer. Proc. **13**, 43–50 (1949). — DE BOODT, M.: Het beoordelen van de bodemstructuur door laboratoriumonderzoek. Mededelingen van de Landbouwhogeschool, Gent, **23**, 465–548 (1958). — Number of soil sample replications needed in soil structure research. Proceedings of the International Symposium on Soil Structure. Mededelingen van de Landbouwhogeschool, Gent, **24**, 267–276 (1959). — DE BOODT, M., and L. DE LEENHEER: Soil structure index and plant growth. Proceedings of the International Symposium on Soil Structure. Mededelingen van de Landbouwhogeschool, Gent, **24**, 312–323 (1959a). — Proposition d'évaluation de la porosité totale du sol sur le terrain. Proceedings of the International Symposium on Soil Structure. Mededelingen van de Landbouwhogeschool, Gent, **24**, 222–233 (1959b). — Proposition pour l'évaluation de la stabilité des agrégats sur le terrain. Proceedings of the International Symposium on Soil Structure. Mededelingen van de Landbouwhogeschool, Gent, **24**, 234–243 (1959c). — DE BOODT, M., L. DE

Leenheer and D. Kirkham: Soil aggregate stability indexes and crop yields. Soil Sci. **91**, 138–146 (1961). — De Boodt, M., L. De Leenheer en P. Mortier: De praktische bruikbaarheid in de bodenkunde van de methode van de neutronendiffusie. Mededelingen van de Landbouwhogeschool, Gent, **28**, 85–132 (1963). — De Leenheer, L., and M. De Boodt: Soil sampling and storage for soil structure research. Proceedings of the International Symposium on Soil Structure. Mededelingen van de Landbouwhogeschool, Gent, **24**, 257–266 (1959 a). — Determination of aggregate stability by the change in mean weight diameter. Proceedings of the International Symposium on Soil Structure. Mededelingen van de Landbouwhogeschool, Gent, **24**, 290–300 (1959 b). — De Leenheer, L., and M. Van Ruymbeke: Monografie der Zeepolders; Repertorium van de bodemkundige eigenschappen der belangrijke bodemtypen in de Belgische Zeepolders. Mededelingen van de Landbouwhogeschool, Gent, **25**, 1–416 (1960). — Is it possible to predict some physical soil characteristics, knowing the soil components. Trans. 7th Intern. Congr. Soil Sci. **1**, 272–278 (1960). — Dettmann, M. G., and W. W. Emmerson: A modified permeability test for measuring the cohesion of soil crumbs. J. Soil Sci. **10**, 215–226 (1959). — De Vries, D. A.: Some results of field determinations from thermal conductivity measurements. Netherlands J. Agricult. Sci. **1**, 115–121 (1953). — Doblies, W., and W. Faust: Aggregatzerstörung durch Verwendung eines Rührwerkes in Verbindung mit chemischer Dispergierung. Inst. f. Bodenkunde, Universität Giessen (W. Germany). 1963.

Ebert, D.: Messung der Krümelstabilität nach der Spülmethode. Tagungsberichte **13**. Probleme der Krümelstabilitätsmessung und der Krümelbildung. Deutsche Akademie der Landwirtschaftswissenschaften zu Berlin, 57–66 (1958). — Emerson, W. W.: The determination of the stability of soil crumbs. J. Soil Sci. **5**, 233–250 (1954).

Faust, W.: Untersuchungen über die Ursachen und Verhütung von Erosionsschäden in Verbindung mit Flurbereinigung. Jahresbericht 1959 an das Minist. f. Ern., Landw. u. Forsten des Landes Nordrhein-Westfalen (1959). — Feng, C. L., and G. M. Browning: Aggregate stability in relation to pore size distribution. Soil Sci. Soc. Amer. Proc. **11**, 67–73 (1947). — Flaig, W., K. Scharrer and G. K. Judel: Zur Methodik der Bestimmung des Redoxpotentials im Boden. Z. Pflanzenernähr., Düng., Bodenkde. **68**, 97–122 (1955). — Flannery, R., and D. Kirkham: A soil core water permeameter for field use. Submitted for publication in Soil Sci. (1963). — Fountaine, E. R.: Investigations into the mechanism of soil adhesion. J. Soil Sci. **5**, 251–263 (1954). — Frese, H., W. Czeratzki and H. Schladerbusch: Krümelzerstörung unter dem Einfluß verschiedener Beregnungsweisen. Z. Pflanzenernähr., Düng., Bodenkde. **73** (**118**), 210–223 (1956).

Gardner, W. R.: Calculation of capillary conductivity from pressure plate outflow data. Soil Sci. Soc. Amer. Proc. **20**, 317–320 (1956). — Gardner, W. R., and M. Fireman: Laboratory studies of evaporation from soil columns in the presence of a water table. Soil Sci. **85**, 244–249 (1958). — Gardner, W., and D. Kirkham: Determination of soil moisture by neutron scattering. Soil Sci. **73**, 391–402 (1952). — Gardner, W. R., and M. S. Mayhugh: Solutions and tests of the diffusion equation for the movement of water in soil. Soil Sci. Soc. Amer. Proc. **22**, 197–201 (1958). — Gätke, C. R.: Über die Untersuchung von Bodenproben aus Stechzylindern nach der Porosimetermethode. Deutsche Landwirtschaft **9**, 81–82 (1958). — Gliemeroth, G.: Die Wirkungen ackerbaulicher Maßnahmen auf die Aggregatstabilität nach Untersuchungen mit der Durchlauf- und Schlämmethode. Tagungsberichte **13**. Probleme der Krümelstabilitätsmessung und der Krümelbildung. Deutsche Akademie der Landwirtschaftswissenschaften zu Berlin, 39–56 (1958). — Greacen, E. L.: Water content and soil strength. J. Soil Sci. **11**, 313–333 (1960a). — Aggregate strength and soil consistence. Trans. 7th Intern. Congr. Soil Sci. **1**, 256–264 (1960b). — Greacen, E. L., and A. N. Huon: Microscopic changes in soil aggregates during peremability tests. Austral. J. Agricult. Res. **4**, 184–192 (1953).

Haines, W. B.: The volume changes associated with variations of water content in soils. J. Agricult. Sci. **13**, 296–310 (1923). — Hallaire, M., et S. Henin: Sur la non validité de l'équation de conductivité pour exprimer le mouvement de l'eau non saturante dans le sol. C. R. Acad. Sci., Paris, **246**, 1720–1722 (1958). — Hardy, F.: The maximum water retaining capacity of colloidal soils, the interpretation of this and of certain other soil moisture constants. J. Agricult. Sci. **13**, 340–351 (1923). — Hartge, K.: Die Messung der Wasserpermeabilität an Stechzylinderproben. Z. Kulturtechn. **2**, 103–114 (1961). — Henin, S.: Etude sur la régénération des éléments structuraux des sols de limon. Ann. Agronom. **6**, 455–472 (1936). — Henin, S., O. Robrichet, et A. Jongerius: Principes pour l'évaluation de la stabilité structurale du sol. Ann. Agronom. 6ème Année, n° 4, Sér. A, 537–557 (1955). — Hooghoudt, S. B.: Een gecombineerde zeef- en pipetmethode voor de bepaling van de granulaire samenstelling van gronden. Verslagen van Landbouwkundige Onderzoekingen, nr **50** (13) A., p. 683. 's Gravenhage: Algemene drukkerij, 1945.

Jacob, H.: Zur Bestimmung der mechanischen Festigkeit trockner Bodenproben. Z. Bodenkde. u. Pflanzenernähr., Berlin, **27** (**70**), 240–250 (1941). — Jamison, V. C.: The significance of air adsorption by soil colloids in picnometric measurements. Soil Sci. Soc. Amer. Proc. **17**, 17–19 (1953). — Janert, H.: Die Anwendung von Methanol bei physikalischen Bodenuntersuchungen. Z. Pflanzenernähr., Düng., Bodenkde. **83**, 124–134 (1958). — Janse, A. R. P.: Measurement of sound absorption in soils. Proceedings of the International Symposium on Soil Structure. Mededelingen van de Landbouwhogeschool, Gent, **24**, 398–410 (1959).

Keen, B. A.: Experimental methods for the study of soil cultivation. Empire J. Exper. Agricult. **1**, 97–102 (1933). — Keen, B. A. and J. Coutts: Single value soil properties; a study of the significance of certain soil constants. J. Agricult. Sci. **18**, 740–765 (1928). — Keen, B. A., and W. Scott Blair: Plastometric studies of soil and clay pastes. J. Agricult. Sci. **19**, 684–700 (1929). — Kirkham, D.: Measurement of the hydraulic conductivity of soil in place. Symposium on permeability of soils; American Society of Testing Materials. Special Technical Publication **163**, 80–97 (1955). — Kirkham, D., and G. S. Taylor: Some tests of a four electrode probe for soil moisture measurements. Soil Sci. Soc. Amer. Proc. **14**, 42–46 (1950). — Kirkham, D., M. De Boodt and L. De Leenheer: Modulus of rupture determination on cylindrical soil core samples. Proceedings of the International Symposium on Soil Structure. Mededelingen van de Landbouwhogeschool, Gent, **24**, 369–376 (1959). — Kirkham, D., and R. J. Kunze: Isotopes methods and uses in soil physics research. Adv. Agronomy **14**, 325–358 (1962). — Koepf, H.: Beregnungsmethode zur Beurteilung der Stabilität von Bodenkrümeln. Tagungsberichte **13.** Probleme der Krümelstabilitätsmessung und der Krümelbildung. Deutsche Akademie der Landwirtschaftswissenschaften zu Berlin, 67–78 (1958). — Köhnke, H.: The practical use of the energy concept of soil moisture. Soil Sci. Soc. Amer. Proc. **11**, 64–66 (1947). — Korobockhin, I., V.: The possibility of investigating the dynamics of soil moisture and of compaction by gamma rays of different energies. Written in Russian, English summary in Soils and Fertilizers **22**, 101 (1959). — Kubiëna, W.: Micropedology. Ames, Iowa, U.S.A.: Collegiate Press. 1938. — Kullmann, A.: Zur Problematik der Krümelstabilitätsmessungen und zur Methodik des Durchflußverfahrens. Tagungsberichte **13.** Probleme der Krümelstabilitätsmessung und der Krümelbildung. Deutsche Akademie der Landwirtschaftswissenschaften zu Berlin, 7–34 (1958). — Kunze, R. J., and D. Kirkham: Simplified accounting for membrane impedance in capillary conductivity determinations. Soil Sci. Soc. Amer. Proc. **26**, 421–426 (1962). — Kuron, J.: Zur Frage der Plastizität der Böden und ihrer praktischen Bedeutung. Z. Pflanzenernähr., Düng., Bodenkde. **42**, 140–147 (1948).

Lambe, T. W.: Soil Testing for Engineers, p. 22–29. New York: J. Wiley. 1951. — Lemon, E. R., and A. E. Erickson: The measurement of oxygen diffusion in the soil with a platinum microelectrode. Soil Sci. Soc. Amer. Proc. **16**, 160–163 (1952). — Lemon, E., and J. Kirstensen: An edaphic expression of soil structure. Trans. 7th Intern. Congr. Soil Sci. **1**, 232–240 (1960). — Leyton, L., and J. Yadav: Effect of drainage on certain physical properties of a heavy clay soil. Soil Sci. **11**, 305–312 (1960). — Loebell, R.: Barometerfreie Luftpyknometer. Z. Pflanzenernähr., Düng., Bodenkde. **60**, 172–181 (1953). — Lutz, J. F.: Apparatus for collecting undisturbed soil samples. Soil Sci. **64**, 399–401 (1947).

Maiboroda, N. M.: Determination of soil moisture by the water method. Pochvovedenie 8, 101–103 (1957). Russian paper; summary in Soils and Fertilizers **21**, 22 (1958). — Mangelsdorff, E. G.: Reference is made to Russell, E. W.: Soil Structure. Imperial Bureau of Soil Science, Techn. Comm. **37,** 7 (1929). Marshall, T. J.: Relations between water and soil. Technical Communication no. 50. Commonwealth Bureau of Soils, Commonwealth Agricultural Bureau, Farnham Royal-Brucks, England 1959. — Meyer, L., und U. von Rennenkampff: Neuer Apparat und Methode zur automatischen Durchführung der Krümelanalyse nach Tiulin und Vorschläge zur Bestimmung des Krümelanteils, der Krümelgroßen und der Krümelfähigkeit des Bodens. Z. Pflanzenernähr. u. Düng. **43**, 268–280 (1936). — Mortier, P., M. De Boodt and L. De Leenheer: Über das Auflösungsvermögen der Neutronendiffusionsmethode für Feuchtigkeitsbestimmungen im Boden. Z. Pflanzenernähr., Düng., Bodenkde. **87** (**133**), 244–250 (1959). — Mortier, P., M. De Boodt, W. Dansercoer, and L. De Leenheer: The resolution of the neutron scattering method for soil moisture determinations. Trans. 7th Intern. Congr. Soil Sci. **1**, 321–329 (1960). — MORTIER, P., and M. DE BOODT: Determination of soil moisture by neutron scattering. Netherlands J. Agricult. Sci. **4**, 111–113 (1956).

NIELSEN, D. R., and R. E. PHILLIPS: Small fritted glass bead plates for determination of moisture retention. Soil Sci. Soc. Amer. Proc. **22**, 574–575 (1958). — NITZSCH, W. V. VON: Porengrößen im Boden, ihre Beziehungen zur Bodenbearbeitung und zum Wasserhaushalt. Heft 85 der R.K.T.L. Schriften, Berlin, 1938.

O'NEAL, A. M.: A key for evaluating soil permeability by means of certain field clues. Soil Sci. Soc. Amer. Proc. **16**, 312–315 (1952).

PASCHALL, A. H., R. T. A. BURKE and L. D. BAVER: Aggregation studies on Muskingum, Chester and Landsdale silt loams. Soil Survey Bull. 16–44, 1935. — PAYNE, D.: The measurement of the breakdown of soil aggregates from determinations of surface areas. Proceedings of the International Symposium on Soil Structure. Mededelingen van de Landbouwhogeschool, Gent, **24**, 335–340 (1959). — PEERLKAMP, P. K.: A visual method of soil structure evaluation. Proceedings of the International Symposium on Soil Structure. Mededelingen van de Landbouwhogeschool, Gent, **24**, 216–221 (1959). — PFEFFERKORN, C.: Zur Kenntnis der Plastizität der Tone und Kaoline. Ziegelwelt **73**, 243 (1942). — PERRIER, E. R., D. R. NIELSEN and J. E. DOAN: Adjustment of bulk density to an oven-dry volume basis. Soil Sci. **88**, 291–293 (1959). — PUCHNER, H. (1911). Reference is made to RUSSELL, E. W.: Soil Structure. Imperial Bureau of Soil Science, Techn. Comm. **37**, 7 (1938).

REDLICH, G. C.: Determination of soil structure by microscopical investigation. Soil Sci. **50**, 3–13 (1940). — REEVE, R. C.: Drainage of irrigated lands. Soil Conservation **23**, 12–15 (1957). — RICHARDS, L. A.: A pressure membrane extraction apparatus for soil solution. Soil Sci. **51**, 377–386 (1941). — Pressure membrane apparatus, construction and use. Agricult. Engng. **28**, 451–455 (1947). — Methods of measuring soil moisture tension. Soil Sci. **68**, 95–112 (1949). — Modulus of rupture as an index of crusting of soil. Soil Sci. Soc. Amer. Proc. **17**, 321–323 (1953). — RICHARDS, L. A., and M. FIREMAN: Pressure plate apparatus for studying moisture sorption and transmission by soil. Soil Sci. **56**, 395–404 (1943). — RICHARDS, L. A., and C. H. WADLEIGH: Soil water and plant growth. See Soil Physical Conditions and Plant Growth, Vol. II of Agronomy, p. 73–225. New York: Academic Press. 1952. — RICHARDS, S. J.: Soil moisture content calculations from capillary tension records. Soil Sci. Soc. Amer. Proc. **3**, 57–64 (1939). — RHODES, E. O.: A machine for testing highway subgrade soils. Amer. Soc. Testing Mat. Proc. **38**, 551 (1938). — ROEMER, TH., und H. FRESE: Bodenbearbeitung. Handbuch der Landwirtschaft, Bd. I, Ackerbaulehre p. 237–309. Berlin und Hamburg: P. Parey. 1952. — RUSSELL, E. W.: Soil Structure. Imperial Bureau of Soil Science, Techn. Comm. **37**, 8 (1938). — RUSSELL, M. B.: Methods of measuring soil structure and aeration. Soil Sci. **68**, 25–35 (1949). — A simplified Air-Picnometer for field use. Soil Sci. Soc. Amer. Proc. **14**, 73–76 (1949).

SCHAFFER, G.: Eine Methode der Abscherwiderstandsmessung bei Ackerböden zur Beurteilung ihrer Strukturfestigkeit im Felde. Landwirtsch. Forsch. **13**, 24–33 (1960). — SCHOFIELD, R. K.: The pF of water in soil. Transactions of the 3rd International Congress of Soil Science, **2**, 37–48 (1935). — SCHUMACKER, W.: Die Physik, Vol. I, Die Physik des Bodens, p. 81–102. Berlin: Wiegandt und Hempel. 1864. — SIDERI, D. I.: Determination of the volume weight of soils. Transactions Soviet Section, International Society of Soil Science, **5**, 266–273 (1936). (Reference by RUSSELL, E. W.: Soil Structure, Imperial Bureau of Soil Science, Techn. Comm. **37**, 7, 1938.) — Soil Science Society of America, Committee on Terminology; Subcommittee for section I: Soil Physics. Soil Sci. Soc. Amer. Proc. **10**, 473–478 (1946). — Soil Survey Manual. Soil Permeability, 167–172. Agricultural Research Administration. United States Department of Agriculture, Washington, D. C., 1951. — STONE, J. F., D. KIRKHAM, and A. A. READ: Soil moisture determination by a portable neutron scattering moisture meter. Soil Sci. Soc. Amer. Proc. **19**, 419–423 (1955). — STRICKLING, E.: Relationship of porosity to water stability in Beltsville soil aggregates. Soil Sci. **80**, 449–457 (1955).

TAYLOR, S. A.: Oxygen diffusion in porous media as affected by compaction and moisture content. Soil Sci. Soc. Amer. Proc. **14**, 55–61 (1949). — TERZAGHI, C.: Simplified soil test for subgrades and their physical significance. Public Roads **7**, 153–162 (1926). — TIULIN, A. F.: Certain considerations on the general soil structure and on methods for its determination. Soviet Section, International Soil Science Society, A1, 111–132 (1933). — TORSTENSSON, G., and S. A. ERICKSON: A new method for determination of the porosity of the soil. Soil Sci. **42**, 405–417 (1936).

UHLAND, R. E., and A. M. O'NEAL: Soil permeability determinations for use in soil and water conservation. United States Department of Agriculture, S.C.S. — T.P., 101, Washington, D.C., 1951.

VAN BAVEL, C. H. M.: Gaseous diffusion and porosity in porous media. Soil Sci. **73**, 91–104 (1952). — Simple diffusion well for measuring soil specific diffusion impedance and soil air composition. Soil Sci. Soc. Amer. Proc. **18**, 229–234 (1954). — Soil densiometry by gamma transmission. Soil Sci. **87**, 50–58 (1959). — VAN BAVEL, C. H. M., and D. KIRKHAM: Field measurement of soil permeability using auger holes. Soil Sci. Soc. Amer. Proc. **13**, 90–96 (1949). — VAN BAVEL, C. H. M., NEWTON UNDERWOOD and R. W. SWANSON: Soil moisture measurement by neutron moderation. Soil Sci. **82**, 29–42 (1956). — VAN BEERS, W.: The auger-hole method. A field measurement of the hydraulic conductivity of soil below the water table. International Institute for Land Reclamation and Improvement, Bull. **1**, Wageningen (Netherlands), 1958. — VAN DER MAREL, H. W.: Rapid determination of soil water by dielectric measurement of dioxane extract. Soil Sci. **87**, 105–119 (1959). — VAN DUIN, R. H. A., and D. A. DE VRIES: A recording apparatus for measuring thermal conductivity, and some results obtained with it in soil. Netherlands J. Agricult. Sci. **2**, 168–175 (1954). — VEIHMEYER, F. J.: An improved soil sampling tube. Soil Sci. **27**, 147 (1929). — VEIHMEYER, F. J., and A. H. HENDRICKSON: The relation of soil moisture to cultivation and plant growth. Proceedings of the 1st International Congress of Soil Science, Washington, **3**, 498–513 (1927). — VILENSKY, D. G.: Methodology for soil tenacity and soil erosion studies. Agricult. Engng. **26**, 465–467 (1945). — VISSER, W. C.: Pore space determination as a field method. Soil Sci. **44**, 467–477 (1937). — VOLKOV, M. I. (1933): Reference is made to RUSSELL, E. W.: Soil Structure. Imperial Bureau of Soil Science, Techn. Comm. **37**, 8 (1938). — VOMOCIL, J. A.: In situ measurements of soil bulk density. Agricult. Engng. **35**, 651 (1954).

WILDE, S. A., and D. M. SPYRIDAKIS: Determination of soil moisture by the immersion method. Soil Sci. **94**, 132–133 (1962). — WOODRUFF, C. M.: Soil moisture and plant growth in relation to pF. Soil Sci. Soc. Amer. Proc. **5**, 36–41 (1940).

YODER, R. E.: A direct method of aggregate analysis of soils and a study of the physical nature of erosion losses. J. Amer. Soc. Agronomy **28**, 337–351 (1936).

c) Die morphologische Untersuchung des Bodengefüges

Von

H.-J. Altemüller

1. Einleitung

Das Gefüge des Bodens kann auf zwei voneinander grundsätzlich verschiedenen Wegen untersucht werden. Der erste Weg führt über die bodenphysikalischen Meßmethoden zu einer vorwiegend *funktionellen*, in Zahlen ausgedrückten Kennzeichnung des Bodens. Ihrem Wesen nach sind diese Methoden stets auf ein bestimmtes *Detail* des Ganzen, d. h. auf eng begrenzte Eigenschaften ausgerichtet, wie z. B. Raumgewicht, Permeabilität, Festigkeit u. a., wobei der Boden rein rechnerisch als *homogener* Körper behandelt oder als vereinfachtes Gefügemodell *vorgestellt* wird. Verschiedene Methoden machen auch eine Veränderung der natürlichen Lagerung zur Voraussetzung.

Der zweite Weg führt über die Untersuchung der *gestaltlich* wahrnehmbaren Erscheinungsformen zu einer Betrachtung des Bodens im *ganzen*, wobei „ganz", wie SANDER (1948) betont, niemals im Sinne von „restlos" zu verstehen ist. Alle erkennbaren Merkmale werden jedoch unmittelbar als Teile des Ganzen *gesehen*. Der Boden erscheint dabei niemals als homogener Körper, sondern stets in seiner natürlichen *Heterogenität*, und die Untersuchung wird stets am ungestörten Objekt vorgenommen. Der Morphologe arbeitet zudem nicht in erster Linie quantitativ, sondern vergleichend und typisierend.

Je nach Aufgabe und Zielsetzung einer Arbeit werden von Fall zu Fall mehr die physikalischen oder mehr die morphologischen Methoden im Vordergrund des Interesses stehen müssen. Wie aber jede Gestalt als ein Träger von Funktionen anzusehen ist, so sind auch die Funktionen des Bodengefüges auf seine gestaltlichen Merkmale zu beziehen, um dort ihren gemeinsamen Angelpunkt zu finden.

Das Ziel der folgenden Ausführungen ist es, einen Einblick in makro- und mikromorphologische Gefügeuntersuchungen zu vermitteln. Mit Rücksicht auf den zur Verfügung stehenden Raum kann dies allerdings nur in kurzer, gedrängter Form geschehen, wobei die Erläuterung einiger Grundbegriffe und eine Auswahl von Gefügeschilderungen und Bildbeispielen die wichtigsten Stützen bilden müssen. Auch in der Auswahl der Literatur war Beschränkung erforderlich; es wurde jedoch versucht, die Hinweise etwas zahlreicher einzustreuen, um die weitere Information zu erleichtern.

Die morphologische Gefügeuntersuchung hat schon seit den Anfängen der Bodenkunde vor allem im Rahmen der Feldarbeiten ihren festen Platz. Als beschreibende Methode blieb sie aber in der folgenden Entwicklung hinter den physikalischen und chemischen Untersuchungsverfahren zurück. Von verschiedenen Seiten wurden deshalb zahlreiche Anregungen zu einem intensiveren Studium des Bodengefüges gegeben. Die nachhaltigste Wirkung ist dabei ohne Zweifel von den Arbeiten KUBIËNAS ausgegangen, auf welche sich die folgenden Ausführungen in weitem Umfange gründen. Mit der Begründung der *Mikromorphologie des Bodens* hat KUBIËNA ein Forschungsgebiet geschaffen, das neue Einblicke in den Boden vermittelt und vor allem für die Untersuchung des Bodengefüges außerordentlich vielseitige Möglichkeiten bietet. Die Mikromorphologie ist auch in besonderem Maße geeignet, eine enge Verbindung zur physikalischen und chemischen Bodenuntersuchung und zur Bodenbiologie herzustellen und dadurch neue Wege der Forschung aufzuzeigen.

Als eine weitere grundlegende Quelle ist die „Einführung in die Gefügekunde der geologischen Körper“ von SANDER (1948 und 1950) zu nennen, die vor allem für die allgemeine begriffliche Erfassung von Gefügen wertvolle Hilfe leistet (s. ALTEMÜLLER 1962 d). Eine Anzahl der hier gebrauchten Begriffe und Einteilungen gründet sich auf die Beschäftigung mit diesem Werke. Sie können einer weiteren Entwicklung der Gefügekunde des Bodens nützlich sein und helfen, auch über die Bodenkunde hinaus Grenzen zu überwinden.

2. Allgemeine Grundbegriffe

A. Gefüge und Struktur

Der Ausdruck Gefüge war in der Bodenkunde schon früh im Gebrauch (z. B. SCHUMACHER 1864), er wurde jedoch später nur wenig benutzt (z. B. SCHUCHT 1929). Erst KUBIËNA gab ihm in Anlehnung an SANDER (1930) einen dem Boden angepaßten, umfassenden Begriffsinhalt und definierte das *Bodengefüge* als „die räumliche Anordnung der Bodengemengteile, und zwar nicht nur in dem engeren Sinne der Aggregatbildung..., sondern im allumfassenden, allgemeinen Sinne, so auch das mannigfaltige Innengefüge dichter Bodenmassen, die Auswirkungen der verschiedensten Ausscheidungs- und Lösungsvorgänge, die Umwandlungen durch den Einfluß von Lebewesen, Stoffbewegungen u. a. betreffend“ (1953, S. 365). Der Begriff Bodengefüge erhielt damit übergeordnete Bedeutung und wurde unabhängig von den Begriffen Struktur, Textur, usw. gehandhabt (KUBIËNA 1938).

Unter *Struktur* versteht KUBIËNA „eine dem Ausdruck Gefüge ähnliche Bezeichnung, doch mehr in dem Sinne der Art der Bildung von Gefügekomplexen (Krümel, Schöllchen, Bröckel oder sonstige Aggregate, ferner auch Art der Absonderung)... Im Schrifttum mancher Länder wird Struktur gleichbedeutend mit Gefüge (s. dort) gebraucht" (1953, S. 370).

Der Begriff „Struktur" hat sich im praktischen landwirtschaftlichen Gebrauch allgemein erhalten (z. B. Krümelstruktur, strukturloser Boden u. a.) und ist vor allem auch in der wissenschaftlichen Literatur weit verbreitet, wogegen die Begriffe „Textur" und „Absonderung" an Bedeutung zurücktreten. Beim näheren Studium wird jedoch deutlich, daß die zahlreichen vorliegenden Definitionen sich selten völlig decken. Der Begriff wird nicht nur auf die Aggregatbildung, sondern zuweilen allgemein auf makromorphologische Merkmale bezogen und von anderen Autoren auch auf den Gesamtboden ausgedehnt, wobei einerseits morphologische, andererseits auch chemisch-physikalische Betrachtungsweisen im Vordergrund stehen. Es wird daher immer schwieriger, ein klares Bild des Strukturbegriffes zu gewinnen. Für den wissenschaftlichen Gebrauch ist es deshalb ratsam, das Gewicht mehr auf den Ausdruck und Begriff *Gefüge* zu legen. Folgende Literatur ist zu diesem Problem besonders zu erwähnen:

Rüger 1930; Kubiëna 1935, 1938, 1953; Baver 1948; Frei 1948; Bach und Deuel 1953; Laatsch 1954; Hoenes 1955; Jongerius 1957; Blümel, Janik und Schiller 1959; Frese 1960; Brewer 1960; Franz 1960; Brewer und Sleeman 1960; Mückenhausen 1958, 1961, 1963; Parfenova und Jarilova 1962; Altemüller 1962d; Koepf 1963.

Um die verschiedenen Aspekte des inneren Baues des Bodens (z. B. morphologische und physikalische Betrachtung, Größenordnungen usw.) in ihren gegenseitigen Beziehungen unter *einem* Oberbegriff allgemein deutlich zu machen, hat der Verfasser auf der Grundlage der Arbeiten von Kubiëna und Sander eine *neue Definition des Bodengefüges* vorgeschlagen, die gegenüber der ersten Veröffentlichung (1962d) hier mit geringer Abänderung wie folgt wiedergegeben wird:

Das Bodengefüge ist zu verstehen als die räumliche Ordnung der Bodengemengteile, betrachtet als Bauelemente des Bodens in gestaltlicher, in funktioneller und in genetischer Hinsicht, unabhängig von den Größen- und Rangordnungen und unabhängig von der stofflichen Beschaffenheit.

B. Gefügeelement

Ganz allgemein werden die unterscheidbaren Bestandteile des Bodens als *Bodengemengteile* bezeichnet. In der Gefügekunde sind sie nicht nur als Einzelteile anzusehen, sondern vor allem in ihrem Anteil am Zusammenbau, als „Bauelemente" (Kubiëna 1962a) und somit als *Gefügeelemente* zu betrachten. Gefügeelemente sind sowohl feste Einzelteile (z. B. ein Sandkorn) oder zusammengesetzte Teile (z. B. Aggregate, Konkretionen, das Gefügeplasma — s. S. 240), als auch flüssige und gasförmige Anteile (Bodenwasser, Bodenluft), die zwar meist unabhängig untersucht werden, aber doch stets auch als Bestandteile des Gefüges gesehen werden müssen.

C. Gestalt (Form)

Zur ausführlichen Erläuterung des Gestaltbegriffes sei auf Sander (1948) und Kubiëna (1948) verwiesen. Der Begriff ist in weiter Definition für die allgemeine Morphologie und Gefügekunde von grundlegender Bedeutung. Im Rahmen dieser Abhandlung wird er nur in enger Fassung für die mittelbar oder unmittelbar sichtbare körperliche Form oder Hohlform eines Gefüges oder Gefügeelementes angewandt. Die Gestalt wird bestimmt:

Durch *Beschreibung*

a) des Gesamtbildes: z. B. kugelförmig, plattig, schwammartig (Nikiforoff 1941);
b) der Oberflächen: z. B. glatt, rauh, schorfig, kantig;
c) der Konturenschärfe: diffus, unscharf, scharf (Brewer und Sleeman 1960).

Durch *Messung* (vgl. REDLICH 1940, HOENES 1955, KUBIËNA, BECKMANN und GEYGER 1961, 1962)

a) von Rundungsgraden (KÖSTER 1960, BREWER und SLEEMAN 1960);
b) von Lappungsquotienten (BECKMANN 1962);
c) der Achsenverhältnisse (ANDERSON und BINNE 1961).

Nach *genetischen Gesichtspunkten.*

D. Größe und Größenordnungen

Größenmessungen werden oft unabhängig von der Gefügeuntersuchung durchgeführt (z. B. Korngrößenmessungen nach Sieb- und Schlämmverfahren). Die Größenverhältnisse der Gefügeelemente eines Bodens sind jedoch stets auch als Gefügemerkmal anzusehen, das in jeder Gefügeuntersuchung zumindest qualitativ anzusprechen ist.

Absolute Größe. Sie wird für ein Einzelelement (Korn, Aggregat, Hohlraum usw.) durch den mittleren Durchmesser oder die Länge der Achsen angegeben; für bestimmte Gruppen von Elementen als statistische Größe. Über die Einteilung von Korngrößen siehe SCHÖNHALS 1955, DE LEENHEER und VAN HOVE 1956, KÖSTER 1960, KATSCHINSKI 1960. Über Aggregat- und Hohlraumgrößen siehe u. a. REDLICH 1940, M. GRAÇANIN 1950, KUBIËNA und Mitarbeiter 1961, ANDERSON und BINNE 1961, BECKMANN 1962, GEYGER 1962.

Relative Größe (Größenverteilung). Die Größenverteilung wird im Bodengefüge vor allem für Körner, Aggregate und Hohlräume bestimmt. Sie kann *qualitativ* z. B. mit Bezug auf die Körner ähnlich wie bei Gesteinen als

gleichkörnig (alle Hauptgemengteile haben ähnliche Größe),
wechselkörnig (stärkere Schwankungen der Korngröße),
ungleichkörnig (große Körner in feiner Grundmasse)

angegeben werden (vgl. HOENES 1955). Für andere Gefügeelemente sind andere Bezeichnungen möglich, z. B. gleichporig, wechselporig usw. Als *quantitative* Angabe ist aus den absoluten Größenmessungen die Verteilungskurve aufzustellen, deren Verlauf auch geschätzt werden kann (deutlicher Gipfel, undeutlicher Gipfel, ungleichmäßige Verteilung — BREWER und SLEEMAN 1960).

Größenordnungen. Die Gefügebegriffe gelten laut Definition in allen Größenbereichen. Grobe Riß- und Polyederbildungen werden ebenso wie die feinsten mikroskopischen Details der Anordnung von Tonteilchen als Gefüge behandelt. Für den jeweils untersuchten Bereich sind deshalb stets die Größenverhältnisse anzugeben. Das geschieht entweder namentlich (Profil, Horizont, Handstück, usw.) oder besser durch die allgemeine Angabe der Maßverhältnisse (z. B. cm-, mm-, oder μ-Bereich). Man kann danach auch einen makro- und mikro- und submikroskopischen Bereich unterscheiden und dann vereinfacht von Makro- und Mikrogefüge usw. sprechen (vgl. JONGERIUS 1957, ALTEMÜLLER 1962d).

E. Rangordnungen

In der gleichen Weise, wie einfache und zusammengesetzte Gefüge-*Elemente* unterschieden werden, sind auch einfacher und komplizierter gebaute *Gefüge* zu unterscheiden. Nach der Organisation ihres inneren Baues können Rangordnungen aufgestellt werden.

KUBIËNA (1935) nennt als wichtige Grundeinheit das *Elementargefüge* (aufgebaut aus Grundelementen) und bezeichnet höher organisierte Gefüge (aufgebaut aus Aggregaten u. dgl.) allgemein als *Gefüge höherer Ordnung.* BREWER (1960b) hat eine ganze Reihe vom „Grundgefüge“ (basic structure) über mehrere Ordnungen bis zum „Gesamtgefüge“ (complete structure) aufgestellt. Diese Ordnungen sind rein morphologisch aufzufassen. Im Gegensatz hierzu kann eine Rangordnung auch nach genetischen Gesichtspunkten bestimmt werden, z. B.

nach dem Entwicklungsgrad eines Gefüges. So unterscheidet Frei (1948) wenig entwickelte, höher und höchstentwickelte Gefüge und nennt als entsprechende Beispiele Primitivgefüge, Feinkoagulatgefüge, Schwammgefüge u. a. Die Festlegung einer Rangfolge ist oft schwierig und nicht immer notwendig.

F. Verteilungs- und Richtungsgefüge

Bei der Untersuchung der Gefügeelemente nach ihrer Lage im Raum wird von zwei Seiten ausgegangen:

Von der *räumlichen Lage der Gefügeelemente*, so wie sie durch Koordinaten ermittelt wird, wobei jedes Element theoretisch durch einen Punkt ersetzt werden kann *(Verteilungsgefüge, Ortslage)*;

Von der *Richtung der Gefügeelemente* im Raum *(Richtungsgefüge, Drehlage)*.

Das *Verteilungsgefüge* kann für bestimmte Gefügeelemente in Auswahl nach der Gestalt, Größe oder anderen Merkmalen bestimmt werden. Verschiedene Arten der Packung gleich großer Kugeln (Ramann 1914) geben einfachste Beispiele. Die Elementargefüge Kubiënas (1935) sind ebenfalls Beispiele für Verteilungsgefüge (wenn auch nicht ausschließlich), die man sich noch weiter abgewandelt vorstellen kann. Einfache sprachliche Kennzeichnungen für verschiedene Ortslage sind mit Worten wie: regulär, gehäuft, gebändert, konzentrisch usw. zu geben (Brewer und Sleeman 1960).

Die Einordnung von Gefügeelementen in ein mehr oder weniger hervortretendes *Richtungsgefüge* ist im Boden in allen Größenbereichen und bei verschiedenstem Material zu beobachten. Diese Regelung (oder Orientierung) steht im Boden im allgemeinen zur Gestalt der Gefügeelemente in Beziehung. Hierher gehört die ganze Mannigfaltigkeit der in Böden — vor allem in Dünnschliffen unter dem Polarisationsmikroskop — beobachtbaren Erscheinungen der Teilchenregelung im Gefügeplasma (doppelbrechende Schlieren, Höfe, Anlagerungen, Schichtungen usw.), die in zahlreichen Arbeiten ausführlich beschrieben sind. Regelungen sind auch in größeren Dimensionen häufig, z. B. bei plattenartigen Aggregaten, in der Richtung von Röhren und Spalten, in der Laub- und Nadelstreu der Waldböden, in Torfen usw.

3. Merkmale und Kennzeichnung des Bodengefüges bei Untersuchungen im Feld

A. Allgemeines

Die Untersuchung des Bodengefüges ist infolge der großen Mannigfaltigkeit der Böden und ihres vielschichtigen und formenreichen Aufbaues eine umfangreiche Arbeit, die kaum je vollständig ausgeführt werden kann. Folglich werden niemals *alle* Daten eines Gefüges erfaßt, sondern es wird eine Auswahl getroffen, und diese Auswahl ist abhängig von dem Interesse, nach welchem ein Gefüge untersucht wird. Allerdings darf die Einheit des Bodens nicht zerrissen werden. Auch eine stark spezialisierte Untersuchung des Gefüges wird um so besser (also nicht unbedingt vollständiger im engeren Sinne), je mehr der Untersuchende vom Boden als Ganzem Kenntnis hat. Die stetige Ausweitung des eigenen Erfahrungsschatzes ist daher wichtigste Aufgabe für den Gefügekundler.

B. Ackerbauliche und bodenkundliche Arbeitsrichtung

Die in der Literatur vorliegenden Beschreibungen und Anleitungen zur Untersuchung des Gefüges unterscheiden sich nach Inhalt und Zielsetzung oft erheblich voneinander. Vor allem lassen sich zwei Hauptrichtungen erkennen, die hier vereinfacht als ackerbauliche und als bodenkundliche Richtung bezeichnet werden können. Die erste ist mehr auf die Untersuchung der bearbeiteten Krumenschicht der Ackerböden, die zweite auf die Böden insgesamt und immer auf die Betrachtung des gesamten Profils eingestellt. Durch die fruchtbare gegenseitige Beeinflussung der Fachgebiete Ackerbau und Bodenkunde sind die Grenzen heute allerdings verwischt und lassen sich nur noch aus der historischen Entwicklung erkennen.

Die *ackerbauliche* Bodenbeurteilung ist schon in der Mitte des 18. Jahrhunderts vornehmlich auf die stoffliche Beschaffenheit und die Körnungsart der Böden ausgerichtet (SPRENGER 1764). Die Bezeichnung Sand-, Ton-, Mergel-, Kalk- oder Humusboden geben davon Zeugnis. Daneben erwähnten SPRENGER (1764) und ein Anonymus (1770) aber auch die Einteilung in leichtes und schweres Land und meinten damit nicht nur die Zusammensetzung nach Sand- und Tonanteilen, sondern auch die Unterschiede in der Art der Bindung der Bodenteile und der Verteilung der Hohlräume. Damit waren also Gefügemerkmale angesprochen. Als SCHUMACHER (1864) später ein schüttiges und bündiges Gefüge unterschied, lag dem ein ähnlicher Inhalt zugrunde.

Zur Kennzeichnung des Verhaltens von Bodensuspensionen prägte HILGARD (1879) die Begriffe Einzelkornstruktur und Flockenstruktur. Diese übertrug WOLLNY (1882) in eine andere Größenordnung und brachte sie als *Einzelkorn-* und *Krümelstruktur* auf den natürlich gelagerten bzw. bearbeiteten Boden zur Anwendung. Nach WOLLNY war die Einzelkornstruktur nicht immer bindungslos, sondern häufig auch kohärent. Er verstand darunter allgemein die dichteste Packung der Bodenbestandteile, also kurz denjenigen Zustand, der von GEDROIZ (1926, zit. nach KRAUSE 1931) später „strukturlos" genannt wurde. Ebenso war die Krümelstruktur nicht unbedingt aus „echten Krümeln" aufgebaut (wie es später verstanden wurde), sondern ebenso aus kleinen Schollen und Bröckchen, die, wie es WOLLNY selbst beschrieb, im Innern Einzelkornstruktur aufweisen. Diese Ansicht wurde auch später noch von MITSCHERLICH (1954) vertreten.

In den Erläuterungen RAMANNS (1914) wird vollkommen deutlich, daß mit diesen Begriffen nicht die *Gestalt* der Gefügeelemente zum Ausdruck gebracht werden sollte; dieser wurde sogar nur geringe Bedeutung zugemessen. Das Interesse war allein auf die Unterschiede der räumlichen *Lagerung*, auf das Verteilungsgefüge gerichtet, auf den stetigen Wechsel zwischen „locker" und „dicht" in der Ackerkrume. Die Beschreibung der Einzelkorn- und Krümelstruktur hat besonders in den Lehrbüchern des Ackerbaues und der Bodenbearbeitung bis in die jüngste Zeit breiten Raum eingenommen (HEUSER 1928, ROEMER 1929, SCHEFFER und TORNAU 1953, KLAPP 1954).

Nach der vorwiegend mechanischen Betrachtung WOLLNYS wurde die Krümelbildung später mehr im Zusammenhang mit Koagulationsprozessen gesehen (EHRENBERG 1915, GEDROIZ 1933), und es entwickelte sich eine umfangreiche Krümelforschung, die in Deutschland besonders auch im Hinblick auf die „Bodengare" betrieben wurde (MEYER und v. RENNENKAMPF 1936, v. NITZSCH und CZERATZKI 1940, SEKERA und BRUNNER 1943, GÖRBING 1947, ROEMER und FRESE 1952).

Diese Arbeiten brachten ganz allgemein starke Anregungen zu einer gründlicheren morphologischen Untersuchung des Bodens. Das Interesse richtet sich dabei nicht mehr allein auf den Krümelzustand, sondern immer mehr auch auf die *Gestalt* der Gefügeelemente. UNDRITZ (1951) betont sogar, daß die rein morphologische Beschreibung am besten zur Charakterisierung des Bodengefüges geeignet sei, während die Beurteilung der Krümelfähigkeit oder die Erkennung von „lebendverbauten" Krümeln im Felde jeder gesicherten Grundlage entbehre. Auch der *Unterboden* mit den verbreitet angetroffenen Verdichtungen („Pflugsohlen") wurde ein wichtiges Untersuchungsobjekt. In einer eingehenden, später zu Unrecht weniger beachteten Studie zur Frage der Sohlenbildung hatte SCHLACHT bereits 1933 auf das Vorkommen plattiger Gefügeformen in den verdichteten Horizonten hingewiesen. Ihre Entstehungsbedingungen wurden später von CZERATZKI (1956) experimentell untersucht. In diesem Zusammenhang sei auf weitere Literatur zur Bildung von Gefügeformen durch äußere Einflüsse verwiesen: WOLKEWITZ 1953, CZERATZKI und FRESE 1956, CZERATZKI 1957, FRESE 1958, KOEPF 1960 und 1961, RID 1961/62.

Von *bodenkundlicher* Seite wurde den gestaltlichen Merkmalen des Bodengefüges schon frühzeitig besondere Aufmerksamkeit geschenkt (GLINKA 1914, STREMME 1926, ZACHAROV 1927, STEBUTT 1930). Sie wurden stets im Zusammenhang mit der bodentypologischen Entwicklung gesehen und dienten nicht nur einer feineren Unterteilung von Horizonten, sondern wurden zum Teil sogar zu wesentlichen Kennzeichen bestimmter Bodentypen. Eine besondere Bedeutung erlangten hierbei die oft charakteristisch geformten *Aggregate*, die vielfach so augenfällig in Erscheinung treten, daß zahlreiche Gefüge auch heute noch nach ihnen benannt werden.

MÜCKENHAUSEN hat in mehreren Veröffentlichungen (1958, 1961, 1963) zusammenfassende und übersichtliche Beschreibungen der wichtigsten Aggregatformen und der daraus gebildeten Gefüge, verbunden mit Hinweisen auf deren Entstehungsbedingungen und Funktionen, gegeben. Auch seine ausführlichen Darstellungen über den makromorphologischen Bau und das Gefüge wichtiger Bodentypen (1962) sind an dieser Stelle zu nennen.

C. Systematische Gliederung

Eine erste *systematische Gliederung der Aggregatformen* hat ZACHAROV (1927, s. BAVER 1948) vorgeschlagen. Er verstand unter Struktur allein die Eigenschaft des Bodens, in Aggregate von verschiedener Form und Größe zu zerfallen. Zwar nannte er für die Beschreibung eines Bodenprofils insgesamt sieben Hauptteile (Farbe, Struktur, Konstitution, Mächtigkeit, Neubildungen und Einlagerungen, Bodenskelett, Horizonte; zit. nach STREMME 1933), von denen auch andere als Gefügemerkmale in unserem Sinne anzusehen sind — so war z.B. die Konstitution eine von der Körnung und Hohlraumgestaltung abhängige Größe – ; die Beurteilung des Bodengefüges nach den Aggregatformen trat jedoch immer mehr in den Vordergrund. Dies wird besonders auch in einer Klassifikation von NIKIFOROFF (1941) deutlich, deren Hauptmerkmale kurz anzuführen sind. Die Einteilung der Aggregate erfolgt dabei in *Typen, Klassen* und *Grade,* wobei in jeder Stufe bestimmte Symbole zur Kurzbezeichnung Anwendung finden.

Aus den geometrischen Hauptformen und der Richtung der Längsachsen ergeben sich die *Typen*:

1. *platy* (vorwiegend horizontale Richtung der Achsen)	=
2. *prismatic* (vorwiegend vertikale Richtung der Achsen)	||
3. *blocky* (etwa gleiche Achsen, kantig)	+
4. *granular* (etwa gleiche Achsen, gerundet)	0

Die *Klassen* richten sich nach der Größe:

sehr fein	aa
fein	a
mittel	b
grob	c
sehr grob	cc

Die *Grade* werden in fünf Stufen angegeben und sind abhängig von der Formausprägung, der Festigkeit und Zerfallsneigung (1. gering, 2. schwach, 3. mäßig, 4. gut, 5. stark entwickelt oder ausgeprägt).

Je nach den angetroffenen Merkmalen werden die Zeichen verschieden kombiniert und bedeuten dann z.B,:
0 a^4 gut entwickelte (oder ausgeprägte) feinkörnige Struktur
+ b^2 stark entwickelte (oder ausgeprägte) mittel-blockige Struktur usw.

Die *Porosität* geht unter der Festigkeit und Zerfallsneigung mit in den Strukturgrad ein. Zuletzt folgen noch individuelle Kennzeichnungen (Überzüge, Konturenschärfe usw.) und unter Umständen auch besondere Namen, soweit charakteristische Ausprägungen dies erfordern (columnar, pencillike, pealike usw.).

Eine solche Klassifikation erscheint durch die zahlreichen Angaben zur Gestalt, zur Richtung usw. ziemlich vollständig. Es wird jedoch leicht übersehen, daß sie sich nur auf einen bestimmten Aspekt des Bodengefüges bezieht und deshalb nicht allgemein anzuwenden ist. Sehr viele Böden oder Bodenhorizonte erweisen sich nämlich als so weitgehend durch andere, ebenfalls am Profil wahrnehmbare Merkmale differenziert (vgl. KUBIËNA 1936/37), daß diese nicht vernachlässigt werden sollten.

In dem Vorschlag von BAVER (1948) sind nicht nur aggregierte Gefüge (compound structure), sondern auch die nicht oder undeutlich in Aggregate gegliederten Gefüge (simple structure) aufgenommen. Dabei wendet BAVER (1948) den Begriff „structure" nicht mehr allein auf die Ausbildung der *Aggregate* an, sondern schließt auch die Art der Anordnung von *Primärteilen* mit ein.

JONGERIUS (1957) geht noch einen Schritt weiter und nimmt außerdem auch die Gestalt und Anordnung der *Hohlräume* mit unter den Strukturbegriff. Er steht damit in Übereinstimmung mit der Definition des Gefüges nach FREI (1948) und kommt auch dem Gefügebegriff KUBIËNAS und der hier gegebenen Definition nahe.

D. Anleitung zur Gefügebeschreibung im Felde

Für die praktische Beschreibung des Bodengefüges im Felde ist es nützlich, die verschiedenen systematischen Gliederungen zu kennen. Es ist jedoch nicht ratsam, von vornherein nach bestimmten Typen zu suchen, denn diese sind nur in den selteneren Fällen vollkommen ausgeprägt. Im allgemeinen existieren — wie Abb. 44 zeigt — die verschiedensten Übergänge und Durchdringungen, so daß die meisten Beschreibungen unbefriedigend sein müßten. Ein Vorschlag von KOEPF (1961, 1963) geht deshalb in Anlehnung an BRYSSINE (1954) von drei „Grundzuständen" des Bodengefüges aus:

1. kohärenzfreie, nicht aggregierte Gefüge,
2. kohärente Gefüge,
3. aggregierte Gefüge.

Diese dienen lediglich zur ersten, orientierenden Einteilung am Profil und sind in den verschiedensten Übergängen denkbar, welche man sich am besten in einer Dreiecksdarstellung deutlich macht (s. MÜCKENHAUSEN 1963).

Kohärenzfreie, nicht aggregierte Gefüge (Abb. 41). Zur Kennzeichnung dienen vor allem die Gestalt und relative Größe der Körner (Körnung), die relative Ortslage (Schichtung nach verschiedenen Korngrößen, Packung), ferner die Art der Intergranularräume und eventuell vorhandener größerer Poren. Da es sich meist um lose, für das unbewaffnete Auge wenig gegliederte Sande handelt, sind die Angaben oft nicht reichhaltig (Einzelkorngefüge).

Kohärente Gefüge (Abb. 42). Diese werden ebenfalls nach der Körnung und Packung, dann aber besonders nach den vorhandenen Hohlräumen (partielle Spalten, Röhren, Poren, Blasenräume usw.) nach deren Gestalt und Anordnung (Verteilung und Richtung) beschrieben. Anlagerungen in Leitbahnen, Bänderungen, Schichtungen, Aussehen von Bruchflächen usw. bringen weitere Kennzeichnungen. Grobkörnige Massen zeigen unter der Lupe häufig die Bindungsart der Elemente (Überzüge, Brückenbildungen von Tonanteilen usw.); feinkörnige, tonige enthalten oft Einschlüsse oder sind aus miteinander „verwachsenen" Aggregaten aufgebaut.

Aggregierte Gefüge (Abb. 43). Für die aggregierten Gefüge steht die schon erwähnte Beschreibung der Gestalt der Aggregate mit den weiteren Merkmalen der Ausprägung

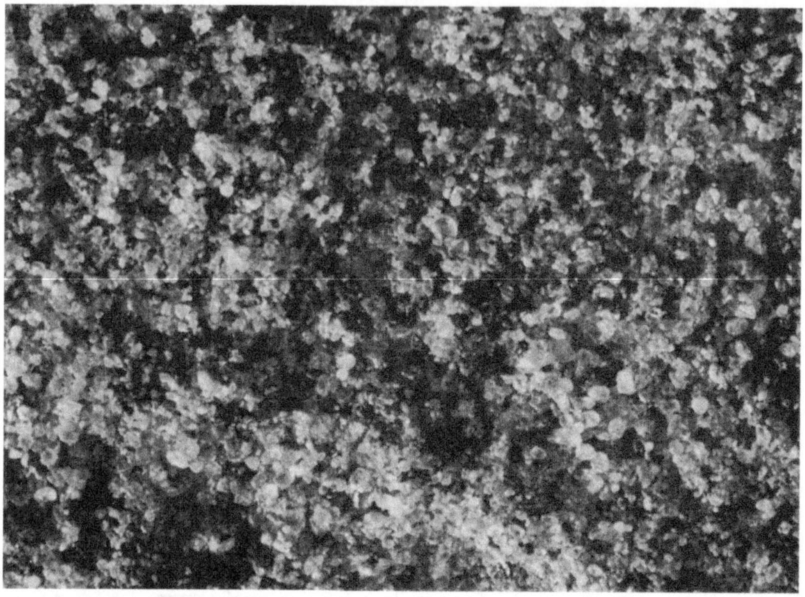

Abb. 41. Einzelkorngefüge aus annähernd gleichkörnigem fluvioglazialem Sand (Hauptgemengteile Quarz, Schwerminerale und Feldspat). Ansicht einer Bruchfläche in natürlicher Lagerung. Abbildungsmaßstab (M) = 3,2 : 1. Alle Aufnahmen vom Verfasser

Abb. 42. Kohärentes Gefüge aus dicht gelagerter, schluffreicher Bodenmasse mit gut ausgeprägtem Hohlraumsystem aus Röhren (Tiergängen) und blasenförmigen Großporen. Probe aus einem Schlammteich. Ansicht einer Bruchfläche. M = 3,2 : 1

Abb. 43. Ideale Ausbildung eines in prismatische Formen aggregierten Gefüges aus montmorillonitischem Ton. Bildung der horizontalen und vertikalen Spalten durch langsames Gefrieren der feuchten Masse (Modellversuche von CZERATZKI, nicht veröffentlicht). Schrägaufsicht, M = 2:1

Abb. 44. Vorwiegend kohärentes Gefüge im A_3-Horizont einer durchschlämmten Parabraunerde aus Löß mit unvollständig ausgebildeten horizontalen und vertikalen Spalten, senkrecht verlaufenden Wurzelröhren und Tiergängen (teilweise erfüllt mit Feinstaggregaten), mit nesterartig verteilten Regenwurmexkrementen und mit nadelstichartigen Feinporen in der dichten Grundmasse. Vertikaler Bruch, M = 1:2,5

im Vordergrund (Mückenhausen 1963). Dazu kommt die Anordnung (nach räumlicher Verteilung und Richtung), die gegenseitige mehr oder weniger schwache Bindung. Die Gestaltung von Oberflächen und das Innengefüge, für welches die bei den kohärenten Gefügen genannten Merkmale beachtet werden müssen, also die Gliederung der Hohlräume usw.

E. Direkte mikroskopische Untersuchung

Ein großer Teil der Merkmale eines Bodengefüges ist bei einfacher Profilbeobachtung nicht zu erkennen, weil die feineren Differenzierungen dem bloßen Auge nicht mehr zugänglich sind. Kubiëna (1932, 1938) hat deshalb den Weg der direkten mikroskopischen Untersuchung vorgeschlagen und dargelegt, daß dieses Verfahren nicht nur für rein gefügekundliche, sondern vor allem auch für bodenbiologische Arbeiten eine wertvolle Hilfe bedeutet. Wo die äußeren Umstände nicht die Untersuchung am Profil erlauben, kann die direkte Beobachtung auch an Bodenmonolithen, Handstücken oder an Aufbereitungen im Labor erfolgen. Diese Art der Untersuchung vermag Einblicke zu geben, die mit anderen Verfahren nicht oder nicht sofort zu erlangen sind. Die direkte mikroskopische Beobachtung ist deshalb stets als unentbehrliches Bindeglied zwischen der Gefügeuntersuchung am Profil und der Untersuchung an Dünnschliffen anzusehen und mit beiden in gleicher Weise eng verbunden (Day 1948, de Craene und Laruelle 1955, Altemüller 1957, Diez 1959, Brand 1960, F. Meyer 1962, Zachariae 1962).

4. Merkmale und Kennzeichnung des Bodengefüges bei Untersuchungen an Dünnschliffen

A. Allgemeines

Rein größenmäßig gesehen besteht zwischen der Beobachtung am Profil oder Handstück und der Untersuchung eines Dünnschliffes keine strenge Grenze mehr. Die Technik der Schliffherstellung ist soweit vervollkommnet, daß auch größere Bodenstücke zu Schliffen verarbeitet werden und somit alle Betrachtungsarten lückenlos ineinander übergehen können. Für die Untersuchungen von Dünnschliffen gelten auch im wesentlichen die gleichen Grundsätze wie für die Untersuchung am Profil. Die Arbeitsweise ist wie am Profil in erster Linie beschreibend, wird aber in stärkerem Maße durch Messungen und Untersuchungen der stofflichen Beschaffenheit, unter Ausnutzung der mikroskopischen Methoden, ergänzt. Die zur Gliederung des Gefüges dienenden Gefügeelemente erscheinen im Dünnschliff zumeist deutlicher ausgeprägt. Einige wichtige Elemente werden im folgenden kurz erläutert.

B. Gefügeelemente

Gefügekorn. Das wichtigste Grundelement eines Gefüges ist für Gesteine und Böden gleichermaßen das „durch eine wahrnehmbare Unstetigkeit des Stoffes oder der Orientierung rings begrenzte“ Gefügekorn (Sander 1950). Im Boden sind in erster Linie alle sogenannten Primärteile, vom Sandkorn bis zu den Teilchen von kolloidaler Größenordnung, als Gefügekorn anzusehen. Nach der Definition gilt der Begriff jedoch auch für zusammengesetzte Elemente, wenn sie sich nach ihrer stofflichen Beschaffenheit oder inneren Regelung gegenüber ihrer Umgebung unterscheiden. So sind z.B. in scheinbar homogenen Tonen auch die Partien gleicher Teilchenregelung als Gefügekorn zu betrachten (Lagenkorn); ebenso Gesteine oder organische Anteile (Stoffkorn) usw.

Aggregate. Im weiteren Sinne sind alle aus Primärteilen zusammengesetzten, im Boden mehr oder weniger deutlich gegenüber ihrer Umgebung abgegrenzten

Komplexe, unabhängig von ihrer Größe, als Aggregate zu bezeichnen. Verschiedene Autoren rechnen nur die durch natürliche Aggregation entstandenen Elemente (Krümel, Exkremente von Bodentieren u. a.) zu den Aggregaten und bezeichnen z. B. die durch Aufspaltung dichter Massen gebildeten Körper als *Absonderungen* (z. B. prismatische, plattige Absonderungen u. a., s. RÜGER 1930, KUBIËNA 1935, 1938, 1953; FRESE 1958). Diese Unterteilung ist in manchen Fällen nützlich, und man kann, um den Unterschied zu betonen, auch von *Aggregaten im engeren Sinne* sprechen. Da jedoch häufig keine eindeutigen Zuordnungen möglich sind und der Gebrauch des Aggregatbegriffes verschieden ist, wird hier an der weiteren Fassung festgehalten. Im englischen Schrifttum erscheint neuerdings für Aggregate der Ausdruck *Ped* (OSMOND 1958, BREWER 1960b).

Andere zusammengesetzte Elemente. Eine weitere Gruppe von zusammengesetzten, aber nicht eigentlich zu den Aggregaten zählenden Elementen faßt BREWER (1960b) unter dem Begriff „*pedological features*" zusammen. Hierzu gehören Fleckenbildungen, Konkretionen, nesterartige Ansammlungen von Korngruppen, Filmbildungen u. a. Sie unterscheiden sich durch die Art der Entstehung, durch stoffliche oder Konzentrationsunterschiede oder andersartige Anordnung deutlich von ihrer Umgebung und verdienen besondere Beachtung.

Hohlräume. Der nicht von festen Anteilen eingenommene Bodenraum ist im ganzen wie im einzelnen ebenfalls als Gefügeelement zu betrachten. Die Einteilung erfolgt in Beziehung zu den festen Gefügeelementen oder nach der Gestalt in Intergranularporen (Kornzwischenräume), Interaggregatporen, Intraaggregatporen usw., ferner Risse, Spalten, Röhren, blasenförmige Poren usw.

Skelett (Gefügeskelett). Nach KUBIËNA (1947) „die Gesamtheit der formbeständigen, mehr oder minder unbeweglichen und chemisch schwer angreifbaren Anteile im Bodengefüge (Mineralkörner der Sandfraktion, Rohhumusteile); sie können bei gegenseitiger Berührung ein mehr oder minder stabiles Gerüst bilden, in dessen Maschen sich die Vorgänge der Stoffveränderung, Stoffumformung und Stoffneubildung vollziehen, die zur Entstehung neuer Gefügeverhältnisse führen".

KUBIËNA gab auch die Bezeichnung „Gefügeskelett", um deutlich zu machen, daß damit nicht der Stein- und Kiesgehalt des Bodens gemeint ist. BRAND (1961) sieht trotzdem eine Gefahr für Mißverständnisse und bevorzugt deshalb den Ausdruck „Feinbodengerüst", BORCHERT (1961b) nennt es „Korngerüst". Im wesentlichen liegt aber beiden Skelettbegriffen die gleiche Vorstellung zugrunde (s. STEBUTT 1930), und bei einer Vereinheitlichung gefügekundlicher Grundbegriffe über die Größenordnungen hinweg, die unbedingt erforderlich ist, lassen sich beide Auffassungen leicht ineinander führen. Die Möglichkeit quantitativer Angaben (z. B. Stein- und Kiesgehalt) wird dadurch in keiner Weise beeinträchtigt. — Der Begriff Skelett sollte im gefügekundlichen Gebrauch nicht streng größenmäßig oder stofflich abgegrenzt werden, da die Skelett*funktion* von sehr verschiedenen Teilen geleistet werden kann.

Plasma (Gefügeplasma). Nach KUBIËNA (1948) „die Gesamtheit der formveränderlichen beweglichen und chemisch leicht umwandelbaren Anteile im Bodengefüge". Die Eigenart des Bodens verlangt eine Zusammenfassung der feineren Anteile gegenüber dem Skelett. Jeder Einblick in den Boden zeigt, daß in den feineren Größenbereichen andere Wirkungsmechanismen ablaufen, die sich im Gefüge deutlich ausprägen. Eine strenge größenmäßige Grenze sollte dabei ebenfalls nicht gezogen werden. Der Ausdruck Plasma wurde nicht allgemein übernommen. JONGERIUS (1957) gebraucht in fast gleichem Sinne das Wort Matrix. BRAND (1961) spricht von Bindesubstanzen, BORCHERT (1961b) von Bindesubstanz oder Matrix. Der Ausdruck *Matrix* wird allerdings von RICHARDS (1959) und BREWER (1960b) für die Gesamtheit der festen Bodenbestandteile benutzt.

C. Teilgefüge

Eine Schar von Gefügeelementen, die sich durch irgendeine Gleichartigkeit, z. B. der Gestalt, Drehrichtung oder sonstigen Beschaffenheit, innerhalb von anderen Scharen von Gefügeelementen auszeichnet und dadurch von diesen unterscheidet, wird *Teilgefüge* genannt (SANDER 1948). Im Boden ist eine gegenseitige Durchdringung verschiedener, unter sich gleicher Scharen von Gefügeelementen fast immer in irgendeiner Form gegeben. Der Begriff dient häufig zur Gegenüberstellung z. B. von verlagerten gegenüber nicht verlagerten, aggregierten gegenüber nicht aggregierten, orientierten gegenüber nicht orientierten Gefügeelementen usw. Eine Überlagerung von zwei Gefügen (also Teilgefügen) bebezeichnete LAATSCH (1954) als *Gefügeinterferenz*.

D. Beispiele für Gefüge in grobkörnigen Substraten

Sandreiche Bodenmassen erweisen sich in Dünnschliffen stets viel stärker differenziert, als man bei flüchtiger Betrachtung im Felde anzunehmen geneigt ist. Vor allem die verschiedenen Merkmale der Skelettkörner treten sofort auffällig in Erscheinung. Der Verlauf einer Größenverteilungskurve ist meist leicht abzuschätzen (z. B. enge Verteilung, breite Streuung), und die Zusammensetzung aus Korngemischen verschiedener Herkunft ist unschwer zu erkennen. So bestehen z. B. manche Flugsande aus ziemlich gleichkörnigen Anteilen, wogegen der sogenannte Sandlöß oft deutlich aus Löß- und Flugsandteilen ohne verbindende Mittelgruppe aufgebaut ist, so daß die Korngrößenanalyse eine klare Zweigipfligkeit der Verteilungskurve ergibt. Die Gestalt der Körner schwankt in starker Abhängigkeit von den beteiligten Mineral- und Gesteinsarten, dem Verwitterungszustand und der mechanischen Beanspruchung. Verwitterungsvorgänge verursachen die verschiedenartigsten Korrosionsformen — entweder durch langsame Auflösung vom Rand her oder durch Skelettierung (PROISSL 1940). Die mechanische Beanspruchung verursacht Abrundung durch gegenseitigen Abrieb oder auch Zertrümmerung von weicheren Elementen. Sofern kein weiterer Transport erfolgt ist, lassen sich die Trümmer bisweilen noch als zusammengehörig erkennen. Die Gestalt der Körner beeinflußt vielfach auch die Ausprägung eines Richtungsgefüges. Körner mit hervortretenden Längsachsen sind entweder sperrig und ungeregelt oder mit einer bestimmten Orientierung eingelagert. Dabei ist die Packung oft verschieden und kann als offen, halboffen oder geschlossen angesprochen werden.

Die Gefüge sind jedoch selten völlig rein aus groben Körnungen aufgebaut, sondern enthalten fast immer auch feinere Anteile aus Schluff oder Ton oder organischen Stoffen. Daraus ergibt sich — vor allem, wenn die Schluffanteile zurücktreten — eine oft ideale Zweiteilung in Gefügeskelett und Gefügeplasma. In dem weiten und gleichmäßig leitfähigen Raum innerhalb des Kornverbandes können sich Stoffbewegungen und Stoffumwandlungen ziemlich ungestört vollziehen. Infolgedessen sind zahlreiche Formen von *Elementargefügen* oft in typischer, sehr reiner Ausprägung zu finden.

In gebänderten Horizonten, vor allem unter Podsolen, Braunerde-Podsolen, Fahlerden, Parabraunerden u. a. finden sich oft sehr vollkommen ausgebildete *Hüllengefüge* (KUBIËNA 1935 und FREI 1948). Sie entstehen dort, wo das Gefügeplasma zeitweise im Bodenwasser peptisiert ist und bei der Austrocknung auf den Kornoberflächen angelagert wird (Abb. 45 und 46). In den Kornwinkeln sind meniskenartige Ausrundungen zu beobachten. Die Hüllen sind verschieden zusammengesetzt und gefärbt und, sofern sie aus Tonteilchen aufgebaut sind, infolge Teilchenregelung häufig doppelbrechend. Auch schorfige, weniger gleichmäßig angelagerte Hüllen z. B. von Eisenhydroxyden sind anzutreffen. In den Ortsteinhorizonten von Humuspodsolen sind zum Teil mehrfach geschichtete Hüllen zu finden, aus welchen der Gang der Profilentwicklung sichtbar werden kann. Dabei treten oft charakteristische Schrumpfungsrisse auf, die sich wie ein Netz über die Kornoberflächen ziehen. (KUBIËNA 1953, MÜCKENHAUSEN 1962, ALTEMÜLLER 1962b, SCHLICHTING 1962). Modellversuche zur Darstellung von Hüllengefügen wurden von KUBIËNA (1938) und BREWER (1956, 1960a), BREWER und HALDANE (1957) angestellt.

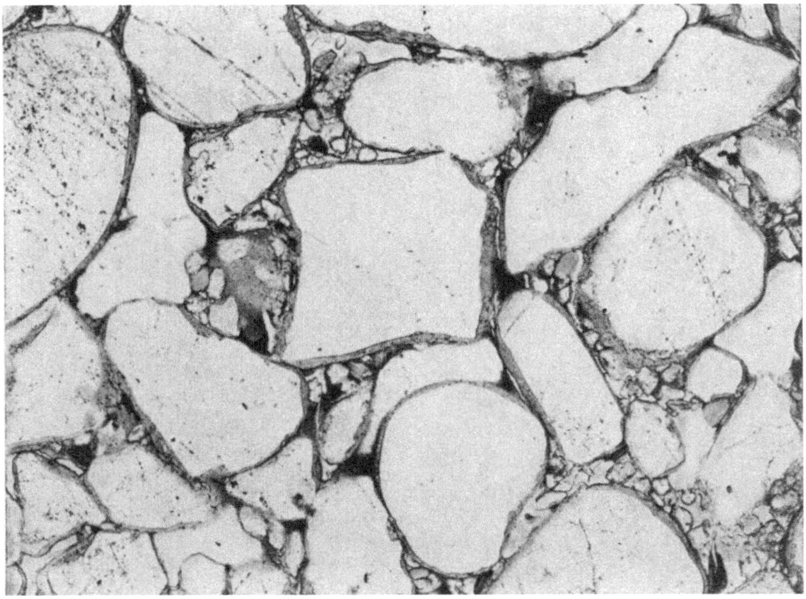

Abb. 45. Hüllengefüge im B-Horizont einer durchschlämmten Parabraunerde aus sandiger Grundmoräne. Die braungelben, eisenreichen Tonanteile sind den Skelettkörnern in dünner Schichtung als zusammenhängender Film aufgelagert. Dünnschliff, M = 32 : 1

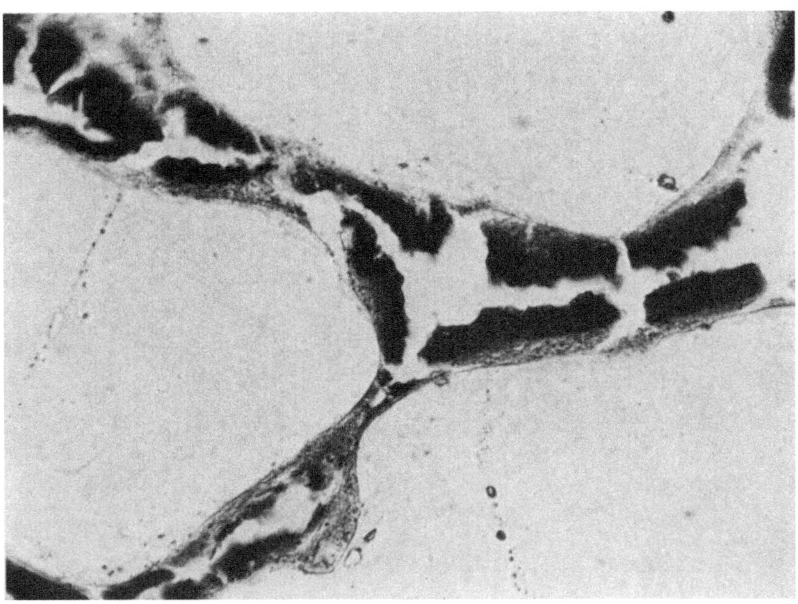

Abb. 46. Stark vergrößerter Ausschnitt aus einem Humusortstein. Hohlraum zwischen drei Quarzkörnern mit geschichteten Anlagerungen von mineralischen Tonteilchen und dunkelbraunen, stark geschrumpften Huminstoffen. M = 360 : 1

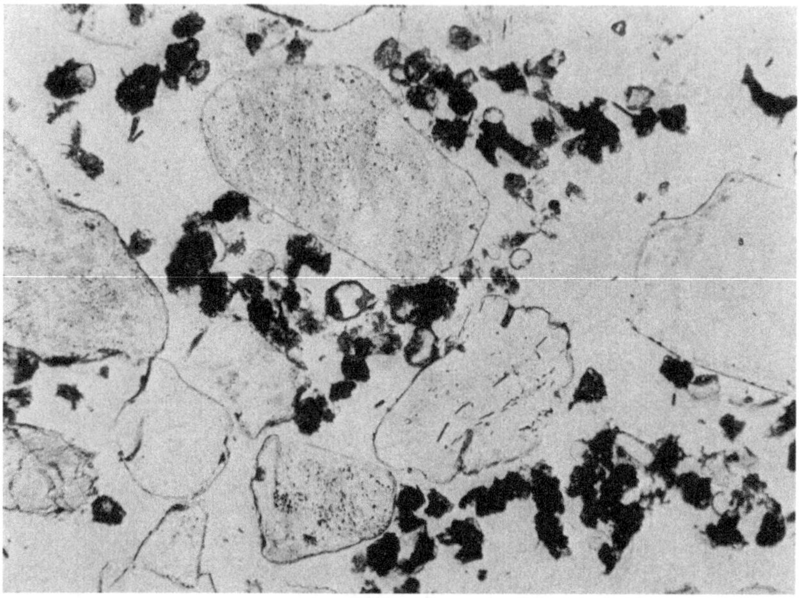

Abb. 47. Intertextisches Gefüge im (B)-Horizont einer Braunerde aus Sand. Die Tonanteile bilden eine graubraune, gut geflockte, fein aggregierte Bindemasse zwischen den Sandkörnern, haften aber nur unregelmäßig auf deren Oberflächen. M = 32 : 1

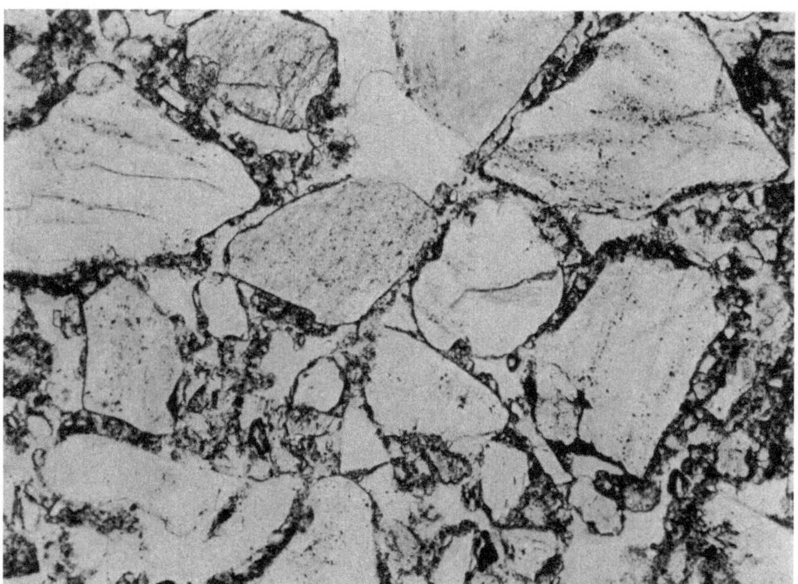

Abb. 48. Agglomeratisches Gefüge im A_1-Horizont (Feinmoder) einer Podsol-Braunerde aus Sand. Lose Ansammlung von organischen koprogenen Aggregaten in einem offenen Grobsandskelett ohne sichtbare gegenseitige Bindung. Die Oberflächen der Sandkörner sind völlig frei von Belägen. M = 32 : 1

In braunen Verwitterungshorizonten von Braunerden, Podsol-Braunerden und verwandten Bodentypen ist das Gefügeplasma zumeist in flockigen, wolkigen Gebilden den Skelettkörnern anhaftend und bildet brückenartige Verbindungen von Korn zu Korn (Abb. 47). Dieses *intertextische Gefüge* (Kubiëna 1935) ist vor allem an trockeneren Standorten und in Böden aus vorwiegend silikatischem Material gut ausgebildet, ist aber auch ganz allgemein im gemäßigten Klima als sogenanntes Braunerdegefüge verbreitet (Kubiëna 1953, Kundler 1961a, Mückenhausen 1962, Reuter 1962, Altemüller 1962b). Hohe Gehalte an Eisenhydroxyden wirken oft verstärkend auf den Flockungszustand, ebenso feindisperse organische Anteile der Mull-Humusform. Stark geflockt erscheinendes Gefügeplasma kann auch in Ortsteinhorizonten von Podsolen auftreten, die sich dann wesentlich von solchen mit Hüllengefüge unterscheiden.

Weitere Unterscheidungen ergeben sich vor allem aus den beteiligten Elementen. *Bleicherdegefüge* (Kubiëna 1935) sind z. B. durch weitgehende Auflösung und Abfuhr des Gefügeplasmas entstanden und bilden ein Skelett völlig gebleichter Sandkörner, in dem nur wenige organische oder mineralische Anteile verstreut enthalten sind.

Aggregatbildungen sind nicht selten, aber nicht so charakteristisch wie in den tonreichen Massen. Kleine, in Kornzwischenräume lose eingelagerte Aggregate von organischem Material treten jedoch in Moderhorizonten in charakteristischer Weise auf (*agglomeratische Gefüge*, Abb. 48, Kubiëna 1935, 1938).

E. Beispiele für Gefüge in feinkörnigen Substraten

In feinkörnigen Böden werden, zum Teil rein physikalisch bedingt, andere Gefügemerkmale vorherrschend. An Böden aus Löß wird z. B. das engere Zusammengedrängtsein der Gemengteile besonders deutlich. Die vorherrschenden Korngruppen aus Grobschluff (20 bis 60 μ) bilden ein Feinskelett, das bei stärkerer Vergrößerung unter dem Mikroskop zwar ein dem Sandskelett ähnliches Aussehen hat, es wird aber deutlich, daß sich die Wasserbewegung und der Stofftransport nicht in so gleichmäßiger Weise in den Intergranularräumen vollziehen wie in Sanden. Die Bodenmasse ist durch Poren und Gänge (Abb. 51) und, je nach den Tonanteilen, auch durch Spalten (Abb. 49) mehr oder weniger gegliedert und besitzt dadurch ein oft gut ausgeprägtes Leitbahnsystem.

Unterschiede in der Körnung, z. B. auch Einmischungen von Fremdanteilen in lößartigem Material, sind immer leicht erkennbar. Die Korngestalt ist zum Teil weniger differenziert als in Sanden, da mit abnehmender Größe gerundete Formen gegenüber kantigen und splitterartigen zurücktreten. Glimmerartige und leistenförmige Anteile im Feinskelett sind durch ihre gute Orientierungsfähigkeit feine Indikatoren für Umlagerungsvorgänge.

Das Gefügeplasma ist mehr oder weniger gleichmäßig in den Kornzwischenräumen verteilt und kann geflockt oder verschieden stark dispergiert erscheinen. Das *intertextische Gefüge* ist bisweilen gut ausgeprägt, aber seltener so vollkommen geflockt wie in durchlässigeren Böden. Im allgemeinen ist eine stärkere Aggregatbildung festzustellen, deren Entstehung vornehmlich auf die Tätigkeit von Bodentieren zurückzuführen ist. Stark ausgeprägte Gefüge dieser Art sind in den Mull-Horizonten von Schwarzerden und Braunerden (auch von Parabraunerden) anzutreffen (Abb. 50). Ferner in alpinen Rasenbraunerden, Lockerbraunerden, Auenböden usw. Das Gesamtgefüge wirkt durch seinen Hohlraumreichtum und die geflockten Elemente oft schwammähnlich und wird dann *Schwammgefüge* genannt.

In zahlreichen Böden tritt die *Verlagerung von Schlämmstoffen* als auf-

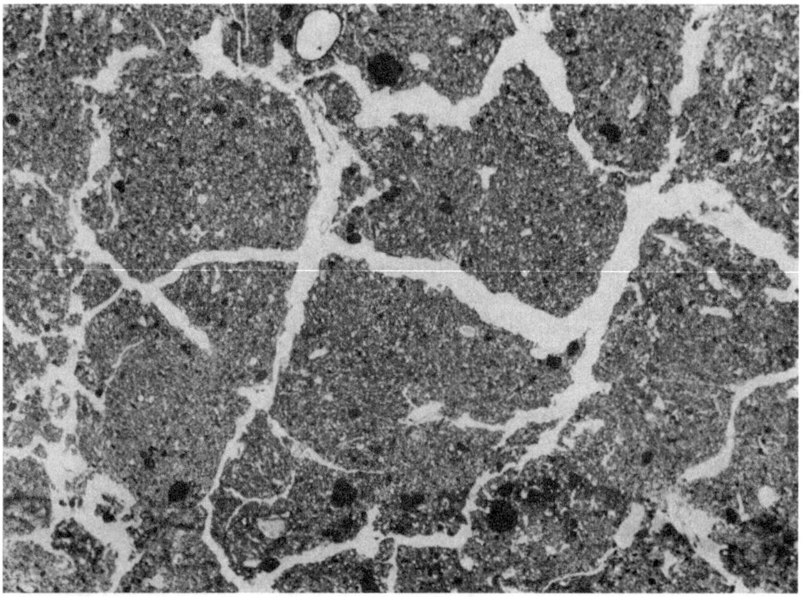

Abb. 49. Polyedrisches Gefüge im Ap-Horizont einer tonreichen Braunerde. Spaltenbildung **durch** Schrumpfung im Verlauf mehrfacher Befeuchtung und Trocknung. M = 10 : 1

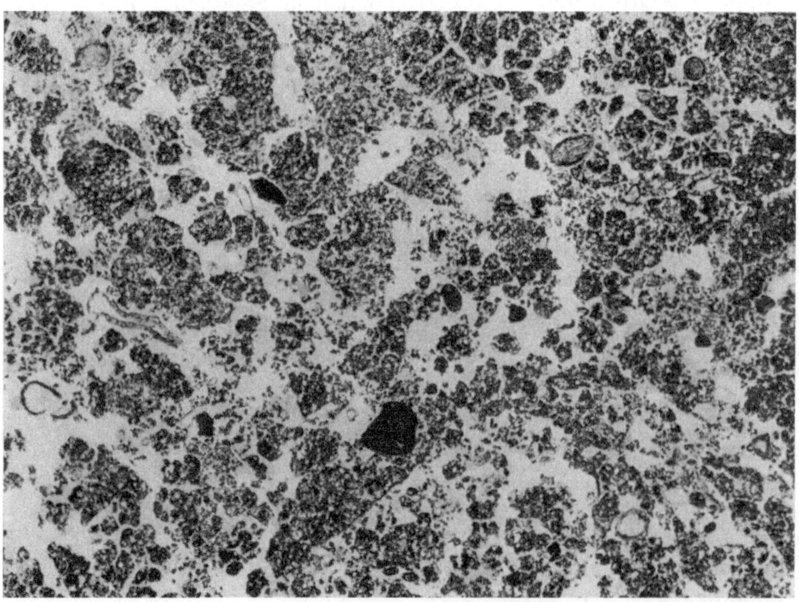

Abb. 50. Koprogenes Krümelgefüge im A_1/(B)-Horizont einer Lockerbraunerde aus spätglazialem Löß. M = 10 : 1

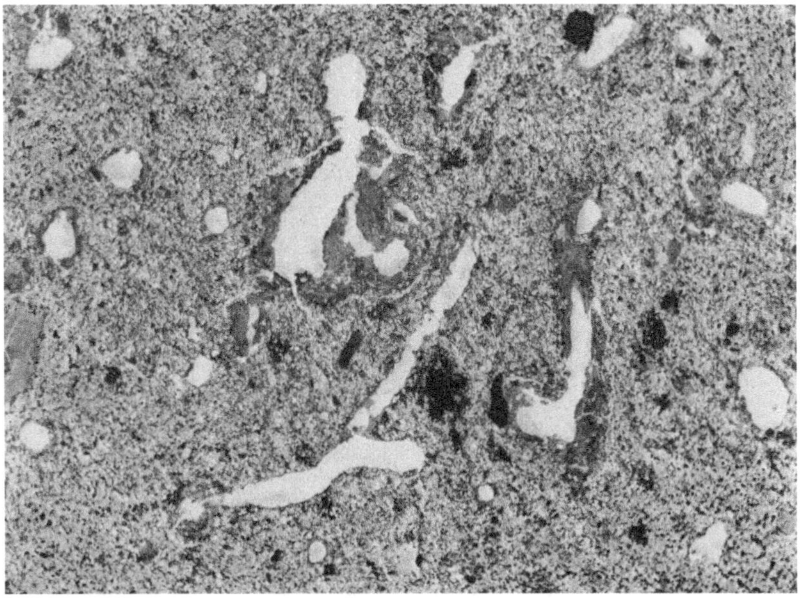

Abb. 51. Kohärentes, feinporiges Gefüge im B/C-Horizont einer Parabraunerde aus Löß mit Anlagerungen von kolloidalem Ton an den Porenwandungen. M = 14 : 1

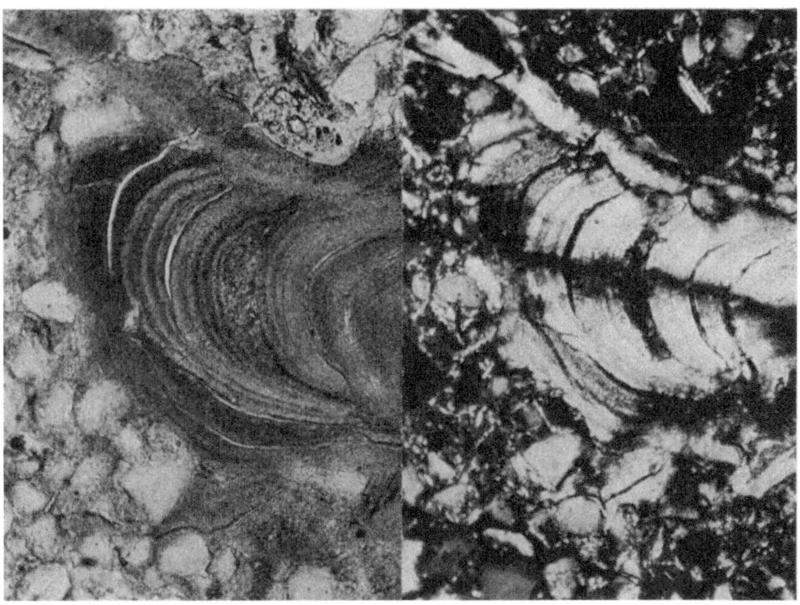

a *b*

Abb. 52a. Feinschichtig angelagerter Ton bei stärkerer Vergrößerung im B-Horizont einer Parabraunerde aus Löß. Man erkennt deutlich die meniskenartige, konkave Form der einzelnen Schichten. Die linke Bildseite ist im Profil unten. M = 100 : 1

Abb. 52b. Derselbe Dünnschliff bei gekreuzten Polarisatoren. Die Schichten sind infolge gleichartiger Teilchenregelung doppelbrechend und leuchten im dunklen Umfeld hell auf fälliges Merkmal in Erscheinung. Diese Schlämmstoffe sind, ähnlich wie manche hüllenbildenden Plasmaformen der sandigen Substrate, meist feindisperse orientiert angelagerte Tonmassen mit ausgeprägter Doppelbrechung. Sie sind von den nicht verlagerten Plasmateilen vielfach deutlich zu unterscheiden (Abb. 51 und 52). Die Verlagerung von Schlämmstoffen in Böden war der Bodenkunde nicht unbekannt. Sie konnte aber erst durch die mikromorphologischen Untersuchungen als bedeutsamer Prozeß voll erkannt und von chemischen Verlagerungsvorgängen getrennt werden. Die Ursachen sind bis heute nur zum Teil geklärt.

Die Beschreibungen von Gefügen mit Anlagerungen von durchschlämmten Tonen sind inzwischen sehr zahlreich. Von KUBIËNA zuerst (1936/37 und 1938) mikroskopisch untersucht, folgten Darstellungen von PETERSON (1937), JOHNSTON (1939), SWANSON und PETERSON (1940), FREI und CLINE (1949). Später wurden u. a. Arbeiten von KUBIËNA (1956), ALTEMÜLLER (1956, 1957, 1962a und b), MINASCHINA (1958), BLUME und SCHLICHTING (1959), KUNDLER (1959, 1961b), NORDMEYER (1959), BRAND (1961), DALRYMPLE (1962), MANIL (1962), REUTER (1962), LIEBEROTH (1962), PARFENOVA und JARILOVA (1962a und b) mit mikromorphologischen Untersuchungen veröffentlicht. Darin werden in der Regel auch braunerdeartige Gefüge behandelt. BARGON (1960) beschreibt mikroskopische Untersuchungen von sogenannten Lockerbraunerden, die durch eine auffällig starke Flockung und völlige Unbeweglichkeit der Teilchen ausgezeichnet sind. Erste Untersuchungen von Lockerbraunerden hat SCHÖNHALS (1957) durchgeführt.

BLÜMEL, JANIK und SCHILLER (1959) haben vergleichende Untersuchungen von Gefügemerkmalen und dem kolloidchemischen Verhalten von braunlehmartigem und von erdigem, geflocktem Boden durchgeführt und dabei enge Beziehungen zwischen dem mikroskopischen Gefügebild und der Kolloidbeweglichkeit festgestellt.

F. Beispiele für Gefüge in tonreichen Substraten

In tonreichen Substraten sind die Gefüge vor allem durch das Zurücktreten der Skelettanteile ausgezeichnet, die bisweilen ohne gegenseitige Berührung wie schwimmend eingebettet sind oder auch völlig fehlen. Die Gefüge sind jedoch deshalb nicht weniger gegliedert, sondern es können oft außerordentlich vielseitige und vielgestaltige Innengefüge festgestellt werden. Die Beziehungen zwischen Skelett und Plasma sind dabei nicht mehr allein bestimmend. Die Kennzeichen zur Gefügegliederung werden mehr aus der Beschaffenheit und Anordnung der Tonanteile selbst gesucht. Einige wesentliche Grundzüge seien hier kurz aufgeführt:

KUBIËNA hat vor allem auf die große Bedeutung des Flockungszustandes der Tonmassen hingewiesen. Zwischen extremer Flockung und extremer Dispergierung existieren alle Übergänge. Als typisches Beispiel für hochdispergierte Tonmassen ist der *Braunlehm* der tropischen Regenwaldzone zu nennen. Das dichte, für eine rasche Wasserbewegung undurchlässige, aber immer mehr oder weniger feuchte Material ist in besonderem Maße geeignet, *Diffusionsbewegungen* von gelösten Stoffen deutlich werden zu lassen. Diese Diffusionsbewegungen bestimmen charakteristische Gefügebildungen, die besonders lebhaft in der Bildung von Eisenanreicherungen zum Ausdruck kommen.

Im wechseltrockenen Savannengebiet werden die Braunlehmmerkmale durch Flockung und Ausscheidung von hämatitreichen Verbindungen mehr und mehr verändert und intensiv rot verfärbt. Dadurch wird in einem ursprünglich dichten Gefüge ein stabiles Sekundärskelett aufgebaut, aus dem schließlich die leichter beweglichen eisenarmen und kieselsäurereichen Tonanteile ausgewaschen werden, so daß im letzten Stadium stark geflockte, rote, erdige Gefüge zurückbleiben. Mit zunehmender Höhenlage treten andere Veränderungen auf, in deren Verlauf sich braunerdeartige Gefüge ausbilden (s. unten).

Im gemäßigten, mitteleuropäischen Klima finden sich braunlehmähnliche Gefüge vor allem auf Kalkgesteinen, und zwar in bester Ausbildung auf möglichst reinen Kalken. Mergelartige Ausgangsgesteine liefern weniger gut ausgeprägte Braunlehm-Merkmale (Kubiëna 1948, 1953; Schmidt-Lorenz 1957, Werner 1958; Kalk und Meyer 1960; Mückenhausen 1962; Smolikova und Ložek 1962). Im ganzen ähnlich verhalten sich auch zahlreiche Verwitterungsmassen aus Tonschiefern und Schieferton verschiedener geologischer Herkunft (Diez 1958, Mückenhausen, Gerkhausen und Kerpen 1959, Altemüller 1960, 1962a, Kerpen 1962, Mückenhausen 1962).

In den wassergesättigten, gequollenen Tonen können sich verschiedenartige Bewegungsvorgänge (innere Gleitbewegungen, Fließvorgänge u. a.) in charakteristischer Weise im Gefüge einprägen. Die Tonteilchen werden dabei in bevorzugte Richtungen eingeregelt und können im Schliff bei gekreuzten Polarisatoren als doppelbrechende, längliche Schlieren oder als doppelbrechende Höfe um festere Elemente erkannt werden (Abb. 53). Die Regelung der Tonteilchen ist im ganzen weniger vollkommen als z. B. in den Hüllengefügen der Sande, da im allgemeinen nicht die völlig freie Beweglichkeit und Orientierungsfähigkeit gegeben ist. Aus den Gefügen ist deutlich die gegenseitige Behinderung und Beeinflussung der Gefügeelemente abzulesen. Nur dort, wo der dichte Verband durch Hohlräume aufgelockert ist, die als Leitbahnen wirksam werden, treten auch die Anlagerungen aus Suspensionen als feingeschichtete Beläge in Erscheinung. Diese Grundvorgänge durchdringen sich gegenseitig und werden von anderen Prozessen vielfältig überlagert, so daß sehr kompliziert aufgebaute Gefüge aus den Tonen entstehen können. Neben den neugebildeten Elementen bleiben oft auch geologisch bedingte Teilgefüge über lange Zeiträume der Bodenbildung erhalten (Kubiëna 1954b, 1955b; Dalrymple 1958). Sehr aufschlußreiche Gefüge, die weitgehende Differenzierungen erlauben, sind z. B. in Böden aus periglacialen Fließerden oder in älteren Verwitterungsdecken erhalten (Abb. 54), in denen durch mechanische Aufarbeitung und Umlagerungen Gefügeelemente von verschiedenster Herkunft und Entstehung nebeneinander erkennbar sind (Altemüller 1960, 1962a).

In den Marschböden spielt die Dispersität der Tonanteile für die Ausprägung der Gefüge ebenfalls eine große Rolle. Deshalb hat Borchert (1961b) vorgeschlagen, die Menge der Ton- und Feinschluffanteile ($>6\,\mu$) und das geschätzte Ausmaß ihrer Orientierung bzw. Aggregation im Gefüge für eine Einteilung der Böden zu verwenden.

G. Eisenausscheidungen und andere Neubildungen im Bodengefüge

Als wichtigster Farbträger treten *Eisenverbindungen* in Böden oft in auffälliger Form in Erscheinung. Durch ihre verschiedene Beschaffenheit und die Art der Ausscheidung wirken sie stark auf die Ausprägung des Bodengefüges ein und verdienen deshalb in jeder Untersuchung besondere Beachtung.

In Braunlehmen und braunlehmartigen Massen sind die Eisenverbindungen vornehmlich als Oxydhydrate in feinstdisperser und homogener Verteilung enthalten und verursachen eine gelbbraune, im Dünnschliff braungelbe bis dottergelbe gleichmäßige Färbung. Aus diesem Zustand lassen sich vor allem in Tropenböden verschiedene Veränderungen ableiten, die nach Kubiëna (1943a, 1953, 1954c, 1956b, 1957, 1962b) als braune Vererdung, Rubefizierung, rote Vererdung und Laterisierung bezeichnet werden. Auch die Pseudogleybildung ist hier zu nennen.

Wie oben schon kurz vermerkt, ist die *braune Vererdung* für wechselfeuchte Lagen mit zeitweiser Abkühlung bezeichnend (Höhenlagen der Tropen und Subtropen) und hat eine Ausflockung des Eisenoxydhydrats und die Bildung

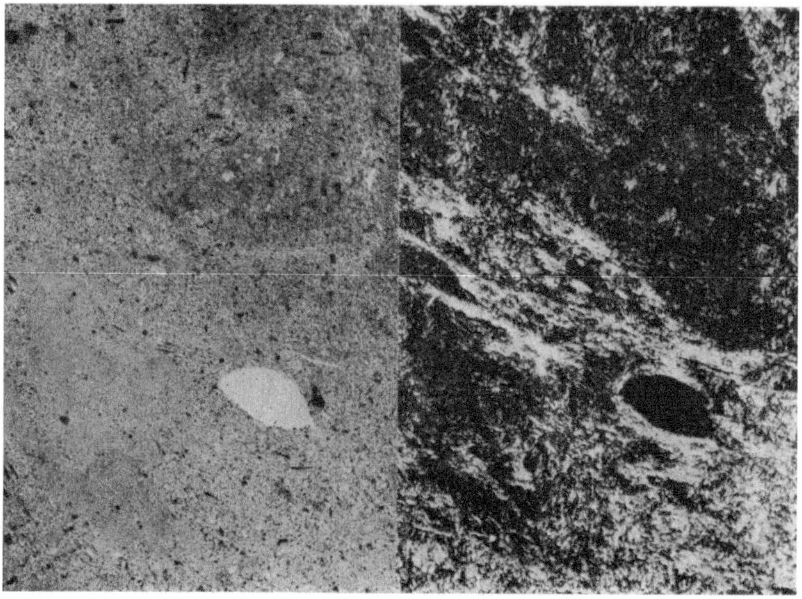

a b

Abb. 53a. Stark verwitterter, hochplastischer Ton aus dem Gipskeuper. Im Bild nur geringe Gliederung durch schwache Fleckung und ungleichmäßige Kornverteilung. Außerdem ist eine rundliche Pore sichtbar. M = 80 : 1

Abb. 53b. Derselbe Dünnschliff bei gekreuzten Polarisatoren. Man erkennt hell aufleuchtende, doppelbrechende Schlieren, die eine gleichartige Teilchenregelung entlang von inneren Gleitflächen verraten

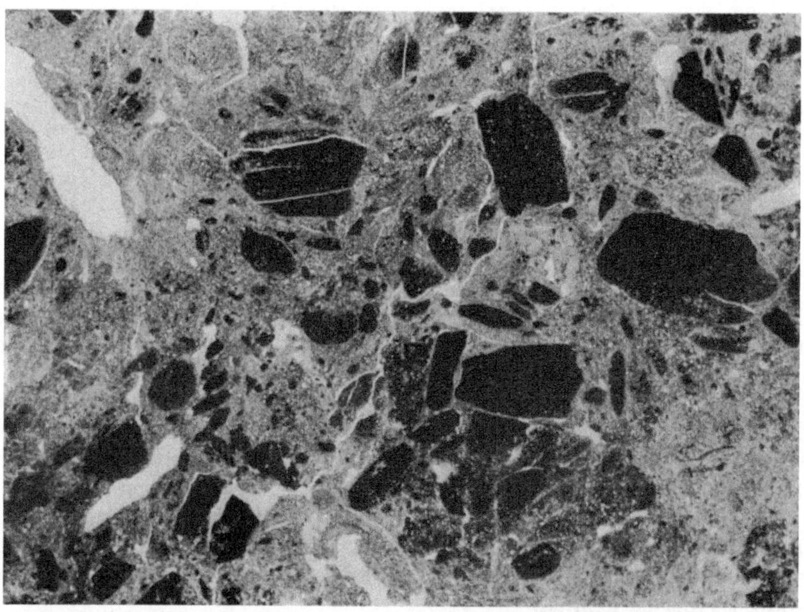

Abb. 54. Brecciöses Gefüge mit verschieden stark aufgearbeiteten Trümmeraggregaten aus Tonschiefer, die in einer dichten plastischen Tongrundmasse eingebettet sind. Lettenkeuper-Fließerde, M = 10 : 1

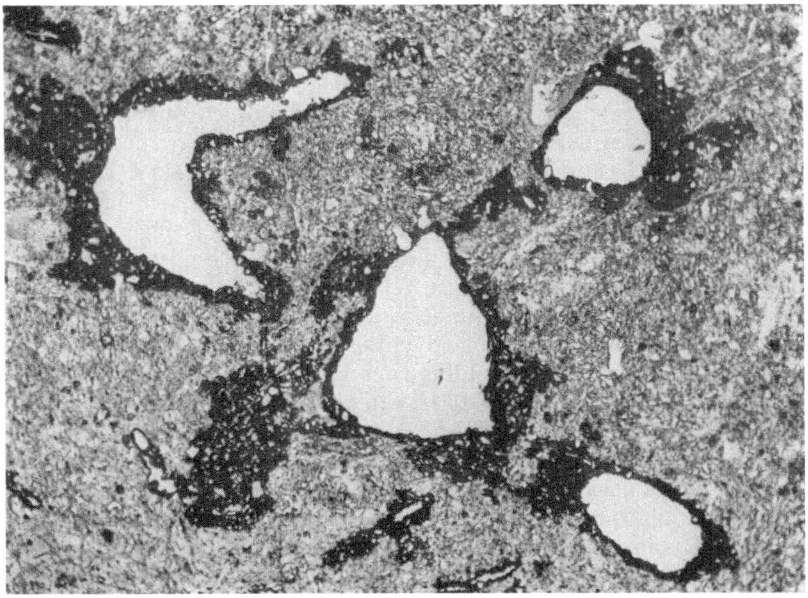

Abb. 55. Saumartige Anlagerung von dunkelbraunen Eisenhydroxydsolen an Porenwandungen im g_2-Horizont eines Pseudogleys aus Löß. M = 27:1

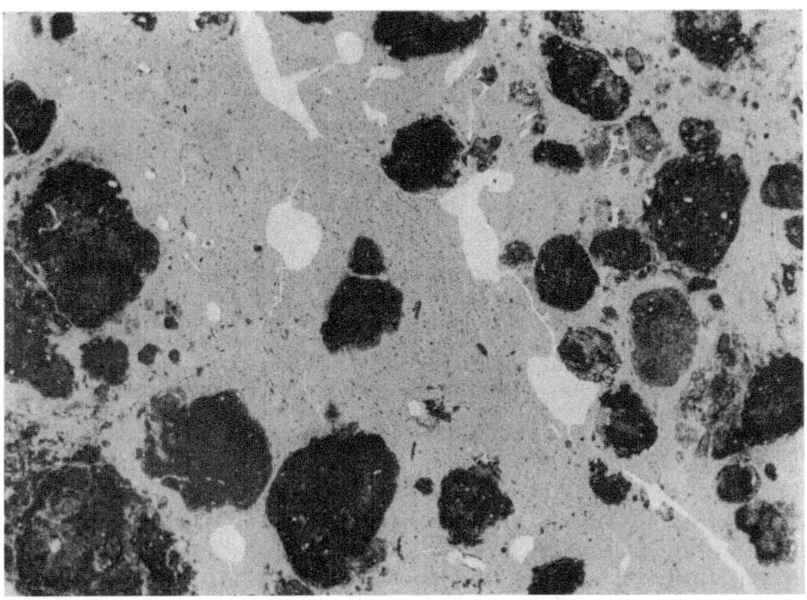

Abb. 56. Gerundete Eisenhydroxydkonkretionen in hellgrauer, eisenverarmter Ton-Grundmasse im g_1-Horizont eines Pelosol-Pseudogleyes aus Turneri-Ton (Lias β). M = 14:1

mehr oder weniger aggregierter, krümeliger Gefüge zur Folge. In wechseltrockenen Gebieten mit zeitweise starker Erwärmung (Savannen) beginnt im Inneren der Bodenmassen eine feinkristalline, mehr oder weniger gleichmäßige Ausscheidung von Goethit und Hämatit, die mit auffallender Rotfärbung verbunden ist *(Rubefizierung)*. Durch weitere Ausflockung entstehen rote erdige Gefüge (*rote Vererdung*). Hierzu siehe auch die Arbeiten von de Craene und Laruelle (1959), Laruelle und Denisoff (1956), Laruelle 1956, Osmond und Stephen (1955). Im Verlauf der *Laterisierung* bilden sich nach Kubiëna sehr formenreiche und mikroskopisch stark differenzierte Gefüge. Im Gegensatz zur Rubefizierung und Vererdung ist dieser Prozeß vornehmlich an tieferliegende und gleichmäßig feuchte Bodenzonen gebunden und läßt kräftige, oft an bestimmten Orten konzentrierte Ausscheidungen von Eisenoxydhydraten und Oxyden entstehen. Die Formen sind entlang von Spalten und Gängen usw. oft saumartig, im Bodeninneren diffus oder konkretionsartig, andere sind hohlraumfüllend und häufig mit dem Mikroskop als kristallin erkennbar (vorwiegend Hämatit und Goethit). Die Ausscheidungen sind anfangs weich und erhärten in fortgeschrittenen Stadien und vor allem dann, wenn die laterisierten Horizonte infolge Erosion des darüberliegenden Bodens an die Oberfläche kommen und der Austrocknung unterliegen (Krusten und Panzerbildung). Alexander und Cady (1962) beschreiben weitere mikromorphologische Merkmale — vor allem in Beziehung zu bestimmten Gesteinen und im Hinblick auf die Herkunft der Eisenverbindungen.

In gemäßigten Klimaten sind *Pseudogley-* und *Gleybildung* neben der Podsolierung die am stärksten auf die Lösung und Ausscheidung von Eisenverbindungen wirkenden Prozesse. Die Ausscheidungsformen sind ebenfalls mannigfaltig, obgleich die mikroskopisch als kristallin erkennbaren Bildungen hinter solchen, die amorph erscheinen, zurücktreten. Diffuse, fleckenartige Ausscheidungen, konzentrische Ringe um Wurzelbahnen, wandständige, im Dünnschliff als Säume sichtbare Anlagerungen um Hohlräume (Abb. 55) oder Aggregate, Bänderungen, mehr oder weniger scharfrandige Konkretionen mit konzentrischem oder andersartigem Innenbau und runder (Abb. 56) oder gelappter Außengestalt, gelartige oder kristalline Hohlraumfüllungen, Pseudomorphosen u. a. sind die wichtigsten Grundformen. Weitere Unterscheidungen ergaben sich durch Farbe und Konzentration und durch gegenseitige Durchdringung oder Überlagerungen. Zahlreiche Eisenausscheidungen gehen auch auf ältere Bodenbildungsprozesse zurück und haben deshalb als Relikte besonderen diagnostischen Wert.

Zur Deutung der Formen stehen vergleichende morphologische Untersuchungen im Vordergrund und haben Zusammenhänge mit bestimmten Prozessen erkennen lassen. Trotz der großen Zahl rein analytischer Untersuchungen zur Frage der Bewegung von Eisenverbindungen im Boden fehlen jedoch experimentelle, auf die mikromorphologischen Merkmale ausgerichtete Arbeiten fast völlig. Beschreibungen von Pseudogley- und Gleymerkmalen anhand von Dünnschliffen wurden von Kubiëna (1938, 1953, 1956a, 1962b), Altemüller (1957, 1960, 1962a), M. Müller (1958), Diez (1959), Kerpen (1960), Blümel (1962), Parfenova und Jarilova (1962b), Zimmermann (1962) u. a. gegeben.

Neben den Eisenverbindungen spielen Ausscheidungen von *Ca-Carbonaten*, *Gips* und anderen leichter löslichen *Salzen* eine bedeutende Rolle als empfindliche Anzeiger für Stoffbewegungen im Boden. Sie sind vielfach einer direkten mikroskopischen Mineralbestimmung im Dünnschliff zugänglich und zum Teil auch leicht erkennbar. Die Neubildungen sind teilweise an feuchte, in der Mehrzahl jedoch an zeitweise trockene Bedingungen gebunden und sehr formenreich. Gelartig erscheinende oder kristalline Ausblühungen *(Effloreszenzen)* an äußeren oder inneren Oberflächen des Bodens, feinkristalline Ausscheidungen im Inneren dichter Bodenmassen *(Interfloreszenzen)*, Kristallbildungen in Hohlräumen (Kristallröhren, Kristallkammern, KUBIËNA 1933, 1936/37), ferner diffuse oder scharfrandige Konkretionen seien als Beispiele erwähnt. Die hierbei sich bildenden Mineralarten nehmen vor allem in Trockengebieten an Zahl zu. So bestimmte FEOFAROVA (1958a und b) in versalzten Böden verschiedene Formen von Carbonaten als Calcit, Aragonit, Dolomit und Ankerit, verschiedene Sulfate als Gips, Tenardit, Glauberit u. a. Außerdem geben JARILOVA (1958), DOBROVOL'SKIJ (1960) und (1961), KERPEN, GEWEHR und SCHARPENSEHL (1960) und PARFENOVA und JARILOVA (1962b) detaillierte mikromorphologische Beschreibungen von Salz-Ausscheidungen in Trockengebieten.

Die vielfach im gemäßigten mitteleuropäischen Klima an Bodenaufschlußwänden feststellbaren weißlichen Effloreszenzen sind nach WERNER (1961) vorwiegend aus Gips gebildet. Auch in Modellversuchen mit Bodenlysimetern wurde bei starker Verdunstung an der Oberfläche als Krustenbildner Gips gefunden (CZERATZKI und ALTEMÜLLER 1960).

In diesem Zusammenhang sei ergänzend auch auf einige Arbeiten zur Frage der Bildung kieseliger Minerale aus Pflanzen (PARFENOVA 1949, PARFENOVA und JARILOVA 1962b) und der Umbildung von Opal in Quarz in der Terra fusca (KALK und B. MEYER 1960) hingewiesen.

H. Organische Stoffe und Humusformen

Die Kenntnisse über die Veränderung organischer Stoffe im Boden und die Bildung der Humusformen sind durch die Einführung der Dünnschliffuntersuchungen wesentlich bereichert worden. Die Pflanzenteile in ihren verschiedenen Stadien der Zersetzung, die Bildung von Abbauprodukten und ihre Verbindung mit dem mineralischen Bodenanteil, das Wachstum von Pilzen, die Arbeit von Bodentieren und andere Gegebenheiten bilden einen vielgestaltigen Zusammenhang, der bis in feinste Details aus dem Schliffbild abzulesen ist. KUBIËNA hat in mehreren Arbeiten auch für diesen Untersuchungszweig die Grundlagen geschaffen und damit Anregungen für weitere Bearbeitungen gegeben (1932a, 1938, 1943b, 1948, 1953, 1955a).

Der Ablauf des biologischen Geschehens kann zwar im fixierten Schliffbild nicht unmittelbar verfolgt werden, wie das bei direkter mikroskopischer Beobachtung am Boden möglich ist — auch lassen sich manche Einzelheiten der chemischen Zusammensetzung nur bedingt, andere auch gar nicht feststellen, doch mindert das den Wert der Untersuchung keinesfalls. Aus der Verbindung mit speziellen Methoden und der Rückspiegelung der gewonnen Ergebnisse in den Gesamtverband gewinnt die Schliffuntersuchung noch mehr an Bedeutung, indem sie für viele Einzeluntersuchungen eine engere und schärfer gefaßte Beziehung zum Boden herzustellen vermag. Zudem sei darauf hingewiesen, daß zahlreiche mikroskopische Untersuchungsverfahren, die in anderen Fachgebieten längst ihren unbestrittenen Platz einnehmen, z. B. Polarisations- und Fluoreszenzmikroskopie, Phasenkontrastverfahren, Mikrophotometrie, Interferenzmikroskopie u. a., in der Bodenmikroskopie erst in den Anfängen oder überhaupt noch nicht genutzt werden. Dazu ist es vorteilhaft, auch andere Präparationsverfahren (z. B. Mikrotomschnitte, die eine differenzierte Vorarbeit ermöglichen) mit heranzuziehen, doch kann man überall dort, wo organisches und anorganisches Material miteinander vermischt ist, nicht auf den Dünnschliff verzichten und wird deshalb die Untersuchungsmethodik dieser Präpariertechnik ständig weiter anpassen.

Die Untersuchungen sind zunächst vor allem auf die Bestimmung der Einzelelemente gerichtet, die in sehr verschiedenen Formen angetroffen werden. Die allgemeine Erkennung von *Pflanzenteilen* ist bei wenig zersetztem Material nach dem Zellverband und den Umrissen einfach. Auch die Zuordnung zu bestimmten Pflanzenarten und -organen (Nadeln, Blätter, Stengel, Rinden, Holzteile u. a.) ist oft leicht möglich, besonders, wenn bei der Probeentnahme die nötigen Bestimmungen vorgenommen wurden. Je mehr die örtlichen Gegebenheiten untersucht werden, um so sicherer wird auch die Ansprache im Schliff, so daß oft kleinste Reste noch bestimmt werden können.

Im Laufe der Zersetzung wechseln vor allem die *Farben* der Pflanzenteile in der Regel zu immer dunkleren Brauntönen, die sich sowohl an wenig zerteiltem Material als auch nach der Aufarbeitung durch Bodentiere infolge der Bildung feinst verteilter Humusstoffe als Ergebnis bakterieller und pilzlicher Tätigkeit einstellen, und die Gewebeteile gleich einem Farbstoff imprägnieren (s. hierzu Abbildungen bei KUBIËNA 1943b, 1948, JONGERIUS 1961, RONDE 1958, MOHTADI 1962). Im Gegensatz dazu kann auch völlige *Ausbleichung* stattfinden, so daß stärker zerteilte und mit dem Mineralboden vermengte Gewebsreste selbst mikroskopisch schwer erkennbar sind, sofern nicht besondere Hilfsmittel angewandt werden. In fahlen g_1-Horizonten von stark vernäßten Pseudogleyen lassen sich z. B. überraschend hohe Anteile von pflanzlichen Geweben im Fluoreszenzlicht feststellen (ALTEMÜLLER, nicht veröffentlicht). In Rohhumus- und Moderformen sind auch charakteristisch *rote Färbungen* häufig, die jedoch bis heute nicht untersucht wurden.

Bei gekreuzten Polarisatoren zeigen frische, zellulosereiche Pflanzenteile meist auffällige *Doppelbrechung*, die sich im Laufe der Zersetzung — oft mit der zunehmenden Braunfärbung parallelgehend — mehr und mehr verliert. KUBIËNA (1943b) machte darauf aufmerksam, daß in Gewebepartien, die stark von Pilzmyzelien durchsetzt sind, die Doppelbrechung oft deutlich schwächer ist als in benachbarten, nicht befallenen Teilen, eine Tatsache, die den Stoffabbau durch Pilzbefall sehr deutlich macht. Die Beurteilung der Doppelbrechung ist folglich für jede Untersuchung sehr wesentlich. Bei gekreuzten Polarisatoren lassen sich überdies manche Feinheiten des Abbaues (z. B. zerteilte Gewebereste in Exkrementen von Bodentieren, die teilweise noch doppelbrechend sind oder bei der Verdauung diese Eigenschaft verlieren) verfolgen, die einer Beurteilung im Hellfeld nicht mehr zugänglich sind.

Bodenpilze zeigen im Dünnschliff zum Unterschied gegen Plattenkulturen nicht farblose, sondern braune Hyphen, welche in Querschnitten oder verschieden langen Teilstücken mit Zellscheidewänden und Verzweigungen entweder einzeln oder zu dichten Knäueln gepackt gut erkennbar sind. Da Fruchtkörper kaum je erfaßt werden, ist eine nähere Bestimmung allein im Schliff kaum möglich. Ruhende oder auskeimende Dauerformen (Sklerotien) sind jedoch häufig anzutreffen (s. SAUERLANDT 1960, Abb. S. 51). Ganz allgemein tritt das Pilzwachstum in sauren Humusformen stark in den Vordergrund und beschränkt sich dabei nicht allein auf die oberen Humushorizonte. Auch Ortsteinhorizonte in Podsolen — um ein Beispiel zu nennen — sind oft stark von Pilzen durchsetzt.

Die Aufarbeitung von Pflanzengeweben wird vor allem von Bodentieren durch Verbiß, Aufnahme und Verdauung und Ausscheidung als Exkremente besorgt, wobei zum Teil charakteristische Produkte entstehen. So sind in Moder- und Rohhumusformen die nesterartigen Ansammlungen der eiförmigen Exkremente einiger Milbenarten (vorwiegend Oribatiden, s. KUBIËNA 1943b, 1948) leicht erkennbar, die bisweilen nur im Inneren der Pflanzenteile (Kavernenfraß) erhalten bleiben und unter Regenwirkung leicht zerfließen. JONGERIUS und Pons (1962), PUFFE und GROSSE-BRAUCKMANN (1963) haben diese Art Exkremente auch in Torfen beschrieben. Auch andere Tierarten, z. B. Asseln und Insektenlarven (KUBIËNA 1943b, JONGERIUS 1961), Enchytraeen (SAUERLANDT und MARZUSCH-TRAPPMANN 1962) und Regenwürmer bilden bestimmte, wiedererkennbare Exkrementformen. Es können jedoch, wie ZACHARIAE (1962) zeigt, allein durch verschiedene Nahrungsaufnahmen verschiedene Abwandlungen entstehen. Dazu kommt, daß in reich von Bodentieren besiedelten Standorten mehrere Artengruppen aufeinanderfolgend gleiches Material durcharbeiten, so daß im Endeffekt sehr komplizierte Verhältnisse vorliegen und die Zuordnung der Exkrementformen zu bestimmten Tiergruppen nicht mehr ohne weiteres möglich ist. — Während die oben erwähnten Milben sich auf rein pflanzliche Nahrung beschränken, nehmen Enchytraeen und Regenwürmer sowie manche Insektenlarven auch Bodenmaterial auf und schaffen damit die Voraussetzungen zur Bildung von Mull-Humusformen. Weitere Angaben hierzu bringt KÜHNELT (1958).

Als *neugebildete Humusstoffe* können bereits die Exkremente der Bodentiere bezeichnet werden, in denen sich die Zellstrukturen immer mehr verlieren und dafür dunkle, oft flockig-wolkige Gebilde entstehen (Abb. 48). Soweit keine mineralischen Anteile damit verbunden sind, können sie sogar verhältnismäßig reine Humusstoffe darstellen. Das trifft noch mehr für solförmig verlagerte Anteile zu, die z. B. in Ortsteinhorizonten deutlich als Beläge über Mineralkörnern erkennbar sind (Abb. 46). Mikroskopisch ist jedoch nicht immer zu entscheiden, ob auch Verbindungen des Eisens gegenwärtig sind, was durch Glühversuche aber qualitativ bestimmbar ist. Schwieriger ist der Nachweis von Aluminium- und anderen Verbindungen. Die früher vorgeschlagenen Färbemethoden und Tüpfelreaktionen (s. KUBIËNA 1938) wurden nicht weiter erprobt. Eine Bearbeitung und Anwendung dieser qualitativen Methoden könnte in manchen Fällen eine große Hilfe sein — um so mehr noch die Anwendung quantitativer mikrochemischer Verfahren.

Für Mull und mullartige Humusformen ist, wie schon erwähnt, eine enge Verbindung von mineralischem und organischem Material typisch. Zwar lassen sich bei starker Vergrößerung unter dem Mikroskop vor allem in den A_1-Horizonten von Schwarzerden, Smonitzen und anderen humusreicheren Böden unzählige feinste, dunkle Partikel von wenigen μ Größe feststellen, die als letzte Reste von Pflanzenzellen anzusehen sind (ALTEMÜLLER 1960) — die feineren kolloidalen Anteile sind jedoch mikroskopisch nicht sichtbar und nicht vom mineralischen Bodenanteil zu trennen.

Bei der Bestimmung der *Humusformen* ist es unbedingt notwendig, die Gesamtheit der Merkmale ins Auge zu fassen und sich nicht auf Einzelbeobachtungen zu beschränken, denn besonders bei den mikroskopischen Untersuchungen wird man häufig feststellen, daß bestimmte Einzelmerkmale in verschiedenen Humusformen wiederkehren können. Die mikromorphologische Beschreibung der wichtigsten Humusformen wurde von KUBIËNA (1953) in zusammengefaßter Darstellung gegeben und soll deshalb hier übergangen werden. In den zitierten Arbeiten und in weiteren Veröffentlichungen von JONGERIUS (1957), KUROTORI und MATSUMOTO (1958), KROLL und B. MEYER (1959), KLINGE (1960), KOWALINSKI (1960), SCHEFFER, FÖLSTER und B. MEYER (1960), RONDE (1958), GRAÇANIN (1962), SZABÓ, MARTON, VARGA und SCHÖNFELD (1962) u. a. sind ebenfalls mikromorphologische Kennzeichnungen von Humusformen gegeben. Außerdem enthält das schon zitierte Werk MÜCKENHAUSENS (1962) im Zusammenhang mit den entsprechenden Bodentypen zahlreiche Angaben.

Um die Ausführungen abschließend wieder auf ihren Ausgangspunkt zu orientieren, sind noch einige Bemerkungen über die Bedeutung der organischen Stoffe als *Gefügeelemente* und der Humusformen als *Gefüge* erforderlich. Soweit mineralische Anteile in den Humusbildungen überwiegen, wie z. B. im Mull, ergibt sich die gefügekundliche Betrachtung aus den Merkmalen, welche in den Abschnitten über die Untersuchung grobkörniger, feinkörniger und toniger Substrate genannt wurden unter Berücksichtigung der neu hinzukommenden Elemente in zwangloser Erweiterung. Besonders hervorzuheben sind die krümeligen und schwammartigen Gefüge, die in humusreichen, biologisch tätigen Horizonten ihre beste Ausbildung finden. Unter bestimmten Verhältnissen sind Gefügeformen zu beobachten, die auch in humusfreiem Material auftreten, wie z. B. plattige, polyedrische oder prismatische Gefüge usw. Auch völlig kompakte Bildungen, wie sie als Verdichtungszonen unter Ackernutzung entstehen können, sind nicht selten.

Einige Formen des *Elementargefüges* erhalten durch feinkolloidale organische Stoffe sehr charakteristische Ausbildungen, wie z. B. das Hüllengefüge der Humusortsteine oder das agglomeratische Gefüge mancher Feinmoder-Horizonte (s. S. 242). Allgemein können die organischen Stoffe sowohl Flockungen hervorrufen oder verstärken als auch in feinstdisperser Verteilung im Bodenprofil verlagert werden und entsprechend an der Gefügebildung beteiligt sein.

Überall dort, wo organische Anteile vorherrschen, sei es in einzelnen Horizonten (z. B. Rohhumus) oder in ganzen Profilen (z. B. Torf) entsteht ein völlig neues, vom Mineralboden abweichendes Bild. Jedoch handelt es sich auch hierbei um typisierbare Gefüge, die aus verschiedenartigen Gefügeelementen gebildet sind. Folglich sind auch alle Methoden der Gefügeuntersuchung in entsprechender Anpassung grundsätzlich anwendbar. Wenn auch in manchen Fällen mit der Beschreibung der stofflichen Beschaffenheit und der Herkunft der Elemente schon Wesentliches ausgesagt ist, kann doch aus der Beantwortung der Fragen nach der räumlichen Verteilung dieser Elemente, ihrer Einordnung in bevorzugte Richtungen, ihrer Rangfolge usw., d. h. aus der Abhandlung gefügekundlicher Fragestellungen, ein vollständigeres Ergebnis erwartet werden, nach dem besonders auch feine Unterteilungen und Abstufungen möglich sind.

5. Verfahren zur Herstellung von Bodendünnschliffen

Das Bodenmaterial für morphologische und gefügekundliche Untersuchungen kann je nach den gestellten Aufgaben in verschiedener Weise vorbereitet und präpariert werden. Allgemeine Voraussetzung ist dabei die Entnahme von Proben in ungestörter Lagerung, entweder als *Profilmonolithe* (GLINKA 1914) oder als kleinere *Rahmenproben* nach den Empfehlungen KUBIËNAS (1938, 1953). Die sogenannten „*Lackprofile*" und „*Klebplattenmonolithe*", wie sie besonders von SCHLACHT (1929), VOIGT (1936), RÖMER (1952), VOGEL und KOHL (1952), BRAND (1959) u. a. vorgeschlagen wurden, dienen vornehmlich als Anschauungsmaterial, sollen aber trotzdem erwähnt werden. Die hierzu benötigten Lacke und Klebstoffe werden nur zur Befestigung des Bodens auf einer Unterlage benötigt und sollen möglichst nicht zu sehen sein. Ähnlich verhält es sich auch bei der Herstellung von *Reliefgüssen* nach REINECK (1962), mit denen vor allem Schichtungsarten in sandigen Sedimenten sichtbar gemacht werden.

Zur Herstellung von Schnitten, Anschliffen und Dünnschliffen sind dagegen möglichst *vollständige Tränkungen* der Proben mit flüssigen Substanzen erforderlich, welche durch nachfolgendes Erhärten alle Bodenteile fest einschließen. Dadurch wird es möglich, die Proben zu schneiden oder zu sägen, ohne das Gefüge des Bodens zu verändern. Von den älteren Verfahren ist vor allem die Methode von PIGULEVSKIJ (1914) zu erwähnen, nach welcher aus paraffingetränkten Bodenproben sehr anschauliche Querschnitte vom Gefüge der Ackerkrume in Größen von etwa 30×60 cm gewonnen wurden. Die Untersuchung erfolgte bei schwacher Vergrößerung im *auffallenden Licht,* wobei die Hohlräume je nach der Art des Einbettungsmittels weiß oder dunkel erschienen. Auch einige später entwickelte Verfahren sind auf die Beobachtung im Auflicht zugeschnitten. So empfahl DAY (1948) polierte Anschliffe für mikroskopische Untersuchungen, wie sie ähnlich auch in der Erz- und Zementmikroskopie und Metallographie benutzt werden. HOMRIGHAUSEN und LANGE (1962) härten größere Bodenmonolithe für makroskopische Untersuchungen am Anschliff und WERNER (1962) verwendet mit fluoreszierenden Farbstoffen versehene Einbettungsmittel, um die Porenverteilung in den Anschliffproben bei UV-Licht sichtbar zu machen. Die Einbettungstechnik ist praktisch dieselbe wie bei der Dünnschliffherstellung, so daß nach Bedarf auch eine weitere Verarbeitung der Proben möglich ist.

Zur *Herstellung von Dünnschliffen* wird eine vollkommen durchtränkte und erhärtete Probe mittels einer Trennscheibe in zwei Hälften gesägt und eine der so geschaffenen Schnittflächen mit Schleifpulver oder Schleifpapieren glattgeschliffen und anschließend möglichst relieffrei poliert. Die sauber gewaschene Schleiffläche wird auf einen Objektträger aus Glas aufgekittet. Nach dem Erhärten der Kittstelle wird der Bodenblock parallel zur Glasfläche abgesägt, so daß ein dünnes, schon durchscheinendes Plättchen auf dem Objektträger verbleibt. Dieses wird maschinell oder von Hand bis zu einer Dicke von 40, 20 oder 15 μ weiter abgeschliffen und wie im ersten Schleifgang sauber poliert und gewaschen. Nach dem Aufkitten eines Deckgläschens ist der Schliff fertig und kann zur mikroskopischen Untersuchung benutzt werden.

Der im Grunde sehr einfache Arbeitsgang wird dadurch erschwert, daß es kaum Tränkungsmittel gibt, die alle gestellten Anforderungen vollkommen erfüllen. Im Laufe der Zeit wurden deshalb schon zahlreiche Vorschläge zur Einbettung und Schliffherstellung gemacht, die entweder nicht befriedigten oder aufgegeben werden mußten, weil die entsprechenden Substanzen nicht mehr hergestellt wurden. Zur Orientierung über die Entwicklung der Methodik sei besonders auf die Arbeiten KUBIËNAS (1937, 1938, 1942) hingewiesen. Eine kurze Übersicht über weitere Verfahren gibt BORCHERT (1961a).

Heute werden zumeist polymerisationsfähige Produkte der Kunststoffindustrie (sogenannte Gießharze u. a.) benutzt, die nach Zugabe von katalytisch wirkenden Stoffen aus dem flüssigen in einen glasartig festen Zustand übergehen. Der Vorteil dieser Mittel liegt vor allem darin, daß sie nach der Polymerisation gegen Erwärmung wenig empfindlich sind und auch maschinelle Verarbeitung ohne weiteres ertragen. Bei guter Durchtränkung sind die Bodenteile so fest eingeschlossen, daß auch dünnste Schliffe hergestellt werden können. In jedem Falle ist allerdings eine sorgfältige Einarbeitung in die Einbettungs- und Schleiftechnik erforderlich.

Die wichtigsten Einbettungsmittel sind derzeit:

1. Methacrylsäureester
2. Aethoxylin- oder Epoxy-Harze
3. Polyesterharze

Zu den *Methacrylsäureestern* gehört das von BORCHERT (1961, 1962) und von KURON und HOMRIGHAUSEN (1959) vorgeschlagene „*Plexigum*“ von Röhm & Haas, Darmstadt. Das monomere Produkt ist sehr dünnflüssig und dringt gut in die Bodenproben ein. Die Verarbeitungszeit kann kurz bemessen werden. Nach einer Tränkungszeit von 15 bis 25 Minuten im Vakuum erfolgt die Polymerisation in 3 bis 4 Stunden bei Zimmertemperatur. Der Brechungsindex liegt bei $n_D = 1{,}49$. Da eine Kittung auf Glas nicht möglich ist, werden Objektträger aus Plexiglas benutzt. BORCHERT benutzt die Präparate ohne Deckglas.

Aethoxylin- oder Epoxy-Harze wurden von REINECK (1958, 1962) und WELLS (1962) vorgeschlagen. Sie werden von den Firmen CIBA in Wehr/Baden bzw. Duxford/Cambridge unter dem Namen „*Araldit*“ hergestellt. Die Harze sind bei Zimmertemperatur ziemlich hoch viskos, werden aber bei Erwärmung über 80° C rasch dünnflüssig und sind dann gut zur Tränkung im Vakuum geeignet. Die Polymerisation erfolgt bei etwa 120° C. Der Brechungsindex wird von WELLS mit $n_D = 1{,}54$ bis 1,55 angegeben. Araldit besitzt eine sehr hohe Haftfähigkeit auf Glas.

Von den zahlreichen im Handel befindlichen *Polyester-Harzen* hat sich vor allem das „*Vestopal H*“ (Chemische Werke Hüls, Marl, Kreis Recklinghausen) bewährt (ALTEMÜLLER 1956a, 1962c). Das Harz wird zur Verarbeitung mit Monostyrol verdünnt und bei Zimmertemperatur verarbeitet. Der Styrolanteil kann vollständig mit einpolymerisiert werden. Um bessere optische und mechanische Eigenschaften zu erzielen, ist es jedoch ratsam, denselben möglichst weitgehend vor der Polymerisation verdampfen zu lassen. Der Brechungsindex liegt bei styrolärmeren Polymerisaten zwischen $n_D = 1{,}55$ und 1,56. Die Haftfähigkeit auf Glas ist gut. Es empfiehlt sich jedoch, in möglichst gleichmäßig temperierten Räumen zu arbeiten. JONGERIUS und HEINTZBERGER (1963) schlagen vor, die Verarbeitungszeit des Vestopals durch entsprechende Verringerung der Katalysatormengen auf mehrere Wochen einzustellen, um damit auch schwerste Tonböden vollständig durchtränken zu können. Auf diese Weise wurden Großschliffpräparate von 8 × 15 cm Größe und normaler Dünnschliffdicke von bester Qualität hergestellt.

Bei stark quellfähigem Bodenmaterial ist die der Präparation vorausgehende Trocknung und Schrumpfung oft von Nachteil. HAARLØV und WEIS-FOGH (1953) haben deshalb *feuchte Bodenproben* gefroren, in passende Stücke zugeschnitten, nach dem Auftauen in Agarlösung überführt und gehärtet, um danach Schnitte anzufertigen. Da die Mineralkörner nicht geschnitten werden können, richtet sich die Schnittdicke nach den jeweils größten Körnern. MITCHELL (1956) hat naturfeuchte Tone in einer wasserverträglichen Polyäthylenglykol-Verbindung (Carbowachs 6000) eingebettet und zu Dünnschliffen verarbeitet.

Das gleiche Verfahren haben MACKENZIE und DAWSON (1961) für feuchte Torfe angewandt und gegenüber der vergleichsweise geprüften Gefriertrocknung mit nachfolgender Einbettung in Harze als vorteilhafter befunden. Auch PUFFE und GROSSE-BRAUCKMANN (1963) berichten über ein Verfahren nach HOLLING zur Einbettung naturfeuchter Torfe in wasserverträgliche Substanzen (Cremolan) und konnten hochwertige Mikrotomschnitte anfertigen.

Im allgemeinen ist zu sagen, daß für Mineralböden noch keine befriedigende Methode vorliegt, um die Gefügemerkmale des naturfeuchten Bodens in gleicher Weise wie an Dünnschliffen aus getrocknetem Material untersuchen zu können. Hierfür sind noch manche Aufgaben zu lösen.

Literatur

Anmerkung: Die mit * bezeichneten Arbeiten sind als Sonderdruck in dem Band „Arbeiten aus dem Gebiet der Mikromorphologie des Bodens“, herausgegeben von H.-J. ALTEMÜLLER und H. FRESE, Weinheim/Bergstr.: Verlag Chemie, 1962, enthalten.

ALEXANDER, L. T., und J. G. CADY: Genesis and hardening of laterite in soils. U.S.D.A. Soil Cons. Serv. Techn. Bull. Nr. 1282. 90 S. Washington 1962. — ALTEMÜLLER, H.-J.: Neue Möglichkeiten zur Herstellung von Bodendünnschliffen. Z. Pflanzenernähr., Düng., Bodenkde. **72** (117), 56–62 (1956a). — Mikroskopische Untersuchungen einiger Löß-Bodentypen mit Hilfe von Dünnschliffen. Z. Pflanzenernähr., Düng., Bodenkde. **72** (117), 152–167 (1956b). — Bodentypen aus Löß im Raume Braunschweig und ihre Veränderungen unter dem Einfluß des Ackerbaues. 250 S. Diss. Bonn 1957. — Mikromorphologische Untersuchungen an einigen Gipskeuperböden im Raum Iphofen. Bayer. Landwirtsch. Jb. **37**, Sonderh. 4, 70–85 (1960). — Mikromorphologische Untersuchungen an Bodenprofilen des mittleren fränkischen Jura. Bayer. Landwirtsch. Jb. **39**, Sonderh. 1, 98–109 (1962a). — Beitrag zur mikromorphologischen Differenzierung von durchschlämmter Parabraunerde, Podsol-Braunerde und Humus-Podsol. Z. Pflanzenernähr., Düng., Bodenkde. **98** (143), 247–258 (1962b)*. — Verbesserung der Einbettungs- und Schleiftechnik bei der Herstellung von Dünnschliffen mit Vestopal. Z. Pflanzenernähr., Düng., Bodenkde. **99** (144), 164–177 (1962c)*. — Gedanken zum Aufbau des Bodens und seiner begrifflichen Erfassung. Z. Kulturtechn. **3**, 323–336 (1962d). — ANDERSON, D. M., und R. R. BINNE: Modal analysis of soils. Soil Sci. Soc. Amer. Proc. **25**, 499–503 (1961). — Anonymus: Von Erkenntnis und Anwendung der verschiedenen Erdarten. 64 S. Leipzig: Christ. Gottl. Hilscher. 1770. — ANTIPOV-KARATAEV, J. N., und V. V. KELLERMAN: Die Bodenaggregate und ihre kolloidchemische Analyse. Z. Pflanzenernähr., Düng., Bodenkde. **98**, 193–204 (1962)*.

BACH, R., und H. DEUEL: Über das Bodengefüge. Schweiz. Z. Forstwesen **104**, 192–206 (1953). — BARGON, E.: Über die Entwicklung von Lockerbraunerden aus Solifluktionsmaterial im vorderen Odenwald. Z. Pflanzenernähr., Düng., Bodenkde. **90** (135), 229–244 (1960). — BAVER, L. D.: Soil physics, 2. Aufl. 398 S. New York: Wiley; London: Chapman and Hall. 1948. — BECKMANN, W.: Zur Mikromorphometrie von Hohlräumen und Aggregaten im Boden. Z. Pflanzenernähr., Düng., Bodenkde. **99** (144), 129–139 (1962)*. — BLUME, H. P., und E. SCHLICHTING: Nachweis einer Tonverlagerung in Bodenprofilen. Z. Pflanzenernähr., Düng., Bodenkde. **85** (130), 227–244 (1959). — BLÜMEL, F.: Formen der Eisenoxydhydrat-Ausscheidungen in Gleyen und Pseudogleyen. Z. Pflanzenernähr., Düng., Bodenkde. **98**, 258–264 (1962)*. — BLÜMEL, F., V. JANIK und H. SCHILLER: Die Mikromorphologie und der Kolloidzustand unterschiedlicher Bodentypen. Landwirtsch.-chem. Bundesversuchsanst. Linz/D., Festschrift LX/4, 89–120 (1959). — BORCHERT, H.: Einfluß der Bodenerosion auf die Bodenstruktur und Methoden zu ihrer Kennzeichnung. Geol. Jb. **78**, 439–502 (1961a). — Entwurf einer Mikrogefüge-Einteilung von Marschböden. Geol. Rdsch. **51**, 643–647 (1961b). — Die Herstellung von Bodendünnschliffen mit dem Festlegungsmittel Plexigum M 7466. Z. Pflanzenernähr., Düng., Bodenkde. **99** (144), 159–164 (1962)*. — BRAND, FR. C.: Die Herstellung von Lackfilm-Profilen erdfeuchter Böden. Z. Pflanzenernähr., Düng., Bodenkde. **86** (131), 123–131 (1959). — Die Ergebnisse von Bodenstrukturuntersuchungen einiger aus der Riß-Grundmoräne und aus Löß entwickelter Bodentypen. Z. Acker- u. Pflanzenbau **112**, 341–376 (1961). — BREWER, R.: Cutans: their definition, recognition and interpretation. J. Soil Sci. **11**, 280–292 (1960a). — The petrographic approach to the study of soils. Trans. 7th Int. Congr. Soil Sci. Madison, Wisc. **1**, 1–11 (1960b). — BREWER, R., und A. D. HALDANE: Preliminary experiments in the development of clay orientation. Soil Sci. **84**, 301–309 (1957). — BREWER, R., und J. R. SLEEMAN: Soil structure and fabric. J. Soil Sci. **11**, 172–185 (1960). — BRYSSINE, G.: Contribution à l'étude des propriétés physiques du sol, IV, Note sur l'appréciation de la structure du sol. Trav. Section Pédologie **8** et **9**, **33** (1954). — Etude de la structure du sol. Mededel. Landbouwhogeschool **24**, 301–306 (1959).

CZERATZKI, W.: Zur Wirkung des Frostes auf die Struktur des Bodens. Z. Pflanzenernähr., Düng., Bodenkde. **72**, 15–32 (1956). — Bodenstrukturbildung bei wechselnder Befeuchtung und Trocknung. Film C 725/1956. Inst. f. d. wiss. Film, Göttingen 1957. — CZERATZKI, W., und H. FRESE: Kinematographische Untersuchungen zur Strukturbildung. Rapp. 6e Congr. Int. Sci. Sol Paris B, 173–178 (1956). — CZERATZKI, W., und H.-J. ALTEMÜLLER: Salzanreicherung durch Bodenverdunstung im Trockenjahr 1959. Landbauforschung Völkenrode **10**, 31–35 (1960).

DALRYMPLE, J. B.: The application of soil micromorphology to fossil soils and other deposits from archaeological sites. J. Soil Sci. **9**, 199–209 (1958). — Some micromorphological implications of time as a soil forming factor, illustrated from sites in south-eastern England. Z. Pflanzenernähr., Düng., Bodenkde. **98** (143), 232–239 (1962)*. — DAY, P. R.: Experiments in the use of the microscope for the study of soil structure. Soil Sci. Soc. Amer. Proc. **13**, 43–50 (1948). — DE CRAENE, A., und J. LARUELLE: Genèse et altération des Latosols équatoriaux et tropicaux humides. Bull. Agric. Congo Belge **46**, 1113–1243 (Bruxelles 1955). — DE LEENHEER, L., und J. VAN HOVE: Werkwijze voor de Mechanische Analyse met de Kettinghydrometer. Mededel. Landbouwhogeschool **21**, 249–274 (1956). — DOBROVOL'SKIJ, V. V.: Zur Mineral-Hypergenese der Quartär-Sedimente der Zentralrussischen Waldsteppe. Bjulleten' M. O-va Isp. prirody, otd. geologii **35**, 83–99 (Moskau 1960) (russ.). — Typomorphe Neubildungen in Quartär-Sedimenten der Wüstenzone der UdSSR. Počvovedenie **1961**, Nr. 10, 44–58 (russ.). — DIEZ, TH.: Entwicklung und Eigenschaften von Böden aus tonigen Substraten. 145 S. Diss. Bonn 1959.

EHRENBERG, P.: Die Bodenkolloide. 563 S. Dresden und Leipzig: Steinkopff. 1915.

FEOFAROVA, J. J.: Bestimmung der Carbonate in versalzten Böden mit mikroskopischen Methoden. Trudy počvenn. inst. im. V. V. Dokučaeva, Moskva, **53**, 75–88 (1958a) (russ.). — Sulfate in versalzten Böden. Trudy počvenn. inst. im. V. V. Dokučaeva, Moskva, **53**, 89–103 (1958b) (russ.). — FRANZ, H.: Feldbodenkunde. Wien und München: Fromme. 1960. — FREI, E.: Gefügeuntersuchungen an landwirtschaftlichen Kulturböden. Landwirtsch. Jb. Schweiz **62**, 1–17 (1948). — Genesis of various types of soil structure. Trans. 4th Int. Congr. Soil Sci. Amsterdam **1**, 46–47 (1950). — FRESE, H.: Zur Bildung von Makro-Gefüge-Typen im Ackerboden durch atmosphärische Einflüsse. Dtsch. Akad. d. Landwirtsch. Wiss. Berlin, Tagungsber. **13**, 117–127 (1958). — Neue Gesichtspunkte für die Beurteilung und Bewirtschaftung unserer Ackerböden. Landwirtsch. Forsch. Sonderh. 12, 49–59 (1959). — Zur Nomenklatur, Definition und Messung ackerbaulich wichtiger, physikalischer Bodeneigenschaften. Landwirtsch. Forsch. Sonderh. 14, 1–6 (1960a). — Über Voraussetzungen und Zweckmäßigkeit einer Vertiefung der Ackerkrume. Dtsch. Akad. d. Landwirtsch. Wiss. Berlin, Tagungsber. **28**, 51–60 (1960b). — FRESE, H., und W. CZERATZKI: Der wissenschaftliche Film in der landwirtschaftlichen Forschung. Strukturbildung im Ackerboden. Umschau **57**, 495–498 (1957).

GEDROIZ, K. K.: Zur Frage der Bodenstruktur und ihrer Bedeutung für die Landwirtschaft. Leningrad 1926. Siehe in: KRAUSE 1931. — Materialien zur Klärung der Frage über die Bodenstruktur. Trudy sovet. sekcii M.A.P. **1** (1933) (russ.). Siehe in: ANTIPOV-KARATAEV und KELLERMAN 1962. — GEYGER, E.: Zur Methodik der mikromorphometrischen Bodenuntersuchung. Z. Pflanzenernähr., Düng., Bodenkde. **99** (144), 118–129 (1962)*. — GLINKA, K.: Die Typen der Bodenbildung, ihre Klassifikation und geographische Verbreitung. 365 S. Berlin: Borntraeger. 1914. — GÖRBING, J.: Die Grundlagen der Gare im praktischen Ackerbau, 2 Bde. 206 und 97 S. Hannover: Landbuch-Verlag. 1947. — GRAÇANIN, M.: To the problem of evaluation of soil structure. Trans. 4th Int. Congr. Soil Sci. Amsterdam **1**, 40–43 (1950). — GRAÇANIN, Z.: Zur Genese, Morphologie und Mikromorphologie der Hangtorfbildung auf Kalksteinen in Kroatien. Z. Pflanzenernähr., Düng., Bodenkde. **98** (143), 264–272 (1962)*.

HAARLØV, N., und T. WEIS-FOGH: A microscopical technique for studying the undisturbed texture of soils. Oikos **4**, 44–57 (1953)*. — HEUSER, O.: Grundzüge der praktischen Bodenbearbeitung auf bodenkundlicher Grundlage. 228 S. Berlin: Parey. 1928. — HILGARD, E. W.: Über die Flockung kleiner Teilchen und die physikalischen und technischen Beziehungen dieser Erscheinung. Forsch. Geb. Agrikult.-Physik **2**, 441–454 (1879). — HOENES, D.: Mikroskopische Grundlagen der technischen Gesteinskunde. In: FREUND, H.: Handbuch der Mikroskopie in der Technik **4**/1, 323–695, Frankfurt/M. 1955. — HOMRIGHAUSEN, E., und E. LANGE: Zur Präparierung und Härtung von großen Bodenmonolithen. Mitt. Inst. f. Bodenkde. u. Bodenerhaltung d. Univ. Gießen **1** (1962).

JARILOVA, E. A.: Mineralogische Charakteristik der Salze in der Černozem-Zone. Trudy počvenn. inst. im. V. V. Dokučaeva, Moskva, **53**, 131–142 (1958) (russ.). — JOHNSTON, J. R.: Micromorphology of Marshall silty clay loam, a Webster clay and a Nacodoches sandy loam profile. Soil Sci. Amer. Proc. **4**, 360–363 (1939). — JONGERIUS, A.: Morfologische Onderzoekingen over de Bodemstructuur. 93 S. Wageningen 1957. — The morphological soil structure classification of the Dutch Soil Survey Institute. Mededel. Landbouwhogeschool **24**, 206–213 (1959). — De Micromorphologie van de Organische Stof. Bodemkunde, 's-Gravenhage **1961**, 43–58. — JONGERIUS, A., und J. SCHELLING: Micromorphology of organic matter formed under the influence of soil organisms, especially soil fauna. Trans. 7th Int. Congr. Soil Sci. Madison, Wisc. **2**, 702–710 (1960). — JONGERIUS, A., und L. J. PONS: Soil genesis in organic soils. Boor en Spade **12**, 156–168 (1962a). — Einige mikromorphologische Bemerkungen über den Vererdungsvorgang im niederländischen Moor. Z. Pflanzenernähr., Düng., Bodenkde. **97** (142), 243–255 (1962b)*. — JONGERIUS, A., und G. HEINTZBERGER: The preparation of mammoth-sized thin sections (im Druck).

KALK, E., und B. MEYER: Progressive Bio-Opal→Quarz-Umwandlung bei der Terra Fusca-Entstehung aus Kalkgestein. Zeiss-Mitt. **2**, 103–112 (1960). — KATSCHINSKI, N. A.: Klassifikation der Böden nach ihrer Textur. Trans. 7th Int. Congr. Soil Sci. Madison, Wisc. **1**, 130–137 (1960). — KERPEN, W.: Die Böden des Versuchsgutes Rengen, Kartierung, Eigenschaften und Standortswert. Wiss. Ber. Landwirtsch. Fak. d. Univ. Bonn 1960, H. 5, 316 S. — KERPEN, W., H. GEWEHR und H. W. SCHARFENSEEL: Zur Kenntnis der ariden Irrigationsböden des Sudan, II, Mikroskopische Untersuchungen. Pedologie **10**/2, 303–323 (1960). — KLAPP, E.: Lehrbuch des Acker- und Pflanzenbaues, 4. Aufl. 407 S. Berlin und Hamburg: Parey. 1954. — KLINGE, H.: Beiträge zur Kenntnis tropischer Böden, I, Die schwarzerdeartigen Böden auf der oberen Serie der Schichten von San Salvador, El Salvador, Zentralamerika. Z. Pflanzenernähr., Düng., Bodenkde. **89** (134), 102–114 (1960). — KOEPF, H.: Vorschlag für eine einheitliche Beurteilung und Ansprache von Böden für die praktische Feldversuchsanstellung und Beratung (Bodenbeurteilungsrahmen). Landwirtsch. Forsch. Sonderh. **11**, 53–61 (1958). — Zur Strukturbildung verschiedener Böden durch Schrumpfung. Z. Pflanzenernähr., Düng., Bodenkde. **89**, 159–170 (1960). — Beitrag zur Strukturbildung bei verschiedenen Bodentypen. Landwirtsch. Forsch. **14**, 1–9 (1961). — Die Kennzeichnung des Bodengefüges im Felde. Z. Kulturtechn. u. Flurbereinigung **4**, 93–101 (1963). — KÖSTER, E.: Mechanische Gesteins- und Bodenanalyse. 171 S. München: Hanser. 1960. — KOWALINSKI, ST.: Differentiation of morphological, physical and chemical properties of black earths as an effect of change of their utilization. Zeszyty naukowe Wyszej Szkoly Rolniczej we Wroclaviu **10**, Nr. 29, 103–117 (1960) (poln.). — KRAUSE, M.: Russische Forschungen auf dem Gebiete der Bodenstruktur. Landwirtsch. Jb. **73**, 603–690 (1931). — KROLL, W., und B. MEYER: Die Charakterisierung von Humusformen durch Siebanalyse und mikromorphologische Untersuchung am Beispiel einer Bodenserie auf mittlerem Buntsandstein bei Göttingen. Landwirtsch. Forsch. **12**, 12–17 (1959). — Kubiëna, W. L.: Mikropedologie. Neue Wege bodenkundlicher und bodenbiologischer Forschung. Biol. Gen. **8**, 513–546 (1932a). — Ein Bodenmikroskop für Freiland- und Laboratoriumsgebrauch. Bodenkundl. Forsch. **3**, 91–102 (1932b). — Mikropedologische Untersuchungen über Kristall-Neubildungen in Bodenhohlräumen. Z. Pflanzenernähr., Düng., Bodenkde. A **31**, 255–278 (1933). — Über das Elementargefüge des Bodens. Bodenkundl. Forsch. **4**, 380–412 (1935). — Beiträge zur Kenntnis des Gefüges kohärenter Bodenmassen. Bodenkde. u. Pflanzenernähr. **2** (47) 1–23 (1936/37). — Verfahren zur Herstellung von Dünnschliffen von Böden in ungestörter Lagerung. Zeiss-Nachr. 2. Folge, 81–91 (1937). — Mikropodology. 243 S. Ames, Iowa: Collegiate Press. 1938. — Die Dünnschlifftechnik in der Bodenuntersuchung. Forschungsdienst Sonderh. 16, 91–97 (1942). — Gefügeuntersuchungen an tropischen und subtropischen Rotlehmen. Beitr. Kolonialforsch. **3**, 48–58 (1943a). — Die mikroskopische Humusuntersuchung. Z. Weltforstwirtsch. **10**, 387–410 (1943b). — Entwicklungslehre des Bodens. 215 S. Wien: Springer. 1948. — Bestimmungsbuch und Systematik der Böden Europas. 392 S. Stuttgart: Enke. 1953. — Neue Wege und Ziele der allgemeinen und angewandten Bodenkunde. Gedenkschrift zur Verleihung des Justus-von-Liebig-Preises 1954, 11–25, Stiftung F.V.S. zu Hamburg. 1954a. — Sobre el Metodo de la Palaeoedafologia. An. Edafologia y Fisiologia Vegetal **13**, 523–543 (1954b). — Micromorphology of laterite formation in Rio Muni (Spanish Guinea). Trans. 5th Int. Congr. Soil Sci. Léopoldville **4**, 77–84 (1954c). — Die Bedeutung des Begriffes „Humusform" für die Bodenkunde und Humusforschung. Z. Pflanzenernähr., Düng., Bodenkde. **69** (114), 3–7 (1955a). — Über die Braunlehmrelikte des Atakor (Hoggar-Geb., Zentral-Sahara). Erdkunde **9**, 115–132 (1955b). — Zur Mikromorphologie, Systematik und Entwicklung der rezenten und fossilen Lößböden. Eiszeitalter u. Gegenwart **7**, 102–112 (1956a). — Rubefizierung und Laterisierung. 6e Congr. Int. Sci. Sol Paris **5**, 247–249 (1956b). — Neue Beiträge zur Kenntnis des planetarischen und hypsometrischen Formenwandels der Böden Afrikas. Stuttgarter Geogr. Studien **69** (Lautensach-Festschrift), 50–64 (1957). — The classification of soils. J. Soil Sci. **9**, 9–19 (1958). — Wesen, Ziele und Anwendungsgebiete der mikromorphologischen Bodenforschung. Z. Pflanzenernähr., Düng., Bodenkde. **97** (142), 193–205 (1962a)*. — Die taxonomische Bedeutung der Art und Ausbildung von Eisenoxydhydratmineralien in Tropenböden. Z. Pflanzenernähr., Düng., Bodenkde. **98**, 205–213 (1962b)*. — Kubiëna, W. L., W. Beckmann und E. Geyger: Zur Methodik der photogrammetrischen Strukturanalyse des Bodens. Z. Pflanzenernähr., Düng., Bodenkde. **92** (137), 116–126 (1961). — Zur Untersuchung der Feinstruktur von Bodenaggregaten mit Hilfe von Strukturphotogrammen. Zeiss-Mitt. **2**, 256–273 (1962). — Kühnelt, W.: Zoogene Krümelbildung in ungestörten Böden. Dtsch. Akad. d. Landwirtsch. Wiss. Berlin, Tagungsber. **13**, 193–199 (1958). — Kundler, P.: Zur Kenntnis der Rasenpodsole und Grauen Waldböden Mittelrußlands im Vergleich zu den Sols lessivés des westlichen Europas. Z. Pflanzenernähr., Düng., Bodenkde. **86** (131), 16–36 (1959). — Zur Charakteristik und Systematik der Waldböden aus Sand im nordostdeutschen Flachland. Arch. Forstwes. **10**, 417–429 (1961a). — Lessivés (Parabraunerden, Fahlerden) aus Geschiebemergel der Würm-Eiszeit im norddeutschen Tiefland. Z. Pflanzenernähr., Düng., Bodenkde. **95** (140), 97–100 (1961b). — Kuron, H., und E. Homrighausen: Die Herstellung von Bodendünnschliffen mit dem Festlegungsmittel Plexigum 7466. Mitt. Inst. f. Bodenkde. u. Bodenerhaltung d. Univ. Gießen **1** (1959). — Kurotori, T., und H. Matsumoto: Microscopical study of forest soils (I), On the method of making thin sections of soils for microscopic observation, and the application of this technique to the study of forest soils. Forest Soils of Japan, Report 9. 11 S., 49 Abb. Tokyo 1958.

Laatsch, W.: Dynamik der mitteleuropäischen Mineralböden, 3. Aufl. 277 S. Dresden und Leipzig: Steinkopff. 1954. — Laruelle, J.: Quelques aspects de la microstructure des sols du nord-est du Congo Belge. Pedologie **6**, 38–57 (1956). — Laruelle, J., und I. Denisoff: Etude microscopique des sols du parc national de la Garamba. 6e Congr. Int. Sci. Sol Paris **5**, 103–109 (1956). — Lieberoth, J.: Über den Einfluß der Ackerkultur auf die Bodenentwicklung im sächsischen Lößgebiet. Albrecht Thaer Arch. **6**, 3–30 (1962).

Mackenzie, A. F., und J. E. Dawson: The preparation and study of thin sections of wet organic soils materials. J. Soil Sci. **12**, 142–144 (1961). — Manil, G.: Diskussion über den Ausdruck „Lessivé" auf Grund mikromorphologischer Beobachtungen. Z. Pflanzenernähr., Düng., Bodenkde. **98** (143), 214–218 (1962)*. — Meyer, L., und U. Rennenkampf: Neuer Apparat und Methode zur automatischen Durchführung der Krümelanalyse nach Tjulin. Z. Pflanzenernähr., Düng., Bodenkde. A **43**, 268–280 (1936). — MINASCHINA, N. G.: Optisch orientierte Tone in Böden. Počvovedenie **1958**, H. 4, 90–96 (russ.; Engl. Übers. in: Soviet Soil Sci.). — MITCHELL, J. K.: The fabric of natural clays and its relation to engineering properties. Highway Res. Board Meeting 1956. — MITSCHERLICH, E. A.: Bodenkunde für Landwirte, Forstwirte und Gärtner, 7. Aufl. 327 S. Berlin und Hamburg: Parey. 1957. — MOHTADI, S.: Beitrag zur Isolierung von pflanzenphysiologisch wirksamen Stoffen während der Rotte von Stroh. 117 S. Diss. Gießen 1962. — MÜCKENHAUSEN, E.: Die Beurteilung des Bodengefüges bei der forstlichen Standortsaufnahme. In: Landwirtschaft — Angewandte Wissenschaft. Heft: Forstliche Standortsaufnahme, Begriffe und Fachausdrücke, S. 44–57. Hiltrup: Landwirtsch. Verlag. 1958. — Bodenkundliche Untersuchungsmethoden. In: BENTZ, A.: Lehrbuch der angewandten Geologie, Bd. 1, S. 957–1057. Stuttgart: Enke. 1961. — Entstehung, Eigenschaften und Systematik der Böden der Bundesrepublik Deutschland. 148 S., 60 Taf. Frankfurt/M.: DLG-Verlags GmbH. 1962. — Form, Entstehung und Funktion des Bodengefüges. Z. Kulturtechn. u. Flurbereinigung **4**, 102–114 (1963). — MÜCKENHAUSEN, E., W. GERKHAUSEN und W. KERPEN: Entstehung und Eigenschaften der Böden auf den fossilen Verwitterungsdecken der Eifel. Z. Acker- u. Pflanzenbau **108**, 201–222 (1959). — MÜLLER, M.: Auewaldböden des schweizerischen Mittellandes. Mitt. Schweiz. Anst. Forstl. Versuchswes. **34**, 37–86 (1958).

NIKIFOROFF, C. C.: Morphological classification of soil structure. Soil Sci. **52**, 193–211 (1941). — NITZSCH, W. v., und W. CZERATZKI: Die Bodenbeschaffenheit und Beeinflußbarkeit der Bodenkolloide und ihre Bedeutung für die Krümelbildung und Krümeleigenschaften. Bodenkde. u. Pflanzenernähr. **18**, 1–50 (1940). — NORDMEYER, H.: Mineralogische Untersuchungen an drei Bodenprofilen: auf Zechsteinletten, Unterem Buntsandstein und Zechsteindolomit. Beitr. Mineral. u. Petrogr. **7**, 1–44 (1959).

OSMOND, D. A.: Micropedology. Soils a. Fertilizers **21**, 1–6 (1958). — OSMOND, D. A., und I. STEPHEN: The micropedology of some red soils from Cyprus. J. Soil Sci. **8**, 19–26 (1957).

PARFENOVA, F. J.: Einige sekundäre Mineralbildungen in Pflanzen und Böden. Problemy sovetskogo počvovedenija, Sb. **15**, 71–79 (1949); (russ.). — PARFENOVA, F. J., und E. A. JARILOVA: Ziele und Methoden der bodenmineralogischen Untersuchung mit dem Polarisationsmikroskop. Počvovedenie **1958**, Nr. 12, 28–35 (russ.; Engl. Übers. in: Soviet Soil Sci.). — Mikromorphologie des Bodens als Methode zur Untersuchung der Geochemischen Landschaften. Z. Pflanzenernähr., Düng., Bodenkde. **97** (142), 265–273 (1962a)*. — Mineralogische Untersuchungen in der Bodenkunde. 206 S. Moskau: Izd. Akad. Nauk SSSR. 1962b (russ.). — PETERSON, J. B.: The micromorphology of some loessial soils of Iowa. Soil Sci. Soc. Amer. Proc. **2**, 9–13 (1957). — PIGULEVSKIJ, M.: Eine Methode zur Fixierung des Bodens. Aus: Izvestij Bjuro po Sel'skoxozjajstvennoj mexanike, Ausg. 2, 1–56, St. Petersburg 1914. — PROISSL, A.: Mikroskopische Untersuchungen über Bodenbildung im niederösterreichischen Waldviertel. Bodenkde. u. Pflanzenernähr. **21/22** (66/67), 112–141 (1940). — PUFFE, D., und G. GROSSE-BRAUCKMANN: Mikromorphologische Untersuchungen an Torfen. Z. Kulturtechn. u. Flurbereinigung **4**, 159–188 (1963).

RAMANN, E.: Bodenkunde, 3. Aufl. Berlin: Springer. 1911. — REDLICH, G. C.: Determination of soil structure by microscopical investigation. Soil Sci. **50**, 3–12 (1940). — REINECK, H. E.: Über das Härten und Schleifen von Lockersedimenten. Senck. leth. **39**, 49–56 (1958). — Reliefgüsse ungestörter Sandproben. Z. Pflanzenernähr., Düng., Bodenkde. **99** (144), 151–153 (1962)*. — REUTER, G.: Lessivé-Braunerde-Interferenzen auf Geschiebemergel. Z. Pflanzenernähr., Düng., Bodenkde. **98** (143), 240–246 (1962)*. — RICHARDS, L. A.: Availability of water to crops on saline soil. Agric. Inform. Bull. Nr. 210, 1–10. Washington 1959. — RID, H.: Strukturveränderungen des Ackerbodens im Bereich der Krume, der Pflugsohle und des Untergrundes, nachgewiesen durch Strukturdiagnosen und Messungen der Grobstruktur. Z. Acker- u. Pflanzenbau **114**, 58–86 (1961/62). — ROEMER, TH.: Bodenbearbeitung. In: Handbuch der Landwirtschaft **2**, 209–273. Berlin: Parey. 1929. — ROEMER, TH., und H. FRESE: Bodenbearbeitung. In: Handbuch der Landwirtschaft, 2. Aufl., **1**, 237–309. Berlin und Hamburg: Parey. 1952. — RÖMER, A.: Die Herstellung von transportablen Bodenprofilen. Landwirtsch. Forsch. **3**, 212–216 (1952). — RONDE, G.: Bodenzoologische Untersuchungen von Stickstoff-Meliorationsflächen im Bayer. Staatsforstamt Schwabach (Mittelfranken). In: „Auswertung von Düngungs- und Meliorationsversuchen in der Forstwirtschaft“, S. 49–127. Bochum: Ruhrstickstoff A.G. 1958.

SANDER, B.: Gefügekunde der Gesteine. 352 S. Wien: Springer. 1930. — Einführung in die Gefügekunde der Geologischen Körper, 2 Bde. 215 und 409 S. Wien und Innsbruck: Springer. 1948 und 1950. — SAUERLANDT, W.: Zum Aufbau eines gärtnerischen Kulturbodens mittels Strohkompost. Organischer Landbau **1960**, H. 3, 49–52. — SAUERLANDT, W., und M. MARZUSCH-TRAPPMANN: Einige Probleme der biologischen Untersuchung der Ackerböden. Z. Pflanzenernähr., Düng., Bodenkde. **97** (142), 216–224 (1962)*. — SCHEFFER, F., und O. TORNAU: Lehrbuch des Ackerbaues (ROEMER-SCHEFFER), 4. Aufl. 457 S. Berlin: Parey. 1953. — SCHEFFER, F., H. FÖLSTER und B. MEYER: Zur Entstehung von Schwarzerden und Schwarzerdeartigen Böden. Chemie d. Erde **20**, 302–330 (1960). — SCHLACHT, K.: Eine neue Methode zur Konservierung von Bodenprofilen. Z. Pflanzenernähr., Düng., Bodenkde. A **13**, 426–431 (1929). — Bodenstruktur und Sohlenbildung, mit Kartierungsbeispielen aus der Vorderpfalz. Z. Pflanzenernähr., Düng., Bodenkde. A **27**, 303–385 (1933). — SCHLICHTING, E.: Zur Deutung von „Ortstein"-Böden im subarktisch-alpinen Gebiet. Z. Pflanzenernähr., Düng., Bodenkde. **100** (145), 121–126 (1963). — SCHMIDT-LORENZ, R.: Ein Beitrag zur Kenntnis der Terra fusca im Südniedersächsischen Bergland. 107 S. Diss. Göttingen 1957. — SCHÖNHALS, E.: Kennzahlen für den Feinheitsgrad des Lösses. Eiszeitalter u. Gegenwart **6**, 133–147 (1955). — Spätglaciale äolische Ablagerungen in einigen Mittelgebirgen Hessens. Eiszeitalter u. Gegenwart **8**, 5–17 (1957). — SCHUMACHER, W.: Die Physik des Bodens in ihren theoretischen und praktischen Beziehungen zur Landwirtschaft. 505 S. Berlin: Wiegandt u. Hempel. 1864. — SCHUCHT, F.: Entstehung, Eigenschaften und Einteilung der Böden. In: Handbuch der Landwirtschaft **2**, 1–40. Berlin: Parey. 1929. — SEKERA, F.: Der allgemeine Bauplan der Bodenstruktur und die Dynamik der Bodengare. Z. Pflanzenernähr., Düng., Bodenkde. **52**, 57–60 (1951a). — SEKERA, F., und A. BRUNNER: Beiträge zur Methodik der Gareforschung. Bodenkde u. Pflanzenernähr. **29**, 169–212 (1943). — SMOLIKOVA, L., und V. LOŽEK: Zur Altersfrage der mitteleuropäischen Terra calcis. Eiszeitalter u. Gegenwart **13**, 157–177 (1962). — SPRENGER, BALTH.: Kurzer, doch gründlicher Begriff des gesamten Feldbaues. 328 S. mit Anhang. Stuttgart: Mezler & Co. 1764. — STEBUTT, A.: Lehrbuch der allgemeinen Bodenkunde. Der Boden als dynamisches System. 518 S. Berlin: Borntraeger. 1930. — STREMME, H.: Grundzüge der praktischen Bodenkunde. Berlin: Borntraeger. 1926. — Bodentypen und Bodenkarten. Handwörterbuch der Naturwissenschaften, 2. Aufl., **2**, 127–135. Jena: Fischer. 1933. — SWANSON, C. L. W., und J. B. PETERSON: Differences in the microstructure of profiles of the Marshall and the Shelby silt loam. Soil Sci. Soc. Amer. Proc. **5**, 297–303 (1940). — SZABÓ, J., M. MARTON, L. VARGA und S. SCHÖNFELD: Komplexe bodenbiologische Untersuchungen an Rendzinen. Počvovedenie **1962**, Nr. 10, 85–95 (russ.).

UNDRITZ, R.-D.: Kritische Beurteilung der verschiedenen „Strukturstufen" bei der Spatendiagnose nach Görbing. 2 Teile, 120 und 58 S. Diss. Hohenheim 1951.

VOGEL, F., und F. KOHL: Das Bodenprofil. Verfahren zu seiner Entnahme und Präparation für Lehr- und Sammlungszwecke. 12 S. München: Bayer. Geol. Landesamt. 1952. — VOIGT, E.: Ein neues Verfahren zur Konservierung von Bodenprofilen. Z. Pflanzenernähr., Düng., Bodenkde. **45**, 111–115 (1936).

WELLS, C. B.: Resin impregnation of soil samples. Nature **193**, Nr. 4817, S. 804. — WERNER, J.: Zur Kenntnis der Braunen Karbonatböden (Terra fusca) auf der Schwäbischen Alb. Arb. a. d. Geol. Paläontol. Inst. d. T. H. Stuttgart Nr. 16, 94 S. (1958). — Gips-Ausblühungen an Bodenaufschlußwänden. Z. Pflanzenernähr., Düng., Bodenkde. **97** (142), 3–9 (1962a). — Über die Herstellung fluoreszierender Bodenanschliffe. Z. Pflanzenernähr., Düng., Bodenkde. **99** (144), 144–150 (1962b)*. — WOLKEWITZ, H.: Über nichtbiologische Faktoren der Strukturbildung von Böden. Die Bröckelung von feuchten Bodenkolloiden beim Trocknen. Diss. Gießen 1953. — WOLLNY, E.: Untersuchungen über den Einfluß der Struktur des Bodens auf dessen Feuchtigkeits- und Temperaturverhältnisse. Forsch. Geb. Agrikult.-Physik **5**, 145–209 (1882).

ZACHARIAE, G.: Zur Methodik bei Geländeuntersuchungen in der Bodenzoologie. Z. Pflanzenernähr., Düng., Bodenkde. **97** (142), 224–233 (1962)*. — ZACHAROV, S. A.: Fortschritte der russischen Wissenschaft in der Morphologie der Böden. Akad. Nauk SSSR, Počvenn. issled. 2. Leningrad 1927. — ZIMMERMANN, K.: Ein Beitrag zur Methodik der Beschreibung und Ausdeutung von Bodendünnschliffen. Diss. Bonn 1962. Zit. in: MÜCKENHAUSEN, E.: Entstehung, Eigenschaften und Systematik der Böden der Bundesrepublik Deutschland. 1962.

d) Die Beeinflussung des Bodengefüges durch Bodenbearbeitung

Von

H. Frese

1. Einleitung

Jede Bearbeitung des Bodens bedeutet eine Veränderung des Bodengefüges. Diese Feststellung wird im allgemeinen unwidersprochen hingenommen, denn tatsächlich ist eine solche Veränderung des Bodengefüges in allen Fällen die unmittelbare Aufgabe jeder Bearbeitungsmaßnahme (Czeratzki 1960a, Frese

Abb. 57. Senkrechter Schnitt durch eine Bearbeitungsgrenzschicht in der Völkenroder Bodenrinne (Löß, Parzelle 7). Maßstab 1:1

1959a, Rid 1962, Schaffer 1961) (Abb. 57). Zweck und Erfolg werden im praktischen Betrieb durchweg von der Erfahrung und der persönlichen Ansicht des Landwirtes bestimmt bzw. beurteilt.

Den exakten Nachweis über das Ergebnis zu erbringen, ist jedoch selbst im wissenschaftlichen Versuch äußerst schwierig, und zwar aus folgenden Gründen:

1. Ausgangs- und Endzustand des Bodens bei einem bestimmten Arbeitsgang lassen sich bisher wegen des Mangels an geeigneten Methoden, vor allem unter schwierigen Bodenverhältnissen, kaum ausreichend genau kennzeichnen.

2. Das Bodengefüge wird dabei zunächst in gestaltlicher Hinsicht (Form und Größe der Schollen, Bruchstücke, Aggregate, Porengröße und -verteilung) verändert.

3. Auch in funktioneller Hinsicht (physikalisches Verhalten in bezug auf Luft-, Wasser- und Wärmehaushalt, Festigkeit u. a.) erfolgt eine Umformung des Bodengefüges.

4. Abhängig hiervon werden außerdem Umfang, Intensität und Ablauf zahlreicher biologischer und chemischer Prozesse beeinflußt.

5. Die rein mechanisch geschaffene Neuordnung des Bodengefüges ist aber kein statischer Zustand, sondern unterliegt einer ständigen Dynamik.

6. Die Dynamik dieser Vorgänge wird sowohl von endogenen (Bodeneigenschaften) wie exogenen Faktoren (Witterung, Bewirtschaftungseinflüsse) entscheidend mitbestimmt.

Aus all dem ergibt sich, daß wir es mit einem schwierigen und nicht leicht erfaßbaren Komplex zu tun haben, wenn wir vom Bodengefüge und seiner Beeinflussung durch Bodenbearbeitungsmaßnahmen sprechen. Dies gilt besonders im Hinblick auf seine Bedeutung für die Pflanzenernährung und Düngung. Wegen dieses weitgespannten Einflusses aller Bearbeitungsmaßnahmen auf den Boden wird deshalb hier anstelle des meist gebrauchten Wortes „Bodenstruktur" der umfassendere Begriff des „*Bodengefüges*" verwendet, wie ihn ALTEMÜLLER (1962 und Beitrag in diesem Handbuch, S. 230) unter Zugrundelegung der Arbeiten von KUBIENA (1938, 1953) und SANDER (1948/1950) neu definiert hat:

Das Bodengefüge ist zu verstehen als die Ordnung der Bodengemengteile, betrachtet als Bauelemente des Bodens in gestaltlicher, funktioneller und genetischer Hinsicht, unabhängig von den Größen- und Rangordnungen und unabhängig von der stofflichen Beschaffenheit.

2. Funktion und Bedeutung des Bodengefüges

Funktion und Bedeutung des Bodengefüges für die Pflanzenproduktion werden am besten durch einen Vergleich deutlich. Es ist bekannt, daß Pflanzen in Wasserkulturen herangezogen werden können. Bei diesem als „Hydroponik" bezeichneten Verfahren können praktisch alle wesentlichen Wachstumsfaktoren wunschgemäß gesteuert und laufend den Bedürfnissen der Pflanzen angepaßt werden. Die Nährstoffe werden gewissermaßen in vorfabrizierter und konzentrierter Form dargeboten, die Klimaverhältnisse können geregelt werden. Bodenmaterial als Standort und Halt für die Pflanze ist nicht erforderlich, wenn es auch manchmal in Form von Ziegelsplitt oder Steingrus verwendet wird.

Unter den natürlichen Produktionsbedingungen des Ackers muß dagegen das Bodenmaterial nicht nur den Pflanzen den Standort bieten, sondern auch die als Rohmaterial vorhandenen oder verabfolgten Stoffe mineralischer oder organischer Art in pflanzenaufnehmbare Form umwandeln sowie die von der Natur zur Verfügung gestellte Energie an Wärme, Licht und Wasser speichern und umsetzen. Der Boden dient also als Akkumulator und Transformator und steuert so zahllose für das Pflanzenwachstum wichtige Prozesse (SCHEFFER 1963). Sie laufen in starker Abhängigkeit von dem gegebenen Gefüge ab, das sich nur bedingt und empirisch beeinflussen läßt. Ein Teil der wichtigsten Faktoren, wie die genetischen Bodeneigenschaften und atmosphärischen Einflüsse, sind zudem der Kontrolle durch den Menschen fast völlig entzogen.

Wegen dieser sehr komplexen Funktionen sind die Ziele, das Bodengefüge durch Bearbeitung in bestimmter Richtung zu verändern, mannigfaltig („Probleme der Bodenbearbeitung" 1960, FRESE 1954, 1956a, 1959a). Der Erfolg jeder Bodenbearbeitung ist von einer sehr großen Zahl von Faktoren abhängig. Alle Maßnahmen greifen außerdem in ihrer Wirkung mehr oder weniger ineinander über. Da auch die Ausgangsbedingungen ständig wechseln, darf man deshalb ohne Übertreibung sagen, daß der Arbeitserfolg in jedem einzelnen Fall etwas Einmaliges ist und sich darum auch nicht mit genügender Sicherheit voraussagen und reproduzieren läßt (FRESE 1956b, 1958a).

3. Die Beeinflussung des Bodengefüges durch Bodenbearbeitung

Der Versuch, die Ziele der Bodenbearbeitung mit dem Zweck, das Bodengefüge zu beeinflussen, in einer übersichtlichen Gliederung darzustellen, muß also immer willkürlich bleiben. Aus der Sicht des Landwirts lassen sie sich wohl am besten wie folgt ordnen:

A. Beabsichtigte Wirkungen

Direktwirkungen einzelner Maßnahmen und Geräte. 1. *Mechanisch-technologische Eingriffe.* Das *Wenden,* meist durch den Pflug, soll Bewuchs, Erntereste, Unkraut oder organische und mineralische Dünger in den Boden einbringen. Der Wendevorgang selbst ist dabei stark abhängig vom gegebenen Bodenzustand, der Form des Pfluges und seiner Geschwindigkeit. Art und Güte dieser Wendearbeit haben nicht nur unmittelbaren Einfluß auf manche im Boden ablaufenden Prozesse (so z. B. Verrottung!), sondern bestimmen auch wesentlich Art und Zahl der für die Saatbettbereitung zusätzlich erforderlichen Arbeitsgänge.

Erst in neuerer Zeit ist es durch die Entwicklung entsprechender Verfahren gelungen, die Bodenwendung durch den Pflug zu messen und hieraus wichtige Schlußfolgerungen für die zweckmäßigste Art der Pflugarbeit im Hinblick auf den angestrebten Endzustand des Bodens in gestaltlicher und funktioneller Hinsicht zu ziehen (FEUERLEIN 1960, FRESE 1963).

Gleichzeitig wird mit dem Pflügen aber auch ein *Mischen* angestrebt, und im allgemeinen wurde dies auch bisher als selbstverständlich vorausgesetzt.

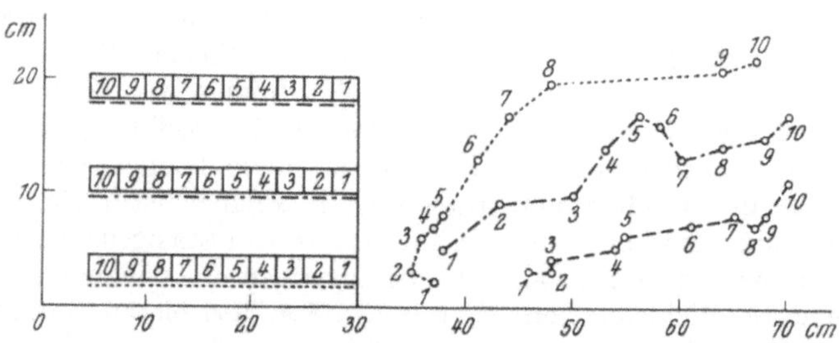

Abb. 58. Der Mischeffekt des Pfluges. Anlehmiger Sand, Schraubenform, 6 km/h. Nach dem Pflügen liegen die Meßkörperchen in der gleichen Reihenfolge wie bisher. Die in den ungepflügten Boden eingebrachten Meßkörperchen sind im gewendeten Furchenbalken nur in der Schichtfolge, nicht aber in der Reihenfolge verlagert. Eine Mischwirkung durch den Pflug ist also nicht erkennbar

Auch hier haben erst neuere Forschungen ergeben, daß die vermutete Mischwirkung des Pfluges oft geringer als angenommen oder gar nicht vorhanden ist (Abb. 58). Während dies für die Einbringung organischer Substanz schon früher mehrfach gezeigt wurde, erscheinen die neueren Ergebnisse vor allem im Hinblick auf die Einmischung mineralischer Düngestoffe interessant und wichtig (RID und SÜSS 1960). Das Einmischen aufgebrachter mineralischer oder organischer Stoffe in den Boden erfolgt sehr viel wirkungsvoller durch rotierende Geräte (Fräse, Krümler, Kreisel-Pflug und -Egge; CZERATZKI, FEUERLEIN und KLÜGEL 1963) oder durch die Nachbearbeitungsgeräte wie Eggen, Drahtwälzeggen u. a.

Eine dem Mischen entgegengesetzte Gefügebeeinflussung stellt das *Sortieren* oder Entmischen gröberer und feinerer Bodenbestandteile dar, wie es beim Einsatz bestimmter Geräte auftreten kann (Kultivatoren, Eggen, rotierende Geräte). Diese Anhäufung unterschiedlich großer Bodenbestandteile in getrennten Schichten kann den Ablauf mancher Prozesse im Boden in sehr unterschiedlicher Weise nachhaltig beeinflussen (z. B. sehr feine Oberflächenschicht durch Walzen, grobkrümlige Oberfläche als Verschlämmungsschutz, Ansammlung von Feinerde durch sogenanntes „Toteggen").

Maßnahmen zum *Aufrauhen* oder *Einebnen* der Oberfläche sind meist gleichbedeutend mit der Schaffung eines *Grobgefüges* bzw. der *Zerkleinerung* gröberer Schollen. Sie wirken sich vor allem auf Wasseraufnahmefähigkeit und -leitung, Verdunstung, Durchlüftung und Erwärmung und dadurch häufig besonders

Abb. 59. Starke Verschlämmung und anschließende Verkrustung durch zu feine Bearbeitung des Saatbettes

schnell und nachhaltig auf das Pflanzenwachstum, vor allem im Keimstadium, aus (KOHLER 1960); aber auch Verschlämmung und Verkrustung können durch Einebnung und Zerkleinerung der Krume gefördert werden (Abb. 59).

Wohl die wichtigste Aufgabe der Bodenbearbeitung ist die Beeinflussung des Bodengefüges hinsichtlich der Dichte der Lagerung, sei es durch *Lockerung* oder bewußte *Verdichtung*. Ohne eine ständig wiederholte Auflockerung würden unsere intensiv genutzten Böden unter dem Einfluß der mechanisierten Wirtschaftsweisen sehr bald so dicht gelagert sein, daß sie in vieler Hinsicht nicht mehr die notwendigen Voraussetzungen für eine ungestörte und volle Entwicklung der angebauten Pflanzen bieten könnten.

Beide Maßnahmen dienen daher im besonderen Maße dem Ziel, das Gefüge in einem möglichst günstigen Sinne zu beeinflussen, ohne daß der Landwirt —

aber auch der Wissenschaftler! — in der Lage wäre, im Einzelfall vorauszusagen oder zu beurteilen, ob die beabsichtigte und erzielte Gefügeveränderung den gewünschten Erfolg haben wird. Grund hierfür ist neben den zahlreichen Lücken in unserem Wissen über die Zusammenhänge zwischen dem Bodengefüge und dem Pflanzenwachstum, dem bisherigen Mangel an geeigneten Methoden zur Erfassung der wirksamen Faktoren und ihrer Wechselwirkungen vor allem die Unmöglichkeit, bei der Bearbeitung des Bodens die kommenden Witterungseinflüsse mit einiger Sicherheit einzukalkulieren.

2. Alle mechanisch-technologischen Eingriffe in das Bodengefüge erfolgen zunächst einmal mit dem Ziel, *physikalische Faktoren* zu steuern. Hierbei geht

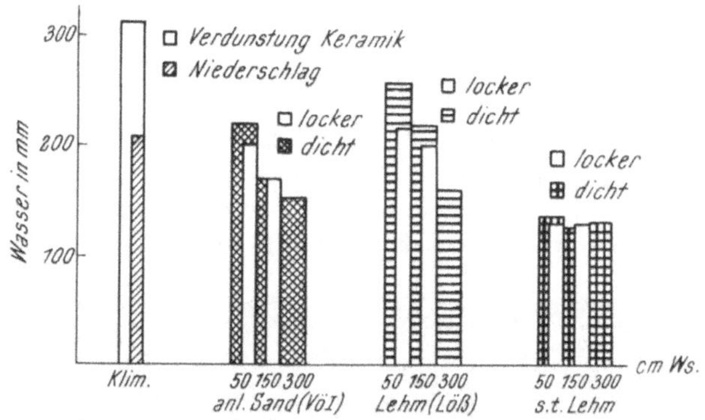

Abb. 60. Verdunstung verschiedener Bodenarten in Unterdrucklysimetern (25 cm Bodensäule) in Abhängigkeit von Saugspannung und Bodendichte. 1. Juli bis 31. Oktober 1961

es vor allem um die Einflußnahme auf den Wasser-, Luft- und Wärmehaushalt des Bodens und die integrierenden und Komplementär-Wirkungen dieser Faktoren.

Ein grobstückig-lockeres Gefüge, insbesondere in schwereren Böden, bedingt bessere Eindringmöglichkeiten für das Wasser, jedoch schlechtere Wasserhaltung und kapillare Leitung als eine dichtere und feinere Lagerung des Bodens (Abb. 60). Der geringeren Wasserkapazität des grob-lockeren Materials entspricht eine hohe Luftkapazität, eine größere Temperatur-Amplitude, niedrigere Wärmekapazität und geringere Wärmeleitfähigkeit.

Demgegenüber finden wir im feinen und dicht gelagerten Boden geminderte Versickerungsmöglichkeit für das Wasser, höhere Wasserkapazität und bessere kapillare Leitfähigkeit (beides jedoch nur bis zu einem gewissen Grade). Abhängig davon ergibt sich eine höhere Wärmekapazität und bessere Wärmeleitung, dagegen geringeres Temperaturspiel, geringere Luftkapazität und unter Umständen behinderten Gasaustausch. Innerhalb dieser hier sehr summarisch dargestellten Extreme finden sich in Abhängigkeit von Witterungsbedingungen, Bearbeitungseinflüssen, Bodenart, Gefüge, Bodentyp und anderen Gegebenheiten des Standortes (z. B. Grundwasserstand) alle Übergänge. Soweit eine bewußte Steuerung dieser Faktoren und ihrer Wechselwirkungen durch Bearbeitungsmaßnahmen angestrebt wird, beziehen sie sich z. B. auf eine Auflockerung zum besseren Abtrocknen, die Herstellung einer Hackschicht zur Verminderung von Verdunstungsverlusten bzw. zur schnelleren Erwärmung des Keimbettes oder aber auf eine gezielte Verdichtung, um die Wasserhaltung und Kapillarität zu verbessern, zu schnelle Erwärmung und Austrocknung der Oberflächen zu verhindern oder stark unterschiedliche Gefüge zweier aneinandergrenzender Schichten vor allem im Hinblick auf die Wasserführung einander anzugleichen (Czeratzki 1962, Gliemeroth 1950a).

Eine der wesentlichen Aufgaben von Bodenbearbeitungsmaßnahmen hierbei ist es, alle physikalischen Extrem-Zustände zu verhindern oder zu beseitigen, weil sie aus vielerlei Gründen das Pflanzen-, vor allem das Wurzelwachstum, sehr nachhaltig beeinträchtigen können. Sie werden deshalb abgestimmt sein müssen auf Bodenart und -typ, den erfahrungsgemäß zu erwartenden Witterungsablauf und die sich daraus ergebenden vorwiegenden Mangelfaktoren (in schweren Böden meist Luftmangel, in leichten Böden Wassermangel).

3. Im Hinblick auf im Boden ablaufende *chemische Prozesse* sind die besprochenen Gefügeeigenschaften von nicht geringerer Bedeutung. Abhängig von den angeführten physikalischen Faktoren werden die chemischen Umsetzungen sehr unterschiedlich verlaufen. Hierbei spielen Wassergehalt und Wasserführung für die Mobilisierung und Leitung der mineralischen Nährstoffe, Temperatur, Feuchtigkeit und Gasaustausch für die Umsetzung des organischen Materials eine entscheidende Rolle. Bekannt ist, daß eine starke Durchlüftung und Erwärmung den Abbau organischer Substanzen stark beschleunigen, Luftmangel und daher Zersetzung unter anaeroben Bedingungen zu Reduktionserscheinungen führen kann, die erhebliche Wachstumsstörungen der Pflanzen zur Folge haben können. In welchem Maße die wechselseitige Konstellation der physikalischen Faktoren die Umsetzung und Bereitstellung mineralischer und organischer Nährstoffe für die Pflanze tatsächlich beeinflußt, ist noch immer zu wenig bekannt. Alle Untersuchungen über die Menge der pflanzenverfügbaren Nährstoffe im Boden erfolgen ja durchweg im Laboratorium unter „unnatürlichen" Bedingungen. Der auf dem Acker wirksame Einfluß des Gefüges bleibt unberücksichtigt. Es muß aber vermutet werden, daß das wirkliche Nährstoffangebot auf dem Feld unter dem Einfluß der sich ständig ändernden Gefügebedingungen in mehr oder weniger weiten Grenzen schwankt.

4. Über viele wichtige *biologische Vorgänge* im Boden besteht eine nicht geringere Unsicherheit in unserem Wissen. Zwar ist aus exakten Laborversuchen bekannt, welch maßgeblichen Einfluß innerhalb verhältnismäßig enger Grenzen Temperatur, Durchlüftung und Wassergehalt auf die Intensität der Entwicklung und die Aktivität der Mikro- und Makro-Fauna haben (Dawson 19€3, Hubbell und Gardner 1948, Koepf 1954, McCalla 1945, Müller 1960). Wieweit sie aber durch einen Eingriff in das gestaltliche und funktionelle Bodengefüge positiv oder negativ beeinflußt werden und mit welchen Folgewirkungen für den Einfluß auf mancherlei Gefügeeigenschaften sowie die Mobilisierung der Nährstoffe, ist bisher noch wenig erforscht.

Stufenweises Erreichen des angestrebten Endzustandes. 1. *Planmäßige Folge mehrerer Bearbeitungsgänge.* Die gesamte, innerhalb einer Vegetationszeit als Vorbereitung zum Anbau einer bestimmten Frucht durchzuführende Bearbeitung des Bodens wird im allgemeinen in mehrere Arbeitsgänge aufgelöst. Dies trifft vor allem zu, wenn sich der Bearbeitungszeitraum vom Herbst bis zum Frühjahr hinzieht. Zwar wird besonders bei schneller Aufeinanderfolge der Früchte (etwa Winterung nach Sommerung) sowohl aus technologischen wie ökonomischen Gründen immer stärker versucht, durch Gerätekombinationen oder Anwendung moderner Geräte mehrere Arbeitsgänge zu vereinigen (Czeratzki und Ruhm 1961). Dies gelingt ohne Schwierigkeiten allerdings meist nur auf leichteren Böden oder dann, wenn die Bearbeitung in einem günstigen Konsistenzzustand erfolgen kann.

Je ungünstiger der Bodenzustand im Augenblick der Bearbeitung für den gedachten Zweck ist und je schwerer die Böden sind, desto mehr Arbeitsgänge sind erforderlich, um das angestrebte Saatbett zu schaffen. Dabei verläuft die Entwicklung meist von einer mehr oder weniger grobscholligen Pflugarbeit mit rauher Oberfläche und relativ hohem Lockerheitsgrad über eine Zerkleinerung der groben Schollen, verbunden mit einer Einebnung der Oberfläche sowie anschließenden Maßnahmen zur Verringerung der ursprünglich überhöhten Lockerung. Dies geschieht durch systematisches Verfestigen des Bodens bis zu einem Grad, der die Bedingungen für die Aussaat günstig erscheinen läßt. Art und Zahl der Maßnahmen und der angestrebte Endzustand sind jeweils lokal und empirisch bedingt. Dabei müssen sie oft gleichzeitig der mechanischen Unkrautbekämpfung dienen und auch auf die Erfordernisse späterer Pflegemaßnahmen in den Beständen abgestimmt sein. Die angewandten Verfahren und die eingesetzten Geräte wechseln deshalb von Fall zu Fall nicht unerheblich.

2. Wesentlich mitbestimmend dabei ist das Bestreben nach einer *Nutzung förderlicher bzw. Beseitigung schädlicher Witterungseinflüsse.* Da diese Einflüsse

Abb. 61. Bodenzustand im gepflügten Land (Hintergrund) und im ungepflügten Land (Vordergrund). Unmittelbar nach dem Auftauen vermochte der ungepflügte Boden die gefallenen Niederschläge nicht aufzunehmen

unübersehbar sind und in die Planung von Bodenbearbeitungsmaßnahmen nur bedingt einbezogen werden können, wird der Landwirt bestrebt sein, den für die Aussaat und die weitere Vegetation als günstig angesehenen Bodenzustand erst möglichst spät herzustellen. Je länger die gesamte Bearbeitungsspanne dauert, um so vorsichtiger wird er seine einzelnen Bearbeitungsmaßnahmen ansetzen müssen, weil er in der Zwischenzeit mit den Einwirkungen der Witterung auf das Bodengefüge zu rechnen hat (Abb. 61). Diese Einflüsse können förderlich sein, ihn also in seinen Bemühungen unterstützen und ihm gewisse Arbeitsgänge abnehmen. Ebenso aber können sie Rückschläge bringen und ihn zwingen, ursprünglich nicht beabsichtigte Maßnahmen einzuschalten oder andere zu wiederholen, um unerwünschte Witterungseinflüsse auf den Boden wieder zu beseitigen (etwa Verschlämmung und Verkrustung, CZERATZKI 1957).

3. Während aller Zwischenperioden zwischen einzelnen Arbeitsgängen, besonders aber dann, wenn für die gesamten Bearbeitungsmaßnahmen nur kurze

Zeit zur Verfügung steht, wird sich die *natürliche Dynamik des Bodengefüges* wiederum teils hemmend, teils fördernd auswirken. Ausmaß und Intensität dieser Dynamik sind vorwiegend von den genetisch bedingten Eigenschaften des Bodenmaterials abhängig. Von der Korngrößenzusammensetzung, der Art und Menge des Tonmaterials sowie der Dichte und Stabilität der Aggregate hängt z. B. der Umfang der *Quellung* und *Schrumpfung* ab, der die Böden je nach Wassergehalt unterworfen sind (CZERATZKI und FRESE 1956).

Unter dem Einfluß des *Regens* erfährt jeder mechanisch aufgelockerte Boden eine ständige *Setzung* und laufende *Verdichtung* (Abb. 62). Dieser Prozeß kommt theoretisch erst zum Stillstand, wenn die dichtestmögliche Lagerung der Boden-

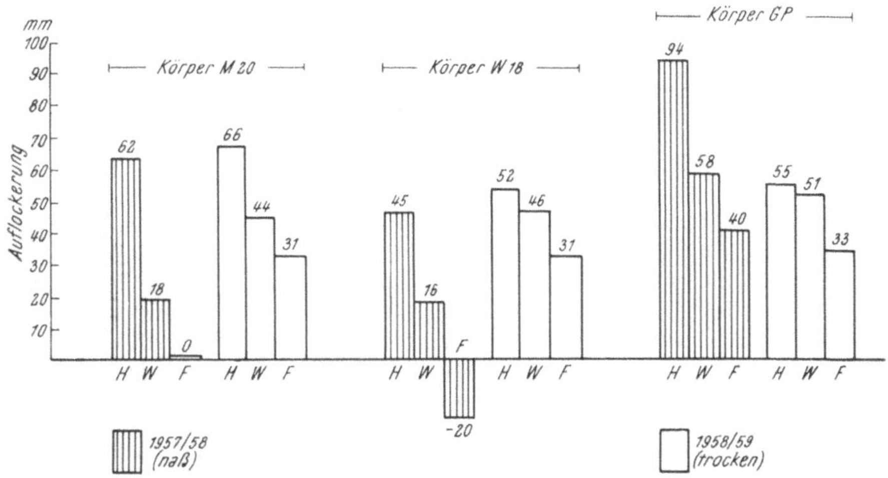

Abb. 62. Auflockerung und Setzung des gepflügten Bodens durch drei verschiedene Pflugkörper in zwei Jahren. Bei trockener Witterung unterscheiden sich Auflockerung und Setzung auf den drei Pflugparzellen nicht sehr stark. Bei feuchter Witterung dagegen ist die im Herbst (*H*) erzielte Auflockerung durch den Pflug sehr unterschiedlich, die Setzung bis zum Ende des Winters (*W*) in zwei Fällen sehr stark. Nach der Frühjahrsbestellung (*F*) hat sich der Boden in der ersten Parzelle wieder auf den Ausgangszustand gesetzt, liegt in der zweiten sogar dichter als ursprünglich und hat nur auf der dritten Parzelle einen Teil der Auflockerung bewahrt

gemengteile erreicht ist, was allerdings nur in extremen Fällen tatsächlich eintritt (hohe Niederschläge, starke Verschlämmung). Zumeist wird sich ein stark vom Kulturzustand, den Bodeneigenschaften und der Stärke der Niederschläge abhängiger Gleichgewichtszustand als mehr oder minder stabiles Stadium einstellen (KÖNNECKE 1935). Besonders einschneidende und nachhaltige dynamische Veränderungen im Ackerboden verursacht der *Frost*. Je nach Stärke, Dauer und Wechsel und abhängig von Tongehalt, augenblicklicher Lagerung und Wassergehalt des Bodens übt er erhebliche Einflüsse auf das Bodengefüge aus (CZERATZKI 1956, CZERATZKI und FRESE 1956, FRESE 1958 b). Im allgemeinen bewirkt der Frost dabei eine *Auflockerung* und (als eine der wichtigsten ackerbaulichen Wirkungen) eine Zerkleinerung der Schollen und Bröckel in der Ackerkrume. Die dabei entstehenden, oft aus sehr feinen, splitterartigen Aggregaten aufgebauten Gefüge (häufig nicht sehr glücklich als „Frostgare“ bezeichnet), sind jedoch im allgemeinen sehr labil. Wird deshalb, besonders im Frühjahr, diese auflockernde und zerkleinernde Arbeit des Frostes durch Regeneinwirkung überlagert, verwandelt sie sich auf verschlämmenden Böden in eine sehr schnelle und intensive Setzung, die zu starken Verdichtungen führen kann.

Diese laufenden Veränderungen in der Lagerung des Bodens bedingen auch eine ständige Dynamik der Gefüge-Funktionen. Aber auch die Bedeutung eines *bestimmten* Gefüges für das Pflanzenwachstum kann sich im Laufe der Vegetationszeit ändern. Ein zunächst und unter gegebenen Witterungsbedingungen ungünstiges Gefüge kann sich zu einem späteren Zeitpunkt bei anderer Witterung oder veränderten Ansprüchen der Pflanzen als förderlich erweisen und umgekehrt (Frese 1959b).

Der Umfang der voraussichtlichen dynamischen Veränderungen im Boden müßte also in die Planung der Bearbeitungsmaßnahmen so weit wie möglich einbezogen werden können, wenn einige Aussicht bestehen soll, die Gefügeeigenschaften des Bodens in gewissem Umfang in einer gewünschten Richtung zu beeinflussen. Daß dies nach dem bisher Gesagten äußerst schwierig, ja beinahe unmöglich ist, ist die Ursache dafür, daß der Erfolg aller Bodenbearbeitungsmaßnahmen stets unsicher bleibt und von Jahr zu Jahr je nach Witterungsablauf in verhältnismäßig weiten Grenzen schwanken muß.

B. Unbeabsichtigte Nebenwirkungen

Bei fast allen Maßnahmen der Bodenbearbeitung müssen neben den erwünschten auch unerwünschte Nebenwirkungen in Kauf genommen werden. So ist eine Wendung meist mit einer Lockerung und Zerkleinerung, eine Verdichtung mit der Zerstörung an sich erwünschter Schöllchen, eine Einebnung oft mit einer partiellen Verdichtung, eine Auflockerung mit der Bildung gröberer Schollen oder einer zu starken Zerkleinerung des Bodens verbunden. Diese Nebenwirkungen können also sowohl förderlich wie unerwünscht sein. Soweit sie nützlich sind oder bei den Folgemaßnahmen ohne größeren Aufwand wieder beseitigt werden können, brauchen sie uns hier nicht zu beschäftigen. Von Bedeutung dagegen sind

Gefügestörungen durch die Bearbeitungsgeräte. Hierzu gehören: 1. *Schollenbildungen*, wie sie meist in Verbindung mit Druck- oder Schmierstellen bei der Bearbeitung zu feuchter Böden entstehen. Ihr Auftreten in größerem Umfang kann die weitere Bearbeitung erheblich erschweren oder das Abwarten längerer Witterungseinwirkungen (Mürbung) erforderlich machen. Starke Schollenbildungen können sich so ungünstig auswirken, daß unter Umständen das Pflanzenwachstum völlig gehemmt wird („Der Pflug in internationaler Sicht" 1959). Eine besonders unerwünschte Folge solcher Schollenbildung ist, daß ein Saatbett erst durch eine größere Zahl ziemlich scharf angreifender Bearbeitungsgänge hergerichtet werden kann (Abb. 63), wobei (s. unten) weitere nachteilige Folgeerscheinungen in Kauf genommen werden müssen.

2. *Scharf abgegrenzte Gefügezonen.* Bei einer Bodenbearbeitung in ungünstigem Feuchtigkeitszustand treten oft scharf abgegrenzte Unterschiede im Gefüge der bearbeiteten und der nicht bearbeiteten Zone auf. Derartige Erscheinungen können in Verbindung mit ungünstigen Witterungsbedingungen die Entwicklung der heranwachsenden Pflanze ebenfalls stark beeinträchtigen. Die Wurzeln vermögen die Grenze zweier derartig unterschiedlich strukturierter Schichten oft nur schwer oder gar nicht zu überwinden. Ferner können an dieser Stelle Wasserstauungen und damit Luftmangel auftreten, die zu Wurzeldeformationen führen, oder aber die Wasserführung wird an der oberen Grenze der dichter lagernden Schicht gestört. Sofern derartige Grenzen innerhalb der üblicherweise bearbeiteten Ackerkrume liegen, lassen sie sich spätestens mit der nächsten Pflugfurche wieder beseitigen. Schwerwiegender und nachhaltiger wirken solche Zonen,

wenn sie an der Grenze der Krume zum unbearbeiteten Untergrund liegen (ALTEMÜLLER 1959). Dort führen sie zu Störungen, die als

3. *Pflugsohlen* bezeichnet werden. Über Ursachen und Entstehung dieser Störzonen liegen noch immer zu wenig klare Erkenntnisse vor. Nach v. NITZSCH (1936) sind sie dadurch charakterisiert, daß sie dichter lagern als die darüberliegende Krume und der darunterliegende ungestörte Untergrund. Je nach Bodenentwicklung und Profilaufbau können aber durch Bearbeitungsmaßnahmen auch Verdichtungen hervorgerufen werden, die durch diese Definition nicht erfaßt werden. Neue Untersuchungen lassen vermuten, daß echte Pflugsohlen tatsächlich durch das Schar,

Abb. 63. Senkrechter Schnitt (Dünnschliff) quer zur Arbeitsrichtung des Pfluges durch zwei übereinanderliegende Bearbeitungsgrenzen mit Eindrücken der Schleifsohlen (Pfeile). Unter den Schleifsohlen-Spuren (Bildmitte und links oben) deutlich verdichtete Zonen (Bodenrinne Völkenrode: Löß, Parzelle 4)

vor allem aber die Schleifsohle des Pfluges verursacht werden, von der häufig erhebliche Kräfte abgestützt werden müssen. Die zunächst noch sehr dünnen Verdichtungszonen werden dann zum Ausgangspunkt zunehmend mächtigerer und stärker verdichteter Zonen (FRESE und ALTEMÜLLER 1962) (Abb. 64). Ähnliche Erscheinungen ließen sich auch unter bestimmten Bedingungen bei der Fräsarbeit feststellen (FEUERLEIN 1957). In jedem Fall bedeuten derartig verdichtete Schichten einen Engpaß für Wasserführung und Durchlüftung und können dadurch, oder auch rein mechanisch, das Eindringen der Wurzeln in die tieferen Schichten behindern.

Einwirkung durch die Zugmittel. Außerdem aber muß bei der Entstehung dieser Störzonen der Einwirkung durch die Zugmittel eine ebenso große Bedeutung zugemessen werden. Die in der Furche gehenden Zugtiere oder die Furchenräder der Schlepper üben auf die freiliegende Bearbeitungsgrenze erhebliche Druckwirkungen aus (Abb. 65). Hierüber liegen fundierte Untersuchungen vor, die diesen Einfluß der Zugmittel auf die Entstehung von Gefügestörungen ziemlich weitgehend klarstellen (Domsch 1955, Feuerlein 1961, Frese 1955, Frese und Feuerlein 1956, Gliemeroth 1948, 1953; Rid 1958, Söhne 1951, 1952, 1953). Dagegen sind die bis jetzt erarbeiteten Ergebnisse über deren

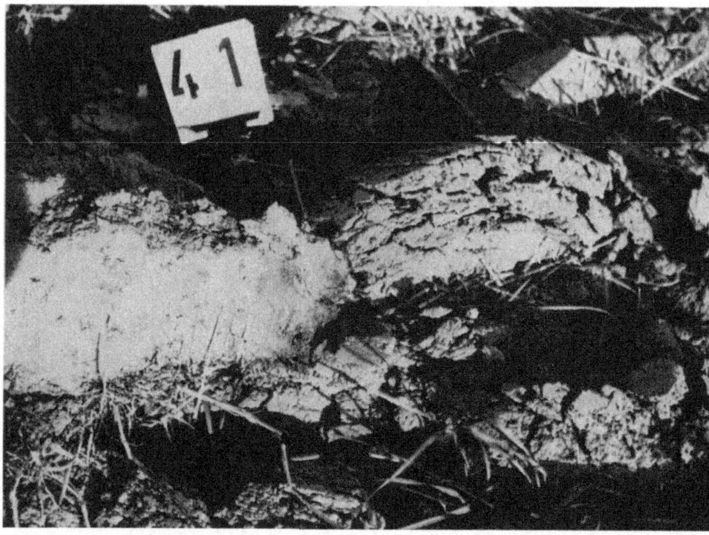

Abb. 64. Die Scholle des Scharpfluges (rechts) ist durch die Scherwirkung des Pflugkörpers aufgesplittert und gemürbt. Die Scholle des Scheibenpfluges (links) zeigt eine glatte, verhärtete und nicht aufgemürbte Schnittfläche

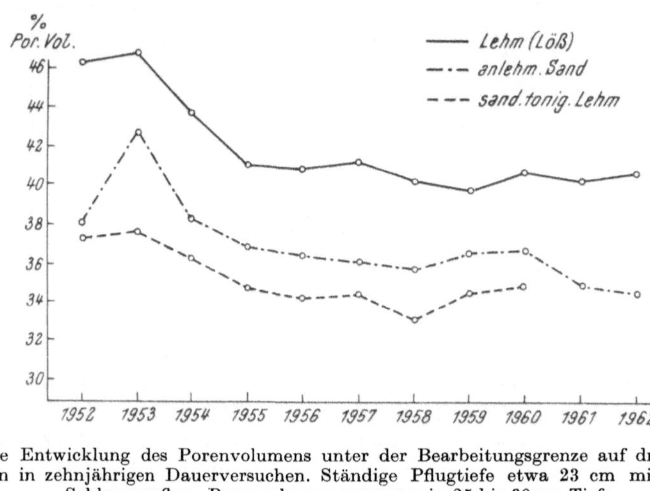

Abb. 65. Die Entwicklung des Porenvolumens unter der Bearbeitungsgrenze auf drei unterschiedlichen Böden in zehnjährigen Dauerversuchen. Ständige Pflugtiefe etwa 23 cm mit einscharigem Schlepperpflug; Porenvolumen gemessen in 25 bis 30 cm Tiefe

Auswirkungen auf das Pflanzenwachstum noch bemerkenswert widersprechend. Während teilweise durch Auflockerung dieser verdichteten Zonen merkliche Mehrerträge erzielt wurden, zeigten sich diese in anderen Versuchen (so in eigenen langjährigen Dauerversuchen) nur für einzelne Früchte und in verhältnismäßig engen Grenzen (Czeratzki 1958, 1963; Frese 1963, Gliemeroth 1950 b, c; Rauhe 1962, Rid 1952, 1960).

Offensichtlicher und deshalb stärker beachtet sind die Einwirkungen der Zugmittel und Fahrzeuge auf Bodengefüge und Pflanzenwachstum bei allen Arbeiten *auf* dem Acker nach dem Pflügen. Infolge der partiell auftretenden Gefügeveränderungen zeichnen sich die Fahrspuren häufig stark ab und lassen dabei sowohl Wachstumshemmungen wie Wachstumsförderungen erkennen. Eine Begünstigung des Pflanzenwachstums tritt dabei dann ein, wenn eine ausreichende Setzung des Bodens nicht schon durch entsprechend langes Ablagern oder Bearbeitungsmaßnahmen erreicht ist, sondern erst durch den Druck der Schlepperräder erzielt wird. Aber auch durch mangelhaftes Wachstum sich abzeichnende Spuren sprechen nicht in allen Fällen für eine nachteilige Druckwirkung. Bei näherer Untersuchung sind sie oft als Folge unzweckmäßiger, nicht richtig auf die verdichtende Wirkung der Schlepperräder abgestimmter Folgemaßnahmen zu erkennen.

Auch die unbeabsichtigten Nebenwirkungen der Bodenbearbeitung auf das Bodengefüge stellen also einen nicht unwichtigen Faktor im modernen, technisierten Ackerbau dar und werfen mancherlei schwierige ackerbautechnische und bodenkundliche Fragen auf, die noch weiterer Untersuchung bedürfen.

Wie dieser knappe Abriß erkennen läßt, stellt die Erforschung des Einflusses der Bodenbearbeitung auf das Bodengefüge und der sich daraus ergebenden Auswirkungen auf das Pflanzenwachstum (die ja immer das eigentliche Ziel jeder Bearbeitungsmaßnahme sind) ein noch sehr junges Wissensgebiet dar (FRESE 1959 c, 1960, 1962). Es erschien deshalb sinnvoll, ja notwendig, besonders die Problematik und die Vielfalt der bestehenden, aber noch unzureichend erforschten Wechselbeziehungen auf diesem Gebiet in den Vordergrund zu stellen. Das Herausgreifen einzelner Ergebnisse, die jeweils unter ganz bestimmten Bedingungen gewonnen wurden und auch nur dort Gültigkeit haben, könnte leicht zu dem Fehlschluß führen, daß mit ihnen allgemeingültige und übertragbare Erkenntnisse gewonnen seien. Wenn also mehr offene Fragen aufgeworfen werden mußten, als beantwortet werden können, sollte dies Anreiz dazu sein, diese für Wissenschaft und Praxis gleichermaßen bedeutungsvollen Fragen aufgreifen und klären zu helfen.

Literatur

ALTEMÜLLER, H.-J.: Verlassene Krumenschichten — eine Frage der Pflugtiefe. Mitt. DLG **74** (14), 403–406 (1959). — Gedanken zum Aufbau des Bodens und seiner begrifflichen Erfassung. Z. Kulturtechn. **3** (6), 323–336 (1962). — Die morphologische Untersuchung des Bodengefüges. Dieses Handbuch, S. 230–263.

CZERATZKI, W.: Zur Wirkung des Frostes auf die Struktur des Bodens. Z. Pflanzenernähr., Düng., Bodenkde. **72** (1), 15–32 (1956). — Bodenverschlämmung — Bodenschutz. Landb.-Forsch. **7** (4), 83–87 (1957). — Ergebnisse über Bodenbearbeitung mit Zweischichtpflügen. Proc. Int. Symp. Soil Structure, Gent. 1958, 137–145. — Die Erhaltung des günstigsten Bodenzustandes durch zweckmäßige Bodenbearbeitung. Arch. DLG **25**, 33–49 (1960 a). — Möglichkeiten zur Messung der Wasserbewegung im Boden mit Unterdrucklysimetern. Ber. Landesanst. Bodennutzungsschutz **1962** (3), 23–33. — Einige Ergebnisse über den Einfluß von Untergrundlockerung, Herbstdüngung und Saatbettbereitung auf den Rübenertrag. Zucker **16** (14), (1963). — CZERATZKI, W., und H. FRESE: Kinematographische Untersuchungen zur Strukturbildung, I und II. 6. Int. Bodenkundl. Kongr., Paris, 1956, Rapp. Vol. B, 173–178. — CZERATZKI, W., und E. RUHM: Saatbettversuche mit Zuckerrüben. Zucker **14** (5), (1961). — CZERATZKI, W., W. FEUERLEIN und H. KLÜGEL: Geräte zur Stroheinbringung. Landb.-Forsch. **13** (1), 1–10 (1963).

DAWSON, R. C.: Soil micro-organisms and agricultural production. In: Food for Peace. ASA Special Publication Nr. 1, 77–84. Madison/Wisc. 1963. — DOMSCH, M.: Probleme der Bodenbearbeitung, Berlin: Deutscher Bauernverlag. 1955.

FEUERLEIN, W.: Die Fräse im landwirtschaftlichen Einsatz. In: 14. Konstrukteurh. 1957, 88–98 (Grundl. Landtechn. H. 9). — Die Pflugarbeit und ihre Beurteilung. In:

18. Konstrukteurh. 1960, 44–50 (Grundl. Landtechn. H. 12). — Zum Einfluß des Schlepperrads auf den Acker. Landb.-Forsch. **11** (3), 69–72 (1961). — Minimale und optimale Aufwendungen für den Boden. Broschüre „Wintertagung 1962". Verb. landw. Gutsbetr. in Österreich. — FRESE, H.: Bodenfruchtbarkeit und Bodenbearbeitung. In: Der Bodenfruchtbarkeitsdienst im Spiegel von Wissenschaft und Wirtschaftsberatung. Z. Landwirtschaft—Angewandte Wissenschaft, Nr. 26, 46–63 (1954). — Bodendruckfragen beim Einsatz luftbereifter Fahrzeuge auf dem Acker. Landb.-Forsch. **5** (2), 31–34 (1955). — Bodenbearbeitung — Brücke zwischen Boden und Pflanze. Förderungsdienst **4**, (9) 257–262; (10), 297–300; (11), 332–336 (1956 a). — Aussichten für eine exakte Beurteilung des Arbeitserfolges von Bodenbearbeitungsgeräten. Grundl. Landtechn. **1956** b (7), 5–10. — Grundsätzliche Fragen der Bodenbearbeitung. Grundl. Landtechn. **1958** a (10), 48–54. — Zur Bildung von Makro-Gefüge-Typen im Ackerboden durch atmosphärische Einflüsse. In: Probleme der Krümelstabilitätsmessung und der Krümelbildung. Dtsch. Akad. Landwirtsch. Wiss. Berlin, Tag. Ber. Nr. 13, 117–127 (1958 b). — Die Beeinflussung der Bodenstruktur durch Bearbeitungsmaßnahmen. Arb. DLG **54**, 68–85 (1959 a). — Neue Gesichtspunkte für die Beurteilung und Bewirtschaftung unserer Ackerböden. Landwirtsch. Forsch. Sonderh. 1959 b (12), 49–59. — Bodenbearbeitung — uraltes Tun, junges Forschungsgebiet. Landb.-Forsch. **1959** c (3/4), 61–64. — Haben wir ein Konzept für eine Wissenschaft von der Bodenbearbeitung? Trans. VIIth Int. Congr. Soil Sci. **1**, 54–66 (1960). — Zur Situation der Forschung auf dem Gebiet der Bodenbearbeitung. Kühn-Arch. **76**, 19–25 (1962). — Methodische Grundlagen der Bodenbearbeitungsforschung. Grundförbättring (Uppsala) **16** (3), (1963). — FRESE, H., und W. FEUERLEIN: Bodendruckschäden — unvermeidlich? Mitt. DLG **71** (13), 301–303 (1956). — FRESE, H., und H.-J. ALTEMÜLLER: Über einige morphologische Beobachtungen an Pflugsohlen. Grundl. Landtechn. **1962** (15), 10–14.

GLIEMEROTH, G.: Selbstverschuldete Strukturstörungen des Bodens unter besonderer Berücksichtigung des Schlepperraddruckes. Ber. Landtechn. **1948** (2), 19–54. — Beziehungen zwischen Struktur und Wasserhaltung auf Lehm- und Lößboden und ihre Bewertung bei der Bearbeitung. Z. Acker- u. Pflanzenbau **93** (3), 319–338 (1950 a). — Die Aufbruchsarbeit des Lockerungsschars des Zweischichtenpflugs. Kuratorium f. Techn. i. d. Landwirtsch. Ber. ü. Landtechn. **1950** b, 1158–1160. — Vergleich von Tiefpflügen und Zweischichtenbearbeitung zu Zuckerrüben auf Lößboden. Zucker **1950** c (6), 109. — Untersuchungen über die Verfestigungs- und Verlagerungsvorgänge im Ackerboden unter Rad- und Raupenfahrzeugen. Z. Acker- u. Pflanzenbau **96** (2), 219–234 (1953).

HUBBELL, D. S., und J. L. GARDNER: Effects of aeration, compaction and waterlogging on soil structure and microflora. J. Amer. Soc. Agron. **1948** (40), 932–940.

KOEPF, H.: Experimenteller Beitrag zur Bodenbeurteilung mittels biochemischer Reaktionen. Z. Pflanzenernähr., Düng., Bodenkde. **67** (112), (3), 262–277 (1954). — KOHLER, M.: Untersuchungen über die Saatbettherstellung zu Winterweizen. Diss. ETH Zürich 1960. — KÖNNECKE, G.: Untersuchungen über das sogenannte Sichsetzen des Bodens. Kühn-Arch. **39**, 296–322 (1935). — KUBIËNA, W. L.: Micropedology. Ames/Iowa: Collegiate Press. 1938. — Bestimmungsbuch und Systematik der Böden Europas. Stuttgart: Enke. 1953.

MCCALLA, T. M.: Influence of microorganisms and some organic substances on soil structure. Soil Sci. **59**, 287–297 (1945). — MÜLLER, G.: Über die bodenbiologische Dynamik tiefbearbeiteter und tiefgedüngter leichter Sandböden. Dtsch. Akad. Landwirtsch. Wiss. Berlin, Tag. Ber. Nr. 28, 167–184 (1960).

NITZSCH, H. v.: Bessere Bodenbearbeitung. RKTL-Schrift 1936, H. 70.

RAUHE, K.: Die Wirkung verschiedener im Frühjahr und Herbst durchgeführter Bodenbearbeitungsmethoden auf lehmigem Sand. Thaer-Arch. **6** (3), 221–238 (1962). — RID, H.: Ergebnisse mehrjähriger Arbeiten auf dem Gebiete der Bodenphysik und Bodenbearbeitung in Bayern. Z. Pflanzenbau **1952**, Sonderh. 3. — Bodenbearbeitung und Bodenpflege. München: Bayer. Landwirtschaftsverlag. 1958. — Versuche mit Untergrundlockerung. Dtsch. Akad. Landwirtsch. Wiss. Berlin, Tag. Ber. Nr. 28, 253–261 (1960). — Einfluß der Bodenbearbeitung auf die Bodenstruktur. Landwirtsch. Forsch. **15** (2), 105–117 (1962). — RID, H., und A. SÜSS: Zur Methodik der Prüfung des Effekts von Bodenbearbeitungsgeräten. Landtechn. Forsch. **10** (3), 62–70 (1960).

SANDER, B.: Einführung in die Gefügekunde der geologischen Körper, 2 Bde. Wien und Innsbruck: Springer. 1948/1950. — SCHAFFER, G.: Veränderungen der Bodenstruktur als Folge ackerbaulicher Maßnahmen. Arbeiten der Landwirtschaftlichen Hochschule Hohenheim, Bd. 6. Stuttgart: Ulmer. 1961. — SCHEFFER, F.: Das Transformationsvermögen der Böden als Grundlage ihrer Bewertung. Landwirtsch.

Forsch. **1963**, Sonderh. 17. — SÖHNE, W.: Das mechanische Verhalten des Ackerbodens bei Belastungen, unter rollenden Rädern sowie bei der Bodenbearbeitung. In: 9. Konstrukteurh. 1951, 87–94 (Grundl. Landtechn. H. 1). — Die Verformbarkeit des Ackerbodens. In: 10. Konstrukteurh. 1952, 51–59 (Grundl. Landtechn. H. 3). — Druckverteilung im Boden und Bodenverformung unter Schlepperreifen. In: 11. Konstrukteurh. 1953, 49–63 (Grundl. Landtechn. H. 5).

Verschiedene Verfasser: Der Pflug in internationaler Sicht. Erfahrungen und Forschungsergebnisse aus Dänemark, Deutschland, England, Niederlande, Norwegen, Schweden. Mitt. DLG **74** (13), (1959). — Probleme der Bodenbearbeitung. Dtsch. Akad. Landwirtsch. Wiss. Berlin, Tag. Ber. Nr. 28 (1960).

e) The Influence of Climate and Weather upon Soil Structure

By

P. K. Peerlkamp

1. Introduction

The genesis of soil structure is very complex. Soil structure, defined as the spatial arrangement of elementary soil particles, is the result of the co-operation of several factors, the activities of which vary with time generally. Among these factors weather is an important one. It acts in different ways. On one hand the mean weather conditions (climate) have a long-term influence upon soil structure, on the other the present weather and its fluctuations affect soil structure more in the short run. In the latter effects weather can act either more directly by means of rainfall (surface clogging and crusting, water erosion, alternate wetting and drying), temperature (alternate freezing and thawing) and wind (wind erosion) or indirectly via the soil moisture conditions under which cultivations are carried out and via plant growth and microbial activities in the soil.

2. Long-term Effects of Climate

BAVER (1934) studied the relation between aggregation and climate on a large number of aggregate analyses of various soils. He found aggregation related to clay content and organic matter percentage of the main soil groups, which in turn are correlated with the climatic factors rainfall and temperature (JENNY 1930, JENNY and LEONARD 1934).

When temperature is kept constant, aggregation, measured as well by the total percentage of aggregates as by the percentage of silt and clay joined into water-stable aggregates, appears to be small in the desert soils, but increasing with increasing rainfall it reaches a maximum at the rainfall in the regions of the chernozems, further showing a decrease until it again reaches low values at the high rainfall of the climate in which podsols are formed. BAVER (1956) explained the course of the total percentage of aggregates with increasing rainfall from the small clay content, caused by the slight weathering under arid conditions, the high clay and organic matter percentages of the chernozems and dark humid-prairie soils and the low percentage of clay in the top soil of podsols, where, in consequence of the great rainfall, clay is eluviated to a lower horizon. The small percentage of silt and clay, which is aggregated in desert soils, is, according to BAVER, due to the low organic matter content and the presence of sodium. In the region of high rainfall this percentage is small too, since the top soil has low organic matter, alumina and iron content, whereas at medium rainfall it reaches a maximum in consequence of a relatively high percentage of organic matter and the presence of divalent bases.

When rainfall was kept constant BAVER found an increasing percentage of clay and silt aggregated with increasing temperature for soils from humid regions, whereas in semi-arid regions the opposite was true. He explained the latter from the decreased organic matter content, the cause in the humid regions, however, from an increased influence of iron and alumina going from podsols to laterites.

It will be clear that, owing to the great number of factors affecting soil structure, the correlations between climate and aggregation are not highly significant, but give only tendencies. These tendencies, however, show the broad effects on soil structure of climate in its capacity as a soil developing factor.

3. Surface Clogging and Crusting

The destructive influence of rainfall on the structure of the surface-layer of the soil was recognised a long time ago. The well-known pioneer in soil-physics WOLLNY (1877) was the first who studied the effect of vegetation in protecting soil structure against deterioration by rain. He found a decrease of 34 to 53% in non-capillary pore-volume if the soil was bare and explained this from a slaking of the aggregates in the soil surface, loading the infiltrating rain-water with fine soil material, which clogs the large pores. In more recent times WOLLNY's conclusions were confirmed generally by the work of LOWDERMILK (1930), HENDRICKSON (1934), DULEY (1939) and others who were studying the problem from the point of view of water erosion. From DULEY's experiments with artificial rain it appears that the compact surface layer, which may be several mm thick, not only has a greater volume-weight, but often contains coarser particles and less organic material than the underlying soil. It seems that this compact structure is formed by fitting some of the finer particles in the pores around the larger ones and that clay particles and organic material suspended in the run-off are transported to lower parts of the soil surface. This accumulation of clay and organic material in depressions and the appearance of a more sandy surface on the ridges is a common picture of bare sandy, sandy loam and sandy clay soils after a heavy rain. It was shown by DULEY that the influence on the infiltration-rate of the shallow compact surface-layer was much greater than that of soil type, slope, moisture percentage or profile. Therefore avoidance of the formation of a dense surface-layer, for instance by means of a crop or a mulch, is very important in soil and water conservation.

Recently DULEY's results regarding the clogging of the top layer were confirmed in a direct way by labelling the clay particles with Rb^{86} (SOR 1961, BERTRAND and SOR 1962) or P^{32} (KAZÓ and GRUBER 1962). The migration of fine particles and the clogging of the topsoil by them mainly occurred in the upper 3 cm. As a result of this displacement by the run-off a decrease of aggregate stability, clay, and organic matter percentage in the upper 2 cm of the topsoil was found.

The mechanism of slaking of the aggregates in the soil surface by rain is rather complex. There are at least four different causes: the compression of entrapped air, dispersion of the soil colloids, irregular swelling of the clay throughout the aggregates and the mechanical energy of raindrop impact. At present the first or last is considered to be most important, depending on the moisture content.

The slaking of aggregates which are immersed in water is a well-known phenomenon, utilized in wet-sieving techniques (TIULIN 1928; YODER 1936) to determine aggregate stability. YODER observed that slaking was only complete if the soil clods were almost air-dry and that it did not occur if they were previously saturated carefully by capillarity or if they were immersed in a non-polar fluid.

He concluded from this that the slaking effect was caused by water entering the clod from the entire surface by capillary tension, entrapping and compressing soil air until its pressure was exceeding the cohesion of the soil and the clod was broken down with a series of miniature explosions. HÉNIN (1938) has given a theoretical treatment of aggregate destruction by entrapped air, based on the cohesion and the wettability of the soil. The latter is evaluated by ROBINSON and PAGE (1950) who were determining the wetting angle of water on clay surfaces and found that organic matter, which is absorbed by the clay-fraction gives a considerable decrease of the wettability. Hence the capillary forces, which compress the entrapped air are smaller and the stability of the aggregates appears to be greater. This increase in stability, however, seems to be caused not only by a decrease in capillary suction but also by an increase in cohesion since a smaller wettability probably means a more difficult hydration of the cementing bonds between the soil particles too. GIUDICI (1954) recognised the importance of the rate with which water enters an aggregate. The greater this rate is, the smaller are the possibilities for the entrapped air to expand and the sooner exceeds its pressure that of the cohesion of the soil. Then a kind of micro-explosion occurs, the soil particles in a shallow outer layer of the aggregate are loosened and the process is repeated within a second layer. GIUDICI has derived formulas for the thickness of these layers and for the slaking velocity. The latter appeared to be about 32 mm per hour for a clay soil, which was slaking step by step in layers of about 1.25 mm.

Deterioration of soil structure under influence of an excess of water by a dispersion of the colloids, which cement the greater soil particles, is a common experimental fact. HÉNIN (1938, 1955) examined it. Dispersion under natural conditions occurs especially in soils in which clay colloids are the most important cementing agents. For a certain soil the rate of dispersion depends on the initial moisture percentage and on the period during which the soil has been immersed (PURI and KEEN 1925), since the rehydration of the clay, so that every particle is again surrounded by a waterfilm, is a rather slow process.

YODER (1936), HÉNIN (1938), ROBINSON and PAGE (1950) and others considered the swelling of the clay colloids as a cause of the destruction of the soil aggregates. If aggregates in the soil surface are wetted by rain this wetting does not occur homogeniously throughout the entire aggregate generally, but takes place very locally at the beginning. This causes unequal swelling of the clay colloids and tensions within the aggregate, which result in a fragmentation along cleavage boundaries formed during a foregoing dry period. ROBINSON and PAGE demonstrated this slaking effect by experiments with aggregates of different clay minerals, which were moistened in a vacuum before testing. Kaolinite-aggregates appeared to be water stable, montmorillonite-aggregates, however, slaked slowly. Illite-aggregates showed some cracks. Organic matter, absorbed by the clay, decreases its swelling properties as was shown by GIESEKING and others (1939, 1941). Therefore this seems to be one way by which organic matter increases aggregate stability.

A very important mechanism of soil structure deterioration by rainfall is the raindrop impact, which produces the so-called splash-erosion (ELLISON 1944). The kinetic energy of a falling raindrop is $^1/_2\, mv^2$ (m = mass, depending on drop size; v = velocity). The velocity is nearly constant in stagnant air, but somewhat varying under natural conditions, due to wind and turbulence.

The first raindrop measurements were made by German scientists around 1900 (see BENNETT and others 1951). More recently LAWS and his associates (1941, 1943) have studied drop-sizes and fall-velocities of raindrops. They found that there is a maximum diameter of about 7 mm. Larger drops break up into smaller ones. The fall-velocity depends on drop-size and wind, but for drops of about 6 mm it equals about 9 m/sec. (GUNN and KINCER 1949). This means that a shower of 10 mm strikes the soil surface with a total kinetic energy of about 4.10^{13} erg/ha, that is to say an energy necessary to raise a cultivated soil layer with a surface of a hectare and a depth of 15 cm up to a height of 20 cm. The energy, however, with which a single drop supplies the soil surface per unit area of impact is more important for the deteriorating effect on soil structure than the total energy per hectare of the entire shower, since the kinetic energy of one drop is passed on the small area of contact with the soil in a very short time. This energy per unit area of impact can be estimated easily by assuming the raindrop spherical and the area of impact equals the maximal cross-section of the drop. For drops of 6 mm diameter and a fall-velocity (see above) of 9 m/sec this energy appears to be about 16.10^4 erg/cm^2. Comparing this with the mean energy a 10 mm-shower applies to unit soil area (4.10^{13} erg/ha or 40.10^4 erg/cm^2) we see the difference to be rather small. Consequently the energy which is concentrated by a raindrop impact on a small part of the soil surface is relatively high and therefore an important source of structure deterioration. A falling raindrop, however, has not a constant spherical shape but an oscillating spheroidal one, a well-known effect in fluid physics (see e.g. GRIMSEHL 1929). Hence EKERN (1951) found in his experiments with artificial rain on sand of which the surface had a certain slope that there were alternating zones with greater and smaller sand transport, coinciding respectively with areas of impact of high thin drops and of broad flattened ones.

The reaction of a bare soil surface to raindrop impact has been studied by ELLISON *et al.* (1944, 1945, 1950), EKERN (1951), ROSE (1960) and BISAL (1960). ELLISON (1944) was the first who measured raindrop splash. He found that raindrops with a diameter of about 5 mm and a fall velocity of $5^1/_2$ m/sec produced splashes up to a distance of 152 cm. Aggregates and particles of 2 mm diameter were thrown as far as 40 cm. The splashing parts of the raindrop were found to be loaded with clay, silt and sand after impact with the surface of a silt loam. The sand clogged the surface as the turbid splash water infiltrated into the soil during the first minutes, taking the clay and silt with it. After 2 or 3 minutes the surface appeared to be sealed and a maximal splash occurs due to a water-film which has been formed on the soil surface. If there is a certain slope this water moves downhill and its eroding effect is increased greatly by turbulences caused by raindrops falling in the shallow water layer (STALLINGS 1957). ELLISON (1944) found the percentage of smaller aggregates (<0.105 mm) in the splash and in the run-off as well as in the $1^1/_2$ cm top layer after rainfall to be greater than in the original top soil. Apparently the bigger aggregates had been broken up by the splash process. ELLISON and SLATER (1945) studied the infiltration rate in relation to the quantity of splashed soil. They found the infiltration to be nearly inversely proportional to the logarithm of the quantity of soil transported per time unit by splash. This quantity (S) in turn is dependent on the impact velocity of the raindrops (V), the drop diameter (d) and the rainfall intensity (I). Up to now, however, it doesn't seem to be sure what on one hand the relation is between S and V, d and I on the other. According to ELLISON (1944) S should be proportional to $V^{4.33}$, $d^{1.07}$, $I^{0.65}$. ROSE (1960), however, for a constant d, states a dependence of S on $V \cdot I$, more than on $V^2 \cdot I$, and according to BISAL (1960) S should be proportional to $V^{1.4} \cdot d$. Nevertheless can be concluded that at the moment of impact the fall velocity of raindrops is an important factor determining soil splash and surface clogging, which is confirmed by the experiments of EKERN (1951). SCHUFFELEN and VAN SCHUYLENBORGH (1950) tested the mechanical influence of raindrop impact in a more direct way by comparing the effects of a raindrop with those of a little metallic ball. They found in both cases curves of the same shape for the relation between aggregate stability and fall height and concluded from this that the mechanical effect of raindrop impact is the most important one. It seems, however, that the parts which raindrop impact and air-compression play in structure deterioration depends on the moisture percentage of the soil (KOEPF 1958). At moisture percentages below about 20% the influence of compressed air decreases with increasing moisture content. Above 25% moisture the kinetic energy of the raindrop is the principal factor.

If a soil which has been exposed to rainfall is desiccated, the sealed surface forms a crust. This is a common experience, even for a sandy soil poor in clay, silt and organic matter, where such a crust may give some protection against wind erosion. The hardest rainfall induced crusts are found on sandy loams, especially if the clay:silt:sand ratio is so that the smaller particles can be fitted in the pores between the greater ones (DULEY 1939). CARNES (1934) introduced a modulus of rupture (R), which is related with rainfall by $R=a\,e^{bx}$, wherein x=the precipitation and a and b are constants. In relation to the great importance of fall velocity of raindrops in splash effects as mentioned above it seems probable, however, that not only the total quantity of precipitation determines the hardness of the soil crust, but that drop velocity should be an important factor too. Surface crusts hinder the water and gas exchange between soil and atmosphere and may prevent the emergence of seedlings (LUTZ 1952, RICHARDS 1953). The rate of clogging and crusting of a bare soil surface depends on rainfall and soil properties. The influence of the latter on marsh silt soils in the northern Netherlands was recently studied by PELGRUM (1963). Clogging and crusting was only small if the percentage of clay ($<2\,\mu$) was more than 20 but became severe if this percentage decreased below 15. To prevent the soil surface from this structure deterioration apparently pH—KCl should be somewhat above 7, the quantity of Ca^{++}-ions in the soil solution higher than 2 maeq/100 g dry soil, and the organic matter percentage 2 or more. The length of the period during which the soil surface was exposed to the influence of weather also appeared to be an important factor. The later in autumn the soil was ploughed the less clogging and crusting could be observed in the next spring.

4. Water Erosion

Water erosion is the transport of soil downwards on the surface of sloping land under influence of rainfall or melting snow. It occurs generally when the intensity of precipitation is greater than the infiltration rate of the soil and if the surface is bare or only partly covered by a crop (f.i. a row-crop). The problem of water erosion is a very broad one and has drawn much attention from agricultural engineers and soil scientists during the last 25 years. References to the following manuals (e.g. STALLINGS 1957) and summarizing papers (e.g. KURON 1954, BLAKELY *et al.* 1957) can be made for a general survey of their work. We have here to restrict ourselves to the influence of weather upon soil structure in the erosion process. Erosion requires a loosening and a transport of soil particles, that means a structure deterioration. The principal mechanisms for soil loosening in water erosion are raindrop splash (splash erosion) and a scouring action of flowing water (scour erosion). The first is discussed in the foregoing section. Experiments (EKERN 1951) have shown that the percentage of the splashed soil which is moving downhill equals 50+S, if S means the slope expressed as a percentage. Raindrop splash decreases the infiltration rate of the soil surface very rapidly, inducing in this way a surface flow of water downhill (run-off), which gives scour erosion. This scouring effect is increased greatly by soil particles suspended in the run-off due to the turbulent action of the waterfilm on the soil surface when raindrops fall on to it. The erosiveness is maximal if the loosening and transporting power of the surface stream are balanced, that is to say when this stream contains abrasive material just enough to loosen as much soil as the stream can carry together with this abrasive material (STALLINGS 1957).

Erodibility of soils for water decreases according to the following sequence: loess loam, sandy loam, clayey loam and clay soils (HEMPEL 1954). Clay soils are eroded only by very high rainfall intensities. When the eroded soil in the run-off, which may be aggregated partly (WEAKLY 1962), settles in depressions the soils thus formed generally have poor structures (KURON 1958).

It will be clear that structure deterioration by precipitation, directly by raindrop impact or more indirectly by a surface flow of rain or melting water, is a fundamental process in water erosion. Therefore maintenance or improvement of soil structure and especially of structure stability (QUENTIN and COMBEAU 1962) is one of the principal measures in soil and water conservation, beside surface covering and provisions for an orderly disposal of run-off water. According to RAM *et al.* (1960) the effectiveness of a cover crop in preventing splash erosion depends on the spatial volume of the crop (=mean height × density).

5. Alternate Wetting and Drying

It is a common experience that a clod of a clay soil or a dried mass of a clayey soil after thoroughly puddling, shows a fragmentation into small pieces if exposed to a few cycles of alternate wetting and drying. Several research workers used this phenomenon in making artificial aggregates for stability measurements (f.i. PETERSON 1943, McHENRY 1945).

The mechanisms by which disruption of clods and aggregates can occur when they are wetted are discussed in section 3. We will rule out of court here the raindrop impact, being a typical surface phenomenon. Considering the topsoil, unequal swelling and compression of entrapped air are the most important causes for a certain fragmentation of the larger structure elements if rewetted at a rather low moisture percentage. KOEPF (1958) has shown that aggregate stability depends on the moisture content and has a maximum at 20—25% of weight moisture for the three different clay soils he was investigating. Apparently the air content at this moisture percentage is so low that disruption of the aggregates by air compression becomes negligible, while on the other hand moisture content is not high enough for mechanical forces to produce a plastic deformation easily. Consequently soil structure has the lowest stability if rewetted in a dry or in a wet state. In the last case however fragmentation into smaller aggregates does not generally occur, but the moisture state of the soil will come into the neighbourhood of the upper plastic limit and soil will collapse into a non-aggregated state due to gravity and other mechanical forces (BOEKEL and PEERLKAMP 1956).

If a very wet clay soil is desiccating, it will shrink. This shrinkage equals the loss of water as long as the soil remains saturated (HAINES 1923). When air enters the soil, however, shrinkage decreases and air content increases strongly. HAINES has shown that this air remains partly occluded in the soil during a subsequent rewetting and causes an increase of the total soil volume.

It will be clear that these changes in volume give rise to unequal strains throughout the soil mass, since wetting as well as drying begins at the exterior, thus bringing about an unequal moisture distribution, which is advanced by the entrapped air. These unequal strains and stresses due to alternate wetting and drying, together with the disruptive action of air compressed in the pores on wetting, produce the fragmentation which has been generally perceived.

KOEPF (1960) studied the shrinkage patterns of different puddled soils when drying. He found a positive correlation between shrinkage and clay and organic matter percentages and a negative one between shrinkage and volume of solid material in undisturbed samples. In general the top soil was showing a stronger shrinkage as compared with a subsoil of the same clay content. These two lastly mentioned effects in connection with the more often occurring variations in moisture percentage of the top soil will be important for the development of the top layer's structure.

6. Alternate Freezing and Thawing

Practical farmers are all familiar with the idea that freezing and thawing cycles during winter may be very helpful in obtaining a good structure of clay soils in the next spring. Therefore it will be clear that the effect of alternate freezing and thawing has drawn the attention of soil scientists during the last century. SCHUMACHER (1864) and WOLLNY (1897) and later JUNG (1931), gave important contributions to our knowledge about this problem. The older work has been summarized in a paper by CZERATZKI (1956).

When a wet soil freezes soil water crystallises first in the greater pores due to the positive correlation between moisture tension and freezing-point depression. The difference in vapour tension above the solid and the liquid state of soil water causes a growth of the ice crystals at the cost of the moisture content of the soil between these crystals, which is removed, and further, what is most important generally, by moisture transport from lower lying zones or from the ground water as is shown by CZERATZKI (1956). Moisture supply from above (atmosphere, precipitation, melting snow) is sometimes possible too. The growing ice mass, which can appear either as layers or lenses of ice or be distributed more homogeneously throughout the soil, depending on the circumstances, exerts pressure on the adjacent soil. Thus these parts of the soil become denser and more stable (VAN SCHUYLENBORGH 1947). Apparently desiccation also contributes to this stabilization, owing to certain physicochemical effects. In this way, freezing of a moist soil brings about a frost structure, which can be homogeneous or heterogeneous (stratified), depending on texture, moisture percentage, water-supply and freezing velocity. Fine textured moist sandy soils (grain size <50—$70\ \mu$, TABER 1930) with a wet subsoil give strongly stratified frost structures when frozen slowly. If capillary water transport is low, as in coarse textures or in very dense soils, if the subsoil is rather dry or if the freezing temperature is low, ice lenses become very thin and small and under suitable conditions there arises a homogeneous distribution of the ice mass throughout the soil. The large numbers of small crystals which are formed then, cause a breaking up of the aggregates and a dispersion of the soil.

In a clay soil with a heterogeneous frost structure and a level surface the ice lenses are nearly horizontal. Together with vertical cracks due to shrinkage in the desiccated zone below frost depth, cracks in which, by continuing frost, ice is formed too, the ice lenses divide the soil into polyhedric clods. In ploughed land with a rough surface frost can penetrate into the soil clods from different directions. This generally gives a more irregular pattern of ice layers, dividing big clods into small ones, which are more dense and stable. This means: freezing improves macro-structure, but not micro-structure (VAN SCHUYLENBORGH 1947). If this granulating action of frost is to be of any importance for soil structure in spring however, it is necessary that the obtained frost structure withstands thawing. Since freezing can increase moisture content considerably (POST and DREIBELBIS 1942 report moisture percentages up to 213%), it is possible that the soil becomes very wet during thawing. Depending on the rate of thawing, water discharge, evaporation and rainfall during the thawing period and also depending on soil characteristics; the soil will remain relatively dry and preserve its frost induced granulation or it will be soaked with water and structure will collapse. The latter occurs f.i. in silty soils and in sodium clay soils. Several authors (WOLLNY 1897, GÜNTHER 1931, GARDNER 1945, GRIM 1954) emphasise the importance of calcium in stabilizing frost structures. Organic matter is considered to be an important stabilizing agent too (BAVER 1956). SCHUFFELEN and VAN SCHUYLENBORGH (1950) have studied the influence of the size and moisture content of aggregates of clay soils on the increase of their stability due to one or more freezing-thawing cycles. They found that this stability-increase increased with increasing moisture percentage for 2—4 mm aggregates, but was constant or decreasing for sizes of 1—2 mm. This can be explained by the co-operation of the stabilizing effect owing to mechanical compression of the aggregates by growing ice crystals (JUNG 1931) and the lowering of stability caused by the drying and wetting that accompanies freezing and thawing. According to JUNG (1931) freezing has an aggregating effect, which is greater for slow than for quick freezing and is maximal for a slow frozen soil with a moisture percentage of about 50 per cent saturation.

The effect of freezing-thawing cycles on the size of soil aggregates depends on size and moisture content of the aggregates before the treatment and on the number of cycles. LOGSDAIL and WEBBER (1959) observed an increase of the fragmentation of aggregates by 3 freezing-thawing cycles with an increasing moisture percentage. If aggregates with a diameter of about 2.5 mm and different moisture percentages are subjected to freezing-thawing cycles the mean weight diameter is decreasing and this effect was found the stronger the greater the number of cycles was. Very small aggregates (<0.25 mm) are showing an increase of the mean weight diameter, increasing with the number of cycles (SILLANPÄÄ 1961, SILLANPÄÄ and WEBBER 1961). Generally, the dispersion of soil rapidly increases with an increasing number of cycles, especially within about the first 10 cycles.

Summarizing, it can be said that the effect of freezing and thawing on soil structure is rather complex due to the great number of factors and their interactions influencing it. Most important is the breaking up of big clods of clay soils into rather small aggregates of an increased stability. It must be kept in mind however that generally these aggregates are not crumbs, but a kind of dense micro-clods, which under very wet conditions can easily be compacted again into bigger units, especially if the lime and organic matter content of the soil is low.

7. Wind Erosion

Wind erosion is the displacement of soil particles on the surface by wind. Fundamental research on soil drifting has been made by BAGNOLD (1941), MALINA (1941), CHEPIL (f.i. 1945/1946) and ZINGG (f.i. 1950).

Wind erosion begins according to MALINA (1941) with an initiation of soil movement, followed by a transport of the soil either in the air or along the surface, and ends with the deposition of the material in a new location. In each of these three steps the structure of the surface soil can be disturbed.

Wind erosion starts generally with the movement of a few loose grains due to wind pressure or to a slight mechanical disturbance of the soil surface by foot steps or a landing bird. These grains often roll on the surface, thus obtaining a rotational momentum, which helps to give the particle an uplift due to the pressure difference between the upper and lower side of a body rotating in an air stream (*Magnus*-effect). In consequence of the great wind velocity gradient near the soil surface the particle obtains an increased kinetic energy, since it is exposed to greater wind velocities at some distance from the surface. The grain will move downwards after a few moments, due to gravity. When it collides with the soil surface its energy is distributed among two or more grains, that begin to leap in turn. In this way a cumulative bouncing movement arises, which is introduced by FREE (1911) as "saltation" to distinguish it from the movement whereby soil particles are carried by wind at greater distances more or less parallel to the soil surface ("suspension movement"). A third type of movement is the rolling or sliding of grains along the surface, named "surface creep" (BAGNOLD 1941). According to CHEPIL (1941, 1945, 1946) type of movement and erodibility depend on the size of the particles (loose grains and aggregates) in the soil surface. The grain fraction 100 — 500 μ or the aggregate fraction 180 — 1200 μ moves in saltation, smaller particles move in suspension. Surface creep occurs with the grain fraction 500-1000 μ and the aggregate fraction 1200 — 2500 μ. Greater particles are not blown by wind generally. The grain fraction 100-150 μ shows maximal erodibility. The greater part of a wind blown soil usually moves in saltation (about 55-75%).

It will be clear from these results that a well structured soil with a high percentage of aggregates greater than 2.5 mm in its surface will have a considerable resistance to wind erosion. A high rate of stability of these aggregates is a self-evident condition for a more durable resistance. Therefore clayey soils are not eroded by wind generally if the clay content is higher than 15% (v. d. SPEK 1950). On the other hand wind erosion on most sandy soils is introduced by a breaking up of the aggregates due to rainfall (see section 3). When in this way a sufficient number of erodible particles is obtained, soil blowing will start, if wind velocity is high enough (generally more than about 7-8 m/sec on the reference height of 6 m). As well by the impacts of the particles moving in saltation as by the scouring action of the surface creep, destruction of soil surface structure can occur, thus increasing soil drifting. Often this destruction is irreversible as the greater part of the soil organic matter and the finest grains are blown far away in suspension. Sometimes a deposition of the particles moving in saltation and surface creep occurs on adjacent fields, which were not eroded before, due to a plant cover or a good structure. Then it is often seen that the vegetation dies or soil structure is disturbed by abrasion and the soil of these fields starts to drift too. Summarizing it can be said that, under suitable conditions, wind can set in motion particles of the soil surface, which then disturb the soil structure by impact and abrasion. Wind erosion can be combatted by measures which increase the size and stability of the soil aggregates or decrease the wind velocity near the soil surface.

8. Indirect Influences of Weather upon Soil Structure

In the sections 3—7 the more direct effects of precipitation, temperature and wind are discussed. Meteorological conditions, however, influence soil structure in an indirect way too, especially via soil cultivations and plant growth.

The influence of ploughing and other cultivations upon the structure of several soils is affected strongly by their moisture content, which in turn is determined to a great extent by weather during the foregoing period. It is e. g. a common experience that a very wet autumn often gives poor structures, especially at present, as it is possible now to drive on the land and to plough it under wet soil conditions, due to motorization. Results of LOW (1955) suggest that cultivations under dry conditions of the soil give a stable tilth.

Plant growth as well as the microbial population of the soil are influenced by soil moisture content and soil temperature. Both are determined by a complex of soil and meteorological factors among which radiation, precipitation, air temperature and wind velocity play an important part. Plant growth brings organic matter into the soil periodically by means of the roots and of above-ground parts which are buried in the soil. This organic material decays by microbial activity, thus giving a temporary improvement of soil structure (PEERLKAMP 1950), which depends as well on the total mass of roots as on microbial activity. It will be clear that in this way meteorological conditions can influence soil structure indirectly.

9. Seasonal Variability of Soil Structure

The different meteorological factors are not constant and therefore their direct and indirect influences vary strongly. Several authors (ALDERFER 1946, 1950, ROWLES 1948, HÉNIN and TURC 1949, KÄMPF 1952, LOW 1955, KULLMANN, 1958, and others) have shown the great variability of soil structure and especially of aggregation, within rather short periods of e. g. a few weeks. HEINONEN and PUKKALA (1954) report great and frequent variations of total pore volume too. Due to the complexity of the entire problem of structure variability in which not only meteorological factors play a part but also measures of soil cultivation, crop rotation, fertilizing, manuring and other agircultural treatments, it has been not possible till now to explain the observed effects fully and to derive rules with a general applicability. Although some of the published variabilities of soil aggregation may be due to the analysing method (KULLMANN and KOITZSCH 1956) it seems to be sure that meteorological conditions and their variability have important effects, which are often much greater than the influences of soil treatments.

Literature

ALDERFER, R. B.: Seasonal variability in the aggregation of Hagerstown silt loam. Soil Sci. **62**, 151–168 (1946). — Influence of seasonal and cultural conditions on aggregation of Hagerstown soil. Soil Sci. **69**, 193–203 (1950).

BAGNOLD, R. A.: The physics of blown sand and desert dunes, 1st ed. London: Methuen. 1941 (reprinted 1954). — BAVER, L. D.: A classification of soil structure and its relation to the main soil groups. Amer. Soil Survey Ass. Bull. **15**, 107–109 (1934). — Soil physics, 3rd ed., pp. 123–198. New York: Wiley. 1956. — BENNETT, H. H., F. G. BELL and B. D. ROBINSON: Raindrops and erosion. U.S.D.A. Circ. no. 895 (1951). — BERTRAND, A. R., and K. SOR: The effects of rainfall intensity on soil structure and irrigation of colloidal materials in soils. Soil Sci. Soc. Amer.

Proc. **26**, 297–300 (1962). — BISAL, F.: The effect of raindrop size and impact velocity on sand-splash. Canad. J. Soil Sci. **40**, 242–245 (1960). — BLAKELY, B. D., J. J. COYLE and J. G. STEELE: Soil. U.S.D.A. Yearbook of Agric. **1957**, 290–307. — BOEKEL, P., and P. K. PEERLKAMP: Soil consistency as a factor determining the soil structure of clay soils. Neth. J. Agric. Sci. **4**, 122–125 (1956).

CARNES, A.: Soil crusts. Methods of study, their strength and a method of overcoming their injury to cotton stand. Agric. Engng. **15**, 167–169, 171 (1934). — CHEPIL, W. S., and R. A. MILNE: Wind erosion of soil in relation to roughness of surface. Soil Sci. **52**, 417–433 (1941). — CHEPIL, W. S.: Dynamics of wind erosion I/VI. Soil Sci. **60**, 305–320, 397–411, 475–480 (1945), **61**, 167–177, 257–263, 331–340 (1946). — CZERATZKI, W.: Zur Wirkung des Frostes auf die Struktur des Bodens. Z. Pflanzenernähr., Düng., Bodenkde. **72**, 15–32 (1956).

DULEY, F. L.: Surface factors affecting the rate of intake of water by soils. Soil Sci. Soc. Amer. Proc. **4**, 60–64 (1939).

EKERN, P. C.: Raindrop impact as the force initiating soil erosion. Soil Sci. Soc. Amer. Proc. **15**, 7–10 (1951). — ELLISON, W. D.: Studies of raindrop erosion. Agric. Engng. **25**, 131–136, 181–182 (1944). — ELLISON W. D., and C. S. SLATER: Factors that affect surface sealing and infiltration of exposed soil surfaces. Agric. Engng. **26**, 156–157, 162 (1945). — Soil erosion by rainstorms. Nav. Docks P-42, 5 p.p. Dept. of the Navy, Washington, D. C., 1950.

FREE, E. E.: The movement of soil material by the wind. U.S.D.A. Bur. Soils Bull. no. 68 (1911).

GARDNER, R.: Some effects of freezing and thawing on the aggregation and permeability of dispersed soils. Soil Sci. **60**, 437–443 (1945). — GIESEKING, J. E.: The mechanism of cation exchange in the montmorillonite-beidellite-nontronite type of clay minerals. Soil Sci. **47**, 1–13 (1939). — ENSMINGER, L. E., and J. E. GIESEKING: The absorption of proteins by montmorillonitic clays and its effect on base exchange capacity. Soil Sci. **51**, 125–132 (1941). — GIUDICI, P.: Évaluation des facteurs déterminant la désintégration hydromoléculaire des agrégats terreux. Trans. 5th Int. Congr. Soil Sci. **1945**, Part II, 219–228. — GRIM, R. E.: Relation of frost action to the clay-mineral composition of soil materials. Highway Res. Board Spec. Rep. **2**, 167–172 (1954). — GRIMSEHL: Lehrbuch der Physik, 7. Aufl., Bd. I, S. 288. Leipzig: Teubner. 1929. — GUNN, R., and G. D. KINCER: The terminal velocity of fall for water droplets in stagnant air. J. Meteor. **6**, 243–248 (1949). — GÜNTHER, E.: Untersuchungen über die Einwirkung des Frostes auf verschiedene Ackerböden und den dadurch bedingten Einfluß auf das Wachstum von Hafer und Gerste. Landw. Jahrb. **73**, 893–922 (1931).

HAINES, W. B.: The volume-changes associated with variations of water content in soil. J. Agric. Sci. **13**, 296–310 (1923). — HEINONEN, R., and R. PUKKALA: Seasonal variation in soil structure. Maataloust. Aikak. **26**, 116–119 (1954). — HEMPEL, L.: Beobachtungen über die Empfindlichkeit von Ackerböden gegenüber der Bodenerosion. Z. Pflanzenernähr., Düng., Bodenkde. **64**, 42–54 (1954). — HENDRICKSON, B. H.: The choking of pore space in the soil and its relation to runoff and erosion. Trans. Amer. Geophys. Union **2**, 500–505 (1934). — HÉNIN, S.: Étude physico-chimique de la stabilité structurale des terres. Monographie des Stat. et Lab. de Rech. Agron. Paris 1938. — HÉNIN, S., et L. TURC: Nouvelles observations sur la variation annuelle de la stabilité structurale du sol. C. R. Acad. Agric. France **36.14**, 587–599 (1950). — HÉNIN, S., O. ROBICHET et A. JONGERIUS: Principes pour l'évaluation de la stabilité de la structure du sol. Ann. Agron. **A 6**, 537–557 (1955).

JENNY, H.: A study on the influence of climate upon the nitrogen and organic matter content of soil. Miss. Agric. Exper. Stat. Res. Bull. no. 152 (1930). — JENNY, H., and C. D. LEONARD: Functional relationships between soil properties and rainfall. Soil Sci. **38**, 363–381 (1934). — JUNG, E.: Untersuchungen über die Einwirkung des Frostes auf den Erdboden. Z. Pflanzenernähr., Düng., Bodenkde. **A 19**, 326–333 (1931).

KÄMPF, R.: Untersuchungen über die Strukturwandlungen der Ackerkrume bei Lößlehmböden. Z. Acker- u. Pflanzenbau **95**, 303–330 (1952). — KAZÓ, B., and L. GRUBER: The investigation of microsolifluction with the aid of tagged isotopes. Int. Ass. Sci. Hydrol. Publ. **59**, 62–66 (1962). — KOEPF, H.: Über den Einfluß der Vorbefeuchtung auf die Stabilität von Bodenaggregaten. Proc. Int. Symp. Soil Structure Ghent 1958, 283–289 (Med. Landbouwhogeschool, Gent, **24**, nr. 1, 1959). — Zur Strukturbildung verschiedener Böden durch Schrumpfung. Z. Pflanzenernähr., Düng., Bodenkde. **89**, 159–170 (1960). — KULLMANN, A., and R. KOITZSCH: Zur Krümelstabilitätsmessung auf leichtem Boden. Z. Pflanzenernähr., Düng., Bodenkde. **73**,

224–235 (1956). — KULLMANN, A.: Zur Problematik der Krümelstabilitätsmessungen und zur Methodik des Durchflußverfahrens. Tagungsber. Deutsch. Akad. Landw. Wiss. Berlin **13**, 7–34 (1958). — KURON, H.: Ergebnisse von 15-jährigen Untersuchungen über Bodenerosion durch Wasser in Deutschland. Veröff. Int. Ass. Hydrol. **36**, 8 S. (1954). — Strukturschäden durch Bodenerosion. Proc. Int. Symp. Soil Structure Ghent 1958, 24–29 (Med. Landbouwhogeschool, Gent **24**, nr. 1, 1959).

LAWS, J. O.: Measurements of the fall-velocity of waterdrops and raindrops. Trans. Amer. Geophys. Union **3**, 709–721 (1941). — LAWS, J. O., and D. A. PARSONS: The relation of raindrop-size to intensity. Trans. Amer. Geophys. Union **2**, 452–459 (1943). — LOGSDAIL, D. E., and L. R. WEBBER: Effect of frost action on structure of Haldimand clay. Canad. J. Soil Sci. **39**, 103–106 (1959). — LOW, A. J.: Improvements in the structural state of soils under leys. J. Soil Sci. **6**, 179–199 (1955). — LOWDERMILK, W. C.: Influence of forest litter on runoff, percolation and erosion. J. Forestry **28**, 474–491 (1930). — LUTZ, J. F.: Mechanical impedance and plant growth. Agronomy, Vol. II, pp. 43–71. New York: Academic Press. 1952.

MALINA, F. J.: Recent developments in the dynamics of wind erosion. Trans. Amer. Geophys. Union **3**, 262–284 (1941). — MCHENRY, J. R.: Mechanics in the formation of water-stable soil aggregates. Iowa St. Coll. J. Sci. **20**, 25–27 (1945).

PEERLKAMP, P. K.: The influence on soil structure of the "natural organic manuring" by roots and stubbles of crops. Trans. 4th Int. Congr. Soil Sci. **1950**, Vol. I, 50–54. — PELGRUM, A.: De gevoeligheid voor verslemping van lichte klei- en zavelgronden. Landbouwvoorlichting **20**, 637–645 (1963). — PETERSON, J. B.: Formation of water-stable structure in puddled soils. Soil Sci. **55**, 289–300 (1943). — POST, F. H., and F. R. DREIBELBIS: Some influences of frost penetration and microclimate on the water relationships of woodland, pasture and cultivated soils. Soil Sci. Soc. Amer. Proc. **7**, 95–105 (1942). — PURI, A. N., and B. A. KEEN: The dispersion of soil in water under various conditions. J. Agric. Sci. **15**, 147–161 (1925).

QUANTIN, P., and A. COMBEAU: L'érosion et la stabilité structurale du sol. Int. Ass. Sci. Hydrol. Publ. **59**, 124–130 (1962).

RAM, D. N., M. T. VITTUM and P. J. ZWERMAN: An evaluation of certain winter cover crops for the control of splash erosion. Agron. J. **52**, 479–482 (1960). — RICHARDS, L. A.: Modulus of rupture as an index of crusting of soil. Soil Sci. Soc. Amer. Proc. **17**, 321–323 (1953). — ROBINSON, D. O., and J. B. PAGE: Soil aggregate stability. Soil Sci. Soc. Amer. Proc. **15**, 25–29 (1950). — ROSE, C. W.: Soil detachment caused by rainfall. Soil Sci. **89**, 28–35 (1960). — ROWLES, W.: Physical properties related to structure of Ste. Rosalie clay and their seasonal variation. Sci. Agric. **28**, 558–573 (1948).

SCHUFFELEN, A. C., and J. VAN SCHUYLENBORGH: Het gebruik van de methode Vilenski ter bestudering van de invloed van de vorst en van de micro-organismen op de structuur van de grond. Landb. Tijdschr. **62**, 339–347 (1950). — SCHUMACHER, W.: Die Physik des Bodens in ihren theoretischen und praktischen Beziehungen zur Landwirtschaft, S. 275. Berlin 1864. — SCHUYLENBORGH, J. VAN: A study on soil structure. Thesis Wageningen. 1947. — SILLANPÄÄ, M.: The dynamic nature of soil aggregation as affected by cycles of freezing and thawing. Acta Agron. Scand. **11**, 87–94 (1961). — SILLANPÄÄ, M., and L. R. WEBBER: The effect of freezing-thawing and wetting-drying cycles on soil aggregation. Canad. J. Soil Sci. **41**, 182–187 (1961). — SOR, K. M.: Effects of rainfall energy on some physical properties of soils. Diss. Abstr. **22**, 4–5 (1961). — SPEK, JAC. VAN DER: Stuivende mariene gronden (Drifting marine soils). Versl. Landb. Ond. **56.11**, 28 pp. (1950). — STALLINGS, J. H.: Soil conservation. Englewood Cliffs, N.Y., 1957.

TABER, S.: Freezing and thawing of soils as factors in the destruction of road pavements. Publ. Roads **11**, 113–132 (1930). — TIULIN, A. F.: Questions on soil structure. II. Aggregate analysis as a method for determining soil structure. Perm. Agric. Exper. Sta. (Div. Agric. Chem.) Rep. **2**, 77–122 (1928).

WEAKLY, H. E.: Aggregation of soil carried in run-off from simulated rainfall. Soil Sci. Soc. Amer. Proc. **26**, 511–512 (1962). — WOLLNY, R. E.: Der Einfluß der Pflanzendecke und Beschattung auf die physikalischen Eigenschaften und die Fruchtbarkeit des Bodens. Berlin 1877. — Untersuchungen über den Einfluß des Frostes auf die physikalischen Eigenschaften des Bodens. Forsch. a. d. Geb. d. Agrikulturphysik **20**, 439–469 (1897/98).

YODER, R. E.: A direct method of aggregate analysis and a study of the physical nature of erosion losses. J. Amer. Soc. Agron. **28**, 337–351 (1936).

ZINGG, A. W., and W. S. CHEPIL: Aerodynamics of wind erosion. Agric. Engng. **31**, 279–282, 284 (1950).

f) Effects of Crop and Rotation on Soil Structure

By

P. Boekel and P. K. Peerlkamp

1. Introduction

The interrelation between crop and soil structure is very complex. Soil structure affects plant growth, but conversely, the crop and the sequence of crops (rotation) have again an influence on soil structure. The effects in both directions depend on many factors, among which those related to soil and climate are very important. Most of these factors vary with time more or less rapidly.

The influences that crop and rotation exert upon soil structure are to be divided roughly into direct and indirect ones. The ways in which a crop affects soil structure directly are: a) the plant cover protects the soil surface from raindrop splash and wind erosion and b) the penetrating and branching roots affect soil structure mechanically. Indirect influences of crop growth and rotation upon soil structure are: c) their effects on moisture content and temperature of the soil, which in turn affect soil structure in several ways, d) the influence which soil organic matter, due to roots and plant residues, has upon soil structure, and e) the effects of tillage operations which are necessary in agriculture.

2. Protection against Raindrop Splash and Wind Erosion

The phenomenon of soil structure deterioration by means of the kinetic energy of falling raindrops, briefly named "raindrop splash", has been discussed in detail in the preceding chapter. A plant cover can absorb this energy more or less, depending on the density of the vegetation. This has been shown already by WOLLNY (1877), who found that, the denser the plant cover, the smaller was the decrease of soil porosity during the vegetation period. From his results it appears that especially the bigger pores are preserved by a vegetative canopy (Table 7)

Table 7. *Effect of a plant cover on the porosity of a humiferous, calcareous sandy soil* (after WOLLNY 1877)

crop	total porosity (% of vol.)		non capillary porosity (in % of total porosity)	
	cover	no cover	cover	no cover
rye	62.2	58.3	54	38
peas	63.6	59.2	51	35
vetch	69.2	65.5	56	44

and that only 45 to 88% of the total rainfall reaches the soil surface directly, depending on the type of crop and on the number of plants per unit area. Important factors determined by the type of the crop are the total surface of the leaves per plant and the shape and position of the leaves. According to KÖRNICKE and WERNER (1885) lucerne has a very large surface of leaves per unit area of soil surface (Table 8), corn and beets have small leave-surfaces and clover and cereals have an intermediate position. With the exception of corn this sequence agrees roughly with that which WOLLNY, HAYNES and SMITH *et al.* (cf. BAVER 1956) obtained for the interception of rainfall by various crops. It will be obvious that the figures given in Table 8 depend not only on the type of crop but on the number of plants per unit area too. Differences in this number seem to be the cause of the different results for corn, which is illustrated by the findings of Wollny (see Baver 1956) that the percentage of total rainfall penetrating a corn crop with

Table 8. *Total surface of leaves* (m^2) *per* m^2 *soil surface*
(after Körnicke and Werner 1885)

crop	leave-surface	crop	leave-surface
lucerne	85.6	barley	14.6
clover	24.5	summer-wheat	13.7
oats	21.1	corn	2.7
winter-wheat	17.5	mangolds	1.6
winter-rye	15.6		

9 or with 36 plants per m^2 is 63 and 44 respectively. It is self-evident that not only the density of the vegetative canopy at a certain moment is important for soil structure but also its changes in the course of time, especially in relation to the fluctuations in rainfall. Sowing or planting date, vegetation period, type of crop and all factors affecting its development (physical and chemical soil fertility, soil and crop treatment, plant diseases a.s.o.) play an important part with regard to the sheltering effect that a crop has upon the soil and therefore they can influence soil structure.

To protect a soil from wind erosion a crop needs to be not so dense as to prevent the surface from raindrop splash. Strips of crops perpendicular to the predominant wind direction are generally sufficient. Several scientists have studied the sheltering effect of wind-breaks (see e. g.: van der Linde and Woudenberg 1951 and Kreutz 1952). It appears that this effect can be shown up to a distance of 10 to 30 times the height of the windscreen on its lee-side and as far as about 10 times this height on the windward side. For practical purposes the shelterbelts are planted generally at distances of 20 to 30 times their height.

3. Mechanical Effects of Plant Roots

It is a common experience that a soil containing a well-branched root system, such as a grass sod, generally has a good structure. How this favourable influence is exerted by the roots is a question, which still cannot be answered completely at present. Roots can affect soil structure in several ways: they can exert pressures upon the soil, the branched root system can keep loose soil particles together, dessicates the soil, thus influencing soil structure, the different types of organic matter arising from a microbial attack of the dead roots affect the structure and its stability, and perhaps root excretions play a part. In this section we will restrict ourselves to the mechanical effects, in which the roots are acting more directly. The indirect influences via moisture content and organic matter will be treated in the sections 4 and 5 respectively.

The growing roots develop a net-work of fine channels in the soil, which is left more or less complete after the roots have been decayed. The volume of this network can be estimated from determinations of root-weights (see e.g. Schuurman 1958), supposing the moisture percentage of the fresh roots is 78 and their density is 1.04 g/cm^3. From the results mentioned in Table 9 it appears that in comparison with the total pore volume of the soil (of say 50%) the root volume is rather small and that in the top layer the latter volume is much greater for grass than for wheat. It must be emphasized however that this effect of increasing the porosity is to a certain extent a cumulative one, since in the next season the roots grow partly on other places and most root holes have lifetimes longer than one year. Therefore, undoubtedly the observed effect will be much greater

Table 9. *Root volumes for perennial ray grass and spring wheat on a clay soil*

crop	layer (cm below surface)	weight (kg/ha) of the air-dry roots[1]	root volume	
			m^3/ha	% of volume
perennial ray grass	0—10	6000	26.2	2.6
	10—40	3000	13.1	0.4
spring wheat	0—10	1500	6.6	0.7
	10—40	800	3.5	0.1

[1] After SCHUURMAN (1958).

than is suggested by the figures in Table 9. It must be noticed that this effect is not only confined to the topsoil, but is especially significant for the structure of the subsoil. In this respect deep rooting crops such as lucerne are important.

Roots are thought to loosen the soil by the pressure they exert on the adjacent soil, due to their transversal growth. Especially when roots penetrate into a clod along a former cleavage plane or another point of weakness they can split up the clod and can introduce or increase the granulation. Mainly this occurs in clayey soils under moderate moisture conditions. If such a soil is wet root pressure gives a plastic deformation, in the dry soil roots are usually unable to penetrate clods and are growing and branching in cracks already present. On the other hand the transversal root pressure can press the particles of a loose soil together into small aggregates and perhaps may increase the stability of such aggregates, if present, by means of a certain compression.

Roots and root hairs often literally enmesh soil granules, thus enlarging their stability. According to Low (1955) this occurs especially in sandy soils, where the relatively large pores enable the roots to penetrate into the soil comparatively easy. The aggregates formed in this way are fairly water-stable, but are generally very weak toward mechanical forces. In clay soils this aggregate building by roots seems to be of little importance because of the strong binding forces in these soils.

Much more experimental results, however, must be available to obtain a complete understanding of the mechanical root effects upon soil structure.

4. Influence of Soil Moisture and Temperature as Affected by Crop Growth

The ways in which soil moisture and soil temperature can affect the structure of the soil were described in the preceding chapter. Here we will report briefly on the influence a crop has on soil moisture content and soil temperature in relation to structure.

A crop diminishes the quantity of precipitation that reaches the soil surface (interception) and decreases the evaporation from this surface. On the other hand it increases the removal of moisture from the soil by transpiration.

The quantity of intercepted rainwater that evaporates directly from the plants varies with the kind of crop and with the kind and size of precipitation.

Although only very few data for the interception of agricultural crops are available, its annual magnitude seems to be generally between 5 and 15 percent of the annual rainfall (Ackermann *et al.* 1955).

Under a dense vegetation the evaporation from the soil surface as compared to that of a bare soil will be small, owing to the strongly diminished air movement and the lower maximal temperatures under the vegetative canopy. The total evaporation (evapotranspiration) from a soil covered with a crop, however, is much greater than the evaporation from a bare soil. Under the circumstances in N.W. Europe the ratio of their annual magnitudes is about 17 as can be derived from the lysimeter results of Maschhaupt (1948) and others. Therefore the total effect of a crop generally will be a lessened moisture content of the soil. On clay or clayey soils this can give cracks, especially in the neighbourhood of the roots where desiccation is greatest, a certain stabilisation of tilth (Low 1955) and easier cultivations after harvest in a wet autumn, resulting in a better structure. Further, keeping a clay soil covered with a crop in autumn will lessen its field capacity during winter owing to the desiccation and the retarded remoistening (Fig. 66) and according to Boekel and Peerlkamp (1956) this will decrease structure deterioration during winter.

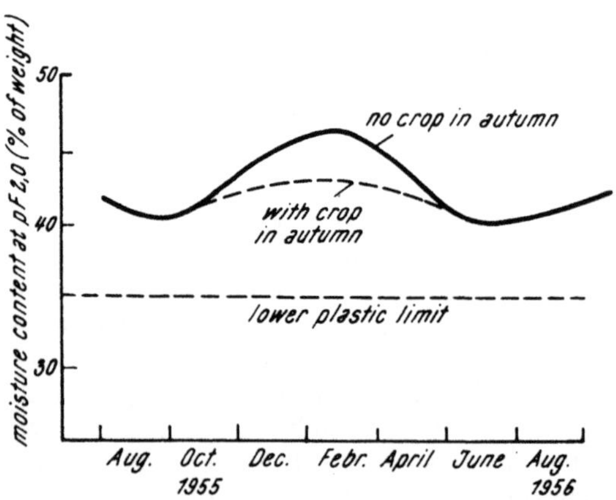

Fig. 66. Variations within a year of the moisture content at pF 2 (field capacity) of a clay soil with or without a crop in autumn (from unpublished data by Boekel)

A plant cover diminishes the diurnal as well as the annual fluctuations of soil temperature, depending on the type of crop and the climate. The smaller diurnal variations and the lower temperature of the top soil in summer, as well as its lower moisture content, may have any influence on the microbial activity in the soil and can affect in this way soil structure (cf. section 5). Covering a clay soil with a crop during winter prevents partly or entirely the beneficial effect of freezing (see the preceding chapter, section 6).

5. Effects of Organic Matter Originating from Vegetation

Generally there will be no difference of opinion about the fact that organic material brought into the soil has a favourable influence upon soil structure. Roots and stubbles of harvested crops and other plant residues contribute to a considerable extent to the organic matter supply of the soil (Table 10, Goedewaagen and Schuurman 1950). This more or less fresh organic material undergoes a series of chemical transformations in the soil owing to microbial activity and depending on weather, soil, crop and other factors that influence this activity. In the course of this transformation various intermediate products arise which can be more or less microbial resistant. It appears that some of these products play a part in respect to soil structure.

Several authors have shown that certain polysaccharides and polyuronides as well as the soil bacteria and fungi themselves affect soil structure stability (see e.g. Rennie *et al.* 1954). All these agents, however, are not microbial resistant but decay rather easily. Therefore fresh organic material must frequently be brought into the soil if it is to affect soil structure in this way. Roots and stubbles of crops give such a periodical supply (PEERLKAMP 1950).

A regular organic dressing of the soil, however, also leads to a certain percentage of organic matter in the soil when a state of equilibrium is attained. This percentage depends on climate, crop, soil and soil management. KÖNEKAMP (1957) emphasized in this respect the intensity of cultivation and the nitrogen and albumen content of the organic material. If the latters were high and the cultivation intensity low the increase of the organic matter percentage was high. According to KORTLEVEN (1959) an annual supply S (dry weight) to the soil with organic materials in the Netherlands resulted eventually in an increase of the organic matter content of the soil of 1/20 S.

Soil organic matter, consisting for the greater part of materials, which are not so easy assailable by the soil microbes as the temporary cementing agents mentioned above, affect soil structure as is shown by several authors (see e.g. PEERLKAMP 1950, BAVER 1956, BOEKEL 1959). The way, however, in which the more stable soil organic matter influences structure is still somewhat obscure at present. BEUTELSPACHER (1955) and FLAIG (1958) report that the humic acids themselves are not able to cement clay particles together, but that sesquioxides must be present too. EGAWA and SEKIYA (1956) found the fraction of very small aggregates (<0.1 mm) to have the highest percentage of colloidal clay and humic acids. Therefore these materials seem to play a part, chiefly in the formation of very small aggregates. Extraction of soil organic matter with various solvents indicates the importance of the easy soluble, so-called raw humus, accumulated in the rhizosphere, in cementing soil particles and small aggregates into big ones. According to FÖRGETEG (1957) aggregation in the top layer of 12 cm occurs especially in the period from spring till harvesting time. He concludes that root growth and microbial activity are the main causes for this aggregation, whereas soil moisture is only a secondary factor.

It is obvious that a crop not only affects soil structure by bringing fresh organic material into the soil but also more indirectly by its effects on soil moisture and temperature, which in turn influence the microbial activity. Generally speaking, however, the micro-organisms in the soil can grow within a wide range of moisture percentages and temperatures, so that the effects of the latter on microbial activities are mainly restricted to extreme circumstances.

Summarizing it can be said that a crop as well by the "natural organic manuring" due to its roots and stubbles, which can be transformed microbially into various structure improving agents, as by its effects on the microbial activity via moisture percentage and temperature can affect soil structure. A complete picture of this complex occurence, however, cannot be given just yet.

6. Effects of Tillage Operations as Related to Crop and Rotation

Soil cultivations and other mechanical soil treatments as e.g. the mechanical harvesting of potatoes, greatly affect soil structure. The result depends mainly on the type of soil and its moisture content. For a certain soil the latter is determined mainly by the climatological conditions. Generally, mechanical soil treatments affect principally the actual structure and not the structure stability. Therefore the result of such a treatment in the course of time mainly depends on the weather in the previous period (see the preceding chapter).

Since every crop has its specific requirements with regard to soil cultivations, etc., different crops and rotations can affect the structure of a soil in different ways, mainly owing to the different times at which the soil is treated and the frequency and intensity with which this is done. The seedbed preparation of a winter crop usually takes place earlier in autumn and therefore generally under somewhat drier conditions and also more intensively than autumn ploughing for a spring crop. At the beginning of the winter, the soil of the winter crop will be finer granulated. What happens during winter depends on the soil and weather. On a sandy clay soil the wintercrop can give a certain protection against structural deterioration, so that the structure in spring is worse on the bare soil. On a heavy clay soil, however, the crop prevents the soil, to a certain extent, from the beneficial freezing, resulting in a better structure on the bare soil in spring.

Fine seeds require fine granulated seedbeds. Low row crops make possible a more intensive treatment of the top soil by hoeing and earthing up. Harvesting potatoes and sugar-beets with lifting-machines gives an intensive soil treatment too. On heavy clay soils these intensive mechanical soil treatments can have a favourable effect upon structure if the soil is not very wet. On light soils, however, granulation becomes too fine and structure will deteriorate very easily.

Crops such as potatoes, sugar-beets and cover crops that must be harvested late in the season increase the chances that harvesting and tillage operations must be done under wet conditions and that, therefore, on clay soils structure will deteriorate.

Generally, however, these effects of mechanical soil treatments are only temporary. Usually they disappear within one year.

7. The Total Effects of Different Crops upon Soil Structure

In the preceding sections the ways in which a crop can affect soil structure have been briefly discussed. What are, however, the total effects of the different crops? Although literature gives a lot of data that could help to answer this

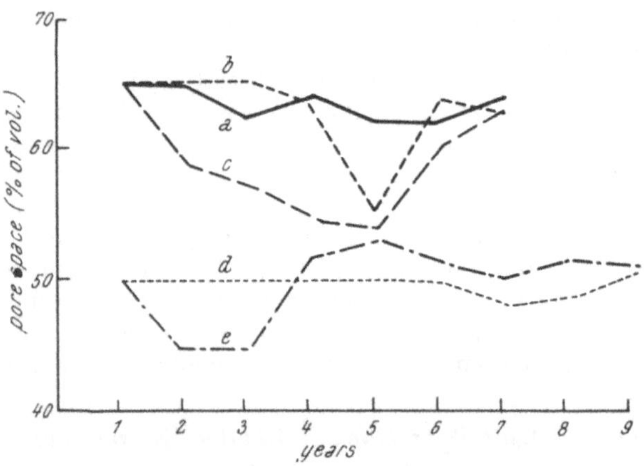

Fig. 67. Effect of ley farming on the pore space with a 2- and 4-year ley-period on a sandy soil and with a 2-year ley-period on a silty loam (according to Boekel)

a Sandy soil, continuous arable land; *b* Sandy soil, ley farming (4th and 5th year grass); *c* Sandy soil, ley farming (2nd, 3rd, 4th and 5th year grass); *d* Silty loam, continuous arable land; *e* Silty loam, ley farming (2nd and 3rd year grass)

question, most of them characterize soil structure in an insufficient way. There is a lack of data especially concerning the actual structure and its changes in the course of time. Further it is often not quite certain that the observed structural differences are entirely due to differences in crops and that no other factors play a part. Some general remarks however can be made.

It is commonly believed that grass generally has a favourable effect upon soil structure (see e.g. EBERT 1953). But it must be emphasized that this is especially true for a sod which has been ploughed up and that often the actual structure of grassland is a relatively dense one, owing to a compression of the soil by cattle. This is a common experience in ley farming, as is illustrated for two different soils in the Netherlands by Fig. 67.

On the sandy soil with a relatively large pore volume this volume is strongly reduced during the ley-period and it tends more to the level of arable land after the ley has been ploughed. On the silty loam, however, which has a much smaller pore space, the depression in the grassland years is not so great and an increased pore space level remains after ploughing the ley. WILLIAMS and COOKE (1961) report that keeping a soil three years under grass stabilizes the structure, mainly during the first year after plowing up the sod. The latter agrees with a result obtained by KOBLET and WEHRLI (1959), who furthermore found that aggregate stability was higher under grass than under legume five months after the establishment of the ley and that this stability increased with an increasing proportion of grass in grass-clover mixtures.

Several authors (EBERT 1953, PRINGLE and COUTTS 1956, WISNIEWSKI *et al.* 1958 and others) have reported on the influences of different species of grass on soil structure. Generally, however, their results are not concordant enough

Table 10. *Air dry weights of roots in the topsoil (0-20 cm), including stubbles if present, of different crops in 100 kg/ha*
(after GOEDEWAAGEN and SCHUURMAN 1950)

crop		weight of roots (incl. stubbles)
grass	old grassland	70 (23 to 110)
	ley (7 years old)	78
	ley (3 years old)	89
cereals	spring barley	14
	spring wheat	15
	winter barley	21
	winter wheat	26
	oats	28
	rye	31
root crops	potatoes	3
	sugar-beets	6
	mangolds	8
legumes and pulses	serradella	3
	peas	4
	vetch	7
	lupins	9
	trefoil	10
	horse beans	12
	white clover	14
	red clover	22
	lucerne	57

for making a sequence of the species regarding their effects on soil structure. This seems to be due to other factors such as type and fertility of the soil, moisture conditions, climate, etc. that affect grass growth and in this way soil structure too. Therefore an improvement of soil structure by growing grasses requires the use of a mixture of grass species that is adapted to the local conditions. Differences in grass species appeared to have a greater effect on the water stable aggregation of a heavy clay soil than differences in herbage utilization (Troughton 1961).

The same difficulties as with the grass species are met if one compares the effects of different crops on soil structure. Generally, however, it is found that grasses have a more favourable influence upon soil structure than cereals and that cereals are better in this respect than root crops (Ebert 1953). This seems to be caused chiefly by differences in the extent of the root-system (see Table 10). It must be emphasized that the differences between grass and other crops are somewhat smaller than Table 10 would suggest at first sight. Grass renews only a part of the root system (say 1/3) each year. A 3-years ley that was ploughed has brought during these 3 years a quantity of organic material into the top soil of $1^2/_3$ of its root mass or about 14500 kg/ha. For cereals this is three times about 2200 or 6600 kg/ha, which is still a big difference. Root crops contribute very little to the organic matter supply of the soil and have therefore not a favourable influence upon soil structure in the long run. They can have a certain favourable temporary influence on soil structure owing to the more intensive tillage that is associated with them (see section 6). Legumes are generally prefered with respect to the effects on soil structure than their root weights suggest. It seems that this is due to their covering action and their moisture uptake (see sections 2 and 4), perhaps sometimes also to their nitrogen effect. They seem however to have only a transitory effect on soil structure (Mortensen and Young 1960).

8. Crop Rotation and Soil Structure

From the results mentioned in the preceding section it follows that crop rotation enables us to influence soil structure. Since, however, the effects of rotation on soil structure depend on several other factors related to soil, soil fertility, cultivation methods, climate and weather, etc. they can only be broadly outlined.

Several authors (Peerlkamp 1950, Mazurak *et al.* 1954, Förgeteg 1957) report that a continuous cropping of potatoes causes a poor soil structure. Others (Woodruff 1939, Wilson *et al.* 1947) have observed the same for corn and (Jones 1961) soybeans. Generally, the structure could be improved by adding cereals, clover as catch crop, lucerne and ley in the rotation. Williamson *et al.* (1959) obtained a very good granulation with an 8 years rotation in which 4 years ley were included. Könekamp (1957) emphasized also the necessity of cropping arable crops alternately with legumes and grasses. Low (1955) has stated, however, that an improvement of a poor structure of arable land by means of ley farming often takes a long time, e.g. 50 years. It is commonly believed that the effects on soil structure of green manuring are smaller than of ley farming. Recently this is borne out by the results of a study made by one of the authors (Boekel 1963) on the influence of three different rotations upon the properties of a clay soil. These rotations were:

a) potatoes, peas, wheat, sugar beets, barley and flax, without any organic manure;

b) the same rotation with legumes or other green manuring crops after wheat and flax;

c) as rotation *a*, but with two years ley between flax and potatoes, with green manuring crops after wheat and flax and with about 20 tons/ha farmyard manure, applied in three out of the eight years, which is one cycle's duration.

After 10 years the organic matter content and the pore and air volume of the soil under rotation *b* appeared to be 0.1% of weight and 0.5—1.0% of volume higher than those of the soil with rotation *a*. For the soil under rotation *c* these differences were 0.5% of weight and 1.0—2.0% of volume respectively.

Different rotations with nothing but arable crops such as cereals, root crops, peas, flax, etc. in various frequencies seem to affect soil structure in about the same way.

Literature

ACKERMANN, W. C., E. A. COLMAN and H. O. OGROSKY: From ocean to sky to land to ocean. "Water", U.S.D.A. Yearbook of Agric. 1955, 41–51.

BAVER, L. D.: Soil physics, 3rd ed. New York: Wiley. 1956. — BEUTELSPACHER, H.: Natürliche Fadenkolloide und Krümelbildung. Landbauforsch. Völkenrode **5**, 90–92 (1955). — BOEKEL, P., and P. K. PEERLKAMP: Soil consistency as a factor determining the soil structure of clay soils. Neth. J. Agr. Sci. **4**, 122–125 (1956). — BOEKEL, P.: Evaluation of the structure of clay soils by means of soil consistency. Proc. Int. Symp. Soil Structure Ghent 1958, 363–368 (Med. Landbouwhogesch. Gent **24**, nr. 1, 1959). — The effect of organic matter on the structure of clay soils. Neth. J. Agric. Sci. **11**, 250–263 (1963).

EBERT, D.: Die Krümelbeständigkeit des Bodens in Abhängigkeit vom Pflanzenbestand. Deutsche Landwirtsch. **4**, 39–42, 90–93 (1953). — EGAWA, T., and K. SEKIYA: Studies on humus and aggregate formation. Soil and Plantfood **2**, 75–82 (1956).

FLAIG, W.: Zur Chemie der anorganischen und organischen Komponenten der Krümelbildung. Tagungsber. Deutsch. Akad. Landw. Wiss. Berlin **13**, 225–244 (1958). — FÖRGETEG, S.: Adatok a vizálló talajmorzsák dinamikájáról (Data on the dynamics of water stable aggregates). Növénytermelés **6**, 3–16 (1957).

GOEDEWAAGEN, M. A. J., and J. J. SCHUURMAN: Wortelproductie op bouw- en grasland als bron van organische stof in de grond. Landb. Tijdschr. **62**, 469–482 (1950).

JONES, J. H.: The effect of cropping systems on aggregate stability, organic matter content and moisture retention in soil. Diss. Abstr. **21**, 2844–2845 (1961).

KOBLET, R., and A. WEHRLI: Über die Struktur von Ackerböden bei Anbau von Kleegras. Z. Acker- u. Pflanzenbau **109**, 337–354 (1959). — KÖHNEKAMP, A. H.: Die Rolle von Klee und Gras bei der Humusversorgung der Böden. Z. Acker- u. Pflanzenbau **104**, 89–102 (1957). — KÖRNICKE, F., and H. WERNER: Handbuch des Getreidebaues, Bd. II, S. 34. Bonn: Strauß. 1885. — KORTLEVEN, J.: The long-term effects of urban refuse compost as an organic fertilizer. Int. Congr. Disposal and Utilization Town Refuse, Apr. 27—May 1, 1959. Scheveningen (Neth.). Publ. in Dutch in: Med. blad V.A.M. **14**, no. 4, 3 pp. (1959). — Kwantitatieve aspecten van humusopbouw en humusafbraak. Versl. Landbouwk. Onderz. **69.1**, 1–109 (1963). — KREUTZ, W.: Der Windschutz. Dortmund: Ardey. 1952.

LINDE, R. J., VAN DER, and J. P. M. WOUDENBERG: On the microclimatic properties of sheltered areas. The oak-coppice sheltered area. Med. I.B.T.O.N. n. 10, Oosterbeek 1951. — LOW, A. J.: Improvements in the structural state of soils under leys. J. Soil Sci. **6**, 179–199 (1955).

MASCHHAUPT, J. G.: Lysimeter-onderzoekingen aan het Rijkslandbouwproefstation te Groningen en elders I. Versl. Landb. Ond. **44**, 1–184 (1938). — MAZURAK, A. P., V. T. VALASSIS and L. C. HARRIS: Water-stability of aggregates from potato plots as affected by different rotation systems under irrigation in Western Nebraska. Soil Sci. Soc. Amer. Proc. **18**, 243–247 (1954). — MORTENSEN, J. L., and J. L. YOUNG: Effect of summerseeded green manures on soil structure and availability of nitrogen. Ohio Agric. Exper. Stat. Res. Bull. **865**, 31 pp. (1960).

PEERLKAMP, P. K.: The influence on soil structure of the "natural organic manuring" by roots and stubbles of crops. Trans. 4th Int. Congr. Soil Sci. **1950**, Vol. 1, 50–54. — PLOTNIKOV, A. A.: The role of leys of annuals and perennials in the formation of water-stable soil structure. Sborn. Nauch. Trud. Ivanov. S.-Kh. Inst. **18,** 110–117 (1960). — Pringle, J., and J. R. Coutts: The effects of grasses on aggregation in light soils. J. Brit. Grassl. Soc. **11,** 185–189 (1956).

Rennie, D. A., E. Truog and O. N. Allen: Soil aggregation as influenced by microbial gums, level of fertility and kind of crop. Soil Sci. Soc. Amer. Proc. **18,** 399–403 (1954).

Schuurman, J. J.: Root development, water uptake and growth of spring wheat and perennial ray grass on three profiles. Rep. Conf. Suppl. Irrigation Comm. VI. I.S.S.S., 11 pp. Copenhagen 1958.

Troughton, A.: The effect on the soil of the components and management of a ley. J. Brit. Grassl. Soc. **16,** 1–5 (1961).

Wiklert, P.: Grassland and soil structure. Grundförbättring **15,** 15–49 (1962). — Williams, R. J. B., and G. W. Cooke: Some effects of farmyard manure and of grass residues on soil structure. Soil Sci. **92,** 30–39 (1961). — Williamson, W. T. H., J. Pringle and J. R. H. Coutts: Short and long ley rotations and soil structure in the North-East of Scotland. Proc. Int. Symp. Soil Structure Ghent 1958, 176–187 (Med. Landbouwhogesch. Gent **24,** nr. 1, 1959). — Wilson, A. H., R. Gish and G. M. Browning: Cropping systems and season as factors affecting aggregate stability. Soil Sci. Soc. Amer. Proc. **12,** 36–38 (1947). — Wisniewski, A. J., M. Salomon and J. B. Smith: Restoration of soil structure with redtop and other sod crops. Soil Sci. Soc. Amer. Proc. **22,** 320–322 (1958). — Wollny, R. E.: Der Einfluß der Pflanzendecke und Beschattung auf die physikalischen Eigenschaften und die Fruchtbarkeit des Bodens. Berlin 1877. — Woodruff, C. M.: Variations in the state and stability of aggregation as a result of different methods of cropping. Soil Sci. Soc. Amer. Proc. **4,** 13–18 (1939).

g) Soil Structure and Soil Fertility

By

L. De Leenheer

1. Introduction

From a practical point of view a good soil structure is represented by that group of soil physical properties which provides the crop with suitable physical conditions for growing during the whole growing season. It is evident that the terms "good" and "bad" structure have only a relative meaning, changing with the kind and variety of the crops, the growing period (germination, stooling, the full vegetative development, ripening) of the plant, and the air and water economy in the soil at each growing period. The air and water economy, on the other hand, change within the soil profile, and change with depth, in the root zone as well as in the arable layer.

Nevertheless, a bad structure can be recognized qualitatively in the field, and much data is available proving that on obviously bad structured fields soil fertility is rather low; the higher the clay content of a soil, the lower the germination coefficient and the slower the vegetative development of wheat through structure deterioration (De Leenheer and De Caestecker 1959). In the Netherlands (Boekel 1959) an experimental field was laid out in 1929 on a heavy clay soil and soil structural differences were caused by varied applications of lime. From 1929 to 1954 the average of the crop production was 36 % higher on the good structured plots than on the other (Fig. 68).

In Belgium the average crop production of wheat in 1953 (a relatively wet year), was 51 % higher on soils of good rather than bad structure (De Leenheer and De Caestecker 1959), the average surplus production on soils of good structure in the same year was 23 % for mangolds (De Caestecker and De Leenheer 1959). These two average differences were found each time on

18 fields of current practice (more than 20 different farmers) in marine alluvial soils. Both studies (BOEKEL; DE LEENHEER and DE CAESTECKER) proved the changing influence of water and air economy during the crop growth.

Many soil scientists, knowing the complexity of the problem, try to correlate crop yield with one or another physical property. Even when a high correlation is found, there is no proof that the results of such a method can be generalized. Physical conditions are so interrelated with chemical and biological properties

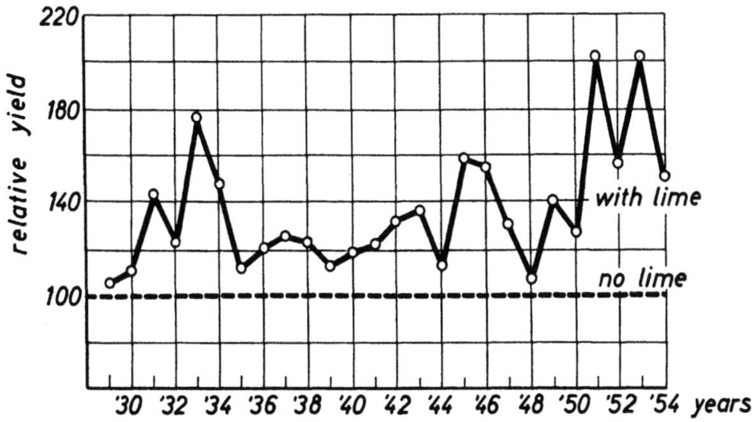

Fig. 68. Relative yields on an experimental field with bad and good structure, caused by varied applications of lime (BOEKEL 1959)

of the soil, that the most important factor of the fertility of a given soil is difficult to detect and changes with circumstances. Only when the exact physical requirements of the plant are known, can satisfactory tests be devised (FOUNTAINE 1959). From this point of view attention should be given to the following factors:

— the capacity of plant roots for mechanical work,
— heat and soil temperature,
— the soil aeration and soil air composition,
— the water requirements of plants.

These factors are briefly discussed, insofar as they were considered suitable for the investigation of soil structure and its influence on soil fertility. The study of the water requirements of plants, however, will be made in a separate chapter by another author. Instead of this aspect, the problem of soil structure stability and its importance for plant growth will be discussed here.

2. Mechanical Soil Properties and Crop Production

We can say that a fair amount of knowledge has been already gathered on the water requirements of plants and on the influence of soil temperature on plant growth, but that rather little is known on soil aeration and on the rates of gas diffusion needed for good plant growth. Nearly nothing is known on the mechanical work capacity of plant roots. We all recognize that a plant root which enters a crack in a soil clod can exert a high pressure on the crack sides, but is unable to enter into the same clod at its surface. For the last reason soil strength tests are not very useful in predicting plant performance or crop production (FOUNTAINE 1959). Nevertheless, in a Belgian silt loam soil the modulus of rupture determination on cylindrical soil core samples (3 to 8 cm depth cores) taken in the fields was well correlated with yield of winter wheat and with the number of blows to drive in the sampling can, acting as a penetrometer, to 8 cm depth (Fig. 69), (Kirkham, De Boodt and De Leenheer 1959 a). In a tillage study, corn yields could be correlated significantly with the number of blows to drive the soil sampling tubes to the desired depth (Swanson and Jacobson 1956).

Other authors found a close relationship between the measure of resistance to penetration and bulk density (Parker and Jenny 1945). Although in certain

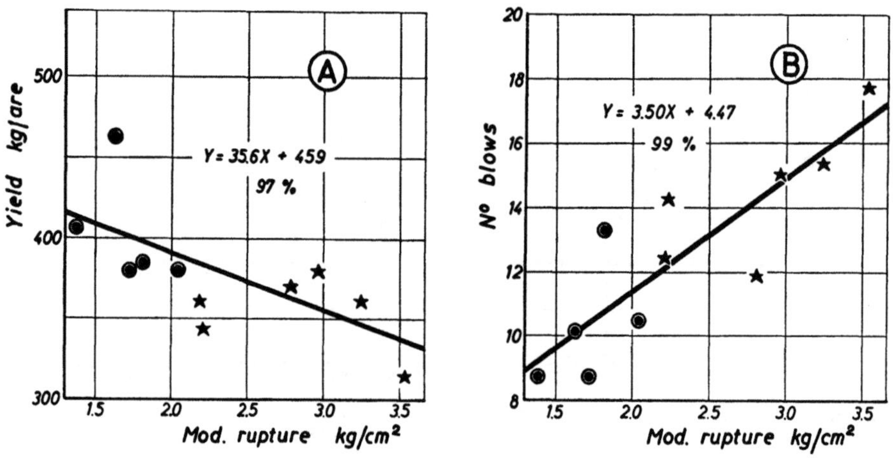

Fig. 69. Correlation of the modulus of rupture with the number of blows to drive in a sampling can (to 8 cm depth), acting as a penetrometer, and with the yield of winter wheat on a Belgian silt loam soil (Kirkham, De Boodt and De Leenheer 1959)

circumstances compaction of loose soil may be desirable for good water economy, it is evident that no general critical values can be stated as they change with texture.

So Veihmeyer and Hendrickson (1948) found that no roots can penetrate any soil with an apparent density higher than 1.9; in other soils the limiting value was 1.8 to 1.7; in clays the limiting value was 1.6 to 1.7. In clay soils root development was markedly decreased as soon as the apparent density reached 1.4, the value being 1.6 in sandy soils.

Mechanized farming and negligence in soil organic matter supply are important factors in soil structure deterioration. Mechanical farming as such is often made responsible for soil compaction and the decrease in soil production resulting from it.

Soil compaction under the arable layer is often a consequence of bad soil management, and soil compaction does not always mean an increase in clay content through percolation in the compacted layer. Some soils are very sensitive in this respect; the pores between sand particles can be filled up to a certain extent with smaller silt particles; the maximum diameter d_1 of the pores between larger particles with diameter d is well known, $d_1 = 0.232\ d$ (supposing tetrahedral packing) (Lutz 1952). So the most sensitive soils for compaction through mechanical influences (gravity and water movement, pressure of machinery, vibration of the tractor) are those soils where enough small particles are present to fill up the pores and cavities between the larger ones. Therefore sandy loam soils are very sensitive in this respect.

Table 11a. *Influence of deep plowing and plowing with subsoiling on crop yields*

A—Crop yields on heavy soil (30-35% clay content). Field: Tr. Re.

	1954: wheat			1955: barley			1956: oats		
	yield in kg/ha	ratio		yield in kg/ha	ratio		yield in kg/ha	ratio	
		for the objects	for the way of plowing		for the objects	for the way of plowing		for the objects	for the way of plowing
Deep plowing (45 cm)									
poor structure	2488	102	96	4191	99	92	3659	78	83
good structure	2507	103		4075	97		4118	88	
Ordinary plowing (20 cm)									
poor structure	2440	100	100	4216	100	100	4659	100	100
good structure	2754	113		4764	113		4753	102	
Plowing with subsoiling (20/45 cm)									
poor structure	2297	94	91	3779	90	90	3703	79	83
good structure	2449	100		4266	101		4086	88	

B—Crop yields on light clay soil (±15% clay content). Field: Va. Lo.

	1954: wheat			1955: barley			1956: oats		
	yield in kg/ha	ratio		yield in kg/ha	ratio		yield in kg/ha	ratio	
		for the objects	for the way of plowing		for the objects	for the way of plowing		for the objects	for the way of plowing
Deep plowing (45 cm)									
poor structure	4474	208	187	4049	90	91	4512	108	101
good structure	4357	203		3923	87		4449	106	
Ordinary plowing (20 cm)									
poor structure	2150	100	100	4502	100	100	4176	100	100
good structure	2558	119		4271	95		4658	112	
Plowing with subsoiling (20/45 cm)									
poor structure	3099	144	136	4287	95	97	4597	110	103
good structure	3306	154		4184	93		4497	108	

Table 11b. *Influence of kind of plowing on physical soil properties and crop yield*

A—1954. Sampling at harvest time. Field: Tr. Re.

	Bulk density		Relative value of the aggregate stability		Relative value of the volume of drainage pores		Relative value of the volume of useful water retention pores		Relative amount of the wheat yields	
Deep plowing										
poor structure	1.207 (0.01)	1.329	70.0 (0.24)	80.6	156.3 (0.001)	84.1	119.7 (0.01)	85.1	102.0	96.2
good structure	1.451		78.8		62.4		75.5		102.7	
Ordinary plowing										
poor structure	1.301 (0.05)	1.230	100.0 (0.04)	100.0	100.0 (0.01)	100.0	100.0 (0.03)	100.0	100.0 (0.26)	100.0
good structure	1.159		85.1		159.9		129.3		112.9	
Plowing with subsoiling										
poor structure	1.244	1.249	73.4	80.2	117.8	95.3	105.4	93.9	94.1	91.4
good structure	1.253		74.5		129.9		109.9		100.4	
Deep vs. ordinary plowing		0.17		0.01				0.19		
Ordinary plowing vs. plowing with subsoiling				0.01						0.20
Deep plowing vs. plowing with subsoiling		0.25								

B—1956. Sampling at harvest time. Field: Tr. Re.

	Bulk density		Relative value of the aggregate stability		Relative value of the volume of drainage pores		Relative value of the volume of useful water retention pores		Relative amount of the oats yields	
Deep plowing										
poor structure	1.291	1.299	106.3	106.9	97.3	87.0	92.6	87.1	99.4	92.0
good structure	1.306		100.7		88.8		90.5		96.6	
Ordinary plowing										
poor structure	1.239	1.210	100.0 (0.16)	100.0	100.0	100.0	100.0	100.0	100.0 (0.03)	100.0
good structure	1.181		92.5		113.7		110.2		113.0	
Plowing with subsoiling										
poor structure	1.224	1.241	89.4	94.0	99.3	91.7	97.4	93.9	89.6 (0.08)	89.6
good structure	1.257		91.2		96.6		100.0		101.1	
Deep vs. ordinary plowing		0.03		0.21		0.14		0.04		0.09
Ordinary plowing vs. plowing with subsoiling				0.21		0.30		0.27		0.02
Deep plowing vs. plowing with subsoiling		0.04		0.04				0.09		

Figures written vertically are showing the probability levels of the observed differences.

Table 11c. *Influence of kind of plowing on physical soil properties and crop yield*

A—1954. Sampling at harvest time. Field: Va. Lo.

	Bulk density			Relative value of the aggregate stability				Relative value of the volume of drainage pores			Relative value of the volume of useful water retention pores			Relative amount of the wheat yields			
Deep plowing																	
poor structure	1.340		1.337	64.0			76.6	104.2		102.2	100.4		100.9	208.0			187.1
good structure	1.334			66.0		0.01		106.6			100.1			202.6		0.001	
Ordinary plowing																	
poor structure	1.334		1.334	100.0	0.09		100.0	100.0		100.0	100.0		100.0	100.0	0.01	0.001	100.0
good structure	1.334	0.02		74.0				106.2	0.02		98.8	0.001		119.0			
Plowing with subsoiling		0.05				0.01			0.10			0.04				0.001	
poor structure	1.385		1.378	65.0			77.4	91.4		91.5	92.6		93.5	144.1	0.24		136.0
good structure	1.373			66.7				96.6			93.2			153.7			

B—1956. Sampling at harvest time. Field: Va. Lo.

	Bulk density		Relative value of the aggregate stability				Relative value of the volume of drainage pores			Relative value of the volume of useful water retention pores			Relative amount of the oats yields			
Deep plowing																
poor structure	1.239	1.249	92.9	0.02		88.2	101.5		96.8	105.8		101.0	90.0			90.9
good structure	1.259		81.4		0.02		95.8			103.6			87.2		0.01	
Ordinary plowing																
poor structure	1.256	1.243	100.0		0.001	100.0	100.0		100.0	100.0	0.20	100.0	100.0	0.11	0.10	100.0
good structure	1.229		96.3				103.8	0.07		107.3			94.9			
Plowing with subsoiling					0.05										0.27	
poor structure	1.239	1.250	115.7	0.29		112.0	96.5		93.6	104.2		99.6	95.2			96.6
good structure	1.262		104.5				94.4			102.3			93.1			

Figures written vertically are showing the probability levels of the observed differences.

It was long believed that deep plowing and subsoiling were the best methods to improve compacted soils, but since 1915 a large amount of data of experimental fields is available, proving the contrary. In these experiments the yields are compared, but very rarely a comparison of the physical properties is simultaneously made. In the tables on pp. 301—303 some data of this kind are given for deep plowed and subsoiled fields (De Leenheer and De Boodt 1959b) (Table 11a, b and c).

These data show very clearly for the soils investigated that[1]:

— deep plowing and subsoiling combined with ordinary plowing, on soils with structure deterioration, decreased crop production (the loss of production progressing with time) when nothing is done to improve simultaneously soil flocculation;

— when nothing was done to improve the crumb stability, deep plowing and subsoiling had no lasting effect on pore size distribution and aggregate stability;

— an accidental improvement shown for one year (see yield 1954 on field Va.Lo.) may be very misleading, just as it was to King (1914), whose conclusion however had an important influence on agricultural practices for many years;

— it is difficult to correlate crop production with physical soil properties in the deep plowed and subsoiled fields.

Surface crust formation is another consequence of soil structure deterioration; it occurs meanly in light clay and loam soils, in which the crumbs and aggregates are not sufficiently water stable. The mechanical effect of the falling raindrops disaggregates the crumbs, leaching out the clay and partly the finer silt fraction. The consequence is that the sandy fraction is left at the surface, together with the silt particles, which fit within the pores of the sandy grains. The beating effect of the rain, moreover, produces an orientation of the particles, so that a thin hard sandy loam crust will be formed. This surface crust prevents a normal germination as the crust is too hard to be perforated. Heavy clay soil also can make germination difficult, when the arable layer loses its aggregation. In a few fields, on light clay soil (15 % clay) and with evident soil structure deterioration, the average germination on the part of the field with crust formation was only 60 % of the germination observed on the parts with better structure and without crust formation. In the same area (marine alluvial soils), on the heavy clay soils (35 % clay or more) the average germination, however, on the bad structured parts of the fields was only 35 % of the germination (expressed as plant density) on the relatively good structured parts of the same fields (De Leenheer and De Caestecker 1959).

Soil consistency measurements, compression tests, or other laboratory methods, are satisfactory to study the work done neither by roots to penetrate a compacted subsoil nor by seedlings to break through a surface crust. Neither is it known how much a plant suffers in exercising different levels of root pressure. But it is evident there are critical values which highly influence crop production.

Gill and Bolt (1955) call attention to Pfeffer's studies, made in 1893, of the root growth pressures exerted by plants against external resistances en-

[1] The influence of deep plowing depends markedly on soil type and plowing depth (texture and carbonate content of the subsoil). The technique of deep plowing is seldom used to improve soils with deteriorated structure, but is economically often justified to improve the agricultural value of soils with heterogeneous profiles. In this respect the Dutch agriculture has an important experience: on the occasion of the reclamation of the Zuider Zee (543.620 acres or 220.000 ha) and coastal areas deep plowing is a current practice to make the best of the reclaimed new land.

countered in the soil. These values of 1893 are the only data available at this time. The root, which has plastic properties, will follow in its growth the line of least resistance. The pressure exerted by roots is both radial and axial, but is different in the two directions, the last being the highest.

If the root development is confined in its radial direction, the value of the axial pressure increases. So tick-beans (faba vulgaris) with free root tip developed a radial pressure of 3.9 atm. and an axial pressure of 10.4 atm.; with radial confined root tips the radial pressure was 5.5 atm. and the axial pressure went up to 19.4 atm. Roots seem therefore to be able to develop enough pressure to surmount in general soil resistance.

FRESE (1956) in Germany, proved that horizontal deviation in the growth of sugarbeet roots was observed in the case of a vertical change from a loose into a compact soil layer, as well as in the case of change from a compact into a loose soil layer. It is not so much the increase in soil resistance, as it is the abrupt change in physical environment, as the air and water economy, which appears to be the major factor hindering normal vertical root growth.

3. Thermal Properties of Soil

Data are available in the literature on the determination of soil temperature at the soil surface and at different depths. But the determination of thermal properties of soil, such as heat capacity and thermal conductivity, is not often made in relation to soil structure.

It is well known that soil temperature influences not only soil physical properties, such as moisture movement and aeration, but also chemical properties, such as solubility, reaction and equilibrium constants, as well as biological activities and organic matter decomposition. Moreover, the influence of soil temperature on the germination, the vegetative growth and ripening and on the quality of the crop is known by every farmer. This explains the large use of certain cultural practices, as shading and mulching, in warmer climates.

The measurement of soil temperature is not a difficult determination, and for the interpretation of the data collected in terms of plant growth and crop production a large number of documents are available (see RICHARDS, HAGAN and Mc CALLA 1952).

As a recent example of the study of thermal soil properties in relation to soil structure, we can mention the work of GUPALA (1959), who found that the coefficient of temperature conductivity and of heat conductivity increase with increasing compaction, the speed of the increase depending on soil moisture. One might conclude that good structure is at a disadvantage compared with bad structure. This may be temporarily true, because it depends on the air and water relationship in the soil pores, the heat conductivity of water at 0° C being 0.0012 and that of air 0.000057 at the same temperature (HODGMAN, Handbook of Chemistry and Physics, 1947).

In this respect the question arises as to why farmers do not like a loose soil in which to sow the seeds of a grain crop. They know by experience that a more or less compacted soil is preferable. The soil should not be warmed up only at the surface, but also a satisfactory heat conductivity is needed to accelerate the growth of seedlings and young plants. This was stressed by WILLIS, LARSON and KIRKHAM (1957), who found an optimum growth for corn at a temperature of 24° C (75° F) at 10 cm (4 inches) depth.

4. Soil Aeration and Plant Growth

M. B. Russell (1952) writes:

"Much of the information relating soil aeration to crop growth is based on inference rather than quantitative data. For example many authors have attributed the observed effects of drainage, tillage and other soil management practices on crop growth to the influence of those factors on soil aeration. Only in a limited number of cases, however, has the validity of that assumption been established by means of direct measurements of soil aeration. This situation undoubtedly is a consequence of the lack of satisfactory methods of characterizing soil aeration in terms of parameters that are of significance in plant growth".

It is surprising that rather little quantitative proof is given in literature that total pore space and pore size distribution are in direct and causal relationship to soil aeration. The knowledge of the total air content in a soil sample as it is studied with an air-picnometer is a first step to study soil aeration.

That soil aeration must be important is well understood. Several chemical and biological reactions occur in the soil (decomposition of organic matter, activities of plant root and micro-organisms), which increase the carbon dioxide and decrease the oxygen content; only through interchange of soil air with atmospheric air can the soil remain in good condition. Therefore the rate of gas diffusion, which is directly related to the macroporosity or the air-filled pore space, is a more important property than the total air content.

Rommel (1935) studied the exchange between air and soil atmosphere and found that the gas diffusion process accounted for 90 % or more of the normal aeration.

Since the rate of diffusion is a linear function of the macroporosity (Penman 1940, Taylor 1949) one can take the air-filled soil volume as a measure of diffusion, but only under the condition that the air-filled pores are really "active" in the diffusion; that is, the air-filled pores considered must be in direct connection with the atmosphere. When soil is moist many pores become "inactive" in the diffusion process, since they become isolated by water necks.

Van Bavel (1952) and Collis-George (1953) proved that the idea of Penman and Taylor (that diffusivity was proportional to active air space) was right from a practical standpoint, because the free movement of a gas molecule is normally small in comparison with the diameter of the pores. When the pores are very small, as in compact heavy soils, there exists no longer a linear relationship between rate of diffusion and air-filled porosity. Information on the openness of the soil air pores may be obtained by a determination of air permeability, which is the rate of a mass flow under unit pressure gradient. This physical property is also proportional to the macroporosity, but to a higher power of it (Dalla Valle 1948).

A rough but easy method of studying soil aeration in the field in relation to plant growth consists in using a blood-tensiometer as used by physicians (Wilde and Steinbrenner 1950, De Leenheer 1953). The air pocket normally put around the arm of the patient is connected with a cylinder steel tube of well-known section, which is put to a constant depth into the soil. Mathematic calculation of the data collected to determine the air permeability is difficult if not impossible. Much better is the air permeability apparatus proposed by Grover (1955), of which an improved model was used in the field by Kirkham, De Boodt and De Leenheer (1959 b).

In the equipment air is delivered at a constant small pressure through a vacuum-cleaner type flexible hose to a seamless tube sampling can driven into the soil. The air comes from a gasometer type float reservoir which falls at a constant speed, depending on the soil permeability, as the air flows through the hose to the soil. The measured rate of fall of the permeameter float to values of air permeability of the soil is given by a simple formula: k (in μ^2) $=c\,x$, where x is the rate of fall and c a constant, depending e.g. on the volume and section of the circular tube put into the soil and on the size of the float reservoir.

The variability in the measurements may be of a logarithmic normal type, but it may be also of another type (Mc Intyre and Tanner 1959), so that only rank analysis (which does not assume the existence of a homogeneous population) can be applied to allow a conclusion. It was concluded that for the loamy and clay soils studied, the air permeability, when measured at field capacity, is a measure of soil structure which can be related to crop growth.

It must be stressed here that many of the air permeability studies since the last World War have been conducted by Kirkham, who published a first field method in 1947, together with the theoretical basis for the mathematical calculation of the permeability, independent of the equipment used. The study of air permeability of thin soil crusts became possible with an equipment described by Evans and Kirkham (1950), and improved later by Kirkham's cooperator, Grover (1955), and by Tanner and Wengel (1957)[1].

A quite different method to characterize soil aeration is the determination of the oxygen and carbon dioxide concentration in the soil. The correlation, however, between plant growth and soil aeration measured this way is not entirely satisfactory. There are of course critical oxygen and carbon dioxide concentrations, but they differ greatly with the plant species, the variety, the level of root activity and the age of the plant. Even in a 0.5 percent oxygen soil atmosphere, some plants slowly grow, as well as in a 30 % carbon dioxide atmosphere. (The normal soil air contains 20.6 percent oxygen and 0.25 percent carbon dioxide.)

Much information concerning oxygen and carbon dioxide concentration in soils in relation to plant growth is given by Russell (1952), who made a review of the literature on this subject. Nowadays the analysis of soil atmosphere has been made easy. The modern equipment called "oxygen analyzer" makes use of the paramagnetic property of oxygen and allows a direct reading of the oxygen concentration.

Temperature strongly influences soil air toxicity. On one hand, increasing temperature means a higher respirational demand of the roots for oxygen; on the other hand, the increase of temperature causes a decreased solubility of oxygen in the soil water. Thus, Cannon (1925) proved that a normal root growth in a 10 % oxygen-atmosphere was impossible at 30° C, although it was at 18° C.

It is well known that a good aeration is important for the uptake of nutritive elements. In this respect the study of Lawton (1945) pointed out the fact that the uptake of nutritive elements is made difficult when the oxygen content is lower than 10 vol. percent and the carbon dioxide content higher than 5 vol. percent. A difficulty in the study of the soil air composition is the fact that the gas analyzed is withdrawn, at the moment of soil air sampling, from the larger soil porosities and hence the oxygen and carbon dioxide concentration in these pores

[1] In using the latter equipment a point made with regard to calibration by Kirkham, De Boodt and De Leenheer (1959 b) should be observed. A device for measuring air permeability, as well as other diffusion coefficients in soil, invented by Kirkham and Adams has recently (1959) been patented in the U.S.A. Patent Office (under N° 2,913,897 Nov. 24, 1959: Apparatus for measuring diffusion coefficients of soil in place).

are entirely different from those at the root-soil interface. Schuffelen, Janse and Wiersema (1954) tried to improve the soil air sampling using an injection syringe of a veterinarian.

The quantitative determination of several effects of soil aeration on plant growth is not sufficiently established, although everybody agrees that aeration is a major factor of plant growth, as it influences not only the uptake of nutritive elements, but also the chemical composition of soil water (e.g. ferrous iron and soluble manganese content, solubility of calcium and magnesium and of phosphorous), the aerobic and anaerobic and other biological processes, etc.

The measurement of the oxidation-reduction (redox) potential on soils has been tried by many authors to characterize soil aeration. The possible oxidation-reduction reactions in soils, however, are numerous; some of them are of a complex nature and the potential measured in the soil has only a relative value, with only occasional relationship to plant growth. (Peech and Batjer 1935). The method has been improved since the redox-potential was combined with the pH into one single reduction value, with the symbol rH. The problem of the practical determination of the redox-potential in the field has been studied with satisfactory results by Starkey and Wight (1945), although these authors did not examine the relationship with plant growth. Recently Flaig, Scharrer and Judel (1955) made a critical investigation of the errors in redox-potential determinations and recommanded a procedure for laboratory determinations which seemed to be reliable. But in this study neither the results with field observation on root development or plant growth was compared.

In the Netherlands interesting work was done by Tendeloo, Vervelde and Zwart-Voorspuy (1944), who studied the electrical potential between the plant roots and the soil. The measurement of this bio-electrical potential changed nicely with the changes in aeration conditions in the root-zone, so the method seemed to be more interesting than the redox-potential determination. A critical study of Taylor (1949) showed, however, that the bio-electrical potential is strongly dependent upon all factors that influence the respiration of root cells.

Also should be mentioned the reducing influence of high carbon dioxide and low oxygen concentrations on root permeability, and the effect of soil aeration on plant diseases, poor aeration increasing normally the activity of pathogens and the susceptibility of the plant.

5. The Aggregate Stability and Soil Fertility

As already said in the chapter on „Soil Structure Measurements", the wet sieving method is the most used in Europe (Kullmann 1958) and also in the U.S.A. (Gardner 1956).

There is no general agreement about the way the results should be expressed. The results of an aggregate sieving are mostly expressed as a mean aggregate diameter; commonly used are the mean weight diameter (MWD) as described by Van Bavel (1950), the geometric mean (GM) as described by Gardner (1956), and the arithmetic mean size (AMS) as described by Stirk (1958). As a better value, De Leenheer and De Boodt (1959 a) proposed the change in mean weight diameter (CMWD), being the difference in MWD between a dry and a wet aggregate sieving, the latter being preceded by an artificial rainfall application and an incubation.

A critical study has very recently been made by De Boodt, De Leenheer and Kirkham (1960). Six aggregate stability indices used for characterizing the water stability of aggregates were compared and tested for their correlation with wheat yields on Belgian loam soil. The following table gives these results.

The inclusion of the fine aggregate fraction 0-2 mm in the material used for sieving sometimes causes, upon wet sieving, a gluing of the larger particles together. The authors have found, as have BRYANT, BENDIXEN and SLATER (1948) and also LAWS and EVANS (1950), that omission of the fine material improves reproducibility of results (lower coefficient of variation). Moreover, the correlation coefficient with wheat yield was highly significant (P=0.001) in this case (cf. Table 12). A critical examination of the MWD index, the 0-2 mm fraction being omitted, allowed the conclusion that the high correlation of this value with the yield was fortuitous. Indeed, the initial size of dry aggregates was made the same for all plots tested, making thus the MWD a measure of the CMWD. The index CMWD (the 0-2 mm fraction omitted) was also correlated with yields of several other crops, growing on different soil types, ranging in texture from a clay to a sand.

The yield correlations showed that yield may decrease or may increase or may not change at all due to increased water stability. The kind of correlation is determined by the weather.

Table 12. *Reproducibility, for a loam soil, of six ways (indices) of determining aggregate stability; and correlations, with yield of wheat, as obtained for the several ways*

Index no	Index (way of calculating the aggregate stability)	Depth of sampling	Number of samples	Average value of index	Standard deviation	Coefficient of variation	Correlation coefficient with wheat yield [1]
		cm		mm	mm	%	
1	CMWD, by dry and wet sieving, all size fractions included	1—3	20	1.463	0.446	30.5	0.040
		3—5	20	1.487	0.351	23.6	0.284
		1—5	40	1.475	0.391	26.5	0.207
2	MWD, by wet sieving only, all size fractions included	1—3	20	2.758	0.443	16.1	0.287
		3—5	20	2.855	0.376	13.2	0.404
		1—5	40	2.806	0.400	14.3	0.459
3	CMWD[3], by dry and wet sieving, 0—2 mm fraction omitted	1—3	20	2.119	0.265	12.5	0.868[2]
		3—5	20	2.395	0.326	13.6	0.684[2]
		1—5	40	2.257	0.289	12.8	0.818[2]
4	MWD, by wet sieving only, 0—2 mm fraction omitted	1—3	20	2.426	0.264	10.9	0.819[2]
		3—5	20	2.228	0.346	15.5	0.706[2]
		1—5	40	2.327	0.300	12.9	0.840[2]
5	Change log normal mean, by dry and wet sieving, all size fractions included	1—3	20	3.485	1.158	33.2	0.294
		3—5	20	3.475	0.811	23.4	0.344
		1—5	40	3.480	0.970	27.9	0.401
6	Log normal mean, by wet sieving only, all size fractions included	1—3	20	1.705	0.553	32.4	0.068
		3—5	20	1.945	0.557	28.6	0.363
		1—5	40	1.825	0.540	29.6	0.278

[1] Correlation coefficients are for wheat yield vs. aggregate stability for the way of calculating the aggregate stabilities shown in the second column.

[2] Significant at probability level $P = 0.001$: entries in the last column not marked with [2] are not significant at $P = 0.300$ or greater, except that for the 40 samples under index no 2, $P = 0.01$.

[3] Index generally used in Belgium.

6. Soil Structure Indices

A. A Laboratory Soil Structure Index

A soil structure index, if it is to reflect increased yield conditions, should be developed to include more than a measure of the water stability of the aggregates. This was attempted by De Boodt and De Leenheer (1959). Such an index cannot be rigid, as the influence of soil structure on the yields will depend very much on the climatical conditions. Besides this, for the same place, the composition of the index should be different following the type of growing season. In a temperate humid region, a different kind of pores will be of critical value

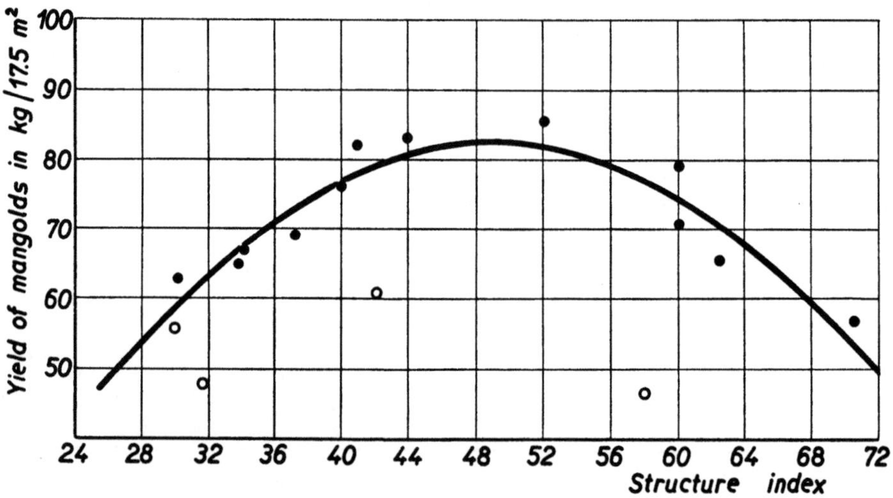

Fig. 70. Correlation of yields of mangolds with soil structure index (product of useful pores with stability index of aggregates) of a sandy soil (Drongen). Formula of the line: $y = -0.0628\,x^2 + 6.1608\,x - 68.62$. Correlation coefficient: $R = 0.91$. Significant at the level 0.002 (99.8 percent). On four plots (white circles in the figure) accidental crop damage was evident: these yields were not taken into account for the calculation of the correlation (De Boodt and De Leenheer 1959)

when the growing season is wet or dry. In the first case the water drainage pores will be of vital value to provide the necessary aeration in the root zone, while in the second case the water storing pores which have to provide water in the root zone will be of main importance. The type of soil also should be involved, as it makes a difference whether or not a soil has a great percent of colloidal material in it. In the first case shrinkage or swelling will change soil physical conditions in one or another sense during the growing season, depending on the moisture content, while in the second case the soil behaviour is much more simple. Finally, the soil structure stability should also be brought into the picture. As an example Fig. 70 is given for a sandy soil where the useful pores were taken as the critical pore space, while Fig. 71 gives the results for a clay soil with the swellable drainage pores as critical pore space.

De Boodt and De Leenheer (1959) also experienced that in some years climatological conditions can be nearly ideal for plant growth: it rains at just the right moment and it stays dry when needed. During such years soil structure has nearly no influence on plant production. They calculated that such ideal cases are present in a humid temperate region (Belgium) on an average once out of every four to five years. This was also found in the Netherlands by Boekel (1959), as is illustrated in Fig. 68.

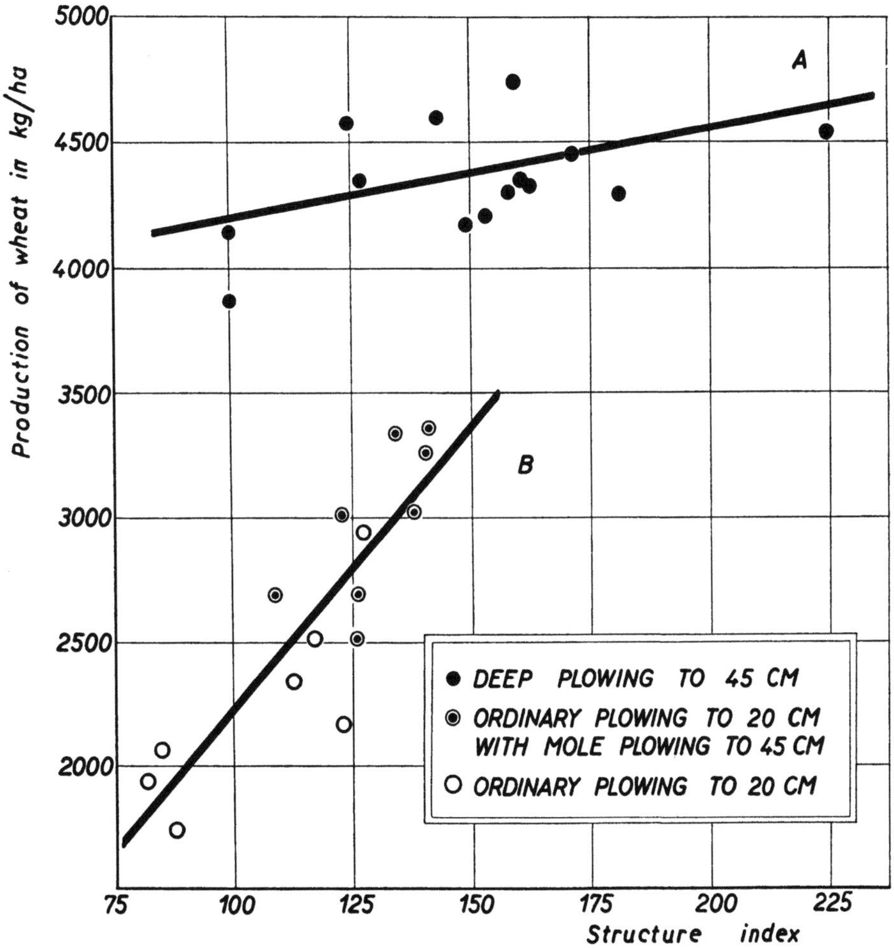

Fig. 71. Correlation of structure index (product of swellable drainage pores and aggregate stability index) with production of wheat (clay soil) (DE BOODT and DE LEENHEER 1959). *A* Correlation on the plots with deep plowing to 45 cm. Formula of the line: $y = 3.35\,x + 3872$. Correlation coefficient: $r = 0.482$. Significant at the level 0.08 (92 percent). *B* Correlation on the plots with ordinary plowing to 20 cm and on the plots with ordinary plowing combined to mole plowing (total depth 45 cm). Formula of the line: $y = 23.25\,x—76.51$. Correlation coefficient: $r = 0.865$. Significant at the level 0.001 (99.9 percent).

B. Field Indices of Soil Structure and Soil Management

It is generally known that mechanized farming and soil management can be done in very different ways.

But, what do we know about the real practical value of several recommendations which are made to maintain or improve soil structure under mechanized farming? Therefore, it is the author's conviction that more regional investigations ought to be made, which are based on field evaluations of soil structure, to study the practical influence of soil management on soil structure. Otherwise it will be very hard to detect the regional importance and interaction of the numerous soil management factors.

Such a regional investigation has been made in Belgium in the loessloam area; the first condition, however, is a good procedure for soil structure evaluation in the field.

In a regional field study of soil structure as a consequence of soil management practices, the distinction of soil compaction and aggregate stability is very important. A given factor (liming e.g.) will have a different influence on stability and compaction. Therefore De Leenheer and co-operators (1960, 1961, 1962, 1963) distinguish a soil compaction index and an aggregate deterioration index, both being evaluated in the field. Such an index represents a frequency combined with the evaluation of the degree of deterioration. Example: 50 fields or plots are visited and 12 times an evident soil compaction at plowsole depth is observed, 20 times the soil compaction is less visible and 18 times no compaction is detected. The frequency percent of evident compaction in this case is 24 and the frequency percent of weak compaction is 40. Half the value of the frequency percent of the less visible deterioration (40:2=20) is added to the frequency percent of the evident deterioration (24) and the total (44) is called the field index of soil compaction. In the same way the field index of aggregate deterioration is calculated (Vandamme, De Leenheer and De Boodt 1963).

To detect soil compaction a simple penetrometer is used, and simultaneously the total soil porosity is evaluated in the field; preference is given to an evaluation at field capacity using a soil sample ring 3 inches or 7,62 cm in diameter (see this handbook, chapter on Methods for Soil Structure Determination).

The total soil porosity is evaluated at three depths: in the arable layer at about 4 inches depth (10 cm), in the plow-sole and immediately below the plow-sole. The conclusion is made in terms of evident plow-sole, beginning or no visible plow-sole. A similar evaluation of aggregate deterioration is made with three gradations: evident, beginning or not visible. It is based on the observation of the crust formation at the soil surface and on the observation of a vertical wall, about 10 inches deep (the profile of the arable layer and its plow-sole). In the loam region of Belgium attention is paid to detecting yellow-white spots of fine sandy material in the arable layer; they are typical in loam soils with deterriorated aggregate stability; the spots have a diameter of the order of magnitude of the centimeter. The more spots appear, the more evidence there is for stability deterioration. At the same time much care is bestowed on detecting the presence of white-gray spots of calciumcarbonate (chalk, sugar sludge, marl, etc.).

This is necessary to classify the arable soils in soils with or without a reserve of visible calciumcarbonate. In the loam soils of Belgium these spots are very small (with a diameter of the order of magnitude of a millimeter), but the presence of such a visible carbonate reserve is now considered as the very first factor (in Belgian loam soils) for a good aggregate stability... and seems to be more important than any other crumb stability factor, including organic matter (providing a minimum of the latter is present). It is evident that the conclusions of a regional soil structure survey are only valid for the region which has been studied; but it is also a fact that the real value of some soil management recommendations can only be proved by field observations, taking into consideration the whole problem of regional mechanised farming (soil type and slope, texture and subsoil influence on permeability, crop rotation, organic and green manure, winter or spring plowing, light or heavy tractor, etc.).

The following results of the soil structure survey of the Belgium loam soils are very instructive in this respect.

1. Influence of the Soil Association (Sublandscape) within the Loam Region. As the influences of soil management and crop rotation are different in the various sublandscapes of the loam region, the field indices of soil compaction and of stability deterioration are different too and seem to be mutually independent (Table 13).

Table 13. *Field indices of different soil associations*

Soil association or sublandscape	Nature of geological substratum	Number of field evaluations of structure	Cattle density per Ha arable land	Field index of plow-sole compaction	Field index of stability deterioration
Normal loam area, West	tertiary fine sand	201	2,9	22	67
Normal loam area, East	tertiary fine sand	284	1,1	45	73
Loam area of Brabant	tertiary coarse sand	146	1,0	40	68
Loam area of Hesbaye	secondary chalk	153	0,7	32	59
Humid loam area	tertiary clay	156	2,0	34	38

2. Influence of Soil Type and Slope. The following comparison (Table 14) has been made on large rolling fields, the three mentioned soil types forming each time the normal toposequence of the loam region: in this case the most frequent loam profile on the plateau, the normal loam soil on the lower part of the slope (partly colluvium), and the typical colluvial profile in the dry depression. The whole toposequence is well-drained (drainage class b).

Table 14. *Field indices of different soils*

Normal soil type (well-drained) on	Normal loam area, West		Normal loam area, East		Brabant		Hesbaye	
	Pl. s.	St. d.	Pl. s.	St. d.	Pl. s.	St. d.	Pl. s.	St. d.
Plateau	14	74	41	75	36	68	29	51
Slope	28	80	51	75	55	73	33	66
Depression	18	68	40	72	50	75	36	66

Pl. s.: field index of plow sole compaction.
St. d.: field index of stability deterioration.

There is no doubt that the slope soil is really sensitive to plow-sole compaction, but not so sensitive to stability deterioration. Probably tire slipping and vibration of a light tractor on slopes are important factors of plow-sole compaction. The small variation in soil compaction in the toposequence of Hesbaye may be explained by the general use of heavy tractors on the mechanized farms.

3. Influence of Traction. Three kinds of traction are distinguished: exclusively horses, exclusively tractors, or mixed. A mixed traction means that only the plowing is done with a tractor, which is normally a light tractor (less than 2000 kg weight); the other soil management operations are done by horses. If all operations are done with a tractor, the latter is normally a heavy one (more than 2000 kg weight).

The influence of traction is illustrated by the following data (Table 15).

It is evident that soil structure on farms with only horse traction is always better than on mechanized farms: this is true for soil compaction as well as for aggregate stability. The comparison of the mixed traction with exclusively tractor traction shows no real difference as far as aggregate stability is concerned. As far as soil compaction goes, the highest compaction index would be expected for tractor traction. However plow-sole compaction under mixed traction is the highest in 2 landscapes out of 5, and these 2 landscapes represent the extremes in cattle density, which does not facilitate the explanation of the frequency found.

Table 15. *Influence of traction*

Kind of Traction	Normal loam area, West		Normal loam area, East		Brabant		Hesbaye		Humid loam area	
	Pl. s.	St. d.	Pl. s.	St. d.	Pl. s.	St. d.	Pl. s.	St. d.	Pl. s.	St. d.
Horse	21	58	18	63	28	55	13	25	29	16
Mixed	23	71	41	75	38	71	39	60	34	35
Tractor	15	60	55	75	45	72	30	61	42	52

4. Influence of the Weight of the Tractor. In the eastern normal loam area special attention has been paid to the influence of the weight of the tractor, taking into account the amount of soil organic matter (both the cattle density and the presence of alfalfa in the crop rotation).

The following data (Table 16) show that the bad influence of the light tractor (less than 2000 kg weight) is evident, especially on plow-sole compaction, and that the structure deteriorating influence is decreasing with an increasing amount of soil organic matter.

Table 16. *Influence of light and heavy tractor*

Weight of tractor	Fields with alfalfa		Fields without alfalfa in the rotation					
			Cattle density in units per ha arable land					
			0—0.8		0.9—1.2		⩾1.3	
	Pl. s.	St. d.	Pl. s.	St. d.	Pl. s.	St. d.	Pl. s.	St. d.
Light tractor	36	68	81	86	55	84	37	77
Heavy tractor	29	32	52	74	45	68	36	76
Caterpillar	—	—	81	95	—	—	—	—

The very bad effect of tractor vibration on soil structure (compaction and stability) is made evident by the influence of the caterpillar traction, which is in use on some larger farms with an extreme low cattle density. It is clear that the structure deteriorating influence of the caterpillar is exactly the contrary effect to the one the farmer expected when buying this expensive machinery. The bad influence of wheel slipping and vibration of the light tractor is also proved by the increase of plow sole compaction on slopes versus plateau. The increase in com-

Table 17. *Influence of tractor on slopes*

Kind of traction	Changes in plow-sole compaction					
	Normal loam area, East			Hesbaye		
	Plateau	Slope	Difference	Plateau	Slope	Difference
Mixed traction	33	53	20	29	47	18
Only tractor	52	66	14	31	28	3

paction frequency is more important under mixed traction (ordinarely made with a light tractor, although horses are also used) than under tractor traction exclusively (which is generally made with a heavy tractor) (Table 17).

5. Influence of Soil Organic Matter. The amount of soil organic matter, which is necessary to maintain a good soil structure, is very different in the 5 sublandscapes of the loam region of Belgium, and, of course, changes with cattle density and the kind of crop rotation (including green manure or alfalfa). In the Western normal loam area, where the number of small farms (less than 10 ha) is high, the cattle density is high too, which is accompanied by a general practice of producing forage crop as second crop in the same year, such as rapes. Here a value of about 3 units of cattle density per ha arable land seems to be necessary; the average cattle density on the farms with evident stability deterioration of the soils was 2.6; on the farms with only a beginning of stability deterioration the cattle density was 3.5. In the eastern normal loam area an important difference was found, both in soil compaction and in aggregate stability, at a cattle density higher or lower than a critical value of 1.7. In the Brabant landscape this critical cattle density is 1.3 and in the humid loam area this value is 2.1. Therefore the amount of soil organic matter which ought to be given to a soil of a known textural class (e.g. the loam soils) cannot be generalised or given in terms of cattle density; the amount changes too much with the kind of soil management and soil use.

The influence of soil organic matter however can nicely be illustrated by comparing the soil structure which results if soil organic matter is given only once or if it is given twice or more times during the crop rotation.

As example of only one application we mention either stable manure or green manure, as example of more applications we mention the gift of both, or the gift of green manure combined with the gift of the crop residues (straw of cereals and leaves of sugar beets).

The following data (Table 18) show the improving influence of a double gift of soil organic matter during a rotation.

Table 18. *Influence of soil organic matter*

Application of soil organic matter per crop rotation	Brabant				Hesbaye			
	Cattle density				All soils		Soils in the depression	
	<1.3		$\geqslant 1.3$					
	Pl. s.	St. d.	Pl. s.	St. d.	Pl. s.	St. d.	Pl. s.	St. d.
Once	50	79	45	60	32	69	42	74
Twice	39	72	28	64	31	39	23	50

6. Influence of a Visible Reserve of Carbonate. This factor seems to be the most important, for the simple reason that its effects are the best visible and the most evident on both field indices of soil structure (Table 19).

Very striking is the improving soil structure effect of carbonate on soils poor in organic matter; if the cattle density is higher than 1.3 per ha arable land the improving effect of the reserve of $CaCO_3$ on plow sole compaction is less marked or even inexistent, but still visible for aggregate stability.

The evident improving effect of a reserve of $CaCO_3$ on loam soils has led to an important practical conclusion in Belgium.

On mechanized farms suffering from soil structure deterioration (assuming no $CaCO_3$-reserve is visible, although the pH may be about 7) the application of 30—40 tons/ha of $CaCO_3$ is recommended; in Belgium the most economical form of $CaCO_3$ is sugar sludge (lime sludge from sugar factories), because the value of

Table 19. *Influence of reserve of carbonate*

Reserve of calcium-carbonate	Normal loam area West; all soils		Normal loam area, East				Brabant				Hesbaye			
			Cattle density				Cattle density				All soils		Soils on slope	
			0.9—1.2		⩾1.3		<1.3		⩾1.3					
	Pl. s.	St. d.	Pl. s.	St. d.	Pl. s.	St. d.	Pl. s.	St. d.	Pl. s.	St. d.	Pl. s.	St. d.	Pl. s.	St. d.
Visible	10	59	38	60	33	71	26	63	37	59	29	41	21	41
Not visible .	28	72	60	88	37	80	63	88	35	65	36	82	46	91

its nitrogen and phosphorous content covers the cost of the product. After two crop rotations a second application of about 18 tons is recommended. The soil structure survey described in the previous pages made also possible the study of the influence of other factors, such as plowing at constant or varying depths, plowing before or after winter time, regular use of sodium-rich fertilizers, specific influence of alfalfa in the rotation, the deteriorating effect of an harvest of sugar beets in a rainy autumn, etc.

Details about these less important factors, however, cannot be given here, but we refer for them to the original papers.

Literature

Appelmans, F., L. De Leenheer and M. Van Ruymbeke: Invloed van de uitbatingswyze van de landbouwbedryven in de natte leemstreek op de structuur van de bovengrond. Mededelingen van de Landbouwhogeschool, Gent, **26**, 173–1754 (1961). — Appelmans, F., M. Van Ruymbeke and L. De Leenheer: Verband tussen uitbatingswyze van de landbouwbedryven in het westelyke deel van de normale leemstreek en de structuur van de bovengrond. Mededelingen van de Landbouwhogeschool, Gent, **27**, 1615–1630 (1962).

Boekel, P.: Some remarks on the influence of soil structure on plant growth. Proceedings of the International Symposium on Soil Structure. Mededelingen van de Landbouwhogeschool, Gent, **24**, 52–57 (1959). — The amelioration of the structure of clay soils by liming. Proceedings of the International Symposium on Soil Structure. Mededelingen van de Landbouwhogeschool, Gent, **24**, 122–127 (1959). — Bryant, J. C., T. W. Bendixen and C. S. Slater: Measurement of the water stability of soils. Soil Sci. **65**, 341–345 (1948).

Cannon, W. A.: Physiological features of roots with special reference to the relation of roots to the aeration of soil. Carnegie Institute Publ., Washington, **368**, 1–168 (1925). — Collis-George, N.: Relationship between air and water permeabilities in porous media. Soil Sci. **76**, 239–250 (1953).

Dalla Valle, J. M.: Micromeritics, 2nd Ed. New York: Pitman. 1948. — De Boodt, M., and L. De Leenheer: Soil structure index and plant growth. Proceedings of the International Symposium on Soil Structure. Mededelingen van de Landbouwhogeschool, Gent, **24**, 312–323 (1959). — De Boodt, M., L. De Leenheer and D. Kirkham: Soil aggregate stability indexes and crop yields. Soil Sci. **91**, 138–146 (1961). — De Caestecker, K., and L. De Leenheer: Influence de la dégradation de la structure du sol sur le développement et la production de betteraves fourragères dans les sols argileux des Polders. Proceedings of the International Symposium on Soil Structure. Mededelingen van de Landbouwhogeschool, Gent, **24**, 44–51

(1959). — De Leenheer, L.: La perméabilité du sol à l'air et la production en betteraves sucrières. Rev. Agriculture, Bruxelles, **6**, 2–8 (1953). — De Leenheer, L., and M. De Boodt: Determination of aggregate stability by the change in mean weight diameter. Proceedings of the International Symposium on Soil Structure. Mededelingen van de Landbouwhogeschool, Gent, **24**, 290–300 (1959a). — Est-il indiqué d'améliorer la structure du sol par un labour profond ou par un labour normal avec sous-solage? Proceedings of the International Symposium on Soil Structure. Mededelingen van de Landbouwhogeschool, Gent, **24**, 154–165 (1959b). — De Leenheer, L., and K. De Caestecker: The influence of poor soil structure on growth and final yield of wheat. Proceedings of the International Symposium on Soil Structure. Mededelingen van de Landbouwhogeschool, Gent, **24**, 36–43 (1959). — De Leenheer, L., M. De Boodt en J. Vandamme: Oefent de mechanisatie en uitbatingswyze een invloed uit op de bodemstructuur in de Leemstreek? Mededelingen van de Landbouwhogeschool, Gent, **25**, 773–802 (1960).

Evans, D. D., and D. Kirkham: Measurement of the air permeability of soil in situ. Soil Sci. Soc. Amer. Proc. **14**, 65–73 (1950).

Flaig, W., K. Scharrer and G. K. Judel: Zur Methodik der Bestimmung des Redoxpotentials im Boden. Z. Pflanzenernähr., Düng., Bodenkde. **68** (113), 97–122 (1955). — Fountaine, E. R.: The physical requirements of plants as criteria for soil structure. Proceedings of the International Symposium on Soil Structure. Mededelingen van de Landbouwhogeschool, Gent, **24**, 30–35 (1959). — Frese, H.: Non published data (1956) from the "Institut für Bodenbearbeitung". Braunschweig.

Gardner, W. R.: Representation of soil aggregate size distribution by a logarithmic-normal distribution. Soil Sci. Soc. Amer. Proc. **20**, 151–153 (1956). — Gill, W. R., and G. H. Bolt: Pfeffer's studies of the root growth pressures exerted by plants. Agronomy J. **47**, 166–168 (1955). — Grover, B. L.: Simplified air permeameters for soil in place. Soil Sci. Soc. Amer. Proc. **19**, 414–418 (1955). — Gupala, A. I.: Thermal properties of soil in relation to its moisture content and compaction. Pochvovedenie **4**, 40–45 (1959). Written in Russian, English summary in Soils and Fertilizers **22**, 360 (1959).

Hodgman, C. D.: Handbook of Chemistry and Physics, 30th Ed. Cleveland, Ohio: Chemical Rubber Publishing Co. 1867–1868, 1947.

King, F. A.: A textbook of the physics of agriculture, 6th Ed. Published by Mrs. F. H. King. Madison, Wisconsin, 1914. — Kirkham, D.: Field methods for determination of air permeability of soil in its undisturbed state. Soil Sci. Soc. Amer. Proc. **11**, 93–99 (1947). — Kirkham, D., M. De Boodt and L. De Leenheer: Modulus of rupture determination on undisturbed soil core samples. Soil Sci. **87**, 141–144 (1959a). — Air permeability at the field capacity as related to soil structure and yields. Proceedings of the International Symposium on Soil Structure. Mededelingen van de Landbouwhogeschool, Gent, **24**, 377–391 (1959b). — Kulmann, A.: Zur Problematik der Krümelstabilitätsmessungen und zur Methodik des Durchflußverfahrens. Tagungsberichte nr. 13. Probleme der Krümelstabilitätsmessung und der Krümelbildung. Deutsche Akademie der Landwirtschaftswissenschaften zu Berlin, 7–34 (1958).

Laws, W., and D. D. Evans: The effects of longtime cultivation on some physical and chemical properties of two rendzina soils. Soil Sci. Soc. Amer. Proc. **14**, 15–19 (1950). — Lawton, K.: The influence of soil aeration on the growth and absorption of nutrients by corn plants. Soil Sci. Soc. Amer. Proc. **10**, 263–268 (1945). — Lutz, J. F.: Mechanical impedance and plant growth. Soil physical conditions and plant growth, Vol. II of Agronomy, pp. 44–67. New York: Academic Press. 1952.

Mc Intyre, D. S., and C. B. Tanner: Anormally distributed soil physical measurements and nonparametric statistics. Soil Sci. **88**, 133–137 (1959).

Parker, E. R., and H. Jenny: Water infiltration and related soil properties as affected by cultivation and organic fertilization. Soil Sci. **60**, 353–376 (1945). — Peech, M., and L. P. Batjer: A critical study of the methods for measuring oxidation-reduction potentials of soils, with special reference to orchard soils. Agricult. Exper. Station Bull., New York State, **625**, 1–23 (1935). — Penman, H. L.: Gas and vapour movements in the soil. II. The diffusion of carbon dioxide through porous solids. J. Agricult. Sci. **30**, 570–581 (1940).

Richards, S. J., R. M. Hagan and T. M. Mc Calla: Soil temperature and plant growth. Soil physical conditions and plant growth. Vol. II of Agronomy, pp. 303–460. New York: Academic Press. 1952. — Rommel, L. G.: Mécanisme de l'aération du sol. Ann. Agronom. **5**, 373–384 (1935). — Russell, M. B.: Soil aeration and plant growth. Soil physical conditions and plant growth. Vol. II of Agronomy, pp. 254–290. New York: Academic Press. 1952.

Schuffelen, A. C., A. R. P. Janse and G. P. Wiersema: Een eenvoudige methode om het koolzuurgehalte van bodemlucht te bepalen. Landbouwkundig Tijdschr., Den Haag, **66**, 36–39 (1954). — Starkey, R. L., and K. M. Wight: Anaerobic corrosion of iron in soil. New York: American Gas Association. 1945. — Stirk, G. B.: Expression of soil aggregate distribution. Soil Sci. **86**, 133–136 (1958). — Swanson, C. L. W., and H. G. M. Jacobson: Effect of soil hardness and compaction on corn growth. Soil Sci. Soc. Amer. Proc. **20**, 161–167 (1956).

Tanner, C. B., and R. W. Wengel: An air permeameter for field and laboratory use. Soil Sci. Soc. Amer. Proc. **21**, 663–664 (1957). — Taylor, S. A.: Oxygen diffusion in porous media as affected by compaction and moisture content. Soil Sci. Soc. Amer. Proc. **14**, 55–61 (1949). — Tendeloo, H. J. C., G. J. Vervelde and A. T. Z. Zwart-Voorspuy: Electro-chemical behaviour of ion exchanging substances. Potential measurements of plant roots. Rec. trav. chim. Pays-Bas **63**, 97–104 (1944).

Van Bavel, C. H. M.: Mean weight diameter of soil aggregates as a statistical index of aggregation. Soil Sci. Soc. Amer. Proc. **14**, 20–23 (1950). — Gaseous diffusion and porosity in porous media. Soil Sci. **73**, 91–104 (1952). — Vandamme, J., L. De Leenheer and F. Appelmans: Invloed van de uitbatingswyze van de landbouwbedryven in de Brabantse leemstreek op de structuur van de bovengrond. Mededelingen van de Landbouwhogeschool, Gent, **28**, 133–164 (1963). — Vandamme, J., L. De Leenheer and M. De Boodt: Verband tussen uitbatingswyze van de landbouwbedryven in het oostelyk gedeelte van de normale leemstreek en de structuur van de bovengrond. Mededelingen van de Landbouwhogeschool, Gent, **28**, 41–79 (1963). — Veihmeyer, F. J., and A. H. Hendrickson: Soil density and root penetration. Soil Sci. **65**, 487–493 (1948).

Wilde, S. A., and E. C. Steinbrenner: Determination of air permeability of soil by means of a sphygmomanometer. J. Forestry **48**, 840–841 (1950). — Willis, W. C., W. E. Larson and D. Kirkham: Corn growth as affected by soil temperature and mulch. Agronomy J. **49**, 323–328 (1957).

C. Der chemische Aufbau des Bodens[1]

a) Die festen anorganischen Bestandteile des Bodens

Von

U. Schwertmann

Die festen anorganischen Bestandteile des Bodens sind Minerale. Sie lassen sich unterteilen in primäre und sekundäre Minerale. Unter primären Mineralen sollen im folgenden alle solche verstanden werden, die magmatischen oder metamorphen Ursprungs sind, während zu den sekundären Mineralen alle bei der Verwitterung entstandenen Minerale gezählt werden sollen, gleichgültig ob sie direkt bei der Verwitterung magmatischer oder metamorpher Gesteine oder auf dem Umweg über das Sediment in den Boden gelangten. Tonminerale des Bodens, die dem Sedimentgestein entstammen, sind daher in gleicher Weise sekundär wie solche, die sich bei der Verwitterung magmatischer Gesteine zum Boden bildeten.

Die zu besprechenden Minerale werden nach folgender Einteilung abgehandelt:

Minerale des Bodens
↙ ↘
Primäre Minerale — Sekundäre Minerale
↙ ↓ ↘
Silikate (Tonminerale) — Oxide und Hydroxide — Carbonate, Sulfate, Phosphate, Sulfide u. a.

[1] Dieser Beitrag wurde 1959 verfaßt und 1963 mit einigen Ergänzungen versehen; er stellt somit im wesentlichen den Stand des Wissens von 1959 dar.

1. Primäre Minerale

Ein wechselnder Anteil der Minerale des Bodens ist magmatischer oder metamorpher Herkunft und mehr oder weniger unverändert aus dem Ausgangsgestein in den Boden gelangt. Da der Boden durch Verwitterungsprozesse entsteht, ist die Wahrscheinlichkeit, daß ein Mineral bei der Entwicklung des Bodens unverändert bleibt, um so größer, je höher dessen Verwitterungsresistenz ist. Daher finden sich unter den primären Mineralen des Bodens hauptsächlich solche mit mittlerer und hoher Verwitterungsresistenz. Der Anteil an schwer verwitterbaren primären Mineralen steigt mit zunehmendem Verwitterungsgrad des Bodens (z. B. Rendzina < Braunerden < Podsol). Da die Verwitterungsgeschwindigkeit mit wachsender Oberfläche ansteigt und die Verwitterungsneubildungen vorwiegend den feineren Fraktionen angehören, steigt andererseits häufig der Anteil primärer Minerale von der Tonfraktion zur Sandfraktion an.

Die wichtigsten primären Minerale des Bodens sind der Quarz, verschiedene Silikate, vor allem Feldspäte, Glimmer, Augite und Hornblenden, sowie einige Schwerminerale mit hoher Verwitterungsresistenz.

Der *Quarz* SiO_2 tritt als häufigstes Mineral in fast allen Böden auf. Sein Anteil hängt vor allem vom Ausgangsgestein und vom Verwitterungsgrad ab. Böden aus Si-reichen magmatischen Gesteinen und aus Sandsteinen und Sanden bestehen meist zu über 50% aus Quarz, während solche aus Ca- und Mg-reichen magmatischen Gesteinen, Kalk- oder Tongesteinen weniger Quarz (25 bis 50%) enthalten (Tab. 20). Da der Quarz wegen der geringen Löslichkeit, der hohen Härte und der fehlenden Spaltbarkeit eines der resistentesten Minerale ist, nimmt sein Anteil am Gesamtboden im Laufe der Verwitterung zu. Stark ausgeprägte Podsole können daher bis >95% Quarz enthalten (VAN DER MAREL 1949). Das Maximum des Quarzgehalts liegt, abgesehen von den Gesteinsresten, meistens in der Grobsandfraktion, doch enthält auch die Tonfraktion noch geringe Mengen an Quarz. Wegen der hohen Verwitterungsresistenz des Quarzes wird dieser in Böden des gemäßigt humiden Klimas als Bezugsgröße für das Ausmaß der stofflichen Veränderungen durch die Bodenbildung (Profilbilanzen) verwendet (BARSHAD 1955, KUNDLER 1959, KÖSTER 1961, BLUME 1963). Aus dem gleichen Grund läßt sich aus der Konstanz des Quarzverhältnisses zweier Sandfraktionen (z. B. 100 bis 200 μ und 200 bis 300 μ) im Profil die Homogenität des Ausgangsmaterials der Bodenbildung messen (KUNDLER 1961).

Von den primären Silikaten sind die Feldspäte und Glimmer am häufigsten vertreten. Der Gesamt-*Feldspat*-Gehalt der Böden liegt häufig zwischen etwa 5 und 30% des Gesamtbodens (Tab. 20) und hängt wie beim Quarz vor allem von der mineralogischen Zusammensetzung des Ausgangsgesteins und dem Verwitterungsgrad des Bodens ab. Böden aus altpleistozänen Sanden sind häufig arm an Feldspäten (< 10%) (VAN DER MAREL 1949, SALGER 1954, KUNDLER 1956), während solche aus jungpleistozänen Sanden im Mittel 15 bis 25% (KUNDLER 1956) und aus feintexturierter Grundmoräne (CANN und WHITESIDE 1955, BAILEY, WHITESIDE und ERICKSON 1957) meist mehr als 20% Feldspäte enthalten. Ein hoher Feldspat-Gehalt findet sich auch in Böden aus Buntsandstein (SCHACHTSCHABEL 1937, HUFFMANN 1954) sowie aus Basalt (HUFFMANN 1954, SMITH 1957), während Böden aus Löß, Ton und Tonschiefer mittlere Werte von 10 bis 15% aufweisen (SCHACHTSCHABEL 1937, SCHROEDER 1954, SCHEFFER, MEYER und KALK 1958). Der Gesamtgehalt an Feldspäten ist im Gegensatz zum Quarz in den feinen Fraktionen (Feinsand, Schluff) höher als in den gröberen Fraktionen (JOHNSON und BEAVERS 1959, WURMAN u. a. 1959).

Die einzelnen Feldspäte verhalten sich entsprechend ihrer verschiedenen Verwitterungsresistenz unterschiedlich. Der relativ stabile *Orthoklas* (K-Feldspat $KAlSi_3O_8$) reichert sich im Laufe der Verwitterung gegenüber den leichter verwitternden *Plagioklasen* (Kalknatronfeldspäte $NaAlSi_3O_8$ bzw. $CaAl_2Si_2O_8$) relativ an (vgl. CANN und WHITESIDE 1955). Aus dem gleichen Grunde ist der Orthoklas-Anteil am Gesamt-Feldspatgehalt in feineren Fraktionen häufig höher als der Plagioklas-Anteil.

Da sich im Laufe der Verwitterung der stabile Quarz gegenüber leichter verwitterbaren Mineralen wie z. B. Feldspäten relativ anreichert, wird zur Kennzeichnung des Verwitterungsgrades häufig das Quarz-Feldspat-Verhältnis verwendet (SCHROEDER 1954, ANDERSON und WHITE 1958, JOHNSON und BEAVERS 1959).

In ähnlicher oder geringerer Menge als die Feldspäte treten die *Glimmer* im Boden auf; ihr Anteil liegt jedoch meist unter 10% (Tab. 20; vgl. die bei den Feldspäten angegebene Literatur). Wie bei den Feldspäten erfolgt auch bei den Glimmern eine Differenzierung der verschiedenen Glimmer nach ihrer Verwitterungsresistenz. Während der dunkle *Biotit* $K(Mg, Fe, Mn)_3(AlSi_3)O_{10}(OH, F)_2$ relativ leicht verwittert, reichert sich der helle *Muskovit* $KAl_2(AlSi_3)O_{10}(OH, F)_2$ meist relativ an. Sein Maximum liegt wegen der guten Spaltbarkeit häufig in der Schlufffraktion. Da die Verwitterung der Glimmer zum Teil in der Abgabe des zwischen den Schichten gebundenen Kaliums besteht, stellen sie mit dem Orthoklas die wichtigsten natürlichen K-Reserven für die Pflanze dar. Außerdem sind sie wie dieser ein wesentliches Ausgangsmaterial für die Tonmineralbildung (S. 337). Der Biotit verliert bei der Verwitterung außer K einen Teil seines Fe-Gehaltes und bleicht hierdurch. Da das Eisen meist unmittelbar nach Austritt aus dem Gitter oxydiert und als Fe(III)-oxid ausgefällt wird, ist der gebleichte Biotit teilweise mit einer braunen Kruste umgeben.

Von den zu den *Schwermineralen* (spezifisches Gewicht > 2,9) gehörenden primären Mineralen, deren Menge meist einige Prozente beträgt, treten in Böden solche auf, die sehr leicht verwittern, und solche, die sehr widerstandsfähig sind. Von den *leicht verwitterbaren*, meist Fe-haltigen Schwermineralen verwittert der *Olivin* $(Mg, Fe)_2SiO_4$ besonders rasch und setzt dabei das Eisen als braunes Oxid frei. Selbst in Böden aus Basalt findet sich daher der Olivin meist nur noch in Gesteinsresten oder in stark angewitterten Körnern (HUFFMANN 1954, SMITH 1957). Ebenfalls relativ leicht, jedoch schwerer als der Olivin, werden *Augite* (Ca, Fe, Na) (Mg, Al, Fe, Ti) $(Si, Al)_2O_6$ und *Hornblenden* $(Ca, Fe, Na, K)_2(Mg, Al, Fe, Ti)_5$ $(Si, Al)_8(O, OH)_2O_{22}$ zersetzt, so daß ihr Anteil meist gering ist und mit der Tiefe zunimmt. Hornblenden wurden jedoch auch in Podsolen des feuchtkühlen Klimas gefunden (VAN DER MAREL 1949).

Die *schwer verwitterbaren* Schwerminerale, zu denen *Turmalin* $Na(Al, Mg, Li)_9$ $B_3[O, OH, F/SiO_4]_6(OH)_{0-2}$, *Zirkon* $ZrSiO_4$, *Rutil* TiO_2, *Titanit* $CaTiSiO_5$, *Ilmenit* $FeTiO_3$, *Apatit* $Ca_5(PO_4)_3(OH, F)$, *Magnetit* Fe_3O_4 und einige andere zu zählen sind, werden wegen dieser Eigenschaft herangezogen, um festzustellen, ob das Ausgangsmaterial der Bodenbildung einheitlich zusammengesetzt ist (BARSHAD 1955). Sie reichern sich im Laufe der Verwitterung relativ an, so daß ihre Menge als Bezugsbasis für die durch Verwitterung verursachten stofflichen Veränderungen dienen kann. So reicherte sich z. B. in tropischen Böden aus rhyolithischem Tuff auf Sumatra das Mineral Zirkon im Laufe zunehmender Verwitterung von 6 auf 40% der Schwermineralfraktion an, während innerhalb der gleichen Verwitterungsserie der Hornblende-Anteil von 42% auf 0% absank (VAN DER MAREL 1947). Als einfache Bezugsgröße wird auch der Zr-Gehalt (KHAN 1959a) oder der Ti-Gehalt (BLUME 1963) herangezogen.

2. Tonminerale[1]

A. Definition und Einteilung der Tonminerale

Den Hauptanteil der sekundären Minerale des Bodens bilden die Tonminerale. Es sind dies vorwiegend kristalline wasserhaltige Alumosilikate mit wechselnden Anteilen von Fe, Ti, Mg, Ca, K und Na, bisweilen auch Li, Zn und Cr, die hauptsächlich der Tonfraktion der Böden angehören. Sie kristallisieren monoklin oder triklin und sind meist blättchenförmig (Phyllosilikate), zum Teil auch röhren-, leisten- oder kettenförmig (Attapulgit, Sepiolit) ausgebildet. Einige sind amorph.

In Tab. 21 sind die einzelnen Gruppen der Tonminerale und deren wichtigste Vertreter sowie deren Strukturformeln aufgeführt. Die Benennung der einzelnen Minerale entspricht den Gepflogenheiten der Fachliteratur. Die hier wiedergegebene Unterteilung erfolgt in erster Linie nach der Struktur der Minerale (S. 321 ff.). Innerhalb der einzelnen Gruppen lassen sich weitere Unterteilungen, z. B. nach Art und Anordnung der kationischen Einzelbausteine (S. 333), vornehmen (vgl. MACKENZIE 1959). Da letztere in weiten Grenzen variieren können, gibt es zwischen den genannten typischen Vertretern zahlreiche Übergänge. Die angegebene Formel entspricht daher jeweils nur der Zusammensetzung des Minerals der ebenfalls angeführten Herkunft.

Den Tonmineralen gemeinsam ist eine Reihe von Eigenschaften, die eng mit ihrem Kristallaufbau und ihrer zum Teil besonders gearteten, meist sehr großen spezifischen Oberfläche verknüpft sind und nicht nur den Tonen, sondern auch vielen tonhaltigen Böden typische und für deren Fruchtbarkeit wichtige Eigenschaften verleihen. Der Besprechung der einzelnen Minerale sollen daher einige Erläuterungen über die Struktur und die Oberfläche der Tonminerale vorangestellt werden.

B. Allgemeines über die Struktur der Tonminerale

Die mit Hilfe einer chemischen Analyse bestimmbaren Bausteine der Minerale, die man sich in vereinfachter Weise als Kugeln vorstellen kann, sind nach einem bestimmten Bauplan in Form einer mehr oder weniger dichten Kugelpackung zu dem Kristallgitter des Minerals zusammengefügt. Die Kugeln dieser Kugelpackung sind im Falle der Tonminerale die großen O- und OH-Ionen, in deren Lücken kleinere Kationen, vor allem Si, Al, Fe und Mg, eingelagert sind (*Zentralkationen*).

---

[1] Bei der Abfassung dieses Kapitels wurden vor allem folgende Zusammenstellungen oder Monographien benutzt:

MARSHALL, C. E.: The colloid chemistry of the silicate minerals. New York: Academic Press Inc. 1949.

Reference clay minerals. A.P.I. Research Project 49. New York: Columbia University. 1951.

GRIM, C. E.: Clay mineralogy. London: McGraw-Hill. 1953.

JASMUND, K.: Die silikatischen Tonminerale, 2. Aufl. Weinheim: Verlag Chemie. 1955.

RICH, C. I., und G. W. THOMAS: The clay fraction of soils. Adv. Agron. **12**, 1 (1960).

SCHEFFER, F., und P. SCHACHTSCHABEL: Bodenkunde. Stuttgart: Enke. 1960.

Darüber hinaus wurden vor allem neuere Original-Arbeiten zitiert. Da das Interesse an der Tonmineralogie auf zahlreiche Wissenschaftszweige verstreut ist, sind die Spezialarbeiten über eine große Zahl verschiedener Zeitschriften mineralogischer, geologischer, keramischer, bodenkundlicher u. a. Art verteilt. Als ausschließlich den Tonmineralen gewidmete Zeitschriften seien hier genannt: „Clay Mineral Bulletin", London, und Berichte der jährlichen „National Conference of Clays and Clay Minerals", USA.

Tabelle 20. *Mineralzusammensetzung einiger Böden des gemäßigt humiden Klimabereiches (Oberböden)*

Bodentyp und/ oder Ausgangsgestein	Ort	Bodenart	Gesteinsreste	Mineralzusammensetzung (%)								Autor
				Quarz	Orthoklas	Plagioklas	Glimmer	Schwerminerale und Opake	Kaolinit und Halloysit	Illit	Montmorillonit	
Podsol (Sand)[1]	Drenthe (Holland)	S	—	84,6	5,5		0,1	0,1		3,9		VAN DER MAREL 1949
Podsol (Sand)	Oldenburg (Deutschland)	hS	—	75	9		3	—	—	2	—	SCHACHTSCHABEL unveröffentlicht
Sandstein	Jena (Deutschland)	S	4	62,5	28,5		1	—	—	2—3	2—3	SCHACHTSCHABEL 1937
Schotter	Voralpen (Deutschland)	sL	4,5	62,5	1,4	3,5	5,0	1,7	0,8	7,9	—	SALGER 1954
Schwarzerde, degrad. (Löß)	Hannover (Deutschland)	sL	—	53	12		8	etwa 10	Sp.	17	—	SCHROEDER 1954
Basalt	Bushmills (Nordirland)	L	22,6	—	—	16,0	16,9[2]	28,0[3]	3,1	—	—	SMITH 1957
Tonschiefer	Thüringen (Deutschland)	tL	4	38	8		5	—	—	8	4	SCHACHTSCHABEL 1937
Ton	Jena (Deutschland)	T	—	31	13		4	—	—	22	11	SCHACHTSCHABEL 1937
Rendzina[4] (Kalkstein)	Göttingen (Deutschland)	tL	—	27,4	8,0	—	34,6	10		19,6		HUFFMANN 1954

[1] Fraktion 16 bis 500μ (= 96,1% des Gesamtbodens) [2] Als Vermikulit angegeben [3] Augit [4] Auf $CaCO_3$-freie Substanz bezogen

Tabelle 21. *Einteilung und Strukturformeln (Beispiele für typische Vertreter) der verbreitetsten Tonminerale*

Bezeichnung und Herkunft	Oktaederschicht	Tetraederschicht	Kationen und H_2O der Zwischenschichten[1]	Autor
1. Kaolinit-Minerale				
Kaolinit	$Al_2\,(OH)_4]$	$[Si_2\,O_5]$		
Halloysit	$[Al_2\,(OH)_4]$	$[Si_2\,O_5]$	0—2,0 H_2O	
2. Montmorillonit-Minerale				
Montmorillonit Wyoming, USA	$[Al_{1,55}\,Fe^{3+}{}_{0,15}\,Mg_{0,33}\,(OH)_2]$	$[Si_{3,92}\,Al_{0,08}\,O_{10}]$	$X_{0,20}\,nH_2O$	Ross und Hendricks 1945
Beidellit Beidell, Colo., USA	$[Al_{1,40}\,Fe^{3+}{}_{0,50}\,Mg_{0,08}\,(OH)_2]$	$[Si_{3,58}\,Al_{0,42}\,O_{10}]$	$X_{0,59}\,nH_2O$	Foster 1954
Nontronit Untergrießbach (Obb.), Deutschland	$[Fe^{3+}{}_{1,76}\,Al_{0,15}\,Mg_{0,11}\,(OH)_2]$	$[Si_{3,50}\,Al_{0,50}\,O_{10}]$	$X_{0,53}\,nH_2O$	Hofmann 1956
3. Illit- (Glimmerartige) Minerale				
Hydromuskovit Pretoria, Südafrika	$[Al_{1,90}\,Fe^{3+}{}_{0,04}\,Mg_{0,06}\,(OH)_2]$	$[Si_{3,06}\,Al_{0,94}\,O_{10}]$	$K_{0,68}\,Na_{0,16}\,X_{0,16}\,nH_2O$	Heystek 1955
Illit Fithian, Ill., USA	$[Al_{1,52}\,Fe_{0,30}\,Mg_{0,19}\,(OH)_2]$	$[Si_{3,54}\,Al_{0,46}\,O_{10}]$	$K_{0,50}\,X_{0,18}\,nH_2O$	Foster 1954
Glaukonit Mittel aus ~40 Analysen	$[Fe^{3+}{}_{0,97}\,Fe^{2+}{}_{0,19}\,Al_{0,47}\,Mg_{0,40}\,(OH)_2]$	$[Si_{3,65}\,Al_{0,35}\,O_{10}]$	(K, Ca, Na)$_{0,84}$ nH_2O	Hendricks und Ross 1941
4. Vermikulit-Minerale				
Vermikulit Kropfmühl (Obb.), Deutschland	$[Mg_{2,61}\,Fe^{3+}{}_{0,10}\,Al_{0,29}\,(OH)_2]$	$[Si_{2,95}\,Al_{1,05}\,O_{10}]$	$X_{0,65}\,nH_2O$	Hofmann 1956
5. Chlorit-Minerale				
Chlorit Rimpfischwäng (Schweiz)	$[Mg_{4,65}\,Fe^{2+}{}_{0,40}\,Al_{0,90}\,(OH)_8]$	$[Si_{3,20}\,Al_{0,80}\,O_{10}]$		Eckhardt 1958
6. Wechsellagerungs-Minerale (Formel entspricht den Anteilen der beteiligten Elementarschichten der einzelnen Minerale)				
7. Allophan-Minerale				

[1] X = wechselnde Anteile verschiedener Kationen (ausgedrückt in Äquivalenten), z. B. Ca, Mg, K, Na, H, H_3O, die austauschbar sind

Einen Überblick über die Größe dieser Ionen gibt Tab. 22. Zwischen den O- bzw. OH-Ionen stehen für die Einlagerung der Zentralkationen zwei Arten von Lücken zur Verfügung, deren Durchmesser 0,55 bzw. 0,61 A betragen. In der einen sind

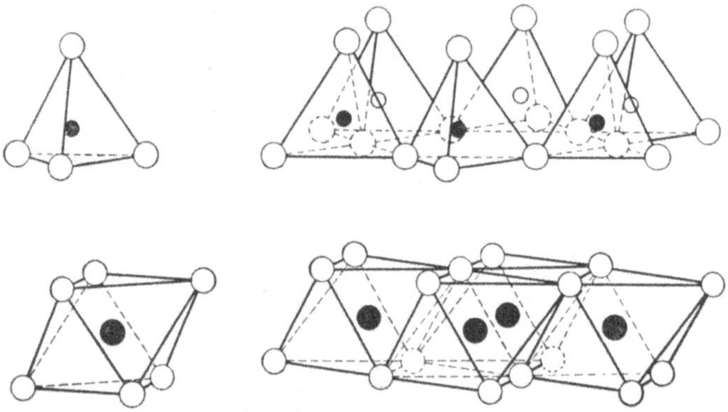

Abb. 72. Bauelemente der Tonminerale und ihre Verknüpfung (schematische Darstellung). Oben: Tetraeder. Unten: Oktaeder. Die größeren Kreise stellen die O- bzw. OH-Ionen, die Punkte bzw. kleineren Kreise die Zentralionen (Si, Al, Mg, Fe u. a.) dar (nach GRIM 1953)

die Zentralkationen tetraedrisch, in der anderen oktaedrisch von den O- bzw. O- und OH-Ionen umgeben (Abb. 72). Der Abstand O-O beträgt im Tetraeder 2,55 A, im Oktaeder 2,60 A.

Tabelle 22. *Ionenradien* (*r*)

Ion	Li	Na	K	NH_4	Mg	Ca	Ba	B	Al	Si	Ti	P
r [A]	0,78	0,98	1,33	1,43	0,78	1,06	1,43	0,20	0,57	0,39	0,64	~0,35

Ion	Fe^{2+}	Fe^{3+}	Mn^{2+}	Mn^{3+}	Mn^{4+}	Co^{2+}	Co^{3+}	Ni^{2+}	Cu^{2+}	Zn
r [A]	0,83	0,67	0,91	0,70	0,52	0,82	0,64	0,78	0,96	0,88

Ion	O, OH	H_3O	F	Cl
r [A]	1,32	1,45	1,33	1,81

Die untereinander durch O-Brücken verknüpften O-*Tetraeder* und O-OH-*Oktaeder* sind in der in Abb. 73 dargestellten Weise schichtförmig angeordnet. Mehrere der so entstehenden Tetraeder- und Oktaederschichten sind wiederum — ebenfalls über O-Brücken — miteinander zur sogenannten *Elementarschicht* verbunden. Dabei bilden die O-Ionen der Basisflächen aller Tetraeder einer Schicht die eine Seite dieser Schicht und sind hier so miteinander verknüpft, daß ein hexagonales Netzwerk von O-Ionen entsteht (Abb. 72 und 73). Die zur anderen Seite zeigenden O-Ionen der Tetraederspitzen stellen die Verknüpfung mit der benachbarten Oktaederschicht her (Abb. 73). Je nach der Aufeinanderfolge von Tetraeder- und Oktaederschichten spricht man von *1:1-* bzw. *2:1-Mineralen.* Während bei den 1:1-Mineralen jeweils eine Oktaederschicht mit einer Tetraederschicht zu einer Elementarschicht verknüpft ist, besteht die Elementarschicht der 2:1-Minerale aus einer Oktaederschicht, die beiderseitig von je einer Tetraederschicht umgeben ist (Abb. 73). Zu den 1:1-Mineralen gehören die Kaolinit-Minerale, zu den 2:1-Mineralen die Montmorillonit-, Vermikulit- und Illit-Minerale sowie die Glimmer, der Talk und der Pyrophyllit. Bei den Chlorit-Mineralen ist zwischen den 2:1-Elementarschichten jeweils eine oktaedrische Brucit-Schicht eingelagert *(2:2-Minerale)*, so daß hier wiederum ein 1:1-Wechsel von oktaedrischen und tetraedrischen Schichten entsteht (Abb. 74).

Art und Ausmaß der Besetzung der Tetraeder- und Oktaederzentren mit den verschiedenen Kationen (Tab. 21) bestimmen die Ladungsverhältnisse der Minerale und bedingen, vor allem bei den 2:1-Mineralen, die große Verschiedenheit der

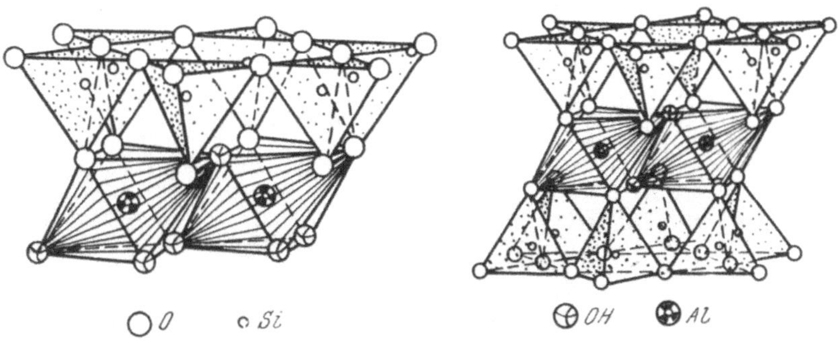

Abb. 73. Schichtförmige Anordnung der Tetraeder und Oktaeder im Gitter der Tonminerale: links 1:1-Minerale, rechts 2:1-Minerale (schematische Darstellung nach JASMUND 1955)

chemischen Zusammensetzung, während die 1:1-Minerale relativ einheitlich zusammengesetzt sind. Sämtliche Tetraederzentren sind mit Zentralionen besetzt, und zwar in erster Linie mit Si und außerdem mit Al. Der Anteil des Al beträgt bei den dioktaedrischen (s. unten) 1:1-Mineralen 0 oder < 1% (S. 329). Bei den 2:1-Mineralen liegt er zwischen 0% (z. B. bei Pyrophyllit und Talk) und 50% (bei Margarit $CaAl_4Si_4Al_4O_{20}(OH)_8$), beträgt bei den Tonmineralen in der Regel jedoch < 15%. Bei Biotit und Muskovit steigt er dagegen auf maximal 25% an.

Im Gegensatz zu den Tetraederzentren schwankt die Oktaederbesetzung zwischen $^2/_3$ und $^3/_3$ aller möglichen Plätze. Bei $^2/_3$-Besetzung spricht man von *dioktaedrischen*, bei $^3/_3$-Besetzung von *trioktaedrischen* Mineralen, die beide innerhalb aller in Tab. 21 genannten Mineralgruppen auftreten. Die Oktaederzentren sind hauptsächlich durch Al, Mg, Fe, in untergeordnetem Maße auch durch Mn, Ti, Cr, Zn und Li besetzt. Bei dioktaedrischer Besetzung mit Al besitzt die Oktaederschicht die Struktur des Hydrargillits mit der Struktureinheit $Al_2(OH)_6$, bei trioktaedrischer Besetzung mit Mg diejenige des Brucits mit der Einheit $Mg_3(OH)_6$. Unter den Glimmern ist der Muskovit dioktaedrisch, der Biotit trioktaedrisch, unter den als Modellminerale geltenden Tonmineralen der Pyrophyllit di- und der Talk trioktaedrisch.

Da bei den letztgenannten Modellmineralen innerhalb der Elementarschicht die Summe der positiven Ladungen der Kationen in den Oktaeder- und Tetraederzentren gleich der Summe der negativen Ladungen der sie umgebenden O- und OH-Ionen (bezogen z. B. auf die Formeleinheit) ist, sind diese Minerale nach außen elektrisch neutral. Sind die genannten Summen jedoch ungleich, so resultiert eine positive oder negative Überschußladung, die entweder durch eine gleich hohe, aber entgegengesetzte in den Nachbarschichten (z. B. beim Chlorit), durch Einlagerung von Gitterbausteinen zwischen die Elementarschichten (z. B. K bei den Glimmern und Illit-Mineralen) oder durch Sorption austauschbarer Ionen (z. B. bei den Montmorillonit-Mineralen) ausgeglichen wird. Eine negative Überschußladung entsteht bei den Tonmineralen in erster Linie dadurch, daß bei gleichbleibender Zahl der Zentralkationen Oktaederzentren nicht mit Al oder Fe^{3+}, sondern mit Mg oder Fe^{2+}, und Tetraederzentren nicht mit Si, sondern mit

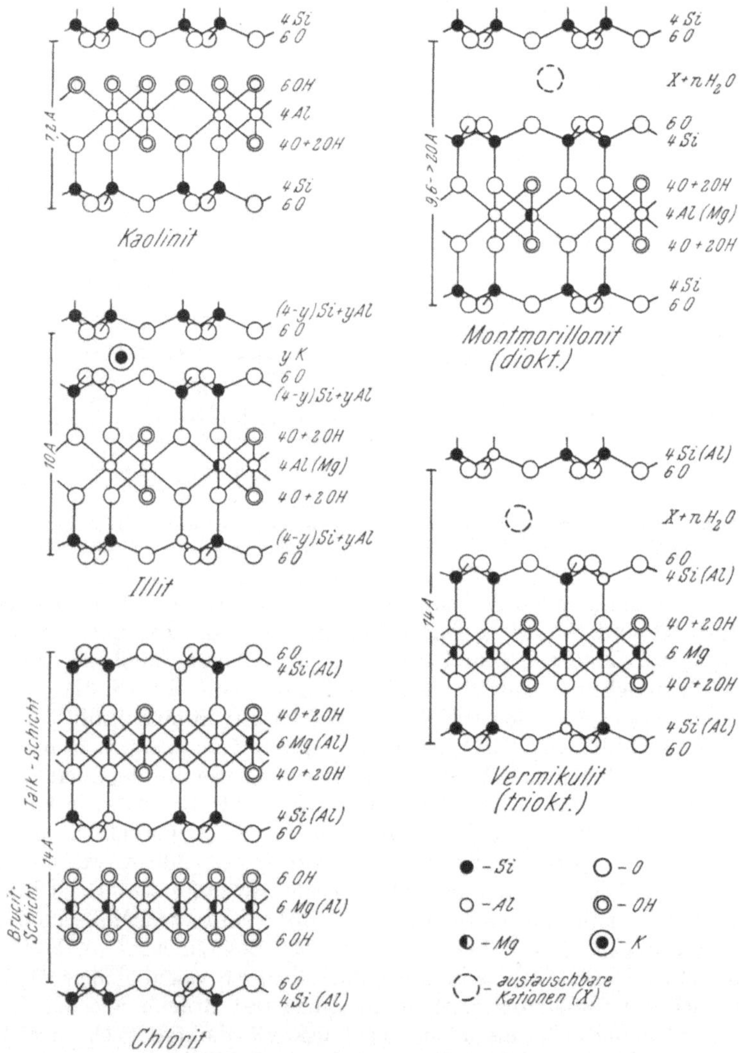

Abb. 74. Strukturschema von Kaolinit, Montmorillonit, Illit, Vermikulit und Chlorit

Al, allgemein also mit niedrigerwertigen Kationen, besetzt sind. Umgekehrt entsteht eine positive Überschußladung z. B. beim Chlorit dadurch, daß Oktaederzentren nicht mit Mg oder Fe^{2+}, sondern mit Al oder Fe^{3+}, also mit höherwertigen Kationen, besetzt sind. Da die Überschußladung fest im Mineral verankert ist und meist bereits bei seiner Entstehung gebildet wird, bezeichnet man sie auch als die *permanente* Ladung. Bei den Tonmineralen ist sie negativ und eine Teilursache des Kationenaustausches dieser Minerale, die wegen dieser Eigenschaft als *Austauscher* bezeichnet werden.

C. Oberfläche der Tonminerale[1]

Die *spezifische Oberfläche* (= Oberfläche pro Gewichtseinheit in m^2/g) ist bei den meisten Mineralen der Tonfraktion auf die äußere Umgrenzung der Teilchen beschränkt (äußere Oberfläche). Lediglich die aufweitbaren Minerale (Montmorillonit-Minerale, Vermikulit-Minerale) besitzen auch zwischen den Elementarschichten eine für die Sorptionsvorgänge zugängliche Oberfläche, die als innere Oberfläche bezeichnet wird.

Die Größe der *äußeren Oberfläche* hängt in erster Linie von der Korngröße ab und wächst, wenn diese abnimmt. So steigt z. B. die Oberfläche eines Würfels mit 1 cm Kantenlänge durch Aufteilung in 10^{15} Würfel mit 0,1 μ Kantenlänge von 6 cm^2 auf 60 m^2. Bei den Tonmineralen besteht die äußere Oberfläche aus den äußeren Basisflächen (parallel zu den Schichten) und den Prismenflächen (senkrecht zu den Schichten).

Im Gegensatz zur äußeren Oberfläche ist die Größe der *inneren Oberfläche* unabhängig von der Teilchengröße, so daß auch das Ausmaß der an ihr stattfindenden Reaktionen bei nicht zu großen Teilchen in keiner Beziehung zur Teilchengröße steht. Die innere Oberfläche der Tonminerale besteht ausschließlich aus Basisflächen zwischen den Tonmineralschichten. Ihre Zugänglichkeit für Wasser und andere Flüssigkeiten mit Dipolcharakter hat die sogenannte innerkristalline Quellung der aufweitbaren Tonminerale zur Folge.

Die Bestimmung der spezifischen Oberfläche (Literatur s. unten) erfolgt in der Regel durch Gasadsorption bei tiefen Temperaturen (z. B. N_2, n-Butan, Äthan) oder durch Adsorption von Flüssigkeiten (z. B. Glykol, Glycerin, Wasser). Aus der Adsorptionsisotherme läßt sich z. B. mit Hilfe der sogenannten BET-Gleichung (Brunauer, Emmet und Teller 1938) die adsorbierte Menge bei Belegung mit einer monomolekularen Schicht errechnen und aus dieser bei Kenntnis des Platzbedarfs eines Moleküls des Adsorptivs die spezifische Oberfläche. Bei aufweitbaren Tonmineralen ist die Dicke der Adsorptionsschicht auch röntgenographisch feststellbar. Aus der Differenz zwischen der Oberfläche bei Anwendung eines Adsorptivs, dem auch die innere Oberfläche zugänglich ist (z. B. Glykol oder Glycerin), und der Oberfläche bei Anwendung eines nur an der äußeren Oberfläche angelagerten Adsorptivs (z. B. N_2, n-Butan), bzw. nach Beseitigung der inneren Oberfläche durch Erhitzen, kann man die innere Oberfläche des Sorbens bestimmen. Schließlich läßt sich die Oberflächengröße bei bekannten Teilchenabmessungen, die z. B. mit Hilfe des Elektronenmikroskopes ermittelt werden, auch aus diesen berechnen.

Die mit den verschiedenen Adsorptiva bestimmten Oberflächenwerte stimmen zwar nicht vollständig überein, doch geben sie die Größenordnung sowie die Relation zwischen den einzelnen Mineralen gut wieder.

Die größte spezifische Oberfläche besitzen die Montmorillonit-Minerale. Die Summe aller Basisflächen beträgt theoretisch je nach der Zusammensetzung des Montmorillonits 760 bis 808 m^2/g. Die mit den verschiedenen Methoden gewonnenen experimentellen Werte liegen je nach der Reinheit und Korngröße etwa zwischen 400 und 900 m^2/g, wobei Werte zwischen 600 und 800 m^2 am häufigsten auftreten. Bei der Fraktion $< 0{,}08\,\mu$ wurden Werte von über 1000 m^2 ermittelt (Sawhney und Jackson 1958). Der Anteil der äußeren Oberfläche beträgt im Mittel nur etwa 5 bis 20% (Orchiston 1954). Die wenigen für den Vermikulit vorliegenden Werte liegen je nach der Korngröße zwischen 200 und 600 m^2 (Diamond und Kinter 1958, Mehra und Jackson 1959); auch hier ist der Anteil der äußeren Oberfläche gering. Die Oberflächengröße von Illiten steigt mit wachsendem Anteil an aufweitbaren Schichten bzw. mit sinkendem K-Gehalt. Die äußere Oberfläche liegt meist zwischen 50 und 100 m^2/g (Brooks 1955), während die Gesamtoberfläche bis auf > 300 m^2 ansteigen kann (Mehra und Jackson 1959, Dyal und Hendricks 1950). Die geringsten Werte findet man bei Kaolinit-

[1] Einschließlich freier Oxide.

Mineralen, die lediglich eine äußere Oberfläche besitzen; je nach der Korngröße beträgt sie 5 bis 30 m^2/g (WEISS u. a. 1956). Eine sehr hohe äußere Oberfläche weisen dagegen wegen der geringen Teilchengröße die Allophan-Minerale auf (etwa 300 m^2/g; DIAMOND und KINTER 1958, ORCHISTON 1959a).

Da die spezifische Oberfläche der Oxide und Hydroxide des Si, Al, Fe und Mn in erster Linie äußere Oberfläche und daher in starkem Maße von der Teilchengröße abhängig ist, schwanken die angegebenen Werte in weiten Grenzen zwischen einigen und mehreren hundert m^2/g. Angaben über die spezifische Oberfläche dieser Substanzen und ihre Abhängigkeit von verschiedenen Faktoren (Herstellungsart, Korngröße, Erhitzungstemperatur) finden sich u. a. bei folgenden Autoren: Kieselsäure: ILER 1955; Quarz: STÖBER 1956; Aluminiumoxide: PRETTRE u. a. 1953; Eisenoxide: ROYEN u. a. 1955; Manganoxide: BUSER und GRAF 1955. Auf Grund der starken Zunahme der spezifischen Oberfläche mit abnehmender Teilchengröße können die im Boden vorliegenden amorphen Anteile dieser Oxide bei nicht zu geringem Gehalt merklich zur Größe der Gesamtoberfläche eines Bodens beitragen (KUN-HUANG und TSEN-TUO 1959).

Die Größe der zugänglichen Oberfläche von Tonmineralen hängt in gewissem Ausmaß auch von der Art der austauschbaren Kationen ab. So wird die Oberfläche von Montmorillonit z. B. durch Sättigung mit K und NH_4 gegenüber der mit Na und zweiwertigen Kationen deutlich erniedrigt (BOWER und GSCHWEND 1952, ORCHISTON 1955, DIAMOND und KINTER 1958), weil vermutlich geringe Anteile der aufweitbaren Schichten unter dem Einfluß dieser Kationen kontrahieren und die innere Oberfläche an diesen Stellen nicht mehr zugänglich ist (DYAL und HENDRICKS 1952).

Durch starkes Erhitzen nimmt die Größe der Oberfläche der Minerale häufig ab, da hierdurch die innere Oberfläche durch Kontraktion der Elementarschichten beseitigt wird oder die Einzelpartikel zu größeren Teilchen aggregieren. Durch Behandlung mit Säuren wird die Oberfläche dagegen vergrößert, da hierdurch Gitterbausteine wie Al, Fe usw. herausgelöst werden und fein verteilte Kieselsäure zurückbleibt. Auch beim Mahlen wird die Oberfläche vergrößert, weil neue Oberflächen freigelegt oder Abbauprodukte mit geringerer Teilchengröße gebildet werden.

Die Beschaffenheit der Oberfläche kann sehr verschieden sein. Neben glatten Kristallflächen treten unregelmäßige, von Spalten und Poren durchsetzte Flächen auf, wodurch die Ladungsdichte der Oberfläche geladener Teilchen oft erheblich verändert wird. Basisflächen sind meist glatter ausgebildet als Prismenflächen. Der Anteil homogener Flächen nimmt mit dem Kristallisationsgrad zu (Kaolinit > Montmorillonit). Außerdem sind die Oberflächen von Tonmineralen häufig mit dünnen Häutchen amorpher freier Oxide des Si, Al und Fe sowie organischer Stoffe umhüllt, die deren Eigenschaften deutlich verändern können.

D. Beschreibung, Vorkommen, chemische Zusammensetzung und Struktur der Tonminerale

Kaolinit-Minerale. Die wichtigsten dioktaedrischen Vertreter der Kaolinit-Minerale sind der *Kaolinit*, der *Nakrit*, der *Dickit*, der *Halloysit* und die *Fireclay*-Minerale. Zu den trioktaedrischen Kaolinit-Mineralen gehören die Mg-silikate *Antigorit* und *Chrysotil* $Mg_6Si_4O_{10}(OH)_8$ sowie die Fe-haltigen bzw. Fe-silikate *Amesit* $(Mg, Fe^{2+})_4Al_4Si_2O_{10}(OH)_8$, *Cronstedtit* $Fe^{3+}{}_4Fe^{2+}{}_4Si_2O_{10}(OH)_8$ und *Berthierin* $(Fe^{2+}, Fe^{3+}, Mg, Al)_6Si_3AlO_{10}(OH)_8$.

Der Kaolinit, das häufigste Mineral dieser Gruppe, ist der wesentliche Bestandteil des Kaolins (Porzellanerde) und tritt als hydrothermales Verwitterungsprodukt Alkalifeldspat-reicher Gesteine sowie in größerer Menge in tropischen und in geringerer Menge auch in vielen Böden der gemäßigten Zone auf. Er ist meist gut kristallisiert und bildet Plättchen von 0,3 bis 4 μ Breite und 0,05 bis 2 μ Dicke mit sechsseitiger Umgrenzung (Abb. 75a). Die Dichte des Kaolinits beträgt 2,6 bis 2,7. Seine Löslichkeit in starken Mineralsäuren ist sehr gering.

Im Gegensatz zum Kaolinit ist beim Halloysit zwischen den Plättchen 1 H_2O-Schicht eingelagert (Endellit[1]), die jedoch schon bei Temperaturen unter 100° C irreversibel abgegeben wird. Außerdem sind die Plättchen zu langgestreckten, röhrenförmigen Teilchen mit einem Durchmesser von 0,04 bis 0,1 μ, einer Länge von 2 bis 8 μ (BATES und COMER 1959) und einer Wandstärke von etwa 100 bis 200 Å (HOFMANN u. a. 1962) aufgebogen (Abb. 75b), die bei der Entwässerung zum Teil entrollt werden. Der Halloysit tritt in größeren Mengen vor allem in Böden auf Si-reichen vulkanischen Tuffen mit Allophan und Kaolinit vergesellschaftet (SUDO und TAKAHASHI 1956) sowie in tropischen Böden (BEUTELSPACHER und VAN DER MAREL 1961) auf. Die Dichte des Halloysits liegt wegen des hohen Wassergehalts nur bei 2,0 bis 2,2.

Abb. 75a bis g. Elektronenmikroskopische Aufnahmen der Tonminerale
Abb. 75a. Kaolinit, Georgia
(Aufnahme WOODWARD 1955)

Der Kaolinit ist der typische Vertreter der 1:1-Minerale; seine Elementarschicht besteht also aus je einer Al-Oktaeder- und Si-Tetraederschicht (Abb. 73 und 74). Die Oktaederseite einer Elementarschicht ist durch eine OH-Schicht, die Tetraederseite durch eine O-Schicht begrenzt. Die Elementarschichten sind durch Wasserstoffbrücken und daher relativ fest miteinander verbunden. Der Abstand zwischen den Elementarschichten ist so klein, daß keine Ionen zwischen ihnen gebunden werden können. Die Oktaederzentren sind ausschließlich mit Al, die Tetraederzentren ausschließlich oder nahezu ausschließlich (vgl. ROBERTSON u. a. 1954, CASHEN 1959) mit Si besetzt, so daß Kaolinite verschiedener Herkunft chemisch relativ einheitlich zusammengesetzt sind. Die Ladung ist daher innerhalb des Gitters ausgeglichen, wie folgende Aufstellung ergibt:

	3 O^{2-}	6 —
Tetraederschicht	2 Si^{4+}	8 +
	2 O^{2-}, $(OH)^-$	5 —
Oktaederschicht	2 Al^{3+}	6 +
	3 $(OH)^-$	3 —

Die einzelnen dioktaedrischen Kaolinit-Minerale, wie z. B. Dickit und Nakrit, unterscheiden sich vor allem durch die Art der Übereinanderlagerung der ein-

[1] Bezüglich der Verwendung der Namen Halloysit, Metahalloysit und Endellit vgl. HOFMANN u. a. (1962).

zelnen Schichten. Beim Halloysit wird der Abstand zwischen den aufgebogenen Elementarschichten darüber hinaus durch Einlagerung von H_2O von etwa 7,1

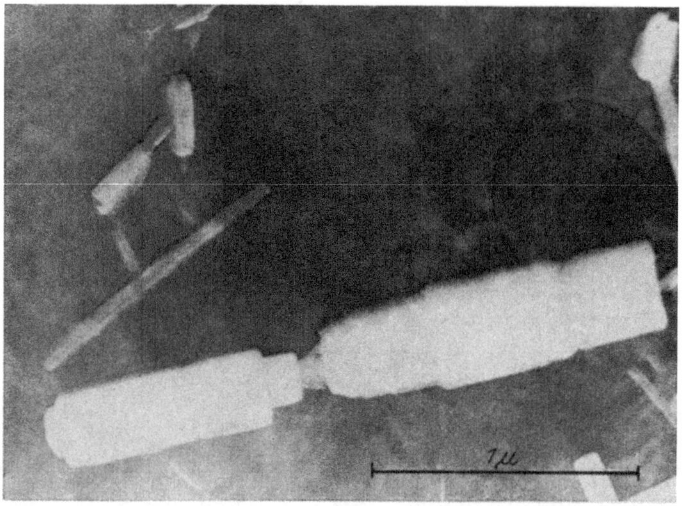

Abb. 75b. Halloysit, Silver City, New Mexico (Aufnahme TAGGART u. a. 1955)

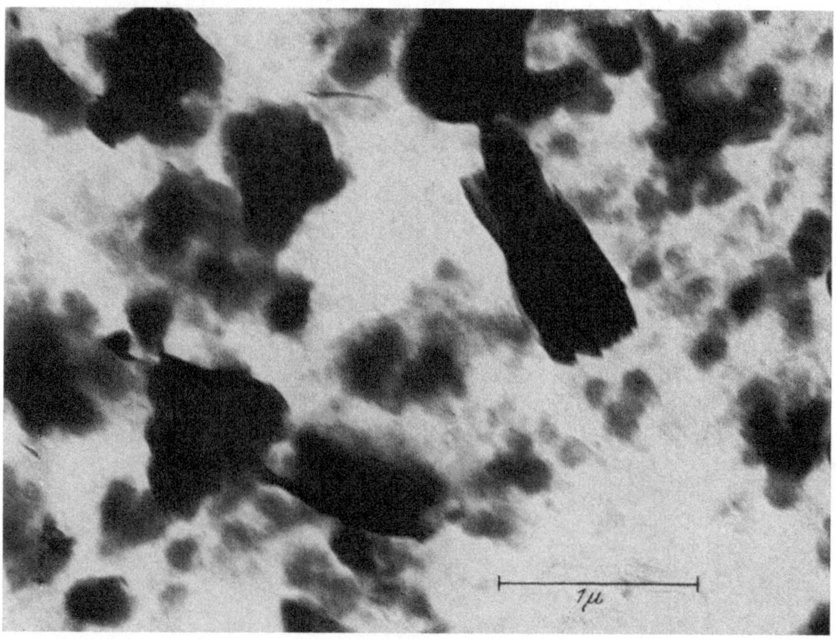

Abb. 75c. Montmorillonit, Polkville, Mississippi (Aufnahme FLAIG und BEUTELSPACHER 1953)

auf etwa 10 Å erweitert. In den trioktaedrischen Kaolinit-Mineralen sind die Oktaederzentren außer bzw. statt mit Al mit Mg, Fe^{2+} und Fe^{3+}, die Tetraederzentren außer mit Si und Fe^{3+} auch mit Al besetzt.

Montmorillonit-Minerale. Zu der großen Gruppe der Montmorillonit-Minerale zählt man die durch ihre chemische Zusammensetzung sich unterscheidenden Minerale *Montmorillonit*, *Beidellit*, *Nontronit*, *Hektorit*, *Saukonit*, *Saponit* und

Abb. 75d. Nontronit, Sandy Ridge, North Carolina (Aufnahme BATES, aus GRIM 1953)

Wolkonskit. Da sie gegenüber dem Kaolinit einen höheren Gehalt an Mg, Ca, Fe und Kieselsäure aufweisen, sind sie als Neubildung besonders in solchen Böden anzutreffen, in denen bei der Tonmineralbildung über längere Zeiten eine nicht zu

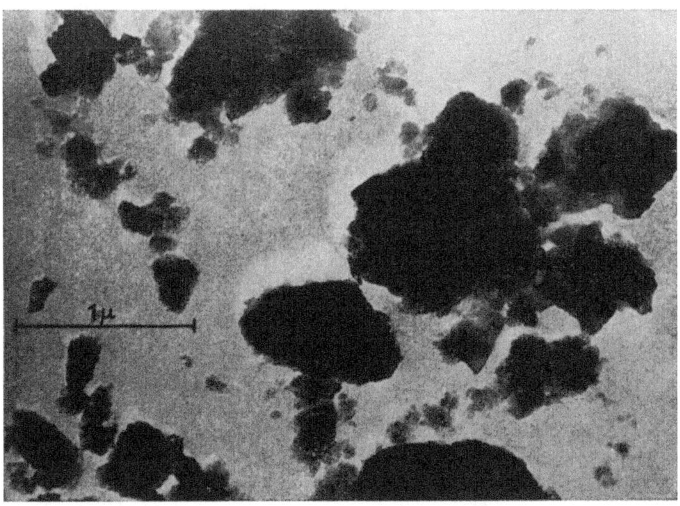

Abb. 75e. Illit, Fithian, Illinois (Aufnahme BATES, aus GRIM 1953)

geringe Konzentration an Mg, Ca, Fe und Kieselsäure vorlag. Dies trifft vor allem für Böden aus basischen magmatischen Gesteinen (z. B. Basalt) und für Böden mit gehemmtem Sickerwasserabfluß und daher geringer Stoffauswaschung

(z. B. in Muldenlagen) zu. In Böden aus Sedimenten, in denen der Montmorillonit häufig durch Abbau von Glimmern entstanden ist (S. 337), ist er besonders im Mittel- und Feinton (0,2 bis 0,08 und <0,08 μ) angereichert (SCHWERTMANN 1961a).

Das elektronenmikroskopische Bild des Montmorillonits zeigt unregelmäßig begrenzte Blättchen oder Blättchenaggregate von 0,3 bis 1,4 μ Durchmesser und 2 bis 6 mμ Dicke (KAHN 1959; Abb. 75c). Der Nontronit ist im Gegensatz zum Montmorillonit und Beidellit leistenförmig ausgebildet (Abb. 75d); die Länge der Leisten beträgt bis zu 4 μ. Die Dichte der Montmorillonit-Minerale liegt je nach ihrem Fe-Gehalt zwischen etwa 2,2 und 2,7. Ihre Löslichkeit in starken Säuren ist wesentlich größer als die des Kaolinits, wobei sich zuerst das Mg, Fe und Al der Oktaederzentren löst (OSTHAUS 1956, GRANQUIST und SUMNER 1959), ohne daß es dabei schon zur Zerstörung des Gitters kommt. Die Mg- und Fe-reichen Glieder lösen sich schneller als die Al-reichen. Bei Behandlung mit Laugen wird sowohl Kieselsäure als auch Al gelöst (STUBIČAN u. a. 1957, HASHIMOTO und JACKSON 1960).

Abb. 75f. Vermikulit (Aufnahme BARBARAS, aus ILER 1955)

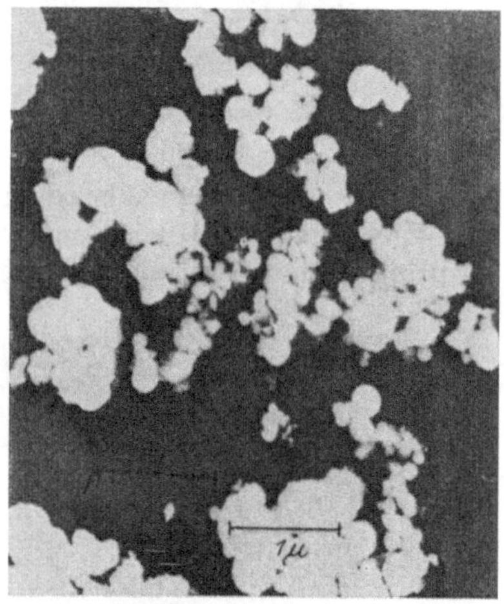

Abb. 75g. Allophan, Gardner Mine Ridge, Indiana (Aufnahme HAUSER 1955)

Die Montmorillonit-Minerale gehören zu den 2:1-Mineralen; jede Oktaederschicht ist demnach von je einer Tetraederschicht umgeben (Abb. 73). Die Elementarschicht ist beiderseitig von O-Schichten begrenzt, so daß eine Verbindung der Elementarschichten durch H-Brücken wie bei den Kaolinit-Mineralen nicht möglich ist. Die Kräfte zwischen den Elementarschichten sind daher wesentlich geringer als bei jenen.

Die einzelnen Montmorillonit-Minerale unterscheiden sich hauptsächlich durch Art, Verteilung und Menge der Zentralkationen, die sich in der chemischen Zusammensetzung widerspiegeln. Die dioktaedrischen Vertreter sind die häufigeren (Abb. 73); sie leiten sich vom Pyrophyllit ab, dessen Oktaederzentren mit Al, und dessen Tetraederzentren sämtlich mit Si besetzt sind; das Gitter ist demnach nach außen neutral. Beim Montmorillonit ist ein Teil der Oktaederzentren mit Mg und ein geringer Teil der Tetraederzentren mit Al besetzt (Tab. 21). Wesentlich höher ist dagegen der tetraedrische Al-Anteil beim Beidellit. Demnach ist die negative Ladung des Montmorillonits hauptsächlich oktaedrisch, die des Beidellits in erster Linie tetraedrisch lokalisiert. Da diese unterschiedliche Ladungsverteilung auch bei anderen 2:1-Mineralen auftritt, unterscheidet U. HOFMANN Minerale montmorillonitischen von solchen beidellitischen Typs. Zu den dioktaedrischen Montmorillonit-Mineralen gehört außerdem der Nontronit, bei dem der Hauptanteil der Oktaederzentren mit Fe, vor allem dreiwertigem Fe, besetzt ist (KORNFELD 1958). Die negative Ladung wird wie beim Beidellit hauptsächlich durch tetraedrisches Al hervorgerufen.

Die wesentlich selteneren trioktaedrischen Montmorillonit-Minerale leiten sich strukturell vom Talk $Mg_3Si_4O_{10}(OH)_2$ ab, bei dem alle Oktaederzentren mit Mg besetzt sind. Das Mg-reiche, zum beidellitischen Typ gehörende Glied wird als Saponit (WEISS u. a. 1955) bezeichnet, das Li- bzw. Cr-haltige, zum montmorillonitischen Typ gehörende als Hektorit bzw. Wolkonskit (WEISS u. a. 1954), das Zn-haltige als Saukonit.

Zwischen dem di- und trioktaedrischen Typ sowie zwischen dem montmorillonitischen und beidellitischen Typ gibt es zahlreiche Übergänge, die in der Regel häufiger sind als die erwähnten Endglieder. Die im Boden auftretenden Montmorillonite sind meist „beidellitischer“ und daher Al-reicher als diejenigen reiner Vorkommen (SAWHNEY und JACKSON 1958), da sie häufig aus Glimmermineralen entstanden sind (SCHWERTMANN 1962).

Wegen der geringen Bindungskräfte zwischen den Elementarschichten sind die Montmorillonit-Minerale aufweitbar und die Zwischenräume für die Einlagerung von Kationen und Quellungsflüssigkeiten zugänglich. Je nach Art und Ausmaß der Einlagerung schwankt der Abstand der Elementarschichten zwischen etwa 12 und 15 Å im lufttrockenen Zustand und ∞ (völlige Aufblätterung in Elementarschichten). Diese Eigenschaften sind von besonderer Bedeutung für den Kationenaustausch und den Wasserhaushalt von Böden, in denen solche Minerale vorkommen.

Illit-Minerale. Von den in Böden vorkommenden Tonmineralen sind die wegen ihrer Verwandtschaft zu den Glimmern auch als *glimmerartige Minerale* bezeichneten Illit-Minerale besonders weit verbreitet. Man rechnet zu ihnen den *Illit*[1], den *Hydromuskovit*, den *Glaukonit* und gelegentlich auch den *Serizit*. Die Häufigkeit — sie wurden in fast allen Böden, besonders in denen des gemäßigten humiden Klimabereiches, und in zahlreichen Sedimenten gefunden — erklärt sich zum Teil aus der großen Stabilität des Illits sowie aus der Tatsache, daß der Illit sowohl durch Aufbau aus amorphen Oxiden als auch durch Abbau von Glimmern entstanden sein kann.

Gemeinsam ist den Illit-Mineralen ein relativ hoher K- (4 bis 7% K_2O) und Wassergehalt sowie ein gegenüber den Montmorillonit-Mineralen geringeres SiO_2/Al_2O_3-Verhältnis. Im elektronenmikroskopischen Bild erscheint der Illit als unregelmäßig begrenzte Blättchen (Abb. 75e), die dem Montmorillonit ähnlich, jedoch meist dicker und größer sind.

Wie die Montmorillonit-Minerale gehören auch die Illit-Minerale strukturell zu den 2:1-Mineralen (Abb. 74), stehen jedoch darüber hinaus den Glimmern weit näher. So weisen sie wie diese einen hohen Al-Anteil in den Tetraedern auf. Die hierdurch entstehende relativ hohe und oberflächennahe Ladung wird zum großen Teil durch Einlagerung von K-Ionen zwischen die Elementarschichten in nicht

[1] Bei der röntgenographischen Untersuchung der Tonfraktion werden nach GRIM alle diejenigen Minerale als Illit bezeichnet, die einen 10 Å-Basisreflex zeigen und nicht aufweitbar sind, wobei man allerdings Minerale sehr unterschiedlicher Zusammensetzung und Struktur erfaßt.

austauschbarer Form neutralisiert. Hieraus erklärt sich der im Vergleich zu den Montmorillonit-Mineralen hohe K-Gehalt der Illit-Minerale, der allerdings stets unter dem der Glimmer liegt. Dies wird entweder mit einem geringeren Al-Gehalt in den Tetraedern (FOSTER 1954), mit dem Einbau von H_3O^+ statt K (NORRISH und BROWN 1952) oder mit der Anwesenheit einzelner aufweitbarer, K-freier Schichten erklärt (MEHRA und JACKSON 1959).

Die K-Ionen stellen eine relativ feste Bindung zwischen den Elementarschichten her, so daß die Illit-Minerale nicht aufweitbar sind. Sie können allerdings im Laufe der Verwitterung durch Verlust des Zwischenschicht-Kaliums und damit des festen Zusammenhalts zwischen den Schichten aufweitbar werden (auch als aufweitbare oder offene Illite bezeichnet). Dabei wird K durch H_3O^+ (bzw. Al) ersetzt und möglicherweise die Ladung erniedrigt, weil eine Aufweitbarkeit nur bei nicht zu hoher Ladung möglich ist. Der Mechanismus dieser Ladungsabnahme ist, abgesehen von den Fe(II)-haltigen Illiten, in denen sie durch Oxydation des Eisens entsteht, bislang nicht geklärt. Außer durch die Aufweitbarkeit sind diese Abbauprodukte dementsprechend durch geringeren K- und höheren H_2O-Gehalt gekennzeichnet. Der aus dem Muskovit entstandene Illit sowie der als Übergangsglied zwischen dem Muskovit und dem Illit stehende Hydromuskovit sind wie der Muskovit selbst dioktaedrisch, während die Abbauprodukte des Biotits wie dieser trioktaedrisch sind[1]. Im ersten Fall sind die Oktaederzentren mit Al, im zweiten mit Mg und Fe^{2+} besetzt. Der Glaukonit (BURST 1958), in dem ein großer Teil der Oktaederzentren mit Fe^{3+} (HOEBEKE und DEKEYSER 1955) besetzt ist, ist dioktaedrisch.

Die Säurelöslichkeit der Glimmer und Illit-Minerale ist für deren Umwandlung im Boden sowie für den K-Haushalt von Bedeutung, weil der Lösungsprozeß ohne wesentliche Änderung des Gitterbaus mit der Abgabe des Kaliums beginnt. Die Entkalisierung läßt sich bei genügend langer Behandlung bereits mit CO_2-haltigem Wasser durchführen und verläuft um so schneller, je geringer die Teilchengröße ist. Unter den Glimmern ist der Biotit wesentlich instabiler als der Muskovit, vermutlich zum Teil wegen der leichten Oxydierbarkeit des in ihm vorhandenen zweiwertigen Eisens. Die Stabilität des Illits gegenüber H-Ionen ist relativ hoch; sie liegt etwa zwischen der des Kaolinits und der des Montmorillonits.

Vermikulit-Minerale. Der *Vermikulit* ist das häufigste der Minerale dieser Gruppe. Außer in reinen Vorkommen kommt er in Böden vor allem als Abbauprodukt von Biotiten, insbesondere von Phlogopiten und Mg-reichen Biotiten (FOSTER 1963), und des Chlorits vor, kann aber auch aus Muskovit entstehen. Er ähnelt in seinem Aussehen sehr dem Biotit (Abb. 75f) und kommt nicht nur in der Tonfraktion, sondern auch in der Schluffraktion vor (bezüglich der Säurelöslichkeit vgl. LOPEZ-GONZALES und CANO-RUIZ 1959).

Der Vermikulit ist entsprechend seiner Herkunft aus Glimmern ebenfalls ein 2:1-Mineral und je nach der Entstehung aus Muskovit oder Biotit di- oder trioktaedrisch. Die hohe, vor allem durch tetraedrisches Al bedingte Gesamtladung ist meistens durch Bindung von Mg und in untergeordnetem Maße von Ca neutralisiert, die im Gegensatz zum K des Illits austauschbar sind. Hieraus erklärt sich zum Teil der hohe Mg-Gehalt der Vermikulit-Minerale. Der Vermikulit ist durch Wasser aufweitbar, in Gegenwart von Mg als austauschbarem Ion bis auf einen Basisabstand von 14 bis 15 Å. Bei Einlagerung von K kontrahieren die Schichten, und die Aufweitbarkeit z. B. durch Wasser geht verloren. Dies kann für den K-Haushalt des Bodens von Bedeutung sein. Ein Verlust der Aufweit-

[1] Der sehr seltene trioktaedrische Illit wird auch als *Ledikit* bezeichnet (MACKENZIE 1959).

barkeit, jedoch ohne Kontraktion, erfolgt auch, wenn das austauschbare Mg und Ca durch Al und Fe ersetzt werden. Hierdurch entstehen sekundäre Chlorite (S. 335).

Chlorit-Minerale. Die im Boden vorkommenden Chlorit-Minerale stammen nur zum Teil aus Metamorphiten und Sedimenten (*primäre Chlorite*), in denen sie sehr häufig (ECKHARDT 1958, FÜCHTBAUER und GOLDSCHMIDT 1959) und in großer Mannigfaltigkeit als meist grün gefärbte blättchenförmige Minerale vorkommen. Nicht selten sind sie hier, wie auch in Böden, aus Hornblenden entstanden (STEPHEN 1952, AOMINE und KAWASAKI 1963). Sie können außerdem aus aufweitbaren 2:1-Mineralen durch Einlagerung von (Al—OH)-Schichten entstehen. Diese Chlorite werden auch als *Bodenchlorite* (VAN DER MAREL 1959) oder *sekundäre Chlorite* (JACKSON 1959) bezeichnet.

Bei den Chlorit-Mineralen metamorpher Gesteine ist zwischen zwei 2:1-Elementarschichten, die die Konstitution von Talk-Schichten besitzen, jeweils eine trioktaedrische Brucit-Schicht ($Mg_3(OH)_6$) eingelagert (2:2-Minerale, s. Abb. 74), deren Mg im Gegensatz zum Vermikulit nicht austauschbar ist. Der Chlorit ist also nicht aufweitbar. Sein Basisabstand beträgt etwa 14 Å. Er wird jedoch aufweitbar, wenn durch Herauslösen der Brucit-Schicht durch Verwitterung der Zusammenhalt der Talk-Schichten verlorengeht, und dann als *aufweitbarer* (*quellbarer*) *Chlorit* bezeichnet.

Die Besetzung der Oktaeder- und Tetraederzentren und die Höhe der dadurch entstehenden Ladung sowie der Ort ihres Ausgleichs innerhalb des Gitters sind sehr variabel, so daß eine große Zahl verschiedener Chlorit-Minerale existiert (vgl. NELSON und ROY 1958). Häufig findet sich in den Oktaederzentren der trioktaedrischen Brucit-Schicht außer Mg auch Al. Die dadurch entstehende positive Überschußladung ist durch eine entsprechende Anzahl von Al in den Tetraederzentren der Talk-Schichten ausgeglichen.

In den sekundären Chloriten liegt zwischen den Elementarschichten statt der trioktaedrischen Brucit-Schicht eine mehr oder weniger vollständige dioktaedrische Hydrargillit-Schicht ($Al_2(OH)_6$) (KLAGES und WHITE 1957, TAMURA 1958, JACKSON 1963). Vermutlich beteiligt sich auch dreiwertiges Fe an dieser Einlagerung (QUIGLEY und MARTIN 1963). Sie enthalten daher in der Regel mehr Al und weniger Mg als die primären Chlorite. Die aufweitbaren Ausgangsprodukte der sekundären Chlorite verlieren durch die Einlagerung der hydrargillitischen Schichten die Fähigkeit zur Kontraktion und Aufweitung.

Wechsellagerungsminerale. Neben den strukturell einheitlich aufgebauten Tonmineralen treten in Böden eine große Anzahl verschiedener Minerale auf, die aus einer regelmäßigen oder unregelmäßigen Wechsellagerung von Elementarschichten anderer Tonminerale bestehen. Sie werden zusammenfassend als *Wechsellagerungsminerale* (mixed layer-, interstratified minerals) bezeichnet. Da sie in Böden häufig bei der Verwitterung der aus dem Ausgangsgestein stammenden nicht aufweitbaren Minerale gebildet werden, bestehen sie häufig aus einer Wechsellagerung von aufweitbaren („verwitterten“) und nicht aufweitbaren Schichten. Die einzelnen Elementarschichten besitzen annähernd die Struktur und die chemische Zusammensetzung derjenigen Schichten, die das Ausgangsmineral aufbauten, da bei der Bildung der Wechsellagerungsminerale in erster Linie die Besetzung zwischen den Elementarschichten verändert wird.

In der Regel werden die Wechsellagerungsminerale nicht mit einem eigenen Namen belegt; nur in Fällen mit regelmäßiger Wechsellagerung wurden gelegentlich neue Namen geprägt, z. B. *Hydrobiotit* für ein Mineral mit regelmäßiger Wechsellagerung von Biotit- und Vermikulit-Schichten und *Corrensit* für eine solche zwischen Chloritschichten und aufweitbaren Schichten (Vermikulit, Montmorillonit) (LIPPMANN 1959, MARTIN-VIVALDI und MCEWAN 1960).

Abb. 76. *Schematische Darstellung wichtiger Bildungs- und Umbildungsabfolgen der Tonminerale in Böden*[1]

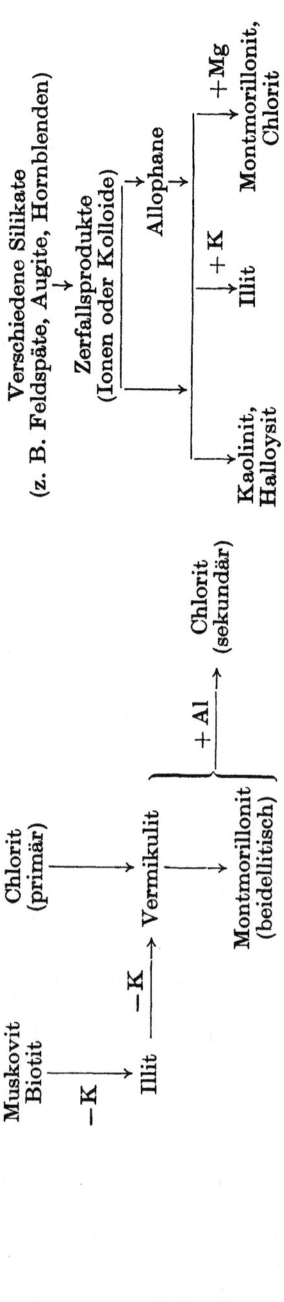

[1] Zwischen den einzelnen Mineralgliedern treten häufig Wechsellagerungsminerale auf.

Allophan-Minerale. Die Allophan-Minerale sind röntgenamorphe Aluminium- oder Eisensilikate mit einem engen SiO_2/R_2O_3-Verhältnis (~ 1) und einem hohen Wassergehalt (bis 30%). Das Al-reiche Glied wird als *Allophan* ($SiO_2 : Al_2O_3 \sim 1$), das Fe-reiche als *Hisingerit* bezeichnet (MACKENZIE 1959). Allophan wurde in Böden auf sauren vulkanischen Aschen Japans (SUDO 1954, AOMINE und YOSHINAGA 1955, AOMINE und JACKSON 1959), Neuseelands (BIRRELL und FIELDES 1952), Chiles (BESOAIN und GARCIA-VICENTE 1962) und Hawaiis (TAMURA u. a. 1953a) zum Teil als vorherrschendes Mineral der Tonfraktion gefunden und kommt dort mit Halloysit, Kaolinit und Hydrargillit vergesellschaftet vor. Wegen ihres röntgenamorphen Charakters stößt die Strukturaufklärung der Allophan-Minerale auf Schwierigkeiten. Man kann jedoch annehmen, daß auch hier die gleichen Bauelemente vorliegen wie bei den anderen Tonmineralen, nämlich Si-O-Tetraeder und Al(Fe)-O-OH-Oktaeder, im Unterschied zu diesen jedoch im lockeren Verband ohne größere Fernordnung. Diese Vermutung wird durch die geringe Größe (zum Teil nur etwa 20 mμ) der Teilchen und deren unregelmäßige, zum Teil abgerundete Form gestützt (s. Abb. 75g). Die Allophan-Minerale besitzen daher eine hohe Oberflächenaktivität und eine relativ große Säurelöslichkeit (BIRRELL und FIELDES 1952). Da es sich bei den erwähnten allophanreichen Böden um sehr junge Bildungen handelt, stellen die Allophan-Minerale vermutlich ein Zwischenstadium bei der Bildung definierter, gut kristallisierter Tonminerale (z. B. Halloysit) dar.

E. Entstehung der Tonminerale

Unsere Kenntnis über die Tonmineralentstehung im Boden gründet sich vor allem auf Vergleiche des Tonmineralgehaltes in verschieden stark bzw. unverwitterten Horizonten innerhalb eines Profils und auf Laboratoriumsversuche unter Bedingungen, die denen der Natur nachgeahmt wurden. Danach können sich Tonminerale entweder durch Umwandlung von Schichtsilikaten (JACKSON u. a. 1952), bei der die Grundstruktur der Silikate erhalten bleibt, oder aus ionaren und kolloiden Abbauprodukten zahlreicher Silikate und anderer Minerale bilden, bei der die Gitter der Tonminerale mehr oder weniger neu entstehen (vgl. Abb. 76).

Das Ausgangsprodukt beim *Abbau von Schichtsilikaten* sind in erster Linie die verschiedenen Glimmer und Chlorite. Bei den Glimmern besteht die Umwandlung unter dem Einfluß saurer Verwitterungslösungen vor allem in einem Ersatz des Zwischenschicht-Kaliums durch H_3O-Ionen und damit einer Aufnahme von Wasser, in einer Teilchenverkleinerung, die sowohl der K-Abgabe folgt als auch sie fördert, und vermutlich auch in einer Ladungsabnahme, die beim Biotit in einer Oxydation des im Gitter enthaltenen zweiwertigen Eisens besteht. Die K-Abgabe erfolgt von den Prismenflächen her, schreitet allmählich weiter nach innen fort und führt schließlich zum Verlust des Schichtzusammenhaltes an einigen Stellen, an denen der Glimmer dann aufblättert. Hinzu kommt ein Zerbrechen der Mineralblättchen senkrecht zur Blättchenebene vor allem durch mechanische Beanspruchung (MEYER u. a. 1963). Auf diese Weise entstehen — beim Muskovit über den Hydromuskovit — der Illit und schließlich über Wechsellagerungsminerale aufweitbare Minerale, deren Aufweitbarkeit gleich der des Vermikulits oder sogar des Montmorillonits sein kann. Im Unterschied zu letzterem ist jedoch nicht nur seine Gesamtladung höher, sondern auch — wie bei den Ausgangsglimmern — in erster Linie tetraedrisch lokalisiert. Dies kommt in einer Reihe von Eigenschaften zum Ausdruck (z. B. Kontrahierbarkeit, Sorptionsverhalten), die den so entstandenen Montmorillonit vom typischen Montmorillonit unterscheiden (SCHWERTMANN 1962). Aus Biotit entsteht durch Ersatz des Zwischenschicht-Kaliums durch Magnesium der Vermikulit, wobei als Zwischenstufe Hydrobiotit auftreten kann.

Die Umwandlung des Chlorits besteht in einem Abbau der Brucit-Schichten, deren Mg zunächst austauschbar (Vermikulit) (DROSTE u. a. 1962) und schließlich völlig gegen andere Kationen ausgetauscht wird.

Die geschilderten Vorgänge sind zum Teil reversibel, d. h. bei genügendem Angebot der durch Verwitterung entfernten Bausteine werden diese wieder in das Gitter eingebaut, so daß das Mineral zum Teil regeneriert wird. Die Zufuhr dieser Kationen, vor allem von K und Mg, kann in der Natur z. B. durch erneute Einwirkung mariner Wässer (WEAVER 1958) oder durch Düngung erfolgen.

Eine besondere Rolle bei den geschilderten Prozessen spielen die *Wechsellagerungsminerale*, weil sie bei fast allen bisher festgestellten Umwandlungen von Schichtsilikaten auftreten und oft den einzigen Anzeiger für eine Mineralverwitterung im Boden darstellen. Ihre Bestimmung durch die Röntgenanalyse ist daher mit besonderer Sorgfalt ausgearbeitet worden (MCEWAN 1956).

Die *Entstehung* von Tonmineralen *aus den ionaren und kolloiden Zerfallsprodukten* von Silikaten tritt vor allem bei nicht schichtförmig gebauten Silikaten (Feldspäte, Olivin, Augite, Hornblenden, silikatische Gläser usw.) hervor und ist besonders von CORRENS (1963) und von MATTSON diskutiert worden. Die genannten Silikate zerfallen durch Hydrolyse und Oxydation in Ionen und kolloide Oxide und Hydroxide, aus denen sich die Tonminerale neu aufbauen. Dabei liegen die Alkali- und Erdalkali-Metalle als Ionen, Kieselsäure und Al-hydroxide je nach dem pH-Wert als Ionen oder Kolloide vor. Beim Aufbau der Tonminerale werden vermutlich zunächst die OH-Ionen der oktaedrisch gebauten Al-hydroxide bzw. (Al, Mg)-hydroxide (TURNER und BRYDON 1962) gegen die tetraedrisch gebauten Silikat-Ionen ausgetauscht (TAMURA und JACKSON 1953). Diese Adsorptionsverbindungen, deren Gehalt an tetraedrischem Al und damit an tetraedrischer Ladung mit steigendem pH des Bildungsmilieus zunimmt (DE KIMPE, GASTUCHE und BRINDLEY 1961), ordnen sich dann, eventuell unter Einbau von Alkali- und Erdalkali-Ionen, zu den Schichten und Schichtverbänden der Tonminerale,

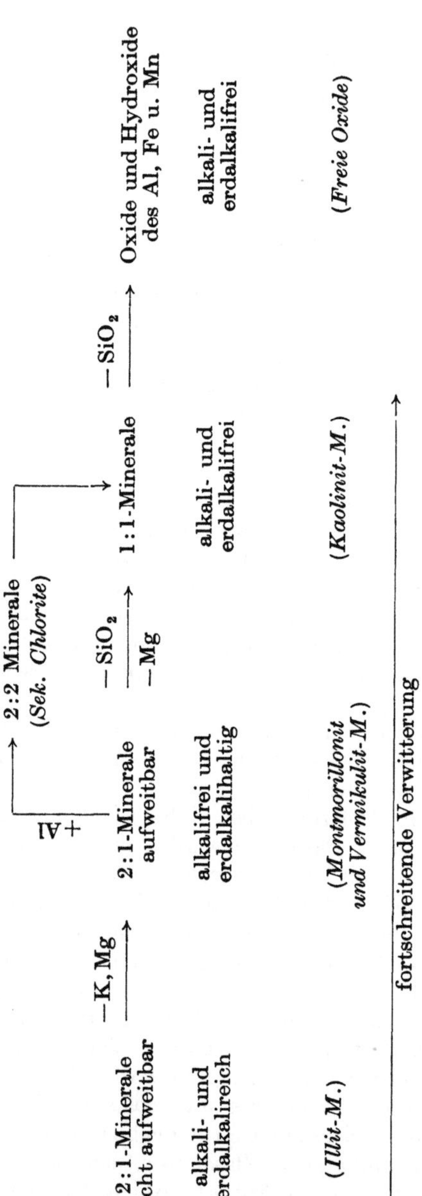

Abb. 77. *Verwitterung und Mineralumbildung im Boden*

ein Alterungsvorgang, der durch häufiges Austrocknen gefördert wird. Als Zwischenprodukte treten hierbei Allophan-Minerale auf. Die Endprodukte dieser Entwicklung sind, wie auch aus zahlreichen Synthese-versuchen hervorgeht (vgl. HENIN 1956), bei genügendem Angebot an Mg und nicht zu rascher Abfuhr der Kieselsäure Montmorillonit oder auch Chlorit-Minerale, und bei genügendem K-Angebot Illit-Minerale, während bei rascher Abfuhr dieser Kationen und der Kieselsäure Kaolinit-Minerale entstehen. Ersteres ist vor allem bei Ca- und Mg-reichen silikatischen Gesteinen bei nicht zu intensiver Verwitterung gegeben, letzteres auf Ca- und Mg-armen Gesteinen, vor allem bei intensiver Verwitterung und starker Auswaschung. Unter extremen Bedingungen kommt es schließlich nicht mehr zur Tonmineralbildung, sondern als Verwitterungsprodukte verbleiben lediglich die freien Oxide (Podsole, Latosole). Es ist allerdings möglich, daß der Abbau der primären, nicht-schichtförmigen Silikate nicht so vollständig erfolgt wie beschrieben, sondern daß z. B. bei der Umwandlung des Orthoklas in ein Schichtsilikat, z. B. Serizit, die bereits im Orthoklas vorliegenden Tetraederschichten erhalten bleiben und als solche zum Aufbau des Schichtsilikats verwendet werden (DE VORE 1959).

Die geschilderten Vorgänge können bei der Verwitterung und der Bildung der Sedimentgesteine nach- und nebeneinander ablaufen. Die große Mannigfaltigkeit der Tonmineralassoziationen im Boden ist daher bei Böden aus Sedimentgesteinen schon zum Teil durch eine ähnliche Mannigfaltigkeit im Ausgangsgestein bedingt; diese wird vor allem durch die stark wechselnden Herkunfts-, Ablagerungs- und Diagenesebedingungen der Sedimente verursacht (vgl. WEAVER 1959).

Junge Böden, z. B. Lößböden und Auenlehme, weisen daher oft den gleichen Tonmineralbestand auf wie die Ausgangssubstrate. Dagegen sind die Tonminerale der aus magmatischen Gesteinen gebildeten Böden ausschließlich in situ entstanden. Kalknatronfeldspäte, Biotit, Augite und Hornblenden gehen dabei vorwiegend in Vermikulit, Illit, Chlorit und Montmorillonit über, Muskovit in Illit und Vermikulit, und Alkalifeldspäte, insbesondere K-Feldspäte, in Illit und Kaolinit. Diesen oft im Laboratoriumsversuch an Proben der genannten Ausgangsminerale rekonstruierten Mineralumwandlungen entspricht der Tonmineralbestand der Böden aus den verschiedenen Gesteinen.

Art und Geschwindigkeit der weiteren Umwandlung der aus den magmatischen oder metamorphen Gesteinen gebildeten oder vom Sediment ererbten Tonminerale im Boden hängen vor allem von der Verwitterungsintensität ab. Von dieser Umwandlung werden sowohl die aus Schichtsilikaten als auch die aus Nichtschichtsilikaten gebildeten Tonminerale (s. Abb. 76) erfaßt. Im allgemeinen besteht sie zunächst in einer Fortführung des Zwischenschicht-Kaliums bzw. -Magnesiums, bei der Illite bzw. Chlorite aufweitbar werden. Da die Verwitterungsintensität im Bodenprofil von unten nach oben zunimmt, steigt oft in gleicher Richtung der Anteil aufweitbarer Minerale bzw. aufweitbarer Schichten innerhalb der einzelnen Mineralteilchen. Bei fortschreitender Versauerung und Verwitterung kommt es dann zum allmählichen Abbau des Gitters, bei dem 2:1-Minerale in 1:1-Minerale umgewandelt werden oder völlig zerfallen können. Das dabei frei werdende Al und Fe kann wiederum in andere noch vorhandene aufweitbare Minerale (z. B. Vermikulit) eingelagert werden und diese in sekundären Chlorit überführen. Auf diese Weise wird die Bildung freier Al-Hydroxide, vor allem Hydrargillit, verhindert, solange noch aufweitbare 2:1-Minerale vorhanden sind (Jackson 1963a). Schließlich erfolgt ein Abbau zu den freien Oxiden des Si, Al und Fe, die demzufolge in extrem verwitterten Böden (z. B. Latosole) erheblich angereichert sind. Hiernach ergibt sich die in Abb. 77 angegebene allgemeine Verwitterungsfolge der Bodenminerale (vgl. auch Jackson und Sherman 1953).

Da die verschiedenen Bodentypen zum Teil unterschiedliche Verwitterungsstadien repräsentieren, herrschen in ihnen charakteristische, dem jeweiligen Verwitterungsstadium entsprechende sekundäre Minerale vor, durch die sich der Verwitterungsgrad kennzeichnen läßt (Jackson 1959).

Die geschilderte Mineralumwandlung im Zuge fortschreitender Verwitterung hat in der Regel auf längere Sicht gesehen eine Abnahme der Bodenfruchtbarkeit zur Folge, da sowohl die Nährstoffnachlieferung aus den sich zersetzenden Mineralen als auch alle günstigen mit den Tonmineralen verbundenen Eigenschaften, wie Nährstoffsorption, Wasserhaltefähigkeit u. a., allmählich reduziert werden. Kulturmaßnahmen, die diesen Versauerungs- und Zersetzungsprozeß verlangsamen oder verhindern, sind daher eine wertvolle Hilfe zur Erhaltung der Bodenfruchtbarkeit.

Das Ausmaß der Bildung sekundärer Minerale der Tonfraktion im Boden hängt von zahlreichen Faktoren wie Klima, Topographie, Ausgangsmaterial und Vegetation ab, wofür Barshad (1959) zahlreiche Beispiele liefert. Die Tonbildungsrate beträgt etwa 0.01 bis 2 mg Ton/Jahr/100 g Ausgangsmaterial. Das Maximum der Tonbildung liegt meistens in der Tiefe von 5 bis 25 cm. Steigende Temperatur und Feuchtigkeitsgehalte erhöhen die Tonbildung. Sie ist auf Ca- und Mg-reichen und feinkörnigen Gesteinen größer als auf Ca- und Mg-armen und grobkörnigen. White (1962) berechnete den K-Verlust in illitischen Böden aus Geschiebemergel Indianas (USA) zu 0,122% K_2O/100 Jahre.

3. Oxide und Hydroxide des Si, Al, Fe und Mn (freie Oxide)

Von den sekundären Mineralen des Bodens sind neben den silikatischen Tonmineralen die Oxide und Hydroxide des Si, Al, Fe und Mn häufig stark vertreten und von besonderer Bedeutung für zahlreiche Bodeneigenschaften. Diese Verwitterungsneubildungen werden häufig und auch im folgenden als *freie Oxide* bezeichnet. Da es jedoch nicht immer möglich ist, auf analytischem Wege die durch Verwitterung gebildeten (sekundären) Oxide und Hydroxide von den primären Verbindungen dieser Art (z. B. Quarz, Magnetit) zu trennen, wird der

Tabelle 23. *Oxidische und hydroxidische Minerale des Si, Al, Fe und Mn-Eigenschaften und Vorkommen in Böden*

Name	Formel	Kristallklasse	Farbe	Vorkommen in Böden
Quarz	SiO_2	trigonal		sehr häufig
Chalcedon im weiteren Sinn	SiO_2	(kryptokristallinischer Quarz)		aus Muttergestein stammend z. B. Flint
Opal	$SiO_2 \cdot nH_2O$	amorph	in der Regel farblos	häufig, zum Teil als Bio-Opal
Cristobalit	SiO_2	tetragonal		aus Muttergestein stammend
Tridymit......	SiO_2	rhombisch		bisher in Böden nicht nachgewiesen
Böhmit	γ–AlOOH	rhombisch		Latosole, Bauxite, Podsole
Diaspor	α–AlOOH	rhombisch	farblos, gelblich, bräunlich	Bauxite
Hydrargillit .. (Gibbsit)	γ–$Al(OH)_3$	monoklin		Latosole, Bauxite, Terra rossa, häufigstes Al-oxid in Böden
Magnetit	Fe_3O_4	kubisch	schwarz	in der Schwermineralfraktion
Hämatit	α–Fe_2O_3	trigonal	rot	Latosole, Terra rossa
Maghemit	γ–Fe_2O_3	kubisch	rotbraun	an Brandstellen, Latosole
Goethit	α–FeOOH	rhombisch	ocker, rostfarben, braun	häufigstes Fe-oxid in Böden des humiden, gemäßigten Bereiches
Lepidokrokit ..	γ–FeOOH	rhombisch	orange	Staunässeböden, Niedermoorböden
Hausmannit ..	Mn_3O_4	tetragonal		bisher als Neubildung in Böden nicht nachgewiesen
Manganit	MnOOH	rhombisch	schwarz bis schwarzbraun	
Pyrolusit	β–MnO_2 δ–MnO_2	rhombisch		häufigste Mn-oxide in Böden und Sedimenten
Mn(Fe)-Manganit	z. B. $3MnO_3 \cdot Mn(Fe)(O, OH)_2 \cdot xH_2O$			

Begriff „freie Oxide" praktisch meist nur auf diejenigen Anteile dieser Stoffe angewandt, die durch die jeweils zur Bestimmung benutzte Methode erfaßt werden. Allerdings decken sich die Anteile der Oxide und Hydroxide des Eisens und Mangans, die durch Extraktion mit einem Reduktionsmittel (z. B. $Na_2S_2O_4$ für Fe, MITCHELL und MACKENZIE 1954, bzw. $NaHSO_3$ für Mn, SCHACHTSCHABEL 1957) erfaßt werden, relativ gut mit den durch Verwitterung neu gebildeten Oxiden und Hydroxiden dieser Elemente (SCHWERTMANN 1959). Bei der freien Kieselsäure und den Al-Oxiden werden dagegen durch vierstündige Behandlung mit heißer 0,5 n-NaOH (FOSTER 1953) nur die amorphen Anteile dieser Verbindungen sowie Allophan-Minerale, nicht dagegen Quarz und gut kristallisierte Al-Oxide und -Hydroxide extrahiert. Da hierbei jedoch auch erhebliche Anteile der Tonminerale gelöst werden können, wird neuerdings eine nur kurze (2,5 Minuten) Behandlung mit 0,5 n-NaOH vorgeschlagen (HASHIMOTO und JACKSON 1960).

In Böden oder Sedimenten entstehen die freien Oxide immer dann, wenn die bei der Verwitterung freigesetzten Anteile dieser Elemente weder ausgewaschen noch bei der Bildung von Tonmineralen verbraucht werden. Dieses ist oft dann der Fall, wenn die jeweiligen freien Oxide stabiler (schwerer löslich) sind als die Tonminerale oder wenn es wegen Fehlens einzelner Bestandteile nicht zur Bildung von Tonmineralen kommt.

Innerhalb der kristallographisch definierten Oxide und Hydroxide, deren wichtigste in Tab. 23 aufgeführt sind, unterscheidet man die Oxide im engeren Sinne (z. B. Al_2O_3) von den reinen Hydroxiden (z. B. $Al(OH)_3$). Dazwischen stehen die Oxidhydroxide (z. B. AlOOH). Außer diesen definierten Formen existieren amorphe Oxide mit wechselndem Gehalt an Wasser, das in der Regel als Adsorptionswasser und daher nicht in stöchiometrischem Verhältnis gebunden ist. Oxide mit definiertem, als H_2O vorliegendem Wassergehalt (Oxidhydrate) sind bei den Oxiden und Hydroxiden des Al, Fe und Mn bisher nicht nachgewiesen.

Die freien Oxide des Bodens gehören in der Regel zur Tonfraktion. Ihre Teilchengröße liegt häufig weit unter $2\,\mu$; ein großer Anteil ist röntgenamorph.

Der Gehalt an freien Oxiden läßt sich oft zur Kennzeichnung bodenbildender (pedogenetischer) Vorgänge (Verwitterung, Verlagerung) heranziehen. So ist häufig eine weit fortgeschrittene Verwitterung mit höherem Gehalt an freien Oxiden verbunden. JACKSON und SHERMAN (1953) sehen die Verwitterungsstadien, in denen freie Oxide gegenüber anderen sekundären Mineralen vorherrschen, als die höchsten an (S. 339). Da sie von der Verlagerung innerhalb des Bodenprofils ebenso wie andere Tonteilchen sehr stark betroffen werden, sind Eluvialhorizonte oft Horizonte, die an freien Oxiden verarmt sind, Illuvialhorizonte dagegen solche, in denen diese angereichert sind.

A. Silicium

Der größte Anteil des Gesamt-Siliciums in Böden ist entweder in Form von Quarz oder primären bzw. sekundären Silikaten gebunden. Nur geringe Mengen liegen als freie Kieselsäure vor. Sie hinterbleibt als Restprodukt bei der Silikatverwitterung oder bildet sich durch Ausfällung aus den bei der Verwitterung entstehenden kieselsäurehaltigen Lösungen.

Der Gehalt an *NaOH-löslicher Kieselsäure* (Methode: FOSTER 1954) in Böden schwankt in weiten Grenzen[1]. Er lag z. B. bei deutschen Marschen zwischen 1 und 10% SiO_2 (SCHACHTSCHABEL u. a. 1958, 1958a), bei amerikanischen Para-

[1] Auf die Grenzen der Verwertbarkeit der nach dieser Methode gewonnenen Werte wurde auf S. 341 hingewiesen.

braunerden zwischen 8 und 12% SiO_2 (Whittig u. a. 1957). Grundwasserbeeinflußte und stark versauerte Schichten zeigten höhere Werte, da hier die Auswaschung und Alterung gering, die Verwitterungsintensität jedoch hoch sind. Tonarme Böden sind dagegen auch bei hohem Verwitterungsgrad ärmer an freier Kieselsäure. Im Gebiet des älteren Pleistozäns Nordwestdeutschlands wurden in sandigen Podsolen und Podsol-Braunerden Werte von 1 bis 3% SiO_2 gefunden, wobei die Podsole ein Maximum im A_e- und ein Minimum im B-Horizont zeigen (Heinemann, unveröffentlicht). Geringere Gehalte (<1%) finden sich in ähnlichen Böden des Jungpleistozäns Schleswig-Holsteins (Schlichting und Blume 1962). Die Werte von Lößböden bei Hannover (degradierte, pseudovergleyte Schwarzerden) lagen bei 1 bis 5% (Schroeder 1954); ihre deutliche Abnahme mit der Tiefe spiegelt die in gleicher Richtung abnehmende Verwitterungsintensität wider. Besonders große Mengen (bis über 20% SiO_2) enthielten Böden auf vulkanischen Aschen (Sasaki und Ishizuka 1957, Whittig u. a. 1957), die jedoch Allophan-Mineralen zuzuschreiben sind.

In der Bodenlösung liegt das Silicium vermutlich als monomere oder doch oligomere Kieselsäure (vermutlich Orthokieselsäure $Si(OH)_4$) vor. Ihre Konzentration ist im neutralen bis schwach alkalischen Bereich sehr gering (einige ppm), steigt allerdings zum sauren Bereich erheblich an (McKeague und Cline 1963). Im stark alkalischen Bereich ($pH > 9$) wird die Kieselsäure in lösliches Silikat übergeführt, so daß der Si-Gehalt der Bodenlösung erneut ansteigt. Eine pH-Erniedrigung solcher Lösungen führt zur Ausfällung von amorphen Kieselgelen, die bereits aus SiO_4-Tetraedern bestehen, jedoch ohne Fernordnung, und durch Wasserverlust allmählich zu *Opal* $SiO_2 \cdot n\,H_2O$, der amorphen Form des SiO_2, verhärten. Der Opal geht dann durch sehr langsame Alterung in kryptokristallinen „*sekundären*" *Quarz* (*Chalcedon* im weiteren Sinne, Smithson 1956) unter weiterer Abnahme seiner Reaktionsfähigkeit und Löslichkeit über. Der so gebildete Quarz unterscheidet sich vom primären Quarz nur durch die sehr kleine Kristallgröße und eine höhere Reaktionsfähigkeit. Eine direkte Ausfällung von Quarz ist trotz dessen sehr geringer Löslichkeit (etwa 10 ppm SiO_2) sehr unwahrscheinlich, obwohl die Bodenlösung hinsichtlich Quarz übersättigt ist (vgl. Siever 1957). Dagegen scheint ein Wachsen vorhandener Quarzkristalle durch Ausfällung von SiO_2 an deren Oberfläche möglich zu sein (Millot u. a. 1963).

Von den verschiedenen Formen des kryptokristallinen Quarzes tritt in Böden auf Kalksteinen (z. B. Rendzinen) der *Flint* häufig auf, der jedoch bereits im Kalkstein enthalten ist (Khan 1959).

Außer als Quarz liegt SiO_2 in zwei weiteren kristallisierten Formen vor, dem *Cristobalit* und dem *Tridymit*. Der Cristobalit wurde in Podsolen und Latosolen aus jungvulkanischen Gesteinen Hawaiis (Tamura, Jackson und Sherman 1953a), Chiles (Besoain und Garcia-Vicente 1962) und Neuseelands (Swindale und Jackson 1956) gefunden und entstammt dem Ausgangsgestein (Hardjosoesastro 1956).

Auffallend geformte Opalkörner treten vor allem in der Grob- und Feinsandfraktion von Böden auf; sie werden wegen ihrer pflanzlichen oder tierischen Entstehung als *Bio-Opal* bezeichnet (Beavers und Stephen 1958, Smithson 1958, 1959). Der Bio-Opal entstammt entweder dem Ausgangsgestein (z. B. Kieselskelette von Meeresorganismen) oder der Vegetationsdecke des Bodens (z. B. verkieselte Epidermiszellen). Die oft gut erhaltene artspezifische Form der Bio-Opalkörner erlaubt ihre Zuordnung zu bestimmten Tier- oder Pflanzenarten (Kann und Arimura 1958). Im Laufe der Bodenentwicklung geht der Bio-Opal zunehmend in Quarz über (Kalk und Meyer 1960).

B. Aluminium

Freie Aluminiumoxiden entstehen vor allem in Böden, in denen infolge intensiver Verwitterung ein großer Teil der Kieselsäure der verwitternden Silikate abgeführt wird. Das dabei ebenfalls in Freiheit gesetzte Aluminium wird dann wie das Eisen zum Teil ausgewaschen, zum Teil in situ oder nach erfolgter Verlagerung als freies Oxid abgeschieden.

In Böden des gemäßigten humiden Klimas, also bei mittlerer Verwitterungsintensität, ist die Menge an neu gebildeten freien Aluminiumoxiden jedoch im allgemeinen gering ($<1\%$ Al_2O_3), da sich hier das bei der Silikatverwitterung freigesetzte Al an der Neubildung von Tonmineralen oder an der Bildung hydrargillitischer Zwischenschichten in aufweitbaren Tonmineralen beteiligt. Ein anderer Teil wird unter stark sauren Bedingungen ausgewaschen. So betrug z. B. die Menge an NaOH-löslichem Al (Methode FOSTER, vgl. Fußnote S. 341) in Podsol-Braunerden bzw. in Podsolen des älteren Pleistozäns Nordwestdeutschlands 0,2 bis 1% bzw. 0,3 bis 0,6% Al_2O_3. Etwa gleich hohe Gehalte wiesen ähnliche Böden des Jungpleistozäns Schleswig-Holsteins auf (vgl. die beim Silicium angegebene Literatur). In Podsolen erfolgt im B-Horizont häufig eine erhebliche relative Al-Anreicherung. Auch Marschen enthalten im Gegensatz zu den hohen Anteilen an freier Kieselsäure nur 0,2 bis 0,9% Al_2O_3. Im gleichen Bereich liegen die in degradierten Schwarzerden aus Löß ermittelten Werte. Besonders große Mengen NaOH-löslicher Aluminiumoxide enthalten Allophanhaltige Böden.

Im Gegensatz zu den Böden des gemäßigten humiden Klimas treten in solchen des tropischen und subtropischen humiden Klimas (z. B. Latosolen) größere Mengen freier Aluminiumoxide auf, da in diesen Böden bei intensiver Silikatverwitterung durch rasche Abfuhr der Kieselsäure die Tonmineralbildung zurücktritt. Das Al wird daher in Form freier Oxide ausgeschieden, die jedoch oft bereits so stark gealtert und durchkristallisiert sind, daß sie sich nicht mehr in verdünnter Alkalilauge lösen.

Die Kenntnisse über die Entstehungsbedingungen der verschiedenen Mineralformen der freien Aluminiumoxide in Böden sind sehr lückenhaft. Laboratoriumsversuche zeigen, daß bei Fällung saurer Al-Salzlösungen mit NH_3 in der Kälte amorphes, wasserhaltiges Al-Oxid, in der Hitze dagegen Böhmit (γ-AlOOH) entsteht, während sich aus Aluminatlösungen bei langsamer Fällung (z. B. durch Einleiten von CO_2) Hydrargillit (γ-$Al(OH)_3$), bei schneller dagegen Bayerit (α-$Al(OH)_3$) bildet, der ebenfalls zu Hydrargillit altert; doch sind die Ergebnisse nicht einheitlich (vgl. GINSBERG u. a. 1961). Die geringen Mengen an alkalilöslichem Aluminium in Böden des gemäßigten humiden Klimas (s. oben) lassen eine Bestimmung der Mineralform nicht zu; vermutlich liegt das so extrahierte Al in Form amorpher Hydroxide vor oder entstammt der Oberfläche von Aluminiumsilikaten (s. S. 341). In Böden des tropischen humiden Klimas mit größeren Anreicherungen an Aluminiumoxiden (z. B. Bauxite, vgl. Amer. Inst. Min. Met. Eng. 1952) altern die wasserhaltigen Gele dagegen relativ schnell zu definierten Oxiden, von denen der *Hydrargillit* (Gibbsit) das häufigste ist.

Hydrargillit wurde in einer großen Zahl von Böden des humiden tropischen Bereichs aus den verschiedensten Gesteinen (vgl. JACKSON und SHERMAN 1953, SWINDALE und JACKSON 1956), vor allem in Latosolen, festgestellt und tritt hier zum Teil in verhärteten Krusten zusammen mit kristallisierten freien Eisenoxiden auf (ALEXANDER u. a. 1956, SIVARAJASINGHAM u. a. 1962). Hydrargillit bildet sich vor allem in Böden mit hoher Permeabilität, die eine starke Fortfuhr von Ca, Mg, K, Na und Kieselsäure ermöglichen (BATES 1962). Auch in fossilen Terra-rossa-Bildungen Jugoslawiens wurde er nachgewiesen.

In Bauxiten und Lateriten tritt neben Hydrargillit in geringerer Menge *Böhmit* auf (BRINDLEY und SHITTON 1957). Er entsteht hier vermutlich aus dem Hydrargillit durch Entwässerung oder auch direkt aus amorphen Aluminiumhydroxiden. Böhmit wurde auch im B-Horizont von Podsolen festgestellt (v. ENGELHARDT 1940, VAN DER MAREL 1949). *Diaspor* wurde außer in Bauxiten (IBERG 1957) in Böden extrem feuchtwarmer Gebiete Mittelamerikas gefunden (LJUNGGREN 1958).

In sämtlichen angeführten Mineralen ist das Aluminium wie in den Tonmineralen oktaedrisch von OH- und O-Ionen umgeben; unterschiedlich ist vor allem die Anordnung und Bindung der Oktaeder untereinander (MILLIGAN und MCATEE 1952).

C. Eisen

Die Oxide und Hydroxide des Eisens und auch die des Mangans werden wegen ihrer auffallenden Färbung, durch die sich Verlagerungen, Anreicherungen und Verarmungen innerhalb des Bodenprofils dem Auge oft leicht zu erkennen geben, zur Klärung vieler, vor allem bodengenetischer Prozesse, herangezogen. Da sich das bei der Verwitterung Fe-haltiger Silikate frei werdende Eisen in wesentlich geringerem Ausmaß am Aufbau von silikatischen Tonmineralen beteiligt als das Aluminium und statt dessen als schwer lösliches Eisen(III)-oxid oder -hydroxid ausgefällt wird, treten Eisenoxide im Gegensatz zu Aluminiumoxiden auch in Böden des gemäßigten Klimas und daher sehr häufig auf. Darüber hinaus wird das Eisen wegen seiner Fähigkeit, in zweiwertiger Form aufzutreten, in der es auch bei höherem pH noch in Lösung und daher wanderungsfähig bleibt, häufig in größeren Mengen dort angereichert, wo eine erneute Oxydation zum dreiwertigen Eisen eintritt. Das dreiwertige Eisen wird hier im Gegensatz zum zweiwertigen wegen der Schwerlöslichkeit seiner Oxide und Hydroxide — und zwar auch bereits im sauren Bereich (ab etwa pH 3) — ausgefällt. Bekannte Erscheinungsformen solcher Anreicherungen sind Konkretionen, Raseneisenbildungen, lateritische Krusten, Flecken in Pseudogleyen u. a. (SCHWERTMANN 1959).

Aus den genannten Gründen schwankt der Gehalt an freien Eisenoxiden in den verschiedenen Böden sehr stark. In Braunerden (KOHL 1958), degradierten Schwarzerden, Gelb-Braunerden und anderen Böden, in denen das bei der Verwitterung frei werdende Eisen an Ort und Stelle verbleibt, liegt der Gehalt meist um etwa 1% Fe_2O_3; er sinkt häufig mit abnehmendem Tongehalt und — wie der Verwitterungsgrad — mit zunehmender Tiefe. In zahlreichen Böden ist das Eisenoxid jedoch im Zuge der Bodenentwicklung verlagert worden und zeigt dann eine ungleichmäßige Verteilung im Profil. Solche Verlagerungen treten z. B. in Parabraunerden (KOHL 1958, WHITE und RIECKEN 1955) auf, deren Gehalt an freien Eisenoxiden im Anreicherungshorizont bis auf 2,5% Fe_2O_3 und mehr ansteigt (MEYER u. a. 1962). Da die Verlagerung hier gemeinsam mit dem Ton erfolgt, geht der Anstieg des Gehaltes an freien Eisenoxiden dem des Tongehaltes parallel. In podsolierten Böden wandert das Eisen dagegen auch unabhängig von den Tonteilchen, zum Teil unter dem Einfluß komplexierender organischer Substanzen aus der Streu, und wird als Oxid im B-Horizont erneut ausgefällt. Der Gehalt der B-Horizonte (Ortstein und -erde) an freien Eisenoxiden ist zwar meist gering ($<1\%$ Fe_2O_3), kann jedoch ein Vielfaches desjenigen der Eluvialhorizonte betragen. Stärkere absolute Anreicherungen findet man in hydromorphen Böden. So können Oxidflecken in Pseudogleyen mehr als 10% Fe_2O_3 enthalten, während die unmittelbar benachbarten gebleichten Stellen weniger als 1% aufweisen. Abbauwürdige Anreicherungen in meist horizontartiger Ausbildung findet man in Gleyböden in Form von Raseneisenerzen mit bis zu mehr als

70% Fe_2O_3 sowie in lateritischen Krusten (vgl. MOHR und VAN BAREN 1954, SIVARAJASINGHAM u. a. 1962).

Die Mineralform, in der sich die freien Eisenoxide im Boden vorfinden, richtet sich nach den Fällungs- und Alterungsbedingungen (vgl. OADES 1963). Werden Ionen des dreiwertigen Eisens z. B. durch pH-Erhöhung ausgefällt, so

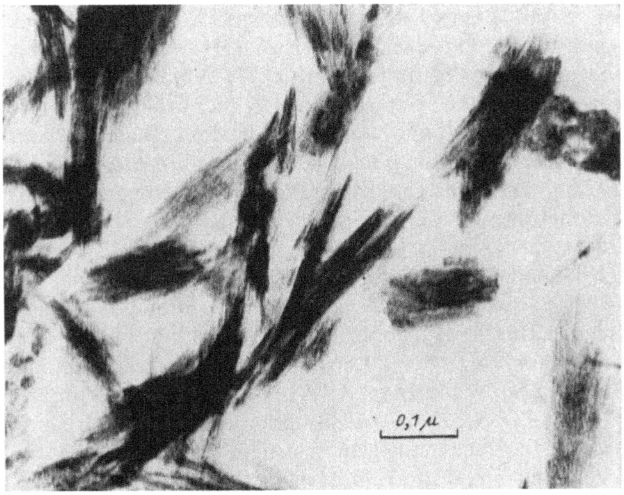

a

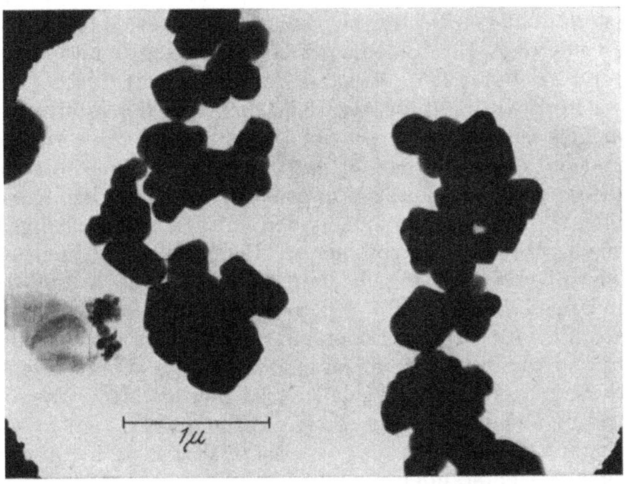

b

Abb. 78a und b. Elektronenmikroskopische Aufnahmen von a) Goethit (Borneo) (Aufn. W. SCHELLMANN, s. SCHELLMANN 1963) und b) Hämatit (Iran) (Aufn. H.-J. ECKHARDT)

entsteht zunächst ein röntgenamorphes Eisen(III)-oxidgel[1], das im Laufe der Zeit entweder zum gelb-braunen Goethit oder zum leuchtend roten Hämatit altert. Hohe Temperatur, niedriges pH und Anwesenheit von Ca und Mg fördern die Bildung des Hämatits, niedrigere Temperatur, hohes pH und die Anwesenheit

[1] Der Gebrauch der Bezeichnung *Limonit* sollte zweckmäßigerweise auf diese wasserhaltige amorphe Form beschränkt werden.

verschiedener Anionen, insbesondere SO_4 und HCO_3, die des Goethits (Schwertmann 1959a, Schellmann 1959). Die langsame Hydrolyse sehr verdünnter Fe(III)-Salzlösungen führt bereits im sauren Bereich unmittelbar zur Bildung von Goethit (Feitknecht und Michaelis 1962). Der seltenere, orange gefärbte Lepidokrokit bildet sich im Gegensatz zum Goethit ausschließlich durch Oxydation von Fe(II)-Verbindungen bei nicht zu hohem CO_2-Gehalt im Oxydationsmilieu (Schwertmann 1959a; vgl. jedoch Feitknecht und Michaelis 1962), wobei dunkelblaugrün gefärbte Zwischenprodukte mit zwei- und dreiwertigem Eisen auftreten (vgl. Feitknecht 1959, Bernal u. a. 1959). Bei höherem CO_2-Gehalt wie auch bei der Oxydation von Eisen(II)-carbonat (Siderit, Weißeisenerz) entsteht dagegen meist ebenfalls Goethit.

Das Kristallgitter von Goethit und Lepidokrokit besteht aus Fe-O-OH-Oktaedern, deren Anordnung beim Goethit dem Diaspor, beim Lepidokrokit dem Böhmit gleich ist. Unter dem Elektronenmikroskop sind beim Goethit aus Lateriten häufig nadelförmige Kristalle (Nadeleisenerz) (s. Abb. 78a), beim Hämatit plättchenförmige Kristalle mit hexagonaler Umgrenzung zu erkennen (Abb. 78b).

Die beschriebenen Bildungsbedingungen bestimmen das Vorkommen der verschiedenen Eisenoxidminerale im Boden. Im humiden Klima des gemäßigten Bereiches ist der durch Hydrolyse bzw. Ausfällung Fe(III)-haltiger Lösungen direkt oder auf dem Wege der Alterung amorpher Gele entstehende *Goethit* das häufigste Mineral (Schwertmann 1959b). Da diese Alterung verschieden weit fortgeschritten ist, besteht ein wechselnder Anteil der freien Eisenoxide aus röntgenamorphen Formen mit sehr geringer Teilchengröße und hoher Oberflächenaktivität. Dieser Anteil läßt sich durch milde Extraktion (z. B. mit Ammoniumoxalat-Lösung von pH 3 im Dunklen) annähernd bestimmen (Schwertmann 1959). Der Anteil des Goethits und dessen Kristallisationsgrad nehmen im Verlauf der Alterung zu. In Latosolen tritt Goethit bevorzugt in schlecht dränenden, z. B. Muldenlagen auf (Jackson und Sherman 1953, Simonett und Bauleke 1963), *Hämatit* dagegen in gut dränenden, z. B. Hang- und Kuppenlagen. Er entsteht hier entweder direkt bei der Alterung des amorphen Eisen(III)-oxidgels aus diesem, nicht dagegen durch Entwässerung des Goethits. Die Kristallisation führt zu harten lateritischen Krusten (Sivarajasingham u. a. 1962). Eine Hämatitbildung im gemäßigten, humiden Klima ist bisher nicht beobachtet worden und nach den obigen Ausführungen auch sehr unwahrscheinlich. Hämatit tritt daher in Böden dieses Klimaraumes nur als fossile Bildung auf und erweist sich dann als sehr stabil, so daß die rote Farbe solcher Böden lange erhalten bleibt (vgl. z. B. Roterden aus Basalt im Vogelsberg, Hessen, aus Granit im Harz, und Böden aus rotem Sandstein des Buntsandsteins).

Lepidokrokit wurde bisher vor allem in Böden mit stauender Nässe beobachtet (Brown 1953, Schwertmann 1959b). Er tritt hier als Mineralneubildung häufig neben Goethit auf und bedingt die orange Farbe der Flecken in solchen Böden. Im Gegensatz zum Goethit ist er gut kristallisiert und nicht durch Alterung amorpher Vorstufen entstanden.

Der Nachweis des ferromagnetischen rotbraunen *Maghemits* in Böden ist bisher nur vereinzelt erbracht worden. Er bedingt den Ferromagnetismus mancher Böden (Henin und Le Borgne 1954, Sherman und Matsusaka 1961) und wurde in größerer Menge in Konkretionen holländischer Auenböden (Van der Marel 1951) und in Niedermoorböden Nordwestdeutschlands (Schwertmann und Heinemann 1959) gefunden. Er entsteht entweder durch Oxydation von im Boden vorhandenen Magnetit, durch Entwässerung von Lepidokrokit oder aus anderen Eisen(III)-Oxiden durch Einwirkung höherer Temperatur unter Mitwirkung der organischen Substanz. Diese bewirkt vermutlich eine Reduktion der Eisenoxide zum Magnetit, der dann unmittelbar zum Maghemit oxydiert wird. Dabei bleibt das kubische Gitter und der Ferromagnetismus des Magnetits erhalten.

D. Mangan

Bei der Verwitterung und Wanderung im Boden verhält sich das Mangan ähnlich wie das Eisen. Da es wie dieses in reduzierter Form auch bei höherem pH in Lösung und daher wanderungsfähig bleibt, reichert es sich mit den Eisenoxiden an Stellen des Luftzutritts an. Bedingt durch eine geringere Oxydierbarkeit ist es jedoch etwas wanderungsfähiger und kann sich dadurch vom Eisen trennen, so daß in horizontartigen Oxidanreicherungen rostbraune Eisenoxidbänder oft gegen schwarze Manganoxidbänder abgesetzt sind, in denen der Mn-Gehalt relativ stark ansteigen kann.

Der Gehalt an freien Manganoxiden in Böden ist im allgemeinen sehr viel geringer als der an freien Eisenoxiden. Die hierfür angegebenen Werte schwanken in den verschiedenen Böden zwischen < 1 mg (z. B. Podsole) und mehreren hundert mg Mn/100 g Boden (z. B. tonreiche Marschen). Durchschnittlich höhere Anteile findet man in Konkretionen (0,05 bis über 1% Mn), in denen mit der Eisen- eine Mn-Anreicherung stattgefunden hat.

Hinsichtlich der Formen der Manganoxide in Böden kann man auf Grund von Untersuchungsergebnissen an sedimentär entstandenen jungen Manganoxidanreicherungen ableiten, daß wasserhaltiges MnO_n mit $n = 1{,}7$ bis 2,0 verschiedener Struktur, vor allem der *Pyrolusit* (vgl. Wadsley 1950) vorherrscht (Finck 1956). Die amorphen Formen dieser Oxide bezeichnet man auch zusammenfassend als *Wad*. Bei der Alterung verlieren sie einen Teil ihres Wassers und gehen in kristallisierte, weniger reaktionsfähige Oxide über.

Von den kristallisierten Formen wurden bisher außer Pyrolusit, vor allem das δ-MnO_2 und OH-haltige Manganoxide gefunden, die außer vierwertigem Mangan auch zweiwertiges Mangan und dreiwertiges Eisen (*Mangan-* bzw. *Eisen-Manganit*, z. B. $3\,MnO_2 \cdot Mn(OH)_2 \cdot x\,H_2O$; Buser u. a. 1954), gelegentlich auch Kalium und Barium enthalten können (Ljunggren 1955, Buser und Grütter 1956). *Hausmannit* und *Manganit* wurden bisher nur in Gesteinen, nicht dagegen in Böden nachgewiesen.

4. Weitere anorganische Bestandteile

Von den bisher nicht erwähnten anorganischen Bodenbestandteilen spielen lediglich die *Carbonate* mengenmäßig eine größere Rolle. Sulfate, Sulfide, Chloride, Phosphate und andere Verbindungen sind zwar oft von erheblicher Bedeutung für die Ernährung der Pflanzen, jedoch nur in einigen Böden (z. B. Salzböden) in größerer Menge vorhanden. So besteht z. B. der häufig in Marschböden fleckenweise auftretende hellgelbe, stark pflanzenschädliche sogenannte Maibolt hauptsächlich aus *Jarosit* $K_2Fe_6(OH)_{12}(SO_4)_4$ (Schwertmann 1961).

Unter den Carbonaten ist das *Calciumcarbonat* (vor allem *Calcit*) im Boden am stärksten vertreten. Es tritt in solchen Böden auf, deren Ausgangsgestein $CaCO_3$ enthält, wie z. B. Kalksteine, Mergel (z. B. Geschiebemergel) und andere kalkhaltige Sedimente (z. B. Löß, Marschen [Verhoeven 1962], Auenlehme). Außerdem kann es durch Düngemittel in ehemals $CaCO_3$-freien Böden bzw. durch Zufuhr $Ca(HCO_3)_2$-haltiger Wässer, z. B. Grundwasser, angereichert werden. Beispiele für das letztere sind die Calciumcarbonat-Solontschaks (Redmont und McClelland 1959) und die sogenannten Wiesenkalke.

Da $CaCO_3$ bei der Verwitterung, abgesehen von den leicht löslichen Salzen, am schnellsten gelöst oder zersetzt wird, geht seine Entfernung aus dem Boden meist der Silikatverwitterung voraus (MEYER u. a. 1962). Der $CaCO_3$-Gehalt eines Bodens ist also außer vom Ausgangsgestein vom Verwitterungsgrad des Bodens abhängig. In Böden des humiden Klimas steigt der $CaCO_3$-Gehalt daher oft mit zunehmender Tiefe an. Das aus dem Oberboden ausgewaschene $CaCO_3$ kann im Unterboden erneut ausgeschieden werden und sich diffus verteilt, mycelartig oder in Form von Konkretionen und Krusten wieder anreichern. Solche Ca-Horizonte findet man z. B. in Böden aus Kalkstein (Rendzinen) und in Lößböden. Die Geschwindigkeit der $CaCO_3$-Auswaschung ist außer vom Grad der Durchfeuchtung von der Korngröße der $CaCO_3$-Teilchen abhängig. Während feinkörniges $CaCO_3$ z. B. im Löß unter stark humidem Klima relativ schnell ausgewaschen wird, können Kalksteinbröckchen (z. B. in Rendzinen aus harten Kalksteinen oder kalksteinhaltigen Schottern) und Muschelschalen (z. B. in Marschen) im Boden noch vorliegen, wenn dessen pH bereits unter 7 liegt.

Außer Calcit kommt in Böden auch *Dolomit* vor, der entweder dem Ausgangsgestein oder dolomitischen Düngemitteln entstammt oder auch im Boden aus Ca-Salzen unter dem Einfluß Mg-haltiger Wässer in situ gebildet sein kann (SHERMAN u. a. 1962). Der Dolomit-Gehalt betrug z. B. im Löß 5 bis 15% (JOHNSON und BEAVERS 1959) und bedingt dessen hohe Werte für das nachlieferbare Magnesium (SCHACHTSCHABEL 1956). Auch in Marschen bestand ein wesentlicher Anteil der Carbonate aus Dolomit (VERHOEVEN 1962). Wegen seiner im Vergleich zum Calcit geringeren Löslichkeit reichert er sich bei der Verwitterung gegenüber diesem zunächst etwas an.

Von den anderen Carbonaten tritt gelegentlich das Eisen(II)-carbonat stärker hervor. Es kommt insbesondere in Nieder- und Übergangsmooren als *Siderit* (Weißeisenerz) lagen- oder linsenförmig als schmutzig-weiße, schmierige Masse vor (VAN HEUVELN 1958), wo es durch überlagernde Torfschichten vor der Oxydation geschützt wird. Bei Zutritt von Luftsauerstoff gehen diese Lagen in sehr feinverteiltes braunes FeOOH über, das als Gasreinigungsmasse benutzt wird. Siderit tritt außerdem als Auskleidung ehemaliger Wurzelröhren in Raseneisenbildungen in Form schwarz glänzender Beläge auf und ist hier gegenüber Sauerstoff wesentlich beständiger.

5. Bestimmung der Minerale der Tonfraktion

Wegen der großen allgemeinen Bedeutung der Minerale der Tonfraktion hat man sich seit langem und mit Erfolg bemüht, Methoden auszuarbeiten, die eine Identifizierung und Strukturbestimmung dieser Minerale, insbesondere der Tonminerale, gestatten.

Das *Lichtmikroskop* ist wegen der geringen Teilchengröße der Tonminerale nur sehr begrenzt verwendbar. Lediglich das *Phasenkontrast-* und *Grenzdunkelfeldverfahren* (MEYER und KALK 1959) läßt die Erkennung feinster Teilchen bei geringen Unterschieden der Lichtbrechung in gewissem Rahmen zu. Am verbreitetsten ist die *röntgenographische* Untersuchung (BROWN 1961). Hierbei werden die auf ein Tonmineralpulver-Präparat (Methode DEBYE-SCHERRER) fallenden monochromatischen Röntgenstrahlen von den Netzebenen (Flächen der Bausteine) abgebeugt bzw. reflektiert und auf einem Film oder mit Hilfe eines Zählrohrs registriert. Die Ausmessung von Lage und Intensität der Reflexe ermöglicht unter Einbeziehung verschiedener Zusatzbehandlungen, wie z. B. Erhitzen, Sättigung mit bestimmten Kationen und Quellflüssigkeiten, die Erkennung und Strukturaufklärung von Tonmineralen.

Mit Hilfe von *Elektronenstrahlen* lassen sich sowohl optische Bilder als auch Beugungsaufnahmen (RADCZEWSKI 1953) herstellen. Erstere ermöglichen wegen des hohen Auflösungsvermögens (50 A) die Erkennung der Einzelteilchen und dienen vornehmlich der Bestimmung ihrer Größe und Form. Im Falle des Halloysits und Kaolinits bilden sie die einzige Unterscheidungsmöglichkeit dieser beiden sonst sehr ähnlichen Minerale. Elektronenbeugungsaufnahmen enthalten im Gegensatz zu Röntgenaufnahmen keine Basisinterferenzen, da die Mineralplättchen senkrecht zur Basis durchstrahlt werden, geben jedoch Aufschluß über einige zur Identifizierung wertvolle Nichtbasisinterferenzen (z. B. 060-Interferenz).

Von ähnlich großer Verbreitung wie die Röntgenanalyse ist die *Differentialthermoanalyse* (DTA) (MACKENZIE 1957), bei der die Probe kontinuierlich erhitzt wird und die Wärmetönungen der dabei auftretenden Reaktionen im Vergleich zu einer inerten Probe mit Hilfe eines Differenzthermoelementes registriert werden. Lage und Ausmaß dieser Effekte sind dabei zum Teil mineralspezifisch. Die gleichzeitig auftretenden Gewichtsveränderungen (Wasserabgabe, CO_2-Abgabe, O_2-Aufnahme) werden bei der *thermogravimetrischen* Analyse registriert (MIELENZ u. a. 1954).

Neuerdings wird auch die für die Konstitutionsaufklärung organischer Stoffe viel verwendete *Ultrarotabsorptionsspektrographie* für Tonminerale (besonders im Wellenlängenbereich von 2 bis 20 μ) angewendet. Sie ermöglicht z. B. spezielle Aussagen über die Bindung der OH-Gruppen in der Substanz (vgl. BEUTELSPACHER 1956, VAN DER MAREL und ZWIERS 1959).

Auch die *chemische* Analyse liefert wertvolle Einzelaussagen, wenn sie auch wegen der ähnlichen Zusammensetzung der Tonminerale eine Identifizierung der einzelnen Minerale nur dann erlaubt, wenn ein einheitlicher Mineralbestand vorliegt. Aus dem SiO_2/Al_2O_3-Verhältnis, dem K-Gehalt u. a. Daten lassen sich jedoch oft wichtige Rückschlüsse auf die Genese der Tonminerale ziehen.

Eine quantitative Bestimmung der Tonminerale ist selbst bei Kombination mehrerer Methoden nur sehr begrenzt möglich. Dies ist vor allem auf die mangelnde Vergleichbarkeit der zur Eichung benutzten Standardminerale zurückzuführen, die z. B. in der wechselnden Zusammensetzung und Kristallausbildung (Teilchengröße, Kristallisationsgrad) begründet ist. Es ist daher sinnvoll, sich auf eine halbquantitative Abschätzung der Mengenanteile zu beschränken. Dies gelingt um so sicherer, je mehr der genannten Methoden zur Bestimmung herangezogen werden.

Literatur

ALEXANDER, L. T., J. G. CADY, L. D. WHITTIG und R. F. DEVER: Mineralogical and chemical changes in the hardening of laterite. Rapp. 6. Congr. Int. Sci. Sol Paris E, 67 (1956). — *Amer. Inst. Min. Met. Eng.*: Problems of clay and laterite genesis. New York 1952. — ANDERSON, J. V., und J. L. WHITE: A study of fragipans in some southern Indiana soils. Proc. Soil Sci. Soc. Amer. **22**, 450 (1958). — AOMINE, S., und M. L. JACKSON: Allophane determination in Ando soils by cation-exchange capacity delta value. Proc. Soil Sci. Soc. Amer. **23**, 210 (1959). — AOMINE, S., und H. KAWASAKI: Vermiculite in some Japanese soils. Soil Sci. a. Plant Nutr. **9**, 18 (1963). — AOMINE, S., und N. YOSHINAGA: Clay minerals in some welldrained volcanic ash soils in Japan. Soil Sci. **79**, 349 (1955).

BAILEY, H. H., E. P. WHITESIDE und A. E. ERICKSON: Mineralogical composition of glacial material as a factor in the morphology and genesis of some podsol, gray wooded, gray-brown podsolic and humic gley soils in Michigan. Proc. Soil Sci. Soc. Amer. **21**, 433 (1957). — BARSHAD, J.: Soil development. In: F. E. BEAR (Ed.), Chemistry of the soil. New York: Reinhold. 1955. — Factors affecting clay formation. Clays and Clay Min. Proc. 6. Nat. Conf. Clays and Clay Min. 1957, 110 (1959). — BATES, TH. F.: Halloysite and gibbsite formation in Hawaii. Clays and Clay Min.

Proc. 9. Nat. Conf. Clays and Clay Min. 1960, 315 (1962). — BATES, TH. F., und J. J. COMER: Further observations on the morphology of chrysotile and halloysite. Clays and Clay Min. Proc. 6. Nat. Conf. Clays and Clay Min. 1957, 237 (1959). — BEAVERS, A. H., und I. STEPHEN: Some features of the distribution of plant-opal in Illinois soils. Soil Sci. **86**, 1 (1958). — BERNAL, J. D., D. R. DASGUPTA und A. L. MACKAY: The oxides and hydroxides of iron and their structural interrelationships. Clay Min. Bull. **4**, 15 (1959). — BESOAIN, M. E., und J. GARCIA-VICENTE: Clay mineralogy of some volcanic ash soils of Chile. N. Jb. Min. Abh. **98**, 349 (1962). — BEUTELSPACHER, H.: Infrarotuntersuchungen an Bodenkolloiden. Rapp. 6. Congr. Int. Sci. Sol Paris **B**, 329 (1956). — BEUTELSPACHER, H., und H. W. VAN DER MAREL: Kennzeichen zur Identifizierung von Kaolinit, "Fireclay"-Mineral und Halloysit, ihre Verbreitung und Bildung. Tonind. Ztg. **85**, 570 (1961). — BIRRELL, K. S., und M. FIELDES: Allophane in volcanic ash soils. J. Soil Sci. **3**, 156 (1952). — BLUME, H. P.: Die Deutung der Tiefenfunktion des Tonmineralbestandes von Böden. Beitr. Min. Petr. **9**, 13 (1963). — BOWER, C. A., und F. B. GSCHWEND: Ethylene glycol retention in soils as a measure of surface area. Proc. Soil Sci. Soc. Amer. **16**, 342 (1952). — BRINDLEY, G. W., und W. H. SHITTON: An x-ray mineralogical study of some British Guayana bauxite deposits. Econ. Geol. **52**, 391 (1957). — BROOKS, C. S.: Nitrogen adsorption experiments on several clay minerals. Soil Sci. **79**, 331 (1955). — BROWN, G.: On the occurrence of lepidocrocite in some British soils. J. Soil Sci. **4**, 220 (1953). — BROWN, G. (Ed.): The x-ray identification and crystal structures of clay minerals. London: Min. Soc. 1961. — BRUNAUER, S., P. H. EMMET und E. TELLER: Adsorption of gases in multimolecular layers. J. Amer. Chem. Soc. **60**, 309 (1938). — BURST, J. F.: Mineral heterogeneity in "glauconite pellets". Amer. Min. **43**, 481 (1958). — BUSER, W., und P. GRAF: Radiochemische Untersuchungen an Festkörpern III. Ionen- und Isotopenaustauschreaktionen an Mangandioxyden und Manganiten. Helv. chim. Acta **38**, 810 (1955). — BUSER, W., P. GRAF und W. FEITKNECHT: Beitrag zur Kenntnis der Mangan(II)-manganite und des δ-MnO_2. Helv. chim. Acta **37**, 2322 (1954). — BUSER, W., und A. GRÜTTER: Über die Natur der Manganknollen. Schweiz. Ann. petr. Mitt. **36**, 49 (1956).

CANN, D. B., und E. P. WHITESIDE: A study of genesis of a podzol-gray brown podzolic intergrade soil profile in Michigan. Proc. Soil Sci. Soc. Amer. **19**, 497 (1955). — CASHEN, G. H.: Electric charges of kaolin. Trans. Farad. Soc. **53**, 477 (1959). — CORRENS, C. W.: Experiments on the decomposition of silicates and discussion of chemical weathering. Clays and Clay Min. Proc. 10. Nat. Conf. Clays and Clay Min. 1961, 443 (1963).

DE KIMPE, C., M. C. GASTUCHE und G. W. BRINDLEY: Ionic coordination in alumo-silicic gels in relation to clay mineral formation. Amer. Min. **46**, 1370 (1961). — DE VORE, G. V.: The surface chemistry of feldspars as an influence on their decomposition products. Clays and Clay Min. Proc. 6. Nat. Conf. Clays and Clay Min. 1957, 26 (1959). — DIAMOND, S., und E. B. KINTER: Surface areas of clay minerals as derived from measurements of glycerol retention. Clays and Clay Minerals, Nat. Acad. Sci., Nat. Res. Council Publ. **566**, 334 (1958). — DROSTE, J. B., N. BHATTACHARYA und J. A. SUNDERMAN: Clay mineral alteration in some Indiana soils. Clays and Clay Min. Proc. 9. Nat. Conf. Clays and Clay Min. 1960, 329 (1962). — DYAL, R. S., und S. B. HENDRICKS: Total surface of clays in polar liquids as a characteristic index. Soil Sci. **69**, 421 (1950). — Formation of mixed layer minerals by potassium fixation. Proc. Soil Sci. Soc. Amer. **16**, 45 (1952).

ECKHARDT, H. J.: Über Chlorite in Sedimenten. Geol. Jahrb. **75**, 437 (1958). — ENGELHARDT, W. v.: Mineralogische Beschreibung eines mecklenburgischen Bodenprofils. Chem. d. Erde **13**, 1 (1940).

FEITKNECHT, W.: Über die Oxydation von festen Hydroxyverbindungen des Eisens in wässrigen Lösungen. Z. Elektrochem. **63**, 34 (1959). — FEITKNECHT, W., und W. MICHAELIS: Über die Hydrolyse von Eisen-(III)-perchlorat-Lösungen. Helv. chim. Acta **45**, 212 (1962). — FINCK, A.: Die Bindungsformen des Mangans im Boden. Z. Pflanzenernähr., Düng., Bodenkde. **73**, 59 (1956). — FLAIG, W., und H. BEUTELSPACHER: Elektronenmikroskopische Untersuchungen über die Gestalt der Tonminerale in Böden. Grundl. Landtechn. 87 (1953). — FOSTER, M. D.: Geochemical studies of clay minerals. III. The determination of free silica and free aluminium in montmorillonites. Geochim. Cosmochim. Acta **3**, 143 (1953). — The relation between "illite", beidellite, and montmorillonite. Clays and Clay Minerals, Nat. Acad. Sci., Nat. Res. Council Publ. **327**, 386 (1954). — Interpretation of the composition of vermiculites and hydrobiotites. Clays and Clay Min. Proc. 10. Nat. Conf. Clays and Clay Min. 1961, 70 (1963). — FÜCHTBAUER, H., und H. GOLDSCHMIDT: Die Tonminerale der Zechsteinformation. Beitr. Min. Petr. **6**, 320 (1959).

GINSBERG, H., W. HÜTTIG und H. STIEHL: Ein Beitrag zum System H_2O/Al_2O_3. Z. anorg. allg. Chem. **309**, 233 (1961). — GRANQUIST, W. T., und G. G. SUMNER: Acid dissolution of a Texas bentonite. Clays and Clay Min. Proc. 6. Nat. Conf. Clays and Clay Min. 1957, 292 (1959). — GRIM, C. E.: Clay mineralogy. London: McGraw-Hill. 1953.

HARDJOSOESASTRO, R. R.: Preliminary note on cristobalite in clay fractions of volcanic ashes. J. Soil Sci. **7**, 185 (1956). — HASHIMOTO, J., und M. L. JACKSON: Rapid dissolution of allophane and kaolinite-halloysite after dehydration. Clays and Clay Min. Proc. 7. Nat. Conf. Clays and Clay Min. 1958, 102 (1960). — HAUSER, E. A.: The colloid science of important clay minerals. Clays and Clay Minerals, Nat. Acad. Sci., Nat. Res. Council Publ. **395**, 442 (1955). — HENDRICKS, S. B., und C. S. ROSS: The chemical composition and genesis of glauconite and celadonite. Amer. Min. **26**, 683 (1941). — HÉNIN, S.: Synthesis of clay minerals at low temperature. Clays and Clay Minerals, Nat. Acad. Sci., Nat. Res. Council Publ. **456**, 54 (1956). — HÉNIN, S., und E. LE BORGNE: Sur les propriétées magnetiques des sols et leurs interprétations pédologiques. Trans. 5. Int. Congr. Soil Sci. Léopoldville **2**, 13 (1954). — HEUVELN, B. VAN: Minerale afzettnigen in het Smeulveen. Boor en Spade **9**, 38 (1958). — HEYSTEK, H.: Some hydrous micas in South African clays and shales. Clays and Clay Minerals, Nat. Acad. Sci., Nat. Res. Council Publ. **395**, 337 (1955). — HOEBEKE, F., und W. DEKEYSER: La glauconite. C. R. Rech. J.R.S.J.A. Nr. 14, 102 (1955). — HOFMANN, U.: Aus der Chemie der hochquellfähigen Tone. Angew. Chem. **68**, 53 (1956). — HOFMANN, U., S. MARCOS und F. W. SCHEINBRA: Das sonderbarste Tonmineral, der Halloysit. Ber. dtsch. keram. Ges. **39**, 474 (1962). — HUFFMANN, H.: Mineralogische Untersuchungen an fünf Bodenprofilen über Basalt, Muschelkalk und Buntsandstein. Heidelb. Beitr. Min. Petr. **4**, 67 (1954).

IBERG, R.: Zur Kenntnis von Böhmit und Diaspor in Bauxit. Chimia **11**, 13 (1957). — ILER, R. K.: The colloid chemistry of silica and silicates. Ithaka, New York: Cornell University Press. 1955.

JACKSON, M. L.: Frequency distributions of clay minerals in major great soil groups as related to the factors of soil formation. Clays and Clay Min. Proc. 6. Nat. Conf. Clays and Clay Min. 1957, 133 (1959). — Aluminium bonding in soils. A unifying principle in soil science. Proc. Soil Sci. Soc. Amer. **27**, 1 (1963). — Interlayering of expansible layer silicates by chemical weathering. Clays and Clay Min. Proc. 11. Nat. Conf. Clays and Clay Min. 29 (1963). — JACKSON, M. L., Y. HSEUNG, R. B. COREY, E. J. EVANS und R. C. VAN DEN HEUVEL: Weathering sequence of clay size minerals in soils and sediments. II. Chemical weathering of layer silicates. Proc. Soil Sci. Soc. Amer. **16**, 3 (1952). — JACKSON, M. L., und G. D. SHERMAN: Chemical weathering of minerals in soils. Adv. Agron. **5**, 213 (1953). — JASMUND, K.: Die silicatischen Tonminerale, 2. Aufl. Weinheim: Verlag Chemie. 1955. — JOHNSON, P. R., und A. H. BEAVERS: A mineralogical characterization of some loess-derived soils in Illinois. Proc. Soil Sci. Soc. Amer. **23**, 143 (1959).

KAHN, A.: Studies on the site and shape of clay particles in aqueous suspensions. Clays and Clay Min. Proc. 6. Nat. Conf. Clays and Clay Min. 1957, 220 (1959). — — KALK, E., und B. MEYER: Progressive Bio-Opal → Quarz-Umwandlung bei der Terra-fusca-Entstehung aus Kalkgestein. Zeiss-Mitt. **2**, 103 (1960). — KANNO, J., und S. ARIMURA: Plant opal in Japanese soils. Soil a. Plant Food **4**, 62 (1958). — KHAN, D. H.: Profile distribution of the sand minerals in some rendzinas, red-brown soils and terra rossas. Soil Sci. **88**, 67 (1959). — Studies on translocation of chemical constituents in some red-brown soils, terra rossas, and rendzinas, using zirconium as a weathering index. Soil Sci. **88**, 196 (1959a). — KLAGES, M. G., und J. L. WHITE: A chlorite-like mineral in Indiana soils. Proc. Soil Sci. Soc. Amer. **21**, 16 (1957). — KÖSTER, H. M.: Vergleich einiger Methoden zur Untersuchung von geochemischen Vorgängen bei der Verwitterung. Beitr. Min. Petr. **8**, 69 (1961). — KOHL, F.: Durchwaschungs- und Durchschlämmungserscheinungen an Böden aus Bayern. Z. Pflanzenernähr., Düng., Bodenkde. **80**, 237 (1958). — KORNFELD, J. A.: Statistical relationships of minor constituents of some nontronites. Clays and Clay Minerals, Nat. Acad. Sci., Nat. Res. Council Publ. **576**, 174 (1958). — KUNDLER, P.: Beurteilung forstlich genutzter Sandböden im nordwestdeutschen Tiefland. Arch. Forstw. **5**, 585 (1956). — Zur Methodik der Bilanzierung der Ergebnisse von Bodenbildungsprozessen (Profilbilanzierung, dargestellt am Beispiel eines Texturprofils auf Geschiebemergel in Norddeutschland). Z. Pflanzenernähr., Düng., Bodenkde. **86**, 215 (1959). — Lessivés (Parabraunerden, Fahlerden) aus Geschiebemergel der Würm-Eiszeit im norddeutschen Tiefland. Z. Pflanzenernähr., Düng., Bodenkde. **95**, 97 (1961). — KUN-HUANG, H., und Ch. TSEN-TUO: Certain physico-chemical properties of the Pinchen clay and the function of its free ferric oxide and organic matter. Proc. Soil Sci. Soc. Amer. **23**, 270 (1959).

LIPPMANN, F.: Corrensit. In: HINTZE-CHUDOBA, Handbuch der Mineralogie, Erg.-Bd. II, S. 688. Berlin: W. de Gruyter. 1959. — LJUNGGREN, P.: Differential thermal analysis and x-ray examination of Fe- and Mn-bog ores. Geol. För. Stockholm Förh. 135 (1955). — A mineralogical examination of some soil samples from southern and central Honduras. Kungl. Fysiograf. Sällsk. Lund Förh. 28, 125 (1958). — LOPEZ-GONZALES, J. DE, und J. CANO-RUIZ: Surface area changes of a vermiculite by acid and thermal treatment. Clays and Clay Min. Proc. 6. Nat. Conf. Clays and Clay Min. 1957, 399 (1959).

MACEWAN, D. M. C.: Fourier transform methods for studying scattering from lamellar systems. I. A direct method for analyzing interstratified mixtures. Koll. Z. **149**, 96 (1956). — MACKENZIE, R. C.: (Ed.): Differential thermal analysis of clays. London: Min. Soc. 1957. — The classification and nomenclature of clay minerals. Clay Min. Bull. **4**, 52 (1959). — MAREL, H. W. VAN DER: Tropical soils in relation to plant nutrition. Soil Sci. **64**, 445 (1947). — Mineralogical composition of a heath podsol profile. Soil Sci. **67**, 193 (1949). — Gamma ferric oxide in sediments. J. Sed. Petrol. **21**, 12 (1951). — Potassium fixation a beneficial soil characteristic for crop production. Z. Pflanzenernähr., Düng., Bodenkde. **84**, 51 (1959). — MAREL, H. W. VAN DER, und J. H. L. ZWIERS: O-H-stretching bonds of the kaolin minerals. Silicates Industr. **24**, 359 (1959). — MARTIN-VIVALDI, J. L., und D. M. C. MACEWAN: Corrensite and swelling chlorite. Clay Min. Bull. **4**, 173 (1960). — MCKEAGUE, J. A., und M. G. CLINE: Silica in soil solutions. Canad. J. Soil Sci. **43**, 70 (1963). — MEHRA, O. P., und M. L. JACKSON: Constancy of the sum of mica unit cell potassium surface and interlayer sorption surface in vermiculite-illite clays. Proc. Soil Sci. Soc. Amer. **23**, 101 (1959). — MEYER, B., und E. KALK: Die Mineralbestimmung in Tonfraktionen des Bodens mit Hilfe von Phasenkontrast- und Grenzdunkelfeldmikroskopie. Landwirtsch. Forsch. **12**, 3 (1959). — — MEYER, B., E. KALK und H. FÖLSTER: Parabraunerden aus primär carbonathaltigem Würm-Löß in Niedersachsen I, II. Z. Pflanzenernähr., Düng., Bodenkde. **99**, 37 (1962), **100**, 1 (1963). — MIELENZ, R. C., N. C. SCHIELTZ und M. E. KNIG: Thermogravimetric analysis of clay and clay-like minerals. Clays and Clay Minerals, Nat Acad. Sci., Nat. Res. Council Publ. **327**, 285 (1954). — MILLIGAN, W. O., und J. L. MCATEE: The structure of hydrous aluminium oxides and hydroxides. Probl. Clay Laterite Genesis 94, Amer. Inst. Min. Met. Eng., New York 1952. — MILLOT, G., J. LUCAS und R. WEY: Research on evolution of clay minerals and argillacous and silicous neoformation. Clays and Clay Min. Proc. 10. Nat. Conf. Clays and Clay Min. 1961, 399 (1963). — MITCHELL, B. D., und R. C. MACKENZIE: Removal of free iron oxide from clay. Soil Sci. **77**, 173 (1954). — MOHR, E. C. J., und F. A. VAN BAREN: Tropical soils. London: Interscience. 1954.

NELSON, B. W., und R. ROY: Synthesis of the chlorites and their structural and chemical constitution. Amer. Min. **43**, 707 (1958). — NORRISH, K., und C. BROWN: Hydrous micas. Min. Mag. **29**, 929 (1952).

OADES, M.: The nature and distribution of iron compounds in soils. Soils a. Fert. **26**, 69 (1963). — ORCHISTON, H. D.: Adsorption of water vapor II. Soil Sci. **78**, 463 (1954). — Adsorption of water vapor III. Soil Sci. **79**, 71 (1955). — Adsorption of water vapor V. Soil Sci. **87**, 276 (1959). — Adsorption of water vapor VII. Soil Sci. **88**, 159 (1959a). — OSTHAUS, B.: Kinetic studies on montmorillonites and nontronite by the acid dissolution technique. Clays and Clay Minerals, Nat. Acad. Sci., Nat. Res. Council Publ. **456**, 301 (1956).

PRETTRE, M., B. IMERLIK, L. BLAUDIM und M. PETITJEAN: Zum Bildungsmechanismus adsorbierter und katalytisch wirksamer Aluminiumoxyde. Angew. Chem. **65**, 549 (1953).

QUIGLEY, R. M., und R. T. MARTIN: Chloritized weathering products of a New England glacial till. Clays and Clay Min. Proc. 10. Nat. Conf. Clays and Clay Min. 1961, 107 (1963).

RADCZEWSKI, O. E.: Über die Bestimmung von Mineralen mittels Beugung im Elektronenmikroskop. Opt. **10**, 163 (1953). — REDMONT, C. E., und J. E. MCCLELLAND: The occurrence and distribution of lime in calcium carbonate solonchak and associated soils of eastern North Dakota. Proc. Soil Sci. Soc. Amer. **23**, 61 (1959). — ROBERTSON, R. H. S., G. W. BRINDLEY und R. C. MACKENZIE: Mineralogy of kaolin clays from Pugu, Tanganyika. Amer. Min. **39**, 118 (1954). — ROSS, C. S., und S. B. HENDRICKS: Minerals of the montmorillonite group. U. S. Geol. Survey Bull. 205-B, 41 (1945). — ROYEN, P., A. ORTH und K. RUTHS: B-Punktsdrucke als charakteristische Konstanten. Z. anorg. allg. Chem. **281**, 1 (1955).

SALGER, M.: Verwitterung und Bodenbildung auf diluvialen Schotterter assen. Heidelb. Beitr. Min. Petr. **4**, 288 (1954). — SASAKI, S., und Y. ISHIZUKA: On the volcanic ash soil in Tokachi district, Hokkaido island, Japan. Soil a. Plant Food **2**,

177 (1957). — SAWHNEY, B. L., und M. L. JACKSON: Soil montmorillonite formulas. Proc. Soil Sci. Soc. Amer. **22**, 115 (1958). — SCHACHTSCHABEL, P.: Mikroskopische und röntgenographische Untersuchungen von Böden. Bodenkunde und Pflanzenernähr. **5**, 375 (1937). — Der Magnesiumversorgungsgrad nordwestdeutscher Böden und seine Beziehungen zum Auftreten von Mangelsymptomen an Kartoffeln. Z. Pflanzenernähr., Düng., Bodendke. **74**, 202 (1956). — Die Bestimmung des Manganversorgungsgrades von Böden und seine Beziehung zum Auftreten der Dörrfleckenkrankheit bei Hafer. Z. Pflanzenernähr., Düng., Bodenkde. **78**, 147 (1957). — SCHACHTSCHABEL, P., R. HAAR und W. KÖSTER: Chemische Untersuchungen an Marschen I. Z. Pflanzenernähr., Düng., Bodenkde. **83**, 113 (1958). — SCHACHTSCHABEL, P., und W. KÖSTER: Chemische Untersuchungen an Marschen II. Z. Pflanzenernähr., Düng., Bodenkde. **85**, 140 (1958a). — SCHEFFER, F., B. MEYER und E. KALK: Mineraluntersuchungen an Würm-Löß südniedersächsischer Lößfluren als Voraussetzung für die Mineralanalyse verschiedener Lößbodentypen. Chem. d. Erde **19**, 338 (1958). — SCHELLMANN, W.: Experimentelle Untersuchungen über die sedimentäre Bildung von Goethit und Hämatit. Chem. d. Erde **20**, 104 (1959). — Zur lateritischen Verwitterung von Serpentinit. Geol. Jb. **81**, 645 (1963). — SCHLICHTING, E., und H.-P. BLUME: Art und Ausmaß der Veränderung des Bestandes mobiler Oxyde in Böden aus jungpleistozänem Geschiebemergel und ihren Horizonten. Z. Pflanzenernähr., Düng., Bodenkde. **96**, 144 (1962). — SCHROEDER, D.: Mineralogische Untersuchungen an Lößprofilen. Heidelb. Beitr. Min. Petr. **4**, 443 (1954). — SCHWERTMANN, U.: Die fraktionierte Extraktion der freien Eisenoxyde in Böden, ihre mineralogischen Formen und ihre Entstehungsweisen. Z. Pflanzenernähr., Düng., Bodenkde. **84**, 194 (1959). — Über die Synthese definierter Eisenoxyde unter verschiedenen Bedingungen. Z. allg. anorg. Chem. **298**, 337 (1959a). — Mineralogische und chemische Untersuchungen an Eisenoxyden in Böden und Sedimenten. N. Jb. Min. Abh. **93**, 67 (1959b). — Über das Vorkommen und die Entstehung von Jarosit in Marschböden (Maibolt). Naturwiss. **48**, 159 (1961). — Der Mineralbestand der Fraktion $<2\mu$ einiger Böden aus Sedimenten und seine Eigenschaften. Z. Pflanzenernähr., Düng., Bodenkde. **95**, 209 (1961a). — Eigenschaften und Bildung aufweitbarer (quellbarer) Dreischicht-Tonminerale in Böden aus Sedimenten. Beitr. Min. Petr. **8**, 199 (1962). — SCHWERTMANN, U., und B. HEINEMANN: Über das Vorkommen und die Entstehung von Maghemit in nordwestdeutschen Böden. N. Jb. Min. Mh. 174 (1959). — SHERMAN, G. D., und Y. MATSUSAKA: Magnetism of iron oxide in Hawaiian soils. Soil Sci. **91**, 239 (1961). — SHERMAN, G. D., F. SCHULTZ und F. J. ALWAY: Dolomitization in soils of the Red River Valley, Minnesota. Soil Sci. **94**, 304 (1962). — SIEVER, R.: The silica budget in the sedimentary cycle. Amer. Min. **42**, 821 (1957). — SIMONETT, D. S., und M. P. BAULEKE: Mineralogy of soils on basalt in North Queensland. Proc. Soil Sci. Soc. Amer. **27**, 205 (1963). — SIVARAJASINGHAM, S., L. T. ALEXANDER, J. G. CADY und M. G. CLINE: Laterite. Adv. Agron. **14**, 1 (1962). — SMITH, J.: A mineralogical study of weathering and soil formation from olivine basalt in northern Ireland. J. Soil Sci. **8**, 225 (1957). — SMITHSON, F.: Silica particles in some British soils. J. Soil Sci. **7**, 122 (1956). — Grass opal in British soils. J. Soil Sci. **9**, 148 (1958). — Opal sponge spicules in soils. J. Soil Sci. **10**, 105 (1959). — STEPHEN, I.: A study of rock weathering with reference to the soils of the Malvern Hills II. J. Soil Sci. **3**, 219 (1952). — STÖBER, W.: Adsorptionseigenschaften und Oberflächenstruktur von Quarzpulvern. Koll. Z. **145**, 17 (1956). — STUBIČAN, V., S. TEZAK und M. WRISCHER: Über einige Eigenschaften der Kieselsäure in Bentoniten. Koll. Z. **151**, 33 (1957). — SUDO, T.: Clay mineralogical aspects of the alteration of volcanic glass in Japan. Clay Min. Bull. **2**, 96 (1954). — SUDO, T., und H. TAKAHASHI: Shapes of halloysite particles in Japanese clays. Clays and Clay Minerals, Nat. Acad. Sci., Nat. Res. Council Publ. **456**, 67 (1956). — SWINDALE, L. D., und M. L. JACKSON: Genetic processes in some residual podsolized soils of New Zealand. Rapp. 6. Congr. Int. Sci. Sol Paris E, 233 (1956).

TAGGART, M. S. JR., W. O. MILLIGAN und H. P. STUDER: Electron micrographic studies of clays. Clays and Clay Minerals, Nat. Acad. Sci., Nat. Res. Council Publ. **395**, 31 (1955). — TAMURA, T.: Identification of clay minerals from acid soils. J. Soil Sci. **9**, 148 (1958). — TAMURA, T., und M. L. JACKSON: Structural and energy relationship in the formation of iron and aluminium oxides, hydroxides and silicates. Science **117**, 381 (1953). — TAMURA, T., M. L. JACKSON und G. D. SHERMAN: Mineral content of low humic, humic and hydrol humic latosols of Hawaii. Proc. Soil Sci. Soc. Amer. **17**, 343 (1953a). — TURNER, R. C., und J. E. BRYDON: Formation of double hydroxides and the titration of clays. Science **136**, 1052 (1962).

VERHOEVEN, B.: On the calcium carbonate content of young marine sediments. Neth. J. Agric. Sci. **10**, 58 (1962).

WADSLEY, A. D.: Synthesis of some hydrated manganese minerals. Amer. Min. 35, 485 (1950). S. a. Rev. Pure Appl. Chem. 1, 203 (1951). — WEAVER, Ch. E.: The effects and geological significance of potassium "fixation" by expandable clay minerals derived from muscovite, biotite, chlorite, and volcanic materials. Amer. Min. 43, 839 (1958). — The clay petrology of sediments. Clays and Clay Min. Proc. 6. Nat. Conf. Clays and Clay Min. 1957, 154 (1959). Dort zahlreiche Literatur. — WEISS, A., G. KOCH und U. HOFMANN: Zur Kenntnis von Wolkonskit. Ber. dtsch. keram. Ges. 31, 301 (1954). — Zur Kenntnis von Saponit. Ber. dtsch. keram. Ges. 32, 12 (1955). — WEISS, A., A. MEHLER, G. KOCH und U. HOFMANN: Über das Anionenaustauschvermögen der Tonmineralien. Z. anorg. allg. Chem. 284, 247 (1956). — WHITE, J. L.: X-ray diffraction studies on weathering of muscovite. Soil Sci. 93, 16 (1962). — WHITE, E. M., und F. F. RIECKEN: Brunizem-gray-brown podsolic soil biosequences. Proc. Soil Sci. Soc. Amer. 19, 504 (1955). — WHITTIG, L. D., V. J. KILMER, R. C. ROBERTS und J. G. CADY: Characteristics and genesis of Cascade and Powell soils of north-western Oregon. Proc. Soil Sci. Soc. Amer. 21, 226 (1957). — WOODWARD, L. A.: Variations in viscosity of clay-water suspensions of Georgia kaolins. Clays and Clay Minerals, Nat. Acad. Sci., Nat. Res. Council Publ. 395, 246 (1955). — WURMAN, E., E. P. WHITESIDE und M. M. MORTLAND: Properties and genesis of finer textured subsoil bands in some sandy Michigan soils. Proc. Soil Sci. Soc. Amer. 23, 135 (1959).

b) Die anorganischen Bestandteile der Bodenlösung

Von

H. Linser

Als Bodenlösung kann man jene wässerige Flüssigkeit bezeichnen, die das Porenvolumen eines Bodens teilweise (im Extremfalle ganz) erfüllt und innerhalb der Hohlräume des Bodens frei beweglich ist, jenen Anteil des Bodenwassers also, den man als Gravitationswasser bezeichnet. Da jenes Wasser, das an den Oberflächen der festen Bodenbestandteile durch elektrostatische oder andere zwischenmolekulare Kräfte festgehalten wird, nicht die gleichen Eigenschaften als Lösungsmittel besitzt wie das frei bewegliche Wasser, kann das Schwarmwasser, das zum Teil noch von der Pflanze genutzt werden kann, eine andere Zusammensetzung hinsichtlich gelöster Stoffe aufweisen als das Gravitationswasser, auf das praktisch keine von den Bodenteilchen ausgehenden Kräfte mehr wirksam sind.

Bei der Gewinnung von Bodenpreßsäften erwies sich der Gehalt der Bodenlösung als abhängig vom Druck und der Menge der abgepreßten Bodenlösung. So wurden folgende Werte beobachtet (Limburgerhof 1939):

Druck (kg/cm^2)	Preßsaft-Menge (cm^3)	mg CaO/l
100	26	383
200	42	463
300	53	502

Dies läßt vermuten, daß die mit verschiedenen Drucken an die Bodenteilchen gebundenen Wasserschichten sehr verschiedene, von der Stärke ihrer Bindung abhängige Gehalte an gelösten Stoffen besitzen.

Der Hauptbestandteil der Bodenlösung ist Wasser. In ihm sind anorganische Salze, Salze organischer Säuren und andere organische Stoffe gelöst; letztere bleiben hier außer Betracht (s. Abschn. I, C, c, S. 382ff. und Abschn. I, C, d, S. 458ff.).

1. Das Wasser

Wasser enthält neben H_2O etwa 0,015% D_2O bzw. HDO (WACHSMANN 1954) und Spuren an HTO, 10^{-18} T je H in norwegischen Wässern (JOHNSTON, WOLFGANG und LIBBY 1951) und besteht aus Wassermolekülen und den aus diesen durch Dissoziation hervorgehenden Wasserstoff- und Hydroxylionen H^+ und

OH^-. Die Wassermoleküle sind Dipole, das heißt, daß sie zwar elektroneutral sind, weil sie eine positive und eine negative elektrische Ladung enthalten, deren Lokalisationen zusammenfallen, unter der Einwirkung eines elektrischen Feldes aber auseinandertreten, so daß das Molekül einen positiven und einen negativen Pol erhält. Als Dipol steht es mit anderen Dipolen in elektrostatischer Wechselwirkung, wobei Wechselwirkungsenergien von etwa 0,9 bis 1,8 kcal/Mol auftreten, die Ordnungskräfte darstellen. Das Dipolmoment μ liegt für Wasser bei $1{,}84 \cdot 10^{-18}$ und bedingt die starke Dielektrizitätskonstante $\Sigma = 81$.

Der Dipolcharakter der Wassermoleküle bedingt, daß sie elektrovalent gebundene Moleküle zu spalten vermögen. Bringt man ein heteropolares Molekül, dessen Form im Gasraum ungeteilt ist, in Wasser, so lagern sich die Wasserdipole an die Ionen an und trennen sie durch eine Schicht von Wassermolekülen, die „Solvat"- oder „Hydrat"-Hülle genannt wird (Abb. 79).

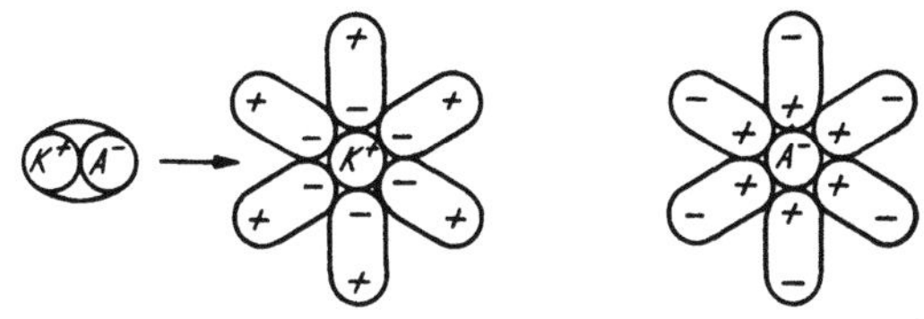

Abb. 79. Solvatation bzw. Hydratation der Komponenten eines heteropolar gebundenen Moleküls durch Wasser-Dipole

Auch das Wassermolekül selbst dissoziiert, wobei die elektrovalent gebundenen H^+- und OH^--Ionen auseinandertreten. Im Wassermolekül hat das Sauerstoffatom seine äußere Elektronenhülle, die sechs Elektronen enthält, durch Hinzunehmen der zwei Elektronen von zwei Wasserstoffatomen zur Edelgasstruktur komplettiert. Von diesem Gesamtkomplex H_2O dissoziiert nun ein Proton H^+ ab, so daß der restliche Komplex mit acht Elektronen nun ein unabgesättigtes Elektron enthält, das ihm eine negative Ladung erteilt:

$$H_2O \rightarrow H^+ + OH^-.$$

Dieser Vorgang ist reversibel, es stellt sich ein Gleichgewichtszustand ein. Das Ion H^+ trennt sich nicht als solches vom Wassermolekül ab, sondern zusammen mit einem Wasserdipol, also hydratisiert in Form von Hydroxonium H_3O^+, so daß folgende Formel gültig sein kann:

$$2\,H_2O \rightleftharpoons H_3O^+ + OH^-,$$

wobei auch höhere Hydratationsstufen möglich erscheinen.

Sind H^+- und OH^--Ionen in gleicher Zahl vorhanden, so zeigt das Wasser „neutrale" Reaktion. Da jedoch H^+-Ionen einerseits und OH^--Ionen andererseits recht kräftige Reaktionspartner für chemische Umsetzungen darstellen, erhält Wasser, in dem ein Überschuß der einen Ionenart über die andere besteht, besondere chemische Reaktionsfähigkeit:

ein Überschuß von H^+ gibt die Eigenschaften einer Säure (bewirkt „saure Reaktion"),

ein Überschuß von OH^- bewirkt die Eigenschaften einer Base („alkalische Reaktion").

Die Messung der Konzentration der H^+-Ionen gibt ein Maß für die Größen dieser Eigenschaften in einer wässerigen Lösung.

Allgemein werden Konzentrationen in Mol/Liter (oder Mol/1000 g) angegeben; die Wasserstoffionen-Konzentration wird in Gramm-Molen je Liter angegeben. Bei $C_{H^+} = 1$ liegen $6{,}023 \cdot 10^{23}$ Wasserstoffionen in einem Liter Flüssigkeit vor. Meistens handelt es sich jedoch um viel kleinere Konzentrationen: $C_{H^+} = 10^{-3}$ bis 10^{-8} mol/l.

Zur Vereinfachung der Schreibweise wurde der Begriff des *pH-Wertes* eingeführt (SÖRENSEN 1909), der den negativen dekadischen Logarithmus der Wasserstoffionen-Konzentration C_{H^+} angibt:

$$pH = -\log C_{H^+}.$$

Kleine pH-Werte bedeuten hohe Wasserstoffionenkonzentration bzw. stark saure Reaktion, hohe pH-Werte bedeuten eine niedrige Wasserstoffionenkonzentration, ein pH > 7 (bei 25° C) ein Überwiegen von OH^--Ionen und eine basische (alkalische) Reaktion. Nach dem Massenwirkungsgesetz gilt nämlich:

$$K = \frac{C_{H^+} \cdot C_{OH^-}}{C_{H_2O}}$$

und da das Wasser in großem Überschuß vorliegt und seine Konzentration daher als konstant gelten darf, ist das Produkt aus Wasserstoffionen und Hydroxylionen konstant (vgl. Abb. 80):

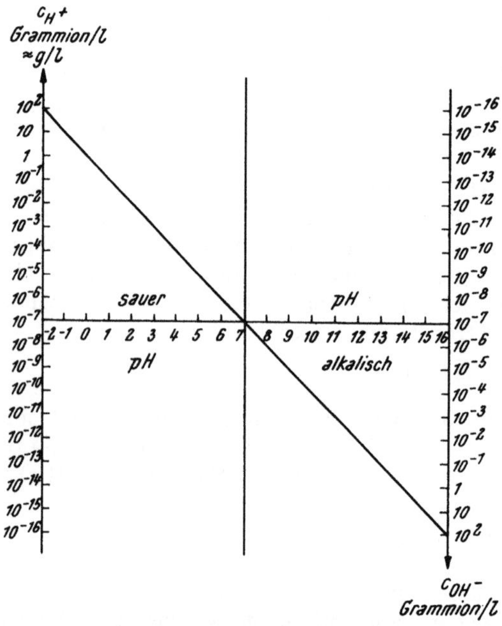

Abb. 80. Zusammenhang zwischen Wasserstoffionenkonzentration bzw. Hydroxylionenkonzentration und pH-Wert. Abszisse: pH-Wert; Ordinaten: links Wasserstoffionenkonzentration, rechts Hydroxylionenkonzentration

$$K = C_{H^+} \cdot C_{OH^-}.$$

Die Konstante K hat dabei einen temperaturunabhängigen Wert, z. B.:

$$K_{0^0} = 0{,}114 \cdot 10^{-14}\ (\text{mol/l})^2$$
$$K_{25^0} = 1{,}0 \cdot 10^{-14}\ (\text{mol/l})^2$$
$$K_{100^0} = 65 \cdot 10^{-14}\ (\text{mol/l})^2$$

Wenn ein Salz KA in Wasser gelöst und dissoziiert vorliegt:

$$\frac{C_{K^+} \cdot C_{A^-}}{C_{KA}} = \text{konst.},$$

so liegen gleich viele Kationen und Anionen vor, die elektrostatischen Ladungen sind einander äquivalent, es liegt also (theoretisch) weder freie Azidität noch freie Basizität vor: Wir haben es mit einem Neutralsalz zu tun. Die Azidität einer Mischung einer starken Säure (z. B. HCl) mit einer starken Base (z. B. KOH) muß daher die des reinen Wassers sein, also $C_{H^+} = 10^{-7}$ (bei 25° C = pH 7). Ist jedoch eines der Ionen einer „schwachen" Säure oder Base zugehörig, die dadurch gekennzeichnet sind, daß sie (als solche) nur wenig dissoziieren, so tritt eine Bindung von Molekülen der Säure (oder Base) mit Wasser ein: wegen der unvollständigen Dissoziation der schwachen Säure bleibt die H^+-Konzentration am Äquivalenzpunkt niedriger als die der (vollständig dissoziierten) Hydroxylionen, weshalb die Lösung einer *schwachen Säure* mit einer *starken Base* alkalisch reagiert und einen pH-Wert größer als 7 zeigt. Umgekehrt reagiert die Lösung einer starken Säure mit einer schwachen Base (z. B. $NH_4 \cdot Cl$) sauer, der pH-Wert ist kleiner als 7.

Bei dem Salz einer schwachen Säure mit einer starken Base wird die pH-Verschiebung der Lösung um so größer, je höher die Salzkonzentration ansteigt. Hydroxylionenkonzentration und Konzentration der undissoziierten Säure sind nicht gleich. Bezeichnet man die Dissoziationskonstante der Säure mit k_s, so gilt

$$pH = pk_s - \lg \frac{C_{HA}}{C_{A^-}}.$$

Setzt man einer solchen Lösung eine kleine Menge Säure zu, so wird die Dissoziation der schwachen Säure in ihre Ionen zurückgedrängt, d. h. es wird die Zahl der H^+-Ionen vermindert, so daß der Zusatz von Säure (H^+-Ionen) keine Veränderung der H-Ionenkonzentration (des pH-Wertes) bewirkt. Eine solche Lösung besitzt deshalb ein *Pufferungsvermögen.* Da auch in der Bodenlösung schwache Säuren vorkommen, ist auch dort mit einem (beschränkten) Pufferungsvermögen zu rechnen.

2. Gelöste Stoffe

Sowohl Gase als auch feste Stoffe besitzen in unterschiedlichem Maße die Fähigkeit, in Wasser gelöst zu werden.

A. Gase in der Bodenlösung

Die Löslichkeit von Gasen in Flüssigkeiten ist bei konstanter Temperatur dem Druck (*P*) proportional (HENRYsches Löslichkeitsgesetz) und außerdem von der Temperatur abhängig. Hierbei hängt die Löslichkeit eines Gases nicht davon ab, ob schon ein anderes Gas in der Flüssigkeit gelöst ist. Aus einem Gasgemisch werden die einzelnen Gase (gemäß dem DALTONschen Gesetz) nach Maßgabe ihrer Teildrucke gelöst, sofern sie nicht untereinander chemische Reaktionen eingehen.

Über die Gehalte von Wasser, das bei 760 Torr Druck mit den entsprechenden Gasen in Gleichgewicht steht, an CO_2, O_2 und N_2 nebst Edelgasen gibt Tab. 24 für verschiedene Temperaturen Auskunft. Die Löslichkeiten nehmen mit steigender Temperatur ab.

Tabelle 24. *Gehalt von Wasser an CO_2, O_2 und N_2 nebst Edelgasen bei verschiedenen Temperaturen (bei Sättigung unter 760 Torr)*
(vgl. VOGEL 1956)

Temperatur °C	0°	10°	20°	30°	40°	50°	60°
CO_2 (g in 100 g Wasser) in H_2O	0,3346	0,2318	0,1688	0,1257	0,0973	0,0761	0,0576
NH_3 (g in 100 g Wasser)..	89,9	68,4	51,8	40,8	33,8	28,4	23,8
O_2 (cm³ im Liter Wasser, CO_2-NH_3-frei) .	10,19	7,87	6,36	5,26	—	—	—
N_2 + übrige Gase (cm³ im Liter Wasser, CO_2-, NH_3-frei)	18,99	14,97	12,32	10,38	—	—	—
Luft (cm³ im Liter Wasser, CO_2-NH_3-frei)	29,18	22,84	18,68	15,64	—	—	—

Da die Bodenlösung bei guter Bodendurchlüftung mit der Atmosphäre im Gleichgewicht stehen kann, wird sie entsprechende Mengen der in der Atmosphäre vorhandenen Gase gelöst enthalten. Da die Untersuchung des Gasgehaltes der Bodenlösung in situ methodisch großen Schwierigkeiten begegnet, sind keine zuverlässigen Messungen verfügbar. Die Anreicherung der von der Wurzelatmung und der Atmung der Bodenmikroflora und -mikrofauna entstammenden CO_2 in der Bodenluft entspricht einer zu erwartenden Erhöhung des Kohlensäuregehaltes der Bodenlösung um das Vielfache, da die Bodenluft gegenüber einem Normalgehalt der Atmosphäre von 0,03 Vol.-% einen Co_2-Gehalt bis zu 10% annehmen kann.

B. In der Bodenlösung gelöste feste Stoffe

Die in der Bodenlösung vorhandenen, gelösten Feststoffe kommen aus verschiedenartigen Quellen in die Bodenlösung:

1. durch Auflösung fester Bodenbestandteile,
2. durch Austausch von festen Bodenbestandteilen,
3. durch Niederschläge, Staub usw.,
4. durch Düngemittel,
5. durch Abgabe aus den Wurzeln lebender Pflanzen,
6. durch Zersetzung abgestorbenen Pflanzenmaterials bzw.
7. durch Abgabe aus lebenden oder im Abbau befindlichen Bodenmikroorganismen.

Die jeweilige Zusammensetzung der Bodenlösung ist zwar weitgehend von der Anlieferung bestimmter Stoffe aus diesen genannten, vielfältigen Quellen abhängig, sie unterliegt jedoch in noch stärkerem Maße anderen, sie beeinflussenden Faktoren:

1. dem Entzug durch Pflanzenbestand und Mikrobenbestand des Bodens,
2. der Auswaschung gelöster Stoffe in tiefere Bodenschichten bzw. dem Aufsteigen von Bodenlösung aus tieferen Bodenschichten,
3. dem Festlegungsvermögen der Böden für bestimmte Stoffe,
4. der gegenseitigen Ausfällung unlöslicher Reaktionsprodukte bei Umsetzungen innerhalb der gelösten Stoffe in der Bodenlösung,
5. durch Verlust gelöster Stoffe an Sorptionsträger des Bodens.

Schließlich unterliegt ihre aktuelle Zusammensetzung den die Löslichkeit allgemein bestimmenden Faktoren physikalisch chemischer Art, vor allem dem Druck und der Temperatur.

Wenn mithin sehr vielfältige Faktoren auf die Zusammensetzung der Bodenlösung von Einfluß sind, so muß ihre Kenntnis doch besonders wünschenswert sein, da die Pflanzenwurzel ja aus ihr den Hauptteil ihres Bedarfes an löslichen Nährstoffen und einen Teil ihres Bedarfes an Gasen zu decken genötigt ist.

Auch im lufthaltigen Boden, dessen Porenvolumen nur zum Teil mit Flüssigkeit erfüllt ist, kann man bei entsprechender mikroskopischer Betrachtung feststellen, daß sich zwischen Wurzelhaaren und Bodenteilchen Flüssigkeitströpfchen und Flüssigkeitsschichten befinden, die große Teile der Wurzelhaaroberflächen bzw. der Wurzelepidermiszellen bedecken. Die Bodenlösung bzw. ihre Zusammensetzung muß daher bei der Stoffaufnahme durch die Wurzel von entscheidender Bedeutung sein.

Da das Gravitationswasser der Krume und der anderen der Pflanze zugänglichen Bodenschichten in Lysimetern nach unten abrinnen und aufgefangen werden kann, sagt auch die Zusammensetzung dieser Wassermengen etwas über die Gehalte der Bodenlösung aus, wenngleich sie nicht mit jener der Bodenlösung in den der Pflanze erreichbaren Bodenbereichen selbst identifiziert werden darf. Sie stellt jene, bei hohen Niederschlägen im Boden bewegliche Lösung dar, die beim Passieren verschiedener Bodenschichten sich mit deren festen Bestandteilen in Lösungs- und Sorptionsgleichgewichte gebracht hat und nach Durchlaufen aller dieser Schichten übrigblieb. Wir sind gewohnt, sie als kennzeichnend für die *Auswaschung* von Nährstoffen aus dem Boden zu betrachten, doch enthält diese Lösung zweifellos die typischen Bestandteile der Bodenlösung, vor allem dann, wenn sie in nicht zu großer Tiefe unterhalb des der Pflanze zugänglichen Bodenraumes abgefangen wird.

Aus Lysimeterversuchen von PFAFF (1950), die sich über 12 Jahre und auf drei Böden erstreckten, geht folgende Zusammensetzung des Sickerwassers hervor (Tab. 25).

Tabelle 25.
Durchschnittsgehalte (mg/l) von Sickerwässern aus langjährigen Lysimeterversuchen
(nach Zahlen von PFAFF 1950)

Boden	Versuchs-dauer	jährliche Sicker-wasser-menge, l	N	SO_4	P_2O_5	MgO	CaO	K_2O	Cl
„gewachsener Boden" Limburgerhof ...	10 Jahre	119	25,2	185,0	3,5	—	210,0	16,8	92,5
saurer Sandboden	12 Jahre	226	16,4	106,2	5,5	17,7	132,9	—	13,3
alkalischer Sandboden	12 Jahre	231	20,8	134,1	5,3	16,5	160,1	—	13,1

Eine 1 m tiefe Bodenschicht enthält bei einem Wassergehalt von (im Mittel bei 100%iger Erfüllung der Wasserkapazität) etwa 35 Vol.-% (HIELSCHER 1938) 350 l Wasser je m², entsprechend einer Regenmenge von 350 mm, also etwa die Hälfte bis ein Drittel einer mitteleuropäischen jährlichen Niederschlagsmenge. Aus der gleichen Bodenschicht werden (in Abhängigkeit von der Bodenart) pro Jahr 5,3 bis 9,0 g N, 0,7 bis 1,5 g P_2O_5, 5,7 bis 13,4 g K_2O und 22,6 bis 64,5 g CaO „ausgewaschen", also an Sickerwasser abgegeben (Limburgerhof 1939), wobei die Sickerwässer etwa 40% der Regenmenge ausmachen.

Tabelle 26. *Analysendaten von Drainagewässern von seit etwa 80 Jahren mit Weizen bebauten Flächen in ppm*
(nach VOELCKER 1871)

	ungedüngt	Stallmist	mineralische Volldüngung
NH_4 — N	0,14	0,20	0,24
NO_3 — N	15,0	62,0	32,9
SO_3	24,7	106,1	73,3
P_2O_5	0,6	—	1,54
Fe_2O_3	5,7	2,6	2,7
MgO	5,1	4,9	7,9
CaO	98,1	147,4	143,9
K_2O	1,7	5,4	4,4
Na_2O	6,0	13,7	10,7
Cl	10,7	20,7	20,7
SiO_2	10,9	35,7	24,7
Organ. Substanz, CO_2 usw.	67,7	77,3	84,6
Summe:	246,4	476,0	407,6

Auch die Zusammensetzung von Drainagewasser mag einige Rückschlüsse auf die Zusammensetzung der Bodenlösung gestatten. Vergleichende Daten für Drainagewasser von ungedüngten, von mit Stallmist gedüngten und von mit Handelsdüngemitteln gedüngten, bebauten Flächen wurden von VOELCKER (1871) mitgeteilt (vgl. Tab. 26).

Die Zahlen zeigen als Hauptbestandteile Calcium und Schwefel und verhältnismäßig mehr Na als K.

Vergleicht man die Verhältnisse, in welchen die einzelnen Bauelemente der Pflanze in diesen Wässern vorliegen, so ergeben sich folgende Zahlen (Tab. 27).

Tabelle 27. *Verhältniszahlen zu Tab. 25 und 26*

N	S	P	Fe	Mg	Ca	K	Na	Cl	wie
21,6	14,2	0,4	5,7	4,4	100	2,0	6,4	15,3	bei „ungedüngt"
59,0	40,3	—	1,7	2,9	100	4,3	9,7	19,6	bei „Stallmistgabe"
32,2	28,5	0,7	1,8	4,7	100	3,6	7,7	20,2	bei „mineralischer Volldüngung"
16,8	41,1	1,0	—	—	100	9,3	—	61,6	bei Sickerwasser „gewachsener Boden Limburgerhof"
17,3	37,2	2,5	—	11,2	100	—	—	14,1	bei Sickerwasser „saurer Sandboden
18,2	39,0	2,0	—	8,7	100	—	—	11,4	bei Sickerwasser „alkalischer Sandboden
27,6	33,5	1,3	3,1	6,4	100	4,8	7,9	23,7	Mittelwert

Mengenmäßig steht somit das Calcium vor dem Schwefel, dann folgen Stickstoff und Chlor sowie in kleineren Mengen Natrium, Magnesium, Kalium und Eisen und, in noch kleinerer Menge, Phosphor.

Die Zusammensetzung jener Bodenlösung, mit der die Wurzel im Boden natürlicherweise in Berührung kommt und aus der sie einen großen Teil der zum Wachstum benötigten Nährstoffe aufnimmt, ist nicht einfach feststellbar. Da der Boden im normalen Fruchtbarkeitszustand einen Wassergehalt besitzt, der seine Kapazität (Festhaltevermögen) für Wasser unterschreitet, kann die Bodenlösung nur schwierig und unter Anwendung besonderer Verfahren in unveränderter Form gewonnen werden (SCHUMACHER 1863, ISCHERIKOW 1907, SIGMOND 1910, HESSELINK VAN SUCHTELEN 1912, BOUYOUCOS und MACCOOL 1915, MORGAN 1916, RAMANN, MÄRZ und BAUER 1916, ZYL 1916, SCHUSTER 1917, LIPMAN 1918, PARKER 1921, BURGESS 1922, GROH und EHRENBERGER 1923, TULAIKOW und KUZMIN 1923, HIBBARD 1923, DOJARENKO und SCHMUCK 1924, VESTERBERG 1924, PIERRE und PARKER 1927, OSUGI und SANO 1928, CONRAD, PROEBSTING und MACKINNON 1930, WRANGELL 1930, 1931; REITEMEIER und RICHARDS 1944, WAGNER 1962).

Da sich ihre Zusammensetzung infolge der recht verschiedenartigen Löslichkeitsverhältnisse der in den festen Bodenteilchen enthaltenen Stoffe bei Zuführung weiterer, den Normalzustand wesentlich überschreitender Wassermengen stark verändert (zugefügtes Wasser verdünnt nicht nur die bisher gelöst gewesenen Stoffe, sondern löst auch bisher ungelöste), muß eine Verdünnung der Bodenlösung mit Wasser bei ihrer Gewinnung vermieden werden. Die methodische Schwierigkeit der Gewinnung der Bodenlösung hat bewirkt, daß nur relativ wenige Forscher sich mit der Untersuchung ihrer Zusammensetzung befaßt haben. Da auch die gewinnbaren Mengen an Bodenlösung nicht groß sind, hat sich ihre analytische Untersuchung zumeist auf ihre Hauptbestandteile beschränkt, so daß fast keine Kenntnisse über die Gehalte der Bodenlösungen aus verschiedenen Bodenarten und Bodentypen an Spurenelementen und akzessorischen Elementen vorliegen.

Eine vergleichende Untersuchung an verschiedenen Bodenarten führten REITEMEIER und RICHARDS (1944) im Hinblick auf einige Hauptnährstoffe durch (vgl. Tab. 28).

Tabelle 28. *Ergebnisse von Analysen an Bodenlösungen in Milliäquivalenten je Liter Bodenlösung (Membran-Preßsaft-Methode)*
(nach REITEMEIER und RICHARDS 1944)

Boden	I Salzboden von feiner Textur: Imperial clay	II nicht salziger „Boden" von feiner Textur: Gile adobe clay	III „nicht salziger Boden" von mittlerer Textur: Millville loam	IV „nicht salziger Boden" von mittelgrober Textur: Hesperia sandy loam	V „gipshaltiger Boden" von feiner Textur: Reogan clay loam
Wassergehalt (%)	27,2	28,9	13,6	10,4	14,2
HCO_3	3,8	3,8	5,4	3,5	11,0
NO_3	89	—	—	17,6	—
SO_4	19,4	26,1	1,3	2,9	47,8
Mg	319,0	6,0	6,8	4,0	63,7
Ca	786,8	23,6	13,3	18,2	47,7
K	2,9	1,0	0,2	1,7	11,4
Na	563,8	22,7	1,3	3,3	30,1
Cl	155,4	15,2	3,1	3,3	30,0
Summe der Kationen (Normalitäten =)	1,67	0,053	0,022	0,027	0,153

In Tab. 28 sind nicht *alle* in der Bodenlösung vorkommenden Stoffe angeführt, sondern nur einige der mengenmäßig bedeutendsten. Neben diesen liegen kleinere Mengen sehr vieler Elemente des gesamten Systems der Elemente vor, über die keine genaueren analytischen Daten erarbeitet worden sind.

Die Zahlen in Tab. 28 zeigen, daß recht unterschiedliche Mengen an Salzen sich in den Bodenlösungen verschiedener Böden gelöst vorfinden und daß es sich auch um recht verschiedenartige Zusammensetzung dieser Salzmengen handelt. Als Hauptbestandteil der gelösten, anorganischen Stoffe der Bodenlösung kann das Calcium betrachtet werden, neben ihm spielt das Magnesium eine bedeutende Rolle. Natrium und Chlor treten nur in besonderen Fällen, mitunter jedoch sehr stark in den Vordergrund und sind mengenmäßig meist bedeutender als das Kalium; Nitrate und Sulfate wechseln stark in ihrer Menge.

An Phosphorsäure fand WRANGELL (1931) in Bodenlösungen (Preßsäften) verschiedener sandiger Lehme 0,016 bis 0,431 mg P_2O_5/l, bei schweren Lehmen 0,019 bis 0,181, bei einem Kalkboden 0,06, bei einem Lauberdeboden 1,74 und bei Komposterde 3,64 mg P_2O_5/l. An Kali fand sie bei sandigem Lehm 2,9 bis 8,7 mg K_2O/l, bei schwerem Lehm 7,2 bis 12,7, bei humosem Waldboden 3,0 bis 13,1, bei einem vorher mit Kali gedüngten Kompostboden jedoch 140 bis 2800 mg K_2O/l. Für Ammoniakstickstoff fand sie bei verschiedenen Lehmböden 1,66 bis 4,69 g mg N/l, bei einem humosen Waldboden 1,24 und bei Komposterde 6,86 mg N/l. Die Nitrat-N-Gehalte betrugen dagegen bei verschiedenen Lehmböden 0,043 bis 78,5 mg N/l, bei humosem Waldboden 21,4 und bei Komposterde 217,0 mg N/l. SCHLÖSING (1870) erhielt für die Bodenlösung bei etwa 20% Wassergehalt des Bodens folgende analytische Daten in ppm:

CO_2	118	MgO	13,5	Na_2O	7,8
N_2O_5	305	CaO	264	Cl	7,4
SO_3	57,9	K_2O	6,9	SiO_2	291

PFAFF fand bei verschiedenen Böden und einem Wassergehalt von 60% der Wasserkapazität 10,5 bis 43,0 mg K_2O/l Preßsaft und 0,02 bis 0,09 mg P_2O_5/l Preßsaft (Limburgerhof 1939).

Wie auch andere Untersuchungen zeigen, liegt der Gehalt der Bodenlösung an Phosphaten weit unterhalb jenem an Sulfaten oder Nitraten (BURD und MARTIN 1924), doch ist es aus prinzipiellen Gründen nicht möglich, über die Zusammensetzung der Bodenlösung *generelle* Aussagen zu machen; ist doch zu erwarten, daß sie außerordentlich große Unterschiede aufweisen muß, je nach der Art der schon genannten vielfältigen Faktoren, welche ihre Zusammensetzung beeinflussen.

Selbstverständlich zeigt die Zusammensetzung der Bodenlösung Veränderungen je nach dem Zeitpunkt innerhalb der Vegetationsperiode, zu welchem die Untersuchung erfolgt. Die für die Pflanze notwendigen Nährstoffe, die sie dem Boden im Verlauf ihres Wachstums entzieht, werden begreiflicherweise am Ende der Vegetationsperiode in geringerer Menge gefunden als am Beginn, sofern nicht eine besondere Zufuhr erfolgt ist. Die Befunde, die in Tab. 29 wiedergegeben sind, zeigen dies besonders im Hinblick auf den Gehalt an NO_3^-, PO_4^{---} und SO_4^{--}.

Tabelle 29. *Ergebnisse von Analysen zu verschiedenen Zeitpunkten an durch Verdrängung unter Druck gewonnenen Bodenlösungen eines kalifornischen Bodens bei einem Wassergehalt von 18,7%. Ionengehalte in mä*
(nach BURD und MARTIN 1924)

Zeitpunkt	Beginn der Vegetationsperiode (30. IV. 1923)	Ende der Vegetationsperiode (4. IX. 1923)	Beginn der Vegetationsperiode (28. IV. 1924)
HCO_3	1,75	2,88	2,33
NO_3	2,28	0,21	2,71
SO_4	7,10	4,36	9,56
PO_4	0,05	0,008	0,04
Mg	3,70	1,80	4,52
Ca	6,75	4,00	8,45
K	1,10	0,59	1,26
Na	2,35	1,13	2,39
Summe der Kationen (Normalität n =)	0,014	0,008	0,017

Ähnliche Befunde stammen von HOAGLAND (1948), von denen einige in Tab. 30 wiedergegeben sind.

Die Zahlen zeigen die relativ starken Entzüge an NO_3^-, SO_4^{--}, PO_4^{---}, Mg^{++}, Ca^{++} und K^+ an, wogegen die Werte für HCO_3^- nach der Vegetationsperiode beträchtlich erhöht sind. Der starke Entzug an Anionen wird offensichtlich durch Aufnahme von CO_2 (zum Hauptteil aus der Atmung der Wurzeln bzw. der Bodenmikrolebewelt stammend) kompensiert. Der Gesamtgehalt an den acht analytisch bestimmten Ionen betrug (im Durchschnitt aller drei Böden) vor Beginn der Vegetation 1272 ppm (1,272 g/l), nach der Vegetationsperiode hingegen nur 756 ppm (0,756 g/l), also nur etwa 60%. Eine genauere Untersuchung der

Tabelle 30. *Ionengehalte der Bodenlösung vor und nach der Vegetation (in ppm)* (nach HOAGLAND 1948)

	Boden A		Boden B		Boden C	
	vor	nach	vor	nach	vor	nach
	der Vegetation		der Vegetation		der Vegetation	
Ca^{++}	283	193	276	153	187	108
Mg^{++}	106	47	93	56	145	38
Na^{+}	49	40	31	28	23	16
K^{+}	24	9	20	11	33	23
HCO_3^-	83	155	93	143	40	112
NO_3^-	174	58	274	88	230	40
SO_4^{--}	655	432	633	275	360	241
PO_4^{---}	1,1	0,6	2,5	1,4	1,2	0,7
Summe	1375	935	1422	755	1019	579

Schwankungen des Gehaltes der Bodenlösung an Stickstoff während der Vegetationsperiode wurde von WRANGELL (1931) durchgeführt. Sie fand bei einem schweren Lehmboden Anfang Juli 11,0 mg N/l an Nitrat, im August nur noch

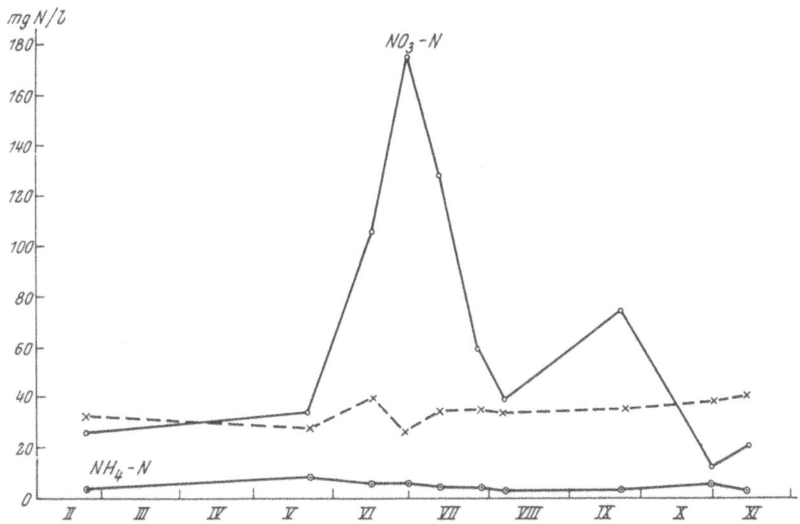

Abb. 81. Jahresschwankungen des Stickstoffgehaltes der Bodenlösung (Komposterde Hohenheim: nach Zahlen von WRANGELL 1931)

0,69, einen Monat später 0,043 und Anfang Dezember 0,088 mg N/l an Nitrat. Dagegen blieb die NH_4-N-Konzentration ziemlich unverändert: 3,4—1,7—2,2——4,3 mg N/l. Ein analoger „Jahresgang" der beiden N-Formen in der Bodenlösung wurde bei einer Hohenheimer Komposterde festgestellt. Er ist aus Abb. 81 ersichtlich.

Die strichlierte Kurve gibt den Wassergehalt des Bodens in % an. Man sieht, daß auch hier der Gehalt an NH_4—N wesentlich geringeren Schwankungen unterlag als der stets vielfach höhere Gehalt an NO_3—N, der im Juni-Juli ein ausgesprochenes Maximum zeigt (das temperaturbedingt sein dürfte). Nach der Ernte ist ein zweiter Gipfel des Nitrat-N-Gehaltes zu beobachten, dessen Minimum zum Beginn des Winters liegt.

Dagegen zeigt die Komposterde das Maximum des Kaligehaltes (nach Kalidüngung) im November, ein Minimum im Februar und nach einem Wiederanstieg auf das Doppelte ein absolutes Minimum im Oktober (Abb. 82). Der Jahresgang des Kaliums erwies sich bei verschiedenen Böden als ziemlich verschieden und zeigte keine prinzipiellen Übereinstimmungen.

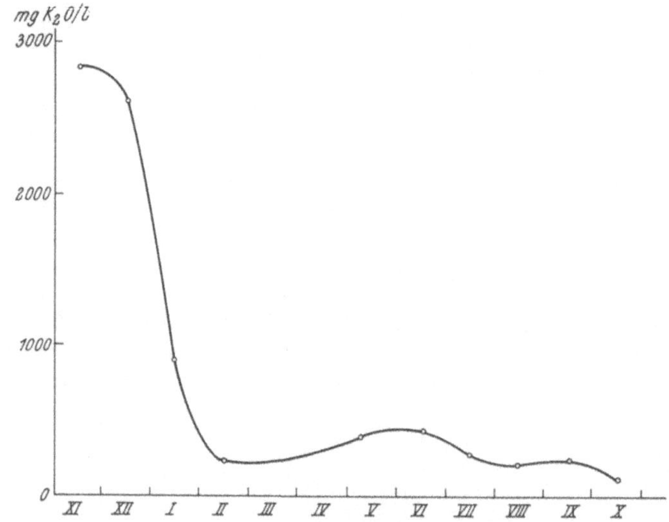

Abb. 82. Kaligehalt der Bodenlösung bei Komposterde, die im Sommer vorher eine (mineralische) Kaligabe erhalten hatte (nach Zahlen von WRANGELL 1931)

Solche Befunde sind allerdings nur dann vergleichbar, wenn die Gewinnung der Bodenlösung bei *gleichen* Gehalten der Böden an Wasser bzw. an Bodenlösung durchgeführt würde. Eine Umrechnung von Befunden bei verschiedenen Wassergehalten auf gleiche Gehalte an Bodenlösung würde fehlerhaft sein, weil eine Veränderung der Wassermenge (pro Bodenvolumen oder Bodengewicht) Verschiebungen im Verhältnis der gelösten Stoffe zueinander zur Folge haben kann bzw. haben muß.

Bei Wassergehalten von 12,5 bis 20,0% fand PFAFF Preßsaftausbeuten von 32 bis 110 cm^3, wobei der Gehalt des Preßsaftes an CaO mit zunehmender Menge abnahm (560 mg/l bis 358 mg/l), während der Gehalt an P_2O_5 auf fast das Doppelte anstieg (0,08 mg/l bis 0,15 mg/l) (Limburgerhof 1939). Während der Gehalt an CaO annähernd der Verdünnung durch den höheren Wassergehalt entsprach, stieg die vom Boden an die Lösung abgegebene Phosphatmenge auf mehr als das Dreifache an.

Die Höhe des Wassergehaltes eines bestimmten Bodens ist daher auch auf die Relationen der Mengen an gelösten Stoffen zueinander von starkem Einfluß. Dies zeigen z. B. Befunde von REITEMEIER und RICHARDS (1944), die in Abb. 83 graphisch dargestellt sind.

Im rechten Teil der Abb. 83 ist als gestrichelte Linie der Verlauf des Sulfatgehaltes der Bodenlösung eingezeichnet, der zu erwarten wäre, wenn der Bodenlösung zugeführte Wassermengen nur verdünnend auf den Gehalt an löslichen

Stoffen wirken würden. Eine Erhöhung des Wassergehaltes des Bodens verdünnt jedoch nicht nur die bereits vorhandene Bodenlösung, sondern bringt neues Lösungsmittel in das System und verändert die Lösungsverhältnisse: während größere Wassermengen zwar nicht mehr Chlor lösen und dessen Konzentration daher nur zu verdünnen scheinen, wird vom Sulfat um so mehr in Lösung gebracht, je mehr Wasser vorhanden ist, so daß dessen Konzentration in der Bodenlösung mit zunehmendem Wassergehalt des Bodens ansteigt.

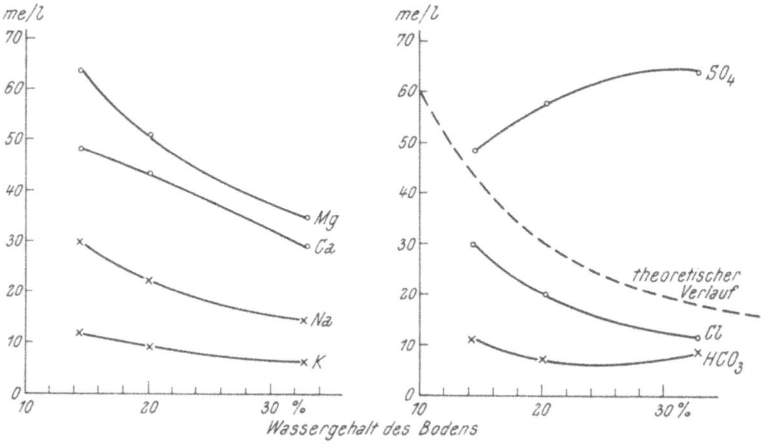

Abb. 83. Die Abhängigkeit des Gehaltes der Bodenlösung eines gipshaltigen tonigen Lehms (Reogan clay loam) vom Wassergehalt des Bodens (nach Zahlen von REITEMEIER und RICHARDS 1944). Die gestrichelte Kurve im rechten Teil der Abbildung deutet den theoretisch zu erwartenden Kurvenverlauf an, wie sie bei einfacher Verdünnung durch höheren Wassergehalt des Bodens (etwa für SO_4) zu erwarten wäre

Die Gleichgewichte, welche sich zwischen den festen (teilweise löslichen) Bodenbestandteilen und der Bodenlösung einstellen, sind von recht komplizierter Art und polyfaktoriell beeinflußt, deshalb schwer durchschaubar. Einige einfache Überlegungen können dies deutlich machen.

Wenn man zu trockenem Boden reines Wasser hinzufügt, so werden die wasserlöslichen Anteile des Bodens in Lösung gehen. Das im Boden befindliche Wasser enthält dann eine Reihe verschiedener gelöster Stoffe. Eine *einzige* bestimmte Substanz würde sich im Wasser so lösen, wie dies durch ihre Löslichkeit im Wasser vorgeschrieben wird. Das Wasser würde sich, wenn genügend von dieser einen Substanz vorhanden ist, mit gelöster Substanz anreichern, bis es zur Sättigung der Lösung kommt, d. h. bis zu einer Konzentration der betreffenden Substanz im Wasser, bei der keine weiteren Substanzmengen mehr gelöst werden können. Der ungelöste Rest der Substanz steht nun „als Bodenkörper" mit einer gesättigten Lösung „im Gleichgewicht". Der Bodenkörper befindet sich in diesem Falle in einer gesättigten Lösung, in der er sich als unlöslich erweist. Für jedes Molekül, das aus dem Bodenkörper herausgelöst wird, wird ein anderes, gleichartiges aus der Lösung an den Bodenkörper abgegeben werden. Dieses Beispiel mag zeigen, daß die Lösungsfähigkeit des Wassers für eine bestimmte Substanz eine andere ist als die Lösungsfähigkeit einer wässerigen Lösung einer gleichen oder anderen Substanz. Im Gleichgewichtszustand, der sich nach längerem Belassen zum trockenen Boden zugefügten reinen Wassers eingestellt hat, bildet das „Lösungsmittel", mit dem die feste Substanz des Bodens nun in Kontakt steht, nicht reines Wasser, sondern die aus dem Gleichgewicht resultierende

Bodenlösung, die eine ganze Reihe verschiedener gelöster Stoffe enthält. Die Löslichkeit neuer (etwa in Form von Düngemitteln) zugefügter Stoffe in dieser Bodenlösung muß nicht gleich ihrer Löslichkeit in Wasser sein. Als Beispiel für die Löslichkeit eines Salzes in Lösungen eines anderen Salzes mag die Löslichkeit von KCl in Lösungen von $MgCl_2$ dienen, über die Tab. 31 einen Überblick für verschiedene Temperaturen gibt. Je höher die Konzentration an $MgCl_2$ in der Lösung ansteigt, desto geringer wird die Löslichkeit für KCl. Auch Tab. 32 gibt Werte für den Gehalt gesättigter Lösungen von K_2SO_4, KCl, NaCl und $NaNO_3$ in Wasser für sich allein bzw. in Gemischen von je zwei Salzen. Man sieht dabei z. B., daß das Verhältnis von KCl zu NaCl sich beim Ansteigen der Temperatur von 10° C auf 60° C von 8,8:20,9 auf 16,2:17,9 verändert, während in einem Gemisch von K_2SO_4 und KCl das Verhältnis der beiden Komponenten zueinander keine so große Änderung erfährt. Bei NaCl und $NaNO_3$ ändert sich das Verhältnis beim Temperaturanstieg von 0° C auf 50° C von 17,0:26,2 auf 10,1:41,5, also von 1:1,54 auf 1:4,10 (Tab. 32). Es mag dies veranschaulichen, wie kompliziert die gegenseitige Beeinflussung der Löslichkeit beim Vorliegen vieler verschiedenartiger Komponenten in einer Lösung werden kann.

Wenn die gewählten Beispiele sich auch mit dem Verhalten gesättigter Lösungen befassen, also sich auf sehr hohe Konzentrationen beziehen, so muß doch berücksichtigt werden, daß auch im Boden ähnlich hohe Konzentrationen erreicht werden können, wenn die Austrocknung genügend weit fortschreitet.

Wenn Salze in Wasser gelöst werden, so werden durch die hydratisierende Wirkung der Wasserdipole Ionen gebildet (vgl. S. 335), also Kationen (K^+) und Anionen (A^-), wobei das Dissoziationsgleichgewicht lautet

$$\frac{C_{K^+} \cdot C_{A^-}}{C_{KA}} = \text{konstant},$$

und $C_{K^+} \cdot C_{A^-} = P$ als Löslichkeitsprodukt bezeichnet wird.

Bei sehr verdünnten Lösungen werden die Entfernungen zwischen den einzelnen Ionen sehr groß, so daß die elektrostatischen Anziehungskräfte der einzelnen Ionen, deren Reichweite beschränkt ist, gegenseitig nicht zur Wirkung gelangen. Ist die Konzentration eines Ionenpaares (K^+, A^-) sehr gering, befinden sich jedoch zahlreiche andersartige Ionenpaare in Lösung (K_1^+, A_1^-), so machen sich die elektrostatischen Anziehungskräfte zwischen ungleichsinnig geladenen Ionen dadurch bemerkbar, daß bei der Messung der H^+-Ionenkonzentration (mit den üblichen Verfahren der H-Ionen-Messung) höhere pH-Werte erhalten werden, als sie der tatsächlichen Ionenkonzentration entsprechen. Gemessen wird daher die *Ionenaktivität* als die Konzentration jener Ionen, deren elektrostatische Anziehung von anderen Ionen der Lösung nicht beansprucht ist. Bezeichnet man diese Ionenaktivität mit dem Symbol a, so bezeichnet der gemessene pH-Wert

$$pH = -\log a_{H^+}$$

eigentlich die *Ionenaktivität*, die nur im Falle einer sehr niedrigen *Gesamt*-Ionenkonzentration der Lösung gleich der tatsächlichen Ionenkonzentration ist. Die gleichzeitige Gegenwart beliebiger andersartiger Ionen in der Lösung, mit der die betrachteten Ionen in elektrostatische Wechselwirkung (Anziehung) treten können, setzt somit die Ionenaktivität herab. Dies ist z. B. die Ursache dafür, daß eine 0,001-molare HCl-Lösung, die einen pH-Wert von etwa 3,0 aufweist, bei gleichzeitigem Vorliegen von 1m Na_2SO_4, obwohl dieses Salz selbst elektroneutral ist, einen pH-Wert von 3,7 aufweist: d. h., daß die wirksame H^+-Ionenkonzentration von 0,001 auf etwa 0,0002 absinkt.

Auch die Abgabe von Ionen aus sorptiven Bindungen an Bodenteilchen an die

Bodenlösung ist abhängig von der Salzkonzentration der Bodenlösung. Während (nach SCHACHTSCHABEL 1940) bei Erhöhung der Wassermenge für 1 mol $CaCl_2$, das mit NH_4-Montmorillonit ins Gleichgewicht gebracht wurde, der %-Gehalt der Lösung an Ca absinkt, steigt die Menge der NH_4-Ionen reziprok dazu an.

Befinden sich mehrere verschiedene Arten von Ionen in einer Lösung, so bestehen, da sie sich beliebig treffen und zu Molekülen zusammentreten können, Möglichkeiten für die Bildung sehr verschiedenartiger Moleküle, deren Löslichkeit

Tabelle 31. *Gehalt (g je 100 g Lösung) Mg-haltiger Lösungen, die mit KCl gesättigt wurden, an KCl bei verschiedenen Temperaturen*
(nach PRECHT und NITTGEN)

Temperatur °C	0°	10°	20°	30°	40°	50°	60°
11% $MgCl_2$ ..	—	14,3	15,9	17,5	19,0	20,5	21,9
15% $MgCl_2$ ..	—	9,9	11,3	12,7	14,2	15,6	17,0
21,2% $MgCl_2$	—	5,3	6,5	7,6	8,8	10,0	11,2
30% $MgCl_2$ ..	—	1,9	2,6	3,4	4,2	5,0	5,8

Tabelle 32. *Bei verschiedenen Temperaturen gesättigte Lösungen von einzelnen Salzen und Salzgemischen enthalten (je 100 g Lösung) folgende Mengen der einzelnen Salze (in g)*
(nach PERCHT und NITTGEN)

Temperatur °C		0°	10°	20°	30°	40°	50°	60°
K_2SO_4	in Wasser	6,87	8,47	10,03	11,49	13,1	14,2	15,4
K_2SO_4	in Wasser ...	—	1,00	1,06	1,14	1,20	1,27	1,33
KCl		—	23,4	24,8	26,2	27,6	28,9	30,1
KCl	in Wasser	22,2	23,8	25,5	27,2	28,7	30,1	31,3
KCl	in Wasser ...	—	8,8	10,2	11,7	13,2	14,7	16,2
NaCl		—	20,9	20,3	19,7	19,1	18,5	17,9
NaCl	in Wasser	26,3	26,3	26,4	26,5	26,7	26,9	27,1
NaCl	in Wasser ...	17,0	15,1	13,3		—	10,1	—
$NaNO_3$		26,2	28,6	32,0		—	41,5	—
$NaNO_3$	in Wasser	42,2	44,6	46,8	49,0	51,2	53,3	55,5

im Wasser sehr verschieden groß sein kann. Bei fehlender (bzw. sehr geringer) Löslichkeit des gebildeten Moleküls im Wasser wird die Substanz, die es bildet, als Niederschlag ausfallen und damit die Ionen, aus denen es sich aufgebaut hat, der Ionenkonkurrenz in der Lösung (weitgehend) entziehen. Diese Ausfällung unlöslicher Stoffe aus der Bodenlösung kann durch das Hinzukommen jener Ionen veranlaßt werden, die zur Bildung unlöslicher Stoffe bzw. Komplexe führen. Eine solche Substanz ist beispielsweise Oxalsäure, deren Anion COO^-/COO^- von Pflanzen abgegeben werden und in die Bodenlösung gelangen kann. Mit den dort vorhandenen, in Lösung befindlichen Ca^{++}-Ionen bildet es das in Wasser fast unlösliche Calciumoxalat $Ca(COO)_2$ (dessen gesättigte Lösung im Wasser je 100 g bei 18° C nur 0,0006 g Oxalsäure enthält). Ähnlich reagieren Ba^{++}-Ionen mit SO_4^{--}-Ionen zu $BaSO_4$ (dessen gesättigte Lösung im Wasser je 100 g bei 18° C nur 0,00023 g $BaSO_4$ enthält).

Die meisten als Pflanzennährstoffe bzw. als pflanzliche Bauelemente bedeutenden Elemente bilden, sofern sie in den häufigsten Ionenformen vorliegen, keine bzw. nur sehr wenige schwerlösliche Verbindungen, da sie Salze bilden, deren Löslichkeiten im Wasser meist so groß sind, daß der Wasserverlust der Bodenlösung ganz extreme Größen annehmen müßte, um zu einer Ausfällung bzw. Kristallisation oberer Salze zum festen Zustand zu führen.

3. Die Löslichkeitseigenschaften der Nährstoffe und der wichtigeren akzessorischen Elemente

Hier seien zunächst die als Pflanzennährstoffe fungierenden bzw. Bauelemente enthaltenden anorganischen Verbindungen im Hinblick auf ihre Löslichkeit im Wasser und einige typische Wechselwirkungen mit anderen, in der Bodenlösung möglicherweise vorkommenden Stoffen hin näher betrachtet. Da verschiedene Bodenschichten verschiedene Zusammensetzung der Bodenlösung besitzen können und Mischungen verschieden zusammengesetzter Bodenlösungen möglich sind, kommt der Bildung unlöslicher Stoffe aus in den Lösungen vorhandenen Ionen eine gewisse Bedeutung zu, weshalb auch die möglicherweise entstehenden unlöslichen Stoffe mit berücksichtigt werden.

Als *Löslichkeit* wird die bei einer bestimmten Temperatur in einer bestimmten Menge (hier 100 g) einer gesättigten Lösung[1] des betreffenden Stoffes (Salzes) gelöste Menge an wasserfreier Substanz (hier g) bezeichnet und zahlenmäßig angegeben. Als *Löslichkeitsprodukt* gilt das Ionenprodukt im Massenwirkungsgesetz:

$$(K^+)\cdot(A^-) = \text{konst.}\,(KA) = L_{(KA)}$$

(vgl. S. 356). Für die schwerlöslichen Salze ist dieses Löslichkeitsprodukt für bestimmte Temperaturen in den folgenden Löslichkeitstabellen (Tab. 33 bis 43) angeführt. Die Zahlen beziehen sich auf Grammoleküle pro Liter.

Anschließend soll auf einige häufige akzessorische Stoffe eingegangen werden. Es ist dabei zweckmäßig, bei der Reihung der Stoffe von den jeweiligen Kationen auszugehen.

A. Ammoniumsalze

Die meisten Ammoniumsalze zeigen eine hohe Wasserlöslichkeit. Tab. 33 stellt die Gehalte gesättigter Lösungen bei verschiedenen Temperaturen einander gegenüber.

Tabelle 33. *Löslichkeit von Ammoniumsalzen*
(vgl. VOGEL 1956, LANGE und FORKER 1956)
Es ist wasserfreie Substanz in g je 100 g Wasser gelöst:

bei °C:	0°	10°	20°	30°	40°	50°	60°
$NH_4\cdot HCO_3$	11,9	16,1	21,7	27,0	36,6	—	59,2
$NH_4\cdot NO_3$	118,5	150,3	178,7	232	(283)	(344)	(415)
$(NH_4)_2\cdot SO_4$	70,4	72,7	75,4	78,1	81,2	84,3	87,4
$NH_4\cdot H_2PO_4$	22,7	—	36,8	—	56,7	—	82,9
$(NH_4)_2\cdot HPO_4$	57,5	—	68,6	—	81,8	—	97,6
$(NH_4)_3\cdot PO_4\cdot 3\,H_2O$	9,4	—	(20,3)	—	—	37,7	—
NH_4Cl	29,7	33,5	37,4	41,4	46,0	50,4	55,3
NH_4–OC·CO–$NH_4\cdot 1\,H_2O$	2,5	3,1	4,4	5,9	8	10,3	—
NH_4VO_3	—	—	—	0,48	0,84	1,32	1,78

[1] Die in den folgenden Tabellen angeführten Löslichkeitswerte beziehen sich auf reine wässerige Lösungen der angegebenen Stoffe. In der Bodenlösung liegen Begleitstoffe bzw. Rechtstromverhältnisse vor, die auf die Löslichkeiten von großem Einfluß sein können.

Alle genannten Ammoniumsalze zeigen somit mit ansteigender Temperatur steigende Löslichkeit im Wasser. Den geringsten Einfluß dieser Art läßt das Ammoniumsulfat erkennen. Als schwerlösliches Ammoniumsalz ist in vorliegendem Zusammenhang das Magnesiumammoniumphosphat zu nennen (vgl. S. 370).

B. Eisensalze

Eisen kann durch Reaktion mit Phosphationen unlösliche Verbindungen bilden. Ferrophosphat $FePO_4 \cdot 2\,H_2O$ (Strengit) ist in Wasser fast unlöslich, löst sich aber bei 100° C zu 0,67 Teilen in 100 Teilen Wasser. $FePO_4 \cdot 4\,H_2O$ verhält sich ähnlich. Eisenpyrophosphat $Fe_4(P_2O_7)_3 \cdot 9\,H_2O$ ist als unlöslich zu bezeichnen. Auch $Fe_3(PO_4)_2 \cdot 8\,H_2O$ (Vivianit) ist wasserunlöslich.

Eisen-Kali-Oxalat $FeK(COO)_4 \cdot 3\,H_2O$ (bzw. $2^1/_2\,H_2O$) lösen sich zu 4,7 (0° C) bzw. 92 (21° C) Teilen in 100 Teilen Wasser.

Über die gewöhnlichen mit Nährstoffionen gebildeten Eisensalze gibt Tab. 34 Auskunft.

Tabelle 34. *Löslichkeit von Eisensalzen*
(vgl. Vogel 1956, Lange und Forker 1956)
Es ist wasserfreie Substanz in g je 100 g Wasser gelöst:

bei °C:	0°	10°	20°	30°	40°	50°	60°	Löslichkeitsprodukt	°C
$Fe(OH)_2$								$1{,}64 \cdot 10^{-14}$	18
$Fe(OH)_3$								$1{,}1 \cdot 10^{-36}$	18
$Fe(HCO_3)_2$								—	—
$Fe(NO_3)_2 \cdot 6\,H_2O$	71,0	—	83,8	—	—	—	165,6		
$Fe(SO_4) \cdot 7\,H_2O$	15,6	10,5	26,6	33,0	40,3	48,6	—		
$Fe(SO_4)_2 \cdot (NH_4)_2 \cdot 6\,H_2O$	17,8	22,1	26,9	32,5	38,5	45,4	53,4		
FeS_2	—	—	0,0006	—	—	—	—	$3{,}7 \cdot 10^{-19}$	18
$(II)FeCl_2 \cdot 6\,H_2O$	—	64,5	—	73,0	77,3	82,5	88,7		
$(III)FeCl_3 \cdot 6\,H_2O$	74,5	81,8	91,9	106,6	—	—	—		

Tabelle 35. *Löslichkeit von Magnesiumsalzen*
(vgl. Vogel 1956, Lange und Forker 1956)
Es ist wasserfreie Substanz in g je 100 g Wasser gelöst:

bei °C:	0°	10°	20°	30°	40°	50°	60°	Löslichkeitsprodukt	°C
$Mg(OH)_2$	—	—	0,0009	—	—	—	—	$1{,}2 \cdot 10^{-11}$	18
$MgCO_3 \cdot 3\,H_2O$	—	—	0,152	—	—	—	—	$2{,}6 \cdot 10^{-5}$	12
$Mg(NO_3)_2 \cdot 6\,H_2O$	63,9	—	70,5	—	81,8	—	93,7		
$Mg(NO_3)_2 \cdot 2\,H_2O$	—	—	—	—	—	—	214,5		
$MgSO_4 \cdot 7\,H_2O$	—	30,0	35,6	40,8	45,4	—	—		
$MgSO_4 \cdot 6\,H_2O$	—	—	—	—	—	—	54,7		
$MgHPO_4 \cdot 7\,H_2O$	—	—	0,31	—	—	—	—		
$NH_4 \cdot MgPO_4 \cdot 6\,H_2O$	0,023	—	0,052	—	0,036	0,030	0,040	$2{,}5 \cdot 10^{-13}$	25
$MgCl_2 \cdot 6\,H_2O$	—	53,5	54,25	55,3	56,5	58,7	60,7		
$MgJ_2 \cdot 8\,H_2O$	120,8	—	139,8	—	173,2	—	—		

C. Magnesiumsalze

Die meisten Magnesiumsalze sind im Wasser leicht löslich. Auch Doppelsalze mit Kl oder K_2SO_4 sind leicht löslich. Magnesiumammoniumphosphat kann als schwerlöslich gelten. Auch Magnesiumstearat löst sich bei 15° C nur zu 0,003 Teilen in 100 Teilen Wasser. Unlöslich ist ferner Magnesiumpyrophosphat, $Mg_2P_2O_7$ bzw. $Mg_2P_2O_7 \cdot 3\,H_2O$. Magnesiumaluminat $MgO \cdot Al_2O_3$ ist wasserunlöslich, ebenso Magnesiummetaborat $Mg(BO_2)_2 \cdot 8\,H_2O$ und $Mg_3(PO_4)_2 \cdot 8\,H_2O$. Auch Magnesiumsilikat $Mg_3Si_4O_{11} \cdot H_2O$ ist wasserunlöslich.

D. Calciumsalze

Die Löslichkeiten verschiedener Calciumsalze sind in Tab. 36 zusammengestellt. Sie zeigt, daß die meisten Calciumsalze im Wasser in beträchtlichen Mengen löslich sind. Eine Ausnahme bilden die Phosphate.

Tabelle 36. *Löslichkeit von Ca-Salzen*
(vgl. VOGEL 1956)
Es ist wasserfreie Substanz in g je 100 g Wasser gelöst:

bei °C:	0°	10°	20°	30°	40°	50°	60°	Löslichkeitsprodukte	°C
$CaCO_3$	—	—	0,0014		0,0015		—	$0{,}99 \cdot 10^{-8}$	15
$Ca(HCO_3)_2$	16,3	—	16,6	—	17,0	—	17,5		
$Ca(OH)_2$	0,130	0,125	0,118	0,109	0,100	—	0,083		
$Ca(NO_3)_2 \cdot 4\,H_2O$	101,0	115,3	1127,0	—	196	—	—		
$Ca(NO_3)_2 \cdot 3\,H_2O$	—	—	—	—	237,5	281,5	—		
$Ca(SO_4)_2 \cdot 2\,H_2O$	0,176	0,192	0,204	0,210	0,212	(0,201)	(0,197)	$6{,}1 \cdot 10^{-5}$	10
$CaHPO_4 \cdot 2\,H_2O$	—	—	0,02	—	0,038	—	—		
$CaHPO_4 \cdot OH_2O$	—	—	—	—	0,014	—	0,015		
$Ca_3(PO_4)_2$	—	—	0,0025	—	—	—	—		
$CaCl_2 \cdot 6\,H_2O$	60,3	65,0	74,5	102	—	—	—		
$CaCl_2 \cdot 2\,H_2O$	—	—	—	—	128,1	132,2	136,8		
CaJ_2	181,9	194,1	208,8	222,6	242,5	—	284,6		

Das im Wasser kaum lösliche Calciumcarbonat wird durch Säuren unter CO_2-Abspaltung leicht gelöst. Durch die Einwirkung von Kohlensäure im Wasser wird Calciumcarbonat gemäß folgendem Vorgang

$$CaCO_3 + H_2CO_3 = Ca(HCO_3)_2$$

in das leichtlösliche Calciumbicarbonat umgewandelt, wobei die Wechselwirkungen zwischen Wasserstoffionenkonzentration, Kohlensäuredruck und Ca-Ionen, welche von BJERRUM und GJALDBAEK (1919) sowie von YAALON (1954) untersucht wurden, in folgender Beziehung ausgedrückt werden können:

$$pCa^{++} = 2\,pH + \log P_{CO_2} - 9{,}85 - 1{,}425\,\sqrt{\mu}$$

wobei μ die Ionenstärke, P den Druck bedeuten und die Konstanten für 25° C gelten.

Nach YAALON (1954) findet man den Logarithmus (log) der gelösten Menge an $CaCO_3$ in mg/l durch die Ausdrücke

$$—0{,}54\ pH + 6{,}23 \text{ und } 0{,}35 \log P_{CO_2} + 2{,}97.$$

Ein Beispiel für die Löslichkeit von $CaCO_3$ in Abhängigkeit vom CO_2-Druck geben die folgenden Zahlen (EPHRAIM 1948):

gelöste Menge $CaCO_3$ mg/l	13	223	533	1086
CO_2-Druck in Atm.	0,00	0,139	0,142	0,984

In einem mit CO_2 unter Atmosphärendruck gesättigten Wasser lösen sich bei 0° C 0,156, bei 15° C nur 0,117 g $CaCO_3$ (auf 100 g Lösung) in Form von $Ca(HCO_3)_2$ auf.

Für die Abhängigkeit der Löslichkeit des Bodenkörpers verschiedener Calciumphosphate gibt BJERRUM folgende Beziehungen an, wobei μ die Ionenstärke bedeutet und die Konstanten für 37° C gelten:

$CaHPO_4 \cdot 2\ H_2O$	$pCa^{++} + pHPO_4^{--} = 6{,}62—4{,}12\sqrt{\mu}$
$Ca_4H(PO_4)_3 \cdot 3\ H_2O$	$4\ p\ Ca^{++} + 3\ pHPO_4^{--} — 2\ pH = 12{,}1—14{,}4\sqrt{\mu}$
$Ca_5(PO_4)_3 \cdot OH \cdot H_2O$	$5\ p\ Ca^{++} + 3\ pHPO_4^{--} — 4\ pH = 7{,}5—16{,}5\sqrt{\mu}$

Ca-Silikate bzw. Ca-Al-Silikate kommen als gesteinbildende Minerale im Boden vor. Ihre Löslichkeiten sind gering, zahlenmäßig nicht näher bekannt. Calciumsilikat $CaSiO_2$ löst sich zu 0,0059 Teilen in 100 Teilen Wasser bei 17° C. Die Plagioklase sind nach GOLDRICH (1938) um so leichter verwitternd, je mehr Ca sie enthalten.

Calciumammoniumphosphat $CaNH_4 \cdot PO_4 \cdot 7\ H_2O$ ist wasserunlöslich. Schwerlösliche Verbindungen liefert Calcium mit der Oxalsäure (im Wasser unlöslich) und mit Huminsäuren.

E. Kaliumsalze

Die Salze des Kaliums sind fast durchweg leicht wasserlöslich. Eine Ausfällung von Kalium aus der Bodenlösung in unlöslicher Form durch in der Bodenlösung natürlicherweise vorkommende Ionen liegt praktisch nicht vor. Auch Kaliumtetraborat ist wasserlöslich. Ebenso Kaliumoxalat, Kaliumsilikat und K-Aluminat.

Tabelle 37. *Löslichkeit von Kaliumsalzen*
(vgl. VOGEL 1956, LANGE und FORKER 1956)
Es ist wasserfreie Substanz in g je 100 g Wasser gelöst:

bei °C	0°	10°	20°	30°	40°	50°	60°
$KOH \cdot 2\ H_2O$	97	103	112	126	—	—	—
$KHCO_3$	22,4	27,7	33,2	39,1	45,4	—	60,0
$K_2CO_3 \cdot 2\ H_2O$	105,5	108	110,5	113,7	116,9	121,2	126,8
KNO_3	13,3	20,9	31,6	45,8	63,9	85,5	110,0
KNO_2	278,8	—	298,4	—	334,9	—	—
$KHSO_4$	36,3	—	51,4	—	67,3	—	—
K_2SO_4	7,35	9,22	11,11	12,97	14,76	16,50	18,17
KH_2PO_4	14,3	18,0	22,7	—	33,9	—	48,6
$K_2HPO_4 \cdot 6\ H_2O$	85,6	122	—	—	—	—	—
$K_2HPO_4 \cdot 3\ H_2O$	—	—	139	—	212,5	—	—
$K_3PO_4 \cdot 7\ H_2O$	77,6	—	—	—	122	—	—
KCl	27,6	31,0	34,0	37,0	40,0	42,6	45,5
KJ	127,5	136	144	152	160	168	176
$KMnO_4$	2,83	4,4	6,38	9,0	12,56	16,89	—

F. Kupfersalze

Die Carbonate des Kupfers, $CuCO_3 \cdot Cu(OH)_2$ und $2\,CuCO_3 \cdot Cu(OH)_2$ sind wasserunlöslich. Cu-Oxalat ist ebenfalls wasserunlöslich, ebenso das Kupriphosphat $Cu_3(PO_4)_2 \cdot 3\,H_2O$. Über die löslichen Kupfersalze vgl. Tab. 38.

Tabelle 38. *Löslichkeit von Kupfersalzen*
(LANGE und FORKER 1956)
Es ist wasserfreie Substanz in g je 100 g Wasser gelöst:

bei °C	0°	10°	20°	30°	40°	50°	60°	Löslichkeitsprodukt	°C
$Cu(OH)_2$									
$Cu(HCO_3)$									
$CuNO_3 \cdot 6\,H_2O$ ..	81,8	95,3	125,1	—	—	—	—		
$CuSO_4 \cdot 5\,H_2O$ ..	14,3	17,4	20,7	25	28,5	33,3	40,0		
CuS	—	—	$3{,}3 \cdot 10^{-5}$ [1]	—	—	—	—	$8{,}5 \cdot 10^{-45}$	18
Cu_2S	—	—	—	—	—	—	—	$2\text{–}10^{-47}$	17
CuCl	—	—	—	—	—	—	—	$1{,}0 \cdot 10^{-6}$	18
$CuCl_2 \cdot 2\,H_2O$ ...	70,7	73,8	77,0	80,3	83,8	87,4	91,2		
CuJ_2	—	—	1,107	—	—	—	—		

[1] Bei 18° C.

Ein Kupferphosphat (vorwiegend $CuHPO_4$) zeigte folgende Löslichkeiten: 0,0108 bei 5°, 0,0103 bei 25°, 0,0085 bei 60°, 0,0101 bei 90° C.

G. Mangansalze

Die Löslichkeiten der für die Bodenlösung in Frage kommenden Salze des Mangans sind in Tab. 39 enthalten. Die Manganphosphate $MnHPO_4 \cdot 3\,H_2O$ und $Mn_3(PO_4)_2 \cdot 7\,H_2O$ sind schwer wasserlöslich, Manganoxalat ist sehr schwach wasserlöslich, Mangansilikat $MnSiO_3$ ist wasserunlöslich.

Tabelle 39. *Löslichkeit von Mangansalzen*
(VOGEL 1956, LANGE und FORKER 1956)
Es ist wasserfreie Substanz in g je 100 g Wasser gelöst:

bei °C:	0°	10°	20°	30°	40°	50°	60°	Löslichkeitsprodukt	°C
$Mn(OH)_2$	—	—	0,002	—	—	—	—	$4 \cdot 10^{-14}$	18
$MnCO_3$	—	—	0,0065	—	—	—	—		
$Mn(NO_3)_2 \cdot 6\,H_2O$	101,98	118,0	142,7	—	—	—	—		
$Mn(NO_3)_2 \cdot 3\,H_2O$	—	—	—	206,6	—	—	—		
$MnSO_4 \cdot 7\,H_2O$	53,23	60,01	—	—	—	—	—		
$MnSO_4 \cdot 5\,H_2O$	—	59,5	62,9	67,8	—	—	—		
$MnSO_4 \cdot 4\,H_2O$	—	—	64,5	66,4	68,8	72,6	—		
MnS	—	—	0,00062[1]	—	—	—	—	$1{,}4\text{–}10^{-15}$	18
$MnNH_4PO_4 \cdot H_2O$	—	—	0,0031	—	—	—	—		
$Mn(H_2PO_4)_2 \cdot 2\,H_2O$	löslich								
$MnCl_2 \cdot 4\,H_2O$	63,4	68,1	79,9	80,7	88,6	98,2	—		

[1] Bei 18° C.

Manganoxyde (MnO, Mn_3O_4, Mn_2O_3, MnO_2) sind wasserunlöslich, dagegen ist MnO_3 und Mn_2O_7 löslich.

H. Zinksalze

Zinkoxyd, ZnO, löst sich zu 0,00042 Teilen in 100 Teilen Wasser bei 18° C. Zinkphosphat $Zn_3(PO_4)_2$ ist wasserunlöslich, ebenso $Zn_3(PO_4)_2 \cdot 4\,H_2O$, $Zn_3(PO_4)_2 \cdot 8\,H_2O$ und Zinkpyrophosphat $Zn_2P_2O_7$. Unlöslich sind ferner Zinksilikat $ZnSiO_4$ und Zinkstearat.

Tabelle 40. *Löslichkeit von Zinksalzen*
(Lange und Forker 1956, Vogel 1956)
Es ist wasserfreie Substanz in g je 100 g Wasser gelöst:

bei °C	0°	10°	20°	30°	40°	50°	60°	Löslichkeitsprodukt	°C
$Zn(OH_2)$	—	—	—	—	—	—	—	$1{,}8 \cdot 10^{-14}$	19
$ZnCO_3$	—	0,001		—	—	—	—		
$Zn(NO_3)_2 \cdot 6\,H_2O$	94,78	—	118,3	—	—	—	—		
$ZnSO_4 \cdot 7\,H_2O$	41,6	47,1	53,8	61,3	—	—	—		
ZnS	—	—	—	—	—	—	—	$1{,}2 \cdot 10^{-23}$	18
Zinkphosphate									
$ZnCl_2 \cdot 3\,H_2O$	208	—	—	—	—	—	—		
$ZnCl_2 \cdot 1{,}5\,H_2O$	—	—	367	—	—	—	—		
$ZnCl_2 \cdot 0\,H_2O$	—	—	—	—	453	—	488		
$ZnJ_2 \cdot 2\,H_2O$	450,9	457,4	484,8	—	—	—	—		

I. Kobaltsalze

Neben löslichen Kobaltsalzen (vgl. Tab. 41) gibt es eine Reihe wasserunlöslicher: CoO, $Co(OH)_2$, $CoCO_3$, $Co_3(PO_4)_2$, $Co_3(PO_4)_2 \cdot 8\,H_2O$, Kobaltsilikat Co_2SiO_4.

Tabelle 41. *Löslichkeit von Kobaltsalzen*
(Lange und Forker 1956)
Es ist wasserfreie Substanz in g je 100 g Wasser gelöst:

bei °C	0°	10°	20°	30°	40°	50°	60°	Löslichkeitsprodukt	°C
$Co(NO_3)_2 \cdot 6\,H_2O$	84,0	—	100,0	—	126,8	—	—		
$Co(SO_4)_2 \cdot 7\,H_2O$	25,6	30,6	36,2	42,3	48,9	55,2	60,4		
CoS	—	—	0,00038[1]	—	—	—	—	$3 \cdot 10^{-26}$	18
$CoCl_2 \cdot 6\,H_2O$	41,6	46,0	50,4	53,5	—	—	—		
CoJ_2	—	159	—	—	—	—	—		

[1] Bei 18° C.

K. Natriumsalze

Fast alle Natriumsalze, die im Boden eine Rolle spielen können, sind wasserlöslich (vgl. Tab. 42), auch alle Natriumsilikate sowie Natriumaluminat. Wasserunlöslich ist lediglich Natriumpyrophosphat $Na_4P_2O_7$ und Natriumoctamolybdat $Na_2Mo_8O_{25} \cdot 4\,H_2O$. Auch Natriumuranat Na_2UO_4 ist unlöslich.

Tabelle 42. *Löslichkeit von Natriumsalzen*
(VOGEL 1956)
Es ist wasserfreie Substanz in g je 100 g Wasser gelöst:

bei °C	0°	10°	20°	30°	40°	50°	60°
$NaOH \cdot 1\,H_2O$	(43,2)	—	107	115	126	145	178
$NaHCO_3 \cdot 0\,H_2O$	6,89	8,20	8,57	11,09	12,7	14,45	16,0
$Na_2CO_3 \cdot 10\,H_2O$	6,86	11,98	21,58	39,70	—	—	—
$NaNO_3 \cdot 0\,H_2O$	70,7	—	88,0	—	104,9	—	124,7
$NaNO_2 \cdot 0\,H_2O$	73,0	—	81,8	—	93,7	—	112,3
$Na_2SO_4 \cdot 10\,H_2O$	4,56	9,14	19,08	40,87	—	—	—
$Na_2SO_4 \cdot 0\,H_2O$	—	—	—	—	48,1	46,6	45,26
$NaAl(SO_4)_2 \cdot 12\,H_2O$	37,4	—	39,7	—	44,3	—	—
$Na_2SO_3 \cdot 7\,H_2O$	13,9	20,0	26,9	36	—	—	—
$Na_2S \cdot 9\,H_2O$	12,4	15,4	18,6	22,5	29,9	(36,4)	(39,1)
$Na_2HPO_4 \cdot 2\,H_2O$	57,7	69,9	85,2	106,4	138,2	—	—
$Na_3PO_4 \cdot 12\,H_2O$	1,4	4,1	11	—	31	43	55
$Na_4P_2O_7 \cdot 10\,H_2O$	2,7	—	5,48	—	12,5	—	21,9
$Na_2MoO_4 \cdot 10\,H_2O$	44,3	—	—	—	—	—	—
$Na_2MoO_4 \cdot 2\,H_2O$	—	—	65,0	—	68	—	73
$NaVO_3 \cdot 2\,H_2O$	—	—	15,3	—	30,2	—	68,4
$NaCl \cdot 0\,H_2O$	35,6	35,70	35,85	36,15	36,42	36,72	37,05
$NaJ \cdot 2\,H_2O$	158,7	168,6	178,7	190,3	204,9	227,8	275,1
$Na_2B_4O_7 \cdot 10\,H_2O$	(1,04)	1,62	2,52	3,90	6,30	9,78	(17,33)
Na_2SIF_6	0,43	0,73	—	—	1,03	—	—

L. Aluminiumsalze

Al_2O_3 ist wasserunlöslich, ebenso $Al_2O_3 \cdot H_2O$, $Al_2O_3 \cdot 2\,H_2O$ und $Al_2O_3 \cdot 3\,H_2O$. Auch Aluminiumphosphat $AlPO_4$ ist wasserunlöslich. Unlöslich sind ferner Kaliumaluminiumsilikat $3\,Al_2O_3 \cdot K_2O \cdot 6\,SiO_2 \cdot 2\,H_2O$ sowie Orthoklas $Al_2O_3 \cdot K_2O \cdot 6\,SiO_2$ und Microclin sowie $Al_2O_3 \cdot Na_2O \cdot 6\,SiO_2$.

Tabelle 43. *Löslichkeit von Aluminiumsalzen*
(VOGEL 1956)
Es ist wasserfreie Substanz in g je 100 g Wasser gelöst:

bei °C	0°	10°	20°	30°	40°	50°	60°	Löslichkeitsprodukt	°C
$Al(OH)_3$	—	—	0,00014	—	—	—	—	$1{,}9 \cdot 10^{-33}$	20
$Al(OH)_3 \cdot 3\,H_2O$ (Säure)	—	—	—	—	—	—	—	$4 \cdot 10^{-13}$	20
$Al(NO_3)_3 \cdot 9\,H_2O$	61	—	73	—	89	—	108		
$Al_2(SO_4)_3 \cdot 18\,H_2O$	31,2	33,5	36,3	40,4	45,6	52,2	58		
$Al(NH_4)(SO_4)_2 \cdot 12\,H_2O$	2,6	—	6,6	—	12,4	—	21,1		
$AlCl_3 \cdot 6\,H_2O$	44,9	—	45,6	—	46,3	—	47,0		
AlJ_3	löslich								

M. Anionen

Es ist nicht erforderlich, die einzelnen Anionen als Komponenten von Salzen gesondert zu besprechen, da diese bereits in den besprochenen Salzen enthalten waren. Einige zusätzliche Notizen mögen jedoch als Ergänzung dienen.

Kohlensäure. Die freie Kohlensäure ist im Wasser nur bei einer Wasserstoffionen-Konzentration vorhanden, die einem pH-Wert von weniger als 8 entspricht. Bei pH 8 liegt die gesamte Kohlensäure im Wasser in Form von HCO_3^--Ionen vor, bei pH-Werten von mehr als 8 steigt der Anteil an CO_3^{--}-Ionen, der bei pH 12,5 fast 100% erreicht. Der Anteil an freier CO_2 wächst ab pH 8 mit fallenden pH-Werten gemäß den Kurven in Abb. 84 und erreicht 100% etwa bei pH 4. Die freie CO_2 diffundiert im Wasser um vieles langsamer als in der Luft

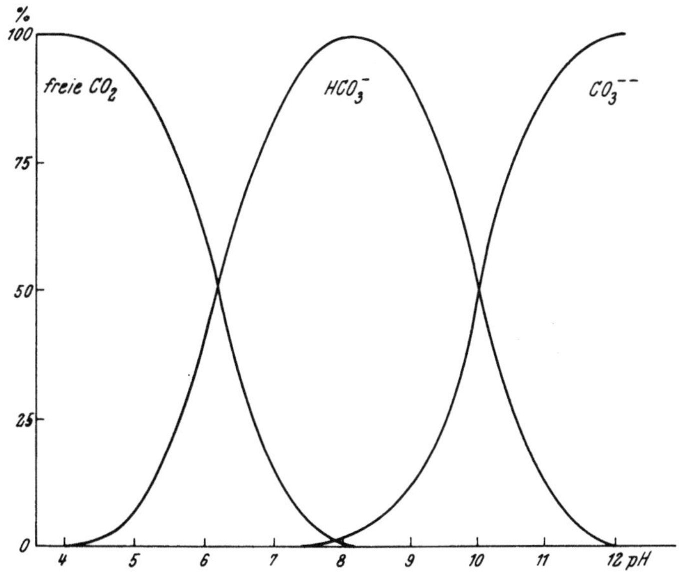

Abb. 84. Anteile der Ionen der Kohlensäure und der freien CO_2 in wässerigen Lösungen von verschiedenen pH-Werten (nach EMERSON und GREEN 1938)

(die Diffusionsgeschwindigkeit beträgt etwa 1/10000), ihre Ionen diffundieren jedoch noch langsamer (mit einer Geschwindigkeit von nur etwa $^1/_8$ bis $^1/_9$ jener der CO_2). Nicht alle Pflanzen vermögen Bicarbonationen zu verwerten. Während z. B. *Scenedesmus*-Arten dazu in der Lage sind, vermag *Chlorella* sie nicht zu verwerten (ÖSTERLIND 1950). Wurzeln höherer Pflanzen können Bicarbonationen verwerten (HÖFNER 1963).

Phosphorsäure. Die Ionen der Phosphorsäure, $H_2PO_4^-$, HPO_4^{--} (und PO_4^{---}) bilden bei verschiedenen pH-Werten der wässerigen Lösung, in der sie sich befinden, mit einigen Kationen verschiedene unlösliche Verbindungen.

Im sauren bis stark sauren pH-Bereich sind es vor allem Al^{+++}- und Fe^{+++}-Ionen, mit denen die Phosphationen sich in unlöslicher bzw. schwer löslicher Form festlegen.

K-, Na- und NH_4-Palmerite, z. B. $H_{1,3}K_{1,1}Al_{1,8}Fe_{0,4}(PO_4)_3 \cdot 5\ H_2O$, sind um so schwerer löslich, je niedriger ihr Phosphatanteil ist. Die Bildung von $Al_3(OH)_3(PO_4)_2 \cdot 5\ H_2O$, Wavellit, sowie $FePO_4 \cdot Fe(OH)_3$, Dufrenit, und der Sesquioxydphosphate, $FePO_4 \cdot 2\ H_2O$, Strengit, und $AlPO_4 \cdot 2\ H_2O$, Variscit, setzt den Gehalt der Lösung an Phosphationen im sauren Bereich herab und schränkt deren Verfügbarkeit für die Pflanze ein. Nach LEFÈVRE (1961) lösen sich im Wasser mg/l von Dicalciumphosphat hydratisiert 5 bis 15, Tricalciumphosphat ~1, Phosphate aus Böden 0,2 bis 0,5, Rohphosphate aus Nordafrika 0,03 und Hydroxylapatit $[Ca_5(PO_4)_3 \cdot OH \cdot H_2O]$ 0,013 (bei 35° C).

Im alkalischen Bereich reagieren Phosphationen mit Ca^{++}- und F^--Ionen unter Bildung von schwerlöslichem bzw. unlöslichem Tricalciumphosphat $Ca_3(PO_4)_2 \cdot CaCO_3$ und Apatiten: $Ca_{10}(PO_4)_6 \cdot F_2$; $Ca_{10}(PO_4)_6 \cdot FOH$; $Ca_{10}(PO_4)_6 \cdot (OH)_2$ (vgl. KURMIES 1953). Mit Zn^{++}-Ionen entsteht unlösliches Zinkphosphat. Infolge dieser und einiger weiterer, hier nicht genannter Reaktionen ist die Phosphorsäure für die Pflanzen nur bei einem schwach sauren bis neutralen pH-Wert der Bodenlösung gut verfügbar.

Borate. B_2O_3 löst sich zu 1,1 (bei 0° C) Teilen in 120 Teilen Wasser. Die Borsäure H_3BO_3 zeigt im Wasser bei verschiedenen Temperaturen die folgenden Sättigungskonzentrationen (in g je 100 g Wasser):

0°	10°	20°	30°	40°	50°	60°
2,66	3,57	4,95	6,6	8,72	11,54	14,81

Molybdate. Molybdänsäure $H_2MoO_4 \cdot H_2O$ ist zu 0,138 (bei 18° C) löslich im Wasser. Phosphormolybdänsäure $20\,MoO_3 \cdot 2\,H_3PO_4 \cdot 48\,H_2O$ löst sich wie die meisten Molybdate leicht im Wasser. $(NH_4)_6Mo_7O_{24} \cdot 4\,H_2O$ löst sich zu 44 (25° C) in 100 Teilen Wasser.

MoO_2 und Mo_2O_3 sind wasserunlöslich, MoO_3 ist zu 0,107 (bei 18° C) löslich. MoS_2 und MoS_4 sind ebenfalls wasserunlöslich, dagegen ist MoS_3 schwer löslich.

Nach BRINTZINGER (1931) sind bei folgenden pH-Werten die folgenden Molybdationen stabil:

pH-Wert:	1,25	1,5—4,5	5,65	6,5—14
Bezeichnung:	Dodekamolybdat	Hexamolybdat	Trimolybdat	Molybdat
Formel:	$Mo_{12}O_{41}^{10-}$	$Mo_6O_{21}^{6-}$	$Mo_3O_{11}^{4-}$	MoO_4^{2-}

Fluoride. CuF_2, Flußspat, löst sich in 100 Teilen Wasser zu 0,0017 g bzw. 0,004 g.

$Na_3[AlF_6]$, Kryolith, natürlich zu 0,039 g, synthetisch zu 0,061 bis 0,063 g, K_2SrF_6 zu 0,176 g, $BaSrF_6$ zu 0,024 g und NaF zu 4,210 bzw. 4,054 g. In schwachen Säuren ist die Löslichkeit für Kryolith bereits wesentlich höher als im Wasser.

N. Natürliche Mineralien

Ein Teil der natürlich vorkommenden Mineralien ist wasserlöslich; sie wurden im Verlauf der langwierigen Vorgänge der Bodenbildung, wo sie in bodenbildenden Gesteinen enthalten waren, bereits der Bodenlösung einverleibt. Andere Mineralien sind unter der Einwirkung von H^+-Ionen löslich. Es ist nicht möglich, hier eine vollständige Aufzählung der natürlichen Mineralien und ihrer Löslichkeitseigenschaften einzufügen. Eine ausführliche Tabelle, in der neben den international bekannten Bezeichnungen von 814 verschiedenen Mineralen auch deren Zusammensetzung angegeben ist, stellt qualitative Angaben über Löslichkeiten und Lösungsmittel zusammen (LANGE und FORKER 1956). Da sie leicht zugänglich ist, soll hier nur darauf verwiesen werden.

4. Nährstoffzufuhr von Quellen außerhalb des Systems Boden—Pflanze

Als gewöhnlich zu berücksichtigende Quellen einer Nährstoffzufuhr zur Bodenlösung kommen in Betracht: Atmosphärische Niederschläge in Form von Staub oder Regen, Düngemittel und Laub- bzw. Nadelstreu.

A. Zufuhr durch atmosphärische Niederschläge

Die Niederschlagswässer enthalten wechselnde Mengen an gelösten Stoffen. Über die Mengen an diesen Stoffen, die, unterschiedlich je nach der geographischen Lage der entnommenen Regenwasserproben, den Feldern jährlich zugeführt werden, vergleiche man den diesbezüglichen Abschnitt dieses Handbuches (Bd. II, ERIKSSON) sowie Ergebnisse von RIEHM und QUELLMALZ (1959). Nach SCHARRER und FAST (1951) wurden Gehalte an Nährstoffen gefunden, die je nach Monat bzw. Jahreszeit beträchtlich schwankten (vgl. Tab. 44).

Tabelle 44. *Gehalte (mg/l) des Regenwassers (Monatsmittel) an Nährstoffen* (nach SCHARRER und FAST 1951)

Nährstoff	Minimalwert	Maximalwert	Sommermittel	Wintermittel	Jahresmittel
HCO_3	15	37	23	26	24,5
N	3,8	14,2	9,0	5,2	6,9
SO_4	50,5	161	83	106,5	95
Ca + Mg[1]	19	78	31	57	41,5
Cl	0,3	6,5	2,1	3,9	3,0

[1] Als Ca berechnet.

Es kommen somit schon im Regenwasser beachtliche Mengen an gelösten Stoffen in den Boden bzw. in die Bodenlösung, teilweise von gleicher Größenordnung, wie sie in den Sickerwässern bzw. der Bodenlösung selbst vorliegen (beispielsweise beim Schwefel).

B. Zufuhr durch Düngemittel

Die Zufuhr von Düngemitteln, bei denen ja darauf geachtet wird, die Nährstoffe nach Möglichkeit in pflanzenverfügbarer, also löslicher Form zu bieten, bringt Salzmengen in den Boden, die sich zum großen Teil im Wasser des Bodens lösen und daher Konzentration und Zusammensetzung der Bodenlösung entscheidend beeinflussen.

Setzt man an, daß 100 kg/ha Ammoniumnitrat verabreicht werden, die sich in dem Wasser, das im Boden bis zu 1 m Tiefe vorhanden ist, auflösen sollen (von den zu erwartenden Sorptionsvorgängen sei hier zunächst abgesehen), so wird die je m² etwa 350 l betragende Bodenlösung um 10 g Ammoniumnitrat angereichert, also um etwa 30 mg/l. Da die Verteilung auf eine Bodentiefe bis zu 1 m ziemlich lange Zeit in Anspruch nimmt und zunächst eine Lösung in nur wenigen mm Niederschlagswasser erfolgt, liegt in den oberen Bodenschichten zunächst eine viel höhere Salzkonzentration vor, nämlich

bei 1 mm Niederschlag	10000 mg/l
bei 5 mm Niederschlag	2000 mg/l
bei 10 mm Niederschlag	1000 mg/l
bei 100 mm Niederschlag	100 mg/l

In den oberen Bodenschichten werden also nach Verabreichung höherer Düngergaben und bei Vorliegen relativ geringer Niederschläge wenigstens *vorübergehend* relativ hohe Salzkonzentrationen hergestellt und auf den Wurzelbereich (vor allem von Jungpflanzen) einwirken. Man darf dabei nicht vergessen, daß nicht nur die Nährstoffe der Düngemittel selbst die Salzkonzentration der Bodenlösung erhöhen.

Die verschiedenen Düngemittel unterscheiden sich nämlich voneinander unter anderen Eigenschaften auch dadurch, daß sie je Nährstoffeinheit verschiedene Mengen an löslichen *Ballaststoffen* mitführen, also die Bodenlösung mit verschiedenen Mengen davon anreichern. So wird z. B. der Bodenlösung mit Superphosphat nicht nur Monocalciumphosphat, sondern auch Gips zugeführt. Zum Vergleich der durch eine Nährstoffeinheit in die Bodenlösung gebrachten Gesamt-Salzmengen kann man den sogenannten *Salzindex* eines Düngemittels heranziehen, der in Abschn. 5 (S. 397) näher besprochen wird.

C. Zufuhr durch Laub- und Nadelstreu

Die vor allem in Wäldern und auf natürlichen Pflanzenbeständen eine Rolle spielende Zufuhr von Laubstreu oder Nadelstreu bedeutet zugleich eine Zufuhr von wasserlöslichen Stoffen. Diese werden durch die Einwirkung des Regens aus der Streu ausgewaschen und in den Boden gebracht. Es ist daher von Interesse, festzustellen, welche Mengen an löslichen, anorganischen Stoffen aus diesen Materialien in den Boden gelangen können. Für eine Reihe verschiedener Pflanzenarten, die Streu liefern, sind die Gehalte an wasserlöslichen Nährstoffen in Tab. 45 zusammengestellt.

Tabelle 45. *Wasserlösliche Mengen an pflanzlichen Bauelementen in verschiedenen organischen Materialien (in % der Trockensubstanz)*

	Eschenstreu	Buchenstreu	Erbsenstroh	Eichenstreu	Fichtenstreu
	(nach MATTSON und KOUTLER-ANDERSSON 1946)			(nach BASILEVICH 1955)	
S	—	—	—	—	0,216
P	0,141	0,133	0,258	0,126	0,009
Fe	—	—	—	0,328	0,720
Mg	0,385	0,170	0,395	0,350	0,060
Ca	2,64	0,23	1,22	3,38	0,236
K	0,622	0,450	2,30	0,149	0,066
Na	—	—	—	0,015	0,074
Mn	0,0014	0,0145	0,0069	0,077	0,240
Si	0,131	0,436	0,075	1,074	8,41

5. Der osmotische Druck der Bodenlösung

Der osmotische Druck der Bodenlösung ist von entscheidendem Einfluß auf die Nährstoffaufnahme und auf den Transport aufgenommener Nährstoffe mit Hilfe des Transpirationsstromes in die oberirdischen Pflanzenorgane (LINSER, MAYR und CHWALA 1962, LINSER und HERWIG 1963). Auf Salzböden wie auch durch hohe Gaben an Düngemitteln, die wasserlösliche Nährstoffe oder bzw. und wasserlösliche Begleitstoffe enthalten, können die osmotischen Drucke der Bodenlösungen so stark erhöht werden, daß Minderungen der Nährstoffaufnahme bzw. des Nährstofftransportes in der Pflanze eintreten können, die von praktischer Bedeutung für die Ertragsbildung sind.

Der osmotische Druck der Bodenlösung kann unmittelbar durch kryoskopische Messungen an Bodenpreßsäften bestimmt werden. Um der komplizierten Methodik der kryoskopischen Messungen von Preßsäften auszuweichen, hat man bei Salzböden aus Leitfähigkeitsmessungen Rückschlüsse auf die osmotischen Drucke der Bodenlösung zu ziehen versucht.

Indem man den Boden in den Zustand der Wassersättigung versetzt und anschließend die Bodenlösung durch Absaugen mittels einer Saugflasche gewinnt, erhält man genügend Flüssigkeit, um eine Leitfähigkeitsmessung durchzuführen. Die so erhaltenen Werte mit dem Faktor 0,36 multipliziert ergeben ein annäherndes Maß für den osmotischen Druck der Bodenlösung in Atm. An Hand solcher Untersuchungen konnten fünf Zustandsformen der Böden hinsichtlich ihres osmotischen Verhaltens unter besonderer Berücksichtigung des Pflanzenertrages herausgestellt werden, wie folgende Übersicht zeigt[1]:

$< 0{,}72$ Atm	Der osmotische bzw. Salzeffekt ist meist zu vernachlässigen
0,72—1,44 Atm	Ertragsdepressionen bei empfindlichen Kulturpflanzen
1,44—2,88 Atm	Stärkere Ertragsdepressionen bei vielen Kulturpflanzen
2,88—5,76 Atm	Nur salztolerante Pflanzen bringen befriedigende Erträge
$> 5{,}76$ Atm	Nur Pflanzen mit sehr hoher Salztoleranz bringen befriedigende Erträge

Der Einfluß verschiedener Düngemittel auf den osmotischen Druck der Bodenlösung kann durch den sogenannten *Salzindex* ausgedrückt werden (RADER, WHITE und WHITTAKER 1943), der den Anstieg des osmotischen Druckes der Bodenlösung angibt, und zwar als Relativzahl für das in Frage stehende Düngemittel, bezogen auf den durch das gleiche Gewicht $NaNO_3$ hervorgerufenen Anstieg des osmotischen Druckes, der gleich 100 gesetzt wird (Salzindex des Düngemittels $S_D = 100$).

Tabelle 46a. *Die relativen Salzindizes pro Einheit eines Pflanzennährstoffes für einige Handelsdüngemittel*
(nach RADER, WHITE und WHITTAKER 1943)

Nährstoff (Berechnungsgrundlage)	Düngemittel	Nährstoffgehalt %	Salzindex je Nährstoffeinheit (S_N)	Salzindex des Düngemittels (S_D)
N	Wasserfreier Ammoniak	82,2	0,572	47,02
	Ammoniumnitrat	35,0	2,990	104,65
	Ammoniumsulfat	21,2	3,253	68,96
	Monoammoniumphosphat	12,2	2,453	29,93
	Diammoniumphosphat	21,2	1,614	34,25
	Stickstofflösung 2 A	40,6	1,930	78,36
	Kaliumnitrat	13,8	5,336	73,64
	Natriumnitrat	16,5	6,060	100,00
	Harnstoff	46,6	1,618	75,40
P_2O_5	Superphosphat	20,0	0,390	7,80
	Tripelsuperphosphat	48,0	0,210	10,08
	Monoammoniumphosphat	61,7	0,485	29,93
	Diammoniumphosphat	53,8	0,637	34,25
K_2O	Kalidüngesalz	20,0	5,636	
	Kaliumchlorid	50,0	2,189	109,45
	Kaliumchlorid	60,0	1,936	116,16
	Kaliumnitrat	46,0	1,580	73,64
	Kaliumsulfat	54,0	0,853	46,06
	Kalimagnesia	21,9	1,971	43,16

[1] Spezialliteratur: Diagnosis and improvement of saline and alkaline soils. Agriculture Handbook Nr. 60; US-Department of Agriculture, Washington.

Die relativen Salzindizes, bezogen auf die Einheit des jeweils betrachteten Nährstoffes (N, P_2O_5 oder K_2O), sind für verschiedene Düngemittel (aus amerikanischer Erzeugung) in Tab. 46a zusammengestellt.

Diese Zahlen ermöglichen den Gesamt-Salzindex für ein aus Komponenten, die in dieser Tabelle genannt sind, hergestelltes Mischdüngemittel zu errechnen. *Für jedes Prozent* eines Nährstoffes ist die aus Tab. 46b ersichtliche Zahl für den Salzindex (S_N) einzusetzen und die Summe zu ermitteln. Für unlösliche Komponenten (oder Füllstoffe, die wasserunlöslich sind) ist der Salzindex gleich Null anzusetzen. In Tab. 46b sind vergleichsweise die Salzindizes (S_D) für zwei verschiedene Mischdüngemittel errechnet, die jedoch gleiche Nährstoffverhältnisse (1:2:2) besitzen. Das Beispiel zeigt, daß das Mischdüngemittel 10:20:20, welches doppelt so viel Nährstoffe enthält wie das Düngemittel 5:10:10, nicht einen doppelt so hohen Salzindex aufweist wie dieses, sondern nur 74,5% davon, daß es also im Hinblick auf die osmotische Beanspruchung der Wurzeln wesentlich schonendere Eigenschaften besitzt als das niedrigprozentigere, von dem bei gleicher Dosis der Nährstoffe ja die doppelte Menge verabreicht werden muß.

Tabelle 46b. *Beispiele für die Berechnung der Salzindizes für Mischdüngemittel* (nach TISDALE und NELSON 1960)

Verwendete Komponente	Düngemittel 5:10:10			Düngemittel 10:20:20		
	Gewicht lbs	Einheiten (%)	Salzindex (anteilig)	Gewicht lbs	Einheiten (%)	Salzindex (anteilig)
Ammonisierungslösung	148	3	5,79	—	—	—
Diammoniumphosphat	—	—	—	373	10 (4)	6,37
Harnstoff	—	—	—	260	6	9,71
Ammmoniumsulfat ...	195	2	6,51	—	—	—
Tripelsuperphosphat ..	—	—	—	417	10	2,10
Superphosphat	1000	10	3,90	—	—	—
50er Kalisalz	400	10	21,89	—	—	—
60er Kalisalz	—	—	—	667	20	38,72
Konditionierungsmittel	100	—	—	110	—	—
Füllstoffe	157	—	—	180	—	—
Summe	2000	25	38,09	2000	50	56,90

Literatur

BASILEVICH, N. J.: Peculiarities of the cycles of ash elements and nitrogen in some soil-plant regions of SSSR. Pedology (Russ.) **1955** (4), 1–33. — BIEDERMANN, G.: Biochemie und Physiologie des Fluors. Berlin: Akademie-Verlag. 1951. — BJERRUM, N.: Investigations on the solubility of calciumphosphates. Nord. Naturforskarmotet i Helsingfors, 1936, 344–348. — BOUYOUCOS, G. J., und M. M. MACCOOL: Die Gefrierpunktmethode als neues Mittel, die Konzentration der Bodenlösung direkt im Boden zu bestimmen. Mich. Str. Techn. Bull. 24 (1915). Exper. Stat. Rec. **34**, 721 (1915). — BURD, J. S., und J. C. MARTIN: Secular and seasonal changes in the soil solution. Soil Sci. **18**, 151–167 (1924). — BURGESS, P. S.: The soil solution, extracted by Lipman's direct-pressure method, compared with 1:4 water extract. Soil Sci. **14**, 191–215 (1922).

CONRAD, J. P., E. L. PROEBSTING und L. MCKINNON: Apparatur und Methodik zur Gewinnung von Bodenpreßlösung. Soil Sci. **29**, 323–328 (1930).

D'ANS, J., und E. LAX: Taschenbuch für Chemiker und Physiker, 2. Aufl. (1896 S.). Berlin-Göttingen-Heidelberg: Springer. 1949. — DOJARENKO, A. G., und A. A. SCHMUCK: Zur Kenntnis der Bodenlösung, I, Methode der Gewinnung und Untersuchung. J. landwirtsch. Wiss. **1**, 577–586 (1924).

EMERSON, R., und L. GREEN: Effect of hydrogen-ion concentration on *Chlorella* photosynthesis. Plant Physiol. **13**, 157–168 (1938). — EPHRAIM, F.: Inorganic chemistry. London: Thome & Roberts. 1948.

GROH, A. W., und P. EHRENBERG: Die Anwendung des Auspressens von Erde zur Ermittlung des Gehaltes des Bodens an gelösten Kalksalzen in verschiedenen Jahreszeiten und unter Berücksichtigung der Beziehung zur Strukturbildung des Bodens. Int. Mitt. Bodenkde. **13**, 107–115 (1923).

HESSELING VAN SUCHTELEN, F. H.: Methode zur Gewinnung der natürlichen Bodenlösung. J. Landwirtsch. **60**, 369–370 (1912). — HIBBARD, L. P.: Vergleich zwischen der durch die Verdrängungsmethode erhaltenen Bodenlösung und dem Wasserauszug alkalischer Böden. Soil Sci. **16**, 465–471 (1923). — HIELSCHER, M.: Z. Pflanzenernähr., Düng., Bodenkde. **7** (52), 257–278 (1938). — HOAGLAND, D. R.: Lectures on the inorganic nutrition of plants. Waltham, Mass.: Chronica Botanica Comp. 1948. — HÖFNER, W.: Einfluß von Zellgiften auf die ^{14}C-Aufnahme isolierter Gerstenwurzeln aus $NaH^{14}CO_3$. Atompraxis **9**, 82–86 (1963).

ISCHERIKOW, W.: Die Gewinnung der Bodenlösung im unveränderten Zustande. Opitn. Agron. (Russ. J. exper. Landwirtsch.) **8**, 147–166 (1907).

JOHNSTON, W. M., R. L. WOLFGANG und W. F. LIBBY: Tritium in nature. Science **113**, 1–2 (1951).

KURMIES, B.: Über die Bildung von Calciumphosphaten in wässeriger Lösung und den Reaktionsablauf bei der Bildung von Hydroxyl- und Fluorapatit. Phosphorsäure **13**, 57–100 (1953). — Über die Bildung von Calciumphosphaten in wässeriger Lösung (48 S.). Essen: Tellus. 1953.

LANGE, N. A., und G. M. FORKER: Handbook of chemistry, 9th ed. (1969 S.). Sandusky, Ohio: Handbook Publ. 1956. — LEFEVRE: Des scories de déphosphoration et leur action fertilisante. Ann. Gembloux **67**, 43–49 (1961). — LEO, M. W. M.: A new suction plate apparatus for extraction of soil solution in conductivity determination. Soil Sci. **95** (2), 142–143 (1963). — *Limburgerhof:* Arbeiten der landwirtschaftlichen Versuchsstation Limburgerhof, 1939, S. 103ff. — LINSER, H., und K. HERWIG: Untersuchungen zur Abhängigkeit der Nährstoffaufnahme vom osmotischen Druck der Außenlösung. Protoplasma **57**, 588–60 (1963). — LINSER, H., H. MAYR und CH. CHWALA: Der Einfluß des osmotischen Drucks einer Nährlösung auf die Phosphoraufnahme durch Wurzeln von *Lycopersicon esculentum.* Atompraxis **8**, 4–6 (1962). — LIPMAN, C. B.: Eine neue Methode zur Gewinnung der Bodenlösung. Univ. of Calif. Publ. Agric. Sci. **3** (7), 131–134 (1918).

MATTSON, S., und E. KOUTLER-ANDERSSON: The acid-base condition in vegetation, litter and humus, IX. Kungl. Lantbrukshögsk. Ann. **13**, 153–178 (1946). — MCGEORGE, W. T.: Iron, aluminium and manganese in the soil solution of Hawaiian soils. Soil Sci. **18**, 1–11 (1924). — MORGAN, J. F.: The soil solution obtained by the soil pressure method. Mich. Agric. Exper. Stat. Techn. Bull. 28 (38 S.), 1916. — MOSS, P.: Some aspects of the cation status of soil moisture, I, The ratio law and soil moisture content. Plant a. Soil **18** (1), 99–113 (1963).

ÖSTERLIND, S.: Inorganic carbon sources of green algae. Physiol. Plant. **3**, 353–360 (1950). — OSUGI, S., und Y. SANO: On the method of obtaining soil solution and the change of nitrata and Ammonia nitrogen in it. J. Sci. Agric. Soc. (307), 223 (1928).

PARKER, F. W.: Methods of studying the concentration and composition of the soil solution. Soil Sci. **12**, 209–232 (1921). — PFAFF, C.: Lysimeter-Versuche. Z. Pflanzenernähr., Düng., Bodenkde. **48** (93), 93–118 (1950). — PIERRE, W. H., und F. W. PARKER: Die Verwendung von Collodiumsäckchen zur Gewinnung klarer Bodenextrakte bei Bestimmung der wasserlöslichen Bestandteile. Soil Sci. **23**, 13–32 (1927).

RADER, L. F., L. M. WHITE und C. W. WHITTAKER: The salt index of fertilizers. Soil Sci. **55**, 201–218 (1943). — RAMANN, E., S. MÄRZ und H. BAUER: Über Boden-Preßsäfte. Int. Mitt. Bodenkde. **6**, 1–26 (1916). — REITEMEIER, R. F., und L. A. RICHARDS: Reliability of the pressure-membrane method for extraction of soil solution. Soil Sci. **57**, 119–135 (1944). — RIEHM, H., und E. QUELLMALZ: Die Bestimmung der Pflanzennährstoffe im Regenwasser und in der Luft und ihre Bedeutung für die Landwirtschaft. In: 100 Jahre Augustenberg, Karlsruhe, 1959.

SCHACHTSCHABEL, P.: Kolloidbeih. **51**, 199 (1940). — SCHARRER, K., und H. FAST: Untersuchungen über die dem Boden durch die Niederschläge zugeführten Pflanzennährstoffe. Z. Pflanzenernähr., Düng., Bodenkde. **55** (100), 97–106 (1951). — SCHLÖSING, TH.: Analysen von in Ackererden befindlichen Bodenlösungen. Compt. Rend. **70**, 98 (1870). — SCHUMACHER, W.: Über die Bodenlösung und solche betreffende Vegetationsversuche. Landwirtsch. Vers.-Stat. **5**, 210 (1863). — SCHUSTER, G. L.: A study of soil solutions by means of a semipermeable membrane supported on a porous clay plate. J. Amer. Soc. Agron. **9**, 333–340 (1917). — SIGMOND, A. v.:

Über die Grundfragen in der Zubereitung der Bodenlösungen für die chemische Analyse. Verh. 2. int. Agrogeologenkonferenz, 1910, Stockholm 1911, S. 71–92. — Stewart, G. R.: Effect of season and crops growth in melyfysing the soil extracted. J. Agricult. Res. **12**, 311–368 (1918).

Tisdale, S. L., und W. L. Nelson: Soil fertility and fertilizers (430 S.). New York: Macmillan. 1960. — Tulaikow, N. M., und M. S. Kuzmin: Über die Frage der Gewinnung von Bodenlösung. Soil Sci. **15**, 235–239 (1923).

Vesterberg, A.: Über eine Methode, die Zusammensetzung des natürlichen Bodenwassers zu ermitteln. Ber. IV. Int. Kongr. Bodenkde., Rom 1924, **2**, 552–556. Voelcker, A.: On the productive power of soils in relation to the loss of plant food by drainage. Trans. Chem. Soc. **24**, 276–297 (1871). — Vogel, H. U. v.: Chemiker-Kalender (560 S.). Berlin-Göttingen-Heidelberg: Springer. 1956.

Wachsmann, F.: Die radioaktiven Isotope. München: Lehmann. 1954. — Wagner, G. H.: Use of porous ceramic cups to sample soil water within the profile. Soil Sci. **94**, 379–386 (1962). — Wiklander, L.: The Soil, in Ruhland, W.: Handbuch der Pflanzenphysiologie, Bd. IV, S. 118–169. Berlin-Göttingen-Heidelberg: Springer. 1958. — Wrangell, M. v.: Die Bestimmung der pflanzenzugänglichen Nährstoffe des Bodens. Landwirtsch. Jahrb. **71**, 149–169 (1930). — Die Untersuchung des Bodens und der Nährstoffversorgung der Pflanzen nach M. von Wrangell, in Honcamp, F.: Handbuch der Pflanzenernährung und Düngerlehre, Bd. I, S. 914–929. Berlin: Springer. 1931.

Yaalon, D. H.: Physico-chemical relationships of $CaCO_3$, pH and CO_2 in calcareous soils. Fifth Int. Congr. Soil Sci., Léopoldville, **2**, 356–363 (1954).

Zyl, J. P. van: Über die Bodenlösung, ihre Gewinnung, Zusammensetzung und Anwendung bei der Schlämmanalyse. Diss. Göttingen: J. Landwirtsch. **64**, 201–275 (1916).

c) Humusstoffe

Von

W. Flaig

1. Entwicklung der Humusforschung

Seit den Anfängen des Ackerbaues ist dem Humusgehalt der Böden eine große Bedeutung zugeschrieben worden. Es hat daher nicht an Bemühungen gefehlt, dessen Bildung, Zusammensetzung und Veränderungen zu studieren.

Unter „Humus“, „Soil Organic Matter“, Matière Noire”, „Peregnoi“ versteht man meist die organischen Stoffe des Bodens mit Ausnahme der lebenden Organismen (Waksman 1936). In den vorliegenden Ausführungen wird der Begriff in diesem Sinne verwendet. Der Humus besteht aus den in den abgestorbenen Organismen enthaltenen Stoffen und deren Umwandlungsprodukten.

Über den Begriff Humus besteht nicht bei allen Autoren eine einheitliche Auffassung; er wird zum Teil insofern eingeengt (Robinson 1939, Maiwald 1931, 1939, Laatsch 1957), als nur der umgesetzte strukturlose, organische Substanzanteil darunter verstanden werden soll. Ferner wird vom genetischen Standpunkt nur derjenige Anteil der organischen Stoffe des Bodens als Humus angesprochen, der sich unter den jeweiligen Bedingungen durch seine geringe Zersetzungsgeschwindigkeit charakteristisch anhäuft (Kubiëna 1948).

Die große Anzahl von Veröffentlichungen über Humus und die darin zahlreich vorhandenen Widersprüche, die jeweils aus der Sicht der verschiedenen Autoren entstanden sind, würden einen historischen Überblick über die Humusforschung sehr umfangreich werden lassen. Aus diesem Grunde wird im Rahmen dieses Beitrages auch keine vollständige Übersicht aller Arbeiten gegeben werden können.

Im Schrifttum gibt es folgende zusammenfassende Darstellungen: Sven Oden (1919), Maiwald (1931), Waksman (1936), Kononowa (1958), Scheffer-Ulrich (1960), Prat (1960), Bremner (1951, 1954, 1956), Flaig (1955a).

In dieser kurzen Zusammenfassung sollen die wesentlichen Erkenntnisse der Chemie der organischen Stoffe im Boden herausgestellt und damit auch Anregungen für die weitere Arbeit gegeben werden.

In einem Handbuch über Agrikulturchemie berichtete WALLERIUS (1761) über einige Tatsachen der Bildung von Humus durch die Zersetzung von Pflanzen und beschrieb einige Eigenschaften, wie die Fähigkeit, Wasser und andere Stoffe zu sorbieren. WALLERIUS betrachtete Humus als einen Pflanzennährstoff. Die Vorstellungen über Stoffe aus dem Humus wurden in dem Maße erweitert, als es möglich war, diese aus natürlichen Produkten zu isolieren. Wahrscheinlich war es ACHARD (1786), der zum ersten Male mit Hilfe alkalischer Lösungen Humussubstanzen aus Torf extrahierte und diese mit Säuren als dunkelbraune Substanzen fällte. Er stellte fest, daß aus den tiefen Schichten mehr Humussubstanzen isoliert werden konnten als aus den oberen.

Zunächst mag als bemerkenswerte Tatsache hervorgehoben werden, daß zu Beginn des Studiums über den Einfluß der verschiedenen Komponenten im Boden auf das Pflanzenwachstum die Wissenschaft den organischen Stoffen im Boden eine größere Bedeutung zumaß als den anorganischen (THAER 1821).

Erst seit den späteren Arbeiten von SPRENGEL (1828, 1837, 1839) und von v. LIEBIG (1846) sind die anorganischen Stoffe in ihrer vollen Bedeutung für die Ernährung der Pflanze erkannt worden und führten im weiteren Verlauf zur Anwendung der Mineraldünger. Die zunehmende Entwicklung auf den Gebieten der Chemie, der Physik und der Biologie im Laufe des 19. und 20. Jahrhunderts hat der Humusforschung viele Anregungen gegeben. Die zu verschiedenen Zeiten unterschiedlichen Fortschritte der Zweige der Naturwissenschaften haben die Richtung der Humusforschung jeweils stark beeinflußt.

2. Einführung und Begriffe

A. Faktoren der Humusbildung

In Abhängigkeit von den klimatischen Bedingungen unterliegen die Bestandteile abgestorbener Organismen verschiedenen Umwandlungen und gelangen mit den anorganischen in der Oberfläche des Bodens mehr oder minder in Wechselwirkung. Ein Teil der abgestorbenen Organismen verbleibt auf dem Boden, ein Teil der Organismen stirbt im Boden ab, während wiederum ein Teil z. B. von Tieren mit dem Boden vermischt wird.

Die nach dem Absterben erfolgenden chemischen Umsetzungen kann man in drei Arten aufteilen:

a) Umsetzungen, an denen sich lebende Organismen beteiligen,

b) Umsetzungen, die durch die noch aktiven Enzyme abgestorbener Organismen bewirkt werden,

c) Umsetzungen, wie sie als chemische Reaktionen ablaufen.

Alle drei Umsetzungen laufen je nach den Bedingungen in verschiedenem Ausmaß nebeneinander ab. Die Umsetzung durch lebende Organismen wird z. B. durch Trockenheit zurückgedrängt, während chemische Reaktionen, wie z. B. die Oxydation, in den Vordergrund treten. Auch die physikalische Beschaffenheit der sich umsetzenden Materialien, wie der Zerteilungsgrad der Stoffe oder Reaktionen in Lösung haben einen Einfluß auf die Geschwindigkeit der Umwandlung der zugeführten Substanzen.

Die Ausgangsmaterialien bestehen aus Verbindungen, deren weiterer Abbau sich teils rascher, teils langsamer vollzieht. Dasselbe trifft für die Umwandlungsprodukte zu. Schließlich werden die organischen Substanzen im Boden bis zu den Endstufen der Oxydation, zu Kohlendioxyd und Wasser abgebaut. Aus den stickstoffhaltigen organischen Verbindungen bildet sich hauptsächlich Ammoniak. In weitaus geringerer Menge bildet sich elementarer Stickstoff und sauerstoffhaltige Stickstoffverbindungen sowie unter reduzierenden Bedingungen Methan und andere aliphatische Kohlenwasserstoffe. Der in den organischen Verbindungen gebundene Schwefel wird unter aeroben Bedingungen zu sauerstoffhaltigen Säuren und unter anaeroben Bedingungen zu Schwefelwasserstoff umgesetzt.

Die Umwandlung der Humusbestandteile wird auch durch die chemische Zusammensetzung und die physikalischen Eigenschaften der anorganischen Bestandteile des Bodens beeinflußt. Dadurch können sich unter gleichen klimatischen Bedingungen unterschiedliche Umwandlungen vollziehen.

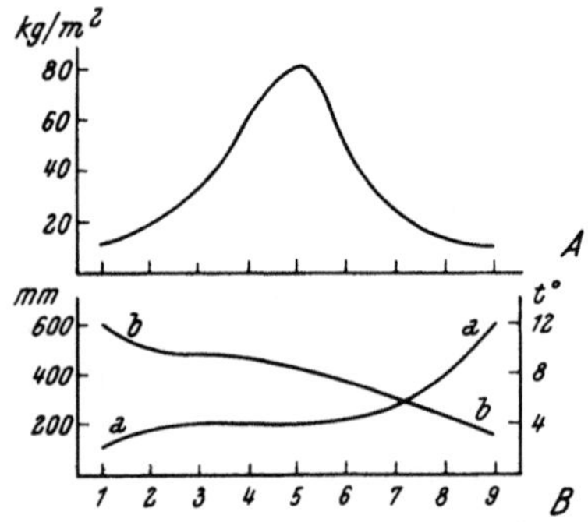

Abb. 85. Humusgehalte russischer Böden nach J. W. Tjurin

1 podsolierte Böden
2 helle, *3* dunkle } graue Waldböden
4 ausgelaugte, *5* mächtige, *6* gewöhnliche } Schwarzerde
7 dunkle, *8* helle } kastanienfarbige Böden
9 Grauerden

A Humusgehalt für eine Tiefe von 100 bis 120 cm in kg/m²
B a jährliche Durchschnittstemperatur
b jährliche Niederschlagsmenge in mm

Die Vielfalt der in den Ausgangsmaterialien enthaltenen Verbindungen wird im Verlauf von deren Umwandlung durch die dabei entstehenden vermehrt. Da die organischen Stoffe im Boden durch das Absterben von Organismen nachgeliefert werden, anschließend einer Umwandlung unterworfen sind und schließlich durch vollständige Oxydation teilweise immer wieder entfernt werden, spricht man auch von einer Dynamik der organischen Stoffe im Boden. Die Menge der organischen Stoffe im Boden stellt sich durch Nachlieferung und Zersetzung zu den Endprodukten der Oxydation je nach den klimatischen Bedingungen unter natürlichen Verhältnissen auf eine bestimmte Höhe ein (Köhnlein 1957, 1959, Tjurin 1937, Jenny 1941).

Die im Boden vorhandene Menge an Humus hängt von den verschiedensten Faktoren ab. Einen wesentlichen Einfluß hat das Klima. Tjurin (1937) hat die Mittelwerte aus zahlreichen Untersuchungen über den Humusgehalt angeführt. Die Menge des Humus ist abhängig von dem Jahresdurchschnitt an Niederschlägen und der Temperatur. Bei einem Temperaturgefälle von Süden nach Norden und einer Zunahme an Niederschlägen bildet sich ein Maximum der Humusbildung ungefähr bei den mittleren Werten aus. Dieses Maximum dürfte jedoch nicht nur ausschließlich durch das Mittel von Temperatur und Niederschlägen bedingt sein, sondern es haben auch noch andere Faktoren eine Wirkung gehabt, wie zum Beispiel die Verteilung der höchsten und tiefsten Temperaturen und die Verteilung der Niederschläge im Laufe des Jahres. Ferner haben die anorganischen Bestandteile des Bodens ebenfalls einen nicht zu vernachlässigenden Einfluß auf die Bildung und Erhaltung von Humusstoffen.

B. Einteilung der Humusstoffe

Aus dem Humus lassen sich einige Verbindungen abtrennen, die in ihrer chemischen Konstitution bekannt sind. Diese werden im nächsten Abschnitt im einzelnen besprochen.

Während der Verrottung nimmt das Ausgangsmaterial mehr und mehr eine dunkle Farbe an. Böden, die mit organischen Stoffen angereichert sind, haben meist eine dunkle Farbe, die, wie z. B. im Falle der Schwarzerde, zu ihrer Namensgebung beigetragen haben.

Durch die Umwandlung der Ausgangsstoffe entstehen Verbindungen, die in ihrem Verhalten nur geringe Unterschiede zeigen. Die Auftrennung macht daher Schwierigkeiten, die bis heute noch nicht gelöst sind. Die bisher angewandten Trennungsmethoden geben noch keine Gewähr dafür, daß die erhaltenen Fraktionen der dunkelgefärbten Stoffe chemisch einheitliche Stoffe sind.

Das Interesse der Bearbeiter des Humus gilt seit jeher vornehmlich den dunkelgefärbten Stoffen, die im Verlaufe der Umwandlung der Ausgangsmaterialien entstehen. Die dunkelgefärbten Stoffe sind meist saurer Natur und werden mit der besten Ausbeute mit alkalischen Lösungen aus den Böden und verrottetem pflanzlichem Material herausgelöst. Die Bedingungen der Extraktion mit anderen Lösungen werden näher beschrieben.

Als erster führte SPRENGEL (1826) systematische Untersuchungen zur Auftrennung und Charakterisierung der Humusstoffe durch. Seine Ergebnisse haben noch heute Gültigkeit. Später haben BERZELIUS (1833) und nach ihm ODEN (1912, 1919, 1922) die Auftrennung der Humusstoffe in einzelne Fraktionen intensiv bearbeitet. Nach deren Vorschlägen erfolgt die Einteilung der Humusstoffe bis zum heutigen Tage.

Beim Ansäuern der alkalischen Extrakte fällt ein Teil der sauren Bestandteile aus. Nach dem Abtrennen des Niederschlages bleibt eine gelbe bis orangerote Lösung zurück. In der sauren wässerigen Lösung befinden sich vorwiegend niedermolekulare Stoffe, die als Fulvosäuren bezeichnet werden. BERZELIUS (1833) bezeichnet die von ihm isolierten Fulvosäuren aus Torf als Krensäuren und die bei der Autoxydation dieser Krensäuren ausfallenden Produkte als Apokrensäuren. Aus dem dunkelbraungefärbten Niederschlag des angesäuerten Extraktes läßt sich mit Alkohol eine Fraktion abtrennen, die nach Vorschlag von HOPPE-SEYLER (1889) als Hymatomelansäuren bezeichnet wird. Der in Alkohol unlös-

Tabelle 47. *Fraktionierung der Humusstoffe*

Fraktionen	Alkali	Säure	Alkohol
Fulvosäuren	löslich	löslich	—
Hymatomelansäuren ...	löslich	unlöslich	löslich
Huminsäuren	löslich	unlöslich	unlöslich
Humine	unlöslich	unlöslich	unlöslich

liche Rückstand wird Huminsäuren genannt; auch diese Stoffgruppe besitzt keine einheitliche chemische Zusammensetzung. Durch Ausfällung mit Neutralsalzen oder starken Laugen lassen sich die Huminsäuren in die Fraktionen der Grau- und Braunhuminsäuren auftrennen (SPRINGER 1938, 1955, FLAIG, SCHEFFER und KLAMROTH 1955).

Die nicht mit Lauge aus dem Boden extrahierbare Fraktion wird als Humine bezeichnet.

Alle diese Namen bezeichnen Fraktionen von Stoffen, deren Einheitlichkeit nicht bewiesen ist, und deren Zusammenhang je nach Herkommen (Kononowa 1958, S. 361) und Extraktionsverfahren Unterschiede aufwiesen. Deshalb sind auch von anderen Autoren noch weitergehende Unterteilungen vorgenommen worden, die sich durch die Variation der Extraktionsverfahren ergeben haben.

Aus dem Bestreben heraus, das Stoffgemisch der entstehenden dunkelgefärbten Stoffe von dem des Ausgangsmaterials und den Abbauprodukten der dunkelgefärbten Substanzen zu unterscheiden, hat Kononowa die Bestandteile des Humus — die Humusstoffe — in spezifische und unspezifische organische Bodenstoffe und Scheffer auf Grund fast gleicher Überlegungen in Huminstoffe und Nichthuminstoffe unterteilt.

Nach Scheffer und Ulrich (1960) sind die organischen Stoffe pflanzlicher und tierischer Herkunft Nichthuminstoffe; Huminstoffe sind aus Nichthuminstoffen gebildete, dunkelgefärbte, bodeneigene Aufbaustoffe.

Sie besitzen eine unterschiedliche Löslichkeit in Acetylbromid, worin die Nichthuminstoffe löslich sind. Da auch Fulvosäuren und Hymatomelansäuren, zwei Fraktionen der Huminstoffe, in dem gleichen Lösungsmittel löslich sind, ist eine Unterscheidung von Nichthuminstoffen und Huminstoffen mit diesem Reagenz nicht möglich. Die für die Beschreibung von Vorgängen in der Bodenkunde geschaffenen Begriffe „Nichthuminstoffe“ und „Huminstoffe“ sind weder durch chemische noch durch physikalische Merkmale eindeutig charakterisiert und daher für eine chemische Systematik des Humus nicht geeignet.

Die unspezifischen organischen Bodenstoffe sind nach Kononowa eine außerordentliche uneinheitliche Gruppe von Substanzen, welche Stoffe aus den zersetzten Pflanzenresten und deren Umwandlungsprodukte sowie Stoffe, die aus der Lebenstätigkeit der Mikroorganismen herrühren, mit einschließt. Sie führt dabei höher- und niedermolekulare Stoffe aus Organismen sowie Verbindungen bekannter Konstitution an, die aus dem Boden extrahiert worden sind.

Unter spezifisch organischen Bodenstoffen versteht sie denjenigen Anteil der Humusstoffe, die sich neu gebildet haben und dunkelgefärbt sind, und deren „Eigenschaften es nicht erlauben, sie den in der organischen Chemie bisher bestehenden Klassen zuzuordnen“ (Kononowa 1958, S. 28).

Als erste ausführliche Anfänge für eine Einteilung der Humusstoffe nach chemischen Gesichtspunkten sind die Isolierungen der in ihrer Konstitution bekannten organischen Verbindungen aufzufassen, wie die durch Schreiner und Shorey (1910) und Shorey (1913) erfolgten; den Untersuchungen dieser Autoren schlossen sich später noch weitere an.

Auch die Auftrennung der Humusstoffe in die mit historischen Namen belegten Fraktionen Fulvosäuren, Hymatomelansäuren, Huminsäuren und Humine ist ein erster Schritt zu einer chemischen Systematik der Humusstoffe. Entsprechende Untersuchungen müssen ergeben, inwieweit diese Fraktionen aus verschiedenen Verbindungen zusammengesetzt sind, und welche Konstitution diesen zukommt.

C. Abgrenzung des Begriffs „Huminsäuren“

In der Literatur findet man die Bezeichnungen Humine, Huminsäuren, humusartige Substanzen auch für dunkelgefärbte Stoffe, die nicht im Boden, sondern bei anderen chemischen Reaktionen gebildet worden sind, die als Rückstände auftreten, und bei denen es bisher nicht möglich war, sie näher zu charakterisieren. Diese Bezeichnung kann man vielleicht historisch damit begründen, daß man im Laufe der Humusforschung versuchte, die Bildung der aus den Böden isolierten Huminsäuren durch Umsetzung von Zuckern und Aminosäuren oder auch anderen pflanzlichen Materialien mit Säuren und Laugen unter mehr oder weniger extremen Bedingungen nachzuahmen.

Der Brauch, dunkelgefärbte Rückstände als „Huminsäure“ oder „humusartige Stoffe“ zu bezeichnen, bürgerte sich mit der Zeit in der chemischen Literatur immer mehr ein. Die Bezeichnung „Huminsäuren“ soll jedoch hauptsächlich derjenigen Stoffgruppe vorbehalten bleiben, die sich bei den Umwandlungen der in den abgestorbenen Organismen enthaltenen Verbindungen bildet (DAVIES und COULSON 1959).

Bei der Verallgemeinerung der Bezeichnung „Huminsäuren“ für auf verschiedene Weise entstandene und in ihrer Zusammensetzung noch unbekannte Stoffe sollte man Vorsicht walten lassen. Es soll daher der Versuch gemacht werden, eine Abtrennung derjenigen Stoffe vorzunehmen, die zweckmäßigerweise als „Huminsäuren“ bezeichnet werden sollen.

Hierher gehören zunächst diejenigen, die aus den Böden isoliert werden können. Dabei wäre es zu ihrer näheren Bezeichnung angebracht, sie als „Bodenhuminsäuren“ bzw. schließlich als „Huminsäuren aus Schwarzerde“ oder aber auch mit der Angabe des Horizontes des betreffenden Bodens zu bezeichnen.

Zu Beginn der Umwandlung der pflanzlichen Substanz lassen sich „Huminsäuren“ extrahieren, wie von KONONOWA (1958, S. 99) in ausführlichen Untersuchungen gezeigt worden ist (vgl. auch KULLMANN 1959). Ferner lassen sich „Huminsäuren“ aus den Produkten der „Inkohlungsreihe“ (ZOLCINSKI 1930), so aus Torf, Braunkohlen und selbst aus Steinkohlen extrahieren. Auch hier wäre es zur genauen Bezeichnung angebracht, die Herkunft vor das Wort Huminsäuren zu setzen.

Später wird noch davon die Rede sein, daß man sich darum bemüht, mit Hilfe von bekannten Verbindungen, die mit möglichen Ausgangsstoffen der natürlichen Huminsäuren in der Natur eine chemische Ähnlichkeit haben, den Vorgang der Bildung der organischen Substanzen des Bodens zu verfolgen (ELLER 1921, 1923, 1925a, 1925b, ELLER und KOCH 1920, ELLER, HERDICKERHOFF und SAENGER 1923, ELLER, MEYER und SAENGER 1923, ERDTMAN 1933a, 1933b, 1933c, 1933d, 1955, ERDTMAN und GRANATH 1954, FLAIG 1950, 1960, SCHEFFER und ZIECHMANN 1958, SCHEFFER, WELTE und ZIECHMANN 1955, SCHEFFER, ZIECHMANN und SCHOLZ 1958; WELTE, SCHATZ und ZIECHMANN 1954). Dabei war man sich von Anfang an bewußt, daß nur ein Teil der in der Natur sich abspielenden Vorgänge abläuft, und die bei diesen Synthesen erhaltenen Produkte auch nur teilweise die Eigenschaften der natürlichen haben. Diese haben zur Abkürzung die Laboratoriumsbezeichnung „Synthesehuminsäuren“.

Um weitere Erkenntnisse über die Bildung von Huminsäuren zu gewinnen, geht man auch dazu über, natürliche Reaktionsbedingungen nachzuahmen, unter denen bekannte Verbindungen Huminsäuren ergeben, die den natürlichen in ihrer Entstehung und Zusammensetzung soweit als möglich ähnlich sind (FLAIG 1955b, 1956, 1958, FLAIG, SCHOBINGER, STEINMETZ 1957, STEINMETZ 1956, SWABY 1956/58).

Die Beziehungen der Untersuchungen zwischen den natürlichen Stoffen und synthetisierten Produkten gehen aus der folgenden Gegenüberstellung hervor.

Natürliche Huminsäuren	*Synthetische Produkte*
Veränderungen des Ausgangsmaterials beim Übergang in Humusstoffe in Abhängigkeit von dessen Zusammensetzung und den Reaktionsbedingungen; Charakterisierung der entstehenden Stoffe.	a) Aufklärung einzelner Reaktionsstufen durch Einsetzen bekannter Verbindungen. b) Herstellung von Substanzen, die den in der Natur entstehenden weitgehend ähnlich sind, unter Verwendung bekannter natürlicher Ausgangsprodukte und unter Angleichung der in der Natur gegebenen Bedingungen.

Tabelle 48. *Zusammensetzung in % wasserfreier Substanz verschiedener Pflanzenteile und tierischer Organismen*

	wasserlösliche Stoffe	Rohprotein	Cellulose	Hemicellulose	Stärke	Lignin	Asche	in organischen Lösungsmitteln löslich	Literatur
Streu der Kiefer	9	16	17	18	—	42	2,5	14	(SCHEFFER und ULRICH 1960)
Streu der Birke	7	6	14	26	—	39	8	14	(SCHEFFER und ULRICH 1960)
Winterweizenstroh	2,1	2,4	36,1	31,3	—	17,9	9,5	—	(SCHOBINGER 1957)
Kleeblätter	—	21,7	15,4	8,1	3,0	4,3	—	23,1	(KONONOWA 1958)
Luzernewurzeln	—	13,3	21,0	11,9	17,8	8,6	—	11,2	(KONONOWA 1958)
Timotheewurzeln	—	7,8	25,9	20,4	2,3	19,5	—	5,3	(KONONOWA 1958)

	Rohprotein	Rohfett	N-freie Extraktstoffe	Rohfaser	Rohasche	Literatur
Sommerroggenstroh	3,6	1,9	48,2	42,7	3,5	(STÄHLIN 1957)
Regenwürmer	29,4	2,5	10,5	0,6	26,3[1]	(STÄHLIN 1957)
Maikäfer[2]	67.7	11,4	0,06	15,7	5,3	(STÄHLIN 1957)

[1] Zum größten Teil Mageninhalt (STÄHLIN 1957).
[2] Als Beispiel einer Analyse für die Zusammensetzung von Arthropoden.

3. Verrottung organischer Substanz

A. Ausgangsmaterial

Die im Boden zur Zersetzung kommenden organischen Substanzen setzen sich aus Inhaltsstoffen der abgestorbenen pflanzlichen und tierischen Organismen zusammen. Man kann hierbei in erster Linie drei Hauptstoffgruppen unterscheiden:

1. Kohlenhydrate,
2. Eiweißstoffe,
3. Lignin.

Während durch die tierischen Organismen fast nur die Komponenten 1. und 2. in den Boden gelangen, setzen sich die pflanzlichen Rückstände, die die Hauptmasse der abgestorbenen organischen Substanz ausmachen, aus allen drei Komponenten zusammen (vgl. Tab. 48).

Neben den genannten Hauptkomponenten spielen Substanzen, wie Fette, Harze, Wachse, Gerb- und Farbstoffe, sowie spezifische Stoffwechselprodukte, wie z. B. Antibiotika, in der prozentualen Zusammensetzung der Ausgangsmaterialien nur eine untergeordnete Rolle.

B. Abbaugeschwindigkeit der organischen Substanz

Der Einfluß der Zusammensetzung des sich zersetzenden Materials und der Einfluß der Umweltbedingungen auf die Abbaugeschwindigkeit ist schon erwähnt worden (s. hierzu S. 384). Über die Verluste an organischen Substanzen ver-

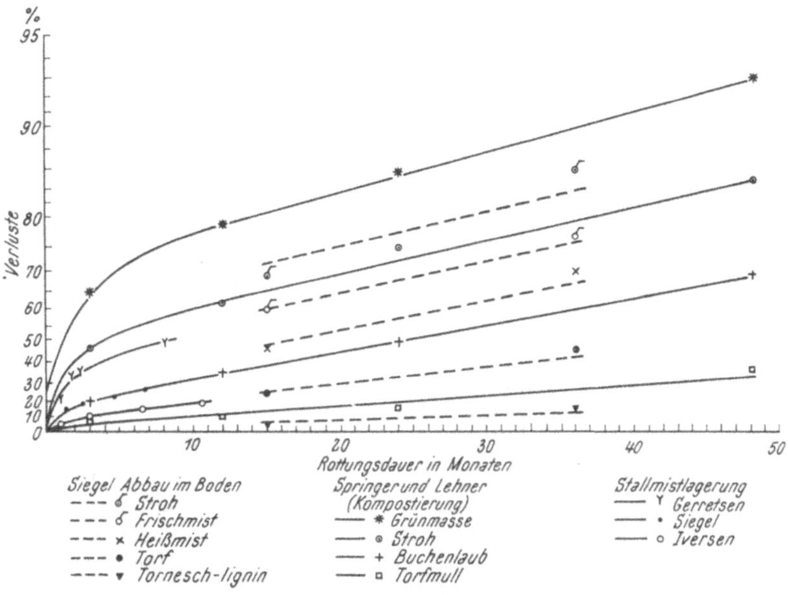

Abb. 86. Die Verluste organischer Düngemittel an organischer Substanz und Trockenmasse während der Kompostierung und Abbau im Boden

schiedenen Ausgangsmaterials gibt z. B. Kolenbrander (1955) eine Zusammenstellung. Die Abbaugeschwindigkeit der einzelnen pflanzlichen Bestandteile wird in den meisten Fällen mit der Methode der annähernden Bestimmung der Pflanzenbestandteile nach Waksman und Stevens (1930) festgestellt.

Die Schwierigkeit der Bestimmung der Veränderungen von Lignin, Cellulose, Hemicellulose und Eiweißstoffen besteht darin, daß diese Methode für noch nicht verrottetes Pflanzenmaterial ausgearbeitet worden ist und die bei der Verrottung entstehenden Produkte nicht mehr scharf abtrennt. So ist es möglich, daß bei der Abtrennung des Lignins auch Humusstoffe mitbestimmt werden. Diese Methode ist daher verschiedentlich variiert worden, wobei vor allem die ausführlichen Untersuchungen von SPRINGER und SEISCHAB (1961), SPRINGER und LEHNER (1952a, 1952b) zu erwähnen sind.

Als Beispiel für die Zersetzung von lignin- und cellulosereichen, aber proteinarmen Pflanzen werden die Vorgänge bei der Verrottung von Stroh angeführt, die bisher am ausführlichsten untersucht worden sind. Diese Verrottung ist unter verschiedenen Bedingungen, bei unterschiedlichen Temperaturen (KAILA 1952) und mit verschiedenen Stickstoffzusätzen (SCHOBINGER 1957, FLAIG, SCHOBINGER und DEUEL 1959, MAEDER 1959) bearbeitet worden. Charakteristisch für alle diese Untersuchungen ist die in Tab. 49 wiedergegebene stoffliche Zusammensetzung der Rotteprodukte, die ein Maß für die unterschiedliche Abbaugeschwindigkeit der einzelnen Komponenten während der Rottezeit ist. Bei anderem pflanzlichen Material, wie Kieferstreu, Laub, Gräser, Leguminosen und unter anderen Bedingungen treten nur graduelle Unterschiede auf.

Tabelle 49. *Veränderungen der Mengen an Bestandteilen im Weizenstroh im Laufe der Verrottung*

Rottezeit in Tagen	Organische Substanz		H_2SO_4-Lignin			Cellulose					
						Holo-			α-		
	Abn. g	% Abn.	g Geh.	%	% Abn.	g Geh.	%	% Abn.	g Geh.	%	% Abn.
0	265		53	20		198	75		106	40	
		53			33			69			67
70	124		35	28		60	43		35	28	
		81			68			94			95
180	50		17	34		12	25		5	10	
		81			64			95			98
260	49		20	38		10	20		2	5	
		81			64			96			97
340	49		19	38		8	15		3	6	
		83			63			96			99
410	47		20	41		10	20		1	1	

g Geh. = Gehalt in g nach den Tagen der Rotte.
Geh. % = Gehalt in % in dem jeweiligen Rotteprodukt.
% Abn. = Abnahme in % bezogen auf ursprüngliche Menge.

Die Elementarzusammensetzung der Strohproben ergibt, daß der Kohlenstoffgehalt im Verlauf der Rotte zunächst ansteigt und dann wieder abfällt, während der Sauerstoffgehalt ab- und der Stickstoffgehalt zunimmt (vgl. Tab. 50).

Diese Feststellung steht mit den gewichtsanalytischen Befunden im Einklang, die ebenfalls besagen, daß zunächst die sauerstoffreiche Cellulose gleich zu Anfang der Rotte stark abnimmt, während das sauerstoffärmere Lignin prozentual zunimmt. Der steigende Gehalt an organisch gebundenem Stickstoff weist auf die zunehmende Bildung von Proteinen durch die Mikroorganismen hin. Der Verlauf der Verengung des C/N-Verhältnisses ist hauptsächlich ein Maß für die Menge der abgebauten Kohlenstoffverbindungen.

Tabelle 50. *Elementarzusammensetzung der Proben von Weizenstroh, bezogen auf organische Substanzen*

Rottezeit in Tagen	C	H	O	N	OCH_3	C:N total
0	46,36	6,47	46,46	0,71	3,80	60,2
70	48,34	6,21	43,57	1,80	3,56	24,2
180	50,25	6,30	40,08	3,37	3,38	12,6
260	49,72	6,01	40,31	3,96	3,30	10,6
340	48,24	6,30	41,64	3,82	3,40	10,7
410	48,32	5,87	41,52	4,29	3,01	9,6

Bei Ansätzen mit Zusatz von verschieden großen Mengen an Stickstoff ist die prozentuale Verteilung des Stickstoffs im Rotteprodukt ohne Stickstoffzusatz und mit verschieden hohen Stickstoffgaben während der ganzen Rottezeit (240 Tage) ungefähr gleich. Der Stickstoff liegt in dem Rotteprodukt zu ungefähr 10% in anorganischer Form vor. Von den 90% organisch gebundenen Stickstoffs sind in der Ligninfraktion bei der Isolierung mit Schwefelsäure nach KLASON ungefähr 30% enthalten.

Abbau der Kohlenhydrate. Die Kohlenhydrate werden relativ schnell sowohl durch Mikroorganismen unter Verwertung als Energiequelle im Stoffwechsel wie auch durch Ectoenzyme (HOFMANN und SEEGERER 1951, HOFMANN und SCHMIDT 1953, SEEGERER 1953) und durch die von den abgestorbenen Pflanzen freigesetzten Enzyme (HOFMANN 1955a) abgebaut (TENNEY und WAKSMAN 1930, WAKSMAN und GERRETSEN 1931, SPRINGER 1944/45, APENITIS, ERDTMAN und LEOPOLD 1951). Ihre Zersetzungsgeschwindigkeit ist dabei auch von der Art der Kohlenhydrate (SCHMALFUSS 1940) und von deren Einbau in die pflanzliche Substanz abhängig.

Die Endprodukte des Abbaus sind Kohlendioxyd und Wasser. Bei unzureichender Sauerstoffnachlieferung entstehen niedermolekulare organische Säuren (DYER und WRENSHALL 1941a, 1941b, ROGERS 1942). Über den Stoffwechsel der Mikroorganismen werden aus den Kohlenhydraten Substanzen gebildet, die durch das Absterben der Mikroorganismen in den Boden gelangen und den verschiedenen Verbindungsklassen angehören. Ein Teil davon beteiligt sich am Aufbau der dunkelgefärbten Stoffe im Humus.

Abbau der Eiweißstoffe. Auch der Abbau der mit den abgestorbenen Pflanzen in den Boden gelangenden Eiweißstoffe vollzieht sich verhältnismäßig rasch. Diese werden zu Peptiden und Aminosäuren hydrolysiert. Letztere erleiden verschiedene Abbaureaktionen, wie sie von WAKSMAN (1952) zusammengestellt worden sind. Dabei entstehen neben Carbonsäure auch Ammoniak und Amine, die bei der Bildung der Huminsäuren eine wichtige Rolle spielen. Neben dem Abbau des pflanzlichen Proteins entstehen durch das Wachstum der Mikroorganismen deren eigene Proteine, die nach Absterben der Mikroorganismen durch Autolyse ebenfalls wieder Zersetzungsvorgängen unterliegen.

Der Abbau der Eiweißstoffe wird in stärkerem Maße als im Falle der Kohlenhydrate durch die Sorption an oberflächenaktiven Bodenkolloiden vermindert. Hierzu sind Untersuchungen über die Abbaugeschwindigkeit von Proteinen durchgeführt worden (ENSMINGER und GIESEKING 1939), die an Tonmineralen sorbiert waren.

Nicht mit der gleichen Geschwindigkeit wie die Proteine werden die Nukleinsäuren im Boden abgebaut (PEARSON, NORMAN und HO 1941, DYER und WRENSHALL 1941a, 1941b). Deren Abbau kann durch die Sorption der Nukleasen an

Tonmineralen verlangsamt werden (FLAIG und KAUL 1956a, 1956b); es ist ferner festgestellt worden, daß die Nukleinsäuren auch nach einer Sterilisation des Bodens abgebaut werden (ROGERS 1942).

Abbau des Lignins. Die bisherigen Untersuchungen auf dem Gebiet der Bildung der dunkelgefärbten Humusstoffe verdichten sich immer mehr zu der Anschauung, daß diese, jedoch speziell die Huminsäuren, durch Reaktionen von aromatischen Ringsystemen mit stickstoffhaltigen Verbindungen (FLAIG 1955b, KONONOWA und ALEXANDROWA 1959) entstehen. Die aromatischen Verbindungen können aus dem Lignin sowie anderen phenolischen Pflanzeninhaltsstoffen und auch aus den Stoffwechselprodukten der Mikroorganismen stammen. Nach den bisherigen Erkenntnissen ist das Lignin als eine der wichtigsten Quellen für die Zurverfügungstellung der aromatischen Ringsysteme anzusehen.

Die hier aufgeführten Untersuchungen und die aus ihnen gezogenen Rückschlüsse werden durch die Untersuchungen von SIMONART und MAYAUDON (1958a, 1958b), MAYAUDON und SIMONART (1958a, 1958b, 1959, 1961) und SIMONART, MAYAUDON und BATISTIC (1958) mit ^{14}C-markiertem pflanzlichem Material bestätigt. Aus der Verteilung der Aktivität von verschieden markiertem pflanzlichem Material nach unterschiedlicher Rottezeit im Boden und Aufteilung der organischen Stoffe nach der in der Humuschemie gebräuchlichen Methode ergibt sich, daß die von der Huminsäurefraktion hydrolysierbaren Anteile, zum größten Teil aus den Aminosäuren und den Kohlenhydraten, die nicht hydrolysierbaren Anteile aus dem Lignin kommen.

Lignin kann durch eine Reihe von Mikroorganismen weitgehend abgebaut werden (STÖCKLI 1952, LINDEBERG 1944, FÅHRAEUS, NILSSON und NILSSON 1949). Hierbei spielen vor allem die im Stoffwechsel dieser Organismen enthaltenen Phenoloxydasen (HOFMANN 1955b, LINDEBERG 1948, LINDEBERG und FÅHRAEUS 1952, DION 1952, GOTTLIEB und GELLER 1949, VLIET 1954) eine wichtige Rolle. Dabei laufen Reaktionen ab, auf die im folgenden noch eingegangen werden soll.

Für die Isolierung von Lignin gibt es verschiedene Methoden. Die Anwendung von Schwefelsäure schließt nicht aus, daß sich das Lignin während der Isolierung

Tabelle 51. *Elementarzusammensetzung der zu verschiedenen Zeitpunkten isolierten Schwefelsäure- und Björkman-Lignine*

	Rottezeit in Tagen	C	H	O	N	S	OCH_3
		%					
H_2SO_4-Lignin	0	58,59	5,60	31,37	0,54	3,9	15,33
	70	58,78	5,93	29,48	2,01	3,8	11,14
	180	58,67	5,91	29,15	2,37	3,9	9,06
	260	58,00	5,83	29,93	3,14	3,9	7,84
	340	54,97	5,96	32,09	3,08	3,9	8,57
	410	54,29	5,86	32,88	2,97	4,0	7,32
Björkman-Lignin	0	60,68	5,79	33,11	0,42		16,76
	70	59,75	7,35	30,93	1,97		10,84
	180	58,81	6,56	31,27	3,36		8,19
	340	57,37	6,95	31,14	4,54		7,50

verändert. Schonende Methoden für die Isolierung des Lignins sind von BRAUNS (1939) und von BJÖRKMAN (1954, 1956) angegeben worden. In beiden Fällen werden organische Lösungsmittel angewandt.

Die Elementarzusammensetzung der Ligninfraktionen, welche einerseits mit Schwefelsäure und zum anderen nach der Methode von BJÖRKMAN mit Dioxan isoliert worden sind, ergibt in beiden Fällen, daß die Kohlenstoffwerte (Broadbent 1954, Flaig, Schobinger, Deuel 1959) ab- und die Sauerstoffwerte zunehmen. Die Zunahme der Stickstoffwerte und die Abnahme der Werte für die Methoxylgruppen läßt auf einen chemischen Zusammenhang zwischen diesen beiden Befunden schließen (Bartlett und Norman 1939, Flaig, Schobinger und Deuel 1959, Flaig 1960). Die Abspaltung der Methoxylgruppen wird auf die Einwirkung der Enzyme der ligninabbauenden Mikroorganismen zurückgeführt (Apenitis, Erdtman und Leopold 1951, Fernandez und Regeiro 1946, Broadbent 1954, Bartlett 1939). Im Verlauf der Rotte nimmt die Austauschkapazität zu (Bartlett und Norman 1939, Broadbent 1954).

Tabelle 52. *Aufteilung und Funktion des Stickstoffgehaltes in den Schwefelsäureligninen aus Weizenstroh nach verschiedener Rottezeit*
(in % der aschefreien Trockensubstanz)

Rottezeit in Tagen	Ursprüngliches Lignin Gesamt-N[1] in %	Hydrolysat α-Aminostickstoff		Hydrolyserückstand		
		in % des ursprünglichen Lignins	in % des Gesamt-N des ursprünglichen Lignins	Gesamt-N in %	NH_3-N in %	„Rest-N" in %
0	0,55	0,32	58,1	0,22	0,22	0
70	2,09	0,82	39,2	0,97	0,58	0,39
180	2,46	0,52	21,1	1,08	0,62	0,46
260	3,26	0,68	20,8	1,38	0,56	0,82
340	3,20	0,65	20,3	1,67	0,82	0,85
410[2]	2,97	0,52	17,5	1,77	0,37	1,50

[1] N % der organischen Substanz, bestimmt nach Dumas.
[2] Mit zusätzlichen Nährsalzen verrottet.

Nach Hydrolyse mit 6-n-Salzsäure ist ein Teil des Stickstoffs als α-Aminostickstoff bestimmbar. Dieser nimmt im Laufe der Rotte zu, jedoch in Prozent zu dem Gesamtstickstoff ab. Dieser Befund bedeutet, daß der als Aminosäure oder als Peptide bei der oxydativen Veränderung des Lignins ankondensierte Stickstoff entweder immer mehr in Kondensationsreaktionen einbezogen wird, oder daß Peptidketten durch die Mikroorganismen im Laufe der Rotte abgebaut werden. Der Stickstoffgehalt der Hydrolyserückstände nimmt im Laufe der Rotte zu. Wird der Hydrolyserückstand unter reduzierenden Bedingungen in einem Gemisch von Kaliumhydroxyd und Natriumacetat in einem Eisentiegel auf 250 bis 270° erhitzt (Flaig, Schobinger und Deuel 1959, Breyhan 1956); so wird mit der Zunahme der Rottezeit in zunehmender Menge Ammoniak abgespalten. Dieses stammt vermutlich aus schwer hydrolysierbaren Anteilen, wie aus Vergleichsuntersuchungen zu schließen ist. Nach dem Ausäthern der wässerigen Lösung der Schmelze lassen sich verschiedene Phenol- und Indolverbindungen nachweisen (Gorham 1959, Flaig, Schobinger und Deuel 1959, Flaig und Breyhan 1956). Der in dem Rückstand der Schmelze vorhandene Reststickstoff ist vermutlich heterocyclisch gebunden. Dieser Anteil des Stickstoffs nimmt im Verlauf der Rotte absolut und prozentual zu. Er scheint zu einer gewissen Stabilisierung des Moleküls beizutragen, da die Verluste an Gesamtsubstanz bei der Hydrolyse mit dessen Ansteigen immer geringer werden.

Die aus den Rotteprodukten als Ligninfraktionen abgetrennten Anteile können nicht mehr als Lignin bezeichnet werden, da ihr Stickstoffgehalt zu hoch und ihr

Methoxylgehalt zu niedrig ist. Sie nehmen mit dem Verlauf der Rotte eine immer dunklere Färbung an und werden auch in anderen Eigenschaften immer mehr den Huminsäuren ähnlich (FLAIG, SCHOBINGER und DEUEL 1959).

Die Zunahme der Sorptionskapazität der Ligninfraktionen besagt, daß im Laufe der Rotte immer mehr saure Gruppen auftreten. Ferner findet man eine Vermehrung der decarboxylierbaren Gruppen (DEUEL und DUBACH 1958a, 1958b, DEUEL, DUBACH und BACH 1958).

Tabelle 53. *Saure und decarboxylierbare Gruppen in den Björkman-Ligninen aus Stroh nach verschiedener Rottezeit*

(Titration mit 0,014 n Na-Colaminat und Decarboxylierung mit 12prozentiger Salzsäure; Werte bezogen auf aschefreie Trockensubstanz)

Rottezeit in Tagen	Saure Gruppen m Äquiv./100 g	Decarboxylierbare Gruppen m Mol CO_2/100 g	CO_2-Abspaltung in % der sauren Gruppen
0	178	10,6	6
70	173	14,5	8
180	233	30,4	13
340	243	—	—

Die abgespaltene Menge CO_2 kann aus dem Abbau der Propanseitenketten der Ligninbausteine oder aus den durch oxydative Aufspaltung des Ringes entstehenden Carbonsäuren stammen. Nach den Untersuchungen von DEUEL, DUBACH und BACH (1958), DEUEL und DUBACH (1958a, 1958b) handelt es sich bei den decarboxylierbaren Verbindungen vermutlich um Ketocarbonsäuren, ungesättigte, heterocyclische und aromatische Hydroxycarbonsäuren.

Tabelle 54. *Methylierung von Lignin-Fraktionen nach Björkman*

Lignin aus frischem Stroh	16,76% OCH_3
Lignin aus frischem Stroh methyliert	21,76% OCH_3
Lignin 70 Tage verrottet	10,84% OCH_3
Lignin 70 Tage verrottet methyliert	13,97% OCH_3

Wie schon erwähnt, nimmt der Methoxylgehalt der Ligninfraktionen im Laufe der Rotte ab. Dabei könnte man zunächst daran denken, daß durch die Methylätherspaltung zusätzliche, phenolische OH-Gruppen entstehen. Remethyliert man Ligninfraktionen aus gerottetem Stroh, so nimmt der Methoxylgehalt zwar zu, aber das Verhältnis der Menge an Methoxylgruppen von dem jeweiligen Ausgangsprodukt zu dem entsprechenden Produkt bleibt immer das gleiche, in dem angegebenen Beispiel ungefähr 4:3. Damit ist erwiesen, daß sich die Anzahl der phenolischen OH-Gruppen durch die Spaltung der Phenolmethyläther nicht erhöht. Die entstandenen Phenole sind sehr reaktionsfähig und reagieren u. a. unter Kondensation mit stickstoffhaltigen Verbindungen weiter. In dem angegebenen Beispiel ist der Gesamtstickstoffgehalt der Ligninfraktionen im Verlauf von 70 Tagen von 0,55% auf 2,01% angestiegen.

Spektrographische Untersuchungen an Ligninfraktionen. Die für das native Lignin charakteristischen Maxima (für Fichtenholz bei 276 und 350 mμ; JONES 1949) nehmen im Verlauf der Rotte ab. Die Abnahme der Absorption im Bereich von 350 mμ dürfte auf die oxydative Veränderung der Phenylpropan-Seitenketten zurückgeführt werden. Der Verlauf der Absorption wird mit der Zunahme der Rottezeit dem von Huminsäuren immer ähnlicher.

Die Infrarotspektren der aus den Rotteprodukten isolierten Ligninfraktionen verändern sich insbesondere im Spektralbereich von 5,5 bis 9,0 μ (BREGER 1951), auch im längerwelligen Bereich treten noch einige Veränderungen auf.

Die Erhöhung der Untergrundabsorption deutet darauf hin, daß die erhaltenen Produkte im Verlauf der Rotte immer mehr aus unterschiedlich substituierten phenolischen Substanzen zusammengesetzt sind (FLAIG 1962).

Im Verlauf der Rotte nimmt die Menge an Carboxylgruppen zu. Hierfür spricht auch die Erhöhung der Absorption bei 5,85 μ. Dieser Befund steht im Einklang mit der Erhöhung der Austauschkapazität der Ligninfraktionen im Verlauf der Rotte.

Die Banden im Bereich von 6,25 bis 6,65 μ verbreitern sich mit zunehmender Rottezeit. Die Stärke der Aromatenbanden ändert sich jedoch wenig. Demnach scheinen die Produkte aus immer mehr unterschiedlich substituierten aromatischen Verbindungen zusammengesetzt zu sein.

Die Änderungen der Banden von 6,85 und 7,05 μ, die den CH-Deformationsschwingungen zugeordnet werden, kann mit dem Abbau der Seitenketten im Lignin erklärt werden. Die erhöhte Absorption bei 8,9 μ (Ätherbindungen) kann darauf zurückgeführt werden, daß die Syringylkomponente sich gegenüber der Guajacylkomponente anreichert, wie mit Hilfe von Modelluntersuchungen wahrscheinlich gemacht werden konnte.

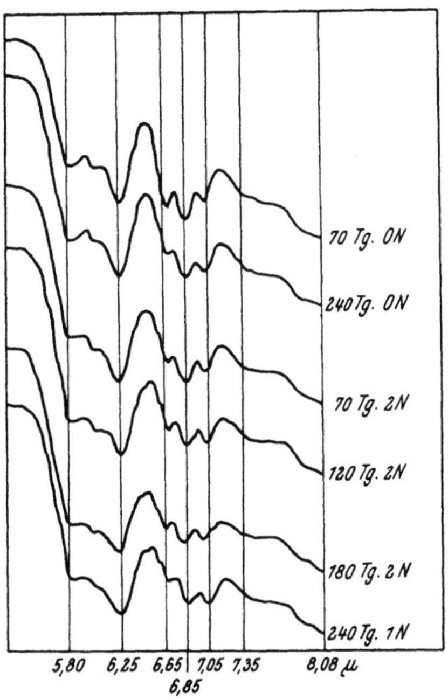

Abb. 87. IR-Spektren von Ligninfraktionen verschiedener Rottezeit im Bereich von 5,25 bis 8,00 μ

Werden den Rotteansätzen unterschiedliche Mengen an anorganischen Stickstoffverbindungen zugesetzt, so verbreitern sich die Banden der Ligninfraktionen im Bereich von 5,85 bis 6,00 μ und von 6,10 bis 6,20 μ mit zunehmender Rottezeit und mit zunehmendem Stickstoffgehalt. Die Banden im Bereich von 5,85 bis 6,00 μ können durch verschiedenartige Carbonylgruppen verursacht werden. Die Banden im Bereich von 6,10 bis 6,20 lassen sich NH-Deformationsschwingungen zuordnen. Die Verbreiterung der Banden steht in einem Zusammenhang mit dem Stickstoffgehalt der Ligninfraktionen. Vergleicht man zwei Ligninfraktionen mit gleichem Stickstoffgehalt, wie 180 Rottetage/2 N (1,6% N) und 240 Tage/1 N (1,70%), so sind die Kurven nahezu gleich. (1 N und 2 N bedeutet 0,5 bzw. 1,0% N als NH_4NO_3 auf Strohtrockensubstanz zu dem Rotteansatz von Roggenstroh.)

Bei der Isolierung des Lignins mit Schwefelsäure ist die Möglichkeit nicht auszuschließen, daß ein Gemisch von Proteinen und anoxydiertem Lignin isoliert wird. Zur Untersuchung dieser Fragen sind Mischungen von Lignin aus frischem Stroh mit Albumin mit dem gleichen Stickstoffgehalt wie die Ligninfraktionen zu den verschiedenen Rottezeiten hergestellt worden.

Der Stickstoffgehalt der Ligninfraktion von 240 Rottetagen/2 N entspricht einem Proteingehalt von maximal 11%. Der Vergleich des Spektrums der Ligninfraktionen mit dem des Gemisches von Lignin mit 11% Albumin läßt eine stärkere

Untergrundabsorption im Spektrum des Gemisches erkennen und ergibt keine Veränderung wie im Falle von stickstoffhaltigen Ligninfraktionen. Dieser Befund kann damit gedeutet werden, daß sich der Stickstoffgehalt der Ligninfraktionen nicht aus einem Gemisch von Lignin und Protein zusammensetzt. Daß eine Reaktion zwischen dem im Laufe der Verrottung veränderten Lignin und stickstoffhaltigen Verbindungen stattfindet, kann auch damit begründet werden, daß sich die Infrarotspektren der Ligninfraktionen mit der Zunahme von deren Stickstoffgehalt im Laufe der Rotte in anderer Weise verändern als die von Gemischen von Lignin des Ausgangsmaterials mit verschiedenen Mengen an Albumin. In dem Gebiet von 6,10 bis 6,20 μ liegen auch C = N-Schwingungen. Die Veränderungen in den Spektren können somit auch auf heterocyclisch gebundenen Stickstoff zurückgeführt werden, wie Vergleichsuntersuchungen mit Hydrolyserückständen der Ligninfraktionen ergeben.

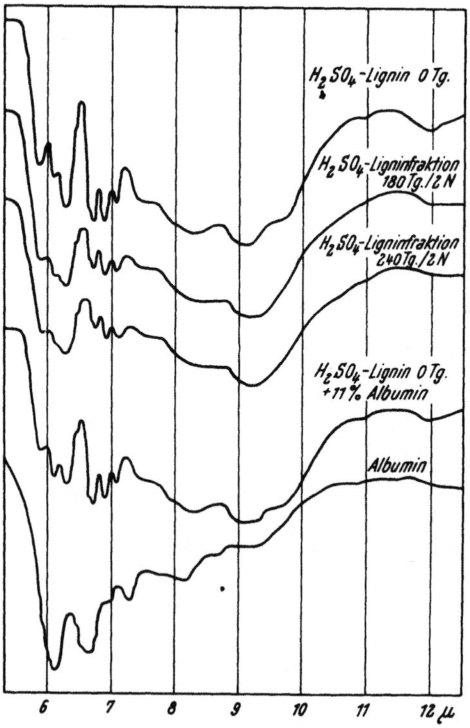

Abb. 88. Vergleich der Spektren von Ligninfraktionen verschiedener Rottezeit und Gemischen aus Lignin und Albumin

In letzter Zeit haben sich auch JENKINSON und TINSLEY (1960) mit dem Lignin-Protein-Komplex befaßt und diesen ausführlich untersucht. Sie finden je nach Herkunft unterschiedliche Lignin-Protein-Komplexe, die aber die gleiche Zusammensetzung an α-Aminosäuren besitzen.

Der Abbau anderer pflanzlicher Inhaltsstoffe. Das Chlorophyll wird nur wenig abgebaut und läßt sich in unterschiedlichen Mengen aus verschiedenen Herkommen extrahieren (GORHAM 1959, BENADE 1954). Fette, Harze und Wachse, wenn sie auch nur in geringen Mengen vorkommen, sollen sich in gewissen Grenzen anreichern können, wenn sich auch anscheinend die Oberflächenwachse der Coniferennadeln (HÄRTEL und PAPESCH 1955) verhältnismäßig rasch zersetzen lassen.

Von den organischen Phosphorverbindungen ist das Phytin das am schwersten zersetzliche (PEARSON, NORMAN und HO 1941, DYER und WRENSHALL 1941a, 1941b, WRENSHALL und DYER 1941). Die Zersetzung schwefelhaltiger organischer Verbindungen wurde von FREDERICK, STARKEY und SEGAL (1957) untersucht.

C. Regelmäßigkeiten bei der Veränderung der Pflanzeninhaltsstoffe im Verlauf der Rotte

Bestimmt man die Elementarzusammensetzung von gerottetem Roggenstroh, dem 1% Ammoniumnitrat als Stickstoffquelle zugegeben worden ist, im Laufe der Rotte und errechnet daraus die Abnahme der Elemente C, H, O in Gramm-Atomen, so bleibt nach 70 Tagen das Verhältnis Kohlenstoff : Sauerstoff weitgehend konstant. Auf 6 Kohlenstoffatome kommen ungefähr 3,3 Sauerstoffatome. Das Verhältnis von Kohlenstoff: Wasserstoff nimmt von 6:9,5 im Verlauf der Rotte auf 6:7,9 ab (FLAIG 1962).

BROADBENT (1954) stellte fest, daß die Anzahl der Carboxylgruppen und die Austauschkapazität in mval/100 g zunimmt.

Extrahiert man die Rotteprodukte mit heißem Wasser, so kann man auf Grund des Methoxylgehaltes und des organischen Stickstoffgehaltes des Extraktes Aussagen über die Menge an Ligninabbauprodukten sowie an wasserlöslichen Peptiden oder Aminosäuren machen.

In Ansätzen, die ohne Stickstoffgabe verrottet worden sind, besteht der organische Anteil der Heißwasserextrakte zu ungefähr 3% aus Aminosäuren und zu 30% aus Ligninabbauprodukten, die noch Methoxylgruppen tragen. Da jedoch papierchromatographisch gezeigt werden konnte, daß sich unter den Ligninabbauprodukten im Wasserextrakt auch solche befinden, die keine Methoxylgruppe tragen, wie z. B. p-Hydroxy-benzaldehyd, muß die Gesamtmenge an Ligninabbauprodukten im Extrakt noch größer sein als 30%. Fügt man den Rotteansätzen etwa 1% vom Strohtrockengewicht an Stickstoff in Form von Ammonnitrat zu, so ist der Gehalt an wasserlöslichen Peptiden oder Aminosäuren, bezogen auf den organischen Anteil der Wasserextrakte, beachtlich höher und ist mit 25 bis 30% einzusetzen. Der Gehalt an Ligninabbauprodukten mit Methoxylgruppen ist ungefähr 7%. Der größere Gehalt an Aminosäuren bei Zufuhr von Stickstoff zu den Rotteansätzen kann seine Deutung darin finden, daß ein stärkeres Mikroorganismenwachstum vorliegt. Dieses geht auch aus dem stärkeren Abbau der organischen Substanz hervor. Die geringere Menge an Ligninabbauprodukten läßt sich damit erklären, daß durch das größere Angebot an stickstoffhaltigen Verbindungen als Peptide oder Aminosäuren sowie deren Abbauprodukte Kondensationsreaktionen stattgefunden haben, die zu wasserunlöslichen Produkten führen.

Eine Stütze für diese Befunde ist die Tatsache, daß in den Rückständen der wässerigen Extraktion der Methoxylgehalt größer ist, als der Menge an isolierten Ligninfraktionen entspricht.

Die Rückstände der Rotteprodukte, die mit Äther und Wasser extrahiert worden sind, enthalten hauptsächlich höhermolekulare Anteile.

Addiert man die daraus isolierten Mengen an Holocellulose, Lignin, Asche und das aus den Stickstoffwerten berechnete Protein, so beträgt deren Summe während der ganzen Rottezeit 95 bis 97%. Die Berechnung des Proteins aus dem Stickstoffgehalt ist jedoch nicht völlig korrekt, da auch „heterocyclisch“ gebundener Stickstoff vorliegt.

Errechnet man aus den Elementaranalysen der Ligninfraktionen die prozentuale Abnahme von Kohlenstoff, Wasserstoff, Sauerstoff und Methoxyl in Gramm-Atomen, so ist die von Kohlenstoff und Wasserstoff nahezu gleich, während sich die von Sauerstoff gegen Ende der Rotte vermindert.

Die stärkere Abnahme von Kohlenstoff und Wasserstoff gegenüber der von Sauerstoff deutet darauf hin, daß zunächst die Seitenketten abgebaut werden. Die jeweilig auftretenden Differenzen im Laufe der Rotte zwischen Kohlenstoff, Wasserstoff und Sauerstoff setzen sich zusammen aus

a) den abgespaltenen Kohlenstoff-Wasserstoff- und Sauerstoffatomen und

b) den Sauerstoffatomen, die im Verlaufe der Oxydation in Ligninfraktionen aufgenommen werden, und

c) den Kohlenstoff-, Wasserstoff-, Sauerstoffatomen, die durch die Kondensation stickstoffhaltiger Verbindungen bei den Ligninfraktionen hinzukommen.

Errechnet man aus den Stickstoffwerten der Ligninfraktionen die Menge an hinzukommenden stickstoffhaltigen Verbindungen in Form von Proteinen, so stellt man fest, daß diese „Proteinwerte" keinen großen Einfluß auf die Differenzen an Kohlenstoff-, Wasserstoff- und Sauerstoffatomen haben, die bei den einzelnen Rottestufen auftreten.

Überschlagsmäßig läßt sich errechnen, daß der Gewichtsverlust an Lignin zu $^2/_3$ auf die Abspaltung der Seitenketten und zu $^1/_3$ auf die Ringspaltung zurückgeführt werden kann, wobei die aus den Methoxylwerten sich ergebende Entmethylierung berücksichtigt ist. Diese Aussagen bedeuten, daß sich die Ligninfraktionen im Verlaufe der Rotte prozentual immer mehr aus Verbindungen mit aromatischen Ringen zusammensetzen. Diese Überlegungen besagen jedoch nicht, daß die im Verlaufe der Rotte entstehenden Produkte in ihrer Zusammensetzung immer mehr aus einheitlichen Polymeren bestehen. Breger (1951) hat mittels physikalisch-chemischer Methode zum erstenmal auf die strukturelle und chemische Verwandtschaft zwischen Lignin und Huminsäuren hingewiesen.

D. Beteiligung der Mikroorganismen bei der Bildung von Huminsäuren

Nicht nur bei dem Abbau der organischen Substanz im Boden, sondern auch bei der Bildung von Humusstoffen kommt der Mitwirkung der Mikroorganismen eine große Bedeutung zu. Bei der Autolyse von Mikroorganismenkulturen entstehen in vielen Fällen dunkelgefärbte Stoffe, die ein ähnliches Verhalten wie Huminsäuren zeigen (Küster 1952). Ein Teil dieser Stoffe wird aus gefärbten Stoffwechselprodukten gebildet (v. Plotho 1951, Laatsch, Hoops und Bauer 1951); die aus diesen Stoffen entstehenden Verbindungen sind stickstoffhaltig (Scheffer 1953, Kumada und Hotta 1958). Der größte Teil des Stickstoffs liegt als Aminosäuren vor. So wurden von verschiedenen Seiten Untersuchungen über die dunkelgefärbten Stoffe angestellt (Thom und Phillips 1932), denn diese traten bei verschiedenen Mikroorganismen, wie z. B. bei Azotobacter chroococcum (Rippel und Ludwig 1925), Aspergillus niger (Bortels 1927) und verschiedenen anderen Pilzen, auf. Der Gehalt dieser dunkelgefärbten Produkte kann bis zu 50% des Mycelgewichtes betragen (Thom und Phillips 1932, Pinck und Allison 1944, Springer und Wagner 1952). Derartige Beobachtungen sind von verschiedenen Seiten gemacht worden (Laatsch, Bauer und Bieneck 1950). Die braunen Stoffe entstehen auch dann, wenn keine aromatischen Verbindungen oder Lignin in den Nährmedien vorhanden sind. Sie bilden sich im Laufe der Autolyse (Behr 1930, Enders 1943, Scheffer, v. Plotho und Welte 1950). Durch mikroskopische Beobachtung ist festgestellt worden, daß auch bei der Zersetzung junger grüner Pflanzenteile ebenfalls Humusstoffe gebildet werden (Kononowa 1958). Waksman (1928) hatte schon vor langer Zeit geschildert, daß auch aus isolierter Cellulose dunkelgefärbte, den Humusstoffen ähnliche Verbindungen entstehen.

Verschiedene Untersuchungen über die Bildung der braungefärbten Stoffe, die eine Ähnlichkeit mit den Huminsäuren haben, sind auch in Abhängigkeit von gefärbten Stoffwechselprodukten von Mikroorganismen gemacht worden. So fand man z. B., daß bei Aspergillus fumigatus, dessen gefärbtes Stoffwechselprodukt ein Chinon, Fumigatin, ist, in dessen Kulturlösungen im Laufe der Autolyse schwarzbraune Produkte entstehen. Mittels der Tannin-Reaktion nach Bavendamm (1928) konnten Phenoloxydasen nachgewiesen werden. Im Falle von Penicillium spinulosum, dessen gefärbtes Stoffwechselprodukt das Spinulosin, ein 2,5-Dihydroxychinon ist, fand weder die Bildung der braunen Produkte statt, noch konnten mittels der Tannin-Reaktion Phenoloxydasen nachgewiesen werden (KÜSTER 1953). Ähnliche Beobachtungen über die Bildung von gefärbten Stoffwechselprodukten sind auch bei verschiedenen Streptomycetenarten gemacht worden (KÜSTER 1955). In einigen Fällen (KÜSTER 1958) konnte nachgewiesen werden, daß die Braunfärbung der Kulturlösungen von der Aktivität der Phenoloxydasen abhängig war. Da nur Phenoloxydasen in der Lage sind aromatische Verbindungen zu oxydieren, muß auch in den Fällen, in denen keine aromatischen Verbindungen als Stickstoff- oder Kohlenstoffquelle den Kulturlösungen der Mikroorganismen zugesetzt waren, die Entstehung der dunkelgefärbten Stoffe über eine aromatische Zwischenstufe verlaufen. Diese Tatsache verdient insofern festgehalten zu werden, als damit die Möglichkeit besteht, daß die in ihrem Verhalten den Huminsäuren ähnlichen Stoffe aus Mikroorganismenkulturen bei Anwesenheit von aromatischen oder nichtaromatischen Kohlenstoff- oder Stickstoffquellen gebildet werden können. Diese Untersuchungen bedürfen jedoch noch weiterer Bearbeitung, da es bisher noch nicht versucht worden ist, die aromatischen Anteile der Huminsäuren, die durch die Mikroorganismentätigkeit in den Kulturlösungen gebildet werden, nachzuweisen.

Diese aromatischen Zwischenstufen können einmal aromatische Aminosäuren sein, wie sie im Stoffwechsel der Mikroorganismen entstehen, zum anderen wäre es möglich, daß die aromatischen Verbindungen über Shikimisäure (BROWN und NEISH 1955) im Stoffwechsel gebildet werden. Ein weiterer Weg der Bildung aromatischer Verbindungen geht über die aktive Essigsäure aus Acetyl-Coenzym-A, wie GRISEBACH (1957) im Falle von Cyanidin gezeigt hat. Ferner ist bekannt, daß einige Mikroorganismen (BRUCKER und DREHMANN 1958) in ihrem Stoffwechsel Gallussäure und andere Phenolcarbonsäuren bilden (GROSS 1958, METZENBERG und MITCHELL 1958), wie Phycomyces blakesleanus.

4. Extraktion und Fraktionierung von Humusstoffen

Zur Extraktion von Humusstoffen wurden zahlreiche Lösungsmittel angewandt (SCHEFFER und ULRICH 1960, TINSLEY und SALAM 1961). Seit der ersten Isolierung von Huminsäuren aus einem Torfboden mit Natronlauge durch ACHARD (1786) war es das Ziel der Humuschemiker, diese schwarze, für den Boden offenbar typische Substanz rein darzustellen, ihre Eigenschaften zu untersuchen und ihre chemische Konstitution aufzuklären. Gegen die Verwendung von Natronlauge als Extraktionsmittel wurden später verschiedene Einwände erhoben: 1. Stärker alkalische Lösungsmittel lösen auch noch nicht umgesetztes pflanzliches Material, z. B. Hemicellulose und Lignine. 2. Unter alkalischen Bedingungen wird das gelöste Material in Gegenwart von Luft oxydiert. 3. Auch andere chemische Veränderungen könnten in alkalischer Lösung stattfinden, z. B. hydrolytische Spaltungen, aber auch Kondensationen. Aus diesen Gründen wurden schonendere Extraktionsmittel herangezogen. SIMON (1929) zeigte, daß die Alkalisalze kalkfällender Säuren, wie Natriumfluorid und Natriumoxalat, die im Humus vorliegenden unlöslichen Humate in wasserlösliche Alkalihumate überführen. BREMNER und LEES (1949) fanden bei ihrem systematischen Vergleich der Lösungen von 18 verschiedenen Natriumsalzen, darunter auch die von SIMON angewandten, daß 0,1 m Natriumpyrophosphatlösung am wirksamsten war. Nach TINSLEY und SALAM (1961) ist Natriumsulfit bei vergleichbarer Konzentration und pH-Wert ebenso wirksam wie Pyrophosphat.

Unter dem Gesichtspunkt, daß die sauren Humussubstanzen nicht nur als Calcium-, sondern auch als Eisen- und Aluminiumhumate festgelegt sein können, wurden, speziell für die Extraktion von Podsolen, Lösungen von Chelatbildnern angewandt. So konnten SCHNITZER, WRIGHT und DESJARDIN (1958) mit ÄDTE-

Na_2-Lösungen die Hauptmenge der organischen Substanz aus dem B_{21}-Horizont eines Podsols extrahieren, während mit dem gleichen Lösungsmittel aus dem A_0-Horizont nur geringe Mengen herausgelöst wurden. Entsprechend extrahierten MARTIN und REEVE (1957a, 1957b) beträchtliche Mengen an organischem Material aus Podsol-B-Horizonten mit wässerigen oder wässerig-acetonischen Lösungen von Cupferron, 8-Hydroxy-chinolin und Acetylaceton.

Es sind zahlreiche Versuche unternommen worden, um die nach einem der angegebenen Verfahren erhaltenen Humusstoffextrakte in charakteristische Fraktionen aufzuteilen. So erfolgt in der Regel durch Ansäuern der alkalischen oder neutralen Extrakte eine Auftrennung in einen säurefällbaren (Huminsäuren) und einen säurelöslichen (Fulvosäuren) Anteil. Durch fraktionierte Extraktion mit Pufferlösungen (SIMON 1934) oder fraktionierte Fällung mit Elektrolyten (SIMON 1935, SPRINGER 1938, FLAIG, SCHEFFER und KLAMROTH 1955) kann die säurefällbare Fraktion weiter unterteilt werden. Durch Behandeln einer schwach essigsauren Huminsäurelösung mit Magnesiumsulfat werden die beiden Fraktionen A-Typ- und B-Typ-Huminsäuren erhalten, durch Fällen schwach alkalischer Huminsäurelösungen mit Natriumchlorid oder Natriumhydroxid die Fraktionen „Grauhuminsäuren" und „Braunhuminsäuren".

Nach HOPPE-SEYLER (1889) läßt sich aus dem säurefällbaren Anteil alkalischer Humusextrakte mit Alkohol eine alkohollösliche Fraktion (Hymatomelansäuren) abtrennen. ZETZSCHE und REINHART (1937) zeigten, daß auch höhere Alkohole Anteile von Huminsäuren lösen. Dabei nimmt die extrahierte Menge mit wachsender Anzahl der Kohlenstoffatome im Alkohol ab. Das Lösungsvermögen der höheren Alkohole wird durch Zusatz geringer Mengen einer Säure erhöht. Die mit diesen Lösungsmitteln erhaltenen Fraktionen werden ebenfalls als Hymatomelansäuren bezeichnet. Auch andere mit Wasser mischbare organische Lösungsmittel wie Aceton, Dioxan oder Tetrahydrofuran stellen mehr oder weniger wirksame Lösungsmittel für Anteile von Huminsäurepräparaten dar, wobei das Lösungsvermögen vom Wassergehalt des Lösungsmittels abhängt (BURKAT 1959).

Die verschiedenen Humusfraktionen, die nach den beschriebenen Methoden erhalten und zum Teil in der Literatur mit besonderen Namen bezeichnet werden, sind im allgemeinen Stoffgemische. Es ist daher kaum zu erwarten, daß nach gleicher Methode aus verschiedenen Böden hergestellte Fraktionen die gleiche chemische Zusammensetzung besitzen (SCHLICHTING 1953). Eine vergleichende Charakterisierung von Böden auf Grund einer Fraktionierung nach einer einheitlichen Methode bleibt solange problematisch, wie die chemische Zusammensetzung dieser Fraktionen nicht bekannt ist. Aus diesem Grunde wird versucht, Unterschiede zwischen solchen nach gleicher Methode aus verschiedenen Böden erhaltenen Fraktionen zahlenmäßig zu erfassen. Dieses Verfahren ist nur sinnvoll, solange qualitativ gleich zusammengesetzte Gemische verglichen werden. Es hat daher nicht an Versuchen gefehlt, durch weitere Fraktionierung einen Einblick in die chemische Zusammensetzung derartiger Gemische zu gewinnen.

Die Fulvosäurenfraktionen aus verschiedenen Böden wurden von FORSYTH (1947a) durch selektive Adsorption an Tierkohle in vier Fraktionen aufgetrennt. Die eine Fraktion enthielt wasserlösliche organische Substanzen wie Zucker und Aminosäuren, die zweite phenolische Glykoside oder Tannine und die dritte Polyuronide. Die vierte Fraktion enthielt stickstoffreiche Verbindungen, Pentosane und organische Phosphorverbindungen. KONONOWA und BELCIKOWA (1960) konnten dialysierte Fulvosäuren aus Rasenpodsol papierchromatographisch mit Butanol-Eisessig-Wasser in zwei Fraktionen auftrennen, die Unterschiede in der Elementaranalyse zeigten.

Hymatomelansäuren aus Braunkohle wurden von Scheele (1937) in einen acetonlöslichen und einen acetonunlöslichen Anteil aufgetrennt. Die beiden Präparate zeigten Unterschiede in der Elementaranalyse. Zetzsche und Reinhart (1937) konnten aus Torfhuminsäuren extrahierte Hymatomelansäuren durch Kombination von Extraktionen mit 1%iger Natriumacetatlösung, Benzol-Alkohol-Gemisch in zwei Haupt- und zwei Nebenfraktionen zerlegen. Die Hauptfraktionen zeigten Unterschiede in der Elementaranalyse. Trojanowski (1957) erhielt bei der Papierchromatographie von Hymatomelansäuren aus Komposten und aus Niedermoortorf mit Butanol-Aceton-0,1 n HCl zwei Komponenten, die als α- und β-Hymatomelansäuren bezeichnet werden und Unterschiede in der Elementaranalyse sowie im potentiometrisch und konduktometrisch ermittelten Äquivalentgewicht zeigen.

Eine präparative Auftrennung von A-Typ- und B-Typ-Huminsäuren aus verschiedenen Böden in drei bzw. vier Fraktionen führten Hayashi und Nagai (1960) mittels selektiver Adsorption an Aluminiumoxid durch. Kononowa (1958) trennte Huminsäuren an Cellulose- und Aluminiumoxidsäulen in drei Fraktionen und papierchromatographisch in zwei Fraktionen auf. Diese Fraktionen zeigen Unterschiede in der Elementaranalyse und im optischen Verhalten.

Auch mit Hilfe der Papierelektrophorese unter Verwendung verschiedener Pufferlösungen lassen sich Huminsäuren in verschiedene Zonen auftrennen. Burges (1960) konnte zeigen, daß die schneller laufende Fraktion eines Huminsäurepräparates aus einem Podsol-B-Horizont nach Ausfällen und Wiederauflösen bei einer erneuten Elektrophorese wieder wie das Ausgangsmaterial in einen langsam und einen schnell laufenden Anteil aufgetrennt wird. Daraus wird der Schluß gezogen, daß die Zonenbildung bei der Papierelektrophorese mehr eine Trennung nach der Teilchengröße als nach chemisch verschiedenen Substanzen ist. Zu einem ähnlichen Schluß kommt Schlüter (1959) bei Sedimentationsmessungen an von Scholz (1959) bei der Chromatographie einer Hydrochinonhuminsäure an Aluminiumoxid erhaltenen Fraktionen.

5. Untersuchungen an natürlichen Huminsäuren

A. Chemische Untersuchungen

Unter Huminsäuren werden in diesem Abschnitt diejenigen Produkte verstanden, die mit Säuren aus alkalischen Bodenextrakten gefällt werden, oder diesen ähnliche, nach anderen Methoden aus Böden isolierte Produkte. Die Ähnlichkeit bezieht sich dabei auf die dunkle Färbung und Löslichkeitseigenschaften. Sie kann entsprechend dem Stand unserer Kenntnis dieser Stoffe nicht näher definiert werden. Es ist das Ziel der chemischen Untersuchungen an den Huminsäuren, zu klären, wie weit die Ähnlichkeit dieser Produkte auf einer ähnlichen chemischen Konstitution beruht und innerhalb welcher Grenzen Unterschiede im chemischen Aufbau bestehen.

Elementarzusammensetzung. Die Elementarzusammensetzung der Huminsäuren ist verschieden; sie scheint von den Bildungsbedingungen abzuhängen. Von den auf S. 384 angeführten Böden ist die Elementarzusammensetzung der Huminsäuren angegeben worden und wird in Tab. 55 aufgeführt.

Im allgemeinen haben die Fulvosäurenfraktionen geringere Kohlenstoff- und höhere Sauerstoffwerte (Kononowa 1956).

Funktionelle Gruppen. Die Huminsäuren enthalten Kohlenstoff, Wasserstoff, Sauerstoff und Stickstoff, gelegentlich auch Schwefel.

Der Sauerstoff kann in Carboxyl-, Carbonyl-, Hydroxy-, Äther- und Estergruppen vorliegen. Für die sauren Eigenschaften sind Carboxyl-Gruppen und

Tabelle 55. *Die Eelementarzusammensetzung von Huminsäuren aus verschiedenen Böden*

Böden, aus denen die Huminsäuren isoliert worden sind	C	H	N	O
Podsoliger Boden	52,39	4,82	3,74	39,05
Rendzina	54,90	4,36	4,07	36,67
Degradierte Schwarzerde	56,34	3,54	3,58	36,65
Tiefe Schwarzerde	57,47	3,38	3,78	35,37
Gewöhnliche Schwarzerde	58,37	3,26	3,70	34,67
Kastanienfarbige Böden	58,56	3,40	4,09	33,95
Podsoliger Boden	56,67	4,79	5,14	33,40
Gewöhnliche Schwarzerde	62,55	2,78	3,32	31,35
Säulenförmige Solonez	59,21	3,83	4,28	32,68

phenolische bzw. enolische Hydroxylgruppen verantwortlich. Im allgemeinen sind Carboxylgruppen saurer als enolische Hydroxylgruppen. Es muß aber bei unbekannten Gemischen, wie sie die Huminsäuren darstellen, auch mit einer Überschneidung der Acidität schwach dissoziierender Carboxylgruppen und stark dissoziierender enolischer Hydroxylgruppen, wie etwa in Hydroxychinon, gerechnet werden. Daher ist eine differenzierende Bestimmung der Carboxyl- und sauren Hydroxylgruppen mit einer auf der Acidität beruhenden Methode nicht möglich. Dazu gehören alle Methoden, die auf Kationenaustausch, Veresterung oder Verätherung beruhen. Es ist nach diesen Methoden möglich, zwischen stärker und schwächer sauren Gruppen zu unterscheiden und diese einzeln zu bestimmen. Die Grenzen zwischen diesen beiden Gruppen ist dabei willkürlich durch die angewandte Methode festgelegt. So werden beim Umsatz mit Calciumacetat die stärker sauren Gruppen erfaßt, beim Umsatz mit Bariumhydroxyd zusätzlich schwächer saure (Stadnikoff 1937). Das gleiche gilt für die Titration mit Natriumaminoaethylat in Dimethylformamid einerseits, in Äthylendiamin andererseits (Wright und Schnitzer 1960). Für eine vergleichende Charakterisierung von Huminsäuren ist das ausreichend, Rückschlüsse auf die chemische Struktur dieser sauren Gruppen lassen derartige Bestimmungen jedoch nicht zu. Dafür bedarf es spezieller Methoden. Eine solche ist für die Bestimmung von Carboxylgruppen die Decarboxylierung mit Bestimmung des gebildeten CO_2. So stellten Broadbent und Bradford (1952) fest, daß 10 bis über 50% der durch Methylierung gefundenen „Carboxylgruppen" durch siedende 12%ige HCl decarboxyliert werden. Wright und Schnitzer (1960) decarboxylierten mit basischem Kupfercarbonat in Chinolin bei 245 bis 250° und erhielten dabei 80 bis 140% der mit Calciumacetat bestimmten sauren Funktionen.

Oxydation und Hydrierung. In zahlreichen Untersuchungen ist versucht worden, durch einen oxydativen Abbau von Huminsäuren einen Einblick in die Konstitution dieser Produkte zu gewinnen. Die Oxydation natürlicher Huminsäuren mit Brom (Feustel und Byers 1936) oder Chlordioxid (Schmidt und Atterer 1927) ergab entweder Chinonderivate, wie Tetrabromchinon, oder Spaltstücke von Chinonen, wie Maleinsäure.

In der Alkalischmelze allein wurden Protocatechusäure und das durch Decarboxylierung daraus hervorgegangene Brenzcatechin (Dragunow, Schelochowtzewa und Strelkowa 1948) sowie Resorcin (Steelink, Berry, Ho und Nordby 1960) erhalten. Die Oxydation mit Nitrobenzol-Alkali liefert in wechselnden Mengen die gleichen Abbauprodukte wie Lignin: p-Hydroxybenzaldehyd, Vanillin und Syringaaldehyd. In Podsol-B-Horizonten sind in der Kaliumhydroxidschmelze und bei der alkalischen Oxydation mit Kupferoxid ebenfalls

Ligninabbauprodukte papierchromatographisch nachgewiesen worden. Bei den von MORRISON (1958) untersuchten Präparaten schwankte der Gehalt an diesen Phenolaldehyden, bezogen auf den gesamten organischen Kohlenstoff je nach Herkunft und Darstellungsmethodik zwischen 0,6% bei einer Huminsäure aus Waldboden und 4% bei einer Huminsäure aus Phragmitestorf.

Eine andere Möglichkeit, durch Oxydation zu definierten Abbauprodukten zu gelangen, geht über eine primäre Nitrierung der Huminsäure und anschließender Oxydation mit Salpetersäure. Auf diesem Wege konnten SCHEFFER und KICKUTH (1961 a, 1961 b) als Oxydationsprodukte des Hydrolyserückstandes einer Zwischenmoorhuminsäure 7,1% aromatische Nitroverbindungen und 18% Phthalsäuren isolieren. Aus Braunkohlenhuminsäuren erhielten FUCHS und STENGEL (1930) weitere Benzolpolycarbonsäuren.

Alle diese Oxydationsversuche zeigen, daß die Huminsäuren aromatische Bestandteile enthalten, von denen ein Teil als unveränderter Ligninbaustein vorliegt. Über die Verknüpfung dieser Bausteine untereinander und mit den aliphatischen Bausteinen können diese Befunde nichts aussagen, zumal sie alle an undefinierten Gemischen durchgeführt wurden.

Bei der Hydrierung mit Nickelkatalysatoren werden Phenole erhalten (KUCHARENKO und SSAWELKEW 1951). Bei Anwendung von Molybdänsulfid findet auch eine gleichzeitige Hydrierung des aromatischen Kerns zu Cyclohexanolderivaten statt (GANZ 1944).

Hydrolyse. Huminsäuren werden bei der Behandlung mit Mineralsäuren in säurelösliche und säureunlösliche Anteile aufgespalten. Im säurelöslichen Anteil finden sich Kohlenhydrate, Aminosäuren und Aminozucker. Der säureunlösliche Anteil zeigt, soweit bisher untersucht, noch die wesentlichen Eigenschaften der Huminsäuren.

FORSYTH (1947b) konnte aus Torfhuminsäuren durch vierstündiges Kochen mit 2%iger HCl 35% des Materials in Lösung bringen. In diesem Hydrolysat befanden sich reduzierende Zucker und ein Drittel des Gesamtstickstoffs. Aus dem Befund, daß nach verschiedenen Methoden gewonnene Fraktionen dieses Huminsäurepräparates bei der Hydrolyse Unterschiede in der Menge des in Lösung gegangenen Hydrolysates zeigten (14 bis 49%), zieht FORSYTH den Schluß, daß diese abhydrolysierbaren Anteile nicht chemisch mit dem Hydrolyserückstand verbunden waren, sondern nur mitgefällt wurden und Adsorptionskomplexe bilden. Ein solcher Schluß wäre begründet, wenn durch die Fraktionierung eine quantitative Trennung in eine hydrolysierbare und eine nichthydrolysierbare Komponente erreicht wäre. Dies ist jedoch auch nicht annähernd der Fall. So bleiben als weitere Möglichkeit offen: Ein Teil des hydrolysierbaren Anteils kann adsorptiv, ein anderer Teil chemisch gebunden sein, dabei kann der chemisch gebundene Anteil gleichmäßig oder ungleichmäßig auf die Huminsäuremoleküle verteilt sein. Im letzteren Falle, der der allgemeinere und wahrscheinlichere ist, lassen sich die Befunde von FORSYTH auch ohne Annahme eines adsorptiv gebundenen Anteils deuten. SCHEFFER und KICKUTH (1961a, 1961b) differenzierten die abhydrolysierbaren Kohlenhydrate einer Huminsäure aus oligotrophem Zwischenmoor durch Extraktion mit Kupferoxid-Äthylendiamin in 16% freie Cellulose, 9% gebundene Cellulose und 16% gebundene Einfachzucker. Das Hydrolysat enthielt in diesem Falle keine Aminosäuren.

Im allgemeinen enthalten Huminsäuren aus Böden bis zu 5% Stickstoff. Etwa die Hälfte des Gesamtstickstoffs befindet sich im sauren Hydrolysat. 30 bis 40% dieses Stickstoffs im Hydrolysat entfielen bei den von BREMNER (1955b) bzw. YOUNG und MORTENSEN (1958) untersuchten Huminsäuren auf Aminosäuren, weitere 3 bis 10% auf Aminozucker. Systematische Untersuchungen über die aus

Huminsäuren abhydrolysierbaren Aminosäuren wurden von Bremner, Flaig und Küster (1955), Bremner (1955c), Kononowa und Alexandrowa (1956, 1958), Okuda und Hori (1954, 1955), Sowden (1955, 1958) und Scharpenseel und Krausse (1962) durchgeführt. Eindeutige Zusammenhänge zwischen Zusammensetzung bzw. Menge der abhydrolysierbaren Aminosäuren und der Herkunft der Huminsäurenpräparate ergeben sich aus diesen Arbeiten nicht. Ebensowenig zeigen einzelne Huminsäurenfraktionen, wie Grauhuminsäuren und Braunhuminsäuren, charakteristische Unterschiede im quantitativen und qualitativen Aminosäuregehalt (Scharpenseel und Krausse 1962). Der Vergleich der Ergebnisse verschiedener Autoren wird dadurch erschwert, daß der Gehalt an abhydrolysierbaren Aminosäuren auch von der Methodik der Herstellung der Huminsäurenpräparate abhängt.

B. Physikalische Eigenschaften

Nach den bisher vorliegenden physikalisch-chemischen Untersuchungen an den Substanzen in der Huminsäurefraktion sind diese höhermolekular. Die bisher angewandten Methoden haben noch keine Beweise erbracht, daß die Abtrennung dieser Fraktion aus den organischen Stoffen des Bodens nach den bisherigen Verfahren und ihre weitere Auftrennung durch Elektrophorese, chromatographische Verfahren und mittels der Ultrazentrifuge zu chemisch einheitlichen Verbindungen auch im Sinne der hochmolekularen Chemie führt. Außerdem lassen die Methoden noch keine genügende Auftrennung der polymolekularen Humusstoffe in Fraktionen mit möglichst einheitlichem Teilchengewicht zu. Die am Aufbau beteiligten Monomeren der Huminsäuren sind in ihrer chemischen Konstitution noch nicht bekannt. So ist man auch zur Zeit noch nicht in der Lage, darüber zu entscheiden, in welchem Ausmaß der in den als Huminsäuren bezeichneten Fraktionen enthaltene Stickstoff als integrierender Bestandteil am Aufbau des Moleküls beteiligt ist oder in welchem Ausmaß er zufällig in Form von polymeren Aminosäuren in die Reaktion einbezogen worden ist. Die Unsicherheit in den Aussagen besteht auch bei den auf synthetischem Wege hergestellten Huminsäuren. Bei der oxydativen Polymerisation von Polyphenolen können Oxydationprodukte der Ausgangsverbindungen entstehen, die sich je nach der Menge an verbrauchtem Sauerstoff in ihrem Oxydationsgrad unterscheiden.

Die bekannt gewordenen Strukturelemente, wie saure Gruppen als Carboxyl- oder phenolische Gruppen, und die verschiedenen Funktionen des Stickstoffs im Molekül der Huminsäuren können in Analogie zu anderen Naturstoffen in unterschiedlichem Maße Bindungen durch Nebenvalenzen bedingen, so daß die Bestimmung der Molekülgröße und der Molekulargewichte dadurch beeinträchtigt wird. Es entspricht daher mehr den Tatsachen, zunächst noch von Teilchengrößen und -gewichten zu sprechen. Die Ergebnisse der bisher bekannten physikalisch-chemischen Messungen sind aus den genannten Gründen in vielen Fällen nicht übereinstimmend.

Elektronenspektren von Huminsäuren. Im Gegensatz zu den Spektren der meisten höhermolekularen Stoffe zeigen die Elektronenspektren von Huminsäuren (Huminstoffe) keine Maxima, sondern eine nach kürzeren Wellen monoton ansteigende Kurve, die, als Wellenlängen gegen den Logarithmus der Extinktion aufgetragen, annähernd linear verläuft.

Direkte Schlüsse auf die Konstitution der Huminsäuren können aus diesen Spektren nicht gezogen werden. Immerhin liefern sie Meßwerte, die zur Charakterisierung und quantitativen Bestimmung herangezogen werden: Die Extinktion bei einer beliebig gewählten Wellenlänge und die Steigung der logarithmischen

Extinktionskurve. Die eine ist ein Maß für die Farbtiefe, die andere für den Farbton einer Huminsäurelösung. Die Spektren von Huminsäuren verschiedener Darstellung und Herkunft unterschieden sich im allgemeinen durch Farbtiefe und Farbton. Es scheint jedoch ein Zusammenhang zwischen beiden Faktoren zu bestehen, nach dem einem steileren Verlauf der Absorptionskurve, zumindest im sichtbaren Gebiet, eine geringere Extinktion entspricht (KUMADA 1955). Nach dem bisher vorliegenden Versuchsmaterial gilt diese Besprechung aber wohl nicht so exakt, daß man darauf eine quantitative Huminsäurebestimmung durch Messung der Extinktion bei zwei Wellenlängen gründen könnte (WELTE 1956). SPRINGER unterschied auf Grund der Sichtbarspektren zwei Typen von Huminsäuren; diejenigen mit hoher Extinktion und kleiner Steigung der Absorptionskurve bezeichnete er als Grauhuminsäuren, diejenigen mit niedriger Extinktion und großer Steigung als Braunhuminsäuren. Auch Unterschiede in anderen Eigenschaften, z. B. der Fällbarkeit durch Elektrolyte, gehen parallel mit diesen spektralen Unterschieden. So zeigen die durch Fällung mit Natriumchlorid erhaltenen Fraktionen einer Schwarzerdehuminsäure die den Braun- bzw. Grauhuminsäuren entsprechenden spektralen Unterschiede (FLAIG, SCHEFFER und KLAMROTH 1955).

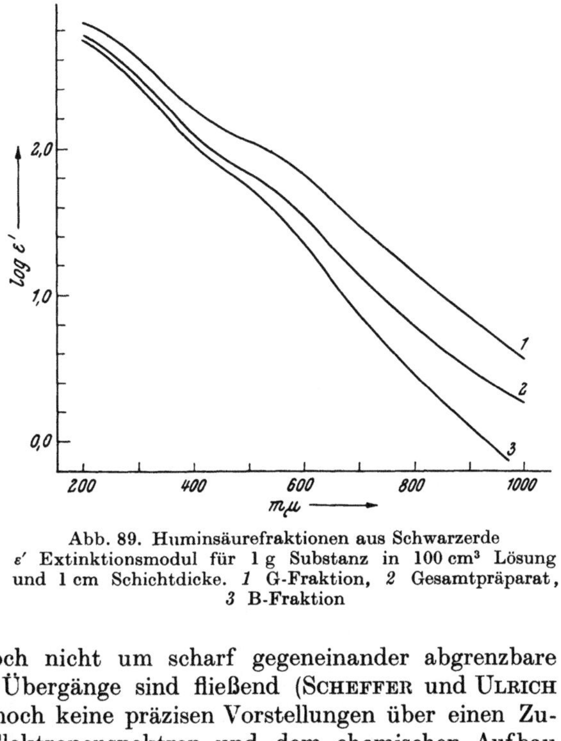

Abb. 89. Huminsäurefraktionen aus Schwarzerde
ε' Extinktionsmodul für 1 g Substanz in 100 cm³ Lösung und 1 cm Schichtdicke. *1* G-Fraktion, *2* Gesamtpräparat, *3* B-Fraktion

Es handelt sich hierbei jedoch nicht um scharf gegeneinander abgrenzbare chemische Stoffgruppen. Die Übergänge sind fließend (SCHEFFER und ULRICH 1960, S. 69). Zur Zeit gibt es noch keine präzisen Vorstellungen über einen Zusammenhang zwischen den Elektronenspektren und dem chemischen Aufbau der Huminsäuren.

Infrarotspektren von Huminsäuren. Die Infrarotspektren von Huminsäuren im Gebiet 2 bis 14 μ, die in der Regel in KBr-Presslingen oder auch in Nujol aufgenommen werden, zeigen eine starke Untergrundabsorption und nur wenige, verhältnismäßig breite Absorptionsbanden. Über die Deutung dieser Spektren liegen in der Literatur bisher wenige systematische Untersuchungen vor (KUMADA und AIZAWA 1958, SCHNITZER, SHEARER und WRIGHT 1959, ZIECHMANN und SCHOLZ 1960, ORLOV, ROZANOVA und MATJUCHINA 1962). Der allgemeine Charakter der Spektren aller bisher untersuchten Huminsäuren ist sehr ähnlich. Sie zeigen starke Absorption bei 3 μ und bei 6 μ sowie im Gebiet 8 bis 10 μ.

Es sind zahlreiche Versuche unternommen worden, die Absorptionsbanden bestimmten Strukturelementen zuzuordnen und so Rückschlüsse auf den chemischen Aufbau der Huminsäuren zu ziehen. So kommen ORLOV, ROZANOVA und MATJUCHINA (1962) auf Grund vergleichender Untersuchungen an zahlreichen Spektren zu dem Schluß, daß Huminsäuren aus verschiedensten Bodentypen durch die gleichen Strukturelemente charakterisiert sind, so daß Zuordnung zu einer einheitlichen Stoffklasse berechtigt erscheint. Bei aller Ähnlichkeit weisen jedoch KUMADA und AIZAWA (1958) auf charakteristische Unterschiede in den Spektren von Huminsäurepräparaten aus verschiedenen japanischen Böden hin. Auf Grund der relativen Intensität einzelner Banden ordnen die Autoren ihre Präparate

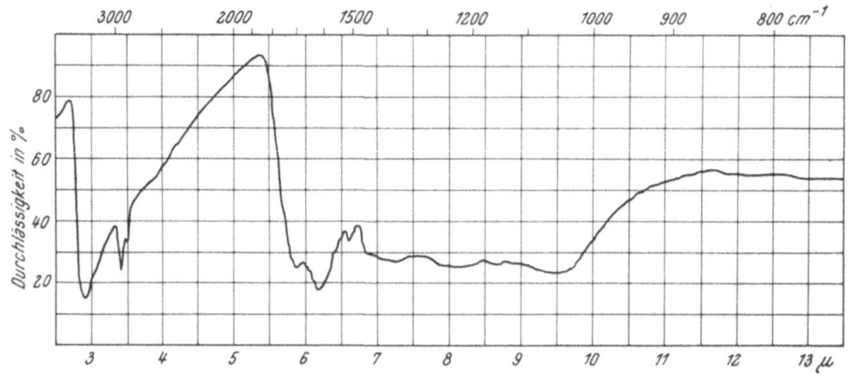

Abb. 90. IR-Spektren einer Huminsäure aus Schwarzerde

vier verschiedenen Gruppen zu. Die zu einer solchen Gruppe gehörenden Präparate zeigen auch andere gemeinsame Eigenschaften, z. B. ähnliche Absorption im

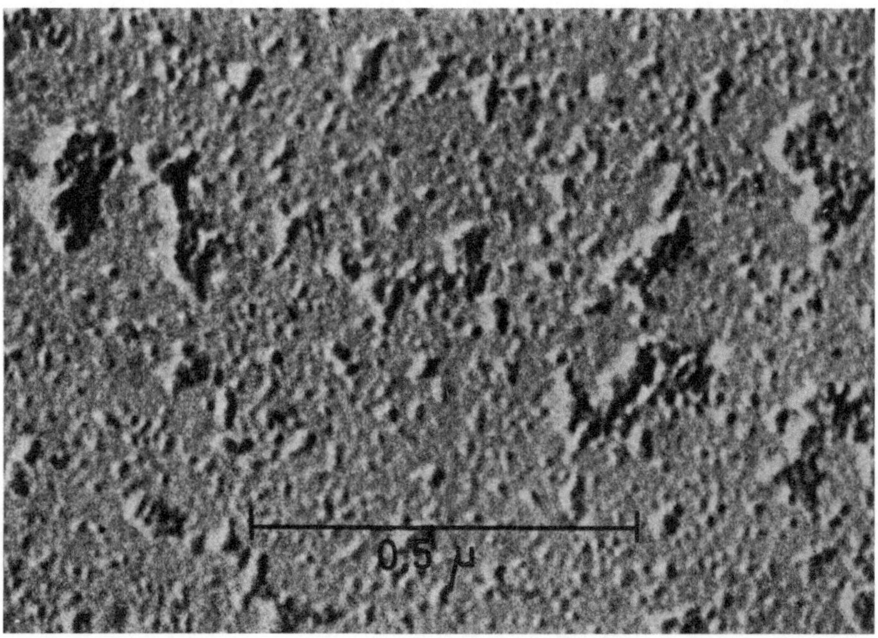

Abb. 91. Natürliche Huminsäuren bei pH 8, Dispergierung in Einzelteilchen (zur Kontraststeigerung mit Platin bedampft)

ultravioletten und sichtbaren Gebiet. SCHNITZER, SHEARER und WRIGHT (1959) studierten insbesondere den Einfluß mineralischer Bestandteile der Huminsäuren auf das IR-Spektrum.

Die Infrarotspektroskopie ist eine zweckmäßige Methode, mit der Huminsäurepräparate in weitgehend unveränderter Form und rasch charakterisiert werden können. Die Infrarotspektren sind daher in besonderem Maße zum Vergleich mit anderen Stoffen und Verbindungen, wie möglichen natürlichen Ausgangsprodukten für die Huminsäurebildung oder Modellsubstanzen geeignet. So treten bei den von KINNEY und DOUCETTE (1958) zusammengestellten Infrarotspektren einer Inkohlungsreihe von Cellulose und Lignin bis zum Anthrazit Spektren auf, die denjenigen von Huminsäuren aus Böden sehr ähnlich sind.

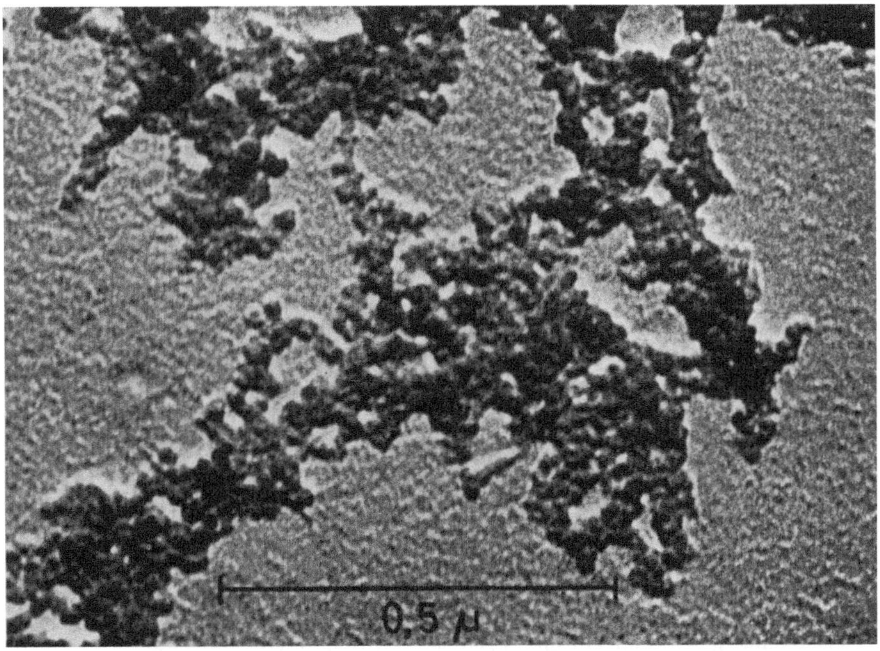

Abb. 92. Huminsäuren aus Schwarzerde bei pH = 3,5

FLAIG, SCHOBINGER und DEUEL (1959) konnten die Bildung von Huminsäuren bei der Verrottung von Stroh infrarotspektrographisch verfolgen (vgl. auch S. 395), ZIECHMANN und PAWELKE (1959) die Umwandlung von wenig humifiziertem Material aus durch Rotfäule zersetztem Eichenholz.

Wenn auch die IR-Spektren zur Zeit noch keine eindeutigen Rückschlüsse auf die Konstitution von Huminsäuren gestatten, so bilden sie doch ein wichtiges Hilfsmittel zur Charakterisierung von Huminsäuren und zur Klärung von Humifizierungsabläufen.

Elektronenmikroskopie. Die elektronenmikroskopische Methode, die in letzter Zeit mehrfach eine wertvolle Hilfe zur Aufklärung der Teilchenform und -größe der Kolloide und deren Verteilung gewesen ist, ergibt, daß sehr oft umgefällte Fraktionen von Huminsäuren aus den verschiedensten natürlichen Herkommen und von Synthesehuminsäuren der verschiedensten Darstellung in alkalischem Medium dispergierte Sphärokolloide sind (FLAIG 1950, FLAIG und BEUTELSPACHER 1951, 1954, BEUTELSPACHER 1952, 1955).

Der Teilchendurchmesser dieser Fraktion beträgt etwa 80 Å, dieser Wert stimmt mit dem aus den Diffusionsmessungen von SCHEELE (1937 errechneten Wert (BEUTELSPACHER 1952) in der Größenordnung gut überein. Mit Hilfe des

Elektronenmikroskops kann auch das Verhalten dieser Kolloide unter verschiedenen Bedingungen untersucht werden.

Im sauren Medium aggregieren die Huminsäuren unter Bildung von Coacervaten im Sinne von Bungenberg de Jong (1949). Mit dieser Methode läßt sich ebenfalls die Dispergierung und Aggregierung in Abhängigkeit vom pH-Wert ähnlich wie mit Hilfe der viskosimetrischen Messungen und der Diffusion (Scheele 1937) nachweisen. Die Befunde dieser drei Methoden deuten darauf hin, daß die Huminsäuren in alkalischem Medium als Molekülkolloide und in saurem als Micellkolloide vorliegen, die über Wasserstoff-Brückenbindungen entstehen (Beutelspacher 1952, Flaig und Beutelspacher 1954, Halla und Ruston 1955).

Beim Vergleich von Teilchen gleicher Größe aus Kohlenstoff wie Gasruß mit Teilchen der Huminsäuren sind letztere elektronenoptisch weniger dicht, d. h. sie bestehen nicht aus kompakten Teilchen, sondern haben eine lockere Struktur mit Poren und Zwischenräumen.

Diese Vorstellung geht auch aus der Berechnung der Volumbeanspruchung hervor. Berechnet man diese aus dem Quotienten der gefundenen Viskositätszahl durch die theoretisch für Sphärokolloide von 0,0025, so erhält man Werte, die unter der Voraussetzung, daß die Huminsäuren hydrophile Kolloide sind, und die Solvatation ähnlich wie bei Glykogen in verdünnten Lösungen keinen Einfluß auf die Viskosität hat (Staudinger und Husemann 1937, Husemann 1941), ebenfalls für eine poröse Struktur der Huminsäuren sprechen. Mit steigendem Stickstoffgehalt nehmen die Werte für die Volumbeanspruchung (V) zu (stickstofffreie Hydrochinonhuminsäure V = 3,9, stickstoffhaltige Hydrochinonhuminsäure 5,56% N, V = 6,9). Durch den Gehalt an Stickstoff werden bei gleichbleibender Größe der Teilchen die Poren enger, und es kann nach Ostwald und Steiner (1925) eine größere Menge an Wasser immobilisiert werden, und damit steigt auch die Viskosität an.

Die elektronenmikroskopische Methode gibt ferner Möglichkeiten der Untersuchung der Salzbildung der Huminsäuren und deren Wechselwirkung mit anorganischen Kolloiden (Beutelspacher 1955).

Viskosimetrie. Die spezifische Viskosität (η spez.) von mehrfach umgefällten dialysierten natürlichen Huminsäuren und stickstofffreien und stickstoffhaltigen Synthesehuminsäuren ergeben bei einem pH-Wert von 8 in Abhängigkeit von der Konzentration Geraden mit verschiedenen Neigungen. Die spezifische Viskosität steigt mit dem Stickstoffgehalt an. Die Viskositätszahl ($Z\,\eta$) ist daher bei stickstoffhaltigen Huminsäuren größer als bei stickstofffreien und jeweils konstant bezogen auf die Konzentration.

$$\eta\,\text{spez.} = \eta\,\text{rel} - 1 \qquad Z\,\eta = \frac{\eta\,\text{spez.}}{c}$$

Die Viskositätszahlen steigen bei Hydrochinonhuminsäuren stickstofffrei ($Z\,\eta = 0{,}0097$) über Huminsäuren aus Schwarzerde (3,8% N; $Z\,\eta = 0{,}0172$) zu Hydrochinonhuminsäuren stickstoffhaltig (5,56% N; $Z\,\eta = 0{,}0185$) an. In wässeriger Suspension beträgt $Z\,\eta = 0{,}0099$ bei der stickstofffreien und 0,0227 bei den stickstoffhaltigen Hydrochinonhuminsäuren. In 1n-Natronlauge liegen die Werte für $Z\,\eta = 0{,}0027$ sehr nahe bei dem von Einstein (1906, 1911) für Sphärokolloide errechneten Wert, der auf Grund von Überlegungen von Einstein und nach der Umformung der Gleichung von Staudinger und Husemann (1937) 0,0025 beträgt. In alkalischem Medium ist keine Abhängigkeit der Viskositätszahl vom Stickstoffgehalt der Huminsäuren zu beobachten.

Aus den Viskositätszahlen ist zu schließen, daß die Substanzen in der Huminsäurefraktion mehr oder weniger Sphärokolloide sind aber keinesfalls Linearkolloide (Beutelspacher 1952).

Bei Erhöhung der Wasserstoffionen-Konzentration steigt im Falle der stickstofffreien Hydrochinonhuminsäuren bis zu einer Konzentration von 1,0 η die relative Viskosität von η rel = 1,0270 nur um 10% an, während die von stickstoffhaltigen Hydrochinonhuminsäuren η rel = 1,128 um das Fünffache n rel = = 5,670 erhöht wird und damit Werte erreicht, die denen von Linearkolloiden entsprechen. Die Erhöhung wird durch Aggregierung der Huminsäureteilchen hervorgerufen. Der Wert geht durch die Verminderung der Wasserstoffionen-Konzentration wieder auf den anfänglichen zurück. Die Durchlaufzeiten gegen den pH-Wert aufgetragen, ergeben für die Huminsäuren charakteristische Kurven, die bei einem bestimmten pH-Wert durch Aggregierung einen Knick aufweisen.

Nach Untersuchungen von Sven Oden (1919) haben Huminsäuren eine spezifische Viskosität von 7,86; diese ist sehr hoch und rührt vermutlich von einer hohen Waserstoffionen-Konzentration oder von Veränderungen durch die durchgeführte Trocknung der Präparate her. Ostwald und Steiner (1925) sind bei Torfhuminsäuren zu ähnlichen Ergebnissen gekommen. Piret, White, Walther und Madden (1960) bestimmen bei einmal umgefällten Torfhuminsäuren die „intrinsic" $\left((\eta) = \frac{\eta\,\text{spez.}}{c} \text{ extrapoliert nach } 0\right)$ Viskosität in Kochsalzlösungen zu $(\eta) = 0{,}115\ (\text{g/ml})^{-1}$ und bestimmten daraus das Reibungsverhältnis zu 1,60 bis 1,86, sie sind der Ansicht, daß die Huminsäuren Ellipsoide mit einem Achsenverhältnis 1:10 sind. Auch Mukherjee, Bhowmik und Lahiri (1957) finden bei Huminsäuren aus oxydierten Kohlen ebenfalls Kurven für die spezifische Viskosität, die mit denen der Torfpräparate vergleichbar waren.

Diffusion und Dialyse. Zur Teilchengewichtsbestimmung mit diesen Methoden liegen noch wenige Messungen vor. Die Werte sind sehr unterschiedlich, da die untersuchten Substanzen verschiedener Herkunft und verschiedenen Vorbehandlungen unterworfen sind (vgl. Tab. 56).

Der von Scheele mitgeteilte Befund, daß die Huminsäuren in alkalischem Medium ein kleineres Teilchengewicht ergeben als in saurem Medium wird durch die Angaben der beiden letzteren Autoren bestätigt. Die Erhöhung des Teilchengewichtes in saurem Medium wird, ähnlich wie bei Eiweißkolloiden (Svedberg 1934) auf die Bildung von Micellkolloiden zurückgeführt.

Ultrazentrifuge. Die Teilchengewichte von Huminsäuren aus Schwarzerde, die durch mehrfache Umfällung einer gewissen Fraktionierung unterworfen worden sind, sind mittels der Messung der Sedimentationsgeschwindigkeit bei pH-Werten von 7 oder 8 bei der Hauptfraktion zu 3 bis $5 \cdot 10^4$ einer weiteren Fraktion $1 \cdot 10^4$ bestimmt worden. Die Untersuchungen ergaben ferner, daß die Huminsäuren Sphärokolloide und polydispers sind (Flaig und Beutelspacher[1]). Stevenson und Mitarbeiter (1953) bestimmen an Huminsäuren aus einem neutralen Pyrophosphatextrakt aus einem Miami clay soil in neutraler Lösung mittels der gleichen Methode ein Teilchengewicht von $5{,}3 \cdot 10^4$ (Sedimentationskonstante $2{,}8 \cdot 10^{-13}\,\text{cm} \cdot \text{sec}^{-1} \cdot \text{dyn}^{-1}$). Auf Grund des Reibungskoeffizienten $f/fo = 2{,}3$ kommen die Autoren zu dem Ergebnis, daß die Huminsäuren ein Achsenverhältnis von 1:30 haben.

Piret und Mitarbeiter (1960) geben ein durchschnittliches Sedimentationsteilchengewicht von $2{,}5 \cdot 10^4$ (Sedimentationskonstante $1{,}83 \cdot 10^{-3}\,\text{cm} \cdot \text{sec}^{-1} \cdot \text{dyn}^{-1}$)

---

[1] Vortrag, gehalten anläßlich der Hauptversammlung der landwirtschaftlichen Untersuchungs- und Forschungsanstalten in Marburg/Lahn am 20. September 1950, vgl. Kononowa, 1958 S. 72.

Tabelle 56. *Teilchengewichte verschiedener Huminsäurepräparate*

Diffusion		Teilchengewicht	Diffusionskoeffizient	
WELTE und Mitarbeiter (1954)	N-haltige Chinonhuminsäuren	750	0,360 cm²/d	
	N-freie Chinonhuminsäuren	600	0,530 cm²/d d = Tag	
SCHEFFER und Mitarbeiter (1955)	natürliche und synthetische Huminsäuren. Herkunft und Darstellung sowie Extraktion nicht angegeben	600–800	nicht angegeben	
ZEILE (1935)	Extrakte aus Torf (Hexamethylentetramin)	1060	0,1345 cm²/d	
	(Harnstoff)	1650	0,1075 cm²/d	
SCHEELE (1937)	a) Huminsäuren aus Kassler Braun (NaF-Extrakt) sauer	11000	0,060 cm²/d	
	alkalisch	5500	0,090 cm²/d	
	b) zweites Präparat wie a) schwach sauer	12600	0,056 cm²/d	
	alkalisch	5000	0,088 cm²/d	
	Humussäure aus Dachauer Moor (NaF-Extrakt) sauer	26000	0,039 cm²/d	
	alkalisch	11000	0,060 cm²/d	
STEVENSON und Mitarbeiter (1953)	Huminsäuren aus Miami clay soil (neutrale Phyrophosphatlösung) neutral	53000	Diffusionskonstante $D_{20} = 3{,}9 \cdot 10^{-7}$ cm²/sec	
SCHEFFER und Mitarbeiter (1958)	Pseudoschwarzerde-Huminsäuren, elektroaktivere Fraktion	—	Diffusionskoeffizient $D_{20} = 14{,}5 \cdot 10^{-6}$ cm²/sec	
	elektroinaktivere Fraktion	—	$= 8{,}6 \cdot 10^{-6}$ cm²/sec	
	N-haltige Hydrochinonhuminsäuren, elektroaktivere Fraktion	—	$= 16{,}3 \cdot 10^{-6}$ cm²/sec	
	elektroinaktivere Fraktion	—	$= 9{,}5 \cdot 10^{-6}$ cm²/sec	
Dialyse FLAIG und BEUTELSPACHER (1954)			Dialysekoeffizient	K
	Schwarztorf (Na_2CO_3-Extrakt) alkalisch	6900	0,0718	5,968
	Schwarzerde (NaOH-Extrakt) alkalisch	4700	0,0643	4,441
	Braunterf	6400	0,0778	5,988
	(NaOH-Extrakt) alkalisch	6350	0,0555	4,441

an, zu dessen Berechnung sie die Reibungskonstante unter Berücksichtigung des Reibungsverhältnisses aus den Viskositätsmessungen von $f/f_0 = 1{,}60$ zu 1,86 und eine ellipsoide Form heranziehen.

Außerdem sind Sedimentationskonstanten mit unterschiedlichen Werten für Huminsäuren aus Böden, Fraktionen von verrottendem Pflanzenmaterial und aus Synthesehuminsäuren, die zum Teil durch Säulenchromatographie oder Elektrophorese aufgetrennt sind, bestimmt worden (WELTE, NIEMANN und ZIECHMANN 1954, SCHEFFER, ZIECHMANN und SCHLÜTER 1958, 1959, SCHLÜTER 1959, ZIECHMANN und KICKUTH 1961).

Die genannten Autoren sind der Ansicht, daß „wegen der fehlerhaften Bestimmung der Diffusionskonstanten die zusätzliche Berechnung des Teilchengewichtes nicht notwendig wäre und die Sedimentationskonstante als relatives Maß für die Teilchengröße zu verwerten sei". Die Begründung, die Bestimmung des Diffusionskoeffizienten wäre nur sinnvoll, wenn es gelänge, durch Herstellung von monomolekularen Huminsäuren die Genauigkeit der Bestimmung zu erhöhen, trifft auch für die Messungen mit der Ultrazentrifuge zu.

Weitere Eigenschaften. Die Huminsäuren wandern bei der Elektrophorese zum positiven Pol und sind demnach negativ geladen. Der ξ-Potential der Huminsäuren aus Schwarzerde liegt bei etwa 20 Millivolt, das der Alkalihumate ist Null (FLAIG und BEUTELSPACHER 1954, BEUTELSPACHER 1952).

Die Bestimmung des Redoxpotentials von Böden hat von verschiedenen Autoren eine Bearbeitung erfahren, während Angaben für Huminsäuren nur wenig zu finden sind.

Die Huminsäuren zeigen rH-Werte von etwa 20. Bei schwach alkalischer Reaktion liegen sie im Bereich des Neutralpunktes (FLAIG und BEUTELSPACHER 1954, BEUTELSPACHER 1952, FLAIG, SCHARRER und JUDEL 1955a, 1955b).

Die Differentialthermoanalyse wird ebenfalls zur Charakterisierung der Huminsäuren und des Humifizierungsgrades vorgeschlagen. Ihre Kurven sind unter Stickstoffatmosphäre uncharakteristisch (MITCHELL 1960). Die Kurven zeigen unter Sauerstoffzuleitung bei etwa 300° ein schwaches Maximum und zwischen 430 und 470° sowie zwischen 480 und 510° zwei starke exotherme Maxima. In der Regel zeigt sich ein umgekehrtes Bild (MITCHELL 1960, VALEK 1962). Über den thermischen Reaktionsverlauf liegen widersprechende Ansichten vor.

Bisher wurde immer wieder versucht, die Huminsäuren mit Hilfe der Elektronenbeugung und auf röntgenographischem Wege zu kennzeichnen. Eine Trennung der Huminsäuren von den anorganischen Bestandteilen ist auch durch zahlreiche Umfällung nicht zu erzielen. Die gelegentlich beobachteten Interferenzlinien können daher auch auf die beigemengten anorganischen Bestandteile zurückgeführt werden. Sorgfältig gereinigte Huminsäuren zeigen keine Interferenzen und müssen als röntgenamorphe Substanzen angesehen werden.

6. Modelluntersuchungen zur Bildung von Stoffen aus dem Humus

Nachdem die bisher bekannten Beiträge über die Zusammensetzung der Fraktionen der Humusstoffe insoweit besprochen worden sind, als die dabei gewonnenen Erkenntnisse eine Vorstellung über deren Bildung geben können, werden nun diejenigen Reaktionen von Substanzen mit ähnlichen chemischen Verhalten mit der Absicht angeführt, für unbekannte Vorgänge noch weitere Aufklärung zu geben. Dieses Vorgehen ist das gleiche, wie es bei der Konstitutionsaufklärung anderer Naturstoffe üblich ist.

Die Huminsäuren lassen sich durch Hydrolyse nicht in einheitliche Bausteine wie im Falle der Cellulose oder der Eiweißstoffe zerlegen. Während bei synthetischem Kautschuk durch oxydativen Abbau über die Verknüpfung der Monomeren Aussagen gemacht werden können, ergeben die bisher auf verschiedenen Wegen erhaltenen Oxydationsprodukte der Huminsäuren noch kein Bild über die Verknüpfung von deren Bausteinen.

In den meisten Böden sind überwiegend aerobe Verhältnisse vorhanden, so daß den Oxydationsreaktionen eine primäre Bedeutung zukommt. Die dabei ablaufenden Kondensationsreaktionen sind insofern von Interesse, als durch diese verschiedene Bausteine bis zu höhermolekularen Stoffen verknüpft werden können.

A. Ligninabbauprodukte

Geht man bei der Bildung der Humusstoffe, insbesondere bei der der Huminsäuren, von Lignin oder anderen pflanzlichen Phenolen, den Pflanzenbestandteilen mit der kleinsten Zersetzungsgeschwindigkeit, aus, so müssen diese bei saurer und neutraler Reaktion gegen schwache Oxydationsmittel verhältnismäßig stabilen Stoffe zunächst eine Änderung erfahren.

Entmethylierung. Bei den Untersuchungen über die Veränderung des Lignins während der Verrottung pflanzlicher Rückstände ist schon mitgeteilt worden, daß dessen Methoxylgehalt abnimmt. Die Entmethylierung erfolgt vermutlich zur Hauptsache durch Enzyme verschiedener Bodenmikroorganismen.

Es bestehen somit zwei Möglichkeiten der Entmethylierung. Die erstere, die am Lignin selbst, die zweite, die an den Spaltstücken des Lignins stattfindet.

Bisher ist die Entmethylierung von Lignin durch Mikroorganismen noch nicht näher untersucht worden, sondern nur an einem Lignan, dem α-Conidendrin (s. unten, KONETZKA, WOODINGS und STOVE 1957). Untersuchungen an Ligninspaltstücken, wie sie in Rotteansätzen festgestellt worden sind, und an solchen, die möglicherweise entstehen können, haben ergeben, daß zunächst die Seitenkette bis zu einer Carboxylgruppe abgebaut wird, bevor eine Entmethylierung stattfindet (HENDERSON 1957).

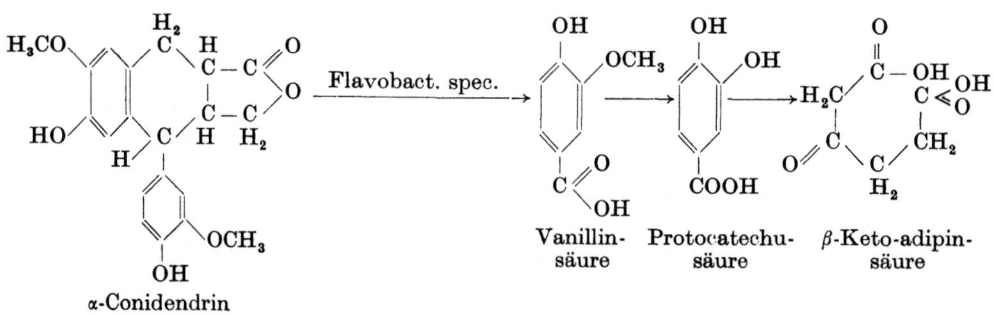

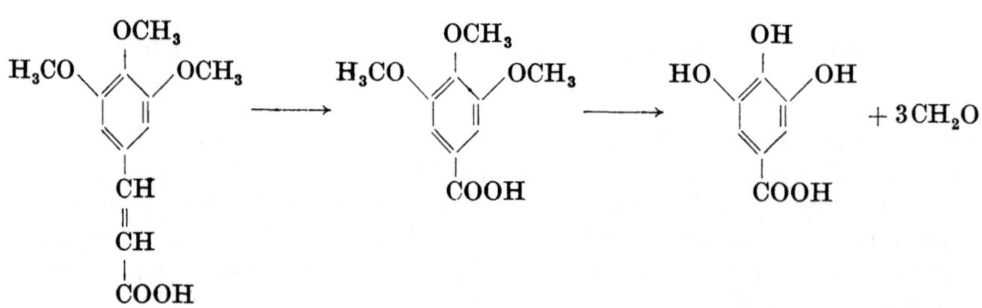

Bei der Oxydation von Vanillinsäure, deren Methoxylgruppe mit ^{14}C markiert war, konnte in Gegenwart von Homogenaten von Weißfäulepilzen ebenfalls eine Entmethylierung beobachtet werden (HAIDER, LIM und FLAIG 1962). Die gebildeten Produkte sind hydrolysiert worden. In den freigesetzten Aminosäuren verbleiben etwa 25% der anfänglich eingesetzten Aktivität, wobei sich in den Aminosäuren Methionin und Serin der Hauptanteil der Aktivität befindet.

Oxydation. Die durch die Entmethylierung entstehenden Phenolverbindungen sind verhältnismäßig leicht oxydierbar. Die dabei entstehenden chinoiden Verbindungen müssen als reaktionsfähige Zwischenstufen bei der Bildung der Humusstoffe angenommen werden. Auf Grund ihrer Reaktionsfähigkeit ist es daher auch in weitem Maße wenig wahrscheinlich, sie in ähnlichen Mengen aus Böden isolieren zu können wie die als Abbauprodukte des Lignins bekannten Phenolverbindungen, die entweder aus Böden oder verrottetem Stroh isoliert worden sind (FLAIG 1950, BÖRNER 1955, BREMNER 1955a, MORRISON 1958, MAEDER 1959). Da anzunehmen ist, daß die chinoiden Zwischenstufen bei der Bildung der Humusstoffe eine Schlüsselstellung einnehmen, ist es von besonderem Interesse, deren Reaktionen insoweit zu studieren, als sie mit der Bildung von Humusstoffen, insbesondere die der Huminsäuren, in Zusammenhang stehen können.

Geht man von dem Strukturschema des Lignins aus, wie es von ADLER (1961) für Coniferenlignin beschrieben ist, so können in Anlehnung an die Ergebnisse, die die Isolierung verschiedener während der Rotte von pflanzlichem Material auftretenden Verbindungen betreffen, die folgenden Verbindungen als Abbauprodukte von Lignin der verschiedenen Pflanzenarten in Betracht gezogen werden:

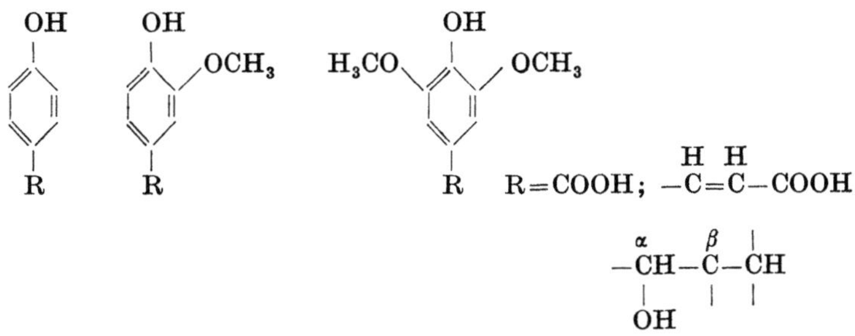

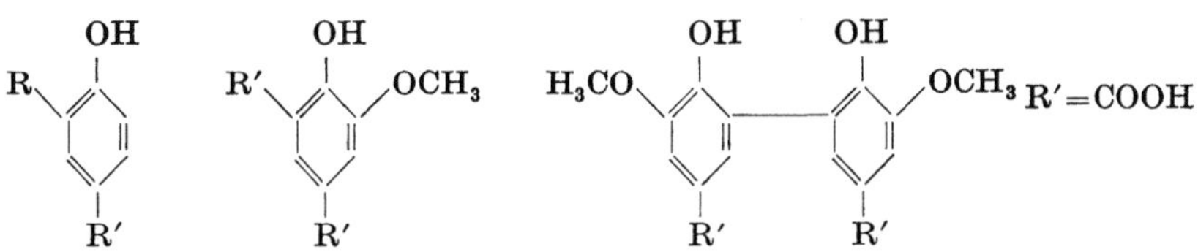

Berücksichtigt man, daß bei einigen Verbindungen eine Entmethylierung stattfinden kann, so kommen noch folgende weitere Möglichkeiten hinzu:

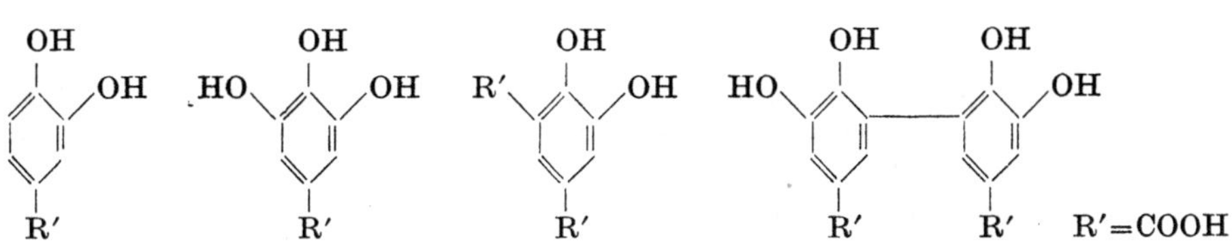

Oxydiert man diese Verbindungen mit Kaliumnitrosodisulfonat (NDS), so findet je nach der Art der Seitenkette eine unterschiedliche Reaktion statt.

a) Bei den Derivaten der Benzoesäure (TEUBER und JELLINEK 1952, TEUBER und RAU 1953, FLAIG 1960) und bei dem Vorhandensein einer OH-Gruppe am α-Kohlenstoffatom der Seitenkette findet durch Decarboxylierung bzw. Abspaltung der Seitenkette (ADLER und LUNDQUIST 1961) die Bildung von p-Benzochinonen statt.

b) Im Falle einer gesättigten oder ungesättigten Seitenkette entstehen zum Teil unter Einführung einer weiteren OH-Gruppe o-Benzochinone (HORNER und DÜRCKHEIMER 1959a).

$NDS = ON(SO_3K)_2$

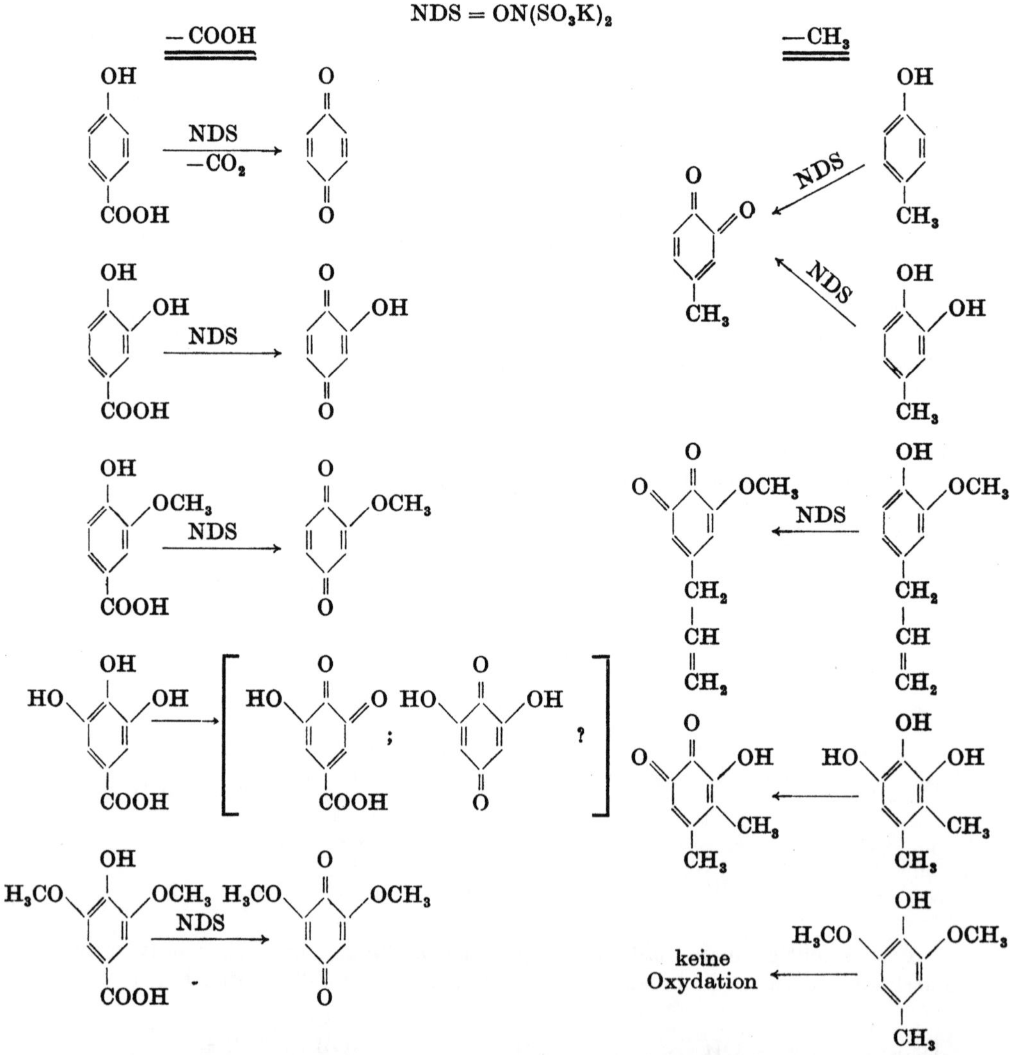

Die Einwirkung von Kaliumnitrosodisulfonat auf diese Phenole bzw. partiell methylierten Phenole ist mit der von Tyrosinase vergleichbar; es findet eine Dehydrierung von o-Diphenolen zu den entsprechenden Chinonen oder die Einführung einer o-ständigen zweiten OH-Gruppe statt. So entsteht u. a. aus p-Hydroxybenzoesäure p-Benzochinon und aus Protocatechusäure Hydroxy-p-benzochinon.

Ebenso ist bei der durch Catecholase aus Polystictus versicolor katalysierten Oxydation von Protocatechusäure Hydroxy-p-benzochinon nach Hydrierung und Acetylierung als Triacetat von Hydroxy-hydrochinon nachgewiesen worden (FLAIG und HAIDER 1961a).

p-Benzochinon und Hydroxy-p-benzochinon spielen bei den Untersuchungen zur Herstellung von „Synthese"-Huminsäuren eine große Rolle. Die Entstehung dieser beiden Verbindungen aus Ligninabbauprodukten ist daher für Überlegungen wichtig, inwieweit die Erkenntnisse aus der Bildung von „Synthese"-Huminsäuren auf die von natürlichen übertragen werden können.

B. Synthesehuminsäuren aus Polyphenolen

Im folgenden soll nun die Oxydation der Polyphenole der Benzolreihe besprochen werden, um einen Einblick in die mannigfaltigen Möglichkeiten bei der Bildung der Humusstoffe in der Natur zu bekommen. Dabei sollen vor allem diejenigen berücksichtigt werden, die unter den Bedingungen im Boden möglicherweise ablaufen können. Diese Betrachtungen dienen ferner dazu, Reaktionen auszuschließen, die für die Bildung der Humusstoffe nicht von Bedeutung sein können.

Für die gefärbten Stoffe der einzelnen nach konventionellen Methoden abtrennbaren Stoffgruppen des Humus wird es zunächst nicht möglich sein, genaue Konstitutionsformeln anzugeben, sondern man wird sich, ähnlich wie bei Lignin, mit der Aufstellung von Strukturschemata vorerst begnügen müssen. Dies trifft insbesondere für die höhermolekularen Vertreter zu. Die Untersuchungen werden demnach darauf hinauslaufen, die Verknüpfung der Bausteine soweit als möglich aufzuklären. Dazu ist es erforderlich, die an natürlichen Humusstoffen ablaufenden Reaktionen mit denen an den in Frage kommenden Ausgangsmaterialien zu vergleichen.

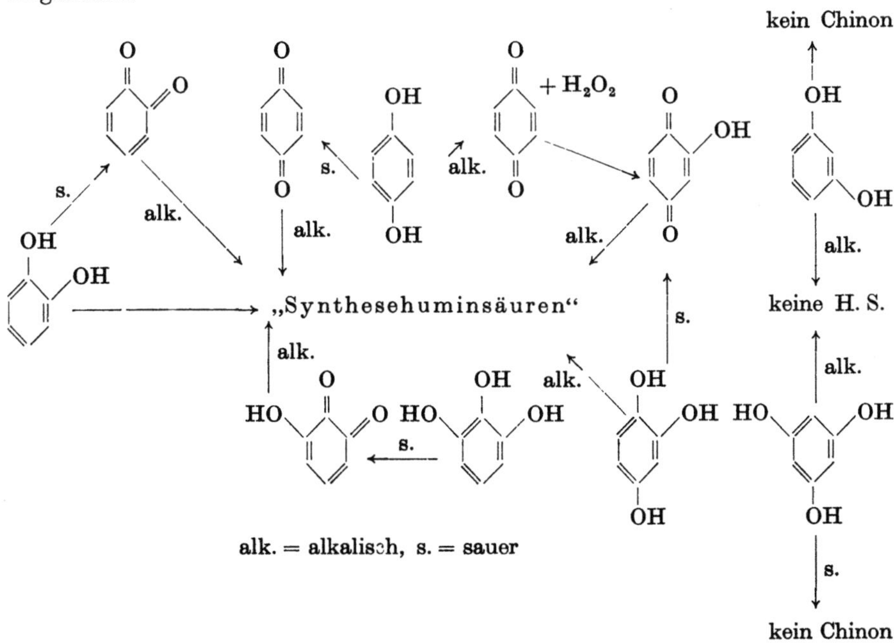

alk. = alkalisch, s. = sauer

Die meisten Phenole und Chinone reagieren in alkalischer Lösung mit Sauerstoff zu schwarzbraunen Produkten, die in der Literatur als „Synthese"-Huminsäuren bezeichnet werden und sehr oft zum Vergleich mit den natürlichen gedient haben.

In einer Übersicht werden diejenigen Di- und Tri-hydroxybenzole zusammengestellt, deren Oxydation in alkalischer Lösung zu „Synthese"-Huminsäuren führt. Die Oxydationsreaktionen verlaufen sehr unübersichtlich und sind im einzelnen noch nicht völlig aufgeklärt.

Aus diesem Schema geht hervor:

1. Nur solche Polyhydroxybenzole, die ein Chinon bilden, gehen in alkalischer Lösung durch Sauerstoffeinwirkung in Huminsäuren über.
2. Aus Chinonen werden unter diesen Bedingungen ebenfalls Huminsäuren gebildet. Im Falle des unsubstituierten Pyrogallols ist 3-Hydroxy-o-benzochinon noch nicht gefaßt worden.
3. Resorcin und Phloroglucin werden nicht zu Chinonen oxydiert; es entstehen in alkalischen Lösungen keine Huminsäuren.

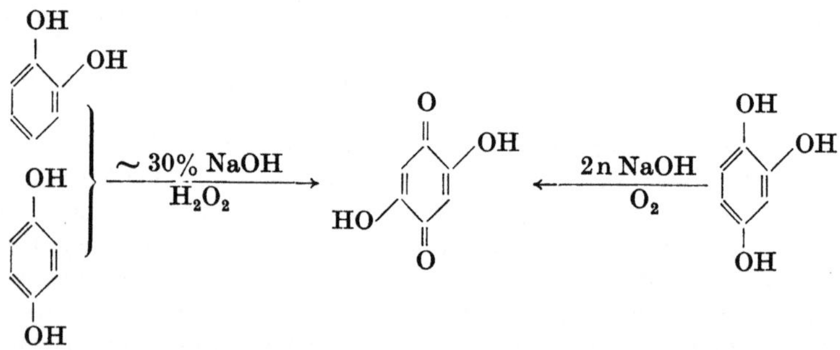

In stärker alkalischen Lösungen mit einem pH-Wert über 9 können aus Brenzcatechin und Hydrochinon (Jones und Shonle 1945, Flaig, Ploetz und Biergans 1955) mit Oxydationsmitteln wie Wasserstoffsuperoxyd in guter Ausbeute 2,5-Dihydroxy-p-benzochinon isoliert werden. Auch aus Hydroxyhydrochinon läßt sich neben anderen Produkten in 2 n Natronlauge bei der Oxydation mit Sauerstoff die gleiche Verbindung isolieren. In neutraler bis saurer Lösung finden andere Reaktionen statt, auf die später noch eingegangen werden soll.

1,4-Diphenole. 1. *Oxydation.* Die Oxydation von Hydrochinon in alkalischer Lösung ist mehrfach untersucht worden. Schon Eller (1921) hat die dabei entstehenden dunkelgefärbten Substanzen als Modelle zum Vergleich mit den aus natürlichen Vorkommen isolierten Huminsäuren herangezogen. Über den Reaktionsablauf der Oxydation von Hydrochinon in alkalischer Lösung ist wenig bekannt. Er ist stark abhängig vom pH-Wert der Reaktionslösung. Die Geschwindigkeit der Sauerstoffaufnahme ist in der ersten Phase der Reaktion dem Quadrat der OH-Ionenkonzentration proportional (Reinders und Dingemans 1934, James und Weissberger 1938, James, Snell und Weissberger 1938). Auch die Folgereaktionen der primären Oxydationsprodukte sind vom pH-Wert der Reaktionslösung abhängig. So erhielten Reinders und Dingemans bei dem pH-Wert 6,93 als Hauptprodukt der Oxydation p-Benzochinon, bei dem pH-Wert 7,88 Huminsäuren.

Die Oxydation von Hydrochinonen in alkalischer Lösung zu Chinonen erfolgt in zwei Stufen über das semichinoide Radikal (Michaelis 1951). Bei der Oxyda-

Tabelle 57. *Substanzen, die eine Rotfärbung während der Oxydation von Hydrochinon bewirken*

	Substanz	Lösungsmittel	Maxima in mμ
I	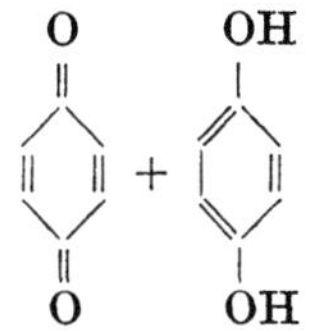 im Überschuß	Wasser pH 7	433
II	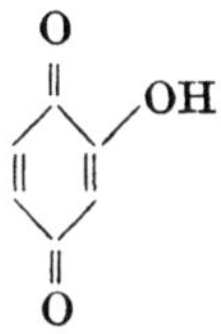	0,1n NaAc	483 261
III	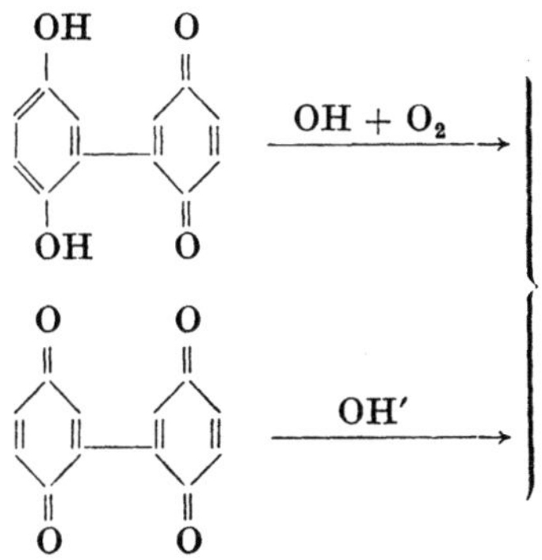	1n NaAc	501 244
IV	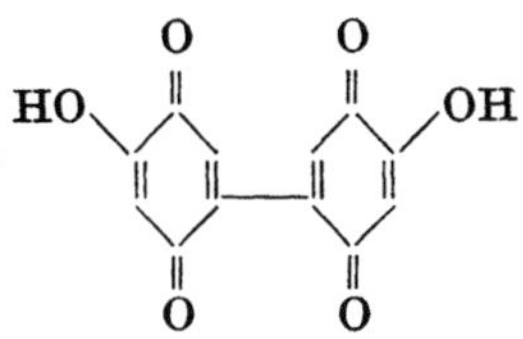	0,1n NaAc	499 273
V	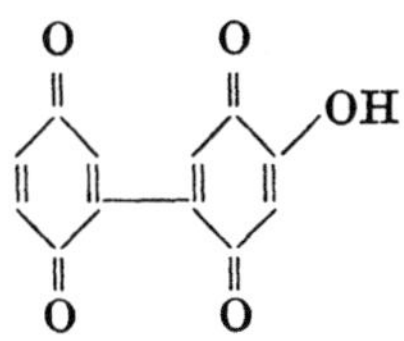	1n NaAc bei ~ pH 6	459[1] 472 244
VI	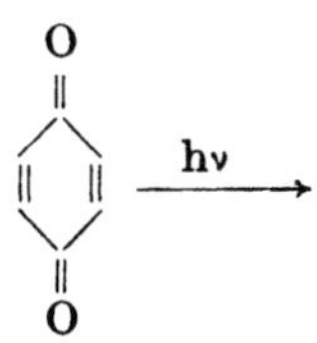	pH 7	496 249

[1] Verändert sich rasch.

tion von Hydrochinon ist die Konzentration des semichinoiden Radikals sehr gering. Es konnte durch sein Elektronen-Spin-Resonanz-Spektrum nachgewiesen werden (VENKATARAMAN und FRAENKEL 1955, BLOIS 1955). Die bei der Autoxydation von Hydrochinon in schwach alkalischer Lösung zunächst auftretende Gelbfärbung dürfte ebenfalls auf das semichinoide Radikal zurückzuführen sein, denn das Spektrum dieser Lösung (FLAIG und SALFELD 1960) zeigt im Gebiet 400 bis 450 mμ die gleiche Doppelbande wie das aus Hydrochinon und p-Benzochinon gebildete p-Benzosemichinon-anion (EIGEN und MATTHIES 1961). Im weiteren Verlauf der Oxydation färbte sich die Reaktionslösung rot, und im

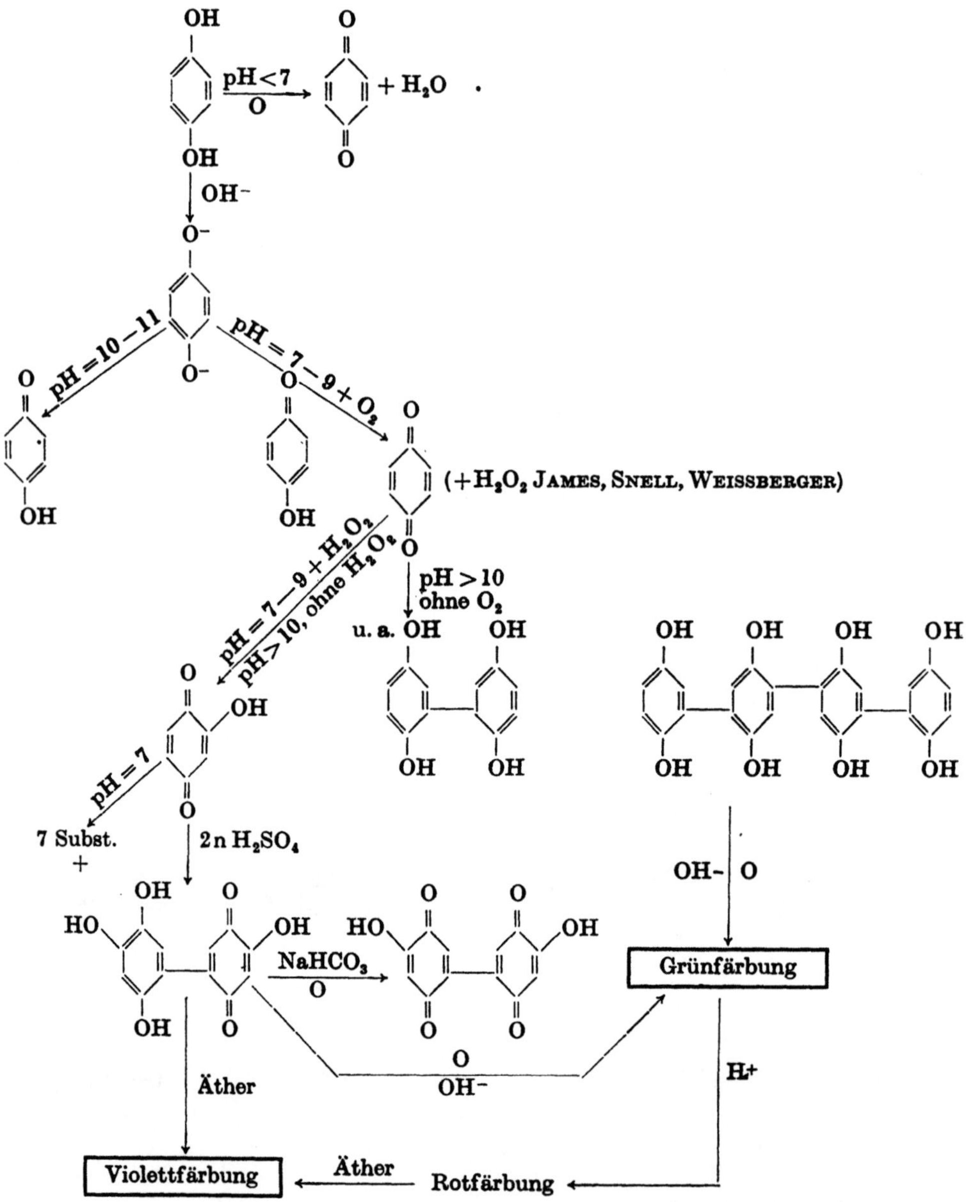

Spektrum tritt die breite Bande vom Anion von Hydroxy-p-benzochinon bei 488 mμ auf. Das Spektrum wird jedoch rasch uncharakteristisch, so daß anzunehmen ist, daß noch weitere Verbindungen, die in diesem Gebiet absorbieren, als Zwischenprodukte gebildet werden. Es sind daher die Oxydationsstufen, insbesondere die die Rotfärbung verursachenden Verbindungen (FLAIG und SALFELD 1960), näher untersucht worden (vgl. auch Tab. 57).

In saurer bis neutraler Lösung rührt die Rotfärbung zu Beginn der Oxydation von einem Assoziat aus p-Benzochinon-Hydrochinon (I) her, bei dem Hydrochinon im Überschuß noch vorhanden ist. Bei der Oxydation in schwach alkalischer, verdünnter Lösung konnte Hydroxy-hydrochinon nach Abtrennung der beiden noch in der Lösung vorhandenen Komponenten, Hydrochinon und p-Benzochinon spektrographisch nachgewiesen werden (FLAIG und SALFELD 1960). Wie alle anderen Hydroxy-p-benzochinone, die in o-Stellung zur OH-Gruppe nicht substituiert sind, so ist auch Hydroxy-p-benzochinon in alkalischer Lösung rot gefärbt (FLAIG und SALFELD 1958).

Die ersten monomeren Reaktionsprodukte der alkalischen Oxydation von Hydrochinon zur Bildung der „Synthese"-Huminsäuren sind demnach die Bildung eines semichinoiden Radikals, die Bildung von p-Benzochinon und die von Hydroxy-p-benzochinon. Die jeweils gebildete Menge hängt von der Wasserstoffionen-Konzentration ab.

In einem Schema sollen diese Befunde zusammengefaßt und noch einige Reaktionen von Hydrochinon und p-Benzochinon erwähnt werden, die für die Bildung der „Synthese"-Huminsäuren weiterhin von Interesse sind.

Wie JAMES, SNELL und WEISSBERGER (1938) zeigen konnten, entsteht bei der alkalischen Oxydation von Hydrochinon p-Benzochinon und gleichzeitig Wasserstoffsuperoxyd. Die verdünnte Lösung von p-Benzochinon bleibt in schwach alkalischer Lösung (0,1 n Natriumacetat) über längere Zeit gelb. Setzt man jedoch dieser Lösung von p-Benzochinon Wasserstoffsuperoxyd zu, so wird die Lösung sofort rot, und Hydroxy-p-benzochinon (FLAIG und SALFELD 1960) kann nach Aufarbeitung des Ansatzes spektrographisch nachgewiesen werden. Bei höheren pH-Werten findet die Bildung von Hydroxy-p-benzochinon aus p-Benzochinon über eine Anlagerung von OH-Ionen auch in Abwesenheit von Wasserstoffsuperoxyd statt; diese Reaktion ist pH-abhängig (EIGEN und MATTHIES 1961).

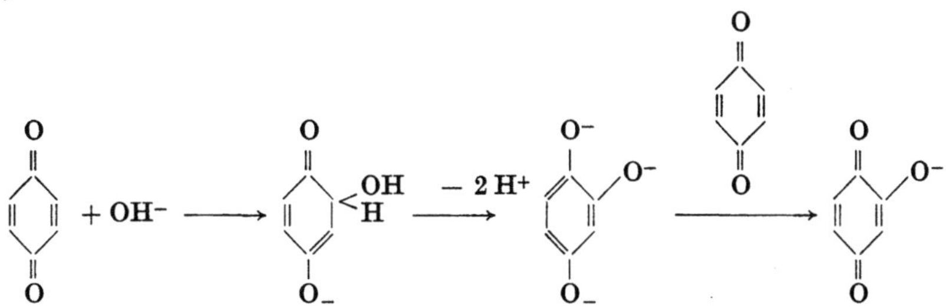

Nach Besprechung der Reaktionen der möglichen, monomeren Zwischenprodukte bei der Bildung der „Synthese"-Huminsäuren soll auch untersucht werden, inwieweit diese polymerisieren. Schon ERDTMAN und GRANATH (1954) beobachteten, daß in alkalischer Lösung von p-Benzochinon in Abwesenheit von Sauerstoff neben vielen dunkelgefärbten Produkten als dimere Verbindung Dihydrochinon erhalten wird. Als Folgeprodukt der Dimerisierung ist die Bildung des intramolekularen Chinhydrons von Di-p-benzochinon, und die von Di-p-benzochinon in Betracht zu ziehen. Das intramolekulare Chinhydron (III, Tab. 57)

autoxydiert in alkalischer Lösung zu einer roten Verbindung, deren Spektrum die gleichen Maxima hat, wie eine alkalische Lösung von Di-p-Benzochinon. Die Maxima liegen jedoch an einer anderen Stelle als die der bisher unter I und II (Tab. 57) erwähnten Verbindungen.

Die Belichtung von wässerigen Lösungen von p-Benzochinon beschleunigt deren Verfärbung nach Rotbraun wesentlich. Aus diesen Ansätzen lassen sich in alkalischer Lösung rot gefärbte Produkte (VI, Tab. 57) abtrennen, deren Spektrum dem der Oxydationsprodukte des intramolekularen Chinhydrons von 4,4'-Dihydroxy-di-p-benzochinon-2,5,2',5' und dessen Chinon in alkalischer Lösung ähnlich ist.

Hydroxy-p-benzochinon dimerisiert in saurer Lösung in guter Ausbeute zu dem intramolekularen Chinhydron von 4,4'-Dihydroxy-di-p-benzochinon-2,5,2'5'; in neutraler Lösung ist diese Verbindung neben sieben anderen papierchromatographisch nachweisbar. In alkalischer Lösung entstehen zur Hauptsache dunkelgefärbte Substanzen, „Synthese"-Huminsäuren. Dieses intramolekulare Chinhydron wird leicht zu 4,4'-Dihydroxy-di-p-benzochinon-2,5,2',5' oxydiert, das sich in alkalischer Lösung ebenfalls mit roter Farbe löst (IV, Tab. 57). Das Spektrum ist von den bisher genannten verschieden. Die Bildung von 4,4'-Dihydroxy-di-p-benzochinon-2,5,2',5' ist auch bei der Oxydation von Protocatechusäure in Gegenwart von Laccase (FLAIG und HAIDER 1961b) als Zwischenprodukt von schwarzbraunen polymeren Produkten möglich.

Zur Untersuchung einer weiteren Möglichkeit der Dimerisierung der Reaktionsprodukte, bei der alkalischen Oxydation von Hydrochinon, ist 2,3,5,2',5'-Pentahydroxydiphenyl synthetisiert worden (FLAIG und SALFELD, unveröffentlicht). Dessen intramolekulares Chinhydron ist ebenfalls stark dunkel gefärbt. Das entsprechende 3-Hydroxy-di-p-benzochinon-2,5,2',5' (V, Tab. 57) löst sich in verdünntem Alkali mit roter Farbe. Das Spektrum ändert sich schneller; auf Grund seines Spektrums zeigt das 3-Hydroxy-di-p-benzochinon-2,5,2',5' ein anderes Verhalten als die Substanzen II, III, IV und VI in Tab. 57. Die Rotfärbung der Oxydationslösungen von Hydrochinon beruht auf der Bildung von Hydroxy-p-benzochinon und eventuell anderen Hydroxychinonen als Folgeprodukte. Wie das Absorptionsspektrum dieser Oxydationslösung in der ersten Phase zeigt, tritt neben dem für die Rotfärbung verantwortlichen Maximum bei Hydroxy-p-benzochinon von 488 mμ eine Doppelbande bei 407 und 429 mμ auf, die nach den Messungen von EIGEN und MATTHIES (1961) auf das semi-p-benzochinoide Radikal zurückzuführen ist.

Das intramolekulare Chinhydron von 4,4'-Dihydroxy-di-p-benzochinon-2,5,2',5' wird in natronalkalischer Lösung zu einer Verbindung oxydiert, deren Färbung ziemlich beständig ist. Eine in ihrem Farbton ähnliche grüne Lösung ergibt Oxydationsprodukte von Tetrahydrochinon in alkalischer Lösung (ERDTMAN, GRANATH und SCHULTZ 1954); da die Identität der jeweils entstehenden Grünfärbung noch nicht feststeht, ist diese in der Abbildung mit einem Strich umrandet.

Die grüne alkalische Lösung des intramolekularen Chinhydrons von 4,4'-Dihydroxy-di-p-benzochinon-2,5,2',5' wird durch Ansäuern rot. Beim Ausschütteln geht diese mit violetter Farbe in Äther; das intramolekulare Chinhydron von 4,4'-Dihydroxy-di-p-benzochinon-2,5,2',5' löst sich in Äther ebenfalls mit violetter Farbe. Es steht auch in diesem Falle noch nicht fest, ob die Färbungen jeweils durch die gleiche Verbindung bewirkt werden. Diese Beobachtungen sind jedoch der Anfang, um die Bildung der „Synthese"-Huminsäuren durch Oxydation von Hydrochinon über den Grad der Dimeren hinaus zu untersuchen; hierfür gibt es bisher nur hypothetische Annahmen.

2. *Dimerisierung.* Das semi-p-benzochinoide Radikal kann in zwei mesomeren Grenzformen formuliert werden.

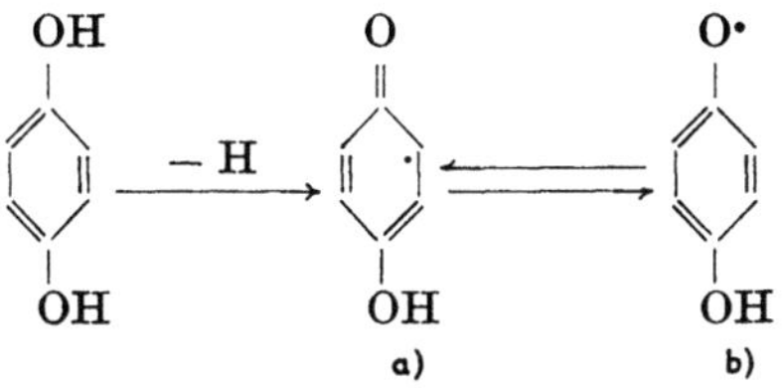

Bei einer Kupplung zwischen a) und b) ist auch die Bildung von Diphenyläthern zu erwarten. Diese sind bisher bei der alkalischen Oxydation von Hydrochinon noch nicht gefunden worden. Ihre Bildung ist sowohl bei den Ansätzen zur Herstellung von „Synthese"-Huminsäuren als auch bei der Bildung der Huminsäuren in der Natur deshalb sehr unwahrscheinlich, da nach GÜNDEL und PUMMERER (1937) 4,4'-Dihydroxy-diphenyläther durch dreiwertige Eisensalze leicht in zwei Moleküle p-Benzochinon gespalten wird. Diese Überlegung gilt ebenfalls für die Bildung von „Synthese"-Huminsäuren aus Brenzcatechin und Pyrogallol.

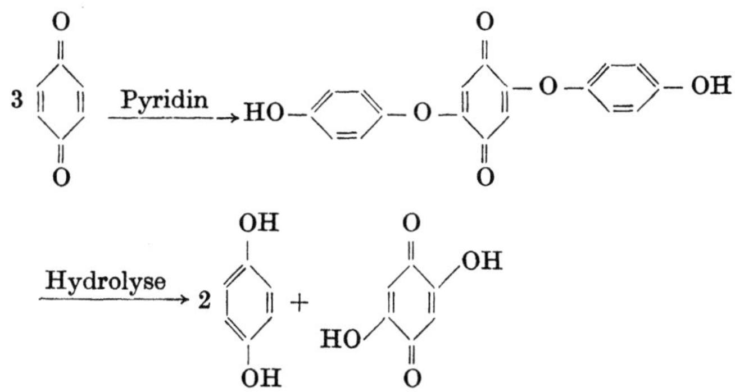

Eine Reaktion, die man vermutlich teilweise mit der Grenzformel eines Sauerstoffradikals von p-Benzochinon formulieren kann, ist die Trimerisierung von Chinon in Pyridin (DIELS und KASSEBART 1937). Doch auch derartige Reaktionsprodukte sind in den Ansätzen zur Herstellung von „Synthese"-Huminsäuren kaum zu erwarten, da das trimere Chinon in Alkali wenig beständig ist und zu zwei Molekülen Hydrochinon und zu 2,5-Dihydroxy-p-benzochinon hydrolisiert wird.

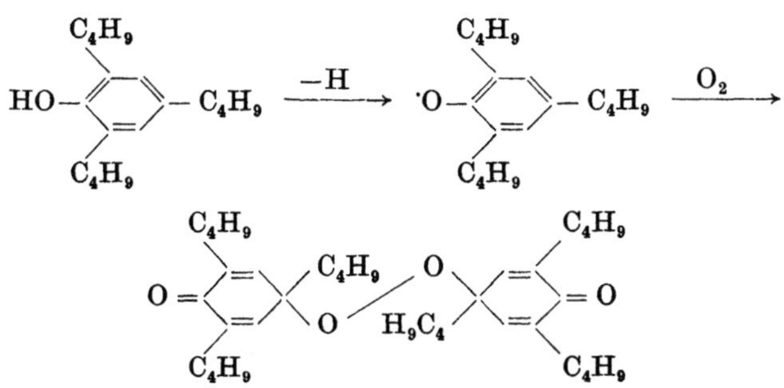

Bei mit tert. Butylgruppen substituierten Monophenolen gelang es, durch Dehydrierung stabile Sauerstoffradikale zu erhalten. Diese reagieren mit Sauerstoff zu chinoiden Peroxyden. Derartige Sauerstoffradikale reagieren mit Phenolen zu Chinoläthern (Müller, Ley und Schiechte 1957).

Durch Substitution mit aliphatischen Resten hat man bei der Oxydation in alkalischer Lösung die Reaktionsmöglichkeiten von Hydrochinon einzuschränken versucht.

Außerdem sollen diese Untersuchungen dazu dienen, Erkenntnisse über die Bildungsmöglichkeiten von Huminsäuren aus verschiedenen phenolischen Verbindungen aus Pflanzen und Mikroorganismen zu gewinnen.

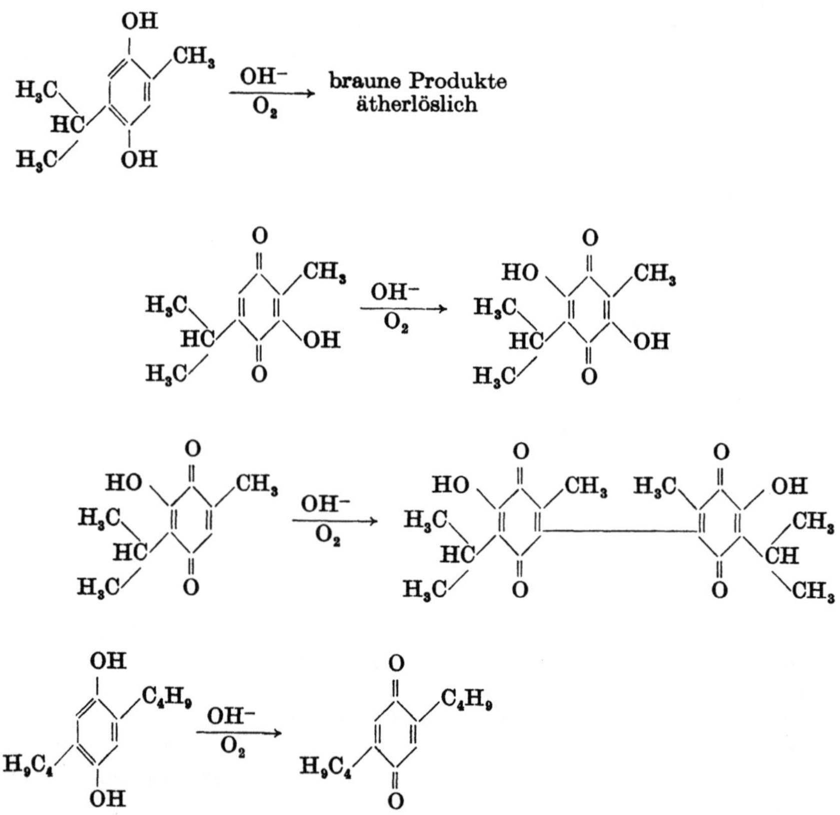

Die Oxydation von Thymohydrochinon ergibt bei pH-Werten zwischen 8 und 9 beim Schütteln mit Sauerstoff wie Hydrochinon ebenfalls dunkelgefärbte Produkte. Diese sind jedoch im Gegensatz zu denen aus Hydrochinon in Äther löslich. Als Oxydationsprodukte entstehen die beiden isomeren Hydroxy-thymochinone. Diese reagieren unter den angegebenen Bedingungen je nach der Stellung der OH-Gruppen unterschiedlich. 3-Hydroxy-thymochinon wird zu 3,6-Dihydroxythymochinon oxydiert. Aus dem 6-Hydroxy-chinon entsteht das entsprechende Dichinon. Die freie Drehbarkeit der beiden Chinonringe ist behindert, wie aus den UV-Spektren hervorgeht (Flaig und Salfeld 1958).

2,5-Ditert. butyl-hydrochinon wird in wässeriger alkalischer Lösung nur bis zum entsprechenden Chinon oxydiert und fällt aus (FLAIG und SCHULZE 1952). Es wird in diesem Falle also weder eine OH-Gruppe eingeführt noch tritt eine Dimerisierung ein.

Von ERDTMAN und STJERNSTRÖM (1959) konnte nachgewiesen werden, daß p-Benzochinon in verdünnter Schwefelsäure zu einem Trimeren polymerisiert, dessen Konstitution durch Untersuchungen an trimerem Tolu- und trimerem Äthylchinon festgelegt wurde.

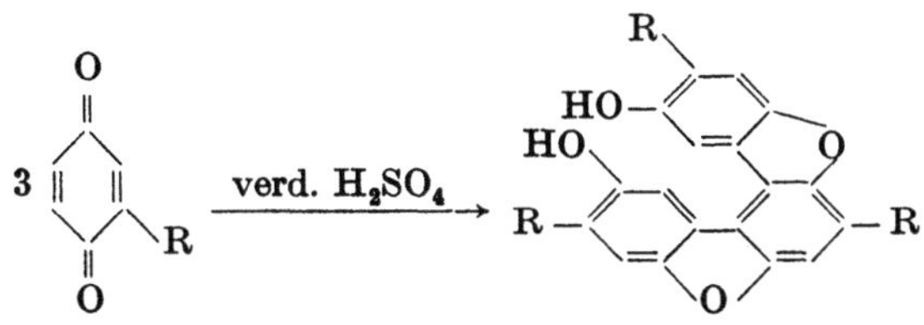

$R = H;\ CH_3;\ C_2H_5$

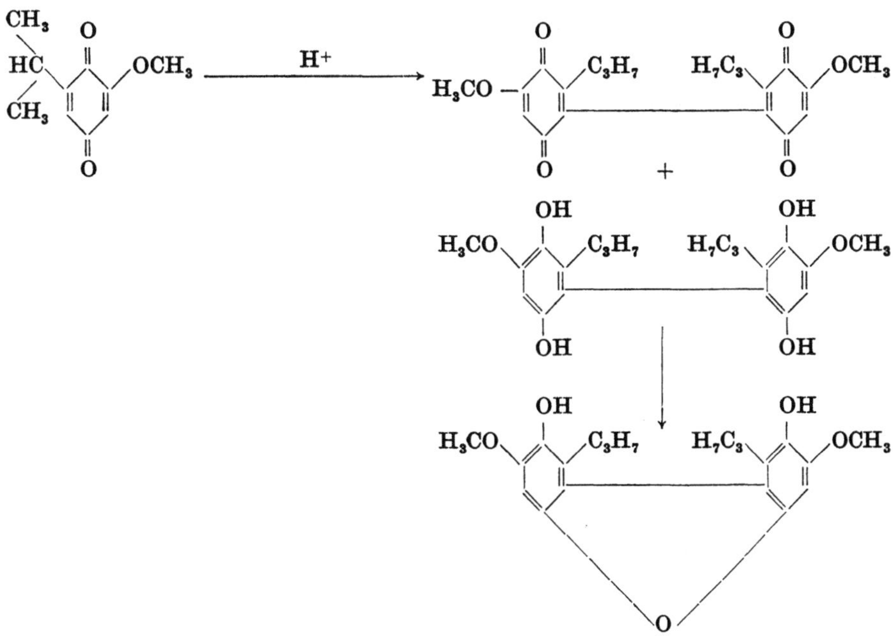

Methoxy-p-benzochinon polymerisiert schon in stark verdünnter Schwefelsäure zu 4,4′-Dimethoxy-p-benzochinon-2,5,2′,5′. Aus 2-Methoxy-6-isopropyl-p-benzochinon bildet sich das entsprechende Dichinon und Dihydrochinon, letzteres geht unter Wasserabspaltung in ein Diphenylenoxid über. Die Polymerisation erfolgt bei beiden Chinonen jeweils in p-Stellung zur Methoxylgruppe:

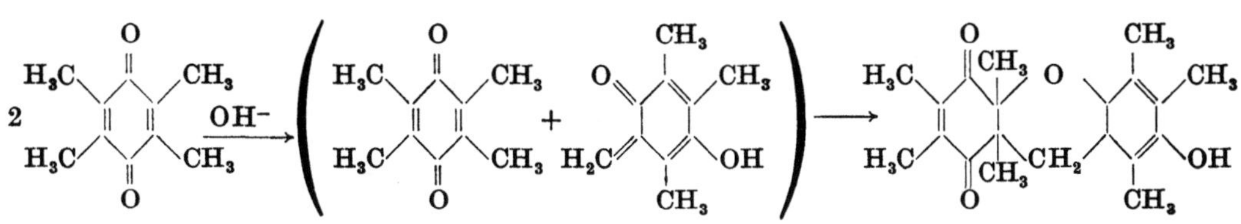

Bei Durochinon findet in alkalischer Lösung eine Dimerisierung durch Diensynthese statt. So reagiert ein Molekül Durochinon mit einem zweiten Molekül, das tautomer als o-Chinonmethid reagiert. Der Verlauf des Absorptionsspektrums zwischen 280 bis 380 mμ von substituierten p-Benzochinonen weist darauf hin, daß eine Verschiebung der Elektronen nach dem o-chinoiden System erfolgen kann (FLAIG, PLOETZ und KÜLLMER 1955, FLAIG und SALFELD 1958).

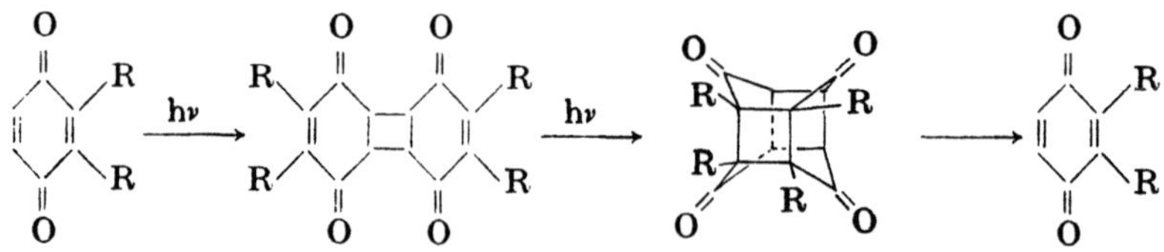

Auch durch die Wirkung von Licht findet eine Dimerisierung verschiedener Chinone statt. 2,3-Dimethyl- (YATES, ARDAO und FIESER 1956, FLAIG, SALFELD und LLANOS 1960) und 2,5-Dimethyl-p-benzochinon (COUKSON, COX und HUDEC 1961) gehen in festem Zustand stufenweise in zwei Dimerisierungsprodukte über. Das zweite Dimerisierungsprodukt von 2,3-Dimethyl-p-benzochinon läßt sich thermisch zu 2,3-Dimethyl-p-benzochinon spalten. Dieser Befund ist ein wichtiger Hinweis auf die Konstitution der beiden entstehenden Dimeren des o-substituierten Chinons. In wässeriger Lösung oder in Essigester ist bei 2,3-Dimethyl-p-benzochinon durch Belichtung ebenfalls eine Dimerisierung festgestellt worden. Auch Naphthochinon-1,4 dimerisiert unter der Einwirkung von Licht; die Konstitution des Dimeren ist analog der des ersten Dimerisierungsproduktes von 2,3-Dimethyl-p-benzochinon. Die Beispiele für die Dimerisierung durch Licht sind voraussichtlich nicht in dem Maße für die Bildung von Humusstoffen als für die Einwirkung derartiger Chinone auf den Stoffwechsel von Pflanzen und der damit veränderten physiologischen Wirksamkeit von Bedeutung.

Diese Modelluntersuchungen zeigen recht anschaulich, inwieweit die Reaktionen der 1,4-Diphenole mit Sauerstoff in alkalischer Lösung von den Substituenten abhängig sein können.

1,2-Diphenole. 1. *Oxydation.* Bei der Oxydation von Brenzcatechin entsteht o-Benzochinon. Bisher gelang es bei keiner Untersuchung, 4-Hydroxy-o-benzochinon aufzufinden. Es würde immer nur das energieärmere Hydroxy-p-benzochinon und nie das 4-Hydroxy-o-benzochinon entstehen. Schon WILLSTÄTTER und MÜLLER (1911) konnten zeigen, daß bei der Oxydation von Hydroxyhydrochinon als einziges Reaktionsprodukt Hydroxy-p-benzochinon gebildet wird.

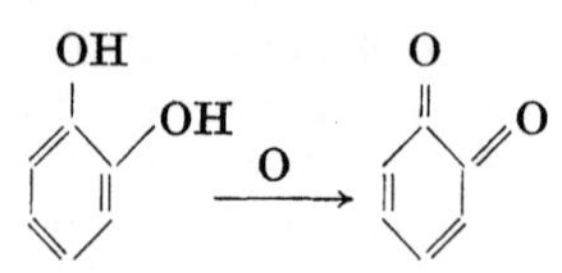

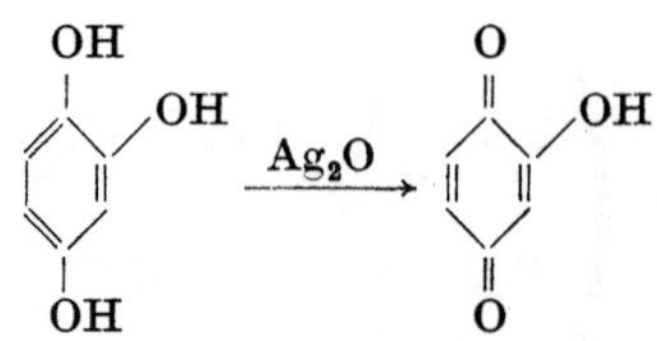

2. *Dimerisierung.* Das semichinoide Radikal, das erste Oxydationsprodukt von Brenzcatechin, kann in drei Grenzzuständen formuliert werden. LE CLERC, MONDY, DOUZOU und LISSITZKY (1959) konnten das Vorhandensein eines Radikals bei der Einwirkung von Polyphenoloxydasen aus Champignon nachweisen. Es ergibt sich aus der Kombination der Reaktionsmöglichkeiten der als Kohlenstoffradikal formulierten Grenzzustände die Bildung von drei isomeren Tetrahydroxydiphenylen.

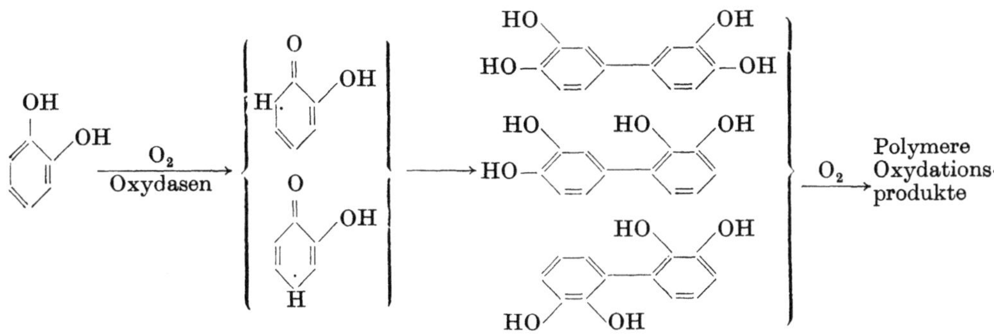

Es gelang, alle drei verschiedenen Tetrahydroxydiphenyle bei der Oxydation von Brenzcatechin in Gegenwart von sauerstoffübertragenden Enzymen darzustellen (FORSYTH und QUESNEL 1957). Die weitere Oxydation führt zu dunkelbraunen, polymeren Produkten mit Eigenschaften, die denen von Huminsäuren ähnlich sind. Bisher war es noch nicht möglich, bei dieser Reaktion Diphenyläther zu isolieren. Die Bildung von Diphenylendioxid-2,3-chinon spricht jedoch dafür, daß bei der enzymatischen Oxydation von Brenzcatechin auch Kohlenstoff-Sauerstoffbindungen auftreten (FORSYTH, QUESNEL und ROBERTS 1960).

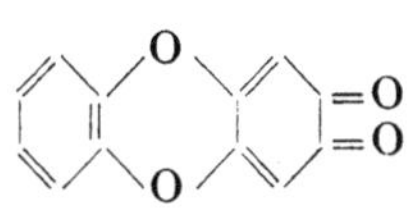

In Gegenwart von Kupfer-amin-komplexen als Modell für sauerstoffübertragende Enzyme wird ein Diamin-o-chinonderivat gebildet (BRACKMAN und HAVINGA 1955a, 1955b, 1955c, 1955d). Diese Untersuchungen sind auch auf andere Phenole ausgedehnt worden; aus Phenol entsteht die gleiche Verbindung. (Siehe Formel auf S. 426 oben.)

Partiell methoxylierte Polyphenole sind für Modelluntersuchungen in bezug auf die Veränderungen von Ligninbestandteilen oder Abbauprodukten von Interesse. So gelang es im Falle der Oxydation von Guajakol sowohl unter physiologischen Bedingungen in Gegenwart einer Polyphenolase vom Laccasetyp wie auch mit Eisen-III-chlorid die drei möglichen, isomeren Dihydroxy-dimethoxydiphenyle zu erhalten. Die entsprechenden Chinone konnten ebenfalls identifiziert werden (WALTER 1960). Diese sind sehr reaktionsfähig und unterliegen der weiteren Polymerisation. Dabei entstehen polymere, dunkelgefärbte Produkte.

Bei der Oxydation von Guajakol mit Phenolasen ist die Bildung von 2,2'-Dihydroxy-3,3'-dimethoxy-diphenyl, die Verknüpfung der Ringe in o-Stellung zur OH-Gruppe bevorzugt. In Gegenwart von Eisen-III-chlorid bilden sich alle

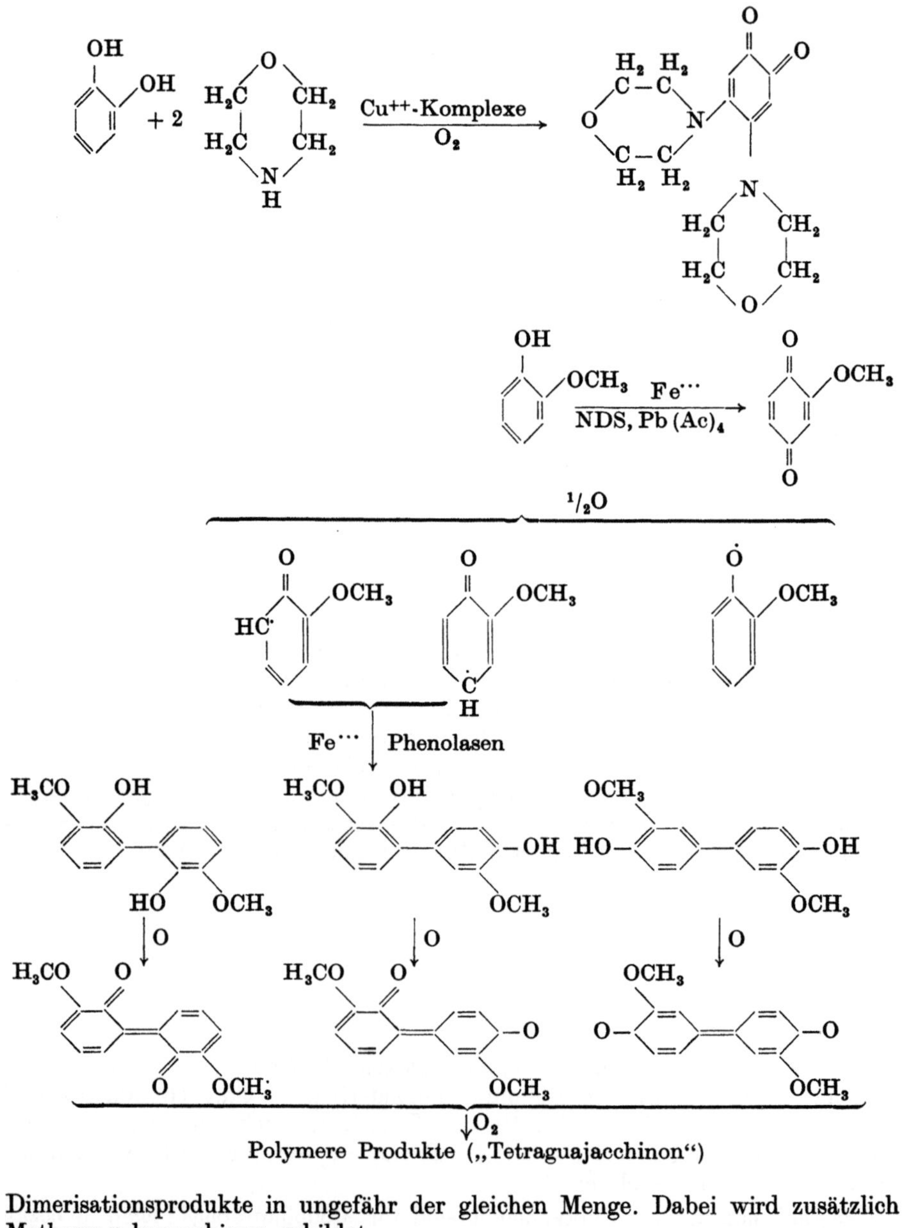

Dimerisationsprodukte in ungefähr der gleichen Menge. Dabei wird zusätzlich Methoxy-p-benzochinon gebildet.

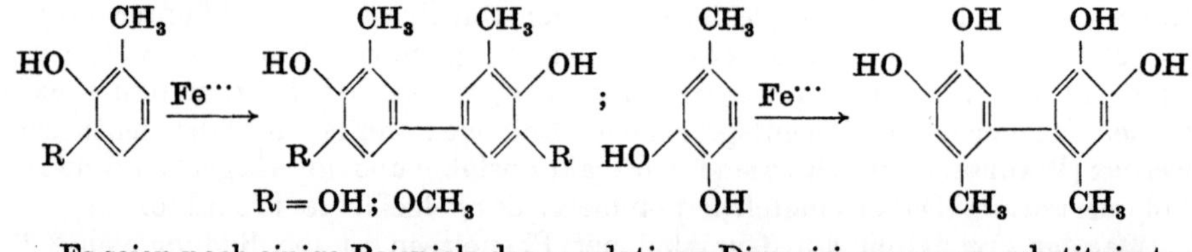

Es seien noch einige Beispiele der oxydativen Dimerisierung von substituierten 1,2-Diphenolen bzw. deren Monomethyläther aufgeführt. So dimerisiert 3-Methylbrenzcatechin in Gegenwart von dreiwertigen Eisenionen zu 3,3'-Dimethyl-4,5,4'5'-tetrahydroxy-diphenyl (MAJIMA und TAKAYAMA 1920). Die Dimerisierung von

1-Methoxy-3-methylbrenzcatechin verläuft in der gleichen Weise. 4-Methylbrenzcatechin dimerisiert zu 2,2'-Dimethyl-4,4',5,5'-tetrahydroxy-diphenyl (Burton und Hopkins 1952):

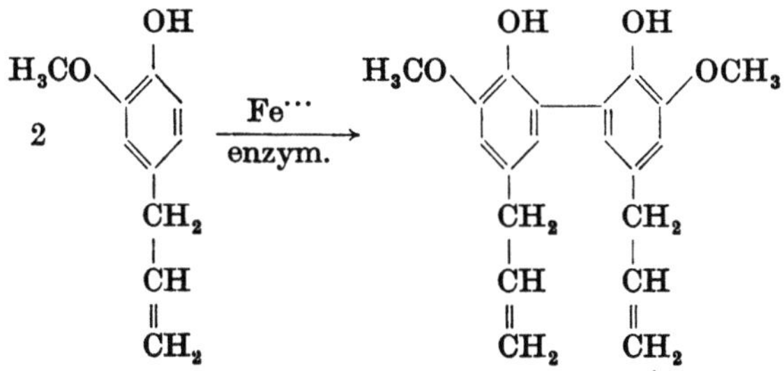

a + a

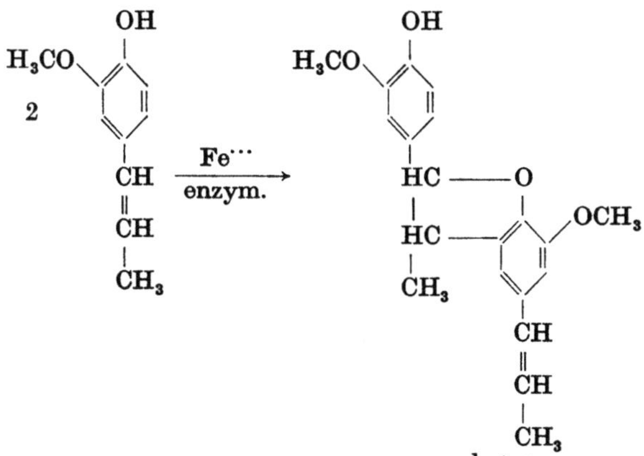

b + a

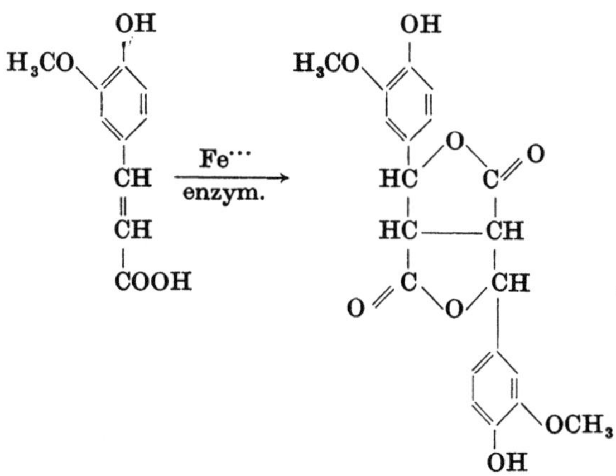

b + b

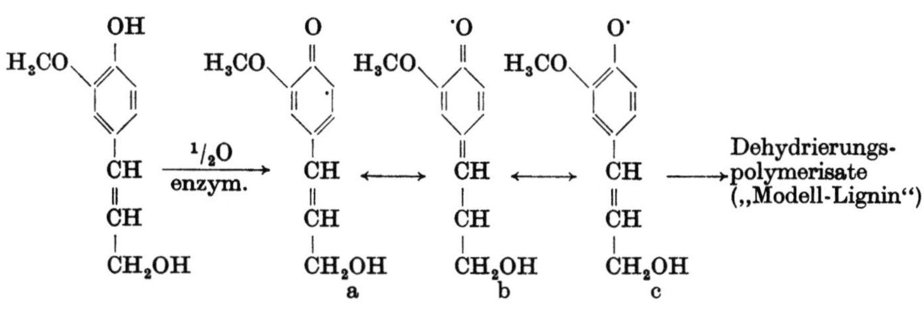

Die Polymerisation kann durch die Lage einer Doppelbindung in der Seitenkette unterschiedlich beeinflußt werden. So dimerisiert Eugenol unter Bildung von Dehydrodieugenol. Die Polymerisation verläuft (ERDTMAN 1933e) unter Verknüpfung beider Ringe in o-Stellung zur OH-Gruppe. Isoeugenol dimerisiert zu Dehydro-di-isoeugenol. Die Polymerisation erfolgt sowohl am β-ständigen Kohlenstoffatom der Seitenkette als auch in o-Stellung zur Hydroxylgruppe am Phenylring (ERDTMAN 1933f, 1935). Im Falle von Ferulasäure findet die Verknüpfung der Monomeren jeweils am β-ständigen Kohlenstoffatom der Seitenkette zu Dehydrodiferulasäure statt. Diese Polymerisation ist im Zusammenhang mit der Bildung von Humusstoffen insofern von Interesse, als sowohl aus frischem Stroh (BÖRNER 1956) wie aber auch aus Stroh, das verschieden lange verrottet worden ist, Ferulasäure isoliert wird (MAEDER 1959).

Zum Vergleich mit den Vorgängen bei der Bildung von Lignin sind die mesomeren Grenzformen der ersten Oxydationsprodukte von Coniferylalkohol angegeben worden, wie sie FREUDENBERG (1956) formuliert. Durch die oxydative Polymerisation in Gegenwart von sauerstoffübertragenden Enzymen entstehen Dehydrierungspolymerisate, die eine große Ähnlichkeit mit dem in der Natur entstehenden Lignin haben. Die verschiedenen mesomeren Grenzformen beteiligen sich in unterschiedlicher Weise.

Im Falle der Dimerisierung von Eugenol reagiert dieses nach der Formulierung a, bei Isoeugenol ein Molekül nach a und eines nach b und bei Ferulasäure beide nach b.

FREUDENBERG (1959) machte bei seinen Untersuchungen über die Dehydrierungspolymerisate die interessante Beobachtung, daß Sinapinalkohol für sich allein bei der Oxydation in Gegenwart von sauerstoffübertragenden Enzymen nicht polymerisiert, aber im Gemisch mit Coniferylalkohol in die Dehydrierungspolymerisation eingebaut wird. Auf die oxydative Dimerisierung von partiell methylierten Pyrogallolderivaten soll später nochmals näher eingegangen werden.

3. *Diensynthese von o-Benzochinonen.*

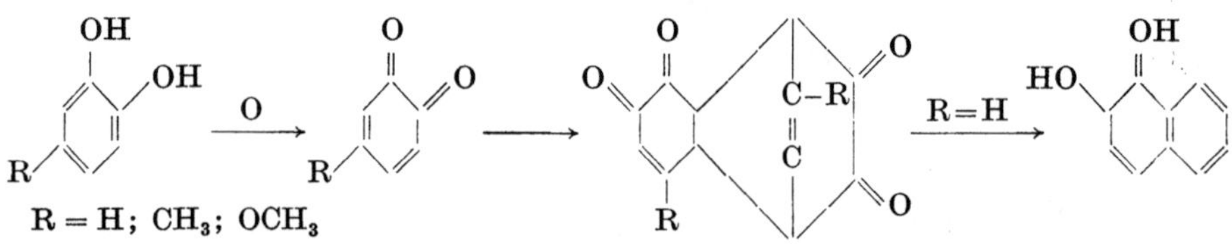

Eine weitere Möglichkeit der Dimerisierung besteht darin, daß o-Benzochinone eine Diensynthese eingehen. Dieser Reaktionsablauf wurde erstmalig von HORNER und STURM (1957) am Beispiel des 4,5-Dimethyl-o-benzochinons aufgefunden und von PATCHETT und WITKOP (1957) und HARLEY-MASON und LAIRD (1958) auf o-Benzochinon übertragen. Die Struktur des dimeren o-Benzochinons wurde durch den Abbau zum 1,2-Dihydroxy-naphthalin bewiesen (HORNER und DÜRCKHEIMER 1958). Entsprechende Dimere wurden aus 4-Methyl-o-benzochinon (HORNER und DÜRCKHEIMER 1959b) und 3-Methoxy-o-benzochinon (ADLER und MAGNUSSON 1959) erhalten. Besonders leicht dimerisieren 3-Hydroxy-o-benzochinone nach diesem Schema (SALFELD 1960).

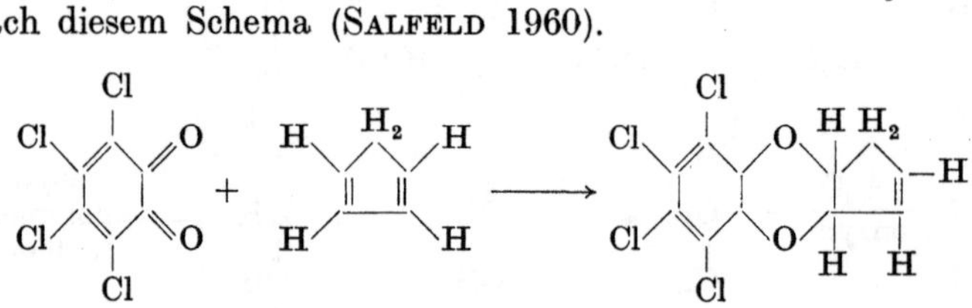

In anderer Weise reagiert Tetrachlor-o-benzochinon nach dem Prinzip einer Diensynthese mit Philodienen z. B. mit Cyclopentadien unter Bildung von Phenylendioxiden.

Triphenole. Im Falle von Trihydroxybenzolen, so von Pyrogallol, lassen sich ebenfalls verschiedene mesomere Grenzzustände für das Semichinon formulieren. Es könnten bei der Kondensation drei isomere Hexahydroxy-diphenyle entstehen. Aus Pyrogallol wird in barytalkalischer Reaktion durch Oxydation 2,3,4,2′,3′,4′-Hexahydroxy-diphenyl gebildet, das bei weiterer Oxydation in dunkelbraun gefärbte Stoffe übergeht, die als „Synthese"-Huminsäuren für die natürlichen als Vergleichssubstanzen gedient haben.

Die entsprechende Dicarbonsäure, die 2,3,4,2′,3′,4′-Hexahydroxy-diphensäure aus Gallussäure, findet sich in der Natur in den Ellagtanninen. Corilagin ist ein Mol Glucopyranose, bei dem in 1-Stellung β-glucosidisch 1 Mol Gallussäure und in 3- und 6-Stellung 1 Mol Hexahydroxy-diphensäure gebunden ist (SCHMIDT und LADEMANN 1951, O. TH. SCHMIDT und D. M. SCHMIDT 1952).

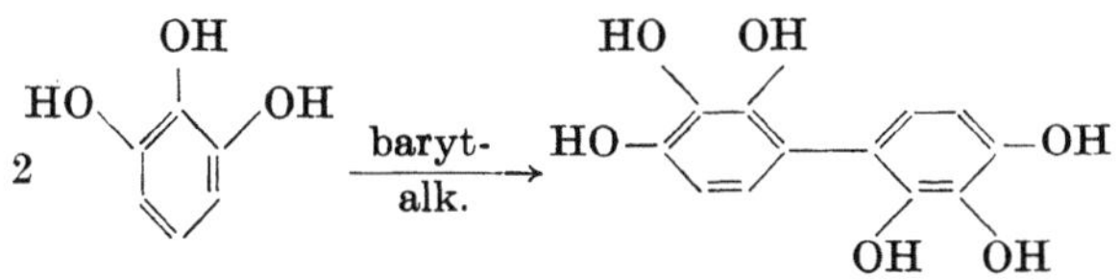

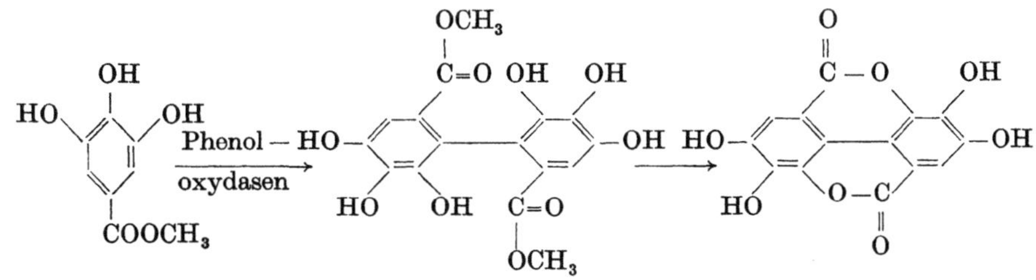

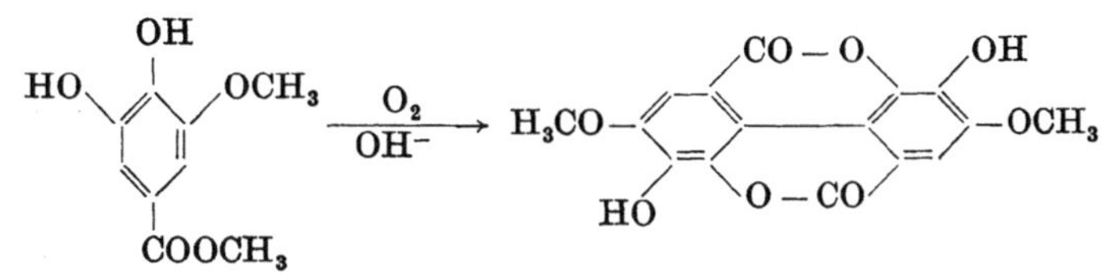

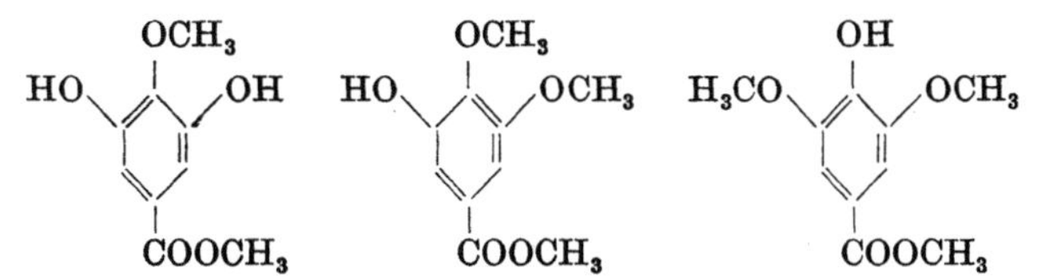

keine Dimerisierung

Im Zusammenhang mit der Diskussion über die Bildung von Huminsäuren aus Abbauprodukten des Lignins sind auch die Reaktionen partiell methylierter Gallussäureester von Interesse. Der Methylester der Gallussäure geht bei der Oxydation in Gegenwart von Polyphenoloxydasen in Ellagsäure (STEINMETZ 1956) über, wobei in geringen Mengen der Diester der 4,5,6,4′,5′,6′-Hexahydroxy-diphensäure als Zwischenprodukt (HATHWAY 1958) isoliert werden konnte. Aus dem in 3-Stellung methylierten Gallussäuremethylester entsteht bei der Oxydation in verdünnten Laugen die in 5,5'-Stellung methylierte Ellagsäure. In 4- oder in 4,5-Stellung methylierte Gallussäureester und der Ester der Syringasäure bilden keine entsprechenden Diphensäuren (HATHWAY 1958). Nach Auffassung des Autors ist damit erwiesen, daß nur bei zwei o-ständigen Hydroxylgruppen die Dimerisierung stattfindet, und diese über eine o-chinoide Zwischenstufe verläuft (HATHWAY 1957).

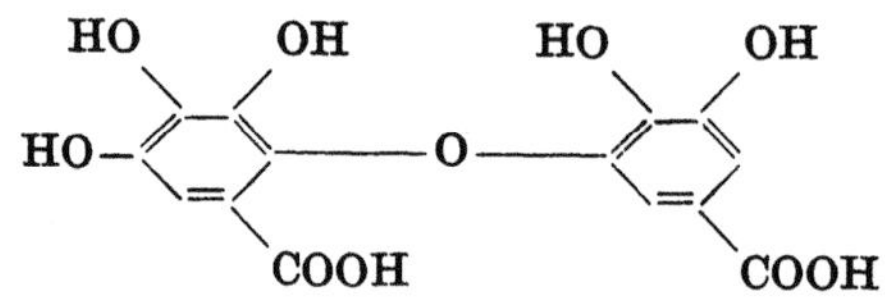

Von den vier möglichen Isomeren der Pentahydroxy-diphenyläther aus Pyrogallol ist als Naturstoff die Dehydrodigallussäure bekannt (MAYER 1952).

Diese disproportioniert in alkalischer Lösung zu Gallussäurechinon und Gallussäure. Das Chinon geht auch in Abwesenheit von Sauerstoff in dunkelgefärbte Produkte über (SCHMIDT und MAYER 1956, MAYER, FIKENTSCHER, J. SCHMIDT und O. TH. SCHMIDT 1960).

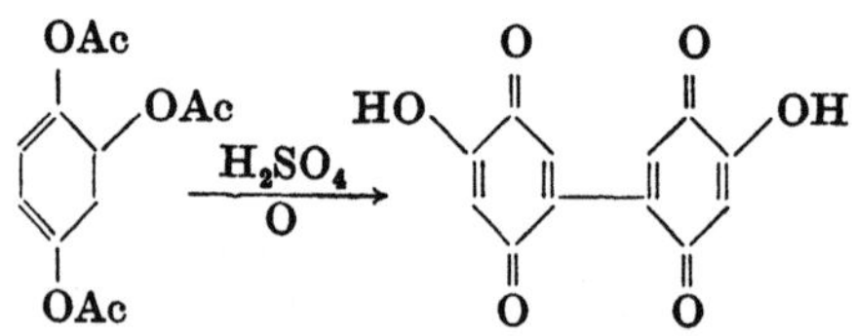

Im Falle von Hydroxychinon läßt sich die Bildung von 6 Isomeren Hexahydroxydiphenylen formulieren. 2,4,5,2',4',5'-Hexahydroxydiphenyl wird intermediär durch Verseifen von Hydroxyhydrochinontriacetat in verdünnter Schwefelsäure gebildet und kann bei gleichzeitiger Oxydation als entsprechendes Dichinon isoliert werden. Das intermolekulare Chinhydron entsteht schon in Gegenwart von Luft beim Versetzen des Diacetats mit verdünnter Schwefelsäure.

Von den neun möglichen Pentahydroxydiphenyläthern aus Hydroxyhydrochinon sind in der Natur bisher noch keine entsprechenden Derivate gefunden worden.

Diensynthese von 3-Hydroxy-o-benzochinonen; Bildung von Tropolonen.

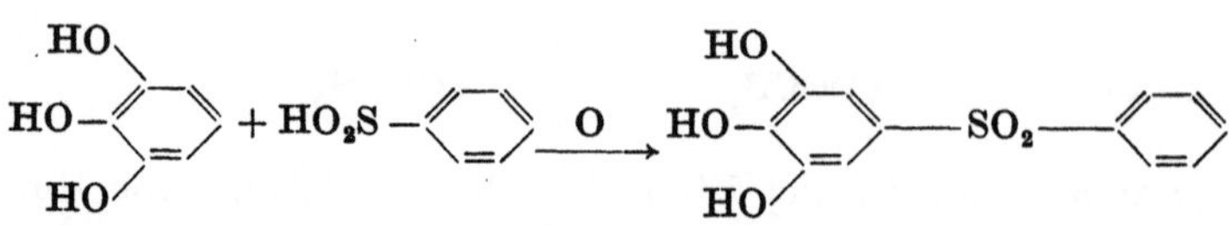

Wird Pyrogallol in wässeriger Kaliumjodatlösung in Gegenwart von Benzolsulfonsäure dehydriert, so bildet sich Trihydroxy-diphenylsulfon (HORNER und DÜRCKHEIMER 1959c). Dieser Befund spricht für die Bildung von 3-Hydroxy-o-benzochinon als erstem Reaktionsprodukt (HINSBERG und HIMMELSCHEIN 1896).

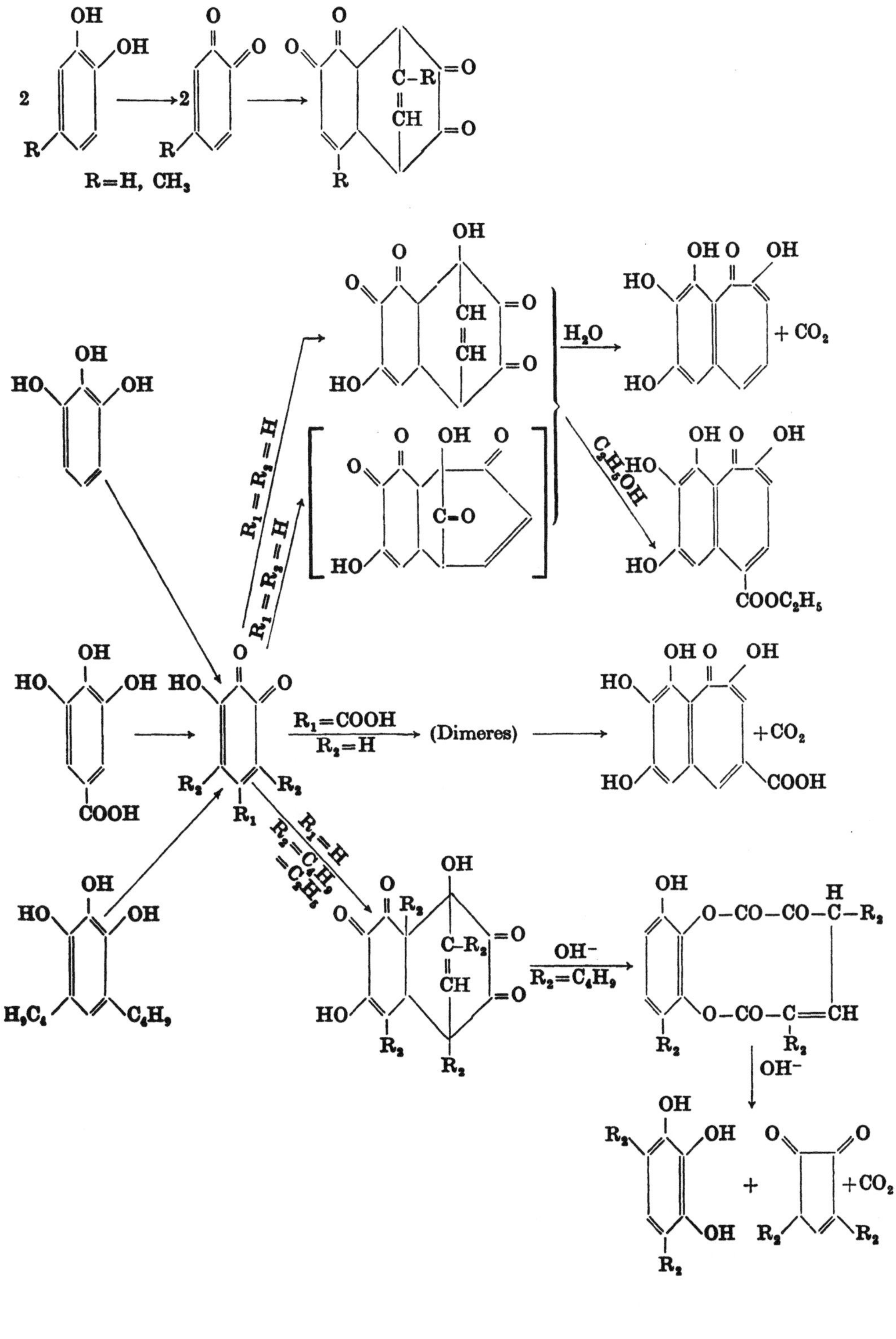

OH
OH
2
R
O
O
2
R
R=H, CH3
O
O
C-R
=O
CH
=O
R
OH
HO
OH
O
O
CH
=O
CH
=O
HO
H2O
OH O OH
HO
+ CO2
HO
O OH O
O
C=O
HO
C2H5OH
OH O OH
HO
HO
COOC2H5
R1=R2=H
R1=R2=H
OH
HO
OH
COOH
HO
O
O
R2
R2
R1
R1=COOH
R2=H
(Dimeres)
OH O OH
HO
+CO2
HO
COOH
R1=H
R2=C4H9
=C6H5
OH
HO
OH
H9C4
C4H9
O
O
R2
OH
C-R2
=O
CH
=O
HO
R2
R2
OH-
R2=C4H9
OH
O-CO-CO-C-R2
H
O-CO-C=CH
R2
R2
OH-
OH
R2
OH
OH
R2
O
O
+
+CO2
R2
R2

Je nach der Art der Substitution bilden sich aus Pyrogallol und dessen Derivaten unterschiedliche Reaktionsprodukte, die jedoch alle über ein 3-Hydroxy-o-benzochinonderivat als Zwischenverbindung und in weiterer Reaktion teilweise über Dienaddukte entstehen. Im Falle von Pyrogallol wird über ein in seiner Konstitution noch nicht bekanntes dimeres Reaktionsprodukt je nach der Art des Lösungsmittels Purpurogallin oder Purpurogallin-γ-carbonsäureester (SALFELD 1957) gebildet. Die isomere Purpurogallin-β-carbonsäure entsteht aus Gallussäure (PERKIN und STEVEN 1903, CROW und HAWORTH 1951) und enzymatisch (HERMANN 1954, STEINMETZ 1956). Durch Substitution von Pyrogallol mit tert. Butyl- oder Äthylgruppen in 4,6-Stellung verläuft die Reaktion nur bis zu einem Dienaddukt (SALFELD 1960). In alkalischen Lösungen wird der zweite Ring aufgespalten, und der Diester von 2,4-Di-tert.butyl-4-oxalocrotonsäure mit 4,6-Di-tert.butyl-pyrogallol gebildet (SALFELD und BAUME 1960). Bei der Behandlung dieses Esters mit Alkali entsteht 4,6-Di-tert. butyl-pyrogallol und Cyclopentandion, das mit tert. Butylresten substituiert ist. Einige weitere Abbauprodukte sind noch von CAMPBELL (1951) und SCHULZE und FLAIG (1952) beschrieben worden.

HORNER und DÜRCKHEIMER (1959c) geben eine Vorschrift für die Darstellung von Benzotropolonen.

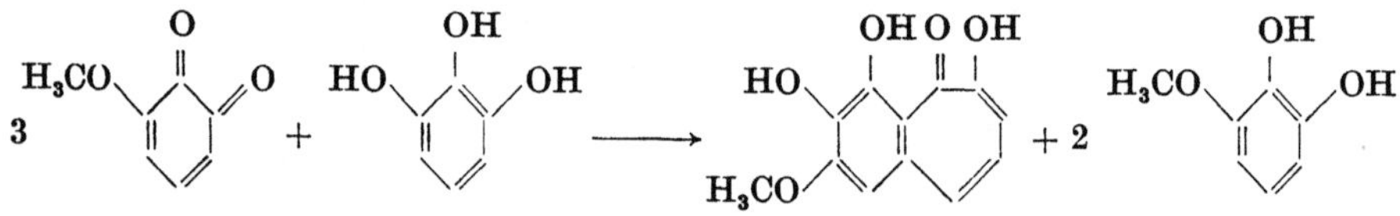

Drei Moleküle eines o-Benzochinons, das eine unbesetzte Doppelbindung haben muß, und ein Mol Pyrogallol, das in 4-, 5- oder 6-Stellung einfach oder mehrfach substituiert sein kann, setzen sich in guter Ausbeute zu Benzotropolonen um.

So ist auch die Bildung des in 3'-Stellung methylierten Purpurogallins durch die Umsetzung des dem Pyrogallol-1-methyläther entsprechenden Chinons mit Pyrogallol bekannt geworden (WILLSTÄTTER und HEISS 1923).

In einer Zusammenfassung über pflanzliche Gerbstoffe macht HATHWAY (1957, 1958, HATHWAY und SEKINS 1957) den Vorschlag, daß die Polymerisation von Catechin zu Phlobatanninen über o-chinoide Zwischenstufen verläuft.

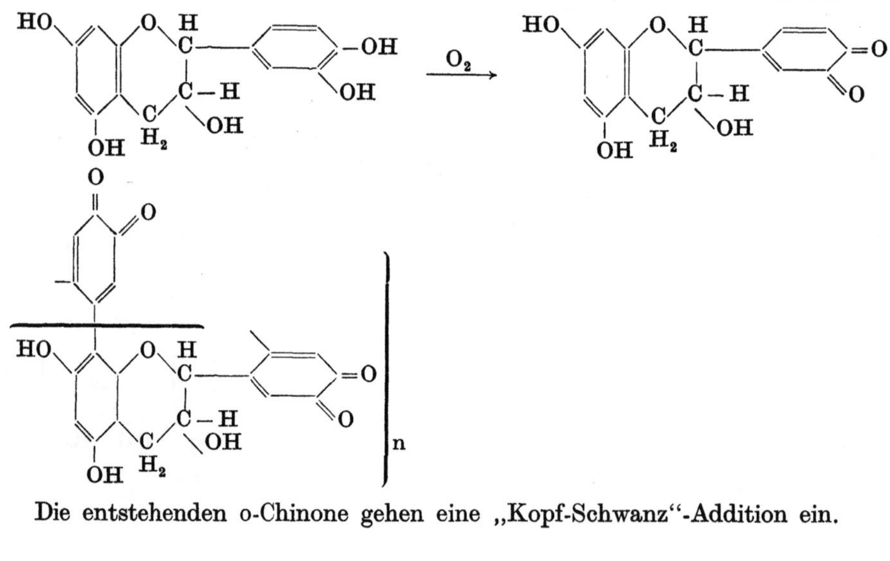

Die entstehenden o-Chinone gehen eine „Kopf-Schwanz"-Addition ein.

Oxydative Kupplung von Polyphenolen.

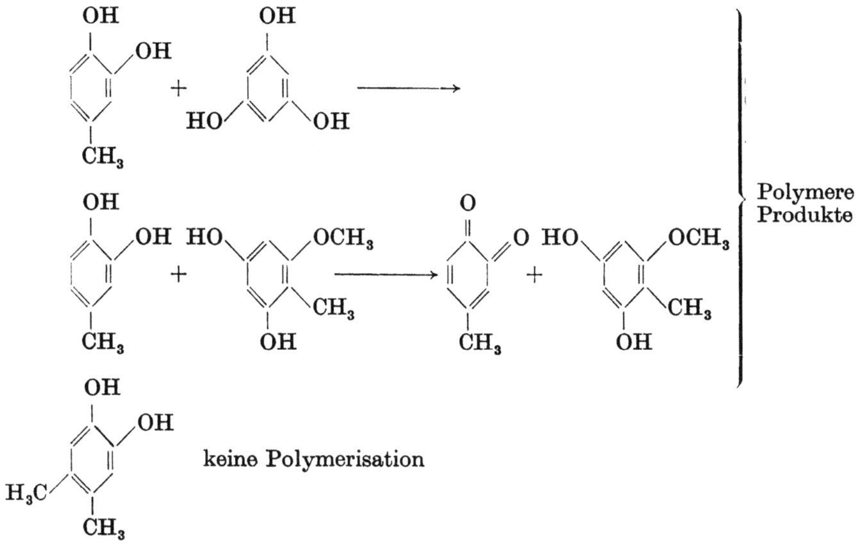

Äquimolare Mischungen von 4-Methylbrenzcatechin mit Phloroglucin oder 4-Methyl-5-methoxy-resorcin geben bei der Luftoxydation Polymere, deren Lichtabsorption denen der oxydierten Catechine sehr ähnlich ist. Der Reaktionsverlauf wird als eine oxydative Addition von 4-Methyl-5-methoxy-resorcin an das gebildete 4-Methyl-o-benzochinon beschrieben. Da eine Addition im Falle von 4,5-Dimethyl-brenzcatechin nicht stattfindet, ist es wahrscheinlich, daß bei den aus oxydiertem Catechin entstandenen Polymeren die Verknüpfung der Bausteine über den o-chinoiden Anteil verläuft. Diese Vorstellung konnte durch weitere Untersuchungen gestützt werden (HATHWAY 1958). Diese Beispiele zeigen, daß auch Polyphenole, die in 1,3- oder 1,3,5-Stellung mit OH-Gruppen substituiert sind, sich an der Bildung dunkelgefärbter polymerer Produkte beteiligen können.

Verknüpfung von Phenolen bzw. Chinonen durch aliphatische Brücken.

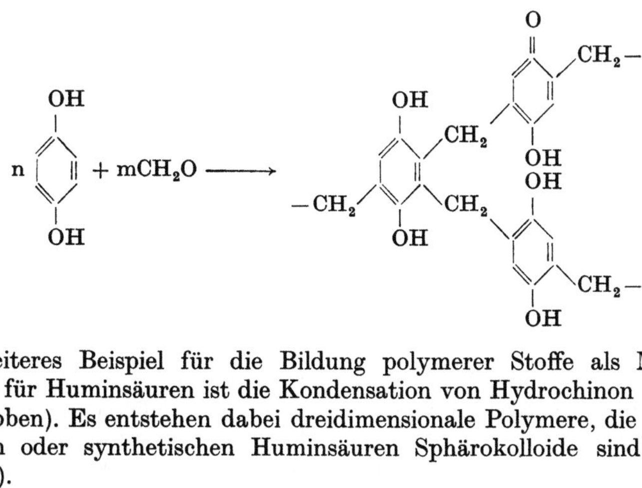

Ein weiteres Beispiel für die Bildung polymerer Stoffe als Modelluntersuchungen für Huminsäuren ist die Kondensation von Hydrochinon mit Formaldehyd (s. oben). Es entstehen dabei dreidimensionale Polymere, die ähnlich den natürlichen oder synthetischen Huminsäuren Sphärokolloide sind (HULTZSCH 1950, 1962).

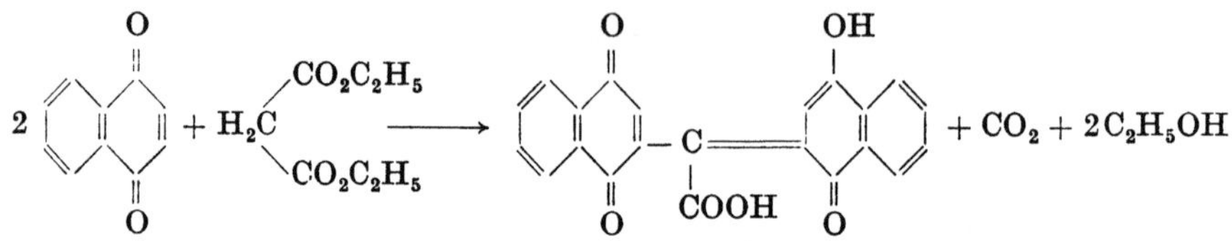

Eine weitere Möglichkeit der Dimerisierung besteht in der Verknüpfung der Ringe über Kohlenstoffbrücken. So reagieren Verbindungen mit aktiven Methylengruppen mit Chinonen, wobei Dichinoylmethylenverbindungen entstehen. Ein Beispiel hierfür ist die Reaktion von Malonsäure mit Naphthochinon (PRATT und BOEHME 1951).

Nach Verseifung und Decarboxylierung entsteht eine Verbindung, bei der auf Grund der Aromatisierungstendenz ein Teil des Chinons als o-Chinon vorliegt.

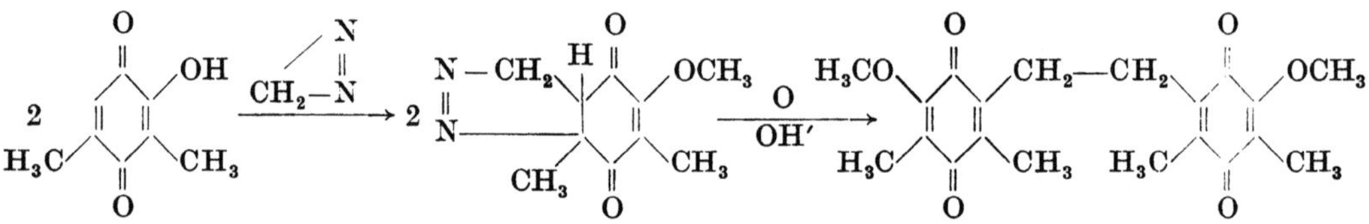

In diesem Zusammenhang sind auch die Umsetzungsprodukte von Diazomethan mit verschiedenen Chinonen zu erwähnen. Behandelt man diese mit verdünntem Alkali, so entsteht z. B. im Fall von 2-Hydroxy-3,5-dimethyl-p-benzochinon ein Dichinon, bei dem die beiden Chinongruppen über eine Aethylenbrücke miteinander verbunden sind (EISTERT 1949, MARINI-BETTOLO und PAOLONI 1954, EISTERT und BOCK 1959, RABIEN 1959). Diese Reaktion ist nicht so sehr für die Bildung von Humusstoffen interessant als für deren Untersuchung selbst, da es bei der Methylierung von Huminsäuren mit Diazomethan und nachfolgender Behandlung mit Alkali auf diese Weise zu Teilchenvergrößerungen kommen könnte.

Polyphenole bzw. Chinone als Elektronenaustauscher.

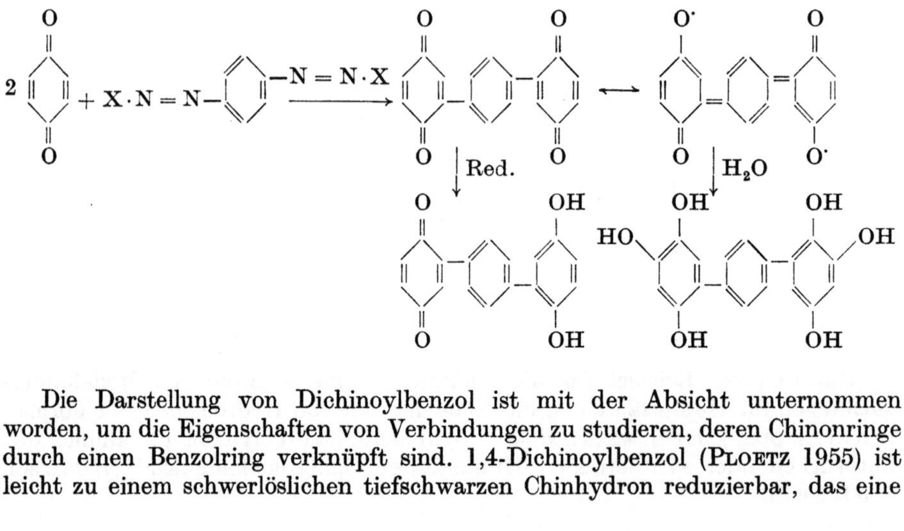

Die Darstellung von Dichinoylbenzol ist mit der Absicht unternommen worden, um die Eigenschaften von Verbindungen zu studieren, deren Chinonringe durch einen Benzolring verknüpft sind. 1,4-Dichinoylbenzol (PLOETZ 1955) ist leicht zu einem schwerlöslichen tiefschwarzen Chinhydron reduzierbar, das eine große Neigung besitzt, auf der Chinhydronstufe zu verharren. Das Chinhydron ist nicht wie in üblicher Weise intermolekular, sondern intramolekular.

Da die natürlichen Huminsäuren nicht nur durch ihre Säurennatur als Kationenaustauscher, sondern zweifellos auch als Elektronenaustauscher für die Vorgänge im Boden von Bedeutung sind, so sind Untersuchungen über inter- und intramolekulare Chinhydrone von Interesse. Auf Grund seines chemischen Verhaltens und seiner Lichtabsorption kann 1,3-Dichinoylbenzol auch als o-chinoides Diradikal formuliert werden. Sollten in Huminsäuren Bauelemente durch Systeme konjugierter Doppelbindungen miteinander verknüpft sein, so ist der rein chinoide Zustand sehr unwahrscheinlich, da das Redoxpotential zu hoch ist. Die tautomeren o-chinoiden Radikale können durch einfache Anlagerung von Wasser in Hydroxyhydrochinone übergehen. Diese sind wiederum sehr reaktionsfähig. Durch Dehydrierung können Hydroxychinone entstehen. Es konnte gezeigt werden, daß ein Gemisch dieser Stoffe ein kontinuierliches UV-Spektrum ergibt, wie es bei natürlichen und synthetischen Huminsäuren in den Arbeiten von Frömel (1938a, 1938b, 1941) und Scheffer und Welte (1950a, 1950b) beschrieben worden ist.

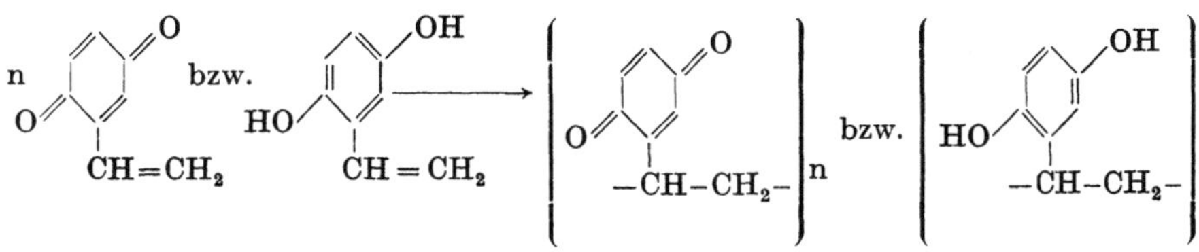

Um über die möglichen Eigenschaften der Huminsäuren als Elektronenaustauscher Erfahrungen zu sammeln, können die Untersuchungen von Cassidy (1949) und Updegraff und Cassidy (1949) mit Vinylchinon bzw. Vinylhydrochinon herangezogen werden. Das polymere Hydrochinon zeigt eine stärkere Neigung zur Semichinonbildung als das monomere.

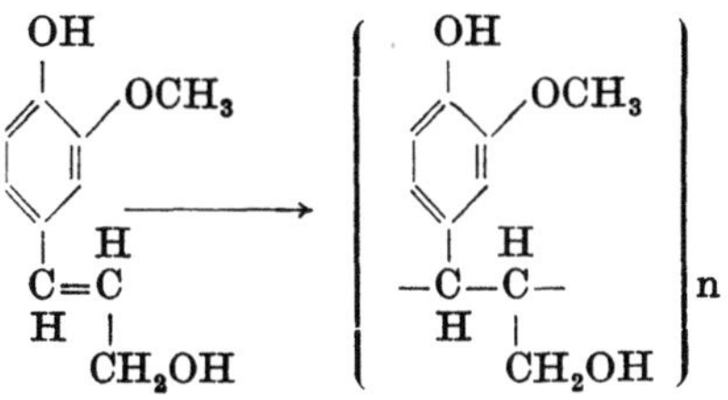

Auch im Falle des Lignins besteht die Möglichkeit, daß ein Anteil der monomeren Bausteine durch Äthylenpolymerisation verknüpft ist (Adler 1957).

Oxydative Ringspaltung von Polyphenolen bzw. Benzochinonen. Die Oxydation von Polyphenolen, die durch den Abbau von Ligninen entstehen oder die zur Herstellung von Synthesehuminsäuren herangezogen werden, führt nicht nur zur Einführung einer weiteren OH-Gruppe, zur Bildung von Chinonen und deren Dimeren, sondern in weiterer Folge auch zur Spaltung des Ringes. Dabei sind Chinone nicht immer Zwischenstufen. Bei enzymatisch katalysierter Reaktion kann der Sauerstoff direkt an den Ring angelagert werden.

In diesem Zusammenhang sollen vor allem die enzymatisch katalysierten Oxydationsreaktionen erwähnt werden und die nichtkatalysierten, soweit sie im Rahmen der Chemie der Humusstoffe von Interesse sind.

Da die an diesen Oxydationsreaktionen beteiligten Enzyme oft sehr spezifisch sind, sollen zunächst deren Funktionen und die danach erfolgte Einteilung Erwähnung finden.

Nach den von MASON (1957) entwickelten Vorstellungen kann man die Enzyme, die die Umsetzung von Sauerstoff mit phenolischen Verbindungen katalysieren, in drei Gruppen einteilen:

Gruppe 1: Sauerstofftransferasen: $A + O_2 \rightarrow AO_2$. Diese übertragen ein Molekül Sauerstoff direkt auf das Substrat (Brenzcatechinoxydase, Protocatechusäureoxydase, Tryptophanoxydase u. a.).

Gruppe 2: „Mixed function oxydases" („Oxydasen mit gemischten Funktionen"). $AH + 2\,\theta + O_2 \rightarrow AOH + O^{--}$. Diese haben zwei voneinander unabhängige enzymatische Aktivitäten, einmal wird ein Atom Sauerstoff reduziert und zum anderen ein zweites Sauerstoffatom als Hydroxylgruppe im Kern eingeführt; so vermag der Phenolasekomplex eine zweite OH-Gruppe zu einer vorhandenen, meist o-ständig, einzuführen und das entstandene Diphenol zum Chinon dehydrieren.

Gruppe 3: Elektronentransferasen: Diese beschleunigen

a) die Reduktion von Sauerstoff zu Wasser:
$4\,\theta + O_2 \rightarrow 2\,O'^{--}$ $(2\,H_2O)$,

b) die Reduktion von Sauerstoff zu Wasserstoffsuperoxyd:
$2\,\theta + O_2 \rightarrow O_2^{--}$ (H_2O_2).

Zu den ersteren gehören die Laccase und die Catecholaseaktivität der Phenolase (Gruppe 2); diese führen Diphenole in Chinone über und sind bei der Dimerisierung von Phenolen beteiligt. Ferner gehören hierher die Enzyme des Cytochromsystems.

Zu den Enzymen, die den Sauerstoff zu Wasserstoffsuperoxyd reduzieren, gehören die Glucoseoxydase und verschiedene Aldehydoxydasen.

Wie durch Untersuchungen mit dem Sauerstoffisotop ^{18}O gezeigt werden konnte, ist nur das Sauerstoffmolekül und nicht der Sauerstoff des Wassers beteiligt (HAYAISHI, KATAGIRI und ROTHBERG 1955, MASON, FOWLKS und PETERSON 1955).

Bei Monophenolen tritt eine Ringspaltung nur nach Einführung einer zweiten OH-Gruppe ein. Diese erfolgt durch die Einwirkung der Enzyme der Gruppe 2 meist in o-Stellung zu der vorhandenen OH-Gruppe.

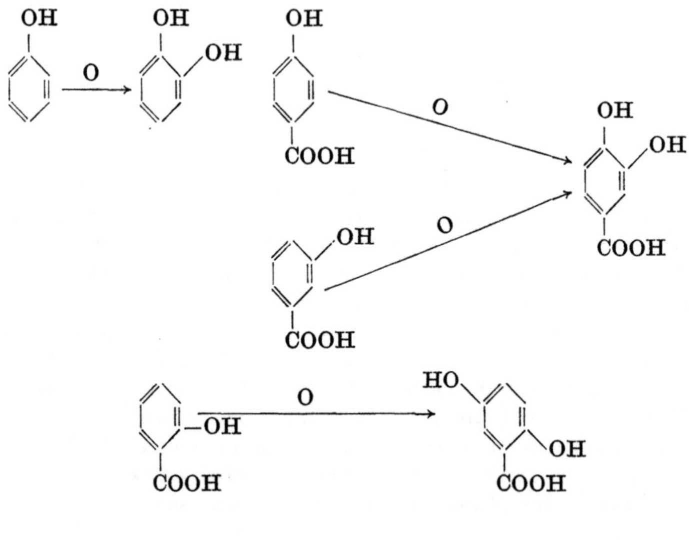

Während m- und p-Hydroxybenzoesäure in Protocatechusäure (EVANS 1947) übergehen, entsteht aus Salicylsäure Hydrochinoncarbonsäure (WALKER und EVANS 1952).

Mit den Reaktionen der enzymatischen Spaltung der Ringe bei Phenolen und Phenolcarbonsäuren haben sich STANIER (1947, 1950), KILBY (1948, 1951), EVANS, SMITH, LINSTEAD und ELVIDGE (1951) und STANIER und INGRAHAM (1954) befaßt.

Sie setzten den Kulturlösungen von Bodenbakterien, wie Pseudomonas fluorescens, Vibrioarten und Fluorobakterien, die erwähnten Substanzen zu und isolierten die Umwandlungsprodukte. Bei Untersuchungen über den Ligninabbau in der darauffolgenden Zeit (FLAIG und HAIDER 1961a, HAIDER, LIM und FLAIG 1961), die mit Weißfäulepilzen hauptsächlich mit Polystictus versicolor und dessen Enzymsystemen durchgeführt wurden, konnten ähnliche Umwandlungsprodukte festgestellt werden.

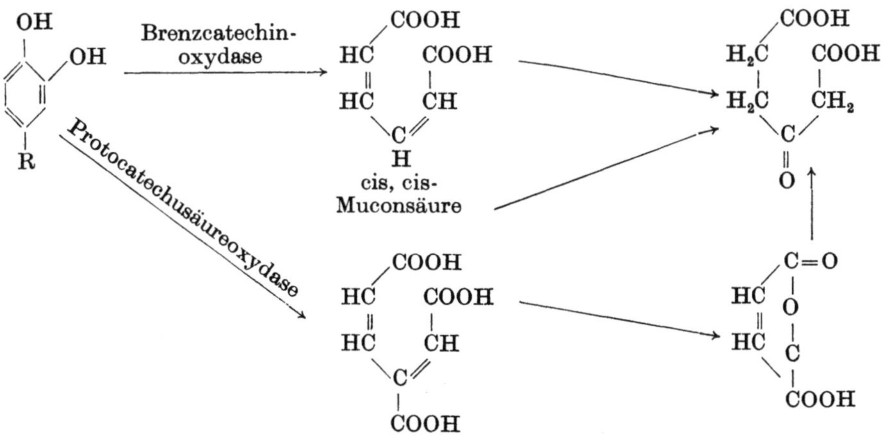

Der Ring wird zwischen den beiden OH-Gruppen durch direkte Anlagerung von Sauerstoff gespalten. Wie Untersuchungen mit ^{18}O zeigen, kann eine Hydratisierung der Doppelbindung oder o-Benzochinon als Zwischenprodukte ausgeschlossen werden. Dies geht auch aus Untersuchungen von FLAIG und HAIDER (1961a) hervor, die zu diesem Zweck an der Carboxylgruppe mit ^{14}C markierte Protocatechusäure einsetzten. Aus der entstandenen Muconsäure bzw. β-Carboxymuconsäure entsteht durch Wasseranlagerung und durch Abspaltung von CO_2 im Falle der Tricarbonsäure β-Ketoadipinsäure. Das Zwischenprodukt bei der Spaltung von Protocatechusäure, das β-Carboxy-γ-methyl-$\Delta\alpha$-butenolid, läßt sich aus Kulturlösungen des Pilzes Neurospora isolieren (GROSS, GAFFORD und TATUM 1956). Die aus Pseudomonas fluorescenz isolierte Protocatechusäureoxydase katalysiert die Oxydation von Brenzcatechin, Hydroxy-p-benzochinon und 2-, 3-, 4-, 2,3- und 2,4-Hydroxybenzoesäuren nicht und ist damit sehr spezifisch (CAIN 1961).

HAYAISHI, KATAGIRI und ROTHBERG (1955) schlagen als Mechanismus der Ringspaltung eine intermediäre Bildung von Peroxiden vor.

TRIPPETT, DAGLEY und STOPHER (1960) sowie DAGLEY, EVANS und RIBBONS (1960) beschreiben eine andere oxydative Spaltung des Brenzcatechins und der Protocatechusäure, bei der die oxydative Spaltung nicht zwischen den beiden Hydroxylgruppen, sondern zwischen dem C-Atom 5 und 6 erfolgt und ein Halbaldehyd der Carboxymuconsäure entsteht; dieser kann mit Ammoniak zu einem Pyridinderivat cyclisieren.

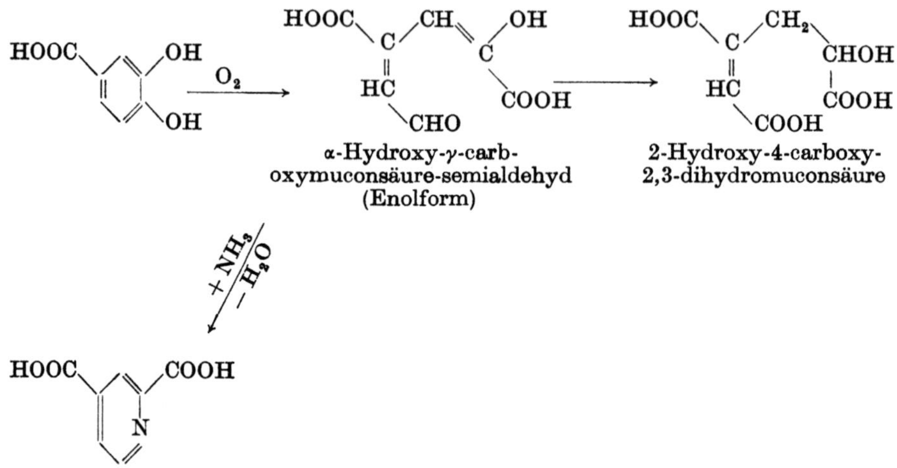

α-Hydroxy-γ-carboxymuconsäure-semialdehyd (Enolform)

2-Hydroxy-4-carboxy-2,3-dihydromuconsäure

Pyridin-α, γ-dicarbonsäure

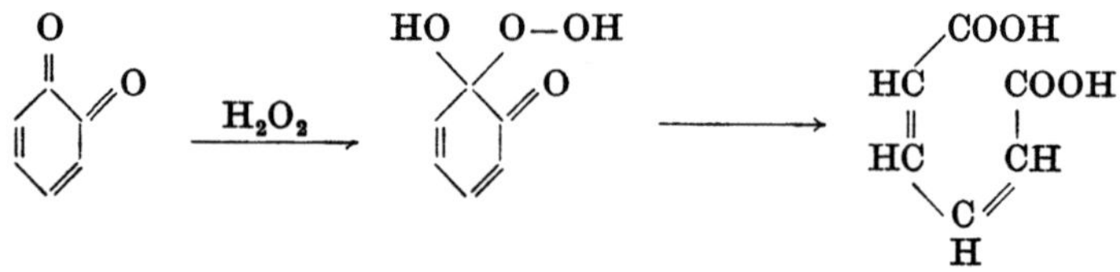

Von Patchett und Witkop (1957) wird beschrieben, daß die Oxydation von o-Benzochinon mit Wasserstoffsuperoxyd in Äther zu Muconsäure über ein Peroxid als Zwischenstufe verläuft.

Brassard und Karrer (1960) beschreiben die Oxydation von 4-Methyl-o-benzochinon und 3,4-Dimethyl-o-benzochinon zu den entsprechenden Derivaten. Die Oxydation von Guajacol (Adler und Magnusson 1959) und Veratrol (Friess, Soloway, Morse und Ingersoll 1952) mit Natriumperjodat bzw. Peressigsäure verläuft unter Ätherspaltung ebenfalls zu Muconsäure; unter solchen Bedingungen kann im Falle von Veratrol auch der Dimethylester der Muconsäue gefaßt werden.

Über die enzymatische Spaltung des Ringes von Hydrochinon ist bisher nichts bekannt. Aus den Lösungen der alkalischen Oxydation von Hydrochinon lassen sich Essigsäure, Oxalsäure und Kohlendioxid (Welte, Schatz und Ziechmann 1954) isolieren. Diese unspezifischen Oxydationsprodukte lassen keine Schlüsse über den Verlauf der Reaktion zu und besagen nur, daß bei dieser Reaktion Ringspaltungen stattgefunden haben müssen.

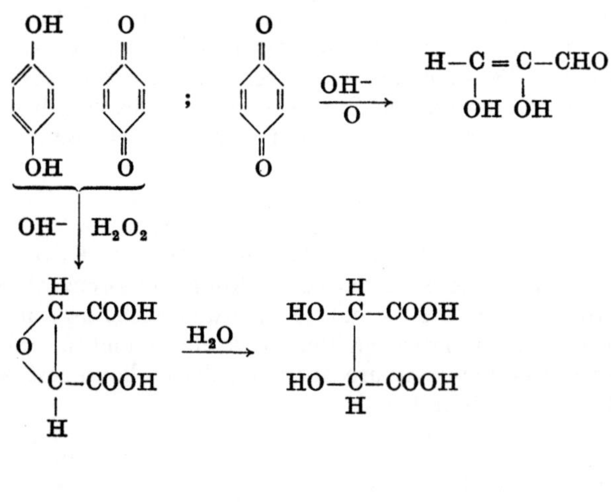

Bei der Oxydation von Hydrochinon oder Chinon mit Wasserstoffsuperoxid in alkalischer Lösung kann Traubensäure isoliert werden, die aus Äthylenoxiddicarbonsäure entsteht (WEITZ, SCHOBERT und SEIBERT 1935). EULER (1949) fand, daß Chinon in zwei Moleküle Redukton, Dihydroxyallylaldehyd, aufgespalten wird.

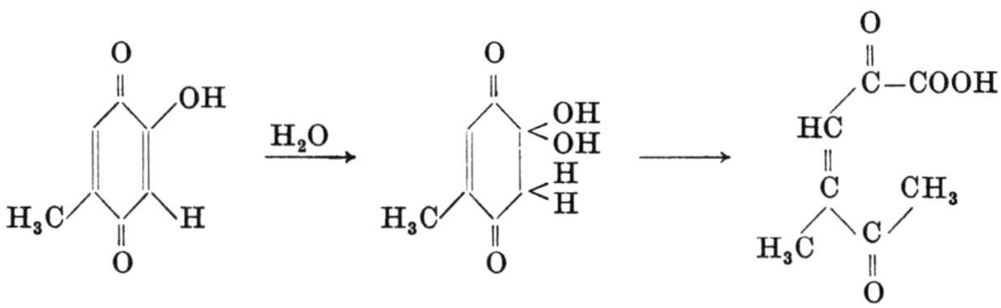

In wässeriger Lösung könnte die Ringspaltung auch über Hydratisierungsreaktionen verlaufen. Derartige Reaktionen hat SCHTSCHUKINA (1952) bei 2-Hydroxy-5-methyl-p-benzochinon und anderen festgestellt. In alkalischer Lösung werden die Hydratisierungsprodukte in ungesättigte Diketomonocarbonsäuren aufgespalten.

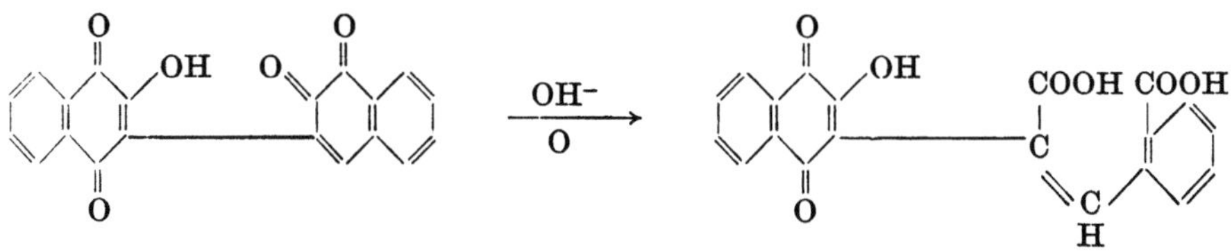

Als ein Beispiel dafür, daß o-Chinone rascher aufgespalten werden als p-Chinone, haben HOOKER und FIESER (1936) die Oxydation von 2-Hydroxy-3,3′-dinaphthyl-1,4,-1′,2′-dichinon in alkalischer Lösung beschrieben. Der ersten angegebenen Reaktionsstufe folgen noch weitere.

Auch im Falle von Pyrogallol und Gallussäure sind bisher bei der durch Enzyme katalysierten Oxydation keine Verbindungen der Ringspaltung gefaßt worden; man ist nur durch die Bildung von Kohlendioxyd bei der Oxydation von Pyrogallol in alkalischer Lösung davon unterrichtet, daß auch eine Ringspaltung stattgefunden haben muß.

RIBBONS und EVANS (1960) beschreiben eine Pseudomonasart, die auf Phthalsäure als einziger Kohlenstoffquelle wächst.

Oxydiert man jedoch 2,4-Di-tert.butyl-pyrogallol mit Phenolasen aus Champignon, so erhält man in gleicher Weise wie bei der Oxydation in schwach alkalischer Lösung 2,4-Di-tert.butyl-4-oxalocrotonsäure (CAMPBELL 1951, FLAIG und SCHULZE 1952, STEINMETZ 1956).

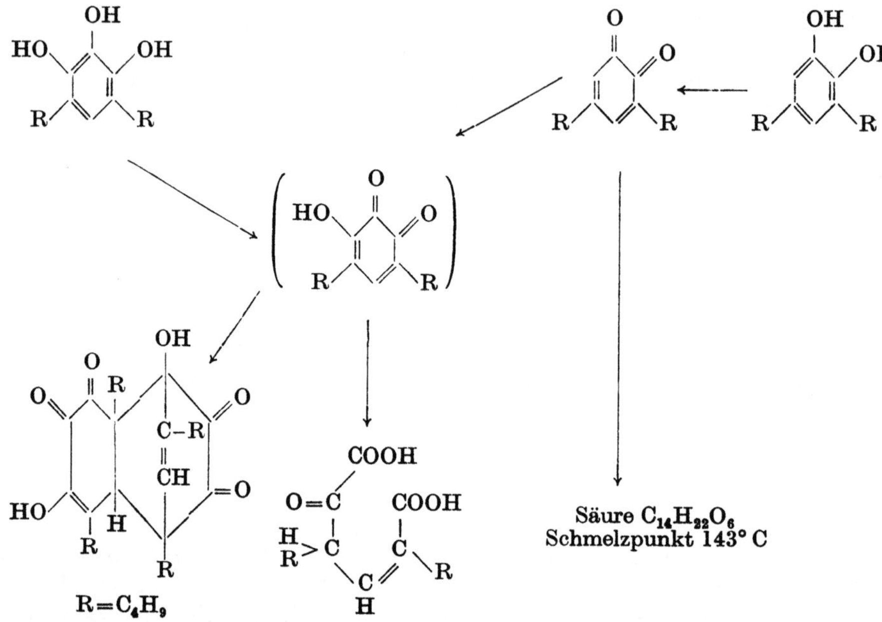

Diese entsteht voraussichtlich aus dem unbeständigen 3-Hydroxy-2,4-di-tert.butyl-o-benzochinon, dessen Bildung durch die Isolierung seines Dimeren festgestellt worden ist. Es ist noch nicht erwiesen, ob die Gegenwart eines Enzyms auch für die Ringspaltung oder nur für die Oxydation zum Hydroxy-o-chinon erforderlich ist.

Interessanterweise erhält man bei der Oxydation von 3,5-Di-tert.butyl-brenzcatechin ebenfalls 2,4-Di-tert.butyl-4-oxalocrotonsäure; neben dieser wird jedoch noch in ungefähr gleicher Menge eine zweite Säure mit der gleichen Summenformel isoliert, deren Konstitution noch nicht bekannt ist. Demnach geht 3,5-Di-tert.butyl-o-benzochinon vermutlich in das 3-Hydroxy-derivat über, aus dem durch Oxydation 2,4-Di-tert.butyl-4-oxalocrotonsäure entsteht. Es bleibt abzuwarten, wie weit sich diese Reaktion auch auf die entmethylierten Ligninabbauprodukte des Guajacyl- und Syringyltyps übertragen läßt.

Diese Reaktion auf Ligninabbauprodukte der beiden verschiedenen Typen angewandt, wäre insofern interessant, als bei beiden das gleiche Endprodukt der Oxydation entsteht.

C. Mögliche Reaktionen zur Bildung von Humusstoffen am Beispiel von Vanillin

Die bisher besprochenen Reaktionen der Phenole, die zu vergleichenden Untersuchungen herangezogen worden sind, sollen nun in einen Zusammenhang mit den möglichen Umwandlungen von Ligninabbauprodukten gebracht werden, wie sie aus verrottendem Material isoliert werden können. Zu diesem Zweck werden die Reaktionen von Vanillin angeführt. In dem angegebenen Schema sind nur solche Reaktionen aufgenommen, die in der Natur und in Kulturlösungen von Mikroorganismen ablaufen und zum Verständnis für die große Anzahl an Reaktionsmöglichkeiten bei der Bildung von Humusstoffen beitragen. Die meisten Reaktionen werden von oxydierenden Enzymen aus Mikroorganismen katalysiert.

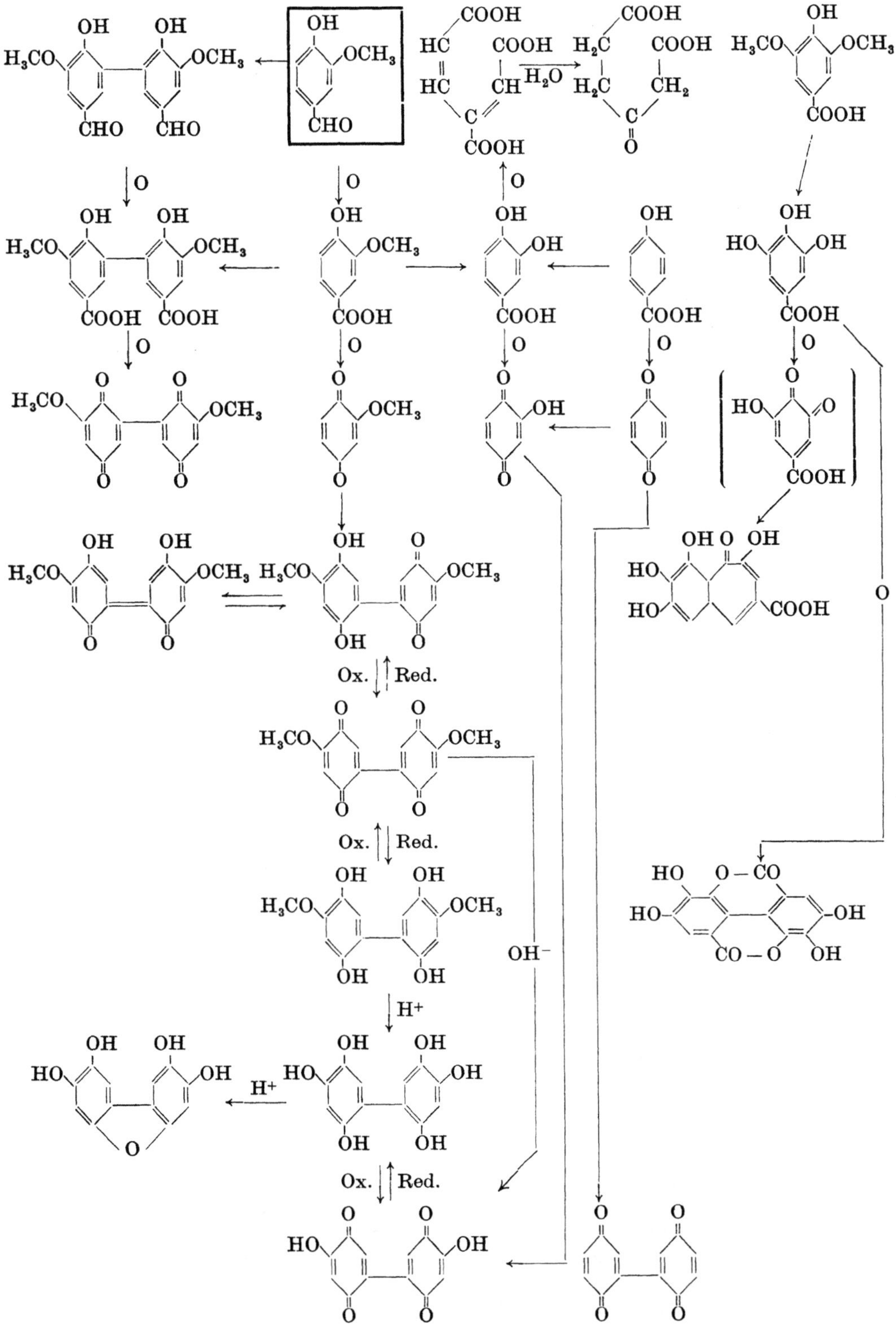

In diesem Schema sind die Reaktionen, die durch die Einwirkung stickstoffhaltiger Verbindungen möglich sind, nicht enthalten, auf die später zurückgekommen wird.

Aus Vanillin entsteht durch oxydative Dimerisierung Dehydrodivanillin. Daraus wird durch weitere Oxydation die entsprechende Dicarbonsäure gebildet, die auch durch Dimerisierung von Vanillinsäure entsteht. Letztere wird bei gleichzeitiger Abspaltung von Kohlendioxid zu Methoxy-p-benzochinon und Dehydrodivanillinsäure zu 3,3'-Dimethoxy-di-p-benzochinon-2,5,2',5' oxydiert.

Methoxy-p-benzochinon dimerisiert zu dem Chinhydron des isomeren 4,4'-Dimethoxy-di-p-benzochinon-2,5,2',5', das Erdtman (1933b) auf Grund seiner kräftigen violetten Farbe mesomer auch als ein o-Chinon formuliert. Aus dem entsprechenden Chinon entsteht durch Hydrolyse in schwach alkalischer und dem entsprechenden Hydrochinon in schwach saurer Lösung 2,4,5,2',4',5'-Hexahydroxy-diphenyl. Durch Einwirkung von Wasserstoffionen geht dieses nach Abspaltung eines Moleküls Wasser in das entsprechende Diphenylenoxid über. 2,4,5,2',4',5'-Hexahydroxydiphenyl läßt sich zu 4,4'-Dihydroxy-di-p-benzochinon oxydieren, einer Verbindung, deren Bildung bei der Oxydation von Protocatechusäure, p-Hydroxybenzoesäure über mehrere Oxydationsstufen ebenfalls möglich ist.

Protocatechusäure kann sowohl durch Entmethylierung aus Vanillinsäure wie auch durch Einführung einer OH-Gruppe in p-Hydroxy-benzoesäure entstehen. Der Weg von Protocatechusäure zu 4,4'-Dihydroxy-di-p-benzochinon-2,5,2',5' ist bisher nur über Hydroxy-p-benzochinon und nicht über eine Dimerisierung der Protocatechusäure bekannt. Die Bildung dieses Dichinons aus p-Hydroxybenzoesäure kann einmal über die Protocatechusäure auf die gleiche Weise und zum anderen über p-Benzochinon und dessen Oxydation zu Hydroxy-p-benzochinon erfolgen. Bisher ist es noch nicht gelungen, von dem Dimerisierungsprodukt von p-Benzochinon, dem Di-p-benzochinon, zu entsprechenden Dihydroxydichinonen zu gelangen (vgl. Tab. 57).

Aus Protocatechusäure entsteht durch Ringaufspaltung durch enzymatische Oxydation β-Carboxy-muconsäure, die über ein weiteres Zwischenprodukt in β-Ketoadipinsäure übergeht.

Zum Vergleich sind die Reaktionen im Falle der Syringasäure angeführt. Aus dieser entsteht durch Entmethylierung Gallussäure, die über verschiedene Zwischenprodukte in Purpurogallin bzw. Purpurogallin-β-carbonsäure (vgl. S. 437) und Ellagsäure übergeht. Weitere Verbindungen, die bei der enzymatischen Oxydation entstehen können, sind bisher nicht bekannt geworden.

Bei der enzymatischen Oxydation, wie auch bei der Oxydation in alkalischer Lösung, entstehen aus den in diesem Schema angeführten Phenolen und Chinonen dunkelgefärbte, in ihren Eigenschaften den Huminsäuren ähnliche Produkte. In Gegenwart von Enzymen enthalten diese Stickstoff.

D. Umsetzungen von Stickstoffverbindungen mit Polyphenolen bei gleichzeitiger Einwirkung von Sauerstoff

Die in der Natur vorhandenen Huminsäuren enthalten alle einen mehr oder minder großen Anteil an Stickstoff, der in den meisten Fällen zwischen 2 bis 4% beträgt. Ein Teil dieses Stickstoffs ist durch Säurehydrolyse als Aminosäuren hydrolysierbar. Es soll daher nur darauf eingegangen werden, welche Möglichkeiten sowohl bei der Bildung natürlicher als auch synthetischer Huminsäuren für den Einbau stickstoffhaltiger Verbindungen bestehen.

Niedermolekulare Verbindungen. Bei der Bildung von Lignin aus dessen Monomeren, z. B. Coniferylalkohol, ist es möglich, daß diese zu einem p-Benzochinonmethid dehydriert werden. In ähnlicher Weise, wie die Anlagerung von Wasser möglich ist, kann auch die Addition von stickstoffhaltigen Verbindungen, so z. B. die von Aminosäuren oder Peptiden, erfolgen.

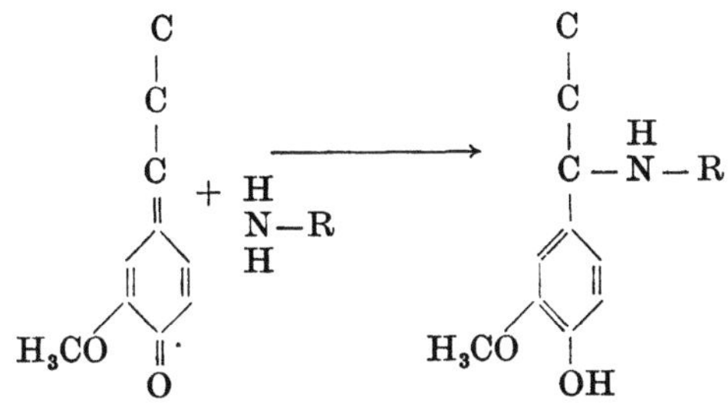

R = Aminosäure- oder Peptidrest

Abgesehen von den bekannten Reaktionen der Addition von primären Aminen an Chinonen zu stickstoffhaltigen Verbindungen besteht eine weitere Möglichkeit durch die Substitution von OH- oder Alkoxylgruppen durch Aminogruppen.

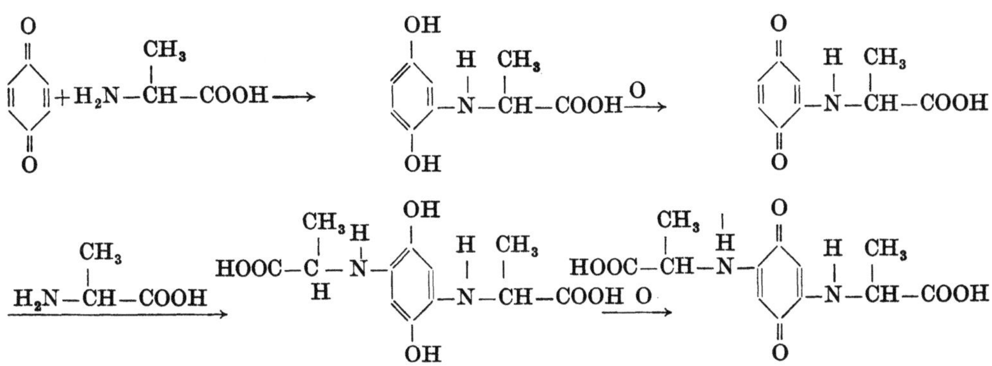

So reagiert Äthanolamin mit 2,6-Dimethoxy-p-benzochinon unter Abspaltung einer Methoxylgruppe. Bei der Einführung eines weiteren Moleküls Äthanolamin bleibt die zweite Methoxylgruppe erhalten (Huisman 1950).

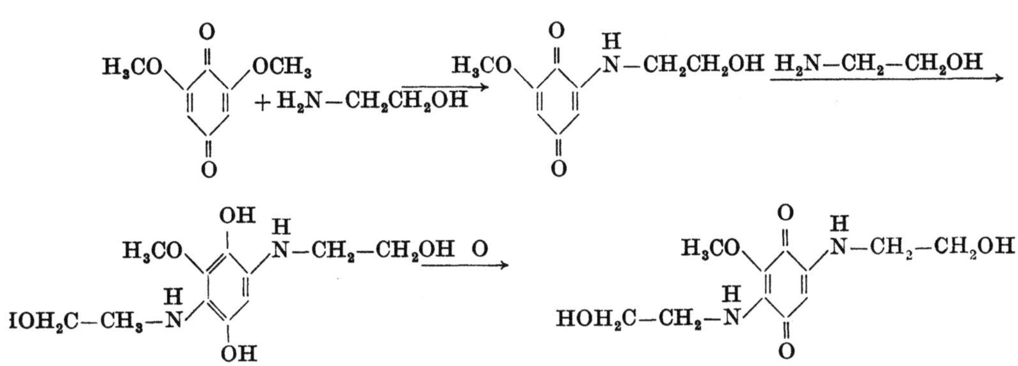

Oft entstehen durch den Einbau stickstoffhaltiger Verbindungen heterocyclische Ringe; so reagiert Orcin mit Ammoniak zu Phenoxazinderivaten (Musso, Matthies, Beecken und Krämer 1957, Beecken, Musso u. a. 1961).

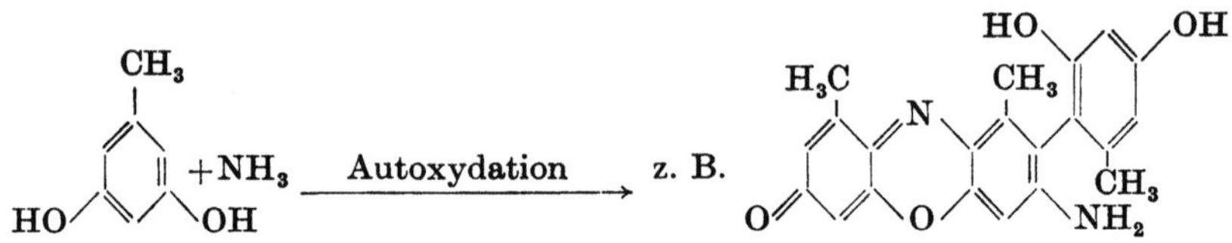

Durch Umsetzung von Cysteinesterchlorhydrat mit α-Naphthochinon oder 2,5-Dimethyl-p-benzochinon entstehen ebenfalls heterocyclische Verbindungen, die ein Stickstoff- und ein Schwefelatom im Molekül enthalten (Kuhn und Beinert 1944).

Horner und Sturm (1957) fanden, daß 4-Aminobrenzcatechin bei der Oxydation unter Polymerisation mit sich selbst reagiert und ein Phenazinderivat bildet.

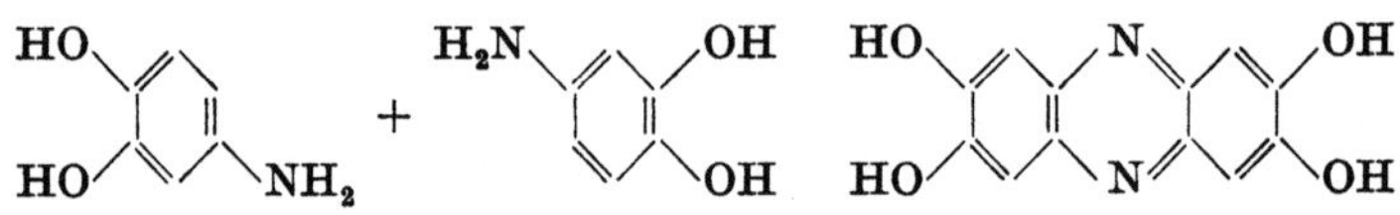

In letzter Zeit haben Michel und Lantz (1961) bekanntgegeben, daß im Hydroxyhydrochinon unter Ausschluß von Luft eine OH-Gruppe in Gegenwart von Ammoniak durch eine NH_2-Gruppe ersetzt wird.

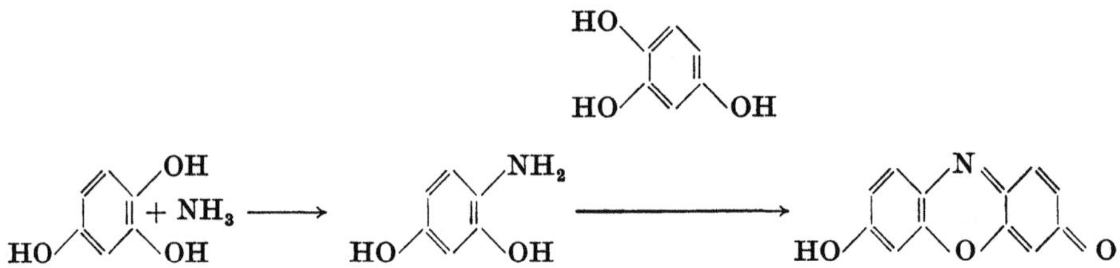

Bei der weiteren Oxydation entsteht dadurch ein 7-Hydroxy-phenoxazen-2. Diese Reaktion könnte eine Bedeutung zur Aufklärung der Bildung von Synthesehuminsäuren in ammoniakalischer Lösung haben.

Höhermolekulare Verbindungen. Bei den Reaktionen zwischen Phenolen bzw. Chinonen und stickstoffhaltigen Verbindungen in Gegenwart von Sauerstoff treten des öfteren polymere Verbindungen auf, deren Konstitution in vielen Fällen noch nicht aufgeklärt ist. Das bekannteste Beispiel im Zusammenhang mit der Chemie der organischen Stoffe des Bodens sind die „Synthese“-Huminsäuren aus Polyphenolen und Ammoniak durch alkalische Oxydation.

Einige Beispiele für die Kondensation von Aminophenolen zu höhermolekularen Produkten seien angeführt.

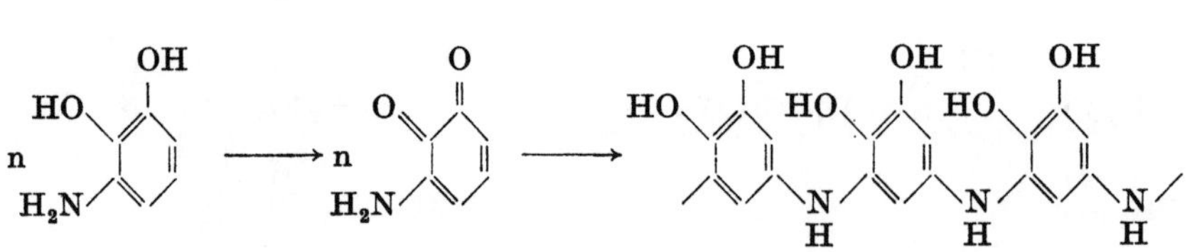

HORNER und STURM (1957) oxydieren 3-Amino-brenzcatechin zu einem dem Melanin ähnlichen Produkt, bei dem sie vermuten, daß die Verknüpfung des Monomeren in der angegebenen Weise erfolgt. Diese Kondensationsreaktion verläuft voraussichtlich über das entsprechende o-Chinon.

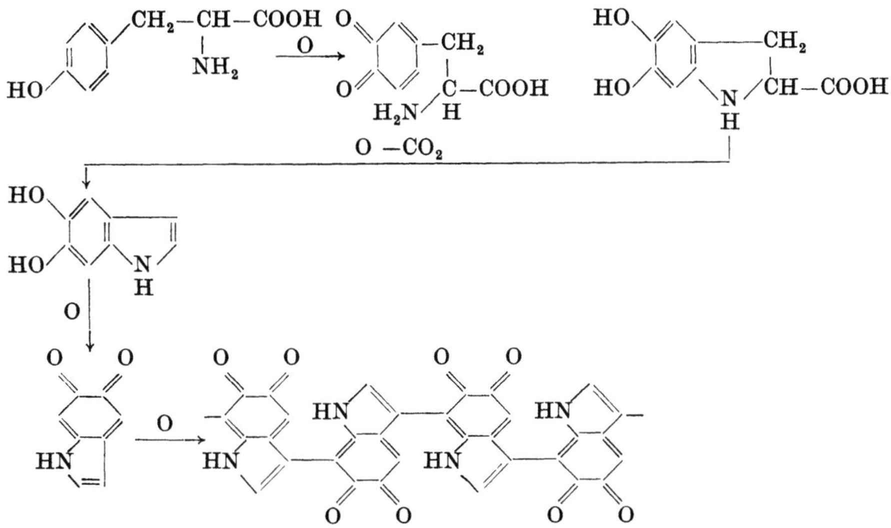

Die Kondensation von Tyrosin unter der Einwirkung oxydierender Enzyme verläuft zu dunkelgefärbten Stoffen. Die Zwischenprodukte dieser Reaktion sind weitgehend in ihrer Konstitution aufgeklärt (EVANS und RAPER 1937a, 1937b, HARLEY-MASON und LAIRD 1958, RAPER 1927).

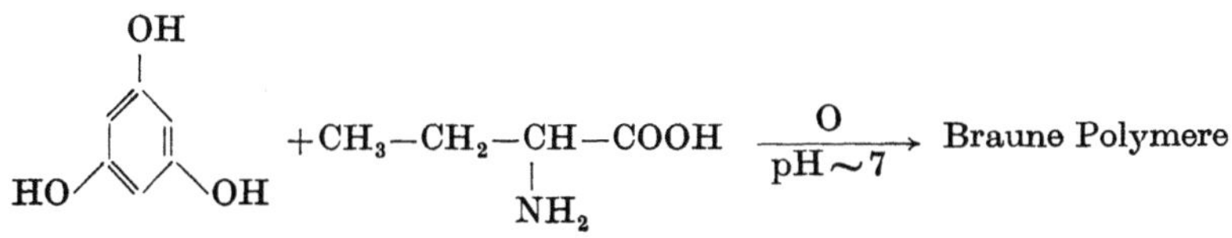

Von STEIN und TENDELOO (1955, 1959) ist die Oxydation von Phloroglucin mit schwachen Oxydationsmitteln wie Kaliumferricyanid in Gegenwart von Aminosäuren in nahezu neutralem Medium untersucht worden. Dabei entstehen ebenfalls dunkelgefärbte Polymere. Phloroglucin kann nicht zu einem Chinon oxydiert werden. Die Autoren sind der Ansicht, daß als erste Reaktionsstufe ein freies Radikal auftritt (STEIN und TENDELOO 1956a, 1956b, 1958).

Untersuchungen über die Bindung des Stickstoffs in den Humussubstanzen sind insofern von Interesse, als dieser Stickstoff bei der Zersetzung im Boden freigesetzt und für die Pflanzen aufnehmbar wird. Die im jährlichen Kreislauf in den Boden gelangenden abgestorbenen Organismen werden zersetzt, und es stellte sich je nach klimatischen Bedingungen eine bestimmte Höhe des Kohlenstoffgehaltes des Bodens in Form von Humussubstanzen ein. Bei diesem Abbau der organischen Substanzen des Bodens werden auch die stickstoffhaltigen heterocyclischen Verbindungen abgebaut, da sie sich sonst im Boden anhäufen würden. Der in den Humussubstanzen enthaltene Stickstoff dient den Organismen, Pflanzen und Mikroorganismen als Nährstoff.

E. Vergleich der Eigenschaften von stickstofffreien und stickstoffhaltigen Synthesehuminsäuren

Die Synthesehuminsäuren werden im allgemeinen dargestellt, indem Polyphenole, wie Hydrochinon, Brenzcatechin und Pyrogallol in alkalischer Lösung oxydiert werden. Als Oxydationsmittel finden Luft, Sauerstoff oder andere, wie z. B. Natriumpersulfat, Verwendung. Die zunächst tiefrote Farbe geht in Abhängigkeit von Konzentration der Lauge und der Phenolverbindung bald in eine tiefbraune bis schwarzbraune Färbung über, die — insbesondere bei Abwesenheit von Stickstoffverbindungen — bei einem Überschuß von Sauerstoff wieder rotbraun wird. Die pH-Abhängigkeit der Oxydation ist schon besprochen worden. Untersuchungen haben ergeben, daß die größten Ausbeuten an Synthesehuminsäuren, d. h. an mit Säure fällbaren Produkten bei pH-Werten zwischen 8 und 10 erhalten werden. Die Synthesehuminsäuren werden für weitere Untersuchungen mit Lauge und Säure mehrfach umgefällt und dann — in manchen Fällen auch direkt — dialysiert und elektrodialysiert. Nach Gefriertrocknung haben die Phenolhuminsäuren eine schwarzbraune Farbe.

Im Falle von Hydrochinon hat sich ergeben, daß die langsame Zugabe der Natronlauge zu der Lösung von Hydrochinon während der Oxydation sich auf die Höhe der Ausbeute günstig auswirkt; durch die laufende Zugabe kann man weitgehend ein Ansteigen des pH-Wertes über 10 vermeiden. Auf diese Weise erhält man nach einem Umsatz von etwa 2,5 Atomen Sauerstoff pro Mol Hydrochinon Ausbeuten von etwa 40%. Bei einem größeren Umsatz an Sauerstoff werden die Ausbeuten wieder kleiner, und nach dem Ansäuern werden beträchtliche Mengen Kohlendioxid und eine harzartige dunkelgefärbte Masse ausgeschieden. Aus der überstehenden weinroten Lösung können mit Erdalkalisalzen schwerlösliche Verbindungen ausgefällt werden (Flaig 1950).

Die aufgenommene Menge Sauerstoff von 2,5 Atomen pro Mol Hydrochinon entspricht summarisch folgender Reaktion:

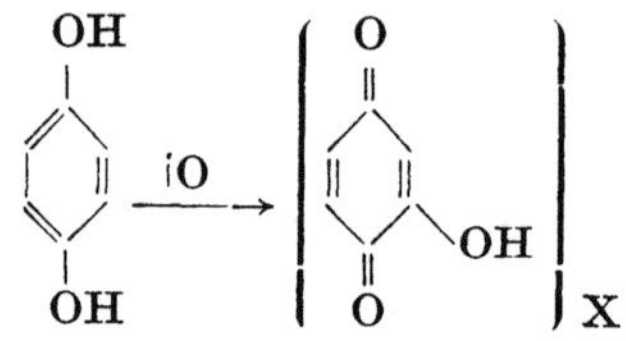

Die von Eller (1923) als erstem aus den Elementaranalysen bestimmte Summenformel entspricht einem polymeren Hydroxy-p-benzochinon.

Bei den Untersuchungen der Bildung von Synthesehuminsäuren aus Hydrochinon ging man auch dazu über, in saurem Medium mit Kieselsäurepräparaten als Katalysatoren zu arbeiten (Ziechmann 1960a, 1960b). Man wollte damit auch den Bedingungen in der Natur näherkommen; im Boden sind neben kristallisierten Mineralen auch je nach Bodenart unterschiedliche Mengen an amorphen Anteilen vorhanden.

Die Reaktionen von Hydrochinon in saurer Lösung sind über längere Zeiträume (Tage statt Stunden) als bei den Oxydationen in alkalischem Medium beobachtet worden. Die Ausbeute an Synthesehuminsäuren ist bei dieser Versuchsanstellung nur wenige Prozent.

Vergleichende Untersuchungen in saurem Medium in Abwesenheit von Kieselsäure haben gezeigt, daß in verdünnten Lösungen die Oxydation von Hydrochinon durch Luftsauerstoff in den untersuchten pH-Bereichen von 5,4 bis 7,0 in langsameren Reaktionen als im alkalischen Medium stattfindet (Salfeld 1961).

Die unterschiedliche beschleunigende Wirkung verschiedener Kieselsäuregele beruht wohl nur auf deren Gehalt an Eisenverbindungen, die als Eisenionen die Oxydation beschleunigen, wie auch mit im Boden vorhandenen Eisenverbindungen gezeigt worden ist (SCHEFFER, MEYER und NIEDERBUDDE 1959). Nach Auswaschen der die Reaktionen beschleunigenden Kieselsäuregele mit Salzsäure ließ deren Wirkung nach (SALFELD 1961). Als weitere Erklärungsmöglichkeit wäre die einer Oberflächenreaktion von Hydrochinon mit Sauerstoff zu nennen.

Aus Ansätzen, denen Kieselsäure zugesetzt worden ist, ließen sich nach Zusatz von o-Aminophenol durch Extraktion mit Äther und heißem Xylol Substanzen isolieren, unter denen sich 3-Hydroxy-phenoxazon (2) befand (ZIECHMANN 1961). Daraus ist geschlossen worden, daß in den Ansätzen 2,5-Dihydroxy-p-benzochinon enthalten ist.

2,5-Dihydroxy-p-benzochinon ist ein stabiles Endprodukt bei der Oxydation von Hydrochinon und spielt bei der Bildung der Synthesehuminsäuren keine wesentliche Rolle.

Oxydiert man unter den oben angegebenen Bedingungen Hydrochinon in Gegenwart von Ammoniak, so betragen die Ausbeuten bei einem Umsatz von 4 Atomen Sauerstoff pro Mol Hydrochinon ungefähr 90%. Beim Ansäuern tritt auch eine Entwicklung von Kohlendioxid ein, die jedoch geringer ist als in natronalkalischer Lösung. Es hat den Anschein, als ob die Abbaugeschwindigkeit der stickstoffhaltigen Synthesehuminsäuren geringer ist als die der stickstofffreien. Im Falle der Herstellung von Hydrochinonhuminsäuren ohne Ammoniak beträgt die Ausbeute bei einem Umsatz von 4 Gramm-Atomen nur noch 3 bis 5%.

Ammoniak wird nur während der Reaktion eingebaut, nachträglich läßt sich dieser nicht mehr in die Synthesehuminsäuren einbauen.

Die in Gegenwart von Ammoniak erhaltenen Produkte haben einen Stickstoffgehalt unter 6%, der sich mit Magnesium nicht abdestillieren läßt. Dieser wird als sogenannter „Kern"-Stickstoff bezeichnet. Die Menge dieses Stickstoffs hängt von der umgesetzten Menge an Sauerstoff ab.

Durch Zusatz geringer Mengen an Schwermetallionen, wie Kupfer- oder Eisensalze, sowie durch Zusatz von Dispergierungsmitteln wird die Geschwindigkeit der Oxydation beschleunigt.

In Abb. 93 sind die Verhältnisse schematisch angegeben, wie die Bildung von Synthesehuminsäuren aus Hydrochinon und Ammoniak bei der Oxydation mit Sauerstoff verläuft. Der Ammoniak ist im Verlaufe der Reaktion zugegeben worden, so daß der pH-Wert nie über 10 angestiegen ist (FLAIG 1950).

Mit steigender Zugabe von Ammoniak nimmt die Geschwindigkeit der Sauerstoffaufnahme und die Menge an Gesamtstickstoff sowie die Bildung an Synthesehuminsäuren zu. Die Menge an „Kern"-Stickstoff steigt bis nahezu 6% an. Nach Aufnahme von etwa 2 Gramm-Atomen Sauerstoff wird die Geschwindigkeit der Sauerstoffaufnahme vermindert. Bei der Oxydation mit mehr als 4 Gramm-Atomen Sauerstoff findet eine Abnahme der Menge an Trockensubstanz — „Synthese"-Huminsäuren - und des Gehaltes an „Kern"-Stickstoff statt, während die Gesamtmenge an Stickstoff noch zunimmt. Dieses deutet darauf hin, daß aus den „Synthese"-Huminsäuren Abbauprodukte mit sauren Gruppen entstehen. Der Gehalt von 6% „Kern"-Stickstoff in den höhermolekularen Syntheseprodukten spricht für eine Umsetzung von 2 Mol Hydrochinon mit einem Mol Ammoniak.

Um die Funktion der Wasserstoffatome der Stickstoffverbindung näher zu charakterisieren, ist auch mit Verbindungen gearbeitet worden, bei denen diese sukzessiv durch aliphatische Gruppen ersetzt worden sind. Während Dimethylamin von p-Benzochinon zu Bis-Dimethylaminochinon (BALTZLY und LORZ 1948) addiert wird, entstehen mit Monomethylamin nur Verharzungsprodukte. Bei

Trimethylamin findet keine Reaktion zu höhermolekularen stickstoffhaltigen Produkten statt. Nach diesen Befunden scheinen sich zwei Wasserstoffatome des Ammoniaks an der Reaktion zu beteiligen.

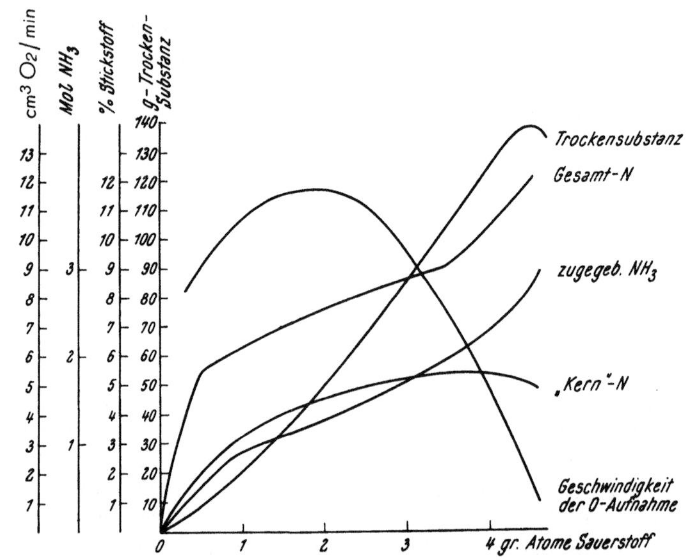

Abb. 93. Umsetzung von Hydrochinon mit Sauerstoff in Gegenwart von Ammoniak

Im Verlaufe dieser Untersuchung ist auch die Beobachtung gemacht worden, daß in Gegenwart von Formaldehyd im Laufe der Bildung der Synthesehuminsäuren in Abwesenheit von Ammoniak Kondensationsreaktionen ablaufen, während diese in Gegenwart von Ammoniak nicht stattfinden. Die isolierten stickstoffhaltigen und stickstofffreien Synthesehuminsäuren kondensieren weder sauer noch alkalisch mit Formaldehyd. Dieser Befund macht es wahrscheinlich, daß Ammoniak eine zusätzliche Vernetzung und — wie aus den Oxydationsversuchen hervorgeht — eine gleichzeitige Stabilisierung der Syntheseprodukte durch Blockierung besonders reaktiver Stellen im Molekül bewirkt (FLAIG 1950).

Derartige Untersuchungen an Synthesehuminsäuren sollen über die möglichen Reaktionen, die bei der Bildung der natürlichen Huminsäuren ablaufen können, Aufschluß geben.

Literatur

ACHARD, F.: Chemische Untersuchung des Torfs. Grell's Chem. Ann. **2**, 391 (1786). — ADLER, E.: Newer views of lignin formation. Tappi **40**, 294–301 (1957). — Über den Stand der Ligninforschung. Z. Papier **15**, 604–609 (1961). — ADLER, E., und K. LUNDQUIST: Estimation of uncondensed phenolic units in spruce lignin. Acta chem. Scand. **15**, 223–224 (1961). — ADLER, E., und R. MAGNUSSON: Periodate oxidation of phenols, I, Monoethers of pyrocatechol and hydroquinones. Acta chem. Scand. **13**, 505–519 (1959). — APENITIS, A., H. ERDTMAN und B. LEOPOLD: Studies on lignin, V, The decay of spruce wood by brown-rotting fungi. Svensk kem. Tidskr. **63**, 195–207 (1951).

BALTZLY, R., und E. LORZ: Addition of dimethylamine to benzoquinone. J. Amer. Chem. Soc. **70**, 861–862 (1948). — BARTLETT, J. B.: The effect of decomposition of the lignin of plant materials. Iowa State Coll. J. Sci. **14**, 11–13 (1939). — BARTLETT, J. B., und A. G. NORMAN: Changes in the lignin of some plant materials as a result of decomposition. Proc. Soil. Sci. Soc. Amer. **3**, 210–216 (1939). — BAVENDAMM, W.: Über das Vorkommen und den Nachweis von Oxydasen bei holzzerstörenden Pilzen. Z. Pflanzenkrankheiten u. Pflanzenschutz **38**, 257–276 (1928). — BEECKEN, H., E. M. GOTTSCHALK, U. v. GIZYCKI, H. KRÄMER, D. MAASSEN, H. G. MATTHIES, H. MUSSO, C. RATHJEN und U. I. ZAHORSZKY: Orcein und Lackmus. Angew. Chem. **73**, 665–688 (1961). — BEHR, G.: Arch. Mikrobiol. **1**, 418 (1930). — BENADE, W.: Mikrobiologie der Moore. Linz 1954. — BERZELIUS, J.: Undersökning of vattnet i Porla källa. Ann. Phys. Chem. **105**, 3, 238 (1833). — BEUTELSPACHER, H.: Physikalische Beiträge zur Humusforschung. Z. Pflanzenernähr., Düng., Bodenkde. **57**, 57–65 (1952). — Wechselwirkung zwischen anorganischen und organischen Kolloiden des Bodens. Z. Pflanzenernähr., Düng., Bodenkde. **69**, 108–115 (1955). — BJÖRKMAN, A.: Isolation of lignin from finely divided wood with neutral solvents. Nature **174**, 1057 (1954). — Studies on finely divided wood, I, Extraction of lignin with neutral solvents. Svensk Papperstidning **59**, 477–485 (1956). — BLOIS, S.: Observation of the semiquinone of benzoquinone. J. Chem. Phys. **23**, 1351 (1955). — BÖRNER, H.: Untersuchungen über phenolische Verbindungen aus Getreidestroh und Getreiderückständen. Naturwiss. **42**, 583–584 (1955). — Der papierchromatographische Nachweis von Ferulasäure in wässerigen Extrakten von Getreidestroh und Getreiderückständen. Naturwiss. **43**, 129 (1956). — BORTELS, H.: Biochem. Z. **182**, 301 (1927). — BRACKMAN, W., und E. HAVINGA: The oxidation of phenols with copper amine catalysts and its relation to the mode of action of tyrosinase, I, The catalytic oxidation of monohydric phenols to orthoquinone derivates. Rec. Trav. chim Pays-Bas **74**, 937–955 (1955a). — The oxidation of phenols with copper amine catalysts and its relation to the mode of action of tyrosinase, II, Oxidation of naphthols with tertiary amines as catalysts. Rec. Trav. chim. Pays-Bas **74**, 1021–1039 (1955b). — The oxidation of phenols with copper amine catalysts and its relation to the mode of action of tyrosinase, III, Kinetics of oxidation of monohydric phenols with copper-morpholine complex as homogeneous catalysts. Rec. Trav. chim. Pays-Bas **74**, 1070–1080 (1955c). — The oxidation of phenols with copper amine catalysts and its relation to the mode of action of tyrosinase, IV, Relation between hydrogen peroxide and the catalytic oxidation of phenol. Rec. Trav. chim. Pays-Bas **74**, 1100–1118 (1955d). — BRASSARD, P., und P. KARRER: Cis, cis-ß, ß'-Dimethylmuconsäureanhydrid. Helv. chim. Acta **43**, 262–264 (1960). — BRAUNS, F. E.: Native Lignin, I, Its isolation and methylation. J. Amer. Chem. Soc. **61**, 2120–2127 (1939). — BREGER, I. A.: Chemical and structural relationship of lignin to humic substances. Fuel **30**, 204–208 (1951). — BREMNER, J. M.: A review of recent work on soil organic matter, I. J. Soil. Sci. **2**, 67–82 (1951). — A review of recent work on soil organic matter, II. J. Soil Sci. **5**, 214–232 (1954). — Recent work on soil organic matter at Rothamsted. Z. Pflanzenernähr., Düng., Bodenkde. **69**, 32–38 (1955a). — Studies on soil humic acids, I, The chemical nature of humic nitrogen. J. Agric. Sci. **46**, 247–256 (1955b). — Nitrogen distribution and amino acid composition of fractions of a humic acid from a chernozem soil (Hildesheimer Schwarzerde). Z. Pflanzenernähr., Düng., Bodenkde. **71**, 63–66 (1955c). — Some soil organic matter problems. Soils a. Fertilizers **19**, 116–125 (1956). — BREMNER, J. M., und H. LEES: Studies on soil organic matter, II, The extraction of organic matter from soil by neutral reagents. J. Agric. Sci. **39**, 274–279 (1949). — BREMNER, J. M., W. FLAIG und E. KÜSTER: Zur Kenntnis der Huminsäuren, IX, Der Gehalt an Aminosäuren in Streptomyceten-Huminsäuren. Z. Pflanzenernähr., Düng., Bodenkde. **71**, 58-63 (1955). — BREYHAN, Th.: Eine Mikromethode zur Stickstoffbestimmung. Z. analyt. Chem. **152**, 412–417 (1956). — BROADBENT, F. E.: Modification in chemical properties of straw during decomposition. Proc. Soil Sci. Soc. Amer. **18**, 165–169 (1954). — BROADBENT, F. E., und G. R. BRADFORD: Cation-exchange groupings in the soil organic fractions. Soil Sci. **74**, 447–457 (1952). — BROWN, S. A., und A. C. NEISH: Shikimic acid as a precursor in lignin biosynthesis. Nature **175**, 688-689 (1955). — BRUCKER, W., und U. DREHMANN: Zur Phenolcarbonsäurebildung von Phycomyces aus ^{14}C-markierten Substraten. Arch. Mikrobiol. **30**, 396–408 (1958). — BUNGENBERG DE JONG, H. G.: Crystallisation—Coacervation—Flocculation. In: KRUYT, H. R.: Colloid Science **2**, 232–258. New York: Elsevier. 1949. — BURGES, A.: Physico-chemical investigations of humic acid. 7th Int. Congr. Soil Sci., Madison, Wisc., USA, 128–133 (1960). — BURKAT, S. E.: Solubility of peat humic acids. J. Appl. Chem. USSR **32**, 227–230 (1959). — BURTON, H., und H. B. HOPKINS: The oxidation of 4-methylcatechol in aqueous solution by ferric ion. J. Chem. Soc. **1952**, 2445–2447.

CAIN, R. B.: The metabolism of protocatechuic acid by a vibrio. Biochem. J. **79**, 298–312 (1961). — CAMPBELL, T. W.: The oxidation of di-tert. butyl. pyrogallol by oxygen in alkaline solution. J. Amer. Chem. Soc. **73** 4190–4195 (1951). — CASSIDY, H. G.: Electron exchange-polymers. J. Amer. Chem. Soc. **71**, 402–406 (1949). — CROW, W. D., und R. D. HAWORTH: Purpurogallin, VII, Constitution of purpurogallin-carboxylic acid. J. Chem. Soc. **1951**, 1325–1327. — COUKSON, R. C., D. A. COX und J. HUDEC: Photodimers of alkylbenzoquinones. J. Chem. Soc. **1961**, 4499.

DAGLEY, S., W. C. EVANS und D. W. RIBBONS: Metabolism of aromatic compounds by microorganisms. Nature **188**, 560 (1960). — DAVIES, R. T., und C. B. COULSON: Humic acid. Soils a. Fertilizers **22**, 159 (1959). — DEUEL, H., und P. DUBACH: Decarboxylierung der organischen Substanz des Bodens, II, Nachweis von Uronsäuren. Z. Pflanzenernähr., Düng., Bodenkde. **82**, 97–106 (1958a). — Decarboxylierung der organischen Substanz des Bodens, III, Decarboxylierung und Fraktionierung decarboxylierbarer Humusstoffe. Helv. chim. Acta **41**, 1310–1321 (1958b). — DEUEL, H., P. DUBACH und R. BACH: Decarboxylierung der organischen Substanz des Bodens, I, Decarboxylierung der gesamten Humusstoffe. Z. Pflanzenernähr., Düng., Bodenkde. **81**, 189 (1958). — DIELS, O., und R. KASSEBART: Zur Kenntnis der durch Pyridin bewirkten Polymerisationsvorgänge, I, Polymerisation des p-Chinons. Liebigs Ann. Chem. **530**, 51–67 (1937). — DION, W. M.: Production and properties of a polyphenol oxidase from the fungus Polyporus versicolor. Canad. J. Bot. **30**, 9–21 (1952). — DRAGUNOW, S. S., N. N. SCHELOCHOWTZEWA und E. I. STRELKOWA: Eine vergleichende Untersuchung von Boden- und Torfhuminsäuren. Pedologija **1948**, 409 (Russ.). — DYER, W. J., und C. L. WRENSHALL: Organic phosphorus in soils, I, Extraction and separation of organic phosphorus compounds from soil. Soil Sci. **51**, 159–170 (1941a). — Organic phosphorus in soils, III, The decomposition of some organic phosphorus compounds in soil cultures. Soil Sci. **51**, 223–239 (1941b).

EIGEN, M., und P. MATTHIES: Über Kinetik und Mechanismus der Primärreaktionen der Zersetzung von Chinon in alkalischer Lösung. Chem. Ber. **94**, 3309–3317 (1961). — EINSTEIN, A.: Eine neue Bestimmung der Moleküldimensionen. Ann. Phys. **19**, 289–306 (1906). — Berichtigung zu: Eine neue Bestimmung der Moleküldimensionen. Ann. Phys. **34**, 591–592 (1911). — EISTERT, B., in: Neuere Methoden der präparativen organischen Chemie, 3. Aufl., S. 359 ff. Weinheim: Verlag Chemie. 1949. — EISTERT, B., und G. BOCK: Reaktionen von α-Dicarbonylverbindungen und Chinonen mit Diazoalkanen, II, Die Umsetzung einiger substituierter p-Benzochinone mit Diazomethan zu p-Epoxymethyl-cyclohexadienonen. Chem. Ber. **92**, 1247–1257 (1959). — ELLER, W.: Künstliche und natürliche Huminsäuren. Brennstoffchem. **2**, 129–133 (1921). — Studien über Huminsäuren, IV, Darstellung und Eigenschaften künstlicher und natürlicher Huminsäuren. Liebigs Ann. Chem. **431**, 133–161 (1923). — Zur Frage der Konstitution der natürlichen Huminsäuren. Brennstoffchem. **6**, 54 (1925a). — Studien über Huminsäuren. Einige Eigenschaften und Reaktionen der Huminsäuren und Humine. Liebigs Ann. Chem. **442**, 160–180 (1925b). — ELLER, W., und K. KOCH: Synthetische Darstellung von Huminsäuren. Ber. dtsch. chem. Ges. **53**, 1469–1476 (1920). — ELLER, W., E. HERDICKERHOFF und H. SAENGER: Einwirkung von Chlor auf Huminsäuren. Liebigs Ann. Chem. **431**, 177–186 (1923). — ELLER, W., H. MEYER und H. SAENGER: Einwirkung von Salpetersäure auf Huminsäuren. Liebigs Ann. Chem. **431**, 162–175 (1923). — ENDERS, C.: Über den Chemismus der Huminsäurebildung unter physiologischen Bedingungen, IV, Die Rolle der Mikroorganismen bei den Humifizierungsvorgängen. Biochem. Z. **315**, 259–292 (1943). — ENSMINGER, L. E., und J. E. GIESEKING: The absorption of proteins by montmorillonitic clays. Soil Sci. **48**, 467–474 (1939). — ERDTMAN, H. G. H.: Studies on the formation of complex oxidation and condensation products of phenols, A contribution to the investigation of the origin and nature of humic acid, I, Studies of the reactivity of simple monocyclic quinones. Proc. Roy. Soc. (London) A **143**, 177–191 (1933a). — Studies on the formation of complex oxidation and condensation products of phenols, II, Coupling of simple phenols and quinones to diphenyl derivatives. Proc. Roy. Soc. (London) A **143**, 191–222 (1933b). — Studies on the formation of complex oxidation and condensation products of phenols, III, Re-arrangements of oxidation-reduction type in the diquinone group. Proc. Roy. Soc. (London) A **143**, 223–228 (1933c). — Studies on the formation of complex oxidation and condensation products of phenols, IV, Termolecular polymerisation products of p-benzoquinone, and α-naphthoquinone. Proc. Roy. Soc. (London) **143**, 228–241 (1933d). — Dehydrierungen in der Coniferylreihe, I, Dehydrodieugenol und Dehydrodiisoeugenol. Biochem. Z. **258**, 172–180 (1933e). — Dehydrierungen in der Coniferylreihe, II, Dehydrodiisoeugenol. Liebigs Ann. Chem. **503**, 283–294 (1933f). — Phenoldehydrierungen, V, Notiz über die Dehydrierung von Trimethylphloroglucin. Die Struktur des Cedrons. Svensk kem. Tidskr. **47**, 223–230 (1935). — Die Reaktion von Benzochinon mit Alkali. Z. Pflanzenernähr., Düng., Bodenkde. **69**, 38–43 (1955). — ERDTMAN, H. G. H., und M. GRANATH: Studies on humic acid, V, The reaction of p-benzoquinone with alkali. Acta chem. Scand. **8**, 811–816 (1954). — ERDTMAN, H. G. H., und N. E. STJERNSTRÖM: Studies on humic acids, VII, The structure of the dihydroxy-dioxidoterphenyl obtained by the action of sulphuric acid on p-benzoquinone. Acta chem. Scand. 13, 653–658 (1959). — ERDTMAN, H. G. H., M. GRANATH und G. SCHULTZ: Studies on humic acids, VI, Triquinone and tetraquinone. Acta chem. Scand. 8, 1442–1450 (1954). — EULER, H. v.: Eigenschaften und Wirkungen der Reduktone. Chimia (Zürich) 3, 209–213 (1949). — EVANS, W. C.: Oxidation of phenol and benzoic acid by some soil bacteria. Biochem. J. 41, 373–382 (1947). — EVANS, W. C., und H. S. RAPER: Eine vergleichende Studie über die Bildung von l-3,4-Dioxyphenylalanin aus Tyrosin durch Tyrosinase von verschiedener Herkunft. Biochem. J. 31, 2155–2161 (1937a). — Die Anhäufung von l-3,4-Dioxyphenylalanin bei der Tyrosinase-Tyrosinreaktion. Biochem. J. 31, 2162–2170 (1937b). — EVANS, W. C., B. S. W. SMITH, R. P. LINSTEAD und J. A. ELVIDGE: Chemistry of the oxidative metabolism of certain aromatic compounds by micro-organisms. Nature 168, 772–775 (1951).

FÅHRAEUS, G., R. NILSSON und G. NILSSON: Studies in the decomposition of wood by means of some white rot fungi. Svensk bot. Tidskr. 43, 343–356 (1949). — FERNANDEZ, O., und B. REGEIRO: Fragmentación enzímica de las ligninas. Farm. Nueva 11, 57–62, 111–115, 169–176 (1946). — FEUSTEL, I. C., und H. G. BYERS: The behavior of lignin and humic acid preparations toward a bromination treatment. Soil Sci. 42, 11–21 (1936). — FLAIG, W.: Zur Kenntnis der Huminsäuren, I, Zur chemischen Konstitution der Huminsäuren. Z. Pflanzenernähr., Düng., Bodenkde. 51, 193–212 (1950). — Contribución al estudio de los ácidos humínicos. Monografías de Ciencia Moderna Nr. 46, Instituto de Edafología y Fisiología Vegetal Madrid (1955a). — Zur Bildungsmöglichkeit von Huminsäuren aus Lignin. Holzforschung 9, 1–4 (1955b). — Zur Chemie der Huminsäuren und deren Modellsubstanzen. VIe Congr. Int. Sci. Sol., Paris, 2, 471–478 (1956). — Die Chemie organischer Stoffe im Boden und deren physiologische Wirkung. Verh. II. u. IV. Komm. Int. Bodenkdl. Ges., Hamburg 1958, Vol. II. — Chemie der Humusstoffe, Suomen Kemistilehti A 33, 229–251 (1960). — Zur Umwandlung von Lignin in Humusstoffe. Freiberger Forschungshefte, Heft A 254, 39–56 1962. — FLAIG, W., und H. BEUTELSPACHER: Zur Kenntnis der Huminsäuren II, Elektronenmikroskopische Untersuchungen an natürlichen und synthetischen Huminsäuren. Z. Pflanzenernähr., Düng., Bodenkde. 52, 1–21 (1951). — Physikalische Chemie der Huminsäuren. Landbouwkund. Tijdschr. 66, 306–336 (1954). — FLAIG, W., und Th. BREYHAN: Über das Vorkommen von Indolverbindungen in Schwarzerde-Huminsäuren. Z. Pflanzenernähr., Düng., Bodenkde. 75, 132–135 (1956). — FLAIG, W., und K. HAIDER: Die Verwertung phenolischer Verbindungen durch Weißfäulepilze. Arch. Mikrobiol. 40, 212–223 (1961a). — Reaktionen mit oxydierenden Enzymen aus Mikroorganismen. Planta Medica Z. Arzneipflanzenforsch. 9, 123–139 (1961b). — FLAIG, W., und R. KAUL: Untersuchungen über die Beeinflussung der Phosphataufnahme von Haferkeimpflanzen durch Phosphatasegifte und Phosphataseaktivatoren. Z. Pflanzenernähr., Düng., Bodenkde. 72, 97–103 (1956a). — Über die Einwirkung von Ribonucleinsäure auf das Pflanzenwachstum. Z. Pflanzenernähr., Düng., Bodenkde. 72, 103–113 (1956b). — FLAIG, W., und J. CHR. SALFELD: UV-Spektren und Konstitution von p-Benzochinonen. Liebigs Ann. Chem. 618, 117–139 (1958). — Zwischenstufen bei der Bildung von Huminsäuren aus Phenolen. Trans. Vol. II, 7th Int. Congr. Soil Sci., Madison, Wisc., USA, 648–656 (1960). — FLAIG, W., und H. SCHULZE: Über den Bildungsmechanismus der Synthesehuminsäuren. Z. Pflanzenernähr., Düng., Bodenkde. 58, 59–67 (1952). — FLAIG, W., TH. PLOETZ und H. BIERGANS: Zur Kenntnis der Huminsäuren XIV, Bildung und Reaktionen einiger Hydroxychinone. Liebigs Ann. Chem. 597, 196–213 (1955). — FLAIG, W., TH. PLOETZ und A. KÜLLMER: Über Ultraviolettspektren einiger Benzochinone. Z. Naturforsch. 10b, 668–676 (1955). — FLAIG, W., J. CHR. SALFELD und A. LLANOS: Bildung von 2,5-Dimethyl-p-benzochinon aus 2,3-Dimethyl-p-benzochinon über dimere Photoprodukte. Angew. Chem. 72, 110 (1960). — FLAIG, W., K. SCHARRER und G. K. JUDEL: Zur Methodik der Bestimmung des Redoxpotentials im Boden. Z. Pflanzenernähr., Düng., Bodenkde. 68, 97–122 (1955a). — Bestimmung des Redoxpotentials von Bodenprofilen. Z. Pflanzenernähr., Düng., Bodenkde. 68, 203–218 (1955b). — FLAIG, W., F. SCHEFFER und B. KLAMROTH: Zur Kenntnis der Huminsäuren VIII, Zur Charakterisierung der Huminsäuren des Bodens. Z. Pflanzenernähr., Düng., Bodenkde. 71, 33–57 (1955). — FLAIG, W., U. SCHOBINGER und H. DEUEL: Umwandlung von Lignin in Huminsäuren bei der Verrottung von Weizenstroh. Chem. Ber. 92, 1973 (1959). — FLAIG, W., U. SCHOBINGER und A. STEINMETZ: Über die Bildung der Huminstoffe. Landbauforsch. 7, 78–80 (1957). — FORSYTH, W. G. C.: Studies on the more soluble complexes of soil organic matter, I, A method of fractionation. Biochem. J. 41, 176–181 (1947a). — The characterization of the humic complexes of soil organic matter. J. Agric. Sci. 37, P. 2, 132–138 (1947b). —

FORSYTH, W. G. C., und V. C. QUESNEL: Intermediates in the enzymic oxidation of catechol. Biochim. biophys. Acta **25**, 155–160 (1957). — FORSYTH, W. G. C., V. C. QUESNEL und J. B. ROBERTS: Diphenylenedioxide-2,3-quinone: An intermediate in the enzymic oxidation of catechol. Biochim. biophys. Acta **37**, 322–26 (1960). — FREDERICK, L. R., R. L. STARKEY und W. SEGAL: Decomposability of some organic sulfur compounds in soil. Proc. Soil Sci. Soc. Amer. **21**, 287–292 (1957). — FREUDENBERG, K.: Lignin im Rahmen der polymeren Naturstoffe. Angew. Chem. **68**, 84–92 (1956). — Über die Biosynthese und Konstitution des Lignins. Chem. Ber. **92**, LXXXIX-XCVIII (1959). — FRIESS, S. L., A. H. SOLOWAY, B. K. MORSE und W. C. INGERSOLL: Reactions of peracids, VII, Nuclear oxidation of aromatic ethers with perbenzoic acid. J. Amer. Chem. Soc. **74**, 1305 (1952). — FRÖMEL, W.: Über Absorptionsspektren von Huminsäuren in Lösungen. Bodenkde. u. Pflanzenernähr. **6**, 93–119 (1938a). — Über UV-Absorptionsspektren von Huminsäuren in Lösungen. Bodenkde. u. Pflanzenernähr. **11**, 129–144 (1938b). — Über Fulvosäuren. Bodenkde. u. Pflanzenernähr. **25**, 345–358 (1941). — FUCHS, W., und W. STENGEL: Über den Abbau der Huminsäuren zu Benzolcarbonsäuren und Nitrophenolen. Liebigs Ann. Chem. **478**, 267–283 (1930).

GANZ, E.: Contribution à l'étude de la structure des acides humiques naturels. Ann. Chim. **19**, 202 (1944). — GORHAM, E.: Chlorophyll derivatives in woodland soils. Soil Sci. **87**, 258–261 (1959). — GOTTLIEB, S., und J. H. GELLER: Enzymatic decomposition of lignin. Science **110**, 189–190 (1949). — GRISEBACH, H.: Zur Biogenese des Cyanidins, I, Versuche mit Acetat-(1-C^{14}) und Acetat (2-C^{14}). Z. Naturforsch. **12b**, 227–231 (1957). — GROSS, S. R.: The enzymatic conversion of 5-dehydroshikimic acid to protocatechuic acid. J. Biol. Chemistry **232**, 5 (1958). — GROSS, S. R., R. D. GAFFORD und E. L. TATUM: The metabolism of protocatechuic acid by Neurospora. J. Biol. Chemistry **219**, 781–796 (1956). — GÜNDEL, W., und R. PUMMERER: Über die Einwirkung aromatischer Nitrosoverbindungen auf Chinone. Liebigs Ann. Chem. **529**, 11–32 (1937).

HÄRTEL, O., und E. PAPESCH: Über die Wachsausscheidungen von Coniferen. Ber. dtsch. bot. Ges. **68**, 133–142 (1955). — HAIDER, K., SOUN-UK LIM und W. FLAIG: Untersuchungen über die Einwirkung von Mikroorganismen auf ^{14}C-markierte phenolische Verbindungen. Landwirtsch. Forsch. **15**, 196 (1962). — HALLA, F., und W. RUSTON: Leitfähigkeitsanomalien bei Huminsäuresolen. Z. Elektrochem. **59**, 525–530 (1955). — HARLEY-MASON, J., und A. H. LAIRD: Dimer of o-benzoquinone. J. Chem. Soc. **1958**, 1718–1719. — HATHWAY, D. E.: Autoxidation of polyphenols, I, Autoxidation of methyl gallate and its o-methyl ethers in aqueous ammonia. J. Chem. Soc. **1957**, 519–523. — An approach to the study of vegetable tannins by the oxidation of plant phenolics. J. Soc. Leather Trudes Chemists **42**, 109–121 (1958). — HATHWAY, D. E., und J. W. SEKINS: Enzymic oxidation of catechin to a polymer structurally related to some phlobatannins. Biochem. J. **67**, 239–245 (1957). — HAYASHI, T., und T. NAGAI: On the components of soil humic acids, VII, Quantitative determination and distribution of components. Soil a. Plant Food **5**, 153–160 (1960). — HAYAISHI, O., M. KATAGIRI und S. ROTHBERG: Mechanism of the pyrocatecase reaction. J. Amer. chem. Soc. **77**, 5450–5451 (1955). — HENDERSON, M. E. K.: Metabolism of methoxylated aromatic compounds by soil fungi. J. Gen. Microbiol. **16**, 686–695 (1957). — HERRMANN, K.: Über die Bildung der Purpurogallincarbonsäure durch fermentative Oxydation der Gallussäure, II, Über gerbstoffartige Substanzen. Arch. Pharmazie **287**, 497–503 (1954). — HINSBERG, O., und A. HIMMELSCHEIN: Über Oxy- und Aminoderivate des Diphenylsulfons. Ber. dtsch. chem. Ges. **29**, 2023–2029 (1896). — HOFMANN, E.: Die Enzyme im Boden und ihre Bedeutung für seine Biologie und Fruchtbarkeit. Z. Acker- u. Pflanzenbau **100**, 31–35 (1955a). — Über die Rolle der Enzyme bei der Humusbildung. Z. Pflanzenernähr., Düng., Bodenkde. **69**, 165–172 (1955b). — HOFMANN, E., und A. SEEGERER: Über das Enzymsystem unserer Kulturböden, I, Saccharase. Biochem. Z. **322**, 174–179 (1951). — HOFMANN, E., und W. SCHMIDT: Über das Enzymsystem unserer Kulturböden, II, Urease. Biochem. Z. **324**, 125–127 (1953). — HOOKER, S. C., und L. F. FIESER: Concerning Wichelhaus' "Di-β-napthoquinone oxide". J. Amer. Chem. Soc. **58**, 1216–1223 (1936). — HOPPE-SEYLER, F.: Über Huminsubstanzen, ihre Entstehung und ihre Eigenschaften. Z. physiol. Chem. **13**, 66–121 (1889). — HORNER, L., und W. DÜRCKHEIMER: Zur Kenntnis der o-Chinone, XI, Die Struktur des dimeren o-Benzochinons. Chem. Ber. **91**, 2532–2534 (1958). — Zur Kenntnis der o-Chinone, XII, o-Chinone aus Brenzcatechin-Derivaten. Z. Naturforsch. **14b**, 741 (1959a). — Zur Kenntnis der o-Chinone, XIII, Dimensierungs-Prinzipien von o-Benzochinonen Z. Naturforsch. **14b**, 742–743 (1959b). — Zur Kenntnis der o-Chinone, XIV, Benzotropolone aus o-Chinonen. Z. Naturforsch. **14b**, 743/44

(1959c). — HORNER, L., und K. STURM: Zur Kenntnis der o-Chinone, X, Modellreaktionen zur Melaninbildung. Liebigs Ann. Chem. **608**, 128–139 (1957). — HULTZSCH, K.: Chemie der Phenolharze. Berlin-Göttingen-Heidelberg: Springer. 1950. — Ring-Kondensate in Alkylphenolharzen. Z. Kunststoffe **52**, 20–24 (1962). — HUISMAN, H. O.: Investigations on quinones and quinone-derivatives. Preparation and antibiotic properties of some substituted p-benzo- and p-toluquinones, hydroquinones and hydroquinone diesters. Rec. Trav. Chim. Pays-Bas **69**, 1133–1156 (1950). — HUSEMANN, E.: Molekulargewichtsbestimmungen an hydrolytisch abgebauten Glykogenen durch Fällungstitration. J. Prakt. Chem. **158**, 163–175 (1941).

JAMES, T. H., und A. WEISSBERGER: Oxidation processes, XI, The autoxidation of durohydroquinone. J. Amer. Chem. Soc. **60**, 98–104 (1938). — JAMES, T. H., J. M. SNELL und A. WEISSBERGER: Oxidation processes, XII, The autoxidation of hydroquinone and of the mono-, di- and trimethylhydroquinone. J. Amer. Chem. Soc. **60**, 2084–2093 (1938). — JENKINSON, D. S., und G. TINSLEY: A comparison of the ligno-protein isolated from a mineral soil and from a straw compost. Sci. Proc. Roy. Dublin Soc. A **1**, 141–147 (1960). — JENNY, H.: Factors of soil formations. New York und London: McGraw Hill. 1941. — JONES, E. J. JR.: The ultraviolet absorption spectra of complex hydroxyaromatic compounds and derivatives, with particular reference to lignin. Tappi **32**, 311–315 (1949). — JONES, R. G., und H. A. SHONLE: The preparation of 2,5-dihydroxyquinone. J. Amer. Chem. Soc. **67**, 1034–1035 (1945).

KAILA, A.: Humification of straw at various temperatures. Acta agr fenn. Helsinki **78**, 1–32 (1952). — KILBY, B. A.: The bacterial oxidation of phenol to β-ketoadipic acid. Biochem. J. **43**, V (1948). — The formation of β-ketoadipic acid by bacterial fission of aromatic rings. Biochem. J. **49**, 671–674 (1951). — KINNEY, C. R., und E. I. DOUCETTE: Infrared spectra of a coalification series from cellulose and lignin to anthracite. Nature **182**, 785–786 (1958). — KÖHNLEIN, J.: Futterbau und Bodenfruchtbarkeit. 9. Sonderheft Landwirtsch. Forsch., 201–212 (1957). — Über die Zusammenhänge der relativen Lage des Humusspiegels und der Nährstoffanlieferung des Bodens. Z. Acker- u. Pflanzenbau **108**, 1/2 (1959). — KOLENBRANDER, G. J.: Die Verluste an organischer Substanz im Stalldünger. Z. Pflanzenernähr., Düng., Bodenkde. **69**, 125–134 (1955). — KONETZKA, W. A., E. T. WOODINGS und J. STOVE: Microbial dissimilation of methoxylated aromatic compounds. Bact. Proc. **1957**, 85. — KONONOWA, M. M.: Humus der Hauptbodentypen der UdSSR, seine Natur und Bildungsweisen. Rapp. VI. Congr. Int. Sci. Sol, Chemie du Sol, S. 5, Moskau 1956. — Die Humusstoffe des Bodens, übersetzt und bearbeitet von H. BEUTELSPACHER. Berlin: VEB Deutscher Verlag der Wissenschaften. 1958. — KONONOWA, M. M., und I. V. ALEXANDROWA: La biochimie des processus de formation des matières humiques. VI. Congr. Int. Sci. Sol, Paris, Vol. C, 133–137 (1956). — Distributive paper chromatography for a study of the forms of nitrogen in organic substances. Pochvovedenie **5**, 86–92 (1958). — The biochemistry of humus formation and some problems of plant nutrition. Soils a. Fertilizers **22**, 2 (1959). — KONONOWA, M. M., und N. P. BELCIKOWA: Zur Erforschung des Wesens der Bodenhumusstoffe unter Anwendung der Fraktionierung. Pochvovedenie 11, 1 (1960). — KUBIËNA, W. L.: Entwicklungslehre des Bodens. Wien: Springer. 1948. — KUCHARENKO, T. A., und A. SSAWELKEW: Hydrierung von Huminsäuren am Nickelkatalysator. Ber. Akad. Wiss. UdSSR **76**, 77 (1951). — KÜSTER, E.: Umwandlung von Mikroorganismen-Farbstoffen in Humusstoffe. Z. Pflanzenernähr., Düng., Bodenkde. **57**, 51–57 (1952). — Mikrobielle Polyphenoloxydasen und Humusbildung. Zbl. Bakt. I. Orig. **160**, 207–213 (1953). — Humusbildung und Phenoloxydasen bei Streptomyceten. Z. Pflanzenernähr., Düng., Bodenkde. **69**, 137–142 (1955). — Der Einfluß der C- und N-Quelle auf die Peptonbraunfärbung durch Streptomyceten. Zbl. Bakt. II. **111**, 227–234 (1958). — KUHN, R., und H. BEINERT: Über die Umsetzung von Cystein mit Chinon. Ber. dtsch. chem. Ges. **77**, 606 (1944). — KULLMANN, A.: Über die Verrottung von Futterpflanzenwurzeln an Hand von Modellversuchen. Z. Pflanzenernähr., Düng., Bodenkde. **84**, 127–132 (1959). — KUMADA, K.: Absorptionspectra of humic acids. Soil a. Plant Food **1**, 29–30 (1955). — KUMADA, K., und K. AIZAWA: The infra-red spectra of humic acids. Soil a. Plant Food **3**, 152–159 (1958). — KUMADA, K., und S. HOTTA: Artificial humic acids prepared from coal and microbial pigments. Soil a. Plant Food (Tokyo) **3**, 148–151 (1958).

LAATSCH, W.: Dynamik der deutschen Acker- und Waldböden, 4. Aufl. Dresden 1957. — LAATSCH, W., I. BAUER und O. BIENECK: Die Bildungsweisen der Huminsäuren. Landwirtsch. Forsch. **2**, 38 (1950). — LAATSCH, W., L. HOOPS und I. BAUER: Über Huminsäuren mit Aminostickstoff. Z. Pflanzenernähr., Düng., Bodenkde. **53**, 20–29 (1951). — LE CLERC, A. M., J. MONDY, P. DOUZOU und S. LISSITZKY: Mise en évidence par oxydation photochimique ou enzymatique du radical libre du catechol. Biochem. biophys. Acta 32, 499–504 (1959). — LIEBIG, J. v.: Die Chemie in ihrer Anwendung auf Agrikultur und Physiologie. Braunschweig: Vieweg. 1846. — LINDEBERG, G.: Über die Physiologie ligninabbauender Bodenhymenomyceten. Symbolae Botanicae Upsalienses 8, 2, 1–183 (1944). — Occurence of polyphenol oxidases in soil inhabiting basidiomycetes. Physiol. Plant. 1, 196–205 (1948). — LINDEBERG, G., und G. FÅHRAEUS: Nature and formation of phenol oxidases in Polyporus zonatus and P. versicolor. Physiol. Pla–283nt. 5, 277 (1952).

MAEDER, H.: Chemische und pflanzenphysiologische Untersuchungen mit Rottestroh. Diss. Gießen 1959. — MAIWALD, K.: Organische Bestandteile des Bodens. In: BLANCK, E.: Handbuch der Bodenlehre 7, 113, Berlin 1931, und Erg. Band 1, 178, Berlin 1939. — MAJIMA, P., und G. TAKAYAMA: Über den Hauptbestandteil des Japanlacks VII, Der Urushiolmonomethyläther und der Mechanismus der Oxydation des Urushiols. Ber. dtsch. chem. Ges. 53, 1907–1916 (1920). — MARINI-BETTOLO, G. B., und L. PAOLONI: Action of diazo compounds on quinones, IV, Action of diazomethane on 2,6-dimethoxyquinone. Gazz. chim. ital. 84, 327–339 (1954). — MARTIN, A. E.. und R. REEVE: Chemical studies on podzolic illuvial horizons, I, The extraction of organic matter by organic chelating agents. J. Soil Sci. 8, 268–278 (1957a). — Chemical studies of podzolic illuvial horizon, II, The use of acetylacetone as extractant of translocated organic matter. J. Soil Sci. 8, 279–86 (1957b). — MASON, H. S.: Mechanism of oxygen metabolism. Adv. Enzymol. 19, 79–233 (1957). — MASON, H. S., W. L. FOWLKS und E. W. Y. PETERSON: Oxygen transfer and electron transport by the phenolase complex. J. Amer. Chem. Soc. 77, 2914–2915 (1955). — MAYAUDON, J., und P. SIMONART: Etude de la décomposition de la matière organique dans le sol au moyen de carbone radioactif, II, Décomposition du glucose radioactif dans le sol, A, Répartition de la radioactivité dans fractions humique du sol. Plant a. Soil 9, 376–380 (1958a). — Etude de la décomposition de la matière organique dans le sol au moyen de carbone radioactif, III, Décomposition des substances solubles dialysables des protéines et des hémicelluloses. Plant a. Soil 11, 170–175 (1958b). — Etude de la décomposition de la matière organique dans le sol au moyen de carbone radioactif, V, Décomposition de cellulose et de lignine. Plant a. Soil 11, 181–192 (1959). — Decomposition of cellulose C^{14} and lignin C^{14} in the soil. Ecology of Soil Fungi (1961). — MAYER, W.: Dehydrodigallussäure. Liebigs Ann. Chem. 578, 34–44 (1952). — MAYER, W., R. FIKENTSCHER, J. SCHMIDT und O. TH. SCHMIDT: Über eine ungewöhnliche Spaltung von Diarylläthern. Chem. Ber. 93, 2761–2776 (1960). — METZENBERG, R. L., und H. K. MITCHELL: The biosynthesis of aromatic compounds by neurospora crassa. Biochem. J. 68, 168 (1958). — MICHAELIS, L.: Theory of oxidation-reduction. The Enzymes, Vol. II, Part I, New York 1951. — MICHEL, E., und R. LANTZ: Action de l'ammoniaque ou des amines primaires sur le benzène substitué en 1, 2, et 4 par des groupes amino ou hydroxyles, non substitués ou substitués, I, Les trois groupes sont des hydroxyles: reactions sur l'hydroxy hydroquinone. Bull. Soc. chim. France 12, 2402–2408 (1961). — MITCHELL, B. D.: The differential thermal analysis of humic substances and related materials. Sci. Proc. Roy. Dublin Soc. A 1, 105–114 (1960). — MORRISON, R. I.: The alkaline nitrobenzene oxidation of soil organic matter. Soil Sci. 9, 130–140 (1958). — MÜLLER, E., K. LEY und G. SCHIECHTE: Zur Dehydrierung von Phenolen. Angew. Chem. 69, 204 (1957). — MUKHERJEE, P. N., J. N. BHOWMIK und A. LAHIRI: Mechanism of the oxidation of coal with special reference to the products of oxidation. Fuel 36, 417–422 (1957). — MUSSO, H., H. G. MATTHIES, H. BEECKEN und H. KRÄMER: Zur Konstitution der Orcein-Farbstoffe. Angew. Chem. 69, 178 (1957).

ODEN, S.: Zur Kenntnis der Huminsäuren des Sphagnum-Torfes. Ber. dtsch. chem. Ges. 35, 651 (1912). — Die Huminsäuren, chemische, physikalische und bodenkundliche Forschungen. Kolloidchem. Beihefte 11, 75–260 (1919). — Die Huminsäuren. Leipzig: Steinkopff. 1922. — OKUDA, A., und S. HORI: II, 1, Chromatographic investigation of amino acids in humic acid and alkaline alcohol lignin. Trans. V. Int. Congr. Soil Sci. Léopoldville, 2, 255–258 (1954). — Identification of amino acids in humic acid. Soil a. Plant Food (Tokyo) 1, 39–40 (1955). — ORLOV, D. S., O. H. ROZANOVA und S. G. MATJUCHINA: Infrarotabsorptionsspektren der Huminsäuren. Pochvovedenie 1962, 1, 17–25. — OSTWALD, W., und A. STEINER: Beiträge zur Kolloidchemie von Humussäure und Torf. Kolloidchem. Beihefte 21, 97–170 (1925).

PATCHETT, A. A., und B. WITKOP: On the mechanism of oxidation of o-quinone by peroxide. J. Org. Chemistry 22, 1477–1484 (1957). — PEARSON, R. W., A. G. NORMAN und CH. HO: The mineralization of the phosphorus of various compounds in soil. Proc. Soil Sci. Soc. Amer. 6, 168–175 (1941). — PERKIN, A. G., und A. B. STEVEN: Purpurogallin, I. J. Chem. Soc. 83, 192–201 (1903). — PINCK, L. A., und F. E. ALLISON:

The synthesis of ligninlike complexes by fungi. Soil Sci. 57, 155–61 (1944). — PIRET, E. L., G. WHITE, H. C. WALTHER JR., und A. J. MADDEN JR.: Some physicochemical properties of peat humic acids. Sci. Proc. Roy. Dublin Soc. A 1, 69–79 (1960). — PLOETZ, TH.: Polymere Chinone als Huminsäuremodelle. Z. Pflanzenernähr., Düng., Bodenkde. 69, 50–58 (1955). — PLOTHO, O. v.: Weitere Untersuchungen zur Humusbildung der Mikroorganismen. Z. Pflanzenernähr., Düng., Bodenkde. 55, 151–169 (1951). — PRAT, S.: Literatura o Humusu. Fysiologie rostlin prirodovedecko fakulty university Karlovy (1960). — PRATT, E. F., und W. E. BOEHME: Reactions of naphthoquinones with malonic ester and its analogs, I, Reactions with malonic ester. J. Amer. Chem. Soc. 73, 444–448 (1951).

RABIEN, G.: Zur Reaktion von Hydroxy-p-benzochinon mit Diazomethan. Diplomarbeit Braunschweig 1959. — RAPER, H. S.: Die Einwirkung von Tyrosinase auf Tyrosin. Fermentforsch. 9, 206 (1927). — REINDERS, W., und P. DINGEMANS: Die Oxydationsgeschwindigkeit von Hydrochinon mit Luftsauerstoff, I. Rec. Trav. chim. Pays-Bas 53, 209–230 (1934). — RIBBONS, D. W., und W. C. EVANS: Oxidative metabolism of phthalic acid by soil pseudomonads. Biochem. J. 76, 310–318 (1960). — RIPPEL, A., und O. LUDWIG: Die Schwarzfärbung von Azotobacter chroococcum Bei. als Melaninbildung. Zbl. Bakt. II. 64, 161 (1925). — ROBINSON, G. W.: Die Böden. Berlin 1939. — ROGERS, H. T.: Dephosphorylation of organic phosphorus compounds by soil catalysts. Soil Sci. 54, 439–446 (1942).

SALFELD, J. CHR.: Zum Reaktionsmechanismus der Purpurogallinbildung. Angew. Chem. 69, 723/24 (1957). — Über die Oxydation von Pyrogallol und Pyrogallolderivaten, II, Die Konstitution dimerer 3-Hydroxy-o-benzochinone. Chem. Ber. 93, 737–745 (1960). — Zur Chemie der Huminsäurebildung. Landbauforsch. Völkenrode 11, 22–24 (1961). — SALFELD, J. CHR., und E. BAUME: Über die Oxydation von Pyrogallol und Pyrogallolderivaten, III, Die Umlagerung von dimerem 3-Hydroxy-4,6-di-tert.butyl-o-benzochinon durch Alkali. Chem. Ber. 93, 745–751 (1960). — SCHARPENSEEL, H. W., und R. KRAUSSE: Aminosäureuntersuchungen an verschiedenen organischen Sedimenten, besonders Grau- und Braunhuminsäurefraktionen verschiedener Bodentypen (einschließlich C^{14}-markierter Huminsäuren). Z. Pflanzenernähr., Düng., Bodenkde. 96, 11–34 (1962). — SCHEELE, W.: Beitrag zur Charakterisierung natürlicher Humusstoffe. Kolloid-Beihefte 46, 368–424 (1937). — SCHEFFER, F.: Humus und dessen Bildung durch die Mikroorganismen. Trans. VI. Congr. Microbiol. 6, 265–278 (1953). — SCHEFFER, F., und R. KICKUTH: Chemische Abbauversuche an einer natürlichen Huminsäure, I. Z. Pflanzenernähr., Düng., Bodenkde. 94, 180–188 (1961a). — Chemische Abbauversuche an einer natürlichen Huminsäure, II. Z. Pflanzenernähr., Düng., Bodenkde. 94, 189–198 (1961b). — SCHEFFER, F., und B. ULRICH: Lehrbuch der Agrikulturchemie und Bodenkunde, III. Teil, Humus und Humusdüngung 1. Stuttgart: Enke. 1960. — SCHEFFER, F., und E. WELTE: Über absorptionsspektrographische Untersuchungen an natürlichen Huminsäuren. Landwirtsch. Forsch. 1, 190–220 (1950a). — Die Ultraviolett-Absorption von Huminsäuren aus Hochmoortorf, Braun- und Steinkohlen. Landwirtsch. Forsch. 3, 1–6 (1950b). — SCHEFFER, F., und W. ZIECHMANN: Über Synthesehuminsäuren aus Polyoxybenzolen. Z. Pflanzenernähr., Düng., Bodenkde. 80, 126–127 (1958). — SCHEFFER, F., B. MEYER und A. NIEDERBUDDE: Huminstoffbildung unter katalytischer Einwirkung natürlich vorkommender Eisenverbindungen im Modellversuch. Z. Pflanzenernähr., Düng., Bodenkde. 87, 26–44 (1959). — SCHEFFER, F., O. v. PLOTHO und E. WELTE: Untersuchungen über die Bildung von Humusstoffen durch Actinomyceten. Landwirtsch. Forsch. 1, 81–92 (1950). — SCHEFFER, F., E. WELTE und W. ZIECHMANN: Über Untersuchungen an künstlichen Huminsäuren. Z. Pflanzenernähr., Düng., Bodenkde. 69, 58–65 (1955). — SCHEFFER, F., W. ZIECHMANN und H. SCHLÜTER: Über die Sedimentation von Huminsäuren in der Ultrazentrifuge. Kolloid-Z. 160, 1–7 (1958). — Die präparative Trennung von Huminsäuren im Schwerefeld der Ultrazentrifuge. Naturwiss. 46, 143–144 (1959). — SCHEFFER, F., W. ZIECHMANN und H. SCHOLZ: Über Synthesehuminsäuren aus Polyoxybenzolen. Z. Pflanzenernähr., Düng., Bodenkde. 80, 134–149 (1958). — SCHLICHTING, E.: Zur Kenntnis des Heidehumus, I, Fraktionierung und Untersuchung des ganzen Humuskörpers. Z. Pflanzenernähr., Düng., Bodenkde. 61, 1–12 (1953). — SCHLÜTER, H.: Über Sedimentationsmessungen an Huminsäuren. Z. Pflanzenernähr., Düng., Bodenkde. 84, 169–174 (1959). — SCHMALFUSS, K.: Über Bodenatmung. Bodenkde. u. Pflanzenernähr. 21–22, 442–454 (1940). — SCHMIDT, E., und M. ATTERER: Zur Kenntnis der Huminsubstanzen. Ber. dtsch. chem. Ges. 60, 1671 (1927). — SCHMIDT, O. TH., und R. LADEMANN: Corilagin, ein weiterer kristallisierter Gerbstoff aus Dividivi, X. Mitt. über natürliche Gerbstoffe. Liebigs Ann. Chem. 571, 232–237 (1951). — SCHMIDT, O. TH., und W. MAYER: Natürliche Gerbstoffe. Angew. Chem. **68**, 103–114 (1956). — Schmidt, O. Th., und D. M. Schmidt: Über das Vorkommen von Corilagin in Myrobalanen, XV. Mitt. über natürliche Gerbstoffe. Liebigs Ann. Chem. **578**, 31–33 (1952). — Schnitzer, M., D. A. Shearer und J. R. Wright: A study in the infrared of high molecular weight organic matter extracted by various reagents from a podzolic B horizon. Soil Sci. **87**, 252–257 (1959). — Schnitzer, M., J. R. Wright und J. G. Desjardin: A comparison of the effectiveness of various extractants for organic matter from two horizons of a podzol profile. Canad. J. Soil Sci. **38**, 49–53 (1958). — Schobinger, U. Chemische Untersuchungen über die Umwandlung von Weizenstrohlignin im Laufe der Verrottung. Diss. ETH Zürich 1957. — Scholz, H.: Die adsorptionschromatographische Auftrennung von Huminsäuren. Z. Pflanzenernähr., Düng., Bodenkde. **84**, 159–169 (1959). — Schreiner, O., und E. Shorey: Chemical nature of soil organic matter. U. S. Dept. Agric. Bureau of Soils Bull. **74**, 5–48 (1910). — Schtschukina, L. A.: Oxydative und oxydativ hydrolytische Umwandlung organischer Moleküle, 21, Hydrolytische Spaltung von 6-Methyl-3-hydroxy-p-benzochinon. J. allg. Chem. **22**, (84) 668–675 (1952). — Schulze, H., und W. Flaig: Zur Kenntnis der Huminsäuren IV, Über die Ringsprengung mehrwertiger Phenole mit Sauerstoff in alkalischem Medium. Liebigs Ann. Chem. **575**, 231–241 (1952). — Seegerer, A.: Der Saccharasegehalt des Bodens als Maßstab seiner biologischen Aktivität. Z. Pflanzenernähr., Düng., Bodenkde. **61**, 251–260 (1953). — Shorey, E.: Some organic soil constituents. U. S. Dept. Agric. Bureau of Soils Bull. **88**, 5–41 (1913). — Simon, K.: Über die Herstellung von Humusextrakten mit neutralen Mitteln. Pflanzenernähr. u. Düng. **14**, 252–257 (1929). — Über die unterschiedlichen Eigenschaften extrahierbarer Humussubstanzen. Z. Pflanzenernähr., Düng., Bodenkde. **34**, 144–162 (1934). — Färbungen und Oxydationswerte von Humusextrakten in gegenseitiger Beziehung. Z. Pflanzenernähr., Düng., Bodenkde. **39**, 1–14 (1935). — Simonart, P., und J. Mayaudon: Etude de la décomposition de la matière organique dans le sol au moyen de carbone radioactif, I, Cinétique de l'oxydation en CO_2 de divers substrats radioactifs. Plant a. Soil **9**, 367–375 (1958a). — Etude de la décomposition de la matière organique dans le sol au moyen de carbone radioactif, II, Décomposition du glucose radioactif dans le sol, B, Répartition de la radioactivité dans l'α-humus. Plant a. Soil **9**, 381–384 (1958b). — Simonart, P., J. Mayaudon und L. Batistic: Etude de la décomposition de la matière organique dans le sol au moyen de carbone radioactif, IV, Décomposition des pigments foliaires. Plant a. Soil **11**, 176–180 (1958). — Sowden, F. J.: Estimation of amino acids in soil. Hydrolysates by the Moore and Stein method. Soil Sci. **80**, 181–188 (1955). — Forms of nitrogen in the organic matter of different horizons of soil profiles. Canad. J. Soil Sci. **38**, 147–154 (1958). — Sprengel, C.: Über Pflanzenhumus, Humussäure und humussaure Salze. Kastners Arch. Ges. Naturlehre **8**, 145 (1826). — Von den Substanzen der Ackerkrume und des Untergrundes, insbesondere, wie solche durch die chemische Analyse entdeckt und voneinander geschieden werden können. Erdmanns J. Techn. Ökonom. Chem. **2**, 423–474, **3**, 42–99, 313–351, 397–421 (1828). — Die Bodenkunde oder die Lehre vom Boden. Leipzig: Immanuel Müller. 1837. — Die Lehre vom Dünger oder Beschreibung aller bei der Landwirtschaft gebräuchlicher vegetabilischer, animalischer und mineralischer Düngermaterialien nebst Erklärung ihrer Wirkungsart. Leipzig: Immanuel Müller. 1839. — Springer, U.: Der heutige Stand der Humusuntersuchungsmethodik mit besonderer Berücksichtigung der Trennung, Bestimmung und Charakterisierung der Huminsäuretypen und ihre Anwendung auf charakteristische Humusformen. Z. Bodenkde. u. Pflanzenernähr. **6**, 312–373 (1938). — Stoffabbau und Humusaufbau untersucht an einem Strohmist und Strohligninmist (Laboratoriumsversuch). Prakt. Bl. Pflanzenbau u. Pflanzenschutz **21/22**, 1–57 (1944/45). — Über Komposthuminsäuren aus verschiedenen pflanzlichen Ausgangsstoffen. Z. Pflanzenernähr., Düng., Bodenkde. **69**, 66–71 (1955). — Springer, U., und A. Lehner: Stoffabbau und Humusaufbau bei der aeroben und anaeroben Zersetzung landwirtschaftlich und forstwirtschaftlich wichtiger organischer Stoffe I. Z. Pflanzenernähr., Düng., Bodenkde. **58**, 193–231 (1952a). — Stoffabbau und Humusaufbau bei der aeroben und anaeroben Zersetzung landwirtschaftlich und forstwirtschaftlich wichtiger organischer Stoffe II. Z. Pflanzenernähr., Düng., Bodenkde. **59**, 1–27 (1952b). — Springer, U., und F. Seischab: Zur Kenntnis der bei der Kompostbereitung auftretenden stofflichen Veränderungen. Mit Ergebnissen eines dreijährigen Gefäßversuches. Bayr. Landw. Jb. **38**, 250–300 (1961). — Springer, U., und A. Wagner: Zur Kenntnis der von Pilzen gebildeten huminsäureähnlichen Substanzen unter besonderer Berücksichtigung von Stachybotris lobulata. Z. Pflanzenbau. u. Pflanzenschutz **3**, 145–172 (1952). — Stadnikoff, G.: Über Huminsäuren. Kolloid-Z. **80**, 60–68 (1937). — Stählin, A.: Methodenbuch, B. XII, Die Beurteilung der Futtermittel. Radebeul-

Berlin: Neumann. 1957. — STANIER, R. Y.: Simultaneous adaption: a new technique for the study of metabolic pathways. J. Bact. 54, 339–348 (1947). — The bacterial oxidation of aromatic compounds, IV, Studies on the mechanism of enzymatic degradation of protocatechuic acid. J. Bact. 59, 527–532 (1950). — STANIER, R. Y., und J. L. INGRAHAM: Protocatechuic oxidase: J. Biol. Chemistry 210, 799–808 (1954). — STAUDINGER, H., und E. HUSEMANN: Über hochpolymere Verbindungen, 155, Über die Konstitution des Glykogens. Liebigs Ann. Chem. 530, 1–20 (1937). — STEELINK, C., J. W. BERRY, A. HO und H. E. NORDBY: Alkaline degradation products of soil humic acid. Sci. Proc. Roy. Dublin Soc. A 1, 59–67 (1960). — STEIN, H. N., und H. J. C. TENDELOO: On the kinetics of the oxidation of phloroglucinol. Rec. Trav. chim. Pays-Bas 74, 905–918 (1955). — Some observations on the oxidation products of phloroglucinol. Rec. Trav. chim. Pays-Bas 75, 1233–1239 (1956a). — Evidence for free radicals as intermediates in the oxidation of phloroglucinol. Rec. Trav. chim. Pays-Bas 75, 1240–1242 (1956b). — The kinetics of the oxidation of 2,4-dimethyl-phloroglucinol. Rec. Trav. chim. Pays-Bas 77, 487–490 (1958). — The oxidation of phloroglucinol as a model for humification processes. Plant a. Soil 11, 131–138 (1959). — STEINMETZ, A.: Modellversuche zur Beteiligung von Polyphenoloxydasen bei der Bildung natürlicher Huminsäuren. Diss. TH Braunschweig 1956. — STEVENSON, F. J.: Isolation and identification of some amino compounds in soils. Proc. Soc. Soil Sci. Amer. 20, 201–204 (1956a). — Effect of some long-time rotation on the amino acid composition of the soil. Proc. Soc. Soil Sci. Amer. 20, 204–208 (1956b). — STEVENSON, F. J., Q. VAN WINKLE und W. P. MARTIN: Physiochemical investigations of clay adsorbed organic colloids, II. Proc. Soil Sci. Soc. Amer. 17, 31–34 (1953). — STÖCKLI, A.: Über den Abbau von Lignin, Cellulose und Hemicellulose durch Pilze. Versuche mit Streumaterialien, Holz und Ligninsulfonsäuren. Promotion ETH Zürich 1952. — SVEDBERG, T.: Die Molekulargewichtsanalyse im Zentrifugalfeld. Kolloid-Z. 67, 2–16 (1934). — SWABY, R.: Soil organic matter. 8th, 9th and 10th C.S.I.R.O. Ann. Rep. Government Printers, Sydney, 1956–58.

TENNEY, F. G., und S. A. WAKSMAN: Composition of natural organic materials and their decomposition in the soil, V, Decomposition of various chemical constituents in plant materials under anaerobic conditions. Soil Sci. 30, 143–160 (1930). — TEUBER, H. J., und C. JELLINEK: Reaktionen mit Nitrosodisulfonat, I, Einwirkung auf Phenole und aromatische Stickstoffverbindungen. Chem. Ber. 85, 95–103 (1952). — TEUBER, H. J., und W. RAU: Reaktionen mit Nitrosodisulfonat, II, Über die Bildung von Chinonen aus einwertigen Phenolen. Chem. Ber. 86, 1036–1047 (1953). — THAER, A.: Die Grundzüge der rationellen Landwirtschaft, Bd. 1–4. Berlin: Reiner. 1821. — THOM, C., und M. PHILLIPS: Ligninartige Komplexe in Pilzen. J. Wash. Acad. Sci. 22, 237–239 (1932). — TINSLEY, J., und A. SALAM: Extraction of soil organic matter with aqueous solvents. Soils a. Fertilizers 24, 81–84 (1961). — TJURIN, I. W.: Die organischen Stoffe der Böden. Moskau: Selchosgis. 1937. — TRIPPETT, S., S. DAGLEY und D. A. STOPHER: Bacterial oxidation of protocatechuic acid. Biochem. J. 76, 9 P (1960). — TROJANOWSKI, J.: Analiza konduktometryczna i potencjometryczna niektorych frakcji humusu. Ann. Univ. Mariae-Curie-Sklodowska (Lublin-Polen) 12, 183–204 (1957).

UPDEGRAFF, I. H., und H. C. CASSIDY: Electron-exchange-polymers, II. J. Amer. Chem. Soc. 71, 407–410 (1949).

VALEK, B.: Das Humusmaximum der Differentialthermokurven aus Bodenproben von verschiedenen Pflanzenbeständen. Studies about Humus. Symposium Humus and Plant, Praha and Brno 28. IX.—6. X. 1961 341–345 (1962). — VENKATARAMAN, B., und G. K. FRAENKEL: Proton hyperfine interactions in paramagnetic-resonance of semiquinones. J. Amer. Chem. Soc. 77, 2707–2713 (1955), Abstr., 126th Meeting Amer. Chem. Soc. Sept. 1954, 22. R. — VLIET, W. F.: Enzymatic oxidation of lignin. Biochim. biophys. Acta 15, 211–216 (1954).

WAKSMAN, S. A.: I. Int. Congr. Soil Sci., Washington, 3, 690 (1928). — Humus Baltimore: Williams and Wilkins. 1936. — Soil Microbiology. New York: Wiley. 1952. — WAKSMAN, S. A., und F. GERRETSEN: Influence of temperature and moisture upon the nature and extent of decomposition of plant residues by microorganisms. Ecology 12, 33–60 (1931). — WAKSMAN, S. A., und K. R. STEVENS: A system of proximate chemical analysis of plant materials. Ind. Engng. Chem. 2, 167–173 (1930). — WALKER, N., und W. C. EVANS: Pathways in the metabolism of the monohydroxybenzoic acid by soil bacteria. Biochem. J. 52, XXIII (1952). — WALLERIUS J.: Agricultural fundamenta chemica. Uppsala 1761. — WALTER, H.: Beitrag zur Kenntnis der Oxydation des Guajakols unter physiologischen Bedingungen. Diss. TH Braunschweig 1960. — WEITZ, E., H. SCHOBERT und H. SEIBERT: Über die Einwirkung von alkalischem Wasserstoffsuperoxyd auf Chinone und auf mehrwertige

Phenole. Ber. dtsch. chem. Ges. 68, 1163–1168 (1935). — Welte, E.: Zur Konzentrationsmessung von Huminsäuren. Z. Pflanzenernähr., Düng., Bodenkde. 74, 219–227 (1956). — Welte, E., A. Niemann und W. Ziechmann: Über Huminsäuren II (Molekular- und Äquivalentgewicht). Naturwiss. 41, 334 (1954). — Welte, E., M. Schatz und W. Ziechmann: Über Synthesehuminsäuren (Oxydationsmessungen). Naturwiss. 41, 213–214 (1954). — Willstätter, R., und H. Heiss: Über die Konstitution des Purpurogallins. Liebigs Ann. Chem. 433, 17–33 (1923). — Willstätter, R., und F. Müller: Zur Kenntnis der o-Chinone, XXVI. Mitt. über Chinoide, Ber. dtsch. chem. Ges. 44, 2171–2181 (1911). — Wrenshall, C. L., und W. J. Dyer: Organic phosphorus in soils, II, The nature of the organic phosphorus compounds. Nucleic acid derivates. Phytin. Soil Sci. 51, 235–248 (1941). — Wright, J. R., und M. Schnitzer: Oxygen-containing functional groups in the organic matter of the Ao and Bh horizons of a podzol. 7th Int. Congr. Soil Sci., Madison, Wisc., USA, 120–127 (1960).

Yates, P., M. I. Ardao und L. E. Fieser: The infrared spectra of p-benzoquinones. J. Amer. Chem. Soc. 78, 650–652 (1956). — Young, J. L., und J. L. Mortensen: Soil nitrogen complexes, I, Chromatography of amino compounds in soil hydrolysates. Ohio Agric. Exper. Stat. Res. Circ. 61, 18 (1958).

Zeile, K.: Zur Kenntnis der Huminsäuren. Kolloid-Z. 72, 211–212 (1935). — Zetzsche, F., und H. Reinhart: Über die Huminsäuren des Torfes. Brennstoff-Chem. 18, 393–95 (1937). — Ziechmann, W.: Über Modellreaktionen zur Bildung synthetischer Huminsäuren, I, Die Synthese von Huminsäuren im alkalischen Milieu. Brennstoffchem. 41, 289–320 (1960a). — Über Modellreaktionen zur Bildung synthetischer Huminsäuren, II, Die Synthese von Huminsäuren im neutralen Milieu. Brennstoffchem. 41, 334–340 (1960b). — Die Bildung von 2,5-Dioxy-p-benzochinon bei der Autoxydation des Hydrochinons im dreiphasigen System. Naturwiss. 48, 456–457 (1961). — Ziechmann, W., und R. Kickuth: Die Struktur der Huminsäuren (Zur Theorie der Huminstoffe). Kolloid-Z. 174, 38–46 (1961). — Ziechmann, W., und G. Pawelke: Zum Vergleich natürlicher und synthetischer Huminsäuren und ihre Vorstufen. Z. Pflanzenernähr., Düng., Bodenkde. 84, 174–184 (1959). — Ziechmann, W., und H. Scholz: Spektroskopische Untersuchungen an Huminsäuren. Naturwiss. 47, 193–196 (1960). — Zolcinski, J.: Eine neue genetische physikalisch-chemische Theorie der Bildung des Humus, Torfs und der Kohle. Die Rolle und Bedeutung der biologischen Faktoren bei diesen Vorgängen. Wiss. Arch. Landw. A 4, 196 (1930).

d) Sonstige organische Bodenbestandteile

Von

W. Flaig, J. Chr. Salfeld und H. Söchtig

Die organischen Stoffe im Boden stammen hauptsächlich aus den abgestorbenen Organismen; in geringem Umfang werden sie von lebenden Organismen ausgeschieden. Jährlich werden unter mitteleuropäischen Verhältnissen ungefähr folgende Mengen an organischer Masse pro Hektar gebildet und umgesetzt:

	Pflanzen 4000 bis 5000 kg	Tiere im Boden 70 bis 400 kg	Mikroorganismen 2000 kg
Allgemeine Stoffwechselprodukte	Proteine, aliphatische und aromatische Aminosäuren Kohlenhydrate Fette, aliphatische Säuren, organische Phosphorverbindungen		
Spezielle Stoffwechselprodukte	Lignin phenolische Verbindungen	Aminozucker Chitin	z. B. Verbindungen von chinoiden Strukturen Antibiotika

Einige chemische Verbindungen sind in den Pflanzen, Tieren und Mikroorganismen die gleichen. Alle Organismen enthalten Proteine, aliphatische und aromatische Aminosäuren, Kohlenhydrate, Fette, aliphatische Säuren und verschiedene organische Phosphorverbindungen.

Eine Reihe weiterer Substanzen sind nur in einigen Arten der Organismen enthalten. Spezielle Stoffwechselprodukte der Pflanzen sind das bei der Humifizierung eine wichtige Rolle spielende Lignin und andere phenolische Verbindungen. In vielen Böden ist auch das Chitin der Insektencuticula und der Pilzhyphen von Bedeutung. Die Mikroorganismen bilden ebenfalls spezielle Stoffwechselprodukte, von denen jedoch nur einige erwähnt werden sollen, die für den Humifizierungsprozeß von Bedeutung sein können. Eine Reihe von Mikroorganismen bildet Pigmente, die eine chinoide Struktur besitzen (ANSLOW und RAISSTRICK 1938a, 1938b, 1938c, RAISSTRICK, ANSLOW und BREEN 1940, QUILLICO 1951).

In neueren Untersuchungen ist die Wirkung von Antibiotika auf das Pflanzenwachstum überprüft worden. Dabei wurde gefunden, daß sie, dem Boden zugesetzt, verhältnismäßig rasch abgebaut werden[1].

Die hochmolekularen Substanzen werden bei dem Umwandlungsprozeß im Boden nahezu alle durch Mikroorganismen abgebaut. Die Geschwindigkeit des Abbaus zu niedermolekularen Anteilen und deren Mineralisierung ist dabei verschieden. Aliphatische Aminosäuren und sonstige aliphatische Säuren, die monomeren Zucker und ihre Abkömmlinge, werden rascher abgebaut als die aromatischen Aminosäuren und heterocyclischen Verbindungen, wie Purin und Pyrimidin aus den Nukleinsäuren. Lignin und einige phenolische Verbindungen sind zumindest bei Versuchen in Reinkulturen verhältnismäßig resistent gegenüber dem Angriff durch Mikroorganismen. Bei diesen Umwandlungsprozessen sollen auch Fermente, die frei im Boden vorliegen, eine Rolle spielen (HOFMANN und HOFFMANN 1961, CLAUS und MECHSNER 1960).

Kohlenhydrate und Proteine machen einen wesentlichen Anteil der abgestorbenen Organismen aus. Der Ligningehalt der Pflanzen ist unterschiedlich und beträgt bei Wiesengräsern etwa 10%, bei Getreidestroh etwa 20% und bei den verschiedenen Holzarten zwischen 25 und 35% des Trockengewichtes.

Der Zerfall der organischen Substanz im Boden ist von der Bildung verschiedener organischer Säuren und Kohlendioxid begleitet (GREENWOOD und LEES 1956, LEES und PORTEOUS 1950, TOMBESI 1953). Ferner ist bei Untersuchungen über die Zusammensetzung der organischen Substanzen im Boden zu berücksichtigen, daß die Mikroorganismen die Bestandteile der abgestorbenen Organismen nicht nur zersetzen, sondern auch körpereigene Proteine und Stoffwechselprodukte erzeugen; hierbei ist im Zusammenhang mit der Humusbildung vor allem die Bildung von Polysacchariden, speziell Polyuroniden, von Bedeutung.

Solange keine ausreichenden Untersuchungen an verschiedenen Bodentypen zu verschiedenen Jahreszeiten und unter unterschiedlicher Vegetation vorliegen, ist es nicht möglich, Abhängigkeiten des Gehaltes an definierten Verbindungen vom Bodentyp, Klima oder Bestand zu erkennen.

Daher muß man sich bei der Isolierung von definierten organischen Verbindungen aus dem Boden darüber im klaren sein, daß die Befunde nur jeweils den augenblicklichen Zustand des Bodens betreffen können und nicht verallgemeinert werden dürfen.

Die Isolierung definierter organischer Substanzen aus Böden ist erstmals von SCHREINER und SHOREY und ihren Mitarbeitern in den Jahren von 1906 bis 1938 durchgeführt worden.

Zu jener Zeit waren die analytischen Hilfsmittel wie Spektroskopie, Elektrophorese und die verschiedenen Arten der Chromatographie noch nicht bekannt

---

[1] Vergleiche in diesem Handbuch Band I, W. FLAIG, F. HEILINGER und G. SCHMID.

oder in dem Maße ausgearbeitet wie heute; die Autoren haben alle Verbindungen isoliert und präparativ dargestellt. Diese Arbeiten gewinnen in der letzten Zeit wieder mehr an Interesse und bedürfen daher einer besonderen Würdigung.

Das Schema der Aufarbeitung war wie folgt:

Boden (organisch gebundener Kohlenstoff in den Huminstoffen = 100)
Extraktion mit 2prozentiger Natronlauge

unlöslich	alkalische Lösung
24 (S u. S)	76 (S u. S)
32 (B)	68 (B)
20 (S)	80 (S)
(Huminstoffe unbekannter Konstitution)	angesäuert und filtriert

Niederschlag)	saures Filtrat
37 (S u. S)	39 (S u. S)
28 (B)	40 (B)
60 (S)	20 (S)
Extrakt mit kochendem Alkohol	(Fulvosäuren) enthält niedermolekulare Verbindungen, deren Konstitution teilweise bekannt ist

unlöslich	alkoholische Lösung
16 (S u. S)	21 (S u. S)
40 (S)	15 (S)
(Huminsäuren, Stoffe unbekannter Konstitution)	(Hymatomelansäuren)
	Extraktion mit Petroläther

unlöslich	Petrolätherlösung
19	2
(kann Harzsäuren und Harze enthalten)	(kann höhere Fettsäuren enthalten)

(S u. S) = Schreiner und Shorey (1910c)
(B) = Berthelot und Andre (1892)
(S) = Schmuk (1930)

Das Ausgangsmaterial bei den Untersuchungen von Schreiner und Shorey (1910c) war ein Mineralboden mit 0,96% organischem Kohlenstoff in den Huminstoffen, der in dem Arbeitsschema = 100 gesetzt worden ist. Zum Vergleich sind Ergebnisse von Berthelot und Andre (1892) sowie von Schmuk (1930) angegeben, aus denen hervorgeht, daß die Verteilung des Kohlenstoffs in den Böden sehr unterschiedlich sein kann.

Schreiner und Shorey haben nach diesem Schema den organischen Anteil des Bodens mit verschiedenen Fraktionierungsmethoden aufgetrennt. Ein Teil der dabei gefundenen Verbindungen entstammt Mikroorganismen. Man muß damit rechnen, daß die eine oder andere Substanz durch die Art der Isolierung entstanden sein kann. Im einzelnen sind die in Tab. 58 aufgeführten Substanzen identifiziert worden.

Diese von Schreiner und Shorey im Jahre 1906 begonnenen Untersuchungen beschränkten sich auf die Isolierung niedermolekularer Verbindungen. Bei der Untersuchung der höhermolekularen Substanzen begnügte man sich zunächst mit den nach der Stoffgruppenanalyse von Waksman oder ähnlichen Methoden erhaltenen chemisch nicht definierten Fraktionen. Die Unzulänglichkeit dieser Methodik führte dazu, daß man sich in neuerer Zeit mehr der Isolierung chemisch definierter Substanzen zuwandte. Diese Arbeitsrichtung wurde durch die Entwicklung neuer Methoden zur Trennung und Identifizierung organischer Substanzen begünstigt.

Tabelle 58. *Zusammenstellung der vom Arbeitskreis Schreiner und Shorey aus Boden isolierten chemischen Verbindungen*

Verbindung	Bemerkung	Literatur
	Alkohole, Zucker und Fettsäuren	
Mannit	zuweilen in größerer Menge gefunden	SHOREY (1913)
Rhamnose		SHOREY (1913)
Oxalsäure	selten gefunden, aber dann zuweilen in größerer Menge	SHOREY (1913)
Bernsteinsäure		SHOREY (1913)
Zuckersäure		SHOREY (1913)
Acrylsäure		SHOREY (1913)
α-Crotonsäure		WALTERS und WISE (1916)
Dioxystearinsäure	häufig gefunden	SCHREINER und SHOREY (1909)
Trioxystearinsäure		SHOREY (1928)
Oxybehensäure		SHOREY (1928)
Ölsäure		SHOREY (1928)
	Stickstoffhaltige aliphatische und heterocyclische Verbindungen	
Cyanursäure	häufig und mitunter auch in größerer Menge gefunden	SHOREY und WALTERS (1914) WISE und WALTERS (1917a, 1917b)
Histidin		SCHREINER und SHOREY (1910a, 1910b)
Arginin		SCHREINER und SHOREY (1910a, 1910b)
Lysin		SHOREY (1913)
Xanthin		SCHREINER und SHOREY (1910a, 1910b)
Hypoxanthin		SCHREINER und SHOREY (1910a, 1910b)
Cytosin		SCHREINER und SHOREY (1910a, 1910b)
Adenin		SHOREY (1913)
Allantoin		SHOREY (1938)
Cholin		SHOREY (1913)
Kreatinin		SHOREY (1912a), SULLIVAN (1911)
Pikolincarbonsäure		SCHREINER und SHOREY (1909) SHOREY (1906) SCHREINER und SHOREY (1908)
	Aromatische Verbindungen	
Benzoesäure	selten gefunden, aber dann mitunter in größerer Menge	SHOREY (1916), WALTERS (1917)
p-Hydroxybenzoesäure		SHOREY (1916), WALTERS (1917)
Salizylaldehyd		SHOREY (1913)
Trithiobenzaldehyd	selten gefunden	SHOREY (1913)
Vanillin		(SHOREY 1913)

Im folgenden werden die einzelnen Verbindungsgruppen näher beschrieben:

1. Kohlenhydrate

A. Monosaccharide

Der Gehalt an freien Monosacchariden im Boden scheint nicht sehr groß zu sein, da sie einerseits rasch von den Bodenorganismen abgebaut, andererseits von anderen Bodenbestandteilen sorbiert werden können. ALVSAKER (1948) isolierte in Kaltwasserextrakten 1% der organischen Substanz norwegischer Böden als reduzierende Zucker. Diese enthielten in Prozent der organischen Substanz des Bodens 0,22% Glucose, 0,02% Galactose, 0,035% Fruktose, 0,03% Xylose, 0,04% Arabinose und weniger als 0,001% Ribose (ALVSAKER und MICHELSON 1957). NAGAR (1962) fand in verschiedenen Böden Glucose. Die monomeren Zucker können von Tonmineralen sehr stark sorbiert werden (GREENLAND 1956, KOHL und TAYLOR 1961). ENDERS und SIGURDSSON (1947) fanden außerdem im Boden Methylglyoxal.

B. Polysaccharide

Die Hauptmasse der Polysaccharide abgestorbener Pflanzenteile gelangt als Cellulose in den Boden. Diese wird von den Mikroorganismen als Kohlenstoffquelle zur Energiegewinnung verhältnismäßig leicht verwertet. In neuerer Zeit wurde die Cellulosezersetzung in Böden durch Einbettung und Bebrütung von Baumwollgewebe in Bodenproben untersucht („burial test"). Dabei wurde einerseits die Kohlendioxidentwicklung und andererseits die Änderung der Zugfestigkeit des Baumwollgewebes als Maß für die Aktivität der cellulosezersetzenden Mikroflora verwendet (SCHMIDT und RUSCHMEYER 1958, RUSCHMEYER und SCHMIDT 1958). Die Korrelationskoeffizienten zwischen den nach dieser Methode bestimmten Werten und verschiedenen Bodeneigenschaften sind in Tab. 59 zusammengefaßt. Der Korrelationskoeffizient zwischen der CO_2-Entwicklung und der Änderung der Zugfestigkeit beträgt —0,73.

Tabelle 59. *Korrelationskoeffizienten zwischen verschiedenen Bodeneigenschaften und der cellulosezersetzenden Aktivität bestimmt durch Messung der CO_2-Entwicklung (A) und der Abnahme der Zugfestigkeit (B) nach dem „burial test"*

Bodenfaktor	*A*	*B*
Sandgehalt	—0,47	—0,41
Feinfraktion	+0,47	+0,41
Tongehalt	+0,41	+0,50
Feuchtigkeitsäquivalent	+0,57[1]	+0,40
pH	+0,76[1]	+0,67[1]
Gehalt an organischem Kohlenstoff	+0,50	+0,31
Gehalt an Gesamtstickstoff	+0,63[1]	+0,44
C:N-Verhältnis	—0,51[2]	—0,50
Nitratstickstoff	+0,57[1]	+0,59[1]
Nitrifizierungsvermögen	+0,69[1]	+0,58[1]
Verwertbarer Phosphor	—0,38	—0,13

[1] signifikant bei GD 1%.
[2] signifikant bei GD 2%.

Aus Tab. 59 geht hervor, daß die Bodenreaktion und die Stickstoffverhältnisse im Boden die wichtigsten Faktoren sind, welche die Aktivität der cellulosezersetzenden Mikroorganismen beeinflussen. Diese Resultate stimmen mit Untersuchungen von WAKSMAN und HUTCHINGS (1935) sowie von WHITE, HOLBEN und JEFFRIES (1949) überein.

Polysaccharidfraktion. In den Polysacchariden des Bodens ist das Verhältnis von Glucose zu Xylose niedriger als in denen der Pflanzen (WHISTLER und KIRBY 1956). Daraus wird geschlossen, daß die Polysaccharide des Bodens hauptsächlich mikrobiellen und nicht pflanzlichen Ursprungs sind (DUFF 1952a, 1952b, 1954, THEANDER 1952, 1954, WHISTLER und KIRBY 1956, CLAPP 1957, FORSYTH 1950). Hierauf weisen auch die Befunde von FULLER (1946, 1947a) hin, nach denen die Zusammensetzung und die Decarboxylierbarkeit der Bodenpolysaccharide mehr der von mikrobiellen als der von pflanzlichen Polysacchariden entsprechen. In dieselbe Richtung weist auch die Beobachtung, daß pflanzliche Polysaccharide wie Cellulose (SCHMALFUSS 1940) verhältnismäßig rasch, dagegen aus Böden isolierte Polysaccharide langsam abgebaut werden (WHISTLER und KIRBY 1956).

Mit Hilfe der Stoffgruppenanalyse bestimmte FORSYTH (1948) 0 bis 5% Cellulose und 5 bis 20% Hemicellulose im Boden. Alle Untersuchungen über die Isolierung definierter Polysaccharidkomplexe führen aber zu erheblich geringeren Ausbeuten (Tab. 60).

Tabelle 60. *Isolierung von Polysaccharidfraktionen aus verschiedenen Böden* (nach MEHTA, DUBACH und DEUEL 1961)

Böden	Extraktionsmethode	Menge an Polysaccharidfraktion % vom Boden	Menge an Polysaccharidfraktion % vom Humus	Literatur
Britische Böden	Puffer	0,05–0,15	—	HAWORTH, PINKARD und STRACEY (1946)
Britische Waldböden	Phosphatpuffer (Zimmertemperatur)	0,01–0,04	—	BERNIER (1958) OGSTON (1958)
Britische und tropische Böden	0,5 n NaOH (Zimmertemperatur)	—	1,32–1,94	FORSYTH (1947, 1950)
Schweizer Böden	do.	0,02–0,13	0,10–1,00	DUBACH, ZWEIFEL, BACH und DEUEL (1955)
Böden aus Wisconsin	do.	0,02–0,25	—	RENNIE, TRUOG und ALLEN (1954), CHESTERS, ATTOE und ALLEN (1957)
Böden aus Schottland	Wasser, 4 Stunden 85°	0,3	—	DUFF (1952a, 1952b, 1954)
Böden aus Indiana	Wasser, 6 Stunden 85°	0,05	1,45	WHISTLER und KIRBY (1956)
Muk-Soil von New York	Wasser, 24 Stunden Soxhlet	—	1,8	CLAPP (1957)
Britische Mineralböden, Torfe und Komposte	98%ige Ameisensäure, 30 Minuten Rückfluß	—	3,5–11,5	PARSON (1958)

Die Polysaccharide des Bodens scheinen zu einem großen Teil an andere höhermolekulare organische oder mineralische Bodenbestandteile sorbiert zu sein. Sie können daher erst nach einer Aufhebung der Sorption erfaßt werden. Wahrscheinlich führt aber nur eine Hydrolyse der Bodenfraktion zu einer vollständigen

Desorption. Daher wird durch keine der erwähnten Bestimmungen eine Erfassung der gesamten Bodenpolysaccharide gewährleistet.

In Tab. 61 sind analytische Daten von Polysaccharidfraktionen aus verschiedenen Böden zusammengestellt.

Tabelle 61. *Kennzeichnung von Polysaccharidpräparaten, die aus verschiedenen Böden isoliert worden sind* (nach Mehta, Dubach und Deuel 1961)

	Britische Böden Forsyth (1947, 1950)	Schottische Böden Duff (1952a, b, 1954)	Böden aus Indiana Whistler und Kirby (1956)	Muk Soil aus New York Clapp (1957)	Britische Mineralböden, Torfe und Komposte Parson (1958)
Äquivalentgewicht	1185	1000	1945	–	–
N %	0,34	1,6	0,34	–	–
OCH_3 %	0	2,0	2,4	–	–
Reduzierende Zucker %	80	–	–	–	37,8–15,5
Uronanhydrid %	15,8	20,1	9,1	–	17,9–6,7
Aminozucker %	0	0	Spuren	–	über 5
Zuckerkomponenten in % des Gesamtzuckers in den Präparaten					
Glucose	20,8	36,0	21,2	27,8–36,4	26,6–38,0
Galactone	20,0		16,6	22,2–22,6	17,8–23,2
Mannose	21,9	29,8	18,5	28,6–29,8	16,3–21,0
Arabinose	11,7		10,4	–	6,8–8,2
Xylose	23,6	10,3	12,6	–	7,5–9,0
Ribose	1,5	4,5	Spuren	–	–
Rhamnose	0	11,0	14,2	7,7–13,6	–
Fucose	0	0,7	0	–	0
Unbekannte Zucker	0	7,0	6,5	3,5–6,6	0

Uronsäuren und Pektine. Die Uronsäuren und Pektine werden häufig nach Lefevre-Tollens durch saure Decarboxylierung bestimmt. Die hiermit ermittelten Gehalte des Bodens an Uronsäuren von bis zu 40% (Shorey und Martin 1930, Springer 1940, Norman und Bartholomew 1943, Fuller 1946, 1947a, 1947b) sind aber nach Deuel und Dubach (1958) als zu hoch zu betrachten, da die angewandte Methode nicht spezifisch ist.

Mit Hilfe der Carbazolmethode sind Uronsäuren in Bodenextrakten in der Größenordnung von 0,07 bis 0,16% des Bodengewichtes bestimmt worden (Lynch, Hearns und Cotnoir 1957, Dubach und Lynch 1959). In anderen Untersuchungen ist der Gehalt an Uronsäuren mit 1 bis 4% bestimmt worden (Lynch, Hearns und Cotnoir 1957, Graveland und Lynch 1961). Als definierte Verbindungen sind von Whistler und Kirby (1956) Glucuron- und Galacturonsäure nachgewiesen worden.

Bei der Isolierung der Polyuronsäuren können Huminsäuren mit ausgefällt werden (Lynch, Wright und Olney 1957). Aus diesem Grunde sind auch die Werte, die mit der Carbazolmethode erhalten werden, unsicher.

Aminozucker. In Böden sind 1 bis 20% des Stickstoffs in Form von Aminozuckern nachgewiesen worden (Bremner 1949, Bremner und Shaw 1954, Stevenson 1954, 1957a, 1957b). Ihre Menge in den Böden ist sehr unterschiedlich. Im Durchschnitt liegen die Werte zwischen 5 und 10%.

Bei verschiedenen Böden lag das Verhältnis von Glucosamin zu Galactosamin in den polymeren Aminozuckern zwischen 1,2 und 4,6 (Bremner 1958, Sowden 1959), die höchsten Werte sind in Waldböden mit saurer Streu gefunden worden.

2. Proteine und Aminosäuren

Nach den bisher vorliegenden Untersuchungen sind 30 bis 70% des organisch gebundenen Stickstoffs im Boden Bestandteil von Aminosäuren (ENSMINGER und PEARSON 1950, BREMNER 1951). Freie Aminosäuren wurden nur in geringem Umfange nachgewiesen (GROV und ALVSAKER 1963, BREMNER 1952, DADD, FOWDEN und PEARSALL 1953). Der weitaus größte Anteil liegt in sorptiv oder chemisch gebundener Form vor.

Ein Teil der Aminosäuren entstammt abgestorbenem Pflanzenmaterial. Ein anderer muß mikrobiellen Ursprungs sein, wie aus der Bestimmung von α-ε-Diaminopimelinsäure hervorgeht. Darüber hinaus können Aminosäuren auch als Wurzelausscheidungen in den Boden gelangen (KATZNELSON, ROUATT und PAYNE 1955). Die Herkunft der Aminosäuren scheint aber auf die Zusammensetzung der Bodenproteine ohne großen Einfluß zu sein (STEVENSON 1956a), da sich in den Analysen nur geringe Differenzen ergeben. Charakteristische Unterschiede ergeben sich jedoch, wenn bestimmte organische Substanzen durch definierte Organismengruppen zersetzt werden (SIMONART und BUYSSE 1954, BREMNER 1955b).

Mit der Entwicklung der Papierchromatographie wurden umfangreichere analytische Untersuchungen an Aminosäuregemischen möglich. So wurden systematische Untersuchungen durchgeführt über Aminosäuren:

in Bodenlösungen (BOSE, STEVENSON und KATZNELSON 1959, DAVIDSON, SOWDEN und ATKINSON 1951, KOJIMA 1947, OKUDA und HORI 1954, PARKER, SOWDEN und ATKINSON 1952, PAUL 1959, PAYNE, ROUATT und KATZNELSON 1956, PUTNAM und SCHMIDT 1956, STEVENSON 1954),

in Hydrolysaten von Böden (BREMNER 1950a, 1950b, 1952, CARLES, SOUBIES und GADET 1958, YOUNG und MORTENSEN 1958, PAVEL, KOLOUSEK und SMATLAK 1954, SCHARPENSEEL und KRAUSSE 1962, SOWDEN und PARKER 1953, SOWDEN 1957, 1959, STEVENSON 1956a, 1956b, 1957a, 1957b, 1957c), von Kohlen (DEGENS und BAJOR 1960, LEHMANN und PRASHNOWSKY 1959), von Torf (COULSON, DAVIES und KHAN 1959), von Komposten (BREMNER 1955a, OKUDA und HORI 1954, LUTWICK, CALDWELL und DE LONG 1952) und von Huminsäuren (BREMNER, FLAIG und KÜSTER 1955, BREMNER 1955a, 1955b, 1955c), KONONOWA und ALEXANDROWA 1956, OKUDA und HORI 1954, OKUDA und HORI 1955, SOWDEN 1955, 1958).

Aus einer Arbeit, in der die Aminosäuren von Bodenhydrolysaten mittels Säulenchromatographie bestimmt worden sind, werden Ergebnisse in Tab. 62 angegeben (YOUNG und MORTENSEN 1958).

Die Zusammensetzung der Aminosäuren ist in den fünf Böden qualitativ nahezu gleich, zeigt aber quantitativ erhebliche Unterschiede.

Nach SCHARPENSEEL und KRAUSSE (1962) sind zumeist in Bodenhydrolysaten Leucin und Isoleucin, Lysin, Alanin und Glycin am stärksten vertreten, die schwefelhaltigen Aminosäuren am schwächsten. Tryptophan wurde auch unter veränderten Hydrolysebedingungen nicht gefunden (BREMNER 1955).

Nach CARLES, SOUBIES und GADET (1958), BREMNER (1955b) und CARLES und DECAU (1960a, 1960b) besteht ein beträchtlicher Einfluß des Klimas auf die Zusammensetzung der Aminosäuren im Boden. Durch stärkere biologische Aktivität im heißeren Klima ist eine Zunahme von Lysin, Äthanolamin, γ-Aminobuttersäure und β-Alanin (Decarboxylierungsprodukte von α-ε-Diaminopimelinsäure, Serin, Glutaminsäure bzw. Asparaginsäure) zu finden. Als Prinzip läßt sich erkennen, daß vom temperierten zum tropischen Gürtel hin der Anteil der basischen

Tabelle 62. *Aminosäuren im Hydrolysat einiger Böden*

Aminosäuren	mg/100 g Boden					% der Gesamtaminosäuren				
	A[1]	B[1]	C[1]	D[1]	E[1]	A[1]	B[1]	C[1]	D[1]	E[1]
Cysteinsäure ...	3,5	5	11	5,3	15,9	2,0	2,3	1,0	1,7	1,7
Asparaginsäure	32[2]	43[2]	182[2]	26,0	66,7	18,2[2]	19,5[2]	17,2[2]	8,3	7,2
Serin				14,0	73,4				4,6	8,0
Threonin	13	16	80	18,2	81,0	7,4	7,2	7,5	5,9	8,8
Glutaminsäure .	26	33	137	20,0	94,4	14,8	14,9	12,9	6,3	10,2
Methioninsulfon	2,0	1,1	10	Spuren	Spuren	1,1	0,5	0,9		
Hydroxyprolin .	5	4	33	–	–	2,8	1,8	3,1	–	
Glycin.........	10	12	80	21,8	73,7	5,7	5,4	7,5	7,1	8,0
Alanin	18	22	109	31,7	76,4	10,2	10,0	10,3	10,3	8,3
β-Alanin.......	2	2	10	5,0	17,3	1,1	0,9	0,9	1,6	1,9
γ-Aminobuttersäure	0,3	0,7	1,7	3,5	12,0	0,2	0,3	0,2	1,1	1,3
Valin..........	11	16	81	21,3	62,0	6,2	7,2	7,6	6,9	6,7
Methioninsulfoxid	0,6	1,0	6,4	11,3	22,7	0,3	0,5	0,6	3,6	2,5
Prolin	10	12	48	10,3	51,7	5,7	5,4	4,5	3,4	5,6
Isoleucin	8	9	33	16,1	47,7	4,5	4,1	3,1	5,2	5,2
Leucin	7	9	58	19,5	63,4	4,0	4,1	5,5	6,3	6,9
Ornithin	8	5	35	8,5	8,6	4,5	2,3	3,3	2,7	0,9
Lysin	13	17	69	44,4	70,0	7,4	7,7	6,5	14,1	7,6
Histidin	0,8	3,5	13	5,8	6,9	0,5	1,6	1,2	1,8	0,8
Tyrosin........	Spuren	1,5	17	8,0	23,3		0,7	1,6	2,5	2,5
Arginin	1	1	12	6,4	17,0	0,6	0,5	1,1	2,0	1,8
Phenylalanin ..	5	7	34	7,4	23,6	2,8	3,2	3,2	2,5	2,6
α-Amino-n-buttersäure ..	–	–	–	1,0	3,5	–	–	–	0,3	0,4
α-ε-Diaminopimelinsäure .	–	–	–	2,5	9,9	–	–	–	0,8	1,0
Summe	176,2	220,8	1060,1	308,0	921,1					

	pH	Humus %	Gesamt-C%	Gesamt-N%	C/N
[1] A Wooster silt loam: Kultivierter Boden	6,8	1,5	0,79	0,084	9,3
B Wooster silt loam: Jungfräulicher Boden	5,2	2,3	1,15	0,108	10,6
C Hoytville clay loam: Laubwald	6,0	10,1	6,65	0,550	12,1
D Flavagan silt loam: Dauermaisversuch	5,0	—	1,65	0,138	12,0
E Flavagan silt loam: Randfläche unter Gras	5,6	—	4,60	0,323	14,2

A bis C nach Young und Mortensen (1958)
D und E nach Stevenson (1956)

[2] Asparaginsäure und Serin.

auf Kosten der anderen Aminosäuren ständig zunimmt (Scharpenseel und Krausse 1962).

Alle diese Ergebnisse können anscheinend nur verallgemeinert werden, solange die den Böden eigenen Zersetzungsbedingungen nicht gestört sind. Greenwood und Lees (1960a, 1960b) stellten eine Empfindlichkeitsskala der Aminosäuren gegenüber dem Abbau unter aeroben und anaeroben Bedingungen auf. Die gelenkte Zersetzung durch Beimpfung mit bestimmten Mikroorganismen führt zu einer qualitativen Veränderung des Aminosäurespektrums (Simonart und

BUYSSE 1954, BREMNER, FLAIG und KÜSTER 1955). CARLES, SOUBIES und GADET (1958) fanden nach Zusatz von Stallmist im Boden eine Zunahme von Leucin, Valin, Prolin, Tyrosin und Phenylalanin, dagegen eine Abnahme von Alanin, Glycin, Asparaginsäure, Glutaminsäure und Serin. Bei der Verrottung von Stroh ändert sich die Zusammensetzung der Aminosäuren nicht, abgesehen von der Zunahme des Gehaltes an durch Mikroorganismen speziell aufgebauten Aminosäuren, wie der α-ε-Diaminopimelinsäure, dem β-Alanin und dem Hydroxyprolin (BREMNER 1955a).

3. Nukleinsäuren, Purine, Pyrimidine, Inositphosphate

Auf das Vorkommen von Nukleinsäuren im Boden wird aus dem Auftreten ihrer hydrolytischen bzw. mikrobiellen Abbauprodukte geschlossen. Als solche konnten Pentosen, Purin- und Pyrimidinbasen wie Adenin, Cytosin, Guanin, Uracil, Thymin sowie Xanthin und Hypoxanthin nachgewiesen werden (SCHREINER und SHOREY 1910b, SCHREINER und LANTHROP 1912, SCHREINER und SKINNER 1912, SHOREY 1912a, 1912b, 1913, BOTTOMLEY 1919, WRENSHALL und McKIBBIN 1937, WRENSHALL und DYER 1941, BOWER 1949, ANDERSON 1957, 1958, 1961a, 1961b). Diese Abbauprodukte konnten jedoch nicht in allen Böden nachgewiesen werden (YOSHIDA 1940); zum Teil sind sie nur in Spuren festgestellt worden (ADAMS, BARTHOLOMEW und CLARK 1954).

Die meisten Nachweise werden über die Isolierung einer Nukleinsäurefraktion geführt. Die Menge dieser Fraktion variiert mit der Methodik und der Probenahme (BARROW 1961). So erhielten DYER und WRENSHALL (1941a, 1941b) in Nukleinsäurefraktionen je nach der angewandten Methode entweder 5 bis 13% oder 35 bis 46% des organisch gebundenen Phosphors.

DYER, WRENSHALL und SMITH (1940) nehmen wegen des zu engen NP-Verhältnisses der Nukleinsäurefraktion an, daß in dieser Fraktion auch Phytin vorliegt. BOWER (1949) versuchte die Trennung der Inositphosphate von der Nukleinsäurefraktion durch Fällung des Phytins als Ca-Salz im alkalischen Medium. Die Hydrolyse der Nukleinsäurefraktion erfordert schärfere Bedingungen, als sie für Nukleinsäuren benötigt werden. Auch ihr Abbau im Boden verläuft langsamer als der von dem Boden zugesetzten Nukleinsäuren (DYER und WRENSHALL 1941a, 1941b, BOWER 1949, PEARSON, NORMAN und HO 1941). DYER und WRENSHALL wie auch ADAMS, BARTHOLOMEW und CLARK (1954) nehmen deshalb an, daß die als Spaltstücke von Nukleinsäuren gefundenen und als solche angesehenen Verbindungen auch in anderer Bindungsform im Boden vorliegen. Diese kann eine Sorption an Tonminerale sein (BARTHOLOMEW und GORING 1948, GORING und BARTHOLOMEW 1950, 1952). STEVENSON, MARKS, VARNER und MARTIN (1952) fanden Nukleinsäuren neben Aminosäuren und Zuckern an Tonminerale sorbiert. Diese Bindung soll nach FLAIG, KURON und KAUL (1955) und FLAIG und KAUL (1955) über die basischen Gruppen der Nukleinsäuren erfolgen.

ANDERSON (1958) fand nach der Hydrolyse von Huminsäuren verschiedener Böden 0,88% des Stickstoffs als Purin und Pyrimidinbasen. Auf Grund des Mengenverhältnisses der einzelnen Purin- und Pyrimidinbasen in den Hydrolisaten schließt ANDERSON auf den mikrobiellen Ursprung der Nukleinsäuren. Deren Anteil macht nur 0,6% des organisch gebundenen Phosphats aus.

FABRY (1960) bestimmte in 13 ungarischen Böden den Anteil des Phytin- und Nukleinphosphors zu 0,5 bis 43,7% des Gesamtphosphats der Böden. BURANGULOWA (1959) fand in Baschkirböden unter den Nukleinsäuren besonders Desoxyribonukleinsäure, neben Spuren von Lipoidphosphor und Phytinphosphor. Weitere Untersuchungen über die Nukleinsäurefraktion mit verschiedenen Methoden wurden durchgeführt von MEHTA, LEGG, GORING und BLACK (1954), MICHAEL und HEIDECKER (1940), WEEKS und WALTERS (1947) und KAILA (1948).

CALDWELL und BLACK (1958a, 1958b, 1958c) bestimmten verschiedene Inosithexaphosphate. Sie kommen zu dem Schluß, daß die isomeren Inositphosphate zum größten Teil mikrobiellen Ursprungs sind (1958b). Das Phytin soll nicht widerstandsfähiger gegen mikrobiellen Abbau sein als die anderen organischen Phosphorverbindungen des Bodens (1958c). COSGROVE (1962a, 1962b) fand bei seinen Untersuchungen der sogenannten Phytatfraktion australischer Böden neben einigen niederen Phosphatestern verschiedene isomere Inosithexaphosphate, darunter d,l-Inosit- und scyllo-Inosithexaphosphat. Da letztere bisher nur in Mikroorganismen nachgewiesen worden sind, deutet auch dieser Befund auf den mikrobiellen Ursprung eines Teils der Phytatfraktion hin.

4. Polyphenole

Mit dem pflanzlichen Material werden größere Mengen phenolischer Substanzen in den Boden eingebracht, außerdem können solche als Stoffwechselprodukte der Mikroorganismen gebildet werden. Niedermolekulare wasserlösliche Polyphenole können aus Blattstreu oder als Ausscheidungen aus Wurzeln in den Boden gelangen. In jedem Falle werden die niedermolekularen Polyphenole, wie sie in lebenden Pflanzen und frischem Pflanzenmaterial als Flavone, Flavonole, Catechine, Leucoanthocyanine und Phenolcarbonsäuren vorliegen (HERRMANN 1958, BATE-SMITH und LERNER 1954, COULSON, DAVIES und LEWIS 1960) oder beim Abbau der höhermolekularen Polyphenole im Boden gebildet werden (MAEDER 1960), im Boden relativ rasch verändert (SÖCHTIG 1962a, 1962b). Dabei ist im einzelnen die Bodenreaktion für die Abbau- und Umwandlungsgeschwindigkeit bestimmend (COULSON, DAVIES und LEWIS 1960). Die Isolierung niedermolekularer Polyphenole aus dem Boden ist daher nur unter besonders günstigen Bedingungen möglich.

WALTERS (1917) isolierte p-Hydroxy-benzoesäure. FORSYTH (1947) spricht eine Fraktion bei der Auftrennung seiner Fulvosäuren als phenolische Glykoside an. In Extrakten aus Ernterückständen von Getreide wurde eine Reihe von Phenolcarbonsäuren papierchromatographisch nachgewiesen (BÖRNER 1955, 1956, MAEDER 1960). Nach der gleichen Methode konnten COULSON, DAVIES und LEWIS (1960) in Extrakten aus Mull- und Moorvorkommen 20 verschiedene Polyphenole nachweisen, von denen Epi-Catechin und D-Catechin identifiziert werden.

Die Hauptmasse der pflanzlichen Polyphenole gelangt mit dem abgestorbenen Pflanzenmaterial als höhermolekulare Produkte wie Gerbstoffe oder Lignine in den Boden und kann nur als humifiziertes Material wieder daraus gewonnen werden. Selbst Lignin, das von allen höhermolekularen Pflanzenstoffen am langsamsten abgebaut wird, konnte bisher nicht in reiner Form aus dem Boden isoliert werden. Es wurden stets nur mehr oder weniger ligninhaltige Präparate erhalten, in denen spektrographisch oder durch oxydativen Abbau Ligninstrukturen nachgewiesen wurden (JENKINSON und TINSLEY 1959, 1960, FLAIG, SCHOBINGER und DEUEL 1959, MORRISON 1958).

5. Sonstige Verbindungen

In Wasserextrakten von verschiedenen Böden wurden die normalerweise bei oxydativem Abbau organischer Substanzen zu erwartenden aliphatischen Säuren gefunden, wie Ameisensäure, Essigsäure, Brenztraubensäure, Milchsäure, Bernsteinsäure, Oxalsäure, Citronensäure und Fumarsäure (SCHREINER und SHOREY 1909, SHOREY 1913, TAKAI 1958, TAKIJIMA 1961, TAKIJIMA und SAKUMA 1961, TAKIJIMA, SAKUMA und CHIBA 1961, JORGENSEN 1963, KAURICHEV, IVANOVA und NOZDRUNOVA 1963).

Aus den Benzol-Methanolextrakten von Böden wurden Wachse isoliert, die aus normalen aliphatischen Säuren und normalen primären aliphatischen Alkoholen und Sterinen zusammengesetzt sind. Außerdem enthalten diese Bodenextrakte pentacyclische und hexacyclische Verbindungen, bei denen es sich um Triterpene handeln dürfte (MEINSCHEIN und KENNY 1957). Benzpyrene scheinen, wenn auch nur in geringen Mengen, normalerweise in Böden vorzukommen. Das cancerogene 3,4-Benzpyren und das inaktive 1,2-Isomere wurden spektroskopisch von BLUMER (1961) in Böden nachgewiesen.

Literatur

ADAMS, A. P., W. V. BARTHOLOMEW und F. E. CLARK: Measurement of nucleic acid components in soil. Soil Sci. Soc. Amer. Proc. **18**, 40–46 (1954). — ALVSAKER, E.: Univ. Bergen Skrifter **23**, 1 (1948). — ALVSAKER, E., und K. MICHELSON: Carbohydrates in cold water extract of a pine forest soil. Acta Chem. Scand. **11**, 1794–1795 (1957). — ANDERSON, G.: Nucleic acid derivatives in soils. Nature (London) **180**, 287–288 (1957). — Identification of derivatives of deoxyribonucleic acid in humic acid. Soil Sci. **86**, 169–174 (1958). — A partial fractionation of alkali-soluble soil organic phosphat. J. Soil Sci. **12**, 276–285 (1961a). — Estimation of purines and pyrimidines in soil humic acid. Soil Sci. **91**, 156–161 (1961b). — ANSLOW, W. K., und H. RAISTRICK: Studies in the biochemistry of microorganisms, LIX, Spinulosin (3,6-dihydroxy-4-methoxy-2,5-toluquinone), a metabolic product of a strain of aspergillus fumigatus fresenius. Biochem. J. **32**, 228 (1938a). — Studies in the biochemistry of microorganisms, LVIII, Fumigatin (3-hydroxy-4-methoxy-2,5-toluqinone) and spinulosin (3,6-dihydroxy-4-methoxy-2,5-toluquinone), metabolic products respectively of aspergillus fumigatus fresenius and penicillium spinulosum Thom. Biochem. J. **32**, 687 (1938b). — Studies in the biochemistry of microorganisms, LVIII, Synthesis of spinulosin (3,6-dihydroxy-4-methoxy-2,5-toluquinone), a metabolic product of penicillium spinulosum Thom. Biochem. J. **32**, 803 (1938c).

BARROW, N. J.: Phosphorus in soil organic matter. Commonwealth Bur. Soils, Soils Fertilizers **24**, 169–173 (1961). — BARTHOLOMEW, W. V., und C. A. I. GORING: Microbial products and soil organic matter, I, Some characteristics of the organic phosphorus of microorganisms. Proc. Soil Sci. Soc. Amer. **13**, 238–241 (1948). — BATE-SMITH, E. C., und N. H. LERNER: Systematic distribution of leucoanthocyanins in leaves. Biochem. J. **58**, 126–132 (1954). — BERNIER, B.: Characterization of polysaccharides isolated from forest soils. Biochem. J. **70**, 590–599 (1958). — BERTHELOT, M., und G. ANDRE: Ann. Chim. Phys. **25**, Ser. 6, 362 (1892). — BLUMER, M.: Benzpyrenes in soil. Science **134**, 474–475 (1961). — BÖRNER, H.: Untersuchungen über phenolische Verbindungen aus Getreidestroh und Getreiderückständen. Naturwiss. **42**, 583 (1955). — Der papierchromatographische Nachweis von Ferulasäure in wäßrigen Extrakten von Getreidestroh und Getreiderückständen. Naturwiss. **43**, 129–130 (1956). — BOSE, B., I. L. STEVENSON und H. KATZNELSON: Metabolic activity of the soil microflora. Proc. Nat. Inst. Sci. India **25B**, 223–9 (1959). — BOTTOMLEY, W. B.: The isolation from peat of certain nucleic acid derivatives. Proc. Roy. Soc. (London) B **90**, 39–44 (1919). — BOWER, C. A.: Studies on the forms and availability of soil organic phosphorus. Iowa Agricult. Exper. Stat. Res. Bull. **362** (1949). — BREMNER, J. M.: Studies on soil organic matter, I, The chemical nature of soil organic nitrogen. J. Agric. Sci. **39**, 183 (1949). — Amino-acids in soil. Nature (London) **165**, 367 (1950a). — The amino-acid composition of the protein material in soil. Biochem. J. **47**, 538–542 (1950b). — A review of recent work on soil organic matter, I. J. Soil Sci. **2**, 67–82 (1951). — The nature of soil-nitrogen complexes. J. Sci. Food Agric. **3**, 497–500 (1952). — Nitrogen transformations during the biological decomposition of straw composted with inorganic nitrogen. J. Agric. Sci. **45**, 469–475 (1955a). — Studies in soil humic acids, I, The chemical nature of humic nitrogen. J. Agric. Sci. **46**, 247–256 (1955b). — Nitrogen distribution and amino-acid composition of fractions of a humic acid from a chernozem soil („Hildesheimer Schwarzerde"). Z. Pflanzenernähr., Düng., Bodenkde. **71**, 63–66 (1955c). — Amino sugars in soil. J. Sci. Food Agric. **9**, 528–532 (1958). — Bremner, J. M., und K. Shaw: Studies on the estimation and decomposition of amino sugars in soil. J. Agric. Sci. **44**, 152 (1954). — Bremner, J. M., W. Flaig und E. Küster: Zur Kenntnis der Huminsäuren, IX, Der Gehalt an Aminosäuren in Streptomyceten-Huminsäuren. Z. Pflanzenernähr., Düng., Bodenkde. **71**, 58–63 (1955). — Burangulova, M. N.: Nucleic acids of soils. Biol. Nuclein. Obmen. Tast. 177–182 (1959).

Caldwell, A. G., und C. A. Black: Inositol hexaphosphate, I, Quantitative determination in extracts of soils and manures. Proc. Soil Sci. Soc. Amer. **22**, 290–293 (1958a). — Inositol hexaphosphate, II, Synthesis by soil microorganisms. Proc. Soil Sci. Soc. Amer. **22**, 293–296 (1958b). — Inositol hexaphosphate, III, Content in soils. Proc. Soil Sci. Soc. Amer. **22**, 296–298 (1958c). — Carles, J., und J. Deceau: Variations in the amino acids of soil hydrolysates. Sci. Proc. Roy. Dublin Soc. A **1**, 177–182 (1960a). — De quelques conditions susceptibles de modifier les proportions des acides anrinés du sol. Ann. agron. Paris **11**, 557–575 (1960b). — Carles, J., L. Soubies und R. Gadet: Les acides aminés du sol et leurs variations. C. R. Acad. Sci. Paris **247**, 1229–1232 (1958). — Chesters, G., O. J. Attoe und O. N. Allen: Soil aggregation to various soil constituents. Soil Sci. Soc. Amer. Proc. **21**, 272 (1957). — Clapp, C. E.: High molecular weight water soluble Muck: Isolation and determination of constituent sugars of a borate complex forming polysacharide, employing electrophoretic technique. Ph. D. Thesis, Cornell University, New York 1957. — Claus, D., und Kl. Mechsner: Über die Brauchbarkeit der von Ed. Hofmann ausgearbeiteten Methoden zur Bestimmung der Enzyme im Boden. Plant a. Soil **12**, 195 (1960). — Cosgrove, D. J.: Phosphorus availability. The chemical nature of organic phosphorus. CSIRO Div. Plant Industr. Rep. **1961–62**, 54–55 (1962a). — Forms of inositol hexaphosphate in soils. Nature (London) **194**, 1265–1266 (1962b). — Coulson, C. B., R. I. Davies und E. J. A. Khan: Chemical studies on upland peat in North Wales. J. Sci. Food Agric. **10**, 209–217 (1959). — Coulson, C. B., R. I. Davies und D. A. Lewis: Polyphenols in plant, humus and soil, 1, Polyphenols of leaves, litter and superficial humus from mull and moor sites. J. Soil Sci. **11**, 20–29 (1960).

Dadd, C. C., L. Fowden und W. H. Pearsall: An investigation of the free amino-acids in organic soil types using paper partition chromatography. J. Soil Sci. **4**, 69–71 (1953). — Davidson, D. I., F. J. Sowden und H. J. Atkinson: Application of paper chromatography to identification and quantitative estimation of amino acids in soil organic matter fractions. Soil Sci. **71**, 347–52 (1951). — Degens, E. T., und M. Bajor: Die Verteilung von Aminosäuren in bituminösen Sedimenten und ihre Bedeutung für die Kohlen- und Erdölgeologie. Glückauf **96**, 1525 (1960). — Deuel, H., und P. Dubach: Decarboxylierung der organischen Substanz des Bodens, II, Nachweis von Uronsäuren. Z. Pflanzenernähr., Düng., Bodenkde. **82**, 97 (1958). — Dubach, P., und D. L. Lynch: Comparison of the determination of "Uronic acids" in soil extracts with carbazole and by decarboxylation. Soil Sci. **87**, 273–275 (1959). — Dubach, P., G. Zweifel, R. Bach und H. Deuel: Untersuchung an der Fulvosäurefraktion einiger schweizerischer Böden. Z. Pflanzenernähr., Düng., Bodenkde. **69**, 97–108 (1955). — Duff, R. B.: The occurrence of methylated carbohydrates and rhamnose as components of soil polysaccharides. J. Sci. Food Agric. **3**, 140–144 (1952a). — The occurrence of L-fucose in soil, peat, and in a polysaccharide synthesized by soil bacteria. Chem. a. Ind. **1952**, 1104–5. — The partial identification of o-methyl-sugars occurring in soil peat and compost. Chem. a. Ind. **1954**, 1513. — Dyer, W. J.: und C. L. Wrenshall: Organic phosphorus in soils, I, The extraction and separation of organic phosphorus compounds from soil. Soil Sci. **51**, 159–170 (1941a). — Organic phosphorus in soils, III, The decomposition of some organic phosphorus compounds in soil cultures. Soil Sci. **51**, 323–329 (1941b). — Dyer, W. J., C. L. Wrenshall und G. R. Smith: The isolation of phytin from soil. Science (Lancaster, Pa.) **91**, 319–320 (1940).

Enders, C., und S. Sigurdsson: Chemie der Huminsäurebildung unter physiologischen Bedingungen. Biochem. Z. **318**, 44–46 (1947). — Ensminger, L. E., und R. W. Pearson: Soil nitrogen. Adv. Agron. **2**, 81–111 (1950).

Fabry, G.: Organic fractions of phosphorus in Hungarian soils. Agrokém. Talajt. **9**, 245–260 (1960). — Flaig, W., und R. Kaul: Zur Aufnahme der Phosphorsäure aus Nukleinsäuren durch die Pflanze. Phosphorsäure **15**, 208–217 (1955). — Flaig, W., H. Kuron und R. Kaul: Über die Sorption von Nukleinstoffen an Tonkolloiden. Z. Pflanzenernähr., Düng., Bodenkde. **71**, 141–154 (1955). — Flaig, W., U. Schobinger und H. Deuel: Umwandlung von Lignin in Huminsäuren bei der Verrottung von Weizenstroh. Chem. Ber. **92**, 1973–1982 (1959). — Forsyth, W. G. C.: Studies on the more soluble complexes of soil organic matter, I, A method of fractionation. Biochem. J. **41**, 176 (1947). — Carbohydrate metabolism in the soil. Chem. a. Ind. **1948**, 515–519.

— Studies on the more soluble complexes of soil organic matter, II, The composition of the soluble polysaccharide fraction. Biochem. J. **46**, 141–146 (1950). — FULLER, J. E.: Influence of purified lignin on nitrification in soil. Science **104**, 313–315 (1946). — Decarboxylation rate of uronic groups contained in soil organic matter, plant gums of known constitution, plant materials and microbial products. Soil Sci. **64**, 183–197 (1947a). — FULLER, W. H.: Investigations on the separation of uronides from soils. Soil Sci. **64**, 403–11 (1947b).

GORING, C. A. I., und W. V. BARTHOLOMEW: Microbial products and soil organic matter, II, The effect of clay on the decomposition and separation of the phosphorus compounds in microorganisms. Proc. Soil Sci. Soc. Amer. **14**, 152–156 (1950). — Adsorptions of mononucleotides, nucleic acids and nucleoproteins by clays. Soil Sci. **74**, 149–164 (1952). — GRAVELAND, D. N., und D. L. LYNCH: Distribution of uronides and polysaccharides in the profiles of a soil catena. Soil Sci. **91**, 162–165 (1961). — GREENLAND, D. J.: The adsorption of sugars by montmorillonite, I, X-ray studies. Soil Sci. **7**, 319–328 (1956). — GREENWOOD, D. J., und H. LEES: Studies on the decomposition of amino acids in soils. Plant a. Soil **7**, 253 (1956). — Studies on the decomposition of amino acids in soils, II, The anaerobic metabolism. Plant a. Soil **12**, 69–80 (1960a). — Studies on the decomposition of amino acids in soils, III, The process of amino acid aerobic decomposition and some properties of amino acid-oxidizing organisms. Plant a. Soil **12**, 175–194 (1960b). — GROV, A., und E. ALVSAKER: Amino acids in soil. Water soluble acids. Acta Chem. Scand. **17**, 2307–2315 (1963).

HAWORTH, W. N., F. W. PINKARD und M. STACEY: Function of bacterial polysaccharides in the soil. Nature (London) **158**, 836 (1946). — HERRMANN, K.: Über Oxyzimtsäuren in Getreideblättern. Naturwiss. **45**, 111–112 (1958). — HOFMANN, ED., und GG. HOFFMANN: Über die Zuverlässigkeit der Methoden zur Bestimmung der Aktivität von Bodenenzymen nach Ed. Hofmann. Plant a. Soil **14**, 96–99 (1961).

JENKINSON, D. S., und J. TINSLEY: Studies on the organic material extracted from soils and composts, I, The isolation and characterization of ligno-proteins from compost. J. Soil Sci. **10**, 245–263 (1959). — A comparison of the ligno-protein isolated from a mineral soil and from a straw compost. Sci. Proc. Roy. Dublin Soc. A **1**, 4, 149 (1960). — JORGENSEN, J. R.: Studies on the nature and occurrence of organic acids in soils. Diss. Abstr. **23**, 2272–2273 (1963).

KAILA, A.: On the organic phosphorus in cultivated soils. Valt. Maatkoet. Julk. **129** (1948). — KATZNELSON, H., J. W. ROUATT und T. M. B. PAYNE: The liberation of amino acids and reducing compounds by plant roots. Plant a. Soil **7**, 35–48 (1955). — KAURICHEV, I. S., T. N. IVANOVA und E. M. NOZDRUNOVA: Content of low-molecular organic acids in the composition of watersoluble organic matter of soils. Pochvovedenie **1963**, No. 3, 27–35. — KOHL, R. A., und S. A. TAYLOR: Hydrogen bonding between the carbonyl group and Wyoming bentonite. Soil Sci. **91**, 223–237 (1961). — KOJIMA, R. T.: Soil organic nitrogen, II, Some studies on the amino acids of protein material in a muck soil from Geneva, New York. Soil Sci. **64**, 245–252 (1947). — KONONOWA, M. M., und I. V. ALEXANDROWA: Distributive paper chromatography for a study of the forms of nitrogen in organic substances. Pochvovedenie **1956**, No. 5, 86–92.

LEES, H., und J. W. PORTEOUS: The release of carbon dioxide from soils percolated with various organic materials. Plant a. Soil **2**, 231 (1950). — LEHMANN, W. M., und A. PRASHNOWSKY: Paläobiogeochemische Untersuchungen an Fauna und Flora in verschiedenen geologischen Formationen. Naturwiss. **46**, 479–80 (1959). — LUTWICK, L. E., B. B. CALDWELL und W. A. DELONG: Leachates from decomposing leaves, I, Some general characteristics. Sci. Agric. **32**, 603–613 (1952). — LYNCH, D. L., E. E. HEARNS und L. J. COTNOIR: Determination of polyuronides in soils with carbazole. Soil Sci. Soc. Amer. Proc. **21**, 160–162 (1957). — LYNCH, D. L., L. M. WRIGHT und H. O. OLNEY: Qualitative and quantitative chromatographic analyses of the carbohydrate constituents of the acid-insoluble fraction of soil organic matter. Soil Sci. **84**, 405–411 (1957).

MAEDER, H.: Chemische und pflanzenphysiologische Untersuchungen mit Rottestroh. Diss. Gießen 1960. — MEHTA, N. C., J. O. LEGG, C. A. I. GORING und C. A. BLACK: Determination of organic phosphorus in soils, I, Extraction method. Proc. Soil Sci. Soc. Amer. **18**, 443–449 (1954). — MEHTA, N. C., P. DUBACH und H. DEUEL: Carbohydrates in the soil. Advances in Carbohydrate Chemistry **16**, 335–355 (1961). — MEINSCHEIN, W. G., und G. S. KENNY: Analyses of a chromatographic fraction of organic extracts of soils. Analytic. Chem. **29**, 1153–1161 (1957). — MICHAEL, G., und L. HEIDECKER: Ein weiterer Beitrag zur Frage der Veränderlichkeit der Phosphorsäurefraktionen in Spinat in Abhängigkeit von verschiedener Phosphorsäuredüngung.

Bodenkde. u. Pflanzenernähr. **17**, 358–372 (1940). — MORRISON, R. I.: The alkaline nitrobenzene oxidation of soil organic matter. Soil Sci. **9**, 130–140 (1958).

NAGAR, B. R.: Free monosaccharides in soil organic matter. Nature (London) **194**, 896–897 (1962). — NORMAN, A. G., und W. V. BARTHOLOMEW: The chemistry of soil organic matter, I, Distribution of uronic carbon in some soil profiles. Soil Sci. **56**, 143 (1943).

OGSTON, A. G.: Sedimentation in the ultracentrifuge. Biochem. J. **70**, 598 (1958). — OKUDA, A., und S. HORI: Chromatographic investigation of amino acids in humic acid and alkaline alcohol lignin. Trans. Fifth Int. Congr. Soil Sci. **2**, 255–258 (1954). — Identification of amino acids in humic acid. Soil a. Plant Food **1**, 39–40 (1955).

PARKER, D. I., F. J. SOWDEN und H. J. ATKINSON: Nitrogen distribution and amino acid content of certain organic matter fractions. Sci. Agric. **32**, 163–169 (1952). — PARSON, J. W.: A chemical study of polysaccharide material isolated from soil. Ph. Thesis, University of Reading, England, 1958. — PAUL, E. A.: The extraction and quantitative estimation of free amino acids in soils. Diss. Abstr. **19**, 2419 (1959). — PAVEL, L., J. KOLOUSEK und V. SMATLAK: Soil humus problem, II, Amino-acid composition of the hydrolysates of humic acids in different soil types. Sbornik Ceskoslov. Akad. Zemedel. Ved. **27** A, 207–212 (1954). — PAYNE, T. M. B., J. W. ROUATT und H. KATZNELSON: Determination of free amino acids in soil. Soil Sci. **82**, 521–524 (1956). — PEARSON, R. W., A. G. NORMAN und C. H. HO: The mineralization of the organic phosphorus of various compounds in soil. Proc. Soil Sci. Soc. Amer. **6**, 168–175 (1941). — PUTNAM, H. D., und E. L. SCHMIDT: Free amino acids in soil. Bact. Proc. Baltimore **1956**, 24.

QUILLICO, A.: Récents développements de la chimie des produits naturels: La chimie des moisissures et des bactéries. Chim. et Ind. **66**, 205 (1951).

RAISTRICK, H., W. K. ANSLOW und J. BREEN: Studies on the biochemistry of microorganisms, 64, Emodic acid (4,5,7-Trihydroxyanthraquinone-2-carboxylic acid) and ω-Hydroxyemodin (4,5,7-trihydroxy-2-hydroxymethyl-anthraquinone), metabolic products of a strain of penicillium cyclopium Westling. Biochem. J. **34**, 159 (1940). — RENNIE, D. A., E. TRUOG und O. N. ALLEN: Soil aggregation as influenced by microbial gums, level of fertility and kind of crop. Soil Sci. Soc. Amer. Proc. **18**, 399 (1954). — RUSCHMEYER, O. R., und E. L. SCHMIDT: Decomposition of cellulose in soil burial beds, II, Cellulolytic activity as influences by alteration of soil properties. Appl. Microbiol. **6**, 115–120 (1958).

SCHARPENSEEL, H. W., und R. KRAUSSE: Aminosäureuntersuchungen an verschiedenen organischen Sedimenten, besonders Grau- und Braunhuminsäurefraktionen verschiedener Bodentypen (einschließlich C^{14}-markierter Huminsäuren). Z. Pflanzenernähr., Düng., Bodenkde. **96**, 11–34 (1962). — SCHMALFUSS, K.: Bodenatmung. Bodenkde. u. Pflanzenernähr. **21/22**, 442 (1940). — SCHMIDT, E. L., und O. R. RUSCHMEYER: Decomposition of cellulose in soil burial beds, I, Soil properties in relation to cellulose degradation. Appl. Microbiol. **6**, 108–114 (1958). — SCHMUK, A.: Pedology No. 3, 5 (1930). — SCHREINER, O., und E. C. SHOREY: The isolation of picoline carboxylic acid from soils and its relation to soil fertility. J. Amer. Chem. Soc. **39** (1908). — The isolation of harmful organic substances from soils. U.S. Dept. Agric. Bur. Soils Bull. **53**, 5–53 (1909). — The presence of arginine and histidine in soils. J. Biol. Chem. **8** (1910a). — Pyrimidine derivatives and purine bases in soils. J. Biol. Chem. **8** (1910b). — Chemical nature of soil organic matter. U.S. Dept. Agric. Bur. Soils Bull. **74** (1910c). — SCHREINER, O., und E. C. LANTHROP: The chemistry of steam-heated soils. U.S. Dept. Agric. Bur. Soils Bull. **89** (1912). — SCHREINER, O., und J. J. SKINNER: The action of nucleic acid and its decomposition product on soils and plants. Science (Lancaster, Pa.) **35**, 390 (1912). — SHOREY, E. C.: Organic nitrogen in Hawaiian soils. Ann. Rep. Hawaii Agric. Exper. Stat. 1906. — The isolation of creatinine from soils. J. Amer. Chem. Soc. **34** (1912a). — Nucleic acids in soils. Science (Lancaster, Pa.) **35**, 390 (1912b.). — Some organic soil constituents. U.S. Dept. Agric. Bur. Soils Bull. **88** (1913). — SHOREY, E. C., und J. B. MARTIN: The presence of uronic acids in soils. J. Amer. Chem. Soc. **52**, 4907–4915 (1930). — SHOREY, E. C., und E. H. WALTERS: A nitrogeneous soil constituent tetracarbonimid. J. Agric. Res. **2** (1914). — The presence of some benzene derivatives in soils. J. Agric. Res. **5** (1916). — Proc. Pap. First Int. Congr. Soil Sci. Washington **3**, 264 (1928). — The presence of allantoin in soils. Soil Sci. **45**, 177–183 (1938). — SIMONART, P., und R. BUYSSE: Bacteria and free amino acids in the soil. Trans. Fifth Int. Congr. Soil Sci. **3**, 136–140 (1954). — SÖCHTIG, H.: Zur Erfassung der bei der Strohrotte gebildeten Ligninabbauprodukte, Kennzeichnung durch UV-Spektren. Landbauforsch. **11**, 13–15 (1961a). — Inwieweit können Strohrotteprodukte das Pflanzenwachstum beeinflussen? Über Veränderungen von Ligninabbauprodukten. Landbauforsch. **11**, 16–18 (1961b). — Sowden, F. J., und D. I. Parker: Amino nitrogen of soils and of certain fractions isolated from them. Soil Sci. **76**, 201–208 (1953). — Sowden, F. J.: Estimation of amino acids in soil hydrolysates by the Moore and Stein method. Soil Sci. **80**, 181–188 (1955). — Note on the occurrence of amino groups in soil organic matter. Canad. J. Soil Sci. **37**, 143–144 (1957). — Forms of nitrogen in the organic matter of different horizons of soil profiles. Canad. J. Soil Sci. **38**, 147–154 (1958). — Investigations on the amounts of hexosamines found in various soils and methods for their determination. Soil Sci. **88**, 138–143 (1959). — Springer, U.: Humifizierung und Zersetzung und ihre Bestimmung in Torfen, Stallmisten und anderen organischen Bildungen. Bodenkde. und Pflanzenernähr. **18**, 129 (1940). — Stevenson, F. J.: Ion exchange chromatography of the amino acids in soil hydrolysates. Soil Sci. Soc. Amer. Proc. **18**, 373–377 (1954). — Isolation and identification of some amino compounds in soils. Soil Sci. Soc. Amer. Proc. **20**, 201–204 (1956a). — Effect of some long-time relations on the amino acid composition of the soil. Soil Sci. Soc. Amer. Proc. **20**, 204–208 (1956b). — Distribution of the forms of nitrogen in some soil profiles. Soil Sci. Soc. Amer. Proc. **21**, 283–287 (1957a). — Investigations of aminopolysaccharides in soils, 1, Colorimetric determinations of hexosamines in soil hydrolysates. Soil Sci. **83**, 113–122 (1957b). — Investigations of aminopolysaccharides in soils, 2, Distribution of hexosamines in some soil profiles. Soil Sci. **84**, 99–106 (1957c). — Stevenson, F. J., J. D. Marks, J. E. Varner und W. P. Martin: Electrophoretic and chromatographic investigations of clay adsorbed organic colloids, I, Preliminary investigations. Soil Sci. Soc. Amer. Proc. **16**, 69–73 (1952). — Sullivan, M. X.: The origin of creatinine from soils. J. Amer. Chem. Soc. **33**, 2035 (1911).

Takai, Y.: The determination of organic acids contained in paddy soil, I, Silica-gel partition chromatography of organic acids and separation of organic acids by extracting with other. J. Sci. Soil Tokyo **28**, 435–438 (1958). — Takajima, Y.: Metabolism of organic acids in soils of paddy fields and their inhibitory effects on rice growth, 5, Growth inhibition of rice seedlings as affected by waterlogged soil and by organic-acid concentration in the leachate. J. Sci. Soil Tokyo **32**, 386–389 (1961). — Takajima, Y., und H. Sakuma: Metabolism of organic acids in soils of paddy fields and their inhibitory effect on rice growth, 7, Production of organic acids in waterlogged soil treated with green manure (Chinese milk vetch) and its relationship to growth inhibition of rice seedlings. J. Sci. Soil Tokyo **32**, 560–564 (1961). — Takajima, Y., H. Sakuma und M. Chiba: Metabolism of organic acids in soils of paddy fields and their inhibitory effects on rice growth, 6, The accumulation of organic acids in soil in presence of sucrose and its relationship to growth inhibition of rice seedlings. J. Sci. Soil Tokyo **32**, 390–394 (1961). — Theander, O.: Sphagnum peat, I, Preliminary studies on the carbohydrate constituents. Svensk Kem. Tidskr. **64**, 197–199 (1952). — Sphagnum peat, III, A quantitative study on the carbohydrate constituents of sphagnum mosses and sphagnum peat. Acta Chem. Scand. 8, 989–1000 (1954). — Tombesi, L.: Anwendung der Papierchromatographie und des Perkolators von Lees auf die Untersuchung der Umwandlungen der Aminosäuren im Boden. Ann. Sperimentaz. agrar. (N.S.) **7**, 1219 (1953) (ital.).

Wajsman, S. A; und I. J. Hutchings: Chemical nature of organic matter in different soil types. Soil Sci. **40**, 347–363 (1935). — Walters, E. H.: The isolation of p-hydroxybenzoic acid from soils. J. Amer. Chem. Soc. **39** (1917). — Walters, E. H., und L. E. Wise: α-Crotonic acid a soil constituent. J. Agric. Res. **6** (1916). — Weeks, M. E., und A. Walters: The effect of phosphorus fertilization on chemical composition and forms of phosphorus in nature corn crops. Proc. Soil Sci. Soc. Amer. **11**, 189–194 (1947). — Whistler, R. L., und K. W. Kirby: Composition and behaviour of soil polysaccharides. J. Amer. Chem. Soc. **78**, 1755–1759 (1956). — White, J. W., F. J. Holben und C. D. Jeffries: Cellulose decomposing power in relation to reaction of soils. Soil Sci. **68**, 229–235 (1949). — Wise, L. E., und E. H. Walters: Isolation of cyanuric acid from soil. J. Agric. Res. **10** (1917a). — The identity of cyanuric acid with so-called "tetracarbonimid". J. Amer. Chem. Soc. **39** (1917b). — Wrenshall, C. L., und R. R. McKibbin: Pasture studies, XII, The nature of the organic phosphorus in soils. Canad. J. Res. B **15**, 475–479 (1937). — Wrenshall, C. R., und W. J. Dyer: Organic phosphorus in soils, II, The nature of the organic phosphorus compounds, A, nucleic acid derivatives, B, Phytin. Soil Sci. **51**, 235–248 (1941).

Yoshida, R. K.: Studies on organic phosphorus compounds in soils, isolation of inositol. Soil Sci. **50**, 81–90 (1940). — Young, J. L., und J. L. Mortensen: Soil nitrogen complexes, I, Chromatography of amino compounds in soil hydrolysates. Ohio Agric. Exper. Stat. Res. Circ. **61**, 18 (1958).

D. Le sol, réservoir de principes nutritifs pour la plante

Par

J. Baeyens

a) Introduction

Un schéma complet du système sol-plante comporte la chaine suivante: *sol*: ions non-échangeables ⇄ ions échangeables ⇄ ions solubles ⇄ racines: *plante*.

Les flèches indiquent l'existence d'un équilibre dynamique dans les deux sens entre les différents états sous lesquels se trouvent les principes nutritifs, équilibre ne signifiant pas nécessairement égalité (BARTLETT et MCLEAN 1959).

Cet équilibre a pu être démontré expérimentalement grâce à l'emploi des ions isotopiques radioactifs suivant le schéma général: $A^*Y + AX \rightleftarrows AY + A^*X$. A^* désigne l'ion radioactif et A l'ion ordinaire.

Ainsi, on a pu prouver que le Ca^{45} s'équilibre rapidement et complètement avec celui absorbé par la kaolinite, l'halloysite, la beidellite et la montmorillonite. Il en est de même du P^{32} dont la partie absorbée s'équilibre rapidement avec la très minime partie en solution. Cette rapidité d'échange explique comment les plantes peuvent soutirer la quantité totale du P nécessaire, malgré la très faible partie actuellement en solution (SINGLETON 1958, ROBERTSON 1958).

D'ailleurs, la somme des ions échangeables ne constitue généralement que 1% de la réserve nutritive totale, ce qui serait insuffisant pour nourrir la végétation, si cette réserve ne fournissait de nouvelles quantités d'ions échangeables, à mesure de leur absorption par les racines.

Les rapports substances solubles ⇄ racines sont du domaine de la physiologie végétale et sont traités dans la 1e partie de cet ouvrage (III et IV). Un aperçu de nos connaissances actuelles sur le rôle dans la plante, des éléments majeurs du sol: N, Ca, Mg, K, P, a été publié récemment par LINSER (1959).

La composition chimique du sol lui-même et sa nature cristalline en rapport avec les ions non-échangeables sont expliquées dans une autre part de ce volume (p. 318).

Dans le chapitre suivant nous nous occuperons des ions non-échangeables, échangeables et solubles ainsi que de leur détermination.

La fertilité d'un sol depend de trois facteurs: la richesse chimique et physique du sol même, son aménagement et le climat (BOGUSLAWSKI 1965).

Vu la grande variété des facteurs de croissance, il serait utile d'admettre pour la fertilité d'un sol une unité conventionelle de fertilité. (LINSER 1965).

La microbiologie du sol est un facteur important de sa fertilité, qui cependant jusqu'ici n'est pas quantitativement mesurable (FRANZ 1965).

Selon KÖHNLEIN (1965), les facteurs de fertilité d'un sol peuvent se grouper en quatre classes: le climat, le sol, la spécificité des plantes, et les soins appliqués à la fumure.

$$E = f \text{ (climat, sol, plante, soins appliqués).}$$

b) Les différents états sous lesquels se trouvent les principes nutritifs dans le sol

1. Les ions non-échangeables

A. Les constituants minéraux des sols

La teneur du sol en principes non-échangeables dépend de la composition chimique de la roche-mère dont le sol dérive, de sa structure cristalline et de son degré d'altération climatique.

Cette teneur représente la grande majorité des principes nutritifs, souvent 99% (Wiklander 1959) et constitue donc la "réserve" nutritive pour la plante. Le degré de mobilisation de cette réserve, à l'état échangeable et soluble, dépend du diamètre de la particule, du pH, de la force de liaison des ions au complexe, de leur position dans ou autour du réseau cristallin, et enfin de l'intensité de l'altération climatique.

Diamètre de la particule en rapport avec sa surface spécifique. Toutes choses égales, plus la particule est petite, plus la surface qu'elle présente aux agents corrosifs (eau — acides) est grande: une argile s'altère donc plus facilement qu'un limon et celui-ci plus facilement qu'un sable grossier. Cette altérabilité dépend cependant plus des facteurs suivants que du diamètre de la particule elle-même:

Le pH du milieu. Un pH bas (acides organiques sécrétés par les racines ou les microbes, le CO_3H-) intensifie la solubilisation et donc l'altération du sol. Les phosphates de Fe et Al font exception.

Un sol pauvre en bases produira une végétation pauvre qui elle donnera un humus acide à pouvoir d'altération prononcé; podzol, humus acide forestier.

Les environs immédiats des racines peuvent présenter un pH < 3, suffisant pour créer une podzolisation locale dans les sols tropicaux.

La composition chimique et la structure cristalline de la particule. Les minéraux silicatés avec leur réseau cristallin compliqué s'altèrent moins facilement que les minéraux de constitution plus simple: $CaCO_3$, $MgCO_3$, $CaSO_4$, qui s'altèrent par simple dissolution.

Les ions à l'intérieur du réseau cristallin ne sortent de celui-ci que par substitution avec un autre ion. Ceux des surfaces extérieures planes, et surtout des bords cassés, sont plus échangeables. Ceux situés entre les feuillets sont échangeables ou non suivant la distance entre les feuillets et le diamètre des ions (voir détails au chapitre de la constitution du réseau cristallin, p. 318).

L'intensité de l'altération climatique, en rapport avec la végétation, dépend des facteurs d'altération du climat: température et pluviosité.

Aux tropiques humides, la température et la pluie élevées détachent de la particule minérale une quantité de principes nutritifs suffisante pour permettre une végétation exubérante (forêt vierge) même sur sol pauvre. Ces sols atteignent rapidement un niveau nutritif très bas (latérites), niveau qui n'est maintenu que par la forêt (Baeyens 1938).

Toutefois, même aux tropiques, les minéraux chimiquement riches peuvent livrer aux plantes une quantité suffisante de principes nutritifs pour permettre une culture sans engrais pendant plusieurs années (sols volcaniques jeunes). Dans ce cas, l'analyse minéralogique du sol peut fournir des indications précieuses sur sa fertilité actuelle ou potentielle.

Au contraire, sous climat tempéré, l'altération minéralogique est si lente que cette analyse ne suffit pas pour prévoir sa fertilité, du moins en agriculture *intensive* où une récolte élevée est impossible sans ajouter d'engrais. L'agriculture *extensive* qui se contente d'une récolte moyenne sur une grande étendue, peut tenir compte de cette réserve non-échangeable.

Les différences d'intensité dans l'altération climatique font que les sols des tropiques humides sont généralement plus pauvres en silice et en principes nutritifs, et plus riches en eau de constitution et en Fe et Al que la roche-mère dont ils dérivent. Sous climat humide, les caractéristiques chimiques des sols se rapprochent plus de celles de la roche-mère (Wiklander 1958).

B. Les ions non-échangeables de la matière organique du sol

Les ions non-échangeables font partie des constituants chimiques de cette matière: phytine, nucléo-protéines sulfurées, et complexes où les cations métalliques, surtout polyvalents: Fe, Al, Ca, Mg, Cu etc. sont chelatés. Cette chélation acquiert une importance particulière dans le cas d'un danger possible de fixation du phosphate par le Fe: dans ces cas les chélates préviennent cette fixation (cfr. les paragraphes du P et du Fe) — (BASTISSE 1957).

Toutefois, vu l'instabilité de la matière organique dans beaucoup de sols, le caractère non-échangeable de ces ions n'est que temporaire et dépend de l'activité microbienne du milieu ainsi que de la nature plus ou moins réfractaire de la matière organique. Dans l'humus ligneux et tourbeux, les ions restent plus longtemps à l'état non échangeable que dans l'humus "doux" d'une terre de culture en bon état.

Toutefois même dans ces formes d'humus défavorables, les arbres forestiers trouvent les matières nutritives grâce à l'activité des mycorhizes de leur système radiculaire (WILDE 1958).

C. Le problème de l'assimilabilité et de la valeur nutritive des ions non-échangeables

Ce problème est compliqué et incomplètement résolu. Nous verrons plus loin que des cas existent où la plante soutire au sol une somme d'ions plus grande que celle révélée par l'analyse des ions échangeables; elles doivent se nourrir donc d'une partie de la réserve non-échangeable. Les minéraux riches en ions non-échangeables (illite pour le K) auront donc une importance plus grande que ceux qui sont pauvres (MORITA 1958). Le minéral le plus riche en K: la leucite (18%), à l'état finement broyé, présente une certaine valeur comme engrais (NICOLIC 1959).

Plusieurs explications sont possibles: nous avons vu qu'aux tropiques humides, l'altération chimique de la roche-mère peut être telle qu'elle puisse fournir à la plante une nourriture non-échangeable. De fait on a constaté une corrélation négative entre la pluie totale aux tropiques et la teneur en bases échangeables des sols.

Ensuite nous verrons que certains ions: K^+ et NH_4^+ peuvent devenir temporairement non-échangeables par pénétration entre les feuillets qui peuvent se dilater après par gonflement (montmorillonite) et remettre en liberté ces ions. En outre, les ions temporairement non-échangeables de la matière organique peuvent être mis à la disposition des plantes par minéralisation de cette matière sous l'action de microbes.

Enfin la végétation permanente (prairie, forêt) dispose d'un temps pratiquement illimité pour dissoudre lentement la partie non-échangeable, ce qui explique que la forêt peut vivre sur un sol trop pauvre pour des cultures annuelles (LAEF 1959).

Expérimentalement on a constaté que la teneur en K échangeable diminue en cours de la saison de végétation et est revenue à son niveau primitif la saison suivante, et cela sans fumure (BLACK 1957).

Ce renouvellement peut se répéter mais diminue d'année en année au fur et à mesure que la réserve non-échangeable s'épuise.

En résumé, on peut dire que la réserve non-échangeable des sols acquiert une importance pratique seulement dans des cas déterminés: dans le cas d'altération intense de la roche-mère (tropiques humides), dans le cas de présence de minéraux riches en un élément nutritif déterminé ainsi qu'en général dans les régions à culture peu intensive. Sous climat tempéré et en agriculture intensive cette livraison est trop lente pour fournir des récoltes élevées.

Cette livraison n'est cependant pas négligeable du moment que la solution du sol est saturée en cations (ce qui arrive rarement dans la pratique). Après une année de saturation (en H), jusque 0,1—0,2% du poids du sol (11—22% de C.E.C.) peut devenir échangeable en provenance du réseau cristallin. Si le liquide est saturé en NH_4^+, la mobilisation du non-échangeable est encore plus grande: 5,3—7,9% du sol superficiel et 8,5—15,5% dans le sous-sol de la C.E.C. En outre, il existe une action compétitive ou synergique entre les ions de la solution: entre H et Mg, NH_4^+ et Ca, Mg et Ca, etc. (WIKLANDER et al. 1963).

En sol forestier couvert d'arbres adultes, il est aventageux de combiner les bases changeables avec les non-échangéables et en % de l'analyse pétrographique (LENTSCHIG 1964 et NEBE 1964).

2. Les ions échangeables

A l'encontre des ions non-échangeables qui se trouvent à l'intérieur du réseau cristallin ou font corps avec la molécule organique, les ions échangeables se trouvent à la surface, extérieure ou intérieure, de la particule adsorbante chargée négativement, positivement ou à caractère amphotère. Cette charge électrique attire un nuage d'ions de charge contraire et constitue donc une couche double électrique dont le potentiel est exprimé par la formule de HELMHOLZ ou par une de ses corrections.

Les argiles à charge négative attireront les cations positifs: Ca, Mg, K, Na, NH_4, H etc. et les argiles positives les anions: PO_4^{---}, SO_4^{--}, Cl^-, NO^-_3 etc.

L'échange se fait stéchiométriquement: c.à.d. suivant la loi des masses et est exprimé en milli-équivalents-grammes (m.e.). Cette expression en équivalents semble dire que l'échange se fait d'une manière équivalente pour tous

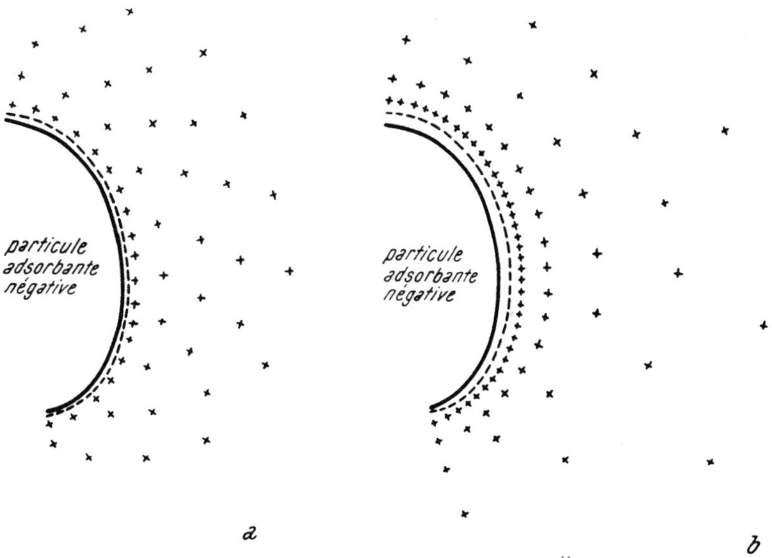

Fig. 94.
Répartition et densité des ions autour de la particule adsorbante d'après HELMHOLZ (a) et GOUY–STERN (b)

les ions. De fait, il existe d'assez nombreuses exceptions à cette règle. Les exceptions sont expliquées soit en admettant que la valence de fait des ions n'est pas celle admise dans les calculs (formation d'ions basiques p.ex. $CaOH^+$ qui diminue la valence réelle surtout des polyvalents), soit en admettant une liaison de nature différente entre l'adsorbant et le cation adsorbé. Cette dernière explication a été admise pour l'adsorption de la matière organique du sol dont la capacité d'adsorption pour le K était dans une expérience de 67 m. e., de 134 m. e. pour le bore et de 287 m. e. pour le cuivre (Broadbent 1955). On met ces grandes différences sur le compte de la formation de chélates qui sont fréquents entre la matière organique du sol et le cation (Black 1957).

Cet échange peut se faire dans les deux sens; sorption et désorption mais est souvent limité par le phénomène d'hysterisis d'échange: un ion une fois adsorbé ne se laisse plus libérer totalement du complexe adsorbant, par pénétration entre les feuillets (K^+, NH_4^+), par substitution dans le réseau cristallin, par remplissage des cavités vides (Al-Si-Mg), par simple précipitation chimique ($Al\,PO_4$-$Fe\,PO_4$) ou enfin par chélation avec la matière organique adsorbante.

La répartition et la densité des ions autour de la particule n'est pas homogène: elle est moins dense aux surfaces qu'aux arêtes et côtés "brisés". Elle est plus dense près de la particule qu'à une certaine distance avec changement brusque (Helmholz) ou graduel (Gouy-Stern): Fig. 94.

La composition chimique d'une garniture ionique autour d'une particule adsorbante est la résultante de plusieurs facteurs agissant à la fois et parfois en sens contraire: la nature de la particule adsorbante, l'énergie de sorption de l'ion, son degré d'hydratation, sa valence, sa concentration et son coefficient d'activité, les changements dans la composition chimique de la solution du sol, son pH etc. (Lebedew 1958).

A. La nature de la particule adsorbante

Capacité de sorption ou d'échange: *T*. La capacité de sorption d'un sol (exprimée en m. e./100 gr.) dépend de sa teneur en matière colloïdale, minérale et organique et de la composition minéralogique: kaolinite: 3–10; illite: 20–40; chlorite: 10–40; montmorillonite: 80–120; matière organique: 150–300; sable: 5—10; limon: 20—30; argile: 50—60 m.e./100 gr. Cette nature minéralogique a aussi répercussion sur la capacité d'échange vis-à-vis des substances organiques (acide acétique et butyrique): Kaolinite < Illite < Halloysite < Montmorillonite < Bentonite. Toutefois ces différences de capacité d'échange subissent des changements d'après la nature du cation adsorbé (Aleixandre et al. 1957). Cette différence est causée par la différence de fixabilité du cation: le K très fixable fournit des chiffres de capacité d'échange inférieurs à ceux obtenus avec un cation peu fixable comme le Ca dans les argiles du type vermiculite alors que pour les argiles à monmorillonite cette différence est moindre vu la nature élastique de ses feuillets (Sawhney et al. 1959).

L'adsorption de polyacrylnitrile par la kaolinite change avec la nature des cations et anions adsorbés au cristal:

$$Th > Ca > Ba > H > NH_4 > K > Na$$

$$F > OH > H_2PO_4 > Cl > CH_3COO > NO_3$$ (Mortensen 1957).

La capacité de sorption dépend de la surface totale, elle augmente donc avec la finesse de texture du sol. Elle dépend en outre de la charge électrique de cette surface et sera donc en rapport avec le pH; pour les cations, elle augmente avec augmentation du pH, et pour obtenir des résultats comparables, toute détermination de la valeur *T* doit s'exécuter à un pH déterminé. La courbe qui représente la relation entre l'augmentation du pH et l'augmentation de la capacité d'échange est plus ou moins sigmoïdale: aux pH bas, ce sont les groupements SiOH qui entrent en jeu, tandis qu'aux pH élevés, ce seraient les groupements AlOH (Wey et al. 1959).

La capacité de sorption est une caractéristique de chaque sol et est assez indépendante de la quantité d'ions ajoutée ou extraite (DE MUMBRUM 1959).

La constitution minéralogique de la particule influe par la structure spéciale de son réseau cristallin: la montmorillonite, avec des charges négatives des deux côtés du feuillet, aura, en principe, une valeur T double de celle de la kaolinite. Il faut en outre tenir compte des charges spécifiques des cassures latérales.

La matière organique du sol avec sa structure "poreuse" (granulaire ou linéaire au microscope électronique) aura une valeur T supérieure à celle des sols minéraux. La capacité d'échange de l'humus "doux" chimiquement riche, est plus élevée que celle de l'humus "brut", plus acide et chimiquement pauvre. Vu le caractère labile de l'humus du sol, une valeur T élevée n'est pas synonyme de forte *rétention* des principes nutritifs: ainsi le phosphore organique est plus assimilable que le phosphore minéral. L'humus du sol, en livrant plus facilement ses éléments mineurs, favorise la guérison des maladies de carence dues à la pénurie de ces éléments (VERMEULEN 1958).

B. La nature de l'ion adsorbé: sa valence, son rayon, son degré d'hydratation et son coefficient d'activité

L'énergie d'adsorption augmente avec la valence et diminue avec son degré d'hydratation, les deux facteurs agissant en sens contraire puisqu'une couche épaisse d'eau d'hydratation empêche le rapprochement de l'ion de la surface adsorbante. Les ions bivalents seront donc plus énergiquement adsorbés que les monovalents (excepté l'H^+).

Ceci vaut pour la matière adsorbante minérale.

Si le complexe adsorbant est surtout organique, la règle de l'énergie de sorption en rapport avec la valence peut être perturbée du fait du pouvoir complexant de cette matière organique qui "bloque" partiellement la valence des ions. Ainsi, l'affinité de la kaolinite pour les cations croît avec leur valence. En ajoutant un corps complexant (citrate), on obtient l'effet inverse.

De même l'humus empêche la fixation de K par les argiles (SCHUFLE 1957). L'effet de la valence sur l'énergie de sorption des ions fait qu'il existe une certaine proportion entre les cations adsorbés: les bivalents étant généralement plus abondants que les monovalents. Toutefois les exceptions à cette règle sont assez nombreuses.

Au point de vue *hydration des ions*, il existe une certaine confusion dans la littérature pédologique; les uns assignent une forte hydratation au Ca^{++} et une hydratation faible au Na^+, les autres admettent le contraire.

L'hydratation des ions a cependant des conséquences agricoles importantes puisqu'une argile sodique humide est pâteuse, à mauvaise structure, alors que cette même argile calcique, à même teneur en eau est bien structurée sans excès d'eau. Ce qui semble indiquer que le Na est plus hydraté que le Ca.

D'ailleurs, la détermination de l'hydratation des ions n'est pas chose aisée et les résultats obtenus varient avec les conditions d'expérimentation.

Les chiffres fournis par les différents auteurs varient entre de larges limites: (m.e.) H^+:1,06—10; Li^+: 3,3—62; Na^+: 1,6—44,5; K^+: 1—29,3; Mg^{++}: 6—33; Ca^{++}: 5,2—22 (GRIESBACH 1957).

Il est difficile de dire si le *rayon ionique* a une influence spécifique sur son adsorption; il est probablement d'ordre secondaire vis-à-vis de la valence et du degré d'hydratation. D'ailleurs les rayons réels ne diffèrent pas tellement d'un ion à l'autre, (K^+: 1,33; NH_4^+: 1,43; Na^+: 0,98; Mg^{++}: 0,78; Ca^{++}: 1,06), alors que les rayons de ces mêmes ions à l'état hydraté sont très différents.

C. La concentration et l'activité des ions dans le liquide du sol

L'effet de dilution. En général, la dilution favorise l'adsorption des bivalents (Ca-Mg) au dépens des monovalents (K-Na). L'augmentation de la concentration, p. ex. par dessication du sol, produit l'effet inverse. Un sol sous climat humide aura donc une tendance à perdre d'abord ses ions monovalents (Na) et par après seulement ses bivalents (Ca).

Ce fait explique l'assainissement rapide des polders marins submergés temporairement par l'eau de mer si le retrait des eaux marines est suivi d'une période de pluie et d'une application d'une bonne dose de sel calcareux *soluble* (SO_4-Ca, $CaCl_2$): transformation rapide de l'argile sodique en argile calcique.

D. La composition minéralogique de la particule adsorbante

Les sols à capacité d'échange élevée ont une tendance à concentrer à leur surface surtout les ions polyvalents (WIKLANDER 1958):

	Kaolinite	*Illite*	*Montmorillonite*
$\frac{Ca}{K}$:	5,0	8,1	22,1

Teneur des sols en ions échangeables. Cette teneur est très variable et est la résultante de plusieurs facteurs. Citons: La richesse de la roche-mère ou sa teneur en principes non-échangeables, la vitesse de livraison de ces principes sous l'influence de l'intensité des facteurs d'altération: climat (pluviosité — température), en rapport avec la végétation et la faune, et enfin, pour les sols cultivés, l'intervention humaine: exportation par les récoltes, importation par les engrais.

La roche-mère imprime son caractère sur le chimisme du sol aussi longtemps que celui-ci reste jeune (Tableau 63).

Tableau 63. *Influence de la roche-mère sur la teneur en ions échangeables des sols développés sous même climat et végétation* (*28 sols forestiers*) (DECKERS et VAN STALLEN 1955)

Assise géologique	cations Ca échangeables: m.e./100 gr.			
	Ca	K	Na	Mg
Salmien	70,3	4,3	6,7	17,4
Gedinnien	43,2	5,2	37,4	13,5
Siegénien inférieur	41,2	8,2	33,5	19,8
Siegénien moyen	59,2	5,4	8,9	23,5
Siegénois supérieur	63,2	7,7	15,8	13,4
Emsien inférieur	70,9	3,9	5,3	19,8
Emsien moyen	56,4	4,6	4,5	34,4

Au fur et à mesure que le sol vieillit, son chimisme s'éloigne de celui de la roche, sous climat humide, dans le sens d'un appauvrissement. Des cas extrêmes: un sol vieux dérivé d'une roche calcaire peut être acide et doit ses caractéristiques pédologiques aux *impuretés* de la roche calcaire: sable, argile. De même une latérite vraie n'a chimiquement plus rien de commun avec la roche dont elle dérive: les bases ont disparu, l'aluminosilicate de la roche a été disloqué, les silicates sont lixiviés et il ne reste qu'un mélange d'hydroxyde de Fe et d'Al.

Le climat agit par ses deux facteurs principaux: pluviosité et température et détermine le caractère plus ou moins altéré du sol.

Les sols des tropiques humides seront donc généralement chimiquement pauvres, même s'ils dérivent de roches chimiquement riches. Les sols des régions arides seront plus riches et les sols des régions tempérées humides seront plus ou moins riches suivant la richesse de la roche-mère et leur degré d'altération.

La végétation: Suivant la composition botanique (feuillus ou conifères), sa densité, son aménagement, son adaptation au milieu (climax), une végétation naturelle (forêt) peut dégrader le sol ou le conserver: climax positif ou négatif (Duchaufour 1958).

Le Tableau 64 montre que la belle forêt conserve mieux la richesse du sol alors que la forêt dégradée l'appauvrit. Cet appauvrissement est surtout marqué dans les horizons superficiels. Les chiffres ne nous disent pas si cet appauvrissement est la conséquence de la végétation dégradée ou si cette végétation maigre est la conséquence du sol pauvre. Comme ces forêts se trouvent sur une même roche-mère, on peut cependant admettre l'action dégradante de la fôrêt. Nous discuterons par après l'effet de l'intervention humaine sur ces mêmes sols.

E. Le climat comme facteur de la composition de la garniture ionique de la particule du sol

Ce facteur est tellement important que certains chercheurs y voyent *le* facteur principal de la formation du sol.

Songeons seulement à son effet sur l'altération du sol: latérites. Je ne fais que mentionner ce facteur puisqu'il appartient au domaine de la pédologie.

Tableau 64. *Relation entre la végétation (belle-forêt et forêt dégradée) et la richesse chimique du sol (valeur V et pH) de quelques sols belges* (Deckers et Van Stallen 1955)

Roche-mère	sols forestiers cultivés			forêt optimale			forêt dégradée		
	horizons	valeur *V*	pH	horizons	valeur *V*	pH	horizons	valeur *V*	pH
Gedinnien	A_p	32,4	5,8	A_1	6,4	4,2	A_1	6,2	3,9
	A_2	21,5	9,4	A_2	12,6	4,8	B_2	27,8	5,1
	DG	42,6	4,9	BD	16,8	4,9	D	29,5	5,1
Siegénien inférieur	A_p	34,1	5,9	A_1	16,1	4,6	A_1	9,2	3,8
	B	23,1	5,4	B_2	16,5	4,6	AB	10,1	4,2
	D	21,2	4,7	D	19,6	5,6	(B)	25,2	4,6
Siegénien supérieur	A_p	28,3	5,7	A_1	25,9	4,8	A_1	4,7	4,3
	AB	18,6	5,6	(B)	33,9	4,6	B	12,9	4,8
	D_2	29,1	5,0	D	33,2	4,7	D	19,3	4,7
Emsien inférieur	A_p	25,2	5,5	A_0	22,1	4	A_0	11,3	4,3
	AB	24,4	5,5	A_1	9,2	4,5	B	9,4	4,6
	Bd	26,3	4,8	BD	12,0	4,7	(B)	22,4	4,7
Emsien moyen	A_p	52,1	6,3	A_1	24,0	5,4	A_1	8,8	4,4
	B	67,2	6,7	(B)	15,9	4,9	B_1	15,1	4,4
	C	65,7	5,8	DG	35,5	5,1	B_2	9,1	4,5
Frasnien	A_p	64,0	5,6	A_1	76,3	7,1	—	—	—
	B	70,0	5,9	B	67,6	7,9	—	—	—
	B–C	82,0	6,2	BC	89,7	8,2	—	—	—
Calcaire	A_1	86,5	7,5	A_{11}	60,8	7,0	—	—	—
	BD_1	89,3	8,2	A_{12}	86,1	7,2	—	—	—
	BD_2	96,3	8,3	AB_D	79,7	7,8	—	—	—

Il s'ensuit que toute étude sur les principes nutritifs du sol doit tenir compte des facteurs du climat: température, régime pluvial et propriétés chimiques et physiques de l'atmosphère (Van Wijk 1963).

Des études intéressantes ont été faites dans le but de mettre en corrélation les cycles de fluctuations de la pluie avec la fertilité du sol. On a trouvé que, après une série d'années sèches (accumulant dans le sol les engrais non-utilisés), la récolte de froment est de 1,50 fois plus élevée qu'après une série d'années humides, et, pour les cultures plus sensibles (-pois), jusqu'à 3 fois. Dans les contrées où la périodicité de ces cycles atmosphériques est assez régulière, l'agriculteur peut prévoir la hauteur de sa récolte et y adapter la quantité de fumure (Van Der Paauw 1962).

La pluie est un facteur de croissance comme les principes nutritifs avec interréactions entre eux. Une fumure potassique doit être copieuse par année de pluie. Le K rend les plantes plus résistantes vis-à-vis de la sécheresse (Studer et Blanchet 1963).

La répartition des ions échangeables à travers le profil (Fig. 95). Cette répartition est une caractéristique de chaque espèce de sol et est une mesure de son âge et degré d'évolution. Dans une séquence de sols forestiers développés sur mêmes matériaux-mère (loess), Livens (1954) a pu démontrer que le degré de satura-

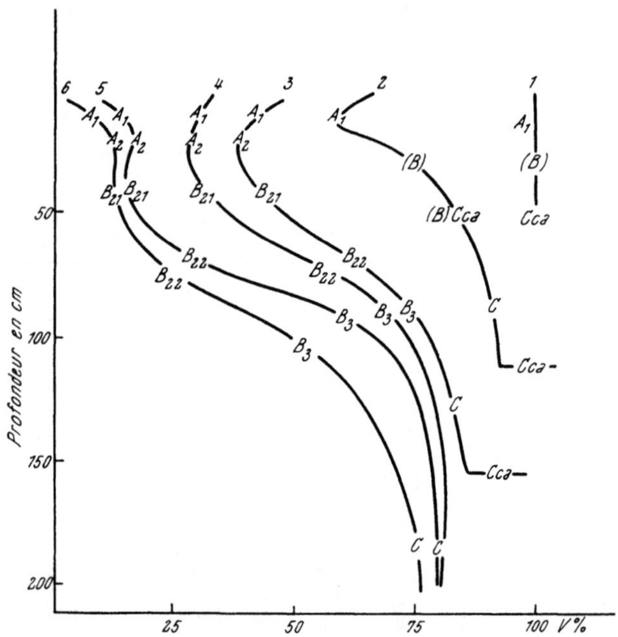

Fig. 95. **Répartition des ions échangeables à travers les profils d'une séquence de sols sur limon loessique (Livens 1954).** *1* Sol brun calcaire, *2* sol brun, *3* sol brun lessivé, *4* sol lessivé, *5* sol podzolique, *6* sol podzolique à gley

tion *V* (et donc la teneur en bases échangeables) diminue avec l'âge du profil dans l'ordre génétique suivant: sol brun calcaire → (*1*) → sol brun (*2*) → sol brun lessivé (*3*) → sol lessivé (*4*) → sol podzolique (*5*) → sol podzolique à gley (*6*).

Ce n'est que dans le sol jeune (*1*) que la valeur *V* est la même à travers tout le profil. Elle diminue de bas en haut dans les sols vieux et cela d'une manière d'autant plus accentuée que le sol est plus dégradé.

L'intervention humaine peut augmenter (engrais) ou diminuer (exportation) la teneur en ions échangeables d'un sol suivant que l'importation l'emporte sur l'exportation ou inversement. Généralement les sols cultivés de longue date seront chimiquement plus riches que la terre vierge du voisinage (Tableau 64). Cette influence humaine affecte directement la couche arable (horizon A), mais peut transformer indirectement tout le profil de sorte que la culture peut changer un profil oligotrophe en un profil eutrophe et le faire "remonter" dans la séquence indiquée à la Fig. 95. Cette influence humaine de longue date a une répercussion sérieuse sur la classification des sols qui jusqu'ici est basée principalement sur les sols "naturels". A l'avenir il faudra faire intervenir sous une rubrique spéciale ces "man-made-soils", du moins dans les régions à agriculture vieille (DELECOUR 1957).

Tableau 65. *Composition chimique des différentes parcelles d'une même exploitation agricole en Belgique* (BAEYENS et STENUIT 1948)

N° de la parcelle	pH	P_2O_5	K_2O	N total	C organique %	
1	5,7	9,0	5,3	104,3	1,29	P_2O_5-K_2O-N : mgr/100 gr
2	5,5	10,2	21,2	85,3	0,91	
3	6,1	1,6	18,5	403,5	8,20	P_2O_5 et K_2O: méthode EGNER-RIEHM (lactate).
4	7,5	19,4	41,3	141,5	1,44	
5	5,9	6,4	15,3	317,5	5,63	
6	7,3	32,8	29,6	128,3	1,59	

Cette influence humaine est cependant très locale et peut varier d'une contrée à l'autre, même d'une parcelle à l'autre à l'intérieur d'une même ferme, du fait, qu'en absence d'analyse régulière de son sol, le fermier distribue un peu au hasard ses engrais sur les différentes parcelles de son exploitation (Tableau 65).

Comme toutes ces parcelles font partie d'une même petite ferme (6 ha) située sur même roche-mère à sol à même texture, il faut admettre que ces différences marquées dans l'état chimique sont dues à des différences de traitement (engrais) pendant de nombreuses années consécutives.

Tableau 66. *Teneurs moyennes en Ca et K échangeables et solubles (ME 100 gr.) de 214 sols loessiques à réactions différentes* (LIVENS 1956)

pH	Nombre d'échantillons	Ca et K échangeables			Ca et K solubles			Rapport: cat. échang. / cat. soluble	
		Ca	K	Ca/K	Ca	K	Ca/K	Ca	K
< 6,5	34	6,72	0,42	16,0	1,31	0,21	6,2	5,1	2,0
6,5—6,9	48	7,68	0,44	17,4	1,77	0,24	7,4	4,3	1,8
7,0—7,4	60	9,28	0,49	18,9	2,00	0,25	8,0	4,9	1,9
7,5—7,9	50	10,82	0,49	22,1	2,24	0,23	9,7[1]	4,8	2,2
> 7,9	22	14,88	0,56	26,5	2,16	0,29	7,5	6,9	2,0

[1] Les rapports Ca/K des sols alcalins sont exagérés du fait que l'extraction solubilise une partie du Ca des carbonates libres de ces sols.

Cette influence dans une contrée à vieille agriculture intensive peut être telle qu'elle a modifié radicalement la vocation agricole naturelle des terrains.

En Belgique, les sols naturellement les plus pauvres (podzols de bruyère) sont devenus après 50 années d'application massive d'engrais, les sols chimiquement les plus riches du pays.

Si la teneur en ions échangeables des sols cultivés peut être très variable, *le rapport* entre les deux ions principaux, Ca et K, semble être assez constant malgré les différences de traitement individuel (Tableau 66).

Il en est de même du rapport entre les cations échangeables et solubles, et ces rapports confirment l'équilibre qui doit exister dans le sol entre ces deux états. Alors que pour le Ca, ce rapport est aux environs de 5, il est beaucoup plus étroit pour le K (± 2).

Comme dans le sol naturel, le K soluble est généralement moins abondant que le Ca et Mg solubles, il faut admettre que dans ces sols cultivés ce K soluble provient principalement, non du sol lui-même, mais bien des engrais.

L'état nutritif d'un sol cultivé de longue date nous apparaît ainsi comme étant sensiblement différent de celui du sol vierge. Malheureusement des études comparatives entre ces deux états sont encore assez rares dans la littérature pédologique.

L'interréaction entre les ions échangeables du sol. C'est peut être le problème le plus compliqué et le moins résolu de la pédologie moderne. Il est cependant à la base de la détermination de la meilleure proportion entre les éléments nutritifs pour la plante et l'antagonisme éventuel entre les ions: l'excès relatif d'un ion déterminé diminuant l'assimilabilité d'un autre ion.

Alors que jadis on admettait la nécessité de rapports définis entre les éléments nutritifs du sol, il paraît maintenant que la plante peut s'adapter à une grande variabilité dans ces rapports du moment qu'un excès prononcé d'un élément vis-à-vis d'un autre est évité (SCHEYS et al. 1963).

Tableau 67. *Rapports entre les éléments N P K d'une même région agricole en Belgique*

P	K	N
10	4,6	41,9
10	17,6	194,2
10	23,9	120,2
10	2,5	59,2
10	9,9	100,6
10	23,3	166,3
10	4,1	102,8
10	3,0	37,2
10	4,0	77,0
10	18,4	85,1
10	89,5	194,8
10	1,4	28,4
10	5,0	135,7
10	67,6	197,6
10	29,5	307,4
10	33,0	179,7
10	116,2	257,1
10	37,8	375,9
10	3,1	82,1
10	79,4	155,0
10	31,1	507,7

La difficulté de la recherche dans ce domaine réside dans les multiples interréactions possibles entre les différents ions: alors qu'un rapport entre deux éléments seulement s'exprime facilement dans un diagramme plan, celui entre plusieurs éléments à la fois exige un système pluri-dimensionnel.

L'interréaction entre les ions du sol aboutit à un stade d'équilibre signalé dans l'introduction de cette étude. L'étude des rapports entre éléments nutritifs du sol se complique du fait que ce rapport n'est pas le même dans le sol que dans la plante. L'analyse de la plante ne peut donc pas servir de base à la fabrication des engrais composés. D'ailleurs, les rapports dans le sol sont tellement variables, même à l'intérieur d'une même région agricole, qu'il faudrait fabriquer un engrais composé presque pour chaque parcelle. Cette grande variabilité est le résultat du morcellement des parcelles en agriculture très avancée.

Le rôle de chaque ion dans cet ensemble peut être suivi commodément par l'emploi, dans l'expérimentation, d'ions radioactifs marqués comme nous le signalerons dans la 2e partie traitant des méthodes employées.

Nous allons résumer l'état actuel de nos connaissances dans ce domaine.

1) L'interréaction entre les ions a introduit en chimie colloïdale du sol la notion importante *d'ion complémentaire:* c'est un autre ion (B) présent dans la solution qui, par son énergie d'adsorption supérieure, chasse un ion (A) donné du complexe et le fait entrer en solution à la place de l'ion B.

L'effet complémentaire d'un ion dépend donc de son *énergie* de sorption propre qui est en général: $Na < K < Mg < Ca < H$.

Plusieurs conséquences importantes agricoles découlent de cette différence d'énergie de sorption.

De tous les cations l'H^+ se laisse le plus énergiquement adsorber et libérera tous les autres ions. Puisque l'eau de pluie, surtout chargée de CO_2, est faiblement mais constamment hydraulysée, tout sol sous climat humide s'acidifie automatiquement et exige un chaulage régulier.

Après l'H vient le calcium qui peut déplacer tous les autres cations: un chaulage libère d'un sol la réserve potassique dont les plantes peuvent profiter. L'énergie de sorption du Na est relativement faible: une argile sodique se laisse facilement désacidifier par un engrais calcique suffisamment soluble: $CaSO_4$, $CaCl_2$, moins bien par le calcaire ordinaire: $CaCO_3$ qui se transforme lentement en $CaH_2(CO_3)_2$.

L'équilibre entre les ions de la solution et ceux adsorbés est régi par la loi de Donnan.

2) Les rapports entre le K et le Na ont été étudiés dans le but, non seulement d'établir l'éventuelle nécessité ou l'utilité du dernier élement, mais aussi pour voir dans quelle mesure le K peut remplacer le Na dans les plantes.

A ce point de vue il semble exister des différences spécifiques entre les différents végétaux, spécificité plus importante que le rapport K/Na dans le sol (Huffaker et Wallace 1959). Toutefois, ce rapport K/Na dans le sol n'est pas négligeable puisque la présence de Na dans la solution libère des minéraux une plus grande quantité de K qui est assimilé par les plantes (Hutcheson et al. 1959).

Les lois de l'échange des ions dans le sol. — Le mécanisme. On peut classer les expressions mathématiques des lois de l'échange des ions en 4 catégories: les expressions purement empiriques et expérimentales, les formules supposant que les phénomènes d'échange suivent la loi des masses, celles se basant sur le côté thermo-dynamique du phénomène et enfin celles qui se basent sur l'équilibre de Donnan et la théorie de Gouy concernant la couche double électrique. La formule de Gapon montre que, dans le cas de la montmorillonite, l'influence de la valence des ions prédomine alors que dans le cas de l'illite ce sont des forces d'adsorption spécifiques (Langerwerff et Bolt 1959).

L'application de la formule de Gouy aux mesures d'absorption négative des chlorides, prouve qu'aux environs du $pH = 10$, les cations alcalino-terreux se conduisent comme des ions monovalents. Une explication possible de ce fait curieux réside dans la supposition de la formation d'ions complexes du type $(MOH)^+$ monovalent (Chaussidon 1963).

Les formules de Vanselov, Davis et Gapon donnent des résultats semblables dans l'échange du Na en terrain calcareux (Davis 1958).

L'équilibre Na-Ca est caractérisé par l'équation différentielle de la couche double électrique de Poisson-Boltzmann (Bower 1959).

L'isotherme d'adsorption de Freundlich, une des premières utilisées en pédologie, continue à susciter l'intérêt des chercheurs qui la trouvent en concordance avec les faits observés (Mitra et Panda 1957).

SCHUFFELEN a soumis ces différentes formules à un examen critique et arrive à la conclusion que la théorie de la couche double électrique représente le mieux la réalité des faits. Il propose de déterminer l'activité des ions dans la solution du sol et de calculer à partir de ces données le "niveau d'énergie" de chaque ion nutritif pour arriver ainsi à une définition physico-chimique de la fertilité du sol (SCHUFFELEN 1958).

Quant au *mécanisme de l'échange* des ions, on suppose qu'il se fait à la suite de l'agitation thermique des ions à la surface de la particule. L'ion entrant profite du moment où l'ion adsorbé est le plus éloigné de la surface pour se glisser entre cette surface et l'ion est ainsi refoulé dans la solution. Une preuve de cette conception thermique du mécanisme réside dans le fait que la témperature augmente la quantité d'ions échangés et absorbés par la plante (ROBINSON et al. 1959).

Nous ne parlons ici que de l'échange des ions entre le sol et sa solution. L'échange entre les ions de la solution et la racine est beaucoup plus compliqué vue l'intervention de la plante "vivante". Ces recherches appartiennent au domaine de la physiologie végétale et sont traités dans une autre partie de cet ouvrage.

Résumons cependant les faits acquis à ce jour.

Les ions entrent dans la plante:

a) Par transport actif au moyen de "carriers". Ce transport est lié à la vie physiologique de la plante, il est sélectif et irréversible même si l'activité des ions dans la plante surpasse celle de la solution du sol.

b) Par transport passif par simple diffusion dans l'"apparent free space" de la racine. Ce transport est non sélectif et réversible.

c) Transport d'échange entre les zones de la plante et ceux de la solution, comparable à l'échange des ions de la surface du sol vers sa solution (MENGEL et MARSCHER 1963).

Ions échangeables et ions assimilables. Longtemps on a considéré ces deux notions comme étant synonymes. Depuis, on a reconnu des cas de plus en plus nombreux de non identité; certaines plantes pouvant assimiler jusqu'au double de la quantité réputée échangeable, alors que dans d'autres cas cette quantité assimilable a été nettement inférieure à la réserve échangeable (MORTLAND et al. 1957).

Par conséquent, il n'existe souvent pas de corrélation entre les teneurs en ions de la plante et celle des ions échangeables dans le sol (PERKINS et STELLY 1958).

De fait, ces deux notions représentent des réalités différentes et déterminées différemment. La détermination des ions échangeables est une opération physico-chimique de laboratoire qui porte un caractère essentiellement relatif comme nous le verrons au chapitre des méthodes appliqueés. La détermination des ions assimilables au contraire est une opération chimico-physiologique puisqu'elle doit tenir compte de l'interréaction entre la solution du sol et l'activité du système radiculaire compte tenu des variantes du climat. La détermination des ions échangeables fournit des chiffres relatifs, c. à. d. dépendant du mode opératoire p. ex. de l'ion déplaçant, mais comparables entre eux.

Pour que les chiffres des ions "assimilables" au laboratoire obtiennent une signification pratique, ils doivent être comparés à un grand nombre de récoltes de champs d'expériences dont le sol a été analysé au laboratoire.

Il faut un grand nombre de champs accompagnés d'analyses de sol et cela pendant plusieurs années consécutives pour pouvoir neutraliser les variations des facteurs climatiques. En agissant de la sorte, on peut obtenir une véritable corrélation significative entre le procédé analytique employé et le résultat de la récolte comme le montrent les diagrammes des Fig. 96 et 97 (BAEYENS et STENUIT 1958, RICHARD 1959, DIEDRICH 1964).

Du moment que la corrélation entre le chiffre des ions assimilables au laboratoire et le résultat des récoltes sur les champs a été établie d'une manière statistique en se basant sur des milliers de cas, le choix de la méthode d'extraction au laboratoire est d'importance secondaire, puisque *tout* réactif faible est capable d'extraire du sol une *certaine* quantité de principes nutritifs qui peut être mise en rapport avec la récolte produite.

Les méthodes de détermination du besoin en engrais des sols sont exposés au chapitre I — H de cet ouvrage.

En se basant ainsi sur un grand nombre de cas, on tient compte aussi de *l'interréaction* entre les différents facteurs de croissance du sol.

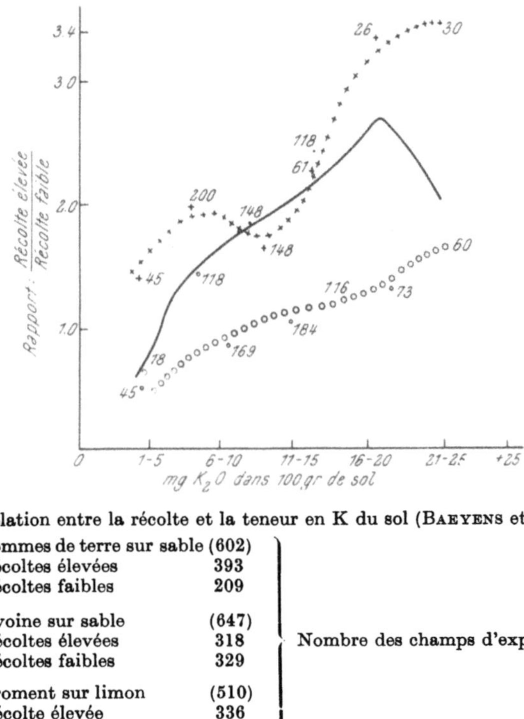

Fig. 96. Relation entre la récolte et la teneur en K du sol (Baeyens et Stenuit 1958)

————————	Pommes de terre sur sable	(602)	Nombre des champs d'expériences analysés
	Récoltes élevées	393	
	Récoltes faibles	209	
ooooooooooo	Avoine sur sable	(647)	
	Récoltes élevées	318	
	Récoltes faibles	329	
+++++++++	Froment sur limon	(510)	
	Récolte élevée	336	
	Récolte faible	174	

Ainsi, la corrélation entre le K assimilable du sol et son effet sur la récolte est faible (0,65) si l'on ne tient compte que du K seul, alors que cette même corrélation devient très significative (0,82) si l'on englobe en même temps dans le système d'appréciation la capacité totale d'échange du sol, sa teneur en Ca et sa matière organique (Jaworski et Barber 1959).

Jusqu'ici toutes déterminations de principes nutritifs assimilables devaient être contrôlées en dernière analyse par le champ d'expérience.

Il semble maintenant que l'emploi d'isotopes radioactifs permettra de prévoir le degré d'assimilabilité d'un ion indépendamment de sa réaction sur la plante.

Ceci est important surtout pour l'ion P dont l'assimilabilité est régie en premier lieu par son degré de fixation au sol. Cette fixabilité peut être déterminée maintenant directement par l'emploi simultané du P^{31} et P^{32} (Russell 1959).

Il permet en outre de déterminer l'assimilabilité des différentes formes d'engrais phosphatés: superphosphate, phosphate bicalcique, tricalcique etc. (Ensminger et Pearson 1957).

L'échange des anions dans le sol. Cet échange semble régi par les mêmes lois que celles des cations (Weiss 1958). Il a été cependant moins étudié excepté pour le PO_4^{-3} dont la grande fixabilité dans tous les sols offre des difficultés spéciales en agriculture.

A l'encontre des cations, ce sont les charges positives du complexe adsorbant qui régissent la fixabilité des anions qui se laissent adsorber dans l'ordre croissant suivant: $Cl = NO_3^- < SO_4^{--} \leq PO_4^{---}$.

Du fait que les nitrates ne se laissent pas adsorber colloïdalement par le complexe négatif du sol, il ne faut pas conclure qu'ils lixivient dans tous les sols:

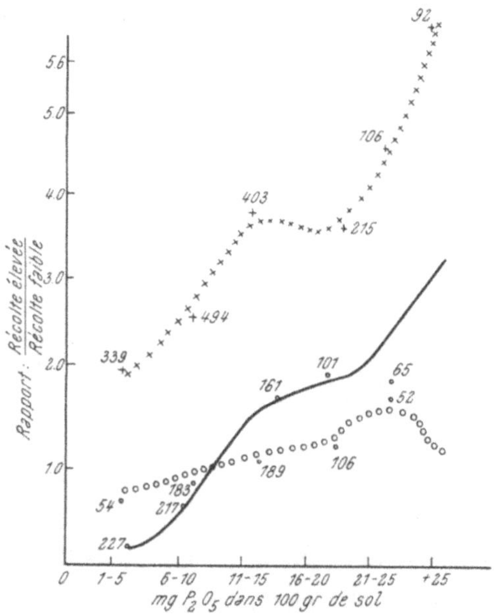

Fig. 97. Relation entre la récolte et la teneur en P du sol (Baeyens et Stenuit 1958)

————————	Navets sur limon	(857)	Nombre des champs d'expériences analysés
	Récoltes élevées	381	
	Récoltes faibles	476	
∘∘∘∘∘∘∘∘∘∘∘	Avoine sur sable	(647)	
	Récoltes élevées	318	
	Récoltes faibles	329	
++++++++++	Froment sur limon	(1649)	
	Récoltes élevées	1198	
	Récoltes faibles	451	

ils peuvent être fixés biologiquement par le phénomène de la néo-protéinisation (cfr. chapitre de la microbiologie du sol).

Cette différence de fixabilité peut s'expliquer par la différence de solubilité: les chlorures et les nitrates sont solubles, les phosphates se combinent au Fe et Al.

Les facteurs de fixation des phosphates seront étudiés au chapitre du P.

Puisque le coefficient de solubilité d'un sel est sous la dépendance du pH, la réaction du sol est importante dans le phénomène d'échange des anions.

Comme pour les cations, la capacité d'échange anionique est en rapport avec la surface totale de la particule adsorbante, externe et interne: la bentonite a une capacité d'échange élevée alors que la pyrophyllite avec sa surface réduite a une capacité d'échange moindre.

L'électrodialyse montre que les anions sont en partie fortement fixés avec une fraction de fixation faible.

La fixation énergique se fait par substitution dans le réseau avec Al et Si, la fraction faiblement fixée est adsorbée extérieurement aux endroits d'échange (SCHELL et JORDAN 1959). En outre la vitesse de fixation de ces deux phases est différente: pour le PO_4^{-3} la forte fixation se fait rapidement: 4 — 7 jours, la fixation faible exige plusieurs semaines (RUSSELL 1959).

Récemment l'attention s'est portée sur *l'aluminium échangeable* sous forme anionique: d'après le pH, de 20 à 90 % de la capacité totale d'échange d'un sol peut être occupé par l'aluminium dont la détermination devient nécessaire dans certaines évaluations du besoin en chaux (McLEAN et al. 1958).

3. Les ions solubles dans le sol

Nous avons étudié au chapitre précédent les facteurs de la grande variabilité en bases échangeables du sol. Puisqu'il existe un équilibre entre ces bases et les ions de la solution du sol, celle-ci aussi sera très variable et sous la dépendance des mêmes facteurs.

A ces causes "internes" de variabilité s'ajoutent encore des causes externes, notamment l'exportation par les plantes ou l'importation par les engrais.

Les Tableaux 66 et 68 montrent qu'il existe un équilibre presque constant entre les cations échangeables et solubles malgré des teneurs très variables en principes nutritifs (LIVENS 1956).

Tableau 68. *Rapport entre K échangeable et soluble de sols à teneurs en K échangeable très différentes (m.e./100 gr. — moyennes de 214 sols différents)*

Teneurs du sol en K échangeable	Nombre de sols analysés	K échangeable	K soluble	Rapport K éch./ K sol.
< 0,30	40	0,26	0,10	2,6
0,30—0,39	46	0,35	0,14	2,5
0,40—0,47	45	0,43	0,21	2,0
0,48—0,55	27	0,51	0,25	2,0
0,56—0,61	16	0,58	0,29	2,0
> 0,62	40	0,82	0,46	1,8

Une vieille terre de culture semble donc "tamponner" son état nutritif vis-à-vis des exportations par la plante et les importations par les engrais. Remarquons toutefois qu'avec l'enrichissement du sol en K, le rapport $\frac{\text{K échangeable}}{\text{K soluble}}$ a une tendance au rétrécissement; à mesure que le sol sature en K, son K soluble devient plus abondant. C'est le cas des vieilles terres de cultures fumées abondamment depuis plusieurs années.

Le même équilibre entre échangeable et soluble a pu être observé pour les anions au moyen de l'expérimentation au phosphore radioactif: P^{32} soluble $+ P^{31}$ échangeable $\rightleftarrows P^{32}$ échangeable $+ P^{31}$ soluble.

En outre cet équilibre est assez indépendant de la quantité du P ajouté au sol (FRIED 1956).

Dans les régions à agriculture intensive, l'état nutritif en principes solubles dans les sols est régi par l'apport d'engrais. Ils seront donc plus riches que les terres vierges de la même région et les variations en principes solubles seront sous la dépendance des mêmes facteurs que ceux qui régissent la variabilité en ions échangeables.

Notons cependant quelques particularités propres aux ions solubles.

La réaction. Une acidité prononcée augmente la concentration en ions solubles à la suite de l'action dissolvante des acides carboniques et organiques sécrétés par les racines et les microbes. Cet effet est surtout marqué en sol calcareux et magnésien. Ceci vaut pour un sol vierge.

En terre cultivée une acidification est toujours un signe d'appauvrissement en bases tant solubles qu'échangeables.

Nous verrons par après qu'il existe une relation générale entre le pH et le degré de saturation d'un sol, donc de sa richesse en ions (Fig. 98).

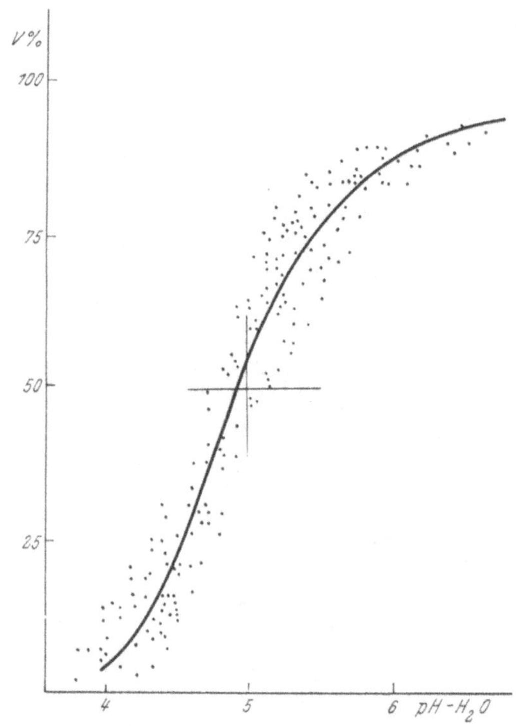

Fig. 98. Relation entre le pH et le degré de saturation d'un sol (Livens 1954)

La saison. En rapport avec l'activité microbienne, la concentration de la solution du sol en ions nutritifs subit des fluctuations saisonnières avec deux sommets: un après l'hiver et un deuxième en automne. Si le sol cultivé porte une végétation, il s'appauvrit au cours de l'été à la suite du pompage des racines. Après l'enlèvement de la récolte, la concentration de la solution augmente de nouveau à la suite de l'activité dissolvante des microbes. Un sol à réserve nutritive élevée peut ainsi produire sans engrais une récolte pendant plusieurs années: le rendement diminue rapidement les premières années pour rester assez longtemps à un niveau plus bas les années qui suivent.

Variations de la pluviosité. En général une abondance de pluie dilue la solution du sol, excepté pour les sulfates dont l'augmentation peut s'expliquer par la solubilité du gypse.

Influence mutuelle des ions en solution. Cette influence affecte surtout l'ion PO_4^{-3} dont la concentration en solution est modifiée par la présence d'autres ions en solution: Ca, Fe, Al, K, qui diminuent ou augmentent la solubilité des phosphates. L'assimilabilité de l'ion PO_4^{-3} est compliquée et sera discutée au chapitre du P.

Conclusion. La variabilité de la composition de la solution du sol est encore plus grande que celle des ions échangeables. Voilà pourquoi peu de recherches systématiques ont été effectuées dans ce domaine. D'ailleurs, hormis les sols salins, la teneur en ions solubles ne représente que quelques pourcents des ions échangeables qui eux ne représentent que 1 à 2 % de la réserve nutritive totale du sol. C'est la raison pourquoi, hormis quelques cas exceptionnels p. ex. terre de jardin ou de serres, les ions solubles ne servent pas de base pour la détermination de l'état nutritif ou le besoin en engrais d'un sol, qui se fonde surtout sur les bases échangeables, ou mieux, assimilables.

Bibliographie

voir p. 522 et suivantes.

c) Les ions nutritifs étudiés séparément

1. Les cations

A. L'hydrogène dans le sol — son acidité

L'hydrogène occupe dans le sol une place à part dans la rangée des cations échangeables puisqu'il régit l'acidité du sol et qu'il n'est pas un principe nutritif pour les plantes. Les différentes espèces d'acidité, actuelle, potentielle, hydraulytique, d'échange ainsi que les problèmes de sa neutralisation par chaulage sont traités ailleurs dans cet ouvrage.

Nous mentionnons seulement ici quelques problèmes spéciaux qui ont fait l'objet de recherches récentes.

Le mécanisme de l'échange de l'ion H^+ avec le complexe du sol est encore incomplètement connu.

L'acidité d'échange est due probablement à la formation d'acide aluminique dans la solution, l'Al provenant du réseau cristallin qui peut se disloquer par pénétration du H^+ dans le réseau : l'ion H_3O^+, porteur de protons, frappe la surface du réseau, l'O de cette surface se combine avec le proton de l'ion d'hydroxonium pour se transformer en OH^- tandis que H_3O^+ devient H_2O. Ce proton confère aux tétraèdres de SiO une fonction acide (ALESCHIN 1958).

La supposition que l'acidité d'échange est liée à l'aluminium se trouve confirmée par le fait qu'il existe un rapport entre les activités de Al^{+++} et celles de H^+ dans les extraits de sol (LINDSAY et al. 1959).

Non seulement l'Al, mais aussi le Ca et le Mg peuvent être remplacés par H dans le réseau et cela en quantités équivalentes. Cet H pénétrant ainsi le réseau est devenu "non-échangeable" (TSCHERNOW et MAXIMOWA 1958).

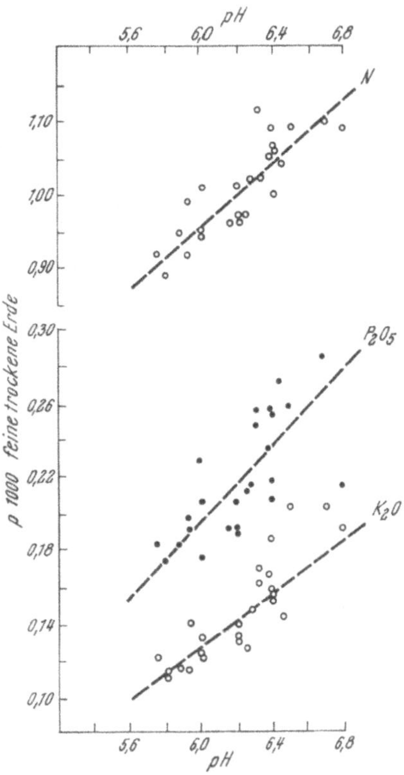

Fig. 99. Teneurs en N, P_2O_5, K_2O, en fonction du pH (LIWERANT 1963)

L'hydrogène suit, comme les autres cations, les lois de l'échange des bases. Par conséquent, le chaulage d'un sol commencera par relever rapidement le pH de celui-ci suivi d'une augmentation lente pour atteindre un équilibre final (NIELSEN 1958).

La capacité totale d'échange T est une fonction de la quantité totale de cations nutritifs présents (S) et la quantité d'ions H: $T = f(S + \mathrm{H})$.

Le degré de saturation d'un sol, tenant compte de la présence éventuelle de cet hydrogène, sera donc: $V = \frac{S \times 100}{T}$.

De ces définitions, il suit qu'il doit exister une relation générale entre pH et V (Fig. 98) avec cependant de nombreux écarts individuels dus à la différence de capacité de sorption des sols et aussi à cause des différents états d'acidité possibles dans un même sol: H représente la quantité totale d'hydrogène alors que le pH ne donne que la partie dissociée (Livens 1954).

En outre, il existe une relation entre le pH et l'état nutritif général du sol: sa teneur en N, P_2O_5 et K_2O. La mesure du pH est une mesure, non seulement de l'acidité du sol, mais non de sa richesse ou pauvreté chimique. Ainsi, pour les sols de la Garonne, un abaissement du pH jusque 5,75 signifie une perte de 30% du Ca et Mg initial échangeable, de 50% pour le K et de 40% de P assimilable. On a constaté en même temps une dégradation de la structure (voir Fig. 99) (Liwerant 1963).

B. Le calcium

A l'état *non-échangeable* il se trouve dans les minéraux et les roches calcaires qu'on peut classer en alumino-silicates calciques (anorthite), en phosphates calciques (apatite), en carbonates (calcite-dolomite), en minéraux argileux calcaires (montmorillonite) et enfin sous forme de gypse.

L'altérabilité, et donc la possibilité de céder des principes nutritifs aux plantes, varie beaucoup d'un minéral à l'autre suivant sa composition chimique et la constitution de son réseau cristallin.

Nos connaissances à ce sujet sont encore fragmentaires. Pourtant cette étude est importante pour les cultures qui doivent vivre de la réserve non-échangeable du sol: forêts, arbres fruitiers.

L'altération des minéraux calcaires *carbonatés* a été mieux étudiée vu l'importance pédologique du calcaire comme facteur neutralisant des sols acides et améliorant leur structure.

Le $CaCO_3$, pratiquement insoluble dans l'eau pure (10–15 mgr par litre) le devient dans l'eau chargée de CO_2, formation de bicarbonate: $CaH_2(CO_3)_2$.

Les facteurs qui régissent l'équilibre de cette transformation sont: le pH, la pression de CO_2(P) et la puissance ionique (mu) (Yaalon 1954)

$$\mathrm{pCa}^{++} = 2\,\mathrm{pH} + \log \mathrm{P_{CO_2}} - 9{,}85 - 1{,}425\,\sqrt{\mathrm{mu}}\,.$$

Yaalon a trouvé en outre:

$$\log S = -0{,}54\,\mathrm{pH} + 6{,}23 \quad \text{et} \quad \log S = 0{,}35\,\mathrm{P} + 2{,}97.$$

S = mgr de $CaCO_3$ dissous par litre.

On voit donc que la quantité de calcaire dissous croit avec la racine carrée de la pression de l'acide carbonique.

Notons cependant que ces relations fournissent des résultats différents suivant qu'elles sont appliquées au calcite pur ou à des sols calcareux, le calcaire des sols étant plus soluble que celui du calcite avec comme conséquence un pH, CO_3H^- et Ca^{++} plus élevés dans la solution du sol (Sterling et al. 1959).

Par conséquent, l'altérabilité des minéraux calcaires est favorisée par un climat humide et une intense activité microbienne du sol, source principale de CO_2.

En région à agriculture intensive, un sol calcareux peut ainsi se décalcifier et même devenir acide dans la couche arable après quelques années de culture.

Par contre, sous climat aride, le bicarbonate peut précipiter en carbonate neutre avec dégagement de CO_2.

Ces principes sont à la base de l'emploi rationnel du calcaire comme chaux agricole:

$$CaCO_3 \rightarrow CaO \rightarrow Ca(OH)_2 \rightarrow CaH_2(CO_3)_2 \rightarrow CaCO_3.$$

La chaux vive ou éteinte doit être enfouie et mélangée aussi intimement que possible avec le sol pour qu'elle puisse se transformer en bicarbonate, seule forme suffisamment soluble pour servir de nourriture aux plantes, désacidifier le sol et améliorer sa structure.

Exposée à l'air sec, elle retourne au stade $CaCO_3$ insoluble.

L'altérabilité du *calcaire phosphaté* a été assez bien étudié, non pas pour le Ca lui-même, mais à cause de la présence du P et sa fixation trop facile dans tous les sols.

Nous étudierons ce phénomène au chapitre du phosphore.

Du côté Ca, on peut dire que sa solubilité dans le phosphate est influencée par le pH du sol ainsi que par la forme chimique du phosphate dont il existe de nombreuses variétés dans la nature: $Ca_3(PO_4)_2$, $CaH_4(PO_4)_2$, $Ca_2H_2(PO_4)_2$, $Ca_4H(PO_4)_3$, 3 H_2O, $Ca_5(PO_4)_3OH \cdot H_2O$ etc.

Le comportement dans le sol de ces différentes formes est compliqué et incomplètement connu.

Enfin, l'altération des *minéraux calcaires silicatés* dans le sol est peu élucidée malgré leur importance probable en sylviculture. Plusieurs auteurs ont établi des échelles qualitatives d'altérabilité de ces minéraux en rapport avec leur importance dans les sols forestiers (AALTONEN 1948).

Le manque d'intérêt s'explique du fait de leur très faible solubilité, quoiqu'aux tropiques et dans les régions à agriculture très extensive, leur rôle ne semble pas négligeable (VAN DER MAREL 1947).

Le calcium échangeable représente jusqu'à 90% des cations échangeables totaux mais ces 90% ne représentent que 1–2% du Ca non-échangeable.

Le Ca échangeable est attaché à la surface des colloïdes du sol en rapport avec la capacité de sorption de ces colloïdes: les colloïdes organiques absorberont donc plus de Ca que les colloïdes minéraux, et parmi ceux-ci les argiles du type 2:1 (montmorillonite) sont plus absorbants que ceux du type 1:1 (Kaolinite).

La teneur globale des sols en Ca est très variable localement. Les sols jeunes sur roche calcaire (Rendsina's) ainsi que les polders jeunes (coquillages marins) sont riches en Ca. Les sols cultivés sont plus riches que les terrains incultes à la suite d'apport d'engrais.

Les sols des tropiques humides (latérites) sont très pauvres en Ca. Au contraire certains sols des régions arides peuvent contenir beaucoup de Ca (carbonates, sulfates).

L'importance du calcium est surtout d'ordre pédologique (réaction du sol — sa structure). Physiologiquement il n'est qu'un élément mineur nécessaire. L'exportation du Ca par les graminées est ± 15 Kgr. (Ha) année, alors que chaulage moyen est de 1.000—4.000 Kgr. par Ha. La carence en Ca se manifeste dans la plante par une analyse des cellules, une déformation de la paroi (gonflement, ramollissement ,coloration) et la production de substances brunes qui occasionnent la coloration des feuilles (BUSSLER 1963).

C. Le magnésium

A l'état non-échangeable il se trouve dans plusieurs minéraux à altérabilité différente: les sulfates sont les plus solubles, puis viennent les carbonates (dolomites) qui sont même plus solubles que le $CaCO_3$.

L'altérabilité des minéraux magnésiens silicatés est minime et d'ailleurs mal connue. Wiklander (1958) cite la série suivante: olivine > hyperstène > augite > > hornblende > biotite > chlorite > illite > vermiculite > montmorillonite.

Après le Ca, le *Mg échangeable* est le mieux représenté dans les sols normaux: 10–30% des cations échangeables totaux. La distribution du Mg à travers le profil est parfois une caractéristique du type de sol et peut servir à déterminer son besoin en Mg (Van der Marel 1959).

Le Mg se trouve surtout dans la fraction fine du sol; il existe une corrélation étroite ($r = +0{,}90$) entre la texture et la réserve en Mg.

La teneur en Mg assimilable diffère avec l'espèce de sol:

Sols tourbeux:	22,7 mgr./100 gr. de sol (Schroeder et al. 1963).
Sols bruns limoneux:	12,9 mgr./100 gr. de sol
Sols bruns sabloneux:	6,8 mgr./100 gr. de sol
Sols bruns podzoliques:	6,3 mgr./100 gr. de sol
Podzols:	4,3 mgr./100 gr. de sol

La carence en Mg dans le sol est liée à une zone de pH (KCl) bien déterminée: < 4,2 et 6,0 >; elle est donc en rapport avec la teneur en Ca du sol, probablement à la suite d'un antagonisme ionique: Ca/Mg.

La teneur en K du sol a aussi son influence (Stenuit et Piot 1958).

La carence n'est pas à craindre si dans la garniture ionique le Mg représente 10–15% et le K 5% (De Haas et Gruppe 1959, Sluijsmans 1964).

Ce même antagonisme semble jouer un rôle dans la formation des nodosités des légumineuses où le Mg est nécessaire et le Ca pas (Norris 1958).

La carence en Mg se manifeste surtout par années pluvieuses et dans les sols à faible capacité adsorptive (sable) et est donc probablement une question de lixiviation.

En Europe occidentale, les précipitations atmosphériques apportent au sol 3–10 Kgr de Mg par Ha par année (Struys 1958).

Le Mg, considéré longtemps comme étant un élément-trace est en réalité un élément majeur, comme le Ca. Les engrais qui apportent le Mg au sol sont les sels potassiques bruts et les scories de déphosphoration. Quoique le SO_4Mg soit plus soluble que le Mg des scories, leur effet agricole est pratiquement le même: 100:95. Une fumure régulière de scories suffit pour couvrir les besoins des plantes en cet élément (Gericke 1962).

D. Le potassium

La teneur des sols en K total varie considérablement d'après les conditions locales.

Les sols les plus pauvres sont les latérites et les podzols, les plus riches les argiles cultivées de longue date dans les régions tempérées.

En général le K se trouve dans les particules les plus fines, de sorte que les sables sont d'ordinaire plus pauvres que les argiles.

A *l'état non-échangeable*, le K se rencontre dans beaucoup de minéraux potassiques: la leucite (18% de K), les feldspaths (7–12%), la muscovite (7–9%), la biotite (5–7%), l'illite (4–5%), la glauconite (5%), l'hornblende (0–2%).

L'assimilabilité, par la plante, de ces minéraux est en général faible. Toutefois la leucite finement moulue a pu être utilisée avec un certain succès comme engrais potassique sur tchernosium et podsol (Nicolic 1959).

Le *K échangeable* se trouvera aussi de préférence dans les fractions fines du sol, où il ne représente cependant que 1% du K total, alors que dans les sols organiques la majorité du K est échangeable vu le caractère labile de l'humus qui se minéralise sous l'action des microbes.

Le *K soluble* dans les sols est généralement moins abondant que le Ca et Mg solubles. Il ne représente que 1 à 10% du K échangeable et est le plus abondant dans les sols cultivés, fortement enrichis par des engrais potassiques.

Dans les terres vierges, la teneur en K soluble dépend, outre du climat (pluvieux ou sec), de l'altérabilité du minéral. Parmi les micas, la biotite est altérable alors que l'illite et la muscovite le sont moins.

Dans la catégorie des feldspaths, les plagioclases sont altérables et l'orthoclase l'est peu.

A l'encontre du phosphore, le K est un élément "mobile" tant dans le sol que dans la plante, qui l'absorbe le mieux par année humide et moins bien par année sèche, de sorte qu'en cette dernière une fumure potassique supplémentaire a un effet marqué (VAN DER PAAUW 1958).

Le K échangeable et soluble peut devenir non-échangeable par *fixation* entre les lamelles des minéraux élémentaires où il remplit les cavités du réseau cristallin occupées antérieurement par d'autres cations: Ca, Mg.

Ceux-ci ont eux-mêmes remplacé le K présent primitivement dans le minéral potassique.

On admet donc des fixations et des défixations successives qui font que les plantes, dans certaines conditions, peuvent assimiler une quantité de K plus grande que celle indiquée par les méthodes habituelles de détermination du K échangeable. K échangeable n'est donc pas synonyme de assimilable. Le K non-échangeable, mais assimilable par la plante, semble lié à la teneur en limon du sol (SMITH et MATHEWS 1957, GARAUDEAUX et al. 1959).

Alors que généralement la fixation du K (à partir des engrais potassiques) est rapide, sa défixation et sa livraison à la plante se font plus lentement de sorte que cette réserve de K, devenue non-échangeable ne peut suffir pour nourrir les plantes de cultures (SCHROEDER 1956).

Par conséquent, après absorption par les plantes du K non-échangeable du sol, le pouvoir fixateur de celui-ci se trouve de nouveau augmenté (VON REICHENBACH 1958).

Les *facteurs* qui régissent cette fixation et défixation de K, sont: la texture du sol, son pH, sa composition minéralogique, l'alternance de sécheresse et d'humidité qui font gonfler ou se rapprocher les feuillets, l'espèce et la concentration des cations présents, spécialement du Ca et du K lui-même ajouté sous forme d'engrais.

Les ions Al dérangent le mécanisme de contraction des feuillets des minéraux du type de 3 couches: illite — montmorillonite, et diminuent ainsi le % de K non échangeable, avec augmentation subséquente de son assimilabilité par la plante (WELTE et NIEDERBUDDE 1963).

Le problème de la fixation des K dans le sol peut s'aborder de deux manières: d'une manière théorique en étudiant les facteurs qui régissent le glissement des ions potassiques entre les feuillets du cristal argileux; d'une manière pratique en établissant pendant plusieurs années des champs d'expériences avec K* marqué de manière à établir le bilan potassique du sol et de la plante (PETERBURGSKY 1961).

Pour la discussion de tous ces facteurs, voir: SCHUFFELEN (1955), DE MUMBRUM et HOOVER (1958), BLACK (1957), HANWAY et SCOTT (1959).

La fixation du K, phénomène indésirable du point de vue effet immédiat, a cependant l'avantage de prévenir la lixiviation et de procurer aux engrais potassiques un arrière-effet parfois pendant plusieurs années.

Cet arrière-effet diminue fortement les premières années après l'application de l'engrais potassique pour se stabiliser par après à un niveau plus bas (Rouse 1958, Van der Marel 1959).

Du point de vue agronomique on peut donc distinguer dans le sol deux "états" potassiques; la partie immédiatement disponible et la réserve plus lente.

Des procédés de détermination de ces deux états ont été mis au point (Fink 1957).

Le phénomène de fixation et de défixation du K dans le sol, rend difficile la détermination du K "assimilable".

D'ordinaire la simple extraction du K au moyen d'un réactif choisi ne suffit pas pour prévoir le besoin en engrais potassique du sol: il faut tenir compte en outre de sa composition minéralogique et de sa valeur T et S (Strassman et al. 1958).

Les autres éléments de la série alcaline: Rb et Cs n'ont pas d'importance pratique. Il faut toutefois tenir compte de la toxicité spéciale du Cs^{137} (voir Strontium), sous produit des usines nucléaires.

Rubidium et potassium sont physiologiquement antagonistes, comme le Cs. Le Rubidium diminue l'assimilation du K et vice-versa. L'emploi du Rb^{86} peut donc servir à étudier le jeu du K, tant dans le sol que dans la plante (Walker et al. 1962).

E. L'ammonium

L'ammonium occupe une place spéciale parmi les ions nutritifs pour la plante du fait de son caractère labile: il est nitrifié, dénitrifié ou neo-protéiné rapidement suivant l'activité microbienne spéciale du sol.

En outre, il est volatilisable par addition de bases fortes (chaulage), et enfin il peut être fixé par le réseau cristallin puisque son rayon est de même grandeur que celui du potasium.

En sol organique cette fixation se fait cependant surtout par la matière organique (Sohn et Peech 1958).

L'illite fixe plus d'NH_4 que la montmorillonite (Stevenson et al. 1958).

Il s'ensuit que les différents états de l'NH_4^+ dans le sol sont assez difficiles à suivre: à l'état échangeable et soluble il s'y trouve d'ailleurs en quantités minimes vu sa transformation continuelle. Toutefois, il paraît qu'à l'état non-échangeable, entre les lamelles des minéraux, des quantités appréciables peuvent être emprisonnées: 5% de l'azote total dans la couche arable et jusqu'à 60% dans le sous-sol.

Le mécanisme de cette fixation est le même que celui du K et dépend donc des mêmes facteurs: illite > montmorillonite > kaolinite (Stevenson et Dhariwal 1959, Van Schreven 1963). Une autre partie se dénitrifie ou se volatilise.

Le fait que le sous-sol fixe beaucoup plus d'NH_3 que la couche arable n'implique pas nécessairement qu'il est moins assimilable (Legg et Allison 1959).

La quantité d'NH_3 fixée croît avec la finesse de la texture du sol et diminue avec un abaissement du pH (Nommik 1957).

Sa détermination offre cependant un intérêt spécial pour la nutrition des plantes puisque l'azote "soluble" du sol semble être constitué en partie appréciable par de l'N ammoniacal (Livens 1959).

Les données au sujet de la teneur en N ammoniacal des sols sont assez rares:

En sol à activité microbienne intense on n'en trouve que des traces; l'ammoniaque étant nitrifié au fur et à mesure de sa production.

En sol acide (forestier) ou submergé (rizières) les quantités sont plus abondantes et les plantes croissant dans ces conditions sont adaptées à l'assimilation ammoniacale plutôt que nitratée, (MATERNA 1964).

Le rétrécissement en profondeur du C/N dans le profil semble dû principalement à l'ammoniaque fixé (DHARIWAL et STEVENSON 1958), ainsi qu'à la teneur plus élevée en N de la matière organique du sous-sol; cet N est un constituant non-protéique de l'acide fulvonique (STEVENSON 1959).

Chez les sols du Ruhr, à surface spécifique, et une fixation de K relativement moinde, la fixation plus forte d'NH_4 est vue à la fixation relativent plus élevée des fractions plus grandes v.a.v. des fractions plus petites (BROEK 1964).

F. Le sodium

Le sodium se trouve à l'état *non-échangeable* dans tous les minéraux sodiques: albite: 4–8%, certains feldspaths potassiques: 3–7%, hornblende: 1–2%, muscovite: 0–2%.

A l'état *échangeable* et *soluble*, sa teneur dans les sols est généralement moindre que celle du potassium.

De tous les cations échangeables, l'énergie d'absorption du Na est la plus faible. Le Na disparaît donc facilement par lixiviation. Ce fait se trouve à la base de l'assainissement des terres submergées accidentellement par l'eau de mer: si le retrait des eaux marines est suivi par une période de pluie, la structure du sol se rétablit en l'espace de quelques mois, surtout après application d'un sel calcique soluble: $CaSO_4$, $CaCl_2$ (BAEYENS 1958).

4.000—8.000 Kgr. de gypse par Ha suffisent pour faire disparaître 30—70% du sodium d'un sol salin. La lixiviation du sodium peut être accélérée par la pluie ou par irrigation (DE BOIXO 1961).

La disparition du Na d'un sol sodique en régions arides et semi-arides, est un problème presque insoluble par manque de pluie ou d'eau d'irrigation. Le moyen classique: Ca SO_4 n'a que peu d'effet vu sa faible solubilité. Les seuls moyens, indirects, sont le mulch, le fumier organique et une forte fumure minérale azotée et phosphorique (LUKEN 1962).

Un autre moyen consiste dans l'introduction de variétés de plantes adaptées aux sols salins et alcalins et résistant à la sécheresse (BAEYENS 1963).

Vu la grande solubilité des sels sodiques, une accumulation de cet élément n'a lieu qu'en climat semi- et désertique: Na_2CO_3, Na_2SO_4, NaCl, ainsi que dans les sols submergés par l'eau de mer ou pulvérisés d'eau de mer par le vent: sel "cyclique" (EGNER et ERIKSON 1955).

Les sables contiennent moins de Na que les argiles.

A l'encontre du calcium, le sodium en excès peut détériorer la *structure* du sol, le rendant boueux à l'état humide et dur à l'état sec. Cet effet peptisant peut même se manifester après une ou plusieurs applications copieuses d'engrais sodiques comme le montre le Tableau 69.

Dans les plantes le Na peut remplacer une partie du potassium (coton) et provoque une activité accrue des protéines et des nitrates solubles dans les feuilles. D'autre part une carence en potassium en présence d'un excès de sodium provoque une accumulation d'acides ammivées dans les feuilles (JOHAM et AMIN 1965).

Le remède contre cette détérioration de la structure est la chaux mais le problème est assez compliqué du fait qu'une application répétée d'engrais sodiques basifie le sol, basicité qui est encore renforcée par l'emploi d'engrais calcaires.

Malgré son affinité chimique avec le potassium, le Na n'est pas un élément majeur pour les plantes. Sa nécessité comme élément mineur est affirmée par plusieurs auteurs quoiqu'il soit difficile d'expérimenter en absence totale de Na. Il est probablement un élément utile en remplaçant partiellement le K, aussi bien dans le sol que dans la plante. Son caractère d'élément mineur est mis en évidence par le fait qu'à hautes doses il devient facilement toxique, toxicité qui varie cependant avec l'espèce de plante (Lehr et Wybenga 1958, Bhardway 1960).

Il est utile à la betterave, moins au tabac (Service Pédologique de Belgique 1963).

Tableau 69. *Effet des engrais sodiques sur la diminution de la structure des sols cultivés*
Quantités d'engrais appliqués par are:

faible dose: 0,629 kg Na dans 1,600 kg NaCl-2,440 $NaNO_3$- 2,584 kg sylvinite
dose moyenne: 1,258 kg Na dans 3,200 kg NaCl-4,880 $NaNO_3$- 5,168 kg sylvinite
dose élevée: 2,515 kg Na dans 6,400 kg NaCl-9,760 $NaNO_3$-10,336 kg sylvinite

L'état structural du sol mesuré une année après l'application des engrais au moyen de la perméabilité (1) et par néphélométrie (2) (chiffres relatifs)

Engrais	Doses	Etat structural	
		(1)	(2)
NaCl	témoin	100	100
	dose faible	96	95
	dose moyenne	85	117
	forte dose	82	115
$NaNO_3$	témoin	100	100
	dose faible	89	123
	dose moyenne	93	129
	forte dose	79	158
sylvinite	témoin	100	100
	dose faible	98	122
	dose moyenne	93	153
	forte dose	87	171

2. Les anions

A. L'azote dans le sol

L'étude de l'azote dans le sol est un des problèmes les plus compliqués de la pédologie, du fait que ses tranformations sont essentiellement sous la dépendance de l'activité microbienne du sol qui, elle, dépend de son état physique et chimique: structure, texture, régime hydraulique, ainsi que du climat local: température et pluviosité pouvant changer de jour en jour et même d'heure en heure.

Comme ces transformations des formes chimiques de l'N: organique, semi-organique, ammoniacal, nitraté, se font principalement sous l'action des microbes, elles sont décrites dans le chapitre concernant la bactériologie du sol.

La migration à travers le sol, des ions NO_3 et NH_4^+ sous l'influence de la pluviosité a été étudiée par H. Linser (1959) qui a trouvé une corrélation intéressante entre cette migration et la valeur T du sol.

Origines et formes chimiques de l'N du sol. Les roches ne contiennent que des traces d'azote qui arrive dans le sol à partir de l'air, soit dissous dans l'eau de pluie (NH_3-nitrates) soit par voie microbienne *(Azotobacter-Rhizobium)*.

Les 99% de l'N du sol sont sous forme organique: azote hétérocyclique ou aminé des matières humiques.

Les 1–2% d'N minéral s'y trouve sous forme ammoniacale ou nitrique.

Cette faible teneur du sol en azote minéral — le seul directement assimilable—semble être en harmonie avec la physiologie azotée des plantes qui exigent une dose d'azote tout le long de leur période de végétation: cette dose doit être fournie par l'activité microbienne du sol ou par des applications successives de doses fractionnées d'engrais azotés (INGVAR et NILSSON 1959).

En outre, la pluis amène de l'N de l'air au sol surtout sous forme d'NH_3. En régions tempérées cette dose est faible (3—4 Kgr/Ha en Russie), plus élevée aux tropiques (BOBRITSKAYA 1962).

Facteurs de la variabilité de la teneur en N des sols. Cette teneur oscille entre 0,02 et 0,4% dans les sols cultivés. Elle est déterminée en premier lieu par le *climat local*: température et humidité.

La teneur diminue avec l'augmentation de la température et donc, toutes choses égales, diminue du pôle à l'équateur, à la suite de l'augmentation de l'activité minéralisante des microbes avec la température.

La teneur augmente avec l'humidité qui stimule la production de matière végétale jusqu'à un optimum au delà duquel l'excès d'humidité diminue l'activité minéralisante des microbes avec accumulation d'N organique: les sols organiques tourbeux contiennent jusqu'à 15% de N.

Comme les deux facteurs, température et humidité, agissent en sens contraire, il y aura interréaction entre les deux, exprimée par JENNY dans son équation classique; $N = 0{,}55\ e^{-0{,}08t})\ (I - e^{-0.005H})$ ou N est le % d'azote dans le sol superficiel, t la température moyenne annuelle et H le facteur d'humidité c.a.d. le rapport de la pluviosité annuelle au déficit de saturation absolue de l'air. Les constantes varient avec la végétation et l'espèce du sol.

La végétation, comme produit du climat local, influence la teneur en N du sol: une végétation permanente (prairie-forêt) produit plus d'humus et donc plus d'N dans le sol qu'une végétation temporaire (cultures annuelles), ceci à la suite de la constance relative du rapport C/N dont nous parlerons tout à l'heure.

Des particularités pédologiques et topographiques ont leur influence: les versants nord contiennent plus d'N que les versants sud, plus chauds.

L'érosion, en enlevant la terre humifère, appauvrit le haut de la pente en N qui s'accumule vers le bas (SCHEYS 1963).

L'argile est plus riche en N que le sable qui retient moins d'eau et de matière organique.

Si l'N total est assez variable, l'N minéral NH_4^+ et NO_3 l'est encore plus, puisque la minéralisation est conditionnée par le climat et suit la variabilité saisonnière de la température et de la pluie: à la fin de l'été, des sols nus, initialement pauvres en N, s'enrichissent (DELAS 1963).

La stabilité relative de l'N dans le sol — rapport C/N. Malgré la grande variabilité de la teneur en azote dans les sols en général, cette teneur dans un sol et sous un climat et végétation déterminés reste relativement stable.

La preuve en est que 1–2% seulement de cet N total se trouve sous forme minérale lixiviable ou assimilable par les plantes, alors que l'N organique ne l'est pas.

Une autre preuve est la constance relative du rapport C/N dans un sol déterminé, ce qui prouve en même temps que le cycle azoté d'un sol est intimement lié à celui du carbone organique (PRITCHETT et al. 1959).

La stabilité relative du rapport C/N explique le fait curieux que des applications répétées et à doses massives d'engrais, même organique, n'ont que peu d'influence sur ce rapport (EBERHARDT et KRZYSCH 1959, DOBRZANSKI 1959).

Les opinions des chercheurs au sujet de cette stabilité peuvent se classer en deux catégories: les thèses physico-chimiques et les thèses microbiologiques.

Suivant les physico-chimistes, l'N se fixe à la lignine sous forme de ligno-protéine stable (WAKSMAN) ou de combinaison entre l'ammoniaque et la lignine (MATTSON). D'autres supposent une absorption entre les minéraux argileux et les protéines (DYAL et ALLISON) ou entre les feuillets des minéraux et l'NH_3; une fixation semblable à celle du potassium, les deux ions ayant un rayon assez voisin.

Les microbiologues prétendent que cette stabilité apparente de l'N dans le sol est en réalité un équilibre dynamique entre la minéralisation de l'N organique avec formation d'NH_3 et nitrates et la "réorganisation" de ces produits par les microbes.

Ces problèmes sont exposés dans le chapitre de la microbiologie des sols.

Remarquons seulement ici que cette minéralisation et cette réorganisation sont liées au rapport C/N du sol. Ce rapport est relativement stable dans un sol, climat et végétation déterminés et explique donc la stabilité de la teneur en N organique de ce sol: si ce rapport a une tendance à s'élargir p. ex. par incorporation de matière glucidique, les microbes fixateurs d'N de l'air trouvent dans cette matière glucidique, si elle est facilement décomposable, un surcroît d'énergie qui stimule leur activité.

Au contraire, le rétrécissement de ce rapport stimule les minéralisateurs: production d'NH_3 et de nitrates lixiviés ou assimilés par les plantes.

Une autre preuve de la liaison entre le C et C/N du sol se trouve dans les expériences d'enfouissement de paille fraîche. Pendant les deux premiers mois, la végétation souffre de carence azotée. Après cette époque, au fur et à mesure que la paille s'humifie, son effet devient favorable pour atteindre un maximum vers six mois (CHAMINADE 1963).

Nous devons insister sur le caractère relatif et local de cette stabilité du rapport C/N:

En sol cultivé des régions tempérées il est de 10 — 12. — Au tropiques humides il est sensiblement moindre (4 — 8) vu la rapidité de décomposition de la matière glucidique (BAEYENS 1938). En sol à matière organique peu décomposable (lignine) il peut monter à 20 — 30: podzols de bruyères (COPPENET et HELIAS 1958).

Le bilan azoté du sol. La stabilité relative de l'N total d'un sol en conditions déterminées nous amène à parler de la question très controversée du bilan azoté d'un sol: ses pertes en N par exportation des récoltes, par lixiviation, par évaporation gazeuse et par érosion: ses gains par fixation microbienne symbiotique et non-symbiotique, par précipitations atmosphériques et par engrais.

Le cadre de cette étude ne permet pas l'exposé de cette question très compliquée et nous devons renvoyer le lecteur aux auteurs qui ont étudié ce domaine, p. ex. BLACK 1957.

Relevons seulement ici le fait curieux que toutes les études dans ce domaine ont abouti à un bilan nettement négatif et on ignore où est allé cet N perdu.

On peut incriminer l'imperfection du bilan qui ne permet pas de détecter toutes les "fuites" mais ces fuites sont telles (jusqu'à 150 kg N par Ha) qu'elles sont probablement réelles. Chose curieuse c'est que plus les apports d'N au sol sont élevés (engrais), plus le bilan est négatif. On suppose généralement que la perte principale se fait sous forme gazeuse: N, N_2O, NO, NH_3 difficile à mesurer (DEIWICH 1956, GREENWOOD 1962, VINE 1963). L'emploi de l'N^{15} a jeté quelque lumière sur les causes du bilan négatif azoté. Aussi l'avoine prend sensiblement plus d'N du N^*O_3, que celui de $N^* H_4^+$ qui reste dans le sol. La perte est due probablement à la dénitrification (JANSSON 1963).

L'assimilation de l'N du sol — et des engrais — se fait par la plante aux différents stades "critiques" de sa croissance. Ces stades critiques sont, pour les céréales, le tallage, la levée, l'épiaison et la floraison. Une application très tardive d'N à ces céréales peut doubler leur teneur en protéine qui se réalise surtout dans les cotylédons. L'allure de l'assimilation de l'N par les céréales peut donc avoir leur répercussion sur la qualité des produits, par exemple la valeur de panification de la farine (MICHALE 1962).

En forêt, une fumure moyenne, de préférence ammoniacole possède une haute valeur nutritive, puisque les conifères se nourissent en grande partie de l'azote organique de la litière moyennant les mychorrises (HOFFMAN 1964).

B. Le phosphore

A l'état *non-échangeable* il se trouve dans les minéraux et roches phosphatés: apatite, phosphorite, vivianite, ainsi que sous forme de phosphate de Fe et Al.

Sous forme organique il est présent dans les lécithines, acides nucléiques, phytines et dérivés.

Vu le caractère labile de ces composés, le P y est partiellement échangeable: il se minéralise sous l'influence de l'activité décomposante des microbes du sol.

Le P organique prédomine (parfois jusqu'à 90 % du P total) dans les sols cultivés ou portant une végétation herbacée (VAN DIEST et BLACK 1959).

Ce P organique se compose principalement de phytine combinée au Fe et Al de l'humus (KOSAKA et KAZUO ABE 1957). Il peut contenir jusque 17 % d'inosite-hexaphosphate (CALDWELL et BLACK 1958).

La teneur du sol en *P soluble* est d'ordinaire très minime vu sa grande fixabilité dans tous les sols mêmes perméables et sableux.

Donc un phosphate, même soluble, appliqué en couverture, reste dans les quelques centimètres superficiels du sol.

Si dans les sous-sols on trouve du P soluble il provient de l'excrétion des racines (KLOKE 1958).

Dans le podzol, le P s'accumule dans l'horizon B, riche en Fe et se transforme en $FePO_4$ peu soluble et peu assimilable par la plante. D'ailleurs cet horizon est souvent compact et empêche la pénétration des racines (SCHLICHTING 1962).

Dans les sols acides, il se trouve sous forme $H_2PO_4^-$ et dans les sols alcalins sous forme HPO_4^{--}. Le produit ionique: $Ca^{++} \times HPO_4^{--}$ en équilibre avec une solution, demeure constant sur toute une série de pH, de concentrations en ions Ca et d'activité ionique (COLE et OLSEN 1959).

Pour le même motif, le *P échangeable* ne représente qu'une faible fraction du P total. Son mécanisme d'échange est très compliqué et dépend, entr'autre, de l'activité des ions Al et Fe présents: précipitation du PO_4 Al et PO_4Fe insolubles. Aux faibles concentrations en P du liquide du sol, l'échange suit l'isotherme de FREUNDLICH. Aux concentrations élevées, l'adsorption est plus forte surtout si le rapport SiO_2/R_2O_3 est large.

Cette anomalie est expliquée par le remplacement superficiel des ions silicatés par les ions PO_4^{---} ainsi que par la formation de $(PO_4)_2$ Ca_3, le Ca sortant du complexe (KURON et GRÜNEBERG 1959).

Les recherches à l'aide du PO_4^{---} radioactif (P^{32}) ont prouvé que cet échange se fait en deux stades: un stade initial rapide et un stade final atteint plus lentement (WIKLANDER 1958).

Il y aurait donc 2 espèces de P échangeable: une forme facilement échangeable qui diminue rapidement par extraction, et une forme moins échangeable mais à échange plus constant (SHAPIRO et FRIED 1958).

On peut même sous-diviser ce phosphore "labile" en plusieurs sous-fractions: très facilement, moyennement et moins facilement échangeable, ces fractions étant influencées par la présence ou absence de substances organiques chélatantes ou non chélatantes: citrate, barbiturate (Arambarri et Talibudeen 1959).

La phase rapide correspond à la dilution isotopique des ions PO_4 retenue dans l'atmosphère ionique qui entoure les particules d'argile, la phase lente à la dilution des ions PO_4 adsorbés à la surface même des particules (Blanchet 1959).

La fixation du P dans le sol se fait de plusieurs manières: par précipitation chimique directe, en sol acide, sous forme de PO_4 Fe et Al; en sol basique sous forme de $(PO_4)_2 Ca_3$; par réactions d'échange ou de substitution à la surface et à l'intérieur du réseau cristallin; par immobilisation biologique sous forme organique; et enfin par le fluor présent qui précipite le P sous forme d'hydroxy-fluor-apatite. Il semble que tous les sols contiennent sufisamment de fluor actif pour réaliser, du moins en partie, cette précipitation (Jahn-Deesbach 1959).

L'emploi du P_{32} a prouvé que le phosphore précipité, probablement en couche mono-moléculaire à la surface des particules du sol, du calcaire et de l'apatite, reste en grande partie assimilable par les plantes.

Ce P "de surface" ou P "labile" provient surtout des engrais. Il augmente avec la dose appliquée et est élevé dans les sols contenant naturellement beaucoup de P.

Ce P de surface s'échange facilement avec P^{32} ajouté et maintient un potentiel d'équilibre constant dans la solution du sol (Singleton 1958, Nye et Forster 1958, Scott et Russell 1959).

Le P superficiel, en partie soluble dans l'eau, intervient pour 85 % dans le P assimilé par les plantes (Bouldin et Sample 1959).

L'assimilabilité du P du sol par les plantes dépend de plusieurs facteurs; du pH du sol en rapport avec la présence d'ions Ca, Fe et Al, de la présence de sels neutres, de l'état hydraulique du sol, de la matière organique, de l'activité du système radiculaire et des microbes et enfin de l'espèce d'engrais appliquée.

L'interréaction de tous ces facteurs est compliquée et encore imparfaitement connue (Lindsay et Stephenson 1959).

La précipitation du superphosphate dans le sol ne serait pas due en premier lieu à la présence du Fe et au Al, mais bien au K qui formerait le composé insoluble: $H_6 K_3 Al_5 (PO_4)_8 \cdot 18 H_2O$ (Lindsay et al. 1959).

Le pH a un effet différent suivant qu'il s'agit de sol ferrugineux ou calcareux.

En sol ferrugineux, la solubilité du P est minime au pH bas: 3—4: formation de PO_4 Fe et PO_4 Al insolubles; l'assimilabilité du P augmente avec l'augmentation du pH.

En sol calcareux cette assimilabilité est minimale aux environs de la neutralité [formation de $(PO_4)_2 Ca_3$] et augmente aussi bien du côté légèrement acide que basique.

Cette immobilisation du P serait due à la formation de cristaux de $Ca_3 (PO_4)_2$ qui servent de germes pour des précipitations ultérieures (Blanchet et al. 1963).

En sols contenant aussi bien du Fe que du Ca, le phénomène se complique du fait de l'interférence des courbes (Wiklander 1958).

Ces courbes expliquent aussi l'effet différent du chaulage: en sols neutres, il diminue l'assimilabilité du P présent: formation de $Ca_3 (PO_4)_2$ et l'augmente en sols acides: précipitation de $Fe(OH)_3$ et $Al(OH)_3$ insolubles.

Les latérites et podsols acides et contenant beaucoup de fer, seront donc des fixateurs énergiques du P: les premières applications d'engrais phosphatés seront si fortement retenues que les plantes n'en profiteront pas (Kauritschew 1958, Blanchet 1959).

Les sels neutres présents dans le sol diminuent l'assimilabilité des phosphates au dessus du point isoélectrique. MATTSON a expliqué cet effet déprimant en se basant sur la distribution des phosphates entre la couche électrique double et la solution du sol d'après l'équilibre DONNAN.

Ces sels neutres agissent d'une manière différente suivant la nature de leur cation: $Fe^{+++} > Ca^{++} > Mg^{++} > K^{+} > Na^{+}$. Les sols sodiques fixent moins le phosphate que les sols calciques (ARAMBARRI et TALIBUDEEN 1959).

La dessication prolongée du sol diminue l'assimilabilité de son P, probablement par transformation du Fe^{++} en Fe^{+++} avec précipitation du PO_4Fe. Son humectation permanente, p. ex. par submersion produit l'effet contraire.

Le drainage et l'irrigation d'un terrain peuvent donc avoir une répercussion sur son état phosphaté.

En sols tourbeux la dessication augmente ou diminue le P assimilable suivant le degré de décomposition de la matière organique (HOFFMANN et STEINFATT 1959).

La matière organique du sol a une influence favorable sur l'assimilabilité de son P, soit par complexation de cette matière avec Fe et Al qui ainsi ne peuvent précipiter le phosphate, soit par minéralisation du P organique sous l'activité microbienne.

L'incorporation d'engrais organiques augmentera donc le phosphore assimilable du sol (RADET 1957, WILLIAMS et al. 1958) et même un engrais phosphaté minéral, incorporé à une substance organique voit son effet utile augmenté (SCHARRER et KÜHN 1959). Même la tourbe brute et acide minéralise son P en l'espace de quelques mois (ARMI KAILA 1958).

Cette minéralisation du P organique semble être en relation avec le degré d'altération du sol (WALKER et ADAMS 1959).

Les chélates de fer, tout en empêchant le fer de se combiner aux phosphates, sont cependant assez labiles pour prévenir la maladie des plantes due à la carence en fer assimilable (MAJUNDER et DUNN 1959).

La preuve que le P organique est utilisable par les plantes réside dans le fait que sa simple extraction aqueuse fournit une quantité de P qui est en corrélation avec le niveau de la récolte (LARSEN et al. 1958).

L'auteur a trouvé que l'assimilabilité du P varie au cours de la période de végétation et est favorablement influencée par l'activité microbienne du sol (BAEYENS 1948).

Le système radiculaire exerce une action solubilisante sur le P du sol, aussi bien dans la rhizosphère que dans le sous-sol (KLOKE 1958) mais avec des différences marquées d'après les espèces: les légumineuses ont un pouvoir solubilisant prononcé et l'esparcette plus que la luzerne (SSASANOW et SSERGEJEMA 1958).

Les prairies avec leur rhizosphère très active (production de CO_2) utilisent mieux les phosphates bruts que les champs de culture (ULRICH 1959).

Enfin *les différents engrais phosphatés* se comportent différemment dans un même sol. Les recherches avec P^{32} ont établi l'ordre décroissant suivant: phosphate d' Am ou Na > superphosphate > métaphosphate de Ca > phosphate bicalcique > phosphate tricalcique.

La transformation d'une de ces formes dans l'autre peut être prévue en se basant sur le potentiel Ca (pH—0,5 p Ca) et en calculant sur lui l'enthalpie des transformations des phosphates en fonction du potentiel Ca. Si le potentiel Ca dépasse 4,5, il se forme des phosphates de Ca même en présence de $Fe(OH)_3$.

L'hydroxylapatite se forme de préférence, puis le phosphate octocalcique et le phosphate dicalcique. En absence de Fe, et aux potentiels Ca entre 3,4 et 4,5, il se forme surtout du phosphate dicalcique; en-dessous de 3,4 c'est presque exclusivement lui qui se forme (ULRICH 1963).

Cette série peut cependant différer suivant l'espèce de sol, surtout d'après son pH (Hagin 1958).

Puisque la durée de la demie-vie du P^{32} n'est que de 14 jours environ, il est difficile d'étudier au moyen des isotopes radioactifs l'*arrière-effet* des engrais phosphatés qui peut s'étendre sur plusieurs années.

On peut cependant mesurer indirectement cet arrière-effet, en comparant l'action du P ordinaire appliqué les années précédentes à celle du P^{32} appliqué l'année même.

Cet arrière-effet des phosphates diffère suivant la nature de l'engrais et le pH du sol mais généralement s'étend sur plusieurs années: Dans un essai 2 à 35 % seulement du P appliqué était utilisé la 1e année (Singleton 1958). Généralement il n'est que de 5 à 10 %, mais peut atteindre 20 % en vase de végétation où les conditions pédologiques sont optimales (Baeyens 1948).

Le cycle du P dans la nature: les roches phosphatées (apatites) s'altèrent lentement et leur P, moyennant les rivières, s'accumule dans l'eau de mer et y nourrit le plancton dont se nourrissent les poissons, qui eux sont mangés par les oiseaux marins dont les déjections forment le guano, qui par altération produit les phosphorites propres à l'industrie du superphosphate. Celui-ci retourne au sol pour nourrir les plantes dont se nourrissent les animaux et l'homme. L'apatite, contenant du Fl, se transforme ainsi en phosphorite qui n'en contient pas (Schreiber 1962).

L'usage du P^{32} s'est révélé intéressant dans le problème de la meilleure *façon d'épandre les engrais phosphatés:* mélangés au sol d'une manière homogène ou placés "en nids" à des distances et profondeurs déterminées des racines.

Des résultats surprenants ont été obtenus: le P^{32} enfoui superficiellement (10 cm) est assimilé au taux de 72 à 85 % du P total (Burton 1957).

Celui enfoui profondément (30 — 40 cm) ne l'est plus qu'au taux de 1 — 3 %.

De même le palmier à huile réagit le plus favorablement au P^{32} appliqué superficiellement (Broeshart 1959).

La solubilité du phosphate influence le developpement du système radiculaire qui est le plus devellopé dans les régions du sol où l'engrais phosphaté est le plus soluble (superphosphate) (Engelstad et Moreno 1965).

Vu la faible diffusibilité des phosphates en général, on avait conseillé jusqu'ici l'enfouissement profond pour obtenir un système radiculaire bien développé en profondeur. Les résultats récents sont intéressants pour la fumure des prairies où il n'y a pas moyen d'enfouir les engrais. D'ailleurs la prairie vit surtout de l'activité microbienne des quelques centimètres superficiels: la rhizosphère.

Dans d'autres essais avec P^{32}, la diffusion du superphosphate augmente avec la profondeur du labour, probablement à la suite de l'amélioration de la structure du sol et de son économie d'eau en rapport avec l'étalement des racines (Rid et Süss 1958, Süss 1958).

Le superphosphate "localisé" près des racines est plus actif que celui mélangé uniformement au sol (Mattingly et Wissowson 1958, Barbier 1957).

L'usage du P^{32} a jeté une certaine lumière sur le problème compliqué de *l'utilisation relative par les plantes du P du sol et du P des engrais*, problème insoluble par la chimie ordinaire.

Au début de leur période végétative, les jeunes plantes prennent plus de P des engrais que du sol; par après c'est l'inverse à la suite de l'étalement du système radiculaire exploitant le sol.

Cette allure générale varie cependant suivant l'espèce de plante et est en relation avec les différences de pénétration des racines.

Parfois les engrais phosphatés augmentent l'utilisation du P du sol, mais généralement la diminuent, les racines assimilant plus facilement le P des engrais que celui du sol (Mattingly 1957, Maleina 1958, Machold 1958).

Signalons en passant que l'emploi de Mg^{28} a montré que le pourcentage de Mg de l'engrais absorbé par la plante croît avec la dose d'engrais appliquée, mais que ce pourcentage diminue avec l'augmentation du Mg échangeable du sol (Brown et Wells 1958).

Par "*état phosphorique*" d'un sol on entend le rapport entre le P absorbé par la plante et celui de l'engrais appliqué au sol. Cet état est une expression du degré d'assimilabilité de l'engrais ou encore son coefficient de rendement utile.

Plusieurs formules ont été proposées p. ex.: $A = B\left(\frac{Sf-1}{Sp}\right)$ où A et B indiquent respectivement les quantités du P du sol et de l'engrais et Sf et Sp l'activité spécifique respectivement du P de l'engrais et de la plante.

Cet état phosphorique a été utilisé pour "calibrer" les procédés de détermination du besoin en P d'un sol, mais le P^{32} a prouvé que cet emploi doit se faire avec prudence et implique des restrictions assez nombreuses (Singleton 1958). Ce calibrage ne se justifie que pour une même catégorie de sols mais ne s'applique plus à des sols à caractéristiques pédologiques très différentes (Russell 1959).

L'emploi du P^{32} a permis de distinguer deux espèces de P assimilable dans le sol: une partie facilement, et une autre partie difficilement assimilable, celle-ci constituant l'"arrière-effet" des engrais phosphatés.

La partie facilement assimilable est aussi la plus diffusible surtout en sols organiques, tandis que l'autre partie reste dans les quelques centimètres superficiels (Sutton 1963).

Vu la forte fixation du P dans le sol et donc le faible coefficient utile de rendement des engrais phosphatés en général [5 — 10 % seulement l'année d'application (Baeyens 1948)], les essais d'applications directes du P sur les feuilles par pulvérisations semblent prometteurs: ce P a, d'après certains, une action au moins égale à celui du sol (Takahashi et Yoshida 1958, Naundorf 1959), alors que d'après d'autres il est moins efficace. Les résultats contradictoires semblent dus à une différence de technique expérimentale: pulvérisations sur les feuilles avec perte ou badigeonnage par pinceau qui évite les pertes (Kick et Hellwig 1959).

Alors que la granulation est généralement favorable pour les engrais azotés et potassiques, elle l'est moins pour les engrais phosphatés composés, si ce phosphore n'y est pas sous forme soluble (Kloke et Przemeck 1959).

Le P est un des facteurs qui détermine la *qualité* des produits agricoles: la teneur en protéine augmente avec la fumure phosphatée et cette augmentation suit la loi de Mitscherlich (voir Fig. 100) (Jürgens-Gschwind 1962).

De multiples recherches ont été exécutées dans le but de connaître la meilleure forme d'engrais phosphaté dans les différents sols. La tendance générale qui se dégage de ces recherches est que, si le sol est pédologiquement en ordre, tous les engrais phosphatés se valent: si le résultat immédiat n'est pas égal, celui à longue échéance (arrière-effet) est le même (Neergard 1962, Strobl 1962).

C. Le soufre

La teneur en soufre du sol varie de 0,01 à 1 %. Ce dernier chiffre se rencontre dans les régions arides où le S se concentre à la surface, souvent sous forme de sulfates. Aux environs des centres industriels jusque 260 kg par Ha et par année peuvent atteindre le sol avec la pluie (Gericke et Kurmies 1957). Il s'ensuit que dans ces mêmes régions les plantes sont plus riches en cet élément (McLEAN et VOLK 1958). Aux E.U. les chiffres oscillent entre 15 et 22 kgr par Ha et par année (JORDAN et BARDSLY 1958). En général l'hémisphère nord en contient plus que l'hémisphère sud. D'autre part, les pertes par lixiviation sont plus élevées (13 Kgr/Ha) que dans l'hémisphère sud (4 Kgr/Ha) (FRENEY et al. 1962).

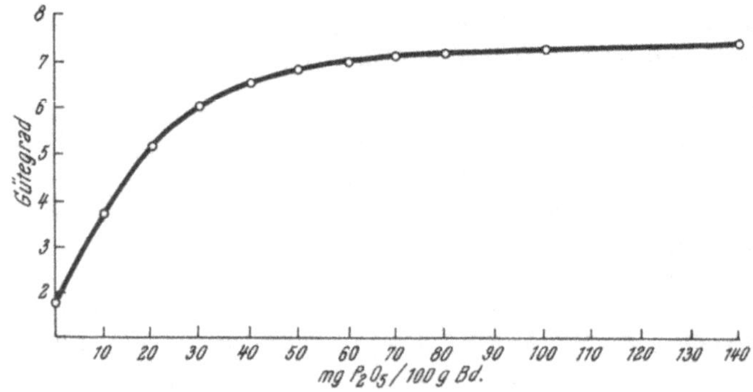

Fig. 100. Abhängigkeit des Futterwertes von Heu vom P-Zustand des Bodens (JÜRGENS-GSCHWIND 1962)

A l'état *non-échangeable* le S se trouve dans les minéraux sulfureux (pyrite-marcacite: FeS, chalcopyrite: $CuFeS_2$ et pyrotite) ainsi que sous forme organique (cystine-cystéine et protéines sulfurées) surtout dans les sols cultivés ou organiques où il constitue la majeure partie du S présent.

Le soufre *échangeable et soluble* du sol est le plus souvent sous forme de SO_4^{--}. Il y est faiblement absorbé et donc facilement lixiviable, mais son absorption croît avec abaissement du pH, ainsi que dans les sols riches en hydroxydes de Fe et Al.

Par année sèche, le gypse se solubilise difficilement mais présente alors un arrière-effet les années suivantes (WALKER et ADAMS 1958).

Vu sa lixiviabilité, le sulphate se trouve surtout dans le sous-sol et dans les sols argileux (NELLER 1959).

Le cycle du S dans le sol est sous l'influence de l'activité microbienne: oxydante (formation de SO_4^{--} à partir des minéraux ou soufre organique) et réductrice (formation de sulfures: FeS_nH_2O, H_2S, toxiques pour la plante, ou même de S élémentaire).

Puisque la racine ne prend le S que sous forme oxydée, la rapidité et les conditions de cette *oxydation* sont importantes pour la fertilité du sol puisque le soufre doit être considéré comme un élément nutritif majeur.

Suivant WIKLANDER l'oxydation du FeS dans le sol se passe en deux stades: un premier, rapide, se fait chimiquement:

$$FeS + H_2O + 1/2\, O_2 = Fe\,(OH)_2 + S - 36 \text{ cal};$$

le 2e stade se fait lentement et sous l'action des microbes sulfureux oxydants:

$$2\,S + 2\,H_2O + 3\,SO_2 = 2\,H_2SO_4 - 238 \text{ cal}.$$

Cette oxydation est fortement endothermique et, comme les bactéries sulfureuses n'ont pas de pigment chlorophyllien, l'énergie nécessaire à cette transformation doit venir d'une source extérieure, probablement des microbes hétérotrophes concomitants décomposant les glucides du sol.

Une autre caractéristique de cette oxydation est la production de H_2SO_4 avec abaissement marqué du pH.

L'oxydation sulfureuse, tout comme la nitrification, a une tendance à acidifier le sol qui donc doit être bien pourvu en cations échangeables, surtout en Ca, sinon l'activité microbienne se ralentit.

Au laboratoire on peut même atteindre des acidités toxiques: pH 2,9.

D'ailleurs entre le cycle biologique du S dans le sol et celui de l'N, il existe un parallélisme, non seulement en ce qui concerne les produits formés, mais aussi au point de vue des conditions du milieu : de ce parallélisme, il résulte qu'il n'est pas étonnant que dans le sol, il faut un rapport entre l'N et le S. En excès d'N et pénurie de S, le NO_3' n'est pas organisé et reste comme tel dans la plante : goût de "salpètre" des légumes (Johansson 1962).

D'autre part, une forte fumure en S pour abaisser les sols alcalins, diminue la minéralisation de l'N organique et favorise la lixiviation des bases: K, Ca, Mg et aussi des sulphates. L'acidification du sol par le S entraîne la fixation du P, sous forme de PO_4Fe et Al (Simon—Sylvestre et Boischot 1962).

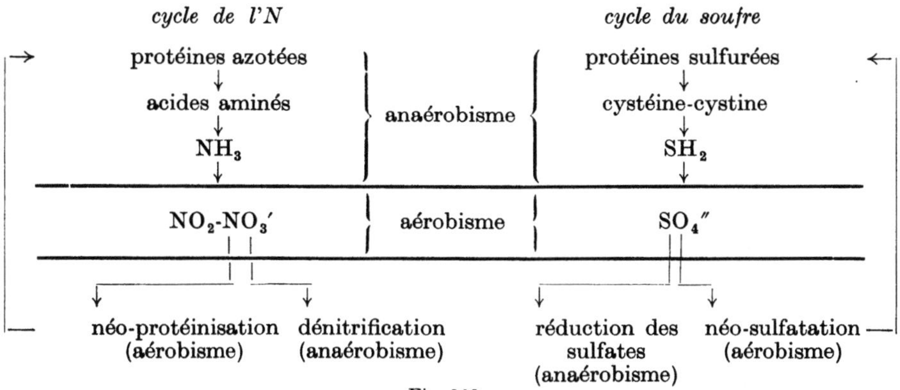

Fig. 101

On voit par ce tableau que le cycle du soufre a sensiblement la même allure que celui de l'N (Baeyens 1965).

Comme pour l'N, le bilan du S est difficile à établir avec exactitude: alors que l'exportation par la plante se détermine par son analyse, la lixiviation du S par

Tableau 70. *Le soufre de l'air et des précipitations en Europe* (Johansson 1962)

	Atmosphère	précipitations	Total
Suède	28	35	63
Norvège	4	12	16
Finlande	6	6	12
Danemark	3	13	16
Islande	1	1	2
Angleterre	7	16	23
Europe Centrale	29	29	58

On voit que le S arrivant au sol est très variable suivant les pays.

les eaux de drainage, et surtout l'apport de S par les eaux des pluies, sont presque impossibles à mesurer (voir Tableau 70). Voici un essai de bilan pour les sols hollandais:

Gains		*Pertes*	
Pluie:	16 Kgr./Ha/année	Exportation:	15 Kgr./Ha/année
Engrais:	15 Kgr./Ha/année	Lixiviation:	30 Kgr./Ha/année
Atmosphère:	19 Kgr./Ha/année		
	50 Kgr./Ha/année		45 Kgr./Ha/année

On voit donc que, contrairement au bilan de l'N qui est toujours négatif, celui du S est positif. Ceci explique que dans la pratique, le besoin en S des sols est moins élevé que par exemple celui du phosphore (Rosanow 1960).

Puisque l'oxydation du soufre acidifie le sol, le S peut être employé pour abaisser le pH des sols trop alcalins. Les conditions de cet abaissement, importantes pour les contrées arides et les sols sujets à certaines maladies de carence liées au pH (Mn) ont été bien étudiées.

Ce sont en général celles qui favorisent l'oxydation du soufre, chimiquement ou microbiologiquement.

Comme cette oxydation se fait, du moins en partie, sous l'influence de microbes autotrophes et aérobies (en symbiose avec certains microbes hétérotrophes, décomposant l'humus avec libération d'énergie), il faut dans le sol une humidité, une aération, un pH, une température et une teneur en matière organique appropriés.

Le soufre dans le sol y est sous forme de polysulphides qui, sous l'influence de bactéries oxydantes, se transforme en S élémentaire, puis en SO_4'' (Hart 1963).

La dose de S nécessaire pour obtenir un abaissement de pH déterminé variera donc avec l'importance relative des facteurs.

Le soufre est le remède indiqué pour guérir une carence en Mn due à l'excès d'alcalinité du sol (Stenuit et Piot 1958).

Le cycle du S dans le sol est en rapport avec son cycle azoté: en pénurie de S, la teneur en amides des plantes croît anormalement alors que l'élaboration des glucides diminue (Rendig et McComb 1959).

La réduction du sulfate sous l'influence des bactéries désulfurantes qui sont hétérotrophes, n'a lieu que dans un sol à potentiel d'oxydo-réduction peu élevé, en général impropre à l'agriculture: excès d'humidité, pH trop bas, anaérobisme.

Dans ces conditions, le S peut s'y accumuler sous forme de FS ou H_2S, qui, après drainage et labour, fournit le H_2SO_4 très toxique pour les plantes avec abaissement anormal du pH.

Un chaulage préalable s'impose avant tout essai de culture. H_2S peut cependant dans ces sols solubiliser les phosphates minéraux présents (Sperber 1958).

Le soufre doit être considéré comme un élément majeur pour les plantes au même titre que le P: certaines familles (légumineuses) contiennent autant et même plus de S que de P. Il est surtout important dans l'association de graminées et trèfles où, en cas de pénurie de soufre, l'engrais azoté fait pousser l'herbe au détriment des trèfles (Walker et Adams 1958).

Malgré son caractère d'élément majeur, les engrais à base de soufre ont pour beaucoup de contrées une importance moindre que les engrais phosphatés, ceci probablement à la suite de l'enrichissement du sol en soufre d'origine atmosphérique: 40–45 Kgr S./Ha en Allemagne occidentale alors que dans cette contrée la lixiviation n'est que de 30 Kgr et l'exportation par les récoltes de 13 Kgr (Buchner 1958).

Dans les cas d'inondations des polders par l'eau de mer, par exemple rupture des digues, le $CaSO_4$ s'est montré très efficace pour remettre en ordre ces sols sodiques (VAN BEEKOM 1961).

Le soufre du sol et des engrais présente une importance particulière pour les plantes oléagineuses. Sans aborder le rôle du soufre dans ces plantes, domaine de la physiologie végétale, nous pouvons noter cependant que pour ces plantes, le S est aussi important que l'N et le P surtout en sols pauvres en Magnésium. Le rôle de celui-ci en rapport avec les autres éléments est peu connu (VENEMA 1963).

D. Le carbone

Le carbone, sous forme de CO_2 ou CO_3H^-, est un anion directement assimilable par les racines comme le SO_4^{--} et le Cl^-. Cette assimilation se fait au moyen d'un enzyme qui carboxylise le phospho-enolpyruvate avec formation d'acide oxalo-acétique et libération du phosphate minéral dont la quantité produite est liée à la quantité de CO_2 absorbé (JACKSON et COLEMAN 1959).

Signalons à ce propos l'intérêt que pourrait présenter l'emploi d'engrais carbonatés (CO_3KH) à réaction alcaline qui dégagent dans l'atmosphère du sol le CO_2 avec solubilisation des principes minéraux (AUDIDIER 1959).

Toutefois, d'après les recherches récentes, la teneur en CO_2 du sol, beaucoup plus grande que celle de l'air, suffirait pour nourrir les racines (STEMMET et al. 1962).

3. Les éléménts mineurs, éléments-trace, micro-éléments

La *définition* de ces éléments, tant cations qu'anions est assez difficile à formuler: alors que les éléments majeurs: N, P, S, K, Mg, Ca, etc., peuvent être définis comme étant indispensables à la plante en quantités considérables, l'élément mineur peut être nécessaire ou non ou simplement utile et présent dans la plante à l'état de trace seulement ou en quantités parfois fortes (Al, Si).

Une caractéristique assez constante des éléments mineurs, même pour ceux réputés nécessaires, est leur toxicité dès qu'une dose, généralement minime, est dépassée, alors que les éléments majeurs peuvent être fournis à la plante en doses massives sans préjudice pour elle.

Les deux espèces d'éléments répondent à la loi de croissance de MITSCHERLICH mais, vue la toxicité des éléments mineurs déjà à dose faible, l'allure des courbes est différente comme le montre le diagramme Fig. 102 (CHARLIERS et CARPENTIER 1959).

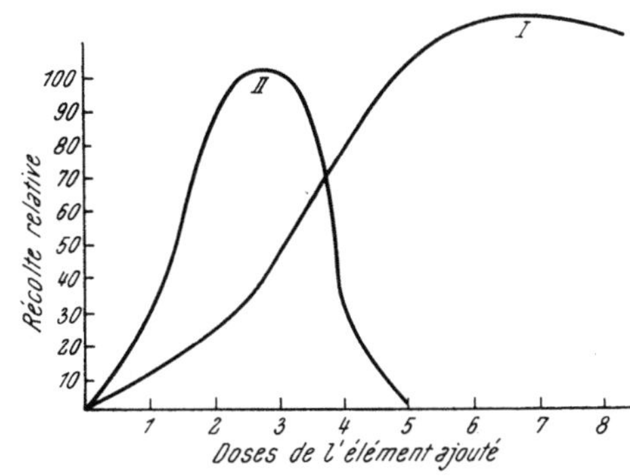

Fig. 102. La loi de croissance de MITSCHERLICH appliquée aux éléments majeurs (*I*) et mineurs (*II*) (CHARLIERS 1959)

Pour que cette toxicité se manifeste, l'élément mineur doit être présent dans le sol à l'état assimilable par la plante, assimilabilité qui dépend des conditions du milieu, comme par ex. du pH, et non de la dose totale présente.

Ainsi, on n'a constaté que peu de cas d'intoxication à la suite de pulvérisations, même répétées, de pesticides cupriques ou arsénicaux qui s'immobilisent facilement dans le sol. D'ailleurs, l'assimilabilité d'un élément mineur semble régie par les mêmes lois d'échange que celles des majeurs.

La toxicité d'un élément mineur sera donc diminuée par la présence, dans le liquide du sol, d'autres éléments mineurs antagonistes.

Des cas de toxicité dus à la présence d'un excès unilatéral d'un élément mineur assimilable ont été constatés juqu'ici avec: Bo, Fe, Cu, Mn, Zn, Mo, Al.

D'autre part, beaucoup de sols peuvent contenir un élément mineur nécessaire en quantité considérable, mais peu assimilable. Dans ce cas les plantes peuvent souffrir d'une carence en cet élément.

Dans ce domaine, l'emploi généralisé de chélates, livrant facilement à la plante l'élément à l'état assimilable, peut rendre des services à l'agriculture.

Les engrais organiques (fumier de ferme) jouent probablement le même rôle (ATKINSON et al. 1958).

L'effet du fumier de ferme est cependant spécifique: Dans une expérience il diminua l'assimilabilité du Zn et de Fe en augmentant celle du Mn (MILLER et OHLROGGE 1958).

La teneur des sols en éléments mineurs est évidemment très variable. Malgré les nombreuses études locales, nous manquons d'une vue d'ensemble dans ce domaine.

En Finlande cette teneur semble liée à la texture du sol: les argiles en contiennent plus que les limons lourds et ceux-ci plus que les limons légers. Les sols tourbeux sont les moins bien pourvus (VUORINEN 1958).

En U.R.S.S., la variabilité de la teneur en éléments mineurs est liée aux grands groupes de sols: podzols, terres noires, sols bruns forestiers et aux régions géographiques (PEIVE 1959).

En Australie du Sud (terra rossa et rendzines), les éléments mineurs sont assez uniformément répartis à travers le profil, sauf le Cu et le Mn qui s'accumulent en surface (MCKENZIE 1959).

En Angleterre, les sols de Rothamsted sont sensiblement plus riches que ceux de Woburn à proximité (Report Rothamsted 1959).

En Espagne, la grande variabilité locale de la teneur en éléments-trace, est due à la variabilité du climat (pluvieux ou aride) et de la roche-mère: calcareuse ou non (DEL RIVIERO 1959).

L'utilité ou la nécessité d'un élément mineur réputé nécessaire n'est que relative et dépend de nos moyens d'investigation: à mesure que nos méthodes de diagnostic se raffinent, des éléments réputés jadis superflus, sont reconnus actuellement nécessaires (Cl): les dernières découvertes dans ce domaine ont été obtenues, non pas par des champs d'essai, mais par des recherches de laboratoire en milieu de culture purifié à l'extrême et moyennant des méthodes de diagnostic raffinées: méthodes microchimiques, chromatographiques et isotopiques (ARNON 1958).

La technique ordinaire du champ d'expérience est trop "grossière" pour révéler le caractère nécessaire ou utile d'un élément mineur, souvent présent dans le sol à doses infinitésimales.

Les cas de maladies de carence dues à l'absence d'un élément mineur nécessaire semblent devenir de plus en plus fréquents.

Il est possible que cette fréquence ait existé de tout temps, mais soit restée inaperçue faute de moyens d'investigation suffisamment précis.

Ces carences peuvent aussi être la conséquence de l'emploi généralisé d'engrais chimiques synthétiques trop "purs", augmentant l'exportation des éléments mineurs mais ne les retournant plus au sol comme le faisaient jadis les engrais plus "impurs": nitrate du Chili, guano, kaïnite, sylvinite, superphosphate et scories de déphosphoration. Une application de 600 Kgr/Ha de ce dernier engrais apporte au sol: 270 Kgr. CaO, 25 Kgr. de Mn, 12 Kgr. de MgO, 2,4 Kgr. de V, 1,8 Kgr. Cr, 1,8 Kgr. Ti, 20 gr. Cu, 20 gr. B, 2 gr. Co.

Par ces doses, le besoin en Co est couvert à 100%, en Cu à 40% et en Bore à 25%.

Aussi longtemps que l'industrie continuera à livrer à l'agriculture des quantités actuelles de scories, les sols seront immunisés contre les maladies de carence (ATANASIU 1962).

Un problème spécial conçernant les éléments mineurs est soulevé par les *éléments mineurs radioactifs*, répandus dans l'atmosphère à la suite de décharges nucléaires et se concentrant dans le sol par précipitations atmosphériques ou par déchets provenant d'usines nucléaires: Sr^{90}, Cs^{137}, Ru^{106}, Ce^{144}, Y^{91}, Zr^{95}, Nb^{95}, etc.

Comme certains de ces isotopes ont une demi-vie assez longue, leur influence sur les plantes et l'organisme humain doit être envisagée.

Le Sr^{90} et le Cs^{137} s'accumulent spécialement dans le sol. Ils peuvent cependant être immobilisés par le Ca, le K et les engrais organiques (GRAHAM 1958, GULJAKIN et JUDINZEMA 1958).

La radioactivité du sol croît d'année en année. A Argonne, elle a augmenté entre 1950 et 1957 de 0,5 M E V pour Ba^{140}, La^{147}, Nd^{174}, Ru^{173} et Rd^{106}, de 0,75 M E V pour $Zr/^{5}$ et Nb^{15} et de 1,6 M E V pour La^{140}.

La conséquence est que, par exemple, la radioactivité de l'herbe augmente d'une façon inquiétante: les premières années à la suite de l'absorption directe par les feuilles des émanations atmosphériques des usines, les années suivantes par accumulation de ces éléments dans le sol (KLOKE et LUDWIEG 1962).

Ces éléments radioactifs sont facilement fixés par le sol et s'accumulent donc dans les couches superficielles (GUSTAFSON et al. 1958).

Cette rétention dépend peu du pH dans les limites ordinaires: 4–8 (RHODES 1957).

Dans un essai, après 9 récoltes successives de trèfle, 24% du Sr^{90} ajouté au sol et seulement 0,7% du Cs^{137} furent assimilés (NISHITA et al. 1958).

Les propriétés d'échange de minéraux argileux sont utilisées actuellement pour immobiliser les déchets radioactifs de l'industrie nucléaire.

Le degré de "désactivation" de ces ions radioactifs dépend du pH et de la mobilité de ces ions dans le sol (AMPHLETT 1958). Il dépend aussi de la quantité d'ion inactif présent dans le liquide du sol: l'addition de traces de Cs entraîneur diminue fortement la fraction de Cs radioactif retenue par le sol (BARBIER et DUVAL 1958).

Les moyens pour se débarrasser de ces substances nocives sont l'application de Ca et K, lixiviation avec des sels calcaires solubles, complexation avec le tartrate de K, le citrate d'Am, éthylène diamine, acétate sodique et autres substances organiques qui par décomposition libèrent le Sr^{90} et favorise sa lixiviation (COHEN et GAILLEDREAU 1961).

Remarquons que *tout* sol présente une faible radioactivité due à la présence d'éléments radioactifs naturels: U, Th, Ra, K^{40}, C^{14} etc. ou provoquée par rayonnement cosmique et solaire. Elle oscille entre 45 et 132 imp/min/Kgr. sol. 10^{-3} (JASTREBOW 1958).

L'influence sur les plantes de ces éléments radioactifs est encore incomplètement connue. Parfois ils stimulent, parfois ils diminuent la croissance (RODRIGUEZ 1959).

Un résumé critique de l'énorme bibliographie des éléments-trace se trouve dans l'ouvrage récent de SWAINE (1962), ainsi que dans le numéro spécial (16) de la "Landwirtschaftliche Forschung" — Frankfurt (1962).

Un aperçu de la teneur en éléments-traces des principales plantes agricoles en rapport avec la fumure se trouve dans SCHMITT (1962).

A. Le fer

A l'état non-échangeable, le Fe, principalement sous forme ferreux, se trouve dans les roches primaires: augite, olivine, hornblende et biotite; dans les minéraux silicatés secondaires: montmorillonite, illite, vermiculite; sous forme d'hydroxydes: limonite, goethite, lépidocrite; sous forme de carbonates: carbonate neutre, $FeCO_3$ peu soluble et bicarbonate $FeH_2(CO_3)_2$ assez soluble; sous forme de sulfures: pyrite (FeS_2), chalcopyrite ($CuFeS_2$), melnikovite ($FeS_2 \cdot nH_2O$) et enfin sous forme de différents oxydes anhydres: FeO, Fe_2O_3, Fe_3O_4 etc.

La solubilité et échangeabilité dans le sol sont régies par l'équation d'équilibre: $Fe^{++} \rightleftarrows Fe^{+++} + e^-$ à laquelle s'applique la loi de NERNST:

$$E = 0{,}77 + 0{,}059 \log\left(\frac{Fe^{+++}}{Fe^{++}}\right).$$

On voit que le sens $\rightarrow$ augmente avec l'élévation du pH et du potentiel d'oxydo-réduction, celui-ci dépendant de la pression d'oxygène dans le sol.

Le chaulage favorisera donc le sens $\rightarrow$ aussi bien en relevant le pH qu'en mettant le sol en meilleure structure grumeleuse, assurant ainsi une meilleure aération.

Par contre, dans les sols compacts, mal aérés et submergés, le sens $\leftarrow$ prédominera.

L'action du calcaire comme agent provocateur de la chlorose en fer est compliquée et imparfaitement élucidée: les récentes recherches au moyen du Fe^{59} et du Ca^{40} ont révélé que ce n'est pas en premier lieu le pH ni la concentration en Ca^{++} qui est en jeu, mais bien l'augmentation des ions CO_3H^- à la suite de l'hydrolyse du calcaire en sol humide. L'augmentation du bicarbonate déprime la respiration des racines et affecte le système de l'oxydase du cytochrome (THORNE et WIEBE 1957).

L'addition au sol de phosphate avec formation de $FePO_4$ insoluble, aussi bien dans le sol que dans la plante, favorise la chlorose en Fe (BROWN et al. 1959).

Les expériences avec du Fe^{59}, par ex. par injection dans la tige, ont prouvé que cet élément réputé "paresseux" peut devenir très mobile si une immobilisation accidentelle par un ion précipitant est évitée (BROWN et HOLMES 1955).

L'incorporation d'un ion Fe^{++} à un sol ayant tendance à s'oxyder, n'a souvent pas d'effet puisqu'il est rapidement transformé en Fe^{+++} avant que les plantes aient pu l'utiliser.

Dans ce cas, il vaut mieux pulvériser une solution de sel ferreux sur les feuilles.

Toutefois, même en sol acide, la chlorose en fer peut se manifester, mais alors pour une cause différente et notamment à la suite d'un antagonisme avec l'ion Mn^{++} qui est parfois en excès dans ces sols acides.

Cet antagonisme est compliqué et imparfaitement élucidé: les expériences au moyen du fer marqué montrent qu'en présence d'un excès de Mn, les plantes absorbent parfois beaucoup de Fe.

Dans d'autres cas c'est le contraire (TWYMAN 1958).

Enfin, un excès de Cu, Ni ou Co peut provoquer la carence en fer (HOLLEY et CAIN 1955).

Cette carence peut avoir une signification pratique dans le cas d'applications répétées de pesticides cupriques quoique, comme nous le verrons au chapitre du Cu, celui-ci est généralement fortement fixé par le sol.

La matière organique du sol immobilise temporairement le fer sous forme de chélates, qui cependant se décomposent facilement sous l'action des microbes.

Les chélates de fer artificiels sont très efficaces pour guérir la carence en fer même en sols calcareux (KROLL 1957).

L'assimilabilité du Fe dans un chélate diffère cependant suivant la nature du chélate: DTTA, ATA ou EDTA (De Kock 1957).

Les expériences avec chélates marqués par C^{14} et N^{15} prouvent que cette différence d'action tient à leur décomposition plus ou moins facile par la plante, l'EDTA étant le moins décomposable et ayant donc l'effet le moins prononcé en sol calcareux (Wallace 1955).

Plus récemment, ce même auteur a essayé un chélate de Fe qui se laisse libérer 100% du sol employé par simple filtration (Wallace 1956).

A la suite des différents résultats obtenus avec les divers chélates, on les a classés en deux catégories: ceux à noyau aliphatique (Fe — RA — 155 ou Fe — ATA, qui soutirent le fer de la solution par sorption de toute la molécule de chélate à l'argile ou par échange du Fe contre le Ca) et ceux à noyau phénolique (Fe — RA — 156) où cet échange avec le Ca n'a pas lieu (Hill et al. 1958).

D'autres admettent un échange du Fe du chélate avec l'Al de l'argile ou encore une précipitation du Fe par les silicates (Hemwall 1958).

Le fer peut être déplacé de ses complexes organiques (Fe EDTA et Fe HEE DTA) par d'autres métaux chélatants: Cu, Zn qui eux deviennent ainsi temporairement inassimilables. Ce fait pourrait expliquer les mobilisations et immobilisations de ces éléments dans le sol (Guinn et Johan 1963).

Notons enfin que les transformations: $Fe^{++} \rightleftarrows Fe^{+++}$ peuvent se faire dans le sol, non seulement par voie simplement chimique, mais aussi par voie microbienne: il existe dans le sol une flore ferrugineuse oxydante (*Leptotrix*, *Gallionella*) et réductrice:

$$FeCO_3 = CO_2 + H_2O \rightleftarrows FeH_2(CO_2)_2.$$

La distribution du Fe à travers les horizons d'un profil dépend des conditions d'oxy-réduction de ce profil: en général le Fe^{+++} diminue en profondeur avec augmentation concomitante de Fe^{++}.

En sols perméables et acides, le Fe migre jusqu'à l'horizon B du podzol. Des concrétions de Fe se trouvent dans l'horizon gley (oscillations de la nappe phréatique) et dans les latérites où il est libéré sous forme d'hydroxydes et d'oxydes à la suite de l'altération du minéral silicaté.

La présence du fer soluble a une influence marquée sur la fixation des phosphates dans le sol (cfr. article du phosphore).

B. Le manganèse

A l'état non-échangeable il se trouve dans la biotite (jusque 1%), dans la hornblende et l'augite (0,3–0,4%), dans la muscovite (0,1%), dans certaines roches calcaires et dolomitiques, mais surtout sous forme de différents oxydes, amorphes ou cristallisés, anhydres ou hydratés, et souvent sous forme manganique: MnO_2, Mn_2O_3, Mn_3O_4, $MnO_2 \cdot nH_2O$.

A l'état échangeable et soluble il se comporte comme le Fe: l'ion Mn^{++} étant soluble et assimilable, les ions Mn^{+++} et Mn^{++++} l'étant beaucoup moins.

A l'état soluble il ne constitue qu'une très faible portion du Mn total dans les sols à pH et potentiel d'oxy-réduction élevés: ce seront ces sols qui seront sujets à la maladie de carence.

On a même pu délimiter statistiquement, pour une catégorie déterminée de sol, des zônes de pH en dessous desquelles ($<5{,}5$) la maladie ne se manifeste pas et au-dessus desquelles ($>7{,}0$) elle est fréquente (Stenuit et al. 1957).

Par contre, en cas de réduction et en conditions très acides, l'accumulation du Mn^{++} peut devenir toxique pour la plante (Bussler 1958).

Les teneurs du sol en Mn échangeable varieront donc non seulement suivant les conditions de ce sol: 1 à 4.000 p.p.m. (WINOGRADOW 1954), mais aussi suivant le réactif employé et le temps de conservation de l'échantillon (STENUIT et VAN HOECK 1957). Du fait que les oxydes de Mn supérieurs: MnO_2 et ses variantes pyrolusite, manganite, hausmannite etc. sont insolubles, on avait conclu à leur non-assimilabilité. Il paraît maintenant que tous sont assimilables à des degrés divers. Ils ne sont pas solubilisés par les microbes puisque le sol manganique stérilisé reste disponible pour la plante (PAGE 1963).

D'ailleurs, les résultats de la détermination du Mn échangeable varient avec le cation déplaçant plus que pour les éléments majeurs échangeables.

Voilà pourquoi la détermination du Mn échangeable ne permet de prévoir une carence en Mn que si celui-ci est inférieur à 1 p. p.m.avec pH>9.

Au dessus de 1 p.p.m. il est préférable de doser le Mn réductible par l'hydroquinone (voir plus loin, détermination du Mn) (COPPENET 1959).

La maladie de carence en Mn est fréquente en sols organiques: ceux-ci sont souvent pauvres en Mn assimilable, probablement par formation de complexes temporairement insolubles.

Il faut cependant distinguer entre la matière organique acide qui augmente la teneur en Mn réductible, et la matière organique alcaline qui la diminue.

Alors que les chélates augmentent l'assimilabilité du Fe, ils semblent diminuer celle du Mn (THORNE et WIEBE 1957).

La présence de beaucoup de fer soluble peut inhiber l'assimilation du Mn (antagonisme ionique).

La relation Fe/Mn est compliquée et incomplètement résolue. Suivant l'état énergétique du rédox:

$$Fe^{++} + Mn^{+++} \rightleftarrows Fe^{+++} + Mn^{++}$$

il peut y avoir antagonisme ou synergisme. En outre, la présence d'autres éléments, P, Ca, Mo, B peuvent influencer ce rédox (DE KOCK et INKSON 1962).

Les échantillons de sols, conservés longtemps à haute température et à l'état sec, voient augmenter leur teneur en Mn échangeable.

Ce fait curieux est assez difficile à expliquer puisque par dessication et aération subséquente, le Mn^{++} doit se transformer en Mn^{+++} (KOUTLER et ANDERSSON 1956).

Cet effet de dessication est le plus prononcé en sol organique calcareux (SANCHEZ et KAMPRATH 1959).

La sensibilité des plantes vis-à-vis d'une carence en Mn diffère avec l'espèce: avoine>froment>orge>seigle (COIC et COPPENET 1958).

Alors qu'une toxicité du sol par excès de Mn^{++} se guérit facilement par élévation du pH et chaulage, sa carence en Mn assimilable est plus difficile à corriger.

Il faut employer une substance acidifiante (soufre qui se transforme en SO_4^{--}) qui ne nuit pas aux plantes ni aux autres propriétés du sol, ou traiter les semences au moyen d'une solution de $MnSO_4$ (STENUIT et PIOT 1957, STENUIT 1962).

Emploi d'engrais contenant du Mn: scories, superphosphate. La comparaison de ces deux engrais, au point de vue remède pour la carence en Mn du sol, est rendu compliquée du fait que les scories ont une réaction basique et le super une réaction acide. Or, nous savons que l'apparition de la carence est favorisée par la basicité et atténuée par la réaction acide.

Le soufre a un effet et arrière-effet, durant quatre ans, beaucoup plus marqué que l'application au sol de $MgSO_4$ ou d'engrais acidifiants (STENUIT 1962, 1963).

Contrairement à ce qui arrive avec le fer, l'addition d'un phosphate soluble semble augmenter le Mn assimilable du sol.

Comme pour le fer, les transformations du Mn dans le sol se font aussi bien par voie simplement chimique que par voie microbienne: il existe dans le sol, suivant les conditions, une flore manganeuse oxydante ou réductrice.

En outre, certains microbes utilisent le Mn pour l'édification de leur protoplasme et le soustraient, du moins temporairement, aux racines des plantes, qui elles semblent excréter des substances solubilisantes pour le Mn: ces substances deviennent inactives avec l'augmentation du pH (BROMFIELD 1958).

Les localisations caractéristiques du Mn dans les tissus des plantes peuvent très bien être mises en évidence au moyen de radiogrammes de feuilles traitées avec Mn^{54}.

A la suite de ce qui précède on peut attendre des déficiences en Mn dans les sols à pH élevé, calcareux, riches en matière organique, donc dans les tourbes calcareuses submergées et podzols fortement chaulés.

Une toxicité peut être attendue dans les sols très acides.

C. Le cuivre

Le cuivre *dans le sol* provient des minéraux à base de cuivre: chalcopyrite: $CuFeS_2$, carbonates, phosphates et sulfures. Puisque le Cu est fortement fixé dans le sol, *sa teneur* dépend de celle de la roche-mère, et varie donc considérablement d'un sol à l'autre: les extrêmes trouvés sont de 2 à 180 p.p.m.

Accidentellement, à la suite de pulvérisations répétées de fongicides cupriques (vignobles), cette teneur peut monter jusqu'à près de 1.000 p.p.m. mais alors seulement dans la couche superficielle. Comme il est fortement fixé par le sol les plantes n'en souffrent pas.

Le cuivre non-échangeable se trouve dans les minéraux cupriques mais surtout dans la matière organique qui fixe le cuivre d'une manière exceptionnellement forte sous forme de complexes organiques (MANSKAJA et al. 1958).

Par conséquent, la teneur en cuivre *échangeable et soluble* sera faible (1% du Cu total) et l'échange se fait difficilement avec les échangeurs ordinaires, alors que les acides et les solutions complexantes (cyanures, acides oxalique et nitrique) en extraient de grandes quantités.

Comme pour les autres cations, la capacité d'échange du sol pour le cuivre augmente avec son pH et l'échange suit l'isotherme de FREUNDLICH (MULLER 1959).

La mobilisation du cuivre dans le sol se fait chimiquement (par oxydation) ou biologiquement par les bactéries sulfureuses et par l'activité solubilisante des racines.

Les facteurs qui influencent la mobilisation du cuivre dans le sol, et donc son assimilabilité, sont en premier lieu sa teneur et la nature de la matière organique qui fixe le Cu, ensuite le pH et enfin les conditions climatiques. Les périodes sèches favorisent la maladie de carence.

L'humus "doux" qui minéralise vite, libère suffisamment de cuivre alors que l'humus "brut" le retient et provoque des maladies de carence (maladie de défrichement dans les sols de bruyères).

La présence de carbonates solubles favorise la libération du Cu de ses complexes organiques (TOBIA et HANNA 1958).

La déficience en cuivre se rencontre surtout en sols sableux, podzoliques et calcareux.

La racine assimile de préférence le Cu chelaté qui peut être absorbé en doses plus grandes que le $CuSO_4$ avant de devenir toxique. Le Ca diminue l'assimilabilité du Cu, puisque le Cu est le plus assimilable en milieu acide (BERINGER 1963).

D. Le zinc

Le zinc présente *dans le sol* certaines ressemblances avec le cuivre. Il s'y trouve sous forme minérale dans les minéraux ferro-magnésiens: augite, hornblende, biotite et aussi sous forme de Zn S.

En général les roches basiques sont plus riches que les roches acides puisque les composés de zinc sont plus solubles en milieu acide qu'alcalin.

Comme le cuivre, le zinc est énergiquement retenu par la matière organique, probablement sous forme de complexes. Il se trouve donc surtout dans la couche arable (Nair et Mehta 1959).

Aux environs des mines de zinc, les sols peuvent contenir des quantités appréciables de cet élément, mais grâce à sa forte fixabilité, les plantes, mêmes sensibles, n'en souffrent pas. Il existe même une flore "zincicole" comme *Armeria Halleri* et *Silene inflata* (Knickmann 1959).

Dans le sol, la grande majorité du Zn se trouve à *l'état non-échangeable.*

Dans le réseau cristallin, il prend des positions non occupées par l'H ou se substitue au Mg dont le rayon (0,78 Å) est assez proche de celui du Zn (0,83 Å).

Les expériences avec Zn^{65} ont cependant révélé que cette substitution diffère suivant l'espèce de minéral: la dolomite et la magnésite, en absorbant le Zn, le privent de son eau d'hydratation alors que la calcite ne le fait pas.

A la suite de sa forte fixation au complexe colloïdal minéral et organique — ici par chélation (Himes et al. 1957, Miller et Ohlrogge 1958) — la *teneur du sol en Zn échangeable et soluble* est minime, mais dépend du pH et de la teneur en matière colloïdale.

Les sols acides contiennent jusqu'à 10 fois plus de Zn que les neutres.

Son énergie d'adsorption, mesurée au moyen du $Zn^{65}Cl_2$ est, d'après Nelson et Melsted (1955): H > Zn > Ca > MG > K. Entre pH 6,5 et 7,5, 90% du Zn sont retenus dans les 5 cm. superficiels tandis qu'il descend jusque 40 cm. à pH 4. L'influence du pH sur la mobilité du Zn du sol tient à son caractère amphotère: en milieu acide le Zn est un cation; en milieu basique il devient l'anion zincate.

Aux environs de la neutralité, il se forme du $Zn(OH)_2$ peu soluble (Scharrer et Höfner 1958).

En pénurie de Zn, la sève des plantes contient des doses anormalement élevées de Mg, d'N et de P. L'antagonisme entre ces éléments et le Zn a été constaté aussi par d'autres chercheurs (Routchenko 1961).

Les maladies de carence dues à la pénurie de Zn assimilable seront plus à craindre en sol basique qu'en sol acide, mais sont assez indépendantes de la teneur totale en Zn du sol. Par conséquent certains chélates de Zn (mais pas tous), qui livrent facilement leur Zn au sol, seront plus actifs en sols calcaires qu'en sols acides (Wallace et Mueller 1959).

En outre il semble exister un certain antagonisme entre le Zn et le Cu: la teneur en Zn assimilable varie beaucoup avec le pH, alors que celle du Cu est relativement indépendante de la réaction du sol (Van Koetsveld et Lehr 1961).

Le même antagonisme existe entre le Zn d'une part et P et Mg d'autre part. La carence en Zn se manifeste dès que son taux dans le sol baisse en-dessous de 3 p.p.m. (Dartiges 1963).

Les récentes études au moyen du Zn^{65} ont été résumées par Thorne et Wiebe (1957).

E. Le cobalt

Le cobalt se trouve de préférence dans les minéraux sulfureux et arsénicaux.

Les sols dérivés de roches basiques sont plus riches en Co que ceux dérivés de roches acides.

Dans le sol, le Co se trouve surtout dans les fractions fines. Un pH élévé diminue son assimilation par la plante. Elle augmente en sol acide. Il se fixe énergiquement au complexe tant minéral qu'organique (ROLVINK 1959, KULIKOW 1961).

L'emploi de Co^{60} a prouvé qu'il existe dans le sol sous forme facilement et difficilement échangeable (SINGLETON 1958).

Le Co est nécessaire aux animaux (vitamine B_{12}) mais il reste à prouver qu'il est indispensable aux plantes quoique certaines espèces (*Nyssa sylvatica*) en absorbent des quantités telles (jusque 253 p.p.m.) qu'elles peuvent servir de plantes indicatrices pour les sols cobaltifères (KUBOTA et LASAR 1958).

Puisque le Co est nécessaire aux animaux qui vivent des prairies, la teneur en Co de l'herbe a une certaine importance. En moyenne on trouve:

herbes:	0,052 mgr. Co par Kgr. de matière sèche
trèfles:	0,102 mgr. Co par Kgr. de matière sèche
autres plantes:	0,146 mgr. Co par Kgr. de matière sèche

Dans un rapport de 1:2:3.

Signalons l'utilité de la présence des soi-disant "mauvaises herbes" dans les prairies (GERICKE 1962).

En Hollande une teneur de 0,1 mgr. par Kgr. de sol (soluble dans l'acide acétique) est estimée comme insuffisante (HENKENS 1963).

En utilisant le Co^{60}, on peut obtenir des solutions nutritives très pures contenant moins de 2×10^{-10} moles de Co^{60} par litre; même à cette extrême dilution, on n'a pas pu provoquer des symptômes de carence chez la tomate (THORNE et WIEBE 1957).

Puisque les ruminants ont besoin de cobalt, il peut être intéressant d'analyser les sols des prairies.

En Hollande, les prairies sur sol sabloneux sont sensiblement plus pauvres en Co que celles sur sol lourd. Une fumure de 2 Kgr. de sel cobaltique suffit pour plusieurs années (HENKENS 1959).

F. Le molybdène

Le molybdène, *à l'état non-échangeable*, est présent dans certains minéraux argileux: l'olivine, la molybdénite MoS_2, le $CaMoO_4$, et dans la matière organique du sol.

Les roches acides contiennent plus de Mo que les basiques.

Les sols en contiennent des quantités variant de 0,3 à 7 p.p.m. principalement localisés dans l'horizon superficiel (WIKLANDER 1958).

Le molybdène échangeable se trouve dans le sol probablement sous forme anionique: MoO_4^{--} ou $HMoO_4^{--}$ et s'échange donc contre les ions ^-OH de l'argile et le PO_4^{--}; celui-ci est un échangeur énergique avec le molybdate et explique que la présence ou l'absence de phosphore dans le sol joue un rôle important dans l'assimilabilité du Mo. Ce fait explique aussi que la fixation du Mo augmente avec l'acidité et diminue avec l'alcalinité du sol, et il explique enfin que c'est surtout dans les sols acides que l'application de Mo guérit la maladie de carence en cet élément (FOY et BARBER 1959).

Suivant les conditions du sol, le *molybdène soluble* est minime ou abondant (0,001 à 3,9 p.p.m.) mais augmente généralement avec le pH.

Le chaulage est donc le remède indiqué contre la carence en molybdène.

Dans les régions (Australie—Nouvelle Zélande) où cette maladie est très répandue, l'application directe d'un molybdate peut être un expédient, mais le chaulage est un moyen plus "naturel" (SAUERBECK 1958).

L'azote nitrique stimule l'assimilation du Mo et l'azote ammoniacal la déprime; cette dépression peut être neutralisée par l'addition d'un engrais à base de Mo (TROBISCH et GERMAR 1959).

La teneur en molybdène soluble dans les acides est en relation avec la teneur du sol en argile, en Fe et en MnO_2 (GALLEGO et JOLIN 1958).

Le molybdène est un élément mineur nécessaire pour les plantes, surtout pour les légumineuses.

Il est indispensable à la formation des nodosités, à la fixation de l'N de l'air, à la réduction des nitrates, ainsi qu'à la formation d'acide glutamique et de l'asparagine (MULDER et al. 1959).

C'est pour cela aussi qu'il existe une relation entre la teneur en Mo et la teneur en N chez les légumineuses (BURKIN 1961).

D'autre part, étant un vrai élément mineur, il peut devenir toxique en sol alcalin et tourbeux.

Les plantes peuvent alors en absorber jusque 1000 p.p.m. et devenir toxiques pour le bétail (BEESON 1958).

G. Le silicium

Le silicium est, avec l'aluminium, l'élément constitutif le plus important de la croûte terrestre: 59,1% dont 12% sous forme de quartz.

Les principaux minéraux siliceux sont: le quartz lui-même, puis les micas, les feldspaths, la hornblende, l'augite et enfin la grande catégorie des minéraux argileux.

Les sols qui dérivent de ces minéraux sont plus riches (horizon A du podzol) ou plus pauvres (latérites) en Si suivant leur degré d'altération.

Les eaux des rivières tropicales sont riches en silicium qui se concentre dans l'eau de mer où il est assimilé par les animaux et plantes "silicicoles": coraux — diatomées — éponges.

Puisque le Si forme le centre des tétraèdres dans le réseau cristallin, la majeure partie du Si se trouve dans le sol sous forme *non-échangeable*.

La minime partie *échangeable* se trouve probablement en couche mono-moléculaire d'acide silicique à la surface des particules de quartz.

Sous forme de H_4SiO_4, il devient *soluble* (en solution vraie ou colloïdale) par altération climatique des sols. Les facteurs de cette altération sont: le climat, la nature de la roche-mère, le pH, la végétation et la présence éventuelle d'agents peptisants ou coagulants.

Alors que sa solubilité varie peu entre pH 1 à 8, elle augmente sensiblement au dessus de pH 9 par ionisation de l'acide silicique:

$$H_4SiO_4 + OH^- = H_3SiO_4^- + H_2O$$

Comme les sols à pH élevés (alcalins) sont riches en Na, ils forment des silicates sodiques solubles.

Puisque le silicium est abondant dans les sols, certaines plantes ou parties de plantes (cypéracées — glumelles des graminées) en contiennent des quantités importantes sans pour cela que soit prouvée sa nécessité.

Récemment on a trouvé que le Si favorise le développement du système radiculaire. Il rend les feuilles plus résistantes aux maladies des piqûres d'insectes, des moisissures et microbes (KNICKMANN 1962).

Le fait qu'une minime partie du phosphore peut être remplacée par le silicium, n'a pas d'importance agricole.

H. Le chlore

Le chlore se trouve dans certains minéraux: sodalite, scapalite et surtout chlorapatite, dans les gisements de sel gemme: kaïnite, sylvinite, dans certains sols salins des régions arides où il s'accumule à la surface, ainsi que dans les polders submergés par l'eau de mer.

Comme la plupart des chlorures sont solubles, il lixivie facilement sous climat humide pour être emporté par les eaux des rivières dans la mer (1,9% de Cl) d'où il retourne cependant à la terre par le vent atmosphérique (sel "cyclique"). Les eaux de pluie peuvent ainsi déposer de 5,1 à 447 Kgr de Cl par Ha-année.

De fait, en tenant compte de la direction dominante des vents et de la topographie, il existe une corrélation négative assez significative entre la teneur en Cl des sols et leur distance de la mer.

Les sols cultivés peuvent contenir des doses appréciables de chlore apporté par les engrais chlorurés potassiques.

A cause de sa monovalence et de son grand rayon ionique, le Cl se fixe peu au sol et est donc *très échangeable et soluble*. Seuls les sols latéritiques à charge positive peuvent en absorber une certaine quantité; comme ces sols sont acides, il semble exister une corrélation positive entre leur pH et leur teneur en Cl, alors que cette corrélation est négative entre le pH et la teneur en soufre et surtout en phosphore (Wiklander 1958).

Malgré son abondance relative dans certaines plantes marines et désertiques, il a été difficile jusqu'ici de prouver sa nécessité physiologique.

Dans l'expérimentation, il doit être lié à un cation K, Na, Mg, etc., et son utilité éventuelle a été confondue avec celle de ces cations (Wigdor et al. 1958).

Récemment cependant, des cas de déficience bien établis en chlore ont été publiés (tomates) de sorte que le chlore peut être considéré maintenant comme un élément nécessaire (Wallace 1959).

Les lemnacées, vu leur croissance rapide, constituent un objet expérimentale intéressant pour les chercheurs. La nécessité du chlore pour elles a été bien établie (Martin 1963).

La grande mobilité de Cl, aussi bien dans le sol que dans la plante, a été confirmée par l'emploi du Cl^{36}: la simple pulvérisation sur les feuilles le fait retrouver rapidemment dans toutes les autres parties de la plante.

Cette mobilité rend en outre difficile l'interprétation des autoradiogrammes avec Cl^{36}.

L'impression que le chlore se concentre dans les pétioles et nervures est simplement due à l'épaisseur plus grande de ces organes (Thorne et Wiebe 1957).

I. Le fluor

Le fluor est présent dans tous les sols qui contiennent des minéraux fluorés par ex. l'apatite.

Il s'y trouve en majorité sous forme insoluble (Jahn-Deesbach 1959). Il s'y accumule à la suite d'applications d'engrais phosphatés à base d'apatite, par l'intermédiaire d'usines émettant dans l'atmosphère des gaz fluorés ou déversant des eaux d'irrigation contenant du fluor.

C'est un élément mineur toxique pour les plantes (Brewer et al. 1959).

Chez les arbres fruitiers, une application de 25 p.p.m. de F provoque une diminution de la croissance, de la surface foliaire, de la quantité et qualité des fruits ainsi qu'une accélération de la chute des feuilles (Brewer et al. 1959). Dans les sols calcareux, le F est assez facilement soluble alors qu'il ne l'est pas dans les terrains argileux.

J. L'iode

Il n'est pas prouvé jusqu'ici que l'iode soit nécessaire à la plante alors qu'il est nécessaire aux animaux qui se nourrissent des plantes.

D'après les recherches récentes, l'iode est utile à l'épinard, le trèfle blanc, la betterave fourragère, les tomates, l'orge, le froment, le lin et autres. En général l'iodate est plus actif que l'iodure. Malgré cela, l'iode reste un vrai élément mineur, toxique ou ineffectif dès qu'on surpasse la dose minime optimale (PAUWELS 1961).

Pour ce motif, beaucoup d'analyses de teneurs en iode des sols ont été exécutées dans différentes parties du monde.

Puisque la mer est le réservoir principal de l'iode (algues marines), les sols côtiers sont plus riches en iode que ceux de l'intérieur (sel cyclique).

Les roches calcaires et les sols alcalins sont les plus riches en iode qui est retenu par les matières adsorbantes: argiles, humus.

Voir aperçu bibliographique de l'iode du sol par GALLEGO et OLIVER (1959).

K. Le bore

Le principal *minéral borique* est la tourmaline (3—4% de B). Ensuite, on le trouve en petites quantités dans beaucoup de minéraux silicatés: muscovite, glauconite, biotite, et dans certains calcaires et dolomites.

Leur teneur oscille entre 2,8 gr et 143 gr de B par Tonne (HEIDE et THIELE 1958).

Comme le bore est un élément essentiel pour les plantes, il retourne avec leurs débris dans le sol (3—100 p.p.m., MITCHELL 1955), d'où il lixivie jusqu'à l'eau de mer qui contient $0{,}027^0/_{00}$ d'acide borique (KALLE 1958).

Voilà pourquoi les sédiments marins argileux sont riches en B (100—160 p.p.m.).

Comme le bore fait partie du sel cyclique, à pluviosité égale, les sols côtiers en sont parfois 50 fois plus riches que ceux de l'intérieur du pays.

Comme le bore est assez lixiviable, les sols des climats humides (podzols) sont plus pauvres que ceux des semi-arides. Le nitrate de Chili, probablement d'origine marine, contient des quantités appréciables de B.

Par conséquent, la teneur en *B échangeable et soluble* des sols sous climat humide est inférieure à celle des sols arides. Le mécanisme de l'échange de bore est peu connu.

La fixation, qui est réelle puisque 50% seulement du bore total se dissout dans l'eau chaude, semble être due aux liaisons de l'H de l'acide borique aux groupements OH de l'Al du minéral argileux.

Le mécanisme de l'échange du bore suit les lois générales de l'échange des ions, et spécialement la formule de LANGMUIR. Cette formule peut servir à prévoir l'effet nocif des eaux boriquées d'irrigation ainsi que la "déborication" des sols boriqués (HATCHER et BOWER 1958).

D'autres chercheurs supposent une substitution de l'Al du réseau cristallin par le B ou une fixation de l'H_3BO_3 dans l'espace interlamellaire des minéraux argileux du type 2:1.

Le bore est un élément nécessaire mais mineur. On peut donc rencontrer des cas de carence en bore aussi bien que des cas de toxicité (irrigation par des eaux boriquées — emploi d'engrais boriqués).

Les doses optimales et toxiques sont assez rapprochées et des intoxications peuvent se produire à la suite d'applications répétées de petites doses de B en vue de guérir les maladies de carence en B (CHABANNES 1958, BRÄUNLICH 1958). En

Hollande, les sols contenant moins de 0,30 p.p.m. de bore soluble dans l'eau chaude, sont sujets à la maladie du cœur de la betterave (LEHR et HENKENS 1959).

Les *facteurs qui favorisent la carence* en bore sont: l'augmentation du pH du sol et sa dessication.

Les cas de maladies de carence se rencontrent donc après chaulage massif et par année sèche, deux facteurs favorisant la fixation du bore.

Comme d'autre part, les sols sableux, perméables, sont plus souvent sujets à cette maladie que les sols argileux, il faut admettre en outre une lixiviation possible (sable) ou une rétention (argile), (QUELLETTE 1958).

Le mode d'action de ces facteurs reste controversé. Le chaulage peut aussi bien provoquer la maladie de carence que prévenir l'effet toxique de doses élevées de B.

D'ailleurs, les maladies de carence se rencontrent aussi bien en sols sous climat humide qu'aride.

En outre, le bore peut être complexé par la matière organique du sol et livré ou non à la plante suivant l'intensité de la minéralisation de cette matière par les microbes (MCLEAN et VOLK 1958).

L'état actuel de nos connaissances sur le bore est condensé dans l'ouvrage de SCROPP (1957).

L. Le sélénium

Le sélénium se trouve dans certaines roches sédimentaires crétacées d'où il est enlevé par érosion et accumulé dans les fonds marins.

De ce fait, les sols dérivés des roches sélénieuses contiennent beaucoup moins de sélénium (2—4 p.p.m.) que ces roches elles-mêmes (10 p.p.m.) (MOXON 1958).

Il se trouve surtout dans le sous-sol sous forme élémentaire, sélénite, sélénate ou organique.

Le sélénium est un véritable élément mineur, toxique pour les animaux.

Les plantes peuvent en supporter des doses minimes et il est même possible qu'il soit pour certaines un élément utile puisqu'il existe une flore indicatrice des sols sélénieux: *Stanleya*, *Onoapsis*, *Xylorrhiza* et *Astragalus*. Les plantes qui ont absorbé des doses de sélénium sans dommage, peuvent devenir toxiques pour le bétail et même pour certains insectes.

M. L'aluminium

L'aluminium *échangeable* est en rapport avec le degré d'acidification et d'altération du sol.

L'acidité d'échange est due aux ions Al.

Dès que la différence entre le pH-eau et pH-KCl dépasse une unité, l'aluminium échangeable atteint des teneurs appréciables (GILLY 1958).

Vu le caractère amphotère de l'aluminium, la notion d'aluminium échangeable est assez imprécise et dépend de la nature de l'échangeur, surtout de son pH.

Si l'échangeur est acide ($< 4{,}8$), une partie de l'aluminium non échangeable entre en solution (MCLEAN 1958). L'addition de KCl à un sol acide augmente son Al échangeable au point de devenir toxique pour la plante, alors que des doses massives d'engrais potassiques en présence de chaux n'ont pas cet effet défavorable (RAGLAND et COLEMAN 1959).

N. Le strontium et cesium

Le strontium *à l'état natif* dans le sol atteint rarement une dose toxique puisque sa teneur à l'état échangeable n'atteint que $1{,}3^0/_{00}$ en moyenne de celle du Ca échangeable (MENZEL et HEALD 1959).

Il en est de même du Sr^{90} qui, à la suite de décharges nucléaires ou comme résidu d'usines atomiques, peut se fixer dans le sol à la dose de 0,5 MC par gramme, alors que sa solution contient moins de 10^{-6} moles par litre (MCHENRY 1958).

A la suite de la longueur de sa demi-vie, le Sr^{90} est un des éléments les plus dangereux de l'ère atomique et son accumulation locale dans le sol a attiré l'attention des pédologues.

Toutefois, on a constaté que dans ces cas d'accumulation, l'ajoute de Sr stable réduit l'assimilation par la plante du Sr^{90} par dilution du carrier (ROMNEY *et al.* 1959). Le maximum d'absorption du Sr se fait aux environs de la neutralité du sol (PROUT 1958).

L'assimilation du Sr^{90} par la plante semble être en relation étroite avec la teneur du sol en Ca échangeable (SCHEFFER et LUDWIG 1959).

La chaulage et les sels potassiques diminuent l'assimilabilité du Sr^{90} (ROMNEY et al. 1959): si celui-ci provient de l'atmosphère, il peut toutefois être absorbé directement par les feuilles et dans ce cas le chaulage n'est d'aucune utilité (SIEGEL 1959).

La majorité du Sr^{90} du sol provient de l'atmosphère à la suite de décharges nucléaires puisqu'il existe une relation entre les précipitations atmosphériques et la teneur en Sr^{90} du sol; il se cantonne dans les 5 cm superficiels et est donc fortement retenu par le complexe absorbant (KNOOP et SCHROEDER 1958).

La plante ne prend que 50% du Sr^{85} superficiel vis-à-vis de 100% de Sr mêlé au sol même (LEE 1962).

Comme le Sr^{90}, le Cs^{137} a une demi-vie suffisamment longue pour contaminer les plantes — pour lesquelles il est toxique — et donc indirectement les animaux et l'homme. Le Cs du sol devient toxique pour la plante dès qu'il attient 8% de capacité d'échange du sol. Cs^{137} et K sont antagonistes. Si le sol contient beaucoup de K assimilable, l'assimilation du Cs^{137} par la plante est déprimée (NISHITA et al. 1962).

L'application de fortes doses de Ca et Mg, sous forme de carbonates (5—10 tonnes par Ha) diminue l'assimilation du Sr^{89}. Au contraire, sous forme de chlorures, ils augmentent le Sr^{89} des plantes. L'N augmente le Sr de la paille, mais diminue celui de la graine (ANDERSEN 1963).

Bibliographie

AALTONEN, V. T.: Boden und Wald. Berlin 1948. — ALEIXANDRE, V., et M. COVABONG RODRIGUEZ: Modificación de las propriedades de los minerales arcillosos mediante tratamiente permico y cation de cambio. An. Edaf. Fis. Veg., Madrid, **16**, 799–856 (1957). — ALESCHIN, S. N.: Adsorption der Wasserstoffionen durch den Boden. Verh. Int. Bod. Ges. **2**, 183–187 (1958). — AMPHLETT, C. B.: Ionenaustausch in Tonmineralien. Endeavour **17**, Nr. 67, 149–155 (1958). — ANDERSEN, A. J.: Influence of liming and mineral fertilization on plant uptake of radiostrontium. Soil Sci. **95**, 52–59 (1963). — ARAMBARRI, P., et O. TALIBUDEEN: Factors influencing the isotopically exchangeable phosphate in soils. Plant a. Soil **11**, 4, 343, 364 (1959). — ARMI KAILIA: Availability for plants of P in some virgin peat samples. J. Agr. Sci. Soc. Finland **30**, 133–142 (1958). — ARNON, D. I.: Trace elements, p. 5–7. New York 1958. — ATANASIU, N.: Die Wirkung der Spurenelemente im Thomasphosphat. Phosphorsäure **22**, 28–47 (1962). — ATKINSON, H. J., G. R. GILES, J. G. DESJARDINS: Farmyard manure and trace-element supply. Plant a. Soil **10**, 1, 32–36 (1958). — AUDIDIER, L.: Un nouvel engrais potassique: le CO_3KH. Congr. mond. Rech. agron. Rome **1959**, 1205–1207.

BAEYENS, J.: Similitude entre le cycle des soufre et celui de l'azote (non édité 1965). — BAEYENS, J., et G. SCHEYS: Etudes sur les polders inondés par l'eau de mer, I–II–III–IV. Agricultura (Louvain) 1948. — BAEYENS, J., G. SCHEYS et F. APPELMANS: Le phosphore dans le sol et dans la plante. Agricultura (Louvain) 1948. — BAEYENS, J.: Les sols de l'Afrique Centrale, p. 70. Bruxelles 1938. — BAEYENS, J.: et D. F. STENUIT: La situation nutritive et les besoins en engrais des sols belges, p. 55. Louvain: Service Pédologique de Belgique 1948. — Beziehung zwischen Ernte und Bodennährstoffen. III° Weltcongr. Düngungsfr., Frankfurt 1958, p. 110. — Les possibilités de culture en sols salins et alcalins en régions semi-arides. Agricultura (Louvain) 1963 (Sous presse). — BARBIER, G.: Untersuchungen über die Wirkungsart von Phosphordünger mit Hilfe radioaktiver Phosphorsäure. Phosphorsäure **17**, 190–193 (1957). — BARBIER, G., et L. DUVAL: Sur l'échange des cations présents en minime proportion dans le sol. Ann. Agron. **11**, 695–712 (1958). — BARLETT, R. J., et E. O. MCLEAN: Effect of K and Ca activities in clay suspensions and solutions on plant-uptake. Proc. Soil. Sci. Soc. Amer. **23**, 285–289 (1959). — BASTISSE, E. M.: Etude de quelques éléments minéraux fixés dans les sols sous forme organique. Ann. Inst. Nat. Rech. Agr. **8**, 585–603 (1957). — BEEKOM, C. W. C. VAN: Les inondations de 1944–1945 et leur conséquence pour l'agriculture. Versl. Landb. Onderz. **64**, 13, 42–44 (1961). — BEESON, K. C.: The relation of soils to the micronutrient element content of plants and to animal nutrition. Trace elements: Proc. Conf. Wooster, Ohio 1957, p. 67–79. U.S.D.A., Soil and Nutr. Lab., Ithaca, N.Y. 1958. — BERINGHER, H.: Aufnahme und Wirkung des Mikronährstoffes Cu in ionoger und chelatisierter Form bei Gerste. Z. Pflanzenernähr., Düng., Bodenkde. **100**, 1, 1–34 (1963). — BHARDWAJ, S. N.: Physiological studies on salt tolerance in crop plants. Agra Univ. J. Res. **9**, Part 2, 195–209 (1960). – BLACK, C. A.: Soil-plant relationships. New-York 1957. — BLANCHET, R.: Pouvoir adsorbant des divers constituants du sol à l'égard des ions phosphoriques et examen des réserves du sol. Compt. rend. Acad. Agric. France **45**, 247 (1959). — BLANCHET, R., et al.: Premières indications sur la dynamique du P dans les sols calcaires de Champagne. Compt. rend. Acad. Agric. France **4**, 305–317 (1963). — BOBRITSKAYA, M. A.: Enrichissement du sol en N provenant de l'atmosphère par les pluies. Pochvovedenie **12**, 53–60 (1962). — Energie d'adsorption des ions minéraux par les colloïdes du sol. Ann. Agron. **10**, 5–52, **2**, 125–154 (1959). — BOGUSLAWSKI, E. VON: Zur Entwicklung des Begriffs Bodenfruchtbarkeit. Z. Pflanzenernähr., Düng., Bodenkde. **108**, 2, 97–115 (1965). — BOIXO, M. DE: Teneur en sel des sols de Cheraia après 25 ans d'irrigation. Compt. rend. Acad. Agric. France **47**, 259–267 (1961). — BOULDIN, D. R., et E. C. SAMPLE: Laboratory and greenhouse studies with phosphates. Proc. Soil. Sci. Soc. Amer. **23**, 338–342 (1959). — BOWER, C. A.: Cation-exchange equilibria in soils affected by sodium salts. Soil Sci. **88**, 32–35 (1959). — BOWER, C. A., et S. O. GOERTZEN: Exchange able sodium in calcareous soils. Proc. Soil Sci. Soc. Amer. **22**, 270 (1958). — BRÄUNLICH, K. VON: Versuchsergebnisse zum Problem der Borüberdüngung. Z. Pflanzenernähr., Düng., Bodenkde. **83**, 1–7 (1958). — BREWER, R. F., et al.: Effect of fluorine additions to substrate on navel orange trees grown in solution cultures. Soil Sci. **84**, 183–188 (1959). — BROADBENT, F. E.: Basic problems in organic matter transformations. Soil Sci. **79**, 107–114 (1955). — BROEK, J. M. M. VAN DER: Les sols alluvionnaires de la Meuse, la Ruhr et le Limbourg. C. R. Rech. Agron. 1964 (Ned.). — BROEK, J. M. M. VAN DER, et H. W. VAN DER MAREL: Magnesium in soils of Limburg. Z. Pflanzenernähr., Düng., Bodenkde. **84**, 237–254 (1959). — BROESHART, H.: Fertilizer placement studies by means of radio-isotopes. Z. Pflanzenernähr., Düng., Bodenkde. **84**, 76–80 (1959). — BROMFIELD, S. M.: The properties of a biologically formed manganese oxyde. Plant a. Soil **9**, 325–337 (1958). — The solution of γ-MnO_2 by substances released from soil and from roots of oaks and vetch in relation to Mn availability. Plant a. Soil **10**, 2, 147–160 (1958). — BROWN, J. C., et R. S. HOLMES: Soil Conservation Mag. **20**, 259 (1955). — BROWN, D. A., et J. P. WELL: Ion exchange in soil-plant root environments, III, The use of Mg^{28} for the Measurement of magnesium absorption by plants. Soil Sci. Soc. Amer. Proc. **22**, 323–326 (1958). — BROWN, J. C., R. S. HOLMES et L. O. TIFFIN: Hypotheses concerning iron chlorosis. Proc. Soil Sci. Soc. Amer. **23**, 231–234 (1959). — BUCHNER, A. VON: Die Schwefelversorgung der Westdeutschen Landwirtschaft. Landw. Forsch. **11**, 79 (1958). — BURKIN, I. A.: Einfluß von Molibdän auf die Qualität und Samenergiebigkeit von Klee und Wicke. Select. u. Samenz. **26**, 38–40 (1961). — BURTON, G.W.: Atomic energy and agriculture, p. 73–76. Washington 1957. — BUSSLER, W. VON: Manganvergiftung bei höheren Pflanzen. Z. Pflanzenernähr., Düng., Bodenkde. **81**, 256–280 (1958). — Entwicklung von Ca-Mangelsymptomen. Z. Pflanzenernähr., Düng., Bodenkde. **100**, 1, 53–58 (1963).

CALDWELL, A. G., et C. A. BLACK: Inositol hexaphosphate content in soils.

Proc. Soil Sci. Soc. Amer. **22**, 296–298 (1958). — Chabannes, J.: Enquêtes sur les conséquences éventuelles d'applications répétées de borax aux terres cultivées. Compt. rend. Acad. Agric. France **44**, 440–445 (1958). — Chaminade, R.: Influence de la paille à différents états de fermentation sur la fertilité d'un sol. Ann. Agron. **14**, 5–11 (1963). — Charliers, N., et C. J. Carpentier: Importance agronomique des éléments autres que N, P, K, en particulier dans le superphosphate. Assoc. Int. fabr. superph. **25**, 5–21 (1959). — Chaussidon, J.: Application of the Gouy-theory to divalent cations. Soil Sci. **95**, 131–133 (1963). — Cohen, P., et C. Gailledreau: La pollution radioactive des sols: Aspect et remèdes. Compt. rend. Acad. Agric. France **47**, 226–228 (1961). — Coic, Y., et M. Coppenet: Sur la carence en Mn des céréales dans les sols humifères de Bretagne. Ann. Agron. **9**, 111–138 (1958). — Cole, C. V., et S. R. Olsen: pH in relation to the solubility in calcareous Soils. I. Proc. Soil Sci. Soc. Amer. **23**, 116–118 (1959). — Coppenet, M., et M. Helias: Valeurs du pH et du rapport carbone-azote de quelques humus forestiers et de landes en Bretagne. Bull. Ass. Franc. Sol. **1958**, 412–420 (1958). — Coppenet, M.: Le Mn dans les sols de Bretagne. Ann. Agron. **10**, 155–218 (1959).

Dartigues, E., et al.: Premiers résultats sur l'étude de la carence en Zn dans le sud-ouest. Compt. rend. Acad. Agric. France **4**, 266–279 (1963). — Davis, L. E.: Exchangeable sodium in calcareous soils. Soil Sci. Soc. Amer. Proc. **22**, 270 (1958). — Deckers, J., et R. van Stallen: Contribution à l'étude de la saturation en bases des sols bruns. Agricultura (Louvain) **3**, 311–340 (1955). — Deiwich, L. C.: Nitrification, p. 218. Baltimore 1956. — De Kock, P. C., et R. L. Mitchell: Uptake of chelated metals by plants. Soil Sci. **84**, 55–62 (1957). — Delas, J.: Evolution de l'N minéral en sol nu, sous climat atlantique, en fonction de diverses fumures organiques. Compt. rend. Acad. Agric. France **3**, 383–392 (1963). — Del Riviero, J. M.: Las enfermidades carenciales de las plantas cultivadas en España. Congr. mond. Rech. agron. Rome **1959**, 1385–1588. — De Mumbrum, L. E.: Exchangeable K levels in vermiculite ... and implications relative to K levels in soils. Proc. Soil Sci. Soc. Amer. **23**, 192–194 (1959). — De Mumbrum, L. E., et C. D. Hoover: Potassium release and fixation related to illite and vermiculite as single minerals and in mixture. Proc. Soil Sci. Amer. **22**, 222–225 (1958). — Dhariwal, A. P. S., et F. J. Stevenson: Determination of fixed ammonium in soils. Soil Sci. **86**, 343–349 (1958). — Diedrich, H.: Zur Methodik der Spurenelementbestimmung in Fichtennadeln. Düng. u. Meliorat. i. d. Forstw. **66**, 58–70 (1964). — Diest, A. van, et C. A. Black: Soil organic phosphorus and plant growth II. Soil Sci. **87**, 154 (1959). — Dobrzanski, B.: Einfluß der Düngung auf den Humusgehalt des Bodens. Z. Pflanzenernähr., Düng., Bodenkde. **84**, 122–127 (1959). — Duchaufour, P. H.: Pédologie forestière. Nancy 1958.

Ebergardt, W. von, et G. Krzysch: Stickstoffgehalt und C/N-Verhältnis eines lehmigen Sandbodens und ihre Abhängigkeit von Pflugarbeit und Düngung. Z. Pflanzenernähr., Düng., Bodenkde. **85**, 117–123 (1959). — Egner, H., et E. Erikson: Current data on the chemical composition of air and precipitation. Tellus **7**, 134–139 (1955). — Engelstad, O. P., et E. C. Moreno: Effect of P concentration and distribution on P uptake and root growth of corn. Soil Sci. **99**, 4, 227–233 (1965). — Ensminger, C. E., et R. W. Pearson: Atomic energy and agriculture, p. 19. Washington 1957.

Fink, J.: Zur Ermittlung des K-Versorgungsgrades lehmiger Braunerden. Kali-Briefe **3**, Nr. 9, 1 (1957). — Foy, C. D., et S. A. Barber: Mo response of alfalfa on Indiana soils in the greenhouse. Proc. Soil Sci. Soc. Amer. **23**, 36–39 (1959). — Franz, H.: Die Bedeutung des bodenbiologischen Geschehens für den Bodenfruchtbarkeitszustand. Z. Pflanzenernähr., Düng., Bodenkde. **108**, 2, 123–128 (1965). — Freney, J. R., et al.: A review of certain aspects of S as a soil constituent and plant nutrient. Plant a. Soil **18**, 3, 295—308 (1962). — Fried, M., et R. E. Shapiro: Phosphate supply power of various soils. Soil Sci. Soc. Amer. Proc. **20**, 471 (1956).

Gallego, R., et T. Jolin: Aportación a la geoquímica del molibdeno. An. Edaf. Fis. Veg., Madrid, **17**, 341–385 (1958). — Gallego, R., et S. Oliver: Estudios sobre yodo en suelos. An. Edaf. Fis. Veg., Madrid, **18**, 3, 207–238 (1959). – Garaudeaux, J. G., de Beaucorps et H. Chevalier: Evolution des ions potassium en sol de limon au cours d'une culture. Z. Pflanzenernähr., Düng., Bodenkde. **84**, 36–42 (1959). — Gericke, S., et B. Kurmies: Pflanzennährstoffe in den atmosphärischen Niederschlägen. Phosphorsäure **17**, 279–300 (1957) — Gericke, S.: Die Wirkung der Magnesia im Thomasphosphat. Phosphorsäure **22**, 162–182 (1962). — Wirkung der Thomasphosphatdüngung auf den Co-Gehalt von Wiesenheu. Phosphorsäure **22**, 48–60 (1962). — Gilly, G.: Acidité et aluminium échangeable. Ann. Agron. A **9**, 679–691 (1958). — Graham, E. R.: Uptake of Sr^{90} and Cs^{137} by soils and vegetation. Soil Sci. **86**, 91–97 (1958). —

GREENWOOD, D. J.: Nitrification and nitrate dissimilation in soil. Plant a. Soil 17, 3, 365–391 (1962). — GRIESSBACH, R.: Austauschadsorption in Theorie und Praxis, p. 252–260. Berlin 1957. — GUINN, G., et H. E. JOHAM: Displacement of iron in Fe EDTA and Fe HEEDTA by Cu and Zn. Soil Sci. 95, 101–104 (1963). — GULJAKIN, I. V., et E. V. JUDINZEWA: Problèmes agro-chimiques de Sr^{90} et Cs^{137} et autres isotopes radioactifs. Compt. rend. Acad. Agr. Timirjasew (russe) 1, 15 (1958). — GUSTAFSON, P. F., L. D. MARINELLI et S. S. BRAR: Natural and fission-products gamma-ray emitting radioactivity of soil. Science 127, 1240–1242 (1958).

DE HAAS, P. G., et W. GRUPPE: Die Bedeutung des Magnesiums für die Ernährung der Obstgehölze. Landw. Forsch., Sonderheft 13, 37–44 (1959). — HAGIN, J.: Availability of dicalcium phosphate to plants, when applied in various forms. Plant a. Soil 10, 2, 101–113 (1958). — HANWAY, J. J., et A. D. SCOTT: Soil potassium moisture relations III. Proc. Soil Sci. Soc. Amer. 23, 22–24 (1959). — HART, M. G. R.: Observations on the source of acid in empoldered mangrove soils. Plant a. Soil 17, 1, 87–98 (1963). — HATCHER, J. T., et C. A. BOWER: Equilibria and dynamics of boron adsorption by soils. Soil Sci. 85, 319–323 (1958). — HEIDE, F., et H. THIELE: Zur Geochemie der Bors. Chem. d. Erde 19, 329–337 (1958). — HEMWALL, J. B.: Reaction of ferric ethyllenediamine tetra-acetate with soil clay minerals. Soil Sci. 86, 126–132 (1958). — HENKENS, C. H.: Sporenelementen op grasland. Kali 51, 42–48 (1963). — HENKENS, L. H.: Kobalt op grasland. Landbouwvoorlichting 16, 642–651 (1959). — HILL, D. G., GOTTINGHAM et C. P. LLOYD-JONES: Behaviour of iron chelates in calcareous soils. Plant a. Soil 10, 2, 194–204 (1958). — HIMES, F. L., et S. A. BARBER: Chelating ability of soil organic matter. Proc. Soil Sci. Soc. Amer. 21, 368–373 (1957). — HOFFMANN, F.: Untersuchungen zur Stickstoffernährung junger Nadelholzpflanzen. Düng. u. Meliorat. i. d. Forstw. 66, 151–166 (1964). — HOFFMAN, W., et K. STEINFATT: Vergleichende Untersuchungen über die Löslichkeit der pflanzenaufnehmbaren Nährstoffe K und P in frischen und lufttrockenen stark humosen und organogenen Böden. Landw. Forsch. 12, 2, 77–87 (1959). — HOLLEY, R. W., et J. C. CAIN: Accumulation of Arginine in plants afflicted with iron-deficiency type chlorosis. Science 121, 172–173 (1999). — HUFFAKER, R. C., et A. WALLACE: Sodium absorption by different plant species at different potassium levels. Soil Sci. 87, 130–134 (1959). — HUTCHESON, T. B., W. G. WOLTZ et S. B. MCCALEB: Potassium-Sodium interrelations ship 1. Soil Sci. 87, 28–36 (1959). — Potassium-Sodium interrelationships 2. Soil Sci. 88, 36–44 (1959).

INGVAR-NILSSON, S., et M. HAHLIN: Synpunkter pä växternas kväveförsörjning (Some aspects of the nitrogen supply of plants). K. Skogs. Lantbr. Akad. Tidskr. 18, 75–96 (1959).

JACKSON, W. H., et N. T. COLEMAN: Fixation of carbon dioxide by plant roots through phosphoenol pyruvate carboxylase. Plant a. Soil 11, 1–16 (1959). — JAHN-DEESBACH, W.: Über die Beziehungen zwischen Fluor und Phosphorsäure in verschiedenen Böden. Phosphorsäure 19, 132–140 (1959). — JANSSON, S. L.: Balance sheet and residual effects of fertilizer N in a 6-year study with N^{15}. Soil Sci. 95, 31–37 (1963). — JASTREBOW, M. T.: Radioactivité naturelle des sols de la Russie d'Europe (russe). Doklady Akad. Nauk. U.R.S.S. 119, 586–589 (1958). — JAWORSKY, C. A., et S. A. BARBER: Soil properties in relation to potassium uptake by Alfalfa. Soil Sci. 87, 37–41 (1959). — JIRO KOSAKA: Organic Ph. in upland soil. Soil and Plant Food, Tokyo 3, 95–99 (1957). — JOHAM, H. E., et J. V. AMIN: Role of Sodium in the potassium nutrition of cotton. Soil Sc. 99, 4, 220–226 (1965). — JOHANSSON, O.: Über die S-Versorgung der landwirtschaftlichen Kulturen in Schweden. Landwirtsch. Forsch. 15, 1–6 (1962). — JORDAN, J. V., G. ORIEN et BAKER: Sulfur studies in North Idaho soils using radiosulfur. Soil Sci. 88, 1–6 (1959). — JORDAN, H. V., et C. E. BARDSLEY: Response of crops to sulfur on south-eastern soils. Proc. Soil Sci. Soc. Amer. 22, 254–256 (1958). — JÜRGENS-GSCHWIND: Phosphatdüngung und Qualität im Futterbau. Phosphorsäure 22, 134–151 (1962).

KALLE, I.: Handb. Pflanzenphysiologie, Vol. IV, p. 171 (W. RUHLAND, éd.). Berlin-Göttingen-Heidelberg: Springer. 1958. — KAURITSCHEW, I. S.: Les transformations de phosphates dans le podzol et l'influence concommitante des processus de réduction. Bul. Acad. Agr. Timirjasew (russe) 2, 109 (1958). — KENNETH, C. BEESON: Soil micronutrients in plants and animals. Trace elements, p. 67–79. New York 1958. — KICK, H., et O. HELLWIG: Vegetationsversuche zur Aufnahme von N, P, K, über das Blatt im Vergleich zur Aufnahme über die Wurzel. Z. Pflanzenernähr., Düng., Bodenkde. 84, 265–271 (1959). — KLETSCHOWSKI, W. M., et I. W. GULJAKIN: Le comportement de petites doses de Strontium, Cesium, Ruthenium et Zirconium dans le sol et les plantes. Pédologie (russe) 3, 1–46 (1958). — KLOKE, A., et F. LUDWIEG: Sr^{90}, Cs^{137}, Pb^{239} in Heuproben von 1953–1961. Naturwiss. 49, 65–66 (1962). —

Kloke, A.: P_2O_5-Transport in Böden. Kali-Briefe 4, Nr. 3, 1 (1958). — Kloke, A., et E. Przemeck: Untersuchungen über die P_2O_5-Düngewirkung von Mehrnährstoffdüngern. Landw. Forsch. 12, 1, 21–29 (1959). — Knickmann, E.: Zur Nützung unfruchtbarer Böden mit hohem Gehalt an Blei und Zink. Z. Pflanzenernähr., Düng., Bodenkde. 84, 255–258 (1959). – Silicium und Pflanze. Landw. Forsch. 15, 130–137 (1962). — Knoop, E., et D. Schroeder: Der Sr^{90}-Gehalt einiger Böden Schleswig-Holsteins. Naturwiss. 45, 436–437 (1958). — Kock, P. C. de, et R. H. E. Inkson: Mn content of mustard leaves in relation to iron and major nutrients supply. Plant a. Soil 17, 2, 183–190 (1962). — Köhnlein, J.: Zur Kennzeichnung und begrifflichen Formulierung der Beziehungen zwischen Ertragsbildung und Bodenfruchtbarkeit. Z. Pflanzenernähr. Düng., Bodenkde. 108, 2, 138–144 (1965). — Koetsveld, E. E. van, et J. J. Lehr: La teneur en Zn des sols des Pays-Bas et leur signification pour le bétail. Landb. Tijdschr. 9, 371–381 (1961). — Kosaka, J., et K. Abe: Organic phosphorus in upland soil. Soil and Plant Food, Tokyo, 3, 95–99 (1957). — Koutler-Andersson, E.: Studies on Gytha soils V. Kungl. Lantbrukshögsk. Ann. 22, 279–295 (1956). — Kroll, H.: The ferric chelate for treatment of lime-induced chloroses. Soil Sci. 84, 51–53 (1957). — Kubota, J., et V. A. Lasar: Co status of soils of Southern U. S. II. Soil Sci. 86, 262–268 (1958). — Kulikow, N. W.: Über die Kobaltaufnahme durch die Pflanzen in Abhängigkeit vom Humusgehalt des Bodens. Bodenkunde, H. 4, 78–81 (1961). — Kullmann, A.: Über die Verrottung von Futterpflanzenwurzeln an Hand von Modellversuchen. Z. Pflanzenernähr., Düng., Bodenkde. 84, 127–132 (1959). — Kuron, H., et F. Grueneberg: Die Phosphatfesthaltung durch die Kolloidfraktion des Bodens. Congr. mond. Rech. Agron. Rome **1959**, 1057–1058.

Lamm, C. G., et al.: Isotopic exchange in soil fertility studies. Soil Sci. 95, 16–23 (1963). — Langewerff, J. V., et G. H. Bolt: Theoretical and experimental analysis of Gapon's equation for ion exchange. Soil Sci. 87, 217–222 (1959). — Larsen, J. E., G. F. Warren et R. Langston: Studies of Ph. availability in organic soils. Proc. Soil Sci. Soc. Amer. 22, 336–339 (1958). — Leafel, A. L.: Release of potassium from feldspatic rock and minerals. Soil Sci. 87, 2–12 (1959). — Lebedew, W. I.: Les facteurs principaux qui régissent l'adsorption et l'échange des ions dans le sol. Pédologie (russe) 6, 21–28 (1958). — Lee, C. C.: Uptake of Sr^{85} by alfalfa. Science 138, 41–42 (1962). — Legg, J. U., et F. F. Allison: Recovery of N^{15} tagged nitrogen from Am.-fixing soils. Proc. Soil Sci. Soc. Amer. 23, 131–134 (1959). — Lehr, J. J., et Ch. H. Henkens: Threshold values of boron contents in Dutch soils. Congr. mond. Rech. Agron. Rome **1959**, 1397–1404. — Lehr, J. J., et J. M. Wybenga: Exploratory pot-experiments on sensitiveness of different crops to sodium. Plant a. Soil 9, 3, 237–253, 9, 4, 385–394 (1958). — Lentschig, S.: Zur Ermittlung der Nährstoffversorgung forstlich genützten Bodens mittels Salzsäureanalyse und Vollanalyse. Düng. u. Meliorat. i. d. Forstw. 66, 71–82 (1964). — Lindsay, W. I., M. Peech et J. S. Clark: Determination of aluminium ion activity in soil extracts. Proc. Soil Sci. Soc. Amer. 23, 266–269 (1959). — Lindsay, W. I., J. R. Lehr et H. F. Stephenson: Nature of monocalcium phosphate monohydrate in soils. Proc. Soil Sci. Soc. Amer. 23, 342–345 (1959). — Lindsay, W. I., et H. F. Stephenson: Nature of the reactions of monocalcium phosphate monohydrate in soils. Proc. Soil Sci. Soc. Amer. 23, 12–18 (1959). — Linser, H.: L'effet physiologique sur les plantes des macroéléments fertilisants à la lumière des expériences scientifiques récentes. Un siècle de succès dans le domaine de la fertilisation, p. 147. Frankfurt am Main 1959. — Fassung und Bedeutung des Begriffs „Bodenfruchtbarkeit". Z. Pflanzenernähr., Düng., Bodenkde. 108, 2, 115–123 (1965). — Linser, H., H. H. Mayr et H. Unzeitig: Untersuchungen über die Wanderung von Ionen in Bodensäulen. Z. Pflanzenernähr., Düng., Bodenkde. 86, 57–65 (1959). — Livens, J.: Le rendement du froment en fonction du pH et la teneur en k. Agricultura Louvain 4, 81–98 (1956). — Degré de saturation d'une séquence de sols forestiers sur limon lœssique. — Livens, J., et R. Vanstallen: Le pH comme indice du degré de saturation. Actes et compt. rend. V. Congrès Science du Sol Léopoldville 1954, II 20, 376–383; II 24, 423–429. — Recherches sur le rôle de la litière dans la désaturation en bases des sols forestiers sur limon lœssique. Agricultura Louvain 4, 113–132 (1956). — Livens, P. J.: Contribution à l'étude de l'azote minéralisable du sol. Agricultura (Louvain) 8, 27 (1959). — Degré de saturation d'une séquence de sols forestiers sur limon. Ve Congr. Int. Sci. Sol. Léopoldville 2, 376–383 (1957). — Livens, P. J., et R. van Stallen: Le pH comme indice du degré de saturation. Ve Congr. Int. Sci. Sol. Léopoldville 2, 423–429 (1957). — Liwerant, J.: Relations entre le degré d'activité et le niveau nutritif des sols d'alluvions anciennes de la Garonne. Ann. Agron. France 14, 2, 227–238 (1963). — Luken, H.: Saline soils under dryland agriculture ... and possibilities for their improvement. Plant a. Soil 17, 1, 26–67 (1962).

Machold, O.: Einfluß des Isotopenaustausches auf die Ermittlung des Boden- und Pflanzenphosphors bei radiometrischen Untersuchungen der P-Aufnahme durch die Pflanzen. Verh. Int. Bod. Ges. **2**, 200–206 (1958). — Majumder, S. K., et Stuart Dunn: Use of organic phosphate in nutrient culture. Plant a. Soil **10**, 3, 266–270 (1959). — Maleina, A. A.: Phosphate Fixation by Various Soils. Dokl. Akad. Nauk **118**, 1031 bis 1033 (1958). — Manskaja, Ss., M. T. W. Drosoowa et M. P. Temelsonawa: Combinations du Cu avec différentes formes de matières organiques. Pédologie (russe) **6**, 41–48 (1958). — Marel, H. W. van der: Potassium fixation, a beneficial soil characteristic for crop production. Z. Pflanzenernähr., Düng., Bodenkde. **84**, 51 (1959). — Tropical soils in relation to plant nutrition. Soil Sci. **64**, 445 (1947). — Martin, G.: Différences spécifiques entre les besoins quantitatifs en Chlore de différentes espèces de Lemna. Plant a. Soil **18**, 2, 258–266 (1963). — Materna, J.: Die Ernährungsverhältnisse in Fichtenbeständen im Gebiet von Manetin. Düng. u. Meliorat. i. d. Forstw. **66**, 36—46 (1964). — Mattingly, G. E. G.: The use of the isotope P^{32} in recent work on soil and fertilizer phosphorus. Soil a. Fertilizers, Bur. Soil. Sci. **20**, 59 (1957). — Mattingly, G. E. G., et F. V. Widdowson: Uptake of P from P^{32} labelled superphosphate by field crops. Plant a. Soil **10**, 2, 161–175 (1958). — Mc Henry, J. R.: Ion-exchange properties of Sr in a calcareous soil. Proc. Soil Sci. Soc. Amer. **22**, 514–518 (1958). — McKenzie, R. M.: Trace-elements in some South Australian terra rossa and rendzina soils. Austr. J. Agric. Res. **10**, 52–57 (1959). — McLean, E. O., et G. W. Volk: Trace-elements, p. 251 et 254. New-York 1958. — McLean, E. O., M. R. Heddleson, Ribartlett et N. Holowaychur: Aluminium in soils. Proc. Soil Sci. Soc. Amer. **22**, 382–387 (1958). — Mengel, K., et H. Marscher: Die aktive Ionenaufnahme durch die Pflanzenzelle. Z. Pflanzenernähr., Düng., Bodenkde. **100**, 3, 193–207 (1963). — Menzel, R. G., et W. R. Heald: Sr and Ca contents of crop plant in relation to exchangeable Sr and Ca of the soil. Proc. Soil Sci. Soc. Amer. **23**, 110–412 (1959). — Michale, B.: Einfluß gestaffelter P-Düngung auf die Eiweißzusammensetzung des Weizenkorns bei unterschiedlicher N-Versorgung der Pflanzen. Phosphorsäure **22**, **1–6** (1962). — Miller, M. H., et A. J. Ohlrogge: Water-soluble chelating agents in organic materials. Proc. Soil Sci. Soc. Amer. **22**, 225–228 (1958). — Mitra, S. P., et N. Panda: Adsorption of glucose by bentonite, Boden und die Oxyde von Eisen und Aluminium. Proc. Nat. Acad. Sci. India A **26**, 67–71 (1957). — Mitchell, W. F.: Trace-elements. Chemistry of the soil **1955**. 253. — Morita, Shuji: Nutrient absorption by rice from media containing different types of clay-minerals. Soil Sci. **86**, 336–342 (1958). — Mortensen, J. L.: Adsorption of hydrolysed polyacrilonitrile on kaolinite. Proc. Soil Sci. Soc. Amer. **21**, 385–388 (1957). — Mortland, M. M., K. Lawton et G. Uehara: Fixation and release of K by some clay-minerals. Proc. Soil Sci. Soc. Amer. **21**, 381–388 (1957). — Moxon, A. L.: Trace-elements, p. 198. New York 1958. — Mulder, E. G., K. Makema et W. L. Van Veen: Molybdenum in symbiotic nitrogen-fixation and in nitrate assimilation. Plant a. Soil **10**, 4, 319–334 (1959). — Muller, J.: Comportement du cuivre dans les sols pauvres en matière organique. Bull. Ass. Fr. Et. Sol. **6**, 295–299 (1959).

Nair, G. G. M., et B. V. Metha: Status of zinc in soils of Western India. Soil Sci. **87**, 155–159 (1959). — Naundorf, F.: Neue Düngungsmethoden beim Cacao. Gordian **58**, 9 (1959). — Nebe, W.: Die Beurteilung der Nährstoffverhältnisse in Waldböden durch Gesteinsanalyse, Düng. u. Meliorat. i. d. Forstw. **66**, 83–92 (1964). — Neergaard, Ch.: Düngungsversuche mit Thomasphosphat in Dänemark. Phosphorsäure **22**, 152–161 (1962). — Neller, J. R.: Extractable sulfate-sulfur in soils of Florida. Proc. Soil Sci. Soc. Amer. **23**, 346–348 (1959). — Nelson et Melsted: Proc. Soil Sci. Soc. Amer. **19**, 439 (1955). — Nicolic, S., D. J. Jelenic, M. Pantovic et A. Vajnberger: Expériences et observations relatives à la fumure potassique — La leucite Yougoslave. Z. Pflanzenernähr., Düng., Bodenkde. **84**, 279–282 (1959). — Nielsen, K. F.: Rate of change of pH in some acid soils due to liming. Canad. J. Soil Sci. **38**, 114–119 (1958). — Nilsson, N.: Synpunkter på västernas kväreförsörgning. Kungl. Skogs. Lantbr. Akad. Tidskr. **98**, 75 (1959). — Nishita, H., A. J. Steen et K. H. Larson: Release of Sr^{90} and Cs^{137} from vina loam upon prolonged cropping. Soil Sci. **86**, 195–204 (1958). — Nishita, H., et al.: Accumulation of Cs and K and growth of bean plants in nutrient solution and soils. Plant a. Soil **17**, 2, 221–242 (1962). — Nommik, H.: Fixation and defixation of ammonium in soils. Acta Agric. Scand. **7**, 395–436 (1957). — Norris, N.: Rhizobium needs magnesium, not calcium. Nature (London) **182**, 734–735 (1958). — Nye, P. H., et W. N. H. Forster: A study of the mechanism of soil-phosphate uptake and plant species. Plant a. Soil **9**, 4, 338–352 (1958).

Page, E. R.: Studies in soil and plant manganese. Plant a. Soil **17**, 1, 99–107 (1963).

— Paauw, F. van der: Potash requirements and meteriological conditions. Plant a. Soil 9, 3, 254–268 (1958). — Periodic fluctuations of soil fertility... by alternating periods of low and high rainfall. Plant a. Soil 17, 2, 155–182 (1962). — Pauwels, G. W. F. H.: Iodine as a micronutrient for plants. Plant a. Soil 14, 4, 377–392 (1961). — Peive, A.: Content of forms of trace-elements that are available to plants in the soils of the USSR. Acad. Sci. Latvia, USSR, Riga, 10, 3 (1959). — Perkins, H. F., et M. Stelly: Ca and Mg content of oats... as affected by Na and K. Soil Sci. 86, 305–309 (1958). — Peterburgski, A. V.: Adsorptionsprozesse im Boden und die Wurzelernährung der Pflanzen. Verh. Int. Bod. Ges. 2, 159–163 (1958). — Peterburgsky, A. V., et al.: Transformation of forms of K in soil during long-time K fertilization. Plant a. Soil 15, 199–210 (1961). — Pritchett, C. F. End et M. N. Malik: The nitrogen status of the mineral soils of Florida. Proc. Soil Sci. Soc. Amer. 23, 127–430 (1959). — Prout, W. E.: Adsorption of radio-active wastes by Savannah River plant. Soil Sci. 86, 13–17 (1958).

Quellette, G. J.: Persistance of borax in sandy soils. Canad. J. Soil Sci. 38, 77–84 (1958).

Radet, E.: Détermination de l'acide phosphorique assimilable dans les sols de craie par le réactif au citrate de Na et évolution des phosphates solubles. Ann. Inst. Nat. Rech. Agr. A 8, 619–632 (1957). — Ragland, J. L., et N. T. Coleman: The effect of Soil Solution Aluminium and Calcium on root growth. Proc. Soil Sci. Soc. Amer. 23, 355–357 (1959). — Reichenbach, H. v.: Über die Beziehungen zwischen K-Nachlieferung und K-Festlegung des Bodens. Verh. Int. Bod. Ges. 2, 188–193 (1958). — Rendig, V. V., et E. A. McComb: Effect of nutrient stress on plant composition: interaction of N with S. Proc. Soil Sci. Soc. Amer. 23, 377–380 (1959). — Rhodes, D. W.: The effect of pH on the uptake of radioactive isotopes from solution by a soil. Proc. Soil Sci. Soc. Amer. 21, 389–392 (1957). — Richard, L.: Interdépendence des éléments de la nutrition des végétaux. C. R. Acad. Paris 248, 3338–3340 (1999). — Rid, H., et A. Suss: Die Verteilung von P^{32} im Boden und dessen Aufnahme durch Sommergerste bei verschiedener Bodenbearbeitung. Landw. Forsch. 11, 40 (1958). — Robinson, R. R., W. G. Sprague et C. F. Gross: The relation of temperature and phosphate placement to growth of clover. Proc. Soil Sci. Soc. Amer. 23, 225–228 (1959). — Rodriguez, J.: Radioactividad en suelos y determinación radiométrica de K^{40}. An. Edaf. Fis. Veg., Madrid, 18, 5, 389–405 (1959). — Rolvink, I. W.: Voorkomen van Co en zijn betekenis voor plant en dier. Landbouwdokumentatie (Nederland) 6, 163–168 (1959). — Romney, E. M., G. V. Alexander, G. M. Le Roy et K. H. Larson: Influence of stable Sr on plant uptake of Sr^{90} from soils. Soil Sci. 87, 42–45 (1959). — Influence of calcium on plant uptake of Sr^{90} and stable Sr. Soil Sci. 87, 160–165 (1959). — Rosanow, M.: Recherches sur le besoin en soufre des prairies. Lab. Rech. du nitrate de Chili 5, 1–7 (1960). — Rouse, R. O.: The K status of Hartsells fine sandy loam as affected by K fertilization. Proc. Soil Sci. Soc. Amer. 22, 334–336 (1958). — Routchenko, W.: Détermination d'une carence en Zn sur maïs. Compt. rend. Acad. Agric. France 47, 739–741 (1961). — Russell, E. W., et R. Scott: The value of measurements of isotopic dilution. Z. Pflanzenernähr., Düng., Bodenkde. 84, 63–75 (1959).

Sanchez, C., et E. J. Kamprath: Effect of liming and organic matter on the availability of native applied Mn. Proc. Soil Sci. Soc. Amer. 23, 302–304 (1959). — Sauerbeck, D.: Auswirkung einer Molybdändüngung auf den Molybdän. Z. Pflanzenernähr., Düng., Bodenkde. 80, 203–217 (1958). — Sawhney, B. L., M. L. Jackson et R. B. Corey: Cation-exchange capacity determination of soils as influenced by cation species. Soil Sci. 87, 243–248 (1959). — Scharrer, K., et W. Höfner: Isotopenversuche über die Aufnahme von Zink durch die Pflanze. Z. Pflanzenernähr., Düng., Bodenkde. 81, 213–224 (1958). — Scharrer, K., et H. Kuhn: Über die Nährstoffwirkung des Humatsdüngers „Wichtel". Landw. Forsch. 12, 2, 111–117 (1959). — Scheffer, F., et F. Ludwieg: Gefäßversuche über die Aufnahme von Radiostrontium aus dem Boden. Z. Pflanzenernähr., Düng., Bodenkde. 85, 244–249 (1959). — Scheys, G., et al.: Expériences de fumures avec engrais cathioniques et anioniques. Agricultura 1, 57–90 (1963). — Schlichting, E.: Über den Haushalt von Raseneisenböden. Phosphorsäure 22, 199–211 (1962). — Schmitt, L.: Beiträge zur Mikronährstoff-Frage III. Landw. Forsch. 16, 93–111 (1962). — Schreiber, R.: Phosphor, eine Lebensfrage für Tier und Mensch. Phosphorsäure 22, 61–73 (1962). — Schreven, D. A. van: N transformations in the former subaqueous soils of polders recently reclaimed from Lake Yssel. Plant a. Soil 18, 2, 143–162 (1963). — Schroeder, D.: Zum Kalihaushalt von Lösböden. Kali-Briefe 5, 1–4 (1956). — Schuffelen, A. C., et H. W. Van der Marel: Potassium fixation in soils. Potassium Symposium, p. 157–201. Bern 1955. — Schuffelen, A. C.: Ionic equilibria in soils. Verh. Int. Bod. Ges. 2, 132 (1958). — Schufle,

J. A.: Effect of complexing agents on the order of cation exchange on kaolinite. Soil Sci. 84, 323–328 (1957). — SCOTT-RUSSELL, R.: The value of measurements of isotopic dilution in the study of soil-plant relationships. Z. Pflanzenernähr., Düng., Bodenkde. 84, 63–75 (1959). — SCROPP, W. H.: Bor und Leguminosen. Arb. Ber. Süd. Vers. Weihenstephan B 2, 5–4 (1957). — Service Pédologique de Belgique: L'importance de l'élément sodium dans le sol. Sol et fumure 5, 1, 11 (1963). — SHAPIRO, R. E., et M. FRIED: Relative release and retentiveness of soil phosphates. Proc. Soil Sci. Soc. Amer. 23, 195–198 (1959). — SHELL, W. R., et J. V. JORDAN: Anionexchange studies of pure clays. Plant a. Soil 10, 4, 303–318 (1959). — SIEGEL, O.: Gehalt der Böden der Bundesrepublik Deutschland an Sr^{90} und Ra^{226}. Stand und Leistung agrikulturchemischer Forschung VI, p. 22–29. Frankfurt 1959. — SIMON-SYLVESTRE, G., et P. BOISCHQT: Etudes sur l'emploi du S élémentaire comme fertilisant du sol. Ann. Agron. 13, 549–574 (1962). — SINGLETON, W. R.: Nuclear radiation in food and agriculture, p. 218, 219 et 223. New-York 1958. — SLUIJSMANS, C. M. S.: L'analyse du sol comme base de la fumure de l'avoine avec Mg. (Ned). Versl. Landbouw. Onderz. 1–37 (1964). — SMITH, J. A., et B. C. MATHEWS: Release of K by 18 Ontario soils during continuous cropping in the greenhouse. Canad. J. Soil Sci. 37, 1–10 (1957). — SOHN, J. B., et M. PEECH: Retention and fixation of ammonia by soils. Soil Sci. 85, 1–9 (1958). — SPERBER, J. I.: Release of phosphate from soil-minerals by Hydrogen sulphide. Nature (London) 181, 934 (1958). — SSASANOW, W. I., et A. G. SSERGEJEWA: L'influence de la luzerne et de l'esparcette sur l'assimilabilité du P des terres noires irriguées. Pédologie (russe) 5, 99–101 (1958). — STEMMET, M. C., et al.: The uptake of CO_2 by plant roots. Plant a. Soil 17, 3, 357–364 (1962). — STENUIT, D. F., et R. PIOT: Magnesium — Hoofdelement voor plantenvoeding. Rev. Agron. 6, 897–920. (1958). — Traitement des semences au sulphate de Mn. Pédologie (Belgique) 7, 266–270 (1957). — STENUIT, D. F., R. PIOT et R. BOON: Relation entre le pH et la teneur en Mn des sols sablonneux et l'apparition de carence en Mn chez l'avoine. Pédologie (Belgique) 7, 259–263 (1957). — STENUIT, D. F., et R. PIOT: La magnésie, un élément essentiel pour l'alimentation des plantes. Rev. Agron. (Bruxelles) 3, 401–428 (1958). — STENUIT, D. F., et G. VAN HOECK: La détermination du Mn dans le sol. Pédologie (Belgique) 7, 251–257 (1957). — STENUIT, D.: Beziehung zwischen der Phosphatform und dem Ertrag auf Sandböden mit Mn-Mangel. Z. Pflanzenernähr., Düng., Bodenkde. 1, 42–52 (1962). — STENUIT, D., et PIOT: La lutte contre la carence en Mn dans les sols sableux. Agricultura 10, 2, 377–420 (1962). — STERLING, R. OLSEN et F. S. WATANABE: Solubility of Ca carbonate in calcareous soils. Soil Sci. 88, 123–129 (1959). — STEVENSON, F. J.: C/N-Relationships in soil. Soil Sci. 88, 201–205 (1959). — STEVENSON, F. J., A. P. S. DHARIWAL et M. B. CHOUDHRI: Further evidence for naturally occuring fixed ammonium in soils. Soil Sci. 85, 42–46 (1958). — STEVENSON, J. F., et A. P. S. DHARIWAL: Distribution of fixed ammonium in soils. Proc. Soil Sci. Soc. Amer. 23, 121–125 (1959). — STRASMAN, A., P. QUIDET et R. BLANCHET: Valeur comparée des divers tests analytiques relatifs au K du sol. Compt. Rend. Acad. Agron. France 44, 639–642 (1958). — STROBEL, G.: Wirkung der Phosphatdüngung auf der Wiese in einem 30jährigen Düngungsversuch. Phosphorsäure 22, 7–27 (1962). — STRUYS, L. C.: De aanvoer van Mg met de neerslag. Kalk 11, 35–38 (1958). — STUDER, R., et R. BLANCHET: Irrigations en région tempérée à influence océanique et interréactions entre l'alimentation potassique et l'alimentation hydrique des plantes. Compt. rend. Acad. Agric. France 5, 339–348 (1963). — SÜSS, A.: Isotopenversuche zur Frage der Nährstoffwanderung im Boden. Verh. Int. Bod. Ges. 2, 235–239 (1958). — SUTTON, C. D., et S. LARSEN: The residual value of fertilizer P applied in two field experiments. Plant a. Soil 18, 2, 267–272 (1963). — SWAINE, D. J.: The trace-elements of fertilizers. Comm. Bur. Soils 52, 305 (1962).

TAKAHASHI, T., et D. YOSHIDA: Fractionation of foliar-adsorbed pH by paper electrophoresis. Soil a. Plant Food 4, 32–36 (1958). — THORNE, P. W., et H. H. WIEBE: Atomic energy and agriculture, p. 50–61–63. Washington 1957. — TOBIA, S. K., et A. S. HANNA: Effect of Cu SO_4 added to irrigation water on copper-status of Egyptian soils. Soil Sci. 85, 302–306 (1958). — TROBISCH, S., et R. GERMAR: Ergebnisse eines Mo-Düngungsversuches zu Blumenkohl. Deutsch. Landw. 10, 189–191 (1959). — TSCHERNOW, W. A., et W. S. MAXIMOWA: Austauschreaktionen zwischen adsorbierten Wasserstoffionen und den Al- und Mg-ionen in Tonen. Verh. Int. Bod. Ges. 2, 179–182 (1958). — TWYMAN, E. S.: Effect of iron supply on tomato. Plant a. Soil 10, 4, 375–388 (1958).

ULRICH, B.: Die Umsetzungen der anorganischen Phosphate als Funktion des Kalkpotentials. Z. Pflanzenernähr., Düng., Bodenkde. 100, 2, 97–102 (1963). —

ULRICH, B.: Theoretische Betrachtungen zur Frage der Rohsphosphatwirkung. Landw. Forsch. **12**, 30–36 (1999).

VENENA, K. C. W.: Some notes regarding the function on the sulphate-anion in the metabolism of oil producing plants. Potash a. Tropic. Agric. **6**, 2, 22–43 (1963). — VERMEULEN, F. H. B.: Stichting Ned. Landb. Kalkbureau — De Bilt (Nederland), Nr. II, 3–7 (1958). — VINE, H.: Some measurements of relase and fixation of N in soil of natural structure. Plant a. Soil **17**, 1, 109–130 (1963). — VUORINEN, J.: On the amounts of minor elements in Finnish Soils. J. Agr. Sci. Soc. Finland **30**, 36–40 (1958).

WALKER, J. M., et S. A. BARVER: Absorption of K and P from the soil by corn roots. Plant a. Soil, XVII, **2**, 243–259 (1962). — WALKER, T. W., et A. F. R. ADAMS: Competition for sulphur in grass-clover association. Plant a. Soil **9**, 4, 353–366 (1958). — Residual effects of calcium sulphate on the yield and composition of a grass-clover association. Plant a. Soil **10**, 2, 176–182 (1958). — Studies on soil organic matter; 2. Soil Sci. **87**, 1–10 (1959). — WALLACE, T.: Symposium on the use of metal chelates in plant nutrition, p. 48 et 72. Teh. nat. Press. California 1956. — WALLACE, A., et R. T. MUELLER: Response of plants to Zinc and Mn-chelates. Proc. Soil Sci. Soc. Amer. **23**, 79 (1959). — WALLACE, T.: Trace-elements. Deficiencies and excesses and their control. Congr. mond. Rech. Agron. Rome **1959**, 511–524. — WEISS, A.: Über das Kationenaustauschvermögen der Tonminerale II. Z. anorg. Chem. **297**, 257–286 (1958). — WELTE, E., et E. A. NIEDERBUDDE: Die unterschiedliche Bindungsfestigkeit von fixiertem K und ihre mineralogische Deutung. Plant a. Soil **18**, 2, 176–190 (1963). — WEY, R., O. SIESKIND et T. P. LIEBENGUTH: L'influence du pH sur l'absorption des cations par les argiles. Bull. Ass. Franc. Et. Sol. **6**, 258–266 (1959). — WIGDOR, S., R. F. STINSOW et W. W. MCCALL: Chloride toxicity of flowering stock and sweet peas. Mich. St. Univ. Agr. Bull. **40**, 468–478 (1958). — WIJK, W. R. VAN: Physics of plant environment, Amsterdam. p. 362, North-Holland Publ. Co. 1963. — WIKLANDER: Handb. Pflanzenphysiologie, Vol. IV, p. 124–128–134–149–153–163 (W. RUHLAND, éd.). Berlin-Göttingen-Heidelberg: Springer. 1958. — WIKLANDER, L., et al.: Influence of exchangeable ions on release of mineral-bound ions. Soil Sc. **95**, 9–15 (1963). — WILDE, S. A.: Forest soils and forest growth. 1958. — WILLIAMS, F., S. M. BROMFIELD et C. U. WILLIAMS: Effect of glucose on phosphate availability. Austr. J. Agric. Res. **9**, 640–663 (1958). — WINOGRADOW, A. P.: Geochemie seltener und nur in Spuren vorhandener chemischer Elemente im Boden. Berlin 1954.

X — Report of the Rothamsted Exper. Station, Harpenden, Engl. (1959).

YAALON, DAN, H.: Physico-chemical relationships of $CaCO_3$, pH and CO_2 in calcareous soils. Trans. V Int. Congr. Soil Sci., Léopoldville **2**, 356–363 (1954). — YUAN, T. L., et J. G. A. FISKEL: 6 Aluminium studies II. Proc. Soil Sci. Soc. Amer. **23**, 202–205 (1959). — YUAN, T. L., et W. K. ROBERTSON: Specific P^{32} sorption by oils. Soil Sci. **86**, 220–225 (1958).

d) Méthodes de détermination des principes nutritifs dans le sol[1, 2]

1. La détermination des ions non-échangeables

Comme ces ions font partie du squelette alumino-silicaté et se trouvent à l'intérieur de la particule du sol, leurs méthodes de détermination coïncident avec celles utilisées en chimie minérale et en pétrochimie.

Nous renvoyons le lecteur aux ouvrages spéciaux parus dans ce domaine. Nous nous contentons ici d'un aperçu bref et critique.

A. Détermination quantitative globale des ions non-échangeables: analyse chimique élémentaire

Notons que cette analyse "totale" englobe aussi les ions échangeables qui toutefois ne représentent souvent qu'une fraction négligeable (moins 1%) de la teneur totale en ions.

---

[1] La détermination des principes nutritifs dans le sol implique la prise préalable des achantillons sur place. Ceci est une opération délicate et compliquée qui dépasse le cadre du présent article (LAMBERT 1964).

[2] L'analyse foliare des arbres n'a de valeur que si elle est combinée avec celle du sol. Elle exige en autre une technique bien définie, variable avec les essences (TAMM 1965, HOINE 1964).

On emploie les macro-méthodes classiques gravimétriques et titrimétriques, les micro-méthodes spectro-photométriques: spectromètre d'émission à flamme et photomètre d'absorption, les méthodes colorimétriques, et enfin les micro-mé thodes au moyen de réactifs organiques: par complexation.

Le choix de la méthode à employer dépend de plusieurs facteurs: notamment de l'exactitude requise, de la simplicité et de la vitesse de la méthode en rapport avec le nombre d'échantillons à analyser, et enfin de la quantité de matière disponible.

Si cette quantité est suffisante, les macro-méthodes gravimétriques et titrimétriques classiques restent les plus exactes, mais sont souvent lentes et ne se prêtent pas à l'analyse en grande série.

Les méthodes spectro-photométriques, tant d'émission que d'absorption, permettent d'analyser plusieurs éléments dans le même extrait mais exigent souvent une préparation laborieuse de l'échantillon en vue d'écarter les substances interférentes.

Enfin, les méthodes à réactifs organiques complexants sont indiquées en micro puisqu'elles sont très sensibles, souvent sélectives, et possèdent un facteur de conversion très favorable (FEIGL 1956).

B. La détermination de la position de l'ion dans le réseau cristallin

Elle se fait au moyen des méthodes de détermination de la structure cristalline des minéraux argileux: röntgenologie, microscope électronique, analyse thermodifférentielle et thermopondérale (RAMACHANDRAN et GARG 1959).

2. La détermination des ions échangeables

Toute détermination d'ions échangeables porte un cartain *caractère conventionnel et relatif:* les résultats des différentes méthodes ne sont comparables qu'à l'intérieur d'une méthode donnée, mais les résultats des différentes méthodes ne sont pas comparables entre eux (BARROWS et DROSDORF 1958).

Le motif est que toute détermination d'ions échangeables est une réaction d'équilibre et non une réaction finale. D'ailleurs, le nuage d'ions autour de la particule est diffus du côté extérieur et passe insensiblement aux ions non-échangeables du côté intérieur (cfr. Fig. 94).

Il est évident que toute détermination d'ions échangeables englobe aussi les ions solubles qui cependant sont négligeables en sol vierge mais peuvent devenir importants en sols salins et alcalins et aussi dans les vieilles terres de culture enrichies par les engrais.

Les résultats dépendent *du choix du liquide déplaçant:* de sa composition chimique, de sa concentration, de son activité ionique et de son pH.

Nous avons vu antérieurement que chaque cation a une énergie d'adsorption — et donc de déplacement — qui lui est propre. Le pouvoir déplaçant dépend en outre de sa présentation chimique: sel neutre ($Ca\ Cl_2$) ou sel hydraulitique ($Ca\ CO_3$), donc de la nature de l'anion présent.

Les méthodes "continues", à lavages répétés, par dialyse ou centrifugation, sont en principe préférables aux méthodes à un seul traitement.

Les premières courent cependant le danger d'extraire, par contact trop prolongé du liquide, une partie des ions non-échangeables et offrent l'inconvénient de peptiser l'échantillon avec colmatation subséquente, surtout par l'emploi d'ions déplaçants monovalents (Na — Li) et de l'eau; l'emploi de solvants organiques anhydres est préférable; employer des membranes dialysantes: cellophane, collodion, etc.

Les sols calcareux et alcalins, contenant des sels hydraulitiques libres donnent des chiffres de bases échangeables supérieurs à leur capacité totale d'échange. Ces sels libres doivent d'abord être éloignés par un lavage au moyen d'un liquide non électrolyte, sinon ils sont comptés comme bases échangeables.

Différents procédés ont été proposés pour contourner cette difficulté: l'emploi de l'$AcNH_4$ alcoolique, introduction d'un facteur de calcul correctif pour la solubilité du $Ca\ CO_3$ dans le filtrat, l'emploi d'un mélange de $Ba\ Cl_2$ + triethanolamine à pH 8,1 qui recouvre les carbonates d'une pellicule insoluble de $Ba\ CO_3$ etc.

Aucune de ces solutions n'est entièrement à l'abri de critiques. En réalité, la détermination des ions échangeables des sols calcareux reste une difficulté.

Toute détermination d'ions échangeables peut se baser, soit sur la *concentration soit sur l'activité:* cette dernière se détermine en principe au moyen d'électrodes sélectives dont la sélectivité diffère avec le système employé. Pour le sodium, l'électrode en verre sodique semble préférable (SYRIN et ORLON 1958).

L'emploi d'isotopes radioactifs comme ions déplaçants, ouvre des perspectives intéressantes dans le domaine des ions échangeables, spécialement pour le P et pour le Ca. Ces procédés sont assez compliqués mais offrent l'avantage de pouvoir suivre l'équilibre qui s'établit entre l'ion absorbé et l'ion entrant en solution et déterminer ainsi le degré de fixabilité d'un ion au sol (FRIED 1957, SCHEFFER et ULRICH 1958).

Il permet aussi de déterminer si l'échange est complet ou partiel ainsi que le pourcentage de l'ion absorbé et de l'ion libéré.

Enfin, il nous donne une idée de l'arrière-effet des engrais et de leur coefficient de rendement: ainsi le S des sulfates est assimilé surtout la première année, tandis que le S élémentaire l'est surtout la deuxième année, quoiqu' à dose moindre que le sulfate, probablement parce qu'il doit d'abord se transformer en SO_4^{--} (JORDAN et BAKER 1959).

Une des difficultés de l'emploi des isotopes réside dans leur différence de masse qui peut fausser l'interprétation de leur réaction chimique (LINSER et KAINDL 1960). Le livre récent de ces auteurs est une excellente mise au point de nos connaissances actuelles dans ce domaine.

Toute détermination d'ions échangeables comporte *deux opérations: l'extraction* de l'ion adsorbé au moyen d'un ion déplaçant approprié et la *détermination* de l'ion libéré au moyen d'un procédé chimique ou physique.

La deuxième opération est du domaine de la chimie analytique générale et ses procédés sont bien mis au point pour chaque ion déterminé.

Le déplacement, au contraire, de l'ion adsorbé est une opération délicate, souvent encore sujette à discussion; nous y reviendrons au paragraphe de la détermination des ions assimilables par la plante.

A. La capacité d'échange ou de sorption d'un sol

Méthodes chimiques. En principe, cette détermination se fait en saturant un sol avec une solution normale d'un sel tamponné au pH 7, par ex. Na Ac et ensuite, par déplacement du cation adsorbé par un autre cation (Ca Ac) à la même normalité. L'excès de l'ion déplaçant est éliminé par lavages au moyen d'un liquide non-électrolyte (acétone).

Comme l'anion acétate est absorbé par certains sols, il vaut mieux saturer le sol avec le $CaCl_2$ qui lui est aussi plus ou moins absorbé, mais qui peut être facilement déterminé. On soustrait alors le Ca correspondant du Ca total et on obtient ainsi la charge négative netto (RICH 1962).

Enfin, le Ca adsorbé est déplacé au moyen du premier solvant (Ac Na) et déterminé dans la solution par titrimétrie ou spectro-photométrie.

Le tamponnement préalable du sol est nécessaire si l'on emploie des sels d'acides forts qui par leur dissociation entravent la sortie de l'ion H^+ du complexe.

Voilà pourquoi, on emploie de préférence un acide organique moins dissocié.

Du côté cationique, on emploie de préférence les cations qui forment des bases fortes: Ca, K, Na, NH_4: ceux-ci donnent des résultats plus constants que les bases faibles: Cu, Zn, Al, Fe qui fournissent souvent des résultats trop élevés (MOSCHEN 1958).

La composition minéralogique du sol doit influencer le choix du cation déplaçant: la montmorillonite est assez indifférente à ce point de vue, mais avec la vermiculite qui fixe le K, celui-ci employé comme déplaçant, fournit des résultats inférieurs à ceux obtenus avec le Ca (SAWHNEY 1959).

Méthode au spectromètre à flamme. L'extraction est en principe la même que dans la méthode chimique, avec cette différence qu'on emploie ici de préférence l'ion NH_4^+ comme ion déplaçant qui permet la lecture photométrique directe de la solution. C'est la méthode la plus sensible et la plus rapide, seulement elle a l'inconvénient de nécessiter une préparation préalable de l'échantillon, dans le but d'éloigner les substances interférentes: Fe, Al, phosphates, etc. Toutefois, pour certains cations, cette interférence suit des règles assez précises et peut donc être neutralisée par un facteur de correction. Ainsi la présence de Ca "exalte" les lignes spectrales du Mg et Mn dont les teneurs réelles peuvent être estimées avec une approximation suffisante en multipliant les résultats obtenus par le facteur de correction 0,4 (BURRIEL MARTI et ALVAREZ HERRERO 1959).

L'élément Mg reste le plus délicat à déterminer; on peut rehausser considérablement l'exactitude des déterminations (0,024—0,27%) en ajoutant au Beckman DU un condensateur électrolytique de 2.000 F, 6 v. (MATHIEU et al. 1961).

Méthode isotopique. Ce procédé présente un intérêt spécial dans la détermination de la capacité d'échange des sols calcareux et alcalins où les méthodes précédentes offrent des difficultés par la présence de sels libres.

L'emploi de déplaçants marqués comme le Sr^{89} a l'avantage de pouvoir déplacer et mesurer de très faibles quantités dans une suspension très diluée. Le Sr^{89} a plusieurs avantages: comme ion divalent il déplace facilement d'autres ions divalents (Ca). La teneur en Sr des sols normaux est extrêmement faible. Il émet des particules β très énergétiques (E_{max}=1,46 MeV), il a une demi-vie favorable (50 jours) et son activité se mesure facilement par un compteur ordinaire (BERGSETH et al. 1963).

En général, les méthodes isotopiques se trouvent à la base de la détermination du besoin en engrais, qui se fait encore actuellement au moyen d'un réactif choisi arbitrairement.

S'il n'y a pas d'effet chimique entre les atomes ordinaires et marqués, les réactions d'échange se font dans des systèmes équilibrés. La désorption au contraire a lieu dans des systèmes non équilibrés (LAMM et al. 1963).

Méthode colorimétrique. En combinaison avec la méthode isotopique. En employant comme ions déplaçants Sr^{90} ou Cs^{137} et le cobalti-hexamine comme colorant, on obtient des résultats aussi précis qu'avec la méthode standard à l'acétate, alors que la méthode est sensiblement plus rapide (AMAVIS 1959).

B. La détermination des cations métalliques (valeur S de Hissink)

On agite le sol avec un acide ou son sel et on mesure exactement les changements de pH pendant l'opération en se basant sur l'équation de BROWN:

$$S \text{ (m/e/100 gr)} = (\text{pH mesuré} - 2{,}31) \cdot 22$$

ou bien on titre le restant de l'acide après la fin de la réaction.

Ce dernier procédé n'est évidemment pas applicable aux sols calcareux dont le calcaire libre neutralise l'acide.

C. Détermination de l'H échangeable (H = T — S d'après Hissink)

En principe, on traite le sol avec un "accepteur" de protons (H^+), donc un sel d'acide faible et on titre l'acide formé au moyen d'une base titrée.

On peut aussi mesurer les changements de pH comme pour la détermination de S.

On peut titrer directement le sol au moyen d'une base forte ou d'un sel hydraulitique.

Enfin, on peut simplement soustraire S de T: cette dernière méthode ne fournit que des résultats approximatifs. La titration de sols très acides fournit des valeurs d'H échangeable trop élevées du fait de l'hydrolyse de l'Al libéré. Dans ce cas on "neutralise" cet Al par complexation avec un fluorure. Après cela, on peut ou bien titrer directement l'H échangeable, ou bien on peut déterminer l'acidité d'échange et celui de l'aluminium; l'H échangeable est alors calculé par différence (YUAN 1959).

D'autres employent le Ba Cl_2 — triehtanol aminé ou l'acétate d'Am. Tous ces procédés fournissent des résultats différents; ils sont seulement comparables à l'intérieur d'une méthode déterminée (PEECH et al. 1962).

Toutes ces méthodes peuvent servir à *déterminer le besoin en chaux d'un sol acide.*

Approximativement, chaque m.e. d'hydrogène échangeable par 100 g de sol exige 500 kg de Ca CO_3 par million de kg de terre arable.

Toute détermination, au laboratoire, du besoin en chaux d'un sol porte un certain caractère de relativité. Les résultats appliqués sur le champ, donnent rarement la neutralité désirée: facteur de conversion, donc erreur possible très élevée, mélange défectueux de la chaux avec le sol, impureté du calcaire, etc.

D'ailleurs, le pouvoir neutralisant des différents calcaires, même purs, n'est pas le même (SHAW et ROBINSON 1959).

La détermination du pouvoir neutralisant des calcaires au moyen de leur solubilité dans l'eau chargée de CO_2, dépend de leur état physique et de l'absence ou présence de Mg (dolomite). En répétant plusieurs fois l'extraction on peut cependant obtenir des résultats qui correspondent à la réalité des choses (GREZES 1959).

Une méthode uniforme et d'application générale à tous les calcaires est encore à trouver (SCHOLLENBERGER et al. 1962).

D. Détermination de chaque cation séparément

Avec le photo-spectographe, on emploie de préférence comme ion déplaçant l'Ac NH_4 normal qui mouille bien l'échantillon, est volatil après extraction, et permet en une seule opération la détermination simultanée des principaux cations échangeables: Ca^{++}, Mg^{++}, K^+, Na^+ (excepté évidemment l'NH_4^+ lui-même).

Pour le Ca, l'extraction à l'acétate normal, contrôlée par l'échange isotopique du Ca^{45} a donné des résultats sensiblement égaux. Cette méthode peut donc servir à déterminer le Ca échangeable du sol. Echangeable ne veut pas dire cependant assimilable, parce que, si la désorption du Ca-sol par le liquide du sol est relativement simple, l'absorption par la plante de ce Ca du liquide-sol est beaucoup plus compliquée (NEWBOULD et SCOTT RUSSELL 1963).

Les cations majeurs échangeables. *Le calcium* se détermine titrimétriquement ou colorimétriquement par la méthode à l'oxalate-cérate, au versénate, au picronolate, à l'oxalate, au chloranilate, etc. (JACKSON 1958), ou mieux par spectrométrie (GETTKANDT 1958).

Cette dernière méthode nécessite cependant l'éloignement préalable de substances interférentes comme les phosphates et les sulfates.

On peut neutraliser ces interférences par un sel de fer ou par le $La\,Cl_3 \cdot 7\,H_2O$ (YOFE et FINKELSTEIN 1958), ou par un pulvérisateur de construction spéciale (FILCEK 1959).

En utilisant un photomètre à flamme avec filtres très sélectifs, on peut déterminer le Ca avec une exactitude de $\pm$ 3%, même en présence d'un grand excès de Na (SCHUHKNECHT 1957).

Le magnésium se détermine par une des méthodes macro classiques au phosphate, gravimétriquement ou titriquement.

Le meilleur réactif d'extraction semble être le $Ca\,Cl_2$. Ce Mg est en corrélation étroite ($r = +0,94$) avec la hauteur de la récolte en vases de végétation (SCHROEDER et al. 1963).

En micro-analyse, il est déterminé colorimétriquement au bleu de phosphomolybdène ou au jaune Titane (SCHILLING 1957), ou bien encore par le 8-hydroxyquinoline, très sensible mais assez compliqué (JACKSON 1958, p. 100) et enfin évidemment par spectrométrie. Dans ce dernier cas, l'emploi de l'alcool isopropylique semble préférable aux autres alcools (SCHACHTSCHABEL et SCHWERTMANN 1958).

La méthode au versénate permet, moyennant précautions spéciales, la détermination du Mg échangeable, même en présence d'un excès de Ca (LEFEVRE 1958).

Le sodium: outre le spectro-photomètre, on emploie le procédé à l'acétate d'uranyle: $Na\,Mg \cdot (UO_2)_2$ Ac. 8 H_2O déterminé gravimétriquement, titrimétriquement, ou colorimétriquement en utilisant la fluorescence jaune de l'acétate: cette dernière méthode est rapide (15 minutes) et peut s'exécuter même sur le champ (MATELSKI et YIEN 1958).

Le Na peut se déterminer aussi au moyen d'une électrode en verre spéciale à l'Ag/Cl Ag dans une solution de Na Cl. Le potentiel mesuré est en relation avec les concentrations de $H^+ + Na^+ + K^+$.

En déterminant séparément K^+ et H^+, on obtient par différence le Na^+ dont les valeurs correspondent suffisamment avec celles obtenues au photomètre à flamme (BOWER 1959).

Le potassium: Les méthodes gravimétriques au chloroplatinate ou au perchlorate sont trop lentes pour pouvoir servir à l'analyse en grande série. D'autres méthodes gravimétriques: — dipicrilamine, hexanitrodiphenylamine et au nitrophenylborate (titration) sont plus expéditives.

Signalons aussi une méthode microbiologique à l'aspergillus niger (FIEDLER 1958) et une méthode complexométrique à l'éthylène-diamine-tétra-acétate disodique qui chélate le K, qui est rapide et très sensible vu le facteur de conversion très favorable dn complexe (PILLERI 1957).

Mais les deux méthodes les plus courantes sont le spectromètre à flamme et la méthode colorimétrique au cobaltinitrite; malgré certains inconvénients, elles restent les méthodes les plus rapides et les plus sensibles.

Le spectro-photomètre est le plus sensible (1—3%) et permet la détermination de plusieurs ions dans le même extrait, non seulement les "grandes" bases échangeables: Ca, Mg, K, Na, mais aussi toute une série d'éléments mineurs: Fe, Cu, Co, Mn, Mo, Zn, etc.

L'emploi du spectromètre à flamme exige de l'analyste une routine et une expérience qui doivent être acquises. Cette méthode si rapide et si sensible est en réalité d'un maniement délicat, même entre les mains d'un chimiste expérimenté.

Signalons quelques points qui méritent une attention soutenue: le remplacement régulier du dessicant, le chauffage préalable, le choix de la cellule photoélectrique, le choix du gaz à brûler (acétylène, hydrogène, gaz de ville) ainsi que sa pression, etc.

La méthode au cobaltinitrite est la plus sensible après celle du photomètre à flamme.

Le fait que la composition chimique du précipité peut dépendre des conditions opératoires, n'empêche pas la parfaite reproductibilité des résultats, du moment que les conditions restent les mêmes: il faut soigner spécialement la standardisation de la solution étalon au KCl. La détermination finale du précipité au cobaltinitrite peut se faire gravimétriquement, titrimétriquement ou colorimétriquement: ce dernier procédé est le plus rapide mais le moins sensible.

Le pouvoir fixateur des sols pour le K: Détermination du K devenu non-échangeable. Plusieurs méthodes ont été essayées: humectations du sol et dessications successives qui sont supposées emprisonner le K entre les feuillets; chauffage du sol à 500° (Legg et Axley 1958); traitement du sol au NaOH bouillant qui dilate les feuillets et libère le K emprisonné.

Détermination du K semi-échangeable, c'est-à-dire, de la partie non-échangeable mais restant assimilable par les plantes (Stahlberg 1958): électrodialyse continue, et enfin extraction du K au moyen d'aspergillus niger (Fiedler 1958).

Tous ces procédés n'ont qu'une valeur relative et ont été essayés pour déterminer le K non échangeable mais assimilable par les plantes, ces deux notions n'étant pas synonymes comme nous l'avons vu plus haut.

L'ammonium échangeable: sa détermination exige certaines précautions, d'une part à cause de son caractère trop labile (transformation microbienne en nitrate, néoprotéinisation ou réduction) ou trop stable: devenant non-échangeable par pénétration entre les feuillets comme le K.

En outre, pendant la conservation de l'échantillon, des composés azotés organiques peuvent s'hydrauliser avec production d'ammoniaque.

L'analyse doit se faire le plus rapidement possible et sur l'échantillon humide (avec détermination concomitante de l'humidité).

L'ion déplaçant peut être H^+ qui permet la distillation directe de NH_4^+ au Kjeldahl, ou l'ion Na^+ ou K^+ qui permet l'extraction de l'NH_4^+ entre les feuillets par leur gonflement (Scott 1958, Dhariwal et Stevenson 1958). L'ammoniaque déplacé peut être déterminé par la méthode Nessler ou par Kjeldahl macro ou micro.

La quantité d'NH_4 échangeable des sols normaux est minime vis-à-vis des autres bases (0,02 — 0,15 m.e.) et peut être négligée dans le calcul de la capacité totale d'échange.

Les anions majeurs assimilables. Quoique les lois de l'échange s'appliquent en principe aussi bien aux anions qu'aux cations, on applique aux premiers plus couramment l'épithète d'assimilable plutôt que d'échangeable. Le motif en est que, plus encore que pour les cations, pour eux assimilable n'est pas du tout synonyme d'échangeable.

D'ailleurs, les procédés d'extraction des anions ont été élaborés en premier lieu en vue de leur disponibilité pour les plantes plus que pour leur caractère échangeable.

La notion "assimilable" est toujours plus ou moins relative et conventionnelle, puisque la quantité extraite du sol dépend du choix du réactif employé.

En principe, ce réactif doit être d'une force telle qu'elle se rapproche du pouvoir d'attaque des racines elles-mêmes. C'est un rapprochement qu'on ne réalisera jamais d'une manière satisfaisante puisque les conditions opératoires du laboratoire ne pourront jamais imiter le jeu souple d'un système radiculaire extrêmement ramifié qui dispose de toute une saison de végétation, ou d'un nombre illimité d'années pour un arbre, pour extraire du sol son capital nutritif nécessaire.

La seule façon rationnelle d'obtenir un réactif à effet "physiologique" valable est de comparer son résultat, moyennant de multiples analyses, aux récoltes de multiples champs d'expériences établis dans ce but (BAEYENS et STENUIT 1958).

L'azote assimilable. On ne peut parler d'N échangeable que sous forme d'ion NH_4^+. Sa détermination a été discutée à propos des cations échangeables.

Restent la détermination de l'azote organique ($\pm 99\%$ de l'N total du sol) et l'azote nitraté qui dérive de l'azote organique par activité microbienne. Comme cette activité est sous la dépendance des facteurs du milieu (humidité, température, aération, etc.) essentiellement variables, la détermination de l'N dit assimilable est une entreprise très compliquée et qui se prête difficilement à un système d'analyses en série.

Voilà pourquoi, on se contente d'ordinaire de la détermination de l'N total et de l'azote nitraté au moyen des procédés classiques au KJELDAHL (N total) ou colorimétriquement avec l'acide nitro-phénolsulfamique et autres (azote nitraté).

Ces méthodes ont subi de multiples variantes dans le but de les adapter aux différentes catégories de sol (minéraux ou organiques), et à la vitesse du travail, dans le but de simplifier les appareils, et surtout dans le but de les rendre *micro* et même *ultra-micro* utilisables.

Nous renvoyons le lecteur aux ouvrages et aux revues spécialisés dans ce domaine.

L'étude des *transformations biologiques* de l'N dans le sol comporte la détermination de l'N total comme point de départ, et des nitrates comme point final: la simple détermination des nitrates n'est pas une mesure de l'intensité de la minéralisation de l'N dans un sol, puisque sa teneur en nitrates à un moment donné est la résultante du bilan azoté de ce sol avec ses *gains* (fixation microbiologique de l'azote de l'air — précipitations atmosphériques) et ses *pertes* (dénitrification, lixiviation, exportation par les plantes).

Toutefois l'intensité de la minéralisation de l'N organique peut être déterminée relativement au laboratoire en incubant les échantillons de sol dans des conditions déterminées de température, d'humidité et d'aération avec déterminations régulières des nitrates formés.

Il est bien entendu que ce procédé ne fournit que des résultats relatifs qui ne représentent pas nécessairement les conditions "in situ naturali".

Toutefois des résultats pratiquement utilisables ont pu être obtenus pour des sols et des endroits écologiquement très semblables (Harmsen et Van Schreven 1955).

Comme la méthode d'incubation est assez compliquée, d'autres auteurs ont cherché et trouvé une corrélation satisfaisante entre cette méthode et la teneur du sol en azote soluble (NH_3 et N organique) dans l'eau bouillante. Cette dernière méthode caractérise en outre assez bien les fluctuations de l'azote minéralisable au cours de la saison (Livens 1959).

Elle permet de contrôler le moment de l'épuisement de la réserve azotée d'un sol et par conséquent de déterminer les époques des applications fractionnées des engrais azotés.

La teneur du sol en N soluble est aussi une indication de la rapidité d'action des différents engrais azotés (Harmsen 1959) et en particulier de la catégorie des produits de condensation de l'urée qui, grâce à leur solubilité assez lente, fournissent à la plante l'N nécessaire pendant toute leur durée de végétation (Atanasiu 1959).

En outre, il faut tenir compte de l'hétérogénéité horizontale étonnante du sol sur place, ce qui nécessite un échantillonnage multiple réparti en plusieurs endroits du champ, avec triage statistique des résultats.

L'N minéralisé descend dans le sous-sol pendant l'hiver et est assimilé par les racines profondes de la plante de l'année suivante. L'échantillonnage du sol, en vue de la détermination de son besoin en N au laboratoire, doit tenir compte de ce fait. Les échantillons pris uniquement dans la couche arable donnent une idée fausse de l'état azoté réel du sol et conduisent à des conseils de fumure erronnés (Gasser 1962).

Dans les sols normaux, la teneur en *nitrites* est négligeable: puisqu'ils sont nitratés au fur et à mesure de leur production. Ils s'accumulent seulement en conditions spéciales, par ex. en sols alcalins après application massive d'engrais ammoniacaux. Leur détermination se fait au moyen de l'acide sulfanilique et (γ)-naphthyaline mais l'opération est délicate vu le caractère changeant de l'anion à déterminer.

Le phosphore échangeable et assimilable. La détermination du P soit-disant échangeable et assimilable se complique du fait de la forte fixabilité de cet élément dans beaucoup de sols.

Cette fixabilité du phosphore dans le sol peut être mesurée sur place par l'emploi du P^{32} et de plaques photographiques spéciales qu'on enfouit dans le sol (Süss 1959)[1]. Puisque l'assimilabilité du phosphore semble dépendre beaucoup du pH du sol, on peut classer les réactifs d'extraction en deux catégories: acides et basiques.

Ces réactifs, comparés entre eux pour un même sol, donnent des résultats qui concordent bien pour un sol riche en P, mais sont peu concordants pour les sols pauvres.

Ce sont cependant surtout ces derniers sols qui exigent une analyse en vue de la détermination de leur besoin en P (Pancorbo et Lucena 1959).

Le choix du réactif doit en outre tenir compte de son influence possible sur les substances interférentes au colorimètre par l'emploi du bleu de phosphomolybdène.

Spécialement, l'extraction à l'acide citrique nécessite la destruction préalable

[1] Signalons ici que le bombardement d'un orthophosphate par des neutrons dans un réacteur nucléaire, produit 50% du P^{32} sous forme non-orthophosphatique qui peut fausser les résultats des expériences avec le P^{32}. Ludwieg (1959) a élaboré une méthode chimique de séparation de ces deux formes de P^{32}.

de celui-ci, destruction qui peut cependant être omise par l'emploi du réactif spécial de LUCENA (1959).

Le phosphore "échangeable". Capacité d'échange des sols pour le P — Pouvoir de fixation des sols par le P.

Cette forme de phosphore semble liée à sa surface spécifique mesurable par échange avec le phosphore isotopique P^{32} (OLSEN et WATONABE 1957).

Il se forme probablement des composés insolubles de P avec le Fe, Al, et Ca que l'on peut déterminer par ex. par le procédé Bass et Sieling, ou par extraction au moyen de l'acide fluorhydrique ou le NH_4F: le fluor forme un ion complexe avec le Fe et l'Al: par ex. Al $(NH_4)_3$ F_6 ou Fe $(NH_4)_3$ F_6 qui libère ainsi le phosphore, complètement du phosphate d'aluminium, et partiellement du phosphate de fer.

Le phosphate "échangeable" ainsi obtenu semble représenter une bonne partie du phosphore "assimilable".

La corrélation entre les chiffres analytiques et le rendement de la récolte obtenue est très significative (CHANG et JACKSON 1957, 1959).

Le phosphore "assimilable". 1) *Méthodes d'extraction:* Il existe toute une série de méthodes allant de la simple digestion de l'échantillon dans l'eau (pure ou carbonatée) à l'extraction du phosphore total par fusion alcaline, dissolution par l'acide perchlorique, HF, etc.

La concordance de ces méthodes est assez significative du moment qu'on classe préalablement ces sols en trois catégories: ceux à dominance de phosphate de fer, ceux à dominance de phosphate de Fe+phosphate de Ca, et ceux où domine le phosphate de Ca seul. Pour qu'une méthode choisie soit comparable à une autre, il faudrait donc déterminer préalablement ces trois formes de phosphates (CHANG et JUO 1963).

Nous ne parlerons pas du phosphore total, puisque actuellement on est à peu près d'accord pour dire que cette forme n'a pas de relation directe avec le phosphore "assimilable" pour les plantes (hormis peut-être certains sols tropicaux).

a) *Phosphore soluble dans l'eau:* ce procédé n'a d'importance pratique que dans les sols très riches en phosphore et dans les eaux de drainage ou d'irrigation;

Il exige l'emploi d'une méthode de détermination très sensible (à l'acide chlorostanneux en milieu sulfurique — cf. plus loin) puisque les sols normaux contiennent souvent moins de 1 p.p.m. de P soluble.

D'ailleurs, la teneur du sol en P soluble dans l'eau n'a aucun rapport avec la quantité d'engrais appliquée à ces sols ni avec leur effet sur la hauteur de la récolte (WEBB et PESEK 1959).

b) *Phosphore soluble dans les acides faibles ou dilués ou leurs sels*: Ces méthodes ont toutes la prétention d'extraire le P "assimilable" des sols, quoique leurs résultats ne sont pas comparables entre eux.

Elles peuvent cependant toutes servir à établir le besoin en phosphore des sols à condition de se baser sur des normes de fertilité (chiffres-limite) établis préalablement au moyen de multiples champs d'expériences aménagés dans ce but.

La corrélation des analyses avec ces champs d'expériences est en général satisfaisante pour les sols acides ou neutres, moins bonne pour les sols alcalins et calcaires (le procédé au fluorure dilué semble convenir aussi pour ces sols).

On a essayé H_2SO_4, HNO_3, HCL, HFl, dilués; les acides organiques: acétique, citrique, lactique ainsi que le $CaCl_2$ (WILD 1959).

Tous ces réactifs sont employés, ou bien à une concentration faible déterminée, supposée imiter le pouvoir d'attaque du système radiculaire, ou bien à une dilution telle que la suspension du sol après agitation donne un pH final déterminé, important surtout pour les sols alcalins et calcaires qui neutralisent une partie de l'acide employé.

c) *Phosphore soluble dans les alcalis dilués:* K_2CO_3, $NaHCO_3$, NaOH dilué, etc.

Ces réactifs sont employés spécialement pour les sols calcareux et alcalins ainsi que pour extraire le phosphore organique.

L'avantage des réactifs carbonatés réside dans le fait qu'ils tiennent l'activité de Ca^{++} suffisamment basse pour prévenir la précipitation (sous forme de $Ca_3(PO_4)_2$) du phosphore libéré pendant l'extraction (CHANG et JACKSON 1957).

N. B. Le P organique peut s'extraire aussi par réactif acide (1 — b) après oxydation de cette matière par H_2O_2, par $HClO_4$ (Metha) ou par incinération.

La différence obtenue entre le P inorganique par colorimétrie avant et après destruction de la matière organique donne le P organique puisque ce n'est que le P minéral qui se laisse déterminer colorimétriquement.

Une variante du procédé Metha (BOHNE et al. 1958) permet d'employer une très petite quantité de $HClO_4$ évitant le danger d'explosion et simplifiant le mode opératoire.

Le procédé Metha fournit des résultats supérieurs de 10% par rapport à celui de Kaila (VAN DIEST et BLACK 1958).

d) *Phosphore échangeable au moyen d'une résine d'échange.* En employant comme base de comparaison le $H_2PO_4^-$ au lieu du P minéral total, cette méthode semble donner une bonne idée du P assimilable du sol (LATHWELL et al. 1958, MOSER et al. 1959).

2) *Détermination du P dans un des extraits mentionnés* (1–a–b–c–d). Les anciennes méthodes chimiques de détermination du P étant trop encombrantes; toutes les analyses se font actuellement photométriquement, soit au moyen du bleu de phospho-molybdène, soit du bleu de phospho-vanado-molybdène (DREWS 1958).

Plusieurs variantes de ces méthodes ont été publiées suivant le réducteur employé: $SnCl_2$ (en milieu chlorhydrique ou sulfurique), acide amino-naphtol-sulfonique, acide ascorbique, etc. Ce dernier réducteur stabilise la couleur bleue (RAUTERBERG et al. 1958).

Avec l'acide ascorbique, en suivant minutieusement le procédé employé, on peut doser le phosphore dans tous les milieux sans interférence des autres substances présentes (DUVAL 1962).

Toutes ces variantes diffèrent d'après leur sensibilité, la rapidité, la durée de stabilité de la couleur, le danger d'interférence d'ions étrangers et les limites optimales de concentration du liquide d'extraction. Voir discussion critique de ces méthodes: JACKSON (1958, p. 138).

La méthode au phospho-vanado-molybdène semble devenir "universelle": non seulement elle donne des résultats satisfaisants pour les sols, mais aussi pour les extraits de végétaux et même pour les engrais phosphatés du commerce (MUNK et LÖSING 1959).

Le phosphore "non-assimilable". Le P non extrait par un des réactifs mentionnés sous la rubrique: "assimilable" est considéré comme non assimilable.

Il fait partie du réseau cristallin (phosphore tétraédrique) dont la sortie est empêchée par les oxydes de fer entourant la particule en fine pellicule.

Ce phosphore non assimilable fait partie de l'arrière-effet de l'engrais P. On a essayé de la déterminer moyennant percolation avec des sels dilués de K. Ce phosphore représente 50% de celui calculé sur le bilan phosphorique (TOMBESI et CALE 1962).

Au sujet de l'assimilabilité du PO_4 Fe et PO_4 Al, plus encore du $(PO_4)_2$ Ca_3, les avis sont partagés et il n'est pas étonnant que des procédés de détermination de ce phosphore ne soient pas entrés dans la pratique courante.

On se contente de faire la différence entre le P total et celui dit assimilable obtenu par un des procédés 1–a–b–c–d. Il est curieux qu'on emploie si peu le spectromètre à flamme dans ce but moyennant une technique spéciale mise au point dans notre Institut de Louvain, nous sommes parvenus à déterminer le Mg avec reproducibilité d'au moins 1,5% (DE LOOSE 1965).

Les procédés classiques de détermination du P étant trop lents et trop compliqués, on applique actuellement à l'analyse des engrais, les procédés colorimétriques actuels élaborés pour l'analyse du sol (FERRARI et al. 1962).

Le soufre assimilable. Aucune tentative de quelque envergure n'a été entreprise jusqu'ici pour trouver un *réactif extrayant* du sol le soufre assimilable pour la plante.

La cause probable de cette carence réside dans le fait que trop longtemps, le soufre a été relégué au rang d'élément mineur alors que son rôle dans la plante est aussi important que celui du phosphore.

Beaucoup de réactifs ont été essayés; l'eau chaude, bicarbonate de Na, incubation, solution de phosphates, $CaCl_2$, aspergillus, etc. Peu de ces méthodes ont été comparées avec la récolte (FRENEY et al. 1962).

Récemment cependant, McCLUNG et al. (1959) ont cru trouver une corrélation entre le rendement de la récolte en vases de végétation et le S extrait du sol au moyen de l'Ac NH_4 (SAALBACH et al. 1962). Ils ont trouvé avec la méthode à chaud de WILLIAMS une bonne corrélation entre le S du sol et celui de la plante, de sorte que ce procédé peut servir à déterminer le S assimilable du sol. La méthode classique de détermination sous forme de Ba SO_4 est lente, compliquée (éloignement des carbonates) et ne se prête pas au micro-dosage.

Voilà pourquoi actuellement, on emploie la méthode titrimétrique au versénate: chélation de l'excès de Ba restant après la précipitation du $BaSO_4$.

Signalons aussi la titration oscillométrique avec $BaSO_4$ et enfin, le chauffage du sol avec $KClO_3$: l'oxygène se combine quantitativement avec le S, et le SO_2 se détermine aisément (AIDINJAN 1957).

Par néphélométrie, le SO_4Ba formé peut être maintenu en suspension fine par adjonction d'un tensio-actif par exemple le Tween 20 (JOUIS 1962).

Les sulfates *solubles* dans le sol se déterminent par la méthode de réduction en H_2S de JOHNSON.

Moyennant précautions, jusqu'à I γ de S dans 3 ml d'extrait peuvent être détectés (FRENEY 1958).

Pour interpréter valablement les chiffres analytiques du S extrait par une des méthodes mentionnées, il faut mettre en rapport ces chiffres avec la concentration de la solution du sol et son pouvoir absorbant (BLANCHET et al. 1963).

La difficulté dans toutes ces méthodes réside dans la volatilisation possible d'une certaine partie du soufre pendant l'incinération des substances végétales. Les meilleures méthodes sont celles qui s'exécutent en vase clos (BUCK 1962).

Les éléments mineurs échangeables et assimilables. Nos connaissances concernant le rôle physiologique des éléments mineurs ne sont pas suffisamment avancées pour permettre aux chimistes la découverte du meilleur *réactif d'extraction* pour chaque élément en particulier. D'ailleurs, ils ont concentré leurs efforts jusqu'ici surtout à l'échafaudage de la meilleure *méthode de détermination* au laboratoire des éléments dans un extrait de sol obtenu plus ou moins arbitrairement, ou choisi de telle façon qu'il présente le moins d'ennui pour l'analyse du laboratoire (absence d'éléments interférents).

Comme il s'agit souvent de détecter des quantités infinitésimales, les macro-méthodes classiques gravi-métriques ou titrimétriques sont de peu d'utilité.

On s'est donc adressé aux micro-méthodes: colorimétriques, complexométriques et surtout au spectroscope d'émission ou d'absorption (photomètre à flamme) qui permet le dosage de plusieurs éléments dans un même extrait, quitte à prendre des précautions pour éviter l'effet des substances interférentes (SCHARRER et MENGEL 1959).

Récemment on a même mis sur pied un laboratoire transportable permettant la détermination rapide sur place de la plupart des éléments mineurs avec une précision de 6 à 10% (PEIVE 1959).

Un aperçu général des méthodes d'analyse des éléments-trace se trouve dans l'ouvrage de PINTA 1962. Il divise les méthodes en spectro-photométriques d'absorption et d'émission et les méthodes polarographiques.

Un aperçu des éléments-trace, spécialement dans les sols, les plantes et les vins des vignobles avec une critique des méthodes employées se trouve dans GARTEL (1962).

Les techniques chimiques mentionnées ici pour le sol, peuvent s'appliquer à *l'analyse des feuilles* dont les teneurs en éléments mineurs, surtout à certains stades de développement, sont en rapport avec les teneurs du sol en éléments assimilables.

On peut appliquer aussi le *diagnostic visuel* des plantes soit en plein champ, soit en vases de végétation.

Des atlas coloriés ont été publiés indiquant les symptômes de carence ou de toxicité caractéristiques pour chaque espèce végétale.

La détermination, pour toute une contrée, de la fréquence d'apparition d'une maladie de carence peut conduire à l'établissement d'une carte régionale de cette carence. Signalons aussi que l'analyse du sol au moyen *d'aspergillus niger* répond assez fidèlement à la présence dans le sol de beaucoup d'éléments mineurs à l'état assimilable pour les plantes supérieures.

Enfin toutes ces méthodes peuvent être contrôlées par *des champs d'essai ou de vases de végétation* avec et sans applications de doses variables de l'élément à étudier.

Connaissant les facteurs de l'apparition d'une carence ou toxicité d'un élément-trace, on peut *trouver le remède à appliquer*.

Cette application est assez facile dans le cas d'une carence puisque souvent une application directe, sur le sol ou sur les feuilles, de l'élément manquant suffit à enrayer le mal.

Plus difficile est la guérison d'un excès: dans ce cas il faut employer souvent un remède indirect: changement des conditions du sol diminuant la teneur, à l'état assimilable, du principe nocif.

1) *Le fer échangeable et assimilable.* a) *Extraction:* Pendant l'extraction à l'Ac NH_4 neutre pour les cations échangeables ordinaires, une partie du Fe^{++} s'oxyde en Fe^{+++} qui précipite.

On peut éviter cette précipitation par l'emploi d'Ac NH_4 ajusté au pH3.

On court cependant alors le danger qu'une partie de fer non-échangeable entre en solution.

b) *Détermination.* a) *Le fer total:* $Fe^{++} + Fe^{+++}$ se détermine spectroscopiquement ou chimiquement avec SCN, par réduction au Zinc amalgamé ou par chélation avec l'orthophénantroline: cette dernière méthode n'est pas gênée par la présence de substances interférentes: même le phosphore ne gêne qu'à des doses qui ne se rencontrent pas dans un sol normal (ROBICHET 1957).

Signalons enfin, la méthode titrimétrique au rédox de fer avec $KMnO_4$ comme oxydant, et Ti Cl_3 comme réducteur (KARBACH 1958).

b) *Fer ferreux*: Fe^{++}. Pour éviter son oxydation, il faut l'extraire du sol très rapidement à l'Ac NH_4 (moins de 5 minutes) et à l'état humide.

Le filtrat, débarrassé de son Ac NH_4 par évaporation et redissous dans l'acide, est traité par l'orthophénanthroline.

c) *Le fer ferrique:* on l'extrait du sol avec Ac NH_4 à pH_3; on obtient ainsi $Fe^{++} + Fe^{+++}$ et on calcule le Fe^{+++} par différence avec le Fe^{++} obtenu par le procédé précédent.

2) *Le manganèse échangeable et assimilable.* Puisqu'on admet que seul le Mn^{++} est échangeable et assimilable, on tâche d'obtenir un réactif qui n'extrait que le Mn^{++} du sol.

On distingue en outre le Mn facilement et difficilement échangeable suivant qu'il est facilement ou difficilement réductible par un réducteur.

Le Mn échangeable total est obtenu par extraction prolongée du sol avec l'Ac NH_4.

Le Mn facilement échangeable est obtenu de la même manière, mais en employant l'hydroquinone comme réducteur. Le Mn est déterminé colorimétriquement dans l'extrait par oxydation au périodate en solution phosphatée. Il peut aussi être transformé en $H_2Mn\ O_3$ qui donne une combinaison colorée avec l'éther méthylé du para-aminophénol (BATALIN 1957).

Outre l'Ac NH_4 comme échangeur, on a essayé une solution de Mg $(NO_3)_2$ dont les résultats semblent être en corrélation avec la maladie de carence en Mn chez les plantes. (BOKEN 1958).

D'autres ont trouvé une corrélation semblable en employant l'Ac Na à pH 4.7 (STENUIT et al. 1957).

D'autres encore ont essayé avec succès la méthode microbiologique à *l'Aspergillus niger* (SCHEFFER 1957).

Enfin le Mn (et le Cu) "assimilables" ont été déterminés par l'analyse de la plante à un stade déterminé de sa croissance, ainsi que par l'essai NEUBAUER, (SCHARRER et al. 1959). HOFF et MEDERSKI (1958) ont comparé les différentes méthodes d'extraction en présence et ont trouvé que l'extraction au phosphate mono-ammoniaque est supérieure à toutes les autres: acétate d'ammonium, de Na et hydroquinone.

3) *Le cuivre échangeable et assimilable.* La plus grande confusion règne encore dans le domaine de la *détermination du Cu soi-disant échangeable:* l'Ac NH_4 n'extrait que des traces insignifiantes de Cu, alors que la digestion du sol dans HFl solubilise pratiquement tout le cuivre.

Entre ces deux extrêmes, toute une série de réactifs a été essayée, chaque auteur proclamant que sa méthode extrait le Cu vraiment assimilable.

JACKSON pense que le Cu qui ne se laisse pas extraire au moyen de $HClO_4$ n'a pas de signification agronomique.

WIKLANDER traite le sol au moyen d'une résine hydrogénée.

ERIKSON extrait tout le Cu absorbé colloïdalement en le transformant en complexe stable: Cu P_2O^{7--}.

STEENBERG emploie le HCl à pH 2 et trouve que l'extraction contenant moins de 0,1 à 0,3 p.p.m., cause la carence en Cu.

ANTIPOV — KARAKAEV utilisent 0,5 N HNO_3 et trouvent qu'un sol est déficient en Cu s'il contient moins de 18 p.p.m.

Enfin, d'autres auteurs ont employé des réactifs biologiques: aspergillus niger.

En Hollande, on extrait le sol avec HNO_3 0,4 N et ici on a trouvé une relation entre ce Cu et la teneur en Cu de l'herbe moyennant des centaines de champs d'expériences. Pour l'avoine, il faut un minimum de 3 mg/Kg et 4 pour le froment, pour l'herbe 7.

Il est probable que tous ces procédés peuvent fournir des indications au sujet du Cu *assimilable* dans les sols, à condition d'établir des chiffres-limite basés sur un grand nombre d'analyses comparées aux résultats de la récolte comme pour les éléments majeurs.

D'ailleurs, il semble que le Cu soit lié au sol par d'autres forces que celles qui sont à la base de l'échange des cations ordinaires.

Il est en rapport avec le pH et l'Al échangeable (BLEVINS et MASSEY 1959).

Si l'extraction du cuivre d'un sol offre des incertitudes, sa *détermination chimique dans l'extrait* a été bien mise au point, par ex. la méthode colorimétrique au diéthyldithio-carbonate qui permet le dosage de très minimes quantités de Cu dans le sol (SCHARRER et SCHAUMLÖFFEL 1958) ou au moyen du Zincon: 2 carboxy — 2′ — Hydroxy — 5′ — sulfoformazylbenzène (PRATT et BRADFORD 1958).

Une méthode rapide du dosage de Cu repose sur la chélation avec la dichinoline (CANNEY et HAWKINS 1958).

4) *Le zinc échangeable et assimilable.* Même confusion que pour le cuivre quant au choix du *réactif d'extraction.*

D'ailleurs, souvent le zinc et le cuivre s'extraient par le même réactif: on a essayé $HClO_4$, 0,1 N HCl, Ac NH_4 à pH 4,6; KCl 1 N à pH 3,2; Mg SO_4 0,1 N etc... (Nair et Metha — 1959) et enfin, aussi la méthode à l'aspergillus niger en utilisant le Zn radioactif (SCHNEIDER et SIEGEL 1958).

Remarquons que tous ces réactifs sont acides ou acidifiés.

Nous avons vu plus haut que la carence en Zn est liée avant tout au pH du sol. D'ailleurs tous ces auteurs ont trouvé une corrélation positive entre la quantité de Zn extraite par leur réactif et les phénomènes de carence en Zn chez les plantes.

La détermination du Zn dans le liquide d'extraction n'offre pas de difficultés.

Elle se fait commodément par colorimétrie au dithizone. Notons aussi la détermination rapide et très sensible du Zn au moyen de son isotope Zn^{65} (THORNE et WIEBE 1957).

5) *Le cobalt échangeable.* Aucune étude systématique n'a été faite au sujet du meilleur réactif d'extraction en rapport avec la maladie de carence en cobalt.

Les chimistes ont concentré leur attention sur la meilleure méthode de détermination du Co dans l'extrait. Cette méthode diffère suivant la quantité de Co disponible: l'emploi du nitroso-naphtol permet de détecter jusque 0,1 γ de Co dans 10 ml (LEWIS 1958).

6) *Le molybdène échangeable.* Comme pour le Co, nous savons peu de chose au sujet de son échangeabilité en rapport avec son assimilabilité par les plantes.

La détermination dans l'extrait se complique du fait de l'interférence possible avec d'autres ions, ce qui nécessite une séparation chimique préalable.

On a longtemps employé la méthode colorimétrique au sulfo-cyanure, en utilisant divers réducteurs: l'acétone, l'acide ascorbique, l'acide chloro-stanneux, etc. (GALLEGO et JOLIN 1958).

Comme cette méthode n'est pas très sensible, on utilise actuellement le complexe coloré au dithiol qui permet l'analyse simultanée du Mo et du Cu.

Par l'emploi de colonnes d'échangeurs ioniques, on peut éliminer les ions interférents, et la méthode devient très sensible (SCHARRER et HÖFNER 1959).

Le complexe $[MoS_4]$. $[Cr(NH_3)\,5\,Cl]$ donne un précipité facilement filtrable, mais nécessite l'élimination préalable des substances interférentes (SPACU et GHEORGHIU 1958).

Une comparaison critique des méthodes employées actuellement se trouve chez RIEHM (1962). L'auteur conclut à la supériorité de la méthode au dithiol vis-à-vis des autres.

7) *L'aluminium échangeable.* Sa détermination offre de l'intérêt en rapport avec les différents degrés d'altération et d'acidification du sol.

La quantité d'Al extraite au moyen d'échangeurs alcalins est souvent sensiblement plus élevée que celle extraite par les échangeurs neutres et parmi ceux-ci le $BaCl_2$, $CaCl_2$, KCl et NH_4Cl extraient plus que l'échangeur "universel" NH_4 Ac.

Vu son caractère amphotère, l'extraction de l'Al échangeable a un caractère conventionnel plus prononcé que celui des autres ions (YUAN et FISKELL 1959).

Toutefois, dans le choix de l'extracteur, il faut toujours tenir compte du pH choisi: ainsi à pH 4,8, l'NH_4 Ac extrait autant d'Al que le NaCl et $BaCl_2$ (MCLEAN et al. 1959).

Sa détermination en très petites quantités se fait commodément au moyen d'une modification de la méthode à l'ério-chrome-cyanine de Hill qui est beaucoup plus rapide et plus sensible que celle à l'aluminon (JONES et THURMAN 1957).

8) *Le strontium.* Il se détermine au moyen de la méthode d'activation des neutrons ou au moyen du spectro-photomètre à flamme. Les deux méthodes fournissent des résultats concordants (HARRISON 1958).

La détermination du Sr^{90} devient un problème d'actualité puisque, de tous les produits radioactifs: émissions atmosphériques ou résidus d'usines nucléaires, le Sr^{90} est le plus dangereux (TIMOFEJEWA et TITLJANOWA 1959).

En principe, la détermination dans le sol du Sr^{90} à côté du Sr^{89} est possible puisque le produit de décomposition du Sr^{90} (Y^{90}) est radioactif alors que celui du Sr ordinaire ne l'est pas.

La difficulté dans l'extraction du Sr^{90} réside dans le fait de ne pas extraire en même temps d'autres substances interférentes. On y parvient cependant par électrodialyse suivie d'échange d'ions au moyen d'échangeurs à base de résine. On obtient un rendement de 96% au moins (GRAHAM 1959).

9) *Le bore assimilable. Pour son extraction,* on a utilisé l'eau froide, l'eau bouillante, et les acides dilués, mais peu de ces procédés ont donné une corrélation avec l'apparition de la maladie de carence en B, de sorte que nous ignorons toujours laquelle de ces méthodes représente le mieux le bore assimilable.

L'extraction, même à l'eau chaude, ne donne qu'une partie du B réellement disponible dans les sols lourds (OELSCHLAGER 1958).

Si les recherches du côté déficient (maladies de carence) sont assez lacunaires, celles du côté excessif (effet toxique) sont plus nombreuses, suite aux accidents survenus par emploi d'eaux d'irrigation boriquées et d'engrais ou de pesticides à base de bore.

En Hollande, les sols avec moins de 0,35 p.p.m. de B présentent dans 73%, la carence en B (LEHR 1962).

La détermination de la capacité d'échange des sols pour le bore, permet actuellement de prévenir l'effet toxique de l'emploi d'eau d'irrigation boriquée (HATCHER et BOWER 1958).

La détermination dans l'extrait se fait macroscopiquement par titration du premier H de H_3BO_3, les autres étant bloqués par complexation avec le mannite.

De petites quantités se déterminent par colorimètre, par ex. par le procédé au curcuma dans une de ses multiples améliorations.

Le procédé au curcuma, après extraction du B à l'eau chaude, donne des résultats reproductibles à condition de maintenir minutieusement les détails du procédé: appareils très purs, essai à blanc du curcuma, température exactement à 55° pour la lecture du rosocyamine, etc. (VAN WESEMAEL 1962).

10) *Le Silicium dans la plante.* Il se dépose le long des parois des cellules et organes de conduction. Par incinération sèche la forme des cellules est conservée. L'examen chimique, microscopique et röntgenologique prouve qu'il s'agit d'opale, α quartz et cristobalite, dont la forme dépend de la présence d'éléments accessoires Na, K, Ca et Mg. Il faut donc se méfier de l'incinération sèche qui probablement ne représente pas le vrai silicium présent dans la plante.

Il faut donc préférer l'incinération humide avec HNO_3 et $HClO_4$ (Jones et Milne 1963).

Les autres procédés colorimétriques à base d'anthraquinone (quinalizarine — coramine — etc) sont très sensibles à de petites variations de température pendant la lecture et exigent la concentration très exacte de 89 % de H_2SO_4.

Toutefois, avec le rouge carmin et un procédé analytique précis, on peut atteindre une exactitude d'environ $2\,\gamma$ B (Burriel Marti 1958).

3. La détermination des ions solubles dans le sol

Comme dans les sols normaux les ions solubles ne représentent qu'un petit pourcentage des ions échangeables, leur détermination ne s'impose que dans des cas exceptionnels d'accumulation de ces ions: sols sous climat aride, submergés par l'eau de mer, terres de jardins et de serres fumées excessivement par engrais chimiques, etc.

La difficulté de détermination dans ces cas réside dans l'obtention d'un extrait liquide représentant fidèlement la composition chimique du liquide du sol.

On a essayé les *méthodes d'extraction* suivantes:

1) *Déplacement ou percolation:* La composition du percolat obtenue dépend de sa nature: eau, huile, alcool, etc., de la quantité employée, de la durée de percolation et du degré de colmatation par peptisation du sol (Kawe 1959).

2) *Par pression:* au moyen d'une presse hydraulique: 500 atm. — La quantité de "suc" obtenue est minime et sa composition diffère avec le caractère sablonneux, argileux ou humique de l'échantillon.

3) *Par centrifugation:* donne peu de résultats puisque la force centrifuge est trop faible vis-à-vis du pF de la plupart des sols.

4) *Par adsorption* au moyen d'une substance adsorbante indifférente: papier-filtre (Schuffelen).

Par succion à travers un papier-filtre, procédé simple, relativement vite, peut suffire pour mesurer la conductivité du filtrat obtenu d'une pâte de sol (Micah 1963).

5) *Par diffusion ou pression à travers une membrane.* Cette méthode exige un temps considérable (2 semaines) pendant lequel l'extrait peut subir des changements à la suite d'activité microbienne.

En outre, les propriétés sélectives de la membrane influencent la composition du filtrat (Griessbach 1957, Coleman et Marsh 1961).

La détermination de la composition ionique de l'extrait peut se faire:

1) *Directement* par évaporation, pesée (ions totaux) et individuellement par les procédés chimiques ou physico-chimiques mentionnés aux "bases échangeables".

2) *Par conductivité électrique:* cette méthode est devenue très populaire depuis l'introduction en pédologie de la notion de pF.

Elle est rapide et sensible, mais ne fournit que des données relatives qui doivent être calibrées au moyen d'une série de solutions de sels à concentration croissante.

En outre, la conductivité électrique diffère d'après la composition chimique de la solution du sol mesurée: nature des cations Ca, Na et des anions: chlorures, sulfates, carbonates.

En principe, la concentration en sels est liée à la conductivité par l'équation:

meq. sels par litre = 12.5 mmOhm/cm (SAYEGH 1958).

L'appareil de mesure consiste, en principe, en un pont de résistance électrique, une cellule à électrodes en platine et accessoires.

Puisque la méthode conductimétrique est essentiellement relative, l'opération doit toujours se faire dans les mêmes conditions.

Exécutée sur le champ même, elle permet de détecter l'hétérogénéité saline locale aussi bien horizontalement que verticalement: tâches salines, horizons salins: en surface sous climat aride, en profondeur sous climat humide.

La conductivité d'une solution de sol est en relation directe avec celle de l'eau d'irrigation et inversement proportionnelle à celle de l'eau de drainage.

Ces mesures peuvent donc servir à constater l'effet des mesures d'amélioration foncière. (PEARSON 1959, LONGENECKER et LYERLY 1959).

3) *Par pression osmotique:* il existe une relation entre la conductivité électrique d'une solution et sa pression osmotique:

$$\text{Atm.} = n\ L\ \text{ohms/cm}.$$

Seulement, comme pour la conductivité, le facteur n varie avec la composition chimique de la solution: il est de 0,36 pour le NaCl et les sols salins et alcalins; 0,28 pour le $Mg\ SO_4$.

4) *La détermination des ions séparément* se fait par une des méthodes déjà signalées à propos des ions échangeables. La détermination des carbonates occupe ici une place spéciale puisqu'elle permet de juger de l'alcalinité totale de l'extrait, ainsi que de déterminer le carbone minéral par opposition au carbone organique de la matière organique.

Signalons enfin les solutions nutritives composées dans le but de déterminer leur influence individuelle et leur interréaction sur les plantes en agriculture, en vases de végétation avec sable et avec sol. Aucune de ces solutions n'est "universelle" et doit être préparée pour une catégorie de plantes déterminées. De là le nombre important de solutions proposées: CRONE, TOLLENS, USERADYA, BELOUSSOW, STOKKOSA, etc. (HOROVITZ 1963).

Bibliographie

AIDINJAN, R. CH.: Méthode de détermination du S total dans le sol. Pédologie (russe) 9, 49–59 (1957). — AMAVIS, R.: Comparaison des méthodes de mesure de la capacité d'échange d'ions d'un sol. Bull. Ass. Franc. Et. Sol. 8, 317–325 (1959). — ANTIPOW, I. N., KARATAJEW et G. M. KADER: Untersuchungen der Austauschreaktionen in Böden mit Hilfe des Ca^{45}. Verh. Int. Bod. Ges. 2, 225–229 (1958). — ATANASIU, N.: Zur Wirkung schwerlöslicher N-Verbindungen. Z. Pflanzenernähr., Düng., Bodenkde. 84, 103–110 (1959).

BARROWS, H. L., et M. DROSDORFF: A comparison of methods for determining the base exchange capacity of soils. Proc. Soil Sci. Soc. Amer. 22, 119–122 (1958). — BATALIN, A. CH.: Méthode rapide de détermination de faibles quantités de Mn dans le sol. Pédologie (russe) 9, 124–125 (1957). — BERGSETH, F. A., et al.: Determination of C.E.C. in colloïdal soil material by means of Sr^{89}. Soil Sci. 95, 97–100 (1963). — BLANCHET, R., et al.: Appréciation détaillée du niveau phosphorique des sels par des techniques chimiques simples. Compt. rend. Acad. Agric. France 3, 348–356 (1963). — BLEVINS, R. L., et H. F. MASSEY: Evaluation of two methods of measuring available soil copper and the effects of soil pH and extractable Al and Cu by plants. Proc. Soil Sci. Soc. Amer. 23, 296–298 (1959). — BOHNE, H., P. GROPLER et M. DITTMER: Beitrag zur Bestimmung der Gesamtphosphorsäure des Bodens mit Überchlorsäure. Z. Pflanzenernähr., Düng., Bodenkde. 82, 42–46 (1958). — BOKEN, E.: Determination of manganese in soil. Plant a. Soil 9, 3, 269–285 (1958). — BOWER, C. A.: Determination of Na in saline solutions with a glass electrode. Proc. Soil Sci. Soc. Amer. 23, 29–31 (1959). — BROWN, D. A., et T. P. WELLS: The use of Mg^{28} for the measurement of Mg absorption by plants. Proc. Soil Sci. Soc. Amer. 22, 323–326 (1959). — BUCK, M.: Geeignete Methoden zur Bestimmung des Gesamt-S-Gehaltes pflanzlicher Substanzen. Landw. Forsch. 15, 135–145 (1962). — BURRIEL MARTI, F., S. J. MENEZGOMEZ et M. RODRIQUEZ DE LA PENA: Determinación de boro en suelos y plantas. An. Edaf. Fis. Veg., Madrid, 17, 565–586 (1958). — BURRIEL MARTI, F., et C. ALVAREZ HERRERO: Determinación de cationes de cambio en suelos por via espectral. An. Edaf. Fis. Veg., Madrid, 18, 1, 11–31 (1959).

CANNEY, J., et A. HAWKINS: Cold acid extraction of copper from soils and sediments. Econ. Geol. 53, 877 (1958). — CHANG, S. C., et S. R. JUO: Available P in relation to phosphates in Soils. Soil. Sci. 95, 91–96 (1963). — CHANG, S. C., et M. L. JACKSON: Fractionation of soil phosphorus. Soil Sci. 84, 133–144 (1957). — Proc. Soil Sci. Soc. Amer. 21, 265–269 (1957). — CLEMENT, C. R., et T. E. WILLIAMS: An incubation technique for assessing the N status of soils newly ploughed from leys. Soil Sci. 13, 82–91 (1962). — COLEMAN, J. D., et A. D. MARSH: An investigation of the pressure-membrane method for measuring the suction properties of soil. Soil Sci. 12, 343–362 (1961).

DELOOSE, R.: Etude critique de quelques méthodes de détermination du Mg dans les graminées dans le cadre de la sélection. Rev. agriculture (Bruxelles) 18, 2, 193–198 (1965). — DHARIWAL A. P. S., et F. T. STEVENSON: Determination of fixed ammonium in soils. Soil Sci. 86, 343–349 (1958). — DREWS, M.: Vergleichende Untersuchungen über die Anwendbarkeit der Photo-Rex-Zinnchlorür und Vanadat-Molybdat-Methode. Z. Pflanzenernähr., Düng., Bodenkde. 83, 64–67 (1958). — DUVAL, L.: Dosage céruléo-molybdique du P dans les sols, végétaux et engrais. Ann. Agron. 13, 5, 469–482 (1962).

FEIGL, F.: Spot tests I/II. 1956. — FIEDLER, H. T.: Kali-Ionen und Bodenorganismen Kali-Briefe 4, 1–4 (1958). — FILCEK, M.: Die Ausschaltung des Phosphateinflusses bei der flammenphotometrischen Ca Bestimmung. Z. Pflanzenernähr., Düng., Bodenkde. 85, 112–117 (1959). — FIVE, C. V.: An evaluation of ammonium fluoride as a selective extrahant for aluminium-bound soil phosphate. Soil Sci. 87, 13–21 et 83–88 (1959). — FRENEY, J. R., et al.: A review of certain aspects of sulphur as a soil constituent and plant nutrition. Plant a. Soil 17, 3, 295–308 (1962). — FRENEY, T. R.: Determination of water-soluble sulphate in soils. Soil Sci. 86, 241–244 (1958). — FRIED, M.: Atomic energy and agriculture, p. 5. Washington 1957.

GASSER, J. K. R.: Effects of long continued treatment of the mineral N content and mineralisable N of soil. Plant a. Soil 17, 2, 209–220 (1962). — GARTEL, W.: Die Bestimmung von Mg, B, Mn, Zn etc. und Mo in Weinbergböden. 1962. — GETTKANDT, G: Eine schnell durchführbare flammenphotometrische Ca-Bestimmung in salzsaurem Bodenauszug. Landw. Forsch. 2, 93–95 (1958). — GRAHAM, E. R.: Sr^{90} in soil by electrodialysis and ion exchange method. Soil Sci. 88, 11–15 (1959). — GREZES, M. G.: Compt. rend. Acad. Agric. France 4, 171–176 (1959).

HARMSEN, G. W.: Was kann uns die Bestimmung des Gehaltes löslichen Stickstoffs im Boden lehren? Z. Pflanzenernähr., Düng., Bodenkde. 84, 98–102 (1959). — HARMSEN, G. W., et D. A. VAN SCHREVEN: Mineralisation of organic nitrogen in soil. Advances in Agronomy 8, 299–398 (1955). — HARRISON, G. E.: Estimation of Sr in biological materials. Nature 182, 792–793 (1958). — HATCHER J. T., et C. A. BOWER: Equilibria and dynamics of boron adsorption by soils. Soil Sci. 85, 319–323 (1958). — HOFF, D. J., et H. J. MEDERSKI: Trace elements, p. 109–116. New York 1958. — HOÏNE, H.: Der Einfluß des Baumalters, der soziologischen Stellung des Baumes (1964). — HOROVITZ, C.: Nährlösung für Zuckerrüben in Wasser und Sandkulturen. Z. Pflanzenernähr., Düng., Bodenkde. 100, 2, 127–129 (1963).

JACKSON, W. A.: Soil chemical analysis. Washington 1958. — JONES, L. H., et D. A. THURMAN: The determination of Al in soil, ash and plant materials using Eriochrome Cyanine R. A. Plant a. Soil 9, 131–142 (1957). — JONES, L. H. P., et A. A. MILNE: Studies of Silica in art plants. Plant a. Soil 18, 2, 207–220 (1963). — JOUIS, E., et al.: Une méthode néphélométrique rapide pour le dosage du soufre dans le sol. Ann. Agron. 11, 5, 483–489 (1962).

KAILA, A.: Effect of superphosphate on the mobilization of nitrogen in a peat-soil. Maataloustieteellinen Aikakauskiya 30, 214–222 (1958). — KARBACH, L.: Potentiometrische Bestimmung von Eisen in Böden. Land. Forsch. 2, 96–108 (1958). — KAWE, A.: Methode zur Bestimmung des Düngerbedürfnisses des Bodens für Kalk, Kalium und Phosphorsäure mit Hilfe der Bodenlösung. Z. Pflanzenernähr., Düng., Bodenkde. 84, 224–231 (1959).

LAMBERT, J.: L'analyse chimique des sols d'Ardenne. Agricultura (Louvain), 12, 2, 247–

261 (1964). – LATHWELL, D. J., N. SAUCUEZ, D. T. LISK et M. PEECH: Availability of soil pH as determined by several chemical methods. Agron. J. 50, 366–369 (1958). – LEFEVRE, P.: Dosage de quantités minimes de Mg en présence de quantités importantes de Ca. Ann. Agron. 9, 481–477 (1958). — LEGG, J. O., et J. H. AXLEY: Investigation of a thermal method for the determination of fixed K in soils. Proc. Soil Sci. Soc. Amer. 22, 287–290 (1958). — LERH, J. J.: Der Borversorgungsgrad holländischer Böden. Landw. Forsch. 16, 72–92 (1962). — LEWIS, J. C.: Cobalt determination in soils and rocks with 2-nitro-1-naphtol. Analyt. Chem. 30, 53–56 (1958). – LINSER, H., et K. KAINDL: Isotope in der Landwirtschaft. Hamburg und Berlin: Parey. 1960. — LIVENS. J.: Contribution à l'étude de l'azote minéralisable du sol. Agricultura (Louvain) 7, 28–44 (1959). — Recherches sur l'N ammoniacal et organique du sol soluble dans l'eau bouillante. Agricultura (Louvain) 7, 519–532 (1959). — LONGENECKER, D. E., et P. J. LYERLY: Chemical characteristics of soils of West Texas as affected by irrigation water quality. Soil Sci. 87, 207–216 (1959). — LUCENA, F., CONDE, et A. MARTINEZ DE PANCORBO: Nuevo reactivo para la determinación colorimetrica del fósforo en suelos. An. Edaf. Fis. Vég., Madrid, 18, 67–72 (1959). — LUDWIEG, F.: Bestimmung der radio-chemischen Verunreinigung der Phosphorsäure in bestrahlten Phosphatdüngern. Z. Pflanzenernähr., Düng., Bodenkde. 84, 81–85 (1959). — LYDERSEN, P. L., et M. OPEM: Zur Wasserdampfdestillation von Ammoniak aus Bodenextrakten. Z. anal. Chem. 159, 339–343 (1958).

MALEINA, A. A.: La méthodique de la détermination du P disponible dans le sol. Pédologie (russe) 4, 113 (1958). — MATELSKI, R. P., et C. H. YIEN: A rapid field and laboratory procedure for determining the sodium content of soil. Proc. Soil Sci. Soc. Amer. 22, 500–502 (1958). — MCCLUNG, A. C., L. M. M. DE FREITAS et W. L. LOTT: Brazilian soils in relation to plant responses to sulphur. Proc. Soil Sci. Soc. Amer. 23, 221–224 (1959). — MCLEAN, E. O., M. R. HEDDLESON et G. J. POST: Comparison of Al extraction methods in soils and clays. Proc. Soil Sci. Soc. Amer. 23, 289–293 (1959). — MATHIEU, A. L., et N. BURTCH: Note an a modification for the Beckman DU flame-spectrophosometer. Canad. J. Soil Sci. 41, 134–135 (1961). — MICAH, W. M. L.: A new suction plate apparatus for extraction of soil solution in conductivity determination. Soil Sci. 95, 142–143 (1963). — MOSCHEW, A. I.: Détermination de la capacité d'échange des colloïdes dans le sol. Pédologie (russe) 6, 80–85 (1958). — MOSER, U. S., W. H. SUTHERLAND et C. A. BLACK: Evaluation of laboratory indexes of adsorption of soil phosphorus by plants. Plant a. Soil 10, 4, 356–374 (1959). — MUNK, H., et LÖSING: Die Anwendung der Vanadat-Molybdat-Methode bei der Bestimmung der Phosphorsäure in Düngemitteln. Landw. Forsch. 12, 2, 125–132 (1959).

NAIR, G. G. M., et B. V. METHA: Status of Zinc in soils of Western India. Soil Sci. 87, 155–159 (1959). — NEWBOULD, P., et R. SCOTT-RUSSELL: Isotopic equilibration of Ca^{45} with labile Soil-Ca. Plant a. Soil 18, 2, 239–257 (1963).

OELSCHLÄGER, W.: Fehlermöglichkeiten bei der Bestimmung des pflanzenaufnehmbaren Bors. Landw. Forsch. 2, 45–52 (1958). — OLSEN, S. R., et F. S. WATONABE: Proc. Soil Sci. Soc. Amer. 21, 144–149 (1957).

PANCORBO, A., DE MARTINEZ, et F. LUCENA, CONDE: Determinación y formas de fósforo en los suelos de la provincia de Salamanca. An. Edaf. Fis. Veg., Madrid, 18, 2, 81–95 (1959). — PEARSON, G. A.: Factors influencing salinity of submerged soils and growth of Caloro Rice. Soil Sci. 87, 198–206 (1959). — PEECH, M., et al.: A critical study of the $BaCl_2$-triethanolamine and the Am. Ac. methods for determining the exchangeable Hydrogen content of soils. Proc. Soil Sci. Soc. Amer. 26, 37–40 (1962). — PEIVE, J. V.: Portable field laboratory for determination of available minor elements in the soil. Z. Pflanzenernähr., Düng., Bodenkde. 84, 231–237 (1959). — PILLERI, R.: Komplexometrische K-Bestimmung. Z. Anal. Chem. 157, 1–2 (1957). — PINTA, M.: Recherche et dosage des éléments trace, p. 726, Paris 1962. — POHLAND, E., et H. FRENZEN: Die Atomtechnik in der Landwirtschaft. Stand und Leistung der agrikulturchemischen Forschung, VI, p. 5–13. Frankfurt 1959. — POLJAKOW, J. A., et W. S. GERMOGENOVA: Anwendung einiger Formen von schwerem Wasser bei bodenkundlichen und agrikulturchemischen Untersuchungen. Verh. Int. Bod. Ges. 2, 240–243 (1958). — PRATT, P. F., et G. R. BRADFORD: Separation and determination of total copper and zinc in soils. Proc. Soil Sci. Soc. Amer. 22, 399–402 (1958). — PUFFELES, M., et N. E. NESSIM: Direct flame-photometric determination of Ca in soil and plant extracts, water and serum. Anal. Chim. Act. (Amsterdam) 20, 38–46 (1959).

RAMACHANDRAN, L. H., et GARG: Differential thermal analysis. Rookee, India. 1959. — RAUTERBERG, F., et H. OSSENBERG-NEUHAUS: Phosphorsäurebestimmung nach der Molybdänblaumethode mit Ascorbinesäure als Reduktionsmittel. Z. Pflanzenernähr., Düng., Bodenkde. 82, 46–49 (1958). — RICH, C. I.: Removal of excess salt in cation-exchange-capacity determinations. Soil Sci. 93, 87–94 (1962). — RICHARD, M. H.:

Etude de l'efficacité et de la rentabilité de la fertilisation basée sur l'expérimentation longue au champ. Compt. rend. Acad. Agric. France 2, 47 (1959). — Riehm, H., et W. Scholl: Chemische Bestimmung von Mo in Böden, Pflanzen und Gesteinen mit Toluol-3,4-Dithiol. Landw. Forsch. 16, 138–147 (1962). — Robichet, O.: Méthodes de dosage du Fer. Ann. Inst. Nat. Rech. Agr. A 8, 257–324 (1957).

Saalbach, E., et al.: Untersuchungen über die Bestimmung des Gehalts an pflanzenverfügbarem S in Böden. Landw. Forsch. 15, 6–15 (1962). — Sawhney, B. L., M. L. Jackson et R. B. Corey: Cation-exchange capacity determination of soils as influenced by the cation species. Soil Sci. 87, 243–248 (1959). — Sayegh, A. H., L. A. Alban et R. G. Petersen: A sampling study in a saline and alkali area. Proc. Soil Sci. Soc. Amer. 22, 252–254 (1958). — Schachtschabel, P., et U. Schwertmann: Der Einfluß von Alkohol bei der flammenspektrometrischen Bestimmung des pflanzenverfügbaren Mg in Böden. Z. Pflanzenernähr., Düng., Bodenkde. 82, 38–42 (1958). — Scharrer, H., E. Russ et K. Mengel: Über die Bestimmung des pflanzenaufnehmbaren Kupfers und Mangans. Z. Pflanzenernähr., Düng., Bodenkde. 85, 1–20 (1959). — Scharrer, H., et W. Höfner: Kationenabtrennung durch Ionenaustauscher bei der photometrischen Mo-Bestimmung mit Dithiol. Z. Pflanzenernähr., Düng., Bodenkde. 86, 49–56 (1959). — Scharrer, K., et H. Mengel: Die Bestimmung von Mikronährstoffen in Mikronährstoffvolldüngern. Landw. Forsch. 12, 2, 118–124 (1959). — Scharrer, K., et E. Schaumlöffel: Die quantitative Bestimmung kleinster Mengen Kupfers mittels Diäthyldithio-carbamat. Landw. Forsch. 2 59–60 (1958). — Scheffer, F., A. Kloke et F. Hunerhoff: Untersuchungen des Bodens auf Mn-Bedürftigkeit nach der mikrobiologischen Testmethode mittels A. Niger. Landw. Forsch. 10, 207–214 (1957). — Scheffer, F., et B. Ulrich: Die Anwendung von Leitisotopen im System: Boden — Bodenlösung — Pflanze. Naturwiss. 45, 49–53 (1958). — Schilling, F.: Über die Mg-Bestimmung mit Titangel b. Z. Landw. Vers. Unters.wesen 3, 236 (1957). — Schroeder, D., et al.: Die Mg-Vorräte schleswig-holsteiner Böden. Z. Pflanzenernähr., Düng. Bodenkde. 100, 3, 261-224 (1963). — Schröel, S. A.: Die Untersuchung niederösterreichischer Weinböden auf deren Mo-Gehalt mit Aspergillus niger. Bodenkultur (Wien) 10, 16 (1958). — Schuhknecht, W.: Die flammenphotometrische Bestimmung geringer Mengen von Ca neben größeren Mengen von Na. Z. Anal. Chem. 157, 338 (1957). — Schneider, K. H., et O. Siegel: Untersuchungen über die Aufnahme von Zink aus kalkhaltigen Böden durch Aspergillus niger mit Hilfe von radioaktivem Zink. Landw. Forsch. 2, 4, 270–274 (1958). — Schollenberger, C. J., et al.: A comparison of methods for evaluating of agricultural limestones. Soil Sci. 93, 161–171 (1962). — Scott, A. D., J. J. Hanway et A. P. Edwariss: Replaceability of ammonium in vermiculite with acid solutions. Proc. Soil Sci. Soc. Amer. 22, 388–392 (1958). — Shaw, W. M., et Brooks Robinson: Chemical evaluation of neutralizing efficiency of agricultural limestone. Soil Sci. 87, 262–272 (1959). — Spacu, P., et C. Gheorghiu: Eine microgravimetrische Methode für die Bestimmung des Molybdäns allein und in Gegenwart anderer Elemente. Z. Anal. Chem. 159, 209–122 (1958). — Stahlberg, S.: Les réserves potassiques des sols — Méthodes de détermination. K. Sokgs. Landbruksak. Tids. 97, 389–401 (1958). — Stenuit, D. F., R. Piot et R. Boon: Relation entre le pH et la teneur en Mn des sols. Pédologie (Belgique) 7, 259–265 (1957). — Süss, A.: Eine Methode zu autoradiographischen Untersuchungen der Phosphatverteilung im Boden. Naturwiss. 46, 146–147 (1959). — Syrin, N. G., et D. S. Orlon: Méthode de détermination de l'activité des ions Na dans les sols et leurs solutions. Compt. rend. Ped. Univ. Moscou 13, 71–80 (1958).

Tamm, C. O.: Die Blattanalyse als Methode zur Ermittlung der Nährstoffversorgung des Waldes. Düng. Meliorat. i. d. Forstw. 66, 7–16 (1965). — Thorne, D. W., et H. H. Wiebe: Atomic energy and agriculture, p. 50. Washington 1957. — Thun, R.: Die Untersuchung von Böden, 4. Aufl. Berlin 1958. — Timofejewa, W. A., et A. A. Titljanowa: La sorption du Sr^{90} dans le sol. Compt. rend. Acad. Sci. U.R.S.S. 24, 3–117 (1959). — Tombesi, L., et M. T. Cale: Studies on the determination of available P in soils. Plant a. Soil 17, 2, 137–145 (1962).

Van Diest, A., et C. A. Black: Determination of organic phosphorus in soils. Proc. Soil Sci. Soc. Amer. 22, 286–287 (1958).

Webb, J. R., et J. T. Pesek: An evaluation of phosphorus fertilizers varying in water solubility. Proc. Soil Sci. Soc. Amer. 23, 381–384 (1959). — Wezemael, J. C. van: Determination of water-soluble Boron in soils. Plant Nutrition Res. 6, 1–11 (1961). — Wild, A.: The determination of $CaCl_2$-soluble phosphate in soils. Z. Pflanzenernähr., Düng., Bodenkde. 84, 220–224 (1959).

Yofe, J., et R. Finkelstein: Elimination of anionic interference in flame photometric determination of Ca in the presence of phosphate and sulphate. Analyt. Chim.

Acta (Amsterdam) 19, 166–173 (1958). — YUAN, T. L.: Determination of exchangeable Hydrogen in soils by a titration method. Soil Sci. 88, 164–167 (1959). — YUAN, T. L., et J. G. A. FISKELL: Aluminium Studies, Sol and plant analysis of Al by modification of the aluminon method. S. Agr. Food. chem. 7, 115–117 (1959). — Aluminon Studies, II, The extraction of Al. from some Florida soils. Soil Sci. Soc. Amer. Proc. 23, 202–205 (1959).

E. Die Mikroorganismen des Bodens und ihre Bedeutung

Von

H. Glathe und **G. Glathe**

a) Quantitative und qualitative Analyse der Mikroflora

Die ältere Literatur der Mikrobiologie des Bodens wurde von LÖHNIS (1910 und 1935) und WAKSMAN (1932) eingehend bearbeitet. Zusammenfassende Darstellungen in Handbüchern gaben RIPPEL (1931, 1939), STAPP (1931) und JENSEN (1959); eine grundsätzliche Bearbeitung der bodenbiologischen Probleme erfolgte durch WAKSMAN und STARKEY (1931), FEHER (1933), CUTLER und CRUMP (1935), POCHON und TCHAN (1948), WAKSMAN (1952), POCHON und BARJAC (1958), BURGES (1958) und ALEXANDER (1961). Sammelreferate über Bodenbiologie wurden von SMITH (1948) sowie THORNTON und MEIKLEJOHN (1957) veröffentlicht. Die spezielle Mikrobiologie der Moore wurde in neuerer Zeit von BENADE (1954) und BURGEFF (1961) bearbeitet. Zu beachten sind ferner die Veröffentlichungen über die Kongresse der Internationalen Mikrobiologischen bzw. Bodenkundlichen Gesellschaft sowie einzelne Arbeiten über methodische Grundlagen (WAKSMAN 1928, NIKLAS 1931, DEMETER 1936 und SIMONART 1957). Als eine Fundgrube für wichtige Ergebnisse bodenbiologischer Arbeiten sind die neun Auflagen des Werkes von RUSSELL, Soil conditions and plant growth (9. Aufl., 1961) zu betrachten. Schließlich sei auf die in verschiedenen Gebieten als klassisch zu bezeichnenden Arbeiten von WINOGRADSKY (1949) verwiesen, die in neuer Form erschienen sind. Im folgenden kann nur ein Teil der neuesten Literatur erfaßt werden.

1. Zahl und Verteilung der im Boden vorkommenden Mikroorganismen

Das Ziel aller Untersuchungen über Zahl und Verteilung von Mikroben im Boden besteht darin, Unterschiede in der Mikroflora von Böden verschiedener Art, Herkunft und Behandlung zu erarbeiten.

Da die Mikroorganismen auf Änderungen der Umweltbedingungen unter Umständen empfindlich reagieren, entsteht bei der Entnahme einer größeren Zahl von Einzelproben das Problem, diese so aufzubewahren, daß möglichst geringe Veränderungen eintreten. JAGNOW (1958) stellte fest, daß die Keimzahlen beim Lagern in der Kühltruhe rasch absanken. Die Beobachtung, daß in vielen Fällen ein gleichmäßiges Maximum der Keimzahl im Herbst eintritt (BURRICHTER 1955, SEIFERT 1960a), veranlaßte NETZSCH-LEHNER und NOWAK (1960), die Proben bei 25° C und möglichst konstanter Feuchtigkeit aufzubewahren. Eine Konstanz der Werte wurde jedoch nicht erreicht.

ABD-EL-MALEK et al. (1961) konnten nach 30- bis 40tägiger Bebrütung bei 30° C in den Gesamtkeimzahlen von Streptomyceten und Bakterien keine signifikante Veränderung feststellen. Erhebliche Veränderungen, besonders der antagonistischen Typen, wurden dagegen nach dem Zusatz von organischen Substanzen wie Glukose, Saccharose bzw. Pepton beobachtet.

Alle Verfahren zur Feststellung des Keimgehalts im Boden (vgl. Thornton 1956, Müller und Steinweg 1958) bedienen sich der räumlichen Trennung sowohl der im Boden verstreuten einzelnen Mikroben als auch der nester- oder kolonieartig angeordneten Organismengruppen, durch Schütteln in sterilen Flüssigkeiten. Mikroorganismen, die in Bodenkrümel eingeschlossen sind, müssen durch Dispergierung der Krümel in Freiheit gesetzt werden. Werden die Krümel nicht dispergiert, dann sinken sie in den Verdünnungen rasch zu Boden und die Mikroorganismen gelangen überhaupt nicht in das Verdünnungsmittel, wie aus nachstehenden Ergebnissen von Kuron, Glathe und Homrighausen (1959; vgl. auch Homrighausen und Rohmer 1958) hervorgeht.

	Keimzahlen je Gesichtsfeld		
	Unbehandelt	Rohagit 0,035%	Rohagit 0,05%
Ohne Dispergierung	3,34 ± 0,254	2,77 ± 0,253	2,25 ± 0,180
Mit Dispergierung	3,30 ± 0,208	3,52 ± 0,259	4,72 ± 0,370
	Keimzahlen in Milliarden je 1 g Boden		
Ohne Dispergierung	1,157	0,950	0,779
Mit Dispergierung	1,143	1,219	1,479

Die Dispergierung erfolgte mit 0,4 n Natriumpyrophosphat. Das im Feldversuch verwendete Rohagit ist ein Bodenstabilisierungsmittel. Die Ergebnisse lassen deutlich erkennen, daß die Keimzahlen bei Anwendung von 0,05% Rohagit um etwa 20% gegenüber „unbehandelt" anstiegen, und zwar ist dies nur deutlich geworden durch die Anwendung des Dispersionsmittels. Bei der kulturellen Zählung von Azotobacter-Zellen im Boden hat sich eine Konzentration von 0,18% Natriumpyrophosphat als zweckmäßige Konzentration erwiesen (Glathe und Ahrens 1963).

Die ersten Angaben über die Zahl der in Böden enthaltenen Keime stammen von R. Koch (1881). Er ermittelte sie durch das Plattengußverfahren, eine Methode, die daher allgemein mit seinem Namen gekennzeichnet wird. Dabei werden Verdünnungen von Bodenaufschwemmungen bekannter Konzentration mit Nährböden vermischt in Petrischalen bebrütet. Die einzelnen Bakterienzellen können sich nun vermehren, zu makroskopisch sichtbaren Kolonien auswachsen und gezählt werden.

Eine Vermehrung ist aber nur möglich, wenn der Nährboden den Anforderungen der Organismen entspricht, und zwar hinsichtlich

1. der Art und Menge der Makro- und Mikronährstoffe (vgl. Starkey 1955),
2. der Qualität der Kohlenstoffquellen,
3. der Reaktion,
4. des Sauerstoffhaushaltes,
5. der Temperatur.

Da die Ansprüche der einzelnen Arten sehr unterschiedlich sind, kann nicht erwartet werden, daß alle im Boden vorhandenen Mikroorganismen unter den jeweils gewählten Bedingungen zur Entwicklung kommen. Der Vorteil der Kochschen Plattenmethode besteht darin, daß sie nicht nur die Ermittlung der Keimzahl gestattet, sondern gleichzeitig auch einen Einblick in die artenmäßige Zusammensetzung der Mikroflora.

Häufig werden Gußplatten von einzelnen Kolonien überwuchert, so daß kleine Kolonien weder gezählt noch isoliert werden können. Dies vermied Unger (1958) dadurch, daß er das von Bucksteeg und Thiele (1958) für die Keimzählung bei Wasserproben oberflächlich angewandte 2, 3, 5 Triphenyl-Tetrazoliumchlorid (TTC) — um Hemmwirkungen zu vermeiden, in kleinen Mengen von 0,01, 0,005 und 0,001% — dem flüssigen Agar zusetzte. Ein Vergleich von überwucherten und nach Verwendung von TTC nicht überwucherten Platten ergab eine sehr gute Übereinstimmung. Das gleiche Resultat erzielte HIRTE (1961) mit dem von ihm für bodenbiologische Arbeiten besonders empfohlenen Glycerin-Pepton-Agar mit Fleischwasserzusatz.

Die KOCHsche Plattenmethode kann durch eine entsprechende Gestaltung von Nährböden und Umweltbedingungen beliebig ausgebaut und den verfolgten Zielen angepaßt werden. Die getrennte Erfassung von Arten mit besonderen physiologischen Eigenschaften ist möglich. Differenzialnährböden für Azotobacter, Escherichia, Zellulosezersetzer, Pilze u. a. m. sind entwickelt worden, um möglichst viele Arten zu erfassen. Die Untersuchung von morphologischen Besonderheiten der Sporen von Actinomyceten und Pilzen sowie der Knöllchenbakterien im Elektronenmikroskop kann nach FLAIG (1952) weitere differenzierende physiologische Untersuchungen ersparen.

WEST und LOCHHEAD (1940b) schlagen zur Bestimmung des „Bakterien-Gleichgewichts" im Boden einen Grundnährboden vor, der durch Zusätze von Aminosäuren und Wuchsstoffen abwandelbar ist. Die prozentuale Häufigkeit von Organismen, die auf den einzelnen Nährböden zur Entwicklung kommen, läßt Einflüsse von Jahreszeit, Pflanzen und Böden erkennen.

Will man aber einer möglichst großen Zahl von Mikroben die erforderlichen Lebensbedingungen schaffen, so kann die Methode durch die übermäßig große Zahl von notwendigen Varianten unhandlich werden.

Bei einzelnen physiologischen Gruppen von Mikroorganismen, die schlecht oder gar nicht auf Platten wachsen, begnügt man sich mit der Bestimmung der Verdünnungsgrenze, bis zu der ein bestimmter Vorgang, etwa die Nitrifikation, die Zersetzung von Zellulose usw., noch nachweisbar ist. Auf diesem Wege erhält man den Titer eines physiologischen Vorganges. SAKAI (1958) hat für die Nitrifikation Sandkulturen in sterilen Röhrchen angewendet, in denen er die Nitrifikationsbakterien anreicherte. MILLBANK (1956) benützt die von SINGH (1946a) zur Zählung von Protozoen empfohlene Methode der Kultur auf Silicagel-Nährböden in etwas abgeänderter Form zum zahlenmäßigen Nachweis von *Nitrosomonas*. Zur Feststellung von zellulosezersetzenden Bakterien hat PUSCHKINSKAJA (1954) den von WINOGRADSKY empfohlenen synthetischen Agar mit Bodensuspensionen beschickt und mit Fließpapier bedeckt. Hierbei sollten gleichzeitig die Begleitorganismen ausgeschaltet werden. Einen für die Züchtung streng aerober Organismen, *Acetobacter*- und *Streptomyces*-Arten, in Submerskultur am Kreisschwingschüttler besonders geeigneten Verschluß der Kulturgefäße beschreibt JANKE (1952).

Schon CONN (1918, 1928) und WINOGRADSKY (1925) haben Methoden entwickelt, die dem Begriff „Gesamtkeimzahl" auch experimentell gerecht zu werden versuchen. Das Prinzip besteht darin, die in den verdünnten Aufschwemmungen des Bodens enthaltenen Keime nicht kulturell zur Entwicklung zu bringen, sondern direkt unter dem Mikroskop zu zählen. Ähnliche Methoden sind für die Zählung von Blutkörperchen oder von Bakterien in der Milch bekannt.

Das erste sich hierbei ergebende Problem bestand darin, ein Färbungsmittel anzuwenden, das geeignet ist, Mikroorganismen von den übrigen Bodenbestandteilen zu trennen. In neuerer Zeit hat STRUGGER (1949; vgl. auch ROUSCHAL und STRUGGER 1943) in dem fluoreszierenden Acridinorange einen Vitalfarbstoff empfohlen, der im Gegensatz zu den in der Mikrobiologie bisher vorwiegend verwendeten giftig wirkenden Farbstoffen von den Organismen ohne schädigende Wirkung aufgenommen werden kann.

Weiterhin mußte die Schwierigkeit überwunden werden, eine eindeutige Beziehung zwischen der Bodenmenge und der unter dem Gesichtsfeld erfaßten Flüssigkeitsmenge herzustellen. STRUGGER verwendet Zählkammern, die nach Abdeckung mit einem Deckglas Flüssigkeitsschichten von 25 μ Dicke aufnehmen. Da mit Hilfe des Mikroskops nur Organismen zu erkennen sind, die in einer etwa 5 μ dicken Schicht enthalten sind, müssen in jedem Gesichtsfeld die mikroskopisch erfaßbaren Ebenen durch Verschiebung in vertikaler Richtung durchgemustert werden. Eine zweite Möglichkeit der quantitativen Untersuchung besteht darin, daß die zu untersuchende Flüssigkeitsmenge auf eine Fläche bekannter Größe verteilt und nach dem Abtrocknen untersucht wird. LEHNER und NOWAK (1957) ziehen die Schätzung vor, weil dabei störende Extremwerte, wie sie die Zählung oft ergibt, ausgeschaltet werden, ohne daß das Gesamtergebnis beeinträchtigt wird (vgl. SEIFERT, 1958, 1960b).

Um die etwa bei der Methode STRUGGER vorhandene Schwierigkeit, daß Bakterien mit rascher Eigenbewegung das Ergebnis fälschen, auszuschalten, haben JONES und MOLLISON (1948) der Aufschwemmung Agar zugesetzt und dadurch feste Agarfilme geschaffen. JAGNOW (1961) benutzte diese Methode in etwas abgewandelter Form und beobachtete die Agarfilme mit Hilfe der Phasenkontrastoptik, bei der die Unterscheidung der einzelnen Mikroorganismen leichter war als bei der Hellfeldbetrachtung. Gegenüber der Arbeitsweise von STRUGGER waren die Präparate beständiger und deshalb für die Auswertung vieler gleichzeitig entnommener Proben und für die Messung von Actinomyceten und Pilzhyphen (angegeben in m/g Boden) besser geeignet. Er ermittelte in Wiesenland je nach Boden und Jahreszeit Actinomycetenmyzel in einer Länge von 3000 bis 25000 m/g, demgegenüber betrug die Gesamtmenge der Pilzhyphen in der obersten Schicht des gleichen Bodens 170 bis 650 m/g. Die Gesamtkeimzahlen bei einzelnen Gräsern und Kräutern stiegen bis zu 350 Mrd./g. Das Gesamtgewicht der Mikroorganismen in der obersten 15 cm mächtigen Schicht betrug 2000 kg/ha (Trockensubstanz). Eine exakte Bestimmung der Keimzahl mit Hilfe der Direktmethode hängt weiterhin davon ab, ob alle Mikroorganismen unter dem Mikroskop einzeln erkennbar werden. Schwierigkeiten bestehen darin, daß besonders in nährstoffreichen, gut durchlüfteten Böden die Vermehrung der Organismen nicht in zwei, sondern drei Dimensionen erfolgt. Die dabei entstehenden Kolonien werden von STRUGGER als Zoogloen bezeichnet und getrennt gezählt. BURRICHTER (1955) beobachtete, daß zwischen den Zahlen der Bakterien und Zoogloen eine gewisse Proportionalität besteht. Die Zahl der Einzelbakterien stieg jedoch langsamer an als die der Zoogloen. Schließlich sind bei allen Methoden der direkten Keimzählung die Mikroorganismen nur in seltenen Fällen direkt in der Flüssigkeit suspendiert. In der Regel ist die Adhäsion an mineralische Bodenbestandteile so groß, daß sie sich während des Anlegens der Verdünnungen nicht von ihnen lösen. Alle auf der blickabgewandten Seite der Partikel haftenden Mikroorganismen entgehen der Zählung. STRUGGER schlägt daher vor, die gefundene Keimzahl um 33% zu erhöhen.

Es sei noch auf die Methode der Keimzählung von GRAY und THORNTON (1928) hingewiesen, bei der eine Bodenaufschwemmung bekannter Konzentration mit einer bestimmten Menge Indigo-Teilchen gemischt wird.

Zur Feststellung der Verteilung der Mikroorganismen im Boden, aber auch zu ihrem quantitativen Nachweis ist die Aufwuchsplattenmethode von ROSSI-CHOLODNY (ROSSI 1921, CHOLODNY 1930, ZIEMICKA 1935, ROSSI 1936, LINFORD 1942, HOPF 1950, ROVIRA 1955 u. a.) herangezogen worden. Sauber polierte Objektträger werden in den Boden eingelegt und können zu jeder beliebigen Zeit entnommen und nach vorsichtiger Reinigung, Fixierung und Färbung (Bengalrosa) untersucht werden. Ihre quantitative Auswertung ist insofern mit einem Unsicherheitsfaktor belastet, als die regionale Verteilung der Organismen stark von den im Boden vorhandenen Partikeln organischer Substanz und besonders von der Entwicklung des Wurzelsystems höherer Pflanzen beeinflußt wird. Nach PFENNIG (1958) führt die Aufwuchsplattenmethode zu einem Einbruch in den Gleichgewichtszustand der Bodenflora. Für die Dauer von etwa zehn Tagen trete eine lebhafte Entwicklung ein, die einem gewissen Reifezustand zustrebe, in dem die Organismen sich größtenteils im Dauerzustand befinden. Diese Bedenken können auf Grund eigener Erfahrungen jedoch nicht bestätigt werden.

RYBALKINA und KONONENKO (1953) erreichten ein intensiveres Wachstum, indem sie die Objektträger vor dem Einlegen mit Nähragar verschiedener Zusammensetzung überzogen und bei 35 bis 45° C trockneten. Zur Färbung verwendeten sie eine 1%ige Carbolerythrosinlösung. Um besonders Schimmelpilze zu erfassen, verwendeten andere Forscher (ROBERTS und BOLLEN 1955, MESSIAEN 1957, J. C. BROWN 1958, KENDRICK und JACKSON 1958) besonders präparierte Objektträger oder anstelle der Objektträger Glasfaserstreifen (LEGGE 1952) bzw. Zellophanblätter oder -taschen (KERR 1958). JENSEN (1935; vgl. auch WINTER 1947, 1949) hat die Befunde der Aufwuchsplatten zum quantitativen Nachweis der Pilze insofern ausgewertet, als er die Länge der Pilzhyphen bestimmte. NICKERSON (1948) und ROBERTS und BOLLEN (1955) wendeten sie zur Beobachtung des Einflusses von Herbiziden auf die Mikroflora an.

LEHNER und ihre Mitarbeiter (LEHNER, NOWAK und SEIBOLD 1958, LEHNER und NOWAK 1958, 1959/60, 1960, NOWAK 1959) entwickelten als weitere Möglichkeit des Direktnachweises von Bakterien im Boden die kombinierte Aufwuchs-Fluorochromierungsmethode.

Eine gewisse Ähnlichkeit mit der Methode CHOLODNY-ROSSI hat auch die von ALEXANDER und JACKSON (1954) empfohlene Dünnschliffmethode. WIGOROW (1955) erreichte eine gewisse Erhaltung der Bodenstruktur dadurch, daß er kleine Bodenkrümel auf Objektträger legte, mit Reagenzien behandelte und dann mikroskopierte. Auf diesem Wege konnte er z. B. nachweisen, daß das Ammoniak sich nicht auf der Oberfläche der Bodenteilchen befand, sondern im Bakterienschleim und in im Abbau befindlichen Pflanzenteilen enthalten war und daß die Humusteilchen mit Bakterien gesättigt waren. Alles deute darauf hin, daß bei der Pflanzenernährung wesentliche Prozesse sich zwischen der Mikroflora und der Pflanzenwurzel abspielen.

Eine Feststellung der Verteilung der Mikroben im Boden und noch mehr die ihrer Zahl muß notwendig das Gefüge des Bodens und damit die Lebensbedingungen der Organismen stören. Je weniger dies geschieht, z. B. bei der Aufwuchsplattenmethode, um so besser ist der Einblick in ihre Verteilung. Die Direktmethoden lassen die Formenfülle leicht erkennen, das Haften an Krümeln und organischen Resten, das Vorherrschen von Stäbchen oder Kokken, von großen Bakterien oder Pilzen, während das Plattengußverfahren, das durch die Zusammensetzung des Nährbodens bereits bestimmte physiologische Gruppen auswählt, die Isolierung einzelner Arten erleichtert. So ist z. B. eine Trennung von echten Bakterien und Actinomyceten nur mit Hilfe der KOCHschen Methode möglich, da, wie M. E. BROWN (1958) feststellt, einzelne Actinomyceten während des größten Teils des Jahres als Stäbchen erscheinen. NOWAK und NETZSCH-LEHNER (1960; vgl. ISHIZAWA et al. 1958) empfehlen, beide Methoden parallel zu verwenden, da besonders für die Direktmethode nicht jeder Boden geeignet ist, andere wieder besonders gute Resultate ergeben.

Es ist zu erwarten, daß die mit den direkten Methoden ermittelten Keimzahlen wesentlich höher liegen als die Ergebnisse des KOCHschen Verfahrens.

Bei Vergleichen der direkten Methoden von CONN (1918) und von WINOGRADSKY (1925) mit der Methode KOCH stellte KÜHLMORGEN-HILLE (1928) fest, daß die KOCHschen Keimzahlen im Mittel 10% der nach CONN und 2,7% der nach WINOGRADSKY gefundenen Keimzahlen betrugen. SKINNER et al. (1952) fanden nach der Methode von JONES und MOLLISON stets 40- bis 50mal so hohe Keimzahlen wie nach der Plattenzählung. THORNTON (1956) ermittelte mit Hilfe der Plattenkulturmethode Keimzahlen zwischen 50 und 100 Mill./g Boden, mit der Direktmethode dagegen mehrere Milliarden, das ergibt ein Verhältnis von etwa 10%. Bei Untersuchungen von SAUERLANDT et al. (1961) und auch in eigenen Versuchen ergaben sich nach KOCH nur etwa 0,5 bis 5% der Keimzahlen, die mit Hilfe der Methode STRUGGER ermittelt wurden.

Zwischen den Ergebnissen des Plattenkulturverfahrens und der direkten Methoden wird im allgemeinen ein Parallelismus bestehen. Eine ständige Korrelation braucht sich jedoch nicht zu ergeben, wenn, wie TAYLOR (1936) annimmt, Verschiebungen in der Zusammensetzung der Mikroflora eintreten können, die nur mit einer von beiden Methoden erfaßbar sind. Das wäre der Fall, wenn eine bestimmte Gruppe von Bakterien eine besonders starke Zu- oder Abnahme erfährt, diese Organismen aber auf dem verwendeten Nährboden überhaupt nicht wachsen oder besonders gut gedeihen, wie SAUERLANDT et al. (1961) bei der Keimzählung nach organischer Düngung feststellten. Ein besonderer Vorteil der Direktmethoden besteht in dem geringeren Zeitbedarf für das Einzelergebnis; denn nur häufige Probenahmen und eine große Zahl von Einzeluntersuchungen lassen Schlüsse auf Einflüsse von Klima, Düngungs- oder Bearbeitungsmaßnahmen zu (vgl. NOWAK und NETZSCH-LEHNER 1961). Zählung bzw. Schätzung sollte bei den Direktmethoden zur Sicherung der Versuchsergebnisse möglichst stets durch den gleichen Beobachter durchgeführt werden.

Beim quantitativen Nachweis von Actinomyceten mit Hilfe der Plattenmethode werden nur die Sporen erfaßt, weil, wie JAGNOW (1958) feststellte, nur diese auskeimen. Beim Direktnachweis bereitet die Unterscheidung der Stäbchenformen von den Bakterien Schwierigkeiten. Für die Plattenkultur zur Keimzahlbestimmung hält HIRTE (1961) Glycerin-Pepton-Agar mit Fleischwasserzusatz für gut geeignet. Zur qualitativen Analyse empfiehlt er wie auch STAPP (1952/54) Na-Asparaginat-Agar nach CONN.

Zum Nachweis von Hefen im Boden setzten MILLER und WEBB (1954) einem Kartoffel-Dextrose-Agar Bengalrosa und Ochsengalle zu. Durchschnittlich enthielten die untersuchten Böden 1000 Hefezellen je g, maximal 50000. DIMENNA (1957) benützte einen Glukose-Pepton-Agar von pH 4 und fand in Lehmboden von nicht gepflügten Schafweiden Neuseelands 6000 bis 240000 Keime je g Boden, wenn die Suspensionsverdünnungen auf die Agar-Oberfläche geimpft wurden. Besondere Verfahren zur getrennten Isolierung von Hefen und Pilzen entwickelten ETCHELLS und seine Mitarbeiter (1954).

Für das spezielle Studium der Pilze ist eine große Zahl von Methoden entwickelt worden. PAPAVIZAS und DAVEY (1959) besprechen die Eignung von 41 Substraten und verschiedenen Hemmzusätzen zur Isolierung von Bodenpilzen. Bei Anwendung der KOCHschen Plattenmethode besteht die Schwierigkeit in erster Linie darin, daß die Bakterien das Wachstum der Pilze hemmen. Deshalb werden Spezialnährböden verwendet, die infolge ihrer hohen Wasserstoffionen-Konzentration (Reaktion pH 5 bis 5,5) die Entwicklung der Bakterien nicht zulassen oder hemmen. WILTGEN und BONNIER (1952; vgl. auch PAHARIA und KOMMEDAHL 1954, MESSIAEN 1957) zerkleinerten den Boden im Mörser und streuten ihn auf Nährböden, die Bengalrosa enthielten, um die Entwicklung der Bakterien zu unterdrücken. Als Hemmstoff für Bakterien verwendeten BAKER-

SPIGEL und MILLER (1953) Ochsengalle, Kristallviolett, Streptomycin oder Penicillin, BUTLER und HINE (1958) Novobiocin als saures Na-Salz. Dabei wurde auch die radiale Entwicklung der Pilzkolonien unterdrückt. WARCUP (1955) wirbelte die Erde in Kolben auf, und nach kurzer Sedimentation wurden die in der Aufschwemmung suspendierten Pilzhyphen auf Dox-Hefeagar kultiviert. Die Bakterienentwicklung wurde durch Verdünnung des Nährmediums auf 1:6 gehemmt. Weiterhin wurde empfohlen (WARCUP 1950, DOMSCH 1955, SCHMIDT 1959), die zu prüfende Erde in Petrischalen zu geben und mit einem dünnen Film von Nähragar zu übersprühen. Pilzfallen wurden in Gestalt von Kammern in Objektträgergröße (LA TOUCHE 1948, SEWELL 1956) oder als durchbohrte Röhrchen (CHESTERS 1948, EVANS 1954, MUELLER und DURRELL 1957) verwendet. Schließlich wurden auch Mais- oder Erbsensamen als Köder für Schimmelpilze ausgelegt (KENDRICK und JACKSON 1958).

Bei der Isolierung von niederen Erdphycomyceten mit Hilfe der Ködermethode stellten HARDER und UEBELMESSER (1958) fest, daß Flechten die Entwicklung der Phycomyceten im Boden wie im künstlichen Substrat hemmten.

Die Entwicklungsphase der Schimmelpilze übt auf die Höhe der Keimzahlen insofern einen Einfluß aus, als von den vegetativen Stadien nur wenige auf Platten zu Kolonien auswachsen, während bei Versporung die Koloniezahl relativ höher ist (HACK 1957).

Zum zahlenmäßigen, kulturellen Nachweis der Protozoen im Boden, besonders der Amöben, hat SINGH (1946a, b) einen Silicagel-Nährboden entwickelt. TCHAN und BUNT (1954) zählten die Protozoen, indem sie die Objektträger mit einer Boden-Wasser-Suspension im Verhältnis 1:5 besprühten und in feuchtem Zustand mit Osmium- oder Formalindämpfen fixierten und mit Erythrosin an- und mit Methylgrün gegenfärbten. Die Protozoen erscheinen blaßrot mit dunklem Kern, Bakterien, Pilze und Algen purpurrot und Bodenpartikel grün.

2. Keimarten und ihre Beeinflussung durch Klima und Düngung

Zu den Mikroorganismen des Bodens gehören Bakterien, Actinomyceten, Pilze (Hefen), Algen als Mikroflora und Protozoen als Mikrofauna, dazu kommen die Kleintiere, die ihrerseits wieder als Wirte von Mikroorganismen aufzufassen sind.

Zu den *echten Bakterien* des Bodens gehören eine Fülle von sporenfreien und sporenbildenden Bakterien, Kokken und gelegentlich Spirillen. Die letzteren finden nur in sehr nassen Böden geeignete Lebensbedingungen. In der älteren und neueren Literatur ist das Verhältnis von Kokken zu Stäbchen eingehend diskutiert worden, und man versuchte auch zu klären, ob die sporenbildenden oder die sporenfreien Mikroorganismen an Zahl im Vordergrund stehen. Berücksichtigt man die Ergebnisse der Plattenkulturmethode, dann muß man den Sporenbildnern eine erhebliche Bedeutung beimessen, da sie in allen Böden in relativ großen Mengen auftreten (vgl. ARISTOVSKAIA und PARINKINA 1958). Für den Vergleich verschiedener Böden ist es besonders wesentlich, Unterlagen über die im Boden auftretenden Formen der Bakterien zu erhalten. Hierfür sind die Methoden der direkten Bodenuntersuchung, die Methode STRUGGER und die Aufwuchsplattenmethode nach CHOLODNY-ROSSI, besonders geeignet. Die Präparate der einzelnen Untersuchungen geben gewissermaßen eine Augenblicksaufnahme der Flora. In zeitlichen Abständen entnommene Präparate gestatten indessen einen Einblick in ihre Dynamik. Bei diesen Untersuchungen hat sich gezeigt, daß sowohl kugel- als auch stäbchenförmige Mikroorganismen im Boden vorkommen. Die mikroskopische Untersuchung kann aber keinen Aufschluß darüber geben, ob es sich hierbei um echte Kokken oder Entwicklungsstadien von sporenbildenden Bakterien oder Actinomyceten handelt. Eine weitere Klärung des Verhältnisses von Kokken zu Stäbchen ist nur mit Hilfe des Mikromanipulators möglich. Ob solche Arbeiten aber infolge der geringen Entwicklungsfähigkeit von Ein-Zell-Kulturen zu dem gewünschten Ziel führen, kann nicht vorausgesagt werden. Nach TAYLOR und LOCHHEAD (1937) zeigten 110 Stämme von *B. globiforme Conn* (eingehendes Studium bei CONN und DIMMICK 1947) = *Achromobacter globiformis Bergey* eine charakteristische Metamorphose von Stäbchen zu Kokken. TAYLOR (1938) fand den Organismus in 89 von 90 kanadischen Böden. In den fruchtbaren Typen kam er vielleicht etwas häufiger vor; in diesen Böden wurden aber auch höhere Gesamtkeimzahlen gefunden. Seine untere Grenze lag bei einer Bodenreaktion von pH 5,0. Bei 45° C war er nach 24 Stunden nicht mehr lebensfähig.

TOPPING (1937/38, 1938) fand bei ihren Untersuchungen überwiegend kleine Stäbchen (Corynebakterien, Proactinomyceten und mit ihnen verwandte bewegliche Organismen) ohne bestimmte physiologische Eigenschaften, mit großer Neigung zu morphologischen Wandlungen. Sie scheinen identisch zu sein mit der von WINOGRADSKY als autochthon bezeichneten Mikroflora. ROUATT und LOCHHEAD (1955) stellten fest, daß wachsende Kulturpflanzen Bakteriengruppen mit einfacheren Nahrungsansprüchen stärker förderten als solche mit komplizierten; aromatische Verbindungen werden von Bakteriengruppen aus wurzelfernem Boden zersetzt (VRANÝ 1960). *Bac. megaterium* kommt nach GROMYKO (1955) häufiger in kultivierten Böden vor als in unkultivierten, mehr in der Krume als im Untergrund. In sauren Hochmoorprofilen haben POSCHENRIEDER und seine Mitarbeiter Mikrokokken gefunden (RÖSCHENTHALER und POSCHENRIEDER 1958a: *Micrococcus candidus;* 1958b, BECK und POSCHENRIEDER 1958), daneben vereinzelt Sarcinen und Achromobacter-Arten, aber wenig Sporenbildner und in tieferen Schichten Pseudomonaden. Sie nehmen an, daß fluoreszierende Bakterien sogar an der Entstehung der Moore beteiligt sind (POSCHENRIEDER und BECK 1958) und daß sie mitwirken bei der Überführung rezenten Pflanzenmaterials über jüngere zu älteren Torfen bis zur Braunkohle.

Bei der Untersuchung harnstoffspaltender Bodenbakterien fanden BORNSIDE und KALLIO (1958) zwei physiologisch verschiedene Gruppen, die einen waren aktiv im pH-Bereich von 7 bis 9, die anderen ab pH 8,5.

Über die Verbreitung der Myxobakterien und ihre Systematik berichtet KÜHLWEIN (1960). Er isolierte verschiedene Arten aus Boden und Wildmist.

Eine Reihe bisher unbekannter Bodenbakterien beschrieben französische Forscher. PRÉVOT und seine Mitarbeiter isolierten aus Böden der Elfenbeinküste *Cl. saturnirubrum* (1946), *Cl. inflabilis mangenoti* n. sp. (PRÉVOT et ZIMMÈS-CHAVEROU 1947) bzw. das nicht sporenbildende anaerobe *Cillobact. Combesi* n. sp. in Guinea (PRÉVOT et LAPLANCHE 1947) und wiesen die Existenz des *Cl. corallinum* P. et R. erstmalig in Böden von Senegal nach (PRÉVOT et SANSONNENS 1947). COUMARTIN (1957) beschrieb eine als *Perabact. spelei* bezeichnete Art, die er in den Kavernen von Ton- und Eisensedimenten fand.

Thermophile Bakterien (Sammelreferat von GAUGHAN 1947) kommen nicht nur in den Böden der heißen und gemäßigten Zone vor, sondern sogar auch in arktischen Böden und Gewässern. MCBEE und GAUGLER (1956) haben 15 aerobe sporenbildende Arten bei 55° C isoliert und 10 davon als *Bac. stearothermophilus*, 3 als *Bac. coagulans* und 2 als *Bac. licheniformis* identifiziert. Es zeigte sich (R. H. MCBEE and V. H. MCBEE 1956), daß die gefundenen Arten bei wesentlich tieferen Temperaturen wachsen, wenn ihnen bestimmte organische Wirkstoffe zur Verfügung stehen, die allerdings noch nicht bekannt sind. Diese Organismen wurden in den Böden unter längst verlassenen Siedlungen gefunden. KOCH und HOFF-

MANN (1911) konnten nachweisen, daß die Temperaturansprüche der thermophilen Bakterien im Boden und in Laboratoriumskulturen unterschiedlich sein können.

Psychrotolerante Bakterien (Sammelreferat von INGRAHAM and STOKES 1959) treten nicht nur als marine Organismen auf (HESS 1934), sondern sind auch immer wieder aus Böden isoliert worden (LOCHHEAD 1926). Auch Gemüse ist von ihnen infiziert worden (HUCKER 1954). Unter ihnen treten auch obligat anaerobe Arten auf, die unter Umständen Störungen in der Konservenindustrie in Gestalt von Bombagen auslösen.

Die *Actinomyceten* (WAKSMAN 1950, 1953, 1959) sind infolge ihres geringen Wasserbedarfs, wie MEIKLEJOHN (1957) nachwies, dazu befähigt, Trockenzeiten im Boden zu überstehen. Ihr Anteil an der Gesamtflora beträgt daher, nach den Ergebnissen von JAGNOW (1956) in Kalktrockenrasen und Trockenwaldstandorten 30 bis 40%, im Maximum sogar 60%. Ihre Entwicklung im Boden, vor allem die Bildung von Chlamydo- bzw. Luftmycelsporen, ist mit Hilfe der Methode CHOLODNY-ROSSI mikroskopisch gut zu verfolgen (PFENNIG 1958). BROWN (1958) impfte *Nocardia cellulans* in partiell sterilisierte Böden des Versuchsfeldes Woburn, die Zahl der Nocardien schwankte am Anfang des Versuchs stark, ihre Entwicklung hemmte die Zahl der echten Bakterien, wie die nicht beimpften Parallelen bewiesen. Die höchste Zahl der Nocardien wurde gefunden, wenn die Stäbchenform vorherrschte, sie sank auf ein Minimum, wenn sie in Mycelform auftraten. Im Boden erscheinen sie vorwiegend als Stäbchen. Über das Vorkommen von *Nocardia uniformis* in Solonetzböden Ungarns berichteten MARTON und SZABÓ (1959); MCCLUNG (1951) fand wenig Actinomyceten in sauren Böden, konnte sie aber bis pH 9 nachweisen. Auch SZABÓ und Mitarbeiter (1959) stellten für die meisten Arten ein pH-Optimum im alkalischen Bereich fest.

In den Hochgebirgsböden Pamirs herrschten nach KUZNETZOW (1960) bestimmte Actinomyceten-Arten vor. Besonders reichlich traten *Mycobacterien* auf. Über ihr Auftreten in Eichenwaldböden berichtet KOSLOWA (1955).

GAERTNER (1955) isolierte aus Ackerböden zwei ungewöhnliche keratinophile Organismen der Gattung *Actinoplanes*.

Auf die Bedeutung der Actinomyceten bei der Bildung von antibiotischen Substanzen (ALESCHINA et al. 1955, KORJENJAKO et al. 1955, SZABÓ 1956, GAUSE et al. 1958, VANĚK et al. 1958, KRASSILNIKOV 1959) und deren Eignung zur Bekämpfung phytopathogener Mikroben (KRASSILNIKOV 1960) sei besonders hingewiesen.

REHM (1961) fand bei Streptomyceten aus Sand- und aus Schwarzerdeböden bei bestimmten Arten (z. B. *Act. chromogenes*-Stämmen) deutliche Unterschiede in der Stärke in der Antibioticabildung, bei anderen (*Strept. griseus* und *Strept. diastaticus*-Stämmen) dagegen gute Übereinstimmungen; die morphologischen Merkmale der Arten blieben konstant. Die antibiotische Tätigkeit dürfte daher für systematische Untersuchungen, wie es KRASSILNIKOV (1959) empfiehlt, nur bedingt geeignet sein.

CAPRIOTTI isolierte *Hefen* aus verschiedenen europäischen Böden in den Niederlanden (1955), in Schweden (*Torulospora nilssoni* n. sp.: 1958a; *Debaryomyces castellii* n. sp.: 1958c; *Candida vanriji* n. sp. 1958d) und in Spanien (1958b). VAN DER WALT und TSCHEUSCHNER beschrieben *Pichia vanriji* n. sp. (1956a) und *Saccharomyces pretoriensis* n. sp. (1956b), VAN DER WALT (1957) drei neue sporenbildende Hefen, die er in südafrikanischen Böden fand. Über das Vorkommen von Hefen in Böden und auf Weidegras in Neuseeland berichtet DIMENNA (1957, 1958).

Die Verbreitung von Lipomyces und seine Beziehung zu Azotobacter in Böden

Jakutiens hat MASILKIN (1957) studiert; es zeigte sich, daß die Bodenart von ausschlaggebender Bedeutung war.

Wegen ihres höheren Sauerstoffbedarfs treten die meisten *Schimmelpilze* (vgl. die Werke von GILMAN 1957 und PARKINSON und WAID 1960) in tieferen Schichten des Bodens seltener auf als in der Krume. Trotzdem konnten HARDER und ÜBELMESSER (1957) nachweisen, daß Chytridineen bis 1,55 m, Pythius bei 3,55 m, Mucorineen bis 1,20 m, septierte Hyphen aber bis 3,55 m und in einem Falle in voreiszeitlichem Keuperton noch bei 5,20 m vorkommen können. Allgemeingültige Angaben werden erst aus einer sehr großen Zahl von Einzeluntersuchungen abgeleitet werden können, weil die natürliche Struktur des Bodens, augenblickliche Höhe des Grundwassers, Einfluß von Wurmröhren usw. die Ergebnisse stark beeinflussen können.

Die Tatsache, daß Pilze in sauren Böden besonders stark vertreten sind, erklärt JACKSON (1957) damit, daß die Fungistasis (vgl. DOBBS und HINSON 1953), die in der Hemmung der Keimung von Pilzsporen besteht, durch Verschiebung der Reaktion nach der sauren Seite herabgesetzt wird. Die isolierten Pilzarten waren gegen eine solche Hemmung verschieden empfindlich.

Aus einer pilzreichen Komposterde isolierte DOMSCH (1960) 133 Arten und gibt eine Übersicht über ihre Systematik, Leistung und Verteilung. In den arktischen Tundren der Nowaja Semlja wurden Schimmelpilze äußerst selten gefunden (SUSCHKINA und RYSCHKOWA 1956), im Norden fehlten sie völlig. Arm an Arten von Schimmelpilzen waren auch die unberührten Gebiete Australiens (ÜBELMESSER 1956, HARDER und GALLWITZ -ÜBELMESSER 1959); im übrigen unterscheiden sich die in diesem Kontinent vorkommenden nicht von denen anderer Kontinente. Am häufigsten, nämlich in 91% der 34 untersuchten Proben, traten folgende Arten auf: *Olpidium luxurians* und *Olpid. pendulum* bzw. *Rhizophydinum sphaerotheca* und *Rh. carpophilum*. GAERTNER (1954) stellte die weltweite Verbreitung der niederen Erdphycomyceten fest. In Böden Nordschwedens waren sie relativ wenig, in Böden Afrikas relativ stark vertreten. Besonders häufig wurde *Karlingia rosea* und *Olpidium pendulum* gefunden.

Einige neue erdbewohnende Mortierella-Arten wies LINNEMANN (1952/54) in Böden des Brockengipfels, in Waldböden und Moor nach. WOLF (1952/54) isolierte 750 Arten von Mycorrhiza-Pilzen aus Torf und fand als Symbionten von Ericaceen überwiegend Verwandte von *Cladosporium*, daneben *Mortierella*, ferner *Penicillium-* und *Verticillium-Arten.*

KOX (1954) stellte bei Penicillien und Verticillien im Hochmoor Pektinasebildung fest; bei der Zellulosezersetzung waren beide Arten passiv; aktiv dabei waren die Symbionten der Ericaceen. Wichtig für die Zersetzung von Taxus-Nadelstreu ist das von SCHNEIDER (1957) isolierte sklerotienbildende *Penicillium taxi*. Über die Verbreitung der Phycomyceten, Ascomyceten und Fungi imperfecti in sehr unterschiedlichen, auch arktischen und tropischen Böden hat NIETHAMMER (1937) eingehende Untersuchungen durchgeführt.

Die vegetativen Formen von Schleimpilzen wachsen als Myxamoeben, die von Bakterienzellen leben. RAPER und QUINLAN (1958) isolierten *Acytostelium leptosomum* aus Boden und Waldstreu. SINGH (1946c, 1947) berichtet von dem Vorkommen von *Dictyostelium mucoroides* und *D. giganteum*, deren Entwicklung wesentlich von der Bodenfeuchtigkeit beeinflußt wird.

Mit der Pilzflora von salzhaltigen Böden befaßte sich SCHOLZ (1958). Die aus den 115 Erdproben des Küsten- und Binnenlandes isolierten Pilze konnten sich größtenteils auch auf Süßwasser entwickeln, nur *Thranstochytrium pachydermum* benötigte einen höheren Salzgehalt. In den Küstenböden war dieser Organismus besonders häufig verbreitet, in den Böden des Binnenlandes dagegen *Rhizophydium halophilum*.

Verschiedene Pilze nehmen im Kampf gegen Pflanzenschädlinge und -krankheiten eine bedeutende Stellung ein. So isolierte DUDDIGTON (1954) aus mit Nematoden verseuchten Böden Englands 20 Pilzarten, die Nematoden zerstörten. REHM (1956) fand in mitteldeutschem Boden 5 *Alternaria tenuis*-Stämme mit unterschiedlicher Fähigkeit, Phytotoxine und Antibiotica zu bilden. Aus australischen Böden isolierten McLENNAN und DUCKER (1954) 26 antibiotisch wirksame Penicillium-Stämme.

NISSEN (1954) andererseits wies bei verschiedenen Bodenpilzen die Fähigkeit zum Abbau von Penicillin, Chloromycetin und Terramycin nach. Über die Beziehungen der Pilze untereinander gibt DEVAY (1956) eine Übersicht.

Algen kommen in allen Böden vor. Ihr Nährstoffbedarf ähnelt dem der höheren Pflanzen. Sie passen sich weitgehend allen Reaktionsbereichen an. TCHAN (1959) hält sie deshalb als biologische Testorganismen zur Prüfung der Bodenfruchtbarkeit für besonders geeignet. Die von Mischkulturen in zwei bis drei Wochen gebildete Chlorophyllmenge wird gemessen. Sie steht in klarer Beziehung zur N-Konzentration des Bodens. Beziehungen zwischen dem Wachstum der blaugrünen Algen und der Bodenfruchtbarkeit waren bereits 1953 von der Sektion für Düngung herausgestellt worden (Sektion für Düngung 1953).

Über Ökologie und Systematik der Algen gibt die Algenkunde von FOTT (1959) Aufschluß. Auf Grund von Angaben der Literatur, ergänzt durch eigene Versuche, hat M. B. ALLEN (1952) einen Überblick über die Myxophyceen gegeben.

SCHTINA (1957) beobachtete eine Förderung der höheren Pflanzen durch Zugabe von Algenkulturen zu Böden im nordosteuropäischen Rußland. Folgende Faktoren begünstigten das Bodenleben: Bildung organischer Substanz, Festhalten von Mineralsalzen, Hemmung der Bodenerosion und nicht zuletzt die Stickstoffbindung (s. S. 588). Grünalgen sind zur Nitratreduktion befähigt, und zwar zeigte KESSLER (1955), daß die Reduktion von Nitrat zu Nitrit nur in Gegenwart von CO_2 oder Glukose stattfindet, während die Nitrit-Reduktion ein photochemisch bedingter Prozeß zu sein scheint.

JOHN (1942) untersuchte die Algenflora britischer Böden und stellte eine Abhängigkeit von Reaktion und Kalkgehalt fest, insofern als auf sauren Böden vorwiegend Chlorophyten (LORENZEN 1958) vorkamen, auf alkalischen eher Cyanophyten, Diatomeen und Heterokonten. Für Kalkböden waren *Anabaena variabilis*, *Achnanthes coarctata*, *Chlorochytrium paradoxum* typisch. FRITSCH (1944) beschrieb *Cladophora calcicola* nov. gen. et sp., die er von Kalkgestein isolierte. LUND (1947) bestätigte die Befunde von JOHN, er bezeichnet Cyanophyten als kennzeichnend für N- und P-reiche Böden. Während JOHN eine Mikroschichtung in der Weise fand, daß von den oberflächlich lebenden Algen die Cyanophyceen oben und die Heterokonten unten wuchsen, beobachtete LUND eine Mikroschichtung nur bei den Xanthophyceen. JOHN stellte auch eine Sukzession der Algenflora fest, bei der zuletzt zumeist die Cyanophyceen dominierten. Entsprechend den Beobachtungen von JOHN, daß die Algenflora am besten in den feuchten, kühlen Monaten ausgebildet ist, stellt DEMOLON (1960) eine starke Vermehrung während des Winters, besonders nach Ammonphosphatdüngung, fest und TITTOR (1956) ein Maximum der Algenentwicklung im Frühjahr. Er fand besonders viele Algen in Lehmböden, die artenreicher waren als die reinen Lößböden. KNAPP und LIETH (1952) fanden auf der Bodenoberfläche reichlich wachsend nur *Nostoc sphaericum* und *Cylindrospermum licheniforme*.

Die *Protozoen* (vgl. auch CUTLER 1923 und KOFFMANN 1934) begrenzen das

Wachstum der Bakterien im Boden. Neben ihnen sind Myxomyceten, Myxophyceen und Antibiotica in der gleichen Richtung wirksam. Erhebliche Abnahmen der Keimzahlen wurden von Cutler (1923) festgestellt, wenn dem Boden Protozoen zugeführt wurden. Die Senkung der Keimzahlen war bei Flagellaten stärker als bei Amöben. Seine Befunde wurden durch Telegdy-Kováts (1932) in Versuchen mit Sand bestätigt.

Nicht alle Bodenbakterien sind als Nahrung für Amöben geeignet (Singh 1947, 1948) und beeinflussen je nach Art den Entwicklungszyklus. Pflanzenpathogene Bakterien erzeugten ein für Amöben toxisches Stoffwechselprodukt (Singh 1942a), ebenso erwiesen sich die Pigmente verschiedener Bodenbakterien als toxisch für Protozoen, d. h. sie bieten den Bakterien Schutz vor den Bodenräubern (Singh 1942b, 1945). So konnte die Flagellate *Cercomonas crassicauda* Rhizobien nicht

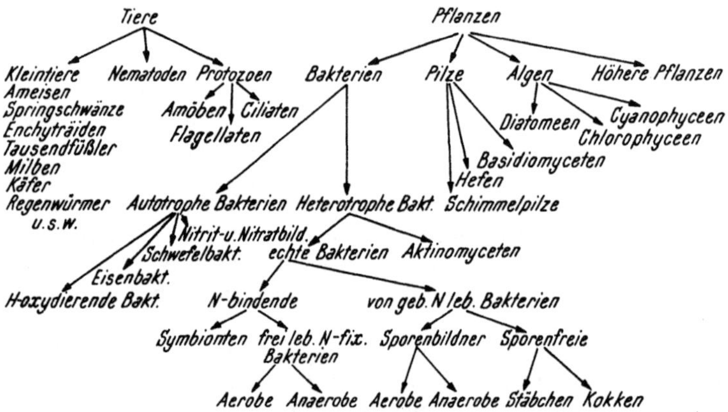

Abb. 103. Übersicht über die im Boden vorhandene Mikroflora und die wichtigsten Kleintierarten

verdauen (Singh 1942a). Szabó und Mitarbeiter (1959) beobachteten, daß sich die Protozoen unter Herausbildung spezifischer physiologischer und morphologischer Eigenschaften den besonderen Verhältnissen der Szikböden anzupassen vermochten.

Mineraldünger wie Stalldunggaben erhöhten die Zahl der Amöben in Böden von Rothamsted (Singh 1949), doch bestand keine Korrelation zwischen ihrer Zahl und dem Gehalt an organischem Kohlenstoff.

Über die Wechselbeziehungen zwischen Mikroorganismen und Kleintieren und deren Verbreitung im Boden sei vor allem auf die zusammenfassenden Darstellungen von Franz (1950), Kühnelt (1950), Graff (1953) u. a. hingewiesen.

Eine schematische Übersicht über die im Boden vorhandene Mikroflora und die wichtigsten Kleintierarten (Abb. 103) gibt Glathe (noch nicht veröffentlicht).

A. Einfluß des Klimas

Umfang der Mikroflora (Arten und Zahl der Individuen) und ihre Aktivität werden durch zahlreiche Faktoren beeinflußt, die in einer schematischen Übersicht (Abb. 104) wiedergegeben sind. Ausschlaggebend für die *Zahl* der im Boden vorkommenden *Mikroorganismen* ist zunächst das Nahrungsangebot, das sie im Boden vorfinden, die Menge des zur Verfügung stehenden Wassers bzw. der Luft und die jeweilige Temperatur. Düngung und Bodenbearbeitung, Reaktion und Struktur, Vorfrucht und der augenblickliche Pflanzenbestand üben ihren Einfluß aus. Wird dem Boden keine Düngung zugeführt, dann kann das Nahrungsangebot als konstant angenommen werden. Da die Biofaktoren Wasser und Luft in umgekehrtem Verhältnis zueinander stehen, können sie als eine Größe betrachtet werden. Wasser und Temperatur bestimmen also in ein und demselben Boden die Aktivität der Mikroorganismen. FEHER (1933a, 1933b, FEHER und FRANK 1937, 1938) und sein Mitarbeiter FRANK haben diese Biofaktoren eingehend im Laboratorium und im Freiland studiert, durch Multiplikation der numerischen Werte beider Komponenten erhielten sie ein Produkt, das sie als R-Faktor bezeichneten. Die numerische Höhe des R-Faktors ist jedoch nicht entscheidend, denn das errechnete Produkt besteht aus zwei physiologisch nicht gleichwertigen Faktoren. Da aber das gleiche Produkt aus einer beliebig großen Zahl von Faktoren gebildet werden kann, ist es als Charakteristikum für einen bestimmten Boden erst dann geeignet, wenn eine größere Zahl von Einzelwerten in einem Raumkoordinatensystem betrachtet wird. Über den Parallelismus zwischen Keimzahl und Wassergehalt geben die in der graphischen Darstellung der Abb. 105 enthaltenen Resultate von GLATHE (noch nicht veröffentlicht) Aufschluß.

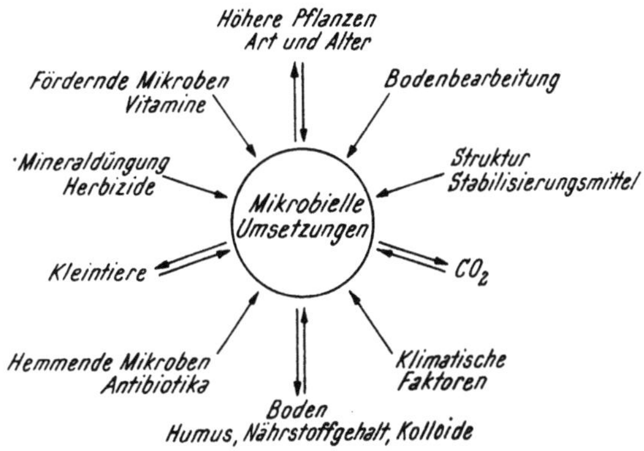

Abb. 104. Faktoren, die Zusammensetzung und Aktivität der Mikroflora beeinflussen

Das Optimum der Mikroorganismentätigkeit liegt etwa bei 60% der Wasserkapazität und 25° C. Bei den einzelnen Arten bestehen jedoch erhebliche Ab-

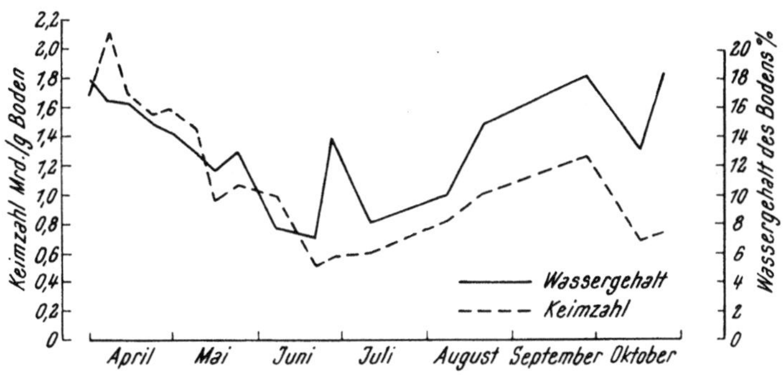

Abb. 105. Keimzahl und Wassergehalt des Bodens

weichungen, so liegt das Optimum für Pilze bereits bei 15% der Wasserkapazität; zwischen nitrifizierenden und denitrifizierenden Bakterien besteht ein gegensätzliches Verhältnis. Bei Temperaturen < 0° C wurden verschiedentlich höhere Keimzahlen festgestellt, hierfür dürfte in erster Linie die krümelsprengende Wirkung des Frosts verantwortlich zu machen sein. Mit zunehmender Tiefe sinkt im Boden sowohl die Menge der Nährstoffe als auch die Menge des zur Verfügung stehenden Sauerstoffs. Der Rückgang der Keimzahl mit zunehmender Tiefe ist daher durchaus verständlich. Nur eine dünne, einige Meter dicke Schicht der Erdkruste ist durchgehend belebt. In der Nähe von Ölfeldern konnte ZOBELL

(1946) noch in einer Tiefe von 2000 m Bakterien nachweisen. Es ist jedoch wahrscheinlich, daß die bakterienführenden Tiefenschichten von mikrobenfreien Deckschichten überlagert werden (BEERSTECHER JR. 1954).

Die Stärke der Abnahme ist abhängig von dem Gehalt an Nährstoffen und Wasser (BLUE et al. 1955). Pilze nehmen rascher ab als Bakterien (WAKSMAN 1944). G. D. THORNTON und ENO (1959) fanden in $2^1/_2$ bis 3 m Tiefe nur mehr 15% der Krumenkeimzahlen, die nitrifizierenden Bakterien waren von 1 Mill./g in der Krume auf 100/g in 120 cm Tiefe zurückgegangen. In einem besonderen Falle stellten BLUE et al. (1955) eine keimarme Zone in 20 bis 25 cm Tiefe fest, die sich daraus erklärte, daß es ihr, verglichen mit der Krume an Nährstoffen, mit den tieferen Schichten an Wasser fehlte. Auch RÖSCHENTHALER und seine Mitarbeiter (1958) fanden in Moorprofilen in tieferen Schichten zunächst einen Anstieg der Keimzahl.

In den Waldböden ist das Verhältnis von Bakterien : Pilzen in den verschiedenen Horizonten sehr unterschiedlich, MEYER (1959) kam zu folgenden Ergebnissen:

Typischer Rohhumus	Bakterien : Pilze = 0,9 : 1
Feuchtrohhumus	Bakterien : Pilze = 8 : 1
Moder	Bakterien : Pilze = 11 : 1
Mull	Bakterien : Pilze = 22 : 1

Danach fällt die Aufgabe der Zersetzung der organischen Substanz des Waldes offenbar zunächst den Schimmelpilzen zu, die nach und nach von Bakterien abgelöst werden.

Nach LOUB (1960) spielt für die Massenentwicklung der Mikroflora im Boden der Wasserhaushalt die maßgebende Rolle; die einzelnen physiologischen Gruppen werden deutlich von chemischen Standorteigenschaften beeinflußt. Entsprechend den periodischen Schwankungen in den Keimzahlen faßt LOUB „ökologische Bodentypengruppen" zusammen.

Beim Vergleich der Pilzfloren verschiedener Erdteile machten sich Einflüsse von Klima und Bodentyp bemerkbar.

NIETHAMMER et al. (1959; KREHL-NIEFFER 1951) stellten einen gewissen Einfluß des geologischen Untergrundes und der Meereshöhe auf die Pilzflora der Böden fest, ein Einfluß der geographischen Breite konnte nicht erkannt werden. Nach ENGHUSEN (1956) ist die Pilzflora charakteristisch für den Bodentyp, doch zeigen landwirtschaftlich genutzte Böden eine andere Flora als Waldböden. So stellten MCLENNAN und DUCKER (1954) eine bemerkenswerte Übereinstimmung zwischen der Pilzflora von Eisenmineralböden in England und Australien fest.

MUSKAT (1955) fand in Bayern Penicillien häufiger, Aspergilli weniger häufig als in Tunis, dort in den feucht-sauren Böden des Nordens Penicillien und Mucorales, im heiß-trockenen Inneren vorwiegend Aspergilli und Fungi imperfecti, NICOT (1955) im Wüstensand in Algerien hauptsächlich Hyphales und Sphaeropsidales neben wenigen Ascomycetes.

Interessant ist die Feststellung von JENNY (1961), daß, obwohl die Regel von der Zunahme des organischem C und N in den Böden bei abnehmenden Temperaturen und steigenden Niederschlägen sowohl für die N-reicheren amerikanischen Böden wie für die N-armen indischen Böden gilt, eine N-Anreicherung in Indien durch organische Düngung und Förderung der photosynthetischen N-Bindung allein nicht erreicht werden kann. Nur zusätzliche Mineraldüngung kann den amerikanischen Verhältnissen entsprechende Erträge bewirken, weil in den indischen Böden die Abbautätigkeit durch hohe Durchschnittstemperaturen zu sehr angeregt wird und sie durch erschöpfende Kulturmaßnahmen verarmt sind (vgl. DHAR 1961). Bodenbearbeitung verändert das Mikroklima und damit das Bodenleben. Deutliche Unterschiede zeigten sich beim Pflügen mit und ohne Wenden. Das Wenden aktivierte die Schicht von 10 bis 20 cm Tiefe (TOROPKINA 1959). Günstige ackerbauliche Maßnahmen fördern die vegetative Entwicklung der Mikroben, ungünstige die Sporenbildung, Drainage die aeroben Organismen (MANZON und DZECINA 1954). Besonders reich an Mikroben sind Regenwurmexkremente und die Wandungen von Regenwurmröhren (BRÜSEWITZ 1959).

POSCHENRIEDER und BECK (1960) glauben, daß in schweren Tonböden durch geeignete Bearbeitungsmaßnahmen die ,,potentielle Abbaukraft" mobilisiert werden kann. Vergleichende mikrobiologische Untersuchungen zeigten, daß ihre Mikroflora und deren Aktivität größenordnungsmäßig durchaus der eines Lößbodens gleichzusetzen ist. Bodenerosion führt in den Anschwemmzonen trotz der verschlämmten Krumenschicht zu lebhaftem Bodenleben, wohl wegen der besseren Versorgung mit abbaufähiger organischer Substanz (KURON et al. 1959).

Unter dem Einfluß der klimatischen Faktoren kommt es zu erheblichen Schwankungen der Keimzahlen im Verlauf eines Jahres, weil der im Minimum befindliche Faktor über die Entwicklung der Mikroflora entscheidet. Maxima sind im Frühjahr und Herbst festzustellen, wobei im Frühjahr im allgemeinen die Temperatur und im Herbst der Wassergehalt des Bodens als Minimum-Faktor zu betrachten sind. Verschiebungen können in den von der klimatischen Norm abweichenden Jahren auftreten. So fand BURRICHTER (1958) bei kaltem Frühjahr und feuchtem Sommer zunächst ein Minimum, während das Maximum erst im Sommer auftrat. TAYLOR (1936) fand eine Abhängigkeit der Keimzahl vom Wassergehalt nur bei heftigen Niederschlägen, während SEIFERT (1960) vor allem der Feuchtigkeit einen starken Einfluß auf die Höhe der Keimzahl zuschreibt, einen Einfluß der Temperatur aber nicht nachweisen konnte. In den Waldgebieten in Südrußland ist nach SAMZJEWITSCH (1955) die Tätigkeit der Mikroorganismen im feuchten Winter intensiver als im trockenen Sommer, sie wird gefördert durch die Wurzelausscheidungen der Bäume, deren Maximum nach dem Laubfall erreicht wird. WITKAMP (1960) stellt eine Abhängigkeit der Keimzahlen für Pilze und des Mycelwachstums bei der Zersetzung von Eichenlaub von den Schwankungen in Temperatur und Feuchtigkeit fest. Das stärkste Mycelwachstum beobachtete er daher im Herbst oder Winter.

Da die einzelnen Organismen in verschiedenem Grade von den Schwankungen in Temperatur und Feuchtigkeit beeinflußt werden, treten in ihrem Verhältnis zueinander Verschiebungen ein. KATZNELSON und CHASE (1944) verfolgten diese Veränderungen für die einzelnen ernährungsphysiologisch verschiedenen Organismengruppen. ISHIZAWA et al. (1958) stellten eine Förderung der Bakterien, aber eine Verminderung der Actinomyceten und Pilze durch stärkere Befeuchtung fest. In einer Dürreperiode bestand die Flora des Podsol nach ARISTOVSKAIA und PARINKINA (1958) vorwiegend aus Actinomyceten. MUSTAFOWA (1959) beobachtete in leichtem, lehmigem Podsol mit Tannenbestand und Sauerklee bzw. Blaubeere die höchsten Keimzahlen in der Streu und der humushaltigen Schicht. Das Maximum erreichten die Bakterien im Sommer, besonders aber im Herbst, die Pilze im Sommer, die Actinomyceten im Frühjahr und Herbst, im Sommer ging ihre Zahl zurück.

Schließlich sei darauf hingewiesen, daß nach THORNTON und GRAY (1934) Schwankungen der Keimzahlen innerhalb weniger Stunden auftreten können. Nach THORNTON und TAYLOR (1935) sind Luftdruck, Licht und Tageszeit nicht die Ursachen der täglichen Schwankungen, die Dynamik ist offenbar in fördernden und hemmenden Wirkungen innerhalb der Flora selbst zu suchen. Die mit Hilfe der Plattenkulturmethode erzielten Abweichungen waren größer als bei der direkten Keimzählung. Daraus ergibt sich, daß die einzelnen auf Platten wachsenden Gruppen der Bakterien sich unterschiedlich verhielten.

B. Einfluß der Düngung

Von den Mineraldüngemitteln wurde vor allem bei Kalk ein besonders starker Einfluß auf die Keimzahl festgestellt. Es darf jedoch nicht vergessen werden, daß sie in allen Fällen mit neutralen bzw. schwach alkalischen Nährböden ermittelt wurde und daß der Kalk die Reaktion des Bodens nach der alkalischen Seite verschoben, also die Mikroorganismen gefördert hatte, die mit den verwendeten neutralen Nährböden mit Sicherheit erfaßt werden konnten. BURRICHTER (1955) konnte mit Hilfe der Methode STRUGGER in dem Reaktionsbereich von schwach-sauer bis alkalisch keinen Einfluß der Reaktion auf die Höhe der Keimzahl nachweisen, und es wäre erwünscht, daß die Untersuchungen mit direkten Methoden auch in das Gebiet stärkerer Azidität ausgedehnt würden. Nach LOUB (1959, FRANZ 1959, SCHNEIDER 1959) wurde in Waldböden die Zahl der Pilzarten, die nennenswerte Säuremengen ausscheiden, durch Ca-Düngung herabgesetzt, durch N-Düngung die Individuenzahl erhöht. Düngung mit Ca und N in Form von Kalkammonsalpeter bewirkte ein biologisches Gleichgewicht. FRANZ und LOUB (1959) konnten die Wirkung des Kalkes auf die Gesamtkeimzahl noch nach 15 Jahren nachweisen: Die Pilzflora war zurückgegangen; die für saure Böden charakteristischen Arten *Penicillium decumbens*, *restrictum* und *purpurogenum* fehlten. Im Schwarzwald wurden 50 Jahre nach der Anlage des Versuchs folgende Keimzahlen ermittelt (s. Tab. 71).

Tabelle 71. *Keimzahlen in Millionen je 1 g Boden*

Schichten von	ungedüngt	gedüngt mit 20 dz $CaCO_3$ und 40 dz Thomasmehl	gedüngt mit 40 dz $CaCO_3$ und 27 dz Thomasmehl
0— 4 cm	1,2	2,5	3,2
4— 8 cm	0,9	2,4	1,5
28—32 cm	1,7	1,6	1,5

Auf den gedüngten Flächen war die Sphagnum-Decke völlig verschwunden. Die Verfasser nehmen an, daß die Umwandlung der Auflagehumusdecke 10 bis 20 Jahre nach der Mineraldüngung einen gewissen Endzustand erreicht hat. FIEDLER und FIEDLER (1961) beobachteten in Waldböden 8 Jahre nach der Kalkung eine deutliche Zunahme von Bakterien und Actinomyceten, dagegen hatte die Keimzahl der Pilze abgenommen. Ähnliche Ergebnisse wurden von FRERCKS, PUFFE und FINDEISEN (1960) bei der Aufkalkung von Moor- und Heidesandböden erzielt. Die Zahl der Pilze ging zurück, die der Bakterien stieg an. Schwache Aufkalkung mit zusätzlicher N-Düngung (23, 25 bzw. 86 kg N/ha) verstärkte bei Hochmoor die Zellulosezersetzung bei sichtbarer Pilzentwicklung. Untersuchungen von WIERINGA (1955) ergaben, daß in sauren Parzellen höchstens 15% der pektinzersetzenden Organismen Streptomyceten, sonst vorwiegend Schimmelpilze waren, in reichlich gekalkten aber 80 bis 90%.

In Alkaliböden konnten MITRA und SHANKER (1957) eine günstige Wirkung von Unkräutern als organische Düngung nur bei gleichzeitigen Gips- oder S-Gaben feststellen.

Bei Versuchen mit *Mineraldüngern* fand MÜLLER (1957a) im Feldversuch keine Unterschiede zur ungedüngten Parzelle. Bei Stickstoffdüngung war die Zahl der Pilze höher als bei Ungedüngt. KOSCHELKOW u. a. (1958) fanden in Grasnarbe-Podsol bei hohen Mineraldüngergaben minimale Mengen von Bakterien, Actinomyceten und Schimmelpilzen, die an der Umsetzung des Stickstoffs beteiligt sind. Vergleichende Untersuchungen über die Wirkung verschiedener Stickstoffdünger im Boden zeigten (SIMON 1960), daß die Mikroorganismen Nitrat-N ebenso gut, aber langsamer aufnehmen als NH_4-N.

Düngung mit P_2O_5 fördert besonders die Stickstoffbindung (s. S. 583). MITRA und SHANKER (1957) nehmen an, daß primär die Oxydation der organischen Substanz im Boden durch Phosphorsäure, besonders durch die Alkalität des Thomasmehls, gefördert wird und damit indirekt die Stickstoffbindung. Auch die Keimzahlen werden durch die Düngung mit Phosphorsäure deutlich gesteigert, wie die graphische Darstellung, Abb. 106, erkennen läßt (GLATHE 1953).

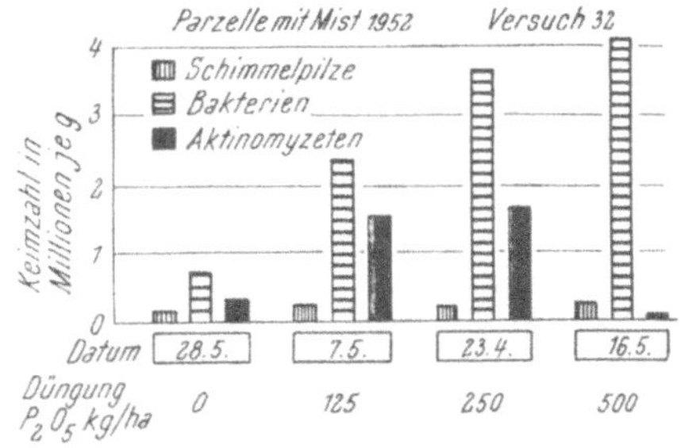

Abb. 106. Wachstum von Pilzen und Bakterien bei verschiedener P-Düngung (nach GLATHE 1953)

Selbst bei der ungewöhnlich hohen Gabe von 500 kg P_2O_5/ha lassen die Bakterien noch eine Förderung erkennen, die Keimzahlen der Actinomyceten beginnen zu sinken. Demgegenüber bleiben die Keimzahlen der Pilze unbeeinflußt.

Düngung mit flüssigem NH_3 (ENO und BLUE 1954) erniedrigte die Zahl der Pilze für 5 bis 7 Wochen, die Zahl der Bakterien nur für 3 Tage; danach schienen sich NH_3-resistente Arten üppig zu entwickeln, die die durch die Düngung erreichte Reaktion von pH 8,5 vertragen. Auf die Dauer störte die NH_3-Gabe jedoch das ökologische Gleichgewicht nicht.

Die Kenntnisse von der Bedeutung der *Mikronährstoffe* sind bei Mikroorganismen weit geringer als bei höheren Pflanzen. Verschiedene Arbeiten haben gezeigt, daß der Bedarf der Bakterien geringer ist als der der Schimmelpilze. Eingehender studiert wurden die Vorgänge bei den Stickstoff bindenden Bakterien; und es hat sich gezeigt, daß Molybdän und Bor für die N-Bindung sowohl bei Azotobacter als auch bei Rhizobien nötig sind. Bei Kupfermangel ließ Azotobacter keine Pigmentierung erkennen. Von den Schimmelpilzen ist vor allem *Aspergillus niger* gründlich untersucht worden. Seine Reaktionen auf Mangel an Mikronährstoffen sind so scharf, daß er als Testorganismus für Kupfer, Magnesium, Molybdän, Zink und Eisen verwendbar ist. Weiterhin konnte festgestellt werden, daß die Mikronährstoffe keine Strukturelemente darstellen, sondern im katalytischen System tätig sind. Weitere Einzelheiten sind dem Sammelreferat von STARKEY (1955) zu entnehmen.

Stärker als die mineralischen wirken sich im allgemeinen die *organischen Düngemittel* in der belebten Welt des Bodens aus. Eine Gegenüberstellung beider Gruppen ist aus den Ergebnissen von RÜBENSAM und STEINBRENNER (1961) ersichtlich, die in Tab. 72 wiedergegeben sind.

Zusatz organischer Stoffe zum Boden beeinflußt die Umsetzungen, die darin ablaufen, wesentlich (betreffend Gründüngung vgl. NOVAK 1958). KOEPF (1956) stellte fest, daß die Schnelligkeit des Abbaus dieser Substanzen, gemessen an der CO_2-Bildung, bei nährstoffarmen Sandböden abhängig ist von dem C/N-Verhältnis der zugegebenen Substanz, bei einem nährstoffreichen Lehmboden dagegen unabhängig von ihrem N-Gehalt. Die Abbauintensität steht in loser Beziehung zu dem Gehalt der Böden an Ton, Kohlenstoff, Gesamt- und Nitrat-Stickstoff. Die Bodenreaktion übt einen wesentlichen Einfluß aus. Der Gehalt an Ammoniak, saccharase- und laktatlöslichen Nährstoffen war ohne Beziehung zur Abbaugeschwindigkeit. Demnach erwies sich die Dynamik des Abbauverlaufs beim

Tabelle 72. *Beeinflussung der Keimzahlen infolge unterschiedlicher Düngung im Bodenfruchtbarkeitsversuch Thyrow*
Keimzahlen in 1000 je cm^3 Boden

Düngung	Gesamtkeimzahl nach Koch	Eiweißzersetzung	N-fix. Org.	Nitrifizierende Bakterien	Zellulosezersetzung	Actinomyceten	Schimmelpilze
Ungedüngt seit 1937	382,6	16,00	280,6	128,3	2,65	218	31,63
N P K Ca	762,4	23,60	474,9	241,8	3,64	334	30,76
N P K Ca + 300 dz/ha Stallmist	1523,9	31,89	694,6	664,9	3,52	603	33,56
N P K Ca + 600 dz/ha Stallmist	1800,6	35,68	674,1	760,4	3,98	702	35,47

Vergleich zweier im N-Gehalt sich unterscheidender organischer Substanzen als kennzeichnend für die Böden. STOTZKY et al. (1956) wiesen nach, daß in Komposterde die mikrobiologische Aktivität der Menge der darin enthaltenen Polyuronide entsprach.

Die durch organische Düngemittel ausgelöste CO_2-Bildung ist abhängig von deren Nährstoffgehalt. Beim Vergleich von Stallmist, Stroh und Mineraldüngung auf gleicher Nährstoffbasis beobachtete SCHAFFER (1954) bei Strohdüngung die intensivere CO_2-Bildung, die aber noch von der Mineraldüngung übertroffen wurde. Bezogen auf gleiche C-Mengen war die CO_2-Entwicklung beim Stallmist um 45% höher. Gemischte Düngung erwies sich als am günstigsten. KICK (1956) stellte beim Vergleich von Stroh, Torf und Stallmist die intensivste CO_2-Bildung beim Stroh fest.

Oberflächendüngung mit Stallmist auf Grasflächen erhöhte die Keimzahl im Boden (ARENS und STUCKMANN 1959), und zwar stärker bei jährlichen geringen Gaben als bei größerer Vorratsgabe (FELDMANN 1957). Nach KOSCHELKOW und OSIPOWA (1956, KOSCHELKOW et al. 1958) förderten Stallmistgaben auf mittelpodsolierten Böden die Nitrifikationstätigkeit und erhöhten den Gehalt an aufnehmbarer Phosphorsäure. AMBERGER et al. (1957) konnten bei Grunddüngung mit N und K und Stallmistgaben feststellen, daß dadurch die Aktivität der Mikroorganismen angeregt wird, daß aber besonders bei P_2O_5-armem Stallmist die Gefahr der Erschöpfung des Bodens an P_2O_5 besteht.

Nach ALDRICH und MARTIN (1954) wird die zahlen- und artenmäßige Zusammensetzung der Mikroflora des Bodens stärker durch die organische Substanz als durch die Zusammensetzung der austauschbaren Kationen beeinflußt. Je nach der Art der Ernterückstände wird ihre Zersetzung durch verschiedene Pilzarten herbeigeführt. Die Verfasser (MARTIN und ALDRICH 1954) beobachteten ihren Abbau während 200 Tagen und stellten fest, daß von den beteiligten Pilzen (*Chaetomium* sp., *Cunninghamella bertholletiae*, *Fusarium* spp., *Monotospora brevis*, *Mucor* spp., *Sepedonium chrysospermum*, *Sporotrichum* sp., *Stachybotrys atra*, *Stysanus stemonites*) einige nur im Anfang intensiv tätig waren, andere während der ganzen Versuchszeit. In besonders sauren Böden entwickelte sich vor allem *Penicillium nigricans*. Nach DAWSON (1947) sind bei der Strohzersetzung oxydasebildende Pilze am aktivsten.

Die Erhöhung der Keimzahlen nach einer Düngung mit Stalldung ist nicht auf die in den Boden eingebrachten Mikroorganismen zurückzuführen, sondern auf eine stärkere Entfaltung der im Boden heimischen Mikroflora infolge Zuführung von Nährstoffen. Der Nachweis wurde durch Vergleichsdüngung mit natürlichem und sterilisiertem Stalldung geführt (TEMPLE 1912).

WLASJUK et al. (1956) beobachteten einzelne Organismengruppen nach der Düngung mit Stallmist, Humus und Torf und bei zusätzlichen Gaben von Superphosphat und Kalk. Es trat eine Vermehrung der nichtsporenbildenden Bakterien ein. Die Zahl der aeroben Zellulosezersetzer und anaeroben N-Bindner wie Nitrifikanten stieg auf das 10- bis 1000fache. Actinomyceten wurden leicht gefördert, die Menge der Pilze durch Torf und Humus etwas reduziert. MEYER und SCHAFFER (1954) stellten in einem Boden, der seit 20 Jahren keine Zufuhr organischer Substanz erhalten hatte, fest, daß harmonische Mineraldüngung die Bodenatmung während einer Vegetationsperiode stärker erhöhte als die gleichen Nährstoffgaben in Form von Stallmist. Die oben erwähnten Ergebnisse von RÜBENSAM und STEINBRENNER (1961) führen dagegen zu dem Schluß, daß mineralische Volldüngung die organische Düngung mit Stallmist in ihrer fördernden Wirkung auf das Bodenleben nicht erreicht und sie nicht ersetzen kann. Besonders deutlich war die Zunahme bei Nitrifikanten, Stickstoffbindern und Actinomyceten (RAUHE und MÜLLER 1960, vgl. auch ABD-EL-MALEK et al. 1961). Bei den Pilzen trat eine Zunahme an Penicillien ein, Dematiaceen nahmen in gleichem Maße zu, wie Trichoderma abnahm. SAUERLANDT et al. (1961) stellten die Förderung einer auf Haferflocken-Bodenextraktagar gut wachsenden Bakteriengruppe durch Mistkompostdüngung fest.

Bei der direkten mikroskopischen Prüfung des Einflusses von organischen Düngemitteln auf die Mikroflora konnte WIGOREW (1958) kein einheitliches Bild über die Entwicklung der Mikroflora nach Art und Zahl gewinnen. Sie wurde durch organische Substanzen besonders gefördert. Schimmelpilze waren stets beteiligt und die sie zersetzenden Bakterien und Actinomyceten.

Über den Einfluß des *Mulchens* auf den Boden geben JACKS et al. (1955) einen Überblick. Die Verbesserung des Wasserhaushalts und die gleichmäßige Temperatur fördern das Leben der Kleintiere und das der Bakterien, besonders auch die Nitrifikation. WHITE et al. (1959) konnten jedoch bei Mulchen mit Kiefernsägemehl auf sandigem bis tonigem Lehm keinen Einfluß auf die Mikroorganismen des Bodens feststellen.

Zugabe von anorganischem Stickstoff zu Strohdüngung förderte die Strohzersetzung; allerdings nahmen die Mikroorganismen den zugegebenen Stickstoff, besonders Ammoniakverbindungen, rascher auf als die Pflanzen, bei günstigen Mineralisationsbedingungen wurde er aber auch rasch wieder verfügbar (JANSSON et al. 1956, STOJANOVIC und BROADBENT 1956). Eine Erhöhung des Humusgehaltes wird jedoch nur auf dem Wege über gesteigertes Pflanzenwachstum erreicht werden (ALLISON 1955, BARBIER und BOISCHOT 1955).

Städtische Abwässer werden an vielen Orten auf landwirtschaftlich genutzte Flächen verrieselt oder verregnet. Nach MITTENDORF (1953) wirkte Verrieselung in dem leichten Sandboden günstiger auf Bodenatmung und Nitrifikation als Verregnung; in beiden Fällen wurde die Keimzahl erhöht. Die Wirkung des Abwassers auf die Mikroflora des Bodens ist nach UNGER (1959/60) abhängig von Bodenart und -nutzung. In einem Ackerboden wurde die Gesamtkeimzahl erhöht, in einem Wiesenboden kaum beeinflußt. Er vermutet einen Abbau der Abwasserflora durch die autochthone Bodenflora. Die Zellulosezersetzung wurde durch Abwassergaben gefördert. Die Zahl der Colikeime ging mehr oder weniger schnell zurück. LEHNER und NOWAK (1956) stellten eine große Widerstandsfähigkeit von *Escherichia coli* nach Grünlandberegnung fest. Häufig zeigt sich nach mehreren Jahren der Berieselung eine Rieselmüdigkeit der Böden. Man hat als Ursache eine Entartung der Mikroflora angenommen. HIRTE (1961a, b, c) kam jedoch auf Grund seiner Untersuchungen zu dem Ergebnis, daß ein negativer Einfluß auf die Mikroflora als Ganzes betrachtet nicht stattfindet. In rieselmüden Böden sind die Keimzahlen für Bakterien, Actinomyceten und Pilze eindeutig höher als in nicht rieselmüden, eine Erscheinung, die ihre Erklärung in höherem Gehalt des Bodens an organischer Substanz und Wasser findet.

Mit der weiteren Zersetzung der beim Abbau von organischen Substanzen entstehenden Stoffe befaßte sich KNÖSEL (1958, 1959). Er isolierte Pilze aus verschiedenen Böden und aus der Rhizosphäre von Lein, Roggen und der geruchlosen Kamille und stellte fest, daß sie fast alle zur Verwertung von p-Oxybenzoesäure, p-Cumarsäure, Ferula- und Vanillinsäure befähigt sind. Für die Zersetzung der p-Oxybenzoesäure fand er etwa zwei Drittel der geprüften Streptomycetenstämme geeignet, und von den Bakterien waren es mehr zymogene Arten als solche der autochthonen Flora. In Lösungskulturen zeigten sich stärkere Einflüsse der zugesetzten Säuren als im Boden, wo nur die Actinomyceten und die autochthonen Bakterien durch die p-Oxybenzoesäure vermindert wurden.

LEMOIGNE et al. (1951) setzten verschiedenen Böden Saccharose zu. Dadurch wurden zunächst *B. megaterium-cereus*-Arten gefördert, dann durch die bei deren Autolyse entstehende β-Oxybuttersäure Pseudomonas-Arten.

Neben den Düngemitteln werden in der Landwirtschaft wie im Gartenbau verhältnismäßig oft *Herbizide*, *Fungizide* und *Insektizide* angewandt, deren Einfluß auf die Mikroorganismen des Bodens wiederholt untersucht wurde (BRAUNS 1956). Übersichten über die Wirkung der Schädlingsbekämpfungsmittel finden sich bei ENO (1958) und ZEUMER (1958). In den meisten Fällen zeigte es sich, daß bei Laboratoriumsversuchen in wässerigen Lösungen wohl Hemmwirkungen beobachtet wurden, daß aber bei Feldversuchen und der Anwendung der üblichen Konzentrationen die Mikroflora entweder unbeeinflußt blieb oder geringe Förderungswirkung erkennen ließ (STEINBRENNER et al. 1960). Die Giftigkeit dieser Mittel scheint durch Adsorption an die kolloidalen Bodensubstanzen herabgesetzt zu werden (IVARSON und PRAMER 1956).

SHEFNER und BURKHARDT (1957) stellten als Maximalkonzentration von Borsäure, die von *Azotobacter chroococcum* noch vertragen wird, 0,12% bei pH 5,6, von *Rhizobium meliloti* 0,06% bei pH 5,2 fest. Bei Zugabe von 4% Mannit erhöhte sich die Maximalgabe auf 0,4% Borsäure. Merkaptoverbindungen, besonders die Merkapto-Aminosäuren, hemmten nach BROWN et al. (1954) die Nitrifikation. NH_3 in Gasform blieb bei Konzentrationen von 300 ppm in Versuchen von ENO et al. (1955) ohne Einfluß, die doppelte Menge reduzierte die Pilzzahl auf 4,9%, die Nitrifikation wurde bei Konzentrationen über 300 ppm stark gehemmt.

Bei Anwendung von Kalkstickstoff zur Unkrautbekämpfung in Tabakbeständen beobachteten WELLS et al. (1957) Schädigungen des Bestandes durch Verschiebung der Reaktion nach der alkalischen Seite und Erhöhung der NH_3-Konzentration. Durch zusätzliche Anwendung von Superphosphat oder organischer Düngung ließen sich beide Werte normalisieren.

Durch das Natriumsalz der 2,4-Dichloressigsäure, das zur Unkrautbekämpfung verwendet wird, kann nach STAPP und FRETER (1952b) eine kurzfristige Schädigung der Bakterienflora in den oberen Bodenschichten eintreten, das Salz wird jedoch von der Mikroflora des Bodens abgebaut, so daß es unter normalen Bedingungen nach spätestens 4 bis 6 Wochen nicht mehr vorhanden ist (STAPP und FRETER 1952a). JOHNSON und COLMER (1955a, 1955b) beobachteten Hemmungen in den in Nährlösungen ablaufenden Umsetzungen bei Konzentrationen von etwa 800 ppm, im Boden wurden jedoch keine Hemmwirkungen festgestellt; nach MAGEE und COLMER (1955) liegen die zur Hemmung erforderlichen Konzentrationen bedeutend höher, als sie auf den Feldern zur Anwendung kommen.

Die Wirtspflanzen erwiesen sich als empfindlicher als die Knöllchenbakterien (CARLYLE und THORPE 1947, ENO und EVERETT 1958, THORNTON und ENO 1959). STAPP und SPICHER (1954/55) reicherten *Flavobacterium peregrinum* an, das in Reinkulturen besonders rasch das normale Herbizid abbaute. Auch WALKER und NEWMAN (1956) isolierten Organismen, die zu seinem Abbau befähigt sind; sie nehmen an, daß es sich um Mutanten handelt. Im Feldboden war 2,4-D innerhalb 3 bis 6 Wochen abgebaut (DE ROSE und NEWMAN 1948). Nach STEENSON und WALKER (1956/57) fördern Temperaturen bis zu 30° C und Belüftung den Abbau.

BRUINSMA (1960) stellte fest, daß das Herbizid DNOC (4,6-Dinitro-o-Kresol) im Boden abgebaut wird, so daß sich die keimungshemmende Wirkung auf Kresse, die sich kurz nach der Anwendung zeigte, später in eine Stimulationswirkung umwandelte. Auf die Bodenorganismen wirkte es zum Teil sterilisierend, z. B. auf solche, die die Kressekeimung hemmten. Insektizide, wie Chlordan, DDT, Toxaphen, Dieldrin, Aldrin, Endrin, Lindan oder Heptachlor, üben in den in der Praxis üblichen Mengen keinen Einfluß auf die Tätigkeit und Zahl der Bakterien und Pilze aus (STOTZKY et al. 1956, MARTIN et al. 1959); gelegentlich wurde eine kurze stimulierende Wirkung je nach Konzentration beobachtet (FLETCHER und BOLLEN 1954, ROBERTS und BOLLEN 1955); in Einzelfällen wurde die Nitrifikation beeinträchtigt (BROWN 1954, ENO und EVERETT 1958). JONES (1952) stellte fest, daß DDT erst im 2. und 3. Jahr nach der Anwendung abgebaut wurde, und zwar um so stärker, je geringer die Konzentration war und je höher der Gehalt des Bodens an organischer Substanz.

BHC (Benzolhexachlorid) hemmte nach GRAY (1954) Nitrifikation und Harnstoffzersetzung nur im Laboratoriumsversuch, nicht im Boden, die Zellulosezersetzung nur bei engem Kontakt mit *Cytophaga*.

Nach POPESCU und CALANCEA (1958) wirkten Asche, Abavit und Hexachloran günstig auf die Mikrobentätigkeit im Boden; Asche wurde erfolgreich gegen *Tilletia* angewandt, Germisan blieb ohne Einfluß. Saatbeizmittel können die Wirkung einer Leguminosenimpfung in Frage stellen, da die Rhizobien, wie HOFER (1958) feststellte, verschieden empfindlich auf die Einwirkung der Fungizide reagierten. MARTIN et al. (1957) beobachteten nach der Behandlung von Böden mit Fungiziden zunächst eine Abnahme der Pilzzahl, dann aber eine starke Vermehrung, aber nur einzelner Arten. Besonders *Trichoderma viride* wurde dominant. Die Behandlung wirkte sich noch nach drei Jahren aus. Eine besonders ungünstige Wirkung der Anwendung von Biociden auf Sandboden vor der Aussaat von *Pinus radiata* auf die Ausbildung der Mycorrhiza der Kieferwurzeln beobachteten WILDE und PERSIDSKY (1956). Sie nehmen an, daß die Biocide die Ausscheidungen der Organismen in der Rhizosphäre verändern.

3. Mikroorganismen der Rhizosphäre

Die Kultivierung eines Bodens und die Art seiner Nutzung wirken sich auf das Mikroklima wesentlich aus. Kulturmaßnahmen, wie Bearbeitung und Düngung, und der Pflanzenbestand je nach Beschattung und Durchwurzelung beeinflussen das Bodenleben. PAKUSIN (1955) fand im bewachsenen Boden ein intensiveres

Bakterienleben als im unbewachsenen, Azotobacter konnte er, wie KATZNELSON und JERSCHOW (1957), nur unter Vegetation nachweisen.

In den landwirtschaftlich nicht genutzten Böden erfolgte der Abbau der Zellulose durch Schimmelpilze (KATZNELSON und JERSCHOW 1957); wurden die Böden in Kultur genommen, dann ging die Zahl der Pilze zurück, an ihre Stelle traten echte Bakterien und Actinomyceten, die Keimzahlen stiegen an (AFANASJEWA und SWATSKAJA 1958). LINNEMANN (1958) kam bei der Untersuchung von Böden aus Mexiko und Lappland zu dem Ergebnis, daß Kultivierung auch eher eine Abnahme als eine Zunahme der Zahl der Pilzarten zur Folge hatte. FLAIG und KUTZNER (1960) fanden, daß unter bestimmten Pflanzen einzelne Streptomyces-Stämme bevorzugt vorkommen, auch unterscheiden sich die Böden in ihrer Streptomycetenflora. Minima und Maxima im bewachsenen und kultivierten Boden werden von der Pflanzenentwicklung beeinflußt (MÜLLER 1959 II). Besonders im Sommer oder Herbst zeigt sich zur Zeit der Wurzelrotte je nach Pflanzenart ein Maximum. Das Maximum für Strahlenpilze in der Rhizosphäre lag im Herbst.

Die Unterschiede in der Mikroflora unter verschiedenen Pflanzen sind im Frühjahr am deutlichsten festzustellen (SEIFERT 1960). Bei einem Vergleich relativ gleichgearteter Bestände — verschiedene Wiesenböden — ist der Einfluß von Bodenart und Reaktion ausschlaggebend (JAGNOW 1958).

Unter Leguminosen wird im allgemeinen ein intensiveres Bakterienleben gefunden als unter Gräsern (u. a. BALICKA 1952, BÖTTCHER 1956). Besonders ausführliche qualitative und quantitative bakteriologische, mykologische und zoologische Bodenuntersuchungen unter Futterpflanzenbeständen in Rein- und Mischsaat führte MÜLLER (1955/56—1958/59, 1956, 1958/59, 1959—1959/60, 1959/60, 1960a, b, MÜLLER und RAUHE 1959) durch. Es ergaben sich enge Beziehungen zwischen R-Faktor und Entwicklung der verschiedenen Organismengruppen. Luzerne förderte besonders die Eiweißzersetzer und Nitrifikanten, Knaulgras zeigte im allgemeinen geringe Unterschiede gegenüber dem nicht durchwurzelten Boden, nur Zellulosezersetzer und anaerobe Arten waren zahlreicher und Pilze, die schwer zersetzbare Stoffe angreifen. Unter Mischkulturen waren beide Förderungswirkungen zu erkennen. Die Vorfrucht beeinflußte das Bodenleben, im 2. Nachbaujahr zeigte sich eine ausgesprochene Neigung zum Ausgleich, nur die Pilze ließen sogar im 3. Jahr noch von der Vorfrucht gesteuerte signifikante Abweichungen erkennen. Bei mehrjährigem Kleebestand trat im 3. Jahr eine Zunahme der Zellulosezersetzer und ammonifizierenden Organismen ein, aber eine Abnahme der nitrifizierenden (SHUKOWA 1958).

Die fördernde Wirkung der Leguminosen geht auch daraus hervor, daß in den mit ihnen bestellten Böden eine sehr hohe Aktivität von Invertase, Katalase, Phosphatase und proteolytischen Fermenten gefunden wurde (WLASSJUK et al. 1957), dagegen zeigten Winterweizen und Hafer geringe Aktivität. Eine größere Keimmenge unter einem Pflanzenbestand führte nach HABER (1959) nicht immer zu entsprechend intensiver Bodenatmung, diese war abhängig von der artmäßigen Zusammensetzung der Flora.

Waldstreifen im Wolgagebiet übten einen günstigen Einfluß auf die Bodenbildung des angrenzenden Graslandes aus (MALJANOW und SARALIDSE 1957), der Humusgehalt des Bodens wurde vermehrt, die Struktur verbessert. WILLIAMSON (1956) stellte beim Vergleich von Wald- und Grünlandboden zwar gleiche Maximalwerte für Actinomyceten, Bakterien und Pilze fest, doch waren die Minimalwerte im Waldboden stets höher. Die Aktivität der Organismen im Nadelwaldboden förderte den Umsatz der organischen Substanz und hielt, wie SONN (1956) feststellte, die Podsolbildung in der Schwarzerde auf.

In Waldböden wurden besonders deutliche Unterschiede festgestellt, und zwar

war der Laubwaldboden reicher an Bakterien, bei Birke besonders auch an nitrifizierenden (SEIFERT 1958), der Nadelwaldboden an Pilzen (CHASE und BAKER 1954).

Sehr günstig für das Bodenleben und die Regeneration des Bodens wirkte sich nach den Untersuchungen von BURRICHTER (1954) die Wiederbewaldung eines Heidepodsols mit Birke und Eiche aus. Während bei Heidebestand und Kiefernaufforstung nur der A-Horizont stärker besiedelt war, konnte eine Zunahme der Mikroorganismen unter den Laubbäumen bis 60 und 70 cm Tiefe nachgewiesen werden.

Da also die höheren Pflanzen aktiv in das mikrobiologische Geschehen des Bodens eingreifen und die Wirkungen der verschiedenen Pflanzenarten unterschiedlich sind, ergibt sich eine mikrobiologische Sphäre des Problems der Fruchtfolge, deren praktische Bedeutung nicht unterstrichen zu werden braucht. Und zwar sind es die Ausscheidungen der Wurzeln höherer Pflanzen, die durch fördernde und hemmende Wirkung zu einer Selektion der Arten führen, so daß von einer spezifischen Flora der Rhizosphäre gesprochen werden kann.

Der Begriff „Rhizosphäre" ist erstmalig von HILTNER im Jahre 1904 angewendet worden. In ihr sind die Keimzahlen, wie bereits HOFFMANN (1914) feststellte, höher als in größerer Entfernung. Weitere Beweise für die Richtigkeit dieser Tatsache wurden von zahlreichen Forschern erbracht. Mit Hilfe der CHOLODNY-ROSSI-Methode (s. S. 554) wurden die entstehenden Mikrobenmäntel erstmals von STARKEY (1938) und STILLE (1938) sichtbar gemacht. Die Stärke dieser Flora schwankt in weiten Grenzen. Sie hängt ab von Art und Alter der Pflanzen.

GLATHE et al. (1952/54) erbrachten mit Hilfe dieser Methode den Nachweis, daß in der Rhizosphärenflora bei zunehmender Austrocknung des Bodens die echten Bakterien in steigendem Maße durch Actinomyceten abgelöst wurden.

Zusammenfassende Übersichten über die Flora der Rhizosphäre wurden von KATZNELSON et al. (1948), KRASSILNIKOV (1957), STARKEY (1958), COOK und LOCHHEAD (1959) und TROLLDENIER (1961) gegeben.

Für die besonders starke Vermehrung der Mikroorganismen in der Rhizosphäre ist die Ausscheidung von Nährstoffen und Wirkstoffen verantwortlich zu machen. Nach einer Zusammenstellung von STARKEY (1958) scheiden die Pflanzenwurzeln Vitamine, Aminosäuren, Zucker, Tannin, Alkaloide und Phosphatide aus, von KANDLER (1951) werden Asparagin, Alanin, Serin, Asparaginsäure, Valin, Glutaminsäure, Leucin und Glutamin genannt. Die Wurzelausscheidungen werden zum Teil angeregt durch das Bakterienwachstum. STOLP (1952) stellte fest, daß sterile Pflanzen keine Aminosäuren ausscheiden. Der Ertrag von sterilen Pflanzen ist nach PÁNTOS (1956) geringer als von solchen, die mit typischen Rhizosphärebakterien beimpft wurden. FEDOROW und PANTOSCH (1958) konnten nachweisen, daß zwischen der Dichte der Bakterienbesiedlung im Wurzelbereich und der neugebildeten organischen Pflanzensubstanz eine direkte Beziehung bestand.

Die Art der Wurzelausscheidungen ist abhängig von der Pflanzenart und wirkt selektiv, hemmend (vgl. BECKER et al. 1951) oder fördernd, auf die Mikroorganismen. So kommen, wie LOCHHEAD und seine Mitarbeiter feststellten, in der Rhizosphäre Arten zur Entwicklung, die Biotin, Thiamin und Aminostickstoff benötigen (WEST und LOCHHEAD 1940a, b), und sie haben bewiesen, daß diese Stoffe von den Wurzeln ausgeschieden werden (WEST 1939, LOCHHEAD und THEXTON 1947, vgl. auch LILLY und LEONIAN 1939, GYLLENBERG 1956). Sie (ROUATT und LOCHHEAD 1955) beobachteten weiter einen hohen Gehalt an Vitamin B_{12} in den Knöllchen von Luzerne und in der Rhizosphäre eine Anreicherung von Mikroorganismen, die dieses Vitamin benötigen. In Lösungskulturen wirkten nach ROVIRA (1956) Zusätze von Wurzelausscheidungen von Erbsen bzw. Hafer günstig auf das Wachstum von aus der entsprechenden Rhizosphäre isolierten Organismen weniger auf das der Bakterien aus dem Kontrollboden. Die Erbsen schieden nach 25 Tagen steriler Sandkultur Aminoverbindungen mit 22 Substanzen aus, Hafer nur 14 und auch quantitativ weniger. Diese Tatsache kann die intensivere Mikrobenentwicklung an den Wurzeln der Leguminosen — verglichen mit der an denen der Gramineen — erklären. METZ (1955) bestätigte frühere Befunde, daß die Mikrobenentwicklung bei Cruciferen intensiver ist als bei Gramineen. Im Wiesenbestand fand JAGNOW (1961) die Wurzeln von *Achillea millefolium* besonders reich besiedelt. Nach GYLLENBURG und HANIOJA (1956) bewirkten Gräser eine größere Steigerung der Bakterienzahl in der Rhizosphäre gegenüber dem Kontrollboden als Bäume (vgl. METZ 1955), Büsche und auch Moos. Verletzte Wurzeln von Waldbäumen und Sträuchern wirkten hemmend auf die Bakterien.

In ihrer Wirkung auf die Mikroorganismen sind nicht alle Wurzelteile als gleichwertig zu betrachten. Ein fördernder Einfluß geht nur von den lebenden Wurzeln und vor allem von den Wurzelhaaren (TULAJKOWA 1954, GYLLENBURG und HANOIJA 1956) aus; außerdem konnten LOCHHEAD und seine Mitarbeiter feststellen, daß das Alter der Pflanze einen Einfluß auf die Stärke der Flora der Rhizosphäre ausübt. Die Ergebnisse wurden wiederholt bestätigt, so u. a. von WALLACE und KING (1954) für Getreide, von TULAJKOWA (1954) für Hanf. Nach ROVIRA (1959) ging mit zunehmendem Alter der in Wasserkulturen gezogenen Tomaten-, Klee- und Phalaris-Pflanzen auch die Menge der ausgeschiedenen Aminosäuren zurück. Nach GOLOWATSCHEWA (1957) erreichte die Rhizosphärenflora während der Bestockungsphase von Mais und Weizen ihren höchsten Stand, ein Rückgang setzte am Ende des Schossens ein. Bei Klee und Gräsern haben FEDOROW und NEPOMILUJEW (1954a, c) eine besonders reiche Rhizosphärenflora während der Blüte gefunden, nach der Mahd gingen die Keimzahlen zurück, die Flora wurde auch artenärmer. In gemischten Wiesenbeständen stellte JAGNOW (1961) eine Beziehung zwischen der Intensität des Pflanzenwachstums und der Besiedlungsdichte der Rhizosphäre der einzelnen Pflanzenarten im Laufe der Vegetationsperiode fest. Das Vorkommen von Azotobacter in der Rhizosphäre der Weinrebe erwies sich nach den Untersuchungen von SSADOWSKAJA (1959) als abhängig von der Entwicklung der Pflanze. Mit der Triebbildung setzte eine Hemmung ein, die zur Zeit der Blüte einen Höhepunkt erreichte und dann abfiel. Über die räumliche Ausdehnung der Rhizosphäre gibt CLARK (1949) folgende Zahlen an (s. Tab 73).

Tabelle 73. *Die räumliche Ausdehnung der Rhizosphäre*

Abstand der entnommenen Probe von der Wurzel	Gesamtkeimz. Mill./g	Sporen Mill./g	Verhältnis Sporen:Gesamtkeimz.
10 —15 cm	5,26	1,7	0,323
5 —10 cm	4,79	1,8	0,376
2,5— 5 cm	4,56	1,7	0,373
0,5— 2,5 cm	5,47	1,1	0,203
0 — 0,5 cm	12,99	1,37	0,105
Wurzeloberfläche	51,00	0,1	<0,001

Danach beträgt die Stärke des Mikrobenmantels der Wurzel etwa 5 mm.

Von den zahlreichen Angaben über das Verhältnis der Keimzahlen in der Rhizosphäre zu denen des nicht durchwurzelten Bodens seien einige herausgegriffen. In der amerikanischen Literatur ist hierfür der Begriff des r/s-Wertes (rhizosphere/soil-ratio) eingeführt. TIMONIN (1940a) fand die Menge der Bakterien und Actinomyceten in der Rhizosphäre 7- bis 71mal so hoch wie im übrigen Boden, bei den Pilzen schwankten die analogen Werte zwischen 0,75 und 3,1, nach TOLLE und RIPPEL-BALDES (1958) bei Getreide zwischen 1,5 und 3,0. NETTE (1955) fand im Wurzelbereich von Eichen 10- bis 100mal so viel denitrifizierende Bakterien wie im wurzelfreien Boden. WELTE und TROLLDENIER (1961) stellten eine Beziehung des r/s-Wertes zur Bodenreaktion fest; sie beobachteten bei abnehmender Wasserstoffionen-Konzentration für Bakterien eine absolute und relative Zunahme des Wertes, bei Pilzen dagegen eine Abnahme, er kann im alkalischen Bereich sogar unter 1 liegen.

Auch die *Düngung* beeinflußt die Flora der Rhizosphäre. Nach FEDOROW und NEPOMILUJEW (1954b) werden die Keimzahlen in der Rhizosphäre durch jede Form der Düngung erhöht. TULAJKOWA (1956) untersuchte den Einfluß von Superphosphat auf die Rhizosphärenflora von Hanf mit Hilfe von Agarkulturen. In der direkten Umgebung des Korns zeigte sich zunächst eine Hemmzone, die aber nach einigen Tagen überwachsen wurde. LOUW und WEBLEY (1959) beobachteten eine Zunahme der Keimzahl bei Zugabe von Superphosphat. In den Gefäßversuchen von MACURA (1958a) war ein Einfluß sowohl der Zeit der N-Düngung als auch des Fehlens von Stickstoff bzw. Phosphorsäure (MACURA 1961) festzustellen.

Die Flora der Rhizosphäre kann bis zu einem gewissen Grad durch die an den *Samen haftenden Mikroorganismen* beeinflußt werden. WALLACE und LOCHHEAD (1951) erbrachten für 2000 Bakterienstämme, die sie von Samen isoliert hatten, den Nachweis, daß diese auch in der jeweiligen Rhizosphäre vorkamen. Auch die Ergebnisse von MACURA (1958b) sprechen dafür, daß die Samenflora Einfluß auf die Rhizosphärenbesiedlung ausübt. Die Versuche zeigten, daß nach Beimpfung von Keimlingen mit Organismen, die am Saatgut hafteten, in der Rhizosphäre Arten vorherrschten, die Aminosäuren zur Entwicklung benötigten, wie dies bei den meisten Pflanzen der Fall ist (LOCHHEAD und ROUATT 1955). Wurde die Bodenflora in Aufschwemmungen zugegeben, dann dominierten Arten, die den Stickstoff in Form von Ammoniak verarbeiteten. Nun wird man aber unter Berücksichtigung der Wechselwirkungen, die von den Wurzeln der verschiedenen Pflanzen im Mischbestand ausgehen (METZ 1955, WINTER 1955, 1959), ebenso wie der Einflüsse der Vorfrucht und der Bodenflora nicht als Regel erwarten können, daß die Samenflora sich durchsetzt. Die Rhizosphärenflora dürfte eine Zwischenstellung zwischen Samen- und Bodenflora einnehmen.

Über die *Zusammensetzung der Flora der Rhizosphäre* liegen zahlreiche Untersuchungen vor. Da sie nicht nur durch die Pflanze selbst, sondern vor allem durch den Boden, ferner aber auch durch die Düngung, vor allem die organischer Natur, und durch die Jahreszeit bestimmt wird, kann den Schlußfolgerungen aus Einzelangaben nur ein vorläufiger Charakter beigemessen werden. Außerdem wurde bereits von THORNTON (1956a) darauf hingewiesen, daß es sich bei allen Angaben nur um Arten handeln kann, die auf den verwendeten Nährböden zu wachsen vermögen. Eine Kontrolle mit der Methode CHOLODNY-ROSSI ist nicht möglich, da sie nur quantitative, aber kaum qualitative Aussagen ermöglicht. Weitere Teilerkenntnisse sind dagegen von Lebendpräparaten zu erwarten, von denen mit Hilfe des Mikromanipulators Ein-Zell-Kulturen angelegt werden. Nach den Untersuchungen von LOCHHEAD und Mitarbeitern (KATZNELSON et al. 1955, LOCHHEAD und ROUATT 1955, ROUATT et al. 1960) kommen in der Rhizosphäre mehr Organismen vor, die einfache Stoffe, wie Zucker und Aminosäuren (vgl. auch STOLP 1952), verarbeiten, im nicht durchwurzelten Boden dagegen Arten, die aromatische Verbindungen abbauen, Hefeautolysat (WEST und LOCHHEAD 1940b, ROUATT und KATZNELSON 1957) oder andere Wachstumsfaktoren benötigen (COOK und LOCHHEAD 1959).

LOCHHEAD (1940) hat ferner nachgewiesen, daß die gramnegativen Stäbchen sich anreichern (vgl. auch ROVIRA 1956), während grampositive Stäbchen, kokkoide Formen und Sporenbildner an Zahl zurücktreten (CLARK 1949). HOLDING (1960) fand bei der Untersuchung von 26 Böden im wurzelfreien Boden 7% gramnegativer Organismen, in der Wurzelnähe dagegen 20%. Zu diesen gehörten in erster Linie fluoreszierende Pseudomonaden (vgl. auch STILLE 1958, FEDOROW und NEPOMILUJEW 1954a) und ihnen nahestehende Arten. Nach KOSLOWA (1955) bilden die Pseudomonaden der Eichenrhizosphäre bis zu 30% der Flora; 50% gehören zu den Mykobakterien; bei älteren Wurzeln traten Pigmentbildner auf. Auch bei Pflanzen, die Mykorrhizen tragen, kommen, wie TRIBUNSKAJA (1955) feststellte, fluoreszierende und eiweißzersetzende Bakterien in größerer Zahl vor. Aerobe stickstoffbindende Mikroorganismen fehlen jedoch. Mykobakterien beobachteten neben den Pseudomonaden auch *Vozniakovskaja* und *Zhil'tsova* (1958) in der Rhizosphäre von Mais, Hafer, Stachelbeeren und Apfelbäumen, FEDOROW und NEPOMILUJEW (1954a, c) bei Klee und als besonders typisch bei Timothegras. HOLDING (1960) nennt ferner *Cytophaga*, *Achromobacter*, *Xanthomonas*, *Chromobact. alcaligenes*, *Agrobacterium*, *Flavobacterium* und fakultativ anaerobe Arten. TULAJKOWA (1954) fand bei Hanf als typische Rhizosphärenorganismen NH_3-bildende Formen vom Typ des *Agrobacterium radiobacter*. Während im wurzelfreien Boden reichlich Corynebakterien vorkommen, sind sie nach LOCHHEAD (1940) in der Rhizosphäre seltener, die ihnen verwandte *Mycoplana* häufiger. Auch eine Vermehrung der chromogenen Formen wurde beobachtet (vgl. auch VOZNIAKOVSKAJA und ZHIL'TSOVA 1958).

Azotobacter ist in der Rhizosphäre zahlreicher Pflanzen nachgewiesen worden. BERNHARD und WORONKOWA (1956) beobachteten eine starke Vermehrung von Azotobacter in Rasenpodsolboden besonders in der Rhizosphäre, CLARK (1948) berichtete von intensivem Azotobacterwachstum an Tomatenwurzeln, während weder KATZNELSON (1946) noch ALLISON (1947, ALLISON et al. 1947) eine Stimulation durch die Wurzel beobachten konnten. Nach STILLE (1958) war Azotobacter im Wurzelbereich stärker vertreten als im wurzelfreien Boden, wenn dieser besonders arm an organischer Substanz war. In sterilen Kulturen stellte er einen Antagonismus zwischen Azotobacter und Fluoreszenten fest, in analoger Weise wie FEDOROW und NEPOMILUJEW (1954b) eine Hemmung von Azotobacter im Wurzelbereich des Timothegrases durch die dort zahlreich vorhandenen Mycobakterien beobachteten. Nur in der frühen Entwicklung des Grases kam er reichlich vor. Nach organischer und PK-Düngung konnte er in späteren Wachstumsphasen noch nachgewiesen werden. Derartige antagonistische Wirkungen können die unterschiedlichen Ergebnisse beim Nachweis dieses Organismus in der Rhizosphäre verschiedener Pflanzen zu verschiedenen Zeiten erklären.

METZ (1955) stellte eine sehr unterschiedliche Wirkung von Wurzelsekreten auf Azotobacter-Kulturen fest. Nach MESCHKOW und CHODAKOWA (1954) wurde Azotobacter in vitro durch Zusatz von Maiswurzelextrakt gefördert. Indessen dürften diese Ergebnisse nur mit Vorsicht auf die Verhältnisse im Wurzelbereich übertragbar sein, denn die Extrakte sowohl von Mais als auch von Erbsen förderten Sporenbildner stärker als Nichtsporenbildner, während im Wurzelbereich die Sporenbildner zurücktraten. Hierbei dürfte die Konzentration der Extrakte eine erhebliche Rolle spielen.

Auch die Actinomyceten kommen, wie TIMONIN (1940a) feststellte, in der Rhizosphäre häufiger vor als in dem nicht durchwurzelten Boden. Nach IVARSON und KATZNELSON (1960) konnte jedoch bei Birkensämlingen in Podsolböden kein stimulierender Einfluß der Pflanze festgestellt werden. In der Rhizosphäre des Klees wurden von FEDOROW und NEPOMILUJEW (1954c) Mycobakterien nachgewiesen. Nach MÜLLER (1957b) ist bei Gräsern eine hohe Strahlenpilzaktivität vorhanden, während das bei Leguminosen nicht der Fall ist. REHM (1961) befaßte sich speziell mit Actinomyceten und stellte fest, daß im Wurzelbereich der Gerste in Sandboden *Str. roseochromogenes* stimuliert worden war. Einige andere Arten, wie *Str. diastaticus, olivochromogenes* und *antibioticus* sowie *Act. cretaceus*, wurden sehr häufig nachgewiesen.

Eine ausgesprochen selektive Wirkung der höheren Pflanzen auf das Vorkommen von Schimmelpilzen im Wurzelbereich wurde wiederholt beobachtet. AGNIHOTHRUDU (1955) fand Pilze in der Rhizosphäre meist in vegetativen Stadien, im Boden aber als Sporen. VIRTANEN et al. (1957) wiesen fungizide Stoffe in der Pflanze nach. ROVIRA (1956) konnte nach Befeuchten eines Bodens mit verdünnten Wurzelausscheidungen in diesem künstlichen Rhizosphärenboden eine Förderung der Bakterien, aber nicht der Pilze feststellen. Dagegen weist GARRETT (1956) auf den fördernden Einfluß besonders der Mykorrhiza bildenden Pflanzen hin. TOLLE und RIPPEL-BALDES (1958) isolierten 40 Pilze aus der Rhizosphäre von Getreide, die bei anderen Pflanzen nicht wuchsen. Die Verfasser vermuten eine Familienspezifität. Die Pilzarten in den Rhizosphären von Baumwolle und Winter- wie Sommerweizen waren nach den Untersuchungen von MOSKOWETZ (1957) deutlich verschieden. Im Wurzelbereich der Luzerne kamen besonders viele Pilze vor, im wurzelfreien Boden nur wenige. TIMONIN (1940a) fand bei Luzerne relativ mehr Bakterien als Pilze, bei Hafer besonders viele Pilze. CHESTERS und PARKINSON (1959) stellten Unterschiede in der Pilzbesiedelung der verschiedenen Wurzelregionen fest. An der Wurzelkrone kamen vorwiegend Zellulose abbauende Arten vor, an den Wurzelspitzen von Anfang bis Ende der Entwicklung Mucoraceen, später Fusarien, Dematien, Ascomyceten, stets gleichbleibend Mortierellen.

Nach MARTIN und ERVIN (1958) tritt beim Bepflanzen von „Zitronen“-Böden mit Orangensämlingen eine Wandlung in der Zusammensetzung der Pilzflora ein. Anscheinend ist jedoch für die Entwicklung der Sämlinge eine angepaßte Pilzflora günstiger. In Buchenwaldboden fanden HARLEY und WAID (1955) jeweils verschiedene Pilzfloren an den Wurzeln, im Boden und in der Laubstreu; die Wurzelspitzen waren pilzfrei. Bei Eichen stellte KOSLOWA (1955) eine ziemlich einförmige Pilzflora fest; im Frühjahr überwog *Penicillium*, im Herbst *Alternaria*.

Die Intensität der Belichtung der Pflanzen im Gewächshaus hatte in den Versuchen von PETERSON (1961) keinen deutlichen Einfluß auf die in der Rhizosphäre von Weizen und Sojabohnen sich entwickelnden Schimmelpilze.

Die *höheren Pflanzen* üben also einen selektiven Einfluß auf die Mikroflora des Bodens aus. Innerhalb der Mikroflora selbst treten ständig hemmende und fördernde Wirkungen auf, so daß die im Augenblick vorhandene Flora das Ergebnis eines Systems von Kräften ist, die das derzeitige Gleichgewicht einem ständigen Wechsel unterwerfen, so daß von einem Gleichgewicht immer nur für einen eng umgrenzten Zeitraum gesprochen werden kann. Diese Kräfte haben also eine ähnliche Funktion wie die Hormone eines einzelnen Organismus (SCHEFFER und KLOKE 1954). Hemmstoffe (WINTER 1942, 1950/51, 1951a), werden sowohl von Mikroorganismen als auch von höheren Pflanzen gebildet. Eine qualitative Trennung von *Wuchs-* und *Hemmstoffen* ist nicht möglich, da germistatische Stoffe, wie LOCHHEAD (1958a) feststellte, bei stärkerer Verdünnung als Wirkstoffe in Erscheinung treten können. Die Wirkung der in den höheren Pflanzen enthaltenen Hemmstoffe ist abhängig von dem Alter der Pflanzen. Von braunen Getreideblättern wurde *Eschericha coli* stets gehemmt. WINTER und WILLECKE

(1952) benutzten zu ihrem Nachweis *Bac. subtilis* als Testorganismus. Nach WINTER (1951b) können sie eine erhebliche Stabilität besitzen. Ihre Zersetzung im Boden erfolgt durch Mikroorganismen (SCHÖNBECK 1956), sie kann durch Stickstoffdüngung beschleunigt werden (WINTER und SCHÖNBECK 1957). Es ist daher durchaus verständlich, daß WALLHÄUSER (1951) einen raschen Abbau in Percolationsversuchen beobachten konnte.

Für die unterschiedliche Stabilität von *Antibioticis* in Böden sind nach JEFFERYS (1952) mehrere Ursachen verantwortlich zu machen. Eine rasche Inaktivierung von Albidin, Frequentin, Gliotoxin, Penicillin und Viridin kann durch ungünstige Reaktion ausgelöst werden. Bei Griseofulvin, Mycophenolsäure und Patulin zeigte sich eine größere Stabilität, wenn der Boden vorher mit Hitze behandelt worden war, also lag nahe, die Aktivität der Mikroben verantwortlich zu machen. Größere Mengen von Streptomycin wurden vom Boden absorbiert, wenn er vor der Prüfung mit Säure behandelt wurde. Schließlich wurde bei Penicillin und Streptomycin eine rein chemische Bindung beobachtet.

Neben diesen bakteriostatischen bzw. bakteriziden Stoffen (Antibiotica) kommen im Boden auch mikrobiogene *Wirkstoffe*, wie Thiamin, Biotin usw. (COOK und LOCHHEAD 1959) vor, und LOCHHEAD (1958b) hat darauf hingewiesen, daß ihnen etwa die gleiche Bedeutung zukommt wie den antagonistischen Phänomenen. Auf den umfangreichen historischen Überblick von LOCHHEAD (1958a, vgl. auch AUDUS 1953) kann verwiesen werden. ROULET und SCHOPFER (1950) schreiben ihr Entstehen den auxiautotrophen Mikroorganismen zu; den Nachweis hatten LOCHHEAD und CHASE (1942, 1943) erbracht. WEST (1939) zeigte, daß meßbare Mengen dieser Stoffe auch von Wurzeln höherer Pflanzen ausgeschieden werden. LOCHHEAD und CHASE (1943) fanden in allen Böden Organismen, die sieben Wirkstoffe nebeneinander benötigten, aber auch solche, deren Entwicklung nur an einen Stoff gebunden war. Zu den vitaminbedürftigen Bakterien gehören vor allem die *Arthrobacter*-Arten.

Als besonders charakteristischen Wirkstoff fanden LOCHHEAD und seine Mitarbeiter (LOCHHEAD et al. 1952, BURTON und LOCHHEAD 1953, BURTON et al. 1954, LOCHHEAD und BURTON 1956, LOCHHEAD 1958b) den „Terreges-Faktor (TF)", der z. B. von *Arthrobacter pascens* erzeugt wird und für das Wachstum von *Arthrobacter terregens* unbedingte Voraussetzung ist.

Die Ansprüche der Organismen eines Ackerbodens gegenüber den einzelnen Wuchsstoffen wurden nach LOCHHEAD und BURTON (1957) in Tab. 74 zusammengestellt.

Tabelle 74. *Wuchsstoffbedarf von Bodenbakterien*

Wuchsstoff	Wuchsstoff beanspruchende Bakterien	
	Prozentualer Anteil an den isolierten Stämmen	Geschätztes Vorkommen in Mill./g Boden
Thiamin	19,4	10,2
Biotin	16,4	8,6
Vitamin B_{12}	7,2	3,8
Pantothensäure	4,6	2,4
Folsäure	3,0	1,6
Nicotinsäure	2,0	1,1
Riboflavin	0,6	0,3
Pyridoxin	<0,2	<0,1
p-Aminobenzoesäure	<0,2	<0,1
Cholin	<0,2	<0,1
Inositol	<0,2	<0,1
Ein oder mehrere Wuchsstoffe	27,1	14,1

Thiamin und Biotin liegen also bei weitem an der Spitze.

Die Bildung von einigen Wuchsstoffen hat LOCHHEAD (1957) bei 316 Stämmen nachgewiesen, in der Rhizosphäre waren sie häufiger vertreten als in nicht durchwurzeltem Boden. Seine Ergebnisse sind in Tab. 75 zusammengestellt.

Tabelle 75. *Wuchsstoffbildung durch Bodenbakterien*

Wuchsstoff	nicht durchwurzelter Boden		Rhizosphäre von Roggen		Rhizosphäre von Gerste	
	Prozentualer Anteil	etwa Mill./g	Prozentualer Anteil	etwa Mill./g	Prozentualer Anteil	etwa Mill./g
Thiamin	35,5	41,1	32,4	374,7	56,7	1487,0
Biotin	19,6	22,7	15,2	175,8	32,7	857,6
Riboflavin	39,2	45,4	35,2	407,1	67,3	1765,0
Vitamin B_{12}	29,9	34,6	33,3	385,5	34,6	907,4
Terregens Faktor	22,4	25,9	19,0	218,8	25,0	655,6
Ein oder mehrere Wuchsstoffe	50,5	58,5	59,0	682,4	81,7	2142,6

Die Wuchsstoffe können von den höheren Pflanzen aufgenommen werden (WINTER und SCHÖNBECK 1958), doch ist noch nicht geklärt, in welchem Umfang das möglich ist.

Weiterhin gilt es festzustellen, welche Rolle diese Wuchsstoffe beim Auftreten von *Wurzelerkrankungen* spielen. Ergebnisse von LOCHHEAD und COOK (1961) lassen vermuten, daß an der Wurzeloberfläche gegen *Fus. oxysporum* anfälliger Pflanzen eine relativ große Anhäufung von Vitaminen und verwandten Substanzen stattfindet. Bakterien aus der Rhizosphäre resistenter Pflanzen waren besser geeignet, Fusarium zu unterdrücken als die anfälliger Arten. Zudem stellten LOCHHEAD et al. (1940) in der Rhizosphäre für Welkekrankheiten anfälliger Pflanzen eine stärkere Anhäufung von Mikroorganismen fest als in der resistenter Individuen. Nach TIMONIN (1940b) änderte zusätzliche Düngung das Verhältnis bei Flachs und Tabak nicht, obwohl sie sich auf die Erträge deutlich auswirkte. WEST und LOCHHEAD (1940a, b) fanden einen besonders deutlichen Unterschied bei Böden, die schlecht mit organischer Substanz versorgt waren (vgl. auch HILDEBRANDT and WEST 1941). Die verschiedene Besiedlung resistenter und anfälliger Pflanzen mit Mikroorganismen steht demnach in enger Beziehung zu Unterschieden in den Wurzelausscheidungen solcher Arten. So schieden gegen *Fus. moniliforme* resistente Reispflanzen nach Untersuchungen von ANDAL et al. (1956) besonders viel Cystin und Asparagin aus, gegen *Fus. vasinfectum* resistente Baumwollpflanzen viel Cystin. BUXTON (1957a, b) stellte fest, daß Exsudate aus gegen *Fus. oxysporum* resistenten Erbsenarten oder aus deren Rhizosphäre die Keimung des Pilzes hinderten, nicht aber Exsudate anfälliger Arten. Die Oosporen von *Phytium mamillatum* keimten nach BARTON (1957) nur bei Wurzelberührung aus.

Wenn nun, wie GYURKO (1959) feststellte, sich zwischen der Flora der Rhizosphäre und der des übrigen Bodens ein relativ resistentes antagonistisches System herausstellt, dann dürfte ein Schutz der Pflanze durch ein Eingreifen der Bodenflora nicht zu erwarten sein. Möglicherweise kann dieser Antagonismus durch bestimmte Gründüngungspflanzen aufgehoben werden; denn WEST und HILDEBRANDT (1941) gelang es, die Wurzelfäule von Stachelbeeren durch Düngung mit Sojabohnen zu bekämpfen. Da dieses Ziel bei Verwendung von Rotklee nicht erreicht wurde, muß geschlossen werden, daß jede Pflanze ihre spezifische Trabantenflora erzeugt. Bei Mischkulturen höherer Pflanzen erfolgt offenbar eine Wechselwirkung zwischen den jeweiligen Floren, die im Falle der Sojabohne zu einer Blockierung oder Vernichtung der pathogenen Art führte. Eine erste Erklärung für ein derartiges Phänomen bringt die Beobachtung von METZ (1955), wonach Wurzeln Hemmungen von Bakterien im Bereich der Wurzeln einer anderen Pflanzenart auszulösen vermögen. Diese Hemmung konnte noch in einer gewissen Entfernung von der Wurzel nachgewiesen werden. Das Vorherrschen oder durch langjährigen Anbau bewirkte Anreichern einzelner Arten kann krankheitsähnliche Erscheinungen hervorrufen. So fand TIMONIN (1946) in dem Wurzelbereich einer Hafersorte, die stark auf Manganmangel reagierte, besonders viel manganoxydierende Bakterien. Wurden diese Organismen durch Bodensterilisation beseitigt, entwickelten sich die Pflanzen gesund und zeigten höhere Erträge.

REHM (1961) glaubte, daß die Gerstenmüdigkeit gewisser Böden entweder durch leichteren Pilzbefall wegen geringerer antifungaler Aktivität der Rhizosphären-Actinomyceten im Frühstadium der Gerstenentwicklung oder durch nachteilige Wirkung von Streptomyceten-Stoffwechselprodukten zu erklären ist.

Diese wenigen Beispiele, die die Vielfalt der Beziehungen zwischen Pflanzen und Mikroben im Wurzelbereich gesunder und kranker Kulturen erkennen lassen, die auch Wege zur Bekämpfung von Pflanzenkrankheiten zu erschließen vermögen, zeigen die besondere Bedeutung des Studiums der Rhizosphärenflora.

4. Stickstoffbindende Mikroorganismen

Eine zusammenfassende Darstellung der biologischen Bindung des atmosphärischen Stickstoffs unter besonderer Berücksichtigung der russischen Literatur ist von FEDOROW (1960) veröffentlicht worden. Über den Mechanismus der N-Bindung berichteten VIRTANEN (1948) sowie WILSON und BURRIS (1953). Weitere Übersichten gaben WILSON (1940), BURK und BURRIS (1941), BURRIS und WILSON (1945), WILSON (1951, 1958).

A. Frei lebende stickstoffbindende Mikroorganismen

Zu den nicht sporenbildenden, stickstoffbindenden Bakterien gehört das Azotobacter (vgl. JENSEN 1954). Im Boden sind verschiedene Arten nachgewiesen worden, und zwar *Azotobacter chroococcum*, *vinelandii*, *macrocytogenes* und *indicum (Beijerinckia)*, während *Azotobacter agile* aus Kanalwasser isoliert wurde.

Azotobacter chroococcum ist am weitesten verbreitet und besonders eingehend studiert worden.

Zur Zählung der Azotobacterzellen in Böden wird die Methode KOCH angewendet. AHRENS (1961) erzielte gute Übereinstimmungen der Parallelen durch Herabsetzung der Menge des Nährbodens auf 5 cm^3. Eine einfachere Methode stellt der von HOLZWEISSIG empfohlene Biotest dar. Er geht von dem Verfahren WINOGRADSKYS aus, Bodenkrümel auf den Nährböden aufzustreuen und die Entwicklung der Azotobacterkolonien zu beobachten. Auch CURIE (1931), KAILA (1954) und KRASSILNIKOW (1957a) hatten sich dieser Methode bedient. Im Biotest werden 50 Krümel aufgelegt und die Zahl der Azotobacterkolonien in % der Gesamtzahl der Krümel ermittelt. AHRENS konnte zeigen, daß bei relativ azotobacterarmen Böden dem Biotest der Vorzug zu geben ist. Bis zu einem Vorkommen von 100 Azotobacterzellen je 1 g Boden stimmen die Resultate befriedigend mit denen der Methode KOCH überein, während ein stärkerer Besatz des Bodens vom Biotest nicht mehr zu erfassen ist, wie die graphische Darstellung Abb. 107 erkennen läßt.

Hierbei wurde die Menge des beim Biotest verwendeten Bodens quantitativ ermittelt. Höhere Keimzahlen von Azotobacter in Böden können nur mit der Methode Koch ermittelt werden. Zusatz von 0,1% Na-Benzoat zum Nährboden unterdrückt nach Aleem (1953) verunreinigende Organismen und vergrößert die Azotobacter-Kolonien.

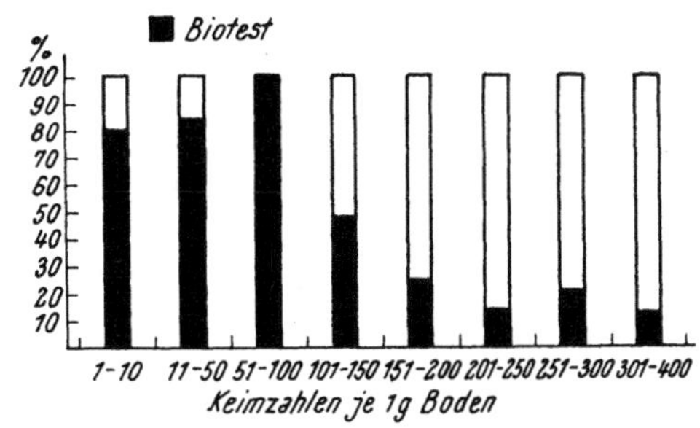

Abb. 107. Keimzahlen nach Biotest je 1 g in % der Keimzahlen nach Koch (Koch = 100%). Bereich der Keimzahlen nach Koch von 5 bis 2895 Keimen je 1 g Boden

Die Sicherheit der Zahlenangaben wird dadurch beeinträchtigt, daß Azotobacter nicht zu allen Jahreszeiten mit gleicher Sicherheit nachweisbar ist. In 40 Böden von Gatersleben, die Rehm (1956a) untersuchte, war der Organismus nur im Frühjahr 1955 nachweisbar, während er im Herbst 1953 und im Frühjahr 1954 nur in 37 Böden gefunden wurde. Van Eseltine (1956) wies Azotobacter im Herbst häufiger als im Frühjahr nach. Kirssanina und Wolkowa (1960) fanden dagegen in Böden des Altaigebirges ein Minimum im Herbst. Neben dem Maximum im Frühjahr stellten Kolker und Dachnowa (1960) bei Regen oder Bewässerung ein zweites Maximum im Herbst fest. Die *Bodenfeuchtigkeit* erwies sich als entscheidender Faktor. So findet Azotobacter nach Tschapsk und Giambiaja (1955) in der Krume keine günstigen Lebensbedingungen, weil sie zu gut durchlüftet und zu trocken ist. Bei geringem Sauerstoffdruck nutzt Azotobacter die Energiequellen besser aus. Anderson (1958a) konnte nachweisen, daß nach einer Regenperiode doppelt so viele Böden Azotobacter enthielten als nach einer Trockenperiode. Höhere Zahlen zeigten nur Böden, die während des ganzen Jahres feucht waren. Bei einzelnen Böden, in denen Azotobacter nicht nachweisbar gewesen war, trat eine starke Entwicklung des Organismus ein, wenn 10% Zucker zugesetzt und der Boden auf einen Feuchtigkeitsgehalt von 18 bis 23% gebracht worden war. Offenbar werden während längerer Trockenheit Cysten gebildet, die eine gewisse Ruhe benötigen, bevor sie wieder vegetative Formen entwickeln. Auch hierdurch können also Schwankungen in dem Auftreten von Azotobacter erklärt werden. Mischustin (1953) hat in den bei Azotobacter bekannten Depressionszeiten das Wachstum durch Zugabe von Kaliumphosphat und Mannit mit anschließender Bebrütung bei 25° C angeregt, so daß der Nachweis erleichtert und gesichert wurde.

Tabelle 76. *Vorkommen von Azotobacter in dänischen Böden*

% der Böden	Zahl der Zellen
42,2	0
26,9	0—100
19,7	100—1000
6,1	1000—10000
1,1	10000—100000

Die Dichte des Azotobactervorkommens ist zweifellos abhängig von der *Bodenart*. Jensen (1950a) kam auf Grund seiner Untersuchungen von dänischen Böden zu den in Tab. 76 wiedergegebenen Werten.

Bei den von Swaby (1939) in Australien untersuchten 21 Böden wurden Schwankungen zwischen 3 und 1600 Zellen/g festgestellt. In schweizerischen Böden fand Stöckli (1944) in 723 Böden Dichten zwischen 10 und 44000 Zellen/g

Boden. In den arktischen Tundren der Insel Nowaja Semlja gelang Suschkina und Ryschkowa (1956) der Nachweis von Azotobacter niemals. Nach Jushina (1958) wirkten die kultivierten, eisenhaltigen Podsole, Hochmoor- und Übergangsböden der Halbinsel Kola toxisch, weniger die humushaltigen Podsole und Niederungsmoorböden. In den toxisch wirkenden Böden hielt sich Azotobacter nur für Stunden, im Höchstfall bis zu drei Tagen. Bei den übrigen Bodentypen wurden Hemmungen nur in bebauten, nicht dagegen in brachliegenden Böden beobachtet. Der Einfluß der höheren Pflanze spielte demnach eine wichtige Rolle. Nach Suschkina und Sjergunina (1955) verschwindet Azotobacter in Böden, die einer intensiven Podsolierung unterworfen sind.

Der Organismus stellt hohe Ansprüche an die *Reaktion* des Bodens. Im allgemeinen ist er in Böden unter pH 5,8 nicht mehr nachzuweisen. So fand Stöckli (1944) bei der Untersuchung von 1000 schweizerischen Böden Azotobacter nur bei Reaktionen über pH 6. McKnight (1949) wies ihn in 77% von 151 Böden, deren Reaktion oberhalb von pH 6 lag, nach, dagegen nur in 14% der Böden mit höherem Säuregrad. In Übereinstimmung damit stellte Jensen (1950b) eine Toleranz von Azotobacter bis pH 5,8 fest, unter 16 Böden mit einem pH unter 5,6 hat er (1959) keinen azotobacterhaltigen gefunden, während Niklas et al. (1926) pH 4,5 als untere Grenze für sein Wachstum angaben. In Lößlehmboden der Universität Idaho fand Anderson (1958a) in Böden um pH 6 wenig, oberhalb von pH 6,3 höhere und ziemlich konstante Zahlen.

Glathe stellte fest, daß einzelne Böden im Bereich von pH 4,6 bis 6,0 wohl von Natur aus Azotobacter enthielten, daß Azotobacter aber aus azotobacterfreien Böden im Bereich von pH 3,8 bis 6,2 nach Impfung nicht reisolierbar war. Auch Ahrens (1961) hat ihn aus Böden mit einer Reaktion unter pH 5,0 isoliert, wie Tab. 77 beweist:

Tabelle 77. *Vorkommen von Azotobacter chroococcum in Böden verschiedener Reaktion* (nach Ahrens)

<table>
<tr><th>Reaktion pH</th><th>Azotobacterhaltig Zahl der Böden</th><th>%</th><th>Azotobacterfrei Zahl der Böden</th><th>%</th></tr>
<tr><td>4,55—5,0</td><td>6</td><td rowspan="2">30</td><td>8</td><td rowspan="2">70</td></tr>
<tr><td>5,05—5,5</td><td>3</td><td>13</td></tr>
<tr><td>5,55—6,0</td><td>10</td><td rowspan="2">56,7</td><td>8</td><td rowspan="2">43,3</td></tr>
<tr><td>6,05—6,5</td><td>7</td><td>5</td></tr>
</table>

Dieses Vorkommen war jedoch selten, denn von den 30 untersuchten Böden zwischen 4,55 und 5,5 enthielten nur 9=30% Azotobacter, von den 30 Böden zwischen 5,55 und 6,5 dagegen 17=56,7%. 20 weitere Böden mit einer Reaktion von pH 3,8 bis 4,5 waren azotobacterfrei. Blinkow (1955b) hat aus sibirischen Böden einen Stamm isoliert, der noch bei pH 5,49 Stickstoff zu binden vermochte. Roy und Mukherjee (1957, Roy 1958) haben aus roten indischen Böden ein Azotobacter isoliert, das sogar unter pH 4,0 zu wachsen vermochte, es bildete aus Kohlenhydraten Säure und wurde daher als *Azotobacter acidum* bezeichnet. Auch Altson (1936) und Martin und seine Mitarbeiter (1937) berichteten über das Vorkommen von Azotobacter in Böden mit einer Reaktion unter pH 5,0.

Die Aufgabe der Zukunft besteht in dem Nachweis, ob es sich um besondere Eigenschaften der betreffenden Böden oder um spezielle Rassen handelt, denn eine Reihe von Angaben deuten darauf hin, daß säureresistente Rassen vorkommen.

Gegen höheren *Salzgehalt* des Bodens zeigte Azotobacter größere Resistenz als die Kulturpflanzen. SUSCHKINA (1956) konnte Azotobacter auch dann noch nachweisen, wenn der Grad der Versalzung eine Nutzung durch Kulturpflanzen überhaupt ausschloß. Das von SUSCHKINA im Jahre 1945 isolierte *Azotobacter halophilum* (1948, BLINKOW 1955a, KRASSILNIKOW 1959a) ist wohl als ökologische Variante aufzufassen, es fixierte Stickstoff intensiver in einem Medium, das 3% Kochsalz enthielt, als in normalem. Im Mittel wurden je g Kohlenstoffquelle allerdings nur 3 mg Stickstoff fixiert. Der Organismus kommt in Salzböden vor.

Da die *Humusstoffe* des Bodens von Azotobacter als C-Quelle verwendet werden, war seine stärkere Verbreitung in organisch gedüngten Böden zu erwarten. Dies ist unter anderem von KATZNELSON (1946) nachgewiesen worden. Nicht alle organischen Stoffe üben jedoch die gleiche Wirkung aus (vgl. WAKSMAN 1952, SZELÉNYI und BALÁZS 1957). In dunklem, mit Dauerhumus angereichertem Rasenpodsol oder Kompost aus Rasenpodsol mit halbverrottetem Pferdedung oder Torf vermehrte sich Azotobacter nach BERNHARD und WORONKOWA (1956) nur langsam, rasch dagegen bei stürmisch sich zersetzenden Wurzelrückständen von Lupine, Wiesenlieschgras oder Stroh. Sie (1960) stellten ferner fest, daß organische Düngung dem Boden häufig nicht genügend leichtlösliche organische Substanz zuführte, die eine Vermehrung von Azotobacter bewirkte. Zusätzliche Stickstoffdüngung zu Stroh erhöhte die Keimzahl von Azotobacter.

Zweifellos kommt es nicht nur auf die Form des organischen Düngers, sondern auch auf den Boden an, in dem die Versuche durchgeführt werden, und ferner darauf, ob es sich um eine einmalige oder mehrmalige Düngung handelt. Eine stärkere Beweiskraft ist langjährigen Versuchen beizumessen. AHRENS (1961) konnte bei der Untersuchung von Bodenproben, die aus dem Versuch „Ewiger Roggenbau“ in Halle entnommen worden waren, nachweisen, daß die Zahl der Azotobacterzellen durch die Stallmistdüngung sogar gesenkt worden war, wie aus Tab. 78 ersichtlich ist.

Tabelle 78. *Zahl der Azotobacterzellen in Bodenproben aus dem Versuch „Ewiger Roggenbau“ in Halle*

Düngung	Zahl der Azotobacterzellen	
	12. 5. 1960	20. 9. 1960
ungedüngt	56	20
PK	53	92
NPK	0	0
N	0	0
Stallmist	13	3

Von den *Mineralstoffen* übt vor allem die Phosphorsäure einen fördernden Einfluß auf Azotobacter aus (GLATHE 1953, MITRA und SHANKER 1957a). Besonders günstig wirkte in Versuchen von FARKASDI (1956) bei einem Sandboden die Düngung mit dem organisch-mineralischen Phosphordünger „Biosuper“ auf die Entwicklung von Azotobacter.

KRIUSCHKOWA und POPOWA (1934, 1935) führten Phosphorsäuredünger in Strangform in den Boden ein und fanden eine Zunahme der Azotobacterzellen bis zu einer Entfernung von 5 cm. Es ist daher durchaus verständlich, daß die Verbreitung von Azotobacter im Boden eine deutliche Abhängigkeit von dessen Gehalt an Phosphorsäure zeigt (MALISZEWSKA 1949, JENSEN 1950a).

HOFMANN und HAMMER (1910) untersuchten die Verwertbarkeit verschiedener Calciumphosphate bereits im Jahre 1910 und kamen zu dem Ergebnis, daß das sekundäre und tertiäre Phosphat besser ausgenutzt wurde als das primäre Salz.

Wenn es sich jedoch um besonders P_2O_5-reiche Böden handelt, wie bei dem Boden des E-Feldes in dem bekannten Dauerdüngungsversuch in Göttingen, kann die Zufuhr von Phosphorsäure sogar zu einem Rückgang der Azotobacterzellen führen, wie aus Tab. 79 ersichtlich ist (AHRENS 1961).

Tabelle 79. *Vorkommen von Azotobacter im Boden des E-Feldes Göttingen*

Zahl der Azotobacterzellen	Düngung					
	K_2O		P_2O_5		N	
	ohne	mit	ohne	mit	ohne	mit
absolut	4969	6685	4969	3347	4969	3450
relativ	100	134,5	100	67,4	100	69,4

Demgegenüber hatte der Organismus auf diesem Boden auf Kalium deutlich positiv reagiert. TORNAU (1959) hat die Ernteerträge dieses Versuches aus den Jahren 1892/1953 zusammengestellt und kam zu dem Ergebnis, daß diese durch Phosphorsäure nicht, wohl aber durch Kali deutlich erhöht worden waren. Azotobacter und Pflanze zeigten also ein analoges Verhalten. Entscheidend für die Wirkung der Düngung auf Azotobacter ist der Vorrat des Bodens an den entsprechenden Nährstoffen.

Anders liegen die Verhältnisse bei der Düngung mit Stickstoff: während die Zahl der Azotobacterzellen in den gedüngten Parzellen sank, wurde der Pflanzenertrag durch Stickstoffdüngung erhöht. In Rothamsted war, wie ZIEMICKA (1932) feststellte, Azotobacter aus den Böden verschwunden, die dauernd schwefelsaures Ammoniak erhalten hatten. Hier spielte allerdings zusätzlich die Senkung der Bodenreaktion eine Rolle.

SCHMALFUSS (1960) hat zu Reinkulturen von Azotobacter Stickstoff in Form von $Ca(NO_3)_2$ zugesetzt, um dessen Einfluß auf die N-Umsetzungen zu verfolgen. In einem Reaktionsbereich von pH 7,3 bis 8,2 erzielte er nach 28 Tagen folgende Ergebnisse (s. Tab. 80):

Tabelle 80

N-Zugabe mg N je 1 g Mannit	Gesamt-N	Zell-N	Filtrat-N	Ammoniak-N	Nitrat-N
2,7	19,5	17,0	2,6	0,4	0
4,7	15,2	13,5	1,8	0,5	0
10,7	10,7	9,0	1,6	0,1	0,5
19,5	18,7	7,5	11,2	2,6	7,6

Bei einem Zusatz von 10,7 bzw. 19,5 mg Nitrat-Stickstoff war also überhaupt kein Stickstoff mehr gebunden worden. Der zugesetzte Stickstoff wurde bei den geringeren Gaben assimiliert, bei den höheren Gaben war die Eiweißbildung gehemmt. Die Zellen wurden zum Teil zersetzt, so daß Ammoniak im Filtrat gefunden wurde, am meisten bei der höchsten N-Gabe.

Die Empfindlichkeit der Azotobacterzellen gegenüber Stickstoff trat auch in den Versuchen von Macura und Kunc (1961) zutage. In Perkolationsversuchen bewirkte die kontinuierliche Zufuhr von Glukose eine erhebliche Vermehrung von Azotobacter, jedoch nur eine geringe Veränderung der Gesamtkeimzahl. Wenn dagegen zunächst eine gewisse Sättigung des Bodens mit Ammonium- bzw. Phosphat-Ionen erfolgt war oder wenn Glukose gemeinsam mit diesen Stoffen zugeführt wurde, stieg zwar die Gesamtkeimzahl, aber nicht die Zahl der Azotobacterzellen.

Von den *Mikronährstoffen* benötigt Azotobacter vor allem Molybdän. Bortels (1936/37, 1937a) stellte weiterhin fest, daß Molybdän zum Teil durch Vanadium bzw. Wolfram ersetzt werden kann. Nach Gerretsen und de Hoop (1954) werden Vermehrung und Schleimbildung von *Azotobacter chroococcum* durch Bor gefördert, so daß auch die Krümelbildung des Bodens günstig beeinflußt wird. Nach Macura und Pavel (1959) erhöht Montmorillonit den Verbrauch von Glukose durch Azotobacter und die Stickstoffbindung. Die Wirkung war um so größer, je kleiner die Montmorillonit-Partikel waren. Der kolloidale Zustand war also Voraussetzung für die Wirkung.

Zwischen Azotobacter und den *höheren Pflanzen* bestehen Wechselwirkungen. In der Rhizosphäre sind sowohl hemmende wie fördernde Einflüsse auf die Entwicklung von Azotobacter festgestellt worden. Stille (1958) beobachtete eine geringe Vermehrung gegenüber dem wurzelfreien Boden, sie kann jedoch erheblich sein, falls der Boden arm an organischen Stoffen ist.

Die einzelnen Pflanzen verhalten sich in ihrer Wirkung recht unterschiedlich. So unterdrückte nach Ssadowskaja (1959) die Weinrebe das Wachstum von Azotobacter, auch nach einer Impfung mit Azotobacterin. Die Hemmung erreichte in der Blütezeit ein Maximum und ging nach der Ernte zurück. Nach Nabijew (1957) nahm der Gehalt an Azotobacter unter Baumwolle stetig ab, er konnte aber durch mineralische und organische Düngung erhöht werden. Besonders günstige Lebensbedingungen fand er für Azotobacter im Wurzelbereich von Wiesenpflanzen bzw. Luzerne.

Vančura und Macura (1961) kamen zu dem Ergebnis, daß Azotobacter befähigt ist, die Wurzelausscheidungen von Gerste und Weizen als C- und Energiequellen zu verwerten. Seine Anreicherung in dem engsten Wurzelbereich wäre daher verständlich. Es hat sich jedoch gezeigt, daß diese Möglichkeit an bestimmte Voraussetzungen hinsichtlich der Kulturbedingungen und des jeweiligen Stammes gebunden ist. Auch als alleinige Energiequelle sind sie für Azotobacter nicht geeignet. Azotobacter seinerseits scheidet nach Vančura und Macura (1960) β-Indolessigsäure (vgl. Bukatsch et al. 1956) aus, die als Wuchsstoff für die höhere Pflanze zu betrachten ist. Ferner wurde festgestellt (Vančura 1960), daß die von den S- bzw. R-Typen von Azotobacter ausgeschiedenen Polysaccharide gewisse Unterschiede erkennen lassen. Die S-Typen enthielten mehr Galacturonsäuren, die R-Typen mehr Galactose.

Derartige Wuchsstoffe (vgl. Rubentschik 1956) und der von Azotobacter gebundene Stickstoff ließen einen günstigen Einfluß auf das Pflanzenwachstum und den Pflanzenertrag vermuten. Es hat daher nahe gelegen, Azotobacter auf dem Wege der *Bodenimpfung* in azotobacterfreie oder -arme Böden zu bringen. Eine Impfung wird aber solange wirkungslos bleiben, als der Boden selbst nicht in den Zustand versetzt wird, in dem er fähig ist, dem zugeführten Organismus günstige Lebensbedingungen zu bieten. Impfungen haben daher zu widersprechenden Ergebnissen geführt (Stapp 1945 bis 1949, 1952, Lehner 1953, vgl.

auch ALLISON 1947). WICHTMANN (1952) hat nach Impfung mit Handelsimpfstoff und Azotobacter-Rohkulturen in Vegetationsversuchen mit Senf während der gesamten Versuchsdauer eine 10- bis 100fache Anreicherung von Azotobacter festgestellt. Im Ertrag zeigte sich jedoch in den beiden Ernten des ersten Jahres kein Impferfolg. In der dritten Ernte im zweiten Jahr war im Durchschnitt ein Mehrertrag von 12,6% festzustellen. Da eine erhöhte N-Aufnahme nicht nachgewiesen werden konnte, wurde eine Stimulationswirkung durch Azotobacter und eine allmähliche Anpassung der eingeimpften Organismen vermutet.

VIRTANEN und LINCOLA (1954) konnten bei Kulturen mit sterilen Pflanzen in Glassand nachweisen, daß nach Impfung mit einem stark stickstoffbindenden *Azotobacter chroococcum* weder bei Erbsen noch bei Weizen eine wachstumsfördernde Wirkung erzielt werden konnte. SPICHER (1954) stellte fest, daß eine Impfung in sterilen, flüssigen Kulturen bei Gerste dann zu einer Förderung des Pflanzenwachstums führte, wenn die Bakterienkulturen ihren Assimilationsvorgang abgeschlossen hatten. Die fördernde Wirkung beruhte nicht auf einer Versorgung der Pflanzen mit gebundenem Stickstoff, sondern auf einem Effekt, der von Azotobacter gebildeten Wirkstoffe. Hemmungen der Azotobacterentwicklung können durch Antagonisten bewirkt werden. Das wurde nachgewiesen für *Aspergillus* und *Penicillium* von ANDERSON (1958b) und für *Pseudomonas fluorescens* von STILLE (1958).

Die im Boden zur Anwendung kommenden Mengen von *Herbiciden* haben nach MAGEE und COLMER (1955) die Atmung von Azotobacter nicht gehemmt, Giberilline haben nach LU und BOLLEN (1957/58, LU et al. 1958) die Entwicklung von Azotobacter gefördert.

Die zu der *Gattung Azotobacter* (Übersicht bei JENSEN 1954) gehörenden Arten unterscheiden sich vor allem auch in physiologischer Hinsicht voneinander. Unterschiede in der Nitrataufnahme von *Azotobacter chroococcum, vinelandii, indicum (Beijerinckia)* und *agile* stellten GREEN und WILSON (1953) fest. Calcium ist für *Azotobacter chroococcum, vinelandii, insigne, macrocytogenes* und *beijerincki* erforderlich, wenn gebundener Stickstoff nicht zur Verfügung steht. *Azotobacter agile* wird durch Calcium zwar gefördert, benötigt ihn aber nicht unbedingt (NORRIS und JENSEN 1957, 1958).

Auch die Art *Azotobacter chroococcum* kann in physiologischer Beziehung nicht als einheitlich betrachtet werden, schon VON METZEN (1936) hat verschiedene Rassen vermutet, SCHLÜTER und BUKATSCH (1959) kamen zu dem gleichen Ergebnis.

Azotobacter indicum wurde später wegen morphologischer (DERX 1950) und physiologischer (JENSEN 1954a) Unterschiede als besondere Gattung *Beijerinckia* bezeichnet, deren geringerer Ca-Bedarf, Säuretoleranz und häufiges Vorkommen in tropischen Böden hervorgehoben wird. Aus lateritischen Roterden und Rotlehmen Südafrikas hat BECKING (1959) eine *Beijerinckia* isoliert, die eine Toleranz im Bereich von pH 4,3 bis 7,0 hatte. Sie war unempfindlich gegen Frost und hohe Temperaturen und zeichnete sich durch einen geringeren Bedarf an P_2O_5 aus. Vermutlich handelt es sich um eine ökologische Variante.

Von den *anaeroben Bakterien* sind *Clostridium pasteurianum* und einige andere Arten zur Stickstoffbindung befähigt. Wenn ihre Leistungen auch geringer sind als die der *Azotobacteraceen* — sie binden nur etwa 4 mg N je 1 g Kohlenstoffquelle — so ist doch ihre Zahl im Boden viel größer als die der aeroben Arten. Zudem gedeihen sie auch in Böden mit saurer Reaktion.

Da KAILA (1954) im Laboratorium Stickstoffgewinne auch in Bodenproben feststellte, die keine klassischen stickstoffbindenden Organismen enthielten, mußte vermutet werden, daß neben diesen auch *andere Arten* im Boden vorkommen, die zur Bindung des atmosphärischen Stickstoffs befähigt sind.

Aus Baumwollkapseln isolierte STAPP (1940) ein stickstoffbindendes Kurzstäbchen, das er *Azotomonas insolita* benannte. In der Fähigkeit, Stickstoff zu binden, war der Organismus leistungsfähiger als *Bacillus asterosporus*, erreichte aber nicht *Azotobacter chroococcum*. Ein weiteres Kurzstäbchen wurde von JENSEN und seinen Mitarbeitern (1959) aus indischen Böden isoliert, das im Bereich von pH 5 bis 9 aktiv ist, es wurde als *Derxia gummosa* bezeichnet.

Unter den Pseudomonaden wurden stickstoffbindende Arten von PROCTOR und WILSON (1958) gefunden, über die bereits von ANDERSON (1955: *Ps. azotocolligens* n. sp.) bzw. VOETS und DEBACKER (1956) berichtet worden war. Sie selbst führten den Nachweis mit radioaktivem Stickstoff N^{15} bei 6 Pseudomonas-Stämmen.

Weiterhin haben PROCTOR und WILSON (1958, 1959) in Boden, Wasser und Urin 8 *Achromobacter* gefunden und ihre Fähigkeit der Stickstoffbindung ebenfalls mit Hilfe von N^{15} bewiesen. Einzelne N-bindende Stämme von *Aerobacter aerogenes* erwähnt JENSEN (1956). Ein sehr schwach bindendes Bakterium (1,6 mg N je 1 g C-Quelle) wurde von ihm (JENSEN 1958) aus einem dänischen Gewässer isoliert, der Nachweis der N-Bindung gelang ebenfalls mit dem radioaktiven N^{15}. Eine geringe N-Bindung wurde ferner von OKUDA et al. (1959) durch *Rhodopseudomonas capsulatus* unter anaeroben Bedingungen bei Photosynthese festgestellt. In Gewässern des Krasnodargebietes wurden von RODINA (1956) verschiedene Stämme von *Spirillum azotocolligens* isoliert, die 4 bis 12 mg Stickstoff je 1 g Mannit zu binden vermochten.

Auch für *Actinomyceten* liegen positive Befunde vor (FEDOROW und KUDRJASCHEWA 1956). METCALFE und BROWN (1957) isolierten die stickstoffbindende *Nocardia calcaria n. sp.*, sie war befähigt, 2 bis 4,5 mg N je g Glukose, Saccharose oder Mannit zu binden. *Nocardia cellulans* n. sp. zersetzt Zellulose in der Kultur und bindet Stickstoff unter Verarbeitung von Glukose, Saccharose, Mannit und Zellulose als C-Quelle. 12 mg Stickstoff je g Zellulose wurden gebunden. Die Organismen wurden aus Kalkverwitterungsboden unter Grünland isoliert. FEDOROW und ILJINA (1959) beobachteten trotz des Zusatzes fördernd wirkender Hefeautolysate nur eine sehr geringe Stickstoffbindung bei Actinomyceten.

Über die Stickstoffbindung durch *Hefen* liegen widersprechende Ergebnisse vor. NÉMETH (1959) schrieb zwei rot pigmentierten Saccharomyceten, die aus einem sauren Sandboden Nordwestungarns isoliert wurden, die Fähigkeit der N-Bindung zu. FREI (1941/42) beobachtete wie SCHANDERL (1939/40) N-Bindung durch verschiedene hautbildende Hefen; die Angaben konnten jedoch von LÖHNIS (1943) für einzelne Stämme von *Pichia*, *Mycoderma* und *Saccharomyces*, von SCHEFFER und KÜSTER (1950) für *Pichia farinosa* und *Mycoderma bispora* und von STAPP (1951) für *Pichia farinosa*, *Pichia membrane faciens* und *Mycoderma bispora* nicht bestätigt werden. METCALFE und CHAYEN (1954) berichten über zwei Hefen, die unter *Calluna vulgaris* isoliert wurden, und BROWN und METCALFE (1957) über eine *Pullularia* aus dem Thallus einer Flechte *(Chladonia unicialis)*, die beide etwa 4 mg N je 1 g Kohlenstoffquelle binden. METCALFE et al. (1954) schreiben auch einem Saccharomyceten, den sie aus einem Heidepodsol isolierten, die Fähigkeit der N-Bindung zu. Die Angabe wird von ROBERTS und WILSON (1954) bestätigt. NÉMETH und MATKOVICS (1958, NÉMETH 1959, MATKOVICS 1960) isolierten aus sterilen Schnitten von Knöllchen von Lupinus luteus Hefen mit rotem Pigment, die unter starker Säurebildung kräftige N-Bindung zeigten.

Von den *Schimmelpilzen* sollen nach SCHANDERL (1939/40) der Kellerschimmel *Chladosporium cellare* und nach FREI (1941/42) *Penicillium glaucum* zu den stickstoffbindenden Organismen zu rechnen sein. Eine schwache N-Bindung durch einen *Aspergillus niger*-Stamm stellte SCHOBER (1930) fest. Sie wurde gehemmt durch Zugabe von gebundenem Stickstoff und wurde durch die bei der Assimilation erhöhte Produktion von Glukon- und Oxalsäure zum Stillstand gebracht.

Für die *Algen* (STOKES 1940, FULLER et al. 1960) *Nostoc, Anabaena* und *Calothrix* wurde der Nachweis der Stickstoffbindung von M. B. ALLEN (1952, 1955) erneut erbracht. BHASKARAN und VENKATARAMAN (1958) vermuten die Fähigkeit der Stickstoffbindung bei der blaugrünen Alge *Nostoc punctiforme* in den Wurzelknöllchen von *Trifolium alexandrinum.* Die Physiologie des Vorgangs der Stickstoffbindung bei blaugrünen Algen wurde durch FOGG (1956) eingehend studiert. Im gemäßigten Klima scheint die N-Bindung durch Algen praktisch eine völlig untergeordnete Rolle zu spielen, wesentlich ist sie jedoch in den Tropen, vor allem in Reisböden (DE 1939, DE und MANDAL 1956, NISHIGAKI und SHIOIRI 1959).

Über die *Mengen des im Boden gebundenen Stickstoffs* gehen die Meinungen begreiflicherweise weit auseinander. Die in der Literatur angegebenen Mengen schwanken zwischen 0 und 67 kg Stickstoff je Hektar und Jahr. Nach HALL (1905) wurden die höchsten Werte auf einem Boden mit natürlicher Vegetation in Broadbalk gefunden, der 22 Jahre lang unbewirtschaftet geblieben war. Wenn für die N-Bindung in erster Linie Azotobacter verantwortlich gemacht wird, dann wäre nach den bisherigen Auffassungen der höchste Stickstoffgewinn auf Böden zu erwarten, deren Reaktion oberhalb von pH 6,0 liegt. Andererseits ist aber bekannt, daß die Stickstoffdüngung eine Hemmung der Stickstoffbindung zur Folge hat. Da der Verbrauch an mineralischen Stickstoffdüngemitteln ständig ansteigt, wäre eine Abnahme der biologischen Stickstoffbindung zu erwarten. Auch die physikalische Beschaffenheit des Bodens beeinflußt die Höhe der N-Bindung. So waren die Bedingungen für die N-Bindung in grobkörnigen westaustralischen Böden offenbar weniger günstig als in feinkörnigen (PARKER 1957). Sie ist weiter abhängig von der Menge der den Mikroorganismen zur Verfügung stehenden leicht zersetzbaren organischen Stoffe (DELWICHE und WIJLER 1956). Auch das Eisen darf nicht fehlen (ALEXANDER and WILSON 1954). Durch anorganische Stickstoffverbindungen wird die N-Bindung gehemmt (LENHARD 1953, EBERT 1959), da die beteiligten Mikroorganismen, vor allem Azotobacter, sie bevorzugt assimilieren. Besonders Ammoniakstickstoff schließt die N-Bindung durch Azotobacter aus, da er leicht aufgenommen wird, weil im Mechanismus der N-Bindung Ammoniak als Ausgangsstoff für die Synthese der Glutaminsäure auftritt, während die Organismen sich an das Nitrat erst anpassen müssen (vgl. VIRTANEN 1948, FEDOROW 1960).

B. Rhizobien

Zusammenfassende Darstellungen der biochemischen und symbiotischen Eigenschaften der Rhizobien wurden von FRED et al. (1932) sowie von E. K. ALLEN und O. N. ALLEN (1950) veröffentlicht, später gaben letztere eine Übersicht über pathologische Formen von Wurzelknöllchen, z. B. die von *Agrobacterium tumefaciens* gebildeten Tumoren, heraus (ALLEN and ALLEN 1954). In einem weiteren Sammelreferat gibt THORNTON (1954) Aufschluß über die Systematik und Dissoziation der Rhizobien und ihr Verhalten im Boden, über Wurzelsekretionen und Wirksamkeit der Knöllchen.

Eine Übersicht über die Ausdehnung der Knöllchenbildung in den einzelnen Familien der Leguminosen, von denen die Papilionaceen bisher am intensivsten untersucht wurden, geben ALLEN und ALLEN (1959, 1961). Danach bilden von den 1285 untersuchten Leguminosenarten 85% Knöllchen. Auf ältere Sammelreferate von WILSON (1935, 1937), WILSON und SARLES (1939), WILSON und BURRIS (1947), ALLEN und ALLEN (1947, 1958), VIRTANEN (1947), L. BOND (1948) und FEDOROW (1960) sei hingewiesen.

Eine *artspezifische Wirksamkeit* der Rhizobien ist erwiesen. Die höhere Pflanze übt dabei deutlich eine selektive Wirkung aus (VINCENT und WATERS 1953). Nach TSCHAILACHJAN und MEGRABJAN (1955) enthalten die Wurzeln von Leguminosen Hemmstoffe, die nicht spezifische Rhizobien unterdrücken. Des weiteren hatte NUTMAN (1945) festgestellt, daß trotz einer sehr hohen Zahl von Knöllchenbakterien die Menge der gebildeten Knöllchen gering blieb; vermutlich infolge von Wurzelausscheidungen, die Knöllchenbildung und Bildung von Seitenwurzeln hemmten. Von ELKAN (1961) wurde der Nachweis geführt, daß ein knöllchenfreier Sojabohnenstamm einen Stoff ausschied, der die Knöllchenbildung bei einem normalerweise knöllchenbildenden Schwesterstamm unterband. Nach KRASSILNIKOW (1954) verlieren die Rhizobien ihre Spezifität bei längerer Kultivierung in Filtraten alter Kulturen. SCHEFFZOWA (1959) stellte fest, daß besonders Filtrate zwei- bis dreiwöchiger Pflanzen bakterizid wirken. Die Wirkung der Hemmstoffe aus Pflanzenwurzeln wurde von ihr experimentell nachgewiesen, indem sie Gewebesäfte aus Rotklee-, Erbsen- bzw. Luzernewurzeln auf Agarplatten brachte, die mit Rhizobien beimpft waren. Besonders resistente Stämme konnten isoliert werden, sie bewirkten eine beschleunigte Bildung und eine erhöhte Zahl von Knöllchen. THOMPSON (1960) extrahierte aus der Samenschale eines Ausläufer treibenden Klees ein besonders aktives Antibioticum, das das Wachstum der Rhizobien hemmte.

Einzelne Autoren stellten verwandte Gruppen innerhalb der Rhizobien fest, so beobachtete HANKE-PACEWICZOWA (1952), daß *Galega officinalis* reichlich Knöllchen bildete, wenn mit Rhizobien von *Galega* selbst oder von *Vicia faba* bzw. *V. sativa* geimpft wurde. BONNIER et al. (1952) erreichten in einem Boden, in dem Sojabohnen knöllchenfrei wuchsen, dann Knöllchenbildung, wenn die Pflanzen mit *Pisum sativum, Vicia faba, V. narbonensis, Trifolium pratense, Lupinus luteus* oder *Medicago sativa* gepfropft wurden, nicht aber bei Pfropfung mit *Lotus corniculatus, Trifolium repens* und *Phaseolus vulgaris*. Bei den Versuchen von WAGENBRETH (1956) mit gepfropften Leguminosen entwickelten sich hingegen nur Knöllchen mit den für die Unterlage spezifischen Rhizobien. VYAS und PRASAD (1959) stellten eine nahe Verwandtschaft der Symbionten von *Zornia diphylla* mit denen der *Pisum*-Gruppe fest. PARKER und ALLEN (1952) impften *Trifolium ambiguum* mit 35 verschiedenen Rhizobien, konnten aber stets nur geringe Knöllchenbildung beobachten und keine echte Beziehung zwischen Symbiont und Wirt.

Die Rhizobien von *Astragalus* sp. stellen nach CHEN und SHU (1944) im Gegensatz zu der Auffassung von BUSHNELL und SARLES (1937) eine spezifische Gruppe dar.

Mit dem *Mechanismus der Infektion* hat sich DÉVAY (1958) eingehend befaßt. Die Rhizobien werden erst 4 bis 5 Tage nach der Keimung wirksam, dabei gehen in den Haarwurzelzellen biochemische Veränderungen vor sich. Anscheinend stehen die Reduktion des Ribo- und Desoxyribonucleinsäure-Spiegels der Zellen, die erhöhte Aktivität der Bernsteinsäure-Hydrogenase und die gleichzeitige Abnahme der Apfelsäure-Hydrogenase in den Wirtszellen mit der Infektiosität der Bakterien in Beziehung.

Nicht alle Knöllchen einer Pflanze enthalten Bakterienstämme gleicher *Wirksamkeit*. BURTON et al. (1952) stellten fest, daß verschiedene aus einem

Knöllchen stammende Stämme mehr Einheitlichkeit in ihren physiologischen Eigenschaften und in ihrer Wirksamkeit (vgl. MELKOUMOWA 1957) zeigen als solche, die aus verschiedenen Knöllchen gewonnen wurden. Nach Untersuchungen von MACCONNELL und BOND (1957 a) enthielten die Knöllchen der wildwachsenden Leguminosen *Ulex europaeus* und *Medicago lupulina* nicht immer wirksame Rhizobien. Die unwirksamen Stämme haben zwar die Fähigkeit, die Knöllchenbildung auszulösen, atmosphärischen Stickstoff zu binden und nach THORNTON (1945) werden auch verzweigte Formen gebildet, aber ihre Lebensdauer ist wesentlich geringer, infolgedessen bleiben die Knöllchen von unwirksamen Stämmen kleiner. Während die Knöllchen wirksamer Stämme 1,5 mm lang waren und eine Bakterienmasse von 0,15 mm^3 enthielten, erreichten die Knöllchen von unwirksamen Stämmen nur eine Länge von 0,5 mm und enthielten 0,03 mm^3 Rhizobienzellen (H. G. THORNTON 1939, vgl. VIRTANEN 1947, BURRIS und WILSON 1952, GY 1957).

Zwischen dem Gewicht der Knöllchen und dem Pflanzenertrag besteht, wie Tab. 81 von FEDOROW (1960) zeigt, eine positive Korrelation (vgl. dagegen

Tabelle 81

Knöllchengewicht je Pflanze in g	Ertrag der grünen Pflanzenteile je Pflanze in g	Zahl der Hülsen je Pflanze
Ohne Knöllchen	162	37
1,0	191	36
4,4	214	63
8,0	252	78
10,0	283	93

MELKOUMOWA 1957). Wie stark die Ertragsdifferenzen nach der Impfung mit wirksamen bzw. unwirksamen Stämmen sein können, zeigte BOND (1958) (zit. nach SCHAEDE 1962) in den Ernten von Erbsen, die in Tab. 82 wiedergegeben sind.

Tabelle 82

Impfung	Gesamt-N in den grünen Teilen von 6 Pflanzen, in mg	N-Gehalt in % des Trockengewichts grüner Teile
Kontrolle	31,14	1,99
unwirksame Stämme	35,97	1,89
	70,80	3,36
	53,36	3,10
wirksame Stämme	114,02	4,28
	117,53	3,68
	131,98	3,90
	154,08	5,15
	225,70	5,45

Eine gleichzeitige Impfung mit wirksamen und unwirksamen Stämmen von Rhizobien hatte nach BURTON et al. (1954) keinen nachteiligen Einfluß auf Wachstum und N-Bindung; der gleiche Impferfolg wurde auch beobachtet, wenn zunächst mit wirksamen und anschließend mit unwirksamen Stämmen geimpft wurde, im umgekehrten Fall war er, wie Tab. 83 zeigt, geringer.

Die *Verbreitung unwirksamer Stämme* ist für die Landwirtschaft durchaus von Bedeutung, denn sie betrug bei 463 Stämmen, die THORNTON (1947) in verschiedenen Bezirken Großbritanniens isolierte, 37%. Eine Pflanze kann sowohl von wirksamen als auch von unwirksamen Stämmen infiziert werden.

Tabelle 83

Impfung	Gesamtmenge von 10 Pflanzen		
	Trocken-gewicht g	Stickstoff %	Gesamt-N mg
Kontrolle	12,3	1,26	154
Wirksamer Stamm D14	17,3	2,85	493
Unwirksamer Stamm D10	14,6	1,18	173
Mischung D14 und D10	17,7	2,84	503
D14 und 21 Tage später D10	18,2	2,82	513
D10 und 21 Tage später D14	17,8	2,57	457
D14 nach 21 Tagen eingeimpft	16,8	2,57	430
D10 nach 21 Tagen eingeimpft	16,2	1,34	226
Mischung D14 und D10 nach 21 Tagen eingeimpft	15,8	2,58	408

Für die Entstehung unwirksamer Stämme können verschiedene Gründe verantwortlich gemacht werden. Eine Schädigung der Rhizobien kann erfolgen: 1. bei für die Arten ungeeigneter Bodenreaktion, 2. bei Anwesenheit antagonistischer Organismen oder 3. unter gewissen Voraussetzungen bei Anwesenheit von Rhizobiophagen (KLECZKOWSKA 1950). Eine verminderte Wirksamkeit trat nach HOLDING et al. (1960) bei der Züchtung von Luzerne-, Klee- und Erbsenrhizobien auf aminosäurehaltigen Medien auf. Gleichzeitig kam es zu morphologischen Veränderungen an den Organismen und an den Kolonieformen. Mutanten der Kolonieform konnten nach JORDAN (1952) auch durch Röntgenstrahlen ausgelöst werden.

In den Knöllchen, die von wirksamen Stämmen gebildet werden, hat VIRTANEN (1945, 1947) ein *hämoglobinähnliches Ferment* nachgewiesen. In sauerstofffreier Atmosphäre wurde das Pigment nicht gebildet, es wurde auch kein Stickstoff gebunden. Die rote Farbe verschwand im alternden Knöllchen. HEUMANN (1952) beobachtete, daß das Knöllchengewebe eine grüne Farbe annahm, wenn die Photosynthese durch Verdunkelung unterbrochen oder das Ferment durch Behandlung mit Knallgas zersetzt wurde. Photosynthese und Leghämoglobinbildung stehen also in direktem Zusammenhang. Die Fermentbildung beschränkte sich auf die Zellen, die intensiv an der Stickstoffbindung beteiligt waren. Wurde das Ferment zerstört, dann unterblieb die N-Bindung, es verschwanden die Bakteroiden, an ihre Stelle traten Stäbchen; außerdem wurde die in den Knöllchen enthaltene Stärke nicht abgebaut. Es kommt also nur zu Bakteroidenbildung und Stickstoffbindung, wenn das Ferment intakt ist. NOWOTNY-MIECZYNSKA (1952) konnte eine Korrelation zwischen der Menge des in den Knöllchen enthaltenen Leghämoglobins und der Menge des gebundenen Stickstoffs nicht feststellen.

Nach EGLE und MUNDIG (1954) steigt der Hämingehalt im Knöllchen im Verlauf der ersten Wachstumsphase stark an, in der zweiten Phase bleibt er konstant (außer bei Erbse) und sinkt bei Beginn der Blühphase.

SPICHER (1952/54) beobachtete, daß in den Knöllchen von Lupine und Serradella überwiegend Stäbchen und nur 1,7 bis 4,5% bzw. 0,3 bis 0,62% Bakteroiden enthalten waren; Bakteroidengehalt und Menge des gebundenen Stickstoffs stiegen gleichsinnig an. Bei Klee überwogen Y-Formen und Keulenformen. Es konnte nachgewiesen werden, daß die Bildung von Bakteroiden von der Pflanze beeinflußt wird. Die Menge der Bakteroiden ist auch von den einzelnen Bakterienstämmen abhängig. Nach NAUNDORF und NILSSON (1942) scheidet *Azotobacter chroococcum* formative Wirkstoffe aus: Bakteroide Formen traten in den Knöllchen zahlreicher auf, wenn ein Gemisch von Rhizobien und Azotobacter zur Impfung verwendet wurde. Auch auf künstlichen Medien traten verzweigte Formen bei Anwesenheit von Azotobacter in größerer Zahl auf. Die Gigasformen der Rhizobien waren sechsmal so lang wie die Normalformen (NAUNDORF und NILSSON 1943).

Die *morphologischen Eigenschaften* der Rhizobien wurden von TSCHISHIK (1959) mit Hilfe der Phasenkontrast- bzw. Lumineszenzmikroskopie untersucht. In den Stäbchen wurden keine Unterschiede festgestellt, die Kernsubstanz war meist in diffusem Zustand vorhanden. Nucleoide traten in Bakteroiden häufiger auf als in den Stäbchen. Je stärker die Verzweigung, um so zahlreicher waren die Nucleoide. Zwischen wirksamen und unwirksamen Stämmen waren Unterschiede nicht festzustellen. Nach GOLEBIOWSKA (1952) können die morphologischen und cytologischen Eigenschaften der Bakterien (*Rh. meliloti*, *Rh. trifolii*) durch die Art der Kohlenstoffquelle beeinflußt werden. Über die Auswirkungen verschiedener Belichtungszeit auf die Knöllchenbildung berichten BONNIER und SIRONVAL (1956, SIRONVAL 1956, 1958). Kurztagspflanzen (Soja) entwickelten weniger Knöllchen und weniger Hämoglobin als Langtagspflanzen, der Einfluß längerer Belichtungszeit wirkt sich also über die Chlorophyllsynthese in den Blättern positiv auf die Knöllchenbildung aus. Nach LINDSTRÖM et al. (1952) wurde durch *Trifolium pratense* bei Dunkelheit nur etwa $^1/_{10}$ der bei Licht gebundenen N_2^{15}-Menge fixiert. Auch die Aufnahme von Ammoniakstickstoff und dessen Einbau in das Zelleiweiß ist vom Licht abhängig. Die N-Bindung scheint also unmittelbar von einem Intermediärprodukt der Photosynthese abhängig zu sein. Bei Verdunkelung über 24 Stunden hinaus hörte die N-Bindung überhaupt auf. Auch VAN SCHREVEN (1959) stellte eine erhöhte Knöllchenbildung bei Belichtung fest. Zusatz von 0,5 bis 2% Glukose bzw. 0,5 bis 1% Saccharose stimulierte die Knöllchenbildung bei Klee. Noch höhere Zuckermengen hatten aber, zumal wenn zusätzlich belichtet wurde, keine oder depressive Wirkung. Ein Besprühen der Pflanzen mit Saccharose bzw. mit Harnstoff hatte eine negative Wirkung auf Zahl und Gewicht der Knöllchen.

Die N-Bindung steht in engem Zusammenhang mit der Menge der vorhandenen *Kohlenstoffverbindungen*. BACH et al. (1958) führten 8 bis 10 Wochen alten Sojapflanzen morgens einige Stunden lang Kohlendioxyd mit markiertem C^{14} zu. Dann assimilierten die Pflanzen normal bis zur Ernte am folgenden Morgen. In den Knöllchen hatten die organischen Säuren am Tage einen niedrigeren prozentualen Anteil an C^{14} als die Aminosäuren, während in den Wurzeln das Umgekehrte der Fall war. Nachts stieg der Gehalt an organischen Säuren in den Knöllchen und fiel in den Wurzeln. In den Knöllchen wurde auch Zucker mit C^{14} nachgewiesen. In abgeschnittenen, zerschnittenen Knöllchen wurde die Bindung von N^{15} durch Zuckerzusatz erhöht. Die Menge des fixierten Stickstoffs ging zurück, wenn die Knöllchen bei Einsetzen der Blüte und zurückgehendem Wachstum abgetrennt wurden (MAGEE und BURRIS 1954). KVASNIKOV et al. (1958) kamen zu dem Ergebnis, daß durch hohen Zuckergehalt der Kulturpflanzen auch die Knöllchenbildung verstärkt wird. Abbauprodukte des Strohs wirkten im gleichen Sinne (BJÄLFVE und NILSSON 1938). In den Versuchen von BONNIER und BROUWERS (1958) wurde die Knöllchenbildung durch Erweiterung des C:N-Verhältnisses gefördert. CO_2 in saurer Kulturlösung erwies sich nach MULDER und VAN VEEN

(1960b) als fördernder Faktor für Knöllchenbildung und N-Bindung, in neutralen und schwachsauren Medien, vor allem bei Zufuhr von Nitrat-N blieb CO_2 ohne Wirkung. Versuche in Böden ergaben keine eindeutigen Resultate.

Die *N-Düngung* fördert das Wachstum der Leguminosen und verbessert dadurch die C-Versorgung der Rhizobien, wie SCHMIDT (1955) bei Versuchen mit *Vicia faba* beobachtete. Eine kleine Stickstoffgabe zur Förderung der Jugendentwicklung der Pflanzen und der Knöllchenbildung hat sich als zweckmäßig erwiesen. Durch höhere Gaben wurde aber, nach den Versuchen von ALLOS und BARTHOLOMEW (1955), die Bindung des atmosphärischen Stickstoffs herabgesetzt, dafür stieg der Anteil des aus der Düngung entnommenen N^{15} in den Pflanzen. Unabhängig von der Stärke der N^{15}-Düngung nahmen die Leguminosen etwa 55% des Dünger-Stickstoffs auf, während diese Menge bei Nicht-Leguminosen etwa 85% betrug. Auf gut mit N versorgten Böden nehmen Sojabohnen wahrscheinlich weniger als die Hälfte des von ihnen benötigten Stickstoffs aus der Luft auf, auf schlecht versorgten mehr als die Hälfte. Wie stark die N-Düngung die N-Bindung herabsetzt, geht besonders deutlich aus den Ergebnissen von ALLOS und BARTHOLOMEW (1955, zit. nach SCHAEDE 1962) hervor, die in Tab. 84 enthalten sind. Die Stickstoffbindung sinkt mit steigenden N-Gaben.

Tabelle 84

Pflanze	Stickstoffgabe während einer 10wöchigen Periode in mg/Topf			
	0	108	216	432
Luzerne				
Ertrag (g/Topf)	3,6	5,3	5,6	8,8
Gesamtaufnahme an N (mg/Topf)	108	172	198	333
Fixierter N (in % der Gesamtaufnahme)	100	59	45	17
Lespedeza				
Ertrag (g/Topf)	5,8	7,5	7,8	8,3
Gesamtaufnahme an N (mg/Topf)	128	198	194	280
Fixierter N (in % der Gesamtaufnahme)	100	71	52	18
Klee				
Ertrag (g/Topf)	1,5	3,5	5,1	5,7
Gesamtaufnahme an N (mg/Topf)	44	112	166	237
Fixierter N (in % der Gesamtaufnahme)	100	35	28	11

Auch WALKER et al. (1956) beobachteten eine negative Korrelation zwischen N-Düngung und N-Bindung. Bei Klee-Gras-Gemischen hatte das Gras 95% des verwerteten Dünger-Stickstoffs aufgenommen. ALLOS und BARTHOLOMEW (1959) stellten ferner fest, daß bei einer bestimmten, die notwendige Menge übersteigenden Gabe die Knöllchenbildung überhaupt aufhört. 50 bis 75% des von der Pflanze verwerteten Stickstoffs wurde durch N-Bindung gewonnen. KNO_3-Düngung hemmte bei Luzerne, Rotklee, Lupine und Soja die Entwicklung von Knöllchen und die Bildung von Hämoglobin, wie Versuche von NOWOTNY-MIECZYNSKA und RUSZKOWSKA (1952) ergaben. Durch Saccharose wurde die Wirkung aufgehoben. Auch HALLSWORTH et al. (1960) stellten fest, daß bei steigenden N-Gaben der Gehalt der Knöllchen an Leghämoglobin abnahm.

Von den *Mineralstoffen* werden Phosphor und Kalium benötigt. Nach NORRIS (1958) können Rhizobien gelegentlich wohl Calcium entbehren, aber nicht Magnesium, letzteres kann aber teilweise durch Calcium bzw. Barium ersetzt werden. Die fördernde Wirkung von Molybdän wurde bereits von BORTELS (1937b) nachgewiesen und von JENSEN (1948) und anderen Forschern bestätigt.

Durch Molybdänmangel wird nach MULDER et al. (1959) die N-Bindung gehemmt, durch Bormangel dagegen das Wachstum des Knöllchengewebes. In Mo-armen Böden wurde bereits 5 Stunden nach einer Mo-Düngung ein deutlicher Zuwachs an Glutaminsäure in den Knöllchen von Luzerne und Rotklee festgestellt. Asparagin trat erst später auf, so daß angenommen werden kann, daß die Glutaminsäure-Dehydrogenase das wichtigste Ferment bei der Aminosäuresynthese ist.

Für *Medicago sativa* wurde von REISENAUER (1960) der Bedarf von Cobalt nachgewiesen, das in dem von den Rhizobien gebildeten Vitamin B_{12} enthalten ist. Nach AHMED und EVANS (1959) gilt das gleiche auch für *Soja hispida*. HALLSWORTH et al. (1960) erbrachten den Nachweis für *Trifolium*, sie konnten gleichzeitig feststellen, daß der optimale Co-Gehalt bei 0,006 bis 0,06 ppm liegt.

Ein günstiger Einfluß verschiedener *Wuchsstoffe* auf die Rhizobien wurde wiederholt beobachtet: Vitamin B_1 (Biotin) erzeugt eine beträchtliche Induktion des Wachstums der Rhizobien (NILSSON, BJÄLFVE und BURSTRÖM 1938) und erhöht das Längenwachstum der Pflanzen (NILSSON, BJÄLFVE und BURSTRÖM 1939). MULDER und VAN VEEN (1960a) stellten eine Förderung der Knöllchenbildung bei Rotklee durch Hefeextrakt, sterilisierte Rhizobienkulturen und Stallmist fest.

Gibberellinsäure (1 bis 1000 ppm.) blieb nach FLETCHER et al. (1958) ohne Einfluß auf die Knöllchenbildung bei *Trifolium repens* sowie auf das Wachstum von *Rh. trifolii*. Dagegen konnten THURBER und Mitarbeiter (1958) durch Besprühen von *Phaseolus vulgaris* in Gefäßversuchen mit K-Gibberellat (0,005%ig) eine Hemmung der Knöllchenbildung beobachten. Auch GALSTON (1959) kam zu dem Ergebnis, daß Gibberellin die Knöllchenbildung hemmt.

Fungizide wirken nach HOFER (1958) auf verschiedene Rhizobienstämme unterschiedlich, auch zeigen nicht alle Wirtspflanzen die gleiche Empfindlichkeit wie die Symbionten. Gegen Ammonium- und Natrium-2,4-dichlorophenoxyazetat sind Rhizobien weniger empfindlich als die Wirtspflanze (CARLYLE and THORPE 1947). Aus Knöllchen, die zwei Jahre lang in gefrorenem Zustand aufbewahrt wurden, konnten Rhizobien isoliert werden (POTTER 1955), jedoch nicht, wenn sie unter Mineralöl tiefgefroren wurden.

Ein vermehrtes Auftreten von *Bakteriophagen* der Rhizobien kann nach BRUCH und ALLEN (1955) zu Kleemüdigkeit führen. Die optimale Temperatur für 4 Phagen der Symbionten von *Lotus corniculatus* lag bei 27 bis 30° C. Durch Temperaturen zwischen 45 und 55° C wurden sie inaktiviert. Bei +2° C wurden die Titer 60 Tage lang gehalten. Die günstigste Reaktion lag zwischen pH 6,5 und 7,5. KATZNELSON (1941) hielt Phagen von Rhizobien jahrelang bei 0° C aktiv. KLECZKOWSKA (1945) entwickelte eine Methode, die es erlaubte, die Bakteriophagen zahlenmäßig zu bestimmen.

Die *Flora der Rhizosphäre* spielt gegenüber den Rhizobien eine verschiedenartige Rolle. Unter den Bakterienarten der Flora wurden Aktivatoren, Antagonisten und indifferente Arten gefunden. THORNTON et al. (1949) isolierten 6 Streptomyceten und 2 Penicillien, die *Rh. meliloti* bei Plattenkultur hemmten. Nach AFRIKJAN und TUMANJAN (1958) gehören zu den Antagonisten unter den Mikroorganismen des Bodens besonders Sporenbildner und Actinomyceten, selten Mycobakterien und nichtsporenbildende Bakterien. Die früheren Arbeiten von WILSON (1939) wurden damit bestätigt.

WILSON hatte festgestellt, daß die antagonistischen Kräfte der Streptomyceten gegenüber Rhizobien sehr veränderlich seien. Auch HARRIS (1953) stellte fest, daß Rhizobien von einigen Bakterien und Pilzen der Rhizosphäre gefördert wurden; andere zeigten eine hemmende Wirkung.

Auch Klima und Boden spielen bei der Knöllchenentwicklung eine Rolle. Nach MEYER und ANDERSON (1959) erreichte die N-Bindung durch Knöllchen von *Trifolium subterraneum* bei 20° C die Wirkung einer Nitratdüngung, während sie bei 30° C unterblieb.

Die Akklimatisation von Rhizobien der Sojabohne auf anderen als Tschernosem-Böden ist nach KLYKOW (1956) nicht gesichert. Eine Knöllchenbildung wurde zwar beobachtet, bei den Organismen konnte es sich indessen um saprophytische Bakterien handeln. Ihre Fähigkeit, Stickstoff zu binden, wurde jedenfalls nicht nachgewiesen. IWANOWSKAJA und FEDOROWA (1956) konnten auf Böden, die noch nie Lupine und Sojabohne getragen hatten, selbst nach viermal wiederholtem Anbau keine Knöllchenbildung beobachten. Schließlich entwickelten sich nichttypische knotenartige Knöllchen. Wenn sie später als Impfstoff verwendet wurden, führten sie jedoch zu typischen Knöllchen. Es wird vermutet, daß sich die Mikroorganismen unter dem Einfluß der Pflanzen änderten.

Über die Mengen des gebundenen Stickstoffs bei Erbsen berichten VIRTANEN und SAUBERT-VON HAUSEN (1953). Insgesamt wurden 67% des Stickstoffs aus der Atmosphäre und 33% aus dem Boden entnommen. Unter Berücksichtigung des als Nachfrucht gebauten Hafers betrug die Gesamtmenge des gebundenen Stickstoffs 100 kg/ha. Nach GEORDETTE (1953) wird in Belgien bei Luzerne mit einer N-Bindung bis zu 400 kg je ha und Jahr gerechnet. In tropischem Klima ermittelte MOORE (1960) beim Vergleich einer Gras- mit einer Leguminosen-Grasweide im fünfjährigen Durchschnitt einen jährlichen Gewinn von 126 kg N/ha. Die Ergebnisse von LYON (1936) in Tab. 85 zeigen die sehr unterschiedlichen Mengen an Stickstoff, die von Leguminosen gebunden werden.

Tabelle 85. *Stickstoffbindung durch Leguminosen und ihr Einfluß auf die folgende Getreideernte*

Mittel aus zwei Versuchen mit fünfmaligem Wechsel von Leguminosen und Getreide (Roggen oder Gerste) in der Fruchtfolge

Frucht	Stickstofferntebei		Gewinn oder Verlust an Stickstoff je Fruchtfolge kg/ha	von Leguminosen gebundener Stickstoff Gesamtmenge kg/ha	Körnerertrag der Getreide dz/ha
	Leguminosen kg/ha	Getreide kg/ha			
Luzerne	334,9	73,9	136,6	504,0	29,1
Klee	140,0	57,1	126,5	291,2	24,3
Süßklee	190,4	57,1	94,1	302,4	23,7
Sojabohnen	197,1	32,5	—8,96	179,2	14,8
Feldbohnen	115,4	28,0	—22,4	78,4	13,3
Getreide, j.J. ...	—	24,6	—11,2	—	10,9

Sie sind bei Grünfutterpflanzen höher als bei solchen, deren Früchte zur Reife kommen. Der Ertrag des Getreides in der Fruchtfolge wird deutlich günstig beeinflußt, auch wenn im Boden kein Stickstoffgewinn festgestellt werden konnte. Nach THORNTON und BROADBENT (1948) sinkt die Menge des gebundenen Stickstoffs mit steigendem N-Gehalt des Bodens.

Über die Möglichkeit der asymbiotischen Stickstoffbindung durch Rhizobien liegen widersprechende Ergebnisse vor. In einer Reihe von älteren Arbeiten (CLARK 1936, RABOTNOWA 1938, WILSON 1939, 1940) wird den Rhizobien diese Fähigkeit nicht zuerkannt. Nach VERNER und KOVALEV (1936) binden sie Stickstoff auf künstlichen Nährböden, wenn diesen Bios zugegeben wird. HEUMANN (1952) beobachtete auf Möhrenagar, dem anstelle des Leghämoglobins 15% frisches Menschenblut zugesetzt wurde, bei Rhizobien von *Pisum* Bildung von Y-Formen und Verwertung des atmosphärischen Stickstoffs. Das Hämoglobin des Blutes konnte durch α-Eisenoxydhydrat ersetzt werden.

C. Knöllchentragende Nichtleguminosen

Zwischen den Leguminosen und den knöllchentragenden Nichtleguminosen bestehen einige interessante Parallelen. BOND (1957b) konnte nachweisen, daß *Casuarina*, *Shepherdia* und *Ceanothus* in Wasserkulturen markierten Stickstoff N^{15} aus einer Atmosphäre von 15% N, 20% O_2 und 65% Ar aufnehmen und in Wurzeln und Zweige ableiten konnten. Bei Erlen wurde von dem gleichen Forscher (BOND 1956) eine N-Fixierung aus einem Gemisch von 20% O_2 und 80% N (davon 18% N^{15}) festgestellt. Nach VIRTANEN (1957) nahmen Fichten in Gemeinschaftskultur mit Erlen einen Teil des von den letzteren gebundenen Stickstoffs auf. GARDNER (1958) stellte in Wasserkulturpflanzen von *Eleagnus angustifolia*, die mit Symbionten von *Hippophaë* geimpft waren, gegenüber ungeimpften einen bis 90fach höheren Gesamtstickstoffgehalt fest. Unter Robinienbeständen (16 bis 20 Jahre alt) stellten IKE und STONE (1958) einen Stickstoffgewinn von 50 kg N/ha fest. Die N-Bindung in N-freier Nährlösung betrug nach BOND und GARDNER (1957) bei *Alnus glutinosa* im 1. Jahr 300 mg, im 2. Jahr 2500 mg, *bei Myrica gale* im 1. Jahr 460 mg, im 2. Jahr 5020 mg, bei *Hippophaë rhamnoides* im 1. Jahr 26 mg, im 2. Jahr 200 mg. In den Versuchen von MACCONNEL und BOND (1957b) stieg das Knöllchengewicht von *Myrica gale* und *Alnus glutinosa* bei steigender Düngung mit Ammonsulfat zwar zunächst an, sank dann aber deutlich ab. Bei der N-Bindung spielt jedoch die Reaktion eine Rolle, und BOND (1957a) konnte feststellen, daß die Knöllchenbildung bei zunehmend saurer Reaktion bei *Casuarina* stärker abfiel als bei anderen Nichtleguminosen. Bei Kulturen von *Hippophaë*, *Myrica* und besonders bei *Alnus* sind die höheren Pflanzen weniger säureempfindlich als die Symbionten. Bei *Alnus* werden die Knöllchen in der Regel durch Actinomyceten gebildet, nur im Jugendstadium wurden Knöllchen beobachtet (POMMER 1956), die die Fadenpilze *Penicillium albidum Sapp.* bzw. *Cylindrocarpon radicicola WR* enthielten. Da die Erle aber gerbstoffhaltige Hemmstoffe bildet, sterben diese letzteren nach spätestens 12 Wochen ab. MACCONNELL (1959) stellte fest, daß der Sauerstoff einen Einfluß auf die Knöllchenbildung ausübt: Wenn der O_2-Gehalt in der zur Durchlüftung der Kulturlösungen verwendeten Luft herabgesetzt wurde, reagierten die knöllchentragenden Pflanzen empfindlicher als die mit N gedüngten. Die N-Düngung hemmte die Knöllchenbildung auch bei *Hippophaë* (BOND et al. 1956). Ohne Knöllchenbildung entwickelten sich die Pflanzen mit NH_4-N besser als bei Düngung mit NO_3-N.

Über den Mechanismus der Infektion bei *Alnus* und die Verdauung der Symbionten berichten KÄPPEL und WARTENBERG (1958) sehr ausführlich.

Literatur

ABD-EL-MALEK, Y., M. MONIB und M. N. ZAYED: Effect of organic matter on the development in soil of antibiotic-producing microorganisms. Plant a. Soil **15**, 25–36 (1961). — AFANASJEWA, A. L., und L. N. SWJATSKAJA: Die mikrobiologischen Prozesse im Boden bei verschiedenen Bearbeitungsmethoden der Neu- und Brachländereien. Agrobiologija **1**, 86–92 (1958); Ref. Z. Pflanzenernähr., Düng., Bodenkde. **83**, 273 (1958). — AFRIKJAN, E. K., und B. T. TUMANJAN: Die antagonistische Wirkung der Mikroorganismen des Bodens auf die Kultur der Knöllchenbakterien. Nachr.

Akad. Wiss. Armen. SSR, H. 2, 37–46 (1958) (russ.); Ref. Z. Pflanzenernähr., Düng., Bodenkde. **85**, 182 (1959). — AGNIHOTHRUDU, V.: State in which fungi occur in the rhizosphere. Naturwiss. **42**, 515–516 (1955). — AHMED, S., und H. J. EVANS: Effect of cobalt on the growth of soybeans in the absence of supplied nitrogen. Biochem. Biophys. Res. Comm. **1**, 271–275 (1959). — AHRENS, E.: Der Einfluß organischer und mineralischer Dünger auf das Verhalten von Azotobacter und die Möglichkeit seines quantitativen Nachweises. Diss. Giessen 1961. — ALDRICH, D. G., und J. P. MARTIN: A chemical and microbiological study of effects of exchangeable cations on soil aggregation. Soil Sci. Soc. Amer. Proc. **18**, 276–281 (1954). — ALEEM, M. J. H.: Counting of Azotobacter in soils. Plant a. Soil **4**, 248–251 (1953). — ALESCHINA, E. N., und L. N. MAKAROWSKAJA: Actinomyceten als Antagonisten. Microbiologija **24**, 309–315 (1955); Ref. Zbl. Bakt. II **109**, 541 (1956). — ALEXANDER, F. E. S., und R. M. JACKSON: Examination of soil microorganisms in their natural environment. Nature (Lond.) **174**, 750–751 (1954); Ref. Zbl. Bakt. II **108**, 717 (1954/55). — ALEXANDER, M.: Introduction to soil microbiology. New York: Wiley. 1961. — ALEXANDER, M., und P. W. WILSON: Large-scale production of the Azotobacter for enzymes. Appl. Microbiol. **2**, 135–140 (1954). — ALLEN, E. K., und O. N. ALLEN: Biochemical and symbiotic properties of the Rhizobia. Bact. Rev. **14**, 273–330 (1950) (mit 415 Literaturangaben). — Biological aspects of symbiotic nitrogen fixation. In: RUHLAND, Handbuch der Pflanzenphysiologie, Bd. 8, S. 48–118 (1958). — The scope of nodulation among members of the family Leguminosae. IX. Int. Bot. Congr., Proc. **2**, 3–4 (1959). — Nitrogen fixation. The scope of nodulation in the Leguminosae. Recent Advances in Botany, S. 585–588. Toronto: University Press, 1961. — ALLEN, M. B.: The cultivation of *Myxophyceae*. Arch. Mikrobiol. **17**, 34–53 (1952). — ALLEN, M. B., und D. J. ARNON: Studies on nitrogen-fixing blue-green algae, II, The sodium requirement of *Anabaena cylindrica*. Physiol. Plant. **8**, 653–660 (1955); Ref. Z. Pflanzenernähr., Düng., Bodenkde. **73**, 163 (1956). — ALLEN, O. N., und E. K. ALLEN: A survey of nodulation among leguminous plants. Soil Sci. Soc. Amer. Proc. **12**, 203–208 (1947). — Morphogenesis of the leguminous root nodule. Abnormal and Pathological Plant Growth **6**, 209–234 (1954) (mit 101 Literaturangaben). — ALLISON, F. E.: Azotobacter inoculation of crops, I, Historical. Soil Sci. **64**, 413–429 (1947). — Does nitrogen applied to crop residues produce more humus? Soil Sci. Soc. Amer. Proc. **19**, 210–211 (1955); Ref. Z. Pflanzenernähr., Düng., Bodenkde. **72**, 185 (1956). — ALLISON, F. E., V. L. GADDY, L. A. PINCK, und W. H. ARMIGER: Azotobacter inoculation of crops, II, Effect of crops under greenhouse conditions. Soil Sci. **64**, 489–497 (1947). — ALLOS, H. F., und W. V. BARTHOLOMEW: Effect of available nitrogen on symbiotic fixation. Soil Sci. Soc. Amer. Proc. **19**, 182–184 (1955); Ref. Z. Pflanzenernähr., Düng., Bodenkde. **72**, 178 (1956). — Replacement of symbiotic fixation by available nitrogen. Soil Sci. **87**, 61–66 (1959). — ALTSON, R. A.: Studies on Azotobacter in Malayan soils. Soil Sci. **26**, 268–280 (1936). — AMBERGER, A., E. WOLF und W. HUNNIUS: Über die Wirkung einer langjährigen Phosphorsäuredüngung mit Handelsdüngern und Stallmist auf Boden und Pflanze. Bayer. Landwirtsch. Jahrb. **34**, 229–248 (1957). — ANDAL, R., K. BHUVANESWARI und N. S. SUBBA-RAO: Root exudation of paddy. Nature **178**, 1063 (1956); Ref. Z. Pflanzenernähr., Düng., Bodenkde. **80**, 89 (1958). — ANDERSON, G.: Nitrogen fixation by a *Pseudomonas* like soil bacterium. J. Bact. **70**, 129–133 (1955). — ANDERSON, G. R.: Ecology of Azotobacter in soils of the Palouse region, I, Occurrence. Soil Sci. **86**, 57–62 (1958a). — Ecology of Azotobacter in soils of the Palouse region, II, Responses to environmental changes. Soil Sci. **86**, 111–116 (1958b). — ARENS, R., und J. STUCKMANN: Zur Frage der Wirkung organischer Dünger auf Grünland. Z. Acker- u. Pflanzenbau **107**, 357–370 (1959). — ARISTOVSKAIA, T. V., und O. M. PARINKINA: Seasonal changes and ecological pecularities of the microflora of some podsol soils. Microbiologija **27** (3), 320–325 (1958). — AUDUS, L. J.: Plant growth substances. London: Leonhard Hill. 1953.

BACH, M. K., W. E. MAGEE und R. H. BURRIS: Translocation of photosynthetic products to soybean nodules and their role in nitrogen fixation. Plant Physiol. **33**, 118–124 (1958); Ref. Z. Pflanzenernähr., Düng., Bodenkde. **83**, 170 (1958). — BAKERSPIGEL, A., und J. J. MILLER: Comparison of oxgall, crystall violet, streptomycin and penicillin as bacterial growth inhibitors in platings of soil fungi. Soil Sci. **76**, 123–126 (1953); Ref. Zbl. Bakt. II **108**, 504 (1954/55). — BALICKA, N.: The influence of mixed culture of red clover and tymothy on the microflora in their rhizosphere. Acta Microbiol. Polon. **1**, 58–64 (1952); Ref. Zbl. Bakt. II **108**, 107 (1954/55). — BARBIER, G., und P. BOISCHOT: Mode d'action de la paille enfouie dans le sol sur la nutrition azotée des cultures. Z. Pflanzenernähr., Düng., Bodenkde. **69**, 8–15 (1955). — BARTON, R.: Germination of oospores of *Phytium mamillatum* in response to exudates from living seedlings. Nature (Lond.) **180**, 613–614 (1957). — BECK, TH., und H. POSCHENrieder: Über die artenmäßige Zusammensetzung der Mikroflora eines sehr sauren Waldmoorprofils. Zbl. Bakt. II **111**, 672–683 (1958). — Becker, Y., J. Guillemat, L. Guyot und D. Lelievre: Sur un aspect phytopathologique du problème des substances racinaires toxiques. C. R. Acad. Sci. **233**, 198–199 (1951); zit. nach Z. Pflanzenernähr., Düng., Bodenkde. **80**, 225–236 (1958). — Becking, J. H.: Nitrogen-fixing bacteria of the genus *Beijerinckia* in South Africa soils. Plant a. Soil **11**, 193–206 (1959). — Beerstecher, E. jr.: Petroleum microbiology, S. 100. Amsterdam: Elsevier. 100, 1954. — Benade, W.: Mikrobiologie der Moore. Linz: Länderverlag Stöber. 1954. — Bernhard, W. W., und E. A. Woronkowa: Die Gewöhnung von Azotobakter an Rasenpodsolboden in Gegenwart verschiedener Quellen organischer Substanz. Bjull. Wsesojusn. Nautschnoissled. Inst. Udobr. i Agropotschwowed. **2**, 15–19 (1956) (russ.); Ref. Z. Pflanzenernähr., Düng., Bodenkde. **77**, 276 (1957). — Die Vermehrung des Azotobacters und die Stickstoffixierung in Gegenwart von organischen Düngern und anderen Quellen der organischen Substanz. Agrobiologia, **1960**, 103–108; Ref. Z. Pflanzenernähr., Düng., Bodenkde. **93**, 88 (1961). — Bhaskaran, S., und G. S. Venkataraman: Occurrence of a blue-green algae in the nodules of *Trifolium alexandrinum*. Nature **181**, 277–278 (1958); Ref. Z. Pflanzenernähr., Düng., Bodenkde. **82**, 254 (1958). — Bjälfve, G., und R. Nilsson: Stroh als Nährboden für Bacterium radicicola. Ann. landwirtsch. Hochsch. Schwedens **5**, 71–88 (1938). — Blinkow, G. N.: Ein säureresistenter Azotobakter. Mikrobiol. **24**, 415–421 (1955b) (russ.); Ref. Z. Pflanzenernähr., Düng., Bodenkde. **72**, 94 (1956). — Geographical distribution of Azotobacter in the soils of Siberia. Uch. Zap. Tomsk. Gos. Ped. Inst. **1955a**, 494–534 (14). — Blue, W. M. G., C. F. Eno und P. J. Westgate: Influence of soil profile characteristics and nutrient concentrations on fungi and bacteria in Leon fine sand. Soil Sci. **80**, 303–308 (1955). — Bond, G.: Evidence for fixation of nitrogen by root nodules of Alder (*Alnus*) under field conditions. New Phytol. **55**, 147–153 (1956); Ref. Z. Pflanzenernähr., Düng., Bodenkde. **75**, 183 (1956). — The development and significance of the root nodules of Casuarina. Ann. Bot. N. S. **21**, 373–380 (1957a); Ref. Z. Pflanzenernähr., Düng., Bodenkde. **80**, 89 (1958). — Isotopic studies of nitrogen fixation in nonlegume root nodules. Ann. Bot. N.S. **21**, 513–521 (1957b); Ref. Z. Pflanzenernähr., Düng., Bodenkde. **83**, 170 (1958). — Symbiotic nitrogen fixation by non-legumes. In: Nutrition of the legumes. S. 216–227, London 1958. Zit. nach Schaede, Die pflanzlichen Symbiosen, S. 69. Stuttgart 1962. — Bond, G., und I. C. Gardner: Nitrogen fixation in non-legume root nodule plants. Nature **179**, 680–681 (1957); Ref. Z. Pflanzenernähr., Düng., Bodenkde. **81**, 91 (1958). — Bond, G., J. T. MacConnell und A. H. MacCallum: The nitrogen-nutrition of *Hippophaë rhamnoides*. Ann. Bot. N.S. **20**, 501–512 (1956); Ref. Z. Pflanzenernähr., Düng., Bodenkde. **77**, 259 (1957). — Bond, L.: Origin and developmental morphology of root nodules of *Pisum sativum*. Bot. Gaz. **109**, 411–434 (1948). — Bonnier, Ch., und L. Brouwers: Nodulation, en tube de verre, de Légumineuses à grosses graines cultivées aseptiquement. Bull. Inst. Agronom. Stat. Rech. Gembloux **26**, 317–321 (1958); Ref. Z. Pflanzenernähr., Düng., Bodenkde. **88**, 191 (1960). — Bonnier, Ch., F. H. Hely und P. Manil: Essai d'adaption à *Soja hispida* de souches de *Rhizobium* non spécifiques. Influence de greffes sur la spécifité d'hôte du genre *Rhizobium*. Bull. Inst. Agronom. Stat. Rech. Gembloux **20**, 137–140 (1952); Ref. Z. Pflanzenernähr., Düng., Bodenkde. **66**, 179 (1954). — Bonnier, Ch., und C. Sironval: Influence of day-length on nodule formation in *Soja hispida* by a specific *Rhizobium* strain. Nature **177**, 93–94 (1956); Ref. Z. Pflanzenernähr., Düng., Bodenkde. **76**, 62 (1957). — Bornside, G. H., und R. E. Kallio: Ecological aspects of ureolytic soil bacteria. Soil Sci. **85**, 38–41 (1958). — Bortels, H.: Über die Wirkung von Molybdän und Vanadiumdünger auf Azotobacterzahl und Stickstoffbindung in Erde. Arch. Mikrobiol. **8**, 1–12 (1937a). — Über die Wirkung von Molybdän und Vanadiumdüngungen auf Leguminosen. Arch. Mikrobiol. **8**, 13–26 (1937b). — Weitere Untersuchungen über die Bedeutung von Molybdän, Vanadium, Wolfram und anderen Erdaschenstoffen für stickstoffbindende und andere Mikroorganismen. Zbl. Bakt. II **95**, 193–218 (1936/37). — Böttcher, K.: Vergleichende Untersuchungen über den Einfluß mehrjähriger Futterpflanzen auf die Mikroflora des Bodens. Dtsch. Landwirtsch. **7**, 400–404 (1956); Ref. Z. Pflanzenernähr., Düng., Bodenkde. **76**, 185 (1957). — Brauns, A.: Angewandte Bodenbiologie und Pflanzenschutz. Nachrichtenbl. dtsch. Pflanzenschutzdienst (Braunschweig) **8**, 10–13 (1956); Ref. Z. Pflanzenernähr., Düng., Bodenkde. **75**, 182 (1956). — Brown, A. L.: Effect of several insecticides on ammonification and nitrification in two neutral alluvial soils. Soil Sci. Soc. Amer. Proc. **18**, 417–420 (1954); Ref. Z. Pflanzenernähr., Düng., Bodenkde. **69**, 313 (1955). — Brown, J. C.: Fungal mycelium in dune soils estimated by a modified impression slide technique. Trans. Brit. Mycol. Soc. **41**, 81–88 (1958). — Brown, M. E.: Preliminary studies on the inoculation of selected microorganisms into partially sterilised soils. J. Gen. Microbiol. **18**, 239–247 (1958). — BROWN, M. E., und G. METCALFE: Nitrogen fixation by a species of *Pullularia*. Nature (Lond.) **180**, 282 (1957). — BROWN, W. T., J. H. QUASTEL und P. G. SCHOLEFIELD: Effects of mercapto compounds on soil nitrification. Appl. Microbiol. **2**, 235–239 (1954). — BRUINSMA, J.: The action of 4,6-dinitro-o-cresol (DNOC) in soil, I, The emergence from cress seed in DNOC-treated soil mixtures. Plant a. Soil **12**, 249–258 (1960). — BRUCH, C. W., und O. N. ALLEN: Description of two bacteriophages active against *Lotus Rhizobia*. Soil Sci. Soc. Amer. Proc. **19**, 175–179 (1955). — BRÜSEWITZ, G.: Untersuchungen über den Einfluß des Regenwurms auf Zahl, Art und Leistungen von Mikroorganismen im Boden. Arch. Mikrobiol. **33**, 52–82 (1959). — BUCKSTEEG, W., und H. THIELE: Verbesserung der Keimzählungstechnik in der Wasserbakteriologie durch Anwendung von 2-, 3-, 5-Triphenyltetrazoliumchlorid (TTC). Zbl. Bakt. II **111**, 235–237 (1958). — BUKATSCH, F., K. BURGER und M. SCHLÜTER: Untersuchungen über Eiweiß und Eiweißstoffwechsel bei Azotobacter mit besonderer Berücksichtigung der Indolkörper. Zbl. Bakt. II **109**, 226–235 (1956). — BURGEFF, H.: Mikrobiologie des Hochmoores mit besonderer Berücksichtigung der Erikazeen-Pilz-Symbiose. Stuttgart: Fischer, 1961. — BURGES, A.: Micro-Organisms in the Soil. London: Hutchinson University Library. 1958. — BURCK, D., und R. H. BURRIS: Biochemical nitrogen fixation. Ann. Rev. Biochem. **10**, 587–618 (1941). — BURRICHTER, E.: Regeneration von Heide-Podsolböden und die Entwicklung des Bodenkeimgehaltes in Abhängigkeit von der Bewaldung. Z. Pflanzenernähr., Düng., Bodenkde. **67**, 150–163 (1954). — Die bakteriologische Kartierung und Beurteilung der Kulturböden des Provinzialgutes Kinderhaus bei Münster/Westf. Landwirtsch. Forsch. **8**, 14–25 (1955); Ref. Z. Pflanzenernähr., Düng., Bodenkde. **76**, 279 (1957). — Untersuchungen über die Massenentwicklung der Bodenbakterien im Laufe des Jahres. Ber. Dtsch. Bot. Ges. **121**, 71–77 (1958); Ref. Zbl. Bakt. II **113**, 176 (1959/60). — BURRIS, R. H., und P. W. WILSON: Biological nitrogen fixation. Ann. Rev. Biochem. **14**, 685–708 (1945). — Effect of haemoglobin and other nitrogeneous compounds in the respiration of the Rhizobia. Biochem. J. **51**, 90–96 (1952); Ref. Zbl. Bakt. II **107**, 196 (1952/54). — BURTON, J. C., O. N. ALLEN und K. C. BERGER: The prevalence of strains of *Rhizobium phaseoli* in some Midwestern soils. Soil Sci. Soc. Amer. Proc. **16**, 167–170 (1952). — Response of beans *(Phaseolus vulgaris L.)* to inoculation with mixtures of effective and ineffective Rhizobia. Soil Sci. Soc. Amer. Proc. **18**, 156–159 (1954). — BURTON, M. O., und A. G. LOCHHEAD: Nutritional requirements of *Arthrobacter terregens*. Canad. J. Bot. **31**, 145–151 (1953). — BURTON, M. O., F. J. SOWDEN und A. G. LOCHHEAD: Studies on the isolation and nature of the "Terregens Factor". Canad. J. Biochem. Physiol. **32**, 400–406 (1954). — BUSHNELL, O. A., und W. B. SARLES: Root nodule bacteria of wild leguminous plants in Wisconsin. Soil Sci. **44**, 409–424 (1937). — BUTLER, E. E., und R. B. HINE: Use of novobiocin for isolation of fungi from soil. Soil Sci. **85**, 250–254 (1958). — BUXTON, E. W.: Some effects of pea root exudates on physiologic races of *Fusarium oxysporum* Fr. f. pisi (Linf.) Snyder and Hansen. Brit. Mycol. Soc. Trans. **40**, 145–154 (1957). — Differential rhizosphere effects of three pea cultivars on physiologic races of *Fusarium oxysporum* Fr. f. pisi (Linf.) Snyder and Hansen. Brit. Mycol. Soc. Trans. **40**, 305–316 (1957).

CAPRIOTTI, A.: Yeasts in some Netherlands soils. A. v. Leeuwenhoek **21**, 145 (1955). — *Torulospora nilssoni* nov. spec., a new yeast from Swedish soils. Arch. Mikrobiol. **28**, 247–254 (1958a). — Neue Hefen, isoliert aus Böden Spaniens. Arch. Mikrobiol. **28**, 338–343 (1958b). — *Debaryomyces castellii* nov. spec. Eine neue Hefe aus Boden Schwedens. Arch. Mikrobiol. **28**, 344–348 (1958c). — *Candida vanriji* nova species, a new yeast from Swedish soils. Arch. Mikrobiol. **30**, 226–230 (1958d). — CARLYLE, R. E., und J. D. THORPE: Some effects of ammonium and sodium 2,4-dichlorophenooxyacetates on legumes and the Rhizobium bacteria. J. Amer. Soc. Agron. **39**, 929–936 (1947); Ref. Z. Pflanzenernähr., Düng., Bodenkde. **44**, 263 (1949). — CHASE, F. E., und G. BAKER: A comparison of microbial activity in an Ontario forest soil under Pine, Hemlock and Naple cover. Canad. J. Microbiol. **1**, 45–54 (1954). — CHEN, H. K., und M. K. SHU: Note on the root nodule bacteria of *Astragalus sinicus* L. Soil Sci. **58**, 291–293 (1944). — CHESTERS, Ch. G. C.: A contribution to the study of fungi in the soil. Trans. Brit. Mycol. Soc. **30**, 100 (1948). — CHESTER, CH. G. C., und D. PARKINSON: On the distribution of fungi in the rhizospheres of oats. Plant a. Soil **11**, 145–156 (1959). — CHOLODNY, N.: Über eine neue Methode zur Untersuchung der Bodenmikroflora. Arch. Mikrobiol. **1**, 620–652 (1930). — CLARK, D. G.: Physiological studies on Rhizobium species. New York (Cornell) Agric. Exper. Stat. Mem. 196 (1936). — CLARK, F. E.: Azotobacter inoculation of crops, III, Recovery of Azotobacter from the rhizosphere. Soil Sci. **65**, 193–202 (1948). — Soil microorganisms and plant roots. Adv. Agronomy **1**, 245 (1949). — CONN, H. J.: The microscopic study of bacteria and fungi in soil. New York Agric. Exper. Stat. (Geneva) Techn. Bull. **64**, 3–20 (1918). — Bacteriological study of a soil type by new methods. Soil Sci. **25**, 263–272 (1928). — CONN, H. J., und J. DIMMICK: Soil bacteria similar in morphology to *Mycobacterium* and *Cornyebacterium*. J. Bact. **54**, 291–303 (1947). — COOK, F. D., und A. G. LOCHHEAD: Growth factor relationships on soil microorganisms as affected by proximity to the plant root. Canad. J. Microbiol. **5** (4), 323–334 (1959). — COUMARTIN, V.: Recherches sur une bactérie des argiles des cavernes et des sédiments ferrugineux. C. R. Acad. Sci. Paris **245**, 1758–1760 (1957). — CURIE, J. H.: Method for the study of Azotobacter and its application to fertility plot soils. Soil Sci. **32**, 9–25 (1931). — CUTLER, D. W.: The action of protozoa on bacteria when inoculated into sterile soil. Ann. Appl. Biol. **10**, 137–141 (1923). — CUTLER, D. W., und CRUMP: Problems in soil microbiology. London: Longmans, Green. 1935.

DAWSON, R. C.: Decomposition of wheat straw by soil fungi. J. Bact. **54**, 35 (1947); Ref. Z. Pflanzenernähr., Düng., Bodenkde. **44**, 234 (1949). — DE, P. K.: The role of blue-green algae in nitrogen fixation in rice fields. Proc. Roy. Soc. Lond. B **127**, 121–139 (1939). — DE, P. K., und N. L. MANDAL: Fixation of nitrogen by algae in rice soils. Soil Sci. **81**, 453–458 (1956). — DELWICHE, C. C., und J. WIJLER: Non symbiontic nitrogen fixation in soil. Plant a. Soil **7**, 113–129 (1956). — DEMETER, J.: Neue Methoden zur mikrobiologischen Untersuchung von Boden. In: ABDERHALDEN, Handbuch der biologischen Arbeitsmethoden, Bd. XII, Teil 2, H. 6, S. 771–887. Berlin-Wien: Urban & Schwarzenberg. 1936. — DEMOLON, A.: Principes d'Agronomie, I, Dynamique du sol. Paris: Dunod, 1960. — DERX, H. G.: *Beijerinckia*, a new genus of nitrogen-fixing bacteria in tropical soils. Meded. Konink. Ned. Akad. Wetens. **53**, 2 (1950). — DE VAY, J. E.: Mutual relationsships in fungi. Ann. Rev. Mikrobiol. **10**, 115–140 (1956). — DÉVAY, M.: Development of rhizobial ineffectibility in the roots of certain *Papilionaceae*. Acta Biol. Suppl. **2**, 28 (Abstr.) (1958); Ref. Zbl. Bakt. II **113**, 289 (1959/60). — DHAR, N. R.: Nitrogen problem. 48. Indian Science Congress, Roorkee 1961. — DIMENNA, M. E.: Isolation of yeasts from Soil. J. Gen. Microbiol. **17**, 678 (1957); Ref. Zbl. Bakt. II **113**, 544 (1959/60). — I. Isolation of yeasts from soil. J. Gen. Microbiol. **17**, 678 (1957). — II. Two new species of yeasts from New Zealand. J. Gen. Microbiol. **18**, 269–272 (1958). — DOBBS, C. G., und W. H. HINSON: A widespread fungistasis in soils. Nature **172**, 197–199 (1953); Ref. Zbl. Bakt. II **108**, 332–333 (1954/55). — DOMSCH, K. H.: Die Kultivierung von Bodenpilzen auf bodenähnlichen Substraten. Arch. Microbiol. **23**, 79–87 (1955). — Das Pilzspektrum einer Bodenprobe. Arch. Microbiol. **35**, 181–196, 229–247, 310–339 (1960). — DUDDIGTON, C. L.: Nematode-destroying fungi in agricultural soils. Nature (Lond.) **173**, 500–501 (1954); Ref. Zbl. Bakt. II **108**, 504 (1954/55).

EBERT, K.: Über die Beeinflussung des Wachstums und der Stickstoffbindung von Azotobakter durch Amid-Ammoniak- und Nitratstickstoff. Diss. Halle. Z. Pflanzenernähr., Düng., Bodenkde. **87** (132), 118–134 (1959). — EGLE, K., und H. MUNDIG: Ausbildung und Funktion von Hämoglobin in den Wurzelknöllchen von Leguminosen. Biol. Zbl. **73**, 577–602 (1954); Ref. Z. Pflanzenernähr., Düng., Bodenkde. **72**, 200 (1956). — ELKAN, G. H.: A nodulation-inhibiting root excretion from a non-nodulating soybean strain. Canad. J. Microbiol. **7**, 851–856 (1961). — ENGHUSEN, H.: Bodenkunde. Mykologische Studie an Stuttgarter Steppenschwarzerden im Vergleich mit Waldböden. Zbl. Bakt. II **109**, 157–177 (1956). — ENO, C. F.: What pesticides do to soils, 2, Insecticides and the soil. Agric. Food Chem. **6** (5), 348 (1958). — ENO, C. F., und W. G. BLUE: The effect of anhydrous ammonia on nitrification and the microbiological population of sandy soils. Soil Sci. Soc. Amer. Proc. **18**, 178–181 (1954). — ENO, C. F., W. G. BLUE und J. M. GOOD: The effect of anhydrous ammonia on nematodes, fungi, bacteria and nitrification in some Florida soils. Soil Sci. Soc. Amer. Proc. **19**, 55–58 (1955). — ENO, C. F., und P. H. EVERETT: Effects of soil applications of 10 chlorinated hydrocarbon insecticides on soil microorganisms and the growth of stringless black Valentine beans. Soil Sci. Soc. Amer. Proc. **22**, 235–238 (1958). — ESELTINE, W. P. VAN: Occurrence of azotobacter in certain western South Carolina soils. Soil Sci. **81**, 29–32 (1956). — ETCHELLS, J. L., R. N. COSTILOW, T. A. BELL und A. L. DEMAIN: Control of molds during the enumeration and isolation of yeasts from soil and plant material. Appl. Microbiol. **2**, 296–300 (1954). — EVANS, E.: Soil recolonization tube for studying recolonization of sterilized soil by micro-organisms. Nature (Lond.) **173**, 1196 (1954); Ref. Zbl. Bakt. II **108**, 501 (1954/55).

FARKASDI, G.: Untersuchungen über die Wirkung der granulierten Phosphordüngemittel auf die Mikroflora des Bodens. Z. Pflanzenernähr., Düng., Bodenkde. **72**, 248–254 (1956). — FEDOROW, M. W.: Biologische Bindung des atmosphärischen Stickstoffs. Dtsch. Übers. von B. FECHER. Berlin: Deutscher Verlag der Wissenschaften. 1960. — Neue Ergebnisse über Ammoniak als Primärprodukt bei der Bindung molekularen Stickstoffs durch Azotobacter. Microbiologija **24**, 626–637 (1955); Ref. Z. Pflanzenernähr., Düng., Bodenkde. **74**, 93 (1956). — FEDOROW, M. W., und T. K. ILJINA: Der stimulierende Einfluß von Hefeautolysat auf die stickstoffbindende Eigenschaft von Actinomyceten des Bodens. Microbiologija **28**, 541–547 (1959); Ref. Zbl. Bakt. II **114**, 221 (1961). — FEDOROW, M. W., und T. K. KUDRJASCHEWA: Stickstoffbindende Aktivität einiger Actinomyceten des Bodens. Ber. Akad. Wiss. UdSSR **108**, 345–348 (1956). — FEDOROW, M. W., und W. F. NEPOMILUJEW: Wesentliche Formen der Rhizosphärenbakterien des Timothegrases (*Phleum pratense*) und die Veränderung ihrer Keimzahl in der Rhizosphäre in Abhängigkeit von den Entwicklungsstufen und dem Lebensalter dieser Pflanzen. Microbiologija **23**, 166–171 (1954a); Ref. Z. Pflanzenernähr., Düng., Bodenkde. **69**, 312 (1955). — Verbreitung und Aktivität der Stickstoffbindung des Azotobacter in der Rhizosphäre mehrjähriger Gräser. Microbiologija **23**, 275–282 (1954b); Ref. Z. Pflanzenernähr., Düng., Bodenkde. **69**, 320 (1955). — Die Grundformen der Rhizosphärenbakterien des Klees und ihr quantitativer Gehalt in der Rhizosphäre nach den Entwicklungsphasen der Pflanze und ihren Lebensjahren. Microbiologija **23**, 431–437 (1954c); Ref. Zbl. Bakt. II **108**, 503 (1954/55). — FEDOROW, M. W., und D. PANTOSCH: Microbiologija **27** (6), 699–704 (1958). — FEHÉR, D.: Untersuchungen über die Mikrobiologie des Waldbodens. Berlin: Springer. 1933a. — Experimentelle Untersuchungen über den Einfluß von Temperatur und Wassergehalt des Bodens auf die Lebenserscheinungen der Bodenbakterien. Arch. Mikrobiol. **4**, 447 (1933b). — FEHÉR, D., und M. FRANK: Experimentelle Untersuchungen über den Einfluß der Temperatur und des Wassergehaltes auf die Tätigkeit des Bodens. Arch. Mikrobiol. **8**, 249–287 (1937). — Untersuchungen über den Einfluß der Temperatur und des Wassergehaltes auf die Tätigkeit der Mikroorganismen des Bodens, II. Die Bestätigung der experimentell abgeleiteten Gesetzmäßigkeiten durch Untersuchung der Wald- und Ackerböden und ihre Übertragung auf den Wärme- und Wasserhaushalt der höheren Pflanzen. Arch. Mikrobiol. **9**, 193 (1938). — FELDMANN, A.: Der quantitative Bakterienbesatz von Dauergrünlandböden und seine Beeinflussung durch organische Düngung. Z. Pflanzenernähr., Düng., Bodenkde. **78**, 54–65 (1957). — FIEDLER, E., und H. J. FIEDLER: Bodenmikrobiologische Untersuchungen an einer Kalkversuchsfläche des Tharandter Waldes. Arch. Forstwesen **10** (7), 733–751 (1961). — FLAIG, W.: Bodenmikroorganismen im Elektronenmikroskop. Landbau-Forschung **1952**, 51–53. — FLAIG, W., und H. KUTZNER: Beitrag zur Ökologie der Gattung *Streptomyces Waksman et Henrici*. Arch. Mikrobiol. **35**, 207–228 (1960). — FLETCHER, D. W., und W. B. BOLLEN: The effects of aldrin on soil microorganisms and some of their activities related to soil fertility. Appl. Microbiol. **2**, 349–354 (1954). — FLETCHER, W., J. W. S. ALCOM und J. C. RAYMOND: Effect of giberellic acid on the nodulation of white clover (*Trif. rep.*). Nature **182**, 1319 (1958); zit. nach GALSTON, A. W.: Nature **183**, 545 (1959). — FOGG, G. E.: Nitrogen fixation by photosynthetic organisms. Ann. Rev. Plant Physiol. **7**, 51–70 (1956); Ref. Z. Pflanzenernähr., Düng., Bodenkde. **75**, 182 (1956) (mit 120 Literaturangaben). — FOTT, B.: Algenkunde VII. Jena: Fischer, 1959; Ref. Zbl. Bakt. II **113**, 614–616 (1959/60). — FRANZ, H.: Bodenzoologie als Grundlage der Bodenpflege. Berlin 1950. — Das biologische Geschehen im Waldboden und seine Beeinflussung durch die Kalkdüngung. Allg. Forstztg. Wien **70**, 178–181 (1959). — FRANZ, H., und W. LOUB: Bodenbiologische Untersuchungen an Walddüngungsversuchen. Zbl. ges. Forstwesen **76**, 129–162 (1959). — FRED, E. B., I. L. BALDWIN und E. McCOY: Root nodule bacteria and leguminous plants. Univ. Wisc. Stud. Sci. **5**, 343 pp. (1932). — FREI, H.: Quantitative Untersuchungen über die Assimilation des elementaren Stickstoffs der Luft durch hautbildende Hefen. Zbl. Bakt. II **104**, 326–365 (1941/42). — FRERCKS, W., D. PUFFE und H. FINDEISEN: Über Keimzahlbestimmungen, Zellulosezersetzung und die Stickstoffwirkung auf den Zersetzungsvorgang in Moorböden und Sandmischkulturen unterschiedlicher Aufkalkung. Z. Pflanzenernähr., Düng., Bodenkde. **89**, 27–42 (1960). — FRITSCH, F. E.: *Cladophorella calcicola nov. gen.* et *sp.*, a terrestrial member of the *Cladophorales*. Ann. Bot. N. S. **8**, 157–171 (1944); Ref. Z. Pflanzenernähr., Düng., Bodenkde. **44**, 257 (1949). — FULLER, W. H., R. E. CAMERON und N. RAICA JR.: Fixation of nitrogen in desert soils by algae. Trans. 7th Int. Congr. Soil Sci. Madison III. Comm. 617–624 (1960).

GALSTON, A. W.: Gibberellins and nodulation. Nature **183**, 545 (1959); Ref. Z. Pflanzenernähr., Düng., Bodenkde. **88**, 69 (1960). — GALSTON, A. W., und Mitarbeiter: Nature **181**. 1082 (1958a); zit. nach GALSTON 1959. — GARDNER, J. C.: Nitrogen fixation in *Elaeagnus* root nodules. Nature **181**, 717–718 (1958); Ref. Z. Pflanzenernähr., Düng., Bodenkde. **82**, 211 (1958). — GARRETT, S. D.: Biology of root-infecting fungi. Cambridge: University Press. 1956; zit. nach MARTIN und ERVIN 1958. — GAERTNER, A.: Über das Vorkommen niederer Erdphycomyceten in Afrika. Schweden und an einigen mitteleuropäischen Standorten. Arch. Mikrobiol. **21**, 4–56 (1954). — Über zwei ungewöhnliche keratinophile Organismen aus Ackerböden. Arch. Mikrobiol. **23**, 28–37 (1955). — GAUGHAN, E. R. L.: The thermophile microorganisms. Bact. Rev. **11**, 189–225 (1947). — GAUSE, G. F., T. P. PREOBRASHENSKAJA, E. S. KUDRINA, N. O. BLINOW, J. D. RJABOWA und M. A. SWESCHNIKOWA: Zur Klassifizierung der Actinomyceten. Dtsch. Übers. von B. FECHER. Jena: Fischer. 1958; Ref. Zbl. Bakt. II **113**, 612 (1959/60). — GEORDETTE, R.: Aperçu de travaux récents consacrés à la fixation symbiotique d'azote chez les légumineuses. Ann. Gembloux **59**, 215–236 (1953). — GERRETSEN, F. C., und H. DE HOOP: Boron, an essential micro-element for *Azotobacter chroococcum*. Plant a. Soil **5**, 349–367 (1954). — GILMAN, J. C.: Manual of soil fungi, 2dn ed. Ames, Iowa: Iowa State College Press. 1957. — GLATHE, H.: Die Bedeutung der Phosphorsäure für die Bodenmikroben. Phosphorsäure **13**, 190–199 (1953). — GLATHE, H., und E. AHRENS: Die Bestimmung des Azotobactergehalts in azotobacterhaltigen Substanzen. Landwirtsch. Forsch. Im Druck (1963). — GLATHE, H., C. VON BERNSTORFF und A. ARNOLD: Lebensgemeinschaft von Mikroorganismen und höheren Pflanzen im Bereich der Rhizosphäre. Zbl. Bakt. II **107**, 481–488 (1952/54). — GOLEBIOWSKA, J.: The effect of the carbon source in the cultures of *Rh. trifolii* and *Rh. meliloti* upon the nitrogen fixation in the inoculated plants. Act. Microbiol. Polon. **1**, 327–337 (1952); Ref. Zbl. Bakt. II **108**, 105 (1954/55). — GOLOWATSCHEWA, R. S.: Der Einfluß von *Clostridium pasteurianum* auf den Ertrag der landw. Kulturpflanzen. Nachr. Akad. Wiss. Estisch. SSR, Biol. Ser. H. **1**, 29–37 (1957); Ref. Z. Pflanzenernähr., Düng., Bodenkde. **86**, 263 (1959). — GRAFF, O.: Die Regenwürmer Deutschlands. Bilderatlas für Bauern, Gärtner, Forstwirte und Bodenkundler. Schriftenreihe der Forsch.-Anst. f. Landwirtsch. Braunschweig-Völkenrode (1953). — GRAY, P. H. H.: Effects of benzenehexachloride on soil microorganisms. Appl. Microbiol. **2**, 37–40 (1954). — GRAY, P. H. H., und H. G. H. THORNTON: The estimation of bacterial numbers in soil by direct count from stained films. Nature **122**, 400–401 (1928). — GREEN, M., und P. W. WILSON: The utilization of nitrate nitrogen by *Azotobacter*. J. Gen. Microbiol. **9**, 89–96 (1953). — GROMYKO, JE. P.: Die Ökologie sowie einige morphologische und physiologische Besonderheiten von *Bac. megaterium*. Microbiologija **24**, 565–572 (1955); Ref. Z. Pflanzenernähr., Düng., Bodenkde. **74**, 92 (1956). — GY, L.: Unterschiede in der N-bindenden Aktivität und Virulenz von Erbsen- und Wicken-Rhizobien, bedingt durch das Entwicklungsstadium der Wirtspflanze. Agrokémia és Talajtan **6**, 177–188 (1957) (ungar.); Ref. Z. Pflanzenernähr., Düng., Bodenkde. **81**, 182 (1958). — GYLLENBERG, H.: The "rhizosphere effect" of graminaceous plants in virgin soils, II. Nutritional characteristics of non-sporogenous bacteria associated with the roots. Physiol. Plant. **9**, 110–129 (1956); Ref. Z. Pflanzenernähr., Düng., Bodenkde. **75**, 184 (1956). — GYLLENBERG, H., und P. HANIOJA: The "rhizosphere-effect" of gramineous plants in virgin soils, III. Comparison with the effect of other plants. Physiol. Plant. **9**, 441–445 (1956); Ref. Z. Pflanzenernähr., Düng., Bodenkde. **76**, 271 (1957). — GYURKO, P.: Die antagonistische Wirkung der Rhizosphären-Bakterien der Zuckerrübe auf die Bodenbakterien. Acta Agron. Hung. **9**, 175–184 (1959); Ref. Z. Pflanzenernähr., Düng., Bodenkde. **89**, 231 (1960).

HABER, W.: Zur Ökologie des Bodenlebens in verschiedenen Pflanzengesellschaften. Ber. Dtsch. Bot. Ges. **71**, 399–410 (1959). — HACK, J. E.: The effect of spore germination and development on plate counts of fungi in soil. J. Gen. Microbiol. **17**, 625–630 (1957). — HALL, A. D.: On the Accumulation of Fertility by Land allowed to run wild. J. Agricult. Sci. **1**, 241 (1905/06). — HALLSWORTH, E. G., S. B. WILSON und E. A. N. GREENWOOD: Copper and cobalt in nitrogen fixation. Nature (Lond.) **187**, 79–80 (1960). — HANKE-PACEWICZOWA, T.: Selection of Rhizobium suitable for inoculation of *Galega officinalis*. Act. Microbiol. Polon. **1**, 36–38 (1952) (Poln. m. russ. u. engl. Zusammenfassung); Ref. Zbl. Bakt. II **108**, 104 (1954/55). — HARDER, R., und E. ÜBELMESSER: Notiz zur Frage des Vorkommens von Chytridineen und anderen Pilzen in tiefen Erdschichten. Arch. Mikrobiol. **26**, 353–357 (1957). — Über die Beeinflussung niederer Erdphycomyceten durch Flechten. Arch. Mikrobiol. **31**, 82–86 (1958). — HARDER, R., und E. GALLWITZ-ÜBELMESSER: Über niedere Erdphycomyceten Australiens. Arch. Mikrobiol. **32**, 115–126 (1959). — HARLEY, L. J., und J. S. WAID: A method of studying active mycelia on living roots and other surfaces in the soil. Trans. Brit. Mycol. Soc. **38**, 104–118 (1955); Ref. Z. Pflanzenernähr., Düng,, Bodenkde. **75**, 184 (1956). — HARRIS, J. R.: Influence of rhizosphere microorganisms on the virulence of *Rhizobium trifolii*. Nature **172**, 507–508 (1953); Ref. Zbl. Bakt. II **108**, 499 (1954/55). — HESS, E.: Effects of low temperatures on the growth of marine bacteria. Contrib. Canad. Biol. and Fisheries Ser. C 8, 491–505 (1934). — HEUMANN, W.: Über Wesen und Bedeutung der Bakteroiden in den Wurzelknöllchen der Erbse. Naturwiss. **39**, 66–67, 239, 287 (1952). — HILDEBRANDT, A. A., und P. M. WEST: Strawberry root rot in relation to microbiological changes induced in root rot soil by the incorporation of certain cover crops. Canad. J. Res. **19**, 183–198 (1941). — HIRTE, W.: Glycerin-Pepton-Agar, ein vorteilhafter Nährboden für bodenbakteriologische Arbeiten. Zbl. Bakt. II **114**, 141–146 (1961). — Vergleichende mikrobiologische Untersuchungen an unterschiedlich rieselmüden Böden der Berliner Rieselfelder, I. Die Mikroorganismentätigkeit (Mikroorganismentätigkeit und biochemische Aktivität). Zbl. Bakt. II **114**, 367–387 (1961a). — Vergleichende mikrobiologische Untersuchungen an rieselmüden und gesunden Böden der Berliner Rieselfelder, II. Qualitative Untersuchungen. Zbl. Bakt. II **114**, 490–519 (1961b). — Mikrobiologische Untersuchungen zum Problem der Rieselmüdigkeit. Arch. Gartenbau **9**, 213–232 (1961c). — HOFER, A. W.: Selective action of fungicides on rhizobium. Soil Sci. **86**, 282–286 (1958). — HOFFMANN, C.: A contribution to the subject of the factors concerned in soil productivity. Univ. Kansas Sci. Bull. **9**, 81–89 (1914). — HOFFMANN, C., und B. W. HAMMER: Some factors concerned in the fixation of nitrogen by Azotobacter. Zbl. Bakt. II **28**, 127–139 (1910). — HOLDING, A. J.: The properties and classification of the predominant gram-negative bacteria occurring in soil. J. Appl. Bact. **23**, 3, 515–525 (1960). — HOLDING, A. J., S. N. TILO und O. N. ALLEN: Modified plant responses induced by Rhizobia cultivated on amino acid media. 7th Int. Congr. Soil Sci., Madison, Wisc., USA **1960**, Trans. II, 608-616. — HOLZWEISSIG: Arbeitsvorschrift. Radicin-Institut, Westerade/Holstein. — HOMRIGHAUSEN, E., und W. ROHMER: Untersuchungen verschiedener Dispergierungsverfahren bei der Vorbehandlung zur fluorescenzmikroskopischen Keimzählung in Böden. Landwirtsch. Forsch. **12**, 253–260 (1958). — HOPF, M.: Untersuchungen über die natürliche Mikroflora des Bodens. Arch. Mikrobiol. **14**, 661-677 (1950). — HUCKER, G. J.: Low temperature organisms in frozen vegetables. Food Technol. **8**, 79–82 (1954).

IKE, A. F., und E. L. STONE: Soil nitrogen accumulation under Black Locust (*Robinia pseudacacia* L.). Soil Sci. Soc. Amer. Proc. **22**, 346–349 (1958). — INGRAHAM, J. L., und J. L. STOKES: Psychrophilic bacteria. Bact. Rev. **23**, 97–108 (1959). — ISHIZAWA, S., T. SUZUKI, T. KODA und O. SATO: Studies on microorganisms and their activities in soil. Bull. Nat. Inst. Agr. Sci. (Japan) Ser. B, Soils 8, 1–211 (1958). — IVARSON, K. C., und D. PRAMER: The persistance and biological effects of surface active agents in soil. Soil Sci. Amer. Proc. **20**, 371–374 (1956); Ref. Z. Pflanzenernähr., Düng., Bodenkde. **76**, 90 (1957). — IVARSON, K. C., und H. KATZNELSON: Studies on the rhizosphere microflora of yellow birch seedlings. Plant a. Soil **12**, 30–40 (1960). — IWANOWSKAJA, T. L., und M. M. FEDOROWA: Die Bildung von Knöllchen an den Wurzeln der Soja im Vegetationsversuch. Agribiologija **6** (102), 29–34 (1956) (russ.); Ref. Z. Pflanzenernähr., Düng., Bodenkde. **78**, 263 (1957).

JACKS, G. V., W. D. BRIND und R. SMITH: Mulching. Commonwealth Bureau of Soil Sci. Techn. Comm. Nr. 49, 1955; Commonwealth Agric. Bur. Farnham Royal, Bucks., England; Ref. Z. Pflanzenernähr., Düng., Bodenkde. **74**, 250 (1956). — JACKSON, R. M.: Fungistasis as a factor in the rhizosphere phenomenon. Nature (Lond.) **180**, 96–97 (1957). — JAGNOW, G.: Untersuchungen über die Verbreitung von Streptomyzeten in Naturböden. Arch. Mikrobiol. **25**, 274–296 (1956). — Untersuchungen über Keimzahl und biologische Aktivität von Wiesenböden. Z. Pflanzenernähr., Düng., Bodenkde. **82**, 50–67 (1958). — Bodenmikroskopische Untersuchungen der engeren Rhizosphäre einiger Grünlandpflanzen auf Wiesenstandorten. Zbl. Bakt. II **114**, 475–489 (1961). — JANKE, A.: Die Wattehaube. Ein Verschluß für Kulturgefäße zur Züchtung streng aerober Mikroorganismen. Arch. Mikrobiol. **17**, 155–159 (1952). — JANSSON, S. L., M. Y. HALLAM und W. V. BARTHOLOMEW: Preferential utilization of ammonium over nitrate by microorganisms in the decomposition of oat straw. Plant a. Soil **6**, 382–390 (1955). — JEFFERYS, E. G.: The stability of antibiotics in soil. J. Gen. Microbiol. **7**, 295–312 (1952). — JENNY, H.: Comparison of soil nitrogen and carbon in tropical and temperate regions, as observed in India and the Americas. Missouri Agric. Exper. Stat. Res. Bull. **765** (1961). — JENSEN, H. L.: Contributions to the microbiology of Australian soils, III. Proc. Linn. Soc. N.S.W. **60**, 145–154 (1935). — Nitrogen fixation in leguminous plants, VII. The nitrogen-fixing activity in root nodule tissue in *Medicago* and *Trifolium*. Proc. Linn. Soc. N.S.W. **72**, 265–291 (1948). — A survey of biological nitrogen fixation in relation to the world supply of nitrogen. Trans. Fourth Int. Congr. Soil Sci. **1**, 165–172 (1950a). — The occurrence of *Azotobacter* in cultivated soils of Denmark. Tidskr. Planteavl. **53**, 622–649 (1950b) (dän. m. engl. Zusammenfassung). — The magnesium requirements of *Azotobacter* and *Beijerinckia* with some additional notes on the latter genus. Acta Agricult. Scand. **4**, 223–236 (1954a). — The *Azotobacteriaceae*. Bact. Rev. **18**, 195–214 (1954) (mit 136 Literaturangaben). — Die Mikrobiologie des Bodens. In: RUHLAND, Handbuch der Pflanzenphysiologie, Bd. 11, S. 707–751 (1959). — JENSEN, H. L., P. K. DE und L. RHATTACHARYA: A new nitrogen-fixing bacterium. Nature (Lond.) **184**, 1743 (1959); Ref. Zbl. Bakt. II **114**, 221 (1961). — JENSEN, V.: Nitrogen fixation by *Aerobacter aerogenes*. Physiol. Plant. **9**, 130–136 (1956); Ref. Z. Pflanzenernähr., Düng., Bodenkde. **75**, 183 (1956). — A new nitrogen fixing bacterium from a Danish watercourse. Arch. Mikrobiol. **29**, 348–353 (1958). — JOHN, R. P.: An ecological and taxonomic study of the Algae of British soils, I. Distribution of the surface growing Algae. Ann. Bot. N.S. **6**, 323–349 (1942); Ref. Z. Pflanzenernähr., Düng., Bodenkde. **44**, 264 (1949). — JOHNSON, E. J., und A. R. COLMER: The effect of herbicides on soil microorganisms, I. The effect of 2,4-dichlorophenoxyacetic acid on some phases of the nitrogen metabolism of *Bacillus cereus*. Appl. Mikrobiol. **3**, 123–126 (1955a). — The effect of herbicides on soil microorganisms, II. The effect of 2,4-dichlorophenoxyacetic acid on some phases of the nitrogen metabolism of *Pseudomonas fluorescens* and the microorganisms of a soil suspension. Appl. Microbiol. **3**, 126–128 (1955b). — JONES, L. W.: Stability of DDT and its effect on microbial activities. Soil Sci. **73**, 237–242 (1952); Ref. Zbl. Bakt. II **107**, 115 (1952/54). — JONES, P. T. C., J. E. MOLLISON und M. H. QUENOUILLE: A technique for the quantitative estimation of soil microorganisms. J. Gen. Microbiol. **2**, 54–69 (1948); Ref. Soils a. Fertilizers **11**, 1757 (1948). — JORDAN, D. C.: Studies on the root nodule bacteria, II, The production and behavior of colonial mutants, produced by X-ray irradiation. Canad. J. Bot. **30**, 125–130 (1952). — JUSHINA, Z. J.: Die Lebensfähigkeit des Azotobacters in den in Kultur stehenden und den brachliegenden Böden der Halbinsel Kola. Mikrobiologija **118**, 201–205 (1958) (russ.); Ref. Z. Pflanzenernähr., Düng., Bodenkde. **85**, 182 (1959).

KAILA, A.: Some observations on the occurrence of Azotobacter in Finnish soils. J. Sci. Agric. Soc. Finnland **26**, 40–49 (1954). — KANDLER, O.: Papierchromatographischer Nachweis der Aminosäureausscheidungen in vitro kultivierter Maiswurzeln. Z. Naturforsch. **6b**, 437–445 (1951). — KÄPPEL, M., und H. WARTENBERG: Der Formenwechsel des *Actinomyces alni Peklo* in den Wurzeln von *Alnus glutinosa Gaertner*. Arch. Mikrobiol. **30**, 46–63 (1958). — KATZNELSON, H.: Studies with the bacteriophage of rhizobia. Sci. Agricult. **22**, 4, 209–215 (1941). — The rhizosphere effect of mangels on certain groups of soil microorganisms. Soil Sci. **62**, 343–354 (1946). — KATZNELSON, H., und F. E. CHASE: Qualitative studies of soil microorganisms, VI. Influence of season and treatment on incidence of nutritional groups of bacteria. Soil Sci. **58**, 473–479 (1944). — KATZNELSON, H. A., A. G. LOCHHEAD und M. J. TIMONIN: Soil microorganisms and the rhizosphere. Bot. Rev. Lancaster **14**, 543–587 (1948). — KATZNELSON, H., J. W. ROUATT und T. M. B. PAYNE: The liberation of amino acids and reducing compounds by plant roots. Plant a. Soil **7**, 35–48 (1955). — KATZNELSON, R. S., und W. W. JERSCHOW: Untersuchung der Mikroflora der brachliegenden und der in Kultur stehenden Böden der Karelischen ASSR. Mikrobiologie **26**, 468–476 (1957) (russ.); Ref. Z. Pflanzenernähr., Düng., Bodenkde. **80**, 272 (1958). — KENDRICK, J. B., und A. R. JACKSON: Factors influencing the isolation of certain soil-borne plant pathogens from soil. Phytopathology **48**, 394 (1958). — KERR, A.: The use of cellophane in growth studies on soil fungi. Trans. Brit. Mycol. Soc. **41**, 14–16 (1958). — KESSLER, E.: Role of photochemical processes in the reduction of nitrate by green algae. Nature **176**, 1069–1070 (1955); Ref. Z. Pflanzenernähr., Düng., Bodenkde. **73**, 174 (1956). — KICK, H.: Beobachtungen über den Abbau organischer Masse im Boden unter dem Einfluß der Kalkung. Z. Pflanzenernähr., Düng., Bodenkde. **74**, 1–10 (1956). — KIRSSANINA, E. F., und W. A. WOLKOWA: Einige Daten über die Verbreitung des Azotobacters in den Böden des Gebirgs-Altai-Gebietes. Mikrobiologie **29**, 551–554 (1960); Ref. Z. Pflanzenernähr., Düng., Bodenkde. **93**, 91 (1961). — KLECZKOWSKA, J.: The production of plaques by Rhizobium Bacteriophage in poured plates and its values as a counting method. J. Bact. **50**, 71–79 (1945). — A study of phage resistant mutants of *Rhizobium trifolii*. J. Gen. Microbiol. **4**, 298–310 (1950). — KLYKOW, A. P.: Das Auftreten von Knöllchenbakterien im Zusammenhang mit der Akklimatisation von Soja in der Nicht-Tschernosem-Zone. Agrobiologija **6** (102), 35–40 (1956) (russ.); Ref. Z. Pflanzenernähr., Düng., Bodenkde. **78**, 263 (1957). — KNAPP, R., und H. LIETH: Über Ursachen des verstärkten Auftretens von erdbewohnenden Cyanophyceen. Arch. Mikrobiol. **17**, 292–299 (1952). — KNÖSEL, D.: Über die Wirkung aus Pflanzenresten freiwerdender, phenolischer Substanzen auf Mikroorganismen des Bodens, I. Der Einfluß von p-Oxybenzoesäure auf die Entwicklung von Pilzen, Actinomyceten und Bakterien. Z. Pflanzenernähr., Düng., Bodenkde. **80**, 225–237 (1958). — Über die Wirkung aus Pflanzenresten freiwerdender, phenolischer Substanzen auf

Mikroorganismen des Bodens, II. Versuche mit p-Cumarsäure, Ferulasäure und Vanillinsäure. Z. Pflanzenernähr., Düng., Bodenkde. **85**, 58–66 (1959). — KOCH, A., und C. HOFFMANN: Über die Verschiedenheit der Temperaturansprüche thermophiler Bakterien im Boden und in künstlichen Nährsubstraten. Zbl. Bakt. II **31**, 433–436 (1911). — KOCH, R.: Zur Untersuchung von pathogenen Organismen. Mitt. Kaiserl. Gesundheitsamts **1**, 34–36 (1881). — KOFFMANN, M.: Die Mikrofauna des Bodens, ihr Verhältnis zu anderen Mikroorganismen und ihre Rolle bei den mikrobiologischen Vorgängen im Boden. Arch. Mikrobiol. **5**, 246–302 (1934). — KOLKER, I. I., und E. A. DACHNOWA: Die Verbreitung von Azotobacter in den Böden der Krim. Mikrobiologija **29**, 555–562 (1960); Ref. Z. Pflanzenernähr., Düng., Bodenkde. **93**, 91 (1961). — KOEPF, H.: Zur Dynamik des Abbaus organischer Substanzen in verschiedenen Böden. Z. Pflanzenernähr., Düng., Bodenkde. **73**, 48–59 (1956). — KORJENJAKO, A. I., O. J. ARTAMONOWA und S. W. LJETUNOWA: Bildung und Beständigkeit antibiotischer Stoffe der Actinomyceten im Boden. Mikrobiologija **24**, 550–557 (1955); Ref. Z. Pflanzenernähr., Düng., Bodenkde. **74**, 93 (1956). — KOSCHEL'KOW, P. N., U. G. OKSSENT'JAN, Z. M. OSSIPOWA und D. W. CHARKOW: Die Bedeutung von Stallmist und mineralischen Düngern für die Erhöhung der Fruchtbarkeit der Grasnarbe-Podsolböden. Bodenkunde **6**, 91–98 (1958) (russ.); Ref. Z. Pflanzenernähr., Düng., Bodenkde. **87**, 165 (1959). — KOSCHEL'KOW, P. N., und Z. M. OSSIPOWA: Die Rolle des Stallmists bei der Verbesserung schwerer podsoliger Böden. Semledjelije, **1956**; 76–79; Ref. Z. Pflanzenernähr., Düng., Bodenkde. **76**, 168 (1957). — KOSLOWA, J. I.: Untersuchungen des Gattungs- u. Artbestandes der Mikroflora in der Eichenrhizosphäre. Mikrobiologija **24**, 558–564 (1955); Ref. Z. Pflanzenernähr., Düng., Bodenkde. **74**, 91 (1956). — KOX, E.: Der durch Pilze und aerobe Bakterien veranlaßte Pektin- und Cellulose-Abbau im Hochmoor unter besonderer Berücksichtigung des Sphagnum-Abbaues. Arch. Mikrobiol. **20**, 111–140 (1954); Ref. Z. Pflanzenernähr., Düng., Bodenkde. **66**, 271 (1954). — KRASSILNIKOV, N. A.: Le principe écologico-taxonomique dans l'étude de la microflore du sol. Sympos. sur les méthodes d'étude microb. du sol, S. 129–137. Soc. Belge de Pédologie Louvain **1957**a. — On the importance of soil microorganisms in plant nutrition (contributions by Soviet microbiologists in 40 years). Mikrobiologija **26**, 637–650 (1957). — Diagnostik der Bakterien und Aktinomyzeten. Jena: Fischer. 1959a. — Über die Bedeutsamkeit antibiotischer Stoffe bei Actinomyceten für die Artbestimmung. Mikrobiologija **28**, 179–186 (1959) (russ. m. engl. Zusammenfassung); Ref. Zbl. Bakt. II **113**, 628 (1959/60). — The biological role of microbes-antagonists, producers of antibiotic substances. Soil a. Plant Food **5**, 184–193 (1960); Ref. Z. Pflanzenernähr., Düng., Bodenkde. **93**, 91 (1961). — KREHL-NIEFFER, R.: Verbreitung und Physiologie mikroskopischer Bodenpilze. Arch. Mikrobiol. **15**, 389 (1951). — KRJUCHKOWA, A. P., und E. P. POPOWA: Zur Untersuchung der ökologischen Variabilität der Mikroorganismen in bezug auf Bodendiagnostik. Mikrobiologia **3**, 232 (1934), **4**, 593 (1935). — KÜHLMORGEN-HILLE, G.: Vergleichende Prüfung der Methoden zur Ermittlung der Keimzahl im Boden. Zbl. Bakt. II **74**, 497–519 (1928). — KÜHLWEIN, H.: Zur Systematik und Verbreitung der Myxobakterien. Zbl. Bakt. II **113**, 480–490 (1960). — KÜHNELT, W.: Bodenbiologie. Mit besonderer Berücksichtigung der Tierwelt. Berlin: Herold. 1950. — KURON, H., H. GLATHE und E. HOMRIGHAUSEN: Bemerkungen zur Durchführung der fluoreszenzmikroskopischen Methode nach Strugger. Zbl. Bakt. II **112**, 204–213 (1959). — KURON, H., E. HOMRIGHAUSEN und W. ROHMER: Fluoreszenzmikroskopische Untersuchungen über die Wirkung der Flächenerosion auf das Bakterienleben in Lößböden. Z. Pflanzenernähr., Düng., Bodenkde. **85**, 44–50 (1959). — KUZNETZOW, W. D.: Aktinomyceten in einigen Böden Pamirs und ihre antagonistischen Eigenschaften. Mikrobiologija **29**, 563–570 (1960); Ref. Z. Pflanzenernähr., Düng., Bodenkde. **93**, 91 (1961). — KVASSNIKOW, B. V., S. T. DOLGIKH und O. N. LEBEDIANTSEVA: Carbohydrate and nitrogen content of varieties of vegetables pea and haricot having different rates of root-nodule formation. Mikrobiologija **27** (5), 587–592 (1958).

LA TOUCHE, C. J.: Slide traps for soil fungi. Trans. Brit. Mycol. Soc. **31**, 281 (1948). — LEGGE, B. J.: Use of glas fibre material in soil mycology. Nature (Lond.) **169**, 759–760 (1952). — LEHNER, A., und W. NOWAK: Ergebnisse zweijähriger bakteriologischer Beobachtungen an Grünlandpflanzen unter Berücksichtigung einiger Abwasserverregnungsgebiete Bayerns. Bayr. Landwirtsch. Jahrb. **33**, 211–223 (1956). — Bakteriologische Bodenuntersuchungen mit Hilfe von Acridinorange nach Strugger. Zbl. Bakt. II **110**, 349–360 (1957). — Eine Schnellmethode zur praktischen Feststellung der Kleinlebewelt des Bodens. Prakt. Bl. Pflanzenbau, Pflanzenschutz **1958**, 149–152; Ref. Zbl. Bakt. II **113**, 163 (1959/60). — Neuere Ergebnisse des Direktnachweises von Bakterien im Boden mittels der kombinierten Aufwuchs-Fluorochromierungsmethode Zbl. Bakt. **113**, 32–34 (1959/60). — LEHNER, A., W. NOWAK und L. SEIBOLD: Eine

Weiterentwicklung des Boden-Fluorochromierungs-Verfahrens mit Acridinorange zur Kombinationsmethode. Landwirtsch. Forsch. **11**, 121–127 (1958). — LEMOIGNE, M., H. GIRARD und G. JACOBELLI: Travaux bactériens écologics. Ann. Agron. **2**, 173–198 (1951); Ref. Z. Pflanzenernähr., Düng., Bodenkde. **66**, 277 (1954). — LENHARD, G.: Die Stickstoffbindung von *Azotobacter chroococcum* in Gegenwart von Ammonium- u. Nitratstickstoff. Z. Pflanzenernähr., Düng., Bodenkde. **63**, 241–247 (1953). — LILLY, V. G., und L. H. LEONIAN: Vitamin B_1 in soil. Science **89**, 292 (1939). — LINDSTROM, E. S., J. W. NEWTON und P. W. WILSON: The relationship between photosynthesis and nitrogen fixation. Proc. Nat. Acad. Sci. U.S.A. **38**, 392–396 (1952); Ref. Zbl. Bakt II **107**, 196 (1952/54). — LINFORD, M. B.: Methods of observing soil flora and fauna associated with roots. Soil Sci. **53**, 93–103 (1942). — LINNEMANN, G.: Einige neue erdbewohnende *Mortierella*-Arten. Zbl. Bakt. II, **107**, 225–230 (1952/54). — Untersuchungen zur Verbreitung und Systematik der Mortierellen. Arch. Mikrobiol. **30**, 256–267 (1958). — LOCHHEAD, A. G.: The bacterial types in frozen soil. Soil Sci. **21**, 225–231 (1926). — Qualitative studies of soil microorganisms, III. Influence of plant growth on the character of the bacterial flora. Canad. J. Res. **18**, 42–53 (1940). — Qualitative studies of soil microorganisms, XV. Capability of the predominant bacterial flora for synthesis of various growth factors. Soil Sci. **84**, 395–403 (1957). — Soil bacteria and growth-promoting substances. Bact. Rev. **22**, 145–153 (1958a). — Two new species of *Arthrobacter* requiring respectively Vitamin B_{12} and the Terregens Factor. Arch. Mikrobiol. **31**, 163–170 (1958b). — LOCHHEAD, A. G., und M. O. BURTON: Incidence in soil of bacteria requiring Vitamin B_{12} and the Terregens Factor. Soil Sci. **82**, 237–245 (1956). — Qualitative studies of soil microorganisms, XIV. Specific vitamin requirements of the predominant bacterial flora. Canad. J. Microbiol. **3**, 35–42 (1957). — LOCHHEAD, A. G., M. O. BURTON und R. H. THEXTON: A bacterial growth factor synthesized by a soil bacterium. Nature **170**, 282 (1952). — LOCHHEAD, A. G., und F. E. CHASE: Bacterial growth factors in soil. Sci. **96**, 387 (1942). — Qualitative studies of soil microorganisms, V. Nutritional requirements of the predominant bacterial flora. Soil Sci. **55**, 185–195 (1943). — LOCHHEAD, A. G., und E. D. COOK: Microbial growth factors in relation to resistance of flax varieties to *Fusarium* wilt. Canad. J. Bot. **39**, 7–19 (1961). — LOCHHEAD, A. G., und J. W. ROUATT: The "rhizosphere effect" on the nutritional groups of soil bacteria. Soil Sci. Soc. Amer. Proc. **19**, 48–49 (1955). — LOCHHEAD, A. G., und R. H. THEXTON: Qualitative studies of soil microorganisms, VII. The rhizosphere effect in relation to the amino acid nutrition of bacteria. Canad. J. Res. C **25**, 20–26 (1947). — LOCHHEAD, A. G., M. J. TIMONIN und P. M. WEST: The microflora of the rhizosphere in relation to resistance of plants to soil-born pathogens. Sci. Agric. **20**, 414–418 (1940). — LÖHNIS, F.: Handbuch der landwirtschaftlichen Bakteriologie, 1. Aufl. Berlin: Borntraeger. 1910. — LÖHNIS, F., und H. GLATHE: Handbuch der landwirtschaftlichen Bakteriologie, 2. Aufl. Berlin: Borntraeger. 1935. — LÖHNIS, M. T.: Sind hautbildende Hefen befähigt, elementaren Stickstoff zu assimilieren? A. v. Leeuwenhoek **9**, 133 (1943). — LORENZEN, H.: pH-Wert und Kohlenstoffversorgung bei Wachstum und Photosynthese von Grünalgen. Flora **146**, 94–108 (1958); Ref. Z. Pflanzenernähr., Düng., Bodenkde. **85**, 264 (1959). — LOUB, W.: Walddüngungsmaßnahmen im Lichte der Bodenmikrobiologie. Forstz. **14**, 369–391 (1959); Ref. Z. Pflanzenernähr., Düng., Bodenkde. **90**, 186 (1960). — Die mikrobiologische Charakterisierung von Bodentypen. Bodenkultur (Wien) Ausg. A **11**, 38–70 (1960); Ref. Landwirtsch. Zbl. II, Pfl. Prod. **6**, 845 (1961). — LOUW, H. A., und D. M. WEBLEY: The bacteriology of the root region of the oat plant grown under controlled plot culture conditions. J. Appl. Bact. **22**, 216 (1959). — LU, K. C., und W. B. BOLLEN: Effect of gibrel, a potassium salt of gibberellic acid, on microbial activities in soil. Plant a. Soil **9**, 318–324 (1957/58). — LU, K. C., C. M. GILMOUR, A. C. ZAGALLO und W. B. BOLLEN: Effects of giberellic acid on soil microorganisms. Nature **181**, 189–190 (1958); Ref. Z. Pflanzenernähr., Düng., Bodenkde. **82**, 255 (1958). — LUND, J. W. G.: Observations on soil Algae, II. Notes on groups other than Diatoms. New Phytologist **46**, 35–60 (1947); Ref. Z. Pflanzenernähr., Düng., Bodenkde. **44**, 255 (1949). — LYON, T. L.: Cornell Agric. Exper. Stat. Bull. **645** (1936).

MACCONNELL, J. T.: The oxygen factor in the development and function of the root nodules of Alder. Ann. Bot. N.S. **23**, 261–268 (1959); Ref. Z. Pflanzenernähr., Düng., Bodenkde. **89**, 69 (1960). — MACCONNELL, J. T., und G. BOND: Nitrogen fixation in wild legumes. Ann. Bot. N.S. **21**, 185–192 (1957a); Ref. Z. Pflanzenernähr., Düng., Bodenkde. **77**, 258 (1957). — A comparison of the effect of combined nitrogen on nodulation in non-legumes and legumes. Plant a. Soil 8, 378–388 (1957b). — MACURA, J.: The rhizosphere microflora of sugar beet. Folia Microbiol. **4**, 3, 129–136 (1958a). — Seed and soil bacteria in relation to the rhizosphere effect. Folia Microbiol. 4, 3, 274–280 (1958b). — Bacterial flora of the root surface of wheat grown in solutions deficient in nitrogen and phosphorus. Folia Microbiol. 6, 279–281 (1961). — MACURA, J., und F. KUNC: Continuous flow method in soil microbiology, II. Observations on glucose metabolism. Folia Microbiol. 6, 398–407 (1961). — MACURA, J., und L. PAVEL: The influence of Montmorillonite on nitrogen fixation by *Azotobacter*. Folia Microbiol. 4, 82–90 (1959). — MAGEE, W. E., und R. H. BURRIS: Fixation of $^{15}N_2$ by excised nodules. Plant Physiol. 29, 199–200 (1954); Ref. Z. Pflanzenernähr., Düng., Bodenkde. 72, 93 (1956). — MAGEE, L. A., und A. R. COLMER: The effect of herbicides on soil microorganisms, II. The effect of some herbicides on the respiration of *Azotobacter*. Appl. Microbiol. 3, 288–292 (1955). — MALISZEWSKA, W.: The microbial characteristics of soils on the farms of the University Mariae Curie- Sklodowska in Lublin. Ann. Univ. Mariae Curie-Sklodowska (E) 4, 37–76 (1949). — MALJANOW, A. P., und T. W. SARALIDSE: Einfluß der Waldstreifen auf die Struktur der dunklen Kastanienböden im Transwolgagebiet. Potschwowedenie Nr. 3, 32–39 (1957) (russ).; Ref. Z. Pflanzenernähr., Düng., Bodenkde. 80, 76 (1958). — MANZON, V. D., und A. V. DZECINA: Die Diagnostik des Aktivitätszustandes der Bodenmikroflora. Mikrobiologija 23, 455–461 (1954); Ref. Zbl. Bakt. II 108, 503 (1954/55). — MARTIN, J. P., und D. G. ALDRICH: Effect of various exchangeable cation ratios on kinds of fungi developing during decomposition of organic residues in soil. Soil Sci. Soc. Amer. Proc. 18, 160–164 (1954). — MARTIN, J. P., R. C. BAINES und J. O. ERVIN: Influence of soil fumigation for citrus replants on the fungus population of the soil. Soil Sci. Soc. Amer. Proc. 21, 163–166 (1957). — MARTIN, J. P., und J. O. ERVIN: Changes in fungus population of California old citrus orchard soils when cropped to orange seedlings in the greenhouse. Soil Sci. 86, 152–155 (1958). — MARTIN, J. P., R. B. HARDING, G. H. CANELL und L. D. ANDERSON: Influence of five annual field applications of organic insecticides on soil biological and physical properties. Soil Sci. 87, 334–338 (1959). — MARTIN, W. P., R. H. WALKER und P. E. BROWN: The occurrence of *Azotobacter* in Iowa soils and factors affecting their distribution. Iowa Stat. Coll. Agric. Exper. Stat. Res. Bull. 217, 225–256 (1937). — MARTON, M., und J. SZABO: *Nocardia uniformis*, a new species from solonetz soil. Acta Microbiol., Budapest 6, 131–134 (1959); Ref. Zbl. Bakt. II 113, 627 (1959/60). — MASILKIN, I. A.: Der Bodenpilz *Lipomyces* und seine Verbreitung in den Böden Jakutiens. Mikrobiologie 26, 477–480 (1957) (russ.); Ref. Z. Pflanzenernähr., Düng., Bodenkde. 80, 272 (1958). — MATKOVICS, B.: Extreme Werte der atmosphärischen N_2-Bindung bei den aus den Wurzelknöllchen von *Lupinus luteus* isolierten Hefen. Naturwiss. 47 (4), 92 (1960). — McBEE, R. H., und L. P. GAUGLER: Identity of thermophilic bacteria isolated from arctic soils and waters. J. Bact. 71, 186–187 (1956). — McBEE, R. H., und V. H. McBEE: The incidence of thermophilic bacteria in arctic soils and waters. J. Bact. 71, 182–185 (1956). — McCLUNG, N. M.: *Actinomycetes* in local soils. Trans. Kansas Acad. Sci. 54, 218–220 (1951). — McKNIGHT, T.: Non-symbiotic nitrogenfixing organisms in Queensland soils. Queensland J. Agric. Aci. 6, 177–195 (1949). — McLENNAN, E. I., und S. C. DUCKER: Micro-fungal population of acid sandy podsols. Nature (Lond.) 174, 1060–1061 (1954). — MEIKLEJOHN, J.: Numbers of bacteria and actinomycetes in a Kenya soil. J. Soil Sci. 8, 240–247 (1957); Ref. Z. Pflanzenernähr., Düng., Bodenkde. 80, 87 (1958). — MELKOUMOWA, T. A.: Besonderheiten der Stämme der Knöllchenbakterien auf verschiedenen Luzernesorten, die unter den Verhältnissen der Aserbajdschanischen SSR angebaut werden. Nachr. Akad. Wiss. UdSSR. Biol. Ser. H. 5, 616–624 (1957) (russ.); Ref. Z. Pflanzenernähr., Düng., Bodenkde. 80, 272 (1958). — MESCHKOW, N. W., und R. N. CHODAKOWA: Der Einfluß von Wurzelausscheidungen der Erbse und des Mais auf die Entwicklung einiger Bodenmikroben in künstlicher Kultur im Rhizosphärenauszug der Pflanze. Mikrobiologija 23, 544–550 (1954); Ref. Z. Zbl. Bakt. 109, 281 (1956). — MESSIAEN, C. M.: L'influence des méthodes d'isolement et des milieux de culture dans l'échantillonage de la flore fongique des sols. C. R. Acad. Agric. Fr. 43, 384–386 (1957). — METCALFE, G., und M. E. BROWN: Nitrogen fixation by new species of *Nocardia*. J. Gen. Microbiol. 17, 567–572 (1957). — METCALFE, G., und S. CHAYEN: Nitrogen fixation by soil yeasts. Nature 174, 841–842 (1954). — METCALFE, G., S. CHAYEN, E. R. ROBERTS und T. G. G. WILSON: Nitrogen fixation by soil yeasts. Nature (Lond.) 174, 841–842 (1954); Ref. Z. Pflanzenernähr., Düng., Bodenkde. 69, 317 (1955). — METZ, H.: Untersuchungen über die Rhizosphäre. Arch. Mikrobiol. 23, 297–326 (1955). — METZEN, O. v.: Vergleichende Betrachtung verschiedener Böden über Keimgehalt, Stickstoffbindung und Salpeterbildung. Zbl. Bakt. II 94. 142–175 (1936). — MEYER, D. R., und A. J. ANDERSON: Temperature and symbiotic nitrogen fixation. Nature 183, 61 (1959); Ref. Z. Pflanzenernähr., Düng., Bodenkde. 88, 191 (1960). — MEYER, F. H.: Untersuchungen über die Aktivität der Mikroorganismen in Mull, Moder und Rohhumus. Arch. Mikrobiol. 33, 149–169

(1959). — MEYER, L., und G. SCHAFFER: Atmungskurven des Bodens unter dem Einfluß von Düngung und Bewachsung. Landwirtsch. Forsch. **6**, 81–95 (1954). — MILLBANK, I. W.: Estimation of numbers of *Nitrosomonas* in soil and culture. Nature (Lond.) **117**, 848–849 (1956). — MILLER, J. J., und N. S. WEBB: Isolation of yeasts from soil with the aid of acid, rose bengal and oxgall. Soil Sci. **77**, 197–204 (1954); Ref. Z. Pflanzenernähr., Düng., Bodenkde. **67**, 277 (1954). — MISCHUSTIN, E. N.: Über die Verbreitung von *Azotobacter* im Boden. Methodisches zum Nachweis des *Azotobacter*. Mikrobiologia **22**, 408–414 (1953); Ref. Zbl. Bakt. II **108**, 312 (1953/54). — MITRA, P. S., und H. SHANKER: Studien über die Wirkung verschiedener Phosphate auf den N-Gehalt alkalischer Böden. Phosphorsäure **17**, 10–16 (1957a). — Amelioration of alkali soil by chemicals in combination with organic-matter-like weeds. Soil Sci. **83**, 471–474 (1957). — MITTENDORF, U.: Die Wirkung der Abwässer auf die biologische Tätigkeit des Bodens. Dtsch. Landwirt **4**, 591–594 (1953); Ref. Z. Pflanzenernähr., Düng., Bodenkde. **66**, 87 (1954). — MOORE, A. W.: Symbiotic nitrogen fixation in a grazed tropical grass-legume pasture. Nature **185**, 638 (1960); Ref. Z. Pflanzenernähr., Düng., Bodenkde. **93**, 90 (1961). — MOSKOWETZ, W. S.: Die Pilzmikroflora in der Rhizosphäre der Luzerne im Süden der Ukrainischen SSR, II, Quantitative und qualitative Zusammensetzung der Pilzmikroflora in der Rhizosphäre der Luzerne. J. Mikrobiol. Akad. Wiss. Ukrain. SSR **19**, 44–50 (1957) (russ.); Ref. Z. Pflanzenernähr., Düng., Bodenkde. **80**, 272 (1958). — MULDER, E. G., K. BAKEMA und W. L. van VEEN: Molybdenum in symbiotic nitrogen fixation and in nitrate assimilation. Plant a. Soil **10**, 319–334 (1959). — MULDER, E. G., und W. L. van VEEN: Effect of pH and organic compounds on nitrogen fixation by red clover. Plant a. Soil **13**, 91–113 (1960a). — The influence of carbon dioxide on symbiote nitrogen fixation. Plant a. Soil **13**, 265–278 (1960b). — MÜLLER, G.: Untersuchungen über die Wechselbeziehungen zwischen Bodenleben und Standortsfaktoren bei Futterpflanzen. VIe Congrès international de la science du sol, Paris III, **6**, 29–37 (1956). — Untersuchung über die Wechselbeziehung zwischen Bodenleben und Standortsfaktoren bei Futterpflanzenarten. Hab.-Schrift, Wiss. Z. Humboldt-Univ. Berlin, Math. naturwiss. Reihe **5**, 191–230 (1955/56a). — Untersuchung über die Wechselbeziehung zwischen Bodenleben, Standortsfaktoren und Ertragsfähigkeit in Kulturpflanzenbeständen nach Futterpflanzen, 2. Wiss. Z. Humboldt-Univ. Berlin, Math. naturwiss. Reihe **5**, 4, 297–320 (1955/56b). — Untersuchung über die Wechselbeziehung zwischen Bodenleben, Standortsfaktoren und Ertragsfähigkeit in Kulturpflanzenbeständen nach Futterpflanzen, 3. Wiss. Z. Humboldt-Univ. Berlin, Math.-naturwiss. Reihe **6**, 5, 563–578 (1956/57). — Untersuchung über die Wechselbeziehung zwischen Bodenleben, Standortsfaktoren und Ertragsfähigkeit in Kulturpflanzenbeständen nach Futterpflanzenarten, 4. Wiss. Z. Humboldt-Univ. Berlin, Math.-naturwiss. Reihe **8**, 4/5, 667–681 (1958/59). — Bodenbiologische Untersuchungen in der engeren und weiteren Rhizosphäre bei lebenden und abgestorbenen Wurzeln von Futterpflanzen. Z. Acker- u. Pflanzenbau **104**, 289–306 (1957b). — Prüfung der Beziehungen zwischen mineralischer Düngung und Bodenleben. Dtsch. Landwirtsch. 8, 24–31 (1957a). — Bodenbiologische Abbauuntersuchungen unter Berücksichtigung der Standortverhältnisse bei Schwarzbrache nach Rein- und Mischsaaten, Zbl. Bakt. II, I **112**, 44–78, II, **112**, 169–203 (1959), III, **113**, 254–269 (1959/60). — Wechselbeziehungen zwischen bodenbiologischer Dynamik und Futterpflanzenbau bei Sandböden. Wiss. Z. Karl-Marx-Univ., Leipzig, Math.-naturwiss. Reihe **9**, H. 5, 837–844 (1959/60). — Beiträge zur Frage der bodenbiologischen Änderungen durch acker- und pflanzenbauliche Maßnahmen. Zbl. Bakt. II **113**, 324–339. (1960). — Über die bodenbiologische Dynamik tiefbearbeiteter und tiefgedüngter leichter Sandböden. Tagungsbericht Nr. 28, Probleme der Bodenbearbeitung d. wiss. Arbeitstagung im Inst. für Acker- und Pflanzenbau. Müncheberg 167–184 (1960). — MÜLLER, G., und K. RAUHE: Zur Tiefkultur auf leichten Böden im besonderen Hinblick auf die Bodenbiologie, II, Bodenbiologischer Teil. Z. Acker- u. Pflanzenbau **109**, 3, 309–332 (1959). — MÜLLER, G., und K. STEINWEG: Varianzanalytische Prüfung bodenbakteriologischer Keimzählmethoden. Zbl. Bakt. II, **111**, 21–25, 625–652 (1958). — MUELLER, K. E., und L. W. DURRELL: Sampling tubes for soil fungi. Phytopathology **47**, 243 (1957). — MUSKAT, J.: Untersuchungen über Schimmelpilze bayrischer und tunesischer Böden, I, Floristisch-ökologischer Teil. Arch. Mikrobiol. **22**, 1–20 (1955). — MUSTAFA, N. N.: Mikrobiologische Beobachtungen in Podsolen im Tannenwald mit einer Bodendecke von Sauerklee und einer solchen von Blaubeere. Nachr. Leningrader Univ., Ser. f. Biol., Folge 3, H. 15, 19–25 (1959) (russ.); Ref. Z. Pflanzenernähr., Düng., Bodenkde. **90**, 186 (1960).

NABIJEW, G. N.: *Azotobacter* in den ungepflügten und bewässerten Wiesenböden der Zone der Grauerde. Nachr. Akad. Wiss. Usbek. SSR, Ser. biol. Wiss. Nr. 3, 21–30 (1957); Ref. Zbl. Bakt. II **112**, 306–307 (1959). — NAUNDORF, G., und R. NILSSON:

Über formbildende Wirkstoffe bei *Azotobacter chroococcum* und der Einfluß dieser formativen Wirkstoffe auf die Bakteroidenbildung von *Bact. radicicola*. Naturwiss. **30**, 50–51 (1942). — Über formbildende Wirkstoffe bei *Azotobacter chroococcum* und der Einfluß dieser formativen Wirkstoffe auf die Bildung von Gigasformen bei *Bact. radicicola*. Naturwiss. **31**, 29–30 (1943). — NÉMETH, G.: A new nitrogen-fixing microorganism producing a red pigment. Nature (Lond.) **183**, 1460–1461 (1959); Ref. Zbl. Bakt. II **113**, 542 (1959/1960). — NÉMETH, GG., und B. MATKOVICS: Über einen neuen, ein rotes Pigment bildenden und den atmosphärischen Stickstoff bindenden Mikroorganismus. Zbl. Bakt. II **111**, 562–565 (1958). — NETTE, I. T.: Denitrifizierende Bakterien in der Eichenrhizosphäre. Mikrobiologija **24**, 429–434 (1955) (russ.); Ref. Z. Pflanzenernähr., Düng., Bodenkde. **72**, 94 (1956). — NETZSCH-LEHNER, A., und W. NOWAK: Verändert Bebrütung den Bakteriengehalt von Böden? Z. Pflanzenernähr., Düng., Bodenkde. **90** (135), 146–153 (1960). — NICKERSON, J. T. R.: A modified little plate method for bacterial counts in vegetable freezing plants. Food Res. **8**, 163–168 (1948). — NICOT, J.: Remarques sur les peuplements des Mycomycetes des sables désertiques. C.R. Acad. Sci. (Paris) **240**, 2082–2084 (1955). — NIETHAMMER, A.: Die mikroskopischen Boden-Pilze. Prag 1937. — NIETHAMMER, A., R. KREHL-NIEFFER und M. HITZLER: Mikroskopische Bodenpilze verschiedener Herkunft unter verschiedenen Kulturbedingungen. Zbl. Bakt. II **112**, 429–439 (1959). — NIKLAS, H.: Die biologischen Untersuchungsmethoden. In: HONCAMP, Handbuch der Pflanzenernährung und Düngerlehre, Bd. 1, S. 807–858 (1931). — NIKLAS, H., H. POSCHENRIEDER und A. HOCK: Über die Verbreitung des Azotobacter in den Böden Bayerns unter Berücksichtigung der Bodenreaktion, des Kalk- und Phosphorsäuregehaltes derselben. Zbl. Bakt. II **66**, 16–28 (1926). — NILSSON, R., G. BJÄLFVE und D. BURSTRÖM: Über Zuwachsfaktoren bei *Bact. radicicola* III. Ann. landwirtsch. Hochsch. Schwedens **7**, 51–61 (1938). — Biotin als Zuwachsfaktor für *Bact. radicicola*. Naturwiss. **27**, 389 (1939). — NISHIGAKI, S., und M. SHIOIRI: Nitrogen cycles in the rice field soil, I, The effect of the blue-green Algae on the fixation of atmospheric nitrogen in the water logged rice soils. Soil a. Plant Food **5**, 36–39 (1959). — NISSEN, T. V.: Effects of antibiotica on carbon dioxide production in soil. Nature (Lond.) **174**, 226–227 (1954); Ref. Z. Pflanzenernähr., Düng., Bodenkde. **69**, 316 (1955). — NORRIS, D.: Rhizobium needs magnesium not calcium. Nature (Lond.) **182**, 734–735 (1958); Ref. Zbl. Bakt. II **113**, 290 (1959/60). — NORRIS, J. R., und H. L. JENSEN: Calcium requirements of *Azotobacter*. Nature **180**, 1493–1494 (1957); Ref. Z. Pflanzenernähr., Düng., Bodenkde. **81**, 181 (1958). — Calcium requirements of Azotobacter. Arch. Mikrobiol. **31**, 198–205 (1958). — NOVÁK, B.: Die Gründüngung vom mikrobiologischen Standpunkte. Sbornik Csl. Akad. tzemed ved. Rostlinná výroba. Rocnik **4** (XXXI), 6/7, 791–798 (1958). — NOWAK, W.: Neuere Ergebnisse des Direktnachweises von Bakterien im Boden mittels der kombinierten Aufwuchs-Fluorochromierungsmethode (Kurzref.). Zbl. Bakt. I, Orig. **176**, 243 (1959). — NOWAK, W., und A. NETZSCH-LEHNER: Vergleichsuntersuchungen zwischen dem Plattengußverfahren und einer fluoreszenzmikroskopischen Direktbetrachtungsmethode des Bodens. Z. Pflanzenernähr., Düng., Bodenkde. **89** (134), 137–148 (1960). — Bodenbakteriologische Untersuchungsergebnisse aus mehrjährigen Feldversuchen unter besonderer Berücksichtigung der Vollstrohdüngung. Bayer. Landwirtsch. Jahrb. **38**, 338–350 (1961). — NOWOTNY-MIECZYNSKA, J.: Some factors influencing the pigmentation nodules of leguminous plants. Act. Microbiol. Polon. **1**, 42–51 (1952) (poln. m. russ. u. engl. Zusammenfassung); Ref. Zbl. Bakt. II **108**, 105 (1954/55). — NOWOTNY-MIECZYNSKA, J., und M. RUSKOWSKA: The influence of mineral nitrogen on the growth of inoculated leguminous plants. Act. Microbiol. Polon. **1**, 205–222 (1952); Ref. Zbl. Bakt. II **108**, 105 (1954/55). — NUTMAN, P. S.: A factor in clover nodule formation associated with the volume of the medium occupied by the roots. Nature **156**, 20 (1945).

OKUDA, A., M. YAMAGUCHI und M. M. KOBAYASHI: Nitrogen fixing microorganisms in paddy soils, IV, Nitrogen fixation by photosynthetic bacteria (*Rhodopseudomonas capsulatus* species) under various conditions. Soil and Plant Food **5**, 73–76 (1959); Ref. Z. Pflanzenernähr., Düng., Bodenkde. **90**, 271 (1960).

PAHARIA, K. D., und T. KOMMEDAHL: A modified planting technique for the study of soil fungi. Phytopathol. **44**, 502 (1954). — PAKUSIN, Z. G.: Microflora of salty soils of the central part of southeastern Shirvania. Dokl. A. N. Azerb. SSR **11** (8), 651–666 (1955). — PÁNTOS, G.: Die Wirkung der Rhizosphäre-Bakterien des Weizens auf die Pflanze unter monobakteriellen Verhältnissen. Agrokémia és Talajtan **3**, 351–358 (1956) (ung.); Ref. Z. Pflanzenernähr., Düng., Bodenkde. **77**, 169 (1957). — PAPAVIZAS, G. C., und CH. B. DAVEY: Evalution of various media and antimicrobial agents for isolation of soil fungi. Soil Sci. **88**, 112–117 (1959). — PARKER, C. A.: Non-symbiotic nitrogen-fixing bacteria in soil, III, Total nitrogen changes in a field soil. J. Soil Sci. 8, 48–59 (1957). — PARKER, D. T., und O. N. ALLEN: The nodulation status of *Trifolium ambiguum*. Soil Sci. Soc. Amer. Proc. **16**, 350–353 (1952); Ref. Zbl. Bakt. II **108**, 104 (1954/55). — PARKINSON, D., und J. S. WAID: The ecology of soil fungi. Liverpool: University Press. 1960. — PETERSON, E. A.: Observations on the influence of plant illumination on the fungal flora of roots. Canad. J. Microbiol. **7**, 1–6 (1961). — PFENNIG, N.: Beobachtungen des Wachstumsverhaltens von Streptomyceten auf Rossi-Cholodny-Aufwuchs-Platten im Boden. Arch. Mikrobiol. **31**, 206–216 (1958). — POCHON, J., und H. DE BARJAC: Traité de Microbiologie des sols. Paris: Dunod. 1958. — POCHON, J., und Y. T. TCHAN: Précis de Microbiologie du Sol. Monographie de l'Inst. Pasteur. Paris: Masson. 1948. — POMMER, E. H.: Beiträge zur Anatomie und Biologie der Wurzelknöllchen von *Alnus glutinosa Gaertner*. Flora **143**, 603–634 (1956); Ref. Z. Pflanzenernähr., Düng., Bodenkde. **76**, 191 (1957). — POPESCU, V., und L. CALANCEA: Der Einfluß einiger Fungizide auf biologische Prozesse im Boden. Lucrari Stiintifice **14**, 79–89 (1958) (rumän. m. engl. Zusammenfassung); Ref. Z. Pflanzenernähr., Düng., Bodenkde. **86**, 172 (1959). — POSCHENRIEDER, H., und TH. BECK: Untersuchungen über die Rolle einiger bei den ersten Stadien des Torfbildungsvorganges beteiligten Bakterienarten. Zbl. Bakt. II **111**, 684–695 (1958). — Mikrobiologische-biochemische Charakterisierung von schweren Tonböden. Bayer. Landwirtsch. Jahrb., Sonderheft 4, 86–98 (1960). — POTTER, L. F.: Isolation of rhizobia from preserved nodules. Appl. Microbiol. **3**, 160–161 (1955). — PRÉVOT, A. R.: Étude d'une bactérie anaérobie chromogène nouvelle: *Clostr. saturnirubrum*. C. R. Séances Acad. Sci. **223**, 1035–1037 (1946). — PRÉVOT, A. R., und J. LAPLANCHE: Étude d'une bactérie anaérobie nouvelle de Guinée francaise *Cillobacterium Combesi* n. sp. Ann. Inst. Past. **73**, 687 (1947). — PRÉVOT, A. R., und R. SANSONNENS: Présence de *Cl. corallinum* P. et R. au Sénégal. Ann. Inst. Past. **73**, 1044 (1947). — PRÉVOT, A. R., und J. ZIMMÉS-CHAVEROU: Études d'une nouvelle espèce anaérobie de Côte d'Ivoire: *Inflabilis mangenoti n. sp.* Ann. Inst. Past. **73**, 602 (1947). — PROCTOR, M. H., und P. W. WILSON: Nitrogen fixation by Gram-negative bacteria. Nature **182**, 891 (1958); Ref. Z. Pflanzenernähr., Düng., Bodenkde. **85**, 94 (1959). — Nitrogen fixation by *Achromobacter spp.* Arch. Mikrobiol. **32**, 254–260 (1959). — PUSCHKINSKAJA, O. J.: Zur Methode der Feststellung von zelluloselösenden Mikroorganismen im Boden. Mikrobiologia **23**, 34–36 (1954); Ref. Zbl. Bakt. II **108**, 482 (1954/55).

RABOTNOWA, I. L.: Are rhizobia capable of nitrogen fixation in pure culture? Microbiologija **7**, 673–682 (1938). — RAPER, K. B., und M. S. QUINLAN: *Acytostelium leptosomum*: A unique cellular slime mould with an acellulae stalk. J. Gen. Microbiol. **18**, 16–32 (1958). — RAUHE, K., und G. MÜLLER: Zur Tiefkultur auf leichten Böden im besonderen Hinblick auf die Bodenbiologie, II, Bodenmikrobiologischer Teil. Z. Acker- u. Pflanzenbau **109**, 309–332 (1960). — REHM, H. J.: Untersuchungen über die Bodenmikroflora von Gatersleben und Umgebung, III, Untersuchungen über *Azotobacter chroococcum Beijerinck*. Kulturpflanze **4**, 208–221 (1956a). — Untersuchungen über die Bodenmikroflora von Gatersleben und Umgebung, IV, Untersuchungen über *Alternaria tenuis Nees*. Kulturpflanze **4**, 222–227 (1956). — Beitrag zur Ökologie der Streptomyceten, 3, Die Streptomycetenarten und ihre antibiotische Aktivität in der Rhizosphäre der Gerste. Zbl. Bakt. II **114** 147–155 (1961a). — Beitrag zur Ökologie der Streptomyceten, 4, Die Streptomycetenarten und ihre antibiotische Aktivität in Sandböden Mitteldeutschlands (im Vergleich mit anderen Böden). Zbl. Bakt. II **114**, 345–355 (1961). — REISENAUER, H. M.: Cobalt in nitrogen fixation by a legume. Nature (Lond.) **186**, 375–376 (1960). — RIPPEL, A.: Bakteriologisch-chemische Methoden zur Bestimmung des Fruchtbarkeitszustandes des Bodens und der Kreislauf der Stoffe. In: BLANCK, Handbuch der Bodenlehre, Bd. VIII, S. 599–671. Berlin: Springer. 1931. — Mikrobiologie des Bodens. In: BLANCK, Handbuch der Bodenlehre, Ergänzungsband I. Berlin: Springer. 1939. — ROBERTS, E. R., und T. G. G. WILSON: Nature **174**, 842 (1954). — ROBERTS, J. E., und W. B. BOLLEN: A micro-plating method for soil molds and its use to detect some effects of certain insecticides and herbicides. Appl. Microbiol. **3**, 190–194 (1955). — RODINA, A. G.: Im Wasser lebende Spirillen mit der Fähigkeit, molekularen Stickstoff zu binden. Mikrobiologija **25** (2), 145–149 (1956). — RÖSCHENTHALER, R., F. BUKATSCH und H. POSCHENRIEDER: Zusammenhänge zwischen Aminosäurebestand und autochthoner Bakterienflora in einem oberbayrischen Moorprofil (Hoch- und Flachmoorschichten). Mitt. f. Landkultur, Moor- und Torfwirtschaft **6**, 25–32 (1958); Ref. Zbl. Bakt. II **113**, 178 (1959/60). — RÖSCHENTHALER, R., und H. POSCHENRIEDER: Untersuchungen über die Bakterienflora eines Hochmoorprofils bei Staltach in Bayern. Zbl. Bakt. II **111**, 653–671 (1958a). — Untersuchungen über die Verwertbarkeit einiger im Moor vorkommender freier Aminosäuren durch die Torfbakterienflora. Mitt. d. Landkultur, Moor- und Torfwirtsch. **6**, 21–24 (1958b); Ref. Zbl. Bakt. II **113**, 177 (1959/60). —

Rose de, H. R., und A. S. Newman: The comparison of the persistence of certain plant growth-regulators when applied to soil. Soil Sci. Soc. Amer. Proc. **12**, 222–226 (1948). — Rossi, G.: Preliminary note on the microbiology of the soil and the possible existence therein of visible germs. Soil Sci. **12**, 409–412 (1921). — Direct microscopic and bacteriological examination of the soil. Soil Sci. **41**, 53–66 (1936). — Rouatt, J. W., and H. Katznelson: The comparative growth of bacterial isolates from rhizosphere and non-rhizosphere soil. Canad. J. Microbiol. **3**, 271–275 (1957). — Rouatt, J. W., H. Katznelson und T. M. B. Payne: Statistical evaluation of the rhizosphere effect. Soil Sci. Soc. Amer. Proc. **24**, 271–273 (1960). — Rouatt, J. W., und A. G. Lochhead: Qualitative studies of soil microorganisms, XIII, Effect of decomposition of various crop plants on the nutritional groups of soil bacteria. Soil Sci. **80**, 147–154 (1955). — Roulet, M. A., und W. H. Schopfer: Les vitamines du sol et leur signification. Trans. 4th Int. Congr. Soil Sci. **1**, 202–203 (1950). — Rouschal, Ch., und S. Strugger: Eine neue Methode zur Vitalbeoachtung der Mikroorganismen im Erdboden. Naturwiss. **31**, 300 (1943). — Rodina, A. G.: Aequous spirilla which fix molecular nitrogen. Microbiologija **25**, (2), 145–149 (1956); Ref. Biol. Abstr. **32**, 31 121 (1958). — Rovira, A. D.: Factors influencing soil metabolism and the relationship between plant roots and soil microorganisms. Thesis University of Sydney 1955. — Plant root excretions in relation to the rhizosphere effect, I, The nature of root exudate from oats and peas; II, A study of the properties of root exudate and its effect on the growth of microorganisms isolated from the rhizosphere and control soil; III, The effect of the root exudate on the numbers and activity of microorganisms in soil. Plant a. Soil **7**, 178–194, 195–208, 209–217 (1956). — Root excretions in relation to the rhizosphere effect, IV, Influence of plant species, age of plant, light, temperature and calcium nutrition on exudation. Plant a. Soil **11**. 53–64 (1959). — Roy. A. B.: A new species of Azotobacter producing heavy slime and acid. Nature (Lond.) **182**, 120-121 (1958); Ref. Zbl. Bakt. II **113**, 164 (1959/60). — Roy, A. B., und M. K. Mukherjee: A new type of nitrogen-fixing bacterium. Nature **180**, 236 (1957); Ref. Z. Pflanzenernähr., Düng., Bodenkde. **80**, 271 (1958). — Rübensam, E., und K. Steinbrenner: Mikrobiologische Veränderungen durch unterschiedliche Düngung im Thyrower Bodenfruchtbarkeitsversuch. Albrecht Thaer-Arch. **5**, 266–281 (1961). — Ruben(t)schik, L. I.: Bakteriendünger. Mitt. Versuchsstat. Gärungsgewerbe (Wien) **10**, 75–79, 95–99 (1956), übers. nach Mikrobiologija **25**, H. 2 (1956). — Vgl. auch L. I. Rubenchik, Azotobacter and its use in agriculture. Übersetzung aus dem Russischen durch Israel Program for Scientific Translation, Jerusalem 1963. — Russell, E. W.: Soil conditions and plant growth, 9th ed. London: Longmans, Green. 1961. — Rybalkina, A. W., und E. W. Kononenko: Die unmittelbare Beobachtung der Bodenmikroflora mit Hilfe der modifizierten Methode nach Cholodny. Mikrobiologia **22**, 439–444 (1953); Ref. Zbl. Bakt. II **108**, 501 (1954/55).

Sakai, H.: On a method for isolation of nitrifying organisms. Soil a. Plant Food **4**, 81–82 (1958); Ref. Z. Pflanzenernähr., Düng., Bodenkde. **86**, 269 (1959). — Samzjewitsch, S. A.: Über die jahreszeitliche und periodische Entwicklung der Bodenmikroorganismen. Mikrobiologija **24**, 615–625 (1955); Ref. Z. Pflanzenernähr., Düng., Bodenkde. **74**, 92 (1956). — Sauerlandt, W., M. Marzusch-Trappmann und C. Tietjen: Der Einfluß der Häufigkeit der organischen Düngung auf den Gehalt des Bodens an organisch gebundenem Kohlenstoff unter besonderer Berücksichtigung der Keimdichte und der Enchytraeiden. Z. Pflanzenernähr., Düng., Bodenkde. **92**, 134–147 (1961). — Schaede, R., und F. H. Meyer: Die pflanzlichen Symbiosen, 3. Aufl. Stuttgart: Fischer. 1962. — Schaffer, G.: Atmungskuryen des Bodens unter dem Einfluß langjähriger, verschiedenartiger Düngung. Z. Pflanzenernähr., Düng., Bodenkde. **67**, 219–230 (1954). — Schanderl, H.: Untersuchungen über die Assimilation des elementaren Stickstoffs der Luft durch hautbildende Hefen und durch *Cladosporium cellare*. Zbl. Bakt. II **101**, 401–408 (1939/40). — Scheffer, F., und A. Kloke: Der Einfluß von Antibiotica auf die Entwicklung und den Nährstoffgehalt von Kulturpflanzen. Z. Pflanzenernähr., Düng., Bodenkde. **66**, 29–38 (1954). — Scheffer, F., und E. Küster: Untersuchungen über die N-Bindung durch hautbildende Hefen. Z. Pflanzenernähr., Düng., Bodenkde. **51**, 224 (1950). — Scheffzowa, J. J.: Der Einfluß von Gewebesäften aus Leguminosenwurzeln auf Knöllchenbakterien. Mikrobiologija **28**, 75–79 (1959) (russ. m. engl. Zusammenfassung); Ref. Zbl. Bakt. II **113**, 542 (1959/60). — Schlüter, M., und F. Bukatsch: Physiologische Untersuchungen zur Rassenfrage von Azotobacter. Zbl. Bakt. II **112**, 509–543 (1959). — Schmalfuss, K.: Über das Ausmaß der biologischen Bindung des elementaren Stickstoffs. Der Einfluß assimilierbarer Stickstoffverbindungen. Ber. Ver. Sächs. Akad. Wiss. Leipzig **104**, H. 1, 1–18 (1960). — Schmidt, G.: Untersuchungen über die Einwirkung von Umweltbedingungen auf die Knöllchenbildung und die N-Bindung von *Vicia faba L.*

Kühn-Arch. **69**, 165–222 (1955); zit. nach EBERT, K.; Diss. Halle 1959. — SCHMIDT, H. L.: Beitrag zur Ermittlung der Pilzbesiedlung bei natürlichen Böden und ihrer Kennzeichnung durch ein besonderes Isolierungsverfahren. Arch. Mikrobiol. **32**, 224–233 (1959). — SCHNEIDER, B.: Chemische und biologische Veränderungen des Rohhumus unter dem Einfluß der Düngung mit Kalk und Phosphorsäure. Forst- u. Holzwirtsch. **14**, 182–186 (1959). — SCHNEIDER, R.: *Penicillium taxi* nov. sp., eine neue sklerotienbildende Art auf Nadelstreu von *Taxus baccata*. Zbl. Bakt. II **110**, 43–49 (1957). — SCHOBER, R.: Luftstickstoffassimilation und Säurebildung bei *Aspergillus niger*. Jahrb. wiss. Bot. **72**, 1–105 (1930). — SCHOLZ, E.: Über niedere Phycomyceten aus Salzböden und ihr Verhalten in Salzlösungen. Arch. Mikrobiol. **30**, 119–146 (1958). — SCHÖNBECK, F.: Untersuchungen über Vorkommen und Bedeutung von Hemmstoffen in Getreiderückständen innerhalb der Fruchtfolge. Z. Pflanzenkrankheiten-Pflanzenschutz **63**, 513–545 (1956); Ref. Z. Pflanzenernähr., Düng., Bodenkde. **77**, 74 (1957). — SCHREVEN, D. A. VAN: Effects of added sugars and nitrogen on nodulation of legumes. Plant a. Soil **11**, 93–112 (1959). — SCHTINA, E. A.: Über die Entwicklung der Bodenalgen in den Wald-Podsol-Böden. Potschwowedenie Nr. 3, 12–18 (1957) (russ.); Ref. Z. Pflanzenernähr., Düng., Bodenkde. **80**, 90 (1958). — SEIFERT, J.: Die Anwendung der Fluoreszenzmikroskopie in der Mikrobiologie des Bodens. Bodenkde., H. 2, 50–54 (1958) (russ.); Ref. Z. Pflanzenernähr., Düng., Bodenkde. **87**, 264 (1959). — The influence of moisture and temperature on the number of microorganisms in the soil. Folia Microbiol. **5**, 176–180 (1960a). — Number of microorganisms, nitrification and biological activity in the soil of plant communities in Koda. Tschechoslov. Akad. Wiss. Sekt. IV (1960b). — Sektion für Düngung, Zusammenstellung der Arbeiten des erweiterten Plenums: Die Bedeutung der Mikroorganismen für die Pflanzenernährung. Moskau: Staatsverlag f. landwirtsch. Literatur. 1953 (russ.); Ref. Z. Pflanzenernähr., Düng., Bodenkde. **69**, 261 (1955). — SEWELL, G. W. F.: A slide-trap method for the isolation of soil fungi. Nature **177**, 708 (1956); Ref. Z. Pflanzenernähr., Düng., Bodenkde. **76**, 93 (1957). — SHEFNER, A. M., und B. J. BURKHARDT: Reversal of boric acid inhibition of growth in certain soil microorganisms. Nature (Lond.) **180**, 1207–1208 (1957). — SHUKOWA, P. A.: Der Einfluß der Kultivierung auf die Mikroflora im humus- und eisenhaltigen Podsolboden auf der Halbinsel Kola. Mikrobiol. **27**, 195–200 (1958) (russ.); Ref. Z. Pflanzenernähr., Düng., Bodenkde. **85**, 181 (1959). — SHUTT, F. T.: Some characteristics of the Western prairie soils of Canada. J. Agric. Sci. **3**, 335–357 (1910). — SIMON, G.: L'enfouissement des pailles dans le sol. Étude générale et répercussions sur la microflore du sol. I. Enfouissement des pailles et chimie du sol. Ann. agron. **11**, 5–53 (1960); Ref. Z. Pflanzenernähr., Düng., Bodenkde. **92**, 256 (1961). — SIMONART, P.: Symposium sur les méthodes d'études microbiologiques du sol. Louvain 1957. — SINGH, B. N.: Selection of bacterial food by soil flaggellates and amoebae. Ann. Appl. Biol. **29**, 18–22 (1942a). — Toxic effects of certain bacterial metabolic products on soil protozoa. Nature **149**, 168 (1942b). — The selection of bacterial food by soil amoebae and the toxic effects of bacterial pigments and other products on soil protozoa. Brit. J. Exper. Path. **26**, 316–325 (1945). — Silica Jelly as a substrate for counting holozoic protozoa. Nature **157**, 302 (1946a). — A method of estimating the numbers of soil protozoa, especially amoebae. Ann. Appl. Biol. **33**, 112–119 (1946b). — Soil *Acrasieae* and their bacterial food supply. Nature **157**, 133 (1946c). — Studies on soil *Acrasieae*, 2, The active life of species of *Dictyostelium* in soil and the influence thereon of soil moisture and bacterial food. J. Gen. Microbiol. **1**, 361–367 (1947). — Limitation of bacteria by micro-predators in soil. Nature **161**, 140 (1948). — The effect of artificial fertilizers and dung on the number of Amoebae in Rothamsted soils. J. Gen. Microbiol. **3**, 204–210 (1949). — SIRONVAL, C.: Influence of day-length on nodule formation in *Soja hispida* by a specific Rhizobium strain. Nature (Lond.) **177**, 93–94 (1956). — Relation between chlorophyll metabolism and nodule formation in Soja bean. Nature **181**, 1272–1273 (1958); Ref. Pflanzenernähr., Düng., Bodenkde. **82**, 230 (1958). — SKINNER, F. A., P. C. T. JONES und J. E. MOLLISON: A comparison of a direct and a plate-counting technique for the quantitative estimation of soil microorganisms. J. Gen. Microbiol. **6**, 261–271 (1952); Ref. Zbl. Bakt. II **107**, 189 (1952/54). — SMITH, N. R.: Microbiology of soil. Ann. Rev. Microbiol. **2**, 453–484 (1948). — SONN, W.: Wechselwirkungen zwischen Wald und Boden. Potschwowedenie Nr. 7, 80–91 (1956) (russ.); Ref. Z. Pflanzenernähr., Düng., Bodenkde. **77**, 83 (1957). — SPICHER, G.: Lebensdauer und Stickstoffbindung der Knöllchenbakterien von Lupine, Serradella und Klee in Abhängigkeit von ihrer Gestalt. Zbl. Bakt. II **107**, 383–418 (1952/54). — Untersuchungen über die Wechselwirkungen zwischen *Azotobacter chroococcum* und höherer Pflanze. Zbl. Bakt. II **107**, 353–383 (1954). — SSADOWSKAJA, P. O.: Der Einfluß des Wurzelsystems vom Wein auf die Entwicklung des

Azotobacters. Mikrobiologija 28, 534–540 (1959); Ref. Zbl. Bakt. II 114, 220 (1961). — STAPP, C.: Der Boden in biologischer Hinsicht. In: HONCAMP, Handbuch der Pflanzenernährung und Düngerlehre, Bd. 1, S. 526–601, Berlin: Springer. 1931. — *Azotomonas insolita*, ein neuer aerober stickstoffbindender Mikroorganismus. Zbl. Bakt. II 102, 1–19 (1940). — Jahresbericht der Biologischen Zentralanstalt für Land- und Forstwirtschaft, Braunschweig 1954–1949, S. 25. — Zur Frage der Bindung atmosphärischen Stickstoffs durch hautbildende Hefen. Arch. Mikrobiol. 16, 48–52 (1951). — Über den Wert von Azotobacter-Impfpräparaten für die deutsche Landwirtschaft. Landwirtsch. Forsch. 3, 176–204 (1952). — Untersuchungen über Actinomyzeten des Bodens, I. Zbl. Bakt. II 107, 129–150 (1952/54). — STAPP, C., und R. FRETER: Untersuchungen über die Wirkung von 2,4-D im Boden, I, Eine verbesserte Nachweismethode und orientierende Versuche über die Haltbarkeitsdauer des Wuchsstoffs im Boden. Phytopathol. Z. 18, 365–375 (1952a). — Untersuchungen über die Wirkung von 2,4-D im Boden, II, Die Reaktion von Bodenbakterien auf den Wuchsstoff. Phytopathol. Z. 19, 20–33 (1952b). — STAPP, C., und G. SPICHER: Untersuchungen über die Wirkung von 2,4-D im Boden, IV, *Flavobacterium peregrinum* n. sp. und seine Fähigkeit zum Abbau des Hormones. Zbl. Bakt. II 108, 113–196 (1954/55). — STARKEY, R. L.: Some influences of the development of higher plants upon the microorganisms in the soil, IV, Microscopic examination of the rhizosphere. Soil Sci. 45, 207–249 (1938). — Relations of micronutrients to development of microorganisms. Soil Sci. 79, 1–14 (1955) (94 Literaturangaben). — Interrelations between microorganisms and plant roots in the rhizosphere. Bact. Rev. Baltimore 22, 154–172 (1958). — STEENSON, T. I., und N. WALKER: Observations on the bacterial oxydation of chlorophenoxyacetic acids. Plant a. Soil 8, 17–32 (1956/57). — STEINBRENNER, K., FR. NAGLITSCH und I. SCHLICHT: Der Einfluß der Herbizide Simazin und W 6658 auf die Bodenmikroorganismen und die Bodenfauna. Albrecht Thaer-Arch. 4, H. 8, 611–631 (1960). — STILLE, B.: Untersuchungen über die Bedeutung der Rhizosphäre. Arch. Mikrobiol. 9, 477–485 (1938). — Beobachtungen über das Verhalten von Azotobacter im Wurzelbereich höherer Pflanzen. Arch. Mikrobiol. 31, 255–261 (1958). — STÖCKLI, A.: Die Verbreitung der Azotobacterorganismen in der Schweiz. Landwirtsch. Jahrb. Schweiz 58, 67–105 (1944). — STOJANOVIC, B. J., und F. E. BROADBENT: Immobilisation and mineralisation rates of nitrogen during decomposition of plant residues in soil. Soil Sci. Soc. Amer. Proc. 20, 213–218 (1956). — STOKES, J. L.: The role of algae in the nitrogen cycle of the soil. Soil Sci. 49, 265–275 (1940). — STOLP, H.: Beiträge zur Frage der Beziehungen zwischen Mikroorganismen und höheren Pflanzen. Arch. Mikrobiol. 17, 1–29 (1952). — STOTZKY, G., W. P. MARTIN und J. L. MORTENSEN: Certain effects of crop residue and fumigant applications on the decomposition of an Ohio muck soil. Soil Sci. Soc. Amer. Proc. 20, 392–396 (1956). — STRUGGER, S.: Fluoreszenzmikroskopie und Mikrobiologie. Hannover: Schaper. 1949. — SUSCHKINA, N. N.: Stickstoffbindung und biologische Besonderheiten eines aus stark salzhaltigem Boden der unteren Wolga isolierten Azotobacter-Stammes. Westn. Mosk. un-ta Nr. 10, 195–209 (1948). — Zur Salzverträglichkeit von Azotobacter. Mikrobiologija 25, 35–40 (1956); Ref. Zbl. Bakt. II 110, 588 (1957). — SUSCHKINA, N. N., und P. S. RYSHKOWA: Über die Mikroflora der Böden an der Westküste der Insel Nowaja Zemlja. Dokl. Akad. Nauk SSSR 106, 914–916 (1956) (russ.); Ref. Z. Pflanzenernähr., Düng., Bodenkde. 75, 185 (1956). — SUSCHKINA, N. N., und L. A. SJERGUNINA: Verbreitung und Entwicklung von Azotobacter in Rasenpodsolböden. Mikrobiologija 24, 408–414 (1955) (russ.); Ref. Zbl. Bakt. II 110, 757 (1957). — SWABY, R. J.: The occurrence and activities of *Azotobacter* and *Clostridium butyricum* in Victorian soils. Austr. J. Exper. Biol. 17, 401–423 (1939). — SZABÓ, I.: Antibiotische Aktivität der Strahlpilzflora in den Böden Ungarns. Agrokémia és Talajtan 5, 433–456 (1956) (ungar.); Ref. Z. Pflanzenernähr., Düng., Bodenkde. 80, 87 (1958). — SZABÓ, I., M. MARTON, J. SZABOLCS und L. VARGA: Anpassung der Mikroflora und Mikrofauna an die Verhältnisse der Szikböden (Alkaliböden) mit besonderer Berücksichtigung eines degradierten Solontschak-Solonetz-Bodens. Acta Agronom. Hung. 9, 9–39 (1959); Ref. Z. Pflanzenernähr., Düng., Bodenkde. 89, 231 (1960). — SZELÉNYI, F., und H. BALÁZS: Forschungsergebnisse über die Stickstoffbindefähigkeit der Azotobacter. A Debreceni Mezögazdaságí Akadémia Tudományos Evkönyve 1957, 159–167 (ungar.); Ref. Z. Pflanzenernähr., Düng., Bodenkde. 82, 254 (1958).

TAYLOR, C. B.: Short period fluctuations in the numbers of bacterial cells in soil. Proc. Roy. Soc. Lond. B 814, 269–295 (1936). — Further studies of *Bact. globiforme* and the incidence of this type of organism in Canadian soils. Soil Sci. 46, 307–320 (1938). — TAYLOR, C. B. und A. G. LOCHHEAD: A study of *Bact. globiforme Conn* in soils differing in fertility. Canad. J. Res. 15, 340–347 (1937). — TCHAN, Y. T.:

Study of soil algae, II, Bioassay of soil fertility by algae. Plant a. Soil **10**, 220–232 (1959). — Tchan, Y. T., und J. S. Bunt: Direct microscopy for study and count of soil protozoa. Nature (Lond.) **174**, 656 (1954); Ref. Zbl. Bakt. II **108**, 717 (1954/55). — de Telegdy-Kovats L.: The growth and respiration of bacteria in sand cultures in the presence and absence of protozoa. Ann. Appl. Biol. **19**, 65 (1932). — Temple, J. C.: The influence of stall manure upon the bacterial flora of the soil. Zbl. Bakt. II **34**, 204–223 (1912). — Thompson, J. A.: Inhibition of nodule bacteria by an antibiotic from legume seed coats. Nature **187**, 619–620 (1960); Ref. Z. Pflanzenernähr., Düng., Bodenkde. **93**, 89 (1961). — Thornton, G. D., J. de Alencar und F. B. Smith: Some effects of *Streptomyces albus* and *Penicillium* spec. on *Rhizobium meliloti*. Soil Sci. Soc. Amer. Proc. **14**, 188 (1949). — Thornton, G. D., und F. E. Broadbent: Preliminary studies of the influence of nitrogen fertilization of peanuts on nodulation, yield and gynophore absorption of this element. J. Amer. Soc. Agron. **40**, 64–69 (1948). — Thornton, G. D., und C. F. Eno: Soil microbiology in Florida, 1939–1959. Soil a. Crop Sci. Soc. Florida Proc. **19**, 39–51 (1959). — Thornton, H. G.: A suggested explanation of the "inefficiency" of certain strains of rhizobia. Trans. IIIrd Com. Int. Soc. Soil Sci., New Brunswick A, 20–23 (1939). — Effective and ineffective strains of legume nodule bacteria. Nature (Lond.) **156**, 654 (1945). — Report of the Department of Soil Microbiology for the years 1939–1945. Rothamsted Exper. Stat. Harpenden, England 1947. — The nodule bacteria and their host legumes; some problems that they still present. Sci. Progr. **166**, 185–204 (1954); Ref. Zbl. Bakt. II **108**, 500 (1954/55). — The development and present problems of soil microbiology. J. Sci. Food Agric. **7**, 93–101 (1956). — The ecology of microorganisms in soil. Proc. Roy. Soc. Lond. B **145**, 364–374 (1956a). — Thornton, H. G., und P. H. H. Gray: The numbers of bacterial cells in field soils as estimated by the ratio method. Proc. Roy. Soc. Lond. B **115**, 522–543 (1934). — Thornton, H. G., und J. Meiklejohn: Soil microbiology. Ann. Rev. Microbiol. **11**, 123–148 (1957). — Thornton, H. G., und C. B. Taylor: Short-period fluctuations in bacterial numbers in soil. Trans. 3rd Int. Cong. Soil Sci., Oxford **1** (3), 175–179 (1935). — Thurber, G. A., J. R. Douglas und A. W. Galston: Inhibitory effect of Gibberellins on nodulization in dwarf beans. *Phaseolus vulgaris*. Nature (Lond.) **181**, 1082–1083 (1958); Ref. Z. Pflanzenernähr., Düng., Bodenkde. 82, 213 (1958). — Timonin, M. J.: The interaction of higher plants and soil microorganisms, I, Microbial population of rhizosphere of seedlings of certain cultivated plants. Canad. J. Res. **18**, 307–317 (1940a). — The interaction of higher plants and soil microorganisms, II, Study of the microbial population of the rhizosphere in relation to resistance of plants to soil-borne diseases. Canad. J. Res. **18**, 444–455 (1940b). — Microflora of the rhizosphere in relation to the manganese-deficiency disease of oats. Soil Sci. Soc. Amer. Proc. **11**, 284–292 (1946). — Tittor, J.: Untersuchungen über die Bodenmikroflora von Gatersleben und Umgebung, II, Untersuchungen über die Verbreitung von Bodenalgen. Kulturpflanze **4**, 195–207 (1956). — Tolle, R., und A. Rippel-Baldes: Untersuchungen über die Rhizosphäre von Gramineen. Zbl. Bakt. II **111**, 204–217 (1958). — Topping, L. E.: The predominant microorganisms in soils. Zbl. Bakt. II **97**, 289–303 (1937/38), **98**, 193–201 (1938). — Tornau, O.: Der Dauerdüngungsversuch des Göttinger E-Feldes. Ein Sammelbericht. Göttingen 1959. — Toropkina, A.: Der Einfluß des Verfahrens der Hauptbearbeitung des Bodens auf die Entwicklung der Mikroflora. Baumwoll-Wirtsch. H. 3, 31–35 (1959) (russ.); Ref. Z. Pflanzenernähr., Düng., Bodenkde. 88, 272 (1960). — Tribunskaja, A. Ja.: Untersuchungen der Rhizosphärenmikroflora der Kiefersämlinge. Mikrobiologija **24**, 188–1952 (1955) (russ.); Ref. Z. Pflanzenernähr., Düng., Bodenkde. **72**, **94** (1956). — Trolldenier, G.: Die Bedeutung der Rhizosphärenorganismen für die Pflanze (Sammelreferat). Landwirtsch. Forsch., 15. Sonderheft, 101–109 (1961). — Tschapek, M., und N. Giambiaja: Nitrogen fixation of Azotobacter in soil. Its inhibition by oxygen. Arch. Mikrobiol. **21**, 376–390 (1955). — Tschishik, G. J.: Über die Struktur der Knöllchenbakterien. Mikrobiologija **28**, 28–33 (1959) (russ. u. engl. Zusammenfassung). Ref. Zbl. Bakt. II **113**, 541 (1959/60). — Tulaykowa, K. P.: Die Wurzelmikroflora des Hanfs, I, Quantitative Charakteristik der Wurzelmikroflora des auf verschiedenen Böden gezüchteten Hanfs. Z. obšč. Biol. **15**, 288–302 (1954); Ref. Zbl. Bakt. II **108**, 503 (1954/55). — Das Wachstum der Wurzelmikroorganismen beim unmittelbaren Kontakt mit granulierten Superphosphat. Mikrobiologija **25**, 299–304 (1956).

Uebelmesser, E. R.: Über einige neue Chytridineen aus Erdboden (Olpidium, Rhizophidium, Phlyctochytrium und Rhizophyctis). Arch. Mikrobiol. **25**, 307–324 (1956). — Unger, H.: 2-, 3-, 5-Triphenyl-Tetrazoliumsalz als Hilfsmittel bei der Keimzählung nach dem Kochschen Gußplattenverfahren. Arch. Mikrobiol. **32**, 20–24 (1958). — Die Änderung der Bakterienzahl abwasserbehandelter Böden mit Berücksichtigung der coliformen Bakterien. Wiss. Z. Karl-Marx-Univ. Leipzig, Math.-Naturwiss. Reihe **9**, H. 4 (1959/60).

VANČURA, V.: Composition of capsular polysaccharides of smooth and rough types of colonies of *Azotobacter chroococcum*. Folia Microbiol. **5**, 100–104 (1960). — VANČURA, V., und J. MACURA: Indole derivates in Azotobacter cultures. Folia Microbiol. **5**, 293–297 (1960). — The effect of root excretions on Azotobacter. Folia Microbiol. **6**, 250–259 (1961). — VANĚK, Z., L. DOLEŽILOVÁ und Z. ŘEHÁČEK: Formation of a mixture of antibiotic substances including antibiotics of a polygene character, by strains of Actinomycetes freshly isolated from soil samples. J. Gen. Microbiol. **18**, 649–657 (1958). — VERNER, A. P., und A. A. KOVALEV: On the problem of nitrogen-fixing power of *Bact. radicicola*. C. R. (Dokl.) Acad. Sci. URSS **13**, 325–329 (1936). — VINCENT, J. M., und L. M. WATERS: The influence of the host on competition amongst clover root-nodule bacteria. J. Gen. Microbiol. **9**, 357–370 (1953). — VIRTANEN, A. I.: Symbiotic nitrogen fixation. Nature (Lond.) **155**, 747–748 (1945). — The biology and chemistry of nitrogen fixation by legume bacteria. Biol. Rev. **22**, 239–269 (1947). — Biological Nitrogen Fixation. Ann. Rev. Microbiol. **2**, 453–484 (1948). — Investigations on nitrogen fixation by the alder, II, Associated culture of spruce and inoculated alder without combined nitrogen. Physiol. Plant. **10**, 164–169 (1957); Ref. Z. Pflanzenernähr., Düng., Bodenkde. **79**, 66 (1957). — VIRTANEN, A. I., P. K. HIETALA und O. WAHLROOS: Antimicrobial substances in cereals and fodder plants. Arch. Biochem. Biophys. **69**, 486–500 (1957); zit. nach STARKEY 1958. — VIRTANEN, A. I., und H. LINCOLA: Über die Wirkung der Azotobakterimpfung auf das Wachstum von Erbsen und Weizen in sterilen Sandkulturen ohne N-Nahrung. Naturwiss. **41**, 70–71 (1954). — VIRTANEN, A. I., und S. SAUBERT-v. HAUSEN: A method for determining in pea-cultures the amount of molecular nitrogen fixed and the amount of combined nitrogen taken up from the soil. Plant a. Soil **4**, 171–177 (1953). — VOETS, J. P., und J. DEBACKER: *Pseudonomonas azotogensis* nov. spec. a new free living nitrogen fixer. Naturwiss. **43**, 40–41 (1956). — VOZNIAKOVSKAJA, IU. M., und G. K. ZHIL'TSOVA: Species composition of the root microflora of some plants. Microbiologija **27**, 599–606 (1958). — VRANÝ, J.: Occurrence of bacteria assimilating benzoic acid and p-hydroxybenzoic acid in cereal rhizosphere and in soil. Folia Microbiol. **5**, 116–119 (1960). — VYAS, S. R., und N. PRASAD: Studies on Rhizobium from Zornia diphylla. Proc. Ind. Acad. Sci. Sect. B **49** (3), 156–160 (1959); Ref. Biol. Abstr. **34**, 11 (1959).

WAGENBRETH, D.: Leguminosenpfropfung und Wirtsspezifität der Knöllchenbakterien. Flora **144**, 84–97 (1956); Ref. Z. Pflanzenernähr., Düng., Bodenkde. **78**, 262 (1957). — WAKSMAN, S. A.: Methoden der mikrobiologischen Bodenforschung. In: ABERDERHALDEN, Handbuch der biologischen Arbeitsmethoden, Bd. 11, Teil 3, H. 5, S. 715–864 (1928). — Principles of Soil Microbiology, 2nd ed. Baltimore: Williams & Wilkins. 1932. — Three decades with soil fungi. Soil Sci. **58**, 89–115 (1944). — The Actinomycetes, their nature, occurrence, activities and importance. Waltham, Mass: Chronic Botanica Co. 1950. — Soil Microbiology. New York: Wiley. 1952. — Guide to the classification and identification of the Actinomycetes and their Antibiotics. Baltimore: Williams & Wilkins. 1953. — The Actinomycetes, Vol. I, Nature, occurrence and activities. Baltimore: Williams & Wilkins. 1959. — WAKSMAN, S. A., und R. L. STARKEY: The soil and the microbe. New York: Wiley. 1931. — WALKER, R. L., und A. S. NEWMAN: Microbial decomposition of 2,4-Di-Chlorophenoxyacetic acid. Appl. Microbiol. **4**, 201–206 (1956). — WALKER, T. W., A. F. R. ADAMS und H. D. ORCHISTON: Fate of labeled nitrate and ammonium nitrogen when applied to grass and clover grown separately and together. Soil Sci. **81**, 339–351 (1956). — WALLACE, R. H., und H. de L. KING: Nutritional groups of soil bacteria on the roots of barley and oats. Soil Sci. Soc. Amer. Proc. **18**, 282–285 (1954). — WALLACE, R. H., und A. G. LOCHHEAD: Bacteria associated with seeds of various crop plants. Soil Sci. **71**, 159–166 (1951). — WALLHÄUSER, K. H.: Untersuchungen über das antagonistische Verhalten von Mikroorganismen am natürlichen Standort. Arch. Mikrobiol. **16**, 237–251 (1951). — WALT, I. P. van der: Three new sporogenous yeasts from soil. A. v. LEEUWENHOEK **23**, 23 (1957). — WALT, I. P. van der, und J. T. TSCHEUSCHNER: *Pichia vanriji* n. sp. isolated from soil. J. Gen. Microbiol. **15**, 459–461 (1956). — *Saccharomyces pretoriensis* n. sp. from South African soil. J. Gen. Microbiol. **14**, 485–486 (1956). — WARCUP, J. H.: The soil-plate method for isolation of fungi from soil. Nature (Lond.) **166**, 117 (1950). — Isolation of fungi from hyphae present in soil. Nature (Lond.) **175**, 953–954 (1955). — WELLS, C. G., W. A. SEAY, C. E. BORTNER und W. D. VALLEAU: The effects of phosphorus and organic matter on the concentration of certain decomposition products of cyanamid in tobacco plant beds. Soil Sci. Soc. Amer. Proc. **21**, 192 (1957). — WELTE, E., und G. TROLLDENIER: Der Einfluß der Wasserstoffionenkonzentration des Bodens auf den Rhizosphäreneffect. Naturwiss. **48**, 14, 509 (1961). — WEST, M. P.:

Excretion of thiamine and biotin by the roots of higher plants. Nature (Lond.) **144**, 1050–1051 (1939). — WEST, M. P., und A. A. HILDEBRAND: The microbiological balance of strawberry root rot soil as related to the rhizosphere and decompsotion of certain cover crops. Canad. J. Res. **19**, 199–210 (1941). — WEST, M. P., und A. G. LOCHHEAD: Qualitative studies of soil-microorganisms, IV, The rhizosphere in relation to the nutritive requirements of soil bacteria. Canad. J. Res. **18**, 129–135 (1940a) — The nutritional requirements of soil bacteria, a basis for determining the bacterial equilibrium of soils. Soil Sci. **50**, 409–420 (1940b). — WHITE, A. W., J. E. GIDDENS und H. D. MORRIS: The effect of sawdust on crop growth and physical and biological properties of Cecil soil. Soil Sci. Soc. Amer. Proc. **23**, 365–368 (1959). — WICHTMANN, H.: Untersuchungen über die Beziehungen zwischen Azotobaktergehalt und Bodenfruchtbarkeit. Arch. Mikrobiol. **17**, 54–78 (1952). — WIERINGA, K. T.: Der Abbau der Pektine; der erste Angriff der organischen Pflanzensubstanz. Z. Pflanzenernähr., Düng., Bodenkde. **69**, 150–155 (1955). — WIGOREW, L. J.: Mikrochemische Untersuchungen von Podsolböden. Agrobiologija H. 5, 93–107 (1955) (russ.); Ref. Z. Pflanzenernähr., Düng., Bodenkde. **72**, 276 (1956). — Ausnutzung des mineralischen Stickstoffs und der organischen Dünger durch die Mikroorganismen in rasenpodsoligen Böden. Agrobiologija **1958**, 88–94 (russ.); Ref. Z. Pflanzenernähr., Düng., Bodenkde. **83**, 186 (1958). — WILDE, S. A., und D. J. PERSIDSKY: Effect of biocides on the development of ectotrophic mycorrhizae in Monterey pine seedlings. Soil Sci. Soc. Amer. Proc. **20**, 107–110 (1956). — WILLIAMSON, O. C.: A study of the microflora of a grassland and a forested soil. Trans. Kansas Acad. Sci. **59**, 32–38 (1956). — WILSON, P. W.: The carbohydrate-nitrogen relation in symbiotic nitrogen fixation. Wis. Agr. Exper. Stat. Res. Bull. **129**, 40 (1935). — Symbiotic nitrogen-fixation by the Leguminosae. Bot. Rev. **3**, 365–399 (1937). — Mechanism of symbiotic nitrogen fixation. Erg. Enzym-Forsch. **8**, 13–54 (1939). — The biochemistry of symbiotic nitrogen-fixation. University of Wisconsin Press 1940. — Biological nitrogen-fixation. In: Bacterial Physiology, Chap. 14. New York: Academic Press. 1951. — Asymbiotic nitrogen fixation. In: RUHLAND, Handbuch der Pflanzenphysiologie, Bd. 8, S. 9–47 (1958). — WILSON, P. W., und R. H. BURRIS: The mechanism of biological nitrogen fixation. Bact. Rev. **11**, 41–73 (1947). — Biological nitrogen fixation—a reappraisal. Ann. Rev. Microbiol. **7**, 415–432 (1953). — WILSON, P. W., und W. B. SARLES: Root nodule bacteria. Tabular Biol. **17**, 338–367 (1939). — WILTGEN, N., und CH. BONNIER: A propos de l'examen mycologique du sol. Description d'une méthode. Bull. Inst. Agron. Stat. Rech. Gembloux, 391–396 (1952); Ref. Z. Pflanzenernähr., Düng., Bodenkde. **66**, 91 (1954). — WINOGRADSKY, S.: Etudes sur la microbiologie du sol, I, Sur la méthode. Ann. Inst. Past. **39**, 299–354 (1925). — Microbiologie du sol. Paris: Masson. 1949. — WINTER, A. G.: Der Einfluß partieller Sterilisation des Bodens auf die Entwicklung der Laufhyphen von *Ophiobolus graminis*. Phytopath. Z. **14**, 204–302 (1942). — Eine Methode zur quantitativen Bestimmung des Wachstums parasitischer oder saprophytischer Pilze im natürlichen Boden. Festschrift Appel, Biol. Zentralanst. Berlin 12—15 (1947). — Untersuchungen über die Beziehungen zwischen *Ophiobolus graminis* und anderen Organismen mit Hilfe der Aufwuchsplattenmethode. Arch. Mikrobiol. **14**, 240–270 (1949). — Untersuchungen über die Ökologie und den Massenwechsel bodenbewohnender mikroskopischer Pilze, I. Arch. Mikrobiol. **15**, 42–71 (1950/51). — Untersuchungen über die Ökologie und den Massenwechsel bodenbewohnender mikroskopischer Pilze, II. Arch. Mikrobiol. **16**, 136–162 (1951a). — Über die Aufnahme von Antibioticis durch höhere Pflanzen und ihre Stabilität in natürlichen Böden. Naturwiss. **38**, 457–458 (1951b). — Untersuchungen über Vorkommen und Bedeutung von antimikrobiellen und antiphytotischen Substanzen in natürlichen Böden. Z. Pflanzenernähr., Düng., Bodenkde. **69**, 224–233 (1955). — Analyse und Synthese in der Bodenbiologie. Landwirtsch. Forsch. **12**, Sonderheft, 97–110 (1959); Ref. Z. Pflanzenernähr., Düng., Bodenkde. **90**, 186 (1960). — WINTER, A. G., und F. SCHÖNBECK: Untersuchungen über den Einfluß von Stickstoff auf den Abbau von Hemmstoffen aus Getreidestroh- und -stoppeln. Z. Pflanzenernähr., Düng., Bodenkde. **79**, 132–142 (1957). — Zum Wirkstoffkreislauf Pflanze-Tier-Boden. Z. Pflanzenernähr., Düng., Bodenkde. **84**, 11–21 (1958). — WINTER, A. G., und L. WILLEKE: Untersuchungen über Antibiotica aus höheren Pflanzen, V, Hemmstoffe in Blättern und Blattstreu der Gramineen. Naturwiss. **39**, 190–191 (1952). — WITKAMP, M.: Seasonal fluctuations of the fungus flora in mull and mor of an oak forest. Inst. Toegepast Biol. Onderzoek in de Natuur (Itbon), Arnhem, Med. **46**, 51 (1960). — WLASJUK, P. A., K. M. DOBROTWORSKAJA und R. N. OLEINIK: Der Einfluß organisch mineralischer Düngung auf die mikrobiologischen Prozesse im Boden. Mikrobiologija **25**, 293–298 (1956). — WLASJUK, P. A., K. M. DOBROTWORSKAJA und S. A. GORDIENKO: Die Intensität der Fermentwirkung in der Rhizosphäre einzelner landw. Kulturpflanzen. Doklady Wssessojuznoj Akademii selskochoz. nau im. Lenina **22**, H. 3, 14–19 (1957) (russ.); Ref. Zbl. Bakt. II **111**, 317 (1958). — WOLF, E.: Beitrag zur Systematik der Gattung Mortierella und Mortierella-Arten als Mycorrhiza-Pilze bei Ericaceen. Zbl. Bakt. II **107**, 523–548 (1952/54).

ZEUMER, H.: Rückstände von Pflanzenschutz- und Vorratsschutzmitteln, von sonstigen Schädlingsbekämpfungs- und Unkrautbekämpfungsmitteln, sowie von Mitteln zur Beeinflussung des Pflanzenwachstums. Literatur-Übersicht. Mitt. Biol. Bundesanst. Berlin-Dahlem, **1958**, 124 S.; Ref. Z. Pflanzenernähr., Düng., Bodenkde. **84**, 309 (1958). — ZIEMICKA, J.: The Azotobacter test of soil fertility applied to the classical fields at Rothamsted. J. Agr. Sci. **22**, 797–810 (1932). — The use of a modified Rossi-Cholodny technique for studying the organisms that decompose certain organic compounds in soil. Zbl. Bakt. II **91**, 379–394 (1935). — ZoBELL, C. E.: Marine microbiology, a monograph on hydrobacteriology. Waltham, Mass. 1946.

b) Bedeutung der Mikroorganismen für die Bodenstruktur

Die Bedeutung der Mikroorganismen für die Bodenstruktur kann sich, wie Abb. 108 zeigt, auf sehr verschiedene Weise manifestieren; denn alle Stoffe bzw. Maßnahmen, die einen Einfluß auf die Mikroorganismen des Bodens ausüben, berühren letztlich auch seine Struktur. Nun wird die Struktur auf primärem Wege durch die Bodenbearbeitung bzw. durch synthetische Bodenstabilisierungs-

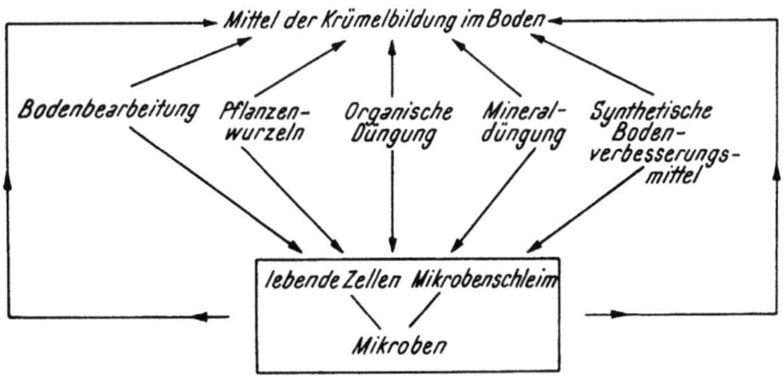

Abb. 108. Schema zur Bedeutung der Mikroorganismen für die Bodenstruktur (nach GLATHE 1958)

mittel verändert; da aber beide Maßnahmen gleichzeitig einen starken Einfluß auf die Mikroflora bzw. auf ihre Aktivität ausüben, müssen sie sich auch auf sekundärem Wege in der Struktur bemerkbar machen. Bisher sind jedoch nur die Einflüsse auf die Zusammensetzung der Mikroflora studiert worden, aber nicht deren zusätzliche Auswirkung auf die Struktur. Im folgenden soll nun die direkte Beeinflussung der Struktur durch die Mikroflora erörtert werden.

MARTIN (1945) gliedert die Faktoren, durch die Mikroorganismen die Struktur des Bodens beeinflussen können, wie folgt:

1. Zellen und Fäden von Mikroorganismen, die organische Stoffe abbauen,
2. Produkte der mikrobiellen Synthese,
3. Abbauprodukte des mikrobiellen Stoffwechsels,
4. Wasserlösliche Stoffe aus der ursprünglichen organischen Substanz.

Darüber hinaus muß auch mit der rein physikalischen Umschließung einzelner Partikel durch Hyphen von Pilzen und Actinomyceten bzw. Fäden von Bakterien gerechnet werden.

Zahlreiche Forscher (KANIVETZ und KORNEEVA 1938, MARTIN und ANDERSON 1943, MARTIN und WAKSMAN 1940, MYERS und MCCALLA 1941, MISCHUSTIN und PUSCHKINSKAJA 1942, PEELE 1940, SWABY 1949, NOVÁK und NOVÁKOWÁ 1960) kamen zu dem Ergebnis, daß die mikrobielle Aktivität die Bodenstruktur verbessere.

MCCALLA (1945b) ging diesem Problem nach, als Bodenerosionen die Ackerkrume des Lößlehms in Peoria entführt hatten und der Untergrund in Kultur genommen werden mußte. In erster Linie sollte nun dessen Struktur so stabilisiert werden, daß der Boden wasseraufnahmefähig bzw. wasserdurchlässig wurde und infolgedessen gegen Erosion geschützt war. Bei Perkolationsversuchen des unbewegten Bodens führte das Wachstum von Mikroorganismen zu einer Verstopfung des Bodens. Wurde der Boden jedoch umgerührt, führte die Aktivität der Mikroorganismen zu einer Krümelstruktur, die durch Wasser nicht aufgehoben wurde. Er (MCCALLA 1944, 1945) führte den Beweis dadurch, daß er die größere Stabilität dieser Krümel gegenüber tropfendem Wasser nachwies (vgl. auch MISONO und SUTO 1958). YODER (1936) entwickelte eine eigene Apparatur. Kontinuierliche Durchflußapparaturen wurden ferner von LEES und MACURA angewendet (LEES und QUASTEL 1946, LEES 1949, MACURA und MALEK 1958, MACURA 1961, 1961a).

Über den Einfluß verschiedener Arten von Mikroorganismen auf die Struktur des Bodens erzielte MCCALLA (1946) mit Hilfe der Tropfenmethode die in Tab. 86 angeführten Ergebnisse:

Tabelle 86. *Einfluß einiger Organismen auf die Stabilität der Bodenstruktur gegen tropfendes Wasser*

(Auszug aus verschiedenen Übersichten von MCCALLA)

Kultur	Tropfen je 0,1 g Boden nach Bebrütung für die Dauer von ... Tagen					
	0	1 Tag	5 Tage	10 Tage	20 Tage	30 Tage
Rhizobien von Soja ...	5,1	5,3	6,7	6,9	13,2	20,3
Rhizobien von Lotus ..	5,1	6,5	14,6	31,7	—	49,3
Azotobacter 2	5,8	7,2	7,9	8,5	9,6	6,0
Sporenbildner	5,1	8,7	7,2	7,4	8,0	8,8
Trichobacterium 1	5,3	—	49,8	49,0	47,5	43,8
Bakterienschleim 2	5,3	9,7	51,8	54,2	52,6	49,2
Mittel von 13 Bakterien	5,3	7,0	18,8	22,9	23,0	22,8
Mittel von 5 Actinomyceten	5,4	6,4	26,9	37,1	44,5	40,4
Mittel von 4 Hefen	5,2	17,6	22,2	19,8	13,9	18,1
Mittel von 7 Pilzen ...	5,1	38,0	45,6	43,2	36,1	33,7

Tab. 86 läßt erkennen, daß einige schleimbildende Bakterien sowie Actinomyceten und Schimmelpilze im Boden stark stabilisierend wirken.

Unter den Bakterien erwiesen sich Stämme von *Agrobacterium radiobacter*, *Bac. subtilis*, *Bac. mesentericus* und *Pseudomonas* als aktiv (MCCALLA 1942, MISCHUSTIN und PUSCHKINSKAJA 1942, SCHUFFELEN und v. SCHUYLENBORGH 1950, LÜTJEHARMS 1952, RENNIE et al. 1954, SCHACHTSCHABEL 1956). GILMOUR et al. (1948) haben verschiedene Pilze, nämlich *Aspergillus niger*, *Penicillium* sp., *Alternaria* sp., *Fusarium* sp., *Helminthosporium* sp., *Curoularia* sp., *Stemphylium* sp. und *Sporotrichium* sp., in drei verschiedenen Böden geprüft. Die Krümelbildung war abhängig von der Wirksamkeit der Pilzart, der organischen Substanz, die gleichzeitig gegeben wurde, und der physikalischen Zusammensetzung des Bodens. Bei Zusatz von 1% gemahlenem Stroh bei einem Wassergehalt von 25 bis 30% bewirkte in Versuchen von MCCALLA und seinen Mitarbeitern (DOWNS et al. 1955) von 12 Zellulose zersetzenden Pilzen die intensivste Krümelung *Stachybotrys atra* (vgl. auch MARTIN et al. 1959). Auch *Alternaria* erwies sich als sehr wirksam (MCCALLA, HASKINS und FROLIK 1957). Es zeigte sich jedoch (MCCALLA,

Haskins and Curley 1958), daß Pilze, wie z. B. *Stachybotrys atra*, erst dann voll zur Wirkung kommen, wenn die natürliche Mikroflora gehemmt wurde. Das geschah durch Dampfbehandlung im Autoklaven bzw. durch Chloropikrin bzw. Telon. Die Zahl der aktiven Mikroorganismen ist wahrscheinlich erheblich höher. So fanden Thornton, Cowie und McDonald (1956), daß unter Kiefern auch *Rhizopogon rubescens*, *Clathrus cibarius* und *Geastrum* sowie Vertreter der Basidiomyceten an der Krümelung beteiligt waren. Martin, Ervin und Shepherd (1959) dehnten die Liste der Pilze aus auf: *Epicoccum purpurascens*, *Aspergillus versicolor*, *Volutella* sp., *Diplodina* sp., *Pyrenochaeta* sp., *Stemphylium consortiale* und *Aspergillus sulfureus* (letzterer mit geringster Wirksamkeit). Müller (1958) erkannte als sehr aktive Krümelbildner je einen Stamm von *Fusarium*, *Trichoderma* und *Penicillium*, ein Gemisch aller geprüften Pilz- und Bakterienstämme sowie eine natürliche Bodenaufschwemmung, Novák und Nováková (1960) Hyphen von Pilzen und Actinomyceten.

Nach McCalla (1945, 1945a und 1950) werden die Krümel durch chemische, physikalische und biologische Kräfte, durch Silicium, Ton, Eisenoxyde und organische Stoffe zusammengehalten. Zu den letzteren gehören Pflanzenwurzeln und Organismen verschiedener Art, Pflanzenharze, Schleimstoffe, Wachs, Fett und Lignin. Bei *Cladosporium* sp. bestanden z. B. die bindenden Kräfte zur Hälfte aus verklebenden Stoffwechselprodukten, zur Hälfte aus Hyphen (McCalla 1942). Auf Grund mikroskopischer Studien mit Hilfe der „Aufwuchsplattenmethode" nach Cholodny-Rossi kam Sekera (1941) ebenfalls zu dem Ergebnis, daß vor allem Pilzhyphen an der Krümelbildung beteiligt sind; ihre Tätigkeit bezeichnete er als „Lebendverbauung" (vgl. auch Kuron et al. 1959). Nicht zu unterschätzen ist die Bedeutung der von den Mikroben gebildeten Kohlensäure. Nach Laatsch (1941) erhöht sie bei genügendem Kalkgehalt die Bikarbonatkonzentration in der Bodenlösung. Dadurch wächst die Phosphationenkonzentration. Die gleichmäßige Verteilung dieser Ionen auf den Kolloidoberflächen fördert Krümelbildung und -stabilität, wodurch wiederum die Mikrobenatmung günstig beeinflußt wird.

Bei den Untersuchungen über die Wirkung der Stoffwechselprodukte der Mikroorganismen wurde der Wirkung des Bakterienschleimes besondere Aufmerksamkeit gewidmet. Rennie et al. (1954) führten ihre Versuche mit einem von *Agrobacterium radiobacter* synthetisierten Polysaccharid durch. Geoghegan und Brian (1946, 1948) konnten nachweisen, daß in geeigneten Nährlösungen auch *Bac. subtilis* Polysaccharide bildete, so daß er einen ähnlichen Bindungseffekt auf Bodenpartikel ausübte wie *Agrob. radiobacter*. Forsyth und Webley (1949) wiesen Bakterien, die befähigt sind, Polysaccharide zu erzeugen, in verschiedenen Acker-, Moor- und Waldböden nach. Ihr Anteil an der mit Hilfe der Kochschen Methode erfaßbaren Bakterienflora schwankte zwischen 5 und 16%. Vier Typen von Polysacchariden wurden ermittelt: Levane, Glukose-Uronsäuren, Glukose-Mannose-Uronsäuren und Glukose-Mannose-Rhamnose-Uronsäuren; die ersten beiden waren am meisten verbreitet. Rennie et al. (1954) verglichen die Wirkung des Agrobacterschleimes mit der des Kriliums. Weiterhin wurde ein Bakterienschleim geprüft, der aus einem schwach krümeligen Boden extrahiert worden war.

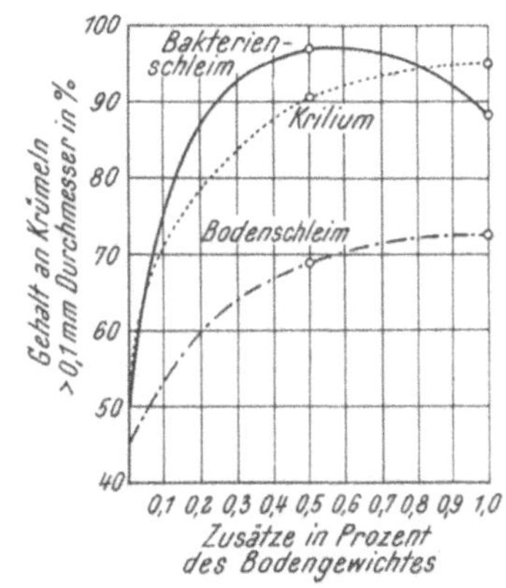

Abb. 109. Wirkung von Bakterienschleim, Krilium und Bodenschleim bei äquivalentem Zusatz zu tonigem Lehmboden auf dessen Gehalt an Krümeln

Die Schleimstoffe waren als Pulver dargestellt, sie wurden dem Boden in wässeriger Lösung zugesetzt. Die Ergebnisse sind in Abb. 109 und Tab. 87 enthalten.

Tabelle 87. *Wirkung von 0,1% Krilium bzw. Bakterienschleim auf die Krümelbildung in Lehmboden* (SPENCER)

Tage nach der Behandlung	Prozentgehalt an wasserunlöslichen Krümeln, bezogen auf das Gewicht des Bodens					
	Kontrolle		Krilium		Bakterienschleim	
	> 0,5 mm ⌀	> 0,1 mm ⌀	> 0,5 mm ⌀	> 0,1 mm ⌀	> 0,5 mm ⌀	> 0,1 mm ⌀
Beginn	17,4	50,3	52,3	62,8	58,5	69,6
7	12,0	33,6	54,8	63,6	49,3	61,2
14	12,6	34,0	53,8	63,3	57.5	66,5
21	10,5	30,6	49,0	61,0	40,0	56,2
42	10,1	31,7	51,9	62,6	31,4	55,0

Alle Produkte führten zu einem Anstieg des Gehalts an Krümeln ⌀ >0,1 mm. Bei einem Zusatz von 0,02 g je 100 g Boden wurde eine Zunahme von etwa 50% an Krümeln erzielt, deren Durchmesser 0,1 mm überstieg. Aus Tab. 87 sind die Gehalte an Krümeln ersichtlich, die zu verschiedenen Entnahmezeiten festgestellt wurden. Die Stabilität der von Bakterienschleim gebildeten Krümel ist nach 42 Tagen bereits wieder zurückgegangen. Daraus kann geschlossen werden, daß der Schleim vermutlich auf mikrobiologischem Wege wieder abgebaut wird. Schon McCALLA (1942) hatte festgestellt, daß die krümelbildenden Polysaccharide durch andere Mikroorganismen abgebaut wurden (vgl. auch MARTIN 1946). Den engen Zusammenhang zwischen dem Gehalt an Krümeln und dem an Bakterienschleim lassen auch die Untersuchungen an einem jungfräulichen Boden (Tab. 88 nach RENNIE et al.) erkennen.

Tabelle 88. *Schleimgehalt und prozentuale Krümelung in drei Horizonten eines jungfräulichen Lehmbodens* (Miami)

Horizont und Tiefe	Beschreibung des Bodens	Schleimgehalt des Bodens %	Gehalt an Krümeln	
			⌀ >0,5 mm	> 0,1 mm
A_1 bis 7,5 cm	krümelige Struktur, dunkelgrau mit zahlreichen Wurzeln	0,23	63,6	82,6
A_2 15 bis 20 cm	plattige Struktur, hellgrau	0,02	2,8	28,6
B_1 30 bis 40 cm	dichte, nußartige Struktur, rotbraun	0,05	14,9	58,3

Die Schleimstoffe sind nicht die alleinige Ursache der Krümelbildung. Nach GEOGHEGAN und BRIAN (1946) stellen sie nur eine Gruppe von den strukturbildenden Stoffwechselprodukten der Mikroorganismen dar, deren Wirkung im Ackerbau jedoch von den jeweiligen Bodeneigenschaften abhängig ist. CHESTERS et al. (1957) bestätigen das und betrachten den Bakterienschleim als den im allgemeinen wichtigsten Faktor, jedoch stand z. B. in Kewaunee-Boden der Gehalt an Eisenoxyd an erster Stelle. Letzteres zeigte eine spezifische Wirkung auf die kleineren Krümel. Demgegenüber war die Wirkung des Tons relativ gering. Die für die Krümelbildung günstigste Reaktion lag bei pH 6,5. Zwischen der CO_2-

Produktion des Bodens und der Krümelbildung bestand keine Korrelation. McCalla (1945b) stellte fest, daß Strohdüngung die Krümelstruktur verbesserte. Er prüfte bei diesen Versuchen die Bindefähigkeit der Krümel mit Hilfe der Perkolationsmethode. Die Ergebnisse sind in Tab. 89 zusammengestellt.

Tabelle 89. *Perkolationsrate von Löß (Peoria) in Abhängigkeit von der Zersetzungsgeschwindigkeit von Weizenstroh* (20 dz Stroh + 90 kg NH_4NO_3/ha)

Tage nach der Strohzugabe	Perkolation Zoll/Stunde						Gesamtmenge in
	1 Std.	2 Std.	3 Std.	4 Std.	5 Std.	6 Std.	6 Std.
0	0,05	0,81	0,76	0,38	0,44	0,37	2,81
10	0,76	2,39	1,84	0,93	0,93	0,87	7,72
20	0,81	3,00	2,21	0,96	1,10	0,76	8,84
26	0,46	2,63	2,76	1,17	1,26	1,26	9,54
46	0,00	1,31	1,53	0,64	0,79	0,81	5,08

Mit zunehmender Dauer der Zersetzung des Strohs stieg die Perkolationsrate, um dann wieder zu sinken. Die größte Wirksamkeit besitzen offenbar die Zersetzungsprodukte (McCalla 1945b). Ein Zusatz von Zucker während der Perkolation senkte infolge der starken Vermehrung der Bakterien den Durchfluß des Wassers ruckartig (McCalla 1951). Das war jedoch nicht der Fall, wenn zusätzlich Sublimat (10 ppm) gegeben wurde. Offenbar unterband das Desinfektionsmittel die Aktivität der Mikroorganismen. Dawson (1946) konnte indessen einen Zusammenhang zwischen Krümelung und Mikroflora unter dem Einfluß der Ernterückstände nicht beobachten. In Versuchen von Stephenson und Schuster (1945) trat nach einer Bedeckung des Bodens mit Stroh eine Zunahme der größeren Bodenkrümel ein. Rennie et al. (1954) stellten eine weitere Erhöhung des Krümelgehaltes im Boden fest, wenn die organische Düngung durch Kalkung ergänzt wurde (vgl. Tab. 90).

Tabelle 90. *Wirkung von organischer Substanz und Kalk auf den Gehalt an Bakterienschleim und Krümeln im Lehmboden* (Spencer)

Düngung	Reaktion pH	Gehalt an Krümeln ø > 0,5 mm in %			Gehalt an Bakterienschleim in kg je 4000000 kg Erde		
		6 Tg.	24 Tg.	58 Tg.	6 Tg.	24 Tg.	58 Tg.
Ungedüngt	5,5	15,2	35,7	31,7	5200	2000	3440
Kalk 100 dz/ha	6,5	50,8	60,5	40,0	5640	4640	4040
Weizenstroh 80 t/ha + Kalk 100 dz/ha	6,5	46,4	76,7	82,1	8800	—	3760
Luzerne 80 t/ha + Kalk 100 dz/ha	6,5	77,4	79,4	78,2	8080	6480	5960

Auch Gromyko (1948) beobachtete eine Erhöhung der Zahl der Makroaggregate nach Kalkung bei gleichzeitigem Anstieg der Keimzahlen von Bakterien und Actinomyceten.

Auf Grund der Untersuchungsergebnisse von Gericke (Tab. 91) stellte Jürgens-Gschwind (1960) fest, daß die Phosphorsäure ebenfalls krümelbildend wirkt.

Da die Phosphorsäure die mikrobielle Aktivität fördert, ist damit zu rechnen, daß neben der Kolloidverklebung, die durch phosphatbeladene Tonteilchen ausgelöst wird, auch eine stärkere Schleimproduktion eintritt, so daß die Krümelbildung durch die Phosphorsäure auch indirekt über die Mikroorganismen gefördert wird.

In einem mehrjährigen Feldversuch konnten RENNIE et al. nachweisen, daß die Kalkgaben sich noch nach sechs Jahren auf die Krümelbildung auswirkten. Ein Maximum an Krümeln wurde im Juli erreicht, ihm folgte ein Rückgang im Herbst. Die Ergebnisse lassen erkennen, daß auch unter natürlichen Verhältnissen die Schleimstoffe im Laufe der Zeit wieder abgebaut werden. Auch AMBROŽ (1956) fand im Sommer und zu Beginn des Herbstes die stärkste Krümelung. Die wasserstabilen Krümel >2 mm Durchmesser enthielten eine geringere Menge freier Huminsäuren als der nicht gekrümelte Boden, aber eine größere Menge an gebundenen Huminsäuren. Die Oberfläche der Krümel enthielt mehr Mikroorganismen, besonders Actinomyceten, als der strukturlose Boden. Es ist zwar nicht Aufgabe dieses Beitrages, über die Wirkung neuer synthetischer Bodenstabilisierungsmittel auf die Bodenstruktur zu referieren, indessen haben diese Stoffe auch einen direkten Einfluß auf die Mikroflora des Bodens. Diesen Zusammenhängen ist von verschiedenen Forschern nachgegangen worden. In welchem Umfang die Stoffe auch eine indirekte Wirkung, nämlich über die Aktivierung der Mikroflora, auf die Struktur auszuüben vermögen, ist zwar zu vermuten, aber noch nicht geklärt worden. Die verschiedenen Möglichkeiten der Einwirkung wurden von FULLER und GAIRAUD (1954) dargelegt, wobei sie insbesondere erörtern, welche Folgen in dem O_2-armen bzw. O_2-freien Inneren der Krümel zu erwarten sein könnten.

Tabelle 91. *Wirkung der Phosphorsäure auf den Krümelgehalt des Bodens nach Gericke*[1]

Düngung kg/ha P_2O_5	Krümelgehalt %	Krümelgehalt rel.
0	45	100
40	50	111
80	62	138
120	64	142
200	66	147

[1] Zitiert nach JÜRGENS-GSCHWIND 1960, S. 227.

Zunächst wurde von BONNIER und HELY (1953) bei einer vergleichenden Prüfung des belgischen Produkts „Aglusol“ mit dem amerikanischen Produkt „Krilium“ (Formel 9) nachgewiesen, daß mit einer toxischen Wirkung auf die Mikroflora des Bodens bei keinem dieser Stoffe zu rechnen ist, dagegen zeigten verschiedene Pilze wie *Mucor*, *Fusarium*, eine stärkere vegetative Vermehrung, während z. B. *Penicillium glaucum*, *Aspergillus niger* u. a. zwar ein geringes Wachstum, aber dafür eine umfangreichere Versporung erkennen ließen. Dagegen hatte das Produkt VAMA (=Copolymerisat aus *V*inyl*a*cetat und dem partiellen Methylester der *Ma*leinsäure) im Sandboden eine, allerdings nur vorübergehende, toxische Wirkung auf die Mikroflora des Bodens, nicht dagegen HPAN (=*h*ydrolisiertes *P*oly*a*cryl*n*itril). Die hemmende Wirkung war jedoch in anderen Böden von Arizona nicht zu beobachten. Nach MORTENSEN und MARTIN (1954, 1956) konnte mit Hilfe der gleichen Stoffe, die aber mit markiertem C^{14} hergestellt worden waren, an Hand des abgegebenen $C^{14}O_2$ nachgewiesen werden, daß innerhalb von 130 Tagen 0,2 bis 2,74% auf mikrobiellem Wege abgebaut wurden. Auch FULLER und GAIRAUD (1954) arbeiteten mit diesen Produkten, sie bestätigten die vorübergehende toxische Wirkung des VAMA und beobachteten eine Intensivierung der mikrobiellen Aktivität des Bodens. Kleine Mengen der Stabilisierungsmittel wurden abgebaut, danach war aber fünf Monate lang offenbar keine weitere Zersetzung eingetreten. Bei Agrylon wurde nach HAGIN (1955) die Nitrifikation auf das Doppelte erhöht. Demgegenüber beobachtete McCALLA (1959) bei HPAN weder einen Einfluß auf den Nitratgehalt, noch im Laboratoriumsversuch auf die Geschwindigkeit der Nitrifikation. HELY et al. (1954) erreichten eine Erhöhung des Krümelbestandes in Löß durch verschiedene Kunstharze, bessere Knöllchenbildung bei Luzerne wurde aber nur durch Aglusol erzielt. HELY und BONNIER (1960) stellten eine Förderung durch die von ihnen geprüften und in praktischen Grenzen angewendeten Stabilisierungsmittel bei Azotobacter und Streptomyceten darin fest, daß die Menge der gebildeten Schleimstoffe viel höher war als in den unbehandelten Böden. Noch fünf Jahre nach der Anwendung war die Vermehrung der Mikroorganismen verbunden mit einer deutlichen Strukturverbesserung durch „Flotal" bei lehmigen und tonigen Böden von SIEGEL und DISSELKAMP (1958) nachgewiesen worden.

Auch die *Bodenbearbeitung* verändert die Struktur des Bodens und damit die Aktivität der Mikroorganismen. So beobachtete FELDMANN (1956) eine Zunahme von Einzelkeimen und von Zoogloen nach dem Rigolen von Gartenboden. Letztere waren vor der Behandlung nicht nachweisbar gewesen. Eine Erhöhung der Keimzahl stellten auch AFANASJEWA und SWATSKAJA (1958) bei dem Einsatz verschiedener Arten von Pflügen fest. Eine sekundäre Beeinflussung der Struktur ist besonders dann zu erwarten, wenn die organische Substanz durch Bearbeitung rascher mineralisiert wird (KIUPERS 1955), etwa durch Freilegung der organischen Substanz aus den Bodenkrümeln (ROVIRA und GREACEN 1957). Eingehende Untersuchungen widmeten MÜLLER und RAUHE (1959) dem Einfluß der Tiefendüngung in Sandboden auf Mikroflora und Kleintiere. Während Bakterien, Actinomyceten und Kleintiere positiv reagierten, ließen die Bodenpilze keine Veränderung erkennen.

Bekannt ist schließlich die durch *Kleintiere des Bodens* bewirkte Verbesserung der Struktur. Die Wirkung ist zum Teil auf ihr Zusammenwirken mit Mikroorganismen zurückzuführen, die sich im Verdauungstrakt der Tiere anreichern, wie es bei der Symbiose von Regenwürmern und Actinomyceten deutlich wird. Durch die mechanische Zerkleinerung wird die Angriffsfläche für die Mikroorganismen erhöht und eine Beschleunigung des Abbaus erzielt. KURTSCHEWA (1960) beobachtete, daß Bakterien und Pilze in 140 Tagen nur 10% des Eichenlaubs zersetzten, während bei Mitwirkung der Kleintiere 50% abgebaut wurden. Gleichzeitig wurde aber der Boden mit Mikroorganismen angereichert. Ihre Bedeutung wird besonders deutlich, wenn berücksichtigt wird, daß die Kleintiere nach GRAFF (1954) im Laufe eines Jahres etwa das 100fache ihres Eigengewichts an Boden (5 bis 6 kg/m^2) umsetzen. Einen zusammenfassenden Überblick über die Leistung der Kleintiere gibt FRANZ (1955, 1960).

Literatur

AFANASJEWA, A. L., und L. N. SWATSKAJA: Die mikrobiologischen Prozesse im Boden bei verschiedenen Bearbeitungsmethoden der Neu- und Brachländereien. Agrobiologija 1, 86–92 (1958); Ref. Z. Pflanzenernähr., Düng., Bodenkde. 83, 273 (1958). — AMBROŽ, Z.: Aggregate formation in relationship to biological activity and different forms of humus in a meadow soil. Sborn. čsl. Akad. zeměd Věd. (Rad. Rostlinná Výroba) 29, 513–524 (1956).

BONNIER, CH., und F. W. HELY: Influence de certaines résines synthétiques, améliorant la structure du sol, sur la population fongique d'un sol. Extrait de l'Institut Agronomique et des Stations de Recherches de Gembloux 21, 14–28 (1953).

CHESTERS, G., O. J. ATTOE und O. N. ALLEN: Soil aggregation in relation to various soil constituents. Soil Sci. Soc. Amer. Proc. 21, 272–277 (1957).

DAWSON, C. R.: Effect of crop residues on soil micro-populations, aggregation and fertility under Maryland conditions. Soil Sci. Soc. Amer. Proc. **10**, 180–184 (1946). — DOWNS, S. C., T. M. MCCALLA und F. A. HASKINS: Stachybotrys atra, an effective aggregator of Peorian loess. Soil Sci. Soc. Amer. Proc. **19**, 179–181 (1955).

FELDMANN, A.: Untersuchungen über den Einfluß tiefer gärtnerischer Bodenbearbeitung auf die Entwicklung des Bodenkeimgehaltes. Z. Pflanzenernähr., Düng., Bodenkde. **74**, 17–24 (1956). — FORSYTH, W. G. C., und D. M. WEBLEY: The synthesis of polysaccharides by bacteria isolated from soil. J. Gen. Microbiol. **3**, 395–399 (1949). — FRANZ, H.: Die Bedeutung der Kleintiere für die Humusbildung. Z. Pflanzenernähr., Düng., Bodenkde. **69**, 176–181 (1955). — Die Zersetzung der organischen Substanz im Boden (mit besonderer Berücksichtigung der Leistungen der Bodentiere). Referat auf der 23. Sitzung des Ausschusses für Ackerbau der DLG, Wiesbaden, 12. I. 1960. — FULLER, W. H., und C. GAIRAUD: The influence of soil aggregate stabilizers on the biological activity of soils. Soil Sci. Soc. Amer. Proc. **18**, 35–40 (1954).

GEOGHEGAN, M. J., und R. C. BRIAN: Influence of bacterial polysaccharides on aggregate formation in soils. Nature **158**, 837 (1946). — Aggregate formation in soil, I, Influence of some bacterial polysaccharides on the binding of soil particles. Biochem. J. **43**, 5–13 (1948). — GILMOUR, C. M., O. N. ALLEN und E. TRUOG: Soil aggregation as influenced by the growth of mold species, kind of soil and organic matter. Soil Sci. Soc. Amer. Proc. **13**, 292–296 (1948). — GLATHE, H.: Bodenstruktur und Bodenleben in Wechselwirkung. Arbeiten der DLG, Bd. 54. Frankfurt/Main: DLG-Verlag. 1958. — GRAFF, O.: Was bedeuten unsere Bodentiere? Landbau-Forsch. **1954**, 45/46. — GROMYKO, E. P.: Die Strukturbildung und die Mikroorganismen im Podsol. Mikrobiologija **27**, 465–471 (1958) (russ.); Ref. Z. Pflanzenernähr., Düng., Bodenkde. **87**, 92 (1959).

HAGIN, J.: Rates of nitrification in natural and "conditioner"-formed soil aggregates of various sizes. Bull. Res. Counc. Israel **5**, 98–104 (1955); Ref. Z. Pflanzenernähr., Düng., Bodenkde. **73**, 276 (1956). — HELY, F. W., und C. BONNIER: The effects of induced soils aggregation on the occurrence of Azotobacter und Streptomyces. A. v. LEEUWENHOEK **20**, 359–373 (1960). — HELY, F. W., CH. BONNIER und P. MANIL: Investigations concerning nodulation and growth of lucerne seedlings in a loess artificially aggregated to various levels. Plant a. Soil **5**, 121–131 (1954); Ref. Z. Pflanzenernähr., Düng., Bodenkde. **67**, 90 (1954).

JÜRGENS-GSCHWIND, S.: Die Wirkung der Phosphatdüngung auf die Bodenstruktur. Phosphorsäure **20**, 224–240 (1960).

KANIVETZ, J. J., und N. P. KORNEEVA: Importance of biochemical structure forms. Pedology **32**, 1429–1441 (1938). — KIUPERS, H.: Eine Untersuchung der Faktoren Bodenstruktur und Stickstoffdüngung. Versl. Landbouk Onderz. **61**, 9 (1955); Ref. Z. Pflanzenernähr., Düng., Bodenkde. **74**, 275 (1956). — KURON, H., H. GLATHE und E. HOMRIGHAUSEN: Bemerkungen zur Durchführung der fluoreszenz-mikroskopischen Methode nach Strugger. Zbl. Bakt. II **112**, 204–213 (1959). — KURTSCHEWA, G. F.: Die Rolle der wirbellosen Tiere bei Zerlegung des Eichenabfalles. Bodenkde. **1960**, 16–23 (russ.); Ref. Z. Pflanzenernähr., Düng., Bodenkde. **92**, 269 (1961).

LAATSCH, W.: Die Gare als kolloidchemisches Problem. Phosphorsäure **10**, 301–317 (1941). — LEES, H.: The soil percolation technique. Plant a. Soil **1**, 221 (1949). — LEES, H., und J. H. QUASTEL: Biochemistry of nitrification in soil. Biochem. J. **40**, 803 (1946). — LÜTJEHARMS, W. J.: Der Einfluß von Bodenpilzen auf die Stabilität von Bodenaggregaten und den Benetzungswiderstand. Pharmac. Weekbl. **87**, 636–640 (1952); Ref. Z. Pflanzenernähr., Düng., Bodenkde. **64**, 88 (1954).

MACURA, J.: Continuous flow method in soil microbiology, I, Apparatus. Folia Microbiol. Praha **6**, 328–334 (1961). — The use of continuous flow method for studying soil metabolism. Trans. VIIth Int. Soil Sci. Congr. 1961a. — MACURA, J., und I. MALEK: Continuous flow method for the study of microbiological processes in soil samples. Nature **182**, 1796 (1958). — MARTIN, J. P.: Microorganisms and soil aggregation, I, Origin and nature of some of the aggregating substances. Soil Sci. **59**, 163–174 (1945). — Microorganisms and soil aggregation, II, Influence of bacterial polysaccharides on soil structure. Soil Sci. **61**, 157–166 (1946). — MARTIN, J. P., J. O. ERVIN und R. A. SHEPHERD: Decomposition and aggregating effect of fungus cell material in soil. Soil Sci. Soc. Amer. Proc. **23**, 217–220 (1959); Ref. Z. Pflanzenernähr., Düng., Bodenkde. **89**, 229 (1960). — MARTIN, J. P., und S. A. WAKSMAN: Influence of microorganisms on soil aggregation and erosion. Soil Sci. 50, 29–47 (1940). — MARTIN, T. L., und D. A. ANDERSON: Organic matter decomposition, mold flora and soil aggregation relationships. Soil Sci. Soc. Amer. Proc. 7, 215–217 (1943). — MCCALLA, T. M.: Influence of biological products on soil structure and infiltration. Soil Sci. Soc. Amer. Proc. 7, 209–214 (1942). — Water drop method of determining stability of soil structure. Soil Sci. 58, 117–123 (1944). — Influence of microorganisms and some organic substances on soil structure. Soil Sci. 59, 287–297 (1945). — Some factors influencing the percolation of water through a layer of loessial soil. Soil Sci. Soc. Amer. Proc. 9, 12–16 (1945a). — The influence of microorganisms and some organic substances on water percolation through a layer of Peorian loess. Soil Sci. Soc. Amer. Proc. 10, 175–179 (1945b). — Influence of some microbial groups on stabilizing soil structure against falling water drops. Soil Sci. Soc. Amer. Proc. 11, 260–263 (1946). — Microorganisms and soil structure. Trans. Kansas Acad. Sci. 53, 91–100 (1950). — Studies on the effect of microorganisms on rate of percolation of water through soils. Soil Sci. Soc. Amer. Proc. 15, 181–186 (1951). — Influence of soil conditioner (HPAN) on nitrification. Trans. Kansas Acad. Sci. 62, 53–57 (1959). — MCCALLA, T. M., F. A. HASKINS und R. D. CURLEY: Soil aggregation by microorganisms following soil fumigation. Soil Sci. Soc. Amer. Proc. 22, 311–314 (1958). — MCCALLA, T. M., F. A. HASKINS und E. F. FROLIK: Influence of various factors on aggregation of Peorian loess by microorganisms. Soil Sci. 84, 155–161 (1957). — MISCHUSTIN, E. N., und O. J. PUSCHKINSKAJA: Role of the microbiological factor in the formation of soil structure. Microbiologija 11, 92–104 (1942); Chem. Abstr. 38, 3066 (1944). — MISONO, S., und S. SUTO: Study on the micro-structure of soil. Bull. Nat. Inst. Agr. Sci. (Japan) Ser. B, Soils 8, 213–259 (1958). — MORTENSEN, J. L., und W. P. MARTIN: Decomposition of the soil conditioning Polyelectrolytes HPAN and VAMA in Ohio soil. Soil Sci. Soc. Amer. Proc. 18, 395–398 (1954). — The microbial decomposition and absorption of synthetic polyelectrolytes in Ohio soils. Atom Energy Com. Rep. Nr. TID — 7512, 235–243 (1956); Ref. Z. Pflanzenernähr., Düng., Bodenkde. 80, 270 (1958). — MÜLLER, G.: Probleme der Krümelstabilitätsmessung und der Krümelbildung. Tagungsberichte der wiss. Arbeitstagung am 10.—11. Okt. 1957 im Institut für Acker- und Pflanzenbau Müncheberg Nr. 13, 167–192 (1958). — MÜLLER, G., und K. RAUHE: Zur Tiefkultur auf leichten Böden im besonderen Hinblick auf die Bodenbiologie, II, Bodenbiologischer Teil. Z. Acker- u. Pflanzenbau 109, 309–332 (1959). — MYERS, H. E., und T. M. MCCALLA: Changes in soil aggregation in relation to bacterial numbers, hydrogen-ion concentration, and length of time soil was kept moist. Soil Sci. 51, 189–200 (1941).

NOVÁK, B., und J. NOVÁKOVÁ: Einfluß der Zugabe organischer Stoffe auf die physikalische, chemische und mikrobiologische Veränderung von Erde unter Berücksichtigung der Humusbildung. Sborn čsl. Akad. Zěměd věd Ročnik 6 (XXXIII), Rostlinná výroba 1399–1420 (1960).

PEELE, T. C.: Microbial activity in relation to soil aggregation. J. Amer. Soc. Agronom. 32, 204–212 (1940).

RENNIE, D. A., E. TRUOG und O. N. ALLEN: Soil aggregation as influenced by microbial gums, level of fertility and kind of crop. Soil Sci. Soc. Amer. Proc. 18, 399–403 (1954). — ROVIRA, A. D., und E. L. GREACEN: The effect of aggregate disruption on the activity of microorganisms in the soil. Austral. J. Agric. Res. 8, 659–673 (1957); Ref. Z. Pflanzenernähr., Düng., Bodenkde. 85, 178 (1959).

SCHACHTSCHABEL, P.: Bodenstruktur und Bodenfruchtbarkeit. Landwirtsch. Forsch. 7, Sonderh., 40–46 (1956); Ref. Z. Pflanzenernähr., Düng., Bodenkde. 78, 258 (1957). — SCHUFFELEN, A. C., und J. v. SCHUYLENBORGH: Die Anwendung von Vilenskis Methode zur Untersuchung des Einflusses von Frost und Mikroorganismen auf die Bodenstruktur. Landb. Tijdschr. 62, 339–347 (1950); Ref. Z. Pflanzenernähr., Düng., Bodenkde. 67, 88 (1954). — SEKERA, F.: Was ist Bodengare ? Phosphorsäure 10, 3–46 (1941). — SIEGEL, O., und CH. DISSELKAMP: Untersuchungen über Wirkung und Bedeutung der Bodenbakterien, 2, Fluoreszenzmikroskopische Untersuchungen über den Einfluß von Flotal auf die Bakteriendichte schwerer Böden. Landwirtsch. Forsch. 11, 265–269 (1958). — STEPHENSON, R. E., und C. E. SCHUSTER: Effect of mulches on soil properties. Soil Sci. 59, 219–230 (1945). — SWABY, R. J.: The relationship between microorganisms and soil aggregation. J. Gen. Microbiol. 3, 2, 236–254 (1949).

THORNTON, R. H., J. D. COWIE und D. C. MCDONALD: Mycelial aggregation of sand soil under Pinus radiata. Nature (Lond.) 177, 231–232 (1956).

YODER, R. E. A.: A direct method of aggregate analysis of soils and a study of the physical nature of erosion losses. J. Amer. Soc. Agronom. 28, 337–351 (1936).

c) Bedeutung der Mikroorganismen für die Nährstoffumsetzungen im Boden

Die Umsetzung von Nährstoffen im Boden ist zum größten Teil das Resultat der Tätigkeit der Mikroorganismen. Bei der Bestimmung dieser Aktivität können verschiedene Wege eingeschlagen werden. Daß eine einzelne Keimzahl nichts über die Aktivität der im Boden enthaltenen Mikroorganismen aussagen kann, wurde von Löhnis bereits betont (vgl. auch Drobnik 1957a). Es kann auch nicht der Fall sein, weil, wie auch Rybalkina und Kononenko (1956) feststellen, keine der Keimzählungsmethoden es erlaubt, die aktive von der potentiellen Mikroflora zu trennen. Wohl aber bietet eine Serie von Keimzählungen in bestimmten Abständen eine von verschiedenen Möglichkeiten, den Erfolg eines technischen Eingriffs in den Boden zu erfassen, weil sie einen Einblick in die Dynamik der Flora gestattet (vgl. Novák 1960). Wenn beispielsweise ein Boden einen Teil seines Wassers zur Verdunstung abgibt, wird, wie z. B. Prusnikiewicz (1959) nachwies, zunächst nur die Tätigkeit der Bakterien gehemmt, während die Aktivität von Actinomyceten und Pilzen noch steigt. In den Untersuchungen von Burrichter (1958) und auch in eigenen Versuchen konnten die jahreszeitlichen Schwankungen der Mikroflora mit Hilfe der fluoreszenzmikroskopischen Methode von Strugger eindeutig nachgewiesen werden (S. 563). Eine weitere Möglichkeit wurde von Meyer (1960) ausgewertet, nämlich die zusätzliche Bestimmung der Atmung; der Quotient Atmung: Keimzahl kennzeichnet die individuelle Leistung. Physiologische Untersuchungen, sei es in Nährlösungen, sei es im Boden selbst, vermögen wohl einen Einblick in einzelne Vorgänge zu gewähren, die aktuelle oder potentielle Fähigkeit eines Bodens in bezug auf bestimmte Umsetzungen erkennen lassen, eine Klärung der Vielfalt der neben-, mit- und gegeneinander ablaufenden Prozesse (Wikén 1955) jedoch nicht herbeizuführen.

Um die Aktivität der Mikroorganismen erfassen zu können, wird im allgemeinen die *CO_2-Produktion* bestimmt. Einen ausführlichen Überblick über die CO_2-Entbindung im Boden und die hierbei angewendeten Methoden geben Haber (1958) sowie Meyer und Koepf (1960). Da das Kohlendioxyd schwerer ist als Luft, sammelt es sich im Untergrund, so daß Konzentrationen bis zu 10% in der Bodenluft gemessen wurden. In Gefäßversuchen ist es daher nötig, eine Ansammlung zu verhindern. Bei der Methode Isermeyer (1952) wird das Kohlendioxyd automatisch in der Lauge absorbiert, während in größeren Gefäßen CO_2-freie Luft eingeleitet wird (Stotzky et al. 1958). Die Bestimmung des Kohlendioxyds kann sowohl titrimetrisch als auch kalorimetrisch (Beck und Röschentaler 1959) erfolgen. Augier und Jacob (1959) benutzten Standflaschen, deren Deckel drei Durchbohrungen aufwiesen (Luftzutritt, Luftaustritt und Probeentnahme). Besonders geeignet ist das Warburg-Gerät (vgl. Stevenson 1956/57 u. a.), weil eine größere Zahl von Gefäßen unter absolut gleichmäßigen Bedingungen gehalten werden kann. Das leistungsfähigste Gerät, das auch bei Freilandversuchen zuverlässig arbeitet, ist zur Zeit der automatisch registrierende Ultrarot-Adsorptionsschreiber (URAS), der kleinste Veränderungen erkennen läßt (Koepf 1952, 1953a, 1955, Horn 1955) und kontinuierliche Resultate liefert. Nuhi (1962) verwendete bei der Isermeyer-Methode mehrere Bodenbeutel in einem Standgefäß, so daß eine Entnahme zu verschiedener Zeit möglich war. Vorteilhaft ist es, wenn gleichzeitig auch der O_2-Verbrauch quantitativ bestimmt wird (Drobnik 1957b).

Die bei der Bodenatmung erfaßte CO_2-Menge stellt nicht die Gesamtproduktion dar. Ein gewisser Anteil wird von den autotrophen Organismen assimiliert. Erfahrungsgemäß ist dieser Teil jedoch sehr gering, er wird von Rippel (1931,

1955) auf etwa 3,6 kg C bzw. 13,2 kg CO_2/ha = 0,1 bis 0,3% des C-Umsatzes geschätzt. Dazu sind noch Mengen zu rechnen, die von Pilzen verarbeitet werden. STOVER und FREIBERG (1958) stellten fest, daß *Fusarium oxysporium* var. *cubense* hierzu befähigt ist. Auch heterotrophe Mikroorganismen fixieren Kohlensäure. SCHLEGEL (1955) beobachtete den günstigen Einfluß von CO_2 auf die Überführung des anorganischen Phosphors in organische Bindung bei *Hydrogenomonas*, *Chlorella pyrenoidosa* und *Rhodospirillum rubrum*. Bei Böden, die mit Pflanzen bestanden sind, muß schließlich auch die bei der Photosynthese verbrauchte Menge CO_2 in Rechnung gestellt werden (WALTER und HABER 1957). Andererseits erhöht die Atmung der höheren Pflanzen (KILBINGER 1953, 1955) die CO_2-Ausbeute von bewachsenen Böden erheblich. Nach MEYER und SCHAFFER (1954) betrug der Anteil der höheren Pflanzen an der Atmungsleistung der untersuchten Böden 37%, wobei sich besonders die Tätigkeit der Pflanzenwurzeln auswirkte.

Wesentliche Einflüsse auf die Bodenatmung schreibt HABER (1959) den örtlichen Verhältnissen zu, der Bodenart, der Struktur, dem Wassergehalt und der Temperatur. KOEPF (1953b) bewies, daß die Menge an CO_2, die von einem Boden bei einer bestimmten Temperatur abgegeben wird, abhängig ist von den vorausgegangenen Temperaturverhältnissen. Jede vorausgegangene Steigerung des Stoffumsatzes minderte die Umsatzleistung.

Über die Zusammenhänge zwischen *Temperatur* und CO_2-Produktion aus Podsol- und Torfböden im Tagesablauf berichtet MAKAROW (1958). In ostafrikanischen Böden, die zunächst getrocknet und dann wieder angefeuchtet wurden, kam es nach BIRCH und FRIEND (1956) zu einem beschleunigten Abbau der organischen Substanz, gemessen an der erhöhten CO_2-Produktion. Nach ihrer Meinung war das nicht die Folge einer Stimulierung der Mikroorganismen bzw. der Zerstörung toxischer Stoffe, sondern der Freilegung der organischen Substanz, die im feuchten Boden an den Ton gebunden und daher biologisch schwer angreifbar war. Die Zersetzung der organischen Substanz kam jedoch nach etwa zehn Tagen zum Stillstand. Das wurde von BIRCH (1958) darauf zurückgeführt, daß die Mikroorganismen zur Ruheform übergehen. Sollte das der Fall sein, dann wäre es notwendig, die Ursache dieser Erscheinung festzustellen. Sie müßte an der Bildung von Sporen zu erkennen sein, diese sind aber in situ nur sehr selten beobachtet worden.

Sowohl *mineralische* wie *organische Düngung* übt einen Einfluß auf die CO_2-Bildung des Bodens aus. Eine besonders starke Erhöhung der CO_2-Abgabe ist nach einer Kalkung des Bodens zu beobachten (FRERCKS 1954, WEISSENBERG 1954, BÉLA 1957, FRERCKS und PUFFE 1959). Da eine Aufkalkung zu einer Erhöhung der pH-Zahl und damit zu einer Verbesserung der Lebensbedingungen des größten Teiles der Bakterien führt, ist diese Tatsache ohne weiteres verständlich. Bei Kalkstickstoff lag die obere Grenze der Gabe, die ohne Hemmung der Atmung ertragen wird, nach NUHI (1962) bei 1400 kg N/ha, also weit über den in der Landwirtschaft angewendeten Gaben. Bis zu dieser Grenze wurde eine starke Erhöhung der CO_2-Produktion gegenüber dem ungedüngten Boden beobachtet. Da der zum Vergleich herangezogene Kalksalpeter eine sehr viel geringere Wirkung zeigte, kann die Ursache nicht in der Menge des zugeführten Kalks liegen, sondern muß der Wirkung des Cyanamids zugeschrieben werden. Bei der Düngung mit Phosphorsäure beobachtete BÉLA (1957) einen Anstieg der CO_2-Produktion.

Organische Düngemittel verbessern in den meisten Böden die Lebensbedingungen für die Mikroorganismen und führen damit zu einer verstärkten Bodenatmung; dies wurde vor allem für Stalldünger wiederholt (u. a. KILBINGER 1953, 1955) festgestellt. Die Atmungsgröße wird als Maßstab für die Geschwindigkeit des Abbaus organischer Stoffe verwendet (vgl. BIRCH 1958); sei es der im Boden enthaltenen, sei es der zugeführten. Bei der Messung der CO_2-Abgabe des gewachsenen Bodens handelt es sich um die *aktuelle* Atmung. Wird in vergleichenden Versuchen die Atmung durch Zusatz leicht verwertbarer Kohlenhydrate (Traubenzucker) stimuliert, so wird die *potentielle* Atmung bestimmt. Bei einem Vergleich der Wirkung organischer Dünger auf die CO_2-Bildung ist die Vergleichsbasis ausschlaggebend: ausgehend von gleichem C-Gehalt der Düngung ergab in Versuchen von SCHAFFER (1954) Stallmist 45% mehr CO_2 als Stroh, bei gleichem Nährstoffgehalt Stroh mehr als Stallmist. Nach KOEPF (1956) ist die Dynamik des Abbauablaufs zweier im N-Gehalt sich unterscheidender organischer Substanzen kennzeichnend für den Boden in der Weise, daß bei einem nährstoffarmen Boden der N-Gehalt des organischen Düngers die Zersetzungsgeschwindigkeit beeinflußt, nicht aber bei einem nährstoffreichen Boden. Eine gewisse Hemmung der CO_2-Bildung beobachtete NISSEN (1954), wenn dem Boden Penicillin, Chloromycetin bzw. Terramycin zugesetzt wurden. Bei Zusatz von Gibberellinsäure konnten LU et al. (1958) mit Hilfe der Warburg-Apparatur keine gesicherte Zunahme der Atmung feststellen, während Gibrel, das K-Salz der Gibberellinsäure, den Abbau der bodeneigenen organischen Substanz förderte (LU und BOLLEN 1957/58). In Wiesenböden beobachtete JAGNOW (1958) die größte CO_2-Abgabe in den obersten Schichten (0 bis 10 cm), bei schweren Böden setzte der Rückgang schon von 5 cm Tiefe abwärts ein. Nach KOSMAT (1959) ist die CO_2-Produktion in der Krume bei Wiesenböden größer als im Ackerboden, doch geht die Atmung in den tieferen Schichten bei Wiesenböden stärker zurück als bei Ackerböden. Auch POSCHENRIEDER und BECK (1960) beobachteten in schweren Tonböden bei Wald- und besonders bei Wiesenbestand verglichen mit Ackerland eine relativ intensive Atmung, die sie mit der starken CO_2-Entwicklung der Pflanzenwurzeln erklärten, die im Feldbestand wesentlich geringer ist. MINA (1957) fand in Waldböden den höchsten CO_2-Gehalt in den tiefen Schichten, die stärkste CO_2-Bildung in der obersten, humushaltigen Schicht, mit Schwankungen je nach Bestand und Jahreszeit.

Der größte Teil der gebildeten Kohlensäure stammt aus der *Zersetzung der Zellulose* im Boden. Zum Nachweis der Zersetzungsintensität verwendete RICHARD (1945) Zellulosetestschnüre, die er mit Stahlnadeln in das Bodenprofil einzog. Die Intensität der Zersetzung wurde in einem Reißapparat anhand der Festigkeitsabnahme geprüft (vgl. auch BERTOLET 1944, MARSH et al. 1945). SCHMIDT und RUSCHMEYER (1958) verwendeten Textilfasern, deren Festigkeit sich parallel mit Bodenreaktion, Nitratgehalt und Nitrifikationskraft änderte. In einer späteren Arbeit (1958) wurden Baumwollstreifen eingelegt. Durch Düngung mit Kalk und Stickstoff wurde die zellulolytische Kraft gesteigert, während Phosphorsäure eine begrenzte Wirkung hatte. UNGER (1960) arbeitete mit 10 g Zellulosewatte, die in PCU-Gazebeuteln eingeschlossen war. Der Abbau wurde durch Veraschung der Restwatte bestimmt. Durch Beregnung mit Abwasser wurde die zellulolytische Kraft in einer Tiefe von 5 cm stärker intensiviert als in 10 cm Tiefe. In Dauergrünland erfolgte die Zersetzung schneller als im Ackerboden unter Winterweizen.

KOZOVÁ (1961) brachte Filtrierpapierstreifen oder -scheiben geschützt durch Silongewebe in den Boden ein. Die unzersetzte Zellulose wurde bestimmt. Bei Feldversuchen wurde festgestellt, daß der Stickstoffverbrauch der wachsenden Kulturpflanzen die Zellulosezersetzung hemmte. Durch Anreicherung des Filtrierpapiers mit Ammon-N konnte der Vorgang gefördert werden.

Die Zersetzung der Zellulose wurde auch in sauren Moorböden (pH 2,48 bis 3,96) durch Erhöhung der CO_2-Produktion nachgewiesen (POSCHENRIEDER et al.

1957). Durch Zusatz von Stickstoff bzw. Kalk wurde sie weiter gesteigert (FRERCKS et al. 1960). Bei schwacher Aufkalkung wurde durch Stickstoffzusatz die Zersetzung unter sichtbarer Pilzentwicklung stark gefördert. Doch wirkte eine normale N-Düngung weniger fördernd als starke Aufkalkung. Stickstoffüberdüngung hemmte die Abbauvorgänge, wenn infolge zu geringer Aufkalkung Pilze überwogen, weniger bei hoher Aufkalkung und Bakterienbesatz. Bakterien setzten den Abbau stetiger fort (FRERCKS und PUFFE 1961). In stärker zersetzten Mooren mit höherem Gehalt an löslichem Stickstoff war die Zersetzungstätigkeit auch ohne N-Gabe schon intensiver als in schwach zersetztem Moorboden.

GARRISON und HARRIS (1954) isolierten aus Kompost einen zellulosezersetzenden Kokkus, wie ihn in ähnlicher Form schon MERKER (1912) beschrieben hatte. HUTCHINSON und CLAYTON (1919) zeigten jedoch später, daß es sich bei dem MERKERschen Organismus um Mikrocysten von Myxobakterien gehandelt hatte, eine Auffassung, die auch STANIER (1942) teilte. Auch von BAKER (1931) wurde ein kokkenartiger Organismus gefunden, aber nicht eingehender beschrieben. Bei dem von GARRISON und HARRIS (1954) isolierten Bakterium handelte es sich um einen unbeweglichen Kokkus pleomorpher Art. Über die anaeroben thermophilen Zellulosezersetzer, die McBEE (1950) in Reinkulturen gewonnen hat, wird eine vergleichende Übersicht gegeben.

Im Gegensatz zur Zellulose werden *Lignin*, *Wachs* und *Fett* langsam abgebaut. GHILDYAL und GUPTA (1959) nehmen an, daß die Zersetzung durch Actinomyceten und Pilze erfolgt. Das Lignin wird nach GOTTLIEB und PELCZAR (1951) mit Sicherheit durch Pilze angegriffen, und zwar das des Holzes durch Basidiomyceten, das der Streu durch Hymenomyceten. LINDEBERG (1955) stellte fest, daß Streupilze Lignin zersetzen, Mykorrhizapilze dagegen äußerst selten. GOTTLIEB und PELCZAR (1951) kommen auf Grund zahlreicher Arbeiten über den Abbau des Holzes zu dem Ergebnis, daß bisher weder ein Bakterium noch ein Enzym isoliert wurde, das befähigt ist, Lignin zu zersetzen.

Auch beim *Kreislauf des Stickstoffs* handelt es sich um ein ständiges Ineinandergreifen von Einzelvorgängen der Mineralisierung und Festlegung. In schematischen Darstellungen versuchten JANSSON (1955) und ATANASIU (1960) den Wirkungsmechanismus deutlich zu machen. Beim Abbau der Eiweißverbindungen werden Ammoniak und Nitrat gebildet, die zusammen als *aktiver* Stickstoff des Bodens bezeichnet werden. Dabei entscheiden klimatische Bedingungen bzw. die Struktur des Bodens, ob es etwa zur Anhäufung von Ammoniak oder Nitrat kommt. In diesen Prozeß greift außerdem die höhere Pflanze ein, indem sie entweder die eine oder die andere Form bzw. beide aufnimmt. Zur Bildung von Ammoniak sind viele Arten von Mikroorganismen befähigt, deren Ansprüche an Nährstoffe und Umweltfaktoren so verschiedenartig sind, daß dieser Prozeß praktisch in allen Böden stattfinden kann. Die Bildung von Nitrat dagegen ist an die Tätigkeit einer Organismengruppe gebunden, deren Ansprüche an Bodenklima und Nährstoffversorgung ihrer Aktivität ziemlich enge Grenzen setzen.

Bei der *Nitrifikation* handelt es sich um einen Oxydationsvorgang, der in zwei Phasen abläuft. Eine Darstellung der biochemischen Grundlagen des Prozesses wird von ENGEL (1958b) gegeben. *Nitrosomonas* bildet aus Ammoniak Nitrit, *Nitrobacter* wandelt das Nitrit in Nitrat um. Ausreichende Belüftung ist die wichtigste Voraussetzung für ihre normale Entwicklung. Da die oberste Bodenschicht meist die größte Sauerstoffmenge enthält, ist dort die Nitrifikation am intensivsten (u. a. PATHAK und SHRIKHANDE 1952, MARTIN und COX 1956). So stellten z. B. KRESGE und MERKLE (1957) fest, daß im Oberboden etwa 10% des Gesamtstickstoffs innerhalb von 12 Wochen nitrifiziert worden waren, in alkalischem Boden 6,5% und in saurem 13%, im Unterboden dagegen nur 5,3 bis 6,1%.

Alle Maßnahmen, die zu einer besseren Sauerstoffversorgung führen, fördern die Nitrifikation. GELLER (1958) und auch MÜLLER (1960a) erzielten eine verstärkte Nitrifikation durch Vertiefung der Pflugfurche von 25 auf 32 cm.

Das *Optimum des Wassergehalts* liegt etwa bei einer Sättigung von 60% der Kapazität. Bei darüber hinaus steigendem Wassergehalt kommt die Nitrifikation allmählich zum Stillstand. ROBINSON (1957) fand im Krumenboden des roten Kikuyu-Lehms einen Stillstand der Nitrifikation, bevor der wilting-point erreicht war. Demgegenüber stellen aber die ammoniakbildenden Mikroorganismen ihre Tätigkeit erst später ein, daher kommt es bei hohem Wassergehalt zu einer Ammoniakanreicherung. In tropischen Böden führt die Austrocknung während der Trockenzeit, wie GREENLAND (1958) beobachtete, zu einer starken Hemmung, aber die Niederschläge bei beginnender Regenzeit bewirken eine rasche Intensivierung der Nitrifikation. In den undurchlässigen Tonböden des Gezira-Gebirges im Sudan war daher zu Beginn der Regenzeit eine Anhäufung von Nitrat festzustellen (JONES 1957).

In Laboratoriumsversuchen können nach FITTS et al. (1955) reproduzierbare Werte nur erzielt werden, wenn die Bodenfeuchtigkeit exakt unter Kontrolle gehalten wird. Auf tensiometrischem Wege ist dies mit größerer Sicherheit zu erreichen als bei laufendem Ersatz des verdunsteten Wassers durch Gewichtskontrollen. Die Konstanz des Feuchtigkeitsgehalts und gleichzeitig eine gute Durchlüftung werden zwar durch Zusatz von Vermiculit erleichtert (STANFORD und HANWAY 1955), bei Nitrifikationsversuchen ist dieser Weg jedoch nicht gangbar, da, wie ALLISON et al. (1953) in vier Wochen lang laufenden Versuchen feststellten, Vermiculit und Illit das Ammoniak stark festhalten. Das in Tonboden festgelegte Ammoniak war im Durchschnitt nur zu 5% (Schwankungen lagen zwischen 0 und 18,4%) für die nitrifizierenden Bakterien zugänglich.

Nach DHARIWAL und STEVENSON (1958) wurde das Ammoniak in Böden mit Illit als vorherrschendem Tonmineral fester gebunden als in Böden mit Montmorillonit. GOLDBERG und GAINEY (1955) beobachteten eine Förderung der Nitrifikation

in K-gesättigtem	Bentonit	um + 9%
in K-gesättigtem	Muscovit	um +19%
in K-gesättigtem	Montmorillonit	um + 7%

bzw. eine Hemmung

in K-gesättigtem	Kaolin	um —21%
in NH_4-gesättigtem	Montmorillonit	um — 7%
in NH_4-gesättigtem	Muscovit	um —70%
in NH_4-gesättigtem	Kaolin	um —69%

Nitrifikation und NH_4-Adsorption standen in direkter Beziehung. Trotzdem konnte JANSSON (1958) mit Hilfe von markiertem Stickstoff feststellen, daß nitrifizierende Bakterien fest sorbiertes Ammonium in stärkerem Maße in Freiheit zu setzen vermögen als chemische Extraktionsmittel.

Die nitrifizierenden Bakterien wachsen auf der Oberfläche der Bodenkrümel an den Stellen, wo Ammoniumsalze vorhanden sind. Nur relativ wenig sind in der Bodenlösung nachweisbar (LEES und QUASTEL 1946). Bei Anwendung von Agrylon zur Strukturverbesserung stellte HAGIN (1955) erhöhte Nitrifikation fest. Eine besonders hohe Nitratausbeute ergab sich bei Versuchen von KUSNETZOWA (1955) im Winter, wohl durch Veränderung des physikalisch-chemischen Zustandes der Bodenkolloide. In Lösungsversuchen beobachteten PATHAK und SHRIKHANDE (1952) eine Überlegenheit des Tonbodens gegenüber Sand- und Schlammboden.

Das *Temperaturoptimum für die Nitrifikation* liegt bei 25° C (vgl. u. a. JENSEN 1950). Doch ist auch von höheren Optimalwerten berichtet worden. Für *Nitrobacter* fanden DEPPE und ENGEL (1960) die Optimaltemperatur bei 34 bis 35° C; die Maximaltemperatur wurde bei 45° C ermittelt. BURANGULOWA und SOLEWJEWA (1956) stellten die höchste Zahl von nitrifizierenden Bakterien in degradierter Schwarzerde in 20 bis 30 cm Tiefe bei 38% Bodenfeuchtigkeit und 30° C fest. FREDERICK (1956) fand, daß die optimale Temperatur für die Nitrifikation zwischen 27 und 35° C lag; unter günstigen Bedingungen wurden bei Temperaturen um den Gefrierpunkt noch 50% des Ammoniakstickstoffs nitrifiziert. SABEY et al. (1956) stellten bei Versuchen in dem Temperaturbereich von 0 bis 30° C einen besonders starken Rückgang der Nitrifikation unter 8° C fest. ANDERSON und PURVIS (1955) fanden, daß bei 8,3° C 30 bis 50% des zugesetzten Ammoniaks innerhalb von 6 Wochen nitrifiziert wurden. Bei 2,8° C wurden minimale Mengen in allen Böden nitrifiziert, ein Befund, den ANDERSON (1960) bei späteren Untersuchungen nicht bestätigte. Daß bei tieferen Temperaturen die Intensität des Prozesses gehemmt ist, geht auch aus den Arbeiten von FISHER und PARKS (1958) hervor. Bei 20 und 30° C wurden auch höhere Mengen Harnstoff als 110 bzw. 120 kg N/ha verarbeitet, bei 10° C dagegen hatte die Höhe der Harnstoffgabe keinen Einfluß auf die Menge des gebildeten Nitrats. Bei den höheren Temperaturen wurden in der ersten Woche bereits 75% des zugesetzten Stickstoffs nitrifiziert. SAKAI (1959) prüfte Böden, die bei 25° C ihre maximale Nitrifikationskraft gezeigt hatten, später bei anderen Temperaturen und erzielte folgende Ergebnisse:

Temperatur	Menge des gebildeten Nitrats
5° C	11 bis 42 %
15° C	64 bis 69 %
25° C	100 %
35° C	60 bis 67 %

Der anfangs bei 25° C gehaltene Boden zeigte bei Bebrütung in anderen Temperaturbereichen eine geringere Nitrifikationskraft als Böden, die ständig bei diesen Temperaturen gehalten worden waren. Die maximale Leistung bei 25° C betrugt 10 mg NO_3-N/100 g Boden/Tag. SABEY et al. (1959) beobachteten bei 25° C in verschiedenen Böden Nitratmengen, die zwischen 44 ppm und 216 ppm Nitrat-N je Woche schwankten. Bei der im Gartenbau üblichen Dämpfung des Bodens trat Nitrifikation erst wieder nach Impfung mit ungedämpfter Erde ein (SCHMIDT 1956). Diese Beobachtung wird man indessen nicht verallgemeinern können, da die Höhe der Dämpfungstemperatur von entscheidender Bedeutung ist. Die nitrifizierenden Bakterien sind sehr hitzeresistent. So überstehen sie die heißen tropischen Trockenperioden und werden bei Befeuchtung sehr rasch wieder aktiv. Ferner ist nach den Ergebnissen von FARKASDI (unveröffentlichte Berichte) die Frage noch nicht entschieden, ob thermotolerante bzw. thermophile Nitrifikationsbakterien in der Natur vorkommen.

Die Nitrifikation verläuft in tropischen Böden, beispielsweise in den Gezira-Böden des Sudan (JEWITT 1945), den Pegasse-Böden Britisch-Guayanas (PAUL und SHARIFF 1954) oder den Hochlandböden Kenias (MILLBANK 1959) ebenso wie in der gemäßigten Zone. Nach PAUL und SHARIFF waren die in Kenia isolierten Stämme in morphologischer und physiologischer Hinsicht identisch mit denen der gemäßigten Zone.

Die Nitrifikation ist möglich in einem *Reaktions*bereich von pH 3,5 bis 10, das Optimum liegt zwischen 6,5 und 8. Die Leistung der Bakterien kann gesteigert werden, wenn die Reaktion in der Richtung des Optimums verschoben wird. In Versuchen von ANDERSON und PURVIS (1955) setzte bei einer Reaktion von pH 4,9 in sandigem Lehm die Nitrifikation von Ammonsulfat wenigstens zwei Wochen später ein als die von Ammoniak. Wenn der Boden zu Beginn des Versuchs eine Reaktion von pH 6,3 zeigte, traten diese Unterschiede nicht auf (vgl. ENO und BLUE 1957). Umgekehrt wurde, wie ENO und BLUE (1954) feststellten, die Nitrifikation von flüssigem Ammoniak gegenüber Ammonsulfat in alkalischem Boden verzögert. MCINTOSH und FREDERICK (1958) begasten einen sandig tonigen Lehmboden mit NH_3; bei einer Reaktion von pH 6,5 erfolgte zunächst eine rasche Nitrifikation, sie wurde aber bereits durch das Absinken der Reaktion auf pH 5,5 bis 5,0 allmählich verzögert. Ammoniak kann die Nitrifikation hemmen, wenn sein Gehalt im Boden die Grenze von 302 ppm übersteigt (ENO et al. 1955). So ist es auch durchaus verständlich, daß THEMLITZ (1958) nach Ammoniak- bzw. Harnstoffdüngung in saurem Fichtenrohhumus eine Nitrifikation erst nach Kalkzugabe beobachtete. Eine Förderung der Nitrifikation in Waldböden wurde durch Kalk (CHASE und BAKER 1954) bzw. durch Gips (HESSE 1957/58) erreicht. Salzschäden traten nach Kaliumsulfat, Magnesiumsulfat und Calciumchlorid bzw. nach Zugabe einwertiger Ionen bei einer Reaktion von pH 4,4 auf, bei pH 6,0 jedoch nicht (MUNK 1958).

Durch alkalische Reaktion und besonders durch freies Ammoniak wird *Nitrobacter* stärker gehemmt als *Nitrosomonas*, so daß es in solchen Fällen zu Nitritanhäufung kommt (SOULIDES und CLARK 1958). STOJANOVIC und ALEXANDER (1958) stellten mit Hilfe der Perfusionsmethode bei hohen Ammoniakgaben Nitritanreicherung fest. Auch in Kulturflüssigkeiten konnten ALEEM und ALEXANDER (1960) eine Schädigung von *Nitrobacter agilis* durch Ammoniumsalze nur im alkalischen Bereich feststellen. Nach TYLER et al. (1959) scheint die Gefahr größer zu sein, wenn zu der alkalischen Reaktion noch niedrige Temperaturen treten, gegen die *Nitrobacter* empfindlicher ist als *Nitrosomonas*.

Ein Bedarf der nitrifizierenden Bakterien an Phosphor, Magnesium und Eisen wurde von ALEEM und ALEXANDER (1960) festgestellt. Von den Mikronährstoffen ist bekannt, daß Bor die Nitrifikation fördert (PINTEA et al. 1958), besonders dann, wenn der Boden, wie der Podsol, arm an organischer Substanz ist. In Schwarzerde wirkte auch Uran günstig auf die Nitrifikation.

Lange Zeit wurde die Meinung vertreten, daß organische Stoffe die Nitrifikation hemmen. WHITING (1926) hat zwar eine große Zahl derartiger Hemmstoffe ermittelt, eine praktische Bedeutung kommt ihnen indessen nicht zu, da sie im Boden im allgemeinen nicht vorkommen. Die *Aminosäuren* Methylamin, Trimethylamin und Cystin werden nicht nitrifiziert, sie müssen erst von anderen Mikroorganismen ammonifiziert werden. Bei der Oxydation des Cystins kommt es zur Bildung von Schwefelsäure und damit zu einer Senkung der Reaktion und Hemmung der Nitrifikation. Durch eine Reihe von Stoffen, z. B. Natriumchlorat, Chloromycetin u. a., wird nur *Nitrobacter* gehemmt; infolgedessen führen sie zu einer Nitritanreicherung (LEES und QUASTEL 1946). Eine zusammenfassende Darstellung über weitere hemmende Aminosäuren und andere, auch anorganische, Hemmstoffe gaben QUASTEL und SCHOLEFIELD (1951). JENSEN (1950) wies Nitrifikation in Gegenwart von 10% Glucose oder 20% Saccharose nach, erst stärkere Konzentrationen wirkten hemmend. Zudem werden Stoffe, die in Reinkulturen hemmend oder stimulierend wirken, in Mischkulturen zersetzt (MEIKLEJOHN 1952). Mit Hilfe der Perkolationsmethode (vgl. LEES und QUASTEL 1944) verfolgten WHEELER und YEMM (1958) die Nitrifikation verschiedener Aminosäuren und stellten bei einer Versuchsdauer von 3 bis 4 Tagen eine Abhängigkeit der Intensität der Nitrifikation von Art und Menge der zugesetzten Substanzen fest.

Selbst in lagernden organischen Düngemitteln findet Nitrifikation statt.

JENSEN (1950) bewies es für den Stalldünger, und bei der Kompostierung von Siedlungsabfällen wurde der Nitratgehalt sogar als Kriterium für den Rottegrad empfohlen (KELLER 1960). Versuche über die Nitrifikation von Blut-, Huf- oder Hornmehl im Boden ließen nicht auf eine unterschiedliche Zersetzbarkeit dieser Stoffe schließen, auch Gaben verschiedener Stärke wirken sich nicht aus, wie OWEN et al. (1950) feststellten. Am 12. bzw. 19. Tag wurden bei 100 ppm 94,2%, bei 200 ppm 93,7% und bei 300 ppm 94,5% des zugegebenen Stickstoffs in Nitrat übergeführt. Die Menge von 300 ppm wurde bereits von WAKSMAN (1923) als zweckmäßig erkannt. Düngung mit Stroh entzieht den nitrifizierenden Bakterien das Ammonium, denn die Strohzersetzung führt zu einer lebhaften NH_4-Assimilation. Bei Bedeckung des Bodens mit Stroh ist die Nitrifikation wohl in erster Linie abhängig von Luftzufuhr und Temperatur. Die Temperatur wird naturgemäß durch die Farbe des Strohs beeinflußt: dunkles Stroh absorbiert mehr Wärme, während helle Farben größere Wärmemengen reflektieren (McCALLA 1947). Der Nitratgehalt des Bodens ist weiterhin abhängig von der Dicke der Strohbedeckung, der Struktur des Bodens und vom Zeitpunkt des Einpflügens bzw. Eineggens des Strohs, von dessen C:N-Verhältnis und von den angebauten Kulturpflanzen (McCALLA and RUSSELL 1943, 1948). Da bei einem zu weiten C:N-Verhältnis auch Nitrat assimiliert und durch eiweißzersetzende Mikroorganismen wieder freigelegt werden kann, wird die Dynamik deutlich, die den jeweiligen Formen des im Boden vorhandenen Stickstoffs eigen ist.

Die Intensität der Nitrifikation wird durch höhere Pflanzen offenbar in unterschiedlichem Maße beeinflußt. Während VANSTALLEN (1960) unter Grünland eine besonders rasche Nitrifikation beobachtete, kam die Nitrifikation in südafrikanischen Lysimeterversuchen nach THERON (1951) unter Gras in der zweiten Vegetationsperiode zum Stillstand. Unter Hirse wurde sie nur kurz vor der Reife unterdrückt, während der Reife setzte sie jedoch wieder ein. Während der Zeit gehemmter Nitrifikation kam es zu einem Anstieg des Ammoniakgehaltes im Boden. Offenbar wurde nur das autotrophe Dehydrogenasesystem gestört, nicht dagegen die NH_3-Bildung. Nach STEVENSON und CHASE (1953) verlief die Nitrifikation unter Pfirsichbäumen bei Luzerne- und Strohbedeckung intensiver als unter Rasen oder Buchweizen in unbedecktem Boden. Unter Berücksichtigung der Ergebnisse von STEPHENSON und SCHUSTER (1945) dürfte in erster Linie die Bodenbedeckung für die Intensivierung der Nitrifikation verantwortlich zu machen sein. In siebenjährigen Versuchen auf Lehmboden (Iowa) fanden FITTS et al. (1953) eine Korrelation zwischen Nitratgehalt und Maiserträgen bei einer Düngung von 45 bis 67 kg N/ha.

Bisher war man der Auffassung, daß die Nitrifikation ohne *Stickstoffverluste* abläuft. Es wurde jedoch nachgewiesen, daß recht beträchtliche Mengen verlorengehen können. Besonders hohe N-Verluste haben GUSSEJNOW et al. (1957) ermittelt. Bei der Düngung von Wiesengrauerde mit Ammoniumsulfat bzw. Ammoniumnitrat stellten sie eine Zunahme der Verluste mit steigenden N-Gaben fest. Im Laboratoriumsversuch waren 6 Stunden nach der Düngung 52 bis 60% verlorengegangen, nach 20 Tagen bei 320 mg Ammoniumsulfat je 1 kg Boden bei Zimmertemperatur 68%, bei 1300 mg Ammoniumsulfat je 1 kg Boden bei 60 bis 65° C 98%; in mittelstark podsolierter Gelberde fanden sie bei 10° C N-Verluste von 11,3 bis 20,5% und bei 60° C von 38,9 bis 42,7%. Im Feldversuch auf Grau-Braun-Erde beobachteten sie bei beiden Düngemitteln nach 18 Tagen Verluste in Höhe von 64 bis 65%. Auch auf Moorböden waren die Verluste hoch. VANSTALLEN (1960) konnte bei einer Düngung mit Ammoniumsulfat oft nur 75% des verabreichten Stickstoffs wiederfinden. Bei einer Reaktion von $pH > 7,5$ gingen 16% verloren; auch in ihren Versuchen bestand eine direkte Korrelation zwischen Düngergabe und Verlusten, bei 100 kg N/ha waren nach 3 Wochen 3%, bei 400 kg N/ha 16% verlorengegangen. Auch Pflanzenbestand und Düngung üben auf die Höhe der Verluste einen Einfluß aus. WALKER et al. (1956) stellten in Versuchen mit markierten N^{15} unabhängig von der Pflanzenart einen Verlust von 30% fest. DHAR (1956) konnte die N-Verluste in Böden Indiens durch Zugabe von Calciumphosphaten herabsetzen. In einem Schlufflehm stellten LOEWENSTEIN et al. (1957) bei einer Reaktion von pH 5,5 nach Zugabe von viermal je 150 ppm N als Ammoniumsulfat nach 11 Monaten im nicht bepflanzten Boden N-Verluste von 35 bis 72% fest. Nach Kalkzufuhr stiegen die Verluste, durch Zusatz von 0,5% Stroh wurden die Verluste herabgesetzt, im bepflanzten Boden gingen nur 7 bis 8% des Stickstoffs verloren. Auch beim Umbruch von Wiesen bzw. bei der Kultivierung von Prärieboden (SHUTT 1910, 1925) treten regelmäßig erhebliche N-Verluste auf.

Die Verluste sind mit dem Entweichen gasförmiger Zwischenprodukte des Nitrifikationsprozesses zu erklären. Nach WAGNER und SMITH (1958) wurden in der Luft über einem gut nitrifizierenden Boden NH_3, NO, NO_2 (in kleinen Mengen) und N_2O (unter Umständen in großen Mengen) gefunden. Es scheint eine Beziehung zwischen Nitritanreicherung und N-Verlusten zu bestehen (SOULIDES und CLARK 1958), und zwar sind die N-Verluste in gut durchlüfteten, sauren Böden möglicherweise auf die Instabilität der HNO_2 zurückzuführen (CLARK et al. 1960). BAUMANN und MAASS (1957), GLIEMEROTH (1959) sowie GERRETSON und DE HOOP (1957) beobachteten N-Verluste in Nitrifikationsversuchen, wenn zu Beginn eine günstige Reaktion herrschte, ein geringes Pufferungsvermögen aber ein Absinken der Reaktion unter pH 5,5 zuließ. Dann können NO-Verbindungen oder auch N_2 entweichen. Um ein Entweichen von NH_3 konnte es sich nicht handeln, da die Verluste durch Anheben der Reaktion über pH 5,5 unterbunden werden konnten. Auch eine Denitrifikation konnte nicht vorliegen, da bei Nitratzusatz Verluste nicht eintraten. Der biologische Charakter des Prozesses wurde aber dadurch nachgewiesen, daß Verluste nicht mehr auftraten, sobald der Boden sterilisiert wurde. N-Verluste durch Auswaschung des gebildeten Nitrats sind unter entsprechenden Voraussetzungen immer möglich, sie stehen jedoch mit dem biologischen Prozeß der Nitrifikation nicht in direkter Beziehung. ANDERSON (1960) empfiehlt, im Herbst nicht mit Ammoniak zu düngen, um N-Verluste durch Nitrat-Auswaschung zu verhindern, es sei denn, die Temperatur sei unter 5° C gesunken.

Die nitrifizierenden Bakterien sind autotroph, aber es hat nicht an Mitteilungen gefehlt, daß auch gewisse *heterotrophe Organismen* befähigt seien, Nitrit bzw. Nitrat zu bilden. FISHER et al. (1952) fanden gramnegative Stäbchen, die heterotroph leben und in Bodenextraktlösung Ammoniak zu Nitrit oxydieren sollen. Weiter beobachteten HORA und IYENGAR (1960) in Peptonlösung eine Bildung von Nitrit und Nitrat durch Penicillien und Aspergilli bei einer Reaktion, die zu Beginn bei pH 7,0 und am Schluß bei pH 7,8 bis 8,4 lag (erste Mitteilung von IYENGAR und HORA 1959). EYLAR und SCHMIDT (1959) bestätigen diese Befunde in einer Pepton und Hefeextrakt enthaltenden Nährlösung, der Zusatz von Ammonsulfat oder Harnstoff förderte aber den Vorgang nicht. Eine enge Verbindung zwischen *Nitrosomonas* und in Mischkulturen auftretenden heterotrophen Bodenbakterien (zwei *Pseudomonas*-Arten und *Hyphomicrobium vulgare*) stellte GUNDERSON (1955) fest. Ob es sich dabei um Symbiose handelt, konnte nicht geklärt werden.

Die Mengen des im Boden enthaltenen Nitrat- und Ammoniakstickstoffs werden, wie bereits erwähnt, als *aktiver Stickstoff* bezeichnet. Sie können nicht als statische Formen angesehen werden (vgl. JANSSON 1955), sondern sind durch die Dynamik von Mineralisierungs- und Festlegungsprozessen bedingt. Die Fraktion des aktiven Stickstoffs kann durch langsamen Abbau der organischen Substanz des Bodens oder durch Düngung mit mineralischem Stickstoff ergänzt werden. Nach LEFÈVRE und HIROUX (1960) scheint die Nachlieferung aus der organischen Substanz des Bodens um so größer zu sein, je größer die N-Verluste des Bodens sind. Bei der Bestimmung des aktiven Stickstoffs können zwei Wege eingeschlagen werden. Einmal können die im Boden vorhandenen Ammoniak- bzw. Nitratmengen in gewissen Zeitabständen bestimmt werden; ihre Summe stellt die „*effektive*" Mineralisation dar. Auf der anderen Seite besteht die Möglichkeit, abgesiebte Teile der entnommenen Bodenproben unter konstanten, optimalen Bedingungen im Laboratorium zu bebrüten und dann die „*potentielle*" Mineralisation zu ermitteln. Der zuletzt genannte Weg wurde in letzter Zeit sowohl für Acker- als auch für Waldboden beschritten. Er hat den Vorzug, daß Auswaschungsverluste, wie sie im Freiland jederzeit die tatsächlichen Verhältnisse verschleiern können, im Laboratoriumsversuch ausgeschaltet werden. ZÖTTL (1960a) führte solche Versuche in Waldboden durch, er betrachtete eine Zeit von 6 Wochen als ausreichend; die gefundenen Werte stellen den „Nettoumsatz" dar, der für die höheren Pflanzen allein von Bedeutung ist. Der „Bruttoumsatz" schließt die von den Mikroben zum Aufbau ihres Körpers verbrauchten Mengen ein. Die durchschnittliche Mineralisationsquote betrug 5% des Gesamtstickstoffs, das entspricht einer Jahreslieferung von mindestens 31 kg N/ha (ZÖTTL 1960d). Hatte das Humusmaterial einen hohen Gehalt an leicht angreifbaren stickstofffreien Substanzen oder wurden rasch verwertbare Kohlenhydrate zugesetzt, so wurden trotz hohen Bruttoumsatzes nur geringe Nettomengen gebildet (ZÖTTL 1960b). Die höchsten Werte von Brutto- und Nettomineralisation traten in dem aus der Streulage entnommenen Material auf (ZÖTTL 1960c).

Bei Grasland basaltischer Herkunft beobachteten MARTIN und COX (1956) im Unterboden einen niedrigeren Ammoniakgehalt als in der Krume. Kultivierung erhöhte den Nitratgehalt. 25 Jahre nachdem das Land in Kultur genommen worden war, hatte sich der Nitratgehalt in der Bodenschicht von 15 bis 30 cm Tiefe erhöht. Durch Perkolationsversuche mit Ammonsulfat wurde wahrscheinlich gemacht, daß das Nitrat aus dem Oberboden durch Einwaschung dahin gelangte. Im Freiland kann Nitrat in warmen trockenen Jahren auch aus tieferen Schichten heraufgehoben werden, so daß es zu einer Anreicherung in der oberen Zone kommt (MAASS 1961). Das Verhältnis, in dem die beiden N-Formen im Boden vorkommen, ist abhängig von seiner Reaktion. SOULIDES und CLARK (1958) beobachteten in Weide- und Ackerland eine NH_4-Anreicherung in Böden bei einer Reaktion von pH 4,7 und 4,9, dagegen nicht bei pH 5,7 bis 7,9. Daraus geht hervor, daß die Ammoniakbildner eine höhere Resistenz gegenüber der Versauerung aufweisen als die Nitrifikanten.

Werden dem Boden große Mengen von Harnstoff zugeführt (bis 800 ppm), wie es BROADBENT et al. (1958) taten, so kann die starke Ammoniakanhäufung zu einer Verschiebung der Reaktion führen. Die Folge ist eine gehemmte Nitratbildung und dafür eine Anreicherung von Nitrit.

Die Intensität der Mineralisation wird naturgemäß von dem Wassergehalt des Bodens beeinflußt. Eine Erhöhung der Menge des aktiven Stickstoffs beobachtete STEVENSON (1956/1957) bei der Befeuchtung eines lufttrockenen Bodens oder eines feldfrischen Bodens dann, wenn seine Wasserkapazität auf 60% abgesättigt wurde. Bei der künstlichen Trocknung von Böden bei 20° bzw. 100° C besteht nach BIRCH (1959) zwischen der Trocknungsdauer und N-Mineralisierung eine einfache logarithmische Beziehung. Wahrscheinlich werden durch den Trocknungsprozeß die Oberflächen der Humuskolloide, die angreifbare Substanzen bieten, vergrößert. Extreme Temperaturen führen zu relativ stärkerer Ammoniakbildung. GREAVES und JONES (1944) hielten 4 Böden 24 Monate bei verschiedenen Temperaturen und kamen zu dem Ergebnis, daß die NH_4-Anhäufung am stärksten bei 10° C und am niedrigsten bei 40° C war. Bei 40° C war die Nitrifikationsfähigkeit in 2 Fällen erloschen. Zu einer Anreicherung von Ammoniak kam es in den Versuchen von DUBOWIK (1956) nach einer Stallmistdüngung dann, wenn die Nitrifikation durch niedrige Temperaturen gehemmt wurde, während die Ammonifikation noch nicht unterbunden war.

Die Nutzung des Landes wirkt sich auf den Gehalt des Bodens an aktivem Stickstoff aus, wie auf den Anteil, den die beiden N-Formen an der Gesamtmenge haben. Hohe Nitratgehalte treten nach BAUMANN und MAASS (1957) sowie GELLER (1958) in Perioden ohne wirksamen Pflanzenstand bzw. nach Lockerung verdichteter Böden auf, während sie zur Zeit intensiver Nährstoffaufnahme durch die Pflanzen stark zurückgehen. Das bedeutet aber keine Hemmung der Nitrifikation; denn MUNSON und STANFORD (1955) konnten sogar eine Korrelation zwischen N-Aufnahme durch die Pflanzen und Nitratbildung beobachten, sobald der Boden im Laboratorium bebrütet wurde. HANWAY und DUMENIL (1955) halten es für möglich, aus dieser Korrelation eine Prognose über die Düngebedürftigkeit des Bodens abzuleiten. ZÖTTL (1958) beobachtete einen Parallelismus zwischen der im Laboratorium im Waldhumus gebildeten Menge des aktiven Stickstoffs und der Wuchsleistung der Baumbestände. Einzelergebnisse von Ammoniak- und Nitratbestimmungen in Böden unter verschiedenem Pflanzenbestand können weder über die Wirkung der Pflanzen auf den Stickstoffhaushalt des Bodens etwas aussagen noch über die Mineralisationsleistung des Bodens. Je nach dem Zeitpunkt der Probenahme entziehen die Pflanzen dem Boden mineralischen Stickstoff. BARNES (1950) stellte bei Probenahmen Ende Dezember bis Anfang Januar in Grünlandboden einen geringeren Gehalt an aktivem Stickstoff fest als im Ackerboden, aber dann in Laboratoriumsversuchen von 4 Monaten Laufzeit im Grünlandboden eine höhere Mineralisationsleistung. SOULIDES und CLARK (1958) wiesen in Weideböden höhere Ammoniak- und niedere Nitratgehalte als in Ackerböden nach.

Düngung, vor allem so weit sie das N/C-Verhältnis im Boden beeinflußt, wirkt sich auf dessen Gehalt an aktivem Stickstoff aus. Durch langjährige Stallmistdüngung konnte, wie KRESKE und MERKLE (1957) feststellten, ein Vorrat an aktivem Stickstoff geschaffen werden. In sauren Böden kommt es nicht zu einem so weitgehenden Abbau, daher wird organisch gebundener Stickstoff angereichert, der allerdings nach Kalkung mineralisiert werden kann. Zu einer Anreicherung des aktiven Stickstoffs kam es in Böden von Tanganjika bei beginnender Regenzeit durch Abbau von Rückständen der letzten Ernte und der Mikrobensubstanz (HAGENZICKER 1957/58). STOJANOVIC und BROADBENT (1957) prüften den Abbau von Weizenstroh und Maisblättern. Zusätze von Ammoniumsulfat oder Kalksalpeter ergaben bei Maisblättern gleich hohe Mengen an Nitrat. Die Kombination von Weizenstroh und Ammoniumsulfat führte nach 6 Wochen zu Stickstoffverlusten; wurde jedoch Nitratstickstoff zu Weizenstroh gegeben, konnte ein geringer Anstieg des aktiven Stickstoffs beobachtet werden. Hieraus geht die geringere biologische Angreifbarkeit der im Stroh enthaltenen Stickstoffverbindungen hervor. Bei der Düngung mit Stallmist, besonders aber mit Stroh, setzt, wie ALTEN et al. (1958) vermittels der CO_2-Produktion nachweisen konnten, sofort eine starke mikrobielle Aktivität ein. In Gefäßen ist die hierbei auftretende N-Assimilation in etwa 14 Tagen, im Freiland nach 8 bis 9 Wochen abgeschlossen. Eine Mineralisation wurde innerhalb einer Vegetationsperiode nicht festgestellt. Mit Hilfe markierten Stickstoffs wurde von JANSSON (1958) eine relativ langsame

Mineralisierung biologisch gebundenen Ammoniakstickstoffs festgestellt. Die Geschwindigkeit der Zersetzung von Bakterieneiweiß hängt von der jeweiligen Art ab (WIGOROW 1955).

Eine Zunahme der Nitrifikationskraft beobachtete BELIAEV (1958) bei langjähriger Anwendung von Ammoniumsulfat in der NPK-Düngung ohne Kalk. Es kam zu einer Anreicherung der Schimmelpilze und zu einer Abnahme der Produktivität des Bodens. Eine Hemmung der NH_3- bzw. NO_2-Bildung beobachtete STUNDL (1959/60) in Kulturlösungen mit Bodenimpfung, wenn Benzin, Gasöl oder Rohbenzol im Verhältnis 1:1000 zugesetzt wurden. Auch die biologische Reinigungswirkung des Bodens wurde durch Treibstoffe herabgesetzt.

Treten im Boden infolge Verdichtung oder Durchnässung anaerobe Verhältnisse auf, kommt es zu einer *Denitrifikation* bzw. zu einer Stickstoffentbindung. Die Vorgänge wurden von JENSEN (1904) und von ENGEL (1958a) dargelegt. Der Nachweis des Vorgangs wurde in der Regel durch eine Stickstoffbilanz des Bodens unter Berücksichtigung des zugeführten und entzogenen Stickstoffs geführt. Angaben über die Höhe der durch Stickstoffentbindung verursachten Verluste sind mit zahlreichen Unsicherheitsfaktoren belastet. Einmal kann im Boden gleichzeitig der entgegengesetzte Prozeß der Stickstoffbindung ablaufen, der nicht quantitativ erfaßbar ist. Im allgemeinen ist jedoch diese Gefahr gering, weil auch die klimatischen Voraussetzungen gegensätzlicher Natur sind. Größere Bedeutung kommt dagegen N-Mengen zu, die dem Boden durch die Niederschläge zugeführt bzw. durch Ammoniakverdunstung oder Auswaschung entzogen werden. Auch die auf dem Wege der Ammon- bzw. Nitratassimilation — wenn auch nur vorübergehend — verschwindenden Mengen wasserlöslichen Stickstoffs haben zu Fehlschlüssen geführt. Hohe Verluste durch Stickstoffentbindung wurden unter anderen von HAUK und MELSTED (1956), HAUK et al. (1958), WALKER et al. (1956) und LOEWENSTEIN et al. (1957) angegeben. In den Versuchen von DE und DIGAR (1954, 1955) in wassergesättigten Böden Westbengalens waren sie in nicht bepflanztem Boden größer als unter Reiskulturen. Um Verlusten vorzubeugen, ist es bei Reisanbau erforderlich, Nitratdünger zu meiden und die Düngung auf die Zeit intensiven Wachstums zu beschränken. Besonders hohe N-Verluste von 143 lbs/acre wurden von RUSSELL (1961) ermittelt, sie sind auf ungewöhnlich hohe Stallmistgaben zurückzuführen. Ein Beweis hierfür wurde von ALLISON et al. (1960) geführt: Bei einem O_2-Partialdruck von 0,46% beobachteten sie eine Reduktion zu N_2 von 10%, wenn keine Energiequelle zugesetzt wurde, dagegen von 50%, wenn 0,5% Glucose beigegeben worden war.

Die zur Denitrifikation befähigten Mikroorganismen gehören ausschließlich der Gruppe der echten Bakterien an, während bei Actinomyceten und Schimmelpilzen diese Fähigkeit nicht nachgewiesen wurde. Häufig wird die Zahl der denitrifizierenden Bakterien angegeben; so fand NETTE (1955) in der Rhizosphäre der Eiche das 10- bis 100fache, verglichen mit dem nicht durchwurzelten Boden. Die Zahl der Organismen berechtigt indessen nicht zu Schlußfolgerungen hinsichtlich der Intensität der Denitrifikation bzw. der Höhe zu erwartender N-Verluste.

Eine wesentliche Erweiterung der Kenntnisse über den Mechanismus der ablaufenden Prozesse ist den Arbeiten von VERHOEVEN (1952), KLUYVER und VERHOEVEN (1954), WIJLER und DELWICHE (1954) und anderen zu verdanken. Sie stellten fest, daß die Vorgänge in hohem Maße auf die Aktivierung des Wasserstoffs zurückzuführen sind. Die Nitrate dienen als H-Acceptoren und anorganische und organische Stoffe als H-Donatoren. Mit Hilfe von N^{15} konnten WIJLER und DELWICHE (1954) nachweisen, daß sowohl N_2O, NO als auch N_2 gebildet wurden. Der größte Teil des Stickstoffs wurde als N_2O abgegeben. Die Denitrifikationsintensität war oberhalb von pH 6 ziemlich konstant. Oberhalb von pH 7 wurde N_2 neben N_2O gebildet. Unterhalb von pH 7 wurde die Reduktion von N_2O stark gehemmt. Die Bildung von NO nahm aber zu, wenn die Reaktion unter pH 6 sank. Der Sauerstoffdruck von 5 mm Hg senkte die Denitrifikation auf 10% gegenüber anaeroben Bedingungen. Eine Reduktion des Nitrats zu Ammoniak trat auch hier nur bei Anwesenheit größerer Mengen organischer Substanz ein. Eine rein chemische Denitrifikation durch Einwirkung von Ammoniak auf salpetrige Säure (VAN-SLYKE-Reaktion) wurde selbst bei niedrigem pH nicht beobachtet.

An der *Nährstoff-Freisetzung* aus den *mineralischen Bestandteilen* des Bodens sind die Mikroorganismen beteiligt. Dabei werden abgeschiedener Bakterienschleim und ausgeschiedene organische und anorganische Säuren wirksam. ARISTOWSKAJA (1956) wies die Fähigkeit, anorganische Substanzen zu assimilieren, bei Bakterien und Pilzen in Podsolböden nach. Die Säurebildung war um so stärker, je weniger aschebildende Substanz im Boden vorhanden war. Die Wirksamkeit von *Aspergillus niger* und anderen Pilzen bei der Aufschließung verschiedener Feldspate wurde u. a. von ENO und REUSZER (1955) und von MÜLLER und FÖRSTER (1961) geprüft. Bis zum 4. Tag wiesen diese Wein-, Oxal- und Bernsteinsäure nach, die im weiteren Verlauf der Versuche in Zitronensäure umgewandelt wurden.

Besonderes Interesse ist der Mitwirkung der Mikroorganismen bei den Umsetzungen der *Phosphorverbindungen* im Boden entgegengebracht worden. Die mikrobielle Aufschließung hat eine Festlegung des Phosphors in Mikrobeneiweiß, dessen Phosphor säurelöslich ist, zum Teil aber auch in Humuskomponenten zur Folge. HARRIS et al. (1951) arbeiteten mit schwachen Lösungen radioaktiver Phosphate (K_2HPO_4). Dabei wurde der größte Teil der Phosphate während der aktiven Zellteilung in den Zellen zurückgehalten, so daß die Rate der P_2O_5-Umwandlung gering war.

In mitteleuropäischen und ukrainischen Schwarzerdeböden wurde die Bodenphosphorsäure nach JAHN-DEESBACH (1956) zu etwa 20% in bestimmten phosphorsäurereichen organischen Stoffen — etwa in Nucleinsäuren — gefunden, die durch ihre feste Bindung an Ton, Eisen o. a. blockiert sind. Bei der biologischen Festlegung der Phosphorsäure kommt es nach KAILA (1954) in erster Linie auf die Höhe des Gesamtphosphorgehalts an. Auf Grund der finnischen Forschungen können organische Stoffe, deren Gesamtphosphorgehalt höher als 0,2% ist, wohl keine erkennbare oder nur eine kurzfristige Festlegung des Bodenphosphors bewirken. Bei Ernterückständen und Stallmist ist das sicher der Fall, bei Stroh dagegen nicht.

Zu einer Festlegung in organischer Form kam es nach JACKMAN (1955) auch bei Superphosphatdüngung auf Weideböden. Die Stärke der Festlegung war abhängig von der Bodenart. Eine starke Festlegung wurde in einem gelbbraunen Lehmboden beobachtet, der einen hohen Gehalt an organischer Phosphorsäure besaß. Dagegen unterblieb sie in einem fruchtbaren Schlufflehmboden, dessen Phosphorsäure nur zu etwa 30% in organischer Form vorlag. Eine Steigerung der Menge löslicher Phosphorsäure durch $CaCO_3$ wurde bereits von FRED und HAAS (1919) nachgewiesen und von MÁTÉ und MOLNÁR (1956) für Wiesenböden bestätigt. Wenn keine Kalkung erfolgte, kam es zur Bildung von Fe- und Al-Phosphaten. Auch THOMPSON et al. (1954) beobachteten bei höheren pH-Werten eine stärkere Mineralisation der Phosphorsäure als bei niedrigen.

TSUBOTA (1959) gibt an, daß Verluste an Phosphorsäure unter anaeroben Bedingungen dann eintreten können, wenn Phosphate durch Buttersäurebakterien zu PH_3 reduziert werden.

Eine dauernde Anreicherung leicht aufnehmbarer Phosphorsäure im Boden ist nicht möglich, sie geht zu einem immer größeren Teil in schwer lösliche Phosphate über. Die Lebenstätigkeit der Mikroorganismen und Pflanzen schützt die Phosphate vor dieser Festlegung (KUDRIN 1952). Die Mikroorganismen können zwei- bis viermal soviel Phosphate aufnehmen, wie in den pflanzlichen Rückständen enthalten sind. Der Umsatz der Phosphorsäure ist mit dem Kreislauf des Kohlenstoffs bzw. Stickstoffs verbunden. Es ist daher durchaus verständlich, daß in den obersten $2^1/_2$ cm eines 13 Jahre lang gemulchten Bodens, der mikrobiologisch besonders aktiv ist, mehr säurelösliche Phosphate enthalten sind als in einem Vergleichsboden, der 15 Jahre lang gepflügt wurde (WITHEE und MCCALLA 1954).

An der Aufschließung der Phosphate beteiligen sich verschiedene Arten von Mikroorganismen. Bei *Escherichia freundii* spielte, wie aus den Ergebnissen von SPERBER (1957) hervorgeht, vor allem Milchsäure eine Rolle, daneben Ameisen-, Essig-, Propion-, Fumar- und Bernsteinsäure. Für die bakterielle Lösung von Mineralphosphaten ist allerdings die ausreichende Versorgung mit leicht löslichen Kohlenhydraten unabdingbare Voraussetzung. MEYER und KÖNIG (1960) stellten fest, daß verschiedene säureliebende Bodenpilze befähigt sind, Rohphosphate aufzuschließen. In Anlehnung an die Methode ROSSI-CHOLODNY wurde ein Verfahren entwickelt, das es ermöglicht, die Lösung der Partikel unter dem Mikroskop zu verfolgen (KÖNIG 1961a). Auch *Aspergillus niger* (vgl. ZIECHMANN und METZLER 1960) sowie *Penicillium* sp. und *Nocardia* sp. hatten eine lösende Wirkung (SPERBER 1958). Daß die Hydrolyse durch die gebildeten Säuren bewirkt wurde, ging daraus hervor, daß sie durch $CaCO_3$ gehemmt werden konnte. Von CASIDA JR. (1959) wurde nachgewiesen, daß Phosphatasen von Aspergillusarten in der Lage waren, besonders resistente P-Verbindungen im Boden anzugreifen. Nach MOREAU (1959) können sich an der Lösung auch *Trichoderma viridis*, *Mortierella vinacea*, *Trisporium nigricans*, *Penicillium* sp. und Azotobacter beteiligen. Auch erwies sich Milchsäure als besonders wirksames Aufschließungsmittel. Nach KÖNIG (1961b) konnten die säurebildenden Pilze *Aspergillus niger* und *Penicillium chrysogenum*, die selbst bei Anwesenheit schwerlöslicher Rohphosphate (Apatite) chemotropisch reagierten, eine große Zahl anderer Pilze veranlassen, sich ebenso zu verhalten. Nach CLAUS et al. (1958) ist die lösende Wirkung bei Azotobacter dem hauptsächlich aus Galaktose und verschiedenen Rhamnose-Bausteinen bestehenden Schleim zuzusprechen. HOFMANN (1958) stellte fest, daß neben verschiedenen Bakterien auch wilde Hefen beim Abbau von Galaktose freie anorganische Phosphorsäure als Hexosemonophosphorsäureester binden. Durch die Enzyme von Schimmelpilzen und Bakterien wird hieraus im Boden wieder P_2O_5 abgespalten und kann von den Pflanzen verwertet werden.

Mikroorganismen, die in der Lage sind, Phosphate zu mineralisieren, sind besonders auch in der Rhizosphäre gefunden worden. SPERBER (1957) führte den Nachweis mit Hilfe von einigen Bakterienstämmen, die er aus dem Wurzelbereich isoliert hatte. Sie konnten die Phosphorsäure aus Apatit lösen. GERRETSEN (1948) verwendete sterile und nichtsterile Kulturen. Mit Hilfe der normalen Rhizosphärenflora wurde eine bessere Ausnutzung der Phosphate und ein intensiveres Wachstum der Pflanzen erzielt. AKHROMEIKO und SHESTAKOWA (1958) führten den Pflanzen P^{32} über das Blatt zu. Dadurch wurde eine verstärkte P-Ausscheidung durch die Wurzel ausgelöst. Die Rhizosphärebakterien legten zwar einen Teil dieser Phosphorsäure fest, aber in einer leicht mineralisierbaren Form.

Phosphordüngung bewirkt eine Verbesserung der *Bodenfruchtbarkeit*, deren verschiedene Möglichkeiten DHAR (1956) folgendermaßen formuliert:

1. Verhinderung von Verlusten an Calcium (Indien).
2. Förderung der Stickstoffbindung, indem sie die Mineralisierung der organischen Stoffe beschleunigt.
3. Erhöhung des Werts von Komposten.
4. Hemmung der Stickstoffverluste bei Düngung mit mineralischem Stickstoff.
5. Verbesserung der alkalischen Böden, besonders durch Rohphosphate und Thomasphosphate in Verbindung mit organischen Substanzen.

KREYBIG (1954) empfahl, Rohphosphate durch Kompostierung mit Stalldung in organische Form überzuführen. Diese Form betrachtete er als „Bodendünger", während er granulierte Düngemittel als „Pflanzendünger" bezeichnete. In den Versuchen von DÖRING (1954) bewirkte Rohphosphat keine Ertragssteigerung, dagegen wirkte Stallmist, in dem Rohphosphate aufgeschlossen worden waren, um 20% besser als der Vergleichsstallmist ohne P. Während der Rotte wurden die Ammoniumcarbonate des Stallmists in Kalkhumate überführt und durch Ca-Entzug wurde das Rohphosphat aufgeschlossen. Weicherdige Rohphosphate können auch mit Erfolg als Düngemittel verwendet werden, Voraussetzung sind jedoch biologisch aktive Böden (HOFMANN und AMBERGER 1957). Sie zeigen zwar eine schwächere Anfangswirkung, haben aber eine bessere Nachwirkung als die handelsüblichen P-Dünger. Eine Erhöhung des Gehalts an zitronensäurelöslicher Phosphorsäure im Boden wurde von GELLER und JUSPE (1954) bei steigenden Mineraldüngergaben beobachtet, die auch Nitrifikation und Zellulosezersetzung intensivierten. Die in den Pflanzenrückständen enthaltene Phosphorsäure steht der nachfolgenden Kultur wieder zur Verfügung, besonders wenn die Pflanzen vor der Blüte geerntet wurden oder wenn es sich um P-reiche Reste handelte (FULLER et al. 1956). In diesem Zustand bieten sie dem mikrobiellen Angriff geringeren Widerstand.

Die höhere Oxydationsstufe der Phosphorsäure ist als Pflanzennährstoff wirksamer; so sind beispielsweise die Phosphate für die Pflanzen leichter verwertbar als Phosphite. ADAMS und CONRAD (1953) konnten zeigen, daß Phosphite im ersten Jahr nicht zur Wirkung kamen, dafür aber in der Nachfrucht. Die Umwandlung war ein mikrobiologischer Prozeß, er konnte durch Zusatz von Toluol unterbunden werden. Beteiligt waren Bakterien, Actinomyceten und Pilze. In Nährlösungen mit Phosphaten und Phosphiten wurden zuerst die Phosphate aufgebraucht.

Auch die *Kieselsäure* und ihre Verbindungen werden von Mikroorganismen angegriffen. HEINEN (1960) stellte fest, daß verschiedene Bakterien (*Serratia marcescens, Bac. subtilis, Escherichia coli* u. a.) anstelle von P auch Si aufzunehmen vermögen. Die Aufnahme war optimal zwischen pH 6,6 und 7,5, stärker aus Silikatlösungen als aus Quarz- oder Silicagelwasser. Durch stärkere Säure inaktivierte Zellen adsorbierten Si.

Die *Eisen- und Manganverbindungen* des Bodens werden nach BROMFIELD (1954) durch die Tätigkeit der Bakterien reduziert, dadurch beweglich und für die höhere Pflanze verfügbar. Voraussetzung ist allerdings, daß den Organismen genügende Mengen an organischer Substanz zur Verfügung stehen. Später (BROMFIELD 1956) stellte er fest, das Corynebakterien- und Chromobakteriengemische das Mangan des Mangansulfats (0,005%) in einer Hefeextraktlösung (0,005%) oxydieren. Die Oxydation ist allerdings auf die stationäre Phase der Entwicklung beschränkt, in der sie sich nicht autotroph ernähren. Bei vergleichenden Untersuchungen in sterilisierten und nicht sterilisierten Böden kam KOSEGARTEN (1957) zu dem Ergebnis, daß die Mikroorganismen im Boden vorwiegend Mangan oxydieren, wobei die organische Substanz reduziert wird.

BAUERNFEIND und POSCHENRIEDER (1960) isolierten aus der Rhizosphäre des Hafers 21 Bakterienstämme, von denen 15 mehr als 20 γ Mangan pro 50 ml Nährlösung festlegten: *Bac. subtilis*, *Pseudomonas convexa*, *Corynebacterium* sp., *Achromobacter delmarvae*, *Bac. pumilus*, *Bac. circulans*, *Flavobacterium lutescens*, *Achromobacter eurydice*. Durch die Tätigkeit dieser Organismen wird die Wirkung einer Mangandüngung erheblich abgeschwächt, sie kann zu Manganmangelerscheinungen führen, wie sie TIMONIN (1946) bei einer Hafersorte beobachtete. Durch Bodensterilisation können sie behoben werden. Ein Schema des Manganumsatzes im Boden geben MANN und QUASTEL (1946). Nach CLAUS et al. (1958) werden Eisen- und Bariumsulfate durch den Schleim von *Azotobacter chroococcum*, *Aerobacter aerogenes* und *Pseudomonas tumefaciens* gelöst, wobei der Schleim von *Aerobacter* die stärkste Wirkung zeigte. Stoffwechselprodukte von *Bac. subtilis*, *Asp. niger* und *Bac. megaterium* enthalten, wie GARIBALDI und NEILANDS (1956) feststellten, Substanzen, die mit den Eisenionen des Kulturmediums eine Komplexverbindung eingehen, die stark färbend wirkt, z. B. bei *Bac. subtilis* tief purpurrot.

Mit der Oxydation von *Schwefelverbindungen* (6 bis 24%) befaßten sich FREDERICK et al. (1957). Cystin, Taurin, Natriumtaurocholat, Sulfosalizylsäure, Formaldehyd-Na-Sulfoxylat wurden schnell zersetzt, während Thiamin und Thioharnstoff langsam oder gar nicht zersetzt wurden. Aus Methionin wird nicht Sulfat, sondern Methylthiol und Dimethyldisulfid gebildet, Stoffe, die die Nitrifikation hemmen. Die Sulfat bildenden Verbindungen vermehren die Zahl der Mikroorganismen im Boden, andere, vor allem Thioharnstoff, Phenylthioharnstoff, Thioäthanol, Sulfonmethan und Natriumdiäthyldithiocarbonat wirken so stark toxisch, daß beispielsweise die Zahl der Pilze um 99% zurückgeht. Die Bildung von Sulfaten erniedrigt die Bodenreaktion. Bei Methionin wird die Reaktion zunächst alkalisch, später wieder sauer. In den Waldböden Ostafrikas wurde von HESSE (1957/58) ebenfalls Sulfat nachgewiesen, und zwar zunehmend mit der Tiefe. In der obersten Schicht wird der organisch gebundene Schwefel oxydiert, in die Tiefe eingewaschen und dort von den Pflanzen aufgenommen. Die Oxydation des Schwefels ist ein extrem langsamer Prozeß, er wird durch $CaSO_4$ und schwefelhaltige Aminosäuren (Cystin und Methionin) gefördert, nicht dagegen durch $CaCO_3$ und schwefelfreie Aminosäuren. Nach BENDA (1957) wird in Hochmooren durch Eiweißzersetzung bzw. Desulfurikation Schwefelwasserstoff durch *Desulfovibrio desulfuricans* in anaeroben Zonen gebildet. Nur Cystin wird abgebaut, während Methionin nicht angegriffen wird. In der Natur läuft der Prozeß bei pH 4,3 ab.

Zur Bestimmung der *mikrobiologischen Aktivität* eines Bodens stehen verschiedene Wege zur Verfügung. Die Gesamtaktivität kommt in der vom Boden abgegebenen CO_2-Menge zum Ausdruck. Sie kann als aktuelle Aktivität bestimmt werden, wenn der Boden lediglich unter konstanten Bedingungen gehalten wird. Erfolgt die CO_2-Bestimmung jedoch nach Zusatz von mineralischen oder organischen Düngemitteln, dann wird die potentielle Aktivität bestimmt. Da die Tätigkeit der Mikroorganismen gleichzeitig zur Bildung von Wärme führt, entwickelte REINAU (1961) die thermokinetische Methode, mit deren Hilfe die Menge der nach Zusatz von Zucker gebildeten Wärme erfaßt wurde. Das Ergebnis stellt eine potentielle Größe dar. Anhand der Keimzahlen kann sowohl die aktuelle als auch die potentielle Aktivität des Bodens bestimmt werden, sowohl mit Hilfe der mikroskopischen als auch nach der KOCHschen Methode. Die letztere kann unter Einschaltung selektiver Nährböden zur Bestimmung der Keimzahl verschiedener physiologischer Gruppen (N-bindende, ammonifizierende, nitrifizierende Bakterien usw.) verwendet werden. HOFMANN und seine Mitarbeiter (1950a, 1950b, 1952, 1953, 1953/54, 1955) empfahlen die Bestimmung des Enzymgehalts des Bodens.

Es ist wiederholt die Frage aufgeworfen worden, ob zwischen *mikrobiologischer Aktivität* und der *Bodenfruchtbarkeit* eine Beziehung besteht. Im allgemeinen wird Azotobacter als Leitorganismus der Bodenfruchtbarkeit angesehen. Das ist zulässig, solange die Intensität seines Wachstums als Maßstab für den Grad der Fruchtbarkeit angesehen wird. Es wäre aber falsch, azotobacterfreie Böden als unfruchtbar zu bezeichnen. Azotobacter kann in geringen Mengen vorhanden sein oder fehlen, wenn

1. die Bodenreaktion für Azotobacter nicht zuträglich ist,
2. der Boden arm an P_2O_5, K_2O oder Mo ist, oder
3. der Boden nicht genug für Azotobacter verwertbare Kohlenhydrate enthält.

Bei den azotobacterfreien Böden muß also geprüft werden, warum der Organismus fehlt.

Weiterhin wurden in neuerer Zeit u. a. Vergleiche angestellt zwischen allgemeiner mikrobiologischer Aktivität und Bodenfruchtbarkeit von MÜLLER (1959b), zwischen Keimzahl und Ertrag von BURRICHTER (1955), zwischen Nitratgehalt und Ertrag von FITTS et al. (1953) und zwischen Gehalt an aktivem Stickstoff und Baumertrag von ZÖTTL (1958). Ferner verwendete KOEPF (1952ff) die CO_2-Bildung als Maßstab der Bodenfruchtbarkeit, JAGNOW (1958) für Wiesenböden und MINA (1957) für Waldböden, während DROBNIK (1957b) den O_2-Verbrauch als Vergleichsbasis empfahl. HOFMANN (l. c.) hält für Vergleichszwecke den Fermentgehalt für geeigneter als die Atmung. Nach den Versuchen von KOEPF (1954) können jedoch Fermente im Boden gespeichert werden, während der tatsächliche Abbau von den Fermenten lebender Organismen bewirkt wird. Der Fermentgehalt stellt also das Ergebnis biologischer Aktivität im Boden dar, während die Atmung die Dynamik der mikrobiologischen Vorgänge wiederzugeben vermag. Er kommt (1955) zu dem Schluß, daß jede der beiden Methoden nur einen Teil der Faktoren der Fruchtbarkeit erfaßt, daß sie sich aber gut ergänzen.

Bei diesen Vergleichen ist zwar eine Beziehung zwischen Bodenfruchtbarkeit, soweit sie an den Erträgen gemessen wurde, verschiedentlich gefunden worden. Trotzdem braucht das nicht grundsätzlich der Fall zu sein, denn die Erträge hängen von den örtlichen klimatischen Verhältnissen ab, während die mikrobiologische Aktivität, solange sie im Laboratorium bestimmt wird, bei konstanter Feuchtigkeit und Temperatur erarbeitet wird. Werden die Böden eines Feldbestandes untersucht, wird das Ergebnis von der Wetterlage am Untersuchungstag beeinflußt. Größere Wahrscheinlichkeit für eine direkte Beziehung besteht dagegen zwischen der Ertragsfähigkeit eines Bodens und seiner potentiellen Aktivität. Grundsätzlich ist jedoch festzustellen, daß für die Ermittlung der Beziehungen zwischen Bodenfruchtbarkeit und mikrobiologischer Aktivität für die letztere noch keine befriedigende Bestimmungsmethode zur Verfügung steht. Es ist daher zweckmäßig, mehrere Methoden nebeneinander anzuwenden.

Literatur

ADAMS, F. A., und J. P. CONRAD: Transition of phosphite to phosphate in soils. Soil Sci. **75**, 361–372 (1953). — AKHROMEIKO, A. J., und V. A. SHESTAKOVA: The influence of rhizosphere microorganisms on the uptake and secretion of phosphorus and sulphur by the roots of arboreal seedlings. Proc. Second Int. Conf. Peaceful Uses of Atomic Energy, Geneva, **27**, 193–199 (1958); Ref. Z. Pflanzenernähr., Düng., Bodenkde. **89**, 71 (1960). — ALEEM, M. J. H., und M. ALEXANDER: Nutrition and physiology of *Nitrobacter agilis*. Appl. Microbiol. **8**, 80–84 (1960). — ALLISON, F. E., J. N. CARTER und L. D. STERLING: The effect of partial pressure of oxygen on denitrification in soil. Soil Sci. Soc. Amer. Proc. **24**, 283–285 (1960). — ALLISON, F. E., J. H. DOETSCH und E. M. ROLLER: Availability of fixed ammonium in soils containing different clay materials. Soil Sci. **75**, 373–382 (1953). — ALTEN, F., C. M. FLORIAN und K. MECHSNER: Abhängigkeit der Wirkung mineralischer Dünger von Höhe und zeitlichem Verlauf der biologischen Aktivität des Bodens. Agrochimica 3, 1–6 (1958); Ref. Z. Pflanzenernähr., Düng., Bodenkde. 88, 74 (1960). — ANDERSON, O. E.: The effect of low temperature on nitrification of ammonia in Cecil sandy loam. Soil Sci. Soc. Amer. Proc. 24, 286–289 (1960). — ANDERSON, O. E., und E. R. PURVIS: Effects of low temperatures on nitrification of ammonia in soils. Soil Sci. 80, 313–317 (1955). — ARISTOWSKAJA, T. W.: Über die Rolle der Mikroorganismen bei der Podsolbildung. Potschwowedenie Nr. 3, 48–62 (1956) (russ.); Ref. Zbl. Bakt. II 111, 731 (1958). — ATANASIU, N.: Düngungsfragen. Erg. landwirtsch. Forsch. H. III, 13–33 (1960). — AUGIER, J., und M. JACOB: Etude de l'action de la microflore des sols sur un substrat donné: principes et apareille. C. R. Acad. Sci. Paris 249, 1157–1159 (1959); Ref. Zbl. Bakt. II, 114, 206 (1961).

BAKER, F.: Preliminary note on the role of coccoid microorganisms in the desintegration of cell-wall substances. Zbl. Bakt. II 83, 452–454 (1931). — BARNES, T. W.: The formation of nitrates in soil following various crop relations. J. Agric. Sci. 40, 166–168 (1950). — BAUERNFEIND, A., und H. POSCHENRIEDER: Untersuchungen über die Manganfestlegung durch Bodenbakterien. Bayer. Landwirtsch. Jahrb. 37, 610–618 (1960). — BAUMANN, H., und G. MAASS: Über den Verlauf des Nitratgehaltes unter verschiedenen Früchten im Ackerboden. Z. Pflanzenernähr., Düng., Bodenkde. 79, 155–167 (1957). — BECK, TH., und R. RÖSCHENTHALER: Eine einfache Methode zur kolorimetrischen Bestimmung der Bodenatmung im Laboratoriumsversuch. Z. Pflanznernähr., Düng., Bodenkde. 87, 143–156 (1959). — BÉLA, M.: Ergebnisse der im Laboratorium durchgeführten Bodenatmungsversuche auf degradiertem Steppenboden. A Debreceni Mezögazda-sagi Akadémia Tudományos Évkönigve 1956, 141–152 (1957) (ung.); Ref. Z. Pflanzenernähr., Düng., Bodenkde. 82, 246 (1958). — BELIAEV, G. N.: Influence of long-term application of fertilizer on microflora of turf-podsol soil containing loam. Mikrobiologija 27, 464–469 (1958). — BENDA, I.: Mikrobiologische Untersuchungen über das Auftreten von Schwefelwasserstoff in den anaeroben Zonen des Hochmoores. Arch. Mikrobiol. 27, 337–374 (1957). — BERTOLET, E. E.: Observations on soil burial procedures. Amer. Dyestuff Rep. 33, 21–24 (1944). — BIRCH, H. F.: The effect of soil drying on humus decomposition and nitrogen availability. Plant a. Soil 10, 9–31 (1958). — Further observations on humus decomposition and nitrification. Plant a. Soil 11, 262–286 (1959). — BIRCH, H. F. und M. T. FRIEND: Humus decomposition in east African soils. Nature 178, 500–501 (1956); Ref. Z. Pflanzenernähr., Düng., Bodenkde. 76, 175 (1957). — BROADBENT, F. E., G. N. HILL und K. B. TYLER: Transformations and movement of urea in soils. Soil Sci. Soc. Amer. Proc. 22, 303–307 (1958). — BROMFIELD, S. M.: Reduction of ferric compounds by soil bacteria. J. Gen. Microbiol. 11, 1–6 (1954). — Oxidation of manganese by soil microorganisms. Austral. J. Biol. Sci. 9, 238–252 (1956); Ref. Z. Pflanzenernähr., Düng., Bodenkde. 78, 261 (1957). — BURANGULOWA, M. N., und E. P. SOLEWJEWA: Nitrifizierende Bakterien in den degradierten Schwarzerden des Voruralgebietes (in der Autonomen Republik Baschkirien). Trudy Baschkir. seljskochos. Inst. 7, 3–15 (1956) (russ.); Ref. Z. Pflanzenernähr., Düng., Bodenkde. 82, 256 (1958). — BURRICHTER, E.: Die bakteriologische Kartierung und Beurteilung der Kulturböden des Provinzialgutes Kinderhaus bei Münster/Westf. Landwirtsch. Forsch. 8, 14–25 (1955). — Untersuchungen über die Massenentwicklung der Bodenbakterien im Laufe des Jahres. Ber. Dtsch. Bot. Ges. 121, 71–77 (1958); Ref. Zbl. Bakt. II 113, 176 (1959/60).

CASIDA, L. E.: Phosphates activity of some common soil fungi. Soil Sci. 87, 305–310 (1959). — CHASE, F. E., und G. BAKER: A comparison of microbial activity in an Ontario forest soil under Pine, Hemlock and Mapel cover. Canad. J. Microbiol. 1, 45–54 (1954). — CLARK, F. E., W. E. BEARD und D. H. SMITH: Dissimilar nitrifying capacities of soils in relation to losses of applied nitrogen. Soil Sci. Soc. Amer. Proc. 24, 50–54 (1960). — CLAUS, D., H. WITTMANN und A. RIPPEL-BALDES: Untersuchungen über die Zusammensetzung von Bakterienschleimen und deren Lösungsvermögen gegenüber schwerlöslichen anorganischen Verbindungen. Arch. Mikrobiol. 29, 169–178 (1958).

DE, P. K. und S. DIGAR: Loss of nitrogen gas from water-logged soils. J. Agric. Sci. 44, 129–132 (1954). — Influence of the rice crop on the loss of nitrogen gas from water-logged soils. J. Agric. Sci. 45, 280–282 (1955). — DEPPE, K., und H. ENGEL: Untersuchungen über die Temperaturabhängigkeit der Nitritbildung durch *Nitrobacter winogradskyi Buch.* bei ungehemmtem und gehemmtem Wachstum. Zbl. Bakt. II 113, 561–568 (1960). — DHAR, N. R.: Calcium phosphates and their importance in nitrogen fixation and alkali soil reclamation. Proc. Nat. Acad. Sci. India 25, A, 211–327 (1956); Ref. Z. Pflanzenernähr., Düng., Bodenkde. 81, 74 (1958). — DHARIWAL, A. P. S., und F. J. STEVENSON: Determination of fixed ammonium in soils. Soil Sci. 86, 343—349 (1958). — DÖRING, H.: Rohphosphataufschluß während der Stallmistrotte. Z. Pflanzenernähr., Düng., Bodenkde. 66, 202–211 (1954). — DROBNIK, J.: Untersuchungen biologischer Umwandlungen organischer Stoffe im Boden. Potschwowedenije Nr. 12, 62–71 (1957a). — Die Atmungsmessung in der Bodenmikrobiologie I (Die Makro-Atmungsmessung). Ceskoslov. Mikrobiol. 2, 116–123 (1957b); Ref. Z. Pflanzenernähr., Düng., Bodenkde. 86. 270 (1959). — Dubowick, J. F.: Einfluß der Bodentemperatur auf die Wirksamkeit der Düngung in der Schwarzerde im nördlichen Kasachstan. Bodenkunde 1956, 62–65 (russ.); Ref. Z. Pflanzenernähr., Düng., Bodenkde. 79, 178 (1957).

Engel, H.: Die Stickstoffentbindung. In: Ruhland, Handbuch der Pflanzenphysiologie, Bd. 8, S. 1083–1106 (1958a). — Die Nitrifikation. In: Ruhland, Handbuch der Pflanzenphysiologie, Bd. 8, S. 1107–1127 (1958b). — Eno, C. F., und W. G. Blue: The effect of anhydrous ammonia on nitrification and the microbiological population of sandy soils. Soil Sci. Soc. Amer. Proc. 18, 178–181 (1954). — The comparative rate of nitrification of anhydrous ammonia, urea and ammonium sulfate in sandy soils. Soil Sci. Soc. Amer. Proc. 21, 392–396 (1957). — Eno, C. F., W. G. Blue und J. M. Good: The effect of anhydrous ammonia on nematodes, fungi, bacteria and nitrification in some Florida soils. Soil Sci. Soc. Amer. Proc. 19, 55–58 (1955). — Eno, C. F., und H. W. Reuszer: Potassium availability from biotite, muscovite, greensand and microcline as determined by growth of Aspergillus niger. Soil Sci. 80, 199–209 (1955). — Eylar, O. R., und E. L. Schmidt: A survey of heterotrophic microorganisms from soil for ability to form nitrite and nitrate. J. Gen. Microbiol. 20, 473–481 (1959).

Fisher, T., E. Fisher und M. Appleman: Nitrification by certain heterotrophic bacteria present in soil. J. Bact. 64, 596 (1952). — Fisher, W. B., und W. L. Parks: Influence of soil temperature on urea hydrolysis and subsequent nitrification. Soil Sci. Soc. Amer. Proc. 22, 247–248 (1958). — Fitts, J. W., W. V. Bartholomew und H. Heidel: Correlation between nitrificable nitrogen and yield response of corn to nitrogen fertilization on Iowa soils. Soil Sci. Soc. Amer. Proc. 17, 119–122 (1953). — Predicting nitrogen fertilizer needs of Iowa soils, I, Evaluation and control of factors in nitrate production and analysis. Soil Sci. Soc. Amer. Proc. 19, 69–73 (1955). — Fred, E. B., und A. R. C. Haas: The etching of marble by roots in the presence and absence of bacteria. J. Gen. Physiol. 1, 631–638 (1919). — Frederick, L. R.: The formation of nitrate from ammonium nitrogen in soils, I, Effect of temperature. Soil Sci. Soc. Amer. Proc. 20, 496–500 (1956). — Frederick, L. R., R. L. Starkey und W. Segal: Decomposability of some organic sulfur compounds in soil. Soil Sci. Soc. Amer. Proc. 21, 287–292 (1957). — Frercks, W.: Die Bodenatmung als Mittel zur Erfassung der Mikroorganismentätigkeit in Moor- und Heidesandböden; ein neues Verfahren zu ihrer Bestimmung und erste Ergebnisse. Z. Pflanzenernähr., Düng., Bodenkde. 66, 39–54 (1954). — Frercks, W., und D. Puffe: Vergleichende Untersuchungen zwischen der „Bodenatmung" und der CO_2-Produktion von Moorböden. Z. Pflanzenernähr., Düng., Bodenkde. 87, 108–118 (1959). — Zur Frage des Pektin- und Zelluloseabbaus in Moorböden. Z. Pflanzenernähr., Düng., Bodenkde. 92, 46–56 (1961). — Frercks, W., D. Puffe und H. Findeisen: Über die Keimzahlbestimmungen, Zellulosezersetzung und die Stickstoffwirkung auf den Zersetzungsvorgang in Moorböden und Sandmischkulturen unterschiedlicher Aufkalkung. Z. Pflanzenernähr., Düng., Bodenkde. 89, 27–42 (1960). — Fuller, W. H., D. R. Nielsen und R. W. Miller: Some factors influencing the utilization of phosphorus from crop residues. Soil Sci. Soc. Amer. Proc. 20, 218–224 (1956).

Garibaldi, A., und J. B. Neilands: Formation of iron-binding compounds by microorganisms. Nature 177, 526–527 (1956); Ref. Z. Pflanzenernähr., Düng., Bodenkde. 76, 59 (1957). — Garrison, R. G., und J. O. Harris: The preliminary description of a cellulose digesting Micrococcus. Trans. Kansas Acad. Sci. 57, 81–84 (1954). — Geller, I. A.: Der Einfluß der Bodenbearbeitung auf die mikrobiologischen Prozesse und den Ertrag der Zuckerrüben. Agrobiologija 1958, 101–107 (russ.); Ref. Z. Pflanzenernähr., Düng., Bodenkde. 88, 189 (1960). — Geller, I. A., und F. B. Juspe: Über den Einfluß mineralischer Düngemittel auf mikrobiologische Bodenprozesse. Mikrobiologija 23, 451–454 (1954); Ref. Z. Pflanzenernähr., Düng., Bodenkde. 69, 319 (1955). — Gerretsen, F. C.: The influence of microorganisms on the phosphate intake by the plant. Plant a. Soil 1, 51–81 (1948). — Gerretsen, F. C., und H. de Hoop: Nitrogen losses during nitrification in solution and in acid sandy soils. Canad. J. Microbiol. 3, 359–380 (1957). — Ghildyal, B. P., und U. C. Gupta: A study of the biochemical and microbiological changes during the decomposition of Crotolaria juncea (sann hemp) at different stages of growth in soil. Plant a. Soil 11, 312–330 (1959). — Gliemeroth, G.: Stickstoffverlagerung über Winter auf einem Lößlehm in Abhängigkeit von Form, Menge, Termin und Verteilung der Herbstdüngung. Z. Pflanzenernähr., Düng., Bodenkde. 85, 20–31 (1959). — Goldberg, S. S., und P. L. Gainey: Role of surface phenomena in nitrification. Soil Sci. 80, 43–53 (1955). — Gottlieb, S., und M. J. Pelczar: Microbiological aspects of lignin degradation. Bact.

Rev. **15**, 55–76 (1951). — Greaves, I. E., und L. W. Jones: The influence of temperature on the microflora of the soil. Soil Sci. **58**, 377–387 (1944). — Greenland, D. J.: Nitrate fluctuations in tropical soils. J. Agric. Sci. **50**, 82–92 (1958). — Gundersen, K.: Observations on mixed cultures of *Nitrosomonas* and heterotrophic soil bacteria. Plant a. Soil **7**, 26–34 (1955). — Gussejnow, D. M., Z. R. Mowssumow und W. W. Seidow: Stickstoffverluste von dem in den Boden eingebrachten Stickstoffdünger. Nachr. Aserbajdschan. Akad. **4**, 111–123 (1957) (russ.); Ref. Z. Pflanzenernähr., Düng., Bodenkde. **82**, 238 (1958).

Haber, W.: Ökologische Untersuchung der Bodenatmung. Flora **146**, 109–157 (1958); Ref. Z. Pflanzenernähr., Düng., Bodenkde. **86**, 269 (1959). — Zur Ökologie des Bodenlebens in verschiedenen Pflanzengesellschaften. Ber. Dtsch. Bot. Ges. **71**, 399–410 (1959). — Hagenzieker, F.: Soil nitrogen studies at Urambo, Tanganjika-Territory, East Africa. Plant a. Soil **9**, 97–113 (1957/58). — Hagin, J.: Rates of nitrification in natural and "conditioner"-formed soil aggregates of various sizes. Bull. Res. Counc. Israel **5**, 98–104 (1955); Ref. Z. Pflanzenernähr., Düng., Bodenkde. **73**, 276 (1956). — Hanway, J., und L. Dumenil: Predicting nitrogen fertilizer needs of Iowa soils, III, Use of nitrate production together with other information as a basis for making nitrogen fertilizer recommendation for corn in Iowa. Soil Sci. Soc. Amer. Proc. **19**, 77–80 (1955). — Harris, J. O., R. E. Hein und R. H. McFarland: Phosphate turnover studies with soil bacteria. Trans. Kansas Acad. Sci. **54**, 560–564 (1951). — Hauck, R. D., und S. W. Melsted: Some aspects of the problem of evaluating denitrification in soils. Soil Sci. Soc. Amer. Proc. **20**, 361–364 (1956). — Hauck, R. D., S. W. Melsted und P. E. Yankwich: Use of N-isotope distribution in nitrogen gas in the study of denitrification. Soil Sci. **86**, 287–291 (1958). — Heinen, W.: Silicium-Stoffwechsel bei Mikroorganismen, I, Aufnahme von Silicium durch Bakterien. Arch. Mikrobiol. **37**, 199–210 (1960). — Hesse, P. R.: Sulphur and nitrogen changes in forest soils of East Africa. Plant a. Soil **9**, 86–96 (1957/58). — Hofmann, Ed.: Enzymreaktionen und ihre Bedeutung für die Bestimmung der Bodenfruchtbarkeit. Z. Pflanzenernähr., Düng., Bodenkde. **56**, 68 (1952). — Die Enzyme im Boden und ihre Bedeutung für seine Biologie und Fruchtbarkeit. Z. Acker- u. Pflanzenbau **100**, 31 (1955). — Die Bedeutung der Phosphorsäure beim Rotteprozeß, ihre Festlegung und Lösung durch biologische Faktoren. Phosphorsäure **18**, 172–176 (1958). — Hofmann, E., und A. Amberger: Zur Phosphorsäurewirkung von Rohphosphaten in Feldversuchen. Z. Pflanzenernähr., Düng., Bodenkde. **76**, 102–110 (1957). — Hofmann, Ed., und Gg. Hoffmann: Über das Enzymsystem unserer Kulturböden, V, α-und β-Galaktosidase und α-Glucosidase. Biochem. Z. **325**, 329–332 (1953/54). — Hofmann, Ed., und J. Niggemann: Über das Enzymsystem unserer Kulturböden, III, Proteinase. Biochem. Z. **324**, 308–310 (1953). — Hofmann, Ed., und H. Scheck: Über den Einfluß des Kaliumions auf Bildung und Aktivität der Fermente bei Hefen und Schimmelpilzen. Biochem. Z. **321**, 98–106 (1950b). — Hofmann, Ed., und A. Seegerer: Der Fermentgehalt des Bodens als Maßstab seiner biologischen Aktivität. Biochem. Z. **321**, 97 (1950a). — Hora, T. S., und M. R. S. Iyengar: Nitrification by soil fungi. Arch. Mikrobiol. **35**, 252–257 (1960). — Horn, S.: Die Bodenatmung und ein kolorimetrisches Verfahren zu ihrer serienmäßigen Bestimmung im Freiland. Z. Acker- u. Pflanzenbau **99**, 1 (1955). — Hutchinson, H. B., und J. Clayton: The decomposition of cellulose by an aerobic organism (*Spirochaeta cytophaga*). J. Agric. Sci. **9**, 143–173 (1919).

Isermeyer, J.: Eine einfache Methode zur Bestimmung der Bodenatmung und der Karbonate im Boden. Z. Pflanzenernähr., Düng., Bodenkde. **56**, 26–38 (1952). — Iyengar, M. R. S., und T. S. Hora: Nitrite oxydation by soil fungi. Naturwiss. **46**, 211 (1959).

Jackmann, R. H.: Organic phosphorus in New Zealand soils under pasture, I, Conversion of applied phosphorus into organic forms. Soil Sci. **79**, 207–213 (1955). — Jagnow, G.: Untersuchungen über Keimzahl und biologische Aktivität von Wiesenböden. Z. Pflanzenernähr., Düng., Bodenkde. **82**, 50–67 (1958). — Jahn-Deesbach, W.: Untersuchungen über den Bindungszustand der Phosphorsäure in einigen europäischen Schwarzerdeböden. Z. Pflanzenernähr., Düng., Bodenkde. **74**, 97–111 (1956). — Jansson, S. L.: Orientierende Studien über den Stickstoffkreislauf im Boden mit Hilfe von ^{15}N als Leitisotop. Z. Pflanzenernähr., Düng., Bodenkde. **69**, 190–198 (1955). — Tracer studies on nitrogen transformation in soil with special attention to mineralisation-immobilization relationships. Kungl. Lantbr. Ann. **24**, 101–361 (1958); Ref. Z. Pflanzenernähr., Düng., Bodenkde. **82**, 82 (1958). — Jensen, H.: Denitrification und Stickstoffentbindung. In: Lafar, Handbuch der teehnischen Mykologie, Bd. 3, S. 182–192 (1904). — Jensen, H. L.: Effect of organic compounds on *Nitrosomonas*. Nature **165**, 974 (1950). — Jewitt, T. N.: Nitrification in Sudan

Gezira Soil. J. Agric. Sci. **35**, 264–271 (1945). — JONES, T. A.: Nitrogen studies on the irrigated soils of the Sudan-gezira. J. Soil Sci. **8**, 211–224 (1957).

KAILA, A.: Über mikrobiologische Festlegung und Mineralisierung des Phosphors bei der Zersetzung organischer Stoffe. Z. Pflanzenernähr., Düng., Bodenkde. **64**, 27–35 (1954). — KELLER, P.: Die Bedeutung und die Methoden zur Beurteilung des Reifegrades von Müll- und Müll-Klärschlamm-Komposten. Int. A. G. f. Müllforschung, Informationsblatt Nr. 10, 6–17 (1960). — KILBINGER, A.: Untersuchungen über bodenbürtige Kohlensäure mit dem „Mikrogas-Analysengerät Wösthoff", 1. Mitt. Schriftenreihe d. Kohlenstoffbiolog. Forschgsstation e.V., Essen-Bredeney **2**, Januar (1953). — Kohlensäure, ein Lebensstoff. Schriftenreihe d. Kohlenstoffbiolog. Forschungsstation e.V., Essen-Bredeney **19** (1955). — KLUYVER, A. J., und W. VERHOEVEN: Studies on true dissimilatory nitrate reduction, II, The mechanism of denitrification. A. v. Leeuwenhoek. **20**, 241–262 (1954). — KÖNIG, E.: Über die Aufschließung von Rohphosphaten in natürlichen Böden durch den direkten Angriff von Bodenpilzen. Landwirtsch. Forsch. **14**, 216–225 (1961a). — Untersuchungen über den Chemotropismus einiger Pilze gegenüber schwerlöslichen Phosphaten. Arch. Mikrobiol. **40**, 395–402 (1961b). — KOEPF, H.: Laufende Messung der Bodenatmung im Freiland. Landwirtsch. Forsch. **4**, 186–194 (1952). — Die Verwendung des Ultrarotabsorptionsschreibers (URAS) für die kontinuierliche Registrierung der Bodenatmung im Freiland. Landwirtsch. Forsch. **5**, 54–62 (1953a). — Die Temperatur-Zeit-Abhängigkeit der Bodenatmung. Z. Pflanzenernähr., Düng., Bodenkde. **61**, 29–48 (1953b). — Die biologische Aktivität des Bodens und ihre experimentelle Kennzeichung. Z. Pflanzenernähr., Düng., Bodenkde. **64**, 138–146 (1954a). — Untersuchungen über die biologische Aktivität des Bodens, I, Atmungskurven des Bodens und Fermentaktivität unter dem Einfluß von Düngung und Pflanzenwachstum. Z. Acker- u. Pflanzenbau **98**, 289–313 (1954b). — Zur Bodenbeurteilung mittels biochemischer Methoden. Z. Acker- u. Pflanzenbau **100**, 36 (1955). — Zur Dynamik des Abbaus organischer Substanzen in verschiedenen Böden. Z. Pflanzenernähr., Düng., Bodenkde. **73**, 48–59 (1956). — KOSEGARTEN, E. K.: Veränderungen des aktiven Mangans von Bodenproben in Abhängigkeit von Wassergehalt und Mikrobentätigkeit. Landwirtsch. Forsch. **10**, 214–222 (1957). — KOSMAT, H.: Kontrolle der Bodenatmung durch akustisch-elektrische Messungen. Bodenkultur A **10**, 130–140 (1959); Ref. Z. Pflanzenernähr., Düng., Bodenkde. **87**, 262 (1959) und **88**, 191 (1960). — KOZOVÁ, J., und Mitarbeiter: Mikrobiologische Zellulosezersetzung bei natürlichen Bodenverhältnissen. Aus dem Forschungsinst. f. pflanzl. Produktion d. Tschechosl. Akad. d. Landwirtsch.-Wissensch. Prag (1961). — KRESGE, C. B., und F. G. MERKLE: A study of the validity of laboratory techniques in appraising the available nitrogen producing capacity of soils. Soil Sci. Soc. Amer. Proc. **21**, 516–521 (1957). — KREYBIG, L.: Neuere biologische Düngemittel und Düngungsmethoden. Dtsch. Landwirtsch. **5**, 601–611 (1954). — KUDRIN, S. A.: Der Kreislauf des Phosphors im Boden und die im Boden stattfindenden Umwandlungen der Phosphatdüngemittel. Agrobiologija **5**, 91–105 (1952); Ref. Z. Pflanzenernähr., Düng., Bodenkde. **64**, 186 (1954). — KUSNETSOWA, A. W.: Der Gehalt an Nitraten in den Böden des Gebietes Tscheljabinsk in verschiedenen Jahreszeiten Agrobiologija **5**, 108–112 (1955) (russ.); Ref. Z. Pflanzenernähr., Düng., Bodenkde. **72**, 277–278 (1956).

LEES, H., und J. H. QUASTEL: A new technique for the study of soil sterilisation. Chem. a. Ind. Nr. 26, 238–239 (1944). — I, Kinetics of, and the effects of poisons on, soil nitrification, as studied by a soil perfusion technique, II, The site of soil nitrification, III, Nitrification of various organic nitrogen compounds. Biochem. J. **40**, 803–814, 815–823, 824–828 (1946). — LEFÈVRE, G., und G. HIROUX: Considérations sur la minéralisation et l'utilisation des réserves azotées du sol. Ann. agron. **11**, 135–162 (1960); Ref. Z. Pflanzenernähr., Düng., Bodenkde. **92**, 259 (1961). — LINDEBERG, G.: Ligninabbau und Phenoloxydasebildung der Bodenhymenomyceten. Z. Pflanzenernähr., Düng., Bodenkde. **69**, 142–150 (1955). — LOEWENSTEIN, H., L. E. ENGELBERT, O. J. ATTOE und O. N. ALLEN: Nitrogen loss in gaseous form from soils as influenced by fertilizers and management. Soil Sci. Soc. Amer. Proc. **21**, 397–400 (1957). — LU, K. C., und W. B. BOLLEN: Effect of gibrel, a potassium salt of gibberellic acid, on microbial activities in soil. Plant. a. Soil **9**, 318–324 (1957/58). — LU, K. C., C. M. GILMOUR, A. C. ZAGALLO und W. B. BOLLEN: Effects of gibberellic acid on soil microorganisms. Nature **181**, 189–190 (1958); Ref. Z. Pflanzenernähr., Düng., Bodenkde. **82**, 255 (1958).

MAASS, G.: Untersuchungen über den Einfluß von Ammoniak- und Nitratdünger auf den Gehalt an mineralischem Stickstoff im Boden. Z. Pflanzenernähr., Düng., Bodenkde. **93**, 26–38 (1961). — MAKAROW, B. N.: Die Änderung der Bodenatmung im Tagesablauf und die CO_2-Gehalte der bodennahen Luftschicht. Ber. Akad. Wiss.

SSR 118, 389–391 (1958) (russ.); Ref. Z. Pflanzenernähr., Düng., Bodenkde. 83, 185 (1958). — MANN, P. J. G., und J. H. QUASTEL: Manganese metabolism in soils. Nature (Lond.) 158, 151 (1946). — MARSH, P. B., G. A. GREATHOUSE, M. L. BUTLER und K. BOLLENBACHER: Testing fabrics for resistence to mildew and rot. USDA Techn. Bull. 892, 22 S. (1945). — MARTIN, A. E., und J. E. COX: Nitrogen studies on black soils from the Darling Downs, Queensland, I, Seasonal variations in moisture and mineral nitrogen fractions; II, The nitrifying activity of subsurface horizons. Austral. J. Agric. Res. 7, 169–193 (1956); Ref. Z. Pflanzenernähr., Düng., Bodenkde. 74, 273 (1956). — MÁTÉ, F., und F. MOLNÁR: Phosphatbindungsversuche auf Wiesenboden mittels radioaktivem Phosphorisotop (P^{32}). Agrokémia és Talajtan 5, 165–170 (1956) (ung.); Ref. Z. Pflanzenernähr., Düng., Bodenkde. 78, 251 (1957). — MCBEE, R. H.: The anaerobic thermophilic cellulolytic bacteria. Bact. Rev. 14, 51–63 (1950). — MCCALLA, T. M.: Light reflection from stubble mulch. J. Amer. Soc. Agron. 39, 690–696 (1947). — MCCALLA, T. M., und J. C. RUSSEL: Nitrate production as affected by graincrop residues on the surface of the soil. Univ. Nebraska Coll. Agric. Exper. Stat. Res. Bull. 131 (1943). — Nitrate production as affected by sweet-clover residues left on the surface of the soil. J. Amer. Soc. Agron. 40, 411–421 (1948). — MCINTOSH, T. H., und L. R. FREDERICK: Distribution and nitrification of anhydrous ammonia in a Nicollet sandy clay loam. Soil Sci. Soc. Amer. Proc. 22, 402–405 (1958); Ref. Z. Pflanzenernähr., Düng., Bodenkde. 85, 179 (1959). — MEIKLEJOHN, J.: Some organic substances and the nitrifying bacteria Proc. Soc. Appl. Bact. 15, 77–81 (1952); Ref. Zbl. Bakt. II 108, 316 (1954/55). — MERKER, E.: Parasitic bacteria from the leaves of *Elodea*. Zbl. Bakt. II 31, 578–609 (1912). — MEYER, F. H.: Vergleich des mikrobiellen Abbaus von Fichten- und Buchenstreu auf verschiedenen Bodentypen. Arch. Mikrobiol. 35, 340–360 (1960). — MEYER, L., und E. KÖNIG: Versuchsergebnisse zur mikrobiellen Aufschließung schwerlöslicher Phosphate durch Bodenpilze. Landwirtsch. Forsch. 13, 7–24 (1960). — MEYER, L., und H. KOEPF: Das Kohlendioxyd und die Kohlensäure im Boden. In: RUHLAND, Handbuch der Pflanzenphysiologie, Bd. 5, S. 24–46 (1960). — MEYER, L., und G. SCHAFFER: Atmungskurven des Bodens unter dem Einfluß von Düngung und Bewachsung. Landwirtsch. Forsch. 6, 81–95 (1954). — MILLBANK, J. W.: The physiology of nitrification in Kenya highland soil. Plant a. Soil 11, 293–311 (1959); Ref. Z. Pflanzenernähr., Düng., Bodenkde. 91, 87 (1960). — MINA, B. N.: Die biologische Aktivität der Waldböden und ihre Abhängigkeit von den physikalisch-geographischen Verhältnissen und von der Zusammensetzung der Anpflanzungen. Bodenkde. 10, 73–79 (1957) (russ.); Ref. Z. Pflanzenernähr., Düng., Bodenkde. 82, 255 (1958). — MOREAU, R.: Sur la libération biologique dans les sols de phosphates solubles à partir des composées phosphorées insolubles. C. R. Acad. Sci. Paris 249, 1804–1806 (1959); Ref. Zbl. Bakt. II 114, 221 (1961). — MÜLLER, G.: Bodenbiologische Abbauuntersuchungen unter Berücksichtigung der Standortsfaktoren bei Schwarzbrache nach Rein- und Mischsaaten, II. Zbl. Bakt. II 112, 169–203 (1959). — Über die bodenbiologische Dynamik tiefbearbeiteter und tiefgedüngter leichter Sandböden. Tagungsber. Nr. 28, Probleme der Bodenbearbeitung, d. wiss. Arbeitstagung im Inst. f. Acker- u. Pflanzenbau Müncheberg 167–184 (1960). — MÜLLER, G., und I. FÖRSTER: Einige methodische Versuche zum Problem der Nährstofffreisetzung aus Mineralien durch Bodenpilze. Zbl. Bakt. II 114, 1–10 (1961). — MUNK, H.: Die Nitrifikation von Ammoniumsalzen in sauren Böden. Landwirtsch. Forsch. 11, 150–156 (1958); Ref. Z. Pflanzenernähr., Düng., Bodenkde. 86, 263 (1959). — MUNSON, R. D., und G. STANFORD: Predicting nitrogen fertilizer needs of Iowa Soils, IV, Evaluation of nitrate production as a criterion of nitrogen availability. Soil Sci. Soc. Amer. Proc. 19, 464–468 (1955); Ref. Z. Pflanzenernähr., Düng., Bodenkde. 73, 177 (1956).

NETTE, J. T.: Denitrifizierende Bakterien in der Rhizosphäre der Eiche. Mikrobiologija 24, 429–434 (1955). — NISSEN, T. V.: Effects of antibiotics on carbondioxide production in soil. Nature (Lond.) 174, 226–227 (1954); Ref. Zbl. Bakt. II 108, 727 (1954/55). — NOVÁK, B.: Mikrobiologie im Studium der Bodenbildung. Sborník čsl. Akad. zeměd věd. Rostlinná výroba. Ročnik 6 (XXXIII), 1029–1032 (1960). — NUHI, A.: Einfluß des Calciumcyanamids auf die Mikroflora des Bodens. Diss. Giessen 1962.

OWEN, O., D. W. ROGERS und G. W. WINSOR: The nitrogen status of soils, I, The nitrification of some nitrogenous fertilizers, II, The effect of levels of application on the nitrification of some nitrogenous fertilizers. J. Agric. Sci. 40, 185–197 (1950).

PATHAK, A. N., und J. G. SHRIKHANDE: Die Wirkung verschiedener mechanischer Fraktionen auf die Leistungsfähigkeit nitrifizierender Organismen. Current Sci. 21 (1952), Kaupur. Goot. Agric. Coll., Chem. Sect. [aus Chem. Zbl. 14, 2180 (1953)], Ref. Z. Pflanzenernähr., Düng., Bodenkde. 64, 88 (1954). — PAUL, H., und M. A. A. SHARIFF: The rates of nitrification of peat soils in British Guinea. J. Agric. Sci. 44,

377–382 (1954). — Pintea, C., A. Leonte und A. Untu: Der Einfluß der Mikroelemente Bor und Uran auf die Leistungsfähigkeit der Nitrifikation des Bodens. Lucrări Stintifice (Bukarest XIV), 111–116 (1958); Ref. Z. Pflanzenernähr., Düng., Bodenkde. **86**, 258 (1959). — Poschenrieder, H., und Th. Beck: Mikrobiologisch-biochemische Charakterisierung von schweren Tonböden. Bayer. Landwirtsch. Jahrb., Sonderheft **4**, 86–98 (1960). — Poschenrieder, H., O. Kraemer und Th. Beck: Untersuchungen über die mikrobiologischen Verhältnisse in einem sehr sauren Waldmoorprofil. Mitt. Moor- u. Torfwirtsch. **4**, 28–37 (1957). — Prusnikiewicz, Z.: Untersuchungen über die biologische Ausnutzbarkeit des Wassers im Boden. Pozn. Tow. Przy. Nauk **5**, 1–55 (1959); Ref. Z. Pflanzenernähr., Düng., Bodenkde. **87**, 264 (1959).

Quastel, J. H., und P. G. Scholefield: Biochemistry of nitrification in soil. Bact. Rev. **15**, 1–53 (1951).

Reinau, E. H.: 50 Jahre Kohlensäuredüngung. Privatdruck, Lörrach 1961. — Richard, F.: Der biologische Abbau von Zellulose- und Eiweiß-Testschnüren im Boden von Wald- und Rasengesellschaften. Mitt. Schweiz. Anst. Forstl. Versuchswes. **24**, 297–307 (1945); Ref. Z. Pflanzenernähr., Düng., Bodenkde. **44**, 87 (1949). — Rippel, A.: Bakteriologisch-chemische Methoden der Fruchtbarkeitsbestimmung. In: Blanck, Handbuch der Bodenlehre, Bd. 8, S. 599–671 (1931). — Rippel-Baldes, A.: Grundriß der Mikrobiologie, 3. Aufl., 292 S. (1955). — Robinson, J. B. D.: The critical relationship between soil moisture content in the region of wilting point and the mineralisation of natural soil nitrogen. J. Agric. Sci. **49**, 100–105 (1957). — Ruschmeyer, O. R., und E. L. Schmidt: Cellulose decomposition in soil burial beds, II, Cellulolytic activity as influenced by alternation of soil properties. Appl. Microbiol. **6**, 115–120 (1958). — Russell, E. W.: Soil conditions and plant growth, 9th ed. London: Longmans, Green. 1961. — Rybalkina, A. W., und E. W. Kononenko: Die aktive Mikroflora der Böden. Potschwowedenije, H. 3, 63–71 (1956); Ref. Zbl. Bakt. II **111**, 732 (1958).

Sabey, B. R., W. V. Bartholomew, R. Shaw und J. Pesek: Influence of temperature on nitrification in soils. Soil Sci. Soc. Amer. Proc. **20**, 357–360 (1956); Ref. Z. Pflanzenernähr., Düng., Bodenkde. **76**, 185 (1957). — Sabey, B. R., L. R. Frederick, und W. V. Bartholomew: The formation of nitrate from ammonium nitrogen in soils, III, Influence of temperature and initial population of nitrifying organisms on the maximum rate and delay period. Soil Sci. Soc. Amer. Proc. **23**, 462–465 (1959); Ref. Z. Pflanzenernähr., Düng., Bodenkde. **90**, 181 (1960). — Sakai, H.: Effect of temperature on nitrification in soils. Soil a. Plant Food **4**, 159–162 (1959); Ref. Z. Pflanzenernähr., Düng., Bodenkde. **89**, 229 (1960). — Schaffer, G.: Atmungskurven des Bodens unter dem Einfluß langjähriger verschiedenartiger Düngung. Z. Pflanzenernähr., Düng., Bodenkde. **67**, 219–230 (1954). — Schlegel, H. G.: Die Beeinflussung des Phosphathaushaltes von Mikroorganismen durch Kohlendioxyd. Arch. Mikrobiol. **23**, 195–206 (1955). — Schmidt, E. L., und O. R. Ruschmeyer: Cellulose decomposition in soil burial beds, I, Soil properties in relation to cellulose degradation Appl. Microbiol. **6**, 108–114 (1958). — Schmidt, G.: Beitrag über die Wirkung der Bodendämpfung auf den Stickstoffhaushalt und die biologische Aktivität eines Bodens. Z. Pflanzenernähr., Düng., Bodenkde. **72**, 167–176 (1956). — Seifert, J.: Number of microorganisms, nitrification and biological activity in the soil of plant communities in Koda, near Srbsko. Tschechosl. Akad. Wiss. Sect. IV (1960). — Shutt, F. T.: Some characteristics of the Western prairie soils of Canada. J. Agric. Sci. **3**, 335–357 (1910). — The influence of grain growing on the nitrogen and organic matter content of the Western prairie soils of Canada. J. Agric. Sci. **15**, 162–177 (1925). — Soulides, D. A., und F. E. Clark: Nitrification in grassland soils. Soil Sci. Soc. Amer. Proc. **22**, 308–311 (1958); Ref. Z. Pflanzenernähr., Düng., Bodenkde. **83**, 275 (1958). — Sperber, J. I.: Solution of mineral phosphates by soil bacteria. Nature **180**, 994–995 (1957); Ref. Z. Pflanzenernähr., Düng., Bodenkde. **80**, 273 (1958). — Solution of apatite by soil microorganisms producing organic acids. Austral. Agric. Res. **9**, 782–787 (1958); Ref. Z. Pflanzenernähr., Düng., Bodenkde. **86**, 262 (1959). — Stanford, G., und J. Hanway: Predicting nitrogen fertilizer needs of Iowa soils, II, A simplified technique for determining relative nitrate production in soils. Soil Sci. Soc. Amer. Proc. **19**, 74–77 (1955); Ref. Z. Pflanzenernähr., Düng., Bodenkde. **74**, 187 (1956). — Stanier, R. Y.: The Cytophaga group: a contribution to the biology of the Myxobacteria. Bact. Rev. **6**, 1–143 (1942). — Stephenson, R. E., und C. E. Schuster: Effect of mulches on soil properties. Soil Sci. **59**, 219–230 (1954). — Stevenson, J. I.: Some observations on the microbial activity in remoistened air-dried soils. Plant a. Soil **8**, 170–182 (1956/57). — Stevenson, J. I., und F. E. Chase: Nitrification in an orchard soil under three cultural practices. Soil Sci. **76**, 107–114 (1953). — Stojanovic, B. J., und M. Alexander: Effect of inorganic nitrogen on nitrification. Soil

Sci. **86**, 208–215 (1958). — Stojanovic, B. J., und F. E. Broadbent: Influence of low temperature on nitrogen transformations in Honeoye silt loam. Soil Sci. **84**, 243–248 (1957). — Stotzky, G., T. Ryan und J. L. Mortensen: Apparatus for studying biochemical transformations in incubated soils. Soil Sci. Soc. Amer. Proc. **22**, 270–271 (1958). — Stover, R. H., und S. R. Freiberg: Effect of carbon dioxide on multiplication of Fusarium in soil. Nature (Lond.) **181**, 788–789 (1958); Ref. Zbl. Bakt. II **113**, 291 (1959/60). — Stundl, K.: Zur Frage des Einflusses von Treibstoffen auf die Abbauleistung von Bodenbakterien. Zbl. Bakt. II **113**, 35–41 (1959/60).

Themlitz, R.: Umsetzung verschiedener N-Dünger mit einem durch voraufgegangene Bestandskalkung bzw. -stickstoffdüngung umgewandelten Fichtenrohhumus. Z. Pflanzenernähr., Düng., Bodenkde. **82**, 165–174 (1958). — Theron, I. I.: The influence of plants on the mineralization of nitrogen and the maintenance of organic matter in the soil. J. Agric. Sci. **41**, 289–296 (1951). — Thompson, L. M., C. A. Black und J. A. Zoellner: Occurrence and mineralization of organic phosphorus in soils, with particular reference to associations with nitrogen, carbon and pH. Soil Sci. **77**, 185–196 (1954); Ref. Z. Pflanzenernähr., Düng., Bodenkde. **67**, 186 (1954). — Timonin, M. J.: Microflora of the rhizosphere in relation to the manganese-deficiency disease of oats. Soil Sci. Soc. Amer. Proc. **11**, 284–292 (1946). — Tsubota, G.: Phosphate reduction in the paddy field I. Soil a. Plant Food **5**, 10–15 (1959); Ref. Z. Pflanzenernähr., Düng., Bodenkde. **89**, 229 (1960). — Tyler, K. B., F. E. Broadbent und G. N. Hill: Low temperature effects on nitrification in four California soils. Soil Sci. **87**, 123–129 (1959).

Unger, H.: Der Zellulosetest, eine Methode zur Ermittlung der zellulolytischen Aktivität des Bodens in Feldversuchen. Z. Pflanzenernähr., Düng., Bodenkde. **91**, 44–52 (1960).

Vanstallen, R.: Fixierung und Verflüchtigung von Ammoniak-Stickstoff. Pedologie **10**, 15–21 (1960) (fläm.); Ref. Z. Pflanzenernähr., Düng., Bodenkde. **92**, 257 (1961). — Verhoeven, W.: Aerobic sporeforming nitrate reducing bacteria. Diss. Delft 1952.

Wagner, G. H., und G. E. Smith: Nitrogen losses from soils fertilizerd with different nitrogen carriers. Soil Sci. **85**, 125–129 (1958). — Waksman, S. A.: Microbiological analysis of soil as an index of soil fertility, V, Methods for the study of nitrification. Soil Sci. **15**, 241–261 (1923). — Walker, T. W., A. F. R. Adams und H. D. Orchiston: Fate of labeled nitrate and ammonium nitrogen when applied to grass and clover grown separately and together. Soil Sci. **81**, 339–351 (1956). — Walter, H., und W. Haber: Über die Intensität der Bodenatmung mit Bemerkungen zu den Lundegårdhschen Werten. Ber. Dtsch. Bot. Ges. **70** (120), 275–282 (1957); Ref. Z. Pflanzenernähr., Düng., Bodenkde. **80**, 271 (1958). — Weissenberg, H.: Die Mikroorganismentätigkeit auf rohhumushaltigem Heidesand, gemessen an der CO_2-Produktion. Z. Pflanzenernähr., Düng., Bodenkde. **66**, 227–238 (1954). — Wheeler, B. E. J., and E. W. Yemm: The conversion of amino acids in soils, I, Amino-acid breakdown and nitrification in cultivated and natural soils. Plant a. Soil **10**, 49–77 (1958). — Whiting, A. L.: Some important factors controlling the rate of nitrification of organic materials. J. Amer. Soc. Agron. **18**, 854–876 (1926). — Wigorow, L. I.: Die Mineralisation der Eiweißstoffe von Bodenbakterien. Mikrobiologija **24**, 422–428 (1955) (russ.); Ref. Z. Pflanzenernähr., Düng., Bodenkde. **72**, 94 (1956). — Wijler, J., und C. C. Delwiche: Investigations on the denitrifying process in soil. Plant a. Soil **5**, 155–169 (1954). — Wikén, T.: Über die Physiologie einiger Mikroorganismen mit besonderer Berücksichtigung der Bedeutung der experimentellen Mikrobiologie für das Verständnis der Umsetzung der organischen Stoffe im Boden. Z. Pflanzenernähr., Düng., Bodenkde. **69**, 155–165 (1955). — Withee, L. V., und T. M. McCalla: The effect of stubble mulching on the acid soluble and absorbed phosphorus in the soil. Soil Sci. Soc. Amer. Proc. **18**, 285–286 (1954).

Ziechmann, W., und K. Metzler: Nebenwirkungen verschiedener Harnstoff-Acetaldehyd-Kondensate im chemischen und mikrobiologischen Test. Z. Pflanzenernähr., Düng., Bodenkde. **89**, 190–201 (1960). — Zöttl, H.: Die Bestimmung der Stickstoffmineralisation im Waldhumus durch den Brutversuch. Z. Pflanzenernähr., Düng., Bodenkde. **81**, 35–50 (1958). — Methodische Untersuchungen zur Bestimmung der Mineralstickstoffnachlieferung des Waldbodens. Forst. Cbl. **79**, 72–90 (1960a). — Dynamik der Stickstoffmineralisation im organischen Waldbodenmaterial, I, Beziehung zwischen Bruttomineralisation und Nettomineralisation. Plant a. Soil **13**, 166–182 (1960b). — Dynamik der Stickstoffmineralisation im organischen Waldbodenmaterial, II, Einfluß des Stickstoffgehaltes auf die Mineralstoffnachlieferung. Plant a. Soil **13**, 183–206 (1960c). — Die Mineralstickstoffanlieferung in Fichten- und Kiefernbeständen Bayerns. Forstw. Cbl. **79**, 221–236 (1960d).

F. Der Wasser-, Luft- und Wärmehaushalt des Bodens

a) Die Bindung von Wasser im Boden

Von

H. Linser

Die Pflanze findet das für ihr Wachstum nötige Wasser zum allergrößten Teil im Boden vor, dem sie es mit Hilfe ihres Wurzelsystems entnimmt. Das im Boden vorhandene Wasser steht ihr jedoch nicht zur Gänze zur Verfügung, da es

1. teilweise an Bestandteile des Bodens mehr oder weniger fest gebunden ist oder wird und da es

2. durch Bewegungsvorgänge aus dem Bereich der Pflanzenwurzel entfernt wird.

Wir haben uns deshalb sowohl mit den Ursachen, Bedingungen und Folgen der Wasserbindung im Boden zu befassen als auch mit der Beweglichkeit des Wassers im Boden (vgl. S. 672 ff.).

1. Die Bindung von Wasser durch zwischenmolekulare Wechselwirkungen

Die atomare Struktur des Sauerstoffatoms bedingt, daß im Molekül des Wassers (H_2O) die Verbindungslinien der Mittelpunkte der beiden Wasserstoffatome zum Mittelpunkt des Sauerstoffatoms einen „Valenzwinkel" von 105° bilden. Dies hat zur Folge, daß der Schwerpunkt der positiven und jener der negativen Ladungen im Molekül nicht zusammenfallen, sondern einen bestimmten Abstand voneinander haben. Das Wassermolekül besitzt deshalb Dipolcharakter und verhält sich so, als ob es einen negativ geladenen und in gewissem Abstand davon einen positiv geladenen Pol besäße. Die Maßzahl für diese Polarität, die im dampfförmigen Zustand noch stärker ist als im flüssigen, ist das Dipolmoment $\mu = 1{,}86 \cdot 10^{-18}$ elektrostatische Einheiten $\times$ cm (Debye); ihre Größe ist beträchtlich und bewirkt, daß Wassermoleküle im elektrischen Feld eine bestimmte Ordnung, d. h. Einstellung ihrer elektrischen Achse in eine bestimmte Richtung erfahren. Es hat weiterhin die Folge, daß sowohl Wassermoleküle untereinander als auch Wassermoleküle und Ionen aufeinander elektrostatische Wechselwirkungen ausüben, die zur elektrostatischen zwischenmolekularen Bindung von Wassermolekülen untereinander oder von Wasser an Ionen oder andere geladene Teilchen führen, wobei den Wassermolekülen auf Grund ihres Dipolmoments eine bestimmte Lage bzw. geordnete Ausrichtung gegeben wird.

Würden sich Wassermoleküle nicht wie Dipole verhalten, sondern wie inerte Kugeln von 1,38 Å Durchmesser, so würden sie in dichtester Kugelpackung dem Wasser eine Dichte von etwa 1,84 verleihen (Frizman 1935). Tatsächlich weist es jedoch die Dichte 1 auf. Auch das Volumen eines Mols Wasser beträgt nicht, wie bei würfelförmiger oder hexagonaler Packung rechnerisch zu erwarten wäre, 9,0 oder 12,74 cm³ (Rode 1959), sondern 18 cm³, was zeigt, daß die Moleküle des flüssigen Wassers weniger direkt angeordnet sind. Dies kommt dadurch zustande, daß sich auf Grund der Dipolmomente und ihrer gegenseitigen Wechselverbindungen lockere Aggregate von Wassermolekülen bilden, so daß wir es mit Mono- (H_2O), Di- (H_4O_2), Tri- (H_6O_3) oder Pentahydrolen ($H_{10}O_5$) zu tun haben, über deren räumliche Struktur wir keine nähere Kenntnis besitzen, von der wir aber annehmen müssen, daß sie weitgehend von Temperaturstößen gestört wird und Komplexe betrifft, die immer wieder neu entstehen und immer wieder durch Temperaturstöße zerstört werden. Demnach würden bei niedrigsten Temperaturen mehr größere Hydrole (Polyhydrole) vorliegen, bei höheren Temperaturen jedoch die Monohydrole überwiegen. Im festen Wasser (Eis) liegen wahrscheinlich 59% in Form von Trihydrolen, 41% als Dihydrole vor, während Monohydrole fehlen (Frizman 1935).

Die Bindungs- bzw. Wechselwirkungsenergie zwischen Dipolen ist wesentlich kleiner als jene zwischen Dipolen und Ionen, und diese wiederum ist kleiner als jene zwischen Ion und Ion. Unter bestimmten rechnerischen Voraussetzungen und bei einem Mittelpunktabstand von 4 Å beträgt beispielsweise die Wechselwirkungsenergie zwischen Dipol und Dipol (je nach gegenwärtiger Lage) 0,9 bis 1,8 kcal/Mol (gegenüber einer kinetischen Energie der Wärmebewegung von etwa 1 kcal/Mol), bei der Bindung von Dipol und Ion aber 8,7 und bei der Bindung von Ion und Ion 83 kcal/Mol (vgl. Hölzel 1956). Die Wassermoleküle sind also (gleiche Mittelpunktabstände vorausgesetzt) untereinander wesentlich schwächer gebunden und gegen Wärmestöße empfindlicher als bei Bindung an Ionen.

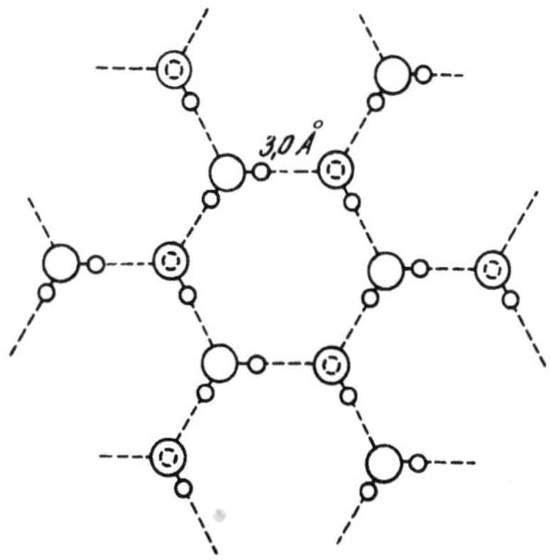

Abb. 110. Hexagonales Netz von Wasseratomen als Schichtebene in Schichtmineralien nach der Hypothese von Hendricks und Jefferson (1938). Große Kreise: Sauerstoffionen; kleine Kreise: Wasserstoffionen

Ionen ordnen auf Grund ihrer Ladung mehrere Wassermoleküle (als Dipole) um sich herum an und halten sie fest. Es liegt keine sehr genaue Kenntnis über die Zahl von Wassermolekülen vor, welche bestimmte Ionen zum Aufbau ihrer Wasserhülle (Hydratation) heranziehen. Die von verschiedenen Autoren angegebenen Zahlen schwanken erheblich (vgl. Gruner und Vogel 1950), nur für das Wasserstoffion lautet die Zahl übereinstimmend 1 Wassermolekül. Es werden Zahlen genannt für Li-Ionen von 11 bis 14 (68 bis 158), Na-Ionen 4 bis 11 (60 bis 70), K-Ionen 2 bis 6 (22), Mg-Ionen 8 bis 23 (40), Ca-Ionen 8 bis 22 (28) und Ba-Ionen 4 bis 22. Auch Anionen können hydratisiert sein, z. B. lauten Zahlen für Cl-Ionen 3 und NO_3-Ionen 2, bei F-Ionen dagegen 4 bis 6 (Hölzel 1956). Hydratisierte Ionen können aus entsprechend konzentrierten Lösungen auskristallisieren, wobei sie Hydratwasser in die Ordnung des Kristallgitters mit einbauen, so daß es in stöchiometrischen Relationen zu den Ionen als *Kristallwasser* vorliegt. Zahlreiche im Boden vorkommende Mineralstoffe enthalten beträchtliche Mengen an Kristallwasser. Dieses ist freilich durch Gitterenergie viel fester gebunden als das Hydratwasser an den Ionen und nur unter Aufwendung beträchtlicher Energiezufuhr zu entfernen.

Quellbare Tonminerale wie z. B. Montmorillonit können Wasser (Wasserdampf) in ihre Kristallgitter-Schichten schichtweise einlagern. Bradley, Grimm und Clark (1937) fanden durch Auswertung von Beugungsspektren, daß der Abstand der Montmorillonitlamellen mit steigendem Feuchtigkeitsgehalt nicht kontinuierlich, sondern sprungweise ansteigt, was zeigt, daß jeweils eine vollständige neue Wasserschicht zwischen die Lamellen eingeschoben wird. Sie konnten die Aufnahme von vier solcher Schichten feststellen, deren jede sechs Wassermolekülen je Elementarlamelle entsprach. Jede dieser Schichten muß die Wassermoleküle in einer dem Kristallgitter des Minerals zugeordneten Anordnung enthalten. Eine von Hendricks und Jefferson (1938) entwickelte Hypothese hierüber ist in Abb. 110 veranschaulicht.

Bindungen, wie sie zwischen Wassermolekül und Ion bestehen, können natürlich auch zwischen an Oberflächen großer Moleküle oder an anderen Teilchen fixierten Ladungen und Wassermolekülen bestehen bzw. hergestellt werden. So sind zahlreiche organische, hochmolekulare Stoffe in der Lage, Hydratwasser festzuhalten bzw. in tertiäre Hyperstrukturen mit einzubauen.

Bei allen diesen Vorgängen erhält sozusagen jedes eingebaute Wassermolekül seine bestimmte strukturelle Funktion und einen bestimmten Platz zugewiesen, an dem es gebunden bleibt, bis es durch entsprechende, geeignete Energiezufuhr davon getrennt wird.

Wasserfilme solcher Art, die sich an Oberflächen ladungstragender Kristallgitter oder anderer Teilchen bilden, haften um so weniger stark und werden um so leichter entfernt, je mehr der Anteil der Bindung Wasserdipol an Wasserdipol gegenüber dem Anteil der Bindung Wasserdipol an Ladungsträger hervortritt, d. h. daß die unmittelbar an den Ladungsträger angrenzende Wasserschicht am festesten gebunden ist und in strengerer Ordnung vorliegt als die nächsten, darüberliegenden Schichten.

Die feste Bindung und Lage solcher Hydratwassermoleküle entziehen sie weitgehend der Möglichkeit, mit Ionen, die durch Diffusionsvorgänge in die Nähe gelangen können, zu reagieren bzw. ihren Platz zu verlassen und sich mit Ionen zu verbinden, es sei denn, daß ein einfacher Austausch eines Hydratwassermoleküls am Ladungsträger gegen ein solches am Ion stattfindet, wodurch das Gesamtsystem ja nicht verändert wird.

Als Hydratwasser gebundenes Wasser befindet sich sozusagen im festen Zustand und kann daher diejenigen Funktionen, die es im flüssigen Zustand löslichen Stoffen gegenüber einnimmt, nämlich jene der Ionisierung löslicher Stoffe und der Hydratation neugebildeter Ionen, nicht mehr zusätzlich übernehmen. Dies führt zu einer Erscheinung, die besonders von TROFIMOW (1925, 1927) festgestellt und untersucht worden ist, nämlich zur Erscheinung des „nicht lösenden Raumes“ oder der „salzfreien Wasserhaut“, die man auch als eine Erscheinung „negativer Adsorption“ charakterisiert hat. Dieses fest gebundene Wasser, das so stark gebunden und geordnet ist, daß es durch Temperaturstöße nicht abgelöst und in flüssigen Zustand gebracht werden kann, ist daher salzfrei und darf bei der Berechnung der Ionenkonzentration der Bodenlösung nicht als Lösungsmittel in Rechnung gestellt werden. Das Volumen des salzfreien Wasserfilms in cm^3 je 100 g Boden liegt nach Untersuchungen von TROFIMOW (1927) je nach Wasserbindungsvermögen des Bodens zwischen 0,70 und 9,95 und macht somit etwa das 1- bis 2fache der maximalen Hygroskopizität aus. Der salzfreie Wasserfilm besitzt auch keine elektrische Leitfähigkeit. DOLGOW (1948) untersuchte den Rückgang der Leitfähigkeit mit abnehmendem Feuchtigkeitsgehalt von Böden und fand, daß die Leitfähigkeit bereits bei Wassergehalten der Böden gleich Null war, die wenig kleiner als deren maximale Hygroskopizität waren.

Kommt trockener Boden mit Wasserdampf in Berührung, so wird auf Grund der eben geschilderten zwischenmolekularen Wechselwirkung Wasser an den Oberflächen der Bodenteilchen sorbiert. Durch den Übergang des in Wärmebewegung befindlichen Wassermoleküls aus der Dampfform in den fest gebundenen und geordneten Zustand der Hydrathülle des Teilchens verliert das Wassermolekül seine kinetische Energie, die nach DUMANSKI (1950) $10{,}04 \cdot 10^{-14}$ erg beträgt. Es tritt daher sogenannte *Benetzungswärme* auf, die (bei Stärke z. B. je Molekül Wasser) ebenfalls mit $10{,}4 \cdot 10^{-14}$ erg gefunden wurde, also praktisch gleich groß ist. Die Benetzungswärme, die bei der Sorption von Wasser an ladungstragende Oberflächen frei wird, ist unabhängig von der Art bzw. chemischen Beschaffenheit der Teilchen und beträgt nach Messungen an verschiedenen Sorbenten 72 bis 80 cal je Gramm gebundenen Wassers. Dieser Wert entspricht etwa der Schmelzwärme von Eis (76,6 g cal/g Wasser), ein Beweis dafür, daß das sorbierte Wasser tatsächlich als im „festen" Zustand befindlich betrachtet werden darf. Die Benetzungswärme eines bestimmten Materials ist bestimmt durch die Zahl der sorbierten Wassermoleküle, daher von der Zahl der an den Oberflächen vorhandenen Ladungsstellen, damit aber auch von der Größe der Oberfläche, welche eine Gewichtseinheit des Materials aufzuweisen hat. Sie ist somit abhängig vom Dispersitätsgrad und ist danach bei kleinen Korngrößen pro Gewichtseinheit verschiedener Bodenfraktionen größer als bei Teilchen mit größerem Durchmesser. Bestimmungen der Benetzungswärme verschiedener Bodenfraktionen wurden von KAPP (1930) an 12 verschiedenen Böden durchgeführt, wobei für die einzelnen Teilchengrößen folgende Benetzungswärmen in cal/g Boden gefunden wurden (Tab. 92):

Tabelle 92. *Benetzungswärme (cal/g Boden) in Abhängigkeit von der Korngröße*

Korngröße	Benetzungswärme	Mittelwert
< 0,00125 mm	2,13 bis 18,22	7,26
0,00125 bis 0,005	1,32 bis 8,43	3,36
0,005 bis 0,010	0,32 bis 0,85	0,55
0,010 bis 0,020	0,00 bis 0,45	0,21
0,020 bis 0,035	0,00 bis 0,54	0,09
0,035 bis 0,050	0,00 bis 0,36	0,04
0,050 bis 0,100	0,00 bis 0,36	0,04

Mitunter zeigen auch größere Bodenteilchen (Korngrößen von mehr als 0,02 mm) beachtliche Werte der Benetzungswärme, und zwar dann, wenn sich an ihren Oberflächen kolloide Schichten als Überzüge befinden, die relativ hohe Benetzungswärmen verursachen können. Daran ist vor allem Humus beteiligt, da organische Stoffe prinzipiell ebenso an der Bildung von Benetzungswärme beteiligt sind wie anorganische Materialien.

Nach Untersuchungen von DIMO (1946) an Schwarzerdeprofilen wurde bei praktisch gleichem Tongehalt eine Abhängigkeit der Benetzungswärme vom Humusgehalt insofern festgestellt, als bei einer Abnahme des Humusgehaltes von 6,25% auf 0,47% ein Absinken der Benetzungswärme von 799 cal auf 576 cal/100 g Boden zu beobachten war.

Daß Benetzungswärme nicht nur durch die Bindung einer einzigen Molekularschicht von Wasser an den sorbierenden Oberflächen zustande kommt, sondern daß über der ersten noch weitere Molekülschichten Wasser sorbiert werden, zeigen Untersuchungen von MITSCHERLICH (1901), der bei stufenweise erhöhtem Wasserzusatz zu Böden fand, daß der Zuwachs an Benetzungswärme je Einheit zugegebener Wassermenge von Stufe zu Stufe exponentiell abnahm. Daraus geht hervor, daß die Benetzungswärme der ersten sorbierten Molekularschicht Wasser am größten ist und daß der Ordnungszustand (und damit die Bindungsfestigkeit) der darüberliegenden Wasserschichten von Schicht zu Schicht kleiner wird und exponentiell abfällt (vgl. RODE 1959). Immerhin kann die Zahl der Schichten Wasser, die mit nach außen abnehmender Bindungsfestigkeit angelagert sind und zur Benetzungswärme meßbar beitragen, auf einige Hundert geschätzt werden. DERJAGIN (1934) fand in einem speziellen Fall die Zahl 270. Bei Versuchen, die Zahl der Wasserschichten nach Sorption von Wasser an Quarzsand zu ermitteln, stellte Tschapek (1938) fest, daß die Anzahl der Schichten stark durch die Korngröße beeinflußt war; sie war um so größer, je größer die Korngröße war (Tab. 93):

Tabelle 93. *Anzahl der sorbierten Wasserschichten in Abhängigheit von der Korngröße*

Korngröße, in 10^{-4} cm	Anzahl der Schichten (bei einer Schichtdicke von 2,76 Å)
1000—250	23
250—100	12
100— 50	11
50— 10	5
10— 5	4
5— 1	1

Die hohe Bindungsfestigkeit der unmittelbar an die ladungstragenden Oberflächen der Teilchen angrenzenden Wasserschichten bewirkt eine ziemlich dichte Packung der Wassermoleküle, das heißt, eine gegenüber flüssigem Wasser *erhöhte Dichte* des Wassers. Nach Angaben von Rode (1959) wurden an Schwarzerdeböden und alluvialen Lehmböden je nach dem Feuchtigkeitsgehalt der Böden (der zwischen 1,6 und 13,9% lag) Werte für die Dichte des Wassers zwischen 1,74 und 1,13 gefunden. Das bedeutet, daß die am stärksten gebundenen Wasserschichten eine Dichte aufweisen, die einem Druck von einigen zehntausend Atmosphären entspricht. Mit zunehmender Entfernung der Wasserschichten und abnehmender Bindungsfestigkeit fällt dieser Druck (exponentiell) bis gegen den Normaldruck, unter dem das flüssige Wasser steht, ab.

Man kann somit die Bindungsfestigkeit des Wassers im Boden auch in Atmosphären berechnen. Dies ist insofern wichtig, als das Aneignungsvermögen der Pflanzenzelle für Wasser ebenfalls in Atmosphären gemessen bzw. angegeben werden kann, nämlich als jener Saugdruck, der sich aus der Differenz zwischen osmotischem Druck des Zellsaftes und dem Wanddruck (Druck der Zellwand gegen den Plasmaschlauch) ergibt. Jenes Wasser, das unter einem höheren Atmosphärendruck festgehalten wird als dem Saugdruck der Pflanze entspricht, kann von der Pflanze dem Boden nicht entzogen werden und wird deshalb auch als „totes Wasser" bezeichnet. Diese Definition besagt, daß die Menge an „totem Wasser" nicht eine den Boden speziell kennzeichnende Größe sein kann, da sie zwar vom Boden abhängig ist, aber entscheidend durch die von der Pflanze bestimmte Größe des Saugdrucks begrenzt wird. Der hohe Druck, den die mit hoher Bindungsfestigkeit unmittelbar an die Bodenteilchen anliegenden Wasserschichten aufweisen, bewirkt eine starke Erniedrigung des Gefrierpunktes dieser Wasserschichten, der selbst bis zu Temperaturen von —78° C absinken kann. Wasser, das selbst bei —78° C nicht gefriert, bezeichneten Bouyoucos und McCool (1916) als „gebundenes" Wasser, das zusammen mit dem zwischen —4° C und —78° C gefrierenden „kapillar-adsorbierten" Wasser das „nicht freie" Wasser ergibt.

2. Die Sorption dampfförmigen Wassers durch den Boden (Hygroskopizität)

In der Natur steht jeder Boden mit Wasser, sei es mit dem Wasserdampf der Atmosphäre bzw. der Bodenluft, sei es mit dem flüssigen Wasser der Bodenlösung bzw. der Niederschläge oder des Grundwassers, in Berührung. Er enthält daher stets Wasser; auch im lufttrockenen Zustand hält er sich im Gleichgewicht mit dem Wasserdampfgehalt der Luft, und es bedarf ganz besonderer Trocknungsmethoden, um dem Boden, ohne ihn in seinem chemisch-physikalischen Bau zu verändern, wasserfrei zu erhalten. Bringen wir so wasserfrei gemachten Boden mit einer wasserdampfhaltigen Atmosphäre in Berührung, so wird der Boden begierig Wasser aufnehmen und sich mit der Atmosphäre in einen Gleichgewichtszustand setzen. Dieses Verhalten des Bodens bezeichnet man als *Hygroskopizität.*

Die Luft über dem Boden besitzt einen Dampfdruck für Wasser gleich P, während der maximale Dampfdruck bei gleicher Temperatur als P_0 bezeichnet wird. Der relative Dampfdruck der Luft über dem Boden ist dann P/P_0. Dieser Wert wird experimentell (vgl. KURON 1930 bis 1936) zweckmäßig so eingestellt, daß man Luft mit Schwefelsäure verschiedener Konzentration ins Gleichgewicht stellt. Man erhält folgende Werte:

H_2SO_4 in Prozenten:	10	20	30	39	48,5	54	60	70	74
P/P_0:	0,942	0,868	0,748	0,582	0,383	0,286	0,177	0,069	0,034

Die vom Boden in Abhängigkeit vom relativen Dampfdruck der Luft sorbierte Wassermenge (a) soll nach SPERANSKI und KRASCHENINIKOW (1907) nach der Gleichung

$$a = a_0 + K\left(\frac{P}{P_0}\right)^2$$

berechenbar sein (wobei a_0 und K Konstanten darstellen), doch fand KURON, daß dies nur für einen mittleren Bereich der Werte von P/P_0 gilt, während bei

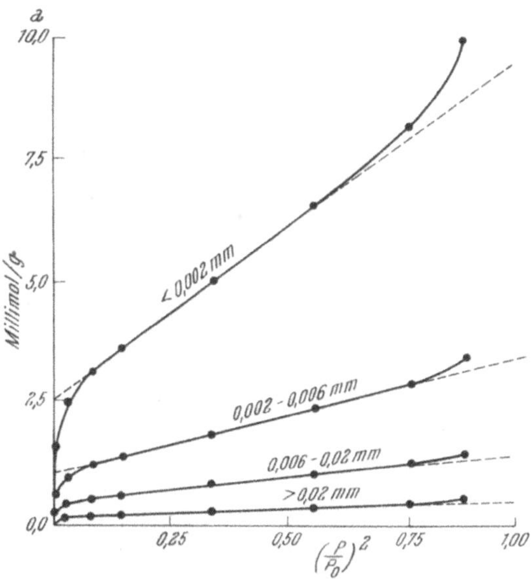

Abb. 111. Abhängigkeit des sorbierten Wasserdampfes von der relativen Luftfeuchte. Bestimmungen KURONS, nach der Gleichung von SPERANSKI ausgewertet

kleinen und großen Werten Abweichungen bestehen. Abb. 111 gibt Werte von KURON für die Abhängigkeit der sorbierten Wassermenge von der relativen

Luftfeuchtigkeit bei vier verschiedenen Bodenfraktionen (Teilchengrößen <0,002 mm, 0,002 bis 0,006 mm, 0,006 bis 0,02 mm, >0,02 mm) und zeigt die

Tabelle 94. *Die Additivität der Sorptionswerte für den Boden aus Wegnersau nach Kuron* (1930)

$\frac{P}{P_0}$	Werte der Wasserdampfsorption in g/100 g trockenen Boden					
	Fraktion < 0,002	Fraktion 0,002—0,006	Fraktion 0,006—0,020	Fraktion > 0,020	Sorption aus der Summe der Fraktionen berechnet	Sorption des Gesamtbodens direkt bestimmt
0,942	2,514	2,090	1,280	1,273	7,057	7,61
0,868	2,055	1,750	1,102	1,012	5,919	6,24
0,748	1,661	1,440	0,911	0,834	4,846	5,03
0,582	1,290	1,140	0,730	0,665	3,825	3,96
0,383	0,975	0,850	0,544	0,509	2,878	2,91
0,177	0,665	0,595	0,370	0,335	1,965	1,97
0,069	0,425	0,384	0,235	0,217	1,261	1,21
0,034	0,345	0,276	0,164	0,155	0,940	0,88
Gehalt an Fraktionen in Prozent	12,19	12,65	13,04	62,12	100,0	

Abweichungen gegenüber der Gleichung von SPERANSKI und KRASCHENINIKOW. Die sorbierte Wassermenge stieg mit fallender Teilchengröße an, und KURON

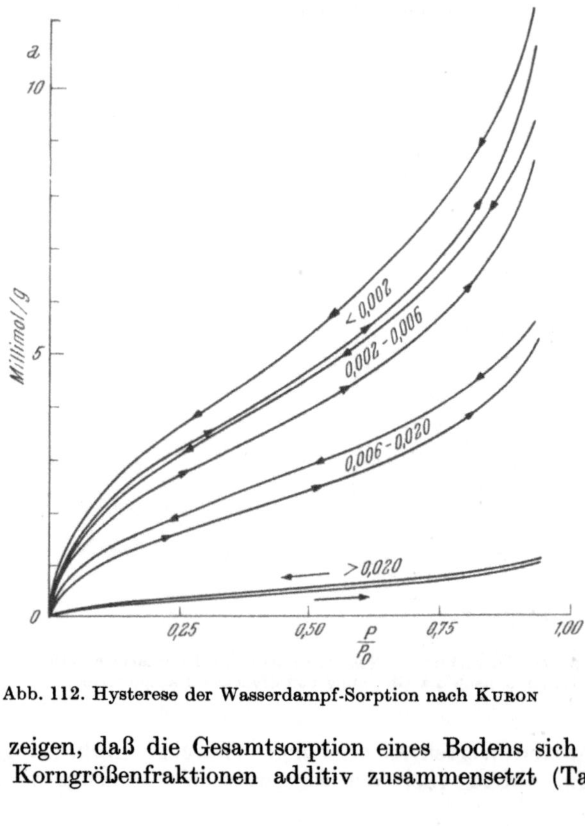

Abb. 112. Hysterese der Wasserdampf-Sorption nach KURON

(1930) konnte zeigen, daß die Gesamtsorption eines Bodens sich aus jener der anteilmäßigen Korngrößenfraktionen additiv zusammensetzt (Tab. 94).

Die Gleichgewichte, die sich zwischen Bodenfraktion und wasserdampfhaltiger Luft im Hinblick auf die Sorption des Wassers einstellen, liegen erfahrungsgemäß verschieden, je nachdem, ob das Gleichgewicht durch Aufnahme von

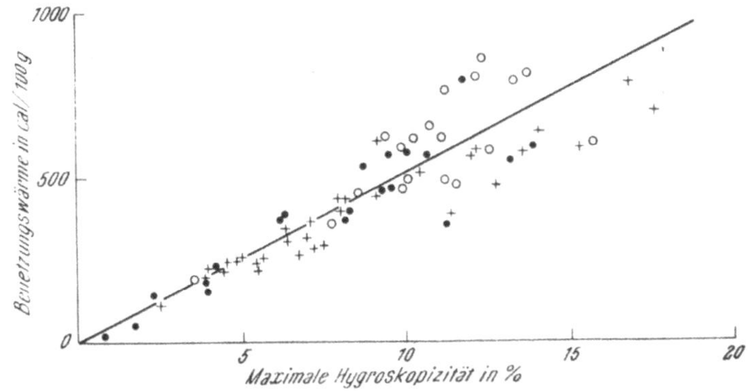

Abb. 113. Abhängigkeit der Benetzungswärme der Böden von der maximalen Hygroskopizität nach ANDRIANOW, DIMO und ZUNKER. Mit Punkten sind die Angaben ANDRIANOWS, mit Kreisen die von DIMO und mit Kreuzen die von ZUNKER bezeichnet (nach RODE 1959)

Wasser aus einer feuchteren Atmosphäre oder aber durch Abgabe von Wasser an eine trockene Atmosphäre hergestellt worden ist. Diese Erscheinung der „Hysterese“ ist im Verlauf der Kurven der Abb. 112 (nach Zahlen von KURON 1936) ersichtlich.

3. Löslichkeit, Solvatation, osmotischer Druck

Haben wir bisher den Fall betrachtet, daß Wasser (bzw. Wasserdampf) mit in Wasser unlöslichen (festen) Bodenteilchen in Wechselwirkung tritt, so müssen wir kurz auch auf jenen Fall eingehen, bei dem die Hydratation der Oberflächen fester Teilchen dazu führt, daß durch sie Bestandteile dieser Teilchen aus ihren bisher bestehenden Bindungen gelöst und aus dem Strukturverband des festen Teilchens herausgedrängt werden. Wenn Ionen oder Moleküle im Verband eines Kristallgitters feste Teilchen bilden, so können sie durch Wechselwirkung mit Wasser, indem sie sich mit Hydrathüllen umgeben, „aufgelöst“ werden. Wir sprechen dann von Solvatation und erhalten aus der Wechselwirkung zwischen Bodenteilchen und Wasser eine wässerige Lösung. Sie besitzt einen „osmotischen Druck“ (P), der nach VAN'T HOFF von der molaren Konzentration des gelösten Stoffes (C) im Lösungsmittel (Wasser), der Gaskonstante (R) und der absoluten Temperatur (T) bestimmt wird:

$$P = R \cdot C \cdot T$$

Nach einer Gleichung von OSTWALD und MÜNDLER (1919) für quellende Systeme setzt sich der beobachtete osmotische Druck (π) aus dem echten osmotischen Druck RTC_1 plus dem Hydratationsdruck des quellenden Systems kC_2^n zusammen, wobei C_1 die Molkonzentration des gelösten Stoffes, C_2 die Konzentration des quellenden Systems (g/1000 cm³) und k sowie n Konstanten bedeuten. Der Hydratationsdruck macht bei Zuckerlösungen bis zu 63% des beobachteten Gesamtdrucks aus. Der osmotische Druck eines umgrenzten Volumens einer Lösung ist ein Maß für den Saugdruck, den er auf umgebendes Wasser ausübt, sofern seine Umgrenzung keinen entgegengerichteten Wanddruck ausübt. Lösungsvolumina mit hohem osmotischem Druck entziehen daher benachbarten

Lösungsvolumina mit niedrigerem osmotischem Druck Wasser bis zum beiderseitigen Ausgleich der osmotischen Drucke, wobei das Volumen des ursprünglich höheren Druckes sich vergrößert, jenes des ursprünglich kleineren Druckes sich dagegen verkleinert. Der osmotische Druck stellt daher einen Faktor dar, der auch für die Wasserbewegungen im Boden Bedeutung besitzt (vgl. S. 673 ff.). Sofern geringe Anteile löslicher Stoffe im trockenen Boden vorhanden sind, treten sie mit Wasser im Sinne der Bildung kleiner Bereiche von Lösungen mit hohem osmotischem Druck in Wechselwirkung, die ihrerseits das Bestreben haben, sich durch weitere Wasseranziehung zu verdünnen. Sie haben dadurch Anteil an der Hygroskopizität des Bodens, die somit nicht ausschließlich und allein aus der Wechselwirkung zwischen festen, wasserunlöslichen Bodenteilchen und dem Wasser ihre Größe ableiten.

Einen weiteren Einfluß auf die Wasserbindung des Bodens üben, neben der Sorption an den Kolloidgrenzflächen, die an den Bodenkolloiden sorbierten Ionen, vor allem die Kationen, aus. Bei geringen relativen Feuchtegehalten der Bodenluft weisen mit mehrwertigen Ionen beladene Bodenkolloide wegen der größeren Hydrathülle dieser Ionen einen höheren H_2O-Gehalt auf als mit 1-wertigen Ionen beladene Kolloide. Mit zunehmender relativen Feuchte der Bodenluft nimmt die Wasseranlagerung der letztgenannten Kolloide aber stärker zu (SCHEFFER und SCHACHTSCHABEL 1960). Während die Wasserbindung bei $<40\%$ relative Feuchte der Hydratation der sorbierten Ionen und der der Kolloidgrenzflächen zugeschrieben werden muß, beruht diese bei höheren Feuchtegehalten neben der Kapillarkondensation auf osmotischen Erscheinungen. Die osmotische (dynamische) Wasserbindung ist, in Abhängigkeit von der Haftfestigkeit der sorbierten Ionen, bei den mit 1-wertigen Kationen beladenen Bodenkolloiden größer als bei den mit höherwertigen Ionen beladenen Kolloiden (KURON 1930, 1931).

4. Bindung von Wasser durch kapillare Erscheinungen

Die innerhalb der Flüssigkeit zwischen ihren einzelnen Molekülen auftretenden Anziehungskräfte (VAN DER WAALSsche Kräfte) heben sich, da jedes Molekül allseitig von Nachbarmolekülen umgeben ist, gegenseitig auf. An der Grenzfläche zu einer gasförmigen Phase aber, also an der Flüssigkeitsoberfläche, sind die Flüssigkeitsmoleküle jedoch nur einseitig von gleichen Molekülen umgeben. In der Gasphase finden sich keine bzw. nur sehr wenige, VAN DER WAALSsche Kräfte ausübende Moleküle vor. Die an der Wasseroberfläche befindlichen Moleküle stehen daher unter der einseitigen Wirkung der Anziehungskraft aller unterhalb befindlichen Wassermoleküle, die nun einen Druck auf die Gesamtmasse der Wassermoleküle ausübt. Es handelt sich um den Oberflächen- oder Kohäsionsdruck des Wassers, der nach VAN DER WAALS etwa 10.700 Atmosphären beträgt. Das Wasser steht dadurch unter einem sehr hohen Druck und kann deshalb durch zusätzlichen Druck kaum noch wesentlich komprimiert werden. Da nun auf Grund der Anziehung in Richtung Inneres der Flüssigkeit soviel als möglich Moleküle in das Innere zu gelangen versuchen, muß das gegebene Volumen bestrebt sein, die kleinstmögliche Oberfläche auszubilden, also Kugelform anzunehmen. Da ferner der Druck der Oberflächenmoleküle ganz analog wie eine gespannte elastische Haut diese Kugelform herbeiführt und daher die Oberflächenmoleküle wie eine gespannte elastische Haut in Erscheinung treten, spricht man auch von „Oberflächenspannung". Sie wird in $\text{dyn}\cdot\text{cm}^{-1}$ gleich $\text{erg}\cdot\text{cm}^{-2}$ gemessen und sinkt mit steigender Temperatur bei Wasser nach der Gleichung

$$\alpha_t = 75{,}6\,(1 - 0{,}002\,t)$$

Zwischen der an der Oberfläche vorhandenen Schicht von Wassermolekülen und jenen im Flüssigkeitsinneren besteht natürlich ein ständiger Wechsel bzw. Austausch. Nach Adam (1938) beträgt die Verweildauer eines Moleküls an der Oberfläche bei 20° C nur etwa den achtmillionsten Teil einer Sekunde, wenn die Flüssigkeit an ein Vakuum grenzt. Es handelt sich also keineswegs um ein stabiles „Oberflächenhäutchen".

Die zwischen Oberflächen fester Teilchen und Wassermolekülen wirksamen zwischenmolekularen Anziehungskräfte zeigen, wie wir gesehen haben, das Bestreben, die Oberflächen mit einem Wasserfilm zu überziehen, der festgehalten wird. Dies ist nur möglich, wenn zwischen die Oberflächen und das Wasser nicht Schichten eingelagert sind, die selbst keine Ladungen tragen bzw. nicht Dipolcharakter besitzen und mehrere Å dick sind, so daß der mögliche Abstand der Wassermoleküle, die herantreten können, größer bleibt, als die Reichweite der

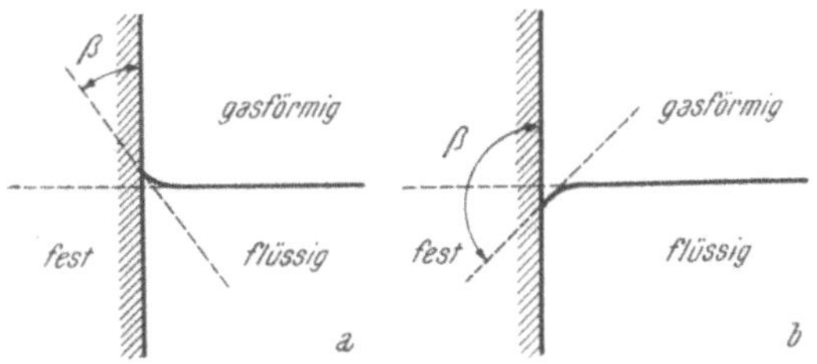

Abb. 114. Randwinkel α und Benetzungswinkel β bei (a) großer und (b) geringer Benetzbarkeit

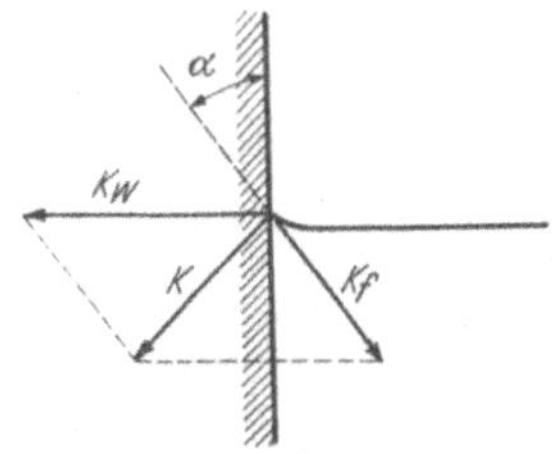

Abb. 115. Kräfteparallelogramm des Zusammenwirkens von zwischenmolekularer Anziehung der Wassermoleküle untereinander (k_w) und der zwischenmolekularen Anziehung zwischen fester Oberfläche und Wassermolekülen (k_f). k Richtung der auf das Wassermolekül wirksamen resultierenden Kraft

zwischenmolekularen Kräfte ausmacht. Die feste Oberfläche ist nur *benetzbar*, wenn sich die Wassermoleküle ihr bis in den Bereich der zwischenmolekularen Anziehungskräfte nähern können. Die Benetzbarkeit ist Voraussetzung für die Bildung von Wasserfilmen an festen Oberflächen. Sie kann gemessen werden durch den Winkel, der beim teilweisen Eintauchen einer festen Oberfläche in eine Flüssigkeit zwischen beiden gebildet wird. Bei guter Benetzbarkeit wird die Flüssigkeit an der festen Oberfläche etwas angehoben und es bildet sich ein Winkel von weniger als 90°, bei schlechter Benetzbarkeit wird die Flüssigkeit etwas hinuntergedrückt, so daß ein Winkel von mehr als 90° entsteht (Abb. 114).

Die Erscheinung kommt dadurch zustande, daß nicht nur zwischen der festen Oberfläche und den Molekülen der Flüssigkeit eine Anziehung (Kraftvektor K_w in Abb. 115) wirkt, sondern daß auch zwischen den einzelnen Molekülen der Flüssigkeit Anziehungskräfte (Kraftvektor k_f in Abb. 115) bestehen.

Die von der Krümmung der Flüssigkeitsoberfläche an der festen Oberfläche betroffene Fläche hat eine bestimmte Größe, die nur von der Natur der Flüssigkeit und von der Natur des Festmaterials abhängig ist. Je kleiner ihre gesamte Flüssigkeitsoberfläche ist, desto größer ist der von der Krümmung betroffene Anteil. Für Wasser bildet sich bei einem Abstand zweier fester Oberflächen von nur wenigen Millimetern bereits keine ebene Wasseroberfläche mehr aus, so daß der Anteil an ebener Fläche bereits 0% der Gesamtoberfläche ausmacht; in einem engen Rohr entsteht dadurch ein Meniskus, der bei guter Benetzung konkav, bei schlechter Benetzung konvex ist.

Je mehr der vom Wassermolekül an der Oberfläche ausgeübte Druck auf die darunterliegende Flüssigkeit unter dem Einfluß der von der festen Oberfläche ausgeübten Anziehung teilweise kompensiert, das heißt in seiner Richtung verändert wird, je mehr also der Einfluß der festen Oberfläche ansteigt, desto kleiner wird der auf die Flüssigkeit ausgeübte Druck, der Oberflächendruck. Er muß in großen Gefäßen (Rohren mit großem Durchmesser) größer sein als in kleinen Gefäßen (Rohren mit sehr kleinem Durchmesser, also Kapillarrohren). Taucht man daher ein offenes Kapillarrohr in eine offene Flüssigkeit ein, so drückt der größere Oberflächendruck der offenen Fläche gegen den geringeren Druck der Meniskusoberfläche in der Kapillare die Flüssigkeit in der Kapillare so hoch, bis der dadurch erwirkte hydrostatische Druck der Flüssigkeitssäule den Betrag der Oberflächendruckdifferenz zwischen Kapillarmeniskus und freier Flüssigkeitsoberfläche kompensiert hat. Die maximale Steighöhe (h) der Flüssigkeitssäule in der Kapillare beträgt

$$h = \frac{2\delta}{\varrho r g}$$

wobei ϱ die Dichte der Flüssigkeit, r den Radius der Kapillare, g die Erdbeschleunigung (981 cm/sec²) und δ die Kapillaritätskonstante darstellen. Bei nicht vollständiger Benetzung multipliziert sich h noch mit $\cos\alpha$, wobei α den Randwinkel (vgl. Abb. 115) bedeutet. Die Höhe des Wasseraufstieges ist damit umgekehrt proportional zum Radius der Kapillare und abhängig von der Benetzbarkeit. Bei gleicher Benetzbarkeit steigt das Wasser (für das $\delta = 74$ dyn/cm beträgt) um so höher, je enger die Kapillare ist, so daß

$$h = \frac{0{,}15}{r}$$

gesetzt werden kann.

Die Geschwindigkeit des Wasseraufstieges in Kapillaren ist um so kleiner, je kleiner die Durchmesser der Kapillaren sind, weil sich die Rand-Reibung vergrößert. Nach einer von WOSHBURN angegebenen und von AGAPOWA (1937) verwendeten Formel kann die Geschwindigkeit des kapillaren Aufstieges ermittelt werden, doch kann sie wegen verschiedener Querschnitte und Formen des Kapillarsystems im Boden nicht unmittelbar angewandt werden.

Bei sehr feinen Kapillaren ist zweifellos die zwischen fester Oberfläche (Wand) und Flüssigkeit (Wasser) bestehende Anziehung groß genug, um die im Gewicht des Wassers gegebene Schwerkraftwirkung zu kompensieren und somit beim Herausnehmen der wassergefüllten Kapillaren aus dem Wassergefäß ein Ausrinnen der Kapillare zu vermeiden. Bei weniger feinen Kapillaren bleibt eine Wassersäule darin nur dann hängen („hängendes Wasser"), wenn der obere Meniskus einen kleineren Krümmungsradius aufweist als der untere. Der Oberflächendruck des unteren Meniskus weist ja nach oben, jener des oberen nach unten; das Gewicht der Wassersäule kann nur durch die Differenz dieser beiden Oberflächendrucke kompensiert werden.

Im Boden spielen diese kapillaren Erscheinungen eine wichtige Rolle für das Bodenwasser bzw. für die Bodenlösung. Zwischen den Bodenskelett-Teilchen, im Inneren der Bodenkrümel, aber auch zwischen kleineren Bodenkrümeln, entstehen zahlreiche Kapillarräume, allerdings von sehr unregelmäßiger Gestalt. Um die Verhältnisse, die sich bei unregelmäßig geformten Kapillaren ergeben, genauer kennenzulernen, haben KOSSOWITSCH (1904) und LEBEDEV (1936) sich mit JAMINschen „Paternosterkapillaren" befaßt, das sind Kapillarrohre mit periodisch der Länge nach verschieden großen Durchmessern. Die Steighöhe des Wassers ist in solchen Kapillaren abhängig davon, an welcher Stelle (bei welchem Durchmesser) der Meniskus eingestellt wird; steht er an einer Stelle mit größerem Durchmesser, ist die Steighöhe geringer, steht er an einer engeren Stelle, ist die Steighöhe entsprechend höher. Es gibt daher nicht einen, sondern *mehrere* Gleichgewichtszustände in solchen Kapillaren (bei verschiedenen Steighöhen je nach Lage des Meniskus); man spricht bei dieser Erscheinung von kapillarer Hysterese. Die Entleerung der Paternosterkapillaren erfolgt ruckweise.

Im Boden haben wir es jedoch mit noch vielgestaltigeren Kapillarräumen zu tun, die nicht Röhrencharakter haben, sondern ein räumliches Netz vielfach verzweigter Hohlräume darstellen, das sich zwischen den Bodenteilchen und Aggregaten erstreckt und alle Übergänge zwischen kapillaren und größeren Hohlräumen aufweist. Diese Verhältnisse hat man dadurch modellmäßig zu idealisieren bzw. zu erfassen versucht, daß man die Bodenpartikelchen als kugelförmig und in Würfel- oder aber in hexagonaler Packung vorliegend betrachtet hat und nun Berechnungen über die sich dabei ergebenden Hohlräume und ihr kapillares Verhalten anstellte (vgl. RODE 1959). Zwischen zwei kugelförmigen Partikelchen (vgl. Abb. 116) wird Wasser in Form einer „Manschette" festgehalten. Die Menge dieses Wassers kann als „Porenwinkelwasser", „Fugenwasser" oder „penduläres Wasser" bezeichnet werden. Ihr Volumen beträgt nach FISHER (1930)

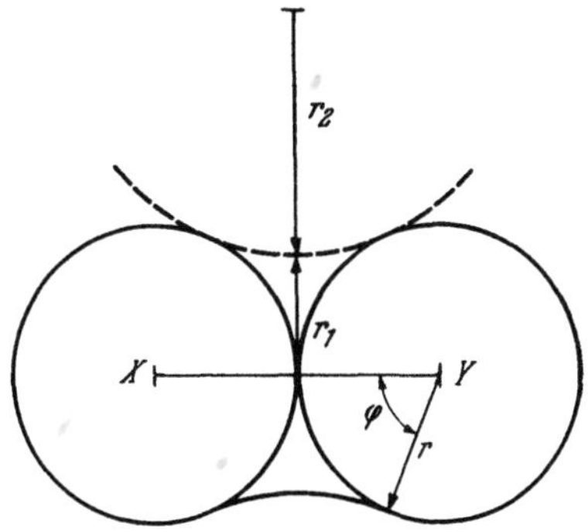

Abb. 116. Die Wassermanschette zwischen gleichen kugelförmigen Partikelchen eines „Idealbodens" (nach RODE 1959)

$$V = 2\pi r^3 (\sec \cdot \varphi - 1)^2 \cdot \left[1 - \left(\frac{\pi}{2} - f\right) \cdot \operatorname{tg} \varphi\right]$$

wobei der Winkel zwischen der Verbindungslinie der Zentren der Partikel und dem Radius vom Zentrum des Partikels bis an den Rand der Wassermanschette und r der Radius der Teilchen ist.

Da die kapillaren Kräfte eine gewisse Größe haben, die das Gewicht der Manschette kompensieren müssen, wenn das Wasser festgehalten und nicht abrinnen soll, besitzt die festgehaltene Wassermanschette ein Maximalgewicht, dessen Größe von dem Radius der kugelförmigen Teilchen abhängig ist. Bei großem r bleiben die Manschetten (bzw. die Winkel) so klein, daß keine gegenseitige Berührung der kapillar festgehaltenen Wassermanschetten untereinander stattfindet und die einzelnen Wassermanschetten voneinander isoliert bleiben, so daß Diffusionsvorgänge von einer zur anderen nicht stattfinden können; bei kleinem r dagegen können die Manschetten relativ so groß werden, daß sie miteinander in Verbindung treten und ein Wasserkontinuum bilden, das viele Teilchen gemeinsam umgibt und Wege für Stofftransporte durch Diffusion bildet. Das Volumen des Fugenwassers beträgt im Augenblick des Zusammenfließens der Wassermanschetten bei würfelförmiger Packung der Teilchen 18,3% und bei hexagonaler Packung 22,6% des gesamten Porenvolumens. Auch beim Zusammenfließen der Wassermanschetten bleiben aber die Lufträume zwischen den Teilchen untereinander (raumnetzartig) verbunden. Befindet sich das Wasser in diesem Zustand der Kontinuität, so spricht man mit VERSLUYS (1917) nicht mehr von pendulärem, sondern von „funikulärem" Wasser. Verdrängt man jedoch ein noch größeres Volumen der Luft durch Wasser, so daß nur noch voneinander isolierte Luftblasen im Kontinuum des Wassers vorliegen, so wird die Wassermenge als „Kapillarwasser" bezeichnet[1].

---

[1] Die untere Grenze der Bindungsfestigkeit des Kapillarwassers im Boden liegt etwa zwischen 0,05 und 0,3 atm. Kapillar adsorbiertes Wasser gefriert in einem Temperaturintervall von –4° C bis –78° C.

Diese häufig verwendeten Begriffe definieren sich somit wie folgt:

Isolierte Wassermanschetten im Luftkontinuum	= penduläres Wasser
Wasserkontinuum neben Luftkontinuum	= funikuläres Wasser
Isolierte Luftblasen in Wasserkontinuum	= kapillares Wasser

In gewisser Hinsicht hat man es im Idealboden also mit einer sehr verzweigten Jaminschen Paternosterkapillare zu tun, und man kann berechnen, daß das Verhältnis zwischen dem Durchmesser der engsten und der breitesten Abschnitte dieses zwischen den festen Teilchen sich erstreckenden räumlichen Kapillarnetzes bei

würfelförmiger Packung der Teilchen 0,73 :0,41 =1,78 beträgt,
bei hexagonaler Packung aber 0,288:0,155=1,86 bis 0,414:0,155=2,66.

Da die Höhe des kapillaren Aufstiegs durch die Querschnitte der engsten Verbindungskanäle bestimmt wird, der 0,42 des Durchmessers des Teilchens beträgt, ist sie von dem Teilchenradius abhängig, und Dolgow (1948) fand bei Messungen des kapillaren Wasseraufstieges in Ziegelsplitt verschiedener Korngrößen tatsächlich gute Übereinstimmung der experimentellen mit den idealen (berechneten) Werten (Tab. 95).

Tabelle 95. *Kapillarer Wasseraufstieg (cm) bei Ziegelsplitt verschiedener Korngrößen (mm) verglichen mit berechneten Werten*
(nach Dolgow 1948)

Korngrößen:	2—1	1—0,5	0,5—0,25	0,25—0,10
im Mittel	1,5	0,75	0,375	0,175
Höhe des Aufstiegs:				
gefunden	5,0	10,8	16,3	41,0
berechnet	4,9	9,8	19,6	42,0

Es zeigt sich somit, daß die tatsächlichen Abweichungen, die durch das Abweichen der Teilchen des Ziegelsplitts von der (idealen) Kugelform zu erwarten sind, im Mittel einander aufheben und die „idealen" Verhältnisse nahezu gültig belassen können. Zweifellos aber sind die Verhältnisse im natürlichen Boden so vielfältig und vom Idealfall so verschieden, daß eine Übertragung der Idealverhältnisse auf den Boden nicht in allen Fällen erfolgversprechend sein kann.

In den Räumen zwischen den Bodenteilchen befinden sich bei Gegenwart von Wasser somit Menisken unterschiedlicher Krümmung, an deren Oberfläche Wasser verdunstet. Der Luftraum wird somit schnell mit Wasserdampf gesättigt, dessen Spannung einerseits von der Temperatur, andererseits aber auch von dem Krümmungsradius (und dessen Vorzeichen) der Menisken abhängig ist. Über konkaven Oberflächen ist nämlich die Dampfspannung kleiner und über konvexen größer als über einer ebenen Wasserfläche. Es besteht folgende Beziehung:

$$RT \cdot \ln \frac{P}{P_0} = \frac{2M \cdot \alpha}{d \cdot r}$$

wobei R die Gaskonstante, T die absolute Temperatur, P_0 den Dampfdruck über einer ebenen Oberfläche der Flüssigkeit bei T^0, P den Dampfdruck über einer gekrümmten Flüssigkeitsoberfläche vom Radius r bei T^0, M das Molekulargewicht der Flüssigkeit, α die Oberflächenspannung der Flüssigkeit bei T^0 und d die Dichte der Flüssigkeit bei T^0 bedeuten. Weil das Wasser in den Kapillaren des Bodens vorwiegend konkave Oberflächen bildet, muß das in der Bodenluft befindliche Wasser (falls Sättigung mit Wasserdampf vorliegt) in den Kapillaren kondensieren, da wie erwähnt der Dampfdruck über diesen herabgesetzt ist.

Diese als „Kapillarkondensation" des Wassers bezeichnete Erscheinung ist nur dann von Bedeutung, wenn der Feuchtigkeitsgehalt des Bodens seiner maximalen Hygroskopizität annähernd gleich wird.

5. Klassifikation des Wassers im Boden nach Bindungsfestigkeiten

Im natürlichen Boden liegen sehr unterschiedlich geformte und unterschiedlich große Hohlräume vor, die in Abhängigkeit von den Witterungsbedingungen recht unterschiedlich mit Wasser erfüllt sind. Nach Einwirkung größerer Niederschlagsmengen versucht oft viel mehr Wasser in den Boden einzudringen, als von ihm aufgenommen bzw. festgehalten werden kann. Im Zustand voller Wassersättigung enthält daher der Boden Wasser in allen möglichen Bindungsformen neben freiem, keiner speziellen Bindungsform unterworfenem Wasser. Man muß daher, um dem Boden das in ihm vorhandene Wasser zu entziehen, verschieden starke Kräfte aufwenden, um die verschieden starken Kräfte zu überwinden, mit denen es gebunden vorliegt. Die Anteile der gesamten im Boden vorhandenen Wassermenge, die mit Kräften bestimmter Größe festgehalten werden, sind verschieden groß. Man kann daher einen Boden hinsichtlich seines Wasserbindungsvermögens dadurch charakterisieren, daß man angibt, welche Wassermengen mit jeweils anzugebenden Bindungsfestigkeiten von ihm festgehalten werden. Zwei Methoden bieten sich hierbei besonders an:

1. Man betrachtet den Boden als hygroskopisches (bzw. osmotisches) System, versucht, ihm das Wasser durch mechanischen oder osmotischen Druck zu entziehen, und berechnet die Bindungsfestigkeiten des Wassers in kg/cm^2 oder in Atmosphären (atm).

2. Man versucht, dem Boden Wasser durch Zentrifugieren zu entziehen, und berechnet die Bindungsfestigkeiten des Wassers als die zu ihrer Lösung erforderliche Zentrifugalkraft (in g).

Da man auch durch eine Zentrifugalkraft von 70.000 g noch nicht alles Wasser dem Boden entziehen kann, da es ferner einfacher ist, den Druck in atm zu messen, und da man bei Kenntnis der Bindungsfestigkeiten in Atmosphären direkt mit dem ebenfalls in Atmosphären angebbaren Saugdruck der Pflanzen vergleichen kann, der sie zur Wasseraufnahme befähigt, gibt man im allgemeinen der erstgenannten Methode den Vorzug.

Da man ferner bestrebt ist, das Wasser im Boden nicht nur nach seiner Bindungsfestigkeit zu klassifizieren, sondern auch nach den Ursachen seiner Bindung und ihrer Festigkeit, hat man Bezeichnungen für einzelne Wasserfraktionen des Bodens gewählt, die jene Ursachen kennzeichnen sollen.

Rodewald und Mitscherlich (1904) fanden, daß die Wassermenge WH, welche die Teilchen eines Materials zur Gänze bedeckt und benetzt, so daß die Benetzungswärme bei ihrer Erreichung gleich Null wird, für verschiedene Materialien in einem Raum über genau 10%iger Schwefelsäure gleiche Größe hat. Sie nannten diese Wassermenge WH (hygroskopisches Wasser) und berechneten sie als

$$WH = \frac{\text{Gewicht des Bodens mit hygroskopischem Wasser}}{\text{Gewicht des trockenen Bodens}}$$

An Stelle von 10%iger H_2SO_4 kann nach Beutelspacher (1939) zur Bestimmung der Hygroskopizität auch eine gesättigte wässerige Lösung von Natriumsulfat dienen, in der sich noch feste Kristalle befinden. Nach Kuron (1930) verwendet man, um Fehler durch Kondensationen auszuschalten, vor allem in den Tropen den Ausgleich über 50%iger H_2SO_4, erhält aber entsprechend kleinere Werte für die Hygroskopizität, was wiederum Fehler bedingt.

Dieses „hygroskopische Wasser" ist an die Bodenteilchen mit einem Druck von 50 atm oder mehr gebunden. Seine Menge kann auch als 1 Hy bezeichnet werden (vgl. SCHARRER 1953). Die doppelte Menge (2 Hy) ist immer noch bis zu 20 atm stark gebunden. Die 1,6- bis 1,7fache Menge (1,6 bis 1,7 Hy) kennzeichnet den „Welkepunkt" der meisten Kulturpflanzen, weil sie hier wegen der Un-

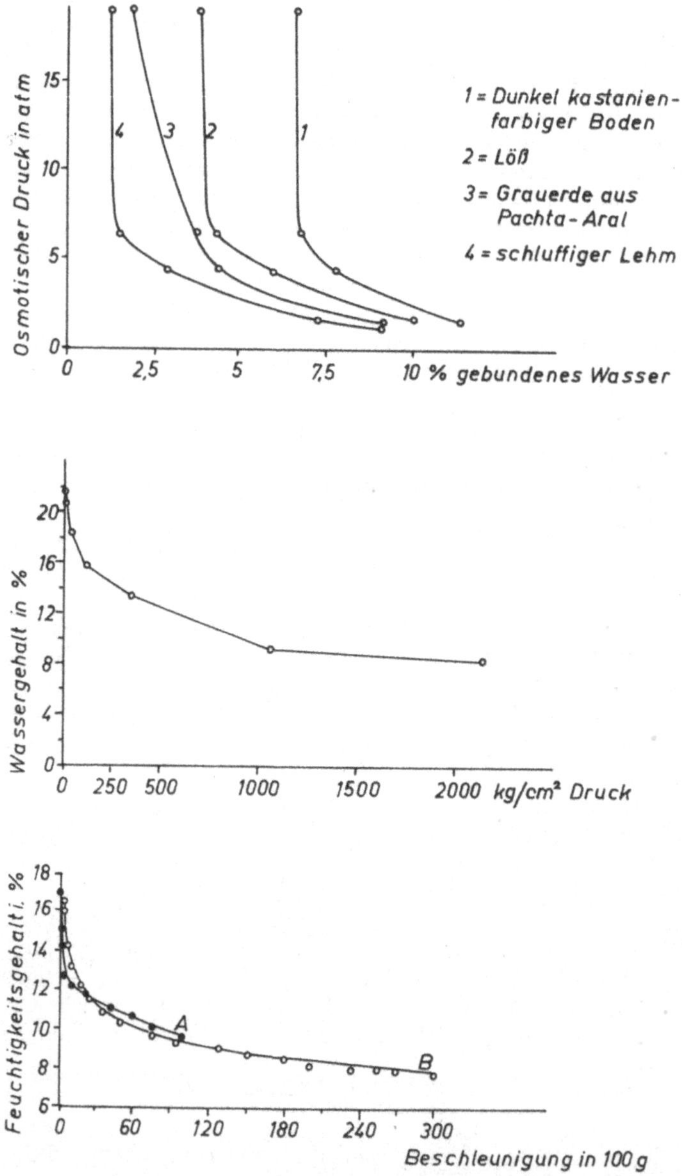

Abb. 117. *Oben:* Menge des vier verschiedenen Bodenproben durch Zuckerlösungen hoher osmotischer Drucke entziehbaren Wassers (DOLGOW 1948)

Mitte: Wassergehalt eines Bodens nach Anwendung von Druck in kg/cm² (KOLJASEV 1941)

Unten: Wassergehalt zweier Böden (*A*, *B*) nach Zentrifugieren bei Beschleunigungen von 1 bis 30000 g (OLMSTEAD 1937) möglichkeit, mit ihrer Saugkraft von etwa 20 atm (Weizen etwa 10 atm, Rüben etwa 15, Kartoffeln etwa 20, Mais etwa 25 atm) aus dem Boden noch Wasser zu erhalten, zu welken beginnen. 2 Hy bedeuten daher etwa die Grenze der Verfügbarkeit des Wassers für den Durchschnitt der Kulturpflanzen, weshalb man die Wassermenge 2 Hy auch als „totes Wasser" bezeichnet hat. Bis zu 4,5 Hy nennt man die Wasserhülle der Bodenteilchen, die dann bis herab zu nur 6,5 atm gebunden ist, auch „Schwarmwasser". Die Menge von etwa 4,5 bis 2,0 = 2,5 Hy ist „biologisch nutzbar", wobei die Grenzwerte natürlich vom Saugdruck der Pflanzenwurzel abhängig sind.

Das 4- bis 5fache der Hygroskopizität nach MITSCHERLICH kann von einem Boden im allgemeinen noch gegen die Wirkung der Schwerkraft festgehalten werden. Wasser, das nicht mehr gegen die Wirkung der Schwerkraft festgehalten wird, wurde von mehreren Autoren als „Gravitationswasser" oder auch als „Senkwasser" bezeichnet.

Jene Wassermenge, die ein Boden aus einer fast mit Wasserdampf gesättigten Luft (relative Feuchtigkeit = 94%) aufnimmt und im Gleichgewicht mit ihr festhält, wird als „maximale Hygroskopizität" (= MH) bezeichnet.

Diejenige Wassermenge, die im Boden durch die Wirkung zwischenmolekularer Anziehungskräfte zurückgehalten werden kann, wird in Gewichtsprozenten des trockenen Bodens als „maximale molekulare Wasserkapazität" angegeben und ist theoretisch identisch mit der auf Grund andersartiger Methodik bestimmbaren „maximalen Adsorptionskapazität".

Die genannte, maximal in einem Boden festgehaltene Menge an Wasser (ausschließlich Gravitationswasser) ohne Berücksichtigung der Größe und der Art der wirksamen Bindungskräfte wird als „Haftwasser", „wasserhaltende Kraft", „Wasserkapazität", „Gesamtwasserkapazität" oder „Grenzwasserkapazität" bezeichnet.

6. Die Saugkraft des Bodens

Wenngleich es wichtig ist, die verschiedenen Ursachen zu kennen, aus welchen Wasser im Boden gebunden wird, und die jeweiligen Anteile zu kennen, mit welchen sich die einzelnen bindenden Komponenten am Gesamtphänomen der Wasserbindung beteiligen, ist doch für das praktisch wichtige Verhalten des Wassers im Boden das Gesamtphänomen selbst von besonderer Bedeutung. Man kann die Summe aller wasserbindenden Faktoren eines Bodens als die „Saugkraft des Bodens" bezeichnen (KORNEW 1924). Sie zu messen ist auf verschiedene Weise möglich. Die sogenannten „Kapillarimeter" benutzen eine aus feinporigem gebranntem Ton bestehende Zelle (ähnlich einer CHAMBERLAINschen Kerze), die mit Wasser gesättigt in den Boden versenkt wird, dessen Saugkraft bestimmt werden soll. Die Zelle ist mit einem Manometer verbunden, welches anzeigt, daß bzw. wieviel der Boden den Zellen Wasser entzieht. Im Manometer geht der Druck zurück, bis ein Gleichgewichtszustand erreicht ist. Dieser Druck ist gleich der Saugkraft des Bodens. (Betreffend die Meßmethodik vergleiche man SEKERA 1931 und DOLGOW 1948.) Diese Methoden gestatten, die Saugkraft des Bodens im Bereich von 0 bis etwa 0,9 atm zu messen. Höhere Drucke lassen sich mit der Zentrifugenmethode nach GARDNER (1920) und SCHOFIELD (1935) ermitteln; sie ist für Saugkräfte im Bereich von 200 bis 10.000 cm Wassersäule bzw. von 0,2 bis 10 atm einsetzbar und gestattet, die Saugkraft des Bodens H (in cm Wassersäule) wie folgt zu berechnen:

$$H = \frac{h \cdot N^2 \cdot r}{2 \cdot 300^2}$$

wobei h die Stärke der Bodenschicht im Zentrifugenbecher in cm, N die Umdrehungszahl der Zentrifuge in der Minute und r der Abstand (Drehradius) der Bodenprobe, gemessen von der Zentrifugenachse bis zur Mitte der Bodenschicht (in cm), ist. Die bei verschiedenen Umdrehungszahlen ermittelten Saugkraftwerte werden mit den Wassergehalten der zentrifugierten Bodenproben in Kurven graphisch dargestellt. Es ergeben sich Kurven der in Abb. 118 dargestellten Art.

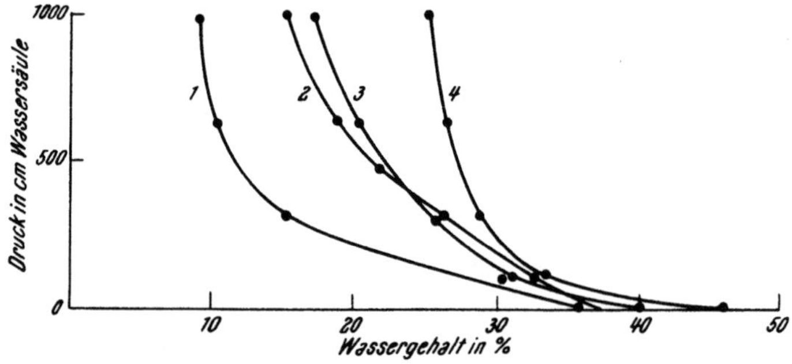

Abb. 118. Wassergehalt des Bodens in Abhängigkeit vom Saugdruck des Bodens (Dolgow 1948a). *1* Schluffiger Lehm, *2* Löß, *3* Grauerde, *4* Andijansker Grauerde

Noch höhere Saugkräfte (im Bereich von 30000 bis 1000000 cm Wassersäule oder 30 bis 1000 atm können mit Hilfe einer Einstellung eines Gleichgewichtszustandes zwischen Bodenprobe und Wasserdampfspannung einer entsprechend eingestellten Atmosphäre gemessen werden (Thomas 1921, Schofield 1935). Die Berechnung der Saugkraft erfolgt nach der Formel

$$H = -\frac{R \cdot T}{M \cdot g} \cdot \ln\left(\frac{h}{100}\right)$$

wobei M das Molekulargewicht des Wassers und h die relative Luftfeuchtigkeit bedeuten. Bei 20° C und der Logarithmenbasis 10 gilt somit

$$H = 3{,}32 \cdot 10^6 \cdot \log \frac{100}{h}$$

Die Saugkraft des Bodens ergibt sich bei voller Wassersättigung mit 0, bei verschiedenen Wassergehalten mit verschieden hohen Werten, bis zu einem Wert von etwa 10^7 cm Wassersäule oder 10^4 atm bei fast völlig trockenem Boden. Sie sinkt also mit steigendem Wassergehalt des Bodens ab.

Um innerhalb des weiten Zahlenintervalls einfachere Angaben machen zu können, wird nach Schofield (1935) die als „freie Energie des Bodens" aufzufassende, in cm Wassersäule gemessene Saugkraft mit F in Analogie zu den pH-Werten als Logarithmus, also in Form von pF, angegeben.

Eine Saugkraft von 1 atm (= 1033 cm Wassersäule) entspricht somit einem pF-Wert gleich 3, bei trockenem Boden erreicht pF den Maximalwert von 7.

Für den Bereich von pF 4,0 bis 4,3 kann der Saugdruck zweckmäßig (nach Schofield 1935) aus der Gefrierpunktserniedrigung t (in Grad Celsius) der Bodenlösung nach folgender Formel bestimmt werden:

$$\mathrm{pF} = 4{,}1 + \log t$$

Die Abhängigkeit des pF-Wertes vom Wassergehalt liegt im Sinne der Erscheinung der Hysterese bei Befeuchtung etwas anders als bei Austrocknung eines Bodens (vgl. Abb. 119), was bei vergleichenden Untersuchungen berücksichtigt werden muß.

Böden mit gleichen pF-Werten besitzen energetisch vergleichbare Wassergehalte, sie können in diesem Sinne als äquivalent betrachtet werden. Wie stark sich hierbei die Bodenart auswirkt, geht aus Abb. 120 hervor, die die pF-Kurven in Abhängigkeit vom Wassergehalt des Bodens für je einen Sand-, Lehm- und Tonboden zeigt.

Die minimale Wasserkapazität von Böden liegt im allgemeinen bei Saugkraftwerten von 0,14 bis 0,65 atm oder pF 2,15 bis 2,81, die maximale molekulare Wasserkapazität bei Saugkraftwerten von 4,1 bis 7,9 atm bzw. pF 3,61 bis 3,90.

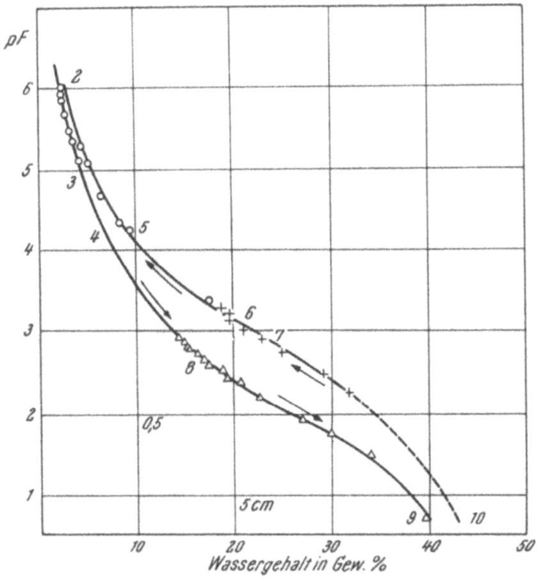

Abb. 119. Abhängigkeit des pF vom Wassergehalt bei Befeuchtung (untere Kurve) und bei Austrocknung (obere Kurve) nach SCHOFIELD. *1* Bei 105° getrocknet, *2* bei 50 % Luftfeuchte, *3* Hysterese über Schwefelsäure von 33 %, *4* Hysterese über Schwefelsäure von 10 %, *5* permanenter Welkepunkt, *6* annähernde Field Capacity, *7* Moisture Equivalent bei Erdbeschleunigung von 1000 g, *8* bei einer Saugkraft von 0,5 atm, *9* bei einer Saugkraft von 5 cm Wassersäule, *10* HILGARDS „Maximum Moisture". ○ Absorptionsmethode von Wasserdampf bei verschiedener relativer Feuchte, + Zentrifugenmethode (nach THOMAS), △ poröse Platten-Methode (nach RICHARDS)

7. Klassifikation des Wassers im Boden im Hinblick auf seine Menge und Verteilung

Diejenige Höchstmenge an Wasser, die der Boden bei vollständiger Auffüllung aller Poren und Hohlräume enthält (gleichgültig, ob er dieses Wasser auch gegen Gravitations- oder andere Einflüsse festzuhalten vermag oder nicht), wird als „volle" bzw. „maximale Wasserkapazität" bezeichnet. Sie kann im Laboratorium leicht bestimmt werden und ist z. B. dort als tatsächlich erfüllt gegeben, wo das Grundwasser in die betrachtete Bodenschicht unmittelbar eingedrungen ist, sofern nicht Hohlräume von ihm unerfüllt geblieben sind. Sie kann ferner dort als tatsächlich erfüllt vorliegen, wo starke Niederschläge den Boden von oben her völlig mit Wasser sättigen.

Bei natürlichen Böden kann der Fall vorkommen, daß aus den Niederschlägen stammendes Wasser, das in die oberen Bodenschichten eindringt, in diesen festgehalten wird und gar nicht so tief einsickern kann, daß es zur Verbindung mit dem (tiefer liegenden) Grundwasser kommt. Dieses „hängende Wasser", das kapillar festgehalten wird und daher auch als „kapillar-hängendes" Wasser bezeichnet wurde (ZUNKER 1930), wird mitunter auch „getragenes Wasser" benannt. Die Höchstmenge an hängendem Wasser, die im Bodenprofil gehalten werden kann, wird als „minimale Wasserkapazität" („Moisture Equivalent", „absolute Wasserkapazität" im Sinne von TJUREMNOV, „Grenz-Feldwasserkapazität" bei ULASEVIC und „kapillare Wasserkapazität" bei KATSCHINSKI, GW bei RODE 1959) bezeichnet. Man kann auch von „kapillarer Wasserkapazität" sprechen.

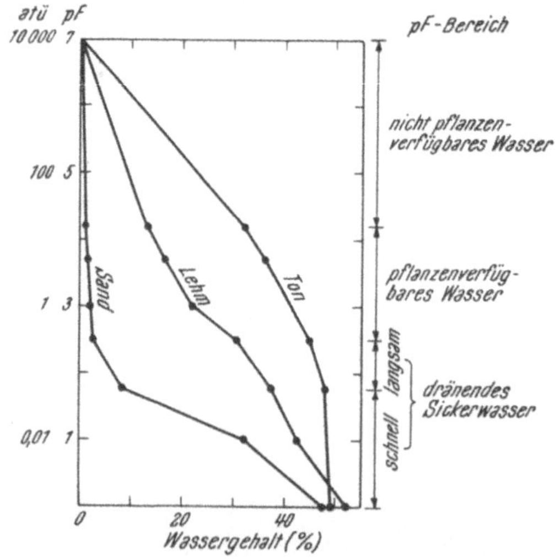

Abb. 120. pF-Kurve eines Sandbodens (pleistozäner Sand), eines Lehmbodens (Schwarzerde) und eines Tonbodens (Knickmarsch) (HARTGE, unveröffentlicht; nach SCHEFFER und SCHACHTSCHABEL 1960)

Infolge von Verdunstung kann hängendes Wasser seinen Zusammenhang verlieren und aufhören, zur verdunstenden Oberfläche aufzusteigen. Die Feuchtigkeit, bei der die Kapillarverbindung abreißt, wird als „lentokapillarer Punkt" oder als „Feuchtigkeit des Abrisses der Kapillarverbindung" bezeichnet.

8. Klassifikation des Wassers im Boden im Hinblick auf seine Verfügbarkeit für die Pflanze

Die Pflanze vermag sich vom Bodenwasser nur jene Anteile anzueignen, welche mit geringeren Kräften vom Boden festgehalten werden, als sie der Saugkraft der Pflanzenwurzeln entsprechen. Es ist somit die Saugkraftdifferenz zwischen der Saugkraft des Bodens (F_B) und der Saugkraft der Pflanze (F_P) für die Wasserversorgung der Pflanze ausschlaggebend; denn nur die dem Wert $F_P—F_B$ entsprechende Wassermenge des Bodens ist für die Pflanze verfügbar. Die Pflanze hat nur geringfügige Möglichkeiten, ihre F_P dem Wert von F_B anzupassen. Steigt F_B (bzw. der pF-Wert des Bodens) an, so wird ein Punkt erreicht, bei dem $F_P—F_B=0$ wird und die Pflanze kein Wasser mehr aufnehmen kann.

Bei Wassermangel zeigt die Pflanze Erscheinungen, die man als drei aufeinanderfolgende Stadien charakterisieren kann:

1. Beginn des Welkens; Pflanzen lassen nur die oberen Blätter hängen.
2. Starkes Welken; die Pflanzen lassen alle Blätter hängen, die Blätter vergilben.
3. Verdorren; die Pflanzen beleben sich nach dem Begießen nicht mehr.

Das Stadium 3 tritt bei einem Boden ein, der nur noch festgebundenes Wasser enthält (60 bis 90% der MH). Dieser Wassergehalt des Bodens, bei dem die Pflanzen absterben, wird als „Grenzwelkepunkt" bezeichnet. Das Stadium 2 entspricht dem „Welkekoeffizienten". Der Bereich zwischen Grenzwelkepunkt und Welkekoeffizient kann als „Welkeintervall" gekennzeichnet werden. Nach FURR und REEVE (1945) liegt dieses Welkeintervall zwischen 9,1 und 22,2 atm bzw. zwischen pF 3,97 und 4,36.

BRIGGS und SHANTZ (1914) stellten bei 1318 Welkepunktbestimmungen fest, daß der Welkepunkt nicht sehr stark von der Pflanzenart bestimmt wird, sondern daß fast alle Pflanzen bei gleichem Wassergehalt des Bodens verwelken (vgl. auch

VEIHMEYER und HENDRICKSON 1928 sowie BOTELHO DA COSTA 1938). Das permanente Welken, bei dem eine Wiedererholung der Pflanze nicht mehr eintritt, erfolgt beim „Welkepunktkoeffizienten" („Welkepunkt", „permanenter Welkepunkt"). Beim Welkekoeffizienten hört (nach RASUMOWA 1950) die Gewichtszunahme der Pflanzen auf; der „produktive Wassergehalt" eines Bodens muß also oberhalb des Welkekoeffizienten liegen.

BOTELHO DA COSTA (1938) stellte fest, daß der Welkepunkt bei allen Böden, die er untersuchte, bei einem pF-Wert von 4,2 (entsprechend 16 atm) liegt. RUSSELL (1939) fand für den „Wilting Point" pF = 4,18 (zugleich für den „Hygroscopic Coefficient pF = 4,54[1] und für das „Moisture Equivalent" pF = 2,70).

Auf das „tote Wasser" des Bodens (VAGELER 1932) wurde bereits S. 665 eingegangen. Es kann im Gefäßversuch auf physiologischem Wege auch indirekt ermittelt werden (v. BOGUSLAWSKI 1941). Der um das tote Bodenwasser verminderte Wassergehalt des Bodens, das „physiologisch wirksame Bodenwasser", entspricht etwa dem kapillargebundenen Bodenwasser. Häufig kommt auch ein Teil davon nicht mehr voll zur Ertragswirkung. Dieser Teil wird zusammen mit dem hygroskopischen Bodenwasser als „ungenutztes Bodenwasser" bezeichnet (SEKERA 1931, 1937, 1938).

9. Klassifikation des Wassers im Hinblick auf seine Beweglichkeit

Neben der Klassifizierung des Bodenwassers nach seinen Bindungsfestigkeiten, seinen Wirkungen auf die Saugkraft des Bodens und nach seiner Verteilung im Boden sowie nach seiner Verfügbarkeit für die Pflanze spielt auch seine Klassifikation im Hinblick auf seine Beweglichkeit eine große Rolle. RODE (1959) unterscheidet nach diesem Gesichtspunkt folgende Wasserformen im Boden:

I. Kristallwasser
II. Festes Wasser
III. Dampfförmiges Wasser
IV. Fest gebundenes Wasser
V. Labil gebundenes Wasser
VI. Freies Wasser
   1. Fugenwasser
   2. Kapillar hängendes Wasser
   3. Sorptionsmäßig eingeschlossenes Wasser
   4. Versickerndes Gravitationswasser
   5. Gestütztes Gravitationswasser
      a) kapillar gestütztes
      b) abfließendes
      c) stehendes
      d) kapillar aufsteigendes

Es gibt im Vergleich zu den Klassifikationen des Bodenwassers bei statischer Betrachtung (im unbewegten Zustand) eine graphische Darstellung für die Klassifikation im bewegten bzw. im beweglichen Zustand (Abb. 121, rechte Hälfte der Darstellung).

Hierbei unterscheidet er:

1. Unbewegliches Wasser (Sorptionskräfte bis zum permanenten Welkepunkt).

2. Schwer bewegliches Wasser (zwischen Welkepunkt und Feuchtigkeit des kapillaren Abrisses).

[1] Bestimmt über etwa 3,3%iger H_2SO_4.

3. Mittel bewegliches Wasser (zwischen Feuchtigkeit des kapillaren Abrisses und minimaler Wasserkapazität).

4. Leicht bewegliches Wasser (zwischen minimaler und voller Wasserkapazität).

Etwas mehr mit den Bindungsformen des Wassers im Boden befaßt sich die

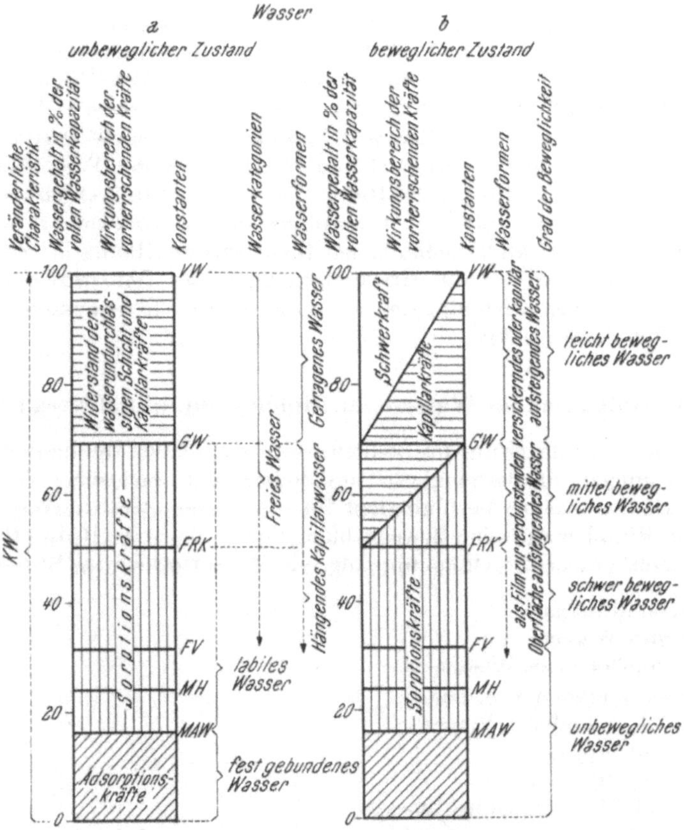

Abb. 121. Schema der Bodenfeuchtigkeit in feinkörnigen Böden nach Rode (*FV—WF*). *VW* Volle Wasserkapazität, *KW* kapillare Wasserkapazität, *GW* minimale Wasserkapazität, *FRK* Feuchtigkeit des kapillaren Abrisses, *FV* Feuchtigkeit des permanenten Welkepunktes, *MH* maximale Hygroskopizität, *MAW* maximale Sorptionskapazität

von Lebedev (1927) vorgeschlagene Einteilung des Bodenwassers nach seiner Beweglichkeit, die auch den Aggregatzustand des zur Bewegung gelangenden Wassers berücksichtigt:

1. Wasserdampf: in den Poren in Dampfform vorliegendes Wasser, das in dieser Form durch Dampfdruckgradienten im Boden verlagert werden kann;

2. Hygroskopisches Wasser: seine Bewegung ist in Dampfform von feuchteren zu trockeneren Stellen des Bodens möglich;

3. Filmwasser in flüssiger Phase: es kann von dickeren Filmen zu dünneren Stellen der Filme transportiert werden;

4. Gravitationswasser: es bewegt sich unter dem Einfluß der Schwerkraft in flüssigem Aggregatzustand innerhalb des Systems der Bodenporen.

Literatur

ADAM, N. K.: The physics and chemistry of surfaces. London 1938 (Moskau 1947).

BEUTELSPACHER, H.: Verhandl. IV. Komm. Int. Bodenkundl. Ges., Stockholm, 5. bis 8. 7. 1939, S. 50. Stockholm 1939. — BOTELHO DA COSTA, J. W.: The indirect determination of the wilting coefficient by the freezing point method. J. Agric. Sci. 28 (1938). — A critical survey of investigations on the wilting coefficient of soils. J. Agric. Res. 28 (1938). — BOGUSLAWSKI, E. v.: Untersuchungen über den Einfluß des Feinerdegehaltes, der Pflanzenbestandsdichte und der Pflanzenart auf die Aufnahme und Verwertung des Bodenwassers. Z. Bodenkde. u. Pflanzenernähr. **24** (69), 265–303 (1941). — BRADLEY, W. F., R. E. GRIMM und G. L. CLARK: A study of the behaviour of Montmorillonite upon wetting. Z. Krist. A **97**, 216–222 (1937). — BRIGGS, L. J., und H. L. SHANTZ: The wilting coefficient for different plants. U.S. Dept. Agric. Bur. Pl. Ind. Bull. 230. Washington 1912. — Relative water requirement of plants. J. Agric. Res. **3** (1914).

DERJAGIN, B. W.: Die mechanischen Eigenschaften dünner Flüssigkeitsschichten. J. Phys. Chem., Moskau-Leningrad, **5** (1934). — DIMO, W. N.: Zur Frage der Benetzungswärme von Böden, Mineralien und Gesteinen und ihre Beziehung zu den Oberflächen. Počvovedenie, H. 5 (1946). — DOLGOW, S. L.: Untersuchungen über die Beweglichkeit des Bodenwassers und seine Zugänglichkeit für die Pflanze. Moskau-Leningrad 1948. — Über die Kräfte, die das Wasser im Boden zurückhalten. Ber. d. Universität Moskau 1948, Ausg. 2. — Über die Kräfte, die das Wasser im Boden zurückhalten. Ber. d. Universität Moskau 1948. — DUMANSKI, A. W.: Benetzungswärme und Hydrophilie von Dispersionssystemen. Kolloid-J., Moskau, **12**, Nr. 5 (1950).

FISHER, R. A.: The Haines formula improving. J. Agric. Sci. **20** (1930). — FRIZMAN, E. CH.: Die Natur des Wassers. Moskau 1935. — FURR, S., und K. REEVE: Range of soil moisture percentage through which plants undergo permanent wilting. J. Agric. Res. **71** (1945).

GARDNER, W.: The capillary transmission constant and methods determinating it experimentally. Soil Sci. **10**, 103–126 (1920). — The capillary potential and its relation to soil moisture constants. Soil Sci. **10**, 357–359 (1920). — A new soil. Soil Sci. **9**, 191–195 (1920). — GRUNER, E., und R. F. VOGEL: Das Quellungsverhalten von Kalzium- und Natrium-Bentonit und seine Deutung. Kolloid-Z. **118**, 100–110 (1950).

HAINES, W. B.: Studies in the physical properties of soil, V, The hysteresis effect in capillar properties and the modes of moisture distribution associated therewith. J. Agric. Sci. **20**, 97–116 (1930). — HENDRICKS, S. B., und M. E. JEFFERSON: The structure of kaolin and talc-pyrophyllite hydrates and their bearing on water sorption. Amer. Min., N. Y., **23** (1938). — HÖLZEL, ST.: Die chemische Reaktion. In: LINSER, H. (Herausg.), Grundlagen der allgemeinen Vitalchemie, II, S. 126. Wien: Urban & Schwarzenberg. 1956.

KAPP, L. C.: The approximate size of soil particles at which the heat of wetting is manifested. Soil Sci. **29**, 401–412 (1930). — KOLJASEV, F. E.: Über die Wasserbewegung im Boden. Ber. d. Universität Moskau, **3** (1941). — KORNEW, W. G.: Die aufsaugende Kraft des Bodens. J. Exper. Agron., Leningrad, **22** (1924). — KOSSOWITSCH, P. S.: Hydrologie der Böden. J. Exper. Agron., St. Petersburg, **5** (1904). — KURON, H.: Versuche zur Feststellung der Gesamtoberfläche an Erdböden, Tonen und verwandten Stoffen. Z. Pflanzenernähr., Düng., Bodenkde. A **18**, 179–203 (1930). — Versuche zur Feststellung der Gesamtoberfläche an Erdböden, Tonen und verwandten Stoffen, III, Der Einfluß des Salzgehaltes auf die Wasserbindung an Böden und Tonen. Z. Pflanzenernähr., Düng., Bodenkde. A **21**, 271–300 (1931). — Versuche zur Feststellung der Gesamtoberfläche an Erdböden, Tonen und verwandten Stoffen, IV, Die Wasserbindung in den einzelnen Horizonten verschiedener Bodenprofile. Z. Pflanzenernähr., Düng., Bodenkde. A **24**, 257–277 (1932). — Versuche zur Feststellung der Gesamtoberfläche an Erdböden, Tonen und verwandten Stoffen, V, Einige Bemerkungen zur Hysteresis der Wasseradsorption. Z. Pflanzenernähr., Düng., Bodenkde. A **25**, 179–187 (1932a). — Versuche zur Feststellung der Gesamtoberfläche an Erdböden, Tonen und verwandten Stoffen, VI, Der Einfluß von Natriumkarbonatzusätzen auf die Wasserbindung von Wasserstoffton. Z. Pflanzenernähr., Düng., Bodenkde. A **33**, 298–308 (1934). — Versuche zur Feststellung der Gesamtoberfläche an Erdböden, Tonen und verwandten Stoffen, VII, Die Wasserbindung an einem mit nicht dissoziierenden Kationen gesättigten Ton. Z. Pflanzenernähr., Düng., Bodenkde., A **36**, 282–287 (1934). — Die Abgrenzung der Formen des vom Boden gebundenen Wassers durch den Einfluß auf die Umsetzung zwischen Boden und Salzen. Z. Pflanzenernähr., Düng., Bodenkde. A **45**, 352–363 (1936).

LEBEDEV, A. F.: Boden- und Grundwasser, 4. Aufl. Moskau-Leningrad 1936.

Mitscherlich, E. A.: Untersuchungen über die physikalischen Bodeneigenschaften. Landw. Jahrb. **31** B (1901).

Olmstead, L. B.: Some moisture relations of the soils from the erosion experiment stations. U.S. Dept. Agr. Techn. Bull. **562** (1937). — Ostwald, W., und K. Mündler: Über Osmose und Quellung disperser Systeme. Kolloid-Z. **24**, 7–27 (1919).

Rasumowa, L. A.: Die wichtigsten Ergebnisse der agrohydrologischen Untersuchungen des hydro-meteorologischen Dienstes. Ausg. 18 (45). Leningrad 1950. — Rode, A. A.: Das Wasser im Boden, 464 S. Berlin: Akademie-Verlag. 1959. — Rodewald, H., und E. A. Mitscherlich: Die Bestimmung der Hygroskopizität. Landw. Vers. Stat. **59**, 433–441 (1904). — Russell, M. B.: Soil moisture sorption curves for four Iowa soils. Soil Sci. Soc. Amer. Proc. **4**, 51–54 (1939).

Scharrer, K.: Agrikulturchemie I, Pflanzenernährung. Sammlung Göschen, Bd. 329, 143 S. Berlin: W. de Gruyter. 1953. — Schofield, R. K.: The pF of the water in Soil. Trans. 3rd Int. Congr. Soil Sci. II (1935). — Sekera, F.: Die nutzbare Wasserkapazität und die Wasserbeweglichkeit im Boden. Z. Pflanzenernähr., Düng., Bodenkde. **22**, 87–111 (1931). — Die Strukturanalyse des Bodens. Statik und Dynamik des Bodenwassers. Z. Bodenkde. u. Pflanzenernähr. **6**, 259–311 (1938). — Forschungsdienst, Sonderheft **7**, 17–27 (1937). — Speranski, K. W., und M. N. Kraschenininkow: Das hygroskopische Wasser im Boden, Bd. 8. Sankt Petersburg 1907.

Thomas, M. D.: Aqueous vapor pressure of soils. Soil Sci. **11**, 409–434 (1921); **17**, 1–18 (1924); **25**, 409–418 (1928); **25**, 419–427 (1928). — Trofimow, A.: Über das Filmwasser im Boden. Bd. IV, Moskau 1927.

Vageler, P.: Der Kationen- und Wasserhaushalt des Mineralbodens, S. 260ff. Berlin: Springer. 1932. — Veihmeyer, F. J., und A. H. Hendrickson: Soil moisture at permanent wilting of plants. Plant Physiol. **3** (1928). — Versluys, J.: Die Kapillarität der Böden. Int. Mitt. Bodenkde. **7** B (1917).

Zunker, H.: Das Verhalten des Bodens zum Wasser. In: Blanck (Herausg.), Handbuch der Bodenlehre, Bd. 6. Berlin: Springer. 1930.

b) Die Bewegungen von Wasser im Boden

Von

H. Linser[1]

1. Allgemeines

Die Beweglichkeit des Wassers in einem Stoff mit kontinuierlichem Porenraum — als solchen dürfen wir den Boden bei eben noch zulässiger Vereinfachung der Verhältnisse betrachten (Childs 1964) — hängt ab von dem Aggregatzustand, in welchem das Wasser sich befindet, oder aber von den Bindungszuständen, in welchem Wasser an bestimmte Strukturteile gebunden vorliegt. Sie hängt ferner ab von jenen Faktoren, welche diese Zustände beeinflussen. Schließlich hängt sie von jenen Eigenschaften des Stoffes ab, die den Bewegungen von Wasser in verschiedenen Formen Widerstand entgegensetzen, vor allem von den Querschnittsflächen der Poren bzw. von der Permeabilität des Materials für die verschiedenen beweglichen Formen des Wassers.

Wie aus dem vorhergehenden Abschnitt hervorgeht, liegt Wasser im Boden teils mehr oder minder fest an feste Bodenteilchen gebunden vor, teils aber „frei“, das heißt in einem der drei Aggregatzustände: gasförmig, flüssig oder fest. Über den Anteil jedes dieser Aggregatzustände am Gesamtwasser des Bodens entscheidet vor allem die Temperatur.

Um Bodenwasser durch Gefrieren in festen Aggregatzustand zu überführen, sind um so tiefere Bodentemperaturen erforderlich, je weniger Wasser im Boden vorhanden ist bzw. je größer sein strukturgebundener Anteil bzw. je höher der

---

[1] Der Autor dankt Herrn Dr. H. U. Preusse, Gießen, für seine Mithilfe bei der Abfassung dieses Beitrages.

pF-Wert des Bodens liegt. Das hygroskopische Wasser wird selbst bei Temperaturen von —78° C noch nicht in Eisform bzw. in festen Zustand überführt, sondern bleibt in seinen lokalisierten Bindungsorten. Je niedriger die Bindungsenergien sind, desto leichter gefriert das Wasser im Boden. Bei Bodentemperaturen von etwa —4 bis 0° C kann nur das Gravitationswasser in festem Zustand vorliegen. Zwischen dem Eis und der umgebenden Atmosphäre findet ein Austausch von Wassermolekülen statt, wobei sich (in einem geschlossenen System) ein stationärer Zustand, ein bestimmter Sättigungsdruck einstellt (Dampfdruck). Ebenso wie flüssiges Wasser steht auch Eis unter bestimmtem Dampfdruck und kann Wassermoleküle unmittelbar in die Gasphase überführen (verdampfen). Auch in gefrorenem Boden liegt daher, ebenso wie neben den flüssigen Wasseranteilen normal temperierter Böden, Wasser in der Gasphase (Wasserdampf) vor. Der Dampfdruck einer Flüssigkeit steigt mit der Temperatur an. Er beträgt für Wasser:

bei	—60°	—40°	—20°	0°	+20°	+40°	+60°	+80°C
cm Hg	0,0007	0,0093	0,077	0,46	1,75	5,53	14,94	35,51

Da eine Bewegung von Wasser in fester Form im Boden wegen der Kleinheit der Bodenporen keinerlei praktische Bedeutung besitzen kann, ist ein Transport von Wasser aus der gefrorenen Form nur über die Gasphase möglich. Die auf diesem Wege pro Zeiteinheit transportierbare Menge steigt mit steigender Temperatur bzw. mit dem steigenden Dampfdruck an.

Die flüssigen Anteile des im Boden befindlichen Wassers können sich als solche innerhalb des Porensystems des Bodens bewegen und gehorchen dabei weitgehend den hydrodynamischen Gesetzlichkeiten. Aber auch aus den flüssigen Anteilen treten Wassermoleküle in die Dampfphase über. Die Transportformen des Wassers im Boden sind somit die Gasphase und die flüssige Phase, wobei diese wechselseitig ineinander überführt werden können. Mit steigenden Temperaturen gewinnt die Gasphase, mit sinkenden die flüssige Phase an Bedeutung. Der Anteil der flüssigen Form des Wassers am Gesamtwassergehalt des Bodens bestimmt die relative Bedeutung des gasförmigen Wassertransports für die Gesamtwasserbewegung im Boden: je trockener ein Boden ist, desto bedeutender ist der Anteil des Wassertransports über die Gasphase am Gesamtwassertransport des Bodens.

Während die Viskosität von Gasen mit steigenden Temperaturen zunimmt, sinkt die Viskosität von Flüssigkeiten mit steigenden Temperaturen ab. Die Viskosität flüssigen Wassers beträgt:

bei	0° C	10° C	20° C	30° C	40° C	50° C	60° C
Centipoise	1,792	1,308	1,005	0,801	0,656	0,549	0,469

Steigende Temperaturen erleichtern also den Durchfluß von Wasser durch enge Öffnungen, wie sie im Porensystem des Bodens gegeben sind.

Die Kräfte, die Wasser im Boden zur Bewegung bringen, sind verschiedenartige:

1. Die Schwerkraft;

2. Saugdruckgradienten zwischen trockeneren und feuchteren Bodenpartien (osmotische Kräfte; Kapillarkräfte, die entlang von Kapillargradienten, infolge von Evaporation oder andersartigem Wasserentzug entstehend, auftreten);

3. Die Molekularbewegung des Wasserdampfs, welche für die Diffusion und das Nachdiffundieren von Wasserdampf an jene Orte verantwortlich ist, an denen Wasserdampf entzogen (gebunden oder kondensiert) wurde;

4. Der mechanische Anstoß bzw. Druck durch die Bodenluft oder das Bodenwasser bzw. durch das Grundwasser oder überstehendes Niederschlags- bzw. Inundationswasser;

5. Die Evaporation an der Bodenoberfläche in Verbindung mit der Kohäsion kapillarer Wassersäulen.

Diese Kräfte wirken sich im Boden aus unter dem Einfluß der durch die strukturellen Faktoren des Bodens, aber auch durch seinen Wasserzustand selbst bedingten Randbedingungen, so daß ein stark verflochtenes Wechselspiel der einzelnen Kräfte die Wasserbewegung im Boden als ein recht komplexes Phänomen erscheinen läßt.

Auf einen trockenen Boden von oben her aufgebrachtes (Niederschlags-, Inundations-, Bewässerungs- oder Beregnungs-) Wasser wird zunächst besonders durch den Saugdruckgradienten zwischen der befeuchteten Bodenoberfläche und den noch trockenen bzw. noch trockeneren Bodenschichten unterhalb des eindringenden Wassers in den Boden gesaugt, wobei daneben zugleich Wasser auf Grund der Gravitation einsickert. Der Saugdruckgradient hat hier an der Sickerbewegung entscheidend großen Anteil (VEIHMEYER und EDLEFSEN 1937). Er wird unterstützt vom hydrostatischen Druck der eventuell überstehenden Wassermenge. Je mehr Wasser in den Boden eindringt, desto größer wird der Anteil der Gravitation an der Wasserbewegung im Boden. Ist auf Grund der Gravitationswirkungen das Gravitationswasser abgeflossen, so bleibt das gegenüber den Gravitationskräften gebundene Wasser zurück. Während das Gravitationswasser definitionsgemäß der Schwerkraft folgend, nach unten hin absickert, kann sich das „Haftwasser“ auf Grund von Saugkraftgradienten nach allen Richtungen, je nach dem, wo und in welchen Richtungen solche Gradienten bestehen oder auftreten, bewegen. Diese Bewegung kann auch nach oben gerichtet sein und

Tabelle 96. *Mittelpodsolierter Waldboden auf leichtem schluffigem Decklehm, in 85 cm Tiefe von sandigem Lehm unterlagert, nach Wassiljew (1950)*

Horizont	Tiefe der entnommenen Probe in cm	Spezifisches Gewicht	Volumengewicht in g/cm³	In Vol.-%			
				MH[1]	*WF*[1]	*GW*[1]	*P(VW)*[1]
A_0	0— 4	1,87	0,09	3,1	5,8	17	95,2
A_1'	4— 9	2,53	0,68	8,5	16,0	34	73,1
A_1''	9— 16	2,62	1,06	6,1	11,5	33	59,6
A_2	16— 25	2,67	1,34	5,3	8,2	30	49,8
A_2B	25— 35	2,68	1,44	4,4	6,8	30	45,3
B	35— 45	2,70	1,54	6,8	10,5	31	43,0
BC_1	45— 55	2,74	1,57	6,1	9,5	31	42,7
	55— 65	2,74	1,61	5,6	8,7	30	41,3
C_1	65— 75	2,74	1,63	5,2	8,1	28	40,5
C_2	85— 95	2,75	1,67	7,0	10,8	25	39,3
	100—110	2,74	1,73	8,3	12,9	26	36,8
	120—130	2,70	1,72	8,5	13,2	25	36,3
	140—150	2,70	1,73	10,5	16,3	24	36,0
C_3	200—210	2,75	1,76	13,4			36,0
	240—250	2,75	1,79	14,9			35,0

[1] Siehe S. 670.

somit einen Aufstieg in Richtung zur Bodenoberfläche hin bewirken. Eine nach oben gerichtete Wasserbewegung kann auch durch die Saugkraft der Evaporation von Wasser an der Bodenoberfläche bewirkt werden, wobei kapillare Wassersäulen auf Grund ihrer Kohäsion zur Bodenoberfläche hin bewegt werden können. Die Bewegungen des Wassers können somit im Boden in allen Richtungen verlaufen, wobei die jeweiligen Verhältnisse innerhalb des Bodens darüber entscheiden, welche Transportrichtung jeweils gerade bevorzugt ist und überwiegt.

Für die Bewegungen des Wassers sind jene Eigenschaften des Bodens ausschlaggebend, die wir als seine „Wasserführung“ kennzeichnend zusammen-

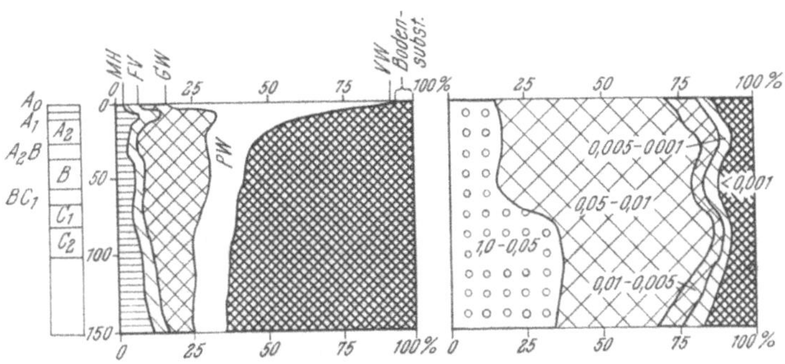

Abb. 122. Wasserführung, Struktur und Textur eines mittelpodsolierten Waldbodens auf leichtem Decklehm nach WASSILJEW (1950)

MH Maximale Hygroskopizität
FV Feuchtigkeit des permanenten Welkepunktes
GW Minimale Wasserkapazität
} Wasservolumen
VW Volle Wasserkapazität
PW Luftvolumen

fassen. Diese kann ihrerseits durch die „hydrologischen Konstanten“ charakterisiert werden, die für ein bestimmtes Bodenprofil in der Weise dargestellt werden können, wie dies in Tab. 96 für das konkrete Beispiel eines Waldbodens gezeigt wird.

Eine graphische Darstellung der Wasserführung im unmittelbaren Vergleich mit Struktur und Textur des Bodens in der Art, wie dies am gleichen Beispiel die Abb. 122 darstellt, ermöglicht eine eindrucksvolle und übersichtliche Charakterisierung des Bodenprofils.

2. Die Bewegung von Wasser in flüssiger Form

Die Bewegungen von Wasser in flüssiger Form, beispielsweise ω einer Mengeneinheit von Punkt A zu Punkt B, erfolgt auf Grund des Vorliegens einer Potentialdifferenz, welche die beim Transport zu leistende Arbeit angibt. Gilt für den Punkt A das hydraulische Potential φ_A und für den Punkt B das hydraulische Potential φ_B, so ermöglicht die Potentialdifferenz

$$U_{A \to B} = \varphi_B - \varphi_A$$

die Bewegung der Wassermenge von A nach B, wenn zwischen beiden Punkten nicht besondere Widerstände die Bewegung beeinträchtigen. Das Wasser steht unter dem Einfluß von dreierlei Komponenten, welche die Potentiale bestimmen:

1. unter dem Einfluß der Gravitation,

2. unter dem Druck der überstehenden, auf der betrachteten Wassermenge lastenden Wassersäule, und

3. unter dem Einfluß etwa seitlich angreifenden hydrostatischen Drucks (z. B. Auswirkungen von Sorptionskräften und osmotischen Kräften).

Betrachten wir zunächst nur die unter 1. und 2. genannten, stets wirksamen Komponenten, so beträgt das Potential

$$\varphi = Z + p$$

wobei Z den Gravitationsanteil (die Höhe des Punktes mit dem Potential über einen gegebenen Wert) und p den hydrostatischen Druck (die Höhe der Wassersäule in einem Manometer, das an den Punkt angesetzt wurde) bedeutet.

Das vertikale Einsickern von Wasser in Sand wurde schon von DARCY (1856) eingehend untersucht. Dem von ihm abgeleiteten und nach ihm benannten Gesetz entsprechend gilt für die Strömungsgeschwindigkeit v:

$$v = -K \cdot \operatorname{grad} \varphi$$

wobei der Gradient des Potentials die Potentialzunahme pro Abstandseinheit in der Richtung des steilsten Potentialanstiegs ist und K die hydraulische Leitfähigkeit des Materials bedeuten.

Der Boden ist ein anisotropes Material, in welchem die Richtung der Wasserbewegung nicht immer mit jener des Potentialgradienten übereinstimmen kann; deshalb wird hier K zu einem Tensor, doch kann ein Fall von anisotroper Leitfähigkeit mathematisch auch in einen aequivalenten Fall isotroper Leitfähigkeit umgewandelt werden, wobei K einen skalaren Faktor darstellt.

Nach GARNER (1920) wird das die Wasserbewegung bewirkende Potential φ wie folgt formuliert:

$$\varphi = -\int_A^B f \cdot \cos\vartheta \cdot ds$$

wobei f die Feldkraft, die pro Gewichtseinheit des Wassers wirksam ist, bedeutet, A den Punkt, auf den sich das Potential bezieht (gewöhnlich die freie Wasseroberfläche angenommen), B den in Betracht gezogenen Punkt, ds ein elementares Wegstück und ϑ den Winkel zwischen der Feldkraft und dem Wegstück.

Die in der Zeiteinheit versickernde Wassermenge Q ist nach DARCY (1856)

$$Q = K \frac{h}{l} \cdot \omega$$

wobei K den Filtrationskoeffizient, h das Druckgefälle, ω den Querschnitt und l die Länge der Filtersäule bedeuten.

Wenn auch hierbei der Filtrationskoeffizient K die Dimension einer Geschwindigkeit (Weg:Zeit)[1] hat, so gibt er doch nur eine scheinbare, nicht die wahre Sickergeschwindigkeit an, da noch durch die Größe des wirksamen Porenvolumens (LEBEDEV 1936, aktive Porosität $= P_0$ nach KATSCHINSKI 1945) dividiert werden muß, um die echte Sickergeschwindigkeit zu erhalten: scheinbare Sickergeschwindigkeit v, echte Sickergeschwindigkeit u.

$$v = \frac{Q}{\omega} = K \frac{h}{l} \qquad u = \frac{V}{P_0}$$

---

[1] $K = \frac{Q \cdot l}{h \cdot w}$; $Q = \frac{\text{Länge}^3}{\text{Zeit}}$; l und H sind Längen; $w = \text{Länge}^2$;

$$K = \frac{\text{Länge}^3 \cdot \text{Länge}}{\text{Zeit} \cdot \text{Länge} \cdot \text{Länge}^2} = \frac{\text{Länge}}{\text{Zeit}}.$$

Der Filtrationskoeffizient ist, wie man sieht, bei einem Druckgefälle $\frac{h}{l} = 1$, also bei einem Druck von 1 cm Wassersäule und bei einer Höhe der Filterschicht von 1 cm gleich der scheinbaren Sickergeschwindigkeit.

Das Hohlraumvolumen eines Bodens ist auch bei einem bestimmten Boden keine konstante Größe, sondern abhängig von dessen Strukturzustand. Das „gesamte Hohlraumvolumen" (P) beträgt bei Böden mit guter Krümelstruktur bis zu etwa 60%, bei verdichtetem Boden dagegen nur 35 bis 40% (APSITS 1937). Es errechnet sich für 100 cm³ gewachsenen Boden aus dem Volumgewicht (VG) des bei 105° C getrockneten und dem spezifischen Gewicht (S) dieses Bodens nach der Formel:

$$P = 100 - \frac{VG}{S} \cdot 100$$

Das Sickerwasser bewegt sich nicht im gesamten, sondern nur im aktiven oder sogenannten „spannungsfreien Porenvolumen" (P_0) (ZUNKER 1928). Es wird ermittelt nach der Formel:

$$P_0 = P - \frac{\mathrm{Hy}}{100} \cdot (1 - P) \cdot S$$

wobei vorausgesetzt wird, daß bei der Benetzung des Bodens durch Wasser die gleiche Wassermenge hygroskopisch gebunden wird wie über 10prozentiger H_2SO_4 (Hy).

Ist neben Wasser noch Luft sorbiert, was vor allem bei Feuchtegehalten der Bodenluft <100% der Fall ist, errechnet sich das spannungsfreie Porenvolumen (P_{01}) wie folgt:

$$P_{01} = P - \frac{\mathrm{Hy} + \mathrm{lg}}{100} \cdot (1 - P) \cdot S$$

(lg = cm³ Luft je 100 g trockenen Boden).

Bei gleichen P ist das spannungsfreie Hohlraumvolumen P_0 um so kleiner, je schwerer ein Boden ist; deshalb ist die Beweglichkeit des Bodenwassers am höchsten bei Sandböden und am niedrigsten bei Tonböden (vgl. die Zahlen der Tab. 97).

Tabelle 97. *Das spannungsfreie Hohlraumvolumen P_0 (in %) im Vergleich zum gesamten Hohlraumvolumen P (in %) und in Abhängigkeit von der Bodenart* (nach SCHEFFER-SCHACHTSCHABEL 1952)

Bodenart	Spannungsfreies Hohlraumvolumen bei		
	$P = 50\%$	$P = 45\%$	$P = 40\%$
Sandboden	40	34	28
Sandiger Lehmboden	30	23	16
Tonboden	10	1	0

Nach VAGELER läßt sich P_0 angenähert ermitteln, indem man das 4,5fache der Hygroskopizität von P subtrahiert.

Wasser fließt im Boden, indem es eine bestimmte Zahl von Kanälen ausfüllt; diese Zahl nimmt mit abnehmendem Feuchtigkeitsgehalt des Bodens ab, ebenso nimmt die Größe der ausfüllbaren Kanäle mit zunehmendem Feuchtigkeitsverlust ab. Die hydraulische Leitfähigkeit steigt daher mit einem Ansteigen des Feuchtigkeitsgehaltes des Bodens nach einer Kurve an, die in Abb. 123 dargestellt ist.

Die Unterteilung des Porenraums in viele feine oder wenige weite Kanäle ist von ausschlaggebendem Einfluß auf die hydraulische Leitfähigkeit eines Bodens, weshalb diese sehr stark von der Bodenstruktur bzw. von den sie beeinflussenden Faktoren abhängig ist[1]. Textur und Struktur der einzelnen Bodenhorizonte sind daher von großem Einfluß auf die Filtrationskoeffizienten der verschiedenen Schichten eines Bodenprofils. Einige extreme Zahlen von Materialien mit extremen Eigenschaften sind in Tab. 98 wiedergegeben.

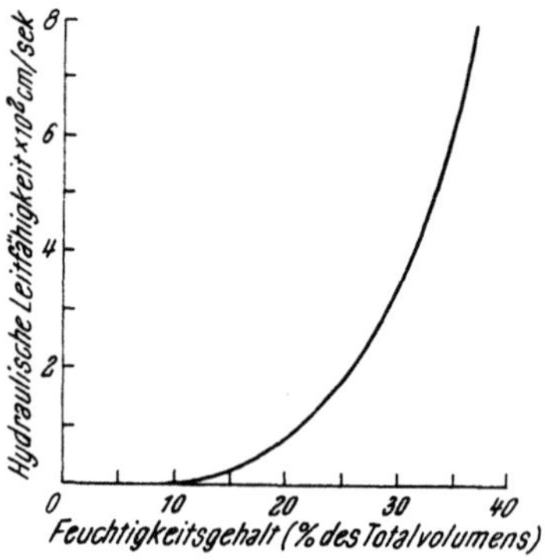

Abb. 123. Die Abhängigkeit der hydraulischen Leitfähigkeit vom Feuchtigkeitsgehalt (nach Chils 1964)

Die Wasserdurchlässigkeit geschichteter Materialien ist somit wesentlich geringer, als die Wasserdurchlässigkeit der ungeschichteten, homogenen Materialien allein. Die Zahl der Schichten macht sich insofern bemerkbar, als ihre Vermehrung die Durchlässigkeit herabsetzt, ja mitunter gleich Null werden lassen kann. An dem Phänomen scheinen ursächlich beteiligt zu sein: 1. Das Vorliegen von Kapillaren mit sehr ungleich großen Durchmessern bzw. ungleichartigen Formen; 2. eine Erhöhung der Viskosität des flüssigen Wassers durch hohe Luftgehalte; 3. eine Vermehrung von Zwischenschichten aus eingeschlossener Luft bzw. die vermehrte Bildung von Grenzflächen (Wasser-Luft) mit Erhöhung des Kapillardruckes gegenüber dem fließenden Wasser. In nativen Bodenprofilen liegen Bodenschichten vor, und es ist daher zu erwarten, daß die Wasserdurchlässigkeit in ihnen durch die geschilderten Wirkungen herabgesetzt ist. Ergebnisse von Dolgopolowa (1948) und Wassiljew (1950) zeigen, daß die maximale Wasserdurchlässigkeit bei Bodenprofilen zwar in der obersten Schicht liegen kann, daß sie mitunter aber auch erst im unteren Teil des „Humushorizontes" oder im „Karbonat-Illuvial-Horizont" erreicht ist. In allen Fällen aber nahm der Filtrationskoeffizient in tieferen Bodenschichten stark ab (Abb. 124).

Tabelle 98. *Filtrationskoeffizienten* (K) *von je 30 cm starken Schichten aus Sand und aus Torf*
(nach Lebedev 1936)

Material	K in cm/sec
Sand allein	$79000 \cdot 10^{-7}$
Torf allein	$1400 \cdot 10^{-7}$
Sand und Torf geschichtet	$48 \cdot 10^{-7}$

Auf die Bodenoberfläche gebrachtes, flüssiges Wasser kann in den Boden durch Risse und größere Spalten durch „Influktuation" eindringen oder aber bei homogenen Bodenoberflächen durch „Infiltration" (vgl. Wyssozki 1933). Im letzteren

[1] Nach De Boodt und De Leenheer (1955) werden Poren mit einem Durchmesser von $> 50\ \mu$ als schnelldränend, solche mit einem Durchmesser von 50 bis $10\ \mu$ als langsam dränend bezeichnet.

Falle kann das Wasser die Poren vollständig ausfüllen und sich sozusagen als kontinuierliche Wassersäule bewegen, wobei auch „kapillar hängendes" Wasser mit einbezogen werden kann, das schon durch 30 cm Wassersäule verdrängt bzw. bewegt werden kann; es kann aber auch keine vollständige Erfüllung der Poren mit Wasser vorliegen, weshalb dann die Bodenpartikelchen als von Wasserfilmen

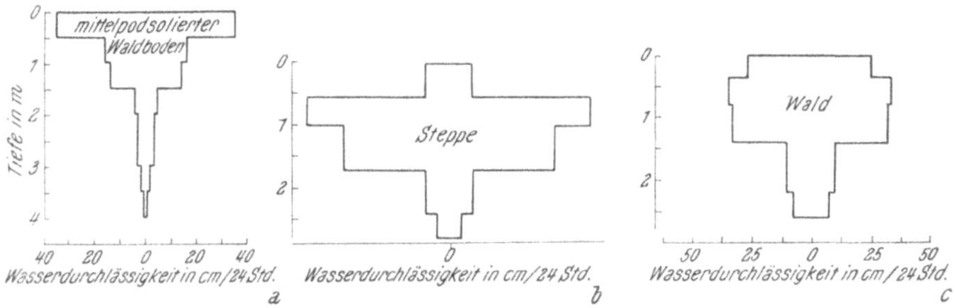

Abb. 124. Wasserdurchlässigkeit von Bodenprofilen

a) mittelpodsolierter Waldboden nach WASSILJEW (1950);
b) tiefgründig lehmiger Tschernosjomboden unter Steppe nach DOLGOPOLOWA (1948), umgerechnet nach RODE (1959);
c) gleicher Boden wie *b*) unter Wald

umgeben zu denken sind, die als solche ebenfalls durch Gravitationskräfte bewegt werden können und die deshalb als „Gravitationsfilmwasser" bezeichnet wurden (PROSKURNIKOW 1948); da „Filmwasser" jedoch in anderem Zusammenhang gebundenes Wasser bedeutet, schlug RODE (1959) vor, von „umfließendem" Gravitationswasser im Gegensatz zu „filtrierendem" Gravitationswasser zu sprechen. In Abhängigkeit von der Ausfüllung der Poren mit Wasser und in Abhängigkeit von der Korngröße des Bodenmaterials liegen bei Sand die Infiltrationsgeschwindigkeiten bei 10^{-3} bis 10^{-2} cm/sec. Sie steigen mit ansteigendem Wassergehalt im Sinne des in Abb. 125 gezeigten Kurvenverlaufes an.

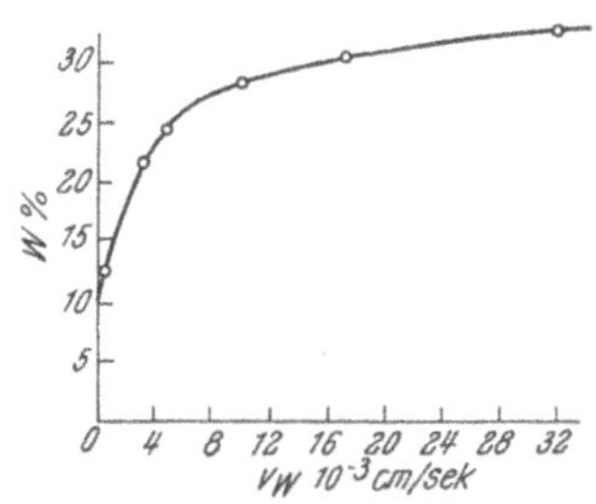

Abb. 125. Die Abhängigkeit der Infiltrationsgeschwindigkeit (Vw) vom Wassergehalt (W) einer Sandfraktion (0,5 bis 0,25 mm) (nach PROSKURNIKOW 1948)

Die Bewegung von Wasser im Bodenprofil nach abwärts kann in den meisten Fällen mit Hilfe der Kontinuitätsgleichung erfaßt werden, die sich von der Überlegung ableitet, daß der absolute Überschuß an abfließendem Wasser aus einem Bodenbereich dessen Feuchtigkeitsverlust gleich ist. Nach DARCYS Gesetz lassen sich die Strömungsgeschwindigkeiten als Potentialgradienten ausdrücken, so daß folgende Beziehung zwischen Feuchtigkeitsgehalt und Potentialverteilung gilt:

$$\mathrm{d}c/\mathrm{d}t = \mathrm{div}\ (K \cdot \mathrm{grad}\ \varphi)$$

Sie beschreibt das Strömen im allgemeinen. Betrachtet man die Veränderung des Feuchtigkeitsprofils mit der Zeit (t) und berücksichtigt man nur die Richtung im Sinne des Potentials bzw. Z (vgl. S 676), so erhält die Kontinuitätsgleichung die Form

$$\mathrm{d}c/\mathrm{d}t = \mathrm{d}\,[K\,(1 + \mathrm{d}p/\mathrm{d}z)]/\mathrm{d}z$$

Wandert die Feuchtigkeit am Profil mit stetiger Geschwindigkeit auf- und abwärts (vom oder zum Grundwasserspiegel), und eliminiert man die Zeit als unabhängige Veränderliche, so geht die Kontinuitätsgleichung in eine integrierbare Differentialgleichung mit zwei Veränderlichen über, vorausgesetzt, daß K als Funktion von Feuchtigkeitsgehalt und Sog bekannt ist. Es wird so ein Feuchtigkeitsprofil erhalten, das die Verteilung des Feuchtigkeitsgehaltes oder des Sogs oberhalb des Grundwasserspiegels darstellt (Childs 1964).

Auch die zeitliche Entwicklung des Feuchtigkeitsprofils nach Überschwemmung der Bodenoberfläche mit Wasser, das nun in einen völlig trockenen Boden eindringt, kann mit Hilfe der Kontinuitätsgleichung verfolgt werden, indem man das Darcysche Gesetz durch das (feuchtigkeitsunabhängige) Diffusionsvermögen D ausdrückt; sie lautet dann:

$$\mathrm{d}c/\mathrm{d}t = \mathrm{d}\,[K + D\,(\mathrm{d}c/\mathrm{d}z)]/\mathrm{d}z$$

Da der Gravitationsanteil K keiner Wärmegröße entspricht, lassen sich übliche Lösungsverfahren für diese Gleichung nicht anwenden. Nach Irmay (1956) wird auch hier die Zeit als Veränderliche eliminiert, so daß die Lösung ein Feuchtigkeitsgehaltsprofil berechenbarer Gestalt bietet, das mit berechenbarer Geschwindigkeit fortschreitet. Youngs (1957) beobachtete eine befriedigende Übereinstimmung mit experimentellen Befunden. Einen Teil seiner Ergebnisse zeigt Abb. 126.

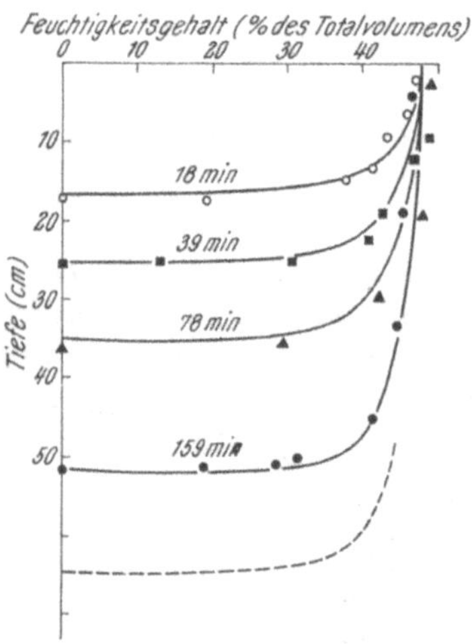

Abb. 126. Fortschreiten der Infiltration aus einer überfluteten Oberfläche nach Youngs (1957). Die Linien sind berechnet, die Einzelpunkte experimentell bestimmt

Da derartige Berechnungen bisher nur für homogene Materialien und Laboratoriumsbedingungen gelten, können sie zwar der Aufklärung der physikalischen Vorgänge der Wasserbewegung im einzelnen dienen, nicht aber für Freilanduntersuchungen herangezogen werden. Hier kann eher ein älteres und einfacheres Modell von Nutzen sein, das Green und Ampt (1912) einführten. Sie betrachteten die Entwicklung eines Feuchtigkeitsprofils als Ergebnis des Fortschreitens einer Wasserfront, die den völlig gesättigten Boden scharf von dem noch nicht wassergesättigten Boden trennt, wobei an dieser Linie ein charakteristischer Sog existieren soll. Danach besteht in allen Stadien, auch bei geschichteten Böden, eine bekannte Potentialdifferenz an den Enden einer Bodensäule bekannten Strömungswiderstandes, so daß sich die Strömungsrate (= Infiltrationsrate) nach dem Darcyschen Gesetz direkt berechnen läßt.

Das Studium der Vorgänge, die mit dem Eindringen von Wasser in den Boden verbunden sind (Perkolationsvorgänge), ist neuerdings durch die Möglichkeit des Einsatzes von Tritium-Wasser um einen methodischen Weg bereichert worden (Scharpenseel und Gewehr 1960). Auch auf die Feuchtigkeitsmessung im Boden mit Hilfe der Neutronen- und der Gamma-Strahlung sei hier kurz hingewiesen (Delcher und Mitarbeiter; vgl. Linser und Kaindl 1960).

Die Verteilung von Wasser in einem Bodenprofil im Verlauf der Zeit läßt sich in einer Weise graphisch darstellen, für die Abb. 127 ein Beispiel gibt. Die Wassergehalte in Gewichtsprozenten sind dabei als Zahlen angegeben, welche Linien („Chronoisoplethen") des Verlaufs dieser Wassergehalte in Abhängigkeit von der Jahreszeit markieren. Es ergibt sich ein anschauliches Bild der jahreszeitlichen

Entwicklung des Feuchtigkeitsprofils eines Bodens, das noch durch die Angabe der Niederschlagsmengen erweitert ist.

Die vom Boden nicht festgehaltenen und in tiefere Bodenschichten abfließenden Wassermengen werden als „Sickerwasser" bezeichnet. Sie können in sogenannten „Lysimeter"-Anlagen gemessen werden, indem man unterhalb der zu prüfen-

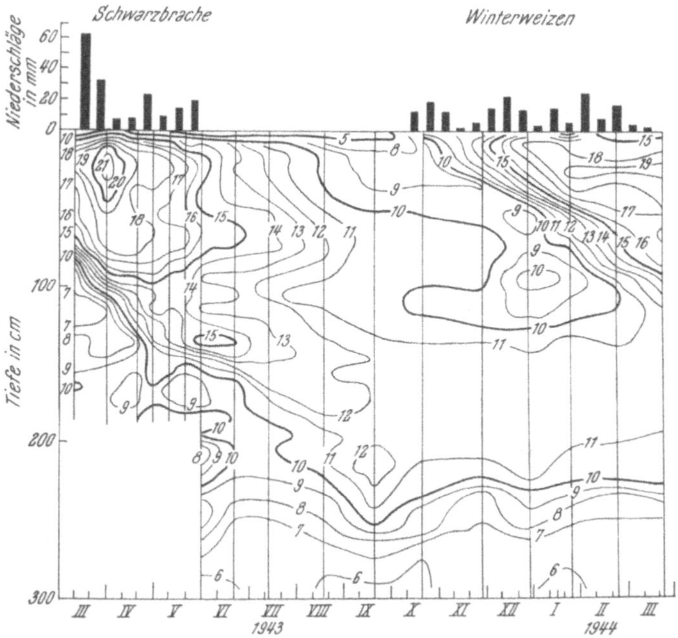

Abb. 127. Wassergehalt einer nicht bewässerten Grauerde unter Schwarzbrache und Winterweizen nach BOLSCHAKOW (1950). Die Zahlen geben den Wassergehalt in Gew.-% an

den Bodenschichten Auffangflächen aus undurchlässigem Material anbringt und das an ihnen ankommende Sickerwasser trichterartig ableitet und sammelt. Wenngleich hierdurch der Kontakt der Bodenprobe mit dem darunterliegenden Boden und dadurch ein eventueller aufsteigender Wasserstrom abgeschnitten wird und unnatürliche Verhältnisse geschaffen werden, geben die gewonnenen Zahlen doch ein Maß für die ohne diesen Kontakt auftretenden Sickerwassermengen. Nach Schätzungen von ENGLER (1921) verteilt sich das Niederschlagswasser, das auf einen Boden auffällt, unter mitteleuropäischen Verhältnissen grob wie folgt (Tab. 99):

Tabelle 99. *Verteilung des Niederschlagswassers bei der Verdunstung*

	im Freiland	im Wald
Verdunstung auf den Bodenoberflächen	24%	5%
Verdunstung auf den Pflanzenoberflächen	10%	15%
Verdunstung durch pflanzliche Transpiration	6%	20%
Gesamtverdunstung	40%	
abfließendes Sickerwasser	60%	
Niederschlagsmenge	100%	

Selbstverständlich sind diese Zahlen stark beeinflußbar und abhängig von Wasserführung, Struktur und Textur des Bodens wie auch von der Art des Pflanzenbestandes.

MITSCHERLICH (1950) nennt nach Ergebnissen von BÜCHLER folgende Zahlen für die Sickerwassermengen in Prozent der Niederschlagsmengen in Abhängigkeit von Bodenart und Pflanzenbestand:

	kahle Fläche	Buchenbestand	Fichtenbestand	Grasfläche
Sandböden	85	74	77	55
Tonböden	77	29	29	18

Wenn einmal eine bestimmte Niederschlagsmenge auf den Boden gelangt ist, die dessen Bedarf an gebundenem Wasser abgesättigt und daher kein Sickerwasser zur Folge gehabt hat (x_0), dann werden nach LUEDECKE (1906, 1912, 1914) Sickerwassermengen (y) gefunden, die den Niederschlagsmengen proportional sind:

$$y = (x - x_0) \cdot b$$

Er fand dabei, daß der Proportionalitätsfaktor b im Winter größer ist als im Sommer und im Durchschnitt des ganzen Jahres Werte im Bereich von etwa 0,36 bis etwa 0,83 annimmt.

3. Wasserbewegung in Kapillargradienten

Wenn man eine Kapillare vom Radius r in eine Flüssigkeit taucht, die die innere Wandung der Kapillare bereits benetzt hat, so steigt sie auf Grund der Kapillarkräfte innerhalb der Kapillare höher an, als es dem äußeren Flüssigkeitsniveau entspricht. Die zu ihrem Anstieg um eine Höhe dh erforderliche Arbeit ist gleich jener, die beim Anheben eines Flüssigkeitsvolumens $\pi \cdot r^2 \cdot dh$ von der Masse $dm = \varrho \cdot \pi \cdot r^2 \cdot dh$ um die Höhe h geleistet würde, wobei ϱ die Dichte der Flüssigkeit bedeutet. Sie beträgt $dm \cdot gh = \varrho \cdot \pi \cdot r^2 \cdot dh \cdot gh$. Zugleich verschwindet ein Stück benetzter Flüssigkeitshaut mit der freien Oberfläche $2\pi \cdot r \cdot dh$, so daß die Oberflächenenergie $2\,\vartheta \cdot \pi \cdot r \cdot dh$ frei wird (wobei ϑ die Oberflächenspannung bzw. die Kapillaritätskonstante ist). Zunächst ist diese freie Energie größer als die zum Anheben erforderliche Arbeit, die Flüssigkeit steigt somit in der Kapillare weiter an. Bei einer bestimmten Steighöhe wird die zum Anheben erforderliche Arbeit größer bzw. gleich groß wie die durch ihren Effekt freiwerdende Oberflächenenergie. Damit ist die maximale Steighöhe nach der Bedingung $\varrho \cdot \pi \cdot r^2 \cdot gh \cdot dh = 2\pi \cdot r \cdot \vartheta \cdot dh$ gegeben als

$$h = \frac{2\vartheta}{\varrho \cdot r \cdot g}$$

Bei nicht vollständiger Benetzung und einem Randwinkel α gilt allgemein

$$h = \frac{2\vartheta}{\varrho \cdot r \cdot g} \cdot \cos\alpha$$

Für Wasser ist $\vartheta = 74$ dyn/cm; bei $\varrho = 1$ und $g = 981$ cm/sec^2 erhält man für die Steighöhe h

$$h = \frac{0{,}15}{r} \text{ oder } h = \frac{0{,}3}{d}$$

Die Aufstiegshöhe von Wasser in Kapillaren ist somit umgekehrt proportional dem Radius (oder Durchmesser d) der Kapillaren; es wird demnach in Kapillarsystemen um so höher aufsteigen, je feiner die Kapillarsysteme sind.

Aus den geschilderten Berechnungen geht hervor, daß bei waagrecht liegenden Kapillaren durch Ansetzen einer noch engeren Kapillare deren Inhalt in die engere Kapillare gesaugt werden kann; das bedeutet, daß in kapillaren Systemen Wasser in Richtung eines Gradienten, der die Veränderung des Radius eines Kapillarsystems darstellt, in Richtung gegen die kleineren Radien transportiert wird. Sie haben dabei den Widerstand zu überwinden, der durch zwischenmolekulare Kräfte (Wasserbindung) der Wasserbewegung entgegengesetzt wird. Diese entgegengesetzt wirkenden Kräfte ersetzen in der obigen Berechnung die Gravitation. Auch hierbei kommt es zur Einstellung eines Gleichgewichtszustandes. Der Aufstieg entgegen der Wirkung der Gravitation ist dabei besonders wichtig.

Die Geschwindigkeit des Wasseraufstieges steht ebenfalls im umgekehrten Verhältnis zum Durchmesser der Kapillaren. Da im Boden die Querschnitte und die Formen der Poren des Kapillarsystems sehr unterschiedlich sind, ist es nicht möglich, die für sie gültige theoretische Berechnungsweise anzuwenden.

Trotzdem wurde von GÖTTSCH (1922) für die Zeitfunktion des Ansteigens von Wasser in senkrecht stehenden Kapillaren folgender Ausdruck angegeben:

$$t = \frac{8 \cdot \eta}{r^2 \cdot g} \left(2{,}302 \cdot H \cdot \log \frac{H}{H-h} - h\right)$$

wobei η die innere Reibung des Wassers, bei $+15°$ C$=0{,}0115$, r der Radius der Kapillare, H die Steighöhe, h die zur Zeit gegebene Steighöhe und g die Schwerebeschleunigung ist. Da danach die innere Reibung des Wassers beim Anheben mit dem Quadrat des Radius des Kapillarrohres abnimmt, wird der Wasser-

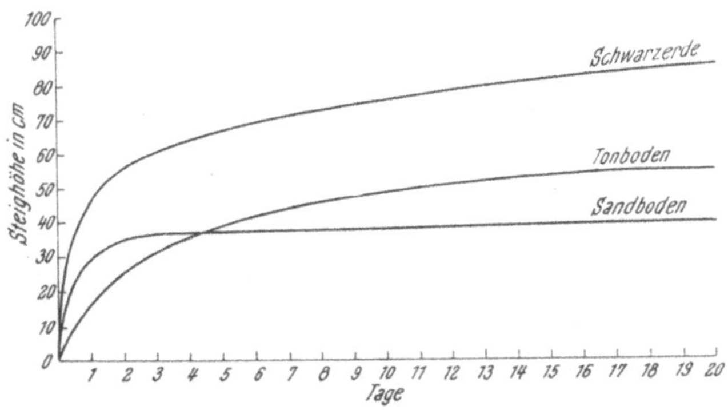

Abb. 128. Der kapillare Aufstieg bei drei Bodenarten in Abhängigkeit von der Zeit (nach SCHEFFER und SCHACHTSCHABEL 1960)

anstieg zugleich beschleunigt. Jedoch nimmt die innere Reibung des Wassers mit der Höhe h des bereits erreichten Anstieges zu, während das Ansteigen selbst nur von der Differenz H—h gefördert wird; bei sehr engen Kapillarröhren kann nun die innere Reibung so groß werden, daß die Zeit, die für den Wasseranstieg benötigt wird, sehr groß, ja unendlich groß werden und daher ein Wasseranstieg oder eine kapillare Wasserleitung überhaupt nicht mehr erfolgen kann (MITSCHERLICH 1950).

Für den Modellfall idealer Böden hat KEEN (1927) folgende theoretische Hubhöhen berechnet:

Feiner Kies	$^1/_2$ Fuß	Schluff	31,25 Fuß
Grobsand	$1^1/_2$ Fuß	Feinschluff bis	150 Fuß
Feinsand	$7^1/_2$ Fuß	Ton	150 Fuß

Der kapillare Aufstieg verläuft in der Zeit in Form einer Sättigungskurve, die erst nach mehreren Tagen ihren Maximalwert bzw. den Gleichgewichtszustand der End-Steighöhe erreicht.

Abb. 128 stellt an drei verschiedenen Bodenarten Steigverläufe einander gegenüber. Hohe Tongehalte verlangsamen die Steiggeschwindigkeit (MITSCHERLICH 1923).

Aufstiegsgeschwindigkeit und erreichbare Aufstiegshöhe stehen in keiner unmittelbaren Relation zueinander. Bei Sand z. B. ist die Aufstiegsgeschwindigkeit relativ hoch, die erreichte Aufstiegshöhe jedoch geringer als bei Tonboden, bei dem die Aufstiegsgeschwindigkeit relativ gering ist.

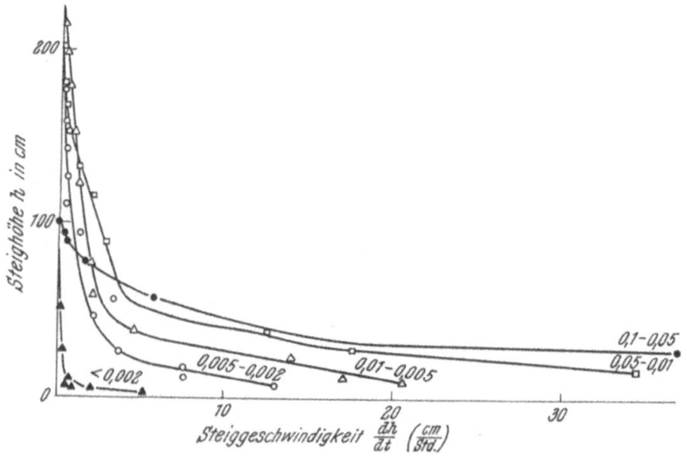

Abb. 129. Steiggeschwindigkeit in verschiedenen Korngrößen in Abhängigkeit von der Steighöhe (nach NOVÁK und PECHÁNEK)

NOVÁK und PECHÁNEK (1943) fanden bei Quarzsand verschiedener Korngröße die aus Abb. 129 entnehmbaren Steiggeschwindigkeiten, woraus sich für das Erreichen bestimmter Steighöhen bei den einzelnen Korngrößen die in Tab. 100 genannten Zahlen ergeben.

Tabelle 100. *Steiggeschwindigkeit in Säulen aus Quarzsand verschiedener Korngröße, erhalten durch Interpolation der Kurven von Novák und Pechánek (1943)*

Fraktionen mm	Steiggeschwindigkeit in cm/Std. auf eine Höhe von				
	20 cm	30 cm	40 cm	50 cm	60 cm
0,1 —0,05	31	17	14	9	0,1
0,05 —0,01	26	16	12	5	2,8
0,01 —0,005	12,5	6,5	4	2,5	2,2
0,005—0,002	4,6	3	2,5	2,0	1,7
<0,002	0,4	0,2	0,2	0,1	0,08

Nach DOLGOW und RYSHOW (1933) gilt für den kapillaren Aufstieg von Wasser in gefüllten Rohren

$$vh = K\,(H + h)$$

wobei v die Steiggeschwindigkeit, h die Steighöhe, H die maximale Steighöhe und K eine Konstante ist. Hieraus kann die maximale Steighöhe ermittelt werden.

In Bodensäulen verteilt sich Wasser zunächst nur langsam fortschreitend vom Ort der Zufuhr aus in die nichtbefeuchteten Teile des Bodens, so daß es zu einer differenten Verteilung des Wassers kommt. Abb. 130 zeigt die unterschiedliche Verteilung des Wassers nach Zufuhr von oben bzw. von unten her.

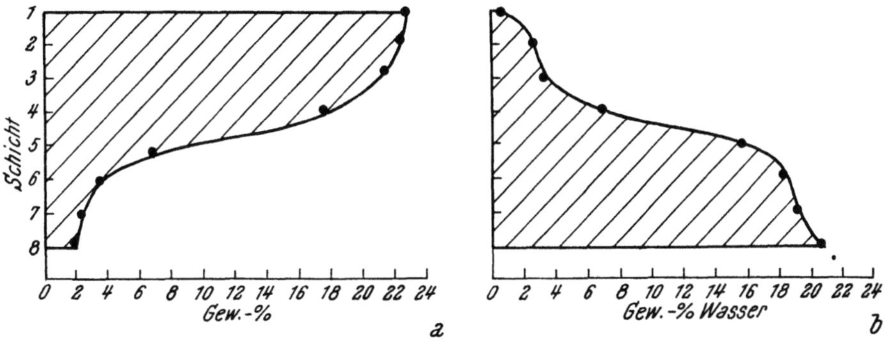

Abb. 130. Wasserverteilung in 36 cm hohen Löß-Monolithen
a) nach 12tägiger Wassersättigung durch Zufuhr des Wassers von oben her (auf Schicht 1); *b*) nach 7tägiger Kapillarsättigung mit Wasser nach Zufuhr von Wasser von unten her (durch Schicht 8). Nach Ergebnissen von ISMAILSKI (1894)

Von oben zugeführtes Wasser sickerte nicht völlig nach unten ab, sondern blieb in Kapillarräumen der oberen Bodenschichten zum größten Teil hängen („hängendes Wasser"). Ebensowenig stieg von unten zugeführtes Wasser nicht bis zur gleichmäßigen Verteilung nach oben, sondern nur zu einem kleinen Teil bis zur obersten Schicht auf; es blieb zum größten Teil in den basalen Schichten kapillar hängen.

4. Die Mechanik der Wasserbewegung in Dampfform

Wasser geht, wie alle flüssigen Stoffe, bei den gewöhnlichen Temperaturen unserer Umwelt allmählich in den gasförmigen Zustand über, d. h. es verdampft. Dies geschieht um so schneller, je höher die Temperatur ist. In einem geschlossenen Raum, der zum Teil mit Wasser gefüllt ist, verdampft Wasser in den freien Raum, bis sich ein stationärer Zustand einstellt, der dadurch zustande kommt, daß nun gleichviele Moleküle Wasser pro Zeiteinheit aus der flüssigen in die Gasphase übertreten, als umgekehrt aus der Gasphase in die flüssige Phase zurückkehren, d. h. kondensieren. Diese Kondensation erfolgt durch molekulare Anziehungskräfte, die mit der Oberflächenspannung zusammenhängen; sie sind einseitig gerichtet und trachten die an der Oberfläche befindlichen Moleküle in das Innere der Flüssigkeit hineinzuziehen. Diese Oberflächenspannungskräfte sind an ebenen Flüssigkeitsoberflächen größer als an gekrümmten, weshalb die Kondensation an ebenen Wasseroberflächen leichter vor sich geht als an den gekrümmten Oberflächen kleiner Wassertröpfchen. Die Kondensation wird gefördert, wenn zu den zwischenmolekularen Wechselwirkungen der Wassermoleküle untereinander noch weitere Anziehungskräfte, z. B. elektrostatische Ladungen, hinzukommen. So fungieren feste, elektrische Ladungen tragende Partikelchen in Wechselwirkung mit Wasserdampf als Kondensationskerne, sie fördern damit die Kondensation von Wasserdampf.

Verdampfung und Kondensation können an verschiedenen Stellen eines Bodens, entsprechend deren jeweiligen Eigenschaften sowie bedingt durch ihren Temperaturzustand, in völlig verschiedenen Verhältnissen zueinander stehen. Von Stellen, an denen die Verdampfung größer ist als die Kondensation, wird somit Wasser in Dampfform zu jenen Stellen transportiert, an welchen die Kondensation überwiegt. Die pro Zeiteinheit transportierte Menge an Wasser ist wiederum der Potentialdifferenz zwischen den Punkten A und B proportional; sie ist um so größer, je größer die Dampfdruckdifferenz zwischen beiden Punkten ist. In Schluff- und Feinsandfraktionen fanden JONES und KOHNKE (1952), daß eine Verdoppelung des Dampfdruckes auch eine Verdoppelung der Dampfbewegung zur Folge hatte. Für die Bewegung von Wasserdampf in Böden wurde von PENMAN (1940) eine Diffusionsgleichung aufgestellt. GURR, MARSHALL und HUTTON (1952) fanden, daß der Wassergehalt eines Bodens für die Geschwindigkeit des Wassertransportes in Dampfform bestimmend ist, der seinerseits das für den Dampf zugängliche Porenvolumen beeinflußt. In Lehmböden fanden sie die Dampfbewegung bei Wassergehalten von 5,3 bis 7,9% maximal, sie war sowohl in feuchteren als auch in trockeneren Böden kleiner. Die Permeabilität des Bodens für die dampfförmige Phase steigt, jene für die flüssige Phase fällt mit sinkendem Wassergehalt des Bodens. Die Dampfbewegung kann von einer entgegengesetzten Flüssigkeitsbewegung im Boden begleitet werden, weshalb die gesamte Wasserbewegung im Bodenraum von den unterschiedlichen Bodenpermeabilitätswerten für die flüssige und für die dampfförmige Phase abhängig ist. Die maximale Dampfbewegung wird in verschiedenen Böden bei verschiedenen Bodenwasserspannungen erreicht. JONES und KOHNKE (1952) stellten fest, daß mit abnehmender Teilchengröße (vom Sand bis Schluff) die Wasserspannung, bei der Bewegung in Dampfform begann, ebenso wie jene, bei der die Dampfbewegung maximal war, anstieg. Die Porosität des Bodens bestimmt die Dampfbewegung nicht so sehr durch die Porengröße als vielmehr durch das Volumen der jeweils ungesättigten Bodenporen. Die Dampfbewegung ist im allgemeinen maximal unterhalb des permanenten Welkepunktes; dies war bei einem Schlufflehm z. B. bei pF = 3,70, bei einem Lehm bei pF 3,85 der Fall.

Wenn an warmen, sonnigen Tagen die oberen Bodenschichten unter ihre Hygroskopizität austrocknen, sorbieren sie nachts, wenn die relative Luftfeuchtigkeit infolge der Abkühlung ansteigt, Wasserdampf aus der Atmosphäre. Dies geschieht auch dann, wenn die Bodenoberfläche wärmer ist als die bodennahe Luftschicht (TRENEL und LINDNER 1960). Die Sorption erfolgt auf Grund der Saugkraft des Bodens und ist der relativen Luftfeuchtigkeit direkt, dem Wassergehalt des Bodens umgekehrt proportional. Die so erzielten Sorptionsgewinne des Bodens an Wasser machen im Frühjahr täglich etwa 0,023 mm, im Sommer 0,013 bis 0,033 mm und im Herbst 0,055 mm je Nacht aus.

Entsprechend dem Temperatur- und Dampfdruckgefälle bewegt sich Wasser nachts aus tieferen Schichten zur Bodenoberfläche empor, wobei 0,13 bis 0,20 mm je Nacht transportiert werden (TRENEL und LINDNER 1960). Nach Beginn der Sonneneinstrahlung steigt der Dampfdruck in den Oberflächenschichten des Bodens schnell an, so daß sich das Dampfdruckgefälle nun umkehrt, wobei 55 bis 80% des abdestillierenden Wassers an die Atmosphäre und etwa 20 bis 45% in die tieferen Bodenschichten abgeführt werden. Dieser in den Untergrund strömende Wasserdampf kondensiert zunächst in den kühleren Schichten und diffundiert bei fortschreitender Erwärmung in immer tiefere Bodenschichten. Nach TERNEL und LINDNER (1960) bewegen sich so an Schönwettertagen im Frühjahr 0,18 bis 0,35 mm, im Sommer 0,28 bis 0,70 mm und im Herbst 0,07 bis 0,13 mm Wasser aus den oberen 5 bis 10 cm des Bodens in tiefere Bodenschichten. Tropische Klima- und Bodenbedingungen führen zu stärkeren Wasserdampfbewegungen im Bodenprofil, die etwa 2,3 bis 6,7 mm je Tag ausmachen, wobei über den aus der Atmosphäre stammenden Anteil an Wasser dabei keine Angaben vorliegen (PAGEL 1960).

5. Wasserbewegung entlang Temperaturgradienten

Nach den Untersuchungen von BOUYOUKOS (1915) und SMITH (1943) spielt bei der Wasserbewegung von wärmeren zu kälteren Bodenpartien die dampfförmige Phase als Transportform nur eine untergeordnete Rolle. Bei einer Temperaturdifferenz von 20° C zwischen feuchtem, warmem und kaltem, trockenem Boden und Trennung beider Bodenproben durch einen dünnen Luftring (so daß keine Kapillarverbindung bestand und der Transport flüssigen Wassers ausgeschaltet war) war auf der kalten Seite nach 8 Stunden eine nur 0,035%ige Wasseranreicherung beobachtet worden.

Neben der Wasserbewegung in Dampfform tritt beim Anlegen eines Temperaturgradienten an den Boden auch eine Wasserbewegung in Kapillarräumen in Erscheinung. Bei Wassersättigung des Bodens tritt keine Wasserbewegung entlang von Temperaturgradienten auf, die Dampfbewegung allein aber genügt nicht, um die Wasserbewegungsraten verschieden feuchter Böden allein zu erklären (KUZMAK und SEREDA 1957). Es muß somit wenigstens ein Teil des Wassers in Kapillarräumen oder in dünnen Filmen fließen. BOUYOUKOS (1915) und SMITH (1943) wiesen darauf hin, daß bei einem Temperaturgefälle im Boden die Menisken einzelner Kapillaren unter verschiedenen Temperaturen stehen.

In einem ungesättigten Boden ist der hydrostatische Druck P, unter dem das flüssige Bodenwasser in Kapillaren steht (ein negativer Wert bezogen auf jenen einer freien Wasserfläche), nach EDLEFSEN und ANDERSON (1943):

$$P = \frac{2S}{r}$$

wobei S die Oberflächenspannung des Wassers und r den Radius der Wasseroberflächen in den Bodenporen bedeuten. Da S mit steigender Temperatur abfällt und von r angenommen werden darf, daß er einen konstanten negativen Wert hat, muß P mit steigender Temperatur ansteigen. In einem gleichmäßigen Boden von gleichmäßigem Wassergehalt muß daher ein Temperaturgradient einen Druckgradienten hervorrufen, der einen Transport von flüssigem Wasser in Richtung zur kälteren Stelle zur Folge hat (GURR, MARSHALL und HUTTON 1952), was auch experimentell gefunden wurde (BOUYOUKOS 1915, SMITH 1943, Mac LEAN und GWATKIN 1946).

Diese erste Phase des Wassertransportes entlang eines Temperaturgradienten wird jedoch von einer zweiten Phase gefolgt, die einen Rücktransport von Wasser in umgekehrter Richtung zur Folge hat. GURR und Mitarbeiter (1952) erklären dies wie folgt:

Wenn genügend Wasser transportiert worden ist, um das Druckgefälle (den Gradienten von P in der flüssigen Phase) zu beseitigen, so wird sich der Dampfdruck des Bodenwassers (p) in Relation zur absoluten Temperatur T wie folgt verhalten:

$$P = \varrho RT \log \frac{p}{P_0} \text{ oder } \log p = \log P_0 + \frac{P}{\varrho RT}$$

wobei R die Gaskonstante je Gramm, ϱ die Dichte des Wassers und p_0 der Dampfdruck einer freien Wasseroberfläche ist. (Bei Böden, deren Wassergehalt den permanenten Welkepunkt nicht unterschreitet, liegt p/p_0 zwischen 1 und 0,98.) Jeder Anstieg von T steigert p_0. Bei konstantem P (und vernachlässigenswert kleinen Änderungen von P) wird dabei $\frac{P}{\varrho RT}$ eine kleinere negative Größe. Infolge der oben erwähnten Beziehung muß p mit steigender Temperatur ansteigen, so daß der Wasserdampf sich im System noch immer bewegen muß, auch wenn die flüssige Phase dies nicht mehr tut. Bewegt sich nur der Wasserdampf im Sinne von Verdampfung und Kondensation so lange, bis ein Dampfdruckgleichgewicht erreicht ist, so wird in dem oben für P gegebenen Ausdruck p zu einer Konstanten. Da p_0 mit T ansteigt und $\log \frac{p}{p_0}$ ein negativer Wert ist, muß nunmehr P mit ansteigendem T kleiner werden. Es muß nunmehr daher Wasser von der kühleren zur wärmeren Seite des Gradienten fließen, also im umgekehrten Sinne wie während der ersten Phase des Gesamtvorganges. Ein Gleichgewichtszustand zwischen beiden Phasen kann aus thermodynamischen Gründen nicht erreicht werden. Praktisch stellt sich im genügend feuchten Boden ein Zirkulationssystem ein, in welchem Wasserdampf in Richtung sinkender Temperatur und flüssiges Wasser in umgekehrter Richtung transportiert wird.

6. Grundwasser und Bewegungen des Grundwasserspiegels

Als „Grundwasser" oder „Untergrundwasser" wird jenes Wasser bezeichnet, das in tieferen Bodenschichten in flüssiger Phase vorliegt und nicht weiter absickern kann, weil es daran gehindert wird, entweder durch das Auftreffen auf wasserundurchlässige Bodenschichten oder durch Kommunikation mit größeren Wasserreservoiren (Seen, Untergrundseen usw.). Für die Pflanzen hat vor allem jenes Grundwasser Bedeutung, das sich noch im Bereich ihres Wurzelsystems befindet. Schon Mitscherlich (1950) wies darauf hin, daß hinsichtlich der Bestimmung des Grundwasserstandes manche Unklarheit besteht und daß keineswegs eine Identität zwischen der Höhe des Wasserspiegels in einem Brunnen mit jenem Grundwasserstand gegeben ist, der für die Pflanze eine Rolle spielt. Der Wasserstand in einem Brunnen liegt gewöhnlich viel tiefer als die grundwasserstauenden wasserundurchlässigen Bodenschichten, die in verschiedenen Bodentiefen verschiedene Grundwasserniveaus („Wasseradern", die übereinander liegen können) herstellen. Solche wasserführende, auf Tonschichten ruhende Sandadern werden beim Brunnengraben durchschlagen, weil ihre Zuflußleistung zu klein ist, um den Wasserbedarf zu decken, der als Brunnenleistung verlangt wird. Das Grundwasser der durchschlagenen Schichten rinnt dabei in den Brunnen, d. h. in die tiefere Grundwasserschicht, welche durch den Wasserspiegel im Brunnen gekennzeichnet ist.

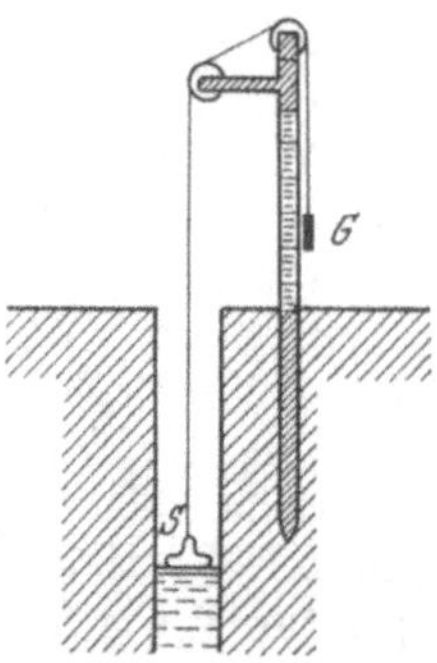

Abb. 131. Der Untergrund-Wasserstandsmesser von Pause nach Mitscherlich (1949)

Für die Pflanze spielt jedoch vor allem die oberste wasserführende Schicht („Wasserader") die Rolle des Grundwassers. Dieser Wasserstand (Abstand von der Bodenoberfläche) kann nach Pause (vgl. Ramann 1911) in einem bis zur Tiefe der ersten Grundwasser führenden Schicht vorgetriebenen Schacht mit Hilfe eines Schwimmers gemessen bzw. laufend kontrolliert werden (Abb. 131).

Dieser Grundwasserstand zeigt je nach den eindringenden Sickerwassermengen, d. h. also in Abhängigkeit von den Niederschlägen, bedeutende Schwankungen. Bei Messungen, die 1850 bis 1886 in München durchgeführt wurden, war der niedrigste Grundwasserstand bei Niederschlagshöhen von 29 bis 112 mm um 0 bis 268 mm erhöht, bei Messungen in Berlin 1870 bis 1885 bei Niederschlägen von 32 bis 66 mm sogar um 0 bis 580 mm (in einem Jahr sogar bis 798 mm) erhöht worden. Dies zeigt, daß Wasseradern offenbar nicht nur von direkt eingedrungenem Sickerwasser, sondern außerdem auch noch von seitlich zufließendem

Sickerwasser gespeist werden. Die Schwankungen des Grundwasserspiegels zeigen eine gewisse Abhängigkeit auch vom Wasserverbrauch der Vegetation bzw. von der Jahreszeit, wie beispielsweise aus den Darstellungen in Abb. 132 hervorgeht.

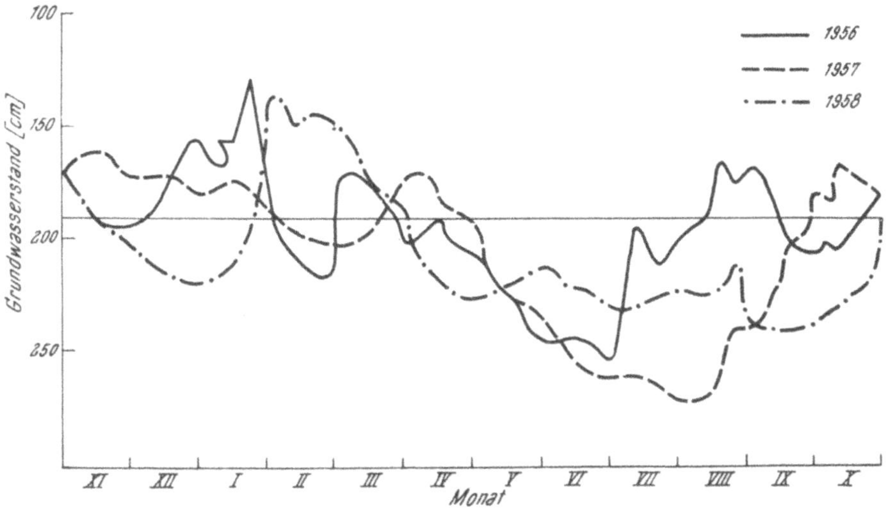

Abb. 132. Jahresgang des Grundwasserspiegels im „Wesuwer Moor" nach LAUENSTEIN (1960)

Der Jahresgang des Grundwasserspiegels kann in eine graphische Darstellung der Chronoisoplethen der Feuchtigkeit eines Bodenprofils mit eingezeichnet werden. Ein Beispiel hierfür gibt Abb. 133.

Der Grundwasserspiegel stellt natürlich nur in dem zu seiner Messung gegrabenen Schacht tatsächlich einen „Spiegel", d. h. eine ebene Wasserfläche dar, die sich scharf als Grenze zwischen flüssigem Wasser und wasserdampfgesättigter Luft darstellt. Innerhalb des Bodens geht die volle Wassersättigung des Bodens, die unterhalb des „Grundwasserspiegels" praktisch gegeben ist, durch das Phänomen des kapillaren Wasseraufstiegs allmählich zu niedrigeren Werten der Wassersättigung über. Der Grundwasserspiegel schiebt somit eine Zone des kapillaren Wasseraufstieges sozusagen vor sich her. MITSCHERLICH (1950) schätzt, daß der kapillare Anstieg des Bodenwassers praktisch etwa 1,5 m betragen dürfte. Bei einer Tiefe des Wurzelsystems der meisten Kulturpflanzen von 2,5 bis 2,7 m (SCHULZE 1906) wäre somit ein Grundwasserstand von 4 m unterhalb der Bodenoberfläche für die Pflanze günstig. MITSCHERLICH (1901) fand bei Erhebungen in verschiedenen Gegenden Deutschlands, daß Ackerböden bei trockenem Wetter Maximalerträge lieferten, wenn der Grundwasserstand höher als in 4 m Tiefe lag,

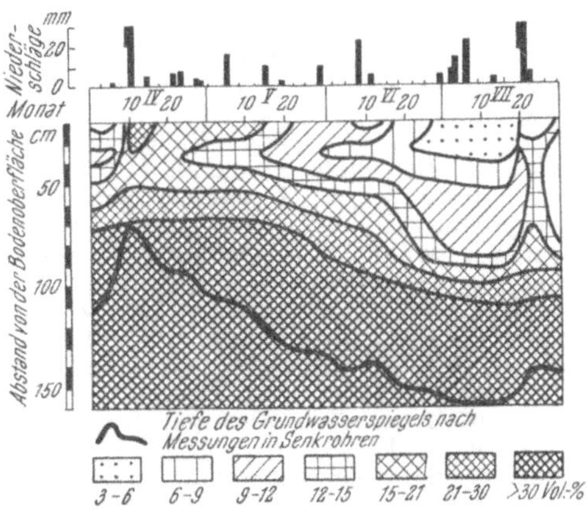

Abb. 133. Chronoisoplethen der Bodenfeuchtigkeit und Grundwasserspiegel eines Sandbodens nach EHWALD und Mitarbeitern (1961)

bei nasser Witterung aber dann, wenn er tiefer als in 4 m Tiefe lag. Daß man für diesen Grenzwert von 4 m aber nicht allgemeine Gültigkeit beanspruchen darf, zeigen Ergebnisse von Visser (1959), der fand, daß jener Grundwasserstand, bei dem optimaler Pflanzenertrag erzielt wird, stark von der Bodenart abhängt und daß er um so tiefer liegt, je schwerer der Mineralboden ist (vgl. Abb. 142, S. 711).

Die Wassermenge, die vom Grundwasser her nach oben in trockene Bodenschichten aufsteigt, nimmt zudem mit der Höhe über dem Grundwasserspiegel sehr schnell ab. v. Nitzsch (1933) fand an natürlich gelagertem Lehmboden einen täglichen Aufstieg von 35 mm (Niederschlagsäquivalent) in den Bereich 2 cm über dem Grundwasser, aber von nur 3 mm in den Bereich 8 cm über dem Grundwasserspiegel. Wenn bei Sandböden zwar der Aufstieg schneller verläuft, so wird dort doch eine geringere Maximalsteighöhe erreicht. In beiden Fällen müssen die Pflanzenwurzeln also relativ dicht an den Grundwasserspiegel heranwachsen, wenn ihre Wasserversorgung aus ihm sichergestellt werden soll.

Literatur

Apsits, J.: Bodenstruktur und Pflanzenwachstum. Z. Bodenkde. u. Pflanzenernähr. 3, 336 (1937).

Belcher, D. J., T. R. Cuykendall und H. S. Sack: The movement of soil moisture and density by neutron and γ-ray scattering. Techn. Developm. Rep. No. 127, Indianapolis, Indiana, 1950. — Bodman, G. B., und N. E. Edlefsen: The soil-moisture system. Soil Sci. 38, 425–444 (1934). — Bolschakow, A. F.: Die Wasserführung der unbewässerten Böden Usbekistans. Тр. Ин-та почв АН СССР Moskau-Leningrad 32 (1950). — Bouyoukos, G. B.: Effect of temperature on the movement of water vapor and capillary moisture in soils. J. Agric. Res. 5, 141–172 (1915).

Childs, E. C.: Die Bewegungen des Wassers im Boden. Endeavour 23, 81–84 (1964).

Darcy, H.: Les fontaines publiques de la ville de Dijon. Paris: Dalmont. 1856. — De Boodt, M., und L. De Leenheer: Studie van de Porienverdeling in de Bodem. Meded. Landbowhogeschoolen en Opzoekingsstations Gent 20, 98–130 (1955). — Dolgow, S. I., und S. N. Ryshow: Zur Frage der Bestimmung des maximalen kapillaren Wasseraufstiegs in Böden und Untergrund. Тр. ВИЧАА, Moskau 1933, Ausg. 2.

Edlefsen, N. E., und A. B. C. Anderson: Thermodynamics of soil moisture. Hilgardia 15, 31–298 (1943). — Ehwald, E., E. Vetterlein und F. Buchholz: Das Eindringen von Niederschlägen und Wasserbewegungen in sandigen Waldböden. Z. Pflanzenernähr., Düng., Bodenkde. 93 (138), 202–209 (1961). — Engler, A.: Int. Mitt. Bodenkde. 11, 12–25 (1921).

Gardner, W. R.: A capillary transmission constant and methods of determining it experimentally. Soil Sci. 10, 357 (1920). — Göttsch, H.: Dissertation Kiel 1922; nach Mitscherlich: Bodenkunde, 70, Berlin u. Hamburg: Parey. — Green, H., und G. A. Ampt: Studien über Bodenphysik, 2. Teil, Die Durchdringlichkeit eines Bodens für Luft und Wasser. J. Agric. Sci. 5, 1–26 (1912). — Gurr, G., T. J. Marshall und J. T. Hutton: Movement of water in soil due to a temperature gradient. Soil Sci. 74, 335–345 (1952).

Irmay, S.: Proc. Symp. Darcey Int. Assoc. Sci. Hydrol. Putl. Ser. 41, 57 (1956). — Ismailski, A. A.: Der Wassergehalt des Bodens und das Grundwasser. 1894.

Jones, H. E., und H. Kohnke: Einfluß der Bodenwasserspannung auf die Dampfbewegung des Bodenwassers. Proc. Soil Sci. Soc. Amer. 16, 245 (1952).

Katschinski, N. A.: Neues in der Theorie von wasserundurchlässigen Abdichtungen im Boden. Poćvovedenie (1945). — Keen, B. A.: System soil — soil moisture. Trans. Faraday Soc. 17, 228–243 (1922). — Kuzmak, J. M., und P. J. Sereda: Flow in liquid phase. Soil Sci. 84, 419–422 (1957).

Lauenstein, H. J.: Untersuchungen zur Boden- und Nährstoffdynamik der neuartigen „Deutschen Sandmischkulturen". Mitt. Arb. staatl. Moor-Vers. Stat. Bremen 8, 133–183 (1960). — Lebedev, A. F.: Boden- und Grundwasser, 4. Aufl. Moskau-Leningrad 1936. — Linser, H., und K. Kaindl: Isotope in der Landwirtschaft, 442 S. Berlin u. Hamburg: Parey. 1960. — Luedecke, C.: Kulturtechniker 9, 101 (1906); 15, 88 (1912); 17, 319 (1914).

McLEAN, D. J., und P. M. GWATKIN: Moisture movement occurring in soil due to the existence of a temperatur gradient. Road Res. Lab., D.S.I.R. England, Note RN/761, 1946. — MITSCHERLICH, E. A.: Bodenkunde, 4. Aufl. 1923, 5. Aufl. 1949, 6. Aufl. 1950, 371 S. Berlin u. Hamburg: Parey.

NITZSCH, W. V.: RKTL-Schriften, H. 41. Berlin: Parey. 1923.

PAGEL, H.: Über die Bewegung von dampfförmigem Wasser im Boden der wechselfeuchten Tropen. Albrecht Thaer Arch. **4**, 325–345 (1960). — PENMAN, H. L.: Gas and vapor movements in the soil I. J. Agric. Sci. **30**, 437–462 (1940). — PROSKURNIKOW, S. M.: Versuchsergebnisse über die Bewegung des Gravitations-Filmwassers in homogenen Sanden. Тр. ГГИ, Leningrad, 1948. — Experimentelle Untersuchungen zum Studium der kapillaren Wasserbewegung in Sanden. Тр. ГГИ, Leningrad, 1948.

RAMANN, E.: Bodenkunde, 3. Aufl., S. 364. Berlin: Springer. 1911.

SCHARPENSEEL, H. W., und H. GEWEHR: Studien zur Wasserbewegung im Boden mit Tritium-Wasser. Z. Pflanzenernähr., Düng., Bodenkde. **88** (133), 35–49 (1960). — SCHEFFER, F., und P. SCHACHTSCHABEL: Bodenkunde, 3. Aufl., 239 S. Stuttgart: F. Enke. 1952, 5. Aufl. 1960. — SCHULZE, B.: Studien über die Bewurzelung unserer Kulturpflanzen. Festschr. d. Landwirtsch. Vers. Stat. Breslau, 1906, S. 77ff. — SMITH, W. O.: Thermal transfer of moisture in soil. Trans. Amer. Geophys. Union **24**, 511–523 (1943).

TRENEL, M., und H. LINDNER: Über die Herkunft des im Boden kondensierten Wassers. Albrecht Thaer Arch. **4**, 175–194 (1960).

VEIHMEYER, F. J., und N. E. EDLEFSEN: Die Interpretation von Bodenwasserproblemen mittels Energiewechsel. Trans. Amer. Geophys. Union **18**, 302 (1937). — VISSER, W. C.: Crop growth and availability of moisture. J. Sci. Food Agric. **11**, 1–11 (1959) sowie Techn. Bull. (Wageningen) **6** (1959).

WASSILJEW, I. S.: Die Wasserführung des Podsols. Moskau-Leningrad 1950. — WYSSOZKI, G. N.: Die Wasserführung der Böden, Grundwasser und Salze in den Steppen und Waldsteppen. Ausg. 6, Leningrad 1933.

YOUNGS, E. G.: Moisture profiles during vertical infiltration. Soil Sci. **84**, 283–290 (1957).

ZUNKER, F.: Über das Schwinden und Quellen der Böden und ein neues Bodenuntersuchungsverfahren. Kulturtechniker **31**, 527 (1928). — Verhalten des Bodens zum Wasser. In: BLANCK, Handbuch der Bodenlehre, Bd. 6, S. 66–220. Berlin: Springer. 1930.

c) Grundwasser und Pflanzenertrag

Von

B. Wohlrab

Unter Grundwasser versteht man nach DIN 4049 (1954) „Wasser, das Hohlräume der Erdrinde zusammenhängend ausfüllt und nur mehr der Schwere (hydrostatischer Druck) unterliegt.“ Der nach Angaben von KLAPP (1958) und KOEHNE (1948) entworfene schematische Geländequerschnitt (Abb. 134) enthält einige wichtige Begriffe der Grundwasserkunde (Deutscher Grundwasserausschuß 1961, DIN 4049/1954).

Mit *Grundwassersohle* wird die Oberfläche eines undurchlässigen oder schwerdurchlässigen Gesteinskörpers bezeichnet. Sie ist die Basis des *Grundwasserleiters*, der im dargestellten Schema (Abb. 134) aus grobem Lockergestein besteht und hier die oberste Schicht der Erdrinde bildet. Der aus dem *Sickerwasseranteil* der Niederschläge gespeiste *Grundwasserstrom* fließt in eine Richtung, die vor allem von der Neigung der Grundwassersohle und der Lage der Hauptvorflut abhängt. Als *Grundwasseroberfläche* bezeichnet man die oberste Grenzfläche des ungespannten Grundwassers, an welcher der Wasserdruck dem Druck der freien Luft entspricht. In nur wenig in das Grundwasser eintauchenden Bohrlöchern stellt sich ein *Grundwasserspiegel* in Höhe dieser Oberfläche ein. Im Boden existiert die Oberfläche nicht wirklich, denn Kapillarkräfte verlegen die tatsächliche Grenze der Wassersättigung je nach den Porendurchmessern und Porenformen des Lockergesteins mehr oder weniger weit nach oben. Dieser Übergangsbereich wird *Saugraum* (DIN 4049/1954) genannt (frühere Bezeichnung: Kapillar- oder Saugsaum). Man unterscheidet einen geschlossenen und einen offenen

Saugraum (KOEHNE 1948). Im geschlossenen sind, abgesehen von kleinen Luftbläschen, sämtliche Poren mit Wasser gefüllt, im offenen nur die mit dem Grundwasser in Verbindung stehenden Kapillaren, während die gröberen Poren Luft und/oder Sickerwasser enthalten.

Im 1. Abschnitt sind die Voraussetzungen, Zusammenhänge und Grenzen zu umreißen, innerhalb deren das Grundwasser Einfluß auf die Vegetation nehmen kann. Es folgt im 2. Abschnitt ein Überblick über die ertragsfördernden und ertragsbegrenzenden Wirkungen des Grundwassers.

1. Über den Grundwassereinfluß auf die Vegetation

Außer den hydropedologischen Kennzeichen und Begriffen erscheinen in dem schematischen Geländequerschnitt (Abb. 134) noch die Kulturen mit ihrem Wurzelraum. Damit werden gleichzeitig die Voraussetzungen und Grenzen des Grundwassereinflusses auf die Vegetation veranschaulicht.

A. Voraussetzungen des Grundwassereinflusses

Unmittelbare Verbindung Saugraum—Wurzelraum. Eine unmittelbare Wirkung des Grundwassers auf den Wasserhaushalt des Wurzelraumes und damit

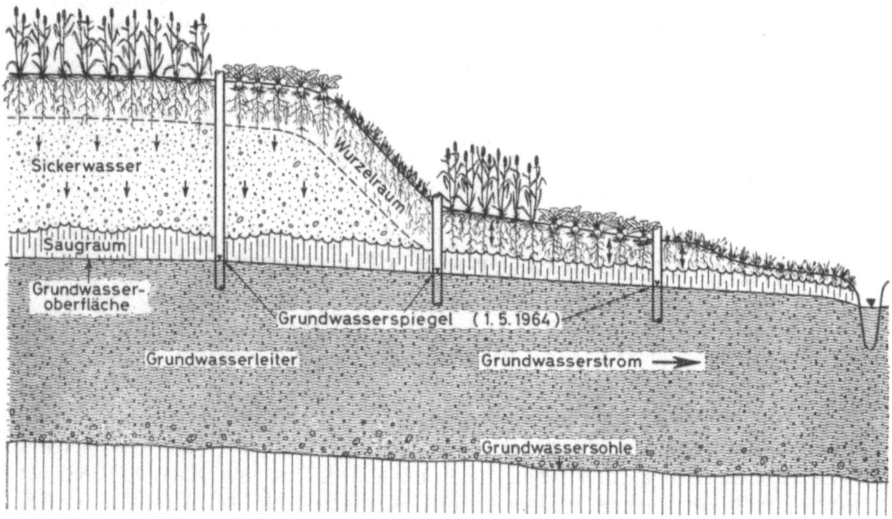

Abb. 134. Schematischer Geländequerschnitt mit Grundwasserstrom

auf den Pflanzenertrag besteht nur dort, wo Saugraum und Wurzelraum miteinander verbunden sind (Abb. 134, rechter Teil). Am günstigsten ist dabei eine Durchdringung von Rhizosphäre und offenem Saugraum, denn die meisten landwirtschaftlichen Kulturpflanzen meiden mit ihrem Wurzelwerk den völlig von Wasser durchtränkten Bodenraum. KÖNEKAMP und KÖNIG (1929) haben jedenfalls eindeutig festgestellt, daß nur Wurzeln von Gräsern, und zwar hauptsächlich des Rohrglanzgrases, in den vom Grundwasser erfüllten Bodenbereich eingedrungen waren.

Liegt der Grundwasserspiegel zu nahe unter der Bodenoberfläche und steht damit für den Luftaustausch gar kein oder ein zu geringer Bodenraum zur Verfügung (Uferbereich des Baches in Abb. 134), dann finden sich mit diesen Verhältnissen lediglich sehr flachwurzelnde oder Nässe ertragende Pflanzenarten ab. Die Nutzungsmöglichkeiten und die Ertragsleistung solcher Standorte sind eng begrenzt und lassen sich nur durch eine Entwässerung erweitern bzw. verbessern (Olbertz und Press 1963, Schroeder 1958, Uhden 1964).

Dampfförmiger Wasseraufstieg. Fehlt die Verbindung von Saugraum und Wurzelraum, wie aus der linken Bildhälfte (Abb. 134) ersichtlich wird, dann nimmt das Grundwasser keinen unmittelbaren Einfluß auf die Vegetation. Sie ist dann im wesentlichen auf das den Niederschlägen entstammende Wasserdargebot angewiesen. Mittelbar besteht jedoch noch die Möglichkeit des dampfförmigen Aufstieges aus dem Grundwasser. Die Kondensation eines solchen Dampfstromes kann vornehmlich im Übergangsbereich von grob- zu feinporenreicherem Boden erfolgen. Heide (1960) und Wohlrab (1959 b) haben für diesen Fall die geologisch-bodenkundlichen Voraussetzungen beschrieben. Nach Kokkonen (zitiert von Trénel 1949) vollzieht sich der dampfförmige Aufstieg vor allem bei Bodenfrost. Lebedeff (1927) hatte bei seinen Untersuchungen eines Tschernosems bei Odessa während der Wintermonate eine Bodenfeuchtezunahme von 66 mm ermittelt, die er nach der aufgestellten Bilanz mit dampfförmigem Aufstieg aus tieferen Schichten als Folge des Temperaturgefälles begründete. Ohne den Vorgang der dampfförmigen Wasserbewegung grundsätzlich zu bestreiten, haben Keilhack (1935) und Zunker (1930) jedoch die Verlagerung und Kondensation derartiger, für den Pflanzenwuchs beachtlicher Wassermengen bezweifelt. Trénel (1949/1950 und 1954) zog aus den Ergebnissen seiner ertragskundlichen Untersuchungen von Lößböden im mitteldeutschen Braunkohlengebiet den Schluß, daß die Kondensationsvorgänge weit über den optisch erkennbaren kapillaren Anstieg des Grundwassers hinausreichen. Er stützte sich dabei auf Untersuchungsbefunde von Weber (1954), gegen dessen Methodik jedoch Olbertz (1957) Bedenken äußerte. Dörter (1954) kam nach seinen Experimenten zu dem Schluß, daß die wirklichen Grundwassereffekte über die erkennbaren Steighöhen hinausreichen.

Diese widersprüchlichen Auffassungen über die Bedeutung des dampfförmigen Wasseraufstieges aus tieferem Grundwasser für den Pflanzenwuchs, wie sie vor allem für die Verhältnisse unter gemäßigt-humidem Klima bestehen, müssen noch durch weitere Untersuchungen geklärt werden. Am zuverlässigsten erscheinen zu diesem Zweck Lysimetermessungen in Verbindung mit Bodenfeuchtebestimmungen[1].

B. Grenzen des unmittelbaren Grundwassereinflusses

Wenn man von dem in seiner Bedeutung für das Pflanzenwasserdargebot — zumindest unter humiden Klimaten — noch umstrittenen dampfförmigen Wasseraufstieg absieht, dann sind für das Wirksamwerden des Grundwassers auf die Vegetation folgende drei Faktoren von wesentlicher Bedeutung:

Mächtigkeit des Saugraumes, d. h. kapillare Steighöhe,
Steiggeschwindigkeit und
Durchwurzelungstiefe.

Kapillare Steighöhe. Sie ergibt sich vor allem aus den physikalischen Bodeneigenschaften. So ist z. B. ihre Zunahme vom Sand zum Lehm und ihre Abnahme wiederum zum Ton aus zahlreichen Untersuchungen an sortierten

[1] Derartige Untersuchungen laufen seit Herbst 1963 bei der Landesanstalt für Immissions- und Bodennutzungsschutz, Essen.

Korngrößengruppen und entsprechenden Gemischen sowie an verschiedenen natürlichen Böden bekannt. An diesen Untersuchungen haben sich vor allem Atterberg (zitiert von Zunker 1930 und Trénel 1949), Dhein und Mott (1955), Hilgard und Loughridge (zitiert von Trénel 1949), Kozeny (1924), Rode (1959), Schucht (zitiert von Weisbrod 1937), Luedecke/Wollny (beide zitiert von Zunker 1930) und Vageler (1932) beteiligt. Ihre Ergebnisse weichen jedoch auch bei gleichen Korngrößenfraktionen zum Teil beträchtlich voneinander ab.

Atterberg hatte die größte Steighöhe von 275 cm in einem Substrat mit der Korngröße 0,02 bis 0,01 mm (Mittelschluff=0,02 bis 0,006 mm) nach 30 Tagen festgestellt. Diese Angabe kommt den Beobachtungen kapillaren Hubs von 295 cm in einem Schluff (Korngröße 0,016 mm) nahe, den Hilgard und Loughridge — jedoch erst nach 475 Tagen — gemessen hatten. Rode beobachtete Steighöhen in gewachsenem Lößlehm von 170 bis 355 cm, im Mittel 237 cm. Nach den Untersuchungen von Dhein und Mott an drei verschiedenen Böden — schluffiger Lehm, feinsandiger Lehm, anlehmiger Sand — war selbst im leichtesten Boden bei 50 cm noch nicht die volle Steighöhe erreicht. Für Mittel- und Grobsande ohne nennenswerten Ton- und Schluffgehalt kann man jedoch nach den vorliegenden Messungen und Geländebeobachtungen für die Saugraummächtigkeit Werte unter 50 cm anhalten (Wohlrab 1961). In diesem Zusammenhang muß noch erwähnt werden, daß die Temperatur (Viskosität des Wassers) und der Luftdruck Einfluß auf die Steighöhe nehmen, der jedoch im gemäßigten Klima vernachlässigt werden kann (Koehne 1948 und Zunker 1930).

Steiggeschwindigkeit. Die Förderleistung im Saugraum, d. h. die sogenannte Steiggeschwindigkeit läßt mit zunehmender Entfernung von der Grundwasseroberfläche nach. Zu diesem Ergebnis kam schon Atterberg (zitiert von Zunker 1930) auf Grund von Untersuchungen mit verschiedenen Korngrößenfraktionen, ferner Oehler (1937) durch Messungen in gestörten und Olbertz (1957) in ungestörten Böden. v. Nitzsch (1933) hatte aus seinen Untersuchungsbefunden — erstmalig in natürlich gelagerten Böden — den Schluß gezogen, daß die Förderleistung schon wenige Zentimeter über der Grundwasseroberfläche unbedeutend sei. Im Gegensatz hierzu stellten Dhein und Mott (1955), ebenfalls in ungestörten Bodensäulen (siehe oben), bei anlehmigem Sand und Lößlehm in 50 cm Höhe eine Förderleistung von rund 1 mm/Tag fest. Sie zogen daraus für diese Böden den Schluß, daß „der Einfluß des Grundwassers auf den Wasserhaushalt des Bodens in einer Höhe von 50 cm und wahrscheinlich noch darüber hinaus sowohl in dynamischer (Grundwassernachlieferung) als auch in statischer Hinsicht (Wasserreserve) recht beträchtlich" ist. Hanus (1962) fand bei einem Lößlehm ähnliche Werte und konnte aus seinen Untersuchungsergebnissen gleichzeitig eine Begründung ableiten für die beachtlichen Unterschiede zwischen den Befunden von v. Nitsch (1937) und denen von Dhein und Mott (1955). Er führte sie darauf zurück, daß die tägliche Nachlieferungsmenge in ausgetrockneten Bodenkernen zunächst sehr gering ist und mit zunehmender Durchfeuchtung zunimmt. Olbertz (1937) kam nach seinen Resorptionsmessungen noch zu folgendem bemerkenswerten Schluß: „Das Nachlieferungsvermögen ist damit weitgehend abhängig von der Intensität, mit der die Feuchte dem Bodenabschnitt durch Pflanzenwurzeln oder übergeordnete Bodenschichten entnommen wird."

Abschließend bleibt zu dieser Frage noch zu sagen, daß die Steiggeschwindigkeit in gemäßigt humiden Klimaten in ihrer Bedeutung nicht überschätzt werden darf. Sie tritt erst dann wesentlich in Erscheinung, wenn im Laufe einer längerwährenden Trockenperiode der vorhandene Vorrat verbraucht ist. Erst dann hängt die Bedarfsdeckung der Kulturen von der Schnelligkeit der Nachlieferung ab (Wohlrab 1961). Unter dieser Voraussetzung wird sie allerdings um so mehr entscheidend sein, als sich die Kulturen auf solchen Standorten gegebenenfalls konstitutionell auf reichliche Wasservorräte eingestellt haben können (Baumann 1949).

Durchwurzelungstiefe. Abgesehen von der arteigenen Ausbildung der Wurzeln ist ihr Vordringen in den Untergrund und ihre Ausbreitung in den einzelnen Schichten sowie ihre Aktivität vor allem von den physikalischen und chemischen Eigenschaften und dem Feuchtegang des Bodens abhängig (u. a. Gliemeroth 1953 und 1954/55, Kmoch 1952 und 1962, van Lieshout 1960). Außerdem wird die Gesamtwurzelmasse und ihr Tiefgang beim Dauergrünland sehr stark von der Nutzung beeinflußt (Klapp 1956). Die in Übersichten von Hahne (1938), Koehne (1948), Kutschera (1960), Trénel (1949), Weisbrod (1937) und Weaver (1926) zusammengestellten eigenen oder fremden Untersuchungsergebnisse umfassen deshalb zum Teil sehr verschiedene Durchwurzelungstiefen. Sie liegen bei den wichtigsten Getreidearten, Hack- und Futterfrüchten zwischen einigen Dezimetern und mehr als zwei Metern. Das Wurzelbild zeigt überall eine deutliche Abnahme der Wurzelmasse mit der Tiefe, die aber je nach Boden und Witterungsverlauf sehr verschiedene Ausmaße aufweist. Die nach Wurzelauswaschungen ermittelte und vielfach in Diagrammen dargestellte Quantität gibt jedoch, worauf u. a. Stocker (1954) und Hanus (1962) hinwiesen, keinen unmittelbaren Maßstab für die Aktivität und das Leistungsvermögen bei der Wasseraufnahme. Klapp (1961) kommt speziell für Lößlehmboden zu dem Schluß: „Die Bedeutung der tiefer als 1 m hinabreichenden Wurzeln für die Ertragssicherung und generell für Notlagen der Wasserversorgung kann jedenfalls nicht bestritten werden.“

Grenzflurabstand. Aus der kapillaren Steighöhe und der Durchwurzelungstiefe ergibt sich der sogenannte Grenzflurabstand, bei dem ein unmittelbarer Einfluß des Grundwassers auf die Vegetation noch möglich ist. Dieser Grenzflurabstand hat besonders im Mittelpunkt von Auseinandersetzungen über Fragen des Grundwasserentzuges und seiner Folgen gestanden (Brüne 1960, Hahne 1938, Heide 1962, Klapp 1961, Trénel 1949/50, Weisbrod 1937, Wohlrab 1959 b, 1961 a und b).

Wenn man von sehr günstigen Bodenverhältnissen nach Art, Mächtigkeit, Zustand und Gefüge ausgeht, wie sie in Schwarzerden, Pararendzinen und nur schwach lessivierten Parabraunerden aus mächtigem Löß gegeben sind, dann kann die Grenze des unmittelbaren Grundwassereinflusses auf einjährige Ackerkulturen und einzelne Arten des Dauergrünlandes bei einem Flurabstand von etwa 4 m gezogen werden (Wohlrab 1961 a). Sie ergibt sich aus der kapillaren Steighöhe in diesem Boden von etwa 2,5 m und einer maximalen Durchwurzelungstiefe der bezeichneten Pflanzenarten im Bestand von durchschnittlich 2,0 m, unter Berücksichtigung einer für den Pflanzenwasserbedarf nicht völlig unbedeutenden Nachlieferung. Mehrjährige Kulturen mit entsprechend ausgeprägtem Wurzeltiefgang, beispielsweise die Luzerne (Könekamp 1934), werden unter den gleichen Bedingungen vielleicht noch Grundwasser mit größerem Flurabstand nutzen können. In gemäßigt humiden Klimaten sollte man aber diesem bei bestem Boden anzuhaltenden Grenzflurabstand keine größere Bedeutung beimessen, als ihm praktisch wirklich zukommt. Diese Böden zeichnen sich bekanntlich durch eine hohe nutzbare Wasserkapazität und damit, von Trockenjahren abgesehen, durch ausreichende Wasservorräte aus.

Bei Sandböden ist der für einjährige Ackerkulturen und für das Grünland anzuhaltende Grenzflurabstand wesentlich geringer. In mittelkörnigen Sanden liegt er nach Beobachtungen im Jahre 1949 von Schmidt und Klapp (1949) etwa bei

1,0 m für das Grünland und bei rund 1,4 m für einjährige Feldfrüchte. In Ton- und Moorböden hängt der kapillare Aufstieg aus dem Grundwasser weitgehend vom Anteil der für diese Bewegung in Frage kommenden Poren ab. Bei voll durchwurzelten Ton- oder Moorüberdeckungen über durchlässigem wasserführendem Untergrund (Sande bis Kiese) besteht eine Verbindung zwischen der Rhizosphäre und dem Grundwasser, wenn dieses mit seinem Saugraum in die Deckschicht hineinreicht. Das gilt auch für Überdeckungen aus Lehm bis lehmigem Sand. Solche Bodenverhältnisse herrschen vor allem in den Flußauen, in denen Niederterrassenschotter, -kiese und -sande von mehr oder weniger sandigen Hochflutlehmen und -tonen in einer Mächtigkeit meist um etwa 1 m überdeckt werden.

C. Grundwasserspiegelschwankung

Bei der bisherigen Behandlung der Voraussetzungen und Grenzen des Grundwassereinflusses war von bestimmten Flurabständen (Abb. 134, Messung am 1. Mai 1964) die Rede. Nun schwankt bekanntlich der Grundwasserspiegel jahreszeitlich und über längere Zeiträume mehr oder weniger stark. Diese Schwankung oder, anders ausgedrückt, die Grundwasserspiegelamplitude, hängt insbesondere von den örtlichen morphologischen und hydrographischen Gegebenheiten, von den hydrogeologischen Eigenschaften des Grundwasserleiters und seiner Sohle sowie von der klima- und witterungsbedingten Grundwasserneubildung (Versickerung) ab (Abb. 134). Weil die Grundwasserspiegelamplitude ökologisch und auch bodendynamisch von wesentlicher Bedeutung ist, hat MÜCKENHAUSEN

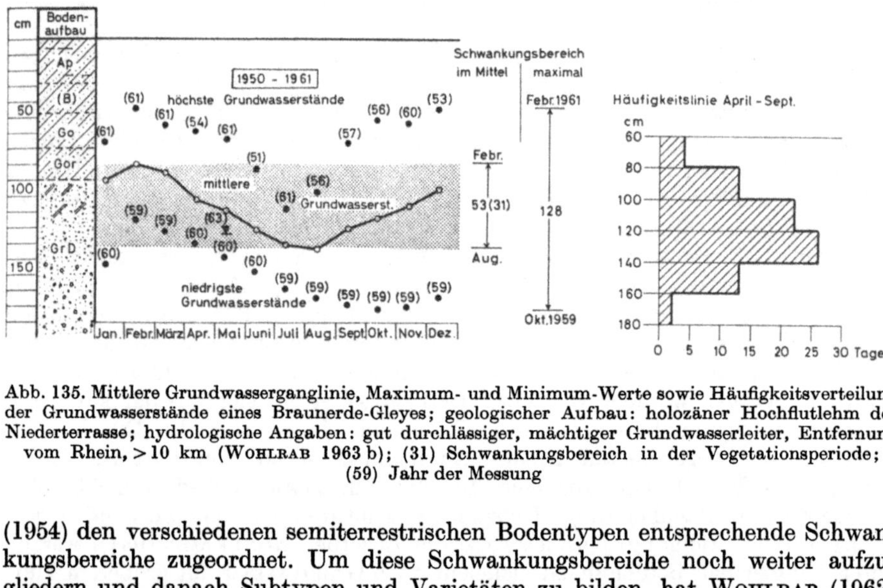

Abb. 135. Mittlere Grundwasserganglinie, Maximum- und Minimum-Werte sowie Häufigkeitsverteilung der Grundwasserstände eines Braunerde-Gleyes; geologischer Aufbau: holozäner Hochflutlehm der Niederterrasse; hydrologische Angaben: gut durchlässiger, mächtiger Grundwasserleiter, Entfernung vom Rhein, >10 km (WOHLRAB 1963 b); (31) Schwankungsbereich in der Vegetationsperiode; (59) Jahr der Messung

(1954) den verschiedenen semiterrestrischen Bodentypen entsprechende Schwankungsbereiche zugeordnet. Um diese Schwankungsbereiche noch weiter aufzugliedern und danach Subtypen und Varietäten zu bilden, hat WOHLRAB (1962) die „Auswertung und Darstellung langjähriger Grundwasserbeobachtungsergebnisse in bodenkundlicher Sicht" vorgeschlagen. Für die untere Rheinniederung wurden einige charakteristische, dort weit verbreitet vorkommende Grundwasserböden — Naßgley, Typischer Gley, Braunerde-Gley, Gley-Braunerde und Auenboden — zusammengestellt und die ihnen eigenen Spiegelamplituden als besonderes Unterscheidungsmerkmal hervorgehoben (WOHLRAB 1963b). Abb. 135 zeigt den Braunerde-Gley aus dieser Folge.

Ähnliche Profilserien lassen sich in Gebieten mit mächtigen wasserführenden Decksanden zusammenstellen, die in Nordwestdeutschland einen bedeutenden Flächenumfang einnehmen (WOHLRAB 1964 b). Von einer solchen Serie der oberen Emssandebene (Senne), bestehend aus Anmoorgley, gebleichtem Gley, Gley-Podsol und Podsol, sind in Abb. 136 die Häufigkeitsverteilungen zusammengestellt.

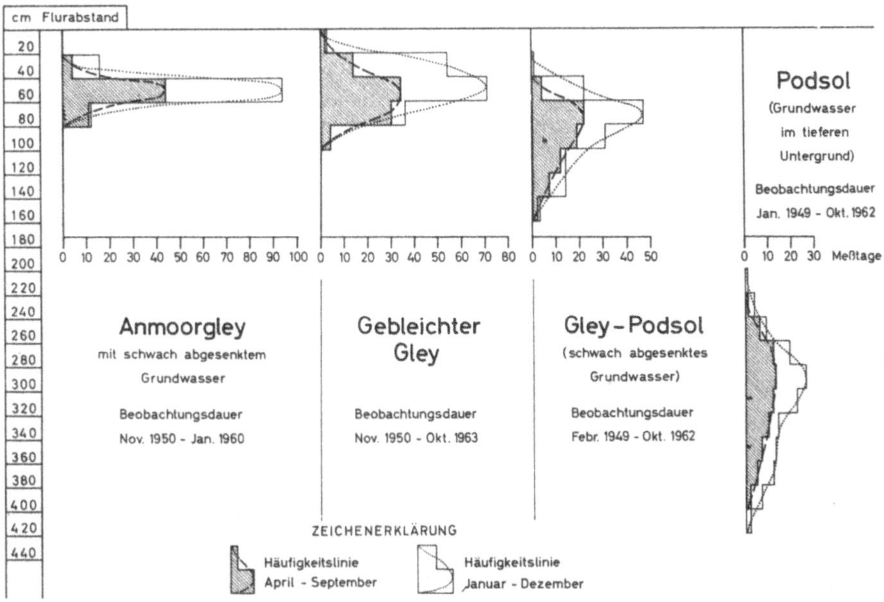

Abb. 136. Häufigkeitsverteilung der Grundwasserstände bei verschiedenen Böden

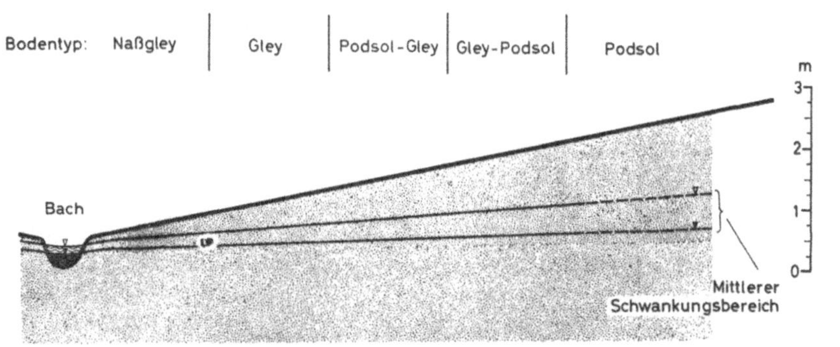

Abb. 137. Schematischer Geländequerschnitt mit verschiedenen grundwasserbeeinflußten Böden

Wenn man diese Reihenfolge nach geomorphologischen Gesichtspunkten einzuordnen versucht, dann ergibt sich etwa das in Abb. 137 dargestellte Schema. Mächtige, relativ gleichmäßig durchlässige Grundwasserleiter vorausgesetzt, nehmen auf einer geneigten größeren Ebene die Flurabstände und der Schwankungsbereich des Grundwasserspiegels in Richtung zur Wasserscheide zu (s. auch STREMME 1960). Die vom Grundwasser mehr oder weniger geprägten Böden gehen danach ohne klare Grenzen ineinander über. Und zwar folgt auf den Naßgley der Gley, bei armen, glazial-fluviatilen Sanden hauptsächlich in der Ausbildung des gebleichten Gleyes. Ihm schließt sich der Podsol-Gley, der Gley-Podsol und der Podsol mit Grundwasserhorizonten im Untergrund an. Für die angegebene Reihenfolge der Bodentypen bzw. Subtypen ergibt sich daraus eine zunehmende Amplitude.

Wie zahlreiche Auswertungen der geschilderten Art beweisen, besteht eine solche Regel jedoch nur unter den angegebenen morphologischen und hydrogeologischen Verhältnissen. Ist hingegen das Relief sehr kleinflächig geformt, beispielsweise in einer Niederung mit einzelnen, sich nur wenige Dezimeter heraushebenden Inseln, dann sind trotz der auf ihnen entwickelten Semigleye oder terrestrischen Bodenbildungen *keine* wesentlich verschiedenen Grundwasserschwankungsbereiche zu erwarten.

Erheblich verwickelter sind die Verhältnisse dort, wo der Grundwasserleiter auf engem Raum wechselnd mächtig und durchlässig ist. Das trifft zu für Gebiete mit ausgedehnten Grundmoränenlinsen oder tonig-schluffigem Untergrund in verschiedener Tiefe. Welche entscheidende Wirkung die Mächtigkeit und Durchlässigkeit des Grundwasserleiters auf die Grundwasserspiegelamplitude hat, zeigte WOHLRAB (1965) an einer anderen Serie von sieben Profilen mit Decksanden von abnehmender Mächtigkeit über Kreidesandmergeln und Grundmoräne.

BUCHHOLZ (1960) stellte bei seinen Untersuchungen in grundwasserbeeinflußten Sandböden folgendes fest: „Nach Niederschlägen steigt der Grundwasserspiegel innerhalb des wassergesättigten Bodenbereiches auf und engt den geschlossenen Kapillarsaum ein. In den Intervallen einzelner Niederschlagsperioden sinkt der Grundwasserspiegel innerhalb des wassergesättigten Bodenbereiches ab und erweitert den geschlossenen Kapillarsaum.“ Die von BUCHHOLZ aus diesen Beobachtungen gezogene Folgerung, daß sich die Schwankungen des Grundwasserspiegels nicht oder höchstens in ganz abgeschwächter Form auf die Oberfläche des geschlossenen Saugraumes übertragen, kann jedoch nur für kurzfristige Schwankungen, aber nicht für den in der Regel vorhandenen jahreszeitlichen Trend gelten.

D. Einflüsse der chemischen Grundwassereigenschaften

Nach den bisherigen Ausführungen vollzieht sich die Grundwasserwirkung auf die Vegetation in erster Linie über den Wasser- und Lufthaushalt des Bodens. Die Durchfeuchtungs- und Durchlüftungsverhältnisse im Boden haben jedoch gleichzeitig wesentlichen Einfluß auf die Verfügbarkeit von Nährstoffen, insbesondere — worauf noch eingegangen wird — auf das Dargebot bodenbürtigen Stickstoffs. Neben diesem mittelbaren Zusammenhang gehen vom Grundwasser noch unmittelbare Wirkungen auf die Nährstoffverlagerung aus. Sie stehen in enger Verbindung mit den jeweiligen chemischen Grundwassereigenschaften und ihrer Dynamik, zu deren Kenntnis u. a. BUCHHOLZ (1960) und v. ZEZSCHWITZ (1964) Untersuchungsergebnisse beigetragen haben. BUCHHOLZ beschäftigte sich mit den Redoxverhältnissen im Grundwasser, und zwar im ständig wassergesättigten Bodenbereich (Reduktionszone) und im Grundwasserschwankungsbereich (Oxydationszone). v. ZEZSCHWITZ behandelte die verwickelten Ursachen der Calciumcarbonatausscheidung in Gleyen und Semigleyen aus Lößlehm und Löß über kalkreichem Untergrund. Kalkreiche Grundwässer haben besonders bei semiterrestrischen Böden aus armen Deck- und Talsanden wesentliche Wirkung auf die Ertragsfähigkeit.

Mit den Einflüssen der Mineraldüngung (Auswaschung) auf den Chemismus des Grundwassers hat sich NÖRING (1951) auseinandergesetzt. SCHWILLE (1953 und 1962) befaßte sich mit den Chloriden und Nitraten in den Grundwässern Rheinhessens. Die Verunreinigung des Grundwassers mit versickernden Abwässern kann örtlich von einiger Bedeutung für den Pflanzenwuchs sein.

Zur ständigen Anreicherung leicht löslicher Salze im Grundwasserschwankungsbereich kam es in den stark alkalischen Gleyböden (Solontschak, Solonez) des semiariden und semihumiden Klimas (FRANZ 1960, RAMSAUER 1964). Mit den besonderen Schwierigkeiten der Kultivierung solcher Böden, die eine tiefgreifende Absenkung des Grundwassers zur Unterbrechung der kapillaren Verbindung mit dem Wurzelraum voraussetzt, befaßten sich außer FRANZ und RAMSAUER u. a. KOSTJAKOW (1959) sowie OLBERTZ und PRESS (1963).

2. Das Grundwasser in seiner Wirkung auf den Pflanzenertrag

Das zuverlässigste Maß für die Beurteilung der leistungsfördernden oder -begrenzenden Wirkungen des Grundwassers ist der Pflanzenertrag.

A. Versuchseinrichtungen

Um die Einflüsse des Grundwassers auf den Pflanzenertrag festzustellen, sind schon die verschiedensten Versuche eingeleitet und durchgeführt worden. Dabei lassen sich im wesentlichen stationäre Anlagen und vergleichende Ertragsfeststellungen unterscheiden.

Stationäre Anlagen. Hierzu sind zunächst alle Gefäßversuche zu rechnen. U. a. experimentierten BERTRAM (1931) und neuerdings HANUS (1962) mit Gefäßen von relativ kleiner Versuchsfläche. HANUS benutzt zu diesem Zweck mit Paraffin und Aluminiumfolie ummantelte ungestörte Bodensäulen, die er bewußt unter bestimmten, von den natürlichen Gegebenheiten abweichenden Gewächshausbedingungen hielt. Eine Versuchsanlage aus größeren Kästen mit eingefülltem Boden und verschiedenen Grundwasserständen (40, 60 und 80 cm unter Flur) hatte wohl als erster PITSCH (1913) eingerichtet. FRECKMANN (1926), KÖNEKAMP und KÖNIG (1929) berichteten über die Landsberger Großgefäßstation, die Betonkästen mit rund 0,8 m^2 Versuchsfläche in einer ersten Reihe mit sechs verschiedenen Böden (Lehm, Ton, Niederungsmoor, lehmiger, feiner und grobkörniger Sand) bei vier konstanten Grundwasserstufen (40, 70, 100 und 130 cm unter Flur) und in einer zweiten Reihe mit lehmigem Sand und wechselnden Grundwasserständen umfaßte. SCHWARZ (1932) hatte diese Versuche fortgesetzt. In den Niederlanden wurden von FRANKENA und GOEDEWAAGEN (1942) neue Versuche mit Kästen durchgeführt, die ungestörte Lehm-, Moor- und Sandprofile mit der zugehörigen Grasnarbe enthielten. 2 m^2 große Vegetationsgefäße aus Eisenbeton mit eingefülltem Boden bei automatisch regulierbaren Grundwasserständen beschrieben COLLIER und ROBELIN (1959). SCHOLL (1962 a und b) berichtete über einen Senkkasten aus Stahlblech, in den der Boden in ungestörter Lagerung eingebracht wurde, und zwar in Anlehnung an das von FRIEDRICH und FRANZEN (1960) entwickelte Verfahren.

Größere Freilandversuchsfelder mit umfangreichen technischen Anlagen zur Haltung bestimmter, zum Teil jahreszeitlich zu verändernder Grundwasserflurabstände sind vor allem in den Niederlanden, und zwar hauptsächlich auf Fluß- und Seemarsch sowie auf Niedermoor eingerichtet worden. Sie wurden u. a. von HOOGHOUDT (1952), VAN HOORN (1958) und MINDERHOUD (1960) beschrieben.

Abb. 138 zeigt im Grundriß die Anordnung und technische Einrichtung des südlich von Utrecht gelegenen Grundwasserstandsversuchsfeldes „De Lucht“ (Minderhoud 1960). Wäre (1955) berichtet über ein finnisches Grundwasserstandsversuchsfeld auf Lehm und Moor. Zu nennen ist ferner ein Grundwasserstandsversuchsfeld auf Niedermoor in England (Eden et al. 1951, Nicholson 1952, Nicholson und Firth 1953, Nicholson und Firth 1958). Einige Versuchsfelder dieser Art wurden außerdem in verschiedenen Staaten der USA eingerichtet (Roe 1936, Clayton et al. 1952, Ellis und Morris 1946). Seit einigen Jahren be-

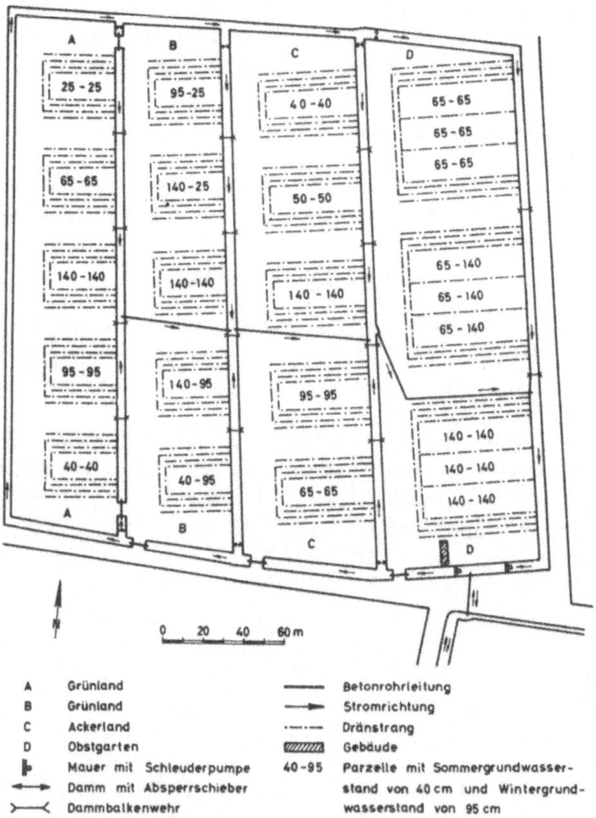

Abb. 138. Grundwasserstandsversuchsfeld „De Lucht“ (Minderhoud 1960)

steht schließlich eine derartige Anlage auf Sandboden in der Boker Heide, Westfalen, über deren Einrichtung und bisherige Ergebnisse Boeker und Kmoch (1963) berichteten.

Zu erwähnen sind darüber hinaus noch die Dränversuche, bei denen es zum großen Teil im Haupt- oder Nebenzweck um die Erzielung und Feststellung optimaler Grundwasserstände geht, wobei optimal gleichzeitig unter dräntechnischen und kostenwirtschaftlichen Gesichtspunkten verstanden wird. Mit dem Dränversuch und seinen Problemen hat sich Bellin (1964) unter Heranziehung und Auswertung der vorliegenden Literatur näher befaßt.

Stationäre Anlagen sind entweder mit sehr hohem technischem und damit auch finanziellem Aufwand verbunden, oder ihre Ergebnisse können, wenn es

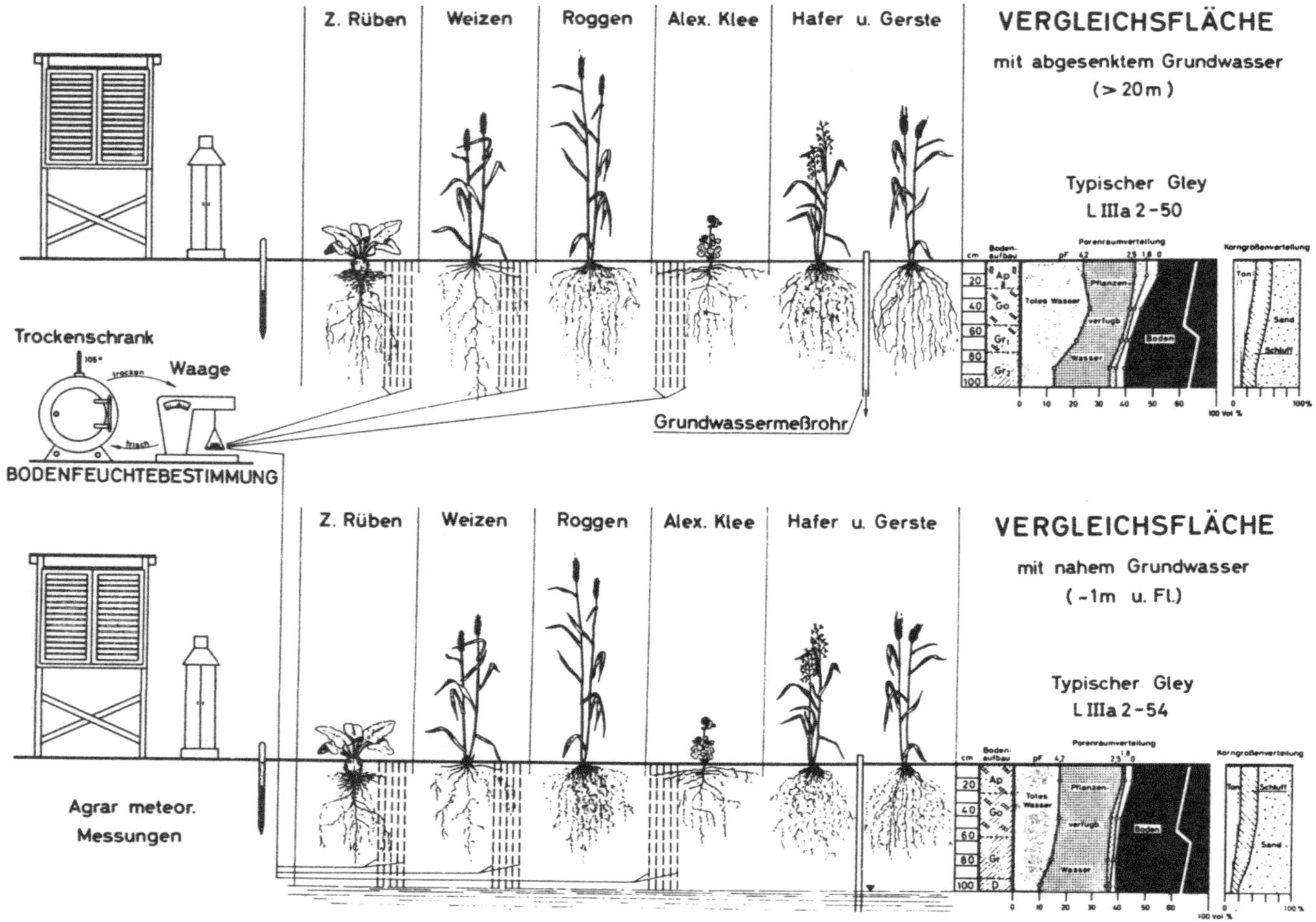

Abb. 139. Vergleichende Ertragsfeststellungen und Untersuchungen über die Grundwasserwirkung (WOHLRAB 1965)

sich um Gefäßversuche handelt, nur bedingt auf Freilandverhältnisse übertragen werden.

Vergleichende Ertragsfeststellungen. Ein anderer Weg, die Wirkung des Grundwassers auf den Ertrag zu erfassen, besteht im Vergleich der Erträge von Flächen mit verschiedenen ungestörten oder durch wasserbauliche Maßnahmen veränderten Grundwasserverhältnissen. Solche vergleichenden Ertragsfeststellungen werden sowohl zur Ermittlung des optimalen Flurabstandes als auch zum Nachweis der Auswirkungen von Grundwasserentzug bzw. -anstieg auf die Ertragsleistung herangezogen (Bergerhoff, Wohlrab und Bohn 1955, Kalisvaart 1949, Makkink 1955, Sieben und Smits 1951, Smits 1962, Trénel 1949/50, van der Woerdt 1953, Wohlrab 1959 a und 1965). Das Prinzip der von Wohlrab beschriebenen Versuchsdurchführung, die neben den ertragskundlichen Feststellungen und Grundwasserstandsmessungen noch agrarmeteorologische Beobachtungen und Bestimmungen der statischen und dynamischen Bodenwasserverhältnisse einschließt, zeigt die Abb. 139.

B. Ursachen, Zusammenhänge und Ausmaß der Ertragsbeeinflussung durch das Grundwasser

Die aufgezählten, zum Teil recht umfangreichen Versuche zur Klärung der Grundwasserwirkung auf den Ertrag gingen von verschiedenen Fragestellungen aus. Im Vordergrund — vor allem bei den niederländischen Versuchen — stand die wirtschaftlich tragbare Entwässerungstiefe (Schöpfwerke), und zwar vor allem bei schweren mineralischen und organogenen Niederungsböden. Die zu diesem Zweck gehaltenen tiefsten Grundwasserstände — 1,4 m auf dem Versuchsfeld „De Lucht" (Abb. 138; Minderhoud 1960) und 1,5 m auf dem Versuchsfeld „Nieuw Beerta" (van Hoorn 1958) — lagen bei den schweren Böden noch so nahe beim Optimum, daß lediglich über die Ertragsbegrenzung infolge beginnenden Wasserüberschusses durch zu oberflächennahes Grundwasser ein abschließendes Urteil möglich war, nicht aber über Ertragsdepressionen als Folge von Wassermangel durch zu tiefe Grundwasserstände. Ein Teil der Versuche lief nur wenige Jahre. Die aus ihren Ergebnissen gezogenen Schlußfolgerungen gelten daher nur für Verhältnisse, wie sie in den Versuchsjahren, vor allem witterungsmäßig, aufgetreten waren. Im übrigen konnte man so deutliche Ergebnisse, wie sie Hanus (1962) im Gewächshaus (ohne Niederschlag) feststellte, unter Freilandbedingungen im gemäßigt humiden Klima nicht erwarten. Es ist daher zu verstehen, daß die Befunde der verschiedenen Versuche und Untersuchungen bei überschläglicher Betrachtung teilweise widersprüchlich erscheinen. Berücksichtigt man jedoch die jeweilige Fragestellung, die verschiedenen Bedingungen und die Dauer der Versuche, dann schälen sich einige wesentliche Erkenntnisse heraus.

Grundwasserwirkung bei verschiedenem Boden und Witterungsverlauf. Schon Pitsch (1913) hatte festgestellt, daß die Reaktion des Graswuchses auf den Grundwasserstand im Trockenjahr 1911 anders war als im nassen Jahr 1910. Könekamp und König (1929) kamen für Kleegras zu dem Schluß, „daß, je schwerer der Boden und je höher die Niederschläge sind, desto günstiger die tieferen Grundwasserstände auf das Pflanzenwachstum wirken, während die hohen Grundwasserstände auf den leichten Böden und besonders in den trockenen Jahren überlegen sind" (Tab. 101).

Sieben und Smits (1951) fanden bei ihren vergleichenden Ertragsermittlungen von Grünland auf Sand, Moor und Lehmboden die Erkenntnisse von Könekamp und König bestätigt. Dieser Feststellung widersprechen auch nicht die in Tab. 102 zusammengefaßten Ergebnisse von Kalisvaart (1949), nach denen in den Jahren 1946 bis 1948 der höchste Ertrag beim geringsten Grundwasserflurabstand (40 cm) auftrat, weil es sich beim Standort seiner Infiltrationsversuche (Grabeneinstau) um einen groben Sandboden handelt.

Tabelle 101. *Heuerträge (dz/ha) bei verschiedenen Grundwasserständen* (KÖNEKAMP und KÖNIG 1929)

Bodenart und Grundwasserstand (unter Flur)	1924	1925	1926	1927	1928	Durchschnitt 1924 bis 1928
Lehm........40 cm	98,1	222,9	145,6	110,0	149,5	145,2
70 cm	80,6	170,4	138,8	112,0	128,9	126,1
100 cm	75,1	137,5	132,6	110,4	125,8	116,3
130 cm	71,0	118,1	100,0	96,6	116,8	100,5
Ton.........40 cm	91,5	180,1	124,3	102,1	144,3	128,5
70 cm	88,9	161,8	137,5	136,5	133,0	131,5
100 cm	75,8	141,8	157,9	148,3	128,9	130,5
130 cm	75,7	140,6	175,4	180,0	127,5	139,8
Niederungsmoor.......40 cm	81,6	172,5	154,8	161,6	173,4	148,8
70 cm	90,1	161,8	148,6	131,4	152,5	136,9
100 cm	96,6	124,4	154,9	158,1	162,6	139,3
130 cm	110,3	123,0	139,5	184,1	162,1	143,8
Grobkörniger Sand.......40 cm	79,3	118,0	85,0	90,0	120,4	98,5
70 cm	29,8	50,4	67,5	86,6	98,6	66,6
100 cm	26,2	42,8	55,6	85,1	83,3	58,6
130 cm	22,6	33,1	50,0	73,3	77,5	51,5
Feinkörniger Sand.......40 cm	82,0	174,4	109,6	96,4	152,3	122,9
70 cm	63,9	151,6	97,5	89,5	138,1	108,0
100 cm	71,1	113,0	92,5	88,3	125,3	98,0
130 cm	41,5	72,3	74,1	68,1	100,3	71,2
Lehmiger Sand.......40 cm	117,3	259,9	165,2	143,9	190,9	175,4
70 cm	108,1	178,5	156,9	156,0	163,1	152,5
100 cm	100,6	163,6	154,1	155,4	158,3	146,4
130 cm	89,6	109,1	130,0	155,9	145,3	126,9
Monatsmittel (mm) der Niederschläge März bis September.	51,3	46,0	71,4	79,2	31,9	
Monatsmittel (° C) der mittleren Tagestemperaturen Mai bis August.....	16,7	18,2	16,9	15,3	14,6	

NICHOLSON und FIRTH (1953 und 1958) ermittelten auf ihrem Versuchsfeld mit kalkreichem Niedermoorboden sowohl in trockenen als auch in nassen Perioden die höchsten Trockensubstanzerträge des als Versuchsfrucht dienenden Welschen Weidelgrases bei den tiefsten Grundwasserständen von 100 cm. Auf dem gleichen Versuchsfeld haben jedoch Ackerfrüchte in trockenen Sommern auf den höheren Grundwasserstufen in ihrem Aufwuchs günstig reagiert.

BERGERHOFF, WOHLRAB und BOHN (1955) wiesen ebenfalls darauf hin, daß die Wirkung des Grundwassers auf den Ertrag bei Flurabständen, die eine Verbindung zwischen Saugraum und Wurzelraum garantierten, in ihrem Ausmaß von der Niederschlagsmenge und -verteilung während der Vegetationszeit und vom Boden, namentlich von seiner nutzbaren Wasserkapazität, abhängt. Die größten Ertragsminderungen waren daher während der Untersuchungsdauer (1949 bis 1953) im Trockenjahr 1949 auf den Vergleichsflächen mit tiefem, von den Pflanzenwurzeln nicht erreichbarem Grundwasser zu verzeichnen.

WÄRE (1955) hatte auf dem finnischen Grundwasserstandsversuchsfeld bei Lehm und Moor keine deutlichen Ergebnisse für die drei verschiedenen Grundwasserstufen

60, 30 und 20 cm feststellen können. Nach VAN HOORN (1958) nahm der Ertrag besonders des Getreides, aber auch anderer Feldfrüchte im Mittel der Jahre 1946 bis 1955 auf schwerer Seemarsch mit der Tiefe des Grundwasserstandes (Stufen 40, 60, 90, 120 und 150 cm unter Flur) zu. Bei Erbsen war der Höchstertrag jedoch schon auf der 90-cm-Grundwasserstufe erreicht. Ein deutlicher Unterschied zwischen den trockenen und nassen Jahren wurde von VAN HOORN nicht beobachtet.

MINDERHOUD (1960) stellte fest, daß der Ertrag von Dauergrünland auf dem Flußmarschboden in Trockenperioden mit zunehmender Entwässerungstiefe abnahm, in nassen Sommern hingegen anstieg. Auf Moorboden traten jedoch bei den tiefsten Sommerwasserständen von 60 cm unter Flur immer die höchsten Erträge auf. Die meisten Ackerfrüchte auf Flußmarschboden, mit Ausnahme einiger Leguminosen, reagierten ebenfalls jederzeit während der Versuchsdauer von 1954 bis 1957 mit einer Zunahme des Ertrages auf größere Grundwasserstandstiefe (140 cm unter Flur als tiefste Stufe).

Tabelle 102. *Grünlanderträge bei verschiedenen Grundwasserflurabständen* (entnommen aus KLAPP 1956)

Pegelstand	Grünlandertrag relativ (rund)
40	100
55	82
70	78
110	61

HOOGHOUDT, VAN DER WOERDT, BENNEMA und VAN DIJK (1960) haben die Zusammenhänge und Folgen irreversibler Austrocknung von mehr oder weniger tonigem Niedermoor in Westniederland durch zu tiefgreifende Entwässerung ausführlich behandelt und durch Untersuchungen belegt. Mit Grundwasserfragen bei Moorböden befaßten sich außerdem vor allem BADEN (1962), BADEN und EGGELSMANN (1955 und 1958), BADEN und SEGEBERG (1960) und SEGEBERG (1962).

VAN DER WOERDT (1953) hatte aus den Resultaten verschiedener Untersuchungen optimale Grundwasserstandsbereiche ermittelt. Er gab sie an für Moorböden zwischen 20 cm (bei trockenen Sommern) und 60 cm (bei nassen Sommern) sowie für Sandboden entsprechend zwischen 40 und 70 cm.

MAKKINK (1955) kam nach Auswertung zahlreicher vergleichender Ertragsfeststellungen von verschiedenen Versuchsfeldern zu dem Ergebnis, daß eine bessere Beziehung zwischen Grundwasserstand und Ertrag abzuleiten ist, wenn man bei ersterem die pF-Werte des jeweiligen Bodens als Maßstab für das kapillare Leitungsvermögen berücksichtigt. WESSELING (1957) betonte nochmals: „Der optimale Grundwasserstand hängt ab von dem Maße, in welchem ein Feuchtdefizit während der Vegetationszeit vorkommt.“

Zusammenfassend läßt sich über die Abhängigkeit der Grundwasserwirkung von Boden und Witterungsverlauf folgendes sagen:

Das Grundwasser kommt unter den im Abschnitt 1 behandelten Voraussetzungen als zusätzliches Wasserdargebot in Frage. Dieses zusätzliche Wasserdargebot ist für die Pflanze dann von Nutzen, wenn die aus dem Niederschlag stammenden leicht verfügbaren Wasservorräte im Boden ihren Bedarf nicht decken. Je größer dieses Defizit im Wasserdargebot ist, sei es infolge von Trockenperioden mit hoher Verdunstung (Witterungsverlauf), sei es infolge geringer nutzbarer Wasserkapazität des Wurzelraumes (Boden), um so mehr kann die verfügbare Reserve Grundwasser den Minimumfaktor ausgleichen und damit den Pflanzenertrag steigern. Ist der Pflanzenwasserbedarf jedoch allein aus den Niederschlagswasservorräten gedeckt, gleichgültig, ob über eine große nutzbare Wasserkapazität (mittlere bis schwere Böden) bei rechtzeitigen ausreichend hohen Niederschlägen oder bei geringer nutzbarer Wasserkapazität (leichte Böden) durch sehr häufige Regenfälle, dann kann durch Grundwasser keine Ertragssteigerung mehr eintreten. In solchem Falle machen sich im Gegenteil mit dem Wasserüberschuß vermehrt nachteilige Einflüsse — Verschlechterung der Bodenstruktur, Verminderung der Wasserdurchlässigkeit und mangelnder Luftaustausch, geringere Verfügbarkeit an Nährstoffen, insbesondere an Stickstoff — bemerkbar, die zunehmend Ertragsdepressionen hervorrufen.

Grundwasserwirkung bei verschiedenen Kulturarten und Feldfrüchten. Da der Wasserbedarf der Pflanzen in annähernder Analogie zum Nährstoffbedarf arteigen und je nach der Entwicklungsphase sehr verschieden ist, zeigen die in den Versuchen eingestellten konstanten oder wechselnden Grundwasserstände bzw. die in den verschiedenen Untersuchungsjahren aufgetretenen natürlichen Grundwasserschwankungen bei den einzelnen Kulturarten und Feldfrüchten verschiedene Wirkung. BERGERHOFF, WOHLRAB und BOHN (1955) kamen zu dem Schluß, daß im gemäßigt humiden Klimabereich vor allem beim Grünland, und zwar in der Regel beim zweiten und weiteren Schnitt, beziehungsweise vom dritten Auftrieb an, ferner in ähnlicher Weise bei den Hauptfruchtfutterpflanzen sowie beim Zweit- und Zwischenfruchtbau, schließlich bei den späten Hackfrüchten ertragsfördernde Wirkungen vom Grundwasser ausgingen, sofern es zu diesem Zeitabschnitt in erreichbarer Tiefe anstand. Getreide, insbesondere Wintergetreide, ist mit Ausnahme von außergewöhnlich trockenem Frühjahr und von leichten Böden nicht auf die Reserve Grundwasser im Wasserdargebot angewiesen (WOHLRAB 1961 a). BAUMANN (1961) hat in einer Übersicht die Rangordnung der Empfindlichkeit der wichtigsten Kulturarten und Feldfrüchte gegen Wassermangel aufgestellt und daraus die Grundwasserwirkung abgeleitet. Er stützte sich dabei in erster Linie auf Gefäßversuche mit verschieden hoher Wassersättigung und auf Beregnungsversuche (BAUMANN und KLAUSS 1956, ZILLMANN 1957, BAUMANN und KLATT 1957, BAUMANN und ZILLMANN 1959, BAUMANN 1960).

BELLIN (1964) kommt unter besonderer Berücksichtigung der Dränung zu dem Schluß: „Auch unter sonst gleichen Witterungsbedingungen wird jede Pflanzenart — physiologisch bedingt — andere optimale Grundwasserverhältnisse fordern.“

Auf Verschiebungen in der Bestandeszusammensetzung der natürlichen Grünlandnarbe oder eingesäter Kleegrasgemische unter dem Einfluß verschieden hoher Grundwasserstände haben schon KÖNEKAMP und KÖNIG (1929) und später u. a. MINDERHOUD (1960) sowie BOEKER und KMOCH (1963) hingewiesen. Diese Beziehung zwischen Artenzusammensetzung und Grundwasserständen hat man sich in der pflanzensoziologischen Standortansprache zunutze gemacht (u. a. DANCAU 1961, ELLENBERG 1952, FOERSTER 1962, KLAPP 1954 und 1959, MEISEL 1960, STÄHLIN und SCHWEIGHART 1960, TÜXEN 1954 a und b, WEISER 1954).

Grundwasserwirkung auf Trockensubstanz- und Rohproteingehalt, Stickstoffeffekt. Daß der Trockensubstanzgehalt der geernteten Grünmasse mit der Tiefe des Grundwasserstandes zunimmt, ist bei allen Versuchen, bei denen man dieser Frage nachgegangen war (u. a. KÖNEKAMP und KÖNIG 1929, MINDERHOUD 1960 sowie BOEKER und KMOCH 1963), eindeutig festgestellt worden und bedarf auch keiner weiteren Klärung. Ähnliches gilt für den Rohproteingehalt. Schon PITSCH (1913) stellte eine Zunahme mit der Entwässerungstiefe fest. KÖNEKAMP und KÖNIG (1929) kamen zu dem Schluß, daß das Heu im allgemeinen um so eiweißreicher ist, je tiefer der Grundwasserstand liegt. Ihre Ergebnisse aus den Jahren 1927 und 1928 (Tab. 103) zeigen deutlich, daß der Rohproteingehalt den Ausfall an Masse im Eiweißertrag verringert, in nassen Jahren sogar ausgleicht.

Auch MINDERHOUD (1960) stellte in mehreren Jahren eine Zunahme des Rohproteingehaltes beim Grünland mit der Entwässerungstiefe fest. Dieser Zusammenhang machte sich vor allem im Spätsommer (4. und 5. Schnitt) von nassen Jahren stark bemerkbar. BOEKER und KMOCH (1963) wiesen ebenfalls auf die Tendenz hoher Rohproteinerträge auf den Stufen mit tiefem Grundwasserstand hin. Sie machten darauf aufmerksam, daß dies u. a. mit der Bestandeszusammensetzung in Verbindung zu bringen ist.

Tabelle 103. *Rohproteinertrag (dz/ha) bei verschiedenen Grundwasserständen* (KÖNEKAMP und KÖNIG 1929)

		1927				1928			
	Grundwasser-flurabstand (cm)	1. Schnitt (9. Juni)	2. Schnitt (1. August)	3. Schnitt (23. Septemb.)	Insgesamt	1. Schnitt (8. Juni)	2. Schnitt (24. Juli)	3. Schnitt (20. Septemb.)	Insgesamt
Lehm	40	2,33	4,61	1,38	8,32	3,89	8,10	1,63	13,62
	70	3,46	3,25	1,44	8,15	4,56	7,25	1,66	13,47
	100	3,18	3,78	2,21	9,17	5,35	5,72	1,46	12,53
	130	3,03	3,47	1,67	8,17	4,85	5,76	1,54	11,15
Ton	40	2,74	3,27	1,56	7,57	4,77	7,05	2,19	14,01
	70	4,58	4,15	3,37	12,10	5,43	6,83	2,03	14,29
	100	4,96	4,55	3,35	12,86	5,21	6,20	1,21	12,62
	130	5,99	5,70	4,68	16,37	7,10	7,65	1,29	16,04
Niederungsmoor	40	3,70	5,96	3,35	13,01	5,75	11,10	2,66	19,51
	70	3,00	4,89	3,73	11,62	5,87	7,01	2,99	15,87
	100	4,83	4,75	4,23	13,81	7,65	5,88	2,92	16,45
	130	7,30	5,25	5,98	18,53	8,10	7,55	3,82	19,47
Grobkörniger Sand ...	40	3,12	2,45	0,80	6,37	3,49	6,18	1,18	10,85
	70	2,27	2,99	0,63	5,89	3,68	5,46	0,73	9,87
	100	1,43	4,01	0,70	6,14	2,62	6,39	0,78	9,79
	130	1,71	3,19	0,74	5,64	2,04	6,06	0,76	8,86
Feinkörniger Sand ...	40	2,58	3,90	1,22	7,70	4,37	7,19	2,00	13,56
	70	1,92	3,52	0,78	6,22	4,18	6,77	1,11	12,06
	100	2,53	2,49	0,75	5,77	3,54	7,51	0,95	11,00
	130	1,53	2,85	0,34	4,72	3,32	5,00	1,06	9,45
Lehmiger Sand	40	4,29	4,32	2,35	10,96	6,35	8,50	2,41	17,26
	70	4,51	4,46	2,42	11,39	4,43	8,40	1,99	14,82
	100	5,21	4,56	2,62	12,39	5,70	6,99	2,06	14,75
	130	5,35	3,71	2,24	11,30	5,87	6,65	1,65	14,17

Vor allem FERRARI (1949), VAN HOORN (1958), MINDERHOUD (1960) und SIEBEN (1962) haben den Stickstoffhaushalt im Zusammenhang mit den Grundwasserflurabständen bzw. mit der Entwässerungstiefe näher untersucht und behandelt. VAN HOORN stellte auf schwerer Seemarsch anhand der verschiedenen Versuchsglieder mit gestaffelten Stickstoffgaben fest, daß der Einfluß der Stickstoffdüngung bei größerer Grundwassertiefe abnimmt. Er konstruierte aus seinen Ergebnissen die in Abb. 140 dargestellten Funktionskurven für drei Grundwasserstufen (40, 90 und 150 cm Flurabstand).

Nach der Auffassung von VAN HOORN ist der abnehmende Einfluß der Stickstoffdüngung bei größeren Wasserstandstiefen demnach anscheinend dem größeren Aufkommen an bodenbürtigem Stickstoff zuzuschreiben. So ist es auch zu erklären, daß der durch hohe Grundwasserstände verursachte Ertragsrückgang besonders bei Getreide fast vollständig durch verstärkte Stickstoffdüngung aufgehoben werden konnte. Dieser Effekt war bei den Kartoffel- und Rübenerträgen nicht so groß wie beim Getreide. Bei den Hülsenfrüchten trat er nicht mehr eindeutig hervor. VAN HOORN führte dies auf die Tätigkeit stickstoffsammelnder Bakterien zurück, die zwar bei zu hohen Grundwasserständen abnimmt, aber immer noch für ein ausreichendes Stickstoffaufkommen sorgen dürfte. Zu ähnlichen Ergebnissen kam MINDERHOUD (1960) für schwere Flußmarsch. Er folgerte daraus, „daß das Pflanzenwachstum auf zweierlei Weise durch den Grundwasserstand beeinflußt wird: über die Wasserversorgung und über den Stickstoffhaushalt. Diese Faktoren haben ihren Ursprung im Wasser- und Luftgehalt des Bodens, der vom Grundwasserstand abhängig ist. Bei Gewächsen, die verhältnismäßig gut mit Wasser versorgt sind, wie beispielsweise tiefwurzelnde Getreidearten auf Flußmarsch, ist der Faktor Stickstoffhaushalt entscheidend für den verschiedenen Aufwuchs. Bei auf demselben Boden aufwachsenden Feld-

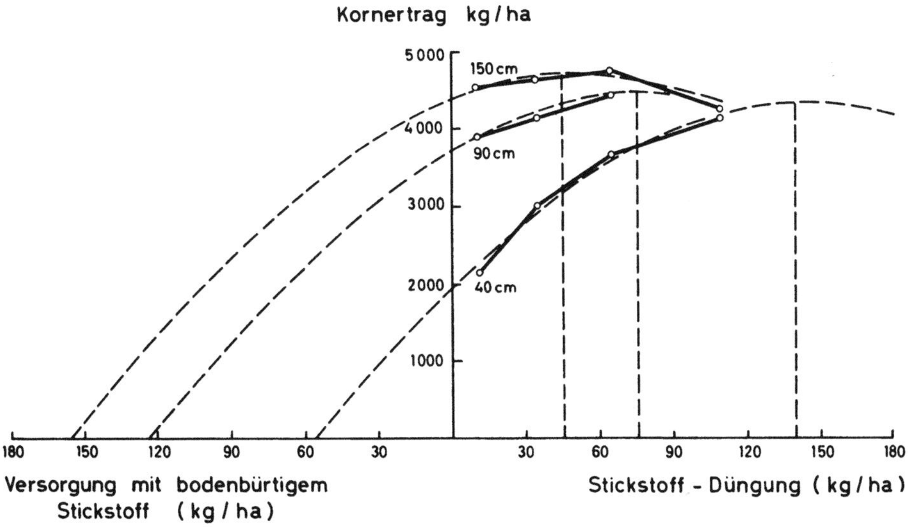

Abb. 140. Grundwasserflurabstand (cm) und Nachlieferung bodenbürtigen Stickstoffs (VAN HOORN 1958)

früchten mit einer guten Stickstoffversorgung, wie Leguminosen, die eine weniger gute Wasserversorgung haben, kann durch eine flache Bewurzelung der Faktor Wasserversorgung das größere Gewicht haben.“

SIEBEN (1962) befaßte sich in diesem Zusammenhang näher mit der Stickstoffbildung in ungedüngtem Boden bei verschiedener Entwässerungstiefe. Er kommt u. a. zu dem Schluß, daß bei schlechter Entwässerung (zu hohe Grundwasserstände) nicht nur die Stickstoffaufnahme gehemmt ist, sondern auch wirkliche Stickstoffverluste auftreten, nach MINDERHOUD (1960) als Folge von Auswaschung und Denitrifikation.

Grundwasserschwankung und Ertrag. Um die Wirkung jahreszeitlich wechselnder Grundwasserstände festzustellen, hatte bereits FRECKMANN (1925) auf seiner Landsberger Station eine Serie Versuchskästen angelegt, über deren Ergebnisse KÖNEKAMP und KÖNIG (1929) berichtet haben (Tab. 104).

Abgesehen vom Anlaufjahr (1924) zeigen alle drei Grundwasserstufen (40, 70 und 100 cm Flurabstand) bei dem untersuchten Boden — lehmiger Sand — nachstehende gleichbleibende Reihenfolge:

Höchstertrag: im Winter trocken, im Sommer dauernd angestautes Grundwasser,
im Winter und Sommer dauernd angestautes Grundwasser,
im Winter trocken, im Sommer alle zwei Wochen wechselweise angestautes und abgelassenes Grundwasser,
im Winter trocken, im Sommer alle vier Wochen wechselweise angestautes und abgelassenes Grundwasser.
Geringster Ertrag: im Winter angestaut, im Sommer trocken.

Tabelle 104. *Heuerträge (relativ) bei wechselnden Grundwasserständen (Winter trocken, Sommer 40 cm = 100)*
(KÖNEKAMP und KÖNIG 1929)

Art der Wasserhaltung	1924	1925	1926	1927	1928
Winter und Sommer 40 cm	117	98	86	92	95
Winter trocken, Sommer 40 cm	100	100	100	100	100
Winter 40 cm, Sommer trocken	85	48	74	90	71
Winter trocken, Sommer 40 cm und trocken, alle 2 Wochen wechselnd	100	66	87	95	92
Winter trocken, Sommer 40 cm und trocken, alle 4 Wochen wechselnd	104	62	78	91	88
Winter und Sommer 70 cm	115	85	91	99	82
Winter trocken, Sommer 70 cm	100	100	100	100	100
Winter 70 cm, Sommer trocken	108	56	77	85	72
Winter trocken, Sommer 70 cm und trocken, alle 2 Wochen wechselnd	103	78	94	90	92
Winter trocken, Sommer 70 cm und trocken, alle 4 Wochen wechselnd	102	77	84	89	88
Winter und Sommer 100 cm	98	83	92	98	104
Winter trocken, Sommer 100 cm	100	100	100	100	100
Winter 100 cm, Sommer trocken	76	53	71	72	98
Winter trocken, Sommer 100 cm, alle 2 Wochen wechselnd	88	77	93	91	121
Winter trocken, Sommer 100 cm, alle 4 Wochen wechselnd	97	68	84	86	104

Zu einem ähnlichen Befund kommt WÄRE (1955). Er hatte den Heuertrag bei einer ein- bis zweijährigen Wechselweide auf Lehm und Moor ebenfalls fast jedes Jahr durch Anhebung des Sommerwasserstandes gesteigert.

Demgegenüber haben sich nach MINDERHOUD (1960) auf dem schweren Flußmarschboden die Erwartungen nicht bestätigt, daß wechselnde Grundwasserstände (Winter tief, Sommer hoch) die Vorteile von hohen und die von konstant tiefen Wasserständen miteinander verbinden. MINDERHOUD weist noch darauf hin, daß bei Dauergrünland die Gefahr des Durchtretens und damit eine Einschränkung der Beweidungsfähigkeit besteht, wenn der Grundwasserstand gehoben wird, ehe die Verdunstung die Niederschläge übertrifft.

Die Ergebnisse von MINDERHOUD (1960) stehen nur scheinbar im Widerspruch insbesondere zu den Befunden von KÖNEKAMP und KÖNIG (1929). Bei letzteren handelt es sich um einen relativ leichten Boden. Außerdem liegt Landsberg/Warthe im kontinentalen Klimabereich.

Auf der schweren Flußmarsch unter ozeanischem Einfluß hingegen entsprach der tiefste der eingestellten Grundwasserstände (140 cm unter Flur) offensichtlich annähernd dem Optimum, so daß allenfalls in außergewöhnlich trockenen Sommern von höheren Grundwasserständen noch ein Erfolg zu erwarten war.

BAUMANN (1961) ging an Hand der Spiegelamplitude näher auf die Wasserversorgung der Pflanzen aus dem Grundwasser ein. Er kam zu dem Schluß, daß bei einer mittleren Grundwasserganglinie mit Höchststand im frühen Frühjahr und Tiefststand im Juli „Arten, deren Wurzelwachstum später einsetzt und deren Hauptwasserbedarf im Juli und August liegt, stärker gefördert werden".

Ganz beträchtliche Ertragsdepressionen stellte BERTRAM (1931) bei einem Versuch auf ummantelten und mit Sommergerste bestellten Bodensäulen (schwerer Weichselniederungsboden) fest, in denen der Grundwasserstand vom 10. Juni bis 10. Juli von 80 cm unter Flur in drei Parallelreihen auf 50, 25 und 10 cm angehoben und danach wieder auf 80 cm gesenkt wurde. Es ist daraus analog zu den Feststellungen von MINDERHOUD (1960) zu schließen, daß längere Zeit anhaltende hohe Grundwasserstände während der Vegetationszeit, wie sie etwa unter dem Einfluß eines ausgedehnten Sommerhochwassers im Nachbarbereich von Flüssen auftreten können, sich zumindest bei schweren Böden nachteilig für den Aufwuchs und die Ertragsbildung auswirken. Sonst aber sind im Winter tiefe und im Sommer nahe unter Flur liegende Grundwasserstände, die aber den Wurzelraum nicht einengen, vor allem für die überwinternden Früchte besonders förderlich. Solche Voraussetzungen bestehen, mit Ausnahme von Tallagen der Hochgebirgsflüsse (Sommer: Hochstand), in der Regel nur bei künstlich vorgenommenem Einstau. Unter natürlichen Verhältnissen verläuft die jahreszeitliche Ganglinie des Grundwasserspiegels im allgemeinen umgekehrt. Es kann also daraus gefolgert werden, daß günstige Wirkungen vom Grundwasser dann zu erwarten sind, wenn die Amplitude der Ganglinie seines Spiegels nicht zu groß ist.

C. Funktionskurven Grundwasser—Pflanzenertrag

In Anbetracht der aufgeführten vielfältigen Abhängigkeit der Grundwasserwirkung (Boden, Witterung, Pflanzenart usw.) ist es verständlich, daß KLAPP (1956) äußerte: „Es gibt keinen allgemein gültigen ‚besten' Grundwasserstand, seine günstigste Lage hängt von den verschiedensten Einflüssen ab, sie ist selbst

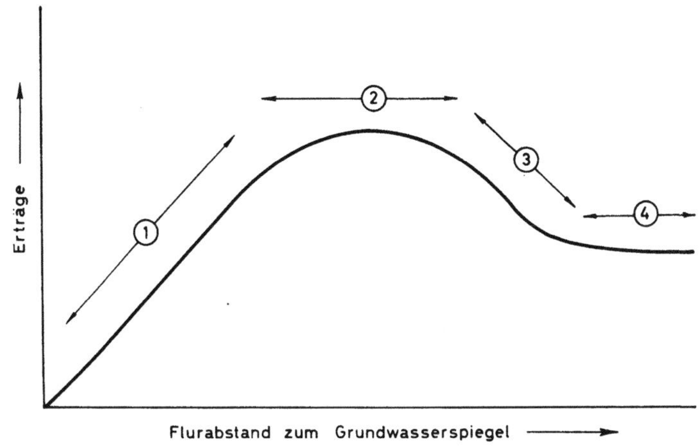

Abb. 141. Idealfunktion zwischen Grundwasserstand und Pflanzenertrag (nach VISSER 1960)

für eine bestimmte Fläche nie für längere Zeit dieselbe." WESSELING (1957) ist ähnlicher Auffassung, wenn er erklärt: „Es besteht keine allgemeine Ertrags-Grundwasserstands-Kurve, die für jedes Jahr gilt." Dennoch fehlt es nicht an Versuchen, gerade von niederländischen Fachleuten, rechnerisch und vor allem graphisch Funktionen zwischen Grundwasserstand und Pflanzenertrag abzuleiten.

Grundwasserstand—Pflanzenertrag. Mit obiger Funktionskurve (Abb. 141) hat VISSER (1960) ganz allgemein und grundsätzlich eine Beziehung zwischen Grundwasserstand und Pflanzenertrag abgeleitet.

Diese ideelle Bezugslinie setzt sich aus vier Abschnitten zusammen, die im wesentlichen die Reaktion der Pflanze auf den Luftvorrat bzw. -austausch einerseits sowie auf das Wasserdargebot andererseits widerspiegeln (BELLIN 1964, VISSER 1960, WESSELING 1957):

Das Ansteigen der Kurve (Abschnitt 1) ist auf die Verbesserung des Luftaustausches im Wurzelraum zurückzuführen.

Im Gipfelbereich (Abschnitt 2) herrschen ausgewogene Verhältnisse.

Der zunehmende Wassermangel macht sich im 3. Abschnitt bemerkbar.

Bei optimalem Luftaustausch unter den gegebenen Bodenverhältnissen beschränkt sich im 4. Abschnitt die Wasserversorgung auf die aus den Niederschlägen stammende und im Boden gehaltene pflanzenverfügbare Feuchte.

Nach der von Mitscherlich (1928) abgeleiteten zweiten Annäherung des Wirkungsgesetzes der Wachstumsfaktoren kann der Luftvorrat bzw. -austausch als „Wirkungsfaktor" und der Wassermangel als „Schädigungsfaktor" aufgefaßt werden. Geht man jedoch davon aus, daß der Flurabstand zum Grundwasserspiegel das Maß für eine gemeinsam wirkende Faktorengruppe, nämlich den „Wasser- und Lufthaushalt des Wurzelraumes", bildet, dann entspricht die Funktionskurve der von v. Boguslawski und Schneider (1962/1964) behandelten dritten Annäherung des Ertragsgesetzes mit folgender Gleichung:

$$y = M \cdot 10^{-z\left[\log \frac{x+i}{m+i}\right]^n}$$

Hierin bedeuten:

y = Ertrag (Mittelwert)

x = verabreichte Menge des Wachstumsfaktors

M = Maximalertrag, welcher bei der gegebenen Konstellation aller Wachstumsfaktoren und ausreichender Zufuhr des variierten Faktors erreicht wird

m = diejenige Menge des Wachstumsfaktors als Teil von x, welche zur Erreichung von M erforderlich ist

i = Abstand des Anfangspunktes der Kurve vom Nullpunkt, die im System ohne Zufuhr von x wirksame Menge des variierten Wachstumsfaktors

z = Konstante, welche die Breite des Maximumgebietes und zusammen mit $m+i$ Anstieg und Abfall der Kurve bestimmt.

Für verschiedene niederländische Böden (Gruppe I bis VII) hat Visser (1959) die in Abb. 142 dargestellten Funktionskurven abgeleitet.

Die Ergebnisse stammen von Grundwasserstandsversuchsfeldern und von vergleichenden Ertragsfeststellungen und schließen Acker sowie Grünland ein. Den Einfluß der anderen Wachstumsfaktoren, welche die Wirkung des Feuchtigkeitseinflusses verdecken könnten, hat Visser durch eine graphisch-statistische Methode eliminiert (s. auch Baumann 1961). Bei den Kurven I bis VI handelt es sich um Mineralböden mit abnehmender nutzbarer Wasserkapazität, bei Bezugslinie VII um Niedermoor.

Legt man die von v. Boguslawski und Schneider (1962/1964) behandelte Formel von der 3. Annäherung des Ertragsgesetzes zugrunde, dann stellen die von Visser abgeleiteten Bezugslinien (I bis VII) Variationen der Parameter „m", „i" und „z" dar. Der Parameter „i", d. h. die im System ohne Zufuhr von „x" wirksame Menge des variierten Wachstumsfaktors „Wasser- und Lufthaushalt" (= Abstand des Anfangspunktes der Kurve vom Nullpunkt), hat in diesem Falle ein umgekehrtes Vorzeichen. Die obere Grenzfläche des völlig mit Wasser gesättigten Bodenraumes liegt nämlich je nach der Mächtigkeit des geschlossenen Saugraumes mehr oder weniger (kapillares Leitvermögen des jeweiligen Bodens) über der eigentlichen Grundwasseroberfläche ($\approx$ Grundwasserspiegel).

An Hand der Bezugslinien VI und I sollen die in die Funktionskurve und die Formel von Schneider (1960) eingegangenen Faktoren bzw. Parameter kurz erläutert werden. Bezugslinie VI entspricht einem Boden (mittel- bis grobkörniger Sand) mit großem grobem Porenraum (pF 0 bis 2,5[1]) und sehr geringer nutzbarer

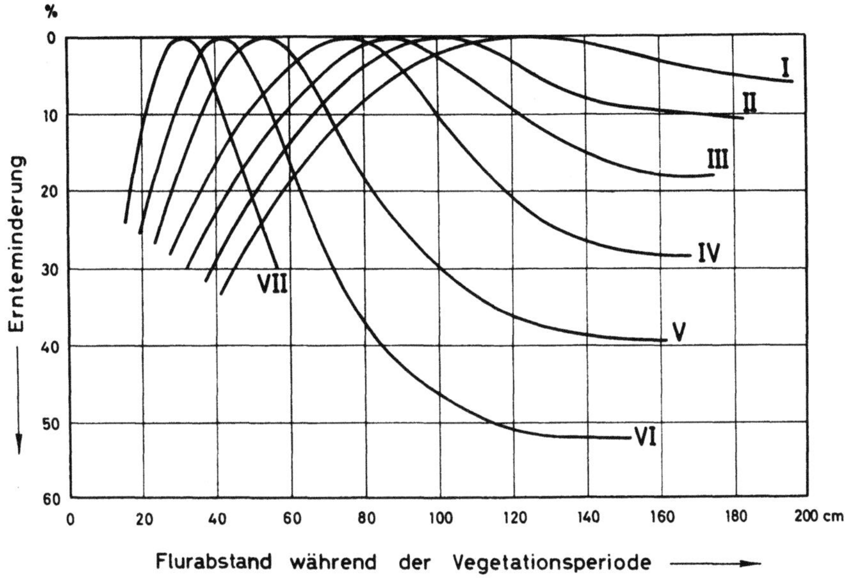

Abb. 142. Beziehungen zwischen Sommergrundwasserstand und Pflanzenertrag bei verschiedenen niederländischen Böden. I bis VI Mineralböden mit abnehmender nutzbarer Wasserkapazität, VII Niedermoor (VISSER 1959)

Wasserkapazität (pF 2,5 bis 4,2), demzufolge nur schwacher Mächtigkeit des geschlossenen Saugraumes. Die Kurve setzt deshalb auf der x-Achse relativ nahe am 0-Punkt an. Wegen der geringen kapillaren Steighöhe und der im allgemeinen nicht so tief reichenden Durchwurzelungsneigung steigt die Funktionskurve verhältnismäßig steil an. Sie fällt außerdem wegen des bei diesem Boden leicht ins Minimum geratenden, allein aus Niederschlägen stammenden Wasserdargebotes relativ stark und tief (50%) ab. Der Maximumbereich ist aus den gleichen Gründen ziemlich schmal.

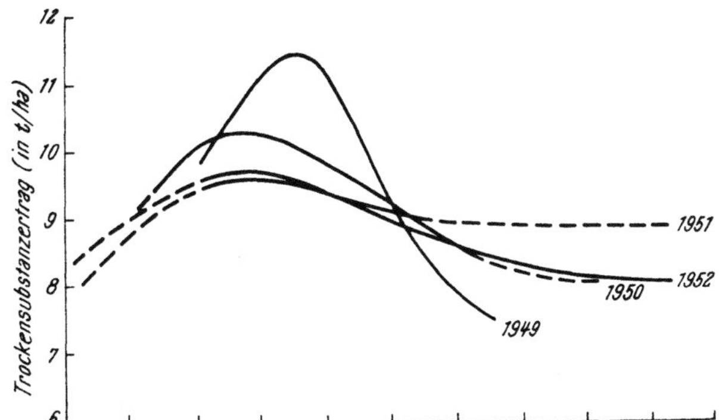

Abb. 143. Beziehungen zwischen Sommergrundwasserstand und Grünlandertrag auf Moormarsch bei verschiedenem Witterungsverlauf (SMITS 1962)

Bei mittelschweren Böden (Bezugslinie I) hingegen reicht die Kapillarwirkung (Mächtigkeit des Saugraumes) wesentlich weiter ($i_{I} > i_{VI}$) und bewirkt gemeinsam

[1] Näheres über Porenvolumen, Porenraumverteilung und pF-Wert ist unter anderem bei SCHACHTSCHABEL und HARTGE (1959), WOHLRAB (1965), WOHLRAB und LANGNER (1964) sowie WOLKEWITZ (1959 und 1962) zu finden.

mit der geringen Durchlüftung (niedrigerer Anteil grober Poren) einen deutlich flacheren Anstieg der Kurve ($m_I > m_{VI}$) sowie in Verbindung mit der größeren Durchwurzelungsneigung einen ausgedehnten Gipfelbereich ($z_I > z_{VI}$). Der Abfall ist wegen der auf diesen Böden seltener geschwächten Wasserversorgung infolge der großen Speicherung von Niederschlagswasser wesentlich geringer.

Jede dieser von Visser (1959) für einen Boden mit bestimmten Porenraumverhältnissen (Wasserbindungs- und kapillares Leitungsvermögen) aufgestellten Funktionskurven kann ihrerseits in verschiedenen Variationen auftreten. Solche Variationen ergeben sich vor allem aus dem Witterungsverlauf, den Feuchteansprüchen der einzelnen Kulturarten und Feldfrüchte und aus der Düngung, insbesondere der Stickstoffdüngung. Smits (1962) hat derartige Bezugslinien abgeleitet, und zwar von den 1949 bis 1952 festgestellten Erträgen auf Acker und Grünland bei verschiedenen Böden und bestimmten Sommergrundwasserständen. Herausgegriffen ist hier eine besonders deutliche Darstellung von witterungsbedingten Variationen der Funktionskurve des Grünlandertrages auf Moormarschböden (Abb. 143).

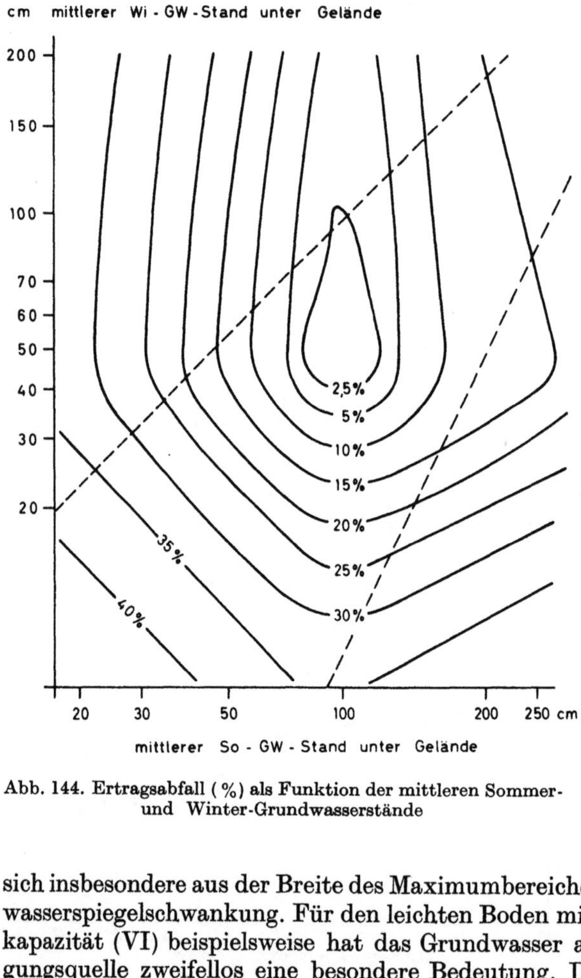

Abb. 144. Ertragsabfall (%) als Funktion der mittleren Sommer- und Winter-Grundwasserstände

Grundwasserschwankung—Pflanzenertrag. Die von Visser (1959) für verschiedene Böden aufgestellten Funktionslinien (Abb. 142) beziehen sich auf einen mittleren Grundwasserstand während der Vegetationszeit. Dabei ergibt sich insbesondere aus der Breite des Maximumbereiches ein Hinweis auf die Grundwasserspiegelschwankung. Für den leichten Boden mit geringer nutzbarer Wasserkapazität (VI) beispielsweise hat das Grundwasser als zusätzliche Wasserversorgungsquelle zweifellos eine besondere Bedeutung. Diese positive Wirkung setzt jedoch neben einem optimalen Flurabstand gleichzeitig analog zum schmalen Maximumbereich der Kurve eine geringe Grundwasserspiegelschwankung während der Vegetationszeit voraus.

Mit der Ertragsbeeinflussung verschiedener Winter- und Sommergrundwasserstände (unter Flur) haben sich vor allem Bloemen (1951) und Visser (1959) befaßt und diese in entsprechenden Zeichnungen dargestellt (s. auch Bellin 1964).

Abb. 144 zeigt derartige Bezugslinien für die unter III (Abb. 142) zusammengefaßten Böden. Auffällig ist bei diesen Böden die größere Optimumbreite der

Wintergrundwasserstände im Vergleich mit der der Sommergrundwasserstände. Wenn man von dem im allgemeinen innerhalb der Fehlergrenze liegenden Ertragsabfall von 2,5% absieht, dann ergibt sich im übrigen, daß tiefe Wintergrundwasserstände offensichtlich keine Ertragsdepressionen hervorrufen. — Auch für die anderen sechs nach der nutzbaren Wasserkapazität geordneten niederländischen Bodenklassen hat VISSER (1959 und 1961) derartige Funktionen abgeleitet und in graphische Darstellungen transponiert.

Literatur

BADEN, W.: Klärung von Grundwasserentzugsfragen in Moorgebieten. Ber. a. d. Landesanst. f. Bodennutzungsschutz, Bochum, **3**, 207–212 (1962). — BADEN, W., und R. EGGELSMANN: Zur Frage der Wasserreserven unserer Moorböden im Frühjahr. Wasserwirtsch. **45**, 228–231 (1955). — Über die Regelung des Wasserhaushaltes bei Moormeliorationen und die dafür notwendigen Vor- und Folgearbeiten. Wasser u. Boden **10**, 29–36 (1958). — BADEN, W., und H. SEGEBERG: Untersuchungen über die Einwirkungen von Grundwasserabsenkungen auf den Wasserhaushalt von Moor- und Anmoorböden der Erftniederung. Wasser u. Boden **10**, 360–366 (1960). — BAUMANN, H.: Die konstitutionelle Anpassung der Kulturpflanzen an die Wasserversorgung. Z. Pflanzenernähr., Düng., Bodenkde. **46** (91), 1/3, 176 (1949). — Wasserversorgung und Ertrag. Nach einem Vortrag, gehalten auf der Hochschultagung der Landwirtschaftlichen Fakultät der Christian-Albrechts-Universität Kiel am 25. Oktober 1960. — Witterungslehre für die Landwirtschaft. Berlin-Hamburg: Parey. 1961a. — Pflanzen-Wasserversorgung und Ertragsbilanz an Standorten mit oberflächennahem Grundwasser. Wasser u. Boden **13** (12), 411 (1961b). — BAUMANN, H., und M. L. KLAUSS: Gefäßversuche mit Sommergetreide zur Frage der konstitutionellen Anpassung der Pflanzen an die Wasserversorgung. Z. Acker- u. Pflanzenbau **101**, 35–52 (1956). — BAUMANN, H., und F. KLATT: Beregnungszeitpunkt mit Hafer- und Winterweizensorten. Z. Acker- u. Pflanzenbau **103**, 380–396 (1957). — BAUMANN, H., und K. H. ZILLMANN: Zum Wasserverbrauch der Sommergerste in mehrjährigen Gefäßversuchen. Z. Acker- u. Pflanzenbau **108**, 62–76 (1959). — BELLIN, K.: Entwicklung und Probleme des Dränversuches unter besonderer Berücksichtigung des Ertragsversuches. Mitt. a. d. Inst. f. Wasserwirtsch. u. landwirtsch. Wasserbau d. T. H. Hannover **4** (1954). — BERGERHOFF, H., B. WOHLRAB und S. BOHN: Grenzen und Möglichkeiten der natürlichen Wasserversorgung von landwirtschaftlichen Kulturen aus oberflächennahen Grundwasservorkommen unter besonderer Berücksichtigung bisheriger Beobachtungen am linken Niederrhein. Z. Acker- u. Pflanzenbau **99** (1), 19–44 (1955). — BERTRAM: Die Auswirkung von Grundwasserstandsschwankungen in künstlich entwässerten Poldern. Kulturtechniker **34**, 194–213 (1931). — BLOEMEN, G.: Twee aspecten van de grondwaterdiepte. Maandbl. Landbouwvoorlichtingsd. **8**, 387–390 (1951). — BOEKER, P., und H. G. KMOCH: Grundwasserstandsversuche in der Boker Heide, Kreis Paderborn. Z. Kulturtechn. u. Flurbereinigung **4**, 142–156 (1963). — BOGUSLAWSKI, E. v.: Was ist Bodenfruchtbarkeit? Bodenfruchtbarkeit, Beiträge, herausgegeben von der Fachgruppe zur Förderung der Bodenfruchtbarkeit im Verband DLUFA. Praxis u. Forschung, Oldenburg, **1955**, 7–15. — BOGUSLAWSKI, E. v., und B. SCHNEIDER: Die dritte Annäherung des Ertragsgesetzes (1.—3. Mitt.). Z. Acker- u. Pflanzenbau **114** (3), 221–236; **116** (2), 113–128; **119** (1), 1–28 (1962–64). — BRÜNE, F.: Die Grundwasserabsenkung im linksrheinischen Erftgebiet und ihr Einfluß auf die Ertragsfähigkeit der dortigen Lößlehmböden. Wasser u. Boden **10**, 347–360 (1960). — BUCHHOLZ, F.: Beiträge zur Dynamik chemischer Grundwassereigenschaften und zum Bodenfeuchtigkeitsverlauf in grundwasserbeeinflußten Sandböden unter Waldbestockung im Gebiet des nordostdeutschen Pleistozäns. Diss. Eberswalde 1960.

CLAYTON, B. S., et al.: Water control in the peat and muck soils of the Florida Everglades. Bull. Nr. 378, Agricult. Exper. Stat. Gainesville, Florida (1942). — COLLIER, D., und M. ROBELIN: Contribution à l'étude de la dynamique de l'eau et des substances chimiques dans les sols argilo-calcaires de Limagne. Ann. Agronom. **4**, 415–453 (1959).

DANCAU, B.: Auswertung der Vegetationskarte für Planung von wasserwirtschaftlichen Maßnahmen. Bayer. Landwirtsch. Jb. **38** (3), 373–380 (1961). — Deutscher Grundwasserausschuß: Grundwasser, Richtlinien für Beobachtung und Auswertung 1961. — DHEIN, A., und N. MOTT: Untersuchungen über den Einfluß des Grundwassers auf den Wasserhaushalt des Bodens in dynamischer und statischer Hinsicht. Z. Acker- u. Pflanzenbau **100** (1), 41–52 (1955). — DIN 4049: Gewässerkunde, Fachausdrücke und Begriffsbestimmungen; Blatt 1 (Teil I: quantitativ). Berlin-Köln: Beuth. 1954. — Dörter, K.: Die Steighöhe des Wassers in humusfreien Sanden in Abhängigkeit von der Hygroskopizität. Kühn-Arch. **68**, 285–332 (1954).

Eden, A., et al.: The effect of ground water-level upon productivity and composition of Fenland grass. J. Agricult. Sci. **41**, part 3, 191–202 (1951). — Ellenberg, H.: Auswirkungen der Grundwasserabsenkung auf die Wiesengesellschaften am Seitenkanal westlich Braunschweig. Angew. Pflanzensoziologie G, Stolzenau, 1952b. — Ellis, N. K., und R. Morris: Preliminary observations on the relation of yield of crops grown on organic soil with controlled water table and the area of aeration in the soil and subsidence of the soil. Proc. 1945 Soil Sci. Soc. Amer. **10**, 282–283 (1946).

Ferrari, Th. J.: Stikstofbemesting en bodemfactoren. Landbouwkundig Tijdschr. **61** (2), 111–120 (1949). — Foerster, E.: Wasserbedingte Bodentypen und Grünland-Pflanzengesellschaften. Ber. a. d. Landesanst. f. Bodennutzungsschutz, Bochum, **3**, 87–90 (1962). — Frankena, H. J., und M. A. J. Goedewaagen: Een vakkenproef over den invloed van verschillende waterstanden op den grasgroei bij drie grondsoorten. Versl. Landbouwkund. Onderz., Wageningen, **48** (6), (1942). — Franz, H.: Feldbodenkunde. Wien: Fromme. 1960. — Freckmann, W.: Versuche über den Einfluß verschieden hoher Grundwasserstände auf das Gedeihen eines Kleegrasbestandes. Landwirtsch. J., Erg. Bd. **1**, 38 (1926).—Friedrich, W., und H. Franzen: Ein neuer Versickerungsmesser (Lysimeter). Dtsch. Gewässerkdl. Mitt. **4**, 105–111 (1960).

Gliemeroth, G.: Wasserhaushalt des Bodens in Abhängigkeit von der Wurzelausbildung einiger Kulturpflanzen. Z. Acker- u. Pflanzenbau, Berlin u. Hamburg, **95**, 1 (1952).

Hahne: Die Bedeutung des Bodenwassers für die Bodenschätzung. Braunkohle **1938**, 698–703. — Hanus, H.: Wurzelprofil und Wasserversorgung der Grasnarbe bei verschiedenen Grundwasserständen. Diss. Bonn 1962. — Heide, G.: Bodenkundliche Fragen zur Grundwasserabsenkung im Erftgebiet. Braunkohle, Wärme, Energie **60** (3), 97–104 (1960). — Hooghoudt, S. B.: Resultaten van het grondwaterstandsproefveld te Nieuw-Beerta. Verslagen Technische Bijjeenkomsten van de Commissie voor Hydrologisch Onderzoek T.N.O. **1–6**, 56–63 (1952). — Hooghoudt, S. B., B. van der Woerdt, J. Bennema und H. van Dijk: Vertrogende veengronden in West-Nederland. Versl. Landbouwkund. Onderz., Wageningen, **6623** (1960). — Hoorn, H. W. van: Results of a ground water-level experimental field with arable crops on clay soil. Netherl. J. Agricult. Sci. **6** (1), 1–10 (1958).

Kalisvaart, C.: Influence of sub-irrigation on grassland; Report of the Fifth International Grassland Congress 1949, 47–50. — Keilhack, K.: Lehrbuch der Grundwasser- und Quellenkunde für Geologen, Hydrologen, Bohrunternehmer usw., 3. Aufl. Berlin 1935. — Klapp, E.: Erträge von Pflanzengesellschaften in Beziehung zu Grundwasser und Nährstoffversorgung. Arb. a. d. Zentralst. f. Vegetationskartierung, Stolzenau, **8** (1954). — Wiesen und Weiden, 3. Aufl. Berlin-Hamburg: Parey. 1956. — Lehrbuch des Acker- und Pflanzenbaues. 5. Aufl., 93–96. Berlin-Hamburg: Parey. 1958. — Die Vegetationskartierung als Mittel zur Beurteilung des Wasserhaushaltes für die Bodennutzung, insbesondere zur Planung und Überwachung wasserwirtschaftlicher Maßnahmen. CEA Brugg/Schweiz 1959. — Zu den Grundwasserabsenkungen im linksrheinischen Erftgebiet und ihrem Einfluß auf die Ertragsfähigkeit der dortigen Lößlehmböden. Wasser u. Boden **13** (4), (1961). — Kmoch, H. G.: Über den Umfang und einige Gesetzmäßigkeiten der Wurzelmassenbildung unter Grasnarben. Z. Acker- u. Pflanzenbau **95**, 363–380 (1952). — Die Luftdurchlässigkeit des Bodens, ihre Bestimmung und ihre Bedeutung für einige ackerbauliche Probleme. Berlin: Borntraeger. 1962. — Koehne, W.: Grundwasserkunde, 2. Aufl. Stuttgart: Schweizerbart. 1948. — Könekamp, A.: Beitrag zur Kenntnis des Wurzelwachstums einiger Klee- und Gräserarten. Landwirtsch. Jb. **80** (1934). — Könekamp, A., und F. König: Untersuchungen über den Einfluß des Grundwassers auf die Entwicklung eines Kleegrasgemisches. Landwirtsch. Jb. **69**, 209–252 (1929). — Kostjakow, A. N.: Grundlagen der Meliorationen. Berlin: VEB-Verlag Technik. 1959. — Kozeny, J. S.: Über den kapillaren Aufstieg des Wassers im Boden. Kulturtechniker **27** (1), (1924). — Kutschera, L.: Wurzelatlas mitteleuropäischer Ackerunkräuter und Kulturpflanzen. Frankfurt/M.: DLG-Verlag. 1960.

LEBEDEFF, A. F.: Die Bewegung des Wassers im Boden und im Untergrund. Z. Pflanzenernähr., Düng., Bodenkde. A 10, 1–36 (1927). — LIESHOUT, J. W. VAN: Invloed van het bodenmilieu op ontwikkeling en activiteit van het wortelstelsel. Versl. Landbouwkund. Onderz., Wageningen, 6618 (1960).

MAKKINK, G. F.: De grasopbrengst en de grondwaterstand. Landbouwkundig Tijdschr. 67 (10), 713–731 (1955c). — MEISEL, K.: Die Auswirkungen der Grundwasserabsenkung auf die Pflanzengesellschaften im Gebiete um Moers (Niederrhein). Arb. a. d. Bundesanst. f. Vegetationskartierung, Stolzenau. — MINDERHOUD, J. W.: Grasgroei en grondwaterstand, onderzoekingen over de betekenis van de grondwaterstand voor komkleigrasland. Proefstation voor de Akker- en Weidebouw, Wageningen, 15 (1960). — MITSCHERLICH, E. A.: Die zweite Annäherung des Wirkungsgesetzes der Wachstumsfaktoren. Z. Pflanzenernähr., Düng., Bodenkde. A 12, 273–282 (1928). — MÜCKENHAUSEN, E.: Die Beurteilung des Grundwassers. Forstliche Standortaufnahmen; Begriffe u. Fachausdrücke; Landwirtsch. — Angew. Wissenschaft. Hiltrup: Landwirtsch. Verlag. 1958.

NICHOLSON, H. H.: The control of ground water-level in crop production. Reprint from the „Report of the Thirteenth International Horticultural Congress, London 1952" (1952). — NICHOLSON, H. H., und D. H. FIRTH: The effect of ground water-level on the performance and yield of some common crops. J. Agricult. Sci. 43, part 1, 95–104 (1953). — The effect of ground water-level on the yield of some common crops on a Fen peat soil. J. Agricult. Sci. 50, part 3, 243–252 (1958). — NITZSCH, W. v.: Über Wasserbewegung und Porosität des Bodens und ihre Beziehungen zur Bodenbearbeitung. RKTL-Schriften 41 (1933). — NÖRING, F.: Einflüsse der Kunstdüngung auf den Chemismus des Grundwassers. Gesundheits-Ing., München, 72, 190–191 (1951).

OEHLER, TH.: Versuche über die Wasserverteilung im Kapillarsaum. Verhandl. VI. Komm. Int. Ges. Zürich, Teil B, 84 (1927). — OLBERTZ, M. H.: Über die am Standort des Kulturbodens erfaßbaren Größen des Wasserhaushaltes. Wiss. Abh. 23. Berlin: Dtsch. Akad. Landwirtsch. Wiss. 1957. — OLBERTZ, M. H., und H. PRESS: Landwirtschaftlicher Wasserbau. Taschenbuch der Wasserwirtschaft, 3. Aufl. Hamburg: Wasser und Boden. 1963.

PITSCH, O.: Proefneming over den invloed van de hoogt van den grondwaterstand op de groote van den oogst aan gras of hooi. Meded. Rijks Hoogere Land-, Tuin- en Boschbouwschool, deel 6, 1–41 (1913).

RAMSAUER, B.: Der Boden. Taschenbuch Landwirtschaftlicher Wasserbau (Herausgeber O. UHDEN). Stuttgart: Franckh. 1964. — RODE, A. A.: Das Wasser im Boden. Neubearbeitet und in deutscher Sprache von M. TRÉNEL. Berlin: Akademie-Verlag. 1959. — ROE, H. B.: A study of influence of depth of groundwater-level on yields of crops grown on peat lands. Bull. No. 330, Agricult. Exper. Stat. Univ. Minnesota (1936).

SCHACHTSCHABEL, P., und K. HARTGE: Die Messung der Bodenstruktur durch Bestimmung der pF-Kurve und der Strukturstabilität. Landwirtsch. Forsch. 12. Sonderh., 30–36 (1959). — SCHMIDT, H., und E. KLAPP: Landwirtschaftlich-wasserwirtschaftliches Gutachten zur Errichtung des Wasserwerkes V in der Senne (unveröffentlicht). 1949. — SCHOLL, G.: Verfahren zur Prüfung des Grundwassereinflusses auf das Pflanzenwachstum. Ber. a. d. Landesanst. f. Bodennutzungsschutz, Bochum, 3, 83–86 (1962a). — Eine Anlage zur Einrichtung künstlicher Grundwasserstände im gewachsenen Boden. Z. f. Kulturtechn. 3, 147 (1962b). — SCHROEDER, G.: Landwirtschaftlicher Wasserbau, 3. Aufl. Berlin-Göttingen-Heidelberg: Springer. 1958. — SCHWILLE, F.: Chloride und Nitrate in den Grundwässern Rheinhessens und des Rheingaues. Gas- u. Wasserfach 94, 410–414 (1953). — Nitrate im Grundwasser. Dtsch. Gewässerkdl. Mitt. 6 (2), 25–32 (1962). — SEGEBERG, H.: Eine Methode zum Nachweis von Grundwasserentzugsschäden in Moor- und Mineralböden. Ber. a. d. Landesanst. f. Bodennutzungsschutz, Bochum, 3, 107–112 (1962). — SIEBEN, W. H.: Einfluß von Lehmmächtigkeit und Sommergrundwasserstand auf die Erträge von „Lehm auf Sand"-Boden. Ber. a. d. Landesanst. f. Bodennutzungsschutz, Bochum, 3, 161–162 (1962). — SIEBEN, W. H., und H. SMITS: De invloed van het weer in 1949 en 1950 op de opbrengsten van grasland. Maandbl. voor de Landbouwvoorlichtingsdienst 8 (2), 71–79 (1951). — SMITS, H.: Der Zusammenhang zwischen Grundwasserstand und Ertrag von Roggen, Hafer und Futterrüben auf humosen diluvialen Sandböden, Moorböden und Moormarschböden in nassen und trockenen Jahren. Ber. a. d. Landesanst. f. Bodennutzungsschutz, Bochum, 3, 147-154 (1962). — STÄHLIN, A., und O. Schweighart: Verbreitete Pflanzengesellschaften des Dauergrünlandes, der Äcker, Gärten und Weinberge. München: BLV-Verlagsges. 1960. — Stocker, O.: Diskussionsbemerkung in: Pflanzensoziologie als Brücke zwischen Land- und Wasserwirtschaft. Angew. Pflanzensoziologie, Stolzenau, 8, 129 (1954). — Stremme, H. E.: Grundwasserschwankung in Feuchtböden auf Geschiebelehm. 7th Int. Congr. Soil Sci., Madison, Wisc., USA. **26**, 184–194 (1960).

Trénel, M.: Zur gutachtlichen Beurteilung des Einflusses der Grundwasserabsenkung auf den Ernteertrag im Löß. Z. Pflanzenernähr., Düng., Bodenkde. **45** (30), 133–147 (1949); **49** (94), 25–59 u. 224–237 (1950). — Über Kondensationsvorgänge in Böden. Sitzungsber. DAL, **III**, H. 6 (1954). — Tüxen, R.: Pflanzengesellschaften und Grundwasserganglinien. Angew. Pflanzensoziologie, Stolzenau, 8 (1954a). — Die Wasserstufenkarte und ihre Bedeutung für die nachträgliche Feststellung von Änderungen im Wasserhaushalt einer Landschaft. Angew. Pflanzensoziologie, Stolzenau, 8 (1954b).

Uhden, O.: Entwässerung. Taschenbuch Landwirtschaftlicher Wasserbau. Stuttgart: Franckh. 1964.

Vageler, P.: Der Kationen- und Wasserhaushalt des Mineralbodens. Berlin: Springer. 1932. — Visser, W. C.: Crop growth and availability of moisture. Inst. for Land and Water Management Research, Wageningen, Techn. Bull. **6** (1959). — Size and shape of sub-divisions in drainage areas. Inst. for Land and Water Management Research, Wageningen, Techn. Bull. **17** (1960). — The agro-hydrological survey of the Netherlands. Inst. for Land and Water Management Research, Wageningen, Techn. Bull. **19** (1961).

Wäre, M.: Pohjavedenkorkeuden vaikutusta selvittäviä kokeita nurmikasveilla vuosina 1939–1954. Maatalous ja Koetoiminta IX, Helsinki (1955). — Weaver, J. E.: Root development of field crops. New York: Mc-Graw-Hill. 1926. — Weber, H.: Über Kondensationsvorgänge im gewachsenen Boden. Diss. Berlin 1954. — Wesseling, J.: Enige aspecten van de waterbeheersing in landbouwgronden. Versl. Landbouwkund. Onderz., Wageningen, **63**, 5 (1957). — Weiser, F.: Die Pflanzensoziologie als Beweismittel bei Grünlandschädigungen durch Grundwasserentzug. Z. Acker- u. Pflanzenbau **97** (4), 393–398 (1954). — Weisbrod, W.: Die Wasserfrage bei der Bodenschätzung. Braunkohle **49**, 883–887 (1937). — Woerdt, B. van der: Nadelen van het overschrijden van de optimale watervoorziening bij blijvend grasland. Landbouwvoorlichting **10** (4), 137–142 (1953). — Wohlrab, B.: Grundwasserfragen im Rheinischen Braunkohlengebiet. Mitt. DLG **74** (41), 1191–93 (1959a). — Untersuchungen über die Auswirkungen des Grundwasserentzuges auf die Ertragsleistung landwirtschaftlicher Kulturen. Kulturtechniker **47**, 230–250 (1959b). — Das Grundwasser als Standortfaktor für die landwirtschaftliche Bodennutzung unter besonderer Berücksichtigung seiner Rolle bei der Wasserversorgung der Kulturen. Wasser u. Boden **13** (1), (1961a). — Grundwasserentzug als Folge wasserwirtschaftlicher oder bergbaulicher Eingriffe; seine Auswirkungen auf die landwirtschaftliche Bodennutzung. Mitt. Geol. Ges. Essen **4** (1961b). — Auswertung und Darstellung langjähriger Grundwasserbeobachtungsergebnisse in bodenkundlicher Sicht. Ber. a. d. Landesanst. f. Bodennutzungsschutz, Bochum, **3**, 45–51 (1962). — Beiträge zur Kennzeichnung des Wasserhaushaltes semiterrestrischer Böden großer Flußniederungen. Ber. a. d. Landesanst. f. Bodennutzungsschutz, Bochum, **4**, 97–121 (1963). — Über den Grundwasserschwankungsbereich in verschiedenen semiterrestrischen Böden. Mitt. Dtsch. Bodenkundl. Ges. **2** (1964); im Druck. — Auswirkungen wasser- und bergbaulicher Eingriffe auf die Landeskultur, Untersuchungen zu ihrer Klärung und für ihren Ausgleich. Habil.-Schrift Gießen 1965. — Wohlrab, B., und Chr. Langner: Zum landwirtschaftlichen Teil des wasserwirtschaftlichen Rahmenplanes. Z. Kulturtechn. u. Flurbereinigung **5** (2), 65–84 (1964). — Wolkewitz, H.: Die Weiterentwicklung des Verfahrens der pF-Untersuchung zur Feststellung der Bindungsintensität des Wassers im Boden. Kulturtechniker **47**, 37–50 (1959). — Methode zur Bestimmung der Bindungsintensität des Wassers im Boden. Diskussion der Kennwerte. Z. Kulturtechn. **3**, 143–146 (1962).

Zezschwitz, E. v.: Grundwasser und Bodenbildung in der Soester Niederbörde. Geol. Jb. **82**, 143–176 (1964). — Zillmann, K. H.: Über den Wasserverbrauch einiger Kulturpflanzen am Standort in Abhängigkeit von den meteorologischen Faktoren und von der Bodenfeuchtigkeit. Z. Acker- u. Pflanzenbau **104**, 257–274 (1957). — Zunker, F.: Das Verhalten des Bodens zum Wasser. Blancks Handbuch der Bodenlehre VI, 66–202, Berlin 1930.

d) Die Bodenluft

Von

H. Koepf

1. Einleitung

Eine ausreichende *Luftführung* des Bodens ist die Voraussetzung für den Fortgang der Bodenbildungsprozesse (wenn man von subhydrischen Böden und den tiefer gelegenen Horizonten der semiterrestrischen Böden absieht), für die Funktion des Bodens innerhalb der Stoffkreisläufe der Biosphäre und das Gedeihen der meisten Kultur- und Wildpflanzen.

Der Luftsauerstoff wird bei der Verwitterung und Verwesung der festen Bodensubstanz benötigt. Bei der Oxydation der organischen Substanz entsteht CO_2. Nach Stöckli (1949) gibt ein Hektar Boden jährlich 5000 bis 10000 kg CO_2 an die Atmosphäre ab, durch die etwa 80% des durch Assimilation aus der Luft entnommenen Kohlenstoffs wieder ergänzt werden. Atmung, Nitrifikation und Stickstoffbindung erfolgen bei ausreichender Sauerstoffzufuhr zum Boden, ist diese ungenügend, so werden diese Prozesse gehemmt, die Zersetzung der organischen Rückstände wird in ungünstige Bahnen gelenkt und Denitrifikation tritt ein. Schließlich benötigen die Wurzeln der höheren Pflanzen für ihre Lebenstätigkeit Sauerstoff und Baver (1956, S. 199) ist der Ansicht, daß bei einer mangelhaften Ausbildung des Wurzelsystems vornehmlich Sauerstoffmangel als begrenzender Faktor auftritt.

Für die angewandte Bodenkunde sind von Interesse: der *Gehalt* des Bodens an Luft, die *Zusammensetzung* der Bodenluft und die *Erneuerung* der Bodenluft oder die *Durchlüftung*. Diese in ihrer Bedingtheit durch die Eigenschaften des Bodens, seine biologischen Prozesse, die meteorologischen Vorgänge und den Pflanzenwuchs darzustellen, ist die Aufgabe dieses Abschnitts.

2. Der Gehalt des Bodens an Luft

Die nicht mit Wasser erfüllten Bodenhohlräume enthalten Luft. Kopecky (1914) definierte die *Luftkapazität* des Bodens als diejenige Menge an Luft, die der Boden enthält, wenn er bis zur absoluten Wasserkapazität mit Wasser gesättigt ist. Anstelle der Wasserkapazität wird heute vielfach die Feldkapazität angegeben. Mit Luftkapazität wird daher besser diejenige Menge an Luft bezeichnet, die eine in ihrer Lagerung ungestörte Bodenprobe dann enthält, wenn ihr Wassergehalt der Feldkapazität entspricht. Allgemein kann man sagen, daß die Luftkapazität etwa gleich dem nicht-kapillaren Porenraum ist. Der tatsächliche oder *aktuelle Luftgehalt* entspricht im Laufe des Jahres nur zeitweilig dem Werte für die Luftkapazität, sonst liegt er darüber oder darunter. Er ist geringer, wenn in den Bodenhohlräumen vorübergehend Senk- oder Sickerwasser anwesend ist, er ist höher, wenn die kapillaren Poren nur teilweise mit Wasser erfüllt sind. Mit anderen Worten: Der Luftgehalt eines Bodens hängt ab von dem *Porenvolumen* (PV) des Bodens, der *Porengrößenverteilung* und dem *Wassergehalt*.

Die Größe des *Gesamt*-PV ist abhängig von der *Textur* und *Struktur* des Bodens. Bei Mineralböden liegt es im Bereich von 35 bis 65% des Bodenvolumens. Bei organogenen Böden ist es wesentlich größer. Gordon (1955) gibt für das Porenvolumen von Torfmull einen Wert von 93 bis 94% an, wird Torfmull zu 90,5% mit Wasser gesättigt, so sind noch rund 40% des Porenvolumens mit Luft erfüllt. Die Beziehungen zwischen der Textur und dem Gesamt-PV sind eher von theoretischem Interesse, da die räumliche Anordnung der festen Bodenpartikel und damit auch das Porenvolumen und seine Gestalt von der im Augenblick vorhandenen Struktur abhängen. Tab. 105 nach Mitscherlich (1949) enthält einige Zahlenwerte für die Beziehungen zwischen Textur und Hohlraumvolumen des Bodens.

Tabelle 105. *Hohlraumvolumenprozente im Boden* (Mitscherlich 1949)

	lufttrocken 1	naß eingeschlämmt 2
Tieflandmoorboden	78	72
humusreicher Sandboden	59	56
humoser Sandboden	55	37
Sandboden	48	33
lehmiger Sandboden	50	36
sandiger Lehmboden	56	35
strenger Tonboden	59	69

In Spalte 2 der Tab. 105 sind die Werte für die dichtere Lagerung nach dem Einschlämmen des Bodens angegeben. Der Tonboden fällt, vermutlich infolge Quellungserscheinungen, vollkommen aus der Reihe. Konta (1959) bestimmte an 10 unverfestigten Tonen vom Pleistozän bis zum Tertiär das Gesamt-PV und das PV bei durchschnittlicher Zimmertemperatur und Luftfeuchtigkeit (effektives PV), er erhielt folgende Werte:

	PV	effektives PV
Kaolinite und Illite	48 — 32%	43 — 31%
Montmorillonit	32 — 27%	19 — 11%

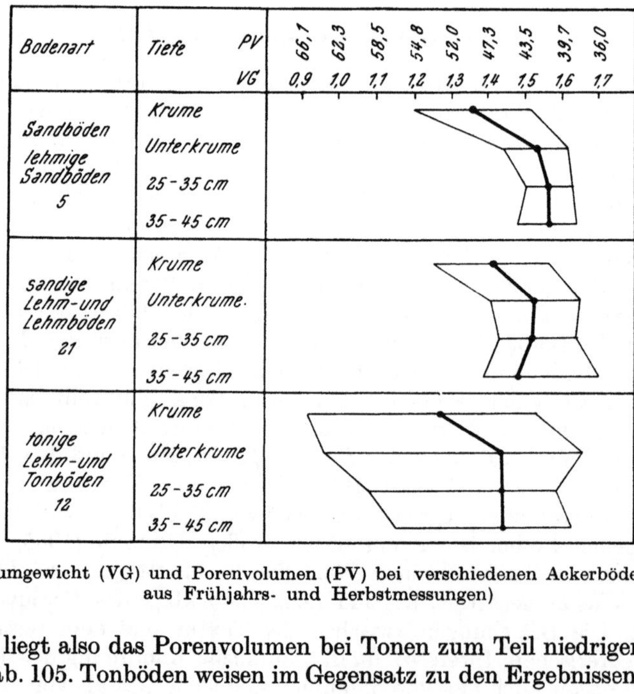

Abb. 145. Volumgewicht (VG) und Porenvolumen (PV) bei verschiedenen Ackerböden (Mittelwerte aus Frühjahrs- und Herbstmessungen)

Danach liegt also das Porenvolumen bei Tonen zum Teil niedriger als bei den Böden in Tab. 105. Tonböden weisen im Gegensatz zu den Ergebnissen von Konta oftmals ein sehr hohes Porenvolumen auf. In Abb. 145 sind die Porenvolumina bei 5 Sandböden, 21 Lehmböden und 12 Tonböden in verschiedenen Bodentiefen dargestellt. (Die Zahlenwerte, die dieser Darstellung zugrunde liegen, verdanken wir Professor H. Frese und Mitarbeitern, stark ausgezogene Linien: Mittelwerte, dünn ausgezogene: Extremwerte.) Man erkennt in dieser Darstellung die höheren Porenvolumina in der Krumenschicht, die Neigung zur Pflugsohlenbildung bei sandigen Lehm- und Lehmböden, die zum Teil extrem hohen PV-Werte bei Tonböden.

Von entscheidender Bedeutung für den Luftgehalt und die Luftführung des Bodens ist aber nicht so sehr der Gesamtporenraum, sondern die *Porengrößenverteilung*, so daß die Bestimmung des Gesamt-PV unter Umständen zu irrtümlichen Vorstellungen über den Lufthaushalt veranlaßt. Dies ergibt sich aus den verschiedenen Funktionen, die den Bodenporen zuzuschreiben sind. De Boodt (1957) unterscheidet zwischen groben Poren, $>10\,\mu$, in denen sich das Sickerwasser bewegt bzw. die mit Luft erfüllt sind, mittleren Poren, 10 bis $0{,}2\,\mu$, in denen das Wasser in pflanzenverfügbarer Form gespeichert wird und feinen Poren, $<0{,}2\,\mu$, aus denen das Wasser weder durch die Schwerkraft noch durch Pflanzenwurzeln entfernt wird. Etwas anders ist die Einteilung von S. Odén (1957):

Tabelle 106. *Einteilung der Porengrößen nach Odén* (1957)

Durchmesser μ		pF	Verhalten zum Wasser	
Grobporen	> 200	< 1,2	dränbar	leicht pflanzenverfügbar
Mittelporen	200—20	1,2—2,2	dränbar	leicht pflanzenverfügbar
Feinporen	20— 2	2,2—3,2	nicht dränbar	leicht pflanzenverfügbar
Mikroporen	2— 0,2	3,2—4,2	nicht dränbar	schwer pflanzenverfügbar
Ultraporen	< 0,2	> 4,2	nicht dränbar	nicht pflanzenverfügbar

Nach dieser Einteilung würde also der von Grob- und Mittelporen eingenommene Raum der Luftkapazität entsprechen. Für die Bestimmung der Porengrößenverteilung wird die pF-Kurve benutzt, diese kann nach dem Gesagten also auch für die Charakterisierung des Lufthaushaltes eines Bodens ausgewertet werden. Abb. 146 nach F. Richard (1955) zeigt an einem Beispiel die Porengrößenverteilung bei einem Sand- und einem Staubboden. Richard gibt die Untergrenze der Grobporen, die also normalerweise Luft führen, mit $8\,\mu$, die Größe der Mittelporen mit 0,2 bis $8\,\mu$ an. Die erheblichen Unterschiede im Gehalt an luftführenden Grobporen treten bei den beiden Böden deutlich hervor.

Nach Scheffer und Schachtschabel (1960) kann man einen Boden, der bei Feldkapazität 15 bis 25% Luft enthält, als gut durchlüftet bezeichnen. Für die einzelnen Bodenarten geben die Autoren folgende mittlere Luftgehalte bei Feldkapazität an: bei Sandböden 30 bis 40%, bei Lehmböden 10 bis 25%, bei Tonböden 5 bis 15%. Da gerade bei Pelosolen der *Luftmangel* besonders häufig als begrenzender Faktor für die Ausbildung des Wurzelsystems auftritt, folgen in Tab. 107 noch einige Angaben über den Luftgehalt verschiedener schwerer Böden bei Feldkapazität nach eigenen Untersuchungen (Feuerstack 1960). Diese Ergebnisse decken sich mit den Befunden von Th. Dietz (1959). Die Luftgehalte liegen bei diesen schweren Böden allgemein sehr niedrig, sie können in einzelnen Horizonten bis auf Null absinken. Dabei ist zu berücksichtigen, daß die Wassergehalte bei diesen Böden wegen deren eigentümlichem Wasserhaushalt (Feuerstack 1960) während langen Zeiten im Jahre die Feldkapazität erreichen.

Als optimal für den Pflanzenwuchs ist der Bodenaufbau dann zu bezeichnen, wenn kapillare und nicht-kapillare Hohlräume ausreichend und im richtigen Verhältnis vorhanden sind. Bei grob texturierten Böden überwiegt das nicht-

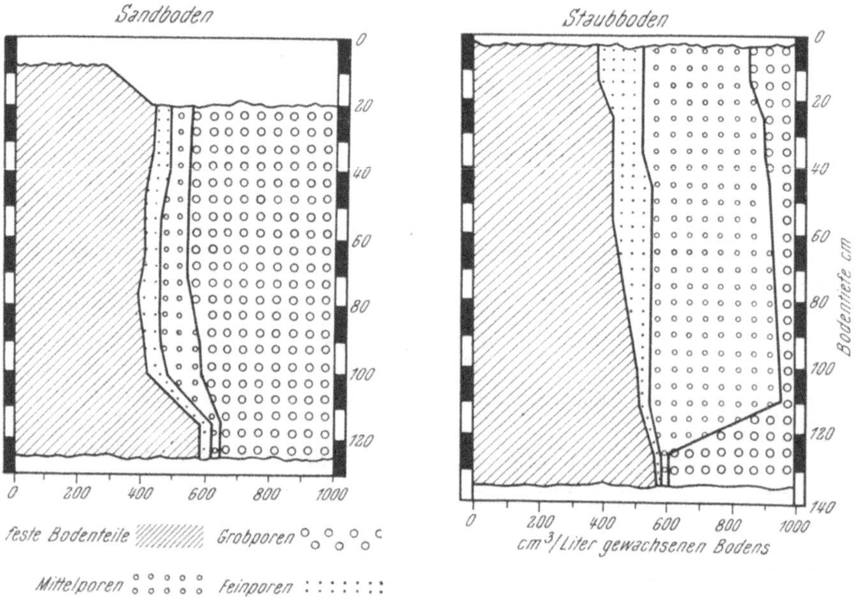

Abb. 146. Substanzvolumen und Porengrößenverteilung bei einem Sand- und einem Staubboden (RICHARD 1955)

kapillare Hohlraumvolumen, bei fein texturierten ist es meist zu klein. *Aggregierung* und *Krümelung* des Bodens verbessern bei diesen Böden das Verhältnis

Tabelle 107. *Luftgehalt einiger Tonböden bei Wassersättigung in Höhe der Feldkapazität* (FEUERSTACK 1960)

Bezeichnung des Bodens	Tiefe cm	Luftgehalt bei F.K. Vol.-%	Bezeichnung des Bodens	Tiefe cm	Luftgehalt bei F.K. Vol.-%
kalkhaltige	0—20	11,9	verbraunter	0—15	7,8
Tonbraunerde	20—47	3,5	Pelosol auf	15—33	7,4
auf Letten-	47—65	0,0	Gipskeuper	33—46	1,9
keuper	65—80	1,9		46—54	9,1
				54—70	10,0
humusreiche,	0—20	18,9	Pelosol–	0—16	5,9
kalkhaltige	20—35	9,4	Pseudogley	16—28	0,5
Tonbraunerde	35—47	10,3	auf Opalinus-	28—47	0,9
auf Posidonien-	47—57	10,8	ton		
schiefer	57—80	11,8			

zwischen kapillarem und nicht-kapillarem Porenraum. Der Zusammenhang zwischen Luftgehalt, nicht-kapillarem Hohlraumvolumen des Bodens und der Aggregatgröße möge durch die Tab. 108 nach KRAUSE (1931) veranschaulicht werden.

Tabelle 108. *Der Luft- und Sauerstoffgehalt des Bodens bei verschiedenem relativen nichtkapillarem Hohlraumvolumen*
(KRAUSE 1931)

Durchmesser der Bodenteilchen in mm	1/2	1/2—1	1—2	2—3	3—5
Nichtkapillares Hohlraumvolumen	8%	49%	54%	59%	63%
Luftgehalt in Volumprozent des Bodens	2,7	24,5	29,6	35,1	38,7
Sauerstoffgehalt der Bodenluft in %	5,4	18,6	19,3	19,4	—
Sauerstoffgehalt in % des Bodens	0,14	4,55	5,74	6,70	7,51

Bei einer Aggregatgröße oberhalb 0,5 mm nimmt nach diesen Zahlen der Luft- und Sauerstoffgehalt des Bodens sprunghaft zu, um dann mit zunehmender Aggregatgröße weiter anzusteigen. Doch weist BAVER (1956, S. 200) auf die Ergebnisse YODERS (1937) hin, der feststellte, daß bei großen Aggregaten (>1/4 Zoll) die Luftkapazität des Bodens wieder geringer wurde.

Aggregierung des Bodens und *Aggregatstabilität* sind deshalb für den Bodenluftgehalt und, wie unten dargestellt wird, auch den Gasaustausch zwischen Boden und Atmosphäre von ausschlaggebender Bedeutung. Aggregatbildung, Lockerung und Durchporung des Bodens hängen ab von den natürlichen Faktoren der Strukturbildung wie Tonsubstanz nach Menge und Art, Humussubstanz nach Menge und Art, Anwesenheit anderer Kleb- und Bindesubstanzen (Oxyde des Fe und Al, $CaCO_3$, organische Linearkolloide), Tätigkeit der Pflanzenwurzeln, Bodentiere, Mikroorganismen, Quellen und Schrumpfen, Bodenfrost usw. Sie werden ferner beeinflußt durch die Düngung, insbesondere die organische Düngung, Fruchtfolge und die mechanische Bodenbearbeitung. Da die Strukturbildung

Tabelle 109. *Einfluß des Regenwurmes auf die Luftkapazität des Bodens*
(STÖCKLI 1928)

Nr.	Bezeichnung der Böden	Wahrscheinliche Luftkapazität[1] der Böden pro dm³ cm³	Volumen der pro dm² ausgestoßenen Wurmexkremente[2] cm³	Volumen der Wurmexkremente in % der Luftkapazität des Bodens
1	Gartenerde	74,9	7,0	9,4
2	Dauerwiese	67,6	32,6	49,6
3	Dauerwiese	103,9	31,0	29,9
4	Waldwiese	143,9	60,7	42,4
5	Obstgarten	148,2	28,8	16,2
6	Wiesenböschung	125,3	13,1	10,5
7	Golfwiese	100,9	67,2	64,0
8	Mischwald	134,9	12,6	9,3
9	Fichtenwald	203,5	15,3	7,5
10	Fichtenwald	106,2	52,4	49,4

[1] Wahrscheinliche Luftkapazität = Bestimmung nach BURGER + 5%.
[2] Volumen der an der Bodenoberfläche abgesetzten Wurmexkremente, von denen für die Berechnung angenommen wird, daß ihre Substanz aus den oberen 10 cm des Bodens stammt.

in anderen Abschnitten dieses Handbuches ausführlich besprochen wird, sollen an dieser Stelle nur noch einige Zahlenwerte über die Beeinflussung des Bodenluftgehaltes durch Faktoren der Strukturbildung mitgeteilt werden. Daß gerade auch die *Bodentiere* zur Lockerung des Bodens und der Ausbildung des nicht kapillaren Hohlraumvolumens beitragen, ist seit den Untersuchungen Wollnys (1890), später von Blank und Giesecke (1924) bekannt. In Tab. 109 werden die Ergebnisse einer Untersuchung Stöcklis (1928) mitgeteilt über den Zusammenhang zwischen Regenwurmtätigkeit und Luftgehalt, aus der der große Anteil der vom Regenwurm geschaffenen luftführenden Bodenhohlräume an der gesamten Luftkapazität hervorgeht.

Tabelle 110. *Organische Düngung und nicht-kapillares Hohlraumvolumen* (*nach* Baver 1956, S. 201)

	Tiefe in Zoll	nicht-kapillare Porosität in Vol.-% Juni 1940	Sept. 1940
Im Herbst 1939 gepflügt	0—3	10,3	3,5
	3—6	3,5	3,0
Im Herbst 1939 10 t Stallmist untergepflügt	0—3	15,2	11,8
	3—6	6,3	4,0
Im Herbst 1939 20 t Stallmist untergepflügt	0—3	10,4	8,7
	3—6	17,5	8,7

Zur Vergrößerung des nicht-kapillaren Porenraums bei Ackerböden trägt insbesondere der Anbau solcher Pflanzen bei, die den Boden reichlich und tief durchwurzeln. Der Einfluß der *organischen Düngung* auf das nicht-kapillare Hohlraumvolumen ist aus den Zahlen für einen „Brookston-Tonboden" zu ersehen (Tab. 110).

Tabelle 111. *Porenvolumen und Luftgehalt* (*in Vol.-%*) *nach verschieden tiefer Bearbeitung bei drei verschiedenen Bodenarten* (Kuipers 1959)

Pflügen auf cm:	15	15	25	5	15	15	25	25	35	12	20	28
Lockern auf cm:		10				10		10				
Porenvolumen in												
7—12 cm Tiefe	49,5	48,9	48,4	44,8	45,6	45,1	45,2	43,6	43,7	47,1	46,3	46,8
17—22 cm Tiefe	43,3	45,9	47,1	42,2	42,8	44,8	44,8	44,6	44,3	52,0	47,5	48,3
27—32 cm Tiefe	—	—	—	43,8	43,5	45,6	42,9	44,5	44,0	56,1	54,6	50,9
Luftgehalt bei pF 1,9 in												
7—12 cm Tiefe	23,3	22,4	21,7	9,6	10,6	9,8	10,1	8,0	4,4	5,6	4,3	5,1
17—22 cm Tiefe	13,9	17,9	19,7	6,1	6,6	9,6	9,6	9,6	4,8	4,7	6,2	7,4
27—32 cm Tiefe	—	—	—	8,2	7,9	10,6	7,1	9,1	4,6	6,9	5,9	4,3
	Sandboden 5% Humus			junger, kalkhaltiger, mariner Tonboden						mariner Tonboden		

—— nicht bearbeitete Schicht
----- gelockerte Schicht

Tab. 111 zeigt den Einfluß verschieden tiefer *Pflugarbeit* und Untergrundlockerung auf das Porenvolumen und den Luftgehalt bei drei verschiedenen Böden nach einer Untersuchung von KUIPERS (1959).

Die Bodenhohlräume sind entweder mit Luft oder mit Wasser erfüllt. Bei dieser engen Wechselbeziehung beeinflußt deshalb jede Änderung des Porenvolumens durch Faktoren der Strukturbildung, Düngung und Bearbeitung den Luftgehalt und die Luftführung in doppelter Hinsicht: 1. Durch die Größe des vorhandenen Porenvolumens, 2. durch die Veränderung der *Wasserführung*. Je höher der Gehalt an Mittel- und Grobporen ist, um so rascher sinkt das Sickerwasser nach Niederschlägen in den Untergrund und kann die Luft wieder in den Boden eindringen. Je höher die Durchfeuchtung eines Bodens ist, um so geringer ist im allgemeinen auch sein Luftgehalt.

3. Die Sorption von Luft und Luftbestandteilen durch die feste Bodensubstanz

Der Boden als poröser Körper kann Luft und Luftbestandteile sorbieren. In Betracht kommen hierbei *polare* und *apolare Sorption*. Bei apolarer Sorption ist die aufgenommene Menge gering. ALTEN und LOOFMANN (1941) zeigten, daß je 100 g trockenem Bentonit nur etwa 0,0014 Mole festgehalten wurden, während dieselbe Substanz aus wasserdampfgesättigter Atmosphäre 1,717 Mole Wasser durch polare Sorption zu binden vermochte. Für die Luftgase gibt MANEGOLD (1955, S. 595) folgende Dipolmomente an:

Gas (trocken)	Dipolmoment Debye-Einheiten
H_2	0
NH_3	1,49
N_2	0
O_2	0
H_2S	0,95
CO_2	0

Bei den im Boden sorbierten Luftbestandteilen liegt also überwiegend apolare Sorption vor. Kapillarkondensation ist bei den Gasen der Luft, auch bei CO_2, dessen kritische Temperatur bei 31,1° C liegt, nicht wahrscheinlich. Sehr wohl kommt dagegen *Chemosorption* in Betracht, da Reaktionen zwischen O_2 und CO_2 und der Bodensubstanz möglich sind. Enthält der Boden Feuchtigkeit, so ist von der festgehaltenen Gasmenge der *gelöste* Anteil abzuziehen, um die eigentliche Sorption zu erhalten. Die sorbierte Gasmenge hängt ab von der *aktiven Oberfläche* des Adsorbens und der Natur des Adsorptivs. Je höher der Siedepunkt eines Gases liegt, um so mehr wird bei sonst gleichen Bedingungen adsorbiert. Ferner steigt die sorbierte Menge an bei schärferer Vortrocknung des Adsorbens und bei wachsendem Druck, sie wird geringer bei zunehmender Sorptionstemperatur.

ALTEN und LOOFMANN (1941) geben für einige Bodenbestandteile folgende Werte für adsorbierte Luft an (Mol · 10^6 bei ca 20° C, in Klammern der Druck in mm Hg): Kieselgel: 4368 (708,8), Bentonit: 1371 (725,5), Sand: 36,2 (704,7), Quarzmehl $< 100\,\mu$: 48,7 (723,2), Kaolin: 125,2 (726,6), glimmerartiges Tonmineral: 244,7 (704,8) je 100 g Substanz.

WIESSMANN und NEUMANN (1935) ermittelten für verschiedene Gase und Bodenbestandteile die in Tab. 112 enthaltenen Werte.

Man erkennt aus Tab. 112 den Einfluß der Sorptionstemperatur und der Entwässerung vor Beginn der Messung. Je stärker vor der Messung entwässert wird, um so höher ist die sorbierte Menge. Wurde allerdings bei Eisenoxydhydrat die Entwässerung bei 250° C vorgenommen, so sank die sorbierte Menge infolge

Alterung des Adsorbens um 4 bis 8%. Das Verhältnis von $N_2:O_2$ wird durch unterschiedliche Wassergehalte entsprechend der Löslichkeit dieser Gase verändert. Bei einem Wassergehalt des Bodens von 9% sinkt nach diesen Autoren die Sorption von N_2 auf Null ab. NAKAYAMA (1958) gibt ebenfalls an, daß bei Boden- und Tonproben, die mehr als 10 bis 15% Feuchtigkeit enthalten, die von der Probe aufgenommene Menge O_2 der Löslichkeit von O_2 in Wasser entspricht (bei Bentonit ist sie allerdings größer). Die CO_2-Sorption ist im allgemeinen größer, als nach der Löslichkeit anzunehmen wäre. NAKAYAMA nimmt auch an, daß beim Trocknen des Bodens organische Moleküle ein vollständiges Schließen der Schichtebenen von Tonmineralien verhindern, so daß ihre Oberflächen für die Gassorption zur Verfügung bleiben. Weitere Angaben über den Einfluß der Tonminerale, den Tongehalt des Bodens und den Wassergehalt des Bodens werden von RUNKLES (1957), SCOTT und NAKAYAMA (1958) sowie JAMISON (1953) gemacht. Abb. 147 nach JAMISON zeigt den Einfluß des Feuchtigkeitsgehaltes auf die adsorbierte Luftmenge bei einem Hurricane- und einem Lloyd-Ton.

Tabelle 112. *Sorption von Gasen an Bodenkomponenten und Böden* (WIESSMANN und NEUMANN 1935)
A = Mole je 100 g Substanz

Präparat	Entwässerungstemperatur	Wassergehalt	Sorptionstemperatur	p mm Hg	N_2 $A \cdot 10^6$	p mm Hg	O_2 $A \cdot 10^6$
Eisenoxydhydrat	45° C	15,7%	0° C	618,1	1480	619,5	1282
	45° C	15,7%	25° C	662,8	1051	639,4	893
	150° C	11,8%	0° C	662,1	2135	575,5	1518
	150° C	11,8%	25° C	653,6	1443	595,3	1091
Hydrargillit ..	45° C	3 Mole	25° C	557,2	0,1	540,4	1,3
$CaCO_3$ gefällt .	45° C	0,0%	25° C	735,1	— 0,9	695,8	1,7
Buntsandsteinboden	45° C	7,1%	25° C	678,9	119	698,5	127
Lößlehm	45° C	2,9%	25° C	692,5	127	700,1	132

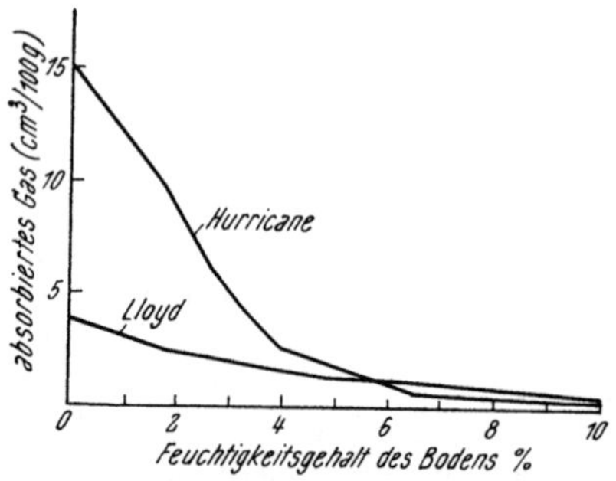

Abb. 147. Luftadsorption von Hurricane- und Lloyd-Ton bei 23° C in Beziehung zum Feuchtigkeitsgehalt des Bodens (JAMISON 1953)

Wesentlich stärker als N_2 und O_2 werden CO_2 und NH_3 absorbiert. DIEKMANN (1937) untersuchte die Gassorption an Braunkohlen und anderen organischen Stoffen. Nach Desorption dieser Stoffe bei 45° C und bei einem Drucke von 1,2—1,4 mm Hg, erhielt er u. a. folgende Zahlen für die sorbierten Gasmengen:

	N_2 Mole/100 g	CO_2 Mole/100 g	NH_3 Mole/100 g
Kassler Braun	$67 \cdot 10^{-5}$	$335 \cdot 10^{-4}$	$810 \cdot 10^{-3}$
Braunkohle, getrocknet	$170 \cdot 10^{-5}$	$370 \cdot 10^{-5}$	$800 \cdot 10^{-3}$

Allerdings schließt er aus dem Vergleich mit den Werken anderer Autoren, daß bei CO_2 1,2 bis 28%, bei NH_3 dagegen nur 1 bis 3,7% auf Adsorption zurückzuführen ist, der Rest auf Chemosorption und Lösung.

Abb. 148 nach Makower, Shaw und Alexander (1937) zeigt die Adsorptionsisothermen von CO_2, N_2 und O_2. Bei einem Druck von 500 mm Hg wird rund 50 mal mehr CO_2 adsorbiert als N_2 und O_2.

Der praktisch geradlinige Kurvenverlauf bei N_2 und O_2 in Abb. 148 beweist keine verschiedenartige Sorption gegenüber CO_2. Würde man diese Gase bei tieferen Temperaturen untersuchen, so wäre der Kurvenverlauf anders. Die von zahlreichen Autoren ermittelten Adsorptionsisothermen folgen der Freundlichschen Adsorptionsisotherme von der Form $a = \alpha \cdot p^{1/n}$. Nach Kuron (1932) sind die Adsorptionsisothermen von nahe verwandten Böden des gleichen Typus affine Kurven. Kuron untersuchte ferner den Einfluß der Ionenbelegung auf die Adsorption von CO_2 an Ton. Die sorbierte Menge war von der Ionenbelegung abhängig und zwar in der Reihenfolge: Mg > Ca > H > K > Ba > Na > Li.

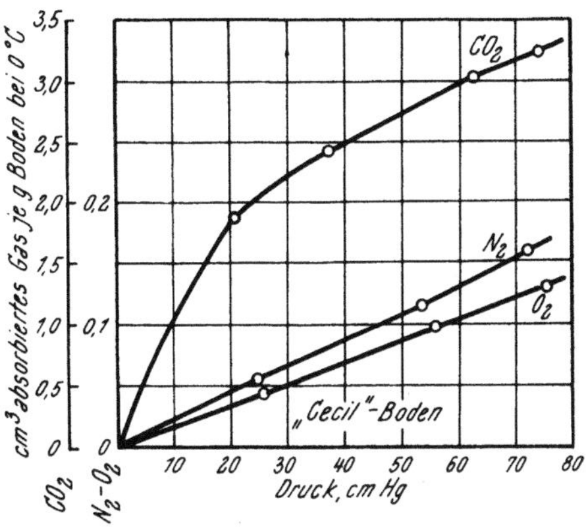

Abb. 148. Adsorptionsisothermen für CO_2, N_2 und O_2 (Makower, Shaw und Alexander 1937)

Nach Wiessmann und Neumann (1935) spielen für die Sorption in erster Linie die *gelförmig-kolloidalen* Bodenbestandteile wie Kieselsäure, Eisenoxyd und Tonerde eine Rolle, wobei weniger deren chemische Individualität als ihre physikalische Beschaffenheit von Einfluß ist. In Trockenzeiten können beachtenswerte Gasmengen sorbiert werden, die Wiederbenetzung nach scharfer Trocknung erfolgt bei manchen Böden bekanntlich langsam. Durch selektive Sorption von Wasserdampf wird die sorbierte Gasschicht allmählich verdrängt. Bei feuchtem Boden ist die Sorption ziemlich bedeutungslos, doch enthält dann das Bodenwasser die Gase der Luft in gelöster Form.

4. Die Zusammensetzung der Bodenluft

Die Wurzeln der höheren Pflanzen und die meisten Bodenmikroorganismen verbrauchen bei ihrer *Lebenstätigkeit* Sauerstoff und geben Kohlendioxyd ab. Der Sauerstoffgehalt der Bodenluft kann deshalb niedriger sein als der der Atmosphäre, der Kohlendioxydgehalt ist in der Regel höher. In welchem Ausmaße dies der Fall ist, hängt von der Intensität der Lebensvorgänge im Boden und der Geschwindigkeit des *Gasaustausches* zwischen Boden und atmosphärischer Luft ab. Die Zusammensetzung der Bodenluft unterliegt zeitlichen und örtlichen Schwankungen, sie ist in den einzelnen Schichten des Bodens verschieden.

Nach Ramann (1911, S. 390) ist die Bodenluft in tieferen Schichten stets, in den oberen meist zu 100% mit *Wasserdampf* gesättigt. Shurawlew (1955) untersuchte die Wasserdampfsättigung der Bodenluft in der oberen Schicht eines unbestellten Tschernosems. Wenn der Feuchtigkeitsgehalt dieses Bodens über der normalen Hygroskopizität liegt, ist die Luft mit Wasserdampf gesättigt. Geiger (1952) stellte bei der Registrierung der Bodenluftfeuchtigkeit an einem Sommertag während der Mittagsstunden in den oberen 20 mm des Bodens ein Absinken bis unter 90% fest.

Die natürliche *Radioaktivität* der Bodenluft pro m³ Boden wird von M. J. Jasstrebow (1958) für einen Wiesenboden mit 0,049 μ C, für einen Rasenpodsolboden mit 0,054 μ C angegeben. Wurde die Luft über dem Boden mit $^{14}CO_2$ markiert, so war bei der Entnahme von Bodenluftproben aus 20 cm Tiefe die Radioaktivität in der Bodenluft und der atmosphärischen Luft dieselbe. Bei den durch Kernreaktionen erzeugten Spaltprodukten in der bodennahen Luftschicht handelt es sich um Aerosolteilchen von 0,05 bis 5 μ Durchmesser (Morgenstern und Rentschler 1959). Diese werden, wenn sie überhaupt auf den Boden niedergeschlagen werden, vermutlich rasch zu Bestandteilen der festen Bodensubstanz.

Die *bodenkundlich* und *pflanzenphysiologisch* wichtigsten Eigenschaften der Bodenluft sind ihr *Sauerstoff-* und *Kohlendioxydgehalt*. Grenzwerte für die Mindestkonzentration an O_2 bzw. die noch tolerierten CO_2-Mengen, bei denen ein normales Wachstum erfolgt, können nur ungefähr angegeben werden, da die Ansprüche der Pflanzen verschieden sind und außerdem die Dauer einer schädlichen Konzentration von Einfluß ist.

Wiegand und Lemon (1958) untersuchten die O_2-Konzentration an der Wurzeloberfläche von Hirse und Baumwolle. Sie fanden, daß diese bei Feldkapazität bei einem Lehmboden optimal, bei einem Tonboden geringer als optimal war. Wichtig für die O_2-Versorgung der Pflanzenwurzeln ist nach Ansicht dieser Autoren die Länge des Diffusionsweges durch die Wasserfilme. Sie soll die Wurzelatmung öfters hemmen als O_2-Mangel in der Bodenluft. Nach Stolwijk und Thimann (1957) stellten Pisum sativum, Vicia faba, Phaseolus vulgaris und Helianthus annuus das Wurzelwachstum bei CO_2-Konzentrationen von 6,5% ein, während Avena sativa und Hordeum vulgare von derselben Konzentration nicht beeinflußt wurden. Erbsen wurden bereits bei einer CO_2-Konzentration von 1,5% geschädigt. Peterson (1950) stellte einige Angaben aus der Literatur zusammen. Bei 5 bis 10% O_2 in der Bodenluft sollen die Pflanzen in der Regel eine normale Entwicklung zeigen und auch Konzentrationen von 1 bis 2% überstehen können. Obstbäume brauchen für ein gutes Gedeihen 10% O_2 in der Bodenluft, bei 0,1 bis 3% wachsen sie langsam. Da sich die O_2-, und CO_2-Konzentrationen im Boden normalerweise reziprok verhalten, ist es oft schwierig die Einflüsse zu trennen, doch scheinen hohe CO_2-Konzentrationen die Ertragsbildung öfters zu beeinträchtigen als geringe O_2-Gehalte. Bei den Versuchen von Shanks und Laurie (1950) zeigten Rosen bei 9, 13 und 17% O_2 in der Bodenluft (angewandt wurden 1 bis 29%) das beste Wachstum, während sie bei 1% O_2 und geringer Feuchtigkeit abstarben. 10% CO_2 inhibierten bei den Untersuchungen von Norris, Wiegand und Johanson (1959) die Atmung von abgetrennten Zwiebelwurzeln. Gill und Miller (1956) stellten bei Maiswurzeln, die gleichzeitig noch in ihrem Wachstum mechanisch behindert wurden, fest, daß bei O_2-Konzentrationen unter 10% die Zuwachsrate abnahm. Bei den Versuchen von Danielson und Russel (1957) mit Mais erfolgte bei 10% O_2 in der Bodenluft noch eine normale Aufnahme von Rb^{86}. Bei belüfteten Sandkulturen von Tabakpflanzen beobachteten Harris und van Bavel (1957) bei einem $CO_2 : O_2$-Verhältnis > 1 eine signifikante Erniedrigung von Blattwachstum, Ertrag und Wasseraufnahme. Tabakpflanzen, die 1 Tag mit Wasser überstaut wurden, erholten sich wieder, 2 oder mehrere Tage langes Überstauen bewirkte eine nachhaltige Beeinträchtigung des Wachstums. Bei den Untersuchungen von Kemper und Amemiya (1957) schien eine 2- bis 3tägige Überstauung von Luzerne mit periodischem Abfall des O_2-Gehaltes in der Bodenluft auf 10 bis 13% nicht nachteilig zu wirken. Wie sehr die erste Wurzelentwicklung von Keimpflanzen durch CO_2-Stau in der Bodenluft beeinträchtigt werden kann, wurde von Michael und Bergmann (1954) gezeigt.

Überblickt man diese Angaben und die älterer Autoren, so kann man allgemein sagen, daß Konzentrationen von weniger als 10% O_2 und mehr als 5% CO_2 etwa die Grenze bezeichnen, bei denen zumindest mit einer Verlangsamung des Wachstums zu rechnen ist, während noch extremere Konzentrationen letal wirken können. Die genannten Konzentrationen können bei schlecht durchlüfteten, nassen Böden durchaus vorkommen. Die Entwicklung der Pflanze wird dann beeinträchtigt, zumal wenn man bedenkt, daß unter diesen Verhältnissen die Gesamtmenge an O_2, die im Boden vorhanden ist, in der Regel ebenfalls gering ist.

Die *atmosphärische* Luft enthält im Mittel 79,0% N_2, 20,96% O_2 und 0,03% CO_2. Bei normal durchlüfteten Böden liegt der O_2-Gehalt nur wenig unter 21%, der CO_2-Gehalt erreicht im Mittel das Zehnfache der CO_2-Konzentration in der Atmosphäre, etwa 0,1 bis 1%. In tieferen Schichten nehmen die O_2-Gehalte ab, die CO_2-Gehalte zu, das $CO_2 : O_2$-Verhältnis in der Bodenluft wird also enger. Schlecht durchlüftete, nasse Böden weisen allerdings erheblich niedrigere O_2- und entsprechend höhere CO_2-Gehalte auf. Im allgemeinen verhalten sich die O_2- und CO_2-Gehalte in der Bodenluft *reziprok*. M. J. Jasstrebow (1958a) ordnete einige Böden nach zunehmendem $CO_2 : O_2$-Verhältnis in der Reihenfolge Rasenpodsol-, Wiesenrasen-, Schlamm- und Sumpfboden. Das weiteste $CO_2:O_2$-Verhältnis mit 0,08 wurde bei dieser Untersuchung in 15 bis 60 cm auf Rasenpodsolböden gefunden, das engste mit 0,39 in der 15 bis 40 cm Schicht von Wiesenrasen und von 1,2 in der 15 bis 30 cm Schicht von Sumpfboden. PENNINGSFELD (1950) bestimmte den aus den Differenzen zum Normalgehalt berechneten Quotienten $CO_2:O_2$ bei einem Waldboden zu 0 bis 2,76 und bei einem Ackerboden zu 0,03 bis 0,92, jeweils von der Bodenoberfläche bis in 120 cm Tiefe.

Tabelle 113. *Sauerstoff- und Kohlendioxydgehalt der Bodenluft in verschiedenen Tiefen bei einem sandigen Lehmboden*
(BOYNTON und REUTHER 1938)

Datum	1 Fuß (= 30,48 cm)		2 Fuß		3 Fuß		4 Fuß		5 Fuß		6 Fuß	
	CO_2	O_2	CO_2	O_2	CO_2	O_2	CO_2	O_2	CO_2	O_2	CO_2	O_2
14. 11. 37	1,20	19,40	2,40	11,60	6,60	3,50	9,60	0,70	10,40	2,40	15,50	0,20
23. 3. 38	0,15	20,15	2,10	0,15	5,60	0,30	—	—	—	—	—	—
21. 4. 38	1,90	18,65	3,20	0,35	6,85	0,25	—	—	—	—	—	—
24. 5. 38	3,70	16,20	3,95	13,95	5,60	13,35	—	—	—	—	—	—
21. 6. 38	1,70	19,25	4,15	17,10	5,35	16,40	9,60	9,45	10,95	2,10	—	—
25. 7. 38	2,00	19,80	3,10	19,10	5,20	17,50	9,10	14,50	11,70	12,40	12,60	9,80
19. 8. 38	2,40	19,00	3,70	17,40	5,00	16,70	8,55	15,25	11,85	12,95	11,90	11,85
22. 9. 38	3,00	15,30	4,80	11,00	5,80	9,95	8,45	8,95	10,55	8,40	10,60	9,00

Tab. 113 nach BOYNTON und REUTHER (1938) zeigt die Zunahme der CO_2- und Abnahme der O_2-Konzentration mit der Tiefe. An diesen Zahlen wird auch das reziproke Verhalten der CO_2- und O_2-Konzentrationen deutlich, ferner die erheblichen *jahreszeitlichen* Schwankungen der Werte.

Abb. 149 nach TAYLOR und ABRAHAMS (1953) veranschaulicht die erheblichen jahreszeitlichen Schwankungen im *Sauerstoffgehalt* der Bodenluft in verschiedenen Tiefen als Funktion des Wassergehaltes im Boden. Daß in tieferen Bodenschichten auch das Oxydations- und Reduktionspotential erniedrigt ist, wurde u. a. von BERTRAND und KOHNKE (1957) gezeigt. Sie stellten ferner fest, daß O_2-Mangel das Pflanzenwachstum begrenzt, wenn die O_2-Diffusion im Boden <16% der Diffusion in freier Luft ist.

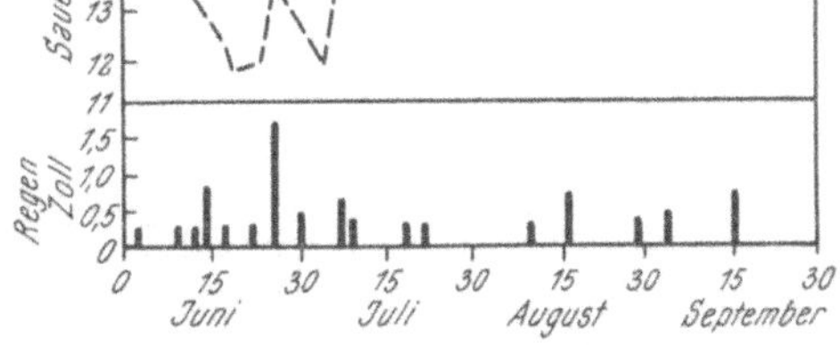

Abb. 149. Sauerstoffgehalt der Bodenluft als Funktion des Feuchtigkeitsgehaltes (TAYLOR und ABRAHAMS 1953)

EPSTEIN und KOHNKE (1957) untersuchten den Einfluß *organischer Substanz* auf den O_2-Gehalt der Bodenluft. Die Verabfolgung von 2 t/ha (Trockensubstanz) frischer Luzerne drückte den O_2-Gehalt der Bodenluft in 1, 3 und 8 Zoll Tiefe, und zwar stärker, wenn die organische Substanz auf der Bodenoberfläche belassen wurde, weniger stark, wenn sie in den Oberboden eingearbeitet wurde. Im letzteren Falle wurde offenbar die Durchlüftung des Bodens verbessert.

Tabelle 114. *Sauerstoffgehalt der Bodenluft und Diffusionsrate in verschiedenen Tiefen des Bodens bei unterschiedlicher Bodenbearbeitung und zwei Fruchtfolgen* (Raney 1950)

A / B Fruchtfolge, die mehr / weniger organische Substanz hinterläßt.

Rotation	Sauerstoff-Partialdruck in der Bodenluft A	B	Raumgewicht des Bodens A	B	Feuchtigkeitsgehalt des Bodens A	B	Diffusionsrate A	B
	mm Hg		g/cm³		%			
Tiefe Bearbeitung Zoll								
4 Untergrundlockerung	141	145	1,27	1,28	19,8	20,2	0,546	0,436
4 Übliche Pflugarbeit	141	145	1,28	1,32	20,0	19,9	0,639	0,568
4 Scheibenpflug	138	143	1,36	1,33	20,0	19,3	0,377	0,327
4 Rotierende Bearbeitung	134	141	1,39	1,29	19,6	20,1	0,356	0,358
Mittel	139	144	1,33	1,31	20,0	19,9	0,479	0,422
8 Untergrundlockerung	137	142	1,32	1,34	21,0	19,7	0,428	0,308
8 Übliche Pflugarbeit	138	143	1,31	1,42	22,2	19,9	0,417	0,269
8 Scheibenpflug	129	133	1,40	1,33	21,3	10,3	0,279	0,215
8 Rotierende Bearbeitung	121	131	1,48	1,45	21,3	20,0	0,209	0,216
Mittel	131	137	1,38	1,39	21,5	20,0	0,333	0,252
12 Untergrundlockerung	133	104	1,42	1,49	22,1	21,6	0,172	0,229
12 Übliche Pflugarbeit	133	134	1,45	1,47	22,6	22,5	0,176	0,149
12 Scheibenpflug	128	129	1,51	1,48	21,0	22,2	0,166	0,253
12 Rotierende Bearbeitung	126	125	1,50	1,52	21,2	18,9	0,154	0,147
Mittel	130	132	1,47	1,49	21,7	21,3	0,167	0,195

Durch Tab. 114 wird der Einfluß verschiedener *Bodenbearbeitung* auf den O_2-Gehalt der Bodenluft illustriert (Raney 1950). Im Vergleich zu normaler Pflugarbeit und Untergrundlockerung lag bei diesem Boden, einem Schlufflehm, der O_2-Partialdruck bei den mit Scheibenpflug und rotierenden Geräten bearbeiteten Parzellen tiefer, desgleichen wurde bei der Fruchtfolge A, die größere Mengen organischer Substanz in den Boden brachte, die O_2-Konzentration verringert. Diese Erscheinung ist offenbar auf verstärkten *Sauerstoffverbrauch* in der Bodenluft zurückzuführen, denn die Diffusionsrate ist bei den Teilstücken der Fruchtfolge A höher als bei der Fruchtfolge B.

Der *CO_2-Gehalt der Bodenluft* ist gleich der Differenz zwischen der *Kohlensäureproduktion* im Boden und der *Ableitung* von CO_2 in die Atmosphäre sowie in die feste und flüssige Phase des Bodens. Diese beiden Vorgänge hängen von einem vielfältigen Komplex gleichsinnig und gegensinnig einwirkender Faktoren ab, die häufig in gegenseitiger Abhängigkeit voneinander variieren. Es ist bei Freilandmessungen daher oftmals nicht einfach, die Ursachen für das Zustandekommen der gemessenen Werte eindeutig zu interpretieren. Obwohl die *biologische Aktivität* in den oberen Bodenschichten am größten ist, nehmen die CO_2-Konzentrationen nach der Tiefe hin zu (Tab. 113). Die Ableitung des Kohlendioxyds in die Atmosphäre erfolgt im gasförmigen Zustand, dazu ist in jedem Niveau ein *Konzentrations-*

gefälle von unten nach oben erforderlich. Doch scheint, wie auf S. 736 erwähnt wird, auch eine CO_2-Anreicherung entgegen einem Konzentrationsgefälle möglich zu sein. Gelegentlich wird allerdings in größeren Tiefen eine Abnahme der CO_2-Konzentration festgestellt. PENNINGSFELD (1950) erklärt diese Beobachtung mit einer besseren Durchlüftung dieser Schichten über feine Risse bzw. mit O_2-Zufuhr und CO_2-Absorption durch ziehendes Grundwasser.

Faktoren, die eine hohe *Luftkapazität* und einen hohen *Luftgehalt* bedingen, wie grobe Textur, lockere Struktur und geringer Wassergehalt, verbessern die Durchlüftung des Bodens und erniedrigen die CO_2-Konzentration in der Bodenluft. LUNDEGÅRDH (1924) zitiert hierzu die in Tab. 115 angegebenen Werte. Schwere Böden weisen demnach durchschnittlich höhere CO_2-Gehalte auf als Sandböden.

Tabelle 115. *CO_2-Gehalt der Bodenluft in verschiedenen Bodenarten* (LUNDEGÅRDH 1924)

	Sandboden %	Lehmboden %	Moorboden %
CO_2-Gehalt in 15 cm Tiefe	0,09—0,19	0,05—0,27	0,10—0,45
CO_2-Gehalt in 30 cm Tiefe	0,06—0,24	0,09—0,47	0,34—1,12
CO_2-Gehalt in 60 cm Tiefe	0,11—0,57	0,20—1,13	1,01—3,77

Bearbeitungsmaßnahmen, durch die der Boden gelockert wird, tragen ebenfalls dazu bei, die CO_2-Konzentration in der Bodenluft zu verringern. Aus Tab. 116 nach BLOHM und MAGERS (1930), die hierzu einige Zahlenwerte bringt, wird gleichzeitig deutlich, daß bei höheren CO_2-Konzentrationen in der Bodenluft die Bodenatmung geringer ist. Man kann daraus schließen, daß bei dieser Untersuchung die höheren CO_2-Konzentrationen vornehmlich auf langsameren Gasaustausch zwischen Bodenluft und Atmosphäre zurückzuführen sind.

Tabelle 116. *Einfluß der Bodenbearbeitung auf den CO_2-Gehalt der Bodenluft* (BLOHM und MAGERS 1930)

Datum	Bearbeitung	Bodenatmung mg $CO_2/m^2 \cdot$ Std.	CO_2-Gehalt der Bodenluft	Luftvolumen %	Wassergehalt Gew. %
28. 6.	gegraben	189	$0{,}395 \pm 0{,}072$	$35{,}5 \pm 2{,}2$	15,5
	unbearbeitet	171	$0{,}464 \pm 0{,}108$	$23{,}9 \pm 0{,}4$	15,8
2. 7.	gegraben	261	$0{,}577 \pm 0{,}030$	$36{,}0 \pm 0{,}7$	15,0
	unbearbeitet	189	$0{,}395 \pm 0{,}072$	$24{,}8 \pm 1{,}3$	14,1
15. 8.	gefräst, geeggt	205	$0{,}379 \pm 0{,}029$	$29{,}7 \pm 0{,}9$	13,9
	gefräst, gewalzt	201	$0{,}417 \pm 0{,}032$	$28{,}3 \pm 0{,}8$	13,6
22. 8.	frisch gegraben	240	$0{,}359 \pm 0{,}019$	45,2	10,7
	frisch gehackt	170	$0{,}598 \pm 0{,}100$	$29{,}8 \pm 0{,}8$	13,9

Die *höheren Pflanzen* beeinflussen den CO_2-Gehalt der Bodenluft aus verschiedenen Gründen: Sie verändern die Struktur, den Wasser- und Lufthaushalt des Bodens, damit aber auch seine Durchlüftung und biologische Aktivität, und sie sind durch die *Wurzelatmung* selbst an der CO_2-Produktion im Boden beteiligt. Vgl. hierzu S. 733. WOLLNY (1880, 1886) fand zwar, daß durch Pflanzen beschatteter Boden, der sich weniger erwärmt, in der warmen Jahreszeit weniger CO_2 enthielt als brachliegender. Doch zeigten andere Untersuchungen, daß der CO_2-Gehalt der Bodenluft unter Pflanzenwuchs meistens erhöht ist. Er ändert sich auch mit dem Entwicklungszustand der Pflanzen, wie aus Tab. 117 nach Parker (1924) hervorgeht.

Tabelle 117. *CO_2-Gehalt der Bodenluft unter verschiedenen Pflanzenbeständen* (Parker 1924)

Alter der Pflanzen Tage	Sorghum %	Erbsen %	Buchweizen %	Bohnen %	Baumwolle %	Unbestellt %
15—21	0,79	1,26	0,82	0,88	0,84	0,65
22—28	0,86	1,83	0,84	1,15	1,01	0,46
29—35	1,26	3,20	0,79	1,40	1,25	0,39
36—42	1,68	5,11	—	2,34	1,85	0,25
43—49	1,60	3,27	—	1,76	1,38	0,14
50—56	2,59	3,43	—	2,20	1,85	0,20
57—63	2,51	2,69	—	2,04	1,85	0,13

Die täglichen und jahreszeitlichen *Schwankungen* im CO_2-Gehalt der Bodenluft hängen eng mit der Durchlüftung des Bodens und der Intensität seiner biologischen Vorgänge zusammen. Nach Romell (zit. nach Giesecke 1930) liegt

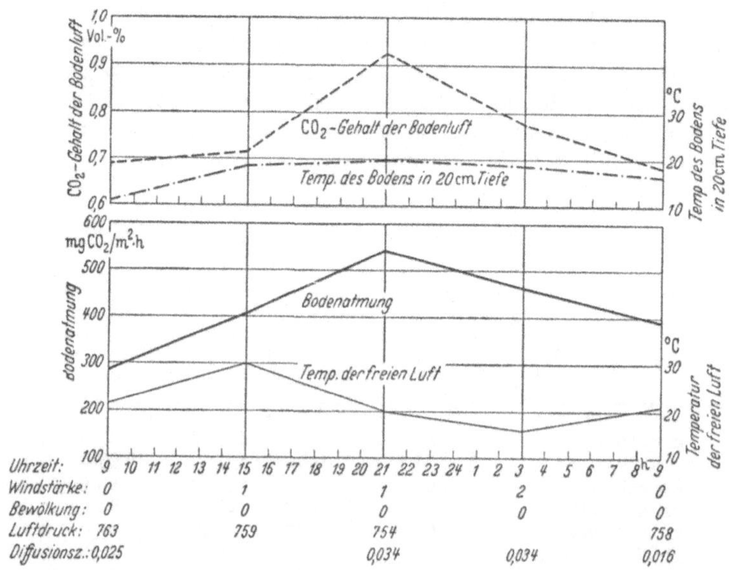

Abb. 150. Tagesgang der CO_2-Konzentration in der Bodenluft in Beziehung zur Bodenatmung, Boden- und Lufttemperatur (Magers 1929)

das Minimum des CO_2-Gehaltes meistens in der kalten Jahreszeit, ein ein- oder zweigipfliges Maximum in der wärmeren Periode. Dieser Verlauf entspricht, wie auf S. 732 beschrieben wird, der Jahreskurve der Bodenatmung. Auch die täglichen Schwankungen des CO_2-Gehaltes sind, wie aus Abb. 150 (Magers 1929) hervorgeht, mit dem Tagesgang der Bodenatmung und der Bodentemperatur gekoppelt.

LEHMANN (1931) fand nach einem *Regen* eine Abnahme der CO_2-Konzentration in 10 cm Tiefe, die er auf die Lösung von CO_2 in Regenwasser sowie verstärkte Luftdruckschwankungen in der Schlechtwetterperiode zurückführte. MINAR (1935) dagegen stellte eine Erhöhung der CO_2-Konzentration nach Regen fest, die er mit einer Verschlämmung der Bodenoberfläche, d. h. also einer Behinderung der Durchlüftung erklärt.

Die Bodendurchlüftung wird im nächsten Abschnitt behandelt. An dieser Stelle muß die *CO_2-Produktion des Bodens*, als ein wesentlicher Faktor, von dem die Zusammensetzung der Bodenluft abhängt, noch etwas eingehender besprochen werden. Die Begriffe *CO_2-Produktion* des Bodens und *Bodenatmung* sind nicht synonym. Unter CO_2-Produktion wird die in einer bestimmten Zeit von der Gewichtseinheit des Bodens erzeugte Menge an Kohlendioxyd verstanden, unter Bodenatmung diejenige CO_2-Menge, die von 1 m^2 Bodenoberfläche innerhalb einer Stunde an die Atmosphäre abgegeben wird (mg (g) $CO_2/m^2 \cdot h$, LUNDEGÅRDH 1924).

Die verschiedene Bezugsbasis bei den CO_2-Produktions- und den Bodenatmungswerten ist bei der Deutung von Untersuchungsbefunden zu berücksichtigen. So stellten z. B. FRERCKS und PUFFE (1959) fest, daß sich bei schwach zersetztem Hochmoortorf das Maximum der Bodenatmung im Laufe einiger Jahre von der am höchsten zu der am wenigsten aufgekalkten Parzelle verlagerte. Im Laboratoriumsversuch dagegen wurde nach 5 Jahren Versuchsdauer auf der am höchsten aufgekalkten Parzelle immer noch die größte CO_2-Produktion festgestellt. Der Grund für dieses unterschiedliche Ergebnis liegt in der verschiedenen Bezugsbasis bei den beiden Messungen. Wenn bei Mineralböden der Gasaustausch nicht durch Bodenverdichtung und hohe Wassergehalte gestört ist, entspricht jedoch die Bodenatmung im großen und ganzen der CO_2-Produktion im ganzen Profil (MAGERS 1929).

Die *CO_2-Bildung im Boden* hängt ab von der Zahl und Art der aktiven Mikroorganismen, ihren Ernährungsbedingungen, dem Pflanzenwuchs und dem Witterungsverlauf. Doch wird bei einem bestimmten Boden die Größe der CO_2-Entbindung nicht allein durch die im Augenblick der Messung bestehende exogene und die mit analytischen Mitteln faßbare endogene Faktorenkonstellation, sondern auch durch die *vorausgegangenen* Einflüsse auf die Bodendynamik bestimmt. KOEPF (1953) faßt das Ergebnis von Untersuchungen über die Temperatur — Zeit — Abhängigkeit der Bodenatmung dahin zusammen: Im Bereich von 10 bis 40° C folgt die CO_2-Produktion des Bodens der VAN'T HOFFschen-Regel. Die Q_{10}-Werte nehmen mit steigenden Temperaturen ab. Auch bei Gleichstellung der übrigen exogenen Bedingungen läßt sich einer bestimmten Versuchstemperatur kein eindeutig bestimmter Wert der CO_2-Produktion eines und desselben Bodens zuordnen. Jede durch eine Temperaturerhöhung erzielte Steigerung des Stoffumsatzes wirkt sich nachträglich in einem proportionalen Nachlassen der Umsatzleistung aus. Der Effekt einer bestimmten Temperaturerhöhung wird maßgeblich von den vorausgegangenen Temperaturen und ihrer Einwirkungsdauer beeinflußt. Für den Wechsel von Feuchtigkeit und Trockenheit und ihren Einfluß auf die CO_2-Produktion lassen sich entsprechende Erscheinungen zeigen (KOEPF 1952). Jede Trockenheitsperiode wirkt wie eine partielle Sterilisation des Bodens. Wird der Boden wieder angefeuchtet, so erfolgt vorübergehend eine beträchtliche Steigerung der Bodenatmungswerte (SCHMALFUSS 1940).

Zwischen der *Zahl* der *Mikroorganismen* im Boden und der Bodenatmung besteht nur eine lockere Beziehung (RIPPEL 1931). Für einige Bodeneigenschaften und die CO_2-Produktion wurden von KOEPF (1954 a, b) folgende Korrelationskoeffizienten errechnet:

Bei 30 verschiedenen Ackerböden:

CO_2-Produktion und Gehalt an abschlämmbaren Teilen	(< 0,01 mm)	$r = 0{,}52$
Kohlenstoffgehalt des Bodens		$r = 0{,}54$

Bei 37 verschiedenen Böden von Acker, Dauergrünland und Wald:

Gehalt an abschlämmbaren Teilen	$r = 0{,}43$
C-Gehalt des Bodens	$r = 0{,}77$
C-Gehalt des Bodens und C/N-Verhältnis	$R = 0{,}81$.

Dabei besteht zwischen der CO_2-Produktion und dem C-Gehalt eine positive, zwischen der CO_2-Produktion und dem C/N-Verhältnis eine negative Korrelation. Zufuhr *organischer Substanz* zum Boden erhöht die CO_2-Produktion und die Bodenatmung, doch ist die Wirkung sehr von der Natur der organischen Substanz und den Eigenschaften des Bodens abhängig (Schmalfuss 1940, Lees und Porteous 1950, Pink und Allison 1951, Sokolow 1954, Weissenberg 1954, Koepf 1956, Frercks und Puffe 1957, Gaarder 1957 u. a.). Die Wirkung der Mineraldünger auf die Bodenatmung wurde schon frühzeitig untersucht, Suchtelen (1910 u.a.). Lundegårdh (1924) und Schmalfuss (1940) betonten dabei insbesondere die Bedeutung des *Nährstoffverhältnisses*, Joshi und Joshi (1953) sowie Schaffer (1954) wiesen darauf hin, daß bei sehr hohen und einseitigen Mineraldüngergaben bereits schon eine Abnahme bzw. keine weitere Steigerung der Bodenatmungswerte eintreten kann, wenn die Pflanzenerträge durch diese Düngung noch gesteigert werden. Daß bei Mangelböden auch die Zufuhr der Spurenelemente Cu und B die Bodenatmung beeinflußt, wurde von Weissenberg (1954) festgestellt.

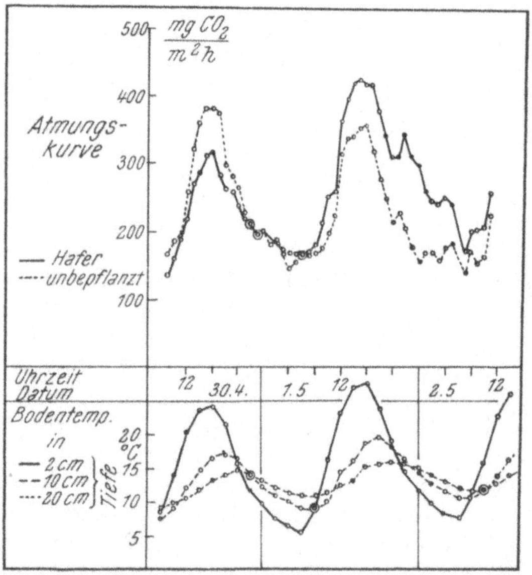

Abb. 151. Atmungskurven einer bepflanzten und einer unbepflanzten Parzelle an zwei Sonnenscheintagen (Koepf 1954c)

Schließlich sei noch die Beeinflussung der CO_2-Produktion durch die *meteorologischen Faktoren* — Feuchtigkeit und Temperatur — sowie durch die höhere Pflanze kurz erörtert. Da sich die meteorologischen Faktoren im Freiland täglich und im Laufe des Jahres ändern, zeigen die CO_2-Produktion und die Bodenatmung eine ausgesprochene tägliche und jahreszeitliche Periodizität mit einem ein- oder zweigipfligen Maximum im Frühjahr und Sommer und den geringsten Werten im Winter. Russel und Appleyard (1915) wiesen bereits darauf hin, daß die Bodenatmung im Sommerhalbjahr vorwiegend in Abhängigkeit vom Feuchtigkeitsgehalt des Bodens, im Winter dagegen in der Hauptsache mit den Außentemperaturen variiert. Feher und Frank (1937) bildeten aus den Temperatur- und Feuchtigkeitswerten ein Produkt, den R-Faktor, und betrachteten die Intensität der Lebenstätigkeit im Boden als eine Funktion dieses Wertes. Die täglichen Schwankungen der Bodenatmung sind, wie aus Abb. 151 nach Koepf (1954c) hervorgeht, vornehmlich eine Folge der täglichen Temperaturschwankungen im Boden.

DOENHOFF (1927), KOEPF (1954c) und MAKAROW (1958) fanden, daß das tägliche Maximum der Bodenatmung in den Mittags- und frühen Nachmittagsstunden liegt, MAGERS (1929) stellte dagegen die höchsten Werte erst gegen 21 Uhr fest. Diese Unterschiede sind als Folge verschiedener Erwärmung und Temperaturausbreitung im Boden zu erklären. Das Zusammenspiel von Temperatur und Feuchtigkeit in ihrem Einfluß auf die Bodenatmung während einer ganzen Vegetationsperiode wird durch Abb. 152 nach H. KOEPF (1952) veranschaulicht, aus der gleichzeitig die Wirkung einer organischen Düngung hervorgeht.

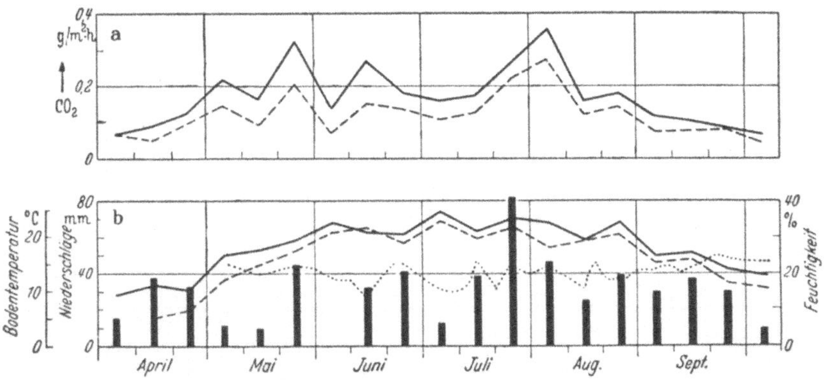

Abb. 152. *a* Atmungskurven eines unbewachsenen Ackerbodens, —— = mit Stallmistdüngung, – – – – = ungedüngt. *b* Der Witterungsverlauf, Säulen = Niederschlagssumme je Dekade, —— = Bodentemperatur in 10 cm Tiefe auf der Stallmistparzelle, – – – = Bodentemperatur auf der ungedüngten Parzelle, ···· = Bodenfeuchte in Gew.-% H_2O (KOEPF 1952)

Daß die *höheren Pflanzen* durch O_2-Verbrauch und CO_2-Ausscheidung ihrer Wurzeln die Zusammensetzung der Bodenluft verändern, wurde bereits erwähnt. Nach den Angaben von LUNDEGÅRDH (1924) wird der Anteil der Wurzelatmung an der gesamten CO_2-Produktion des Bodens meist auf $^1/_3$ geschätzt. Bei der Untersuchung von MEYER und SCHAFFER (1954) wurde die Bodenatmung durch den Faktor Bewachsung um 37% erhöht. M. J. JASSTREBOW (1958) gibt den Anteil der Wurzelatmung an der CO_2-Produktion im Boden mit 29 bis 37% an. Doch liegt auch beim Einfluß der höheren Pflanzen keine konstante Wirkung vor, diese ändert sich mit der Jahreszeit und dem Entwicklungszustand der Pflanzen.

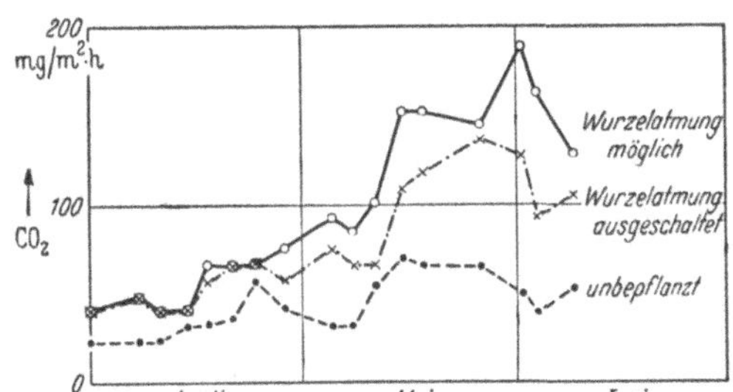

Abb. 153. Atmungskurven des Bodens in einem Ackerbohnenbestand mit und ohne Ausschaltung der Wurzelatmung (MEYER und SCHAFFER 1954)

MEYER und SCHAFFER (1954) trennten unter den Bedingungen des Freilandes die Wurzelatmung von den übrigen Wirkungen des Pflanzenwuchses (Struktur, Wassergehalt usw.) und erhielten die in Abb. 153 graphisch dargestellten Werte.

MAGERS (1929) ordnete die Bodenatmungswerte unter verschiedenen Pflanzenbeständen in der Reihenfolge: Erbsen $>$ Zuckerrüben $>$ Kartoffeln, Hafer, Gerste $>$ unbestellt, WURMBACH (1934) dagegen: Dauergrünland $>$ Leguminosen $>$ Getreide $>$ Zuckerrüben $>$ Brache. Tab. 118 nach STANGEN (1955) veranschaulicht die unterschiedliche CO_2-Produktion eines und desselben Bodens unter verschiedener Nutzung und zugleich die Aktivitätsverteilung im Profil.

Überblickt man diese Angaben über die CO_2-Produktion, so zeigt sich, daß die biologische Aktivität des Bodens, die darin zum Ausdruck kommt, eine zeitlich und örtlich veränderliche Größe ist. Ihre Änderungen hängen von einem komplexen Faktorengefüge ab. Dieses ist zum Teil durch die Beschaffenheit des Bodens selbst und den Witterungsverlauf gegeben. Es wird aber auch durch die Bewirtschaftung beeinflußt. Bei den CO_2-erzeugenden bodenbiologischen Vorgängen wird zugleich Sauerstoff verbraucht. Die Partialdrucke dieser Gase in der Bodenluft werden dadurch einseitig verändert. Schädliche Über- bzw. Unterkonzentrationen an CO_2 und O_2 würden im Boden die Lebenstätigkeit rasch hemmen, wenn nicht ein ständiger Gasaustausch zwischen dem Boden und der Atmosphäre stattfinden würde. Dieser soll im nächsten Abschnitt besprochen werden.

Tabelle 118. *CO_2-Produktion in verschiedenen Bodentiefen (Milligramm CO_2 je 100 g Boden)*
(STANGEN 1955)

Kulturart	Profilstufe cm	Nach 48 Std.	% Humus (Gesamt-C)
Dauergrünland	0—4	95,2	3,0
	6—10	14,4	2,1
	12—16	7,9	1,7
	18—22	4,5	1,7
	24—28	2,4	1,2
	30—34	2,3	0,9
Mischwald	0—4	40,2	6,9
	6—10	12,8	1,7
	12—16	9,4	1,0
	18—22	4,4	1,4
	24—28	3,0	1,0
	30—34	3,4	1,0
Fichtenhochwald	0—4	51,3	8,1
	6—10	10,8	4,0
	12—16	5,8	2,7
	18—22	4,0	3,4
	24—28	3,8	2,8
	30—34	3,8	0,9
Rübenacker	0—4	14,0	2,0
	6—10	10,4	2,2
	12—16	15,9	2,7
	18—22	12,7	2,2
	24—28	7,4	1,3
	30—34	6,9	0,8

5. Der Gasaustausch zwischen dem Boden und der Atmosphäre

Der einseitig gerichteten Veränderung der Partialdrucke von O_2 und CO_2 durch biologische Prozesse wirkt der Gasaustausch zwischen Bodenluft und atmosphärischer Luft entgegen. Dieser *Gasaustausch*, oft auch als *Erneuerung der Bodenluft* bezeichnet, wird *Durchlüftung* genannt. Unter *Luftdurchlässigkeit* des Bodens oder *Permeabilität* für Luft soll dagegen diejenige Luftmenge verstanden werden, die bei der Verwendung einer entsprechenden Meßanordnung die Bodenhohlräume unter der Wirkung eines Druckgradienten passiert.

Die *Bodendurchlüftung* vollzieht sich fast ausschließlich in der Gasphase. Andere Faktoren spielen beim O_2- und CO_2-Haushalt des Bodens nur eine untergeordnete Rolle. Es ist wahrscheinlich, daß sauerstoffhaltiges, ziehendes Grundwasser die O_2-Versorgung tieferer Bodenschichten verbessern kann. Ferner ist zu berücksichtigen, daß Temperaturänderungen des Bodens das Gleichgewicht zwischen Gasen in der Bodenluft und gelösten Gasen im Bodenwasser verändern. So weist TURLJUN (1957) darauf hin, daß bei einer intensiven Erwärmung des Bodens die entbundene CO_2-Menge 3 bis 4 mal größer sein kann, als nach der VAN'T HOFFschen Regel für biologische CO_2-Entbindung zu erwarten wäre. Bei trockenem Boden sind Sorptions- und Desorptionsvorgänge als Folge von Temperaturänderungen und Befeuchtung in Betracht zu ziehen. Der Abtransport von CO_2 aus dem Boden durch Lösung im Sickerwasser ist gering. Legt man durchschnittliche Verhältnisse zugrunde, nämlich eine Bodenatmung von 100 mg/m² · h (entsprechend 876 g/m² · Jahr), 600 mm Jahresniederschläge, wovon rund 50% als Sickerwasser in den Untergrund wandern und einen CO_2-Gehalt der Bodenluft von 0,3 Vol.-%, so werden durch das Sickerwasser bei einer Durchschnittstemperatur von 10° C rund 2,12 g CO_2/m² im Jahr ausgewaschen. Diese Zahl stimmt in der Größenordnung mit der Angabe ROMELLS (1922) überein, der auf Grund des CO_2-Gehaltes von Gewässern errechnete, daß durch Wasser im Laufe des Jahres soviel CO_2 weggeführt wird, wie der Boden in 1 bis 2 Tagen an die Atmosphäre abgibt. Wenn das Sickerwasser Gelegenheit hat, im Boden Kalk zu lösen, so ist sein CO_2-Gehalt allerdings größer. Bei kalkreichen Aueböden mit ziehendem Grundwasser und hoher CO_2-Konzentration in der Bodenluft dürfte deshalb die Menge an gelöstem CO_2, die aus dem Boden fortgeführt wird, größer sein. Schließlich sind noch CO_2-Verluste des Bodens als Folge der Wasserverdunstung möglich. Den Angaben von HELBIG (1930, S. 221), LAATSCH (1954) und SCHUBACH (1952) sowie mündlichen Mitteilungen der Agrarmeteorologischen Forschungsstelle Hohenheim läßt sich entnehmen, daß eine tägliche Verdunstungsrate von 3 bis 5 mm an einzelnen Tagen schon sehr hoch ist. Das entspricht einer Verdunstung während der heißesten Tageszeit von etwa 0,5 Liter/m² · h, eine Menge, durch die nicht viel CO_2 in Freiheit gesetzt wird. Die Ansicht WURMBACHS (1934), wonach durch hohe Wasserverdunstung an warmen Tagen die Bodenatmungswerte erhöht werden können, dürfte demnach nur in eingeschränktem Maße zutreffen. Die von TURLJUN (1957) genannten Effekte müssen dabei berücksichtigt werden. Die genannten Vorgänge beeinflussen den Gashaushalt des Bodens nach dem Gesagten nur geringfügig und vorübergehend. Doch können Temperatur- und Feuchtigkeitswechsel des Bodens eine gewisse Phasenverschiebung zwischen CO_2-Produktion und Bodenatmung bewirken.

Nach dieser Abgrenzung des Problems soll nunmehr die Bodendurchlüftung besprochen werden. Die grundlegende Arbeit hierzu stammt von ROMELL (1922), ältere Untersuchungen von HANNEN (1892), HENSELE (1893), BUCKINGHAM (1904) und LUNDEGÅRDH (1924). Neuerdings hat die Bodendurchlüftung eine vielseitige experimentelle und theoretische Bearbeitung erfahren, so durch BLOHM und MAGERS (1930), SCHMIDT (1930), ROMELL (1935), PENMAN (1940, 1940a), BLAKE und PAGE (1948), WERSCHININ und KINIBENKO (1948), TAYLOR (1949), RANEY (1950), DE VRIES (1950), VAN BAVEL (1951, 1952a, 1952b), COLLIS-GEORGE (1953), EVANS (1953), RUST, KLUTE und GIESEKING (1957), MARSHALL (1959) u. a. Eine neuere zusammenfassende Darstellung stammt von BAVER (1956, S. 209—218).

Bei der *Durchlüftung des Bodens* lassen sich nach BAVER (1956) zwei Faktorengruppen unterscheiden:

a) *die Diffusion*, die stetig wirkt, wenn ein Konzentrationsgefälle zwischen Bodenluft und Atmosphäre besteht,

b) die *meteorologischen Faktoren* wie Temperaturschwankungen, Luftdruckschwankungen, Luftbewegung und Veränderung der Bodenluft durch eindringendes Regenwasser. Deren gemeinsames Kennzeichen ist, daß sie unstetig wirken.

Schwankungen des Luftdrucks: BOUYOUCOS und MCCOOL (1924) stellten fest, daß der Luftdruck im Boden in 10 Fuß Tiefe stets derselbe ist, wie in der Atmosphäre. Die Bodenluft macht die Druckschwankungen der Atmosphäre mit. Bei zunehmendem Luftdruck muß also atmosphärische Luft in die Bodenhohlräume eintreten und umgekehrt. SCHMIDT (1930) wies auf solche Massenverschiebungen bei raschem Luftdruckwechsel hin. Nach seiner Ansicht wird die Erneuerung der Bodenluft durch die Unruhe in der Luft erheblich beeinflußt. Nach den Berechnungen von BUCKINGHAM (1904) ist der Einfluß von Barometerschwankungen auf die Durchlüftung nur gering, ROMELL (1922) schätzt ihn auf maximal $^1/_{100}$ der normalen Durchlüftung.

Temperaturschwankungen: Temperaturschwankungen der Bodenluft können wegen der damit verbundenen Ausdehnung und Zusammenziehung einen Massenwechsel zwischen Bodenluft und Atmosphäre veranlassen. ROMELL (1922) veranschlagt diese als Folge der täglichen Temperaturschwankungen eintretende Lufterneuerung im Boden auf $^1/_{100}$ der Normaldurchlüftung. Ferner besteht zwischen der Bodenluft und der Atmosphäre ein wechselnder Temperaturgradient, der einen Luftaustausch bewirkt. Dieser Effekt wird von ROMELL auf $^1/_{240}$ bis $^1/_{400}$ der normalen Durchlüftung geschätzt.

ONTSCHUKOW (1956) bestimmte gleichzeitig die Temperaturen der festen Bodensubstanz und der Bodenluft. Die größten Differenzen von 10,6° C wurden an der Bodenoberfläche gemessen, in 20 cm Tiefe bestanden keine Unterschiede mehr. Er schließt aus diesen Messungen, daß *konvektive* Strömungen innerhalb der Bodenluft auf die obersten Schichten beschränkt bleiben.

Wind: ROMELL (1922) schätzt die Wirkung der Luftbewegungen in der Atmosphäre auf $^1/_{1000}$ der normalen Durchlüftung bei bewachsenem Boden. FUKUDA (1955) leitete auf Grund des DARCYschen Gesetzes eine theoretische Formel für das Eindringen des Windes in den Boden ab und prüfte diese auch experimentell. Selbst bei einem Sandboden mit 0,5 bis 0,25 mm mittlerem Teilchendurchmesser dringt bei böigem Wind die Luft nur etwa 5 mm unter die Bodenoberfläche. Das durch Windstöße ausgetauschte Luftvolumen entspricht je nach deren Frequenz einer Säule von 1 bis 20 m pro Tag. Die auf diese Weise aus dem Boden abtransportierte Wasserdampfmenge ist gering, sie beträgt beispielsweise bei 0° C Bodentemperatur 10 mg/cm^2 · Tag, bei 20° C 35 mm. Aus dieser Untersuchung kann man schließen, daß bei böigem Wind die Grenzfläche zwischen Bodenluft und Atmosphäre einige mm unter die Bodenoberfläche verlegt wird.

Regenfall: Niederschlagswasser verdrängt beim Eindringen in den Boden die Bodenluft und löst darin vorhandene Gase, vornehmlich CO_2. Je nach Häufigkeit und Intensität der Niederschläge erfolgt also durch Sickerwasser periodisch eine partielle oder nahezu völlige Erneuerung der Bodenluft, da nach dem Verschwinden des Sickerwassers die atmosphärische Luft wieder in den Boden eindringt. Eine völlige Erneuerung der Bodenluft auf diesem Wege ist auch bei starken Niederschlägen nicht anzunehmen, da ein gewisser Anteil an eingeschlossener Luft in den Poren verbleiben wird. ROMELL (1922) schätzt die Wirkung des Regens auf $^1/_{12}$ bis $^1/_{16}$ der gesamten Durchlüftung.

Nach dem Gesagten verbleibt also die *Diffusion* als derjenige stetig wirkende Vorgang, durch den die Erneuerung der Bodenluft in der Hauptsache erfolgt. Diffusion findet statt, sobald zwischen den Bodenhohlräumen und der Atmosphäre ein *Konzentrationsgefälle* besteht und *freie Verbindungswege* vorhanden sind. Doch macht TURLJUN (1958) darauf aufmerksam, daß in den tieferen Schichten des Bodens mit sehr engen Hohlräumen sich die Gase in Ultraverteilung befinden und für ihre Bewegung andere Gesetze gelten, nämlich *Sorption* durch die feste und flüssige Phase sowie das Prinzip der *Effusionsströmung*. Bei der Effusionsgegenströmung nehmen die Gase die Temperatur ihrer Umgebung an und bewegen sich entlang eines *Temperaturgradienten*, gegebenenfalls auch in Richtung auf höhere Konzentrationen. Die Verlagerung erfolgt also selektiv, je nachdem ob die Gase einen Energiezuwachs oder -verlust erfahren. Doch gilt dies nur für Schichten mit feinsten gasführenden Poren. Der Gasaustausch zwischen den Poren, die normalerweise Luft enthalten, und der Atmosphäre erfolgt durch Diffusion entlang eines *Konzentrationsgefälles.* Für den Stofftransport durch Diffusion gilt die FICKsche Gleichung:

$$dS = -\ kF \frac{dc}{dx} \cdot dt$$

Die transportierte Stoffmenge dS, die in der Zeiteinheit dt durch den Querschnitt F unter dem Einfluß des Konzentrationsgefälles dc/dx hindurchwandert, ist also proportional zum *Diffusionskoeffizienten* k. Dieser hat die Dimension cm^2/sek und liegt nach MANEGOLD (1955, S. 179) bei Gasen unter gewöhnlichen Bedingungen im Bereich von 10^{-1} bis 10^{-2} cm^2/sek. Für k gilt ferner $k = \text{const.}\ T^{1,5}/p$. Der Diffusionskoeffizient ist also proportional zur absoluten Temperatur und umgekehrt proportional zum Druck sowie abhängig von der Natur der Gase. Für die stationäre Diffusion gilt:

$$S = \frac{k\,F\,t\,(Ce - Ca)}{d}$$

wobei Ce die Eintrittskonzentration, Ca die Austrittskonzentration ist und d der Diffusionsweg (MANEGOLD 1955, S. 178). Für den Gasaustausch im Boden folgt also, daß dieser vom *freien Porenquerschnitt,* der *Weglänge* der Gase und dem *Konzentrationsgefälle* zwischen zwei betrachteten Niveaus abhängt.

Bei der theoretischen Bearbeitung dieses Diffusionsvorganges wurden verschiedene Wege eingeschlagen. Die Berechnungen von ROMELL (1922) sowie von VAN BAVEL (1951, 1952b) erfolgten unter Berücksichtigung der biologischen Aktivität, d. h. der O_2 verbrauchenden und CO_2 produzierenden Prozesse im Boden. PENMAN (1940, 1940a) untersuchte die Diffusion durch ein poröses Medium. Er führt die verringerte Diffusion durch einen porösen Körper auf zwei Umstände zurück: Den geringeren freien Porenquerschnitt und die größere Weglänge der Gase, da die Hohlräume nicht geradlinig in Richtung auf die Bodenoberfläche verlaufen. Diese Beziehungen kommen in der Gleichung

$$D = D_0 S \frac{l}{le}$$

zum Ausdruck. Dabei ist D die Diffusionskonstante in dem porösen Medium, D_0 der Diffusionskoeffizient in Luft, S der freie Porenquerschnitt, l die Länge des porösen Körpers, le die tatsächliche Weglänge der Gase durch die gewundenen Hohlräume des Bodens. Das Verhältnis D/D_0 wurde von ihm zu 0,66 bestimmt. VAN BAVEL (1951) gibt folgende experimentelle Daten an (Tab. 119):

Tabelle 119. *Verhältnis D/D_0 für die Diffusion von Alkohol durch die Luft von zwei verschiedenen porösen Medien*
(VAN BAVEL 1951)

Zusammensetzung des Mediums	Porosität	D/D_0	Mittl. Fehler
Quarzsand 4 % > 1 mm, 70 % 0,5—1,0 mm 15 % 0,25—0,5 mm, 11 % < 0,25 mm	0,415	0,589	0,003
50 % Boden < 0,1 mm, 50 % Quarzsand 0,25—0,5 mm	0,355	0,577	0,003

Er schließt aus seinen Versuchen und denjenigen Penmanns, daß in einem *Porositätsbereich* von 0,1 bis 0,60 praktisch mit einem Verhältnis D/D_0 von 0,6 gerechnet werden kann. De Vries (1950) wies allerdings darauf hin, daß diese durch eine Gerade darstellbare Beziehung von der Form der festen Partikel abhängt. Je mehr diese von der Kugelform abweicht, umso mehr weicht auch die Kurve für die Beziehung zwischen D/D_0 und S von einer Geraden ab. Ferner wurde von Blake und Page (1948) bereits darauf hingewiesen, daß diese Kurve nicht durch den Nullpunkt zu gehen braucht. Rust, Klute und Gieseking (1957),

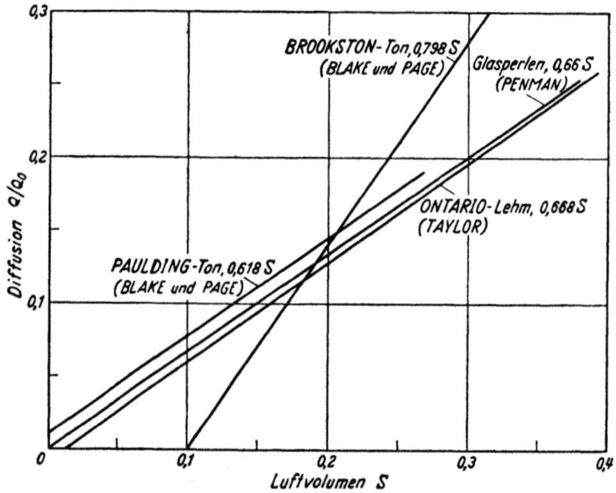

Abb. 154. Die Beziehung zwischen der Diffusion und dem lufterfüllten Hohlraumvolumen des Bodens (Baver 1956, S. 216)

die die Diffusion unter Verwendung von $C^{14}O_2$ prüften, fanden als lineare Annäherung die Formel $D/D_0 = 0{,}04 + 0{,}60\ S$. Marshall (1959) findet, daß die Formel $D/D_0 = \varepsilon^{3/2}$ eine bessere Annäherung ergibt als $D/D_0 = 0{,}66\ \varepsilon$. Dabei ist ε die Porenfläche, ausgedrückt in cm^2/cm^2. In Abb. 154 nach Baver (1956) sind die Ergebnisse verschiedener Autoren zusammengefaßt. Dabei ist bemerkenswert, daß bei dem schweren Brookston-Ton die Diffusion bereits bei einem PV von 10% den Wert 0 erreicht. Daraus ist zu schließen, daß dieser Boden *blockierte Hohlräume* ohne Verbindung zur Außenluft enthält.

Aus den mitgeteilten Daten geht hervor, daß der Diffusionskoeffizient unabhängig ist vom Porenraum und, wie von Collis-George (1953) betont wird, auch von der Permeabilität. Doch treten bei einem Porenvolumen unter 10% Abweichungen ein. Die geförderte Gasmenge hängt demnach von zwei Faktoren ab, dem Konzentrationsgefälle und dem freien Porenraum. Dieser wiederum richtet sich nach dem Gesamtporenvolumen und dem Wassergehalt. Diese Beziehungen werden am besten durch die Ergebnisse Taylors (1949) illustriert, die in Abb. 155a bis d wiedergegeben sind. Taylor führt anstelle des Ausdruckes D/D_0 den Parameter λ, die "*äquivalente Diffusionsdistanz*" ein, die die Dimension einer Länge hat. λ gibt die Länge einer Röhre mit bestimmtem Querschnitt an, durch die bei unbehinderter Diffusion die gleiche Gasmenge diffundieren würde wie durch einen porösen Körper mit demselben Querschnitt.

Textur, Struktur und *Wassergehalt* des Bodens sind diejenigen Faktoren, von denen die *Gestalt* und *Größe* des *luftführenden Porenraums* abhängen, sie bestimmen damit auch den Gasaustausch zwischen Boden und Luft. Das Konzentrationsgefälle zwischen Bodenluft und Atmosphäre hängt weiterhin ab von der *Intensität der Lebensvorgänge* im Boden, d. h. von den O_2 verbrauchenden und CO_2 produzierenden Prozessen. Besondere Berücksichtigung verdient bei der Durchlüftung

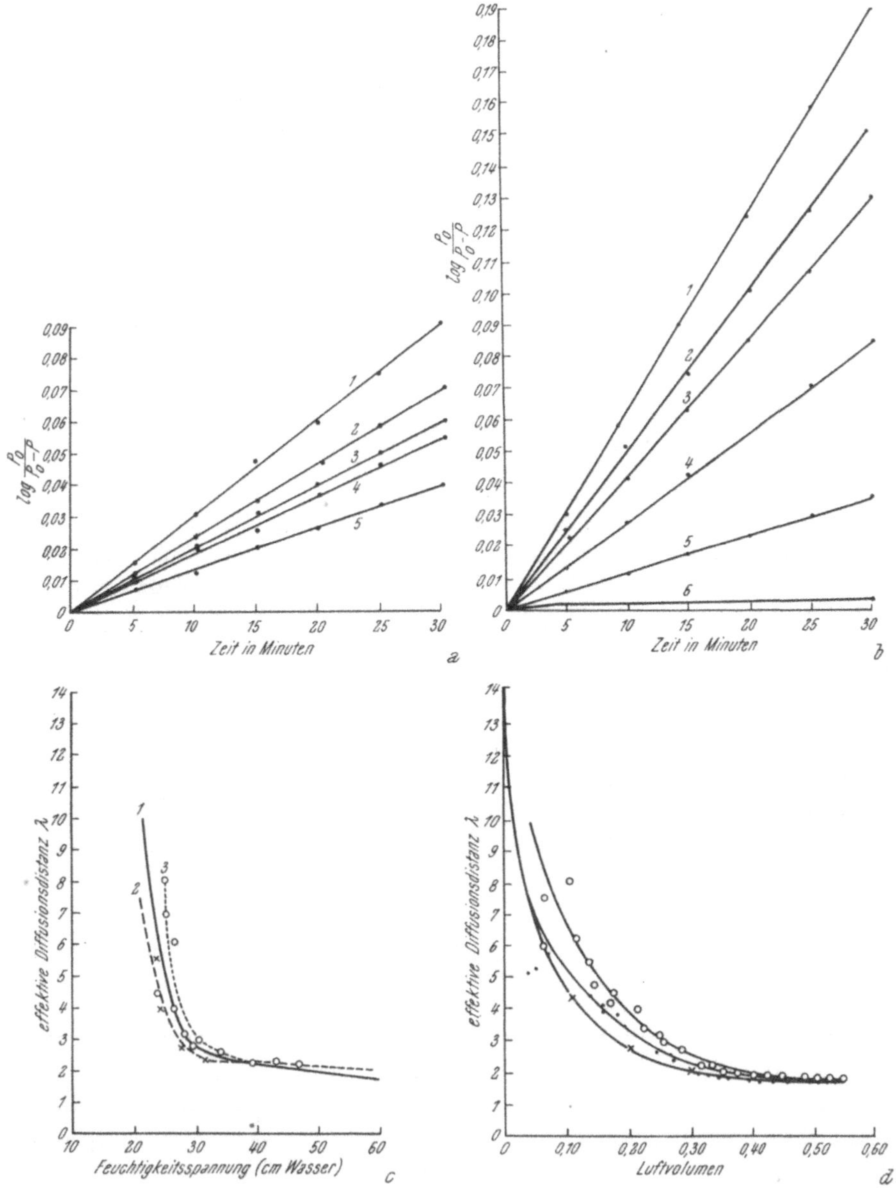

Abb. 155 *a* bis *d*. Die Diffusion als Funktion des Porenvolumens und des Wassergehaltes von porösen Körpern (TAYLOR 1949)

der *Wasserhaushalt.* Wie die Abb. 154 und 155 zeigen, ist eine ausreichende Durchlüftung auch bei dicht gelagertem und verkrustetem Boden durchaus möglich. Dies trifft, wie aus Abb. 155b und c hervorgeht nicht mehr zu, sobald ein bestimmter

Feuchtigkeitsgehalt überschritten wird. DOMBY und KOHNKE (1956) wiesen deshalb auch darauf hin, daß sich die Wirkung einer *Bodenverkrustung* auf die Durchlüftung erst bei feuchtem Boden bemerkbar macht. Wassergesättigte Krusten können einen Gasaustausch ganz verhindern, trockene, verkrustete Böden können eine höhere Diffusion aufweisen als feuchte, nicht verkrustete. Die Beeinflussung des Wassergehaltes im Boden durch Bewässerung (ROBINSON 1957) oder Drainage (JAMISON und DOMBY 1956) wirkt sich deshalb erheblich auf die O_2-Versorgung der Kulturpflanzen aus. BLAKE und PAGE (1948) untersuchten die Porosität und die Diffusionsrate bei verschiedenen *Fruchtfolgen* und verschiedener *Bodenbearbeitung*. Ihre Ergebnisse sind in Abb. 156 auszugsweise wiedergegeben.

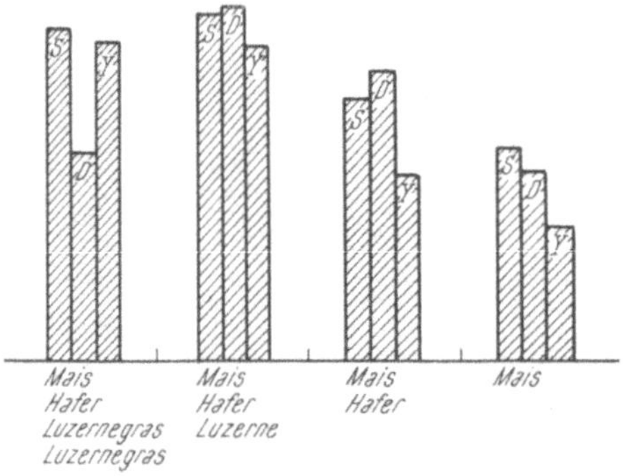

Abb. 156. Porosität des Bodens, Diffusionsrate und Ertrag an Mais bei vier verschiedenen Fruchtfolgen. S = Porosität, D = Diffusion, Y = Ertrag. (BLAKE und PAGE 1948)

Diese Darstellung zeigt, daß eine vielseitige Fruchtfolge die Porosität des Bodens verbessert und die Diffusionsrate erhöht. Auch die Erträge an Mais liegen bei den Fruchtfolgen mit Luzerne und Luzernegras erwartungsgemäß höher als bei der reinen Getreidefruchtfolge. Doch spielen bei der Ertragsbildung neben der Bodendurchlüftung zweifellos noch andere Faktoren eine Rolle. Die Sauerstoffgehalte in der Bodenluft, die hier ebenfalls bestimmt wurden, sanken in keinem Fall unter 18,5% in 1 Fuß Tiefe und 12,5% in 3 Fuß Tiefe ab. Die Verwendung der verschiedenen Bodenbearbeitungsgeräte bedingte Unterschiede im O_2-Gehalt der Bodenluft von nicht mehr als 0,2 bis 0,3%.

Die Permeabilität des Bodens für Luft. Zahlenwerte für die Bodendurchlüftung können auf verschiedene Weise erhalten werden. Die Diffusion und die Zusammensetzung der Bodenluft können direkt bestimmt werden. Auch die Luftkapazität des Bodens ist ein Maß für seine Durchlüftung. Häufig wird jedoch auch die *Permeabilität für Luft* gemessen. Man preßt einen Luftstrom durch den Boden und mißt den hierzu notwendigen Druck. Vorrichtungen hierzu wurden u.a. von RENK (1879), WOLLNY (1893), JANERT (1937), EVANS und KIRKHAM (1949) und RID (1960) beschrieben.

Tabelle 120. *Zeit in Sekunden für den Abfall des Manometers beim Durchpressen von Luft durch eine Scholle*
(DE BOODT und KIRKHAM 1953)

Ida Schlufflehm	Vertikal		Horizontal	
	von oben	von unten	von Seite 1	von Seite 2
A Horizont	344,6	349,4	153,5	182,7
B Horizont	214,8	245,9	180,8	229,4

Die Permeabilität des Bodens für Luft hängt in erster Linie ab von seinem Gehalt an groben Poren. Die Durchströmungsgeschwindigkeit nimmt mit dem Quadrat des Porendurchmessers zu. Grobe Textur, Vorhandensein von Aggregaten, zwischen denen sich größere Hohlräume befinden, Spalten und Röhren, geringe Wassergehalte des Bodens erhöhen deshalb die Permeabilitätswerte sehr rasch.

Da die Durchströmbarkeit für Luft ungleich höher ist als die für Wasser, spricht die Methode auf geringe Unterschiede im Gehalt an groben Bodenporen an. Eine genügende Homogenität in der Lagerung des Bodens ist für die Reproduzierbarkeit der Meßwerte wichtig. De Boodt und Kirkham (1953), ferner Maasland und Kirkham (1955) machten darauf aufmerksam, daß sich mit Permeabilitätsmessungen häufig eine *Anisotropie* von Schollen nachweisen läßt. Böden, die zu einer schieferigen Struktur neigen, haben, wie aus Tab. 120 hervorgeht, in horizontaler Richtung eine höhere Permeabilität als in der vertikalen.

Literatur

Alten, F., und H. Loofmann: Die Luftadsorption an Böden und Bodenbestandteilen und ihr Einfluß auf die Bestimmung des Porenvolumens und des spezifischen Gewichtes von Ackerböden mit dem Druckluftpyknometer. Z. Bodenkde. Pflanzenernähr. **26**, 1 — 13 (1941).

Van Bavel, C. H. M.: A soil aeration theory based on diffusion. Soil Sci. **72**, 33 — 46 (1951). — Gaseous diffusion and porosity in porous media. Soil Sci. **73**, 91 — 104 (1952a); Ref.: Soils a. Fertil. **15**, 187 (1952a). — A theory of the soil atmosphere in and around a hemisphere in which soil gases are used or released. Proc. Soil. Sci. Soc. Amer. **16**, 150 — 153 (1952b); Ref.: Soils a. Fertil. **15**, 323 (1952b). — Baver, L. D.: Soil Physics, 3. Aufl. New York: Wiley, und London: Chapman and Hall 1956. — Bertrand, A. R., and H. Kohnke: Subsoil conditions and their effects on oxygen supply and the growth of corn roots. Proc. Soil. Sci. Soc. Amer. **21**, 135 — 140 (1957). — Blake, G. R., and J. B. Page: Direct measurement of gaseous diffusion in soils. Proc. Soil Sci. Soc. Amer. **13**, 37 — 42 (1948). — Blank, E., und F. Giesecke: Über den Einfluß der Regenwürmer auf die physikalischen und biologischen Eigenschaften des Bodens. Z. Pflanzenernähr., Düng., Bodenkde. B **3**, 198 (1924). — Blohm, G., und H. Magers: Die Durchlüftung des Bodens als physikalische Eigenschaft. Fortschr. Landwirtsch. **5**, 713 - 719 (1930). — De Boodt, M. F., and D. Kirkham: Anisotropy and measurement of air permeability of soil clods. Soil Sci. **76**, 127 - 133 (1953). — De Boodt, M.: Bijdrage tot de kennis van de bodemstruktur. Diss. Gent 1957. — Bouyoucos, G. J., and M. M. McCool: The aeration of soils as influenced by air-barometric pressure changes. Soil Sci. **18**, 53—63 (1924). — Boynton, D., and W. Reuther: A way of sampling soil gases in dense subsoil and some of its advantages and limitations. Proc. Soil Sci. Soc. Amer. **3**, 37 — 42 (1938). — Buckingham, E.: Contribution to our knowledge of the aeration of soils. U.S. Dep. Agric. Bur. Soils Bull. **25**, 52 (1904).

Collis-George, N.: Relationship between air and water permeabilities in porous media. Soil Sci. **76**, 239 - 250 (1953).

Danielson, R. E., and M. B. Russel: Ion absorption by corn roots as influenced by moisture and aeration. Proc. Soil Sci. Soc. Amer. **21**, 2 - 6 (1957); Ref.: Z. Pflanzenernähr., Düng., Bodenkde. **78**, 86 (1957). — Diekmann, J.: Gassorption an Braunkohlen. Bodenkde. u. Pflanzenernähr. **2**, 315 — 327 (1937). — Dietz, Th.: Entwicklung und Eigenschaften von Böden aus tonigen Substraten. Diss. Bonn 1959. — Domby, C. W., and H. Kohnke: The influence of soil crusts on gaseous diffusion. Proc. Soil Sci. Soc. Amer. **20**, 1 - 5 (1956); Ref.: Z. Pflanzenernähr., Düng., Bodenkde. **75**, 93 (1956); LZ. II, 1106 (1957). — Doenhoff, G.: Untersuchungen über die Größe und Bedeutung der Bodenatmung auf landwirtschaftlich kultivierten Flächen. Kühn-Arch. **15**, 457 - 510 (1927).

Epstein, E., and H. Kohnke: Soil aeration as affected by organic matter application. Proc. Soil Sci. Soc. Amer. **21**, 585–588 (1957); Ref.: Soils a. Fertil. **21**, 99–100 (1958). — Evans, D. D., and D. Kirkham: Measurement of the air permeability of soil in situ. Proc. Soil Sci. Soc. Amer. **14**, 65–73 (1949). — Evans, D. D.: Effect of combined pressure and concentration gradients on gaseous flow through soils. Jowa St. Coll. J. Sci. **27**, 165–166 (1953); Ref.: Soils a. Fertil. **17**, 25 (1954).

Feher, D., und M. Frank: Experimentelle Untersuchungen über den Einfluß der Temperatur und des Wassergehaltes auf die Tätigkeit der Mikroorganismen des Bodens. Arch. Microbiol. **8**, 249–287 (1937). — Feuerstack, W.: Bodenkundliche Kennzeichnung schwerer Böden zum Zwecke der Beurteilung ihrer Bewirtschaftungsunterschiede. Diss. Hohenheim 1960. — Frercks, W., und D. Puffe: Der Einfluß der Bodentemperaturen und -feuchten auf den Verlauf der Bodenatmung bei Moor- und Heidesandböden, sowie Dampfpflug- und Fehnkulturen. Z. Pflanzenernähr., Düng., Bodenkde. **78**, 107–121 (1957). — Vergleichende Untersuchungen zwischen der „Bodenatmung“ und der CO_2-Produktion von Moorböden. Z. Pflanzenernähr., Düng., Bodenkde. **87**, 108–118 (1959). — FUKUDA, H.: Air and vapor movement in soil due to wind gustiness. Soil Sci. **79**, 249–256 (1955); Ref.: Soils a. Fertil. **18**, 315 (1955); LZ II, 24–25, **394** (1956).

GAARDER, T.: Studies in soil respiration in Western Norway, the Bergen district. Univ. Bergen Arb. No. **3, 24** (1957); Ref.: Soils a. Fertil. **21**, 100 (1958). — GEIGER, R.: Bericht über ein von G. Hofman entwickeltes Gerät zur Registrierung der Bodenluftfeuchtigkeit. Ber. Dtsch. Wetterdienst US-Zone Nr. 32, 24–27 (1952). — GIESECKE, F.: Das Verhalten des Bodens gegen Luft. Handbuch der Bodenlehre, Bd. VI, S. 253–342. Berlin: Springer. 1930. — GILL, W. R., and R. P. MILLER: A method for study of the influence of mechanical impedance and aeration on the growth of seedling roots. Proc. Soil Sci. Soc. Amer. **20**, 154–157 (1956). — GORDON, M: Das Porenvolumen von Torfmull. Torfnachrichten **6**, 5–6 (1955); Ref.: Soils a. Fertil. **18**, 315 (1955).

HANNEN, F.: Untersuchungen über den Einfluß der physikalischen Beschaffenheit des Bodens auf die Diffusion der Kohlensäure. Forsch. Geb. Agr. Phys. **15**, 6 – 25 (1892). — HARRIS, D. G., and C. H. M. VAN BAVEL: Growth, yield and water absorption of tobacco plants as affected by the composition of the root atmosphere. Agron. J. **49**, 11–14 (1957); Ref.: Soils a. Fertil. **20**, 87 (1957). — HELBIG, M.: Die Verdunstung des Wassers aus dem Boden. Handbuch der Bodenlehre, Bd. 6. Berlin: Springer. 1930. — HENSELE, G. A.: Untersuchungen über den Einfluß des Windes auf den Boden. Forsch. Geb. Agr. Phys. **16**, 311–364 (1893).

JAMISON, V. C.: The significance of air adsorption by soil colloids in picnometric measurements. Proc. Soil Sci. Soc. Amer. **17**, 17–19 (1953); Ref.: Z. Pflanzenernähr., Düng., Bodenkde. **65**, 274 (1954). — JAMISON, V. C., and C. W. DOMBY: The effect of a dense soil layer and varying air—water relations on the growth, root development and nutrient uptake of cotton in Commerce silt loam. Proc. Soil Sci. Soc. Amer. **20**, 447–453 (1956). — JANERT, H.: Die Durchlüftbarkeit des Bodens. Verh. VI. Kommiss. Int. Bodenkundl. Ges. B, Zürich **1937**; zit. nach JANERT, Bodenkundliches Praktikum, S. 77. Berlin: VEB Verlag Technik. 1953. — JASSTREBOW, M. J.: Der Einfluß der wichtigsten biologischen Faktoren auf die Zusammensetzung der Bodenluft in den Schwemmlandböden des Flusses Kljasma. Bodenkunde, Nr. **10**, 81–88 (1958a) (russ.); Ref.: LZ. II, 44 (1960); Soils a. Fertil. **22**, 32 (1959). — Bestimmung der Radioaktivität von Bodenluft und atmosphärischer Luft, die letztere ist mit $C^{14}O_2$ markiert, um die Reinheit von Bodenluftproben festzustellen. Dokl. Akad. Nauk. **119**, 792–794 (1958) (russ.); Ref.: Soils a. Fertil. **21**, 296 (1958). — JOSHI, N. V., and S. G. JOSHI: Studies in carbon dioxide evolution in soils. The depression in the production of carbon dioxide in the soils observed after the addition of mineral fertilizers. J. Indian. Soc. Soil Sci. **1**, 115–119 (1953); Ref.: Soils a. Fertil. **17**, 342 (1954).

KEMPER, W. D., and M. AMEMIYA: Alfalfa growth as affected by aeration and soil moisture stress under flood irrigation. Proc. Soil. Sci. Soc. Amer. **21**, 657–660 (1957); Ref.: Z. Pflanzenernähr., Düng., Bodenkde. **81**, 84–85 (1958). — KOEPF, H.: Bodenklima und Bodenatmung. Ber. Dtsch. Wetterdienst US-Zone **32**, 34–37 (1952). — Die Temperatur/Zeit-Abhängigkeit der Bodenatmung. Z. Pflanzenernähr., Düng., Bodenkde. **61**, 29–48 (1953). — Experimenteller Beitrag zur Bodenbeurteilung mittels biochemischer Methoden. I. Enzymreaktionen und CO_2-Ausscheidung bei verschiedenen Böden. Z. Pflanzenernähr., Düng., Bodenkde. **67**, 262–270 (1954 a). — Experimenteller Beitrag zur Bodenbeurteilung mittels biochemischer Methoden. II. Enzymreaktionen und CO_2-Ausscheidung des Bodens bei einem statischen Düngungsversuch und den drei Hauptkulturarten. Z. Pflanzenernähr., Düng., Bodenkde. **67**, 271–277 (1954 b). — Untersuchungen über die biologische Aktivität des Bodens, Teil I: Atmungskurven des Bodens und Fermentaktivität unter dem Einfluß von Düngung und Pflanzenwachstum. Z. Acker- u. Pflanzenbau **98**, 289–312 (1954 c). — Zur Dynamik des Abbaues organischer Substanzen in verschiedenen Böden. Z. Pflanzenernähr., Düng., Bodenkde. **73**, 48–59 (1956). — KONTA, J.: Porosity of clay rocks and their imbibition capacity. Silicates industr. **24**, 133–136 (1959); Ref.: Z. Pflanzenernähr., Düng., Bodenkde. **86**, 188 (1959). — KOPECKY, I.: Die physikalischen Eigenschaften des Bodens. Intern. Mitt. Bodenkde. **4**, 138–198 (1914). — KRAUSE, M.: Russische Forschungen auf dem Gebiete der Bodenstruktur. Landwirtsch. Jahrb. **73**, 603–690 (1931). — KUIPERS, H.: Pore space on three experimental fields with different plowing depth. Meded. Landbouwhogesch. Gent **24**, 146–151 (1959). — KURON, H.: Adsorption von Dämpfen und Gasen an Böden und Tonen und ihre Verwendung zur Oberflächenermittlung dieser Stoffe. Koll. Beih. **36**, 179 – 256 (1932).

LAATSCH, W.: Dynamik der mitteleuropäischen Mineralböden, S. 179. Dresden und Leipzig: Steinkopff. 1954. — LEES, H., and J. W. PORTEOUS: The release of carbon dioxide from soils percolated with various organic materials. Plant a. Soil 2, 231 - 241 (1950); Ref.: Z. Pflanzenernähr., Düng., Bodenkde. 53, 82 (1951). — LEHMANN: Messungen der freien Kohlensäure in und über dem Boden der bioklimatischen Stationen des Lunzer Gebietes. Österr. Bot. Z. 80, 98 (1931); zit. nach R. MELVILLE, Handbuch der Bodenlehre, Erg.-Bd. I, S. 245. Berlin: Springer. 1939. — LUNDEGÅRDH, H.: Der Kreislauf der Kohlensäure in der Natur. Jena: Fischer. 1924.

MAASLAND, M., and D. KIRKHAM: Theory and measurement of anisotropic air permeability in soil. Proc. Soil Sci. Soc. Amer. 19, 395–400 (1955); Ref.: Soils a. Fertil. 19, 40 (1956). — MAGERS, H.: Untersuchungen über die Produktion der Kohlensäure im Ackerboden. Wiss. Arch. Landwirtsch. A 2, 472–544 (1929). — MAKAROW, B. N.: Veränderungen in der Bodenatmung und im CO_2-Gehalt der bodennahen Luftschicht während des Tagesverlaufes. Ber. Akad. Wiss. UdSSR 118, 389–391 (1958) (russ.); Ref.: LZ. II, 1608–1609 (1959); Soils a. Fertil. 21, 168 (1958). — MAKOWER, B., T. M. SHAW and L. T. ALEXANDER: The specific surface and density of some soils and their colloids. Proc. Soil Sci. Soc. Amer. 2, 101–109 (1937). — MANEGOLD, E.: Kapillarsysteme, 1. Bd., Straßenbau, Chemie u. Technik, Heidelberg 1955. — MARSHALL, T. J.: The diffusion of gases through porous media. J. Soil Sci. 10, 79–82 (1959); Ref.: LZ. II, 2051 (1959). — MEYER, L., und G. SCHAFFER: Atmungskurven des Bodens unter dem Einfluß von Düngung und Bewachsung. Landwirtsch. Forsch. 6, 81–95 (1954). — MICHAEL, G., und W. BERGMANN: Bodenkohlensäure und Wurzelwachstum. Z. Pflanzenernähr., Düng., Bodenkde. 65, 180–194 (1954). — MINAR, M.: Einfluß der atmosphärischen Niederschläge auf den Kohlensäuregehalt der Bodenluft und der Atmosphäre. Sborn. Českosl. Akad. Zemed. 10, 106 (1935). — MITSCHERLICH, E. A.: Bodenkunde, 5. Aufl. Berlin: Parey. 1949. — MORGENSTERN, W., und W. RENTSCHLER: Anteil „heißer Teilchen" an der Aktivität der bodennahen Atmosphäre. Naturwiss. 46, 472–473 (1959).

NAKAYAMA, F. S.: Gas sorption by soils and clay minerals. Diss. Abstr. 19, 1883 (1958); Ref.: Soils a. Fertil. 22, 273 (1959). — NORRIS, W. E., J. D. WIEGAND, and L. JOHANSON: Effect of carbon dioxyde on respiration of exised onion root tips in high oxygen atmospheres. Soil Sci. 88, 144–149 (1959); Ref.: Soils a. Fertil. 23, 29 (1960).

ODÉN, S.: Förslag till klassifikation av markporer. (Ein Vorschlag für die Klassifizierung der Bodenporen.) Kungl. Skogs Lantbr. Acad. Tidskr. 96, 297–313 (1957); Ref.: Z. Pflanzenernähr., Düng., Bodenkde. 81, 279 (1958); LZ. II, 34 (1959). — ONTSCHUKOW, D. N.: Gesetzmäßigkeiten in der Verlagerung der Wärme und Feuchtigkeit im Boden im Laufe eines Tages und einer Nacht. Bodenkunde, Nr. 5, 25–30 (1956) (russ.); Ref.: LZ. II, 255–256 (1958).

PARKER, F. W.: Carbon dioxyde production of plant roots as a factor in the feeding power of plants. Soil Sci. 17, 229–247 (1924). — PENMAN, H. L.: Gas and vapor movements in the soil I. J. Agric. Sci. 30, 437–462 (1940). — Gas and vapor movements in the soil II. J. Agric. Sci. 30, 570–581 (1940 a). — PENNINGSFELD, F.: Über Atmungsuntersuchungen an einigen Bodenprofilen. Z. Pflanzenernähr., Düng., Bodenkde. 50, 135–164 (1950). — PETERSON, J. B.: Relations of soil air to roots as factors in plant growth. Soil Sci. 70, 175–186 (1950). — PINK, L. A., and F. E. ALLISON: Maintenance of soil organic matter. III. Influence of green manures on the release of native soil carbon. Soil Sci. 71, 67–76 (1951).

RAMANN, E.: Wasserdampfsättigung der Bodenluft. Bodenkunde. Berlin: Springer. 1911. — RANEY, W. A.: Field measurements of oxygen diffusion through soil. Proc. Soil Sci. Soc. Amer. 14, 61–65 (1950); Ref.: Soils a. Fertil. 14, 32 (1951). — RENK, F.: Über die Permeabilität des Bodens für Luft. Forsch. Geb. Agr. Phys. 2, 339–347 (1879). — RICHARD, F.: Über Fragen des Wasserhaushaltes im Boden. Schweiz. Forstw. 4, 1–23 (1955). — RID, H.: Über eine Feldmethode zur Messung der Bodendurchlüftung. Z. Pflanzenernähr., Düng., Bodenkde. 88, 227–231 (1960).— RIPPEL, A.: Bakteriologisch-chemische Methoden zur Bestimmung des Fruchtbarkeitszustandes des Bodens und des Kreislaufes der Stoffe. Handbuch der Bodenlehre, Bd. 8, S. 599–671. Berlin: Springer. 1931. — ROBINSON, F. E.: A diffusion chamber for studying soil atmosphere. Soil Sci. 83, 465–469 (1957); Ref.: Z. Pflanzenernähr., Düng., Bodenkde. 80, 85–86 (1958). — ROMELL, L. G.: Die Bodenventilation als ökologischer Faktor. Medd. Stat. Skogsförsöksanst. 18, 282–339 (1922). — Mechanismus der Durchlüftung des Bodens. Ann. agron. 5, 373 (1935); Ref.: Z. Pflanzenernähr., Düng., Bodenkde. 42, 110 (1936). — RUNKLES, I. R.: Diffusion, sorption and depth distribution of oxygen in soils. Jowa St. Coll. J. Sci. 31, 513–515 (1957); Ref.: Soils a. Fertil. 20, 264 (1957). — RUNKLES I. R.,

A. D. SCOTT and F. S. NAKAYAMA: Oxygen sorption by moist soils and vermiculite. Proc. Soil Sci. Soc. Amer. 22, 15–18 (1958); Ref.: Soils a. Fertil. 21, 234–235 (1958). — RUSSEL, E. J., and A. APPLEYARD: The atmosphere of the soil, its composition and the causes of variation. J. Agric. Sci. 7, 1–45 (1915). — RUST, R. H., A. KLUTE, and I. E. GIESEKING: Diffusion-porosity measurements using a nonsteady state system. Soil Sci. 84, 453–463 (1957); Ref.: Z. Pflanzenernähr., Düng., Bodenkde. 81, 180 (1958).

SCHAFFER, G.: Atmungskurven des Bodens unter dem Einfluß langjähriger verschiedenartiger Düngung. Z. Pflanzenernähr., Düng., Bodenkde. 67, 219–230 (1954). — SCHEFFER, F., und P. SCHACHTSCHABEL: Lehrbuch der Agrikulturchemie und Bodenkunde, I. Teil, Bodenkunde, 5. Aufl. Stuttgart: Enke. 1960. — SCHMALFUSS, K.: Über Bodenatmung. Bodenkde. u. Pflanzenernähr. 21/22, 442–454 (1940). — SCHMIDT, W.: Versuche zur Bodenatmung. Forsch. u. Fortschr. 6, 198–199 (1930). — SCHUBACH, K.: Wasserhaushaltsuntersuchungen an verschiedenen Bodenarten unter besonderer Berücksichtigung der Verdunstung. Ber. Dtsch. Wetterdienst US-Zone 40, 3–14 (1952). — SHANKS, I. B., and L. LAURIE: Rose root studies: the optimum oxygen concentration of the soil atmosphere. Proc. Amer. Soc. Horticult. Sci. 55, 499–503 (1950); Ref.: Soils a. Fertil. 15, 29 (1952). — SHURAWLEW, M. S.: Die Kondensation des atmosphärischen Wasserdampfes in der oberen Schicht des Schwarzerdebodens. Bodenkunde Nr. 9, 37–44 (1955) (russ.); Ref.: LZ. II, 428 (1957). — SOKOLOW, D. F.: The influence of moistening the soil on the intensity of evolution of CO_2 in the decomposition of forest litter. Dokl. Akad. Nauk SSSR 95, 1317–1320 (1954) (russ.); Ref.: Soils a. Fertil. 17, 428 (1954). — STANGEN, G.: Atmungskurven des Bodens unter dem Einfluß verschiedener Kulturarten. Diss. Hohenheim 1955. — STOLWIJK, I. A. I., and K. V. THIMANN: On the uptake of carbon dioxyde and bicarbonate by roots, and its influence on growth. Plant Physiol. 32, 513–520 (1957); Ref.: Soils a. Fertil. 21, 168 (1958). — STÖCKLI, A.: Studien über den Einfluß des Regenwurms auf die Beschaffenheit des Bodens. Landwirtsch. Jb. Schweiz 42, 5–121 (1928). — Der Einfluß der Mikroflora und -fauna auf die Beschaffenheit des Bodens. Z. Pflanzenernähr., Düng., Bodenkde. 45, 41–53 (1949). — VAN SUCHTELEN, H.: Über die Messung der Lebenstätigkeit der aerobiotischen Bakterien im Boden durch die Kohlensäureproduktion. Zbl. Bakt. II. Abt. 28, 45 (1910).

TAYLOR, S. A.: Oxygen diffusion in porous media as a measure of soil aeration. Proc. Soil Sci. Soc. Amer. 14, 55–61 (1949); Ref.: Soils a. Fertil. 14, 32 (1951). — TAYLOR, G. S., and J. H. ABRAHAMS: A diffusion-equilibrium method for obtaining soil gases under field conditions. Proc. Soil Sci. Soc. Amer. 17, 201–206 (1953); Ref.: Z. Pflanzenernähr., Düng., Bodenkde. 66, 84 (1954). — TURLJUN, I. A.: Zur Theorie des Gasaustausches in Böden. Bodenkunde 7, 22–30 (1957) (russ.); Ref.: Z. Pflanzenernähr., Düng., Bodenkde. 81, 89 (1958); LZ. II, 1365 (1958). — Über die Natur der Wanderung der Gase und Dämpfe im Bodenuntergrund. Bodenkunde 9, 89–100 (1958) (russ.); Ref.: LZ. II, 2503–2504 (1959); Soils a. Fertil. 22, 106 (1959).

DE VRIES, D. A.: Some remarks on gaseous diffusion in soils. Trans. IVth Int. Congr. Soil Sci. 2, 41–43 (1950).

WERSCHININ, P. V., und N. V. KINIBENKO: Über die Diffusion von CO_2 durch den Boden. Bodenkunde 325–328 (1948) (russ.). — WEISSENBERG, H.: Die Mikroorganismentätigkeit auf rohhumushaltigem Heidesand, gemessen an der CO_2-Produktion. Z. Pflanzenernähr., Düng., Bodenkde. 66, 227–239 (1954); Ref.: Soils a. Fertil. 18, 30 (1955). — WIEGAND, C. L., and E. R. LEMON: A field study of some plant-soil relations in aeration. Proc. Soil Sci. Soc. Amer. 22, 216–221 (1958); Ref.: Z. Pflanzenernähr., Düng., Bodenkde, 85, 172 (1959); LZ. II, 706–707 (1959); Soils a. Fertil. 21, 296 (1958). — WIESSMANN, H., und W. NEUMANN: Gassorption an Bodenkomponenten und Böden I. Z. Pflanzenernähr., Düng., Bodenkde. 40, 49–81 (1935). — WOLLNY, E.: Untersuchungen über den Einfluß der Pflanzendecken und der Beschattung auf den Kohlensäuregehalt der Bodenluft. Forsch. Geb. Agr. Phys. 3, 1–14 (1880). — Untersuchungen über den Einfluß der physikalischen Eigenschaften des Bodens auf dessen Gehalt an freier Kohlensäure. Forsch. Geb. Agr. Phys. 9, 65–194 (1886). — Untersuchungen über die Beeinflussung der Fruchtbarkeit der Ackerkrume durch die Tätigkeit der Regenwürmer. Forsch. Geb. Agr. Phys. 13, 381–395 (1890). — Untersuchungen über die Permeabilität des Bodens für Luft. Forsch. Geb. Agr. Phys. 16, 193–222 (1893). — WURMBACH, H.: Beiträge zur Kenntnis der Bodenatmung und der Kohlensäurekonzentration der Bodenluft unter landwirtschaftlich genutzten Flächen. Wiss. Arch. Landw. Abt. A 10, 484–532 (1934).

YODER, R. E.: The significance of soil structure in relation to the tilth problem. Proc. Soil Sci. Soc. Amer. 2, 21–33 (1937).

e) Die Bodentemperatur

Von

H. Koepf

1. Einleitung

Die im Erdboden herrschenden Temperaturen sind in erster Linie durch seine klimatische Lage bedingt. Durch Maßnahmen des Menschen können sie nur innerhalb enger Grenzen beeinflußt werden. Doch ist ihre Kenntnis wichtig, weil die Bodentemperatur für die Bodengenese, das Bodenleben und das Pflanzenwachstum ein Faktor von *umfassender Bedeutung* ist. Die mechanische Verwitterung wird durch häufigen Temperaturwechsel begünstigt. Die chemische Verwitterung wird durch höhere Temperaturen beschleunigt. So konnte JENNY (1941) zeigen, daß — bei Gleichheit der übrigen Faktoren — der Tongehalt von Böden parallel zur durchschnittlichen Jahrestemperatur ansteigt. Die Intensität der Lebensvorgänge wächst nach der VAN 'T HOFFschen Regel mit der Temperatur bis zu einem Optimum. Die Zersetzung der organischen Substanz erfolgt bei höherer Temperatur rascher. Die Böden der wärmeren Klimate sind deshalb in der Regel humusärmer als die der kühleren Regionen, in denen hohe Kohlenstoffgehalte, ungenügende Zersetzung und Rohhumusbildung häufig beobachtet werden. Die höheren Pflanzen stellen für die Keimung und ein optimales Wachstum artcharakteristische Ansprüche an die Bodentemperatur. Bei Temperaturen um den Nullpunkt und darunter ruht die Lebenstätigkeit im Boden oder ist sehr verlangsamt.

Der Boden macht als *oberste Verwitterungsschicht* der festen Erdoberfläche die *Temperaturschwankungen in der Atmosphäre* mit, jedoch mit anderen *Amplituden* und *Phasenverschiebungen* gegenüber den Perioden in der Luft. Wärmezufuhr und Wärmeverlust des Bodens sind nahezu ausschließlich durch *Energiezufuhr* und *-verlust* an der *Bodenoberfläche* bestimmt. Andere Faktoren sind demgegenüber bedeutungslos. Die beim *Abbau der organischen Substanz* im Boden frei werdende Energie beeinflußt die Bodentemperatur nicht meßbar. Der aus dem *Erdinneren* kommende Wärmestrom beträgt nach GEIGER (1950, S. 29) 0,0001 cal/cm^2 · min, er kann ebenfalls vernachlässigt werden. Die ausschlaggebende Wärmequelle für den Boden ist also die von der *Sonne* zugestrahlte Wärmeenergie. Wärmeverluste erfolgen durch *Ausstrahlung* in die Atmosphäre und *Verdunstung*. Es soll nun zunächst der Wärmeumsatz an der Bodenoberfläche betrachtet werden.

2. Der Wärmeumsatz an der Bodenoberfläche

Eine eingehende Darstellung der Glieder des Wärmeumsatzes an der Erdoberfläche stammt von R. GEIGER (1950). GEIGER unterscheidet zwischen einem *Einstrahlungstypus*, der am deutlichsten an heiteren Sonnenscheintagen verwirklicht ist, und dem *Ausstrahlungstypus*, der in klaren Nächten in Erscheinung tritt. Hoher Wasserdampfgehalt der Luft, Bewölkung und Verunreinigung der Luft mit Schwebeteilchen schaffen kompliziertere Verhältnisse.

A. Die Wärmezufuhr zur Bodenoberfläche

Von der *Sonnenstrahlung*, die 1,94 cal/cm^2 · min beträgt, wird beim Durchgang durch die *Erdatmosphäre* ein wechselnder Anteil von den Wolken reflektiert; rund 8% werden diffus zerstreut und gelangen teils auf die Erdoberfläche, teils zurück in den Weltenraum. Wasserdampf und Kohlensäure in der Luft absorbieren ebenfalls Strahlung. *Direkte Sonnenbestrahlung* und *diffuse Himmelsstrahlung* bilden in der Hauptsache den Energiestrom, der auf die Erdoberfläche gelangt und dort teils absorbiert, teils reflektiert wird. Die Summe von Sonnenstrahlung und Himmelsstrahlung wird als *Globalstrahlung* bezeichnet. Die dem Boden zugeführte Wärme hängt ab von

1. den Verlusten der Sonnenstrahlung beim Durchgang durch die Erdatmosphäre;
2. der Erhebung der Bodenoberfläche über den Meeresspiegel;
3. dem Einfallswinkel der Sonnenstrahlen, der sich nach der geographischen Breite, der Inklination und Exposition eines Ortes richtet, ferner nach dem Sonnenstand im Jahres- und Tageslauf;
4. dem Absorptions- und Reflektionsvermögen (der Albedo) der Erdoberfläche, die nicht nur von der Bodenbeschaffenheit, sondern insbesondere auch von der Pflanzendecke bestimmt werden.

In Tab. 121 werden einige Zahlenwerte nach J. SCHUBERT (1928) für die Strahlungsbeträge in den einzelnen Monaten des Jahres an heiteren Tagen und bei mittlerer Bedeckung mitgeteilt.

Tabelle 121. *Sonnenstrahlung in cal/cm² · min um die Mittagszeit an heiteren Tagen und bei mittlerer Bewölkung in Potsdam 1907 bis 1923* (nach J. SCHUBERT 1928)

	Jan.	Feb.	März	Apr.	Mai	Juni	Juli	Aug.	Sept.	Okt.	Nov.	Dez.
Strahlung an heiteren Tagen												
Horizontalfläche	0,287	0,471	0,649	0,937	1,065	1,113	1,027	0,918	0,796	0,569	0,324	0,240
Nordhang 30° .	—	—	0,105	0,379	0,569	0,654	0,583	0,433	0,228	—	—	—
Ost-Westhang 30°	0,249	0,408	0,562	0,811	0,922	0,964	0,890	0,495	0,690	0,493	0,281	0,208
Südhang 30°	0,735	0,916	1,019	1,244	1,276	1,274	1,196	1,157	1,152	1,002	0,447	0,676
Strahlung bei mittlerer Bewölkung												
Horizontalfläche	0,075	0,137	0,253	0,394	0,522	0,568	0,493	0,450	0,374	0,228	0,087	0,060
Nordhang 30° .	—	—	0,041	0,159	0,279	0,334	0,280	0,212	0,107	—	—	—
Ost-Westhang 30°	0,065	0,118	0,219	0,341	0,452	0,492	0,427	0,390	0,324	0,197	0,076	0,052
Südhang 30° ..	0,191	0,266	0,397	0,522	0,625	0,650	0,574	0,567	0,541	0,401	0,202	0,169

Diese Zahlen zeigen, daß der die Erdoberfläche erreichende Strahlungsbetrag durch *Bewölkung* beträchtlich vermindert wird. Im gleichen Sinne wirkt der *Wasserdampfgehalt der Luft* und deren Verunreinigung mit Schwebeteilchen, so daß in Gebieten mit durchschnittlich hoher Luftfeuchtigkeit und häufiger Bewölkung die Einstrahlung auf die Erdoberfläche verringert ist. Allerdings sind dies gleichzeitig Gebiete mit verringerter Ausstrahlung. Die Nähe großer *Wasserflächen*, Meere und Seen beeinflußt auf diese Weise den Strahlungsumsatz an der Erdoberfläche. Gleichzeitig stellen größere Wassermassen bzw. warme Meeresströmungen bedeutende Wärmespeicher dar, die ausgleichend auf den Temperaturverlauf benachbarter Landflächen wirken.

Daß die Einstrahlung mit der *Erhebung eines Ortes über NN* zunimmt, möge durch die folgenden Zahlen nach MÖRIKOFER (zit. nach GEIGER 1950, S. 10) belegt werden:

Mittagsstrahlung im Januar an wolkenlosen Tagen nach Messungen in Davos, St. Blasien und Karlsruhe

Seehöhe	100	500	1500	4000 m
Strahlung	0,8	1,2	1,4	1,6 cal/cm² · min

Die starken täglichen und jahreszeitlichen Temperaturextreme in Höhenlagen sind durch diesen Umstand mitbedingt. Expositionsunterschiede wirken sich im Gebirge stärker aus als im Flachland.

Wenn der Strahlungsstrom S_0 eine Fläche unter dem Einfallswinkel i trifft, so empfängt diese die Strahlung $S = S_0 \cdot \cos i$. Der einer Bodenfläche zugeführte Energiebetrag hängt deshalb ab von der *geographischen Breite*, der *Inklination* und der *Exposition* eines Ortes, ferner vom *Sonnenstand*, d. h. der *Jahres-* und *Tageszeit*. In Tab. 122 nach COLLMANN (1958) sind die Bestrahlungs-

Tabelle 122. *Durchschnittliche Monats- und Jahresmittel der Globalstrahlung in cal/cm² · Tag*
(nach COLLMANN 1958)

Station	Monatsmittel												Jahresmittel			
	Jan.	Feb.	März	April	Mai	Juni	Juli	Aug.	Sep.	Okt.	Nov.	Dez.	beob.	extr.	beob./extr. %	Zeitraum
Stockholm .	29	78	201	308	467	517	500	392	243	112	32	18	241	492	49	1946—1953
Hamburg Fuhlsbüttel	50	99	197	326	408	426	402	349	248	134	57	34	228	547	42	1950—1955
Potsdam ..	59	105	217	327	420	479	446	370	274	142	64	42	245	560	44	1937—1952
Königsstein Taunus ..	64	122	236	354	437	467	453	395	289	170	71	43	259	582	45	1941—1954
Tübingen ..	59	119	218	277	354	375	379	323	227	164	86	48	220	598	37	1950—1954
Oberstdorf .	87	138	232	318	361	377	389	342	261	174	95	71	237	608	39	1940—1946
Wien	70	134	226	337	442	467	465	396	277	158	80	52	259	599	43	1938—1954
Athen	186	264	341	459	477	590	624	554	468	287	200	166	385	692	56	1953—1954

stärken für Orte auf verschiedenen geographischen Breiten angegeben. Bei den Jahresmitteln sind die beobachteten Beträge und theoretischen Höchstwerte (extraterrestrische Strahlung) angegeben. Letztere würde erreicht, wenn keine Erdatmosphäre vorhanden wäre. Die Entfernung von Stockholm nach Athen entspricht etwa 20 Breitegraden. Das Jahresmittel in Stockholm beträgt rund 60% der Strahlung in Athen, in den Sommermonaten ist der Unterschied nur gering, in den Wintermonaten dagegen hoch. Der Beobachtungszeitraum ist für die einzelnen Stationen nicht gleich, die beobachteten Werte sind durch den unterschiedlichen Witterungsverlauf im Beobachtungszeitraum mit beeinflußt. Trotzdem ist unschwer zu erkennen, daß die *klimatische Lage* der einzelnen Orte, insbesondere die durchschnittliche Bedeckung und Luftfeuchtigkeit, gewisse Abweichungen bedingt.

Aus den Zahlen in der Tab. 121 ist bereits ersichtlich, daß *Hangneigung* und *Hangrichtung* den Strahlungsgenuß der Bodenoberfläche erheblich verändern. Als weitere unabhängige Variable kommt noch der jahres- und tageszeitlich wechselnde Sonnenstand hinzu. Abb. 157 nach J. SCHUBERT (1928) zeigt in anschaulicher Weise das Zusammenspiel dieser verschiedenen Faktoren. In den Morgen- und Abendstunden ist die Bestrahlung der Horizontalflächen wegen des schrägen Auftreffens der Sonnenstrahlen geringer; die Höchstwerte im Tagesgang werden um die Mittagsstunden, im Jahresgang im Juni ermittelt.

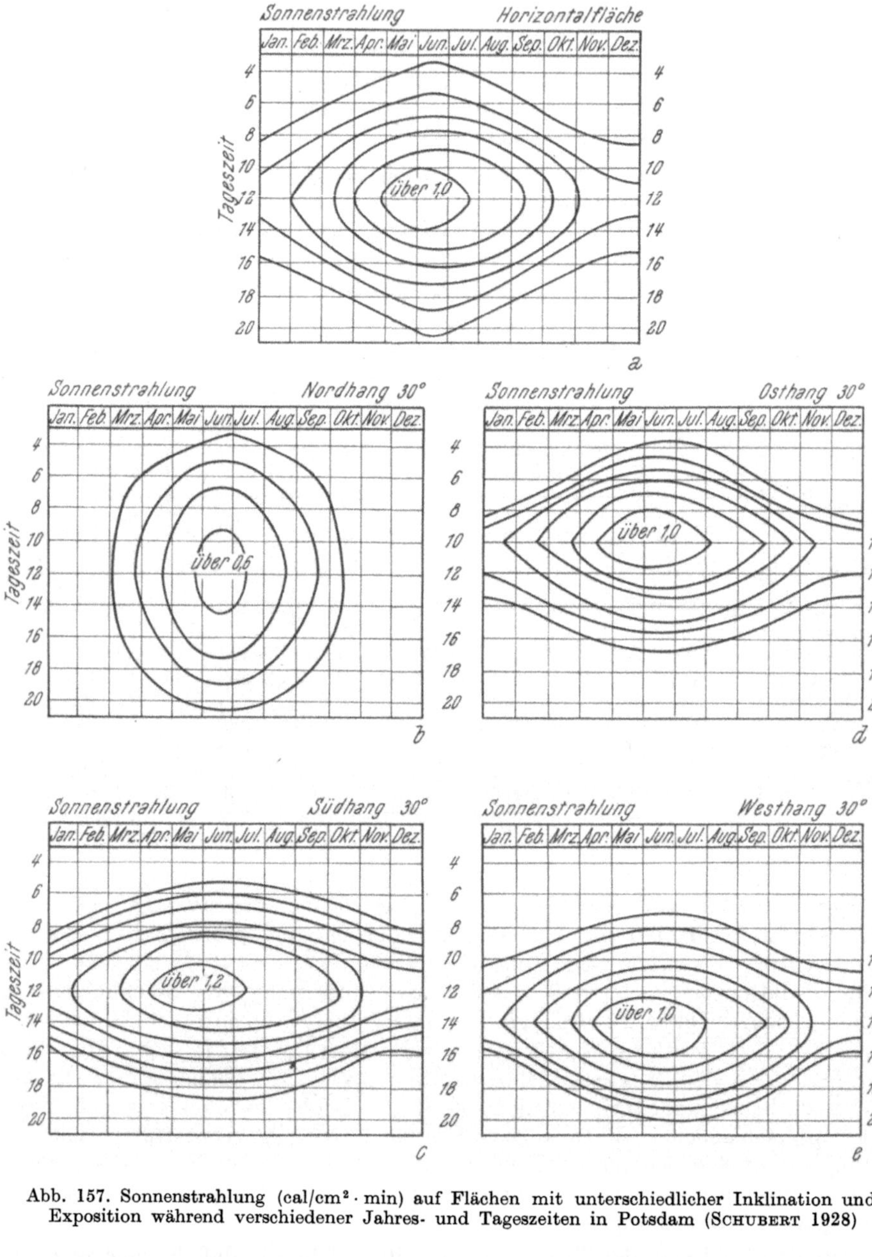

Abb. 157. Sonnenstrahlung (cal/cm² · min) auf Flächen mit unterschiedlicher Inklination und Exposition während verschiedener Jahres- und Tageszeiten in Potsdam (SCHUBERT 1928)

Bei der 30° nach Süden geneigten Fläche ist die ankommende Strahlung am höchsten, beim Nordhang am geringsten, dieser empfängt vom Oktober bis Februar überhaupt keine direkte Bestrahlung. Beim Osthang ist das Tagesmaximum in die Vormittagsstunden, beim Westhang in die Nachmittagsstunden verlegt.

Die der Bodenoberfläche zugeführte Globalstrahlung setzt sich zusammen aus der *direkten Sonnenbestrahlung* und der *diffusen Himmelsstrahlung*. Abb. 158 nach GRUNOW (1952) zeigt die Ergebnisse von Messungen am Hohenpeißenberg, einer freistehenden Kuppe im Alpenvorland. Die Neigung der Hänge in den Richtungen 345° und 165° beträgt rund 30°. Im oberen Teil dieser Darstellung sind die Monatsmittel der Globalstrahlung angegeben, der schraffierte Teil der Säulen zeigt die Werte für den Nordhang. In den Sommermonaten sind die

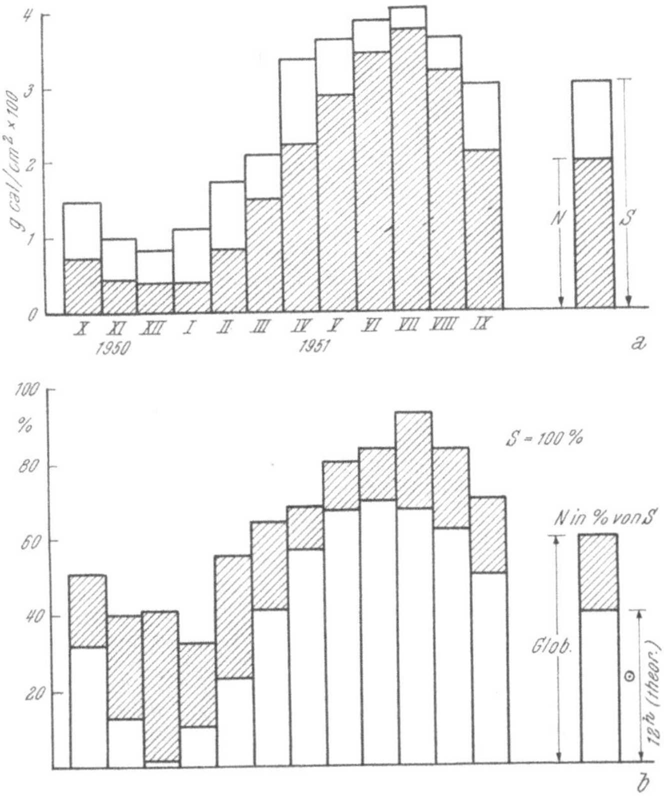

Abb. 158. Messung der Globalstrahlung am Nord- und Südhang (30°) des Hohenpeißenbergs (GRUNOW 1952); *a*) Monatsmittel der Tagesmessung, ganze Säulen: Südhang, schraffiert: Nordhang, *b*) Strahlungsgenuß des Nordhanges in Prozenten des Südhanges, schraffiert: Anteil der Himmelsstrahlung an der Globalstrahlung

Unterschiede zwischen Nord- und Südhang nur gering, in den Wintermonaten dagegen hoch. Die Säulen im unteren Teil der Zeichnung zeigen den Strahlungsgenuß des Nordhanges in Prozent des Südhanges, wobei der Anteil der Himmelsstrahlung schraffiert ist. Die Wärmezufuhr zum Nordhang während der Wintermonate erfolgt fast ausschließlich durch Himmelsstrahlung.

Die Energieaufnahme durch die diffuse Himmelsstrahlung hängt zwar von der Hangneigung, nicht aber von der Exposition ab. Je höher der Anteil der diffusen Himmelsstrahlung an der Globalstrahlung ist, um so weniger fallen Expositionsunterschiede ins Gewicht. Es folgt daraus, daß bei *trübem Wetter* geringere Expositionsunterschiede festgestellt werden als bei heiterem.

B. Die Wärmeaufnahme und Wärmeabgabe durch die Bodenoberfläche

Die Bodenoberfläche stellt für die ankommende Strahlung eine Grenzschicht dar, in die sie nur wenig eindringen kann. Sie wird, wie von verschiedenen Autoren gezeigt wurde, innerhalb einer wenige Millimeter dicken Oberflächenschicht absorbiert bzw. reflektiert. Frischmann (1939) verwendete auf 0,5 mm Korngröße gesiebte Böden. Diese ließen bei einer Schichtdicke von 8 mm keine sichtbaren Strahlen mehr durch, bei 2 bis 4 mm war zum Teil ein schwacher Durchgang zu verzeichnen. Nach Feher und Frank (1936) sowie Baumgartner (1953) verschiebt sich das Spektrum des eindringenden Lichtes zugunsten des Rotanteils. Baumgartner gibt an, daß die relative Lichtintensität von etwa 1% der Außenhelligkeit für verschiedene Quarzsande in einer Tiefe herrscht, die etwa dem 6- bis 8fachen des Korndurchmessers entspricht.

Die *Reflexionszahl* oder die *Albedo* ist derjenige Anteil der Strahlung, der reflektiert wird. Ein Körper mit einer Albedo von 0,5 oder 50% absorbiert also 50% und reflektiert 50% der ankommenden Strahlung. Die Albedo verschiedener Oberflächen hat nach Landsberg und Blanc (1958) die folgenden Werte für die *kurzwellige* Strahlung (Tab. 123):

Tabelle 123. *Albedo von Oberflächen für ankommende kurzwellige Strahlung* (Landsberg und Blanc 1958)

Oberfläche	%	Oberfläche	%
Frischer Schnee	83	Grünes Gras	20
Eis	70	Bepflanzte Felder	15
Alter Schnee	50	Feuchter Sand	9
Hohes, trockenes Gras	32	Wald	7
Wüstensand	25	Wasseroberfläche	6
Trockenes Feld, gepflügt	22		

Die Albedo für die *Infrarotstrahlung* ist erheblich geringer. Falkenberg (1928) nennt hierfür folgende Absorptionskonstanten, in Klammern die Albedo:

Sand, heller	0,89	(11%)
Kalkstein, hellgrauer	0,92	(8%)
Kies, grober	0,92	(8%)
Pelargonie, zonale Blätter	0,96	(4%)
Erdscholle mit Rasen	0,98	(2%)
Tannennadeln	0,96	(4%)
Schnee	über 0,995	(0,5%)

Schnee ist also überraschenderweise für langwellige Strahlung ein fast idealer „schwarzer Körper“. Aus den Zahlen der Tab. 123 geht hervor, daß der größte Teil der sichtbaren Strahlung absorbiert wird, im Mittel etwa 4/5. *Dunkle* Oberflächen absorbieren mehr, helle weniger. Für die langwellige Strahlung liegen die Absorptionskoeffizienten noch höher als für die sichtbare, nämlich um 90% und darüber. Bei einem mit *Pflanzen* bestandenen Boden ist die Albedo für kurz- und langwellige Strahlung verringert. Eine Ausnahme hiervon macht, was ohne weiteres verständlich ist, der Bestand an abgestorbenem, trockenem Gras, dieser reflektiert ein Drittel der sichtbaren Strahlung. Bei bewachsenem Boden ist allerdings zu berücksichtigen, daß je nach Höhe des Bestandes und *Bedeckungsgrad* der Bodenoberfläche nur ein Teil der Strahlung bis zur Bodenoberfläche gelangt, der größte Teil wird in der Blattregion absorbiert oder reflektiert.

Bodenfarbe und *Feuchtigkeitsgehalt* des Bodens bestimmen in erster Linie das Absorptionsvermögen der Bodenoberfläche. Die folgenden Zahlen für die Reflexion *R* sind einer Untersuchung von FRISCHMANN (1939) entnommen:

	R
weißer Meeressand	17,4%
licht rotgelber Dünensand	14,9%
licht rotgelber Flußsand	13,6%
lichtgelber, mit Sand vermischter Lehmboden	10,6%
roter Lateritboden	9,6%
brauner Lehmboden	9,2%
schwarzgrauer Lehmboden	8,5%
grauschwarzer Lehmboden	8,1%
Tierkohlenstaub	0,7%

Je dunkler der Boden ist, um so mehr Strahlung wird absorbiert. Feuchte Böden erscheinen dunkler als trockene, sie weisen eine kleinere Reflexionszahl auf als trockene. Der Wasserfilm an der Oberfläche von Bodenpartikeln und Blättern hält einen Teil der Strahlung zurück. BAUMGARTNER (1953) stellte hierzu nach den Ergebnissen verschiedener Autoren die folgenden Zahlen zusammen. Die Reflexion *R'* der *feuchten Oberfläche* ist erheblich geringer als die Reflexion *R* am trockenen Körper:

	R	*R'*
Schwarze Humuserde	14	9
Grauer Sandboden	18	9
Wellsand	24	13
Helles, frisches, hohes Gras	32	20
Dünensand	37	24

Es muß nunmehr der *Ausstrahlungstypus* R. GEIGERS (1950, S. 15—25) noch etwas näher betrachtet werden. Die Energiezufuhr zur Bodenoberfläche erfolgt während des Tages überwiegend durch kurzwellige Strahlung. Diese Energiequelle fehlt in den *Nachtstunden*, es erfolgt *Wärmeausstrahlung* von der Bodenoberfläche. Nach dem STEPHAN-BOLTZMANNschen Strahlungsgesetz wächst die ausgestrahlte Gesamtenergie mit der 4. Potenz der absoluten Temperatur eines strahlenden Körpers.

$$S = a \cdot 8{,}26 \cdot 10^{-11} \cdot T^4 \text{ [cal/cm}^2\text{ min]}$$

a ist die Absorptionskonstante, sie hat bei einem schwarzen Körper den Wert 1. Der Boden kann annähernd als schwarzer Körper angesehen werden. Der Betrag an ausgestrahlter Wärme wächst also sehr rasch mit der Temperatur. Gleichzeitig ändert sich die Qualität der Ausstrahlung vom Boden. Nach dem WIENschen Verschiebungsgesetz ist die Strahlung eines Körpers um so kurzwelliger, je heißer er ist. Das Maximum der Sonnenstrahlung liegt demzufolge im *kurzwelligen*, das der Erde im *langwelligen* Strahlungsbereich. Die Erdoberfläche empfängt vorwiegend kurzwellige Strahlung und gibt Wärmestrahlung wieder ab.

Dieser Unterschied ist für den *Wärmehaushalt* des Bodens und der unteren Atmosphärenschichten von weittragender Bedeutung. Die Atmosphäre läßt die kurzwelligen Strahlen durch. Wasserdampf und Kohlensäure absorbieren jedoch langwellige Strahlung. Die bekannte *Glashauswirkung* der Atmosphäre beruht darauf, daß die kurzwellige Strahlung bis zur Bodenoberfläche hereindringt, die langwellige Ausstrahlung dagegen wegen des Wasserdampf- und Kohlensäuregehaltes der Luft zurückgehalten wird. Nun stehen weiterhin Absorption und Emission bei einer gegebenen Temperatur in einem konstanten Verhältnis zueinander. Wasserdampf und Kohlensäure der Luft, die die Wärmestrahlung von der Bodenoberfläche aufnehmen, senden deshalb eine *langwellige*

Gegenstrahlung zur Erde, deren Größe vom Gehalt der Atmosphäre an Wasserdampf und Kohlensäure, sowie deren Temperatur abhängt. Soweit der Wärmeverlust des Bodens während der Nachtstunden durch Strahlung erfolgt, wird er also bestimmt durch die Temperatur des Bodens und die *Differenz* zwischen Wärmeausstrahlung und Gegenstrahlung. Diese Differenz wird als effektive Ausstrahlung bezeichnet. Dazu muß noch die Absorption durch die Wolken und die Reflexion an der Wolkenunterseite in Rechnung gestellt werden, durch die die Gegenstrahlung erhöht und die effektive Ausstrahlung verringert wird. Abb. 159 nach Geiger (1950, S. 20) zeigt den Zusammenhang zwischen *Bewölkung* und effektiver Ausstrahlung. Die Geraden geben die errechneten Werte an, wobei *k* eine von Wolkenart, Wolkenhöhe und Temperatur der Wolkenuntergrenze abhängige Konstante ist, die Kurve zeigt die gemessenen Werte.

Allgemein läßt sich also sagen, daß bei hohem *Wasserdampfgehalt* der Luft und bei *Bewölkung* sowohl die auf die Erdoberfläche gelangenden Strahlungsbeträge, wie auch die effektive Ausstrahlung verringert ist. Der Boden erwärmt sich am Tage weniger, er erleidet aber, da die Gegenstrahlung erhöht ist, in der Nacht auch geringere Wärmeverluste. Mit anderen Worten, die Amplitude der Temperaturschwankungen am Boden wird kleiner.

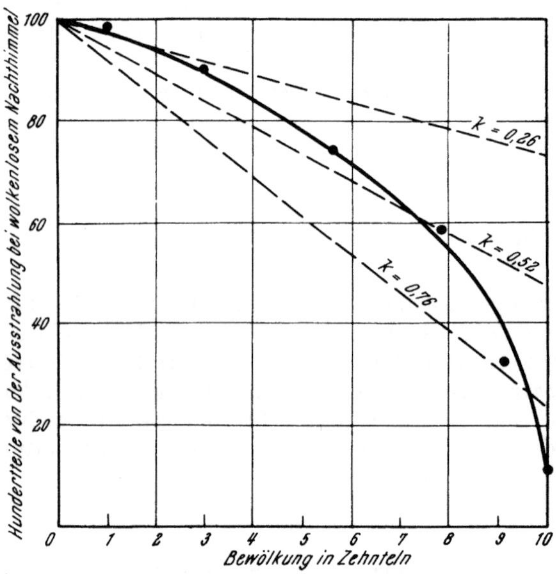

Abb. 159. Abhängigkeit der effektiven Ausstrahlung von der Bewölkung (Geiger 1950, S. 20)

Ergänzend sei darauf hingewiesen, daß die effektive Ausstrahlung in *Höhenlagen* etwas höher ist als in tiefer gelegenen Gebieten. Da die Bodentemperatur in Höhenlagen niedriger ist, ist auch die Wärmeausstrahlung kleiner, doch nimmt die Gegenstrahlung mit der Erhebung über dem Meeresspiegel rascher ab, so daß in Höhenlagen insgesamt eine größere effektive Ausstrahlung zu verzeichnen ist. Ferner ist zu bemerken, daß die Geländeform die nächtliche Ausstrahlung beeinflußt. Wird in welligem Gelände und in Tallagen an den tiefer gelegenen Stellen ein Teil der Himmelshalbkugel abgeschirmt, so wird die Ausstrahlung verringert.

Die durch Strahlung erfolgende Wärmeaufnahme und -abgabe macht den Hauptanteil des Wärmeumsatzes an der Bodenoberfläche aus. Globalstrahlung und Gegenstrahlung beeinflussen den Wärmehaushalt *positiv*, die Ausstrahlung *negativ*. Als weiteres Glied im Wärmeumsatz an der Bodenoberfläche sind noch die Zustandsänderungen des Wassers zu betrachten. Durch die *Verdunstung* werden dem Boden erhebliche Wärmemengen entzogen. Die Verdampfungswärme des Wassers ist

bei	0° C	595 cal/g
bei	20° C	585 cal/g
bei	40° C	574,5 cal/g.

Für die Verdunstung von 1 mm Wasser aus dem Boden werden also rund 60 cal/cm² benötigt. SCHUBERT (1928) rechnet für das mittlere Norddeutschland in der Vegetationszeit mit einer täglichen Verdunstung von nahezu 2 mm. Dafür sind 120 cal/cm² notwendig, während die durchschnittlich zugeführte Wärme 257 cal/cm² beträgt. Wird bei der Verdunstung Wärme entzogen, so wird bei der *Tau-* und *Reifbildung* Wärme frei. Die auf diese Weise dem Boden zugeführte Wärme macht jedoch nur einen geringen Anteil am Wärmeumsatz aus. Die auf den Boden fallenden *Niederschläge* können wärmer oder kälter sein als dieser, also Wärme zuführen oder den Boden abkühlen. Ferner wird durch *Austausch* und *Leitung* Wärme vom Boden an die Luft abgegeben oder aus dieser aufgenommen. Die Wärmekapazität der Luft ist jedoch gering, dieser Umsatz spielt deshalb im Wärmehaushalt des Bodens ebenfalls nur eine untergeordnete Rolle.

Schließlich sei noch auf die *Wärmebilanz* der Bodenoberfläche während eines ganzen Jahres hingewiesen. GEIGER (1950, S. 193) zitiert hierzu die folgenden von ALBRECHT errechneten Werte für Potsdam, denen Messungen im Jahre 1903 zugrunde liegen.

Wärmebilanz eines Jahres in 1000 cal/cm²:

Zur Bodenoberfläche zugeführte Strahlung	19,8
Von der Bodenoberfläche an die Luft abgegeben	—0,4
Von der Bodenoberfläche an den Boden abgegeben	0,2
Für die Verdunstung abgegeben	20,0
Gesamter jährlicher Wärmeumsatz	63,5

Die letztgenannte Zahl ist die Summe aller Wärmebewegungen ohne Rücksicht darauf, ob es sich um Zufuhr oder Verlust handelt. Die der Bodenoberfläche zugeführte Strahlung erreicht annähernd denselben Betrag wie die zur Verdunstung verbrauchte Wärmemenge. Wärmeaufnahme durch den Boden und Wärmeabgabe halten sich, wenn man die Werte für das ganze Jahr zusammenfaßt, annähernd die Waage. Gliedert man nach einzelnen Jahreszeiten auf, so ändert sich naturgemäß das Bild. In den Frühjahrs- und Sommermonaten ist die Wärmebilanz des Bodens positiv, im Spätherbst und Winter negativ.

Die bisherigen Ausführungen bezogen sich ausschließlich auf die *Bodenoberfläche.* Wärmezufuhr und -verlust des Bodens erfolgen hier. Mit dem Übergang von der *Grenzschicht* zwischen Boden und Luft zum *Bodenprofil* treten grundsätzlich andersartige Verhältnisse auf. Es hat sich gezeigt, daß der Wärmeumsatz zwischen Bodenoberfläche und Atmosphäre überwiegend durch Strahlung und Verdunstung erfolgt. Die erstere spielt im Boden selbst nur eine untergeordnete Rolle. Zunächst ist nun die Wärmekapazität des Bodens zu besprechen, dann die Wärmeausbreitung im Boden und schließlich die Bodentemperaturen selbst und deren Änderungen als Funktion der Bodeneigenschaften und des Wärmeumsatzes an der Bodenoberfläche.

3. Die Wärmekapazität des Bodens

Die Wärmeübertragung von der Bodenoberfläche in tiefere Schichten richtet sich nach der *Wärmeleitfähigkeit* des Bodens, die Wärmespeicherung dagegen nach dessen *Wärmekapazität.* Wenn zwei Böden den gleichen Strahlungsbetrag an ihrer Oberfläche absorbieren und die gleiche Wärmeleitfähigkeit besitzen, so weist der Boden, der die höhere Wärmekapazität besitzt, niedrigere Temperaturen auf als der Vergleichsboden mit kleinerer Wärmekapazität. Sinngemäß gilt für Wärmeverluste die folgende Überlegung: Wenn zwei Böden in den Nachtstunden die gleiche Wärmemenge ausstrahlen, die bei beiden Böden aus derselben Schicht, z. B. den oberen 20 cm, entnommen werden soll, so wird sich der Boden mit höherer Wärmekapazität weniger abkühlen als der mit kleinerer Wärmekapazität. Die in den einzelnen Schichten eines Bodenprofils gemessenen Temperaturen und Temperaturänderungen richten sich folglich nach dem Wärmeumsatz an der Bodenoberfläche durch Strahlung, Verdunstung usw., der Wärmeleitfähigkeit und der Wärmekapazität des Bodens. *Temperaturausbreitung* und *Wärmeausbreitung* im Boden sind verschiedenartige Erscheinungen.

Die *Wärmekapazität* eines Körpers von m Gramm Masse ist $m \cdot c$. Dabei ist c die *spezifische Wärme*, sie wird gemessen in cal/g · Grad und bezeichnet diejenige Wärmemenge, die notwendig ist, um 1 g Substanz um 1° C zu er-

Tabelle 124. *Spezifische Wärme verschiedener Böden und Bodenkonstituenten*

Stoff	spez. Wärme cal/g·Grad	Autor
Quarz	0,188	Biedermann (1937)
Magnesiaglimmer	0,206	Biedermann (1937)
Kaliglimmer	0,208	Biedermann (1937)
Orthoklas	0,188	Biedermann (1937)
Augit	0,193	Biedermann (1937)
Hornblende	0,195	Biedermann (1937)
Apatit	0,190	Biedermann (1937)
Kalkspat	0,200	Biedermann (1937)
Dolomit	0,222	Biedermann (1937)
Fe_2O_3	0,165	Ulrich (1894) zit. nach Bayer (1956)
Al_2O_3	0,217	Lang (1878) zit. nach Bayer (1956)
$Fe(OH)_3$	0,226	Ulrich (1894)
Granit	0,197	Biedermann (1937)
Basalt	0,205	Biedermann (1937)
Sandstein	0,174	Biedermann (1937)
Marmor	0,212	Biedermann (1937)
Tuffstein	0,331	Biedermann (1937)
Quarzsand	0,191	Ulrich (1894)
Kaolin	0,224	Ulrich (1894)
sand. Lehmboden	0,22	Kostjakow (1959, S. 147)
Lehm	0,22	Kostjakow (1959, S. 147)
Torf	0,25	Kostjakow (1959, S. 147)
Humus	0,443	Ulrich (1894)
kalkhaltiger Sandboden	0,249	Lang (1878)
humoser kalkhaltiger Sandboden	0,257	Lang (1878)
Gartenboden	0,267	Lang (1878)
ausgeglühte, mineralische Bodensubstanz	0,184	Dmitrijew (1958)
organische Substanz	0,3	Dmitrijew (1958)
chemisch gebundenes Wasser im Boden	0,45	Dmitrijew (1958)
Wasser	1,00	—
Luft	0,237	—

wärmen (bei bestimmter Temperatur). In Tab. 124 sind für einige Bodenkonstituenten und Böden die spezifischen Wärmen nach den Angaben verschiedener Autoren zusammengestellt.

Dieser Übersicht läßt sich entnehmen, daß die spezifische Wärme der *anorganischen* festen Bodensubstanz um 0,18 bis 0,22 cal/g · Grad liegt. Die *organischen* festen Bodenbestandteile haben eine spezifische Wärme von 0,3 bis 0,4 cal/g · Grad. Schwerere Böden und solche, die höhere Gehalte an organischer Substanz aufweisen, besitzen auch in trockenem Zustand eine etwas höhere spezifische Wärme als leichte und humusarme Böden (DIMO 1952).

Der Boden besteht aus *fester Substanz, Luft und Wasser*, seine spezifische Wärme setzt sich *additiv* aus den spezifischen Wärmen seiner Konstituenten zusammen. Sind $c_1, c_2, c_3 \ldots$ die spezifischen Wärmen der Bodenkonstituenten, die mit der Masse $m_1, m_2, m_3 \ldots$ an seinem Aufbau beteiligt sind, so ist die spezifische Wärme c_B des Gesamtbodens:

$$c_B = \frac{c_1 m_1 + c_2 m_2 + c_3 m_3 \ldots}{m_B}$$

Sie ist, da die spezifische Wärme der festen Bodenbestandteile nur $^1/_4$ bis $^1/_5$ von derjenigen des Wassers beträgt, weitgehend vom Wassergehalt des Bodens abhängig. Abb. 160 nach KOSTJAKOW (1959) zeigt hierzu ein Beispiel. Der gleichen Bodenmenge (100 kg) wird dieselbe Wärmemenge (300 kcal) zugeführt. Man erhält dann je nach dem Wassergehalt eine verschiedene Temperaturerhöhung, die durch die beiden Kurven für eine spezifische Wärme der festen Bodensubstanz von 0,20 bzw. 0,25 cal/g · Grad angegeben wird. Trockener Boden z. B. erwärmt sich in dem angegebenen Beispiel um 15° C, solcher mit 10% Wassergehalt um 10°, solcher mit 40% Wassergehalt um 5° C.

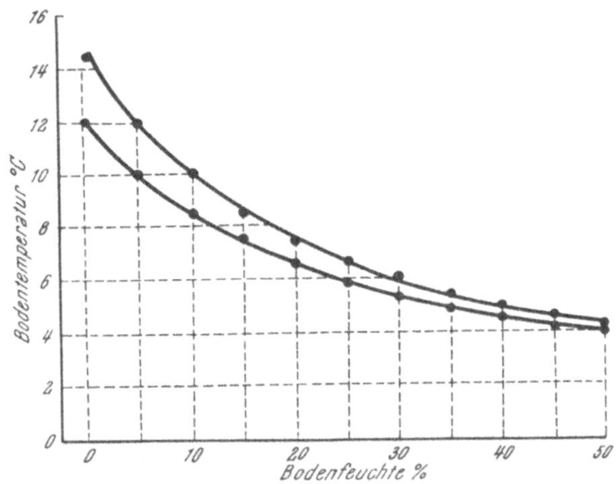

Abb. 160. Temperaturerhöhung zweier Böden mit unterschiedlichem Wassergehalt bei gleicher Wärmezufuhr (KOSTJAKOW 1959)

Feuchter Boden benötigt also für seine Erwärmung auf eine bestimmte Temperatur mehr Wärme als trockener Boden mit derselben Ausgangstemperatur. Böden mit durchschnittlich *hohem Wassergehalt* erwärmen sich in der Regel langsamer als durchschnittlich *trockenere* Böden. Andererseits speichern sie aber auch mehr Wärme, sie kühlen nicht so leicht ab.

GEIGER (1950), KERSTEN (1952) und BAVER (1956) weisen mit Recht darauf hin, daß das *Bodenvolumen* als Bezugsbasis für die Betrachtung der Wärmekapazität besser geeignet ist als die Gewichtseinheit des Bodens. Die von der Bodenoberfläche ausgehende Wärme erwärmt den darunter befindlichen Bodenraum. Dessen Wärmekapazität in cal/cm³ richtet sich nach seinem Wassergehalt (*WG*), dem Substanzvolumen (*SV*) und der spezifischen Wärme *c* der festen Substanz. Die Wärmekapazität von 1 cm³ Boden ist dann

$$VG \cdot c + WG \cdot 10^{-2} \text{ [cal/cm}^3 \cdot \text{Grad]}$$

Dabei ist *VG* das Volumengewicht des Bodens, das bei Mineralböden zwischen 1,2 und 1,6 g/cm³ liegt. Es kann aus dem Substanzvolumen *SV* (%) errechnet werden. Das *VG* von 1 cm³ trockenem Boden ist $VG = SV \cdot s \cdot 10^{-2}$, das spezifische Gewicht *s* von Mineralböden liegt bei 2,65. Der Wassergehalt *WG* wird in Volumprozent ausgedrückt. Die Wärmekapazität der Bodenluft kann vernachlässigt werden, sie beträgt, wie aus den Zahlen der Tab. 124 leicht zu ersehen ist, rund $^1/_{1000}$ der Wärmekapazität der festen Bodensubstanz bezogen auf gleiche Volumina. BAVER (1956) zitiert folgende Werte nach ULRICH für die Wärmekapazität der Volumeinheit Kaolin bzw. Humus:

	trocken	50% wassergesättigt	100% wassergesättigt
Kaolin	0,233	0,539	0,846
Humus	0,165	0,555	0,945

Innerhalb des *Profils von Mineralböden* wechselt die Wärmekapazität des *trockenen* Bodens nach DIMO (1952) nur wenig. Dies dürfte auch zutreffen, wenn man die Wärmekapazität auf das Bodenvolumen bezieht. In tieferen Schichten ist zwar in der Regel das Substanzvolumen höher, doch sind die oberen Horizonte reicher an organischer Substanz, deren spezifische Wärme größer ist als die der anorganischen Komponente. Betrachtet man allerdings die Wärmekapazität des feuchten Bodens, so sind innerhalb des Profils *erhebliche Unterschiede* als Folge verschiedener *Wassergehalte* möglich. Da diese sich im Laufe des Jahres ändern, ist auch die Wärmekapazität des Bodens keine konstante Größe, sondern ändert sich entsprechend dem *Befeuchtungszustand* des Bodens. In Tab. 125 nach BAUMGARTNER (1956) ist als Beispiel die auf die Volumeinheit bezogene Wärmekapazität eines Fichtenwaldbodens angegeben:

Tabelle 125. *Wärmekapazität in cal/cm³ · Grad in einem Profil unter Fichtenwald* (BAUMGARTNER 1956)

Schicht	Bodenart	Wärmekapazität des trockenen Bodens	Wärmekapazität des Wassers	Wärmekapazität des Gesamtbodens
0— 2 cm	Nadelstreu	0,08	0,08	0,16
2— 5 cm	Rohhumus	0,15	0,12	0,27
5— 7 cm	Rohhumus	0,18	0,26	0,44
10—20 cm	Rohhumus und	0,22	0,30	0,52
20—30 cm	Mineralboden	0,25	0,35	0,60

4. Die Wärme- und Temperaturleitfähigkeit des Bodens und sein Wärmegehalt

In Tab. 126 sind die von SCHUBERT (1904) angeführten Durchschnittswerte für den täglichen Verlauf der Bodentemperaturen in Pawlowsk im August 1888 wiedergegeben.

Man erkennt aus dieser Darstellung den *Gang* der Bodentemperaturen in den einzelnen Tiefen. Die *Minima* (einfach unterstrichen) treten in *tieferen* Schichten *später* auf, das gleiche gilt für die tägliche Wanderung der *Maxima* (doppelt unterstrichen). Durch die Richtung und Anzahl der Pfeile wird die Richtung und Größe des *Temperaturgefälles* angedeutet, durch die Punkte werden die Stellen bezeichnet, an denen das Temperaturgefälle seine Richtung wechselt. In tieferen Schichten ist die Amplitude dieser täglichen Temperaturwelle gering, ihre Schwingung ist abgeschwächt und tritt von einer bestimmten Tiefe an nicht mehr meßbar in Erscheinung.

Da die tägliche Durchschnittstemperatur der oberen Bodenschichten in der warmen Jahreszeit aber ansteigt, in der kalten absinkt, existiert darüber hinaus noch die *jährliche Temperaturperiode*, die sich in tiefere Schichten erstreckt als die tägliche. In den Oberflächenschichten des Bodens wird also die jährliche Temperaturwelle von der täglichen überlagert. Allgemein gilt für dieses Temperaturspiel in Böden, daß es nach der Tiefe hin *abgeschwächt* wird, gleichzeitig wird

Tabelle 126. *Bodentemperaturen zu Pawlowsk, August 1888, °C*
(Schubert 1904, S. 3)

Tiefe	Vormittag						Nachmittag					
cm	1 h	3 h	5 h	7 h	9 h	11 h	13 h	15 h	17 h	19 h	21 h	23 h
0	10,7	10,2	10,3	·14,6	19,8	25,0	26,9	24,7	·20,7	16,0	13,1	11,6
	↑ ↑	↑ ↑	↑	· ↓ ↓	↓ ↓	↓↓ ↓	↓↓ ↓	↓ ↓	· ↑	↑ ↑	↑ ↑	↑ ↑
2	11,6	11,1	11,0	·13,5	18,3	22,9	24,6	23,7	·21,0	·17,1	14,3	12,7
	↑ ↑	↑	↑	· ↓	↓ ↓	↓↓ ↓	↓↓ ↓	↓↓ ↓	↓	· ↑	↑	↑ ↑
5	12,9	12,2	11,8	12,9	·16,0	19,2	21,0	21,3	20,2	·17,9	15,5	14,0
	↑	↑	↑		· ↓	↓ ↓	↓ ↓	↓	↑	· ↑	↑	↑
10	14,1	13,4	12,9	13,0	·14,7	16,9	18,7	19,6	19,4	·18,2	·16,5	15,2
	↑	↑	↑	↑	· ↓	↓	↓ ·	↓	↓	↓	·	↑
20	15,4	·14,8	14,2	13,8	·13,9	·14,6	15,7	16,7	17,2	17,3	·16,9	16,1
		·		↑		·	↓	↓	↓	↓	↓	
40	15,3	·15,2	15,1.	14,9	14,7	·14,5	14,5	14,6	14,8	15,1	15,2	15,3

der Eintritt der Maxima und Minima *zeitlich* verzögert. Der Grund für diese *periodischen Temperaturänderungen* des Bodens ist in der *Wärmeübertragung* von einer Bodenschicht zur andern zu suchen. Besteht im Boden ein Temperaturgefälle von oben nach unten, so strömt Wärme in derselben Richtung und umgekehrt. In einem homogenen Körper, der an allen Stellen die gleiche Wärmeleitfähigkeit und Wärmekapazität aufweist, ist die Wärmemenge, die durch eine Fläche senkrecht zur Richtung des Temperaturgefälles hindurchgeht, proportional zu diesem. Anzahl und Richtung der Pfeile in Tab. 126 deuten also zugleich die Größe und Richtung des Wärmestromes an, allerdings nur dann, wenn der Boden bis zu der angegebenen Tiefe von 40 cm als homogener Körper angesehen werden darf. In der Regel trifft das für die einzelnen Schichten eines Bodenprofils aber nicht zu, so daß diese Proportionalität in mannigfacher Weise gestört wird.

Die *Wärmeübertragung* erfolgt durch *Wärmeleitung*, *Wärmestrahlung* und *Massenaustausch*. Die letztgenannten Vorgänge spielen im Boden praktisch keine Rolle, hier erfolgt die Wärmeübertragung vorwiegend durch Leitung. Auf die Bewegung von *Bodenfeuchte* unter dem Einfluß eines Temperaturgradienten, die mit Energieübertragung verbunden ist, wird auf S. 761 hingewiesen.

Für den *stationären Zustand* der *Wärmeleitung* gilt (Manegold 1960, S. 1228)

$$K = \frac{\lambda F (Te - Ta)\, t}{d} \text{ [cal]}$$

Dabei ist K die Wärmemenge in cal, F die Flächeneinheit (cm^2), Te und Ta die Eintritts- und Austrittstemperaturen, t die Zeit in sec, d die Meßstrecke in cm. λ ist der „*Wärmeleitfähigkeitskoeffizient*" oder die „*Wärmeleitzahl*" mit der Dimension cal/cm$^2 \cdot$sec$\cdot$°C/cm. Weiterhin ist die Wärmemenge C, die von einem Volumen V (ccm) bei einer Temperaturerhöhung um (Te—Ta) °C aufgenommen wird

$$C = V \cdot c \cdot \varrho\, (Te - Ta) \text{ [cal]}$$

Dabei ist c die spezifische Wärme des Körpers, ϱ sein spezifisches Gewicht in g/ccm. Wird dieser Ausdruck in die obige Gleichung eingesetzt, so erhält man

$$V \cdot c \cdot \varrho\,(Te - Ta) = \frac{\lambda\,F\,(Te - Ta)\,t}{d}$$

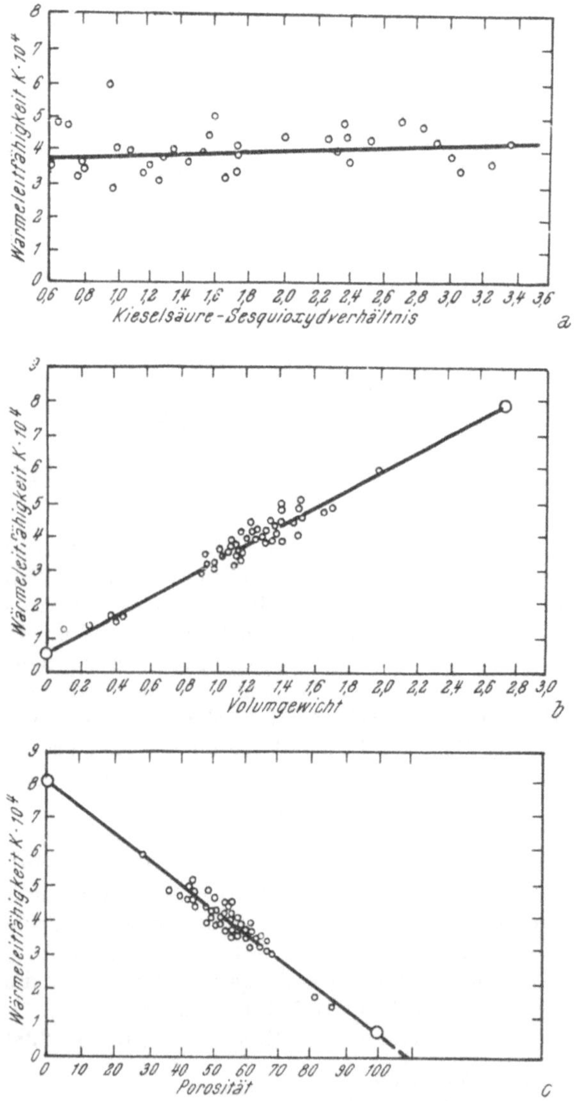

Abb. 161 *a* bis *c*. Die Wärmeleitfähigkeit von trockenem Boden als Funktion des Kieselsäure-Sesquioxydverhältnisses, des Volumgewichtes und der Porosität (Smith und Byers 1938)

In dieser Gleichung ist die Größe $\frac{\lambda}{c \cdot \varrho}$ enthalten. Sie ist die „*Temperaturleitzahl*" und erhält meist die Bezeichnung a, ihre Dimension ist cm^2/sec. Wir entnehmen diesen Gleichungen, daß die Wärmeleitung in einem homogenen Körper zur Wärmeleitzahl und dem Temperaturgefälle innerhalb der beobachteten Strecke proportional ist.

Nun ist der Boden aber ein *poröser Körper*, trockener Boden besteht aus fester Bodensubstanz und Luft. Manegold (1955, S. 470) hat an Hand einer Modellbetrachtung bei Eis dargestellt, daß sich der Wärmedurchgang durch diesen Körper additiv zusammensetzt aus den Werten für die beiden Konstituenten Luft und Eis. Überträgt man dieses Modell auf trokkenen Boden, so erhält man, wenn K_B der Wärmedurchgang durch dieses System ist:

$$\frac{K_B \cdot d}{F\,(Te - Ta)\,t} = \lambda_B \cdot V_B + \lambda_L\,(1 - V_B)$$

Dabei ist λ_B die Wärmeleitzahl für die feste Bodensubstanz, λ_L die Wärmeleitzahl der Luft, V_B ist die Raumerfüllung der festen Bodensubstanz, $1 - V_B$ der Porenanteil des Bodens, $V_B + (1 - V_B) = 1$. Diese Betrachtung gilt, wie von Smith und Byers (1938) experimentell gezeigt wurde, auch beim Boden mit guter Annäherung. Die Wärmeleitfähigkeit eines trockenen Bodens ist um so geringer, je größer sein *Porenvolumen* und je kleiner sein *Raumgewicht* ist. Abb. 161 b und c gibt die Befunde dieser Autoren wieder. Die relativ gute Übereinstimmung zwischen der Messung und der theoretisch zu erwartenden Beziehung ist vermutlich auf die Vorbehandlung zurückzuführen, es wurden nämlich auf 2 mm Korngröße gesiebte Böden verwendet. In der Natur ist diese Übereinstimmung weniger zu erwarten, da die Bedingungen des Wärmeübergangs durch das Substanz- und Porenvolumen nur *annähernd* beschrieben werden. λ ist für die feste Bodensubstanz hoch, für die Luft klein, d. h. der Wärmeübergang erfolgt fast ausschließlich durch die feste Substanz. Besteht zwischen den festen Bodenpartikeln nur eine *punktförmige* Berührung, so muß die Wärmeübertragung gegenüber einem Zustand mit *flächiger* Berührung verlangsamt sein. Es kommt also auch noch auf die innere Geometrie des Systems an.

Es liegt nahe, diese Betrachtung auf das System feste Bodensubstanz, Wasser und Luft auszudehnen. Theoretisch scheint das statthaft zu sein, bei praktischen Messungen tritt aber die oben erwähnte Feuchtigkeitsverlagerung auf, d. h. das System ändert sich während des Meßvorganges.

Nach diesen allgemeinen Bemerkungen sollen nunmehr einige Zahlenangaben für die Wärmeleitfähigkeit folgen. In Tab. 127 sind die Werte für verschiedene Stoffe zusammengestellt.

Tabelle 127. *Wärmeleitzahl für verschiedene Materialien*

Substanz	$\lambda \cdot 10^3$, cal/cm² · sec · Grad/cm	Autor
Luft 0° C	0,056	MANEGOLD 1955, S. 471—475
Luft 20° C	0,060	MANEGOLD 1955, S. 471—475
Wasser 4° C	1,278	MANEGOLD 1955, S. 471—475
Wasser 20° C	1,416	MANEGOLD 1955, S. 471—475
Schnee Raumerfüllung 0,199	0,239	MANEGOLD 1955, S. 471—475
Schnee Raumerfüllung 0,266	0,400	MANEGOLD 1955, S. 471—475
Sandstein	3,0	BIEDERMANN 1937, S. 329
Marmor	5,4—8,4	BIEDERMANN 1937, S. 329
Erde trocken	0,333	MANEGOLD 1955, S. 471—475
Erde feucht	1,610	MANEGOLD 1955, S. 471—475

Aus diesen Zahlen ist zu entnehmen, daß die Wärmeleitzahl für Wasser etwa 25mal höher ist als die von Luft. Die Bedeutung des Porenvolumens, die bereits aus Abb. 161b und c hervorgeht, zeigt sich wiederum bei der Wärmeleitzahl für Schnee. Einer Zunahme der Raumerfüllung von 20 auf 27% entspricht die Erhöhung von λ von $0{,}24 \cdot 10^{-3}$ auf $0{,}40 \cdot 10^{-3}$ cal/cm² · sec · Grad/cm. Der Einfluß der Bodenbeschaffenheit auf den Wert λ geht aus den Ergebnissen von SMITH und BYERS (1938) hervor, die in Tab. 128 auszugsweise wiedergegeben sind.

Aus diesen Zahlen läßt sich eine Reihe von Beziehungen entnehmen. Bei den Mineralböden ist die Wärmeleitzahl im A-Horizont am niedrigsten. Diese Erscheinung ist nach Ansicht der Autoren mitbedingt durch die höheren *Humusgehalte* in diesem Horizont. Ferner ist der Einfluß der *Textur* erkennbar. Sandige Böden besitzen eine höhere Wärmeleitfähigkeit als lehmige. Der höchste Wert für λ wird im A_2-Horizont des Podsols gefunden. *Organogene* Böden besitzen in trockenem Zustande eine niedrigere Wärmeleitfähigkeit als Mineralböden. Einer Zunahme des Volumgewichts, d. h. einer *Verringerung des Porenvolumens*, entsprechen höhere Werte für λ. In der Abb. 161b und c ist diese Beziehung für alle untersuchten Böden graphisch dargestellt. Die *chemische* Zusammensetzung der mineralischen Bodensubstanz scheint die Werte für λ geringfügig zu beeinflussen, wie aus Abb. 161a hervorgeht.

Smith (1939) untersuchte weiterhin die *Wärmeleitfähigkeit von feuchten Böden.* Diese ist in Tab. 129 für einige Beispiele angegeben und zwar für absolut trockenen

Tabelle 128. *Wärmeleitzahl bei trockenen Böden* (Smith und Byers 1938)

Boden	Horizont	$\lambda \cdot 10^4$ $\left(\frac{cal}{cm^2} / \frac{Grad}{cm}\right) \cdot sec^{-1}$	Volumgewicht g/cm³	Dichte g/cm³	Porenvolumen %	$\frac{SiO_2}{R_2O_3}$ in der Kolloidsub.
Podsol	A_1	1,61	0,38	2,10	81,9	—
lehm. Sand	A_2	5,12	1,49	2,64	43,6	—
	B_1	4,32	1,35	2,64	48,9	—
	B_2	4,43	1,46	2,65	44,9	—
	C	4,61	1,52	2,66	42,9	—
grau—brauner	A	3,34	1,14	2,71	57,9	1,72
podsoliger Boden,	B	3,90	1,18	2,78	61,2	1,74
Lehm	C	3,90	1,18	2,80	57,8	1,53
Tschernosem	A	3,44	1,10	2,59	57,6	3,05
Lehm	B_1	3,76	1,10	2,67	58,8	3,00
	B_2	4,08	1,15	2,71	57,6	3,25
	C	4,16	1,27	2,72	53,4	3,36
Sphagnumtorf	—	1,23	0,07	1,54	95,5	—
Schilf—Moor, kultiviert	—	1,72	0,43	1,56	72,4	—

Boden und 4 Feuchtigkeitsstufen: 75% und 99,3% „relativer Feuchtigkeit" entsprechen dem Wassergehalt des Bodens im Gleichgewicht mit dem Wasserdampfgehalt der Luft bei 75- und 99,3- prozentiger Sättigung. Das „moisture equivalent" ist ein im Laboratorium bestimmter Wert, der dem der *Feldkapazität* nahekommt. Proben aus dem A-Horizont der Schwarzerde enthielten bei diesen Feuchtigkeitsstufen die folgenden Wassermengen:

75% rel. Feuchtigk.	0,045 g H_2O/g Boden
99,3% rel. Feuchtigk.	0,086 g H_2O/g Boden
50% moisture equivalent	0,126 g H_2O/g Boden
100% moisture equivalent	0,252 g H_2O/g Boden

Bei der *Anfeuchtung* des Bodens ist also eine bemerkenswerte *Erhöhung* der Wärmeleitfähigkeit zu verzeichnen; der Anstieg ist besonders hoch zwischen 50 und 100% des „moisture equivalent". In dem Maße, wie die isolierende Wirkung der Bodenluft entfällt und das Porenvolumen zum Teil mit Wasser erfüllt ist, wird also die Wärmeleitfähigkeit des Bodens erhöht.

Die Temperaturgradienten, die bei diesen Messungen zwischen der kalten und warmen Seite der Bodenproben bestanden, waren niedrig. Die Zeit, die bis zum Erreichen eines stationären Zustandes erforderlich war, war je nach dem Wassergehalt des Bodens verschieden, weil während der Messung eine *Feuchtigkeitsverlagerung* von der warmen zur kalten Seite der Probe stattfand. Diese war bei 75% relativer Feuchtigkeit und bei 100% des „moisture equivalent" gering, hoch dagegen bei den beiden mittleren Wassergehalten. Diese Wasserbewegung erfolgt offenbar durch Verdunstung auf der warmen und Kondensation auf der kalten Seite der Probe. Es findet also neben der Wärmeleitung eine *Wärmeübertragung durch Massenaustausch* statt. Zu ähnlichen Ergebnissen kamen WOODSIDE und CLIFFE (1959) sowie WOODSIDE und DE BRUYN (1959).

Tabelle 129. *Wärmeleitzahl von Böden bei verschiedenen Feuchtigkeitsstufen* (SMITH 1939)

Boden	Horizont	Wärmeleitzahl $\left(\frac{cal}{cm^2} / \frac{Grad}{cm}\right) \cdot sec^{-1} \cdot 10^4$ bei				
		ofen-getrockn. Boden	75% rel. Feucht.	99,3% rel. Feucht.	50% „moist. equ."	100% „moist. equ."
Podsol, lehm. Sand	A_1	1,61	2,02	2,11	2,49	6,42
	A_2	5,12	5,92	5,99	6,92	12,36
	B_1	4,32	5,31	5,80	6,81	11,54
	B_2	4,43	4,28	5,66	8,64	11,70
	C	4,61	5,71	6,28	8,57	12,89
grau—brauner podsoliger Boden, Lehm	A	3,34	4,53	5,14	8,69	11,51
	B	3,90	4,81	6,01	7,23	10,21
	C	3,90	4,90	6,11	9,33	11,59
Tschernosem Lehm	A	3,44	4,08	4,50	4,90	8,61
	B_1	3,76	4,47	4,64	5,62	8,87
	B_2	4,08	4,90	5,39	6,29	10,40
	C	4,16	5,21	5,73	5,98	12,91

DE VRIES (1952) führte am Boden in situ Kurzzeitmessungen der Wärmeleitfähigkeit durch. Die Werte für die Wärmeleitfähigkeit wiesen im Laufe des Tages *Schwankungen* auf, die als Folge von *Wassergehaltsänderungen*, z. B. auch durch einsickerndes Regenwasser, in Erscheinung traten. Besonders anschaulich tritt der Zusammenhang zwischen der Wärmeleitfähigkeit des Bodens und den *Niederschlägen* bei den Messungen hervor, die ALBRECHT (1937) im Juli 1937 gemacht hat. Sie sind in Abb. 162 wiedergegeben. Der obere Teil der Darstellung zeigt die Niederschläge, der untere die Wärmeleitfähigkeit des Bodens in 1, 10 und 50 cm Tiefe. Der enge Zusammenhang zwischen der Wärmeleitfähigkeit in 1 cm Tiefe und den Niederschlägen ist deutlich erkennbar. An trockenen Tagen weist diese Schicht eine Tagesperiode der Wärmeleitfähigkeit auf. Diese ist vielleicht auf Verdunstungs- und Kondensationsvorgänge zurückzuführen. Die 10- und 50-cm-Schichten sind feuchter als die Oberflächenschicht, weshalb hier die Werte für λ höher sind; außerdem fällt bei den tiefer gelegenen Schichten eine deutliche Phasenverschiebung auf. Die Kurve in 50 cm Tiefe steigt deutlich in den Zeiten, in denen die 10-cm-Kurve sinkt. In dieser Erscheinung dürfte die Bewegung des *Sickerwassers* im Boden zum Ausdruck kommen.

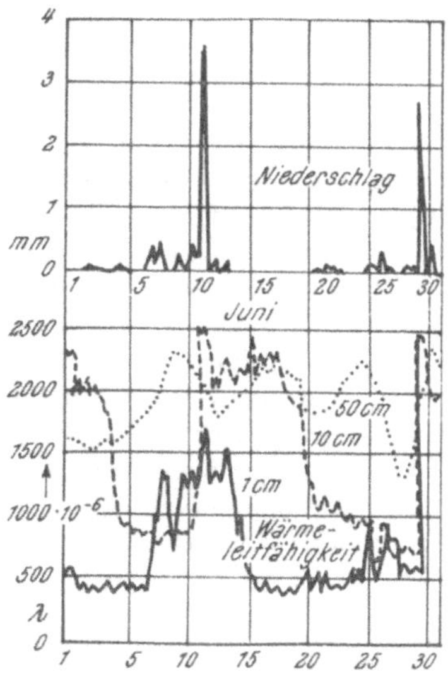

Abb. 162. Niederschläge und Wärmeleitfähigkeit in drei verschiedenen Tiefen während des Juli 1937 (ALBRECHT 1937)

In Tab. 126 wurde ein Beispiel für die *tägliche Temperaturwelle* im Erdboden angeführt. Diese geht aus von der Bodenoberfläche und pflanzt sich mit verringerter Amplitude und zeitlicher Verzögerung nach der Tiefe hin fort. In dieser Temperaturwelle kommt zum Ausdruck, daß sich im Boden ein *Wärmestrom* bewegt, der von der Bodenoberfläche ausgeht, wenn ein Temperaturgefälle von oben nach unten besteht, der in Richtung zur Bodenoberfläche geht, wenn sich das Temperaturgefälle umkehrt. Die Wärmemenge, die sich auf diese Weise durch den Boden bewegt, hängt ab von der Wärmezufuhr zur Bodenoberfläche und deren Wärmeverlusten. Die wichtigsten Faktoren, von denen die Wärmebewegung im Boden abhängt, sind sein Wasser- und Luftgehalt, ferner auch der Humusgehalt des Bodens und seine Textur. Als Folge dieses Wärmestromes tritt im Boden eine *Wärmevorratsänderung* ein. Diese ist in Tab. 130, die der Arbeit von Baumgartner (1956) entnommen ist, für verschiedene Bodenarten und Bodendecken dargestellt.

Tabelle 130. *Tagesgänge des Wärmestromes an heiteren Hochsommertagen in verschiedenen Bodenarten unter verschiedenen Bodendecken* ($mcal/cm^2 \cdot min$) (Baumgartner 1956)

Uhrzeit	0—1	3—4	6—7	9—10	12—13	15—16	18—19	21—22
Potsdam nackter Sand	— 70	— 63	26	157	165	39	— 84	— 84
Leipzig lehmig, Rasen	— 65	— 40	64	155	105	— 3	— 57	— 77
Palkäne Humus, Gras	— 72	— 56	83	127	137	76	— 15	— 67
Eberswalde Sand, Gras	— 95	— 55	15	160	130	25	— 85	— 85
Eberswalde Sand, Kiefernwald	— 50	— 15	5	55	50	5	— 30	— 255

Um die angegebenen Beträge verringert bzw. vergrößert sich der Wärmevorrat des Bodens während der verschiedenen Tageszeiten. Diese Tabelle zeigt den *nächtlichen Wärmeverlust* und die *Zunahme* des Wärmevorrates während der *Tagesstunden*. Was an diesen Zahlen aber besonders interessiert, sind die ausgetauschten Energiebeträge. Sie hängen insbesondere von der *Bodendecke* ab, d. h. dem Schutz der Bodenoberfläche durch eine Pflanzendecke, die sowohl die *Wärmeeinstrahlung* wie den *Wärmeverlust* verringert. Die in Tab. 130 angeführten Beträge für den Wärmestrom sind deshalb bei nacktem Boden hoch; sie sind bei dem unter Kiefernwald befindlichen Sandboden von Eberswalde am geringsten. Ganz gleich liegen die Verhältnisse, wenn man die *jährliche Periode* des Wärmeaustausches im Boden betrachtet. Süring (1927, S. 25) führt hierzu die folgenden Werte für Eberswalde an:

Monat:	I	II	III	IV	V	VI	VII	VIII	IX	X	XI	XII	
Feld:	—300	—166	—9	353	498	469	345	147	—133	—386	—425	—393	cal/cm^2
Wald:	—232	—140	—41	169	294	356	277	165	—16	—232	—298	—302	cal/cm^2

Auch an diesem Beispiel zeigt sich, daß der jährliche Wärmeumsatz im Waldboden erheblich kleiner ist als auf freiem Felde.

Es sind nun die *Beziehungen zwischen dem Wärmestrom und der Bodentemperatur* zu betrachten. In der Regel wird ja die Bodentemperatur als leicht zu bestimmende

Größe gemessen. Doch ist eine *Deutung* der beobachteten *Temperaturen* und *Temperaturänderungen* nur möglich, wenn man berücksichtigt, daß in ihnen der *Wärmevorrat und die Wärmevorratsänderungen zum Ausdruck* kommen. *Die Temperaturleitzahl* des Bodens wurde bereits auf S. 758 erwähnt. Die Größe von $a = \frac{\lambda}{c \cdot \varrho}$ cm²/sec hängt ab von der Wärmeleitfähigkeit λ und der spezifischen Wärme des Systems feste Bodensubstanz, Wasser und Luft. Wird der Bodenoberfläche eine bestimmte Wärmemenge zugeführt, so wird in tieferen Schichten um so eher eine Zunahme der Temperatur zu verzeichnen sein, je größer λ und je kleiner $c \cdot \varrho$ ist. $c \cdot \varrho$ ändert sich, da die Wärmekapazität der Luft vernachlässigt werden kann, in erster Linie mit dem *Wassergehalt* des Bodens. Unter der Voraussetzung, daß die Verlagerung von Wasser während des Meßvorganges bei den gewählten Feuchtigkeitsgehalten vernachlässigt werden darf, wurden die Werte für a auf Grund der Daten von SMITH (1939) für zwei Böden berechnet. Dabei wurde die spezifische Wärme c des Bodens mit 0,22 cal/g·Grad angenommen.

Tabelle 131. *Wärmeleitfähigkeit λ und Temperaturleitzahl a bei verschiedenen Wassergehalten zweier Böden*

	Lehmboden			Sandboden		
Wassergehalt Volumanteile	0	0,059	0,269	0	0,005	0,082
$\lambda \cdot 10^3$ cal/cm² sec °C/cm	0,376	0,447	0,887	0,512	0,592	1,236
$a \cdot 10^2$ cm²/sec	0,155	0,139	0,166	0,156	0,175	0,311

Die Wassergehalte entsprechen 75% relativer Feuchtigkeit und 100% des „moisture equivalent" in der Arbeit von SMITH. Durch die Anfeuchtung auf 100% des „moisture equivalent" wird bei beiden Böden die Wärmeleitfähigkeit etwas mehr als *verdoppelt*. Die Temperaturleitzahl des *Lehmbodens* wird durch die Anfeuchtung nur wenig verändert, bei dem *Sandboden* dagegen durch 0,08 Volumteile Wasser nahezu verdoppelt. Diese Erscheinung wird verständlich, wenn man bedenkt, daß die Wärmeleitfähigkeit λ durch 0,08 Volumteile Wasser beträchtlich erhöht wurde, während andererseits die Wärmekapazität dieser geringen Wassermenge noch nicht erheblich ins Gewicht fällt.

Unter Benutzung der *Temperaturleitzahl* $\frac{\lambda}{c \cdot \varrho}$ ist es möglich, die Fortpflanzung der *Temperaturwelle* im Boden und die Verringerung der *Amplitude* nach der Tiefe hin mathematisch zu erfassen. Die Temperaturänderungen an der Bodenoberfläche pflanzen sich nach der Tiefe hin mit gleichförmiger Geschwindigkeit fort. Für die Temperaturschwankungen gilt, daß die Logarithmen der Amplituden mit der Tiefe linear abnehmen. Entsprechende Formeln werden von SÜRING (1927), GEIGER (1950) und LANDSBERG und BLANC (1958) angegeben. *Die Verzögerung* einer Temperaturänderung ist:

$$\Delta t = (d_2 - d_1) \frac{T}{2\pi} \sqrt{\frac{\pi \cdot \varrho \cdot c}{T \cdot \lambda}} \quad [\text{sec}]$$

wobei Δt = der Verzug in sec ist, d_1 und d_2 = zwei Tiefen in cm, T in sec ist die Periode der Temperaturwelle, $\frac{\lambda}{c \cdot \varrho}$ ist die Temperaturleitzahl.

Für die *Dämpfung* der Temperaturwelle gilt:

$$r_2 = r_1 \cdot e^{(d_2 - d_1)\sqrt{\frac{\pi \cdot \varrho \cdot c}{T \cdot \lambda}}}$$

Dabei sind r_1 und r_2 die Unterschiede zwischen den Maximal- und Minimaltemperaturen in den Tiefen d_1 und d_2 während der Periode T.

In der *Isothermendarstellung der Bodentemperatur* kommen diese Beziehungen anschaulich zum Ausdruck. Ein Beispiel für die tägliche Temperaturwelle wurde bereits in Tab. 126 angeführt. Abb. 163 zeigt die jährlichen Änderungen der Bodentemperaturen in Isothermendarstellung.

Im oberen Teil dieser Zeichnung sind die Isothermen dichter, die Temperaturen ändern sich rascher, die Amplituden der Temperaturwelle sind hier am größten.

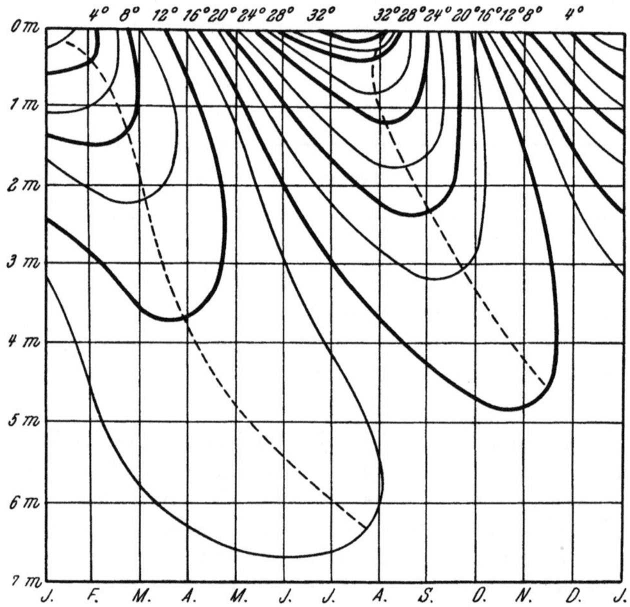

Abb. 163. Jahresgang der Bodentemperaturen in Tiflis (SÜRING 1927)

Deutlich ist die Verzögerung der Temperaturwelle im Laufe des Jahres zu erkennen; die Geschwindigkeit, mit der sich Temperaturänderungen bei diesem Beispiel fortpflanzen, beträgt 4,3 cm je Tag. Die Richtung des Temperaturgefälles im ganzen Profil kehrt sich im Laufe des *Jahres* zweimal um. Diese *Umkehr des „Temperaturprofils“* ist auch während der *Tagesperiode* zu beobachten.

Durch die *Temperaturleitzahl a* werden die *Beziehungen zwischen dem Wärmeumsatz an der Bodenoberfläche und den Bodentemperaturen* hergestellt. Denkt man sich zwei Böden, die im Laufe des Tages an ihrer Oberfläche die gleiche Wärmemenge absorbieren, so folgt für die Temperatur: Ist a groß (d. h. die Wärmeleitfähigkeit λ groß, der Ausdruck $c \cdot \varrho$ klein), so wird die Temperaturwelle rascher und mit einer größeren Amplitude in die Tiefe vordringen als bei dem Vergleichsboden mit kleinerer Temperaturleitzahl a. Der erstgenannte Fall trifft z. B. für Sandböden mit mäßigem Wassergehalt zu, der zweite für Böden mit höherem Wassergehalt wie z. B. Moorböden.

5. Temperaturen und Temperaturänderungen bei verschiedenen Böden

In den ersten drei Abschnitten dieses Beitrags wurde der Wärmehaushalt des Bodens besprochen und die Faktoren, von denen die Bodentemperaturen und Temperaturänderungen im allgemeinen abhängen, dargestellt. Es folgen nunmehr die Ergebnisse von *Temperaturmessungen* bei verschiedenen Böden. Ähnlich

Tabelle 132. *Durchschnittliche Bodentemperaturwerte* (*College Station Texas 1951—1955*) (FLUCKER 1958)

Tiefe unter der Oberfläche cm	durchschn. Max. Temp. °C	durchschn. Min. Temp. °C	durchschn. Jahrestemp. °C	Jahresamplitude der Temp. °C	Verzug gegenüber der Bodenoberfläche Tage
Luft	30,0	10,5	20,8	9,8	0
6	35,2	11,1	24,1	12,1	0
33	33,9	13,9	23,9	10,0	5
66	32,9	15,0	23,8	9,0	15
100	31,9	16,0	23,6	8,0	22
133	30,8	16,6	23,6	7,1	32
166	30,7	17,6	23,6	6,6	39
200	28,7	18,5	23,5	5,1	47
260	27,1	19,5	23,4	3,8	62
300	26,3	20,3	23,4	3,0	74

wie bei anderen Klimadaten geben *Durchschnittswerte* für das ganze Jahr nur sehr bedingt einen Einblick in die Temperaturverhältnisse des Bodens. Als Beispiel für solche Durchschnittswerte soll hier die Untersuchung von FLUCKER (1958) angeführt werden, die sich über einen Beobachtungszeitraum von 5 Jahren erstreckt.

Die oben schon erwähnte *Dämpfung der Amplitude* und die *zeitliche Verzögerung der Temperaturwelle* mit der Tiefe tritt auch bei diesen Werten hervor. Bemerkenswert ist, daß die jährlichen Durchschnittswerte der Bodentemperatur, wenn man von der Oberflächenschicht absieht, im ganzen Profil von bemerkenswerter *Konstanz* sind; sie nehmen mit der Tiefe nur geringfügig ab. Daß die oberflächennahen Schichten durchschnittlich etwas wärmer sind als die tiefer gelegenen, ist durch die klimatische Lage des Beobachtungsortes bedingt. Vgl. hierzu Tab. 133. Durch die Aufgliederung der Temperaturwerte in kürzere Perioden können die Temperaturverhältnisse des Bodens besser charakterisiert werden. Eine für bodenkundliche und pflanzenbauliche Zwecke sehr brauchbare Form der Darstellung wurde von BAIER (1952) gewählt. Sie ist in Abb. 164 wiedergegeben.

Aus dieser Darstellung läßt sich entnehmen, wann im Laufe der Vegetationsperiode die Schichten von 10, 50 und 100 cm eine bestimmte Temperaturschwelle übersteigen. In den tiefer gelegenen Schichten sinken, im Gegensatz zur Krume, die Temperaturen nicht unter 4° C ab. Die Temperaturwelle pflanzt sich in der aufsteigenden Jahreszeit allmählich in die Tiefe fort, im Herbst bleiben die tiefen Schichten länger warm. Derselben Arbeit von BAIER, in der über Bodentemperaturmessungen in Hohenheim berichtet wird, läßt sich ferner entnehmen, daß die absolut höchsten *Monatsschwankungen* im Mai liegen, und zwar sowohl unter unbewachsenem als auch bei grasbewachsenem Boden. Von April bis September sind die Bodentemperaturschwankungen in allen Tiefen größer als in der Zeit von Oktober bis März. Die geringsten Schwankungen wurden im Januar und Februar beobachtet.

Über die *Höchst-* und *Tiefsttemperaturen* sowie die *Ausbreitung der Temperaturwelle* im Boden liegen zahlreiche Angaben vor, von denen hier nur einige erwähnt werden sollen. Nach Chang (1957) ist die Tiefe der Schicht, in der eine Temperaturunruhe feststellbar ist, in den einzelnen *Klimazonen* verschieden. Unveränderliche Temperaturen herrschen in den Tropen ab 10 m, in den mittleren Breiten ab 15 m, im nördlichen Kanada und Alaska ab 20 m Tiefe. Besonders starke Temperaturschwankungen weist naturgemäß die *Oberflächenschicht* des Bodens auf. Yakuva (1958) gibt z. B. für verschiedene Bodenarten an klaren Sommertagen die folgenden Maximalwerte für die Bodentemperaturen an:

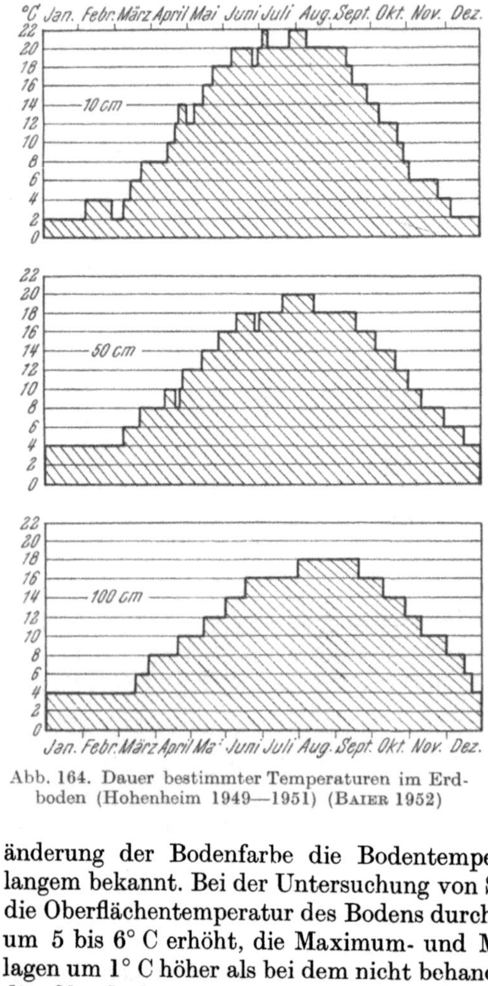

Abb. 164. Dauer bestimmter Temperaturen im Erdboden (Hohenheim 1949—1951) (Baier 1952)

Sandboden	53,5° C
vulk. Sandboden	48,6° C
toniger Lehm	46,7° C
Moorboden	39,0° C
Tonboden	35,9° C

Diese Werte können in manchen Fällen noch erheblich *überschritten* werden. Sie hängen nicht nur vom *Wassergehalt* des Bodens ab, sondern auch von der *Bodenfarbe*. Bei feuchtem Boden erfolgt die Ableitung der absorbierten Wärme rascher, außerdem wird durch die Verdunstung dem Boden Wärme entzogen, die Erwärmung der Oberflächenschicht ist also geringer. Daß durch eine Veränderung der Bodenfarbe die Bodentemperaturen beeinflußt werden, ist seit langem bekannt. Bei der Untersuchung von Suzuki und Maruyama (1957) wurde die Oberflächentemperatur des Bodens durch Bestreuen mit Kohlenstaub am Tage um 5 bis 6° C erhöht, die Maximum- und Minimumtemperaturen in 3 cm Tiefe lagen um 1° C höher als bei dem nicht behandelten Boden. Mallik (1951) konnte die Oberflächentemperatur dunkler Böden durch Bestreuen mit einem weißen Puder um 14° C senken.

Die Angaben über die *Geschwindigkeit*, mit der sich die Temperaturwelle im Boden fortpflanzt, sollen hier noch durch einige Beispiele ergänzt werden. Baumgartner (1956) konnte nach seinen Messungen im Juni und Juli 1952 für die Eintrittszeiten der täglichen Extremwerte folgende Tageszeiten festlegen:

	Luft in Bodennähe	2	5	10	20	30 cm Tiefe
			im Waldboden			
Maximum	15	17	18	20	22,30	11,30 Uhr
Minimum	4,30	6	6,30	8	9	3 Uhr
			auf einer Kahlfläche			
Maximum	14	14	16	18	21	11,30 Uhr
Minimum	4	5	6	8	9	1 Uhr

Die Fortpflanzung der Temperaturwelle bis in 30 cm Tiefe dauert demnach bis in die Nacht- bzw. Vormittagsstunden des nächsten Tages. Diese Zahlen geben naturgemäß nur allgemeine Anhaltspunkte; sie sind von Boden zu Boden verschieden. SHAW (1955) gibt z. B. an, daß die tägliche Maximaltemperatur bei Mineralboden in 30 cm Tiefe um 22 bis 23 Uhr erreicht wird, während sich das Minimum in derselben Tiefe 7 bis 8 Stunden nach Sonnenaufgang einstellt. Sehr geeignet für die anschauliche Wiedergabe des täglichen Temperaturganges im Erdboden ist die Darstellung mittels *Tautochronen*, d. h. den Verbindungslinien gleicher Temperaturen zu einer bestimmten Zeit. In Abb. 165 ist eine solche Tautochronen-Darstellung wiedergegeben.

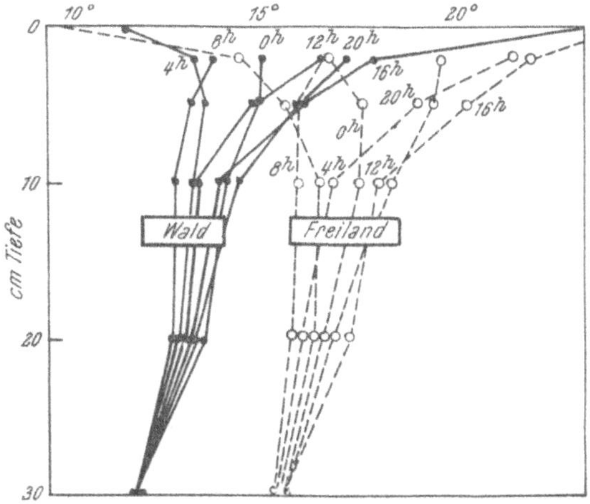

Abb. 165. Tautochronen der Bodentemperatur in Wald und Freiland (BAUMGARTNER 1956)

Dieser Zeichnung liegen die Meßergebnisse während einer Schönwetterperiode nach vorausgegangener kühler Witterung zugrunde. Der Waldboden ist während der ganzen Beobachtungsperiode kühler als der Freilandboden. Die tägliche Amplitude der Temperaturschwankungen ist in den oberen 10 cm am größten; sie nimmt dann mit der Tiefe rasch ab, die tägliche Temperaturwelle reicht bis etwa 30 cm Tiefe. Beide Kurvenbündel verlaufen von links unten nach rechts oben. Hierin kommt zum Ausdruck, daß die Beobachtungszeit in eine Periode der Bodenerwärmung fiel. Ferner zeigt sich an dieser Darstellung die täglich zweimal erfolgende *Umkehrung des „Temperaturprofils“* im Boden. Diese Umkehrung läßt sich auch bei der *jährlichen Temperaturwelle* feststellen, nur mit dem Unterschied, daß hier die Temperaturschwankungen bis in größere Bodentiefen reichen. In den Frühjahrs- und Sommermonaten besteht ein Temperaturgefälle von oben nach unten, in den Herbst- und Wintermonaten ein solches von unten nach oben.

Aus allem, was bisher dargestellt wurde, geht auch hervor, daß aus *klimatischen Daten* nur in grober Annäherung Rückschlüsse auf die Bodentemperatur gezogen werden können. Bei der Untersuchung von DUKER (1957) zeigte sich eine positive Korrelation zwischen der Lufttemperatur und der Bodentemperatur in 3 Zoll Tiefe, eine negative Korrelation dagegen während der warmen Jahreszeit zwischen der Temperatur und der mittleren Windgeschwindigkeit sowie dem Wassergehalt des Bodens. DREIBELBIS (1950) stellte fest, daß die Schwankungen der Bodentemperatur in $^1/_2$ Zoll Tiefe denen der Luft folgen.

Die *Zusammensetzung des Bodens*, seine Textur und Struktur, der augenblickliche *Wassergehalt* usw. bestimmen neben den Gegebenheiten des Wärmeumsatzes an der Bodenoberfläche seine Temperatur. Der Einfluß der Bodenzusammensetzung zeigte sich u. a. bei den Messungen von W. KREUZ (1943) an der Gießener Lysimeteranlage.

Im Jahresmittel liegen die Temperaturen der oberflächennahen Schicht etwas niedriger als die der 50-cm- und 100-cm-Schicht. Man kann nach dem bisher Gesagten versuchen, die Zahlen der Tab. 133 zu deuten. In der Reihenfolge Lehm,

Sand, Humus steigen die Jahresmitteltemperaturen an. Diese Erscheinung dürfte auf die verschiedene Wasserhaltefähigkeit der drei Bodenarten zurückzuführen sein. Sandboden erwärmt sich im Frühjahr rasch, Humusboden vermag dank seiner größeren Wasserhaltefähigkeit mehr Wärme zu speichern, er kühlt im Herbst nicht so rasch aus. Dies wird die Ursache dafür sein, daß die Mittelwerte in beiden Fällen höher liegen als beim Lehmboden. Die Zahlen

Tabelle 133. *Jahresmitteltemperaturen und Abstand der Maxima und Minima bei drei verschiedenen Bodenarten*
(Kreuz 1943)

Mitteltemperaturen °C				Amplituden °C			
Luft 8,2°	Lehm	Sand	Humus	Luft 9,9°	Lehm	Sand	Humus
5 cm Tiefe	9,0	9,3	10,1	5 cm Tiefe	9,9	10,5	11,0
10 cm Tiefe	9,5	9,3	10,4	10 cm Tiefe	9,9	10,3	11,0
20 cm Tiefe	9,4	9,6	10,4	20 cm Tiefe	9,5	10,0	10,5
50 cm Tiefe	10,3	10,7	10,7	50 cm Tiefe	8,2	8,4	9,2
100 cm Tiefe	10,6	11,1	11,3	100 cm Tiefe	7,2	7,1	7,1

für die Amplituden der Temperatur weisen in dieselbe Richtung. Sandboden erwärmt sich rasch, er kühlt, da sein Wärmevorrat gering ist, rasch wieder aus. Die Temperaturschwankungen in der Oberflächenschicht sind deshalb höher als beim Lehmboden. Humusboden absorbiert viel Wärme. Doch wirkt sich die hohe Wärmekapazität der oberen Bodenschichten dahin aus, daß die Temperaturerhöhung im Laufe des Tages nicht in tiefere Schichten vordringt.

Auch der Gehalt der Krume an *Steinen* beeinflußt die Temperaturverhältnisse. Troedsson (1956) untersuchte die Temperaturen bei zwei Böden a) mit wenig, b) mit viel Steinen an der Oberfläche. Die Sommertemperaturen waren bei b) um 1 bis 3° C höher als bei a), die Wintertemperaturen unterschieden sich weniger voneinander. Die lockernde Wirkung des Steingehaltes hatte im Frühjahr an steinigen Stellen einen rascheren Temperaturanstieg zur Folge. Eine *Lockerung in der Krumenschicht* bedingt in der Regel einen höheren Luftgehalt in dieser und begünstigt die Abtrocknung. Dadurch wird nicht nur der Temperaturgang im Boden selbst, sondern auch die Temperatur der Luft in unmittelbarer Nähe der Bodenoberfläche beeinflußt. Messungen von Slater (1954) zeigten, daß die Minimumtemperaturen unmittelbar über *kompaktem* Boden etwas höher liegen als bei lockerem Boden. Milojić (1958) untersuchte bei einem Tschernosem und einem Marschboden die Bodentemperaturen nach verschiedenartiger *Bearbeitung* sowie unter Weizen, Mais und Gras. *Höhe und Bestandesdichte* dieser Pflanzen beeinflußten die Temperaturen erheblich. Nach Herbstfurche lag die Temperatur der Krume um 0,3 bis 0,6° C höher als nach der Frühjahrsfurche, Eggen hatte niedrigere, Walzen des Bodens höhere Bodentemperaturen zur Folge. Potter (1956) stellte fest, daß die *Frostperiode* in gepflügtem Lande kürzer war als unter Gras. Pflanzenbaulich wichtig ist das Verhalten der Krume zur Wärme vor allem in der Zeit des *Vegetationsbeginnes.* Böden, die im Frühjahr zeitig abtrocknen, erwärmen sich rasch, da ihre Wärmekapazität und Wärmeleitung verringert ist. Andererseits kann eine abgetrocknete lockere Oberschicht die *Spätfrostgefahr* verschärfen, sie speichert tagsüber zu wenig Wärme. Wenn dann klare Nächte auftreten, kühlt sie rasch aus, während gleichzeitig der Wärmestrom aus dem Inneren des Bodens verlangsamt ist. Dadurch wird die Wärmemenge, die die Bodenoberfläche in den Nacht- und

Frühstunden ausstrahlt, herabgesetzt. Daß eine oberflächliche Bodenlockerung mit nachfolgenden kalten Nächten junge Kulturen erheblich beeinträchtigen kann, ist hinlänglich bekannt. Auch die *topographische* Lage eines Feldstückes beeinflußt, wie häufig zu beobachten ist, die Spät- und Frühfrostgefahr. Tiefergelegenes Gelände ist im Frühjahr und Herbst stärkeren Temperaturamplituden ausgesetzt als Hänge und Kuppen.

Die *organogenen Böden* zeigen besonders *extreme Temperaturverhältnisse.* Baden und Eggelsmann (1958) untersuchten die Temperaturverhältnisse bei verschiedenen *Hochmoorkulturen.* Der große Torfanteil bei Moorböden bedingt eine geringe Wärmeleitfähigkeit. Ferner hat ihr hoher *Wassergehalt* eine hohe Wärmekapazität zur Folge. Diese bewirkt, daß sich die tieferen Schichten nur schlecht erwärmen. Die Temperaturextreme sind an und über der Bodenoberfläche ausgeprägter als bei Sandmischkulturen. In Strahlungsnächten erfolgt die Wärmeleitung nur langsam. *Nachtfröste* treten mit um so tieferen Temperaturen auf, je mehr die obersten cm des Moores abgetrocknet sind. *Sandmischkulturen* haben bei geringeren Humus- und Wassergehalten eine höhere Temperaturleitfähigkeit und eine geringere Wärmekapazität als Hochmoore. Die eingestrahlte Wärme wird rascher in die Tiefe geleitet, bei Abkühlung der oberen Schichten wird diesen rascher wieder Wärme zugeführt als bei Hochmooren. Im Mittel aus 4 Jahren betrug die Anzahl der Tage mit Nachtfrost über unbewachsenem Boden bei

	im April	Mai	Juni	Sept.	Okt.
Deutscher Hochmoorkultur	14,3	5,5	1,8	4,8	11,1
Handgekuhltem Hochmoor	13,8	5,0	0,8	2,5	10,5
Tiefgepflügtem Hochmoor	12,8	4,8	0,3	2,0	10,8
Tiefgepflügtem Heidesand	11,5	5,0	0,3	2,0	10,5

Die raschere Erwärmung der Sandmischkulturen wirkt sich vor allem im Frühjahr und Frühsommer günstig für das Pflanzenwachstum aus.

Die Verringerung des *Wassergehaltes* im Moorboden hat zur Folge, daß die jahreszeitlich bedingten Temperaturschwankungen rascher und tiefer in den Boden eindringen. Dies zeigt sich an der Eindringtiefe des Frostes und dem Auftauen des Bodens nach den Beobachtungen im Winter 1953/54.

	Bodenfrosttiefe am 4. 3. 1954, cm	vollständiges Auftauen später als D, Tage
A. Deutsche Hochmoorkultur	29	10
B. Handgekuhltes Hochmoor	51	7
C. Tiefgepflügtes Hochmoor	60	4
D. Tiefgepflügter Heidesand	70	–

Zu entsprechenden Ergebnissen kam Pessi (1953, 1957a, 1958) bei Untersuchungen an finnischen Mooren. Wichtig ist es ferner, bei diesen Böden die *Wasserverdunstung* und die damit verbundene Abkühlung zu verringern. Nach De Vries und De Wit (1954) genügt eine Sanddecke von 13 cm Dicke, um die thermischen Verhältnisse von Moorböden denen von vergleichbaren Mineralböden ähnlich zu gestalten.

Der *Wassergehalt* ist selbstverständlich nicht nur bei Moorböden ein entscheidender Faktor für die Temperaturverhältnisse, sondern auch bei Mineralböden. Die hohe Wärmekapazität des Wassers hat bei Böden, die viel Feuchtigkeit enthalten, eine geringere und vor allem langsamere Erwärmung im Frühjahr zur Folge. Dies geht u. a. aus den Untersuchungen Zöttls (1958) hervor. Auf den von ihm untersuchten Lößlehm—Waldstandorten lagen die Temperaturen im Frühjahr bei Wassersättigung um so tiefer, je länger die Sättigung des Bodens mit Wasser anhielt.

Die *Frosttiefe* und die *Wanderungsgeschwindigkeit des Frostes* im Erdboden hängen von den meteorologischen Bedingungen und dem Bodenzustand ab. Unter einer Schneedecke wird die Eindringtiefe des Frostes verringert und das Vordringen des Frostes in tiefere Schichten ganz erheblich verlangsamt. So stellte z. B. Mosienka (1957) unter den extremen Klimabedingungen der Kulundasteppe fest, daß auf Lößlehmböden der Frost auf schneearmen Flächen 240 cm tief in den Boden eindrang, unter einer Schneedecke dagegen nur 72 cm. *Schwere* Böden, die einen hohen Wassergehalt aufweisen, gefrieren weniger tief als leichte und locker gelagerte Böden. Dies geht bereits schon aus den oben angegebenen Beobachtungen von Baden und Eggelsmann (1958) hervor. Czeratzki (1956) weist ferner auf die Feststellung Dukers hin, wonach in wassergesättigten Böden die in der Zeiteinheit gefrierende Wassermenge bei gleichen Frostgraden ungefähr konstant ist. Die Frosteindringtiefe hängt deshalb u. a. auch von der in die Frostzone gehobenen Wassermenge ab, diese wiederum von der Gefriergeschwindigkeit und der Bodenart. Beim Gefrieren des Wassers wird Wärme frei, beim Tauen wird Wärme verbraucht. Der stetige Verlauf der Abkühlung bzw. Erwärmung des Bodens wird darum beim Gefrieren bzw. Auftauen vorübergehend durch einen *stationären* Temperaturzustand unterbrochen (Evans, Horrigan und Chinn 1958). Aniol (1952) hat eine Liste aller erreichbaren Daten für die *größten* beobachteten Frosttiefen in Deutschland zusammengestellt, die in Tab. 134 auszugsweise wiedergegeben sind. Diese Frosttiefen werden häufig nicht erreicht. Da aber der Frost für die Strukturbildung der Böden ein wichtiger Faktor ist, ist es von Interesse, zu wissen, daß gelegentlich Frosttiefen von 70 bis 120 cm vorkommen.

Tabelle 134. *Größte beobachtete Frosttiefen in Deutschland* (Aniol 1952)

Ort	Bodenart	größte Frosttiefe cm	Jahr
Wyk a. Föhr	Sand	92	1941
Stade	leichter Geestboden	78	1942
Oldenburg	Sand	125	1929
Herford	sand. Lehm	85	1929
Hildesheim	Lehm	71	1942
Köln-Butzweilerhof	Lehm	91	1942
Gießen Versuchsfeld	sandiger Lehm	100	1942
Oberlahnstein	Löß	85	1929
Trier—Stadt	mittelschwerer Devonschiefer	80	1929
Kreuznach	kiesiger Lehm	100	1942
Allmenfeld Kr. Groß-Gerau	Sand	79	1940

Abschließend ist noch der Einfluß der *Bodenbedeckung* — durch Pflanzen, Schnee, künstlich aufgebrachte Decken — auf die Bodentemperatur und ihre Änderungen zu besprechen. Dabei ist es nützlich, zunächst wiederum vom Energieumsatz an der *Bodenoberfläche* auszugehen. Baumgartner (1956) hat in seiner mehrfach zitierten Arbeit den Anteil des Wärmestroms im Boden am *Strahlungssaldo* bei verschiedenen Bodendecken berechnet. Für den Strahlungssaldo S gilt $S = S_0 - S_u$. Dabei ist S_0 die zur Erdoberfläche gelangende Strahlung, die sich zusammensetzt aus direkter Sonneneinstrahlung, diffuser

Himmelsstrahlung und langwelliger Gegenstrahlung der Atmosphäre. S_u setzt sich zusammen aus der Reflexion dieser Komponenten an der Erdoberfläche und deren langwelliger Eigenstrahlung. S_B ist der Strahlungssaldo an der Bodenoberfläche; er ist bei bewachsenem Boden kleiner als S, da ein Teil der Strahlung an der Oberfläche des Pflanzenbestandes umgesetzt wird.

Tabelle 135. *Anteil des Wärmestromes im Boden (%) am Strahlungssaldo bei verschiedenen Bodendecken*
(BAUMGARTNER 1956)

Bodendecken	S_B in % von S	Anteil des Bodens in %	
		an S	an S_B
nackter Boden	100	15—30	15—30
Weidewiese	30—80	15	20—35
Getreidefeld	10—30	10	25—35
langes Gras	5—10	5	30—40
Wald	5	2—3	30—45

Diese Tabelle macht auch die Beziehungen deutlich, die zwischen der Bodendecke und den Bodentemperaturen, die ja aus dem Wärmestrom im Boden und der Bodenbeschaffenheit resultieren, bestehen. Zunächst ist zu bemerken, daß der Anteil von S_B an S um so geringer ist, je dichter die Bodendecke ist. Dasselbe gilt für den Anteil des Wärmestromes im Boden an S. Der Nutzeffekt von S_B für den Boden nimmt jedoch, wie die dritte Spalte der Tabelle zeigt, mit dem Überschirmungsgrad zu. Erwärmung und Abkühlung des Bodens verlaufen unter einer Bodendecke *langsamer*, die Temperaturextreme werden gemildert. So beobachtete VAARTAJA (1955) an exponierten Stellen in der dünnen Oberflächenschicht eines Koniferenwaldes Maximaltemperaturen von 50 bis 70° C, während unter Beschattung die Maximumwerte bei 9 bis 13° C bleiben können. Beschattung durch Getreidepflanzen bewirkte nach Beobachtungen von PESSI (1957b) eine *Erniedrigung* der Bodentemperatur in 10 bis 20 cm Tiefe um 3 bis 4° C. KRISTENSEN (1957) bestimmte vom Juni 1955 bis August 1956 die Temperaturen 5 cm unter und 5 cm über der Bodenoberfläche und kam zu folgenden Ergebnissen:

	mittlere wöchentliche Temperaturschwankung (rel.):	
	in der Luft	im Boden
ungelockerter, nackter Boden	1,00	1,00
kurzgehaltenes Gras	1,04	0,87
10 cm Strohmulch	1,20	0,11

FUSS (1958) verwendete Rapsstroh als *Bodenbedeckungsmittel*. 1,5 kg/m² Rapsstroh bewirkten eine erhebliche Erniedrigung der Maximaltemperaturen gegenüber dem unbedeckten Teilstück. Die absolute Jahresschwankung wurde bei Bodenabdeckung in 5 cm Tiefe um 12,1° C, in 50 cm Tiefe um 2,4° C herabgesetzt. Die abgedeckten Teilstücke waren durchschnittlich *feuchter* und zeigten *geringere Schwankungen* des Wassergehaltes.

Umfangreiche Versuche über Bodenbedeckung mit Sägemehl, Torfmull, Kaolinstaub, Kohlenstaub, Schiefer, Aluminiumfolien und Glasfasermatten zu gärtnerischen Kulturen wurden von WEGER (1951) in Geisenheim ausgeführt. Auch bei diesen Untersuchungen zeigte sich ein erheblicher Einfluß der Bedeckung auf die Bodentemperaturen. Ferner werden durch die Bodenbedeckung der

Befeuchtungszustand des Bodens und die *Strahlungsverhältnisse* im Bestand verändert. Mit einer Abdeckung versehener Boden ist in der Regel feuchter als nicht abgedeckter. LANDSBERG und BLANC (1958) haben die Beeinflussung des Bodenklimas durch Bedeckung in einer Tabelle zusammengestellt, die mit einigen Veränderungen hier wiedergegeben wird:

Tabelle 136. *Einfluß einer Bodendecke auf das Bodenklima* (LANDSBERG und BLANC 1958)

Art der Bodenbedeckung	durch diese ändert sich	und beeinflußt die	Richtung der Beeinflussung
weißer Staub	Albedo	Bodentemperatur	tagsüber niedriger
schwarzer Staub	Albedo	Bodentemperatur	nachts höher
Strohdecke	Leitfähigkeit	Bodentemperatur	tagsüber niedriger nachts höher
Strohdecke	Wasserdampfbewegung	Verdunstung	verringert
Aluminiumfolie	Albedo	Bodentemperatur	tagsüber niedriger
Aluminiumfolie	Strahlung	kurzwellige Rückstrahlung	tagsüber höher
Aluminiumfolie	Wasserdampfbewegung	Verdunstung	verringert
Plastikfolie	Wärmeabsorption	Bodentemperatur	tagsüber und nachts höher
Plastikfolie	Wasserdampfbewegung	Verdunstung	verringert
helle Spaliermauer	Reflexion	kurzwellige Strahlung	tagsüber höher
Schutzhecke	Wind	Verdunstung	verringert

Literatur

ALBRECHT, F.: Meßgeräte des Wärmehaushaltes an der Erdoberfläche als Mittel der bioklimatischen Forschung. Meteor. Z. **54**, 471–475 (1937). — ANIOL, R.: Die größten Frosttiefen. Naturwiss. Rdsch. **5**, 493–495 (1952); Ref.: Soils a. Fertil. **17**, 26 (1954).

BADEN, W. und R. EGGELSMANN: Über das Bodenklima verschiedener Hochmoorkulturen und sein Einfluß auf den Pflanzenwuchs. Z. Acker- u. Pflanzenbau **106**, 127–152 (1958); Ref. LZ II, 1382 (1959). — BAIER, W.: Ergebnisse von Bodentemperaturmessungen. Ber. Dtsch. Wetterdienst US-Zone Nr. 38, 189–195 (1952). — BAUMGARTNER, A.: Das Eindringen des Lichtes in den Boden. Forstwiss. Cbl. **72**, 171–184 (1953). — Untersuchungen über den Wärme- und Wasserhaushalt eines jungen Waldes. Ber. Dtsch. Wetterdienst **28** (Bd. 5), (1956). — BAVER, L. D.: Soil Physics, 3. Aufl. New York: Wiley; London: Chapman & Hall 1956. — BIEDERMANN, R., W. A. ROTH und I. KOPPEL: Chemiker-Taschenbuch, III. Teil, S. 312–315. Berlin: Springer. 1937.

CHANG, J. H.: Global distribution of the annual range in soil temperature. Trans. Amer. Geophys. Un. **38**, 718–723 (1957); Ref.: Soils a. Fertil. **21**, 101 (1958). — COLLMANN, W.: Diagramme zum Strahlungsklima Europas. Ber. Dtsch. Wetterdienst Nr. 42, 1–8 (1958). — CZERATZKI, W.: Zur Wirkung des Frostes auf die Struktur des Bodens. Z. Pflanzenernähr., Düng., Bodenkde. **72**, 15–32 (1956).

DIMO, V. N.: Fundamental heat characteristics of some soils of Kutuluk terraces. Trudy Pochv. Inst. Dokuchaeva **37**, 312–328 (1952) (russ.); Ref.: Soils a. Fertil. **16**, 281 (1953). — DMITRIJEW, J. A.: Zur Frage einiger Faktoren, die die spezifische Wärmekapazität der festen Phase einiger Böden bestimmen. Nachr. Moskauer Univ. Ser. Biol., Bodenkde., Geol., Geogr. **13**, Nr. 4, 103–111 (1958) (russ.); Ref.: L. Z. II, 820–821 (1960). — DREIBELBIS, F. R.: A summary of data on soil and air temperatures at the North Appalachian experimental watershed, Coshoeton, Ohio. Proc. Soil Sci. Soc. Amer. **15**, 394–399 (1950); Ref.: Soils a. Fertil. **15**, 115 (1952). — DUKER

W. L.: Determination of soil temperatures from meteorological data. Iowa St. Coll. J. Sci. **31**, 389–390 (1957); Ref.: Soils a. Fertil. **20**, 266 (1957).

Evans, D. D., I. A. Horrigan and C. E. Chinn: Temperature—time curves as soil freezes related to moisture—stress curves. Proc. Soil Sci. Soc. Amer. **22**, 375–378 (1958); Ref.: Soils a. Fertil. **22**, 34 (1959).

Falkenberg, G.: Absorptionskonstanten einiger meteorologisch wichtiger Körper für infrarote Wellen. Meteor. Z. **45**, 334–337 (1928). — Feher, D., und M. Frank: Untersuchungen über die Lichtökologie der Bodenalgen. Arch. Mikrobiol. **7**, 1 (1936). — Flucker, B. J.: Soil temperatures. Soil Sci. **86**, 35–46 (1958); Ref.: Z. Pflanzenernähr., Düng., Bodenkde. **83**, 271 (1958); LZ. II, 486 (1959). — Frischmann, F.: Experimentelle Untersuchungen über das Eindringen der strahlenden Energie in den Boden. Bodenkde. u. Pflanzenernähr. **14**, 257–290 (1939). — Fuss, F.: Veränderungen des Bodenklimas durch Bodenbedeckung bei ausdauernden Kulturen. Angew. Meteor. **3**, 74–76 (1958); Ref.: LZ II, 2257–2258 (1958).

Geiger, R.: Das Klima der bodennahen Luftschicht. 3. Aufl. Braunschweig: Vieweg. 1950. — Grunow, I.: Beiträge zum Hangklima. Ber. Dtsch. Wetterdienst US-Zone **35**, 293–298 (1952).

Jenny, H.: Factors of soil formation, S. 141. New York and London: McGraw-Hill. 1941.

Kersten, M. S.: Thermal properties of soils. Highway Res. Bd. Spec. Rep. **2**, 161–166 (1952); Ref.: Soils a. Fertil. **16**, 109 (1953). — Kostjakow, A. N.: Grundlagen der Meliorationen, S. 147–148. Berlin: VEB Verlag Technik. 1959. — Kreuz, W.: Der Jahresgang der Temperatur in verschiedenen Böden unter gleichen Witterungsverhältnissen. Z. angew. Meteor. **60**, 65–76 (1943). — Kristensen, K. I.: Jordbehandlingens inflydelse på jordens og luftens temperatur. (Der Einfluß einer Bodenbehandlung auf die Temperatur von Boden und Luft.) Tidsskr. Landokon. **4**, 211–222 (1957) (dän.); Ref.: Soils a. Fertil. **22**, 34 (1959).

Landsberg, H. E., und M. L. Blanc: Interaction of soil and weather. Proc. Soil Sci. Soc. Amer. **22**, 491–495 (1958).

Mallik, A. K.: The control of the micro-climate for given practical purposes. Indian. J. Meteor. Geophys. **2**, 165–171 (1951); Ref.: Soils a. Fertil. **16**, 30 (1953). — Manegold, E.: Kapillarsysteme, 1. u. 2. Bd. Heidelberg: Straßenbau, Chemie und Technik, Verl.-Ges. 1955 und 1960. — Milojić, B.: Investigations on temperature changes in the arable layer of soil covered with different field crops. Zborn. Rad. poljopr. Fak. Univ. Beograd **6**, 55–114 (1958) (serb., engl.); Ref.: Soils a. Fertil. **22**, 439 (1959). — Mosienka, N. A.: Gefrieren und Auftauen des Bodens in der Kulunda-Steppe. Bodenkunde Nr. 1, 45–51 (1957) (russ.); Ref.: Z. Pflanzenernähr., Düng., Bodenkde. **79**, 185 (1957); LZ II, 34 (1958).

Pessi, Y.: Effect of mineral soil on soil temperature in the cultivated peat area. Suo. Nr. **6**, 67–70 (1953); (finn.) Ref.: Soilsa. Fertil. **17**, 252 (1954). — Über die Wärmeverhältnisse in Mineral- und Torfböden in Pelsonsuo in den Jahren 1955–1956. Valtion Maatalouskoetoiminnan (State agric. Res. Publ. Finland), Julkaisuja (Rep.) Nr. 159, 1–30 (1957 a) (engl.); Ref.: LZ. II, 1608 (1959); Soils a. Fertil. **21**, 297–298 (1958). — On the influence of plant cover upon soil temperature. Suom. Maataloust. Sen. Aikakausk. **29**, 92–95 (1957 b); Ref.: Soils a. Fertil. **20**, 266 (1957). — On the influence of bog draining upon thermal conditions in the soil and the air near the ground. Acta Agric. Scand. **8**, 359–374 (1958); Ref.: Z. Pflanzenernähr., Düng., Bodenkde. **85**, 177 (1959); Soils a. Fertil. **22**, 107 (1959). — Potter, L. D.: Yearly soil temperatures in eastern North-Dakota. Ecology **37**, 62–70 (1956); Ref.: Soils a. Fertil. **19**, 256 (1956).

Schubert, J.: Der Wärmeaustausch im festen Erdboden in Gewässern und in der Atmosphäre. Berlin: Springer. 1904. — Die Sonnenstrahlung im mittleren Norddeutschland nach den Messungen in Potsdam. Meteor. Z. **45**, 1—16 (1928). — Shaw, R. H.: The measurement of soil temperature. Iowa State Coll. J. Sci. 30, 255–261 (1955); Ref.: LZ. II, 428 (1957); Soils a. Fertil. **19**, 148 (1956). — Slater, C. H. W., and I. P. Ruxton: The effect of soil compaction on incidence of frost. E. Malling Res. Sta. Rep. 88–91 (1954). — Smith, W. O. and H. G. Byers: The thermal conductivity of dry soils of certain of the great soil groups. Proc. Soil Sci. Soc. Amer. **3**, 13–19 (1938). — Smith, W. O.: Thermal conductivities in moist soils. Proc. Soil Sci. Soc. Amer. **4**, 32–40 (1939). — Süring, R.: Leitfaden der Meteorologie. Leipzig: Tauchnitz. 1927. — Suzuki, S., and E. Maruyama: Increase of temperature of soil coates with carbon black and exposed to the sun. Met. Agency Tokyo Mem. Industr. Met. **20**, 2, 39–43 (1957) (jap.); Ref.: Soils a. Fertil. **22**, 107 (1959).

Troedsson, T.: Soil temperatures in soils with stony surface layers. Skr. K. Skogshögsk. Stockh. Nr. 25, 25 (1956) (Schwed.); Ref.: Soils a. Fertil. **20**, 208 (1957).

Vaartaja, O.: Temperature and evaporation at and near ground level on certain forest sites. Canad. J. Bot. 32, 760–783 (1984); Ref.: Soils a. Fertil. 18, 131 (1955). — DE VRIES, D. A.: A nonstationary method for determining thermal conductivity of soils in situ. Soil Sci. 73, 83–90 (1952). — DE VRIES, D. A., und C. T. DE WIT: Die thermischen Eigenschaften der Moorböden und die Beeinflussung der Nachtfrostgefahr dieser Böden durch eine Sanddecke. Meteor. Rdsch. 7, 3/4, 41–45 (1954); Ref.: Soils a. Fertil. 18, 131 (1955).

WEGER, N.: Beiträge zur Frage der Beeinflussung des Bestandsklimas, des Bodenklimas und der Pflanzenentwicklung durch Spaliermauern und Bodenbedeckung. Ber. Dtsch. Wetterdienst US-Zone, Nr. 28 (1951). — WOODSIDE, W., und C. M. A. DE BRUYN: Heat transfer in a moist clay. Soil Sci. 87, 166–173 (1959); Ref.: LZ. II, 2501 (1959); Soils a. Fertil. 22, 273 (1959). — WOODSIDE, W., and I. B. CLIFFE: Heat and moisture transfer in closed systems of two granular materials. Soil Sci. 87, 75–82 (1959); Ref.: Z. Pflanzenernähr., Düng., Bodenkde. 86, 190 (1959).

YAKUWA, R.: On the soil temperature of peat and other mineral soils in Hokkaido (Japan). Reclam. Melior 3, 38–39 (1958); Ref.: Soils a. Fertil. 21, 369 (1958).

ZÖTTL, H.: Die Abhängigkeit der Bodentemperatur vom Wasserhaushalt wechselfeuchter Standorte. Forstwissensch. Cbl. 77, 329–335 (1958); Ref.: LZ. II, 2270 (1959).

G. Air and Precipitation as Sources of Nutrients

By

E. Eriksson

1. Introduction

The atmosphere contains a wide variety of elements and in greatly varying concentrations. The average composition with respect to all known elements and compounds is shown in Table 137. Concentration of gases are expressed in atmospheres, those of particulate matter in microgram per m³ ($\mu g \cdot m^{-3}$). The table is not complete as there is a number of organic compounds present the composition of which are by and large unknown. In precipitation the concentration of organic matter is about the same as of inorganic matter. Organic compounds have been found in precipitation (NEUMANN *et al.* 1959) especially substances behaving like hunus. Also fats or fatty acids have been shown to occur.

In the table the variability in concentration is indicated. The major gases show no detectable variation like the gases which are inert. Those elements or compounds of gaseous nature taking part in the biological circulation show as a rule a great variability and the same is true for particulate matter whether derived from the sea or not.

Elements or compounds which can be regarded as essential for plants are apart from carbon dioxide and molecular nitrogen, nitrogen compounds, sulphur compounds, halogens, calcium, potassium, sodium, magnesium, boron and iodine which all may be referred to as atmospheric micronutrients. There may be others present like copper, manganese, kobolt, zink, etc. but they are not studied yet.

The compounds listed above are all necessary ingredients in plants which may take them up either from the soil by roots or directly from the atmosphere by the green parts of plants. In what way it happens is immaterial in the present context; if they are brought down to the surface of the earth they are at plants' disposal. Particulate matter may fall out on leaves or be swept out of the air by leaves. Once deposited there is but little chance that they are released into the air again expecially if they are hygroscopic. They are either assimilated by the leaves or they are washed down by rain and are taken up by plant roots.

There are thus two processes to consider. One is the transport of nutrients from the atmosphere by falling precipitation, the other is the so called turbulent transport of gaseous or particulate matter from the air to the vegetation or soil. At our present knowledge the second process seems to be at least as important as the first one, and it will be referred to as *dry deposition* as contrasted to *wet deposition* by falling precipitation.

The *wet deposition* of nutrients is, of course, assessed by analysing precipitation and is the part of atmospheric chemistry studied solely until recently.

Table 137. *The chemical composition of the troposphere (1 atm. pressure)*

Element	Compound	Partial pressure in atmospheres	Concentration in microgram per m³	Remarks
Nitrogen	N_2	0.78101		constant
Oxygen	O_2	0.20946		constant
Water	H_2O	~ 0.02		very variable
Argon	Ar	0.00917		constant
Carbon dioxide	CO_2	3.3×10^{-4}		somew. variable
Neon	Ne	1.82×10^{-5}		constant
Helium	He	5.24×10^{-6}		constant
Methane	CH_4	1.5×10^{-6}		constant
Krypton	Kr	1.14×10^{-6}		constant
Nitrous oxide	N_2O	5×10^{-7}		constant
Hydrogen	H_2	5×10^{-7}		constant
Ozon	O_3	$\sim 5 \times 10^{-7}$		very variable
Carbon monoxide	CO	1×10^{-7}		variable
Xenon	Xe	8.6×10^{-8}		constant
Sulphur dioxide	SO_2	1×10^{-8}		somew. variable
Ammonia	H_3N	$\sim 5 \times 10^{-9}$		variable
Hydrochloric acid	HCl	$\sim 10^{-8}$		variable
Nitrogen dioxide	NO_2	$\sim 10^{-9}$		variable
Iodine	J_2	$\sim 2 \times 10^{-13}$		variable
Chloride	Cl		~ 1	very variable
Sulphate-sulphur	SO_4-S		~ 1	very variable
Ammonia nitrogen	H_3N-N		~ 1	very variable
Nitrate nitrogen	NO_3-N		~ 1	very variable
Sodium	Na^+		~ 1	very variable
Calcium	Ca^{2+}		~ 1	very variable
Magnesium	Mg^{2+}		~ 0.5	very variable
Potassium	K^+		~ 0.5	very variable

There are no difficulties other than purely technical ones to carry out such investigation and the rate of deposition per unit area and time is simple to compute.

As to *dry deposition* this is much more difficult to assess. Application of turbulent diffusion theories, wellknown in meteorology, can be done but requires determination of air concentrations at different heights above the ground together with wind velocity measurements at various heights. Even then calculations involve assumptions which are not entirely satisfactory except under certain conditions.

There is, however, another method which seems to be promising. The turbulent mechanism transporting chemical compounds vertically in the atmosphere is the same for all gaseous compounds and, at least for all practical purposes, also the same for gases and particulate matter. The intensity of this transport mechanism can be expressed by a transport velocity which tells at what rate air from one level is brought to another level and vice versa. Chosing the site of one of these levels close to ground or vegetation and the other at say 2 meters height, this transport velocity can be determined experimentally using radioactive tracers and correlated to other meteorological conditions, e.g. wind velocity and air stability. The rate at which a compound is transported to or from the ground is them simply this transport velocity times, the difference in concentration of that compound at the higher level and at the ground level.

The dry deposition will be discussed in some detail later.

The total rate of deposition of plant nutrients from the atmosphere can thus be estimated but will in general vary geographically. The sources of these nutrients are sometimes limited to certain regions. The rate at which they are emitted to the atmosphere may also vary inside the source region. It is therefore clear that the rate at which they are deposited will depend upon prevailing atmospheric motion relative to the source region. And atmospheric motion is not a steady state but rather irregular, causing considerable lateral dispersion from the source region. The rate of deposition will thus depend on factors like geographical position, air circulation and all factors that are determining the vertical transport velocity. A good description of the rate of deposition of nutrients therefore requires regional investigations. The rate of deposition thus becomes a climatic element which may show as large regional variation as any other climatic element.

But air circulation and other meteorological conditions change from year to year causing variations from year to year in the climate. The *chemical climate*, using this expression for the atmospheric environment of plants relevant to its nutrient supply, also changes from year to year in a fashion similar to the *physical climate*. Observations on the rate of deposition of nutrients therefore require several years in order to get a good average picture.

These variations from year to year may partly be of an irregular, random nature but in part secular i.e. slow and systematic. Investigations from one series of years can therefore not be applied immediately to the same regions several decades later. Normally the secular changes are so slow compared to the variation from year to year that they may be neglected in a short time series. There are, however, areas especially in Western Europe where industrial activity and allied human habits have changed so rapidly that these changes may be noticable in atmospheric chemistry also within a shorter period.

Though there are numerous investigations in the past on the chemical composition of precipitation they are not very useful presently as they were carried out in different periods and sampling and analyses were made with different techniques. More recently continental networks of air and precipitation sampling have grown up and their data will be used practically entirely in the following. Most results will be presented as charts with yearly deposition data and so called isolines, i.e. lines connecting places with presumably the same rate of deposition. The unit used for rate of deposition will always be kg · ha^{-1} · year^{-1}.

2. Dry Deposition

As most of the research in this field has been carried out very recently, published in various journals, a review of this research is motivated. In 1932 EGNER discussed the vertical transports of ammonia in connection with ammonia losses from manure and pointed out that the amount of air that came in contact with the surface could be estimated from the relative humidity of air and rate of evaporation. He arrived at a figure of the order of 1000 m · day^{-1} for this vertical exchange corresponding to a vertical transport velocity around 1 cm · sec^{-1}. The usefulness of this idea was not recognized properly until very recently. In connection with studies of the dispersal of radio iodine from atomic power plants CHAMBERLAIN and CHADWICH (1953) made experiments to assess the absorption of radioiodine by grass. They found that the rate of absorption could be expressed as a velocity multiplied with the concentration of radioiodine

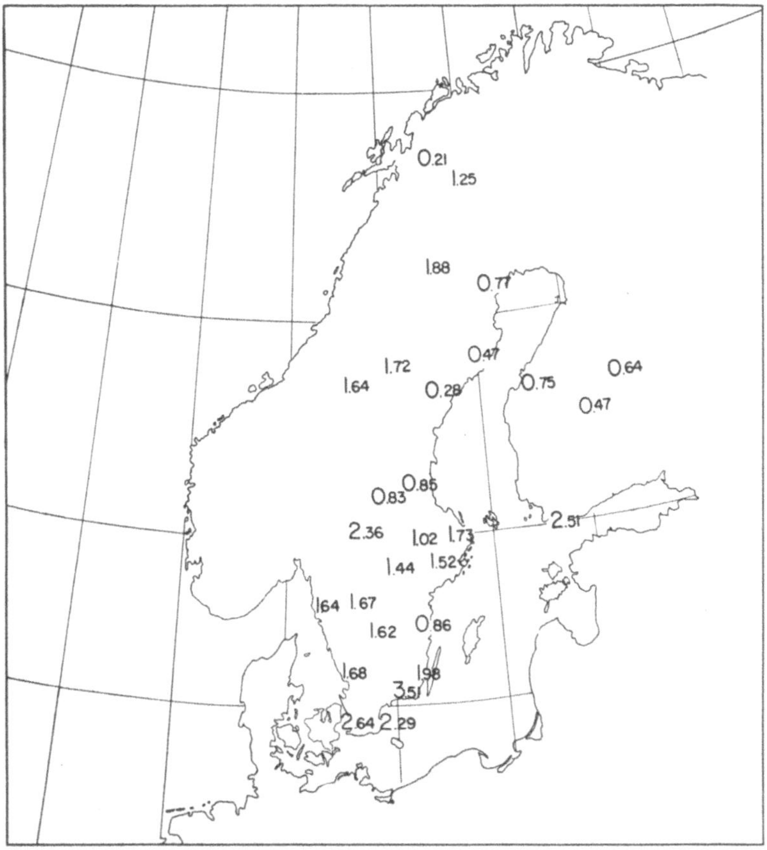

Fig. 166. Deposition velocities for sea salt particles in Scandinavia computed from yearly run-off of chloride in river water, yearly precipitation of chloride and average chloride concentration in the air. Values in cm · sec^{-1} (from ERIKSSON 1959)

in air and arrived at figures around 2 cm · sec^{-1} which again should be the rate at which air from normal observation height is brought in contact with ground, thus a vertical transport velocity.

JOHANSSON (1959) recently studied the absorption by soils and plants of sulphur dioxide in pot experiments and from his data ERIKSSON (1959) computed vertical transport velocities up to 3 cm · sec^{-1} or more. JOHANSSON also found that this vertical transport of sulphur dioxide supplied plants with many times more sulphur than they got from sulphur in precipitation.

ERIKSSON (1959) also concluded from JOHANSSON's (1959) experiments that most of the sulphur supplied to the plant from the atmosphere was released from the soil into the atmosphere, and that even the soils themselves both take up sulphur as sulphur dioxide from the atmosphere and release it again, presumably as hydrogen sulphide. This exchange with the atmosphere was recently verified by Mr. O. GUNNARSSON[1] at the Royal Agricultural College in Sweden who using radioactive sulphur found a rather rapid turnover and exchange with the atmosphere corresponding to a vertical transport velocity in the atmosphere of the order of magnitude mentioned earlier.

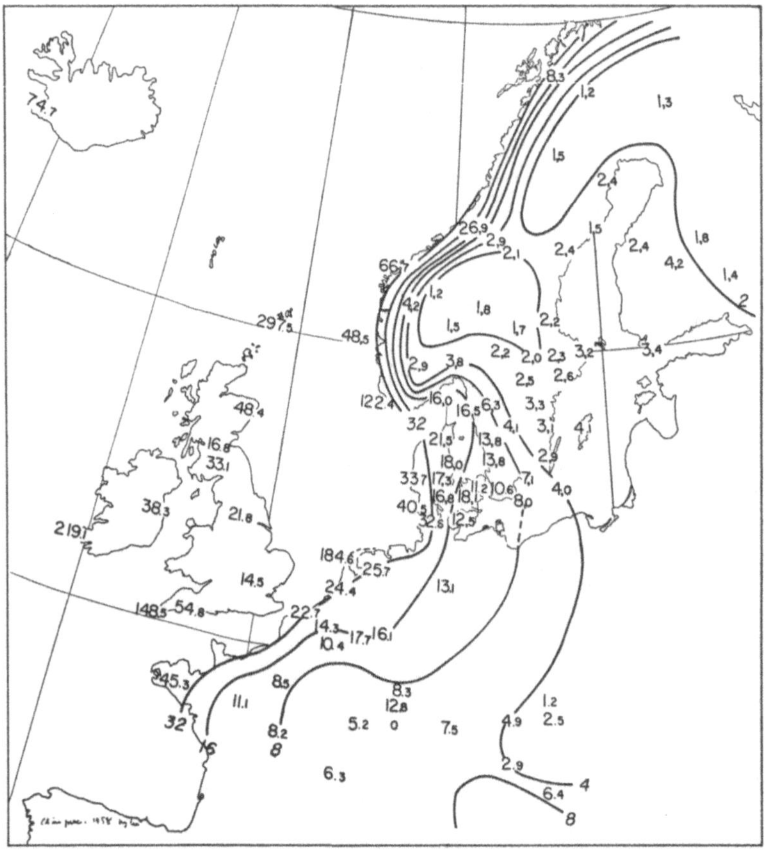

Fig. 167. Deposition by precipitation of chloride in Western Europe during 1958. Values in kg · ha^{-1}

It looks from this that in the case of sulphur the rate of dry deposition of sulphur can be estimated by using a vertical transport velocity of 2 cm · sec^{-1}. The rate is thus this velocity multiplicated with the concentration in the atmosphere.

Considering particulate matter in the atmosphere which may contain an appreciable part of the ammonia and nitrat in the atmosphere its rate of deposition can be estimated in a similar way. ERIKSSON (1959) found good evidence that chloride in particles was deposited at a rate corresponding to a transport velocity between 1 and 2 cm · sec^{-1}. A map over Scandinavia showing the distribution of transport velocities computed from chloride in river water and in precipitation is shown in Fig. 166. It is seen that they increase southward and

[1] I am indebted to Mr. GUNNARSSON for this information.

decrease in general eastwards. The value around 2—3 cm · sec^{-1} in the western part of Europe is likely. However, ERIKSSON (1959) found also that ratios between precipitated chloride and chloride in run-off were fairly constant, amounting to about 0.3 and concluded that with respect to particulate matter the total deposition is about three times the wet deposition. RIEHM and QUELLMALZ (1959) have recently discussed the total deposition of nutrients in various places

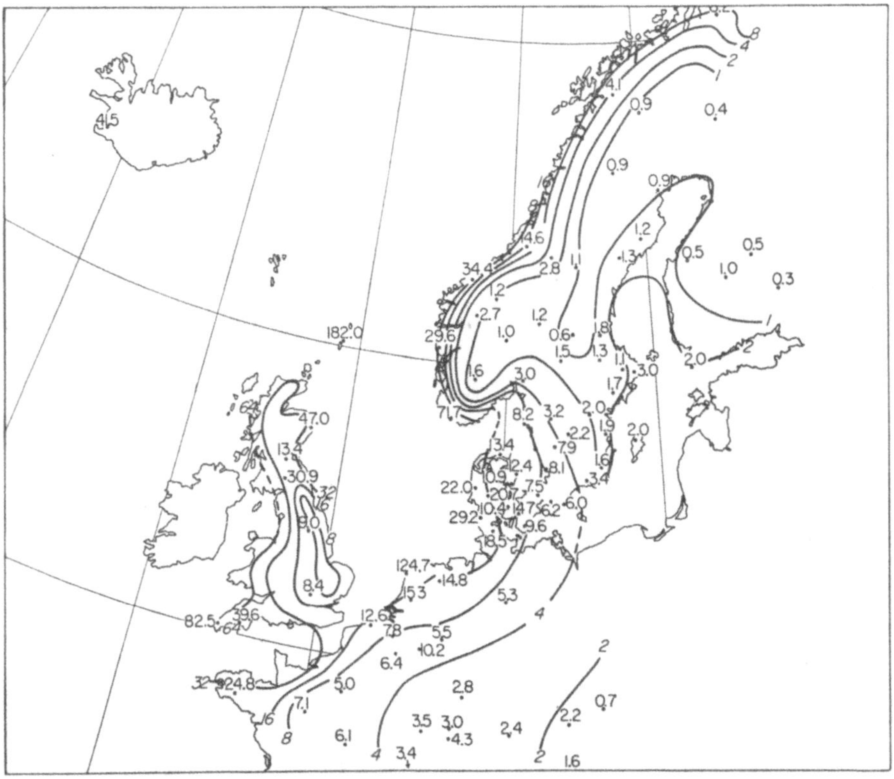

Fig. 168. Deposition by precipitation of sodium in Western Europe during 1958. Values in kg · ha^{-1}

in Germany and they used a vertical transport velocity of about 0.3 cm · sec^{-1} (100 km · year^{-1}) for all nutrients. This is most probably too low for sulphur for reasons discussed above but may be more proper for particulate matter containing nitrate, calcium, magnesium, sodium, potassium and chloride. Thus in estimates done later the total deposition of these will be estimated by multiplying the wet deposition by three and for sulphur the dry deposition will be estimated by using air concentrations and a transport velocity of 2 cm · sec^{-1} for the summer months April to September. Consequently, only air concentrations for those months have been used for computing the dry deposition. A similar procedure will be applied to ammonia which may largely be present in gaseous forms in the atmosphere but one must be somewhat cautious in applying these data.

With the data available presently one can get a rather good picture of the total deposition of various nutrients over western Europe and the United States. The network in Western Europe was initiated in 1954 when a previous Swedish network was reorganized (EGNER and ERIKSSON (1955) and enlarged to cover the whole of Scandinavia. Additions to this network took place later so that for 1958 a fairly large part of Europe is represented. The 1958 data will be used here. All primary data are published previously on a quarterly basis in *Tellus*.

An impressive network was organized by Junge and coworkers (1956, 1958) in the United States and was kept operating for a year. Data from this network were given as concentrations. From this and rainfall data for the same period

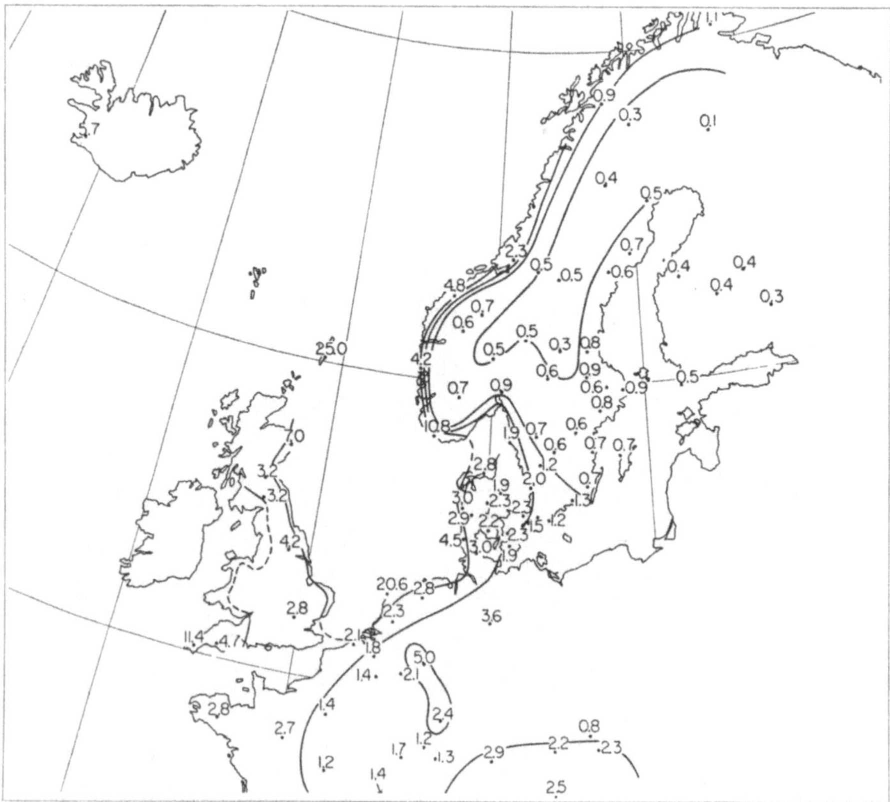

Fig. 169. Deposition by precipitation of magnesium in Western Europe during 1958. Values in kg · ha^{-1}

precipitated amounts have been computed. As there are no data on air concentrations it is not possible to make estimates of the total deposition of sulphur and ammonia nitrogen as in the Western European area.

From the southern hemisphere similar networks have been organized in Australia and data published by Hutton and Leslie (1958) and by Hingston (1958) will be discussed and compared to European data.

In Western Europe some determinations of iodine and boron in precipitation have also been made but are not published quarterly. The results are presently being worked up and published.

3. Yearly Deposition of the Sea Salt Components, Cl, Na, K, Ca, and Mg

In Fig. 167, 168, 169, 170, and 171 the wet deposition of Cl, Na, Mg, K, and Ca during 1958 in Western Europe are shown. Multiplying these figures by three should give a rough estimate of the total yearly deposition of these elements. It is seen that deposition rates of Cl and Na vary geographically in a similar fashion and that their source by and large is the Atlantic Ocean. They decrease rather rapidly inland in places where mountains border the sea but can penetrate rather far in places where the country is flat. Inland the deposition rate seems to attain a practically constant value. This is especially noticable with Mg. As to K and Ca they are much more evenly distributed and Ca shows in addition rather irregular features which indicates local sources.

Table 138. *Yearly deposition by precipitation of chloride, sodium, magnesium, potassium and calcium in various places in Australia*
From HUTTON and LESLIE 1958 and HINGSTON 1958. Values in kg · ha^{-1}

Place	Year	Distance from ocean in km.	Wet deposition kg · ha^{-1} of				
			Cl	Na	Mg	K	Ca
Perth, W. Australia	1950—1956	6	100	55	9	3	16
Coolup, W. Australia	1953	24	76	44	6	2	6
Dwellingup, W. Australia	1951—1956	40	110	83	9	6	7
West Pingaly, W. Australia	1953	105	26	17	2	1	3
Kojunup, W. Australia	1950—1956	153	21	13	2	1	2
Cape Bridgwater, Victoria	1955	1	175	100	12	4	12
Heywood, Victoria	1955	15	73	45	7	2	4
Cavendish, Victoria	1955	100	22	16	2	1	3
Horsham, Victoria	1955	190	6	6	2	1	3
Hopetown, Victoria	1955	260	7	6	2	.5	7
Walpeup, Victoria	1955	260	5	4	1	.5	3
Merbein, Victoria	1955	320	6	6	1	1	12
Warrnambool, Victoria	1955	3	94	60	8	3	7
Warragul, Victoria	1955	65	32	22	3	1	6
Cressy, Victoria	1955	65	20	15	2	1	3
Derrinallum, Victoria	1955	100	16	11	2	1	3
Parwan, Victoria	1955	80	11	9	2	1	2
Beaufort, Victoria	1955	160	8	7	1	1	2
Kynetown, Victoria	1955	130	7	7	1	.5	2
Seymoor, Victoria	1955	160	6	6	1	.5	1
Elmore, Victoria	1955	210	7	6	1	1	2
Charlton, Victoria	1955	310	4	4	1	.5	2
Sale, Victoria	1955	30	16	10	2	1	2
Swifts Creek, Victoria	1955	80	6	6	1	.5	2
Omeo, Victoria	1955	100	2	3	1	.5	2
Mitta Mitta, Victoria	1955	160	4	3	1	1	2
Tallangatta, Victoria	1955	190	4	7	1	1	2
Rookie, Victoria	1955	260	4	3	1	1	1
Kew, Victoria	1955	65	22	17	1	2	8

In the United States the distributions of Cl, Na, K, and Ca are shown in Fig. 172, 173, 174, and 175 constructed from JUNGE's data. Also here high coastal values are evident for Cl and Na. The aridity also enters as seen in the western part where rather low figures for the yearly wet deposition are found. As to K and Ca they are more regular with an interesting maximum in the agricultural region in the middlewest. The total deposition is probably about three times as large even here.

Table 139. *Wet deposition of chloride, sodium, magnesium, potassium, calcium, sulphate-sulphur, ammonia and nitrate-nitrogen and estimated dry deposition of sulphur dioxide—sulphur and ammonia—nitrogen at four places in Africa during 1958*

Place	Wet deposition in kg·ha⁻¹								Dry dep. in kg·ha⁻¹	
	S	Cl	NO_3-N	H_3N-N	Na	K	Mg	Ca	H_3N-N	SO_2-S
Yangambi, B. Congo[1]	7.0	4.0	4.6	1.5	2.1	2.1	1.2	3.7	19	23
Mulunga, B. Congo[1]	4.5	4.0	4.2	5.3	1.6	2.6	1.1	3.4	19	7
Binza, B. Congo[1] ..	2.8	2.2	2.9	.7	1.3	1.4	.9	4.6	23	14
Pretoria, S. A. Union[2] ..	10.0	3.9	1.6	.8	2.7	1.8	3.9	14.4	13	18

[1] Dry deposition computed from average air concentrations for the whole year and 2 cm · sec⁻¹ deposition velocity.

[2] Dry deposition computed from the six summer months' average air concentration and 2 cm · sec⁻¹ deposition velocity.

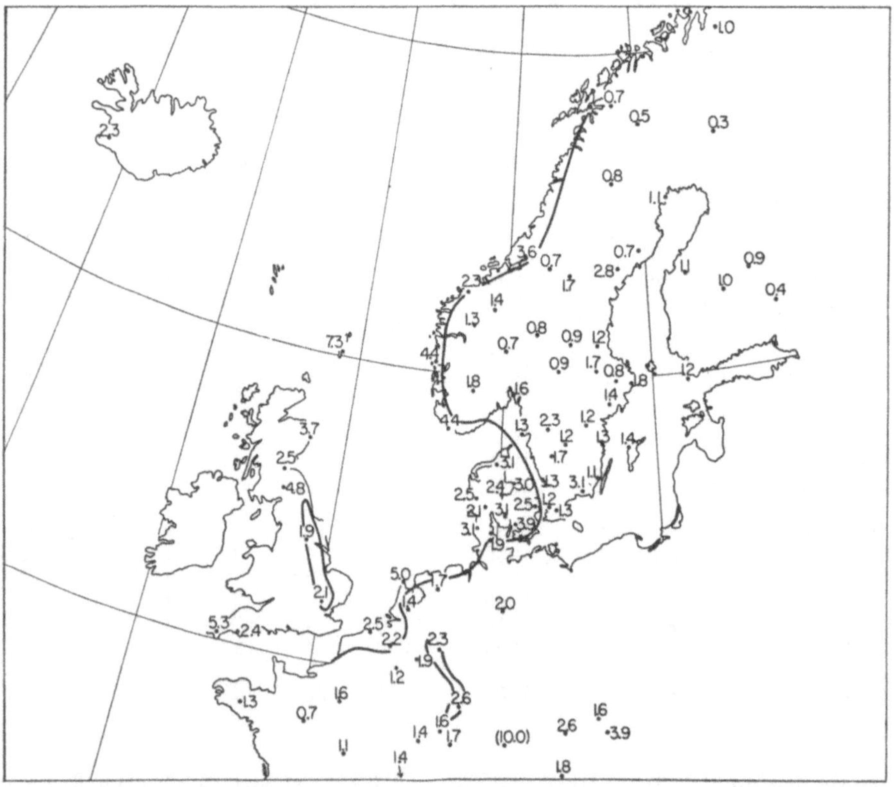

Fig. 170. Deposition by precipitation of potassium in Western Europe during 1958. Values in kg · ha⁻¹

As to Australia Fig. 176 gives the yearly wet deposition of Cl in the southern and western parts. As to the other elements Table 138 gives some data quoted from HUTTON and LESLIE (1958) and HINGSTON (1958) for various places in the same area as represented in Fig. 176.

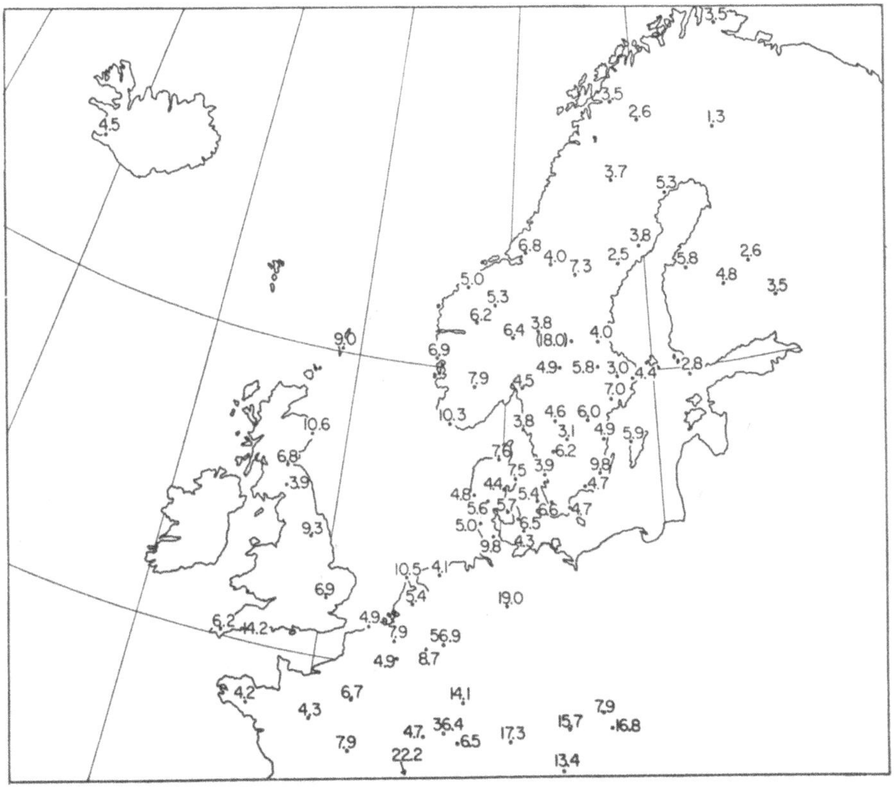

Fig. 171. Deposition by precipitation of calcium in Western Europe during 1958. Values in kg · ha^{-1}

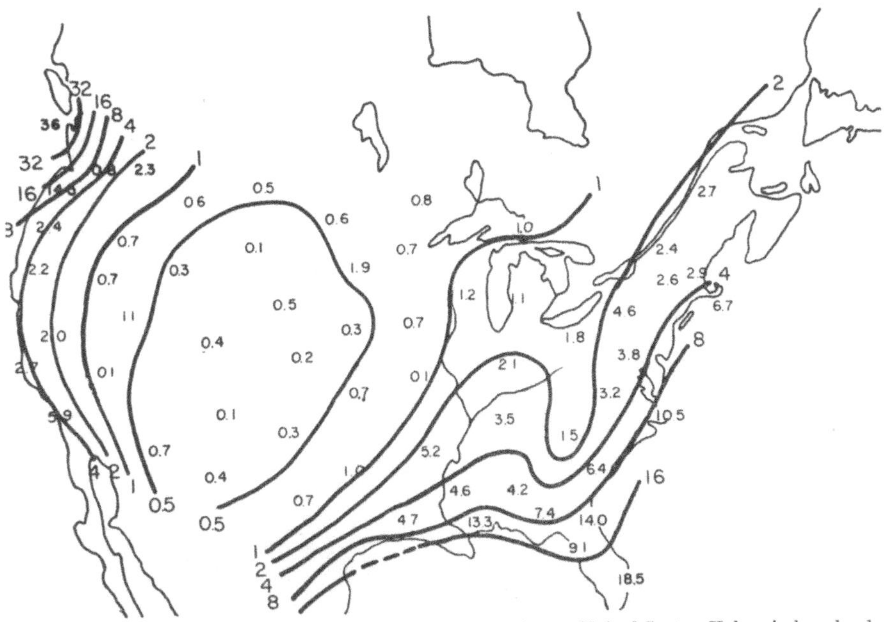

Fig. 172. Yearly deposition by precipitation of chloride in the United States. Values in kg · ha^{-1} computed from data by JUNGE and WERBY 1958

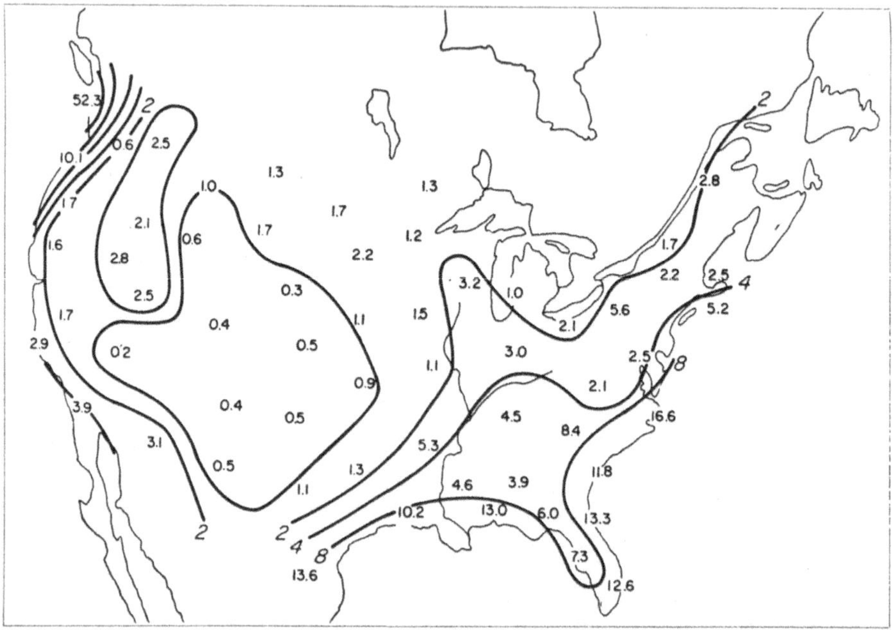

Fig. 173. Yearly deposition by precipitation of sodium in the United States. Values in kg · ha^{-1} computed from data by JUNGE and WERBY 1958

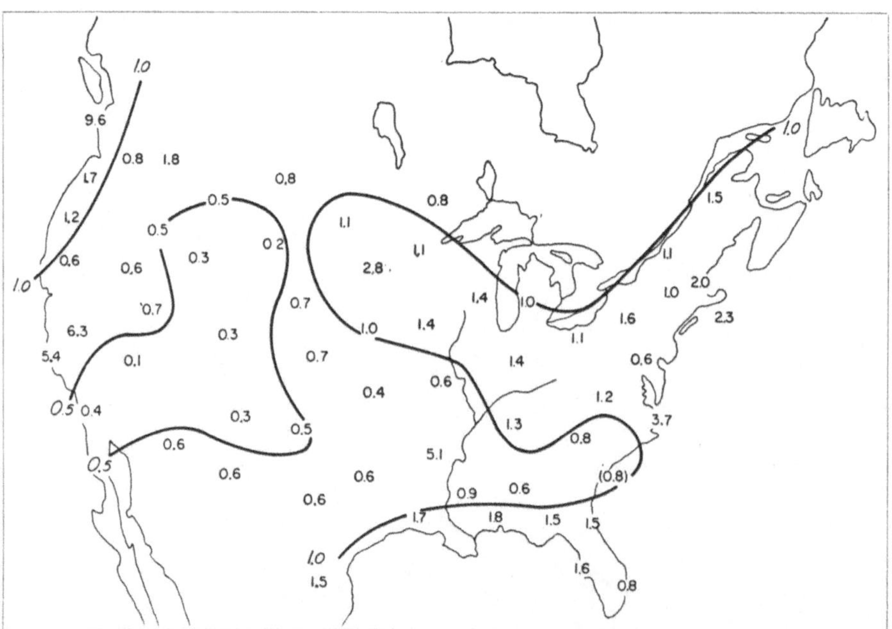

Fig. 174. Yearly deposition by precipitation of potassium in the United States. Values in kg · ha^{-1} computed from data by JUNGE and WERBY 1958

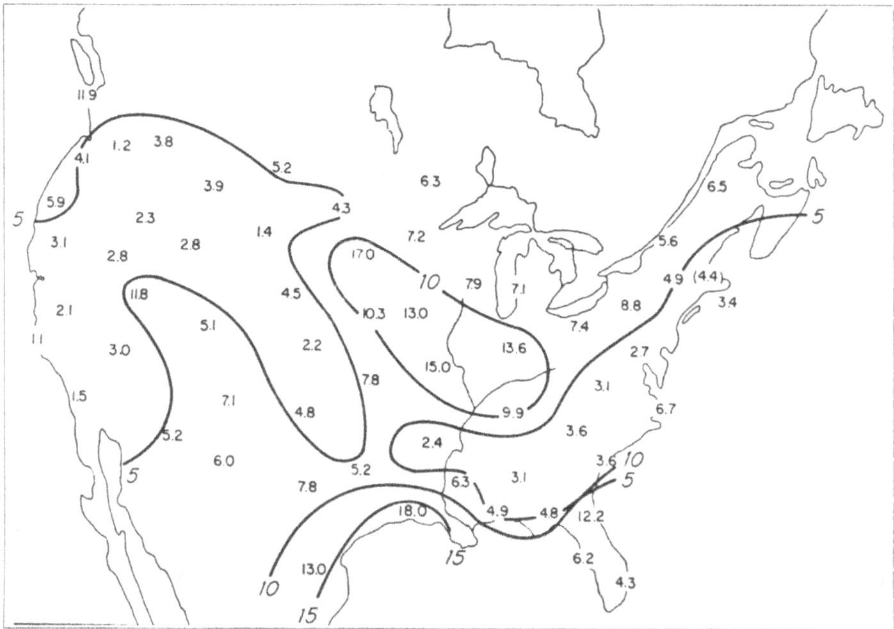

Fig. 175. Yearly deposition by precipitation of calcium in the United States. Values in kg · ha^{-1} computed from data by JUNGE and WERBY 1958

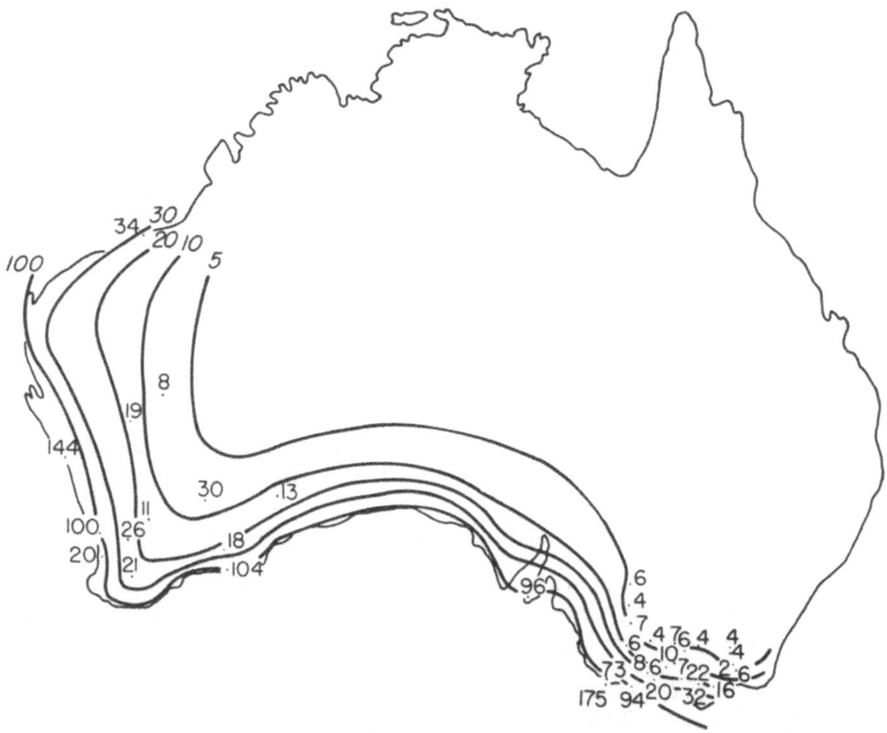

Fig. 176. Yearly deposition by precipitation of chloride in parts of Australia (from ERIKSSON 1959)

As to other parts of the world measurements on the composition of precipitation has been made recently at some places in Africa[1] and average data are shown in Table 139.

In Europe additional investigations were carried out in Soviet during the International Geophysical Year but these data are not yet available. Recently

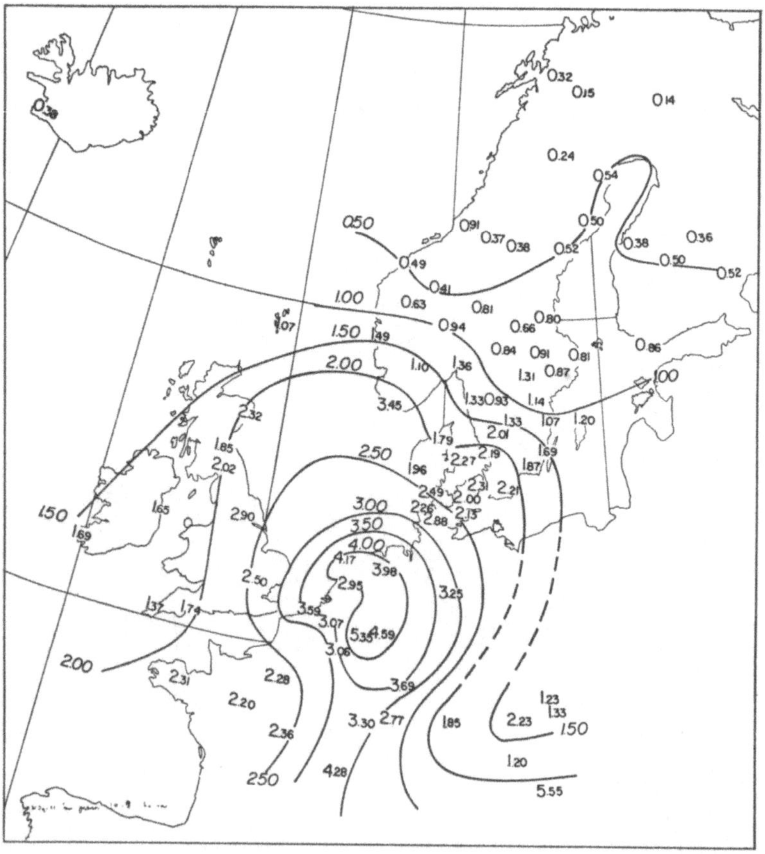

Fig. 177. Deposition by precipitation of nitrate-nitrogen in Western Europe during 1958. Values in kg · ha^{-1}

a network of precipitation sampling stations have been organized in Czechoslovakia (Macku *et al.* 1959) but at present their data do not cover a long enough period to make them strictly comparable to the 1958 data presented here.

4. Yearly Deposition of Nitrogen Compounds

As nitrate occurs in the atmosphere as particulate matter one can anticipate it to be deposited in the same way as sea salts. The total deposition of nitrate nitrogen is therefore probably three times the wet deposition.

[1] For the Belgian Congo stations the samples were analysed by Dr. J. Bouquiaux, Institute de Chemie et Physique, Brussels, and reported to WMO Data Centre for the IGY. The Pretoria samples were analysed in Pretoria. I am indebted to Dr. J. F. Nagel of the Weather Bureau, Pretoria, for these data.

For Western Europe the wet deposition in 1958 is shown in Fig. 177. A maximum in the deposition is found over the most populous area in Europe which indicates that nitrate nitrogen is partly anthropogenic. It is known to be formed in various combustion processes. The rapid decrease northward is also noticeable.

In Fig. 178 the same is shown for the United States computed from JUNGE's data. Again a maximum is seen in the most populous region in the north east,

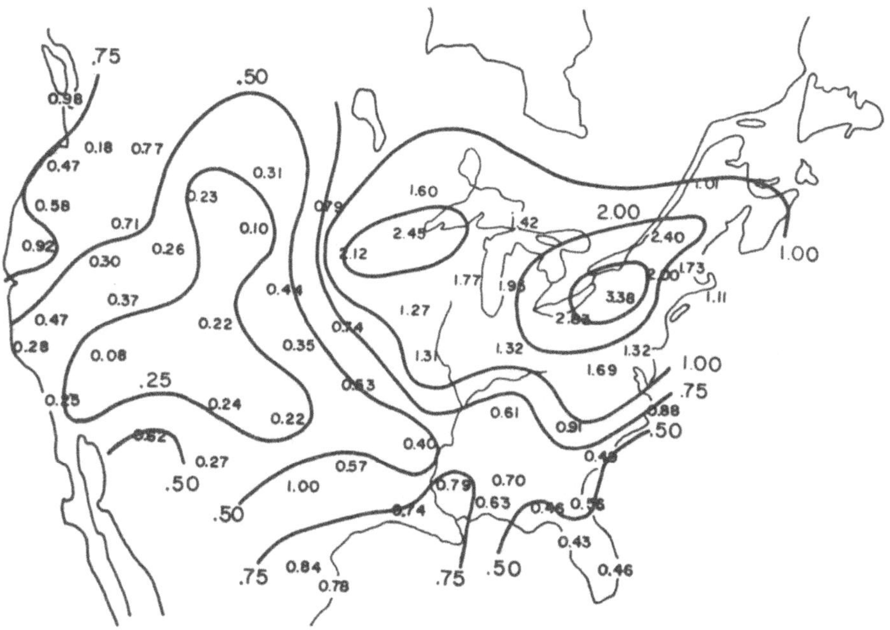

Fig. 178. Yearly deposition by precipitation of nitrate-nitrogen in the United States. Values in kg · ha^{-1} computed from data by JUNGE and GUSTAVSON 1956

other values being rather uniform except in the dry west. On the whole the U.S. data are lower than the European data except in densily populated north east.

As to Africa the four places listed in Table 139 indicate deposited amounts of about the same magnitude as in other places.

Earlier data on nitrate nitrogen in precipitation have been reviewed by ERIKSSON (1952).

Ammonia, being present also in gaseous form in the atmosphere may have a greater chance of dry deposition, i.e. uptake by plant leaves, than nitrate nitrogen. It may be interesting to compare the wet deposition with the estimated dry deposition computed in the way mentioned above. The wet deposition in 1958 for Western Europe is shown in Fig. 179. The values vary as seen between 0.2 and 5.8 kg, the highest being found near industrial centres in England. But it is also remarkable that intensely cultivated extensive agricultural areas like in Denmark also show comparative high values, no doubt being due to release from soil of ammonia. The estimated dry deposition is shown in Fig. 180 and is as seen much higher than the wet deposition. This means that provided plants assimilate ammonia from the air as readily as they assimilate sulphur dioxide, then a substantial contribution by direct absorption takes place in nature. It must be pointed out that this point has not been proved yet and experimental work designed to solve this question is certainly desirable. From Fig. 180 it is seen that the estimated dry deposition is as high as 20 kg · ha^{-1} in some areas. The high Finnish values, however, should not be regarded as representative as such high air concentrations have sofar appeared only in 1958.

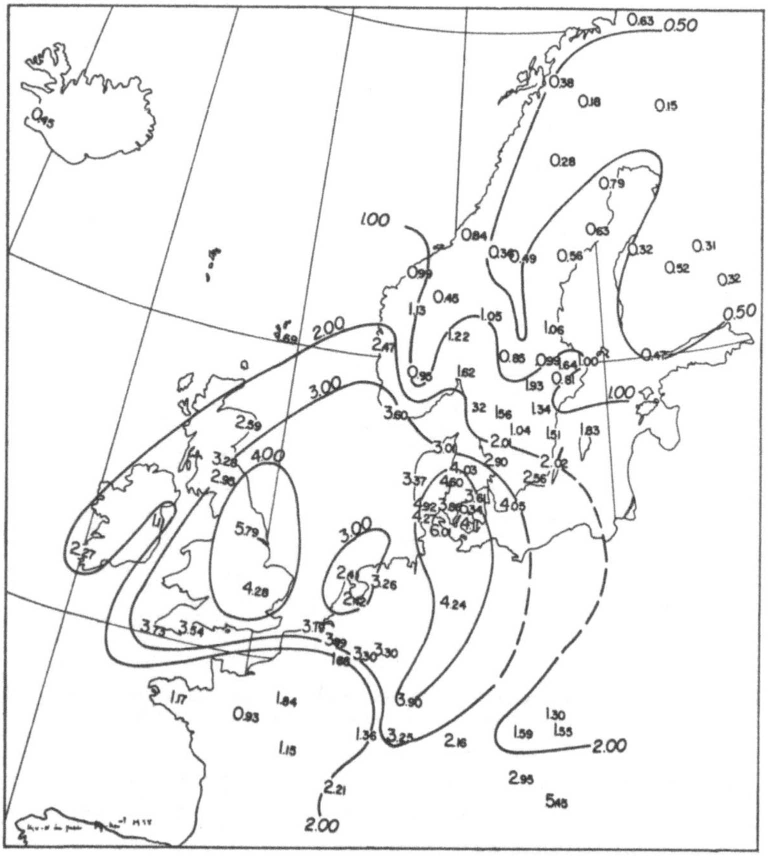

Fig. 179. Deposition by precipitation of ammonia-nitrogen in Western Europe during 1958. Values in kg · ha^{-1}

In the United States Junge's figures on the wet deposition give the result shown in Fig. 181. On the whole these values are lower than the European values but it is interesting to note that they are highest in the agricultural areas in the north. The remarkably low values in the south-east have been explained by Junge as due to absorption of ammonia from the air by the lateritic soils in this region.

As to other continents the four stations in Africa give the yearly amounts shown in Table 139 where also dry deposition has been estimated. The precipitated ammonia is rather moderate when compared to the nitrate nitrogen but air concentrations give dry deposition values of the same size as in Europe.

As to earlier data on the wet deposition in various places Eriksson's (1952) review shows that they vary considerably but more moderate figures are similar to the European data.

5. Yearly Deposition of Sulphur

The wet deposition of sulphur as sulphate in Western Europe during 1958 is shown in Fig. 182. Quite naturally the largest amounts are found in the densily populated areas in the west where it is considerable. The estimated dry deposition in this area computed from the average sulphur concentrations in air during April–September using a deposition velocity of 2 cm · sec^{-1} is shown in Fig. 183.

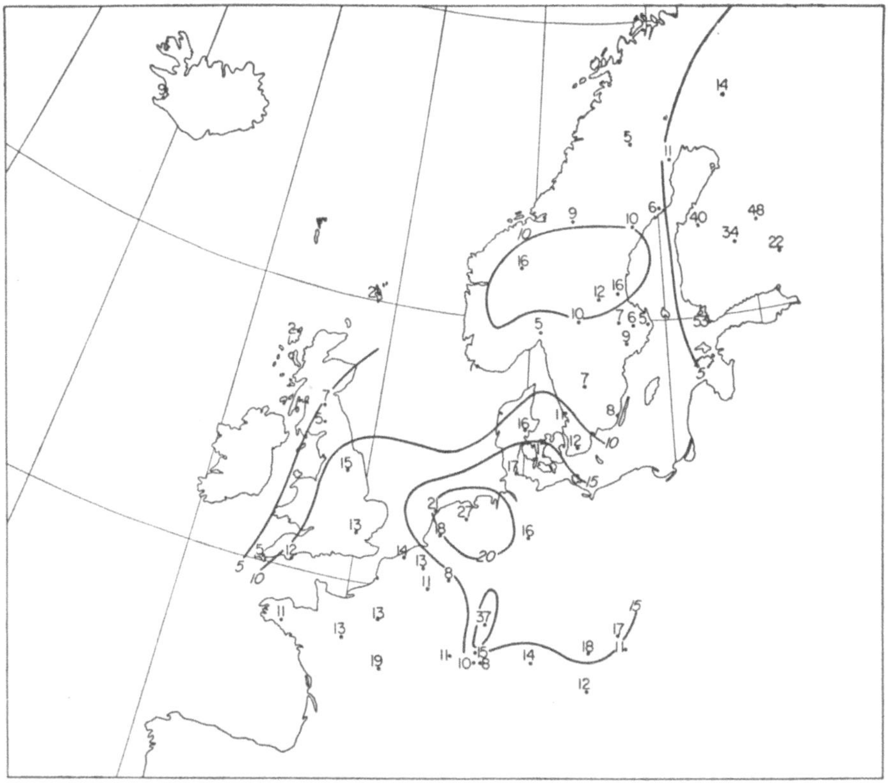

Fig. 180. Estimated yearly dry deposition of ammonia-nitrogen in Western Europe. Values in kg · ha^{-1} computed from average air concentrations April—September in 1958 and a deposition velocity of 2 cm · sec^{-1}

Only the summer months have been considered as this is the active period of the plants. Sulphur in the air (sulphur dioxide) shows a remarkably strong seasonal variation, with high winter values and sometimes extremely low summer values.

It is remarkable to note that the dry deposition estimated in this way is again about twice as high as the wet deposition thus conforming to the same rule as particulate matter. In the figures high values are also found in intensely cultivated areas like Denmark. It is possible that the application of superphosphate in these areas has upset the original sulphur balance of the soils so a large fraction of the sulphur in this fertilizer is released into the atmosphere (as hydrogen sulphide) giving uniform and high values over the whole area.

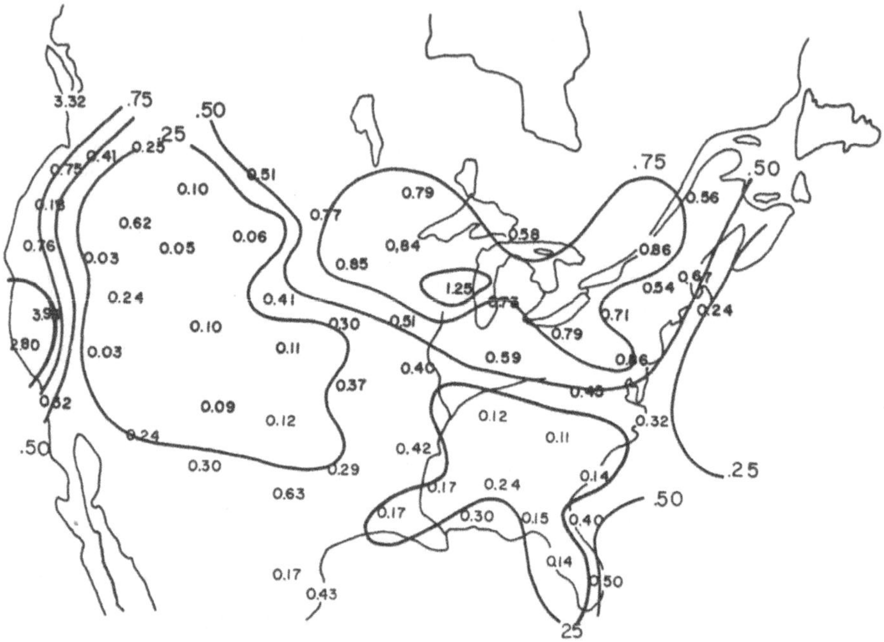

Fig. 181. Yearly deposition by precipitation of ammonia-nitrogen in the United States. Values in kg · ha⁻¹ computed from data by JUNGE and GUSTAVSON 1956

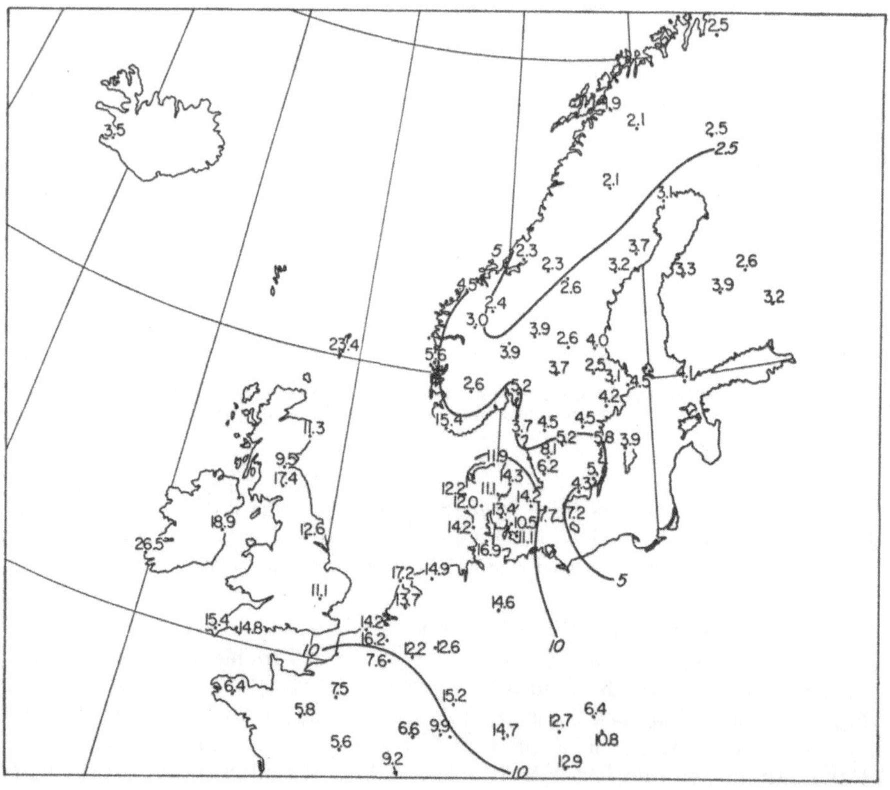

Fig. 182. Deposition by precipitation of sulphate-sulphur in Western Europe during 1958. Values in kg · ha⁻¹

In the United States the yearly wet deposition of sulphate-sulphur is shown in Fig. 184. Noticeable here is the rather high values in the northeast part of the area again showing a correlation to the population density and the industrial activity. Coastal regions also receive more as a rule, though only part of this originates as spray.

The African values are shown in Table 139. In Australia the reported sulphur deposition is fairly low.

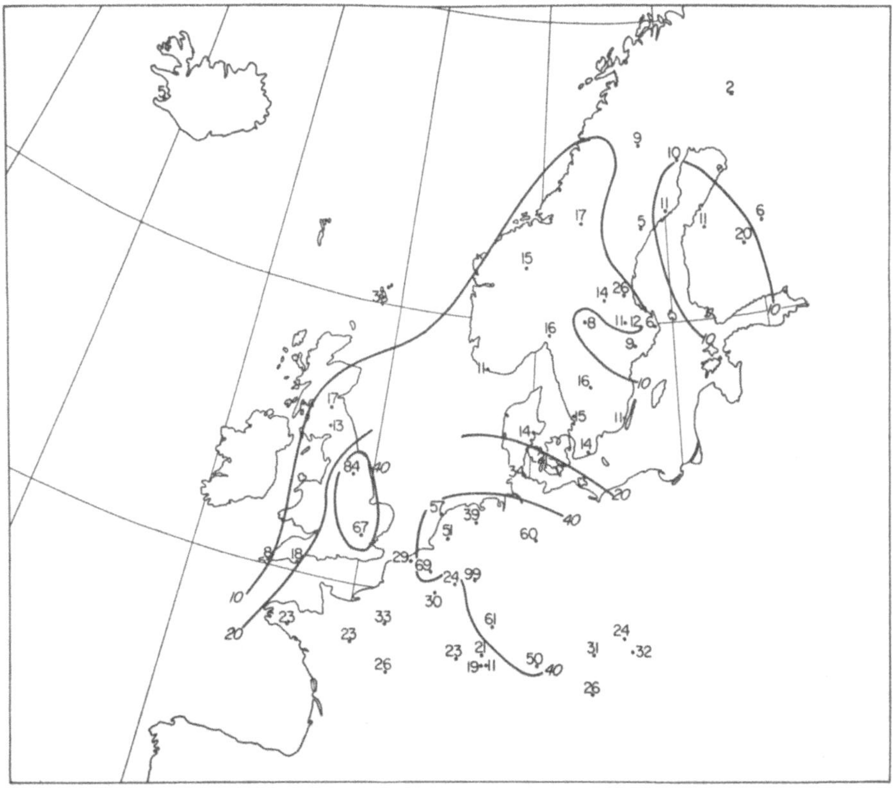

Fig. 183. Estimated dry deposition of sulphur in Western Europe during 1958. Values in kg · ha^{-1}

6. Boron and Iodine

In the Scandinavian network some boron and iodine determinations in precipitation have been made in the past in Sweden. Iodine[1] seems to be deposited fairly uniformly amounting to 3–10 g · ha^{-1} year with somewhat higher values in the south and west. Boron[2] which is mainly derived from the sea is deposited by precipitation in amounts of about 10 g · ha^{-1} · year^{-1}, by and large following the same pattern as iodine.

---

[1] Unpublished data.

[2] In am indebted to Mr. G. Brodin, the Royal Agricultural College, Sweden, for these figures prior to publication.

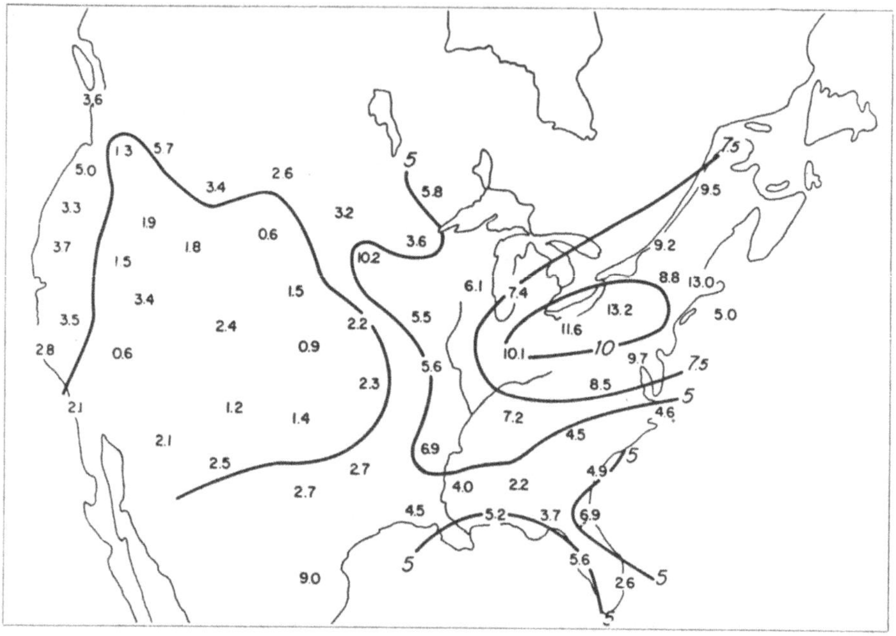

Fig. 184. Yearly deposition by precipitation of sulphate-sulphur in the United States. Values in $kg \cdot ha^{-1}$ computed from data by JUNGE and WERBY 1958

Literature

CHAMBERLAIN, A. C., and R. C. CHADWICK: Deposition of airborne radioiodine vapor. Nucleonics **11**, 8, 22–25 (1953).

EGNER, H.: Stallgödselns kväveförluster genom ammoniakavdunsting. Medd. Centr. anst. f. försöksv. på jordbruksomr. **409**, 37 (1932). — EGNER, H., and E. ERIKSSON: Current data on the chemical composition of air and precipitation. Tellus **7**, 133–139 (1955). — ERIKSSON, E.: Composition of atmospheric precipitation. I and II. Tellus **4**, 215–232, 280–303 (1952).— The yearly circulation of chloride and sulphur in nature; meteorological, geochemical and pedological implications. Tellus **11**, 4 and **12**, 1, 74 pp. (1959).

HINGSTON, F. Y.: The major ions in Western Australian rainwaters. C.S.I.R.O. Div. Rep. Bureau of Soils, No. 1, 11 pp. (1958). — HUTTON, J. T., and T. J. LESLIE: Accession of non-nitrogeneous ions dissolved in rainwater to soils in Victoria. Austr. J. Agr. Res. **9**, 492–507 (1958).

JOHANSSON, O.: On sulphur problems in Swedish agriculture. Ann. Roy. Agr. Coll. Sweden **25**, 57–169 (1959). — JUNGE, CH., and P. E. GUSTAFSON: Precipitation sampling for chemical analyses. Bull. Amer. Met. Soc. **37**, 244 (1956). — JUNGE, CH. E., and R. T. WERBY: The concentration of chloride, sodium, potassium, calcium and sulphate in rain water over the United States. J. Met. **15**, 417–425 (1958).

MACKU, M., J. PODZIMEK, and L. SRAMIK: Distribution of some chemical compositions in atmospheric precipitation over Czechoslovakia. Geof. pura appl. **42**, 24–31 (1959).

NEUMANN, G. H., S. FONSELIUS, and L. WAHLMAN: Measurements on the content of organic material in atmospheric precipitation. Int. J. Air Poll. **2**, 2 (1959).

RIEHM, H., and E. QUELLMALZ: Die Bestimmung der Pflanzennährstoffe im Regenwasser und in der Luft und ihre Bedeutung für die Landwirtschaft. *In*: Hundert Jahre Augustenberg, ed. H. RIEHM, Karlsruhe 1959.

H. Die Auswaschung von Nährstoffen durch Niederschläge

Von

H. Linser

Wenn Niederschläge einem Boden soviel Wasser zuführen, daß es zum Teil als Gravitationswasser in tiefere Bodenschichten absickert, so nimmt dieses Wasser gelöste Stoffe aus der Bodenlösung des Wurzelbereichs mit in jene tieferen Schichten, in denen sie vom Wurzelsystem der Pflanzen nicht mehr erreicht werden. Man spricht von einer Auswaschung von Nährstoffen aus dem den Wurzeln zugänglichen Bodenbereich in Schichten, die für die Wurzeln nicht mehr zugänglich sind. Die Tiefe des betreffenden Bodenbereichs wird dabei durch die Ausbildung des Wurzelsystems der betreffenden Pflanze bestimmt. Die Auswaschung von Nährstoffen kann nur dann erfolgen, wenn diese nicht in bestimmten Bindungen an den Bodenteilchen oder in den Bodenmikroorganismen festgehalten werden. Ihre Größe, die in g/m²/Jahr, aber aus praktischen Gründen zumeist in kg Nährstoff/ha/Jahr angegeben wird, ist demnach von folgenden Parametern abhängig bzw. bestimmt:

zeitliche Verteilung der Niederschläge;
Sickerwassermenge je Flächeneinheit;
Nährstoffzufuhr aus Düngemitteln und Niederschlägen;
Nährstoffgehalt des Bodens;
chemisches und biologisches Festlegungsvermögen des Bodens;
Ionensorptionskapazität des Bodens;
Tiefe des durchwurzelten Bodenraumes;
Art der Pflanze und dem von ihr bewirkten Nährstoffentzug.

Die Größe „Nährstoffauswaschung“ in kg/ha ist deshalb eine von Boden, Klima und Pflanze abhängige und sehr wandelbare Größe, die je nach den gewählten bzw. gegebenen Bedingungen recht verschieden gefunden werden kann. Zur Messung bedient man sich sogenannter Lysimeter-Anlagen (HANAMANN 1901, SEELHORST 1904, LYON und BIZZELL 1921, GERLACH 1926, HIGBY 1927, JOFFE 1929, PFAFF 1936, BAUMANN 1950), die vielfach aus Kleinparzellen von 1, 2 oder 4 m² Größe in Betonkästen bestehen und zumeist 1 m tief sind (vgl. Abb. 185). Die Betonkästen sind in dieser Tiefe durch eine schräge, wasserdicht gemachte Betonfläche nach unten hin abgeschlossen, so daß an ihr das Sickerwasser sich ansammelt bzw. zu einer Abschlußöffnung rinnt, an der es durch ein Glasrohr gesammelt und in ein Auffanggefäß geleitet wird. Mit dessen Hilfe kann seine Menge gemessen und an Hand einer entnommenen Probe von Zeit zu Zeit der Nährstoffgehalt des Sickerwassers analytisch bestimmt werden. Die Bodenproben können entsprechend ihrer natürlichen Lagerung bzw. Schichten-Folge Schicht für Schicht in die Betonkästen eingebracht werden, so daß sie einem natürlichen Bodenprofil ähnlich sind. Der Kontakt mit dem Grundwasserstrom ist durch die Bodenplatte des Lysimeters abgeschnitten. Einmal versickertes

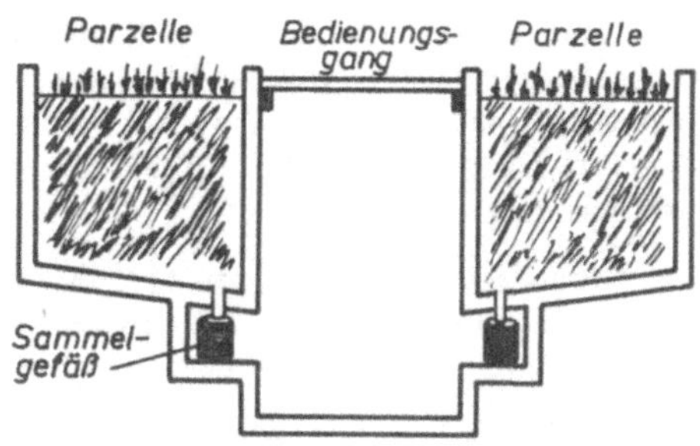

Abb. 185. Querschnitt durch eine Lysimeteranlage mit mittlerem Bedienungsgang

Wasser kann dann auch nicht mehr wieder nach oben steigen und Nährstoffe nach oben transportieren, wenn eine starke Austrocknung der Bodenflächen erfolgt.

Man wird bei Lysimeteranlagen durch schwache Beregnung in Trockenzeiten dafür sorgen, daß eine mittlere Niederschlagsmenge eingehalten wird, die dem zu prüfenden Standort entspricht, ohne daß eine zu starke Austrocknung den Pflanzenbestand, dem meistens unter den natürlichen Verhältnissen eine mehr oder weniger große Menge von aufsteigendem Wasser zur Verfügung steht, gefährdet. An Lysimetern gewonnene Zahlen können deshalb nicht unmittelbar als Maß für die Auswaschungsverhältnisse der natürlich gelagerten Böden dienen, sondern nur als Zahlen, die eine Abschätzung der tatsächlichen Auswaschungsverluste an Nährstoffen ermöglichen können, wenn man die speziellen Verhältnisse berücksichtigt, die bei der Messung geherrscht haben. Die Technik der Lysimeteranlagen wurde von Kohnke und Dreibelbis (1940) diskutiert und ihre Konstruktion in zahlreichen Abbildungen dargestellt. Da die Böden im allgemeinen ein höheres Sorptionsvermögen für Kationen, jedoch ein nur geringes Anionensorptionsvermögen besitzen, spielt bei den in Kationenform befindlichen Nährstoffen die Sorption am Bodenkomplex eine weitaus größere Rolle als für die Nährstoffe, die sich in Anionenform befinden. Auch die Phänomene des Ionenaustausches sind demgemäß bei den Kationen von größerem Einfluß auf die Auswaschung als bei den Anionen. Tonböden zeigen infolge ihrer höheren Sorptionskapazität geringere Auswaschungsverluste an Kationen als Sandböden. Nährstoffe in organischen Bindungsformen liegen in der Bodenlösung meist in geringen Mengen vor und gelangen auch in das Sickerwasser.

Stickstoff

Ammonium-Ionen werden vom Sorptionskomplex des Bodens gebunden und daher wenig ausgewaschen. Im Sickerwasser finden sich nur sehr geringe Mengen an Ammonium-Stickstoff. Dagegen werden Nitrationen stark ausgewaschen.

Sandböden und Lehmböden zeigten bei Gehalten von 70 bis 110 mg N je 100g und durchschnittlichem Niederschlag von 570 mm in langjährigen Versuchen (15 Jahre Versuchsdauer) jährlich eine N-Auswaschung von 51 bis 70 kg/ha N bei Sickerwassermengen von 90 bis 450 l/m². Das entspricht einem jährlichen Verlust von 1 bis 2% des Gesamt-N-Vorrates (Pfaff 1963a). Der Stickstoff verschiedener N-Düngemittel (60 kg N/ha) wurde bei extremen Auswaschungsbedingungen wie folgt ausgewaschen:

Ammoniakdüngemittel zu 23% der zugegebenen Menge
Ammonsalpeterdüngemittel zu 48% der zugegebenen Menge
Salpeterdüngemittel zu 68% der zugegebenen Menge.

Bei Brache werden höhere Auswaschungswerte erhalten als bei guter Bodenbedeckung (bzw. bei hohem Wasserverbrauch durch den Pflanzenbestand); die durchschnittliche jährliche N-Auswaschung betrug in langjährigen Dauerversuchen:

bei Brache	160 kg/ha N
bei Reben	80 kg/ha N
bei Grünland	10 kg/ha N.

Sie kann deshalb auch bei guter N-Düngung, falls diese die Bodenbedeckung erhöht und die Wasserausnutzung durch die Pflanze bzw. die Durchwurzelung des Bodens steigert, herabgesetzt werden. Bei einem Volldüngungs-Steigerungsversuch mit Nitrophoska erhielt PFAFF (1963a) im siebenjährigen Durchschnitt Auswaschungszahlen (kg/ha N), die in Tab. 140 wiedergegeben sind.

Tabelle 140. *N-Auswaschung (kg/ha N) in einem siebenjährigen Düngungsversuch*

Pflanzen	ungedüngt	2 × 40 = 80 kg/ha N	2 × 80 = 160 kg/ha N	2 × 120 = 240 kg/ha N
Frühkartoffeln/Mais	42	43	48	66
Roggen/Sommerraps	19	15	20	25

Unter verschiedenen Beständen wurden folgende Auswaschungswerte für Stickstoff erhalten (Tab. 141):

Tabelle 141. *N-Auswaschung unter verschiedenen Feldbeständen*

Kartoffeln	bei NPK 47 kg/ha N	bei PK 43 kg/ha N
Winterroggen	bei NPK 74 kg/ha N	bei PK 61 kg/ha N
Hafer	bei NPK 60 kg/ha N	bei PK 59 kg/ha N

Aus sauren Böden wird häufig bis zu $^1/_3$ weniger N ausgewaschen als aus alkalischen Böden.

Die größten N-Mengen werden während der Wintermonate ausgewaschen, da in dieser Zeit meist auch die größten Sickerwassermengen anfallen. So betrug die N-Auswaschung in 15 Jahren (nach PFAFF 1963a) im Durchschnitt von mehreren Jahren im Winter ein Vielfaches von jener im Sommer (Tab. 142).

Tabelle 142. *N-Auswaschung im Sommer und im Winter bei verschieden hohen Niederschlägen*

Jahre	bei	im Sommerhalbjahr	im Winterhalbjahr
5	niedrigen Niederschlägen	1 kg/ha N	11 kg/ha N
7	mittleren Niederschlägen	7 kg/ha N	16 kg/ha N
3	hohen Niederschlägen	14 kg/ha N	46 kg/ha N

Mit der Auswaschbarkeit von NH_4- und NO_3-Ionen aus gärtnerischen Erden (Standard-Kultursubstraten) befaßte sich SOUKUP (1962). Zugabe von Bentonit erhöhte den sorptiv-festgehaltenen N-Anteil beträchtlich.

Phosphorsäure

HANAMANN (1901), CREYDT, v. SEELHORST und WILMS (1901) sowie SELIWANOFF (1901), GERLACH (1910) fanden keine Auswaschungsverluste an P_2O_5. NORTON (1908) fand an Hand von Drainagewässern eine P_2O_5-Auswaschung von nur 0,74 kg/ha/Jahr. SCHMITT (1933) fand Auswaschungsverluste von 0,15 bis 0,81 kg/ha/Jahr, GISIGER (1933) bis zu 2,9 kg/ha/Jahr, SCHARRER und KÜHN (1952) 0,16 bis 0,92 kg/ha/Jahr, HIGBY (1927) etwa 9 kg/ha/Jahr und DE VRIES und HETTERSCHIJ (1936) auf moorkolonialen Böden Hollands eine bemerkenswerte P_2O_5-Wanderung, die 15 bis 30 kg/ha/Jahr P_2O_5 mit den Drainagewässern abführte. MACINTIRE und Mitarb. (1947) fanden, daß zwei saure Böden binnen 12 Jahren nur insgesamt etwa 4 kg/ha P_2O_5 durch Auswaschung verloren; bei einem schwach alkalischen, ungekalkten Boden waren es 9,6 kg, beim gekalkten gleichwertigen Boden aber 56 kg/ha und bei Anwendung von Dolomit 73,2 kg/ha P_2O_5-Verlust durch Auswaschung. Gegenüber Phosphorsäure zeigen verschiedene Böden unterschiedliches Reaktionsvermögen. Die ausgewaschenen Mengen an P_2O_5 sind daher von den Bodeneigenschaften besonders abhängig. PFAFF (1963b) fand in langjährigen Versuchen folgende Werte (Tab. 143):

Tabelle 143. *Phosphatauswaschung bei verschiedenen Böden*

Boden	P_2O_5 mg/100 g Boden EGNER-RIEHM	P-Fixierungsvermögen (BASS und SIELING)	pH	P_2O_5-Auswaschung kg/ha/Jahr		
				ohne Düngung	NPK-Düngung	NPK + Stallmist
Lehm C	58	21	7	4,9	—	4,9
Mittl. Lehm DE	28	14	4	5,4	4,0	3,6
Sandboden D	13	5	4	7,6	7,8	7,4
Sandboden E	2	6	7	9,6	12,2	12,3

Böden mit höherem Gehalt an sorbierenden Tonkomplexen liefern geringere P_2O_5-Auswaschungswerte als Sandböden, hohes P-Fixierungsvermögen verringert die P_2O_5-Auswaschung. Phosphatdüngung erhöhte den Gehalt der Sickerwässer an P_2O_5 nur bei dem alkalischen Sandboden. Dem entsprechen auch Befunde von OZANNE (1961). Zwischen NPK-Düngung, Sickerwassermengen und P_2O_5-Auswaschung fand PFAFF (1963b) in siebenjährigen Versuchen mit Haupt- und Zweitfrucht eine Beziehung, die sich aus den Zahlen der Tab. 144 erkennen läßt.

Tabelle 144. *Phosphatauswaschung nach verschieden hoher Volldüngung*

als Nitrophoska	kg/ha P_2O_5	Liter Sickerwasser je m²	kg/ha P_2O_5 im Sickerwasser
	0	239	12
	80	234	10
	160	210	8
	250	194	8

Im Gegensatz zu anderen Nährstoffen steigen mit zunehmender Niederschlagsmenge die Phosphatkonzentrationen im Sickerwasser häufig an, was MATTSON (1950) als eine Folge der DONNAN-Verteilung der Ionen und mit einer Verdünnung der Bodenelektrolyte zu erklären versucht. Im gleichen Sinne wirken sich die Maßnahmen künstlicher Beregnung aus.

Kalium

Die Kali-Auswaschung betrug bei den langjährigen Versuchen von PFAFF (1963 b) auf verschiedenen Böden ohne Mineraldüngung 16 bis 23 kg/ha K_2O, bei Düngung mit K-haltigen Düngemitteln, auch Stallmist, 22 bis 72 kg/ha K_2O. Die Auswaschung von Düngerkali hängt weitgehend vom T-Wert des Bodens ab; bei hohen T-Werten wurden keine K-Verluste vom Düngerkali durch Auswaschung beobachtet. Aus leichten Böden wurde auch das Kali des Stallmistes zu 6 bis 15% ausgewaschen. Durch Kalkung wird die K_2O-Auswaschung nicht erhöht, N-Düngung setzt wie gewöhnlich herab. Nur Natronsalpeter erhöht die K_2O-Auswaschung etwas, wahrscheinlich weil K^+ an seinen Sorptionsstellen durch Na^+ ausgetauscht werden kann. Zugabe organischer Substanz (Kohle, Torf) ließ die K_2O-Auswaschung ziemlich unbeeinflußt. Mit der Auswaschbarkeit der K-Ionen aus gärtnerischen Kultursubstraten befaßte sich SOUKUP (1962).

Natrium

Böden, die Natrium enthaltende Düngemittel erhalten haben, zeigen Na_2O-Auswaschungsverluste, die zahlenmäßig höher liegen als jene an K_2O. Sie betrugen bei den langjährigen Versuchen von PFAFF (1963 b) 32 bis 120 kg/ha. Die Niederschlags- und Sickerwassermengen beeinflußten dabei die Na-Auswaschung stärker, als dies durch verschiedenartigen Pflanzenbestand der Fall war.

Calcium

Das Sickerwasser aus ungedüngten und gedüngten Böden zeigt etwa die gleichen Ca-Gehalte wie die Grundwässer und die Trinkwässer gleicher Gegenden. Die CaO-Gehalte ungedüngter Böden liegen etwas niedriger (z. B. 130 mg/l) als jene mineralisch-organisch gedüngter Böden (z. B. 180 mg/l).

Die Auswaschung an CaO betrug in den langjährigen Versuchen PFAFFS (1963 c) 231 bis 338 kg/ha/Jahr; im Mittel etwa 270 kg/ha/Jahr (bei Sickerwassermengen von 207 bis 221 l/m^2). Die Ca-Verluste steigen mit dem Ca-Vorrat des Bodens einerseits und mit der biologischen Tätigkeit des Bodens andererseits an. Physiologisch saure Düngemittel veranlassen beträchtliche Steigerungen der Kalkauswaschung aus den Böden; Zuführung von Torf setzt die Auswaschung an CaO herab.

Magnesium

Die Auswaschung an MgO beträgt nur etwa $^1/_{10}$ jener an CaO. PFAFF (1963 c) fand Werte von 22 bis 25 kg/ha/Jahr; physiologisch saure Düngung steigert solche Werte noch etwas. JAKOB (1955) gibt durchschnittliche Auswaschungsverluste von 20 bis 40 kg/ha MgO pro Jahr an. Bei PK-Düngung ist die MgO-Auswaschung gegenüber ungedüngt etwas erhöht, wofür wahrscheinlich Austauschvorgänge verantwortlich sind.

Mangan

Die Mangan-Auswaschung wurde mit verschiedenen Böden an Bodensäulen von ZADDACH (1960) bei Niederschlagsmengen entsprechend 700 mm studiert und recht verschieden hoch gefunden, nämlich mit Werten zwischen 0,37 kg/ha bei nicht aufgekalktem, saurem Podsolboden und 112,5 kg/ha bei einem sauren Boden aus Darmstadt. BUCHNER (1952) nennt für die Mn-Auswaschung 0,15 bis 0,30 kg/ha/Jahr, doch muß zweifellos mit einer viel größeren Streubreite der Auswaschungszahlen gerechnet werden. Die Höhe der Auswaschungszahlen ist weniger von der Bodenreaktion als von dem Gehalt des Bodens an reduzierbarem bzw. austauschbarem Mn abhängig. Sie kann durch Aufkalkung der Böden (auch noch bei fast neutralen Böden, pH 6,4) auf pH 7,5 wesentlich vermindert werden (ZADDACH 1960). Gedüngtes Mangan wurde nur in sorptionsschwachen, stark sauren Böden schnell nach unten verlagert und zur Gänze ausgewaschen, nicht jedoch bei pH-Werten von mehr als 5,8. Aufkalkung setzt auch die Auswaschungsverluste an gedüngtem Mangan herab.

Zink

Zink wird in den meisten Böden nur schwer verlagert. HÖFNER (1957) fand bei fünf verschiedenen Böden, die sich hinsichtlich Korngrößenverteilung und Bodenreaktion unterschieden, an Säulenversuchen bei Wassermengen entsprechend 540 mm Niederschlag keine Zinkauswaschung aus dem den Pflanzenwurzeln zugänglichen Bodenbereich.

Kupfer

Kupfer wird ebenfalls in den meisten Böden stark gebunden bzw. in den obersten Bodenschichten festgehalten. SCHAUMLÖFFEL (1958) fand nur geringfügige Kupfermengen in den Sickerwässern saurer und aufgekalkter Böden (2,8 bis 15,08 je Zylinder von 5 cm Durchmesser und 40 cm Höhe) und eine Auswaschung zugedüngter Kupfermengen von maximal 5,7% bei einer 2000 mm Niederschlag entsprechenden Wassermenge.

Sulfat

Sulfationen werden auf sauren und alkalischen Sand- und Lehmböden stark ausgewaschen. In seinen 15jährigen Versuchen fand PFAFF (1963 c) Werte von 200 bis 300 kg/ha SO_4/Jahr. Mit schwefelsaurem Ammoniak zugeführtes Sulfat macht infolge seiner physiologischen Azidität entsprechende Mengen an Ca und Mg löslich. Es wird unter „Mitnahme" von 38% seiner Menge an CaO und etwa 2% an MgO zur Gänze ausgewaschen. Auf durchlässigen Böden mit geringen Kalkreserven sollten deshalb für je 100 kg Ammonsulfat 75 bis 100 kg magnesiumhaltiger Kalkdünger (z. B. mit 85% $CaCO_3$ 10% $MgCO_3$) mit verabreicht werden, um den Kalkspiegel des Bodens aufrechtzuerhalten.

Chlorid

Für Chlor fand PFAFF (1963 c) Auswaschungswerte von 21 bis 32 kg/ha/Jahr. Als Ammoniumchlorid dem Boden zugeführtes Chlor wurde zu etwa 90%, jenes aus 40er Kalisalz zu etwa 70% ausgewaschen.

Borat

KRÜGEL, DREYSPRING, v. LOTTHAMMER (1937) sowie SCHARRER und Mitarbeiter (1955) untersuchten die Auswaschung von Dünger-Borat aus Böden. Im Gegensatz zu den erstgenannten Autoren fanden SCHARRER und Mitarbeiter keine Auswaschung von Dünger-Borat innerhalb eines Jahres im Bodenbereich unterhalb 1 m Tiefe.

Literatur

BAUMANN, H.: Wasserhaushalt der Pflanzen und des Bodens in wasserwirtschaftlicher Betrachtung. Z. Pflanzenernähr., Düng., Bodenkde. **49** (94), 71 (1950). — BUCHNER, A.: Zur Spurenelemente-Bilanz unserer Böden. Mitt. DLG **17**, 275—277 1952).

CREYDT, v. SEELHORST und WILMS: Untersuchungen über Drainagewässer. J. Landw. **49**, 251 (1901).

DE VRIES, O. v., und G. W. G. HETTERSCHIJ: Beweglichkeit der Phosphorsäure im Boden. Z. Bodenkde. u. Pflanzenernähr. **2** (47), 178 (1936).

GERLACH, M.: Über die durch Sickerwässer dem Boden entzogenen Wasser und Nährstoffe. Illust. Landw. Ztg. **95**, 879 (1910). — GISIGER, L.: Die Wanderung der Düngerphosphorsäure im Wiesenboden. Phosphorsäure **3**, 462 (1933).

HANAMANN, J.: Lysimeterversuche des Jahres 1899. Z. landw. Versuchswesen in Österreich **4**, 34 (1901). — HIGBY, W. M.: Lysimeterstudien. Soil Sci. **24**, 51 (1927). — HÖFNER, W.: Untersuchungen über die Auswaschung und Sorption des Zinks im Boden und über die Zinkaufnahme der Pflanzen. Diss. Justus-Liebig-Univ., Gießen, 123 S. (1957).

JAKOB, A.: Magnesia, der fünfte Pflanzenhauptnährstoff. Stuttgart: F. Enke. 1955. — JOFFE, J. S.: A new type of lysimeter at the New Jersey Agricultural Experiment Station. Science **70** (1806), 148 (1929).

KOHNKE, H., und F. R. DREIBELBIS: A survey and discussion of lysimeters and a bibliography on their construction and performance. U. S. Dept. Agric. Miss. Publ. Nr. 372 (1940). — KRÜGEL, C., C. DREYSPRING und R. LOTTHAMMER: Auswaschungsversuche mit Boraten. Das Superphospat **13**, 99–104 (1937).

LYON, T. L., und J. A. BIZZELL: Lysimeterversuche. New York Cornell Stat. Mem. **41**, 51 (1921).

MAC INTIRE, W. H., S. H. WINTERBERG, L. B. CLEMENTS und A. J. STERGES: Phosphate migration in limestoned and dolomited soils as registered by plant response. Soil Sci. Amer. **12** (1947). — MATTSON, S., E. G. WILLIAMS, E. ERIKSON und K. VAHTRAS: Boden-, Bodenlösungsgleichgewichte und Phosphatlöslichkeit. Z. Pflanzenernähr., Düng., Bodenkde. **50**, 39–59 (1950).

NORTON, J. H.: Mengen und Zusammensetzung von Drainagewässern. Amer. Chem. Soc. **30**, 1186 (1908).

OZANNE, P. G.: The loss of phosphorus from sandy soils. Austral. J. Agric. Res. **12**, 409—423 (1961).

PFAFF, C.: Über Lysimeter-Versuche. Forschungsdienst, Sonderh. 6, 102–114 (1936/37). — Sickerwasserverhältnisse in Lysimetern bei Zwischenfruchtfutterbau. Forschungsdienst, Sonderh. 7, 61–66 (1937). — Lysimeter-Versuche mit verschiedenen Stickstoffdüngemitteln. Z. Bodenkde. u. Pflanzenernähr. **12** (57), 45–52 (1939).— Lysimeter-Versuche. Z. Pflanzenernähr., Düng., Bodenkde. **48** (93), 93–118 (1949). — Über Nährstoffauswaschung bei Leguminosen-Gründüngung. Landw. Forsch. **5**, 163–167 (1953). — Nährstoffauswaschung aus dem Boden bei Anbau von Reben. Weinberg u. Keller **7**, 225–229 (1960). — Einfluß der Beregnung auf die Nährstoffauswaschung bei mehrjährigem Gemüsebau. Z. Pflanzenernähr., Düng., Bodenkde. **80**, 93–108 (1958). — Einfluß der Beregnung auf die Nährstoffauswaschung bei mehrjährigem Gemüsebau. Gartenbauwiss. **27** (9), 27–39 (1962). — Das Verhalten des Stickstoffs im Boden nach langjährigen Lysimeterversuchen I. Z. Acker- u. Pflanzenbau **117** (1), 77–99 (1963). — Das Verhalten der Phosphorsäure und der Alkalien im Boden nach langjährigen Lysimeterversuchen II. Z. Acker- u. Pflanzenbau **117** (1), 100–113 (1963). — Über die Auswaschung von Calcium, Magnesium, Chlorid und Sulfat aus dem Boden (Lysimeterversuche) III. Z. Acker- u. Pflanzenbau **117** (2), 117–128 (1963). — PFAFF, C., und F. FRIEDRICH: Die Wasserbilanz des bewachsenen Bodens nach Lysimeter-Versuchen. Wasser u. Boden **9**, 281–287 (1954).

SCHARRER, K., und H. KÜHN: Sickerversuche zur Ermittlung der Auswaschung und Tiefenwirkung verschiedener Phosphorsäuredüngemittel allein und bei Anwesenheit von Stalldünger. Z. Pflanzenernähr., Düng., Bodenkde. **58** (103), 146–162 (1952). — SCHARRER, K., H. KÜHN und J. LÜTTMER: Über die Auswaschung des Bors im Boden. Landw. Forsch. **7** (2), 89–105 (1955). — SCHAUMLÖFFEL, E.: Über die Bestimmung kleinster Mengen Kupfer und dessen Aufnahme durch die Pflanzen. Diss. Justus-Liebig-Univ., Gießen, 154 S. (1958). — SCHMITT, L.: Besteht bei der Herbstdüngung die Gefahr der Auswaschung von Düngerphosphorsäure? Superphosphat **5**, 103 (1933). — SELIWANOFF, TH.: Drainwässer der Salzmoraste und Rieselfelder der Stadt Odessa. Landw. Versuchsstationen **55**, 475 (1901).— SOUKUP, J.: Die Sorption und Auswaschung von Nährstoffionen der gärtnerischen Erden. Mitt. I: NH_4- und NO_3-Ionen. Z. Pflanzenernähr., Düng., Bodenkde. **96** (141), 139–143 (1962). — Die Sorption und Auswaschung von Nährstoffionen in gärtnerischen Erden. Mitt. II: K-Ionen. Z. Pflanzenernähr., Düng., Bodenkde. **98** (143), 55–57 (1962). — STEUBING, L.: Beobachtungen an Lysimetern verschiedener Bauart, Ber. Dtsch. Bot. Ges. **69**, 35–48 (1956).

ZADDACH, M.: Die analytische Bestimmung des Mangans und Untersuchungen über die Verlagerung des Mangans im Boden. Diss. Justus-Liebig-Univ., Gießen, 101 S. (1960).

J. Methoden zur Bestimmung des Düngebedürfnisses der Böden

a) Chemische und physikalisch-chemische Verfahren

Von

E. Rauterberg

1. Entnahme von Bodenproben

Vom Ausschuß für Landwirtschaftliche Bodenuntersuchungen im Deutschen Normenausschuß und vom Reichsverband der Landwirtschaftlichen Untersuchungsämter und Versuchsanstalten sind Anweisungen zur Entnahme von Bodenproben für Nährstoffuntersuchungen ausgearbeitet worden, s. Herrmann 1943 und Methodenbuch Bd. I (S. 1—14). Die wichtigsten Anweisungen zur Entnahme von Bodenproben können wie folgt zusammengefaßt werden:

Mineralböden. *Ackerland:* Die Probenahme soll nach der Ernte, wenn der Boden nicht schmiert, erfolgen (zu empfehlen ist, nach einer abtragenden Frucht). Proben für eine Mischprobe sind nur von einheitlich bewirtschafteten Flächen zu nehmen. 15 bis 20 Einzelproben, gleichmäßig verteilt über die zu untersuchende Fläche genommen, werden zu einer Mischprobe vereinigt. (Ein Randstreifen von 10% der Fläche soll bei der Probenahme unberücksichtigt bleiben.) Die Flächen, von denen eine Mischprobe genommen wird, sollen nicht größer als 1 ha sein. Die Probenahme kann mit einem Spaten oder einem Bohrer erfolgen. Sämtliche Einzelproben müssen bis zur gleichen Tiefe im Boden, etwas weniger tief als gepflügt wird, genommen werden. Bohrer sind dafür in den meisten Fällen am brauchbarsten. Die Bodenproben werden in naturfeuchtem Zustand in Beuteln oder Schachteln an die Untersuchungsstationen geschickt. Die Proben sind kurz auf haltbarem Material mit wasserunempfindlicher Schrift zu bezeichnen. Für jede Durchschnittsprobe ist möglichst ein kurzer, aber wohl überlegter Fragebogen auszufüllen.

Auf *Wiesen und Weiden* sind die Proben ebenso zu nehmen wie auf Ackerland. Sie sollen freilich nur aus der gut durchwurzelten Schicht, bis 10 cm Tiefe, genommen werden. Es wird empfohlen, eine weitere Probe aus der Schicht 10 bis 20 cm zu nehmen. Für die Probenahme auf Wiesen und Weiden ist der Spaten geeigneter als der Bohrstock.

Von *Weinbergböden* sind immer aus der oberen Schicht und aus dem Untergrund Proben zu nehmen.

Vorbereitung der Bodenproben für die Untersuchung. Die in naturfeuchtem Zustand bei den Untersuchungsstationen eingegangenen Bodenproben werden an der Luft oder im Luftstrom bei etwa 30° getrocknet und dann durch ein 2-mm-Sieb — zum Teil auch 1-mm-Sieb — gegeben. Dabei ist darauf zu achten, daß von der Bodenmenge, die auf das Sieb gegeben wird, alles, was kleiner als 2 bzw. 1 mm ist, durch das Sieb fällt.

Moorböden. Bei der Entnahme von Bodenproben auf Moorböden sind die entsprechenden Anweisungen der Versuchsanstalten (s. Methodenbuch Bd. I, S. 5) zu beachten (dort auch weitere Literaturangaben). Moorböden werden vor der Untersuchung nicht getrocknet.

2. Bestimmung der Stickstoff-Bedürftigkeit

Da praktisch alle Mineralböden mit Stickstoff gedüngt werden müssen, damit mittlere und hohe Erträge erzielt werden, ist das Interesse, Methoden zur Bestimmung der Stickstoff-Bedürftigkeit der Böden zu entwickeln, immer gering gewesen. Wenn etwas über den Stickstoffbedarf eines Bodens ausgesagt werden soll, ist nach unseren heutigen Erkenntnissen dafür in erster Linie das C/N-Verhältnis des Bodens und sein Gehalt an Gesamtstickstoff maßgebend.

Über die N-Bedürftigkeit von Niederungsmoorböden, die häufig so gut mit Stickstoff versorgt sind, so daß eine N-Düngung nicht nötig ist, wird wahrscheinlich auch ihr Gehalt an Gesamtstickstoff und ihr C/N-Verhältnis am besten Auskunft geben.

3. Bestimmung der Phosphorsäure-Bedürftigkeit

A. Chemische Methoden

Bestimmung der Gesamtphosphorsäure und der in Mineralsäuren löslichen Phosphorsäure. Zur Bestimmung der Düngebedürftigkeit der Böden für Phosphorsäure wurde zuerst der Gehalt der Böden an Gesamtphosphorsäure und die in starken Mineralsäuren, in erster Linie Salzsäure, lösliche Phosphorsäure ermittelt. Die Böden wurden zum Teil bei Zimmertemperatur und zum Teil bei Siedehitze mit 10 oder 25 %iger Salzsäure behandelt, wie aus den Ausführungen von König (1923, S. 124) hervorgeht. Aus den Untersuchungsergebnissen von Emmerling (1895), der zahlreiche Böden aus Schleswig-Holstein untersucht hat, ist zu ersehen, daß 40 bis 85 % der in heißer Salzsäure löslichen Phosphorsäure von kalter Salzsäure gelöst werden. Der große Unterschied in der Löslichkeit der Bodenphosphorsäure in kalter und heißer Salzsäure ist wahrscheinlich darauf zurückzuführen, daß durch die heiße Salzsäure ein großer Teil der organisch gebundenen Phosphorsäure hydrolysiert wird. Emmerling (1899) ist der Ansicht, daß aus Nucleinen bei Behandlung des Bodens mit Wasserdampf Phosphorsäure abgespalten wird. Die heiße Salzsäure wird natürlich ebenso wirken. Emmerling (1895) erwähnt auf S. 190 einen Satz von Erasmus Darwin, dem Großvater von Charles Darwin, aus seiner „Philosophy of agriculture" um das Jahr 1800: „Die sprichwörtlich fruchtbaren Ackererden zeichnen sich insbesondere durch einen hohen Gehalt an phosphorsaurem Kalk aus." Auch Emmerling (1895, S. 190) ist auf Grund seiner umfangreichen Untersuchungen zu dem Ergebnis gekommen, „daß die Phosphorsäure besser als jeder andere Nährstoff geeignet ist, eine Beziehung zwischen Ertrag und Gehalt des Bodens an Nährstoff erkennen zu lassen". Emmerling (1895) führt S. 190 gleiche Ergebnisse von Schorlemer, Schütze, Thomas, Wagner, Truchat und Weber an. Von verschiedener Seite sind die Böden auf Grund ihres Gehaltes an in Salzsäure löslicher Phosphorsäure auf ihre Düngebedürftigkeit bewertet worden, wie aus Grenzzahlen zu ersehen ist, die von König (1929, S. 124) angegeben werden. Interessant ist, daß Liebscher (1895) Eisenoxyd und Tonerde, die sich zusammen mit der Phosphorsäure in der Salzsäure lösen, bei der Auswertung des ermittelten Phosphorsäuregehalts der Böden berücksichtigt hat. Je weiter das Verhältnis: „Phosphorsäure zu Eisenoxyd plus Tonerde" ist, desto schlechter wird die Phosphorsäure von den Pflanzen ausgenutzt.

Im letzten Drittel des vorigen Jahrhunderts wurde aber auch schon festgestellt, daß aus dem Gehalt der Böden an Gesamtphosphorsäure und an in starker Salzsäure löslichen Phosphorsäure nicht immer geschlossen werden kann, ob eine Phosphorsäuredüngung wirkt.

Verdünnte Salpetersäure wurde von SCHLÖSING JUN. nach VON SIGMOND (1903) und VON SIGMOND (1903) zur Bestimmung der Phosphorsäurebedürftigkeit der Böden verwendet. VON SIGMOND (1903) bestätigt Untersuchungsergebnisse von SCHLÖSING JUN., daß es 2 Arten von Phosphaten im Boden geben soll, solche, die von einer Salpetersäure mit 0,1 bis 0,2 % N_2O_5 und solche, die erst von einer Salpetersäure mit mindestens 1 % N_2O_5 gelöst werden. Nach WEIBULL (1911) haben auch FRAPS und STODDART festgestellt, daß von 1/5 n HNO_3 nur die als Calciumphosphat vorliegende P_2O_5 im Boden gelöst wird, die als Eisenphosphat vorhandene P_2O_5 dagegen nicht. Von VON SIGMOND (1929) ist später eine Methode zur Bestimmung der leicht von den Pflanzen assimilierbaren Phosphorsäure ausgearbeitet worden. Er bestimmt zuerst die Basizität des Bodens und verwendet dann eine derartig konzentrierte Salpetersäure, daß dieselbe nach der Behandlung mit dem Boden 0,005 bis 0,015 n ist. Die Basizität des Bodens wird außerdem bei der Bewertung der gefundenen Phosphorsäuremenge berücksichtigt. BONDORFF und STEENBJERG (1933) benutzen ebenfalls verdünnte Salpetersäure unter Berücksichtigung der Basizität des Bodens.

KIRSANOFF (1931) hat eine Schnellmethode zur Bestimmung der Phosphorsäurebedürftigkeit der Böden mit 0,2 n HCl als Lösungsmittel entwickelt. BRAY (1929) verwendet ebenfalls eine 0,2 n HCl. Als Ergänzung bei Bestimmung der leichtlöslichen Nährstoffe, Phosphorsäure und Kalium, nach der Ammoniumlactat-Essigsäure-Methode (s. S. 807) schlagen EGNER, RIEHM und DOMINGO (1960) vor, die Nährstoffreserven durch Behandlung des Bodens mit 2 n HCl zu bestimmen.

2 g Boden werden mit 50 ml 2n HCl 2 Stunden im siedenden Wasserbad behandelt, dabei wird die salzsaure Bodensuspension alle halbe Stunde gut umgeschüttelt.

MORGAN (1929) berichtet über eine gute Korrelation zwischen der in 0,01 n H_2SO_4 löslichen Phosphorsäure und Tabakerträgen im Gefäßversuch. Die von TRUOG (1930) ausgearbeitete Schnellmethode, die zum Teil modiziert wurde, wird relativ häufig benutzt.

1 g Boden wird 1 Std. mit 200 ml einer 0,002 n H_2SO_4 geschüttelt, die je Liter 3 g $(NH_4)_2SO_4$ oder K_2SO_4 enthält, um einen End-pH-Wert von 3 zu erreichen. P-Versorgung: bis 5 ppm P sehr arm, 5 bis 15 ppm P gering, 15 bis 25 ppm P mittel und über 25 ppm P angemessen hoch.

PEECH et al. (1947) behandeln 1 g Boden mit 100 ml der Lösung nach TRUOG. LOHSE und RUHNKE (1933) verwenden eine verdünnte Kaliumhydrogensulfatlösung, End-pH 2,0. Für kalkhaltige Böden muß die Konzentration der Lösung ausprobiert werden.

BRAY und KURTZ (1945) schütteln 2,85 g Boden 1 Min. mit 20 ml einer Lösung, die für HCl 0,025 n und für NH_4F 0,03 n ist. Im Filtrat bestimmen sie die Phosphorsäure. Ungefähre Grenzzahlen: 3 ppm P sehr geringer; 3 bis 7 ppm P niedriger; 7 bis 20 ppm P mittlerer und 20 ppm P hoher Gehalt, aber unterschiedlich für verschiedene Pflanzen.

Eine saure ammoniumfluoridhaltige Lösung soll besonders geeignet sein, Aluminium- und Eisenphosphate zu lösen, es bilden sich komplexe Aluminium- und Eisenfluorid-Ionen. Die BRAYsche Methode (s. auch BRAY 1948) ist nach Untersuchungen von SMITH, ELLIS und GRAVA (1957), von WELCH, ENSMINGER und WILSON (1957), von THOMPSON und PRATT (1954), von CALDWELL, HUSTRULID und HAMMERS (1956) und von RUSSEL und STANFORD (1954) sehr geeignet zur Bestimmung der P_2O_5-Bedürftigkeit der Böden. Die Salzsäure, in der das Ammoniumfluorid gelöst ist, wird zum Teil auch in höherer Konzentration, 0,1 und 0,2 n benutzt. Eine 0,5 n NH_4F-Lösung wird nach BRAY und DICKMANN (1942) zur Bestimmung der an Aluminium gebundenen Phosphorsäure im Boden verwendet, denn an Eisen und an Calcium gebundene Phosphorsäure wird in neutraler und alkalischer Ammoniumfluoridlösung nicht gelöst.

Lösungen organischer Säuren und ihrer Salze. 1. Die *Zitronensäure* wurde schon im vorigen Jahrhundert benutzt, um die Phosphorsäurebedürftigkeit der Böden zu bestimmen. So berichtet GERLACH (1896), daß er „früher" 100 Böden, deren Reaktion auf eine Düngung mit Phosphorsäure bekannt war, mit 1 %iger Zitronensäure untersucht hat und „daß in 80 % sämtlicher Fälle dieses Reagens seine Schuldigkeit tat". Etwas später wurde von DYER in Rothamsted 1894 festgestellt, daß die Azidität vom Wurzelsaft von 100 verschiedenen Pflanzenarten im Durchschnitt der Azidität einer 1 %igen Zitronensäure entspricht. Auf Grund dieser Feststellung sollte die 1 %ige Zitronensäure ein besonders brauchbares Lösungsmittel zur Bestimmung der pflanzenverfügbaren Nährstoffe im Boden sein. Nach DYER (1894) ist ein Boden phosphorsäurebedürftig, wenn er weniger als 10 mg zitronensäurelösliche Phosphorsäure je 100 g Boden enthält.

Eine 1 %ige Zitronensäure wurde von verschiedenen Seiten als Lösungsmittel zur Bestimmung der Phosphorsäurebedürftigkeit der Böden gewählt, so von LEMMERMANN und FRESENIUS (1923), von KÖNIG und HASENBÄUMER (1926) und v. D. PAAUW (1956). ARRHENIUS (1928) benutzte dagegen eine 2 %ige Zitronensäure und VON DER SPUIJ (1925) eine 1/2 %ige Zitronensäure.

Während in den meisten Fällen die Phosphorsäurebedürftigkeit nur auf Grund der in der Zitronensäure löslichen Phosphorsäuremenge beurteilt wird, bestimmt LEMMERMANN (1931) außerdem die Gesamtphosphorsäure im Boden, s. auch LEMMERMANN und FRESENIUS (1923), und berechnet die „relative Löslichkeit" der Phosphorsäure, das ist die prozentuale Menge der Gesamtphosphorsäure, die in 1 %iger Zitronensäure unter den gewählten Bedingungen löslich ist. Zur Beurteilung der Phosphorsäurebedürftigkeit eines Bodens stehen also 2 Werte zur Verfügung: mg P_2O_5 je 100 g Boden löslich in 1 %iger Zitronensäure und relative Löslichkeit der Phosphorsäure. Erfahrungsgemäß kann die Phosphorsäure in 2 Böden mit gleicher Menge an zitronensäurelöslicher Phosphorsäure aus dem Boden leichter aufgenommen werden, für den die relative Löslichkeit der Phosphorsäure größer ist. Von LEMMERMANN sind gemeinsam mit verschiedenen Mitarbeitern eine Reihe von Arbeiten erschienen, die sich mit der von ihm entwickelten Methode beschäftigen; s. LEMMERMANN, FRESENIUS und LESCH (1927) und LEMMERMANN und FRESENIUS (1930). Die Methode ist von verschiedenen Seiten nachgeprüft worden, so von ENGELS (1926), der vorschlug, die in 10 %iger Salzsäure lösliche Phosphorsäure zur Berechnung der relativen Löslichkeit der Phosphorsäure zu bestimmen, weiter von BAMBERG (1928), der den Einfluß des Trocknens der Böden auf die Phosphorsäuremenge bestimmte, die von der Zitronensäure gelöst wird. Aus lufttrocknen Bodenproben wird eine größere Phosphorsäuremenge gelöst als aus frischen Bodenproben. BAMBERG stellte weiter fest, daß der End-pH-Wert der Lösung einen großen Einfluß auf die gelöste Menge an Phosphorsäure hat.

Die von KÖNIG und HASENBÄUMER und die von LEMMERMANN entwickelten Methoden zur Bestimmung der Phosphorsäurebedürftigkeit der Böden mit Zitronensäure als Lösungsmittel wurden zur Untersuchung großer Serien von Böden relativ wenig benutzt. Hierfür dürfte weniger die Brauchbarkeit dieser Methoden verantwortlich zu machen sein als die Keimpflanzenmethode von NEUBAUER (s. diesen Bd., S. 863), die ungefähr gleichzeitig mit den eben erwähnten Methoden bekanntgegeben wurde. Der Keimpflanzenmethode wurde sofort nach ihrer Bekanntgabe ein größeres Vertrauen entgegengebracht als den chemischen Bodenuntersuchungsmethoden, da die Beurteilung der Böden auf ihren Nährstoffbedarf auf Grund der Nährstoffmenge erfolgte, die von Pflanzen aus dem Boden aufgenommen wird und nicht auf Grund der von chemischen Reagenzien gelösten Nährstoffmenge.

Auf die Zitronensäure als Lösungsmittel zur Bestimmung der Phosphorsäurebedürftigkeit ist man erst wieder zurückgekommen, als die Keimpflanzenmethode für Massenuntersuchungen zu umständlich wurde. Durch Verwendung colorimetrischer Methoden konnte die Phosphorsäure in den Auszügen leicht und schnell bestimmt werden. ARRHENIUS (1928) hat wohl als Erster in großer Zahl Böden auf ihre Phosphorsäurebedürftigkeit mit Zitronensäure als Lösungsmittel untersucht. Bis vor kurzem wurde auch der Phosphorsäuregehalt der Böden in Belgien auf Grund des „P-citr. value" (mg P_2O_5/100 g Boden löslich in 1 %iger Zitronensäure) nach V. D. PAAUW (1956) ermittelt.

Die analytische Bestimmung der Pflanzennährstoffe soll in diesem Abschnitt nicht behandelt werden. Der Verfasser hält es aber für notwendig, hier, wo zum erstenmal colorimetrische Methoden erwähnt werden, darauf hinzuweisen, daß nur solche Analysenmethoden benutzt werden sollten, von denen man genau weiß, daß die Ergebnisse durch zufällig in den Auszügen vorhandenen Stoffen nicht beeinflußt werden. Ein brauchbares Lösungsmittel zur Bestimmung der Phosphorsäurebedürftigkeit der Böden kann nicht eingeführt werden, wenn es nicht möglich ist, den Phosphorsäuregehalt in den Auszügen mit der notwendigen Genauigkeit, die sehr sorgfältig überlegt sein soll, schnell von angelernten Arbeitskräften bestimmen zu lassen.

Lösungen von *Salzen der Zitronensäure* werden relativ selten zur Bestimmung der Phosphorsäurebedürftigkeit der Böden benutzt, obgleich der Phosphorsäuregehalt verschiedener Düngemittel mit Ammoniumcitratlösungen bestimmt wird. KÖNIG (1923, S. 68) erwähnt, daß eine 2 %ige Ammoniumcitratlösung zur Untersuchung von Böden benutzt wurde, er schreibt: „auch dürfte Ammoniumcitrat am geeignetsten sein, diejenige Menge der Bodenphosphorsäure, welche der citratlöslichen des Thomasmehles gleichkommt, zu bestimmen". BAMBERG (1936) hat verschiedene Böden mit einer neutralen Ammoniumcitratlösung behandelt. Eine Bewertung der Böden auf Grund der gelösten Phosphorsäure entsprach einer Beurteilung nach der Methode MITSCHERLICH. Die neutrale 1 %ige Ammoniumcitratlösung hat aber den Nachteil, daß die Bodenauszüge sehr langsam filtrieren, da Ammoniumcitrat ein gutes Peptisationsmittel für Ton ist.

2. *Essigsäure.* DEHÉRAIN (1884) berichtet, daß LECHARTIER an Stelle von Kohlensäure Essigsäure benutzte, um die assimilierbare Phosphorsäure im Boden zu bestimmen.

DEHÉRAIN (1891) war der Ansicht, daß nicht nur der Gehalt an Gesamtphosphorsäure, sondern auch die in Essigsäure lösliche Phosphorsäure bestimmt werden sollte, wenn nach dem Phosphorsäuregehalt die Fruchtbarkeit eines Ackers zu beurteilen ist. DEHÉRAIN und MEYER (1879) war auch bekannt, daß Essigsäure die an Sesquioxyd gebundene Phosphorsäure nicht löst, sondern nur diejenige, die an „Kalk" und andere Basen gebunden ist. Nach GEDROIZ (1903) sagen die in 2 %iger Essigsäure und auch die in 2 %iger Zitronensäure löslichen Mengen an Phosphorsäure nichts aus über die Düngerbedürftigkeit der Böden. Essigsäure als Lösungsmittel zur Bestimmung der Phosphorsäurebedürftigkeit der Böden hat den Nachteil, daß bei einigen Böden einmal von der Essigsäure gelöste Phosphorsäure bei längerer Behandlung des Bodens mit Essigsäure wieder festgelegt wird, wie Untersuchungen von LEMMERMANN, EINECKE und FRESENIUS (1917) und von LEMMERMANN und RAUTERBERG (1949) gezeigt haben. Da Essigsäure und besonders Acetatpufferlösungen, die an Eisen gebundene Phosphorsäure nicht lösen, werden derartige Lösungen wahrscheinlich besonders brauchbar sein, um die in dieser Form vorhandene Phosphorsäure im Boden zu bestimmen, außerdem muß dann natürlich die an Calcium und an Eisen gebundene Phosphorsäure bestimmt werden, vielleicht durch Behandlung der Böden mit Oxalsäure oder

Oxalatpufferlösungen. Aluminiumphosphat wird freilich von der Essigsäure leichter gelöst als Eisenphosphat, über die Löslichkeit dieser beiden Phosphate in Acetatlösungen verschiedener pH-Werte liegen keine Untersuchungsergebnisse vor. Es ist außerdem überhaupt fraglich, ob sich in einem Boden Aluminiumphosphat bildet, wenn schwerer lösliches Eisenphosphat gebildet werden kann.

Acetatpufferlösungen werden bei verschiedenen Schnellmethoden zur Bestimmung der Phosphorsäurebedürftigkeit der Böden verwendet. MORGAN (1937) benutzt z. B. eine Natriumacetatlösung pH 4,8, s. auch Methodenbuch S 254. Nach Untersuchungen von RAUTERBERG und OSSENBERG-NEUHAUS (1950) löst eine Acetatpufferlösung pH 4,88 (freilich Ammoniumacetat) aus sauren Böden sehr kleine Phosphorsäuremengen, während aus gleich behandelten neutralen Böden erheblich größere Mengen an Phosphorsäure gelöst werden, so daß bei Verwendung von einem derartigen Lösungsmittel der pH-Wert der Böden bei Auswertung der Ergebnisse ebenso berücksichtigt werden muß wie bei der Lactatmethode, die später behandelt wird. Die MORGAN-Methode ist zum Teil etwas geändert worden, so werden in Frankreich nach einem Vorschlag von BARBIER, s. KNICKMANN (1955)

5 g Boden mit 10 ml Natriumacetatlösung, pH 4,8, 1 Min. geschüttelt, sofort filtriert und im Filtrat P_2O_5, K und zahlreiche andere Pflanzennährstoffe und für Pflanzen giftige Elemente bestimmt.

Eine Essigsäure-Calciumacetatlösung pH 4 wird von KRUMINŠ (1932) zur Untersuchung von Mineralböden mit pH-Werten über 5,5 und von humusreichen Böden benutzt. Für saure Böden verwendet KRUMINŠ eine Essigsäure-Aluminiumchloridlösung pH 2,6. BURRIEL und HERNANDO (1950) verwenden ein Lösungsmittel, das sie aus Essigsäure, Schwefelsäure, Calciumcarbonat und Magnesiumcarbonat herstellen.

RUBINS und DEAN (1946) haben verschiedene Methoden, die Acetate und Essigsäure als Lösungsmittel benutzen und andere zur Untersuchung einiger Böden verwendet und die Ergebnisse miteinander verglichen. HOFFMANN (1930) hat 42 Mineralböden auf ihren Phosphorsäuregehalt nach verschiedenen Methoden untersucht. Er bestimmte dabei die Löslichkeit der Bodenphosphorsäure in Acetatlösungen pH 3 und pH 4. VUORINEN und MÄKITIE (1955) verwenden eine saure Ammoniumacetatlösung (0,5 n CH_3COOH, 0,5 n CH_3COONH_4, pH 4,65) zur Bestimmung von P_2O_5, K_2O und Ca. Sie haben 70 Böden nach verschiedenen Methoden (1 n CH_3COONH_4, pH 7,0; MORGAN-Lösung; Lactatmethode; 2,5 %iger Essigsäure, 1 %iger Zitronensäure und 0,2 n H_2SO_4 und obige Lösung) untersucht (s. auch MÄKITIE 1956).

MITSCHERLICH (1939) hat gemeinsam mit BEUTELSPACHER etwa 1362 Böden, die nach der von ihm entwickelten Gefäßmethode untersucht waren, mit einer Magnesiumacetat-Essigsäure-Pufferlösung pH 3,5 auf ihren Phosphorsäuregehalt untersucht. MITSCHERLICH behandelt 4 g Boden mit 100 ml der Pufferlösung, die 1 g MgO und 15 ml Eisessig in 1 l enthält, 15 Std. in der Rotiermaschine. Im Filtrat bestimmt er nach Zerstörung der organischen Substanz (H_2O_2) und Abscheidung der Kieselsäure die Phosphorsäure. Die gleichen Böden waren in einer Gemeinschaftsarbeit nach der Lactat-Methode, nach der Magnesiumbicarbonat-Kohlensäure-Methode und zum Teil nach der Aspergillusmethode untersucht worden, außerdem ist die Festlegung der Düngerphosphorsäure in Magnesiumbicarbonat-Kohlensäure bestimmt worden. Die starke Phosphorsäurebedürftigkeit der meisten Böden, die untersucht wurden, zeigt sich auch in der relativ geringen Phosphorsäuremenge, die in den 3 Lösungsmitteln gelöst wurde.

3. *Oxalsäure.* Oxalsäure wurde von GERLACH (1896), EMMERLING (1899), GIGIEL (1924), KUDASHEV (1905), JAFFA (1895), SNYDER (1895) und DAIKUHARA

(1903) zur Bestimmung der Phosphorsäurebedürftigkeit der Böden benutzt. GIGIEL hat festgestellt, daß von den im Boden vorkommenden Phosphaten von einer 0,5 n Oxalsäure Tricalciumphosphat, Aluminiumphosphat, Eisenphosphat, Apatit und Vivianit gelöst werden, Lezithin und Nucleinsäureverbindungen dagegen nicht. GERLACH hat mit Zitronensäure bessere Ergebnisse erzielt als mit Oxalsäure. Nach EMMERLING (1899) muß Oxalsäure geeigneter sein, die „absorbierte Phosphorsäure" im Boden zu bestimmen als Zitronensäure, da Zitronensäure bei einem Eisenoxydüberschuß nur 1/10 der vorhandenen Ferriphosphate löst. Daß Oxalsäure eine größere Menge Bodenphosphorsäure löst als Zitronensäure, Weinsäure und Essigsäure wurde auch von SNYDER und DAIKUHARA festgestellt. Selbst eine $\frac{n}{4}$ Ammoniumoxalatlösung pH 4,88 löst nach RAUTERBERG und OSSENBERG (1950) erheblich mehr Bodenphosphorsäure als eine $\frac{n}{4}$ Ammoniumacetatlösung pH 4,83 und zwar ungefähr die gleiche Menge, die von einer 0,5 n Salzsäure gelöst wird. KUDASHEV (1905) hat bei der Untersuchung von 62 Böden mit 0,5 %iger Oxalsäure brauchbare Angaben über die Phosphorsäurebedürftigkeit der Böden erhalten.

4. *Milchsäure.* Die Milchsäure selber wird als Lösungsmittel der Bodenphosphorsäure nicht verwendet, aber eine Pufferlösung der Milchsäure. Sehr viel benutzt wird zur Zeit eine mit Salzsäure auf pH 3,7 eingestellte Calciumlactatlösung. Dieses Lösungsmittel ist von EGNER (1932) vorgeschlagen worden. EGNER (1940, 1951/52, 1954/55) und RIEHM (1938, 1940, 1946) haben durch zahlreiche Untersuchungen gezeigt, daß mit diesem Lösungsmittel die Düngebedürftigkeit des Bodens bestimmt werden kann, wenn bei der Auswertung der Ergebnisse die pH-Zahl des Bodens und die Bodenart berücksichtigt wird.

5 g Boden werden mit 250 ml Lactatlösung, die für Calciumlactat 0,02 n und für Salzsäure 0,01 n ist und genau den pH-Wert 3,7 haben soll, 2 Std. geschüttelt. Im Filtrat wird Phosphorsäure colorimetrisch und Kalium flammenphotometrisch bestimmt. Die Auswertung der Ergebnisse geht nach RIEHM 1940 auf einen Vorschlag von O. FRANK zurück, der folgende Grenzzahlen aufgestellt hat:

bis 5,5	mg P_2O_5/100 g Boden unbedingt phosphorsäurebedürftig
5,6 — 11	mg P_2O_5/100 g Boden schwach phosphorsäurebedürftig
über 11	mg P_2O_5/100 g Boden nicht phosphorsäurebedürftig.

Mit diesen Grenzzahlen werden nun aber die erhaltenen Untersuchungsergebnisse nicht direkt verglichen, sondern sie werden zur Berücksichtigung der Bodenart und des pH-Wertes der Böden mit Faktoren multipliziert. Diese Faktoren sind:

für reine Sandböden		1,0
für lehmige Sande und sandige Lehme		1,5
für Lehm- und Tonböden		2,0
für Böden mit einem pH-Wert in KCl bis	5,5	1,0
für Böden mit einem pH-Wert in KCl bis	5,6—6,0	0,9
für Böden mit einem pH-Wert in KCl bis	6,1—6,5	0,8
für Böden mit einem pH-Wert in KCl über	6,5	0,7.

Diese Faktoren berücksichtigen, daß die Lactatlösung aus schweren Böden eine relativ geringere Phosphormenge löst als aus leichteren Böden, was auf das starke Adsorptionsvermögen schwerer Böden für Phosphorsäure zurückzuführen ist.

Die Faktoren für die pH-Werte der Böden berücksichtigen, daß die in sauren Böden an Eisen und Aluminium gebundene Phosphorsäure, die auch zum Teil von den Pflanzen aufgenommen wird, nicht in dem Maße von Lactatlösung gelöst wird wie die an Calcium gebundene Phosphorsäure, die in erster Linie in den neutralen und alkalischen Böden enthalten ist. Die geringe Löslichkeit der Bodenphosphorsäure saurer Böden in Calciumlactatlösung im Vergleich zur

Löslichkeit der Phosphorsäure schwach saurer und neutraler Böden im gleichen Lösungsmittel ist aus Untersuchungen von RAUTERBERG und OSSENBERG-NEUHAUS (1950) zu ersehen. Bodenproben, stark unterschiedlich im pH-Wert, auf Parzellen eines Versuchsfeldes genommen, das seit 1929 nur mit Stallmist bzw. mit Stallmist und mineralischer Phosphorsäure gedüngt war, waren nach der Lactatmethode untersucht worden. Die neutralen Böden enthalten, wie folgende Zahlen zeigen, über 2 1/2mal soviel Phosphorsäure wie die sauren Böden. Nur mit Stallmist gedüngte Böden bei pH 6,6 bis 6,7 22 bis 23 mg P_2O_5 und bei pH 4,3 bis 5,0 8,2 bis 8,6,mg P_2O_5. Bei Phosphorsäurezufuhr durch Stallmist und mineralische Düngung bei pH 6,7 und 6,8 28 und 31 mg P_2O_5 und bei pH 4,5 11 bis 12 mg P_2O_5. RAUTERBERG und OSSENBERG-NEUHAUS (1950) haben auf Grund dieser Ergebnisse vorgeschlagen, zu prüfen, ob der Einfluß der Bodenreaktion auf den Gehalt der Böden an lactatlöslicher Phosphorsäure durch weitere Faktoren berücksichtigt werden müßte. Die Auswertung der bei der Lactatmethode erhaltenen Ergebnisse läßt sich einfacher nach einer Tabelle mit Grenzzahlen durchführen, die RIEHM (1940) unter Berücksichtigung der oben angegebenen Grenzzahlen und Faktoren für die Böden verschiedener Bodenarten und pH-Werte aufgestellt hat.

Da bei der Behandlung karbonathaltiger Böden mit der Lactatlösung der pH-Wert des Extraktionsmittels verändert wird, ist diese Methode zur Bestimmung der P_2O_5-Bedürftigkeit karbonathaltiger Böden unbrauchbar, wie auch von SCHEFFER, WELTE und HEINEMANN (1952) festgestellt wurde. RIEHM (1958) hat daher zur Bestimmung der leichtlöslichen Phosphorsäure dieser Böden vorgeschlagen, sie mit einer Ammoniumlactat-Essigsäure (AL-Lösung) zu behandeln.

5 g Boden werden mit 100 ml der AL-Lösung (0,1 n Ammoniumlactat, 0,4 n Essigsäure, pH 3,7) 4 Stdn. bei $20 \pm 1°$ C in einer Rotiermaschine geschüttelt. Im Filtrat wird die gelöste Phosphorsäure bestimmt.

Von EGNER, RIEHM und DOMINGO (1960) sind Analysenvorschriften ausgearbeitet worden, um die gelösten Nährstoffe, Phosphorsäure und Kalium mit wirklich genügender Zuverlässigkeit bestimmen zu können. Zur Zeit werden von ihnen eingehende Untersuchungen darüber durchgeführt, ob mit dieser Methode die P_2O_5- und K-Bedürftigkeit der Böden mit größerer Sicherheit bestimmt werden kann als mit anderen Methoden, die zur Zeit benutzt werden. RIEHM (1958) gibt folgende vorläufige Grenzzahlen an, die noch überprüft werden müssen: I (gut versorgt) über 20 mg, II (mäßig versorgt) 10 bis 20 mg, III (schlecht versorgt) 0 bis 10 mg P_2O_5 je 100 g Boden.

Zur Bestimmung der Phosphorsäurebedürftigkeit der Böden kann auch an eine Behandlung der Böden mit 2 verschiedenen Lösungsmitteln nacheinander, ohne die erste Lösung abzutrennen, gedacht werden. RAUTERBERG und OSSENBERG-NEUHAUS (1951/52) hatten Böden (4 g) mit 100 ml Calciumhydroxyd (mit 10 mg CaO) 1 Std. geschüttelt und dann 100 ml einer doppelt konzentrierten Calciumlactatlösung dazu gegeben und eine weitere Stunde geschüttelt. Im Filtrat wurde die gelöste Phosphorsäure bestimmt. Die gleichen Böden waren auch nach der üblichen Vorschrift auf ihren Gehalt an lactatlöslicher Phosphorsäure untersucht worden. Die unterschiedliche Versorgung der Böden mit mineralischer Phosphorsäure in etwa 15 Jahren war aus den Untersuchungsergebnissen, die bei Vorbehandlung des Bodens mit Calciumhydroxydlösung erhalten waren, erheblich besser zu erkennen als aus den Untersuchungsergebnissen der normalen Lactatmethode.

Auch eine 1%ige Asparaginsäurelösung wurde von MAXWELL (1899) zur Bestimmung der P_2O_5-Bedürftigkeit von Zuckerrohrböden benutzt.

Untersuchung der Bodenlösung. Die Bodenlösung kann aus dem angefeuchteten Boden durch Auspressen oder durch Verdrängen mit einem Lösungsmittel erhalten werden. RAMANN (1918) hat festgestellt, daß zu verschiedenen Zeiten im Jahr genommene Bodenproben im Preßsaft unterschiedliche Mengen an Phosphorsäure enthalten. WRANGELL und HAASE (1926) berichten, wie von verschiedenen Seiten Bodenlösungen gewonnen werden, und haben selber den Einfluß verschiedener Faktoren auf den Phosphorsäuregehalt der Bodenlösung untersucht. Auch von DRATSCHEW und ALEXANDROWA (1932) wurden entsprechende Untersuchungen durchgeführt. Aus diesen verschiedenen Untersuchungen muß der Schluß gezogen werden, daß aus dem Gehalt der Bodenlösung an Phosphorsäure keine bindenden Schlüsse gezogen werden können, ob ein Boden phosphorsäurebedürftig ist oder nicht. Mitverantwortlich für diese Schlußfolgerung sind die unterschiedlichen Mengen an Anionen in der Bodenlösung, die nicht vom Boden adsorbiert werden. Diese Anionen sorgen dafür, daß die Bodenlösung eine entsprechende Menge an Kationen, in erster Linie Calciumionen, enthält. Die Calciumionen bestimmen dann ihrerseits die Konzentration der Phosphorsäure in der Bodenlösung. Ein Versuchsergebnis von NOLTE (1914) zeigt, daß eine Bodenlösung bei einem engen Bodenwasserverhältnis relativ viel Calcium und relativ wenig Phosphorsäure enthält. Mit steigendem Bodenwasserverhältnis nimmt der Calciumgehalt in der Bodenlösung wenig, der Phosphorsäuregehalt dagegen stark zu.

KAWE (1935, 1936, 1959) verwendet Bodenlösungen zum Ausschütteln von Bodenproben zur Bestimmung der Phosphorsäure und Kalibedürftigkeit der Böden. Die Bodenlösungen werden durch Verdrängen gewonnen und 50 Boden werden mit 125 ml 1:1 mit Wasser verdünnter Bodenlösung, die aus dem gleichen Boden hergestellt wurde, behandelt. Im Bodenauszug wird der Phosphorsäuregehalt bestimmt. (Bei Bestimmung der Kalibedürftigkeit berücksichtigt KAWE den Kaligehalt in der Bodenlösung s. S. 817). Die Methode hat bei Vergleichsuntersuchungen gut abgeschnitten (s. auch KIRSTE, 1952).

Kohlensäurefreies Wasser als Lösungsmittel. Schon 1859 wurde von STÖCKHARDT (1859) festgestellt, daß die Nährstoffaufnahme von Pflanzen aus Böden in einem Vegetationsversuch in Beziehung steht zu der Nährstoffmenge, die sich bei dreimaligem Digerieren dieser Böden mit reinem Wasser lösen. Über andere Untersuchungen mit reinem Wasser berichten STEWART (1918) und GEHRING (1931). Bei der Verwendung von kohlensäurefreiem Wasser als Lösungsmittel für die Bodenphosphorsäure muß bedacht werden, daß die gelöste Phosphorsäuremenge durch die gelöste Calciummenge beeinflußt wird, worauf schon hingewiesen wurde. Den Einfluß des Calciums auf diese Löslichkeit der Bodenphosphorsäure zeigen nach DIRKS und SCHEFFER (1930/2), auch Untersuchungsergebnisse von GREGOIRE, der verschiedene Böden zehnmal hintereinander mit Wasser behandelte und in den Auszügen CaO und P_2O_5 bestimmte. Die ersten Bodenauszüge enthalten mehr CaO als die späteren und verschiedentlich weniger P_2O_5.

Als colorimetrische Methoden zur Bestimmung kleiner Phosphorsäuremengen entwickelt waren, wurde von verschiedener Seite wieder versucht, die Phosphorsäurebedürftigkeit der Böden in Wasserauszügen zu bestimmen. So wurden von NEMEC (1926, 1928) von DIRKS und SCHEFFER (1928, 1930) und von VON WRANGELL (1931) entsprechende Methoden vorgeschlagen.

NEMEC behandelt 30 g Boden 1 Std. mit 100 ml Wasser und bestimmt im Filtrat P_2O_5 und SiO_2. Nach DIRKS (1930) wird kohlensäurefreies Wasser als Lösungsmittel nur zur Untersuchung saurer Böden benutzt, für neutrale und alkalische Böden wird eine mit Kohlensäure gesättigte Calciumbicarbonatlösung benutzt, es wird freilich zum kohlensäurefreien Wasser eine kleine Menge (0,0125%) KCl gegeben, damit klare Filtrate erhalten werden. Es werden 30 g Boden 1 Std. mit 75 ml Lösung geschüttelt. VON WRANGELL behandelt 1 g Boden mit 100 ml Wasser 5 Stdn. in der Rotiermaschine, der Boden wird in einer Zentrifuge vom Auszug getrennt und wieder mit 100 ml Wasser behandelt. In beiden Auszügen wird die gelöste Phosphorsäure bestimmt. Aus den erhaltenen Werten a mg P_2O_5 im ersten Auszug und b mg P_2O_5 im zweiten Auszug wird die Gesamtmenge an löslicher Phosphorsäure x nach der Gleichung $x = \frac{a^2}{a-b}$ berechnet, s. auch MEYER (1931).

Von DIRKS und SCHEFFER (1930) wird darauf hingewiesen, daß die Ergebnisse durch das gelöste Calcium beeinflußt werden. Die wasserlösliche Phosphorsäure muß in den beiden Auszügen mit möglichst großer Genauigkeit bestimmt werden, da die Streuung der berechneten Gesamtmenge an wasserlöslicher Phosphorsäure erheblich größer ist als die Streuung der Einzelwerte, was durch die Differenz der Werte *a* und *b* im Nenner bedingt ist. Die Methode von VON WRANGELL ist von TOMMASI und MARIMPIETRI (1935) noch erweitert worden, sie bestimmen außer der Gesamtmenge an wasserlöslicher Phosphorsäure die Gesamtmenge an Phosphorsäure, die bei steigender Phosphorsäuregabe gelöst bleibt. Die Ergebnisse werden graphisch aufgetragen und aus der erhaltenen Kurve wird die notwendige Phosphorsäuredüngung abgelesen.

ROHDE (1933) bestimmt die in kochendem Wasser lösliche Bodenphosphorsäure.

25 g Boden werden mit 2 l destilliertem Wasser in einem 3 l Rundkolben $^1/_2$ Stunde gekocht, anschließend werden bei mäßigem Weiterkochen mit einer Chamberland-Filterkerze 500 ml Lösung abgesaugt. Ein mittelversorgter Boden enthält nach ROHDE 4 bis 6 mg P_2O_5 je 100 g, also eine relativ große Menge an „wasserlöslicher" Phosphorsäure, das ist wahrscheinlich darauf zurückzuführen, daß ein Teil der organisch gebundenen Phosphorsäure hydrolysiert wird, s. EMMERLING (1899).

THUN (1936/37) hat die Methode ROHDE mit der Keimpflanzen-Methode verglichen und für Phosphorsäure eine relativ gute Übereinstimmung der Werte festgestellt.

ACHROMEIKO (1928) hat festgestellt, daß durch Trocknen des Bodens sein Gehalt an wasserlöslicher Phosphorsäure steigt.

Kohlensäurehaltiges Wasser und Lösungen von Karbonaten und von Bikarbonaten (zum Teil mit einem Überschuß an Kohlensäure). Da die Bodenlösung mehr oder weniger mit Kohlensäure angereichert ist, die sowohl von den Mikroorganismen als auch von den Pflanzenwurzeln ausgeschieden wird, und da diese gelöste Kohlensäure die Löslichkeit der im Boden vorhandenen Salze beeinflußt, also einen Einfluß auf die Konzentration der Pflanzennährstoffe in der Bodenlösung hat, war es naheliegend, den Boden mit kohlensäurehaltigem Wasser zu behandeln, um die vorhandenen Pflanzennährstoffe zu bestimmen. Es sind, wie aus den Ausführungen von GEHRING (1931) hervorgeht, auch schon in der zweiten Hälfte des vorigen Jahrhunderts und später zahlreiche entsprechende Untersuchungen durchgeführt worden. So gibt BIELER (1909) an, daß Böden phosphorsäurebedürftig sind, wenn sie weniger als 0,015% P_2O_5 löslich in CO_2-haltigem Wasser enthalten und wenn der Gesamt-P_2O_5-Gehalt kleiner als 0,25% ist. Eingehend hat sich auch MITSCHERLICH (1954, s. S. 126) damit beschäftigt, die Düngebedürftigkeit der Böden durch Ermittlung der in kohlensäurehaltigem Wasser löslichen Pflanzennährstoffe zu bestimmen, und es ist interessant, daß MITSCHERLICH (1910) überzeugt war, das Problem gelöst zu haben, denn er schreibt:

„In diesen Ergebnissen finden Sie bereits das alles, was wir suchen, nämlich, daß wir in der Tat mit unseren Bodenanalysen den Gehalt des Bodens an aufnehmbaren Pflanzennährstoffen festzustellen vermögen, daß wir also das erreicht haben, wonach wir jahrelang in unserer stillen Werkstätte strebten."

LEMMERMANN (1914) kritisierte die Ausführungen MITSCHERLICHS, MITSCHERLICH hat seine Ansicht auch später geändert. Er vertrat viele Jahre sogar den Standpunkt, daß nur durch Vegetationsversuche in Gefäßen neben Felddüngungsversuchen die Düngebedürftigkeit eines Feldes wirklich bestimmt werden kann.

Kohlensäurehaltiges Wasser in Verbindung mit Calciumcarbonat, also eine gesättigte Lösung von Calciumbicarbonat in überschüssiger Kohlensäure, wird in der Methode von DIRKS und SCHEFFER (1928, 1930) zur Bestimmung der leichtlöslichen Phosphorsäure neutraler und alkalischer Böden benutzt, während saure Böden mit kohlensäurefreiem Wasser (mit 0,0125% KCl) behandelt werden, wie S. 808 ausgeführt. Durch die Behandlung calciumhaltiger Böden mit diesem Lösungsmittel wird ein unterschiedlicher Einfluß der Calciumionen des Bodens auf die Löslichkeit der Bodenphosphorsäure ausgeschaltet, denn nach dem Massenwirkungsgesetz muß bei einem Überschuß an HCO_3-Ionen die Konzentration der Ca-Ionen sehr klein sein. Die Konzentration der Phosphorsäure im Auszug wird also in erster Linie durch die Menge entsprechend löslicher Phosphorsäure und dem Adsorptionsvermögen des Bodens für Phosphorsäure bestimmt. Da die Phosphorsäure in saurem Boden einerseits und in neutralen und basischen Böden andererseits verschieden gebunden ist und da die Löslichkeit dieser adsorbierten Phosphorsäure und der vorhandenen Phosphate verschieden ist, dürfte die Verwendung verschiedener Lösungsmittel für saure Böden und für neutrale und basische Böden wie bei der DIRKS-SCHEFFER-Methode ein Problem sein, das wert ist nochmals eingehend untersucht zu werden. Die gelegentlich festgestellte schlechte Übereinstimmung dieser Bodenuntersuchungsmethode mit den Ergebnissen der Keimpflanzenmethode ist vielleicht zum Teil auf analytische Schwierigkeiten zurückzuführen, denn bei der früher benutzten Molybdänblau-Methode mit Stannochlorid als Reduktionsmittel traten bei nicht sehr erfahrenen Analytikern leicht Fehler auf, die nicht erkannt wurden. Heute stehen Analysenmethoden zur colorimetrischen Bestimmung der Phosphorsäure und sorgfältig entwickelte Photometer zur Verfügung, mit denen kleine Phosphorsäuremengen in Bodenauszügen sehr viel genauer als früher bestimmt werden können. Wahrscheinlich ist auch bei der Auswertung der Untersuchungsergebnisse, die mit der DIRKS-SCHEFFER-Methode erhalten werden, nicht genügend berücksichtigt worden, daß die Bodenphosphorsäure von den Pflanzen in sauren Böden schlechter aufgenommen wird als in neutralen und alkalischen Böden, daß aber andererseits diese unterschiedliche Aufnahme der Bodenphosphorsäure nicht vollkommen durch die Verwendung der verschiedenen Lösungsmittel für saure Böden einerseits und für neutrale und alkalische Böden andererseits ausgeglichen werden kann, denn sonderbarerweise werden für alle Böden, also für die mit verschiedenen Lösungsmitteln erhaltenen Ergebnisse, die gleichen Grenzzahlen angegeben.

DREYSPRING und HEINZ (1934) verwenden ebenso wie DIRKS und SCHEFFER die Lösung eines Erdalkalicarbonats in kohlensäuregesättigtem Wasser zur Bestimmung der Phosphorsäurebedürftigkeit der Böden, nämlich eine 0,1 n Magnesiumbicarbonatlösung mit überschüssiger Kohlensäure, die einen pH-Wert von 6,45 bis 6,55 haben soll. Dieses Lösungsmittel wird aber für alle Böden benutzt. Die Verfasser bestimmen nicht nur die Löslichkeit der Phosphorsäure in der 0,1 n $Mg(HCO_3)_2$-Lösung, sondern auch das Nachlieferungsvermögen und die Festlegung der Düngerphosphorsäure und die Nachlieferung derselben.

20 g Boden werden mit 100 ml des Lösungsmittels behandelt, die Lösung wird durch eine Filterkerze, die sich im Stopfen der Schüttelflasche befindet, abgesaugt und der Boden wird noch einmal mit 100 ml Lösungsmittel behandelt. Zur Bestim-

mung der Festlegung der Düngerphosphorsäure und zur Ermittlung der Nachwirkung der Düngerphosphorsäure wird ebenso verfahren, nur, daß zu den 20 g Boden 1 mg P_2O_5 in leichtlöslicher Form gegeben wird (s. auch DREYSPRING und HEINZ 1934 und 1935, DREYSPRING und KRÜGEL 1929).

Alkalische Lösungsmittel. Nach KÖNIG (1923) (S. 72) glaubte GRANDEAU gefunden zu haben, „daß die Fruchtbarkeit der Ackererde in direkter Beziehung stehe zu der Menge der an organischer Substanz gebundenen, in Ammoniak löslichen Mineralstoffe (Phosphorsäure, Kali und Kalk), daß der Fruchtbarkeitsunterschied verschiedener Bodenarten mehr durch den Phosphorsäuregehalt des Ammoniakauszuges als durch den des Säureauszuges angezeigt wird". KÖNIG bringt (S. 71) auch einige Zahlen über den Phosphorsäuregehalt von 6 Böden in 6 %igem Ammoniak. Außerdem ist der Phosphorsäuregehalt in 10 %iger Salzsäure, in 2 %iger Zitronensäure und in 2 % Ammoniumzitratlösung angegeben.

STODDART (1909) bestimmte die an Eisen und Aluminium gebundene Phosphorsäure durch Behandlung des Bodens mit 1 % NaOH. Auch RAUTERBERG und OSSENBERG-NEUHAUS (1950) haben die in 1 %iger Natronlauge lösliche Phosphorsäure in Böden bestimmt. Der prozentuale Anteil von der Gesamtphosphorsäure, die in Natronlauge löslich ist, steht in direkter Beziehung zum pH-Wert des Bodens.

Nach JACKSON (1958) (S. 162) gibt es zwei Arten von alkalischen Lösungsmitteln zur Extraktion der Bodenphosphorsäure: schwach-alkalische Lösungen, wie Natriumbicarbonat (von OLSEN und Mitarbeiter) und Kaliumcarbonat einerseits, und stark-alkalische Lösungen wie Natronlauge andererseits. Die in schwachalkalischen Lösungen lösliche Phosphorsäure steht in Beziehung zu der pflanzenverfügbaren Phosphorsäure, während von der Natronlauge die an Eisen und Aluminium gebundene und die in organischen Verbindungen enthaltene Phosphorsäure gelöst wird.

Bestimmung der austauschbaren Phosphorsäure unter Verwendung von P^{32}. Wenn ein Boden mit Wasser oder einem Lösungsmittel, in dem nur ein kleiner Teil der Bodenphosphorsäure löslich ist, behandelt wird, dann hat sich nach einiger Zeit ein Gleichgewicht eingestellt zwischen der Phosphorsäure in der Lösung und der Phosphorsäure gebunden vom Boden. Durch Zugabe einer bestimmten Menge P^{32} ist versucht worden, die Menge der Bodenphosphorsäure zu bestimmen, die mit der Phosphorsäure in der Lösung im Gleichgewicht steht. Da sich die P^{32} enthaltenden Phosphationen ebenso verhalten wie die P^{31} Phosphationen, muß das Verhältnis P^{32} gebunden am Boden zu P^{32} in der Lösung gleich dem Verhältnis P^{31} am Boden zu P^{31} in der Lösung sein, wenn keine weiteren Umsetzungen erfolgen. Da P^{32} Gesamt und P^{32} in der Lösung durch die Intensität der β-Strahlung gemessen werden kann und P^{31} in der Lösung analytisch zu bestimmen ist, kann P^{31} am Boden berechnet werden. Diese Berechnung von P^{31} am Boden ist aber nur möglich, wenn es wirklich zur Einstellung eines Gleichgewichts kommt, d. h. wenn in der Zeiteinheit ebenso viele Phosphationen sowohl mit P^{31} als auch mit P^{32} gelöst und wieder vom Boden gebunden werden. Ein derartiges Gleichgewicht wird wahrscheinlich nur möglich sein, wenn die Phosphorsäure adsorptiv im Boden gebunden ist. Enthält der Boden dagegen schwer lösliche Phosphate, dann werden die zugegebenen P^{32} Phosphationen nach und nach aus der Lösung verschwinden. In der gesättigten Lösung eines Salzes, das noch als Bodenkörper vorhanden ist, lösen sich die kleinen Kristalle nach und nach auf und die großen Kristalle wachsen. Beim Wachsen der großen Kristalle werden P^{31} und P^{32} Phosphationen im vorhandenen Mengenverhältnis abgeschieden. Da einmal auf den großen Kristallen abgeschiedene P^{32} Ionen nicht wieder gelöst werden, denn gelöst werden Phosphationen aus den kleinsten Kristallen, müßen die P^{32} Phosphationen nach und nach aus der Lösung verschwinden.

Eine Bestimmung der adsorptiv gebundenen Phosphorsäure im Boden wird also nur möglich sein, wenn der Boden keine schwer löslichen Phosphate enthält, bzw. gebildet werden.

Aus Untersuchungen, die bisher von MCAULIFFE et al. (1947), OLSEN (1953_1, 1953_2) durchgeführt wurden, kann noch nicht bindend geschlossen werden, ob die Menge an austauschbarer Phosphorsäure im Boden unter Verwendung von P^{32} bestimmt werden kann.

B. Physikalisch-chemische Methoden

Untersuchung des Adsorptionsvermögens der Böden für Phosphorsäure. EMMERLING berichtet 1895 (S. 232), daß von zahlreichen auch sonst eingehend untersuchten Böden die „Absorptionskraft" des Bodens für Phosphorsäure bestimmt wurde. Eine Beziehung zwischen Fruchtbarkeit der Böden und ihrer Absorptionskraft für Phosphorsäure konnte EMMERLING freilich nicht feststellen. Aus Untersuchungsergebnissen von RAUTERBERG (1931), der die Berücksichtigung der Bodenart bei den Untersuchungen der Böden auf ihre Phosphorsäurebedürftigkeit untersucht hat, ist dagegen zu ersehen, daß zwischen dem Adsorptionsvermögen des Bodens für Phosphorsäure und dem Gehalt des Bodens an pflanzenverfügbarer Phosphorsäure nach der Keimpflanzenmethode eine Beziehung vorhanden ist, wenn von der Annahme ausgegangen wird, daß die Phosphorsäure nur von den Teilchen $< 1\,\mu$ im Boden gebunden wird.

Bei der oben beschriebenen Methode von DREYSPRING und HEINZ (1934) (s. S. 810) wird die Festlegung der Düngerphosphorsäure im Boden und die Nachwirkung der Düngerphosphorsäure dadurch bestimmt, daß zu der Magnesiumbicarbonatlösung, mit der 20 g Boden behandelt werden, 1 mg P_2O_5 gegeben wird. Im ersten und zweiten Auszug wird dann die nicht adsorbierte bzw. die wieder gelöste Phosphorsäure bestimmt. Siehe auch den Vorschlag von TOMMASI und MARIMPIETRI (1935, S. 809), die bei der Bodenuntersuchung nach VON WRANGELL verschiedene Phosphorsäuremengen zusetzen. PURVIS und BLUME (1941) behandeln 10 g Boden 30 Min. lang mit 25 ml einer KH_2PO_4-KCl-Lösung und bestimmen im Filtrat P_2O_5 und K_2O, außerdem wird von ihnen der Gehalt an Al, Ca, Mg, NH_4 und NO_3 bestimmt. BEATER (1949) verwendet eine Pufferlösung pH 1,5 aus Schwefelsäure und Natriumborat und behandelt 2,5 g Boden a) ohne P_2O_5 Zugabe, b) bei P_2O_5 Zugabe 30 Sek. mit 25 ml Pufferlösung bei 30°. Mit der Methode sollen sehr günstige Resultate erzielt worden sein. JACKSON (1958, S. 178) und DEAN und RUBINS (1947) berichten über Methoden zur Bestimmung der Austauschkapazität des Bodens für Phosphorsäure. Es kann auch daran gedacht werden, aus der Menge der austauschbaren Phosphorsäure und der Austauschkapazität des Bodens für Phosphorsäure auf die Bedürftigkeit des Bodens für Phosphorsäure zu schließen.

RAUTERBERG (1960) hat vorgeschlagen, die von einem Boden austauschbar gebundene Menge an Phosphationen dadurch zu bestimmen, daß der Boden einmal mit einem Extraktionsmittel ohne und ein weiteres Mal mit Zugabe (B_1—P_2O_5) von Phosphationen behandelt wird. Aus den in beiden Auszügen ermittelten P_2O_5-Mengen (P_0 und P_1) wird die Phosphorsäure (P_A) berechnet, die von dem Boden unter den gewählten Bedingungen adsorptiv gebunden ist und bei der Behandlung gelöst wird. Die Berechnung von P_A geht auf die Überlegung zurück, daß für beide Extraktionen die Verteilung der Phosphationen in der Lösung und gebunden vom Boden durch dieselbe Gleichsgewichtskonstante bestimmt wird.

$$K = \frac{P_0}{P_A - P_0} \qquad K = \frac{P_1}{P_A + B_1 - P_1}$$

Aus den beiden Gleichungen können P_A und K berechnet werden:

$$P_A = \frac{B_1 P_0}{P_1 - P_0} \text{ und } K = \frac{P_1 - P_0}{B_1 - (P_1 - P_0)}$$

P_A muß mit der P_2O_5-Menge im Boden in Beziehung stehen, die von den Pflanzen aufgenommen werden kann, und K gibt an, wie stark der Boden Phosphationen bindet. Da in der Gleichung zur Berechnung von P_A die Differenz von P_1 und P_0 im Nenner steht, muß der P_2O_5-Gehalt der Auszüge sehr genau bestimmt werden.

Elektrolyse des Bodens. KÖNIG (1923, S. 74) berichtet über verschiedene Vorschläge, die gemacht wurden, die Nährstoffbedürftigkeit der Böden mit Hilfe von Gleichstrom zu bestimmen. Eingehend ist das Problem aber erst von KÖTTGEN und Mitarbeitern (1929 bis 1941) bearbeitet worden. KÖTTGEN hat ein Elektro-Ultrafiltrationsgerät entwickelt. Der Boden wird in eine Mittelkammer gegeben, die von 2 Seitenkammern, in denen sich die Elektroden befinden, durch Ultrafilter getrennt ist. Bei der Elektrolyse werden die Kathoden- und Anodenflüssigkeiten aus den Seitenkammern abgesaugt, gleichzeitig wird in die Mittelkammer Wasser nachgegeben. In den nach und nach abgesaugten Lösungen werden die Nährstoffionen bestimmt. Da der Adsorptionskomplex negativ geladen ist, wandert er mit zur Anode und kann das Ultrafilter verstopfen, wie weit dieses ausgeschaltet werden kann, ist dem Verfasser nicht bekannt.

HERZNER (1931 und 1932) verwendet in einer Suspension von 1 g Boden in 50 ml Wasser als Anode ein Platinblech und als Kathode Quecksilber. Das Quecksilber befindet sich in einem Glasrohr, das mit Filtrierpapier und einem Leinenlappen verschlossen ist. Das Glasrohr enthält 10 ml 0,1 n HCl. Da das Quecksilber im Glasrohr durch Wasserstoffentwicklung dauernd in Bewegung ist, kommt es nicht zur Amalgambildung und die gesamten Kationen werden in der Salzsäure gelöst. Sie können einzeln bestimmt und ihre Gesamtmenge durch Titration ermittelt werden. Dauer der Elektrolyse bei 90 bis 100 Volt 3 Tage. S. auch GLERIA (1932), BRAWER und RANKIN (1933), DEAN (1934), REIFENBERG (1932) und TRÉNEL und FREY (1932).

Bestimmung der Erwärmung des Bodens im Thermostaten bei Zugabe von Zucker und Pflanzennährstoffen. Von REINAU (1938) ist eine thermokinetische Methode zur Bestimmung der leichtverfügbaren Pflanzennährstoffe entwickelt worden. REINAU bestimmt die Erwärmung der mit einer Zuckerlösung angefeuchteten Bodenprobe bei Zugabe unterschiedlicher Mengen an Stickstoff, Phosphorsäure- und Kaliumverbindungen:

100 g Boden werden unter Zugabe von Wasser nach und nach gut verkrümelt, bis aus der Probe ein Brei zu werden beginnt. Eine Bodenmenge, die unter Zugabe von Wasser beim Krümeln mit 2 Gabeln ein Volumen von 125 ml ergibt, wird mit einer zuckerhaltigen Nährlösung, deren Menge der Krümelversuch ergeben hat, in eine Thermosflasche gegeben, die in einen Thermostaten gestellt wird. Innerhalb von 48 Stdn. wird 8- bis 10mal alle 4 bis 6 Stdn. die Temperatur in den Thermosflaschen gemessen. Für jeden Boden werden von REINAU 8 Versuchsreihen mit unterschiedlichen Mengen an N, P_2O_5 und K_2O angesetzt. Die Temperaturen in den einzelnen Flaschen werden in Beziehung zur Zeit der Temperaturmessung graphisch aufgetragen.

Aus dem Verlauf der Temperaturkurve wird auf die Düngerbedürftigkeit des Bodens für die einzelnen Nährstoffe geschlossen.

Bei der Nachprüfung verschiedener Bodenuntersuchungsmethoden, über die SCHMITT (1938) und HERRMANN (1941) berichten, hat die thermokinetische Methode von REINAU bei der Untersuchung von 150 Böden aus 27 Feldversuchen in 68% der Fälle die Düngebedürftigkeit der Böden für Phosphorsäure und Kali hinreichend genau wiedergegeben. Die sicherste Methode war aber die Keimpflanzenmethode von NEUBAUER. Es kann natürlich auch daran gedacht werden, anstatt der Wärme, die von den Mikroorganismen entwickelt wird, die gebildete Kohlensäure zu bestimmen, zum Beispiel nach der Methode ISERMEYER (1952), vielleicht nach 24 und 48 Stdn. bei einem Wechsel des Barytwassers nach 24 Stdn.

Bindung der Nährstoffionen durch Ionenaustauscher. TEPE (1956) geht von der Überlegung aus, daß die Nährstoffionen im Boden vom Adsorptionskomplex auf Grund eines Konzentrationsgefälles zu den Wurzeln wandern und von den Wurzeln im Austausch gegen H-Ionen bzw. HCO_3-Ionen gebunden werden. Die H- und HCO_3-Ionen wandern ihrerseits wieder auf Grund eines Konzentrationsgefälles in den Boden. Zur Bestimmung der Nährstoffleistung des Bodens, früher auch Bodenleistung genannt, verwenden TEPE und LEIDENFROST (1956, 1958) sogenannte Austauschereinheiten (AE).

Die Austauschereinheit ist ein Dialysierschlauch, der mit einem Gemisch von einem Kationenaustauscher und einen Anionenaustauscher gefüllt ist. Die Austauscher haben 5 mval H- bzw. 5 mval HCO_3-Ionen gebunden. Die Austauschereinheit wird in ein Becherglas gestellt, in das 100 ml Boden und dazu Wasser bis zur Oberfläche des Bodens gegeben wird. Nach 24 Stdn. wird die Austauschereinheit aus dem Boden genommen, abgespült und die gebundenen Ionen werden mit 2 n HCl oder H_2SO_4 eluiert. Im Eluat werden die NH_4, K, Ca, Mg, Mn, NO_3 und Phosphationen bestimmt, auch B kann bestimmt werden.

Nach weiteren Untersuchungen von TEPE und LEIDENFROST (1958) ergaben die ermittelten AE-Werte und pflanzenphysiologisch ermittelten Werte die gleichen Tendenzen. TEPE legt besonderen Wert auf die Wanderung der Nährstoffionen aus dem Boden zur Austauschereinheit, er spricht von kinetischen Bodenwerten im Gegensatz zu statischen. TEPE und LEIDENFROST haben auch die Zone im Umkreis der Austauschereinheit bestimmt, aus der Ionen zur Austauschereinheit wandern. Diese Zone ist für die einzelnen Nährstoffionen verschieden, sie variiert auch von Boden zu Boden. Von MØLLER und MOGENSEN (1953) wurde mit Ionenaustauschern nur die verfügbare Phosphorsäure im Boden bestimmt. LATHWELL, SAUCHEZ, LISK et al. (1958) bestimmen die im Boden verfügbare Phosphorsäure mit Ionenaustauschern und haben bei 30 Böden gute Übereinstimmung mit Feldversuchsergebnissen feststellen können.

Sie behandelten 10 g lufttrockenen Boden mit 10 g Natrium-Zeolit und 250 ml dest. Wasser 6 Stdn. Im Filtrat wurde die P_2O_5 colorimetrisch bestimmt. mg/l PO_4 ist der Phosphatwert, eine Größe, die etwas über die Verfügbarkeit der P_2O_5 aussagt. Da die Ca-Ionen des Bodens von Zeolit gebunden und dafür Na-Ionen ausgetauscht werden, löst sich relativ viel P_2O_5.

4. Bestimmung der Kalium-Bedürftigkeit

A. Chemische Methoden

Starke Salzsäure als Lösungsmittel. Zur Bestimmung der K-Bedürftigkeit wurden die Böden im vorigen Jahrhundert ebenso wie zur Bestimmung der P_2O_5-Bedürftigkeit mit starker Salzsäure behandelt. Beim Kochen des Bodens mit starker Salzsäure muß berücksichtigt werden, daß eine 20%ige HCl bei 110° bei gleichbleibender Konzentration siedet. Bei der Arbeitsvorschrift von VON SIGMOND (1931) wird dieses berücksichtigt. Von verschiedener Seite ist auch angegeben, bei welchem K-Gehalt des Bodens, löslich in 10%iger bzw. 25%iger Salzsäure, die Böden nicht auf eine K-Düngung reagieren (s. KÖNIG (1923), S. 68, und LIEBSCHER 1895). WEIBULL (1906, 1907) hat zur Bestimmung der K-Bedürftigkeit der Böden auf Grund ihres Gehaltes an Kalium, löslich in starker Salzsäure, außerdem den Gehalt der Böden an Aluminiumoxyd, löslich in starker Schwefelsäure, bestimmt. Die Mittelwerte der Ergebnisse für Sandböden, sandige Böden, lehmige Sandböden, sandige Lehmböden und Lehmböden wurden von WEIBULL graphisch aufgetragen. Al_2O_3-Gehalt auf der Abszisse und K_2O-Gehalt auf der Ordinate. Die erhaltene Kurve war fast eine Gerade. Nach WEIBULL ist ein bestimmter Boden K-bedürftig, wenn für seinen K_2O- und Al_2O_3-Gehalt der Punkt unter der Kurve liegt, der Boden ist ausreichend mit Kalium versorgt, wenn der Punkt für seinen K_2O- und Al_2O_3-Gehalt oberhalb der Kurve liegt. Interessant bei dieser Methode ist, daß nicht nur die in Salzsäure lösliche K-Menge des Bodens bestimmt wird, sondern noch eine andere Größe, die etwas über den Tongehalt des Bodens aussagt. Es wird also bei der Auswertung der Untersuchungsergebnisse die Bodenart berücksichtigt.

Verdünnte Mineralsäuren. Da bei Behandlung der Böden mit konzentrierten Mineralsäuren auch Pflanzennährstoffe gelöst werden, die von Pflanzen nicht oder nur langsam ausgenutzt werden, ist man dazu übergegangen, die Böden mit verdünnten Säuren zu behandeln. So wird auch verdünnte Salzsäure in verschiedener Konzentration benutzt, um gerade die Pflanzennährstoffe zu lösen, die von den Pflanzen aufgenommen werden können. BRAY (1929) verwendet zum Beispiel eine 0,2 n HCl, VAN DER PAAUW (1956) eine 0,1 n HCl und LEAF (1958) eine 0,4 n HCl. VAN DER PAAUW bestimmt nach einem Vorschlag von VISSE auch den Calciumgehalt und berücksichtigt denselben bei Auswertung der K-Ergebnisse. Wie bei Bestimmung der P_2O_5-Bedürftigkeit der Böden (S. 802) angegeben, behandeln EGNER, RIEHM und DOMINGO (1960) den Boden zur Ermittlung der P_2O_5- und K-Reserven mit heißer 2 n HCl.

Als Lösungsmittel zur Bestimmung der K-Bedürftigkeit der Böden wird auch verdünnte Salpetersäure in verschiedener Konzentration benutzt. BECKER (1929), BAMBERG (1927/28) verwenden die 0,015 bis 0,025 n HNO_3 nach VON SIGMOND (1929), BAMBERG (1936) bestimmt auch die in 0,1 n und 1 n HNO_3 lösliche Kalimenge und bezeichnet die Differenz der Kalimenge löslich in diesen beiden Lösungsmitteln als Aktivität des passiven Bodenkaliums. SEMB und UHLEN (1955) verwenden eine sogenannte REITEMEIER-Methode, bei der 10 g Boden mit 150 ml 1 n HNO_3 gekocht werden. LEAF (1958) kocht Waldböden ebenfalls mit 1 n HNO_3. Aus der älteren Literatur (KÖNIG 1923) geht ebenfalls hervor, daß Böden zur Bestimmung der K-Bedürftigkeit mit verdünnter HNO_3 behandelt wurden.

Verdünnte Schwefelsäure 0,002 n ist das Lösungsmittel der TRUOG-Methode (1930), mit der auch die K-Bedürftigkeit der Böden bestimmt wird. VUORNIEN und MÄKITIE (1955) haben zur Bestimmung der K-Bedürftigkeit die Böden mit verschiedenen Lösungsmitteln behandelt und auch eine 0,2 n H_2SO_4 benutzt.

HUNTER und PRATT (1957) behandeln den Boden (10 g) mit 25 ml Wasser und geben dazu 1 ml konz. H_2SO_4. Das erhitzte Gemisch bleibt 30 Min. stehen und wird dann filtriert. Der Boden auf dem Filter wird mit 0,1 n H_2SO_4 ausgewaschen, bis 100 ml Filtrat vorhanden sind. Die Korrelation zwischen der gelösten und der von LUZERNE aufgenommenen K-Menge ist gut ($r = 0{,}99$).

Organische Säuren. 1. *Zitronensäure.* Nach DYER (1894), der festgestellt hatte, daß die Azidität des Wurzelsaftes ungefähr der Azidität einer 1%igen Zitronensäure entspricht, ist ein Boden nicht genügend mit Kalium versorgt, wenn in 1%iger Zitronensäure je 100 g Boden weniger als 5 mg K_2O gelöst werden.

HALL und PLYMEN (1902) haben Böden mit 1 %iger Zitronensäure und mit Salzsäure und Essigsäure, äquivalenter Konzentration, außerdem mit kohlensäure-gesättigtem Wasser, mit ammoniakalischer Ammoniumzitratlösung und mit konz. Salzsäure behandelt. Nach ihren Ergebnissen stimmen die mit 1 %iger Zitronensäure erhaltenen Ergebnisse am besten mit den Erfahrungen überein. KÖNIG und HASENBÄUMER (1926) haben ebenfalls gute Erfahrungen mit der 1 %igen Zitronensäure zur Bestimmung der K-Bedürftigkeit gesammelt, sie geben als Grenzzahl 16 mg K_2O je 100 g Boden an, s. auch HASENBÄUMER (1931), KÖNIG (1923), GEHRING (1931, S. 134 bis 143) und BEHRENS (1935, besonders S. 101 bis 118).

Von VUORNIEN und MÄKITIE (1955) wurden Böden zur Bestimmung der K- und P_2O_5-Bedürftigkeit mit verschiedenen Lösungsmitteln behandelt, unter anderem auch mit 1 % Zitronensäure. Nach BARBIER werden in zahlreichen landwirtschaftlichen Versuchsstationen in Frankreich die Böden zur Bestimmung der K- und P_2O_5-Bedürftigkeit mit Zitronensäure behandelt (30 g Boden + 150 ml 2 % Zitronensäure 4 Stdn. geschüttelt und nach einer Pause von 16 Stdn. nochmals 4 Stdn. geschüttelt).

2. *Essigsäure.* Über die Brauchbarkeit der Essigsäure als Lösungsmittel zur Bestimmung der K-Bedürftigkeit der Böden liegen wenig Ergebnisse vor. Wie bei der Besprechung der Methoden zur Bestimmung der P_2O_5-Bedürftigkeit der Böden behandelt, wurde schon von BIRNER (1870) Essigsäure als Lösungsmittel vorgeschlagen. SEMB und UHLEN (1955) haben die K-Bedürftigkeit verschiedener Böden nach mehreren Methoden untersucht und auch die WILLIAMS- und STEWART-Methode benutzt (20 g Boden werden mit 100 ml 2,5 %iger [0,48 n] CH_3COOH behandelt). Die Übereinstimmung der Ergebnisse mit einigen anderen Methoden (Austauschbares K mit CH_3COONH_4, EGNER-Methode $CH_2ClCOOH$ und BRAY-Methode, Boden mit $NaNO_3$ ausgewaschen) war gut. VUORNIEN und MÄKITIE (1955) haben ebenfalls Böden auf K- und P_2O_5-Bedürftigkeit mit verschiedenen Lösungsmitteln untersucht und auch eine 2,5 %ige Essigsäure verwendet.

Kohlensäurefreies Wasser. HOOD, BRADY und LATHWELL (1956) behandeln 10 g mit Quarz gemischten Boden nach und nach mit 10 l H_2O. Das aus dem Boden ausgewaschene Kalium wird von Austauschern gebunden und daraus mit wenig HCl für die Analyse eluiert. Das wasserlösliche Kalium ist nach Ansicht der Verfasser ein besseres Maß für die K-Bedürftigkeit der Böden als das austauschbare Kalium. Die wasserlösliche K-Menge wurde verglichen mit dem K-Gehalt von Ladino-Klee. MATTHEWS (1952) hat einen Apparat entwickelt, mit dem das Wasser nacheinander immer wieder durch eine Säule mit Boden und eine Säule mit einem Kationenaustauscher gepumpt wird, s. auch MATTHEWS und SMITH (1957). VON WRANGELL (1931) hat das wasserlösliche Kalium ebenso wie die wasserlösliche Phosphorsäure durch Behandlung von 1 g Boden mit 100 ml Wasser bestimmt. Auch NEMEC (1926) hat im wässerigen Bodenauszug (30 g Boden + 100 ml H_2O) neben der Phosphorsäure auch das Kalium bestimmt. Ebenso hat ROHDE (1933) in der gleichen wässerigen Lösung, in der von ihm Phosphorsäure bestimmt wurde, auch den K-Gehalt ermittelt (Behandlung des Bodens mit kochendem Wasser). DIRKS und SCHEFFER (1928) bestimmen die K-Bedürftigkeit saurer Böden, ebenso wie die P_2O_5-Bedürftigkeit durch Behandlung der Böden mit Wasser (30 g Boden + 75 ml H_2O).

Kohlensäurehaltiges Wasser. BIELER (1909) und MITSCHERLICH (1954, S. 126 bis 133) haben die K-Bedürftigkeit der Böden ebenso wie die der Phosphorsäure durch Behandlung der Böden mit CO_2-haltigem Wasser bestimmt.

Bodenpreßsaft und Bodenlösung. Über die Gewinnung von Bodenpreßsaft und Bodenlösung s. Abschnitt Phosphorsäure, S. 807. Ebenso wie Phosphorsäure wird auch Kalium in den Lösungen bestimmt. KAWE (1935, 1936, 1959) behandelt, wie unter Phosphorsäure ausgeführt ist, die Böden zur Bestimmung der P_2O_5- und K-Bedürftigkeit mit der 1 zu 1 mit Wasser verdünnten Bodenlösung. KAWE bestimmt dann den K-Gehalt in der Bodenlösung und im Bodenauszug. Für die Bewertung der K-Bedürftigkeit ist die K-Menge maßgebend, die von der K-haltigen Bodenlösung gelöst wird.

Salzlösungen. Die meisten Lösungsmittel, mit denen die P_2O_5-Bedürftigkeit der Böden bestimmt wird, werden auch zur Bestimmung der K-Bedürftigkeit der Böden verwendet. Die größte Anzahl dieser Lösungsmittel ist zuerst auf ihre Brauchbarkeit, die P_2O_5-Bedürftigkeit der Böden festzustellen, geprüft worden. Später wurde dann untersucht, ob im gleichen Auszug auch die K-Bedürftigkeit der Böden ermittelt werden kann. Mit zahlreichen der geprüften Lösungsmitteln kann auch mit einer mehr oder weniger großen Genauigkeit die Düngebedürftigkeit der Böden der beiden Nährstoffe Phosphorsäure und Kalium bestimmt werden.

In diesem Abschnitt werden auch Lösungsmittel besprochen, mit denen in erster Linie die K-Bedürftigkeit der Böden festgestellt werden soll. Der Boden wird mit einer Salzlösung behandelt, dabei kommt es zum Kationenaustausch, ein mehr oder weniger großer Anteil des austauschbaren Kaliums geht in Lösung und dafür wird eine äquivalente Menge des zugesetzten Kations gebunden. Die Lösungen folgender Salze werden zur Bestimmung der K-Bedürftigkeit der Böden mit mehr oder weniger großem Erfolg verwendet.

NH_4Cl, NH_4NO_3, NH_4-Zitrat, NH_4-Azetat, NH_4-Karbonat, NH_4-Laktat,
NaCl, $NaNO_3$, $NaHCO_3$, Na-Azetat, $NaClO_4$
$Mg(HCO_3)_2$
$CaCl_2$, $Ca(HCO_3)_2$, Ca-Laktat

Von den aufgeführten Salzen wurde NH_4Cl zuerst benutzt, um festzustellen, ob eine K-Düngung notwendig ist KELLNER (1923) behandelte 20 g Boden mit 50 ml einer kaltgesättigten NH_4Cl-Lösung 1/2 Std. lang unter häufigem Umrühren auf dem Wasserbad. Die Behandlung wurde mehrfach wiederholt. In den vereinten Auszügen wurde der K-, Ca- und Mg-Gehalt bestimmt. Durch diese relativ langwierige Methode wird die Gesamtmenge an austauschbaren Basen im Boden erfaßt. Nach einer anderen von KELLNER (1897) vorgeschlagenen Methode werden 20 bis 25 g Boden in ein Glasrohr (20 bis 25 mm Durchmesser) gefüllt und mit 20 bis 25 ml 10% NH_4-Lösung unter Zugabe von 1 bis 3 Tr. $(NH_4)_2CO_3$ behandelt, anschließend werden mit 100 bis 250 ml der NH_4Cl-Lösung die gesamten austauschbaren Basen ausgewaschen.

Lösungen von NH_4Cl zur Bestimmung der K-Bedürftigkeit der Böden verwenden weiter WHEETING (1930), s. auch GISIGER (1936/37), PAVLOVSCHI und MAVRODINEANU (1939) und VAGELER und ALTEN (1931) über diese Methode wird später berichtet, S. 820.

Eine NH_4NO_3-Lösung wird von NEHRING (1942) verwendet.

Er behandelt 20 g Boden 1 Std. mit 100 ml $\frac{n}{5}$ NH_4NO_3, fällt das Calcium durch Zugabe von $NH_4H_2PO_4$ aus und bestimmt das Kalium flammenphotometrisch.

Die Übereinstimmung der Ergebnisse mit denen der Keimpflanzenmethode ist nach NEHRING gut. LEAF (1958) berichtet über die Behandlung von Waldböden mit n NH_4NO_3, auf pH 4,0 und pH 6,9 eingestellt.

Das Ammoniumsalz der Essigsäure, zum Teil bei einem bestimmten pH-Wert der Lösung, wird relativ häufig zur Bestimmung der K-Bedürftigkeit der Böden verwendet. BAMBERG (1928, 1929) hat verschiedene Böden nach MITSCHERLICH und Neubauer und durch Behandlung mit verschiedenen Lösungsmitteln auf ihre K-Bedürftigkeit untersucht und stellte fest, daß eine 2 % CH_3COONH_4-Lösung brauchbare Ergebnisse liefert. Vuorinen und Mäkitie (1955) verwenden eine Lösung, die aus gleichen Teilen 0,5 n CH_3COONH_4 und 0,5 n CH_3COOH hergestellt wird und den pH-Wert 4,65 hat. Mervin und Peech (1951) arbeiten mit einer 1 n CH_3COONH_4-Lösung. Leaf (1958) hat bei Vergleichsuntersuchungen (s. oben unter NH_4NO_3) Böden mit CH_3COONH_4, eingestellt auf pH 4,6 und 6,9, behandelt. Semb und Uhlen (1955) haben 20 g Boden mit 100 ml CH_3COONH_4 behandelt und anschließend mit dem gleichen Lösungsmittel den Boden ausgewaschen. Ein verbesserter Morgan-Barbier-Test, besonders zur Bestimmung der K-Reserven im Boden, ist von Blanchet und Perigand (1960) vorgeschlagen worden. Sie haben das Natrium in der essigsauren Pufferlösung durch Ammonium ersetzt. Es ist leicht verständlich, daß durch die kleinen NH_4-Ionen bei schweren Böden eine größere Menge an K-Ionen ausgetauscht wird als durch Na-Ionen.

Da bei der flammenphotometrischen K-Bestimmung die Ca-Ionen bei Verwendung der meisten Flammenphotometer ausgefällt werden müssen, hat Schachtschabel (1941) vorgeschlagen, den Boden mit einer Lösung zu behandeln, die für NH_4-Ionen 1 normal ist, 80 % der Gegenionen sollen Acetat- und 20 % Oxalationen sein. Der Boden wird im Verhältnis 1:2,5 mit dem Lösungsmittel behandelt. Die Methode lieferte zu 85 % von 350 untersuchten Böden die gleiche K-Bedürftigkeit wie die Keimpflanzen-Methode von Neubauer. Auf den geringen Arbeitsaufwand dieser Methode bei alleiniger Bestimmung der K-Bedürftigkeit des Bodens in einem Bodenauszug weist Herrmann (1943/1) hin. Die Treffsicherheit der Methode ist nach Herrmann ebenso gut wie die der Methoden Nehring (NH_4NO_3) und Egner (Calciumlaktat).

Mit 1 %igen und 2 %igen Ammoniumcitratlösungen hat Bamberg (1928) auch die Böden behandelt, die von ihm auf ihre K-Bedürftigkeit mit 2 %iger Ammoniumacetatlösung, nach Mitscherlich und nach Neubauer untersucht waren. Die beiden Ammoniumcitratlösungen waren ebenso gut brauchbar wie die Ammoniumacetatlösung. Außerdem erwähnt König (S. 69) schon 1923, daß auch Ammoniumcitratlösungen zur Bodenuntersuchung verwendet werden. Mit einer $(NH_4)_2CO_3$-Lösung arbeitet Kühn (1935).

Von Egner, Riehm und Domingo (1960) wird zur Zeit geprüft, ob mit einer Ammoniumlactat-Essigsäure pH 3,7 (AL-Lösung) die K-Bedürftigkeit mit größerer Zuverlässigkeit ermittelt werden kann als mit Methoden, die zur Zeit verwendet werden. (Arbeitsvorschrift s. unter Phosphorsäure S. 807.)

Natriumchloridlösungen zur Bestimmung der K-Bedürftigkeit der Böden werden von Penseler (1930), Groetzner (1934), Krauss (1936) und Bondorff (1936) benutzt. Nach Penseler werden mit Natriumchloridlösungen brauchbare Ergebnisse erzielt.

Groetzner (1934) hat dagegen keine Beziehung zwischen der K-Menge, die von 0,1 n NaCl-Lösung bei mehrfacher Behandlung gelöst wurde, und den Bodenuntersuchungsergebnissen nach dem Gefäßversuch von Mitscherlich und der Keimpflanzenmethode von Neubauer feststellen können. Krauss verwendet eine quecksilberchloridhaltige Natriumchloridlösung (100 g NaCl und 10 g $HgCl_2$/1), die durch Magnesiumhydroxyd schwach alkalisch gemacht ist. Durch das Quecksilberchlorid sollen die im Boden vorhandenen NH_4-Ionen als $HgNH_2Cl$ gebunden werden, damit das Kalium im Bodenauszug direkt mit Natriumkobaltinitrit bestimmt werden kann. Bondorff (1936) behandelt 50 g Boden mit 50 ml 2 n und 4 n NaCl ($HgCl_2$ und $Mg(OH)_2$ haltig nach Krauss) und berechnet die Gesamtmenge an austauschbaren Kalium nach einer Formel, die ausführlich begründet wird. Die Formel nach VAGELER ist nach seiner Ansicht für diesen Zweck unbrauchbar. Außerdem berechnet er die während der Vegetationsperiode den Pflanzen zur Verfügung stehende K-Menge in kg/ha.

Mit einer Natriumnitratlösung behandelt KRUMINŠ (1934) die Böden bei seiner sogenannten Kalimetermethode (4 g Boden + 10 ml $HgCl_2$ haltige 2,5 n $NaNO_3$, die $Ca[OH]_2$ enthält). Der K-Gehalt der Auszüge wird an der Intensität der Trübungen vom gebildeten Kaliumkobaltinitrit in alkoholischer Lösung erkannt.

BRAY verwendet nach SEMB und UHLEN (1955) ebenfalls eine Natriumnitratlösung (20 g Boden + 100 ml n $NaNO_3$, nach dem Filtrieren wird der Boden mit $NaNO_3$ ausgewaschen bis 500 ml Filtrat vorhanden sind).

OHLSEN, COLE, WANATATE und DEAN (1954) arbeiteten mit einer Natriumbicarbonatlösung pH 8,5.

MITSCHERLICH (1939) hat die Untersuchungsergebnisse von etwa 1000 Böden, die nach seiner Gefäßmethode und durch Behandlung mit Natriumacetat-Essigsäure untersucht waren, zusammengestellt. Die gleiche Bodenprobe wurde zweimal hintereinander mit 0,2 n CH_3COONa-CH_3COOH-Lösung pH 4,7 im Verhältnis 1 : 25 geschüttelt. Die in beiden Auszügen gelöste K-Menge gab die Düngebedürftigkeit der Böden zu 72,4 % ebenso wieder wie der Gefäßversuch. Auch VUORNIEN und MÄKITIE bestimmen den K-Bedarf der Böden durch Behandlung mit einer sauren Natriumacetatlösung (0,73 n CH_3COONa und 0,52 n CH_2COOH). Die Morgan-Lösung enthält auch CH_3COONa und CH_3COOH (s. unter Phosphorsäure S. 805). BRAY (1948) behandelt 5 g Boden zur schnellen Bestimmung des K-Bedarfs mit 10 ml 25 %iger $NaClO_4$- oder 25 %iger $NaNO_3$-Lösung. Auch RUSSEL und STANFORD (1954) verwenden die 25 %ige $NaClO_4$ während LONG und SEATZ (1953) die Böden mit $NaClO_4$-Lösung behandeln, die mit $HClO_4$ angesäuert ist.

Die 0,1 n $Mg(HCO_3)_2$-Lösung, mit der DREYSPRING und HEINZ die P_2O_5-Bedürftigkeit und P_2O_5-Festlegung der Böden bestimmen, wird nach MITSCHERLICH (1939) auch benutzt, um den K-Bedarf der Böden zu ermitteln. Von 1000 nach verschiedenen Methoden untersuchten Böden wurde der K-Bedarf für 70 % der Böden durch die $Mg(HCO_3)_2$-Methode ebenso erfaßt wie durch den Gefäßversuch nach MITSCHERLICH.

Eine 2 %ige $CaCl_2$-Lösung benutzte RÜMPLER (1901), um die K-Bedürftigkeit der Böden zu bestimmen.

Von DIRKS und SCHEFFER (1928) wird die K-Bedürftigkeit neutraler und alkalischer Böden ebenso wie die P_2O_5-Bedürftigkeit durch Behandlung von 30 g Boden mit 75 ml mit CO_2 gesättigtem Wasser unter Zusatz von 1 g $CaCO_3$, also mit einer $Ca(HCO_3)_2$-Lösung bestimmt. Die Verwendung verschiedener Lösungsmittel zur Bestimmung der P_2O_5-Bedürftigkeit saurer und neutraler bzw. alkalischer Böden hat viel für sich. Die gleichen Lösungsmittel zur Bestimmung der K-Bedürftigkeit zu verwenden, scheint keine ideale Lösung zu sein, besonders, wenn dann noch für die mit den beiden Lösungen erhaltenen Ergebnisse die gleichen Grenzzahlen verwendet wurden.

Das von EGNER (1951/52 und 1954/55) vorgeschlagene Lösungsmittel zur Bestimmung der P_2O_5-Bedürftigkeit der Böden, eine saure Calciumlactatlösung (pH 3,7) (s. Methoden zur Bestimmung der P_2O_5-Bedürftigkeit der Böden unter Milchsäure, S. 806) wurde später auch zur Bestimmung der K-Bedürftigkeit der Böden benutzt. RIEHM (1940, 1943) hat die Brauchbarkeit für diesen Zweck eingehend untersucht und auch Grenzzahlen aufgestellt. Bei den Grenzzahlen wird der pH-Wert der Böden nicht berücksichtigt, was, wie ausgeführt, bei

Auswertung der P_2O_5-Ergebnisse notwendig ist. Folgende Grenzzahlen werden als Richtzahlen angegeben.

Über 20 mg K_2O je 100 g Boden = zur Zeit versorgt, 10 bis 20 mg mäßig versorgt, unter 10 mg schlecht versorgt.

Durch zahlreiche Methoden wird der Gehalt an austauschbarem Kalium im Boden ermittelt. Diese Größe ist nun nicht immer brauchbar, etwas über den K-Bedarf der Böden auszusagen. Von DELONG, MACKAY und STEPPLER (1953) ist nämlich festgestellt worden, daß das Verhältnis mval austauschbares Calcium zu mval austauschbares Kalium im Boden ein besseres Kriterium für den K-Gehalt der Blätter ist als der Gehalt an austauschbarem Kalium. Weiter haben die Untersuchungen von HAUSER (1941) ergeben, daß gewisse Böden Kalium in nichtaustauschbarer Form festlegen und daß dieses Kalium von den Pflanzen aufgenommen werden kann. Es handelt sich dabei um Böden, die das Tonmineral Illit enthalten, von dem K-Ionen wohl eingetauscht, aber nur langsam wieder ausgetauscht werden, s. SCHEFFER und SCHACHTSCHABEL (1960, S. 196 ff.).

Alkalische Lösungsmittel. RÜMPLER (1901) behandelte 50 g Boden nach und nach mit 1200 ml Kalkwasser und bestimmte im Auszug das gelöste Kalium. Bei Behandlung von 10 g Boden mit 200 ml 2prozentiger Calciumchloridlösung in kleinen Portionen erhielt er die gleichen Werte für gelöstes Kalium je 100 g Boden.

NATH PURI (1933) bestimmt das austauschbare Kalium und Natrium mit Bariumhydroxydlösung. 10 bis 20 g Boden werden mit 100 ml 0,2 n $Ba(OH)_2$ 2 Stunden behandelt, anschließend wird filtriert und mit 0,1 n $Ba(OH)_2$ ausgewaschen bis 1 l Filtrat vorliegt. Die Erdalkalien im Filtrat werden in Karbonate übergeführt und die Alkalien titriert.

B. Physikalisch-chemische Methoden

Von den physikalisch-chemischen Methoden, die in dem Abschnitt Phosphorsäure (S. 812) beschrieben sind, werden die Methoden, welche Gleichstrom zur Abtrennung der Nährstoffionen aus dem Boden verwenden, auch zur Bestimmung der K-Bedürftigkeit der Böden verwendet, ebenso die thermokinetische Methode von REINAU (1938) und die Methode TEPE (1956), Behandlung eines Bodenbreis mit Austauschern in einem Dialysierschlauch. Ebenso wie TEPE, der die im Boden vorhandenen Nährstoffionen an Austauscher bindet, verwendet auch MØLLER (1937) Austauscher zur Bindung der Nährstoffionen. MØLLER gibt zu der Bodensuspension künstlich hergestellten grobkörnigen Natriumpermutit, den er durch Siebe vom Boden trennen kann. Die vom Permutit gebundenen K- und NH_4-Ionen werden dann von ihm bestimmt. Auch Phosphorsäure soll im Boden mit entsprechenden Austauschern bestimmt werden können. Wenn die grobkörnigen Austauscher so beständig sind, daß sie bei Behandlung mit dem Boden nicht zerkleinert werden, sie also quantitativ vom Boden getrennt werden können, dürfte ein derartiges Verfahren, die von den Pflanzen aufnehmbaren Pflanzennährstoffe zu bestimmen, wahrscheinlich recht brauchbare Ergebnisse liefern.

Auch die von VAGELER und ALTEN (1931) vorgeschlagene Methode, Behandlung gleicher Bodenmengen mit zwei verschiedenen Mengen an NH_4-Ionen zum Austausch der vorhandenen K-, Ca-, Mg- und Na-Ionen gehört zu dieser Gruppe von Methoden, denn das Wesentliche dieser Methode ist die Berechnung der Gesamtmenge an austauschbaren Kationen.

Nach VAGELER und ALTEN werden 100 g Boden mit 500 ml und weitere 100 g Boden mit 1000 ml $\frac{n}{10}$ NH_4Cl (Sand- und Lehmboden) oder $\frac{n}{5}$ NH_4Cl (Tonboden) behandelt. Im Filtrat werden die ausgetauschten K-, Na-, Ca- und Mg-Ionen und die noch vorhandenen NH_4-Ionen bestimmt. Nach der Gleichung $y = \frac{x \cdot E}{x + qE}$ (Gleichung einer um 90° gedrehten, durch den 0-Punkt gehenden, Hyperbel) wird die Gesamtmenge an ausgetauschten Ionen und die Gesamtmenge der eingetauschten NH_4-Ionen berechnet.

Bei Berechnung der Gesamtmenge an austauschbaren K-Ionen bedeuten x die mval NH_4-Ionen, mit dem der Boden behandelt wurde, y die mval K-Ionen, die bei Einwirkung von x mval NH_4-Ionen auf den Boden ausgetauscht werden. E ist die Gesamtmenge an austauschbaren K-Ionen und q eine weitere Konstante, die von VAGELER und ALTEN Sorptionsmodul genannt wird. Da man sich unter den Sorptionsmodul wenig vorstellen kann, dürfte es besser sein, die Größe $q\,E$ im Nenner als Konstante zu bezeichnen, denn $q\,E$ entspricht gerade der Menge an NH_4-Ionen (in mval) mit der gerade 50% der austauschbaren K-Ionen, bzw. Ca-, Mg- und Na-Ionen, ausgetauscht werden. Das bedeutet, je kleiner $q\,E$, desto leichter sind die vorhandenen Ionen austauschbar. Wird $q \cdot E = h$ gesetzt, so würde die Gleichung lauten: $y = \frac{x\,E}{x + h}$.

VAGELER und ALTEN geben an, wie die Größen E und q aus der reziproken Gleichung berechnet werden können. Einfacher ist die Berechnung nach RAUTERBERG (1939). Es ist $E = \frac{y_1 y_2 (x_2 - x_1)}{x_2 y_1 - x_1 y_2}$ und $q = \frac{x_1 x_2 (y_2 - y_1)}{y_2 y_1 (x_2 - x_1)}$ und für $x_2 = 2\,x_1$ (gleiche Bodenmengen werden a) mit x mval und b) mit $2\,x$ mval NH_4-Ionen behandelt) ist $E = \frac{y_1 y_2}{2\,y_1 - y_2}$ und $q = \frac{x_2 (y_2 - y_1)}{y_1 y_2}$. Die hier vorgeschlagene Größe h wäre: $h = \frac{x_1 x_2 (y_2 - y_1)}{x_2 y_1 - x_1 y_2}$ bzw. $h = \frac{2\,x_1\,(y_2 - y_1)}{2\,x_1 - y_2}$.

Die Größen y_1 und y_2 müssen sehr genau bestimmt werden, da sich kleine Abweichungen von den richtigen Werten bei der Berechnung von E und q bzw. h stark auswirken. Ein weiterer Nachteil der Methode ist nach WONG PO CHUEN (1935) die unterschiedlich starke Hydrolyse bei Behandlung gleicher Bodenmengen mit unterschiedlichen Flüssigkeitsmengen. Dieser Fehler könnte vielleicht dadurch behoben werden, daß gleiche Bodenmengen einmal mit $\frac{n}{10}$ und einmal mit $\frac{n}{5}$ NH_4Cl behandelt werden. Die unterschiedliche Dissoziation der $\frac{n}{10}$ und $\frac{n}{5}$ NH_4Cl-Lösungen kann die Ergebnisse freilich auch beeinflussen. Nach ARMERO (1935) befriedigen die Ergebnisse auch nur zum Teil.

5. Bestimmung der Kalk-Bedürftigkeit

Calcium wird den Böden nur in seltenen Fällen zugeführt, um den Pflanzen Calcium als Pflanzennährstoff in ausreichender Menge zur Verfügung zu stellen. Die Zufuhr von Calcium als Hauptbestandteil eines Düngers erfolgt dagegen, um den pH-Wert eines sauren Bodens zu erhöhen und um die physikalischen Eigenschaften besonders schwerer Böden zu verbessern. Hauptbestandteil ist Calcium in den Düngemitteln $CaCO_3$, CaO und $Ca(OH)_2$, dazu kommt Hochofenschlacke mit Calcium als Silikat. Um festzustellen, ob eine Kalkung notwendig ist, wird der pH-Wert des Bodens bestimmt, und zwar entweder in einer Suspension des Bodens in Wasser oder in 1 n KCl-Lösung. Hat die Untersuchung ergeben, daß der pH-Wert des Bodens durch Kalkung erhöht werden muß, ist durch eine weitere Untersuchung festzustellen, wieviel Kalk zu geben ist. Bevor der Begriff „pH-Wert" von SÖRENSEN geschaffen und in die Bodenkunde eingeführt war, wurden Böden, die etwas Calciumcarbonat enthalten müssen, durch Bestimmung der mit Salzsäure entwickelten Kohlensäure auf ihre Kalkbedürftigkeit untersucht (s. Apparat nach Passon in WIEGNER 1926, S. 180—185).

Nachdem erkannt war, daß durch Zufuhr einer zu großen Kalkmenge zum Boden der Humus zu schnell abgebaut wird, wurden zahlreiche Methoden entwickelt, um die Kalkmenge zu bestimmen, die notwendig ist.

A. Elektrometrische Neutralisation und Titration

Die brauchbarste dieser Methoden ist wohl die elektrometrische Neutralisation nach JENSEN (1924).

10 g Boden werden in Erlenmeyerkolben mit 50 ml verdünnter Calciumhydroxydlösung (3 bis 30 ml $\frac{n}{30}$ $Ca(OH)_2$) 3 Tage lang verschlossen unter häufigem Umschütteln stehen gelassen. Anschließend wird der pH-Wert der Bodensuspensionen gemessen, und die ermittelten pH-Werte werden abhängig von der zugesetzten $Ca(OH)_2$-Menge graphisch aufgetragen. Auf der Kurve kann abgelesen werden, welche $Ca(OH)_2$-Menge notwendig ist, um 10 g Boden auf einen bestimmten pH-Wert zu bringen. Die je ha (etwa 3000000 kg Boden) notwendige Kalkmenge wird aus diesem Wert berechnet.

Mit dieser Methode werden sehr brauchbare Ergebnisse erzielt, sie hat aber den Nachteil, sehr aufwendig zu sein. Die Methode ist zum Teil als Standardmethode bezeichnet worden, denn zahlreiche andere Methoden sind durch Vergleich mit dieser Methode auf ihre Brauchbarkeit geprüft worden.

Zur schnellen Bestimmung der Kalkbedürftigkeit der Böden ist von GOY und ROOS (1937) die elektrometrische Titration des Bodens entwickelt worden.

Eine Suspension von 10 g Boden in 50 ml 0,1 n KCl wird kräftig gerührt und mit 0,2 n NaOH elektrometrisch titriert. Die NaOH-Lösung wird dabei in Abständen von 1 Min. zugegeben, dabei wird vor jeder Zugabe der pH-Wert der Lösung gemessen. Aus der NaOH-Menge, die nötig ist, um den Boden auf pH 6,5 bzw. 7,7 zu bringen, wird die je ha notwendige Kalkmenge berechnet.

Diese Methode hat aber den Nachteil, daß die ermittelten Kalkmengen zum Teil sehr viel kleiner sind, als die Kalkmengen, die bei Verwendung der Methode JENSEN ermittelt werden, denn verschiedene Böden können die zugesetzte Natronlauge nicht in einigen Minuten binden. Wahrscheinlich werden mit dieser Methode bessere Ergebnisse erzielt, wenn ebenso wie bei der konduktometrischen T—S-Wert-Bestimmung nach SANDHOFF (s. S. 824) der Boden mit KOH vorbehandelt und die zugesetzte Kalilauge mit einer äquivalenten Menge HCl neutralisiert wird.

B. Kalkbedürftigkeit und pH-Wert von Bodensuspensionen

SCHACHTSCHABEL (1951/2) ermittelt die Kalkbedürftigkeit der Böden durch Bestimmung vom pH-Wert der Suspension des Bodens in einer Calcium-acetatlösung.

10 g Boden + 25 ml $(CH_3COO)_2Ca$ (84 g/l, pH 7,0). Auf einer Tabelle wird die Kalkmenge abgelesen, die notwendig ist, um den Boden auf den pH-Wert 7,0 zu bringen. Soll der Boden auf einen pH-Wert unter 7 z. B. 5,5; 6,0 oder 6,5 gebracht werden, ist zusätzlich der pH-Wert der Suspension des Bodens in einer 1 n KCl-Lösung zu bestimmen. Aus Tabellen wird die notwendige Kalkmenge abgelesen.

Diese Methode ist zur Zeit Verbandsmethode vom Verband Deutscher Landwirtschaftlicher Untersuchungs- und Forschungsanstalten.

C. Ermittlung der Kalkbedürftigkeit aus dem T- und S-Wert des Bodens

Der T-Wert ist nach HISSINK (1925) die Basenmenge in mval, die 100 g Boden austauschbar binden können und der S-Wert die Basenmenge, die 100 g Boden gerade gebunden haben. Aus dem T- und S-Wert wird der V-Wert, die prozentuale Absättigung des Bodens mit Basen, berechnet. Ist der V-Wert klein, müssen dem Boden Basen zugeführt werden, d. h. der Boden muß gekalkt werden. Ist der V-Wert z. B. 43 und sollte auf Grund praktischer Erfahrungen der Boden einen V-Wert von 70 haben, dann muß der V-Wert um 27 erhöht werden. Aus dem T-Wert des Bodens kann berechnet werden, wieviel mval Kalk einem V-Wert des Bodens von 27 entsprechen.

Zur Bestimmung vom T-Wert und S-Wert und außerdem zur Bestimmung vom (T-S)-Wert, also der Basenmenge, die ein Boden noch binden kann, sind zahlreiche Methoden vorgeschlagen worden.

Die nach verschiedenen Methoden ermittelten Größen für den T-, den S- und den (T-S)-Wert eines Bodens sind nicht gleich. Die Vorbereitung des Bodens, die benutzten Reagenzien und die Arbeitsweise haben einen Einfluß auf das Ergebnis. Es sind daher nur Werte, die nach der gleichen Methode ermittelt werden, vergleichbar.

Zur Bestimmung des S-Wertes behandelt HISSINK (1925) den Boden mit 1 n NaCl-Lösung.

Der Auszug wird abfiltriert und der Boden so lange mit 1 n NaCl-Lösung ausgewaschen, bis zweimal je 1 l Filtrat aufgefangen sind. In den Filtraten wird nur das Calcium bestimmt. Im ersten Liter ist die gesamte Menge an austauschbarem Calcium des Bodens und Calcium, das durch Umsetzung von $CaCO_3$ des Bodens mit der NaCl-Lösung gelöst wurde, enthalten. Im zweiten Liter ist dagegen nur Calcium gelöst, das durch Umsetzung von $CaCO_3$ des Bodens mit der NaCl-Lösung gelöst wurde. Die Differenz der Ca-Mengen in den beiden Litern Filtrat ergibt die Menge an austauschbarem Calcium des Bodens.

Mit dieser Methode wird nur das austauschbare Calcium des Bodens bestimmt, während die austauschbaren Mg, K und Na-Ionen nicht erfaßt werden. Schon bevor von HISSINK (1925) die Begriffe T-, S- und V-Wert geschaffen waren, sind auch schon die Calciummengen ermittelt worden, die durch Behandlung des Bodens mit einer Ammoniumsalz-Lösung (NH_4Cl oder NH_4NO_3) gelöst werden, und aus dieser Menge wurde geschlossen, ob ein Boden gekalkt werden muß oder nicht. KELLNER (1897, 1923) behandelte z. B. die Böden mit 10%iger NH_4Cl-Lösung. GEHRING, PEGGAU und WEHRMANN (1926/27) bestimmen den T-Wert des Bodens ähnlich wie Hissink den S-Wert durch Auswaschen des Bodens mit 1 n NaCl-Lösung. Der Boden wird nur, bevor er mit der NaCl-Lösung behandelt wird, mit Ca-Ionen gesättigt.

Zur schnellen Umsetzung der austauschbar gebundenen Kationen des Bodens mit zugesetzten Na-Ionen einer 2 n NaCl-Lösung leitet VASILIADIS (1934, 1937) Wasserdampf in die Suspension. Um die löslichen Salze aus dem Boden zu entfernen, wäscht VASILIADIS den Boden vor der Behandlung mit der NaCl-Lösung mit auf 80° erwärmtem Alkohol aus (200 bis 250 ml je 10 g Boden). Da die austauschbaren Ionen im Boden in erster Linie aus Ca- und Mg-Ionen bestehen (nach HISSINK sind im Durchschnitt am Adsorptionskomplex an metallischen Kationen 79% Ca-, 13% Mg-, 6% Na- und 2% K-Ionen gebunden), kann heute durch komplexometrische Titration, bei welcher Ca- und Mg-Ionen zusammen erfaßt werden, der S-Wert des Bodens in Bodenauszügen mit Neutralsalzen relativ einfach bestimmt werden (s. Jackson, S. 64). Der S-Wert des Bodens wird häufig nach einem Vorschlag von Kappen (1929, S. 64—67) durch Behandlung des Bodens mit einer 0,1 n HCl bestimmt, die vom Boden neutralisierte Salzsäure entspricht dem S-Wert. Diese Methode ist natürlich nur zur Untersuchung karbonatfreier Böden verwendbar, oder es muß der $CaCO_3$ Gehalt des Bodens gesondert bestimmt werden.

Zur Bestimmung vom T- und S-Wert des Bodens wird heute besonders eine von Mehlich (1948) vorgeschlagene und von Schachtschabel (1951_1) verbesserte Methode empfohlen.

Der Boden wird durch Behandlung mit einer Bariumchlorid-Triäthanolamin-Pufferlösung (pH 8,1) mit Ba-Ionen gesättigt, die überschüssigen Ba-Ionen werden ausgewaschen, und die eingetauschten Ba-Ionen (T-Wert) werden durch Mg-Ionen ausgetauscht und bestimmt. Der S-Wert des Bodens wird durch Titration der Pufferlösung, mit der der Boden behandelt wurde, ermittelt.

Riehm, Ulrich und Ulrich (1954) bestimmen den T-Wert durch Behandlung des Bodens mit $CaCO_3$, $BaSO_4$ und einer 0,1n $(COONH_4)_2$-Lösung. Durch das $CaCO_3$ in der Suspension wird dafür gesorgt, daß der Boden mit Basen gesättigt wird. Vom Boden gebunden werden freilich NH_4-Ionen, denn austauschbare Ca-Ionen werden als Oxalat ausgefällt, wodurch wieder erreicht wird, daß eine Sättigung des Bodens mit NH_4-Ionen erfolgt. $BaSO_4$ sorgt für eine gute Klärung des Bodenauszuges. Die Menge der adsorbierten NH_4-Ionen wird durch Bestimmung der im Bodenauszug verbliebenen NH_4-Ionen ermittelt.

Das einfachste Verfahren, den S-Wert und den T-Wert eines Bodens angenähert zu bestimmen, dürfte wohl die Bestimmung der pH-Zahl der Suspension des Bodens in verdünnter Essigsäure und in einer Calciumacetatlösung sein.

Brown (1943) schüttelt 2,5 g Boden mit 25 ml 1 n CH_3COOH (pH 2,31) 1 Std. und bestimmt dann den pH-Wert der Suspension. Aus dem pH-Wert (auf 0,01 bis 0,02 pH genau bestimmt) wird der S-Wert des Bodens wie folgt berechnet: gemessener pH-Wert vermindert um 2,31 und multipliziert mit 22 ist der S-Wert. Bei Behandlung von 2,5 g Boden mit 10 ml 1 n CH_3COOH können nur Böden mit einem S-Wert bis 10 untersucht werden. Mit der Methode können aber auch Böden mit S-Werten bis 20 untersucht werden. Anstatt 2,5 g sind dann 1,25 g Boden mit 25mal n CH_3COOH zu behandeln, und der nach oben berechnete S-Wert ist mit 2 zu multiplizieren.

Nach Schachtschabel (s. Methodenbuch S. 63) kann aus dem pH-Wert einer Bodensuspension mit Calciumacetat als Lösungsmittel (pH 7,16) auf den (T-S)-Wert des Bodens geschlossen werden. Ein pH-Wert der Suspension von 6,87 entspricht einem T-S-Wert von 1, pH-Werte von 6,33 und 6,11 entsprechen T-S-Werten von 5 und 8.

Der T-Wert und auch der T—S-Wert des Bodens können außerdem durch konduktometrische Titration des Bodens nach einem Vorschlag von Sandhoff (1954) bestimmt werden.

0,25 bis 3 g Boden je nach Humusgehalt werden mit 0,01 n KOH geschüttelt, bei höherem Humusgehalt im Wasserbad auf 50° erwärmt. Die Suspension wird mit Salzsäure angesäuert, wodurch die Humusstoffe ausgefällt werden. Der Boden wird abfiltriert, Cl′ frei gewaschen und in ein Leitfähigkeitsgefäß gegeben, in dem sich außerdem eine Glaselektrode befindet. Unter Einleiten von Stickstoff wird mit 0,2 n Ba $(OH)_2$-Lösung, bei Zugabe von je 0,1 oder 0,2 ml titriert. Aus der konduktometrischen Titrationskurve kann der T-Wert abgelesen werden, die potentiometrische Kurve zeigt, bei welchem pH-Wert der Boden mit Basen gesättigt ist.

Soll der T—S-Wert bestimmt werden, ist die KOH-Menge genau abzumessen (100 ml). Vor dem Filtrieren wird der Suspension eine der Kalilauge äquivalente Menge Salzsäure zugesetzt und weiter verfahren wie bei der T-Wertbestimmung. S. auch Mortland und Mellor (1954).

Peter und Mitarbeiter (1959_1, 1959_2) bestimmen den Kalkbedarf des Bodens aus dem pH-Wert des Bodens in einer 1 n KCl Suspension und dem sogenannten M.B-Wert (Methylenblau-Wert).

Der Boden wird mit einer 0,25 %igen MB-Lösung behandelt und die nicht adsorbierte Menge MB photometrisch bestimmt. In Tabellen wird für die gemessenen pH- und MB-Werte die notwendige Kalkmenge des untersuchten Bodens abgelesen (Peter hat über 100000 Böden nach dieser Methode untersucht).

Die Wirkung einer Kalkdüngung durch Untersuchung des Bodens kann nach Mitchell (1936) frühestens nach sechs Monaten festgestellt werden. 50 bis 70% der in den Boden gebrachten Ca-Menge waren nach sechs Monaten austauschbar vom Boden gebunden. Nach weiteren sechs Monaten waren nur geringe Änderungen eingetreten.

D. Methoden für Moorböden

Die Kalkbedürftigkeit der Moorböden wird nach Brüne und Arndt (1938) durch Zugabe von $\frac{n}{30}$ $Ca(OH)_2$-Lösung zur Suspension des Bodens in n $BaCl_2$-Lösung bestimmt. Je nach Nutzung sind Moorböden bis zu pH-Werten 3,4 bzw. 4,0 zu halten. Schachtschabel (1953) hat vorgeschlagen, den Kalkbedarf der Moorböden durch Bestimmung vom pH-Wert der Suspension des Moorbodens in: a) 1 n KCl-Lösung und b) 1 n KCl-Lösung, die 50 ml $\frac{n}{30}$ $Ca(OH)_2$ enthält, zu ermitteln. Die notwendige Kalkmenge zur Bewirtschaftung der Böden als Grünland und als Ackerland sind auf Tabellen abzulesen.

6. Bestimmung der Magnesium-Bedürftigkeit

A. Chemische Methoden

Während Calciumverbindungen, wie im letzten Abschnitt ausgeführt wurde, dem Boden in erster Linie zugeführt werden, um den Boden in einen für ihn günstigen Reaktionszustand zu bringen und um die physikalischen Eigenschaften des Bodens zu verbessern, wird mit Magnesiumverbindungen gedüngt, damit die Pflanzen Magnesium in genügender Menge aufnehmen können. Zur Bestimmung der Gesamtmenge an adsorbiertem Magnesium im Boden bestimmte Kellner (1897) das Magnesium ebenso wie das Calcium und Kalium durch mehrfache Behandlung des Bodens mit gesättigter NH_4Cl-Lösung. Ramann (1927) hat nach Angaben von Jacob (1955, S. 69) die Böden mit zehnprozentiger NH_4NO_3-Lösung behandelt. Aus der gelösten Magnesiummenge hat er auf den Mg-Bedarf von Forstböden geschlossen. Ramann gibt auch Grenzzahlen an. Für die Untersuchung landwirtschaftlich genutzter Böden konnte sich die Methode nicht durchsetzen. Wahrscheinlich spielt dabei mit eine Rolle, daß mehr oder wenig regelmäßig mit Handelsdüngern versorgte Böden schwerer zu beurteilen sind als wenig gedüngte Böden, wie die Forstböden, denn in Böden, denen Handelsdünger zugeführt werden, können die Pflanzennährstoffe bei weitem nicht so gleichmäßig im Boden verteilt sein wie in Böden, in denen die Pflanzennährstoffe durch Verwitterung der Bodenmineralien nach und nach pflanzenaufnehmbar geworden sind und die Pflanzennährstoffe durch Zersetzung der Pflanzenrückstände in den Boden zurückkommen. Je gleichmäßiger verteilt die Pflanzennährstoffe im Boden sind, desto schlechter werden sie aber von den Pflanzen ausgenutzt, wie Rauterberg (1937) zeigen konnte.

Bei Bestimmung der Pflanzennährstoffe in den Auszügen mit Salzsäure verschiedener Konzentration wurde zum Teil auch die gelöste Mg-Menge bestimmt, z. B. von KÖNIG, HASENBÄUMER und COPPENRATH (1907), die 10% HCl verwendeten und GROUVEN (1868), der die Böden mit einer etwa 0,6 n HCl behandelte.

Zur Bestimmung der Mg-Bedürftigkeit der Böden sind verdünnte Säuren nach BRIOUX und JOUIS (1932) nicht geeignet; denn die Beziehung zwischen der in verdünnter HNO_3 löslichen Mg-Menge der Böden und der Wirkung einer Mg-Düngung dieser Böden in Vegetationsgefäßen war schlecht. Da zahlreiche Mineralien Magnesium enthalten, das wahrscheinlich von verdünnten Säuren in mehr oder weniger großer Menge gelöst wird, von den Pflanzen dagegen schlechter verwertet wird als austauschbares Magnesium, z. B. auch Dolomit, ist dieses Ergebnis von BRIOUX und JOUIS leicht verständlich.

Für die Bestimmung der Magnesiumbedürftigkeit der Böden durch eine chemische Bodenuntersuchung hat sich schon frühzeitig LOEW (1909, 1914) eingesetzt. LOEW war der Ansicht, daß das Ca/Mg-Verhältnis im Boden für die Entwicklung der Pflanzen von großer Bedeutung ist. Auf keinen Fall darf dieses Verhältnis kleiner als 1 sein, es soll je nach Pflanzenart zwischen 1 und 3 liegen. Obgleich verschiedene Untersuchungen, die durchgeführt wurden, nachdem LOEW auf die Bedeutung des Ca/Mg-Faktors hingewiesen hatte, ergeben haben, daß der Ca/Mg-Faktor für die Entwicklung der Pflanzen nicht die Bedeutung hat, die LOEW ihm zuschrieb, scheint nach neueren Untersuchungsergebnissen das Nährstoffverhältnis der Kationen zueinander und auch das der Anionen zueinander für die Entwicklung der Pflanzen und für die Produktion von Ernteprodukten hoher Qualität sehr wichtig zu sein, s. HOMÈS (1953) und RAUTERBERG und BUSSLER (1960) und weiter Arbeiten, in denen auf eine Beziehung zwischen dem Auftreten der Weidetetanie und dem Mineralstoffgehalt der Gräser hingewiesen wird.

Interessant sind in diesem Zusammenhang die Untersuchungsergebnisse von JOHNSON, DAVIS und BENNE (1957), die feststellten, daß auf einem organischen Boden der Ertrag an Sellerie durch Mg-Düngung nicht verbessert wurde, aber die Qualität der Ernteprodukte. Die Mg-Mangelsymptome wurden bei einem Ca/Mg-Verhältnis von 3:1 auf ein Minimum reduziert. Der V-Wert für Mg-Ionen soll ungefähr 22 sein. Der Mg-Mangel wird durch starke Kalium- und Natriumgaben vergrößert. Auch SEMB und FRAGETHON (1958) weisen darauf hin, daß das Auftreten von Mg-Mangelerscheinungen an Pflanzen vom K-Gehalt des Bodens abhängig ist. Bei relativ hohem Mg-Gehalt der Böden (10 mg MgO/100 g Boden) traten nur bei sehr hohem K-Gehalt der Böden Mg-Mangelerscheinungen auf. An anderer Stelle schreiben SEMB und TORSTEINSSON (1955), ihre Untersuchungen hätten ergeben, daß der Mg-Mangel bei Pflanzen mehr auf ein ungünstiges Mg/K-Verhältnis als auf niedrigen Mg-Gehalt des Bodens zurückzuführen sei. Auch STENUIT (1959) hat festgestellt, daß Mg-Mangel an Apfelbäumen häufig auftritt, wenn das K_2O/MgO-Verhältnis des Bodens (Na-acetat-Methode) > 3 ist.

MCCOLLOCH, BINGHAM und ALDRICH (1957) haben ebenfalls festgestellt, daß die Kaliversorgung der Pflanzen das Auftreten von Mg-Mangelerscheinungen beeinflußt. Das Verhältnis der austauschbaren Mengen K:Mg im Boden, besonders in 45 bis 75 cm Tiefe, war stark korreliert zum Mg-Gehalt der Blätter von Citrus.

Um aus der Menge an austauschbarem Magnesium im Boden auf die Düngebedürftigkeit der Böden für Magnesium schließen zu können, sind die Böden mit verschiedenen Salzlösungen und verdünnten Säuren behandelt worden. Nach HADDOCK (1933) ist die Beziehung zwischen der von 1 n KCl gelösten Mg-Menge zum Auftreten von Mg-Mangelerscheinungen gut. Nach PEECH (1939) ist von CH_3COONH_4, CH_3COONa und NaCl-Lösungen die 1 n NaCl-Lösung am geeignetsten, die Mg-Bedürftigkeit der Böden zu bestimmen. LINEBERRY und BURKHART (1943) haben darauf hingewiesen, daß bei der Bestimmung der Mg-Bedürftigkeit der dolomithaltigen Böden Magnesium erfaßt wird, daß von den Pflanzen nicht aufgenommen werden kann. Nach SHAW (1941) wird Dolomit-Magnesium bei Behandlung der Böden mit CH_3COONH_4 nicht gelöst. Auch BOVER (1955) verwendet zur Bestimmung der Mg-Bedürftigkeit dolomithaltiger Böden eine Acetatlösung, und zwar n CH_3COONa pH 8,2. Ebenso wird von STENUIT und PIOT (1957, 1958) zur Bestimmung der Mg-Bedürftigkeit der Böden eine CH_3COONa-Lösung verwendet, die freilich auf den pH-Wert 4,8 eingestellt ist, sie geben auch Grenzzahlen an und heben hervor, daß Ertragssteigerungen durch Mg-Düngung bei Hafer nur auftraten, wenn Mg-Mangelsymptome an Hafer beobachtet wurden. Die Mg-Mangelsymptome sind besonders vom 1. bis 10. Mai von ihnen beobachtet worden, s. auch STENUIT (1959). Bei dem Schnellverfahren zur Bestimmung der Mg-Bedürftigkeit der Böden nach MORGAN (1937) wird auch eine Acetatlösung verwendet.

Von den Salzlösungen, die zur Bestimmung der Mg-Bedürftigkeit der Böden verwendet werden, hat sich die von SCHACHTSCHABEL (1954, 1959) vorgeschlagene 0,025 n $CaCl_2$-Lösung als sehr brauchbar erwiesen. Es ist die Methode, die in Deutschland heute in erster Linie benutzt wird.

Es werden 5 g Boden mit 50 ml 0,025 n $CaCl_2$-Lösung 2 Stdn. geschüttelt. Im Filtrat wird das gelöste Magnesium mit Titangelb bestimmt, s. auch SCHACHTSCHABEL und ISERMEYER (1954) und BRUGGER (1958).

Durch die einmalige Behandlung der Böden mit der verdünnten $CaCl_2$-Lösung wird natürlich nur ein Teil des austauschbaren Magnesiums erfaßt, von leichten Böden etwa 85 % und von Lehmböden etwa 65 %, diese relativen Mengen streuen natürlich stark. Als Grenzzahlen mg Mg je 100 g Boden werden angegeben für leichte Böden 5, für Lehmböden 7 und für Tonböden 12.

Nach MICHAEL und SCHILLING (1957) kann der Mg-Versorgungsgrad durch Behandlung der Böden mit Erdalkali- oder Alkalisalzlösungen hinreichend genau bestimmt werden und zwar besser als der K-Versorgungsgrad, da die Nachlieferung aus festgelegtem Nährstoff beim Magnesium eine kleinere Rolle spielt als beim Kalium.

Um zu vermeiden, daß Magnesium aus Dolomit bei Behandlung karbonathaltiger Böden gelöst wird, verwendet TUCKER (1954) 1 n NH_4Cl in 60 %igem Alkohol. 10 g Boden werden mit dieser Lösung ausgewaschen, s. auch PFEFFER (1956).

Mg-Mangel wird in erster Linie auf sauren und auf leichten Böden beobachtet, s. HOFFMANN, RIEHM und SCHROEDER (1959) und weiter STENUIT (1959). Häufig wird der Mg-Mangel durch eine Kalkung beseitigt. Je mehr H-Ionen der Adsorptionskomplex des Bodens gebunden hat, desto schwerer werden die adsorbierten metallischen Kationen gegen H-Ionen, z. B. ausgeschieden von den Pflanzenwurzeln, ausgetauscht.

B. Physikalisch-chemische Methoden

Theoretisch können alle Methoden, nach denen die Düngebedürftigkeit der Böden für Kalium ermittelt wird, auch zur Bestimmung der Düngebedürftigkeit für Magnesium verwendet werden. Da die meisten dieser Methoden entwickelt wurden, als das Interesse für die Düngebedürftigkeit der Böden für Magnesium noch gering war, ist über diese Methoden in bezug auf das Magnesium nur wenig zu sagen. PRINCE und TOTH (1937) haben festgestellt, daß die Mg-Menge, die unter bestimmten Bedingungen bei der Elektrolyse erfaßt wird, ungefähr 60 % vom austauschbaren und 3 bis 4 % vom Gesamt-Magnesium entsprechen.

7. Bestimmung der Düngerbedürftigkeit für Spurennährstoffe

Nach unseren heutigen Erkenntnissen sind für das Wachstum der Pflanzen die Spurennährstoffe: Eisen, Kupfer, Mangan, Zink, Molybdän und Bor notwendig.

Wenn auch bisher schon verschiedene Methoden zur Bestimmung der Düngerbedürftigkeit der Böden für Spurennährstoffe entwickelt worden sind, so muß doch gesagt werden, daß wir uns auf diesem Gebiet der Forschung noch in den Anfängen befinden. Bei Entwicklung derartiger Methoden muß beachtet werden, daß der pH-Wert des Bodens, die Bodenart, der Humusgehalt und die Zusammensetzung der Humusstoffe wahrscheinlich noch stärker berücksichtigt werden müssen als bei den Massennährstoffen, außerdem ist die Oxydationsstufe des Mangans zu beachten. Grenzzahlen gelten bei einigen Spurennährstoffen nur für bestimmte pH-Bereiche. Mangel an einzelnen Spurennährstoffen tritt überhaupt nur unterhalb oder oberhalb bestimmter pH-Werte auf.

Die Bestimmung der Düngerbedürftigkeit der Böden für Spurennährstoffe scheint durch eine Beeinflussung der Spurennährstoffe im Boden oder eventl. auch in der Pflanze besonders erschwert zu sein, denn Delmas, Bats und Remy (1957) berichten über einen durch Phosphorsäureüberschuß hervorgerufenen Manganmangel.

Bei Behandlung der Methoden zur Bestimmung der Spurennährstoffe im Boden werden absichtlich die Analysenmethoden, mit denen die einzelnen Spurennährstoffe bestimmt werden, nicht angegeben, denn eine Analysenmethode, die heute als brauchbar angesehen wird, kann morgen durch eine bessere überholt sein. Die mit den verschiedenen Lösungsmitteln, mit denen die Böden behandelt werden, gemachten Erfahrungen sind dagegen (oder sollen es jedenfalls sein) unabhängig von den Analysenmethoden, die zur Bestimmung der gelösten Spurennährstoffe verwendet werden. Das gleiche gilt natürlich für die Massennährstoffe.

A. Eisen

Methoden zur Bestimmung der Eisenbedürftigkeit der Böden sind nicht entwickelt worden, denn Eisenmangel an Pflanzen tritt relativ selten auf, und wenn er beobachtet wird, dürfte wirklicher Eisenmangel im Boden wohl nur in wenigen Fällen die Ursache sein.

B. Kupfer

Kupfer kommt im Boden anorganisch und organisch gebunden vor. Auf Böden, die aus Ergußgesteinen entstanden sind, tritt nach Mitchell (1956) selten Kupfermangel bei den Pflanzen auf. Das Kupfer ist in diesen Gesteinen sehr gleichmäßig verteilt und in allen Bruchstücken ist Kupfer enthalten. Nach Untersuchungen von Wehrmann (1955) enthalten von den Böden Schleswig-Holsteins die Marschböden 28 bis 48 kg/ha Cu, die Braunerden 19 bis 43 kg/ha Cu, die Niedermoore 11 bis 16 kg/ha Cu und die Podsole 4 bis 16 kg/ha Cu. Kupfermangel tritt im allgemeinen nur auf Böden auf, die arm an Tonmineralien sind, und die bestimmte Humusstoffe enthalten, von denen das Kupfer festgelegt wird. Im Podsol enthält z. B. der A_0 und A_1 Horizont relativ viel Kupfer. Das organisch in den Pflanzen gebundene Kupfer wird wahrscheinlich schwer mineralisiert.

Zur Bestimmung der Kupferbedürftigkeit der Böden wird zum Teil der Gesamtgehalt des Bodens an Kupfer bestimmt z. B. von Menzel und Jackson

(1951) durch Behandlung des Bodens mit H_2F_2 und H_2SO_4. Auch $HClO_4$ wird zur Bestimmung der „Gesamtmenge“ an Kupfer im Boden benutzt. Bei der Untersuchung von Böden mit schwer zersetzbaren Silikaten wird aber mit H_2F_2 und H_2SO_4 mehr Kupfer gefunden als mit $HClO_4$ wie HOLMES (1945) festgestellt hat.

Die in $HClO_4$ lösliche Cu-Menge sagt nach LUNDBLAD, SVANBERG und EKMAN (1945) etwas aus über die potentielle Verfügbarkeit von Kupfer in organischen Böden. Steigt der Kupfergehalt von 1 bis 8 ppm[1], dann nimmt der Cu-Gehalt der Pflanzen von 2 bis 5 ppm zu. Bei weiterem Steigen des Cu-Gehalts der Böden bis auf 100 ppm steigt aber der Cu-Gehalt der Pflanzen nur auf 7 ppm. Nach MITCHELL (1956) ist die Korrelation zwischen dem Gesamt-Cu-Gehalt der Böden und dem Cu-Gehalt im Weidegras schottischer Böden besser als die Korrelation zwischen der in HCl, in CH_3COOH und in H_2O gelösten Cu-Menge der Böden einerseits und dem Cu-Gehalt im Weidegras der betreffenden Böden andererseits. ANTIPOV-KARATAC (1947) hat neben dem Gesamt-Cu-Gehalt auch die in 0,5 n HNO_3 lösliche Cu-Menge bestimmt, nach seinen Ergebnissen sind Böden Cu-bedürftig, wenn weniger als 50% vom Gesamt-Cu in 0,5 n HNO_3 löslich ist oder wenn die Böden weniger als 18 ppm Cu enthalten. Auch WESTERHOFF (1954/55) verwendet zur Bestimmung der Cu-Bedürftigkeit der Böden verdünnte HNO_3. Er behandelt 10 g Boden mit 100 ml HNO_3, die aus 30 ml HNO_3 (D 1,39) durch Verdünnen auf 1 l hergestellt wird. Durch die Ergebnisse dieser HNO_3-Methode wird die Cu-Bedürftigkeit der Böden ebenso beurteilt wie durch die HCl-Methode (10%ige HCl) und nach dem Aspergillus-Test nach STAPP und WETTER. Nach SCHARRER, RUSS und MENGEL (1959) sind folgende Lösungsmittel: 10 %ige HCl, 2 %ige HNO_3, 2 %ige Zitronensäure, Morgan-Lösung und Lösung nach BARON in gleicher Weise geeignet, etwas über die Cu-Bedürftigkeit der Böden auszusagen. SCHARRER und KÜHN (1940) haben neben der in 10%iger HCl gelösten Cu-Menge auch die in Wasser gelöste Cu-Menge bestimmt. STEENBJERG und BOKEN (1948) verwenden eine verdünnte HCl pH 2,0. Werden von diesem Lösungsmittel weniger als 0,1 bis 0,3 ppm Cu gelöst, dann sind die Böden Cu-bedürftig. BARON (1954/55) schlägt vor, zur Bestimmung der Düngerbedürftigkeit der Böden für B, Co, Cu, Mn, Mo Zn und Fe eine Lösung zu verwenden, die 1 n für CH_3COOH und 1 n für $(NH_4)_2SO_4$ ist; durch CH_3COONH_4 wird die Lösung auf pH 4,0 $\pm$0,1 gebracht. ASLESON (1951) verwendet eine 1 n CH_3COONH_4-Lösung. ERIKSSON (1949) hat vorgeschlagen, das Kupfer bei der Extraktion als stabilen Komplex Cu P_2O_7'' zu binden. WIKLANDER (1949) behandelt den Boden mit der Suspension eines Harzaustauschers, später wird der Austauscher von dem Boden durch Spülen auf einem Sieb getrennt. Das vom Austauscher gebundene Cu wird mit HCl ausgewaschen und bestimmt. PRATT und BRADFORD (1958) schmelzen 1 g Boden mit 3 g $NaCO_3$, lösen die Schmelze in 1 n HCl. Nach Abscheidung der SiO_2 wird der Rückstand mit 3 n HCl behandelt. Cu, Zn und Fe werden von einem Anionenaustauscher als Chloroanionen gebunden, nacheinander werden Cu, Fe und Zn mit HCl unterschiedlicher Konzentration ausgewaschen. VERMAAT und VAN DER BIE (1950) haben den Kupfergehalt des Bodens nach MORGAN-VENEMA in 3%iger CH_3COOH bestimmt, außerdem in 2%iger H_2SO_4 und 10%iger CH_3COOH und den Gehalt an löslichen und unlöslichem Kupfer in der Asche des Bodens. Für den MORGAN-Test (Schnelltest auf Pflanzennährstoffe, Methodenbuch Bd. I, S. 254) zur Bestimmung von Cu und ebenfalls Zn, Mo und auch Hg, Pb und As wird in einem 30 ml Becherglas ein gestrichener Teelöffel Boden tropfenweise unter Umrühren mit CH_3COONa Lös. bis zum völligen Durch-

[1] ppm = parts per million (1 ppm = 1 mg/kg).

feuchten versetzt. Dann werden weitere 5 ml der Lösung hinzugefügt, umgerührt und filtriert (CH_3COONa Lösung: 100 g CH_3COONa in einem 1 l Meßkolben in 500 ml H_2O lösen, dazu 30 ml Eisessig geben und auf 1 l verdünnen, pH der Lösung 4,8). Das Filtrat wird mit einer Augentropfpipette durch Pumpen gemischt, und mit dieser Tropfpipette wird die notwendige Anzahl Tropfen für die einzelnen Reaktionen in die Vertiefungen von Porzellantüpfelplatten oder in Glasröhrchen gegeben.

C. Mangan

Nach FINCK (1954) kommt das Mangan in folgenden Formen im Boden vor: 1. freie Mangan-Ionen in der Bodenlösung, 2. als Ion adsorptiv an Austauscher gebunden, 3. in anorganischer Form gebunden, a) schwerlösliche Salze, b) Oxydhydrate, c) in anorganischen Gelen eingeschlossen, 4. an organische Stoffe gebunden, a) komplexgebunden an einzelne organische Stoffe, b) in organischen Gelen eingeschlossen, c) in Organismen gebunden. Von diesem Mangan-Vorkommen können die Pflanzen das in der Bodenlösung vorhandene und das adsorptiv am Austauschkomplex des Bodens gebundene Mangan leicht aufnehmen. Von dem in anderer Form im Boden vorhandenen Mangan ist das in 3- und 4wertiger Form vorliegende, also die Oxydhydrate, zum Teil leicht reduzierbar. Dieses leicht reduzierbare Mangan ist für die Pflanzen auch verwertbar. Bei der Entwicklung von Methoden zur Bestimmung der Düngebedürftigkeit der Böden für Mangan ist also zu beachten, daß nicht nur das austauschbare sondern auch das leicht reduzierbare Mangan erfaßt wird. Austauschbares und leicht reduzierbares Mangan zusammen bestimmt, werden aktives Mangan genannt.

Mit den folgenden Lösungsmitteln wird nur das austauschbare Mangan erfaßt: 1 n neutr. CH_3COONH_4-Lösung nach PIPER (1947), $NH_4H_2PO_4$, HNO_3, H_2SO_4 und H_3PO_4 nach HOFF und MEDERSKI (1958), Citronensäure nach LUNDEGARDH und nach GULLY und HIPOLTSTEINER (s. FINCK, 1954) n $MgSO_4$ nach SCHACHTSCHABEL (1957) und mit der von BARON (1954/55) vorgeschlagenen Lösung (s. unter Kupfer S. 829). Auch beim MORGAN-Test (s. unter Kupfer, S. 829) wird nur das austauschbare Mangan erfaßt. Beim MORGAN-Test zur Bestimmung von Mn, P, K, Mg, NH_4 und NO_3 wird ein gestrichener Teelöffel Boden mit 10 ml der CH_3COONa-Lösung 1 Minute lang kräftig geschüttelt und in ein Röhrchen filtriert.

Zur Bestimmung des aktiven Mangans ist ein Reduktionsmittel notwendig und eine Salzlösung oder eine verdünnte Säure. So werden benutzt: von LEEPER (1934) Hydrochinon in einer CH_3COONH_4-Lösung. JONES und LEEPER (1951) verwenden ebenfalls Hydrochinon. Der Boden wird aber zuerst mit alkoholischer Lösung von Hydrochinon behandelt, dabei wird das leicht reduzierbare Mangan reduziert aber nicht gelöst. Die Hydrochinonlösung wird abgetrennt und anschließend wird der Boden mit $Ca(NO_3)_2$ Lösung extrahiert. FINCK (1954) bestimmt das aktive Mangan einmal mit der Laktatlösung von EGNER-RIEHM, die 0,01% Hydrochinon enthält und außerdem mit einer n CH_3COONH_4-Lösung, in der $Na_2S_2O_4$ (Dithionit) gelöst ist. SCHACHTSCHABEL (1957) hat unter verschiedenen Bedingungen das aktive Mangan bestimmt: 1. mit einer 2 n $MgSO_4$-Lösung (pH 1,5), die Na_2SO_3 und H_2SO_4 enthielt, 2. mit einer n $MgSO_4$-Lösung (pH 5,5), in der Na_2SO_3 und $Na_2S_2O_5$ (Na-pyrosulfit) gelöst war, 3. mit einer n $MgSO_4$-Lösung (pH 8,0) unter Zusatz von Na_2SO_3, 4. mit einer n $MgSO_4$ hydrochinonhaltig (2 g/l), 5. unter reduzierenden Bedingungen im Boden: 30 g Boden wurden in einer Flasche vom Durchmesser 5,5 cm mit 50 ml H_2O überschichtet, ohne den Boden aufzuwirbeln. Nach 3tägigem Stehen wurde nach

Zugabe von 50 ml 2 n $MgSO_4$-Lösung und 0,2 g Aktivkohle das austauschbare Mangan bestimmt. Bei der 1stündigen Behandlung von 100 g Boden mit 500 ml H_2O auf dem Wasserbad bei 6maligem Umschütteln, wie von ROTH und PFAFF (1955) vorgeschlagen, wird wahrscheinlich auch ein Teil vom leicht reduzierbaren Mangan mit erfaßt, nämlich reduziert durch organische Verbindungen im Boden.

Die Aufnahme von leicht reduzierbaren Manganverbindungen durch die Pflanzen ist von JONES und LEEPER (1951) bewiesen worden. Sie haben verschiedene Mn-Oxyde hergestellt und ihre kristalline Struktur durch Röntgenstrahlen bestimmt. Diese Mn-Verbindungen haben sie in Mn-Mangelböden, auf denen sie Hafer anbauten, gegeben. Die Löslichkeit der Verbindungen in Wasser mit 0,05 % Hydrochinon steht in Korrelation zur Mn-Menge, die von den Pflanzen aufgenommen wurde. Von den zahlreichen Methoden zur Bestimmung des aktiven Mangans im Boden sind nach Untersuchungsergebnissen von FINCK (1954) folgende brauchbar, die Mn-Bedürftigkeit der Böden zu bestimmen: 1. Vorbehandlung des Bodens mit einer 0,05 %igen Lösung von Hydrochinon in 50 %igem Alkohol nach JONES und LEEPER und anschließende Bestimmung des austauschbaren Mn's mit Mg $(NO_3)_2$-Lösung (FINCK verwendet anstatt einer Ca $(NO_3)_2$-Lösung eine Mg $(NO_3)_2$-Lösung). 2. Extraktion des Bodens nach SCHACHTSCHABEL (1957) mit einer n Mg $(NO_3)_2$-Lösung, die 0,2 % Hydrochinon enthält. 3. Extraktion des Bodens mit n Mg $(NO_3)_2$, die 0,02 % Hydrochinon und 0,02 % Zitronensäure enthält, von FINCK vorgeschlagen. Unbrauchbar sind nach FINCK die Zitronensäure, die Calciumlaktatlösung mit 0,01 % Hydrochinon, CH_3COONH_4 mit Hydrochinon und CH_3COONH_4 mit Dithionit. Nach Untersuchungen von HOFF und MEDERSKI (1958) ist die Korrelation vom Mn-Gehalt der Blätter von Sojabohnen zur Zeit der Ernte zur Mn-Menge in den Auszügen mit $NH_4H_2PO_4$, mit H_3PO_4 und mit $Ca(NO_3)_2$ nach Vorbehandlung mit alkoholischer Hydrochinonlösung gut. Die von SCHACHTSCHABEL (1957) vorgeschlagene Methode wird in den Landwirtschaftlichen Untersuchungs- und Forschungsanstalten in Deutschland zur Bestimmung der Mn-Bedürftigkeit der Böden verwendet (Methodenbuch 1955, Bd. I, S. 199 und 200). Es werden 5 g Boden unter Zusatz von 0,2 g Aktivkohle mit 50 ml einer 1 n $MgSO_4$-Lösung, die pro 1 2 g $Na_2SO_3 \cdot 8\,H_2O$ und 1 g $NaHSO_3$ enthält, und 5 ml einer 0,1 %igen $AgNO_3$-Lösung 2 Stdn. geschüttelt.

Nach SCHARRER, RUSS und MENGEL (1959) kann aus dem Gesamt-Mn-Gehalt des Bodens, aus den in folgenden Lösungsmitteln löslichen Mn-Mengen, 10 %ige HCl, 2 %ige HNO_3, 2 %ige Zitronensäure, und aus der Menge an „gesamt reduzierbarem und austauschbarem Mn" nicht auf die Mn-Bedürftigkeit der Böden geschlossen werden. Aus der im Heißwasserauszug nach ROTH und PFAFF gelösten Mn-Menge kann angenähert die Mn-Bedürftigkeit der Böden geschätzt werden. Die Mn-Mengen gelöst nach BARON und gelöst nach SCHACHTSCHABEL (austauschbares Mn des lufttrockenen Bodens) sagen noch etwas mehr über die Mn-Bedürftigkeit der Böden aus. Nach den von SCHARRER und Mitarbeitern besprochenen Ergebnissen kann aus der nach MORGAN und aus der nach SCHACHTSCHABEL (Sulfit Mn, pH 8,0) gelösten Mn-Menge am besten auf die Mn-Bedürftigkeit der Böden geschlossen werden.

WEHRMANN (1955) hat den Gesamtmangangehalt zahlreicher Böden Schleswig-Holsteins bestimmt und festgestellt, daß Braunerden und Marschböden mit 1200 bis 1400 kg/ha Mn erheblich manganreicher sind als Podsol- und Niedermoorböden mit 400 bis 700 kg/ha Mn. In Podsolböden ist das Mn zum Teil ausgewaschen und im B-Horizont angereichert.

Bei der Bestimmung der Mn-Bedürftigkeit der Böden ist der pH-Wert der Böden zu berücksichtigen, denn die Dörrfleckenkrankheit, also das äußere Zeichen für Mn-Mangel, tritt praktisch nur auf neutralen und alkalischen Böden auf. Von zahlreichen Böden, die von SCHACHTSCHABEL (1957) untersucht wurden, war die Dörrfleckenkrankheit mit 3 Ausnahmen auf Böden beschränkt, die einen pH-Wert 5,9 und höher hatten. Ausgeschlossen ist natürlich Mn-Mangel auch auf sauren Böden nicht. Nach STENUIT, PIOT und BOON (1956) treten typische Mn-Mangelerscheinungen auf Böden unter pH 5,5 (KCl) nicht auf, dagegen in erster Linie auf Böden über pH 7,0 (KCl).

D. Zink

Zur Bestimmung der Zn-Bedürftigkeit werden die Böden zum Teil mit den gleichen Reagenzien behandelt, die zur Bestimmung der Cu- und Mn-Bedürftigkeit verwendet werden.

PRATT und BRADFORD (1958) schmelzen 1 g Boden mit 3 g Na_2CO_3 und bestimmen in der in HCl gelösten Schmelze die Spurennährstoffe Kupfer und Zink (s. unter Kupfer S. 829). GEILMANN und NEEB (1954) erhitzen 2 bis 5 g Boden im Wasserstoffstrom auf 1000 bis 1100°. Die Zinkverbindungen werden reduziert und das Zink wird in einer Kapillare als Spiegel abgeschieden. Es wird dann gelöst und polarographisch oder colorimetrisch bestimmt.

Der größte Anteil vom Zink im Boden wird, ähnlich wie beim Kupfer nach HOLMES (1945) bei Behandlung des Bodens mit $HClO_4$ erfaßt. Auch WIKLANDER (1949) bestimmt durch die gleiche Behandlung des Bodens mit einem mit H-Ionen gesättigten Harzaustauscher die Zn- und Cu-Bedürftigkeit des Bodens (s. unter Kupfer S. 829). Häufig werden Essigsäure und deren Salze zur Bestimmung der Zn-Bedürftigkeit der Böden verwendet, zum Beispiel die Lösung nach MORGAN (1937) (s. unter Kupfer S. 829) und die MORGAN-Lösung unter Zusatz von $CaCl_2$ von SCHARRER und MUNK (1956), die Lösung von BARON (1954 bis 1955, s. unter Kupfer S. 829) von LYMANN und DEAN (1952) und eine CH_3COOH-KCl-Lösung pH 3,2 von THORNE et al. (1942). Auch SHAW und DEAN (1952) behandeln den Boden mit CH_3COONH_4-Lösung, gleichzeitig aber mit einer Lösung von Dithizon in Tetrachlorkohlenstoff. Das durch die NH_4-Ionen ausgetauschte Zn wird von Dithizon gebunden, dadurch kann durch eine Ausschüttelung des Bodens die Gesamtmenge an austauschbarem Zn erfaßt werden. WEAR und SOMMER (1947) verwenden 0,1 n HCl und BERGH (1948) 0,1 n $MgSO_4$, auf den pH-Wert des Bodens eingestellt, zur Bestimmung der austauschbaren Zn-Menge im Boden. SCHARRER und MUNK (1956) haben in folgenden Lösungsmitteln: 1 n und 10 %iger HCl, 1 und 2 %iger Zitronensäure, Laktatauszug nach EGNER-RIEHM, MORGAN-Lösung, $CaCl_2$-haltige MORGAN-Lösung und Wasser, die löslichen Mengen an Zn bestimmt. Über die Brauchbarkeit der einzelnen Lösungsmittel zur Bestimmung der Zn-Bedürftigkeit der Böden geben sie freilich nichts an.

E. Molybdän

Zur Bestimmung der Mo-Bedürftigkeit der Böden sind auch mit zahlreichen Lösungsmitteln Versuche durchgeführt worden. HALEY und MELSTED (1957) extrahierten Böden mit verschiedenen Lösungsmitteln und verglichen die gelösten Mo-Mengen mit dem Gesamtgehalt der Böden an Molybdän und dem Mo-Gehalt der Pflanzen, die auf dem Boden gewachsen waren. Mit folgenden Reagenzien wurden die Böden behandelt: 1. 0,1 n NaOH; 2. 0,275 m Oxalatlösung nach TAMM; 3. 0,1 n HCl + 0,03 n NH_4F; 4. 0,03 n NH_4F. Der Gehalt an Gesamt-Molybdän im Boden wurde nach der Methode GRIGG und PERRIN bestimmt. Die Pflanzen werden nach JOHNSON und ARKLEY untersucht. Die untersuchten Böden enthielten 1,90 bis 9,20 ppm Mo, die Pflanzen 0,27 bis 10,65 ppm Mo. In 0,1 n NaOH wurde die größte Mo-Menge gelöst, es folgen die anderen Lösungsmittel in der angegebenen Reihenfolge. Der Korrelationskoeffizient zwischen dem Mo-Gehalt der Pflanzen, die auf dem Boden gewachsen waren, und der aus dem Boden gelösten Mo-Menge ist bei Verwendung von 0,1 n NaOH $r = 0{,}849$ und bei der Oxalatlösung $r = 0{,}795$. Die Verfasser schreiben, daß diese beiden Lösungsmittel eine Basis zur Entwicklung einer Methode sind, mit welcher die Mo-Bedürftigkeit der Böden bestimmt werden kann.

SCHARRER und EBERHARDT (1956) haben einen Boden, der gut, und einen Boden, der schlecht mit Molybdän versorgt war, mit verschiedenen Lösungsmitteln behandelt und zwar benutzten sie: 1. Oxalatlösung (angenähert der TAMMschen Lösung); 2. MORGAN-Lösung; 3. die sogenannte Doppellaktatlösung nach EGNER-RIEHM; 4. Wasser; 5. eine 0,4 n Na_2HPO_4-Lösung und 6. eine 0,4 n $MgSO_4$-Lösung. Aus dem Boden, der ausreichend mit Molybdän versorgt war, wurde von den 4 letzten Extraktionsmitteln 11 bis 21mal soviel Molybdän gelöst als aus dem Mo-Mangelboden. Mit dem Extraktionsmittel 1 war das Verhältnis 3,8 und mit Lösung 2 1,6. Bei der Untersuchung weiterer Böden stellten die Verfasser fest, daß mit sinkendem pH-Wert der Böden das Verhältnis: wasserlösliches Molybdän zu oxalatlösliches Molybdän von 29:100 auf 1:100 abnimmt. BARON (1954/55) schlägt vor, zur Bestimmung des Mo-Gehalts der Böden das gleiche Lösungsmittel zu verwenden, mit dem von ihm die anderen Spurennährstoffe im Boden bestimmt werden (s. unter Kupfer, S. 829).

Zur Bestimmung vom Gesamt-Mo-Gehalt des Bodens wird der Boden nach einem Vorschlag von ROBINSON et al. (1951) mit wasserfreiem Na_2CO_3 geschmolzen, und im Wasserauszug der Schmelze wird nach Abscheidung der Kieselsäure die gelöste Mo-Menge bestimmt. Diese Methode ist auch im Methodenbuch (1955, S. 136) aufgenommen worden. Nach MORGAN (1937) werden Molybdän, Zink und Kupfer im gleichen Auszug bestimmt (s. unter Kupfer, S. 829).

Der Gehalt an wasserlöslichem Molybdän des Bodens steht nach Untersuchungen von AMIN und JOHAM (1958) in enger Beziehung zum Gesamt-Mo-Gehalt der Embryonen von Baumwollsamen. Die Verfasser haben aus dieser Beziehung den Schluß gezogen, daß zur Bestimmung der für die Pflanzen verfügbaren Mo-Menge im Boden, der Gehalt der Baumwollembryonen an Molybdän ermittelt werden kann. Umgekehrt muß natürlich auch der Schluß gezogen werden können, daß die in Wasser gelöste Mo-Menge etwas über die Versorgung des Bodens mit pflanzenverfügbarem Molybdän aussagt. Bei Beurteilung der Mo-Bedürftigkeit der Böden muß berücksichtigt werden, daß durch Kalkung der Mo-Mangel zum Teil behoben werden kann, und daß er durch Düngung mit $(NH_4)_2SO_4$ verstärkt wird, was auch TROBISCH und GERMAR (1959) beobachtet haben. Auch GAMMON, VOLK, MCGUBLIN und EDDINS (1954) berichten über Faktoren, welche die Mo-Aufnahme bei Blumenkohl beeinflußen.

F. Bor

Um festzustellen, ob ein Boden den Pflanzen Bor in genügender Menge zur Verfügung stellen kann, wird der Gesamtborgehalt, die in starker Säure lösliche B-Menge, die von Salzlösungen gelöste B-Menge und das in Wasser lösliche Bor bestimmt. WHETSTONE et al. (1942) schmelzen zur Bestimmung der Gesamt-B-Menge ein Gemisch von 0,5 g Boden und 3 g Na_2CO_3. Der Schmelzkuchen wird mit 4 n H_2SO_4 behandelt. Das säurelösliche Bor bestimmen WHETSTONE et al. (1942) durch Behandlung von 50 g Boden mit 75 bis 100 ml 85 %iger H_3PO_4 im Dampfbad. Das gelöste Bor wird im Methylalkohol-Dampfstrom als Methylester abdestilliert. BARON (1954/55) hat, wie unter Kupfer S. 829 angegeben, vorgeschlagen, auf die B-Bedürftigkeit der Böden aus der B-Menge zu schließen, die von einer Lösung, die 1 n für CH_3COOH, 1 n für $(NH_4)_2SO_4$ und durch CH_3COONH_4 auf pH 4,0 gebracht ist, gelöst wird. Beim Morgan-Test (s. unter Kupfer, S. 829) wird zur Bestimmung der B-Bedürftigkeit ein Teelöffel naturpenider Boden durch ein Filter gepreßt und tropfenweise mit 10 ml der CH_3COONa-Lösung behandelt.

In erster Linie wird aber der Boden zur Bestimmung der Borbedürftigkeit mit heißem Wasser behandelt. Die von Berger und Truog (1939) zuerst vorgeschlagene Methode ist von verschiedener Seite etwas abgeändert worden. Das enge Boden-Wasser-Verhältnis 1 : 2 ist aber meistens beibehalten worden. 20 g Boden werden mit 40 ml H_2O 5 Minuten gekocht, zu der Suspension geben Berger und Truog einige Tropfen $CaCl_2$-Lösung, um klare Zentrifugate zu erhalten, s. auch Müller (1957). Oehlschläger (1958) verwendet dagegen einige Tropfen $MgSO_4$-Lösung, um eine Ausfällung von Bor durch die Ca-Ionen auszuschalten. Oehlschläger (1958) hat sich auch eingehend mit den Fehlermöglichkeiten bei der Bestimmung des pflanzenaufnehmbaren Bors im Boden beschäftigt, zum Beispiel Auslaugen vom Glas, Verunreinigungen der Chemikalien, Bindung des Bors beim Eindampfen der wässrigen Extrakte. Er hat auch festgestellt, daß einem Lehmboden zugesetztes Bor im ersten Heißwasserextrakt nur zu 65 bis 72 % wieder gefunden wird. Selbst im fünften Heißwasserauszug wird noch Bor gefunden. Oehlschläger ist der Ansicht, daß der Boden Bor einmal durch Bindung an Ca- und Sr-Ionen, weiter durch Adsorption an Fe- und Al-Silikate und schließlich auch durch Einbau in Silikatgitter als Ersatz für SiO_4-Tetraeder festlegt. Parks und Shaw (1941) haben darauf hingewiesen, daß Bor im Boden so stark festgelegt werden kann, daß mit heißem Wasser kein Bor herauszulösen ist. Besonders gefördert wird die Festlegung von Bor im Boden durch Ca-Ionen, Trockenheit und alkalische Reaktion. Bucher (1957) hat die in Acetat-Sulfat (Lösung nach Baron) und die in heißem Wasser lösliche B-Menge von je 15 Sandböden, mittelschweren Böden und schweren Böden bestimmt. Die Durchschnittswerte der Heißwasserextrakte lagen bei 0,40, 0,47 und 0,55, während die Durchschnittswerte der Auszüge mit der Lösung von Baron 0,42, 0,83 und 1,37 waren. Die Acetat-Sulfatlösung erfaßte also aus den schwereren Böden Bor, das vom heißen Wasser nicht gelöst wurde. Philipson (1953) hat das Boden-Wasser-Verhältnis bei der Bestimmung des in heißem Wasser löslichen Bors erweitert. Er behandelt 20 g Boden mit 200 ml Wasser. Nach seinen Untersuchungen wird durch Behandlung des Bodens mit heißem Wasser nur wenig Bor mehr gelöst als durch Wasser von 18°. Eine Wasserextraktion befriedigt nach seinen Erfahrungen aber nicht, da der Tag der Probenahme also der augenblickliche Bodenzustand, die Ergebnisse stark beeinflußt. Auch Pichotta (1951) hat den Boden im Verhältnis 1 : 10 mit heißem und auch mit kaltem Wasser behandelt. Hasler (1942) hat dagegen den Boden mit kaltem Wasser im Verhältnis 1 : 5 extrahiert. Ein weiterer Vorschlag, das Bor aus dem Boden mit Wasser „auszulaugen", stammt von Clung (1951). Bodenproben werden in eine Soxhlet-Hülse gegeben und 6, zum Teil auch 12 oder 24 Stunden mit Wasser extrahiert. Diese ausgelaugte B-Menge ist ungefähr 24mal so groß wie die mit heißem Wasser bei einmaliger Behandlung extrahierte B-Menge.

G. Kobalt

Kobalt ist zwar kein Pflanzennährstoff, es ist aber doch zu empfehlen, kobaltarme Böden mit Kobaltsalzen zu düngen, da die Rinder sich nur normal entwickeln, wenn die Futtermittel Kobalt in entsprechender Menge enthalten. Baron (1954/55) hat vorgeschlagen, die Düngebedürftigkeit der Böden für

Kobalt mit dem gleichen Lösungsmittel zu bestimmen, mit dem von ihm die Menge an Kupfer, Mangan, Zink, Molybdän und Bor im Boden ermittelt wird (s. unter Kupfer, S. 829).

8. Auswertung der Untersuchungsergebnisse, die bei Bestimmung der Nährstoffbedürftigkeit der Böden erhalten werden

Bei Untersuchung der Böden unter Verwendung von chemischen und physikalisch-chemischen Untersuchungsmethoden wird in den meisten Fällen festgestellt, daß der Boden eine bestimmte Anzahl mg Pflanzennährstoff je 100 g Boden enthält. Das erhaltene Ergebnis wird mit Grenzzahlen verglichen, die angeben, bei welchen Mengen (ausgedrückt in mg je 100 g Boden) der Boden schlecht, mittel und gut mit dem Pflanzennährstoff versorgt ist. Die bei der Untersuchung erhaltenen Ergebnisse werden zum Teil auch auf kg je ha oder pounds per acre umgerechnet. Eine derartige Umrechnung verführt aber leicht zur Aufstellung von Nährstoffbilanzen, das heißt, es werden die erhaltenen Ergebnisse um die mit den Düngemitteln gegebenen Mengen vermehrt und um die mit den Ernten dem Boden entzogenen Mengen vermindert. Solche Nährstoffbilanzen dürfen nicht aufgestellt werden. Die durch Bodenuntersuchungen erhaltenen Ergebnisse geben nicht an, daß nur die gefundene Nährstoffmenge mg/100 g Boden oder kg/ha von den Pflanzen aufgenommen werden kann und der sonst im Boden vorhandene Anteil nicht ausgenutzt wird. Die erhaltenen Ergebnisse sagen nur aus, daß die Pflanzen aus Böden mit hohem Untersuchungsergebnis den Pflanzennährstoff leichter aufnehmen als aus Böden mit kleinem Untersuchungsergebnis. Angaben, ob eine Düngung notwendig ist, können nur in Verbindung mit den oben erwähnten Grenzzahlen gemacht werden.

Literatur

Achromeiko, A.: Der Einfluß des Pulverisierens und Trocknens des Bodens auf dessen Fruchtbarkeit (2. Mitteilung). Z. Pflanzenernähr., Düng., Bodenkde. **A 11**, 65–89 (1928). — Amin, J. V., und H. E. Joham: Molybdän im Baumwoll-Embryo. Soil Sci. **86**, 293–297 (1958); Ref. Z. Pflanzenernähr., Düng., Bodenkde. **85**, 255 (1959). — Antipov-Karataev: Pedology **1947**, 652, nach Jackson, S. 402. — Armero, L. de: Untersuchungen an Böden statischer Düngerversuche nach der Methode Vageler-Alten. Kühn-Arch. **39**, 185–244 (1935). — Arrhenius, O.: Phosphatfrage. Z. Pflanzenernähr., Düng., Bodenkde. **A 14**, 185–194 (1928). — Asleson, J. A.: 1951, nach Jackson, S. 402.

Bamberg, K.: Untersuchungen über die chemische Bestimmung der Bodenfruchtbarkeit I. Z. Pflanzenernähr., Düng., Bodenkde. **A 9**, 161–175 (1927); II, ebenda **11**, 115–141 (1928). — Studien über das austauschbare Kalium. Z. Pflanzenernähr., Düng., Bodenkde. **A 14**, 177–184 (1929). — Aktive und passive Nährstoffe im Boden und deren Bestimmung. Z. Pflanzenernähr., Düng., Bodenkde. **A 45**, 154–188 (1936). — Barbier: Mündliche Mitteilung, s. OECE, S. 53 und 54. — Baron, H.: Gemeinsame Extraktion und chemische Bestimmung des leichtlöslichen Anteils der Mikronährstoffe Bor, Eisen, Kobalt, Kupfer, Mangan, Molybdän und Zink im Boden. Landwirtsch. Forsch. **7**, 82–89 (1954/55). — Beater, B. E.: A rapid method for obtaining readly soluble phosphates and phosphat-fixation in soils. Plant a. Soil **1**, 215–220 (1949). — Becker, E.: Vergleichende Untersuchungen und Versuche mit verschiedenen Methoden für Bestimmung des Nährstoffbedürfnisses der Böden an Phosphorsäure und Kali. Z. Pflanzenernähr., Düng., Bodenkde. **A 13**, 274–294 (1929). — Behrens, W. U.: Die Methoden zur Bestimmung des Kali- und Phosphorsäurebedarfes landwirtschaftlich genutzter Böden. Berlin: Verlag Chemie. 1935. — Berger, K. C., und E. Truog: Ind. Engng. Chem. Anal. Ed. **11**, 540 (1939), nach Oehlschläger, W., Landwirtsch. Forsch. **11**, 45–52 (1958). — Bergh, K.: K. Norske Vidensk. Selskabs Skrifter 1945, Nr. 3 (1948), nach Jackson, S. 408. — Bieler, Th.: Untersuchungen über das aufnehmbare Kali der Böden. Ann. Agricult. Suisse **1909**, 161–184; Ref. Hoffmann Jber. **1909**, 58, s. Blanck,

Bd. 8, S. 132. — BIRNER: Verhandlungen der VI. Wanderversammlung Deutscher Agrikulturchemiker usw. Landwirtsch. Versuchsstat. **13**, 10 (1870), nach BLANCK, Bd. 8, S. 144. — BLACK, C. A.: Soil-plant relationships. New York: J. Wiley. 1957. — BLANCHET, R., und S. PERIGAUD: Sur l'appreviation du niveau des reserves potassique des sols; possibilité d'utilisation d'un test Morgan-Barbier amelioré. Ann. Agronom. **11**, 347-355 (1960). — BLANCK, E.: Handbuch der Bodenlehre, Bd. 8. Berlin: Springer. 1931. — Handbuch der Bodenlehre. Erster Ergänzungsband. Berlin: Springer. 1939. — BONDORFF, K. A., und F. STEENBJERG: Studien over Jordens Fosforsyreinhold. J. Tidsskr. Planteavl. **38**, 273 (1932), und **29**, 549 (1933), nach BLANCK, I. Erg. Bd. S. 366. — BONDORFF, K. A.: Laboratoriumsbestimmungen des Kalibedürfnisses dänischer Böden. Ernährung d. Pflanze **32**, 41–42 (1936). — BOWER, C. A.: Determination of exchangeable magnesium in soils containing dolomite. Proc. Soil Sci. Soc. Amer. **19**, 40–42 (1955). — BRAWER, P. H., und R. B. RANKIN: Elektrodialysis compared with the Neubauer method for determining mineral nutrient deficiencies in soil. J. Amer. Soc. Agronomy **25**, 414 (1933), nach BLANCK, I. Erg. Bd., S. 367. — BRAY, R. H.: A field test for available phosphorus in soils. Illinois Stat. Bull. **337**, 589–602 (1929), nach Diagnostic Techniques for Soil and Crops, S. 86. — Correlation of soil tests with crop. Reponse to added fertilizers and with fertilizers requirement. Diagnostic Techniques for Soil and Crops, S. 53–86. — BRAY, R. H., und N. DICKMANN: Proc. Soil Sci. Soc. Amer. **6**, 312 (1942), nach JACKSON, S. 161. — BRAY, R. H., und L. T. KURTZ: Determination of total, organic, and available forms of phosphorus in soils. Soil Sci. **59**, 39–45 (1945), nach SMITH, E. W., B. G. ELLIS und J. GRAVA, Proc. Soil Sci. Soc. Amer. **21**, 400–404 (1957). — BRIOUX, E., und E. JOUIS: Ann. Agronom. **2**, 146–169 (1932), nach JACOB, S. 69. — BROWN: Soil Sci. **56**, 353 (1943), nach JACKSON, S. 68/69. — BRUGGER, G.: Bemerkung zur Magnesiumbestimmung im Boden nach Schachtschabel. Landwirtsch. Forsch. **11**, 202 (1958). — BRÜNE, FR., und TH. ARND: Ein neues Verfahren zur Bestimmung der Kalkbedürftigkeit von Moorböden. Bodenkde. u. Pflanzenernähr. **9/10**, 51–64 (1938). — BUCHER, R.: Beziehungen zwischen den wasser- und acetatlöslichen Bormengen in Mineralböden. Landwirtsch. Forsch. Sonderheft 9, 114–122 (1957). — BURRIEL, F., und V. HERNANDO: Die Bestimmung des aufnehmbaren Phosphors in spanischen Böden. Bericht d. 4. Intern. Kongreß f. Bodenkunde, Amsterdam 1950, Bd. II, 134–135.

CALDWELL, A. C., A. HUSTRULID und E. L. HAMMERS: Residual availability in the soil of various sources of phosphate as measured by plant absorption P^{32} and by soil test. Proc. Soil Sci. Soc. Amer. **20**, 25–28 (1956). — CLUNG, MC.: Proc. Soil. Sci. Soc. Amer. **15**, 268 (1951), nach JACKSON, S. 381.

DAIKUHARA, G.: Über das Verhalten der Phosphorsäure in den Böden gegen verschiedene organische Säuren. Bull. Coll. Agr. Tokio **5**, 505 (1903); Ref. Hoffmanns Jber. **1903**, 37, nach BLANK, Bd. 8, S. 145. — DEAN, L. A.: Elektrodialyses as a means of studying the nature of soil phosphates. Soil Sci. **30**, 33 (1934), nach BLANK, Erg. Bd., S. 367. — DEAN, L. A., und E. J. RUBINS: Anion exchange in soils. I: Exchangeable phosphorus, the anionexchange capacity. II: Methods of study. III.: Applicability to problems of soil fertility. Soil Sci. **63**, 377–406 (1947). — DEHÉRAIN, P. P., und A. MEYER: Wirkungslosigkeit der Düngung mit Phosphorsäure auf dem Versuchsfeld zu Grignon. Ann. Agronom. **5**, 161–176 (1879); Ref. Biedermanns Zbl. Agrikulturchem. **9**, 567 — 570 (1880). — DEHÉRAIN, P. P.: Über die Wirksamkeit der Superphosphate, nach C. R. **48**, 1286 — 1288 (1884); Ref. Biedermanns Zbl. Agrikulturchem. **13**, 605 (1884). — Die Phosphorsäure im Boden. Ann. Agronom. **17**, 445–454 (1891), nach Biedermanns Zbl. ·Agrikulturchem. **21**, 728–729 (1892). — DELMAS, J., J. BATS und P. REMY: Durch Phosphorsäureüberschuß hervorgerufener Manganmangel. C. R. hebd. Séances Acad. Sci. **244**, 1971–74, 1/4 (1957); Landwirtsch. Zbl. **3**, Abt. II, 1173 (1958). — DELONG, W. A., D. C. MACKAY und H. A. STEPPLER: Coordinated soil-plant analysis: I. Nutrient cations. Proc. Soil Sci. Soc. Amer. **17**, 262–266 (1953). — *Diagnostic Techniques for Soil and Crops*: Edited by H. BROEDEL KITCHEN. Published by the American Potash Institute Washington 6, D. C. (1948). — DIRKS, B.: Der Kohlensäure-Bikarbonatauszug und der Wasserauszug als Grundlage zur Ermittlung der Phosphorsäurebedürftigkeit der Böden. Landwirtsch. Jb. **71**, 73 (1930). — DIRKS, B., und F. SCHEFFER: Vergleichende Untersuchungen über das Nährstoffbedürfnis der Kulturböden. Landwirtsch. Jb. **67**, 779–821 (1928). — Der Kohlensäure-Bikarbonatauszug und der Wasserauszug als Grundlage zur Ermittlung der Phosphorsäurebedürftigkeit der Böden. Landwirtsch. Jb. **71**, 73–99 (1930). — Über die Bestimmung der leichtlöslichen Boden-Phosphorsäure. Wiss. Arch. Landwirtsch. Abt. A **4**, 641–652 (1930). — DRATSCHEW, S. M., und V. P. ALEXANDROWA: Die durch den Feuchtigkeitszustand bewirkten

Änderungen der Zusammensetzung der Konzentration der freien Bodenlösung. Z. Pflanzenernähr., Düng., Bodenkde. **A 25**, 349-365 (1932). — DREYSPRING, C., und W. HEINZ: Über die gleichzeitige Ermittlung des leichtlöslichen Phosphorsäure-Vorrats und der Festlegung der Düngerphosphorsäure im Boden. Z. Pflanzenernähr., Düng., Bodenkde. **A 35**, 362–374 (1934). — Eine neue Saugapparatur zur restlosen Trennung des Bodenauszugs vom Bodenkörper. Z. Pflanzenernähr., Düng., Bodenkde. **A 38**, 213–221 (1935). — DREYSPRING, C., und C. KRÜGEL: Vergleichende Untersuchungen schwedischer Böden auf pflanzenaufnehmbare Phosphorsäure. Superphosphat **2**, 121 (1929). — DYER, B.: On the analytical determination of probably avaible "mineral" plant food in soils. J. Chem. Soc. 1894, 115, nach BEHRENS, S. 181.

EGNÉR, H.: Metod att bestämma lattlöslig fosforsyra i akejord. Medd. Nr. 425 from Centralanstalten Stockholm, 1932, nach BLANCK, I. Erg. Bd., S. 364. — Bestimmung der Kalibedürftigkeit des Bodens auf chemischem Wege. Bodenkde. u. Pflanzenernähr. **21/22**, 270–277 (1940). — Über die Laktatmethode und ihre Anwendung in Schweden. Landwirtsch. Forsch. Sonderheft **4**, 106–109 (1951/52). — Neue Beiträge zur chemischen Bodenuntersuchung unter besonderer Berücksichtigung der Lactatmethode. Landwirtsch. Forsch. Sonderheft **6**, 28–32 (1954/55). — EG-NER, H., H. RIEHM und W. R. DOMINGO: Untersuchungen über die chemische Bodenanalyse als Grundlage für die Beurteilung des Nährstoffzustandes der Böden, II, Chemische Extraktionsmethoden zur Phosphor- und Kaliumbestimmung. Ann. Kgl. Landwirtsch. Hochschule Schwedens **26**, 199–215 (1960). — EMMERLING, A.: Agrikulturchemische Untersuchungen. Eine Festschrift, gewidmet den Schleswig-Holsteinischen Landwirten als ein Rückblick auf die 25jährige Tätigkeit der agrikulturchemischen Versuchsstation Kiel 1895. — Über die verschiedenen Formen der Phosphorsäure im Boden und deren Bestimmung. Landwirtsch. Versuchsstat. **52**, 60–70 (1899). — ENGELS, O.: Weitere Untersuchungen über die Feststellung des Phosphorsäuregehaltes resp. der Phosphorsäurebedürftigkeit des Bodens nach der Neubauerschen und Lemmermannschen Methode. Z. Pflanzenernähr., Düng., Bodenkde. **B 5**, 537-546 (1926). — ERIKSSON: Ann. Roy. Agr. Coll. Sweden 16–72 (1949), nach JACKSON, S. 402.

FINCK, V. A.: Methoden zur Bestimmung des für Hafer verfügbaren Mangans. Z. Pflanzenernähr., Düng., Bodenkde. **67**, 198–211 (1954).

GAMMON, N. JR., G. M. VOLK, E. N. MCGUBBIN und A. H. EDDINS: Soil factors affecting molybdenum uptake by cauliflower. Proc. Soil Sci. Soc. Amer. **18**, 302–305 (1954). — GEDROIZ, K. K.: Die chemischen Methoden zur Bestimmung der Fruchtbarkeit der Böden in bezug auf Phosphorsäure. J. exper. Landw. **4**, 432 (1903); Ref. Biedermanns Zbl. Agrikulturchem. **33**, 145 (1904). — GEHRING, A.: Die Bestimmung des Fruchtbarkeitszustandes des Bodens mit Hilfe chemischer Untersuchungsmethoden (1931). BLANCK, Bd. 8, S. 106–148. — GEHRING, A., A. PEGGAU und O. WEHRMANN: Über die Methode zur Bestimmung des Kalkbedürfnisses des Bodens unter Benutzung des Sättigungszustandes an Kalk. Z. Pflanzenernähr., Düng., Bodenkde. **A 8**, 321–331 (1926/27). — GEILMANN, W., und R. NEEB: Ein einfaches Verfahren zur Bestimmung des Zinkgehaltes in Böden. Z. Pflanzenernähr., Düng., Bodenkde. **67**, 38–47 (1954). — GERLACH, M.: Über das Verhalten der wasserlöslichen Phosphorsäure gegen absorbierende Bestandteile des Bodens. Landwirtsch. Versuchsstat. **46**, 201–219 (1896). — GIGIEL, J. A.: Wirkung der Oxalsäure auf die Phosphorsäureverbindungen des Bodens. Roczniki Nauk Polniczyck **11**, 481 (1924); Ref. Z. Pflanzenernähr., Düng., Bodenkde. **A 10**, 382 (1927/28). — GISIGER, L.: Untersuchungen über die Löslichkeit des Bodenkalis und seine Aufnehmbarkeit durch die Pflanzen. Bodenkde. u. Pflanzenernähr. **2**, 23–34 (1936/37). — GLERIA, J. DI: Die Bestimmung des Nährstoffgehaltes der Böden nach Elektrodialyse. Verh. 2. intern. Kongr. Bodenkde. **4**, 185 (1932), nach BLANCK, I. Erg. Bd., S. 367. — GOY, S., und O. ROOS: Über den Ausbau der elektrometrischen Bodentitration zu einer Methode für die genaue Bestimmung des Kalkbedarfs unter Berücksichtigung des physikalischen Bodenzustandes. Bodenkde. u. Pflanzenernähr. **2**, 166–178 (1937). — GROETZNER, E.: Untersuchungen über die Bestimmung des pflanzenaufnehmbaren Kaliums. Z. Pflanzenernähr., Düng., Bodenkde. **A 35**, 269–296 (1934). — GROUVEN, H.: Über den Zusammenhang zwischen Witterung, Boden und Düngung. Bodenanalyse S. 28–39. Glogau: C. Flemming. 1868.

HADDOCK, J.: Mass. Agr. Exper. Sta. Ann. Rep. **305**, 11 (1933), nach JACOB, S. 70. — HALEY, L. E., und S. W. MELSTED: Preliminary studies of molybdenum in Illinois soils. Proc. Soil Sci. Soc. Amer. **21**, 316–319 (1957). — HALL und PLYMEN: Bestimmung der wirksamen Pflanzennährstoffe im Boden. Biedermanns Zbl. Agrikulturchemie **31**, 298 (1902). — HASENBÄUMER, J.: Die Feststellung der Nährstoffbedürftigkeit des Bodens. Die chemische Bodenanalyse (1931), HONCAMP, Bd. I, S. 771–806. — HASLER: Landwirtschaft. Jb. d. Schweiz **56**, 486 (1942), nach PICHOTTA, G., Z. Pflanzenernähr., Düng., Bodenkde, **54**, 146–49 (1951). — HAUSER, G. F.: Die nichtaustauschbare Festlegung des Kalis im Boden. Wageningen: H. Veenman. 1941. — HERRMANN, R.: Überprüfung der thermokinetischen Bodenuntersuchungsmethode von E. Reinau. Bodenkde. u. Pflanzenernähr. **25**, 358–367 (1941). — Zusammenfassender Berichte über die Überprüfung der Schnellmethoden zur Bestimmung des pflanzenaufnehmbaren Kaliums und der Apparatur von Riehm-Lange, Berlin-Zehlendorf. Bodenkde. u. Pflanzenernähr. **29**, 252–274 (1943). — Normblätter für landwirtschaftliche Bodenuntersuchung. Bodenkde. u. Pflanzenernähr. **30**, 329–342 (1943). — HERZNER, R. A.: Beiträge zur physikalischen Chemie unserer Kulturböden VI. Z. Pflanzenernähr., Düng., Bodenkde. **19**, 93–110 (1931); VI, ebenda **23**, 68–76 (1932). — HISSINK, D. J.: Der Sättigungszustand des Bodens, A: Mineralböden (Tonböden). Z. Pflanzenernähr., Düng., Bodenkde. **4**, 137–158 (1925). — HOFF, D. J., und H. J. MEDERSKI: The chemical estimation of plant available soil manganese. Proc. Soil Sci. Soc. Amer. **22**, 129–132 (1958). — HOFFMANN, E.: Studien über die Löslichkeit der Bodenphosphorsäure und Bodenkieselsäure als Grundlage für die Bewertung des Phosphorsäurebedürfnisses. Landwirtsch. Jb. **1930**, 791–834. — HOFFMANN, W. E., H. RIEHM und D. SCHROEDER: Magnesium-Untersuchungen an deutschen Böden. Magnesium-Boden-Pflanze. Landwirtsch. Forsch. Sonderheft **13**, 9–16 (1959). — HOLMES, R. S.: Determination of copper, zinc, cobalt and lead in soils and soil solutions. Soil Sci. **59**, 77–84 (1945), nach SHAW und DEAN, Soil Sci. **73**, 341–347 (1952). — HOMÈS, M. V.: L'alimentation minérale des plantes et le problème des engrais chimiques. Paris: Masson. 1953. — HONCAMP, F.: Handbuch der Pflanzenernährung und Düngung, I. Bd. Berlin: Springer. 1931. — HOOD, J. T., N. C. BRADY und D. J. LATHWELL: The relationship of water soluble and exchangeable potassium to yield and potassium uptake by ladino clover. Proc. Soil Sci. Soc. Amer. **20**, 228–231 (1956). — HUNTER, A. H., und P. F. PRATT: Extraction of potassium from soils by sulfuric acid. Proc. Soil Sci. Soc. Amer. **21**, 595–598 (1957).

ISERMEYER, H.: Eine einfache Methode zur Bestimmung der Bodenatmung und der Karbonate im Boden. Z. Pflanzenernähr., Düng., Bodenkde. **56**, 26–38 (1952).

JACKSON, M. L.: Soil chemical analysis. Englewood Cliffs, N. J.: Prentice-Hall. 1958. — JACOB, A.: Magnesia, der fünfte Pflanzenhauptnährstoff. Bodenuntersuchung auf Magnesia — Chemische Methoden, S. 68–74. Stuttgart: F. Enke. 1955. — JAFFA, M. E.: Vergleichende Versuche über die Einwirkung der Salzsäure und Oxalsäure auf die Bestandteile des Ackerbodens. Ref. Exper. Stat. Rec. **6**, 792 (1895), Hoffmanns Jber. **1895**, 588, nach BLANCK, Bd. 8, S. 145. — JENSEN, S. T.: Internat. Mitt. Bodenkunde **14**, 112 (1924), nach Methodenbuch Bd. I, S. 149–151. — JOHNSON, K. E. E., J. F. DAVIS und E. J. BENNE: Control of magnesium deficiency in Utah 10B celery grown on organic soil. Proc. Soil Sci. Soc. Amer. **21**, 528–532 (1957). — JONES, L. H. P., und G. W. LEEPER: The availability of various manganese oxydes to plant. Available manganese oxydes in neutral and alkaline soils. Plant a. Soil **3**, 141–159 (1951).

KAPPEN, H.: Die Bodenazidität. Berlin: Springer. 1929. — KAWE, A.: Methode zur Bestimmung des Düngebedürfnisses eines Bodens für Kalium und Phosphorsäure mit Hilfe der Bodenlösung. Kühn-Arch. **39**, 282–293 (1935). — Eine natürliche Methode zur Bestimmung des Düngebedürfnisses des Bodens für Kalium- und Phosphorsäure mit Hilfe der Bodenlösung. Z. Pflanzenernähr., Düng., Bodenkde. **A 43**, 69–83 (1936). — Methode zur Bestimmung des Düngerbedürfnisses des Bodens für Kalk, Kalium und Phosphorsäure mit Hilfe der Bodenlösung. Z. Pflanzenernähr., Düng., Bodenkde. **84**, 224–231 (1959). — KELLNER, O.: Landwirtsch. Versuchsstat. **33**, 358 (1897), nach HONCAMP, I, S. 786. — Originalmitteilung nach KÖNIG (1923), S. 69. — KIRSANOFF, A. T.: Schnellmethode zur Feststellung der Phosphorsäurebedürftigkeit. Bull. d. Leningrader Dep. 38 (1931); Ref. Z. Pflanzenernähr., Düng., Bodenkde. **A 25**, 382 (1932). — KIRSTE, A.: Welches ist die beste Methode zur Untersuchung des Bodens auf pH, Kalk, P_2O_5 und K_2O? Landwirtsch. Forsch. **4**, 51–59 (1952). — KNICKMANN, E.: Der Morgan-Test in Frankreich. Landwirtsch. Forsch. Sonderheft 6, 47–50 (1955). — KÖNIG, J.: Untersuchung landwirtschaftlich und landwirtschaftlich gewerblich wichtiger Stoffe. 5. Aufl. Berlin: P. Parey. 1923. — KÖNIG, J., und J. HASENBÄUMER: Die Ermittlung des Düngebedürfnisses des Bodens. Z. Pflanzenernähr., Düng., Bodenkde. **B 3**, 497–532 (1924), **B 5**, 443–468 (1926). — KÖNIG, J., J. HASENBÄUMER und E. COPPENRATH: Landwirtsch. Versuchsstat. **66**, 401 (1907), nach HONCAMP, Bd. 1, S. 799. — KÖTTGEN, P.: Die Bestimmung der leichtlöslichen Nährstoffe durch elektrischen Gleichstrom zur Ermittlung der Fruchtbarkeitsveranlagung unserer Kulturböden. Z. Pflanzenernähr., Düng., Bodenkde. **A 29**, 275–290 (1933). — Über die Gültigkeit der Vagelerschen Sorptionsgleichung zur Bestimmung der relativen Löslichkeit der Nährstoffe eines Bodens. Z. Pflanzenernähr., Düng., Bodenkde. **A 32**, 320 (1933). — Die Bestimmung der leichtlöslichen und der sorptiv gebundenen Ionen nach der Methode des Gießener Bodenkundlichen Instituts. Bodenkde. u. Pflanzenernähr. **3**, 56–89 (1937). — Köttgen, P., und R. Diehl: Über die Anwendung der Dialyse und Elektroultrafiltration zur Bestimmung des Nährstoffbedürfnisses des Bodens. Z. Pflanzenernähr., Düng., Bodenkde. **A 14**, 65–105 (1929). — Köttgen, P., und L. Jung: Über die Abhängigkeit der Nährstofflieferung von Phosphorsäure, Kalk und Kali von dem Wassergehalt, untersucht an humosen Böden. Bodenkde u. Pflanzenernähr. **24**, 129–150 (1941). — Zur Frage der nachschaffenden Kraft des Bodens. Bodenkde. u. Pflanzenernähr. **24**, 257–265 (1941). — Über den Einfluß der Ca- und K-Ionen auf die Löslichkeit der Phosphorsäure, untersucht an einem Löss-Lehmboden der Wetterau. Bodenkde. u. Pflanzenernähr. **25**, 58–64 (1941). — Krauss, J.: Ein zur billigen Massenuntersuchung geeignetes Verfahren zur Prüfung von Böden auf Kalibedürftigkeit. Bodenkde. u. Pflanzenernähr. **1**, 301–330 (1936). — Krumiņš, K.: Die Abscheidung des Ammoniums bei der Kalimetermethode (Verbesserte Kalimetermethode). Lauksainniecibas Menešrakstz Riga Nr. 11/12, 1933 und Riga 1934, 3–16. — Ein Photometer zur schnellen Bewertung des Phosphorsäurebedarfs. Z. Pflanzenernähr., Düng., Bodenkde. **A 26**, 337 (1932). — Kudashev, A. S.: Available phosphoric acid in chernozem soils. Russ. J. exper. Landw. **6**, 457 (1905), nach Blanck, Bd. 8, S. 145. — Kühn, St.: Untersuchungen zur Bestimmung des leichtaufnehmbaren Kali- und Phosphorsäurevorrates der Böden. Különlenyomat a Kisérletügyi Közlemények **38**, (1935) kötet 5–6 füzetéböl.

Lathwell, D. T., N. Sanchez, D. J. Lisk *et al.*: Availability of soil phosphorus as determined by several chemical methods. Agronom. J. **50**, 366–369 (1958) (Cornell Univ. Ithaca, N. Y.); Ref. Soil Fertilizers **21**, 287 (1958). — Leaf, A. L.: Determination of available potassium in soils of forest plantations. Proc. Soil Sci. Soc. Amer. **22**, 458–459 (1958). — Leeper, G. W.: Relationship of soils to manganese deficiency of plants. Nature **134**, 972–973 (1934). — Lemmermann, O.: Zur Frage der Ermittlung des Düngungsbedürfnisses der Böden mit Hilfe der chemischen Bodenanalyse. Landwirtsch. Versuchsstat. **83**, 345–358 (1914). — Die Bestimmung der relativen Löslichkeit der Phosphorsäure im Boden (1931), s. Blanck, Bd. 8, S. 174–183. — Untersuchungen über das Kalkbedürfnis der Böden durch Laboratoriumsmethoden und Düngungsversuche. Berlin: Verlag Chemie. 1933. — Lemmermann, O., A. Einecke und L. Fresenius: Untersuchungen über die Feststellung des Wirkungswertes der Bodennährstoffe Phosphorsäure und Kali durch den Vegetationsversuch und die Bestimmung ihrer relativen Löslichkeit durch Säuren. Landwirtsch. Versuchsstat. **89**, 81–195 (1917). — Lemmermann, O., und L. Fresenius: Über die Bestimmung der relativen Löslichkeit der Bodenphosphorsäure. Z. Pflanzenernähr., Düng., Bodenkde. **2**, 363–369 (1923). — Die Bestimmung des Düngungsbedürfnisses der Böden für Phosphorsäure nach der Zitronensäuremethode. Z. Pflanzenernähr., Düng., Bodenkde. **A 15**, 249–258 (1930). — Lemmermann, O., L. Fresenius und W. Lesch: Die Ermittlung des Düngungsbedürfnisses der Böden für Phosphorsäure mit Hilfe der Zitratmethode. Z. Pflanzenernähr., Düng., Bodenkde. **B 6**, 163–178 (1927). — Lemmermann, O., und E. Rauterberg: Das verschiedene Verhalten der Phosphorsäureverbindungen des Bodens gegenüber 1%iger Essigsäure. Z. Pflanzenernähr., Düng., Bodenkde. **43**, 193–207 (1949). — Liebscher: Untersuchungen über die Bestimmung des Düngerbedürfnisses der Ackerböden und Kulturpflanzen. J. Landwirtsch. **43**, 49–216 (1895). — Lineberry, R., und L. Burkhart: Plant Physiol. **18**, 324–33 (1943), nach Jacob, S. 69. — Loew, O.: Ist es berechtigt, bei Bodenanalysen die Magnesiumbestimmung außer acht zu lassen? Chem. Ztg. **33**, 118–119 (1909). — Ist die Lehre vom Kalkfaktor eine Hypothese oder eine bewiesene Theorie? Landwirtsch. Jb. **46**, 733–752 (1914). — Lohse, G. W., und G. N. Ruhnke: Studien über leichtlösliche Phosphate in Böden. I: Extraktion der raschlöslichen Phosphate aus Böden mit Hilfe von verdünnter Kaliumbisulfat-Lösung. Soil Sci. **35**, 437 (1933); Ref. Z. Pflanzenernähr., Düng., Bodenkde. **A 35**, 256 (1934). — Long, O. H., und L. F. Seatz: Correlation of soil tests for available phosphorus and potassium with crop yield responses of fertilization. Proc. Soil Sci. Soc. Amer. **17**, 258–262 (1953). — Lundblad, K., O. Svanberg und P. Ekman: The availability and fixation of copper in swedish soils. Plant a. Soil **1**, 277–302 (1949). — Lyman, C., und L. A. Dean: Zinc deficiency of pineapples in relation to soil and plant composition. Soil Sci. **54**, 315–324 (1942), nach Lyman und Dean, Soil Sci. **73**, 341–347 (1952).

Magnesium-Boden-Pflanze. Landwirtsch. Forschung, Sonderheft 13, 16 Mg-Arbeiten, 100 S. (1959). — Mäkitie, O.: Studies on the acid ammoniumacetate extraction method in soil testing. Agrogeologisia Julkaisuja Helsinki Nr. 66, 1–25 (1956). — Matthews, B. C.: The fixation and release of soil potassium. Ph.-D.-Thesis, Cornell University 1952, nach Hood, J. T., N. C. Brady und D. J. Lathwell, Proc. Soil Sci. Soc. Amer. **20**, 228–231 (1956). — Matthews, B. C., und J. A. Smith: Eine Perkolationsmethode zur Messung der Kalinachlieferungskraft von Böden. Canad. J. Soil Sci. **37**, 21–28 (1957); Ref. Landwirtsch. Zbl. **3**, 2492 (1958). — Maxwell, W.: Kalk, Kali und Phosphorsäure im Boden von Hawaii. Chem. Zbl. 1899, II, 1136, nach Blanck, Bd. 8, S. 145. — McAuliffe, C. D., N. S. Hall, L. A. Dean und S. B. Hendricks: Exchange reactions between phosphates and soils by hydroxylic surfaces of soil mineral. Proc. Soil Sci. Soc. Amer. **12**, 119–123 (1947), nach Black, 1957, S. 259 und 284. — McColloch, R. C., E. T. Bingham und D. G. Aldrich: Relation of soil potassium and magnesium to magnesium nutrition of citrus. Proc. Soil Sci. Soc. Amer. **21**, 85–88 (1957). — Mehlich, A.: Determination of cation- and anion-exchange properties of soils. Soil Sci. **66**, 429–445 (1948). — Menzel und Jackson: Anal. Chem. **23**, 1861 (1951), nach Jackson, S. 399. — Merwin und Peech: Proc. Soil Sci. Soc. Amer. **15**, 125 (1951), nach Jackson, 1958, S. 131. — *Methodenbuch*, Bd. I, Die Untersuchung der Böden, von Thun, R., R. Herrmann und E. Knickmann. Verband Deutscher Landwirtschaftlicher Untersuchungs- und Forschungsanstalten. 3. Aufl. Radebeul und Berlin: Neumann. 1955. — Meyer, L.: Vergleichende Untersuchungen zur Beurteilung des Phosphorsäurezustandes nach v. Wrangell. Landwirtsch. Jb. **73**, 119 (1931). — Michael, G., und G. Schilling: Über den Magnesium-Versorgungsgrad mitteldeutscher Ackerböden. Z. Pflanzenernähr., Düng., Bodenkde. **79**, 31–50 (1957). — Mitchell, R. L.: The base status of scottish soils. I. The effects of lime on the five typical soils from northeast Scotland. J. Agricult. Sci. 26. 664 (1936), nach Blanck, I. Erg. Bd., S. 352. — Soil analysis and trace elements. OEEC 1956, S. 144. — Mitcherlich, E. A.: Georgine. Land- und Forstwirtsch. Zg. **1910**, S. 13, nach Lemmermann, Landwirtsch. Versuchsstat. **83**, 364 (1914). — Zweiter Bericht über die Arbeiten und über die Tagung der Arbeitsgemeinschaft zur Prüfung der Laboratoriumsmethoden für die Bestimmung des Kali- und Phosphorsäurebedürfnisses der Böden. Verhandelt in Stockholm 5. 7. 1939. Internationale Bodenkundliche Gesellschaft. — Bodenkunde für Landwirte, Forstwirte und Gärtner, 7. Aufl., S. 126–133. Berlin: P. Parey. 1954. — Møller, J.: Studien über den Ionenaustauschprozeß mit besonderer Berücksichtigung der Agrikulturchemie. Kolloid-Beih. **46**, 1–112 (1937), nach Blanck, Bd. 1, S. 362. — Møller, J., und T. Mogensen: Use of an ion-exchanger for determining available phosphorus in soils. Soil Sci. **76**, 297–306 (1953). — Morgan, M. F.: Tobacco as an indicator plant in studying nutritional deficiencies of soils under greenhouse conditions. J. Amer. Soc. Agronom. **21**, 130 (1929), nach Behrens, S. 96. — Bull. 292, Connecticut Agric. Exper. Stat., New Haven, Connecticut, 1937, nach Methodenbuch, 2. 254. — Mortland, M. M., und J. L. Mellor: Conductometric titration of soil action-exchange capacity. Proc. Soil Sci. Soc. Amer. **18**, 363–364 (1954). — Müller, F. W.: Serienmäßige Borbestimmung in Böden mittels der Heißwasserextraktion nach Berger und Truog. Landwirtsch. Forsch. **10**, 32–35 (1957).

Nath Puri, A.: Eine neue Methode zur Bestimmung des austauschbaren Natriums und Kaliums im Boden. Soil Sci. **36**, 355 (1933). — Nehring, K.: Die Bestimmung des austauschbaren Kaliums als Mittel zur Beurteilung der Kalibedürftigkeit der Böden. Bodenkde. u. Pflanzenernähr. **30**, 36–51 (1942). — Nemec, A.: Ein neues Verfahren zur raschen Bestimmung des Düngebedürfnisses der Ackerböden. Dtsch. Landwirtsch. Presse **53**, 463–464, 629–631, 642–643 (1926). — Ein neues rasches Verfahren zur Ermittlung der Phosphorsäuredüngungsbedürftigkeit der Ackerböden. Fortschr. Landwirtsch. **3**, 919–924 (1928). — Nolte: J. Landwirtsch. **55**. 14 (1914), nach Rauterberg, E., Die Festlegung der Phosphorsäure im Boden, Forschungsdienst Sonderheft 6, 96–102 (1937).

OEEC: The organisation and rationalisation of soil analysis, project No. 156. Published by the European Productivity Agency of the Organisation for European Economic Co-operation, Paris, 1956. — Oelschläger, W.: Fehlermöglichkeiten bei der Bestimmung des pflanzenaufnehmbaren Bors im Boden. Landwirtsch. Forsch. **11**, 45–52 (1958). — Ohlsen, S. R., C. V. Cole, F. S. Wanatabe und L. A. Dean: Estimation of available phosphorus in soils by extraction with sodium bicarbonate. U. S. Dep. Agric. Circ. 939, 1–19 (1954). — Olsen, S., R.: Inorganic phosphorus in alkaline and calcareous soils. Agronomy **4**, 89–122 (1953), nach Blanck, S. 259 und 285. — The measurement of phosphorus on surface of soil particles and its relationship to plant available phosphorus. Kansas Agr. Exper. Sta. Rep. 4, 59–67 (1953), nach BLANCK, S. 259 und 285.

PAAUW, F. v. D.: Calibration of soil test methods for the determination of phosphate and potash status. Plant a. Soil 8, 105–125 (1956).—PARKS, R. Q., und B. T. SHAW: Possible mechanism of boron fixation in soil. J. Chem. Soil Sci. Soc. Amer. Proc. 6, 219 (1941), nach SCHARRER, S. 103. — PAVLOVSCHI, GH., und R. MAVRODINEANU: Beiträge zum Studium des austauschbaren Kaliums. Bodenkde. u. Pflanzenernähr. 15, 250–254 (1939). — PEECH, M.: Determination of exchangeable magnesium in soils by titan yellow with reference to magnesium deficiency in citrus. Proc. Soil Sci. Soc. Amer. 4, 189–195 (1939), nach Diagnostic Techniques for Soil and Crops, S. 9 und 50. — Chemical methods for assessing soil fertility. Diagnostic Techniques for Soil and Crops, S. 1–52 (1948). — PEECH, M., et al.: Methods of soil analysis for soil-fertility investigations. U. S. Dep. Agr. Circ. 757, nach Diagnostic Techniques for Soil and Crops, S. 51. — PENSELER, R. G.: Die Theorie des salzsauren Auszugs im Vergleich mit mehreren Methoden zur Bestimmung der Bodenfruchtbarkeit. Kühn-Arch. 26, 415–445 (1930). — PETER, H., und S. MARKERT: Die Bestimmung des Kalkbedarfs von Böden aus den pH- und MB-Werten. Z. Landwirtsch. Versuchs- u. Untersuchungswesen 5, 148–164 (1959). — PETER, H., S. MARKERT und G. GERICKE: Die Bestimmung der Sorptionseigenschaften von Böden mit Methylenblau. Z. Landwirtsch. Versuchs- u. Untersuchungswesen 5, 165–172 (1959). — PFEFFER, P.: Zur Bestimmung der austauschbaren Basen und des S-Wertes von Böden. Z. Pflanzenernähr., Düng., Bodenkde. 75, 17–32 (1956). — PHILIPSON, T.: Boron in plant and soil with special regard to Swedish agriculture. Acta Agricult. Scand. 3, 121–242 (1953). — PICHOTTA, G.: Bestimmung von Bor-Mikromengen in wäßrigen Auszügen zu landwirtschaftlichen Zwecken. Z. Pflanzenernähr., Düng., Bodenkde. 54, 146–149 (1951). — PIPER, C. S.: Soil and plant analysis. New York: Interscience. 1947. — PRATT, P. F., und G. R. BRADFORD: Separation and determination of total copper and zinc in soils. Proc. Soil Sci. Soc. Amer. 22, 399–402 (1958). — PRINCE, A., und S. TOTH: Proc. Soil Sci. Soc. Amer. 2, 207–14 (1937), nach JACOB, S. 69. — PUCHNER, H.: Bodenkunde für Landwirte. Chemische Bodenanalyse S. 565–576. Stuttgart: F. Enke. 1923. — PURVIS, E. R., und J. M. BLUME: Rapid soil tests for determining the fertilizer requirements of vegetable crops in Eastern Virginia. Bull. 106 der Virginia Truck Experiment Station, Norfolk, Virginia, Jan. 1941; Ref. Z. Pflanzenernähr., Düng., Bodenkde. 43, 88–91 (1949).

RAMANN, E.: Bodenfragen. Dtsch. Landwirtsch. Presse 45, 256 (1918). — Ernährung d. Pflanze 23, 28 (1927). nach JACOB, S. 69. — RAUTERBERG, E.: Die Berücksichtigung der Bodenart bei den Untersuchungen der Böden auf ihre Phosphorsäurebedürftigkeit. Z. Pflanzenernähr., Düng., Bodenkde. A 20, 302–356 (1931). — Die Wirkung der Nährstoffe bei nesterartiger und bei absolut gleichmäßiger Verteilung im Boden. Ernähr. d. Pflanze 33, 201–208 (1937). — Über die Beziehung zwischen Wachstumsfaktor und Ertrag unter besonderer Berücksichtigung der Berechnung der Konstanten A und c in der logarithmischen Gleichung von Mitscherlich. Bodenkde. u. Pflanzenernähr. 14, 10–28 (1939). — Die Bedeutung der mit verschiedenen Methoden erhaltenen Ergebnisse bei Bestimmung der Phosphorsäurebedürftigkeit der Böden. Trans. 7th International Congress of Soil Science, Madison, Wisc., USA, 1960, Vol. II, 422–427, Int. Soc. Soil Sci. — RAUTERBERG, E., und H. OSSENBERG-NEUHAUS: Die Löslichkeit der Phosphorsäure in verschiedenen Lösungsmitteln und die Bewertung der lactatlöslichen Phosphorsäure bei verschiedener Bodenreaktion. Z. Pflanzenernähr., Düng., Bodenkde. 50, 256–278 (1950). — Löslichkeit der Bodenphosphorsäure in Calciumlactatlösung. Landwirtsch. Forsch. Sonderheft 4, 100–106 (1951/52). — RAUTERBERG, E., und W. BUSSLER: Die Ermittlung der optimalen Nährstoffzusammensetzung für die Pflanze nach Homès. Z. Pflanzenernähr., Düng., Bodenkde, 90 5–18 (1960). — REIFENBERG, A.: Die kataphoretische Bestimmung der Düngerbedürftigkeit von Böden. Z. Pflanzenernähr., Düng., Bodenkde. A 24, 20–30 (1932). — REINAU, E. H.: Thermo-kinetische Bodenuntersuchung. Technik i. d. Landwirtschaft 19, 69–74 (1938). — RIEHM, H.: Die Bestimmung der laktatlöslichen Phosphorsäure im Boden unter Verwendung eines lichtelektrischen Colorimeters. Bodenkde. u. Pflanzenernähr. 9/10, 30–50 (1938). — Die Berücksichtigung der Bodenart und der Reaktion bei der Phosphatmethode nach Egnér. Bodenkde. u. Pflanzenernähr. 21/22, 316–327 (1940). — Die Bestimmung des pflanzenaufnehmbaren Kalis nach der neuen Kalimethode von Egnér. Fortschr. d. Landwirtsch.-chem. Forschung 1940. Forschungsdienst Sonderheft 15, 83–92 (1941). — Bestimmung der Kalibedürftigkeit des Bodens aus dem Kaligehalt des Laktatextraktes. Bodenkde. u. Pflanzenernähr. 31, 346–356 (1943). — Bestimmung der Sorptionskapazität des Bodens bei Massenuntersuchungen nach H. Riehm und ihre Bedeutung insbesondere für die Auswertung der Laktatwerte. Z. Pflanzenernähr., Düng., Bodenkde. **31**, 61–74 (1946). — Die Ammoniumlaktat-Essigsäure-Methode zur Bestimmung der leichtlöslichen Phosphorsäure in karbonathaltigen Böden, Agrarkimica **3**, 49–65 (1958). — RIEHM, H., B. ULRICH und M. ULRICH: Schnelle Bestimmung der Kationensorptionskapazität. Landwirtsch. Forsch. **6**, 95–105 (1954). — ROBINSON, W. O., G. EDGINGTON, W. H. ARMINGER und A. V. BREEN: Availability of Molybdenum as influenced by Liming. Soil Sci. **72**, 267–274 (1951). — ROHDE, G.: Einfache Methode zur Bestimmung des Nährstoffzustandes der Böden auf Grund der Löslichkeit ihrer Nährstoffverbindungen in kochendem Wasser. Z. Pflanzenernähr., Düng., Bodenkde. **A 30**, 331–344 (1933). — ROTH, H., und C. PFAFF: Über ein Verfahren zur Bestimmung des pflanzenaufnehmbaren Mangans. Landwirtsch. Forsch. Sonderheft 6, 41–47 (1955). — RUBINS, E. J., und L. A. DEAN: A comparison of certain methods for determining readily soluble phosphorus in soils. J. Amer. Soc. Agronom. **38**, 820–823 (1946); Ref. Soils a. Fertilizers **10**, 25 (1947). — RUHLAND, W.: Handbuch der Pflanzenphysiologie, Bd. 4, Die mineralische Ernährung der Pflanze. BERGMANN, W., Die Ermittlung der Nährstoffbedürftigkeit des Bodens, S. 867–942. Berlin-Göttingen-Heidelberg: Springer. 1958. — RÜMPLER, A.: Auszug des Bodens mit Kalkwasser zur Bestimmung des leichtlöslichen Kalis. Landwirtsch. Versuchsstat. **55** (1901), nach WANNSCHAFFER und SCHUCHT, S. 133. — RUSSEL, D. A., und G. STANFORD: A laboratory manual for soil fertility students, 2. Aufl. Dubuque, Iowa: Brown. 1954.

SANDHOFF, H.: Konduktometrische Bestimmung des T-Wertes von Böden und Bodenbestandteilen. I. Mitt. Z. Pflanzenernähr., Düng., Bodenkde. **62**, 1–19 (1953), II ebenda **67**, 24–38 (1954). — SCHACHTSCHABEL, P.: Die Bestimmung des Kalibedarfs im Boden. Bodenkde. u. Pflanzenernähr. **24**, 371–384 (1941). — Die Bestimmung von S-Wert, T-Wert und Sättigungsgrad. Z. Pflanzenernähr., Düng., Bodenkde. **53**, 7–20 (1951). — Die Methoden zur Bestimmung des Kalkbedarfs im Boden. Z. Pflanzenernähr., Düng., Bodenkde. **54**, 134–145 (1951). — Reaktion und Kalkbedarf von Hochmoorböden. Z. Pflanzenernähr., Düng., Bodenkde. **60**, 21–27 (1953). — Das pflanzenverfügbare Magnesium des Bodens und seine Bestimmung. Z. Pflanzenernähr., Düng., Bodenkde. **67**, 9–23 (1954). — Die Bestimmung des Manganversorgungsgrades von Böden und seine Beziehung zum Auftreten der Dörrfleckenkrankheit bei Hafer. Z. Pflanzenernähr., Düng., Bodenkde. **78**, 147–167 (1957). — Die Bestimmung des Magnesiumversorgungsgrades von Böden mittels chemischer Methoden. Magnesium-Boden-Pflanze. Landwirtsch. Forsch. Sonderheft 13, 88–89 (1959). — SCHACHTSCHABEL, P., und H. ISERMEYER: Die Magnesiumbestimmung mittels Titangelb. Z. Pflanzenernähr., Düng., Bodenkde. **67**, 1–8 (1954). — SCHARRER, K.: Biochemie der Spurenelemente. Berlin-Hamburg: P. Parey. 1955. — SCHARRER, K., und W. EBERHARDT: Zur quantitativen Bestimmung kleinster Mengen Molybdän in der agrikulturchemischen Analyse. Z. Pflanzenernähr., Düng., Bodenkde. **73**, 115–127 (1956). — SCHARRER, K., und H. KÜHN: Die Dithizon-Mischfarbenmethode und ihre Anwendung bei der Bestimmung kleinster Mengen Kupfer in Böden und biochemischen Substanzen. Bodenkde. u. Pflanzenernähr. **21/22**, 344–364 (1940). — SCHARRER, K., und H. MUNK: Die quantitative Bestimmung kleinster Mengen Zink in Böden, pflanzlichen und tierischen Substanzen und Düngemitteln. Z. Pflanzenernähr., Düng., Bodenkde. **74**, 24–42 (1956). — SCHARRER, K., E. RUSS, und K. MENGEL: Die Bestimmung des pflanzenaufnehmbaren Kupfers und Mangans. Z. Pflanzenernähr., Düng., Bodenkde. **85**, 1–20 (1959). — SCHEFFER, F., und P. SCHACHTSCHABEL: Lehrbuch der Agrikulturchemie und Bodenkunde, I. Teil, Bodenkunde, 5. Aufl. Stuttgart: F. Enke. 1960. — SCHEFFER, F., E. WELTE und C. HEINEMANN: Über die Laktatlöslichkeit von bodenbürtiger und Mineraldünger-Phosphorsäure in karbonathaltigen Böden. Z. Pflanzenernähr., Düng., Bodenkde. **58**, 40–59 (1952). — SCHMITT, L.: Ergebnisse der Prüfung von neuen Bodenuntersuchungsmethoden. Forschungsdienst Sonderheft 11, 37–46 (1938). — SEMB, G., und O. TRAGETHON: Undersøkelser over Magnesiummangel J Herredene Rygge og Rade J Østfold. Tiddskript for det Norske Landbruk **65**, 230–243 (1958). — SEMB, G,. und J. TORSTEINSSON: Plantenes magnesium-forsyning og magnesiuminnholdet i en del norske jordprover. Saertykh av Tidsskript for Det Norske Landbruk Nr. 8, 1–17 (1955). — SEMB, G., und G. UHLEN: A comparison of different analytical methods for the determination of potassium and phosphorus in soil based on field experiments. Acta Agricult. Scand. **1**, 44–68 (1955). — SHAW, W.: J. Assoc. Agric. Chem. **24**, 244–49 (1941), nach JACOB, S. 69. — SHAW, E., und L. A. DEAN: Use of dithizone as an extractant to estimate the zinc nutrient status of soils. Soil Sci. **73**, 341–347 (1952). — v. SIGMOND, A.: Ein neues Verfahren zur Bestimmung der leicht assimilierbaren Phosphorsäure im Boden. Wiener Landwirtsch. Ztg. 1902, Nr. 42, 362, nach Biedermanns

Zbl. Agrikulturchem. **32**, 145–146 (1903). — Verhandlungen der II. Kommission der Internat. Bodenkundl. Gesellsch. A 147 (1929), nach RAUTERBERG, Z. Pflanzenernähr., Düng., Bodenkde. **20**, 356 (1931). — Die Bestimmung der in Salzsäure löslichen Mineral- und Nährstoffe des Bodens und die Bewertung der Befunde des Salzsäureauszuges (1931), nach BLANCK, Bd. 8, S. 148–174, s. besonders S. 168. — SMITH, F. W., B. G. ELLIS und J. GRAVA: Use of acid-fluoride solutions of the extraction of available phosphorus in calcareous soils and in soils to which rock phosphate has been added. Proc. Soil Sci. Soc. Amer. **21**, 400–404 (1957). — SNYDER, H.: The action of organic and mineral acids upon soils. J. Amer. Chem. Soc. **17**, 148 (1895), nach BLANCK, Bd. 8, S. 145. — VAN DER SPUIJ, M. J.: Vergleichende Untersuchungen über die Feststellung der Düngebedürftigkeit der Böden an Phosphorsäure. Z. Pflanzenernähr., Düng., Bodenkde. **A 5**, 281–325 (1925). — STEENBJERG und BOKEN: Tids Planteavl. **52**, 375 (1948), nach JACKSON, S. 402. — STENUIT, D., und R. PIOT: La magnésie: element essentiel de l'alimentation des plantes III. Rev. Agricult. **11**, Nr. 6, 897–920. (1958); Ref. Z. Pflanzenernähr., Düng., Bodenkde. **83**, 256 (1958). — La magnésie, un élément essentiel pour l'alimentation des plantes I. Rev. Agricult. **10** (1957); Ref. Z. Pflanzenernähr., Düng., Bodenkde. **83**, 246 (1958). — STENUIT, D., R. PIOT und R. BOON: Relation entre le pH et la teneur en manganese des sols sablonneux et l'apparition des symptomes de carence en manganese chez l'avoine. Rapp. VI/ Congres international de la Science du Sol, Vol. A, 234–235 (1956). — STENUIT, D. F.: Beziehungen zwischen Magnesium-Gehalt und Ertrag belgischer Böden. Magnesium-Boden-Pflanzen. Landwirtsch. Forsch. Sonderheft 13, 23–29 (1959). — STEWART, G. R.: J. Agricult. Res. **12**, 311 (1918), nach DIRKS und SCHEFFER, Wiss. Arch. Land. Abt. A 4, 641–651 (1930). — STÖCKHARDT, A.: Studien über den Boden. Landwirtsch. Versuchsstat. **1**, 21 (1859), nach BLANCK, Bd. 8, S. 107. — STODDART, C. W.: Die Beziehung zwischen leichtlöslicher Phosphorsäure und Bodenazidität. Res. Bull. Nr. 2 (1909), The University of Wisconsin; Ref. Biedermanns Zbl. Agrikulturchem. **39**, 287–288 (1910). — STOCKLASA, J., und E. G. DOERELL: Handbuch der biophysikalischen und biochemischen Durchforschung des Bodens. Berlin: P. Parey. 1926.

TEPE, W.: Die Bestimmung der Bodenleistung mit Ionenaustauschern im Hinblick auf die Ernährung der Pflanzen. Landwirtsch. Forsch. Sonderheft 7, 83–89 (1956). — TEPE, W., und E. LEIDENFROST: Einfluß des H^+/HCO_3-Verhältnisses und der Selektivität bei der Bestimmung der Bodenleistung mit Ionenaustauschern. Z. Pflanzenernähr., Düng., Bodenkde. **75**, 222–227 (1956). — Ein Vergleich zwischen pflanzenphysiologischen, kinetischen und statischen Bodenuntersuchungswerten. Landwirtsch. Forsch. **11**, 217–229 (1958). — THOMPSON, L. F., und P. F. PRATT: Solubility of phosphorus in chemical extractants as indexes to available phosphorus in Ohio soils. Proc. Soil Sci. Soc. Amer. **18**, 467–470 (1954). — THORNE *et al.*: Soil Sci. **54**, 463 (1942), nach JACKSON, S. 408. — THUN, R.: Die Löslichkeit der Nährstoffe Kali und Phosphorsäure in kochendem Wasser. Bodenkde. u. Pflanzenernähr. **2**, 34–44 (1936/37). — THUN, R., R. HERRMANN und E. KNICKMANN: s. Methodenbuch. — TOMMASI, G., und MARIMPIETRI: Neue chemische Methode zur Feststellung der Phosphorsäurebedürftigkeit der Böden. Z. Pflanzenernähr., Düng., Bodenkde. **38**, 161–163 (1935). — TRÉNEL, M., und H. J. FREY: Untersuchungen über die Bestimmung von Kali, Kalk und Phosphorsäure im Boden durch Elektrodialyse in ihrer Beziehung zur Bodenazidität. Z. Pflanzenernähr., Düng., Bodenkde. **24**, 293–308 (1932). — TROBISCH, S., und R. GERMAR: Ergebnisse eines Molybdändüngungsversuchs zu Blumenkohl. Dtsch. Landwirtsch. **10**, 189–91 (1959); Ref. Chem. Zbl. **130**, 12671 (1959). — TRUOG, J.: Amer. Soc. Agronom. **22**, 874 (1930), nach JACKSON, S. 155. — TUCKER, B. M.: The determination of exchangeable calcium and magnesium in carbonate. Soils Austr. J. Agric. Res. **5**, 706–715 (1954), nach PFEFFER, Z. Pflanzenernähr., Düng., Bodenkde. **75**, 17–32 (1956).

VAGELER, P., und F. ALTEN: Böden des Nils und Gash III. Z. Pflanzenernähr., Düng., Bodenkde. A **22**, 21–51 (1931). — VASILIADIS, CH.: Schnellverdrängungsmethode des absorbierten Calciums und Magnesiums aus dem Bodenkomplex mit Hilfe von Wasserdampf. Bestimmung von Calcium und Magnesium. Z. Pflanzenernähr., Düng., Bodenkde. A **33**, 357–362 (1934). — Bemerkungen zu den Wasserdampfmethoden bei Verdrängung der adsorbierten Kationen aus dem Bodenkomplex. Bodenkde. u. Pflanzenernähr. **3**, 334–336 (1937). — VERMAAT, J. G., und G. J. VAN DER BIE: On the occurrence of copper in tropical soils. Plant a Soil **2**, 257–282 (1950). — VUORINEN, J., und O. MAKITIE: The method of soil testing in use in Finland. Agrogeologisia Julkansnja, Helsinki, Nr. 63, 1–44 (1955).

WAHNSCHAFFE, F., und F. SCHUCHT: Anleitung zur wissenschaftlichen Bodenuntersuchung, 3. Aufl. Berlin: P. Parey. 1914. — WEAR, J. J., und A. L. SOMMER: Acid extractable since of soils in relation to the occurrence of zinc-deficiency symptoms of corn (a method of analysis). Proc. Soil Sci. Soc. Amer. **12**, 143–144 (1947). — Wehrmann, J.: Mangan, Kupfer und Kobalt in Pflanzen und Böden schleswig-holsteinischer Weidegebiete. Plant a. Soil **6**, 61–83 (1955). — Weibull, M.: Ein Beitrag zur praktischen Bodenanalyse. Chem. Z. **30**, 722 (1906). — Studien über schwedische Ackerböden. Ein Beitrag zur praktischen Bodenanalyse. J. Landwirtsch. **55**, 215–231 (1907), nach Prjanischnikow, Die Düngerlehre. Nach der 5. russischen Auflage herausgegeben von M. v. Wrangell. S. 305–307. Berlin: P. Parey. 1923. — Die Bedeutung der Reaktion des Ackerbodens für die Untersuchung des Bodens und für dessen Produktionsvermögen. Meddelande frän Alnarps Laboratorium XII. Kongl. Landtbruks. Akademiens Handlinger och Tidskript Stockholm 1911; Ref. Biedermanns Zbl. Agrikulturchem. **41**, 147–150 (1912). — Welch, L. F., L. E. Ensminger und C. M. Wilson: The correlation of soil phosphorus with the yields of Ladino clover. Proc. Soil Sci. Soc. Amer. **21**, 618–620 (1957). — Westerhoff, H.: Beiträge zur Kupferbestimmung im Boden. Landwirtsch. Forsch. **7**, 190–193 (1954/55). — Wheeting, L. C.: A study of methods for the determination of the available potassium of soils. Soil Sci. **29**, 1 (1930), nach Gisiger, Bodenkde. u. Pflanzenernähr. **2**, 34 (1936/37). — Whetstone *et al.*: U. S. D. A. Techn. Bull. 797 (1942), nach Jackson, S. 382. — Wiegner, G.: Anleitung zum quantitativen agrikulturchemischen Praktikum. Berlin: Borntraeger. 1926. — Wiklander: Ann. Roy. Agr. Coll. Sweden **16**, 670 (1949), nach Jackson, S. 402. — Wong Po Chuen: Über die Bestimmung des Kationenumtausches nach Vageler-Alten. Kühn-Arch. **39**, 158–183 (1935). — v. Wrangell, M.: Die Untersuchung des Bodens und der Nährstoffversorgung der Pflanzen s. Honcamp, I. Bd., S. 914–929 (1931). — v. Wrangell, M., und W. Haase: Über den Phosphorsäuregehalt natürlicher Bodenlösungen. Landwirtsch. Jb. **63**, 707–738 (1926).

b) Pflanzenphysiologische Verfahren

Von

N. Atanasiu

1. Einleitung

Unter rein „pflanzenphysiologischen“ Verfahren zur Bestimmung des Düngebedürfnisses der Böden müssen diejenigen Verfahren verstanden werden, bei denen sowohl die „Extraktion“ der Nährstoffe aus dem Boden als auch die Auswertung durch pflanzenphysiologische Vorgänge und ihre Meßergebnisse bedingt sind. So sind dabei z. B. diejenigen Verfahren zu nennen, bei denen die Nährstoffe durch Pflanzen dem Boden entzogen worden sind und die Beurteilung des Versorgungsgrades der Böden aus den gebildeten Pflanzenerträgen vorgenommen wird.

Der *Feldversuch* und der *Gefäßversuch* nach Mitscherlich bilden zwei derartige Methoden, bei denen der gebildete Ertrag bei der Auswertung maßgebend ist.

Die *Anwelkmethode* nach Arland benutzt als Test die relative Transpiration.

Andere Verfahren wie die *Keimpflanzenmethode* nach Neubauer-Schneider, oder die *Blattanalyse* nach Lundegårdh benutzen nur für die „Extraktion“ der Nährstoffe aus dem Boden einen physiologischen Vorgang: Die Aufnahme durch die Pflanzenwurzeln, während die Beurteilung des Versorgungsgrades durch die chemische Analyse der Pflanzensubstanz erfolgt.

In der Fachliteratur werden die letzten zwei Methoden verschieden einklassifiziert. So befindet sich die Neubauer-Methode im *Methodenbuch*, Bd. I: „Die Untersuchung von Böden“ zusammen mit den „chemischen“ Verfahren gruppiert. Im „Handbuch der Pflanzenphysiologie“, Bd. VI, werden diese Methoden von W. Bergmann als „physiologisch-chemische Verfahren“ charakterisiert.

Wie auch durch andere Verfahren wird durch pflanzenphysiologische Methoden in erster Linie die Beschaffenheit der Böden im Hinblick auf ihren Gehalt an verfügbarem *Kali* und *Phosphor* untersucht. Der *Stickstoff* wird weniger berücksichtigt, obwohl im Vergleich zu den anderen Methoden gerade die pflanzenphysiologischen und darunter besonders der Feldversuch die besten Voraussetzungen dafür bieten. Was die Mikronährstoffe anbelangt, sind durch pflanzenphysiologische Verfahren einige brauchbare Ansätze gemacht worden, welche in den weiteren Ausführungen auch Berücksichtigung finden werden.

Bekanntlich wird heute die Bodenuntersuchung auf Nährstoffbedürftigkeit fast ausschließlich mittels chemischer Verfahren serienmäßig durchgeführt. Man stellt sich aus diesem Grunde die berechtigte Frage über den Sinn und Zweck der pflanzenphysiologischen Verfahren zur Bodenuntersuchung, da diese heute eine viel kleinere Verbreitung als die chemischen aufweisen. Mit stärker werdender Anwendung der mineralischen Düngung wurde auch die Notwendigkeit einer gebietsmäßig umfassenderen Bodenuntersuchung auf den Gehalt an Hauptnährstoffen immer akuter. Man war daher bestrebt, Methoden zu entwickeln, welche schnell und ohne große Arbeits- und Materialaufwendungen die gewünschte Auskunft geben konnten. In diesem Bestreben, ein geeignetes Verfahren für Serienuntersuchungen auszuarbeiten, sind in den ersten zwei Jahrzehnten unseres Jahrhunderts pflanzenphysiologische, mikrobiologische und chemische Verfahren zur Bestimmung der Nährstoffbedürftigkeit der Böden ausgearbeitet worden. Darunter konnten der Gefäßversuch nach MITSCHERLICH und die Keimpflanzenmethode von NEUBAUER-SCHNEIDER eine große Bedeutung erlangen. Besonders das Verfahren MITSCHERLICH konnte sich auch im Ausland erfolgreich durchsetzen. Jedoch gewannen die chemischen Verfahren wegen ihrer Schnelligkeit und Einfachheit immer mehr an Bedeutung, so daß heute die Bodenuntersuchung fast ausschließlich durch derartige Schnellverfahren durchgeführt wird.

Die Entwicklung dieser Methoden ist mit einer Fülle von Arbeiten, welche auch weitgehend zur Klärung von Problemen über das Verhalten der Nährstoffe im Boden und über ihre Verfügbarkeit für die Pflanze beigetragen haben, verbunden. Es wurden dabei in Verbindung mit dem Boden und den Nährstoffen Begriffe klargelegt wie: „Gesamtgehalt", „löslicher Anteil", „pflanzenverfügbarer Anteil", „Nährstoffgehalt", Nährstoffbedürfnis", „Düngerbedürfnis" usw.

Während die chemischen Schnellverfahren den *löslichen* Anteil der Bodennährstoffe (*löslich* muß in bezug auf das betreffende Lösungsagens der Methode gesetzt werden!) ermitteln können, geben die pflanzenphysiologischen Verfahren den *pflanzenverfügbaren* Anteil wieder.

MITSCHERLICH kommt das große Verdienst zu, in zahlreichen Arbeiten (darunter: MITSCHERLICH 1923, 1925, 1925, 1926, 1930, 1933, 1936, 1939) diesen Umstand besonders betont zu haben, sowie das Hervorheben der Tatsache, daß eine sinnvolle Ermittlung des Düngungsbedürfnisses des Bodens nur in Verbindung mit der Pflanze möglich ist. Dieser Gedanke wurde auch von LEMMERMANN (1930), wenn auch in etwas anderer Form, hervorgehoben.

Die chemischen oder mikrobiologischen Verfahren können eine Aussage über den Nährstoffgehalt bzw. Düngerbedarf des Bodens nur dann machen, wenn zwischen dem durch sie ermittelten löslichen Anteil und dem mittels Pflanzen bestimmten verfügbaren Anteil eine Korrelation aufgestellt werden kann. Die chemischen oder Schnell-Verfahren besitzen also die Eigenart, daß sie nur als „konventionelle" Verfahren angesehen werden können, welche mittels eines pflanzenphysiologischen Verfahrens *kalibriert* oder *geeicht* werden müssen. Es muß sogar betont werden, daß diese Verfahren *nur dann bestehen und leistungsfähig sein können, wenn sie mit zuverlässigen pflanzenphysiologischen Verfahren nicht nur geeicht, sondern auch ständig überprüft werden.*

Der Gefäßversuch von Mitscherlich ist bereits früher als Standard-Verfahren angenommen worden. Im Rahmen der Internationalen Bodenkundlichen Gesellschaft, Sektion für Bodenfruchtbarkeit, wurden in den Jahren 1935 bis 1938 zahlreiche Schnellverfahren auf ihre Eignung, den verfügbaren Nährstoffgehalt des Bodens zu ermitteln, geprüft und dabei mit den Ergebnissen des Gefäßversuches nach Mitscherlich als Standardmethode verglichen. Das Verfahren hat auch heute diese Bedeutung in vollem Umfang beibehalten. Es gibt zwar Tendenzen, die Schnellverfahren in erster Linie oder sogar ausschließlich mit dem Feldversuch zu eichen. In den weiteren Ausführungen aber werden eingehend die Gründe behandelt, weshalb die Eichung und die Überprüfung mit dem Gefäßversuch unerläßlich ist.

J. W. Fitts und Nelson (1956) teilen mit, daß auf Grund der Arbeiten der „National Soil Test Work Group" (1956) bei der Eichung von chemischen mit pflanzenphysiologischen Verfahren eine höhere Korrelation zu den Gefäß- als zu den Feldversuchen erreicht wird. Diese Auffassung wird auch durch die weiteren Ausführungen dieses Kapitels unterstützt.

Bei den oben erwähnten amerikanischen Studien wurde neben dem Korrelationskoeffizient r auch das „Bestimmtheitsmaß" $= r^2 \times 100$ (coefficient of determination) eingeführt. Bei $r = 0{,}8$ beträgt z. B. das Bestimmtheitsmaß 64 und zeigt eine gute Übereinstimmung der verglichenen Methoden. Bei $r = 0{,}5$, welcher auch statistisch gesichert sein mag, beträgt das Bestimmtheitsmaß nur 25, und dadurch wird ausgedrückt, daß nur in 25% der Fälle, welche durch den chemischen Test vorausgesagt werden, eine tatsächliche Ertragswirkung im Felde eintritt.

Fitts (1955) schlug den Begriff „Wahrscheinlichkeit" (probability) vor: als die Prozentzahl der Fälle, in welchen eine Düngung eine *rentable* Ertragssteigerung gebracht hat, von den durch den chemischen Test vorausgesagten Fällen. Diesem Begriff kommt nur ein qualitatives Maß zu, da die Höhe der erzielten Ertragssteigerung gar nicht berücksichtigt wird.

Im Rahmen dieses Handbuches werden als „pflanzenphysiologische" Verfahren folgende Methoden zusammengruppiert und behandelt:

Verfahren	Autor
1. Felddüngungsversuch	Mitscherlich
2. Gefäßdüngungsversuch	Mitscherlich
3. Anwelkmethode	Arland
4. Keimpflanzmethode	Neubauer-Schneider
5. Blattanalyse	Lundegårdh

Als Ergänzung dazu werden noch einige neuere Verfahren erwähnt, welche für die Bodenuntersuchung an Bedeutung gewinnen können, einer weiteren Prüfung aber noch bedürfen.

2. Der Felddüngungsversuch

Der Feldversuch stellt das älteste Verfahren zur Bestimmung des Düngebedürfnisses der Böden dar. Seine Anfänge sind mit den Namen Lawes und Gilbert und mit der Versuchsstation Rothamsted eng verbunden. In Deutschland haben Grouven (1868), Hanamann (1873), Maerker (1876) und P. Wagner (1920) die ersten Schritte zum exakten Feldversuch getan. Vor allem P. Wagner verdanken wir die Verkleinerung der bis dahin 3 bis 6 ar großen Parzellen auf 1 ar und eine Anlage in dreifacher Wiederholung mit einer klaren Fragestellung.

Bei der modernen Ausgestaltung des Feldversuches hat sich besonders E. A. Mitscherlich (1954) große Verdienste erworben. Er reduzierte die Parzellengröße noch weiter, von 100 auf 25 qm, arbeitete besondere Anlagemethoden und ein Ausgleichsverfahren zur Ausschaltung des Fehlers, welcher durch die Bodenungleichmäßigkeiten bedingt wird, aus und schuf durch seine sinnvolle Formulierung des Ertragsgesetzes die Grundlagen zur quantitativen Auswertung der Ertragsergebnisse bei der Bestimmung des Düngebedürfnisses der Böden.

Lange Zeit herrschte die Meinung, daß der Feldversuch die *exakteste* und *zuverlässigste* Methode zur Bestimmung des Düngebedürfnisses der Böden sei. Es ist heute aber erkannt worden, daß sowohl der Exaktheit als auch der Zuverlässigkeit dieser Methode Grenzen gesetzt sind.

(Eine eingehende Behandlung der Grundlagen zur Auswertung des Feldversuches befindet sich in Bd. I dieses Handbuches im Rahmen des Abschnittes B, „Wachstums- und Ertragsgesetze".)

A. Fragestellung und Anlage

Für die Durchführung von Feldversuchen zur Bestimmung des Düngebedürfnisses der Böden werden von verschiedenen Seiten verschiedene Fragestellungen vorgeschlagen. BEHRENS (1935) unterscheidet zwischen der *direkten* und der *indirekten* Methode. Bei der direkten Methode wird die Wirkung der einzelnen Nährstoffe (N, K, P) allein erfaßt, während sie bei der indirekten Methode in Kombination mit einer Grunddüngung untersucht wird. Es ergeben sich folgende Versuchsglieder:

Direkte Methode:	Indirekte Methode:
1. Ungedüngt	1. Ungedüngt
2. + Kalidüngung	2. + Stickstoff- und Phosphatdüngung
3. + Phosphatdüngung	3. + Stickstoff- und Kalidüngung
4. + Stickstoffdüngung	4. + Kali- und Phosphatdüngung
	5. + Stickstoff-, Kali- und Phosphatdüngung

KNICKMANN (1955) unterscheidet zwischen:

a) *Nährstoffmangelversuch* (oder Differenzdüngungsversuch), welcher in der Fragestellung genau der *indirekten* Methode entspricht, und

b) *Steigerungsversuch*, in dem die einzelnen Nährstoffe in steigenden Gaben verabfolgt werden.

Als Beispiele:

N-Steigerung	K-Steigerung	P-Steigerung
KP	NP	NK
KPN_1	NPK_1	NKP_1
KPN_2	NPK_2	NKP_2
KPN_3	NPK_3	NKP_3

Der *Nährstoffsteigerungsversuch* kann als ein *allgemeiner* Versuch betrachtet werden, aus dessen Ergebnissen zwar auch das Bedürfnis des Bodens ermittelt werden kann. Als *spezieller* Versuch dazu ist aber der Nährstoffmangelversuch zu betrachten.

MITSCHERLICH empfahl zunächst (1950) den normalen Düngungsversuch mit folgender Gestalt:

Teilstück Nr.	1	2	3	4	5	6	7
	—	N	N	N	N	N	—
Düngung:	—	—	K	K	—	—	—
	—	—	—	P	P	—	—

Diese Versuchsgliederung deckt sich mit der des Nährstoffmangelversuches.

Später wurde von ihm (1954) der *exakte* Düngungsversuch ausgearbeitet und empfohlen, welcher in folgender Weise anzulegen ist:

1	2	3	4	5	6	7	8	9	10	
N	N	N	N	N	N	N	N	N	N	
K	K	K	K	K	K	K	K	K	K	P-Reihe
	P		P		P		P		P	
N	N	N	N	N	N	N	N	N	N	
P	P	P	P	P	P	P	P	P	P	K-Reihe
	K		K		K		K		K	

Die von MITSCHERLICH als *normal* und als *exakt* ausgearbeiteten Feldversuche werden wegen ihrer Bedeutung in unseren Ausführungen besondere Berücksichtigung finden.

Zur Wahl der *Anlageart* mögen zunächst einige grundsätzliche Bemerkungen gemacht werden.

Die Anlage muß übersichtlich sein, damit die Bonitierungen während des Wachstums leicht durchführbar sind.

Die Anlage darf nicht zu kompliziert sein, da sonst Verwechselungen bei der Düngerverabreichung entstehen können.

Die Anlage muß eine sinnvolle Verrechnung der Ergebnisse gestatten.

Die hier angeführten Feldversuche lassen sich in zwei „Arten" einteilen:

a) Versuche, welche sinnvoll durch Aufstellung von *Ertrags-Kurven* werden; dazu gehören die Nährstoffsteigerungsversuche.

b) Versuche, bei denen aus Mittelwerten *Differenzen* gebildet werden, das sind die übrigen (normaler und exakter Feldversuch).

Die zur Zeit üblichen Verrechnungsverfahren können ebenfalls in zwei Gruppen eingeteilt werden:

a) Verrechnungen, welche *Mittelwerte* liefern, die *gut von Bodenungleichmäßigkeiten bereinigt* sind. Dazu gehören z. B. die Ausgleichsrechnung von MITSCHERLICH (1934) mittels gleitender Mittel oder neuere Methoden, welche dazu Polynome höheren Grades anwenden (SCHNEIDER 1958). Diese beiden Verfahren setzen eine *Reihenanlage* nach MITSCHERLICH voraus.

b) Verrechnungen, welche eine gute Erfassung des *zufälligen Fehlers* erlauben, wie die Varianzanalyse. Diese Verrechnung kann am sinnvollsten für Anlagen im lateinischen Quadrat angewendet werden, bei denen moderne Gesichtspunkte wie die *gerechte* Verteilung der Versuchsglieder nach BEHRENS (1956) notwendigerweise beachtet werden müssen.

Es ergibt sich aus diesen Ausführungen, daß die zweckmäßigste Anlage für *Nährstoffsteigerungsversuche*, deren Auswertung aus den gebildeten Kurven vorgenommen wird, der *Reihenanlage* nach MITSCHERLICH entspricht. Ebenfalls dazu geeignet ist die Anlage von v. BOGUSLAWSKI, die *Ertragsflächenmethode* (1940), welche einen Ausgleich mittels Gleitmittelverfahren in horizontaler wie auch vertikaler Richtung erlaubt.

Die übrigen Versuche, bei denen die Differenzen zwischen den erhaltenen Erträgen für die Auswertung maßgebend sind und bei deren Erfassung der mittlere Fehler eine wichtige Rolle spielt, können in Anlagen durchgeführt werden, welche varianzanalytische Verrechnungen erlauben (z. B. lateinisches Quadrat mit gerechter Verteilung).

Für den *normalen* Feldversuch empfiehlt MITSCHERLICH ebenfalls die Reihenanlage nach ihm (1950) und die Auswertung mittels Gleitmittelverfahren. Man kann für diesen Versuch auch ohne weiteres eine andere Anlage, wie z. B. lateinisches Quadrat wählen, dessen statistische Verrechnung varianzanalytisch erfolgen kann.

Für den *exakten* Felddüngungsversuch sieht MITSCHERLICH die bereits oben angegebene Anlage vor. Mit dieser Anlage ist auch eine besondere Verrechnungsart verbunden, welche in den weiteren Ausführungen noch im einzelnen gezeigt wird.

B. Die Parzellengröße

Es werden allgemein für Düngungsversuche Parzellengrößen empfohlen, welche den Standraumansprüchen der Versuchspflanze entsprechen: Für Getreideversuche 10 qm (also 2×5 m) und für Hackfrüchte (Kartoffeln, Rüben) 25 qm (also $2{,}5 \times 10$ m). KNICKMANN (1955) und SCHMITT (1959) empfehlen etwas größere Teilstücke, für Halmfrüchte nicht unter 16 qm und für Hackfrüchte zwischen 25 und 50 qm.

Der Parzellen*größe* sind sowohl nach oben als auch nach unten Grenzen gesetzt. Je größer die Parzelle ist, um so weniger machen sich die Einflüsse der technischen Maßnahmen bemerkbar wie: Wägung und Streuung der Düngung, sowie genetischer Art, wie die Individualität der einzelnen Pflanze. Wiederum wird aber der Boden um so ungleichmäßiger, je größer das Teilstück wird. Aus diesen Gründen muß das Teilstück so groß gewählt werden, daß diese Einflüsse ein Minimum darstellen. Wir schließen uns in dieser Hinsicht den Empfehlungen von MITSCHERLICH (1948) an (für Halmfrüchte = 10 qm, für Hackfrüchte = 25 qm) und betrachten als vorteilhafter, wenn die Gesamtversuchsfläche es erlaubt, eine Erhöhung der Wiederholungszahl als eine Vergrößerung des Teilstückes.

C. Die Düngung

Sowohl für den *normalen* als auch für den *exakten* Feldversuch sind dieselben Nährstoffgaben anzuwenden. Für diese Gaben sind von MITSCHERLICH besondere Ertragstafeln ausgearbeitet worden, deren Anwendung weiter erläutert wird.

Es sind folgende Mengen zu verabfolgen:

a) an Stickstoff: für Getreide 50 kg/ha N, für Hackfrüchte 80 kg/ha N in Form von Kalkammonsalpeter;

b) an Phosphatdüngung: für Getreide und für Hackfrüchte in gleicher Weise 80 kg/ha P_2O_5 in Form von Superphosphat;

c) an Kalidüngung: für Getreide und für Hackfrüchte in gleicher Weise 120 kg/ha K_2O als schwefelsaures Kali.

Die Düngung erfolgt nach MITSCHERLICH bei dem exakten Feldversuch: Für Stickstoff und Kali bei der P-Reihe und für Stickstoff und Phosphat bei der K-Reihe derart, daß die Gesamtversuchsfläche mit der entsprechenden Düngermenge gedüngt wird. Parzellenweise wird nur die „Differenzdüngung" verabfolgt.

D. Technische Durchführung des Versuches

Zur technischen Durchführung des Versuches wird besonders auf das entsprechende Kapitel dieses Handbuches, „Technik des Feldversuches" (Bd. I, Abschn. XI C), hingewiesen. Es sei hier nur auf einige besondere Hinweise von MITSCHERLICH (1954) aufmerksam gemacht. „Jede Bodenbearbeitung, die Grunddüngung wie die Saat erfolgt über 12 Teilstücke hinweg *längs der Gesamtfläche*; denn es sollen hierbei eventuell auftretende Versuchsfehler alle Teilstücke in der gleichen Weise treffen." Nach der Verabfolgung der Differenzdüngung macht MITSCHERLICH folgende Empfehlung: „Nunmehr erfolgt, um ein Überschleppen der Düngung zu verhindern, jedwede Bodenbearbeitung *längs der einzelnen Teilstücke*".

Über die Saat, Ernte und weitere Verarbeitung des Erntegutes wird auf Bd. I, Kap. X C dieses Handbuches hingewiesen. Man kann mit gutem Erfolg auch nach den Vorschriften von Mitscherlich vorgehen: „Zur Ernte, die bei trockenem Wetter erfolgen muß, nimmt man am besten eine Dezimalwaage mit aufs Feld, die unter Windschutz aufzustellen ist. Bei Hackfrüchten erntet man nur die Knollen bzw. Wurzeln, bei Getreide u. a., welches mit Sichel oder Sense möglichst dicht am Boden abzuschneiden ist, die ganzen Pflanzen (Korn + Stroh)." Es werden dabei parzellenweise die Knollen- bzw. Wurzelgewichte bei Kartoffeln bzw. Rüben und die Gesamtgewichte bei Getreide festgestellt. Für die weitere Verarbeitung der Ergebnisse können bei Kartoffeln und Rüben die Frischgewichte, für Getreide die lufttrockenen Gewichte herangezogen werden. Wenn aber die Anlagen es erlauben, sind absolut trockene Gewichte zu empfehlen.

E. Auswertung des Feldversuches

Sowohl bei dem normalen als auch bei dem exakten Felddüngungsversuch müssen wir bei der Auswertung zwischen der *statistischen Verrechnung* der Ertragsergebnisse und der *Berechnung des Nährstoffgehaltes* („*b*") des Bodens (vgl. Bd. I, Abschn. VI C, XI C) unterscheiden.

Die *statistische Verrechnung* ist in Bd. I, Kap. XI D eingehend behandelt. Zur allgemeinen Orientierung wird auf diesen Abschnitt hingewiesen. Da aber der *exakte* Feldversuch eine spezielle statistische Auswertung benutzt, sei diese hier kurz erläutert, am besten an Hand eines Beispieles von Mitscherlich (1954) selbst.

Die Ertragsergebnisse des exakten Feldversuches werden in derselben Reihenfolge geschrieben, wie es der Anlage entspricht. Dann werden die Differenzen zwischen „gedüngt" und „ungedüngt" (also zwischen NKP und NK, bzw. zwischen NKP und NP) ermittelt und diese Differenzen als Prozente der ungedüngten (NK- bzw. NP-) Parzelle errechnet. Die Differenzen werden zu den beiden Nachbarparzellen ungedüngt, also links und rechts, gebildet.

In einem Feldversuch zu Hanf sind folgende Erträge erzielt worden:

Teilstück:	1	2	3	4	5	6	7	8	9	10
					P-Reihe					
Düngung	NK	NKP	NK	NKP	NK	NKP	NK	NKP	NK	NKP
kg/25 qm	27,4	46,6	26,8	47,2	26,6	51,0	26,8	50,9	28,4	48,1
Differenz in %	70,0	73,9	76,0	77,4	91,7	91,7	89,9	79,2	69,4	
					K-Reihe					
Düngung	NP	NKP	NP	NKP	NP	NKP	NP	NKP	NP	NKP
kg/25 qm	34,1	41,2	34,5	43,2	37,0	40,7	37,5	40,9	37,0	40,8
Differenz in %	20,8	19,4	24,6	16,8	10,0	8,5	9,1	10,5	10,3	

Man berechnet aus den je neun Differenzen das arithmetische Mittel und den mittleren Fehler und man erhält dann:

Ertragssteigerung durch P-Düngung in % $= 79{,}9 \pm 2{,}5$
Ertragssteigerung durch K-Düngung in % $= 14{,}4 \pm 1{,}4$

Auf der Grundlage seines Wirkungsgesetzes der Wachstumsfaktoren (vgl. Bd. I des Handbuches) hat Mitscherlich Tabellen ausgerechnet, aus denen man für die relative Ertragssteigerung den „*b*"-Wert, also den Gehalt des Bodens an „pflanzenaufnehmbaren Nährstoffen", ablesen kann. Wir geben die Tabellen, welche für eine Düngung von + 80 kg/ha P_2O_5 bzw. + 120 kg/ha K_2O Gültigkeit haben, wieder (Tab. 145).

Tabelle 145. *Relative Ertragssteigerung und „b"-Wert für P und K nach Mitscherlich*

P-Tabelle			K-Tabelle		
Differenzdüngung = 80 kg/ha P_2O_5			Differenzdüngung = 120 kg/ha K_2O		
Relative Ertragssteigerung (Differenz in %)	„b" (dz/ha P_2O_5)	Düngungsbedarf (dz/ha P_2O_5)	Relative Ertragssteigerung (Differenz in %)	„b" (dz/ha K_2O)	Düngungsbedarf (dz/ha K_2O)
211	0,2	2,0	211	0,3	3,0
130	0,3	1,9	150	0,4	2,9
91	0,4	1,8	114	0,5	2,8
67	0,5	1,7	91	0,6	2,7
52	0,6	1,6	74	0,7	2,6
41	0,7	1,5	61	0,8	2,5
33	0,8	1,4	52	0,9	2,4
27	0,9	1,3	44	1,0	2,3
22	1,0	1,2	38	1,1	2,2
19	1,1	1,1	33	1,2	2,1
16	1,2	1,0	29	1,3	2,0
13	1,3	0,9	25	1,4	1,9
10	1,5	0,7	18	1,7	1,6
8,9	1,6	0,6	14	1,9	1,4
7,1	1,7	0,5	10	2,2	1,1
6,0	1,8	0,4	8,3	2,4	0,9
5,2	1,9	0,3	6,6	2,6	0,7
4,4	2,0	0,2	4,4	3,0	0,3
3,4	2,2	—	3,4	3,3	—

Aus den Tabellen werden für unser Beispiel abgelesen:

„b" für $P_2O_5 = 0{,}44 \pm 0{,}01$ dz/ha und
„b" für $K_2O = 1{,}88 \pm 0{,}08$ dz/ha.

Aus dem Nährstoffgehalt des Bodens „b" kann nun auch der Nährstoffbedarf des Bodens errechnet werden. Nach MITSCHERLICH gilt ein Boden als „gut versorgt", wenn er mindestens 3,3 dz/ha K_2O und 2,2 dz/ha P_2O_5 als „pflanzenaufnehmbare" Nährstoffe enthält. Man muß also, um einen „gut versorgten" Boden zu schaffen, die Mengen düngen, welche zu den angegebenen Grenzwerten fehlen.

F. Vor- und Nachteile des Feldversuches

Auf den ersten Blick scheint der Feldversuch die zuverlässigste Methode zur Bestimmung des Düngebedürfnisses der Böden zu sein, da er:
unter *natürlichen Boden-* und *Klimabedingungen* durchgeführt wird und den Einfluß des Untergrundes erfaßt.
Es kann aber als Nachteil entgegengehalten werden, daß

1. die Übertragung der Ergebnisse einer begrenzten Versuchsfläche von z. B. 20 × 25 m auf den gesamten Schlag Schwierigkeiten bereitet, hauptsächlich, wenn der Schlag nicht ganz gleichmäßig beschaffen ist;
2. die Übertragung der Ergebnisse von einem Jahr auf das nächste Jahr ebenfalls schwierig ist, da die Witterungsverhältnisse anders sein können;
3. durch den Feldversuch im Gegensatz zum Gefäßversuch nicht auch der *Düngervorrat* des Bodens festgestellt werden kann.

Die Bestimmung des Düngervorrates des Bodens wird in dem Kapitel „Der Gefäßdüngungsversuch" ausführlich behandelt. Es sei hier vorwegnehmend darauf hingewiesen, daß der Gefäßversuch mit einer Mischung von $^1/_3$ Boden + $^2/_3$ Sand (volumenmäßig) als Gefäßfüllmaterial durchgeführt wird. Man bestimmt also nur $^1/_3$ der Nährstoffmenge, welche einem ganz mit Boden gefüllten Gefäß entspricht. Diese dann mit 3 zu multiplizierende Nährstoffmenge kann bei entsprechender Nährstoffversorgung des Bodens Kali- bzw. Phosphorwerte ergeben, welche für mehr als die eine Ernte des Versuchsjahres ausreichen würden. Daraus ergibt sich der Dünger*vorrat* des Bodens. Im Feldversuch dagegen, wo keine „Bodenverdünnung" möglich ist, kann man nur bestimmen, ob ein Boden für das Versuchsjahr gut versorgt ist oder nicht. Im Falle einer guten Versorgung kann man keine Schlüsse ziehen, ob der Boden für eine oder auch für mehrere Ernten über die für Höchsterträge notwendigen Nährstoffe verfügt.

G. Betrachtungen über die Berechnung des „b"-Wertes

Bekanntlich und wie im Bd. I, Kap. VI C dieses Handbuches besonders betont, ist der Wirkungswert c in der Mitscherlichschen Gleichung nicht konstant, sondern vielmehr von den Umweltsfaktoren bzw. von dem jeweiligen Höchstertrag abhängig. Somit wird eine der wichtigsten Grundlagen für die Anwendung der oben angegebenen Ertragstafel in Frage gestellt. Eine exakte Auswertung würde dann zuerst eine Ermittlung der Konstante „c" für jede Versuchsreihe erforderlich machen. Da dies aber technisch nicht durchführbar ist, wollen wir hier kurz die Frage untersuchen, inwieweit die möglichen Schwankungen von „c" bei Anwendung der konstanten Werte von Mitscherlich die Richtigkeit und Genauigkeit der Auswertung beeinflussen. Ein Rechenbeispiel soll dies erläutern. Es wird angenommen, daß anstelle des c-Wertes für $P_2O_5 = 0{,}6$ ein c-Wert von 0,4 der richtige wäre. Man hat aus den Tabellen ($c = 0{,}6$) einen „b"-Wert von 0,8 dz/ha gefunden, man würde bei Anwendung von Tabellen ($c = 0{,}4$) „b" $= 1{,}2$ dz/ha finden. Man müßte im ersten Falle, um den Boden „gut versorgt" zu gestalten, 1,0 dz/ha P_2O_5 düngen, im zweiten Falle 2,1 dz/ha P_2O_5. Man düngt also 2,1—1,0 = 1,1 dz/ha P_2O_5 weniger, und man erhält dadurch anstelle von 95% (da in diesem Falle nur $1{,}2 + 1{,}0 = 2{,}2$ dz/ha P_2O_5 im Boden sind) bei einem c-Wert $= 0{,}4$ nur 87% des höchstmöglichen Ertrages, also 8% weniger. Bei einer Schwankung des c-Wertes in umgekehrter Richtung (größer als 0,6) würde man bei der Auswertung auf Grund der Mitscherlichschen Tabellen etwas *mehr* als sonst notwendig düngen.

Es muß aber hier bemerkt werden, daß die von Mitscherlich für P_2O_5 und K_2O angegebenen c-Werte als *optimale* Werte angesehen werden können, da sie unter den sehr günstigen Bedingungen des Gefäßversuches ermittelt worden sind. Aus diesem Grunde ist im allgemeinen unter den Feldverhältnissen meist mit Schwankungen der c-Werte nach oben zu rechnen; dadurch fallen die als notwendig ermittelten Nährstoffgaben etwas höher aus, als bei niedrigeren c-Werten notwendig wäre.

Es sei hierzu noch die Tatsache erwähnt, daß das im Feldversuch ermittelte Düngungsbedürfnis für das *nächste* Jahr Gültigkeit haben soll. Die Witterungsschwankungen der verschiedenen Jahre bedingen ebenfalls Schwankungen der „c-Werte". Dadurch werden einer übertriebenen Forderung von *exakter* Berücksichtigung der „c-Werte" Grenzen gesetzt und der Anwendung der Mitscherlichschen Ertragstafel bei der Auswertung von Düngungsversuchen genügend Berechtigung gegeben.

Betrachtungen zum Felddüngungsversuch und zu seiner Weiterentwicklung. Eine Abwandlung der Gedanken von MITSCHERLICH findet man bei BRAY (1948) und BALBA und BRAY (1956), welche eine logarithmische Beziehung zwischen dem prozentischen Nährstoffgehalt der Pflanze und dem Nährstoffgehalt des Bodens annehmen. Es gilt z. B. nach BRAY für den Nährstoff Phosphor in bezug auf Weizenkornertrag folgende Gleichung:

$$y = 0{,}5834(1 - e^{-0{,}03296 \cdot b - 0{,}02697 \cdot x}),$$

worin y = P-Gehalt der Weizenkörner in % bei einer P-Düngung $= x$ und einem P-Gehalt des Bodens $= b$ (beide in pounds ausgedrückt) ist.

Für BRIGGsche Logarithmen und als logarithmische Gleichung ausgedrückt ergibt sich:

$$\log(0{,}5834 - y) = \log 0{,}583 - 0{,}01431 \cdot b - 0{,}00117 \cdot x$$

Bei der Anwendung dieser Methode muß man aber bedenken, daß, wie VANDECAVEYE (1940) in einer umfangreichen Zusammenstellung von Versuchsergebnissen zeigen konnte, der P-Gehalt der Pflanzen bei steigender P-Düngung nicht nur zunehmen, sondern auch konstant bleiben oder sogar abnehmen kann. Diese Schwankungen sind sehr stark von den speziellen Bedingungen der Versuche abhängig. Eine sehr genaue Einhaltung von konstanten Arbeitsbedingungen ist also eine wichtige Voraussetzung dieses Verfahrens.

Eine Neuerung in der Methode des Felddüngungsversuches kann ferner in der Berücksichtigung des „optimalen Nährstoffverhältnisses" erblickt werden. In Deutschland wird besonders von E. v. BOGUSLAWSKI (1954) und in Belgien von HOMÈS (1956) der Standpunkt vertreten, daß man nicht von der Feststellung eines Bedarfes an *einem Nährstoff* sondern von einer *Düngungskombination* sprechen müßte. So entsprechen geringeren N- und K-Gaben auch geringere P-Gaben, während bei höheren N- und K-Gaben auch höhere P-Mengen optimal wirken.

v. BOGUSLAWSKI empfiehlt deshalb zur Feststellung des Düngungsbedürfnisses der Böden den „komplexen" Düngungsversuch, welcher eine Steigerung der einzelnen Nährstoffe N, K und P bei gegenseitiger Abstufung vorsieht.

Ähnliche Gedanken entwickelt HOMÈS (1956) durch die von ihm propagierte „méthode des variantes systematiques" zur Feststellung der optimalen Düngung.

Diese Gedankengänge, deren Richtigkeit nicht zu leugnen ist, sind leider bis jetzt noch nicht in Form von methodischen Vorschriften zur Durchführung des Felddüngungsversuches niedergelegt und veröffentlicht worden. Man kann aus diesem Grunde hier nur auf die zur Zeit vorhandene Fachliteratur hinweisen (BRAUNEWELL 1952, POPP 1958, WÜSTENHAGEN 1959).

Eine Verbindung zwischen der Anlage nach MITSCHERLICH und den von E. v. BOGUSLAWSKI gestellten Forderungen kann darin gesehen werden, daß die verhältnismäßig einfache Versuchsanstellung nach MITSCHERLICH den K- und P-Bedarf des Bodens nur für *eine* vorgeschriebene *mittlere* Düngungsintensität feststellt, während man durch die „komplexe" Versuchsanstellung v. BOGUSLAWSKIS K- und P-Bedarfsmengen z. B. für *niedrige*, *mittlere* und *hohe* Intensitätsstufen ermitteln kann.

Ohne große Schwierigkeiten könnte man auch die Versuchsanstellung nach MITSCHERLICH variieren, und zwar derart, daß die *Grunddüngung* je nach der gewünschten Intensität angepaßt wird. Diese Anpassung betrifft hauptsächlich die von MITSCHERLICH vorgesehene N-Gabe, welche für verschiedene Gegenden Deutschlands und des Auslandes eventuell als zu niedrig zu betrachten ist.

Wenn in der Versuchsanstellung von MITSCHERLICH andere N-Gaben gewählt werden, so können ohne weiteres die oben angegebenen Ertragstafeln für K und P benutzt werden. Für andere als in den obigen Vorschriften vorgesehene Differenzdüngungen an K und P müßten selbstverständlich auch entsprechende Ertragstafeln benutzt werden.

Aus den obigen Ausführungen kann man schließen, daß der Feldversuch nach wie vor eine große Rolle als Methode zur Bestimmung des Düngebedürfnisses des Bodens spielt. Er müßte einerseits viel mehr als bisher ein „praktisches" Verfahren werden und ferner stärker als bis jetzt als Eichmethode der chemischen Schnellverfahren herangezogen werden.

Als Eichmethode kann der Feldversuch in der von MITSCHERLICH vorgeschriebenen Form angewendet werden, obwohl dieser ursprünglich in erster Linie für die praktischen Landwirte gedacht wurde. Bei der Eichung einer Methode mit dem Feldversuch wäre allerdings notwendig, den Vergleich anhand einer *großen* Anzahl von Feldversuchen durchzuführen, wie dies jetzt in Holland (V. D. PAAUW 1956, 1959) gehandhabt wird, wobei allerdings die Versuche nicht nur mit einer Differenzgabe, sondern mit gesteigerten Mengen an K bzw. P angelegt werden.

Zur Zeit ist die Durchführung von Feldversuchen entweder nur bei Versuchs- und Forschungsanstalten oder in der praktischen Landwirtschaft nur in Verbindung mit einem guten Beratungsdienst möglich. Im ersten Falle fehlt ihnen die Bedeutung als praktische Untersuchungsverfahren, im zweiten Falle wird der Feldversuch zu teuer. Diese Umstände lassen zur Zeit die Anwendung des Feldversuches mehr als Eichmethode zu. Es wäre aber anzustreben, daß der Felddüngungsversuch durch entsprechende Beratung und Schulung ein Allgemeingut der praktischen Landwirte wird.

3. Der Gefäßdüngungsversuch

Bereits im vorigen Jahrhundert ist der Vegetationsversuch im Gefäß als Hilfsmittel zur Bestimmung des Nährstoffbedüfnisses der Böden herangezogen worden (WAGNER 1880, PFEIFFER 1918). Die dabei entwickelte sogenannte „Differenzmethode", bei der gesteigerte Nährstoffgaben auf den zu untersuchenden Boden verabfolgt wurden, erlaubte grundsätzlich einen Einblick in den Nährstoffhaushalt der Böden. Als schwierig erwies sich jedoch die praktische Auswertung dieser Versuche und die Übertragung der erzielten Ergebnisse auf den Acker.

MITSCHERLICH (1923, 1930) konnte im Gefäßversuch ein Verfahren zur Bestimmung des Nährstoffgehaltes bzw. -bedarfs der Böden entwickeln, bei dem die Auswertung unter Zugrundelegung des von ihm formulierten Ertragsgesetzes erfolgte. Nach diesem Verfahren ist eine Übertragung der im Gefäßversuch erzielten Ergebnisse auf das Feld ohne weiteres gegeben.

Dieses Verfahren, als Methode MITSCHERLICH bekannt, konnte sich allgemein durchsetzen und wird hier auch näher beschrieben. Andere Verfahren wie: der Gefäßversuch nach WIESSMANN (1927, 1929) oder die „Nährlösungskultur" nach JANERT (1953) sind abgewandelte Formen der Methode MITSCHERLICH, welche etwaige Mängel dieser Methode auszuschalten versuchen. Grundsätzlich ist der Gefäßversuch nach MITSCHERLICH ebenso wie der Feldversuch ein *Nährstoffmangelversuch*, durch den man bei entsprechender Versuchsanstellung das Bedürfnis der Böden an allen uns bekannten Nährstoffen bestimmen kann. Von MITSCHERLICH selbst sind Richtlinien zur Untersuchung der Hauptnährstoffe Kali und Phosphor ausgearbeitet worden, welche in den anschließenden Ausführungen Berücksichtigung finden werden.

Bevor man zur Beschreibung der Gefäßversuchsmethode übergeht, soll hier auf eine sehr ausführliche Darstellung dieses Verfahrens von E. v. BOGUSLAWSKI (1952) hingewiesen werden. Als besonders reichhaltige Literaturquelle auf diesem Gebiet wird der von W. BERGMANN (1958) stammende Beitrag geschätzt. Genauso wie bei dem Feldversuch wird auch bei dem Gefäßversuch in den folgenden Ausführungen in erster Linie darauf Wert gelegt, eine allgemein verständliche Darstellung der Methode MITSCHERLICH wiederzugeben.

A. Versuchsdurchführung

Der Versuch wird in den bekannten, von MITSCHERLICH entwickelten Vegetationsgefäßen durchgeführt. Eine genaue Beschreibung dieser Gefäße befindet sich in Bd. I, Kap. X B, und es kann daher hier darauf verzichtet werden. Die Gefäße werden auf Bänke oder Fahrgestelle aufgestellt. Der gesamte Aufstellungsraum wird gegen Vogelfraß mit einem Netz aus Draht oder neuerdings aus Kunststoff eingezäunt. Derartige Vegetationsanlagen sind auch unter dem Namen „MITSCHERLICH-Station" bekannt.

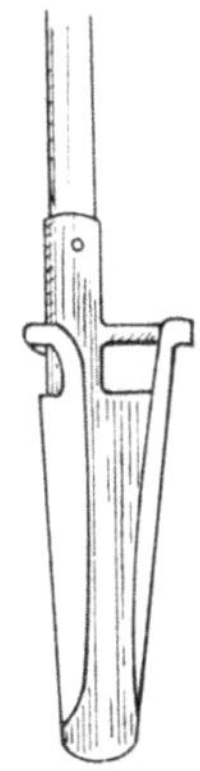

Abb. 186. Probeentnahmespaten nach MITSCHERLICH

Der für den Versuch benutzte Boden muß eine *repräsentative Durchschnittsprobe* der zu untersuchenden Ackerfläche darstellen. *Die Probeentnahme* soll nach MITSCHERLICH an mindestens 50, auf der gesamten untersuchten Fläche gleichmäßig verteilten Stellen, durchgeführt werden. Nach Arbeiten von MITSCHERLICH (1923) und R. FLÖSS (1912) stellte sich heraus, daß diese Mindestzahl unabhängig von der Größe des untersuchten Schlages ist. Auch in den holländischen Versuchsstationen (s. VERMEULEN 1959) werden für die Entnahme einer guten Durchschnittsprobe mindestens 40 Einstiche, unabhängig von der Größe des Schlages, empfohlen. Der Boden wird gewöhnlich aus der Ackerkrume, also aus der Schicht von etwa 0 bis 20 cm Tiefe entnommen. Für besondere Zwecke kann aber auch der Boden aus tieferen Schichten zur Untersuchung herangezogen werden. Zur Probeentnahme hat MITSCHERLICH einen besonderen Spaten konstruiert, der hier in Abb. 186 wiedergegeben ist. Damit werden bei jedem Einstich Bodensäulen von $6{,}5 \times 20$ cm ausgestochen, welche zweckmäßigerweise in einem Eimer gesammelt werden. Bei 50 Einstichen sammelt man eine Bodenmenge von etwa 50 kg. Der entnommene Boden wird gut durchgemischt, durch ein Maurersieb abgesiebt und lufttrocken gemacht.

Die Gefäßfüllung besteht nicht aus reinem Boden sondern aus einem Boden-Sand-Gemisch von einem Volumenteil des zu untersuchenden Bodens und zwei Volumenteilen eines nährstofffreien Sandes. Dazu haben sich in Deutschland z. B. die Sande Hohenbocka bei Dresden oder Dörentrupp a. d. Lippe gut bewährt. Es wird dabei so verfahren, daß ein Gefäß mit Boden in etwa natürlicher Lagerung gefüllt und die Bodenmenge gewichtsmäßig festgestellt wird. Ein Drittel dieses Gewichtes und zwei Drittel des in gleicher Weise festgestellten Sandgewichtes werden dann in einer Wanne gemischt. (Eine genauere Beschreibung des Füllvorganges wird in Bd. I, Kap. X B dieses Handbuches gegeben.)

Die Düngung der Gefäße erfolgt dem Versuchsplan entsprechend. Im allgemeinen werden die Böden auf ihren P- und K-Gehalt untersucht. Dazu werden folgende Versuchsglieder benötigt:

1. Volldüngung $= +N, +K, +P$
2. ohne Kali $= +N, +P$
3. ohne Phosphor $= +N, +K$

Bei je drei Parallelbehandlungen sind für eine solche Untersuchung 9 Gefäße erforderlich. Will man sich auch über den N-Gehalt des Bodens orientieren, so müßte man dann noch ein viertes Versuchsglied: ohne Stickstoff = +K, +P ansetzen. Es kommen folgende Nährstoffmengen und -formen in Anwendung (s. Mitscherlich und Atanasiu 1956):

1,5 g N als NH_4NO_3-Lösung (1 g N = 10 ml Lösung)
1,5 g P_2O_5 als Lösung von $Ca(H_2PO_4)_2$ oder Superphosphat (1,5 g = 40 ml Lösung)
1,5 g K_2O als K_2SO_4 (1,5 g K_2O = 40 ml Lösung).

Als allgemeine Grunddüngung werden allen Gefäßen in gleicher Weise noch folgende Salze gegeben:

1,0 g $MgSO_4 \cdot 7H_2O$ (= 30 ml Lösung)
0,5 g NaCl (= 10 ml Lösung)
1,5 g $CaCO_3$.

(Für Böden, bei denen eine Festlegung der Phosphorsäure durch Kalk vermutet wird, empfiehlt Mitscherlich noch ein Versuchsglied: wie das Versuchsglied 3, aber auch ohne Kalk.)

Das Ansetzen wird von Mitscherlich und Mitarbeiter (1956) in folgender Weise empfohlen: „Der lufttrockne Boden wird in dem angegebenen Verhältnis in einer ‚Mischwanne' gut mit dem Quarzsand gemischt. Es wird dazu der $CaCO_3$ verabfolgt und noch einmal durchgemischt. Jetzt folgen die in Wasser gelösten Nährstoffe, die mittels Pipette zugeführt werden, und 500 cm³ Wasser, nach deren Verabfolgung jedesmal gut durchgemischt wird. Mittels eines Fülltrichters wird diese Mischung unter leichtem Druck (unten fester, nach oben lockerer) eingefüllt."

Das Pflanzen der Gefäße erfolgt mittels eines besonderen *Markiergerätes*, mit dem 21 Pflanzlöcher von etwa 0,5 cm im Durchmesser und 1,5 cm Tiefe je Gefäß hergestellt werden. In diese Löcher werden dann 34 Haferkörner Hochzucht eingesteckt (in die äußeren 4 × 3 Löcher je 2, in die übrigen je 1) und durch leichtes seitliches Drücken zugedeckt.

Solange die Haferpflänzchen noch nicht aufgelaufen sind, bleiben die Gefäße mit den Untersätzen zugedeckt. Nach dem Auflauf werden die Gefäße etwa drei Wochen lang nach Bedarf mit Wasser versorgt, und zwar erhalten alle Gefäße *gleiche* Wassermengen. Es ist zweckmäßig, von Woche zu Woche die einmalige Wassergabe um je 500 ml zu erhöhen, damit nach etwa drei Wochen die Wasserkapazität des Gefäßes erreicht wird. Von nun an werden die Gefäße täglich einmal, an heißen Tagen zweimal auf Durchlauf gegossen. Bezüglich der weiteren Pflegemaßnahmen wird auf das entsprechende Kapitel in Bd. I hingewiesen. Etwa zwei Wochen vor der Vollreife werden die Gefäße nicht mehr gegossen. Bei Erreichung der Vollernte werden sie abgeerntet. Es werden festgestellt die lufttrockenen, oder wenn die Einrichtungen es zulassen, die absolut trocknen Gewichte der Gesamtpflanzenmasse je Gefäß.

B. Versuchsauswertung

Auch bei der Auswertung des Gefäßversuches muß man, wie bei der des Feldversuches, zwischen der statistischen Verrechnung der Erträge und der Ermittlung des Nährstoffgehaltes des Bodens unterscheiden. Die statistische Verrechnung hat als Zweck die Ermittlung des arithmetischen Mittels der Erträge der Parallelbehandlungen und des dazugehörigen mittleren Fehlers. Diese Verrechnung wird eingehend in Bd. I behandelt, so daß wir uns im folgenden nur auf den zweiten

Teil der Auswertung, die Ermittlung des Nährstoffgehaltes des Bodens einschränken können. (Dazu s. auch: MITSCHERLICH 1930, BEHRENS 1935, v. BOGUSLAWSKI 1952).

Diese Auswertung beruht wie bei dem Feldversuch auf der Anwendung des MITSCHERLICHschen Wirkungsgesetzes der Wachstumsfaktoren (s. Bd. I, Kap. VI B). Nach diesem Gesetz existiert folgende Beziehung zwischen der angewandten Nährstoffmenge x und dem gebildeten Ertrag y:

$$\log (A - y) = \log A - cx \,. \tag{1}$$

In dieser Gleichung bedeuten A den Höchstertrag der Versuchsreihe und c einen Proportionalitätsfaktor oder Wirkungswert.

Führt man einen Versuch mit Ackerboden durch, so erhält man auch bei „ungedüngt" einen Ertrag „a", welcher dem Nährstoffgehalt des Bodens „b" entspricht. Für diesen Fall gilt die Gleichung:

$$\log (A - a) = \log A - cb \,. \tag{2}$$

Sind die Größen a, A und c bekannt, so kann man die Gleichung nach b auflösen:

$$b = \frac{\log A - \log (A - a)}{c} \,.$$

Im Gefäßversuch kann man im Gegensatz zum Feldversuch mit Sicherheit annehmen, daß bei entsprechender Volldüngung der jeweilige Höchstertrag A erreicht wird. Aus diesem Grunde kann man hier die Gl. (2) direkt anwenden, was bei dem Feldversuch nicht möglich ist. Unter Annahme von konstanten c-Werten hat MITSCHERLICH Ertragstafeln ausgerechnet, aus denen man direkt den b-Wert ablesen kann. Für diesen Fall braucht man den a-Wert als % von A auszudrücken und den entsprechenden Wert „b" aufzusuchen. Diese Tabellen sind für dz Nährstoff je ha berechnet (Tab. 146).

Bei der Auswertung der Versuchsergebnisse muß zunächst berücksichtigt werden, daß nur $^1/_3$ des Gefäßes mit Boden gefüllt ist und ferner, daß bei der Probeentnahme auf dem Acker nur die Krume, also nur eine Tiefe von 15 cm erfaßt wurde. Die einfache Auswertung der Ertragsergebnisse des Gefäßversuches würde also nur die Nährstoffmengen ausdrücken, welche in dem Boden eines Drittels des Gefäßes waren. Will man das auf das Volumen des ganzen Gefäßes übertragen, so muß man die gefundenen Nährstoffmengen mit 3 multiplizieren. Will man sie ferner statt in g je 300 qcm (= Gefäßoberfläche) in dz/ha ausdrücken, so muß man die entsprechende Umrechnung durchführen. Will man ferner, wie MITSCHERLICH es vorschreibt, die Nährstoffmenge der 15 cm tiefen Gefäße auf eine Ackerkrumentiefe von 30 cm umrechnen, so sind diese Werte noch mit 2 zu multiplizieren.

Die Auswertung sei anhand eines Beispieles erläutert:

In einem Düngungsversuch im Gefäß wurden folgende Erträge erzielt (MITSCHERLICH und Mitarbeiter 1956):

Versuchsglied	Gesamtertrag Hafer g/Gefäß
N K P	234 ± 5 (= A)
N P	103 6 (= a_K)
N K	114 2 (= a_P)

Bekanntlich sind die Wirkungswerte für g und Gefäß nach MITSCHERLICH folgende:

$$c_N = 0{,}64 \qquad c_{K_2O} = 1{,}33 \qquad c_{P_2O_5} = 2{,}0$$

Tabelle 146. *Ertragstafel für die Auswertung des Gefäßversuches*

a % von *A*	„*b*" in dz/ha		*a* % von *A*	„*b*" in dz/ha		*a* % von *A*	„*b*" in dz/ha	
	K_2O	P_2O_5		K_2O	P_2O_5		K_2O	P_2O_5
1	0,01	0,007	34	0,45	0,30	67	1,19	0,79
2	0,02	0,013	35	0,46	0,31	68	1,22	0,81
3	0,03	0,02	36	0,48	0,32	69	1,25	0,83
4	0,04	0,026	37	0,50	0,33	70	1,29	0,86
5	0,05	0,033	38	0,51	0,34	71	1,33	0,89
6	0,065	0,043	39	0,53	0,35	72	1,36	0,91
7	0,075	0,05	40	0,55	0,36	73	1,40	0,94
8	0,085	0,06	41	0,56	0,38	74	1,45	0,96
9	0,100	0,07	42	0,58	0,39	75	1,48	0,99
10	0,11	0,075	43	0,60	0,40	76	1,53	1,02
11	0,12	0,08	44	0,62	0,41	77	1,57	1,05
12	0,13	0,09	45	0,64	0,43	78	1,62	1,08
13	0,14	0,095	46	0,66	0,44	79	1,67	1,11
14	0,16	0,10	47	0,68	0,45	80	1,72	1,15
15	0,17	0,11	48	0,70	0,47	81	1,77	1,18
16	0,18	0,12	49	0,72	0,48	82	1,83	1,22
17	0,20	0,13	50	0,74	0,49	83	1,89	1,26
18	0,21	0,14	51	0,76	0,51	84	1,96	1,30
19	0,22	0,15	52	0,79	0,53	85	2,03	1,35
20	0,23	0,16	53	0,81	0,54	86	2,10	1,40
21	0,25	0,17	54	0,83	0,56	87	2,17	1,45
22	0,26	0,18	55	0,85	0,57	88	2,26	1,50
23	0,28	0,19	56	0,88	0,59	89	2,35	1,57
24	0,29	0,20	57	0,90	0,60	90	2,45	1,63
25	0,30	0,205	58	0,93	0,62	91	2,56	1,70
26	0,32	0,21	59	0,95	0,64	92	2,68	1,78
27	0,33	0,22	60	0,98	0,66	93	2,81	1,88
28	0,35	0,23	61	1,01	0,67	94	2,97	1,98
29	0,36	0,24	62	1,04	0,69	95	3,26	2,17
30	0,38	0,25	63	1,06	0,71	96	3,50	2,33
31	0,39	0,26	64	1,09	0,73	97	3,80	2,54
32	0,41	0,27	65	1,12	0,75	98	4,22	2,81
33	0,43	0,29	66	1,16	0,77	99	5,00	3,33

Aus Gl. (2) werden folgende Nährstoffmengen errechnet:

$$b_K = \frac{\log 234 - \log(234 - 103)}{1{,}33} = 0{,}190 \text{ g } K_2O$$

$$b_P = \frac{\log 234 - \log(234 - 114)}{2{,}0} = 0{,}145 \text{ g } P_2O_5$$

Bei der Umrechnung dieser Mengen auf das gesamte Gefäßvolumen erhalten wir:

$$\text{für } K_2O\text{: } 0{,}190 \times 3 = 0{,}570 \text{ g}$$
$$\text{für } P_2O_5\text{: } 0{,}145 \times 3 = 0{,}435 \text{ g.}$$

Werden diese Mengen in dz/ha ausgedrückt, so müssen folgende Rechnungen durchgeführt werden:

$$\text{für } K_2O\text{: } 0{,}570 \times 10^{-5} \times \frac{10^8}{300} = 0{,}570 \times \frac{10}{3} = 1{,}88 \text{ dz/ha}$$

$$\text{für } P_2O_5\text{: } 0{,}435 \times 10^{-5} \times \frac{10^8}{300} = 0{,}435 \times \frac{10}{3} = 1{,}43 \text{ dz/ha.}$$

Diese Nährstoffe sind nach der Auswertung des Gefäßversuches als „pflanzenaufnehmbar“ im untersuchten Acker in einer Schicht von 0 bis 25 cm vorhanden. Diese Mengen können nach den obigen Ausführungen noch mit 2 multipliziert werden, wenn die Annahme gemacht wird, daß dieselben Mengen auch in der Schicht von 15 bis 30 cm enthalten sind.

Diese Auswertung vereinfacht sich erheblich, wenn man die von MITSCHERLICH vorgeschlagenen Tabellen benutzt. Sie geht in diesem Falle wie folgt vor sich:

Man berechnet die Mangelerträge „a“ als Prozente des Höchstertrages A:

$$a_K = 44{,}0\%$$
$$a_P = 48{,}7\%$$

Es werden in Tab. 146 an $K_2O = 0{,}62$ dz/ha und an $P_2O_5 = 0{,}477$ dz/ha abgelesen. Diese Mengen sind nach den obigen Ausführungen noch mit 3 zu multiplizieren, wodurch man dann folgende Nährstoffmengen erhält:

$$0{,}62 \times 3 = 1{,}86 \text{ dz/ha } K_2O \text{ und}$$
$$0{,}477 \times 3 = 1{,}43 \text{ dz/ha } P_2O_5.$$

Auch diese Werte müßten bei einer Umrechnung auf eine Bodenschicht von 30 cm noch mit 2 multipliziert werden.

Auf russischen Böden konnte KIRSANOW (1930) dagegen beobachten, daß der Nährstoffgehalt im Feld bis etwa 8fach höher als der im Gefäß für eine Krumentiefe von 15 cm gefunden worden war.

Die Umrechnung ist von MITSCHERLICH ursprünglich aus empirischen Gründen eingeführt worden, und zwar mit Rücksicht auf die in Ostpreußen herrschenden Bodenverhältnisse. Spätere Untersuchungen von MITSCHERLICH und ATANASIU (1952) im Raum von Berlin haben gezeigt, daß unter den dortigen Verhältnissen diese weitere Multiplikation für den Nährstoff Phosphor nicht gerechtfertigt ist. Bei eingehenden Untersuchungen der Böden von 0 bis 90 cm Tiefe erhielten sie die in Tab. 147 enthaltenen Ergebnisse:

Tabelle 147. *Nährstoffgehalte in verschiedenen Bodentiefen nach Mitscherlich und Atanasiu (1952) in dz/ha*

Schicht in cm	Bodenprofil I		Bodenprofil II	
	dz/ha P_2O_5	dz/ha K_2O	dz/ha P_2O_5	dz/ha K_2O
0—25	$3{,}64 \pm 0{,}23$	$4{,}68 \pm 0{,}19$	$2{,}05 \pm 0{,}09$	$3{,}34 \pm 0{,}09$
25—50	$0{,}56 \pm 0{,}03$	$3{,}75 \pm 0{,}11$	$1{,}23 \pm 0{,}03$	$3{,}41 \pm 0{,}11$
50—70	$0{,}08 \pm 0{,}00$	$4{,}20 \pm 0{,}20$	$0{,}11 \pm 0{,}01$	$2{,}76 \pm 0{,}13$
70—90	$0{,}30 \pm 0{,}01$	$7{,}71 \pm 0{,}30$	$0{,}56 \pm 0{,}00$	$3{,}48 \pm 0{,}05$

Aus diesen Untersuchungen ist zu entnehmen, daß der P_2O_5-Gehalt des Bodens in tieferen Schichten stark abnehmen kann. Ähnliche Versuche sind in Rauisch-Holzhausen bei Gießen von E. v. BOGUSLAWSKI erzielt worden (noch nicht veröffentlichtes Material).

Dieser Umstand ist sicher einer der wichtigsten Ursachen für die oft nicht erzielbare Übereinstimmung zwischen den Ergebnissen des Gefäß- und des Felddüngungsversuches.

Wir teilen daher die Auffassung von E. v. BOGUSLAWSKI (mündliche Mitteilung), daß die Ergebnisse des Gefäßdüngungsversuches nur auf die bei der

Probeentnahme erfaßte Schicht bezogen werden müssen und daß eine Extrapolation dieser Werte auf tiefere Schichten ohne genaue Kenntnis der vorherrschenden Bodenverhältnisse nicht zulässig ist.

Bei dem Gefäßversuch kann man wie bei dem Feldversuch aus dem festgestellten Nährstoffgehalt des Bodens „*b*" auf seinen Nährstoffbedarf schließen. Es gelten hier wie beim Feldversuch dieselben Grenzwerte für den Versorgungsgrad der Böden: bei „gut versorgt" 2,2 dz/ha P_2O_5 bzw. 3,3 dz/ha K_2O. Will man nun die untersuchte Ackerschicht bzw. die Krume als „gut versorgt" mit Nährstoffen versehen, so sind dann die Differenzen zwischen den Grenzwerten und den gefundenen Gehaltszahlen noch zu düngen.

Aus dem Kapitel „Auswertung" ist es leicht ersichtlich, daß beim Gefäßversuch *b*-Werte auch höher als 2,2 dz/ha für P_2O_5 und 3,3 dz/ha für K_2O errechnet werden können. Mit MITSCHERLICH kann man darüber folgende Überlegung anstellen.

Wird bereits für die Bodenmenge des Gemisches ($^1/_3$ Boden + $^2/_3$ Sand) eine ausreichende Versorgung mit dem betreffenden Nährstoff nachgewiesen (gleicher Ertrag wie bei Volldüngung), so würde es bedeuten, daß bei Umrechnung auf die gesamte Bodenmenge, die ein Gefäß faßt (analog einer 15 cm tiefen Ackerkrume), der Boden mindestens noch zwei Jahre (abzüglich des Untersuchungsjahres) ausreichend mit diesem Nährstoff versorgt ist. Würde man die Verhältnisse auf eine Ackerkrume von 30 cm Tiefe übertragen, so würde man finden, daß der Boden noch mindestens 5 Jahre ($3 \times 2 = 6$ Jahre, abzüglich des Untersuchungsjahres) ausreichend versorgt ist.

Eine Vereinfachung der Auswertung ist durch die Anwendung normographischer Tafeln von M. GÄRTNER (1941) gegeben.

C. Vor- und Nachteile des Gefäßversuches

Im Vergleich zum Feldversuch weist der Gefäßversuch zunächst den Vorteil auf, daß dabei eine Durchschnittsprobe der gesamten zu untersuchenden Fläche berücksichtigt wird.

Ferner kann man, wie oben bereits gezeigt wurde, nicht nur den *Bedarf*, sondern auch den *Vorrat* eines Bodens an Nährstoffen ermitteln.

Auch die Exaktheit des Gefäßversuches ist im Vergleich zum Feldversuch viel größer wegen der besser kontrollierbaren Arbeitsbedingungen.

Als Nachteile erweisen sich zunächst die Schwierigkeiten bei der *Übertragung der Ergebnisse aus den Gefäßen auf die Feldverhältnisse*. Es werden dabei im allgemeinen folgende Einwände geltend gemacht:

1. die *Bodenverdünnung* mit Sand verursacht eine stärkere Ausnutzung der Bodennährstoffe;
2. die *optimale Wasserversorgung* in Gefäßen bedingt ebenfalls einen höheren Löslichkeitsgrad, besonders für Phosphor, als im Feld;
3. die sehr *günstigen* übrigen *Vegetationsbedingungen* des Gefäßversuches haben als Folge viel höhere Erträge als im Felde;
4. die Wirkungswerte „*c*" in der Ertragsgleichung von MITSCHERLICH sind nicht konstant.

Bezüglich der ersten Frage sind von MITSCHERLICH (1934), SCHEFFER (1933) und SCHÄFER (1957) eingehende Untersuchungen bei Mischungen mit verschiedenen Boden-Sand-Verhältnissen durchgeführt worden. Die Ergebnisse von MITSCHERLICH lassen eine direkte Proportionalität zwischen der pflanzenaufnehmbaren Nährstoffmenge und der im Gefäß benutzten Bodenmenge erkennen. Aus den von SCHEFFER mitgeteilten Zahlen ist im Gegensatz dazu eine Zunahme der Ausnutzbarkeit der Bodennährstoffe durch die Pflanzen mit steigender Bodenverdünnung zu entnehmen. Das von SCHÄFER verarbeitete Zahlenmaterial einer großen Reihe von Versuchen von MITSCHERLICH läßt die gleiche Tendenz wie die Ergebnisse von SCHEFFER erkennen. Jedoch sind die Unterschiede zwischen den Mischungen Boden: Sand wie 1:0 bis 1:4 derart gering, daß bei der praktischen Auswertung eine Übertragung der Ergebnisse aus Gefäßversuchen mit Boden-Sand-Gemischen im Verhältnis bis 1:4 auf reinen Boden doch als zulässig zu betrachten ist.

Was die *günstige Wasserversorgung* des Gefäßes anbelangt, so wurde von NEHRING (1955) gezeigt, daß bei steigender Wassergabe die Löslichkeit der Bodenphosphorsäure zunimmt. Man kann daher tatsächlich mit einer höheren Löslichkeit in Gefäßen als unter den Feldbedingungen rechnen. Wie aber von MITSCHERLICH und ATANASIU (1956) gezeigt wurde, wird z. B. im Gefäß für denselben Boden bei voller Versorgung mit voller Wasserkapazität ein höherer Bedarf an P_2O_5 als bei Versorgung mit halber Wasserkapazität errechnet. Dies ist dadurch begründet, daß bei der besseren Wasserversorgung nicht nur die Erträge absolut gesehen höher liegen, sondern daß die relative Ertragssteigerung durch Volldüngung höher ist. Aus diesen Überlegungen kann man ersehen, daß man im allgemeinen durch den Gefäßversuch einen etwas höheren Bedarf ermittelt als dem Feld entspricht.

Die *Inkonstanz der c-Werte* hat theoretisch die Nichtanwendbarkeit der MITSCHERLICHschen Ertragstafeln sowie die Notwendigkeit der c-Bestimmung für jede Versuchsreihe zur Folge. WIESSMANN (1927/28) hat deswegen den Gefäßversuch derart modifiziert, daß für jede Versuchsreihe zusätzlich eine Nährstoffsteigerung in reinem Quarzsand zwecks Bestimmung der jeweiligen Ertragskurven für K und P durchgeführt wird. Die Einwände gegen die Inkonstanz der c-Werte werden dadurch aber nur teilweise behoben, da der Faktor Gefäßfüllung in diesem Falle doch nicht gleich ist (Sand bzw. Boden-Sand-Gemisch).

Da im Gefäßversuch mit günstigeren Vegetationsbedingungen als im Feld gerechnet werden muß, werden auch die c-Werte im Gefäß kleiner als im Feld ausfallen. Dieser Umstand führt zur Feststellung eines höheren Nährstoffbedarfs durch den Gefäß- im Vergleich zum Feldversuch.

Aus diesen Ausführungen kann man schließen, daß man bei der Untersuchung des Bodens auf seine Nährstoffbedürftigkeit stets Bedarfsmengen feststellt, welche für die Feldverhältnisse auch bei *optimaler Witterung* Genüge leisten.

Ferner wird aus diesen Ausführungen klar, warum man bei einem Vergleich zwischen Gefäß- und Feldversuchsergebnissen mit Schwankungen rechnen muß. Es wurde aus diesen Gründen schon wiederholt betont (MITSCHERLICH 1954 und ATANASIU 1954), daß bei der Beratung mittels Gefäßdüngungsversuchen eine Angabe des Nährstoffbedarfes auf 1 kg Genauigkeit als eine zu hohe Anforderung angesehen werden muß. Es genügt, wenn man mit einer Einteilung der Böden in Versorgungsklassen arbeitet, wobei folgende Klassen von den beiden Autoren vorgeschlagen worden sind:

Gruppe I = reichlich versorgt, enthält mehr als 2,2 dz/ha P_2O_5
bzw. mehr als 3,3 dz/ha K_2O

Gruppe II = mittelmäßig versorgt, enthält zwischen 1,1 und 2,2 dz/ha P_2O_5
bzw. zwischen 1,65 und 3,3 dz/ha K_2O

Gruppe III = stark bedürftig, enthält unter 1,1 dz/ha P_2O_5
bzw. unter 1,65 dz/ha K_2O.

D. Betrachtungen zum Gefäßdüngungsversuch und zu seiner Weiterentwicklung

BORDEN (1946) gibt einige Abänderungen der Gefäßmethode von MITSCHERLICH bekannt bei der Untersuchung von Böden mit Rücksicht auf ihr Nährstoffbedürfnis beim Anbau von Zuckerrohr. Diese Methode hat in Hawaii besonders zu einem erfolgreichen Anbau des Zuckerrohrs beigetragen. Die Änderungen betreffen: Versuchspflanze, Boden-Sand-Mischung, P-Gabe zur Erreichung des Höchstertrages sowie die im Gefäß eingefüllte Bodenmenge. Ferner wird von BORDEN für die Auswertung von Ergebnissen neben den Ertragstafeln nach MITSCHERLICH auch noch eine Einteilung der Böden in vier Versorgungsgruppen benutzt: unzureichend versorgt, fraglich versorgt, mittel und hoch versorgt.

Eine Abwandlung der Gedanken MITSCHERLICHS findet man in der Methode von FRIED und DEAN (1952), welche den pflanzenverfügbaren Nährstoffgehalt des Bodens über die Aufnahme markierten Nährstoffs in folgender Weise ermitteln:

$$A = \frac{B(1-y)}{y}$$

wobei: A der verfügbare Nährstoffgehalt des Bodens,
B die zugesetzte Nährstoffgabe,
y die relative Nährstoffaufnahme der Pflanze aus der Düngung ist.

Bei Gefäßversuchen mit 32 P-Düngung unternahm DEAN (1954) einen Vergleich zwischen dem von ihm eingeführten A-Wert und dem b-Wert MITSCHERLICHS. Für die untersuchten Böden ergibt sich $b < A$, und zwar in einem angenähert gleichen Verhältnis. Dieser Umstand kann dadurch erklärt werden, daß der Ermittlung des b-Wertes die Exponential-Funktion zugrunde gelegt ist, während bei der Ermittlung des A-Wertes von DEAN eine lineare Regression angenommen wird.

Diese Methode findet in den USA eine breite Anwendung und wurde auch bei den Arbeiten der National Soil Test Work Group (s. FITTS und NELSON 1956) bei dem Vergleich der chemischen Tests stark berücksichtigt.

JENNY und Mitarbeiter (1950) haben ebenfalls in Anlehnung an die MITSCHERLICH-Methode einen starkvereinfachten Gefäßversuch zur Feststellung der Nährstoffversorgung des Bodens ausgearbeitet. In einfachen Blumentöpfen (Inhalt: 1600 g Trockenboden) werden folgende Versuchsglieder angesetzt:

1. $N_0 P_0 K_0$ = Ungedüngt
2. $N_2 P_3 K_1$ = Volldüngung
3. $N_0 P_3 K_1$ = Mangeldüngung (ohne N)
4. $N_2 P_0 K_1$ = Mangeldüngung (ohne P)
5. $N_2 P_3 K_0$ = Mangeldüngung (ohne K)

Die Versuche werden mit Salat durchgeführt und dauern im Sommer 6 und im Winter 8 Wochen.

Die geernteten Pflanzensubstanzen der übrigen Versuchsglieder werden als Prozente der Volldüngung ausgedrückt. Der relative Ertrag ist dabei ein Maß für das Lieferungsvermögen des Bodens für den betreffenden Nährstoff. Die Autoren betrachten z. B. einen Boden, dessen Mangelertrag 30% des Volldüngungsertrages im Gefäß ausmacht, als ausreichend mit Nährstoff versorgt. In Californien sind die Böden systematisch mittels dieser Methode untersucht und dabei in zwei Gruppen eingeteilt:

„schwach“ (low) und „gut“ (high) versorgt.

Dieselbe Methode von JENNY wurde von F. T. BINGHAM (1949) zur Eichung von chemischen Bodenuntersuchungen auf Phosphor benutzt.

Der Gefäßversuch als Methode zur Bestimmung der Nährstoffbedürftigkeit der Böden hatte besonders in Ost- und Westpreußen in der Zeit zwischen den beiden Weltkriegen eine große, praktische Rolle gespielt, wo mehrere „MITSCHERLICH-Stationen" für diesen Zweck die Böden untersuchten. Auch im Ausland wurde die Methode mit Erfolg eingeführt.

Als Haupthindernisse zur weiteren Ausbreitung dieser Methode können, neben den kostspieligen Anlagen und den damit verbundenen hohen Untersuchungsgebühren, die lange für die Untersuchung benötigte Zeit und die große Bodenprobe — etwa 50 kg —, welche vom Schlag entnommen und an die Station versandt werden muß, angesehen werden. Zwar kann diese Probe bei einer Erweiterung des Mischverhältnisses Boden : Sand von 1:2 auf 1:4 (was ohne weiteres möglich wäre) auf etwa die Hälfte reduziert werden, diese Mengen sind jedoch immer noch zu groß für die heute anzustrebende systematische Bodenuntersuchung in der Landwirtschaft. Diese Schwierigkeiten haben die Suche nach „Schnellverfahren", welche einerseits geringere Bodenmengen beanspruchen und andererseits geringere Unkosten verursachen, veranlaßt und immer mehr intensiviert. Diese Verfahren, die hauptsächlich chemischer Art sind, können naturgemäß aber nur dann eine praktische Bedeutung erlangen, wenn sie gründlich auf ihre Brauchbarkeit bei der Düngungsberatung überprüft werden. Darin liegt eine der größten Bedeutungen des Gefäßversuches, und zwar zur Prüfung, Eichung und Überwachung derartiger Schnellverfahren. Die Meinung, daß der Feldversuch das beste und natürlichste Eichmittel darstellt, ist nur teilweise berechtigt. Die Tatsache, daß einerseits die Probe für die Untersuchung im Labor nur aus einer begrenzten, etwa 20 cm tiefen Schicht gezogen wird, und daß andererseits die Pflanzen im Feld auch weitere Schichten auszunutzen imstande sind, führt allerdings dazu, Ergebnisse vergleichen zu wollen, welche durch zwei *verschiedene* Methoden zu *verschiedenartigen* Böden erzielt werden. Dagegen bietet der Gefäßversuch die Möglichkeit zu prüfen, welche Beziehungen zwischen den Ergebnissen des chemischen und des pflanzenphysiologischen Verfahrens unter Anwendung *gleichen* Bodens bestehen. Dies müßte unseres Erachtens (s. auch ATANASIU 1955) bei jeder Schnellmethode der erste Schritt ihrer systematischen Ausarbeitung sein. Wenn derartige Beziehungen ausgearbeitet und bekannt sind, kann nun der zweite Schritt, die Übertragungen der Ergebnisse auf die Feldbedingungen, also der Vergleich mit dem Feldversuch unternommen werden. In diesem Falle werden Momente wie: Witterung, lokale Beschaffenheit tieferer Schichten als die Krume, Pflanzenart, Intensität der Bewirtschaftung u. a. m. dafür maßgebend sein.

4. Die Bestimmung des Nährstoffgehaltes des Bodens mittels der Keimpflanzenmethode nach Neubauer — Schneider

Die Methode, welche etwa in den Jahren 1921 bis 1923 entwickelt wurde, baut auf der Beobachtung auf, daß junge Pflanzen (im Keimlingstadium) eine rasche Entwicklung des Wurzelsystems und ein großes Aneignungsvermögen für Nährstoffe aufweisen. Diese Eigenschaften der jungen Pflanzen werden von den beiden Autoren dazu ausgenutzt, um mittels der Keimlinge aus dem Boden *rasch* und *vollständig* die „pflanzenaufnehmbaren" Nährstoffmengen zu „extrahieren". Die Verfasser (1923) versuchen dadurch ein *rascheres* und *billigeres* Verfahren, als es die des Gefäß- und Felddüngungsversuches sind, zu schaffen. Die so extrahierten Nährstoffmengen werden dann chemisch-analytisch quantitativ bestimmt.

Ein etwas älteres Verfahren als dieses, welches ebenfalls das starke Aneignungsvermögen der Keimpflanzen für Nährstoffe ausnutzte, wurde von Greisenegger und Vorbuchner (1919) entwickelt. Die Auswertung bei diesem Verfahren erfolgt aber nicht über die Nährstoffanalyse der Pflanzensubstanz, sondern über die gebildeten „Erträge".

Das 1923 von Neubauer und Schneider veröffentlichte Verfahren erfuhr im Laufe der Zeit Änderungen und Verbesserungen. Man unterscheidet zwischen dem Arbeitsverfahren nach Neubauer-Schneider und dem etwas vereinfachten Verfahren nach Lederle (1948, 1949, 1951).

A. Arbeitsverfahren nach Neubauer — Schneider

Der zu untersuchende Boden wird in nicht ganz abgetrocknetem Zustand (lufttrocken) durch ein 2 mm Sieb gegeben. Es werden dann davon 100 g Trockenboden in Neubauer-Schalen (Glasschalen von 11 bis 11,5 cm Durchmesser und 7 cm Höhe) eingefüllt. Auf den Boden wird *eine Deckschicht* von 250 g nährstofffreiem Sand getan, auf die dann 100 Roggenkörner gelegt werden. Die Roggenkörner werden dann wieder mit einer Sandschicht von 0,5 cm bedeckt. Die Körner werden mit der Basis (= Keimling) nach unten in den Sand gesteckt.

Es werden dann 80 g Wasser zugesetzt.

Die Gefäße werden mit einer Glasplatte zugedeckt und am Fenster bei Zimmertemperatur stehen gelassen. Die Deckel werden entfernt, wenn die Keimlinge so hoch sind, daß sie sich sonst krümmen würden. Von diesem Zeitpunkt an wird täglich das verdunstete Wasser zurückgegeben. Nach 17 Tagen erfolgt die Ernte. Dafür wird das Boden-Sand-Gemisch abgespühlt und die gesamte Pflanzensubstanz für die weiteren Untersuchungen herangezogen. Diese Versuche werden in drei Parallelbehandlungen angesetzt und durchgeführt.

Da die Roggenkörner selbst auch Nährstoffe enthalten, werden außer den Bodengefäßen auch Gefäße mit nur 400 g Sand angesetzt, welche als „Blindversuche" dienen. Die bei ihnen bestimmten Nährstoffmengen werden als Blindwerte berücksichtigt.

Die gesamte, abgeerntete Pflanzenmasse wird in Veraschungsschalen gebracht, welche mit Aluminiumdeckel bedeckt werden. Die Substanz wird bei etwa 500° (nicht höher!) verascht, die kohlefreie Asche dann mit heißem Wasser und 10% HCl übergossen und dann auf dem Wasserbad zur Abscheidung von Kieselsäure abgedampft. Nach dem Eintrocknen wird die Substanz mit 10% HCl und Wasser aufgenommen und in 100 ml-Meßkölbchen übergebracht, welche bis zur Marke aufgefüllt werden. In der Flüssigkeit werden in aliquoten Teilen, das Kalium gravimetrisch oder flammenphotometrisch und der Phosphor kolorimetrisch bestimmt (s. Knickmann 1955).

Die Nährstoffe werden nach Abzug der Werte des Blindversuches in mg Nährstoff (K_2O bzw. P_2O_5) je 100 g Trockenboden ausgedrückt.

Besondere Maßnahmen bei dem Ansetzen und bei der Durchführung der Keimpflanzenmethode. Das *Saatgut* muß mit besonderer Sorgfalt ausgesucht werden: gleichmäßiges hohes Tausendkorngewicht von 40 g wird verlangt. Es wird ferner hochwertiges, einheitliches Saatgut für eine gute Reproduzierbarkeit der Methode vorausgesetzt (Pfaff). Vom Verband Deutscher Landwirtschaftlicher Untersuchungs- und Forschungsanstalten wird als Lieferant für Saatgut, Fa. F. von Lochow-Petkus G.m.b.H., Saatzucht Hasselhorst, 3103 Bergen, Kreis Celle, vorgeschlagen. Die sehr sorgfältig ausgesuchten Körner werden mit Ceresan (2 g für 1 kg Saatgut) gebeizt.

Neubauer (1939) hatte die Methode insofern modifiziert, daß er an Stelle des Blindversuches eine sorgfältige Untersuchung der Roggenkörner verlangte und den Abzug der darin festgestellten Nährstoffe empfahl. Da nach seinen Beobachtungen nicht die Gesamtmenge der Nährstoffe der 100 Körner im Sand wiederzufinden war, werden erfahrungsgemäß von dem in 100 Körnern gefundenen Kalium 95% und von dem gefundenen Phosphor 89% als Blindwerte berücksichtigt. Diese Maßnahme hat wesentlich zur Vereinfachung und Sicherung der Methode beigetragen.

Der Vegetationsraum soll möglichst Nordlage und keine direkte Sonneneinstrahlung haben. Die Temperatur des Raumes soll nicht außer den Grenzen 18° bis 22° C schwanken und im Sommer werden notfalls besondere Kühleinrichtungen empfohlen.

B. Arbeitsverfahren nach P. Lederle

Dieses Verfahren unterscheidet sich grundsätzlich von dem von Neubauer-Schneider dadurch, daß nicht die gesamte Pflanzensubstanz, sondern nur die Sprosse zur Analyse herangezogen werden. Dadurch wird eine wichtige Fehlerquelle der Keimpflanzmethode vermieden. Bei diesem Verfahren werden dann als Abzugswerte nur 70% der bei Körneranalyse ermittelten Kali- und 60% der Phosphormenge berücksichtigt.

Ferner hat Lederle noch vorgeschlagen, an Stelle von reinem Wasser beim Anfeuchten eine verdünnte NH_4NO_3-Lösung (40 mg NH_4NO_3 in 1000 ml Wasser) anzuwenden. Auch Schachtschabel (1948) wollte an Stelle von Anmachwasser eine verdünnte Stickstofflösung angewendet wissen, da dadurch zu frühes Nachlassen des Wachstums der Keimlinge und dadurch der K- und P-Aufnahme (wegen eventuellen N-Mangels) vermieden werden.

Außer diesen Maßnahmen unterscheidet sich das Arbeitsverfahren von Lederle von dem von Neubauer-Schneider im wesentlichen nicht.

C. Auswertung der Ergebnisse

Neubauer und Schneider sind bei der Auswertung der Ergebnisse von einer direkten Übertragung der in den Glasschalen gewonnenen Werte auf die Feldverhältnisse ausgegangen. So wurden die mg Nährstoffe in 100 g wasserfreiem Boden in kg/ha umgerechnet: bei der Annahme, daß die Ackerkrume 20 cm und das scheinbare Volumengewicht 1,5 betragen, entspricht 1 mg Nährstoff in 100 g Trockenboden 30 kg Nährstoff je ha.

Unter Berücksichtigung der Nährstoffaufnahme von verschiedenen Kulturpflanzen bei mittlerer und bei hoher Ernte, hat Neubauer (1938) Grenzzahlen ausgerechnet, welche in folgender Tabelle angeführt werden.

Auch Roemer (1924), Haselhoff (1931), Lemmermann (1932) haben Grenzzahlen angegeben, welche teilweise über, teilweise unter denen von Neubauer liegen. Für die Böden und Ernten Nordamerikas wurden von Kohnke (1937) und Moers (1938) Grenzzahlen mitgeteilt. Die Auswertung mittels solcher Grenzzahlen baut naturgemäß auf einer großen Anzahl von Verallgemeinerungen auf wie u. a.: Bodenbeschaffenheit, Krumentiefe, Nährstoffentzug durch Pflanze, ausgenutzte Bodentiefe usf. Diese Momente führten dazu, daß nicht immer eine gute Übereinstimmung der Untersuchungsergebnisse der Keimpflanzmethode mit denen des Feld- oder des Gefäßdüngungsversuches festzustellen war. Vielmehr gewann die Überzeugung mehr an Boden, daß die Keimpflanzenmethode, ähnlich wie die chemischen Schnellmethoden, nur „Anhaltspunkte“ über den Nährstoffgehalt der Böden geben kann. Die Auswertung der Ergebnisse bzw. die Düngerberatung können daher nicht *einfach* und direkt mittels der *mg Nährstoff in 100 g Trockenboden*

	Phosphorsäure (P_2O_5)		Kali (K_2O)	
	mittlere Ernte	hohe Ernte	mittlere Ernte	hohe Ernte
1. Roggen	4	5	13	17
2. Weizen	4	5	15	20
3. Gerste	5	6	19	24
4. Hafer	4	6	16	21
5. Kartoffeln	5	6	28	37
6. Zuckerrüben	5	6	19	25
7. Futterrüben	6	7	29	39
8. Raps	7	9	15	18
9. Rotklee	4	5	19	25
10. Luzerne	7	9	25	35
11. Wiesenheu	4	5	19	25

erhaltenen mg-Zahl vorgenommen werden. Vielmehr müssen dabei auch andere Faktoren, wie: Bodenart (Sorptionskapazität, Tonanteil, Gehalt an organischer Substanz u. a. m.), Untergrundverhältnisse, Bewirtschaftung, berücksichtigt werden, abgesehen von der Witterung, welche bekanntlich jeder Methode eine gewisse Schwankungsbreite der Aussage verleiht.

Diese Überlegungen und Erfahrungen führten dazu, daß zur Interpretation der Ergebnisse der Keimpflanzenmethode Klassen von Grenzwerten gebildet werden, ähnlich wie z. B. bei der Lactatmethode.

So gibt Dietz (1953) Grenzwerte für fünf Klassen an:

mg P_2O_5 : 0 — 3, 3 — 5, 5 — 7, 7 — 9 und > 9 und
mg K_2O : 0 — 12, 12 — 17, 17 — 25, 25 — 35 und > 35.

Selke (1957) empfiehlt für die Auswertung der Keimpflanzenmethode folgende Grenzwerte und Klassen:

	mg P_2O_5	mg K_2O
arm	<4	<15
mittel versorgt	4 bis 7	15 bis 26
gut versorgt	>7	>26

So gesehen, betrachtet man die Keimpflanzenmethode heute nicht mehr als ein „absolutes“ Verfahren sondern als ein „konventionelles“, bei welchem die Keimlinge von Roggen das Extraktionsmittel darstellen.

Auch für Hochmoor- und Heidesandböden sind von W. Hoffmann und Steinfatt (1955) Klassen gebildet worden, und zwar:

	mg P_2O_5	mg K_2O	mg P_2O_5	mg K_2O
normal versorgt	5	8	6	8
gut versorgt	8	12	7	12

Für Niedermoorböden wollten die Verfasser entsprechende Grenzzahlen noch ausarbeiten.

D. Kritische Betrachtungen

Der Gedanke, die *Pflanze* selbst als „Extraktionsmittel" für die *pflanzenverfügbaren* Bodennährstoffe zu benutzen, schien von Anfang an als sehr glücklich, jedoch wurde die Methode von zahlreichen Seiten her kritisch untersucht. Im Rahmen dieser Abhandlung kann man auf die vielen Arbeiten, welche die Keimpflanzmethode als Untersuchungsgegenstand haben, unmöglich eingehen. Es seien deswegen hier nur die grundsätzlichen Punkte der kritischen Betrachtungen erwähnt. Auf eine umfangreiche diesbezügliche Literaturzusammenstellung sei bei W. BERGMANN (1958) hingewiesen.

Eine allgemeine, grundsätzliche Kritik der Keimpflanzenmethode wurde u. a. von LEMMERMANN (1926), MITSCHERLICH (1933) und THUN (1936) ausgeübt. Aus diesen Arbeiten war es zu ersehen, daß die Keimpflanzen, entgegen der von NEUBAUER und SCHNEIDER gehegten Hoffnung, nicht die *gesamten* im Boden als *pflanzenaufnehmbar* vorhandenen Nährstoffe ausziehen, sondern daß die Ausnutzung von der *Menge* der pflanzenaufnehmbaren Nährstoffe abhängt. Um das zu demonstrieren, variierte MITSCHERLICH (1933) auf der einen Seite die Boden- und auf der anderen die Nährstoffmenge. Es mögen hier einige seiner Ergebnisse angeführt werden.

Tabelle 148. *Abhängigkeit der durch die Neubauer-Methode festgestellten mg-Zahl von der Bodenmenge*

angewandte Bodenmenge in g	durch Keimpflanzen entzogene mg P_2O_5
100	$2{,}5 \pm 0{,}2$
50	$1{,}6 \pm 0{,}4$
25	$1{,}3 \pm 0{,}1$

Tabelle 149. *Abhängigkeit der durch die Neubauer-Methode festgestellten mg-Zahl von der pflanzenverfügbaren Nährstoffmenge des Bodens*

100 g Boden + x mg P_2O_5	durch Keimpflanzenmethode entzogene mg P_2O_5
$x = 0$	5,04
10	9,12
20	12,82
40	18,11
80	26,72
160	35,71

Zu sinngemäß ähnlichen Ergebnissen führten auch Untersuchungen von LEMMERMANN (1926), von WIESSMANN (1927), aber auch von NEUBAUER (1928) selbst.

Der Einfluß anderer Faktoren wie: des Lichtes, der Temperatur, der Wassergabe, der zersetzbaren organischen Substanz des Bodens, des Stickstoffs ist ebenfalls Gegenstand zahlreicher Untersuchungen gewesen. Es sei hier auf die wichtigste Literatur hingewiesen (LEMMERMANN 1926, HÄHNE 1926, WIESSMANN 1925, GERICKE 1926, DIRKS 1928, DREWES 1928).

Alle diese Untersuchungen zeigten, daß, wie bereits schon oben erwähnt wurde, die von den Keimpflanzen aus dem Boden entzogenen Nährstoffe von den momentanen Bedingungen abhängen, sie stellen soweit nicht den gesamten Vorrat an wurzellöslichen Nährstoffen des Bodens dar.

Der „konventionelle" Charakter dieser Methode wird auch durch die Vergleiche, welche von verschiedenen Autoren zwischen den Ergebnissen der Keimpflanzenmethode und denen des Gefäß- und Felddüngungsversuches angestellt wurden, demonstriert.

In einer entsprechenden Arbeit stellten MITSCHERLICH und ATANASIU (1952) folgende Übereinstimmungen zwischen den Untersuchungsergebnissen nach der Keimpflanzenmethode und denen des Feldversuches bzw. des Gefäßversuches fest.

Tabelle 150. *Übereinstimmung der Ergebnisse der Neubauer-Schneider-Methode mit dem Feldversuch*

	P_2O_5		K_2O	
	Station Rostock	Station Jena	Station Rostock	Station Jena
1950	33%	33%	40%	42%
1951	40%	62%	30%	33%

Tabelle 151. *Übereinstimmung der Ergebnisse der Neubauer-Schneider-Methode mit dem Gefäßversuch*

	P_2O_5		K_2O	
	Station Rostock	Station Jena	Station Rostock	Station Jena
1950	37%	29%	37%	45%
1951	56%	62%	31%	40%

In einer späteren Arbeit fand ATANASIU (1955) ebenfalls bei zweijährigen Vergleichen (1954 und 1955) für die P-Ergebnisse 60% Übereinstimmuungen zwischen der Keimpflanzenmethode und dem Gefäßversuch.

BEHRENS (1935) fand zwischen der Keimpflanzenmethode und dem Gefäßversuch eine Übereinstimmung von 76 bis 89%.

SKORIC (1953) in Jugoslawien und BROUWER (1949) in Deutschland fanden, daß Proben desselben Feldes zu verschiedenen Terminen (verschiedene Jahreszeiten bzw. vor oder nach Beregnung) entnommen, stark abweichende Ergebnisse im K- und P-Gehalt nach der Keimpflanzenmethode aufwiesen.

Andererseits werden von KIVINEN (1944) gute Übereinstimmungen der Keimpflanzenmethode mit dem Feldversuch gemeldet.

Dadurch ist die Keimpflanzenmethode, ähnlich wie die chemischen Schnellmethoden nur eine „konventionelle" Bodenuntersuchungsmethode. Sie weist aber ihnen gegenüber den Nachteil der größeren Aufwendigkeit sowohl an Material als auch an Zeit auf.

E. Zur Weiterentwicklung der Neubauer — Schneider-Methode

Die anfänglich nur für Kali- und Phosphoruntersuchungen ausgearbeitete Methode ist verschiedentlich auch für andere Nährstoffe benutzt worden.

So versuchten BLANCK und SCHEFFER (1925), THUN (1930), NEUBAUER (1946) und GRAMMS (1955, 1958) den aufnehmbaren Bodenstickstoff mittels der Roggenkeimlinge zu bestimmen. Da aber die ganzen biologischen Vorgänge der N-Umsetzung (Mineralisation usf.), welche für die N-Lieferung im Feld ausschlaggebend sind, in den Glasschalen und in der kurzen Zeit der Keimpflanzenmethode keine Berücksichtigung finden können, sind die dabei ermittelten Werte ohne praktische Bedeutung für den N-Haushalt der Böden. Nach GRAMMS (1958) kann man mittels der Keimpflanzenmethode die Umsetzungen verfolgen, welche bei längerer Lagerung natürlicher, organischer Dünger zu Änderungen des Verhältnisses: Gesamt-N zu Wurzellöslich-N führen.

Eine Weiterentwicklung der Keimpflanzenmethode bedeutet ihre Anwendung zur Bestimmung von pflanzenverfügbaren Mikronährstoffen im Boden. E. RUSS (1958) führte derartige Arbeiten im Agrikulturchemischen Institut Gießen durch und stellte fest, daß bei Cu und Mn prinzipiell die Keimpflanzenmethode zur Bestimmung verfügbarer Spurenelemente herangezogen werden kann. Da aber bei höheren Cu- bzw. Mn-Gaben hauptsächlich eine Anreicherung der Keimlingswurzeln stattfindet, müssen bei der chemischen Analyse sowohl die Sprosse als auch die Wurzeln unbedingt berücksichtigt werden.

In der letzten Zeit wird von K. EHRENDORFER (1959) ein interessanter Vorschlag zur Abänderung der NEUBAUER-SCHNEIDER-Methode in der Weise gemacht, daß an Stelle von Roggenkeimlingen Salatkeimlinge für die Extraktion der Nährstoffe verwendet werden sollen. Der Vorteil soll in dem sehr niedrigen Gehalt der Salatkörner an Nährstoffen liegen, wodurch sich sehr niedrige Blindwerte ergeben.

Ein weiterer Verbesserungsvorschlag ist die Verwendung von Lichtschränken als Vegetationskammern. Es sollen dadurch: Unabhängigkeit von der Jahreszeit und genau definierte Lichtverhältnisse erreicht werden. Die von EHRENDORFER (1959) mitgeteilten Ergebnisse zeigen, daß eine gute Korrelation der neuen „Salatmethode" mit dem EGNER-RIEHM-Verfahren vorhanden ist.

Inwieweit die Verwendung von Lichtschränken in der Tat zu einer besseren Reproduzierbarkeit der Ergebnisse führt, muß noch durch umfangreiche Versuche bewiesen werden. Ebenfalls ist es auch fraglich, ob diese Maßnahme nicht eine gewisse technische Erschwerung des Verfahrens darstellt.

5. Die Anwelkmethode nach Arland

In ganz anderer Weise testet ARLAND (1956) den Nährstoffbedarf der Pflanzen. Nach ihm besteht eine feste Beziehung zwischen der relativen Transpiration $\left(=\frac{\text{absolute Transpiration}}{\text{transpirierende Pflanzenmasse}}\cdot 100\right)$ und der optimalen Düngungskombination in der Weise, daß die letztere durch die niedrigste relative Transpiration gekennzeichnet wird. Ferner nimmt ARLAND an, daß diese Beziehung bereits bei den sehr jungen Pflanzen besteht, so daß Messungen bei erst 18 Tage alten Pflanzen eine Aussage über die Düngungsbedingungen zulassen.

Versuchsdurchführung. Mengen von 2,8 kg des zu untersuchenden Bodens werden in schalenartige Gefäße eingefüllt und mit verschiedenen Düngungskombinationen versetzt. In den Gefäßen werden je 60 Gerstepflanzen 18 Tage (4 bis 5 Blattstadium) angezogen. Die Wasserhaltung beträgt 60% der Wasserkapazität des Bodens (ein- bis zweimal tägliches Gießen).

Es werden folgende Düngergaben verabfolgt:

$$N = 32\ \text{kg/ha}$$
$$P_2O_5 = 54\ \text{kg/ha}$$
$$K_2O = 80\ \text{kg/ha.}$$

Die Mengen werden pro Gefäß umgerechnet unter Zugrundelegung von 3,000.000 kg Bodenmenge je ha.

Eine Reihe von Kombinationen, welche von ARLAND empfohlen wird, ist u. a. folgende:

1. N P K	5. N P —
2. $\frac{N}{2}$ P K	6. N — K
3. N $\frac{P}{2}$ K	7. — P K
4. N P $\frac{K}{2}$	8. — — —

Die über der Bodenoberfläche abgeschnittenen Pflänzchen läßt man in besonderen Kammern bei 25° bis 30° C 30 Minuten transpirieren.

Nach der Berechnung der relativen Transpiration wird die Düngungskombination als optimal angesehen, bei der die niedrigste relative Transpiration gefunden wird. Diese Düngung wird auch im Feld für die laufende Vegetation empfohlen. Die Anwelkmethode muß deswegen rechtzeitig im Frühjahr durchgeführt werden.

Die bestechend einfache Methode hat zu zahlreichen Untersuchungen und kritischen Äußerungen Anlaß gegeben.

U. a. konnten TRÈNEL und FRANKE (1955) experimentell nachweisen, daß diese Beziehung nicht immer zu Recht besteht.

Eine ausführliche Darstellung der Methode und ein reichhaltiger Schrifttumsnachweis darüber sind in der Arbeit von ARLAND, „Ein Beitrag zur Anwelkmethode" (1956), enthalten.

6. Pflanzenphysiologische Methoden zur Bestimmung von Mikronährstoffen

Zu diesen Verfahren zählt in erster Linie das bei der Keimpflanzenmethode besprochene Verfahren zur Cu- und Mn-Bestimmung. Nach den Erfahrungen, welche im Agrikulturchemischen Institut Gießen mit der Methode gemacht wurden, scheint die Keimpflanzenmethode bei entsprechenden Abänderungen auch für die Bestimmung anderer Mikronährstoffe Anwendung finden zu können.

Der physiologische Test auf Bor. Ein eigens ausgearbeiteter Test ist der physiologische Test auf Bor.

Auf die Erfahrungen COLWELLS (1943) mit Sonnenblumenpflanzen als Indikator für Bormangel aufbauend hat BUSSLER (1956) ein entsprechendes Verfahren ausgearbeitet.

Bei diesem Verfahren werden als Kulturgefäße Reagenzgläser aus Jenaer Glas benutzt. Von dem zu untersuchenden Boden werden 100 bis 500 mg eingefüllt und mit 10 ml einer borfreien, aber sonst vollständigen Nährlösung versetzt. Besonders angezogene Sonnenblumenpflänzchen von 6 bis 10 Tagen Alter werden in den Reagenzgläsern (je Glas eine Pflanze) weiter wachsen gelassen.

Zu diesen Boden-Nährlösungskulturen wird eine *Vergleichsreihe* mit bekannten, steigenden Bormengen (0 bis 20 g H_3BO_3 je Gefäß) angesetzt. Während der Kultur werden nach Bedarf alle 2 bis 3 Tage 5 ml Nährlösung ergänzt. Die Mangelerscheinungen, nach denen dann getestet wird, treten in etwa 14 Tagen aber auch früher auf.

Der Boden wird als „borbedürftig" angesehen, wenn nach diesem Test weniger als 1 ppm Bor gefunden wird.

Über die praktischen Auswirkungen des Verfahrens kann infolge Mangel an ausreichenden Erfahrungen noch keine kritische Stellungnahme geäußert werden. Es hat zweifelsohne den Vorteil gegenüber chemischen Verfahren, daß das System Boden + Nährlösung: Pflanze ein dynamisches ist und dadurch eine *kontinuierliche* Erschöpfung des pflanzenverfügbaren Bodenbors ermöglicht wird. Die sehr geringe Bodenmenge, welche für den Test benutzt wird, kann als ein Nachteil empfunden werden, jedoch muß man bedenken, daß mit größeren Bodenmengen das Verfahren selbst im allgemeinen andere Dimensionen einnehmen müßte, ein Umstand, welcher zur Erhöhung sonstiger Fehler beitragen würde.

Literatur

ARLAND, A.: Die Beurteilung des Düngerbedürfnisses des Bodens nach dem Transpirationsvermögen der Pflanzen. Z. Pflanzenernähr., Düng., Bodenkde. A 28 (1933). — Ein Beitrag zur Anwelkmethode. Sitz. Ber. Dtsch. Akad. Landwirtsch. Wiss. Berlin. 5, H. 6. Leipzig: Hirzel. 1956. — ATANASIU, N.: Die Gefäß- und die Felddüngungsversuche und ihre Vergleichbarkeit unter Zugrundelegung des Wirkungsgesetzes. Z. Pflanzenernähr., Düng., Bodenkde. 66, H. 2 (1954). — Ein Vergleich zwischen den Bodenuntersuchungsergebnissen nach den Methoden Mitscherlich, Egnér-Riehm und Neubauer. Z. Landw. Vers. u. Untersuchungswesen 1, H. 6 (1955).

BALBA, A. M., und R. H. BRAY: The application of the Mitscherlich equation for the calculation of plant composition due to fertilizer increments. Proc. Soil Sci. Soc. Amer. 20, 515 (1956). — BALBA, A. M., und L. E. HALEY: Comparison of results obtained by the Balba-Bray-equation and radioactive techniques for the determination of nutrient uptake by plants from different nutrient forms. Soil Sci. 82 (1956). — BEHRENS, W. U.: Die Methoden zur Bestimmung des Kali- und Phosphorsäurebedarfs landwirtschaftlich genutzter Böden. Berlin: Verlag Chemie. 1935. — Die Eignung verschiedener Feldversuchsanordnungen zum Ausgleich der Bodenunterschiede. Z. Acker- u. Pflanzenbau 101 (1956). — BERGMANN, W.: Handbuch der Pflanzenphysiologie, Bd. IV. Berlin-Göttingen-Heidelberg: Springer. 1958. — BINGHAM, F. T.: Soil test of phosphate. Calif. Agricult. 3, 11, 14 (1949). — BLANCK, E., und F. SCHEFFER: Die Neubauer-Methode und die Bestimmung des Stickstoffbedürfnisses der Böden. Z. Pflanzenernähr., Düng., Bodenkde. B 4 (1925). — BOGUSLAWSKI, E. v.: Der Gefäßversuch zur Bestimmung der Bodenfruchtbarkeit. Internat. Soc. Soil Sci. Trans. 2 (1952). — Zur Kenntnis des „b"-Wertes. Z. Pflanzenernähr., Düng., Bodenkde. 66 (1954). — BOGUSLAWSKI, E. v., A. VÖMEL und G. REICHELT: Nährstoffverhältnis in der Düngung und Ertragsbildung. Z. Acker- u. Pflanzenbau 97 (1954). — Das Zusammenwirken der Wachstumsfaktoren bei der Ertragsbildung. Z. Acker- u. Pflanzenbau 98 (1954). — BORDEN, R. J.: Modified Mitscherlichmethode for soil cultures. Soil Sci. 62, 51-60 (1946). — BRAY, R. H.: Diagnostic techniques for soils and crops. (H. D. KITCHEN, ed.) Washington, D. C.: American Potash Institute. 1948. — BRAUNEWELL, W.: Untersuchungen über das Nährstoffverhältnis bei Zucker- und Futterrüben. Diss. Gießen 1956. — BROUWER, W.: Steigerung der Erträge der Hülsenfrüchte durch Beregnung sowie Fragen der Bodenuntersuchung und Düngung. Z. Acker- u. Pflanzenbau 91 (1949). — BUSSLER, W.: Ein physiologischer Test auf Bor. Z. Pflanzenernähr., Düng., Bodenkde. 73 (1956).

COLWELL, W. E.: A biological method for determining the relative Boron content of soils. Soil Sci. 50 (1943).

DEAN, L. A.: Proc. Soil Sci. Soc. Amer. 18, 462 (1954). — DENSCH, A., und C. PFAFF: Versuche zur Bestimmung des Nährstoffbedürfnisses der Böden durch Gefäßversuche nach Mitscherlich nebst einigen Bemerkungen zur Methode Neubauer. Z. Pflanzenernähr., Düng., Bodenkde. B 4, 321 (1925). — DENSCH, A.: Die Bestimmung des Düngerbedürfnisses des Bodens durch Laboratoriumsversuche nach Neubauer. Z. Pflanzenernähr., Düng., Bodenkde B 5, 130 (1926). — Erfahrungen mit der Methode Neubauer. Z. Pflanzenernähr., Düng., Bodenkde B 5, 97 (1926). — DIETZ, R.: Vergleich von Bodenuntersuchungsergebnissen nach den Methoden Neubauer, Egnér und Dirks-Scheffer. Landw. Forsch., Sonderheft 4, 83-87 (1953). — DREWES, K.: Über die Beteiligung von Mikroorganismen an der Aufschließung unlöslicher Phosphate. Zbl. Bakt. II, 76 (1928).

EHRENDORFER, K.: Die Salatmethode. Eine Abänderung der Neubauermethode im Vergleich zu den Methoden Egnér-Riehm und Dirks-Scheffer. Z. Pflanzenernähr., Düng., Bodenkde. (1959).

FITTS, J. W.: Better crops with plant food **39**, No. 3, 17–21 (1955). — FITTS, J. W., und W. L. NELSON: The determination of lime and fertilizer requirements of soils through chemical tests. Advances in Agronomy 8 (1956). — FRIED, M., und L. A. DEAN: A concept concerning the measurement of available soil nutrients. Soil Sci. **73**, 263–271 (1952).

GÄRTNER, M.: Die graphische Auswertung der Gefäßversuche zur Bestimmung des Düngerbedürfnisses des Bodens nach dem Verfahren MITSCHERLICH. Z. Pflanzenernähr., Düng., Bodenkde. **23** (1941). — GERICKE, S.: Die Neubauersche Keimpflanzenmethode und die Citratmethode zur Erkennung des Düngerbedürfnisses des Bodens an Phosphorsäure. III. Landw. Ztg. (1926). — Der Einfluß hoher Kalkgaben auf die Wurzellöslichkeit der Nährstoffe Kali und Phosphorsäure im Boden. Fortschr. Landw. **1** (1926). — GREISENEGGER, I. K., und K. VORBUCHNER: Feststellung des Düngerbedürfnisses durch Bodenerschöpfung. Österr. Ung. Z. Zuckerind. u. Landw. **47** (1918). — GROUVEN: Dritter Bericht über die Arbeiten der Versuchsstation Salzmünde. Halle 1868 (zit. nach ROEMER: Der Feldversuch).

HÄHNE, H.: Die Neubauersche Methode zur Bestimmung der wurzellöslichen Nährstoffe Phosphorsäure und Kali. Z. Pflanzenernähr., Düng., Bodenkde. A **6** (1926). — HANAMANN, J.: Sechsjähriger Vegetations- und Düngungsversuch in Verbindung mit meteorologischen Beobachtungen und Bodenanalyse. Prag 1873 (zit. nach ROEMER: Der Feldversuch). — HASELHOFF, E.: Die Bestimmung des Fruchtbarkeitszustandes des Bodens mit Hilfe biologischer Methoden. In: BLANCK: Handbuch der Bodenlehre, Bd. VIII. Berlin: Springer. 1931. — HOFFMANN, W., und K. STEINFATT: Über die Bestimmung der pflanzenaufnehmbaren Nährstoffe in Hochmoorböden und stark humosen Böden (Heidesand). Landw. Forsch. **7**, H. 3 (1955). — HOMÈS, M.: Recherche d'une formule d'engrais pour cotonnier par la methode des variantes systematiques. VIe Congrès international de la science du sol, Vol. A, IV — 78, Paris (1956). — HOMÈS, M., und G. VAN SCHOOR: Experimentation sur le tabac par la methode des variantes systematiques portant sur huit elements nutritifs. VIe Congrès international de la science du sol, Vol. A, IV — 79, Paris (1956).

JANERT, H.: Bodenkundliches Praktikum. Berlin: VEB Verlag Technik. 1953. — JENNY, H., J. VLAMIS und W. E. MARTIN: Greenhouse assay of fertility of California soils. Hildegardia **20**, May (1950).

KERTSCHER, F.: Grenzen der Auswertung der Nährstoffuntersuchungen der Böden auf Grund der Erfahrungen mit der Keimpflanzenmethode. Z. Pflanzenernähr., Düng., Bodenkde. **29**, 74 (1943). — KIRSANOF, A. F.: Mitscherlich-Theorie, ihre Analyse und praktische Anwendung. Moskau-Leningrad: Staatl. Landw. Verlag. 1930. — KIVINEN, E.: Determination of the fertilizer requirements of soils. Chem. Abstr. **38** (1944). — KNICKMANN, E.: Die Untersuchung von Böden (Handbuch der landwirtschaftlichen Versuchs- und Untersuchungsmethodik, 1. Bd.), 3. Aufl. Radebeul und Berlin: Neumann. 1955. — KOHNKE, H.: Testing the fertility of Alberta soils by the Neubauer and Lemmermann methods. Sci. Agric. **17** (1937).

LEDERLE, P.: Neues über die Neubauer-Methode. Z. Pflanzenernähr., Düng., Bodenkde. **40** (1948). — Vorschläge zur vereinfachten Durchführung des Keimpflanzenverfahrens nach Neubauer und zur Erzielung besserer Reproduzierbarkeit seiner Ergebnisse, Teil I. Landw. Forsch. **1** (1949). — Vorschläge zur vereinfachten Durchführung des Keimpflanzenverfahrens nach Neubauer und zur Erzielung besserer Reproduzierbarkeit seiner Ergebnisse, Teil II. Landw. Forsch. **3** (1951). — LEMMERMANN, O.: Die Bestimmung des Düngungsbedürfnisses der Böden durch Laboratoriumsversuche. Z. Pflanzenernähr., Düng., Bodenkde. B **5** (1926). — Über die heutigen Laboratoriumsmethoden zur Bestimmung des Düngungsbedürfnisses der Böden. Z. Pflanzenernähr., Düng., Bodenkde. **9** (1930). — Methoden für die Untersuchung des Bodens. Berlin: Verlag Chemie. 1932.

MAERKER, M.: Feldversuche über die zweckmäßige Verwendung von künstlichen Düngemitteln für Kartoffeln. Z. landw. Zentral-Versuchsst. Prov. Sachsen (1876-1878). — MC GEORGE, W. T.: Modified Neubauer method for soil cultures. Soil Sci. **62**, 61–70 (1946). — Methodenbuch der Association of Official Agricultural Chemists. Benj. Franklin Station, 8. Ed. Washington 1955. — MITSCHERLICH, E. A.: Die pflanzenphysiologische Lösung der chemischen Bodenanalyse. Landw. Jahrb. **58** (1923). — Bodenkunde für Land- und Forstwirte. Berlin: Parey. 1923. — Zur Bestimmungsmethode des Düngerbedürfnisses des Bodens. Z. Pflanzenernähr., Düng., Bodenkde. B **4** (1925). — Die Bestimmung des Düngerbedürfnisses des Bodens. Berlin: Parey. 1925. — Zur Bestimmung des Nährstoffgehaltes des Bodens. Landw. Jahrb. (1926). — Die Bestimmung des Düngerbedürfnisses des Bodens. Berlin: Parey. 1930. — Die pflanzenphysiologische Bodenanalyse. Abh. 2. Internat. Kongreß für Bodenkunde, Leningrad-Moskau 1933. — Untersuchungen über den pflanzenphysiologischen Wert von chemischen und biologischen Laboratoriumsmethoden zur Bestimmung des Düngerbedürfnisses des Bodens. Schriften der Königl. Königsb. Gel. Ges., Naturw. Klasse **10**, H. 1, Königsberg/Pr. 1933. — Die Bestimmung des Kalivorrates nach MITSCHERLICH. Ernährung d. Pflanze **30** (1934). — Die Verarbeitung landwirtschaftlicher und anderer biologischer Versuchsergebnisse. K.G.G. **10**, H. 4 (1934). — Der Vergleich der Ergebnisse der Kali- und Phosphorsäureuntersuchungen sämtlicher Methoden mit denen des Gefäßversuches sowie sämtlicher Methoden untereinander bei Berücksichtigung der einzelnen Bodenarten. Abhandl. d. Internat. Gesellsch. f. Bodenkunde, Königsberg/Pr. 1936. — Die Ergebnisse der Rundfrage über die Anwendung der Methoden zur Bestimmung des Kali- und Phosphorsäurebedarfes der Böden für praktische Zwecke in den verschiedenen Ländern. Abhandl. d. Internat. Gesellsch. f. Bodenkunde, Königsberg/Pr. 1939. — Über die Größe der Teilstücke bei Feldversuchen. Z. Pflanzenernähr., Düng., Bodenkde. **40** (85), 15–19 (1948a). — Bodenkunde, 6. Aufl. Berlin und Hamburg: Parey. 1950. — Bodenkunde für Landwirte, Forstwirte und Gärtner in pflanzenphysiologischer Ausrichtung und Auswertung, 7. Aufl. Berlin und Hamburg: Parey. 1954. — MITSCHERLICH, E. A., und N. ATANASIU: Zur Bestimmung des Nährstoffgehaltes bzw. des Düngerbedürfnisses der Böden. Z. Acker- u. Pflanzenbau **95**, H. 1 (1952). — Zur Methode des Gefäßdüngungsversuches. Z. Acker- u. Pflanzenbau 35 (1956). — MOOERS, C. A.: An evaluation of the Neubauer, the *Cunninghamells* and *Aspergillus* methods for determination of the fertilizers needs of a soil. Soil Sci. **46**, 211–227 (1938).

National Soil Test Work Group. National Soil and Fertilizer Research Committee North Caroline Agr. Exp. Sta. Techn. Bull. **121** (1956). — NEHRING, K.: Der Einfluß verschiedener Wasserversorgung auf den Mineralstoffgehalt von Grünfuttersorten. Z. Landw. Vers.- u. Untersuchungswesen **1**, H. 2/3 (1955). — NEUBAUER, H., und W. SCHNEIDER: Die Nährstoffaufnahme der Keimpflanzen und ihre Anwendung auf die Bestimmung des Nährstoffgehaltes der Böden. Z. Pflanzenernähr., Düng., Bodenkde. A **2** (1923). — NEUBAUER, H.: Mitteilungen über die Keimpflanzenmethode. Z. Pflanzenernähr., Düng., Bodenkde. B **8** (1929). — Die Keimpflanzenmethode ... nach Neubauer und Schneider. In: HONCAMP: Handbuch der Pflanzenernährung und Düngerlehre. Berlin: Springer. 1931. — Die Keimpflanzenmethode. Berlin: Verlags-Ges. f. Ackerbau. 1939. — Die Ausdehnung der Keimpflanzenmethode auf die Stickstoffbestimmung im Boden und in Komposterden. Z. Pflanzenernähr., Düng., Bodenkde. **37** (1946).

PAAUW, F. VAN DER: Die Auswertung der Bodenuntersuchung auf Phosphorsäure und Kali in den Niederlanden. Landw. Forsch., 12. Sonderheft, 86–94 (1959). — Colibration of soil tests methods for the determination of phosphate and potash status. Plant a. Soil **8**, 2, 105–125 (1956). — PFEIFFER, TH.: Der Vegetationsversuch als Hilfsmittel zur Lösung von Fragen auf dem Gebiete der Pflanzenernährung unter besonderer Berücksichtigung der Sand- und Bodenkulturen in Gefäßen. Berlin: Parey. 1917/18. — POPP, FR.: Untersuchungen über den Einfluß variierter Nährstoffgaben- und -verhältnisse auf Ertragsbildung, Nährstoffaufnahme- und -ausnutzung. Diss. Gießen. 1958.

REED, J. F.: Chemical basis for soiltesting. J. Agric. Food Chem. **4**, 116 (1956). — ROEMER, TH., B. DIRKS und E. GÜNTHER: Neubauer-Analysen im Vergleich zu Felddüngungsversuchen. Landw. Ztg. **44** (1924). — RUSS, E.: Über die Bestimmung von pflanzenaufnehmbarem Kupfer und Mangan in Böden unter besonderer Berücksichtigung der Keimpflanzenmethode. Diss. Gießen 1958.

SCHACHTSCHABEL, P.: Diskussionsbeitrag zu P. LEDERLE. Z. Pflanzenernähr., Düng., Bodenkde. **40** (1948). — SCHÄFER, P.: Die Problematik der Bodenverdünnung durch Sand bei der pflanzenphysiologischen Gefäßmethode zur Bestimmung des im Boden vorhandenen pflanzenaufnehmbaren Nährstoffvorrats. Plant a. Soil **9** (1957). — SCHEFFER, F.: Bestimmung des Nährstoffvorrates im Feldversuch und im Gefäßversuch. Kühn-Arch. **38** (1933). — SCHMITT, L.: Richtlinien für die Anstellung von Düngungsversuchen. Frankfurt/M.: DLG-Verlag. 1959. — SCHNEIDER, B.: Methoden der Zeitreihentheorie in der Biometrie. Biom. Z. **1**, H. 3 (1959). — SKORIC, A.: A critical contribution to the problem of available *P* and *K* determination in the soil. Soil and Plant, Jugoslav. Soil Sci. Soc. **2**, 3 (1953).

THUN, R.: Untersuchungen über die Neubauer-Methode sowie bezüglich des leicht aufnehmbaren Stickstoffs. Z. Pflanzenernähr., Düng., Bodenkde. A **16** (1930). — Die wichtigsten Methoden zur Nährstoffuntersuchung des Bodens und ihr Wert für die praktische Landwirtschaft. Bidermanns Zbl. Agrikult. chem. Abt. A, N.F. **6** (1936). — TRÉNEL, M., und A. FRANCKE: Zur Bestimmung des Nährstoffbedarfs durch die Anwelkmethode nach Arland. Z. Pflanzenernähr., Düng., Bodenkde. **71** (1955).

VALKANOW, V.: Untersuchungen über das Vorhandensein von aufnehmbarer Phosphorsäure auf Wiesenland in verschiedenen Bodenschichten auf Grund der Erfahrungen mit der Keimpflanzenmethode. Z. Pflanzenernähr., Düng., Bodenkde. **31** (1943). — VANDECAVEYE, S. C.: Effect of soil type and fertilizer treatments of the chemical composition of certain forage and small grains. Soil Sci. Amer. Proc. **5** (1940). — VERMEULEN, F. A. B.: Fehlerquellen bei der Bodenuntersuchung. Vortrag auf der Jahreshauptversammlung des Verbandes Deutscher landw. Untersuchungs- und Forschungsanstalten, Karlsruhe, Sept. 1959.

WAGNER, P.: Beiträge zur Begründung und Ausbildung einer exakten Methode der Düngungsversuche. J. Landw. **28** (1880/81). — Anwendung künstlicher Düngemittel. Berlin: Parey. 1920. — WIESSMANN, H.: Über den Einfluß des Lichtes auf die Nährstoffaufnahme der Pflanzen im Jugendstadium. Z. Pflanzenernähr., Düng., Bodenkde. B **4** (1925). — Untersuchungen über die Feststellung des Nährstoffbedürfnisses des Bodens. Z. Pflanzenernähr., Düng., Bodenkde. B **6** (1927). — Die Bestimmung des Nährstoffgehaltes der Böden durch den Gefäßversuch. Z. Pflanzenernähr., Düng., Bodenkde. A **10** (1927/28). — Vergleich zwischen Gefäßmethode nach MITSCHERLICH und WIESSMANN. Z. Pflanzenernähr., Düng., Bodenkde. A **13** (1929). — WÜSTENHAGEN, CHR.: Zur Ertragsbildung und Nährstoffaufnahme in Abhängigkeit von der Düngung und dem Nährstoffverhältnis. Diss. Gießen 1959.

c) Mikrobiologische Verfahren[1]

Von

H. Bortels

1. Einleitung

Schon bald, nachdem man erkannt hatte, daß auch Mikroorganismen die für höhere Pflanzen lebensnotwendigen Elemente zu ihrer Entwicklung benötigen, kam man bei der zunehmenden Intensivierung der Landwirtschaft auf den Gedanken, die den Kulturpflanzen in Böden verfügbaren Mengen an Nährstoffen auf mikrobiologischem Wege zu bestimmen, zumal die chemischen Verfahren damals noch viele Mängel aufwiesen und biologische Methoden mit höheren Pflanzen als Testobjekten wesentlich mehr Zeit beanspruchten. Die ersten Anregungen zu mikrobiologischer Bodenanalyse von LIPMAN (1906), CHRISTENSEN (1907), WAKSMAN und STARKEY (1924), WINOGRADSKY (1925, 1927) sowie BENECKE und SÖDING (1927/28) haben heute, abgesehen vielleicht von dem *Azotobacter*-Plattentest WINOGRADSKYS, nur noch historische Bedeutung. Überhaupt müssen alle biologischen, auch die zunächst wegen ihrer verhältnismäßig schnellen Durchführbarkeit sehr begrüßten mikrobiologischen Verfahren immer mehr an Wert verlieren, je treffsicherer die rein chemischen Schnellmethoden ausgebaut werden; und heute bietet, wie FINCK und SCHLICHTING (1955) ausführen, die quantitative chemische Bestimmung eines Nährelements in einer Bodenprobe — damals noch mit Ausnahme des pflanzenaufnehmbaren Molybdäns — kein wesentliches Problem mehr. Dennoch sollen hier auch die mikrobiologischen Verfahren zur Bestimmung des pflanzenaufnehmbaren Phosphors, Kaliums und Magnesiums wenigstens kurz erwähnt werden. Sie finden zuweilen noch Anwendung,

[1] Vgl. auch BERGMANN (1958).

weil es sich immerhin um biologische Schnellmethoden handelt. Die mikrobiologische Bestimmung der Mikroelemente soll dagegen ausführlich besprochen werden. Wenn es auch schon möglich ist, mit Hilfe moderner chemisch-physikalischer Analysenmethoden solche Elemente im Boden schnell und sicher zu ermitteln, so war doch z. B. die mikrobiologische Bestimmung des Molybdäns bis vor wenigen Jahren noch der chemischen überlegen.

Auf die vielen Versuche, mit Hilfe von Mikroorganismen, nämlich Bakterien, Pilzen und Algen (Benecke, Söding 1927/28, Morgan 1933, Tschan 1956), die Fruchtbarkeit eines Bodens qualitativ zu ermitteln, soll hier nicht weiter eingegangen werden. Als brauchbare Methoden haben sich auch jene quantitativen nicht durchsetzen können, die Bakterien wie *Streptococcus faecalis* (Barton, Wright 1945) oder *Bacterium globiforme* (Conn 1928) als Testorganismen verwendeten (Conn, Darrow 1938). Auch die Mangan-Bestimmungsverfahren mit Hilfe von *Lactobacillus arabinosus* (Bentley, Snell, Phillips 1947) oder *Bacillus subtilis* (Charney, Fisher 1951) sind für Bodenanalysen zu kompliziert bzw. zu unempfindlich. Dagegen scheinen die wegen ihrer stickstoffbindenden Eigenschaft in stickstofffreier Nährlösung besonders konkurrenzfähigen *Azotobacter*-Arten wenigstens für Phosphorbestimmungen geeigneter zu sein. Daß Christensen (1907) den Kalkzustand eines Bodens mit dem sehr säureempfindlichen und calciumbedürftigen *Azotobacter chroococcum* bestimmt hat, was heute durch die pH-Messung längst überholt ist, sei nur als historische Tatsache am Rande vermerkt. Darüberhinaus aber haben Niklas und Mitarbeiter (Niklas, Scharrer, Strobel 1926, Niklas 1931) schon früh auf die Möglichkeit hingewiesen, den Phosphorgehalt eines Bodens mit *Azotobacter* zu ermitteln, und ein entsprechendes Verfahren unter Verwendung von *Azotobacter chroococcum*-Rohkulturen in stickstofffreier Nährlösung ausgearbeitet. Jedoch ergab sich dann bald, daß zu viele bekannte und damals noch unbekannte Faktoren störend wirkten (Dirks, Scheffer 1928, Stöckli 1929, Pittman, Burnham 1932, Scheffer, Kiessling 1934). Wenn man diese heute auch weitgehend ausschließen könnte, z. B. durch Verwendung von Reinkulturen unter Berücksichtigung aller lebensnotwendigen Elemente einschließlich Eisen, Kupfer, Molybdän, Calcium und Bor, so ist die Methode doch allzusehr subjektiven Fehlern unterworfen und kann außerdem, weil keine Säure gebildet wird, allenfalls die leicht löslichen Anteile der Nährstoffe ermitteln, aber nicht mit genügender Genauigkeit. Daran dürften neuere Verfahrensweisen wie diejenige nach Bravo (1952) nichts ändern.

Etwas günstiger liegen die Verhältnisse schon, wenn nicht wie bei der Methode von Niklas die Dicke einer *Azotobacter*-Haut auf einer Nährlösung geschätzt und so als Maß für die Menge des aufnehmbaren Anteils eines zu bestimmenden Nährelements benutzt wird, sondern nach dem Verfahren von Winogradsky (1925, 1927) Zahl und Größe der *Azotobacter*-Kolonien unmittelbar auf einer untersuchten Erdprobe festgestellt werden. Ein Hauptvorteil dieser Methode liegt ferner im Zeitgewinn. Sie dauert nur 48 bis 72 Stunden, wurde zunächst von Sackett und Stewart (1931) und dann auch noch von anderen (Krjutschkowa 1930, Keller 1934) angewandt, verbessert und auf ihre Brauchbarkeit vor allem zur Bestimmung von Phosphor, Kalium und Calcium geprüft. Dabei hat sich ergeben, daß sie ebenfalls nur imstande ist, ungefähre Richtwerte und keine genauen Angaben über die einem Boden fehlenden Nährstoffmengen zu machen (Fuller 1935). Allerdings könnten auch hier durch Berücksichtigung neuerer Erkenntnisse wie z. B. der Molybdänbedürftigkeit von *Azotobacter* (Bortels 1930) noch Verbesserungen angebracht werden. *Azotobacter* benötigt verhältnismäßig viel Phosphor und Molybdän, bildet andererseits keine aufschließend wirkenden Säuren bzw. Alkalien und ist selbst säureempfindlich.

2. Methodik

Eine orientierende Untersuchung solcher Art auf Phosphor und Kalium, gegebenenfalls auch Molybdän (Bortels 1937), nicht auf Calcium, wäre dann folgendermaßen durchzuführen (Allen 1957): 50 g pulverisierten und auf 2 mm gesiebten, lufttrockenen Bodens einer repräsentativen Mischung[1] werden mit 2,5 g löslicher Stärke und 0,5 g $CaCO_3$ gut vermengt, mit 1 ml einer wässerigen Lösung von je nach Art der Bestimmung 15% K_2HPO_4 oder 30% Na_2HPO_4 oder 15% K_2SO_4 und in jedem Falle mit 1 ml von 2% $MgSO_4 \cdot 7H_2O$ sowie 0,02% $Na_2MoO_4 \cdot 2H_2O$ versetzt, mit 1 ml einer Aufschlämmung von *Azotobacter chroococcum* einer 72 Stunden alten Reagenzglas-Agarkultur in 10 ml Wasser beimpft und mit Leitungswasser bis 60% der Kapazität des Bodens angefeuchtet. Ist dieser zu lehmig, kann er mit Quarzsand aufgelockert und im umgekehrten Falle bei zu sandiger Beschaffenheit mit Kaolin knetbarer gemacht werden. Die Mischung wird dann in eine Petrischale eingedrückt, mit angefeuchtetem Spatel oberflächlich geglättet und offen in einer größeren Deckelschale als Feuchtkammer bei 30° C bebrütet. Diese soll an der Innenseite ihres Deckels angefeuchtetes Filtrierpapier enthalten, damit kein Wasser auf die eingestellte Erdplatte abtropfen kann. Nach 2 bis 3 Tagen können Größe und Zahl der auf der glatten Erdoberfläche gewachsenen Kolonien bestimmt und so durch Vergleich Anhaltspunkte für den Gehalt des Bodens an Kalium und Phosphor gewonnen werden. Der Vorteil der Methode liegt in der großen Zeitersparnis.

In etwas anderer Weise gingen Halversen und Hoge (1942) vor, die 50 g trockenen, gesiebten Bodens mit 1 g feinen Quarzsandes bzw. 1 g eines Gemenges von 100 g Quarzsand mit 15 g K_2SO_4 bzw. 30 g $Na_2HPO_4 \cdot 12\,H_2O$ bzw. 15 g K_2HPO_4 mischten und diese Mischungen dann mit jeweils 40 ml heißem 2%igem Wasseragar versetzten. Nach Eingießen dieser Gemenge in Petrischalen und Erstarren des Agars wird die völlig glatte Oberfläche mit je 2 Tropfen einer mehrfach gewaschenen und zentrifugierten *Azotobacter*-Aufschlämmung intensiv beimpft. Wenn den Proben außerdem die oben angegebenen Mengen Magnesium und Molybdän zugesetzt werden, kann diese Methode, bei der ebenfalls die *Azotobacter*-Oberflächenentwicklung nur geschätzt wird, gegenüber derjenigen von Sackett und Stewart (1931) einige weitere Vorteile bieten. Eine modernere, und zwar quantitative Abwandlung der Nährlösungsmethode nach Niklas (1931) hat Wang (1951) vorgeschlagen. Sie beruht auf der kolorimetrischen Messung eines von dem Bakterium in quantitativer Abhängigkeit von einer der notwendigen Aschenelemente aus Benzoesäure gebildeten schwarzen Farbstoffes. Die Grundnährlösung, bestehend aus H_2O 1000,—, $CaCl_2$ 0,375, KH_2PO_4 oder NaH_2PO_4 4,0, $MgSO_4 \cdot 7\,H_2O$ 1,25, NaCl 2,5, $FeSO_4 \cdot 7\,H_2O$ 0,075, $MnSO_4 \cdot 4\,H_2O$ 0,04, wird auf 1/20 verdünnt, mit 2 g Natriumbenzoat auf 1000 ml versetzt, mit NaOH oder KOH auf pH 7,4 eingestellt, 20 Minuten bei 120° C sterilisiert, mit einem geeigneten Stamm von *Azotobacter chroococcum* beimpft und bei 36° C bebrütet. Das Impfmaterial wird einer Kultur entnommen, die so lange ohne das zu bestimmende Element in entsprechenden Nährlösungen vorgezüchtet wurde, bis diese sich nicht mehr schwarz färbten. Die Standardreihen, mit denen die Bodenkulturen verglichen werden, enthalten bei Phosphor-Bestimmungen Phosphat und bei Bestimmung eines anderen Elementes dieses als Sulfat in Konzentrationen von 0, 0,01, 0,05, 0,1, 0,25, 0,5, 0,75, 1,5, 10 mg je 100 ml Nährlösung, die vorher von dem zu bestimmenden Element nicht besonders gereinigt zu werden braucht. Damit dessen Menge in der untersuchten Boden-

---

[1] Bodenproben für mikrobiologische Nährstoffbestimmungen werden ebenso entnommen und aufbewahrt wie für chemische Bestimmungen.

probe im linearen Teil der logarithmischen Ertragskurve der Standardreihe liegt, wird ein Vorversuch mit 0,01, 0,1 und 1 g lufttrockener auf 1 mm gesiebter Erde je 100 ml Nährlösung vorausgeschickt. Im allgemeinen genügen zur Bestimmung von Phosphor, Kalium, Magnesium, Eisen und Mangan 1 g und von Calcium 0,01 g Boden je 100 ml Nährlösung. Zur Auswertung werden die Nährlösungen nach Zugabe von 0,5 ml 1/10 n NaOH zu 100 ml zentrifugiert und photometriert oder kolorimetriert. Das Verfahren wirkt wegen seiner Einfachheit sehr bestechend, dauert aber leider sehr lange wegen der angeblich notwendigen 14 Tage Bebrütungszeit. Vielleicht ließe sich diese bei Berücksichtigung der für *Azotobacter chroococcum* und seine normale Schwarzfärbung notwendigen Elemente Molybdän, Kupfer (Burema, Wieringa 1942) und Bor (Gerretsen, de Hoop 1954) wesentlich abkürzen.

Weil Pilze im allgemeinen mehr organische Substanz bilden, eignen sie sich als Testorganismen zur Bestimmung der Düngebedürftigkeit eines Bodens noch besser als Bakterien. Anregungen hierzu gingen schon 1909 von Buttkewič und Kosceleckii aus. Pantanelli hat schon 1926 eine Methode zur Phosphorsäurebestimmung mit Hilfe von *Aspergillus*-Arten bekanntgegeben. Lumia (1926) verwendete in einem ausgeprägten Schnellverfahren Hefe zur Bestimmung von Kalium und Phosphor und Seidel (1931) außer Hefe auch *Rhizopus*.

Diese Organismen wurden in Nährlösungen gezüchtet, während Mehlich, Truog und Fred (1934) und Mehlich (1938) den Pilz *Cunninghamella* in Anlehnung an die bereits beschriebene Methode Winogradskys auf Erdplatten züchteten und den Durchmesser der gewachsenen Kolonie als Maß für die im Boden verfügbare Nährstoffmenge, in erster Linie Stickstoff, benutzten. In vier Parallelansätzen werden z. B. zur Phosphor-Bestimmung je 50 g lufttrockener, auf 1 mm gesiebter Boden mit folgender Nährlösung bis zur gut knetbaren Konsistenz angefeuchtet: H_2O 1000, Glukose 25,—, NH_4NO_3 2,5, Pepton 0,5 und 5 ml folgender Vorratslösung: H_2O 1000,—, KCl 20,—, $MgSO_4 \cdot 7\,H_2O$ 10,—, $FeSO_4 \cdot 7\,H_2O$ 0,2, $ZnSO_4 \cdot 7\,H_2O$ 0,2. Der pH-Wert wird in üblicher Weise auf 6,8 bis 7,2 eingestellt und der Boden in Schalen von 15 cm Durchmesser und 2 cm Höhe eingedrückt. Wie bei der *Azotobacter*-Plattenmethode wird er oberflächlich geglättet und in der Mitte punktförmig mit in Wasser aufgeschlämmten Sporen einer 3 bis 4 Wochen alten Kultur von *Cunninghamella blakesleeana* oder *C. elegans* auf Malzagar beimpft. Nach 4 bis 5 Tagen (wenn der zu untersuchende Boden reich an organischer Substanz ist, besser nach 7 Tagen) bei 28 bis 30° C wird der Durchmesser der gewachsenen Kolonien bestimmt. Bei weniger als 10 mm ist der Boden sehr stark phosphorbedürftig, bei 11 bis 15 mm mäßig, bei 16 bis 21 mm schwach und bei mehr als 22 mm nicht bedürftig. Sollen andere Elemente bestimmt werden, ist die Nährlösung entsprechend abzuändern und ein anderer Bewertungsschlüssel zu verwenden. Die Brauchbarkeit dieses Verfahrens als einer sichere Richtwerte liefernden Methode wird von Mooers (1938) und Mc Cool (1947) bestätigt und von Allen noch 1957 in seinem Methodenbuch angeführt. Falangue, Martinelli, Neder und De Camargo (1961) haben sie durch Zugabe des Antibioticums Nystatin zum Boden weiter verbessert.

Es hat auch nicht an Versuchen gefehlt, an Hand der von der Gesamtheit der Mikroorganismen in einem Boden gebildeten Kohlensäuremengen Aussagen über dessen Düngebedürftigkeit zu machen (Holben 1932, Andrews 1937, Desai 1946, Schwegler, Poschenrieder und Bukatsch 1959). Jedoch sind alle diese eben erwähnten Verfahren noch zu ungenau oder zu langwierig und entsprechen jedenfalls nicht den Anforderungen, die heute an eine zuverlässige Schnellmethode gestellt werden müssen. Wenngleich auch solche Methoden wie diejenige von Lumia (1926) zur Bestimmung der Phosphorsäure-Bedürftigkeit mit Hilfe von Hefe nach Einbeziehung neuzeitlicher Erkenntnisse und vielleicht auch diejenige von Schwegler et al. (1959) als orientierende Schnellmethoden unter Umständen gute Dienste leisten würden, so haben sie sich doch gegenüber den Verfahren mit *Aspergillus niger* als Testorganismus nicht durchsetzen können. Diese werden ihrerseits heute, wenigstens soweit es die Bestimmung von Makroelementen betrifft, von photometrischen (kolorimetrischen), polarographischen und spektrographischen Schnellmethoden mehr oder weniger verdrängt. Immerhin hat die *Aspergillus*-Methode zur Bestimmung der Kalium- und Phosphor-, weniger der Magnesium-Düngebedürftigkeit eines Bodens, für deren Ausbau sich besonders Niklas mit seinen Mitarbeitern (Niklas 1931, Behrens 1935, Niklas und Poschenrieder 1932, 1936, Niklas und Toursel 1940, 1941a, b, Niklas, Poschenrieder, Trischler 1930 a, b, Niklas, Vilsmeier, Poschenrieder 1932, Niklas, Vilsmeier, Kohl 1933) verdient gemacht hat, längere Zeit eine bedeutende Rolle gespielt und wird auch heute noch vereinzelt angewandt. Sie soll deshalb unter Berücksichtigung neuerer Erkenntnisse kurz geschildert werden, soweit es sich um die Kalium- und Phosphor-Bestimmung handelt:

In Erlenmeyerkolben von 500 ml Inhalt werden 5 g lufttrockenen, auf 1 mm gesiebten Bodens eingefüllt, gleichmäßig verteilt und mit je 50 ml folgender Nährlösung überschichtet: H_2O 1000,—, Rohrzucker 100,—, Zitronensäure 10,— $(NH_4)_2SO_4$ 6,—, Pepton 1,—, $MgSO_4 \cdot 7\,H_2O$ 0,614, $CuSO_4 \cdot 5\,H_2O$ 0,0005, $ZnSO_4 \cdot 7\,H_2O$ 0,005, $FeSO_4 \cdot 7\,H_2O$ 0,005, $MnSO_4 \cdot 4\,H_2O$ 0,002, $Na_2MoO_4 \cdot 2\,H_2O$ 0,001. Dazu kommen bei Kalium-Bestimmungen 1,215 g $(NH_4)H_2PO_4$ und bei Phosphor-Bestimmungen 0,37 g K_2SO_4. Nach Abdecken der Kolben mit Bechergläsern und 10 Minuten Sterilisation bei 120° C werden die Lösungen mit je 1 Tropfen einer dichten Sporenaufschlämmung in sterilem Wasser mit 0,3% Polyviol (Stapp, Wetter 1953) beimpft und 3 Tage bei 35° C bebrütet. Die Sporenaufschlämmung mit Polyviol hat eine Beschleunigung der Pilzentwicklung in geschlossener Decke zur Folge und ist vorteilhafter als die Trockenimpfung mit einem Sporen-Talkum-Gemisch. Mit jedem Boden werden 4 Parallelkulturen angesetzt. Anstelle der Erlenmeyerkolben können aus praktischen Erwägungen auch Bechergläser, mit Schalen abgedeckt, oder Deckelschalen Verwendung finden. Sind die Bodenproben humusreich, dann empfiehlt es sich, sie in aschefreie Filter verpackt in die Kulturgefäße einzulegen, damit nicht Humusteilchen das Mycelgewicht erhöhen können. Nach quantitativem Abernten der gewachsenen Pilzmycelien werden diese mehrfach mit destilliertem Wasser gewaschen, über Nacht bei 100° C getrocknet und nach Abkühlen gewogen. Aus den Gewichten wird im Vergleich mit den Werten einer Testkurve die aufnehmbare Menge Kalium oder Phosphor des Bodens festgestellt. Dabei handelt es sich überwiegend um die in verdünnten Säuren leicht löslichen Mengen. Denn die Nährlösung enthält Zitronensäure, und der Pilz bildet selbst organische Säuren und aus dem Ammonsulfat der Nährlösung Schwefelsäure. Die Testkurve ergibt sich aus den Pilzernten auf obiger Nährlösung ohne Zusatz einer Bodenprobe oder eines Kalium- bzw. Phosphorsäure-Salzes und etwa 8 weiteren Lösungen mit steigenden Kalium- bzw. Phosphor-Gaben, die etwa 2 Dosen mit Maximalwirkung einschließen sollen. Diese Kulturen werden im übrigen ebenso behandelt wie die Bodenkulturen.

Die in der Grundnährlösung enthaltene Zitronensäure soll Bakterienentwicklung und Autolyse des Pilzes verhindern, vor allem, wenn die Sterilisation, die unerwünschte Veränderungen am Boden auslösen könnte, fortfällt. Eine noch bessere Unterdrückung von Fremdkeimen wird in diesem Falle mit etwa 2% Tannin erreicht (Stöckli 1940).

Wenn die zu untersuchenden Böden kalkreich sind, d. h. mehr als 1% $CaCO_3$

enthalten, werden sie vor Zugabe der Nährlösung mit einer dem jeweiligen Kalkgehalt entsprechenden größeren Zitronensäuremenge 10 bis 25 Stunden stehen gelassen (NIKLAS, POSCHENRIEDER, TRISCHLER 1930 b, VILSMEIER 1933 b). NIKLAS, POSCHENRIEDER und TRISCHLER (1930 a) geben hierfür eine Neutralisationstabelle. Aber das Calcium bleibt dennoch ein Störungsfaktor, weil es die Pilzernten auf jeden Fall erhöht, auch als Citrat (NIKLAS, VILSMEIER, POSCHENRIEDER 1932), und die Feststellung geringerer Unterschiede im Magnesiumgehalt angeblich sogar unmöglich macht (NIKLAS, POSCHENRIEDER 1936). NIKLAS und POSCHENRIEDER (1932) haben deshalb vorgeschlagen, bei Untersuchung von Kalkböden von den Pilzgewichten mit zunehmender Größe gleichfalls zunehmende Mengen bis zu 1,25 g für 4 Parallelernten in Abzug zu bringen. Obwohl NIKLAS und TOURSEL (1940) später die Bewertungsgrenzen für den Kalium- und Phosphorgehalt der Böden auf Grund der *Aspergillus*-Ernten neu festgesetzt haben, blieb die Methode wegen weiterer, teils schwer kontrollierbarer Faktoren doch noch mit vielen Mängeln behaftet. So beeinflussen nicht nur das Verhalten des verwendeten Pilzstammes (MEHLICH, TRUOG, FRED 1933) sowie Calcium und Humus (KIESSLING 1931 a) die Höhe der Pilzernten, sondern z. B. auch verschiedene andere Kationen und Anionen der Nährlösung und des Bodens (VAGELER, ALTEN 1931). VARALLYAY (1941) hat deshalb den Bodenbewertungen neben den absoluten Pilzgewichten vor allem Relativwerte, gewonnen aus dem Verhältnis der Ernten mit gedüngtem zu denjenigen mit ungedüngtem Boden, zugrunde gelegt.

SEKERA (GERICKE, PFARRE 1942) ist noch einen Schritt weitergegangen und hat Phosphor und Kalium mit Hilfe von *Aspergillus niger* auf dem Wege eines Mikrodüngungsversuchs bestimmt, wobei die für höhere Kulturpflanzen benötigte Kalium- und Phosphor-Düngemenge nach einem empirisch gewonnenen Auswertungsschlüssel mit den Noten 0 = nicht bis 4 = sehr stark bedürftig aus dem prozentualen Verhältnis der Pilzernten ohne Kalium- bzw. Phosphordüngung zu denjenigen mit Volldüngung (P + K) ermittelt wird. Nach STÖCKLI (1940) sind die Ergebnisse der Phosphorbestimmung, die für phosphorreiche Böden zu hoch liegen, besser, wenn die Grenzprozente (für P 60% und für K 85%), über denen eine Düngung nicht erforderlich ist, von 60 auf 43 erniedrigt werden. Die Methode wird folgendermaßen durchgeführt: in Glasschalen von 7 cm Durchmesser und 4 cm Höhe werden 5 g lufttrockenen, auf 1 mm gesiebten Bodens mit der weiter oben bereits erwähnten Grundnährlösung, die zusätzlich 2% Tannin enthält und nicht sterilisiert wird, überschichtet, und zwar bei Kalium-Bestimmungen mit 25 ml und bei Phosphor-Bestimmungen mit 15 ml. Dazu kommen die Differenzdüngungen. Nach Beimpfung und 2 bis 3 Tagen Bebrütung bei 35° C werden die Pilzernten nach Augenschein oder nach Trockengewicht bestimmt. Wenn immer mit demselben *Aspergillus*-Stamm gearbeitet wird, vermag auch dieses Verfahren in kurzer Zeit gute Richtwerte zu liefern.

Die Urteile über die *Aspergillus*-Methoden zur Kalium- und Phosphor-Bestimmung nach NIKLAS, MEHLICH und Mitarbeitern sowie nach SEKERA sind sehr unterschiedlich. RIPPEL (1931, 1939) und VANDEKAVEYE (1948) haben Vor- und Nachteile verschiedener mikrobiologischer Verfahren kritisch beleuchtet, während KIESSLING und SCHMIDT (1932) auf die Fehler hingewiesen haben, die organische Substanzen der Bodenprobe auslösen können. IHLE und STOLLENWERK (1932), VILSMEIER (1933a) sowie STOCK (1933) lehnen die NIKLAS-Methode als zu ungenau mehr oder weniger ab, und LEAF (1958) zeigt am Beispiel von Forstböden, welche Grenzen hier den mikrobiologischen Verfahren mit *Aspergillus* und *Azotobacter* gesetzt sind. Dagegen hat sich eine noch größere Zahl von Autoren positiv geäußert. NIKLAS und MILLER (1932), NIKLAS, POSCHENRIEDER und

TRISCHLER (1933) sowie NIKLAS und TOURSEL (1941b) betonen die gute Übereinstimmung der Ergebnisse ihrer Methode mit denen des Feldversuchs, der NEUBAUER- und zum Teil auch der MITSCHERLICH-Methode. Gute Übereinstimmung mit NEUBAUER-Zahlen fanden aber z. B. auch GOLLING (1932) und SMITH und DRYBURGH (1943). KIESSLING (1931b) hat ebenfalls festgestellt, daß *Aspergillus*, sofern es sich um humusarme Böden handelt, sehr gut auf alle Düngungsmaßnahmen anspricht, und GERRETSEN (1948) unterstreicht die Empfindlichkeit und den praktischen Wert dieser mikrobiologischen Schnellmethode. TOURSEL (1941) und NIESCHLAG (1943) haben sogar die Magnesium-Düngebedürftigkeit verschiedener Böden mit Hilfe von *Aspergillus niger* recht gut bestimmen können, obwohl das bei den in Frage kommenden geringeren Mengen mit reineren Nährlösungen (s. weiter unten!) noch viel genauer geschehen kann. Vielleicht würde es sich überhaupt empfehlen, auch bei den Kalium- und Phosphor-Bestimmungen, wenn man schon die *Aspergillus*-Methode anwendet, dann wenigstens mit sehr reinem Wasser und analysenreinen Salzen einschließlich aller notwendigen Mikroelemente und einer kristallinen Stickstoffverbindung statt Pepton zu arbeiten. Die Verfahren lassen sich auch vereinfachen, wenn nach Vorschlägen von GUTSCHY (1959) und NJEGOVAN (1960) nicht das Mycelgewicht, sondern der in aliquoten Nährlösungsteilen verbliebene Reststickstoff bestimmt wird. Vergleichsuntersuchungen über verschiedene mikrobiologische und chemische Verfahren einerseits und die biologischen nach NEUBAUER und MITSCHERLICH andererseits haben SANCHEZ-MARROQUIN und TAMOYO (1946), MICHNIEWICZ (1950) und WAHHAB und BOKHARI (1953) davon überzeugt, daß die *Aspergillus*-Methode nach NIKLAS und der Gefäßversuch nach MITSCHERLICH unter bestimmten Voraussetzungen die Düngebedürftigkeit der Böden am besten kennzeichnen. VARRALYAY (1934) hat sogar aus der prozentualen Erhöhung der Mycelgewichte nach einer Bodendüngung zutreffend auf die zu erwartende Erhöhung der Weizenernte im Freiland geschlossen. Selbst in neuerer Zeit wurde die *Aspergillus*-Methode noch verschiedentlich erfolgreich angewandt, z. B. von ROSSELET (1955) und von ALVAHYDO und DÖBEREINER (1955). ENO und REUSZER (1951, 1955) haben darauf aufmerksam gemacht, daß wesentliche Teile des Bodenkalis durch die Tätigkeit von Mikroorganismen den höheren Pflanzen verfügbar werden und daß schon deshalb die *Aspergillus*-Methode als die natürlichere vor den rein chemischen den Vorzug verdiene.

In einem zusammenfassenden Urteil über die erwähnten mikrobiologischen Verfahren zur Bestimmung der Kalium-, Phosphor- und Magnesium-Düngebedürftigkeit verschiedener Böden ist also festzustellen, daß für die Praxis ausreichende Richtwerte am schnellsten, billigsten und treffsichersten mit Hilfe der *Aspergillus*-Methode unter Berücksichtigung aller für diesen Pilz notwendigen Nährstoffe und zur Zeit bekannten sonstigen optimalen Bedingungen zu erreichen sind, sofern es sich nicht um ausgesprochen humus- oder kalkreiche Böden handelt. Die mikrobiologische Bestimmung der Stickstoff-Düngebedürftigkeit ist auf jeden Fall schwieriger wegen der im Boden vorhandenen organischen, zum Teil schwer angreifbaren heterozyklischen Stickstoffverbindungen, andererseits aber auch von untergeordneter Bedeutung, weil Kulturböden wohl stets auf eine Stickstoffdüngung ansprechen werden.

Die modernen chemischen Schnellverfahren haben jedoch die mikrobiologischen fast ganz verdrängt, weil sie viel weniger als diese durch äußere Faktoren beeinflußbar sind. Nur bei Untersuchung der Böden auf Mikroelementbedarf spielen mikrobiologische Methoden, und zwar fast ausschließlich die *Aspergillus*-Methode, noch immer eine gewisse Rolle. Vor allem Molybdän kann auf diese Weise mit der notwendigen Genauigkeit bestimmt werden. *Aspergillus niger* reagiert auf die meisten biologisch wichtigen Schwermetall-Spuren ebenso empfindlich, wie andere Mikroorganismen auf Vitamine und Aminosäuren ansprechen, auf Molybdän-Spuren zum Teil sogar empfindlicher als chemische Analysenverfahren.

Nachdem RAULIN schon 1869 auf die biologische Bedeutung gewisser Elemente in geringsten Spuren aufmerksam gemacht und STEINBERG (1919) sowie BORTELS (1927) die Lebensnotwendigkeit einiger Schwermetalle für *Aspergillus niger* festgestellt hatten, ergab sich bei der fortschreitenden Intensivierung der Landwirtschaft und der sich daraus ergebenden Verarmung der Kulturböden an lebensnotwendigen Mikroelementen auch bald der Wunsch nach der Bestimmung des Mikroelement-Düngebedürfnisses der Böden. Da die chemischen Analysen hierfür anfänglich nicht ausreichten, bot sich als Methode der Wahl das erfahrungs-

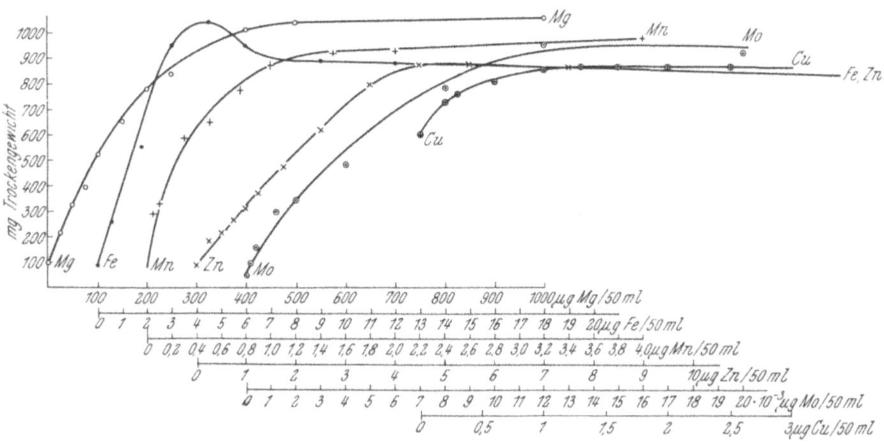

Abb. 187. Testkurven für Mg, Fe, Zn, Mo, Cu nach STAPP und WETTER (1953) und für Mn nach WETTER (1954)

gemäß schnelle, zuverlässige und billige mikrobiologische Verfahren mit *Aspergillus niger* als Testorganismus an. Inzwischen war von STEINBERG (1919) und BORTELS (1927) schon gezeigt worden, daß sich eine synthetische Nährlösung für *Aspergillus* durch Ausfällung und Adsorption der Schwermetallspuren sehr weitgehend davon befreien läßt. Das ist eine notwendige Voraussetzung für das Gelingen des Unentbehrlichkeitsnachweises ebenso wie der mikrobiologischen Mikroelementbestimmung in natürlichen Substraten. Bei der ähnlichen Untersuchung auf Makroelemente genügt es dagegen, die Nährlösungen mit destilliertem Wasser und allenfalls reinem Zucker und analysenreinen Salzen einschließlich einer kristallinen Stickstoffverbindung anzusetzen. Selbst im destillierten Wasser und in chemisch reinen Nährstoffen sind unter Umständen noch so viel Mikroelemente vorhanden, daß sie zur Entwicklung des Pilzes ausreichen.

Das Verfahren ist eine logische Weiterentwicklung der NIKLAS-Methode zur Bestimmung von Kalium und Phosphor und beruht auf der Feststellung, daß *Aspergillus niger* zur normalen Entwicklung außer Wasser, Kohlenstoff, Stickstoff, Kalium, Phosphor, Magnesium und Schwefel auch die Mikroelemente Eisen, Zink, Kupfer, Mangan und Molybdän benötigt. Magnesium, obwohl im quantitativen Sinne kein Mikroelement, soll hier aus praktischen Gründen mit abgehandelt werden. Alle diese Elemente wirken spezifisch, sind also nicht durch andere zu ersetzen, und stehen in einem quantitativen Verhältnis zum Erntegewicht des

Pilzmycels oder zur Konidienmenge und -farbe. Der Pilz vermag ein zu bestimmendes Mikroelement, wenn es ihm anders nicht geboten wird, aus einer der Nährlösung zugesetzten Bodenprobe zu entnehmen, soweit diese ausreichende Mengen in verfügbarer Form enthält.

Durch Vergleich zwischen den Mycelgewichten oder Konidien-Mengen und -farben in Nährlösungen mit einem zu untersuchenden Boden und denjenigen Testlösungen, die das zu bestimmende Mikroelement in nach der Reinigung zugesetzten steigenden Konzentrationen enthalten, wird die Menge des vom Boden an den Pilz abgegebenen Elements bestimmt. Beide, bodenhaltige und Testlösungen, sollen zuvor von dem entsprechenden Mikroelement soweit befreit sein, daß eine Kurve der Erntegewichte als Funktion der gestaffelten Mikroelement-Konzentrationen der Testreihe im Nullpunkt oder unmittelbar darüber beginnt, also der Pilz sich ohne Zugabe des Elements überhaupt nicht oder nur sehr kümmerlich entwickelt (Abb. 187). Solche Testkurven verlaufen mehr oder weniger steil aufwärts, um dann zur Horizontalen umzubiegen oder sehr langsam wieder zu fallen. Nur der ansteigende Ast der Kurven kommt für die Bestimmungen in Betracht.

Als erste vorbereitende Maßnahme zur Mikroelementbestimmung in Böden (Spicher 1954) mit Hilfe von *Aspergillus niger* sind alle Glasgeräte wie Kolben, Schalen, Pipetten usw. aus Pyrex-, Duran- oder Jenaer-Glas Nr. 20, die praktisch keine der in Frage kommenden Schwermetalle mehr an Wasser abgeben und daher nicht fabrikneu sein sollen, gründlich zu reinigen. Das geschieht mit Königswasser, heißer Salpetersäure oder Chromschwefelsäure. Die Gefäße bleiben dann nach kurzem Spülen mit gewöhnlichem destilliertem Wasser 48 Stunden damit stehen und werden schließlich mit 2 × aus Glas destilliertem Wasser nachgespült. Mit der Öffnung nach unten werden sie nun in gereinigte Bechergläser gestellt und die ablaufenden Wassertropfen mit 0,01% Dithizon (Diphenylthiokarbazon) als Komplexbildner in reinem mit frisch destilliertem Ammoniak alkalisiertem Tetrachlorkohlenstoff auf etwa noch vorhandene Metallspuren getestet. Die grüne Dithizonlösung darf sich nicht mehr rot färben.

Mit destilliertem Wasser werden nun folgende Nährlösungen angesetzt:

I. Grundnährlösungen

H_2O	1000.— g	
Glucose	50.— g	(gewöhnlicher Handelszucker ist ebenso brauchbar)
KNO_3	5.— g	(NH_4Cl) } bei Mn-Bestimmungen (Wetter
$Na_2HPO_4 \cdot 12H_2O$	2.5 g	(KH_2PO_4) } 1954, Bortels, Wetter 1954)
$MgSO_4 \cdot 7H_2O$	1.— g	(Na_2SO_4 bei Mg-Bestimmungen)
Nach Nicholas (1952) zusätzlich	0.5 g	$Ca(NO_3)_2$

II. Mikroelementlösungen

In 0.1 ml, die zu 50 ml Grundnährlösung gegeben werden, sind enthalten:

$FeSO_4 \cdot 7H_2O$	2.— mg	
$ZnSO_4 \cdot 7H_2O$	1.— mg	und in weiteren 5 Lösungen dasselbe ohne
$CuSO_4 \cdot 5H_2O$	0.2 mg	ein jeweils anderes der aufgeführten 5 Salze
$MnSO_4 \cdot 4H_2O$	0.1 mg	
$Na_2MoO_4 \cdot 2H_2O$	0.05 mg	

Nach Nicholas (1952) zusätzlich $Ga(No_3)_2 \cdot 8H_2O$ 0.2 mg, $Co(NO_3)_2 \cdot 6H_2O$ 0.1 mg, $Ni\,(NO_3)_2 \cdot 2H_2O$ 0.08 mg, $NaVO_3 \cdot 4H_2O$ 0.08 mg.

Die Wirkung der Mikroelemente ist u. a. auch von der Zusammensetzung der Grundnährlösung abhängig, besonders von der Art der Stickstoffquelle.

Molybdän kann z. B. als aktiver Bestandteil der Nitratreduktase nur zusammen mit Nitratstickstoff zur Wirkung kommen (STEINBERG 1937, 1939, NICHOLAS, NASON 1954).

Die Grundnährlösung wird vor dem Abfüllen in die Kulturkolben nach erprobten Verfahren von den immer vorhandenen, biologisch noch wirksamen Spuren der zu bestimmenden Elemente befreit. Dabei sollte zwar sehr gewissenhaft vorgegangen, aber nicht mit umständlichen Manipulationen eine geringfügige und unnötige Steigerung des Reinheitsgrades angestrebt werden. So haben z. B. MULDER (1939, 1948), SMIT und MULDER (1940) sowie NICHOLAS und FIELDING (1948) lediglich durch Verwendung besonders reinen Wassers und chemisch reiner Substanzen ohne Anwendung eines besonderen Reinigungsverfahrens den *Magnesium*-Gehalt ihrer Nährlösungen so weit erniedrigen können, daß *Aspergillus* hierin ohne Zusatz von Magnesium nicht gewachsen ist. Es ließ sich also hiermit bereits eine ideale Testkurve aufstellen und mit ihrer Hilfe auch der Magnesium-Gehalt von Böden bestimmen. Magnesium wird ja im Vergleich mit den echten Mikroelementen allgemein in verhältnismäßig großen Mengen benötigt. JENSEN und HENRIKSEN (1954) haben dann später ein übriges getan und die Magnesium-Spuren aus der Grundnährlösung einfach durch Ausschütteln mit aktiver Kohle[1] und Filtration entfernt. NICHOLAS und FIELDING (1948, 1949) haben sich zunächst mit Umkristallisieren der Grundnährstoffe begnügt, sich später (NICHOLAS 1949, 1950 a, b, NICHOLAS, FIELDING 1951) aber einer umständlichen Beseitigung der Magnesium-Verunreinigungen bedient, die dann auch STAPP und WETTER (1953) benutzt haben. Dabei werden der doppelt konzentrierten, phosphatfreien Grundnährlösung etwas Magnesiumsulfat und 2 ml einer 2%igen Oxin-(8-Hydroxychinolin-)-Lösung in 2n Essigsäure zugesetzt. Mit frisch destilliertem Ammoniak wird bis pH 8 alkalisiert und nach 24 Stunden (33° C) filtriert.

Selbstverständlich werden bei allen Reinigungen stets aschefreie Papierfilter oder gereinigte Glasfilter verwendet. Ein auf den Pilz toxisch wirkender Oxin-Überschuß wird durch mehrfaches Ausschütteln mit frisch destilliertem Chloroform (jeweils 50 ml) und dann mit frisch destilliertem Äther beseitigt. Schließlich wird zur Entfernung des Äthers kurz auf 70° C erhitzt oder mit 0,1% Kohle filtriert, mit frisch destillierter Salzsäure auf pH 4 bis 5 eingestellt, das umkristallisierte Phosphat zugegeben und mit 2 × aus Glas destilliertem Wasser auf 1000 ml aufgefüllt. Derartig komplizierte Verfahren dürften sich jedenfalls beim Magnesium nicht lohnen (s. auch Abb. 187).

Bei anderen Elementen sind sie dagegen in ähnlicher Form notwendig. Die einfachste Reinigungsmethode, die für *Eisen* ebenso brauchbar ist wie für Zink, Kupfer und Mangan, ist die Fällung und Adsorption mit Calciumcarbonat nach STEINBERG (1919, 1935). Die Grundnährlösung wird mit 15 g $CaCO_3$ 20 Minuten bei 120° C autoklaviert, über Nacht stehen gelassen und dann filtriert. Zwar läßt sich die Fällungs- und Adsorptionszeit mit basischem Magnesiumcarbonat beschleunigen (20 Minuten 100° C), aber dann wird zuviel Phosphat mitgefällt. STEINBERG (1950) hält sein Verfahren wegen der großen Einfachheit auch später noch für allen anderen überlegen, sofern es sich nicht um die Beseitigung von Molybdän handelt. NICHOLAS und FIELDING (1948) betonen aber, daß die nach diesem Verfahren gewonnenen Testkurven nicht gut reproduzierbar seien. BORTELS (1927) setzte der Grundnährlösung einen Tropfen Ammoniumsulfid zu und filtrierte dann das gefällte und an 5% aktive Kohle adsorbierte Eisensulfid ab. NICHOLAS (1949, 1950a, b, 1952) sowie NICHOLAS und FIELDING (1951)

[1] Z. B. Glycerinkohle „Merck".

haben die von Waring und Werkman erstmalig benutzte, beim Magnesium schon beschriebene, umständlichere Bindung an Oxin vorgezogen, und zwar durch 4maliges Ausschütteln der vorher mit etwas zusätzlichem Eisen- oder Aluminiumsalz versehenen Grundnährlösung mit je 30 ml einer 5%igen Oxinlösung in Chloroform bei pH 5 und Einstellen der gereinigten Nährlösung auf pH 4 bis 5. Bei einem experimentellen Vergleich zwischen 15 verschiedenen Reinigungsmethoden kommen Donald, Passey und Swaby (1952a, b) sowie Swaby und Passey (1953) zur Feststellung, daß Eisen ebenso wie Zink, Kupfer und Mangan am einfachsten und zugleich am gründlichsten auf folgende Weise aus der Nährlösung entfernt werden: In einem 1500 ml fassenden Kolben werden zu 1000 ml Nährlösung 5 g Aluminiumoxyd (analysenrein), in etwas Wasser gelöst, gegeben. Der pH-Wert wird dann mit reiner Lauge auf 7,5 eingestellt. Nach kräftigem Schütteln wird 20 Minuten auf 120° C erhitzt, dann umgeschwenkt, damit das Glas innen gut benetzt wird, über Nacht stehen gelassen und durch ein Glasfilter filtriert.

Zink, das von *Aspergillus niger* in verhältnismäßig großen Mengen benötigt wird, kann ebenso wie Eisen aus der Grundnährlösung entfernt werden. Außerdem hat sich nach Nicholas (1949, 1950a, b) und nach Nicholas und Fielding (1948, 1949, 1951) die Bindung des Zinks an den Komplexbildner Dithizon bewährt. Die 5-fach konzentrierte Gundnährlösung (200 ml) wird bei pH 6 mehrfach mit je 10 ml 0,02% Dithizon in reinem Tetrachlorkohlenstoff ausgeschüttelt, bis dieses sich nicht mehr rot färbt. Dithizon-Rückstände müssen durch etwa 2 weitere Behandlungen der Nährlösung mit Tetrachlorkohlenstoff beseitigt werden. Dann wird die Mikroelementlösung zugegeben und mit reinem Wasser auf 1000 ml aufgefüllt. Tucker und Mitarbeitern (1953) ist es gelungen, ihre Nährlösungen lediglich durch Verwendung besonders reiner Salze und durch Reinigung des Wassers und Zuckers mit Ionenaustauschern weitgehend von Zink und anderen Kationen zu befreien. Die Kationenaustauscher müssen zuvor von Mikroelementen, besonders Eisen und Kupfer, mit 5%iger Magnesiumsulfatlösung gereinigt werden. Das Magnesium wird dann wieder mit 2n HCl und danach mit reinem Wasser ausgewaschen.

Wie Eisen und Zink wird nach den bisher beschriebenen Verfahren auch *Kupfer* (mit Dithizon bei pH 4) aus den Lösungen entfernt (Mulder 1939a, b, Acock 1941). Ebenso wirksam wie die Dithizon-Methode ist die Fällung mit Schwefelwasserstoff (Nicholas 1949, 1950a, b, 1952, Nicholas, Fielding 1951, Stapp, Wetter 1953, Dole 1952). Zur 5-fach konzentrierten Grundnährlösung wird 1 ml 5%ige Kupfersulfatlösung (statt Kupfersulfat auch Wismutnitrat, Bleinitrat oder Cadmiumsulfat) und frisch destillierte Salzsäure bis pH 2 zugegeben. Dann wird durch ein Wattefilter reiner Schwefelwasserstoff bis zur vollständigen Fällung des Kupfers (etwa 30 Minuten) eingeleitet, über Nacht stehen gelassen und am besten durch Glasfilter filtriert. Die Schwefelwasserstoff-Behandlung muß unter Umständen wiederholt werden, bis kein Niederschlag mehr ausfällt. Dann wird der überschüssige Schwefelwasserstoff durch Kochen bis zur Geruchlosigkeit entfernt, notfalls mit etwas Kohle filtriert, und der pH-Wert mit reinstem Alkali wieder auf 4 bis 5 eingestellt.

Dieses Verfahren eignet sich am besten für die Nährlösungsreinigung von *Molybdän*-Spuren (Nicholas, Fielding 1950, Hewitt, Hallas 1951), die aber auch mit anderen Methoden, z. B. der Copräzipitation mit Oxin in Gegenwart von 0.01% Eisenionen, wie bei der Beseitigung von Eisenspuren möglich ist.

Zur Entfernung von *Mangan*-Spuren, die wohl auch am besten durch Fällung und Adsorption mit Calciumcarbonat nach Steinberg (1919, 1935) oder mit

Aluminiumoxyd nach DONALD, PASSEY und SWABY (1952a, b) sowie SWABY und PASSEY (1953) erreicht wird, hat NICHOLAS (1952) ebenfalls eine etwas umständlichere Methode empfohlen: Zur Grundnährlösung mit Kaliumnitrat und Kaliumbiphosphat werden 5 ml einer 5%igen Kupfersulfatlösung und 10 ml einer 0.2%igen Natriumdiäthyl-dithiocarbamat-Lösung gegeben. Ein Carbamat-Überschuß wird mit Kupfersulfat gefällt und abfiltriert und dann weiter verfahren wie bei der Kupfer-Reinigung mit Schwefelwasserstoff bei pH 3. Diese Umstände sind aber im Hinblick auf die anderen ebenso wirksamen und einfacheren Methoden kaum erforderlich.

Nach Reinigung erfolgt jedesmal die Korrektur des pH-Wertes. Je weiter Minimum und Maximum einer Vergleichskurve auseinander liegen und je gestreckter, d. h. allmählicher ansteigend sie verläuft, um so ausgedehnter ist der nutzbare Bestimmungsbereich der gestaffelten Mikroelement-Konzentrationen. Da der Kurvenverlauf unter anderem auch vom pH-Wert der Nährlösung abhängig ist, muß dieser auf den jeweils günstigsten Wert eingestellt werden, bei der Bestimmung des Mangans auf 3 (WETTER 1954, BORTELS, WETTER 1954), des Molybdäns auf 2 (NICHOLAS, FIELDING 1951) und der übrigen Elemente auf 4 bis 5.

Dann werden der Grundnährlösung die Mikroelemente wieder zugesetzt bis auf dasjenige Element, dessen Menge in einer zu untersuchenden Bodenprobe bestimmt werden soll und das deshalb den Testnährlösungen in abgestuften Konzentrationen gegeben wird. Auch die Vollnährlösung mit allen für *Aspergillus niger* lebensnotwendigen Elementen wird zunächst gereinigt und dann wieder mit den dadurch entfernten Mikroelementen in optimalen Konzentrationen ergänzt, damit nicht durch toxische Wirkungen von unter Umständen verbliebenen Resten des Reinigungsmittels Fehler entstehen. Die Mikroelementsalze sollen von vornherein einen möglichst hohen chemischen Reinheitsgrad aufweisen, werden aber meistens vor Gebrauch noch ein- bis zweimal umkristallisiert und dann in wässeriger Lösung in Flaschen aus inertem Material, z. B. ,,Lupolen"-Kunststoff vorrätig gehalten, und zwar nach einem Vorschlag von DONALD, PASSEY und SWABY (1952b) am zweckmäßigsten in 6 verschiedenen Lösungen. Von diesen enthält eine alle 5 in Betracht kommenden Elemente Eisen, Zink, Kupfer, Mangan und Molybdän, während den 5 weiteren ein jeweils anderes fehlt. So wird die Gefahr einer erneuten Verunreinigung mit dem beseitigten Element wegen der geringen zusätzlichen Wassermenge möglichst klein gehalten. Im Gegensatz zu den Behauptungen von STEINBERG (1938) und BERTRAND (1941) konnten ertragsteigernde Wirkungen von Calcium und Gallium bzw. Vanadium sowie von 29 weiteren Elementen einschließlich Kobalt, Nickel und Bor später nicht festgestellt werden. Dennoch befinden sich hierunter Elemente von landwirtschaftlicher Bedeutung wie Bor und Kobalt, für deren mikrobiologische Bestimmung in Böden noch keine brauchbaren Methoden vorliegen, obwohl GERRETSEN (1952, 1953) z. B. die Borbestimmung mit Hilfe von *Azotobacter chroococcum* für möglich hält. *Aspergillus niger* ist auch ziemlich unempfindlich gegen höhere Konzentrationen der Mikroelemente mit Ausnahme von Kupfer, weshalb Störungen durch Ionenantagonismen nicht auftreten. Die Reinheit der Mikroelement-Lösungen kann spektrographisch geprüft werden. Sie sind wegen unvermeidbarer Verunreinigungen von Zeit zu Zeit neu anzusetzen. Die fertigen Nährlösungen, die alle notwendigen Elemente mit Ausnahme des einen zu bestimmenden Elements enthalten, werden dann wie bei der Kalium- und Phosphor-Bestimmung abgefüllt und sterilisiert. Die Anzahl der benötigten Gefäße richtet sich nach der Länge der gleich noch zu erwähnenden Testreihen und der Zahl der zu untersuchenden Bodenproben.

Geeignete Testkurven ergeben sich aus folgenden Konzentrationsreihen:

Mg: 0, 25, 50, 100, 150, 200, 300, 400, 500, 1000 μg/50 ml
Fe : 0, 0.5, 1, 2, 3, 4, 5, 10 μg/50 ml
Zn : 0, 0.5, 1, 2, 3, 4, 5, 10 μg/50 ml
Cu : 0, 0.05, 0.1, 0.2, 0.4, 0.6, 0.8, 1, 1.5, 2.5, 5 μg/50 ml
Mn: 0, 0.03, 0.05, 0.1, 0.3, 0.5, 1, 1.5, 2, 5 μg/50 ml
Mo: 0, 0.00005, 0.0001, 0.0003, 0.0005, 0.001, 0.002, 0.003, 0.004, 0.005, 0.006, 0.008, 0.01, 0.02 μg/50 ml

Nach Festlegung der betreffenden Testkurve wird die Bestimmung des Elements in den gut gemischten, lufttrockenen, auf 1 mm gesiebten Bodenproben durchgeführt, die anstelle des betreffenden Elements in die Versuchsgefäße eingefüllt, auf der Grundfläche gleichmäßig verteilt und dann gegebenenfalls nach Alkoholbehandlung mit der vorher sterilisierten Mangelnährlösung überschichtet oder zusammen mit dieser 20 Minuten bei 120° C sterilisiert werden. Die Vergleichsbestimmung der dem Pilz verfügbaren Mikroelementmenge einer Bodenprobe ist nur dann mit der erforderlichen Genauigkeit möglich, wenn sie im Bereich des ansteigenden Astes der Testkurve liegt. Deshalb ist die Bodenprobe entsprechend zu dosieren. Während für die Bestimmung des Kalium- oder Phosphor-Düngebedürfnisses bis zu 10 g je 50 ml benötigt werden, sind es für die Mikroelementbestimmung nur 0,1 bis 1 g. Darin liegt natürlich eine Fehlerquelle, wenn diese kleine Probe für die untersuchte Freilandfläche nicht repräsentativ ist. Roschach (1961) hat auf die Bedeutung der Bodeneinwaage besonders hingewiesen.

Im einzelnen werden gegeben

für die Bestimmung von Mg 0,5 bis 1 g,
für die Bestimmung von Mo 0,5 bis 1 g,
für die Bestimmung von Zn 0,25 bis 0,5 g,
für die Bestimmung von Cu 0,1 bis 0,5 g,
für die Bestimmung von Mn 0,1 bis 0,15 g.

Wetter (1954) empfiehlt, für die Manganbestimmung 2 g Boden mit 10 ml quarzdestilliertem Wasser kräftig zu schütteln, 24 Stunden stehen zu lassen, zu filtrieren und 1 ml des Filtrats zu 50 ml Nährlösung zu geben.

Zuweilen ist es ratsam, zwei verschiedene Mengen zu verwenden, weil die organischen und anorganischen Bestandteile des Bodens nicht ohne Einfluß sind (Spicher 1954). Hemmende, unter Umständen auch fördernde Wirkungen auf *Aspergillus niger*, der allerdings hinsichtlich organischer Wirkstoffe sehr anspruchslos ist, können auch von den Fremdkeimen der Bodenprobe ausgehen. Diese wird deshalb häufig zusammen mit der Nährlösung sterilisiert, was jedoch die Löslichkeit des zu bestimmenden Elements beeinflussen kann. Acock (1941) hat deshalb 0,5 ml 70%igen Alkohols auf 0,5 g Erde eine Stunde in bedeckter Schale einwirken lassen, dann bei gelüftetem Deckel durch kurzes Erwärmen auf dem Wasserbad den Alkohol wieder entfernt und so eine genügende Teilsterilisation erreicht, ohne die Löslichkeitsverhältnisse wesentlich zu ändern. Daß diese ohnehin in einer glukosehaltigen Lösung und unter dem säuernden und reduzierenden Einfluß der Stoffwechselprodukte des Pilzes anders sein werden als im „gewachsenen“ Boden, ist anzunehmen. Deshalb bekommen die mit dieser mikrobiologischen Methode gewonnenen Analysenzahlen auch erst dann einen praktischen Wert, wenn sie an den von den Kulturpflanzen im Freiland gezeigten Symptomen geeicht werden. Nicht sofort verwendete Bodenproben müssen in luftdicht verschlossenen Behältern aus inertem Material („Lupolen“) aufbewahrt werden.

Wegen der oben erwähnten Fehlerquellen und weil sich *Aspergillus niger* wie alle Organismen in seinen Eigenschaften nicht völlig konstant verhält, ist es unbedingt erforderlich, bei jedem Versuchsansatz einige wenige Stufen der Testreihe aus dem ansteigenden Kurventeil und einer Konzentration aus dem Maximumbereich in 4facher Wiederholung mitlaufen zu lassen (ROSCHACH 1956). Ebenso wichtig ist es, den zu testenden Boden nicht nur als Quelle des fehlenden Mikroelements zur Mangelnährlösung zu geben, sondern auch teilweise zur Vollnährlösung, um etwaige fördernde oder hemmende Nebenwirkungen des Bodens erkennen zu können (DONALD, PASSEY, SWABY 1952 b).

Da sich verschiedene Stämme von *Aspergillus niger* sehr unterschiedlich verhalten, ist es unbedingt erforderlich, sich bei laufenden Untersuchungen auf einen für den Nachweis eines bestimmten Mikroelements besonders geeigneten Stamm festzulegen (ACOCK 1941). Das gilt besonders für die Kupferbestimmung, die besser an Hand der Sporenfarbe als des Mycelgewichts vorgenommen wird. Am meisten wurden bisher MULDERS Stamm M und der Stamm Cu 17 von STAPP und WETTER benutzt. Da aber jeder Stamm im Laufe der Zeit seine Eigenschaften

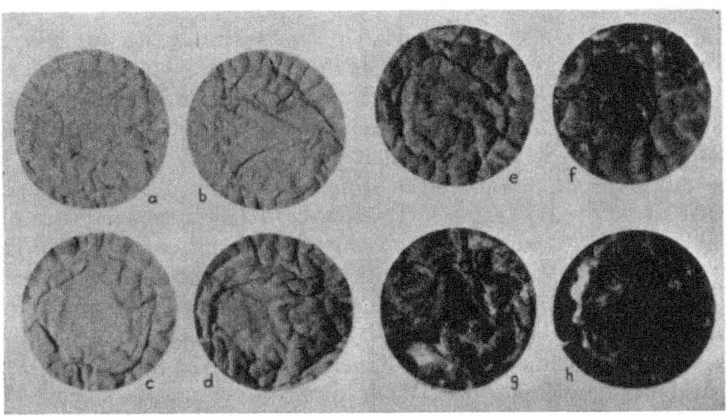

Abb. 188. Testreihe für Cu mit zunehmender Schwärzung der Konidien nach NICHOLAS und FIELDING (1951)

a = 0 μg Cu/50 ml
b = 0,1 μg Cu/50 ml
c = 0,2 μg Cu/50 ml
d = 0,4 μg Cu/50 ml
e = 0,6 μg Cu/50 ml
f = 0,8 μg Cu/50 ml
g = 1,0 μg Cu/50 ml
h = 2,0 μg Cu/50 ml

ändern kann, ist es empfehlenswert, eine größere Menge keimfähiger Sporen eines erprobten Pilzstammes trocken, kühl und geschützt vor Fremdinfektionen aufzubewahren. Mit diesen in doppelt destilliertem Wasser mit 0,3% Polyviol (STAPP, WETTER 1953) bis zur durchscheinend braunen Färbung aufgeschlämmten Sporen werden die Nährlösungen beimpft, und zwar mit 0,1 ml je Gefäß. Wird eine solche Aufschlämmung verdünnt, dann kann unter Umständen das Erntegewicht beträchtlich absinken (ROSCHACH 1956). Das könnte ein Zeichen dafür sein, daß noch weitere Mikroelemente wie Gallium, Vanadium usw. für *Aspergillus niger* lebensnotwendig sind (STEINBERG 1938, BERTRAND 1941). In diesem Falle wäre es besser, solche Elemente der Nährlösung ebenfalls zuzusetzen, wie es z. B. NICHOLAS und FIELDING (1951) getan haben, und dann schwächer zu impfen. Denn mit dem Impfmaterial gelangen in die Lösung auch wieder geringe

Mengen des daraus entfernten Spurenelements, die sich außerdem durch Vorzüchtung des Pilzes in Mangelnährlösung weiter verringern lassen.

Nach 4 Tagen Bebrütung bei 30° C werden die Kulturen durch 10 Minuten Erhitzen auf 100° C im Dampftopf getötet, um eine Verseuchung der Arbeitsräume mit den Sporen zu vermeiden, und dann die Pilzdecken quantitativ durch Filtration geerntet (Roschach 1956). Diese werden mit destilliertem Wasser mehrfach gewaschen, über Nacht bei 100° C getrocknet, abgekühlt und gewogen. Die Gewichte werden mit der Testkurve verglichen und so die dem Pilz verfügbaren Mikroelementmengen in den Bodenproben ermittelt. Die Kupferbestimmung erfolgt besser nach der Sporenfarbe, die sich mit steigenden Kupfer-Konzentrationen von farblos über gelb, hellbraun, dunkelbraun bis schwarz vertieft (Abb. 188). Zum Vergleich dient zweckmäßig eine Photographie der Standardreihe unter Berücksichtigung der mitgeführten Konzentrationen dieser Reihe. Die Konidienfarbe wird durch verschiedene Zink- und Molybdän-Konzentrationen im Nährsubstrat nicht beeinflußt, durch verschiedene Magnesium-, Eisen- und Mangan-Konzentrationen nur die Konidienzahl (Stapp, Wetter 1953). Nach Henkens (1961) verdinen jedoch neuere chemische Verfahren zur Kupferbestimmung den Vorzug vor der *Aspergillus*-Methode.

Die Ergebnisse der bisher vorliegenden *Aspergillus*-Teste auf Mikroelemente sind durchaus zufriedenstellend (Nicholas 1949, 1950 b, Nicholas, Fielding 1949, Stapp, Wetter 1953, Wetter 1954, Tucker, Kurtz 1955, Adams, Elphick 1956), wenn auch hier natürlich Ausnahmen die Regel bestätigen. Schließlich hat jede Methode ihre Schwächen (Russ 1958, Jensen, Lamm 1961). Man muß bedenken, daß *Aspergillus niger* stark säuert und auch reduziert. Daraus ergibt sich, daß z. B. Jensen, Henrikson (1955) in Lehmböden von pH 8 mit *Aspergillus* 2- bis 3mal soviel wie chemisch austauschbares Magnesium gefunden haben und bei Eisen- und Mangan-Mangelsymptomen an höheren Pflanzen oft kein Eisen- oder Manganmangel im Boden mit Hilfe von *Aspergillus* festgestellt werden kann (Donald, Passey, Swaby 1952 b, Scheffer, Kloke, Hünerhoff 1957 I, II, Jensen 1951), weil dieser durch Säuerung und Reduktion von den schwer löslichen Anteilen des Eisens bzw. Mangans noch genügende Mengen löst.

Deshalb hat Wetter (1954) auch nicht den Boden selbst, sondern einen wässerigen Auszug daraus benutzt, und Scheffer, Kloke und Hünerhoff (1957 II) haben nur 5 mg Boden je Gefäß gegeben. Andererseits kann aber auch nach Schneider und Siegel (1958) der Pilz einem kalkhaltigen Boden unter Umständen kein Zink mehr entnehmen, so daß er sich nicht zu entwickeln vermag. In solchen Fällen das Zink mit Mineralsäure in Lösung zu bringen, wäre aber eine wesentliche Einschränkung des biologischen Charakters der Methode. Molybdän wird in sauren Böden häufig ebenfalls durch *Aspergillus* in ausreichenden Mengen nachgewiesen, obwohl höhere Pflanzen auf denselben Böden unter typischen Mangelerscheinungen erkranken. Als Ursache werden antagonistische Wirkungen des Mangans auf höhere Pflanzen und nicht auf *Aspergillus* angenommen (Mulder 1953, 1954). Nach Donald, Passey und Swaby (1952 b) kann bei Eisenmangel-Chlorosen in höheren Pflanzen dennoch kein Eisenmangel im Boden mit Hilfe von *Aspergillus* festgestellt werden, weil entweder Eisenoxyde von *Aspergillus*, aber nicht von höheren Pflanzen verwertet werden können oder aber das Eisen zwar auch von höheren Pflanzen aufgenommen, dann aber in ihnen festgelegt wird. Solche Beispiele ließen sich noch mehr anführen. Sie können aber dem praktischen Wert der *Aspergillus*-Methode keinen wesentlichen Abbruch tun. Biologische Objekte sind immer durch eine Unsumme verschiedener Faktoren individuell beeinflußbar. Darin liegt die Schwäche aller biologischen Methoden. Als ein biologisches, den natürlichen Verhältnissen besser entsprechendes und zugleich schnelles Analysenverfahren, das allerdings eine sehr gewissenhafte, saubere Arbeitsweise erfordert, war die Mikroelementbestimmung mit *Aspergillus niger* den rein chemischen Verfahren vor kurzem noch etwa ebenbürtig, in gewissen Fällen sogar eindeutig überlegen. Neuerdings dürfte es ihnen aber wohl unterlegen sein. Mit anderen Mikroorganismen waren keine besseren Ergebnisse zu erzielen. In Vergleichsuntersuchungen von DONALD, PASSEY und SWABY (1952a, b) mit 2 Hyphenpilzen, 1 Alge, 1 Hefe und 6 Bakterienarten hat sich *Aspergillus niger* doch wieder als der geeignetste Testorganismus erwiesen.

Literatur

ACOCK, A. M.: An examination of Mulder's rapid biological method for estimating the amount of copper in soils. J. Aust. Coun. Sci. Ind. Res. **14**, 288–300 (1941). — ADAMS, A. F. R., und B. L. ELPHICK: The copper content of some soils and pasture species in Canterbury. New Zealand J. Sci. Technol. Ser. A **38**, 345–358 (1956). — ALLEN, O. N.: Experiments in soil bacteriology. Madison, Wisc. 1957. — ALVAHYDO, R., und J. DÖBEREINER: Contribuiçao ao estudo do efeito residual dos fosfatos de Trauira natural e calcinado. Bol. Inst. Ecol. Exper. Agric. **14**, 5–37 (1955); Ref.: Biol. Abstr. **31**, II, 22259 (1957). — ANDREWS, W. B.: Carbon dioxide production in mannitol-treated soils as a measure of crop response. J. Amer. Soc. Agron. **29**, 253–268 (1937).

BARTON WRIGHT, E. C.: The theory and practice of the microbiological assay of vitamin B complex; together with assay of selected amino acids and potassium. Analyst **70**, 283 (1945). — BEHRENS, W. U.: Die Methoden zur Bestimmung des Kali- und Phosphorsäurebedarfs landwirtschaftlich genutzter Böden. Berlin: Verlag Chemie. 1935. — BENECKE, W., und H. SÖDING: Beiträge zum Ausbau der mikrobiologischen Bodenanalyse. Z. Pflanzenernähr., Düng., Bodenkde. A **10**, 129–159 (1927/28). — BENTLEY, O. G., E. E. SNELL und P. H. PHILLIPS: A microbiological method for the determination of manganese. J. Biol. Chem. **170**, 343–350 (1947). — BERGMANN, W.: Die Ermittlung der Nährstoffbedürftigkeit des Bodens. Handb. d. Pflanzenphysiologie von W. RUHLAND, Bd. IV, S. 867–913. Berlin-Göttingen-Heidelberg: Springer. 1958. — BERTRAND, D.: Importance of the trace element vanadium for Aspergillus niger. C. R. Acad. Sci., Paris **213**, 254–257 (1941). — BORTELS, H.: Über die Bedeutung von Eisen, Zink und Kupfer für Mikroorganismen. (Unter besonderer Berücksichtigung von Aspergillus niger.) Biochem. Z. **182**, 301–358 (1927). — Molybdän als Katalysator bei der biologischen Stickstoffbindung. Arch. Mikrobiol. **1**, 333–342 (1930). — Über die Wirkung von Molybdän- und Vanadiumdüngungen auf Azotobacter-Zahl und Stickstoffbindung in Erde. Arch. Mikrobiol. **8**, 1–12 (1937). — BORTELS, H., und C. WETTER: Fortgeführte mikrobiologische Untersuchungen über die Mikronährstoffe Mangan und Molybdän. Nachrichtenbl. d. Dtsch. Pflanzenschutzdienstes **6**, 2–5 (1954). — BRAVO, A. R.: Un método rápido de determinación de nutrientes por medio de Azotobacter. Acta Agronom. (Univers. Palmira-Columbien) **2**, 73–113 (1952); Ref.: Z. Pflanzenernähr., Düng., Bodenkde. **85** (**130**), 182–183 (1959). — BUREMA, S. J., und K. T. WIERINGA: Molybdenum as a growth factor of Azotobacter chroococcum. Antonie van Leeuwenhoek **8**, 123–133 (1942). — BUTKEVIČ, W.: The culture of the fungus Aspergillus niger as a method of soil investigation. Russ. J. Exper. Agric. **10**, 136–141 (1909).

CHARNEY, J., und W. P. FISHER: A microbiological assay method for microgram quantities of manganese in biological material. Science **114**, 687–688 (1951). — CHRISTENSEN, H. R.: Über das Vorkommen und die Verbreitung des Azotobacter chroococcum in verschiedenen Böden. Ein Beitrag zur Methodik der mikrobiologischen Bodenforschung. Zbl. Bakt. II **17**, 109–119 (1907). — CONN, H. J.: Bacterium globiforme and soil fertility levels. N. Y. Agric. Exper. Stat. Techn. Bull. **138** (1928). — CONN, H. J., und M. A. DARROW: The effect of moisture charges on soil as a medium for bacterial growth. Soil Sci. **46**, 365 (1938).

DESAI, S. U.: Environmental requirements of crops and methods of assessing soil fertility. Proc. Indian Sci. Cong. **33**, 34–36 (1946). — DIRKS, B., und F. SCHEFFER: Vergleichende Untersuchungen über das Nährstoffbedürfnis der Kulturböden. Landw. Jb. **67**, 779–821 (1928). — DOLE, M. W.: The Aspergillus niger-method for determining copper in soils. Soil. Sci. **73**, 135–147 (1952). — DONALD, C., B. I. PASSEY und R. J. SWABY: A comparison of methods for removing trace metals from microbiological media. J. Gen. Microbiol. **7**, 211–220 (1952a). — Bioassay of available trace metals from Australian soils. Aust. J. Agric. Res. **3**, 305–325 (1952b).

Eno, Ch. F., und H. W. Reuszer: The availability of potassium in certain minerals to Aspergillus niger. Proc. Soil Sci. Soc. Amer. **15**, 155–159 (1951). — Potassium availability from biotite, muscovite, greensand, and microcline as determined by growth of Aspergillus niger. Soil Sci. **80**, 199–209 (1955).

Falangue, H. A., R. N. Martinelli, N. und R. de Camargo: Modification of the Mehlich method of measuring available phosphorus in soil by the introduction of antibiotic nystatin. Rev. Agric. (Piracicaba) **35**, 269–272 (1960); Ref.: Biol. Abstr. **36**, 77800 (1961). — Finck, A., und E. Schlichting: Möglichkeiten und Grenzen der chemischen Bodenuntersuchung zur Ermittlung des Düngebedarfs. Z. Acker- u. Pflanzenbau **99**, 267–272 (1955). — Fuller, J. E.: The Azotobacter soil plaque test for determining soil nutrient deficiencies. An. Rep. Mass. Agricult. Exper. Stat. Bull. **315**, 21 (1935).

Gericke, S., und E. Pfarre: Bodenuntersuchung und Feldversuch. Bodenkde. u. Pflanzenernähr. **26** (**71**), 180–194 (1942). — Gerretsen, F. C.: On the use of Aspergillus niger for the determination of plant nutrients in the soil. Anal. Chim. Acta **2**, 781–792 (1948). — Some aspects of the microbiological determination of magnesium, zinc, copper, and boron. Trans. Int. Soc. Soil Sci., Dublin **1**, 151–166 (1952). — The influence of boron on the growth of Azotobacter chroococcum. 6. Intern. Kongr. f. Mikrobiologie, Rom **3**, 120–122 (1953). — Gerretsen, F. C., und H. de Hoop: Boron, an essential microelement for Azotobacter chroococcum. Plant a. Soil **5**, 349–367 (1954). — Golling, M.: Vergleichende Bodenuntersuchungen auf Phosphorsäure und Kali mittels Schnellmethoden. Kühn-Arch. **30**, 163 (1932). Gutschy, L.: Eine mikrobiologische Methode zur quantitativen Bestimmung der pflanzenaufnehmbaren Nährstoffe im Boden. Z. Pflanzenernähr., Düng., Bodenkde. **87**, 193–201 (1959).

Halversen, W. V., und W. G. Hoge: The Azotobacter plaque test as applied to the determination of phosphate deficiency in Idaho soils. J. Amer. Soc. Agron. **34**, 503–512 (1942). — Henkens, C. H.: The copper content of the soil determined with biological and chemical methods. Versl. Landbouwk. Onderzoek. **67**, 3–28 (1961); Ref.: Biol. Abstr. **38**, 886 (1962). — Hewitt, E. J., und D. G. Hallas: The use of Aspergillus niger (van Tiegh.) M strain as a test organism in the study of molybdenum as a plant nutrient. Plant a. Soil **3**, 366–408 (1951). Holben, F. J.: Soil respiration in relation to plot yields. Penns. Agr. Exper. Stat. Techn. Bull. **273**, 26–28 (1932).

Ihle, K., und W. Stollenwerk: Vergleichende Untersuchungen zur Kalibestimmung in Böden nach der Keimpflanzen- und Aspergillusmethode. Landw. Jb. **76**, 291–296 (1932).

Jensen, H. L.: Microbiological determination of micro nutrients in soil. Suomen Kemistil., Ser. A **24**, 197–204 (1951). — Jensen, H. L., und A. Henriksen: Microbiological and chemical determination of magnesium in soil. Acta Agr. Scand. **5**, 98–112 (1954). — Zur Bestimmung des Magnesiums in dänischen Böden. Tidskr. Planteavl. **58**, 396–420 (1955); Ref.: Z. Pflanzenernähr., Düng., Bodenkde. **74**, 184 (1956). — Jensen, H. L., und C. G. Lamm: On the zinc content of Danish soils. Agric. Scand. **11**, 63 (1961).

Keller, H.: Zur Bestimmung des Nährstoffbedarfs der Böden mit der Bodenplattenmethode nach Winogradsky. Z. Pflanzenernähr., Düng., Bodenkde. A **33**, 308 (1934). — Kiessling, L. E.: Über die Beeinflussung der Aspergillusmethode durch Huminsäure. Z. Pflanzenernähr., Düng., Bodenkde. A **21**, 86–104 (1941a). — Untersuchungen über den Kalihaushalt eines langjährigen einseitig gedüngten Versuchsfeldbodens mittels der Aspergillusmethode nach Niklas-Poschenrieder-Trischler. Z. Pflanzenernähr., Düng., Bodenkde. A **19**, 241–250 (1931 b). — Kiessling, L. E., und A. Schmidt: Die Beeinflussung des Wachstums von Aspergillus niger durch organische Substanzen. Wiss. Arch. Landw. **9**, 293 (1932). — Koscelеckii, A.: Bodenfruchtbarkeit in Beziehung zur Phosphorsäure. Russ. J. exper. Landwirtsch. **10**, 321 (1909). — Krjutschkowa, A.: Die mikrobiologische Feststellung des Kalk- und Phosphorsäurebedürfnisses eines Bodens. Abh. d. Wiss. Inst. f. Düngemittel, Moskau **76**, 14 (1930); Ref.: Z. Pflanzenernähr., Düng., Bodenkde. A **23**, 120 (1932).

Leaf, A. L.: Determination of available potassium in soils of forest plantations. Soil Sci. Soc. Amer. Proc. **22** (**5**), 458–459 (1958). — Lipman, J. G.: Bacteriological indications of mineral requirements of soils. Ann. Rep. New. Jers. Agricult. Exper. Stat. **27**, 177 (1906). — Lumia, C.: Nuovo metodo di valutazione della assimilabilita di materiali diversi concimanti. Actes 4. Conf. Intern. Pedologie, Rom **3**, 236–240 (1926).

McCool, M. M.: Nitrogen availability in soils as measured by growth response of rye grass and Cunninghamella Blakesleeana. Contrib. Boyce Thompson Inst. **14**, 363–368 (1947). — Mehlich, A.: Use of Cunninghamella Blakesleeana and Aspergillus niger for measuring the fertilizer requirements of plants. Proc. Soil Sci. Soc. Amer. **2**, 279–288 (1938). — Mehlich, A., E. Truog und E. B. Fred: The Aspergillus niger method of measuring available potassium of soil. Soil Sci. **35**, 259–277 (1933). — The Cunninghamella plaque method of measuring available phosphorus in soil. Soil Sci. **38**, 445–458 (1934). — Michniewicz, M.: An estimate of the microbiological methods of determining soil fertility. Ref.: Soils Fertilizer, Commonw. Bur. of Soil Sci. **13**, 179 (1950). — Mooers, C. A.: Evolution of the Neubauer, Cunninghamella, and Aspergillus niger methods for the determination of fertilizer needs of soils. Soil Sci. **46**, 211–227 (1938). — Morgan, M. F.: Choix de méthodes uniformes et appropriées pour la détermination des besoins d'engrais de terrains. Actes II. Conférence Intern. des Engrais Chim. (Rom) **2**, 167–177 (1933). — Mulder, E. G.: Über die Bedeutung des Kupfers für das Wachstum von Mikroorganismen und über eine mikrobiologische Methode zur Bestimmung des pflanzenverfügbaren Bodenkupfers. Arch. Mikrobiol. **10**, 72–86 (1939 a). — On the use of micro-organisms in measuring a deficiency of copper, magnesium, and molybdenum in soils. Antonie van Leeuwenhoek **6**, 99–109 (1939 b). — The microbiological estimation of copper, magnesium, and molybdenum in soil and plant material. Analyt. Chim. Acta **2**, 793–800 (1948). — The essentiallity of trace elements for microorganisms and the microbiological determination of these elements. 6. Cong. Intern. di Microbiological Roma **6**, 293–313 (1953). — Molybdenum in relation to growth of higher plants and microorganisms. Plant a. Soil **5**, 368–415 (1954).

Nicholas, D. J. D.: The use of Aspergillus niger for the determination of mineral nutrients available in soils to crop plants. Chemistry a. Industry **1949**, 809–810. — Use of Aspergillus niger for determining magnesium, copper, zinc and molybdenum in soils. J. Sci. Food Agricult. **1**, 339–344 (1950 a). — Use of Aspergillus niger for determining magnesium, copper, zinc, molybdenum, and manganese in soils and plants. Fertilizer Soc. Proc. **10**, 13–44 (1950 b). — The use of fungi for determining trace metals in biological materials. Analyst **77**, 629–642 (1952). — Nicholas, D. J. D., und A. H. Fielding: The use of Aspergillus niger (M strain) in the bioassay of magnesium, copper, zinc, and molybdenum in soils. I. Ann. Rep. Agric. and Horticult. Res. Stat. Lang Ashton, Bristol, **1947**, 126–137 (1948). — The use of Aspergillus niger for the determination of Mg, Cu, Zn, and Mo available in soils to crop plants. Proc. I. Intern. Congr. Biochem., Cambridge, Abstr. **121**, 302–303 (1949). — Use of Aspergillus niger as a test organism for determining molybdenum available in soils to crop plants. Nature (London) **166**, 342–343 (1950). — The use of Aspergillus niger (M) for the determination of magnesium, zinc, copper, and molybdenum available in soils to crop plants. J. Horticult. Sci. **26**, 125–147 (1951). — Nicholas, D. J. D., und A. Nason: Molybdenum and nitrate reductase II. Molybdenum as a constituent of nitrate reductase. J. Biol. Chem. **207**, 353–360 (1954). — Nieschlag, F.: Über die Magnesia-Düngebedürftigkeit leichter Sand- und Moorböden. Bodenkde. u. Pflanzenernähr. **30** (**75**), 157–173 (1943). — Niklas, H.: Die biologischen Untersuchungsmethoden. Handb. d. Pflanzenernähr. u. Düngerlehre von Honkamp, Bd. I, S. 807–858. Berlin: Springer. 1931. — Niklas, H., und M. Miller: Grundlegende Untersuchungen über den Zusammenhang der Ergebnisse der Keimpflanzen- und der Aspergillusmethode. Ernähr. d. Pfl. **28**, 117–420 (1932). — Niklas, H., und H. Poschenrieder: Die Ausführung der Aspergillusmethode zur Prüfung auf Kali. Ernähr. d. Pfl. **28**, 86–88 (1932). — Zur Feststellung der Magnesia-Düngebedürftigkeit und Magnesia-Düngewirkung im Boden mittels Aspergillus niger. Bodenkde. u. Pflanzenernähr. **1** (**46**), 235–247 (1936). — Niklas, H., H. Poschenrieder und J. Trischler: Eine neue mikrobiologische Methode zur Feststellung der Düngebedürftigkeit der Böden. 1. Mitteilung: Bestimmung der Kalibedürftigkeit der Böden mittels Aspergillus niger. Z. Pflanzenernähr., Düng., Bodenkde. A **18**, 129–157 (1930 a). — Die Kultur des Schimmelpilzes Aspergillus niger zur biochemischen Bestimmung der Kali- und Phosphorsäurebedürftigkeit der Böden. Ernähr. d. Pflanze **26**, 97–103 (1930 b). — Urteile und Erfahrungen über die Verwendung und Brauchbarkeit der Aspergillus-Kalimethode und deren Beurteilung nach dem Stand der bisherigen Forschungsergebnisse. Z. Pflanzenernähr., Düng., Bodenkde. B **12**, 109–130 (1933). — Niklas, H., K. Scharrer und A. Strobel: Phosphatlöslichkeit und Azotobacterwachstum. Landw. Jb. **63**, 387–410 (1926). — Niklas, H., und O. Toursel: Die Bodenuntersuchung mittels Aspergillus niger. Bodenkde. u. Pflanzenernähr. **18**, 79–107 (1940). — Die Ermittlung von Spurenelementen mittels Aspergillus niger. Bodenkde. u. Pflanzenernähr. **23** (**68**), 357–366 (1941). — Ergebnisse aus dem Methodenvergleich zwischen dem Gefäßversuch von Mitscherlich, der Keimpflanzenmethode von Neubauer, der Laktat- und Aspergillusmethode bezüglich der Phosphorsäureergebnisse von insgesamt 1264 Böden. Bodenkde. u. Pflanzenernähr. **24** (**69**), 310–314 (1941). — Niklas H., G. Vilsmeier und F. Kohl: Die Bestimmung der Phosphorsäurebedürftigkeit der Böden mittels Aspergillus niger. Z. Pflanzenernähr., Düng., Bodenkde. A **32**, 50–70 (1933). — Niklas, H., G. Vilsmeier und H. Poschenrieder: Der Einfluß des Kalkgehaltes auf das Aspergilluswachstum bei der Prüfung des Kalibedürfnisses der Böden. Z. Pflanzenernähr., Düng., Bodenkde. **24**, 167–178 (1932). — Njegovan, V. N.: Microbiological method for determining the manurial requirement of soil. Z. Acker- u. Pflanzenbau **110**, 438–450 (1960).

Pantanelli, E.: Determinazione biochemica del fosfato assimilabile del terreno. Actes 4. Conf. Intern. Pedologie, Rom **3**, 232–236 (1926). — Pittmann, D. W., und Burnham: Observations of the use of commercial fertilizers on the acid soils of Utah. Exper. Stat. Rec. **66**, 720 (1932).

Raulin, J.: Etudes chimiques sur la végétation. Ann. Sci. Nat., V. Bot. **11**, 92–299 (1869). — Rippel, A.: Bakteriologisch-chemische Methoden zur Bestimmung des Fruchtbarkeitszustandes des Bodens und der Kreislauf der Stoffe. Handb. d. Bodenlehre von Blanck, Bd. VIII, S. 599–671. Berlin: Springer. 1931. — Mikrobiologie des Bodens. Handb. d. Bodenlehre von Blanck, 1. Ergänzungsbd., S. 440–591. Berlin: Springer. 1939. — Roschach, H.: Die Reproduzierbarkeit der Aspergillusmethode beim quantitativen Nachweis von Zink. Landw. Forsch. **8**, 227–234 (1956). — Die Bedeutung der Bodeneinwaage für die Spurenelementbestimmung mittels Aspergillus niger. Landw. Forsch. 15. Sonderh., 109–116 (1961). — Rosselet, F.: Bio-assay of P, K, and Mg in citrus fertilizer plots using Aspergillus niger. S. African J. Sci. **51**, 331–335 (1955). — Russ, E.: Über die Bestimmung von pflanzenaufnehmbarem Kupfer und Mangan in Böden unter besonderer Berücksichtigung der Keimpflanzenmethode. Landw. Diss. Gießen 1958.

Sackett, W. G., und L. C. Stewart: A bacteriological method for determining mineral soil deficiencies by use of the soil plaque. Colo. Agr. Exper. Stat. Bull. **375**, 1–36 (1931). — Sánchez-Marroquin, A., und M. R. Tamoyo: Aspergillus niger as a measure of the potassium and phosphorus contents of soils. Anales escuela nac. cienc. biol. (Mex.) **4**, 199–212 (1946); Ref.: Chem. Abstr. **41**, 2192 (1947). — Scheffer, F., und L. E. Kiessling: Eignet sich Azotobacter zur Ermittlung des Phosphorsäure- und Kalidüngungsbedürfnisses von Böden? Ernähr. d. Pfl. **30**, 161 (1934). — Scheffer, F., A. Kloke und F. Hünerhoff: Untersuchungen des Bodens auf Manganbedürftigkeit nach der mikrobiologischen Testmethode mittels Aspergillus niger. I. Teil: Mikrobiologischer Nachweis von Mangan in Modellböden. Landw. Forsch. **10**, 176–185 (1957). — Untersuchungen des Bodens auf Manganbedürftigkeit nach der mikrobiologischen Testmethode mittels Aspergillus niger. II. Teil: Mikrobiologischer Nachweis von Mangan in natürlichen Böden. Landw. Forsch. **10**, 207–214 (1957). — Schneider, K. H., und O. Siegel: Untersuchungen über die Aufnahme von Zink aus kalkhaltigen Böden durch Aspergillus niger mit Hilfe von radioaktivem Zink Landw. Forsch. **11**, 270–274 (1958). — Schwegler, E., H. Poschenrieder und F. Bukatsch: Die Atmungsgröße des Bodens als Maßstab für eine Phosphorsäure-Düngebedürftigkeit, bestimmt mit Hilfe des Mikro-Düngungsversuches. Mitt. f. Landkultur, Moor- u. Torfwirtsch. **7**, 134–149 (1959). — Seidel, K.: Eine neue mikrobiologische Methode zur Beurteilung der Nährstoffverhältnisse eines Bodens. Wiss. Arch. Landw. A **6**, 536–576 (1931). — Smit, J., und E. G. Mulder: Biological estimation of Cu and Mg in soils and plants. Rec. Trav. Chim. Pay-Bas **59**, 623 (1940). — Smith, A. M., und A. Dryburgh: The examination of soils by means of Aspergillus niger. J. Soc. Chem. Ind. **53**, 250–254 (1934). Ref.: Z. Pflanzenernähr., Düng., Bodenkde. **39**, 249 (1935). — Spicher, G.: Zur Frage der Zuverlässigkeit mikrobiologischer Methoden beim Nachweis von Spurenelementen und Magnesium im Boden. Zbl. Bakt. II, **108**, 259–267 (1954). — Stapp, C., und C. Wetter: Beiträge zum quantitativen mikrobiologischen Nachweis von Magnesium, Zink, Eisen, Molybdän und Kupfer im Boden. Landw. Forsch. **5**, 167–180 (1953). — Steinberg, R. A.: A study of some factors in the chemical stimulation of growth of Aspergillus niger. Amer. J. Bot. **6**, 330–372 (1919). — Nutrient solution purification for removal of heavy metals in deficiency investigations with Aspergillus niger. J. Agr. Res. **51**, 413–424 (1935). — Role of molybdenum in the utilization of ammonium and nitrate nitrogen by Aspergillus niger. J. Agr. Res. **55**, 891–902 (1937). — The essentially of gallium to growth and reproduction of Aspergillus niger. J. Agr. Res. **57**, 569–574 (1938). — Effects of nitrogen compounds and trace-elements on growth of Aspergillus niger. J. Agr. Res. **59**, 731–748 (1939). — Trace element impurities in nutrient solutions for fungi. Arch. Biochem. **28**,

111 (1950). — STOCK, J.: Kulturversuche mit Aspergillus niger als Indikator für die Düngerbedürftigkeit. Bot. Arch. **35**, 1–76 (1933). — STÖCKLI, A.: Die Azotobactermethode zur Bestimmung der pflanzenaufnehmbaren Phosphorsäure des Bodens. Landw. Jahrb. Schweiz **43**, 811–840 (1929). — Mikrodüngungsversuch nach Sekera. Bodenkde. u. Pflanzenernähr. **21/22** (**66/67**), 287–315 (1940). — SWABY, R. J., und B. I. PASSEY: Availability of trace elements from rocks and minerals. Aust. J. Agricult. Res. **4**, 292–304 (1953).

TOURSEL, O.: Die Bestimmung der Mg-Versorgung von Böden nach der Aspergillusmethode von Niklas. Bodenkde. u. Pflanzenernähr. **24** (**69**), 315–320 (1941). — TSCHAN, Y. T.: Estimation of nutrient elements in soil by algae. Principle of the method. VIe Congr. Int. Sci. Sol. Rapp. C, 249–251, Paris (1956). — TUCKER, T. C., und L. T. KURTZ: A comparison of several chemical methods with the bio-assay procedure for extracting zinc from soils. Soil Sci. Soc. Amer. Proc. **19**, 477–481 (1955). — TUCKER, T. C., L. T. KURTZ und D. L. LYNCH: Zinc status of some Illinois soils as estimated by an Aspergillus niger method. Soil Sci. Soc. Amer. Proc. **17**, 111–114 (1953).

VAGELER, P., und F. ALTEN: Böden des Nils und Gash III, IV. Z. Pflanzenernähr. A **22**, 21–50, 191–267 (1931). — VANDEKAVEYE, S. C.: Biological methods of determining nutrients in soil. Diagn. Techn. f. Soils and Crops **7**, 199–230 (1948). — VARALLYAY, G.: Die Beurteilung der Kali- und Phosphorwirkung durch eine Relativarbeit mit der Aspergillusmethode. Z. Pflanzenernähr., Düng., Bodenkde. A **34**, 218–223 (1934). — Über neuere Laboratoriumsmethoden zur Beurteilung der Düngerwirkung. Bodenkde. u. Pflanzenernähr. **23** (**68**), 98–104 (1941). — VILSMEIER, G.: Beiträge zur Aspergillusmethode. Landw. Jb. **78**, 41–54 (1933 a). — Der Einfluß des Kalkgehaltes der Böden bei der Prüfung auf ihre Phosphorsäurebedürftigkeit mittels Aspergillus niger. Z. Pflanzenernähr., Düng., Bodenkde. A **31**, 279–286 (1933 b).

WAHHAB, A., und A. S. BOKHARI: Comparison of biological tests for assaying fertility of Punjab soils. Soil Sci. **76**, 323–329 (1953). — WAKSMAN, S. A., und R. L. STARKEY: Microbiological analysis of soil as an index of soil fertility. VII. Soil Sci. **17**, 141–161 (1924). — WANG, T. L.: Nouvelle méthode de dosage des éléments minéraux fertilisants du sol par les Azotobacter. Plant a. Soil **3**, 41–46 (1951). — WETTER, C.: Über ein Verfahren zur quantitativen Bestimmung von Mangan im Boden mit Hilfe von Aspergillus niger. Landw. Forsch. **6**, 114–119 (1954). — WINOGRADSKY, S.: Etudes sur la microbiologie du sol. II. Sur les microbes fixateurs d'azote. Ann. Inst. Pasteur **40**, 455–520 (1925). — Etudes sur la microbiologie du sol: III. Sur le pouvoir fixateur des terres. Ann. Inst. Pasteur **42**, 36–62 (1927).

d) Leaf Analysis

By

D. J. D. Nicholas

1. Introduction

The analysis of plants for mineral nutrients is but one method used in a complimentary role with others to determine their mineral status. The aim is to find how the nutrient composition of plants is related to growth and, in particular, to establish an empirical range of values for the mineral nutrients in crop plants, during the growing season, that may result in maximum yields. Any significant differences from these sufficiency values would indicate a deficiency or a toxicity effect for one or more of the nutrients which could then be corrected by proper fertilizer treatment. Should the analysis be done early in the season then an adjustment in fertilizer practice could be made so as to increase the final yields. This is particularly so for correcting trace metal deficiencies (GOODALL 1949, NICHOLAS and FISHER 1951).

The nutrient status of crops may be correlated with (a) the onset of visual symptoms of deficiencies or toxicities, (b) yields of crop, (c) quality factors in crops. For (a) or (b) an empirical approach is often used but information for (c) must be more quantitative since it requires some knowledge of the function of the mineral nutrients in metabolism.

2. Mineral Nutrients

At the present time mineral nutrients found in plants are divided into two main groups, (a) essential elements, N, P, K, Mg, Ca, S, Fe, Cu, Zn, Mn, Mo, B, Cl, Na, (b) other elements present in the ash but not yet shown to be essential for plant growth.

Plants have a regulatory mechanism controlling the intake of mineral nutrients from the soil or solution culture so that healthy plants seem to have characteristic amounts of mineral elements in their ash. The regulatory mechanism, however, breaks down when plants grow under adverse soil conditions, e. gs. acid soils where Mn and Al accumulate to toxic concentrations in plants or in alkaline soils where Fe, Mn or B may be rate limiting for growth. It is often found that some plants accumulate elements, e. gs. Mn in clove, Al and Zn in hickory, Se in *Astragulus*, Ba in brazil nut, Au in *Equisetum palustre* and Hg in *Holosteum umbelliferae*. Injurious effects from toxic amounts of chloride in red currant, boron in citrus, manganese in brassica and heavy metals including Zn, Cu, Cr and Ni in cereals and root crops are well known (WALLACE 1951).

The three criteria of essentiality for a mineral element as defined by ARNON (1950) are as follows:—(a) the plant cannot complete its life cycle without it, (b) the effect on the plant must be direct, (c) its action must be specific and cannot be replaced wholly or in part by any other nutrient. As pointed out previously, NICHOLAS (1957), concept (c) is perhaps too rigid and in the light of recent work on the function of trace elements in plant metabolism, NICHOLAS (1957, 1959), it may be better to substitute the term 'functional element' for 'essential element'. This term would include any element that can function in plant metabolism irrespective of whether the effect is specific or not.

It is convenient to subdivide the essential elements into macro or major nutrients:—N, P, K, Mg, Ca, P, S, Na and the micro, minor or trace elements:— Fe, Cu, Zn, Mn, Mo, B and Cl. If we accept the concept of "functional elements" then Co and possibly Ni and others should be included since they can activate certain plant enzymes non-specifically.

GOODALL and GREGORY (1947) have defined a mineral deficiency in relation to crop yield as follows:— "a plant is deficient in a certain element if supplying that element in a suitable form causes an increase in yield, this effect being specific for the element in question". This definition can be extended to include the effect of excesses of nutrients thus: "when an element either produces no further increases in yield or else leads to a decrease in yield". GOODALL and GREGORY'S concept, however, must now be extended to include the quality factors in crops and these would, of course, vary with the type of crop used, e. gs. high sugar content in sugar beet and cane, colour or vitamin c content in fruit crops, special amino acid, carotene, or high trace metal content (Cu, Co, I) in herbage for animal feeding and fat or oil content of seeds. The quality of plant products depends on the proper functioning of mineral nutrients in enzyme systems and in other non-enzymic cell processes responsible for the production of the desired cell constituents. Although notable advances have been made recently in characterizing important metallo-enzymes in plants, the precise way in which they control metabolism is not understood. Eventually it may be possible not only to get maximum yields, but also to modify enzymic mechanisms in plants, by fertilizer practice so as to produce the desired optimum quality factor or factors. For this purpose chemical and biochemical analyses of the plants would be essential. The other point that needs emphasis here is that maximum yields of product need not necessarily coincide with the level of nutrition required for optimum quality and it may well be that yields below the maximum would result in maximum production of the desired quality factor or even suppress an undesirable factor. Future work with leaf analysis and in plant biochemistry must take into account the effects of nutrients on the production of plant constituents (SCHUPHAN 1957).

Several reviews on plant analysis have appeared recently and should be consulted since only special aspects of the subject can be discussed here because of limited space; LUNDEGÅRDH (1951), STEENJBERG (1951), WALLACE (1951), NICHOLAS (1951), ULRICH (1952). BOYNTON and COMPTON (1945), GOODALL and GREGORY (1957) and PRÉVOT and OLLAGNIER (1957).

3. Appraisal of Methods of Analysis

Three processes control the mineral content of plants (a) the absorptive characteristics of the roots, (b) capacity to accumulate nutrients and the location of the reserves including their mobility, (c) translocation processes from roots to tops and *vice versa*, whether through xylem or phloem whether in ionic or bound form.

There have been two main trends in work with plant analysis, namely the use of either fresh or dried plant materials. The problems associated with sampling plant material for chemical analysis have been recognised from early studies and are common to both methods.

A. Field Sampling Methods

It is often impracticable to determine the mineral nutrients in whole plants and this has led early investigators in particular ATTERBERG (1886) and ALWAY (1928) to select particular parts. The choice of plant tissue is still the subject of much debate but the leaf is usually accepted as a suitable sample and when actively growing is likely to reflect nutrient deficiencies or excesses sooner than other plant portions. Three types of leaf samples are advocated (a) those comparable in age and position on the plants, (b) all leaves from a minimal number of plants in a population, (c) random selection of leaves from all over the plant, from a fixed number of plants. GOSSARD (1943) showed that the error in method (c) was greater than that for (a).

At Long Ashton a study has been made of the occurrence and location of visual symptoms and of the chemical contents of various plant parts grown under optimal, deficiency and toxicity amounts of nutrients (NICHOLAS 1956). Thus K, Mg, P and Mn deficiencies usually appear in older leaves, since those nutrients are readily translocated to younger leaves whereas Ca, B and Fe deficiencies occur usually in meristematic tissues. The time and cost involved in differential sampling of the leaves for chemical analysis would not be justified, thus a compromise is made and one leaf sample is used to determine a range of nutrients. It has been found that mid-stem leaves in most crops reflect the nutrient changes in whole plants satisfactorily and ensure a continuity of sampling throughout the season. The following examples illustrate the methods used for the various crops:—

Potato, tomato, hop, etc.: — leaves comparable in age from the mid-stem position.

Cereals: — the internode adjacent to the growing point. At the young stage, the first mature leaf.

Pasture herbage: — cutting tops above soil level with stainless steel shears.

Onion: — leaf bases.

Cauliflower, sugar beet and other root crops: — the first upright mature leaf outside incurled leaves.

Fruit trees: — leaves from the basal portions of well-exposed new leader growths.

When collecting leaf samples from experimental plots, a randomised procedure is used but for diagnostic purposes only, it is sufficient to collect comparable samples from normal and abnormal plants.

B. Preparation of Test Samples for Analysis

When plant material is to be analysed for major nutrients there is usually no need for special cleaning procedures but for trace metals, especially for Fe, this is essential. A number of workers have washed leaves in water containing a detergent or in dilute hydrochloric acid. JACOBSON (1945), SMITH, REUTHER and SPECHT (1950) and MASON (1950). Iron, shown to be a serious contaminant on leaf surfaces of tomato, could be removed completely by washing the leaves for 30 sec. in water containing a detergent 'teepol', sodium alkyl sulphate (NICHOLAS, LLOYD-JONES and FISHER 1957). The results showed that half the iron present was removed by this washing procedure but there was no loss of Mn. By giving iron labelled with ^{59}Fe via the roots it was shown that there was no loss of iron, during washing, from within the leaves.

MITCHELL (1956) has shown that the possibility of soil contamination affecting the results for P, K, Ca, Zn, Cu, Mn and Mo determinations in pasture herbage can generally be disregarded. For Co, however, soil contamination from trampling by animals is serious and he advises restricting a small area by fencing and cutting the herbage regularly to simulate the conditions of animal grazing. It is possible to assess the extent of soil contamination of a pasture herbage by testing for Ti, the most common element in the earth's crust. The Ti contents of clean herbage is usually between 0.5 and 2 μg/g. dry weight and above 5 μg/g. in contaminated samples. The determination of Ti, however, can only be readily done in a spectrochemical laboratory.

Fungicidal spray residues of Hg and Cu can interfere seriously with trace metal determinations in plants and it is essential to wash the leaves with detergent should spray materials have been applied or else leaf sampling must be done before or very late after the spray applications.

Plant materials which are to be dried for trace metal determinations should be put in polythene bags. Processed or reinforced polythene should not be used since trace metals are often used as catalysts in this process. Fresh plant tissues must be put in a botanical vasculum between damp sheets of bench blotting paper or moss and transported to the laboratory immediately. They should be stored in a refrigerator at 4° C until they are ready for processing. The extraction of the nutrients from fresh plant tissues must be made as soon as possible after sampling. If the material is to be dried then this also must not be delayed since autolysis of leaf protein occurs readily resulting in release of nutrients into the intercellular spaces and then on to the external leaf surfaces. Good quality greaseproof paper is used to line the aluminium trays in the drying oven.

Fresh Plant Tissues. The literature on this topic has been reviewed recently, NICHOLAS (1956), so that only a brief mention of salient points will be made here. Early work was concerned with qualitative staining techniques. HOFFER (1926) showed that Fe accumulated in the nodal tissues of K-deficient maize and McGILLIVRAY, RALEIGH, THUT and VAN OHLEN (1930) detected K in sections of tomato petioles immersed in platinic chloride. A further development was the determination of nutrients in secretions of plant sap from hydathodes or cut stems, LOWRY and TABOR (1931) or in sap squeezed out of plants in special presses up to 1 ton per sq. inch (FONDER 1929, COOK 1930). McCOOL (1926), however, showed that the composition of sap varied with the pressure applied and GASSNER and GOEZE (1932) improved the reproducibility of the results by killing the tissues before pressing. The results obtained by these methods, however, have not been reproducible.

Subsequently solutions, usually buffers, were used to extract nutrients by (a) diffusion method for removal of easily soluble or non-assimilated nutrients, (b) maceration or blendor methods which, in addition, remove some of the bound nutrients. The diffusion method gives empirical results only and is used in the field to supplement visual diagnosis. The blendor methods are more quantitative and are used in the laboratory although it is possible to operate blendors from batteries in a mobile field laboratory. The variety of extractants used include cold and hot water, organic acids, buffered acetate or citrate solutions, hydrochloric acid, sulphuric acid and ethanol. Other extraction solutions include the reagents used to determine mineral nutrients (THORNTON, CONNER and FRASER 1934).

Most of the chemical tests described in early papers were based on microchemical staining techniques for testing tissues *in situ* but later attempts were made to make the tests semi-quantitative and reproducible. For field, work the chemical tests had to be simple involving the minimum of time and apparatus. Thus the tests had to be either specific for the element or be free from interference from other nutrients, at the concentrations found in the extracts. Full details of the Long Ashton tests have been given elsewhere (NICHOLAS 1953) and because of shortage of space will not be discussed here. Sufficient is it to say that the extraction of the nutrients is made with either 2% acetic acid or Morgan's solution (sodium acetate/acetic acid, pH 4.8) and the usual colorimetric tests for the nutrients made on the extracts Since acetic acid alone often gives similar results to those obtained with Morgan's solution, the former is preferable since K, Na and Ca can be determined direct in these extracts using a flame photometer. The following colorimetric methods are used; vanadate for PO_4, titan yellow for Mg and formaldoxime or 'tetrabase' for Mn (NICHOLAS 1953, 1956).

Dried Plant Materials. Plant materials must be dried quickly in aluminium trays covered with good quality greaseproof paper, in a well ventilated electric oven lined with good quality aluminium (except when Ga is to be determined) to avoid metal contamination. The oven should be thermostatically controlled at 110° C and 90° C respectively and dustproof hot air circulated continuously over the samples. The plant material is killed by heating at 110° for 2 hr. and then dried at 90° C for a further period to constant weight.

Plant material for analysis of major nutrients are crushed in various types of hammer mills but for trace metal analysis, milling material while still warm in a toughened pyrex glass pestle and mortar type mill is often desirable. This procedure is not recommended for boron since glass contains the element. PALM and BECKWITH (1956) have modified a Christy Norris mill to avoid iron contamination whereas KRETSCHMER and RANDOLPH (1954) have used a nylon slip roller pulverizer which eliminated trace metal contamination. MITCHELL (1948) found that with a good steel hammer mill, contamination of herbage was only 1 to 2 μg Fe in a total of 50 to 100 μg Fe/gm. drs weight.

Dry ashing is not usually favoured nowadays since losses of some of the more volatile mineral constituents occur. This technique is, however, still in use for boron determination and sodium hydroxide or solid calcium carbonate is mixed in well with the dried material before burning it in a silica lined muffle furnace at 400° C, so as to avoid loss by volatilization.

A wet digestion is often preferred to dry ashing. The usual method is to take 5 g. of dried material in a micro-Kjeldahl flask with 50 ml. Analar or C.P. grade HNO_3 and the material heated on a thermostatically controlled coil heater for about 30 min. in a P.V.C. (polyvinylchloride) lined fume cupboard fitted with injector and extraction fans. After this period when the material is well digested, it is cooled and 5 ml. 60% W/V $HClO_4$ is added and the digest continued until the solution is clear. Excess acid is removed by heating to the "incipient dryness" stage when silica has been dehydrated. Deionised or glass distilled water (50 ml.) is added and the solution boiled for 15 min. to hydrolise the lower phosphates to orthophosphate and then filtered through a No. 41 Whatman (acid washed) paper into 100 ml. volumetric flask and made to volume with distilled water. The following mineral nutrients may be determined in these extracts; P, K, Mg, Ca, S, Mn, Zn, Cu, Na, Cl, Mo. When Fe is to be determined it is preferable to include 2 ml. Analar H_2SO_4 98% W/V in the digest since a better recovery of the element is achieved in some plant materials using this modification, e.g. brassica leaves. Some workers have advocated the use of H_2O_2 instead of $HClO_4$ as an oxidant but evidence that it is superior to perchloric acid remains to be established. NICHOLAS, LLOYD-JONES and FISHER (1957) showed that there was no loss of radioactive iron when added at different stages of digestion of a variety of dried plant materials with HNO_3 and $HClO_4$.

The special procedures used for spectrochemical analysis of plants have been well reviewed by MITCHELL (1948) and LUNDEGÅRDH (1951) and need not be discussed here.

C. Methods for Determining Mineral Nutrients

The chemical methods used for determining mineral nutrients must remain the choice of the individual and the parent laboratory. If expensive spectrochemical equipment is available in a specialist laboratory then this will be the method of choice but usually in laboratories not so equipped, colorimetric techniques involving the use of absorptiometers or spectrophotometers will be used. The last named are often preferable since they do not depend on glass or gelatin filters which deteriorate with time.

Microbiological methods of assay can also be used to determine Fe, Cu, Zn, Mn and Mo either in digests of plant materials or direct in plant extracts (NICHOLAS and FIELDING 1951; NICHOLAS 1952). This is a particularly useful technique when only small amounts of material are available. The determination of Mo by this technique is the most sensitive and specific method known for determining small amounts of the element direct in the sample, without preliminary concentration or separation of the micronutrient. The effective range of assay mineral nutrients is given in Table 152.

Assays for these nutrients can also be made in extracts of fresh or dried plant tissues, and in protein and enzyme fractions of leaves (NICHOLAS 1959).

Neutron activation analysis has been used to determine minute amounts of Cu, Zn, Mn, Sr and B in dried plant materials (BOWEN 1955, 1956, 1959). As for the microbiological assay, the method can be used direct on dried plant materials thus obviating any errors during preparative separations. Dried plant tissues put in clean polythene packets are packed together with nutrient standard

Table 152. *Trace-metal assay series for Aspergillus niger*

Element	Metal in standard, μg per 50 ml. of basal solution	Standard deviation	Method of assessment
Iron	0, 0.1,* 0.5, 1, 2.5, 5,* 10	±0.1	Weight of mycelial felt
Copper	0, 0.05,* 0.1, 0.2, 0.4, 0,6. 0.8,* 1, 1.5, 2, 4	±0.1	Weight of mycelial felt and colour of spores
Zinc	0, 0.5,* 1, 2, 3, 4, 5,* 10	±0.5	Weight of mycelial felt
Molybdenum ..	0, 0.0005,* 0.001, 0.002, 0.004, 0.005, 0.01*	±0.001	Weight of mycelial felt
Manganese ...	0, 0.01,* 0.025, 0.05, 0.1, 0.5, 1,* 5	±0.01	Mycelial characters including intensity of sporulation

* Effective range for assay, coinciding with the maximum differences in mycelial characters, including spore cover and dry-weights of felts.

solutions in similar bags in an aluminium can. The can is then irradiated for various times in a suitable pile, depending on the element to be activated. The B.E.P.O. pile at Harwell has a flux of about 10^{12} thermal neutrons cm^2/sec. and in this the following activation times are given:— Mn, 2.5 hr.; and Ba, Sr, Cu, Zn, 12 hr. After irradiation the material is usually digested in hot HNO_3 and chemical methods are used to separate the element to be radioassayed from the others. This technique, therefore, is limited to those who have access to a pile and it is the author's experience that the present cost of the irradiation service at Harwell, England is so high as to preclude the use of this elegant technique for routine purposes.

4. Expression and Interpretation of Results

Results for macro-elements are preferably expressed as the percentage of the element in dry weight, not as the oxides, e.gs. % K, Ca etc. and those for trace metals or minor elements as μg/g. dry weight and not in parts per million (p.p.m.). The latter term is confusing since it is applicable to either volume or weight; thus μg/g. is more precise and correct and should be adopted. When extracts are prepared from fresh tissues then the results are usually given as μg element extracted per g. fresh weight.

Since the quantitative functions of the elements in plants are not known, it follows that the exact concentration required for maximum growth cannot be estimated. In the absence of this information empirical methods are used to relate nutrient content to symptoms of deficiences or excesses and also to final yield of crop.

A. Mineral Content in Relation to Visual Symptoms

Mineral contents of plants are often correlated first with the onset of visual symptoms, so that a critical range of values may be determined for deficiency or toxicity effects of the various nutrients. (Wallace 1951, Nicholas 1952, 1953, 1956.) Examples of this type of information are given in Table 153.

Table 153. *Healthy and deficiency values of mineral elements in various crops (Leaves except where otherwise stated)*

	Healthy (H) or Deficient (D)	As % Dry Weight				μg/gm. Dry Weight		
		CaO	MgO	K_2O	P_2O_5	Fe	Mn	B
Wheat	H	1.93	0.50	4.93	0.81	140	60	25
(Young Plants: Tops)	D	0.20	0.13	0.71	0.10		15	2
Oat	H	1.60	0.56	5.0	0.78	100	66	30
(Young Plants: Tops)	D	0.58	0.15	0.80	0.20		9	4
Flax	H	0.80	0.20	1.45	0.20	60	50	12
(Mature Plants: Tops)	D	0.40*	0.08*	0.70*	0.12*	24*	6*	5*
Marrow Stemmed ...	H	5.0	0.54	4.64	0.82	120	90	65
Kale	D	0.78	0.32	0.80	0.30		10	9
Brussels Sprout	H	6.60	0.97	4.50	0.60	160	85	70
	D	0.62	0.23	0.92	0.21		8	8
Savoy	H	5.90	0.76	5.0	0.82	110	40	60
	D	0.69	0.24	0.68	0.20		5	10
Cauliflower	H	5.0	0.82	4.3	0.96	145	60	36
	D	1.30	0.24	0.70	0.15		10	13
Swede	H	5.6	0.86	4.6	0.98	87	63	40
	D	1.0	0.30	1.0	0.33		6	15
Turnip	H	5.9	1.20	3.4	1.02	71	75	52
	D	1.3	0.28	1.2	0.36		12	12
Sugar Beet	H	3.7	0.91	5.07	0.80	125	46	29
	D	0.92	0.16	0.97	0.28		17	17
Mangold	H	5.65	1.25	3.32	1.00	110	52	28
	D	0.68	0.21	1.21	0.50		15	11
Potato	H	4.7	0.80	6.5	0.83	88	40	35
	D	1.0	0.34	3.7	0.20		7	
Tomato	H	7.97	0.80	6.0	0.90	269	46	46
	D	0.81	0.22	1.2		20	7	7
Field Bean	H	2.47	0.37	1.82	1.12	75	36	37
	D	0.76		0.80			8	11
Runner Bean	H	1.3	0.58	2.10	0.70	86	68	42
	D	0.25	0.12	0.42			10	7
Apple	H	1.56	0.4	1.97	0.42	90	30	20
	D	0.78	0.25	0.72	0.23	11	15	4
Pear	H	2.96	0.40	3.70	0.56	75	35	26
	D		0.10	0.97			11	
Sweet Cherry	H	2.72	0.61	3.38	0.65	65	41	30
	D		0.21	0.56			7	
Black Currant	H	4.4	0.44	1.68	0.56	200	64	43
	D		0.28	0.65	0.12		15	
Gooseberry	H	3.48	0.51	5.07	0.67	110	45	37
	D		0.30	1.09	0.08			

* Values on soils giving deficiencies in other crops.

Note: On acid soils manganese is often present in plants in toxic amounts. Values exceeding 10,000 μg/gm. dry weight have been recorded.

The results show clearly that the nutrient standards fixed vary with the type of crop and tissue sampled. A detailed knowledge is also required on the effects of various soil types, climatic conditions and season on the nutrient status of crops, before nutrient standards can be fixed with confidence. For diagnostic purposes, however, it may be sufficient to determine the nutrient status of normal and abnormal plants when differences are highly significant for the deficient or toxic elements. On the other hand these differences may be small, the so-called 'good' sample may be near the deficiency level, since there are no apparent symptoms to guide the sampler. Thus a knowledge of the range of nutrient standards expected for normal plants and those deficient in mineral nutrients is a pre-requisite to an accurate diagnosis. Another difficulty is that small stunted plants may contain more of the deficient nutrient than normal plants when the results are expressed on a dry weight basis.

B. Mineral Status in Relation to Yields

The second objective is to correlate the nutrient content of plants with final yields of crop so that the results can serve as a guide to fertilizer practice. This goal is more difficult to achieve since numerous environmental factors other than nutrient supply have a profound effect on final yield. GOODALL and GREGORY (1957) have considered in detail the effects of nutrient supply on mineral uptake by plants and on final yields. The simplest relation is where external supply of a nutrient is increased from a sub-optimal level, resulting in a rise in both nutrient content of the plant and yield of crop. (PFEIFFER 1919, GAS 1937, MACY 1936.) LUNDEGÅRDH (1941) has proposed a formula $y = \frac{a}{x^c}$ where y = yield increment resulting from the addition of a single nutrient and x, concentration of this nutrient in the plant at flowering time; a and c are constants. The second type of relation considered by LUNDEGÅRDH takes into account the interaction with other nutrients. He has shown that the absolute increment in yield resulting from fertilizer treatment is dependent on the internal concentration of nutrients in the plants and on the amounts and ratio of the minerals in the fertilizer applied. The yield increment was represented by him as:—

$$y = \frac{K}{x_1^{c_1} \cdot x_2^{c_2} \cdot x_3^{c_3} \text{ etc,}}$$

so that for factors acting simultaneously, the yield increment is proportional to the product of the effects of single factors. Whether or not this is the correct formula remains to be established. LUNDEGÅRDH's method is based on analysis of plants before the application of fertilizers.

The following difficulties are usually met when correlations between mineral content of plants and yields are attempted:—

(a) The assumption is made that for each set of nutrient concentration there is a unique response in yield to the application of each nutrient. Should the same plant composition be associated with more than one potential yield response to fertilizer treatment, then plant analysis cannot distinguish between them. In practice this difficulty seldom occurs since the effects of fertilizer treatment are directly correlated with the balance of the nutrient content within the plant which in turn is reflected in a growth response.

(b) The same nutrient content of the plant might be situated on both the ascending and descending limb of the S-shaped yield curve. PIPER (1942) and STEENJBERG (1951) have given instances of this phenomenon. When the limiting nutrient is added, e.g. nitrogen to timothy grass or copper to cereals, the yield increased but the N or Cu content, on a concentrational basis, *decreased* initially but later increased. This was due to a marked stimulation in dry weight production on adding the deficient element, so at this stage the concentration of the nutrient per unit dry weight was lower than before it was applied. BOULD, NICHOLAS, TOLHURST, WALLACE and POTTER (1950) have also observed this phenomenon when correcting Cu deficiency in fruit trees.

(c) Nutrients may accumulate, late in growth, when they are no longer required for metabolism. This phenomenon should not be a problem in practice since it is readily recognised. In any event plant analysis, to be of value, must be done during early season when the plants are metabolically active so that corrective measures for trace metal deficiencies in particular, can be applied during the current season. (NICHOLAS and FISHER 1951; GOODALL 1949.)

(d) In experiments with oats, despite highly significant increases in final yields of grain resulting from foliage sprays of $MnSO_4$, symptoms of Mn deficiency appeared in the sprayed plants late in the season after spraying had been discontinued. At this stage the Mn content of the mid-stem leaves was low agreeing with visual symptoms but not with final yields (NICHOLAS and FISHER 1951). The correlation between Mn content of mid-stem leaves and final yields of grain were highly significant ('r' = 0.87) during early growth and at this stage 40 μg Mn/g. dry weight was found to be the optimum level above which an increase in yield as a result of Mn treatment would be unlikely to occur (NICHOLAS and FISHER 1951). GOODALL (1949) has shown that the Mn content of winter wheat can be used to forecast its response to foliage sprays of $MnSO_4$. The Mn contents of wheat above which no further increase in grain can be expected as a result of increasing the element in the plants were determined as follows:— during tillering 34 p.p.m. Mn in dry matter of lower leaf blades and 36 p.p.m. for stems prior to leaf emergence. These values are higher than those associated with the onset of visual symptoms (15 to 20 p.p.m. dry matter) so that plants may be deficient in relation to final yields without showing external symptoms. Thus plant analysis is of use in determining these sub-optimal levels when symptoms are not present to guide the investigator.

C. Methods for Relating Leaf Analysis to Yields

Statistical methods including correlation coefficients ('r' values) and regression data are used to relate mineral content of plants to final yields. Thus correlation coefficients were successfully determined by GOODALL (1948, 1949), NICHOLAS and FISHER (1951) and NICHOLAS (1952) for Mn and K contents of various parts of cereal plants, determined at intervals during growth, and final yields of grain.

A complimentary method involves a plot of the regression of increase in yields following the application of the limiting nutrient, on the nutrient content in plant tissues taken from the non-treated plants. This method has been used by GOODALL (1948, 1949), NICHOLAS and FISHER (1951) and NICHOLAS (1952.) An example of this type of information is given in Table 154.

These critical K levels, determined 5 weeks after seeding, are required for optimum yield of grain. There is evidence, however, that these values fall as the season advances especially in older and in mid-stem leaves presumably because of translocation of K to the ears. Thus the 1.5% dry matter value of tops, 3 weeks before ear emergence, would probably approximate to GOODALL's values of 0.92 and 1.01% K for old leaf blades and stems of barley at leaf emergence. Minimum levels required for development of symptoms are again higher so that

Table 154. *Minimum values for K in various portions of barley required for maximum yields of grain (5 weeks after seeding)*; (ref. NICHOLAS 1952)

μg K extracted in Morgan's solution per g. fresh weight					%K in dry weight
Method of Analysis	Diffusion Method			Waring Blendor	Total analysis
Plant Portion	Young leaves	Mid-stem leaves	Internodes	'Tops'	'Tops'
Minimum values	700	500	600	2000	1.5

chemical methods are of value in determining sub-optimal levels of K in relation to final yields.

Important contributions to the study of nutrient content of plants have been made by numerous investigators including LUNDEGÅRDH (1941, 1951), STEENJBERG (1951), ULRICH (1952), PRÉVOT and OLLAGNIER (1957) and others. Because of limited space it is not possible to discuss these in detail.

D. Interaction of Mineral Nutrients

LAGATU and MAUME (1934) introduced their Foliar Diagnosis method in which three elements were determined in vine leaves, on three or more occasions during the growing season. Thus they determined, N, P, K or Ca, Mg, K. The sum of the three nutrients on a milligram equivalent basis is calculated for each sample and also the percentages of the three individual nutrients. Thus the sum (N + P + K) or (Ca + Mg + K) gives the "quantity" or "intensity" of nutrition and the percentages of each in the total, the "quality" of nutrition, e.g. ratios of N:P:K or Ca:Mg:K. The latter values for the various sampling times are plotted on a triangular diagram and when the points are joined the course of the seasonal trend for a given plant tissue is shown. The diagram for the plant suspected of a deficiency is compared with healthy plants. The foliar diagnosis method has also been used by THOMAS (1937, 1945) in America and by HALAIS (1948) in Mauritius.

NICHOLAS (1957) has used the radii of a circle to represent the mineral content of plant parts. Healthy plants have a symmetrical diagram. Any deviation from the normal shows up the interactions between the various nutrients. Thus K deficiency results in increased Ca, N, P and Mg whereas low Ca results in increased Mn; low N also lowers the Mg content.

5. Conclusions

Future work should be concerned with the following points:—

(a) Search for chemical, physical and microbiological methods that are specific, sensitive and devoid of interference by other factors, for determining trace metals in plants.

(b) Use of statistical methods to correlate nutrient status of plant parts with final yields of crop to determine what plant portions are the best indicators of fertilizer treatments.

(c) Determination of sub-optimal levels of nutrition which occur in the absence of external symptoms of deficiencies or excesses.

(d) Use of biochemical techniques for diagnosis of nutrient disorders since enzyme patterns change markedly in plants deficient in mineral nutrients long before symptoms appear. More quantitative data is required on the role of metals in plant metabolism which in turn govern the quality factors in crops.

Fundamental studies along these lines will result in plant analysis being more reliable and quantitative. It may then be possible to change nutrient patterns by fertilizer practice in such a way as to stimulate rate of reaction of certain enzymes to produce the desired quality factors in economic crops.

Literature

ALWAY, F. J.: Detection of sulphur deficiency of soils by means of plants. lst Int. Conf. Soil Sci. **3**, 590–618 (1928). — ARNON, D. I.: Criteria of essentiality of inorganic micronutrients for plants. Trace Elements in Plant Physiology, p. 31–39 Waltham, Mass.: Chronica Botanica Co. 1950. — ATTERBERG, A.: Die Beurteilung der Bodenkraft nach der Analyse der Haferpflanze. Landw. Jb. **15**, 415–419 (1886).

BOULD, C., D. J. D. NICHOLAS, J. A. H. TOLHURST, T. WALLACE and J. M. S. POTTER: Copper deficiency of fruit trees in Britain. Nature **165**, 920–921 (1950). — BOWEN, H. J. M., and J. A. DYMOND: Strontium and barium in plants and soils. Proc. Roy. Soc. B **144**, 355–368 (1955). — BOWEN, H. J. M.: The determination of manganese in biological materials by activation analysis with a note on gamma spectrum of blood. J. Nuclear Energy 1956, 18–24. — The determination of copper and zinc in biological material by activation analysis. Int. J. App. Radiation and Isotopes **4**, 214–220 (1959). — BOYNTON, D. and O. C. COMPTON: Leaf analysis in estimating the potassium, magnesium and nitrogen needs of fruit trees. Soil Sci. **59**, 339–351 (1945).

COOK, R. L.: Effect of soil type and fertilizers on nitrate content of the expressed sap and the total nitrogen content of the tissue of the small grains. J. Amer. Soc. Agron. **22**, 393–408 (1930).

FONDER, J. F.: The relationship of soil type to the calcium and magnesium contents of green bean stems and leaves and of their expressed juice. Soil Sci. **27**, 415–431 (1929).

GASSNER, G., and G. GOEZE: Zur Frage der Frosthärtebestimmung durch refraktometrische Untersuchung von Pflanzenpreßsäften. Phytopath. Z. **4**, 387–413 (1932). — GAST, P. R.: Studies on the development of conifers in raw humus. III. The growth of *Pinus sylvestris* L seedlings in pot cultures of different soils under varied radiation intensities. Med. Skogsförsökanst. Stockh. **29**, 587–682 (1937). — GOODALL, D. W.: Studies in the diagnosis of mineral deficiency. IV. The mineral content of barley plants in relation to potassium deficiency. Ann. Appl. Biol. **35**, 605–623 (1948). — Studies in the diagnosis of mineral deficiency. V. Manganese deficiency in wheat. Ann. Appl. Biol. **36**, 26–39 (1949). — GOODALL, D. W., and F. G. GREGORY: Chemical composition of plants as an index of nutritional status. Imp. Bur. of Hort. and Plant. Crops. Tech. Com. East Malling, England, No. 17 (1947). — GOSSARD, A. C.: A study of methods of sampling pecan leaves for total nitrogen analysis. Proc. Amer. Soc. Hort. Sci. **42**, 109–114 (1943).

HALAIS, P.: Results of biochemical control of sugar cane plantations for 1948. Rev. agric. Isle Maurice **1948**, 276–295. — HOFFER, G. N.: Testing corn stalks chemically to aid in determining their plant food needs. Bull Ind. Agric. Exp. Sta. **298** 1–31 (1930).

JACOBSON, L.: Iron in the leaves and chloroplasts of some plants in relation to their chlorophyll content. Plant Phys. **20**, 233–245 (1945).

KRETSCHNER, A. E., and J. W. RANDOLPH: Non-contaminating nylon strip-roll pulverizer for grinding dry plant samples. Anal. Chem. **26**, 1862 (1954).

LAGATU, H., and L. MAUME: Examen critique du diagnostic foliare. C. R. Acad. Agric. Fr. **20**, 246–257 (1934). — LOWRY, M. W., and P. TABOR: Sap for analysis by bleeding corn plants. Science (N. S.) **73**, 453–456 (1931). — LUNDEGÅRDH, H.: Die Tripleanalyse. Theoretische und praktische Grundlagen einer pflanzenphysiologischen Methode zur Bestimmung des Düngerbedürfnisses des Ackerbodens. Lantbr. Högsk. Ann. **9**, 127–221 (1941). — *Leaf Analysis*, translated by R. L. MITCHELL. London: Hilger & Watts, Ltd. 1951.

MACY, P.: The quantitative mineral nutrient requirements of plants. Plant Phys. **11**, 749–764 (1936). — MASON, A. C.: Cleaning leaves prior to analysis. Ann. Rep. East Malling Res. Sta. **1952**. 104–107. — MCCOOL, M. M.: Relation of soil to plant cell-sap. Use of fertilizers lowers freezing point of cell-sap, phosphorus content of plant affected. Quart. Bull. Mich. Agric. Exp. Sta. **9**, 60–64 (1926). — MCGILLIVRAY, J. H., G. J. RALEIGH, H. THUT and F. VON OHLEN: A study of plant food deficiencies in tomatoes. Proc. Amer. Soc. Hort. Sci. **26**, 132–136 (1930). — MITCHELL, R. L.:

The spectrographic analysis of soils, plants and related materials. Bur. Soil Sci. Tech. Comm. 44, Harpenden, England (1948).

Nicholas, D. J. D., and D. J. Fisher: The manganese status of cereal crops in relation to yield of grain. 1949 and 1950. Ann. Rep. Long Ashton Res. Sta. **1951**, 77–81. — Nicholas, D. J. D.: A survey of the use of chemical tissue tests for determining the mineral status of crop plants. Soils a. Fertilizers **14**, 1–9 (1951). — Nicholas, D. J. D., and A. H. Fielding: The use of *Aspergillus niger* for the determination of magnesium, zinc, copper, and molybdenum available in soils to crop plants. J. Hort. Sci. **26**, 125–147 (1951). — Nicholas, D. J. D.: The use of fungi for determining trace metals in biological materials. Analyst **77**, 629–642 (1952) and *in*: Proc. Int. Sympos. Microchemistry, p. 205–211. London: Pergamon. 1959. — The potassium status of barley in relation to yields of grain. J. Agr. Sci. **42**, 468–475 (1952). — Chemical tissue tests for determining the mineral status of plants. Mimeographed publication. Long Ashton Research Sta., Univ. Bristol, England, 1–23 (1953). — An appraisal of the use of chemical tissue tests for determining the mineral status of crop plants *in* Analyse des Plantes et Problèmes des Fumures Minerales 119–139. I.R.H.O. Paris (1956). — Nicholas, D. J. D., C. P. Lloyd-Jones and D. J. Fisher: Some problems associated with determining iron in plants. Plant a. Soil **8**, 367–377 (1957). — Nicholas, D. J. D.: Role of trace metals in the nitrogen metabolism of plants with special reference to micro-organisms. J. Sci. Food Agr. **8**, Supp. Issue S 15–S 25 (1957). — Metabolism of inorganic nitrogen and its compounds in plants. 4th Int. Congress of Biochemistry **13**, 307–331 (1959).

Palm, A. W., and R. S. Beckwith: Modification of Christy and Norris mill to avoid iron contamination of ground plant materials. Anal. Chem. **28**, 1637–1638 (1956). — Pfeiffer, T., W. Simmermacher and A. Rippel: Der Gehalt der Haferpflanzen an Stickstoff, Phosphorsäure und Kali unter verschiedenen Bedingungen und seine Beziehungen zur durch eine Nährstoffzufuhr bedingte Ertragserhöhung. J. Landw. **67**, 1–7 (1919). — Piper, C. S.: Investigations on copper deficiency in plants. J. Agr. Sci. **32**, 143–178 (1942). — Prévot, P., and M. Ollagnier: Directions for use of foliar diagnosis. Fertilité **2**, 3–12 (1957).

Schuphan, W.: Subjektive und objektive Qualitätsmerkmale als Grundlage für eine Abgrenzung und Definition des Begriffs „Qualität und Nahrungspflanzen" in Pflanzenqualität Nahrungsgrundlage, p. 8–24. Frankfurt: Sauerländer. 1957. — Smith, P. F., W. Reuther and A. W. Specht: Mineral composition of chlorotic orange leaves and some observations in the relation of sample preparation technique to the interpretation of results. Plant Phys. **25**, 496–506 (1950). — Steenjberg, F.: Yield curves and chemical plant analysis. Plant a. Soil **3**, 97–109 (1951).

Thomas, W.: Foliar diagnosis: application of the concepts of quantity and quality in determining response to fertilizers. P. Amer. Soc. Hort. Sci. **35**, 269–274 (1937). — Present status of diagnosis of mineral requirements of plants by leaf analysis. Soil Sci. **59**, 339–351 (1945). — Thornton, S. F., S. D. Conner and R. R. Fraser: The use of rapid chemical tests on soil and plants as aids in determining fertilizer needs. Circ. Ind. Agric. Exp. Sta. **204**, 1–16 (1939).

Ulrich, A.: Physiological bases for assessing the nutritional requirements of plants. Ann. Rev. Plant Phys. **3**, 207–228 (1952).

Wallace, T.: The diagnosis of mineral deficiencies in plants by visual symptoms. A colour atlas and guide. London: H.M.S.O. 1961. 3rd ed.

GPSR Compliance

The European Union's (EU) General Product Safety Regulation (GPSR) is a set of rules that requires consumer products to be safe and our obligations to ensure this.

If you have any concerns about our products, you can contact us on ProductSafety@springernature.com

In case Publisher is established outside the EU, the EU authorized representative is:

Springer Nature Customer Service Center GmbH
Europaplatz 3
69115 Heidelberg, Germany

Batch number: 10577971

Printed by Printforce, the Netherlands